Pearson New International Edition

Principles of Human Physiology
Cindy L. Stanfield
Fifth Edition

PEARSON

Pearson Education Limited
Edinburgh Gate
Harlow
Essex CM20 2JE
England and Associated Companies throughout the world

Visit us on the World Wide Web at: www.pearsoned.co.uk

ISBN 10: 1-292-02642-1
ISBN 13: 978-1-292-02642-8

British Library Cataloguing-in-Publication Data
A catalogue record for this book is available from the British Library

Printed in the United States of America

Table of Contents

Introduction to Physiology

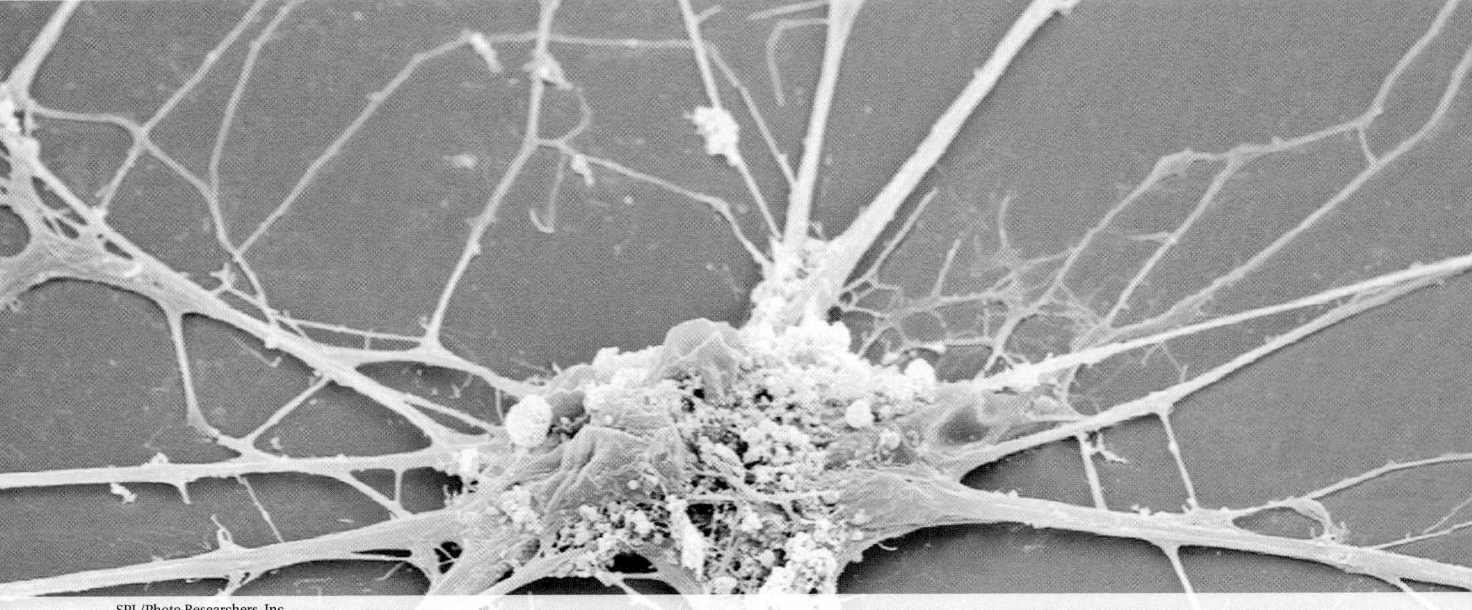

SPL/Photo Researchers, Inc.

Colored scanning electron micrograph (SEM) of a neuron (nerve cell).

The human body is capable of surviving in a dazzling variety of environmental conditions. It can live in jungles, mountains, crowded cities, or deserts. It can withstand the heat of a summer in India or the cold of a New England winter. With proper training, it can acclimate to altitude changes while scaling Mount Everest or survive running a 26.2-mile marathon.

How does the human body do it? As you will learn in this chapter, our bodies have a remarkable ability to adapt to changes in the environment, thereby minimizing internal changes. For example, when a person travels from a cold environment into a warm one (which raises the body's temperature), the body quickly responds by sweating and increasing blood flow to the skin to help bring the temperature back to normal. The body's ability to maintain a normal internal environment, called *homeostasis,* is a primary theme throughout this text.

Welcome to the study of human physiology.

CHAPTER OUTLINE

MasteringA&P®

Go to MasteringA&P for helpful A&P Flix 3-D animations, chapter quizzes, pre-tests, Interactive Physiology tutorials, and more!

LEARNING OUTCOMES *After studying this chapter, you should be able to:*

- Name the four major types of cells in the human body, and describe their defining characteristics.
- Describe the distribution of water in the body, and define the different body fluid compartments.
- Define *homeostasis* and explain its significance to the function of the body.
- Describe the role of negative feedback in homeostasis.
- Explain why diabetes is considered an epidemic.

Physiology, the study of the functions of organisms, comes in many forms—plant physiology, cell physiology, microbial physiology, and animal physiology, to name a few. This text focuses on human physiology, the study of how our bodies work. We emphasize normal physiology, but occasionally describe pathophysiology—what happens when normal body function is disrupted—to better demonstrate typical body function. For example, the effects of diabetes on body function are described throughout the text to illustrate the delicacy of body function and the interdependency of organ systems.

In this text, we take the systems approach to physiology; that is, we study one organ system at a time. An **organ system** is a collection of anatomical structures that work together to carry out a specific function. For example, the cardiovascular system functions to deliver oxygen- and nutrient-rich blood to the various organs of the body. We will learn more about organ systems shortly. As we use the systems approach to studying physiology, you must remember that a single system cannot function alone.

Because nearly everyone is curious about how the human body works, we hope that studying physiology will be one of your most satisfying academic experiences. You will also come to realize that physiology, like the other sciences, is not just a collection of well-worn facts but rather a work in progress. You will recognize that there are significant gaps in our understanding of how the body works, and you will see that much of our current understanding is subject to change as new discoveries are made.

Regardless of your background or current interests, your study of physiology will broaden your scientific outlook. You will begin to see the "big picture," understanding body function not as a collection of unrelated phenomena but rather as a connected whole. You might even discover something else—that physiology is beautiful. Most of us who have decided to make it our life's work think so.

1 Organization of the Body

If you have ever spent time examining a detailed anatomical chart or model of the human body, you have seen that it is an exceedingly complex and intricate structure. Despite the complexity of its structure, however, an underlying simplicity characterizes the function of the human body.

To a student, perhaps the most interesting thing about the body is that its operation can be explained in terms of a relatively small set of principles. For this reason, our approach to describing the body is to strip away all unnecessary details so that the essentials—that is, the unifying themes and principles—can be seen more clearly. To get an idea of what this means, consider **Figure 1**. The brain contains billions of cells that are classified into four groups according to differences in their four general shapes (morphologies). When you consider the function of these cells, however, the similarities among them outweigh the differences, allowing them to be grouped into just one category: All cells in this category are specialized to transmit information in the form of electrical signals from one body location to another. Because of this shared function, all these cells are classified as *neurons* (or *nerve cells*).

Just as the body's underlying simplicity is one of physiology's major themes, so is the degree of interaction among its various parts. Although each of the body's **cells** (cells are the smallest living units) is independently capable of carrying out its own basic life processes, the various types of cells are specialized to perform different functions important to the operation of the body as a whole. For this reason, all the cells ultimately depend on one another for their survival. Similarly, the body's organs are specialized to perform certain tasks vital to the operation of other organs. You know, for example, that your cells need oxygen to live and that oxygen is delivered to your cells by the bloodstream, but consider some of the many things that must occur to ensure that oxygen delivery is sufficient to meet the cells' needs. Oxygen is carried in the bloodstream by cells called *erythrocytes,* which are manufactured by *bone marrow,* a tissue found inside certain bones. To ensure that adequate numbers of erythrocytes are present in the blood, the synthesis of these cells is regulated by a hormone called *erythropoietin,* which is secreted by the kidneys. To ensure adequate blood flow to the body's tissues, the heart must pump a sufficient volume of blood every minute, and for this reason the rate and force of its contractions are regulated by the nervous system. To ensure that the blood carries enough oxygen, the lungs must take in sufficient quantities of air, which requires the control of breathing muscles (such as the diaphragm) by the nervous system. Finally, to provide the energy necessary to drive these and other processes, the gastrointestinal system breaks ingested food down into smaller molecules, which are absorbed into the bloodstream and distributed to cells throughout the body.

This example shows that proper body function requires not only that each part be able to carry out its own particular function, but also that the parts be able to work together in a coordinated manner. To help you better understand how the body's parts work together, the remainder of the chapter outlines broad principles pertaining to body function in general.

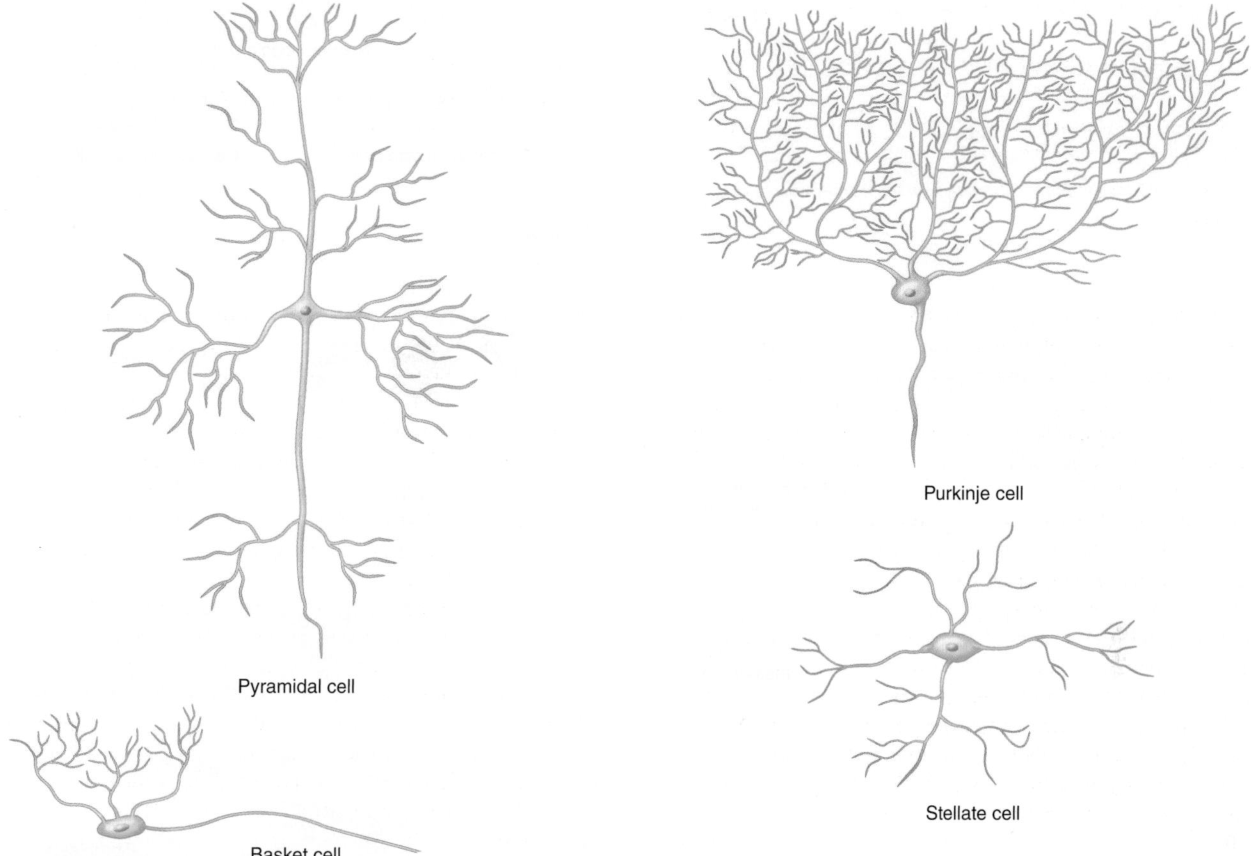

Figure 1 Shapes of cells found in the brain. Each of these four cells is a neuron that transmits electrical and chemical signals.

Cells, Tissues, Organs, and Organ Systems

The human body is a remarkable structure consisting of cells arranged in an orderly fashion. Cells are grouped together to form tissues, which in turn are grouped together to form organs. Organs work together as organ systems. We now describe each of these hierarchical components.

Cells and Tissues

Although more than 200 distinguishable kinds of cells are present in the body, there are only four major classes: (1) *neurons,* (2) *muscle cells,* (3) *epithelial cells,* and (4) *connective tissue cells.* Representative cells belonging to each of these cell types are shown in **Figure 2**. These classifications are very broad and are based primarily on functional differences. Other, more rigorous ways to classify cells have been developed based on anatomical distinctions and embryological origins.

As mentioned previously, nerve cells, or **neurons** (Figure 2a), are specialized to transmit information in the form of electrical signals. For this purpose, neurons typically possess branches that function to receive signals from or transmit signals to other cells. Certain neurons, such as those in the eyes that respond to light or those in the skin that respond to touch, receive information from the outside environment and allow us to perceive the world through our senses. Other neurons relay signals to muscles, glands, and other organs, enabling the control of movement, hormone secretion, and other bodily functions. Still other neurons, such as those in the brain, process information, enabling us to conceptualize, remember, formulate plans of action, and experience emotion.

Muscle cells, or **muscle fibers** (Figure 2b), are specialized to contract, thereby generating mechanical force and movement. These cells are found in the muscles of the arms, legs, and other body parts whose movements are under voluntary control (called *skeletal muscle*), but they are also found in structures not under voluntary control, such as the heart (*cardiac muscle*) and blood vessels (*smooth muscle*). The flexing of an arm, the pumping of blood by the heart, and the mixing of food in the stomach are all examples of muscle cells in action.

Epithelial cells are found in tissues called **epithelia** (singular: *epithelium*), which consist of a continuous, sheetlike layer of cells in combination with a thin underlying layer of noncellular material called a *basement membrane* (Figure 2c). Depending

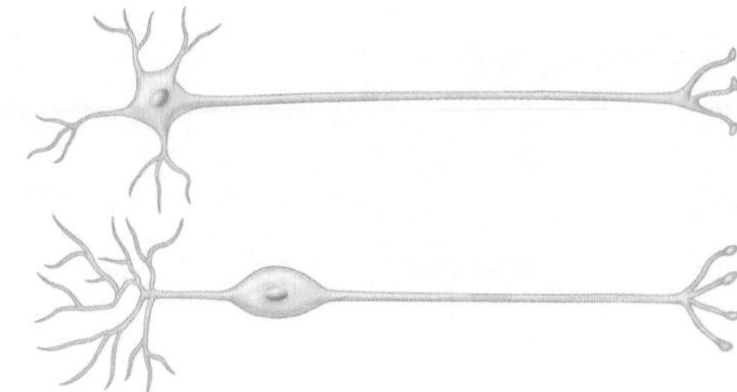

(a) Neurons

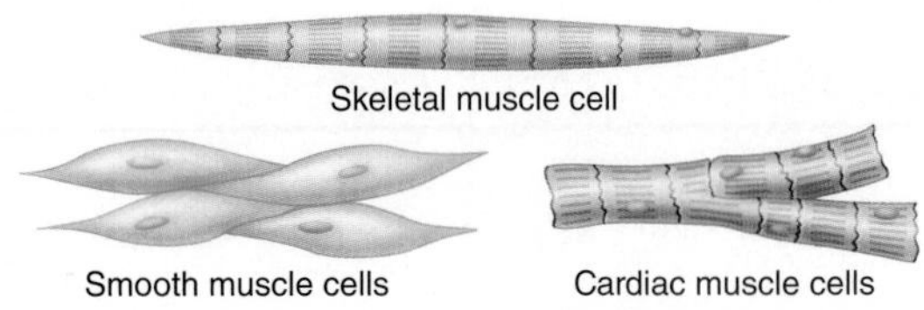

(b) Muscle cells

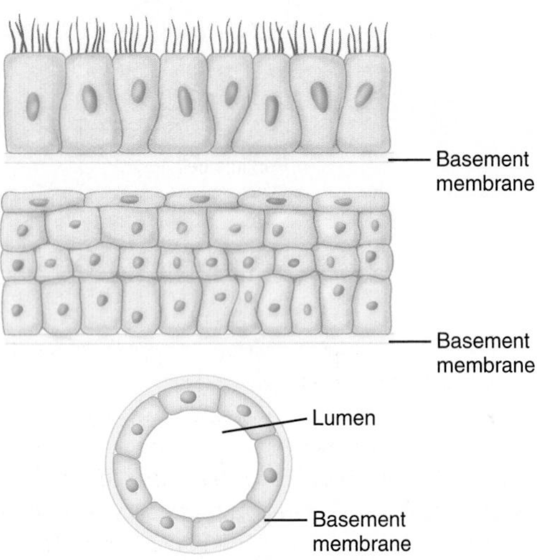

(c) Epithelial cells

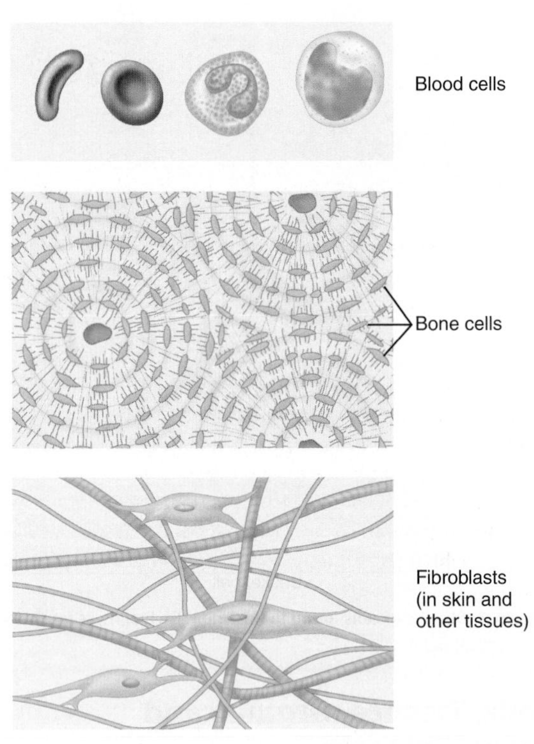

(d) Connective tissue cells

Figure 2 Major cell types in the human body. (a) Neurons. **(b)** Muscle cells. **(c)** Epithelial cells. **(d)** Connective tissue cells.

on the epithelium in question, the cell layer may be one cell thick (*simple*) or several cells thick (*stratified*), and the cells may vary in shape from short and flattened (*squamous*), to regular square-shaped (*cuboidal*), and in some cases to tall and oblong (*columnar*). In all cases, however, cells join closely together to form a barrier that prevents material on one side of the epithelium from mixing freely with material on the other side. Appropriately, epithelia are found wherever body fluids must be kept separate from the external environment, such as the skin surface or the lining of the lungs. Epithelia are also found in the linings of hollow organs such as the stomach, intestines, and blood vessels, where they separate fluids in the interior cavity from the surrounding body fluids. The interior cavity of a hollow organ or vessel is generally referred to as the **lumen.**

Certain epithelial cells are specialized to transport specific materials, such as inorganic ions, organic molecules, or water, from one location to another. For example, cells in the lining of the stomach transport acid (hydrogen ions) into the lumen of the stomach to aid in the digestion of food. Cells in the lining of the intestine, in comparison, transport nutrients and water from the lumen of the intestine into the bloodstream.

Some epithelial cells form **glands,** organs specialized in the synthesis and secretion of a product. Two types of glands are distinguished: exocrine and endocrine. **Exocrine glands** secrete a product into a duct leading to the external environment (**Figure 3**a). Examples of exocrine glands include *sweat glands* and *salivary glands.* **Endocrine glands** secrete **hormones,** chemicals that communicate a message to cells of the body, into the bloodstream (Figure 3b). Examples of endocrine glands include the *pituitary gland* and *adrenal gland.*

The last remaining major cell type, **connective tissue cells,** is the most diverse. This cell type includes blood cells, bone cells, fat cells, and many other kinds of cells that seem to have little in common in terms of structure or function (Figure 2d).

In a narrow sense, the term "connective tissue" refers to any structure whose primary function is to provide physical support for other structures, to anchor them in place, or to link them together. Familiar examples of connective tissue structures are *tendons,* which anchor muscles to bones; *ligaments,* which connect bones together; and the elastic tissue in the skin that gives it its toughness and flexibility. Another example of a connective tissue is the bones themselves, which provide direct or indirect support

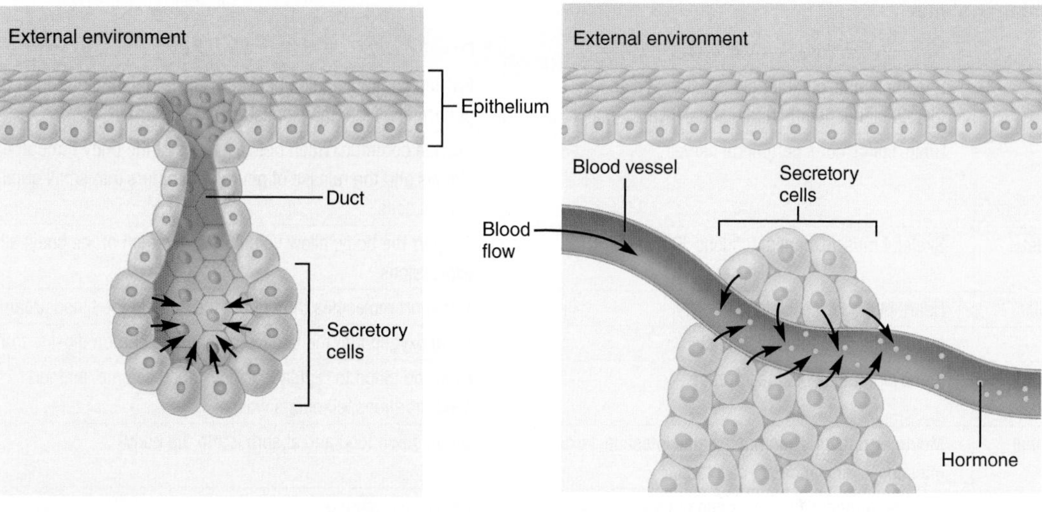

(a) Exocrine gland

(b) Endocrine gland

Figure 3 Glands. (a) Exocrine gland. The secretory cells release their product, which travels via a duct to the external environment. **(b)** Endocrine gland. The secretory cells release their product, a hormone, into the bloodstream, which transports the hormone throughout the internal environment.

for all of the body's structures. In most cases, connective tissue consists of widely scattered cells embedded in a mass of noncellular material called the *extracellular matrix,* which contains a dense meshwork of proteins and other large molecules. Among the most important constituents of the extracellular matrix are the long, fibrous proteins *elastin* (which gives the tissue elasticity) and *collagen* (which gives the tissue tensile strength—that is, the ability to resist stretching).

In a broader sense, the term "connective tissue" encompasses fluids such as the blood and lymph, which do not provide structural support like other connective tissue but instead serve to "connect" the various parts of the body together by providing avenues of communication. The blood, for example, delivers oxygen from the lungs to the rest of the body's tissues and carries hormones from the glands that secrete them to the tissues that respond to them. Similarly, the lymph carries water and other materials that leak out of blood vessels throughout the body and returns them to the blood.

It is a general rule that cells of a given type tend to cluster together in the body with cells of the same type. Nerve cells, for example, are always found in conjunction with other nerve cells, and epithelial cells are always joined with other epithelial cells. Any such collection of cells performing similar functions is referred to as a **tissue.** Thus tissues are also classified into four basic groups: *nervous tissue, muscle tissue, epithelial tissue,* and *connective tissue.* (The term "tissue" is also used more loosely to refer to any of the materials of which the body is composed.)

Organs and Organ Systems

Generally, when two or more tissues combine to make up structures that perform particular functions, those structures are called **organs.** The heart, for example, is an organ whose primary function is to pump blood. Although composed mostly of muscle tissue, it also contains nervous tissue (the endings of nerves that control the heartbeat), epithelial tissue (which lines the heart's chambers), and connective tissue (which makes up the heart's valves and other tissues that hold the muscle fibers together).

The various organs are organized into **organ systems,** collections of organs that work together to perform certain functions. An example is the *cardiovascular system,* whose function is to deliver blood to all the body's tissues. The cardiovascular system includes the heart, blood vessels, and the blood (which is not an organ, but rather a tissue). Another organ system is the *gastrointestinal system,* whose function is to break down food into smaller molecules and then transport these molecules into the bloodstream. This organ system includes the mouth, salivary glands, esophagus, stomach, intestines, liver, gallbladder, and pancreas. In some organ systems (for example, the cardiovascular and gastrointestinal systems), the organs are physically connected. In other cases, the organs are disconnected and more widely scattered. This is true of the *endocrine system,* which encompasses all the glands in the body that secrete hormones, and the *immune system,* which protects the body from invading microorganisms and other foreign materials. The body's organ systems and their primary functions are listed in Table 1.

Although the concept of an organ system is simple in principle, the distinction between one organ system and another is not always clear-cut—many organs perform functions that are integral to more than one organ system. A prime example is the pancreas, which is considered to be part of both the digestive system, because it secretes fluid and digestive enzymes into the intestines, and the endocrine system, because it secretes certain hormones.

Quick Check 1

1. Define *physiology*.
2. Name and describe the four basic types of cells and tissues.
3. Name the ten organ systems, and briefly state the function of each.

TABLE 1 Organ Systems

System	Some organs/tissues within system	Function
Endocrine	Hypothalamus, pituitary gland, adrenal gland, thyroid gland, parathyroid glands, thymus, pancreas	Provide communication between cells of the body through the release of hormones into the bloodstream
Nervous	Brain, spinal cord, peripheral nerves	Provide communication between cells of the body through electrical signals and the release of neurotransmitters into small gaps between certain cells
Musculoskeletal	Skeletal muscle, bones, tendons, ligaments	Support the body; allow voluntary movement of the body; allow facial expressions
Cardiovascular	Heart, blood vessels, blood	Transport molecules throughout the body in the bloodstream
Respiratory	Lungs, pharynx, trachea, bronchi	Bring oxygen into the body and eliminate carbon dioxide from the body
Urinary	Kidneys, ureters, bladder, urethra	Filter the blood to regulate acidity, blood volume, and ion concentrations; eliminate wastes
Gastrointestinal	Mouth, esophagus, stomach, small intestine, large intestine, liver, pancreas, gallbladder	Break down food and absorb it into the body
Reproductive	Gonads, reproductive tracts and glands	Generate offspring
Immune	White blood cells, thymus, lymph nodes, spleen, tonsils, adenoids	Defend the body against pathogens and abnormal cells
Integumentary	Skin	Protects the body from the external environment

The Overall Body Plan: A Simplified View

When physiologists attempt to understand and explain body functions, they usually try to reduce the body's complexity to its essential elements so that unifying themes and principles can be seen more easily. This tendency to simplify is nowhere more apparent than in **Figure 4**, which shows a physiologist's "minimalist" view of the human body. This figure does not look anything like a real body: Not only is it the wrong shape, but it is simplistic and seems to be missing some parts. The gastrointestinal system, for example, is drawn as a straight tube that extends through the body from one end to the other, and the lungs are shown as a single hollow sac. The body's intricate network of blood vessels is depicted as a simple loop, and the heart, which pumps blood around this loop, as just a box. Different cell types, such as nerve, muscle, and connective tissue cells, are drawn to look alike and are given the generic label "cells." Furthermore, the kidneys are shown simply as a single blind-ended tubule that leads to the outside.

The Body's External Environment

The most important concept highlighted by the simplistic rendering of Figure 4 is that a layer of epithelial tissue separates the *external environment* from the interior of the body. This epithelial barrier includes not only the skin, but also the linings of the lungs, gastrointestinal system, and kidney tubules, which are continuous with the external environment. In other words, when air enters the lungs or food enters the stomach, these materials are still actually in the external environment because they are on the external side of this epithelial barrier. Figure 4 also indicates that this barrier is continuous; that is, there is no real separation between the outer surface of the skin and the inside surfaces of the lungs, gastrointestinal system, and kidney tubules. They are all part of the same "fabric," if you will.

The Body's Internal Environment

To live, cells must take in oxygen and nutrients from their surroundings and release carbon dioxide and other waste products into their surroundings. The ultimate source of oxygen and nutrients, and the ultimate repository for discarded waste products (including carbon dioxide), is the external environment. As shown in Figure 4, however, most of the body's cells are not able to exchange materials directly with the external environment because they are not in direct contact with it. Instead, cells receive oxygen and nutrients from the bloodstream, which also carries carbon dioxide and waste products away from cells. Moreover, most cells are not in *direct* contact with the blood, but instead are surrounded by a separate fluid that exchanges materials with the blood. Because this fluid constitutes the immediate environment of most of the body's cells, it is called the **internal environment.** (The term "internal environment" also applies to the fluid in the bloodstream that surrounds blood cells.)

Figure 4 also shows that the blood is contained within epithelium-lined blood vessels. This epithelium differs from that of the gastrointestinal tract, airways, kidneys and skin, in that the epithelium and the blood within it have no connection with the external environment and, therefore, are part of the internal environment. Thus the epithelium that lines the blood vessels is called the endothelium (*endo* = within).

The Exchange of Materials Between the External and Internal Environments To do its job, the blood must obtain oxygen, nutrients, and other needed materials from the

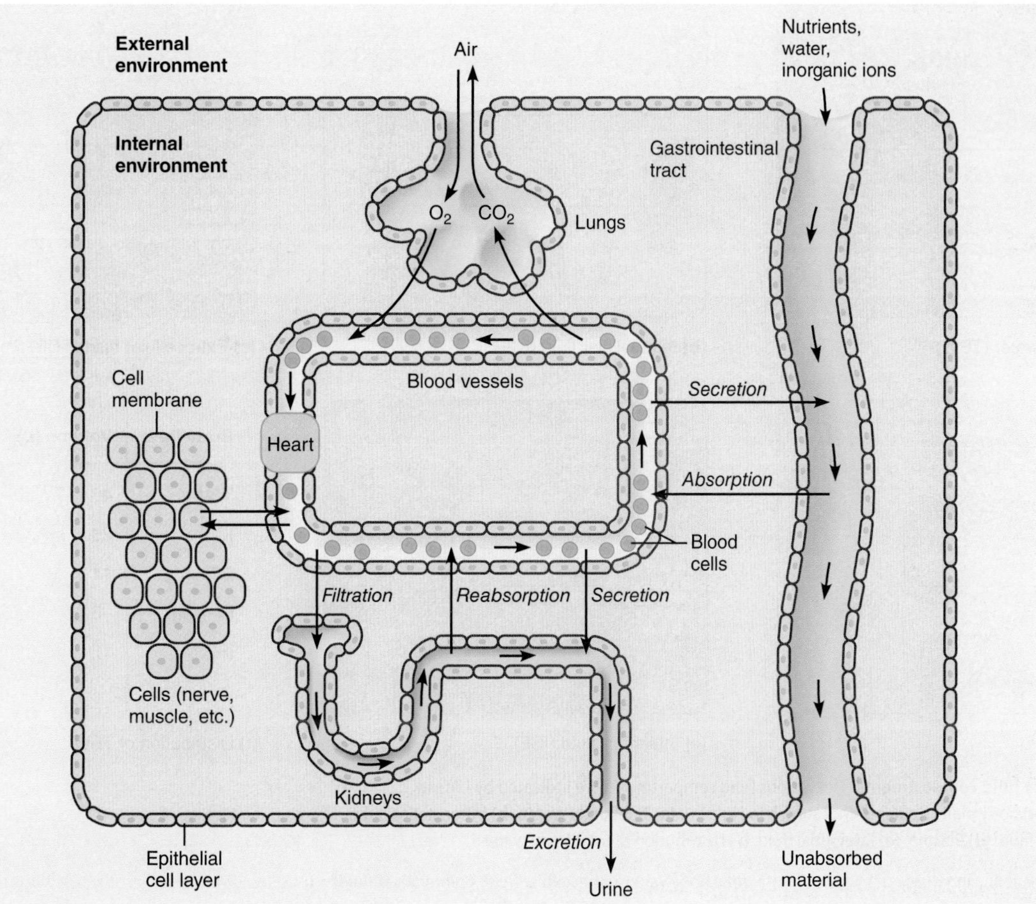

Figure 4 A highly simplified view of the overall plan of the human body. Flows of material are indicated by arrows.

external environment and must release carbon dioxide and other unneeded materials into it. As shown in Figure 4, material is exchanged between the blood and the external environment in a variety of places, including the lungs, the gastrointestinal tract, and the kidneys.

In the lungs, oxygen enters the bloodstream from the air that is breathed in during *inspiration*, whereas carbon dioxide exits the bloodstream and is expelled in the air that is breathed out during *expiration*. In the gastrointestinal tract, the water, inorganic salts, and nutrients obtained from digested food are transported from the lumen to the bloodstream, a process referred to as *absorption*. To aid in the digestion of food, the stomach uses materials from the blood to produce acids and proteins that are then transported into the lumen, a process called *secretion*. Unabsorbed materials (plus bacteria and cellular debris) remain in the gastrointestinal tract and are ultimately eliminated from the body as feces (a process called *excretion*).

In the kidneys, fluid from the bloodstream first enters tubules via a mechanism known as *filtration*. As this fluid travels along the length of the tubules, needed materials (including water, inorganic salts, and nutrients) are selectively transported back into the bloodstream, a process known as *reabsorption*. At the same time, unneeded materials are selectively transported from the bloodstream into the tubules by the secretion process. The fluid that eventually reaches the ends of the tubules constitutes the urine, which is eliminated from the body by excretion. Materials contained in the urine include cellular waste products as well as excess salts and water that are not needed by the body.

Body Fluid Compartments The most abundant substance in the body is *water*, which acts as a solvent for the great variety of solutes found in body fluids. These solutes include small molecules such as inorganic ions, sugars, and amino acids, and large molecules such as proteins. Figure 4 shows that the interior of the body is divided into separate compartments (which are filled with fluid) by barriers of different types, including epithelial tissues and *cell membranes*, which separate the contents of cells from their surroundings. Although these compartments are physically separated, they are still able to exchange materials with each other because the barriers that separate them are *permeable*—that is, they permit molecules to pass through them. These barriers let certain types of molecules through more easily than others, and even exclude

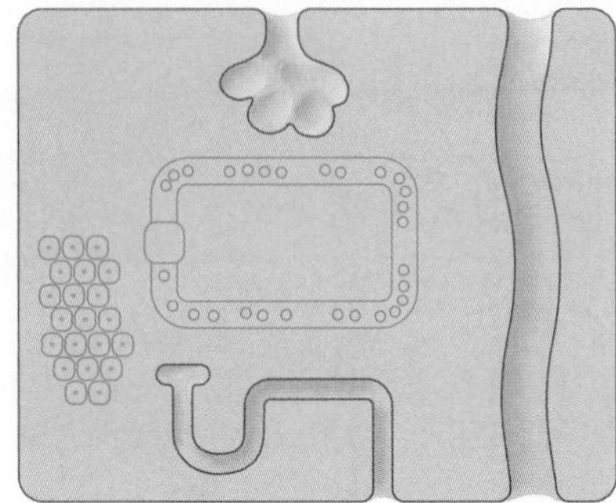

(a) Total body water (TBW)

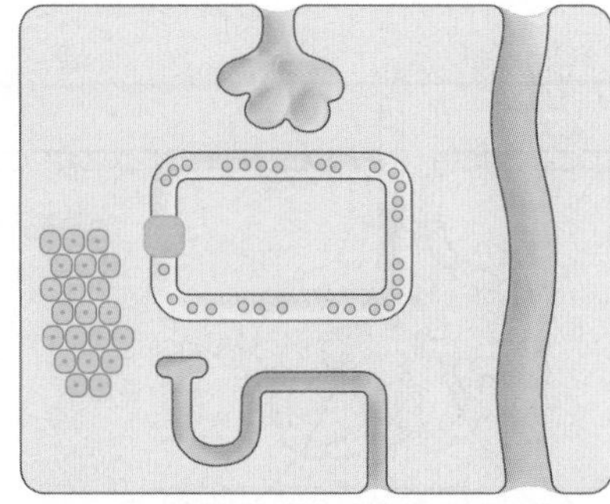

(b) Intracellular fluid (ICF)

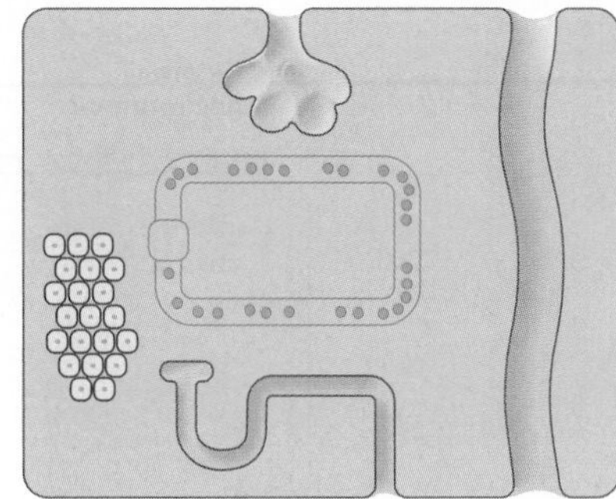

(c) Extracellular fluid (ECF)

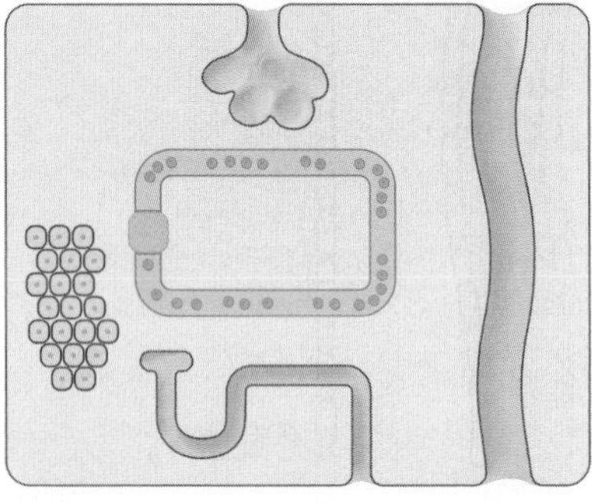

(d) Plasma

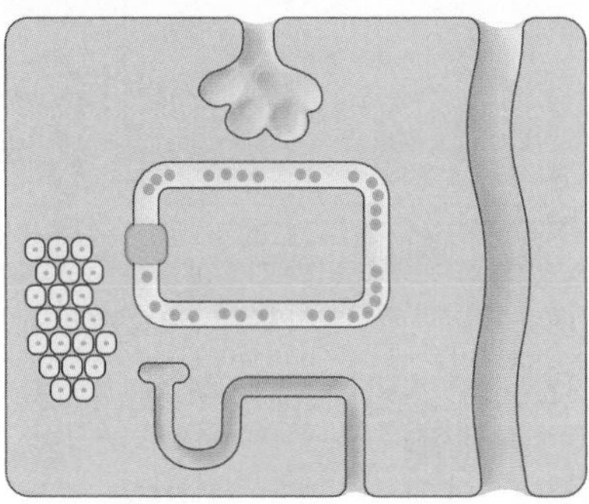

(e) Interstitial fluid (ISF)

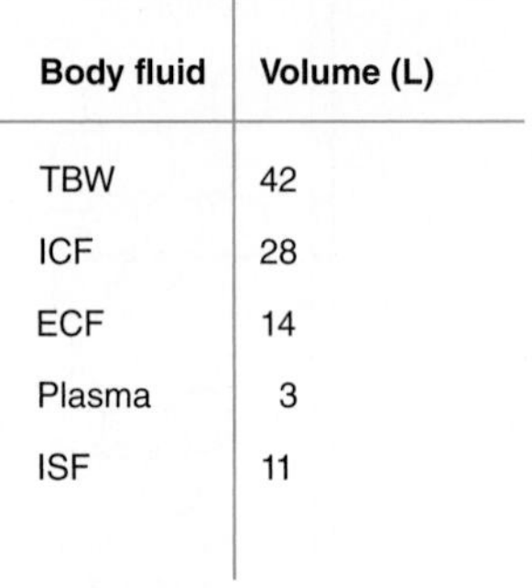

Body fluid	Volume (L)
TBW	42
ICF	28
ECF	14
Plasma	3
ISF	11

(f) Distribution of TBW

Figure 5 Body fluid compartments. The various fluid compartments are indicated by blue in several simplified body plans similar to that shown in Figure 4. **(a)** Total body water. **(b)** Intracellular fluid. **(c)** Extracellular fluid. **(d)** Plasma. **(e)** Interstitial fluid. **(f)** Distribution of total body water.

Q Which of the following does not include plasma: total body water, extracellular fluid, or interstitial fluid?

some molecules from entering certain compartments entirely. Thus it is more accurate to say that cell membranes and epithelial tissues are *selectively permeable* or *semipermeable.*

The volume of water contained in all the body's compartments is termed the **total body water (TBW),** meaning the total volume of fluid enclosed within the outer epithelial layer (**Figure 5**a). For a person weighing 70 kilograms (150 pounds), the volume of TBW is 42 liters, which accounts for approximately 60% of total body weight. Total body water includes water present in fluid located inside cells, called **intracellular fluid (ICF)** (Figure 5b), and fluid located outside cells, called **extracellular fluid (ECF)** (Figure 5c). In the body plan diagram, the volume of ECF relative to ICF is highly exaggerated; in reality, approximately two-thirds of TBW is in the ICF, and only one-third is in the ECF. (Note as well that *extracellular fluid* is synonymous with the *internal environment.*)

ICF and ECF are separated by cell membranes and differ significantly in composition. Intracellular fluid contains many proteins and is relatively rich in potassium, for example, whereas extracellular fluid contains few proteins and is relatively rich in sodium. These differences in composition support the proper functioning of cells and are made possible by the relatively low permeability of cell membranes to many solutes, such that membranes permit the selective exchange of certain solutes.

Of the total volume of extracellular fluid, approximately 20% is found in the blood, and the remainder is found outside the blood. The portion that is present in the blood (that is, the liquid, noncellular part of the blood) is **plasma** (Figure 5d). The portion that is present outside the blood and that bathes most of the cells in the body is called **interstitial fluid (ISF)** (Figure 5e). Plasma and interstitial fluid are very similar in composition; the only major difference between the two is that plasma is relatively rich in proteins, which are scarce in interstitial fluid. The similarity in composition between plasma and interstitial fluid is due to the fact that the walls of the smallest and most numerous blood vessels, called *capillaries,* are highly permeable to most solutes except proteins.

Quick Check 2

1. What is the difference between *absorption* and *reabsorption*?
2. Why is *extra*cellular fluid referred to as the body's *internal* environment?
3. Give the proper term for each of the following: (a) all the water that is contained in the body, (b) fluid that is contained within cells, (c) fluid that is located outside cells, (d) fluid that is located outside cells and found in the blood, and (e) fluid that is located outside cells and found outside the blood.

UNDERSTANDING EXERCISE

Challenging Homeostasis

Maintaining homeostasis can be quite challenging to the organ systems, especially when a person is faced with a stressor—that is, a stimulus that puts strain on the body. Although many stressors are deemed harmful, one stressor—exercise—is actually considered good for the body.

During exercise, the organ systems have to kick into high gear to maintain a normal internal environment. Breathing has to be deeper to bring more oxygen into the body and to eliminate carbon dioxide. The heart has to beat faster and stronger to enhance blood flow, which will rapidly transport oxygen and other nutrients to the cells and carry carbon dioxide and other waste products away from the cells. The liver releases glucose into the bloodstream and adipose tissue releases fatty acids so that exercising muscles can use these substances for energy. Increased muscle and nerve activity also causes extracellular levels of potassium to increase. Blood flow to the skin increases, and glands increase sweat production as the body releases the heat produced by the increased muscle activity.

These are just a handful of the events occurring inside the human body during exercise. With these organs working together correctly, we can lift weights, jump, and even run marathons. Although the changes described in the preceding paragraph are short-term changes that enable the body to maintain homeostasis, the benefits of exercise also persist over the long term. Our bodies adapt to repeated stressors, such as exercise. Marathon runners have slender muscles that are slow to fatigue, whereas weight trainers have large muscles that generate a lot of force but are quick to fatigue. Adaptive benefits of exercise include increased efficiency of the heart and increased metabolism. Exercise helps prevent development of diseases such as type 2 diabetes mellitus. It also burns fat. Thus, while stressors may temporarily take the body out of homeostasis, the body can learn to adapt to some stressors so that they do not seem as harsh the next time around.

2 Homeostasis: A Central Organizing Principle of Physiology

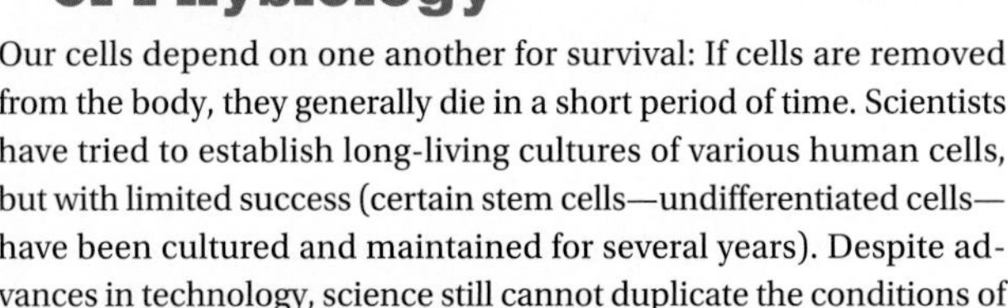

Our cells depend on one another for survival: If cells are removed from the body, they generally die in a short period of time. Scientists have tried to establish long-living cultures of various human cells, but with limited success (certain stem cells—undifferentiated cells—have been cultured and maintained for several years). Despite advances in technology, science still cannot duplicate the conditions of the human body necessary to sustain life.

Given our cells' sensitivity to changing conditions, how can the body tolerate the widely varying conditions it encounters in the external environment? After all, humans can live in both very hot climates, such as the tropics, and much colder climates. We can live at sea level, where oxygen is plentiful, or in the mountains, where the oxygen concentration in air is lower. We can live in the dryness of a desert or in the extreme humidity of a rain forest. How can the body adapt to such a variety of conditions?

The body has all sorts of regulatory mechanisms that work to keep conditions in its internal environment constant despite changes in the external environment. This maintenance of relatively constant conditions in the internal environment is known as **homeostasis.** The concept of homeostasis is a central organizing principle in physiology. In fact, nine of the ten organ systems function to maintain homeostasis (the exception is the reproductive system, which functions to maintain the species, not the individual). Disruption of homeostasis can lead to disease, yet the body is also capable of adapting to mild stressors that disrupts homeostasis (see **Understanding Exercise: Challenging Homeostasis**).

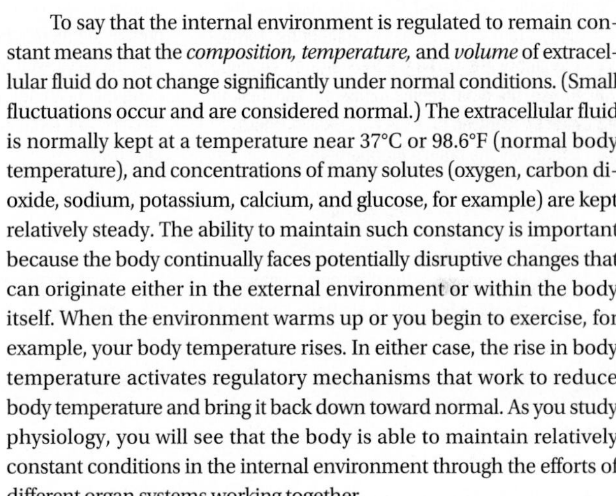

To say that the internal environment is regulated to remain constant means that the *composition, temperature,* and *volume* of extracellular fluid do not change significantly under normal conditions. (Small fluctuations occur and are considered normal.) The extracellular fluid is normally kept at a temperature near 37°C or 98.6°F (normal body temperature), and concentrations of many solutes (oxygen, carbon dioxide, sodium, potassium, calcium, and glucose, for example) are kept relatively steady. The ability to maintain such constancy is important because the body continually faces potentially disruptive changes that can originate either in the external environment or within the body itself. When the environment warms up or you begin to exercise, for example, your body temperature rises. In either case, the rise in body temperature activates regulatory mechanisms that work to reduce body temperature and bring it back down toward normal. As you study physiology, you will see that the body is able to maintain relatively constant conditions in the internal environment through the efforts of different organ systems working together.

Apply Your Knowledge

Hemorrhage is the loss of whole blood, which consists of approximately 55–60% plasma. Without new fluids entering the body, how can plasma volume be elevated toward normal levels to maintain homeostasis?

Even though homeostatic regulatory mechanisms work to resist changes in the internal environment, every regulatory system has its limitations, even when it is undamaged by disease or trauma and is functioning normally. For example, body temperature can be maintained close to normal only so long as environmental temperatures are not too extreme and other stresses

CLINICAL CONNECTIONS

Heat Exhaustion and Heat Stroke 

Heat exhaustion is a consequence of the body's effort to regulate its temperature—in particular, its efforts to get rid of excess heat. When the body must get rid of a large quantity of heat, it produces massive quantities of sweat, leading to a significant reduction in blood volume. In addition, blood flow to the skin increases markedly, which diverts blood from other areas of the body. Together, these changes produce a reduction in blood pressure, which reduces blood flow to the brain and precipitates symptoms such as weakness, dizziness, and even loss of consciousness.

A far more serious condition is *heat stroke,* in which the body's temperature rises out of control because of failure of the thermoregulatory system. Extreme overexertion or high environmental temperatures can overwhelm the body's capacity for getting rid of heat. When this happens, the body's temperature rises in spite of its thermoregulatory efforts. As the temperature continues to rise, the brain begins to malfunction. Delirium sets in, followed by a loss of consciousness. Eventually the brain's thermoregulatory centers begin to fail. The brain then inappropriately stops sending signals to the sweat glands that tell them to secrete fluid. As a result, sweat production comes to a halt, compromising the body's ability to get rid of heat and causing the temperature to rise even faster. If left untreated, this spiral of events leads inexorably to death.

The skin of a person experiencing heat stroke has a flushed appearance (due to increased blood flow) but will also be dry (due to the absence of sweat). These signs make it easy to distinguish heat stroke from heat exhaustion, in which sweating is profuse and the skin is flushed and wet. If someone is experiencing heat stroke or is in danger of doing so, immediate medical attention is of the utmost importance. Often, a person's life can be saved by immersing the body in ice water, which reduces the body temperature quickly to within the range at which normal thermoregulation is possible. Assuming that the elevated temperatures have not caused permanent damage to the brain's thermoregulatory centers, regulatory mechanisms can then take over.

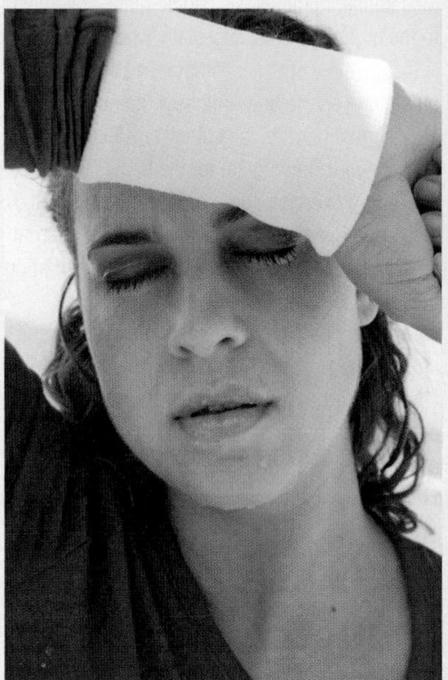

perCut Images/SuperStock.

Critical Thinking Questions

1. In what ways are the differences between heat exhaustion and heat stroke important?
2. What are the roles of blood pressure and sweat production in thermoregulation?
3. What could be alternatives to water immersion for curtailing the effects of heat stroke?

placed on the regulatory system are not too great. Intense exercise or high environmental temperatures, however, can cause body temperature to rise out of control, with potentially fatal consequences **(Clinical Connections: Heat Exhaustion and Heat Stroke)**. In fact, it is generally true that failure of any system to maintain homeostasis ultimately gives rise to signs and symptoms of disease because such failure adversely affects the function of organ systems.

Negative Feedback Control in Homeostasis

Because body temperature is not free to vary but is instead regulated to stay within relatively narrow limits, it is referred to as a **regulated variable.** Plasma concentrations of potassium, sodium, and calcium are also regulated variables because they are kept constant by homeostatic regulatory mechanisms of the organ systems. Most homeostatic regulatory mechanisms follow the same pattern: If a regulated variable increases, the system responds by making it decrease; if it decreases, the system responds by making it increase. Systems behaving in this manner are said to operate by **negative feedback.**

A familiar example of a negative feedback system is the cruise control in a car, which operates to keep the speed of the car steady at a certain desired point (**Figure 6**a and b). If a car running on level ground starts up a hill, the car will begin to slow down. When the control mechanism detects a difference between the actual speed of the car and the desired speed, it feeds more gasoline to the engine, and the car's speed increases. When the car's speed reaches the desired speed, the system "throttles back" to maintain

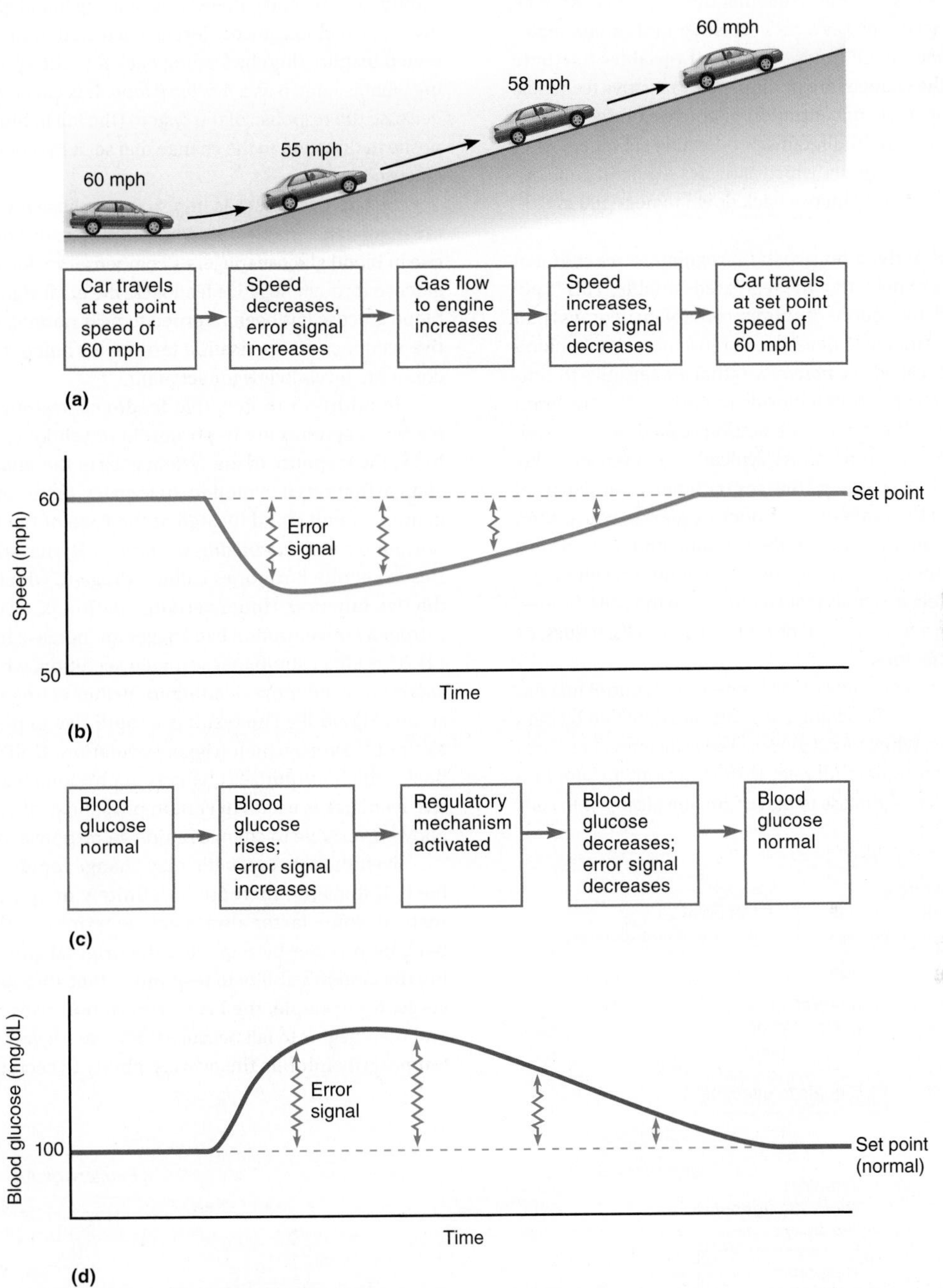

Figure 6 Negative feedback control of a regulated variable. (a) Events occurring as the speed of an automobile is regulated by a cruise control mechanism. **(b)** Graph showing changes in the automobile's speed as it climbs a hill. The dashed line represents the set point speed; vertical arrows indicate the error signal. **(c)** Events occurring as blood glucose is adjusted to normal following an initial rise. **(d)** Graph showing changes in blood glucose.

that speed. As long as the car's actual speed does not differ from the desired speed, the system makes no further adjustments to the flow of gasoline.

Like the cruise control in a car, most homeostatic regulatory mechanisms make adjustments only when they detect a difference between the actual value of the regulated variable and the normal "desired" value, called the **set point.** Any difference between the actual value and the set point constitutes an *error signal.* Because these mechanisms normally work to bring the regulated variable closer to the set point, they ultimately function to make error signals as small as possible. Note that the set point cannot be held absolutely constant, just as a car's cruise control cannot maintain a specified speed at all times. Regulated variables fluctuate continuously, but the changes are minimized by negative feedback. For example, the normal concentration of glucose in the blood is approximately 100 mg/dL (milligrams per deciliter) of blood. After a meal, blood glucose increases, which activates regulatory mechanisms that bring the blood glucose back down to near 100 mg/dL (Figure 6c and d).

To operate properly, a homeostatic regulatory mechanism must have a means of detecting the regulated variable. This is accomplished through the actions of *sensors,* cells (often neurons) that are sensitive to the variable in question. For instance, certain blood vessels contain cells called *chemoreceptors* that are sensitive to concentrations of oxygen and carbon dioxide in the blood; in the brain and other parts of the body, there are neurons called *thermoreceptors* that are sensitive to temperature. Typically, such sensors relay signals (called *input*) to an **integrating center** (often a particular set of neural circuits in the brain or an endocrine gland), which then compares the regulated variable to the set point and orchestrates the appropriate response. In response to the input it receives, the integrating center relays signals (called *output*) to the cells, tissues, or organs that bring about the final response. These cells, tissues, or organs are called **effectors.**

Figure 7 shows an example of the homeostatic control mechanism for blood glucose. Blood glucose levels are detected by beta cells in the pancreas. When blood glucose levels increase, beta cells act as the integration center and release the hormone *insulin* into the blood. Insulin causes glucose to move from the plasma into cells throughout the body, thereby decreasing blood glucose levels. This decrease in blood glucose levels is detected by the beta cells that secreted insulin, thereby feeding back into the system's input, forming what is known as a *feedback loop.* It is called *negative* feedback because the response of the system (the fall in blood glucose) is opposite in direction to the change that set it in motion (a rise in blood glucose).

Negative feedback is important because it triggers changes in the regulated variable only when appropriate. For example, once a rise in blood glucose triggers a compensatory lowering of the blood glucose through negative feedback, the error signal decreases until blood glucose has been returned to near normal. In this way, negative feedback compensation terminates before blood glucose decreases to levels below the set point.

In addition to negative feedback systems, a few positive feedback systems are important in physiology. In **positive feedback,** the response of the system goes in the *same* direction as the change that sets it in motion. In females, for example, the pituitary gland (a small gland located at the base of the brain) secretes a hormone called *luteinizing hormone (LH)* that stimulates the ovaries to secrete hormones called *estrogens,* which regulate reproductive function. Under certain conditions, a rise in the plasma estrogen concentration can trigger an increase in the secretion of LH. This effect stimulates estrogen secretion, which enhances LH secretion even more, leading to further estrogen secretion, and so on (**Figure 8**). The result is a rapid rise in plasma LH, known as the *LH surge,* which triggers ovulation. Unlike negative feedback, which minimizes changes in physiological variables, positive feedback is useful in certain physiological systems because it allows a variable to change rapidly in response to a stimulus.

Even though a variable may change rapidly in positive feedback, it does not increase indefinitely or spiral out of control. Instead, some factor always acts to terminate the positive feedback loop either by removing the original stimulus or by limiting the system's ability to respond to that stimulus. During an LH surge, for example, the LH concentration rises rapidly to a peak and then begins to fall because the surge triggers ovulation, which temporarily inhibits the ovaries' ability to secrete estrogens. The

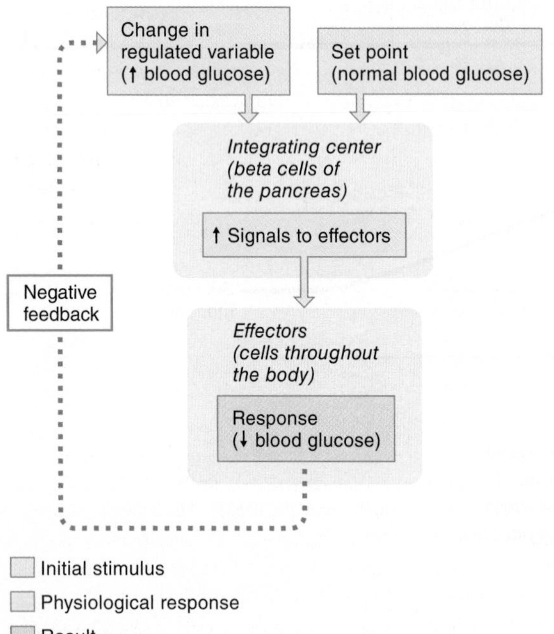

Figure 7 A negative feedback loop. This feedback loop operates in the control of blood glucose, as described in the text. Up and down arrows within boxes indicate increases and decreases, respectively. The dashed line indicates that the response of the system affects the input.

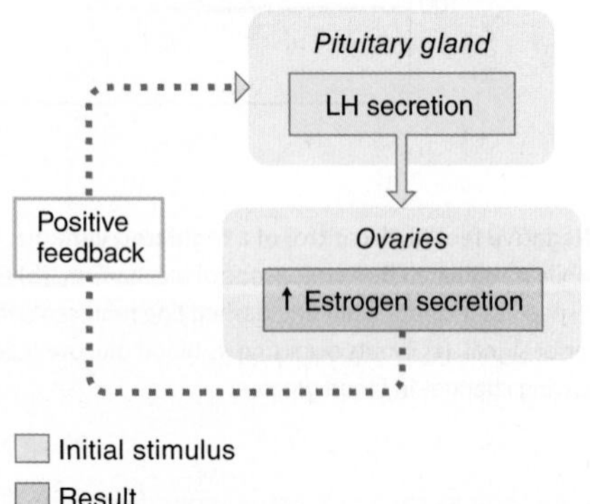

Figure 8 A positive feedback loop. This feedback loop operates in the control of estrogen secretion from the ovaries. LH stimulates estrogen secretion, which stimulates LH release, which stimulates more estrogen secretion, and so on.

resulting fall in plasma estrogen levels removes the stimulus that caused LH secretion to rise in the first place, thereby allowing LH levels to fall.

Quick Check 3

1. Define *homeostasis*, and explain how it enables the body to adapt to changes in its environment.
2. Explain how negative feedback works to maintain homeostasis.
3. In certain forms of diabetes, beta cells of the pancreas fail to secrete insulin at adequate levels. Describe how a lack of insulin would affect blood glucose levels.

3 The Diabetes Epidemic

Diabetes mellitus is a metabolic disease that affects blood glucose levels and plasma volume and causes excessive thirst and fluid loss, among other things. In fact, diabetes affects every system in the human body. The term "diabetes" was first used in the second century A.D. by a Greek physician, Aretus the Cappadocian. *Diabetes* is a Greek word meaning "siphon" or "flow through," which describes the excessive urine flow that characterizes the disease.

Prevalence of Diabetes

Diabetes affects certain individuals more than others. Table 2 shows the prevalence of diabetes among age groups in the United States. Fewer than 1% of people younger than 20 years have diabetes; however, as a person ages, he or she becomes more prone to develop diabetes. More than 10% of people older than 20 years have diabetes; among those persons older than 65 years, 27% have diabetes.

The prevalence of diabetes also varies with race or ethnicity (Table 3). Native Americans are most prone to develop diabetes, followed by African Americans and Hispanics. The prevalence of this disease in the white and Asian American populations is considerably lower.

TABLE 2 Prevalence of Diabetes in the United States Based on Age and Sex Data from the American Diabetes Association

Age group	Number of people with diabetes	Percentage of people with diabetes
Younger than 20 years	215,000	0.26%
20 years and older	25.6 million	11.3%
65 years and older	10.9 million	26.9%
Women	12.6 million	10.8%
Men	13 million	11.8%

TABLE 3 Prevalence of Diabetes in the United States, 2007–2009, Age 20 and Older, Based on Race and Ethnicity

Race/ethnicity	Percentage with diabetes (%)
White	7.1
Asian American	8.4
Hispanic American	11.8
African American	12.6
Native American	14.2*

*Prevalence varies considerably among groups, with the lowest prevalence (5.5%) in Alaskan Natives and the highest prevalence (33.5%) among American Indian adults in southern Arizona.

Source: Data from the Centers for Disease Control and Prevention, *National Diabetes Fact Sheet.* 2011.

The high prevalence of diabetes extends beyond the United States. According to the International Diabetes Association, 285 million people (6.6% of the world population) have diabetes. Only 30 million people worldwide were diagnosed with diabetes in 1985, but by 2000 that number had increased to 150 million (**Figure 9**). By 2030, the number of people with diagnosed diabetes is expected to reach 438 million. By country, diabetes is most prevalent on the Pacific island of Nauru, where 30.9% of the population is diagnosed with diabetes. The country with the next highest level is the United Arab Emirates, with 18.7% of the population diagnosed.

FOCUS ON DIABETES

The prevalence of diabetes mellitus in the world population makes it a major health concern to all. According to the Centers for Disease Control and Prevention's *National Diabetes Fact Sheet, 2011*, an estimated 8.3% of the U.S. population (26 million people) have diabetes mellitus. Of these 26 million cases of diabetes mellitus, 7 million are undiagnosed. In addition, another 79 million people are prediabetic, meaning that there is a high probability that they will develop diabetes mellitus in the future.

One of the most detrimental aspects of diabetes mellitus is its pervasiveness—it affects all organ systems. Diabetes is an example of what happens when homeostasis is disrupted during a disease and how the body responds in an attempt to compensate.

This feature is designed to boost your knowledge of human physiology as it pertains to the pathophysiology of diabetes and its complications—for example, how diabetes can lead to kidney disease or atherosclerosis (vessel disease). We hope you will find it informative and helpful as you advance your study of human physiology.

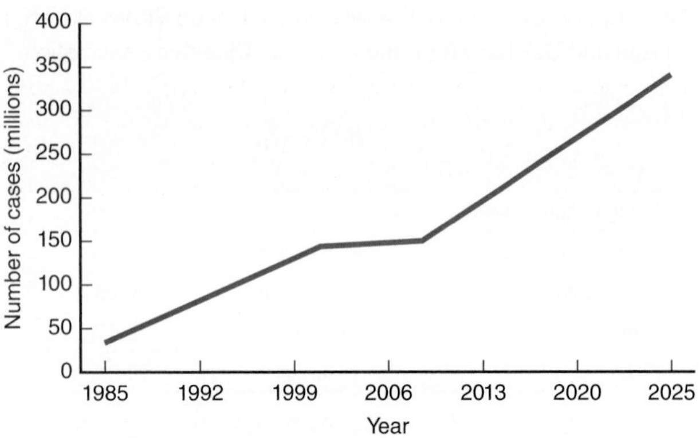

Figure 9 Prevalence of diabetes in the world. The 2025 data are predicted based on current trends.

TABLE 4 Likelihood of Developing Type 2 Diabetes Mellitus Based on BMI

Numbers are relative to the likelihood of developing type 2 diabetes mellitus with a BMI less than 25.*

BMI	Likelihood of developing type 2 diabetes mellitus
<25	1.00*
25–29.9	2.42
30–34.9	3.35
>35	6.16

*Data are given relative to a BMI less than 25, which is assigned a value of 1.00.
Source: Data adapted from the American Heart Association, *A Nation at Risk: Obesity in the United States Statistical Sourcebook.* 2005.

The cost of treating diabetes is high, both in the United States and around the world. Direct medical costs to treat diabetes in the United States in 2007 were estimated at $116 billion. If indirect costs (including time off work, disability, and mortality) are included, this expense increases to $174 billion.

Obesity and Diabetes

Why is the prevalence of diabetes increasing? One major factor is the aging of the population. People now live longer, and a large percentage of the population survives into the higher-risk ages (older than 65 years). In addition, the lifestyles of many young people may lead them to develop diabetes. Diets today incorporate more fats and carbohydrates than in years past. In addition to producing direct metabolic changes, such diets lead to obesity, the number one cause of diabetes. Many young people today also have a sedentary lifestyle, with video games replacing playground time, and physical education classes often absent from the curriculum. Such a sedentary lifestyle can lead to both metabolic changes and obesity—key factors that lead to diabetes.

The prevalence of obesity in the United States is increasing, and obesity itself is becoming an epidemic. *Obesity* is defined as having a high body fat content relative to lean body mass. It is identified using the body mass index (BMI) scale, which measures weight relative to height. The equation that measures BMI is

$$\text{BMI} = \frac{\text{Body weight (kg)}}{\text{Height (m)}^2}$$

A BMI of 25–29 indicates a person is overweight, a BMI of 30–39 indicates obesity, and a BMI greater than 39 indicates severe obesity.

The percentage of adults who are obese has doubled over the last 30 years in the United States. According to the Centers for Disease Control and Prevention, 35% of U.S. adults are overweight and 35% are obese, up from 15% who were obese in 1980. Among children, 10% of preschoolers and 19% of children ages 6–19 years are obese. The significance of these numbers is that obesity and development of type 2 diabetes mellitus (as well as other diseases) are strongly correlated, as shown in Table 4.

Classification of Diabetes

When diabetes is mentioned, most people think of insulin and glucose. Insulin is a hormone released from the pancreas when blood glucose levels are elevated. It promotes glucose uptake into cells, thereby decreasing blood glucose levels—a classic negative feedback system. Once inside the cells, glucose can be used as energy or stored as glycogen or triglycerides to be used as energy at a later time. Although some diabetics must inject themselves with insulin to regulate their blood glucose levels, diabetes is actually much more complicated than that. We will focus on four types of diabetes in this text: type 1 diabetes mellitus, type 2 diabetes mellitus, gestational diabetes, and diabetes insipidus. The prevalence of diabetes described earlier refers to type 2 diabetes mellitus.

Diabetes Mellitus

As mentioned earlier, *diabetes* is Greek for "siphon." *Mellitus* is another Greek word, which means "honey." Honey is sweet, and so is the fluid (urine) excreted by an untreated diabetic. A notable characteristic of diabetes mellitus is high blood glucose levels, such that some glucose is eliminated in the urine, giving it a sweet taste. Conversely, nondiabetics rarely excrete glucose in their urine.

Diabetes mellitus consists of two types: type 1 and type 2. Type 1 diabetes mellitus (type 1 DM) was formerly referred to as insulin-dependent diabetes mellitus or juvenile-onset diabetes mellitus: "insulin-dependent" because damage to the beta cells of the pancreas keeps them from secreting enough insulin in the blood to regulate blood glucose, and "juvenile-onset" because this form of the disease becomes apparent at an early age. Type 1 DM is the type familiar to most people, but it is by far the less common of the two types, accounting for only 5 to 10% of all diabetes mellitus cases.

Type 2 diabetes mellitus (type 2 DM) is characterized by failure to respond to insulin when its levels are normal. Type 2 DM, formerly referred to as non-insulin-dependent diabetes mellitus or adult-onset diabetes mellitus, is the disease associated with the diabetes epidemic. People with type 2 DM have normal beta cells in their pancreas that secrete insulin at appropriate levels to regulate blood glucose. Because the target cells are incapable of responding to insulin, however, they do not take up glucose from the blood.

This failure has two major consequences: (1) The cells do not get the glucose they need for energy and (2) the glucose levels in the blood rise.

Gestational Diabetes

A subclass of type 2 DM is gestational diabetes, which develops in 4% of pregnant women (135,000 cases per year in the United States). Current data suggest that hormones produced in high amounts during pregnancy induce insulin resistance. Of the women who get gestational diabetes, 5–10% develop type 2 DM after giving birth.

Prediabetes

In prediabetes, blood glucose levels are elevated, but not as much as in full diabetes. Prediabetics have fasting blood glucose levels of 100–125 mg/dL. According to the Centers for Disease Control and Prevention, 79 million Americans are considered prediabetic and many are likely to develop type 2 DM within 10 years of developing the signs of prediabetes.

Diabetes Insipidus

Diabetes insipidus differs from diabetes mellitus in that the problem lies not in regulating blood glucose levels, but rather in regulating plasma volume. The word *insipidus* is Latin for "having no flavor," thereby distinguishing the urine of people with this disease from the sweet urine of people with diabetes mellitus. Antidiuretic hormone (ADH) decreases the amount of water lost in the urine. Thus one cause of diabetes insipidus is inadequate ADH secretion. Another cause of this disease is tissue resistance to ADH, which can occur during pregnancy.

The primary symptoms shared by people with diabetes mellitus and those with diabetes insipidus include copious urination and excessive thirst. A person with diabetes insipidus may drink more than 4 gallons of water and excrete more than 3 liters of urine per day. Diagnostic procedures include measuring ADH levels in blood, magnetic resonance imaging of the brain (ADH is secreted by the pituitary gland, which is located in the brain), and a water deprivation test in which the patient is deprived of water and plasma sodium concentration is measured every hour (as plasma volume decreases due to a loss of fluid, plasma sodium concentration increases).

Diabetes insipidus is a rare disease that we will not discuss further here.

Diagnosing Diabetes Mellitus

To test for diabetes, fasting plasma glucose levels are measured and a glucose tolerance test is performed. Random measurements of plasma glucose may give the first indication of prediabetes or diabetes; fasting plasma glucose levels and a glucose tolerance test would then confirm the diagnosis.

The fasting plasma glucose test requires that the patient not eat for 8 hours before a blood sample is taken. Normal fasting glucose levels are 60–100 mg/dL. Plasma glucose levels of 100–125 mg/dL indicate prediabetes, and levels greater than 125 mg/dL indicate diabetes.

The oral glucose tolerance test requires fasting for 8 hours, followed by consumption of a solution containing 75 grams of glucose dissolved in water. A plasma glucose measurement is taken 2 hours after the person consumes the solution. A 2-hour plasma glucose level of less than 139 mg/dL is considered normal, between 140 and 199 mg/dL is prediabetic, and greater than 200 mg/dL indicates diabetes.

Symptoms of Diabetes Mellitus

Diabetes has many symptoms, which can vary depending on the person. The symptoms most commonly identified for diagnosis purposes are elevated blood glucose and glucose in the urine. Glucose in the urine pulls water out of the body with it, resulting in loss of body fluids and dehydration.

As diabetes progresses, it affects virtually every system in the human body. Because their cells cannot utilize glucose normally, diabetics often suffer from fatigue. Fluctuations in blood glucose can lead to lethargy and, with extremely high levels, coma. The disease may affect the eyes, heart, vasculature, kidneys, stomach, and peripheral nerves. Diabetes can also cause erectile dysfunction. In addition, it promotes atherosclerosis, increasing the likelihood that the person will suffer a stroke or heart attack.

Treatment of Diabetes Mellitus

Current therapies used in the treatment of diabetes mellitus help control blood glucose levels. Maintaining fasting blood glucose levels between 70 and 120 mg/dL, for example, has proved effective in decreasing the incidence of eye disease, kidney disease, and nerve damage. Diabetic patients are placed on a strict diet and must test their blood glucose frequently to ensure the levels are not fluctuating too much on a daily basis. A more recently discovered test may be even better at determining blood glucose levels over a period of time—namely, measuring the amount of a certain protein, hemoglobin A_{1c}, present in the blood.

Glucose binds to hemoglobin A_{1c} inside red blood cells. Normally, glucose is transported in the plasma portion of the blood. However, when present at high levels, glucose enters the red blood cells and binds to hemoglobin A_{1c}; the higher the blood glucose levels, the more glucose bound to hemoglobin A_{1c}. Red blood cells have a short life span of 3–4 months. During this period, the cells accumulate glucose until they are removed from the blood by the spleen. Thus measurements of hemoglobin A_{1c} levels indirectly measure the average blood glucose levels for 2–3 months.

Monitoring blood glucose levels does not cure diabetes, but it provides feedback so that the patient can make lifestyle changes (for example, in diet and exercise patterns) to bring blood glucose back toward normal levels. If behavioral changes do not fix the problem, pharmaceutical interventions may be required. Individuals with type 1 DM typically have to administer insulin to control their blood glucose levels. On occasion, persons with type 2 DM may have to do the same to decrease their blood glucose levels, although other pharmaceutical options are also available.

Sulfonylureas and meglitinides are drugs that stimulate beta cells to increase insulin secretion. Thiazolidinediones enhance the action of insulin on muscle and fat cells and decrease the amount of glucose produced by the liver. Biguanides decrease the amount of glucose produced by the liver. Other drugs to treat type 2 DM are also available, but their mechanisms of action require more knowledge of physiology.

As diabetes progresses, treatment must also progress. As secondary problems arise from the diabetes, such as cardiovascular disease, these secondary diseases must be treated. These treatments are beyond the scope of this chapter.

SYSTEMS INTEGRATION In this chapter, we learned that our bodies are made up of cells of various types, and that these cells must work together in a coordinated fashion to maintain homeostasis.

CHAPTER REVIEW

SUMMARY

1 Organization of the Body

- Cells—the smallest living units—are specialized to carry out different functions in the body.
- Specific types of cells are organized into tissues, which are then combined to make organs.
- Organs perform specific functions; organ systems are organs that work together to perform certain tasks.
- Four major types of cells are distinguished: neurons, muscle cells, epithelial cells, and connective tissue cells.
- The internal environment is the fluid surrounding the body's cells.
- The bloodstream delivers and removes materials to and from the internal environment and then exchanges the materials with the external environment.
- The body is divided into several fluid-filled compartments.
- Total body water (TBW) is the total volume of water in all compartments; it includes both intracellular fluid (ICF) and extracellular fluid (ECF).
- ICF is located inside cells; ECF is located outside the cells. The two types of fluids are separated by cell membranes.
- The portion of ECF in the blood is called plasma; that found outside the blood is called interstitial fluid (ISF).
- Plasma and ISF are similar in composition and are separated by the epithelial tissues that line blood vessels.

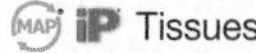

iP Fluid and Electrolytes, Introduction to Body Fluids

2 Homeostasis: A Central Organizing Principle of Physiology

- Homeostasis is the body's ability to maintain constant conditions in the body's internal environment.
- To maintain homeostasis, regulatory mechanisms work to control regulated variables so that variations are minimized.
- Negative feedback occurs when a change in a regulated variable triggers the opposite response.
- Homeostatic regulatory mechanisms include sensors, an integrating center, and effectors.
- Positive feedback occurs in some physiological variables in which a change triggers a similar response.

3 The Diabetes Epidemic

- Obesity increases the likelihood that a person will develop diabetes mellitus.
- There are two types of diabetes mellitus: type 1 DM and type 2 DM. Type 1 DM results from inadequate insulin secretion; type 2 DM results from inadequate response of tissues to insulin.
- Gestational diabetes is a subclass of type 2 DM; it develops in 4% of pregnant women.
- In prediabetes, blood glucose levels are elevated, but not as much as in full diabetes. Prediabetics are likely to develop type 2 DM in the future.
- Diabetes insipidus is associated with inadequate ADH secretion and results in copious urination.
- The major symptom of diabetes mellitus is hyperglycemia. Diagnosis of the disease depends on blood glucose levels and levels of hemoglobin A_{lc}.
- Diabetes mellitus treatment includes regulating blood glucose levels with diet, insulin injections, and/or oral medications.

Check your understanding at MasteringA&P and apply your new knowledge to real-life case studies to ensure that you've mastered the content!

EXERCISES

Multiple-Choice Questions

1. Which of the following best illustrates the concept of positive feedback?
   a) The secretion of acid by cells in the stomach lining is suppressed when the acidity of the stomach contents increases.
   b) A rise in blood pressure stimulates the elimination of water in the urine, which reduces the blood pressure.
   c) An increase in the concentration of H^+ in blood causes an increase in H^+ excretion in urine.
   d) An increase in the carbon dioxide concentration of the blood stimulates breathing, which increases the rate at which carbon dioxide is eliminated from the body.
   e) Contractions of the uterus push the fetus against the cervix, which triggers release of oxytocin into the bloodstream; oxytocin then stimulates stronger contractions of the uterus.
2. The hormone aldosterone stimulates the reabsorption of sodium ions from the lumen of a kidney tubule. Based on your knowledge of the body's cell types, you can surmise that this hormone acts on
   a) Neurons.
   b) Muscle cells.
   c) Epithelial cells.
   d) Connective tissue cells.
3. Hormones are secreted
   a) Into the blood.
   b) From exocrine glands.
   c) From endocrine glands.
   d) Both a and b are true.
   e) Both a and c are true.
4. Which cell type is specialized for contraction and generation of force?
   a) Muscle
   b) Epithelial
   c) Connective tissue
   d) Nerve
5. Normal blood glucose level is
   a) 50 mg/dL.
   b) 100 mg/dL.
   c) 50 mg/mL.
   d) 100 mg/mL.
   e) 50 mg/L.

Objective Questions

6. The body's internal environment is synonymous with (extracellular fluid/intracellular fluid).
7. Maintenance of constant conditions in the internal environment is known as ______.
8. Plasma is extracellular fluid. (true/false)
9. In homeostasis, *all* physiological variables are regulated to stay constant. (true/false)
10. The protein elastin is found in (epithelial/connective) tissue.
11. Plasma and interstitial fluid are identical in composition. (true/false)
12. A hormone that causes movement of glucose from plasma to cells is ______.

Essay Questions

13. Draw a tree diagram illustrating the relationship between total body water, extracellular fluid, intracellular fluid, interstitial fluid, and plasma.
14. Suppose that molecules of glucose are ingested, absorbed into the bloodstream, and then converted inside muscle cells to carbon dioxide, which is eliminated through the lungs. Describe the path of these molecules as they travel through the body, being sure to mention the various barriers (epithelia or cell membranes) that must be crossed.

Critical Thinking

15. Diabetes is a growing health concern in the United States. Compare the different types of diabetes in terms of their prevalence, primary cause, and diagnostics. How might health care workers tackle the growing epidemic of diabetes?

Photo credits for icons in feature boxes: Focus on Diabetes: Tanyae/Dreamstime; Clinical Connections: Iofoto/Dreamstime; Understanding Exercise: Pixattitude/Dreamstime; Discovery: Yuri Arcurs/Cutcaster.

Glossary

effectors cells, tissues, or organs that respond to neural or chemical signals; in homeostatic regulatory systems, cells, tissues, or organs that respond to output signals of the integrating center and bring about the final response; also called *effector organs*

endocrine glands glands that secrete hormones into interstitial fluid

epithelia (ep-ih-thee-lee-ah) continuous, sheetlike layers of cells found in the skin and linings of hollow organs; specialized to regulate exchange; singular, *epithelium*

exocrine glands glands that are specialized for the transport of materials from the body's internal environment to the external environment

extracellular fluid (ECF) fluid located outside cells, accounting for one-third of the total body water; synonymous with *internal environment*

homeostasis (hom-ee-oh-stay-sis) maintenance of relatively constant conditions within the body's internal environment

hormones chemical messengers released from endocrine cells or glands into the interstitial fluid, where they then diffuse into the blood and travel to target cells

integrating center group of cells, usually in the central nervous system (CNS) or an endocrine gland, that use sensory information to determine a response and communicate commands to effectors

internal environment fluid that surrounds the cells inside the body, including the fluid in the bloodstream that surrounds blood cells; synonymous with *extracellular fluid*

interstitial fluid (ISF) extracellular fluid that is present outside the blood, and that bathes most of the cells of the body

intracellular fluid (ICF) fluid located inside cells, accounting for two-thirds of total body water

lumen interior compartment of a hollow organ or vessel

muscle fibers muscle cells

negative feedback a type of feedback commonly employed in homeostatic regulatory systems in which the response of a system goes in a direction opposite to the change that set it in motion

neurons specialized cells in the nervous system that communicate via electrical and chemical signals; also known as *nerve cells*

organ group of different tissues put together to form a structure with a specific function

organ system group of organs that work together to accomplish a task

plasma liquid in the blood made up of water and dissolved solutes, including proteins; represents approximately 20% of the total volume of extracellular fluid

positive feedback a type of feedback in which the response of a system goes in the same direction as the change that set it in motion

set point the normal or desired value of the regulated variable in a homeostatic regulatory system

tissue a group of similar cells that carry out a specific function

total body water (TBW) the volume of water that is contained in all the body's compartments

Answers

Figure Question Answers

Fig 5 Interstitial fluid

Apply Your Knowledge Answers

Hemorrhage causes a decrease in blood volume. Body fluids compensate for this effect immediately by shifting interstitial fluid into the plasma. This influx causes blood volume to increase, which helps maintain a normal blood pressure. However, the increase in total blood volume involves only an increase in the plasma, not the blood cells. It takes several days for the number of blood cells to return to normal.

Answers to End-of-Chapter Multiple Choice and Objective Questions

Multiple Choice Questions

1. e 2. c 3. e 4. a 5. b

Objective Questions

6. extracellular fluid
7. homeostasis
8. true
9. false
10. connective
11. false
12. insulin

The Cell
Structure and Function

The Cell
Structure and Function

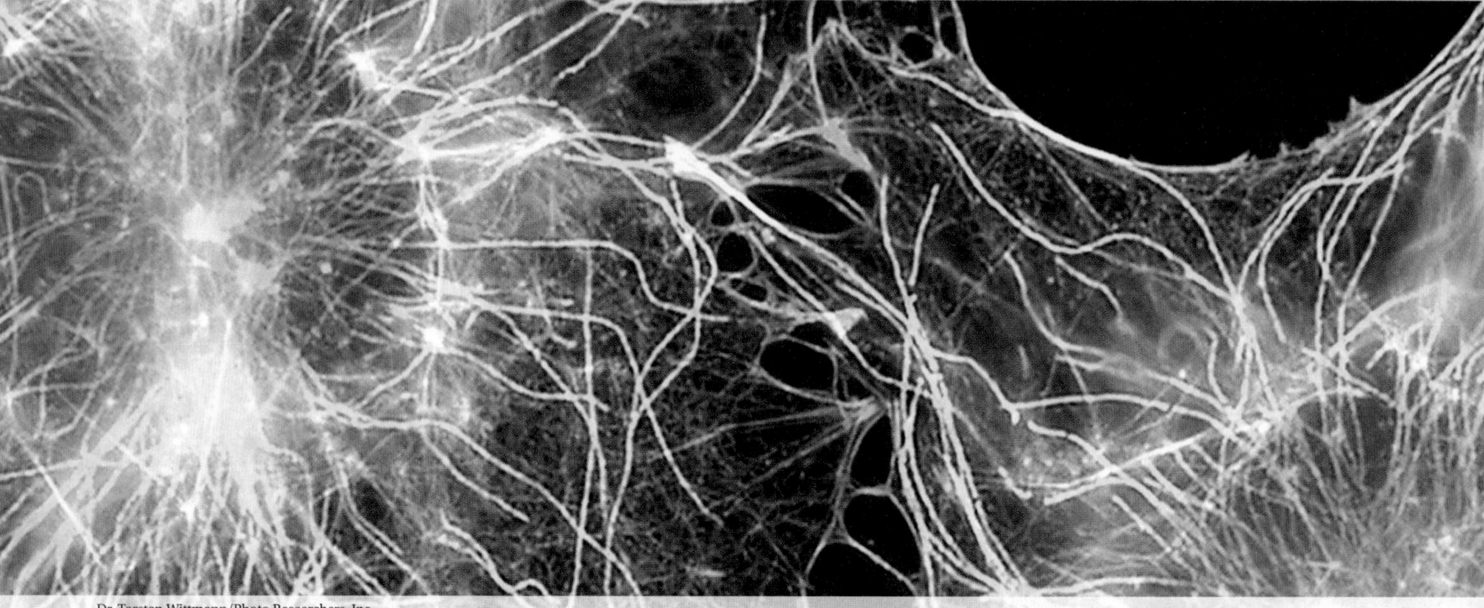

Dr. Torsten Wittmann/Photo Researchers, Inc.

Colorized electron micrograph of two nerve cells.

If you've ever taken a drop of pond water and looked at it under a microscope, then you've probably observed several single-celled organisms. These organisms take up nutrients from their environment, break them down, and convert them to usable energy or to substrates required to synthesize large molecules needed for life. They release waste products into the environment, are motile in water, and respond to various types of stimuli. Lastly, these organisms can grow and reproduce.

The cells of your body are remarkably similar to these unicellular organisms. Your cells require nutrients for energy and for synthesizing needed molecules. Your cells generate waste products that must be eliminated. Some of your cells are motile, and most of your cells can grow and reproduce. However, unlike the unicellular organisms, your cells are part of a multicellular organism—you. As such, your cells must carry out coordinated activities, with different types of cells specialized for certain functions. The ability of organ systems to carry out their functions depends on the cells that form them.

CHAPTER OUTLINE

LEARNING OUTCOMES *After studying this chapter, you should be able to:*

- Know the general characteristics of the four major classes of biomolecules (carbohydrates, proteins, lipids, and nucleic acids), and give a brief description of their functions in cells.
- Describe the structure and major functions of each of the following cellular components: plasma membrane, nucleus, ribosomes, rough endoplasmic reticulum, smooth endoplasmic reticulum, Golgi apparatus, mitochondria, lysosomes, peroxisomes, cytoskeleton.
- Define *transcription* and *translation,* and describe the role of each of the following in protein synthesis: DNA, genes, codons, genetic code, messenger RNA, transfer RNA, ribosomes, anticodons, rough endoplasmic reticulum.
- Explain how genetic information is stored in DNA, and how this information is passed on to daughter cells during mitosis.
- Briefly describe what happens to proteins following their synthesis, taking into account the different fates of membrane proteins, secreted proteins, and cytoplasmic proteins.

Before You Begin

Make sure you have mastered the following topics:

- Cell types
- Body fluid compartments

To understand organ systems, we must understand their basic units: cells. This chapter describes the basic structures and functions of cells. To understand cells and their functions, we must also understand the molecules that form cells and that interact with cells. Most of these molecules are *biomolecules.*

1 Biomolecules

Biomolecules are molecules that are synthesized by living organisms and contain carbon atoms (**Chemistry Review: Atoms and Molecules**). Carbon has four electrons in its outer shell and forms covalent bonds by sharing these electrons with other atoms, including other carbon atoms. Other atoms common to biomolecules include oxygen, hydrogen, and nitrogen. The ability of carbon to share four electrons and to form covalent bonds with other carbon atoms enables carbon-containing molecules to be large and complex. In biomolecules, the component carbon atoms are often arranged in chains or rings.

The four basic types of biomolecules are *carbohydrates, lipids, proteins,* and *nucleotides*. Some biomolecules are *polymers,* which consist of repeated subunits. For example, proteins are polymers of *amino acids*. The following sections discuss each class of biomolecule. Table 1 lists some chemical functional groups commonly found in biomolecules.

TABLE 1 Common Functional Groups Found in Biomolecules

Functional group	Chemical formula	Structure	Chemical property
Hydroxyl	—OH	—O—H	Polar
Sulfhydryl	—SH	—S—H	Polar
Phosphate	$—HPO_4^-$	—O—P(=O)(—O⁻)—OH	Polar
Carboxyl	—COOH	—C(=O)—OH	Acid
Amino	$—NH_2$	—N(—H)—H	Base

Carbohydrates

Carbohydrates are composed of carbon, hydrogen, and oxygen in a ratio of 1:2:1, with the general chemical formula $(CH_2O)_n$. The chemical formula can also be written as $C_n(H_2O)_n$, which might be interpreted as *hydrated* carbons, or carbons surrounded by water—thus the name *carbohydrates.* However, this name can be deceiving because the carbons actually bond to hydroxyl groups (—OH) and hydrogens (—H), not to water molecules. The presence of several hydroxyl groups makes carbohydrates polar molecules, so they readily dissolve in water (**Chemistry Review: Polar Molecules and Hydrogen Bonds**).

Monosaccharides, Disaccharides, and Polysaccharides

Carbohydrates can be further classified into three major groups: (1) monosaccharides, (2) disaccharides, and (3) polysaccharides. **Monosaccharides** are simple sugars, composed of a single unit (**Figure 1a**). The most common monosaccharide is **glucose,** which is an important source of energy for our cells. Glucose has the general formula $C_6H_{12}O_6$. Two other common monosaccharides, *fructose* and *galactose,* also have the general formula $C_6H_{12}O_6$ and can be converted to glucose in our cells. They are distinct molecules, however—the atoms are arranged differently in these monosaccharides, giving each its own chemical properties. *Ribose* and *deoxyribose* are two other common monosaccharides that are important components of *nucleotides,* another class of biomolecules that are described later.

Disaccharides are carbohydrates formed by the covalent bonding of two monosaccharides (Figure 1b). Common examples of disaccharides include *sucrose,* which is composed of a glucose subunit and a fructose subunit joined together, and *lactose,* which

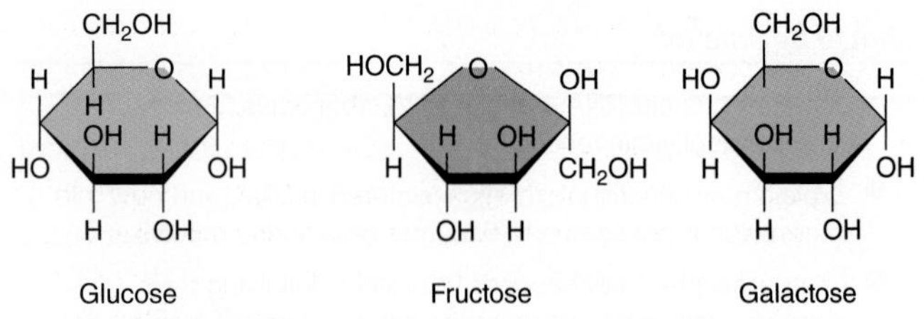

(a) Monosaccharides

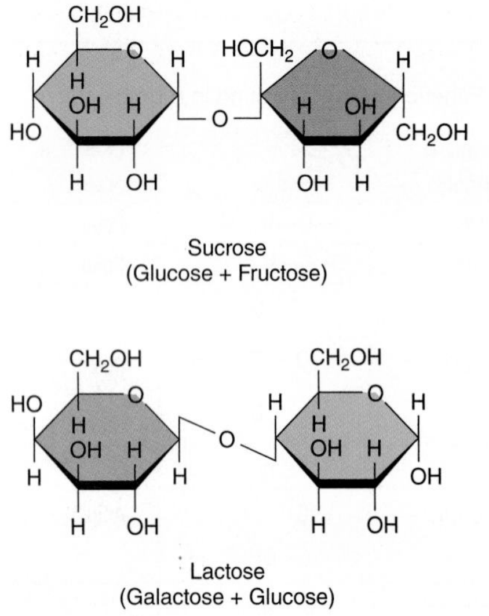

(b) Disaccharides

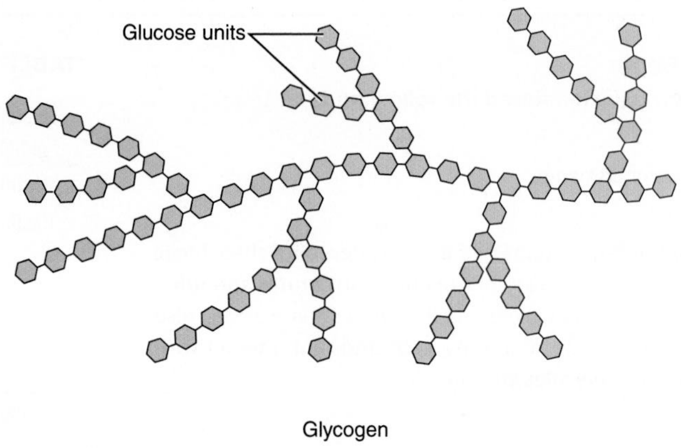

(c) Polysaccharide

Figure 1 Carbohydrates. (a) Examples of three monosaccharides with the same chemical formula: $C_6H_{12}O_6$. **(b)** Examples of two disaccharides. Sucrose, or table sugar, consists of a glucose subunit and a fructose subunit. Lactose, a carbohydrate found in milk, consists of a glucose subunit and a galactose subunit. **(c)** A polysaccharide, which is a polymer of a monosaccharide. Glycogen is a glucose polymer found in animals.

Q How many hydroxyl groups are in a single glucose molecule?

is composed of a glucose subunit and a galactose subunit joined together. Sucrose is also known as table sugar, whereas lactose is a carbohydrate found in milk. Notice that the names of the monosaccharides and disaccharides all end in *-ose.*

Polysaccharides are carbohydrates formed by the covalent bonding of several monosaccharides. **Glycogen** (Figure 1c) is a polymer of glucose subunits and is found in animal cells. Several types of cells in the body can store glucose as glycogen and later break glycogen down when they need glucose for energy. Glycogen is needed for a simple reason: If cells stored energy in the form of monosaccharides, the monosaccharides would exert an osmotic force that would draw water into the cells, making the cells swell and possibly burst. By comparison, a single large storage molecule will not draw in nearly as much water. **Starch** is a polysaccharide found in plants. Humans consume starch in various plant food products; the digestion process makes the glucose subunits of starch available as energy sources. **Cellulose,** another polysaccharide found in plants, is consumed by humans, but we are unable to digest and absorb it. Therefore, cellulose, also known as dietary fiber, passes through our gastrointestinal system. In addition to storing energy, polysaccharides are important components of cell membranes, a topic to be described later.

Synthesis and Breakdown of Saccharides

To synthesize disaccharides and polysaccharides, monosaccharides must be linked together by covalent bonds. The type of reaction that links monosaccharides together is called **condensation,** reflecting the fact that water is produced in the process. For example, when a glucose subunit and a fructose subunit react together to form sucrose, water is released. Similarly, each time a glucose subunit is added to glycogen, a molecule of water is produced.

In contrast, when disaccharides and polysaccharides must be broken down to their monosaccharide subunits, they are reacted with water. The water splits (or lyses) the larger molecule into its smaller components by a process called **hydrolysis.** During the lysis, a hydrogen atom from the water binds to one of the products, while the hydroxyl group binds to the other product of the reaction. We will see that condensation and hydrolysis are used in the synthesis and breakdown of the other biomolecules.

Lipids

Lipids are a diverse group of biomolecules that contain primarily carbon and hydrogen atoms linked together by nonpolar covalent bonds. Therefore, lipids generally are nonpolar molecules and do not dissolve in water. Most lipids also contain some oxygen, and several contain phosphates (—HPO_4^-), which, depending on the structure, may provide areas of the molecule that are polar. A molecule that contains both polar and nonpolar regions is called **amphipathic.** The five main classes of lipids vary both structurally and functionally: (1) *triglycerides,* (2) *ketones,* (3) *phospholipids,* (4) *eicosanoids,* and (5) *steroids.*

Triglycerides

Triglycerides—what we typically call "fats"—are composed of one glycerol molecule and three fatty acid molecules. **Glycerol** is a three-carbon alcohol (**Figure 2a**) that functions as the "backbone" of a triglyceride; **fatty acids** are long chains of carbon atoms with a carboxyl group (—COOH) at one end (Figure 2b and c). **Triglycerides** are formed by linking each of three fatty acids to a different carbon in the glycerol backbone (Figure 2d).

Most fatty acids have an even number of carbon atoms, most commonly 16 or 18. An important feature of fatty acid chains is the number of double bonds between carbons. If there are no double bonds, then each carbon is linked to a maximum number of hydrogen atoms and, therefore, is *saturated* with hydrogen atoms. Thus *saturated fatty acids* contain carbons linked only by single bonds (Figure 2b). In contrast, *unsaturated fatty acids* contain one or more pairs of carbons linked by double bonds and, therefore, have fewer hydrogens per carbon (Figure 2c). A *monounsaturated fatty acid* contains exactly one double-bonded pair of carbons, whereas a *polyunsaturated fatty acid* contains more than one double-bonded

(a) Glycerol

(b) Saturated fatty acid

Oleic acid, a *cis* fatty acid

Eladeic acid, a *trans* fatty acid

(c) Unsaturated fatty acid

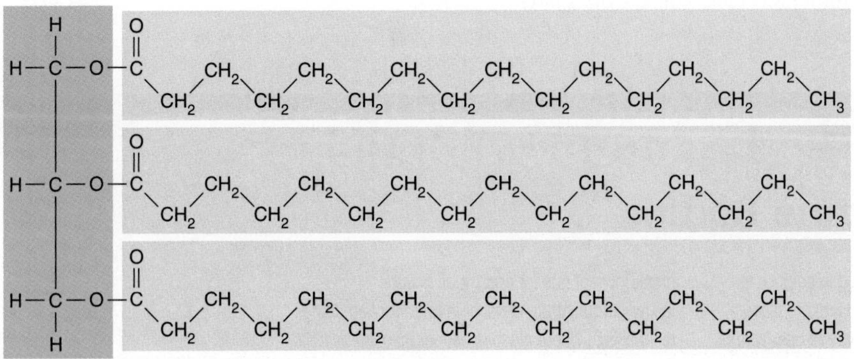

(d) Triglyceride

Figure 2 Triglycerides. (a) Glycerol, a three-carbon alcohol found in both triglycerides and phospholipids. **(b)** A fatty acid chain, which usually consists of an even number of carbon atoms with a carboxyl group (—COOH) on the end. This example is a saturated fatty acid because it has no double bonds between carbon atoms. **(c)** This fatty acid is unsaturated because of the presence of the double bond between carbons 9 and 10. Because it contains only one double bond, it is a monounsaturated fatty acid. **(d)** A triglyceride, which is composed of a glycerol backbone and three fatty acids.

Q Each of the fatty acids in Figure 2(b) and (c) has 16 carbons. Which of the two fatty acids has more hydrogens?

CHEMISTRY REVIEW

Atoms and Molecules

All matter is composed of fundamental units called *atoms,* which are on the order of one-billionth of a centimeter in diameter. Atoms, in turn, are made up of three types of elementary particles: (1) *protons,* which each carry one unit of positive electrical charge; (2) *electrons,* which each carry one unit of negative electrical charge; and (3) *neutrons,* which carry no charge. Normally, an atom possesses equal numbers of protons and electrons, giving it a net charge of zero. The protons and neutrons are densely concentrated at the atom's center in a core called the *nucleus,* while the electrons travel in orbits or *shells* located at various distances from the nucleus.

Atoms are distinguished from one another on the basis of their *atomic number,* which equals the number of protons (and, therefore, also the number of electrons) they possess. The atomic number also determines the chemical properties of an atom. Pure substances consisting entirely of atoms having the same atomic number are referred to as *elements.* Although more than 100 elements are known, just 4 account for more than 99% of all the atoms in the body: *hydrogen (H), carbon (C), nitrogen (N),* and *oxygen (O),* whose atomic numbers are 1, 6, 7, and 8, respectively.

Most substances are composed of two or more atoms linked or *bonded* together to form *molecules.* Molecules of water—the most abundant substance in the body—contain two hydrogen atoms and one oxygen atom and are designated by the formula H_2O. Carbon dioxide, a waste product generated by cells, contains two oxygen atoms and one carbon atom and is designated as CO_2. Most often, a molecule's atoms are held together by *covalent bonds,* which consist of pairs of electrons shared between adjacent atoms. Because all atoms of a given type possess the same number of sharable electrons, each element has a certain capacity for forming bonds. Hydrogen atoms form only a single bond, whereas oxygen, nitrogen, or carbon can form two, three, or four bonds, respectively:

H· ·O· ·N· ·C·

As long as each atom forms the correct number of bonds, the atoms can be combined to form a virtually unlimited variety of molecules. Molecules of water, hydrogen (H_2), and methane or *natural gas* (CH_4) can be represented as follows:

H:O:H	H:H	H:C:H (with H above and below C)
Water	Hydrogen	Methane

Sometimes atoms share two pairs of electrons, forming *double bonds.* This is illustrated in the following representation of a molecule of carbon dioxide:

O::C::O

Carbon dioxide

pair of carbons. The double bonds can also be described as *trans,* indicating that the single hydrogen and the carbon bonded with it are on opposite sides of the chain, or *cis,* indicating that the single hydrogen and the carbon bonded with it are on the same side of the chain (Figure 2d). The degree of saturation and the *trans* or *cis* nature of the double bonds determine important properties of a lipid, including some with significant clinical implications. *Cis* fatty acids are generally produced by natural processes, whereas *trans* fatty acids are produced artificially. For example, saturated fatty acids and *trans* fatty acids are implicated in the development of plaques that can clog arteries, which can lead to stroke or heart attack.

Triglycerides and fatty acids are nonpolar molecules because of the presence of nonpolar carbon-to-carbon and carbon-to-hydrogen bonds. They do not dissolve in water, but they readily dissolve in nonpolar solvents such as oil or benzene.

FOCUS ON DIABETES

Ketone Bodies

When triglycerides are hydrolyzed in adipose tissue, free fatty acids are released into the blood as an immediate alternative source of energy (the primary energy source for cells is glucose, when it is plentiful). Some fatty acids, however, travel to the liver where they are converted to **ketone bodies** or **ketones** (see **Figure 3**). Ketone bodies include *acetoacetic acid, α-hydroxybutyric acid,* and *acetone.* When glucose levels are low, requiring the rapid breakdown of fats for energy, as occurs in low-carbohydrate diets, or in uncontrolled diabetes mellitus, where glucose cannot be used by cells for energy, then ketone bodies are rapidly produced and blood levels of these acids become elevated, producing a condition called *ketosis.* This buildup of acids can result in a *ketoacidosis,* a serious problem in diabetes mellitus. Another common, but not severe consequence of ketosis is "acetone breath."

Figure 3 The general chemical structure of a ketone (a) and an example of a ketone, acetoacetic acid (b).

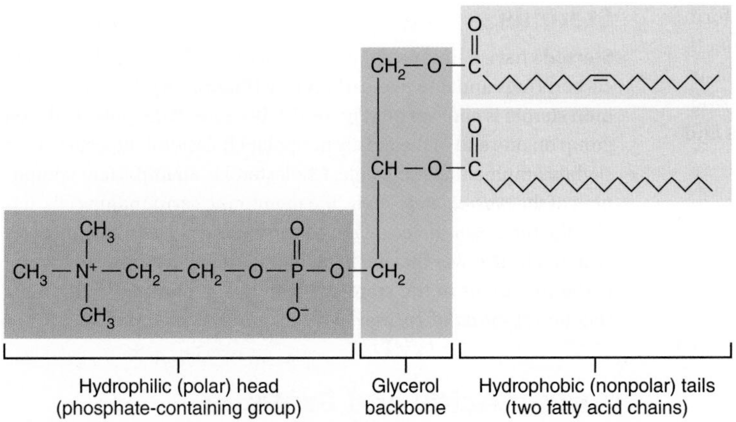

(a) Phospholipid molecule (phosphatidylcholine)

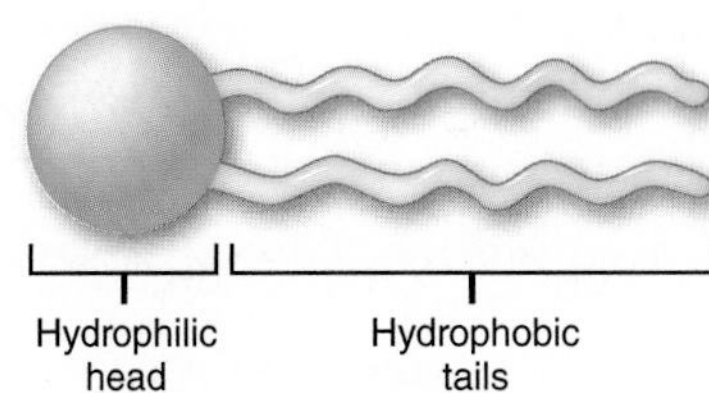

(b) Schematic representation of phospholipid molecule

Figure 4 Phospholipids. (a) A phospholipid, which consists of a glycerol backbone linked to two fatty acid chains and a phosphate-containing group. The phospholipid shown here is *phosphatidylcholine*. **(b)** The standard way to schematically depict a phospholipid, emphasizing the hydrophilic head and hydrophobic tails of the molecule.

Phospholipids

Phospholipids are lipids that contain a phosphate group. They are similar in structure to triglycerides, in that a glycerol forms the backbone. However, instead of three fatty acids, a phospholipid contains two fatty acids and has a phosphate group attached to the third carbon of glycerol (**Figure 4**). The two fatty acids form the *tail* region of the phospholipid, which is nonpolar because of their long chains of carbon atoms. The phosphate group is generally attached to another chemical group, and together they form the *head* region of the phospholipid, which is polar. For example, the phospholipid known as phosphatidylcholine has a choline group attached to the phosphate subunit. Because phospholipids have both a polar region and a nonpolar region, they are amphipathic molecules.

The amphipathic property of phospholipids gives them unique behaviors in an aqueous or watery environment. The polar regions can dissolve in water, but the nonpolar regions cannot. Therefore, when phospholipids are placed in an aqueous environment, the polar regions face the water, and the nonpolar regions face each other. Phospholipids form two physiologically important structures when placed in an aqueous environment: phospholipid bilayers and micelles (**Figure 5**). In a phospholipid bilayer (Figure 5a), which is the core structure of cell membranes, the phospholipids are arranged in two parallel layers: The tails of parallel phospholipids face inward toward each other, and the heads face the outside, where they come into contact with the aqueous environment. A micelle (Figure 5b) is a spherical structure composed of a single layer of phospholipids; it functions in the transport of nonpolar molecules in an aqueous environment. The heads of the phospholipids face outward, where they come into contact with the aqueous environment; the tails face inward, forming a hydrophobic interior.

Another benefit of the amphipathic property of phospholipids is that when applied to water, they decrease the attraction between water molecules. *Surfactants* are a type of phospholipid found in the airways of the lungs, where they decrease the attraction between water molecules lining the lung airways. The water attraction creates surface tension that tends to collapse the lungs—an effect that is prevented by surfactant. In premature babies, surfactant is not yet produced in the lung tissue; thus these babies require mechanical ventilation to keep their lungs open.

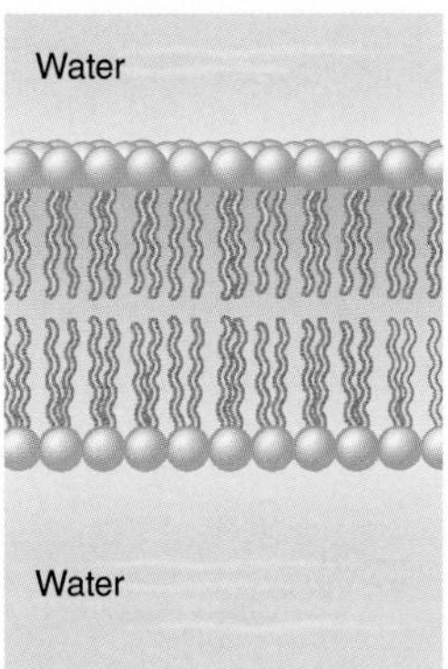

(a) Phospholipid bilayer

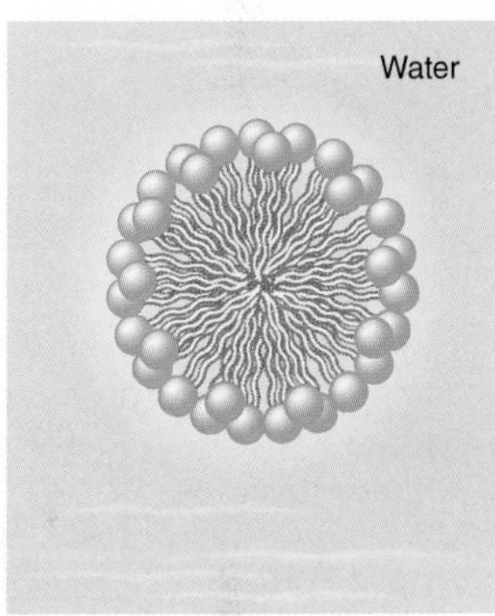

(b) Micelle

Figure 5 Structures formed by phospholipids in an aqueous environment. (a) A lipid bilayer, which consists of two sheets of phospholipids aligned such that the polar heads face the aqueous environment and the nonpolar tails face each other. **(b)** A micelle, a sphere formed by a single layer of phospholipids aligned such that the polar heads face the aqueous environment and the nonpolar tails face each other.

Apply Your Knowledge

Which types of metabolic adjustments occur in a person consuming a high-carbohydrate diet versus a high-fat diet?

Eicosanoids

Eicosanoids are modified fatty acids that function in intercellular communication. An eicosanoid is derived from a 20-carbon polyunsaturated fatty acid, arachidonic acid, that is found in phospholipid components of cell membranes (**Figure 6**). To form eicosanoids, enzymes cause the fatty acid to fold upon itself, forming a five-carbon ring in the middle with two chains of carbon atoms extending parallel to each other away from the ring. Eicosanoids are polar molecules and include *prostaglandins* and *thromboxanes.*

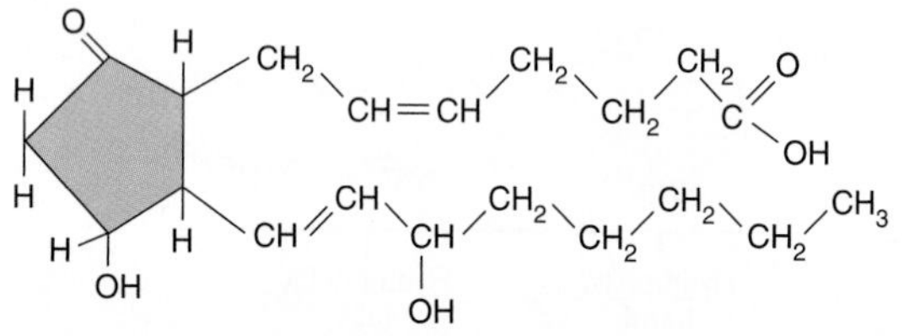

Figure 6 Eicosanoids. In these modified 20-carbon fatty acids, the carbons in the middle of the carbon chain form a ring that causes the molecule to fold upon itself.

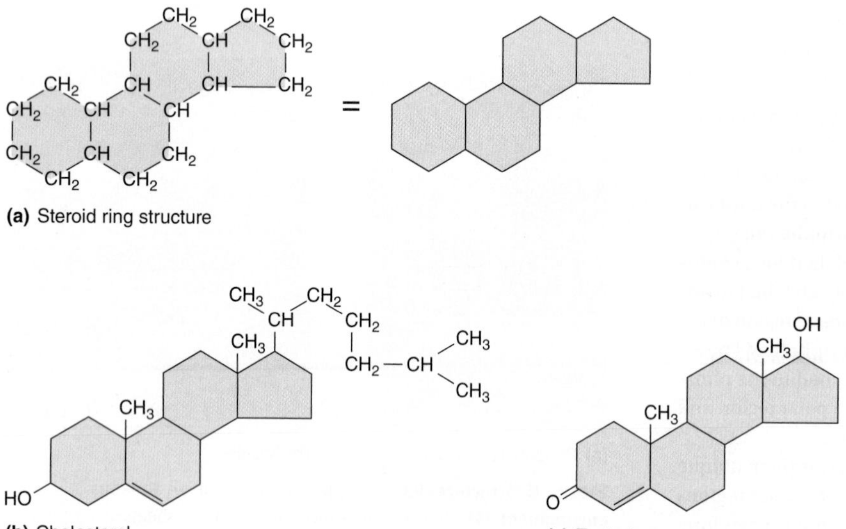

Steroids

Steroids have a unique chemical structure consisting of three six-carbon rings and one five-carbon ring (**Figure 7a**). The most common steroid is cholesterol (Figure 7b). Because of the polar hydroxyl group on one end of the mostly nonpolar cholesterol molecule, it is a slightly amphipathic molecule. Cholesterol is an important component of the *plasma membrane,* the membrane surrounding cells. It is also the precursor molecule for the formation of *bile salts,* molecules secreted by the liver that aid in fat digestion. In addition, cholesterol is the precursor to the steroid hormones, including *testosterone* (Figure 7c), *estradiol, cortisol,* and *1,25-dihydroxyvitamin* D_3.

(a) Steroid ring structure

(b) Cholesterol

(c) Testosterone

Figure 7 Steroids. (a) The basic structure of all steroids, which consists of three six-carbon rings and one five-carbon ring. **(b)** Cholesterol, the most common steroid and the precursor for all other steroids in the body. **(c)** Testosterone, an example of a steroid hormone.

Amino Acids and Proteins

Proteins are polymers of **amino acids,** relatively small biomolecules that contain a central carbon, an amino group, a carboxyl group, a hydrogen, and a residual (R) group (**Figure 8a**). Twenty different R groups exist, so there are 20 different corresponding amino acids. The four amino acids shown in Figure 8b—alanine, tyrosine, glutamate, and lysine—have R groups with different chemical properties. The R group is nonpolar in alanine, a polar aromatic ring structure in tyrosine, acidic in glutamate, and basic in lysine. Although we will discuss amino acids here as components of proteins, they have other functions as well, including playing a role in intercellular communication.

Polymers of amino acids are formed by joining two amino acids together with a *peptide bond;* these polymers are, therefore, called **polypeptides.** A peptide bond forms between the amino group of one amino acid and the carboxyl group of another amino acid by a condensation reaction (**Figure 9**). Polypeptides vary in length from just two to several hundred amino acids; the name given to them

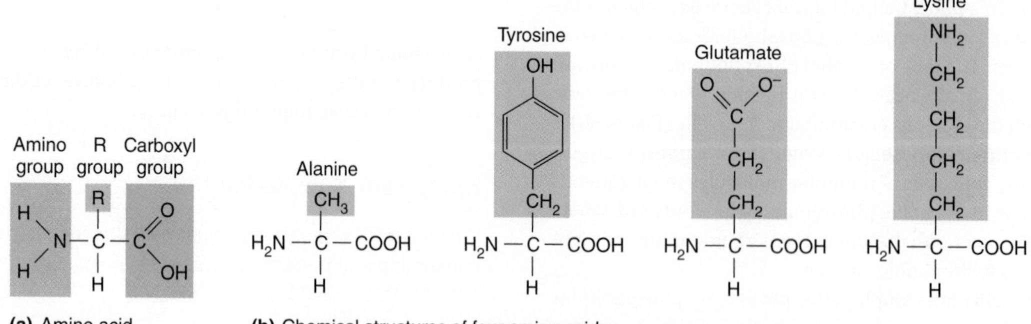

(a) Amino acid

(b) Chemical structures of four amino acids

Figure 8 Amino acids. (a) The basic structure of an amino acid. The central carbon is bonded to an amino group, a carboxyl group, a hydrogen, and an R group. **(b)** Structures of four of the 20 amino acids. The R group is nonpolar in alanine, a polar aromatic structure in tyrosine, acidic in glutamate, and basic in lysine.

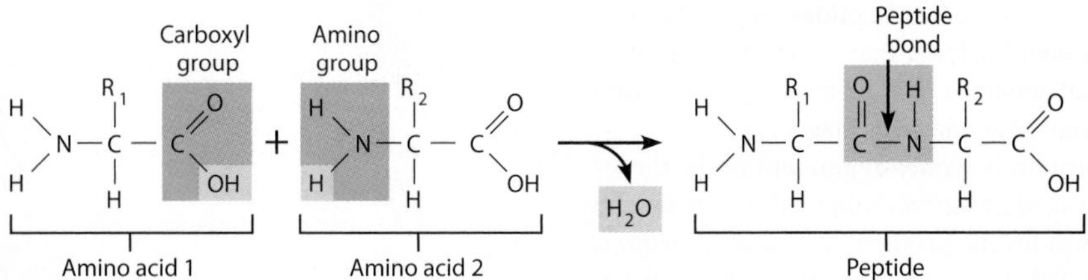

Figure 9 Formation of a peptide bond by a condensation reaction. The peptide bond is formed between two amino acids, and water is released.

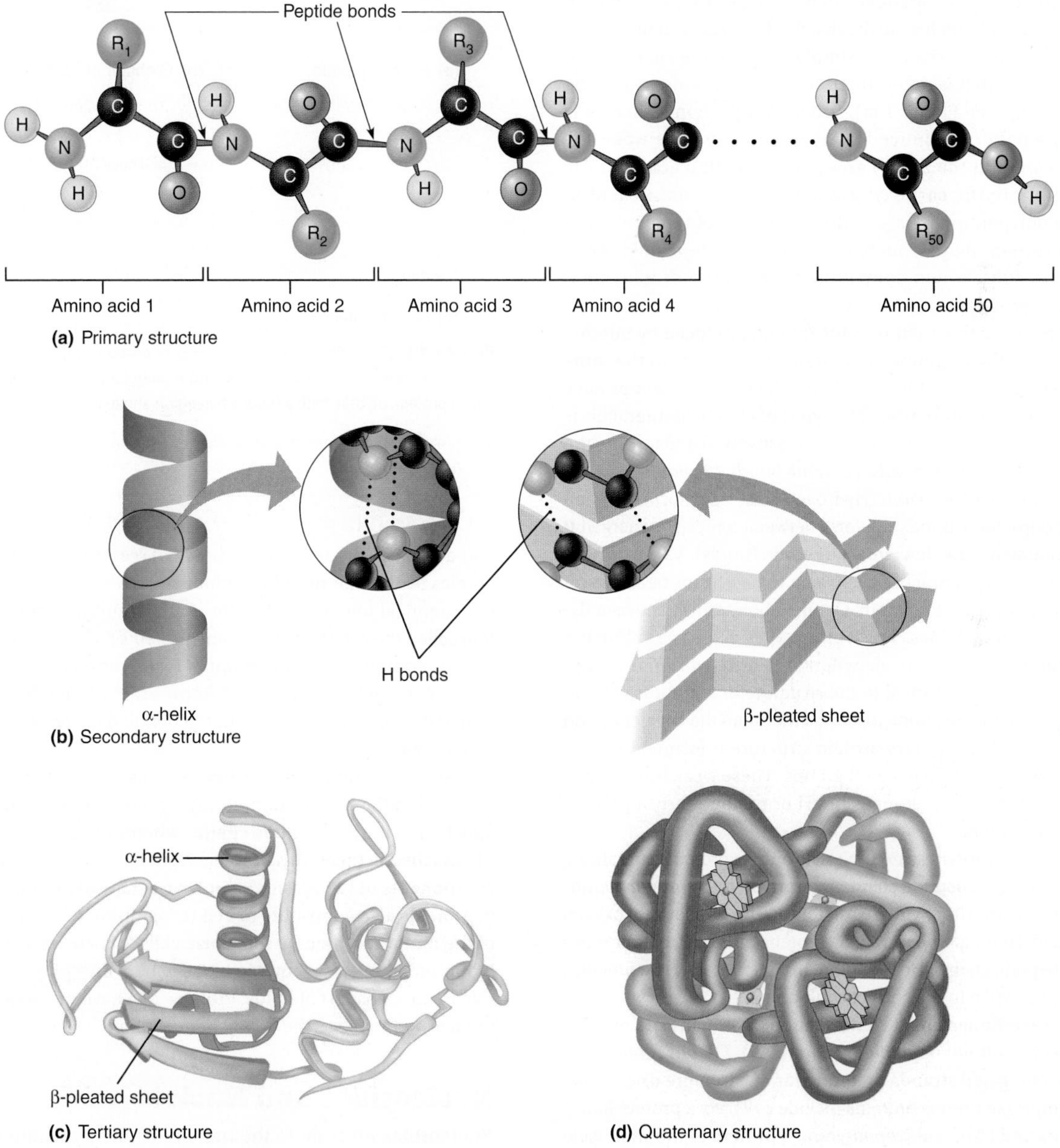

(a) Primary structure

(b) Secondary structure

(c) Tertiary structure

(d) Quaternary structure

Figure 10 Levels of protein structure. (a) Primary protein structure, which is the sequence of amino acids. **(b)** Secondary protein structure, which is caused by hydrogen bonding between the amino hydrogen of one amino acid and the carboxyl oxygen of another amino acid. Common secondary structures include α-helixes and β-pleated sheets. **(c)** Tertiary protein structure, which is the folding pattern produced by interactions between the R groups of amino acids. The protein shown here is the enzyme *lysozyme*. **(d)** Quaternary protein structure, which is the arrangement of more than one polypeptide chain in a single protein. Shown here is *hemoglobin*, which consists of four polypeptide chains.

differs based on their length or function. **Peptides** are short chains of amino acids, usually containing fewer than 50 amino acids. Proteins consist of chains that generally include more than 50 amino acids and often are hundreds of amino acids long.

The function of a protein is highly dependent on its three-dimensional structure or *conformation.* Protein structure can be described at four different levels: *primary, secondary, tertiary,* and *quaternary* (**Figure 10**). The levels of protein structure can be likened to a telephone receiver's cord: Primary structure is the cord stretched out straight, secondary structure is the coiled cord, tertiary structure is the loops and bends in the coiled cord, and quaternary structure is the existence of two cords wrapped together. Now let's relate this analogy to the details of protein structure.

Primary protein structure simply refers to the sequence of amino acids, which is determined by peptide bonds within the peptide chain (see Figure 10a). Secondary protein structure is the folding pattern produced by hydrogen bonds between the hydrogen atom in the amino group of one amino acid and the oxygen atom in the carboxyl group of another amino acid in the same polypeptide. These hydrogen bonds fold the polypeptide into various shapes, such as *α-helixes* and *β-pleated sheets* (see Figure 10b), which contribute to the three-dimensional structure of proteins.

Tertiary protein structure is the folding produced by interactions between the R groups of different amino acids in the same polypeptide (see Figure 10c). Recall that the various R groups have different chemical properties. The types of chemical interactions that can occur depend on the amino acids involved and include the following: (1) *hydrogen bonds,* (2) *ionic bonds,* (3) *van der Waals forces,* and (4) *covalent bonds.* Hydrogen bonds can form between polar R groups. Ionic bonds can form between ionized or charged R groups (**Chemistry Review: Ions and Ionic Bonds**). Van der Waals forces are electrical attractions between the electrons of one atom and the protons of another atom. Covalent bonds form between the R groups of two cysteines. The R group of cysteine is a sulfhydryl group, and the resulting covalent bond is called a *disulfide bridge.* Disulfide bridges are critical to the structure of antibodies and to the structure of the hormone insulin. Other than the covalent bond of disulfide bridges, tertiary protein structure is formed by weak chemical interactions between R groups. These weak interactions are easily disrupted by changes in pH or temperature, a process called *denaturation.*

Quaternary protein structure exists only in proteins containing more than one polypeptide chain. An example is the protein hemoglobin (see Figure 10d), which functions in the transport of oxygen in the blood. Hemoglobin is a single protein consisting of four separate polypeptide chains. Hemoglobin, like all proteins, can function properly only when in the correct conformation.

The three-dimensional conformation of proteins can be classified as either *fibrous* or *globular* (**Figure 11**). *Fibrous proteins* are generally elongated strands that function in structure or contraction. Examples of fibrous proteins include *collagen,* a protein found in tendons and hair, and *tropomyosin,* a protein found in muscle cells. *Globular proteins* are irregular and bulky. Among their many functions, they act as *chemical messengers* for intercellular communication, as *receptors* that bind chemical messengers, as *carrier proteins* that transport substances in blood or across membranes,

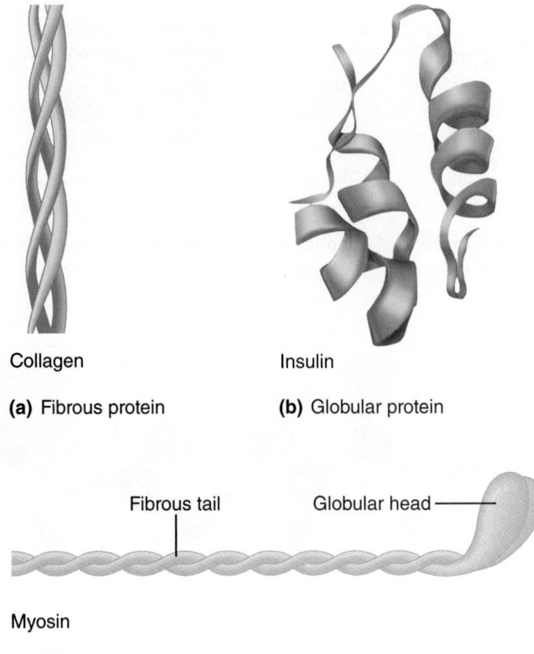

Figure 11 Three-dimensional structures of proteins. (a) A fibrous protein, collagen, which consists of thin strands. **(b)** A globular protein, insulin. **(c)** A mixed protein, myosin, with a fibrous tail region and a globular head region.

Q Which types of chemical forces cause the insulin molecule to fold into its three-dimensional structure?

and as *enzymes* that catalyze chemical reactions in the body. Examples of globular proteins include the oxygen binder *myoglobin,* the chemical messenger *insulin,* and a membrane transport protein called the *Na^+/K^+ pump.* Some proteins have both globular and fibrous components; one example is the muscle protein *myosin,* which has a globular head and fibrous tail. All the proteins mentioned here are discussed further in detailed descriptions of organ systems later in the text.

Some proteins have other classes of organic molecules attached to them. For example, *glycoproteins* have carbohydrates attached to the polypeptide chains, whereas *lipoproteins* have lipids attached to them. Glycoproteins and lipoproteins are important components of the plasma membrane that surrounds cells. Glycoproteins also contribute to cell recognition, which is the ability of the immune system to recognize cells that are part of the body. Lipoproteins are also found in blood, where they play an important role in the transport of lipids. Proteins can have metals attached to them as well, such as the iron in hemoglobin.

Nucleotides and Nucleic Acids

Nucleotides function in the transfer of energy within cells, and they form the genetic material of the cells. The basic structure of a nucleotide is shown in **Figure 12a**. **Nucleotides** contain a five-carbon carbohydrate, a nitrogenous base, and one or more phosphate groups. The carbohydrates found in nucleotides are ribose

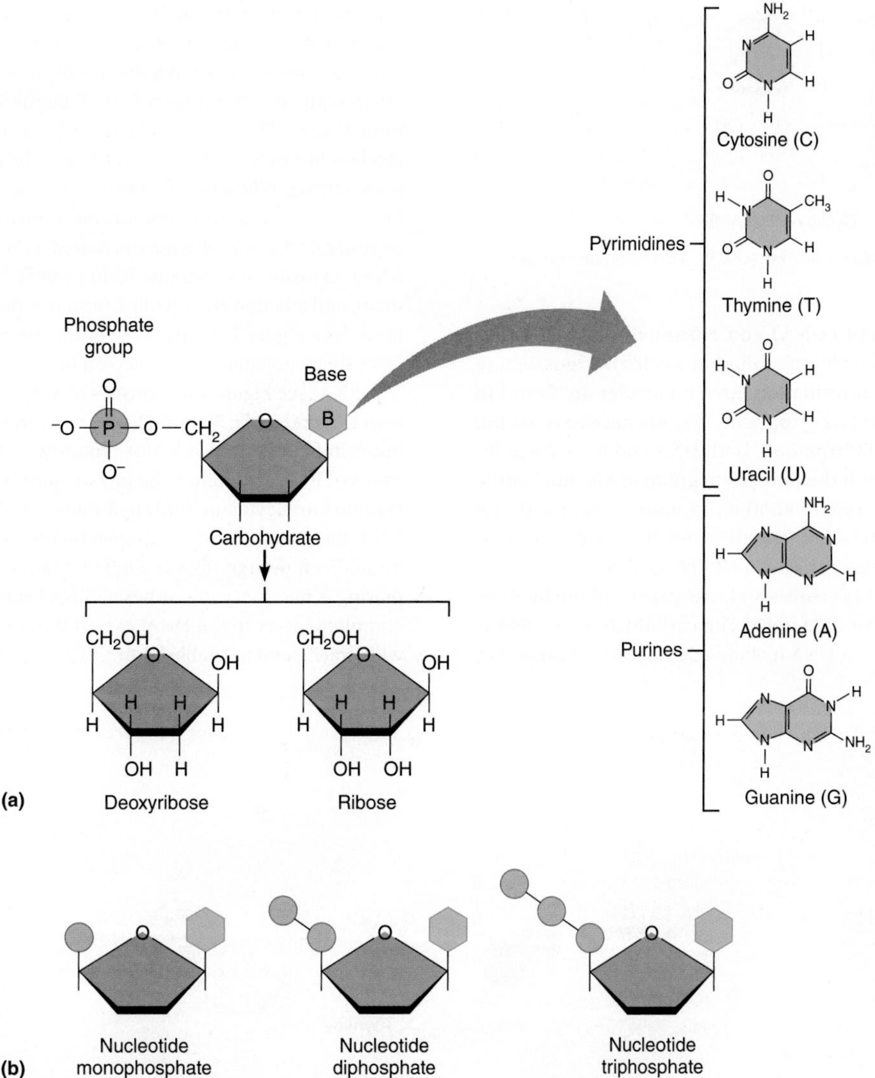

Figure 12 Nucleotides. (a) The basic structure of a nucleotide. The carbohydrate in a nucleotide is either deoxyribose or ribose. The five possible bases include the pyrimidines, which contain a single ring, and the purines, which contain two rings. **(b)** A schematic representation of nucleotides. Nucleotide monophosphates contain a single phosphate group, nucleotide diphosphates contain two phosphate groups, and nucleotide triphosphates contain three phosphate groups.

Q What are the chemical formulas for ribose and deoxyribose?

and deoxyribose. The nitrogenous bases in nucleotides include two classes: (1) *pyrimidines,* which contain a single carbon ring and include *cytosine, thymine,* and *uracil,* and (2) *purines,* which contain a double carbon ring and include *adenine* and *guanine.* A nucleotide that has one phosphate subunit is called a *nucleotide monophosphate* (Figure 12b); similarly, a nucleotide with two phosphate subunits is a *nucleotide diphosphate,* and a nucleotide with three phosphate subunits is a *nucleotide triphosphate.*

Several nucleotides play roles in the exchange of cellular energy. For example, the energy contained in the phosphate bonds of *adenosine triphosphate (ATP)* can drive cell processes when ATP is hydrolyzed to *adenosine diphosphate (ADP). Nicotinamide adenine dinucleotide (NAD)* and *flavin adenine dinucleotide (FAD)* transfer energy in the form of electrons.

Some nucleotides form a ring due to covalent bonding between an oxygen atom of the phosphate group and a carbon atom of the carbohydrate group, as shown in **Figure 13**. These nucleotides, called *cyclic nucleotides,* include chemical messengers (molecules that function as signals for the cell to do something) inside cells, such as *cyclic AMP (cAMP)* and *cyclic GMP (cGMP).*

Polymers of nucleotides include the **nucleic acids** that function in the storage and expression of genetic information:

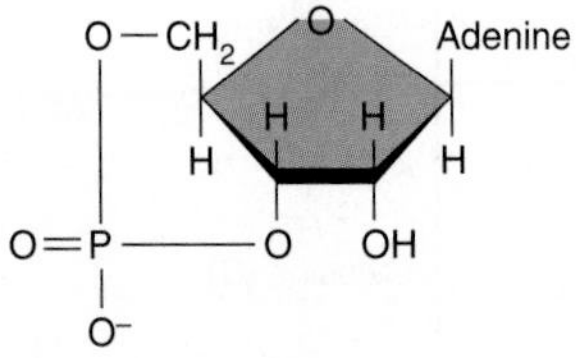

Figure 13 **The nucleotide cyclic AMP (cAMP), a common messenger molecule.**

deoxyribonucleic acid (DNA) and **ribonucleic acid (RNA)** (**Figure 14**). DNA molecules are found in a cell's nucleus, where they store the genetic information. RNA molecules are found in both a cell's nucleus and its cytoplasm; they are necessary for the expression of genetic information. Both DNA and RNA are polymers of nucleotides, with the phosphate group of one nucleotide linked to the carbohydrate of another, forming a chain with the bases sticking out to the sides. Nevertheless, there are important differences in their structures, which we discuss next.

DNA (see Figure 14a) consists of two strands of nucleotides coiled together into a double helix. The carbohydrate in DNA is deoxyribose. The bases in DNA include adenine (A), guanine (G), cytosine (C), and thymine (T). The ends of the strands are labeled as 3′ or 5′, with 3′ at the carbohydrate end and 5′ at the phosphate end. The two strands of DNA in a single molecule run antiparallel to each other, with one strand from 5′ to 3′ aligned with the other strand from 3′ to 5′. The two strands of DNA are held together by hydrogen bonding between bases according to the **law of complementary base pairing**: *Whenever two strands of nucleic acids are held together by hydrogen bonds, G in one strand is always paired with C in the opposite strand, and A is always paired with T in DNA (or with U in RNA).* Cytosine and guanine form three hydrogen bonds between them, and adenine and thymine form two hydrogen bonds between them (see Figure 14b). Because of this pairing, the two strands of DNA are complementary to each other.

RNA (see Figure 14c) consists of a single strand of nucleotides with a 3′ end and a 5′ end. The carbohydrate in RNA is ribose. The bases in RNA include adenine, guanine, cytosine, and uracil. The bases of RNA also follow the law of complementary base pairing: Guanine and cytosine form hydrogen bonds between them, as in DNA, but adenine forms hydrogen bonds with uracil instead of thymine. Even though RNA is single stranded, complementary base pairing is necessary to synthesize RNA from DNA (the RNA will be complementary to the DNA, except that thymine will be replaced with uracil), and to enable folding of a single RNA molecule on itself.

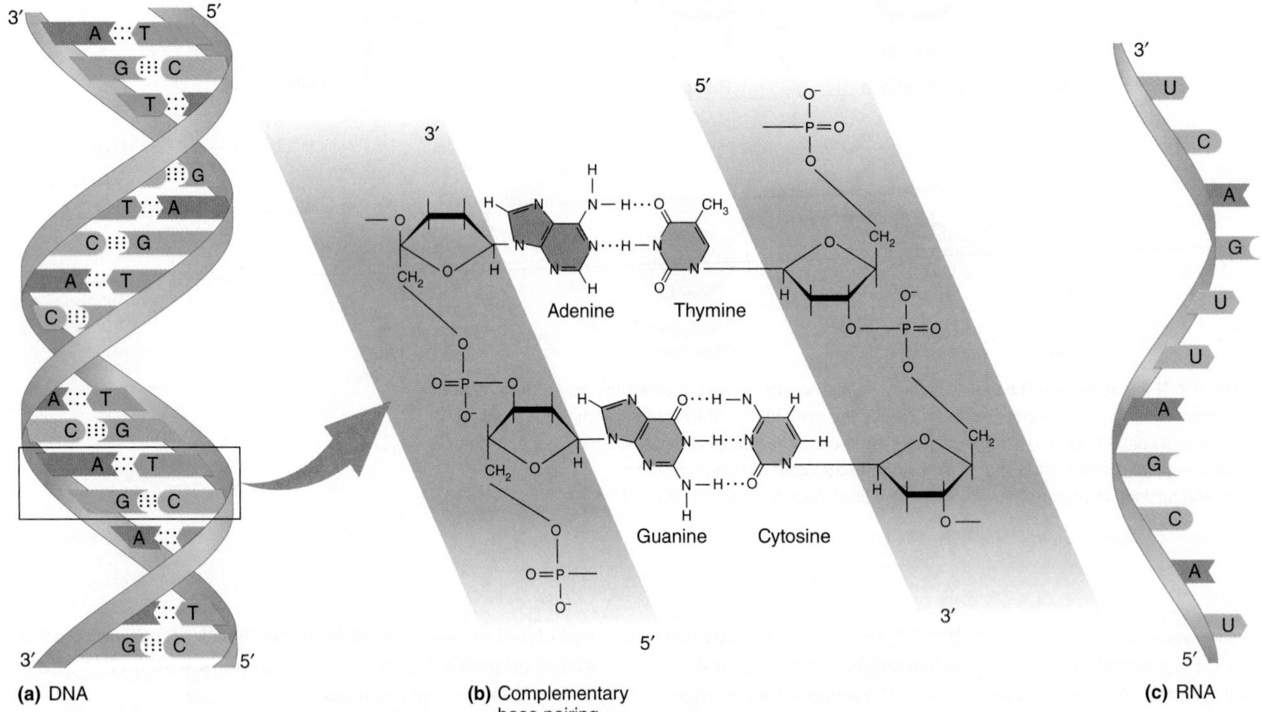

Figure 14 **Nucleic acids. (a)** DNA, which consists of two strands of nucleotides held together by hydrogen bonding according to the law of complementary base pairing. **(b)** Complementary base pairing, with adenine (A) and thymine (T) forming two hydrogen bonds between them, and cytosine (C) and guanine (G) forming three hydrogen bonds between them. **(c)** RNA, which consists of a single strand of nucleotides.

Q Which type of chemical bond links the phosphates and carbohydrates together to form a chain?

Quick Check 1

1. Are carbohydrates generally polar or nonpolar? What about triglycerides?
2. Name the two specialized structures that phospholipids can form in an aqueous environment. Which physiological functions are associated with these structures?
3. Name the subunits that make up the following polymers: glycogen, proteins, and nucleic acids. What are the general functions of each of these polymers?
4. List the four types of chemical interactions responsible for tertiary protein structure in the order of their strength of bonding.

2 Cell Structure

When put together in a certain way, the biomolecules form the basic units of life: cells. The human body contains more than 100 trillion cells that work together to maintain homeostasis. Remarkably, all of these cells are derived from a single fertilized egg. During development, the cells differentiate into more than 200 different types of cells, each with its own specialized function. Although cells can be specialized both anatomically and functionally, they generally have the same basic components. In this section we describe the basic components of a "typical" cell (**Figure 15**).

Each cell is bound by a **plasma membrane,** which separates the cell from the extracellular fluid. Inside the cell are two main components: the **nucleus,** which is a membrane-bound structure

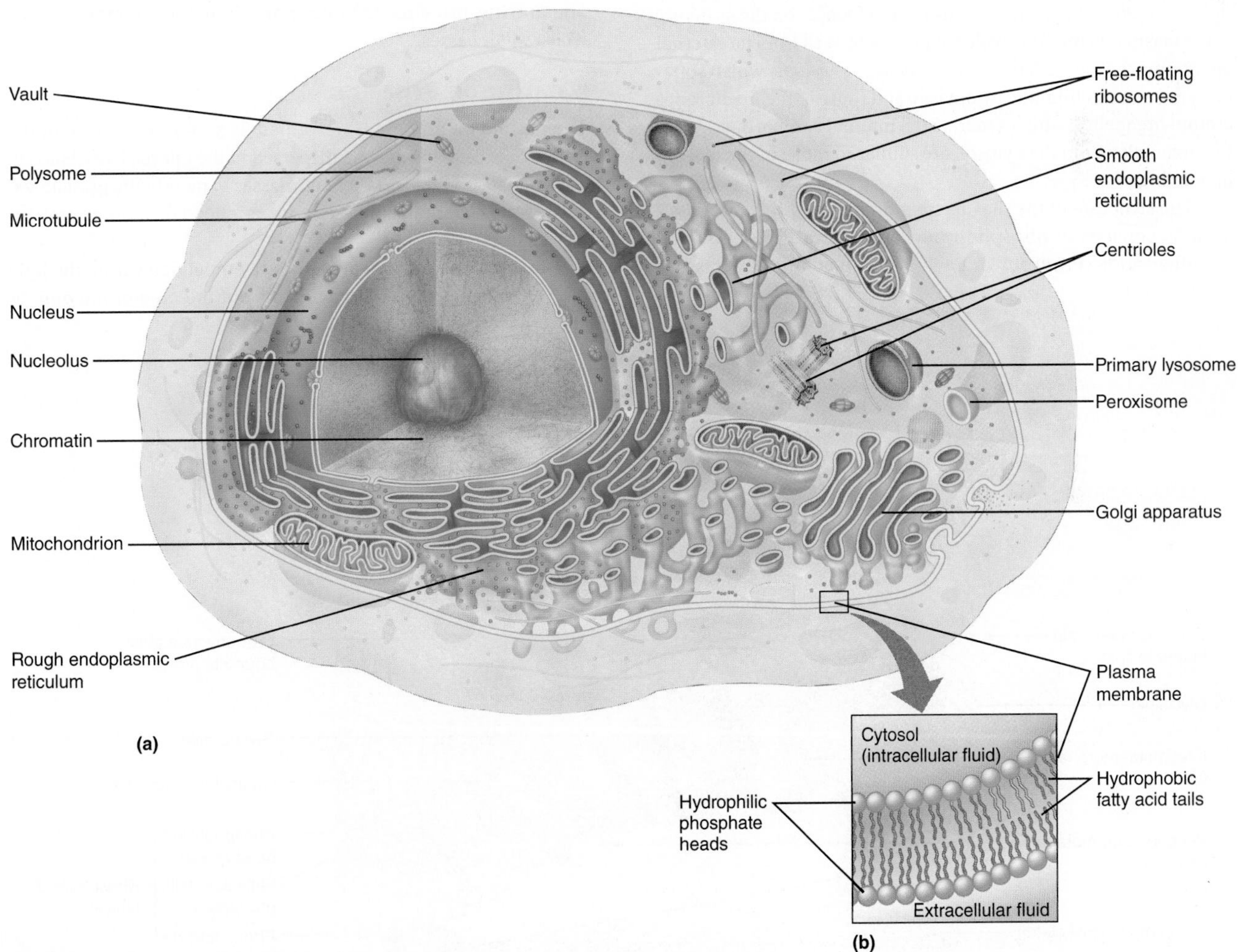

Figure 15 Structures of a typical cell. (a) Three-dimensional view of the cell. The plasma membrane separates the cell from the extracellular fluid. The nucleus contains the genetic information. The cytoplasm contains the organelles and the cytoskeleton, both of which are surrounded by a fluid called cytosol. Organelles include mitochondria, endoplasmic reticulum, Golgi apparatus, lysosomes, peroxisomes, ribosomes, and vaults. **(b)** Enlargement of a portion of the plasma membrane showing the lipid bilayer.

that contains the genetic information for the cell, and the **cytoplasm,** which includes everything inside the cell except the nucleus. The cytoplasm itself consists of two main components: the **cytosol** (the gel-like intracellular fluid) and the **organelles.** The **organelles,** which are structures made up of a variety of biomolecules, carry out specific functions in the cell, much as the organs carry out specific functions in the body. *Membranous organelles* are separated from the cytosol by one or more membranes, and *nonmembranous organelles* have no such boundary with the cytosol.

This section describes the anatomy of a cell; general functions of the various components of a cell are discussed in a subsequent section.

Structure of the Plasma Membrane

The cell contains several types of membranes that function as barriers between compartments. This section focuses on the structure of the plasma membrane, which separates the cell from its external environment. The structures of the *nuclear envelope,* which separates the nucleus from the cytoplasm, and those of the membranes around organelles, which separate the internal compartment(s) of the organelles from the cytosol, are similar to those of the plasma membrane.

The structure of the plasma membrane, described as a *fluid mosaic,* consists of phospholipids, cholesterol, proteins, and carbohydrates (**Figure 16**). Each of these components is described next.

Phospholipid Bilayer

The phospholipids are arranged in a bilayer such that the hydrophilic heads face the aqueous environments inside the cell (the cytosol) and outside the cell (the extracellular fluid), whereas the hydrophobic tails face each other (see Figure 16). This bilayer forms the basic structure of the membrane. The membrane is considered *fluid* because the phospholipids and the other molecules in the membrane are not linked together by chemical bonds and can move about laterally, and they even occasionally move from one side of the bilayer to the other. Cholesterol molecules are found within the lipid bilayer, where they interfere with hydrophobic interactions between phospholipid tails, which could cause crystallization of the bilayer and decrease the fluidity.

The phospholipid bilayer serves as a barrier to the movement of large polar molecules. Although water molecules are polar, they can generally cross the lipid bilayer through water channels called **aquaporins,** proteins that form pores in the membrane through which water passes.

Membrane Proteins

The plasma membrane is described as a *mosaic* because of the presence of proteins that are dispersed in the bilayer, like islands in a sea of phospholipids. Two main classes of membrane proteins exist: integral membrane proteins and peripheral membrane proteins (see Figure 16).

Integral membrane proteins are embedded within the lipid bilayer, so they can be dissociated from the membrane only by

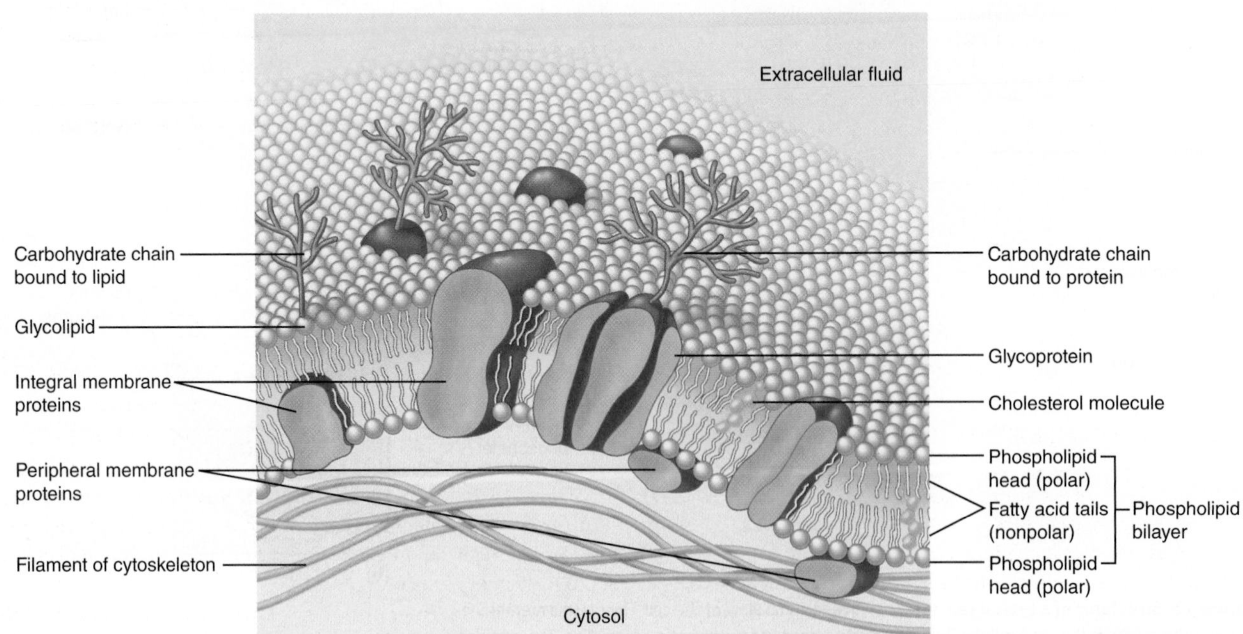

Figure 16 **Plasma membrane.** The flexible plasma membrane is primarily a double layer of phospholipids, with the hydrophobic tails of the phospholipid buried within the bilayer, and the hydrophilic phosphate heads facing the cytosol and extracellular fluid. Integral membrane proteins are scattered throughout the bilayer. Peripheral membrane proteins are associated with the integral membrane proteins, primarily on the side facing the cytosol, where they function as part of the cytoskeleton. Cholesterol molecules are scattered throughout the lipid bilayer. Carbohydrates bound to membrane lipids or membrane proteins face the extracellular fluid. These lipids and proteins are called glycolipids and glycoproteins, respectively.

CHEMISTRY REVIEW

Polar Molecules and Hydrogen Bonds

When two atoms are covalently bonded, they share electrons. However, this sharing may or may not be equal. Certain atoms, such as oxygen (O), nitrogen (N), and sulfur (S), hold on to electrons tightly. Therefore, when bonded to other atoms, they have a tendency to pull electrons away from the other atoms. As a result, each oxygen, nitrogen, or sulfur atom acquires a partial negative charge because it has slightly more than its "fair share" of electrons—that is, it has more electrons than are necessary to balance the positive charge of the protons in its nucleus. In contrast, the atom to which the oxygen, nitrogen, or sulfur is bonded is left with less than its "fair share" of electrons, which gives it a partial positive charge. This situation can be illustrated as follows for molecules in which oxygen, nitrogen, or sulfur is bound to hydrogen, with the R representing the remainder of the molecule and the δ representing a partial charge:

$$\overset{\delta^-}{\text{R:O}}\overset{\delta^+}{\text{:H}} \qquad \text{R:}\overset{\delta^-}{\text{N}}\text{:H} \; (\text{with } \overset{\delta^+}{\text{H}} \text{ below N}) \qquad \overset{\delta^-}{\text{R:S}}\overset{\delta^+}{\text{:H}}$$

Bonds characterized by such unequal electron sharing are known as *polar bonds.* In contrast, in *nonpolar bonds* electrons are shared more or less equally, such that the atoms remain uncharged. Carbon-to-carbon (C—C) and carbon-to-hydrogen bonds (C—H) are common examples of nonpolar bonds found in the molecules of the body.

The presence of electrical charges within a polar molecule produces electrical forces. The direction of electrical forces follows a simple rule: *Opposite charges attract; like charges repel.* Thus the positive region of one polar molecule is electrically attracted to the negative region of another polar molecule (in some cases, the positive and negative regions can be within the same large polar molecule). This electrical attraction holds the two polar molecules together, forming a *hydrogen bond.* The polar bond is called a hydrogen bond because in a polar molecule the hydrogen atom typically contains the partial positive charge. The most common polar molecule is water. Hydrogen bonds between polar water molecules can be represented as shown, above:

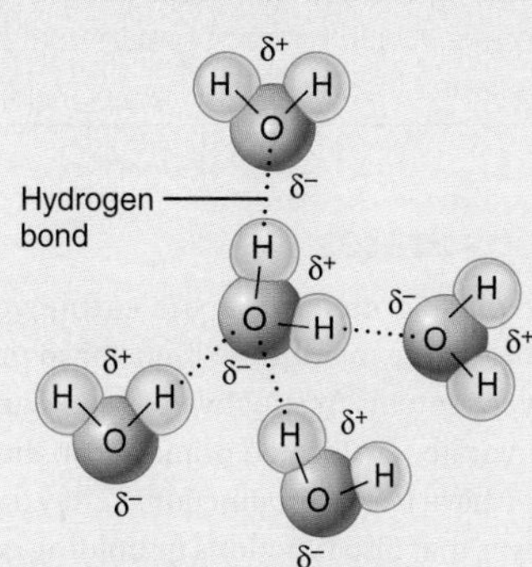

Even though hydrogen bonds are weak bonds that can easily be broken, they are important in determining the structure of large biomolecules such as proteins and nucleic acids, and in establishing the properties of substances in water. How molecules behave in water is critical to cell function. Polar molecules are electrically attracted to the polar water molecules; as a consequence, they dissolve in water. Polar molecules are called hydrophilic ("water-loving") because of their ability to dissolve in water. Nonpolar molecules are called hydrophobic ("water-fearing") because they do not dissolve in water.

Although water is the most common solvent in the body, membranes are composed of lipids, and how a substance behaves in lipids is also important to physiology. Nonpolar molecules dissolve in lipids and can permeate the phospholipid bilayer of membranes; thus they are called lipophilic ("lipid-loving"). Polar molecules can neither dissolve in lipids nor permeate the phospholipid bilayer of membranes; thus they are called lipophobic ("lipid-fearing").

The following table summarizes the nature of molecules held together by covalent bonds. Notice that hydrophilic molecules are lipophobic, and hydrophobic molecules are lipophilic.

Electron sharing in covalent bond	Class of covalent bond	Property in water	Property in lipid
Equal sharing	Nonpolar	Hydrophobic	Lipophilic
Unequal sharing	Polar	Hydrophilic	Lipophobic

physically disrupting the bilayer. These amphipathic molecules maintain contact with both the lipid bilayer and the aqueous environment. The polar surfaces of the protein face the aqueous environment, which could be the cytosol, extracellular fluid, or both. The nonpolar areas are embedded within the lipid bilayer.

Some integral membrane proteins are called **transmembrane proteins,** because they span the lipid bilayer, with surfaces exposed to both the cytosol and the extracellular fluid. Often, a transmembrane protein crosses the membrane at several places. Transmembrane proteins include channels that allow ions to *permeate* (or cross) the membrane, and carrier proteins that transport molecules from one side of the membrane to the other. Other integral membrane proteins are located on only one side of the membrane. Some of the proteins facing the cytosol function as enzymes that catalyze chemical reactions in the cytosol or as a special class of proteins called G proteins (described later). Some of the proteins facing the extracellular fluid function as enzymes that catalyze reactions in the extracellular fluid or as receptor molecules that bind chemical messengers from other cells.

Peripheral membrane proteins are loosely bound to the membrane by associations with integral membrane proteins or phospholipids. These proteins can be dissociated from the membrane and still leave the membrane intact. Most peripheral membrane proteins are located on the cytosolic surface of the plasma membrane and often function as part of a group of proteins that make up the *cytoskeleton.*

Apply Your Knowledge

A scientist wants to determine whether a particular protein is a membrane protein, and if so, whether it is an integral or peripheral membrane protein. Which types of experiments would answer these questions?

Membrane Carbohydrates

Also associated with the plasma membrane are carbohydrates, which are covalently bound to the membrane lipids or to proteins to form glycolipids or glycoproteins, respectively (see Figure 16). Plasma membrane carbohydrates are located primarily on the extracellular surface, where they have two main functions: They form the *glycocalyx,* a protective layer that also functions in holding cells together (**Figure 17**), and they function in cell recognition, labeling the cell as a part of the body or as a distinct type of cell. In fact, certain glycolipids on the membrane of red blood cells determine blood type; if a person receives transfusion of the wrong type of blood, he or she can have an immune reaction to those glycolipids, or antigens. This possibility illustrates the importance of the immune system being able to recognize tissue that is part of the person. Cell recognition is also important in regulating tissue growth.

Structure of the Nucleus

Most cells have a single nucleus that contains the genetic material of the cell, DNA. The DNA exists as thin threads called **chromatin,** except during cell division. The nucleus generally appears as a prominent spherical structure in the cell (**Figure 18**). Surrounding the nucleus is the **nuclear envelope,** which consists of two membranes. These membranes fuse intermittently, leaving gaps called **nuclear pores** that allow selective movement of molecules between the nucleus and the cytoplasm. Within the nucleus is a structure called the *nucleolus,* which is the site of synthesis of a type of RNA called *ribosomal RNA (rRNA).* The nucleus functions in the transmission and expression of genetic information. Encoded in the DNA is the genetic information for RNA and protein synthesis. The very important roles of DNA and RNA in protein synthesis are described later in this chapter.

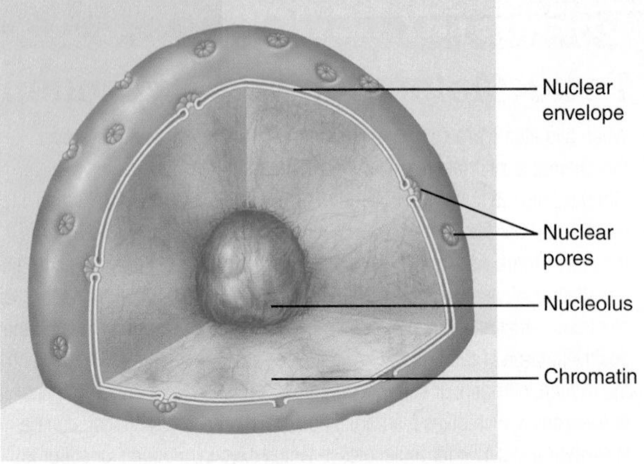

(a)

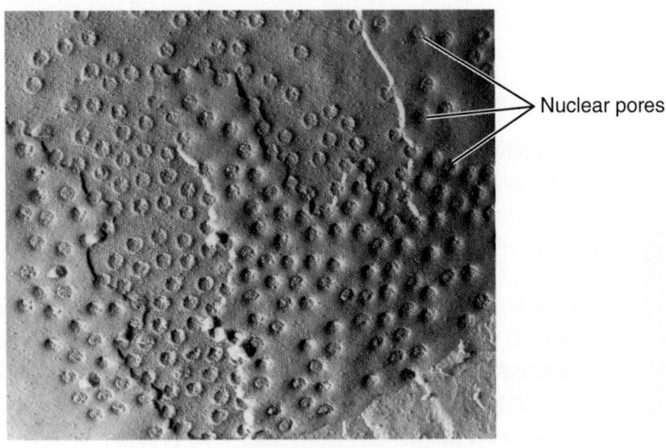

(b)

Don W. Fawcett/Photo Researchers, Inc.

Figure 18 **Nucleus. (a)** The nucleus is separated from the cytoplasm by the nuclear envelope. Pores in the nuclear envelope allow the movement of specific substances between the nucleus and cytoplasm. Located in the nucleus is the DNA, which contains the genetic information of each individual. The DNA exists as thin threads called chromatin. Also within the nucleus is the nucleolus, the site of rRNA synthesis. **(b)** Electron micrograph of a freeze-fractured nuclear envelope showing nuclear pores.

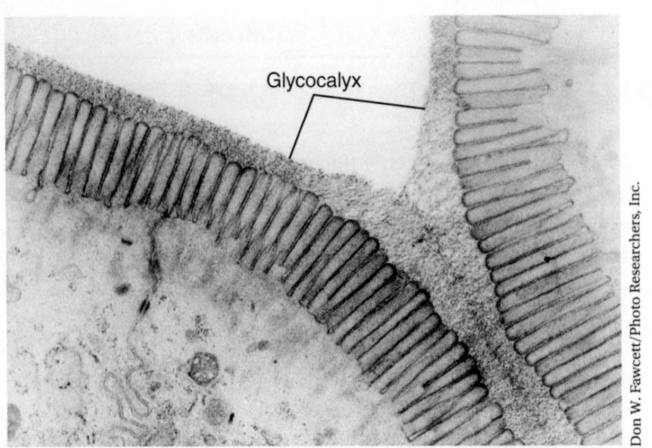

Don W. Fawcett/Photo Researchers, Inc.

Figure 17 **Electron micrograph of the plasma membranes of two cells showing the glycocalyx.**

Contents of the Cytosol

The cytosol is a gel-like fluid that is crucial to many cell processes. A variety of enzymes in the cytosol catalyze specific chemical reactions, such as the ten enzymes that break one glucose molecule into two molecules of pyruvate in the process called *glycolysis.* The cytosol also stores energy in the form of triglycerides or glycogen in masses called **inclusions.** Molecules that are to be released from the cell, or **secreted,** are stored in membrane-bound sacs called **secretory vesicles.** In addition, the molecular composition of the cytosol is critical to the function of cells. For example, the ionic composition of the cytosol plays a crucial role in the activities of nerve and muscle cells.

CHEMISTRY REVIEW

Ions and Ionic Bonds

When atoms form chemical bonds, the electrons of the atoms interact. In **Chemistry Review: Polar Molecules and Hydrogen Bonds** , we saw how atoms could share electrons to form covalent bonds. Some atoms, however, have a tendency to gain or lose electrons completely during a chemical reaction, so that they end up with an excess or a deficit of electrons. When electrons are gained or lost from an atom or molecule, the number of negatively charged electrons no longer equals the number of positively charged protons; in such a case, a charged particle called an *ion* is formed. Atoms that gain an electron acquire a negative charge and are called *anions*, whereas atoms that lose an electron acquire a positive charge and are called *cations*.

When anions and cations are present in solids, they tend to form crystals in which the cations and anions are closely associated. A familiar example is *sodium chloride (NaCl),* also known as table salt, which contains sodium ions (Na^+) and chloride ions (Cl^-). Sodium ions form when sodium atoms lose an electron, producing an ion with 11 protons and 10 electrons. Chloride ions form when chloride atoms gain an electron, producing an ion with 17 protons and 18 electrons. This process occurs as follows:

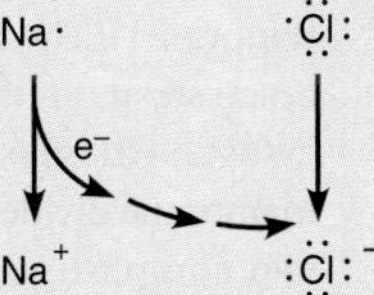

In a crystal of NaCl, the cations (Na^+) and anions (Cl^-) are held together by electrical forces of attraction due to their opposite charges. These forces are sometimes called *ionic bonds.* When ionic solids dissolve in water, ionic bonds are disrupted by electrical attractions between the ions and polar water molecules, leaving cations and anions free to dissociate into separate particles. For sodium chloride, this process can be illustrated as shown at right.

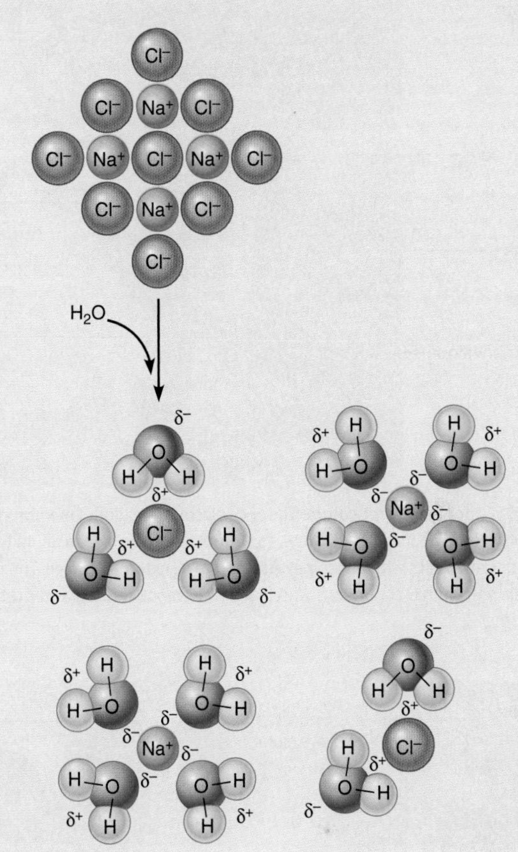

Solutions containing dissolved ions are described as *electrolytic* because they are good conductors of electricity, and ionic substances are referred to as *electrolytes.* Body fluids are electrolytic and contain a number of small ions (known as *inorganic ions*), including sodium, potassium (K^+), calcium (Ca^{2+}), hydrogen (H^+), magnesium (Mg^{2+}), chloride, sulfate (SO_4^{2-}), and bicarbonate (HCO_3^-).

Ionized chemical groups can also be found on certain types of biomolecules. Ions and molecules containing significant numbers of ionized groups are described as *hydrophilic* because they are electrically attracted to water.

Structure of Membranous Organelles

The membranes that surround certain organelles provide a barrier between the cytosol and the interior of the organelle, which in turn creates compartments within the cytoplasm. In some cases, two membranes surround an organelle, creating compartments within the organelle. The structure of each membranous organelle is described next.

Endoplasmic Reticulum

The **endoplasmic reticulum** consists of an elaborate network of membranes enclosing an interior compartment called the lumen (**Figure 19**). Two types of endoplasmic reticulum are distinguished based on their differing appearance and function: *rough endoplasmic reticulum* and *smooth endoplasmic reticulum.* The rough endoplasmic reticulum gets its name from its granular or "rough" appearance under magnification, which is attributable to the presence of **ribosomes**—complexes of rRNA and proteins that function in protein synthesis. In addition, rough endoplasmic reticulum looks like flattened sacs. Smooth endoplasmic reticulum, by contrast, consists of tubules and does not have ribosomes attached to it, giving it a "smooth" appearance. The membrane of the rough endoplasmic reticulum is continuous on one side with the outer membrane of the nuclear envelope, and on the other side with the smooth endoplasmic reticulum (see Figure 19).

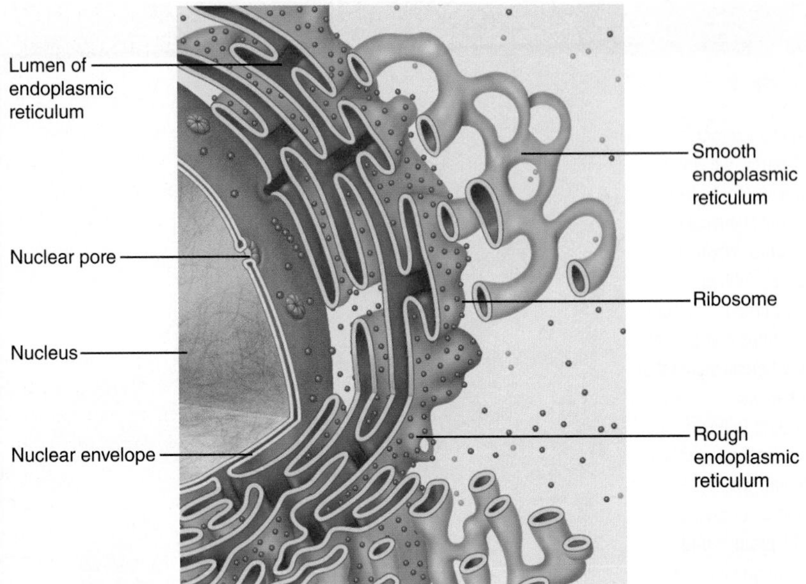

Figure 19 Endoplasmic reticulum. The endoplasmic reticulum's membrane is continuous with the outer membrane of the nuclear envelope. The rough endoplasmic reticulum is closest to the nucleus and consists of flattened sacs with ribosomes attached to its external surface. The membrane of the rough endoplasmic reticulum is continuous with that of the smooth endoplasmic reticulum, which is tubular and lacks ribosomes.

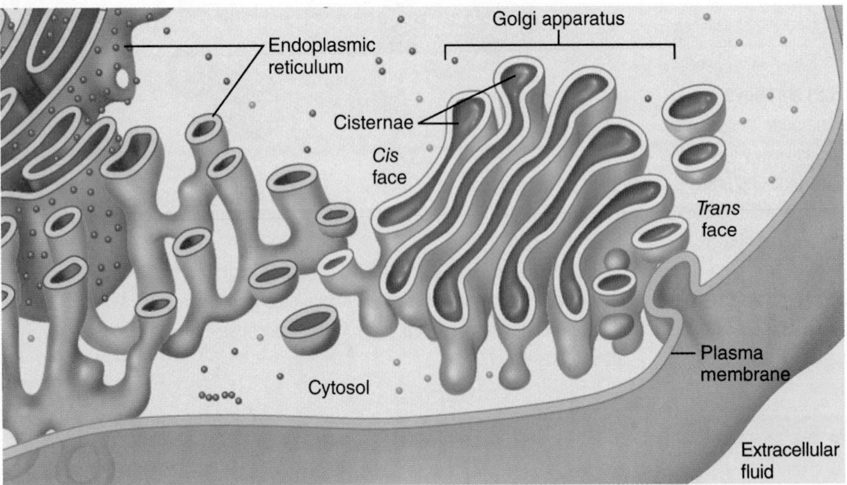

Figure 20 Golgi apparatus. The Golgi apparatus consists of stacks of flattened sacs called cisternae. One side of the Golgi apparatus faces the endoplasmic reticulum; the other side faces the plasma membrane.

The endoplasmic reticulum is important in the synthesis of several types of biomolecules. The rough endoplasmic reticulum is associated with the synthesis of proteins that will be secreted from the cell or incorporated into the plasma membrane, or are destined for another organelle. The smooth endoplasmic reticulum is the site of the synthesis of lipids, including triglycerides and steroids, and is a site for storage of calcium ions. In addition, the smooth endoplasmic reticulum is specialized in certain cells. In liver cells, for example, the smooth endoplasmic reticulum contains detoxification enzymes that break down toxic substances in the blood.

Golgi Apparatus

The **Golgi apparatus** consists of membrane-bound flattened sacs called *cisternae* (**Figure 20**). The Golgi apparatus is closely associated with the endoplasmic reticulum on one side, called the *cis face,* although the membranes of the Golgi apparatus and endoplasmic reticulum are separate. The other side of the Golgi apparatus faces the plasma membrane and is called the *trans face.* The Golgi apparatus processes molecules synthesized in the endoplasmic reticulum, and prepares them for transport to their final location. It packages molecules into vesicles and directs the vesicles to the appropriate location. Some vesicles transport substances to intracellular sites, whereas secretory vesicles transport substances out of the cell.

Mitochondria

Mitochondria are bound by two membranes (**Figure 21**). The outer mitochondrial membrane separates the mitochondrion from the cytosol, whereas the inner mitochondrial membrane divides each mitochondrion into two compartments: the **intermembrane space,** the area between the two membranes, and the **mitochondrial matrix,** the innermost compartment. The inner mitochondrial membrane is a functional compartment in itself because it houses a series of proteins and other molecules called the *electron transport chain.* To increase the area available to the electron transport system, the inner membrane is folded into tubules called **cristae.** Mitochondria are often called the "powerhouse" of the cell because most of the cell's usable energy form, ATP, is produced in these organelles.

The number of mitochondria per cell varies considerably among different cell types based on the particular cell's energy needs. For example, because muscle cells demand large amounts of energy, they contain numerous mitochondria. By contrast, red blood cells, which function primarily in the transport of gases, contain no mitochondria.

Lysosomes

Lysosomes are small spherical organelles surrounded by a single membrane (**Figure 22**). They contain enzymes that degrade intracellular debris and extracellular debris that has been taken into the cell. For example, after lysosomes fuse with old organelles that are no longer functioning, the enzymes inside the lysosomes break

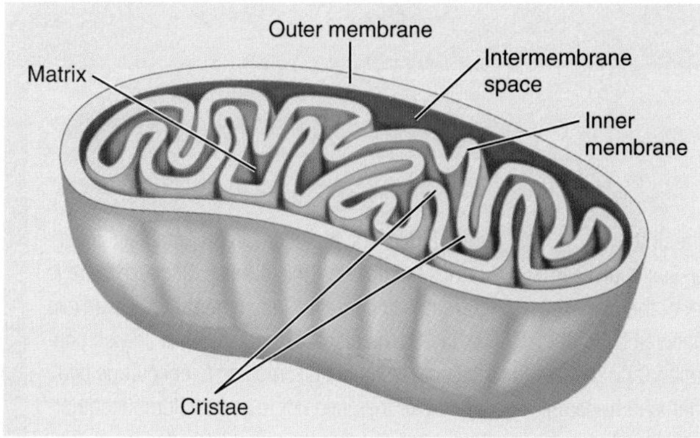

(a)

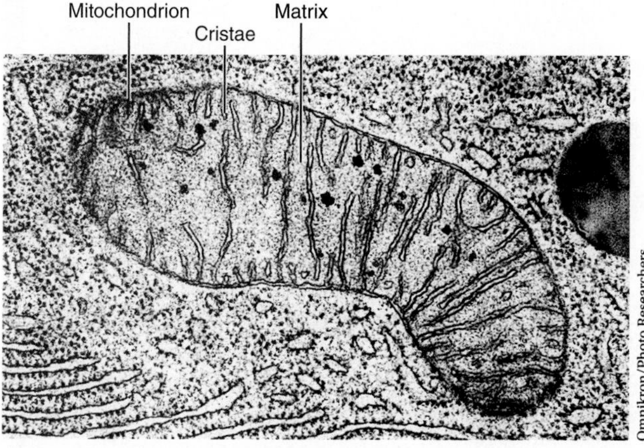

(b)

Figure 21 Mitochondria. (a) Drawing of a mitochondrion. The outer mitochondrial membrane separates the mitochondrion from the cytosol, whereas the inner mitochondrial membrane separates the mitochondrion into an inner matrix and the intermembrane space. The inner membrane has numerous folds called cristae. **(b)** Electron micrograph of a mitochondrion.

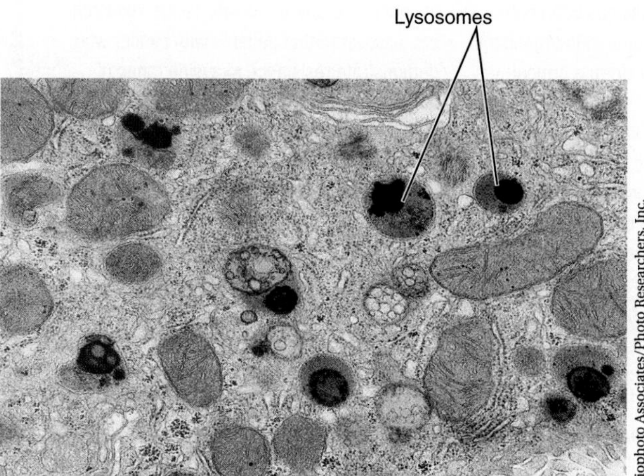

Figure 22 Electron micrograph of lysosomes.

down the organelles, with usable components being recycled and waste products being eliminated from the cell. In the case of extracellular debris, cells can engulf extracellular particles by a process called *endocytosis,* in which the particles are enclosed within a vesicle and brought inside the cell. Lysosomes then fuse with the vesicle, enabling their enzymes to degrade the particles. The importance of lysosomes is illustrated by Tay-Sachs disease, a genetic disease that results in the death of the patient by age 5 (see **Clinical Connections: Tay-Sachs Disease**).

Peroxisomes

Peroxisomes are spherical organelles that are slightly smaller than lysosomes and are surrounded by a single membrane. They function in the oxidation, and thereby degradation, of molecules such as amino acids, fatty acids, and toxic foreign matter. During these oxidation reactions, hydrogen is removed from the molecules being degraded and added to oxygen; this reaction produces hydrogen peroxide (H_2O_2), which in itself is toxic. One of the most notable characteristics of peroxisomes is the presence of the enzyme *catalase,* which catalyzes the breakdown of hydrogen peroxide to form water and oxygen:

$$2\,H_2O_2 \rightarrow H_2O + O_2$$

This reaction prevents the buildup of toxic hydrogen peroxide.

Peroxisomes are most abundant in the liver and kidney cells that remove oxidative waste products from the body. For example, liver peroxisomes contain the enzyme alcohol dehydrogenase, which breaks down ethanol to acetaldehyde.

Structure of Nonmembranous Organelles

Even though nonmembranous organelles have no barrier separating them from the cytosol, they are considered organelles because they consist of biomolecules organized into structures that perform specific functions within the cell.

Ribosomes

Ribosomes, which are very small organelles (25 nm in diameter), are dense granules composed of rRNA and proteins that function in protein synthesis. Each ribosome consists of a small subunit (30S) and a large subunit (50S) that are separate and located in the cytosol when the ribosome is not active (**Figure 23a**). During protein synthesis (discussed later in this chapter), the two subunits come together in the cytosol to form a functional ribosome (Figure 23b). After protein synthesis is initiated, some ribosomes remain free in the cytosol, whereas others become attached to the rough endoplasmic reticulum. Proteins synthesized in association with free ribosomes can remain in the cytosol or enter a mitochondrion, the nucleus, or a peroxisome. Proteins synthesized in association with the rough endoplasmic reticulum will be packaged into vesicles by the Golgi apparatus and directed to their ultimate destination.

CLINICAL CONNECTIONS

Tay-Sachs Disease 

Tay-Sachs disease is a genetic disorder in which mutated genes produce lysosomal enzymes that are less effective than normal. The enzymes most commonly affected break down lipids called gangliosides. When these enzymes do not work properly, the buildup of gangliosides inside cells damages the cells, causing adverse effects. Development of the infant is impaired, with vision loss beginning in the first few months of life. Babies with Tay-Sachs disease are easily startled—a hallmark of the disease. Eventually paralysis sets in, followed by deafness, seizures, inability to swallow, and ultimately breathing difficulties that lead to death.

Tay-Sachs disease is most common among Eastern Europeans, especially those of Ashkenazi Jewish descent. About one in 30 American Jews carries the gene for this disease. Because there is currently no cure and limited treatments available for Tay-Sachs disease, the Ashkenazi Jewish population uses genetic screening to encourage two carriers of the gene not to have children together. This screening has led to a 90% reduction in Tay-Sachs disease in the United States and Canada.

Critical Thinking Questions

1. How does the impact of Tay Sachs on the functionality of cells affect the disease symptoms?
2. In order to eradicate genetic diseases, such as Tay Sachs, should testing be required for all genetic disorders in the U.S.? Why or why not?
3. What would be the advantages and disadvantages of mandatory genetic testing? What other factors should be considered in this debate?

DISCOVERY

Vaults and Chemotherapy

Scientists discovered vaults, illustrated in the accompanying figure, in human liver cells in the 1980s. Vaults are ribonucleoproteins consisting of RNA (called vault RNA [vRNA]) and proteins. They are larger than ribosomes, which are also ribonucleoproteins, and have been found in almost every cell in the body that has been searched for their presence. Most vaults reside within the cytoplasm, but a few have been found in the nucleus; one study even located vaults in the nuclear pores. Three types of proteins appear in vaults, with the most abundant being *major vault protein (MVP)*. The molecular structure of the vault proteins has been conserved in the evolution of eukaryotic organisms, which suggests that they perform a critical function.

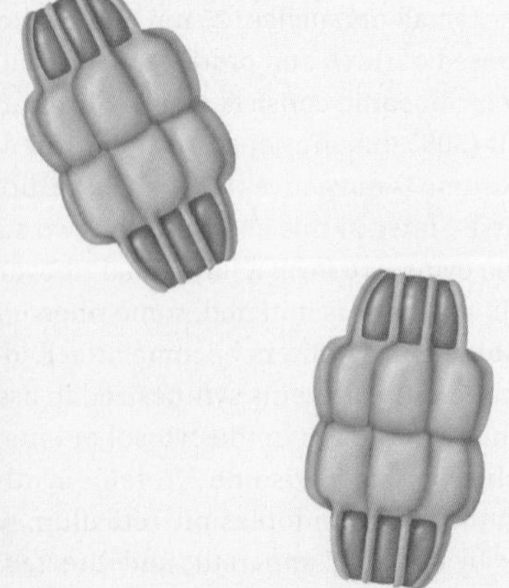

Although the function of vaults remains under investigation, some evidence indicates that they contribute to intracellular transport. Vaults have a large internal cavity capable of holding hundreds of proteins. Several studies suggest that they function as shuttles between the nucleus and the cytoplasm, transporting substances such as ribosomes, RNA, and proteins along this pathway. Vaults and the cytoskeleton may work together in cytoplasmic transport.

Vaults may also function in embryonic development by signaling cell survival or differentiation. As the embryo develops, many cells undergo a process called *apoptosis*, or programmed cell death. Vaults may induce biochemical changes inside cells that oppose the signals responsible for apoptosis.

Although their function remains a mystery, vaults might potentially assume a valuable role in cancer therapy. In the late 1990s, researchers discovered that patients with cancer who demonstrate resistance to a wide range of chemotherapies, called *multidrug resistance (MDR)*, have a greater concentration of vaults in their cells than do persons who lack such resistance. Several studies suggest that MVP levels in cancer patients may be used as a marker to identify those individuals who have resistance to proposed treatments for their disease. Some studies also indicate that MVP contributes to the progression of cancer, possibly through its role as a transporter. Further research is needed to more fully elucidate the function of vault proteins, which might also lead to better treatment of cancer.

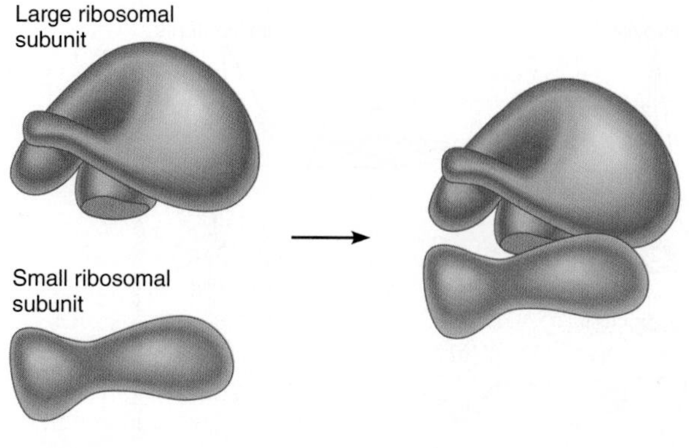

Figure 23 Ribosomes. For protein synthesis to begin, **(a)** the two free subunits in the cytosol must come together to form **(b)** a ribosome.

Vaults

Vaults are the most recently discovered organelle, and their discovery may hold promise for cancer treatment (see **Discovery: Vaults and Chemotherapy**). Vaults are barrel-shaped structures comprising proteins and vault RNA (vRNA). Most vaults are located in the cytoplasm and interact with the cytoskeleton; a small percentage of these organelles are located in the nucleus. Although the function of vaults remains unknown, studies suggest that they assist in intracellular transport and may act as chemical signals for cell survival.

Centrioles

Centrioles are short cylindrical structures consisting of bundles of protein filaments. Each cell has two centrioles oriented perpendicular to each other (see Figure 15). They function in directing the development of a structure called the *mitotic spindle* during cell division.

Cytoskeleton

The **cytoskeleton** is a flexible lattice of fibrous proteins, also called filaments, that gives the cell structure and support, much as the skeleton provides support for the entire body (**Figure 24**). The cytoskeleton is not rigid, nor does it have a fixed structure. The fibrous proteins that form it can disassemble and reassemble as necessary. Functions of the cytoskeleton include mechanical support and structure, intracellular transport of materials, suspension of organelles, formation of adhesions with other cells, contraction, and movement of certain cells. Several types of filaments form the cytoskeleton, including *microfilaments, intermediate filaments,* and *microtubules.* The classification of filaments is based on their diameters.

Microfilaments have the smallest diameter (7 nm) among the fibrous proteins. One microfilament, called *actin* (**Figure 25a**), has several functions in cells, including muscle contraction, "amoeboid-like" movement of cells, and separation of the cytoplasm during cell division. Other actin microfilaments provide structural support for special cell projections called **microvilli** (Figure 25b), which are often found in epithelial cells that are specialized for the exchange of

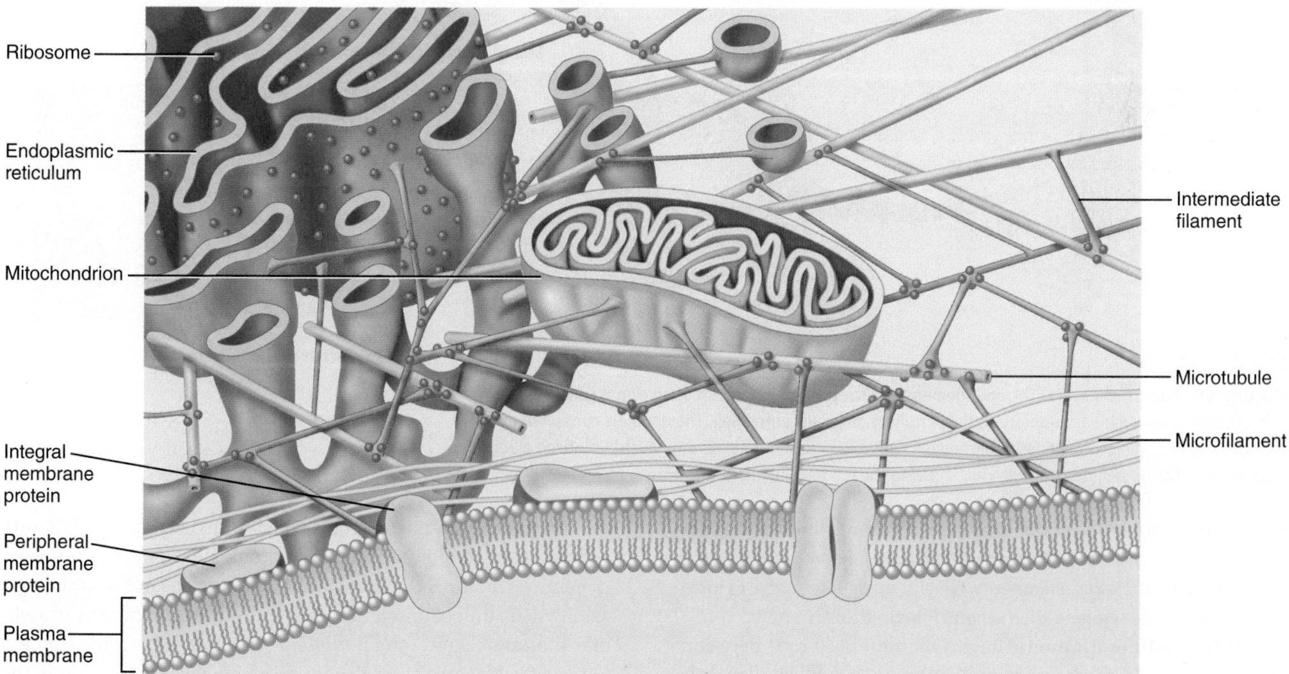

Figure 24 Cytoskeleton. The cytoskeleton consists of three types of filaments that give the cell structural support and enable some degree of motility or contractility. Organelles are suspended in the cytosol by the filaments.

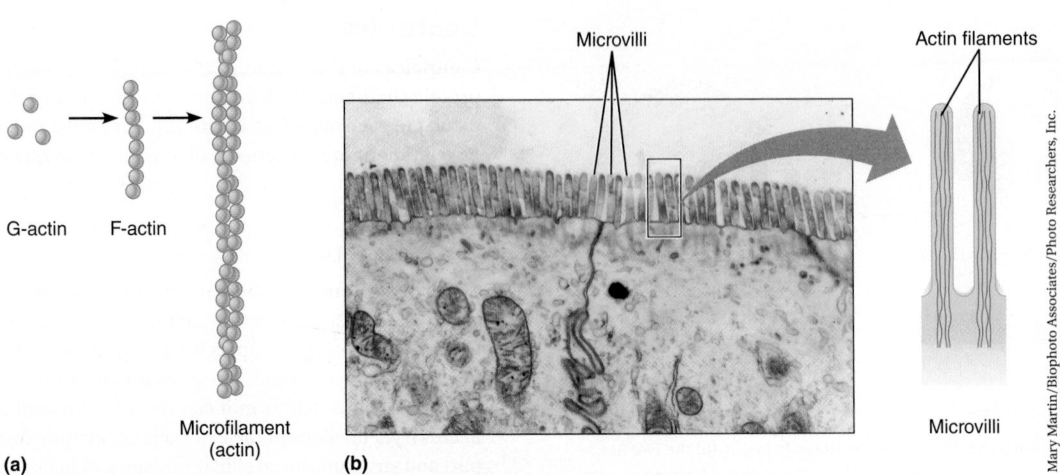

Figure 25 Microfilaments. (a) An actin filament, the most common microfilament. Actin filaments consist of two helically arranged strands of F-actin, a fibrous protein formed of chains of a globular protein, G-actin. **(b)** Electron micrograph of microvilli on the surface of an epithelial cell. The enlarged drawing shows a schematic of the arrangement of microfilaments that maintain the fingerlike microvilli.

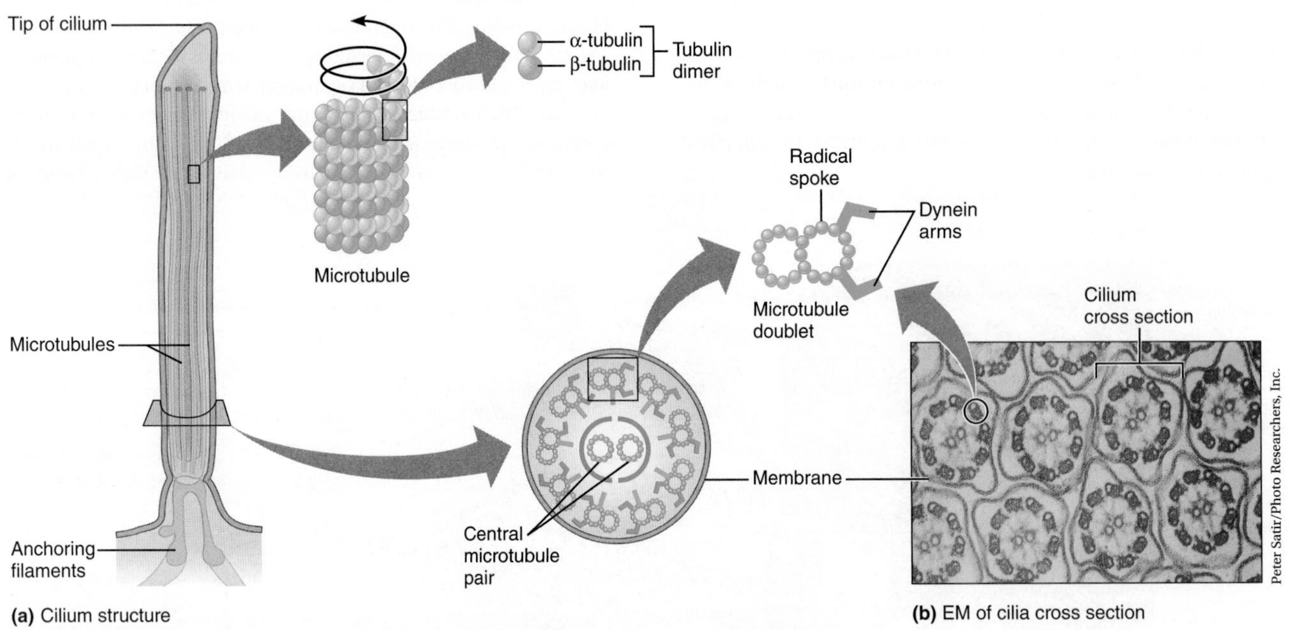

Figure 26 Microtubules. (a) The arrangement of microtubules in a cilium. Each cilium contains ten microtubule pairs—nine as doublets on the periphery and one central pair. The doublets contain dynein arms for motility. Microtubules assemble and disassemble in subunits called dimers. **(b)** Colorized electron micrograph of cross sections through cilia, showing the ten pairs of microtubules.

molecules. Actin is a dynamic structure, constantly undergoing assembly and disassembly. The unit of actin is a globular protein called G-actin. To form the actin filaments, G-actin must polymerize to form F-actin, which then joins with a second F-actin to form a helix.

Intermediate filaments have a diameter (10 nm) between those of microfilaments and microtubules. Intermediate filaments tend to be stronger and more stable than microfilaments. Intermediate filaments called *keratin* are found in skin and hair cells. *Myosin* is an intermediate filament found in muscle cells, where it works with actin to produce contraction.

Microtubules, the filaments with the largest diameter (25 nm), consist of long hollow tubes. Like microfilaments, microtubules are dynamic structures that constantly undergo assembly and disassembly. The unit of microtubule assembly is a heterodimer of globular subunits called α and β *tubulin.* Microtubules provide strength to the cytoskeleton. They also form *spindle fibers,* which aid in the distribution of chromosomes during cell division, and assist with directional movement of vesicles and large biomolecules within the cytosol. In addition, microtubules are components of two motile structures: **cilia** and **flagella** (Figure 26).

Cilia and flagella are hairlike protrusions from the cell surface that can move in a wavelike manner. Movement of cilia propels particles along hollow conduits in the body. In the respiratory tract, for example, cilia move mucus containing trapped inhaled particles away from the lungs and up toward the mouth, where it can be swallowed. A cilium consists of ten pairs of microtubules, with nine of the pairs surrounding a central pair. The paired microtubules are connected by proteins called *dynein arms* that help to generate the force needed for the microtubules to slide past each other; the sliding movement leads to the wavelike motion of a cilium. Flagella are similar to cilia in basic structure, but they are longer and are found in only one cell type in humans, the sperm. Flagella move sperm through the female reproductive tract toward the ovum.

Apply Your Knowledge

Cigarette smoke is known to paralyze and destroy cilia located in the passageways to the lungs. What is the consequence of the loss of the action of cilia?

Table 2 summarizes the different components of the cell and their major functions.

Quick Check 2

1. Where is the genetic information of a cell stored? Which type of molecule stores this information?
2. Describe the basic structure of the plasma membrane. Name four functions of membrane proteins.
3. Describe the basic functions of each of the organelles.
4. What are the three types of filaments found in the cytoskeleton? Name a special function of each type of filament.

3 Cell-to-Cell Adhesions

In many tissues, cells are held together by special membrane proteins called *cell adhesion molecules.* Other proteins function in more specialized cell adhesions, in which the purpose is more than merely holding the cells together. There are three main types of special junctions: tight junctions, desmosomes, and gap junctions.

TABLE 2 Summary of Cell Structures and Functions

Cell part	Structure	Function
Plasma membrane	Lipid bilayer with scattered proteins and cholesterol molecules	Maintains boundary of cell and integrity of cell structure; embedded proteins serve multiple functions
Nucleus	Surrounded by double-layered nuclear envelope	Houses the DNA, which dictates cellular function and protein synthesis
Nucleolus	Dark oval structure inside the nucleus	Synthesis of ribosomal RNA
Cytosol	Gel-like fluid	Cell metabolism, storage
Membranous organelles		
Rough endoplasmic reticulum	Continuous with the nuclear envelope; flattened sacs dotted with ribosomes	Protein synthesis and post-translational processing
Smooth endoplasmic reticulum	Continuous with rough endoplasmic reticulum; tubular structure without ribosomes	Lipid synthesis and post-translational processing of proteins; transport of molecules from endoplasmic reticulum to Golgi apparatus; calcium storage
Golgi apparatus	Series of flattened sacs near the endoplasmic reticulum	Post-translational processing; packaging and sorting of proteins
Mitochondria	Oval-shaped, with an outer membrane and an inner membrane with folds called cristae that project into the matrix	ATP synthesis
Lysosomes	Granular, saclike; scattered throughout the cytoplasm	Breakdown of cellular and extracellular debris
Peroxisomes	Similar in appearance to lysosomes, but smaller	Breakdown of toxic substances, including hydrogen peroxide
Nonmembranous organelles		
Vaults	Small, barrel-shaped	Unknown; possibly transport of molecules between nucleus and cytoplasm
Ribosomes	Granular organelles composed of proteins and rRNA; located in cytosol or on surface of rough endoplasmic reticulum	Translation of mRNA to synthesize proteins
Centrioles	Two cylindrical bundles of protein filaments that are perpendicular to each other	Direction of mitotic spindle development during cell division
Cytoskeleton	Composed of protein filaments, including microfilaments, intermediate filaments, and microtubules	Structural support of cell; cell movement and contraction

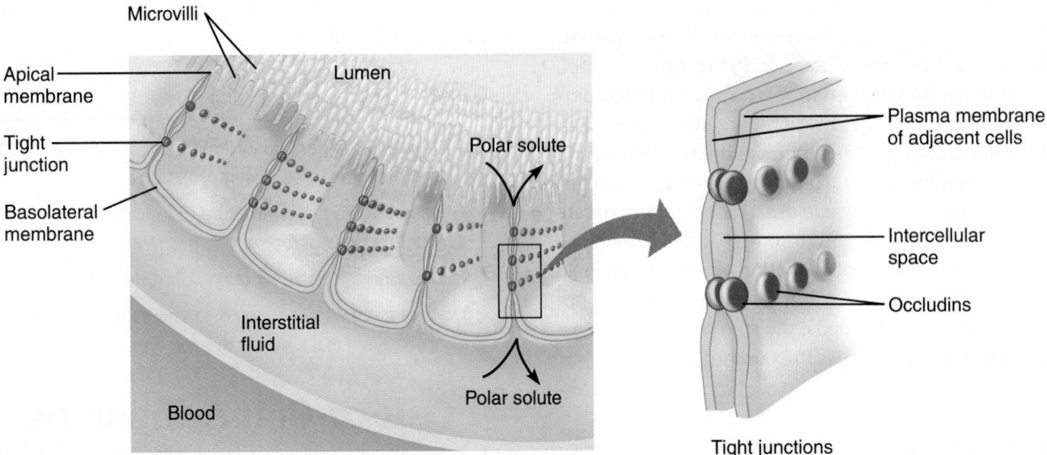

Figure 27 Tight junctions. Tight junctions are commonly found in epithelial tissue, where they limit the movement of molecules through the interstitial space between epithelial cells. Proteins called occludins link two adjacent cells together such that they form nearly impermeable adhesions. Polar molecules that cannot cross lipid bilayers also generally cannot move between cells connected by tight junctions.

Tight Junctions

Tight junctions are commonly found in epithelial tissue that is specialized for molecular transport. In **tight junctions,** integral membrane proteins called *occludins* fuse adjacent cells together to form a nearly *impermeable* barrier to the movement of substances between cells (**Figure 27**). Because of this barrier, solutes generally must cross the epithelial cell layer to go from one side of an epithelium to the other (*transepithelial transport*), rather than going around the cells (*paracellular movement*).

Epithelial cells often line hollow structures, such as the organs of the gastrointestinal tract or the tubules in the kidneys. The internal compartment of a hollow organ, the *lumen,* is separated from the cytosol of each epithelial cell by the *apical membrane.* The membrane facing the extracellular fluid is called the *basolateral membrane.* Epithelial cells lining the organs often have tight junctions that restrict or regulate the movement of molecules from the lumen of the organ into the blood, or from the blood into the lumen. Because of these tight junctions, the composition of the contents of the lumen can differ from that of the blood.

Desmosomes

Desmosomes are found in tissues subject to mechanical stress, such as those in the heart, uterus, and skin. A **desmosome** is a filamentous junction between two adjacent cells that provides strength so that the cells do not tear apart when the tissue is subjected to stress. **Figure 28a** shows the structure of desmosomes in the heart, where they are closely associated with another type of adhesion called gap junctions (described next). At the site of each desmosome is a *plaque* formed by glycoproteins clustered inside each cell. Extending from this plaque are both intracellular protein filaments and protein filaments called *cadherins* that cross the plasma membrane into the extracellular space. The cadherins are linked to intracellular filaments at the plaque.

Gap Junctions

Gap junctions, although found in a variety of tissue types, are most notable for their presence in smooth muscle and in the muscle of the heart, where they allow the muscle to contract as a unit. In these areas, two adjacent cells are connected by membrane proteins called *connexons* (Figure 28b). Each connexon is composed of six smaller membrane proteins called *connexins.* The connexons in the adjacent cells bind to each other, forming small channels that enable ions and small molecules to move between the two cells.

Gap junctions provide direct electrical and metabolic coupling of adjacent cells. Movement of ions between cells electrically couples the cells, such that if the electrical properties of one cell change, ions move from this cell to the adjacent cell through the gap junction, producing the same type of electrical change. This process allows the cells of the heart, for example, to act in a coordinated fashion so that they can contract in unison. In other areas, such as in bone, gap junctions allow nutrients to travel from cells close to the bloodstream to cells deep within the tissue. Gap junctions also allow the passage of some chemical messengers such as the molecule cAMP. In this manner, a signal initiated in one cell can be transmitted to adjacent cells.

4 General Cell Functions

Our cells have specialized functions that contribute to homeostasis. For example, muscle cells contract to generate force, and nerve cells use electrical impulses and chemical messengers for communication. Despite these special functions, most cells of the body carry out several general functions, which we introduce next.

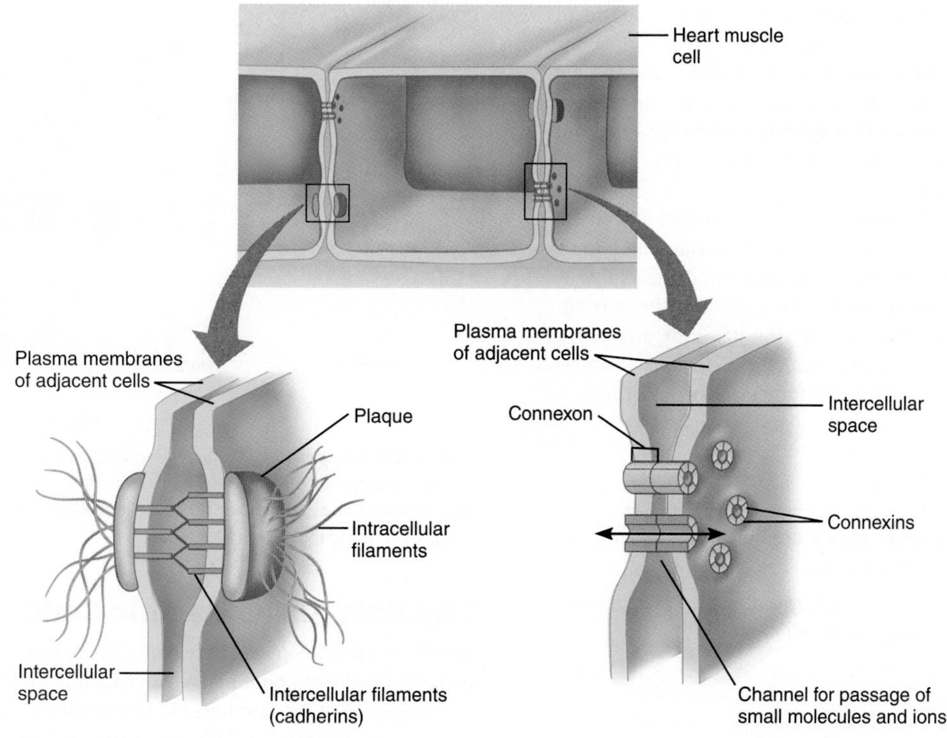

Figure 28 Desmosomes and gap junctions, as shown in heart muscle. **(a)** Desmosomes are strong junctions between cells in tissues subjected to stress. Protein filaments extend from a plaque located near the plasma membrane where the two cells join together. Some of these filaments are located inside the cells (intracellular filaments), whereas other filaments called cadherins (intercellular filaments) extend into the intercellular space, linking the two cells together. **(b)** Gap junctions are areas of membrane containing proteins called connexons that form channels linking the cytoplasm of two adjacent cells. These channels allow ions and small molecules to pass between adjacent cells.

Metabolism

All cells of the body carry out a variety of chemical reactions. **Metabolism** refers to all the chemical reactions that occur in the body. Metabolism can be divided into two classes of chemical reactions: (1) **anabolism,** the synthesis of large molecules from smaller molecules, which requires an input of energy, and (2) **catabolism,** the breakdown of large molecules to smaller molecules, which releases energy. Every second, thousands of anabolic and catabolic reactions occur in virtually every cell of the body. These reactions are highly regulated by the presence of enzymes.

Cellular Transport (MAP)

As noted earlier, the lipid bilayer acts as a barrier to hydrophilic molecules. However, cells require hydrophilic molecules to carry out their functions, and cells also release hydrophilic molecules into the extracellular environment. The process by which molecules move into and out of cells is called *membrane transport.*

Nonpolar molecules can permeate the plasma membrane and will simply move across the membrane by diffusion. The smallest of the polar molecules can also permeate most plasma membranes. In contrast, larger polar molecules cannot permeate the lipid bilayer; some of these molecules can cross plasma membranes only with the assistance of transmembrane proteins or vesicles. For example, some molecules such as glucose and amino acids are transported across the membrane by *carrier proteins*. Other molecules, such as water and ions, can move through *channels*. Proteins and other *macromolecules* are too large to be transported across the plasma membrane by carrier proteins or to move through channels. Instead, movement of these macromolecules relies on the packaging of the molecules into vesicles.

Intercellular Communication

Communication is another function that all cells perform. A single cell in the body cannot exist in isolation; rather, cells must communicate with one another to maintain an environment in which all of them can survive. This communication can be direct, as occurs through gap junctions. Often, however, communication involves the release from one cell of a chemical messenger that acts on a second cell. The two communicating cells can be adjacent to each other, or they can be in different areas of the body. In general, messenger molecules are released from one cell and reach other cells, called *target cells,* which contain proteins that function as *receptors* for that specific messenger. The binding of the messenger molecule to its specific receptor triggers a response in the target cell.

In one example of such intercellular communication, a chemical messenger, antidiuretic hormone (ADH), is released into the blood by exocytosis from an endocrine gland in the brain. When ADH reaches its target cells—certain kidney cells—the hormone binds with its receptors. This binding stimulates the insertion of water channels into the kidney cells, which increases water conservation by removing some of the water from urine being formed.

Quick Check 3

1. Briefly describe the three types of cell adhesions, and give an example of where each is found.
2. Describe the role of transmembrane proteins in the transport of molecules across the plasma membrane.

5 Protein Synthesis

By now you should realize that proteins are very important biomolecules in our cells. Contained within DNA are the codes for the synthesis of all proteins in our cells. This **genetic code** is a universal finding among all animal species. A **gene** is a portion of DNA that codes for a particular protein. A **genome** is the collection of genes in a given species. The mysteries of the human genome were unraveled in 2001, when scientists working on the Human Genome Project identified more than 25,000 different genes that code for even more proteins (one gene can lead to the synthesis of more than one protein). Different cells express different genes (or synthesize different proteins), which is the basis of our different cell types. Next, the methods by which proteins can be synthesized based on the genetic code are described.

Even though DNA is located in the nucleus, protein synthesis takes place in the cytoplasm. Therefore, the code must be transmitted to another molecule, called **messenger RNA (mRNA),** that can move into the cytoplasm. The general steps of protein synthesis are as follows:

1. DNA is *transcribed* according to the genetic code to form a complementary mRNA in the nucleus.
2. mRNA moves from the nucleus to the cytoplasm.
3. mRNA is *translated* by ribosomes to form the correct amino acid sequence of the protein in the cytoplasm.

Mitochondria have their own DNA, and consequently some mitochondrial proteins have a different path for synthesis (see **Clinical Connections: Mitochondrial DNA in Disease, Anthropology, and Forensics**).

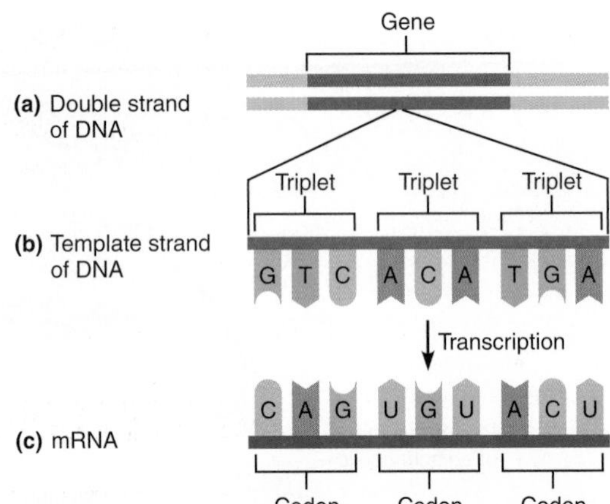

Figure 29 The genetic code. (a) A gene, which is a portion of DNA that codes for a particular protein(s). **(b)** Triplets, the three-base sequences that code for an amino acid sequence. **(c)** Transcribed mRNA codons, which are complementary to the code in DNA triplets.

The Role of the Genetic Code

Recall that DNA consists of two strands of nucleotides held together by hydrogen bonding between complementary base pairs. The sequence of the four bases in DNA codes for the sequence of amino acids in proteins. Although a gene contains the section of DNA that codes for a particular protein or proteins, only one DNA strand, the **template strand,** contains the actual code (**Figure 29a**).

Twenty different amino acids are found in proteins. Thus the sequence of the DNA bases must provide enough information to specify each amino acid in a polypeptide chain. A sequence of three bases, called a **triplet,** codes for a single amino acid (Figure 29b). Each base can be thought of as a letter of an alphabet that is used in words that are all three letters long and indicate a particular amino acid. With four bases that can take up three possible positions in a triplet, there are $4^3 = 64$ possible words in the genetic code. During transcription, which is described shortly, this code is passed on to mRNA, whose base sequence is complementary to the template strand of DNA. Therefore, a DNA triplet is complementary to a three-base sequence in mRNA, called a **codon,** that codes for the same amino acid (Figure 29c).

Note that there are 64 (four bases in three possible positions = 4^3 combinations) possible codons, but only 20 amino acids. Thus, even though each codon codes for only one amino acid, some amino acids are coded for by more than one codon. For example, CCC always codes for the amino acid proline, but CCG also codes for proline. One codon, called an *initiator codon,* is found in every mRNA. This initiator codon, AUG, also codes for the amino acid methionine. Three *termination codons* indicate the end of each protein code. Because these termination codons do not code for any amino acid, only 61 of the 64 codons actually code for amino acids. The genetic code is shown in Table 3.

Transcription

Transcription is the process in which RNA is synthesized using information contained in the DNA; it occurs in the nucleus. Three types of RNA can be transcribed from DNA: (1) mRNA, (2) rRNA,

CLINICAL CONNECTIONS

Mitochondrial DNA in Disease, Anthropology, and Forensics

The nucleus of every somatic cell contains 46 chromosomes consisting of a total of more than 3 billion base pairs with approximately 25,000 genes. The genes code for proteins with a wide range of functions in the body, including functions inside mitochondria. Nuclear DNA codes for 95% of the proteins found in mitochondria. However, each mitochondrion also has its own DNA, which codes for the other 5% of proteins. Mitochondrial DNA (mDNA) consists of a single loop of DNA containing approximately 16,500 base pairs containing 37 genes. All 37 genes are essential for mitochondrial function, and mutations in mitochondrial genes can lead to diseases. Of the 37 genes in mitochondrial DNA, 13 code for enzymes involved in oxidative phosphorylation. The other 24 genes code for transfer RNAs and ribosomal RNAs that assist in protein synthesis inside the mitochondria.

Unlike nuclear DNA, which comes from both the mother's and father's chromosomes, mitochondrial DNA comes only from the mother. During fertilization, an oocyte and a sperm each provide a single set of 23 nuclear chromosomes as they fuse to form a zygote; thus the zygote contains two sets of chromosomes, one donated from the oocyte and one from the sperm. During the fertilization process, however, only the head of the sperm fuses with the oocyte. The head contains a nucleus, but not mitochondria or other organelles. As a consequence, only nuclear DNA is donated to the new cell by the father. The mitochondria of the zygote, and hence the mDNA, come solely from the oocyte, giving mDNA a special role in health, anthropology, and forensics.

Mutations in mitochondrial DNA are generally of somatic origin; that is, they are not hereditary. Although most mutations affecting proteins in the electron transport chain are lethal, some mutations result in disease. One of the more common mitochondrial disorders is Leber hereditary optic neuropathy (LHON). LHON can be caused by a mutation in one of four different mitochondrial genes that code for proteins that are part of complex I of the electron transport chain. These mutations damage the optic nerve, whose role is to transmit signals containing visual information from the eye to the brain. Thus LHON results in blindness. Other mitochondrial DNA mutations cause cancer, encephalopathy (brain damage), lactic acidosis (increased acidity of blood), neuropathy (nerve damage), retinitis pigmentosa (progressive loss of vision due to retinal damage), and deafness. Associations have also been found between mutated mDNA and neurodegenerative diseases including Alzheimer's disease and Parkinson's disease.

The maternal origin of mDNA has also provided anthropologists and forensic scientists with a tool to determine ancestry. When maternal and paternal nuclear DNA are combined during fertilization, the offspring inherits characteristics of both parents. In the germ cells of this offspring, the DNA comprises a combination of maternal and paternal genes. Each generation, therefore, increases the mixing of genes. In contrast, mDNA is conserved through generations, varying only as a result of mutations. The rate of DNA mutations is uncertain, with estimates ranging from 1 in 30 generations to 1 in 600 generations.

Anthropologists use mitochondrial DNA to examine ancestry as part of their research. For example in 1987, scientists reported finding an ancestor common to all of us. According to dating (which may be inaccurate because it is based on an estimated rate of DNA mutation), this ancestor lived in Africa 200,000 years ago. The proposed ancestral lineage is based on the similarity of the mDNA with that of people living today.

Forensic scientists use mDNA primarily in missing persons cases. In those situations, they compare the mDNA of the person in question with the mother's mDNA.

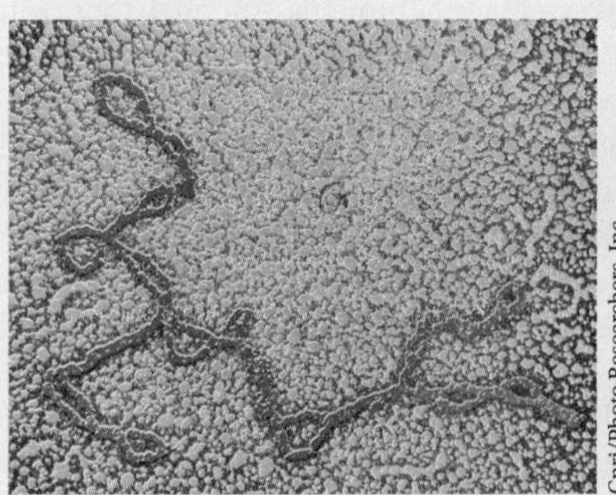

Critical Thinking Questions

1. Why is only maternal mitochondrial DNA passed to the posterity during reproduction?
2. How can mDNA be used to determine ancestry?
3. What are some diseases related to genetic mutations in mDNA, and what do they have in common?

and (3) *tRNA (transfer RNA)*. Each type of RNA is involved in protein synthesis. This section describes the process of transcribing RNA from DNA, using mRNA as the example.

Figure 30 illustrates the process of transcription. The first step is the uncoiling of DNA and its separation into two strands at the site where transcription is occurring. The template strand of DNA contains the gene and functions as a template for the synthesis of mRNA. The section of DNA with the gene is identified by the **promoter sequence,** a specific base sequence to which the enzyme **RNA polymerase** can bind. When RNA polymerase binds to the promoter sequence, it initiates the separation of DNA into two strands. This untwisting of the DNA's double helix allows free ribonucleotides in the triphosphate form (ATP, UTP, GTP, and CTP) to align with the DNA template according to the law of complementary base pairing. (Recall that the base uracil in RNA replaces thymine in DNA.) RNA polymerase also catalyzes the formation of bonds between the aligned ribonucleotides, making a polynucleotide called pre-mRNA that will eventually become the single-stranded mRNA.

The pre-mRNA undergoes *post-transcriptional processing* in the nucleus before the final mRNA product moves through the nuclear

TABLE 3 The Human Genetic Code

First base in codon	Second base in codon: U	C	A	G	Third base in codon
	Phe	Ser	Tyr	Cys	U
	Phe	Ser	Tyr	Cys	C
U	Leu	Ser	*STOP*	*STOP*	A
	Leu	Ser	*STOP*	Trp	G
	Leu	Pro	His	Arg	U
	Leu	Pro	His	Arg	C
C	Leu	Pro	Gln	Arg	A
	Leu	Pro	Gln	Arg	G
	Ile	Thr	Asn	Ser	U
	Ile	Thr	Asn	Ser	C
A	Ile	Thr	Lys	Arg	A
	Met	Thr	Lys	Arg	G
	Val	Ala	Asp	Gly	U
	Val	Ala	Asp	Gly	C
G	Val	Ala	Asp	Gly	A
	Val	Ala	Glu	Gly	G

pores and enters the cytoplasm (**Figure 31**). Within a gene are regions of excess bases, called *introns,* that do not code for the amino acid sequence of proteins. Although these introns are transcribed in the initial mRNA molecule, they must be removed before mRNA leaves the nucleus. After introns are cut out of the initial mRNA molecule, the remaining coding segments, called *exons,* are joined together. Exons can be spliced together in different ways, with each splicing leading to a different code and, therefore, the synthesis of a different polypeptide. Through this mechanism, one gene can code for several proteins. Other aspects of post-transcriptional processing include addition of a chemical group called a *CAP* to the 5′ end, which is necessary for initiating translation, and addition of several adenine nucleotides called a poly A tail to the 3′ end, which protects the mRNA from degradation in the cytoplasm. After processing is complete, the mRNA enters the cytoplasm, and translation of mRNA can occur. As for the introns that were removed, it was once believed that they had no function, but recent studies indicate that they regulate the transcription of RNA by as yet unknown mechanisms.

Translation

Translation is the process in which polypeptides are synthesized using mRNA codons as a template for the assembly of the correct amino acids along the sequence. Another type of RNA molecule, tRNA, carries the appropriate amino acid to the ribosome based on interactions with the mRNA codon. Because of complementary base pairing within its single strand of nucleotides, tRNA is shaped somewhat like a cloverleaf (**Figure 32**). The 3′ end of the tRNA molecule contains a binding site for a specific amino acid. Another region of the tRNA contains a base sequence, called the *anticodon,* that is complementary to the mRNA codon. Ultimately, the correct interaction between the codon on the mRNA and the anticodon on the tRNA ensures that the correct amino acid is added to the elongating polypeptide chain.

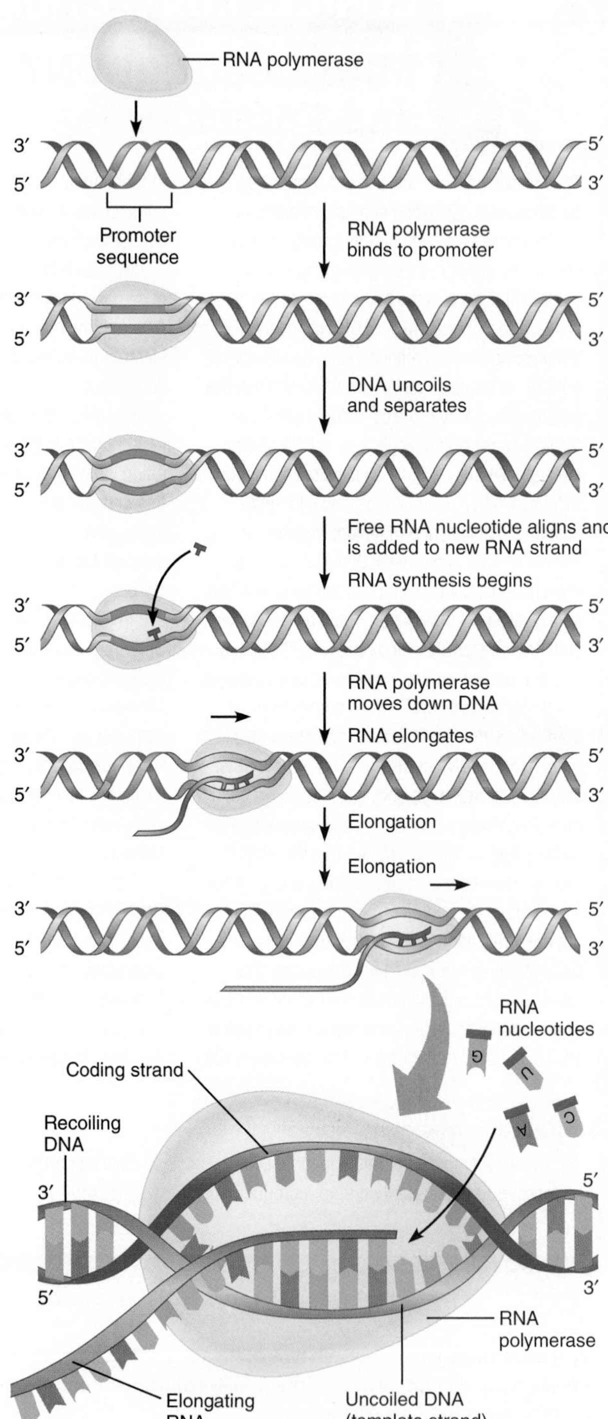

Figure 30 **Transcription.** RNA polymerase binds to the promoter sequence of a DNA molecule, causing the DNA to uncoil. The two strands of DNA separate, exposing the bases for complementary base pairing with free nucleotides. The free nucleotides align, and RNA polymerase catalyzes the formation of a bond between them, generating an elongating strand of mRNA. The enlargement shows that the process of elongation occurs according to the law of complementary base pairing.

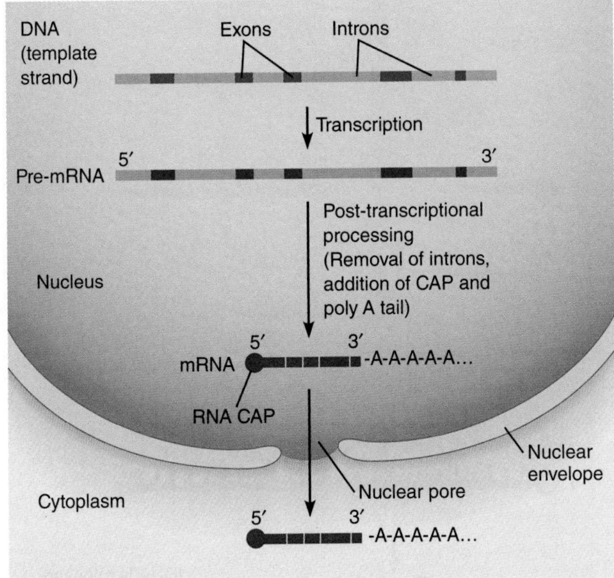

Figure 31 Post-transcriptional processing. Following transcription, pre-mRNA must undergo further processing in the nucleus. Transcribed introns must be removed, and the remaining exons spliced together. A chemical structure called a CAP is added to the 5′ end, and a poly A tail (several nucleotides with the adenine base) is added to the 3′ end. After processing is complete, the mRNA molecule can move from the nucleus to the cytoplasm.

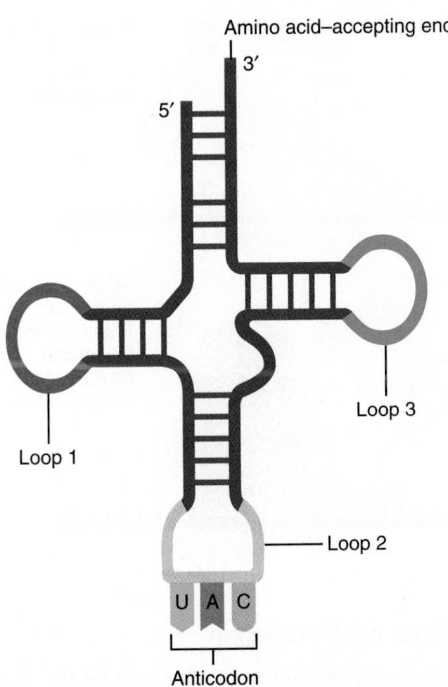

Figure 32 Transfer RNA. Hydrogen bonding between complementary base pairs is responsible for the cloverleaf structure of transfer RNA. The tRNA binds a specific amino acid on one end. At the other end, tRNA contains a base sequence called an anticodon that recognizes the complementary mRNA codon.

Translation occurs in the cytoplasm in association with ribosomes. Recall that ribosomes can be found free in the cytosol or attached to the rough endoplasmic reticulum. The ribosome aligns the mRNA with the tRNAs carrying the appropriately coded amino acids, and contains the enzymes that catalyze the formation of peptide bonds between the amino acids, thereby forming a polypeptide. Ribosomes can hold two tRNA molecules at the same time in two distinct ribosomal regions identified as the *P site* and the *A site*.

Now that we have some knowledge of the components needed for translation, we turn to the process itself. The first step of translation is *initiation* (**Figure 33**). Ribosomes not actively involved in translation are found free in the cytosol as their individual subunits. To start translation, certain *initiation factors* bind to the CAP that was added to the 5′ end of the mRNA molecule during post-transcriptional processing; other initiation factors form a complex with the 30S ribosomal subunit and a *charged tRNA* (tRNA bound to its amino acid) bearing the anticodon complementary to the initiation codon (AUG). Binding of the initiation factors triggers the binding of the 30S ribosomal subunit to the mRNA, such that the first

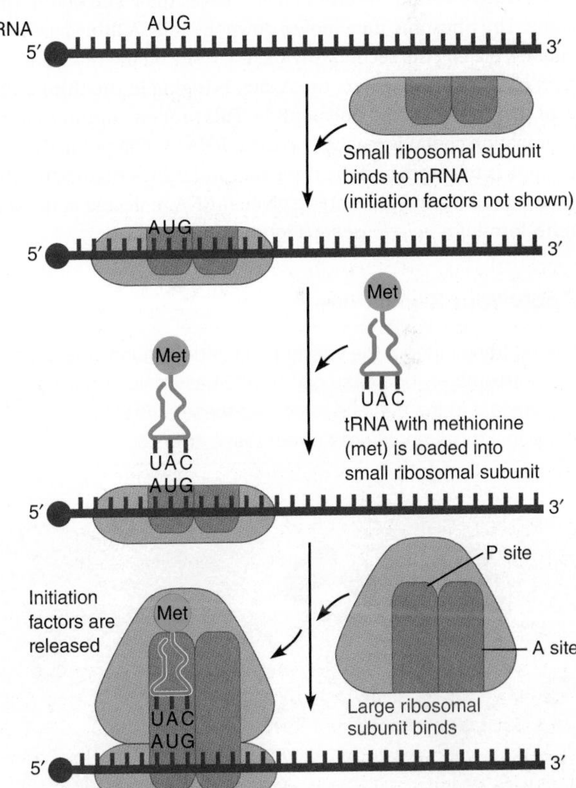

Figure 33 Initiation of translation. In the cytosol, initiation factors bind to the CAP region of an mRNA molecule and to the small ribosomal subunit, triggering the binding of the small ribosomal subunit to the mRNA. A tRNA with the anticodon complementary to the initiation codon of mRNA binds to the mRNA by the law of complementary base pairing. The large ribosomal subunit binds such that the first tRNA is located in the P site of the ribosome, and the initiation factors are released.

Figure 34 Translation. Following initiation, tRNA molecules will occupy sites on the ribosome according to complementary base pairing between the anticodon of the tRNA and the codon of the mRNA. The tRNA carries the appropriate amino acid to be added to the polypeptide chain. Enzymes in the ribosome catalyze the formation of a peptide bond. The first tRNA molecule is released from the P site of the ribosome. The ribosome then moves down the mRNA molecule one codon, putting the second tRNA in the P site. A new tRNA occupies the A site, bringing with it the next amino acid to be added to the polypeptide chain.

Q What is a DNA triplet code for the amino acid serine?

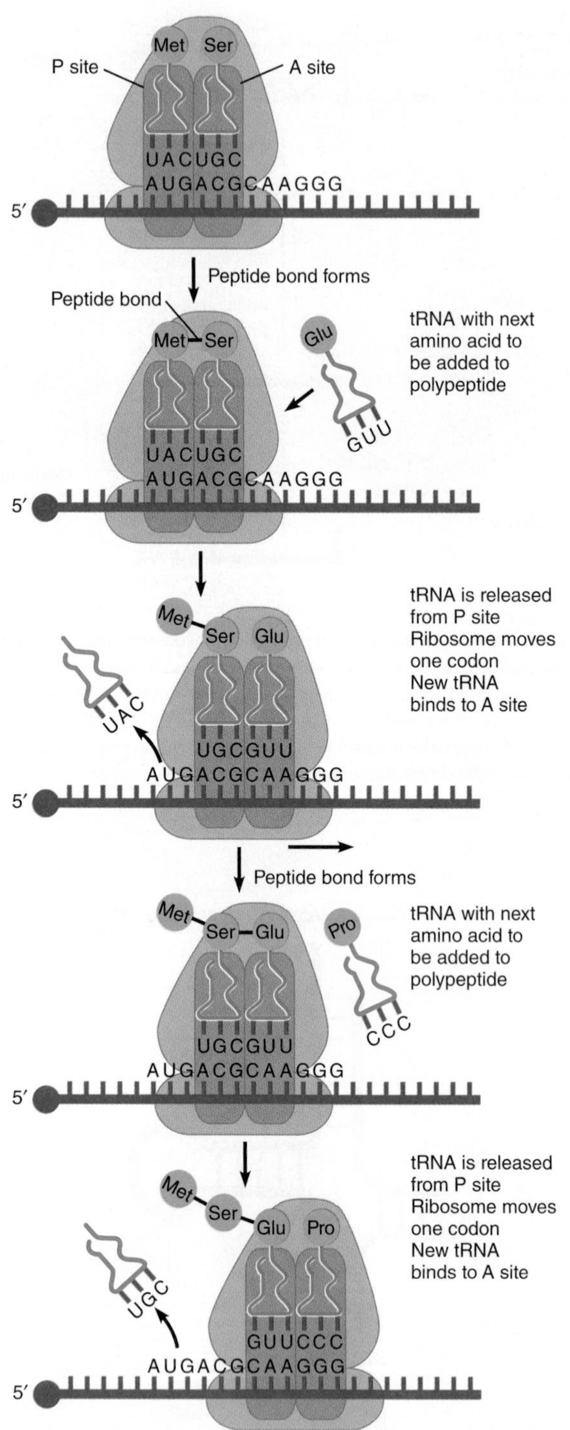

initiation codon (AUG) found within the mRNA is aligned correctly for the start of translation. The 50S ribosomal subunit then binds, causing dissociation of the initiation factors and alignment of the first tRNA in the P site of the ribosome. A second charged tRNA molecule with the appropriate anticodon then enters the A site of the ribosome, and translation begins. (To learn the dangers of chemicals that block translation, see **Clinical Connections: How Ricin Kills**)

During the process of translation (**Figure 34**), an enzyme in the ribosome, *peptidyl transferase,* catalyzes the formation of a peptide bond between the two amino acids. The first amino acid, which is always methionine, is released from the tRNA molecule, and the *free tRNA* (tRNA without its amino acid) leaves the P site on the ribosome. The ribosome then moves down the mRNA three bases (one codon), placing the second tRNA in the P site on the ribosome. The next charged tRNA enters the A site, bringing in the third amino acid to be added to the polypeptide. This process continues until a termination codon is reached on the mRNA. At that point, the polypeptide is released, and the ribosome and mRNA dissociate. More than one ribosome translates a given mRNA molecule at the same time, forming a *polyribosome* (**Figure 35**).

Apply Your Knowledge

During transcription, the wrong nucleotide sequence is attached in approximately one case per 10,000 base pairs. Enzymes are present in the nucleus to correct these errors. What would happen if a cell could not correct these errors?

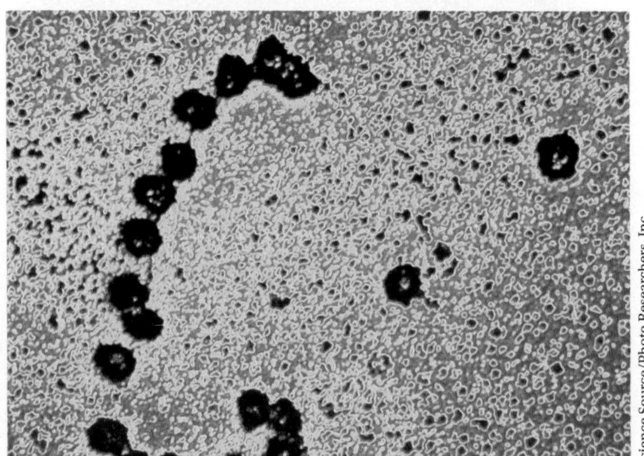

Science Source/Photo Researchers, Inc.

Figure 35 Colorized election micrograph of a polyribosome.

Destination of Proteins

Proteins synthesized in a cell can be destined for the nucleus, cytosol, plasma membrane, or one of the organelles, or they can be secreted from the cell into the extracellular fluid. Unless the final destination of a protein is to remain in the cytosol, the first functioning sequence of amino acids that is translated in a polypeptide chain is the **leader sequence.** The leader sequence is synthesized during translation by free ribosomes within the cytosol and determines the destination of the protein (**Figure 36a**). First, the leader sequence determines whether a protein will be synthesized in the cytosol or in association with the endoplasmic reticulum. Second, it determines whether proteins synthesized in the cytosol will remain there (in which case there is no leader sequence) or enter a mitochondrion, a peroxisome, or the nucleus (Figure 36b). The leader sequence determines the fate of proteins by specific interactions with other proteins on either the membrane of an organelle or the nuclear membrane. How the leader sequence causes ribosomes to attach to the endoplasmic reticulum and where proteins synthesized by those ribosomes are destined are described next.

The process of protein synthesis in association with the endoplasmic reticulum is illustrated in **Figure 37**. Protein synthesis starts in the cytosol, as previously described. If the leader sequence designates that synthesis is to occur in association with the endoplasmic reticulum, then the leader sequence and its associated ribosome bind to *signal recognition proteins* on the endoplasmic reticulum. Once the ribosome is attached to the endoplasmic reticulum,

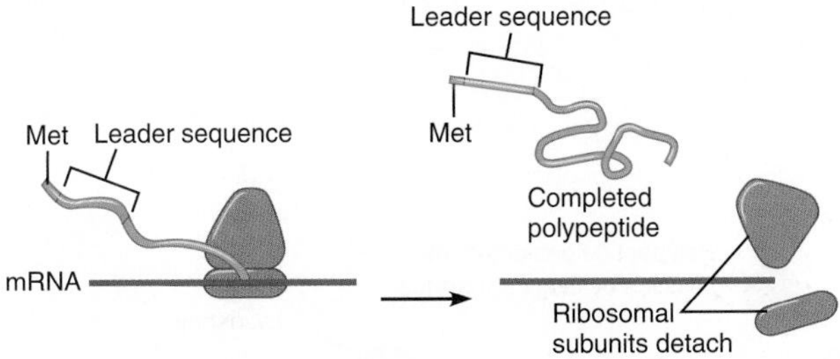

(a) Translation of leader sequence and polypeptide in cytosol

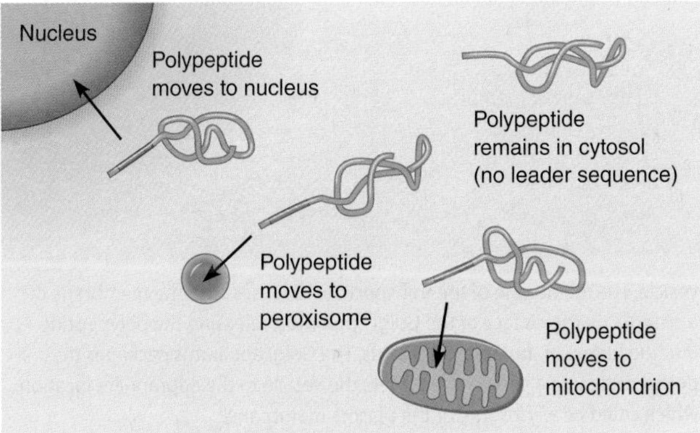

(b) Destination of polypeptides translated in cytosol

Figure 36 Targeting of proteins translated in the cytosol. (a) The first amino acid of a new polypeptide chain is a methionine (met) because its codon is also the initiation codon. A sequence of amino acids synthesized early during translation is the leader sequence. Following translation, the ribosome, mRNA, and polypeptide separate. **(b)** The leader sequence determines the ultimate fate of a protein—whether it stays in the cytosol (in which case no leader sequence would have been translated), moves into a mitochondrion or peroxisome, or enters the nucleus.

protein synthesis continues, with the polypeptide chain passing through the membrane and entering the lumen of the rough endoplasmic reticulum as it is formed. The polypeptide chain travels from the lumen of the rough endoplasmic reticulum to the lumen of the smooth endoplasmic reticulum, and eventually to the Golgi apparatus by mechanisms described shortly. Along this route, another important process also occurs: *post-translational processing.*

Post-translational Processing and Packaging of Proteins

Translation alone generally does not provide a functional protein. Changes in the initial polypeptide chain, called *post-translational processing,* must occur to produce the final protein product. Post-translational processing includes cleavage of some amino acids, which can occur anywhere along the route from the rough endoplasmic reticulum to the Golgi apparatus, and may even occur after packaging by the Golgi apparatus. For example, the leader sequence must be removed from the polypeptide; cleavage of other "excess" amino acids is also common. Other post-translational processing includes the addition of chemical groups such as lipids and carbohydrates to form lipoproteins and glycoproteins, respectively. Carbohydrates are added to the polypeptide in the endoplasmic reticulum and Golgi apparatus by a process called *glycosylation.*

From the lumen of the endoplasmic reticulum, the polypeptide must get to the Golgi apparatus (unless it is destined to remain in the endoplasmic reticulum), whose lumen is not continuous with that of the endoplasmic reticulum. To enable the polypeptide to make this journey, the smooth endoplasmic reticulum packages it into small *transport vesicles* that bud off of the smooth endoplasmic reticulum (see Figure 37). The membrane of the vesicle then fuses with the membrane of the Golgi apparatus, emptying its contents into the lumen of the Golgi apparatus at a Golgi cisterna close to the endoplasmic reticulum.

Once in the Golgi apparatus, the polypeptide undergoes further processing as it travels to the *trans* face on the opposite end of the Golgi apparatus. Because the sacs of the Golgi apparatus are separate membrane-bound structures, the mechanism by which the polypeptide moves from the *cis* sac to the *trans* sac is unclear but probably involves vesicular transport or actual movement of the Golgi sac. The Golgi apparatus sorts and packages proteins into vesicles that are targeted for their appropriate locations. The precise mechanism of how vesicles are directed to their targets is not fully understood, but certain chemicals function as markers for the sorting of proteins. Targets for proteins packaged by the Golgi apparatus include sites within the cell (lysosomes or Golgi apparatus) or the plasma membrane; alternatively, proteins may be secreted from the cell by exocytosis.

Regulation of Protein Synthesis

Cellular function can often be regulated by varying the amount of a certain protein in the cell. For example, increasing the number of carrier proteins enhances the transport of a particular molecule. Protein synthesis can be regulated either at the level of transcription or at the level of translation.

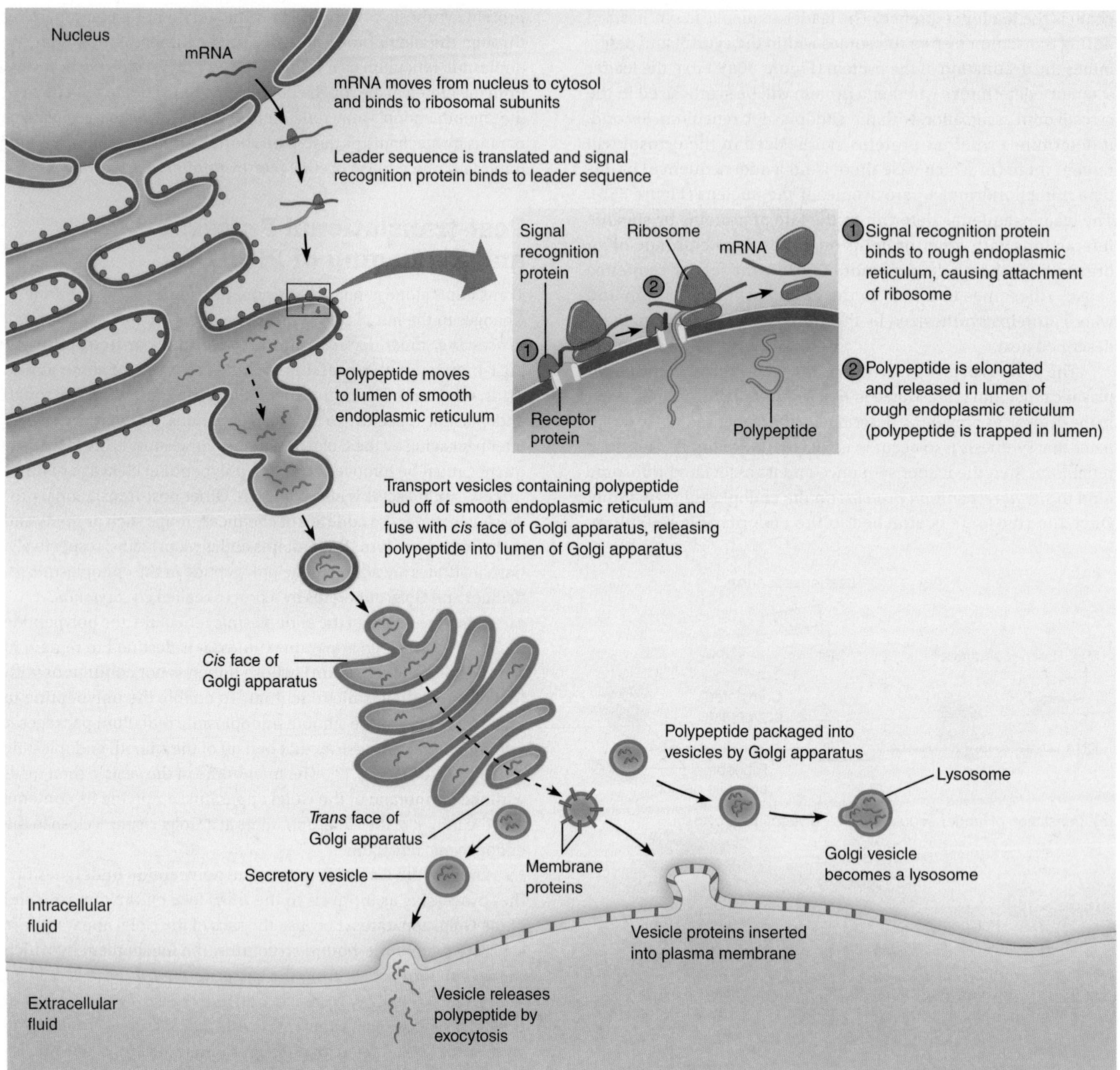

Figure 37 Synthesis of proteins on the endoplasmic reticulum. When the leader sequence causes the ribosome to attach to the endoplasmic reticulum, translation continues with the polypeptide forming in the lumen of the rough endoplasmic reticulum. The polypeptide then moves to the lumen of the smooth endoplasmic reticulum, where it is packaged into a transport vesicle. The membrane of the transport vesicle fuses with the membrane of a cisterna on the *cis* face of the Golgi apparatus, releasing the polypeptide into the lumen of the Golgi apparatus. The Golgi apparatus packages the polypeptide into a vesicle and targets the vesicle to the appropriate location, which could be a lysosome or the plasma membrane.

Regulation of Transcription

Once synthesized and in the cytoplasm, mRNA is present in a functional form only briefly. Some mRNA binds to proteins in the cytoplasm that render it inactive until an appropriate chemical signal is received; other mRNA is rapidly degraded by enzymes in the cytosol. Because mRNA is necessary for protein synthesis, the rate of transcription affects the amount of protein in a cell. Transcription can be either turned on *(induced)* or turned off *(repressed)*, depending on whether the cell requires more or less of the protein. Regulation of transcription typically occurs at the step in which RNA polymerase binds to the promoter sequence on the DNA. At this stage, certain molecules either enhance or decrease the ability of RNA polymerase to bind to the promoter sequence.

FOCUS ON DIABETES

One protein synthesized in the rough endoplasmic reticulum and packaged by the Golgi apparatus is insulin, the hormone that is deficient in 5% of diabetes mellitus cases. Insulin is synthesized in beta cells of the pancreas. Initially, ribosomes translate the mRNA to form a polypeptide of 110 amino acids called preproinsulin (**Figure 38**). The leader sequence is removed in the lumen of the endoplasmic reticulum, thereby forming a chain of 86 amino acids called proinsulin. Proinsulin folds into its tertiary structure, which includes three disulfide bridges, and is transported to the Golgi apparatus, where it is packaged into secretory vesicles. Inside the vesicles, proinsulin is cleaved into insulin, which consists of 51 amino acids and C-peptide. C-peptide is found in the center of the proinsulin (the connecting chain); thus the final insulin product comprises two polypeptide strands held together by two disulfide bridges. Both insulin and C-peptide are secreted by exocytosis, and both have biological activity.

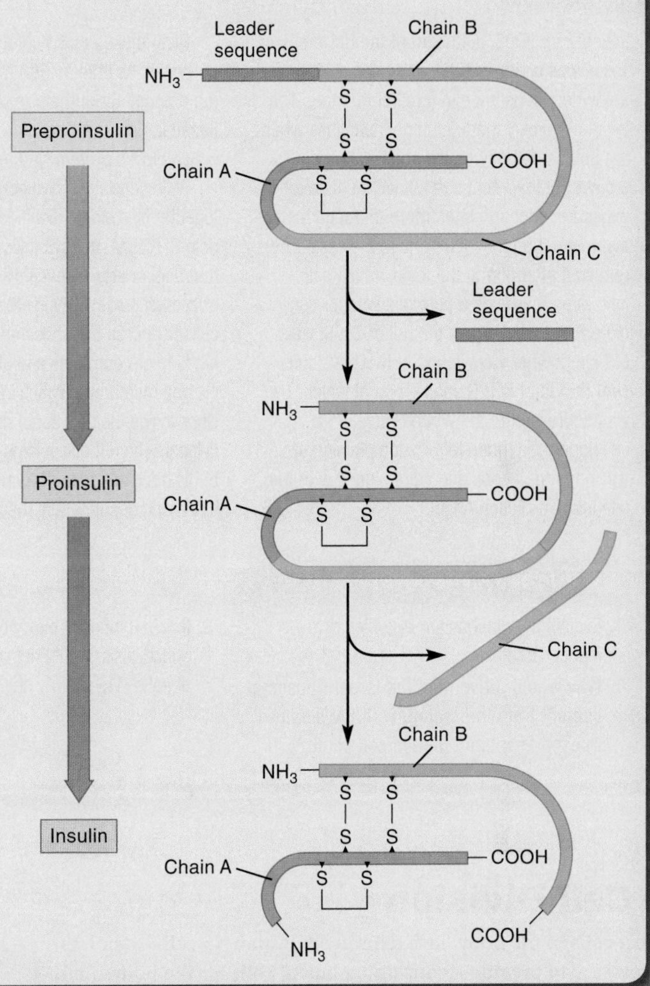

Figure 38 Synthesis of insulin. Insulin is initially synthesized as preproinsulin, which is converted to proinsulin by cleavage of 24 amino acids. Proinsulin is then converted to insulin by cleavage of another 35 amino acids.

Regulation of Translation

Cells can regulate the translation of proteins by mechanisms that remain poorly understood. Regulation of translation typically occurs during initiation. For initiation to occur, at least 11 proteins, including initiation factors, must be present. Some of these proteins can be rendered active or inactive, thereby turning translation on or off.

Protein Degradation

Most proteins in cells are present for a long period of time, but other proteins are short-lived. Most short-lived proteins are marked for rapid degradation by another protein called *ubiquitin.* The binding of several ubiquitin molecules to a protein causes the protein to associate with a *proteasome,* a complex that contains *proteases* to hydrolyze proteins to their amino acid subunits. The ubiquitin molecules will be released from the complex to tag another protein.

Quick Check 4

1. Define *transcription* and *translation*. Where in the cell do these processes occur?
2. Describe the functions of RNA polymerase in transcription.
3. What is the name of the first part of a polypeptide chain that is translated from mRNA? What is its function?

CLINICAL CONNECTIONS

How Ricin Kills 

In February 2005, a student at the University of Texas was doing laundry when she discovered a powdery substance in a roll of quarters. The incident quickly made national headlines when the powder initially tested positive for ricin, a powerful toxin that can kill within hours of exposure. Although later, more thorough tests disproved the presence of ricin, the case drew renewed attention to the toxin, which has long been feared as a potential weapon for bioterrorists. Indeed, in the fall of 2003, vials of ricin powder were found in two envelopes sent through the U.S. mail—one of which was addressed to the White House. Both envelopes contained letters complaining of trucking regulations and were signed with the pseudonym "Fallen Angel."

Ricin makes up 5% of the dry weight of castor beans. As little as 500 micrograms of ricin (an amount about the size of a pinhead) is lethal because ricin is a ribosome-inactivating protein, which blocks protein synthesis.

Ricin consists of two polypeptide chains held together by a single disulfide bridge: ricin A and ricin B. Ricin A, the cytotoxic portion, is an enzyme that depurinates a specific adenine in the large ribosomal subunit. By itself, ricin A cannot enter cells to access the ribosomes. Ricin B, however, binds to glycoproteins and glycolipids of the plasma membrane, which induces endocytosis. Once inside the cell, some ricin escapes into the cytosol, where it dissociates into ricin A and ricin B. Ricin A binds to the 50S ribosomal subunit catalyzing depurination, thereby preventing the 50S subunit from interacting with the 30S subunit during protein synthesis. A single ricin A molecule can inactivate as many as 50,000 ribosomes, thereby preventing protein synthesis and causing cell death. There is no known cure for ricin poisoning, but researchers are currently investigating ways of inactivating the ricin with antibodies that would bind to the ricin B component, preventing ricin from entering cells.

Ted Kinsman/Photo Researchers, Inc.

Castor beans.

Critical Thinking Questions

1. Explain the mechanism by which ricin induces death.
2. Why would preventing the larger ribosomal subunit from interacting with the smaller ribosomal subunit induce cell death?
3. If ricin B induced exocytosis rather than endocytosis, would ricin exposure still be deadly? Why or why not?

6 Cell Division 

Most cells in the body have a limited life span, so cell division is necessary to provide a continuing line of cells for the body. Cell division is also necessary for tissue growth, especially during the early years of life. Two types of cell division take place: **mitosis,** which occurs in most cells, and **meiosis,** which occurs only in reproductive cells. Here we focus on mitosis.

Replication of DNA 

Each cell in the body (except mature red blood cells) contains a copy of all the body's genes in its DNA. Therefore, when a cell divides, DNA must be copied exactly so that each daughter cell receives a complete set of DNA. Before we can understand the process of cell division, we must understand the mechanism of copying DNA, called **replication.**

Together, one complete molecule of DNA carrying a specific set of genetic information, plus its associated proteins, make up a chromosome (**Figure 39**). Most human cells contain 23 pairs of chromosomes. Within each pair, one chromosome was inherited from the father, and one from the mother. Together, these 23 pairs of chromosomes form the *human genome.* Within the nucleus, the chromosomes are coiled around proteins called *histones.* Chromosomes are usually scattered throughout the nucleus in the form of beadlike structures called *chromatin.*

During DNA replication, each strand of DNA serves as a template for the synthesis of a new strand of DNA. This process is termed *semiconservative* because each new DNA molecule consists of a new strand plus an old strand (or conserved strand) that functioned as the template for the new strand. The process starts with the action of **helicase,** an enzyme that causes a section of DNA to unwind and separate forming a *replication fork* (**Figure 40**).

Each strand of DNA acts as a template for adding the appropriate complementary nucleotide to the growing new strand. The nucleotides, in their triphosphate forms (ATP, TTP, CTP, and GTP), are found free within the nucleus. When they align according to the law of complementary base pairing, the enzyme **DNA polymerase** catalyzes a reaction that adds the next nucleotide to

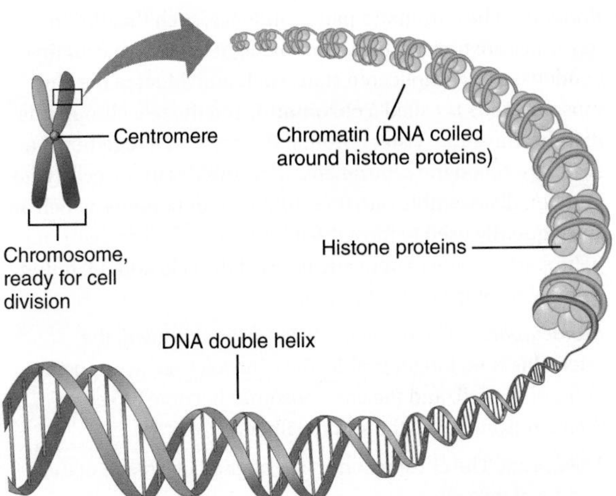

Figure 39 A chromosome. A chromosome is a tightly wound and organized structure present just prior to cell division. At other times, a cell's genetic material is present in the nucleus as chromatin, which consists of the DNA double helix wrapped around histone proteins.

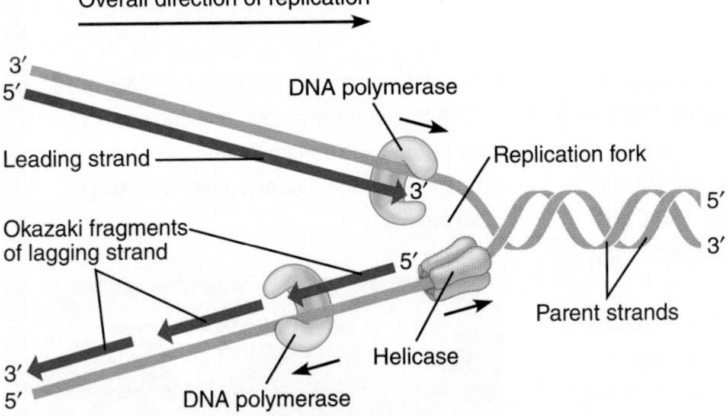

Figure 40 DNA replication. Helicase catalyzes unwinding of DNA, with the two strands separating to expose their component bases. DNA polymerase catalyzes the formation of complementary DNA from both strands, with the leading strand synthesized from the 5′ to 3′ parent strand and the lagging strand from the 3′ to 5′ parent strand.

the growing polynucleotide chain. In the process, two phosphates are cleaved from the nucleotide, providing some of the energy for the reaction.

DNA polymerase catalyzes replication only from 5′ to 3′. This poses a problem because DNA consists of two antiparallel strands. For the 5′ to 3′ strand, DNA polymerase simply starts at one end of the unwound section of DNA and catalyzes formation of the complementary strand from start to finish—this newly synthesized strand is called the *leading strand.* For the 3′ to 5′ strand of DNA, DNA polymerase catalyzes portions of the new DNA from 5′ to 3′, forming a series of segments, called *Okazaki fragments.* Another enzyme, *DNA ligase,* links the Okazaki fragments together to form the *lagging strand.* The end result of the process is two identical copies of the DNA, one to pass on to each of two daughter cells during cell division.

Normally, DNA replication is precise; any errors that may have occurred are corrected by enzymes in the nucleus. Sometimes, however, alterations in the DNA, called *mutations,* may occur. Mutations within a single cell are usually not deleterious to the whole body. In contrast, mutations of genes that control cell growth, called *proto-oncogenes,* can be very serious because growth of cells containing mutated *proto-oncogenes* is no longer regulated. Such mutations are responsible for cancer (**Clinical Connections: Cancer**).

The Cell Cycle

The life cycle of a cell can be separated into two main divisions: interphase and cell division. During **interphase,** the cell carries out its normal physiological functions; during *cell division,* the cell stops performing many of its normal physiological functions. **Figure 41** shows the cell cycle. Notice that interphase is the period between the previous cell division and the beginning of the next cell division. The duration of interphase varies considerably among cell types. For example, cells lining the gastrointestinal tract undergo cell division every few days, whereas skeletal muscle cells never undergo cell division.

Interphase can be divided into four phases: G_0, G_1, *S*, and G_2. The "G" stands for "gap," because little related to cell division is

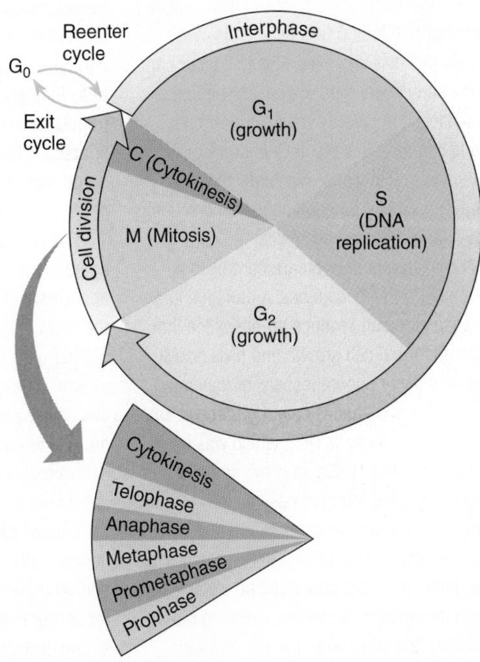

Figure 41 The cell cycle. Most cells move through a continuous cycle that includes cell division. Interphase includes the periods between cell divisions and includes four phases: G_0, G_1, S, and G_2. Mitosis and cytokinesis are the periods of actual cell division. Mitosis includes five phases: prophase, prometaphase, metaphase, anaphase, and telophase. Cytokinesis is the division of the cytoplasm. Some cells leave the cycle and enter G_0, during which no activities related to cell division occur. Some of these cells remain in G_0, whereas others reenter the cycle in response to a chemical signal.

happening in the nucleus. During G_0, the cell carries out its functions that are unrelated to cell division. Cells that do not undergo cell division, such as skeletal muscle cells, remain in G_0 for their entire life. Other cells, such as certain white blood cells, require a signal to move from G_0 to G_1. During G_1, the cell starts showing signs of an upcoming cell division, such as an increase in the rate of protein synthesis. In the S phase, DNA replication occurs, and the chromosomes now exist in pairs. During interphase, the centrioles are duplicated, so that they also exist in pairs at the start of cell division. During G_2, rapid protein synthesis continues as the cell grows to about twice its original size. From G_2, the cell enters cell division.

Cell division can be divided into two phases: mitosis and cytokinesis (see Figure 41). **Mitosis,** the division of the nuclear components, includes five phases: (1) prophase, (2) prometaphase, (3) metaphase, (4) anaphase, and (5) telophase. Although the phases of mitosis are continuous and overlap, they can be identified by certain events:

1. *Prophase:* The chromatin pairs condense, such that the chromosomes are now visible under the light microscope. In this condensed and duplicated state, each individual of the chromosome pairs is called a **chromatid,** and the two chromatids in a chromosome (sister chromatids) are linked together at a structure called the *centromere.* Microtubules in the cell's cytoskeleton disassemble into their tubulin components, which are subsequently used to form the mitotic spindle. The centriole pairs start moving to opposite poles of the cell, and the mitotic spindle develops between them.
2. *Prometaphase:* The nuclear envelope breaks down, the nucleolus is no longer visible, the centrioles are at opposite poles of the cell, and the chromosomes become linked at their centromeres to the spindle fibers.
3. *Metaphase:* The chromosomes are aligned in a plane at the middle of the cell.

CLINICAL CONNECTIONS

Cancer

Each year in the United States, more than 1 million individuals are diagnosed with cancer, and approximately 500,000 people die every year from this disease. Even though cancer can arise in nearly every type of body tissue and leads to illness in diverse ways, at least one theme is common to cancer: Cancerous cells divide uncontrollably and eventually displace the body's normally functioning cells, consuming energy vital to normal body cells.

Cell growth and division are controlled by genes called *proto-oncogenes.* A mutation in a proto-oncogene can produce an *oncogene* that leads to abnormal cell growth and division. Mutations may occur spontaneously, or they may be induced by substances called *mutagens.* One example of a mutagen is DDT, which was used in pesticides in the 1960s. In some cases, including cancer, the mutated cells divide rapidly, passing the mutation on to their daughter cells, which then also divide rapidly. To rid the body of cancer, these mutated cells must be eliminated.

Most therapeutic strategies, including radiation therapy and chemotherapy, are directed at different stages of actively dividing cells. Because cancer cells divide rapidly, these therapies harm cancer cells more than they harm the normal, infrequently dividing cells of the body. Unfortunately, normal body cells that divide frequently are also affected by these therapeutic strategies. For example, because hair cells undergo rapid cell division, hair loss is typically one of the consequences of cancer therapy.

The role of genetics and of environmental agents (such as tobacco) in the initiation of abnormal cell growth and division is currently the topic of intense research. Researchers are learning to manipulate the body's immune system, hoping to halt the growth of cancerous cells by "tricking" the patient's own immunity defenses into destroying the cancer. Sometimes a patient's immune system will not recognize cancer cells as foreign because the cancer cells are not different enough from normal cells to stimulate an immune reaction. At other times the immune system may recognize cancer cells, but its response is not strong enough to destroy the cancer. Various kinds of immunotherapies have been designed to help the immune system recognize cancer cells as distinctly foreign and to strengthen the attack in the hope that malignant cells will be destroyed.

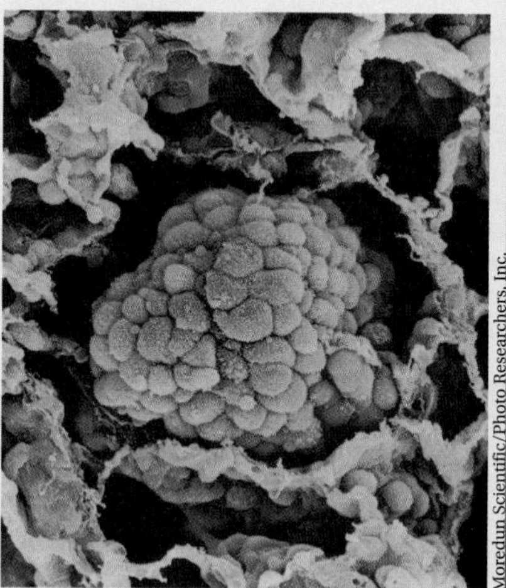

Color-enhanced scanning electron micrograph of cancerous tumor in lung tissue.

Critical Thinking Questions

1. Explain why and how a mutation in a proto-oncogene causes cancer.
2. Which characteristic(s) of cancer cells are therapeutic treatments directed toward, and why?
3. How would strengthening the immune system assist in overcoming cancer?

4. *Anaphase:* The chromatid pairs separate, and the chromosomes move along the mitotic spindle toward opposite poles of the cell.
5. *Telophase:* New nuclear envelopes develop on the two sides of the cell, the chromosomes begin to decondense to chromatin, and the mitotic spindle begins breaking down.

Cytokinesis is the division of the cytoplasm. The cytoplasm divides by cleavage, which starts during the latter phases of mitosis (anaphase or telophase). The cleavage is accomplished by contraction of microfilaments (primarily actin) that form a ring around the midline of the cell, and the mitotic spindle disassembles. The two resulting daughter cells then enter interphase at G_0.

Quick Check 5

1. Define *replication.* How does replication differ from transcription?
2. Name the five stages of mitosis. During which stage does the nuclear envelope start to disintegrate?

SYSTEMS INTEGRATION In this chapter, we learned about cells and the biomolecules found in them. We also learned about proteins in detail; proteins carry out many important functions in the cells.

This chapter also introduced the various organelles inside cells and their unique functions. As we go through the systems, certain organelles will be emphasized, such as smooth endoplasmic reticulum in muscle cells and lysosomes in white blood cells.

CHAPTER REVIEW

SUMMARY

1 Biomolecules

- The four basic types of biomolecules are carbohydrates, lipids, proteins, and nucleotides.
- Carbohydrates, which are polar molecules, include monosaccharides, disaccharides, and polysaccharides.
- Lipids, which are generally nonpolar molecules, include triglycerides, phospholipids, eicosanoids, and steroids.
- Phospholipids are amphipathic molecules.
- Proteins are polymers of amino acids.
- Nucleic acids include DNA and RNA and are polymers of nucleotides.

2 Cell Structure

- The plasma membrane separates the cell from the extracellular fluid; it is composed of a phospholipid bilayer with proteins and cholesterol embedded within the bilayer.
- Carbohydrates are found on the outer surface of the membrane as glycolipids or glycoproteins.
- The nucleus (inside the cell) contains DNA.
- The cytoplasm (also inside the cell) consists of cytosol and organelles.
- The cytosol contains enzymes, glycogen and triglyceride stores, and secretory vesicles.
- The organelles include the rough and smooth endoplasmic reticulum, the Golgi apparatus, mitochondria, lysosomes and peroxisomes, and ribosomes.
- The cytoskeleton gives the cells their shape, provides structural support, transports materials within the cell, suspends organelles, forms adhesions with other cells to form tissues, and causes contraction or movement in certain cells.
- The cytoskeleton consists of protein filaments including microfilaments, intermediate filaments, and microtubules.

MAP iP The Cell

3 Cell-to-Cell Adhesions

- Three types of cell adhesions are found between cells: tight junctions, desmosomes, and gap junctions.

4 General Cell Functions

- General functions carried out by almost all cells include metabolism, cellular transport of molecules across membranes, and intercellular communication.

5 Protein Synthesis

- DNA has genes that contain the codes for protein synthesis, which is transcribed to mRNA in the nucleus.
- The mRNA then moves to the cytoplasm, where it is translated by ribosomes to synthesize proteins.
- When a protein is synthesized in the cytosol, the leader sequence determines whether the protein will remain in the cytosol or enter a mitochondrion, a peroxisome, or the nucleus.
- When a protein is synthesized in association with the endoplasmic reticulum, it will eventually be packaged into vesicles by the Golgi apparatus and directed to the appropriate site in the cell or secreted from the cell.

6 Cell Division

- Interphase is the period between cell divisions; mitosis and cytokinesis make up the process of cell division.
- DNA is replicated during the S phase of interphase so that each daughter cell will receive an exact DNA copy.
- The five phases of mitosis are prophase, prometaphase, metaphase, anaphase, and telophase. Mitosis is followed by cytokinesis.

Check your understanding at MasteringA&P and apply your new knowledge to real-life case studies to ensure that you've mastered the content!

EXERCISES

Multiple-Choice Questions

1. Which of the following biomolecules is *not* a polymer?
   a) Polysaccharide
   b) Phospholipid
   c) Protein
   d) Nucleic acid

2. A fatty acid that has two double bonds between carbon atoms is called
   a) A saturated fatty acid.
   b) A desaturated fatty acid.
   c) A monounsaturated fatty acid.
   d) A polyunsaturated fatty acid.
   e) An eicosanoid.

3. Which of the following molecules is *not* a component of a phospholipid?
   a) Cholesterol
   b) Glycerol
   c) Phosphate
   d) Fatty acid

4. Hydrogen bonding between the amino hydrogen of one amino acid and the carboxyl oxygen of another amino acid in a protein is responsible for
   a) Primary protein structure.
   b) Secondary protein structure.
   c) Tertiary protein structure.
   d) Quaternary protein structure.

5. Which of the following nucleic acids is *not* a pyrimidine?
   a) Cytosine
   b) Thymine
   c) Uracil
   d) Adenine

6. Glycogen and lipids are stored
   a) As inclusions in the cytosol.
   b) In storage vesicles.
   c) In secretory vesicles.
   d) In lysosomes.
   e) In the nucleolus.

7. Which organelle produces most of a cell's ATP?
   a) Nucleus
   b) Peroxisome
   c) Golgi apparatus
   d) Mitochondrion
   e) Smooth endoplasmic reticulum

8. Which cell-to-cell adhesion allows quick transmission of electrical signals between adjacent cells?
   a) Desmosome
   b) Gap junction
   c) Tight junction
   d) All of the above

9. To initiate transcription of DNA, the enzyme RNA polymerase binds to the
   a) Initiator codon.
   b) Termination codon.
   c) Promoter sequence.
   d) Centromere.
   e) Leader sequence.

10. The base sequence on tRNA that recognizes the mRNA codon by the law of complementary base pairing is called the
    a) tRNA codon.
    b) Anticodon.
    c) Amino codon.
    d) Leader codon.
    e) Initiator codon.

11. For proteins to be synthesized in association with the rough endoplasmic reticulum, the leader sequence must bind to a ________ on the membrane of the endoplasmic reticulum.
    a) transport vesicle
    b) secretory vesicle
    c) coated vesicle
    d) signal recognition protein
    e) promoter

12. During replication, which enzyme causes DNA to uncoil?
    a) DNA polymerase
    b) RNA polymerase
    c) DNA kinase
    d) Helicase
    e) DNA ligase

13. What is a lagging strand?
    a) A sequence of nucleotides in DNA that binds RNA polymerase
    b) A sequence of amino acids at the beginning of a polypeptide that directs the polypeptide to its destination
    c) The newly synthesized DNA strand formed by ligation of Okazaki fragments
    d) The protein filaments that extend between two cells to form desmosomes
    e) The protein filaments that form cilia

14. During which phase of mitosis is the mitotic spindle developing?
    a) Prophase
    b) Metaphase
    c) Anaphase
    d) Telophase
    e) Prometaphase

15. During which phase of mitosis do the chromosome pairs move to opposite poles of the cell?
    a) Prophase
    b) Metaphase
    c) Anaphase
    d) Telophase
    e) Prometaphase

Objective Questions

16. Monosaccharides are (polar/nonpolar/amphipathic) molecules.
17. Triglycerides are (polar/nonpolar/amphipathic) molecules.
18. The precursor molecule for all steroids is ________.
19. According to the law of complementary base pairing, in RNA, adenine base pairs with ________.
20. Transmembrane proteins are examples of (integral/peripheral) membrane proteins.
21. The membrane of the smooth endoplasmic reticulum is continuous with the membrane of the Golgi apparatus. (true/false)
22. Microfilaments provide structural support for hairlike projections of the plasma membrane called ________.
23. The proteins that form gap junctions between two adjacent cells are called ________.
24. A section of DNA that codes for a specific protein is called a ________.
25. The mRNA codon that is transcribed from the DNA triplet ATC is ________.

26. More than one codon may code for a single amino acid. (true/false)

27. The first section of a polypeptide to be translated is called the ________; it is important in directing the translated protein to its final destination.

28. Replication of DNA occurs during (interphase/mitosis).

29. Division of the cytoplasm into two daughter cells during cell division is called ________.

30. The chemical that tags proteins for destruction in a cell is called ________.

Essay Questions

31. Name the four major classes of biomolecules. Describe the major function of each class.

32. Define the four levels of protein structure, and explain the chemical interactions responsible for each level.

33. Describe the structure of the plasma membrane. Name at least one function for each of the major components.

34. Certain cells in the adrenal gland synthesize and secrete lots of steroid hormones. Which organelle would be abundant in these cells, and why?

35. Name the three special types of cell adhesions, and list the special functions of each.

36. Define each of the following terms: *transcription, translation,* and *replication.* Where does each of these processes occur in a cell?

37. Describe the anatomical and functional relationship between the rough endoplasmic reticulum, smooth endoplasmic reticulum, and Golgi apparatus in protein synthesis.

38. Describe how the concentration of a certain protein in a cell is regulated.

39. Draw a flowchart showing the synthesis of proteins and targeting of proteins to their final destination in the cell.

Critical Thinking

40. Plasma, which is the fluid portion of blood, is made up of mostly water. The plasma membrane consists of mostly phospholipids. Based on what you know about biomolecules, predict which classes of biomolecules can dissolve in plasma and which can cross the plasma membrane.

41. Name some factors that cause cancer and potential targets of these factors in cells. How can radiation both cause cancer and be used to treat it?

42. What are some of the differences you would expect to find between muscle cells, fat cells, saliva-secreting cells (saliva contains mostly water and proteins), and skin cells?

43. Explain how a mutation of a single base (called a *point mutation*) in a gene may or may not produce a disease. Can you name a disease that is caused by a point mutation in the protein *hemoglobin*? What is the role of mutations in evolution? Are all mutations deleterious?

44. The average human chromosome contains 150×10^6 nucleotide pairs. There are 23 human chromosomes that exist as pairs in all of our cells except the gametes. Errors in replication must occur on occasion. What are the consequences of these errors? Do enzymes in the cell fix all the errors?

Photo credits for icons in feature boxes: Focus on Diabetes: Tanyae/Dreamstime; Clinical Connections: lofoto/Dreamstime; Understanding Exercise: Pixattitude/Dreamstime; Discovery: Yuri Arcurs/Cutcaster.

Glossary

amino acid (ah-meen-oh) biomolecule containing an amine group, a carboxyl group, hydrogen, and a residual (R) group attached to a central carbon; found in proteins

amphipathic (am-fuh-path-ick) a molecule having both polar and nonpolar regions, as in phospholipid molecules

anabolism (an-nab-oh-lizm) synthesis of large molecules from smaller molecules, generally requiring an input of energy

biomolecules molecules synthesized by living organisms that contain carbon

catabolism breakdown of large molecules
into smaller molecules; generally releases energy

cellulose polysaccharide of glucose found in plants that humans are unable to digest or absorb

chromatin the form in which DNA, along with its associated proteins, exists throughout most of the cell cycle; loosely coiled DNA and proteins scattered throughout the nucleus

cilia hairlike processes found on certain epithelial cells in the respiratory tract and oviduct

codon three-base sequence in mRNA that codes for a specific amino acid

cytokinesis phase of cell division during which the cytoplasm divides

cytoplasm everything inside the cell except the nucleus

cytoskeleton flexible lattice of fibrous proteins that gives structure and support to the cell

deoxyribonucleic acid (DNA) (dee-ox-see-ry-boh-noo-klay-ik) a biomolecule consisting of two strands of nucleotides coiled together as a double helix; found in the nucleus and stores genetic information

desmosome (dez-moh-some) filamentous junction between two adjacent cells that provides a strong physical linkage between them; enables tissues to withstand stretching without cells being torn apart from one another

disaccharide (dy-sak-er-ide) carbohydrate consisting of two monosaccharides covalently bonded together

eicosanoids modified fatty acids (all derived from arachidonic acid) that function in intercellular communication

endoplasmic reticulum (en-doh-plas-mik reh-tik-you-lum) elaborate network of membranes inside cells that encloses a single interior compartment; includes rough endoplasmic reticulum and smooth endoplasmic reticulum

fatty acids long hydrocarbon chains with a carboxyl group (i COOH) at one end

genetic code the correspondence between DNA triplets and specific amino acids that governs the expression of all genetic information

glucose most common monosaccharide; provides important source of cellular energy

glycerol (gliss-er-ol) three-carbon alcohol that functions as the "backbone" of a triglyceride or phospholipid

glycogen (gly-coh-jen) a glucose polymer found in animal cells; functions as an energy store

Golgi apparatus (goal-jee) an organelle consisting of membrane-bound flattened sacs called *cisternae* that process molecules synthesized in the endoplasmic reticulum and prepare them for transport

helicase enzyme that causes a section of DNA to unwind during replication

inclusions cytosolic particles composed of triglycerides or glycogen; serve as energy stores for cellular metabolism

integral membrane proteins proteins that are tightly embedded within the lipid bilayer

intermediate filaments fibrous proteins with a diameter between that of microfilaments and microtubules; stronger and more stable than microfilaments

interphase period in the life cycle of the cell during which it is carrying on its normal physiological functions

law of complementary base pairing states that whenever two strands of nucleic acids are held together by hydrogen bonds, G in one strand is always paired with C in the opposite strand, and A is always paired with T in DNA (or U in RNA)

leader sequence initial sequence of amino acids that is present in a newly synthesized polypeptide chain and functions in directing the fate of the final protein

lipids biomolecules that contain primarily carbon and hydrogen atoms linked together by nonpolar covalent bonds

lysosomes small spherical membrane-bounded organelles containing numerous degradative enzymes; involved in the breakdown of unneeded intracellular material or foreign matter that has been taken into the cell

meiosis (my-oh-sis) a series of two cell divisions that generate daughter cells (gametes) with half the normal chromosome number; involved in gametogenesis

messenger RNA (mRNA) molecule that carries genetic information from the nucleus to the cytoplasm; contains the code for the amino acid sequence of a protein

metabolism sum of all the chemical reactions that take place in the body

microfilaments the smallest-diameter protein filaments of the cytoskeleton

microtubules the largest-diameter protein filaments of the cytoskeleton, composed of long, hollow tubes made of tubulin

mitochondria oval-shaped organelles bounded by two membranes; primary site of adenosine triphosphate (ATP) synthesis in the cells

mitochondrial matrix the innermost compartment in mitochondria, bounded by the inner mitochondrial membrane; contains the Krebs cycle enzymes

mitosis (my-toh-sis) the type of cell division that yields two daughter cells containing the normal number of chromosomes; necessary for the growth and replacement of cells

monosaccharide (mah-no-sak-er-ide) simple sugar composed of a single unit

nuclear envelope barrier that separates the nucleus from the cytoplasm; consists of two membranes

nuclear pores pores in the nuclear envelope that allow selective movement of molecules between the nucleus and cytoplasm

nucleic acids polymers of nucleotides that function in the storage and expression of genetic information; see deoxyribonucleic acid (DNA) and ribonucleic acid (RNA)

nucleotide a biomolecule containing one or more phosphate groups, a five-carbon carbohydrate, and a nitrogenous base; involved in energy exchange and in the storage and transmission of genetic information in cells

nucleus in cells, a membrane-bound structure that contains a cell's DNA; in the central nervous system, a collection of neuron cell bodies other than in the cerebral cortex; plural, *nuclei*

organelles structural units of cells consisting of biomolecules; carry out specific functions of the cell

peptides short chains of amino acids (usually less than 50)

peripheral membrane proteins proteins that are loosely bound to the lipid bilayer by associations with integral membrane proteins or phospholipids

peroxisomes spherical membrane-bounded organelles that function in the degradation of molecules such as amino acids, fatty acids, and toxic foreign matter

phospholipid amphipathic lipid molecule consisting of a glycerol backbone to which two fatty acids and a phosphate-containing chemical group are attached

plasma membrane barrier separating a cell from the extracellular fluid; consists of phospholipids, proteins, and cholesterol

polypeptide polymer containing amino acids joined together by peptide bonds

polysaccharide polymer composed of many monosaccharides joined together by covalent bonds

promoter sequence a specific base sequence in DNA to which the enzyme RNA polymerase can bind, thereby initiating transcription

protein a polymer containing amino acids joined together by peptide bonds; usually refers to chains containing more than 50 amino acids

replication duplication of DNA prior to cell division

ribonucleic acid (RNA) polynucleotide molecules found in a cell's nucleus and cytoplasm that are necessary in the expression of genetic information; includes messenger RNA (mRNA), transfer RNA (tRNA), and ribosomal RNA (rRNA)

ribosomes complexes of ribosomal RNA (rRNA) and proteins that function in protein synthesis; found in the cytosol and rough endoplasmic reticulum

secretory vesicles intracellular vesicles containing molecules destined for secretion from the cell

starch polysaccharide found in plants

steroids lipids derived from cholesterol consisting of three six-carbon rings and one five-carbon ring; many function as hormones

tight junction a junction that connects two adjacent cells together, forming a nearly impenetrable barrier that limits the passage of molecules between them

transmembrane proteins integral membrane proteins that span the lipid bilayer, with surfaces exposed to both the cytosol and interstitial fluid

triglyceride a lipid consisting of three fatty acids linked to a glycerol backbone; commonly called *fat*

triplet three-base sequence in DNA that codes for a specific amino acid

vaults barrel-shaped cell organelles that may be involved in the transport of molecules across the nuclear envelope and may have other functions

Answers

Figure Question Answers

Fig 1	5
Fig 2	The fatty acid in Figure 2b
Fig 11	Hydrogen bonds, iconic bonds, and van der Waals forces
Fig 12	Ribose is $C_5H_{10}C_5$; deoxyribose is $C_5H_{10}C_4$
Fig 14	Covalent bonds
Fig 34	TGC

Apply Your Knowledge Answers

When a person consumes a high-carbohydrate diet, the carbohydrates are converted to glucose, which is readily available for the cells to use for energy. Excess glucose is converted to glycogen and fats for storage. With a high-fat diet, triglycerides will be broken down to fatty acids if needed for energy; if not used for energy, the triglycerides can be stored in adipose tissue.

To determine the location of a particular cell protein, scientists must separate the cell into its different components in stepwise fashion and test the various components for the presence of protein. This is done by disrupting the plasma membrane but leaving the organelles and membrane fragments intact. A series of centrifugation steps can then separate the components based on density.

To determine if a membrane protein is an integral membrane protein or a peripheral membrane protein, scientists must determine whether the protein can be easily dissociated from the plasma membrane. This can be done by agitating a solution containing cells and then centrifuging it, which causes the cells to accumulate at the bottom and the solution to remain on the top. The solution is then tested for the presence of protein. If the protein under question is found in the solution, then it was dissociated from the membrane without damaging the membrane. Thus the protein was a peripheral membrane protein.

Cilia act to move debris out of the respiratory airways. When they are damaged, the debris builds up and can make its way to the lungs. To clear the airways of the debris requires coughing—hence the "smoker's cough."

If errors in transcription were not corrected, then the wrong codon would be present in the strand of mRNA. Presence of the wrong codon would most likely mean the wrong amino acid would be translated during protein synthesis (exceptions occur because, in a few cases, two codons may code for the same amino acid). The wrong amino acid in a protein might either have no effect on protein function, decrease protein function, or enhance protein function. For example, in sickle cell anemia, only one nucleotide is wrong in the gene coding for hemoglobin, resulting in one errant amino acid; it causes polymerization of hemoglobin in the red blood cells, distorting the red blood cell into a sickle shape. However, "good" errors can also occur and are the basis of evolution.

Answers to End–of-Chapter Multiple Choice and Objective Questions

Multiple Choice Questions

1. b 2. d 3. a 4. b 5. d 6. a 7. d 8. b
9. c 10. b 11. d 12. d 13. c 14. a 15. c

Objective Questions

16. polar
17. nonpolar
18. cholesterol
19. uracil
20. integral
21. false
22. microvilli
23. connexons
24. gene
25. UAG
26. true
27. leader sequence
28. interphase
29. cytokinesis
30. ubiquitin

Cell Metabolism

Cell Metabolism

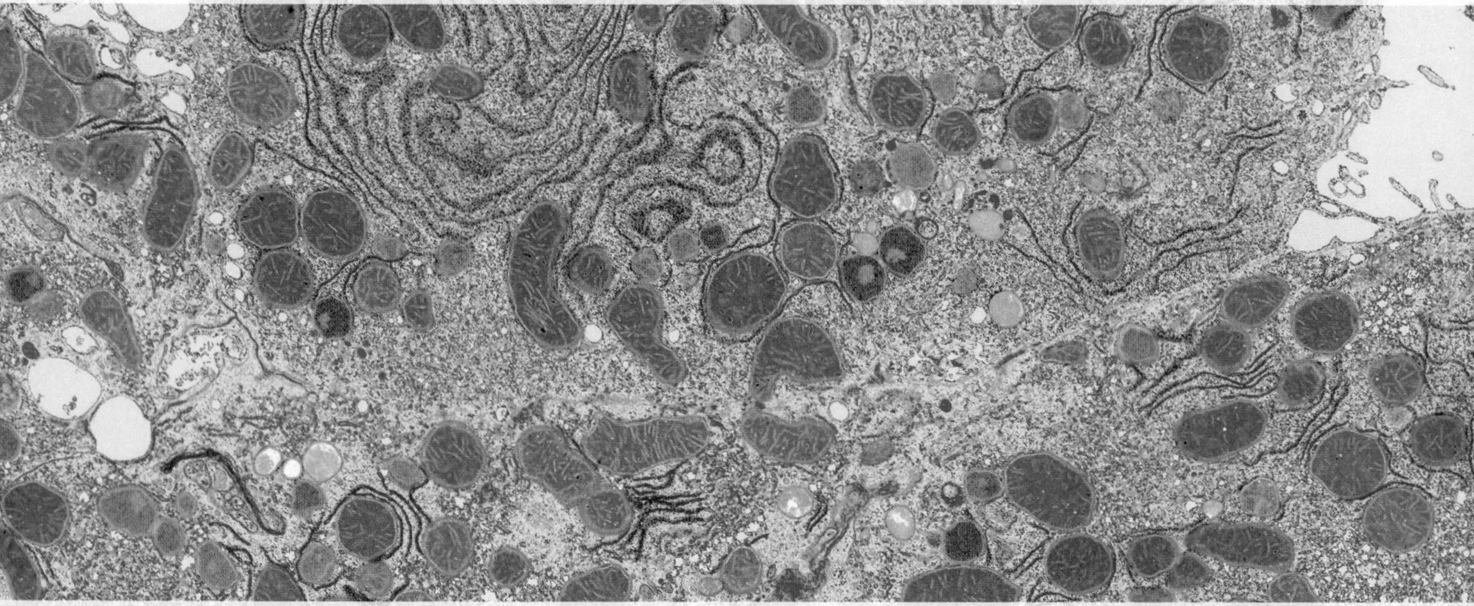

MedImage/Photo Researchers, Inc.

Colorized electron micrograph of liver cells.

Imagine a factory that manufactures hundreds of chemical compounds using only raw materials that can be found in any grocery store. This factory operates 24 hours a day and can speed up or slow down production on a moment's notice according to the demand for its products. The factory generates only environmentally safe waste products and can repair itself when the need arises. Sounds like a utopian dream of the future, right? Well, it actually happens every day right now: You carry billions of chemical factories that operate just like this in your own body—your cells. Each cell is its own "factory," but the cells cannot function alone. They rely on one another for raw materials, waste removal, and energy generation. In this chapter, however, we focus on the reactions that take place within a single cell.

CHAPTER OUTLINE

Go to MasteringA&P for helpful A&P Flix 3-D animations, chapter quizzes, pre-tests, Interactive Physiology tutorials, and more!

LEARNING OUTCOMES ***After studying this chapter, you should be able to:***

- Define *metabolism,* and distinguish catabolic reactions from anabolic reactions.
- Describe what occurs in the following basic types of metabolic reactions: hydrolysis, condensation, phosphorylation, dephosphorylation, oxidation, and reduction.
- Explain the relationship of energy change in a reaction to (1) the energies of the reactants and products and (2) the direction of the reaction.
- Define *activation energy,* and describe its influence on the rates of reactions.
- Describe the general mechanism of enzyme action, and discuss the various factors that influence the rates of enzyme-catalyzed reactions.
- Describe the role of ATP in energy metabolism.
- Describe in general terms the events in the following stages of glucose metabolism: glycolysis, the Krebs cycle, and oxidative phosphorylation.
- Describe in general terms how the body is able to obtain energy from the breakdown of fats, proteins, and glycogen, and how it stores energy by synthesizing these compounds.

Before You Begin

Make sure you have mastered the following topics:

- Biomolecules
- Mitochondria
- Membrane structure

Homeostasis is the fundamental theme of physiology. Homeostasis requires coordination of the organ systems in maintaining the internal environment of the body in a state compatible with life. To carry out their functions, cells that make up the organ systems must work efficiently, which requires many different chemical reactions and energy. The sum total of all chemical reactions that occur in cells is called **metabolism.** In this chapter, we focus on the types of reactions that provide the energy cells need to perform their work. First this energy must be extracted from biomolecules ingested in food; then it must be processed so that it can be either used for work or set aside in storage for future use. The set of reactions involved in energy storage and use (otherwise known as *energy exchange*) is called **energy metabolism.**

1 Types of Metabolic Reactions

Any chemical reaction involves the transformation of materials that enter the reaction, called *reactants,* into different materials, called *products.* In the reaction

$$A + B \rightarrow C + D$$

the reactants A and B are transformed into the products C and D. By convention, reactions are written with reactants on the left and products on the right. A reaction is said to "go forward" when it proceeds from left to right—that is, when reactants are converted to products. Under certain conditions, however, some reactions may go in the opposite direction; in that case, a reaction is said to go "in reverse."

A proper understanding of chemical reactions requires us to realize that even though we say a reaction goes forward or in reverse, it actually is *bidirectional,* meaning that it goes in both directions at the same time. In other words, when reactant molecules are being converted to products, some product molecules are also being converted to reactants. To indicate this bidirectionality, reactions are often drawn with double arrows:

$$A + B \rightleftharpoons C + D$$

When we say that a reaction goes forward or in reverse, we are referring to the direction of the overall (net) reaction. Suppose, for example, that over a certain time interval some product molecules are converted to reactants, but an even greater number of reactant molecules are converted to products. In this case, the net reaction goes forward because more reactant molecules are converted to products than vice versa. Unless noted otherwise, we use the terms *forward* and *reverse* to refer to the direction of the *net* reaction.

Metabolic reactions are broadly classified as being either catabolic or anabolic, depending on whether the product molecules are larger or smaller in size than the reactant molecules. A reaction is **catabolic** if it involves the breakdown of larger molecules into smaller ones. Examples of catabolic reactions include the breakdown of proteins into amino acids and the breakdown of glycogen into individual glucose molecules, both of which occur when food is being digested. A reaction is **anabolic** if it involves the production of larger molecules from smaller reactants. Examples include the synthesis of proteins from amino acids and the synthesis of glycogen from glucose, both of which occur in cells following the absorption of ingested nutrients. The breakdown and synthesis of larger molecules are sometimes referred to as *catabolism* and *anabolism,* respectively.

In most cases, metabolic reactions are linked together in a series of steps such that the products of one reaction serve as the reactants in the next. Such a series of reactions is referred to as a **metabolic pathway,** which can be written as follows:

$$A \rightarrow B \rightarrow C \rightarrow D$$

This particular pathway consists of a series of three reactions. In the first step, the starting material A, called the *initial substrate,* is converted to B. In the second step, B becomes a substrate that is transformed into C. C then becomes the substrate for the third reaction, which yields D, called the *end-product.* Substances in the middle of the pathway (B and C in this example) are called *intermediates.*

Later in this chapter we take a detailed look at a group of metabolic pathways involved in *glucose oxidation,* an important reaction that supplies much of a cell's energy needs. Although each of the individual reactions that make up these metabolic pathways is unique, reactions can be categorized. In the following sections we examine three general types of metabolic reactions, which occur in many metabolic processes including glucose oxidation: (1) *hydrolysis* and *condensation* reactions, (2) *phosphorylation* and *dephosphorylation* reactions, and (3) *oxidation-reduction* reactions.

Hydrolysis and Condensation Reactions

When you eat a meal containing protein, your gastrointestinal tract breaks down the relatively large protein molecules into amino acids, which are considerably smaller and more easily absorbed into the bloodstream. This reaction, like most catabolic reactions that occur in the body, is an example of **hydrolysis,** in which water reacts with molecules, thereby breaking the bonds that link a molecule together. In this case, the bonds that are broken are the peptide bonds that link amino acids together to form proteins. The general form of a hydrolysis reaction is as follows:

$$\text{A—B} + H_2O \rightarrow \text{A—OH} + \text{H—B}$$

where A and B represent chemical groups held together as a single molecule. In this reaction, the molecule A—B is broken down into its constituent parts, A and B, as a result of the breaking of the bond that originally linked them together. In this process, a water molecule (H_2O) splits into two parts, a hydroxyl group (—OH) and a hydrogen (—H), one of which combines with A and the other of which combines with B. As a result, the bond between A and B is broken and replaced with new bonds. The splitting of a disaccharide into two monosaccharides is an example of hydrolysis, as shown for the hydrolysis of sucrose:

$$\text{sucrose} + H_2O \rightarrow \text{glucose} + \text{fructose}$$

The reverse of hydrolysis, called *condensation,* involves the joining together of two or more smaller molecules to form a larger one, as when amino acids are joined together to form proteins. In this process, water is generated as a product:

$$\text{A—OH} + \text{H—B} \rightarrow \text{A—B} + H_2O$$

Sucrose is synthesized from the condensation of glucose and fructose:

$$\text{glucose} + \text{fructose} \rightarrow \text{sucrose} + H_2O$$

Phosphorylation and Dephosphorylation Reactions

In many metabolic reactions, particularly those involved in energy exchange, a phosphate group is either added to a molecule or removed from it. Addition of a phosphate group (abbreviated as P) is known as **phosphorylation** and can be written in an abbreviated fashion as follows:

$$\text{A} + P_i \rightarrow \text{A—P}$$

Here, the free phosphate group is abbreviated as P_i, for inorganic phosphate, which under physiological conditions exists mostly in the ionized forms HPO_4^{2-} and $H_2PO_4^-$. The bond formed in this reaction (that is, the bond linking A to P) is known as a *phosphate bond.* A common example of phosphorylation is the addition of a phosphate group to ADP to synthesize ATP:

$$\text{ADP} + P_i \rightarrow \text{ATP} + H_2O$$

Because water was produced, this reaction is also a condensation.

Removal of a phosphate group, known as *dephosphorylation,* can be written as follows:

$$\text{A—P} \rightarrow \text{A} + P_i$$

An example of dephosphorylation is the removal of a phosphate group from ATP to form ADP:

$$\text{ATP} + H_2O \rightarrow \text{ADP—}P_i$$

Because water was used to split the ATP, this reaction is also considered a hydrolysis reaction.

Oxidation-Reduction Reactions

Central to energy metabolism are **oxidation-reduction reactions.** The concept of oxidation was first developed in the late 1700s after Antoine Lavoisier first defined oxygen (*oxygene*, French for "acid former") as a substance in air that reacts with burning substances. Thus oxidation was first described as any reaction of a substance with oxygen.

For example, cellular metabolism includes the oxidation of glucose to water and carbon dioxide, which occurs as follows:

$$C_6H_{12}O_6 + 6\,O_2 \rightarrow 6\,CO_2 + 6\,H_2O$$

In this reaction, O_2 is the molecular form of oxygen and CO_2 is carbon dioxide. The special feature of oxygen is its ability to pull electrons away from other atoms in a molecule, thereby oxidizing them.

Today, the term **oxidation** is used to describe reactions that remove electrons from an atom or molecule. The electrons removed must be accepted by another atom or molecule in a process called **reduction.** Thus oxidation and reduction reactions are always coupled. An oxidation-reduction reaction can be written as follows:

$$\text{A}\cdot + \text{B} \rightleftharpoons \text{AB}\cdot$$

In this reaction, the dot (·) after A represents an electron. A is oxidized (loses an electron) and B is reduced (gains an electron). Although the product of this reaction is AB, B actually has a greater pull on the electron than A (which is similar to what we just saw for oxygen).

In other oxidation and reduction reactions, electrons are fully removed from and added to molecules. Such reactions are central to *oxidative phosphorylation*, the primary mechanism for synthesizing ATP in our cells. During oxidative phosphorylation, electrons (e^-) are removed from pairs of hydrogen atoms, causing these atoms to become positively charged hydrogen ions (H^+, also called protons). This reaction is an example of oxidation:

$$H_2 \rightarrow 2\,H^+ + 2\,e^-$$

At other points in oxidative phosphorylation, the reverse occurs. That is, electrons combine with hydrogen ions to form uncharged hydrogen atoms, an example of reduction:

$$2\,H^+ + 2\,e^- \rightarrow H_2$$

Another common form of oxidation-reduction reactions in biological systems is the removal or addition of hydrogen ions to a molecule, as shown in the following reaction:

$$HA{-}BH \rightarrow A{=}B + 2\,H$$

In this example, removal of two hydrogen atoms causes the single bond linking A and B to be replaced with a double bond. Even though hydrogen atoms (not electrons) are removed, the reaction is still considered oxidation because each of these atoms carries an electron. As a consequence, removal of these atoms removes electrons that originally belonged to the reactant molecule. (As a reflection of this fact, hydrogen atoms are sometimes referred to as *reducing equivalents*.) Addition of hydrogen atoms to a molecule—the reverse of the previous reaction—is called reduction and is shown as follows:

$$A{=}B + 2\,H \rightarrow HA{-}BH$$

Quick Check 1

1. What is the primary distinction between an anabolic reaction and a catabolic reaction?
2. Describe in general terms what occurs in hydrolysis, phosphorylation, and oxidation-reduction reactions.

2 Metabolic Reactions and Energy

The metabolic reactions that occur in the body perform many functions. One of the most important functions is enabling cells to transform raw materials from the environment into the great variety of substances from which the body is made. Metabolic reactions also provide us with **energy,** broadly defined as the capacity to perform work. The body most obviously performs work during movement. When you push a grocery cart, for example, you exert a certain amount of force against the cart and move it a certain distance. The amount of work you perform is the product of the force exerted and the distance moved (work = force × distance). However, the body performs work in other ways, too—even while at rest—and it must do so on a continual basis just to stay alive. The heart performs work when it pumps blood, the kidneys perform work when they form urine, and cells perform work when they multiply to repair damaged tissues. All of this work relies on energy derived from metabolic reactions. In this section, we explore the nature of the relationship between chemical reactions and energy.

Energy and the Laws of Thermodynamics

To understand how energy can produce work, we need to know the forms of energy and the physical laws governing it. There are two forms of energy: kinetic energy and potential energy. **Kinetic energy** is associated with motion (kinesis = motion) and includes thermal energy, radiant or electromagnetic energy, and electrical energy. Simply stated, molecules possess kinetic energy because they move or vibrate randomly at any temperature above absolute zero. This thermal energy increases in direct proportion with increases in the molecules' temperature. **Potential energy,** in contrast, is stored energy that can eventually be converted to kinetic energy. (In other words, potential energy has the *potential* to become kinetic energy.) Potential energy includes chemical energy, mechanical energy (such as that stored in a stretched rubber band), nuclear energy (the energy that holds the nucleus of an atom together), and gravitational energy. Thus molecules have potential energy in the nucleus and in the bonds between atoms. Nuclear energy is not accessible (except during nuclear fission reactions), but the energy stored in bonds can be released when those bonds break.

The *first law of thermodynamics* states that the energy in a closed system is constant; that is, energy can be neither created nor destroyed. Thus kinetic energy and potential energy are interchangeable, but the overall energy of a system cannot change. Take, for example, the stretching of a rubber band. We apply kinetic energy to the rubber band to stretch it out of its resting length—that is, the length the rubber band has when no external force acts on it. As long as the rubber band is stretched, it is storing energy. Thus the kinetic energy used to stretch the rubber band becomes potential energy. If the stretched rubber band is released, the potential energy is converted back to kinetic energy.

The *second law of thermodynamics* states that natural processes tend to proceed in the direction that spreads out the energy. For example, breaking a large molecule into several small molecules spreads out the energy contained in the original molecule. Another example is the spreading out of molecules from areas where they are more concentrated (have higher energy) to areas where they are less concentrated (have lower energy); this process is called *diffusion*. In the next section, we discuss the exchange of energy in chemical reactions.

Energy Changes in Reactions

Chemical reactions are always accompanied by either the release of energy or the input of energy. In other words, all reactions involve energy exchange of one sort or another. This exchange occurs because different types of molecules possess different amounts of energy. When a reaction releases energy, it does so because the reactant molecules possess more energy than the product molecules; in this case, reactants enter the reaction with a certain amount of energy, and products come out of the reaction with less energy. The "extra" energy must be released in one form or another during the course of the reaction:

$$\text{reactants} \rightarrow \text{products} + \text{energy}$$

The released energy might be heat or take some other form, and under certain conditions it can be used to perform work.

A familiar example of an energy-releasing reaction is the combustion of gasoline. When gasoline is burned in the open air, it generates energy—heat and light—in the form of flames. When it is burned in the internal combustion engine of a car, some of this energy is used to perform the work of moving the car.

The energy change of any reaction (expressed as ΔE) is the difference between the energy of the products (E_{products}) and the energy of the reactants ($E_{\text{reactants}}$): $\Delta E = E_{\text{products}} - E_{\text{reactants}}$. (The symbol Δ, or delta, is frequently used to denote a change in some quantity.) Because the products have less energy than the reactants in an energy-releasing reaction, the energy change ΔE in such a reaction is negative. This situation is illustrated in **Figure 1**a, where distance along the y-axis represents the amount of energy present in the reactant or product molecules, and distance along the x-axis represents the progress of the reaction—that is, the transformation of reactants into products. An energy-releasing reaction is called an **exergonic reaction.**

Figure 1b depicts the energy change that occurs in an energy-requiring reaction, which we can write as follows:

$$\text{reactants} + \text{energy} \rightarrow \text{products}$$

In this case, the products possess more energy than the reactants, so the energy change ΔE is positive. Because energy cannot be created, the "extra" energy acquired by the products must come from some source other than the reactant molecules themselves. An energy-requiring reaction is called an **endergonic reaction.** An example of an endergonic reaction is the synthesis of proteins; energy is required to form each peptide bond in a protein.

The energy change ΔE is typically expressed in units of energy or heat, such as calories (cal) or kilocalories (kcal), but it may also be given in joules (J) or kilojoules (kJ). A **calorie** is the amount of energy that must be put into 1 gram (or 1 mL) of water to raise its temperature by 1 degree centigrade (°C) under a standard set of conditions. One calorie is equivalent to 4.18 joules, and a kilocalorie or kilojoule is 1000 calories or joules, respectively.

The amount of energy released or consumed in a reaction depends not only on the nature of the reactants and products involved, but also on the quantity of reactants consumed or products produced. As this quantity increases, the size of the energy change increases in direct proportion. For this reason, ΔE is commonly expressed in units of kcal/mole or some equivalent.

Chemical energy—that is, the energy stored in the chemical bonds holding molecules together—is a form of potential energy. When we are talking about the energy change in a reaction, we are talking about change in *potential* energy. In energy diagrams such as Figure 1, it is assumed that reactant and product molecules are at the same temperature (that is, have the same kinetic energy), so any difference in energy between reactants and products must reflect differences in potential energy. In the next section, we see that the energy change in a reaction determines not only whether energy is released or required, but also in which direction the reaction proceeds.

How the Energy Change of a Reaction Determines Its Direction

We know from experience that chemical reactions have a natural tendency to run in particular directions. For instance, once you light a piece of paper, it burns until it is reduced to ashes (provided there is sufficient oxygen). Conversely, you never see the combustion of paper stop and go in reverse, such that ashes are reconstituted into paper. The reason for this is that the burning of paper is an exergonic reaction, and *exergonic reactions always proceed spontaneously in the forward direction.* (To say that a process occurs spontaneously means that it occurs on its own—that is, without any energy being put into it.) To reconstitute ashes into paper would require the paper-burning reaction to go in reverse, which would require energy rather than release it. This does not happen because *endergonic reactions do not go forward spontaneously; they go forward only when energy is put into them.* (In fact, endergonic reactions will spontaneously go in reverse if energy is not put into them.)

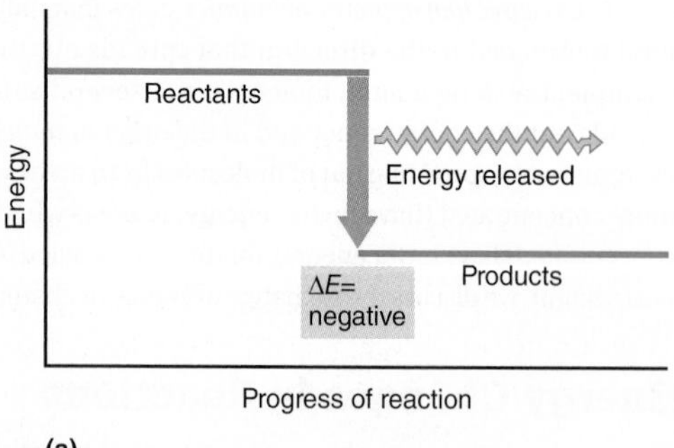

(a)

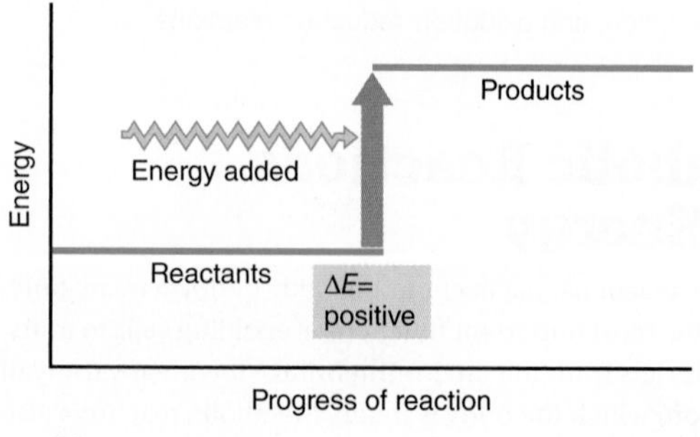

(b)

Figure 1 Energy changes in reactions. Wavy arrows indicate the transfer of energy in the reaction. **(a)** When the energy of the reactants is higher than that of the products, the energy change is negative, and energy is released in the reaction. **(b)** When the energy of the products is higher than that of the reactants, the energy change is positive, and energy is added to the reaction.

These rules reflect another fundamental truth of nature: Systems have a natural tendency to go from states of higher potential energy to states of lower potential energy. We know, for example, that a rock will roll downhill spontaneously but not uphill (not unless it has momentum to begin with), because the potential energy of the rock decreases when it rolls downhill; to roll uphill, its potential energy would have to increase. Likewise, an exergonic reaction (reactants → products + energy) goes forward spontaneously because the potential energy decreases as high-energy reactants are transformed into lower-energy products. In contrast, an endergonic reaction (reactants + energy → products) does not go forward spontaneously because the potential energy increases as low-energy reactants are converted to higher-energy products. (Strictly speaking, the *free energy* change of a reaction, rather than the *potential energy* change, determines its direction. Free energy is the energy that can be extracted from a molecule.)

Given the energy-releasing nature of catabolic reactions and the energy-requiring nature of anabolic reactions, we can surmise that catabolic reactions should occur spontaneously in cells, whereas anabolic reactions should occur only when energy is put into them. Anabolic reactions are able to occur in cells because certain cellular mechanisms *couple* these reactions with catabolic reactions. Such coupling harnesses the energy that is released from catabolic reactions and uses it to drive anabolic reactions. In the following example, the reaction A → B releases energy, which is then used to drive the reaction C → D:

$$A \rightarrow B + \textit{energy}$$

$$C + \textit{energy} \rightarrow D$$

Chemical Equilibrium

A reaction is said to be in **equilibrium** when there is no net reaction direction—that is, when the reactant is converted to the product at the same rate the product is converted to the reactant. Equilibrium occurs when the reactant and the product have the same energy, such that the energy change, ΔE, is zero. Note that equilibrium is reached *not* when the concentration of reactant equals the concentration of product, but rather when the energy levels of reactant and product are equal. For example, when reactant molecules have more energy than product molecules, at equilibrium the concentration of product must be greater than the concentration of reactant for the energy levels of reactant and product to be equal. When the difference in potential energy between the reactants and products is large, then the difference in concentration when the reaction is at equilibrium is also large. As shown in the next section, we can change the energy of either reactants or products by changing their concentration.

The Law of Mass Action

The energy of reactant or product molecules in a solution is determined by many different factors, including the numbers and types of chemical bonds that are present, the nature of the solvent in which the molecules are dissolved, and the molecules' *concentrations*. In general, as the concentration of molecules increases, the energy of the solution increases (see **Chemistry Review: Solutions and Concentrations**). Mathematically, the relationship between concentrations of reactant and product at equilibrium can be described with the *equilibrium constant*. Consider the following reaction:

$$aA + bB \rightleftharpoons cC + dD$$

where a, b, c, and d represent the number of A, B, C, and D molecules, respectively, in a balanced chemical reaction. The equilibrium constant for this reaction at a particular temperature is defined as K, where

$$K = \frac{[C]^c\,[D]^d}{[A]^a\,[B]^b}$$

An equilibrium constant of 1 would mean that the reactants and products contain the same amount of energy; thus there would be

CHEMISTRY REVIEW

Solutions and Concentrations

A mixture of molecules dissolved in a liquid is known as a *solution*. The dissolved substance, which is usually a solid or gas in its pure form, is known as the *solute*, whereas the liquid is referred to as the *solvent*. Solute molecules are said to be *dissolved* when they are completely separate from one another and are surrounded by solvent molecules.

The *concentration* of a solution is a measure of the quantity of solute contained in a unit volume of solution. Solute quantity is most commonly expressed in terms of moles (mol), and volume is most often given in liters (L). If 1 mole of solute is present in 1 liter of solution, the concentration is said to be 1 *molar* (1 M = 1 mol/L). If 1/1000 of a mole (that is, 1 *millimole*) of solute is dissolved in 1 liter of solution, the concentration is 0.001 molar or 1 *millimolar* (1 mM = 1×10^{-3} mol/L). For very dilute solutions, concentrations may be expressed in *micromoles* per liter (μmol/L = 1×10^{-6} mol/L) or *nanomoles* per liter (nmol/L = 1×10^{-6} mol/L). Concentrations of specific substances are often indicated using brackets. For example, $[Na^+]$ represents sodium ion concentration.

One should not confuse molarity with *molality*. Molality is the number of moles per kilogram solvent, which in the case of water is 1 liter at 25°C. Thus, at 25°C, molarity and molality are approximately the same for aqueous solutions.

Occasionally, concentrations are expressed in terms of the *mass* of solute contained in a unit volume of solution—in grams per liter (g/L) or micrograms per liter (μg/L = 1×10^{-6} g/L), for instance. Often in physiology, the unit of volume is a *deciliter* (dL = 0.1 L), which is equivalent to 100 milliliters (100 mL). The concentration of a solution containing 1 gram of solute per deciliter is frequently expressed as 1 *percent* (1%) because physiological solutions are mostly water, 100 mL of which weighs 100 grams.

FOCUS ON DIABETES

The Law of Mass Action

The law of mass action is critical in regulating whole-body metabolism because concentrations of reactants and products in cells and body fluids are constantly changing. For example, following a meal, blood glucose levels increase as glucose is absorbed from the gastrointestinal tract into the blood. In response, the hormone insulin is released into the bloodstream and travels to various cells in the body to increase glucose uptake into these cells. During this *post-absorptive state,* body cells have ample glucose to use it as their primary source of energy. In diabetes mellitus, however, insulin level or activity is low, such that levels of glucose inside cells remain low, even in the post-absorptive state. Thus glucose metabolism is slow and the cell must turn to other molecules as sources of energy, such as lipids and proteins.

no net reaction. An equilibrium constant greater than 1 indicates an exergonic reaction in which the net reaction proceeds to the right. An equilibrium constant less than 1 indicates an endergonic reaction in which the net reaction proceeds to the left.

Because *K* is a constant, an increase in the concentration of reactant would drive the reaction to the right—that is, more product would be produced. Likewise, an increase in the concentration of product would drive the reaction to the left; therefore, less product would be produced or the reaction could go in reverse. This phenomenon is commonly referred to as the **law of mass action.** As it relates to physiology, the law of mass action can be stated more simply as follows: *An increase in the concentration of reactants relative to products tends to push a reaction forward, and an increase in the concentration of products relative to reactants tends to push a reaction in reverse.*

Apply Your Knowledge

Given a chemical reaction, reactant $\rightleftharpoons$ product, the equilibrium constant can be calculated as follows:

$$K = \frac{[\text{product}]}{[\text{reactant}]}$$

If the equilibrium constant is 2, then what is the ratio of reactant to product when the reaction is at equilibrium? Is this an exergonic or endergonic reaction? Answer the same questions for equilibrium constants of 10, 0.1, and 0.01.

Chemical equilibrium and the law of mass action are illustrated in **Figure 2**. The two graphs show the change in the concentration of reactant and product for a simple reaction (reactant $\rightleftharpoons$ product) over time. The slope of the curve represents the rate of reaction, with a greater slope indicating a faster rate. In Figure 2a, the reaction is allowed to proceed until equilibrium. This reaction is exergonic because more product than reactant is present at equilibrium. In Figure 2b, the reaction proceeds to equilibrium, but then more reactant is added. Addition of more reactant drives the formation of more product until equilibrium is again established.

Activation Energy

When reactant molecules enter into a reaction, their conversion into product molecules does not occur abruptly. Instead, the reacting molecules go into a high-energy intermediate form called a *transition state,* which then breaks down into the products. The reverse reaction (the conversion of products to reactants) also must go through this transition state. Thus the transformations that occur in a reaction are continuous and gradual, not sudden, as shown in **Figure 3**a. The "hump" in the middle of the curve, known as an *activation energy barrier,* arises because the potential energy of the transition state is greater than that of either the reactants or the products.

For example, we previously described the burning of paper as an energy-releasing process. However, paper does not spontaneously combust. Instead, energy, such as a flame or heat provided through a magnifying glass, is required to initiate this process by overcoming the activation energy barrier. For reactants to become products, or vice versa, molecules must have sufficient potential energy to surmount the activation energy barrier. To do so, they must acquire some "extra" energy, called **activation energy,** which is the difference between the energy of the transition state and the energy of either the reactants or the products. The activation energy is indicated in Figure 3b as the vertical distance between the initial or final energies and the peak of the curve. Note that for the exergonic reaction shown in the figure, the activation energy for the forward reaction is less than the activation energy for the reverse reaction. For an endergonic reaction, the reverse would be true.

If molecules must acquire extra energy to surmount the activation energy barrier and react, where does this "extra" energy come from? Molecules acquire this energy by colliding with one another, which happens all the time because they are in constant thermal motion. Consider, for example, the reaction A + B $\rightarrow$ C + D. For molecules of A and B to react, they first must collide. When this happens, some of the molecules' thermal kinetic energy is converted to potential energy. If this gain in potential energy is equal to or greater than the activation energy, then the two molecules will enter the transition state and be converted to the products C and D (**Figure 4**a).

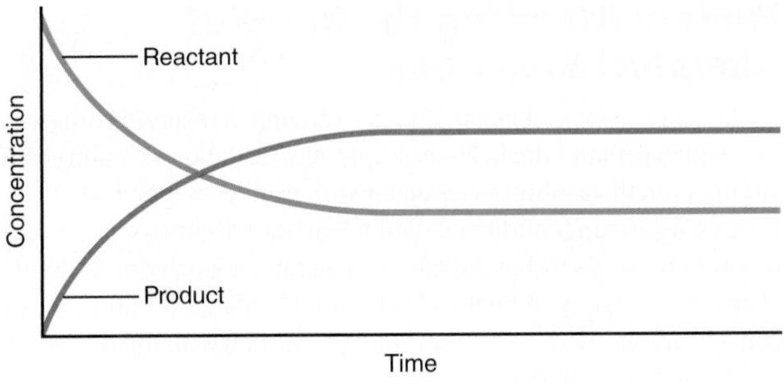

(a)

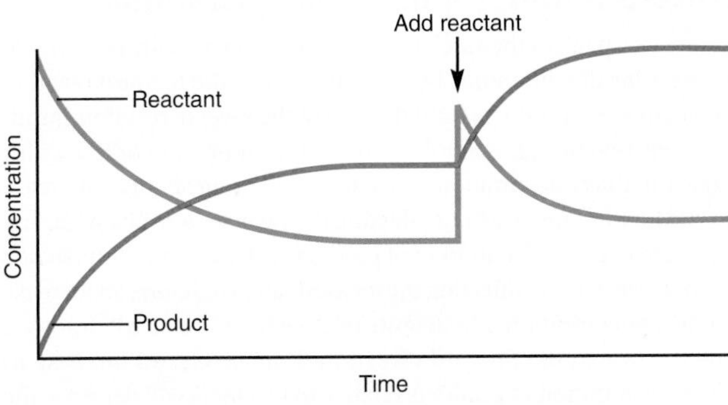

(b)

Figure 2 Chemical equilibrium and the law of mass action. For a simple exergonic reversible reaction with one reactant converted to one product (reactant $\rightleftharpoons$ product), the concentrations of reactant (green line) and product (red line) are plotted over time. **(a)** The reaction is allowed to proceed to equilibrium. **(b)** After reaching equilibrium, more reactant is added, which, by the law of mass action, causes the production of more product.

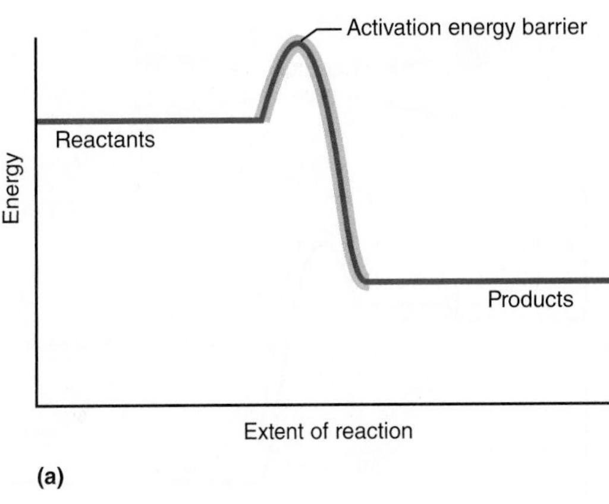

(a)

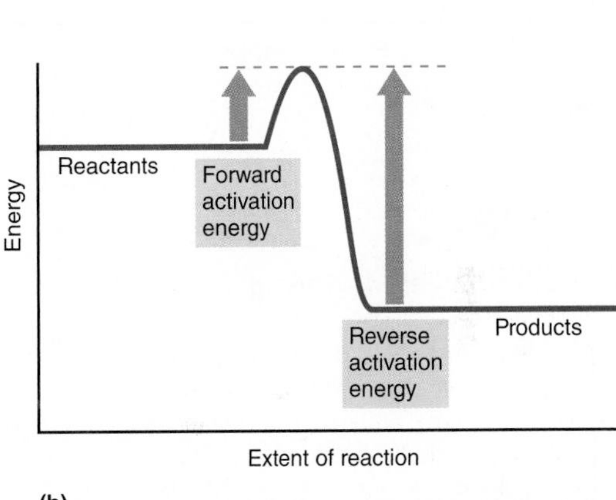

(b)

Figure 3 The activation energy barrier. The reactions depicted are energy-releasing reactions. **(a)** An energy diagram similar to that in Figure 1a, but with the activation energy barrier included. **(b)** An energy diagram in which the forward and reverse activation energies are indicated by the vertical arrows.

The significance of the activation energy barrier is that it limits how fast a reaction can go, a topic that is explored in detail in the next section. Note, however, that not every collision between reactant molecules produces a reaction, because some collisions do not generate enough potential energy to surmount the activation energy barrier. The more quickly two molecules about to collide are moving, the more potential energy they gain in the collision. If the energy gained is less than the activation energy, then the colliding molecules will not enter the transition state and will not react (Figure 4b); instead, they will emerge from the collision unaltered.

Quick Check 2

1. In an exergonic reaction, which has more energy—the reactants or the products?
2. Which factor determines the direction of a reaction? Are reactions that go forward spontaneously exergonic or endergonic?
3. Give a brief description of the law of mass action.
4. What is activation energy? How does it affect a reaction?

3 Reaction Rates

The *rate* of a chemical reaction is a measure of how fast it consumes reactants and generates products; it is usually expressed as a change in concentration per unit time (moles/liter-second, or some equivalent). For example, as the reaction $A + B \rightarrow C + D$ proceeds, the concentrations of A and B diminish while the concentrations of C and D increase. The rate of this reaction can be expressed as the change in the concentrations of C or D per unit time.

The rate of a metabolic reaction is of great physiological significance because proper body function demands that reactions proceed at a rate that matches the body's needs at the moment. Rates higher or lower than required ultimately lead to severe impairment of cellular function. Perhaps the most vivid demonstration of this danger occurs in *hypothermia,* when body temperature falls below normal. Any decline in body temperature causes metabolic reactions to slow down, and a decline in core temperature of only a few degrees can cause a person to become weak and disoriented, and possibly even to lose consciousness. A further drop in

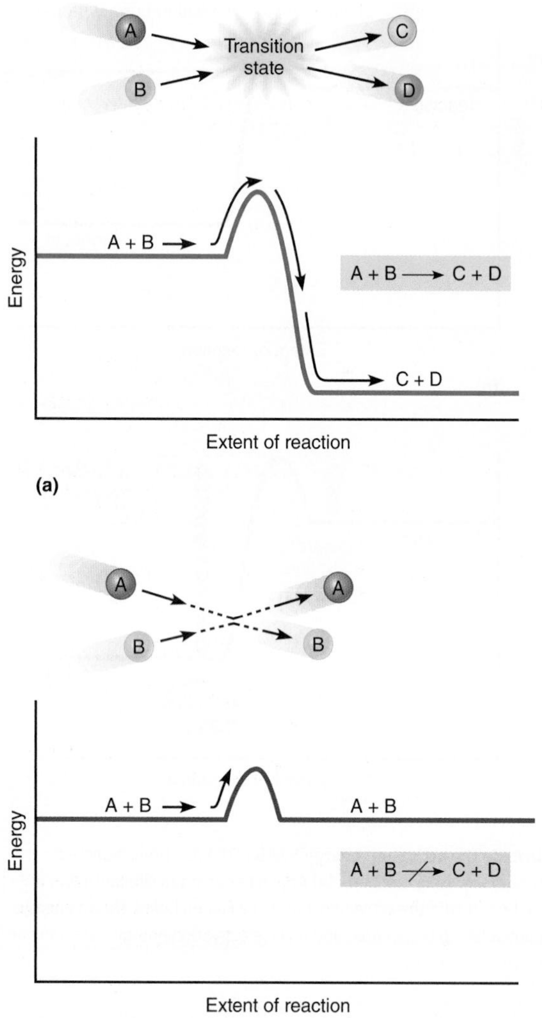

Figure 4 Effect of the activation energy barrier. (a) Collision of two reactant molecules, A and B, with enough energy to enter the transition state. The energy of the collision surmounts the activation energy barrier, resulting in the formation of products C and D. **(b)** Collision of A and B with insufficient energy to enter the transition state. The activation energy barrier is not surmounted, so no reaction occurs.

Q As the activation energy barrier is lowered, does the energy change in the reaction increase, decrease, or stay the same?

temperature can precipitate cardiac arrest (stoppage of the heartbeat) and death.

Later in this section, we will see that metabolic reactions are accelerated or *catalyzed* by special molecules called *enzymes,* and that the rates of metabolic reactions are normally regulated through changes in enzyme activity. First, however, we begin our exploration of reaction rates by examining the factors that affect the rates of chemical reactions in general.

Factors Affecting the Rates of Chemical Reactions

We know from experience that certain chemical reactions progress more quickly than others. For example, one oxidation reaction—the combustion of gasoline—can occur with explosive speed, whereas the rusting of iron (another oxidation reaction) occurs so slowly that it may take weeks to see the effects. The rate of a reaction is determined by a variety of factors, including (1) reactant and product concentrations, (2) temperature, and (3) the height of the reaction's activation energy barrier.

Reactant and Product Concentrations

When we speak of the rate of a reaction, we are usually referring to its *net rate:* the difference between the rate of the forward reaction (reactants ⟶ products) and the rate of the reverse reaction (products ⟶ reactants). According to the law of mass action, any increase in the concentration of reactants will bring about an increase in the forward rate without affecting the reverse rate. Likewise, any increase in the concentration of products will cause the reverse rate to increase without affecting the forward rate. Therefore, an increase in the concentration of reactants relative to products will increase the *net rate* in the forward direction. Conversely, an increase in the concentration of products relative to reactants will decrease the net forward rate and, if the change in concentration is large enough, can even make the reaction go in reverse. The effect of reactant and product concentrations on reaction rates reflects the fact that concentrations affect the frequency of collisions between molecules: As the concentration of molecules increases, the number of collisions occurring in a given time period also increases.

Temperature

In general, the rate of a reaction increases with increasing temperature and decreases as the temperature falls. We refrigerate food, for example, because cooling slows down the rate of decomposition reactions, which helps the food "keep" longer.

Temperature influences reaction rates because it affects the frequency and energy of molecular collisions. As temperature increases, the average kinetic energy of molecules increases, which increases the energy of collisions. The result is an increase in the proportion of collisions having enough energy to surmount the activation energy barrier, which in turn increases the rate of a reaction.

The Height of the Activation Energy Barrier

The height of the activation energy barrier differs among reactions. All else being equal, the rate of a reaction increases as the height of the barrier decreases. The reason for this is that at any given temperature, only a fraction of the collisions between molecules have sufficient energy to surmount the barrier and produce a reaction (**Figure 5**). As the height of the barrier decreases, the proportion of collisions having the requisite energy increases—not because the collisions themselves are any more energetic, but because the minimum energy requirement for a "successful" collision has been reduced. (This situation is analogous to a high-jump competition: If

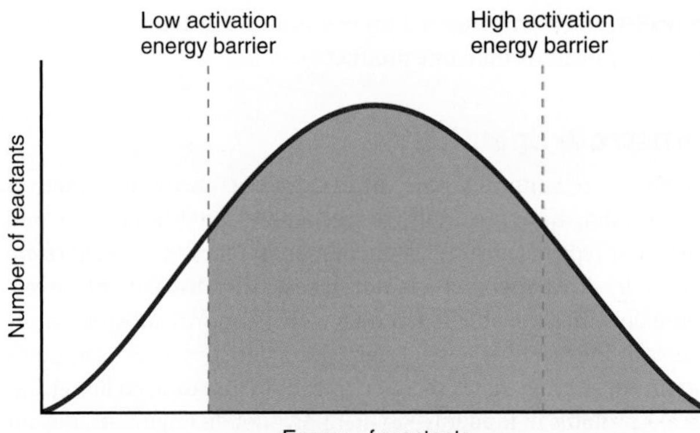

Figure 5 Energy of reactant molecules. The energy of individual reactant molecules varies, but roughly follows a bell-shaped population curve. For reactants to be converted to products, their energy must be greater than the activation energy barrier. If the activation energy barrier is relatively low, a higher proportion of reactants will have enough energy to be converted to product (all reactants to the right of the low activation energy barrier line on the graph). In contrast, when the activation energy barrier is relatively high, few reactants will have enough energy to be converted to product (only those reactants with energy above the high activation energy barrier line on the graph).

the bar is lowered, the number of contestants making a successful jump will increase.)

Note that as the height of the activation energy barrier decreases, rates of both the reverse reaction and the forward reaction increase because the activation energy of both reactions becomes lower. The net rate of the reaction still increases, despite the increase in the rate of the reverse reaction; this trend occurs for reasons beyond the scope of this text.

The Role of Enzymes in Chemical Reactions

If you were to set up a typical metabolic reaction by mixing reactants together in a beaker, it would proceed very slowly. In fact, most metabolic reactions would proceed too slowly to be compatible with life. These reactions are able to proceed at considerably faster rates in cells because of the presence of **enzymes,** biomolecules (almost always proteins) specialized to act as *catalysts,* a general term for substances that increase the rates of chemical reactions. In this section, we examine how enzymes accomplish this task and how the regulation of enzyme activity controls reaction rates according to the body's needs.

Mechanisms of Enzyme Action

Cells contain a wide variety of enzymes, each specialized to catalyze a particular reaction or group of related reactions. (Table 1 lists some of the key enzymes found in the human body, along with their functions.) To catalyze a reaction, an enzyme molecule must first bind to a reactant molecule, which in the context of enzyme-catalyzed reactions is called a **substrate.** The enzyme then acts on

TABLE 1 Sample of Enzymes

Enzyme	Reaction Catalyzed
RNA polymerase	DNA uncoiling and synthesis of RNA
DNA polymerase	DNA uncoiling and synthesis of DNA
Protein kinase	Phosphorylation of a protein
Phosphoprotein phosphatase	Dephosphorylation of a protein
Catalase	Breakdown of hydrogen peroxide to water and oxygen
Hexokinase	Phosphorylation of glucose
ATP synthase	Synthesis of ATP through oxidative phosphorylation
Lactate dehydrogenase	Conversion of lactic acid to pyruvic acid
Glucose-6-phosphatase	Removal of phosphate from glucose-6-phosphate
Carbonic anhydrase	Conversion of carbonic acid to water and carbon dioxide
Amylase	Breakdown of complex carbohydrates
Lipase	Breakdown of triglycerides to monoglycerides and fatty acids
Sucrase	Breakdown of sucrose to glucose and fructose

the substrate to generate a product molecule, which is subsequently released. Thus the action of the enzyme can be written as a two-step reaction:

$$\underset{\text{enzyme}}{E} + \underset{\text{substrate}}{S} \rightleftharpoons \underset{\text{enzyme-substrate complex}}{E \cdot S} \rightleftharpoons \underset{\text{product}}{P} + \underset{\text{enzyme}}{E}$$

The double arrow in the first step (the *binding step*) signifies that it is possible for the substrate to bind to the enzyme and then be released before the enzyme has a chance to act on it (see Figure 6a). Such binding is said to be *reversible* because the substrate interacts with the protein through weak interactions and can readily dissociate from the enzyme after binding to it (**Toolbox: Ligand-Protein Interactions**). If the substrate molecule stays bound to the enzyme long enough, it will eventually be converted to product in the second step (the *catalytic step*).

After the catalytic step, the enzyme molecule emerges from the reaction in the same form as it entered, indicating that it is neither consumed nor otherwise altered in the reaction. The enzyme is then free to act on a new substrate molecule and generate another product. In fact, a single enzyme molecule is capable of catalyzing a reaction repeatedly and theoretically could generate an unlimited quantity of product. As indicated by the double arrow, the enzyme can also catalyze the reaction in reverse, such that the product is converted back to substrate.

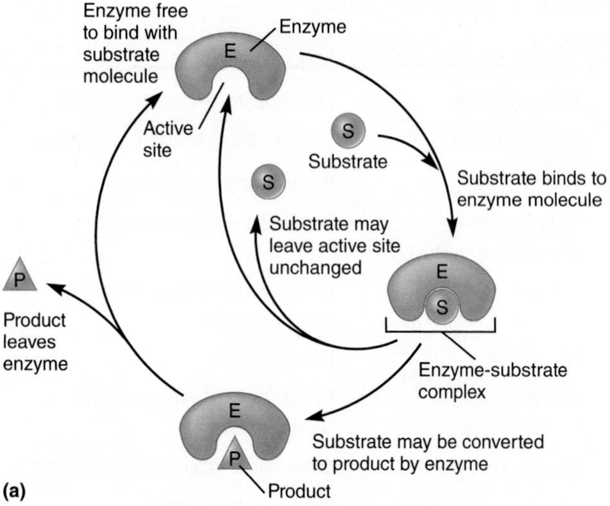

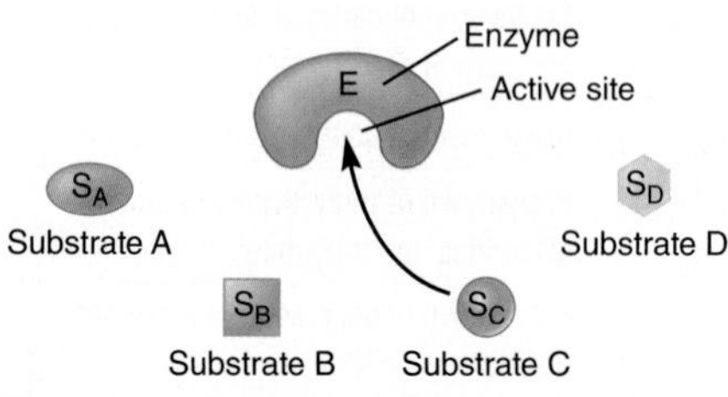

Figure 6 The role of substrate specificity in the mechanism of enzyme action. (a) The mechanism of enzyme action. The closeness of fit between substrate molecule S and the enzyme's active site enables binding and the formation of an enzyme-substrate complex. After binding, the substrate may leave the active site unchanged, or it may be converted to product P by the enzyme. The product then exits the active site, leaving the enzyme free to bind another substrate molecule and catalyze the reaction again. **(b)** Substrate specificity. Because only the shape of candidate substrate S_C is complementary to the enzyme's active site, only it can bind with the enzyme.

Although the preceding reaction illustrates the essentials of enzyme action, it pertains to only the simplest type of enzymatic reaction, in which an enzyme acts on just one substrate molecule at a time to generate a single product, as expressed in the following shorthand notation:

$$S \xrightarrow{E} P$$

In reality, most enzymes act on two or more different substrates and generate more than one product.

Substrate Specificity

Enzymes are generally able to catalyze one particular reaction because they have the ability to "recognize" and bind to only one particular type of substrate, a phenomenon referred to as *substrate specificity*. This specificity is not necessarily absolute, however: Some enzymes are able to act on a wide variety of substrate molecules as long as they have certain characteristics. One example is *pepsin*, an enzyme that is secreted by cells in the stomach lining and breaks proteins in food into smaller polypeptide fragments. Pepsin can act on almost any protein, so long as the protein contains certain amino acids, but it cannot act on other constituents of food, such as fats or carbohydrates.

The basis of substrate specificity relates to the *complementary* shapes of the enzyme and substrate molecules: Only if a substrate molecule closely fits and binds to a particular site on the enzyme molecule, called the *active site,* can the enzyme act on the substrate and catalyze the reaction. As illustrated in Figure 6b, uncomplementary "candidate" substrate molecules cannot fit into the active site very well and, therefore, are not acted upon by the enzyme.

Two models currently describe the mechanism of enzyme binding: the lock-and-key model (Figure 6) and the induced-fit model (**Figure 7**). In the *lock-and-key model*, the substrate matches the active site just as a key matches a lock, with this relationship explaining substrate specificity. The downside of this model is that it does not explain the reversibility of most enzyme-catalyzed reactions—if the enzyme fits the substrate, then it cannot also fit the product for the reverse reaction. The newer *induced-fit model* is more amenable to explaining reversible reactions. In this model, the substrate fits the active site more like a foot fits a sock than like a lock fits a key; the shapes are similar but not precisely complementary. When the substrate binds to the enzyme, it induces a conformational change in the enzyme to produce a better fit, much like a foot inside a sock makes the sock take on the shape of the foot.

Cofactors and Coenzymes

Although enzymes are proteins, many possess additional nonprotein components called **cofactors,** which are necessary for the enzymes to function properly. Many enzymes contain metal ions

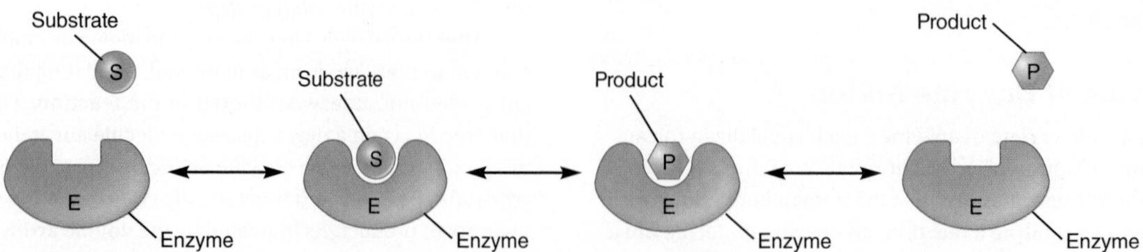

Figure 7 Induced-fit model of enzyme activity. In this model, both the substrate and the product can bind to the active site, allowing the reaction to proceed in reverse.

TOOLBOX

Ligand-Protein Interactions

When a substrate binds to an enzyme, it does so through weak interactions (hydrogen bonds, ionic bonds, and van der Waals forces, but *not* covalent bonds). Molecules that interact with proteins in this manner are called *ligands*. Thus we can generalize the interaction between proteins and ligands as a reversible reaction:

$$\text{protein} + \text{ligand} \rightleftharpoons \text{protein—ligand}$$

Ligands can bind to proteins because they have a shape complementary to a portion of the protein, called the *binding site*, much like a key has a complementary shape to a lock. Ligands are also chemically attracted to protein binding sites. The binding between a ligand and a protein is specific because only certain ligands fit the binding site.

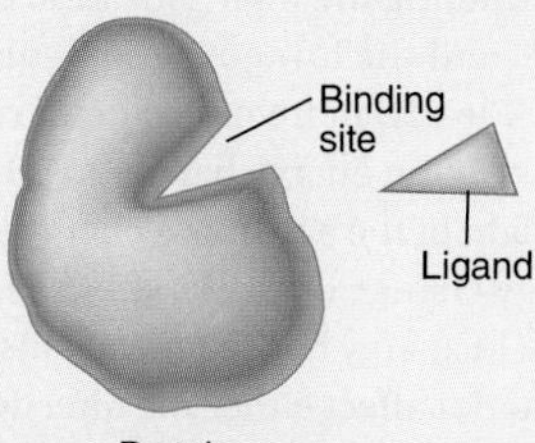

The interaction between proteins and ligands follows the law of mass action—that is, an increase in the amount of either protein or ligand increases the number of proteins with bound ligand. This is illustrated in the following graph, which shows the relationship between the number of proteins with bound ligand and the concentration of ligand. Two curves are shown, one for a low concentration of protein and one for a higher concentration.

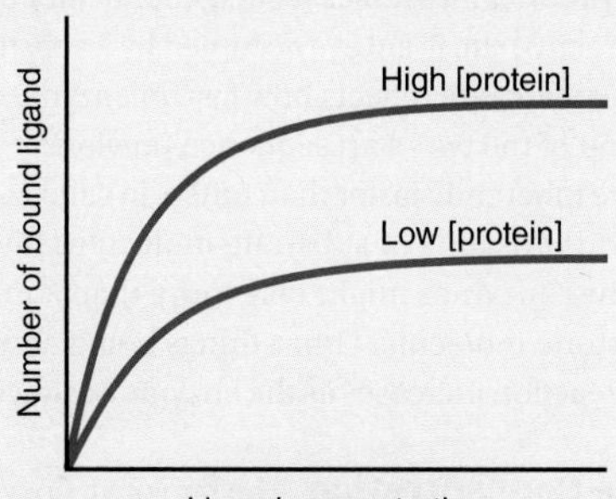

The strength of the interaction between a protein and a ligand, called *affinity,* depends on the degree of complementary shape and the number and strength of the chemical interactions. In the following example, ligand A forms more ionic bonds with the protein than ligand B. Therefore, the protein has a greater affinity for ligand A.

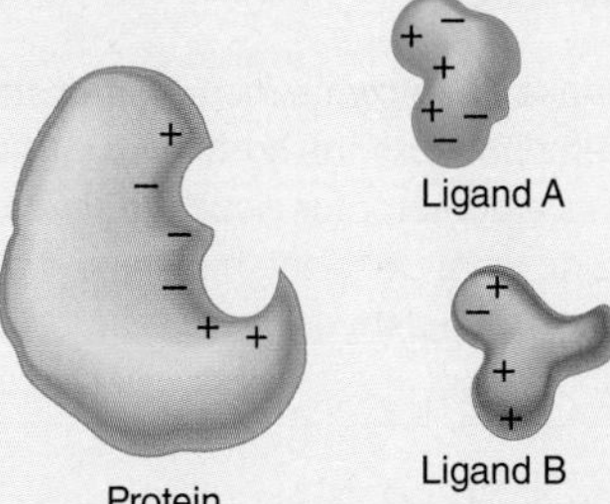

A higher affinity increases the likelihood that a given protein will have a ligand bound to it. In terms of a population of proteins, a higher affinity means more proteins will have ligand bound to them. This situation is shown in the following graph for the binding of ligand A or B to the protein, assuming only one ligand is present at a time.

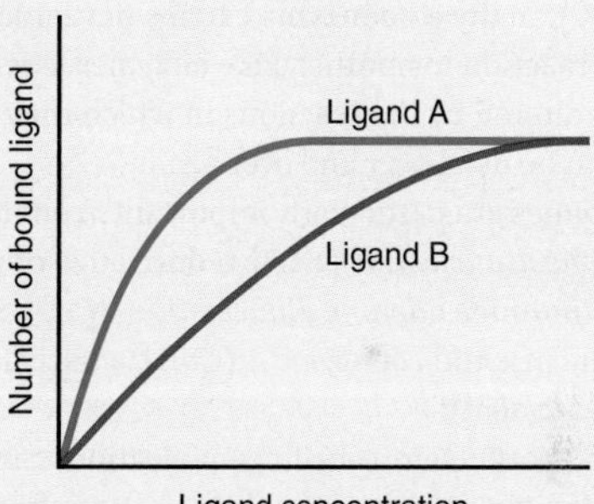

Several types of protein-ligand interactions occur in the body, as shown in the following table and discussed throughout the text.

Protein	Ligand	Function
Enzyme	Substrate	Catalyze chemical reactions in the body
Enzyme	Modulator	Allosteric regulation of enzyme activity
Receptor	Chemical messenger (e.g., neurotransmitters, hormones)	Communicate messages between cells
Carrier	Solute	Transport solutes across plasma membranes
Carrier	Solute	Transport solutes in blood

such as iron, copper, and zinc, which function as cofactors by binding tightly to the side chains of certain amino acids, thereby holding the enzyme in its normal conformation. Without these ions, the enzymes' shapes would be altered, causing them to lose their activity. Other enzymes contain **vitamins** (various organic molecules of which only trace amounts are required for proper functioning of the body, but which must be obtained through the diet) or vitamin derivatives as cofactors.

Apply Your Knowledge

Alcohol dehydrogenase, an enzyme found in the stomach and liver, removes ethanol from the body by oxidizing it to acetaldehyde. Zinc is a cofactor for alcohol dehydrogenase. If a person's intake of zinc were deficient, how would this deficiency affect the activity of alcohol dehydrogenase and the person's ability to tolerate alcohol?

Some vitamin-derived cofactors function as **coenzymes,** molecules that do not themselves have catalytic activity but nonetheless participate directly in the reactions catalyzed by their enzyme partners. A given coenzyme may be used by more than one type of enzyme; often a coenzyme can dissociate from one enzyme and bind to another, enabling it to participate in more than one reaction. In a few cases, coenzymes are permanently bound to their partner enzymes. Usually a coenzyme carries particular chemical groups from one reaction to another. Like enzymes, coenzymes are not permanently altered by the reactions in which they participate and, therefore, can be used over and over again.

Three coenzymes are particularly important in energy metabolism: *flavin adenine dinucleotide (FAD),* a derivative of vitamin B_{12} (riboflavin); *nicotinamide adenine dinucleotide (NAD),* a derivative of vitamin B_3 (niacin); and *coenzyme A (CoA),* a derivative of vitamin B_5 (pantothenic acid).

FAD and NAD participate as hydrogen (electron) carriers in certain oxidation-reduction reactions, shuttling electrons from one place to another inside cells. FAD acquires electrons by picking up pairs of hydrogen atoms and in the process becomes reduced to $FADH_2$:

$$\text{FAD} + 2\,\text{H} \rightarrow \text{FADH}_2$$

Subsequently, $FADH_2$ releases these electrons to other electron acceptors. In so doing, it reverts to its oxidized form, FAD, which is then free to pick up another pair of electrons.

NAD^+ also carries pairs of electrons, but in a different manner: It picks up one electron in a hydrogen atom and the other as a free electron, which it takes from a hydrogen atom, leaving a hydrogen ion (H^+) in solution. Thus the reduction of NAD^+ occurs as follows:

$$\text{NAD}^+ + 2\,\text{H} \rightarrow \text{NADH} + \text{H}^+$$

In this process, NAD^+, which carries a single positive charge, is reduced to NADH, which carries no charge. (The positive charge is canceled by the negative charge of the acquired electron.) Subsequently, NADH releases its pair of electrons to other electron acceptors through the reverse of the previous reaction and is transformed into the oxidized form NAD^+, which is free to pick up another pair of electrons. We will see several examples of NAD^+ and FAD at work later in the chapter.

CoA picks up chemical groups called *acetyl groups* ($—CH_2COOH$) in certain metabolic reactions and carries them to other reactions. In the process, each CoA becomes covalently bound to an acetyl group, forming a compound called *acetylcoenzyme A (acetyl CoA).* The role of CoA in glucose oxidation is described later in this chapter.

Factors Affecting the Rates of Enzyme-Catalyzed Reactions

Enzymes accelerate metabolic reactions by reducing the height of the activation energy barrier. This relationship can be seen in Figure 5, by considering the high activation energy level in the absence of enzyme and the low activation energy level in the presence of enzyme. Clearly, more molecules have sufficient energy to overcome the activation energy barrier in the presence of enzyme. Note that the height of the activation energy barrier does not affect the overall energy change of a reaction (ΔE), which depends only on the difference in energy between reactants and products. Therefore, enzymes cannot affect either the direction of a reaction or the amount of energy it releases or requires; instead, they can affect *only* the *rate* at which a reaction occurs.

The rate at which an enzyme can catalyze a reaction is, in turn, affected by several factors, including (1) the enzyme's catalytic rate, (2) the substrate concentration, (3) the enzyme concentration, and (4) the affinity of the enzyme for the substrate.

Catalytic Rate The **catalytic rate** of an enzyme is a measure of how many product molecules it can generate per unit time, assuming that the active site is always occupied by a substrate molecule. As such, the catalytic rate reflects how fast an enzyme can carry out the catalytic step of the two-step sequence previously described. Some enzymes are inherently faster than others in catalyzing reactions and can convert thousands of substrate molecules into products each second; other enzymes might take more than a minute to act on a single substrate molecule. Other things being equal, the rate of an enzymatic reaction increases as the enzyme's catalytic rate increases.

Substrate Concentration The rate of an enzyme-catalyzed reaction increases as the substrate concentration increases based on the law of mass action. An increase in the number of substrate molecules increases the likelihood that a substrate will be bound to an enzyme—and an enzyme can catalyze a reaction only when substrate is bound to it. In addition, an enzyme can be used over again. Once one substrate has been converted to product and the product released, the enzyme can bind to another substrate and convert it to product.

Thus, when the substrate concentration is low, more time will pass before the next substrate molecule binds to the enzyme, and as a result its active site will remain unoccupied for a greater percentage of the time. Under these conditions, the reaction rate will be low because for much of the time the enzyme is idle. By comparison, at higher substrate concentrations, less time will pass before the next substrate molecule comes along. This means that the enzyme's active site will be

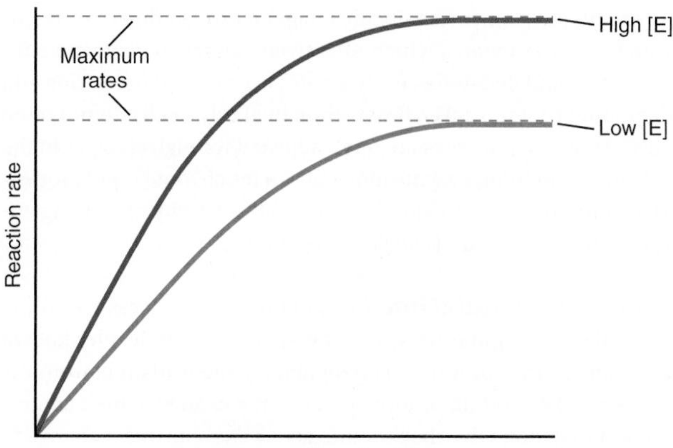

Figure 8 Influence of substrate concentration on the rate of an enzyme-catalyzed reaction. Binding of substrate to the enzyme increases with increasing [S] until high substrate concentrations are reached, in which case all enzyme molecules are bound (100% saturation). At a higher concentration of enzyme, more substrate can bind enzyme and a faster reaction rate will be achieved.

Q As the reaction rate is increasing, is the concentration of free (unbound) enzyme molecules increasing or decreasing?

occupied more of the time, which gives the enzyme more opportunity to catalyze the reaction. Consequently, the reaction rate will be higher.

Figure 8 shows how the rate of an enzyme-catalyzed reaction varies with the substrate concentration, [S], and enzyme concentration. At lower and moderate concentrations, the rate of the reaction increases as [S] increases, as we would expect. At higher concentrations, however, the curve levels off, indicating that further increases in [S] fail to raise the rate appreciably. Given a fixed enzyme concentration, when [S] is very high, the active site of every enzyme molecule is occupied by substrate virtually 100% of the time, and the enzyme is said to be *100% saturated.* (The *percent saturation* of an enzyme indicates the proportion of time that the active site is occupied; thus, at 25% saturation, the active site is occupied 25% of the time.) Raising [S] beyond this point does not cause any further increase in the rate because all of the enzyme molecules are already bound to substrates and are already catalyzing the formation of products. Under these conditions, the rate of the reaction is limited only by the enzyme's catalytic rate and its concentration. When the concentration of enzyme increases, the reaction can occur at a faster rate for any given concentration of substrate, as described next.

Enzyme Concentration The catalytic rate of an enzyme influences the rate of an enzymatic reaction by determining how fast individual enzyme molecules can convert substrates to products. In addition, the rate of an enzymatic reaction increases in direct proportion to the enzyme concentration based on the law of mass action. As described earlier for substrate concentration, an increase in the number of enzyme molecules increases the likelihood that a substrate will be bound to an enzyme and, therefore, converted to product. Because enzymes are proteins, enzyme concentration may be varied by regulating the rate of protein synthesis and degradation.

Figure 8 shows two curves, one for a low concentration of enzyme and the other for a higher concentration. Note that for any given concentration of substrate, the reaction rate is faster for the higher concentration of enzyme. Also note that even though a maximum reaction rate occurs under both concentrations of enzyme, the maximum reaction rate is faster when the concentration of enzyme is greater.

Affinity The **affinity** of an enzyme is a measure of how tightly substrate molecules bind to its active site. (Affinity is a general term referring to the attraction between two objects.) Generally speaking, higher affinities translate into higher rates for enzyme-catalyzed reactions. This makes intuitive sense when you consider that an enzyme's active site must be occupied by a substrate molecule before the enzyme can catalyze any reaction. When substrate molecules are present at a given concentration, an enzyme's active site will be occupied a greater percentage of the time if the enzyme has a high affinity for the substrate. If the affinity is lower, the active site will be occupied for less of the time, all else being equal. Therefore, a high-affinity enzyme is able to generate more product molecules in a given length of time.

A high affinity implies a close fit between an enzyme's active site and a substrate molecule. Such a fit maximizes the area of contact between substrate and enzyme, thereby maximizing any attractive forces that might exist between the two. Affinity is also influenced by other factors that increase this force of attraction, such as the presence of opposite electrical charges on the enzyme and the substrate.

Figure 9 shows how the degree of saturation of an enzyme and, hence, the rate at which it can catalyze a reaction are affected not only by the substrate concentration but also by the affinity of the enzyme for the substrate. At a given substrate concentration, an enzyme with higher affinity will exhibit a higher degree of saturation than will an enzyme with lower affinity, so it reaches its maximum rate at lower substrate concentrations. The maximum rate for a given concentration of enzyme will not be altered, however.

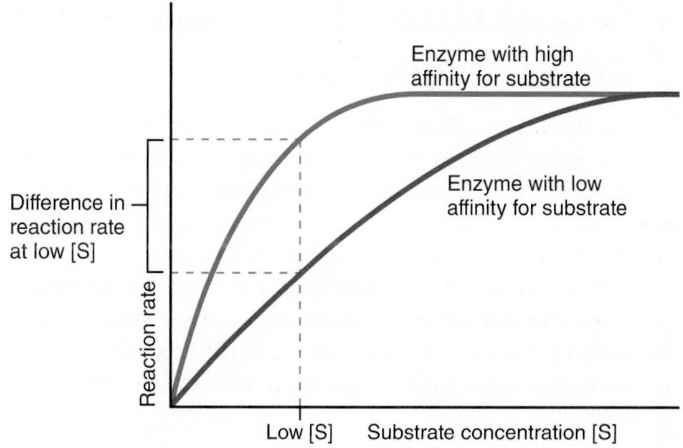

Figure 9 Influence of enzyme-substrate affinity on the rate of an enzyme-catalyzed reaction. Plots of reaction rate versus substrate concentration are shown for two enzymes having different affinities for the substrate. For purposes of comparison, it is assumed that the two enzymes have the same catalytic rate and are present at the same concentration. Note that at lower substrate concentrations, the enzyme with higher affinity is able to catalyze the reaction at a faster rate.

Other Factors Affecting the Rate of Enzyme-Catalyzed Reactions The rates of enzymatic reactions can be affected by other conditions as well, such as the temperature of the reaction mixture and its pH (see **Chemistry Review: Acids, Bases, and pH**). Changes in temperature and pH alter the structure of all proteins; thus such changes alter the shape, and therefore the activity, of enzymes. In general, enzyme activities decline if the temperature or pH becomes significantly higher or lower than normal. Within the body, changes in temperature are rarely significant because body temperature is regulated to stay constant. In intracellular and extracellular fluid, pH is also regulated to stay relatively constant (within the range 7.35 to 7.45), but it varies appreciably in certain locations within the gastrointestinal tract. The lumen of the stomach, for instance, contains fluid that is highly acidic (the pH can be as low as 2); in the intestine, the lumenal fluid is more basic, with a pH greater than 8. Significantly, the activity of *pepsin*, an enzyme that works in the stomach, is highest when the pH is in the acidic range, whereas that of *trypsin*, a similar protein-digesting enzyme that works in the intestine, is highest when the pH is in the basic range.

Regulation of Enzyme Activity

We have just seen how enzymes catalyze reactions and how the rate of an enzyme-catalyzed reaction is influenced by various factors. Now we look at how the body regulates the activity of enzymes to adjust the rates of reactions.

As conditions in the body change, the rates of metabolic reactions are continually adjusted to meet the body's needs. This adjustment is accomplished by means of mechanisms that regulate the activities of certain enzymes. In some cases, cellular enzyme concentrations are adjusted to speed up or slow down reactions. More enzyme molecules may be synthesized, for example, which raises the enzyme concentration and increases reaction rates. Alternatively, enzyme molecules may be broken down or inactivated by other means, which effectively lowers the enzyme concentration and decreases reaction rates. However, synthesizing and degrading proteins takes time—time the body's cells do not often have. Thus reaction rates are often adjusted through changes in the activities of existing enzyme molecules, a much more rapid process. Two common mechanisms for altering the activity of existing enzymes are *allosteric* and *covalent regulation.*

Allosteric Regulation Certain enzymes possess a binding site called the **regulatory site** that is specific for molecules known generally as **modulators.** (The regulatory site is distinct from the active site.) A modulator induces a change in an enzyme's conformation that alters the shape of the active site, causing a change in the enzyme's activity by altering its catalytic rate, its affinity for the substrate, or both. This type of regulatory mechanism is known as **allosteric regulation** (**Figure 10**).

In allosteric regulation, the binding of the modulator to the regulatory site occurs through weak interactions and, therefore, is reversible. Thus the modulator can readily leave the regulatory site after binding to it, and an equilibrium develops between the enzyme and the modulator. The rate of enzyme-catalyzed reactions subject to allosteric regulation varies with the concentration of substrate in a sigmoidal function, as shown in Figure 10b. The sigmoid curve indicates that the enzyme has a greater affinity for the substrate when the substrate is present at greater concentrations than when it is present at low concentrations—a property common to allosteric-regulated enzymes.

Binding of a modulator to an enzyme can bring about either an increase or a decrease in the enzyme's activity, depending on the enzyme in question. A modulator that increases the activity of an enzyme is called an *activator;* a modulator that decreases the activity of an enzyme is called an *inhibitor.* An allosteric activator shifts the curve to the left, indicating that reaction rate increases

CHEMISTRY REVIEW

Acids, Bases, and pH

Certain substances release hydrogen ions or *protons* (H^+) when dissolved in water and are known as *acids.* A familiar example is *hydrochloric acid* (HCl), which dissociates into hydrogen and chloride ions:

$$HCl \rightarrow H^+ + Cl^-$$

HCl is an example of a *strong acid,* an acid that dissociates completely. Certain other acids are *weak acids,* meaning that they dissociate incompletely. For example, *carboxyl groups* (—COOH), which are found on amino acids and other biomolecules, act as weak acids and dissociate as follows:

$$R\text{—}COOH \rightleftharpoons R\text{—}COO^- + H^+$$

Here, the double arrow signifies that protons can not only dissociate from the anion (R—COO^-), but also combine with it.

Substances that combine with protons are called *bases* and are classified as strong or weak depending on whether they combine completely or incompletely. Examples of weak bases include *amino groups* (—NH_2), which are found on amino acids and other compounds and react with protons as follows:

$$R\text{—}NH_2 + H^+ \rightleftharpoons R\text{—}NH_3^+$$

The *acidity* of a solution is determined by its hydrogen ion concentration, which can be expressed in terms of molarity or denoted by a number called the *pH,* which is defined according to the following expression:

$$pH = \log(1/[H^+]) = -\log[H^+]$$

where $[H^+]$ is the hydrogen ion concentration. Note that as $[H^+]$ increases, the pH decreases. Note also that because the pH scale is logarithmic, a change of one pH unit represents a tenfold change in the hydrogen ion concentration.

The hydrogen ion concentration in pure water is 10^{-7} M, giving a pH of 7. Solutions with a pH of 7 are said to be *neutral.* A solution whose pH is less than 7 is said to be *acidic;* if the pH is greater than 7, the solution is *basic.*

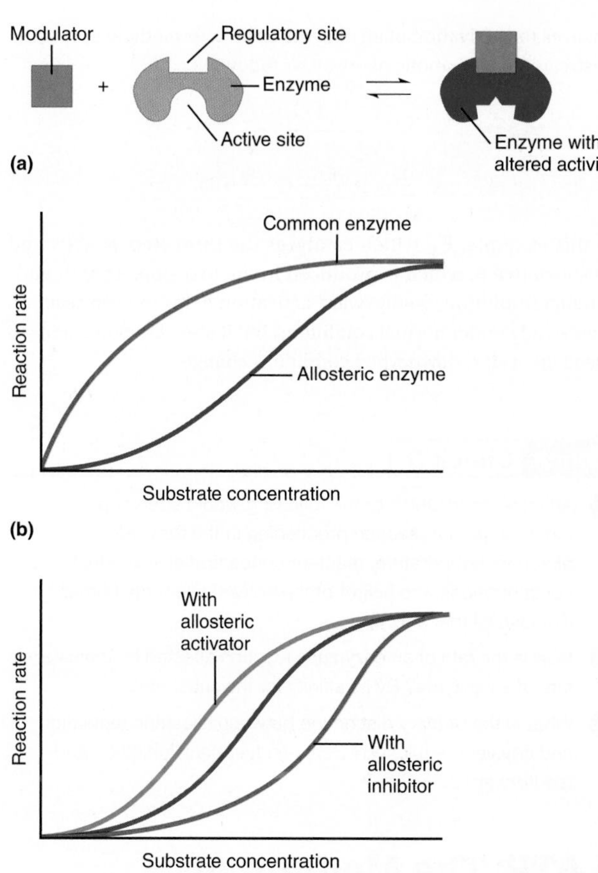

Figure 10 **Allosteric regulation of enzyme activity.** **(a)** Reversible binding of a modulator molecule to a regulatory site on an enzyme. When bound, the modulator alters the active site and, therefore, the activity of the enzyme. **(b)** The effect of modulator concentration on enzyme activity. Rates of allosteric enzyme-catalyzed reactions increase with substrate concentration in a sigmoidal (S-shaped) function. **(c)** Activators increase the reaction rate at any given concentration of substrate, whereas inhibitors decrease the reaction rate.

Q Why does the number of bound enzyme molecules increase as the modulator concentration increases?

for a given concentration of substrate, whereas an allosteric inhibitor shifts the curve to the right, indicating a decreased reaction rate (Figure 10c). Some enzyme molecules have more than one regulatory site and can respond to more than one type of modulator. Sometimes these different modulators affect the enzyme in different ways—some are inhibitory, whereas others are stimulatory.

Covalent Regulation Another type of enzyme regulation is **covalent regulation,** in which changes in an enzyme's activity occur as a result of the covalent bonding of a specific chemical group to a site on the enzyme molecule (**Figure 11**). In this process, formation of the covalent bond between the chemical group and the enzyme protein is catalyzed by another enzyme (enzyme A in Figure 11). Because covalent bonds are relatively strong, the

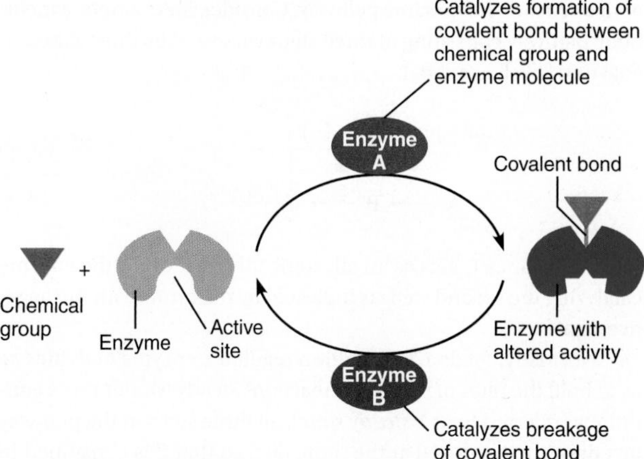

Figure 11 **Covalent regulation of enzyme activity.** Covalent bonding of a chemical group to an enzyme molecule alters the enzyme's activity. Formation of these covalent bonds is catalyzed by another enzyme (A); removal of the chemical group requires the action of yet another enzyme (B), which breaks the covalent bonds.

chemical group remains attached to the enzyme unless these bonds are broken, which requires the action of other enzymes (enzyme B in Figure 11).

A common form of covalent regulation involves the addition or removal of a phosphate group (P_i) to the enzyme molecules. Phosphorylation of a target enzyme molecule (E) is catalyzed by a type of enzyme called a **protein kinase:**

$$E + P_i \xrightarrow{\text{protein kinase}} E—P$$

Conversely, dephosphorylation is catalyzed by an enzyme called a **phosphatase:**

$$E—P \xrightarrow{\text{phosphatase}} E + P_i$$

Protein kinases and phosphatases are usually specific for particular target enzymes and are themselves the targets of other allosteric or covalent regulatory mechanisms.

Feedback Inhibition Through allosteric or covalent regulation, the activity of an enzyme may be increased or decreased, but the reactions in a metabolic pathway are generally catalyzed by different enzymes, with only certain enzymes being regulated. Through regulation of certain enzymes, cells can exert control over entire pathways.

Certain enzymes in metabolic pathways tend to be targets of regulation more than others. For example, within a metabolic pathway the rates of the individual reactions vary. The rate of the entire pathway can be only as fast as the rate of the slowest reaction within the pathway, which is therefore called the *rate-limiting step.* The enzyme that catalyzes the slowest reaction is called the *rate-limiting enzyme,* and is frequently subject to a regulatory process called *feedback inhibition.*

In **feedback inhibition,** an intermediate product of a metabolic pathway allosterically inhibits an enzyme that catalyzes an earlier reaction in the same pathway. Consider, for example, a metabolic pathway consisting of three steps catalyzed by three different enzymes (E_1, E_2, and E_3):

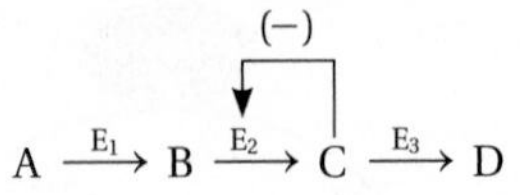

$$A \xrightarrow{E_1} B \xrightarrow{E_2} C \xrightarrow{E_3} D \quad (C \overset{(-)}{\dashrightarrow} E_2)$$

Here, substance C acts as an allosteric inhibitor of E_2, the enzyme catalyzing the second step, as indicated by the arrow with the negative sign.

Normally, feedback inhibition regulates enzyme activities so as to hold the rates of metabolic reactions steady. Under these conditions (referred to as a *steady state*), all three steps in the pathway just described proceed at the same rate, so that C is consumed in the third step as fast as it is generated in the second step, which keeps the concentration of C steady. If the rate of the second step increased suddenly for some reason, then the concentration of C would rise, suppress the activity of E_2, and counteract the change in concentration.

Feedback inhibition can also speed up or slow down metabolic reactions according to changes in the body's needs. Suppose, for instance, that in the previous example the third step speeds up as an appropriate response to an increased cellular demand for the final product, D. The speed-up of the third step will cause C to be consumed at a greater rate, which should cause the concentration of C to begin to fall. This decrease in the concentration of C will reduce the inhibitory action of C on E_2, thereby allowing the second step to speed up.

Often, the allosteric inhibitor in feedback inhibition is the end-product of the metabolic pathway; thus this type of regulation is called *end-product inhibition*. For example, in the following pathway, the end-product of the pathway, D, inhibits E_1:

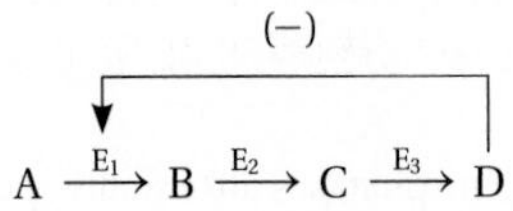

$$A \xrightarrow{E_1} B \xrightarrow{E_2} C \xrightarrow{E_3} D \quad (D \overset{(-)}{\dashrightarrow} E_1)$$

The enzyme allosterically inhibited is often the rate-limiting enzyme of the metabolic pathway, or in some cases, the first enzyme in a metabolic pathway.

Some metabolic pathways have branch points, and it is at the branches where regulation of enzymes occurs. Regulating the first enzymes at branch points determines the relative rates of the branched reactions. Consider the following metabolic pathway:

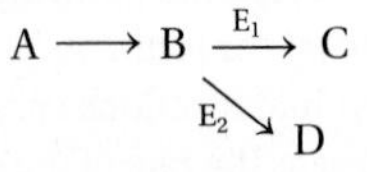

$$A \longrightarrow B \xrightarrow{E_1} C$$
$$B \xrightarrow{E_2} D$$

Intermediate B can be converted to C or D, depending on the relative activity of enzymes 1 and 2.

Feedforward Activation Another, less common mechanism of enzyme regulation is **feedforward activation,** which involves the activation of an enzyme by an intermediate appearing upstream in a metabolic pathway, as follows:

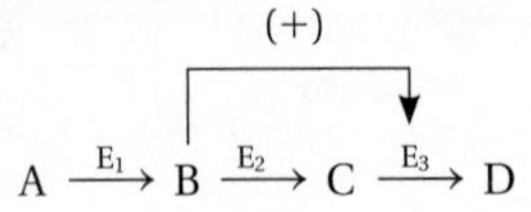

$$A \xrightarrow{E_1} B \xrightarrow{E_2} C \xrightarrow{E_3} D \quad (B \overset{(+)}{\dashrightarrow} E_3)$$

In this example, E_3, which catalyzes the third step, is activated by substance B, which is produced in the first step. As with end-product inhibition, feedforward activation helps to keep reaction rates steady under normal conditions, but it also allows reactions to speed up or slow down when conditions change.

Quick Check 3

1. An increase in which of the following would speed up the net rate of a reaction proceeding in the forward direction: temperature, reactant concentrations, product concentrations, the height of the activation energy barrier? (Choose all that apply.)
2. How is the rate of an enzymatic reaction affected by the catalytic rate of an enzyme? By its affinity for the substrate?
3. What is the primary distinction between allosteric regulation and covalent regulation? Between feedback inhibition and feedforward activation?

4 ATP: The Medium of Energy Exchange

We have seen that energy is released in certain chemical reactions and consumed in others; we have also seen that released energy can be used to perform cellular work. When energy is released in a reaction, it must be "captured" in certain forms before it can be used to do work; otherwise, it is simply dissipated (released into the environment) as heat. When coal burns in a furnace, for example, it releases energy in the form of heat. If this energy is not captured somehow, it just escapes into the atmosphere, without any work being performed. A power plant, in contrast, captures energy released from the burning of coal by using the heat to generate steam under pressure. This steam is then used to drive an electrical generator, and in doing so it performs work. Thus the steam acts as a temporary energy store or "middleman" that facilitates the transfer of energy from the burning coal to the electric generator.

Cells harness the energy released in certain exergonic reactions to synthesize a compound called **adenosine triphosphate (ATP),** which serves as a temporary energy store (**Figure 12**). ATP is synthesized from a nucleotide called *adenosine diphosphate (ADP)* and a phosphate (P_i) according to the following reaction:

$$ADP + P_i + \text{energy} \rightarrow ATP\ (+\ H_2O)$$

ATP synthesis is an example of a condensation reaction because water is generated as a product. (This water is often omitted from the reaction notation for simplicity's sake.) The quantity of energy

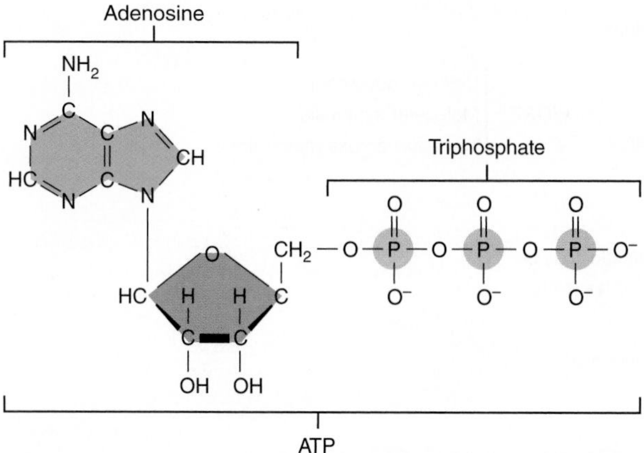

Figure 12 ATP.

required to make one mole of ATP under normal cellular conditions is approximately 7 kcal ($\Delta E = +7$ kcal/mole).

The formation of ATP is a *phosphorylation* reaction because it involves the addition of a phosphate group (P) to another compound, ADP. ATP synthesis occurs through two basic processes: *substrate-level phosphorylation* and *oxidative phosphorylation.* In **substrate-level phosphorylation,** a phosphate group is transferred from a metabolic intermediate (X) to ADP to form ATP:

$$\text{X—P} + \text{ADP} \rightarrow \text{X} + \text{ATP}$$

For example, in muscle cells, *creatine phosphate* can donate its phosphate group to ADP to form ATP and creatine:

$$\text{creatine phosphate} + \text{ADP} \rightarrow \text{creatine} + \text{ATP}$$

In **oxidative phosphorylation,** ADP binds with a free inorganic phosphate (P_i) to form ATP:

$$\text{ADP} + P_i \rightarrow \text{ATP}$$

Oxidative phosphorylation requires the electron transport system in mitochondria and oxygen. Shortly, you will learn specific examples of substrate-level phosphorylation and oxidative phosphorylation. We now look at the breakdown of ATP to release energy.

We noted that steam produced in a power plant is a temporary energy store because it is eventually used for generating electricity; during this process it loses the energy that it gained from the burning coal. Likewise, ATP is a temporary energy store because it is eventually broken down to ADP and P_i, in the process losing the energy that was put into it when it was made:

$$\text{ATP} (+ H_2O) \rightarrow \text{ADP} + P_i + \text{energy}$$

This reaction is often referred to as *ATP hydrolysis* because water is a reactant, even though it is usually omitted from the reaction notation. (Because ATP hydrolysis releases energy and involves the splitting of a single bond—the bond between ATP and one of the attached phosphate groups—that bond is commonly termed a *high-energy phosphate bond.*)

When cells need energy to perform work or to run energy-requiring reactions, they obtain it by hydrolyzing previously synthesized ATP. In the next section, we learn how glucose and other nutrient molecules supply the energy for making ATP, so that these reactions serve as the ultimate sources of cellular energy.

5 Glucose Oxidation: The Central Reaction of Energy Metabolism

We have learned about how enzymes catalyze reactions, so that they can occur quickly enough to meet the body's needs. In this section, we discuss how one type of nutrient—glucose—can be used to produce and store energy to maintain all bodily processes. Later, we show that other nutrient molecules feed into most of the same ATP-producing metabolic pathways as glucose.

Have you ever wondered why we breathe? The simple answer is that breathing enables us to obtain oxygen from the air when we inhale and to get rid of carbon dioxide when we exhale. This exchange of gases is crucial to human survival because we derive most of our energy from the reaction of oxygen with glucose and other nutrient molecules that serve as fuels. When oxygen reacts with these fuels, energy is released and carbon dioxide is generated as a waste product. This outcome is apparent in the reaction of oxygen with glucose ($C_6H_{12}O_6$):

$$C_6H_{12}O_6 + 6\,O_2 \rightarrow 6\,CO_2 + 6\,H_2O + \text{energy}$$

Here we can see that complete oxidation of one molecule of glucose requires six oxygen molecules and generates six molecules each of carbon dioxide and water. We also see that this reaction releases energy, which is why this reaction has such central importance to cells. The energy released amounts to 686 kcal for every mole of glucose that enters the reaction ($\Delta E = -686$ kcal/mole).

Coupling Glucose Oxidation to ATP Synthesis

Cells are able to synthesize ATP using energy derived from glucose oxidation because glucose oxidation involves a negative energy change and, therefore, occurs spontaneously. In contrast, the synthesis of ATP involves a positive energy change and does not occur spontaneously. When cells use glucose oxidation (or any other energy-releasing reaction) to make ATP, some of the energy released in this reaction is used to drive the energy-requiring process of ATP synthesis. This task is accomplished by mechanisms that couple the energy-releasing reaction to the energy-requiring reaction, so that they must occur together (**Figure 13**). When ATP is subsequently broken down, that energy-releasing reaction is coupled to mechanisms that harness the energy to perform work.

We have seen that the oxidation of one mole of glucose releases 686 kcal of energy. Because only 7 kcal is required to synthesize one mole of ATP, the oxidation of one mole of glucose releases enough energy to make 98 moles of ATP. When the actual

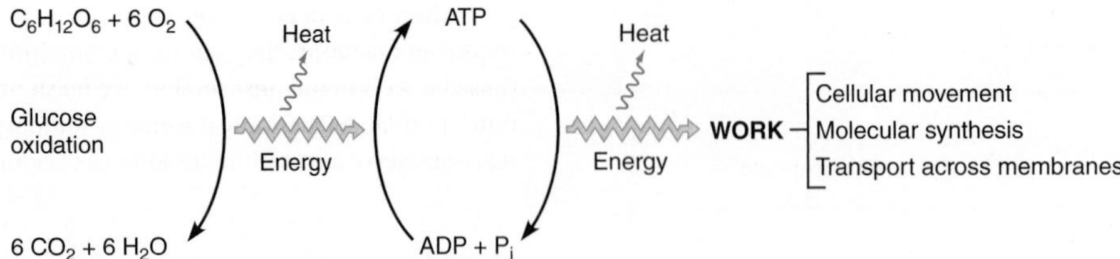

Figure 13 The coupling of reactions in energy transfer. Reactions such as glucose oxidation release energy that is used to synthesize ATP. The subsequent breakdown of ATP releases energy that is used to perform various types of cellular work.

Q As ATP is broken down to ADP and P, does the molecule's potential energy increase or decrease?

amount of ATP is measured, however, it turns out that the quantity of ATP synthesized is only about one-third this amount—approximately 38 moles of ATP per mole of glucose. Why so few? To understand this point, we must consider the energy change for the entire process.

When a mole of glucose is oxidized and 38 moles of ATP is produced, the entire process can be written in the form of a single reaction:

$$C_6H_{12}O_6 + 6\ O_2 + 38\ ADP + 38\ P_i \rightarrow 6\ CO_2 + 6\ H_2O + 38\ ATP$$

The energy change for this reaction is the sum of the energy change for glucose oxidation (−686 kcal) and the energy change for the synthesis of 38 moles of ATP (38 moles $\times$ 7 kcal/mole = 266 kcal). Therefore glucose oxidation releases more energy than is used in ATP synthesis, giving a net energy change that is negative ($\Delta E = -686$ kcal + 266 kcal = −420 kcal). Because the net energy change is negative, the reaction can proceed in the forward direction, yielding ATP as a product.

This example illustrates the general principle that when an energy-releasing reaction is used to drive an energy-requiring process, the released energy cannot be used 100% efficiently; some of the released energy always goes to waste! Because cells normally use approximately 266 kcal of energy for ATP synthesis for each 686 kcal of energy released from the oxidation of one mole of glucose, only about 40% of the released energy is put to use (266 kcal/686 kcal = $0.388 \approx 0.40$). As for the remaining 60% of the released energy, experience tells us at least some of it appears as heat (see Figure 13). In fact, "body heat" is actually heat generated as a by-product of metabolic reactions, including glucose oxidation.

Quick Check 4

1. Is the oxidation of glucose catabolic or anabolic? Does it release energy or require it?
2. Where does the energy for ATP synthesis come from? When ATP is broken down, what is the released energy used for?
3. When energy from glucose oxidation is used to make ATP, only a certain fraction of the released energy is used for this purpose. What happens to the rest of the released energy?

6 Stages of Glucose Oxidation: Glycolysis, the Krebs Cycle, and Oxidative Phosphorylation

We have seen that the oxidation of glucose releases energy and that cells utilize some of this released energy to synthesize ATP. Exactly how cells perform this feat is the topic of this section.

In a cell, the oxidation of glucose is not carried out in a single reaction. Instead, it occurs in three distinct stages, which are carried out in different parts of the cell: (1) *glycolysis* takes place in the cytosol; (2) the *Krebs cycle* (also called the *citric acid cycle, tricarboxylic acid cycle,* or *TCA cycle*) occurs in the mitochondrial matrix; and (3) *oxidative phosphorylation* occurs across the inner mitochondrial membrane. Each of these stages involves a series of chemical reactions that are summarized next, beginning with glycolysis.

Glycolysis

Glycolysis, which means "splitting of sugar," is a metabolic pathway comprising ten reactions, each catalyzed by a different cytosolic enzyme (**Figure 14**). In glycolysis, each glucose molecule (which contains six carbons) is broken down into two molecules of *pyruvate* (the ionized form of *pyruvic acid*) containing three carbons apiece. These molecules of pyruvate are normally broken down further in the subsequent stages of glucose oxidation.

The major results of glycolysis are shown in **Figure 15** and can be summarized as follows:

1. By the end of glycolysis, each glucose molecule has been split into two molecules of pyruvate.
2. During this process, two ATP molecules are consumed (one in reaction 1 and one in reaction 3), but four more are produced by substrate-level phosphorylation (two in reaction 7 and two in reaction 10). This gives a net synthesis of two molecules of ATP for each molecule of glucose consumed.
3. Two molecules of NAD^+ are reduced in step 6, yielding two molecules of NADH for every molecule of glucose.

When all reactants and products are accounted for, the overall reaction of glycolysis is as follows:

$$\text{glucose} + 2\ NAD^+ + 2\ ADP + 2\ P_i \rightarrow 2\ \text{pyruvate} + 2\ NADH + 2\ H^+ + 2\ ATP$$

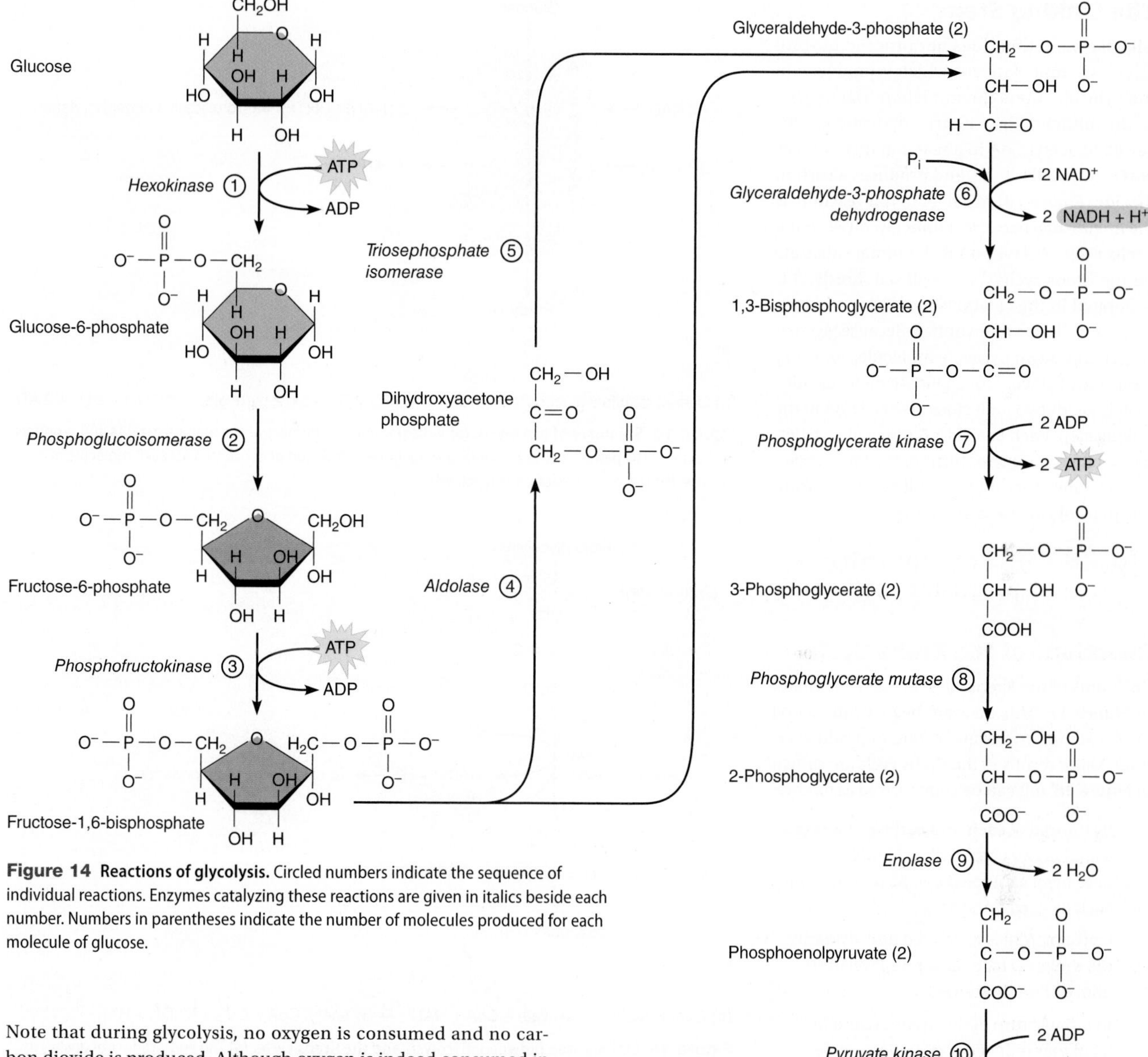

Figure 14 **Reactions of glycolysis.** Circled numbers indicate the sequence of individual reactions. Enzymes catalyzing these reactions are given in italics beside each number. Numbers in parentheses indicate the number of molecules produced for each molecule of glucose.

Note that during glycolysis, no oxygen is consumed and no carbon dioxide is produced. Although oxygen is indeed consumed in glucose oxidation and carbon dioxide is produced, these events do not occur in glycolysis, but rather in the later stages of glucose oxidation.

Although glycolysis is useful to cells because it produces some ATP, its primary importance derives from the fact that it sets the stage for subsequent events that yield even more ATP. The NADH that is produced in glycolysis will eventually give up its electrons, thereby releasing energy that will be used to synthesize more ATP by oxidative phosphorylation. Also, the pyruvate that is generated will eventually be catabolized in the next stage of glucose oxidation, the Krebs cycle.

As we see later in the chapter, pyruvate can proceed through further stages of glucose oxidation only when oxygen is readily available inside the cell. If the availability of oxygen becomes limited, pyruvate is instead converted to a compound called *lactic acid* (or *lactate*), which is not broken down further. As we continue our discussion of glucose metabolism, assume that oxygen is readily available unless otherwise noted.

The Krebs Cycle

Glycolysis resembles most metabolic pathways in that it is a sequence of steps with distinct starting and ending points. In contrast, the **Krebs cycle,** the next stage of glucose oxidation, has no starting or ending points because it is circular or *cyclic,* as its name implies.

The Linking Step

Glucose oxidation does not proceed beyond glycolysis until pyruvate, which is produced in the cytosol, enters the mitochondrial matrix. In the mitochondrial matrix, pyruvate is converted to acetyl CoA in a reaction that reduces NAD^+ to $NADH + H^+$ and produces a carbon dioxide (**Figure 16**). We refer to this reaction as the *linking step* because it links glycolysis to the Krebs cycle. Acetyl CoA is the initial substrate for the Krebs cycle. As we will see shortly, it is generated in the catabolism of lipids and proteins as well as carbohydrates. Because glycolysis generates two pyruvate molecules for every molecule of glucose, one glucose molecule ultimately yields two molecules of acetyl CoA in the linking step. Each acetyl CoA molecule participates in one complete "turn" of the Krebs cycle.

The net reaction for the linking step from one mole of glucose is as follows:

$$2 \text{ pyruvate} + 2 \text{ CoA} + 2 \text{ NAD}^+ + 2 \text{ H}_2\text{O} \rightarrow 2 \text{ acetyl CoA} + 2 \text{ NADH} + 2 \text{ H}^+ + 2 \text{ CO}_2$$

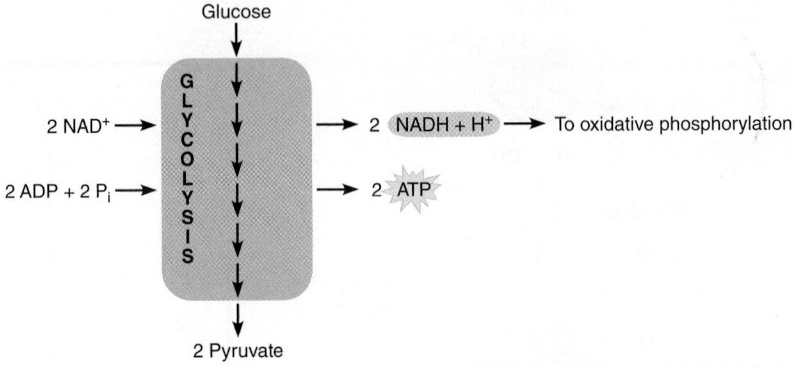

(b) Overall reaction: glucose + 2 NAD^+ + 2 ADP + 2 P_i ⟶ 2 pyruvate + 2 NADH + 2 H^+ + 2 ATP

Figure 15 Summary of glycolysis. (a) A diagrammatic summary of the major reactants and products of the glycolysis pathway. Numbers indicate quantities produced or consumed for each molecule of glucose. **(b)** The overall reaction of glycolysis.

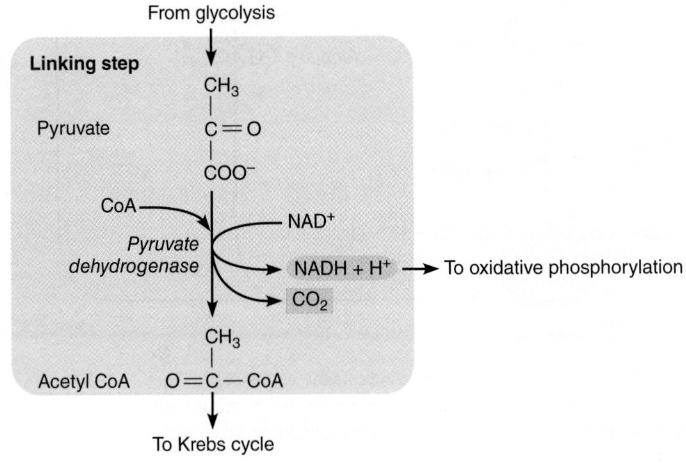

(b) Overall reaction: pyruvate + CoA + NAD^+ ⟶ acetyl CoA + CO_2 + NADH + H^+

Figure 16 Linking step between glycolysis and the Krebs cycle. (a) The end-product of glycolysis, pyruvate, is converted to the initial substrate of the Krebs cycle, acetyl CoA, in the mitochondrial matrix. **(b)** The overall reaction of the linking step.

Reactions of the Krebs Cycle

Reactions of the Krebs cycle are given in detail in **Figure 17**. This diagram begins with acetyl CoA reacting with oxaloacetate to produce citrate. Major results of the Krebs cycle are shown in **Figure 18** and can be summarized as follows:

1. By the end of each turn of the Krebs cycle, a total of two carbon dioxide molecules have been generated as end-products (one each in steps 3 and 4).
2. Only one ATP is generated directly during the Krebs cycle by substrate-level phosphorylation (in step 5).
3. In a single turn of the Krebs cycle, a total of four reduced coenzymes—three $NADH + H^+$ and one $FADH_2$—are produced (in steps 3, 4, 6, and 8).
4. The final product is oxaloacetate, which reacts with acetyl CoA to start the cycle over again, so long as another acetyl CoA molecule is present.

The overall reaction for one turn of the Krebs cycle is as follows:

$$\text{acetyl CoA} + 3 \text{ NAD}^+ + \text{FAD} + \text{ADP} + \text{P}_i + 3 \text{ H}_2\text{O} \rightarrow 2 \text{ CO}_2 + 3 \text{ NADH} + 3 \text{ H}^+ + \text{FADH}_2 + \text{ATP}$$

When combined with the two ATP molecules per glucose molecule that are netted in glycolysis, the additional two ATP molecules that are generated in the Krebs cycle make a running total of four molecules of ATP for every molecule of glucose. As we will see, however, almost all of the ATP that is formed is generated during the final stage of glucose oxidation, oxidative phosphorylation. The Krebs cycle plays an important role in this process because it supplies eight of the 12 reduced coenzyme molecules that eventually go to the electron transport chain to give up their electrons, thereby releasing energy for making ATP.

The Krebs cycle and linking step produce 100% of the six carbon dioxide molecules that result from the complete oxidation of glucose. Note that absolutely no oxygen has been consumed so far, meaning that 100% of the expected oxygen consumption (six molecules of oxygen per molecule of glucose) must occur in oxidative phosphorylation, the final stage of glucose oxidation, if it is to occur at all. As we will see next, oxygen plays a vital role in oxidative phosphorylation,

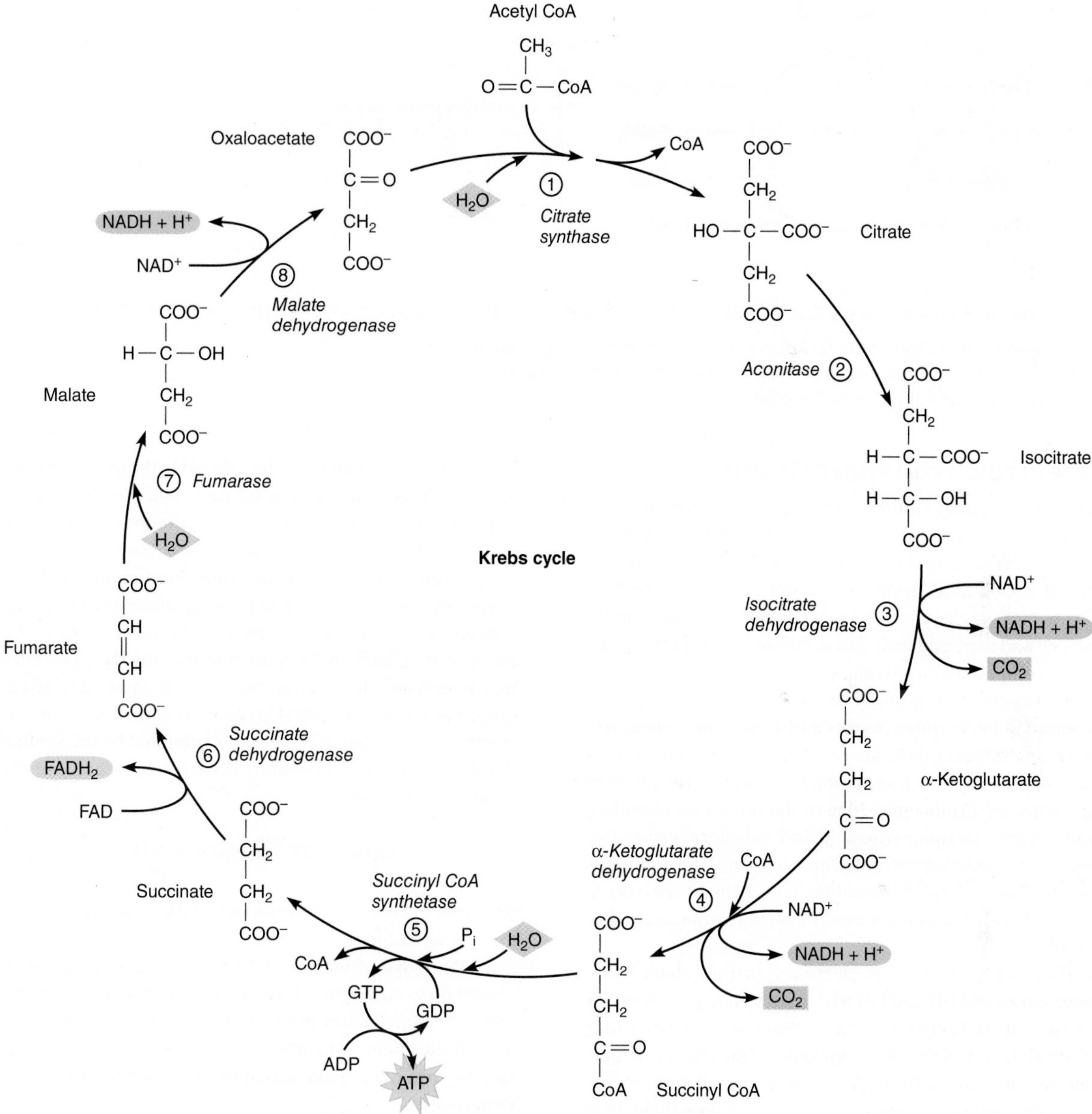

Figure 17 Reactions of the Krebs cycle. Circled numbers indicate the sequence of individual reactions. Enzymes catalyzing these reactions are given in italics beside each number.

Q Step 8 of the Krebs cycle is an example of which type of reaction?

because oxygen is the ultimate acceptor of all the electrons that are given up by NADH + H^+ or $FADH_2$. Without it, the electrons would have nowhere to go, and oxidative phosphorylation would stop.

Oxidative Phosphorylation

Most ATP that is made in cells is made by oxidative phosphorylation, which involves two simultaneous processes: (1) the transport in the inner mitochondrial membrane of hydrogen atoms or electrons through a series of compounds, known as the *electron transport chain,* which releases energy, and (2) the harnessing of this energy to make ATP, which is carried out by a mechanism called *chemiosmotic coupling.*

In oxidative phosphorylation, the reduced coenzymes (NADH + H^+ and $FADH_2$) generated in glycolysis, the linking step, and the Krebs cycle serve as the energy source for making ATP. NADH + H^+ and $FADH_2$ release electrons to the electron transport chain, and as these electrons travel through the chain, energy is released. Much of that energy is captured and harnessed to drive the synthesis of ATP, which is catalyzed by an enzyme called **ATP synthase.**

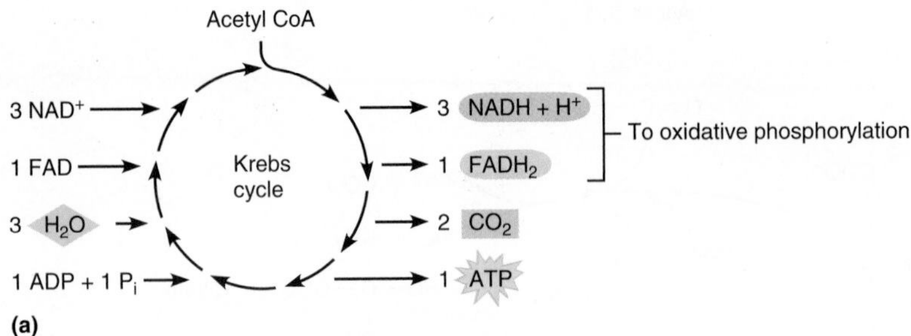

(b) Overall reaction: acetyl CoA + 3 NAD^+ + FAD + ADP + P_i + 3 H_2O $\longrightarrow$ 2 CO_2 + 3 NADH + 3 H^+ + $FADH_2$ + ATP + CoA

Figure 18 Summary of the Krebs cycle. (a) A diagrammatic summary of the major reactants and products of the Krebs cycle. Numbers indicate quantities produced or consumed for each molecule of acetyl CoA that enters the Krebs cycle. **(b)** The overall reaction of the Krebs cycle.

The Electron Transport Chain

The **electron transport chain** comprises a set of diverse compounds located in the inner mitochondrial membrane. Most of these compounds are proteins specialized to function as electron carriers; that is, they bind electrons reversibly and, therefore, have the ability to pick up electrons and subsequently give them up (or to be reduced and then oxidized). Among these electron carriers are a number of compounds called *cytochromes,* which possess special iron-containing chemical groups known as *hemes.* (Heme groups are also found in hemoglobin, as this molecule's name suggests.) Also present are various *iron-sulfur proteins,* which contain iron atoms bound to sulfur. Most of these proteins are aggregated into large complexes that are firmly embedded in the inner mitochondrial membrane. A few electron carriers are individual molecules that move freely in the lipid bilayer and shuttle back and forth, carrying electrons from one complex to another. One, called *coenzyme Q,* is not a protein at all, but rather a small molecule composed mainly of hydrocarbon.

Electrons are carried to the electron transport chain by reduced coenzymes (NADH and $FADH_2$) generated in glycolysis, the linking step, and the Krebs cycle. These reduced coenzymes then release their electrons to certain components of the chain that function as electron acceptors. These electron acceptors, in turn, donate their electrons to other electron acceptors, which pass them on to still other acceptors, and so on. Each time electrons move from one component to the next, they lose some energy—energy that ultimately is used in making ATP.

As electrons pass through the chain, they go from one electron acceptor to the next in the specific sequence shown in **Figure 19**. NADH donates its pair of electrons to the first electron acceptor in the chain, called *flavin mononucleotide (FMN).* In the process, NADH becomes oxidized to NAD^+ (which is then free to pick up more electrons in glycolysis, the linking step, or the Krebs cycle), and FMN is reduced. FMN then passes its electrons to the next component of the chain, an iron-sulfur protein, and becomes oxidized and ready to pick up another pair of electrons from NADH.

Although electrons leave NADH in the form of hydrogens, most components of the electron transport chain do not carry actual hydrogen atoms, only free electrons. Consequently, at some point electrons are stripped from the original hydrogen atoms, which are left as hydrogen ions (H^+) in solution, and the electrons continue to move through the chain. Certain components of the chain *do* carry hydrogen ions, but they transport these ions across the inner mitochondrial membrane. At three components, hydrogen ions move into the intermembrane space, where they become highly concentrated. At the last component, these ions move in the opposite direction, back to the mitochondrial matrix. In essence, then, we can envision the electron transport chain as carrying either hydrogen atoms or an equivalent combination of hydrogen ions and electrons. The release of electrons (denoted by the symbol e^-) to the electron transport chain by NADH is represented at the top of Figure 19 as follows:

$$NADH + H^+ \rightarrow NAD^+ + 2\,H^+ + 2\,e^-$$

Note that the release of electrons by $FADH_2$ is written in Figure 19 in a similar fashion.

After moving from FMN to the iron-sulfur protein, electrons are passed on to coenzyme Q, which passes them on to cytochrome *b,* and so forth. When the electrons eventually reach the final component of the chain, cytochrome a_3, they recombine with hydrogen ions to form hydrogens, which react immediately with oxygen to form water:

$$2\,e^- + 2\,H^+ + {}^1\!/_2\,O_2 \rightarrow H_2O$$

(The single oxygen atom is written as $\frac{1}{2}O_2$ because oxygen enters the reaction in its molecular form, O_2.) By combining this reaction with the previous one, we see that the net result of electron transport is the same as if NADH had simply given up its electrons directly to oxygen to form water, which is an energy-releasing reaction:

$$NADH + H^+ + {}^1\!/_2\,O_2 \rightarrow NAD^+ + H_2O + \text{energy}$$

If this is the net result, then what use is the electron transport chain? The electron transport chain provides a way of harnessing this energy so that it can be used to do something useful—namely, to make ATP. (We will see how this is accomplished shortly.) The precise quantity of ATP made depends on conditions in the cell;

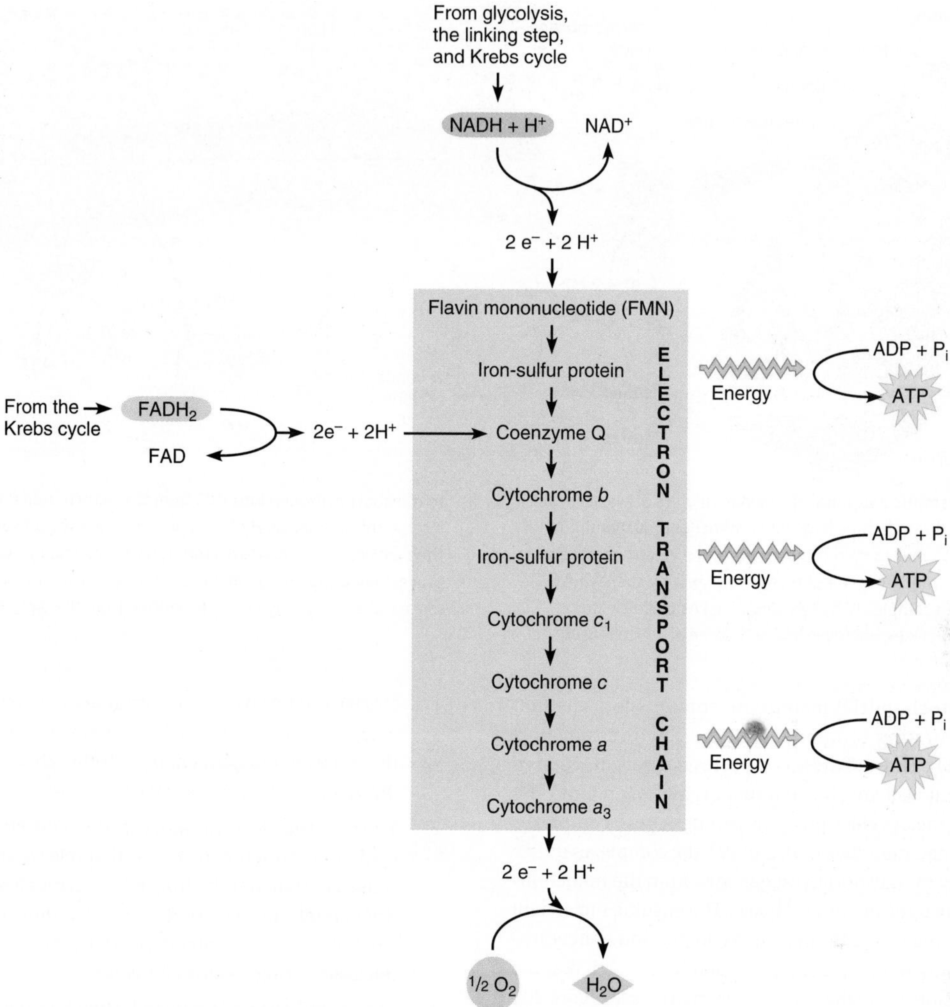

Figure 19 The electron transport chain. Arrows indicate the path of electron flow from initial electron donors (NADH or $FADH_2$) to oxygen, the final electron acceptor. Wavy arrows at right indicate that the movement of electrons through the chain releases energy, which is harnessed to drive ATP synthesis by oxidative phosphorylation.

Q Which component(s) of the electron transport chain contain(s) heme groups?

two or three ATP molecules are synthesized for every pair of electrons released by NADH.

$FADH_2$ also donates electrons to the electron transport chain, but it does not donate its electrons to FMN. FAD is permanently bound to its partner enzyme succinate dehydrogenase (see step 6 of the Krebs cycle, Figure 17), which unlike other Krebs cycle enzymes is embedded in the inner mitochondrial membrane. For this reason, $FADH_2$ is not free to move about, as is NADH. Because of its location, $FADH_2$ is physically removed from FMN and, therefore, cannot donate electrons to it; instead, it donates its electrons to coenzyme Q (see Figure 19). These electrons, like those that come from NADH, eventually combine with oxygen to form water, but because they enter the chain at a point downstream, they release less energy as they move through the chain. As a consequence, only one or two ATP molecules are produced for each pair of electrons released by $FADH_2$, compared to the two or three ATP molecules produced when electrons are released by NADH.

We know that energy is released when electrons move through the electron transport chain and that this energy is used to make ATP. The mechanism by which this occurs is called chemiosmotic coupling.

Chemiosmotic Coupling

The process that couples electron transport to ATP synthesis, known as **chemiosmotic coupling,** first uses the energy released in the electron transport chain to transport hydrogen ions across the inner mitochondrial membrane against their *concentration gradient,* and then uses the energy stored in this gradient to make ATP (**Figure 20**).

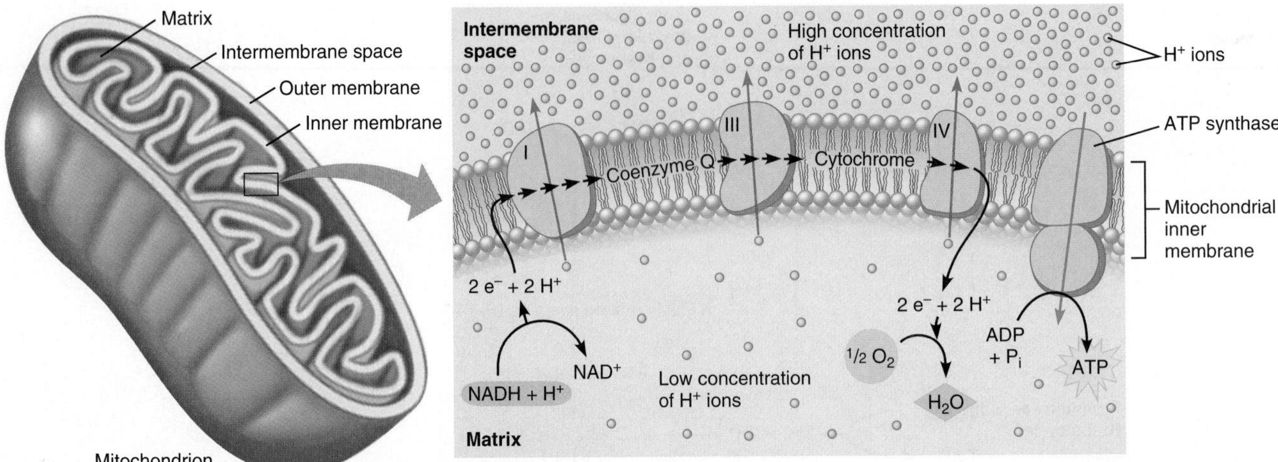

Figure 20 Chemiosmotic coupling. The oxidation of NADH yields a pair of electrons that pass through three large inner mitochondrial membrane complexes (I, III, and IV) that contain components of the electron transport chain. (Electrons are carried from complex I to complex III by coenzyme Q, and from complex III to complex IV by cytochrome *c*.) As these electrons move through the complexes, energy is released, some of which is used to transport hydrogen ions (H^+) from the mitochondrial matrix to the intermembrane space. Hydrogen ions flow in the opposite direction (down their concentration gradient) through the enzyme ATP synthase, in the process releasing energy that is used to synthesize ATP. (For simplicity, the release of electrons by $FADH_2$ is omitted from the diagram.)

The inner mitochondrial membrane contains four distinct complexes incorporating components of the electron transport chain (see Figure 20). Three of these complexes not only function as electron carriers but also are able to transport hydrogen ions (H^+) across the membrane. As electrons pass through these complexes (designated by Roman numerals I, III, and IV), the complexes utilize the released energy to transport hydrogen ions from the mitochondrial matrix into the intermembrane space. The ensuing movement of hydrogen ions creates a difference in hydrogen ion concentration (a concentration gradient) across the membrane such that the concentration outside it is higher than in the matrix (see Figure 20). This concentration gradient represents a store of potential energy. (A difference in electrical potential across the membrane further adds to this stored energy.)

The enzyme ATP synthase, which catalyzes the formation of ATP, resides in the inner mitochondrial membrane, along with components of the electron transport chain, and takes advantage of energy stored in the hydrogen ion gradient to perform its function. Like the complexes just described, it is able to transport hydrogen ions across the membrane. In this case, however, the hydrogen ions move down their concentration gradient, rather than up it (see Figure 20). This flow of hydrogen ions releases energy, which ATP synthase harnesses to make ATP. Sometimes, hydrogen ions leak across the membrane independent of the ATP synthase; this movement of hydrogen ions is *uncoupled* from ATP synthesis and is one of the reasons ATP production per pair of electrons donated from NADH + H^+ or $FADH_2$ is less than optimal (three or two ATP molecules, respectively).

Summary of Oxidative Phosphorylation

We have seen that in oxidative phosphorylation, the final stage of glucose metabolism, the following events occur:

1. NADH and $FADH_2$, which are generated in the earlier stages of glucose metabolism, release their electrons and protons to the electron transport chain, thereby returning to their oxidized forms (NAD^+ and FAD).
2. The released electrons move through the chain by a series of oxidation-reduction reactions that release energy. This energy is used to transport hydrogen ions across the inner mitochondrial membrane, creating a concentration gradient of hydrogen ions across the membrane that stores some of the energy released during electron transport.
3. This stored energy is released when hydrogen ions flow through ATP synthase, which uses the energy to make ATP. At most, three ATP molecules are made for every pair of electrons released from NADH, and two ATP molecules are made for every pair of electrons released from $FADH_2$.
4. The released electrons and protons come together again in the mitochondrial matrix to form water as an end-product.

Given that 10 molecules of NADH and 2 molecules of $FADH_2$ are generated for every molecule of glucose that is oxidized, we expect a maximum of 34 molecules of ATP to be made during oxidative phosphorylation [(10 NADH $\times$ 3 ATP/NADH) + (2 $FADH_2$ $\times$ 2 ATP/$FADH_2$) = 34 ATP]. Because these 12 molecules of reduced coenzymes carry a total of 12 pairs of hydrogens, we also expect 12 molecules of water to be generated, which requires that 6 molecules of oxygen be consumed. Therefore, the overall reaction for oxidative phosphorylation is as follows (**Figure 21**):

$$10\ NADH + 10\ H^+ + 2\ FADH_2 + 34\ ADP + 34\ P_i + 6\ O_2 \rightarrow 10\ NAD^+ + 2\ FAD + 12\ H_2O + 34\ ATP$$

Note that this reaction accounts for 100% of the oxygen consumed in the oxidation of one molecule of glucose. Perhaps surprisingly,

DISCOVERY

Can Uncouplers Aid in Weight Loss?

The movement of electrons down the electron transport chain is coupled to ATP synthesis by chemiosmotic coupling. If this coupling were perfect, then three ATP molecules would be produced for each NADH + H^+ complex that provides electrons and protons (hydrogen ions) to the electron transport chain, and two ATP molecules would be produced for each $FADH_2$ that provides the electrons and protons. However, in actuality, some of the protons that are transported into the mitochondrial intermembrane space leak back across the inner mitochondrial membrane into the matrix independently of the ATP synthase channel. These "leaked" protons do not contribute to ATP synthesis and are said to be "uncoupled" from oxidative phosphorylation. A basal leak exists all the time, and *uncouplers* can increase the amount of proton leak. The greater the uncoupling, the less ATP produced per oxygen consumed by the body.

Uncoupled mitochondria are found in nearly every tissue, but uncoupling agents are particularly prevalent in adipose tissue, the liver, and skeletal muscle. Although the physiological role of uncouplers in humans is unknown, they are of interest to weight loss researchers because uncoupling increases the body's metabolic rate. An increased metabolic rate enables the body to "burn" more energy—and potentially lose weight.

In the 1930s, the uncoupler 2,4-dinitrophenol (DNP) was used to promote weight loss. While patients lost weight with DNP, this agent's toxic effects (e.g., inadequate ATP production leading to lethargy, hyperthermia, and death) outweighed its benefits and DNP was pulled from the market. Pharmaceutical companies continue to search for an uncoupler with a greater difference between the beneficial dose and the toxic dose, such as lipophilic weak acids that bind protons and transport them across the inner mitochondrial membrane from the intermembrane space to the matrix. Until scientists come up with the right uncoupler, we must still rely on exercise and eating right to maintain a healthy weight.

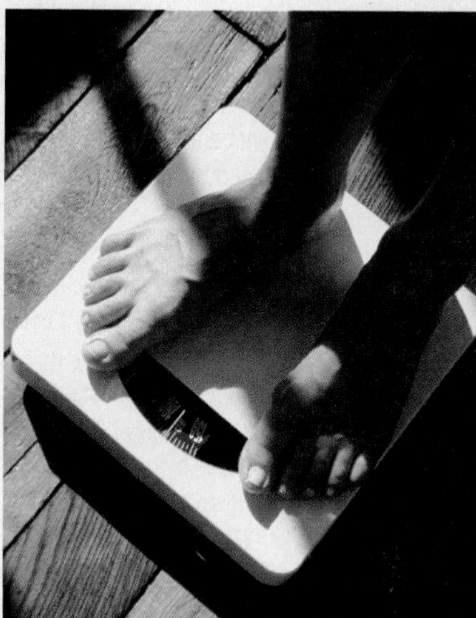

JPL/Anne/Photo Researchers, Inc.

it also shows that more water molecules are produced—12 instead of 6. However, 6 water molecules are consumed in the Krebs cycle, so the net result is 6 molecules of water produced, which is in accord with our expectations.

Apply Your Knowledge

Recall that two of the NADH + H^+ complexes produced during glucose oxidation were generated in the cytosol from glycolysis. These two NADH + H^+ cannot cross the mitochondrial membrane to donate their electrons directly to the electron transport chain. Instead, the electrons are shuttled across the membrane by various molecules. The shuttled electrons can be donated to either NAD^+ or FAD in the mitochondria. Discuss the consequence in terms of ATP production if both electrons are shuttled to mitochondrial NAD^+, both are shuttled to mitochondrial FAD, or one electron is shuttled to NAD^+ and the other to FAD.

When compared to the quantity of ATP generated by substrate-level phosphorylation in glycolysis and the Krebs cycle, clearly the output from oxidative phosphorylation is responsible for most of the ATP that cells produce. Given that substrate-level phosphorylation and oxidative phosphorylation generate 4 and 34 molecules of ATP per glucose, respectively, the latter accounts for nearly 90% of the total of 38 molecules. (Although oxygen is essential to produce most of the energy used by our cells, its high reactivity contributes to several disease states. See **Clinical Connections: Antioxidants and the Problem with Free Radicals.**)

Summary of Glucose Oxidation

Now that we have examined the details of all three stages of glucose metabolism, we can put everything together to see the "big picture." To help us accomplish this integration, the whole of glucose metabolism is summarized in **Figure 22**, a compilation of Figures 15, 16, 18, and 21. Note that when all the reactants and products in all three stages are tallied up in Figure 22b, the result is that for every molecule of glucose consumed, six molecules of oxygen are also consumed. In addition, six molecules each of carbon dioxide and water are produced. These figures are in exact agreement with the generalized reaction for glucose oxidation ($C_6H_{12}O_6 + 6\,O_2 \rightarrow 6\,CO_2 + 6\,H_2O$). Because 38 molecules of ATP are generated, the overall reaction for the three stages of glucose metabolism is as follows:

$$\text{glucose} + 6\,O_2 + 38\,\text{ADP} + 38\,P_i \rightarrow 6\,CO_2 + 6\,H_2O + 38\,\text{ATP}$$

Neither NAD nor FAD appears anywhere in this reaction because these coenzymes are reduced in glycolysis, the linking step, and the Krebs cycle but then oxidized in oxidative phosphorylation and returned to their original forms.

CLINICAL CONNECTIONS

Antioxidants and the Problem with Free Radicals

You see it on juice bottles, on the labels of dietary supplements, and in articles touting the benefits of eating dark chocolate or drinking green tea: "Filled with antioxidants!" What, exactly, are *oxidants*—and how can they be harmful?

The metabolism of oxygen results in the production of *free radicals,* highly reactive molecular species that can damage the body's cells. Free radicals have one or more unpaired ("free") electrons. Because electrons are more stable when paired, free radicals are highly reactive. They achieve stability in cells by reacting with another molecule, gaining an electron in the process by "taking" an electron from the other molecule (that is, oxidizing the other molecule). Free radicals and other oxidizing agents are collectively referred to as oxidants. Oxidants have normal functions in cells—for instance, they help destroy bacteria during inflammation and can induce protective mechanisms within a cell during cell stress. However, when produced in excess of the cell's antioxidant capacity, oxidants may react with critical biochemicals in the cell (including DNA, proteins, and phospholipids) and cause cell damage—a condition known as *oxidative stress.*

Most cells of the body generate oxidants and are subject to oxidative stress. In turn, oxidative stress has been associated with numerous disease states, including arthritis, heart disease, stroke, and degenerative illnesses such as Alzheimer's disease and insulin-dependent diabetes mellitus. Consequently, scientific research has focused much attention on antioxidants—which break down oxidants and molecules that react with free radicals, changing their reactive nature—as potential agents to prevent disease. Some researchers have hypothesized that dietary supplements of vitamins C and E might help prevent heart disease, as might consumption of certain fruits and vegetables, dark chocolate, and red wine. The research is not yet conclusive, but some evidence suggests that antioxidant vitamins can slow down diseases and perhaps even the aging process.

Photos.com/Getty Images.

Dark chocolate is a source of antioxidants.

Critical Thinking Questions

1. Is there a difference between an oxidant and a free radical? If so, what is it?
2. How do oxidants damage tissue and cause disease?
3. What is the nature of oxidative stress, and its association with disease states?

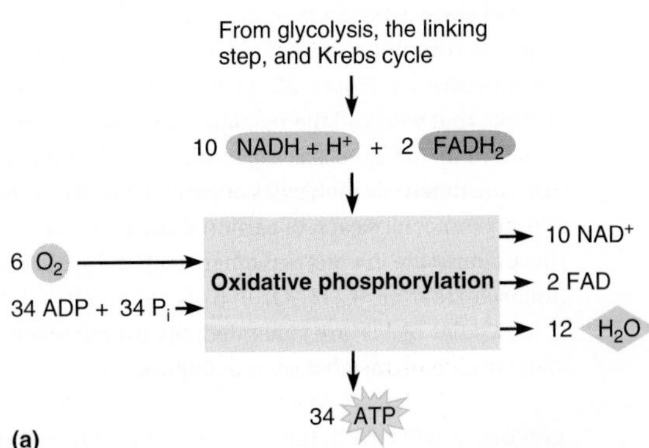

(b) Overall reaction: $10\ NADH + 10\ H^+ + 2\ FADH_2 + 34\ ADP + 34\ P_i + 6\ O_2 \longrightarrow 10\ NAD^+ + 2\ FAD + 12\ H_2O + 34\ ATP$

Figure 21 Summary of oxidative phosphorylation. (a) A diagrammatic summary of the major reactants and products of oxidative phosphorylation. Numbers indicate quantities produced or consumed for each molecule of glucose that enters glycolysis. **(b)** The overall reaction for oxidative phosphorylation.

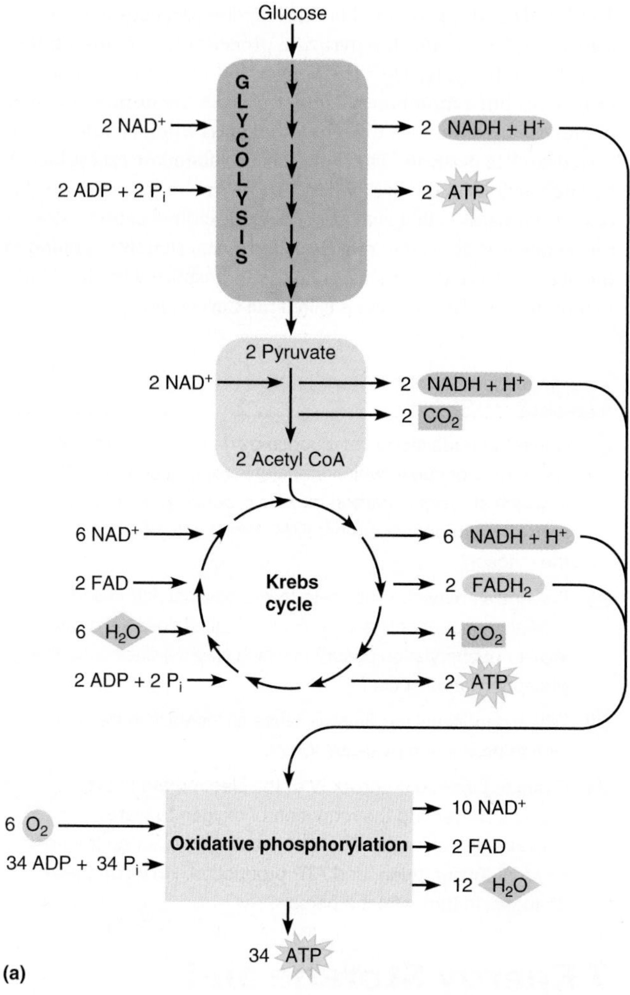

(b) Overall reaction: glucose + 6 O_2 + 38 ADP + 38 P_i ⟶ 6 CO_2 + 6 H_2O + 38 ATP

Figure 22 Summary of glucose oxidation. (a) A compilation of Figures 15a, 16a, 18a, and 21a, showing all three stages of glucose oxidation and their interconnections. **(b)** The overall reaction of glucose oxidation as a sum of the individual reactions. A maximum of 38 ATP molecules can be produced for every molecule of glucose that is oxidized.

Glucose is not the only sugar that can be broken down by cells and used for energy. Cells can also metabolize other monosaccharides, such as fructose, a component of table sugar; galactose, a component of milk sugar; and another monosaccharide called *mannose.* To utilize these sugars, cells first convert them to certain intermediates of glycolysis, such as glucose-6-phosphate and fructose-6-phosphate. These intermediates then are simply "fed into" the glycolysis pathway at the appropriate steps and oxidized in the usual manner.

Glucose Catabolism in the Absence of Oxygen

Because oxygen is the ultimate electron acceptor in oxidative phosphorylation, it must be supplied to tissues on a continual basis by the blood if glucose oxidation is to go to completion. The rate at which oxygen must be supplied depends on how fast the tissues are consuming it; in other words, it depends on the tissues' *metabolic demand.* If oxygen can be delivered to tissues at a rate high enough to match the body's demands, oxygen will always be available to accept electrons, and glucose oxidation will proceed to completion. In the event that oxygen delivery is no longer able to keep pace with demand (because of a reduction in blood flow to a particular organ or tissue, for instance), oxygen concentration in the affected tissue will fall to very low levels, perhaps even approaching zero (called anaerobic conditions).

Under these conditions, fewer oxygen molecules will be available to accept electrons as they reach the end of the electron transport chain, so it becomes difficult to "offload" them from the chain's electron carriers. As a result, the flow of electrons through the chain slows, as does the rate of ATP synthesis by oxidative phosphorylation. Under these conditions, an electron "traffic jam" develops in the electron transport chain, which eventually decreases the number of FMN molecules available to accept electrons from NADH. When this happens, coenzyme molecules become "trapped" in their reduced forms, and levels of the coenzyme NAD^+ decrease.

Any drop in NAD^+ levels in a cell is a potential threat to all ATP production because a supply of NAD^+ is necessary for proper operation of both the glycolysis pathway and the Krebs cycle. Of particular interest here is step 6 of glycolysis, which occurs before any of the steps in which ATP is produced by substrate-level phosphorylation. As a consequence, if oxygen becomes depleted and the supply of NAD^+ is decreased, not only does oxidative phosphorylation slow, but substrate-level phosphorylation slows as well. Thus a reduction in oxygen availability has the potential to slow (or perhaps even stop) all ATP-producing steps, which would be disastrous for a cell, because its supply of ATP would eventually become exhausted.

Fortunately, this disaster can be averted. Most cells contain an enzyme called *lactate dehydrogenase (LDH),* which can convert pyruvate (the end-product of glucose breakdown in glycolysis) to *lactate* according to the following reaction:

$$\text{pyruvate} + \text{NADH} + \text{H}^+ \xrightarrow{LDH} \text{lactate} + \text{NAD}^+$$

The significance of this reaction is as follows: In the event that NADH cannot unload its electrons to the electron transport chain because of a limitation in the availability of oxygen, this reaction provides an alternative pathway that allows electrons to be unloaded in a reaction that generates free NAD^+, which can then pick up electrons in step 6 of glycolysis. Therefore, when the electron transport chain is not available to accept electrons, NADH and NAD^+ can still shuttle back and forth between the conversion of pyruvate to lactate and step 6 of glycolysis (**Figure 23**). This reaction ensures a steady supply of NAD^+, which enables step 6 to run even when oxygen availability is limited. More important, the fact that step 6 can run means that the remaining steps in glycolysis can also run, including steps 7 and 10, which synthesize ATP.

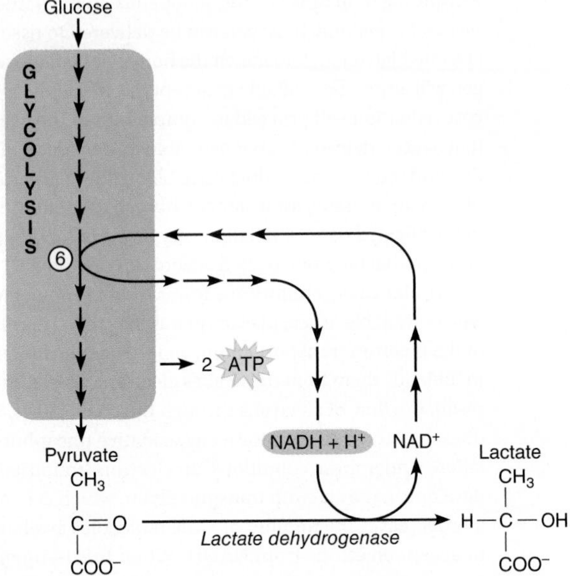

Figure 23 Conversion of pyruvate to lactate. NADH, which is generated in step 6 of glycolysis, can be converted back to NAD^+ in a reaction that generates lactate from pyruvate. This reaction ensures that the cell has a steady supply of NAD^+, which allows for the continued operation of step 6 and the continued production of ATP even under conditions of reduced oxygen availability.

The ability to synthesize ATP in the manner just described allows muscles and other tissues to continue working even when the availability of oxygen is low, but it is important to realize that this does not mean the body's tissues can live indefinitely in the absence of oxygen. For one thing, only two ATP molecules are produced in glycolysis for every molecule of glucose consumed, which is only about 5% of the ATP that is normally produced as a result of complete glucose oxidation. Unless a cell's demand for ATP is very low, it is unlikely that the body will be able to supply enough ATP through glycolysis alone. (Interestingly, red blood cells, which transport oxygen, must obtain all their ATP from glycolysis because they lack mitochondria.) Furthermore, because the conversion of pyruvate to lactate is a "dead end" reaction, lactate tends to accumulate in cells when it is produced at a rapid rate, which can cause acidification of the intracellular fluid. Lactate can also leak into extracellular fluid and "spill over" into the bloodstream, causing acidification of the blood. Acidification of tissues will eventually begin to interfere with proper cellular function and, therefore, cannot continue indefinitely. Even with these negative features, anaerobic glycolysis has a very important role in muscle metabolism (**Understanding Exercise: Sources of Energy for Muscle Cells**).

Because lactate is potentially harmful to cells, they must get rid of it. When oxygen availability to a cell returns to normal, pyruvate begins to proceed to the Krebs cycle as it usually does, and the concentration of pyruvate in cells decreases. As a result of mass action, the lactate dehydrogenase reaction begins to run in reverse, so that lactate is converted back to pyruvate and NAD^+ is converted to NADH:

$$\text{pyruvate} + \text{NADH} + \text{H}^+ \xrightarrow{LDH} \text{lactate} + \text{NAD}^+$$

The NADH that is generated in this reaction proceeds to oxidative phosphorylation, and the pyruvate proceeds to the linking step and the Krebs cycle. Thus the conversion of pyruvate to lactate represents but a momentary "side trip" from the normal pathway of glucose oxidation—a diversion that ends when lactate is converted back to pyruvate. The body can also eliminate excess lactate through actions in the liver, where lactate can be converted to glucose in a process called *gluconeogenesis*, described shortly. Most of the lactate is produced in muscle cells, transported by the blood to the liver, converted to glucose, and then transported in blood back to muscle cells. This process is called the **Cori cycle.**

Quick Check 5

1. Name the three stages of glucose oxidation in the order in which they occur. In which stage(s) is oxygen consumed? In which stage(s) is carbon dioxide produced? Which stage normally produces the most ATP? Which stage(s) occur(s) in the cytosol?
2. Distinguish between substrate-level phosphorylation and oxidative phosphorylation. In which stage(s) does substrate-level phosphorylation occur? In which stage(s) does oxidative phosphorylation occur?
3. Which conditions are likely to cause an increase in the rate of lactate production by cells? Why?
4. Cyanide binds to complex IV of the electron transport system, preventing the reduction of oxygen to water. Describe the ultimate effects of cyanide on electron transport, oxygen consumption, and ATP production. How can these changes, in turn, affect a person?

7 Energy Storage and Use: Metabolism of Carbohydrates, Fats, and Proteins

So far, our examination of energy metabolism has been restricted to one energy source—glucose. Nevertheless, the metabolism of fats and proteins (as well as that of carbohydrates other than glucose) also makes a significant contribution to our energy needs. Although a detailed description of fat and protein metabolism is beyond the scope of this text, a general understanding of the connection between these metabolic pathways and those we have just studied is crucial. Eventually we will need this knowledge to understand the role of the endocrine system in regulating whole body energy metabolism.

The body is able to draw on fats and proteins as alternative sources of energy whenever glucose is scarce and needs to be conserved. When glucose supplies are limited, the body can break down fats and proteins into smaller molecules through catabolic processes that release energy. When the supply of glucose is plentiful, the body can run these reactions in reverse to synthesize fats, proteins, and a large glucose storage molecule called *glycogen*. In this manner, the body stores energy for future use.

UNDERSTANDING EXERCISE

Sources of Energy for Muscle Cells

During exercise, our skeletal muscle cells utilize ATP to contract. How much ATP is used and where the ATP comes from depend on the intensity and type of exercise. Consider a weight trainer who uses close to maximal exertion to lift heavy weights. This type of exercise makes the muscles stronger by increasing muscle cell size. When a person lifts a heavy weight, the muscles need a lot of energy *quickly.* Blood flow to the tissue is not fast enough to provide the needed oxygen for oxidative phosphorylation. Thus the muscle cells depend on anaerobic glycolysis for their energy.

Weight training generally involves a number of different exercises working on different muscle groups. For each exercise, the weight trainer tends to do two or three sets (series of repetitions) of the same movement. For example, to strengthen the biceps, the individual lifts weight while doing biceps curls (flexing of the arm holding the weight). He then performs the exercise with a heavy weight eight to ten times (one set). After a brief pause, the trainer repeats the exercise the same number of times (a second set). He can lift this heavy weight only for a few repetitions, because anaerobic glycolysis is not very efficient and the ATP is utilized rapidly. In addition, lactate builds up in the muscle cell, producing a burning sensation and contributing to fatigue. The second set is harder than the first, because not all the energy used for the first set has been replenished.

Now consider a distance runner. A trained individual who plans on running two miles will not use maximum exertion for most of the distance, and therefore, the runner does not require anaerobic glycolysis for energy, as the weight trainer does. Because of the low to moderate level of exertion, the runner's muscle cells utilize ATP at a slower rate than the weight trainer; thus blood flow can increase to sufficient levels, providing muscle cells with the oxygen needed to sustain a higher rate of oxidative phosphorylation to generate enough ATP. If the runner were in a race, she would probably use maximum exertion at the end to increase her likelihood of winning, knowing that she can generate enough energy through anaerobic glycolysis to finish the race. Should she start the final sprint to the finish line too soon, she will use her energy up too quickly and have to slow down before the race is over.

Figure 24 shows the relationships among the pathways for glucose, fat, and protein metabolism. The three stages of glucose oxidation—glycolysis, the Krebs cycle, and oxidative phosphorylation—form the centerpiece of the figure. (Note that the depiction of these three stages is essentially the same as that shown in Figure 22, except that only the major products of each stage are shown.) In addition, NADH and $FADH_2$ are simply given the abbreviation *coenzyme-2H.* (For simplicity, reduced coenzymes generated in glycolysis or in the linking step are not shown.) Double arrows indicate reactions that can go in both directions; single arrows denote reactions that essentially go one way only. These reactions are described as being *irreversible,* whereas the two-way reactions are said to be reversible.

One-way reactions are not "irreversible" in the absolute sense, but only in the sense that they require large quantities of energy to run in reverse. In many cases, however, one-way reactions do run in reverse under certain circumstances. When this occurs, a reaction does not follow exactly the same course while running in reverse as it does when running in the forward direction, and it is catalyzed by a different set of enzymes. For this reason, the reverse reaction is sometimes referred to as a *bypass reaction*—that is, it follows a "route" that bypasses the enzymes that would normally catalyze the reaction when it is running in the forward direction.

In contrast, reversible reactions feature smaller energy changes and can be made to go in reverse without large amounts of energy. These reactions readily reverse direction in response to changes in reactant or product concentrations in accordance with the law of mass action. Furthermore, these reactions are catalyzed by the same enzymes whether they are going forward or in reverse.

Whenever a metabolic pathway consists of an irreversible reaction in combination with a bypass reaction, the reactions are generally regulated so that they do not run simultaneously, which would be counterproductive. When the irreversible (forward) reaction is running, enzymes catalyzing the bypass reaction are turned off; when the bypass reaction is running, enzymes catalyzing the forward reaction are turned off. In this manner, cells can force the reaction to go in the direction that best suits the body's needs. Thus, when cells need to break down fats for energy, for example, they can turn on the enzymes that perform this function. When they need to synthesize fats to store energy, they can turn these enzymes off and turn on other enzymes that catalyze the necessary bypass reactions. These metabolic adjustments are largely coordinated by hormonal signals that depend on our eating and fasting patterns.

Glycogen Metabolism

Glycogen is a branched-chain molecule found in animal cells; it is similar to starch, a carbohydrate found only in plants. Glycogen is composed of glucose molecules joined together to form a polymer . When glucose is in abundant supply, as after a meal, most of the glucose molecules are not oxidized immediately, but rather are stored as glycogen. This process, called **glycogenesis,** is represented in **Figure 25** by the arrows running from glucose to glycogen. Note that this process requires production of glucose-6-phosphate, an intermediate of glycolysis.

Tissues differ in their capacity to synthesize and store glycogen; liver and skeletal muscle cells are particularly adept at this process. During fasting or when glucose is being used up quickly, the glucose supply is replenished by the breakdown of glycogen into individual glucose molecules, a process known as **glycogenolysis.** The fate of glucose released during glycogenolysis depends on the location where it occurs. When glycogen is broken down, glucose-6-phosphate, an intermediate of glycolysis, is produced. In most cells,

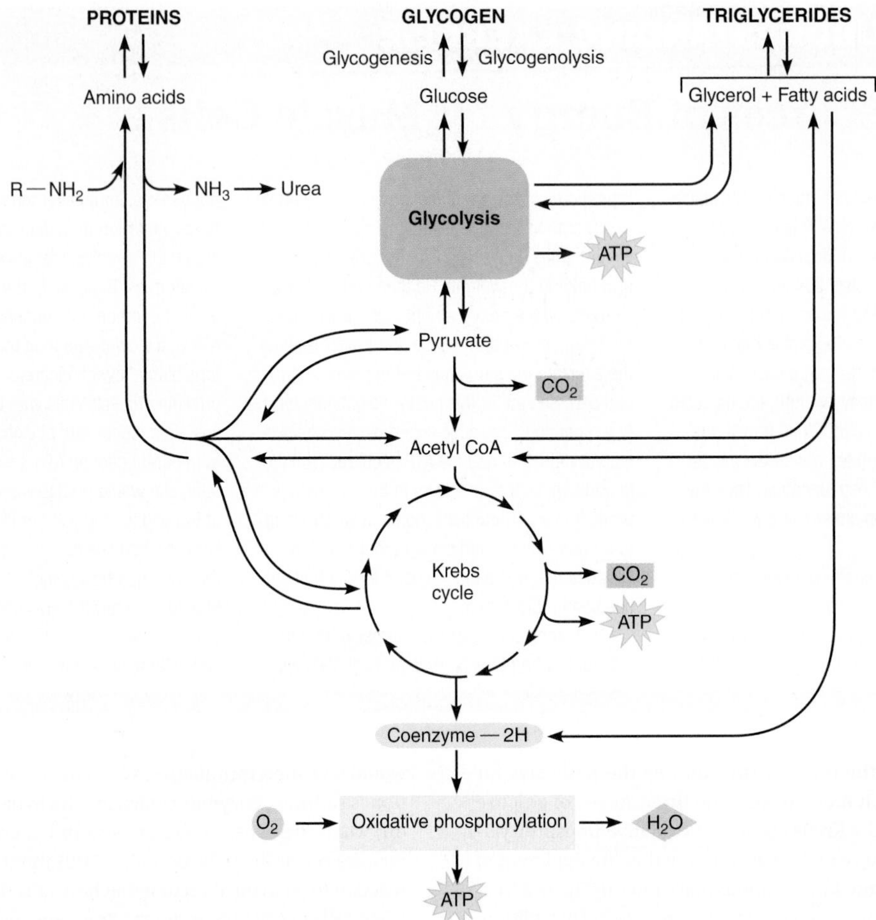

Figure 24 Metabolic pathways involved in protein, glycogen, and fat metabolism. The three stages of glucose metabolism are shown in the center. One-way arrows indicate irreversible reactions; double arrows indicate reversible reactions. Arrows leading into or out of this central pathway indicate points of entry or exit for the breakdown products of protein or triglyceride metabolism.

glycolysis continues and ATP is produced. In liver cells (and, to a smaller extent, kidney cells), however, glucose-6-phosphate can be converted to glucose (only liver and kidney cells have the enzyme, glucose-6-phosphatase, for this reverse reaction; see Figure 25). Glucose, but not glucose-6-phosphate, can be transported across cell membranes. Thus most of the glucose produced by glycogenolysis in liver cells is transported out of the liver and into the bloodstream for uptake and use by other tissues. In other tissue cells, glucose-6-phosphate is trapped inside the cell and can be used either for glycolysis or for glycogenesis.

Gluconeogenesis: Formation of New Glucose

Although fats and proteins can be used by most tissues as energy substitutes for glucose, an adequate supply of glucose must be maintained in the bloodstream at all times because the nervous system—particularly brain tissue—has a lower capacity than other tissues for switching to alternative energy sources. The nervous system requires an uninterrupted glucose supply; otherwise, loss of consciousness and possibly death may result. Nervous tissue is never entirely free of this glucose requirement, although it does have a limited capacity for using other energy sources. Under normal conditions, the liver's glycogen reserves are sufficient to provide energy for a few hours. If a person fasts for a longer period of time, glycogen stores become depleted, which could potentially cause a dangerous fall in blood glucose concentrations. Fortunately, new glucose molecules can be synthesized from noncarbohydrate precursors via a process called **gluconeogenesis** (**Figure 26**). Gluconeogenesis is carried out mainly by the liver, because the liver has all the enzymes necessary to run glycolysis in reverse (see Figure 25).

Glucose can be made via gluconeogenesis from three sources: (1) glycerol, which is produced by breaking down triglycerides; (2) lactate; and (3) amino acids, which can be produced by the breakdown of proteins. In gluconeogenesis, glycerol is first

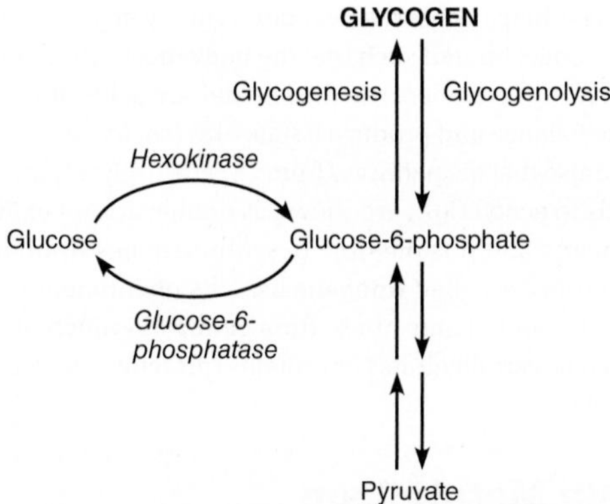

Figure 25 Glycogen metabolism. The relationship between glycogen and glucose is shown with the critical intermediate glucose-6-phosphate. The curved reaction arrows between glucose and glucose-6-phosphate indicate that different enzymes catalyze the forward and reverse reactions. Hexokinase, found in most cells, catalyzes the first step of glycolysis. Glucose-6-phosphatase, found only in liver cells, catalyzes the reverse reaction.

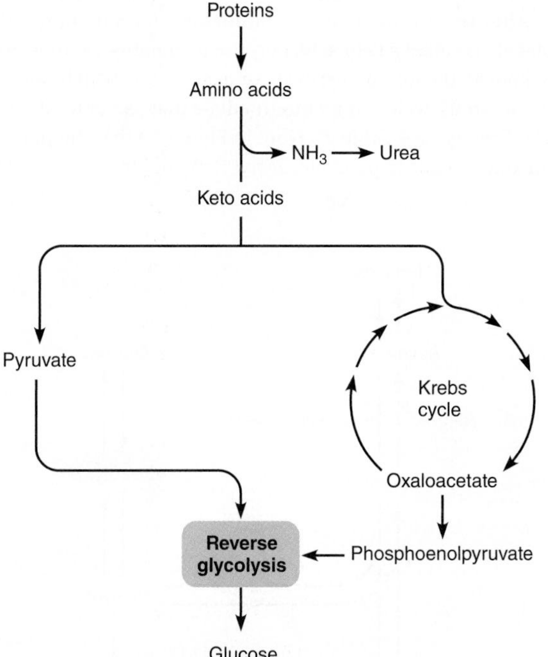

Figure 26 Metabolic pathways involved in gluconeogenesis.

converted to glycerol phosphate, an intermediate of the glycolysis pathway, and lactate is first converted to pyruvate, the normal end-product of glycolysis (see Figure 26). These molecules then proceed through the glycolysis pathway *in reverse,* such that glucose molecules are generated.

Certain amino acids can be converted to glucose after first being converted to pyruvate, which then enters the glycolysis pathway in reverse (**Table 2**). Certain other amino acids, however, are converted to glucose only after entering the Krebs cycle, where they

TABLE 2 Amino Acids That Can Be Used for Gluconeogenesis

Amino acid	Keto acid
Alanine	Pyruvate
Arginine	α-Ketoglutarate
Aspartate	Oxaloacetate
Cysteine	Pyruvate
Glutamate	α-Ketoglutarate
Glutamine	α-Ketoglutarate
Glycine	Pyruvate
Histidine	α-Ketoglutarate
Isoleucine	Succinyl CoA
Methionine	Succinyl CoA
Phenylalanine	Fumarate
Proline	α-Ketoglutarate
Serine	Pyruvate
Threonine	Pyruvate
Tryptophan	Pyruvate
Tyrosine	Fumarate
Valine	Succinyl CoA
Threonine	Pyruvate

are converted to oxaloacetate. Oxaloacetate can be converted into a glycolytic intermediate, phosphoenolpyruvate; which then participates in the glycolysis pathway in the reverse direction. Note, however, that three amino acids cannot be converted to glucose (leucine, tryptophan, isoleucine); these amino acids can be converted to acetyl CoA, but there is no pathway permitting the synthesis of glucose from acetyl CoA. Acetyl CoA can be converted to oxaloacetate in the Krebs cycle, but in the process an equal amount of oxaloacetate is consumed due to the cyclic nature of this series of reactions. Consequently, acetyl CoA cannot be converted to glucose. Fatty acids, which are generated from triglyceride breakdown, cannot be converted to glucose either because they, too, are converted to acetyl CoA.

Fat Metabolism

Like glycogen, fats (lipids) can be broken down and used for energy when other energy supplies are running low. Fats can also be synthesized to store energy when they are in abundant supply in the diet. *Adipose tissue,* which contains fat cells, is the primary storage depot for fats.

Lipids are stored predominantly in the form of *triglycerides,* which consist of three fatty acids joined to a glycerol backbone. **Lipolysis,** the first stage of fat breakdown, entails the separation of fatty acids from the glycerol molecule (**Figure 27**). Glycerol then enters the glycolysis pathway (as dihydroxyacetone phosphate) and proceeds from there to the Krebs cycle and oxidative phosphorylation. Fatty acids are catabolized to acetyl CoA by a

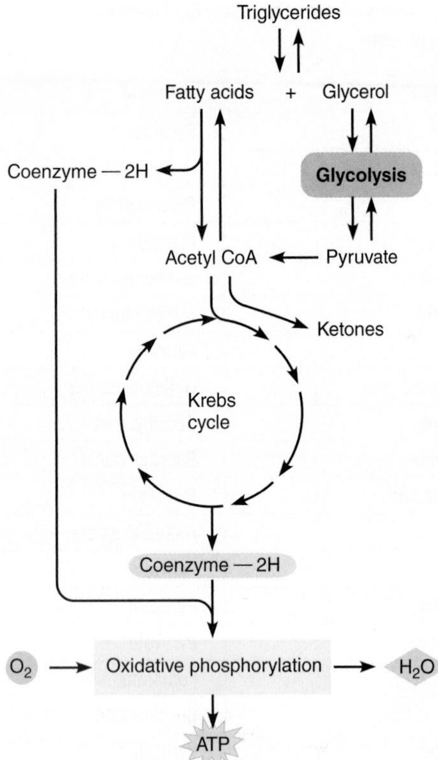

Figure 27 Metabolic pathways involved in lipolysis, the breakdown of fats (triglycerides) for energy. Note that fatty acid breakdown leads to the production of ketones, which are generated from acetyl CoA through a series of reversible reactions.

process called **beta oxidation,** which occurs in the mitochondrial matrix. Fatty acids obtained in the diet are usually 12 to 18 carbons in length, with an even number of carbons. Acetyl CoA molecules contain two carbons. Thus, for a fatty acid with n carbons, beta oxidation generates $n/2$ acetyl CoA molecules—considerably more than the two acetyl CoA molecules produced from each glucose molecule. In addition, each time an acetyl CoA molecule is cleaved from the fatty acid, coenzymes are reduced to generate one $NADH + H^+$ and one $FADH_2$. Thus $n - 1$ $NADH + H^+$ and $FADH_2$ are produced. The acetyl CoA molecules enter the Krebs cycle and oxidative phosphorylation; the reduced coenzymes enter oxidative phosphorylation. Because a gram of fat yields more energy than a gram of glucose or other carbohydrates, fats are known as high-calorie foods.

If triglycerides are broken down for energy at a rate that exceeds demands, compounds called *ketones* are generated as a by-product (see Figure 27). These compounds are synthesized from acetyl CoA in a set of reversible reactions and are manufactured at higher rates when the concentration of acetyl CoA increases. Because many acetyl CoA molecules can be generated from a single fatty acid, the breakdown of fats tends to generate large quantities of ketones—an important effect, because ketones can be utilized by the nervous system as a partial alternative to glucose. This process helps conserve the body's supply of glucose if energy intake becomes limited such that the body must turn to stored fats for energy. However, because the ketones are acids, they can alter acid-base balance and produce a state of *ketoacidosis.*

Note also that the pathways from glycerol to glycolysis, and from fatty acids to acetyl CoA, are shown as double arrows in Figure 27, which means that it is possible to synthesize fats from other nutrients, a process called **lipogenesis.** This phenomenon explains why people can become obese from eating too much of nonlipid foods such as carbohydrates or proteins (proteins can also produce acetyl CoA).

Protein Metabolism

In the metabolic breakdown of proteins for energy (**Figure 28**), proteins are first broken down to amino acids, a process known as **proteolysis.** The amino acids are then *deaminated;* that is, an amino group ($—NH_2$) is removed. This deamination process produces ammonia (NH_3), which is toxic. Fortunately, ammonia is carried by the bloodstream to the liver, where it is converted to *urea,* which is relatively innocuous and is eventually eliminated by the kidneys.

After the amino acid is deaminated, the remainder of the molecule is called a keto acid, because it contains a ketone group. The keto acid could be pyruvate or acetyl CoA (which then goes into the Krebs cycle), or an intermediate that can enter the Krebs cycle directly (see Table 2). Note in Figure 28 that the pathways from amino acids to pyruvate, acetyl CoA, and the Krebs cycle are

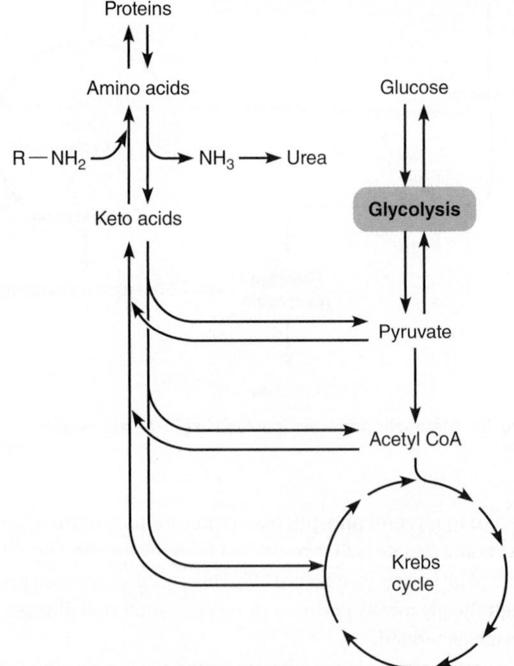

Figure 28 Metabolic pathways involved in the breakdown of proteins for energy. Note that the breakdown of amino acids generates ammonia (NH_3), which is converted to urea in the liver.

FOCUS ON DIABETES

Effects of Insulin on Cell Metabolism

Cells can use carbohydrates, lipids, and proteins for energy, and they can store energy as glycogen or triglycerides when energy is plentiful. What determines whether they use biomolecules for energy or storage? Cell metabolism is regulated by chemical messengers such as insulin. Metabolic changes induced by chemical messengers vary based on the tissue type. The effects of insulin on adipocytes and liver cells demonstrate how metabolic pathways can be regulated. Recall that insulin is released when blood glucose is elevated to higher than normal levels. Thus the responses described next occur when blood glucose is high.

The function of adipocytes is to store energy as triglycerides and to release fatty acids into the blood when they are needed to provide energy for other cells. Insulin promotes energy storage in adipocytes by increasing glucose and fatty acid uptake into the cells, and by activating four enzymes that convert glucose and fatty acids to triglycerides.

In terms of whole-body metabolism, the liver stores carbohydrates (and, to a lesser degree, triglycerides) when blood glucose is plentiful, and releases glucose into the blood for neural tissue uptake when blood glucose is low. In the liver, insulin activates glycogen synthetase, which adds glucose to glycogen, and glucokinase, which converts glucose to glucose-6-phosphate. Insulin also inhibits glucose-6-phosphatase, which converts glucose-6-phosphate to glucose. Thus carbohydrates are stored in the liver as glycogen when insulin is present.

In diabetes mellitus, when insulin production or action is decreased, these metabolic changes are hindered, resulting in excessive lipolysis and decreased glycogenesis. Thus blood levels of fatty acids, ketone bodies, and glucose are all elevated. These metabolic disturbances have detrimental effects on many organ systems.

shown as double arrows, meaning that nonprotein nutrient molecules such as carbohydrates or fats can serve as raw materials for the synthesis of amino acids, which can then be used to make proteins. (As the figure shows, this production process requires the addition of an amino group, which is obtained from another molecule, indicated as $R—NH_2$). Consequently, certain amino acids can be present in cells even when those amino acids are not present in the diet. Other amino acids, however, cannot be synthesized in the body and must be obtained from the diet; these amino acids are referred to as *essential* amino acids. An **essential nutrient** is any biomolecule necessary for proper body function that cannot be synthesized in cells and, therefore, must be obtained from dietary sources. The eight essential amino acids are isoleucine, leucine, lysine, methionine, phenylalanine, threonine, tryptophan, and valine.

Quick Check 6

1. What is *gluconeogenesis,* and why is it important? Which three substances can serve as the raw materials for this process?
2. What happens to the ammonia that is produced when amino acids are broken down for energy?
3. Describe the roles of acetyl CoA in carbohydrate, lipid, and protein metabolism.

SYSTEMS INTEGRATION

In this chapter, we discussed chemical reactions in terms of energy exchange, enzyme-catalyzed reactions, and energy metabolism. Many of the concepts learned in this chapter will be applied to membrane transport. As chemical reactions spontaneously proceed from states of high energy to states of low energy, molecules tend to move spontaneously from areas of higher energy to areas of lower energy. To move a molecule in the opposite direction requires the use of ATP, the energy transfer molecule. We will also learn about proteins that function as transporters, which have many properties similar to enzymes. In fact, some transporters are allosterically or covalently regulated.

Throughout the text, processes requiring energy are described. It is critical to always understand the source of this energy—substrate-level and oxidative phosphorylation. But what happens when the energy produced is insufficient to meet all the body's needs? For this, we need to learn about whole-body metabolism and understand how energy metabolism is regulated to ensure that the brain has sufficient glucose at all times, including in between meals when blood glucose levels tend to decrease. The liver has an important role in maintaining blood glucose levels.

CHAPTER REVIEW

SUMMARY

1 Types of Metabolic Reactions

- Metabolism is the sum total of chemical reactions occurring in the body; energy exchange is the arrangement of reactions involved in energy metabolism.
- Catabolic reactions generate smaller products from larger reactants; anabolic reactions generate larger products from smaller reactants.
- Three types of metabolic reactions are commonly observed in the body: hydrolysis and condensation reactions, phosphorylation and dephosphorylation reactions, and oxidation-reduction reactions.

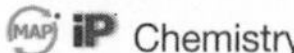

2 Metabolic Reactions and Energy

- Metabolic reactions enable cells to transform raw materials from the environment into structural and functional components; they also provide cells with energy.
- Two forms of energy are distinguished: kinetic energy (associated with motion) and potential energy (stored energy that has the potential to become kinetic energy).
- The energy change of a reaction determines its direction. Catabolic reactions (exergonic) are generally energy releasing and proceed spontaneously; anabolic reactions (endergonic) require energy to proceed.
- Chemical equilibrium is reached when the energy levels of product and reactant are equal.
- A reaction can be made to go in either direction by changing reactant or product concentrations according to the law of mass action.
- The rate of a reaction is limited by its activation energy, reflecting the fact that molecules must go through a high-energy transition state before they can react.

3 Reaction Rates

- Factors affecting the rate of a reaction include reactant and product concentrations, temperature, and the height of the activation energy barrier.
- Metabolic reactions are catalyzed by enzymes, a form of special biomolecule (usually proteins). The enzyme molecule acts on a reactant molecule (a substrate) to generate and release a product molecule.
- Many enzymes require nonprotein cofactors for proper function. Important cofactors include NAD^+ and FAD, which carry electrons between reactions.
- Rates of enzyme-catalyzed reactions are influenced by the enzyme's catalytic rate, enzyme and substrate concentration, and the enzyme-substrate affinity. The temperature of the reaction mixture and its pH can also have an effect on the reaction rate.
- Certain enzymes possess a regulatory site specific for a certain modulator molecule, whose binding alters the enzyme's activity.
- In allosteric regulation, binding of the modulator is reversible. In covalent regulation, an enzyme's activity is altered by the covalent binding of a chemical group (often a phosphate group) to a specific site on the enzyme.
- Metabolic pathways are often regulated by feedback inhibition, in which an enzyme is allosterically inhibited by the product of a reaction occurring downstream. Feedforward activation is a less common form of enzyme regulation, in which enzymes are stimulated by intermediates appearing upstream.

4 ATP: The Medium of Energy Exchange

- Adenosine triphosphate (ATP) is a compound synthesized by cells that serves as a temporary storage place for energy released in certain exergonic reactions.
- There are two basic processes of phosphorylation: substrate-level phosphorylation and oxidative phosphorylation.

5 Glucose Oxidation: The Central Reaction of Energy Metabolism

- Cells obtain much of their energy from the oxidation of glucose: $C_6H_{12}O_6 + 6\,O_2 \rightarrow 6\,CO_2 + 6\,H_2O$. The energy derived from glucose oxidation allows cells to synthesize ATP.
- Later in this process, cells use energy released from the breakdown (hydrolysis) of ATP to perform useful work.

6 Stages of Glucose Oxidation: Glycolysis, the Krebs Cycle, and Oxidative Phosphorylation

- Glucose oxidation occurs in three stages: (1) glycolysis, or the linking step (in the cytosol); (2) the Krebs cycle (in the mitochondrial matrix); and (3) oxidative phosphorylation (in the inner mitochondrial membrane).
- Oxidative phosphorylation involves two processes that occur simultaneously: (1) the movement of electrons through the electron transport chain and (2) chemiosmotic coupling.
- For each glucose molecule, four ATP molecules are synthesized by substrate-level phosphorylation in glycolysis and the Krebs cycle.
- Six CO_2 molecules, ten NADH molecules, and two $FADH_2$ molecules are also produced in glucose oxidation. Later on, NADH and $FADH_2$ release their electrons (hydrogens) to the electron transport chain, and these hydrogens ultimately react with oxygen to form water.
- Energy released in this process is used to make ATP by oxidative phosphorylation, which accounts for most of the ATP molecules that are generated from the complete oxidation of one glucose molecule (34 of the 38 total molecules).

7 Energy Storage and Use: Metabolism of Carbohydrates, Fats, and Proteins

- The body is able to draw upon stored fats and proteins as alternative sources of energy when glucose is scarce and needs to be conserved.
- In certain tissues, energy can be stored by converting glucose to glycogen (by glycogenesis), which can be later broken down to yield glucose (glycogenolysis).
- Glucose can also be synthesized from noncarbohydrate precursors (gluconeogenesis), thereby ensuring a steady supply of glucose in the bloodstream, which is necessary for proper nervous system function.
- Fats and proteins can be broken down to smaller molecules (by lipolysis and proteolysis, respectively); these molecules then enter the pathway for glucose oxidation at various points.
- The body can also store energy by synthesizing fats, which are stored primarily in adipose tissue, and proteins.

MasteringA&P®

Check your understanding at MasteringA&P and apply your new knowledge to real-life case studies to ensure that you've mastered the content!

EXERCISES

Multiple-Choice Questions

1. When glucose is oxidized in cells, oxygen reacts with
   a) Carbon to form CO_2.
   b) Hydrogen to form H_2O.
   c) Components of the electron transport chain.
   d) Inorganic phosphate to form ATP.
   e) Acetyl CoA to form acetate.

2. Which of the following illustrates *substrate-level* phosphorylation? (P = phosphate, X = metabolic intermediate)
   a) $ATP \rightarrow ADP + P_i$
   b) $ADP + P_i \rightarrow ATP$
   c) $X{-}P \rightarrow X + P$
   d) $X{-}P + ADP \rightarrow X + ATP$

3. The enzyme pyruvate dehydrogenase, which converts pyruvate to acetyl CoA, can be activated or inactivated by phosphorylation or dephosphorylation. This is an example of
   a) Allosteric regulation.
   b) Substrate-level phosphorylation.
   c) Oxidative phosphorylation.
   d) Saturation.
   e) Covalent regulation.

4. The following reactions occur in conjunction with step 5 of the Krebs cycle:

   $$GDP + P_i \rightarrow GTP$$
   $$GTP + ADP \rightarrow GDP + ATP$$

   The net reaction in this process is
   a) $P_i + GTP \rightarrow GDP$
   b) $P_i + GDP \rightarrow GTP$
   c) $GTP + ADP \rightarrow GDP + ATP$
   d) $ADP + P_i \rightarrow ATP$
   e) $GDP + P_i \rightarrow ATP$

5. Which of the following reactions is an example of hydrolysis?
   a) Amino acid 1 + amino acid 2 → dipeptide + H_2O
   b) Lactose + H_2O → glucose + galactose
   c) Glucose-1-phosphate → glucose-6-phosphate
   d) Glucose + 6 O_2 → 6 CO_2 + 6 H_2O
   e) Water (liquid) + heat → water (steam)

6. Which of the following classes of organic molecules can be used to synthesize triglycerides?
   a) Carbohydrates only
   b) Lipids only
   c) Proteins only
   d) Carbohydrates and lipids only
   e) Carbohydrates, lipids, and proteins

7. Enzymes increase the rate of an enzyme-catalyzed reaction by
   a) Increasing the activation energy barrier.
   b) Decreasing the activation energy barrier.
   c) Increasing the free energy of the reactant.
   d) Increasing the free energy of the product.
   e) Decreasing the free energy of the reactant.

8. In the following reaction, NAD^+ receives electrons from the substance AH_2:

   $$NAD^+ + AH_2 \xrightarrow{E} NADH + H^+ + A$$

   where E is the enzyme that catalyzes the reaction. In this reaction, which substance undergoes oxidation?
   a) NAD^+
   b) NADH
   c) AH_2
   d) A

9. Which of the following is carried out by enzymes in the cytosol?
   a) Glycolysis
   b) Linking step
   c) Krebs cycle
   d) Oxidative phosphorylation
   e) Chemiosmotic coupling

10. In which of the following is carbon dioxide generated?
    a) Glycolysis only
    b) Linking step only
    c) Krebs cycle only
    d) Glycolysis and Krebs cycle
    e) Linking step and Krebs cycle

Objective Questions

11. The removal of hydrogen atoms from a molecule is an example of (oxidation/reduction).
12. The forward rate of a reaction can be increased by increasing the concentrations of the (reactants/products).
13. For a reaction to proceed spontaneously in the forward direction, the energy change must be (positive/negative).
14. Liver and muscle cells are able to store glucose in the form of a branched-chain molecule called ______.
15. Allosteric regulation involves the attachment of a modulator molecule to an enzyme by means of a covalent bond. (true/false)
16. Glycogenolysis is an example of a catabolic reaction. (true/false)
17. The conversion of noncarbohydrate precursors into glucose is called ______.
18. When ATP is synthesized using energy released in glucose oxidation, most of the ATP is synthesized by oxidative phosphorylation. (true/false)
19. When a substrate molecule binds to an enzyme, it binds to a particular location known as the (active/regulatory) site.
20. Unless a substance is an essential nutrient, it is not necessary for proper cellular function. (true/false)
21. An enzyme catalyzes a reaction at the (lowest/highest) possible rate when it is 100% saturated.
22. Conversion of pyruvate to lactate occurs under (aerobic/anaerobic) conditions.

Essay Questions

23. Explain the significance of the energy change in a reaction (ΔE) with regard to (a) the direction of the reaction and (b) whether the reaction releases energy or requires it.
24. Compare and contrast the mechanisms of substrate-level phosphorylation and oxidative phosphorylation, and describe the role of each in glucose oxidation.
25. Explain how the conversion of pyruvate to lactate enables ATP production to continue even when oxygen is not readily available.
26. Compare and contrast allosteric regulation and covalent regulation of enzyme activity.
27. Explain the concept of activation energy and describe how it influences reaction rates. Include a discussion of how reaction rates are affected by temperature and reactant and product concentrations.

Critical Thinking

28. As you know, fatty acids can be oxidized to provide energy for making ATP. This process begins with a set of reactions known as the fatty acid oxidation cycle, or beta oxidation. Before entering the cycle, a fatty acid reacts with coenzyme A to form a molecule called fatty acyl CoA. This molecule then enters the cycle and continues to go through it until all its carbons have been eliminated.

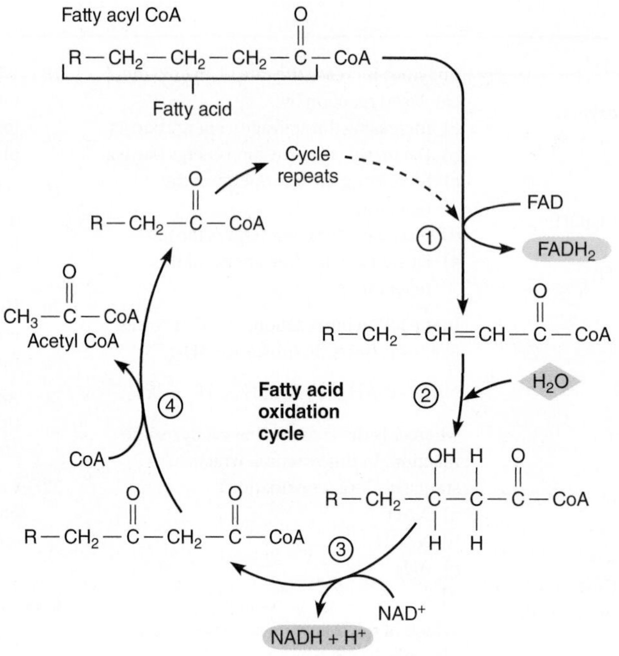

From your examination of this diagram, answer the following questions for a 12-carbon fatty acid.

a) How many acetyl CoA molecules will be produced by beta oxidation of this fatty acid? How many ATP molecules can be produced from each acetyl CoA molecule that goes through the Krebs cycle and oxidative phosphorylation?

b) How many NAD^+ and FAD molecules will be reduced? (Hint: How many bonds must be broken to split a 12-carbon fatty acid into acetyl CoA molecules?)

c) How many ATP molecules will be produced per each NADH + H^+ and each $FADH_2$ molecule that provides electrons to the electron transport system?

d) It takes two ATP molecules to get the cycle started. Based on this information and your answers to the previous questions, how many ATP molecules can be generated from a 12-carbon fatty acid?

29. Carbon monoxide and cyanide are both toxic. Carbon monoxide binds hemoglobin in red blood cells, interfering with the transport of oxygen in cells. Cyanide binds to complex IV of the electron transport chain, interfering with the reduction of oxygen to water (carbon monoxide can also bind to complex IV, but does so to a lesser extent than it binds hemoglobin). Based on what you learned in this chapter, compare the ways these two poisons are lethal, and describe potential treatments.

Photo credits for icons in feature boxes: Focus on Diabetes: Tanyae/Dreamstime; Clinical Connections: Iofoto/Dreamstime; Understanding Exercise: Pixattitude/Dreamstime; Discovery: Yuri Arcurs/Cutcaster.

Glossary

activation energy additional energy that molecules must acquire in order to react; the difference between the energy of the transition state and the energy of the reactants or products

adenosine triphosphate (ATP) (ah-den-oh-seen) compound that serves as the primary direct energy source for cell activities; synthesized from adenosine diphosphate (or ADP) and inorganic phosphate (P_i)

affinity a measure of how tightly ligand molecules bind to proteins

allosteric regulation regulatory mechanism in which a modulator binds reversibly to the regulatory site on an enzyme, inducing a change in its conformation and activity

ATP synthase the enzyme that synthesizes adenosine triphosphate (ATP) during oxidative phosphorylation; located in the inner mitochondrial membrane

beta oxidation metabolic pathway that catabolizes fatty acids to acetyl coenzyme A

calorie amount of energy that must be put into one gram (or one milliliter) of water to raise its temperature by one degree centigrade (°C) under a standard set of conditions

catalytic rate for enzyme-catalyzed reactions, a measure of the number of reactants converted to product per unit time

chemiosmotic coupling the entire process that couples electron transport to adenosine triphosphate (ATP) synthesis; the utilization of energy released during electron transport to transport hydrogen ions across the inner mitochondrial membrane up their concentration gradient

coenzymes (koh-en-zimes) molecules that do not themselves have catalytic activity but that are necessary for proper enzyme function and participate directly in reactions catalyzed by enzymes; often serve to transfer certain chemical groups from one reactant to another

cofactors nonprotein components of some enzymes necessary for them to hold normal conformation during metabolic reactions

covalent regulation regulatory mechanism in which changes in an enzyme's activity are brought about by the covalent bonding of a specific chemical group to a site on the enzyme molecule; usually involves bonding of a phosphate group

electron transport chain a series of electron acceptors and other proteins in the inner mitochondrial membrane; involved in the synthesis of adenosine triphosphate (ATP) by oxidative phosphorylation

energy capacity to do work

energy metabolism set of reactions involved in energy storage and use

enzymes (en-zimes) biomolecules, usually proteins, specialized to act as catalysts in metabolic reactions

equilibrium a condition of no change without requiring energy expenditure; in chemical reactions, when the rates of the forward and reverse reactions are equal

essential nutrient any biomolecule necessary for proper body function that cannot be synthesized in cells and, therefore, must be obtained from dietary sources

feedback inhibition regulatory mechanism in which an enzyme in a metabolic pathway is inhibited by an intermediate appearing downstream

feedforward activation regulatory mechanism in which an enzyme in a metabolic pathway is stimulated by an intermediate appearing upstream

gluconeogenesis (gloo-koh-nee-oh-jen-ih-sis) process during which new glucose molecules can be synthesized from noncarbohydrate precursors by the liver

glycogen (gly-coh-jen) a glucose polymer found in animal cells; functions as an energy store

glycogenesis (gly-koh-jen-eh-sis) synthesis of glycogen from glucose monomers

glycogenolysis (gly-koh-jen-nol-ih-sis) breakdown of glycogen to glucose monomers

glycolysis the first stage of glucose oxidation, occurring in the cytosol, in which each glucose molecule is broken down to two molecules of pyruvate

hydrolysis use of water to split another molecule

kinetic energy energy associated with motion

Krebs cycle a cyclical metabolic pathway occurring in the mitochondrial matrix, in which acetyl coenzyme A is a primary reactant and carbon dioxide and reduced coenzymes are produced; also called the *citric acid cycle, tricarboxylic acid cycle,* or *TCA cycle*

law of mass action states that a reaction can be made to go either forward or in reverse by raising or lowering the concentrations of reactants relative to products

lipogenesis (ly-poh-jen-eh-sis) process by which fat is synthesized from nonlipid nutrients, such as proteins and carbohydrates

lipolysis (ly-pol-ih-sis) the first stage of lipid breakdown; in regard to triglycerides, separation of the fatty acids from the glycerol backbone

metabolic pathway sequence of enzyme-catalyzed reactions

metabolism sum of all the chemical reactions that take place in the body

modulator in allosteric regulation, the specific molecule that binds to a regulatory site on an enzyme molecule and regulates its activity

oxidation removal of electrons from a molecule

oxidative phosphorylation the process of synthesizing adenosine triphosphate (ATP) by harnessing the energy released when hydrogen atoms or electrons are transported through the electron transport chain in the inner mitochondrial membrane

phosphorylation addition of a phosphate group to a molecule

potential energy stored energy

proteolysis (proh-tee-oh-ly-sis) process by which proteins are broken down to amino acids

reduction addition of electrons to a molecule

regulatory site binding site on an enzyme molecule that is specific for molecules known as modulators

substrate a reactant in an enzyme-catalyzed reaction; binds to the active site of an enzyme molecule

substrate-level phosphorylation a mechanism of adenosine triphosphate (ATP) synthesis, in which an enzyme transfers a phosphate group from a substrate to adenosine diphosphate (ADP);occurs in steps 7 and 10 of glycolysis and step 5 of the Krebs cycle

vitamins organic molecules that are required in trace amounts in the body; often act as coenzymes

Answers

Figure Question Answers

Fig 3.4	Stays the same
Fig 3.8	Decreasing
Fig 3.10	Because of the law of mass action
Fig 3.13	Decreases
Fig 3.17	An oxidation-reduction reaction
Fig 3.19	All the cytochromes

Apply Your Knowledge Answers

If the equlibrium constant is 2, then the ratio of reactant to product is 0.5 and the reaction is exergonic. If the equilibrium constant is 10, then the ratio of reactant to product is 0.1 and the reaction is exergonic. If the equilibrium constant is 1, then the ratio of reactant to product is 1.0, and the reaction is neither exergonic or endergonic. If the equilibrium constant is 0.1, then the ratio of reactant to product is 10 and the reaction is endergonic.

A zinc deficiency would decrease the activity of alcohol dehydrogenase, thereby decreasing alcohol tolerance.

Most ATP is produced when the cytosolic NADH donates its electrons to mitochondrial NAD (3 molecules per cytosolic NADH molecule); the least is produced when cytosolic NADH donates its electrons to mitochondrial FAD (2 molecules per cytosolic NADH molecule). In terms of overall ATP production from glucose oxidation, 36 ATP molecules are produced when the cytosolic electrons are shuttled to 2 FAD molecules, 37 ATP molecules when cytosolic electrons are shuttled to 1 NAD molecule and 1 FAD molecule, and 38 ATP molecules when cytosolic electrons are shuttled to 2 NAD molecules.

Answers to End–of-Chapter Multiple Choice and Objective Questions

Multiple Choice Questions

1. b 2. d 3. e 4. d 5. b 6. e 7. b 8. c
9. a 10. e

Objective Questions

11. oxidation
12. reactants
13. negative
14. glycogen
15. false
16. true
17. gluconeogenesis
18. true
19. active
20. false
21. highest
22. anaerobic

Cell Membrane Transport

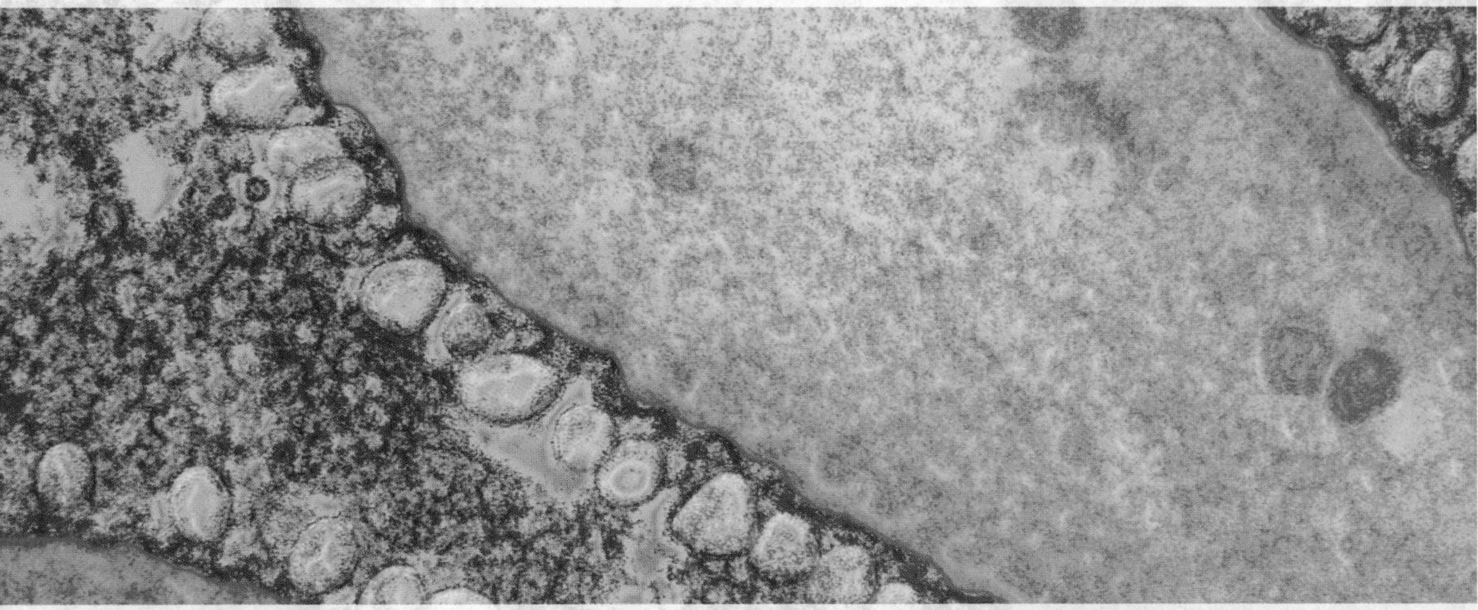

David M. Phillips/Photo Researchers, Inc.

Colorized scanning electron micrograph (SEM) of pinocytosis.

Marcus and Carter are 22-year-old twins. They are both runners, and today they ran 4 miles. Afterward, Marcus drank a liter of pure bottled water, whereas Carter drank a liter of a sports drink that contains solutes such as sodium, potassium, and glucose. Within an hour Marcus had voided approximately 400 mL of urine, whereas Carter had voided only 100 mL. Assuming both men's physical condition was equal and that each lost an equal amount of sweat, can you explain the difference?

Sweat contains water and ions. When Marcus replaced the water, he diluted the ions in his body fluids. Carter, by contrast, replaced both water and ions. Clearly, the difference in urine output is related to the presence or absence of solutes in the drink each man chose. As you will learn in this chapter, our cells have transport mechanisms for water and solutes. The kidneys use these transport mechanisms to excrete water and solute in urine at concentrations that keep the composition of body fluids stable.

The kidney is not the only body system that uses membrane transport mechanisms to maintain homeostasis. *All* body systems use the transport mechanisms described in this chapter.

CHAPTER OUTLINE

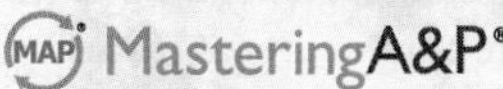

Go to MasteringA&P for helpful A&P Flix 3-D animations, chapter quizzes, pre-tests, Interactive Physiology tutorials, and more!

LEARNING OUTCOMES *After studying this chapter, you should be able to:*

- Explain the role of chemical, electrical, and electrochemical driving forces in the passive transport of substances across a membrane, and distinguish between passive transport and active transport.
- Identify the three general factors that influence the rate at which a substance can be passively transported across a membrane, and identify the two general factors that influence the rate of active transport.
- Identify four factors affecting the permeability of membranes to molecules that cross by simple diffusion, and explain how and why each affects permeability.
- Compare the movement of molecules across the plasma membrane by carriers and through channels.
- Explain the distinction between primary and secondary active transport, and give examples of each.
- Explain how a difference in solute concentration across a membrane can cause the movement of water, and explain the distinction between the osmolarity and tonicity of a solution.
- Define *endocytosis* and *exocytosis*, and explain the primary difference between these and other mechanisms of cellular transport. Distinguish between receptor-mediated endocytosis and pinocytosis.
- Explain in general terms how the polarity of epithelial cells enables them to absorb or secrete materials.

Before You Begin

Make sure you have mastered the following topics:

- Lipid bilayer structure
- Membrane proteins
- ATP

To support life, the cells in your body need to exchange materials with their immediate environment—that is, the fluid that surrounds them. For example, for cells to oxidize glucose for energy, they need glucose and oxygen. As we know, oxygen can permeate the phospholipid bilayer of the plasma membrane, but the polar molecule glucose cannot. How, then, can our cells use glucose for energy?

We also know that membranes contain proteins that carry out a variety of functions, including the transport of specific molecules across the plasma membrane. Depending on the proteins present, the compositions of intracellular fluid and extracellular fluid differ (Table 1). In this chapter, we will learn about which molecules can permeate the plasma membrane and which molecules can cross the membrane only with the assistance of membrane proteins. We will also learn what determines the direction of movement.

In this chapter, we focus on basic cellular and molecular theories of membrane transport, as these ideas are crucial to our current understanding of organ system physiology. These theories are based in large part on the principles of cellular metabolism. The chapter begins with a discussion of the factors most closely related to energy—that is, those factors that determine the *direction* and *rate* of transport.

TABLE 1 Millimolar Concentrations of Selected Solutes in Intracellular Fluid (ICF)* and Extracellular Fluid (ECF)

Solute	ICF (mM)	ECF (mM)
K^+	140.0	4.0
Na^+	15.0	145.0
Mg^{2+}	0.8	1.5
Ca^{2+}	$<0.001^{\dagger}$	1.8
Cl^-	4.0	115.0
HCO_3^-	10.0	25.0
P_i	40.0	2.0
Amino acids	8.0	2.0
Glucose	1.0	5.6
ATP	4.0	0.0
Protein	4.0	0.2

mM = millimolar

*Intracellular fluid composition varies for different cell types.

†Refers to calcium ions free in the cytosol. A significant quantity of intracellular calcium is sequestered in membrane-bound organelles and/or bound to proteins.

1 Factors Affecting the Direction of Transport

We know that the reactant and product molecules in a metabolic reaction have different energies. We also know that the energy change of a reaction—the difference between the reactant and product energies—determines the direction of the reaction and whether it proceeds spontaneously or requires energy. In this section, we see that similar principles govern the transport of molecules across membranes. That is, the energy difference in the molecules on the two sides of the membrane determines the direction of spontaneous transport *(passive transport)* and the need for energy to move the molecule against the energy gradient *(active transport)*.

Passive Transport Versus Active Transport

Transport of molecules across a membrane follows the same principles as chemical reactions: spontaneous (or passive) movement occurs from areas of high energy to areas of low energy; to move in the opposite direction requires an input of energy. The energy of a solution depends on the solute concentration (and charge, if the solute is an ion); energy increases as solute concentration increases (see **Toolbox: Energy of Solutions**). Therefore, solutes move passively from an area where they are in greater concentration to an area where they are in lesser concentration, or down their concentration gradient. To move from low concentration to high

TOOLBOX

Energy of Solutions

Molecules that are transported across membranes are in solution in the intracellular fluid and extracellular fluid. The transport of these molecules depends on the free energy difference of these molecules in solution.

The free energy of a solution of a noncharged solute (S) can be described mathematically as follows:

$$G = RT \ln [S]$$

where G is the free energy, R is the universal gas constant (0.082 liter-atm/mole-K), T is the absolute temperature in kelvins (K), and [S] is the concentration of solute S in moles. The driving force for a molecule to move into the cell depends on the difference in energies (ΔG) from where the molecule ends up (inside) to where the molecule started (outside):

$$\Delta G = \text{energy inside} - \text{energy outside} = RT \ln [S]_i - RT \ln [S]_o$$

where the subscripts "o" and "i" refer to outside and inside the cell, respectively. This equation can be rewritten as follows, in the form known as van't Hoff's equation:

$$\Delta G = RT \ln \frac{[S]_i}{[S]_o}$$

As for chemical reactions, when the change in free energy is negative, the process occurs spontaneously. When the concentration of solute S is greater inside the cell, the ratio of $[S]_i/[S]_o$ is greater than 1 and the natural log of a value greater than 1 is positive. Thus ΔG for moving S into the cell is positive and the process does not occur spontaneously; that is, energy is required to move S from outside the cell to inside. When the concentration of solute S is greater outside the cell, ΔG for moving S into the cell is negative and the process occurs spontaneously; that is, S moves passively into the cell.

concentration requires an input of energy. Transport of molecules across a membrane is called **active transport** if it requires energy, and **passive transport** if it does not. **Simple diffusion**—the movement of a molecule into or out of the cell by its own thermal motion—is one form of passive transport, but certain forms of protein mediated transport are also passive. In contrast, active transport is always mediated by transport proteins referred to as **pumps.**

Consider the movement of glucose into and out of the cell as an example. The concentration of glucose is usually about 1 mM in intracellular fluid, but is nearly 6 mM in extracellular fluid (see Table 1). If glucose is allowed to move spontaneously across the membrane of a typical cell, it will move from higher to lower concentration (higher to lower energy) and, therefore, will move into the cell. In contrast, transport of glucose in the opposite direction (from lower to higher concentration) is active because it does not occur spontaneously and requires energy for its initiation.

Driving Forces Acting on Molecules

Any difference in energy existing across a membrane acts as a *driving force* that tends to push molecules in one direction or another. The direction of this force is always from higher to lower energy, which indicates the direction in which molecules will go if they are allowed to move spontaneously. These driving forces can arise as a result of concentration differences or other factors that affect molecular energies. In the following sections, we see that molecules are generally influenced by three types of driving forces: chemical, electrical, and electrochemical.

Chemical Driving Forces

When a substance is present in different concentrations on either side of a membrane, a **concentration gradient** is said to exist across the membrane. We consider the term "concentration gradient" to be synonymous with the difference in concentration and give it the symbol ΔC. (Strictly speaking, the term "concentration gradient" is used in reference to any difference in concentration between one location and another—not just differences across membranes—and is defined as the rate at which the concentration changes with distance.) When molecules are moving from higher to lower concentration, we can say that they are moving *down a concentration gradient;* movement in the opposite direction is *up a concentration gradient.*

Because molecules will move down a concentration gradient spontaneously, we can think of a concentration gradient as a kind of force that "pushes" molecules in that particular direction. Thus we refer to a concentration gradient as a **chemical driving force,** the direction of which is always down the concentration gradient (**Figure 1**). As we will see, the rate at which a substance is transported varies with the size of the concentration gradient and generally increases as the size of the gradient increases. Therefore, we can say that the magnitude of the chemical driving force increases as ΔC increases. When more than one substance is present, as is the case with real cells, more than one concentration gradient exists. Any chemical driving force that might be acting on a given substance depends only on the concentration gradient of *that particular substance.*

Note that a chemical driving force is fundamentally different from the more familiar types of forces because it does not really act on molecules in the same way that the force of gravity pulls on a rock or a magnetic force pulls on a piece of iron. When molecules move across a membrane down a concentration gradient, they do so simply because more molecules are present on one side of the membrane than on the other. *Individual* molecules are not pushed down the gradient but, in fact, are equally likely to move in either direction. This issue is explored later in this chapter.

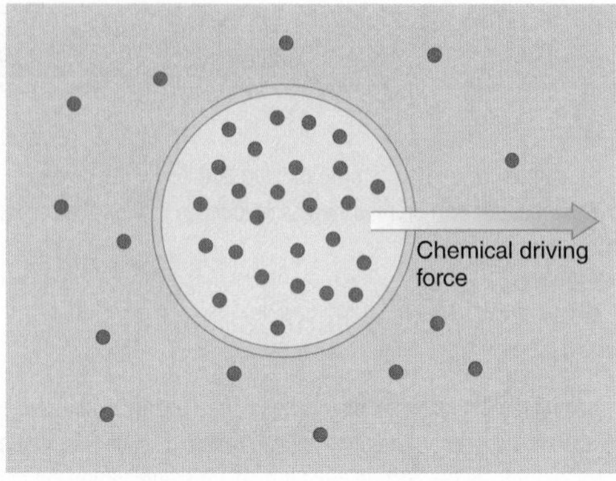

(a)

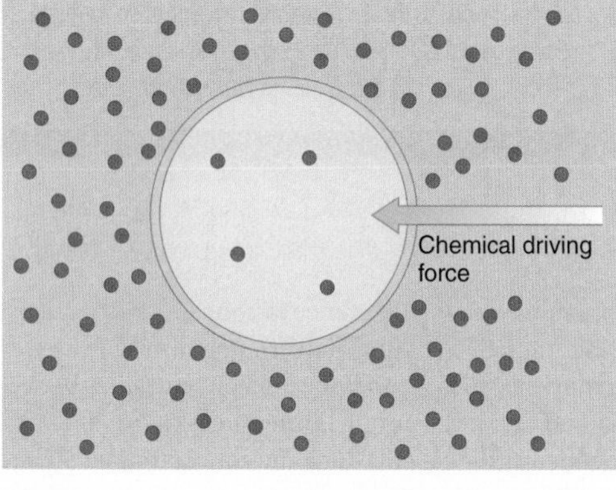

(b)

Figure 1 Chemical driving forces. **(a)** When the concentration of molecules (dots) is higher inside a cell than in extracellular fluid, the direction of the chemical driving force (arrow) is outward. **(b)** When the concentration of molecules is higher in extracellular fluid, the direction of the chemical driving force is inward. In both cases, molecules will move passively in the direction of the driving force, or down the concentration gradient.

Quick Check 1

1. What is the difference between passive transport and active transport?
2. Why is a concentration gradient referred to as a driving force?
3. In which direction (into or out of a cell) does the chemical driving force for amino acids go? (See Table 1.)

Electrical Driving Forces

Molecules moving across membranes can be affected by factors other than the chemical driving force. This is particularly true of ions, which are influenced by *electrical driving forces* in addition to chemical driving forces. Electrical driving forces arise due to the **membrane potential,** a difference in *electrical potential* or *voltage* that exists across the membranes of most cells. In the next section, we see that the existence of a membrane potential reflects an unequal distribution of positively charged ions and negatively charged ions across the plasma membrane. Once we have a firm understanding of what a membrane potential is, we then consider its influence on the forces that affect ion transport.

The Membrane Potential The fluids in the body contain a wide variety of solutes, including many ions (also called *electrolytes*)—that is, substances possessing an electrical charge. Some ions are **cations,** which have a positive charge; others are **anions,** which have a negative charge. Ions are also present in salt solutions, such as seawater, but we normally cannot detect the presence of the ions' electrical charges because the number of positive charges matches the number of negative charges. Such a solution is said to be electrically *neutral* because the positive and negative charges cancel out each other, giving a net (total) electrical charge of zero. Likewise, the total electrical charge of your body is zero because the number of cations in your body equals the number of anions. A person may pick up and emit charges, however, such as when you rub your feet on a carpet (pick up negative charge) and then touch a metal object or other conducting material to release this electrical charge.

In intracellular or extracellular fluid, cations and anions are present in unequal numbers; consequently, these fluids are not electrically neutral. Intracellular fluid contains a slight excess of anions over cations, giving it a net negative charge. Extracellular fluid contains a slight excess of cations over anions, giving it a net positive charge. Because positive and negative charges are distributed unequally between the inside and outside of a cell, a *separation of charge* is said to exist across the membrane (**Figure 2**).

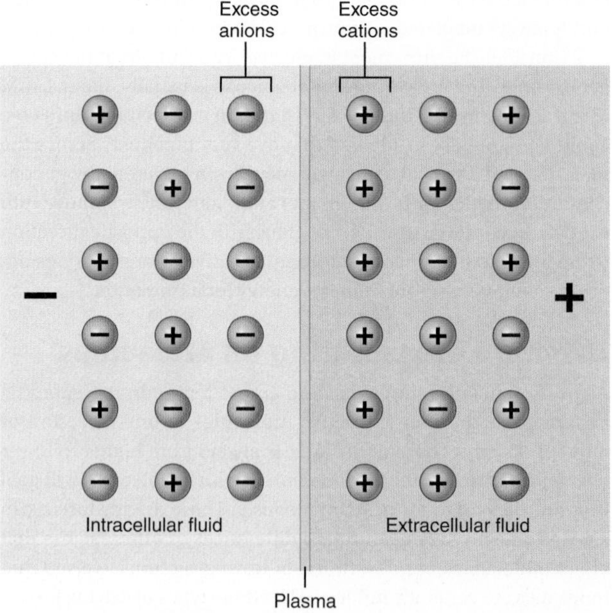

Figure 2 Separation of charge across a cell membrane. Under normal conditions, the fluid inside a cell has a slight excess of anions (negative charges), and the fluid outside the cell has a slight excess of cations (positive charges). These excess charges are clustered in the region near the membrane. Net charges inside and outside the cell are indicated by (+) and (−) signs on either side of the membrane.

When charges are separated, a potential energy (voltage) exists. The excess negative and positive charges of intracellular and extracellular fluid tend to be clustered close to the membrane because the excess negative charges on one side of the membrane are attracted to the excess positive charges on the other side.

A cell's membrane potential reflects this separation of charge and is given in units of electrical potential—in *millivolts* (mV), which are 1/1000 of a volt. The magnitude of the membrane potential (number of millivolts) depends on the degree of charge separation: The greater the difference in charge between the two sides of a membrane, the larger the membrane potential. By convention, the *sign* of the membrane potential (positive or negative) is taken to be the sign of the net charge *inside* the cell relative to outside. Because the inside of a cell is typically more negatively charged than the outside, the membrane potential is usually negative. (Under certain conditions, however, the membrane potential can be positive.) For many cells, the membrane potential, which is denoted by the symbol V_m, is approximately negative 70 millivolts ($V_m = -70$ mV).

How the Membrane Potential Creates an Electrical Driving Force That Acts on Ions An electrical potential is a form of potential energy; that is, it is an electrical force acting on charged particles and has the potential to cause those particles to move. For example, AA batteries separate 1.5 volts of charge across the two terminals. One terminal is positive, the other negative. Should a switch be thrown to provide a connection between the two terminals, then charge will flow as an electrical current from one terminal to another. If a light bulb is placed in between the terminals, then the current can be used to power the bulb.

The membrane potential works in a similar fashion. The separation of charge is a potential energy for current flow across the membrane. Current in biological systems is caused by ion movement. Thus the membrane potential creates an electrical driving force for the movement of ions. To determine the direction of the electrical driving force, we need simply know the *valence* (or charge) of the ion, the sign of the membrane potential (usually negative), and this simple rule: *Opposites attract, likes repel.* Thus cations are attracted by the negative charge inside the cell and have an inward-directed electrical driving force. Anions, by contrast, are repelled by the negative membrane potential and have an outward-directed electrical driving force (**Figure 3**). Uncharged molecules, such as glucose, are not affected by the membrane potential and, therefore, have an electrical driving force of zero.

The *magnitude* of the electrical driving force on an ion depends on the size of the membrane potential and the quantity of charge carried by the ion, and it increases as either of these factors gets larger (**Figure 4**). A larger negative membrane potential, for instance, means a greater number of negative charges inside and positive charges outside, which increases the attractive and repulsive forces acting on an ion. If an ion carries more charge, the attractive and repulsive forces are also increased, which makes the electrical driving force stronger.

Electrochemical Driving Forces

To determine whether ions are being transported passively or actively, a physiologist must identify all the driving forces that might be acting on them. In general, when ions are transported across membranes, two driving forces are influential: (1) a chemical force reflecting the ions' tendency to move down their concentration gradient (from higher to lower concentration) and (2) an electrical force reflecting the ions' tendency to be pushed in one direction or the other by the membrane potential. The total force acting on the ions is the combination of these chemical and electrical driving forces, referred to as the **electrochemical driving force.** (The term "electrochemical driving force" also can be used in reference to uncharged molecules, which are not influenced by electrical driving forces; in this case, the electrochemical driving force is synonymous with the chemical driving force.)

The direction of the electrochemical driving force acting on an ion depends on the net direction of the electrical and chemical driving forces. If both forces go in the same direction, then the electrochemical driving force also acts in that direction. Conversely, if the electrical and chemical forces go in opposite directions, then the electrochemical force acts in the direction of the larger force.

To determine whether the electrical or chemical force is larger, a physiologist must know an ion's **equilibrium potential,** a hypothetical value for the membrane potential at which the electrical driving force is equal and opposite to the chemical driving force,

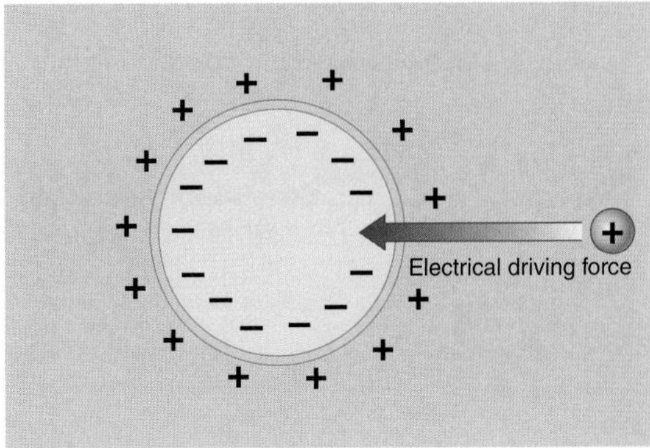

(a)

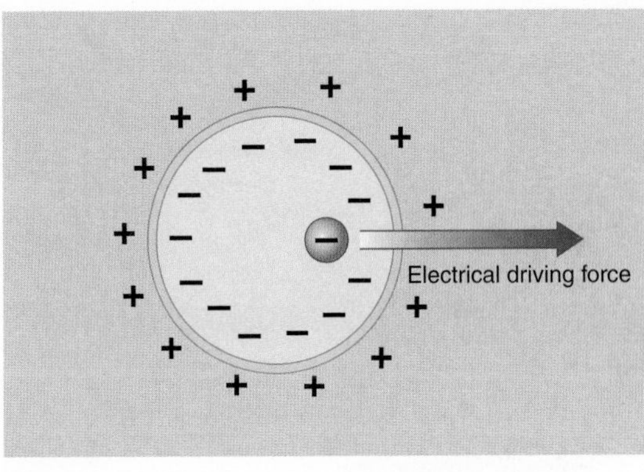

(b)

Figure 3 Electrical driving forces. (a) The direction of the electrical driving force (arrow) on a positively charged ion is inward. **(b)** The direction of the electrical driving force on a negatively charged ion is outward. In both cases, it is assumed that the membrane potential is negative, as indicated by (+) and (–) signs on either side of the membrane.

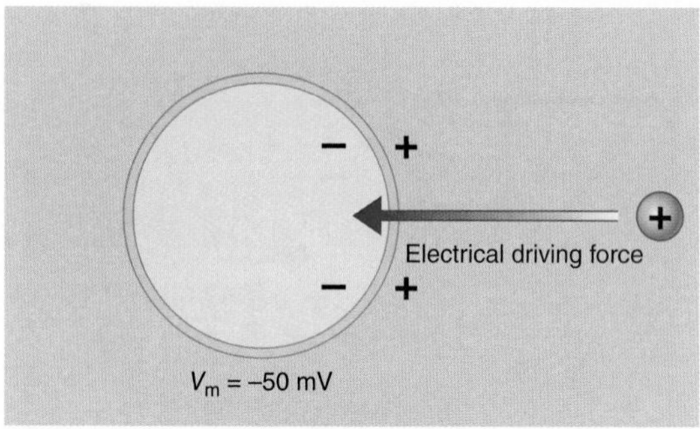

(a)

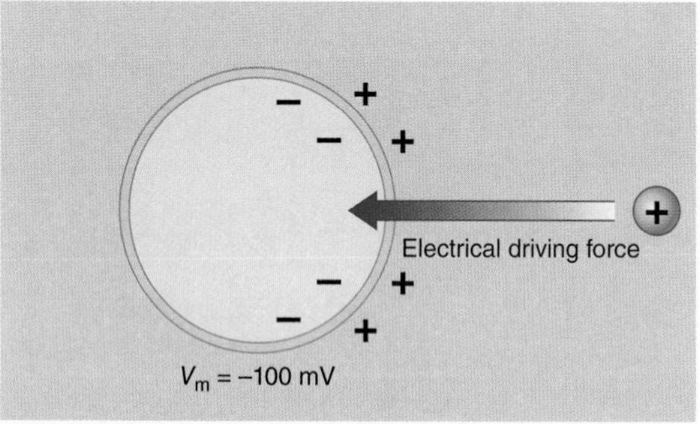

(b)

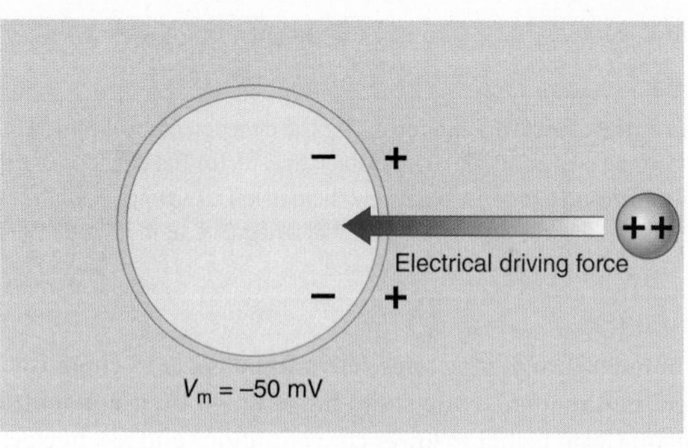

(c)

Figure 4 Effects of membrane potential and ion valence (charge) on electrical driving force. **(a)** In the presence of a given negative membrane potential ($V_m = -50$ mV), an inward electrical driving force (arrow) acts on a positively charged ion crossing the membrane. **(b)** When the magnitude of the membrane potential increases to −100 mV, the magnitude of the electrical driving force also increases, as indicated by the wider arrow. **(c)** For ions of higher valence, the magnitude of the electrical driving force also increases.

producing an electrochemical driving force of zero. If the membrane potential equals the equilibrium potential for an ion, that ion will not move spontaneously in either direction because the total driving force acting on it is zero. In other words, the ion will be at *equilibrium.* (Recall that a chemical reaction is at equilibrium when its net direction is neither forward nor backward.)

The magnitude and sign of an ion's equilibrium potential depend on the size and direction of the ion's concentration gradient, and on the ion's valence. Larger concentration gradients mean larger equilibrium potentials because a greater electrical force is required to equal or "balance" a larger chemical force. The sign of the equilibrium potential is such that the electrical force goes in the direction opposite to the chemical force. In the case of sodium ions (Na^+), for instance, which are found in higher concentrations outside a cell, the chemical force is directed inward. Thus an *outwardly directed* electrical force is required to balance the chemical force. Because Na^+ is positively charged, a *positive* membrane potential will exert an outward electrical force that balances the inward chemical force. This means that the Na^+ equilibrium potential must be positive. If an ion's concentration on either side of a membrane is known, its equilibrium potential can be calculated using the Nernst equation (**Toolbox: Equilibrium Potentials and the Nernst Equation**).

Determining the Direction of the Electrochemical Driving Force The direction of the electrochemical driving force acting on an ion can be determined using the following procedure. To carry it out, we need to know the ion's equilibrium potential (which is given the symbol E_X for ion X) and the cell's membrane potential:

1. Using the principles described in the preceding sections, identify the directions of the chemical and electrical driving forces acting on the ion. If the two driving forces are going in the same direction, the electrochemical force also acts in that direction, and no further analysis is necessary.
2. If the chemical and electrical forces act in opposite directions, compare the sizes of the equilibrium potential and the membrane potential. If they are equal in magnitude, then the electrochemical force is zero and the ion is at equilibrium. If not, proceed to step 3.
3. If the equilibrium potential is larger in magnitude than the membrane potential, then the chemical force is larger than the electrical force; the electrochemical force, therefore, acts in the same direction as the chemical force. If the membrane potential is larger in magnitude than the equilibrium potential, then the electrical force is larger than the chemical force; the electrochemical force, therefore, acts in the same direction as the electrical force.

Application of this procedure is illustrated in **Figure 5** for potassium ion (K^+), whose intracellular and extracellular concentrations are 140 mM and 4 mM, respectively (see Table 1), giving an outwardly directed chemical force and an equilibrium potential of −94 mV ($E_K = -94$ mV). In all three panels in Figure 5, the membrane potential is negative, producing an inwardly directed electrical force that opposes the chemical force. In Figure 5a, the membrane potential and the equilibrium potential are equal, giving an electrochemical force of zero. In Figure 5b, the equilibrium potential is larger than the membrane potential (−70 mV), giving an electrochemical force that is directed out of the cell. In Figure 5c, the membrane potential (−100 mV) is larger than the equilibrium potential, so the electrochemical force is directed into the cell.

Another method to determine the direction of the electrochemical driving force is to compare the equilibrium potential to the actual membrane potential of the cell. The direction of the

TOOLBOX

Equilibrium Potentials and the Nernst Equation

We learned in the last Toolbox feature that the chemical driving force for an uncharged solute to move into a cell is determined by the van't Hoff equation:

$$\Delta G = RT \ln \frac{[S]_i}{[S]_o}$$

where $[S]_o$ is the concentration of solute S outside the cell and $[S]_i$ is the concentration of solute S inside the cell.

When we consider ions, however, there is an electrical force in addition to the chemical force. The following equation describes the electrochemical driving force for an ion (I) to move into the cell:

$$\Delta G = RT \ln \frac{[I]_i}{[I]_o} + zFE$$

where G is the free energy, R is the universal gas constant (0.082 liter-atm/mole-K), T is the absolute temperature (K), z is the valence of the ion, E is the membrane potential, F is Faraday's constant for electrical forces (9.65×10^4 joules/volt-mole), and [I] is the concentration of ion I. Notice that the electrochemical force is a sum of the chemical force and the electrical force (zFE).

With knowledge of an ion's charge and its intracellular and extracellular concentrations, we can find the equilibrium potential of any ion (E_I) by setting ΔG to 0 and solving for E:

$$-zFE_I = RT \ln \frac{[I]_i}{[I]_o} \quad \text{or} \quad E_I = \frac{RT}{zF} \ln \frac{[I]_o}{[I]_i}$$

Plugging in constants and replacing natural log (ln) with $\log_{10}$ (log), we get:

$$E_I = \frac{61 \text{ mV}}{z} \log \frac{[I]_o}{[I]_i}$$

This equation, called the Nernst equation, gives the value of the equilibrium potential in millivolts and assumes that the temperature is at or near the normal body temperature of 37°C.

Using typical intracellular and extracellular concentrations, we can find the equilibrium potential for sodium, E_{Na}, by making the appropriate substitutions, as follows:

$$E_{Na} = \frac{61}{1} \log \frac{145 \text{ mM}}{15 \text{ mM}} = 60.1 \text{ mV} \approx 60 \text{ mV}$$

In similar fashion, we can find the equilibrium potential for potassium, E_K:

$$E_K = \frac{61}{1} \log \frac{4 \text{ mM}}{140 \text{ mM}} = -94.2 \text{ mV} \approx -94 \text{ mV}$$

Note that in both cases the valence is +1, so that the sign of the equilibrium potential depends solely on the direction of the concentration gradient. Note also that E_K is larger in magnitude than E_{Na}. This makes sense because a larger concentration gradient requires a larger membrane potential to balance it, and the K^+ gradient is larger than the Na^+ gradient. If a concentration gradient is small, intracellular and extracellular ion concentrations will be more nearly equal. If the concentrations are identical, the concentration ratio C_o/C_i equals 1, making the equilibrium potential zero. (Recall that the log of 1 is zero). As the concentrations become more and more dissimilar, the ratio becomes either much larger or much smaller than 1, making the log term larger and more positive, or larger and more negative, respectively.

electrochemical driving force is such that the ion will move in the direction that will bring the membrane potential closer to the equilibrium potential for that ion; that is, an ion tends to move in the direction that puts it at equilibrium. Using the example of potassium above, the membrane potential is –70 mV and the potassium equilibrium potential is –94 mV. If potassium moved into the cell, it would bring its positive charge with it, thereby making the membrane potential less negative and taking potassium further from equilibrium. Conversely, if potassium moved out of the cell, it would take its positive charge out of the cell, thereby bringing the membrane potential closer to potassium's equilibrium potential. Thus the direction of the electrochemical driving force on potassium, when the membrane potential is –70 mV, is out of the cell.

Significance of the Electrochemical Driving Force

Because the electrochemical force is the total driving force acting on transported ions, it determines the direction in which the ions move if they are allowed to cross the membrane spontaneously. When ions are transported passively, they always move in the direction of the electrochemical driving force. Put another way, when ions are transported passively, they move *down* their **electrochemical gradient.** When they are being transported actively, they move in the direction opposite to the electrochemical force, or *up* their electrochemical gradient.

Quick Check 2

1. Determine the direction of the electrical driving force for each of the following ions, assuming the cell membrane potential is negative: Na^+, K^+, Cl^-, HCO_3^-, Ca^{2+}.
2. Assume that sodium and calcium ions are being transported through a given cell membrane. For which ion—Na^+ or Ca^{2+}—is the electrical driving force larger? Explain.
3. Refer to Table 1 for intracellular and extracellular ion concentrations, and determine the direction of the chemical driving force on each of the ions in question 1. Determine the direction of the electrochemical driving force on Na^+, assuming that $E_{Na} = +60$ mV and $V_m = -70$ mV.

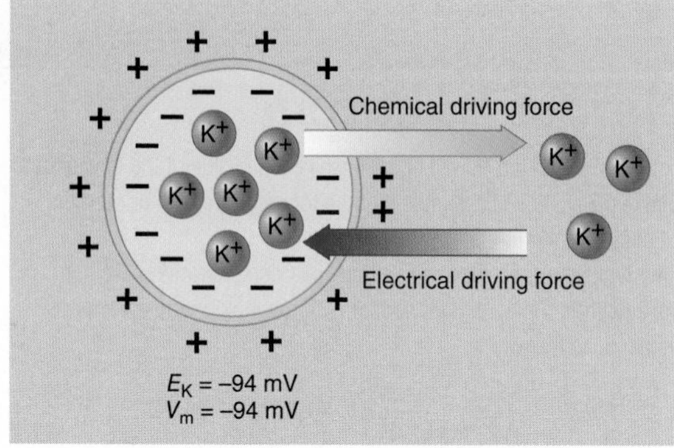

(a)

(b)

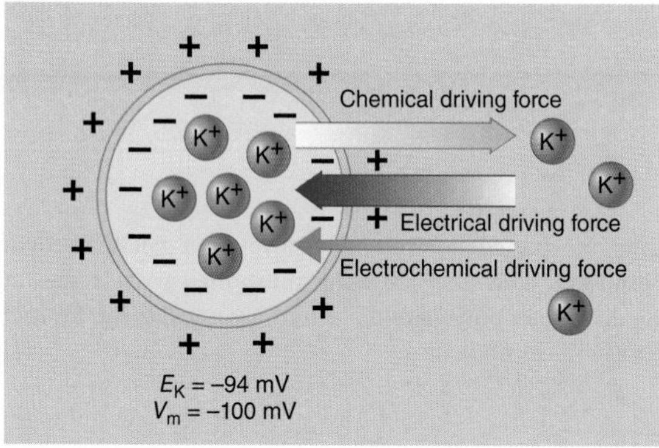

(c)

Figure 5 Electrical, chemical, and electrochemical driving forces acting on potassium ions crossing a cell membrane. The equilibrium potential for potassium (K^+) is −94 mV, and the membrane potential is assumed to be negative. **(a)** When the membrane potential equals the equilibrium potential ($V_m = -94$ mV), the electrochemical force is zero. **(b)** When the membrane potential is −70 mV, the electrical force is smaller than the chemical force, and the electrochemical force is outward. **(c)** When the membrane potential is −100 mV, the chemical force is smaller than the electrical force, and the electrochemical force is inward.

2 Rate of Transport

The rate of a metabolic reaction is important because reactions must proceed at a rate fast enough to meet the body's metabolic demands. For this to occur, molecules must be transported across membranes at sufficient rates. When you are resting, for instance, oxygen is transported across the epithelium lining the lungs (the *pulmonary epithelium*) at a rate that matches the rate of oxygen consumption in your body. When you begin exercising, metabolic reactions consume oxygen at a higher rate; to keep up with this increased demand, the rate of transport across the pulmonary epithelium must also increase, or the cells in your body will eventually run out of oxygen. In certain pathological conditions, such as *pulmonary edema* (excess fluid in the lungs), diffusion of oxygen (and of carbon dioxide) is impeded. Under these conditions, a person has difficulty exercising because oxygen cannot move across the pulmonary epithelium fast enough. If the condition is severe, oxygen delivery may be impaired so much that the individual feels breathless even at rest.

The *rate* at which a substance is transported across a membrane refers to the number of molecules that cross the membrane in a given length of time, which is called the **flux.** Flux is usually expressed in units of moles per second, or some equivalent. (Strictly speaking, flux is the number of molecules crossing a membrane per unit time per unit membrane surface area. For simplicity, we omit area from our definition.)

Note that when we say that molecules are transported in one direction or the other (into or out of a cell, for instance), we are referring to a *net* movement of molecules across the membrane (similar to the *net* direction of a chemical reaction). In virtually every case, individual molecules actually move across a membrane in *both* directions, but more of them may move in one direction than in the other. This bidirectional flow of molecules is illustrated in **Figure 6**, which depicts a membrane separating two solutions in a chamber. In this example, molecules cross the membrane by simple diffusion under two conditions: (1) when their concentration is the same on both sides and (2) when the concentration on the left side is twice that on the right side.

In the first condition (Figure 6a), molecules cross both from side 1 to side 2 and from side 2 to side 1. However, because the concentration of molecules is the same on either side, and each molecule has the same likelihood of crossing to the other side in any given length of time, no net transfer of molecules from one side to the other occurs, the concentrations do not change, and the system is in *diffusional equilibrium.*

In the second condition (Figure 6b), the concentration on side 1 is twice what it originally was, and twice that of side 2. Under these conditions, more molecules move from side 1 to side 2 than from side 2 to side 1. As a result, there is net movement of molecules from side 1 to side 2. The rate of this movement is the *net flux.* Whenever physiologists say that molecules are being transported passively or actively in a certain direction, they are always referring to the direction of the net flux.

As we will see later in this chapter, transport rates are influenced by many variables, some of which are specific to certain transport mechanisms.

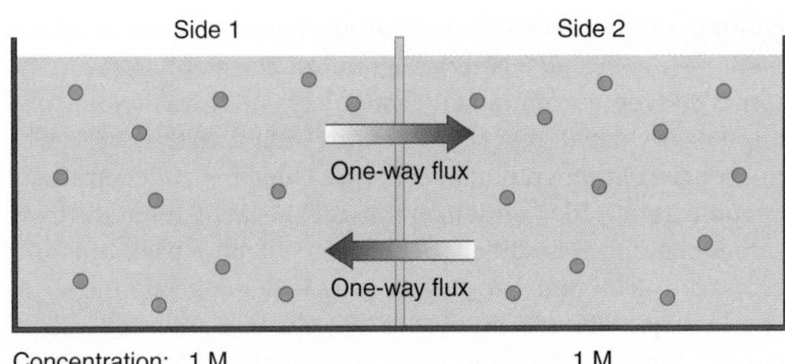

(a)

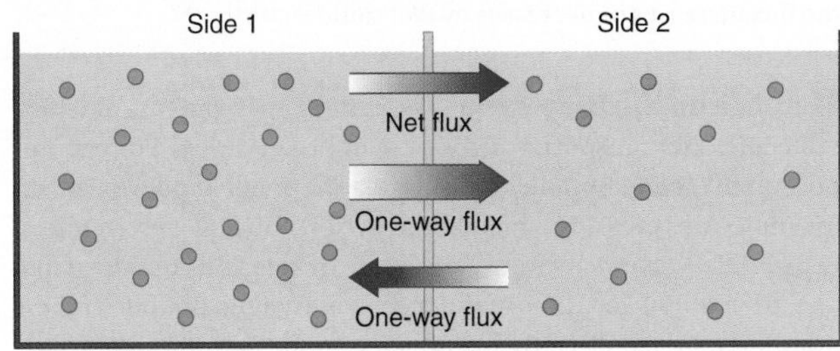

(b)

Figure 6 Directional flow of molecules and net flux. (a) Diffusional equilibrium. When solute molecules (dots) are at the same concentration on both sides of a membrane, the one-way flux from side 1 to side 2 equals that from side 2 to side 1 (arrows). Under these conditions, the net flux across the membrane is zero. **(b)** Net flux. When the concentration on side 1 is increased by a factor of 2, the one-way flux from side 1 to side 2 is twice as large as before, but the one-way flux in the opposite direction is unaffected. The one-way fluxes are now unequal, producing a net flux down the concentration gradient from side 1 to side 2.

3 Passive Transport 

In passive transport, molecules move across the membrane down their chemical or electrochemical gradients. No energy is required. Types of passive transport include *simple diffusion, facilitated diffusion,* and *diffusion through ion channels.*

Simple Diffusion: Passive Transport Through the Lipid Bilayer

Simple diffusion is the least complicated of all transport mechanisms. Although we have already learned about the forces for simple diffusion, here we learn what makes it work. We will also learn about some of the factors that affect the rate of diffusion and the permeability of a membrane to substances moving by simple diffusion.

The Basis for Simple Diffusion

We use the term **simple diffusion** to describe the passive transport of molecules through a biological membrane's lipid bilayer, but in fact the mechanism of simple diffusion is not strictly biological. When someone opens a bottle of perfume nearby, for example, you will eventually smell the fragrance even if there are no air currents to carry it to your nose. This phenomenon occurs because some molecules from the perfume travel from the bottle to your nose by simple diffusion through the air. The molecules travel away from the bottle because they are more concentrated near the bottle and less concentrated farther away, and because these molecules, like any others, move spontaneously down their concentration gradient owing to their state of constant thermal motion.

The movement of molecules from one location to another simply as a result of their own thermal motion is called **diffusion,** a general term that applies to molecules moving through any medium. Thermal motion is often called *random* thermal motion because individual molecules move helter-skelter in many different directions because of collisions with other molecules. As a consequence, molecules do not go in any particular direction for long before changing course, and when this change occurs the direction is unpredictable.

But if thermal motion is random, then how is it that diffusing molecules always move *down* their concentration gradient, which is clearly not a random event? The answer lies in the distinction between *individual molecules,* which move randomly, and a *population of molecules,* which always moves down its concentration gradient. For example, if there are eight molecules on the left and two molecules on the right, all ten individual molecules are just as likely to move right or left. However, when this population of ten molecules is considered, molecules on the left—the side of greater concentration—are four times more likely to move to the right than those on the right are likely to move to the left. Thus there will be more movement from left to right.

Factors Affecting Rates of Simple Diffusion

When a substance is transported passively across a membrane by simple diffusion, the rate at which it is transported depends on three factors: the magnitude of the driving force, the membrane surface area, and the **permeability** of the membrane, a measure of the ease with which molecules are able to move through it. Each of these factors is discussed next.

The Magnitude of the Driving Force When a driving force acts on molecules crossing a membrane, it influences not only the direction in which they move, but also the rate at which they are transported. In most cases, the net flux increases as the magnitude of the driving force increases, though not always.

In simple diffusion, the rate of transport is directly related to the size of the driving force. Consider a situation in which a membrane separates two solutions in a chamber (**Figure 7**). Initially (Figure 7a), side 1 contains solute at a concentration of 1 M, while side 2 contains an equal volume of pure water. The presence of this concentration gradient leads to a net flux of solute molecules from side 1 to side 2. As a result, the concentration on side 1 decreases while the concentration on side 2 increases. In Figure 7b, the concentration on side 1 has decreased to 0.75 M while the concentration on side 2 has increased to 0.25 M. Because the size of the concentration gradient has decreased, the net flux has also decreased, as indicated by the thinner arrow. These changes

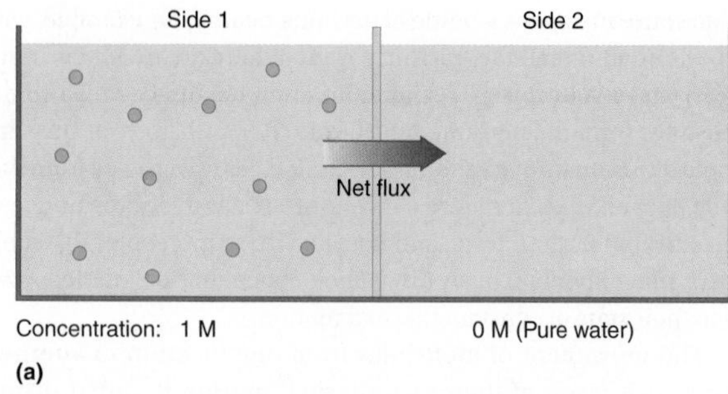

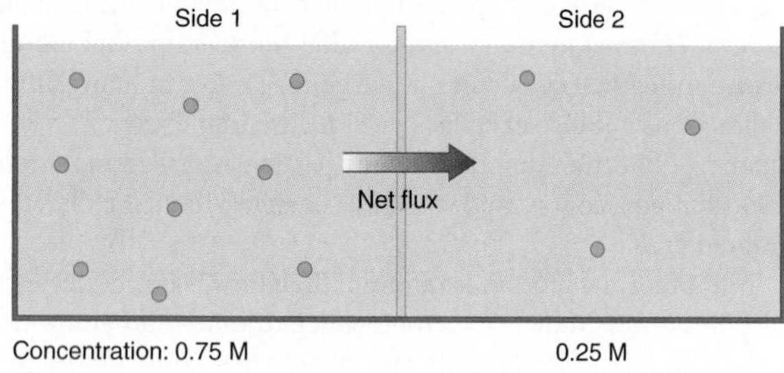

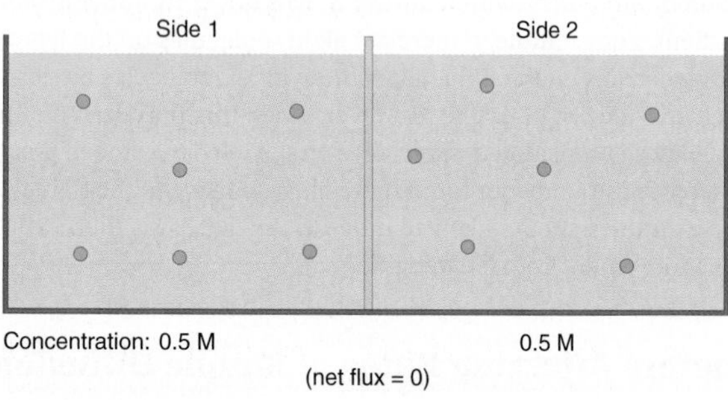

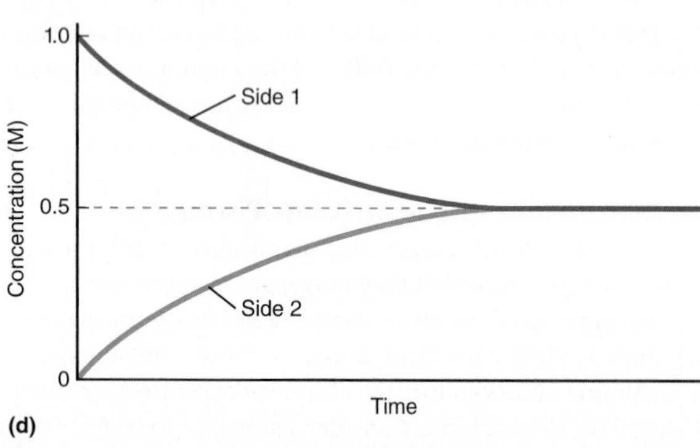

Figure 7 Changes in net flux of molecules across a membrane as the concentration gradient changes. **(a)** A large concentration gradient creates the net flux indicated by the arrow. **(b)** Because of the net flux from side 1 to side 2, the concentration on side 1 has decreased to 0.75 M, while that on side 2 has increased to 0.25 M. The net flux has also decreased. **(c)** Concentrations on both sides of the membrane have become equal, and the net flux is now zero. **(d)** Changes in concentration on the two sides over time.

Q As time passes, does the one-way flux from side 2 to side 1 increase, decrease, or stay the same?

continue until the two concentrations become equal, at which point the net flux equals zero (Figure 7c). The graph in Figure 7d shows that the concentrations change less rapidly as they become more nearly equal; that is, the net flux diminishes as the size of the concentration gradient decreases. Once the concentrations become equal, they no longer change, because the net flux is zero. Under these conditions, molecules still move back and forth across the membrane, but at equal rates in both directions.

In simple diffusion, the net flux of a substance is directly proportional to the size of the concentration gradient—or to the size of the electrochemical gradient, if the substance is charged (**Figure 8**). As the concentration gradient increases or decreases by a given factor, the flux increases or decreases by that same factor.

Membrane Surface Area As a general rule, the rate at which molecules are transported across a membrane varies in direct proportion to the membrane's surface area. To understand why, imagine molecules crossing a membrane in a chamber at a given rate. If two identical chambers were placed side by side and combined into one, then the surface area of the membrane would double, and the number of molecules crossing the membrane per unit time would also double. Note that this would be true regardless of the mechanism of transport.

Various tissues that are specialized for transport—the pulmonary epithelium, the intestinal epithelium, and the walls of capillaries, for instance—have large membrane surface areas, which enhance the ability of these tissues to transport large quantities of material quickly. The surface areas of the pulmonary and intestinal epithelia are increased by their high degree of folding, whereas capillary surface areas are increased by the extensive branching that characterizes these vessels.

Membrane Permeability The permeability of a membrane to a particular substance depends on both the nature of the transported substance and the properties of the membrane (other than surface area) that influence the ease with which molecules are able to penetrate it. For any mechanism of passive transport, a higher permeability translates into a higher rate of transport, other things

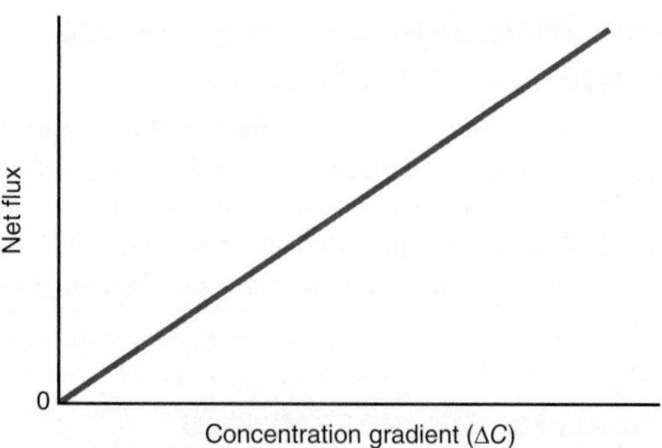

Figure 8 Direct proportionality between net flux and the magnitude of the concentration gradient for an uncharged substance crossing a membrane by simple diffusion.

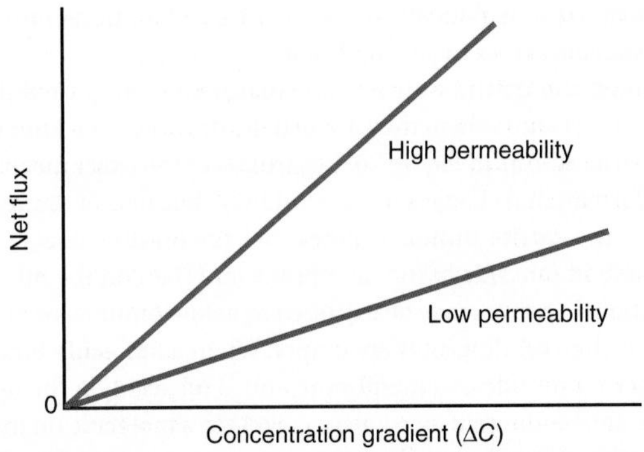

Figure 9 **Effect of permeability on the net flux of an uncharged substance crossing a membrane by simple diffusion.**

being equal. The influence of permeability on transport is illustrated graphically in **Figure 9**, which pertains to a substance crossing a membrane by simple diffusion. When permeability is higher, the net flux is larger for a given concentration gradient (**Toolbox: Fick's Law and Permeability**). We will discuss the various factors that affect membrane permeability shortly.

The permeability of a membrane to a substance X is given the symbol P_X and is often referred to as the "permeability of X." For example, P_{Na} would be the "permeability of sodium" (or alternatively, the "sodium permeability"). Although this usage is very common, note that the term "permeability" applies to membranes, not to transported molecules. A membrane's permeability is not the same for all substances, so permeability of the membrane is generally described in reference to a particular molecule. Strictly speaking, a substance that crosses a membrane is *permeant,* whereas the membrane itself is *permeable* (able to be permeated).

Among the factors influencing the permeability of cell membranes are the following:

1. *The lipid solubility of the diffusing substance.* We have seen that molecules vary in the degree to which they can dissolve in lipids. Hydrophobic (nonpolar) substances are the most lipid soluble, whereas hydrophilic (polar or ionized) substances are the least lipid soluble. Given that molecules must move through a lipid bilayer in simple diffusion across the cell membrane, we would expect lipid-soluble molecules to be able to enter the bilayer more readily. Other things being equal, the more lipid soluble a substance is, the greater a membrane's permeability to that substance. Lipid solubility is also influenced by the lipid composition of the membrane, which varies somewhat from one membrane to another.
2. *The size and shape of diffusing molecules.* Molecules vary considerably in their sizes (physical dimensions and molecular weights) and shapes. Generally speaking, larger molecules and those with more irregular shapes move through the bilayer more slowly, making the membrane permeability lower.
3. *Temperature.* Molecules move faster at higher temperatures, which increases their permeability. This effect is rarely important in human physiology, however, because body temperature is relatively constant.
4. *Membrane thickness.* Theoretically, a thicker membrane would have a lower permeability because molecules must travel farther to penetrate it. Although cell membranes have similar thickness, tissue thickness varies considerably. Tissues specialized for transport, such as capillary walls or the pulmonary epithelium, tend to have relatively thin walls; this thinness increases the permeability and enhances the rate of transport through these tissues.

Of all the factors just mentioned, lipid solubility has the strongest influence on permeability. Differences in lipid solubility are largely responsible for the fact that some substances can cross cell membranes readily by simple diffusion, whereas others cannot. Because most substances in the body are hydrophilic and, therefore, do not penetrate lipid bilayers easily, the list of substances known to be transported by simple diffusion is fairly limited: fatty acids, steroid hormones, thyroid hormones, oxygen, carbon dioxide, and the fat-soluble vitamins (A, D, E, and K).

Even though our focus here has been solely on simple diffusion across *membranes,* a process that applies to relatively few substances, cells ultimately depend on simple diffusion for delivery or removal of *all* solutes, because molecules must diffuse through *water* to move toward and away from membrane surfaces. Recall, for instance, that the accumulation of excess fluid in the lungs impedes oxygen transport, which in turn decreases the lungs' capacity for delivering oxygen to the blood. The reason is that the flux of oxygen across the pulmonary epithelium is affected by the thickness of the fluid layer covering the epithelium. When fluid accumulates, the effect is similar to an increase in tissue thickness, and the diffusion rate slows down.

TOOLBOX

Fick's Law and Permeability

Permeability has a precise mathematical definition. We have seen that when a substance crosses a membrane by simple diffusion, the net flux is proportional to the size of the concentration gradient. This relationship is expressed mathematically by *Fick's law,* which can be written as follows:

$$\text{net flux} = PA(\Delta C)$$

In this expression, A is the membrane surface area, ΔC is the concentration gradient (the difference in concentration across the membrane), and the coefficient P represents permeability. The numerical value of P depends on the particular membrane in question and the identity of the transported substance. Note that for a concentration gradient of a given size, the flux increases as P gets larger.

Quick Check 3

1. In simple diffusion, do individual molecules always go down their concentration gradient? Why or why not?
2. Which properties of transported molecules affect membrane permeability in simple diffusion? Which properties of a membrane affect the permeability?
3. For a substance crossing a tissue by simple diffusion, which of the following (choose all that apply) cause the permeability to increase: increased tissue thickness, decreased size of diffusing molecules, decreased lipid solubility of diffusing molecules, increased temperature.

Facilitated Diffusion: Passive Transport Utilizing Membrane Proteins

We noted earlier that some substances that are transported passively do not cross membranes by simple diffusion but instead cross by way of transport proteins in the membrane—a process known as **mediated transport.** Physiologists use the term **facilitated diffusion** to distinguish passive mediated transport from simple diffusion and from active transport (another type of mediated transport).

Carriers in Facilitated Diffusion

A **carrier** is a transmembrane protein that binds molecules on one side of a membrane and transports them to the other side by means of a conformational change. A carrier possesses one or more binding sites that are usually specific for molecules of certain substances or classes of substances. Examples of substances transported by carriers include monosaccharides and amino acids.

To be transported by a carrier, a molecule must first enter a binding site. Once the molecule is in the binding site, the carrier undergoes a conformational change that exposes the binding site to the fluid on the other side of the membrane. At this point, the molecule is free to dissociate from the carrier and be released into the fluid. **Figure 10** schematically depicts how a carrier transports a glucose molecule across a cell membrane.

Although the nature of the conformational changes exhibited by carriers is unknown, they are usually drawn with a kind of hingelike structure, as in Figure 10. Regardless of the exact mechanism, conformational changes occur randomly (because of thermal agitation of the carrier protein) and expose the binding site(s) to the fluid first on one side of the membrane and then on the other. Conformational changes may be triggered by solute binding, or may occur even when binding sites are empty. When a molecule binds to a carrier on one side of a membrane and is released on the opposite side, the binding site becomes available to a molecule on that side. If a molecule binds to it, the carrier can then transport it in the opposite direction. If a molecule does not bind to it, the carrier can revert to its original conformation with the binding site empty, as shown in Figure 10.

The net flux of facilitated diffusion depends on the frequency of solute binding to the carrier molecule on the two sides of the membrane, so net transport occurs from the side with a greater frequency of binding. Two factors affect the binding of a solute to a carrier: (1) the affinity of the binding site on the carrier and (2) the concentration gradient (or electrochemical gradient, if ions are being transported) of the solute across the membrane. In facilitated diffusion, the affinity of the binding site for the solute is the same when the binding site is exposed to the intracellular fluid as it is when exposed to the extracellular fluid. Therefore, any difference in the binding of solute to carrier on either side of the membrane depends on the concentration gradient.

Solute is most likely to bind the carrier when there is more solute present. Thus, if a concentration gradient is present, more solute binds to carriers when the binding sites are facing the side with the greater solute concentration. When the carrier changes conformation, the solute is released on the other side. The net effect is a flux of solute down its concentration (or electrochemical) gradient. If solute is present at the same concentration on both sides of the membrane, then solute will bind equally when the carrier faces either side of the membrane and the net flux is zero. At this

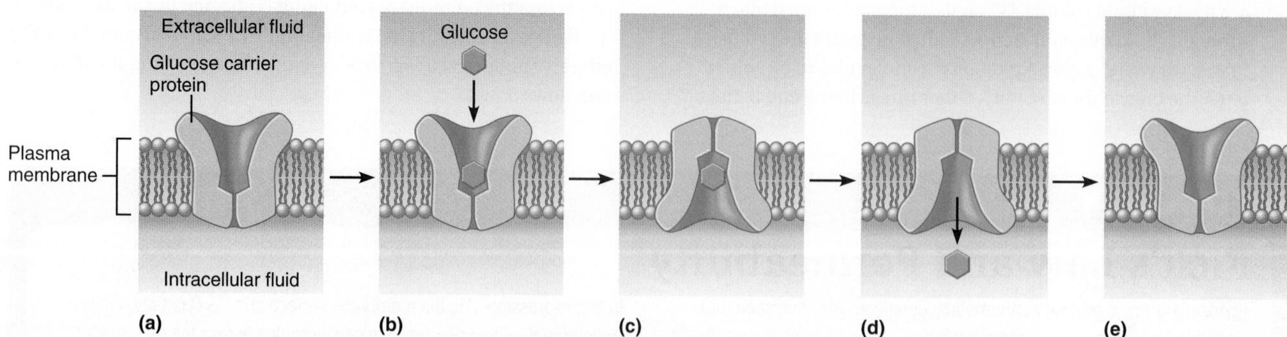

Figure 10 Transport of a glucose molecule across a cell membrane by a carrier protein. **(a)** A carrier protein with an empty binding site. **(b)** Binding of a glucose molecule to the protein's binding site, which faces the outside of the cell. **(c)** Conformational change in the carrier protein, such that the binding site now faces the inside of the cell. **(d)** Release of the glucose molecule. The binding site is once again empty. **(e)** Return of the carrier to its original conformation. The carrier is now ready to bind another glucose molecule.

point, the solute is at diffusional equilibrium across the membrane. This means that a net flux exists only in the presence of a concentration gradient. If a concentration gradient exists, the direction of the net flux is down the gradient. In this regard, facilitated diffusion is like simple diffusion, even though the mechanism is different.

Factors Affecting the Rate of Facilitated Diffusion

The rate of facilitated diffusion is determined entirely by three factors: (1) the transport rates of the individual carriers, (2) the number of carriers in the membrane, and (3) the magnitude of the concentration (or electrochemical) gradient of the transported substance. An increase in any of these factors translates into an increased rate of facilitated diffusion.

Individual carriers transport molecules at different rates, depending on their type. A glucose carrier, for example, typically can transport molecules at a maximum rate of 10,000 glucose molecules per second.

Figure 11 shows how the transport rate of a population of carriers varies as the size of the concentration gradient changes. When the concentration on one side of the membrane increases while that on the other side is kept constant, the net flux increases rapidly at first but then levels off. (Compare the graph in Figure 11 with that in Figure 8, which pertains to a substance crossing a membrane by simple diffusion.) This leveling off occurs because the carrier proteins become 100% saturated at high concentrations.

Saturation of carriers is similar to saturation of enzymes and occurs for the same reason: At a given moment in time, a fixed number of carriers is present, each with its own binding site(s). When the concentration of molecules on one side of a membrane is high, the likelihood that at any given time the binding sites are occupied by molecules is also high. If the concentration is very high, binding sites are occupied virtually 100% of the time. Under these conditions, the carriers are transporting molecules as fast as they can; increasing the solute concentration further will have no effect on the transport rate.

Because the rate of facilitated diffusion depends on the degree of saturation of the carrier proteins, cells can regulate the rate by modifying the number of carriers that are present in the membrane. An increase in the number of carriers will increase the likelihood that solute will bind to carrier and be transported across the membrane (just as an increase in the concentration of enzyme will increase the rate of a chemical reaction). Many hormones exert their actions by altering the number of carriers.

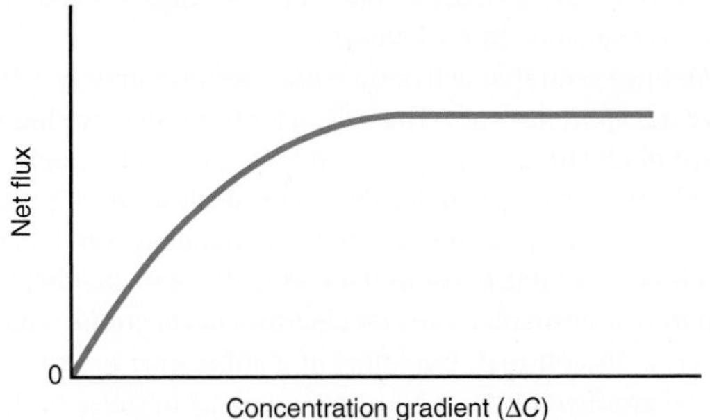

Figure 11 Relation of net flux and concentration gradient for molecules crossing a membrane via facilitated diffusion.

Diffusion Through Channels

A **channel** is a transmembrane protein that transports molecules via a passageway or *pore* that extends from one side of the membrane to the other (**Figure 12**). Like carriers, channels are usually specific for certain substances or classes of substances. Common channel types include water channels (aquaporins) and ion channels.

Diffusion of Water Through Aquaporins

Although some water crosses the lipid bilayer of cell membranes by simple diffusion, most water diffuses across cell membranes through **aquaporins,** highly selective pores that permit water—but no solutes—to move across the membrane by diffusion. Although aquaporins are a fairly recent discovery (the first aquaporin was

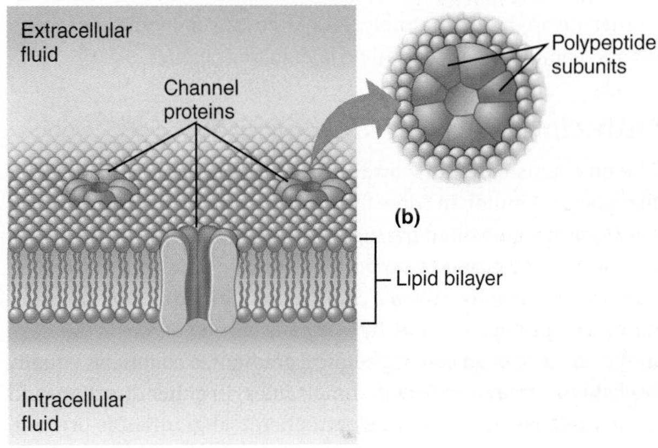

Figure 12 Membrane channels. (a) Channels are transmembrane proteins. **(b)** Top view of an ion channel. **(c)** Structural diagram of a sodium channel. Note the four different domains, each consisting of six units that cross the plasma membrane.

FOCUS ON DIABETES

Glucose Transporter 4

An example of a facilitated diffusion carrier protein is the glucose transporter 4 (GLUT4), which is found in muscle cells and adipocytes. Insulin increases glucose transport into these cells by inducing the insertion of GLUT4 proteins into the plasma membrane. GLUT4 molecules are synthesized in the rough endoplasmic reticulum, packaged in the membrane of secretory vesicles, and stored in the cytosol until insulin triggers their insertion into the plasma membrane.

In diabetes mellitus, the lack of insulin action means fewer GLUT4 transporters in the plasma membrane of these cells, so glucose uptake is decreased, the cells have less energy resources, and glucose stays in the blood where it can have deleterious effects.

discovered in 1992), 13 classes had been identified in cells as of 2009. The various aquaporins differ in their permeability for water, their selectivity, and their location. Water can also cross cell membranes through ion channels. The forces for movement of water across membranes are described later in this chapter.

Diffusion Through Ion Channels

The mechanism of transport through an ion channel depends on the type of channel. In some ion channels, the pore appears to act simply as a water-filled passageway through which ions move by diffusion. In other ion channels, one or more binding sites for ions exist within the pore. When a channel has numerous binding sites, ions move through the pore by "jumping" from one site to the next. In the absence of an electrochemical gradient, a channel is equally likely to transport ions (or other molecules) in either direction, and no net flux will occur. If an electrochemical gradient is present, however, the net flux through the channel will be down the gradient.

Note the distinction between a channel with binding sites and a carrier: An empty channel's binding sites are accessible from both sides of the membrane *at the same time*. In contrast, for a carrier, the sites are accessible from only one side or the other at any given time.

Factors Affecting the Rate of Transport Through Ion Channels

The rate of ion movement through channels depends on the transport rate of individual channels and the number of channels in the membrane. The rate of individual channels varies depending on the type of channel. For those channels that function primarily as pores, ion movement follows the same basic principles as simple diffusion, except that ions move down an electrochemical gradient, not just a chemical gradient. For ion channels with binding sites, transport is generally slower and the channels can be saturated, similar to what occurs with facilitated diffusion.

Ion channels of both types can be regulated to vary the rate of transport. Specifically, most ion channels can exist in two conformations: a closed state and an open state. In the closed state, ions cannot move through the channel. In the open state, ions diffuse through the channel down their electrochemical gradients. Thus the rate of transport through ion channels depends not so much on the total number of ion channels in the membrane but rather on the number of "open" ion channels.

Quick Check 4

1. What is the difference between a carrier and a channel? How are they similar?
2. As the concentration of a substance increases, the flux of molecules transported by carriers increases at first but eventually levels off. Why?
3. When an uncharged substance such as glucose enters a cell by facilitated diffusion, does it go up or down its concentration gradient?

4 Active Transport 

So far we have focused all our attention on passive transport, but cells perform active transport as well. The importance of active transport to the life of a cell is underscored by the fact that some cells expend large amounts of energy (in some cases, as much as 40% of their total ATP production) to actively transport certain molecules across their membranes. Indeed, many vital physiological processes depend on active transport either directly or indirectly; these processes include the generation of electrical signals in neurons and other excitable cells, the regulation of muscle contraction, the absorption of nutrients and water by the digestive system, and body fluid regulation by the kidneys.

We have seen that active transport requires energy, whereas passive transport does not. The reason for this difference lies in the concept of electrochemical gradients: When an electrochemical driving force is acting on molecules, it effectively exerts a "push" in a certain direction. Molecules will move spontaneously in that direction unless something prevents them from doing so. In other words, transport of a substance *down an electrochemical gradient* requires no energy. In contrast, transport of a substance *up an electrochemical gradient* requires an input of energy because molecules are moving against the electrochemical force that is pushing them.

Whether a substance is being transported actively or passively relates to the following rules:

1. If the direction of the net flux is down an electrochemical gradient, the transport is passive.
2. If the direction of the net flux is up an electrochemical gradient, the transport is active.

The two basic forms of active transport—primary and secondary active transport—differ in the nature of the energy source expended. **Primary active transport** uses ATP or some other chemical energy source *directly* to transport substances. Proteins involved in primary active transport are called **pumps.** Secondary **active transport** is powered by a concentration gradient or an electrochemical gradient that was previously created by primary active transport.

The transport proteins that carry out active transport are similar to carriers in many respects, but possess an ability that carriers do not have: Active transporters can harness energy to drive the transport of molecules in a preferred direction across a membrane. In contrast, carriers show no such preference; in the absence of an electrochemical gradient, a carrier is equally likely to transport molecules in either direction. Recall that two factors affect the binding of a solute to a carrier: affinity and the concentration (or electrochemical) gradient. The difference that allows active transporters, and not carriers, to move a molecule in a certain direction relates to the affinity of the transport protein for the molecule being transported. Whereas carrier proteins have equal affinity for the molecule on either side of the membrane, the affinity of active transporters is greater when the binding site is exposed to one side of the membrane than when the binding site is exposed to the other side.

Assume we start with equal concentrations of a solute on both sides of a membrane, as shown in **Figure 13**a. The transporter in the membrane has a greater affinity for the solute when the binding site is facing side A (it took energy, as described shortly, to induce this greater affinity). The greater affinity causes more solute to bind on side A and be transported to side B. The net transport of solute in one direction produces a concentration gradient across the membrane such that the concentration on side B becomes greater than that on side A. Now two factors are affecting binding—greater affinity on side A (favoring transport to side B) and greater concentration on side B (favoring transport to side A). Once the concentration gradient reaches a certain level, solute will be just as likely to move from B to A as to move from A to B (Figure 13b). At this point, the system has reached a *steady state* (a steady state indicates that concentration will not continue to change, but maintaining this state requires energy). The result of this active transport is that the solute becomes concentrated on one side of the membrane. Later, we will learn about the importance of these concentration gradients in other transport processes and in establishing the resting membrane potential.

Like carriers, active transporters are specific for certain molecules and possess fixed numbers of binding sites. Consequently, active transporters, like carriers, become saturated as the concentration of transported molecules increases.

In the following sections we explore the mechanisms of primary and secondary active transport in greater detail.

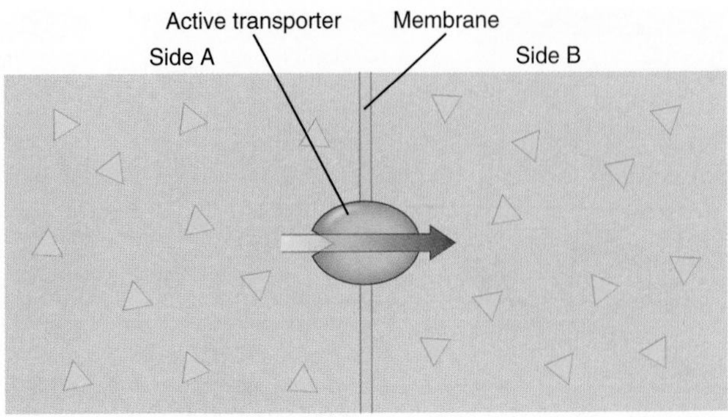

(a) Equal concentration of solute on sides A and B

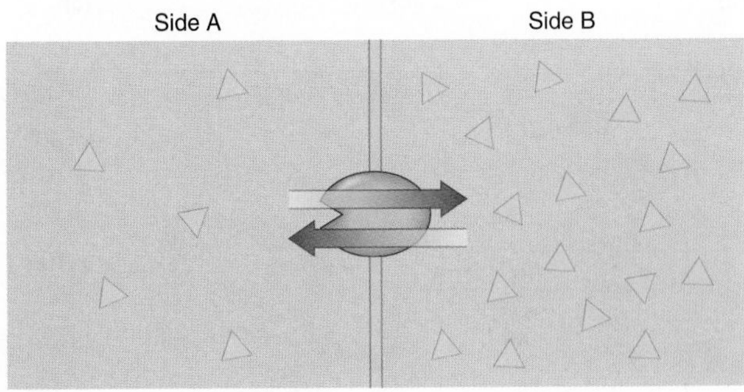

(b) Steady state

Figure 13 Active transport across a membrane. A membrane separates a solute present in equal concentrations on sides A and B. An active transporter is present in the membrane and has a greater affinity for the solute when the binding site faces side A. **(a)** Initially, solute will be transported from side A to side B owing to the greater affinity of the transporter when the binding site faces side A. **(b)** A steady state is reached when solute has an equal likelihood of binding to either side. This occurs when the effect of concentration on the binding of solute offsets the effect of affinity on the binding of solute. The net result of the transporter is to concentrate solute on side B.

Primary Active Transport

The membrane proteins that perform primary active transport function both as transport proteins and as enzymes. In their capacities as enzymes, most of these proteins harness energy from ATP by catalyzing ATP hydrolysis. For this reason, these proteins are frequently referred to as *ATPases.* To understand how primary active transport works, we concentrate on the **sodium-potassium pump** (also called the *Na^+/K^+ pump* or *Na^+/K^+ ATPase*), which is present in nearly every cell and is crucial to several important physiological processes, including electrical signaling in neurons and the absorption of glucose by intestinal epithelial cells. A model explaining the pump's action is depicted in **Figure 14**.

Figure 14 shows that the Na^+/K^+ pump transports Na^+ and K^+ ions in opposite directions across the plasma membrane. For each cycle of the pump, three Na^+ ions are transported out of the cell, and two K^+ ions are transported into the cell. Transport is active in each case because both types of ions move up their electrochemical gradients. In each cycle of the pump, one ATP molecule is hydrolyzed to provide the energy required for this process.

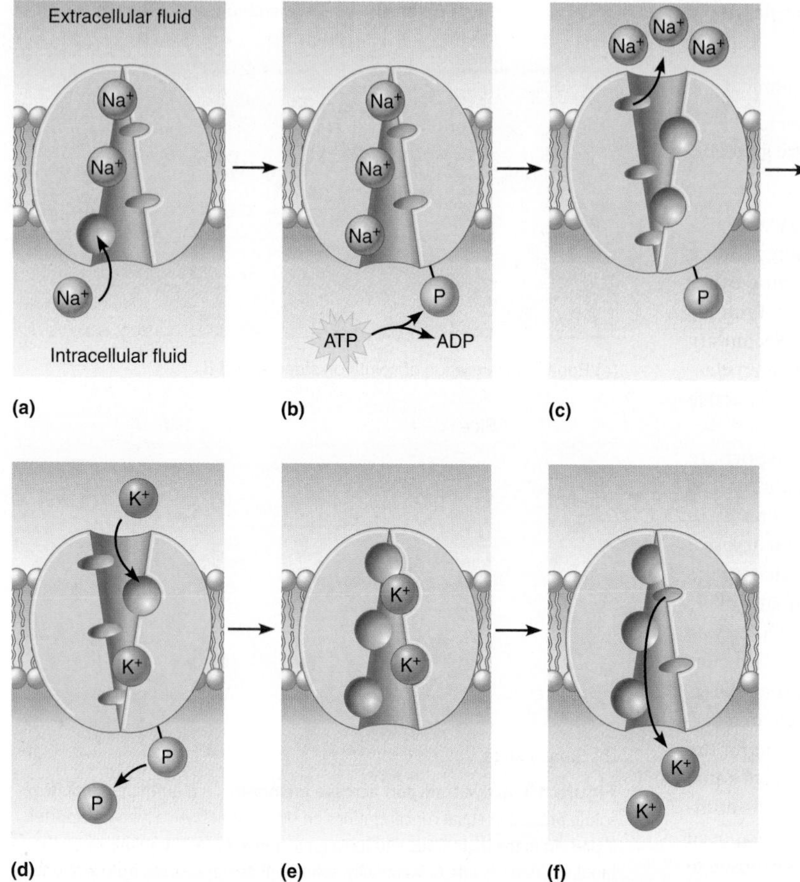

Figure 14 **Primary active transport by the Na^+/K^+ pump.** This pump, which includes three sodium binding sites and two potassium binding sites, uses ATP directly to transport sodium ions out of the cell and potassium ions into the cell up their electrochemical gradients. **(a)** Intracellular Na^+ ions bind to the pump protein. **(b)** The binding of three Na^+ ions triggers phosphorylation of the pump by ATP. **(c)** Phosphorylation induces a conformational change in the protein that allows the release of Na^+ into the extracellular fluid. **(d)** Extracellular K^+ ions bind to the pump protein and trigger release of the phosphate group. **(e)** Loss of the phosphate group allows the protein to return to its original conformation. **(f)** K^+ ions are released to the inside of the cell, and the Na^+ sites become available for binding once again.

Q In Figure 14a, is the affinity of potassium binding sites high or low compared to Figure 14d?

The details of how the Na^+/K^+ pump transports ions in preferred directions are shown in Figure 14. In the first step of the pump cycle, binding sites face inward and are available for Na^+ in the cell to bind to them (see Figure 14a). After three Na^+ ions bind, ATP hydrolysis occurs (see Figure 14b), resulting in phosphorylation of the pump protein. Phosphorylation induces a conformational change in the protein that turns the binding sites outward, so that Na^+ can exit into the extracellular fluid (see Figure 14c). During this conformational change, the binding sites change their shape and decrease their affinity for Na^+, making it very unlikely that Na^+ in the extracellular fluid will bind to the pump protein. For these extracellular Na^+ ions to bind and be carried into the cell would defeat the purpose of the pump. Because the Na^+ binding sites are available on the inside of the membrane but not the outside, the preferred direction of Na^+ transport is out of the cell. In the remaining steps of the pump cycle (see Figure 14d–f), a similar process preferentially transports K^+ into the cell.

The Na^+/K^+ pump is responsible for creating the concentration gradients of Na^+ and K^+ listed in Table 1. Due to continual outward pumping of Na^+ and inward pumping of K^+, intracellular fluid is rich in K^+ but poor in Na^+ relative to extracellular fluid. The concentration gradients of Na^+, K^+, and other ions play a crucial role in generating the membrane potentials essential to transmission of electrical signals in neurons.

The Na^+/K^+ pump is not the only primary active transport system in the cells of your body. For example, pumps in the lining of the stomach transport H^+ out of cells while transporting K^+ into the same cells. These pumps are responsible for the secretion of acid in the stomach—an important component of digestion. Other examples include the Ca^{2+} pumps located in muscle cells, where calcium triggers muscle contraction. Some Ca^{2+} pumps are located on the plasma membrane, where they transport calcium out of the cell. Other Ca^{2+} pumps are located on the smooth endoplasmic reticulum, where they transport calcium from the cytosol into the organelle. In both cases, the transport of calcium out of the cytosol allows the muscle cell to relax.

We can see in Figure 14 that the Na^+/K^+ pump is like an ordinary carrier protein in two respects: It has specific binding sites for Na^+ and K^+, and it transports these ions across the membrane by means of a conformational change. Careful study of the cycle, however, reveals an important difference: An ordinary carrier's binding sites can face either side of the membrane, so that molecules can be transported in both directions (see Figure 10). In the Na^+/K^+ pump, however, affinities of the binding sites for Na^+ or K^+ change as the pump goes through its cycle, such that a binding site's affinity is higher when it faces one side of the membrane and lower when it faces the opposite side. As a consequence, Na^+ or K^+ ions preferentially bind to the pump protein on one side and are released on the other side, which means that Na^+ and K^+ ions are transported in one direction only. As previously mentioned, transport of molecules in a preferred direction is central to the action of any active transporter.

Secondary Active Transport

We know that metabolic reactions can be coupled together, so that an exergonic reaction (glucose oxidation, for example) can be used to drive an endergonic reaction (such as ATP synthesis). In secondary active transport, something similar happens: A transport protein couples the flow of one substance to that of another. One substance moves passively down its electrochemical gradient, in the process releasing energy that is then used to drive the movement of the other substance up its electrochemical gradient.

Figure 15 shows two types of secondary active transport systems in cells. In Figure 15a, the two transported substances are moving in the same direction (in this case, inward across the cell membrane); in Figure 15b, they are moving in opposite directions. The transport of two substances in the same direction is called **cotransport** (or sometimes *symport*). The example of cotransport shown in Figure 15a is *sodium-linked glucose transport,* which couples the inward flow of Na^+ with the inward flow of glucose molecules. In this process, Na^+ moves down its electrochemical gradient and releases energy that then drives the flow of glucose against its concentration gradient. Specifically, Na^+ increases the affinity of the carrier for glucose when the binding site for glucose faces the extracellular fluid. The transport of two substances in opposite directions is called **countertransport** (or sometimes *antiport* or *exchange*). The example of countertransport shown in Figure 15b is *sodium-proton exchange,* in which the inward flow of Na^+ is coupled to the outward flow of protons (H^+). Here, energy released from the flow of Na^+ down its electrochemical gradient is harnessed to drive the flow of H^+ up its electrochemical gradient. Specifically, Na^+ increases the affinity of the carrier for H^+ when the binding site for H^+ is facing the intracellular fluid.

These examples illustrate that secondary active transport may involve either cotransport or countertransport, and that the actively transported substance may move into cells in some cases and out of cells in others. In addition, many secondary active transporters utilize the movement of Na^+ down its electrochemical gradient and into the cell to drive the movement of another ion or molecule against its electrochemical gradient. Note that the direction of transport *alone* does not tell us anything about which substance is being transported actively or which is flowing passively; this status is determined entirely by the directions of the transported substances' electrochemical gradients. The substance that is moving up its electrochemical gradient is the one that is being actively transported, and thus the other substance is moving down its electrochemical gradient and is being passively transported. Energy released from the passive movement of this substance down its gradient is used to drive the flow of the other substance up its gradient.

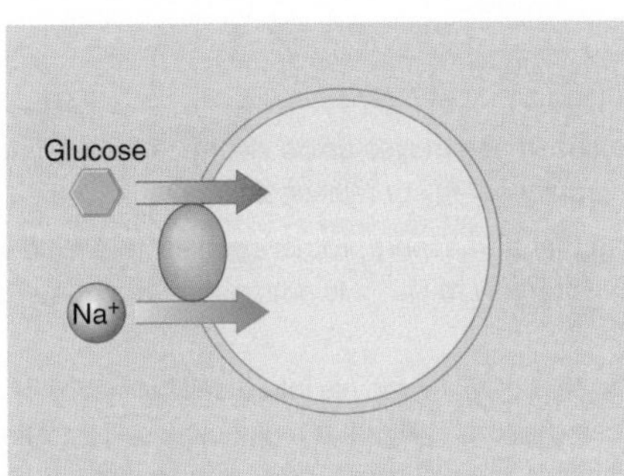

(a) Cotransport

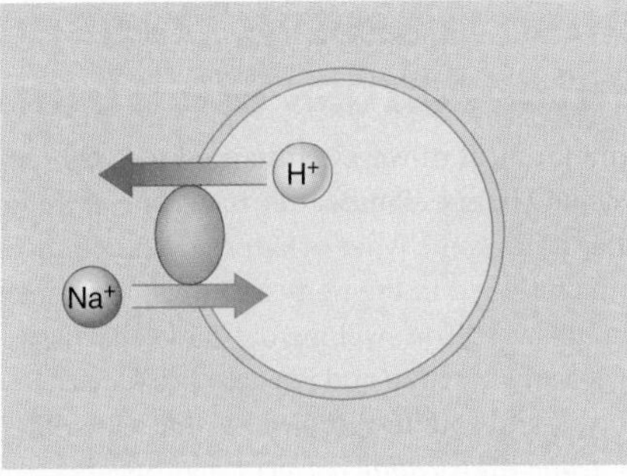

(b) Countertransport

Figure 15 Secondary active transport. Ovals spanning the cell membrane represent transport proteins. **(a)** Sodium-linked glucose transport, an example of cotransport. **(b)** Sodium-proton exchange, an example of countertransport.

Apply Your Knowledge

Compare allosteric and covalent regulation of enzymes to primary and secondary active transport.

Factors Affecting Rates of Active Transport

Because active transport is carried out by pumps, two factors are the sole determinants of the rate at which molecules are actively transported across any membrane: the rate of transport by individual active transporters, and the number of active transporters that are present in the membrane. As either of these variables increases, the rate of transport increases.

The transport rate of an individual active transporter depends on the nature of the transporter. In addition, it is influenced by factors such as the concentrations of the transported substance on either side of the membrane, the size of the electrochemical driving force for that substance, and other variables that affect the conditions under which a transporter must operate. These variables affect different active transporters in different ways, and a more thorough discussion of this topic is beyond the scope of this text.

Table 2 summarizes the major characteristics of the different mechanisms to transport small solutes across cell membranes.

Coexistence of Active and Passive Transport Mechanisms in Cells

We have seen that intracellular and extracellular fluids differ substantially in composition (see Table 1). Now it should be apparent that these differences in composition are created when certain substances are actively transported into or out of cells, increasing or decreasing (respectively) the concentrations of these substances in intracellular fluid with respect to extracellular fluid.

Normally, the composition of intracellular fluid remains fairly steady, which raises a question: If substances are actively transported into or out of cells, why don't intracellular concentrations change? The answer lies in the fact that these substances are simultaneously transported passively (leaked) across the membrane in

TABLE 2 Characteristics of Transport Processes

	SIMPLE DIFFUSION	MEDIATED TRANSPORT			
		Passive transport		Active transport	
		Channel	Facilitated diffusion	Primary	Secondary
Direction of net flux	Down electrochemical gradient	Down electrochemical gradient	Down electrochemical gradient	Up electrochemical gradient	Up electrochemical gradient
Transport protein required?	No	Yes, ion channel	Yes, carrier	Yes, pump	Yes, pump
Requires energy?	No	No	No	Yes	Yes
Energy source	(Not applicable)	(Not applicable)	(Not applicable)	ATP or other chemical energy source	Electrochemical gradient of another solute
Saturation?	No	Sometimes	Yes	Yes	Yes
Specificity?	No	Yes	Yes	Yes	Yes
Character of transported substance	Hydrophobic (nonpolar)	Hydrophilic (ionized or polar)	Hydrophilic (ionized or polar)	Hydrophilic (ionized or polar)	Hydrophilic (ionized or polar)
Examples	Fatty acids, O_2, CO_2	Inorganic ions (Na^+, K^+, Cl^-, Ca^{2+})	Organic molecules (glucose)	Inorganic ions (Na^+, K^+, H^+, Ca^{2+})	Organic molecules and inorganic ions (glucose, amino acids, H^+, Ca^{2+})

the opposite direction but at the same rate, such that the net flux across the membrane (active and passive transport combined) is zero (**Figure 16**). Thus, even though Na^+ ions, for example, are actively transported out of cells by the Na^+/K^+ pump, they leak passively into cells through channels and other proteins. For every Na^+ ion that is transported out under normal conditions, another leaks in passively, such that the intracellular Na^+ concentration does not change. Likewise, for every K^+ ion that is actively transported in, another passively leaks out, such that the intracellular K^+ concentration does not change.

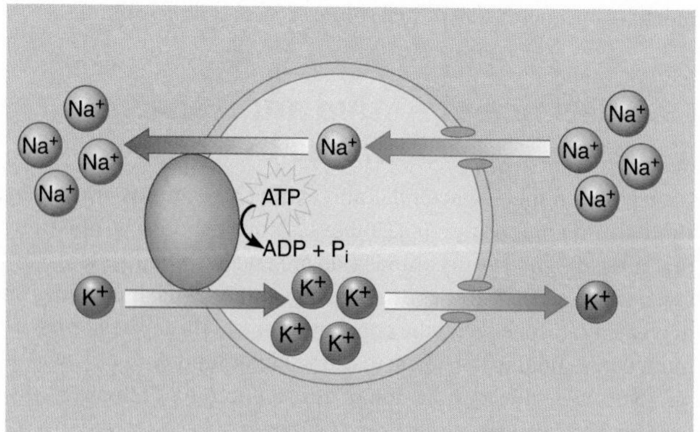

Figure 16 Pumps and leaks in a cell. Active transport of Na^+ and K^+ by the Na^+/K^+ pump (oval at left) is counteracted by passive leaks of these ions through channels (at right).

Quick Check 5

1. When a substance moves up its electrochemical gradient, does it require energy or release energy?
2. Suppose that a transport protein couples inward flow of Na^+ with outward flow of Ca^{2+}. Is cotransport or countertransport occurring?
3. Suppose that glucose moves into a cell by means of sodium-linked cotransport, and that the glucose concentrations are 2.5 mM inside the cell and 1.0 mM outside the cell. Is sodium being transported actively or passively?

5 Osmosis: Passive Transport of Water Across Membranes

Normally, the amount of water contained in cells remains fairly steady, but it can change significantly in certain extreme or pathological conditions. In some types of kidney dysfunction, for instance, excess water is retained in the body and begins to move into cells, causing them to swell. The swelling of brain cells disrupts nervous system function and typically produces headache, nausea, and vomiting, indicating that a condition known as *water intoxication* has occurred. Severe cases can progress to seizures, coma, and death. In contrast, severe dehydration can cause water to move out of cells, such that the cells then shrink. This effect is also detrimental to brain function, and can even be fatal.

Under normal conditions, most cells in the body neither swell nor shrink because there is no net movement of water across membranes; forces that might cause water to cross membranes are absent. In some locations in the body, however, movement of water across cell membranes is a normal occurrence. For glands to secrete fluids such as sweat, tears, or saliva, for instance, water must be transported across specialized epithelial cells. Also, when you drink water, it moves across the epithelial cells lining the intestine to be absorbed into the bloodstream. Such water movement is important because it ultimately affects the volume and composition of all body fluids.

Water transport has always had a "special" status in physiology because of its influence on cell volume, and because its description has its own unique terminology. In reality, water transport is no different from some other types of transport we have already studied; indeed, it is simpler in many respects. Water transport is simple because water flow across membranes is *always* passive, is unaffected by membrane potentials, and is always driven by its own concentration gradient. The flow of water across a membrane down its concentration gradient is called **osmosis.** (Technically, osmosis is the flow of any solvent across a semipermeable membrane in response to a difference in chemical potential across the membrane.)

The effects of osmosis are familiar to nearly everyone who has studied high school biology. When you place red blood cells in pure water, water flows into the cells, which then swell up, burst, and release hemoglobin into the water, a process called *hemolysis.*

Figure 17 depicts the conditions that drive osmosis. As shown in Figure 17a, when a red blood cell is placed in pure water, water molecules flow into the cell because in doing so they are flowing down their concentration gradient. This gradient results from the presence within the cell of various solutes (Na^+, K^+, and proteins, among others) at a combined concentration of about 300 millimolar. The presence of these solute molecules reduces the water concentration inside the cell by taking up a certain amount of space that would have been occupied by water molecules, which means that fewer water molecules are present in a given volume of solution. In other words, the *concentration of water within the cell is lower than outside the cell.*

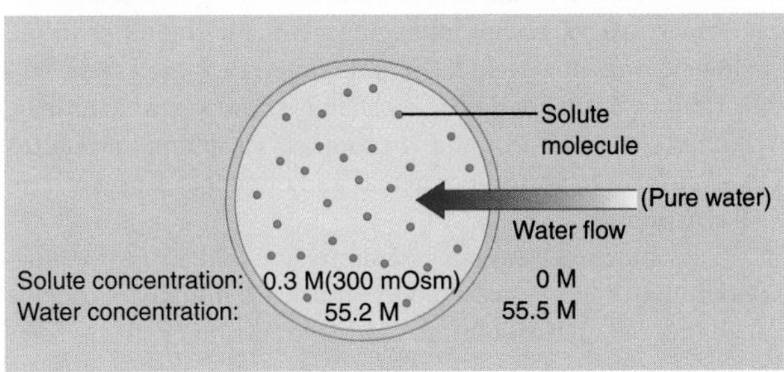

(a)

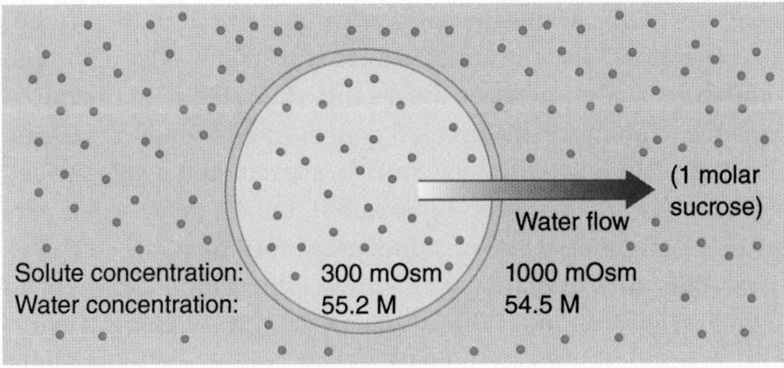

(b)

Figure 17 **Osmosis. (a)** Flow of water into a red blood cell placed in pure water. Water molecules flow passively into the cell down their concentration gradient, thereby moving to an area of higher solute concentration. **(b)** Flow of water out of a red blood cell placed in 1 M sucrose. Note that water again moves from higher to lower water concentration, and from lower to higher solute concentration.

At body temperature, the concentration of pure water is 55.5 molar, and the normal total solute concentration in intracellular fluid is 300 millimolar (0.3 molar). Thus the intracellular water concentration is roughly 55.2 molar (55.5 molar – 0.3 molar; see Figure 17a). Therefore, when a cell is placed in pure water, there is a 0.3 molar smaller water concentration inside it than outside it; as a consequence, water molecules move passively into the cell down their concentration gradient, causing the cell to swell and eventually burst.

The direction of passive water flow into or out of a cell depends on the direction of the water concentration gradient across the plasma membrane. For instance, when a cell is placed in a solution of sucrose at a concentration of 1 molar (1000 millimolar), the cell shrinks because water moves out (Figure 17b). Outward water movement occurs because the total solute concentration is lower inside the cell than it is outside, which makes the water concentration *higher* inside the cell. Therefore, water moves passively out of the cell.

Concentrations of solutions are usually described in terms of the concentration of solute, not solvent. Concentrations of solute and solvent vary inversely. Thus, in osmosis, water movement occurs from a low concentration of solute to a high concentration of solute. The term *solute particle* is often used to reflect that some solutes separate into more than one particle when placed in a solution; each of these resultant particles will have an osmotic influence on movement of the solvent.

Osmolarity

The total solute particle concentration of a solution is known as its **osmolarity.** A solution containing 1 mole of solute particles is said to be at a concentration of 1 *osmolar* (1 Osm). One mole of solute particles is referred to as 1 *osmole.* We use the term *solute particle* to distinguish *osmolar* from *molar* because certain solutes when put into solution dissociate into particles, and each solute particle present decreases the concentration of water. For example, 1 liter of a solution containing 0.1 mole of glucose has an osmolarity of 0.1 osmolar because glucose does not dissociate in solution. However, a 1-liter solution containing 0.1 mole of NaCl has an osmolarity of approximately 0.2 osmolar because NaCl dissociates in solution to give approximately 0.1 mole of sodium ions and 0.1 mole of chloride ions, for a total of 0.2 mole of solute particles. Osmolarity also depends on the total concentration of *all* solute particles. Thus a 1-liter solution containing 0.1 mole of glucose plus 0.1 molar NaCl is a 0.3 osmolar solution.

In physiology, the terms *milliosmole* and *milliosmolar* (mOsm) are frequently used because the solute concentration of body fluids is only a fraction of 1 osmolar. A 1 milliosmolar (1 mOsm) solution contains 1 milliosmole (1/1000 of an osmole) of solute particles per liter. The normal osmolarity of intracellular and extracellular fluid is approximately 300 mOsm (the actual range is 280–296 mOsm), which means that its total solute concentration is 300 milliosmoles per liter.

Two solutions having the same osmolarity are said to be **iso-osmotic.** Thus a 300 millimolar glucose solution is iso-osmotic with intracellular fluid because both solutions are 300 milliosmolar. A solution whose osmolarity is higher than another is said to be **hyperosmotic;** a solution with lower osmolarity is **hypo-osmotic.** When two solutions are iso-osmotic, they have not only the same solute concentration but also the same water concentration. In a hyperosmotic solution, the water concentration is lower because the solute concentration is higher. In a hypo-osmotic solution, the water concentration is higher because the solute concentration is lower.

Apply Your Knowledge

For each of the following solutions, calculate the osmolarity and determine whether the solution is iso-osmotic, hypo-osmotic, or hyperosomotic relative to body fluids: 200 mM solution of KCl, 100 mM solution of $MgCl_2$, and a solution containing 50 mmoles of $CaCl_2$ in 250 mL water.

Osmotic Pressure

Osmotic pressure is another term that describes a solution's total solute concentration; it is often used when comparing other physiological forces described in terms of pressure (**Toolbox: Determining the Osmotic Pressure of a Solution**). The osmotic pressure of a solution (which is indicated by the symbol π) is an indirect measure of its solute concentration and is expressed in ordinary units of pressure, such as atmospheres or millimeters of mercury (mm Hg). As the total solute concentration (osmolarity) increases, the osmotic pressure increases. Therefore, when water moves by osmosis from low solute concentration to high solute concentration, water is also flowing *up* an osmotic pressure gradient (**Figure 18**). In Figure 18, the total solute concentration on the left side of the semipermeable membrane (side 1) is less than that on the right (side 2), making the osmotic pressure on side 2 higher. Thus, as water moves from side 1 (low solute concentration) to side 2 (high solute concentration), water is flowing from lower to higher osmotic pressure, or up the osmotic pressure gradient ($\Delta\pi$).

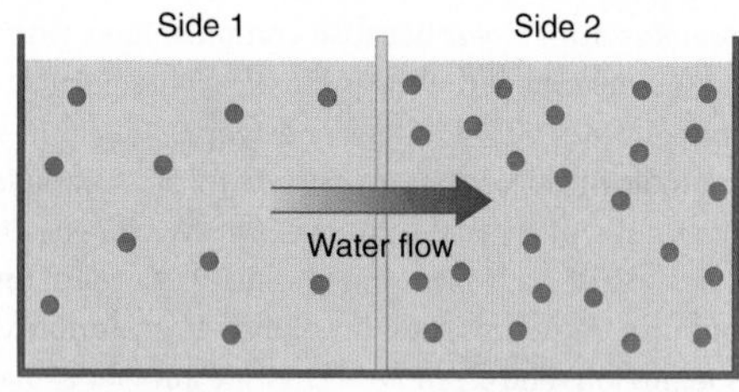

Figure 18 Movement of water up an osmotic pressure gradient. When a semipermeable membrane—which is permeable to water but not to solute—separates two solutions in a chamber, water flows down its concentration gradient toward the side with higher solute concentration. Because osmotic pressure increases as solute concentration increases, water also flows from lower to higher osmotic pressure, or up the osmotic pressure gradient ($\Delta\pi$).

Note that the tendency of water to flow *up* an osmotic pressure gradient does not violate the rules concerning gradients, which state that substances tend to move passively *down* chemical and electrochemical gradients. When water moves up a gradient of osmotic pressure, it is merely moving down a chemical gradient—its own concentration gradient.

Tonicity

When someone suffers massive hemorrhage, it is quite common for emergency medical personnel to administer a saline (sodium chloride) solution intravenously to replace the lost blood volume. This procedure is done to keep the person alive while awaiting a blood transfusion. The administered solution is referred to as *isotonic saline* because its total solute concentration has been carefully formulated to match that of extracellular fluid. Such a solution will not alter cell volumes when injected into the bloodstream. In contrast, a more concentrated saline solution would cause water to flow out of cells, making them shrink; a less concentrated solution would cause water to flow into cells, making them swell.

Whereas a solution's osmolarity is based solely on its total solute concentration, its **tonicity** is a function of the concentration of nonpermeating solutes outside a cell relative to the concentration inside the cell, and it determines the behavior of a cell placed in the solution. A solution is said to be **isotonic** when it does not alter cell volume; when a cell comes into contact with an isotonic solution, it neither shrinks nor swells. (This point refers to the cell's final volume. Under certain conditions a cell may shrink or swell initially, even if the solution is isotonic.) In contrast, a solution that causes cells to shrink is **hypertonic,** whereas a solution that causes cells to swell is **hypotonic**.

The distinction between osmolarity and tonicity is best illustrated in the situation depicted in **Figure 19**. Initially, the cell is placed in a 300 mOsm solution of *urea,* a substance that permeates most cell membranes readily (Figure 19a). The osmolarity of this solution equals the initial osmolarity inside the cell (300 mOsm), making it *iso-osmotic* with the intracellular fluid. Under these conditions, water does not move into or out of the cell because the water concentration is the same on both sides of the cell membrane. Over time, urea moves into the cell due to the initial presence of an inwardly directed concentration gradient (Figure 19b). Intracellular solutes, in contrast, are retained inside the cell because they are relatively *impermeant;* that is, they cannot permeate the membrane. As a result of urea movement, the cell gains solute, and the osmolarity of the intracellular fluid increases so that it exceeds its initial value of 300 mOsm. (The volume of extracellular solution is taken to be very large, so extracellular osmolarity is not affected by solute movement into the cell.) Because the total solute concentration is now higher inside the cell than outside, water moves into the cell, and the cell swells. The cell swells even though the extracellular solution is originally iso-osmotic to the intracellular fluid. Because the cell swells, the extracellular solution is *hypotonic.*

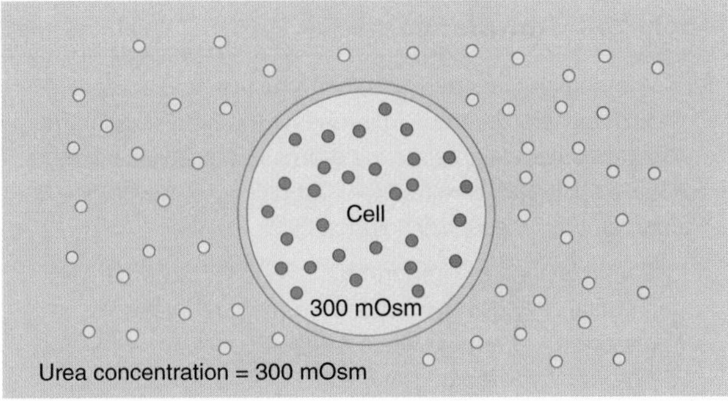

(a)

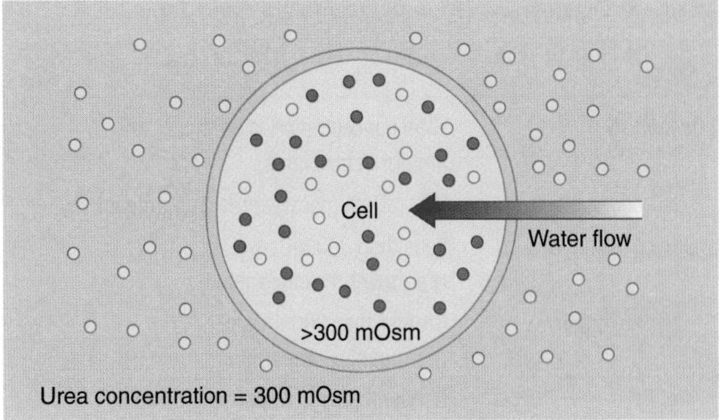

(b)

Figure 19 The distinction between osmolarity and tonicity. (a) A cell containing impermeant solutes (purple dots) at a concentration of 300 mOsm placed in an iso-osmotic solution containing the permeant solute urea (yellow dots) at a concentration of 300 mOsm. **(b)** Swelling that occurs over time as urea moves into the cell down its concentration gradient. Because of the resultant increase in the solute concentration inside the cell, water then flows into the cell down its concentration gradient, indicating that the extracellular solution is hypotonic.

Whether a given solution causes cells to swell, shrink, or stay the same is determined solely by the concentration of impermeant solutes contained in it. A solution is isotonic if it contains impermeant solutes at a concentration of 300 mOsm, the normal concentration of impermeant solutes in intracellular fluid. If the concentration of impermeant solutes is either greater or less than 300 mOsm, the solution will be hypertonic or hypotonic, respectively. A solution's tonicity is not affected by the concentration of any permeant solutes that may or may not be present. The characteristics that determine a solution's osmolarity or tonicity are summarized in Table 3.

When a cell comes into contact with solutions that are hypotonic or hypertonic, the degree to which it swells or shrinks is

TOOLBOX

Determining the Osmotic Pressure of a Solution

It might seem strange that physiologists would use the term *osmotic pressure* to describe a solution's total solute concentration. However, solute concentration does, indeed, exert a pressure that can be countered by other kinds of pressure, such as atmospheric pressure or a piston.

Osmotic pressure can be determined only operationally—that is, by experimental measurement of the pressure required to counter it. Two compartments in a chamber are separated by a *semipermeable* membrane, which is permeable to water but not to any solute. One compartment contains pure water, while the other compartment contains the *test solution* having an unknown osmotic pressure. The compartment containing the test solution is equipped with a piston that allows us to apply pressure to that side. When no pressure is applied, water flows down its concentration gradient into the compartment with the test solution. We can slow this flow of water by applying pressure to the test solution. If we increase the pressure to a sufficient degree, the water flow stops completely. At this point, the applied pressure is equal to the osmotic pressure of the test solution.

As a result of measurements of the type just described, scientists were able to demonstrate that the following three rules apply to the osmotic pressure of a solution:

1. Osmotic pressure increases as the concentration of nonpermeating solute rises.
2. As a general rule, osmotic pressure depends on the total nonpermeating solute concentration rather than the molecular identities of the solutes.
3. If a solution contains solute molecules that ionize into separate particles (such as sodium chloride, NaCl, which dissociates into Na^+ and Cl^- ions), the osmotic pressure depends on the total concentration of particles. Thus a 1 molar solution of NaCl, which contains 1 mole per liter of Na^+ ions and 1 mole per liter of Cl^- ions, has twice the osmotic pressure of a 1 molar solution of glucose, which does not ionize.

The osmotic pressure of a solution is given by the following equation:

$$\pi = CRT$$

where C is the total solute concentration, R is the universal gas constant (0.082 liter-atm/mole-K), and T is the absolute temperature (K). Using this equation, it can be shown that at body temperature, the osmotic pressure of a 1 osmolar solution is surprisingly large—a full 24.6 atmospheres, which is equivalent to more than 18,000 mm Hg!

TABLE 3 Distinctions Between Osmolarity and Tonicity

Terms	Definitions and solute concentrations
Osmolarity	Total concentration of permeant and impermeant solutes
Iso-osmotic*	300 mOsm (permeant + impermeant)
Hypo-osmotic*	Less than 300 mOsm (permeant + impermeant)
Hyperosmotic*	Greater than 300 mOsm (permeant + impermeant)
Tonicity	Concentration of impermeant solutes relative to intracellular fluid
Isotonic*	300 mOsm (impermeant)†
Hypotonic*	Less than 300 mOsm (impermeant)†
Hypertonic*	Greater than 300 mOsm (impermeant)†

mOsm = milliosmolar.
*These designations are relative to a cell containing 300 mOsm solutes, which are assumed to be impermeant.
†Permeant solutes may or may not be present.

determined by the initial concentrations of impermeant solutes in intracellular and extracellular fluid. For example, when a cell containing 300 mOsm impermeant solutes is placed in a large volume of solution containing impermeant solutes at half that concentration (150 mOsm), the cell will swell because water flows in (**Figure 20**a). Because the amount of solute inside the cell is fixed, the increase in cell volume causes the intracellular solute concentration to decrease, but the cell continues to swell until the concentration reaches 150 mM, at which point the cell has expanded to twice its original volume (designated V_o). Swelling stops at this point because there is no longer a concentration gradient of water, and hence no driving force to cause water to flow across the membrane. (In Figure 17a, the red blood cell continues to swell until it bursts because the solute concentration is always higher inside the cell than outside.) By contrast, when an identical cell is placed in a solution containing 600 mOsm impermeant solutes (Figure 20b), it shrinks, which causes the intracellular solute concentration to rise. The cell continues to shrink until the concentration inside the cell reaches 600 mOsm, at which point it has shrunk to half of its original volume. See what happens to the osmolarity of body fluids during exercise in **Understanding Exercise: Sweat Production**.

Apply Your Knowledge

1. Common solutions used clinically include 0.9% saline (NaCl) and 5% dextrose (glucose; $C_6H_{12}O_6$). Following are the molecular weights for the atoms in these molecules: Na = 23, Cl = 35, C = 12, H = 1, and O = 16. Determine the osmolarity of these two solutions.
2. Cell membranes are impermeable to ions, but dextrose moves across cells by mediated transport. Using this information, determine whether each solution in question 1 is isotonic, hypotonic, or hypertonic. Describe clinical situations during which 0.9% saline or 5% dextrose would be administered to a patient.

6 Transport of Material Within Membrane-Bound Compartments

Macromolecules are too large to cross the plasma membrane even with the assistance of proteins; thus they are transported across with the assistance of vesicles. During **endocytosis,** molecules in the extracellular fluid enter the cell through the formation of vesicles, called *endosomes,* from the plasma membrane. During **exocytosis,** molecules within cells are packaged into **secretory vesicles,** which then fuse with the plasma membrane and release their contents into the extracellular fluid. Both of these processes, which require the input of energy, are described next.

Transport of Molecules into Cells by Endocytosis

There are three forms of endocytosis: *phagocytosis, pinocytosis,* and *receptor-mediated endocytosis* (**Figure 21**). In all cases, extracellular fluid and sometimes particulate matter are brought into the cell by the formation of an endosome.

In **phagocytosis** (which means "cell-eating"), a cell uses amoeboid-like movements of its plasma membrane to extend the membrane around particulate matter in the extracellular fluid (see Figure 21a). When the membrane completely surrounds the particle, the two sides of the plasma membrane pinch together to form a large endosome called a *phagosome* in the cytoplasm; the particle and some extracellular fluid are inside the vesicle. In this manner,

FOCUS ON DIABETES

Tonicity is normally not affected by blood glucose levels because glucose permeates the membranes of most cells by mediated transport. In diabetes mellitus, however, insulin-dependent glucose transport is impaired or absent, such that glucose becomes a nonpermeating solute. Hyperglycemia creates a hyperosmotic and hypertonic extracellular fluid, which causes body cells to shrink. This is only one of many reasons why regulating blood glucose is crucial in the management of diabetes.

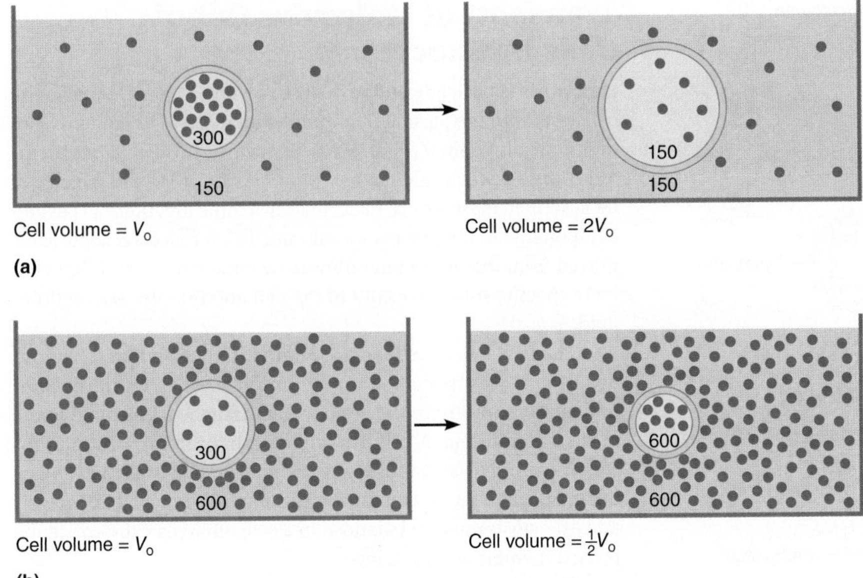

Figure 20 Changes in the volumes of cells placed in hypotonic and hypertonic solutions. All solutes (dots) are assumed to be impermeant; numbers inside and outside cells indicate solute concentrations (mOsm). **(a)** A cell placed in a 150 mOsm solution, which is hypotonic, swells until its volume reaches twice the original volume (V_0). **(b)** A cell placed in a 600 mOsm solution, which is hypertonic, shrinks until its volume is reduced to half of the original volume.

Q If a cell were placed in a solution containing 100 mOsm impermeant solutes, how would its final volume compare to its initial volume?

the cell *engulfs* the particle. Once inside the cell, the membrane of the phagosome fuses with the membrane of a lysosome, forming a *phagolysosome,* which exposes the engulfed particle to the degradative enzymes of the lysosome. The enzymes break down the particle and usable components are recycled. Phagocytosis is common in certain *white blood cells,* which are responsible for removing foreign particles and bacteria from our bodies.

In **pinocytosis** (which means "cell-drinking"), the plasma membrane develops an indentation, and its outer edges pinch together to form an endosome in the cytoplasm (see Figure 21b). Pinocytosis, which occurs continuously in most cells, is a nonspecific process, and the contents of the resulting vesicle comprise extracellular fluid containing dissolved solutes.

Receptor-mediated endocytosis is similar to pinocytosis in that the plasma membrane develops an indentation to form the endosome (see Figure 21c). Unlike pinocytosis, however, receptor-mediated endocytosis is specific. Proteins in the plasma membrane function as *receptors* that recognize and bind specific particles in the extracellular fluid. Binding of particles to receptors concentrates the particles to areas where endocytosis will occur. The area of plasma membrane that forms the vesicle is coated with proteins, called **clathrin,** on its cytosolic surface. The membrane indents in this area, forming what is called a **coated pit.** The coated pit becomes a coated vesicle containing the receptors and the particles bound to them. The protein coat rapidly leaves the vesicle, and the clathrin molecules are recycled. The now uncoated vesicle fuses with a lysosome, forming an *endolysosome.* The enzymes in the lysosome will degrade the particles brought into the cell. The receptors are often recycled by exocytosis, which is described next.

UNDERSTANDING EXERCISE

Sweat Production

We all know that our sweat production increases when we exercise: The harder we exercise, the more we sweat. Sweating is a mechanism for dissipating heat produced by the body (body heat is used to evaporate the sweat), and exercising muscle generates a lot of heat. At the same time, excessive sweating can potentially be dangerous.

Sweat is a fluid, composed of water and solutes (especially sodium and chloride). It is derived from blood plasma, but is less concentrated than plasma. As a result, excessive sweating can increase the osmolarity of plasma, causing cells to shrink. The small amount of cell shrinkage that occurs during exercise is usually not problematic. Conversely, drinking too much fluid after exercise can be dangerous: Drinking too much pure water after exercise can cause water toxicity. An excess of water can decrease plasma osmolarity, causing cells to swell. This effect can lead to neurological problems, such as a headache, because brain cells are contained in a fixed space (within the skull), and swelling of brain cells can increase the pressure exerted on the cells, causing erroneous brain signaling. Too much water can even be deadly, as was sadly reported when a Sacramento, California, radio station held a contest called "Hold your wee for Wii." Contestants competed to see who could drink the most without urinating; the winner would receive a Nintendo Wii player. Unfortunately, one contestant died five hours after the contest from water intoxication.

Sport drinks are popular with both professional and amateur athletes. These drinks contain electrolytes and glucose, so drinking them can maintain normal plasma osmolarity and provide additional nutrients. There is debate about the benefit of nutrients in these drinks, however, because consuming glucose defeats the purpose of exercise for many of us—depletion of our stored energy reserves.

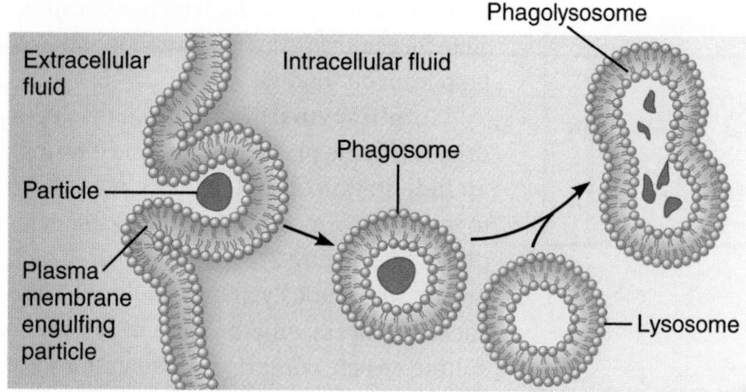

(a) Phagocytosis

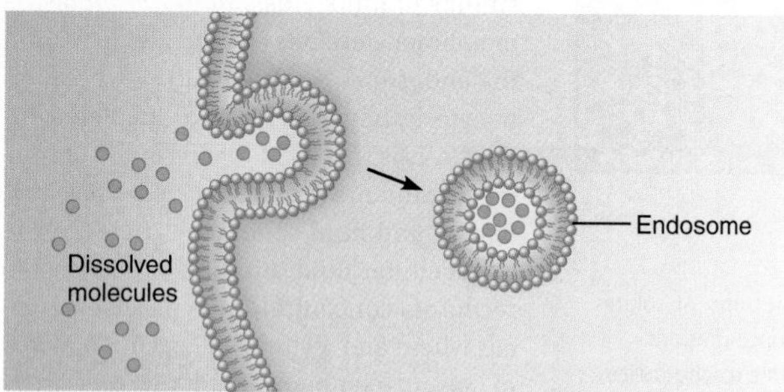

(b) Pinocytosis

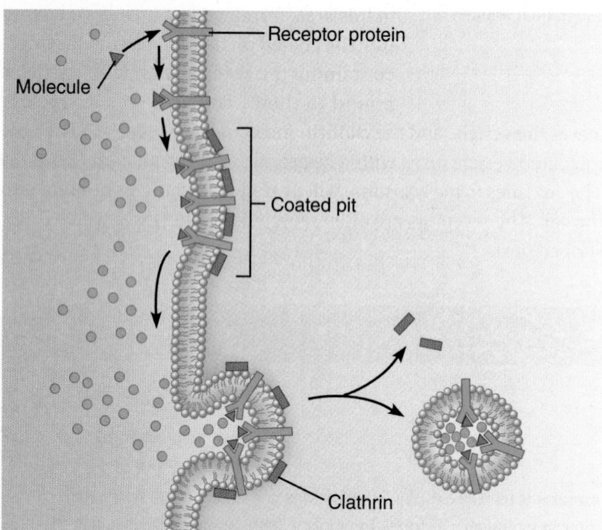

(c) Receptor-mediated endocytosis

Figure 21 Endocytosis. (a) Phagocytosis. After the plasma membrane surrounds a particle in the extracellular fluid, the membrane pinches together to form a phagosome around the particle, which then usually fuses with a lysosome. The enzymes of the lysosome degrade the particle. (b) Pinocytosis. The plasma membrane develops an indentation to form an endosome, containing extracellular fluid and dissolved solutes. (c) Receptor-mediated endocytosis. First, receptor proteins on the plasma membrane bind specific molecules, triggering endocytosis. The plasma membrane then becomes indented around the molecules, forming a vesicle that pinches off of the plasma membrane and enters the cell.

Transport of Molecules Out of Cells by Exocytosis

Exocytosis is basically endocytosis in reverse: A vesicle inside the cell fuses with the plasma membrane and releases its contents into the extracellular fluid (**Figure 22**). Exocytosis involves complex interactions between calcium and proteins, the process of which will be described for neurons. Exocytosis has three functions: (1) to add components to the plasma membrane, (2) to recycle receptors removed from the plasma membrane by endocytosis, and (3) to secrete specific substances out of the cell and into the extracellular fluid.

The first two functions are related in that both add components to the plasma membrane. During exocytosis, whatever components are present in the vesicle membrane will be added to the plasma membrane. A cell can add certain proteins, phospholipids, or carbohydrates to the plasma membrane, or it can replace the membrane that is lost during endocytosis. In fact, endocytosis and exocytosis must be balanced in a cell; otherwise, the size of the plasma membrane will change.

The third function of exocytosis, the secretion of materials, serves a number of functions. Certain white blood cells secrete antibodies to fight infections. Most cells—in particular, neurons and endocrine cells—secrete chemical messengers that communicate with other cells. Cells lining certain hollow ducts or passageways, such as the gastrointestinal tract or respiratory airways, secrete a sticky fluid called mucus, which acts as a protective coating.

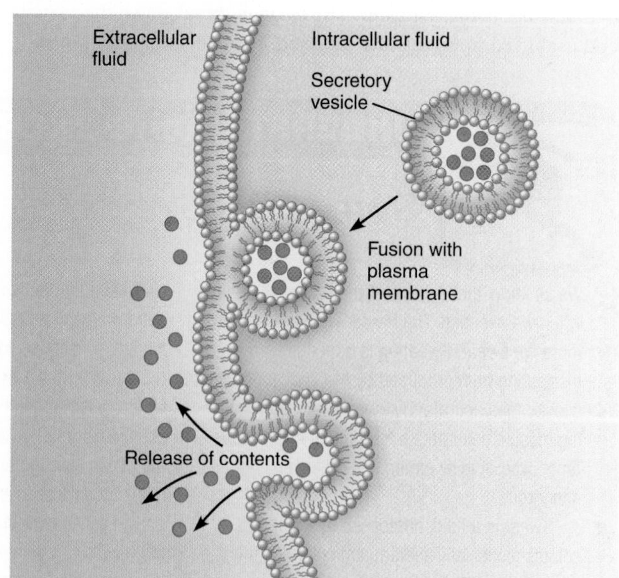

Figure 22 Exocytosis. An intracellular vesicle fuses with the plasma membrane, and the contents of the vesicle are secreted from the cell.

Quick Check 6

1. Name the three types of endocytosis. Which type(s) transport(s) *specific* molecules across the plasma membrane?
2. What are the three functions of exocytosis?

7 Epithelial Transport: Movement of Molecules Across Two Membranes

Up to this point, we have been focusing on cell membrane function as it pertains to the transport of materials into or out of cells. In many epithelial tissues, however, cell membranes function to transport materials *across* cells. This occurs, for instance, when the intestine delivers nutrients to the bloodstream, or when sweat glands produce sweat.

Recall that one function of epithelial tissues is to form barriers between the body's internal environment and the external environment, as well as between different fluid compartments within the body. In addition to acting as barriers, certain epithelia (such as those lining the stomach, intestine, and secretory glands) are able to transport materials into the internal environment from outside (a process called **absorption**) or from the internal environment to the outside (**secretion**).

For an epithelium to absorb or secrete materials, the cells that make up the epithelial tissue must transport substances inward across the membrane on one side of the cell and outward across the membrane on the opposite side of the cell. For this to occur, the membranes on either side must possess different transport systems. Because the membranes on the two sides are distinctly different (in both structure and function), epithelial cells are said to be *polarized.* In this section, we see how the polarity of epithelial cells gives them the ability to absorb or secrete materials. We begin with an overview of epithelial structure.

Epithelial Structure

In an epithelial cell layer of the type that is specialized for absorption or secretion (**Figure 23**), one side of an epithelial cell faces the lumen of a body cavity. The membrane on this side is called the **apical membrane.** The membrane on the opposite side faces the internal environment and is in contact with interstitial fluid, which exchanges materials with the blood. This *blood-facing* membrane or **basolateral membrane** rests on a **basement membrane** consisting of noncellular material that is relatively permeable to most substances. The basement membrane anchors the basolateral membrane and provides physical support for the epithelial layer.

An important feature of epithelial tissues is that adjacent cells are joined by tight junctions that limit the passage of material through the spaces between cells called *paracellular* spaces. (Movement of molecules between the cells is called *paracellular transport.*) These junctions permit the fluids on either side of the cell layer to differ in composition. Tight junctions are important in maintaining homeostasis because even though the composition of fluid in the lumen of an organ may vary widely (consider the variety of foodstuffs that enter the lumen of the stomach), the composition of interstitial fluid should remain consistent because it is part of the body's highly regulated internal environment. The "tightness" of these junctions varies from location to location; *tight epithelia* have junctions with extremely low permeabilities, whereas *leaky epithelia* have junctions that are more permeable.

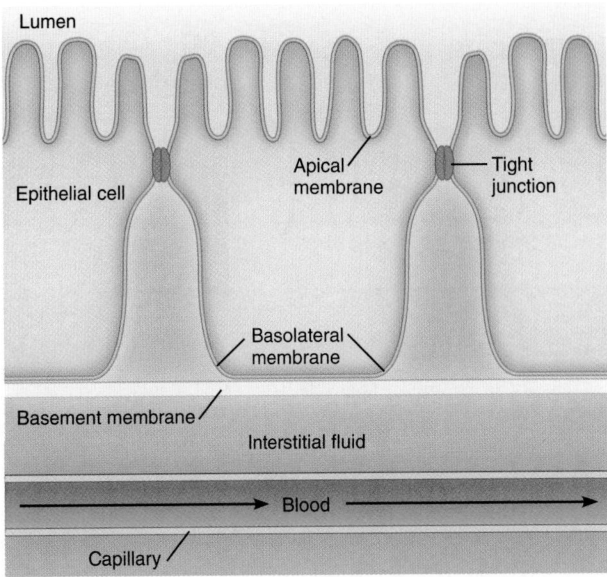

Figure 23 The general structure of an epithelium.

Epithelial Solute Transport

The mechanisms whereby epithelial cells transport solute molecules across an epithelial layer are illustrated in **Figure 24**, which depicts the types of epithelial cells that absorb nutrients in the intestines or that function in body fluid regulation in kidney tubules. Note first that a comparison of the transport systems in the apical and basolateral membranes of both cells reveals the previously mentioned polarity: Only the basolateral membranes of these cells have Na^+/K^+ pumps, which transport Na^+ out of the cell and K^+ into the cell. In addition, only the basolateral membranes possess K^+ channels, which allow K^+ to leak out of the cells down its electrochemical gradient. This pump-leak system maintains a nearly constant concentration of K^+ inside the cells that is higher than the concentration outside the cells. The Na^+/K^+ pump also maintains a low Na^+ concentration inside the cells, which creates an inwardly directed Na^+ gradient. Note as well that in both cells the apical membrane possesses transport systems that the basolateral membrane lacks.

In the cell depicted in Figure 24a, which absorbs Na^+, the Na^+ ion leaks into the cell through Na^+ channels, which are present in the apical membrane only. This leak is counteracted by the active transport of Na^+ out of the cell across the basolateral membrane. Sodium ions enter the cell on one side and exit it on the other side—that is, they are transported completely across the cell from the lumen to the interstitial fluid.

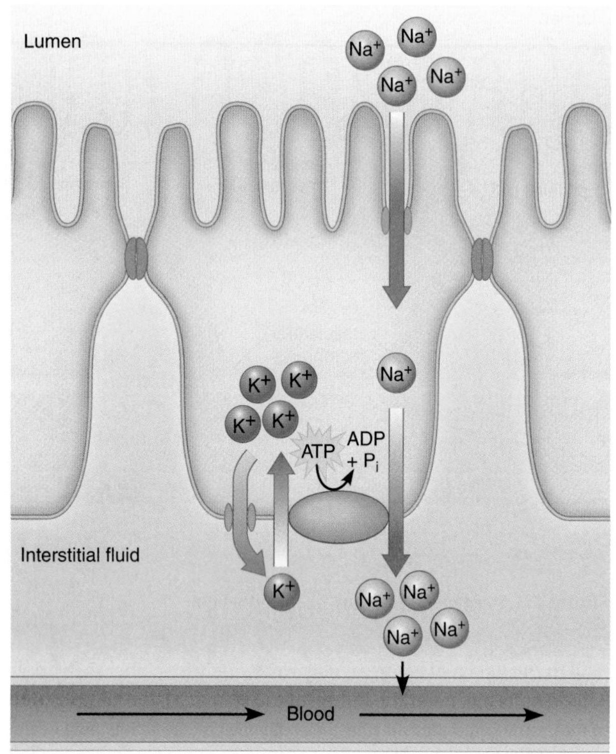

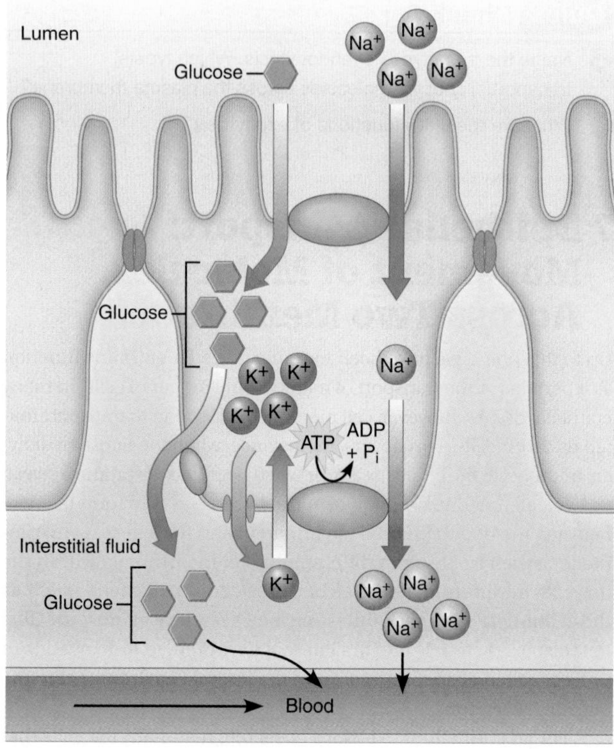

(a) (b)

Figure 24 Mechanisms of epithelial solute transport. (a) Absorption of sodium ions. (b) Absorption of glucose and sodium ions.

Q If glucose is being transported across the basolateral membrane by a carrier (see Figure 24b), is the glucose concentration in the interstitial fluid higher or lower than that inside the cell?

Although Na^+ flows passively across the apical membrane, Na^+ transport across the cell as a whole is *active,* because entry of Na^+ into the cell across the apical membrane depends on the presence of an inwardly directed electrochemical gradient, which in turn depends on the ability of the Na^+/K^+ pump to remove Na^+ from the cell. If the pump were to stop working, the concentration of Na^+ inside the cell would increase; eventually, the Na^+ gradient across the apical membrane would disappear, bringing Na^+ entry to a halt. Because Na^+ transport across the cell is active, Na^+ can be absorbed from lumen to interstitial fluid up an electrochemical gradient.

In the cell depicted in Figure 24b, which absorbs both Na^+ and glucose, Na^+ again enters passively across the apical membrane, but does so by way of a sodium-linked cotransport system, which couples the inward movement of Na^+ with the inward movement of glucose. The Na^+ electrochemical gradient provides the energy that drives the secondary active transport of glucose up its concentration gradient into the cell. Glucose then passively exits the cell by moving down its concentration gradient via a carrier protein in the basolateral membrane. Although glucose exit across the basolateral membrane is passive, glucose transport across the epithelium as a whole is active because it depends on the presence of the inwardly directed Na^+ electrochemical gradient, which in turn depends on the action of the basolateral Na^+/K^+ pump. Glucose can, therefore, be transported from the lumen to interstitial fluid up a concentration gradient.

Epithelial Water Transport

Whenever epithelial cells secrete or absorb fluid, water transport occurs by osmosis. This happens, for instance, when you sweat, when the pancreas secretes pancreatic juice, or when you replenish your body fluids by drinking a glass of water.

Epithelia absorb or secrete water by first using the active transport of solutes to create a difference in osmotic pressure—an osmotic pressure gradient—between the solutions on either side of the cell layer. Water then flows across the epithelium passively by osmosis. Because water flow occurs in response to transport of solutes, water transport is said to be *secondary to* solute transport.

In epithelial water transport (**Figure 25**), an epithelium creates an osmotic pressure gradient (or simply an *osmotic gradient*) to absorb water. First, epithelial cells actively transport solute molecules (shown as dots in the figure) across the basolateral membrane into the interstitial fluid (Figure 25a). As a result, the solute concentration is slightly higher outside the cells than

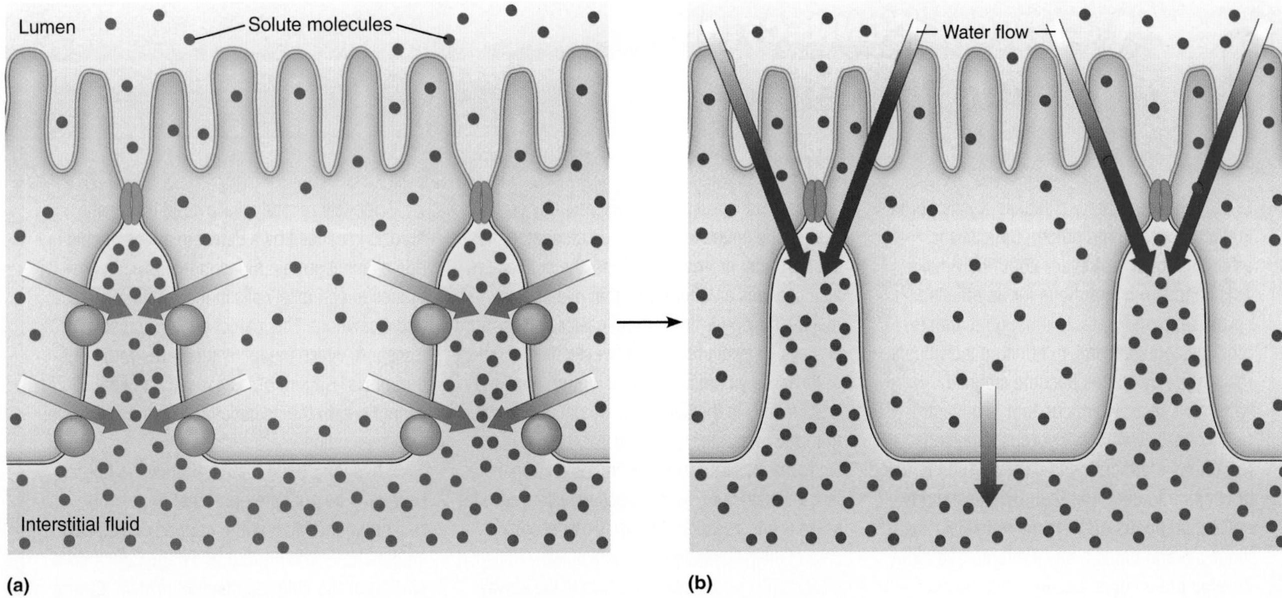

Figure 25 Epithelial water transport in an epithelium that absorbs water and solutes. (a) Pumps (circles) in the basolateral membrane actively transport solute molecules (dots) into interstitial fluid, raising the solute concentration in interstitial fluid, particularly in the paracellular spaces. **(b)** Active solute transport creates a gradient of osmotic pressure across the epithelium that drives passive water flow from the lumen to interstitial fluid (large arrows).

Q In this situation, does solute transport tend to increase or decrease the water concentration in interstitial fluid?

inside—particularly in the paracellular spaces, where the diffusion of solute molecules is hampered by the narrowness of the passages. (This is indicated in the figure by the higher density of dots between the cells.) Because this solute transport creates a difference in osmotic pressure between the solutions on either side of the epithelium, water then flows across the cell layer by osmosis (Figure 25b). Note that the direction of water flow is toward interstitial fluid, or up the osmotic pressure gradient; for epithelia that secrete fluids, the same principle applies, except that solute and water transport go in the opposite direction—away from interstitial fluid.

In certain pathological conditions, epithelial water transport is either excessive or insufficient. In *cystic fibrosis,* for example, the epithelium lining the respiratory airways does not transport enough solute to support adequate fluid production. As a result, the lungs become clogged with mucus, which hampers breathing and can lead to serious respiratory infection (see **Clinical Connections: Cystic Fibrosis**).

Transcytosis

Macromolecules cross epithelial cells by a process called **transcytosis** (**Figure 26**), which involves both endocytosis and exocytosis. During transcytosis, a large molecule is taken into the cell by endocytosis, but the endocytotic vesicle does not fuse with a lysosome. Instead, the vesicle travels to the opposite side of the cell and fuses with the plasma membrane to release its content by exocytosis.

Quick Check 7

1. Why are epithelial cells said to be polarized? How is polarity important to the function of certain epithelia?
2. What does it mean to say that epithelial water transport is secondary to solute transport?

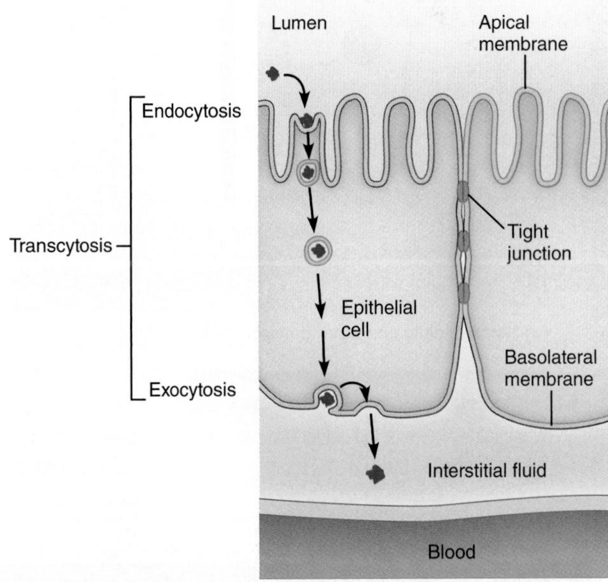

Figure 26 Transcytosis. Macromolecules can be transported across epithelial cells by entering the cell on one side via endocytosis and exiting the cell on the other side via exocytosis.

CLINICAL CONNECTIONS

Cystic Fibrosis 

Of all lethal hereditary diseases, cystic fibrosis is the most common among Caucasians, affecting about 1 in every 2500 individuals. This condition is notorious for its effects on the respiratory tract, although it affects other systems as well. In afflicted individuals, respiratory passages become clogged with thick, viscous mucus that is difficult to dislodge, even with vigorous coughing. Breathing is difficult, and patients run the risk of choking to death on their own secretions unless strenuous effort is made to clear the lungs several times a day. Patients frequently develop pneumonia, because the mucus-clogged airways offer a fertile environment in which bacteria can thrive and multiply. Even though there is no cure for cystic fibrosis, improved treatment methods have increased the average life span for people with this disease from approximately 30 years in 1990 to 50 years today.

In all humans, mucus is secreted by certain cells in the epithelium that lines respiratory passages. Under normal conditions, the epithelium also secretes a watery fluid that dilutes this mucus, making it thinner and easier to clear from airways. In cystic fibrosis, however, the secretion of the watery fluid is impaired. In turn, the undiluted mucus is thicker and quite difficult to clear from the respiratory passages.

In the respiratory epithelium, as in all epithelia that transport fluid, water transport occurs secondary to solute transport. To drive water secretion, cells of the respiratory epithelium actively transport chloride (Cl^-) from interstitial fluid to the airway lumen, creating a negative electrical potential in the lumen that drives the passive flow of sodium (Na^+) in the same direction (figure a). Movement of Na^+ and Cl^- raises the osmotic pressure of fluid bathing the luminal side of the epithelium. As a consequence, water moves passively down the osmotic gradient from interstitial fluid to lumen.

Scientists recently discovered that cystic fibrosis is caused by a defect in a type of chloride channel protein found in the apical membranes of epithelial cells in the respiratory tract and elsewhere. This defect directly impedes Cl^- transport, which indirectly interferes with transport of Na^+ (figure b). As a result, the epithelium cannot create the osmotic gradient necessary for water secretion.

It has long been known that cystic fibrosis is caused by a single abnormal gene, and recent biochemical and electrophysiological studies have shown that this gene codes for a portion of the chloride channel protein. Efforts are underway to develop a therapy that will permit normal genes to be inserted into the DNA of diseased epithelial cells. If this can be accomplished, these cells will then manufacture normal channel proteins, which should allow epithelia to secrete fluid in the normal manner.

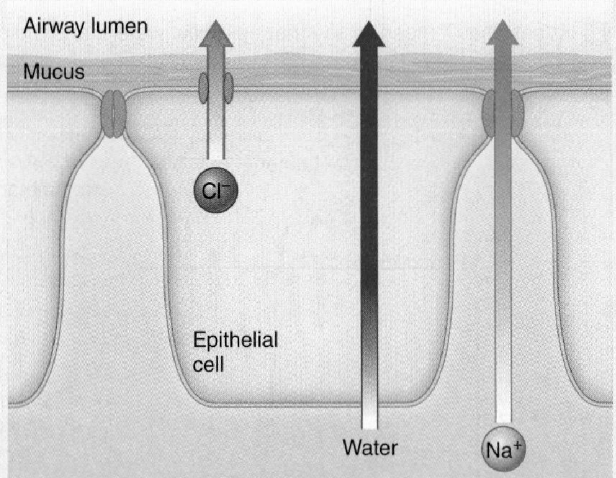

(a) Normal solute and water transport

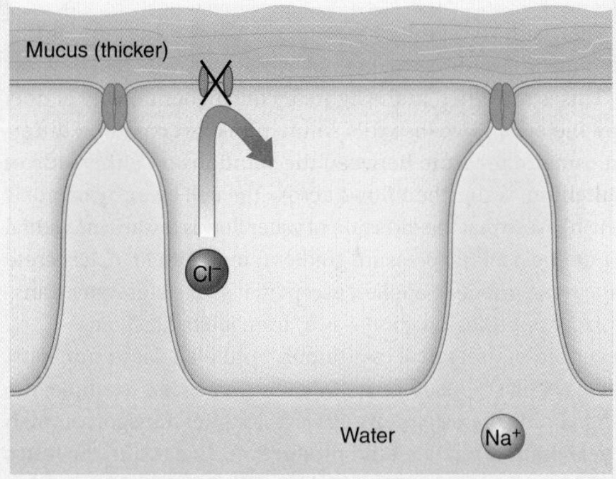

(b) Defective solute and water transport in cystic fibrosis

Critical Thinking Questions

1. What is the nature of the defect that causes cystic fibrosis?
2. Describe the roles that water transport and solute transport play in the development of cystic fibrosis.
3. If successful gene therapy is developed that enables cells to manufacture normal channel proteins, do you think it would cure this disease or just alleviate it?

SYSTEMS INTEGRATION We just learned that membranes and epithelium form selectively permeable barriers that only certain molecules can cross. Membrane proteins are critical to the movement of most molecules across membranes, and we can regulate this movement by varying the concentrations of these transport proteins in the membrane.

Organ systems in which materials are exchanged, such as the respiratory system, digestive system, and urinary system, most obviously utilize the transport mechanisms described in this chapter. Nevertheless, all of the other organ systems use them as well. For example, the nervous system communicates via electrical signals produced by the opening and closing of ion channels. The endocrine system communicates by secreting chemicals called hormones into the bloodstream. Muscle cells contract and relax in response to changes in transport of ions across the plasma membrane and the membrane of the smooth endoplasmic reticulum. Molecules are exchanged between blood and tissue through their movement across the walls of the smallest blood vessels, capillaries. Thus the plasma membrane, with its transport proteins, is a major determinant of organ function.

CHAPTER REVIEW

SUMMARY

1 Factors Affecting the Direction of Transport

- Molecules that cross cell membranes move either by simple diffusion through the lipid bilayer or by mediated transport, which involves specialized transport proteins.
- Active transport requires energy; it is carried out by proteins called pumps.
- Passive transport does not require energy; it includes simple diffusion and some forms of mediated transport.
- Transported molecules are generally influenced by three types of driving forces: chemical driving forces, which are due to the presence of concentration gradients; electrical driving forces, which reflect the influence of a cell's membrane potential on the movement of ions; and electrochemical driving forces, a combination of the chemical and electrical driving forces that represents the net force acting on molecules.
- Passively transported molecules move in the direction of the electrochemical force or down their electrochemical gradient.
- Actively transported molecules go against the electrochemical force or up their electrochemical gradient.

MAP iP Nervous I, The Membrane Potential

MAP iP Fluids and Electrolytes, Introduction to Body Fluids

2 Rate of Transport

- Flux is the rate at which a substance moves across a membrane. The flow of molecules in one direction or the other is the unidirectional flux; the difference between the unidirectional fluxes is the net flux.

3 Passive Transport

- Diffusion is movement of molecules from one location to another as a result of their random thermal motion.
- The rate at which a substance is transported passively by simple diffusion depends on the magnitude of the driving force, the membrane surface area, and the permeability of the membrane.
- Membrane permeability is determined by: the lipid solubility of the diffusing substance (the strongest influence on permeability), the size and shape of diffusing molecules, temperature, and membrane thickness.
- Facilitated diffusion involves carrier proteins, which bind molecules on one side of a membrane and transport them to the other side by means of a conformational change.
- Ion channels are pores that extend from one side of the membrane to the other.
- The rate of facilitated diffusion is determined by the transport rate of individual carriers, the number of carriers in the membrane, and the electrochemical gradient of the transported substance.
- In facilitated diffusion and ion channels, membrane permeability is determined by the transport rate of individual carriers or channels and the number of carriers or channels in the membrane.

4 Active Transport

- The two basic forms of active transport are primary and secondary active transport. Primary active transport uses ATP or some other chemical energy source; secondary active transport uses the electrochemical gradient of one substance as a source of energy to drive the active transport of another substance.
- The rate of transport for an actively transported substance is determined by the rate at which individual active transporters move the substance and the number of active transporters in the membrane.

MAP iP Nervous I, The Membrane Potential

MAP iP Urinary, Early Filtrate Processing

5 Osmosis: Passive Transport of Water Across Membranes

- Water flow across membranes affects cell volumes as well as the secretion and absorption of fluids.
- Osmosis—the flow of water across a membrane down its concentration gradient—is always passive and driven by a concentration gradient of water.
- Because the water concentration of a solution decreases as its solute concentration increases, a difference in osmolarity across a membrane implies the existence of a water concentration gradient.
- Because the osmotic pressure of a solution increases with increasing solute concentration, the flow of water down its concentration gradient is equivalent to flow up an osmotic pressure gradient.
- The volume of a cell is determined by the tonicity of the solution surrounding it, which depends on the solute concentration and the permeability of the membrane to the solutes that are present.

MAP iP Urinary, Early Filtrate Processing

6 Transport of Material Within Membrane-Bound Compartments

- Endocytosis occurs when molecules in the extracellular fluid enter the cell through endosomes from the plasma membrane. Exocytosis occurs when molecules within cells are enclosed within secretory vesicles, fuse with the plasma membrane, and release their contents into the extracellular fluid.
- There are three forms of endocytosis: phagocytosis, pinocytosis, and receptor-mediated endocytosis. In these processes the plasma membrane forms a ring that surrounds the extracellular fluid; the edges of the membrane fuse together and form a vesicle inside the cell.
- Phagocytosis and receptor-mediated endocytosis are specific in that a certain molecule or other matter is brought into the cell.

7 Epithelial Transport: Movement of Molecules Across Two Membranes

- Certain epithelia are specialized to transport materials into (absorption) or out of (secretion) the body's internal environment.
- Epithelial cells are said to be polarized because the membranes on the two sides differ in both structure and function.
- Epithelia form barriers between body fluid compartments or between the internal and external environments.
- The basolateral membrane, which rests on a noncellular basement membrane, faces the internal environment; the opposite membrane is the apical membrane.
- In transcytosis, macromolecules cross epithelial cells by entering via endocytosis and exiting via exocytosis.

MAP iP Urinary, Early Filtrate Processing

Check your understanding at MasteringA&P and apply your new knowledge to real-life case studies to ensure that you've mastered the content!

EXERCISES

Multiple-Choice Questions

1. For a substance crossing a cell membrane, the chemical driving force
   a) Depends only on the concentration gradient, regardless of whether the substance is an ion.
   b) Depends only on the concentration gradient if the substance is uncharged, but also depends on the electrical force if the substance is an ion.
   c) Is the total driving force on the substance, even if it is an ion.
   d) Is the force that pushes molecules across the membrane, but only if the substance is actively transported.
   e) Always favors movement of a molecule into the cell.

2. Which of the following is located in greater concentration inside cells compared to outside?
   a) Potassium ions
   b) Sodium ions
   c) Proteins
   d) Potassium and sodium ions are both located in greater concentration inside cells.
   e) Potassium ions and proteins are both located in greater concentration inside cells.

3. One example of primary active transport is the
   a) Transport of Ca^{2+} up an electrochemical gradient by a protein that hydrolyzes ATP.
   b) Transport of Ca^{2+} up an electrochemical gradient by a protein that couples Ca^{2+} flow to the flow of Na^+ down an electrochemical gradient.
   c) Movement of Ca^{2+} down an electrochemical gradient through channels.
   d) Transport of glucose molecules down a concentration gradient by carriers.
   e) Transport of glucose up a concentration gradient by a protein that couples glucose flow to the flow of Na^+ down an electrochemical gradient.

4. If a certain anion is located in greater concentration inside the cell and a negative membrane potential exists, then which of the following statements is true?
   a) The electrical force on the anion tries to move it into the cell.
   b) The chemical force on the anion tries to move it into the cell.
   c) The equilibrium potential for the anion is a positive value.
   d) Both a and c are correct.
   e) All of the above are correct.

5. Given that the potassium equilibrium potential is −94 mV and the sodium equilibrium potential is +60 mV, which of the following statements is true for forces acting on sodium and potassium when a cell is at −70 mV?
   a) The electrochemical gradient for Na^+ tries to move it into the cell.
   b) The electrochemical gradient for K^+ tries to move it into the cell.
   c) Both a and b are correct.
   d) Neither a nor b is correct.

6. When the membrane potential is equal to the equilibrium potential of Na^+ ($E_{Na} = +60$ mV)
   a) Na^+ moves into a cell down its electrochemical gradient.
   b) Na^+ moves out of a cell down its electrochemical gradient.
   c) The net flux of Na^+ is zero because it is at equilibrium.

7. The osmotic pressure of a solution depends on
   a) The concentrations of all solute particles contained in it.
   b) The concentrations of all permeant solute particles contained in it.
   c) The concentrations of all impermeant solute particles contained in it.
   d) The pressure exerted on the solution by the atmosphere.
   e) The volume of water in which the solute particles are dissolved.

8. Assuming that only impermeant solutes are present, which of the following will occur when a cell is placed in a solution whose osmolarity is 200 mOsm?
   a) Water will move into the cell.
   b) Water will move out of the cell.
   c) Water will not cross the cell membrane.

9. A solution is hypotonic if
   a) The concentration of all solutes contained in it is less than 300 mOsm.
   b) The concentration of all permeant solutes contained in it is less than 300 mOsm.
   c) The concentration of all impermeant solutes contained in it is less than 300 mOsm.
   d) Osmolarity is less than 300 mOsm.

10. Movement of Na^+ in sodium-linked glucose transport, in sodium-proton exchange, and via the sodium-potassium pump are all examples of
    a) Primary active transport.
    b) Passive transport.
    c) Mediated transport.
    d) Simple diffusion.
11. Which of the following molecules would be most likely to cross the lipid bilayer by simple diffusion?
    a) A small polar molecule
    b) A large polar molecule
    c) A small nonpolar molecule
    d) A large nonpolar molecule
12. Assuming that a substance is uncharged and is transported across a membrane by carriers, the net flux of that substance will tend to increase as
    a) The membrane surface area decreases.
    b) The magnitude of the concentration gradient decreases.
    c) The membrane potential becomes more positive.
    d) The number of carriers in the membrane increases.
    e) All of the above.
13. What do active transporters and carriers have in common?
    a) They both transport molecules up electrochemical gradients.
    b) They both transport molecules down electrochemical gradients.
    c) They both transport lipid-soluble substances preferentially.
    d) They both utilize ATP to transport molecules.
    e) They both are specific for certain molecules.
14. A leukocyte—a type of white blood cell—fights bacterial infections by sending out projections of its plasma membrane that surround an invading bacterium. The membrane then fuses together, entrapping the bacterium in a vesicle inside the cell. This is an example of
    a) Exocytosis.
    b) Transcytosis.
    c) Receptor-mediated endocytosis.
    d) Pinocytosis.
    e) Phagocytosis.
15. Which of the following transport mechanisms functions to bring a specific extracellular substance into the cell?
    a) Receptor-mediated endocytosis
    b) Pinocytosis
    c) Phagocytosis
    d) Both a and c
    e) All of the above

Objective Questions

16. Substances that cross cell membranes by simple diffusion are mostly (hydrophilic/hydrophobic).
17. A channel carries out (active/passive) solute transport.
18. In simple diffusion, an uncharged solute always flows from a region of higher concentration to a region of lower concentration. (true/false)
19. In facilitated diffusion, passive flow of an uncharged solute always goes from higher to lower concentration. (true/false)
20. A concentration gradient is also referred to as a(n) ________ driving force.
21. When a membrane potential is positive, there is an excess of cations over anions inside the cell. (true/false)
22. Transport of water from the bloodstream to the lumen of the intestine is an example of (secretion/absorption).
23. A cell will shrink if it is placed in a hypertonic solution. (true/false)
24. When water diffuses across a membrane, it normally flows from a region of higher osmotic pressure to a region of lower osmotic pressure. (true/false)
25. Junctions connecting adjacent epithelial cells are ________ junctions.

Essay Questions

26. To determine whether a substance is being transported actively or passively requires knowledge of only two factors: the direction of the electrochemical gradient and the direction of the net flux. Explain.
27. Describe the various factors that determine membrane permeability in simple diffusion.
28. Explain the mechanism of glucose absorption by intestinal epithelial cells. Include a discussion of the significance of cellular polarity.
29. Compare the diffusion of molecules through the lipid bilayer to the diffusion of ions through channels. Compare simple diffusion to facilitated diffusion.
30. A cell is placed in a saltwater solution of 0.2 mM NaCl. Neither Na^+ nor Cl^- can permeate the plasma membrane. What happens to the cell?
31. In cholera, bacterial toxins stimulate certain epithelial cells in the small intestine to oversecrete solutes into the lumen. What will happen to water transport across the intestinal epithelium? Which clinical sign is a result of this altered transport?

Critical Thinking

32. Assuming that $E_{Na} = +60$ mV, $E_{Cl} = -90$ mV, and $V_m = -70$ mV, find the direction of the electrochemical driving forces acting on Na^+ and Cl^- ions. In which direction will the ions move if they are transported passively? If they are transported actively? (Extra challenge: From the sign of E_{Cl}, determine the direction of the concentration gradient of Cl^- ions.)
33. Repeat question 1 for the case in which $V_m = -100$ mV.
34. Design an epithelium that moves potassium from the lumen to the interstitial fluid with a minimum number of proteins. Include all proteins *required* to accomplish the task.
35. Ion channels can be regulated to open or close, changing the permeability of the membrane to a specific ion. Assume that a cell at a membrane potential of -70 mV has few open sodium channels. Knowing that the equilibrium potential for sodium is $+60$ mV, predict what would happen to the membrane potential if many sodium channels suddenly went from a closed state to an open state. Explain your answer.
36. Oubain is a drug that blocks the Na^+/K^+ pump. Describe and explain the effects you would expect oubain to have on (1) sodium concentration gradients across cell membranes, (2) potassium concentration gradients across cell membranes, and (3) movement of molecules by secondary active transport coupled with sodium.

Photo credits for icons in feature boxes: Focus on Diabetes: Tanyae/Dreamstime; Clinical Connections: Iofoto/Dreamstime; Understanding Exercise: Pixattitude/Dreamstime; Discovery: Yuri Arcurs/Cutcaster.

Glossary

absorption movement of a substance from the external environment to the internal environment by transport across an epithelium; in the cardiovascular system, the bulk flow of fluid from the interstitial space into a capillary

active transport any method of protein-mediated transport of molecules across a membrane that requires the use of energy

apical membrane membrane on the side of an epithelial cell that faces the lumen of a body cavity; the lumen-facing membrane

aquaporins water channels located in the plasma membrane of most cells

basement membrane a layer of noncellular material that is relatively permeable to most substances; anchors the basolateral membrane and provides physical support for the epithelial layer

basolateral membrane membrane on the side of an epithelial cell that faces the internal environment and is in contact with interstitial fluid; blood-facing membrane

channel a transmembrane protein that transports molecules by way of a passageway or pore that extends from one side of the membrane to the other

coated pit an indentation of the plasma membrane that eventually forms an endocytotic vesicle; coated on its inner surface by specific proteins

chemical driving force difference in energy due to a concentration gradient that causes a molecule to move from an area of high concentration to an area of low concentration

concentration gradient difference in the concentration of a substance between regions

cotransport carrier-mediated transport by which two transported substances move in the same direction

countertransport secondary active transport by which two transported substances move in opposite directions

diffusion the passive movement of molecules from one location to another as a result of their own thermal motion

electrochemical driving force the sum of the chemical and electrical driving forces

electrochemical gradient the sum of the electrical and chemical gradients acting on an ion or charged molecule

equilibrium potential the membrane potential at which the electrical driving force is equal and opposite to the chemical driving force, giving an electrochemical driving force of zero

endocytosis (en-doh-sy-toh-sis) uptake of material into a cell via vesicles that pinch off from the plasma membrane; enables macromolecules and larger particles to enter cells

exocytosis (ex-oh-sy-toh-sis) transport of material out of a cell via vesicles that fuse with the plasma membrane; involved in the cellular secretion of hydrophilic molecules

facilitated diffusion the passive movement of molecules across a membrane by way of a transport protein

flux the number of molecules crossing a membrane in a given length of time; a measure of the rate at which a substance is being transported

hyperosmotic a solution having an osmolarity that is higher than another solution's osmolarity

hypertonic a solution that draws water out of a cell, causing the cell to shrink

hypo-osmotic a solution with a lower osmolarity than another solution's osmolarity

iso-osmotic two solutions having the same osmolarity

isotonic a solution that will not alter cell volume

mediated transport transport of molecules across a membrane utilizing transmembrane proteins

simple diffusion passive movement of molecules in response to chemical or electrical forces

membrane potential a difference in electrical potential or voltage that appears across the membranes of most cells

osmosis the passive movement of water across a membrane down its concentration gradient

osmolarity total concentration of solute particles in a solution

osmotic pressure an indirect measure of a solute's concentration, expressed in ordinary units of pressure

passive transport any method of transport of molecules across a membrane that does not require the use of energy

pumps proteins that actively transport molecules across a membrane

phagocytosis (fag-uh-sy-toh-sis) process by which a cell engulfs microorganisms, abnormal cells, and foreign particles present in blood and tissues

primary active transport active transport of molecules utilizing a protein (pump) that uses adenosine triphosphate (ATP) as the energy source

pinocytosis a form of endocytosis in which a cell takes up fluid and dissolved molecules via endocytotic vesicles that pinch off the plasma membrane

pumps proteins that actively transport molecules across a membrane

receptor-mediated endocytosis a form of endocytosis in which endocytotic vesicles contains receptors that recognize and bind specific molecules in the extracellular fluid; enables cells to selectively take up certain molecules

secretory vesicles intracellular vesicles containing molecules destined for secretion from the cell

simple diffusion passive movement of molecules in response to chemical or electrical forces

sodium-potassium pump a protein that utilizes adenosine triphosphate (ATP) to actively transport sodium ions out of the cell and potassium ions into the cell against their electrochemical gradients

secretion movement of substance from the internal environment to the external environment by transport across an epithelium; movement of substance from inside a cell to outside the cell by movement across the plasma membrane

transcytosis the transport of macromolecules across epithelial cells; involves endocytosis at one membrane, followed by exocytosis at the opposite membrane

Answers

Figure Question Answers

Fig 7	Increases
Fig 14	Low
Fig 20	It would be three times larger
Fig 24	Lower
Fig 25	Decreases

Apply Your Knowledge Answers

During allosteric regulation of an enzyme, a modulator binds by weak interactions to a site on the enzyme, thereby altering its affinity for the substrate. In secondary active transport, a similar process occurs.

During covalent regulation of an enzyme, a chemical group—usually a phosphate group—becomes bound to the enzyme by a covalent bond, which alters the activity of the enzyme. In primary active transport, a similar process occurs as the ATPase (pump) catalyzes phosphorylation of itself to increase the affinity of the pump for the solute when the pump faces a certain side of the membrane.

A 200 mM solution of KCl consists of 200 mmoles potassium and 200 mmoles chloride in 1 liter of fluid. The solution is thus $200 + 200 = 400$ mOsm. This solution is hyperosmotic relative to body fluids (which are 300 mOsm).

A 100 mM solution of $MgCl_2$ consists of 100 mmoles of magnesium and 200 mmoles of chloride in 1 liter of fluid. The solution is thus $100 + 200 = 300$ mOsm. This solution is iso-osmotic relative to body fluids.

A solution of 50 mmoles of $CaCl_2$ in 250 mL water is the same concentration as 200 mmoles in 1 liter of water—that is, 200 mM. Thus 1 liter of solution would contain 200 mmoles calcium and 400 mmoles chloride. The solution is $200 + 400 = 600$ mOsm. This solution is hyperosmotic relative to body fluids.

Osmolarity of 0.9% NaCl: A 0.9% solution contains 0.9 g NaCl in 100 mL water, or 9 g per liter. The molecular weight of NaCl $= 23 + 35 = 58$ g/mole. The molarity of a 0.9% solution, therefore, is 9 g/58 g/mole $= 0.155$ moles or 155 mmoles per liter. Because NaCl dissociates into 2 ions in solution, the osmolarity is 310 mOsm. This is an iso-osmotic and isotonic solution.

Osmolarity of 5% dextrose: A 5% solution contains 5 g dextrose in 100 mL water, or 50 g per liter. The molecular weight of dextrose is $(12 \times 6) + (12 \times 1) + (16 \times 6) = 180$ g/mole. The molarity of a 5% solution, therefore, is 50 g/180 g/mole $= 0.278$ mole/liter $= 0.278$ M $= 278$ mM. Because dextrose is one solute particle in solution, 278 mM $= 278$ mOsm. This is an iso-osmotic solution compared to body fluids, but it is hypotonic (similar to the example of urea).

Answers to End-of-Chapter Multiple Choice and Objective Questions

Multiple Choice Questions

1. a 2. e 3. a 4. c 5. a 6. c 7. a 8. a
9. c 10. c 11. c 12. d 13. e 14. e 15. d

Objective Questions

16. hydrophobic
17. passive
18. true
19. true
20. chemical
21. true
22. secretion
23. true
24. false
25. tight

Chemical Messengers

Chemical Messengers

Alfred Pasieka/Photo Researchers, Inc.

Polarized light micrograph of crystals of the female hormone oxytocin.

A community is a group of people who live in the same area or who have a common interest, such as working in the same city. Can you imagine how the city would function without a means of communication between people in the community? Even under normal circumstances, communication is necessary for food distribution, proper waste disposal services, and transportation from one area of the city to another. Now think what could happen if a disaster such as a tornado or earthquake cut off all means of communication. Chaos would ensue.

Our bodies can be thought of as a community of cells. Just like the people living in a city, our body cells must communicate with one another in order to maintain order. When communication is disrupted, a break in homeostasis almost inevitably follows. For example, when a person gets bitten by a black widow spider, a toxin causes certain neurons to communicate excessively to skeletal muscle cells. As a result, muscle cramping occurs, especially in the abdomen. Respiratory muscles can also be affected. As another example, insulin-dependent diabetes mellitus is caused by an insufficient amount of a chemical messenger, insulin. Insulin increases cells' uptake of glucose; thus a deficiency in insulin decreases the amount of glucose available to the cells and increases blood glucose levels. Either too much or too little communication in the body can cause chaos.

CHAPTER OUTLINE

MasteringA&P®

Go to MasteringA&P for helpful A&P Flix 3-D animations, chapter quizzes, pre-tests, Interactive Physiology tutorials, and more!

LEARNING OUTCOMES *After studying this chapter, you should be able to:*

- Compare the three major functional classes of chemical messengers (paracrines, neurotransmitters, and hormones) with respect to the cell that releases them and the ways the messenger gets to the target cell.
- Describe the basic structure and function of each chemical class of messenger (amines, catecholamines, peptides, steroids, and eicosanoids) with regard to mechanisms of synthesis, release, transport, and signal transduction.
- Compare and contrast the signal transduction mechanisms for lipophilic and lipophobic messengers.
- Describe the following functional classes of receptors: channel-linked receptors, enzyme-linked receptors, and G protein–linked receptors.
- Compare and contrast communication mediated through the nervous and endocrine systems.

Before You Begin

Make sure you have mastered the following topics:

- Epithelial cells
- Biomolecules
- Plasma membrane structure
- Plasma membrane proteins

We have already learned about cell structure and basic functions. Our bodies are made up of approximately 10 trillion (10^{14}) cells, however, and they cannot survive in isolation. Instead, our cells depend on one another to maintain the internal environment in a state compatible for life. To achieve this goal, cells must communicate. This chapter describes the mechanisms of intercellular communication. Such communication requires a variety of biomolecule classes, including proteins. Communication requires secretion of messengers by various processes. These messengers often cause a response in other cells through alteration of enzyme activity. We will now discover how cells communicate both with neighboring cells and with cells on opposite ends of the body.

1 Mechanisms of Intercellular Communication

Almost all body functions require communication between cells. Seeing an apple requires communication between the cells in the eyes and the neurons in the brain. Fighting an infection requires communication among several cell types in the blood. Moving your hand requires communication between neurons in the brain and spinal cord and the hand's muscle cells. Maintaining a normal blood glucose level requires the pancreas to release chemicals into the blood that communicate with cells throughout the body. Hundreds of other examples of intercellular communication could be cited. Remarkably, all the cells in the body use just a few mechanisms to communicate with one another. In a relatively small number of instances, cells are physically linked by gap junctions; in the vast majority of instances, cells communicate through chemical messengers.

Direct Communication Through Gap Junctions

Recall that gap junctions link adjacent cells and are formed by *connexins,* plasma membrane proteins that form structures called *connexons* (**Figure 1a**). These connexons form channels that allow ions and small molecules to pass directly from one cell to another. The movement of ions through gap

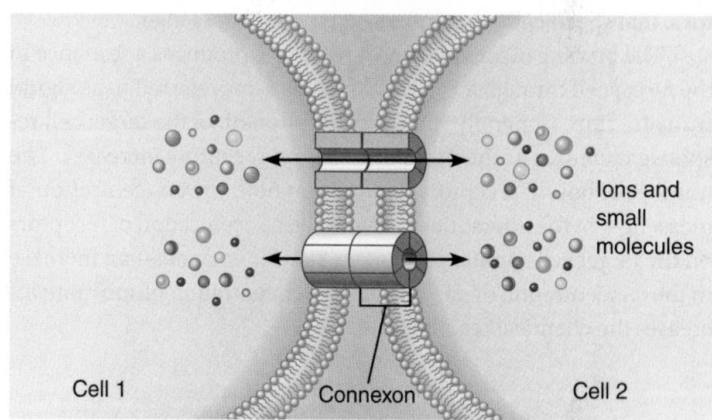

(a) Direct communication through gap junctions

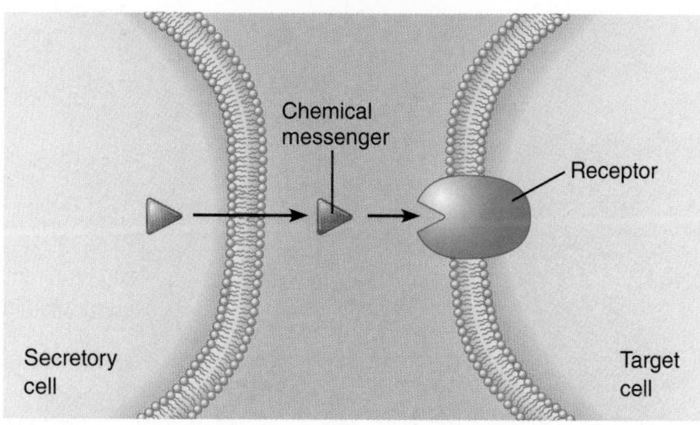

(b) Communication via chemical messengers

Figure 1 Types of intercellular communication. (a) Direct communication through gap junctions. Gap junctions are composed of membrane protein structures called connexons that link cytosols of two adjacent cells, allowing ions and small molecules to move between cells. **(b)** Communication via chemical messengers. After a secretory cell releases a messenger into the extracellular fluid, the messenger binds to receptors on target cells, triggering a response in the target cell.

junctions electrically couples the cells, such that electrical signals in one cell are directly transmitted to the neighboring cells. For example, gap junctions found in heart muscle and the smooth muscle of other internal organs (e.g., the intestines and blood vessels) cause the muscle cells to contract as a unit; that is, the cells contract at the same time. Gap junctions are also found in some glands and between some neurons in the brain and retinas, where they function in synchronized communication.

Indirect Communication Through Chemical Messengers

Most often, cells communicate via chemical messengers (Figure 1b) in the form of *ligands,* molecules that bind to proteins reversibly (see **Toolbox: Ligand-Protein Interactions**). The body has hundreds of these chemical messengers with a multitude of functions.

Communication through chemical messengers occurs when one cell releases a chemical into the interstitial fluid, usually by a process called secretion, and another cell, called the *target cell,* responds to that chemical messenger. The target cell responds to the chemical messenger because it has certain proteins, called **receptors,** that specifically recognize and bind the messenger.

The binding of messengers to receptors produces a response in the target cell through a variety of mechanisms referred to as **signal transduction.** Generally speaking, the strength of the target cell response increases as the number of bound receptors increases. The number of bound receptors depends on both the concentration of messenger in the interstitial fluid and the concentration of receptors on the target cell, similar to what we saw for enzymes—an increase in the concentration of enzyme or the concentration of substrate increases the chemical reaction rate.

2 Chemical Messengers

Chemical messengers can be classified according to their function or chemical structure. First we consider the functional classes of chemical messengers.

Functional Classification of Chemical Messengers

Most chemical messengers can be classified into three main categories: (1) paracrines, (2) neurotransmitters, and (3) hormones (or endocrines) (**Figure 2**). When released into the interstitial fluid, each type of messenger transmits a specific signal by binding to receptors on a target cell.

Paracrines are chemicals that communicate with neighboring cells. The target cell must be close enough that once the paracrine is secreted into the extracellular fluid, it can reach the target cell by simple diffusion (Figure 2a). Paracrines generally include *growth factors, clotting factors,* and *cytokines.* **Growth factors** are proteins that stimulate proliferation and differentiation of cells. For example, nerve growth factor is necessary for embryonic development of the nervous system and for maintenance or regeneration of neurons; vascular endothelial growth factor stimulates proliferation of blood vessels. **Clotting factors** are proteins that stimulate formation of a blood clot. **Cytokines** are peptides; usually released from immune cells, their function is to coordinate the body's defense against infections. Although most cytokines function as paracrines, others travel in the bloodstream to distant target cells and function more like hormones, which are described later in this section.

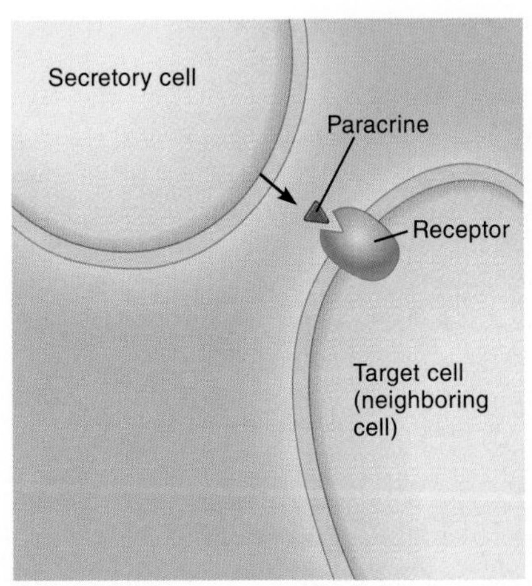

(a) Paracrines

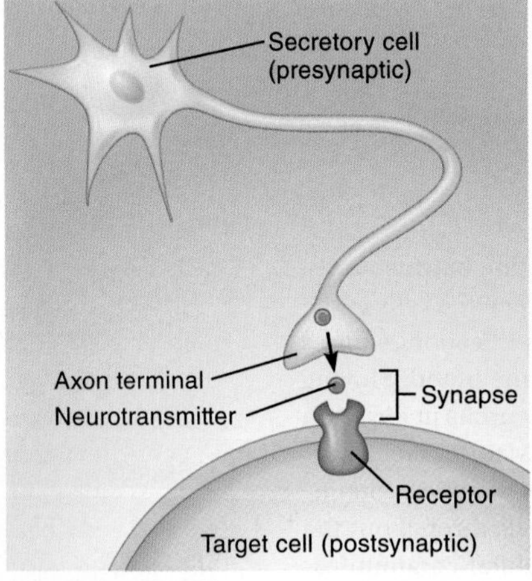

(b) Neurotransmitters

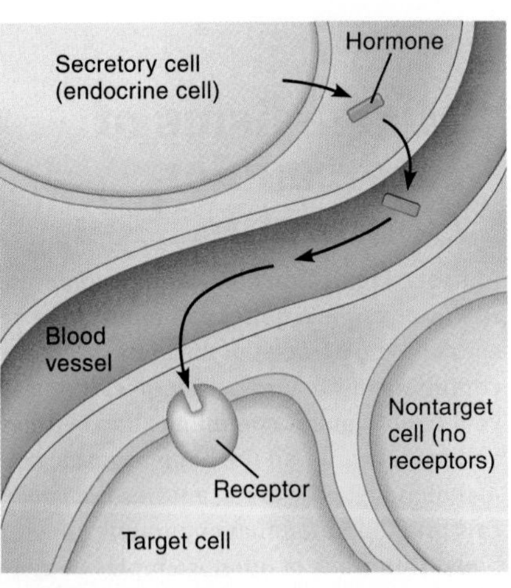

(c) Hormones

Figure 2 Functional classes of chemical messengers.
(a) Paracrines are secreted by one cell and diffuse to a nearby target cell. **(b)** Neurotransmitters are secreted from neurons at functionally specialized structures called synapses. The axon terminal of a presynaptic cell releases the neurotransmitter, which then diffuses a very short distance to bind to receptors on a specific target cell, called the postsynaptic cell. **(c)** Hormones are secreted by endocrine cells into the interstitial fluid. Hormones then diffuse into the bloodstream for transport to target cells in the body. Target cells are identified by the presence of receptors for the specific hormone. Cells without receptors for the hormone cannot respond to the hormone's signal.

DISCOVERY

Antihistamines

Histamine is a biogenic amine with paracrine and neurotransmitter functions. Once released, it acts on target cells by binding to one of three types of histamine receptors: H_1, H_2, or H_3. Histamines binding to H_1 receptors on target cells can produce several symptoms of allergic response, including constriction of the airways, increased fluid secretion in the airways, and dilation of the blood vessels. During severe allergic reactions, blood pressure may drop to a level so low that blood can no longer move through the circulation in adequate amounts, resulting in anaphylactic shock. Another condition associated with anaphylactic shock is a swelling of the airways, which causes respiratory distress.

Treatment of allergic reactions and some ulcers often includes administration of *antihistamines,* which work by blocking H_1 or H_2 receptors on target cells. Classic H_1 receptor antagonists include diphenhydramine hydrochloride (Benadryl) and dimenhydrinate (Dramamine). When receptor antagonists are used to treat an ailment, however, they block receptors throughout the body, including those not involved in the ailment, thereby producing side effects. In the case of H_1 receptor antagonists, the predominant side effect is drowsiness caused by blockage of the H_1 receptors in the brain that are involved in mental alertness. Recently, H_1 receptor antagonists that do not access the brain have been developed that result in fewer sedative side effects; examples include loratidine (Claritin) and fexofenadine (Allegra).

Histamine binding to H_2 receptors stimulates acid secretion by the stomach, which may contribute to the production of *heartburn,* a burning sensation caused by acid irritation of the esophagus. Acid secretion also may produce gastric and intestinal ulcers, perforations of the linings of these organs. As a result, heartburn and some ulcers are treated with administration of H_2 receptor antagonists such as cimetidine (Tagamet) and rantidine (Zantac). Side effects associated with use of H_2 receptor antagonists include headache and drowsiness.

The more specific a drug is for a certain receptor type, the more specific the therapeutic actions of the drug, and the fewer its side effects. Pharmaceutical companies are constantly designing new drugs to bind specifically to receptor subclasses for a certain messenger. In addition, drugs can be modified chemically to limit access to certain structures, again to eliminate unwanted side effects. The combined result is better treatments of ailments.

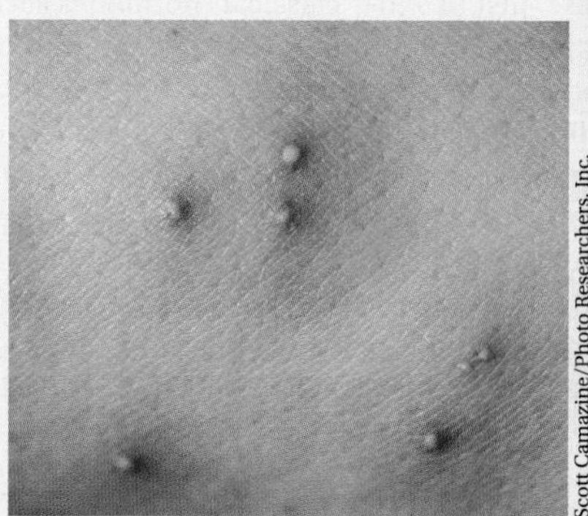

Scott Camazine/Photo Researchers, Inc.

Histamine causes inflammation at the site of insect bites.

Histamine is an example of a paracrine chemical messenger that is important in allergic reactions and inflammation. It is secreted by *mast cells* scattered throughout the body (see **Discovery: Antihistamines**). Histamine is responsible for the runny nose and red, watery eyes associated with allergic reactions. Occurring in response to bacterial infections and various forms of tissue damage, the release of histamine by mast cells is part of a complex cascade of events called *inflammation,* which is characterized in part by redness and swelling. In inflammation, histamine increases blood flow to affected tissues (producing redness) and causes fluid to leak out of the blood vessels and into the tissue (producing swelling).

Autocrines are a subclass of paracrines that act on the same cell that secreted them. Thus the secretory cell is also the target cell. Often an autocrine regulates its own secretion.

Neurotransmitters are chemicals released into interstitial fluid by neurons. These substances are released from a specialized portion of the neuron called the *axon terminal* (Figure 2b), which is very close to the target cell. Because the juncture between the axon terminal and the target cell is called a *synapse,* communication by neurotransmitters is often called *synaptic signaling.* The cell that releases the neurotransmitter is called the **presynaptic neuron,** whereas the target cell (which can be another neuron, a gland cell, or a muscle cell) is called the **postsynaptic cell.** Upon release from the presynaptic neuron, the neurotransmitter quickly diffuses over the short distance from the axon terminal and binds to receptors on the postsynaptic cell, triggering a response. Communication between a neuron and its target cell(s) is very specific because the neurotransmitter is directed only to cells with which the neuron has an anatomical association: the synapse. An example of a neurotransmitter is *acetylcholine,* which is released by the neurons that trigger contraction of skeletal muscles.

Hormones are chemicals released from *endocrine glands* (or occasionally endocrine or other cells dispersed in a tissue) into the interstitial fluid, where they can then diffuse into the blood (Figure 2c). The hormone then travels in the blood to its target cells, which can be distant from the site of hormone release. Although the bloodstream distributes the hormone to almost all cells of the body, only cells possessing receptors specific for the hormone are able to respond and thus serve as its target cells. An example of a hormone is *insulin,* which is secreted by the pancreas

TABLE 1 Functional Classification of Chemical Messengers

Class	Secretory cell type	Distance to target cell	Mode of transport to target cell	Chemical classification of messenger
Paracrine	(Several)	Short	Diffusion	Amines, peptides/proteins, eicosanoids
Neurotransmitter	Neuron	Short*	Diffusion	Amino acids, amines, peptides/proteins
Hormone	Endocrine	Long	Blood	Amines, Steroids, Peptides/Proteins

*Even though neurotransmitters diffuse over only a short distance to the postsynaptic cell, some neurons are involved in long-distance communication because the neuron that releases the neurotransmitter is often quite long (up to 1 meter).

and acts on target cells throughout the body to regulate energy metabolism.

A special group of hormones, called **neurohormones**, are released by a special class of neurons called *neurosecretory cells* through a mechanism similar to that of neurotransmitter release. Just like the "classical" hormones secreted by endocrine glands, neurohormones are released into the interstitial fluid and then diffuse into the blood, which distributes them to target cells throughout the body. An example of a neurohormone is *vasopressin,* or *antidiuretic hormone (ADH),* which is synthesized by neurosecretory cells originating in an area of the brain called the hypothalamus. Once vasopressin is released from the axon terminals of these neurosecretory cells, which are located in the posterior pituitary gland, it travels in the blood to its target cells. The primary target cells are located in the kidneys, where vasopressin affects the volume of urine that is excreted.

Table 1 summarizes some characteristics of the functional classes of chemical messengers. It is important to realize that a single chemical messenger may belong to more than one of these functional classes. For example, serotonin is a neurotransmitter when released from neurons in certain regions of the brain, but it is a paracrine when released from platelets, cell fragments found in blood.

To learn how the different classes of chemical messengers work together, see **Understanding Exercise: Chemical Messengers of Exercise.**

Chemical Classification of Messengers

A messenger's chemical structure determines its mechanisms of synthesis, release, transport, and signal transduction. The most important chemical characteristic is whether the messenger can dissolve in plasma (which is primarily water) or cross the lipid bilayer in the plasma membrane of cells. **Lipophilic** (hydrophobic) molecules are lipid soluble and, therefore, readily cross the plasma membrane—but they do not dissolve in plasma. Hydrophilic (or **lipophobic**) molecules are water soluble and, therefore, dissolve in plasma but do not cross the plasma membrane.

In the following sections, we discuss the five major classes of chemical messengers: (1) amino acids, (2) amines, (3) peptides/proteins, (4) steroids, and (5) eicosanoids (Table 2). Other chemical messengers, such as acetylcholine and nitric oxide, do not fit into any of these classes.

Amino Acid Messengers

Glutamate, aspartate, glycine, and *gamma-aminobutyric acid (GABA)* are amino acids that are classified as chemical messengers because they function as neurotransmitters in the brain and spinal cord. Glutamate, aspartate, and glycine are among the 20 amino acids (alpha amino acids) that are used in protein synthesis, whereas GABA belongs to a different class of amino acids (gamma amino acids). Amino acids are lipophobic; thus they dissolve in water but do not cross plasma membranes.

Amine Messengers

Amines, which are chemical messengers derived from amino acids, are so named because they all possess an amine group ($—NH_2$). The amines include a group of compounds called **catecholamines,** which contain a *catechol* group (a six-carbon ring)

TABLE 2 Chemical Classification of Messengers

Class	Chemical property	Location of receptors on target cell	Functional classification
Amino acids	Lipophobic	Plasma membrane	Neurotransmitters
Amines*	Lipophobic	Plasma membrane	Paracrines, neurotransmitters, hormones
Peptides/proteins	Lipophobic	Plasma membrane	Paracrines, neurotransmitters, hormones
Steroids	Lipophilic	Cytosol†	Hormones
Eicosanoids	Lipophilic	Cytosol	Paracrines

*One exception is the thyroid hormones, which, although amines, are lipophilic and have receptors in the nucleus of target cells.
†A few steroid hormones have receptors on the plasma membrane.

UNDERSTANDING EXERCISE

Chemical Messengers of Exercise

When a person exercises, she contracts *skeletal muscles*—that is, muscles attached to bone or other connective tissue that allow voluntary movement of the body. Skeletal muscle contraction requires the action of all three functional classes of chemical messengers. The voluntary movements of exercise require neural signals from the brain and spinal cord to tell various muscles whether to contract and how strongly to contract. These signals come in the form of a neurotransmitter called acetylcholine. Remarkably, all of our body's muscle contractile activity is controlled by this one chemical—either acetylcholine is released by the nervous system and causes the muscle to contract, or it is not released and the muscle relaxes. The more acetylcholine that communicates with a given muscle cell, the more strongly that muscle cell contracts.

Muscles cannot contract without energy, however. Thus a contracting muscle needs to receive increased blood flow, bringing oxygen and nutrients to it in larger quantities. To increase blood flow, the heart needs to beat faster and stronger; likewise, the blood vessels supplying the muscles need to dilate (become larger in diameter). When the heart receives neural signals and endocrine signals in the form of norepinephrine and epinephrine, respectively, it responds by beating faster and stronger. The blood vessels dilate primarily in response to endocrine and paracrine signals. Increases in the levels of the hormone epinephrine and the paracrine nitric oxide cause blood vessels to dilate. Thus epinephrine affects both the heart and the blood vessels during exercise. Hormones commonly have more than one type of target cell.

Dilation brings more blood to the muscle, but other chemical signals are necessary to ensure uptake of the required nutrients. For example, exercising muscles demonstrate enhanced glucose transport into the cells. The mechanism underlying this increase has not been definitively established, but it is known to require increased glucose transporters expressed in the muscle cell membrane. Unlike most cell membrane glucose transporters, those expressed in exercising muscle are not dependent on insulin. Current theories indicate the involvement of paracrines such as calcium, nitric oxide, or bradykinin in this glucose transporter expression.

and are derived from the amino acid tyrosine. Catecholamines include *dopamine, norepinephrine,* and *epinephrine.* Dopamine and norepinephrine function primarily as neurotransmitters, whereas epinephrine functions primarily as a hormone. Other amines include the neurotransmitter *serotonin,* which is derived from tryptophan; the *thyroid hormones,* which are derived from tyrosine; and the paracrine *histamine,* which is derived from histidine. Except for the thyroid hormones, all of the amines are hydrophilic.

Peptide/Protein Messengers

Most chemical messengers are polypeptides—that is, chains of amino acids linked together by peptide bonds. These messengers are classified as peptides or proteins based on their size, which varies from just 2 amino acids to more than 100 amino acids. The term *peptide* generally refers to chains containing fewer than 50 amino acids, whereas proteins are longer chains of amino acids. Polypeptides are hydrophilic.

Steroid Messengers

Steroids are a class of compounds derived from cholesterol. All of the body's steroid messengers function as hormones. Recall that cholesterol is a lipid with a distinctive four-ring structure. Because steroids are derived from hydrophobic cholesterol, they are also hydrophobic.

Eicosanoid Messengers

Eicosanoids include a variety of paracrines that are produced by almost every cell in the body. Most eicosanoids are derivatives of arachidonic acid, a 20-carbon fatty acid that is found in various plasma membrane phospholipids. Because eicosanoids are lipids, they readily cross the plasma membrane and are hydrophobic. Eicosanoids include the following families of chemically related compounds: prostaglandins, leukotrienes, and thromboxanes.

Quick Check 1

1. Name the three functional classes of messengers. Which messengers are transported in the blood to their target cells?
2. Name the five chemical classes of messengers. Which are hydrophilic, and which are hydrophobic?
3. Which chemical class of messengers is derived from cholesterol? To which functional class do these messengers belong?

Synthesis and Release of Chemical Messengers

The general synthetic pathways and mechanisms of release for chemical messengers are similar within a chemical class. In this section, we examine the synthesis and release of each class of messenger.

Amino Acids

Although amino acids can be obtained from the diet, the four amino acids that function as neurotransmitters must be synthesized within the neuron that will secrete them.

Glutamate and aspartate are synthesized from glucose through a three-step series of reactions. First, glucose is catabolized to pyruvic acid by glycolysis; pyruvic acid is in turn converted to acetyl CoA, which then enters the Krebs cycle; and finally the amine groups are added to certain Krebs cycle intermediates to form glutamate or aspartate. Glycine is synthesized from a glycolytic intermediate, 3-phosphoglycerate, in a series of four reactions. GABA is synthesized from glutamate in a single reaction catalyzed by the enzyme *glutamic acid decarboxylase*.

Following their synthesis in the cytosol, amino acid neurotransmitters are transported into vesicles, where they are stored until they are released by exocytosis.

Amines

All amines are derived from amino acids, and all except thyroid hormones are synthesized in the secretory cell by a series of enzyme-catalyzed reactions. Which amine is produced depends on the enzymes present in a given cell.

Catecholamines Figure 3 shows the pathway for synthesis of the catecholamines, which are derived from the amino acid tyrosine. In this pathway, dopamine acts as the precursor for norepinephrine, which in turn serves as the precursor for epinephrine. Because dopamine is a precursor for the other catecholamines, all catecholamine-secreting cells possess the two enzymes that catalyze its synthesis: *tyrosine β-hydroxylase* and *dopa decarboxylase*. Both of these enzymes are found in the cytosol. After dopamine is synthesized, it is transported into vesicles, where it is stored for later release by exocytosis or used as the substrate for synthesis of norepinephrine. The enzyme that converts dopamine to norepinephrine is *dopamine β-hydroxylase,* which is found inside the vesicle. If norepinephrine is the final product, it is stored in the vesicle until released by exocytosis. For cells that secrete epinephrine, the norepinephrine is transported out of the vesicle into the cytosol, where the enzyme *phenylethanolamine N-methyl transferase (PNMT)* is located. PNMT converts norepinephrine to epinephrine, and the epinephrine is then transported back into a vesicle and stored until exocytosis.

Serotonin Serotonin, or 5-hydroxytryptamine (5-HT), is synthesized from the amino acid tryptophan, one of the essential amino acids. Serotonergic neurons have tryptophan carriers that actively transport tryptophan into the cell. Tryptophan is converted to 5-HT through a pair of enzyme-catalyzed reactions: (1) *tryptophan-5-hydroxylase* converts tryptophan to 5-hydroxytryptophan, and (2) *aromatic L-amino acid decarboxylase* converts 5-hydroxytryptophan to 5-HT. Like the catecholamines, 5-HT is packaged into vesicles and stored until its release by exocytosis.

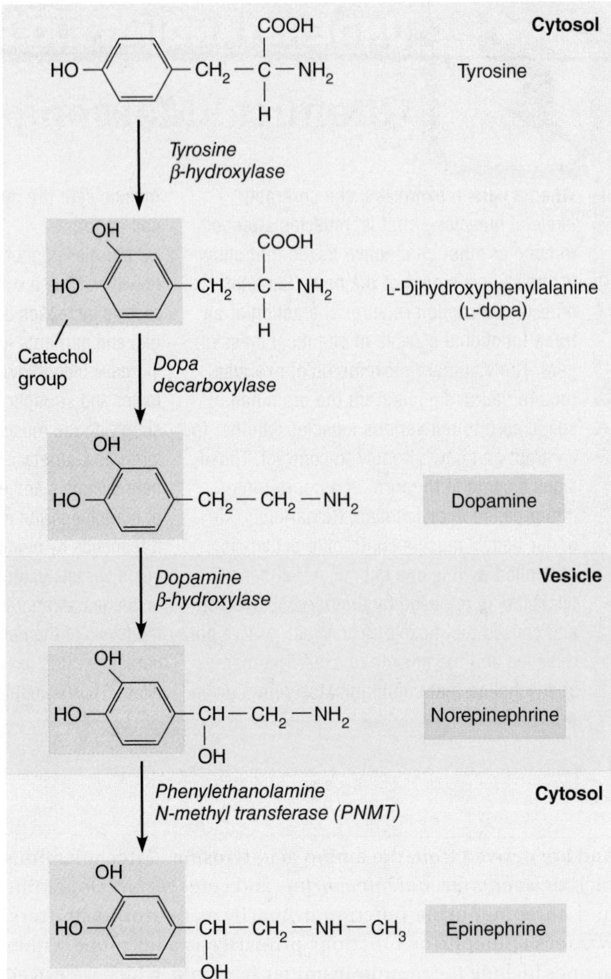

Figure 3 Catecholamine synthesis. Catecholamines are synthesized from the amino acid tyrosine by a sequence of enzyme-catalyzed reactions in which one catecholamine functions as the precursor for the next. The names of catecholamines that function as messengers are highlighted.

Q Which chemical group is removed from L-dopa to form dopamine?

Histamine Histamine is produced from histidine in the cytosol by a one-step reaction that is catalyzed by the enzyme *histidine decarboxylase*. Histamine is then packaged into vesicles and stored until its release by exocytosis.

Peptides and Proteins

Peptides and proteins are synthesized in the same way as other proteins destined for secretion. Briefly, cytosolic mRNA serves as the template that codes for the amino acid sequence in the peptide or protein. Translation of this mRNA begins on ribosomes that are found in a free state in the cytosol. The subsequent steps, shown in **Figure 4a** using a peptide as an example, are the following:

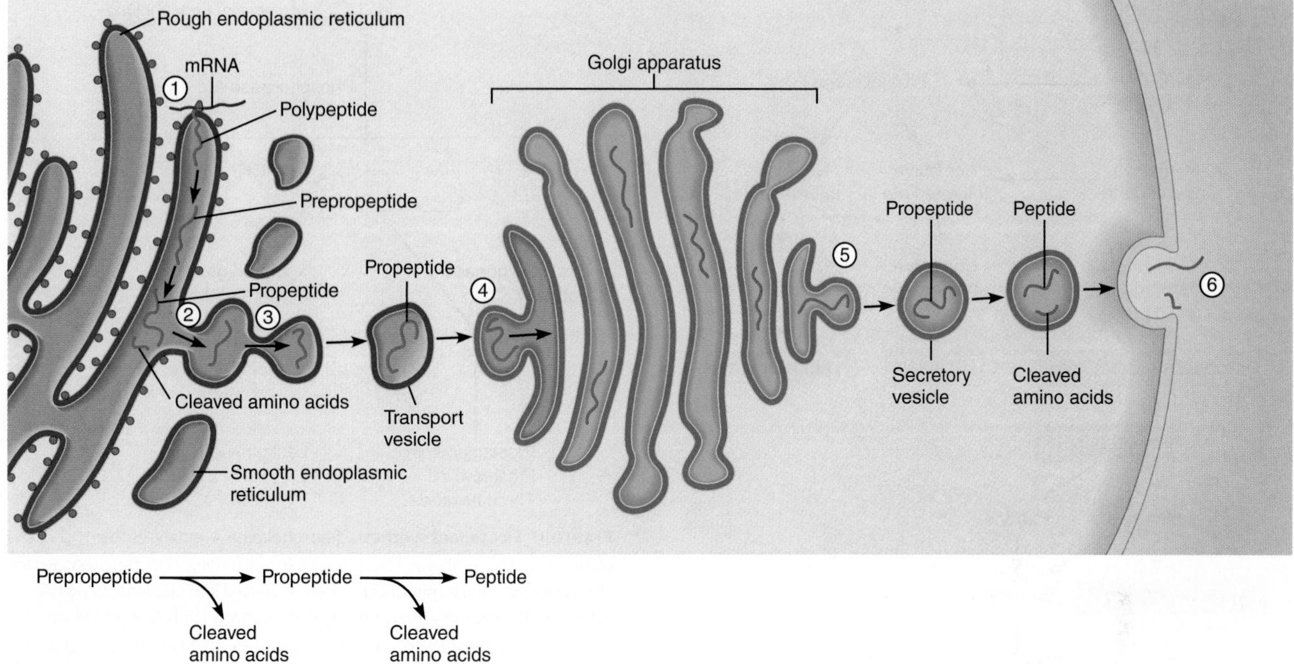

(a) Peptide synthesis

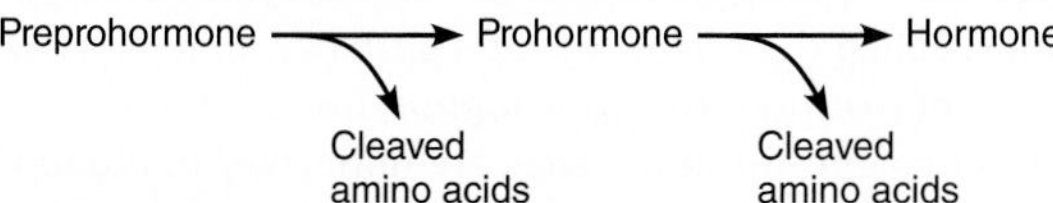

(b) Generalized scheme of hormone synthesis

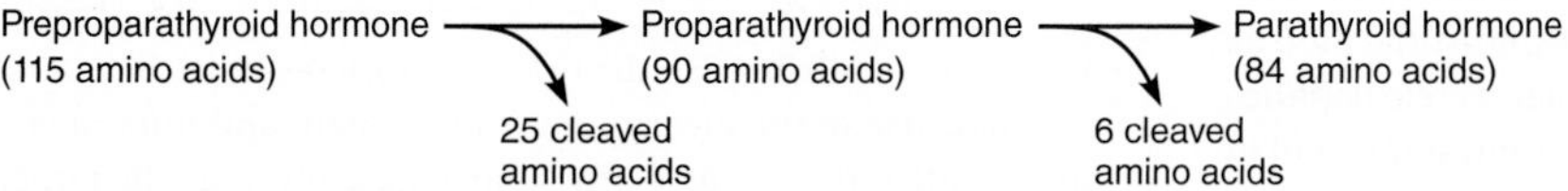

(c) Parathyroid hormone synthesis

Figure 4 Peptide synthesis and release. (a) The steps in peptide synthesis. For a detailed description, see the text. **(b)** The general steps in peptide hormone synthesis. **(c)** The modification of preproparathyroid hormone, which contains 115 amino acids, to the active parathyroid hormone, which contains 84 amino acids.

① Once translation starts, the ribosome attaches itself to the rough endoplasmic reticulum, where the rest of translation occurs. The polypeptide is formed inside the lumen of the rough endoplasmic reticulum, first as a *prepropeptide*.

② In the lumen of the endoplasmic reticulum, *proteolytic enzymes* cleave off some amino acids from the prepropeptide, yielding the *propeptide*.

③ In the smooth endoplasmic reticulum, the propeptide is packaged into transport vesicles.

④ The vesicles transport the propeptide to the Golgi apparatus.

⑤ In the Golgi apparatus, the propeptide is packaged into a secretory vesicle for storage until its release is triggered. More amino acids are cleaved off by proteolytic enzymes in the Golgi apparatus or in the secretory vesicle to give the final product, an active messenger.

⑥ Release occurs by exocytosis.

Typically, peptide fragments generated by proteolysis are released along with the primary messenger, and they may or may not exert their own biological effects. Figure 4c shows a schematic example of the synthesis of a protein hormone—in this case, parathyroid hormone.

Steroids

Steroid messengers are synthesized from cholesterol in a series of reactions (**Figure 5**) catalyzed by enzymes located in the smooth endoplasmic reticulum or mitochondria. During this process, the cholesterol molecule is modified, but its basic ring structure

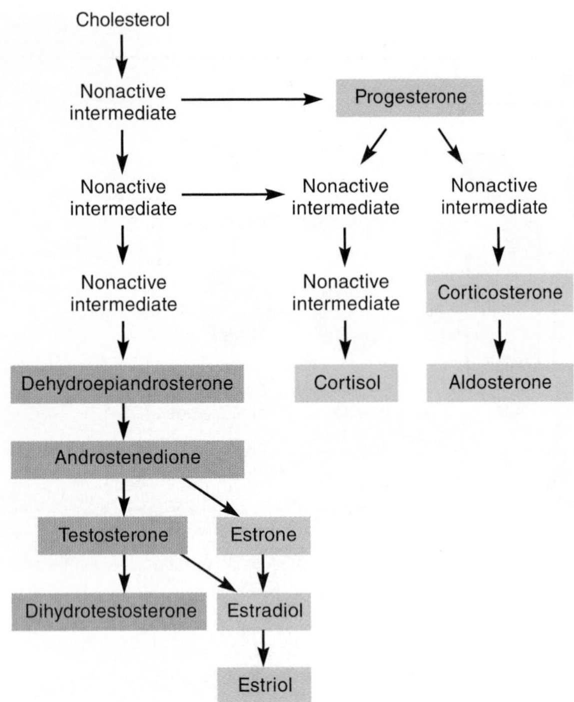

Figure 5 Synthetic pathway for steroids. Each arrow indicates an enzyme-catalyzed reaction. Green boxes indicate hormones produced in the adrenal cortex; blue boxes indicate male sex hormones; orange boxes indicate female sex hormones.

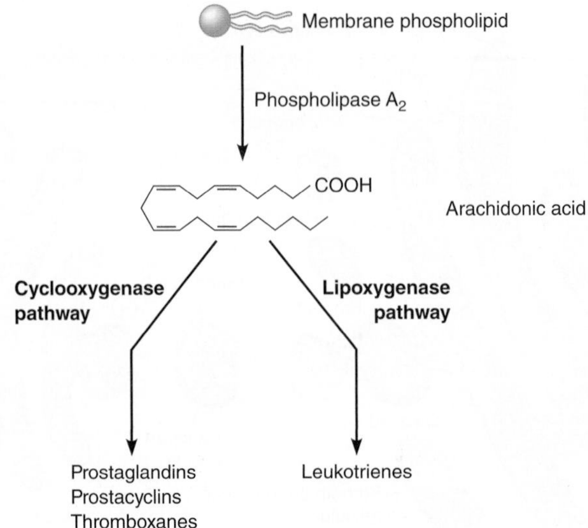

Figure 6 Eicosanoid synthesis. Phospholipase A_2 catalyzes the conversion of a membrane phospholipid to arachidonic acid, the precursor for all eicosanoids. Arachidonic acid is then converted into eicosanoids via two pathways: The cyclooxygenase-dependent pathway leads to the production of prostaglandins, prostacyclins, and thromboxanes, whereas the lipoxygenase-dependent pathway leads to the production of leukotrienes.

remains intact, as does its hydrophobic character. Consequently, all steroids are capable of crossing the plasma membrane. Because they are membrane permeant, steroids cannot be stored prior to release and instead diffuse out of the cell into the interstitial fluid as soon as they are synthesized. Thus, while cells that secrete peptides or amines can synthesize messengers in advance and store them in vesicles to be released on demand, steroid hormones are synthesized on demand and released immediately.

Apply Your Knowledge

Use Figure 5 to explain how certain male sex hormones are required for the synthesis of certain female sex hormones.

Eicosanoids

Like steroids, eicosanoids are synthesized on demand and released immediately because they are hydrophobic and able to pass through plasma membranes easily. The first step in eicosanoid synthesis (**Figure 6**) involves an enzyme called *phospholipase A_2*, which is activated in response to chemical signals of various kinds (paracrines, hormones, neurotransmitters, and even foreign chemicals). When active, this enzyme catalyzes the release of arachidonic acid from membrane phospholipids. Once arachidonic acid is released from the membrane, the final product depends on the complement of enzymes present in the particular cell.

To become an eicosanoid, a molecule of arachidonic acid first binds with one of two enzymes: either *cyclooxygenase* or *lipoxygenase.* Cyclooxygenase is the first enzyme in a series of reactions, called the *cyclooxygenase pathway,* that leads to the synthesis of *prostacyclins, prostaglandins,* or *thromboxanes.* Prostacyclins and thromboxanes are important in blood clotting; prostaglandins play roles in several systems, including inducing pain and the inflammatory response. Lipoxygenase is the first enzyme that participates in a series of reactions, called the *lipoxygenase pathway,* that leads to the synthesis of *leukotrienes,* which also contribute to the inflammatory response.

Because of the eicosanoids' role in pain and inflammation, many anti-inflammatory drugs, such as aspirin, act by targeting enzymes involved in eicosanoid synthesis. By inhibiting the activity of the enzyme cyclooxygenase, aspirin decreases not only inflammation but also blood clotting. It is this latter effect of aspirin that has led physicians to prescribe low doses of aspirin to patients believed to be at risk for a heart attack or stroke; these conditions can be caused by blood clots in the coronary and cerebral arteries, respectively. (See **Clinical Connections: From Aspirin to COX-2 Inhibitors.**)

Transport of Messengers

Once released, a messenger must first reach and then bind to receptors on the target cell for the signal to be transmitted. In many instances, the messenger is released from a cell that is near the target cell, with the messenger then reaching the receptor by simple diffusion. This transport mechanism is used by paracrines and neurotransmitters, for example. Typically these messengers are quickly degraded in the interstitial fluid and become inactive, minimizing the spread of their signaling. In contrast, hormones are transported in the blood, so they have access to most cells in the body.

CLINICAL CONNECTIONS

From Aspirin to COX-2 Inhibitors

Aspirin is one of a number of nonsteroidal anti-inflammatory drugs (NSAIDs) that relieve pain by decreasing the production of certain prostaglandins (PG), eicosanoids that produce pain and inflammation. A critical enzyme in the synthesis of PGs is cyclooxygenase (COX). In the early 1990s, two forms of COX were identified: COX-1 and COX-2. COX-1, which is always present in the body, is generally associated with the synthesis of PGs necessary for maintaining homeostasis. In contrast, COX-2 becomes activated in the presence of chemicals released on tissue damage or infection and leads to production of PGs associated with pain and inflammation. Aspirin nonselectively inhibits the activity of both forms of COX; it is the inhibition of COX-2 that results in pain relief.

Although aspirin is used to treat pain, inflammation, and fever, its inhibition of COX-1 causes several side effects—some favorable, some not. For example, aspirin decreases the production of thromboxane A_2, a chemical involved in the formation of blood clots. However, at higher doses, aspirin also decreases the production of PGI_2 (prostacyclin), a chemical that inhibits the production of blood clots. In recognition of these effects, doctors frequently prescribe 82 mg of aspirin ("baby aspirin") to guard against the formation of blood clots that can trigger heart attack or stroke, whereas 350 mg of aspirin is generally used to treat pain and would actually promote formation of a blood clot. In addition, aspirin inhibits production of PGE_2 in the stomach, an eicosanoid that indirectly protects the stomach lining from stomach acid. Thus major side effects of aspirin may include gastric ulcers and stomach bleeding.

In their search for safer pain relievers, pharmaceutical companies have developed drugs that selectively inhibit COX-2 to treat chronic pain and inflammatory diseases, such as arthritis. The chief advantages of COX-2 inhibitors (such as rofecoxib, also known as Vioxx, or celecoxib, also known as Celebrex) is that they do not inhibit production of PGE_2; thus they cause less harm to the stomach lining than does aspirin. Unfortunately, COX-2 inhibitors produce serious side effects of their own. For reasons that are not understood, these drugs appear to increase the risk of heart attack and stroke in individuals who are already susceptible to cardiovascular disease. Alas, the quest for a perfect "aspirin" still eludes us.

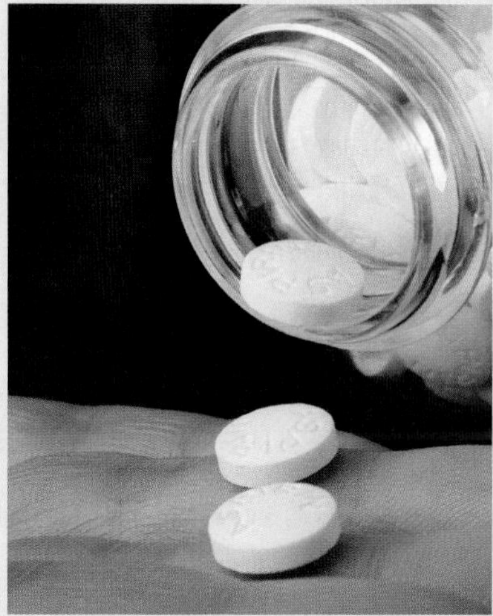

PhotoLink/Photodisc/Getty Images.

Critical Thinking Questions

1. How does inhibiting enzymes prevent the sensation of pain?
2. What negative effects do elevated doses of aspirin have on the body? How would short-term and long-term consequences compare?
3. Describe the advantages and disadvantages of using COX-2 inhibitors, rather than NSAIDs, to treat pain.

Hormones can be transported in the blood either in dissolved form or bound to *carrier proteins.* To be transported in dissolved form, the messenger must be a hydrophilic messenger (**Figure 7a**). Peptides and amines (except thyroid hormones) are generally transported in this manner. Because steroids and the thyroid hormones are hydrophobic and do not dissolve well in blood, these hormones are largely transported while bound to carrier proteins (Figure 7b). Although most of the catecholamines that function as hormones are hydrophilic and are transported in dissolved form, for instance, some are bound to carrier proteins. Some carrier proteins are specific for a particular hormone; one example is *corticosteroid-binding globulin,* which transports the steroid hormone *cortisol.* Other carrier proteins—for example, *albumin*—are not specific and can transport many different hormones.

Even though hydrophobic hormones are transported primarily in bound form, a certain fraction of the hormone molecules dissolve in plasma (generally less than 1%). For each such hormone, an equilibrium develops in the bloodstream between the amount of

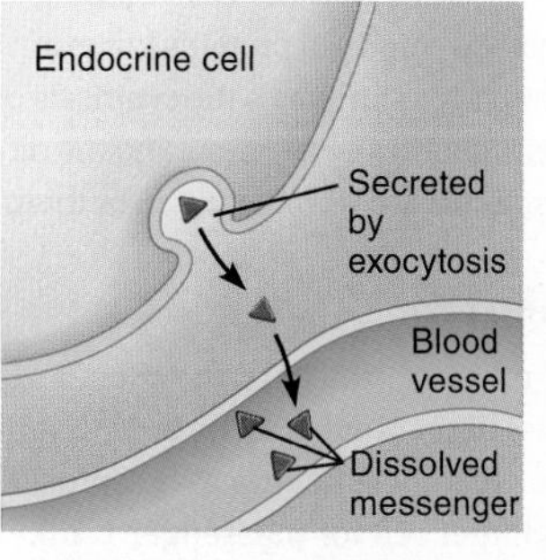

(a) Hydrophilic messenger

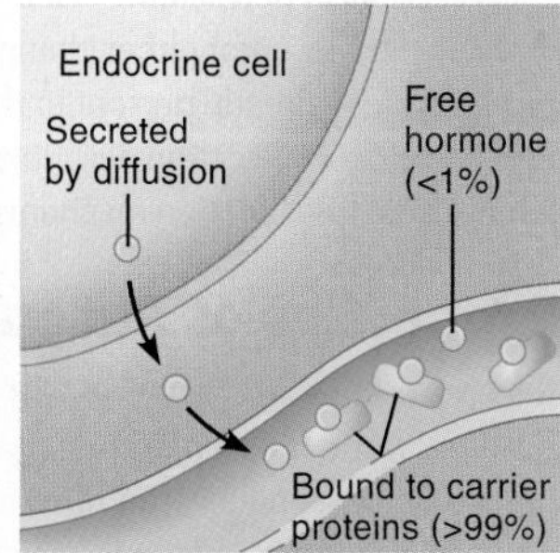

(b) Hydrophobic messenger

Figure 7 Transport of messengers in blood. (a) Hydrophilic messengers are secreted by exocytosis, enter the bloodstream, and dissolve in the plasma. **(b)** Hydrophobic messengers are secreted by simple diffusion and then enter the bloodstream. Most of the messenger molecules are transported bound to carrier proteins. Only the small amount of free hormone in the plasma is immediately available for binding with target cell receptors.

hormone that is bound to a carrier protein (Pr) in the form of a complex (H—Pr) and the amount of free hormone (H) that is dissolved in the plasma:

$$H{-}Pr \rightleftharpoons H + Pr$$

Only free hormone is available to leave the blood and bind to receptors on target cells. However, once the hormone leaves the blood, the equilibrium between bound and free hormone in blood shifts to the right, causing more hormone to be released from the carrier proteins. Likewise, the secretion of hormones into the blood causes the equilibrium to shift to the left, such that more hormone binds to carrier proteins.

Once in the bloodstream, hormones are ultimately degraded. How long a hormone persists in blood is described in terms of **half-life,** the amount of time it takes for half of the hormone in the blood to be degraded. Bloodborne messengers are generally degraded by the liver and excreted by the kidneys. Hormones that are present in dissolved form have relatively short half-lives, usually measured in minutes. By comparison, hormones that are bound to carrier proteins are protected from degradation and have longer half-lives, generally in the range of hours.

Quick Check 2

1. Name the three catecholamines. Which amino acid is the precursor for catecholamines? Which catecholamine generally functions as a hormone?
2. Phospholipase A_2 causes the release of which fatty acid from membrane phospholipids? Which chemical class of messengers is produced from this fatty acid?
3. Which chemical class(es) of hormones is (are) transported in blood while bound to carrier proteins? Does such binding generally *increase* or *decrease* the half-life of a hormone?

3 Signal Transduction Mechanisms

Chemical messengers transmit their signals by binding to target cell receptors located either on the plasma membrane, in the cytosol, or in the nucleus. The location of the receptor depends on whether the messenger is capable of permeating cell membranes. In either case, binding of messenger to receptor either changes the activity of proteins (for example, enzymes) already present in the cell or stimulates the synthesis of new proteins. This section describes properties of receptors and the different signal transduction mechanisms that are set into motion by them.

Properties of Receptors

Receptors show *specificity* for the messenger; that is, they generally bind only one messenger or a class of messengers. Observe that in **Figure 8**, messenger 1 can bind to receptor A but not to receptors B and C. Therefore, only cell A is a target cell for messenger 1. Likewise, cell C is the target cell for messenger 2. The binding between a messenger and a receptor is a brief, reversible chemical interaction that is similar to the enzyme-substrate interactions. The strength of the binding between a messenger and its receptor is termed *affinity.*

A single messenger can often bind to more than one type of receptor, and these receptors may have different affinities for the

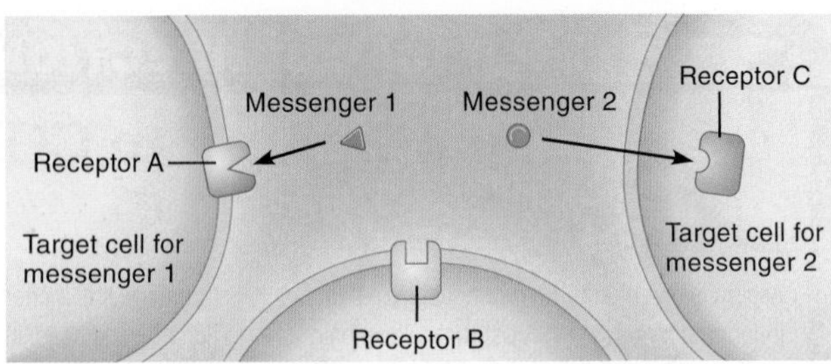

Figure 8 Receptor specificity. Receptor A is specific for messenger 1, receptor C is specific for messenger 2, and neither messenger can bind to receptor B. Note that receptors can be located either on the plasma membrane (receptors A and B) or inside the cell (receptor C).

Q If messenger 2 is to bind receptor C (located inside the target cell), it must cross the plasma membrane of the target cell. Which chemical property must messenger 2 possess to enable its entry into the cell?

messenger. For example, the catecholamine chemical messengers epinephrine and norepinephrine can both bind to *adrenergic* receptors (epinephrine is also called *adrenaline,* and norepinephrine is also called *noradrenaline*). Several different types of alpha (α) and beta (β) adrenergic receptors exist, including α_1, α_2, β_1, β_2, and β_3. α receptors have a greater affinity for norepinephrine than for epinephrine; thus, if norepinephrine and epinephrine are present in equal concentrations, a given α receptor is more likely to bind norepinephrine than epinephrine. β_2 receptors, by contrast, have a greater affinity for epinephrine than for norepinephrine. β_1 and β_3 receptors have approximately equal affinities for norepinephrine and epinephrine.

A single target cell may possess receptors for more than one type of messenger. For example, skeletal muscle cells have receptors for both the neurotransmitter acetylcholine and the hormone insulin. Acetylcholine receptors are directly involved in stimulating muscle contraction, whereas insulin receptors are involved in stimulating glucose uptake and glycogen synthesis in muscle cells.

The Relationship Between Receptor Binding and the Magnitude of the Target Cell Response

As a general rule, the magnitude of a target cell's response to a chemical messenger depends on three factors: (1) the messenger's concentration, (2) the number of receptors present, and (3) the affinity of the receptor for the messenger.

The response of a target cell generally increases as the concentration of messenger increases. This relationship reflects the fact that messengers usually exert their effects by binding reversibly to target cell receptors, as shown in the following reaction:

$$M + R \rightleftharpoons MR \rightarrow \text{response}$$

where M is the messenger, R is the receptor, and MR indicates that the messenger is bound to the receptor. As messenger concentration increases, the reaction is driven to the right. **Figure 9** summarizes the relationship between the concentration of messenger and the number of bound receptors. As the concentration of messenger increases, the proportion of bound receptors increases until all receptors have messengers bound to them—a condition in which the system is said to be *100% saturated.*

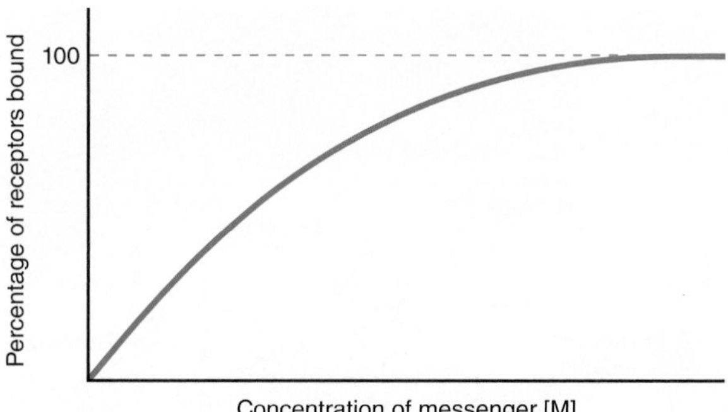

Figure 9 Effect of messenger concentration on messenger-receptor binding. The proportion of receptors bound increases as the concentration of messenger increases. Because the amount of bound receptor determines the magnitude of target cell response, the *y*-axis could also have been labeled "target cell response."

The target cell's response also depends on the number of receptors it possesses. The more receptors there are (the higher their density), the more likely it is that a messenger will bind to a receptor (**Figure 10a**). Thus, at any given concentration of messenger, the number of bound receptors will be greater when more receptors are present, and the response will be stronger.

The number of receptors that a target cell possesses can vary under different circumstances as a result of the synthesis of new receptors or turnover of old receptors. *Up-regulation*—an increase in the number of receptors compared to "normal" conditions—occurs when cells are exposed to low messenger concentrations for a prolonged period. By producing more receptors, target cells adapt to the relative lack of messenger by becoming more responsive to it. For example, in a person with undiagnosed low levels of thyroid hormones, target tissues start expressing more thyroid hormone receptors. Once this condition is diagnosed, the physician will prescribe continuously increasing doses of synthetic thyroid hormones, allowing the target tissues to express fewer and fewer receptors until a dose is reached where the body has a normal response to the synthetic thyroid hormone. Had the physician initially prescribed this final dose, the patient would have had an enhanced and potentially dangerous response to the drug due to the hypersensitivity of the target tissue. While this example demonstrates how up-regulation may be produced through use of medications, in some cases up-regulation of receptors is a normal response that serves to prime the target tissue.

Down-regulation—a decrease in the number of receptors—occurs when messenger concentrations remain higher than normal for a prolonged period. In this case, target cells adapt by producing fewer receptors and becoming less responsive to the messenger. For example, patients may develop *tolerance* to certain drugs because drugs often act by binding to receptors for messengers. The receptors then down-regulate their activity in the presence of high levels of drugs.

The target cell response also depends on the affinity of its receptors for the messenger. When a messenger is present at a given concentration, receptors with higher affinity are more likely to become bound than are receptors with lower affinity, as shown in Figure 10b. Consequently, target cells possessing high-affinity receptors will respond more strongly to a given messenger, all else being equal.

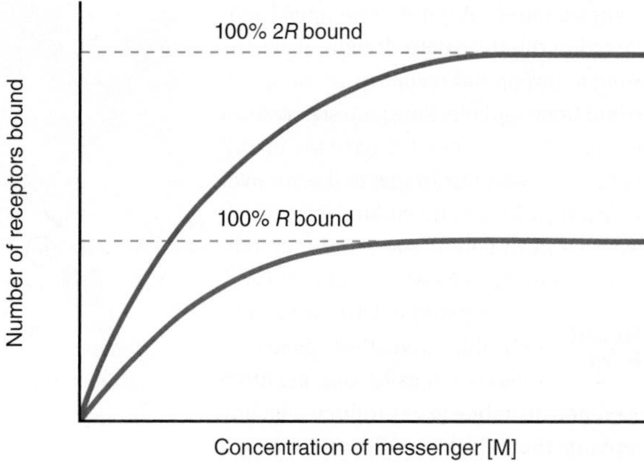

(a) Effects of receptor concentration

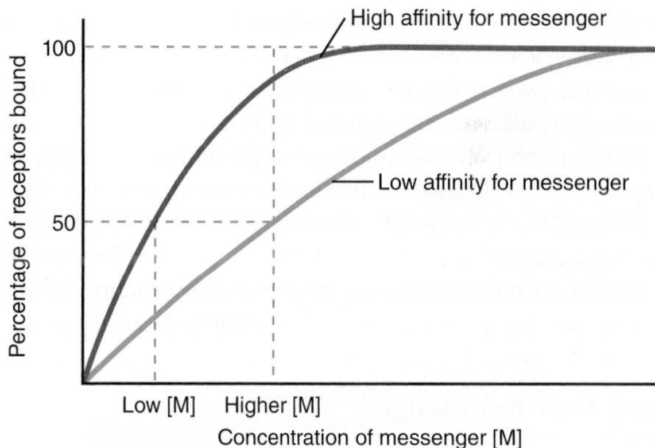

(b) Effects of receptor affinity

Figure 10 Effects of receptor concentration and affinity on messenger-receptor binding. (a) Effects of receptor concentration. The two curves illustrate the effects of doubling the concentration of a given receptor on a target cell. *R* refers to a given concentration of the receptor; 2*R* refers to twice that concentration of that receptor. When the receptor concentration doubles, the maximum number of receptors that can bind with messenger also doubles. **(b)** Effects of receptor affinity. The two curves illustrate the effects of receptor affinity on the proportion of receptors with messenger bound to them. At any concentration of messenger less than 100% saturation, a higher proportion of high-affinity receptors have bound messenger compared to low-affinity receptors. The high-affinity receptors reach saturation at a lower messenger concentration than do the low-affinity receptors. Note that the maximum number of sites that can be bound is independent of receptor affinity.

Receptor Agonists and Antagonists

Many pharmaceutical companies have designed or are designing drugs that interact with messenger receptors (see **Discovery: Antihistamines**). Many of these drugs are ligands that exert their effects by binding to the receptor and acting as an *agonist* or an *antagonist*. Ligands that bind to receptors and produce a biological response are called **agonists. Antagonists,** in contrast, are ligands that bind to receptors but do not produce a response; instead, these agents may actually compete with agonists for the receptor, decreasing the likelihood that the binding of agonist to receptor will occur and bring about a response.

For example, *morphine* is an opioid agonist used in the treatment of pain. It acts by binding to mu opioid receptors in the spinal cord and brain and blocking pain signals from reaching consciousness. We have mu opioid receptors because our brains make our own opioids (enkephalins and endorphins), which prevent us from feeling pain under certain circumstances, such as when a person is extremely stressed. Unfortunately, morphine and other opioid agonists, such as heroin, are often abused because they also produce a feeling of euphoria that is highly addictive. Because a tolerance to morphine develops over time, addicts usually require increasingly higher doses of morphine to produce the same level of euphoria. Opioid overdoses are relatively common in drug addicts and have the potential to be lethal, owing to opioids' suppression of breathing and heart function. If found in time, however, a person who has overdosed on opioids can be treated with *naloxone*, an opioid antagonist that binds to mu opioid receptors with more than 10 times the strength of morphine; in other words, the mu receptors have a 10 times greater affinity for naloxone than for morphine. By binding the mu receptors, naloxone prevents the morphine from binding there, thereby reversing the effects of morphine on the user.

Health Fact

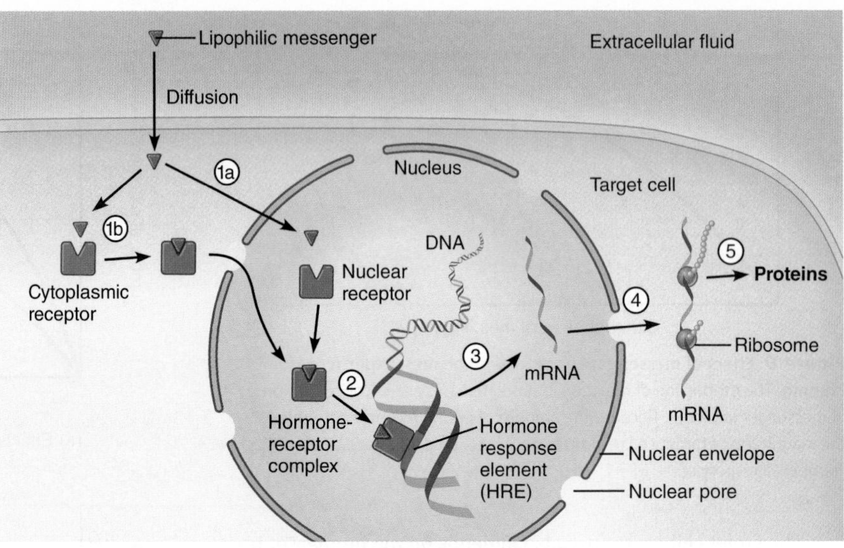

Figure 11 Actions of lipophilic hormones on the target cell. See the text for a detailed description.

Apply Your Knowledge

Figure 9 shows the effects of the concentration of a messenger on the percentage of receptors bound. How would adding an antagonist to the extracellular fluid affect the percentage of receptors with messenger bound? Draw a graph depicting the percentage of receptors bound on the *y*-axis and the concentration of messenger on the *x*-axis. Include three curves representing (1) no antagonist present, (2) low concentration of antagonist, and (3) high concentration of antagonist.

Signal Transduction Mechanisms for Responses Mediated by Intracellular Receptors

Receptors for lipophilic messengers are usually located in the cytosol or nucleus of target cells and are readily accessible because these messengers easily permeate the plasma membrane. The binding of the messenger to the receptor alters the synthesis of a specific protein (or proteins) by the mechanism depicted in **Figure 11**, which shows the action of a lipophilic hormone:

(1a) If a receptor is located in the nucleus, the hormone diffuses into the nucleus and binds to it, forming a hormone-receptor complex.

(1b) If a receptor is located in the cytosol, the hormone binds to it there, forming a hormone-receptor complex that then enters the nucleus.

FOCUS ON DIABETES

Type 1 diabetes mellitus is caused by an insufficient amount of insulin, a hormone produced by beta cells in the pancreas. Insulin is a peptide hormone that is released by exocytosis in response to high blood glucose levels, among other things. Once in the blood, insulin can travel throughout the body, increasing glucose uptake and energy storage in most cells of the body. It exerts its effects by binding to an enzyme-linked receptor, with the associated enzyme being a tyrosine kinase. Thus insulin causes phosphorylation of proteins in the target cell. The ultimate effect of insulin in most target cells is to increase glucose uptake by stimulating greater expression of certain glucose transporters (GLUT-4) in the plasma membrane of the target cells, thereby removing glucose from the blood. Insulin also alters metabolism in its target cells, although the metabolic pathways affected vary for different types of target cells.

When insulin levels are insufficient, glucose uptake into cells decreases, leading to an increase in blood glucose levels. The resulting hyperglycemia causes most of the adverse effects of diabetes mellitus. In addition to special diets, persons with type 1 diabetes mellitus generally require insulin injections to maintain normal blood glucose levels.

② Inside the nucleus, the hormone-receptor complex functions as a transcription factor by binding to a region of DNA called the *hormone response element (HRE)*, which is located at the beginning of a specific gene.

③ Binding of the complex to the HRE activates or deactivates the gene, which affects transcription of mRNA and ultimately increases or decreases synthesis of the protein coded by the gene. In the example in Figure 11, the gene is activated and mRNA synthesis is increased.

④ The mRNA moves into the cytosol.

⑤ The mRNA is translated by ribosomes to yield proteins.

Because changes in protein synthesis can take hours or even days, the effects of lipophilic messengers are generally slow to develop. In addition, because these newly synthesized proteins often remain in the target cells long after the messenger is gone, the effects can persist a long time.

Mechanism of Action for Steroid Hormones

Although many types of steroid hormones are found in the human body, their actions on target cells are quite similar. As described earlier for lipophilic messengers, steroid hormones typically bind to receptors in the cytosol, although some can bind nuclear receptors and others bind receptors on the plasma membrane (actions for plasma membrane receptors are described later in this chapter). The receptors for steroid hormones are known as **nuclear receptors**. For the steroid hormone-receptor complex to activate (or deactivate) a particular gene, two complexes must bind to the HRE in a process called *dimerization* (**Figure 12a**).

Mechanism of Action for Thyroid Hormones and 1,25-Dihydroxyvitamin D_3

Thyroid hormones and 1,25-dihydroxyvitamin D_3 also bind nuclear receptors to alter transcription. Like steroid hormones, dimerization is required for these hormones to alter transcription, but only one of the dimers binds the hormone; the other binds 9-*cis*-retinoic acid, a derivative of vitamin A. Thus two different receptor complexes, or *heterodimers*, must bind the HRE to alter transcription (Figure 12b).

Signal Transduction Mechanisms for Responses Mediated by Membrane-Bound Receptors

Lipophobic messengers cannot permeate the plasma membrane to any significant degree. Thus their receptors are located on the plasma membrane, with the binding site facing the extracellular

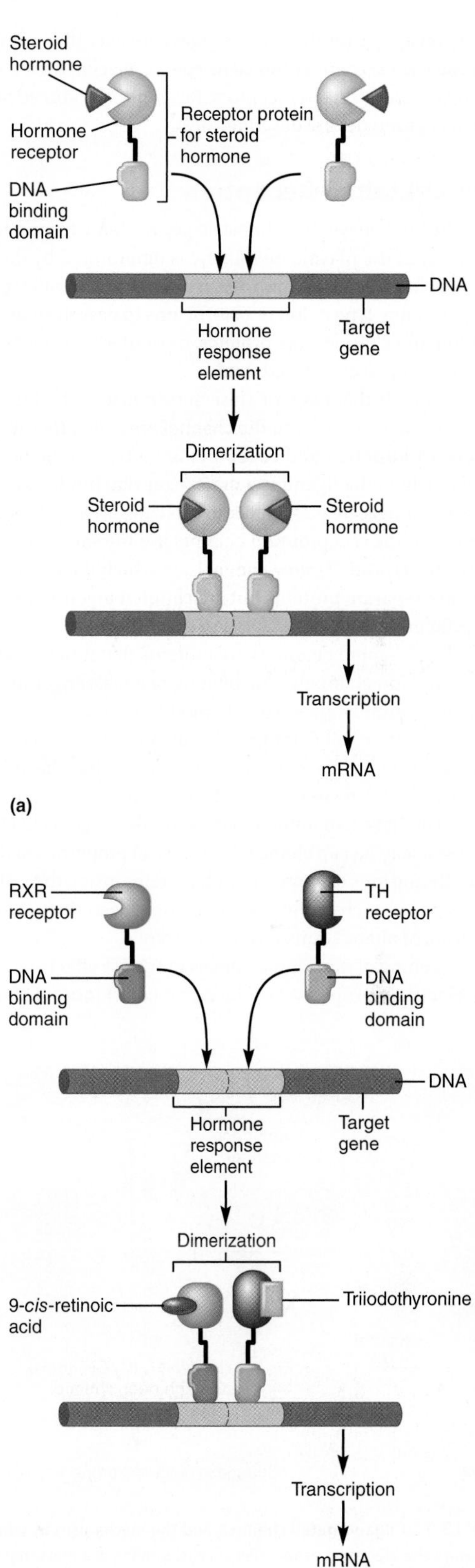

Figure 12 Mechanism of action for steroid and thyroid hormones. (a) Steroid hormones bind to intracellular receptors to form a dimer that alters transcription. **(b)** Thyroid hormones and 9-*cis* retinoic acid bind to nuclear receptors to form a heterodimer that alters transcription.

fluid. The receptors for these messengers are classified into three general categories: channel-linked receptors, enzyme-linked receptors, and G protein–linked receptors. Responses produced at each of these receptor types are described next.

Channel-Linked Receptors

Because the lipid bilayer has almost no permeability to ions, the ion permeability of the plasma membrane is determined by the presence of ion channels in it. These ion channels are generally specific, allowing only one type of ion or class of ions to pass through them. In addition, ion channels are proteins, most of which can be regulated between open and closed states.

Ion channels that open or close in response to the binding of a chemical to a receptor or to the channel are called **ligand-gated channels. Channel-linked receptors** are a type of ligand-gated channel in which the ligand is a messenger that binds to a receptor. These channels are classified into two categories: (1) fast channels, in which the receptor and channel are the same protein (as described next), and (2) slow channels, in which the receptor and channel are separate proteins but are coupled together by a third type of protein, called a *G protein* (described shortly).

Fast ligand-gated channels are proteins that function as both receptors and ion channels. The binding of a messenger to the receptor/ion channel causes the channel to open, increasing the membrane's permeability for that specific ion. Open ion channels allow a specific ion or class of ions to move across the plasma membrane down its electrochemical gradient. Ion movement into or out of the cell can have two different effects on the target cell: (1) Ions entering and leaving can change the electrical properties of the cell and (2) entering ions can interact with proteins inside the cell to induce a response such as muscle contraction, secretion, change in metabolism, or altered transport of a substance.

The opening of most ion channels produces effects by changing the electrical properties of the target cell (**Figure 13**). Recall that a difference in potential exists across the plasma membrane at rest, such that the inside of the cell is negative relative to the outside. Ions that move through an open channel carry with them a charge that changes the membrane potential. For example, the neurotransmitter acetylcholine stimulates skeletal muscle contraction by binding to nicotinic cholinergic receptors on skeletal muscle cells, thereby inducing the opening of ion channels. Sodium ions move into the cell (potassium also moves out, albeit to a lesser degree), carrying a positive charge into the muscle cell. Given that the receptor and ion channel are the same protein molecule, these changes in membrane potential begin rapidly, within a millisecond. Because the binding of messenger to receptor is brief, and the channel remains open only while the messenger is bound, the change in membrane potential usually terminates in a few milliseconds. Changes in membrane potential are crucial to the function of neurons and muscle cells.

In other cases, fast ligand-gated channels exert their effects by opening calcium channels (**Figure 14**). When these channels

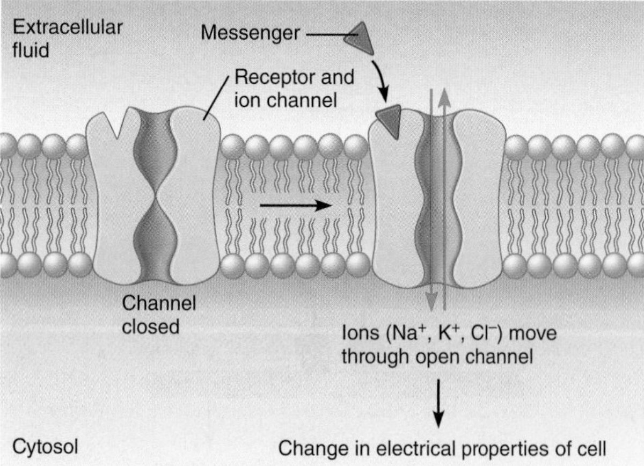

Figure 13 Fast ligand-gated channels and the mechanism by which they change the electrical properties of cells. Binding of a messenger to the receptor/channel opens the ion channel. The opening of most ion channels results in movement of ions into or out of the cell, which in turn changes the cell's electrical properties.

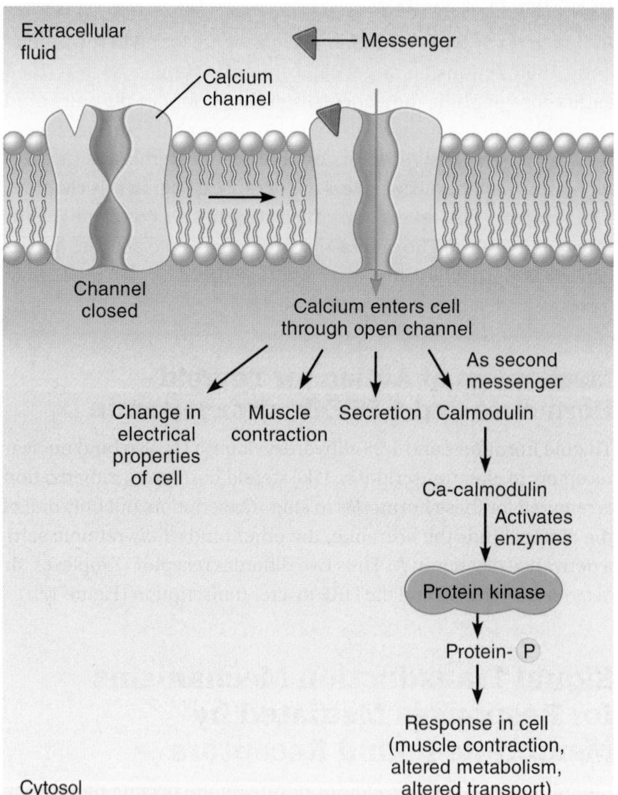

Figure 14 Fast ligand-gated calcium channels. Binding of a messenger to the receptor/channel opens calcium channels, enabling calcium ions to enter the cell. Calcium entry will change the electrical properties of the cell, but can also trigger a variety of responses, such as secretion of some product by exocytosis, muscle contraction, or a change in a protein's activity. In the last instance, calcium acts as a second messenger, binding to the protein calmodulin to form a calcium-calmodulin complex. This complex activates a protein kinase, which phosphorylates a protein that produces a response in the cell.

open, calcium ions enter the cell. Depending on the target cell, the interaction of these calcium ions with intracellular proteins may then trigger a variety of responses, including muscle contraction, secretion of a product by exocytosis, and functioning as a **second messenger**—an intracellular messenger produced by the binding of an extracellular messenger (the *first messenger*) to a receptor. As a second messenger, calcium binds to a cytosolic protein called **calmodulin.** The resultant calcium-calmodulin complex usually activates a protein kinase, an enzyme that catalyzes the phosphorylation of a protein, thereby altering its structure and function through covalent regulation.

Calcium is well suited for its role in intracellular signaling because it is normally present in very low concentrations in the cytosol (10^{-7} to 10^{-6} molar), as compared to its usual concentration of 10^{-3} molar in extracellular fluid. The significance of this low cytosolic concentration of calcium is that entry into a cell of even a small quantity of calcium causes a relatively large percentage change in the concentration, which means that the system is sensitive. Intracellular calcium levels are maintained at their normal low levels by three processes that remove calcium ions from the cytosol: (1) active transport of calcium across the plasma membrane, (2) sequestration of calcium by binding with proteins in the cytosol, and (3) active transport of calcium into certain organelles, such as the smooth endoplasmic reticulum and mitochondria.

Enzyme-Linked Receptors

Certain receptor proteins, known as **enzyme-linked receptors,** function both as enzymes and as receptors. In these transmembrane proteins, the receptor side faces the interstitial fluid and the enzyme side faces the cytosol. The enzymes become activated when a messenger binds to the receptor, which allows them to catalyze intracellular reactions.

Most enzyme-linked receptors are *tyrosine kinases,* which catalyze the addition of a phosphate group to the side chains of the amino acid tyrosine in certain locations in target proteins. The following events occur at the tyrosine kinase receptors (**Figure 15**):

① A messenger binds to the receptor, changing its conformation.

② The conformation change activates the tyrosine kinase.

③ The tyrosine kinase then catalyzes phosphorylation of an intracellular protein.

④ Phosphorylation of the protein changes its activity by covalent regulation, bringing about a response in the target cell.

An example of a messenger that uses the tyrosine kinase signal transduction is the hormone insulin. Other enzyme-linked receptors include *guanylate cyclases,* which catalyze the conversion of *GTP (guanosine triphosphate)* to the second messenger *cGMP (cyclic guanosine monophosphate)*. cGMP then activates a protein kinase, which catalyzes phosphorylation of a protein.

G Protein–Linked Receptors

G protein–linked receptors work by activating special membrane proteins called **G proteins.** (See **Clinical Connections: Cholera and G Proteins**, for an example of the significance of G proteins.) G proteins are located on the intracellular side of the plasma membrane, where they function as links between the G protein–linked

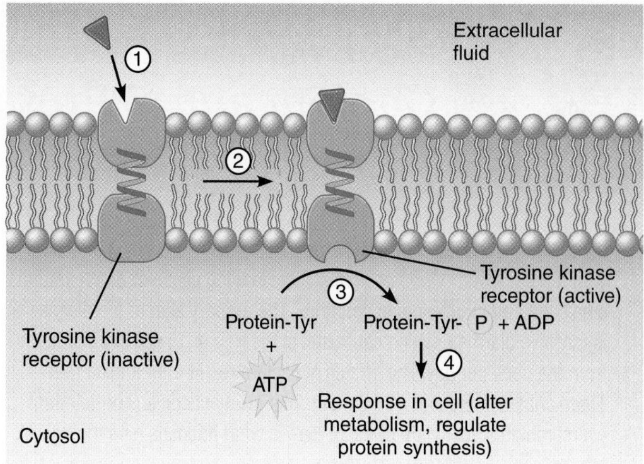

Figure 15 An enzyme-linked receptor. The enzyme-linked receptor depicted here is tyrosine kinase. See the text for a detailed description.

receptor and other proteins in the plasma membrane, called effectors. Effectors include ion channels and enzymes. G proteins, which get their name from their ability to bind guanosine nucleotides, include three subunits: alpha (α), beta (β), and gamma (γ). The guanosine binding site is in the alpha subunit. In its inactive state, a G protein binds GDP (guanosine diphosphate). When a messenger binds to the G protein–linked receptor, the G protein releases the GDP, binds a molecule of GTP (guanosine triphosphate), and becomes active, as shown in the following equation:

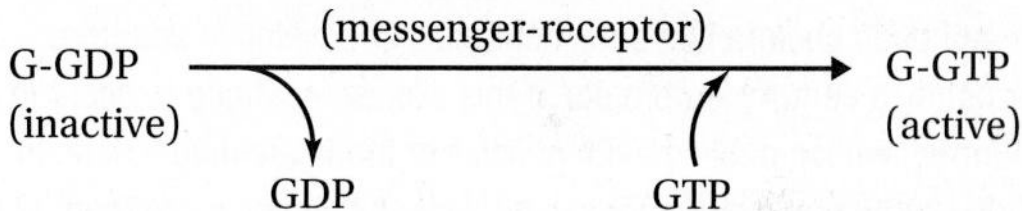

In its active state, the alpha subunit separates from the beta and gamma subunits, leaving a beta-gamma dimer. The alpha subunit, and sometimes the beta-gamma dimer, moves to the effector(s), causing a change in the effector's activity. The G protein does not stay active very long because it also functions as an enzyme that hydrolyzes the GTP, in the process returning itself to the inactive state with bound GDP:

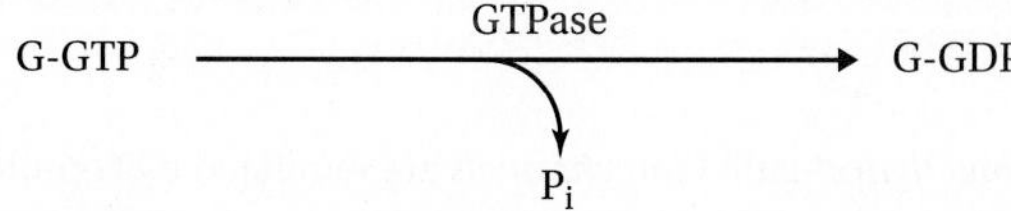

G proteins are a diverse group, formed from at least 20 different alpha subunits, 5 beta subunits, and 13 gamma subunits. Functionally, G proteins can be classified into three basic types: (1) those that affect ion channels, (2) *stimulatory G proteins,* and (3) *inhibitory G proteins.* Stimulatory G proteins (G_s proteins) and inhibitory G proteins (G_i proteins) are associated with the activation and inhibition, respectively, of **amplifier enzymes,** enzymes that catalyze the production of second messengers in the intracellular fluid. (Other types of amplifier enzymes are also found in cells, but they will not be covered in this text.) The functions of some different types of G proteins are discussed next.

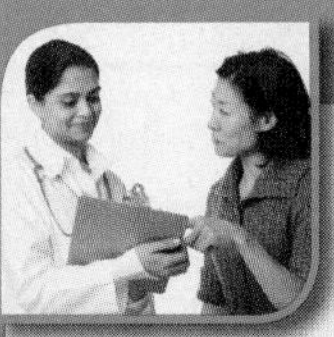

CLINICAL CONNECTIONS

Cholera and G Proteins

G proteins are susceptible to attack by several bacterial toxins, such as the toxin that causes cholera. Cholera, a disease caused by infection with the bacterium *Vibrio cholerae,* is the leading cause of death among small children in developing countries. The primary sign of the disease is massive diarrhea, which can result in the loss of 15–20 liters of fluid from the body per day. The effects of cholera result from a toxin that *Vibrio cholerae* releases into the gut. This toxin acts on a G protein that overstimulates adenylate cyclase. Here is what happens (see the figure):

① The cholera toxin binds to a membrane *ganglioside* (a phospholipid with carbohydrate residues attached) on secretory cells in the small intestine.
② A toxin subunit enters the cell, causing the sustained activation of a G protein.
③ This G protein activates adenylate cyclase.
④ Adenylate cyclase catalyzes the formation of cAMP.
⑤ cAMP activates protein kinases.
⑥ Phosphorylation of proteins enhances the secretion of chloride ions.
⑦ The flow of negatively charged chloride ions out of the cell causes positively charged sodium ions to follow them.
⑧ Water follows the electrolytes into the lumen of the small intestine by osmosis, resulting in severe diarrhea.

Although cholera has been controlled in developed countries for nearly a century, outbreaks of this disease continue to occur in countries whose public health policies or funds are inadequate. In 2008–2009, a cholera epidemic hit Zimbabwe, with more than 30,000 individuals becoming infected and 1500 deaths occurring. Citizens of Zimbabwe, lacking sufficient treated water supplies, had been getting their drinking water from streams contaminated with sewage. *Vibrio cholerae* is generally transmitted through the feces of infected individuals.

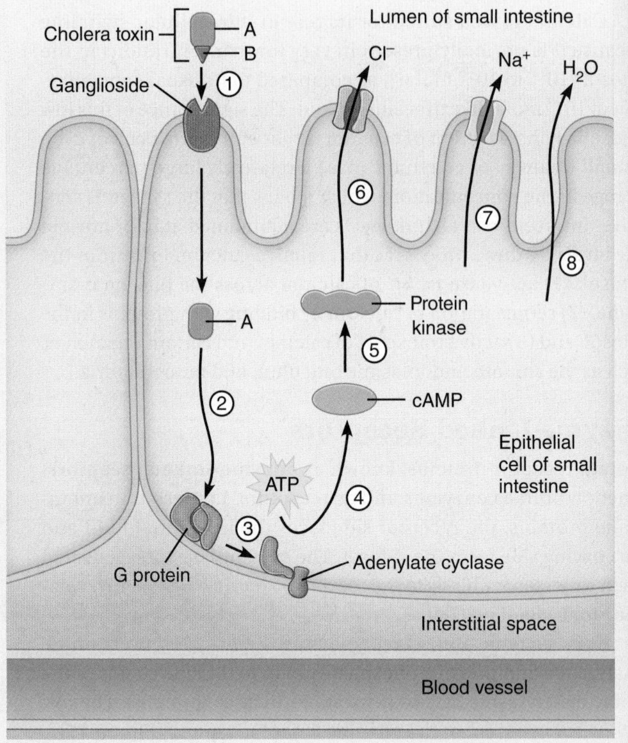

Critical Thinking Questions

1. Describe how the cholera toxin induces excessive water loss from the body.
2. Describe the cycle of exposure to the cholera toxin that perpetuates this disease.
3. What do you think would be the best approach to eradicating cholera worldwide?

Slow ligand-gated ion channels are regulated by G proteins, which cause the channels to open or close in response to a messenger binding to its receptor (**Figure 16**). When the G protein is activated, the alpha subunit moves to the ion channel, causing a conformational change that leads it to open or close. These channels exert effects similar to those exerted by fast ligand-gated channels, albeit with two important differences:

1. At fast ligand-gated channels, messenger binding to receptors only *opens* the channel, so it increases the permeability of the target cell for the specific ion. By contrast, G protein–linked ion channels can be either *opened* or *closed* by messenger binding to the receptor.
2. Binding of a messenger to fast ligand-gated channels produces an immediate and brief (only a few milliseconds) response in the target cell. In contrast, G protein–linked ion channels are slow to open or close in response to receptor binding and stay open or closed for longer periods, often minutes.

G protein–regulated enzymes are associated with the production of second messengers in the cytosol. Five major second messengers account for most of the communication through G protein–regulated enzymes: (1) *cAMP* (cyclic adenosine monophosphate), (2) cGMP, (3) *inositol triphosphate,* (4) *diacylglycerol,* and (5) calcium. These second messenger systems, which are described next, are summarized in Table 3.

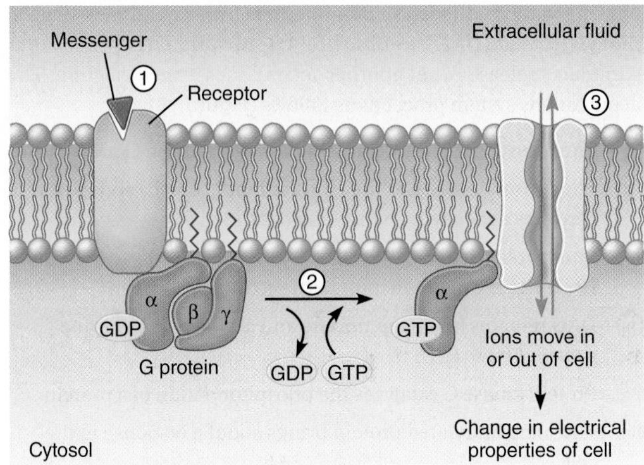

Figure 16 Action of a G protein on a slow ligand-gated ion channel. ① Binding of the messenger to the receptor activates the G protein. ② The alpha subunit moves to an ion channel in the membrane. ③ The ion channel opens or closes, changing the permeability of the membrane to a specific ion. The movement of ions across the plasma membrane is altered, changing the electrical properties of the cell.

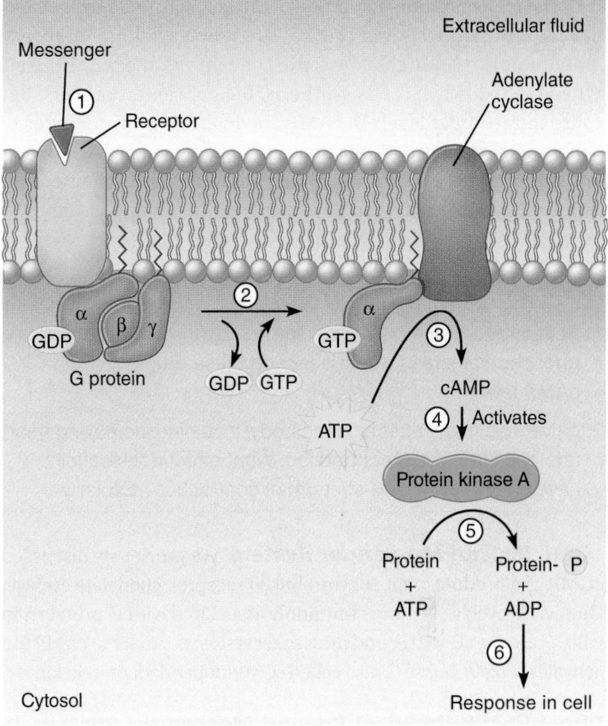

Figure 17 The cAMP second messenger system. See the text for a detailed description.

cAMP Second Messenger System The mechanisms of action of cAMP, the most common second messenger, are as follows (**Figure 17**):

① The first messenger binds to the receptor, activating a G_s protein. (Some messengers inhibit the cAMP second messenger system by activating a G_i protein, which is not shown in Figure 17.)

② The G protein releases the alpha subunit, which binds to and activates the enzyme **adenylate cyclase.**

③ Adenylate cyclase catalyzes the conversion of ATP to cAMP.

④ cAMP activates *protein kinase A,* also called cAMP-dependent protein kinase.

⑤ The protein kinase catalyzes the transfer of a phosphate group from ATP to a protein, thereby altering the protein's activity through covalent regulation.

⑥ Altered protein activity causes a response in the cell.

Termination of the actions of cAMP requires its degradation by the enzyme *cAMP phosphodiesterase.* For the actions of the phosphorylated protein to be terminated, the phosphate group must be removed from the protein by a chemical reaction. The enzymes that dephosphorylate a protein are called *phosphoprotein phosphatases,* and these enzymes are also regulated. Thus the state of a cAMP-regulated protein (whether it is phosphorylated or not) depends on the relative concentration and activity of the phosphoprotein phosphatases and protein kinases, the latter of which depend on cAMP levels in the cell.

TABLE 3 Second Messenger Systems

Second messenger	Precursor	Amplifier enzyme	Usual action	Examples of first messengers in the system
Cyclic adenosine monophosphate (cAMP)	ATP	Adenylate cyclase	Activates protein kinase A	Epinephrine, vasopressin, ACTH, glucagon
Cyclic guanosine monophosphate (cGMP)	GTP	Guanylate cyclase	Activates protein kinase G	Atrial natriuretic peptide, endothelins
Diacylglycerol (DAG)	Inositol-4,5-bisphosphate (PIP_2)	Phospholipase C	Activates protein kinase C	Angiotensin II, histamine, vasopressin
Inositol triphosphate (IP_3)	Inositol-4,5-bisphosphate (PIP_2)	Phospholipase C	Stimulates calcium release from intracellular stores	Angiotensin II, histamine, vasopressin
Calcium*	None	None	Binds to calmodulin, then activates a protein kinase	Angiotensin II, histamine, vasopressin

*Calcium increases in the cytosol in response to opening of ion channels either in the plasma membrane or in certain organelles.

The concentration of cAMP in a cell is determined by the relative rates of synthesis and breakdown. When synthesis proceeds more rapidly than breakdown, the concentration of cAMP rises. When breakdown exceeds synthesis, the concentration falls. Because intracellular levels of cAMP are determined by the rates of two competing enzymes (one that synthesizes it, adenylate cyclase, and one that breaks it down, phosphodiesterase), the effects of stimulating one enzyme can be mimicked by inhibiting the other.

Apply Your Knowledge

1. Several drugs, including caffeine, inhibit the enzyme cAMP phosphodiesterase. What effect do these drugs have on cAMP levels?
2. cAMP has many effects in the body, including increasing heart rate and strength of contraction. What effect does caffeine have on heart rate and strength of contraction? Explain.

cGMP Second Messenger System We previously discussed cGMP as a product of an enzyme-linked receptor, guanylate cyclase. Guanylate cyclase is more commonly associated with G proteins, in which case the cGMP second messenger system is similar to cAMP but activates *protein kinase G,* also called cGMP-dependent protein kinase.

Phosphatidylinositol Second Messenger System In the **phosphatidylinositol system,** a membrane phospholipid, *phosphatidylinositol-4,5-bisphosphate (PIP_2),* undergoes an enzyme-catalyzed reaction that liberates two second messengers, **diacylglycerol (DAG)** and **inositol triphosphate (IP_3);** the latter stimulates release of yet another second messenger, calcium. The action of this system proceeds as follows (**Figure 18**):

(1) The messenger binds to its receptor, activating a G protein.

(2) The G protein releases the alpha subunit, which binds to and activates the enzyme *phospholipase C.*

(3) Phospholipase C catalyzes the conversion of PIP_2 to DAG and IP_3, each of which functions as a second messenger.

(4a) DAG remains in the membrane and activates the enzyme *protein kinase C.*

(5a) Protein kinase C catalyzes the phosphorylation of a protein.

(6a) The phosphorylated protein brings about a response in the cell.

At the same time,

(4b) IP_3 moves into the cytosol.

(5b) IP_3 triggers the release of calcium from the endoplasmic reticulum.

Depending on the cell, calcium then does one of two things:

(6b) It acts on proteins to stimulate contraction or secretion.

(6c) It acts as a second messenger by binding to calmodulin, thereby activating a protein kinase that phosphorylates a protein that produces a response in the cell.

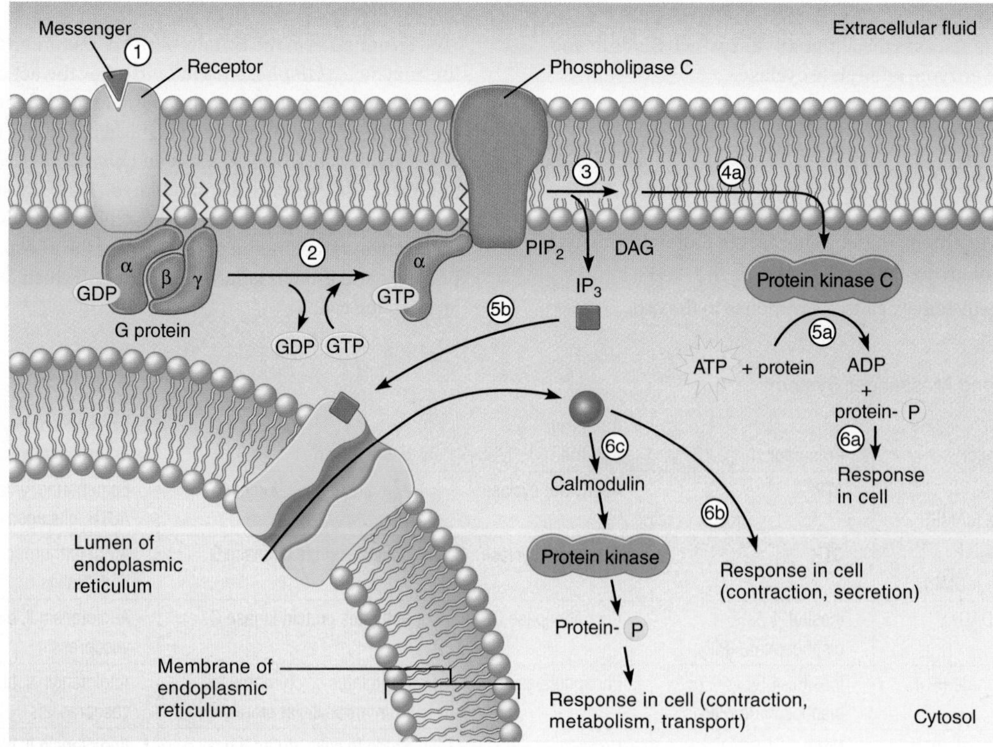

Figure 18 The phosphatidylinositol second messenger system. See the text for a detailed description.

TABLE 4 Signal Transduction Mechanisms for Selected Chemical Messengers

Messenger	Functional class	Chemical class	Signal transduction mechanism
Epinephrine	Hormone, neurotransmitter	Amine	G protein–coupled receptors
Thyroid hormones	Hormone	Amine (lipophilic)	Altered transcription of mRNA
Vasopressin (ADH)	Hormone, neurotransmitter	Peptide	G protein–coupled receptors
Insulin	Hormone, neurotransmitter	Peptide	Enzyme-linked receptors
Estrogen	Hormone	Steroid	Altered transcription of mRNA
Glutamate	Neurotransmitter	Amino acid	Channel-linked receptor, G protein–coupled receptors
Serotonin	Neurotransmitter, paracrine	Amine	Channel-linked receptor, G protein–coupled receptors
Prostaglandins	Paracrine	Eicosanoid	G proteins, unknown for many
Interleukins	Paracrine	Peptide	Enzyme-linked receptors
GABA	Neurotransmitter	Amino acid	Channel-linked receptor

The signal transduction mechanisms for selected chemical messengers in various functional and chemical classes are listed in Table 4.

Signal Amplification in Chemical Messenger Systems

Given that some cells have simple signal transduction mechanisms such as the one-step activation of tyrosine kinase, we might wonder why second messenger systems are needed, when they ultimately have the same basic effect as tyrosine kinase—that is, phosphorylation of a protein. The reason relates to one of the striking features of second messenger systems—the ability of relatively small changes in the concentration of a chemical messenger to elicit marked responses in target cells, a phenomenon known as *signal amplification.*

Figure 19 depicts how signal amplification works, using the cAMP system as an example. The diagram illustrates how a single activated receptor can activate several G proteins, each of which can in turn activate an adenylate cyclase molecule. While active, each adenylate cyclase can generate hundreds of molecules of cAMP, each of which then activates a molecule of protein kinase A. Each protein kinase A molecule can then phosphorylate hundreds of target proteins. The net result is that a large number of end-product molecules can be regulated in response to the binding of a single ligand molecule to its receptor. In the example shown in Figure 19, the action of one first messenger led to the phosphorylation of approximately 2,500,000 proteins.

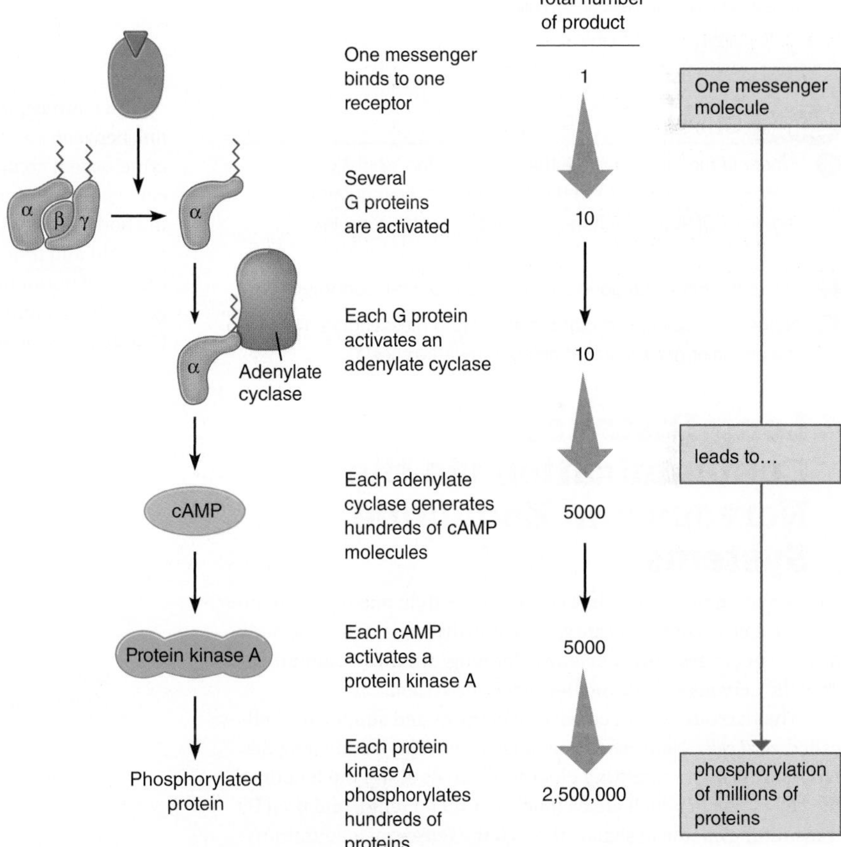

Figure 19 Signal amplification, in this case by the second messenger cAMP.

The sequence of reactions shown in Figure 19 is an example of a **cascade**—a series of sequential steps that progressively increase in magnitude, in much the same manner as numerous tiny brooks, added together, eventually become a major river. Cascades of one type or another are common in chemical messenger systems and account for much of the signal amplification that occurs.

Now that our discussion of the various aspects of chemical messengers is complete, Table 5 summarizes the properties of chemical messengers. We turn next to a discussion of long-distance communication in the body.

TABLE 5 Properties of Chemical Messengers

Property	Lipophobic messenger (hydrophilic)	Lipophilic messenger (hydrophobic)
Chemical classes	Amino acids, amines, peptides	Steroids, eicosanoids, thyroid hormones*
Storage in secretory cell	Secretory vesicles	None
Mechanism of secretion	Exocytosis	Diffusion
Transport in blood†	Dissolved	Bound to carrier protein
Location of receptor	Plasma membrane	Cytosol or nucleus
Signal transduction mechanism	Open/close ion channels‡ Activate membrane-bound enzymes G proteins and second messenger systems	Alter transcription of mRNA (alter protein synthesis)
Relative time to onset of response	Fast	Slow
Relative duration of response	Short	Long
Relative half-life	Short	Long

*Thyroid hormones are amines but are lipophilic.
†Refers to hormones and certain cytokines only.
‡Some of these effects are mediated by G proteins.

Quick Check 3

1. Where in the target cell are the receptors for lipophilic messengers located? What is the name of the regulatory region of DNA to which the hormone-receptor complex binds?
2. Name the three categories of membrane-bound receptors.
3. Name five substances that act as second messengers. Name the amplifier enzymes that catalyze their synthesis.

4 Long-Distance Communication via the Nervous and Endocrine Systems

To maintain homeostasis, it is crucial that cells in one region of the body be able to communicate with cells in distant regions. The body has two organ systems specialized for long-distance communication: the nervous system and the endocrine system.

The nervous system consists of neurons and supporting cells called *glial cells*. Neurons are capable of communicating long distances, first by transmitting electrical signals along the length of the cell (some of which are as much as 1 meter long), and then by transmitting chemical signals through the release of a neurotransmitter from the axon terminal. The neurotransmitter then diffuses over a short distance to the target cell (**Figure 20**). Because of the direct cell-to-cell communication of chemical signals at a synapse, the nervous system is often considered a "wired" system. (One exception involves the neurohormones, which travel via the bloodstream.) Communication in the nervous system generally involves the opening and closing of ion channels, which is very fast and typically of short duration. Signals transmitted by the nervous system travel quickly and have short lives, making the system ideal for controlling movements and monitoring the world around us.

In contrast, the endocrine system lacks any direct anatomical link between secretory cells and their target cells. Instead, the endocrine system communicates through hormones, with these chemical messengers traveling via the bloodstream to virtually all cells in the body. Strictly speaking, hormones are secreted into the interstitial fluid and then diffuse into blood, but we often speak of the secretion of hormones into blood for the sake of simplicity. Hormones generally communicate by altering protein synthesis or activating G proteins—processes that are considerably slower than the electrical

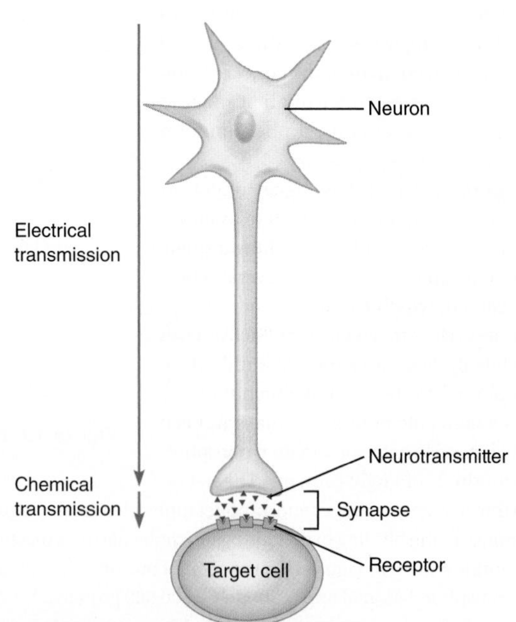

Figure 20 Signal transmission in neurons. Neurons transmit both electrical signals within the cell and chemical signals between cells.

TABLE 6 Characteristics of the Nervous and Endocrine Systems

Characteristic	Nervous system	Endocrine system
Secretory cell	Neuron	Endocrine cell
Target cell	Neuron, muscle, or gland	Most cell types in body
Messenger	Neurotransmitter	Hormone
Pathway for communication	Across synapse	Via bloodstream
Basis of specificity	Receptors on postsynaptic target cell	Receptors on target cells throughout body
Time to onset of effect	Immediate	Delayed
Duration of effect	Brief	Long

and chemical signals used by the nervous system. The relative slowness of the endocrine system and its ability to broadcast signals over wide areas are important in coordinating metabolic activities among organ systems. Various aspects of the nervous and endocrine systems are compared in Table 6.

SYSTEMS INTEGRATION In this chapter, we learned how cells communicate with other cells through chemical messengers. Throughout this text, "communication" is usually assumed to mean "neural or endocrine control of the organ systems to maintain homeostasis."

CHAPTER REVIEW

SUMMARY

1 Mechanisms of Intercellular Communication

- Communication between cells can occur (1) via gap junctions, which allow electrical signals and small molecules to move directly from one cell to adjacent cells, and (2) via the secretion of chemical messengers, which allow signals to be transmitted from one cell to others that may be at distant locations.
- Chemical messengers produce responses in target cells by binding to specific receptors.

MAP iP Nervous II, Anatomy Review

MAP iP Nervous II, Synaptic Transmission

2 Chemical Messengers

- Chemical messengers are classified into three major functional categories: (1) paracrines, (2) neurotransmitters, and (3) hormones.
- Many messengers exert their effects only on cells that are close to the cells that secrete them, but hormones and neurohormones can act at distant sites.
- Messengers are organized into five major classes: amino acids, amines (amino acid derivatives), peptides/proteins, steroids (cholesterol derivatives), and eicosanoids (derivatives of arachidonic acid).
- Steroids, eicosanoids, and some amines (the thyroid hormones) are lipophilic; other messengers are hydrophilic (lipophobic).

MAP iP Nervous II, Synaptic Transmission

MAP iP Nervous II, Anatomy Review

3 Signal Transduction Mechanisms

- The magnitude of a target cell's response to a chemical messenger depends on the messenger's concentration, the number of receptors present, and the affinity of the receptor for the messenger.
- The response of the target cell increases with an increase in messenger concentration. When exposed for long periods to messenger concentrations that are either very low or very high, target cells can alter the number of receptors, bringing about a change in their responsiveness to the messenger.
- A decrease in the number of receptors is called down-regulation; an increase is called up-regulation.
- Lipophilic messengers bind to receptors in the cytosol or nucleus of target cells; the resulting complex binds to DNA to regulate gene transcription and protein synthesis.
- Hydrophilic messengers bind to three types of cell surface receptors: (1) channel-linked receptors, which affect the opening and closing of fast ligand-gated channels; (2) enzyme-linked receptors, which catalyze reactions inside cells; and (3) G protein–linked receptors, which activate specific membrane proteins called G proteins.

- Activated G proteins can activate (or inhibit) a variety of intracellular proteins, including enzymes or channels.
- Many of these enzymes catalyze the formation of second messengers inside the cells, including cyclic AMP (cAMP), cyclic GMP (cGMP), diacylglycerol (DAG), inositol triphosphate (IP_3), and calcium ions (which often work by binding to calmodulin, forming a complex that activates protein kinases).

MAP iP Nervous II, Synaptic Transmission

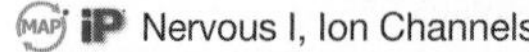

MAP iP Nervous I, Ion Channels

4 Long-Distance Communication via the Nervous and Endocrine Systems

- In the nervous system, neurons send signals to specific groups of target cells, to which they are connected by synapses.
- The endocrine system broadcasts signals to target cells throughout the body, exerting effects that are generally slow and long lasting.
- Hormones are secreted by specialized endocrine cells, which are usually found in endocrine glands.

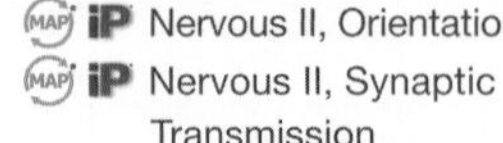

MAP iP Nervous II, Orientation

MAP iP Nervous II, Synaptic Transmission

MAP MasteringA&P®

Check your understanding at MasteringA&P and apply your new knowledge to real-life case studies to ensure that you've mastered the content!

EXERCISES

Multiple-Choice Questions

1. Arachidonic acid is the raw material for the synthesis of
   a) Amines.
   b) Thyroid hormones.
   c) Eicosanoids.
   d) Steroids.
   e) Peptides.
2. Epinephrine is a(n)
   a) Amino acid.
   b) Steroid.
   c) Eicosanoid.
   d) Adrenocorticoid.
   e) Catecholamine.
3. Most chemical messengers fall into which of the following chemical classes?
   a) Amines
   b) Amino acids
   c) Peptides/proteins
   d) Eicosanoids
   e) Steroids
4. All amino acid chemical messengers function as
   a) Paracrines.
   b) Autocrines.
   c) Neurohormones.
   d) Neurotransmitters.
   e) Hormones.
5. All steroid chemical messengers function as
   a) Paracrines.
   b) Autocrines.
   c) Neurohormones.
   d) Neurotransmitters.
   e) Hormones.
6. Which of the following is likely to cause an increase in intracellular cAMP levels?
   a) Stimulation of phosphodiesterase activity
   b) Activation of an inhibitory G protein targeting adenylate cyclase
   c) Binding of chemical messengers to enzyme-linked receptors
   d) Binding of chemical messengers to receptor-operated channels
   e) Stimulation of adenylate cyclase activity
7. Which of the following messenger classes bind to intracellular receptors?
   a) Catecholamines only
   b) Peptides only
   c) Steroids only
   d) Both catecholamines and steroids
   e) Both peptides and steroids
8. The response of a target cell to a messenger depends on which of the following?
   a) Concentration of the messenger
   b) Concentration of receptors on the target cell
   c) Affinity of the receptor for the messenger
   d) Both a and c
   e) All of the above
9. G proteins are involved whenever
   a) Binding of messenger molecules to cell surface receptors triggers a target cell response.
   b) Binding of ligand molecules to cell surface receptors triggers activation or inhibition of enzymes.
   c) Binding of ligand molecules to cell surface receptors triggers synthesis of second messengers.
   d) Binding of ligand molecules to cell surface receptors triggers a change in membrane permeability to ions.
   e) All of the above are true.
10. Which enzyme catalyzes the synthesis of diacylglycerol?
   a) Adenylate cyclase
   b) Tyrosine kinase
   c) Phospholipase C
   d) Protein kinase
   e) Phosphoprotein phosphatase

Objective Questions

11. Cells that secrete a messenger are called ________.
12. An (endocrine/autocrine) agent acts on the same cell that secretes it.
13. Endocrine glands release (neurotransmitter/hormone) into the ________, where it travels to the target cell.
14. When a G protein is activated, it releases a (GDP/GTP) and binds a (GDP/GTP).
15. The enzyme that catalyzes conversion of ATP to cAMP is called ________.
16. Following activation of the phosphatidylinositol system, (IP_3/DAG) liberates calcium from intracellular stores.
17. (Lipophilic/Lipophobic) messengers exert their effect on target cells by activating or inactivating specific genes.
18. Examples of locally acting chemical messengers are (steroids/eicosanoids).
19. Cytosolic calcium often exerts its effect by binding to cytosolic (protein kinase C/calmodulin).
20. Lipophobic messengers are secreted by (exocytosis/diffusion across the cell membrane).
21. Amino acids are (hormones/neurotransmitters).
22. The (nervous system/endocrine system) is a more rapid means of communication.

Essay Questions

23. Describe the different types of chemical signals (paracrine, neurotransmitter, and hormone).
24. Compare lipophilic and lipophobic chemical messengers with respect to the location of their receptors, their transport in blood, and their general signal transduction mechanisms.
25. Describe the effects of activating the enzyme phospholipase C with regard to second messengers produced and their actions in the cell.
26. Describe the cAMP second messenger system, including all the steps from a messenger binding to its receptor to the response in the target cell.
27. Compare long-distance communication by the nervous and endocrine systems with respect to anatomy, speed, and mechanisms of action.
28. Describe the two types of receptor-mediated channels and their mechanisms of action. Which mechanism is faster?
29. Describe the role of protein kinases in signal transduction.

Critical Thinking

30. Describe what would happen in cells that normally secrete epinephrine if a defect occurred in the enzyme phenylethanolamine *N*-methyl transferase (PNMT).
31. The pharmaceutical industry develops many drugs that act as agonists or antagonists at specific receptor types. In designing these drugs, specificity is critical. Explain the benefits of receptor specificity and other chemical properties that must be considered in designing a drug.
32. Although lipophilic messengers almost always alter protein synthesis by binding to internal receptors, lipophobic messengers can alter protein synthesis as well. Explain how.
33. In binding with $alpha_1$ receptors, epinephrine and norepinephrine cause vasoconstriction of blood vessels, which increases blood pressure. By binding to $beta_1$ adrenergic receptors in the heart, epinephrine increases heart rate and the force of contraction, which together cause an increase in blood pressure. Which class(es) of adrenergic agent (alpha or beta, agonist or antagonist) could be used in the treatment of hypertension? Explain your choice(s).

Photo credits for icons in feature boxes: Focus on Diabetes: Tanyae/Dreamstime; Clinical Connections: Iofoto/Dreamstime; Understanding Exercise: Pixattitude/Dreamstime; Discovery: Yuri Arcurs/Cutcaster.

Glossary

adenylate cyclase (ad-den-ah-late sy-klase) intracellular enzyme that catalyzes the conversion of adenosine triphosphate (ATP) to cAMP
calmodulin (kal-mod-you-lin) cytosolic calcium-binding protein; modulates the activity of intracellular proteins
catecholamines amine compounds that contain a catechol group and are derived from the amino acid tyrosine
clotting factors plasma proteins that form blood clots when activated; also called *coagulation factors*
cytokines (sy-toh-kines) peptides or proteins that are usually secreted by immune cells in response to a stimulus
diacylglycerol (DAG) a second messenger released by the phosphatidylinositol system
enzyme-linked receptor plasma membrane protein that functions as both a receptor and an enzyme
growth factors proteins that stimulate the growth and proliferation of a particular type of tissue
half-life time it takes for half the amount of a molecule to be degraded
hormones chemical messengers released from endocrine cells or glands into the interstitial fluid, where they then diffuse into the blood and travel to target cells
inositol triphosphate (IP_3) a second messenger released by the phosphatidylinositol system that stimulates the release of calcium
ligand-gated channels ion channels that open or close in response to a chemical messenger binding to the channel
lipophilic pertaining to a molecule that is lipid soluble, or water insoluble; hydrophobic
lipophobic pertaining to a molecule that is water soluble (not lipid soluble)
neurohormones special class of hormones released by neurosecretory cells
phosphatidylinositol system (fos-fah-tide-il-in-os-ih-tol) a signal transduction system that produces two second messengers, diacylglycerol and inositol triphosphate
presynaptic neuron at a synapse, a neuron that transmits signals to a second neuron or an effector cell
postsynaptic cell cell that receives communication from a neurotransmitter released at a synapse
second messenger an intracellular messenger molecule that is produced in response to the binding of an extracellular messenger (the first messenger) to a receptor
signal transduction the process by which the binding of a chemical messenger to receptors brings about a response in a target cell

Answers

Figure Question Answers

Fig 3 Carboxyl
Fig 8 Messenger 2 must be lipophilic (or hydrophobic)

Apply Your Knowledge Answers

Steroid hormones are synthesized from metabolic pathways, starting with cholesterol. Production of androgens and estrogens follows the same path until it branches with the estrogens being synthesized from androgens. Thus cells in the testes have enzymes for producing the androgens, and the ovaries have enzymes for producing androgens as well as enzymes that convert the androgens to estrogens.

Addition of an antagonist to the extracellular fluid decreases the amount of messenger bound to a receptor. The graph should show the line for the antagonist following a shallower curve but reaching the same maximum bound (but at a higher concentration of messenger). The higher the concentration of antagonist, the shallower the curve.

Inhibition of phosphodiesterase causes cAMP levels to be elevated. Because cAMP increases heart rate and strength of contraction, caffeine would also increase heart rate and the strength of contraction by inhibiting phosphodiesterase.

Answers to End-of-Chapter Multiple Choice and Objective Questions

Multiple Choice Questions

1. c 2. e 3. c 4. d 5. e 6. e 7. c 8. e
9. c 10. c

Objective Questions

11. secretory cells
12. autocrine
13. hormone, blood
14. GDP, GTP
15. adenylate cyclase
16. IP3
17. Lipophilic
18. eicosanoids
19. calmodulin
20. exocytosis
21. neurotransmitters
22. nervous system

The Endocrine System:
Endocrine Glands and Hormone Actions

The Endocrine System: Endocrine Glands and Hormone Actions

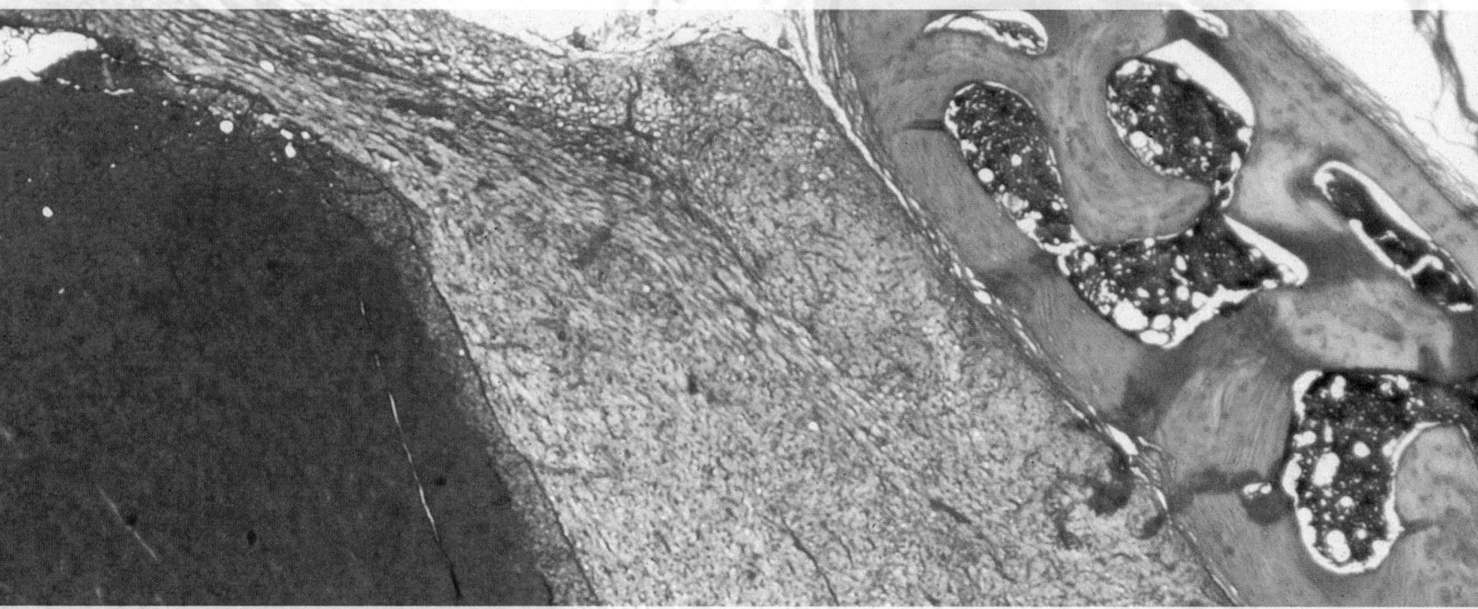

Biophoto Associates/Photo Researchers, Inc.

Colorized photomicrograph of the pituitary gland.

It's Monday evening and you are on your way home from work in rush-hour traffic. The car in front of you brakes suddenly, causing you to slam on your brakes. Even though you didn't hit the car in front of you, your heart is still beating faster and stronger than usual, and you feel a little shaky. You are feeling the effects of *adrenaline,* a hormone released from the adrenal medulla.

Adrenaline, also known as *epinephrine*, is the hormone of the fight-or-flight response that helps the body prepare for quick action. Adrenaline gave early humans the boost needed to hunt food and avoid predators. This "boost" is responsible for what many call the "adrenaline rush"—a feeling of excitement and energy induced by fear, exercise, or other stressors. Adrenaline affects heart rate and energy level by acting on specific target cells spread through the body. In this chapter, you will learn about the glands that secrete hormones and how hormones affect target cells.

CHAPTER OUTLINE

LEARNING OUTCOMES *After studying this chapter, you should be able to:*

- Name the primary and secondary endocrine glands, and the hormones associated with each.
- Describe the links between the hypothalamus with the anterior and posterior pituitary lobes.
- Describe the role of tropic hormones in regulating the release of other hormones, including the feedback loops involved in this process.
- Describe the types of interactions between hormones acting on the same target cell, including additive, synergistic, and permissive interactions.

Before You Begin

Make sure you have mastered the following topics:

- Epithelial cells
- Hormones
- Signal transduction mechanisms

We know how cells communicate with one another. For long-distance communication, the body utilizes the slow endocrine system, which releases hormones, and the fast nervous system, which releases neurotransmitters. In this chapter we focus on the endocrine system.

The organs of the endocrine system consist of endocrine glands are derived from epithelial tissue. Endocrine glands are found in many organs of the body (**Figure 1**). Two types of endocrine organs exist: **primary endocrine organs,** whose primary function is the secretion of hormones, and **secondary endocrine organs,** for which the secretion of hormones occurs secondary to some other function. Some primary endocrine organs are located within the brain, including the hypothalamus, pituitary gland, and pineal gland. Most primary endocrine organs, however, are located outside the nervous system, including the thyroid gland, parathyroid glands, thymus, adrenal glands, pancreas, and gonads (testes in the male and ovaries in the female). The placenta also functions as an endocrine gland in pregnant females. Secondary endocrine glands include organs such as the heart, liver, stomach, small intestine, kidney, and skin.

1 Primary Endocrine Organs

In this section, we examine the functions of the various primary endocrine organs, beginning with the hypothalamus and pituitary gland.

Hypothalamus and Pituitary Gland

Together, the hypothalamus and pituitary gland (**Figure 2**) function to regulate almost every body system. The **hypothalamus** is a part of the brain with many functions in addition to its role as an endocrine gland. It is considered a primary endocrine gland because it secretes several hormones, most of which affect the **pituitary gland** (also called the *hypophysis*), a pea-sized structure that is connected to the hypothalamus by a thin stalk of tissue called the *infundibulum*. The pituitary gland is divided into two structurally and functionally distinct sections called the **anterior lobe** (or *adenohypophysis*), which is derived from epithelial tissue, and the **posterior lobe** (or *neurohypophysis*), which is derived from neural tissue. The different connections between the hypothalamus and the two lobes of the pituitary gland are critical to the function of both endocrine organs.

Neural Connection Between the Hypothalamus and the Posterior Pituitary Gland

The posterior lobe of the pituitary gland contains neural tissue consisting of the axon terminals of neurons originating in the hypothalamus (**Figure 3**). These neural endings in the posterior pituitary gland secrete two peptide hormones: *antidiuretic hormone* (*ADH;* also called *vasopressin*) and *oxytocin*. These hormones are synthesized in neurons originating in different regions of the hypothalamus; ADH is synthesized primarily in the paraventricular nucleus and oxytocin is synthesized primarily in the supraoptic nucleus. Following synthesis, the peptides are packaged into secretory vesicles, which are transported to the axon terminals in the posterior pituitary. The hormones are released by exocytosis when these neurons receive a signal, usually from other neurons; this reaction is called a *neuroendocrine reflex*. The hormones are released into the blood, as are other hormones. Because these hormones are secreted by neurons instead of endocrine glands, they are called *neurohormones*.

The controls of ADH and oxytocin release are completely independent. ADH release is stimulated by an increase in plasma solute concentration, and its target cells are located in the kidneys; they respond by increasing water reabsorption. Oxytocin release is stimulated by pressure in the uterus of a pregnant woman and also by suckling by an infant. This hormone's target cells are found in the uterus, where oxytocin increases uterine contractions, and in the breasts, where oxytocin stimulates milk letdown following childbirth.

Blood Connection Between the Hypothalamus and the Anterior Pituitary Gland

The anterior lobe of the pituitary gland and the cells of the hypothalamus that control it secrete primarily **tropic hormones** (also called *trophic hormones*), which are hormones that regulate the secretion of other hormones. A tropic hormone can be a *stimulating hormone*, which increases the secretion of another hormone, or an *inhibiting hormone*, which decreases the secretion of another hormone. The general signaling pathway is as follows: The hypothalamus releases a tropic hormone that effects the release of another tropic hormone from the anterior pituitary; this tropic hormone then effects the release of a third hormone from another endocrine

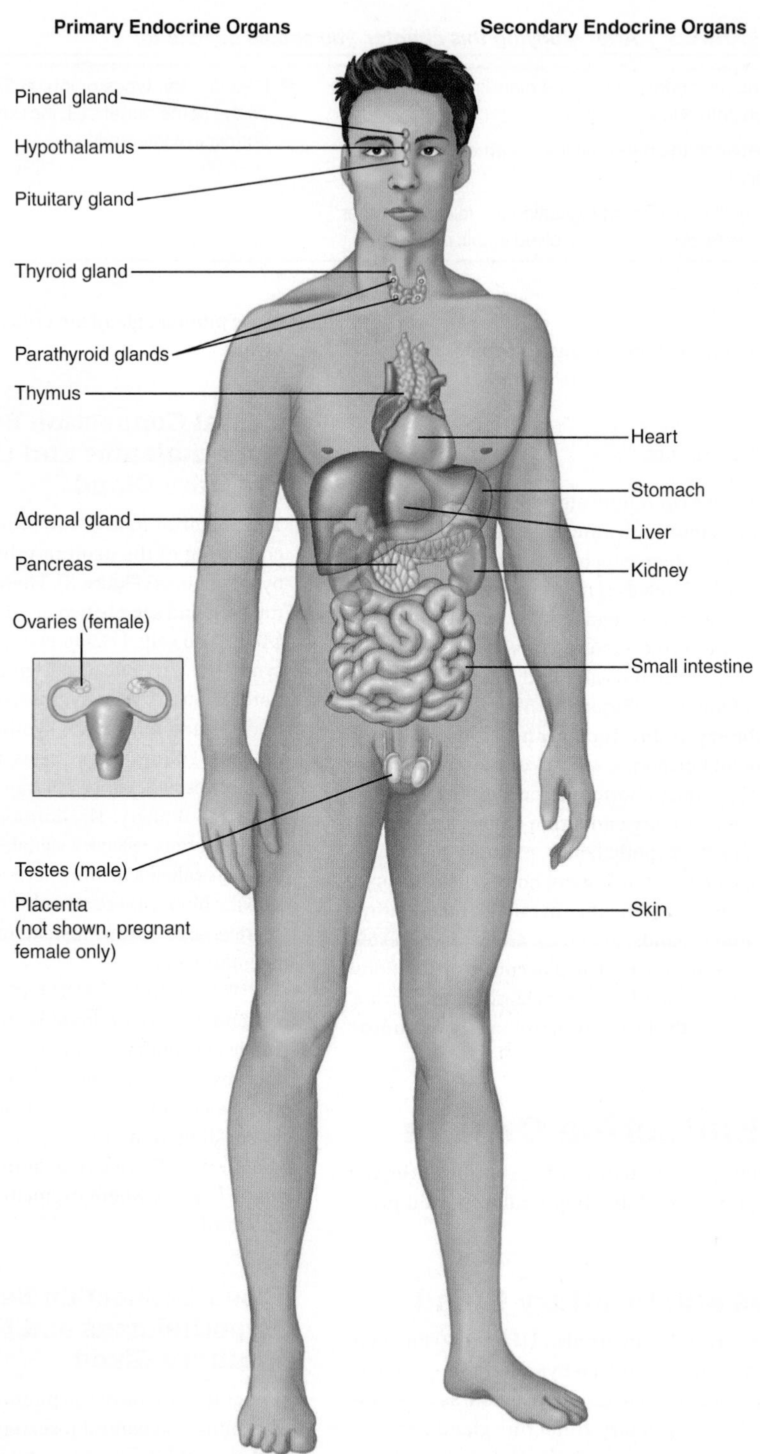

Figure 1 Endocrine organs.

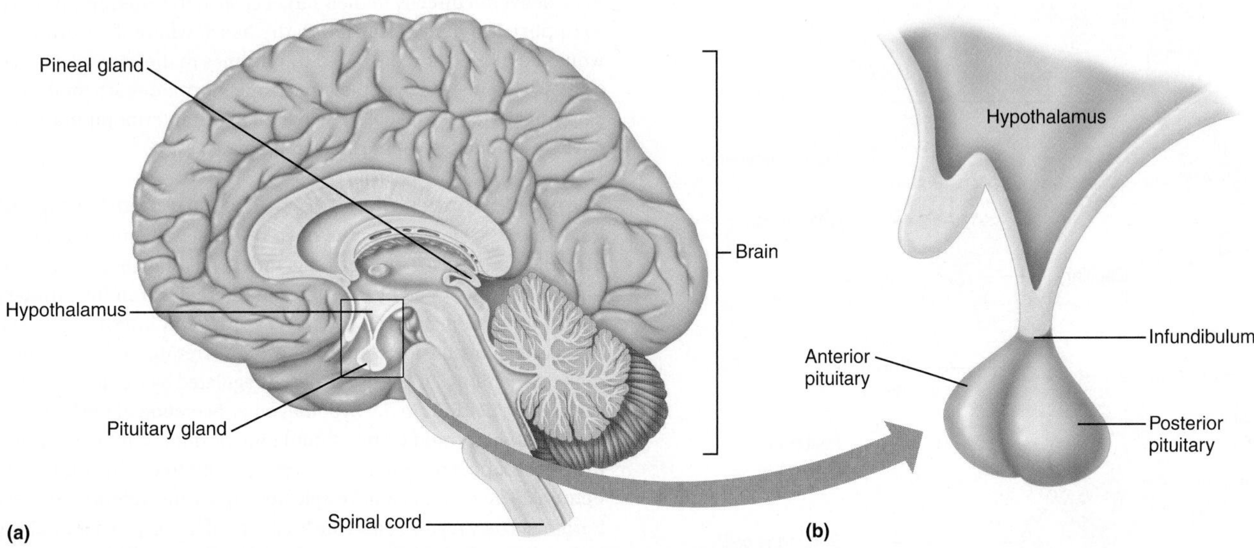

Figure 2 Hypothalamus and pituitary gland. (a) Locations of the hypothalamus and pituitary gland in the brain. **(b)** Connection of the hypothalamus and the pituitary gland by the infundibulum. The pituitary gland is divided into an anterior lobe and a posterior lobe.

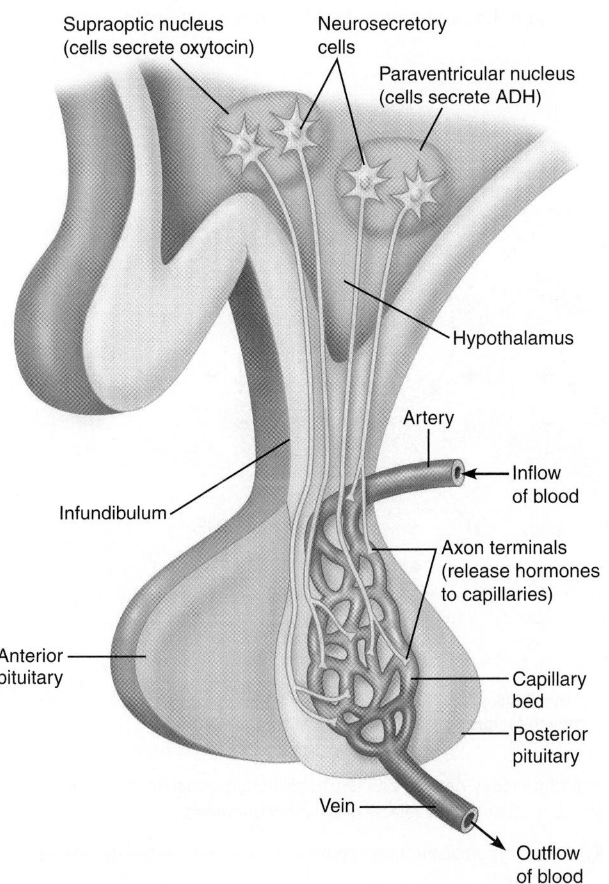

gland, and this third hormone exerts effects on target cells throughout the body.

Anatomy of Hypothalamic-Pituitary Portal System

The hypothalamus and anterior lobe are connected by the **hypothalamic-pituitary portal system** (Figure 4). A *portal system* is a specialized arrangement of blood vessels in which two capillary beds are located in series, one after the other. Capillaries are the smallest of the blood vessels and are the sites where compounds can be exchanged between blood and tissue. Normally, the heart pumps blood that enters many capillary beds in parallel, not in a series, so that each capillary bed receives fresh blood. The blood then returns from each capillary bed via veins back to the heart. Neurons in the hypothalamus secrete tropic hormones into capillary beds located in the *median eminence* of the hypothalamus (see Figure 4). These tropic hormones, which are neurohormones, travel down the infundibulum to the pituitary gland via a portal vein, from which they enter a second capillary bed. The hypothalamic tropic hormones then stimulate or inhibit the release of hormones from the anterior pituitary.

Figure 3 Connection between the hypothalamus and the posterior pituitary gland. Neurons originating in the hypothalamus send projections to the posterior pituitary gland. Neurons originating in the paraventricular nucleus synthesize and secrete ADH; neurons originating in the supraoptic nucleus synthesize and secrete oxytocin. ADH and oxytocin are secreted from nerve terminals into the bloodstream in the posterior pituitary.

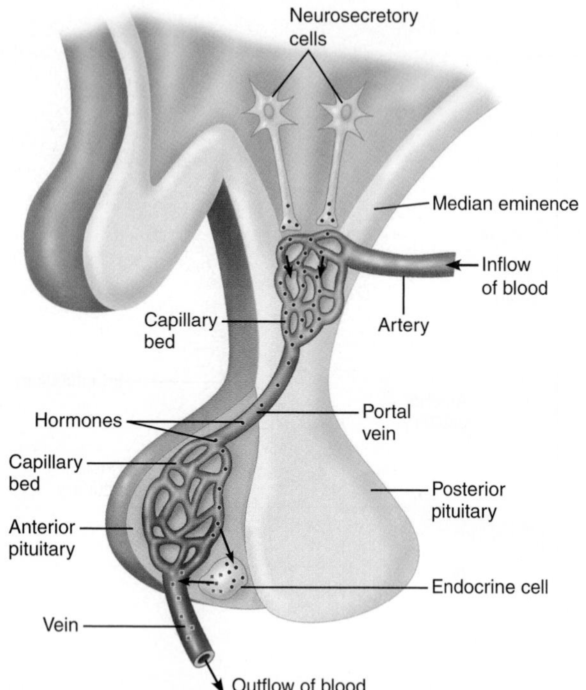

Figure 4 The hypothalamic-pituitary portal system. Neurosecretory cells of the hypothalamus secrete tropic hormones into the hypothalamic-pituitary portal system. The tropic hormones travel to the anterior pituitary, where they effect release of anterior pituitary hormones into the blood.

The portal system enables the hypothalamic tropic hormones to be delivered directly to their target cells in the anterior pituitary, as opposed to being returned to the heart, where the hormones would be diluted and degraded by enzymes in the general circulation. As a result, the hypothalamic tropic hormones are more concentrated in the portal blood delivered to the anterior pituitary and have a greater effect on hormone release.

Tropic Hormones of the Hypothalamus and Anterior Pituitary Compared to the posterior lobe, the anterior lobe is more like a "typical" endocrine gland, in that hormones are synthesized and secreted by cells located entirely within the pituitary gland. Within the anterior lobe, each of several intermingled cell types is responsible for synthesizing and secreting a particular hormone. Secretion of these hormones is regulated by tropic hormones secreted by neurons in the hypothalamus. Secretion of the hypothalamic tropic hormones (five stimulating hormones and two inhibiting hormones) is regulated by neural input to the hypothalamic neurons. The hypothalamic tropic hormones, the anterior pituitary hormones they regulate, and the functions of those anterior pituitary hormones are listed below and shown in **Figure 5**. With one exception, all of these tropic hormones are peptides. The relationship between each hypothalamic tropic hormone with its anterior pituitary tropic hormone and the target endocrine gland is called an *axis*.

1. **Prolactin releasing hormone** *(PRH)* stimulates the anterior pituitary to release **prolactin,** which stimulates mammary gland development and milk secretion in females.

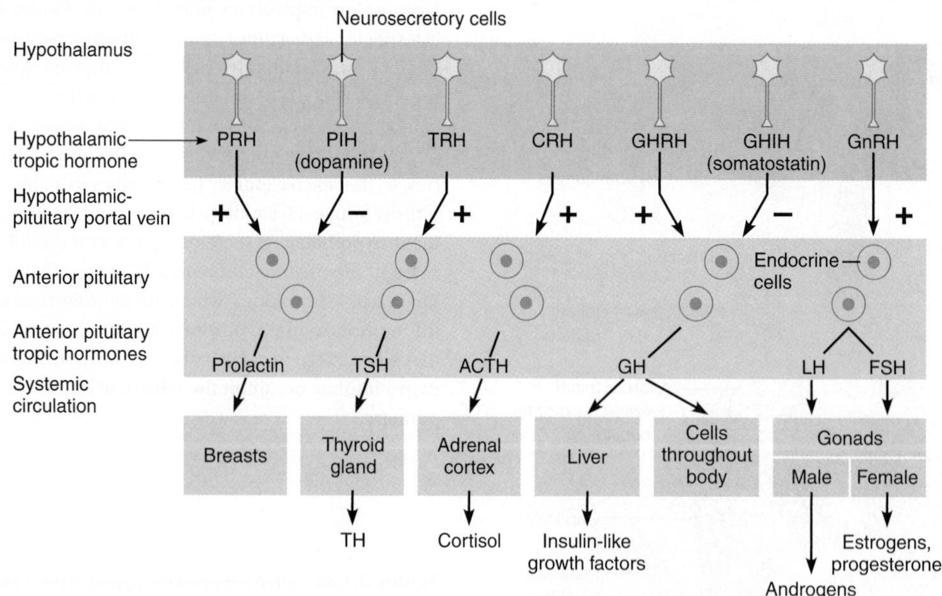

Figure 5 Hypothalamic and anterior pituitary tropic hormones. The hypothalamus secretes into the hypothalamic-pituitary portal system seven tropic hormones that are either releasing hormones (–RH) or inhibiting hormones (–IH). These tropic hormones act on endocrine cells in the anterior pituitary to affect the release of hormones that (except for prolactin) are also tropic hormones. Anterior pituitary tropic hormones act on other endocrine glands to stimulate the release of still other hormones.

Q Which hypothalamic tropic hormone indirectly stimulates the release of cortisol?

2. **Prolactin inhibiting hormone** *(PIH),* or dopamine (a catecholamine), inhibits the release of prolactin.
3. **Thyrotropin releasing hormone** *(TRH,* or *thyrotropin)* stimulates the release of **thyroid stimulating hormone** *(TSH)* from the anterior pituitary. TSH then stimulates the secretion of **thyroid hormones** by the thyroid gland. Thyroid hormones regulate metabolism.
4. **Corticotropin releasing hormone** *(CRH)* stimulates the release of **adrenocorticotropic hormone** *(ACTH)* by the anterior pituitary. ACTH then stimulates the secretion of **glucocorticoids,** such as **cortisol,** from the adrenal cortex, the outer layer of the adrenal gland. Cortisol is the main hormone that regulates metabolism when the body is stressed.
5. **Growth hormone releasing hormone** *(GHRH)* stimulates the secretion of **growth hormone** *(GH)* by the anterior pituitary. GH regulates growth and energy metabolism but also functions as a tropic hormone by stimulating the secretion of **insulin-like growth factors** *(IGFs)* by the liver.
6. **Growth hormone inhibiting hormone** *(GHIH),* or *somatostatin,* inhibits the secretion of growth hormone by the anterior pituitary, thereby decreasing IGF release from the liver.
7. **Gonadotropin releasing hormone** *(GnRH)* stimulates the release of the **gonadotropins, follicle stimulating hormone** *(FSH),* and **luteinizing hormone** *(LH)* by the anterior pituitary. LH stimulates ovulation in females, and it stimulates the secretion of sex hormones (**estrogens** and **progesterone** in females and **androgens** in males) by the gonads. FSH promotes the development of egg cells in females and sperm cells in males, and it stimulates the secretion of estrogens in females and *inhibin* in both sexes.

Feedback loops regulate the multistep pathways by which certain hypothalamic and anterior pituitary tropic hormones (CRH, ACTH, TRH, TSH, GnRH, FSH, and LH) are released (**Figure 6**). The tropic hormone from the anterior pituitary may act through negative feedback on the hypothalamus to decrease its own release. The inhibition of hypothalamic tropic hormones by the anterior pituitary tropic hormone, called *short loop negative feedback,* prevents the buildup of excess anterior pituitary tropic hormone. Additionally, the hormone whose secretion is stimulated by the tropic hormone generally feeds back to the hypothalamus (and often the anterior pituitary as well) to inhibit secretion of the tropic hormone, thereby limiting its own secretion. This phenomenon is called *long loop negative feedback.*

Figure 7 depicts the regulation of cortisol release by the adrenal gland as a specific example of negative feedback regulation. Corticotropin releasing hormone (CRH) stimulates the release of adrenocorticotropic hormone (ACTH), an anterior pituitary tropic hormone. ACTH, in turn, stimulates the release of cortisol from the adrenal gland. If cortisol levels in the blood increase above a certain level, then negative feedback loops cause a decrease in the release of both CRH and ACTH. In both cases, the net result is that the cortisol acts to inhibit its own secretion. However, cortisol provides negative feedback only to its own tropic hormones; it has no effect, for example, on the cells that secrete gonadotropin releasing hormone or luteinizing hormone.

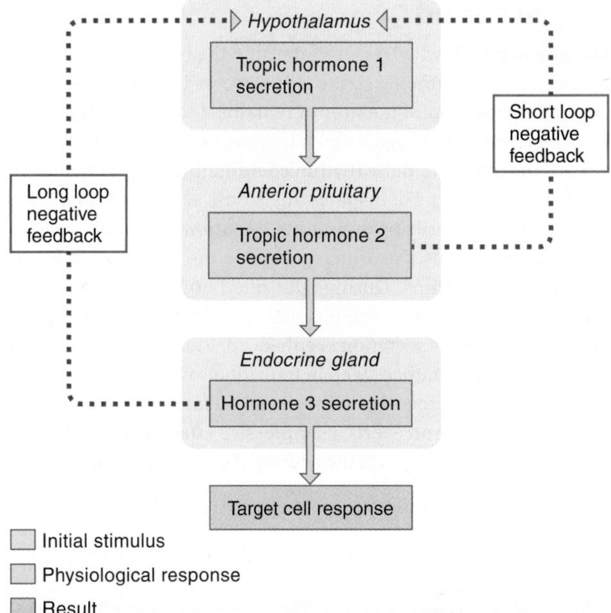

Figure 6 **Negative feedback loops affecting hypothalamic and anterior pituitary tropic hormones.**

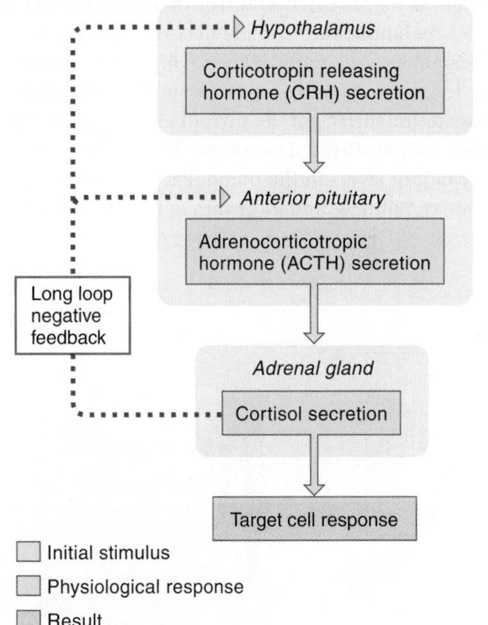

Figure 7 **Regulation of cortisol release.**

Quick Check 1

1. Name the two hormones released in the posterior pituitary gland. Which type of cell releases these hormones?
2. Name each hypothalamic tropic hormone and explain how it affects the release of a hormone in the anterior pituitary gland.
3. What is the difference between short loop and long loop negative feedback?

Pineal Gland

The **pineal gland,** which is located within the brain (see Figure 2), is composed of epithelial tissue that secretes the hormone *melatonin.* The function of melatonin remains a topic of debate, but most recent studies suggest that it is important in establishing the **circadian rhythm** (a daily rhythm coordinating body activities to the day-night cycle). For instance, melatonin secretion is known to be regulated primarily by input from the *suprachiasmatic nucleus* of the hypothalamus, the primary center in the brain for establishing circadian rhythms. During light-filled hours, input from the suprachiasmatic nucleus inhibits melatonin secretion, whereas in the dark, melatonin secretion is enhanced. Melatonin is also used therapeutically to induce sleep in individuals with sleep disorders and to assist in the recovery from jet lag. Melatonin also enhances immune function and exerts a suppressive effect on reproductive function by interfering with the activity of certain hormones.

Thyroid Gland and Parathyroid Glands

The **thyroid gland** is a butterfly-shaped structure located on the ventral surface of the trachea (**Figure 8a**). This gland secretes the two thyroid hormones, **tetraiodothyronine** *(T_4)* and **triiodothyronine** *(T_3),* as well as **calcitonin**. The thyroid hormones regulate the body's metabolic rate and are necessary for normal growth and development; calcitonin decreases blood calcium levels.

The four **parathyroid glands** are smaller structures located on the posterior surface of the thyroid gland (Figure 8b). These glands secrete **parathyroid hormone** *(PTH),* an important regulator of calcium levels in the blood. Parathyroid hormones act on the bones, kidneys, and intestines to increase blood calcium levels.

Thymus

The **thymus** lies close to the heart (see Figure 1) and secretes the hormone **thymosin**. In addition, the thymus is critical for normal immune function because immune cells called *T lymphocytes* (or T cells), which are essential for effective immune responses against invading microorganisms, mature there. Thymosin regulates T-cell function.

Adrenal Glands

The **adrenal glands** are located above the kidneys (see Figure 1); for this reason, they are also called the suprarenal glands. Each adrenal gland consists of an outer layer called the *cortex,* which accounts for approximately 80% of the gland's total mass, and an inner core called the *medulla* (**Figure 9a**). Like the pituitary, each adrenal gland is essentially two glands in one, because the cortex and the medulla are structurally and functionally distinct. The cortex, like the anterior pituitary, is derived from epithelial tissue, whereas the medulla, like the posterior pituitary, is derived from neural tissue.

The **adrenal cortex** is stratified into three distinct layers (Figure 9b): (1) an outer layer called the *zona glomerulosa,* (2) a middle layer called the *zona fasciculata,* and (3) an inner layer called the *zona reticularis.* Because cells in these layers possess different complements of the enzymes involved in the synthesis of adrenal cortex hormones, these cells manufacture and secrete various hormones in different proportions.

The adrenal cortex secretes a number of hormones collectively called **adrenocorticoids,** a term that refers to their site of origin (*adren*al *cort*ex) and the chemical class of compounds to which they belong (ster*oids*). The adrenocorticoids include three types of hormones:

1. **Mineralocorticoids** (primarily *aldosterone*) are secreted exclusively by cells in the zona glomerulosa. Aldosterone regulates sodium reabsorption and potassium secretion by the kidneys.

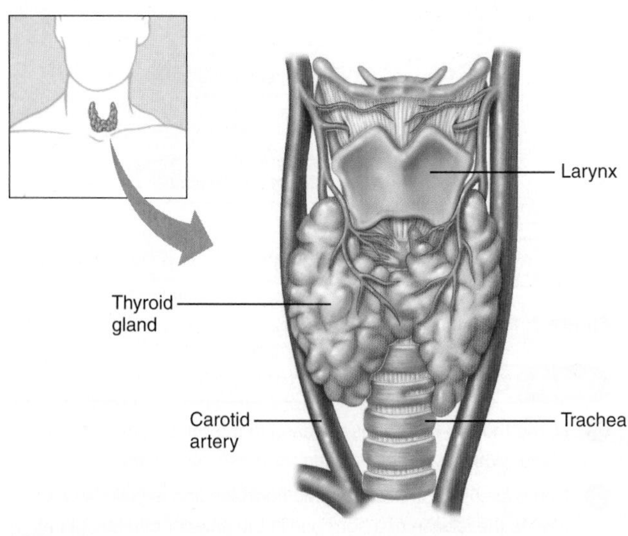

(a) Ventral view

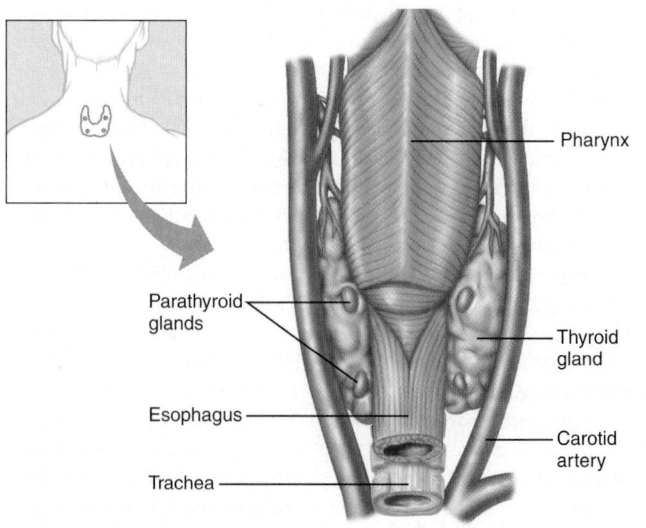

(b) Dorsal view

Biophoto Associates/Photo Researchers, Inc.

Figure 8 Locations of the thyroid and parathyroid glands.

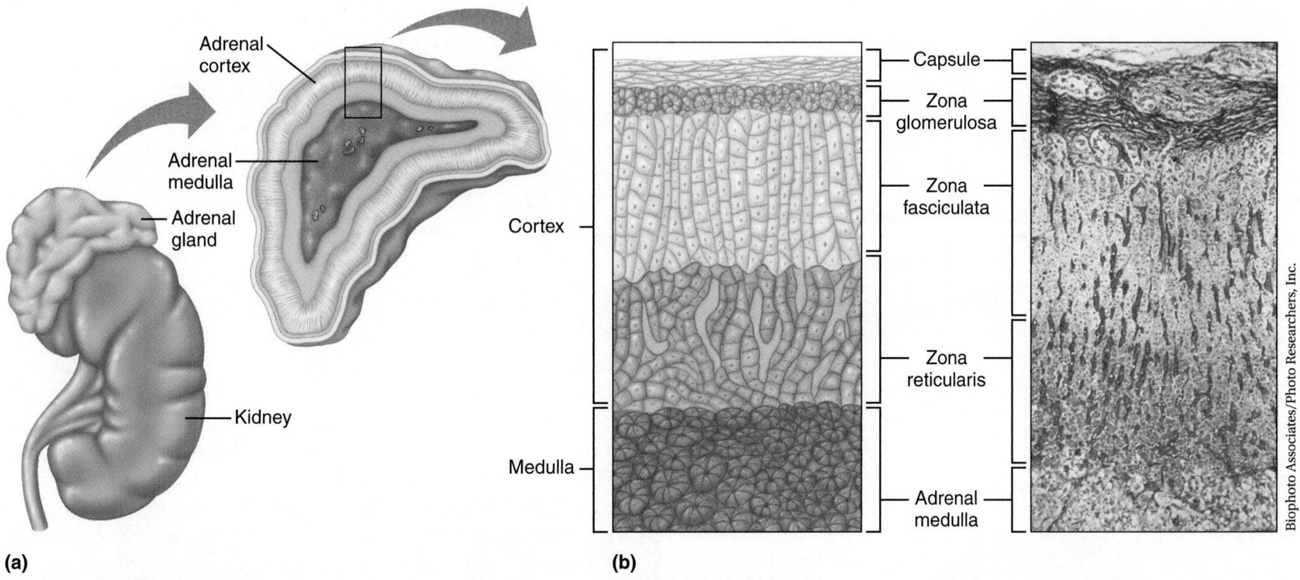

Figure 9 Anatomy of the adrenal gland. (a) The adrenal glands, located just above the kidneys, are divided into an inner medulla and outer cortex. **(b)** The adrenal cortex contains three distinct layers, or zonae, shown in a diagram (left) and photomicrograph (right).

Q Which hormones are secreted from the zona reticularis?

2. **Glucocorticoids** (primarily *cortisol*) are secreted mainly by cells in the zona fasciculata and zona reticularis. Cortisol regulates the body's response to stress; protein, carbohydrate, and lipid metabolism in a variety of tissues; and blood glucose levels. Cortisol is regulated by the CRH-ACTH-cortisol axis.
3. **Sex hormones** (primarily *androgens*) are secreted by cells in the zona fasciculata and zona reticularis (and by the gonads). Sex hormones regulate reproductive function and a variety of other processes. Because androgens are secreted in much larger amounts by the gonads in males, secretion of these hormones by the adrenal cortex is normally of little physiological significance; in females, however, adrenal androgens may stimulate the sex drive.

The **adrenal medulla** contains *chromaffin cells* and secretes catecholamines; approximately 80% of the secreted hormones is epinephrine (also called *adrenaline*), approximately 20% is norepinephrine, and less than 1% is dopamine. Epinephrine is released during times of stress or excitation; among other things, it elicits an increase in heart rate and mobilization of energy stores. The primary stimulus for secretion of epinephrine, and to a lesser extent norepinephrine and dopamine, is neural.

Pancreas

The **pancreas** functions as both an endocrine gland and an exocrine gland. The *exocrine pancreas* includes acinar cells and duct cells that secrete enzymes and fluid into the gastrointestinal tract; the *endocrine pancreas* consists of cell clusters called **islets of Langerhans,** which are scattered throughout the pancreas in spaces between the ducts (**Figure 10**).

The islets of Langerhans are the source of two major hormones, each of which is secreted by a different cell type: **insulin,** which is secreted by **beta cells** *(B cells),* and **glucagon,** which is secreted by **alpha cells** *(A cells).* There are approximately twice as many beta cells as alpha cells in the islets. Both insulin and glucagon regulate energy metabolism and blood glucose levels: Insulin decreases blood glucose levels by promoting its uptake into cells and conversion to glycogen, whereas glucagon increases blood glucose levels by promoting glycogenolysis and gluconeogenesis.

Two other cell types are also located in the islets of Langerhans: D cells and F cells. *D cells,* or *delta cells,* secrete *somatostatin,* which helps to regulate digestion and absorption of nutrients and regulates the secretion of other pancreatic hormones. (Recall that somatostatin is also a hypothalamic tropic hormone that inhibits secretion of growth hormone from the anterior pituitary.) F cells secrete pancreatic polypeptide, which inhibits exocrine secretions of the pancreas and gallbladder contractions.

Gonads

The gonads (ovaries and testes) have both endocrine and nonendocrine functions. In both sexes, they produce the *gametes*—sperm in males and oocytes in females—and secrete sex hormones. In males, the predominant sex hormones are the androgens, testosterone and androstenedione (see **Understanding Exercise: Why Athletes Take Steroids**); in females, the major sex hormones are estradiol (which belongs to a group of related hormones known as estrogens) and progesterone.

The placenta also functions as an endocrine gland in pregnant females, secreting primarily estrogens and progesterone. The placenta also secretes human chorionic gonadotropin (hCG), which can be used to determine pregnancy.

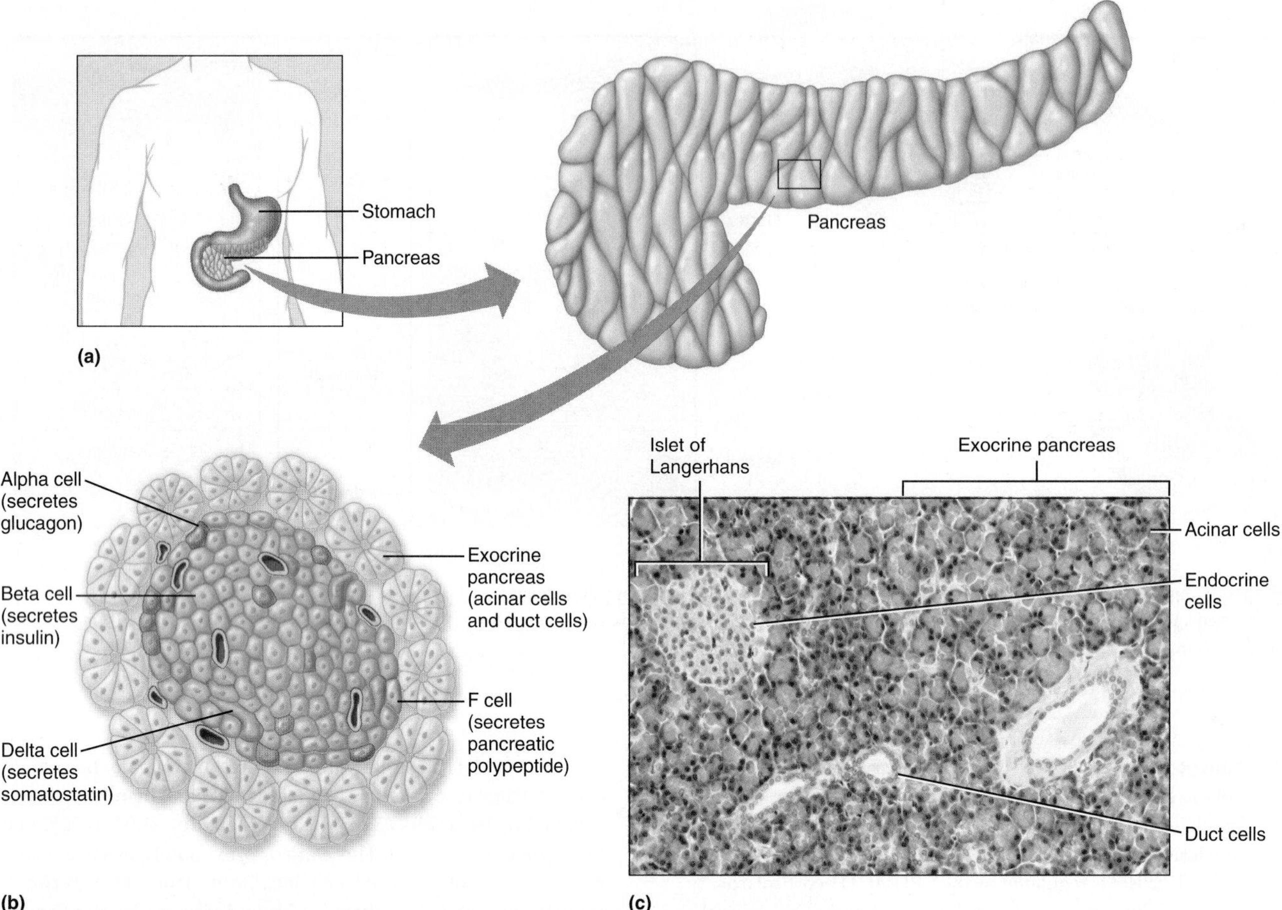

Figure 10 Anatomy of the pancreas. (a) The pancreas is located in the abdominal cavity below the stomach. **(b)** Histology of the pancreas. The pancreas consists of acinar cells and duct cells that secrete exocrine products into ducts, and islets of Langerhans, which contain cells that secrete hormones into the interstitial fluid. Four types of endocrine cells are located in the islets; each type secretes a different hormone. **(c)** Photomicrograph of the islet of Langerhans and the exocrine cells.

2 Secondary Endocrine Organs

In addition to fulfilling their primary function, many organs of the body secrete hormones. These secondary endocrine organs include the heart, kidneys, digestive organs, liver, and skin (see Figure 1). The heart secretes *atrial natriuretic peptide (ANP),* which regulates sodium reabsorption by the kidneys. The kidneys secrete *erythropoietin,* which stimulates production of red blood cells by the bone marrow. Organs in the gastrointestinal tract secrete several hormones that are important in regulating the digestion and absorption of food, including *gastrin, secretin,* and *cholecystokinin.* The liver secretes *insulin-like growth factors (IGFs),* which promote tissue growth. Finally, the skin, liver, and kidneys are involved in the production of *1,25-dihydroxy vitamin* D_3, which regulates blood calcium levels.

The hormones secreted by the primary and secondary endocrine organs, as well as their functions, are summarized in Table 1.

Quick Check 2

1. Name the three hormones released by the thyroid gland. Which hormone regulates blood calcium levels?
2. Which hormones are released from the adrenal cortex? From the adrenal medulla?
3. Which endocrine gland secretes insulin? Estrogens? Melatonin?

UNDERSTANDING EXERCISE

Why Athletes Take Steroids

You hear it in the news often: Steve the athlete tested positive for steroid use. Why would Steve risk his career, and his health, to take illegal drugs?

Today's athletes are stronger and faster than ever. World records continue to be broken, even those records that were thought at one time to be untouchable. Although most of these achievements are attributable to advances in training and technology, some are the result of steroid abuse.

Steroid hormones produced in the body include the adrenocorticosteroids (such as aldosterone and cortisol) and the sex hormones (androgens, estrogens, and progesterones). Steroid drugs include corticosteroids, which are prescribed to fight inflammation, and anabolic steroids, which are used by athletes to make their muscles stronger.

Anabolic steroids are synthetic forms of androgens, the male sex hormones. Athletes use them to increase muscle mass. Excessive androgen levels may cause several side effects, including liver tumors, aggressive and hostile behavior, enlargement of the heart, acne, and high levels of cholesterol and triglycerides in the blood, which predisposes the person to heart disease and stroke. Because androgens are substrates for estrogens in the liver and adipose tissue, men who take synthetic androgens may develop female characteristics, including gynecomastia—abnormal growth of female-like mammary tissue. Conversely, women who take synthetic androgens develop masculinization of the body, including facial hair growth, deepening of the voice, and a decrease in menstrual cycles. Adolescents who take synthetic androgens may also experience stunted growth due to premature closure of the epiphyseal plate.

With so many risks to their health and careers, why do some athletes continue to take anabolic steroids? Androgens can increase their strength by 5–20% and their lean body mass by 2–5 kg. Although these gains might seem small, even hundredths of a second in a 100-meter race may mean the difference between a gold medal (and millions of dollars in endorsements) and a silver medal (and a plane ticket home).

TABLE 1 Endocrine Organs and the Hormones They Secrete

Organ	Hormones	Functions
Primary Endocrine Organs		
Hypothalamus	Releasing and release-inhibiting hormones	Regulate secretion of anterior pituitary hormones
Anterior pituitary gland	Growth hormone (GH)	Essential for growth; stimulates bone and soft-tissue growth; regulates protein, lipid, and carbohydrate metabolism
	Adrenocorticotropic hormone (ACTH)	Stimulates glucocorticoid secretion by the adrenal cortex
	Thyroid stimulating hormone (TSH)	Stimulates secretion of thyroid hormones by the thyroid gland
	Prolactin	Stimulates development of the breasts and milk secretion by the mammary glands
	Follicle stimulating hormone (FSH)	Females: stimulates growth and development of the ovarian follicles, estrogen secretion; males: stimulates sperm production by the testis
	Luteinizing hormone (LH)	Females: stimulates ovulation, transformation of the ovarian follicle into the corpus luteum, and secretion of estrogen and progesterone; males: stimulates testosterone secretion by the testis
Posterior pituitary gland	Antidiuretic hormone (ADH, or vasopressin)	Decreases urine output by the kidneys; promotes constriction of blood vessels (arterioles)
	Oxytocin	Females: stimulates uterine contractions and milk ejection by the mammary glands; males: function unknown
Pineal gland	Melatonin	Regulates biological rhythms according to day-night cycles
Thymus	Thymosin	Stimulates proliferation and function of T lymphocytes
Thyroid gland	Thyroid hormones (triiodothyronine and tetraiodothyronine)	Increase basal metabolic rate; necessary for normal development
	Calcitonin	Promotes calcium deposition in bone; lowers blood calcium levels

continued

TABLE 1 *Continued*

Organ	Hormones	Functions
Parathyroid glands	Parathyroid hormone (PTH)	Promotes calcium release from bone, calcium absorption by the intestine, and calcium reabsorption by the kidney tubules; raises blood calcium levels; stimulates vitamin D_3 synthesis
Adrenal cortex	Mineralocorticoids (aldosterone)	Stimulate sodium reabsorption and potassium secretion by the kidney tubules
	Glucocorticoids (cortisol, corticosterone)	Promote catabolism of proteins and fats; raise blood glucose levels; adapt the body to stress
	Androgens (dehydroepiandrosterone, androstenedione)	Promote sex drive
Adrenal medulla	Epinephrine	Stimulates fight-or-flight response
Pancreas	Insulin	Lowers blood glucose levels; stores energy by promoting protein, triglyceride, and glycogen synthesis
	Glucagon	Raises blood glucose levels; mobilizes energy by promoting glycogenolysis, gluconeogenesis
	Somatostatin	Inhibits secretion of pancreatic hormones; regulates digestion and absorption of nutrients by the gastrointestinal system
Gonads		
Testes	Androgens (testosterone, androstenedione)	Necessary for sperm production by testis; promote sex drive and development of secondary sex characteristics (e.g., facial hair, deep voice)
Ovaries	Estrogens (estradiol)	Necessary for follicular development; promote development of secondary sex characteristics (e.g., breasts, body fat distribution)
	Progestins (progesterone)	Promote endometrial growth to prepare the uterus for pregnancy
Placenta (during pregnancy)	Chorionic gonadotropin, estrogens, progesterone	Maintain the corpus luteum; reinforce the actions of hormones secreted by the corpus luteum
Secondary Endocrine Organs		
Heart	Atrial natriuretic peptide (ANP)	Inhibits sodium reabsorption by the kidney tubules
Kidneys	Erythropoietin	Stimulates production of red blood cells in bone marrow
Gastrointestinal tract		
Stomach	Gastrin	Stimulates acid secretion by the stomach and intestinal motility
Small intestine	Secretin, cholecystokinin (CCK), glucose-dependent insulinotropic peptide (GIP)	Regulate gastrointestinal motility and secretion; regulate exocrine secretion by the liver and pancreas
Liver	Insulin-like growth factors (IGFs)	Promote bone and soft-tissue growth
Skin, liver, kidney*	1,25-dihydroxy vitamin D_3	Promotes absorption of calcium by the intestine

*The skin, liver, and kidney are all necessary for the activation of 1,25-dihydroxy vitamin D_3

3 Hormone Actions at the Target Cell

The first part of this chapter described the glands that secrete hormones. In this section, we discuss in detail the actions of hormones at the target cell. Factors influencing the magnitude of a target cell's response to a hormone include the types of receptors in the target cell and the signal transduction mechanism, as well as the concentration of free hormone in the blood.

Control of Hormone Levels in Blood

The concentration of free hormone in the blood depends on three factors: (1) the rate of hormone secretion, (2) the amount of hormone transported bound to carrier proteins, and (3) the rate at which the hormone is metabolized.

Rate of Hormone Secretion

With few exceptions, endocrine cells and other secretory cells release chemical messengers at variable rates. When these cells receive certain signals, the rate of secretion rises or falls. Faster rates of secretion translate into greater concentrations of hormone in blood and more hormone molecules bound to receptors on target cells. Therefore, when hormone levels in blood increase, they "trigger" changes in the target cells. In a few cases, however, hormones are secreted at a relatively steady rate. Because the concentrations of these messengers do not change appreciably, they do not trigger changes in the target cell in the same way that other hormones do. Instead, they facilitate processes that are normally ongoing. For example, thyroid hormones are usually secreted at nearly constant rates in adults and are necessary for the maintenance of normal metabolism and nervous system function.

In general, endocrine cells alter hormone secretion in response to two types of input: neural signals and **humoral** (bloodborne) signals. Each type of signal may be either stimulatory or inhibitory. Neural signals directly regulate hormone secretion by the hypothalamus (tropic hormones), posterior pituitary gland (ADH and oxytocin), and adrenal medulla (epinephrine), which can affect the secretion of other hormones. For example, stress activates neural signals that stimulate the hypothalamus to release corticotropin releasing hormone, which in turn stimulates the release of ACTH from the anterior pituitary (**Figure 11**). ACTH stimulates the adrenal cortex to secrete cortisol, a hormone that helps the body cope with stress.

Humoral signals are classified into three basic categories: (1) hormones, (2) ions, and (3) metabolites. We have already discussed hormonal control of hormone release with respect to the hypothalamus–anterior pituitary–endocrine gland axes and their negative feedback loops. Many hormones also regulate the blood concentrations of ions or metabolites, which operate through negative feedback to regulate secretion of the hormones. For example, **Figure 12a** depicts the control of blood glucose levels by insulin. High blood glucose levels stimulate the release of insulin from beta cells of the pancreas. Insulin then stimulates uptake of glucose by most cells of the body, causing a decrease in blood glucose. When glucose levels decrease to normal, the stimulus for insulin secretion is reduced and its release slows. Figure 12b depicts another example of negative feedback—the regulation of potassium levels in

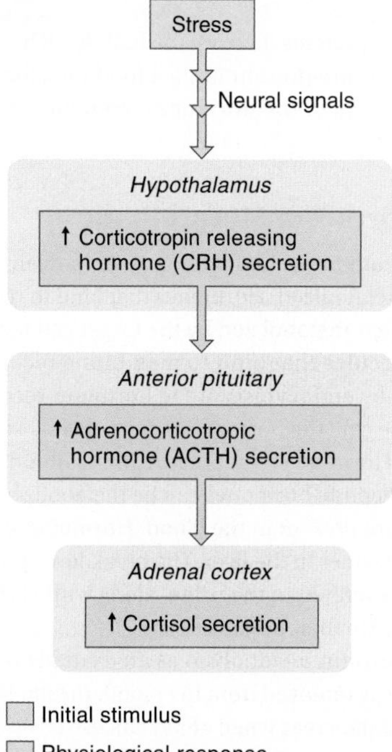

Figure 11 Neural control of hormone release from the hypothalamus. Neural input resulting from stressful stimuli stimulates the release of CRH from the hypothalamus, which eventually leads to the release of cortisol from the adrenal cortex.

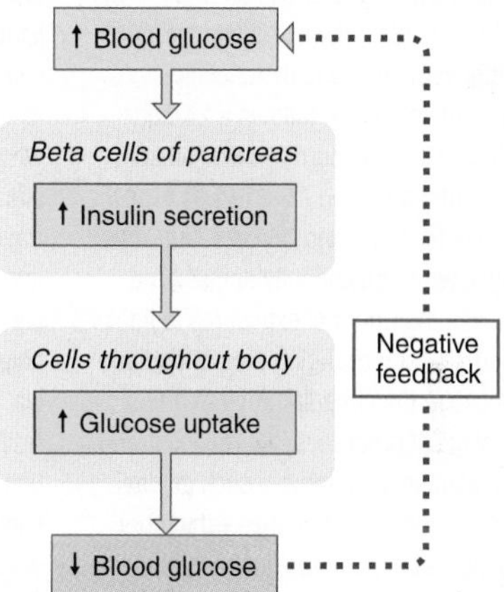

(a) Regulation of blood glucose levels

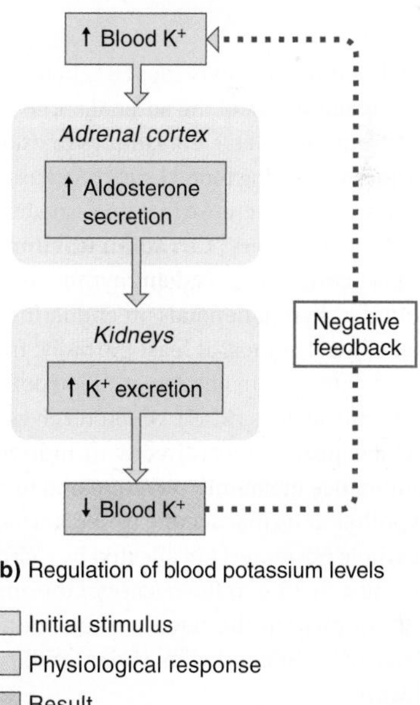

(b) Regulation of blood potassium levels

Initial stimulus
Physiological response
Result

Figure 12 Examples of humoral control of hormone release. **(a)** Control of blood glucose levels by insulin. **(b)** Control of blood potassium levels by aldosterone.

Q Insulin is a peptide hormone, and aldosterone is a steroid hormone. By which mechanism is insulin released from beta cells and aldosterone released from cells of the adrenal cortex?

DISCOVERY

Circadian Rhythms and Jet Lag

The invention of air travel had an unexpected consequence: It created the phenomenon of "jet lag," or symptoms of fatigue, experienced by travelers who fly across time zones. Jet lag occurs because the body's *circadian rhythm* is disrupted. The circadian rhythm is an internal "clock" that governs many body functions. Typically, this internal clock follows a 24-hour cycle and is linked to the normal variations of light and dark that a person experiences over the course of a day (*circa* means "almost" and *dies* means "day"). When travelers fly across time zones, however, their normal exposures to light are disrupted, thereby changing their circadian rhythms. Because the circadian rhythm drives many physiological processes, alterations in the rhythm can result in symptoms—such as daytime sleepiness and loss of energy—that characterize jet lag. Jet lag can last for several days, and its duration tends to increase in proportion with the number of time zones crossed.

Travelers can take steps to help lessen the effects of jet lag. Because a change in the level of light exposure is what disrupts the circadian rhythm, travelers can simulate normal exposures to light with the help of a bright, artificial light while flying. Artificial light at night benefits those traveling westward, while artificial light in the morning benefits those traveling eastward. Medications may also help some travelers. The use of melatonin (a hormone of the pineal gland that is believed to be linked to circadian rhythms) to prevent jet lag has grown considerably over the last several years. However, scientific evidence of its effectiveness is not conclusive.

Exactostock/SuperStock

the blood. High levels of potassium ions (K^+) in the blood stimulate cells in the zona glomerulosa of the adrenal cortex to secrete aldosterone. Aldosterone stimulates the kidneys to excrete potassium ions into the urine, decreasing blood levels of potassium.

Many hormones are secreted in a pattern governed by a circadian rhythm (see **Discovery: Circadian Rhythms and Jet Lag**). The mechanisms behind the circadian rhythm are not completely understood, but the rhythm depends on endocrine and neural input. Hormonal control comes, at least partially, from melatonin. The neural input comes from an area of the hypothalamus called the suprachiasmatic nucleus (SCN), which receives input from the eyes. The SCN has projections (directly or indirectly) to the pineal gland to influence melatonin secretion and to neurosecretory cells of the hypothalamus that secrete tropic hormones. All of the hypothalamic tropic hormones are affected by a circadian rhythm, thereby imparting a rhythm to the release of the anterior pituitary hormones and, in turn, to the release of hormones affected by them. The degree of circadian rhythm varies for the different tropic hormones, however.

Transport of Hormones Bound to Carrier Proteins

Recall that hydrophobic messengers, including the steroid and thyroid hormones, are transported in blood bound to carrier proteins. When hormones are transported in this manner, only the concentration of *unbound* hormones affects the binding of hormone to receptor. Carrier proteins increase the half-life of hormones, which ensures that they are present in the blood for a longer period, by decreasing the rate of the hormones' metabolism, as described next.

Rate of Hormone Metabolism

Hormones remain in the blood for only a relatively short time before they are metabolized. Hormones that bind to receptors on target cells are often metabolized by the target cell itself. Even those hormone molecules that bind to membrane receptors often get "internalized" by endocytosis of the hormone-receptor complex. Once inside the cell, the hormone is degraded by enzymes located in lysosomes. Hormones that are free in the blood can be broken down as well. Peptide hormones can be metabolized by proteolytic enzymes that are present in the blood. Hormones can also be metabolized by enzymes in the liver. The breakdown products of these hormones are excreted in the urine, along with certain hormones that are excreted in their original forms.

Let's use insulin metabolism as an example of how a hydrophilic hormone is removed from the blood. Insulin is released from beta cells of the pancreas when blood glucose levels are high. Insulin then acts on target cells throughout the body to increase glucose transport into the cells. Insulin's half-life is short, with the hormone lasting only 4–6 minutes in blood. In fact, as soon as the blood passes through the liver, much of the insulin is removed from it by receptor-mediated endocytosis. Receptor-mediated endocytosis is also the mechanism by which insulin is taken into its target cells. Once inside the cells, insulin is exposed to insulin-degrading enzyme (IDE), which is located in the cytosol. Some insulin is eliminated from the body by the kidneys. Kidney cells contain IDE; they can break insulin down into smaller peptide fragments and then excrete those fragments from the body.

The hydrophobic hormones—that is, steroids, 1,25-dihydroxyvitamin D_3, and thyroid hormones—are metabolized more slowly (have longer half-lives) than the hydrophilic hormones for two reasons: (1) Hydrophobic hormones are transported in blood bound to carrier proteins, and (2) hydrophobic hormones can be stored temporarily in fatty tissue. Because both of these processes are reversible, hormones can be released from these pools when the free hormone concentration in the plasma begins to fall. The existence of this releasable pool of hormones tends to keep their plasma concentrations elevated long after the rate of secretion has returned to its resting level. For example, when the steroid hormones are metabolized by the liver, they are converted to hydrophilic metabolites that are transported in blood to the kidneys and then excreted in the urine. This process is the basis of the urine test that is used to check for abuse of anabolic steroids by professional athletes. Because of the lipid solubility of the steroids, they persist in the body for a couple of weeks, thereby allowing detection of excess levels for as long as two weeks after their administration.

4 Abnormal Secretion of Hormones

Abnormal secretion of hormones can have serious consequences. Some disease conditions are caused by an excess in the secretion of a hormone, termed *hypersecretion;* others are caused by too little secretion of a hormone, termed *hyposecretion.* An example of hypersecretion is *acromegaly,* a disease caused by an excess of growth hormone secretion in adults, which causes the bones to thicken and organs to grow excessively.

In a primary secretion disorder, the abnormality originates in the endocrine gland that secretes the hormone. For example, in primary hypersecretion of cortisol, the adrenal gland secretes too much cortisol (**Figure 13a**). One possible cause of hypersecretion is tumors of the endocrine cells (see **Clinical Connections: Pituitary Adenomas**). In primary hypersecretion, the blood levels of tropic hormones tend to be lower than normal due to the increased negative feedback from the hormone regulated by the tropic hormones. In the example of cortisol, levels of CRH and ACTH in the blood are low because the excess cortisol inhibits their release through negative feedback.

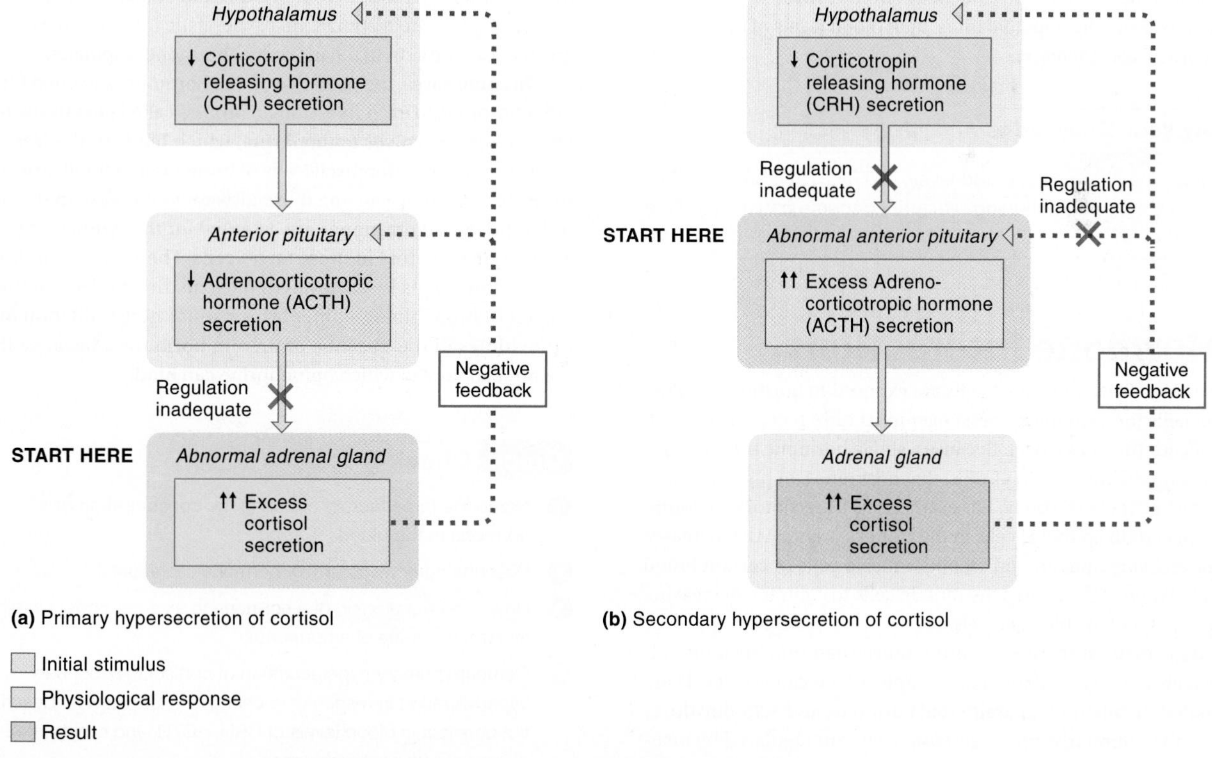

Figure 13 Abnormal secretion of cortisol. (a) Primary hypersecretion of cortisol. An abnormality of the adrenal gland causes it to secrete too much cortisol. Excess cortisol in the blood induces strong negative feedback, decreasing CRH and ACTH release into the blood. **(b)** Secondary hypersecretion of cortisol. An abnormality of the anterior pituitary causes it to secrete too much ACTH. Excess ACTH levels in the blood stimulate the adrenal gland to secrete excess cortisol. Negative feedback to the hypothalamus increases, so CRH levels in the blood decrease.

FOCUS ON DIABETES

An example of hyposecretion of hormones is *insulin-dependent diabetes mellitus*, a disease caused by insufficient secretion of insulin from the beta cells of the pancreas. When too little insulin is present in the blood, cells cannot use glucose adequately for energy. In cases of hypersecretion or hyposecretion, the disease process can be primary (acting directly in the endocrine gland) or secondary (involving a problem with the tropic hormone).

In primary hyposecretion of cortisol, hormone levels follow just the opposite pattern: Cortisol levels are decreased, but CRH and ACTH levels increase as a consequence of reduced negative feedback.

In a secondary secretion disorder, the abnormality originates in the endocrine cells of either the anterior pituitary or the hypothalamus, which secrete the tropic hormone. In secondary hypersecretion of cortisol, for example, blood levels of cortisol increase due to either an excess of ACTH secretion by an abnormal anterior pituitary (Figure 13b) or an excess of CRH secretion by an abnormal hypothalamus. If excess ACTH is secreted from the anterior pituitary, cortisol levels in the blood increase, but levels of CRH fall because of increased negative feedback from cortisol. If excess CRH is secreted from the hypothalamus, then blood levels of both ACTH and cortisol also increase.

Apply Your Knowledge

Describe the changes in blood levels of TRH, TSH, and thyroid hormones if a nonfunctioning pituitary adenoma causes too little TSH secretion.

5 Hormone Interactions

As we saw earlier, almost all cells are exposed to hormones, so the target cells for a chemical messenger must have receptors that are specific for that messenger. Because a single hormone may have receptors on different types of cells, that hormone can produce more than one effect in the body. For example, ADH receptors are found both on certain epithelial cells in the kidneys, where ADH increases water reabsorption, and on smooth muscle cells of certain blood vessels, where ADH causes the muscle cells to contract, decreasing the diameter of the blood vessels.

As is evident in Table 1, often more than one hormone affects a given body function. For example, blood calcium levels are regulated by calcitonin, parathyroid hormone, and 1,25-dihydroxy vitamin D_3. Similarly, blood glucose levels are regulated by insulin, glucagon, epinephrine, cortisol, and growth hormone. In some cases, the effects of the hormones oppose each other, a process called **antagonism.** For example, parathyroid hormone increases blood calcium levels, whereas calcitonin decreases blood calcium levels. Likewise, glucagon increases blood glucose levels, whereas insulin decreases blood glucose levels.

In other cases, the hormones produce the same direction of effect, generally by different means. When two or more hormones produce the same type of response in the body, the effect can be either **additive,** in which case the net effect equals the sum of the individual effects, or **synergistic,** in which case the net effect is greater than the sum of the individual effects. Assume, for example, that hormones A and B produce equal effects when present at a concentration of 1 nanogram per deciliter (ng/dL) of blood. If both hormones were present at 1 ng/dL blood, then the response in an additive interaction would be the sum of that of the individual hormones, but the response in a synergistic interaction would be greater than the sum of the individual hormone responses.

In some cases, the presence of one hormone is needed for another hormone to exert its actions, a process called **permissiveness.** One example involves epinephrine, which binds to β adrenergic receptors on smooth muscle cells of bronchioles (small airways of the lungs), thereby causing these airways to increase in diameter (dilate). Thyroid hormones are essential for the synthesis of the β adrenergic receptors in these target cells. Therefore, even though thyroid hormones by themselves have no direct effect on the diameter of bronchioles, epinephrine cannot trigger dilation of the bronchioles in the absence of thyroid hormones, because there are no receptors to which epinephrine can bind.

Quick Check 3

1. Name the three factors affecting the concentration of a hormone in the blood.
2. Describe humoral control of hormone secretion.
3. How does the binding of a hormone to a carrier protein in blood affect the half-life of a hormone?
4. Compare primary hypersecretion of cortisol to secondary hypersecretion in response to changes in CRH levels. Describe the changes in blood levels of CRH, ACTH, and cortisol that are observed with each imbalance.
5. Compare antagonism, additive effects, synergism, and permissiveness as they relate to hormone interactions.

CLINICAL CONNECTIONS

Pituitary Adenomas

Pituitary adenomas are benign (noncancerous) tumors of epithelial cells in the pituitary gland. Although they are noncancerous, pituitary adenomas are nonetheless potentially dangerous, because they can affect the normal levels of hormone secretion in the pituitary gland.

Pituitary adenomas consisting of inactive tumor cells are called *nonfunctioning adenomas.* Nonfunctioning adenomas tend either to decrease the secretion of certain hormones by compressing normal endocrine cells or to cause other neurological disturbances by compressing neurons near the tumor. Adenomas with tumor cells that actively secrete a certain hormone (thereby increasing the secretion of pituitary hormones) are called *functioning adenomas.* Approximately 50% of functioning pituitary adenomas affect prolactin secreting cells and can result in inappropriate breast discharge *(galactorrhea),* irregular menstrual periods *(amenorrhea),* or sexual dysfunction. Growth hormone secreting cells are affected in approximately 30% of functioning pituitary adenomas, resulting in excessive height *(gigantism)* in children or excessive thickening of bones and enlargement of soft tissues *(acromegaly)* in adults. ACTH secreting cells are affected in approximately 20% of functioning pituitary adenomas, resulting in *Cushing's disease.* Because their cortisol secretion is increased, people with Cushing's disease tend to have elevated blood glucose levels, which can lead to diabetes mellitus. Other symptoms include obesity of the trunk (but not extremities), purple striations ("stretch marks") on the abdomen due to a collagen deficit in the skin, and a rounded face due to accumulation of fluids.

Pituitary adenomas less than 1 cm in diameter are called *microadenomas.* Those greater than 1 cm in diameter are called *macroadenomas.* Most functioning pituitary adenomas are microadenomas, because they cause symptoms that result in early detection of the tumors. Treatment of pituitary adenomas is similar to treatment of other tumors and includes pharmacological treatments (usually effective for hypersecretion of prolactin and growth hormone, but not ACTH), radiation therapy, or surgical removal of the tumor.

Mykhailo Markiv/AP Images

Gigantism.

Critical Thinking Questions

1. Describe the differences between functioning and nonfunctioning adenomas, and their effects.
2. Which hormones are affected by the functioning adenomas, and what are the symptoms of the resulting conditions?
3. Why is it beneficial that most functioning pituitary adenomas are microadenomas?

SYSTEMS INTEGRATION

As we learned in this chapter, the endocrine system involves the release of hormones from many different organs. These hormones, in turn, travel in the blood to affect target cells throughout the body. Thus, as we learn other systems, the endocrine system will appear frequently. In our exploration of the cardiovascular system, we will learn about the role of epinephrine in regulating the amount of blood pumped from the heart and the diameter of blood vessels. We will also learn about the hormones angiotensin II and vasopressin, and their roles in decreasing the diameter of blood vessels. In our study of the renal system, we will learn about hormones that regulate fluid and electrolyte balance. In studying whole-body metabolism and growth, we will learn about the role of numerous hormones that function in these roles. Several hormones are also important in reproduction.

CHAPTER REVIEW

SUMMARY

1 Primary Endocrine Organs

- Primary endocrine organs include the pituitary gland (which is divided into anterior and posterior lobes), pineal gland, thyroid gland, parathyroid glands, thymus, pancreas, and gonads.
- Secretion by the anterior pituitary is regulated by tropic hormones secreted by neurosecretory cells in the hypothalamus.
- Secretion of these and other hormones is regulated by negative feedback.

MAP iP Endocrine, Orientation
MAP iP Endocrine, System Review
MAP iP Endocrine, The Hypothalamic-Pituitary Axis
MAP iP Name the Hormone

2 Secondary Endocrine Organs

- A secondary endocrine organ secretes a hormone in addition to carrying out another primary function.
- Secondary endocrine organs include the heart, liver, and kidneys.

3 Hormone Actions at the Target Cell

- The magnitude of a target cell's response to a hormone varies with the hormone's concentration in the plasma, which itself depends on three factors: the rate of hormone secretion, the amount of hormone that is bound to carrier proteins in the blood, and the rate at which the hormone is metabolized.
- The liver degrades most hormones, and the resulting breakdown products are eventually excreted in the urine.

MAP iP Endocrine, Actions of Hormones on Target Cells
MAP iP Endocrine, Biochemistry, Secretion, and Transport of Hormones
MAP iP Mechanism of Hormone Action

4 Abnormal Secretion of Hormones

- Abnormal hormone secretion includes hyposecretion (too little hormone) and hypersecretion (too much hormone); such a disorder can be either primary or secondary.
- A primary secretion disorder exists when the endocrine organ is abnormal and secretes the wrong amount of hormone.
- A secondary secretion disorder exists when the hypothalamus or anterior pituitary gland is abnormal and secretes the wrong amount of tropic hormone, thereby triggering the inappropriate secretion of hormone from the target endocrine gland.

5 Hormone Interactions

- A single hormone may regulate more than one body function, and a given function may be regulated by two or more hormones, which may exert effects that are additive, antagonistic, synergistic, or permissive.

MAP iP Endocrine, Response to Stress

MAP MasteringA&P®

Check your understanding at MasteringA&P and apply your new knowledge to real-life case studies to ensure that you've mastered the content!

EXERCISES

Multiple-Choice Questions

1. Epinephrine is released from which area of the adrenal gland?
   a) Zona reticularis
   b) Zona fasciculata
   c) Zona glomerulosa
   d) Medulla
2. Which of the following is an accurate statement regarding regulation of pituitary hormone secretion by the hypothalamus?
   a) All pituitary hormones are regulated by tropic hormones from the hypothalamus.
   b) All anterior pituitary hormones are regulated by a releasing hormone and a release-inhibiting hormone from the hypothalamus.
   c) All posterior pituitary hormones are regulated by a releasing hormone from the hypothalamus.
   d) All anterior pituitary hormones are tropic hormones.
   e) None of the above is accurate.
3. Most hypothalamic and pituitary hormones are
   a) Amino acids.
   b) Peptides/proteins.
   c) Steroids.
   d) Eicosanoids.
   e) Catecholamines.
4. Gonadotropin releasing hormone stimulates release of which of the following from the anterior pituitary?
   a) Sex hormones
   b) Follicle stimulating hormone
   c) Luteinizing hormone
   d) Both follicle stimulating hormone and luteinizing hormone
5. Which of the following adrenal hormones is secreted by chromaffin cells?
   a) Cortisol
   b) Aldosterone
   c) Epinephrine
   d) Androgens
6. In primary hyposecretion of thyroid hormones,
   a) Levels of thyroid hormones in the blood decrease.
   b) Levels of TRH in the blood decrease.
   c) Levels of TSH in the blood decrease.
   d) All of the above are true.
   e) None of the above is true.
7. Which of the following organs secretes glucagon?
   a) Liver
   b) Anterior pituitary
   c) Posterior pituitary
   d) Adrenal gland
   e) Pancreas

8. Which of the following is an example of permissiveness?
   a) Glucagon increases blood glucose levels, and insulin decreases blood glucose levels.
   b) Glucagon, epinephrine, and cortisol all increase blood glucose levels.
   c) Estrogen stimulates synthesis of progesterone receptors in the endometrium.
   d) All of the above.
   e) None of the above.

Objective Questions

9. (Endocrine/Exocrine) glands secrete hormones.
10. Neural input to the hypothalamus is involved in regulating secretion of hormones by both lobes of the pituitary. (true/false)
11. Epinephrine is secreted by the adrenal (medulla/cortex).
12. Thyroid hormones are classified as (amines/steroids).
13. Lipophobic messengers are secreted by (exocytosis/diffusion across the cell membrane).
14. Calcitonin is secreted by the (thyroid gland/parathyroid gland).

Essay Questions

15. Describe the role of the hypothalamus in the regulation of anterior pituitary hormone secretion.
16. Describe the various factors that affect the concentration of a hormone in plasma.
17. Describe the anatomy of the adrenal gland. What are its two main subdivisions? What are the minor subdivisions of its outer region? Which hormones are secreted from the different subdivisions?

Critical Thinking

18. Antagonistic hormones have opposing actions on the target tissue. Why might this dual effect be advantageous? What would you predict about factors releasing these hormones?
19. When a substance is first discovered and is *thought* to be a hormone, it is referred to as a *factor*. What evidence would you predict a scientist would need to change the name from *factor* to *hormone*?
20. 1,25-Dihydroxy vitamin D_3 is a hormone that increases calcium absorption by the digestive tract and calcium reabsorption by the kidneys, thereby increasing blood calcium levels. Based on what you know about controls of hormone release, what effect does calcium have on the release of 1,25-dihydroxy vitamin D_3?

Photo credits for icons in feature boxes: Focus on Diabetes: Tanyae/Dreamstime; Clinical Connections: Iofoto/Dreamstime; Understanding Exercise: Pixattitude/Dreamstime; Discovery: Yuri Arcurs/Cutcaster.

Glossary

adrenal cortex outer portion of the adrenal gland; secretes adrenocorticosteroids

adrenal medulla inner portion of the adrenal gland; secretes catecholamines

adrenocorticoids steroid hormones secreted from the adrenal cortex

adrenocorticotropic hormone (ACTH) (ad-ren-oh-kor-tih-koh-troh-pik) tropic hormone secreted from the anterior pituitary that stimulates secretion of glucocorticoids from the adrenal cortex

androgens a class of sex hormones secreted by the testes

antagonism in the endocrine system, when one hormone opposes the actions of another hormone

calcitonin peptide hormone released from C cells of the thyroid gland that regulates plasma calcium levels

circadian rhythm (sir-kay-dee-an) endogenous fluctuations in body functions that occur on a 24-hour cycle

estrogens a class of sex hormones secreted by the ovarie

glucocorticoids steroid hormones secreted from the adrenal cortex that regulate the body's response to stress; regulate protein, carbohydrate, and lipid metabolism in a variety of tissues; and regulate blood glucose levels; the primary glucocorticoid is cortisol

growth hormone (GH) peptide hormone secreted by the anterior pituitary; essential for normal growth

hypothalamus a region at the base of the brain that regulates autonomic functions and secretes several hormones, most of which regulate secretory activity of the pituitary gland

insulin a peptide hormone secreted by beta cells of the pancreas; promotes metabolic processes of the absorptive state

luteinizing hormone (LH) a gonadotropic hormone that stimulates sex hormone secretion and regulates other gonadal functions in either sex

mineralocorticoids steroid hormones secreted from the adrenal cortex that regulate sodium reabsorption and potassium secretion by the kidneys; the primary mineralocorticoid is aldosterone

parathyroid glands primary endocrine glands located on the posterior surface of the thyroid gland; secrete parathyroid hormone (PTH)

permissiveness phenomenon in which one hormone is needed for another hormone to exert its actions

pineal gland primary endocrine gland located in the brain; secretes the hormone melatonin

pituitary gland primary endocrine gland connected to the hypothalamus at the base of the brain; divided into the anterior pituitary and the posterior pituitary

primary endocrine organs organs whose primary function is the secretion of hormones

secondary endocrine organs organs whose secretion of hormones is secondary to another function

sex hormones steroid hormones including estrogens, progesterone, and androgens secreted from the adrenal cortex and gonads, which regulate reproductive function and promote gametogenesis, growth and maintenance of reproductive organs, and development of secondary sex characteristics

synergistic pertaining to a process in which the net effect is greater than the sum of the individual effects

T_4 tetraiodothyronine, one of the thyroid hormones

thymus primary endocrine gland located near the heart; secretes the hormone thymosin; also is the site of T lymphocyte maturation

thyroid gland butterfly-shaped primary endocrine gland located on the ventral surface of the trachea; secretes tetraiodothyronine triiodothyronine and calcitonin

Answers

Figure Question Answers

Fig 6.5 Corticotropin releasing hormone

Fig 6.9 Glucocorticoids and sex hormones

Fig 6.12 Insulin is released by exocytosis; aldosterone is released by diffusion

Apply Your Knowledge Answers

With too little TSH secretion, thyroid hormone secretion will decrease. Thus the person will be hypothyroid (low blood thyroid hormone levels). The low thyroid hormone will have less negative feedback on the secretion of TRH from the hypothalamus. Thus TRH levels in the blood will be elevated. In summary, TRH levels will be high, TSH levels low, and thyroid hormone levels low.

Answers to End–of-Chapter Multiple Choice and Objective Questions

Multiple Choice Questions

1. d 2. a 3. b 4. d 5. c 6. a 7. e 8. c

Objective Questions

9. Endocrine
10. true
11. medulla
12. amines
13. exocytosis
14. thyroid gland

Nerve Cells and Electrical Signaling

Nerve Cells and Electrical Signaling

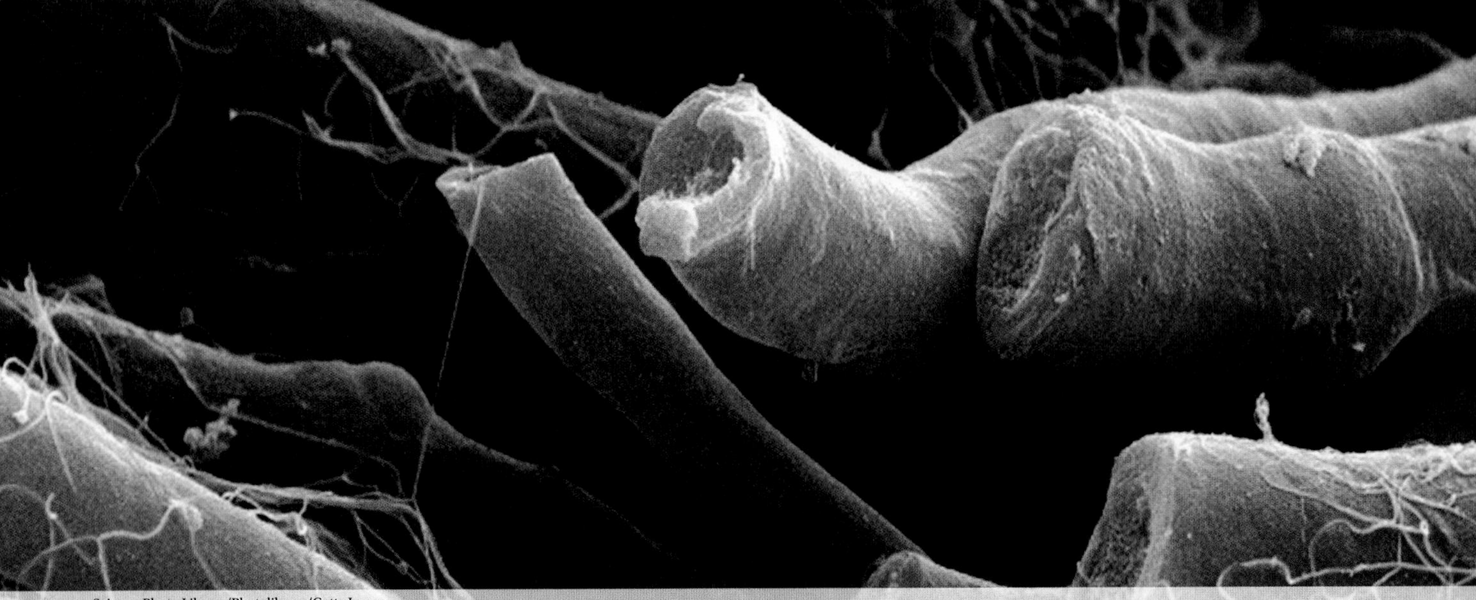

Science Photo Library/Photolibrary/Getty Images

Colorized scanning electron micrograph (SEM) of axons.

Imagine that you are standing at a crosswalk waiting for the light to change. As you cross, out of the corner of your eye, you notice a truck has run the red light and is speeding toward the intersection. Nerve cells in your brain immediately send electrical impulses down your spinal cord, where these impulses are relayed to the nerve cells that control the muscles in your legs. Less than a second after you see the truck, you step backward onto the curb—just in time to avoid being struck.

Every minute of every day, vast networks of nerve cells throughout the body fire off messages. In this example, the messages were transmitted by way of electrical signals smaller than 1/15 of the voltage of a typical flashlight battery! Some nerve cells send messages from the nervous system to organs, while others send messages from the organs to the nervous system. Such back-and-forth signaling is crucial to short-term maintenance of homeostasis because it provides second-by-second feedback, which is necessary for regulation of the body's internal environment.

CHAPTER OUTLINE

LEARNING OUTCOMES *After studying this chapter, you should be able to:*

- Describe the major components of the nervous system and the direction of information flow within and among them.
- Describe the basic anatomy of a neuron. Compare the functions of each part of a neuron, and describe the types of ion channels located in each part. Describe the grouping of neurons within the central nervous system and the peripheral nervous system.
- Describe the structure and function of myelin.
- Explain the ionic basis of the resting membrane potential.
- Describe the various properties of graded potentials, including the direction of change in a potential, the magnitude of change, and temporal and spatial summation. Explain how graded potentials in neurons can trigger an action potential.
- Explain the ionic basis of an action potential. Describe the gating mechanisms for voltage-gated sodium and potassium channels.
- Describe the propagation of action potentials from axon hillock to axon terminal, and compare propagation in myelinated and unmyelinated axons.
- Describe refractory periods, including what causes the absolute and relative refractory periods, and explain their physiological significance.

Before You Begin

Make sure you have mastered the following topics:

- Na^+/K^+ pump
- Ion channels
- Chemical and electrical forces

We know about intercellular communication. Long-distance communication is a function of the endocrine system and the nervous system. To understand how the nervous system functions, we must understand the properties of cell membranes, the properties of energy, ion channels and active transport, and signal transduction mechanisms. In this chapter, we examine how the electrochemical gradients are responsible not only for the resting membrane potential, but also for electrical signaling among nerve cells.

1 Overview of the Nervous System

The nervous system can be divided into two main anatomical parts: the central nervous system and the peripheral nervous system (**Figure 1**). The **central nervous system (CNS)** consists of the brain and spinal cord. It receives and processes information from sensory organs and the viscera to determine the state of the external environment **(sensory information)** and internal environment **(visceral information)**. The CNS integrates this information and makes decisions on appropriate actions, then sends instructions to certain organs (such as muscles and glands) instructing them to perform appropriate tasks. The central nervous system is also the site of learning, memory, emotions, thoughts, language, and other complex functions.

The **peripheral nervous system** consists of neurons that provide communication between the central nervous system and organs throughout the body. The peripheral nervous system can be subdivided into two divisions: afferent and efferent. Neurons of the

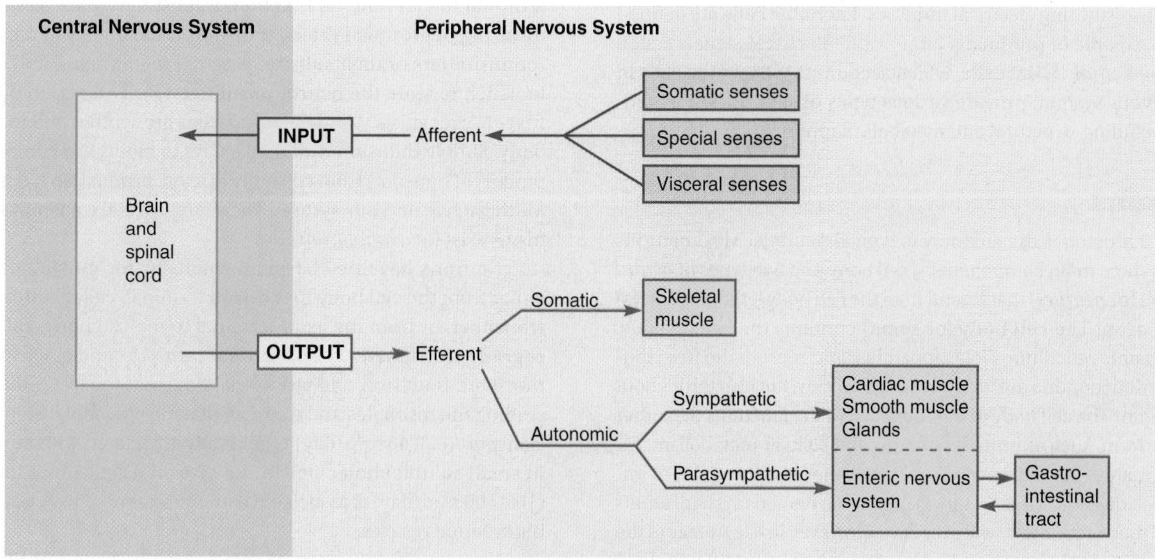

Figure 1 Organization of the nervous system. Arrows indicate the direction of information flow.

afferent division transmit sensory and visceral information from the organs to the central nervous system. Information transmitted to the central nervous system includes the *somatic senses* (associated with the skin, muscles, and joints), the *special senses* (vision, hearing, equilibrium, smell, and taste), and visceral information pertaining to the internal environment (such as fullness of the stomach, blood pressure, and blood pH). Neurons of the *efferent* division transmit information from the central nervous system to organs in the periphery, called **effector organs,** that perform functions in response to commands from neurons. Effector organs are usually muscles and glands. A neuron capable of transmitting messages to an effector organ or receiving information from a sensory organ is said to **innervate** that organ.

The efferent division can be further subdivided into two main branches: the somatic (or voluntary) and autonomic (or involuntary) nervous systems. The **somatic nervous system** consists of the **motor neurons,** which regulate skeletal muscle contractions. The **autonomic nervous system** consists of neurons that regulate the function of internal organs and other structures (such as sweat glands and blood vessels) that are not under voluntary control. The autonomic nervous system can be divided into two branches: the *parasympathetic* and *sympathetic* nervous systems, which tend to have opposite effects on organs.

The body also has an *enteric nervous system,* which comprises an intricate network of neurons in the gastrointestinal tract that can function independently of the rest of the nervous system but communicates with the autonomic nervous system.

2 Cells of the Nervous System

The nervous system contains two main classes of cells: neurons and glial cells. In the nervous system, the neuron is the *functional unit*—that is, the smallest unit of a tissue that can carry out the function of that tissue. Neurons are "excitable cells" that communicate by transmitting electrical impulses. **Excitable cells** are defined as cells capable of producing large, rapid electrical signals called *action potentials.* **Glial cells,** which account for 90% of the cells in the nervous system, provide various types of support to the neurons, including structural and metabolic support.

Neurons

Figure 2 illustrates the anatomy of typical neurons. Most neurons contain three main components: a cell body and two types of *neural processes* (or *neurites*) that extend from the cell body—the dendrite(s) and an axon. The **cell body** (or **soma**) contains the cell nucleus, endoplasmic reticulum, Golgi apparatus, and most of the free ribosomes. Mitochondria are located in the cell body, but also throughout the neuron. The cell body carries out most of the functions that other cells perform, such as protein synthesis and cellular metabolism. Although mature neurons retain their nuclei, they lose the ability to undergo cell division. Thus, in most areas of the nervous system, adults have all the neurons they will ever have. However, in a few areas of the adult human brain new neurons can develop from undifferentiated cells, or **stem cells** (see **Discovery: Neurogenesis**).

Dendrites branch from the cell body and receive input from other neurons at specialized junctions called **synapses.** (Cell bodies themselves can also receive input at synapses.) The branching pattern of dendrites varies between different types of neurons, as shown for the star-shaped stellate cells and the highly branching Purkinje cells in **Figure 3**. The extent of branching is an indication of the number of synapses with the neuron, as the majority of synapses occur on the dendrites.

Neurons have another branch that comes off the cell body called an **axon,** or *nerve fiber*. Unlike a dendrite, whose function is to *receive* information, an axon *sends* information. Generally, a neuron has only one axon, but axons can branch, thereby sending signals to more than one destination. The branches of an axon are called **collaterals,** and as with dendrites, the extent of branching varies among neurons and is indicative of the amount of communication with other cells.

The axon functions in the rapid transmission of information over relatively long distances in the form of electrical signals called **action potentials,** which are brief, large changes in membrane potential during which the inside of the cell becomes positively charged relative to the outside. (Properties of action potentials are described in detail later in this chapter.) The beginning and end of an axon are specialized structures called the axon hillock and the axon terminal, respectively. The **axon hillock**—the site where the axon originates from the cell body—is specialized in most neurons for the *initiation* of action potentials. Once initiated, action potentials are transmitted to the axon terminal by mechanisms we examine shortly. The **axon terminal** is specialized to release neurotransmitter on arrival of an action potential. The released neurotransmitter molecules carry a signal to a *postsynaptic cell,* usually to a dendrite or the cell body of another neuron or to the cells of an effector organ. The neuron whose axon terminal is releasing neurotransmitter is called the *presynaptic cell.*

Axons range in length from 1 millimeter to approximately 1 meter; thus the axon terminal may be located far from the cell body that contains the nucleus and endoplasmic reticulum. For the axon terminal to carry out its function, it must have enzymes for synthesizing neurotransmitters, transporter molecules to move neurotransmitters or their substrates across membranes, and vesicles in which to store the neurotransmitters until an action potential triggers exocytosis. All of these enzymes are synthesized in the cell body. Simple diffusion would take days to months to complete the process of transport from cell body to axon terminal, so it is too slow for the active nervous system. Therefore, special transport mechanisms exist for *axonal transport.*

Neurons have two basic mechanisms for moving products either from the cell body to the axon terminal, called **anterograde transport,** or from the axon terminal to the cell body, called **retrograde transport:** (1) *slow axonal transport* and (2) *fast axonal transport.* Both slow and fast axonal transport involve proteins, including microtubules and a variety of neurofilaments. Slow axonal transport (0.5–40 mm/day) is generally associated with movement of small soluble molecules in the cytosol. Fast axonal transport (100–400 mm/day) is associated with movement of vesicles, including synaptic vesicles.

Fast axonal transport of vesicles is illustrated in **Figure 4**. Microtubules extend the length of the axon and function as "tracks"

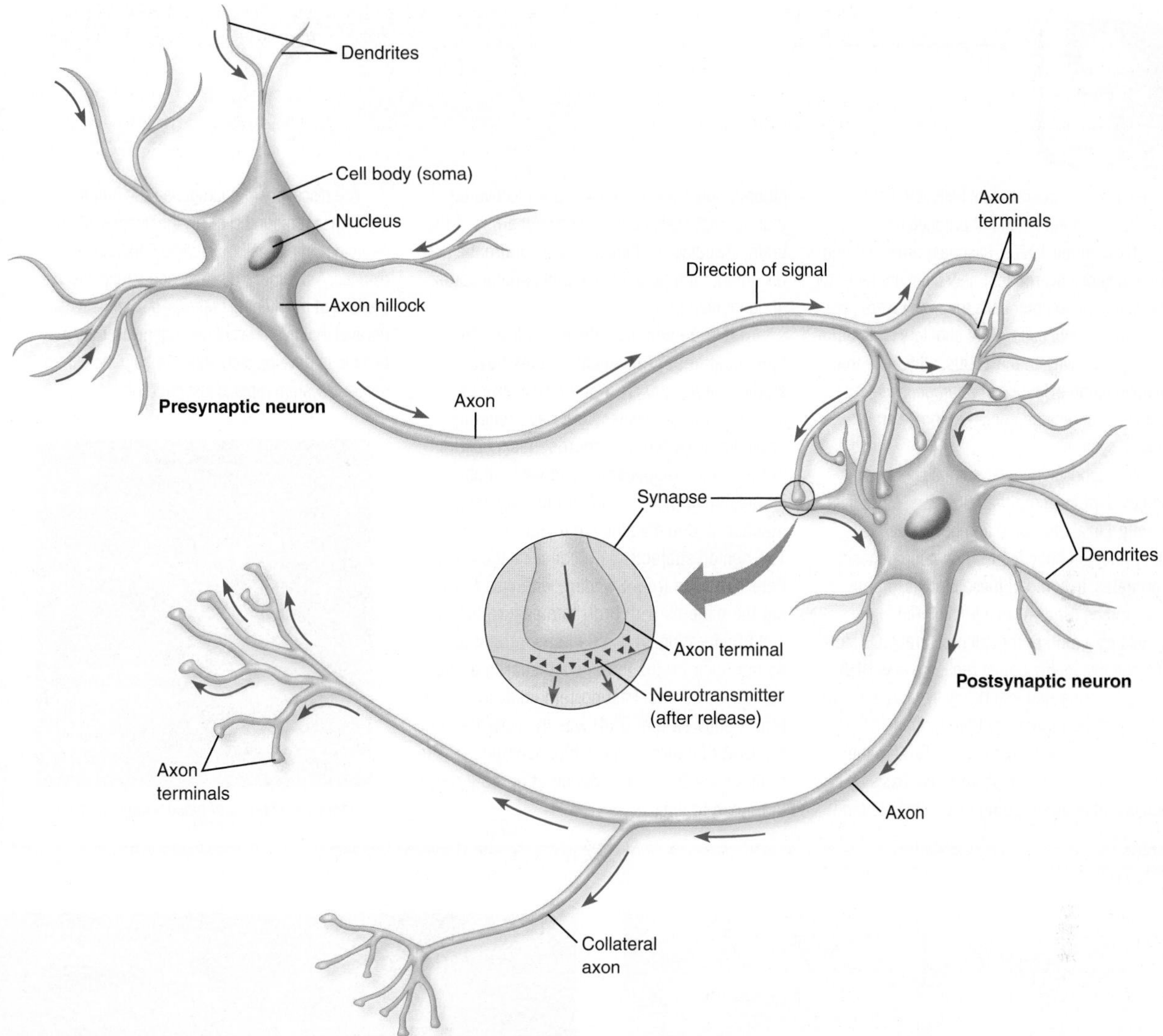

Figure 2 Structure of a typical neuron. Two neurons are shown; the upper neuron (presynaptic neuron) communicates with the lower neuron (postsynaptic neuron), as indicated by the arrows representing information flow. The enlargement shows an axon terminal of one neuron releasing a chemical messenger (neurotransmitter) that communicates with another neuron at a synapse.

for transport molecules. Among the molecules that run on these tracks are *kinesins*, proteins that essentially "walk" down the microtubules, carrying a vesicle with them. The vesicle can be a synaptic vesicle, or the vesicle can be used to transport other products from the site of synthesis in the body to the axon terminal. The process requires adenosine triphosphate (ATP).

Localization of Ion Channels in Neurons

Because different regions of a neuron generally have specialized functions (although some exceptions exist), each region tends to have specific types of ion channels, most of which are gated (that is, they can open or close). The opening or closing of ion channels changes the permeability of the plasma membrane for a specific ion, resulting in a change in the electrical properties of the cell or the release of a neurotransmitter.

Leak channels (or nongated channels), which are found in the plasma membrane throughout a neuron, are always open. They are responsible for the resting membrane potential.

Ligand-gated channels open or close in response to the binding of a chemical (the ligand) to a specific receptor in the plasma membrane. In neurons, ligand-gated channels are most densely located in the dendrites and cell body—areas that receive communication from presynaptic neurons in the form of neurotransmitters.

Voltage-gated channels open or close in response to changes in membrane potential. Voltage-gated sodium channels and voltage-gated potassium channels are located throughout the

DISCOVERY

Neurogenesis

For decades, neuroscientists believed that the adult human brain could not produce new neurons. Then, in the 1970s, Pasquale Graziadei and colleagues discovered that new olfactory receptor cells (sensory receptor cells for smell) were produced from basal cells in the olfactory epithelium of the nasal cavity. In the 1990s, other researchers have found evidence of *neurogenesis*—the production of new neurons—in the central nervous system.

Neurogenesis in lower animal species has been evident for more than 30 years. More recently, Elizabeth Gould and her colleagues from Princeton have focused on neurogenesis in primates. In some of these studies, bromodeoxyuridine (BrdU) was administered to the animals as a marker for cell proliferation. Before cells can divide, they must replicate new DNA. BrdU is incorporated into newly synthesized DNA and, therefore, marks any DNA formed after its administration. Cells that contain BrdU in their DNA are identified on a postmortem basis by histological analysis. Using this technique (and others), Gould and her colleagues discovered that neurogenesis occurs in several areas of the brain, including the hippocampus, prefrontal cortex, inferior temporal cortex, and parietal cortex of nonhuman primates.

Fred Gage and his colleagues from the Salk Institute for Biological Studies have studied neurogenesis in the hippocampus in lower animal species, and have recently provided evidence that neurogenesis occurs in the human hippocampus as well. Their studies in humans used BrdU in the same manner as Gould's study in lower primates. The human subjects were cancer patients of Peter Eriksson from Sweden, who was treating the patients with BrdU to measure tumor cell proliferation. These patients agreed to donate their brains to Gage's research group upon their deaths. Postmortem analyses of the brains showed that BrdU was incorporated into the DNA of neurons in the hippocampus. These findings are the first evidence of neurogenesis in human brains.

The discovery of neurogenesis in human brains has profound clinical significance. Many neurological diseases, including Parkinson's disease, Alzheimer's disease, and stroke, involve the loss of functioning neurons. If scientists can unravel the mysteries of neurogenesis, they may be able to develop techniques to help the central nervous system replace lost neurons.

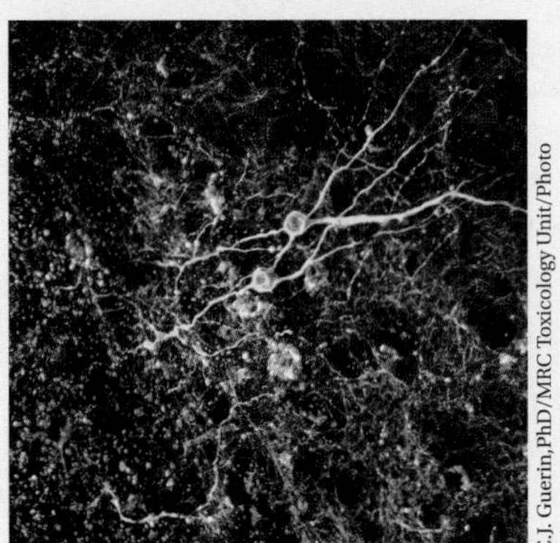
C.J. Guerin,PhD/MRC Toxicology Unit/Photo Researchers, Inc.

Fluorescent staining of hippocampal neurons.

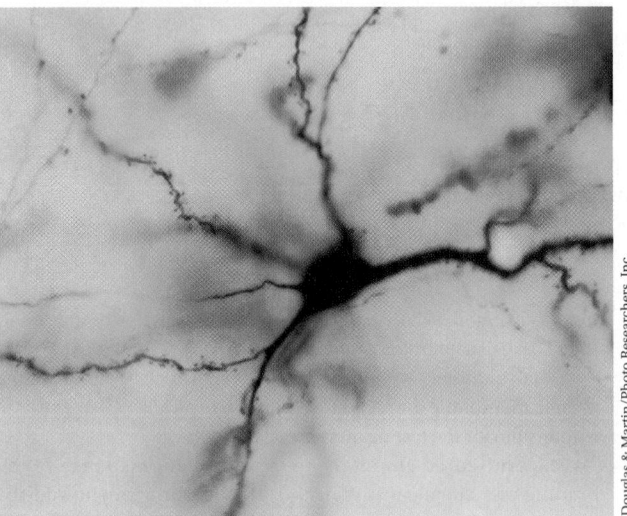
Douglas & Martin/Photo Researchers, Inc.

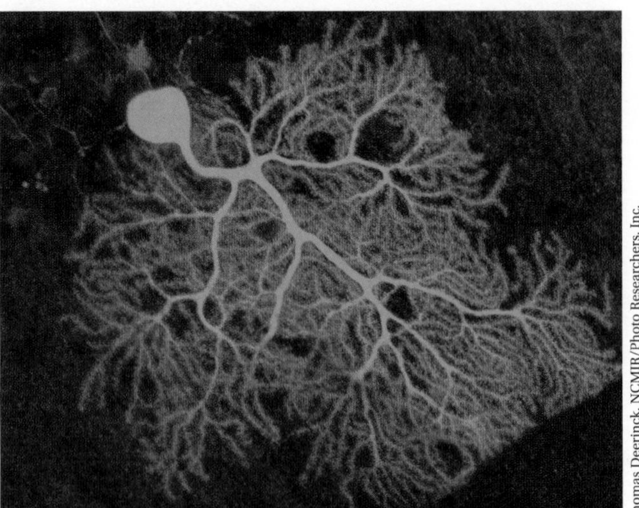
Thomas Deerinck, NCMIR/Photo Researchers, Inc.

Figure 3 Variations in the extent of dendritic branching. (a) Stellate cell. **(b)** Purkinje cell.

neuron, but are more densely clustered in the axon and are present in greatest density in the axon hillock. These channels are necessary for the initiation and propagation of action potentials. The greatest density of voltage-gated calcium channels is found in the axon terminals; these channels open in response to the arrival of an action potential at the axon terminals. When these channels are open, calcium enters the cytosol of the axon terminals and triggers the release of neurotransmitter.

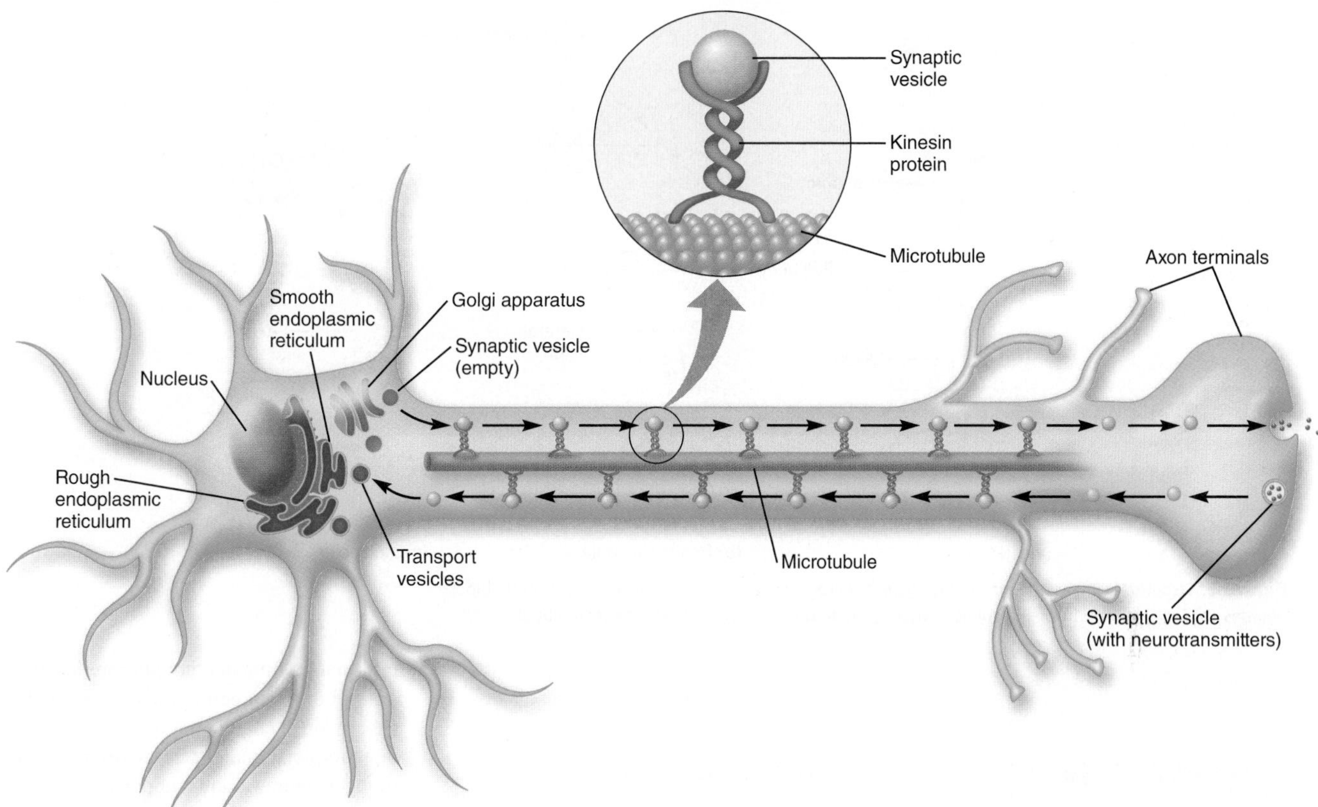

Figure 4 Fast axonal transport of vesicles.

Structural Classification of Neurons

Neurons can be classified structurally according to the number of processes (axons and dendrites) that project from the cell body (**Figure 5**). Neurons commonly found in humans include bipolar neurons and multipolar neurons. **Bipolar neurons** are generally sensory neurons with two projections, an axon and a dendrite, coming off the cell body. Typical bipolar neurons (Figure 5a) function in the senses of olfaction (smell) and vision. Most sensory neurons, however, are a subclass of bipolar neurons called *pseudo-unipolar* neurons (Figure 5b). This name arises because the axon and dendrite projections *appear* as a single process that extends in two directions from the cell body; in actuality, however, the dendrite is modified to function much like an axon, and is a functional continuation of the axon. The modified dendritic process is called the *peripheral axon* because it originates in the periphery with sensory receptors and functions as an axon in that it transmits action potentials. The axon process is called the *central axon* because it ends in the central nervous system, where it forms synapses with other neurons. **Multipolar neurons,** the most common neurons, have multiple projections from the cell body (Figure 5c); one projection is an axon, all the others are dendrites.

Functional Classification of Neurons

Three functional classes of neurons exist: efferent neurons, afferent neurons, and interneurons (**Figure 6**). **Efferent neurons** transmit information from the central nervous system to effector organs. Recall that efferent neurons include the motor neurons that extend to skeletal muscle and the neurons of the autonomic nervous system (see Figure 1). Notice in Figure 6 that the cell body and dendrites of efferent neurons are located in the central nervous system (autonomic postganglionic neurons are an exception). However, the axon leaves the central nervous system and becomes part of the peripheral nervous system as it travels to the effector organ it innervates.

The function of **afferent neurons** is to transmit information from either *sensory receptors* (which detect information pertaining to the outside environment) or *visceral receptors* (which detect information pertaining to conditions in the interior of the body) to the central nervous system for further processing. Most afferent neurons are pseudo-unipolar neurons, with the cell body located outside the central nervous system in a *ganglion* (the general term for a cluster of neural cell bodies located outside the CNS). The endings of the peripheral axon are located in the peripheral organ (sensory organ or visceral organ), where they either are modified into sensory receptors or receive communication from separate sensory receptor cells. The central axon terminates in the central nervous system, where it releases neurotransmitter to communicate with other neurons.

The third functional class of neurons is **interneurons,** which account for 99% of all neurons in the body. They are located entirely in the central nervous system. Interneurons perform all the functions of the central nervous system, including processing sensory

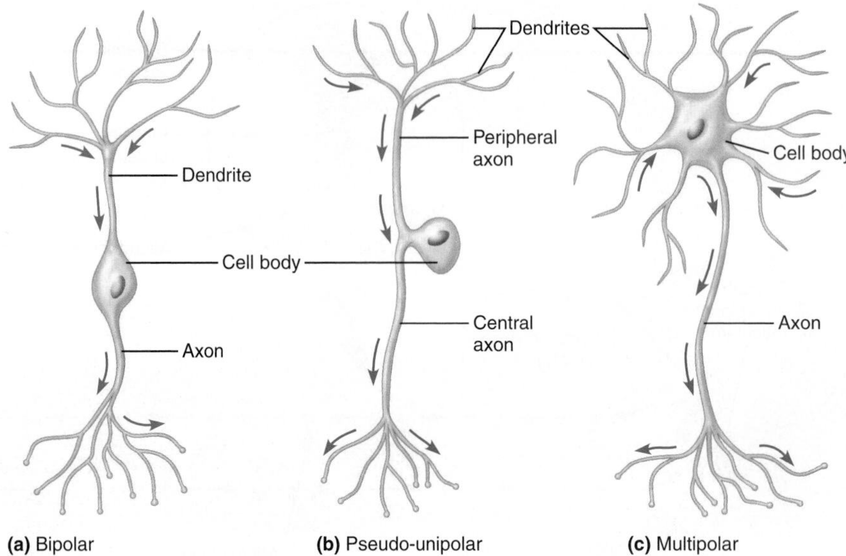

Figure 5 Structural classes of neurons. (a) Afferent neurons associated with vision and olfaction are bipolar neurons. **(b)** The vast majority of afferent neurons are pseudo-unipolar. **(c)** Most neurons are multipolar.

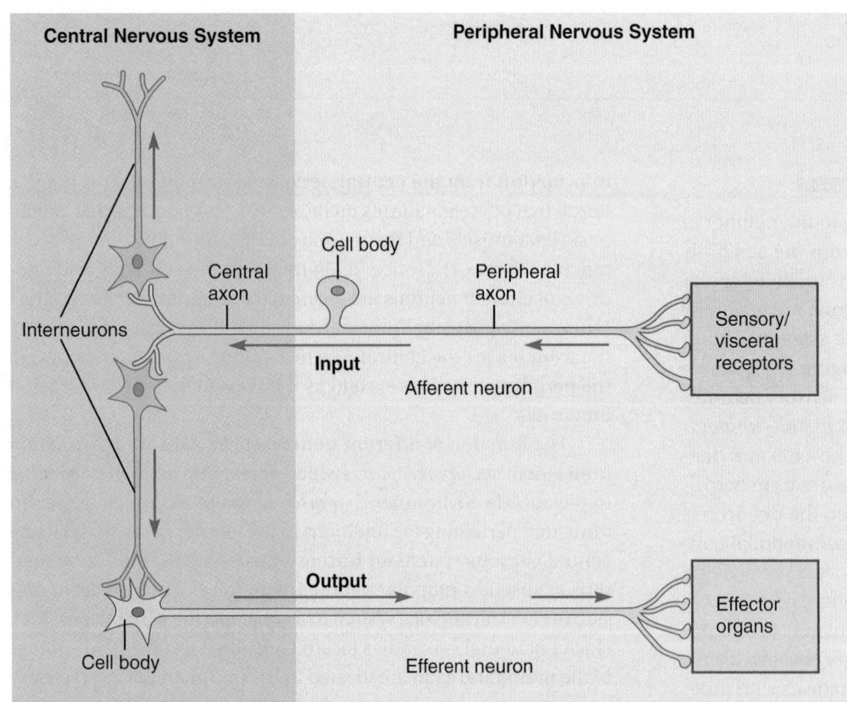

Figure 6 Functional classes of neurons. Afferent neurons originate in the periphery with sensory or visceral receptors. Efferent neurons originate in the central nervous system, where the cell body and dendrites receive synaptic communication from other neurons. Interneurons lie entirely in the central nervous system and can communicate with afferent neurons, efferent neurons, or other interneurons.

Q Name two types of effector organs.

information from afferent neurons, creating and sending out commands to effector organs through efferent neurons, and carrying out complex functions of the brain such as thought, memory, and emotions.

Structural Organization of Neurons in the Nervous System

Neurons are arranged within the nervous system in an orderly fashion, with those having similar functions tending to be grouped together. In addition, neurons are aligned in such a way that cell bodies and dendrites of adjacent cells tend to be grouped together, and axons of adjacent cells tend to be grouped together. In the central nervous system, cell bodies of neurons are often grouped into **nuclei** (singular: *nucleus*), and the axons travel together in bundles called **pathways, tracts,** or **commissures.** In the peripheral nervous system, cell bodies of neurons are clustered together in **ganglia** (singular: *ganglion*), and the axons travel together in bundles called **nerves.**

Glial Cells

Glial cells, the second class of cell found in the nervous system, account for 90% of all cells in the nervous system. Their main functions include providing structural integrity to the nervous system (*glia* is Latin for "glue") and chemical and anatomical support that permits neurons to carry out their functions. Recent studies suggest that glial cells may also play important roles in intercellular communication.

There are four types of glial cells: *astrocytes, microglia, oligodendrocytes,* and *Schwann cells.* Of these glial cells, only Schwann cells are located in the peripheral nervous system; the rest are found in the central nervous system. Because one of the functions of **oligodendrocytes** and **Schwann cells** (or *neurolemmocytes*) is crucial to electrical transmission in neurons, we discuss these cells here.

The primary function of oligodendrocytes and Schwann cells is to form an insulating layer of **myelin** around the axons of neurons. Such insulation enables neurons to transmit action potentials more efficiently and rapidly. **Figure 7a** shows the formation of a myelin sheath by a Schwann cell. Myelin consists of concentric layers of the plasma membranes of either oligodendrocytes or Schwann cells. Oligodendrocytes form myelin around axons in the central nervous system; one oligodendrocyte sends out projections providing the myelin segments for many axons (Figure 7b). In contrast, Schwann cells form myelin around axons in the peripheral nervous system, but each Schwann cell provides myelin for only one axon (Figure 7c). Many oligodendrocytes or Schwann cells are needed to provide the myelin for a single axon.

Because the lipid bilayer of a plasma membrane has low permeability to ions, the several layers of membrane that make up a myelin sheath substantially reduce leakage of ions across the cell membrane. However, in gaps within the myelin, called **nodes of Ranvier** (or simply nodes), the axonal membrane contains voltage-gated sodium and potassium channels that function in the transmission of action potentials by allowing ion movement across the membrane. We discuss the nature of these electrical signals and their origins in the next few sections of this chapter.

Quick Check 1

1. Name the different parts, divisions, and branches of the nervous system and give the basic functions of each.
2. Draw a neuron and label the following structures: cell body, dendrite, axon, and axon terminal. Briefly state the function of these structures and the type of ion channels (ligand-gated or voltage-gated) that can be found in each.
3. Which glial cell forms myelin in the central nervous system? Which forms myelin in the peripheral nervous system?

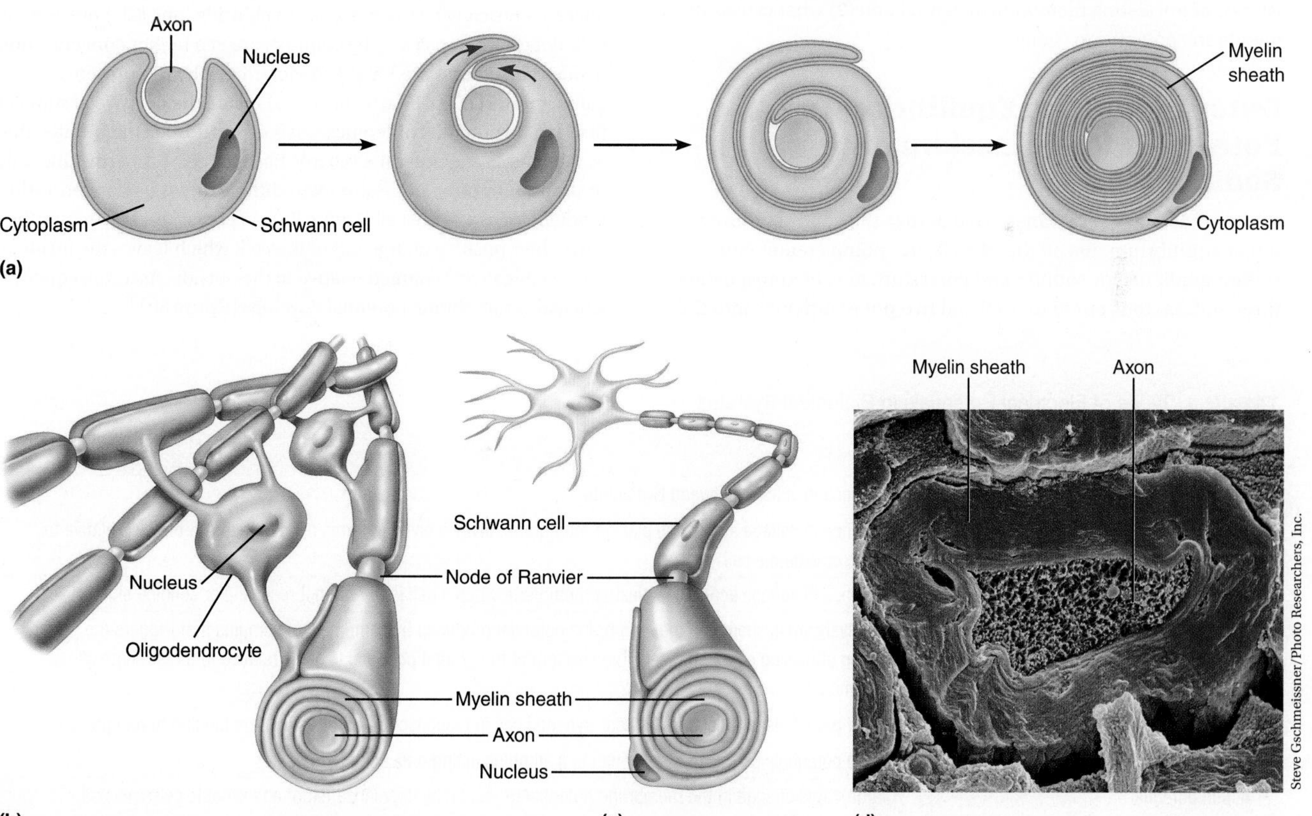

Figure 7 Formation and origins of myelin sheaths. (a) Formation of a myelin sheath by a Schwann cell. **(b)** Arrangement of myelin sheaths formed by oligodendrocytes in the central nervous system. **(c)** Arrangement of myelin sheaths formed by Schwann cells in the peripheral nervous system. **(d)** Electron micrograph showing cross section of a myelinated axon.

3 Establishment of the Resting Membrane Potential

We learned that the membrane potential is an electrical force. In this section, we look at the establishment of the baseline (resting) membrane potential. To further your understanding of electrical forces, see **Toolbox: Electrical Circuits in Biology.**

A cell at rest has a potential difference across its membrane such that the inside of the cell is negatively charged relative to the outside. This difference is called the **resting membrane potential** (resting V_m) because the cell is at rest—it is not receiving or transmitting any signals. The resting membrane potential of neurons is approximately −70 mV. Membrane potentials are *always* described as the potential inside the cell relative to that outside the cell. Therefore, the inside of a typical neuron at rest is 70 mV more negative compared to the outside. (Although we are discussing the resting membrane potential of neurons, all cells in the body have a negative resting membrane potential, ranging from −5 mV to −100 mV.)

Neurons communicate by generating electrical signals in the form of changes in membrane potential. Some of these changes in membrane potential trigger the release of neurotransmitter, which then carries a signal to another cell. Table 1 defines the different types of electrical potentials that are described in this chapter. In the following sections we explore (1) what is responsible for the existence of the resting membrane potential and (2) what causes the membrane potential to change.

Determining the Equilibrium Potentials for Potassium and Sodium Ions

An ion that is actively transported across the cell membrane is not at equilibrium. Recall that the Na^+/K^+ pump creates concentration gradients for sodium and potassium ions by transporting three sodium ions out of the cell and two potassium ions into the cell per ATP molecule hydrolyzed. Sodium ions are more highly concentrated outside the cell, which creates a chemical driving force that tends to push sodium ions into the cell. Potassium ions, in contrast, are more highly concentrated inside the cell, resulting in a chemical driving force that tends to push potassium ions out of the cell. As we will see shortly, the chemical forces for moving sodium and potassium ions across the plasma membrane, and the differences in the permeability of the plasma membrane to these two ions, establish the resting membrane potential.

To understand what causes the resting membrane potential, we first look at two hypothetical situations. First, we consider a cell that is permeable to potassium ions only (that is, it has open potassium channels in its plasma membrane). Next, we consider a cell that is permeable to sodium ions only (that is, it has open sodium channels in its plasma membrane).

Membrane Potential of a Cell Permeable Only to Potassium

Figure 8 illustrates how the potassium equilibrium potential is achieved in a cell freely permeable to only potassium ions. Sodium ions (Na^+) are at a higher concentration outside the cell and are balanced electrically by the presence of chloride ions (Cl^-) outside the cell. Potassium ions (K^+), by contrast, are at a higher concentration inside the cell and are balanced electrically by the presence of organic anions (A^-, primarily proteins) inside the cell. We assume at first that no potential difference exists across the cell membrane; that is, the membrane potential is 0 mV. Because cell 1 is permeable only to potassium ions, potassium will diffuse down its concentration gradient, or out of the cell (Figure 8a). As potassium ions move, they carry their positive charge out of the cell, which leaves the inside of the cell negatively charged relative to the outside. As a consequence, a negative membrane potential develops (Figure 8b).

TABLE 1 Types of Electrical Potentials in Biological Systems

Potential	Definition
Potential difference = E	Difference in voltage between two points
Membrane potential = V_m	Difference in voltage across the plasma membrane; always given in terms of voltage inside the cell relative to voltage outside the cell
Resting V_m	Difference in voltage across the plasma membrane when a cell is at rest (not receiving or sending signals)
Graded potential	A relatively small change in the membrane potential produced by some type of stimulus that triggers the opening or closing of ion channels; the strength of the graded potential is given relative to the strength of the stimulus
Synaptic potential	Graded potentials produced in the post-synaptic cell in response to neurotransmitters binding to receptors
Receptor potential	Graded potentials produced in response to a stimulus acting on a sensory receptor
Action potential	A large, rapid change in the membrane potential produced by depolarization of an excitable cell's plasma membrane to threshold
Equilibrium potential	The membrane potential that counters the chemical forces acting to move an ion across the membrane, thereby putting the ion at equilibrium

TOOLBOX

Electrical Circuits in Biology

Neurons and muscle cells form excitable tissues, meaning that they function through electrical signals in the form of changes in the membrane potential. To better understand these signals, we must first understand the relationship between an electrical potential, current, and resistance.

Electrical forces exist between charged particles. The direction of the electrical force follows a simple rule: Opposite charges attract each other; like charges repel each other. To separate opposite charges requires energy. Once separated, however, the opposite charges store the energy as an *electrical potential* (or voltage). The strength of the potential depends on the amount of charge separation: The greater the separation of charge, the greater the potential.

Electrical potentials are produced in biological systems by separating oppositely charged ions, which are attracted to each other and will move toward each other if possible. When ions move, they carry their charge with them. The movement of electrical charges is called a current (I). In biological systems, currents are typically expressed in units of microamps (10^{-6} amperes). The greater the electrical potential, the greater the *force* for ion movement; however, the presence of a force does not necessitate ion movement.

How easily ions can move depends on the properties of the substance through which they must move. **Resistance** *(R)* is a measurement of the hindrance to charge movement. The greater a substance's resistance, the more difficult it will be for ions to move through it, and the weaker the current will be. A neuron's plasma membrane has high resistance to current flow because its permeability to ions is low. The intracellular and extracellular fluids, by contrast, have low resistance to current flow because these fluids are rich in ions.

The inverse of resistance is **conductance** *(g)*:

$$g = 1/R$$

Because the ability of an ion to cross a plasma membrane depends on the permeability of the plasma membrane to that ion, the conductance of a particular ion increases as the membrane's permeability to that ion increases.

The relationship between potential difference, current, and resistance is defined by **Ohm's law:**

$$I = E/R$$

where E is the potential difference or voltage. Understanding Ohm's law is crucial to understanding neural physiology because plasma membranes have an electrical potential across them, ions present inside and outside the cell are available to carry a charge across the plasma membrane, and resistance to charge movement can be changed by the opening or closing of ion channels.

Once the membrane potential has developed, two forces act on the potassium ions: a *chemical force* due to the concentration gradient and an *electrical force* due to the membrane potential. (Recall that the net force acting on an ion is called the *electrochemical force,* the sum of the electrical and chemical forces.) The direction of the chemical force now present in cell 1 is such that it *pushes* potassium ions out of the cell; the direction of the electrical force is such that it *pulls* potassium ions back into the cell because of the attraction of the positively charged potassium ions for the negative charge inside of the cell.

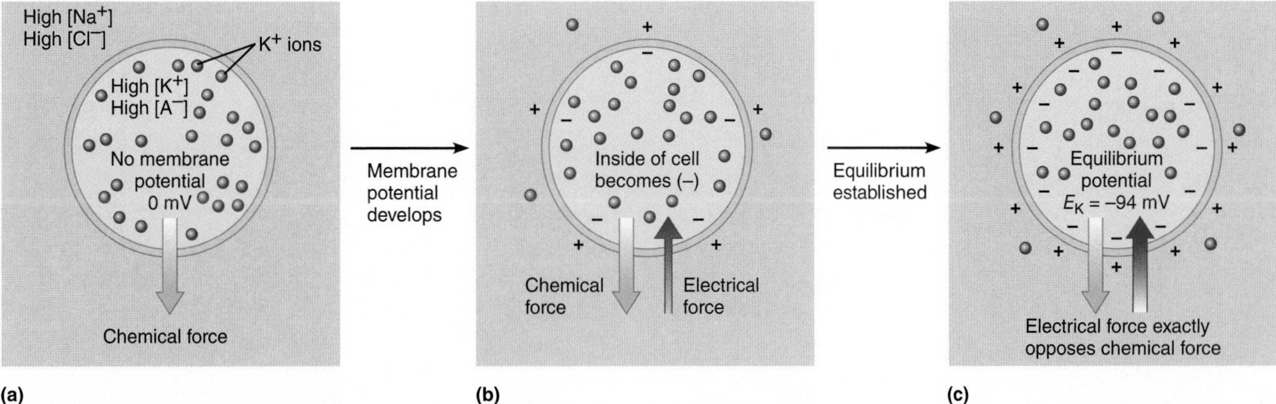

Figure 8 Membrane potential of a cell permeable to potassium only. Potassium (K^+) and organic anions (A^-) are present in greater concentration inside the cell. Sodium (Na^+) and chloride (Cl^-) ions are present in greater concentration outside the cell. The width of an arrow is relative to the strength of ion movement in the direction of the arrow. **(a)** Potassium ions move out of the cell because of a chemical force. **(b)** As some potassium ions leave the cell, the inside of the cell becomes negatively charged relative to the outside. This change in charge distribution creates an electrical force that moves potassium ions into the cell, opposing the chemical force. **(c)** Eventually, enough potassium leaves the cell that the electrical force becomes strong enough to oppose further movement of potassium ions out of the cell because of the chemical force, resulting in no net movement of potassium ions. At this membrane potential, potassium is at equilibrium. This potential is equal to the potassium equilibrium potential and is approximately −94 mV in neurons.

While the chemical force for potassium ions to move across the membrane remains constant (the number of potassium ions that actually move across the membrane is small relative to the total number of potassium ions in the intracellular and extracellular fluids—less than 0.01%), the electrical force will continue to change as potassium ions move across the membrane. Initially the chemical force is greater than the electrical force because the membrane potential is small, so potassium ions continue to move out of the cell (see Figure 8b). However, the more potassium ions that leave the cell, the greater the membrane potential becomes, and consequently the greater the electrical force to pull potassium back into the cell. Thus, as the membrane potential develops, the rate of potassium movement out of the cell becomes slower and slower as the net (electrochemical) force on potassium decreases. Once the electrical force has become just strong enough to exactly balance the opposing chemical force, no net movement of potassium occurs across the membrane (Figure 8c). Under these conditions, potassium is said to be at equilibrium because the electrochemical force is zero. The membrane potential under these conditions is equal to the equilibrium potential for potassium (E_K), which is approximately −94 mV. (The equilibrium potential varies in different neurons based on the concentration gradient for potassium ions.) Recall that the equilibrium potential of any ion (E_x) depends only on that ion's charge and the size of its concentration gradient. Thus the membrane potential of a cell that is permeable only to potassium ions is the potassium equilibrium potential.

Membrane Potential of a Cell Permeable Only to Sodium

Figure 9 illustrates how the sodium equilibrium potential is achieved in a cell freely permeable to only sodium ions. As before, we assume that initially no potential difference exists across the cell membrane. Because sodium can cross the membrane, it diffuses down its concentration gradient into the cell (see Figure 9a). As sodium moves, it carries a positive charge into the cell, which makes the inside of the cell positively charged relative to the outside, creating a positive membrane potential. In the presence of this membrane potential, sodium ions are now acted upon by an electrical force in addition to the chemical force. The direction of the chemical force tends to push sodium into the cell, while the electrical force tends to push sodium out of the cell because of the repulsion between the positively charged sodium ions and the net positive charge inside the cell (see Figure 9b). Although the chemical force for sodium ions to move across the membrane remains constant (due to few sodium ions crossing the membrane), the electrical force will continue to change as sodium ions move across the membrane until equilibrium is established. Sodium continues to flow into the cell, making the membrane potential more positive, until the electrical force becomes just large enough to exactly balance the chemical force (see Figure 9c). At this point, sodium comes to equilibrium, with the membrane potential being equal to the equilibrium potential for sodium (E_{Na}), which is approximately +60 mV. (The actual equilibrium potential for sodium varies based on the concentration gradient for sodium ions across a given neuron.) Thus the membrane potential of a cell that is permeable only to sodium ions is equal to the sodium equilibrium potential.

Resting Membrane Potential of Neurons

Now let's consider the resting membrane potential of neurons. Neurons have the same ion gradients across the cell membrane as the hypothetical cells mentioned earlier, but they have both potassium and sodium channels, making the membrane permeable to both ions (**Figure 10**). The number of open potassium channels far exceeds the number of open sodium channels, however; as a

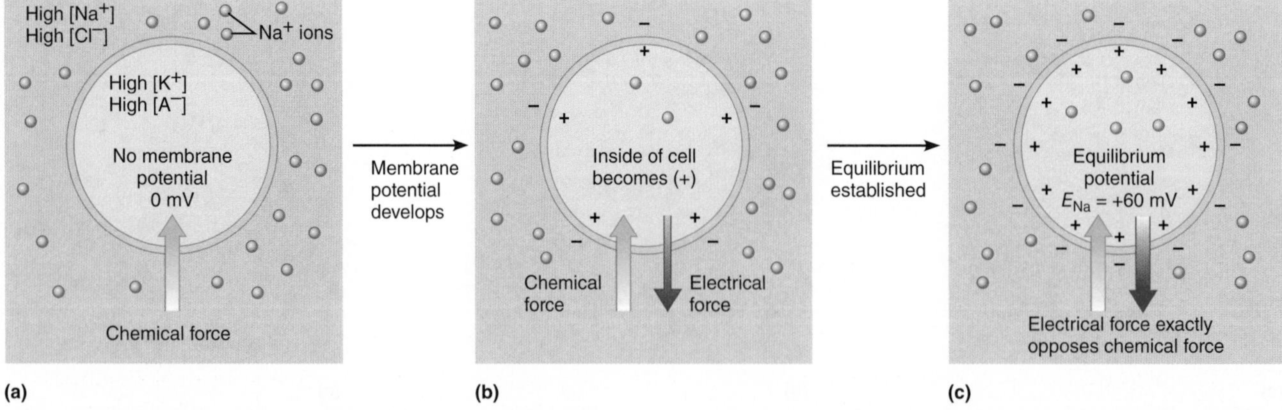

Figure 9 Membrane potential of a cell freely permeable to sodium only. Potassium (K^+) and organic anions (A^-) are present in greater concentration inside the cell. Sodium (Na^+) and chloride (Cl^-) ions are present in greater concentration outside the cell. The width of an arrow is relative to the strength of the force for ion movement in the direction of the arrow. **(a)** Sodium ions move into the cell because of a chemical force. **(b)** As some sodium ions enter the cell, the inside of the cell becomes positively charged relative to the outside. The change in charge distribution creates an electrical force to move sodium ions out of the cell, opposing the chemical force. **(c)** Eventually, enough sodium enters the cell that the electrical force becomes strong enough to oppose further movement of sodium ions into the cell because of the chemical force, resulting in no net movement of sodium ions. At this membrane potential, sodium is at equilibrium. This potential is equal to the sodium equilibrium potential and is approximately +60 mV in neurons.

consequence, the membrane is approximately 25 times more permeable to potassium than to sodium. Assuming that no membrane potential exists initially, let's consider what happens when potassium and sodium ions are both permeant.

Because both potassium and sodium are able to cross the cell membrane, these ions move in the direction of their chemical driving forces—that is, potassium ions move out of the cell and sodium ions move into the cell (Figure 10a). Nevertheless, the outward movement of potassium at this point exceeds the inward movement of sodium because the permeability of the membrane to potassium is greater than the permeability to sodium. Under these conditions, a net outward movement of positive charge occurs, which gives rise to a negative membrane potential (Figure 10b). As the unequal flows of potassium and sodium continue, the membrane potential becomes more negative. It does not increase indefinitely, however, because the negative membrane potential exerts electrical driving forces on potassium and sodium ions that oppose potassium movement and enhance sodium movement (Figure 10c). Therefore, as the membrane potential becomes more negative, outward potassium movement slows down while inward sodium movement speeds up (Figure 10d). Eventually, the flows of the two ions become equal and opposite, so that there is no net movement of positive charge into or out of the cell (Figure 10e). At this point, the membrane potential holds steady at about −70 mV, which is a typical value for the resting membrane potential of a neuron.

In neurons, then, both sodium and potassium are moving across the membrane, and the movement of each ion tends to bring the membrane potential toward its respective equilibrium potential (as depicted in Figure 8 and Figure 9). However, neither ion can

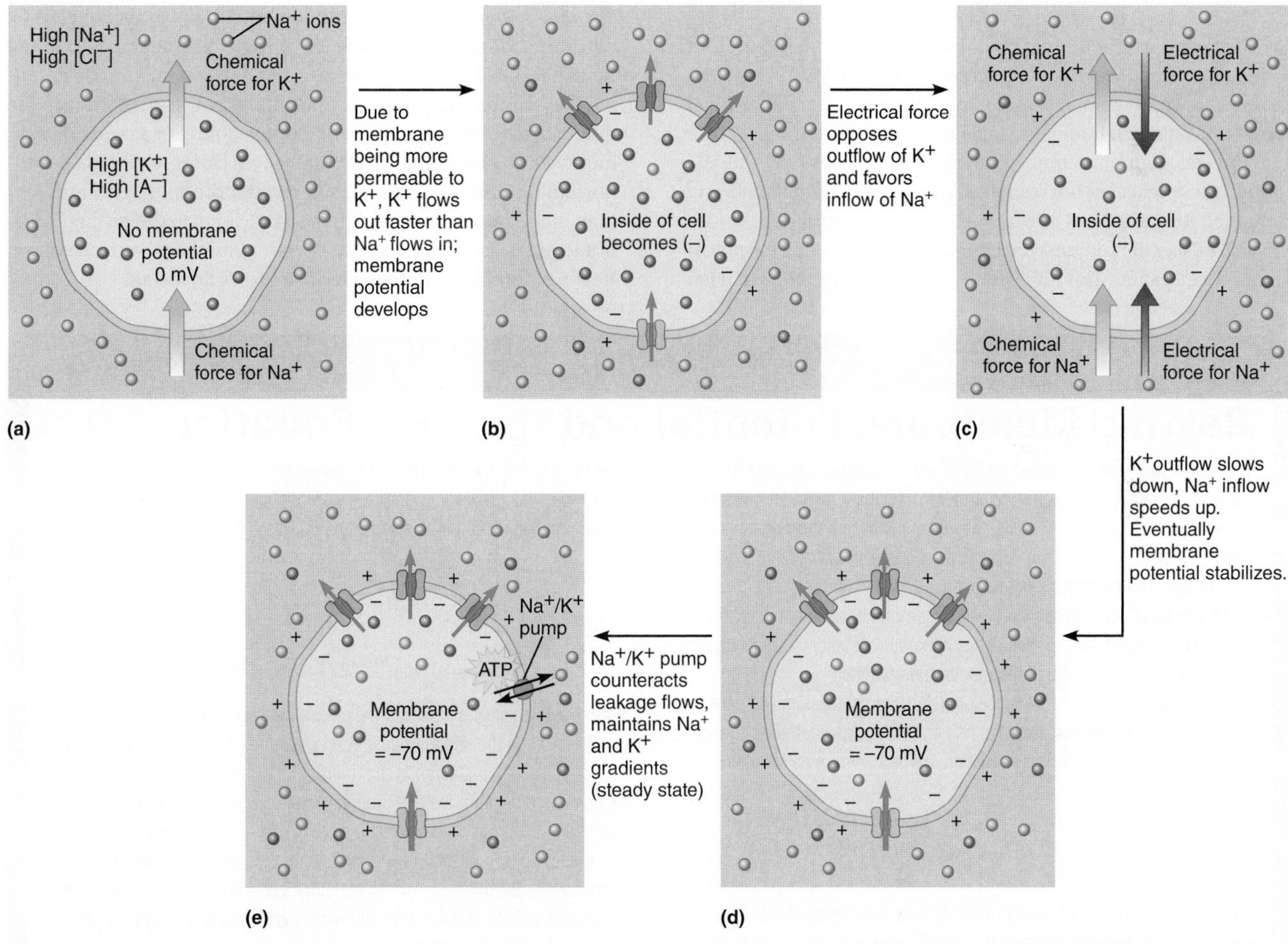

Figure 10 **Establishing a steady-state resting membrane potential.** Potassium and organic anions are present in greater concentration inside the cell. Sodium and chloride ions are present in greater concentration outside the cell. The width of an arrow is relative to the strength of the force for ion movement in the direction of the arrow. The cell is permeable to both sodium and potassium ions, but more permeable to potassium. **(a)** Chemical forces act so that potassium ions leave the cell and sodium ions enter the cell. **(b)** More potassium leaves the cell than sodium enters because of the greater permeability for potassium, creating a negative membrane potential. **(c)** Electrical forces now act on the ions, drawing both sodium and potassium ions into the cell, and increasing sodium movement in and decreasing potassium movement out. **(d)** Eventually, a steady state is established, whereby the movement of sodium into the cell is balanced by the movement of potassium out of the cell, and no net charge movement occurs. This potential is called the resting membrane potential and is approximately −70 mV in neurons. **(e)** To prevent the sodium and potassium concentration gradients from dissipating, the Na^+/K^+ pump moves sodium out of the cell and potassium into the cell, establishing a steady state at −70 mV.

ever come to equilibrium because the movement of each opposes the other. Given these circumstances, the final resting membrane potential is between the equilibrium potentials of sodium and potassium. Because the membrane is much more permeable to potassium, the resting membrane potential, −70 mV, is much closer to the potassium equilibrium potential than to the sodium equilibrium potential. The actual resting membrane potential varies among cells, because the types and numbers of ion channels in the plasma membranes of those cells vary, but the resting membrane potential is always negative.

Apply Your Knowledge

If a neuron had equal permeability to sodium and potassium ions, would the resting membrane potential of that cell be more negative or less negative than −70 mV?

Note that in a neuron at rest, neither sodium nor potassium is at equilibrium, because the membrane potential is not equal to the equilibrium potential of either ion. Therefore, electrochemical forces are acting on both ions, causing sodium to continually *leak into* the cell and potassium to continually *leak out* of the cell. Although these leakages are responsible for creating the membrane potential, they also tend to slowly alter ion concentrations inside the cell, raising the sodium concentration while lowering the potassium concentration. This imbalance would present a problem if left unchecked, because these changes would eventually abolish the concentration gradients of both ions, and the membrane potential would go to zero. However, Na^+/K^+ pumps in the cell membrane avert this problem by actively transporting sodium out of the cell and potassium into the cell using ATP for energy. Normally, sodium is pumped out as fast as it leaks in, and potassium is pumped in as fast as it leaks out. Therefore, the Na^+/K^+ pump not only establishes the concentration gradients, but also maintains them. Furthermore, because the Na^+/K^+ pump is *electrogenic*—that is, it transports a net positive charge out of the cell—it contributes directly to the resting membrane potential, although this effect is minimal and accounts for only a few millivolts of charge separation. Because energy is required to sustain the resting state of a neuron, the cell is not at equilibrium; instead, it is in a *steady state*.

Neurons at Rest

Generally speaking, the membrane potential can be affected by two, three, or more ions. Each ion will try to bring the membrane potential closer to its equilibrium potential. The membrane potential depends on the relative permeabilities of the membrane to the different ions. As the membrane's permeability to a particular ion increases, the membrane potential moves closer to that ion's equilibrium potential. (Those ions at concentration magnitudes less than other ions do not contribute significantly to the resting membrane potential.) This relationship is explored further in **Toolbox: Resting Membrane Potential and the GHK Equation.**

TOOLBOX

Resting Membrane Potential and the GHK Equation

Earlier we saw how the Nernst equation can be used to calculate the equilibrium potential for a specific ion. The Nernst equation cannot be used to calculate the membrane potential, however, because a membrane is permeable to more than one ion, and the membrane's permeability to various ions differs. The membrane potential thus depends on the concentration gradients for all ions across the plasma membrane, and on the permeability of the membrane to those ions. If the permeability for an ion is zero, that ion will not contribute to the membrane potential.

For situations in which only sodium (Na^+) and potassium (K^+) are permeant, the membrane potential (V_m) can be approximated using the Goldman–Hodgkin–Katz (GHK) equation (named after its developers):

$$V_m = 61 \log \frac{P_{Na}[Na^+]_o + P_K[K^+]_o}{P_{Na}[Na^+]_i + P_K[K^+]_i}$$

where the subscripts "o" and "i" indicate concentrations outside and inside the cell, respectively, and P_{Na} and P_K are the membrane's permeabilities to sodium and potassium, respectively.

If we divide the numerator and denominator both by P_K, then this equation becomes

$$V_m = 61 \log \frac{(P_{Na}/P_K)[Na^+]_o + [K^+]_o}{(P_{Na}/P_K)[Na^+]_i + [K^+]_i}$$

This form of the equation gives the membrane potential in millivolts.

We can calculate the membrane potential under resting conditions, assuming that P_K is 25 times as large as P_{Na} ($P_{Na}/P_K = 1/25 = 0.04$):

$$V_m = 61 \log \frac{(0.04)(145\text{ mM}) + 4\text{ mM}}{(0.04)(15\text{ mM}) + 140\text{ mM}}$$

$$= 61 \log(0.0697) = -70.6\text{ mV}$$

Note that this value is closer to the potassium equilibrium potential (−94 mV) than the sodium equilibrium potential (+60 mV), as we would expect. If there is a change in the membrane's permeability to sodium relative to potassium, the GHK equation will give us the new membrane potential, as long as we know the ratio P_{Na}/P_K. If the permeability to either sodium or potassium is equal to zero, then the GHK equation becomes the Nernst equation for the other ion. Thus, if the membrane is permeable to only one ion, then the membrane potential is equal to the equilibrium potential of that ion.

Apply Your Knowledge

Write the GHK equation to calculate the membrane potential for a membrane that is somewhat permeable to sodium, potassium, and chloride ions but impermeable to other ions.

Because neither sodium nor potassium is at equilibrium at the resting membrane potential, a net electrochemical force acts on each ion. The following rule describes electrochemical forces on ions: *The net electrochemical force on an ion tends to move that ion across the membrane in the direction that will move the membrane potential toward that ion's equilibrium potential—that is, bring the ion closer to equilibrium.* Therefore, sodium tends to move into the cell to bring the membrane potential toward +60 mV, and potassium tends to move out of the cell to bring the membrane potential toward −94 mV. The strength of the electrochemical force acting on a specific ion is proportional to the difference between the membrane potential and the equilibrium potential for that ion. Thus, because sodium is 130 mV away from equilibrium (at a resting membrane potential of −70 mV) whereas potassium is only 24 mV away from equilibrium, the electrochemical force moving sodium into the cell is much greater than the electrochemical force moving potassium out of the cell (indicated by the width of the arrows in Figure 10e).

The actual movement of sodium or potassium ions across the membrane is the sodium or potassium current, respectively. Ion currents can be described mathematically as shown in the following equations, which describe the sodium and potassium currents:

$$I_{Na} = g_{Na}(V_m - E_{Na})$$

$$I_K = g_K(V_m - E_K)$$

In these equations, I_I equals the current of a specific ion, g_I equals the conductance of that ion (which is directly related to permeability), E_I equals equilibrium potential of that ion, and V_m equals membrane potential.

The sodium and potassium channels that are responsible for the resting membrane potential are leak channels, which are always open. In addition to these leak channels, neurons have gated ion channels. In the next section we see how the opening and closing of gated channels are responsible for the changes in membrane potential that occur during electrical signaling.

Quick Check 2

1. Describe the concentration gradients for sodium and potassium across the plasma membrane. What establishes these concentration gradients?
2. When a neuron is at rest, what are the directions of the electrochemical forces that drive sodium and potassium movement?
3. If channels that permitted both sodium ions and potassium ions to move through them suddenly opened in the plasma membrane, in which direction would each ion move—into or out of the neuron? Which ion would move more? Why? What change in membrane potential would occur?

4 Electrical Signaling Through Changes in Membrane Potential

Electrical signals occur in neurons via changes in membrane potential that take place when certain ion channels, called *gated channels,* open or close in response to particular stimuli. When gated ion channels open or close, they change the membrane permeability for that specific ion, thereby affecting the movement of that ion across the plasma membrane. For example, if sodium ion channels open, then sodium movement into the cell increases, driving the membrane potential toward the sodium equilibrium potential.

Three types of gated ion channels exist: voltage-gated channels, ligand-gated channels, and mechanically gated channels. Voltage-gated and ligand-gated channels were described previously. Mechanically gated channels open or close in response to a mechanical force on the membrane. These channels are usually found associated with sensory or visceral receptors located at the end of afferent neurons. Gated channels are crucial for normal functioning of the nervous system; in fact, many toxins exert their poisonous effects by interfering with the actions of ion channels (see **Clinical Connections: Neurotoxins**).

Describing Changes in Membrane Potential

Changes in membrane potential are described based on the direction of change relative to the resting membrane potential, as illustrated in **Figure 11**. Because the membrane potential is a

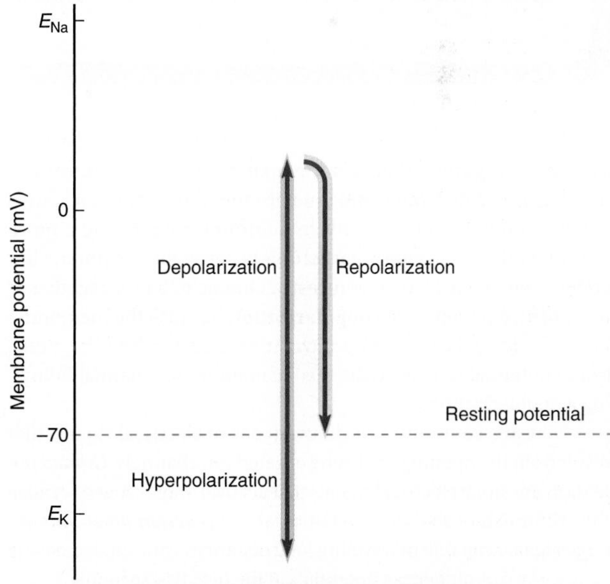

Figure 11 Changes in membrane potential. The membrane potential can change with the opening or closing of ion channels. If gated potassium channels open, then potassium moves out of the cell, bringing the membrane potential toward the potassium equilibrium potential (E_K), or hyperpolarizing the cell. If gated sodium channels open, then sodium moves into the cell, bringing the membrane potential toward the sodium equilibrium potential (E_{Na}), or depolarizing the cell.

CLINICAL CONNECTIONS

Neurotoxins 

The Chinese proverb "To throw away life, eat blowfish" arose because blowfish (also known as puffer fish) contain the neurotoxin tetrodotoxin (TTX). A neurotoxin is a toxin that attacks some aspect of nervous system function. TTX is *very* potent—it is toxic at nanomolar (10^{-9} mole/L) concentrations. This neurotoxin works by blocking voltage-gated sodium channels necessary for producing an action potential.

Blowfish is one of the most prized delicacies in the restaurants of Japan. This fish is prized not only for its taste but also for the tingling sensation one gets around the lips when eating it. In blowfish, TTX is concentrated in certain organs of the blowfish, including the liver and gonads. Because TTX cannot be destroyed by cooking, preparation of the fish can be done only by licensed chefs who are skilled at removing the poison-containing organs without crushing them and thereby possibly contaminating normally edible parts of the fish.

TTX is also present in some kinds of salamanders, octopus, and goby. Species that carry the toxin have sodium channels that differ slightly, rendering them either less susceptible or completely resistant to the toxin.

A closely related toxin that blocks voltage-gated sodium channels is saxitoxin (STX), which is produced by some marine dinoflagellates and by a freshwater cyanobacterium. These microorganisms become a major health threat during certain periods of the year when they multiply rapidly, or "bloom." At high concentrations, they may even impart a reddish color to the water, producing the phenomenon known as "red tide." The primary danger to humans during red tide comes from eating shellfish, which are filter feeders that can accumulate the toxin in their bodies. (Shellfish are resistant to the toxin.) Eating even a single contaminated shellfish can be fatal. Like TTX, STX is not destroyed by cooking. Unfortunately, one cannot always tell whether conditions are hazardous by simple observation; seawater can contain dangerous concentrations of dinoflagellates without appearing red. Fortunately, shellfish are routinely monitored by public health officials for the presence of this toxin, and harvesting of shellfish is banned whenever there is cause for concern.

Stuart Philpott/Alamy

Puffer fish contain a potent neurotoxin.

Critical Thinking Questions

1. Which blowfish organs contain concentrated tetrodotoxin, and how does tetrodotoxin work?
2. What effects would tetrodotoxin have on action potentials?
3. Describe how and why eating organisms that feed on dinoflagellates could be dangerous to humans.

difference in potential across the membrane, the membrane is *polarized*. Because the resting membrane potential is a negative value (approximately −70 mV in neurons), a change to a more negative value is a **hyperpolarization** because the membrane becomes *more polarized*. In contrast, a change to a less negative or to a positive potential is a **depolarization** because the membrane becomes *less polarized*. **Repolarization** occurs when the membrane potential returns to the resting membrane potential following a depolarization.

Neurons communicate via two types of electrical signals that result from the opening or closing of gated ion channels. *Graded potentials* are small electrical signals that act over short ranges because they diminish in size with distance, whereas *action potentials* are large signals capable of traveling long distances without decreasing in size. We examine these potentials in the next two sections.

Graded Potentials

Graded potentials are small changes in membrane potential that occur when ion channels open or close in response to a stimulus acting on the cell (**Figure 12**). Stimuli that produce graded potentials include chemical stimuli, such as neurotransmitter molecules binding to receptors on a dendrite or the cell body of a neuron, and sensory stimuli, such as a touch or light, acting on a sensory receptor at the peripheral ending of an afferent neuron. The magnitude of the change in membrane potential varies

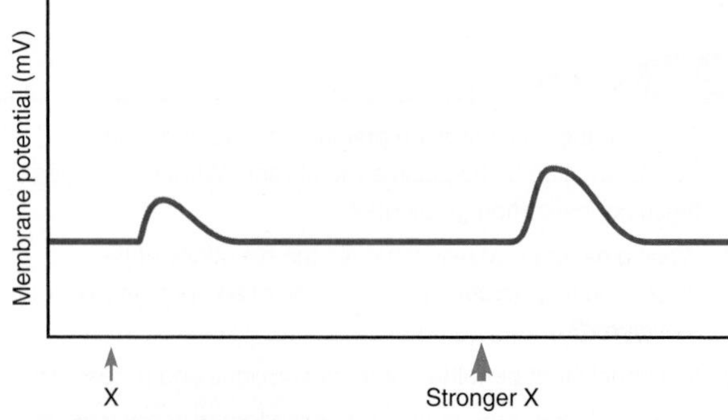

Figure 12 Effect of stimulus strength on size of graded potential. Graded potentials are small changes in the membrane potential (V_m), in response to a stimulus; a stronger stimulus produces a larger change in V_m.

with—is *graded*—according to the strength of the stimulus: A weak stimulus produces a small change in membrane potential, whereas a stronger stimulus produces a greater change in membrane potential (Figure 12). For example, if the presence of 5000 neurotransmitter molecules acting on ligand-gated channels caused a cell to depolarize by 1 mV, then 10,000 neurotransmitter molecules (a stronger stimulus) will cause the cell to depolarize by approximately 2 mV.

Graded Potentials Are Decremental

A graded potential can travel away from the site of stimulation for only a short distance because it is *decremental*; that is, the change in membrane potential decreases in size as it moves along the membrane away from the site of stimulation. To understand why a graded potential is decremental, let's first consider what happens to water flow through a leaky hose. If the hose is intact, the amount of water exiting the hose at one end is the same as that entering the hose at the faucet end. However, if the hose has leaks (much like a membrane experiences ion leakage), then some of the water is lost and the amount of water exiting at the end of the hose is decreased.

Now let's look at the flow of current along a membrane. When a change in potential occurs across a cell membrane at a particular site, this change generates differences in potential within the intracellular and extracellular fluids. Because a separation of charge creates a force for charge to move (current), the graded potential creates charge separation within the intracellular fluid and within the extracellular fluid, which generates currents in these fluids. These currents travel to adjacent areas of the cell membrane, causing voltage changes in these areas. This spread of voltage by passive charge movement is called **electrotonic conduction.** As the graded potential spreads from the site of the stimulation (**Figure 13**), the current is spread over a larger area, and some current leaks across the plasma membrane. As a result, the size of the membrane potential change decreases as it moves from the site of initial stimulation.

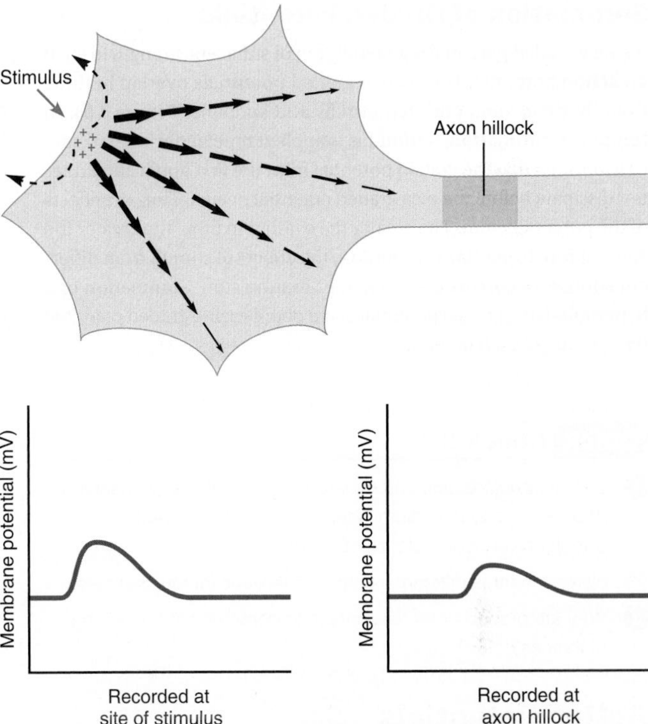

Figure 13 Decremental property of graded potentials. A graded potential dissipates as it moves to adjacent areas of the plasma membrane (solid arrows) and across the plasma membrane (dashed arrows). Therefore, the potential change recorded at the site of stimulus is greater than that recorded distant from the site of stimulus. (Darker arrows indicate a stronger current.)

Graded Potentials Can Be Depolarizations or Hyperpolarizations

Some graded potentials are depolarizations, whereas others are hyperpolarizations (**Figure 14**). The direction of change depends on the particular neuron in question, the stimulus applied to it, and the specific ion channels that open or close in response to the stimulus. For example, if one type of neurotransmitter binding to its receptors caused sodium channels to open, then sodium ions would move into the cell and the resulting graded potential would be a depolarization. If another type of neurotransmitter binding to its receptors caused potassium channels to open, then potassium ions would move out of the cell, and the resulting graded potential would be a hyperpolarization.

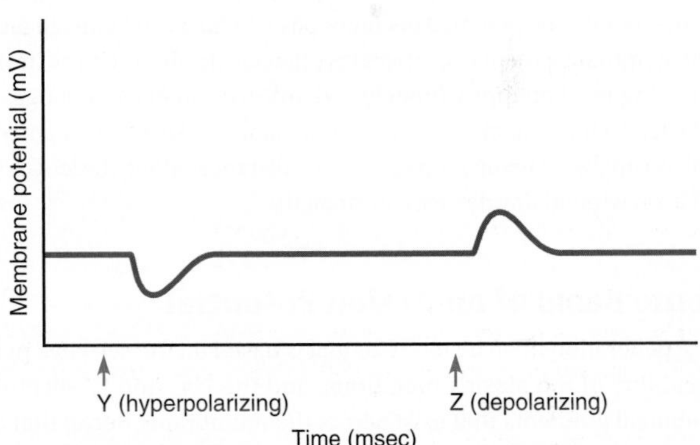

Figure 14 Effect of stimulus type on graded potentials. Depending on the stimulus and the neuron, graded potentials can be either a hyperpolarization or a depolarization.

The primary significance of graded potentials is that they determine whether a cell will generate an action potential. Graded potentials generate action potentials if they depolarize a neuron to a certain level of membrane potential called the **threshold,** a critical value of membrane potential that must be met or exceeded if an action potential is to be generated. Graded potentials that are depolarizations are described as *excitatory,* because they bring the membrane potential closer to the threshold to generate an action potential. In contrast, graded potentials that are hyperpolarizations are described as *inhibitory,* because they take the membrane potential away from the threshold to elicit an action potential.

Summation of Graded Potentials

A single graded potential is generally not of sufficient strength to elicit an action potential. However, if graded potentials overlap in time, then they can sum, both temporally and spatially (**Figure 15**). In **temporal summation,** a stimulus is applied repeatedly in rapid succession such that the graded potential from the first application does not dissipate before the next graded potential occurs. Thus the effects of the potentials sum. The greater the overlap in time, the greater the summation. In **spatial summation,** the effects of stimuli from different sources occurring close together in time sum. Summation of a hyperpolarizing graded potential and a depolarizing graded potential tend to cancel each other out.

Quick Check 3

1. Define *depolarization* and *hyperpolarization.* If the permeability of a membrane to sodium increases, would the membrane depolarize or hyperpolarize? Explain.
2. Name the three different gating mechanisms for ion channels.
3. Why are graded potentials unable to communicate over long distances?

Action Potentials

In most of the remainder of this chapter we discuss action potentials. Action potentials occur in the membranes of excitable tissue (nerve or muscle) in response to graded potentials that reach threshold. Table 2 compares the properties of graded and action potentials. During an action potential a large, rapid depolarization occurs in which the polarity of the membrane potential actually reverses; that is, the membrane potential becomes positive for a brief time. In fact, the membrane potential changes very quickly (in about 1 msec) from a resting level of approximately −70 mV to +30 mV (a change of 100 mV!). Once initiated, an action potential, unlike a graded potential, is capable of being propagated long distances along the length of an axon without any decrease in strength.

Ionic Basis of an Action Potential

The generation of an action potential is based on the selective permeability of the plasma membrane and the Na^+ and K^+ electrochemical gradients that exist across the membrane. Recall that at rest the plasma membrane is approximately 25 times more permeable to potassium ions than to sodium ions because of the presence of many more potassium leak channels than sodium leak channels. In excitable cells, changes in the permeability of the plasma membrane resulting from the opening and closing of gated ion channels can produce action potentials.

An action potential in a neuron consists of three distinct phases (**Figure 16a**):

1. *Rapid Depolarization.* The first phase of an action potential is a **rapid depolarization** during which the membrane potential changes from −70 mV (rest) to +30 mV. This depolarization is caused by a sudden and dramatic increase in permeability to

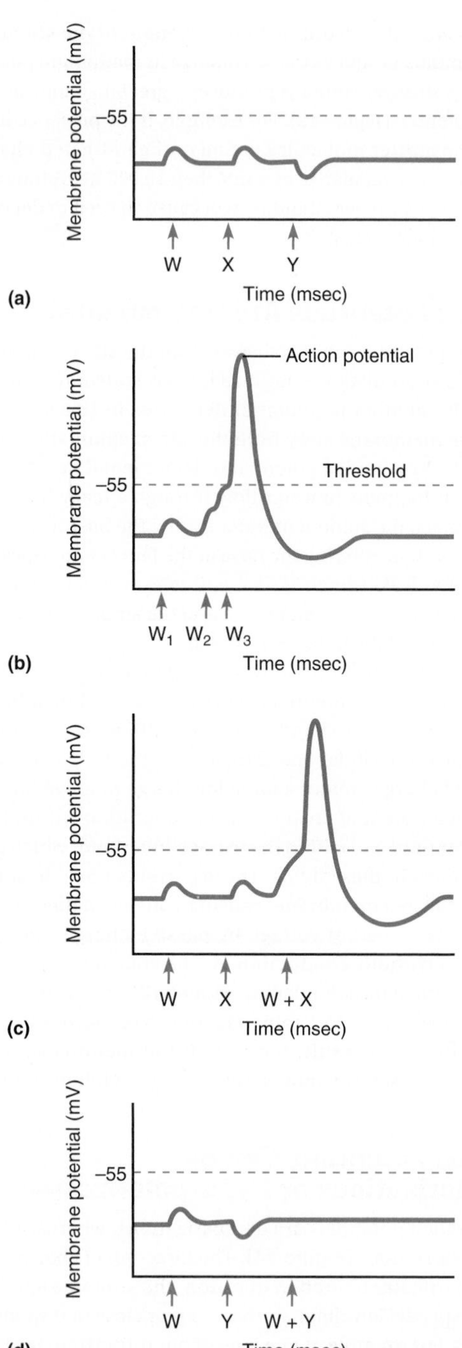

Figure 15 Temporal and spatial summation. (a) Effects of stimuli W, X, and Y on membrane potential. Stimuli W and X depolarize the cell; stimulus Y hyperpolarizes the cell. **(b)** Temporal summation of stimulus W resulting in depolarization to above threshold and generation of an action potential. **(c)** Spatial summation of stimuli W and X resulting in depolarization to above threshold and generation of an action potential. **(d)** Spatial summation of stimuli W and Y resulting in no change in membrane potential and, therefore, no action potential.

TABLE 2 Comparison of Graded Potentials and Action Potentials

Property	Graded potential	Action potential
Location	Dendrites, cell body, sensory receptors	Axon
Strength	Relatively weak, proportional to strength of the stimulus; dissipates with distance from the stimulus	100 mV; all-or-none
Direction of change in membrane potential	Can be depolarizing or hyperpolarizing depending on the stimulus	Depolarizing
Summation	Spatial and temporal	None
Refractory periods	None	Absolute and relative
Channel types involved in producing change in potential	Ligand-gated, mechanically gated	Voltage-gated
Ions involved	Usually Na^+, Cl^-, or K^+	Na^+ and K^+
Duration	Few milliseconds to seconds	1–2 msec (after-hyperpolarization phase may last 15 msec)

sodium (Figure 16b) followed by an increase in the movement of sodium ions into the cell, down sodium's electrochemical gradient. With permeability to sodium now greater than permeability to potassium, the membrane potential approaches the sodium equilibrium potential of +60 mV.

2. *Repolarization.* The second phase of an action potential is a **repolarization** of the membrane potential during which the membrane potential returns from +30 mV back to resting levels (−70 mV). Within 1 msec after the increase in sodium permeability, sodium permeability decreases rapidly, reducing the inflow of sodium. At approximately the same time, potassium permeability increases. Potassium then moves down its electrochemical gradient out of the cell, repolarizing the membrane potential to bring it back to resting levels.
3. *After-hyperpolarization.* The third phase of an action potential is termed **after-hyperpolarization.** Potassium permeability remains elevated for a brief time (5–15 msec) after the membrane potential reaches the resting membrane potential, resulting in an after-hyperpolarization. During this time, the membrane potential is even more negative than at rest as it approaches the potassium equilibrium potential (−94 mV).

Next we discuss the ion channels responsible for the changes in permeability to sodium and potassium ions during an action potential.

The Role of Voltage-Gated Ion Channels in Action Potentials

The changes in permeability associated with the phases of an action potential are due to the time-dependent opening and closing of voltage-gated sodium and potassium channels located primarily in

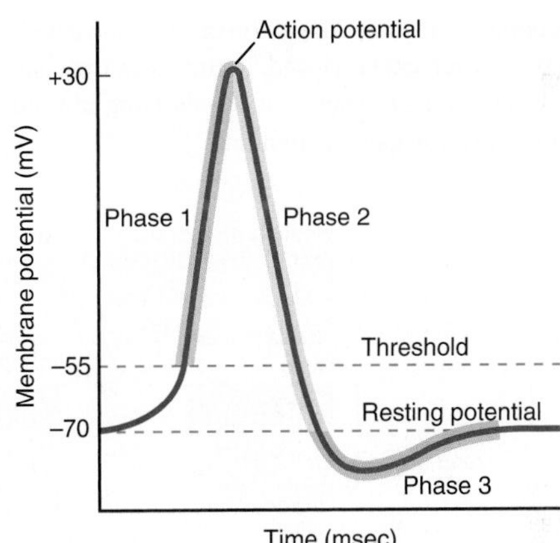

(a) Three phases of an action potential

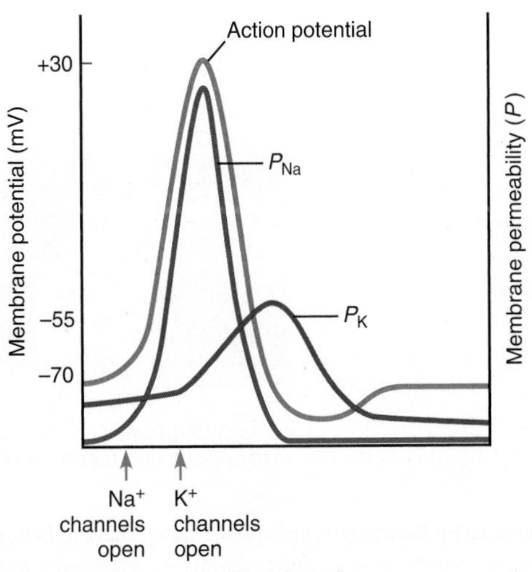

(b) Permeability changes for Na^+ and K^+ during an action potential

Figure 16 The phases and ionic basis of an action potential. (a) The three distinct phases of an action potential. **(b)** The permeability changes for sodium ions (P_{Na}) and potassium ions (P_K) that occur during an action potential.

Q At the beginning of repolarization, permeability to sodium is greater than permeability to potassium. Why does the membrane potential *repolarize*?

the plasma membrane of the axon hillock and axon. (These voltage-gated sodium and potassium channels are also found in the plasma membrane of some muscle cells.) In myelinated axons, these channels are at a greater concentration at the nodes of Ranvier; in unmyelinated axons, these channels are evenly distributed along the entire axon. Because the exact mechanisms of gating in the voltage-gated sodium and potassium channels are not known, we use models to describe their function.

The model for explaining the actions of voltage-gated sodium channels involves two types of gates: activation gates and inactivation gates. **Activation gates** are responsible for the opening of sodium channels during the depolarization phase of an action potential, whereas **inactivation gates** are responsible for the closing of sodium channels during the repolarization phase of an action potential. For a sodium channel to be open, both gates must be open. Both types of gates open and close in response to changes in the membrane potential, and based on the position of these two gates, a sodium channel can exist in three conformations (**Figure 17**).

1. *Closed but capable of opening.* At rest the inactivation gate is open, but the activation gate is closed. In this state, the channel is closed, but it can be opened by a depolarizing stimulus that causes the activation gate to open.
2. *Open.* On depolarization, the activation gate opens and with both gates in their open position, the channel is open and sodium ions move through the channel into the cell; this is what occurs during the depolarization phase of an action potential.
3. *Closed and incapable of opening.* Within approximately 1 msec after the initial stimulus to open the activation gate, the inactivation gate closes. The closing is a delayed response initiated by the same depolarization that caused the activation gate to immediately open. With the inactivation gate closed and the activation gate open, the channel is closed. The inactivation gate remains closed until the membrane potential returns to near its resting value. Until this happens, the channel cannot open in response to a second depolarizing stimulus because the inactivation gate would remain closed; that is, it does not open in response to a depolarization. Once repolarization has occurred, the inactivation gate opens and the activation gate closes, returning the channel to its resting state.

The opening of sodium activation gates is a **regenerative** mechanism; that is, the opening of some sodium activation gates causes more sodium activation gates to open by *regenerating* the stimulus to open the gates (depolarization). This regenerative

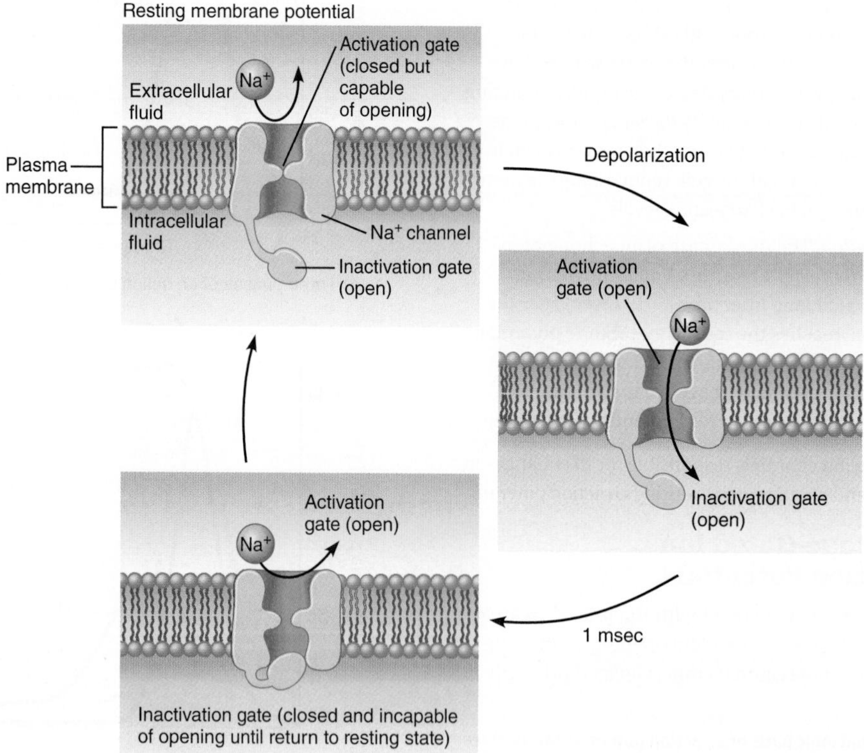

Figure 17 A model for the operation of voltage-gated sodium channels. Voltage-gated sodium channels are schematically represented here by two gates. At rest, the sodium inactivation gate is open and the activation gate is closed but can open in response to a depolarizing stimulus. Following a depolarizing stimulus to threshold, both the activation and inactivation gates are open, and sodium can move through the channel. Approximately 1 msec after a depolarizing stimulus, the inactivation gate closes; it remains closed until the cell has repolarized to the resting state. Before repolarization, the channel cannot open in response to a new depolarizing stimulus.

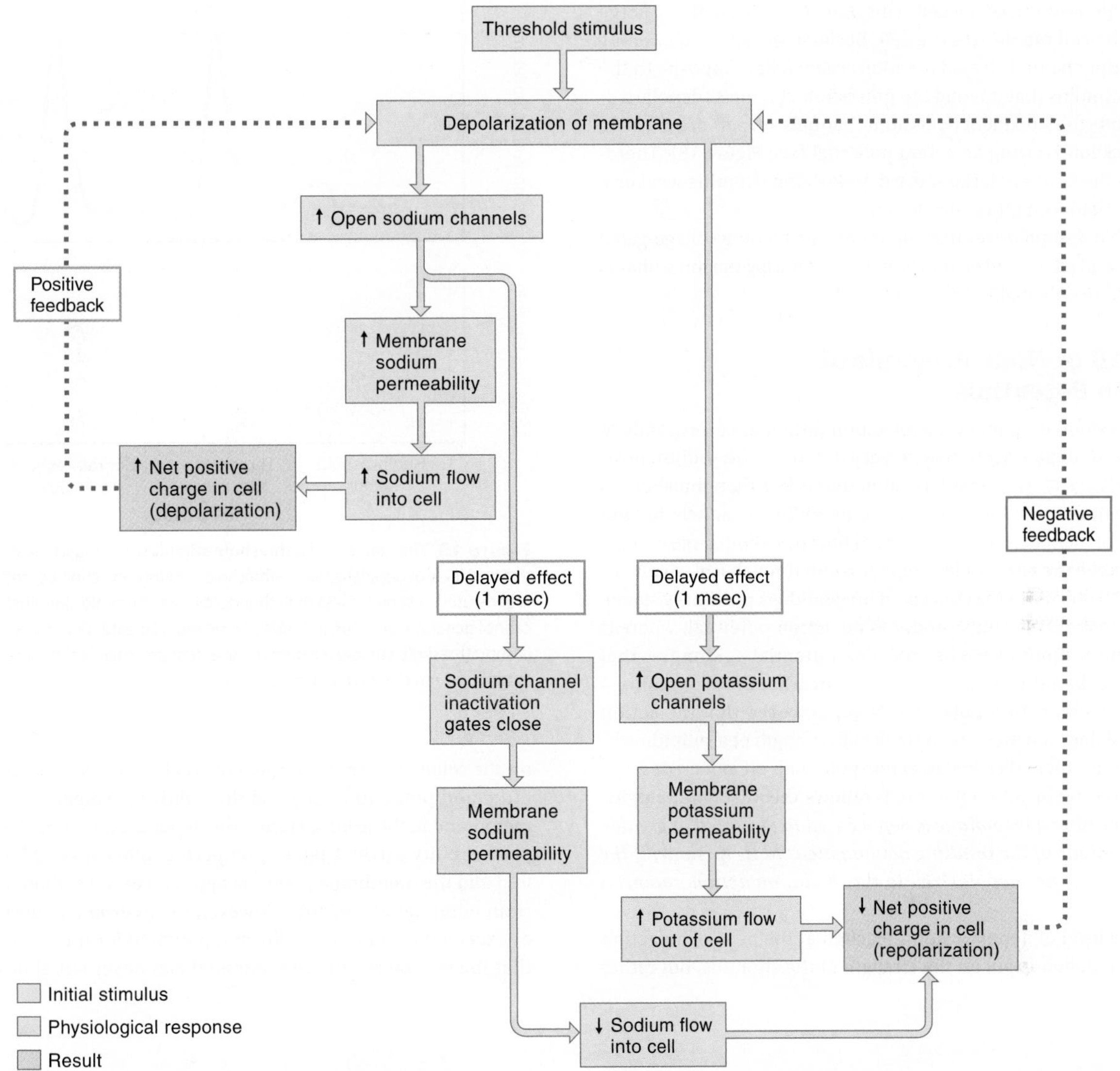

Figure 18 Gating of sodium and potassium channels during an action potential. Sodium channel opening is part of a positive feedback loop that allows for the rapid depolarization of the cell. When the cell is depolarized to threshold, sodium channels open. Opening allows sodium to move into the cell, causing further depolarization and opening more sodium channels. The feedback loop continues until the sodium inactivation gates close, approximately 1 msec after the depolarization to threshold. Potassium channel opening and closing is part of a negative feedback loop. Depolarization stimulates the slow opening of potassium channels, which allows potassium to move out of the cell, repolarizing it. Because repolarization opposes the depolarizing stimulus for opening potassium channels, the potassium channels close.

Q What stops the positive feedback loop for sodium channels to open?

mechanism works as follows: At first, depolarization triggers the opening of a few sodium channels, which allows some movement of sodium ions into the cell, which depolarizes the cell further. The increased depolarization causes still more sodium channels to open, leading to a larger inflow of sodium ions and more depolarization, and so on. This positive feedback loop causes the very rapid (less than 1 msec) depolarization phase of the action potential. The positive feedback loop terminates when the sodium inactivation gates close (**Figure 18**).

The level of depolarization necessary to initiate the regenerative opening of sodium channels is the threshold for initiating an action potential. Threshold occurs when the inward flux of sodium ions exceeds the outward flux of potassium ions.

In contrast to the sodium channels just described, the model for voltage-gated potassium channels describes only a single gate that opens more slowly in response to depolarization. At approximately the same time that sodium inactivation gates are closing (1–2 msec after depolarization to threshold), the potassium channels begin to open. The increased permeability to potassium coupled with the strong electrochemical gradient for potassium to move out of the cell (when the membrane is depolarized, potassium is far from equilibrium) increases the movement of potassium ions out of the cell. This movement of positive charge out of the cell repolarizes the cell. Because the effect of opening potassium channels (repolarization) is an action opposite to the initial stimulus that opened the potassium channels (depolarization), voltage-dependent potassium channels are part of a negative feedback loop during an action potential (see Figure 18). Therefore, as the cell repolarizes, the depolarizing stimulus weakens, and potassium channels slowly close.

Table 3 summarizes the conditions surrounding voltage-gated sodium and potassium channels at rest and during the three phases of an action potential.

The All-or-None Principle of Action Potentials

The threshold for generating an action potential corresponds to the level of depolarization necessary to induce the sodium positive feedback loop. A depolarization that is less than threshold (a **subthreshold** stimulus) may open some sodium channels, but not enough of them to produce an inward flow of sodium large enough to overcome the outward flow of potassium through leak channels. Figure 19 illustrates the concept of threshold. As previously stated, a subthreshold stimulus produces no action potential, whereas a threshold stimulus elicits an action potential. A stimulus that is greater than threshold—that is, a **suprathreshold** stimulus—also elicits an action potential. Note, however, that the action potential does not increase in size as the strength of a suprathreshold stimulus increases; that is, action potentials are not graded.

Initiation of action potentials follows the **all-or-none principle:** *Whether a membrane is depolarized to threshold or greater, the amplitude of the resulting action potential is the same; if the membrane is not depolarized to threshold, no action potential occurs.*

The level of depolarization reached at the peak of an action potential depends *not* on the strength of the stimulus, but rather on the relative strengths of the electrochemical gradients for sodium and potassium ions and the relative permeabilities of the membrane to these ions. During the depolarization phase, sodium permeability exceeds potassium permeability several hundredfold and the membrane potential approaches the sodium equilibrium potential of +60 mV. However, the neuron can never meet or exceed the sodium equilibrium potential for the same reason that the resting membrane potential can never equal or exceed

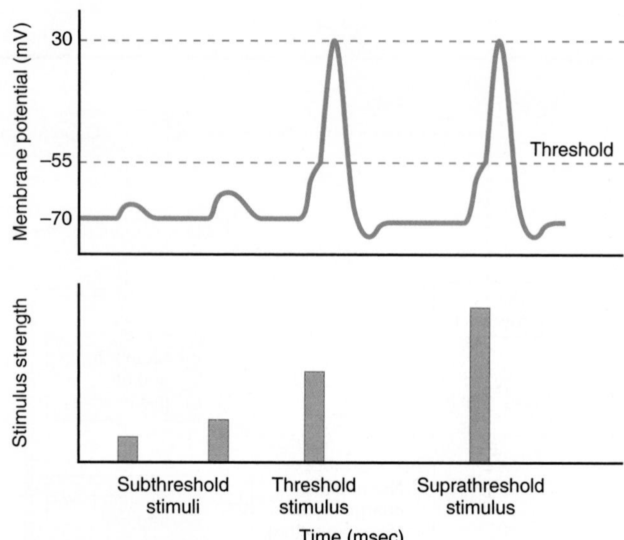

Figure 19 The concept of a threshold stimulus. A stimulus must cause a critical level of depolarization—threshold—before an action potential is generated. A stimulus less than threshold (a subthreshold stimulus) cannot generate an action potential. Any stimulus greater than threshold (a suprathreshold stimulus) generates an action potential of the same magnitude and duration as a threshold stimulus.

TABLE 3 Characteristics of a Neuron at Rest and During the Different Phases of an Action Potential

	Resting	Depolarization	Repolarization	After-hyperpolarization
Membrane potential	−70 mV	−70 mV to +30 mV	+30 mV to −70 mV	−70 mV to −85 mV
Voltage-gated sodium channel	Closed	Open	Closed	Closed
Activation gate	Closed	Open	Open	Closed
Inactivation gate	Open	Open	Closed	Open
Sodium flow	Low inward, through leak channels	High inward, through voltage-gated channels*	Low inward, through leak channels	Low inward, through leak channels
Voltage-gated potassium channel	Closed	Closed	Open	Closing
Potassium flow	Low outward, through leak channels	Low outward, through leak channels	High outward, through voltage-gated channels*	High outward, through voltage-gated channels, but decreasing*

*Even though at any given time ions move through both voltage-gated channels and leak channels, the conductance through the leak channels is negligible compared to that through voltage-gated channels.

the potassium equilibrium potential: Sodium movement into the cell is countered by potassium movement out of the cell, primarily through potassium leak channels and later through the opened voltage-gated potassium channels.

Quick Check 4

1. During the depolarization phase of an action potential, is the membrane more permeable to sodium or to potassium? What about during the repolarization phase?
2. Compare the gating mechanisms of the voltage-gated sodium and potassium channels. Why is the *timing* of events so critical to development of an action potential?
3. During which phase of an action potential are the sodium channels closed but capable of opening? During which phase are they open? During which phase are they closed and incapable of opening?

Refractory Periods

During and immediately after an action potential, the membrane is less excitable than it is at rest. This period of reduced excitability is called the **refractory period.** The refractory period can be divided into two phases: the absolute refractory period and the relative refractory period (**Figure 20**).

The **absolute refractory period** spans all of the depolarization phase plus most of the repolarization phase of an action potential (1–2 msec). During this time, a second action potential cannot be generated in response to a second stimulus, regardless of the strength of that stimulus. The absolute refractory period occurs for two reasons. First, during the rapid depolarization phase of an action potential, the regenerative opening of sodium channels that has been set into motion will proceed to its conclusion and will not be affected by a second stimulus. Second, during the beginning of the repolarization phase, most of the sodium inactivation gates are closed and, therefore, the channels are in the closed and incapable of opening state. A second action potential cannot be generated until the majority of the sodium channels have returned to their resting state, a situation that occurs near the end of the repolarization phase. At that time, the activation gates have closed and the inactivation gates have opened, so the sodium channels are closed but capable of opening in response to depolarization.

The **relative refractory period** occurs immediately after the absolute refractory period and lasts approximately 5–15 msec. During this period, it is possible to generate a second action potential, but only in response to a stimulus stronger than that needed to reach threshold under resting conditions. The relative refractory

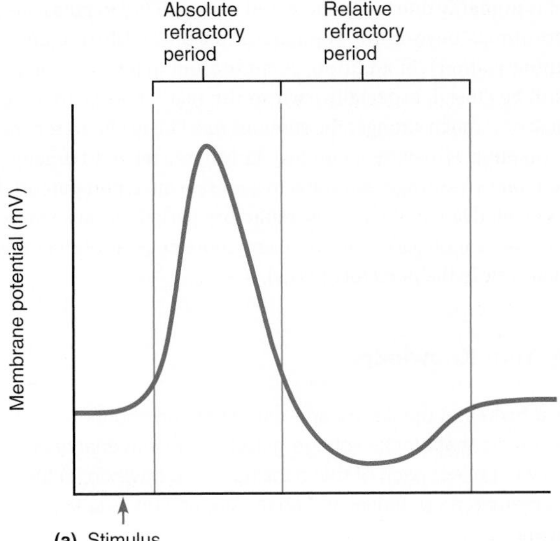

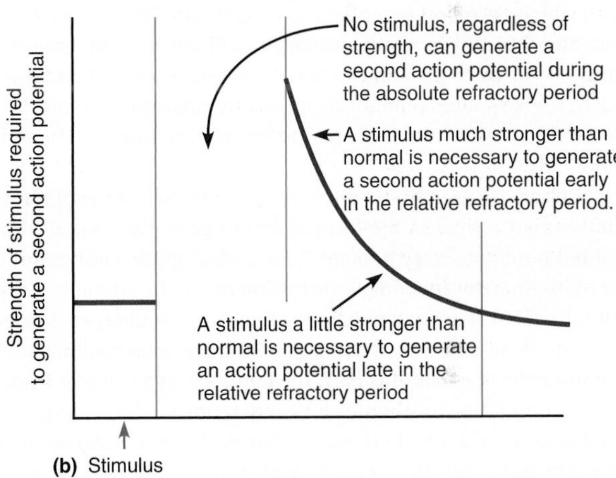

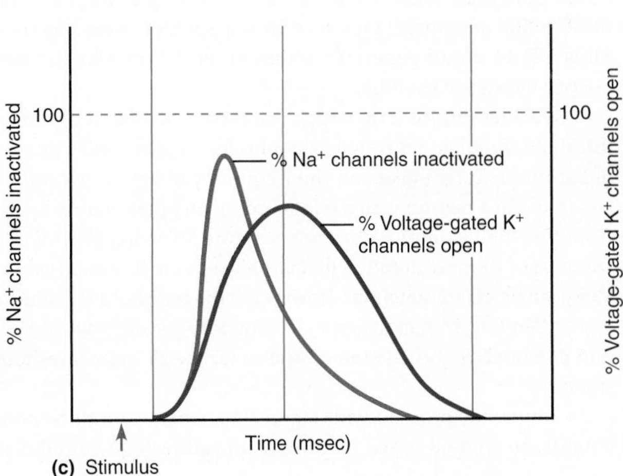

Figure 20 Refractory periods associated with an action potential. The graphs represent **(a)** membrane potential changes occurring during the action potential; **(b)** the stimulus strength needed to elicit a second action potential; and **(c)** the percentages of sodium channels that are inactivated (inactivation gates are closed) and of voltage-gated potassium channels that are open.

period is primarily due to the increased permeability to potassium that continues beyond the repolarization phase (during after-hyperpolarization). In addition, some sodium inactivation gates may still be closed, especially early in the relative refractory period. Just how much stronger the stimulus must be to elicit a second action potential is a matter of timing. Early in the relative refractory period it takes a stronger stimulus to generate an action potential than is needed late in the relative refractory period, because more sodium inactivation gates are closed and more potassium channels are open early in the refractory period.

Apply Your Knowledge

TTX is a toxin that blocks voltage-gated sodium channels, and TEA is a toxin that blocks voltage-gated potassium channels. Predict what effect each of these toxins would have on (1) the resting membrane potential and (2) the generation of action potentials.

The refractory periods establish several properties of action potentials, including their all-or-none property, the frequency with which an action potential can be generated by a single neuron, and the unidirectional propagation along an axon. In terms of the all-or-none principle of action potentials described previously, unlike graded potentials, action potentials cannot sum, because the absolute refractory period prevents an overlap of action potentials.

Refractory periods are also important in the coding of information that arrives in the form of action potentials. Recall that graded potentials vary in magnitude based on the strength of a stimulus, thereby encoding information about the strength of the stimulus. That action potentials are all-or-none would seem to create a problem: How can action potentials relay information about the intensity of a stimulus (say, the loudness of a sound)? Information pertaining to stimulus intensity is encoded by changes in the *frequency* of action potentials—that is, changes in the number of action potentials that occur in a given period of time. Because graded potentials typically last much longer than action potentials, stronger, longer-lasting graded potentials may generate a burst of action potentials. Depending on the size of the graded potential, these action potentials may occur further apart or closer together. Thus a loud sound generates action potentials at a higher frequency than a soft sound.

Consider **Figure 21a**, which compares the effects of a subthreshold stimulus, a threshold stimulus for 10 msec, and a threshold stimulus for 20 msec on the frequency of action potentials generated in a neuron with a relative refractory period of 15 msec. A subthreshold stimulus does not generate an action potential; a stimulus of 10-msec duration that just reaches the threshold generates a single action potential. However, if the threshold stimulus is applied for longer than the relative refractory period, a second action potential can be generated, as seen for the 20-msec threshold stimulus.

Compare Figure 21a with Figure 21b, where suprathreshold stimuli are applied to the neuron. A suprathreshold stimulus is more likely to generate more than one action potential because it can generate a second action potential during the relative refractory period of the first action potential. To generate a second action potential during the relative refractory period, a stimulus must be strong enough to open enough sodium channels such that sodium inflow overcomes the elevated potassium outflow that occurs during the relative refractory period, and the stimulus may have to overcome some sodium inactivation gates that are still closed. Also, a stronger stimulus can produce a second action potential closer in time to the first action potential. However, because a second stimulus cannot generate a second action potential during the absolute refractory period, the absolute refractory period limits the maximum frequency of action potentials in a neuron to approximately 500–1000 per second.

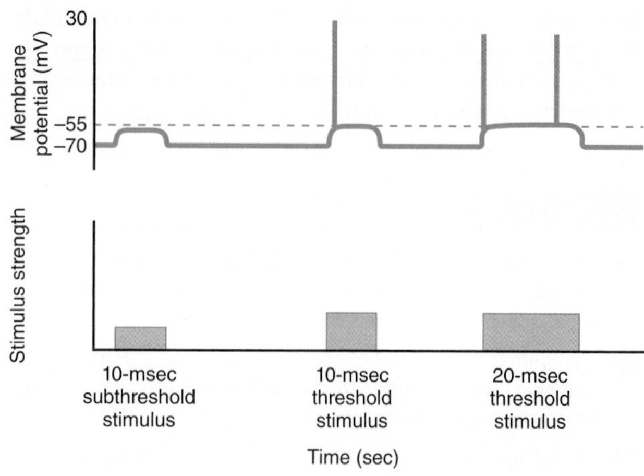

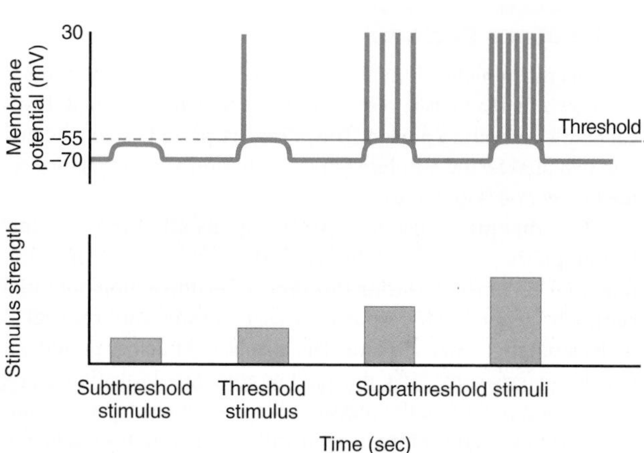

Figure 21 Frequency coding: how action potentials convey intensity of stimuli. The entire refractory period for this neuron is 15 msec. **(a)** Subthreshold stimulus, threshold stimulus for 10 msec, and threshold stimulus for 20 msec. **(b)** Subthreshold, threshold, and suprathreshold stimuli.

Quick Check 5

1. Can an action potential be generated during the absolute refractory period? During the relative refractory period?
2. During the absolute refractory period, are the majority of sodium channels open, closed but capable of opening, or closed and incapable of opening?
3. How do action potentials encode the intensity of a stimulus?

Propagation of Action Potentials

Once an action potential is initiated in an axon, it is propagated down the length of the axon from the trigger zone to the axon terminal without decrement. An action potential does not actually travel down the axon; instead, an action potential sets up electrochemical gradients in the extracellular and intracellular fluids. Because the extracellular and intracellular fluids have low resistance to current flow, positive charges move from the area where the membrane has been depolarized to the adjacent area of membrane, depolarizing it as well. Current flows to adjacent areas of the axon's plasma membrane by electrotonic conduction. The first action potential produced at the trigger zone produces a current that causes a second action potential in the adjacent membrane, which produces a current that causes a third action potential, and so on until an action potential is produced at the axon terminal. In this sense, the propagation of an action potential down an axon is analogous to the sequential falling of a row of dominoes set close together on end: When the first domino falls over, it tips the adjacent one over, and so on all the way to the end of the row of dominoes. The propagation mechanisms differ, however, depending on whether the axon is unmyelinated or myelinated.

Propagation of Action Potentials in Unmyelinated Axons

Recall from our discussion of graded potentials that electrotonic conduction is the passive spread of voltage changes along a neuron, away from the site of origin. Electrotonic conduction is the mechanism by which action potentials are propagated in unmyelinated axons (**Figure 22**). When an action potential occurs in an axon, the entire axon does not depolarize at once; instead, depolarization is restricted to just one region of the axon at any one time. Within this region of depolarization, the sign of the membrane potential is

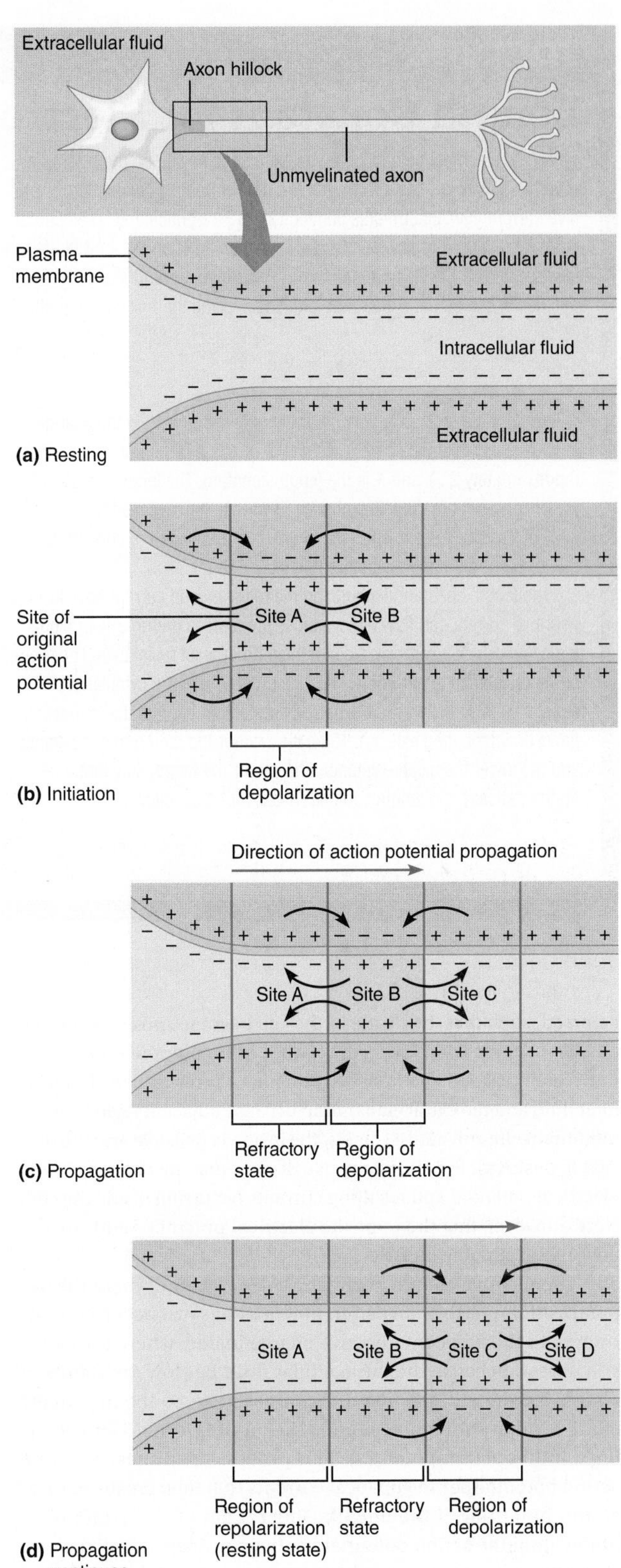

Figure 22 Action potential conduction in unmyelinated axons. **(a)** State of the membrane at the resting membrane potential. **(b)** Initiation of conduction. When an action potential occurs at site A on the membrane, separation of charge occurs in the intracellular fluid and extracellular fluid. The charge separation serves as a force for current to move. The local currents produced by positive ions moving toward the negative regions of the intracellular fluid are shown by arrows. **(c)** Propagation of conduction. The current depolarizes the adjacent region of membrane (site B) to threshold, eliciting an action potential there. **(d)** Continuation of propagation. Depolarization of adjacent regions continues until the action potential has been propagated all the way to the axon terminal. Refractory periods prevent action potentials from traveling in the reverse direction.

TOOLBOX

Length Constant for Electrotonic Conduction

When the electrical potential changes at a specific site on the plasma membrane, current flows from this site by electrotonic conduction, or passive current flow. Electrotonic conduction plays a critical role in electrical signaling. As current flows passively from the site of origin, it dissipates by leakage across the plasma membrane. Assuming a cylindrical structure, this decay can be described mathematically using the following equation:

$$V_x = V_0 e^{-x/\lambda}$$

where V_0 is the change in voltage at the site or *origin*, V_X is the change in voltage at a distance x from the site of origin, e is the natural log base (approximately 2.7), and λ is the length constant. The length constant is the distance where the initial change in voltage has decayed to 1/e or to 37% of its value, as shown in the figure. It is a measure of how far down the axon current can flow before it leaks out of the axon.

The length constant depends on the resistance to current flow. In neurons, current can flow either across the plasma membrane or along the cytoplasm. Current flows down the path of least resistance. The membrane resistance to current flow (r_m) is greater than the cytoplasmic resistance (also called axial resistance, reflecting the fact that current flows down the long axis, r_a). Thus, the greater the membrane resistance and the lower the axial resistance, the longer the length constant. Mathematically, the length constant is described as follows:

$$\lambda = \sqrt{\frac{r_m}{r_a}}$$

Length constants are generally less than 1 mm; that is, by the time a voltage change has spread 1 mm away from the site of origin, the voltage change is less than 37% of the initial change.

The main factor affecting membrane resistance is the degree of myelination. Myelin insulates the axon increasing resistance. The main factor affecting axial resistance is the diameter. Larger-diameter axons have lower resistance to longitudinal current flow and, therefore, a lower axial resistance.

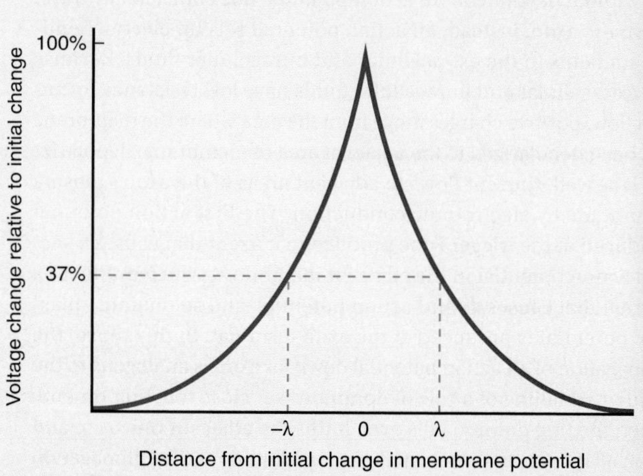

reversed, such that the inside of the cell becomes positive and the outside becomes negative. This change in membrane potential in one region creates a difference in potential within both the intracellular fluid and the extracellular fluid because adjacent regions of the membrane are still at rest (that is, the inside is negative and the outside is positive). For simplicity, we discuss just the changes in the electrical potential and resulting currents occurring inside the cell. Note, however, that the opposite electrical potentials and current flow also occur outside the cell.

The positive charge inside the cell at the site where the depolarization phase of an action potential has just occurred is attracted to the adjacent negative charge inside, where the cell is still at rest. Because the intracellular fluid has low resistance to charge movement, the positive charges move to the negatively charged adjacent region, depolarizing it to threshold and creating another action potential. This process continues, with one action potential setting up local currents that then create another action potential in the neighboring region of the axon membrane, until the action potential has been propagated all the way to the axon terminal. Once the first action potential is generated at the axon hillock, the current spread to the adjacent membrane is always of sufficient magnitude to depolarize the neighboring membrane to threshold and elicit an action potential. This is why action potentials—unlike graded potentials—do not get smaller as they travel.

The diameter of an axon determines how quickly current spreads and, therefore, the conduction velocity of action potentials. The larger the diameter, the less resistance to longitudinal current flow (that is, current flow down the axon). This situation is analogous to a six-lane highway that allows heavy traffic to move faster than a two-lane highway does (there is more space for cars, or ions, to "maneuver"). Therefore, action potentials are propagated more quickly from axon hillock to axon terminal in large-diameter axons (**Toolbox: Length Constant for Electrotonic Conduction**).

When a particular region of an axon is depolarized during an action potential, the resulting currents travel both downstream and upstream to adjacent regions of the membrane. What prevents an action potential from traveling upstream is that the portion of the membrane nearer to the axon hillock has just recently experienced an action potential of its own; therefore, it is in an absolute refractory state. Thus the refractory period prevents action potentials from traveling backward, ensuring unidirectional propagation of action potentials (see Figure 22c and d).

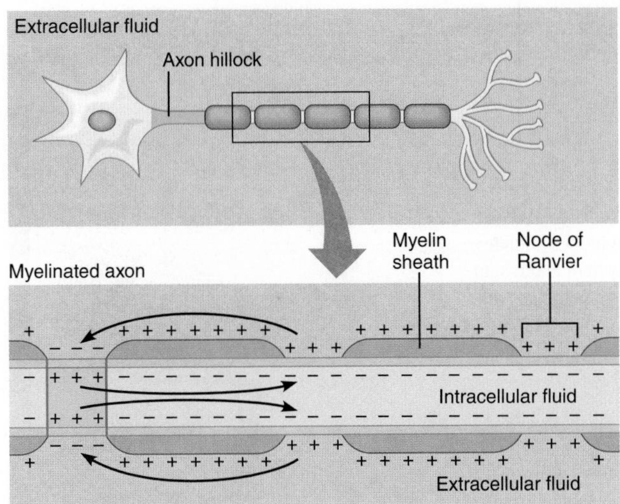

Direction of action potential propagation

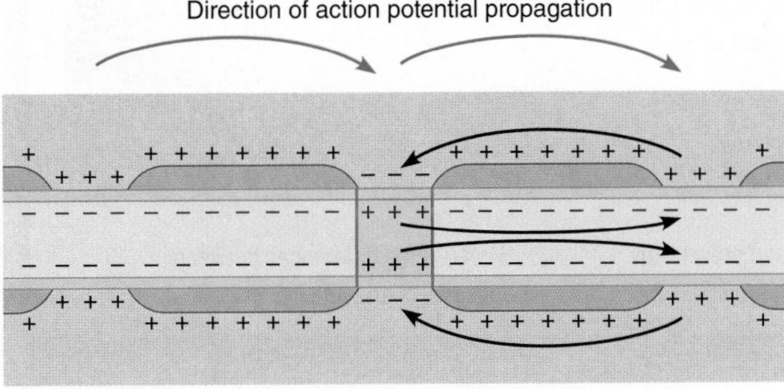

Figure 23 Saltatory conduction in myelinated axons. An action potential in a myelinated axon produces electrical gradients in the intracellular and extracellular fluids that are similar to those observed in unmyelinated axons (see Figure 22). However, because very little current flows across the membrane where myelin insulates it, the current must flow all the way to the next node of Ranvier, where it depolarizes this area of the membrane to threshold and initiates an action potential.

Action Potential Propagation in Myelinated Axons

In axons that are sheathed in myelin, action potentials are propagated by a specialized type of electrotonic conduction called **saltatory conduction** (**Figure 23**). Myelin provides high resistance to ion movement across the plasma membrane, but longitudinal resistance is low. As noted previously, the nodes of Ranvier are gaps in the myelin where the axon membrane lacks insulation, is exposed to the interstitial fluid, and has a high concentration of voltage-gated sodium and potassium channels. In myelinated fibers, action potentials are produced at the nodes of Ranvier. The concept is similar to that described for electrotonic conduction, except that in myelinated axons action potentials cannot be produced where myelin is present. Therefore, the separation of charge in the intracellular fluid causes current to flow from one node of Ranvier to the next. These currents move rapidly under the myelin sheath but diminish in amplitude because some current leaks across the axon membrane. However, because the distance between nodes of Ranvier is short, the current remains strong enough to depolarize the membrane at the node of Ranvier to threshold and to generate an action potential. In this way, action potentials are generated at each node along the axon until an action potential is generated at the axon terminal. The jumping of action potentials from node to node is the basis of the term *saltatory conduction* ("saltatory" comes from *saltare,* Latin for "to leap").

We saw in the previous section that conduction velocity is greater in large-diameter axons. Because action potentials can move in large jumps in myelinated axons, conduction velocities in myelinated axons are greater than those in unmyelinated axons. Think of it this way: Express trains that make few stops (like the action potentials in myelinated axons) take less time to reach a destination than local trains that make many stops (like the action potentials in unmyelinated axons). The fastest-conducting axons, therefore, are both large in diameter and myelinated (**Table 4**). The fastest-conducting axons are generally found in pathways requiring quick action; thus the axons that control skeletal muscle contractions are myelinated and of the largest diameter, and exhibit the fastest conduction velocities. To see what happens when action potential conduction is blocked, see **Clinical Connections: Local Anesthetics**.

Apply Your Knowledge

Multiple sclerosis is an autoimmune disease in which the immune system attacks myelin in the central nervous system. What effect does multiple sclerosis have on the conduction of action potentials in the nervous system?

TABLE 4 Conduction Velocities in Axons of Various Nerve Fiber Types

Fiber type	Myelin present?	Example of function	Fiber diameter (μm)	Conduction velocity (m/sec)
A alpha	Yes	Stimulation of skeletal muscle contraction	12–20	70–120
A beta	Yes	Touch, pressure sensation	5–12	30–70
A gamma	Yes	Stimulation of muscle spindle contractile fibers	3–6	15–30
A delta	Yes, but little	Pain, temperature sensation	2–5	12–30
B	Yes	Visceral afferents, autonomic preganglionics	1–3	3–15
C	No	Pain, temperature sensation, autonomic postganglionics	0.3–1.3	0.7–2.3

CLINICAL CONNECTIONS

Local Anesthetics 

Before drilling on a patient's tooth, a dentist typically gives the patient a shot of a local anesthetic, such as procaine hydrochloride (Novocaine) or lidocaine (Xylocaine). The injection is given in the mouth near the location of nerves that innervate the particular tooth. When the anesthetic reaches the neurons, certain regions of the face, mouth, and tongue lack sensation or become numb. This numbing effect is due to the ability of anesthetics to block the production of action potentials in neurons.

Novocaine and other local anesthetics work by blocking the voltage-gated sodium channels in neurons. If these channels cannot open, then action potentials cannot be produced in the neuron. Action potentials in sensory neurons are necessary for these neurons to communicate to the central nervous system that a stimulus has occurred. When action potentials are blocked, however, the brain (and the patient) is unaware of the presence of painful stimuli in the mouth.

More complicated procedures, such as root canals, may require the administration of analgesics after the procedures are completed. Analgesics differ from anesthetics in that they block only painful stimuli, whereas anesthetics block all stimuli. Analgesics are more specific because they affect synaptic communication of neurons.

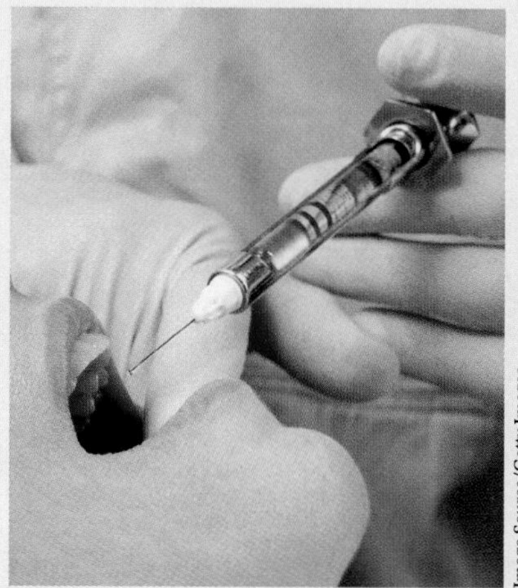

Image Source/Getty Images

Critical Thinking Questions

1. Describe the process by which the application of Novacaine or Xylocaine results in numbness.
2. In what ways does Novacaine prevent the brain from sensing painful stimuli?
3. What are the differences between anesthetics and analgesics?

FOCUS ON DIABETES

Peripheral Neuropathy

Peripheral neuropathy, a disease of the peripheral nervous system, can affect the somatic or autonomic efferent or the afferent branch. Its symptoms vary depending on the site of disease, but the more common symptoms include numbness, tingling, or pain in the hands or feet. In autonomic neuropathy, symptoms are associated with internal organs and include dizziness, diarrhea, indigestion, and impotence. As the disease progresses, symptoms increase.

According to the Neuropathy Association, 20 million Americans suffer from neuropathy. Although many different causes are possible, 30% of neuropathies occur in conjunction with diabetes. How diabetes causes neuropathy is not known, but there are correlations between blood glucose regulation and the development of neuropathy. Thus a person with diabetes who better regulates his or her blood glucose levels is less likely to develop neuropathy. High blood glucose levels have direct effects on the ability of neurons to generate electrical signals, but can also affect blood vessels to the neurons, thereby indirectly causing nerve damage. Recent studies suggest that 30–40% of people with diabetes develop some form of neuropathy. Studies also suggest that another 30% of people with diabetes have asymptomatic neuropathy that has gone undiagnosed.

There is no cure for neuropathy, and the disease tends to progress, especially if blood glucose is not adequately regulated. Several medications can be used to treat the symptoms. Once it has set in, however, the neuropathy cannot be reversed.

5 Maintaining Neural Stability

In the beginning of this chapter, we saw the importance of ion concentration gradients and equilibrium potentials to the establishment of the resting membrane potential; subsequent sections discussed the role of ion movement into or out of the neuron to produce the resting potential, graded potentials, and action potentials. But this raises questions: If ions move across the cell membrane, would not the gradient dissipate, and the equilibrium potentials change? If the gradients dissipate, what happens to the resting membrane potential?

The key to answering these questions lies in the Na^+/K^+ pumps in the neuron membrane. Earlier in this chapter we discussed the Na^+/K^+ pump in the context of its importance in developing concentration gradients and sustaining the gradients when the cell is at rest, when sodium and potassium ions are moving in and out of the cell through leak channels. To compensate for the movement of sodium and potassium ions in an active cell (a cell propagating action potentials), the Na^+/K^+ pump transports even more sodium and potassium ions to compensate for the greater movement of these ions through opened voltage-gated channels. Although an action potential is a relatively large change in the membrane potential, very few ions need to cross the membrane to produce this change. In fact, even without the Na^+/K^+ pump, no measurable change in the resting membrane potential would occur until a neuron has produced more than 1000 action potentials. Therefore, the action of the Na^+/K^+ pump is sufficient to prevent changes in concentration gradients for sodium and potassium in a cell performing at normal levels of activity.

Quick Check 6

1. What is meant by "all-or-none" in reference to action potentials?
2. In which types of axons does saltatory conduction occur?
3. Explain the effects of fiber diameter and myelination on the conduction velocity of action potentials.

SYSTEMS INTEGRATION In this chapter, we learned about how the nervous system generates and transmits electrical signals. These electrical signals are important for communication among all the organ systems and play a critical role in maintaining homeostasis.

CHAPTER REVIEW

SUMMARY

1 Overview of the Nervous System

- The nervous system can be divided into the central nervous system and the peripheral nervous system.
- The central nervous system consists of the brain and spinal cord.
- The peripheral nervous system is subdivided into afferent and efferent divisions.
- The afferent division consists of neurons that transmit information from the periphery to the central nervous system; the efferent division consists of neurons that transmit information from the central nervous system to the periphery.
- The efferent division is divided into two main branches: (1) the somatic nervous system, which communicates to skeletal muscle, and (2) the autonomic nervous system, which communicates to smooth muscle, cardiac muscle, glands, and adipose tissue.
- The autonomic nervous system is divided into the sympathetic and parasympathetic nervous systems.

iP Nervous I and II, Orientation

2 Cells of the Nervous System

- The nervous system contains neurons, cells specialized for transmitting electrical impulses, and glial cells, which provide metabolic and structural support to the neurons.
- Parts of a neuron include the cell body, dendrites, and the axon.
- The dendrites, and to a lesser extent the cell body, receive information from other neurons at synapses.
- The axon includes an axon hillock, where electrical impulses (action potentials) are initiated, and an axon terminal.
- The axon terminal transmits information via neurotransmitters to other neurons at synapses.
- Slow axonal transport or fast axonal transport are used to move products from the cell body to the axon terminal (anterograde transport) or from the axon terminal to the cell body (retrograde transport).
- Neurons are classified functionally into three classes: efferent neurons, afferent neurons, and interneurons.
- Two types of glial cells function in forming myelin around axons: (1) oligodendrocytes, in the central nervous system, and (2) Schwann cells, in the peripheral nervous system.
- Myelin enhances the propagation of electrical impulses by providing insulation to the axon.

iP Nervous I and II, Anatomy Review

iP Nervous I and II, Ion Channels

3 Establishment of the Resting Membrane Potential

- At rest, cells have a membrane potential across them such that the inside of the cell is negatively charged relative to the outside of the cell.
- The membrane potential exists because electrochemical forces

encourage potassium ions to move out of the cell and sodium ions to move into the cell and because the cell membrane is more permeable to potassium ions at rest.

- The resting membrane potential is close to the potassium equilibrium potential and is maintained by the Na^+/K^+ pump.

MAP iP Nervous I, The Membrane Potential
iP Nervous II, Synaptic Potentials and Cellular Integration
MAP iP Resting Membrane Potential

4 Electrical Signaling Through Changes in Membrane Potential

- Changes in the membrane potential can be produced by changing the permeability of the plasma membrane to ions.
- Graded potentials are small changes in membrane potential that occur in response to a stimulus that opens or closes ion channels.
- An action potential is produced when graded potentials result in a depolarization of the neuron to threshold.
- A single graded potential is usually not of sufficient magnitude to depolarize a neuron to threshold, but graded potentials can be temporally and/or spatially additive.
- Action potentials are rapid depolarizations of the plasma membrane that are propagated along axons from the trigger zone to the axon terminal.
- The rapid depolarization phase of an action potential is caused by the opening of sodium channels and ensuing sodium ion movement into the cell.
- The repolarization phase is caused by the closing of sodium channels and the opening of potassium channels, followed by potassium movement out of the cell.
- After-hyperpolarization occurs because potassium channels are slow in closing, allowing continued movement of potassium out of the cell for a brief time.
- Voltage-gated sodium channels have activation gates and inactivation gates, an arrangement that allows the channels to exist in three configurations: closed but capable of opening, open, and closed and incapable of opening.
- Action potentials are all-or-none phenomena, meaning that their size does not vary with the strength of the stimulus eliciting them.
- The strength of a stimulus is coded by the frequency of action potentials, with stronger stimuli producing more action potentials per unit time.
- Absolute and relative refractory periods ensure the unidirectional flow of action potentials and limit the frequency of action potentials.

iP Nervous I, The Action Potential

5 Maintaining Neural Stability

- The Na^+/K^+ pump establishes the concentration gradients for sodium and potassium ions, thereby generating the chemical gradients that establish the resting membrane potential.
- The Na^+/K^+ pump also prevents dissipation of the concentration gradients by returning sodium and potassium ions that have crossed the membrane to their original sides of that membrane.

Check your understanding at MasteringA&P and apply your new knowledge to real-life case studies to ensure that you've mastered the content!

EXERCISES

Multiple-Choice Questions

1. Depolarization of a neuron to threshold stimulates
   a) Opening of sodium channels.
   b) Delayed closing of sodium channels.
   c) Delayed opening of potassium channels.
   d) Both a and c.
   e) All of the above.

2. Neurotransmitters are released most commonly from the
   a) Cell body.
   b) Dendrites.
   c) Axon terminals.
   d) Axon hillock.

3. If a cation is equally distributed across the cell membrane (that is, its concentration inside the cell equals its concentration outside the cell), then which of the following statements is *false*?
   a) At −70 mV, the chemical force on the ion is zero.
   b) At −70 mV, the electrical force on the ion acts to move it into the cell.
   c) At +30 mV, the chemical force on the ion is zero.
   d) The equilibrium potential for the ion is zero.
   e) At −70 mV, the electrochemical force on the ion acts to move it out of the cell.

4. The depolarization phase of an action potential is caused by the
   a) Opening of potassium channels.
   b) Closing of potassium channels.
   c) Opening of sodium channels.
   d) Closing of sodium channels.

5. During the relative refractory period, a second action potential
   a) Cannot be elicited.
   b) Can be elicited by a threshold stimulus.
   c) Can be elicited by a subthreshold stimulus.
   d) Can be elicited by a suprathreshold stimulus.

6. Nerves are found
   a) In the central nervous system.
   b) In the peripheral nervous system.
   c) Both a and b.
   d) Neither a nor b.

7. If the membrane potential of a neuron becomes more negative than it was at rest, then the neuron is ________. In this state, the neuron is ________ excitable.
   a) depolarized; more
   b) hyperpolarized; more
   c) depolarized; less
   d) hyperpolarized; less

8. Oubain is a poison that blocks the Na^+/K^+ pump. If this pump is blocked, then the concentration of potassium inside the cell would
   a) Increase.
   b) Decrease.
   c) Not change.

9. If potassium concentrations in the extracellular fluid of the brain increased, activity in the brain would
   a) Increase.
   b) Decrease.
   c) Not change.

10. Which of the following neurons are part of the peripheral nervous system?
    a) Motor neurons innervating skeletal muscles
    b) Parasympathetic neurons
    c) Sympathetic neurons
    d) All of the above
11. Which of the following axons exhibits the greatest conduction velocity?
    a) An unmyelinated axon with diameter 5 μm
    b) A myelinated axon with diameter 5 μm
    c) An unmyelinated axon with diameter 20 μm
    d) A myelinated axon with diameter 20 μm
12. Which of the following best describes the status of sodium channels at the resting membrane potential?
    a) Activation gates are open and inactivation gates are closed.
    b) Activation gates are closed and inactivation gates are open.
    c) Activation gates and inactivation gates are closed.
    d) Activation gates and inactivation gates are open.
13. Which of the following is not a part of the efferent division of the nervous system?
    a) Parasympathetic nervous system
    b) Sympathetic nervous system
    c) Motor neurons
    d) Sensory receptors
14. Of the following ions, which is (are) located in greater concentration inside the cell?
    a) Sodium only
    b) Potassium only
    c) Chloride only
    d) Sodium and potassium
    e) Potassium and chloride
15. Which of the following statements about graded potentials is *false*?
    a) The magnitude of a graded potential varies with the strength of the stimulus.
    b) Some graded potentials are hyperpolarizations; others are depolarizations.
    c) Graded potentials are produced at ligand-gated ion channels.
    d) Graded potentials can sum over space and time.
    e) Graded potentials are limited in duration by the refractory period.

Objective Questions

16. What are the subdivisions of the peripheral nervous system?
17. Information from the periphery is brought to the central nervous system by (afferent/efferent) pathways.
18. Which cell type is more abundant in the nervous system—glial cells or neurons?
19. Voltage-gated calcium channels are located in which region(s) of a neuron?
20. (Schwann cells/Oligodendrocytes) form myelin in the peripheral nervous system, and (Schwann cells/oligodendrocytes) form myelin in the central nervous system.
21. Myelin (increases/decreases) conduction velocity in axons.
22. If an anion is present in greater concentration outside the cell compared to inside the cell, would the equilibrium potential for that anion be positive, negative, or zero?
23. Which ion is closer to equilibrium at the resting membrane potential of −70 mV—sodium or potassium?
24. In the peripheral nervous system, cell bodies of afferent neurons are located in ________.
25. The electrochemical force for potassium ions when the membrane potential is at the peak of an action potential is (greater than/less than) the electrochemical force for potassium ions when the membrane potential is at rest.
26. Both sodium and potassium channels have inactivation gates that close shortly after the activation gates open. (true/false)
27. When sodium inactivation gates are closed, a second action potential is impossible. (true/false)
28. In myelinated axons, action potentials are propagated by ________ conduction.
29. The Na^+/K^+ pump causes the repolarization phase of an action potential. (true/false)
30. When a neuron is at the peak of an action potential (+30 mV), the direction of the electrical force for potassium ions is (into/out of) the cell.

Essay Questions

31. Draw a typical neuron and label the main structures. Then list the functions of each structure.
32. Compare the chemical and electrical forces acting on sodium and potassium ions when a cell is at rest.
33. List some similarities and some differences between graded potentials and action potentials.
34. Explain the ionic basis of an action potential.

Critical Thinking

35. Predict what the membrane potential would be for a cell with equal permeability to sodium and potassium ions. Assume the concentrations for sodium and potassium ions inside and outside the cell are as given in Table 1.
36. Muscle cells, like neurons, are excitable cells in that they can produce action potentials. The resting membrane potential in muscle is approximately −90 mV. Assuming that only sodium and potassium ions establish the membrane potential, explain how muscle cells can have a more polarized potential than neurons.
37. Many local anesthetics (drugs that block sensory input) work by blocking sodium channels. Explain how sensory input would be lost through this type of action.
38. Motor neurons originate in the central nervous system and terminate with a synapse onto skeletal muscle cells in the periphery. Part of the axon, therefore, is in the central nervous system but most of it is in the peripheral nervous system. The axon is myelinated. Which type of glial cell would form the myelin?
39. Predict what would happen to the resting membrane potential and the ability of a neuron to elicit action potentials if the concentration of potassium in the extracellular fluid was decreased to 50% of its normal value. What would happen if the concentration of sodium in the extracellular fluid was decreased to 50% of its normal value?

Photo credits for icons in feature boxes: Focus on Diabetes: Tanyae/Dreamstime; Clinical Connections: lofoto/Dreamstime; Understanding Exercise: Pixattitude/Dreamstime; Discovery: Yuri Arcurs/Cutcaster.

Glossary

absolute refractory period period during and immediately following an action potential during which a second action potential cannot be generated in response to a second stimulus, regardless of the strength of that stimulus

action potentials large changes in the membrane potential of excitable cells in which the inside of the cell becomes positive relative to the outside; function in transmitting information over long distances in axons

afferent neurons neurons that transmit either sensory information or visceral information to the central nervous system for further processing

after-hyperpolarization third phase of an action potential, during which the membrane potential is more negative than at rest because of high permeability to potassium

all-or-none principle states that action potentials either occur fully or not at all

autonomic nervous system the division of the nervous system that encompasses efferent neurons that synapse with and regulate the function of internal organs and other structures not under voluntary control

axon branch that extends from the cell body of a neuron and sends information to other neurons or effector cells via action potentials and release of neurotransmitter

axon hillock the site where the axon originates from the cell body of a neuron and the point of initiation of action potentials; trigger zone

axon terminal the end of the axon that forms a synapse with another neuron or an effector cell

cell body portion of neuron where nucleus is located; also called *soma*

central nervous system (CNS) the division of the nervous system that includes the brain and spinal cord; consolidates information received from the organs and develops commands to be sent to the organs; is the site of learning, memory, emotions, and cognition

collateral branch of an axon

dendrites branches that extend from the cell body of a neuron and receive information from other neurons

depolarization any change in membrane potential in which the inside of the cell becomes more positive (less negative) than it is at rest

efferent neurons neurons that transmit information from the central nervous system to effector organs

electrotonic conduction the passive spread of a graded potential from the site of origin within a cell

excitable cells cells capable of producing action potentials

oligodendrocytes (oh-lih-goh-den-droh-sites) glial cells that form myelin around axons in the central nervous system; one oligodendrocyte provides myelin segments for many axons

glial cells (glee-al) cells in the nervous system that provide various types of support to the neurons, including structural and metabolic support

graded potentials small changes in membrane potential that occur when ion channels open or close in response to a stimulus, such as the binding of neurotransmitters to receptors; the magnitude of the potential change varies with the strength of the stimulus

hyperpolarization any change in membrane potential in which the inside of the cell becomes more negative than it is at rest

inactivation gates one of two types of gates in voltage-gated sodium channels; close slowly in response to depolarization

innervate to make a synaptic connection with an effector organ

interneurons neurons located entirely in the central nervous system; account for 99% of all neurons in the body

leak channels ion channels that are always open; responsible for the resting membrane potential

ligand-gated channels ion channels that open or close in response to a chemical messenger binding to the channel

myelin (my-uh-lin) layers of plasma membrane from oligodendrocytes and Schwann cells that form insulation around the axons of neurons

node of Ranvier (rahn-vee-ay) gap in the myelin along axons

peripheral nervous system the division of the nervous system that contains nerves that provide communication between the central nervous system and organs of the body; includes afferent and efferent branches

refractory period period of reduced membrane excitability; during and immediately after an action potential, when the membrane is less excitable than it is at rest

regenerative a signal that feeds into itself by positive feedback

relative refractory period period immediately following the absolute refractory period during which it is possible to generate a second action potential, but only with a stimulus stronger than that needed to reach threshold under resting conditions

repolarization return of the membrane potential of a cell to the resting potential following a depolarization

resting membrane potential the voltage that exists across a cell membrane when the cell is not transmitting electrical signals; polarity is such that the inside of the cell is negative with respect to the outside

saltatory conduction type of action potential conduction that occurs in myelinated axons; action potentials "jump" from node to node

Schwann cells glial cells that form myelin around axons in the peripheral nervous system; one Schwann cell provides a segment of myelin for one axon

somatic nervous system the division of the nervous system that encompasses nerve cells that regulate skeletal muscle contractions

spatial summation the addition of graded potentials generated at different locations that occurs when they are stimulated more or less simultaneously

subthreshold in the nervous system, a stimulus too weak to generate an action potential

suprathreshold in the nervous system, a stimulus that exceeds the threshold to generate an action potential

temporal summation the addition of graded potentials generated at a particular site that occurs when it is stimulated at a high frequency

threshold in an excitable cell, the critical value of the membrane potential to which the cell must be depolarized in order to trigger an action potential

voltage-gated channels channels that open or close in response to a change in membrane potential

Answers

Figure Question Answers

Fig 6 Muscle and glands

Fig 16 Potassium is further from equilibrium, so potassium movement out of the cell exceeds sodium movement into the cell

Fig 18 The closing of the sodium inactivation gates and the opening of potassium channels

Apply Your Knowledge Answers

The membrane potential would be less negative.

$$V_m = 61 \log \frac{P_{Na}[Na^+]_o + P_K[K^+]_o + P_{Cl}[Cl^-]_i}{P_{Na}[Na^+]_i + P_K[K^+]_i + P_{Cl}[Cl^-]_o}$$

TTX (tetrodotoxin) blocks the voltage-gated sodium channels responsible for the rapid depolarization phase of the action potential and, therefore, blocks the generation of action potentials. Because some of these channels (albeit only a few) are open at rest, TTX would decrease sodium permeability, thereby causing a hyperpolarized resting membrane potential.

TEA (tetraethylammonium) blocks the voltage-gated potassium channels responsible for the repolarization phase of the action potential. Thus, once an action potential occurs, the depolarization phase will be prolonged. Because some of these channels are open at rest, TEA would decrease potassium permeability, thereby causing a depolarized resting membrane potential.

By decreasing the amount of myelin, multiple sclerosis slows down the conduction of action potentials in the central nervous system. Eventually, the degree of demyelination causes the cessation of action potentials along some axons.

Answers to End-of-Chapter Multiple Choice and Objective Questions

Multiple Choice Questions

1. e 2. c 3. e 4. c 5. d 6. b 7. d 8. b
9. a 10. d 11. d 12. b 13. d 14. b 15. e

Objective Questions

16. afferent and efferent
17. afferent
18. glial cells
19. axon terminal
20. Schwann cells, oligodendrocytes
21. increases
22. negative
23. potassium
24. ganglion
25. greater than
26. false
27. true
28. saltatory
29. false
30. out of

Synaptic Transmission and Neural Integration

Synaptic Transmission and Neural Integration

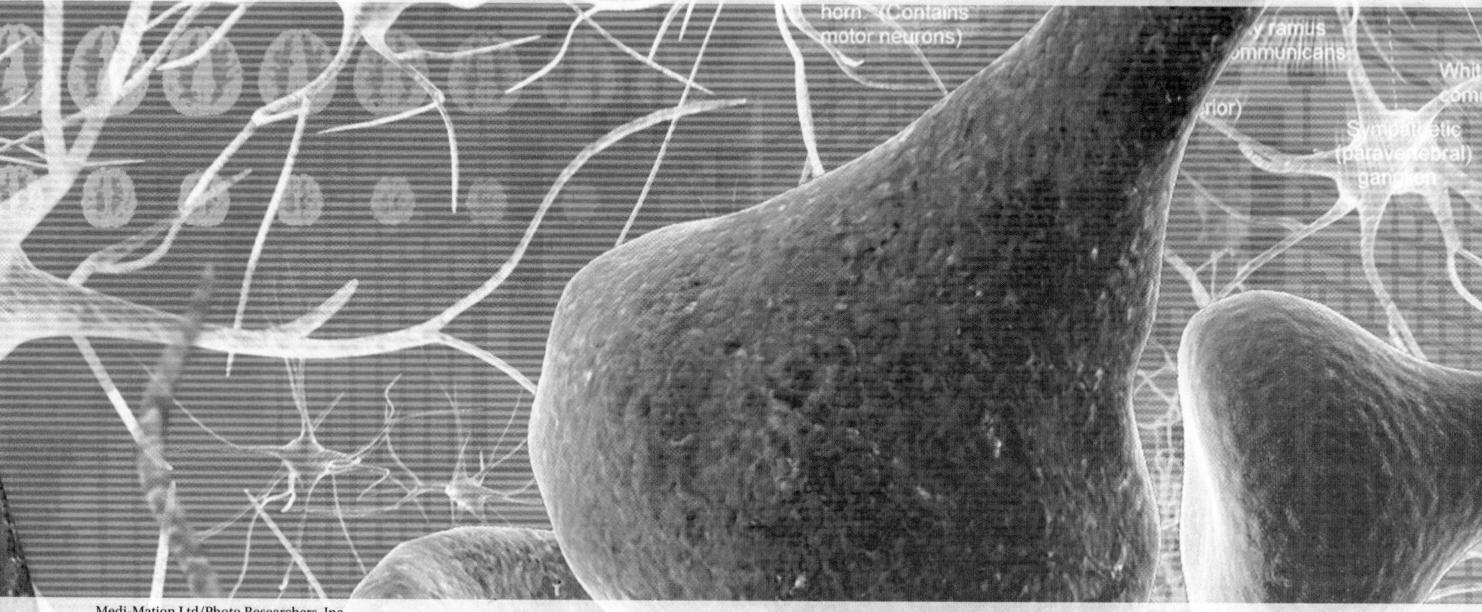

Medi-Mation Ltd/Photo Researchers, Inc.

Composite computer artwork of a junction (or synapse), between two neurons.

Janice is on her college gymnastics team. During one meet, she misses catching the high bar on her dismount from the uneven parallel bars, and falls on her side with her left arm pinned underneath her. The impact breaks her arm—a serious injury that puts Janice in a lot of pain. She is taken from the meet to the hospital, where a doctor gives her morphine. Within minutes, her pain starts to subside.

Why does morphine—a product of the opium poppy—relieve Janice's pain? The answer is related to the ways that neurons communicate via neurotransmitters. Morphine binds to certain neurons in the central nervous system at receptors for naturally occurring morphinelike messenger molecules called *enkephalins* and *endorphins*. As a result, morphine suppresses the transmission of pain signals.

In this chapter, we discuss how enkephalins and other neurotransmitters communicate across synapses by binding to receptors on the postsynaptic neurons. Anywhere in the body that a chemical messenger acts by binding to a receptor represents a potential target for a drug.

CHAPTER OUTLINE

MasteringA&P®

Go to MasteringA&P for helpful A&P Flix 3-D animations, chapter quizzes, pre-tests, Interactive Physiology tutorials, and more!

LEARNING OUTCOMES *After studying this chapter, you should be able to:*

- Understand the differences between chemical and electrical synapses.
- Describe the communication across chemical synapses. Explain how neurotransmitters are released, and describe their actions after release.
- Compare fast and slow responses at synapses.
- Describe the process of neural integration and the role of the axon hillock in this process.
- Describe the major classes of neurotransmitters, including their chemical structures, synthesis, degradation, and signal transduction mechanisms.

Before You Begin

Make sure you have mastered the following topics:

- Gap junctions
- Exocytosis
- Membrane potential
- Graded potentials
- G proteins
- Catecholamine synthesis
- Protein synthesis

We saw that neurons generate electrical signals in the form of graded potentials and action potentials that transmit messages from one area of the cell to another. In this chapter, we learn how those electrical signals lead to communication between neurons at synapses.

Two types of synapses are found in the nervous system: electrical synapses and chemical synapses. **Electrical synapses** operate by allowing electrical signals to be transmitted from one neuron to another through gap junctions. **Chemical synapses** operate through the release of neurotransmitters that activate signal transduction mechanisms in the target cell.

1 Electrical Synapses 

Electrical synapses exist between neurons and either other neurons or glial cells. At electrical synapses, the plasma membranes of adjacent cells are linked together by gap junctions such that when an electrical signal is generated in one cell, it is directly transferred to the adjacent cell by means of ions flowing through the gap junctions. Second messenger molecules can also move through these junctions.

The existence of electrical synapses in the central nervous system has been known for decades, but the functions of these synapses are just now being elucidated. Earlier studies demonstrated the importance of electrical synapses in the development of the nervous system, a topic beyond the scope of this text. More recent studies indicate that these junctions transmit signals in the adult brain.

Electrical synapses allow rapid communication between adjacent neurons that synchronizes the electrical activity in these cells. This communication is often bidirectional, although some gap junctions allow current flow in only one direction. The communication can be excitatory or inhibitory at the same synapse, as either a depolarizing or a hyperpolarizing current can spread through these junctions. Some electrical synapses are always active, whereas others seem to have gating mechanisms that permit current flow only under certain circumstances.

Electrical synapses have been identified in the retina of the eye and certain areas of the cortex, where they are believed to function in transmission of signals. Electrical synapses are also found in areas of the brainstem that regulate breathing, a rhythmical process. The gap junctions are believed to synchronize the neurons responsible for inspiration. The hypothalamic neurons that release tropic hormones are connected to other hypothalamic neurons that release the same tropic hormones by gap junctions, thereby synchronizing their activity and resulting in *bursts* of tropic hormone release. Nevertheless, much remains unknown about the function of electrical synapses.

The vast majority of synapses in the nervous system are chemical synapses, which are much better understood in terms of function. We turn now to chemical synapses, the topic of the remainder of this chapter.

2 Chemical Synapses 

Almost all neurons transmit messages to other cells at chemical synapses. In a chemical synapse, one neuron secretes a neurotransmitter into the extracellular fluid in response to an action potential arriving at its axon terminal. The neurotransmitter then binds to receptors on the plasma membrane of a second cell, triggering in that cell an electrical signal that may or may not initiate an action potential, depending on a number of circumstances.

A neuron can form synapses with other neurons—the situation that is the topic of this chapter—or with effector cells such as muscle or gland cells. Muscles and glands are generally referred to as *effector organs,* and a synapse between a neuron and an effector cell is called a *neuroeffector junction.* Neuroeffector junctions operate according to the same basic principles presented here for neuron-to-neuron synapses. In a few instances, non-neuronal cells can form synapses with neurons, as occurs, for example, with taste receptor cells in taste buds on the tongue.

Functional Anatomy of Chemical Synapses

Figure 1 depicts the possible arrangements in typical neuron-to-neuron synapses. The first neuron, which transmits signals to the second, is designated the **presynaptic neuron;** the second neuron, which receives signals from the first, is referred to as the **postsynaptic neuron.** The narrow space between the presynaptic and

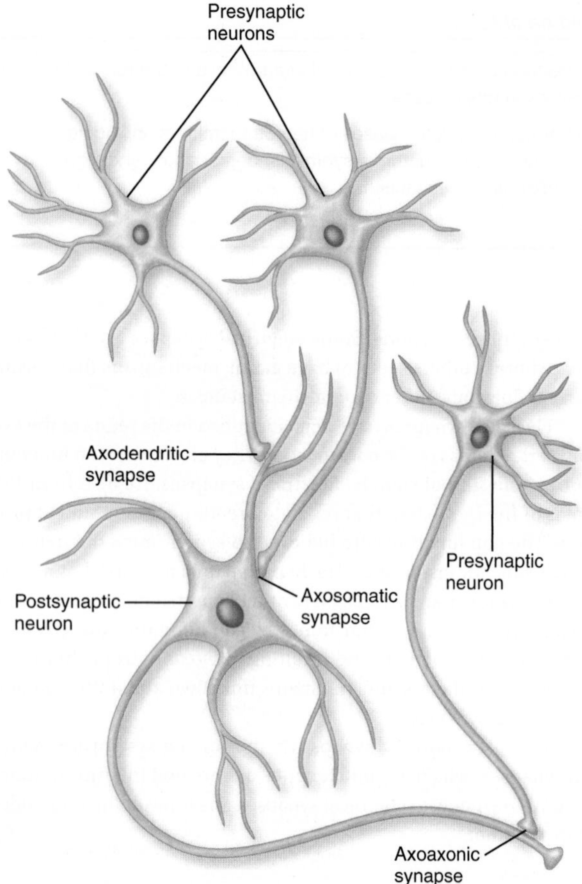

Figure 1 Neuron-to-neuron chemical synapses. Synapses can occur at dendrites (axodendritic synapses), at the cell body (axosomatic synapses), or with another axon (axoaxonic synapses).

postsynaptic neurons is called the **synaptic cleft;** it is generally only 30–50 nm wide (1 nm $= 10^{-9}$ m). Signaling across a chemical synapse is unidirectional—the presynaptic neuron communicates to the postsynaptic neuron.

Most often the presynaptic neuron's axon terminal forms a synapse with a dendrite of the postsynaptic neuron, in which case the synapses are referred to as axodendritic synapses. However, synapses between the axon terminal and soma of the postsynaptic cell, called axosomatic synapses, also occur and have the same function as axodendritic synapses. The presynaptic neuron's axon terminal sometimes forms a synapse with the postsynaptic neuron's axon terminal, in which case the synapse is called an axoaxonic synapse. Axoaxonic synapses have a special function in modulating communication at axodendritic and axosomatic synapses; they will be described later in this chapter. Dendrodendritic synapses have also been identified, but their function is beyond the scope of this text. The following paragraphs describe axodendritic and axosomatic synapses.

Figure 2 shows a close-up of a synapse and the mechanism of transmitting a signal across the synapse. In all cases, the axon terminal of the presynaptic neuron releases neurotransmitters into the synaptic cleft. Once released into the synaptic cleft, the neurotransmitters diffuse rapidly across the cleft and bind to receptors on the postsynaptic neuron. The binding of the neurotransmitter to the receptors produces a response in the postsynaptic neuron by signal transduction mechanisms.

Note that the axon terminal of the presynaptic neuron contains numerous small, membrane-bound compartments called, which store neurotransmitter molecules. Most neuro-transmitters are synthesized in the cytosol of the axon terminal, where the enzymes for their synthesis are located. After synthesis, neurotransmitters are actively transported into synaptic vesicles, where they are stored until their eventual release by exocytosis.

Cytosolic calcium triggers the release of neurotransmitter by exocytosis. Recall that the membrane of a neuron contains ion channels of various types that depend on their location in the neuron. Most abundant in the membrane of the axon terminal are voltage-gated calcium channels. These calcium channels open when the axon terminal is depolarized, which occurs upon arrival of an action potential at the axon terminal (see ① in Figure 2b). When the calcium channels open ②, they allow calcium to flow down its electrochemical gradient into the axon terminal, thereby increasing the concentration of cytosolic calcium in the axon terminal. Calcium then causes the membranes of synaptic vesicles to fuse with vesicle attachment sites on the inner surface of the axon terminal membrane and undergo exocytosis, which releases the neurotransmitters into the synaptic cleft ③.

The amount of neurotransmitter released depends on the concentration of calcium in the cytosol of the axon terminal, which depends on the frequency of action potentials in the presynaptic neuron. Following a single action potential, neurotransmitter release stops within a few milliseconds because the voltage-gated calcium channels close soon after opening, and because calcium ions are actively transported out of the axon terminal on a continual basis, bringing the cytosolic calcium concentration back to its resting level. If a second action potential arrives before neurotransmitter is cleared from the synaptic cleft, however, then the cytosolic calcium levels increase. Increases in cytosolic calcium levels cause more neurotransmitter to be released from the presynaptic cell, thereby increasing the amount of neurotransmitter in the synaptic cleft. When a series of action potentials arrives at an axon terminal in a short time, cytosolic calcium levels increase even more, thereby releasing even more neurotransmitter. As a consequence, the concentration of neurotransmitter in the synaptic cleft increases as the frequency of action potentials increases.

Once in the synaptic cleft, neurotransmitter molecules diffuse away from the axon terminal and toward the postsynaptic neuron, where they bind to receptors (see ④ in Figure 2b), thereby inducing a response in the postsynaptic neuron ⑤. The binding of a neurotransmitter molecule to a receptor is a brief and reversible process. If neurotransmitter molecules were to remain indefinitely in the synaptic cleft following their release, they would bind to receptors over and over again, inducing a continual response in the postsynaptic neuron.

Continual binding of neurotransmitter to receptor does not occur because a number of processes quickly clear the neurotransmitter from the cleft, thereby terminating the signal. Some neurotransmitter molecules are degraded by enzymes, which may be

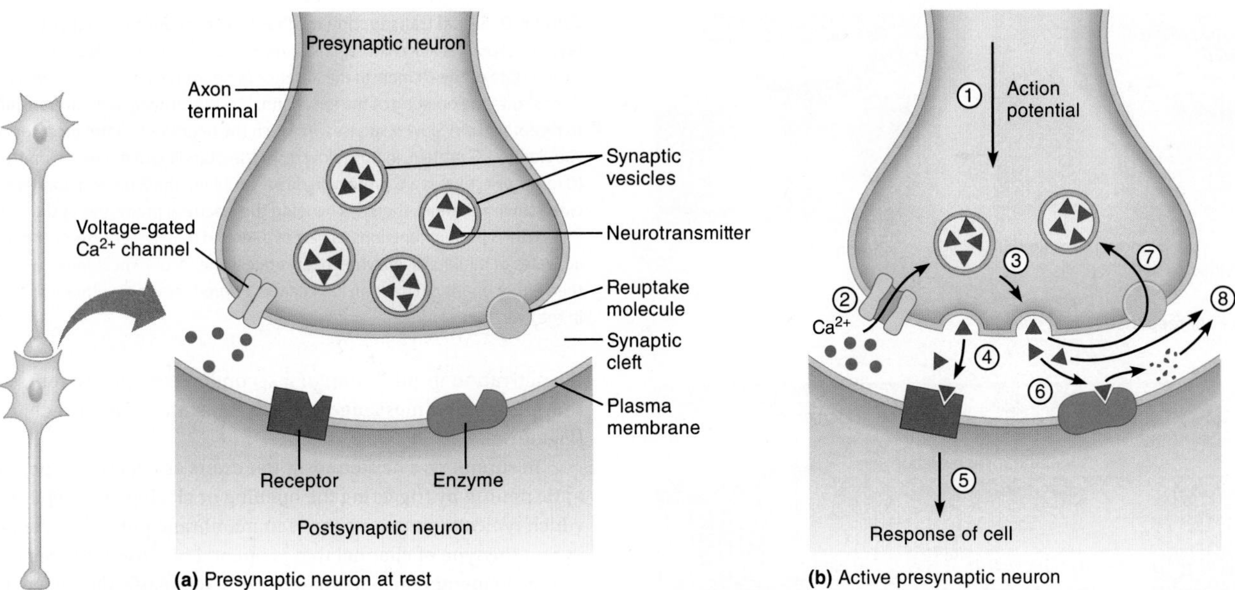

Figure 2 Functional anatomy of a synapse. (a) The presynaptic neuron at rest. Calcium channels are closed, and no neurotransmitter is being released. **(b)** An active presynaptic neuron. See the text for a detailed description of the steps.

located on the postsynaptic neuron's plasma membrane (see ⑥ in Figure 2b), on the presynaptic neuron's plasma membrane, on the plasma membranes of nearby glial cells, in the interstitial fluid of the synaptic cleft, or even in the cytoplasm of the presynaptic neuron or glial cells. Other neurotransmitter molecules can be actively transported back into the presynaptic neuron that released them ⑦, a process known as reuptake. Once inside the neuron, these neurotransmitter molecules are usually degraded and the breakdown products recycled to form new neurotransmitter molecules. Still other neurotransmitter molecules in the synaptic cleft simply diffuse out of the cleft ⑧. So long as neurotransmitter molecules are not bound to receptors, they may be subject to any of the fates just described. As a result, neurotransmitter is usually present in the synaptic cleft for only a few milliseconds after its release from the presynaptic neuron.

Apply Your Knowledge

Several pharmaceutical agents work by decreasing the activity of specific neurotransmitter reuptake proteins or enzymes that degrade neurotransmitters. Describe the effects of each type of inhibitor on the concentration of neurotransmitter in the synaptic cleft and, therefore, on communication across specific synapses.

It takes approximately 0.5–5 msec from the time an action potential arrives at the axon terminal before a response occurs in the postsynaptic cell. This time lag, called the **synaptic delay,** is due mostly to the time required for calcium to trigger the exocytosis of neurotransmitter. Once in the synaptic cleft, diffusion of neurotransmitter to the receptor occurs so rapidly that the time is negligible.

Quick Check 1

1. What are the essential differences between a chemical synapse and an electrical synapse?
2. List the steps involved in the transmission of signals at a chemical synapse.
3. Following its release, a neurotransmitter can be cleared from a synaptic cleft by which three processes?

Signal Transduction Mechanisms at Chemical Synapses

The neurotransmitter released by a presynaptic neuron induces responses in the postsynaptic cell through the signal transduction mechanisms. Neurotransmitter can induce either a fast or slow response in a postsynaptic neuron.

The fast response (**Figure 3a**) occurs whenever a neurotransmitter binds to a **channel-linked receptor,** also called an **ionotropic receptor.** All channel-linked receptors are ligand-gated channels. The binding of the neurotransmitter opens the ion channel, allowing one or more specific ions to permeate the plasma membrane and change the electrical properties of the postsynaptic neuron. The typical response is a change in the membrane potential, called a **postsynaptic potential (PSP),** which occurs very rapidly and terminates rapidly (normally within a few milliseconds) because the channel closes as soon as the neurotransmitter leaves the receptor.

Slow responses, by contrast, are mediated through G protein–linked receptors called **metabotropic receptors.** In the nervous system, G proteins can trigger either the opening or the closing of ion channels, depending on the specific synapse. These G protein–regulated ion channels respond to the binding of neurotransmitter

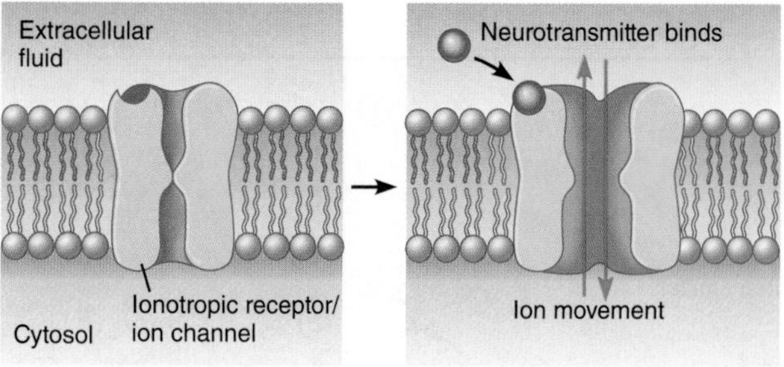

(a) Fast response

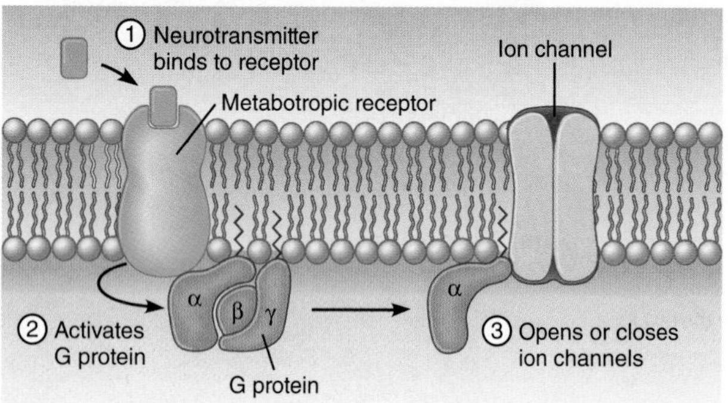

(b) Slow response, direct coupling

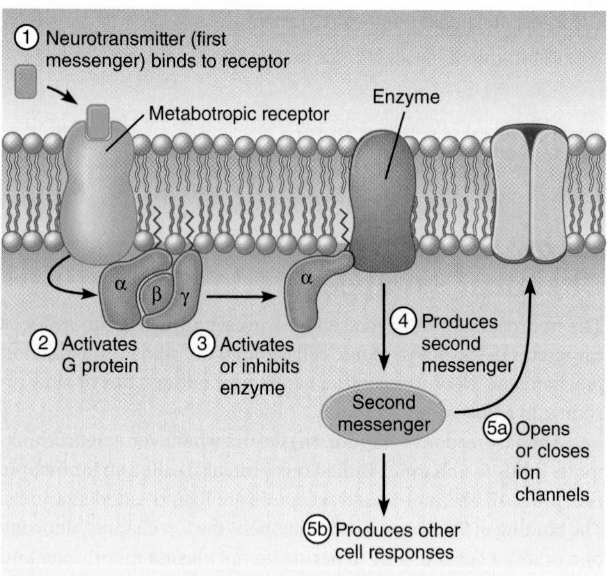

(c) Slow response, second messenger system

Figure 3 Signal transduction mechanisms at chemical synapses. **(a)** Fast responses at a channel-linked receptor. The panel at left shows the closed state of the channel in the absence of neurotransmitter; the panel at right shows the opening of the ion channel by the binding of neurotransmitter to receptor. **(b, c)** Slow responses, in which the neurotransmitter receptor is coupled to a G protein. In both slow responses, binding of neurotransmitter to receptor activates a G protein. In direct coupling, the G protein opens or closes an ion channel, thereby changing the electrical properties of the cell. When the G protein functions as part of a second messenger system, it either activates or inhibits an enzyme that produces the second messenger, which then either opens or closes an ion channel or produces some other response in the cell.

more slowly than the channels that mediate the fast response, with durations ranging from milliseconds to hours, depending on the synapse, but with the same ultimate effect: The opening or closing of ion channels alters the permeability of the postsynaptic neuron to a specific ion or ions, generally resulting in changes in the membrane potential. The G protein can serve as a direct coupling between the receptor and the ion channel (Figure 3b), or it can trigger the activation or inhibition of a second messenger system, such that the second messenger affects the state of the ion channel (Figure 3c).

In summary, a neurotransmitter exerts its effects on a postsynaptic neuron by triggering the opening or closing of ion channels, which typically causes a change in membrane potential. Depending on the type of channel that is opened or closed, the resulting change in membrane potential may be a depolarization or a hyperpolarization. Recall that a depolarization is considered an excitation because it brings the membrane potential closer to the threshold to generate an action potential; likewise, a hyperpolarization is considered an inhibition because it takes the membrane potential further away from the threshold. At any given synapse, the direction of the response is always the same, and the synapse is classified as being either excitatory or inhibitory.

Excitatory Synapses

An **excitatory synapse** is one that brings the membrane potential of the postsynaptic neuron closer to the threshold for generating an action potential; that is, excitatory synapses depolarize the postsynaptic neuron. This depolarization, which is called an **excitatory postsynaptic potential (EPSP),** can occur as either a fast response or a slow response. EPSPs are graded potentials, with the amplitude of depolarization increasing as more neurotransmitter molecules bind to receptors. As with all graded potentials, the depolarization is greatest at the site of origin (usually a dendrite or cell body) and decreases as it moves toward the axon hillock.

Fast EPSPs are generally caused by the binding of neurotransmitter molecules to their receptors on the postsynaptic cell, in the process opening channels that allow small cations (sodium and potassium ions) to move through them (**Figure 4a**). Recall that at the resting membrane potential, potassium is much closer to equilibrium than sodium. Therefore, a much stronger electrochemical force acts to move sodium ions into the cell than acts to move potassium ions out of the cell. Thus, when channels open that allow both sodium and potassium ions to move through them, more sodium ions move in than potassium ions move out, producing a net depolarization. Fast EPSPs can last from just a few to several hundred milliseconds.

Slow EPSPs can be caused by various mechanisms, such as the closing of potassium channels by cAMP (Figure 4b). When the neurotransmitter binds to a receptor ① and activates a G protein ②, the G protein then activates the enzyme adenylate cyclase ③, which in turn catalyzes the reaction converting ATP to cAMP ④.

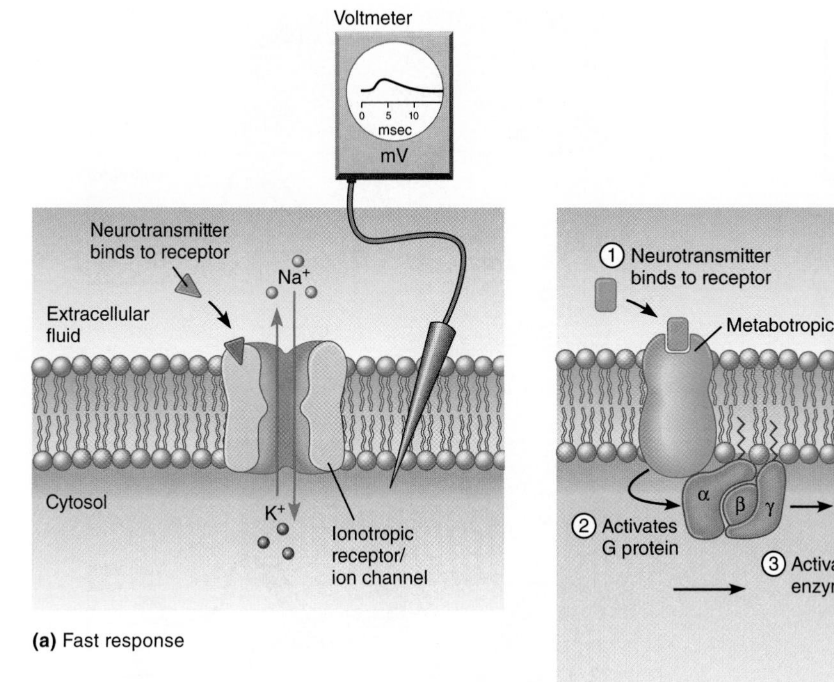

Figure 4 **Excitatory synapses.** An electrode inserted into a neuron and connected to a voltmeter can measure electrical activity in the cell. The type of voltmeter depicted in this and subsequent figures is an oscilloscope. **(a)** A fast excitatory synapse. The neurotransmitter opens ion channels, allowing Na^+ to enter the cell and K^+ to leave it. Sodium movement is greater than potassium movement, so the net effect is a depolarization (an EPSP) lasting several milliseconds. **(b)** A slow excitatory synapse. Activation of a G protein by neurotransmitter leads to the series of events depicted. Phosphorylation of the potassium channel in the final step decreases the leakage of potassium out of the cell, producing a depolarization. Note that this response takes seconds to occur.

cAMP activates protein kinase A ⑤, which catalyzes the addition of a phosphate group to the potassium channel ⑥. Phosphorylation of the potassium channel closes it ⑦.

Before the potassium channel was phosphorylated, the channel was open, and potassium ions were leaking out of the cell while sodium ions were leaking into the cell. No change in membrane potential was occurring because the sodium and potassium leaks balanced each other. When the potassium channel is closed by phosphorylation, however, fewer potassium ions leak out of the cell. The net effect is a depolarization: If permeability to potassium decreases while permeability to sodium remains the same, the membrane potential will move away from the equilibrium potential for potassium and toward the equilibrium potential for sodium—the same effect as if sodium permeability increased.

In comparison to fast EPSPs, slow EPSPs take longer to develop and last longer (seconds to hours). In the case of cAMP-dependent closure of potassium channels, the membrane potential will not return to resting conditions until the potassium channels are dephosphorylated, which will not occur until enough cAMP has been degraded by phosphodiesterase to return the cAMP concentration to its resting level.

Inhibitory Synapses

An **inhibitory synapse** is a synapse that takes the membrane potential of the postsynaptic neuron away from the action potential threshold by hyperpolarizing the neuron or, alternatively, stabilizes the membrane potential at the resting value. In either case, activity at the synapse decreases the likelihood that an action potential will be generated in the postsynaptic neuron—which is why the synapse is said to be *inhibitory*. At inhibitory synapses, the binding of a neurotransmitter to its receptors opens channels for either potassium ions or chloride ions.

When a neurotransmitter causes potassium channels to open (**Figure 5**), potassium will move out of the cell, hyperpolarizing it. This hyperpolarization is called an **inhibitory postsynaptic potential (IPSP).** An IPSP, like an EPSP, is a graded potential: When more neurotransmitter molecules bind to receptors, more potassium channels open and greater hyperpolarization occurs. IPSPs (like EPSPs and all other graded potentials) decrease in size as they move toward the axon hillock, where the hyperpolarization moves the membrane potential away from threshold, thereby decreasing the likelihood that an action potential will be generated in the postsynaptic neuron.

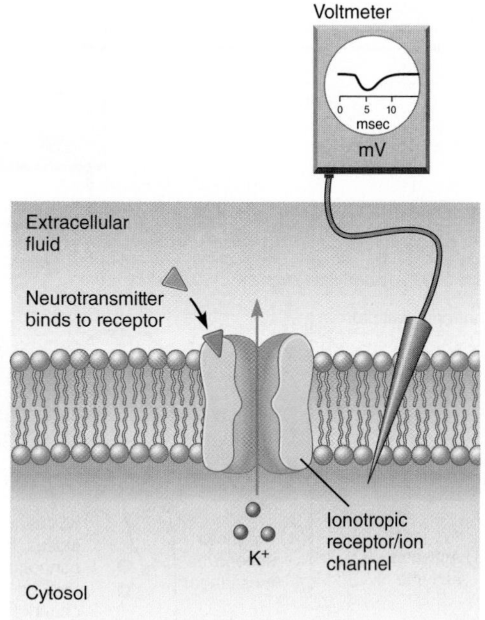

Figure 5 An inhibitory synapse involving potassium channels. At this type of inhibitory synapse, potassium channels are opened by the binding of neurotransmitter to the receptor. Potassium flows out of the cell, hyperpolarizing it and producing an IPSP.

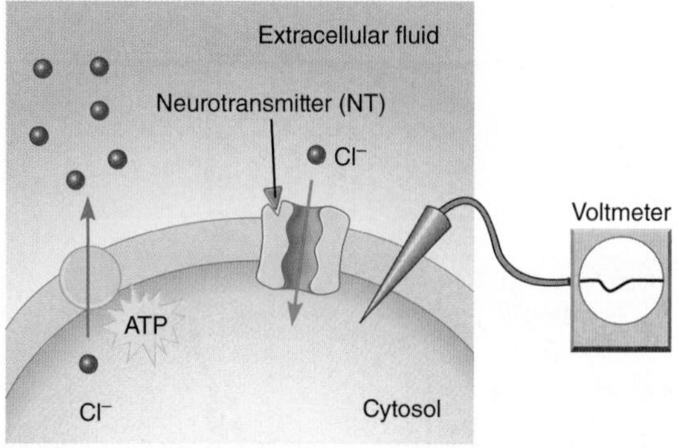

(a) Neuron actively transports chloride out of the cell

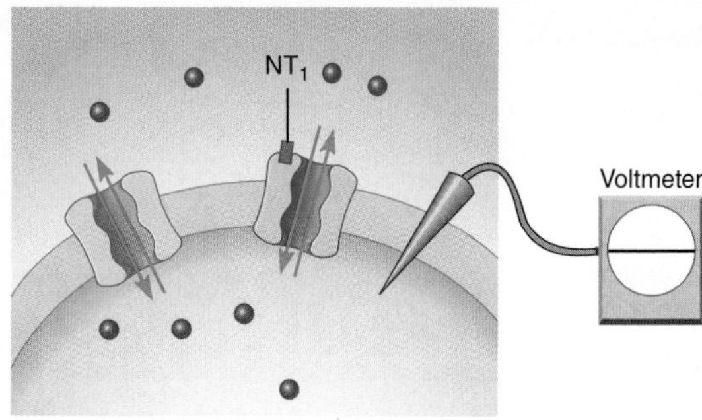

(b) No active transport of chloride, chloride at equilibrium

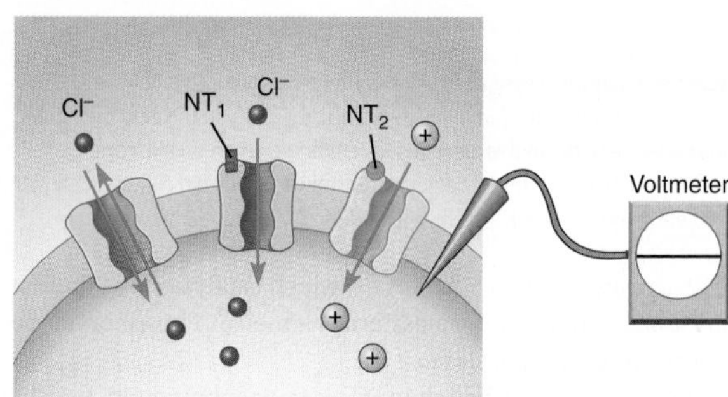

(c) No active transport of chloride, excitatory synapse active

Figure 6 The roles of chloride channels in inhibitory synapses. (a) In neurons that actively transport chloride out of the cell, binding of a neurotransmitter that causes chloride channels to open allows chloride to move into the cell, hyperpolarizing it (inducing an IPSP). **(b)** Cells that lack chloride transporters have chloride leak channels, and chloride is at equilibrium across their membranes. If binding of neurotransmitter 1 (NT_1) causes more chloride channels to open, no net movement of chloride occurs because chloride is already at equilibrium. **(c)** This figure shows the same cell as in panel b except it includes a second synapse. If NT_1 is present at an inhibitory synapse while NT_2 is present at an excitatory synapse, chloride will move into the cell at the same time that positive charges enter the channel opened by NT_2, opposing any change in membrane potential.

What happens when a neurotransmitter causes chloride (Cl^-) channels to open depends on the size and direction of the electrochemical driving force acting to move chloride ions across the postsynaptic neuron membrane. Because the chloride ion has a negative charge (it is an anion), the membrane potential at rest is an electrical force that acts to move chloride out of the cell. However, because chloride ions are found in greater concentration outside the cell than inside, a chemical force acts to move chloride into the cell. The net electrochemical force depends on whether the neuron in question has active chloride transporters.

In some neurons, chloride ions are actively transported out of the cell, making the concentration inside much lower than outside the cell (**Figure 6a**). Therefore, chloride is not at equilibrium; instead, a net electrochemical driving force for moving chloride into the cell exists. When chloride channels are opened by neurotransmitters binding to receptors, chloride moves into the cell, producing a hyperpolarization, or an IPSP. Thus we see the same type of effect that occurs when potassium channels open.

In other neurons, no active transport of chloride occurs, but there are some ion channels through which chloride can diffuse. Under these conditions, the passive movement of chloride coupled with the lack of active transport allows chloride ions to be at equilibrium. In a neuron that lacks active transport of chloride either into or out of the cell (Figure 6b), chloride simply diffuses through chloride leak channels and distributes itself across the membrane at concentrations that will put it at equilibrium. (Note that the membrane potential determines the concentration gradient for chloride ions when they are not actively transported; this behavior stands in contrast to that of sodium and potassium ions, whose concentration gradients, established by active transport, determine the resting membrane potential.) If chloride is at equilibrium, then the opening of chloride channels will not cause a change in the membrane potential (Figure 6b). In this case—perhaps surprisingly—the synapse is still considered to be *inhibitory*. To see why, we must consider how this synapse affects signals that might be generated at other synapses at the same time.

Consider a postsynaptic neuron in which an excitatory synapse and an inhibitory synapse are active at the same time. While the excitatory synapse tends to induce an EPSP (a positive change in the membrane potential) in the postsynaptic neuron, the inhibitory synapse also causes chloride channels to open (Figure 6c). The result is that the EPSP will be diminished, or may even be absent. This outcome occurs because depolarization of the membrane means that chloride is no longer at equilibrium. The more positive potential inside the cell pulls the negative chloride ions into the cell, stabilizing the membrane potential by countering the influence of positive charge moving into the cell. This effect is considered an inhibitory action because it decreases the likelihood that the neuron will reach threshold for an action potential.

As this example illustrates, the activity of a single synapse may influence the likelihood that an action potential will occur, but it usually is not the sole determining factor. Instead, many synapses are normally active at the same time, and it is the particular *combination* of synaptic inputs arriving at any given time that determines whether a postsynaptic cell will produce an action potential. The process that makes this determination, known as *neural integration*, is our next topic of study.

Quick Check 2

1. Activation of which type of receptor—ionotropic or metabotropic—produces the faster response?
2. What is the role of G proteins in the operation of metabotropic receptors?
3. If a G protein activates a second messenger that closes sodium channels, will this activation produce an EPSP or an IPSP? Will the response be fast or slow?

3 Neural Integration

Communication in the nervous system is generally not a process in which one presynaptic cell communicates to one postsynaptic cell. Instead, the axon of one neuron often has several collaterals that communicate to several other neurons, an arrangement called *divergence* (**Figure 7a**). Likewise, a given neuron typically receives communication from many neurons (hundreds or thousands), an arrangement called *convergence* (Figure 7b). Because of convergence, one synapse alone does not determine whether an action potential is generated in the postsynaptic neuron, but rather the axon hillock of the postsynaptic neuron acts as an integrator that, in effect, adds up all the signals arriving from all active synapses. This summation process is called **neural integration,** and it operates according to one simple rule: *An action potential is triggered if the membrane potential at the axon hillock is depolarized to threshold; if the potential is below threshold, no action potential will occur.* Summation is necessary because a single EPSP is rarely of sufficient magnitude to depolarize the postsynaptic neuron to threshold.

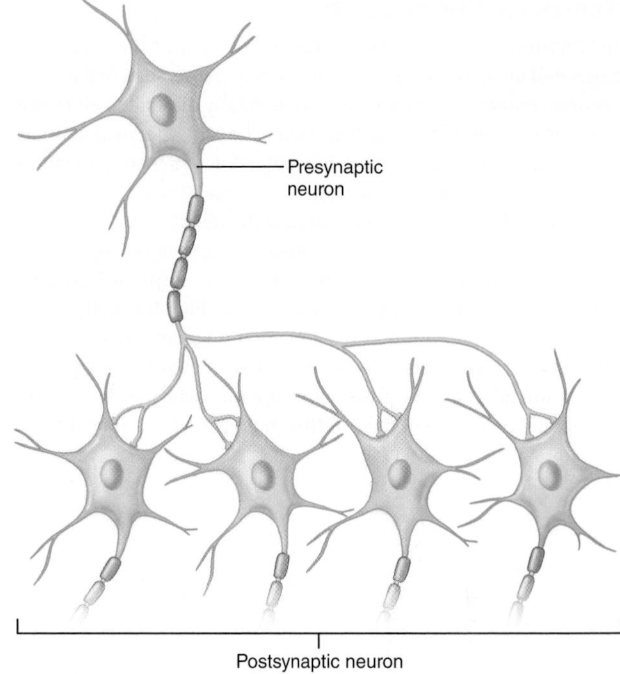

(a) Divergence

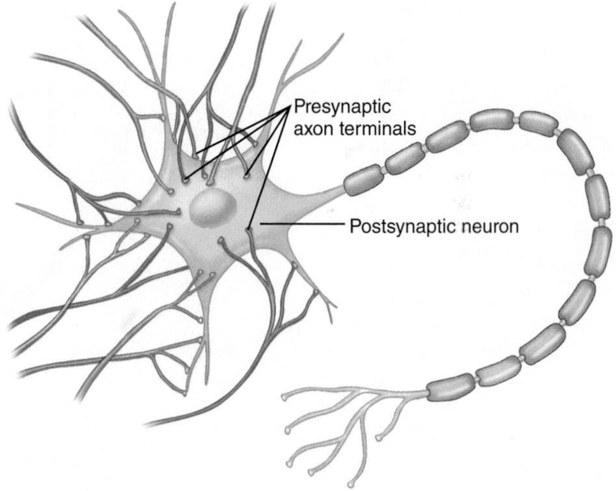

(b) Convergence

Figure 7 Divergence and convergence. (a) In divergence, a single neuron communicates to several other neurons. **(b)** In convergence, a single neuron receives communication from several other neurons.

Summation

Because EPSPs and IPSPs are graded potentials, they undergo the process of summation. This summation can be either *spatial* or *temporal*, depending on whether the postsynaptic potentials being summed arise at the same synapse or at different synapses, respectively.

Temporal Summation

In temporal summation, two or more postsynaptic potentials are generated in rapid succession at the same synapse, such that a postsynaptic potential does not have time to fully dissipate before the next one is generated. When this occurs, the postsynaptic potentials overlap such that the resulting depolarization or hyperpolarization is larger than for a single postsynaptic potential.

Figure 8a depicts measurement of the membrane potential at the axon hillock of a postsynaptic neuron that receives input from three presynaptic neurons (A, B, and C). The graph in Figure 8b shows the events in temporal summation. For simplicity, consider that when neuron A fires a single action potential, it releases enough neurotransmitter to cause an EPSP of approximately 8 mV at the axon hillock in the postsynaptic neuron, which is not enough to depolarize the membrane to threshold (in reality, most EPSPs are less than 1 mV in magnitude). Once this EPSP has terminated and the membrane potential has returned to the resting value, a second action potential in neuron A triggers another EPSP of the same size. However, when two action potentials in neuron A occur closer together in time (within a few milliseconds) such that the second EPSP is produced before the first has had a chance to dissipate, the two EPSPs sum. In this case, the sum of the two EPSPs is large enough to depolarize the postsynaptic neuron's membrane to threshold and generate an action potential.

Note that IPSPs can also sum temporally in a similar manner. When two IPSPs generated at an inhibitory synapse occur close together in time, they sum such that the resulting hyperpolarization is larger than that produced by a single IPSP. The degree of hyperpolarization that can be reached, however, is limited by the equilibrium potential for the ion causing the hyperpolarization.

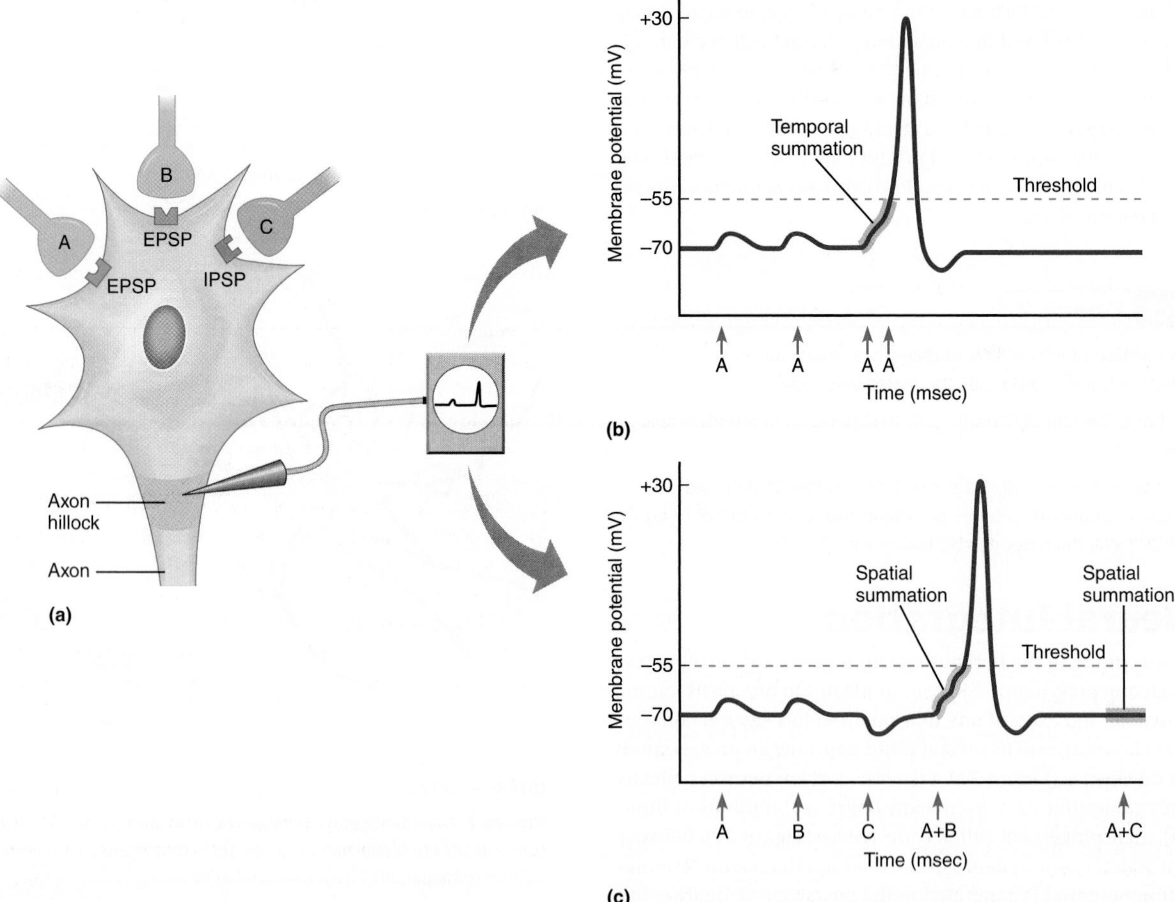

Figure 8 Temporal and spatial summation. (a) A postsynaptic neuron receiving excitatory input from neurons A and B, and inhibitory input from neuron C; three different neurotransmitters are involved. If the sum of all synaptic potentials at the axon hillock results in depolarization to threshold, then an action potential is generated. **(b)** Temporal summation occurs when action potentials arriving from a presynaptic axon terminal of neuron A occur close enough together in time such that the EPSPs produced in response to the binding of neurotransmitter overlap and sum. **(c)** Spatial summation occurs when different synapses are simultaneously active. If EPSPs from synapses with neurons A and B occur at the same time, the EPSPs sum and reach threshold, triggering an action potential. Note that if synapses with neurons A and C were active simultaneously, the resulting IPSP and EPSP would tend to cancel each other out, producing little change in membrane potential.

Q The synapse of neuron A with the postsynaptic cell is closer to the axon hillock than is the synapse of neuron B. If a single action potential in each presynaptic neuron results in an 8-mV depolarization at the axon hillock, which presynaptic neuron had to produce the larger depolarization at the site of the synapse?

When, for example, the opening of potassium channels produces a hyperpolarization, the more potassium channels that are open, the greater the hyperpolarization will be, until the membrane potential approaches −94 mV (the potassium equilibrium potential). The membrane potential can never exceed this value; in fact, it will never reach −94 mV because even when all potassium channels in a membrane are open, sodium leak channels will allow some sodium to leak into the cell, counteracting the movement of potassium.

Temporal summation can occur because postsynaptic potentials last considerably longer than action potentials. This extended duration makes it possible for a presynaptic neuron to fire a second action potential, thereby releasing a second burst of neurotransmitter molecules, while neurotransmitter molecules from the first action potential are still in the synaptic cleft. The higher concentration of neurotransmitter in the synaptic cleft opens more ion channels, producing a greater postsynaptic potential. In contrast, when a presynaptic neuron fires two action potentials farther apart in time, the neurotransmitter released as a result of the first action potential is completely cleared from the synaptic cleft before the next action potential arrives. In this situation, the two postsynaptic potentials cannot sum. (Note, however, that when metabotropic [slow response] receptors are involved, a postsynaptic potential may persist even after the neurotransmitter that initiated it has been cleared from the synaptic cleft.)

Spatial Summation

In spatial summation (Figure 8c), two or more postsynaptic potentials originating from different synapses are generated at approximately the same time, such that when they spread to the axon hillock, they overlap and sum. When neurons A and B have action potentials at different times, each neuron induces an EPSP in the postsynaptic neuron, neither of which is large enough on its own to trigger an action potential. As earlier, for simplicity assume that for a single action potential in neuron A or B, enough neurotransmitter is released to produce an 8-mV depolarization at the axon hillock of the postsynaptic neuron, and an action potential in neuron C releases enough neurotransmitter to cause an 8-mV hyperpolarization. When neurons A and B are activated at the same time, the resulting EPSPs sum to produce a depolarization that *is* large enough to trigger an action potential. Moreover, if neurons A and C were active at the same time, no change in the membrane potential of the postsynaptic neuron would result, because the EPSP that originated from synapse A and the IPSP that originated from synapse C would cancel each other out. Thus spatial summation occurs as postsynaptic potentials originating at different synapses spread to the axon hillock, overlapping along the way.

Figure 8 depicts a highly simplified example of the concept of neural integration. First, the number of synapses with a postsynaptic neuron can reach the hundreds or thousands. Second, because some synapses are excitatory and others are inhibitory, the number of possible combinations of active synaptic inputs is astronomical. Temporal and spatial summation of postsynaptic potentials occurs continually, and the magnitude of the summed potential at the axon hillock determines if and when action potentials occur.

Frequency Coding

Earlier we saw that once a depolarizing stimulus exceeds threshold, the degree of depolarization does not affect the size of action potentials, but rather their frequency. Increases in the strength of such suprathreshold stimuli cause the frequency of action potentials to increase, an effect called **frequency coding.** Because the summation of postsynaptic potentials affects the degree of depolarization at the axon hillock, it follows that summation influences the frequency of action potentials in the postsynaptic neuron. When action potentials occur at a higher frequency, more neurotransmitter is released from the neuron. Therefore, a higher frequency of action potentials corresponds to stronger communication to the next neuron in a neural pathway.

The excitatory and inhibitory synapses described previously occur primarily at axodendritic and axosomatic synapses. We now describe what happens at axoaxonic synapses.

Quick Check 3

1. Compare spatial summation and temporal summation.
2. Explain what is meant when the axon hillock is described as an *integrator*.

4 Presynaptic Modulation

Axoaxonic synapses function as *modulatory synapses*—that is, synapses that regulate the communication across another synapse. Recall that axoaxonic synapses can form between the axon terminal of one neuron and the axon terminal of another neuron (see Figure 1). In axoaxonic synapses, neurotransmitter from the presynaptic neuron (also called a *modulating neuron*) does not generate electrical signals in the postsynaptic neuron. Instead, by binding to receptors on the membrane of the axon terminal of the postsynaptic neuron, the neurotransmitter induces a change in the amount of calcium that enters the axon terminal in response to an action potential, in turn altering the amount of neurotransmitter released from the postsynaptic neuron. In some cases the release of neurotransmitter is enhanced, a phenomenon called **presynaptic facilitation;** in other cases the release of neurotransmitter is decreased, a phenomenon called **presynaptic inhibition.** We consider these phenomena next.

Presynaptic Facilitation

Presynaptic facilitation is illustrated in **Figure 9a**. In the figure, neurons C and D are presynaptic neurons that form excitatory synapses with postsynaptic neuron X; neuron E is a modulating neuron that forms a synapse on the axon terminal of neuron C. In this example, an action potential in neuron C triggers the release of enough neurotransmitter to produce a 10-mV EPSP in neuron X when E is not active. However, when neuron E is active, a single action potential in neuron C will release enough neurotransmitter to produce a 15-mV EPSP and trigger an action potential. Thus the activity of neuron E increases the amount of calcium that enters the cytosol of the axon terminal of neuron C when C is active, thereby increasing the amount of neurotransmitter released from neuron C. As a result, neuron C triggers a larger EPSP in neuron X, compared to when neuron E is inactive. Note that activity in neuron

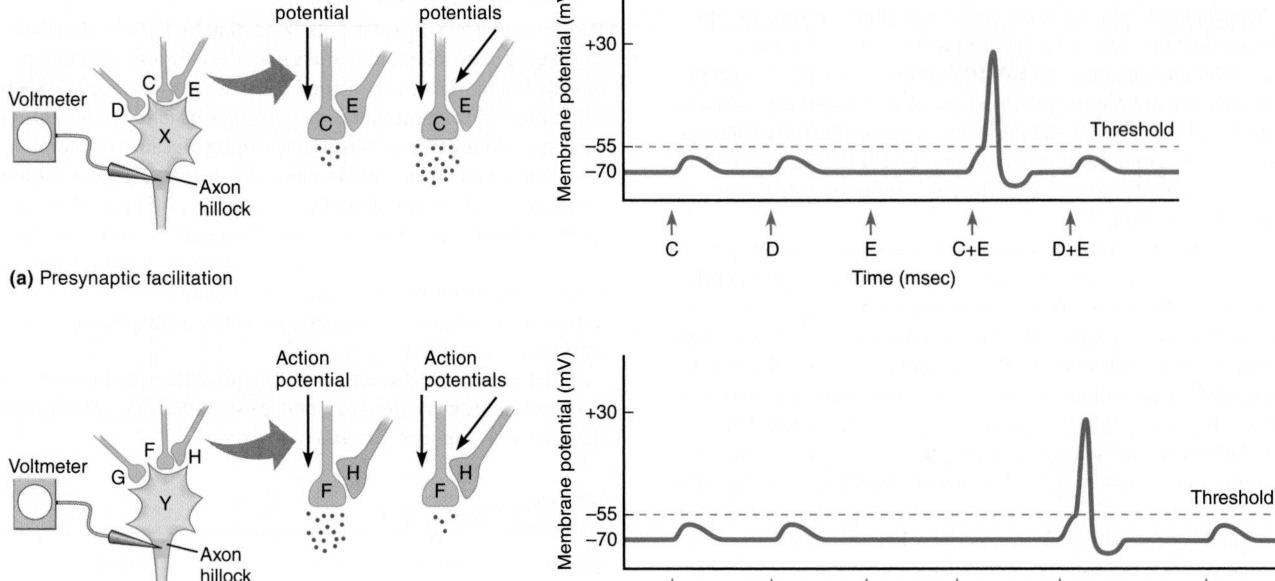

Figure 9 Presynaptic modulation at axoaxonic synapses. (a) Presynaptic facilitation. When neurons C and E are both active, neurotransmitter from neuron E enhances the release of neurotransmitter from neuron C, increasing the strength of the resulting EPSP to threshold and generating an action potential in neuron X. **(b)** Presynaptic inhibition. When neurons F and H are both active, neurotransmitter from neuron H decreases the release of neurotransmitter from neuron F, decreasing the strength of the EPSP in neuron Y. When neurons F and G are both active, their EPSPs sum spatially, and neuron Y is depolarized to threshold, generating an action potential. However, when neurons F, G, and H are all activated, the EPSP from neuron F is diminished due to presynaptic inhibition by neuron H, such that the sum of the EPSPs from neurons F and G is subthreshold, and an action potential is not generated.

E alone has no direct effect on neuron X because neuron E does not form a synapse with neuron X. Note as well that activity in neuron E has no effect on any EPSPs triggered by input from neuron D.

This example illustrates an important characteristic of presynaptic modulation at axoaxonic synapses that differentiates it from axodendritic and axosomatic synapses: Presynaptic modulation affects transmission to the postsynaptic neuron at one specific synapse, thereby altering the ability of that one synapse to excite or inhibit the postsynaptic neuron. In contrast, at axodendritic and axosomatic synapses, presynaptic neurons nonselectively excite or inhibit the postsynaptic neuron by directly producing changes in the membrane potential of the postsynaptic neuron in the form of EPSPs and IPSPs.

Presynaptic Inhibition

Figure 9b shows an example of presynaptic inhibition. In this example, neurons F and G make direct excitatory synaptic connections with neuron Y; neuron H is a modulating neuron that forms an axoaxonic synapse with neuron F. When either neuron F alone or neuron G alone is active, an EPSP is produced in neuron Y. When neuron H is active by itself, no change in the membrane potential of neuron Y occurs. However, when neurons F and H are active at the same time, the response of neuron Y to input from neuron F is decreased; that is, neuron H decreases the amount of calcium that enters the cytosol of the axon terminal of neuron F, thereby suppressing the release of neurotransmitter from neuron F. In contrast, activity in neuron H does not affect the response of neuron Y to input from neuron G.

Apply Your Knowledge

Enkephalins are morphinelike substances produced in the body. They presynaptically inhibit the release of neurotransmitter from neurons that convey signals from pain receptors (called nociceptors) in the skin to the spinal cord. Based on this information, explain what effect, if any, enkephalins would have on the perception of pain, which occurs at the level of the brain.

Note that although Figure 9 illustrates presynaptic facilitation and presynaptic inhibition at excitatory synapses, these phenomena can occur at inhibitory synapses as well.

Quick Check 4

1. At which type of synapse does presynaptic modulation occur?
2. During presynaptic inhibition, does the amount of neurotransmitter released increase or decrease?

5 Neurotransmitters: Structure, Synthesis, and Degradation

Neurotransmitters belong to a variety of chemical classes, including acetylcholine, biogenic amines, amino acids, purines, and neuropeptides. Most are small molecules, with the notable exception of the neuropeptides. A few other neurotransmitters, such as *nitric oxide,* are unlike any others and are grouped together in their own class. Selected members of the different classes of neurotransmitters discussed in the following sections are listed in Table 1.

Acetylcholine

Acetylcholine (ACh) is released from neurons in both the central and peripheral nervous systems. The most abundant neurotransmitter in the peripheral nervous system, it is found in efferent neurons of both the somatic and autonomic branches.

Acetylcholine is synthesized in the cytosol of the axon terminals of neurons (**Figure 10**). The synthesis of acetylcholine from two substrates, *acetyl CoA* and *choline,* is catalyzed by **choline acetyl transferase (CAT),** as follows:

$$\text{acetyl CoA} + \text{choline} \xrightarrow{\text{CAT}} \text{acetylcholine} + \text{CoA}$$

We saw that one of these substrates, acetyl CoA, is a two-carbon molecule produced during the catabolism of lipids, carbohydrates, and proteins, and it is the initial substrate for the Krebs cycle. Therefore, acetyl CoA is found in almost all cells of the body, including those neurons that synthesize and release acetylcholine (called *cholinergic* neurons). In contrast, choline cannot be synthesized by neurons (although some is synthesized in the liver); instead, most choline is obtained from the diet and delivered by the bloodstream to cholinergic neurons, where it is taken in via an active transport system.

Once synthesized, acetylcholine is transported into and stored in synaptic vesicles (see Figure 10) until an action potential triggers its release by exocytosis. After its release, acetylcholine can bind to receptors on the postsynaptic cell (called **cholinergic** receptors) and/or be degraded by an enzyme called **acetylcholinesterase (AChE),** which is found on either the presynaptic neuron membrane or the postsynaptic neuron membrane, or both. Acetylcholinesterase catalyzes the breakdown of acetylcholine into acetate (an acetyl CoA molecule without the CoA attached) and choline, as follows:

$$\text{acetylcholine} \xrightarrow{\text{AChE}} \text{acetate} + \text{choline}$$

Neither of the breakdown products of this reaction has any transmitter activity, but the choline that is produced in this way is taken back into the presynaptic cell by an active transport mechanism and can be used to synthesize more acetylcholine. The acetate diffuses away from the synapse and enters the bloodstream.

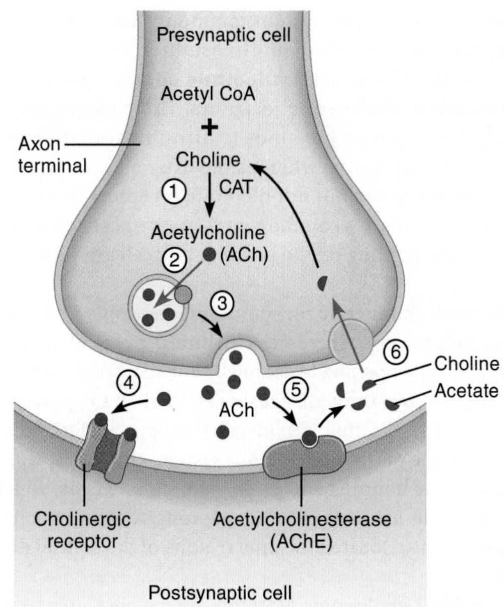

Figure 10 Neurotransmitter synthesis, action, and degradation at cholinergic synapses. The sequence of events at a cholinergic synapse. See the text for details on the steps.

Q What triggers the release of acetylcholine by exocytosis?

TABLE 1 Selected Members of the Classes of Neurotransmitters

Choline derivative	Biogenic amines	Amino acids	Purines	Neuropeptides	Unique molecules
Acetylcholine	Catecholamines	Glutamate	ATP	TRH	Nitric oxide
	Dopamine	Aspartate	ADP	Vasopressin	Endocannabinoids
	Epinephrine	Glycine	Adenosine	Oxytocin	
	Norepinephrine	GABA		Substance P	
	Serotonin			Cholecystokinin	
	Histamine			Endogenous opioids	
				Enkephalins	
				Endorphins	
				Orexin	

Cells in the nervous system commonly have different types of receptors for a specific neurotransmitter. Receptors for acetylcholine are of two types: **nicotinic cholinergic receptors** and **muscarinic cholinergic receptors.** These names reflect the fact that the drug nicotine binds to nicotinic receptors but not to muscarinic receptors, whereas the drug muscarine binds to muscarinic receptors but not nicotinic receptors. Even though these receptors share a common messenger—acetylcholine—the effects of their binding on the postsynaptic cell are very different (**Figure 11**).

Nicotinic cholinergic receptors are ionotropic and have two binding sites for acetylcholine. The binding of acetylcholine to them triggers the opening of channels that allow both sodium and potassium to move through, causing an EPSP in the postsynaptic cell (Figure 11a). Nicotinic cholinergic receptors are located in several areas of the peripheral nervous system—for example, in certain autonomic neurons and on skeletal muscle cells, which are the effector cells for somatic motor neurons. Nicotinic cholinergic receptors are also located in some regions of the central nervous system.

Muscarinic cholinergic receptors are metabotropic receptors that operate through the action of a G protein. Because G proteins can exert a wide range of effects in the postsynaptic cell, the effects of acetylcholine binding to muscarinic cholinergic receptors include opening or closing of ion channels and activation of enzymes (Figure 11b). Muscarinic cholinergic receptors are found on some effector organs for the autonomic nervous system and are the dominant cholinergic receptor type found in the central nervous system.

The different effects of acetylcholine at nicotinic and muscarinic cholinergic receptors illustrate an important physiological concept: *The action of any chemical messenger ultimately depends not on the nature of the messenger, but rather on the signal transduction mechanism activated by the receptor once the messenger binds to it.*

Biogenic Amines

The **biogenic amines** are a class of neurotransmitters derived from amino acids; they are called *amines* because they all possess an amine group ($—NH_2$). (Note that this class does not include amino acids, which also possess amine groups.) The biogenic amines include the catecholamines, serotonin, and histamine. The catecholamines contain a *catechol group*, a six-carbon ring with two hydroxyl groups, and include the neurotransmitters dopamine, norepinephrine, and epinephrine. Following or during synthesis, the biogenic amines are packaged into synaptic vesicles.

Dopamine and norepinephrine are both released primarily by neurons in the central nervous system, but norepinephrine is also released from neurons in the peripheral nervous system. Even though epinephrine is released from some neurons of the central nervous system, it is better known as a hormone that is released from the adrenal medulla in response to commands from the sympathetic nervous system.

The different catecholamines bind to specific receptors. Receptors for epinephrine and norepinephrine are called **adrenergic** receptors, of which there are two main classes: *alpha adrenergic* and *beta adrenergic* receptors. Each of these classes have subclasses designated by numerical subscripts, with the primary receptor types being alpha$_1$, alpha$_2$, beta$_1$, beta$_2$, and beta$_3$. Epinephrine has the highest affinity for beta$_2$ receptors, whereas norepinephrine has higher affinity for alpha and beta$_1$, receptors. Nevertheless, both compounds bind to all of these receptors. Adrenergic receptors are found in the central nervous system and in the effector organs for the sympathetic branch of the autonomic nervous system. These receptors are referred to as being *adrenergic* because alternate terms for epinephrine and norepinephrine are *adrenaline* and *noradrenaline*, respectively. In similar fashion, receptors that bind dopamine are called *dopaminergic*.

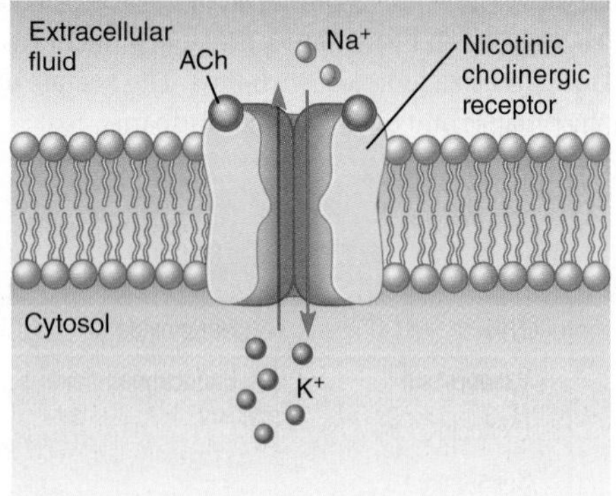

(a) Nicotinic cholinergic receptors **(b)** Muscarinic cholinergic receptor

Figure 11 Signal transduction mechanisms at cholinergic receptors. (a) Nicotinic cholinergic receptors are receptor-operated channels that permit both sodium and potassium ions to move through the cell membrane. When acetylcholine binds to these receptors, the channels open, resulting in an EPSP. **(b)** Muscarinic cholinergic receptors are coupled to G proteins that can either directly open/close ion channels or activate/inhibit an enzyme that catalyzes the production of a second messenger. The second messengers produced can have a variety of effects on the postsynaptic cell, including opening or closing ion channels.

Q Which cholinergic receptor type (nicotinic or muscarinic) produces the faster response?

Catecholamines generally produce slow responses mediated through G proteins and changes in second messenger systems. In addition, they often function as autocrines, binding to receptors (called **autoreceptors**) on the axon terminal of the cell that released them. These receptors enable a neuron to modulate its own release of neurotransmitter, generally by altering the amount of calcium that enters the cell in response to an action potential. Neurons possess autoreceptors for other neurotransmitters as well.

Following their release, catecholamines can be degraded by two enzymes, **monoamine oxidase (MAO)** and **catechol-*O*-methyltransferase (COMT).** MAO is located in the synaptic cleft, in the mitochondria of the axon terminal of the neurons that release the catecholamines, and in some glial cells. COMT is found in the synaptic cleft. To understand the importance of the enzymes that degrade neurotransmitters, see **Clinical Connections: Treating Depression**.

Serotonin and histamine are biogenic amines, but they are not catecholamines. Serotonin is found in the central nervous system, particularly the lower portion of the brain called the *brainstem.* Functions of serotonin include regulating sleep and emotions. Histamine, which is better known for its release from non-neuronal cells during allergic and other types of reactions, can also function as a neurotransmitter. It is found in the central nervous system, primarily in the hypothalamus.

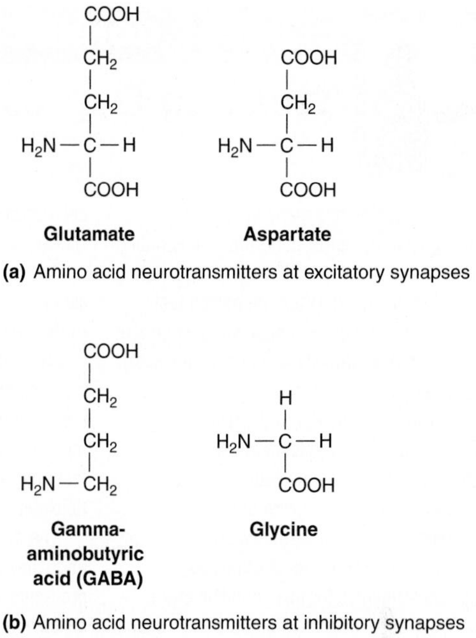

Amino Acid Neurotransmitters

The amino acid neurotransmitters are the most abundant class of neurotransmitters in the central nervous system, where they are widely distributed throughout virtually all areas. **Aspartate** and **glutamate** are released at excitatory synapses, whereas **glycine** and *gamma-aminobutyric acid,* more commonly known as **GABA,** are released at inhibitory synapses (**Figure 12**).

Glutamate is the most commonly released neurotransmitter at excitatory synapses in the central nervous system. It can bind to three main classes of receptors: α-amino-3-hydroxy-5-methyl-4-isoxazole propionate (AMPA) receptors, *N*-methyl-D-aspartate (NMDA) receptors, and kainate receptors. When glutamate binds to AMPA or kainate receptors, a fast EPSP is produced by sodium movement into the cell. When glutamate binds to NMDA receptors, calcium channels open, resulting in an influx of calcium. Calcium acts as a second messenger, producing biochemical changes in the postsynaptic neuron.

GABA is the neurotransmitter most commonly released at inhibitory synapses in the central nervous system. To date, three categories of GABA receptors have been identified, including two ionotropic receptors ($GABA_A$, and $GABA_C$) and one metabotropic receptor ($GABA_B$). The binding of GABA to $GABA_A$ receptors opens chloride channels, producing an IPSP on the postsynaptic cell membrane. $GABA_B$ receptors are found in several locations, including postsynaptic dendrites and somas, where slow IPSPs are produced; presynaptic axon terminal membranes, where they function as autoreceptors; and near axon terminals, where they function in presynaptic inhibition. $GABA_C$ receptors are located in the retina, where they play a role in communication of visual information. Sedatives, such as Valium, bind to $GABA_A$ receptors in the CNS, thereby depressing neural activity (see **Clinical Connections: The Role of GABAergic Agents in Anxiety and Sleep Disorders**).

COOH–CH_2–CH_2–C(H)(H_2N)–COOH

Glutamate

COOH–CH_2–C(H)(H_2N)–COOH

Aspartate

(a) Amino acid neurotransmitters at excitatory synapses

COOH–CH_2–CH_2–CH_2–H_2N

Gamma-aminobutyric acid (GABA)

H–C(H)(H_2N)–COOH

Glycine

(b) Amino acid neurotransmitters at inhibitory synapses

Figure 12 The amino acid neurotransmitters. (a) Amino acid neurotransmitters released at excitatory synapses. **(b)** Amino acid neurotransmitters released at inhibitory synapses.

Purines

ATP is the primary energy source for cellular processes; it is synthesized from ADP and later catabolized to ADP when energy is needed. Thus all of the cells of the body require ATP. When receptors for ATP (and other purines) were first discovered in the brain, it was difficult to fathom that ATP could function as a neurotransmitter. Today, its role as a neurotransmitter is widely accepted, along with the roles of GTP, ADP, and AMP, all of which are stored in synaptic vesicles and released by exocytosis. Adenosine is also considered a neurotransmitter, albeit not in the classical sense. Adenosine is not stored in synaptic vesicles, but rather is generated from ATP released into the extracellular fluid by enzymes located there.

ATP seems to be a major neurotransmitter in the enteric nervous system of the gastrointestinal system. However, it also functions as a neurotransmitter in the peripheral and central nervous systems, where several types of purinergic receptors have been identified. The two primary types of purinergic receptors are P2X and P2Y. P2X is an ionotropic receptor whose channel allows cations—especially sodium and calcium ions—to move through the cell membrane. Both sodium and calcium enter the cell through these channels, depolarizing the cell. Thus communication at P2X receptors is excitatory. Neurotransmitters that can activate the P2X receptors include ATP and ADP. P2Y is a metabotropic receptor coupled to G proteins. Adenosine, ADP, and ATP are all neurotransmitters that can bind to the P2Y receptor. In addition to the purinergic receptors, certain receptors bind only adenosine, the A_1 and A_2 receptors.

CLINICAL CONNECTIONS

Treating Depression 

Clinical depression has many symptoms, including lack of energy, abnormal eating habits (too much or too little), and/or difficulty sleeping or sleeping too much. Often the person feels worthless, may be preoccupied with thoughts of suicide, and has difficulty functioning in society. The cause(s) of depression and its symptoms are not well understood, but depression is an illness associated with biochemical changes in the brain. Much evidence suggests that depression is associated with deficiencies in the biogenic amines serotonin and norepinephrine. Therefore, pharmacological treatment strategies for this condition often seek to increase biogenic amine concentrations in the brain.

One class of antidepressants is *monoamine oxidase inhibitors.* Monoamine oxidase is the enzyme that breaks down biogenic amines, including norepinephrine and serotonin. Because these antidepressants inhibit their degradation, these neurotransmitters remain in the synaptic cleft longer; the effect is similar to an increased release of these neurotransmitters. Monoamine oxidase inhibitors, including phenelzine (brand name: Nardil) and isocarboxazid (Marplan), have been used successfully to treat many cases of clinical depression.

Another class of commonly used antidepressants is *selective serotonin reuptake inhibitors (SSRIs).* By decreasing the reuptake of serotonin into the cell that released it, SSRIs selectively increase the amount of serotonin present at the synaptic cleft. SSRIs are more specific than monoamine oxidase inhibitors because they affect only serotonergic synapses. SSRIs include fluoxetine (Prozac) and paroxetine (Paxil).

The *tricyclics,* named for the three carbon rings they contain, are the oldest of the antidepressants. First used in the 1950s, their mechanisms of action are the least understood. Several hypotheses related to these mechanisms have been proposed, most of which involve alterations of activity at adrenergic or serotonergic synapses. Tricyclic antidepressants include imipramine (Tofranil), amitriptyline (Elavil), and desipramine (Norpramin).

The therapeutic benefits of monoamine oxidase inhibitors and SSRIs for depression are similarly not fully understood. Even though these drugs immediately decrease monoamine oxidase activity or serotonin reuptake, respectively, they do not affect the depression until they have been ingested for several weeks. Much research remains to be done on the mechanisms of action of antidepressants and other medications.

C.I. Aguera/Corbis

Critical Thinking Questions

1. Describe the ways in which monoamine oxidase inhibitors treat or prevent depression.
2. How do selective serotonin inhibitors treat or prevent depression?
3. How would you characterize the limitations of the current treatments for depression? What aspects of treating or preventing this condition seem the most promising?

Once released into the synaptic cleft, purine neurotransmitters can be degraded by enzymes. Nucleotidases break down ADP and ATP, whereas adenosine deaminase breaks down adenosine.

Neuropeptides

The **neuropeptides** are short chains of amino acids that are synthesized in the same manner as proteins. More than 50 of these compounds have been found in neurons, where their function as neurotransmitters is either well established or highly probable. Like other peptides or proteins that are to be secreted from cells, neuropeptides are synthesized in the rough endoplasmic reticulum and packaged into secretory vesicles by the Golgi apparatus. In neurons, these events take place in the cell body—the only part of the cell that contains the necessary machinery. After this synthesis and packaging, secretory vesicles are slowly transported down the axon to the axon terminal, where they are stored. Many of the neuropeptides are more classically known as hormones, including **TRH,** which regulates release of another hormone called TSH; **vasopressin,** which regulates urine output by the kidney; **oxytocin,** which regulates contractions of the uterus and the flow of milk from the breasts; **substance P,** which decreases gastrointestinal motility, and **cholecystokinin,** which regulates gallbladder contraction. Also included among the neuropeptides are the **endogenous opioids,** which exert effects similar to those produced by the drug morphine. The endogenous opioids include the **enkephalins** and **endorphins.**

A more recently discovered neuropeptide is **orexin,** a hypothalamic neurotransmitter that regulates the sleep-wake cycle.

CLINICAL CONNECTIONS

The Role of GABAergic Agents in Anxiety and Sleep Disorders

GABA is the most common inhibitory neurotransmitter in the central nervous system. As such, agents that enhance the activity at GABA synapses tend to depress central nervous system (CNS) activity. Drugs that act on GABAergic systems include benzodiazepines (such as Valium), sleeping aids (such as zolpidem), and alcohol. These drugs interact with $GABA_A$ receptors, which are ionotropic receptors with chloride channels. When GABA binds to the receptor, the chloride channels allow chloride to enter the cell.

Benzodiazepines are used to treat anxiety. Everyone becomes anxious at one time or another, but some people suffer from chronic anxiety or from anxiety attacks. Anxiety occurs in response to a stressful or fearful situation that activates the sympathetic nervous system. This branch of the autonomic nervous system is responsible for the fight-or-flight response, which is characterized by increases in heart rate and strength of heart contraction, respiration, perspiration, and mobilization of energy stores. The fight-or-flight response helps prepare the body for handling the stressor by increasing blood flow and nutrient delivery to cells, especially the heart and skeletal muscle cells. Although it has allowed humans to survive dangerous situations, for some people this response occurs in the face of relatively minor events or even no event at all.

At low doses, benzodiazepines decrease anxiety by allosterically modulating a class of $GABA_A$ receptors called BZ receptors and by increasing the receptor's affinity for GABA. By enhancing GABA transmission, benzodiazepines produce a general inhibitory effect on the central nervous system. At high doses, however, these agents act as stimulants by inhibiting inhibitory circuits in the central nervous system.

Certain sleep aids, such as zolpidem, act similarly to benzodiazepines but bind preferentially to a more limited subclass of $GABA_A$ receptors (BZ-1). Zolpidem has a sedative hypnotic effect, but not the antianxiety effects of benzodiazepines. Both benzodiazepines and zolpidem are potentially addictive, so they should be taken with caution.

Alcohol depresses the central nervous system through several mechanisms, one of which involves $GABA_A$ receptors. Similar to benzodiazepines, alcohol enhances the actions of GABA at $GABA_A$ receptors, although alcohol's actions appear to take place at the cell membrane and not the actual receptor. Because alcohol, benzodiazepines, and zolpidem interact at the same receptors, taking them together can be dangerous: The response in such a case is synergistic (that is, the response of any two drugs together is greater than the response of the sum of each drug alone).

Alcohol depresses the CNS.

Critical Thinking Questions

1. What changes in the body are experienced during a flight-or-fight response?
2. How do low doses of benzodiazepines decrease anxiety?
3. Compare and contrast the differences between alcohol and sleep aids, such as zolpidem, in terms of their effects on the body.

Specifically, orexin induces arousal or wakefulness. Pharmaceutical companies are currently studying synthetic orexin's potential as a treatment for narcolepsy, a disease in which a person rapidly, and often unexpectedly, falls asleep. Other studies suggest orexin could be used to prevent sleep, much as caffeine is used today.

Most neuropeptides are colocalized with other small neurotransmitters in the same neurons; that is, they are released from the same axon terminal. However, neuropeptides are packaged in larger vesicles called **dense core vesicles.** Although they are released from the same axon terminal, neuropeptide release requires a higher frequency of action potentials than does release of its colocalized small neurotransmitter. Neuropeptides often act on metabotropic receptors and modulate the response of the postsynaptic neuron to the colocalized small neurotransmitter.

Unique Neurotransmitters

Several chemical substances that have been identified in neurons appear to function as neurotransmitters, even though they are neither stored in synaptic vesicles nor released by exocytosis. Two such chemicals are the gas nitric oxide and endocannabinoids.

Nitric oxide has a particularly simple structure, one nitrogen atom and one oxygen atom. This molecule does not fit the general pattern of neurotransmitter function, in that it is not synthesized and stored in synaptic vesicles in advance of its release; instead, nitric oxide is released as soon as it is synthesized because it easily crosses the plasma membrane. Thus nitric oxide release is controlled via regulation of the rate of synthesis in a reaction catalyzed by the enzyme **nitric oxide synthetase.** Following its release, nitric oxide diffuses to target cells, which are not necessarily connected to the secretory cell at a synapse. Once nitric oxide reaches the target cell, it enters the cell by diffusion and alters the activity of proteins. Because it degrades quickly on its own, without the assistance of enzymes, its life span is only several seconds.

Nitric oxide may not be the only gas that functions as a neurotransmitter. Recent evidence suggests that carbon monoxide may also be a neurotransmitter.

Endocannabinoids are a family of chemicals that are manufactured in neurons from membrane phospholipids; they include anandamide (AEA) and 2-arachidonylglycerol (2-AG). Their production is stimulated by an increase in cytosolic calcium levels. Once synthesized, endocannabinoids diffuse out of the cell and to receptors on the target cell. One class of endocannabinoid receptors identified in the central nervous system comprises the CB1 receptors, which are metabotropic and the most common type of cannabinoid receptor in the brain. One drug that targets the CB1 receptor is tetrahydrocannabinol (THC)—the active component of *cannabis*, or marijuana. While the function of endocannabinoids is not known, some states now permit the use of THC as a medical treatment for neurological disorders including Parkinson's disease, anxiety disorders, and post-traumatic stress disorder.

Quick Check 5

1. Explain how it is possible for a given neurotransmitter to inhibit one cell and excite another.
2. Name the different classes of neurotransmitters. Which neurotransmitters are dominant in the peripheral nervous system?
3. Most small neurotransmitters are synthesized and stored in the axon terminal before release. How do neuropeptides deviate from this pattern? How do adenosine and endocannabinoids deviate from it?
4. Unlike other neurotransmitters, nitric oxide does not bind to receptors on the surface of a postsynaptic cell. How does it exert its effects?

SYSTEMS INTEGRATION Electrical signals can be in the form of graded potentials and action potentials. In this chapter, we learned how these electrical signals are used for intercellular communication, and through the processes of convergence and divergence, how electrical information becomes integrated into useful signals. Neurons use electrical signals to inform us of both the external environment and the internal environment (detailed in chapters on specific organ systems). Neurons also use electrical signals to cause muscles to contract and glands to secrete product. Thus, electrical signaling is critical for every organ system to function properly in maintaining homeostasis. This is readily apparent with electrolyte imbalances, which can cause a number of malfunctions including muscle spasms and cardiac arrhythmias.

CHAPTER REVIEW

SUMMARY

1 Electrical Synapses

- Neurons communicate with other neurons or with effector organs at synapses.
- Neurons communicate via both electrical synapses and chemical synapses.
- Electrical synapses exist where gap junctions connect two neurons.

iP Nervous I, Anatomy Review

2 Chemical Synapses

- At a chemical synapse, the transmission of a message from a presynaptic neuron to a postsynaptic neuron occurs in response to an action potential in the presynaptic neuron.
- The action potential is propagated from the trigger zone to the axon terminal, where it stimulates the opening of voltage-gated calcium channels.
- The release of neurotransmitter by exocytosis is triggered by cytosolic calcium.
- The neurotransmitter binds to receptors on the postsynaptic neuron, triggering a response in the postsynaptic neuron, usually a change in the electrical properties of the cell.
- Signal transduction refers to the mechanism by which a messenger produces a response in a cell.
- At ionotropic (fast response) receptors, the neurotransmitter causes a rapid opening of ion channels, producing an immediate change in the electrical properties of the cell.
- At metabotropic (slow response) receptors, the neurotransmitter activates a G protein, which either opens or closes ion channels to change the electrical properties of the cell, or, alternatively, activates an enzyme that produces a second messenger, which may then open or close ion channels or produce some other response in the cell.
- At excitatory synapses, the membrane potential of the postsynaptic neuron is depolarized (EPSP), bringing it closer to threshold to generate an action potential.
- At inhibitory synapses, the membrane potential of the postsynaptic neuron is either hyperpolarized (IPSP) or stabilized, making the neuron less likely to fire an action potential.

iP Nervous II, Anatomy Review

MAP iP Nervous II, Synaptic Transmission

iP Nervous II, Synaptic Potentials and Cellular Integration

3 Neural Integration

- Neural integration is the spatial and temporal summation of synaptic potentials at the axon hillock of a postsynaptic neuron.
- An action potential will be generated if the axon hillock is depolarized to threshold, unless the potential is below threshold, in which case no action potential will occur.
- Once depolarization reaches threshold, greater depolarizations will elicit a higher frequency of action potentials.

iP Nervous II, Synaptic Potentials and Cellular Integration

4 Presynaptic Modulation

- The most common type of modulation occurs at axoaxonic synapses and is called presynaptic modulation.
- At the axoaxonic synapse, the presynaptic cell modulates the release of neurotransmitter from the postsynaptic cell.
- In presynaptic facilitation, communication at a specific synapse is enhanced; in presynaptic inhibition, communication at a specific synapse is attenuated.

5 Neurotransmitters: Structure, Synthesis, and Degradation

- A variety of neurotransmitters can be released from neurons.
- Acetylcholine is the most abundant neurotransmitter in the peripheral nervous system; it is also found in the central nervous system.
- The biogenic amines include the catecholamines, serotonin, and histamine.
- The catecholamines include norepinephrine, epinephrine, and dopamine; norepinephrine is a common neurotransmitter in the peripheral nervous system.
- Other classes of neurotransmitters include the amino acid neurotransmitters and neuropeptides.
- Recently discovered neurotransmitters include nitric oxide, ATP, and endocannabinoids.
- Neurotransmitters communicate by binding to receptors specific for them.
- Generally, more than one type of receptor exists for each neurotransmitter, and the response produced in the postsynaptic cell can vary considerably based on the receptor type activated and the coupling mechanism involved.

MAP iP Nervous II, Synaptic Transmission

MAP MasteringA&P®

Check your understanding at MasteringA&P and apply your new knowledge to real-life case studies to ensure that you've mastered the content!

EXERCISES

Multiple-Choice Questions

1. Suppose that the electrochemical force for anion X (X^-) acts to move the anion out of the cell. If a neurotransmitter binding to its receptor opened channels for X^- on the postsynaptic cell, then the response would
   a) Be an EPSP.
   b) Be an IPSP.
   c) Be stabilization of the membrane.
   d) Not occur.
2. Suppose that all the calcium could be removed from the extracellular fluid surrounding a neuron. Such removal would inhibit the ability of a neuron to
   a) Produce action potentials.
   b) Release neurotransmitter.
   c) Respond to the binding of a neurotransmitter to its receptor.
   d) Degrade neurotransmitters.
3. Synaptic vesicles
   a) Store calcium.
   b) Release neurotransmitters by exocytosis.
   c) Degrade neurotransmitters.
   d) Form gap junctions.
   e) Synthesize neurotransmitters.
4. If sodium channels closed in response to a stimulus, then
   a) The neuron would be depolarized.
   b) The neuron would be hyperpolarized.
   c) The membrane potential would be stabilized.
   d) A second messenger would be produced.
   e) The neuron would remain at rest.
5. A fast EPSP is most commonly produced by
   a) The opening of sodium-selective channels.
   b) The opening of potassium-selective channels.
   c) The opening of chloride channels.
   d) The opening of channels selective for both sodium and potassium.
   e) The opening of calcium-selective channels.
6. The enzyme that catalyzes the synthesis of acetylcholine is
   a) Adenylate cyclase.
   b) Choline acetyl transferase.
   c) Monoamine oxidase.
   d) Acetylcholinesterase.
   e) Catechol-*O*-methyltransferase.
7. Which of the following neurotransmitters is a biogenic amine but not a catecholamine?
   a) Norepinephrine
   b) Serotonin
   c) Dopamine
   d) Epinephrine
8. Which of the following is most likely to occur at an axoaxonic synapse?
   a) An EPSP
   b) An IPSP
   c) Stabilization of the membrane potential
   d) Temporal summation
   e) Presynaptic modulation
9. What happens to the concentration of neurotransmitter in the synaptic cleft when the frequency of action potentials increases in the presynaptic neuron?
   a) It increases
   b) It decreases
   c) It remains constant
10. The EPSPs from two different synapses occur at the same time and cause a larger depolarization than either one alone can cause. This is an example of
    a) Membrane stabilization.
    b) Presynaptic inhibition.
    c) Presynaptic facilitation.
    d) Temporal summation.
    e) Spatial summation.

Objective Questions

11. At electrical synapses, which type of junction exists between the two cells?
12. When the opening of ion channels allows both sodium and potassium ions to move through, no change in membrane potential occurs because sodium moves into the cell and potassium moves out of the cell. (true/false)
13. Neurotransmitter receptors are found at (chemical/electrical) synapses.
14. Whether a synapse is excitatory or inhibitory is determined by the mechanism of coupling between the neurotransmitter receptor and ion channels in the postsynaptic cell. (true/false)
15. The synaptic delay includes the time it takes for an action potential to travel from the trigger zone of a presynaptic cell to the axon terminal. (true/false)
16. A given neurotransmitter might be excitatory at one synapse and inhibitory at another synapse. (true/false)

17. Given that release of an inhibitory neurotransmitter is altered by presynaptic facilitation, the response in the postsynaptic cell will be a (larger/smaller) degree of hyperpolarization.

18. The response to a neurotransmitter is faster at (ionotropic/metabotropic) receptors.

19. The enzymes that catalyze the degradation of catecholamines are _______ and _______.

20. Adenylate cyclase catalyzes the formation of _______.

Essay Questions

21. Describe the sequence of events occurring at a chemical synapse, starting with an action potential in the presynaptic cell and ending with a response in the postsynaptic cell.

22. Compare and contrast the events caused by the binding of neurotransmitters to ionotropic and metabotropic receptors.

23. Explain the ionic basis of a fast EPSP.

24. Explain the role of axoaxonic synapses.

25. Describe the steps of the cAMP second messenger system. Explain how cAMP can produce different responses in the many different types of cells that use cAMP as a second messenger.

Critical Thinking

Questions 26–28: The accompanying diagram depicts three neurons in a neural pathway in which neuron 1 synapses on neuron 2, and neuron 2 synapses on neuron 3. Neuron 1 can be artificially activated through a stimulating electrode that passes current across the membrane. The membrane potentials of all three neurons are recorded by an intracellular electrode connected to a voltmeter. The voltage tracings in questions 26 and 28 are of membrane potential (V_{m1} for cell 1, and so on). Action potentials are shown as vertical spikes; the horizontal bar over the voltage tracing indicates the time period during which the stimulating electrode was turned on.

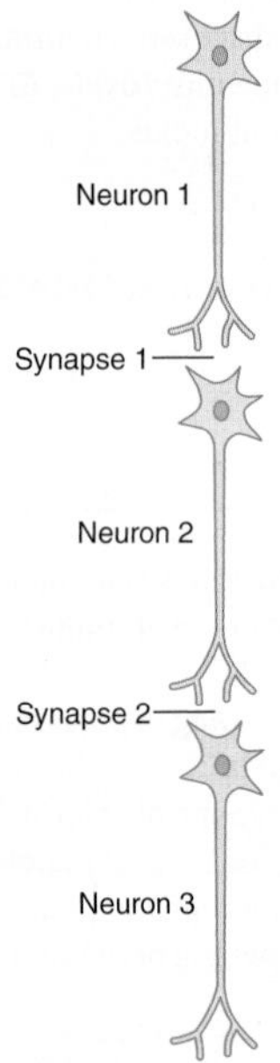

26. From this voltage tracing for cell 1, one can conclude that

a) The stimulation depolarizes the cell membrane.
b) The stimulation hyperpolarizes the cell membrane.
c) The stimulation has no effect on the cell membrane potential.
d) The stimulation mimics the effect of an inhibitory neurotransmitter.

27. If you were to observe an increase in action potential frequency in cell 2 during stimulation of cell 1, you could conclude that
a) Synapse 1 is excitatory.
b) Synapse 1 is inhibitory.
c) The findings are inconclusive.

28. Given the following results, what would your conclusion be?

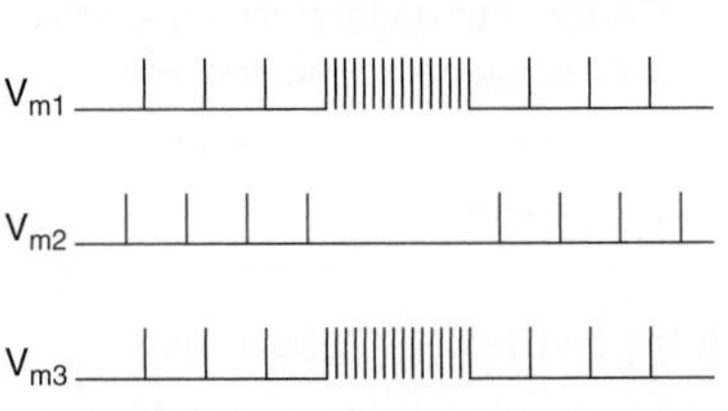

a) Synapse 1 is excitatory, whereas synapse 2 is inhibitory.
b) Synapse 1 is inhibitory, whereas synapse 2 is excitatory.
c) Synapses 1 and 2 are both inhibitory.
d) Synapses 1 and 2 are both excitatory.
e) No definite conclusion is possible.

29. Alzheimer's disease affects memory and cognitive function due to a loss of cholinergic neurons in certain areas of the brain. Propose a treatment strategy based on what you learned in this chapter.

Photo credits for icons in feature boxes: Focus on Diabetes: Tanyae/Dreamstime; Clinical Connections: lofoto/Dreamstime; Understanding Exercise: Pixattitude/Dreamstime; Discovery: Yuri Arcurs/Cutcaster.

Glossary

acetylcholine (ACh) (ass-ih-teel-koh-leen) a neurotransmitter widely employed in both the central and peripheral nervous systems; the most abundant neurotransmitter in the peripheral nervous system, found in efferent neurons of both the somatic and autonomic nervous systems

adrenergic pertaining to epinephrine (adrenaline) or norepinephrine (noradrenaline)

aspartate (ah-spar-tate) amino acid that also functions as a neurotransmitter

autoreceptors receptors on a target cell that bind the same messenger that is released from the secretory cell

catechol-*O*-methyltransferase (COMT) enzyme that degrades catecholamines

channel-linked receptor protein that functions as both an ion channel and a receptor; binding of a messenger to the receptor portion of the protein opens the channels

chemical synapse synapse where a neuron secretes a neurotransmitter into the extracellular fluid in response to an action potential arriving at the axon terminal

choline acetyl transferase (CAT) enzyme that catalyzes the synthesis of acetylcholine

electrical synapse a gap junction that links two cells electrically

endogenous opioids morphinelike substances produced within the body

endorphins (en-dor-fins) morphinelike neuropeptides produced in the nervous system

enkephalins (en-kef-a-lins) morphinelike neuropeptides produced in the nervous system

excitatory postsynaptic potential (EPSP) a graded depolarization caused by neurotransmitter binding to receptors on the postsynaptic neuron

excitatory synapse site of communication at which activity in the presynaptic neuron causes depolarization of the postsynaptic neuron

frequency coding the coding of stimulus intensity by the frequency of action potentials in a neuron, in which a stronger depolarizing stimulus above threshold causes the action potential frequency to increase

GABA gamma-aminobutyric acid; an amino acid neurotransmitter

glutamate amino acid neurotransmitter

glycine (gly-seen) amino acid neurotransmitter

inhibitory postsynaptic potential (IPSP) a graded hyperpolarization caused by neurotransmitter binding to receptors on the postsynaptic neuron

inhibitory synapse site of communication whereby activity in the presynaptic neuron decreases the likelihood of an action potential occurring in the postsynaptic neuron

ionotropic receptor (eye-oh-no-troh-pik) a receptor protein that also functions as an ion channel that opens or closes in response to the binding of a chemical messenger

metabotropic receptor any receptor that works by triggering biochemical changes rather than triggering a direct change in membrane permeability

monoamine oxidase (MAO) (moan-oh-am-ine) enzyme that degrades biogenic amines

neural integration the summation of input to a neuron that determines whether the neuron will fire an action potential

neuropeptides polypeptides that are released from neurons by exocytosis at synapses

nitric oxide gas that functions as a chemical messenger

oxytocin (ox-see-toh-sin) hormone secreted by the posterior pituitary that regulates contractions of the uterus and the flow of milk from the breasts; functions as a neurotransmitter in some neurons

postsynaptic neuron at a synapse, the neuron that receives signals from another neuron

postsynaptic potential (PSP) a change in the membrane potential in a postsynaptic neuron that occurs in response to the binding of neurotransmitters to receptors

presynaptic facilitation a phenomenon occurring at an axo-axonic synapse such that activity in the presynaptic neuron enhances the release of neurotransmitter from the postsynaptic neuron

presynaptic inhibition a phenomenon occurring at an axo-axonic synapse such that activity in the presynaptic neuron decreases the release of neurotransmitter from the postsynaptic neuron

presynaptic neuron at a synapse, a neuron that transmits signals to a second neuron or an effector cell

substance P hormone that decreases gastrointestinal motility; functions as a neurotransmitter in some neurons

synaptic cleft the extracellular space between the axon terminal of the presynaptic cell and the postsynaptic cell at a synapse

synaptic delay time between the arrival of an action potential at the axon terminal of the presynaptic cell and when a response occurs in the postsynaptic cell

vasopressin hormone secreted by the posterior pituitary that regulates urine output by the kidney; stimulates water reabsorption by the kidneys; functions as a neurotransmitter in some neurons; also known as *antidiuretic hormone (ADH)*

Answers

Fig 8 Neuron B, because the resulting graded potential would dissipate to a greater extent by the time it reached the axon hillock

Fig 10 Calcium ions entering the cell through voltage-gated calcium channels

Fig 11 Nicotinic

Apply Your Knowledge Answers

Drugs that block the degradation or reuptake of neurotransmitter increase the concentration of neurotransmitter in the synaptic cleft of active neurons, enhancing the communication. However, if a neuron is not releasing neurotransmitter, then these drugs will have no effect.

For a person to perceive pain, information must be transmitted from the sensory receptor in the skin to the primary somatosensory cortex. Disruption of the communication at any level in between the skin and the cortex will decrease the perception. Thus, by decreasing communication from the afferent to the second-order neuron in the spinal cord, less signal is transmitted to the cortex and the perception of pain is diminished.

Answers to End-of-Chapter Multiple Choice and Objective Questions

Multiple Choice Questions

1. a 2. b 3. b 4. b 5. d 6. b 7. b 8. e 9. a 10. e

Objective Questions

11. gap junction
12. false
13. chemical
14. true
15. false
16. true
17. larger
18. ionotropic
19. monoamine oxidase (MAO), catechol-*O*-methyltransferase (COMT)
20. cAMP

The Nervous System: Central Nervous System

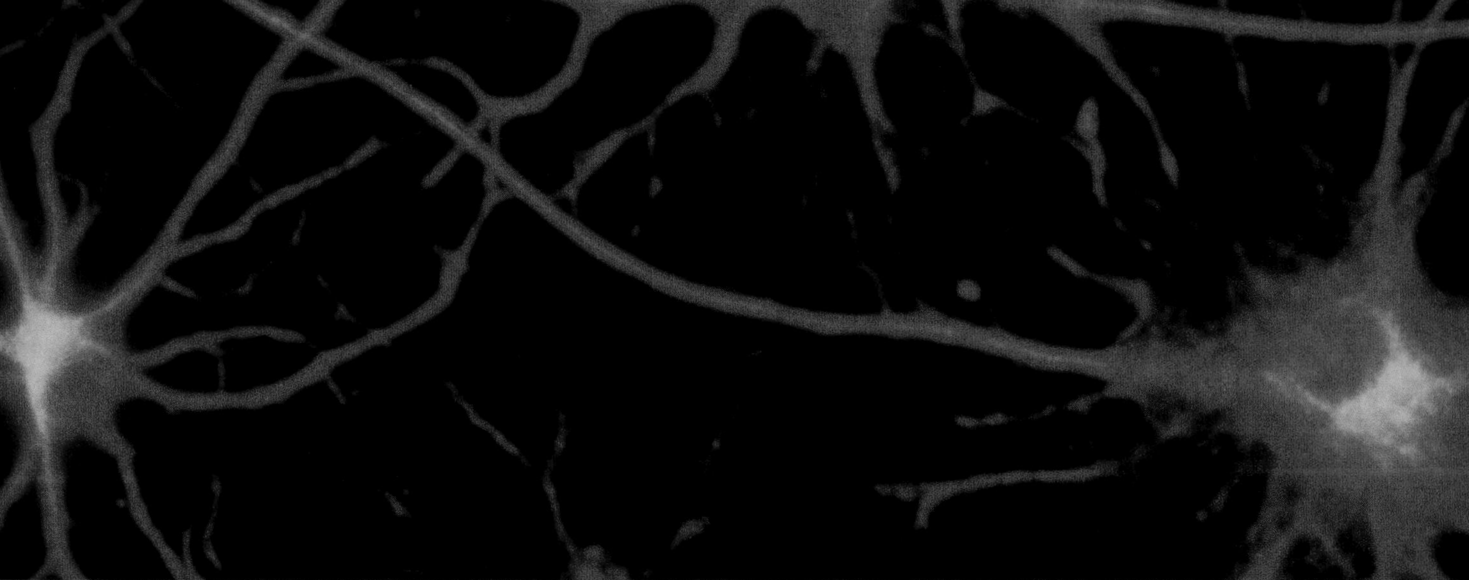

Prof Stephen Waxman, Hank Morgan/Photo Researchers, Inc.

Fluorescent micrograph of neurons.

As you read this, your central nervous system is performing many tasks, including perceiving words on the page, comprehending them, and making judgments about their importance and whether they should be stored in memory. As you read further, you may start integrating information with previous memories or thoughts. However, if you are reading this late at night, you might be sleepy and doze off; tomorrow, you may recall little of what you read. In any case, what happens as you read is a consequence of central nervous system processing.

One of the things that makes the nervous system so fascinating is that most of its functions are carried out by neurons. Since all neurons operate according to a small set of principles, how can they be responsible for all of the nervous system's many complexities? How are neurons involved when you feel angry? How can neurons enable you to remember someone's name or how the heart works? One neuron alone can perform none of these functions, but when billions of neurons (and the glial cells associated with them) are organized to form the nervous system, they can handle these and thousands of other functions.

Go to MasteringA&P for helpful A&P Flix 3-D animations, chapter quizzes, pre-tests, Interactive Physiology tutorials, and more!

CHAPTER OUTLINE

LEARNING OUTCOMES *After studing this chapter, you should be able to:*

- Describe the anatomy of the brain and spinal cord, and relate structure to function.
- Indicate which structures protect the central nervous system and which are involved in neural signaling.
- Describe the anatomy, physiology, and consequences of the blood-brain barrier.
- Describe the energy supplies of the brain, and explain why blood flow is so critical.
- Define *reflex arc*. Describe the following reflex pathways: muscle spindle stretch reflex, withdrawal reflex, and crossed-extensor reflex.
- Describe the areas of the brain that contribute to voluntary control of skeletal muscles, and the basic roles these areas play.
- Describe the different functions of the two language centers: Wernicke's area and Broca's area.
- Describe the different stages of sleep, and explain how the brain shifts from the sleep state to the conscious state.
- Describe the different types of learning and memory. Define *neural plasticity*, and explain how it contributes to learning and memory.

Before You Begin

Make sure you have mastered the following topics:

- Tight junctions
- Organization of the nervous system
- Neurotransmitters
- Neuron structure
- Communication at synapses
- Neural integration

We have learned about electrical signaling in neurons and chemical signaling from one neuron to another at synapses. Amazingly, the nervous system—with all its myriad complexities—functions primarily through these two basic processes. In this chapter, we learn how chains of neuron-to-neuron connections (*neural pathways*) allow the central nervous system to carry out its numerous functions.

The central nervous system is ultimately responsible for everything we perceive, do, feel, and think. It gives each of us our unique personality and sense of self-identity. It also performs many critical functions that typically escape our notice. For example, it coordinates the activities of all our organ systems, a function that is necessary for the maintenance of homeostasis. It is estimated that the central nervous system contains approximately 100 billion (10^{11}) neurons and 100 trillion (10^{14}) synapses, all contained within two remarkable structures, the brain and the spinal cord.

Scientists are continuing to unravel the mysteries of the brain even as you read these words. Because the brain is so complex and its functions so difficult to fathom using current scientific methodology, much of what you will read in this chapter comprises the commonly accepted portions of a constantly developing theory of central nervous system function. Perhaps someday we will understand the brain as well as we understand, say, the lungs or the kidneys. For now, we begin by examining the anatomy of the central nervous system.

1 General Anatomy of the Central Nervous System

The central nervous system (CNS) consists of the brain and the spinal cord. Because it is made up of soft tissue, with a consistency much like Jell-O, it is particularly vulnerable to damage by physical trauma. Fortunately, the CNS is protected by glial cells, bone, connective tissue, and cerebrospinal fluid, all of which are described in this section.

Glial Cells

When we think of the nervous system, we think of neurons. In reality, 75–90% of the central nervous system is composed of glial cells (or *neuroglia*), nonexcitable cells that provide support to neurons. The role of glial cells in neural communication, however, may go beyond simple support. For example, the higher up the organism is on the evolutionary scale, the more glial cells in the brain. Thus humans have more glial cells than any other animal.

There are four types of glial cells: Schwann cells, oligodendrocytes, microglia, and astrocytes (**Figure 1**). Astrocytes are by far the most abundant of the four types. All of the glial cells release growth factors necessary for development and maintenance of the nervous system. In addition, recent studies suggest that glial cells, even though they are nonexcitable cells, communicate with neurons and with each other. Here we discuss the function of astrocytes and microglia.

Astrocytes

Astrocytes, which are so named because of their star-like appearance, are the most diverse and numerous of the glial cells. Although astrocytes surround neurons with gaps of only 20 nm between cells, they do not overlap one another; that is, each cell remains in its own "territory." Astrocytes form a structural and functional link between neurons and non-nervous tissue and extracellular fluid. They direct the development of special capillaries that restrict the movement of certain molecules between blood and the CNS, called the *blood-brain barrier*. Astrocytes guide developing neurons, especially neurites, to their correct destination and regulate the development and maintenance of synapses. They may also support the regeneration of damaged axons.

Astrocytes are critical to the maintenance of the normal extracellular environment surrounding neurons, especially at synapses. They play an especially important role in maintaining normal extracellular potassium levels—a state that is critical to neuron excitability. Astrocytes also remove certain neurotransmitters, such as glutamate and biogenic amines, from the synaptic cleft. Excess

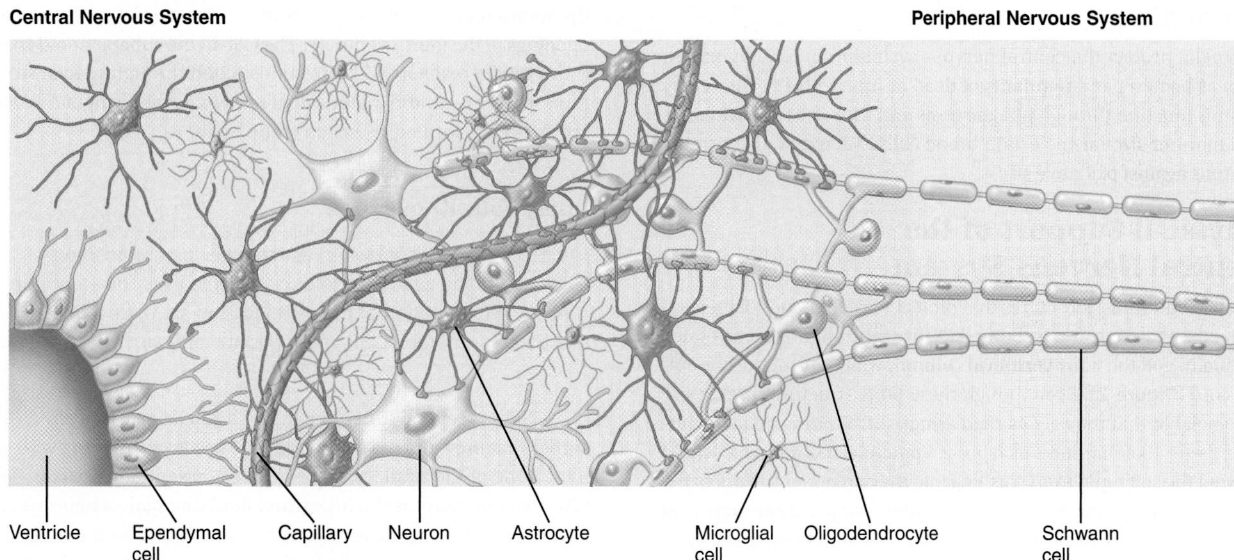

Figure 1 **Glial cells in the nervous system.**

levels of glutamate are toxic and contribute to the spread of neurological damage during a stroke or other types of brain damage.

Astrocytes synthesize and store molecules for use by neurons. For example, they synthesize glutamine, which is then released into the interstitial fluid and picked up by other neurons to form glutamate, an excitatory neurotransmitter. Astrocytes store some glycogen, which can be broken down to lactate; the lactate is then transported to neurons, where it serves as an important energy source for active areas of the brain.

Finally, recent studies suggest that astrocytes function in conjunction with *microglia,* described next, to protect neurons from toxic substances. Specifically, astrocytes protect neurons from oxidative stress and help remove cellular debris (see **Clinical Connections: Glial Cells in Neurodegenerative Disorders**).

CLINICAL CONNECTIONS

Glial Cells in Neurodegenerative Diseases

A growing body of evidence suggests that glial cells may contribute to the development of neurodegenerative diseases such as multiple sclerosis, Alzheimer's disease, and Parkinson's disease.

Multiple sclerosis results from the loss of myelin in the central nervous system. Multiple sclerosis is an autoimmune disease, meaning that the immune system attacks a part of the body—in this case, oligodendrocytes. The loss of myelin (and some axons) in the central nervous system slows down or stops communication along certain neural pathways. Symptoms of multiple sclerosis include blurred vision, muscle weakness, and difficulty maintaining balance.

Alzheimer's disease is caused by the loss of cholinergic neurons in certain brain areas and the replacement of the lost neurons with scar tissue called *plaques.* During the degeneration of cholinergic neurons, astrocytes and microglia become overly active. These glial cells release inflammatory chemicals that enhance further degeneration of cholinergic neurons. Thus a vicious cycle takes place. Early signs of Alzheimer's disease include loss of memory and confusion; later signs include motor dysfunction, loss of communication skills, and decrease in cognitive functions. Treatment options for this disease are limited, chiefly because it is difficult to diagnose the disease before it reaches an advanced stage. Early stages may be treated with acetylcholinesterase inhibitors, but these agents are not effective in advanced Alzheimer's disease.

Parkinson's disease is a degenerative disease involving the loss of dopaminergic neurons. As in Alzheimer's disease, glial cells are thought to enhance neural degeneration through the production of inflammatory agents.

Critical Thinking Questions

1. Describe the symptoms and causes of multiple sclerosis, Alzheimer's disease, and Parkinson's disease.
2. What are the main areas of commonality between these three diseases?
3. What are the key differences between these three diseases?

Microglia

Microglia protect the central nervous system from foreign matter, such as bacteria and remnants of dead or injured cells. They carry out this function through phagocytosis and the release of cytokines in a manner similar to certain blood cells. Microglia also protect neurons against oxidative stress.

Physical Support of the Central Nervous System

The two outermost structures that protect the soft tissues of the central nervous system are the bony skull, or **cranium,** which surrounds the brain, and the bony **vertebral column,** which surrounds the spinal cord (**Figure 2**). Even though these bony structures are clearly beneficial in that they act as rigid armor surrounding delicate nervous tissue, their hardness also poses a potential hazard: What would prevent the soft brain from crashing into the hard inner surface of the skull when, for example, you make a sudden stop in a car moving at freeway speeds? Between the bone and the nervous tissue are a series of three membranes called the meninges and a layer of fluid called cerebrospinal fluid, which provide protection against such impact.

The **meninges** are three connective tissue membranes that separate the soft tissue of the CNS from the surrounding bone (see Figure 2). The three meningeal membranes are the **dura mater,** the **arachnoid mater,** and the **pia mater.** *Mater* is Latin for "mother," indicating the protective nature of the meninges. The dura mater is the outermost layer, closest to the bone. *Dura* is Latin for "hard and durable"—an apt name given that the dura mater is a very tough, fibrous tissue that is almost the consistency of leather. The arachnoid mater is the middle layer. *Arachnoid,* Greek for "spider," appropriately describes the arachnoid mater's weblike structure. Normally, no space exists between the dura and the arachnoid. In fact, if a blood vessel passing through the dura ruptures, then blood can accumulate between the dura and arachnoid maters, forming a *subdural hematoma.* The innermost layer of the meninges, the pia mater, is located immediately adjacent to the nervous tissue; *pia* is Latin for "tender and kind." The space between the pia mater and arachnoid mater, called the **subarachnoid space,** is filled with cerebrospinal fluid.

Cerebrospinal fluid (CSF) is a clear, watery fluid that bathes the CNS; it is similar (but not identical) in composition to plasma (**Table 1**). CSF completely surrounds the CNS and fills a number of cavities located within the brain and spinal cord. The brain contains four such cavities, called **ventricles,** which are continuous with one another. Two C-shaped *lateral ventricles* are connected to a midline *third ventricle* by the *interventricular foramen.* The *cerebral aqueduct* connects the third ventricle to the *fourth ventricle,* which is continuous with the **central canal,** a long thin cylindrical cavity that runs the length of the spinal cord (**Figure 3**). The lining of the ventricles and central canal is composed of epithelial cells called **ependymal cells.**

The vascularized lining of the ventricles forms a tissue called the **choroid plexus,** which consists of the pia mater, capillaries, and ependymal cells, and functions in the synthesis of CSF. The total volume of CSF is only 125–150 mL, but because it is recycled approximately three times per day, the choroid plexus must produce 400–500 mL/day. As CSF is produced, it circulates through the ventricular system and enters the subarachnoid space through openings of the fourth ventricle. The CSF in the subarachnoid space is eventually reabsorbed into venous blood through special structures in the arachnoid mater called *arachnoid villi* (singular: *villus;* see Figure 2b) located at the top of the brain.

Apply Your Knowledge

Hydrocephalus occurs when an increase in cerebrospinal fluid causes the ventricles to become enlarged. Based on the preceding description of the ventricular system, describe a mechanism whereby CSF levels might increase.

CSF has several functions in the brain. It acts as a shock absorber that prevents the soft nervous tissue from colliding with the hard bone of the skull, because the CNS essentially floats in CSF. CSF also functions as the interstitial fluid that bathes neurons and glial cells, providing these cells with essential nutrients and removing waste products from them. The CSF contributes to the maintenance of normal ionic composition around neurons, which is essential for normal excitability of neurons. For the CSF to carry out its metabolic functions, it must be replenished by the blood supply to the CNS, as described next.

Blood Supply to the Central Nervous System

Although the CNS accounts for only some 2% of body weight (the adult brain and spinal cord weigh approximately 3–4 pounds), it receives almost 15% of the blood that the heart pumps to all of the body's organs and tissues under resting conditions. This large blood supply is necessary because CNS tissue has a high rate of metabolic activity compared to most other body tissues and, therefore, has a high demand for glucose and oxygen to meet its energy needs. Under resting conditions, for example, the brain accounts for approximately 20% of all oxygen that the body consumes, and approximately 50% of all glucose consumed. To ensure delivery of these needed materials, adequate blood flow to the CNS must be maintained at all times. In fact, the CNS is so dependent on this blood supply that disruption of blood flow for even a few minutes can result in irreversible damage to CNS tissue. A reduction of blood flow to a particular area of the CNS can cause deficits in certain functions, such as the ability to speak or move an arm. Such changes in function might occur, for example, following a *stroke*—an event in which blood flow becomes interrupted because of a blocked or ruptured blood vessel in the brain (see **Clinical Connections: Stroke**).

The CNS is particularly sensitive to interruptions in blood flow because cells in the CNS contain very little glycogen (compared to muscle and liver) and, therefore, must obtain glucose directly from the blood. Furthermore, most cells in the CNS do not have access to fatty acids for energy, which increases their demand for glucose. (Recall that oxidation of a single fatty acid molecule can yield as much energy as the oxidation of several molecules of glucose.) Finally, whereas many other tissues can obtain energy from anaerobic metabolism during periods of reduced oxygen

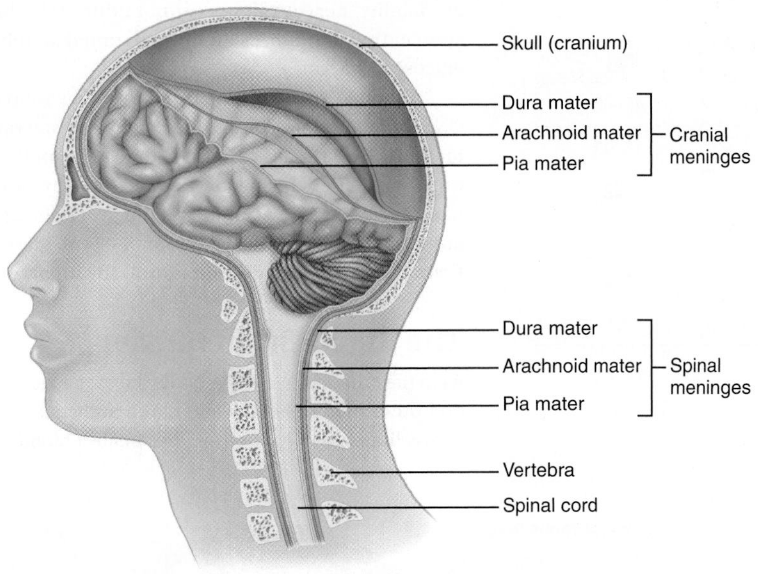

(a)

(b)

Figure 2 Protective structures of the CNS. (a, b) Sections of CNS protective structures. Bony outer structures include the cranium and vertebrae. The meninges, which are located between the bony structures and the soft nervous tissue, are composed of three layers: dura mater, arachnoid mater, and pia mater. The cushioning presence of cerebrospinal fluid within the subarachnoid space provides yet another level of protection.

TABLE 1 Compositions of Plasma and CSF*

	Plasma	CSF
Glucose (mg/dL)	80–100	45–80
Proteins (mg/dL)	7000	15–45
Sodium (mM)	145	149
Potassium (mM)	4	3.1
Chloride (mM)	103	129
Calcium (mM)	2.5	2.1

*CSF levels vary based on sampling site.

availability, nervous tissue cannot utilize this alternative energy source; thus it requires an uninterrupted supply of oxygen and glucose to stay alive.

Although glucose is the primary energy source for cells in the CNS, under certain conditions (such as in starvation or diabetes mellitus) CNS tissues can use ketones to supply as much as two-thirds of their energy needs. Ketones are a by-product of lipid catabolism (produced primarily in the liver) when glucose supplies are limited. However, the CNS cannot rely on these compounds entirely, but rather must depend on a steady supply of glucose.

The Blood-Brain Barrier

As in the other tissues of the body, the exchange of oxygen, glucose, and other materials between blood and cells in the CNS occurs across the walls of *capillaries*, the smallest blood vessels. Capillary

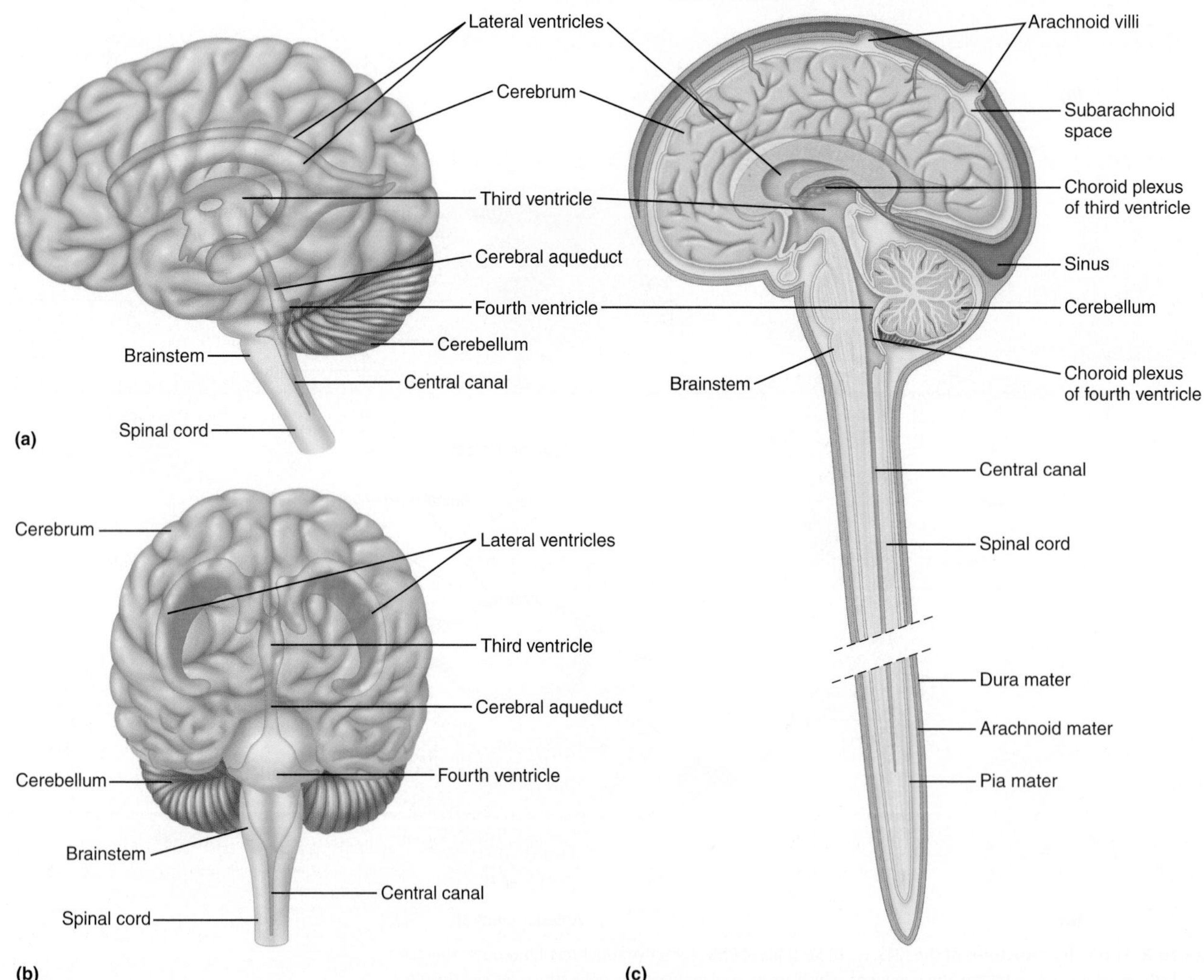

Figure 3 Ventricular system of the CNS. The brain's four ventricles—two lateral ventricles, the third ventricle, and the fourth ventricle—are continuous with the central canal of the spinal cord. **(a)** Lateral view. **(b)** Frontal view. **(c)** Relationships of the ventricles to other CNS structures. Note the presence of the choroid plexus within the ventricles.

CLINICAL CONNECTIONS

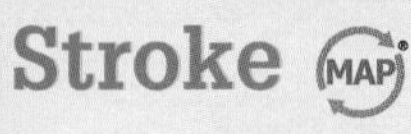

Stroke

Cerebrovascular diseases—those that affect blood vessels in the brain—are the third leading cause of death in the United States today, behind only heart disease and cancer. Cerebrovascular diseases are responsible for impairment of brain function in approximately 2 million Americans each year.

Stroke (also known as *cerebrovascular accident* or *CVA*) is a cerebrovascular disorder characterized by a sudden decrease or stoppage of blood flow to a part of the brain. Decreased blood flow (ischemia) is dangerous to any tissue, but brain tissue is especially vulnerable, in part because of the high rate of this tissue's metabolic reactions, which consume glucose and oxygen quickly. In fact, stoppage of blood flow for no more than three or four minutes may be sufficient to cause death of most brain cells. For this reason, a stroke can kill people within minutes or leave them with permanent brain damage.

Strokes are classified as either *occlusive* or *hemorrhagic* and may occur in the interior of the brain or on its surface. In occlusive stroke, blood flow through a vessel is blocked, usually as a result of a blood clot. In hemorrhagic stroke, a blood vessel ruptures, causing *hemorrhage* (bleeding). Conditions that increase the likelihood of stroke include *arteriosclerosis* (narrowing of arteries due to buildup of fatty deposits on the inner wall), *aneurysm* (bulging of an artery due to weakening of the wall), and *hypertension* (high blood pressure). Stroke symptoms vary with the location and extent of the affected area and are sudden in onset. It is the sudden onset of symptoms that distinguishes stroke from other neurological disorders. The affected individual may experience muscle weakness or paralysis, loss of sight (usually in just a part of the visual field), loss of other senses, tingling or other sensory disturbances, disturbances of language or spatial perception, changes in personality, disorientation, or memory loss. Effects of a stroke may be confined to the initial symptoms, or they may intensify or become more widespread over the course of a few hours or days.

The best treatment for a stroke is prevention. Some people are more prone to strokes due to hypertension or high cholesterol levels. Other people may develop an early warning sign called a *transient ischemic attack (TIA)*, in which temporary blockage of a blood vessel occurs. In either case, the patients may be put on anticoagulants such as aspirin or warfarin, or if there is partial blockage of a cerebral artery, a stent can be surgically implanted to hold the artery open.

Once a stroke occurs, treatment depends on whether the stroke was ischemic or hemorrhagic. Ischemic strokes can be treated with tissue plasminogen activator (tPA), which can break down blood clots. This treatment is effective only if it is administered within the first three hours of the onset of symptoms, however. Hemorrhagic strokes generally require surgical intervention. Full recovery following a major stroke is rare, but some functionality often returns over time.

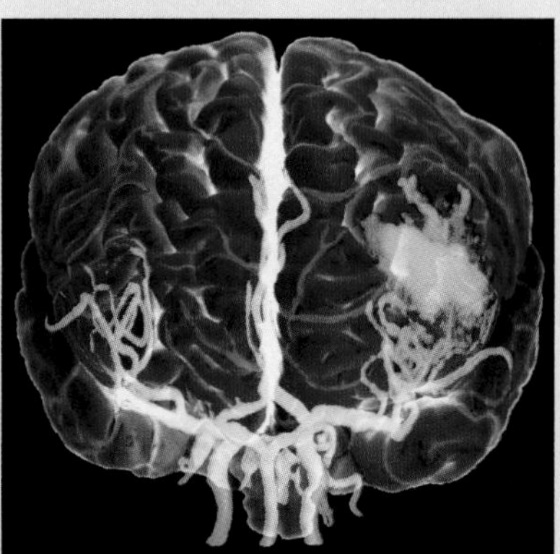

Three-dimensional image showing blood supply to the brain and area of stroke.

Critical Thinking Questions

1. Describe how stroke causes death or permanent brain damage.
2. What conditions can lead to a stroke? What are key stroke symptoms?
3. Identify differences in the treatment and expected outcomes of ischemic and hemorrhagic strokes.

walls are composed of little more than a single layer of endothelial cells (a type of epithelial cell), providing a short diffusion distance for exchange.

In most tissues, small molecules such as gases, inorganic ions, monosaccharides, and amino acids move freely across capillary walls (**Figure 4a**). Hydrophobic molecules diffuse across the membranes of the endothelial cells, whereas hydrophilic molecules diffuse through relatively large gaps (pores) between the endothelial cells. Cells and macromolecules such as proteins, however, are too large to move through these gaps; instead, certain macromolecules can be actively transported across the endothelial cells by transcytosis—the movement of a molecule across an endothelial cell by endocytosis into the cell followed by exocytosis out of the cell.

In the CNS, most hydrophobic molecules can diffuse across the endothelial cells of capillaries, as in other tissues. However, transcytosis does not occur across capillary endothelial cells in the CNS, and the movement of hydrophilic molecules across capillary walls is restricted by the **blood-brain barrier** (Figure 4b), a physical barrier that exists between the blood and CSF, which is the interstitial fluid in the CNS. The existence of this barrier is due to the presence of tight junctions between the capillary endothelial cells, which eliminate capillary pores, thereby restricting the diffusion of hydrophilic molecules between the cells. Astrocytes are critical to the formation of the blood-brain barrier in that they stimulate endothelial cells to develop and maintain tight junctions.

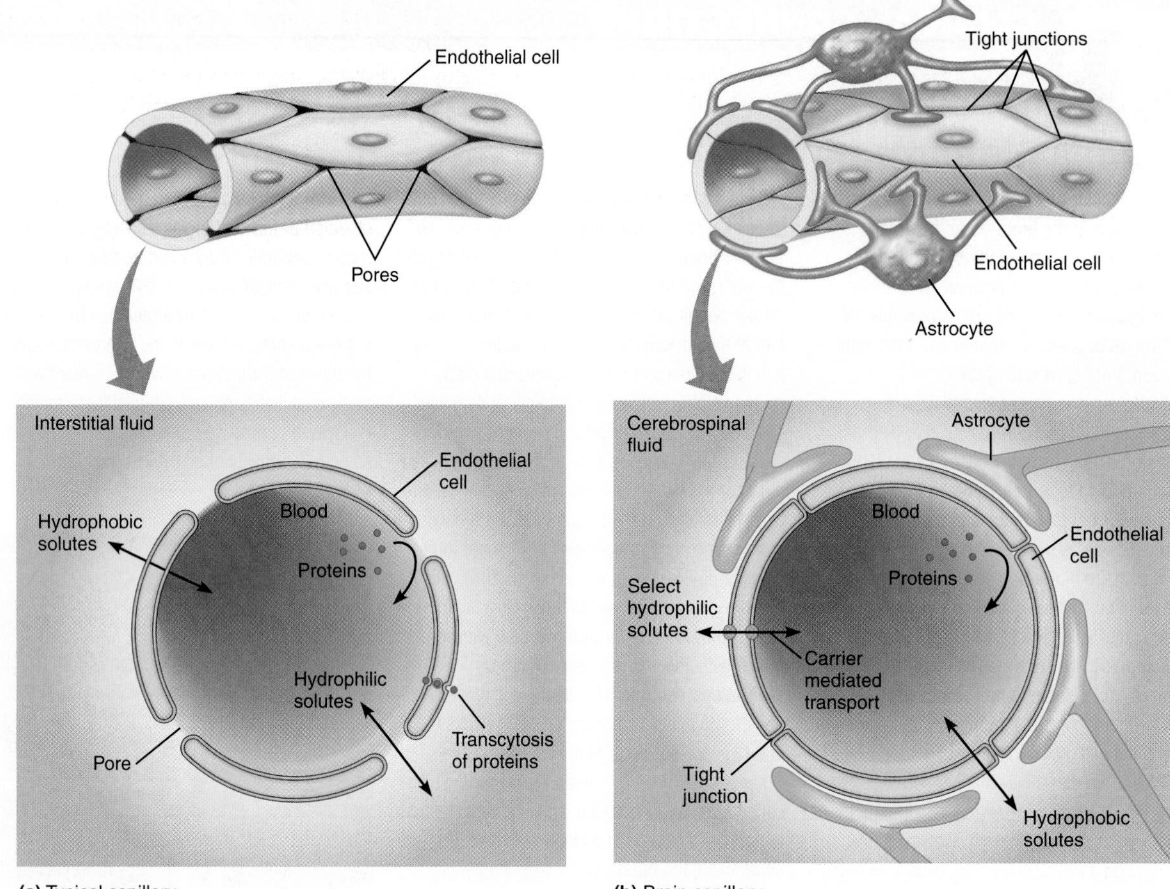

Figure 4 Blood-brain barrier. (a) Typical capillary (found in most regions of the body). **(b)** Brain capillary. Even though astrocytes are found in close association with brain capillaries, they do not constitute a functional barrier.

The blood-brain barrier protects the CNS from harmful substances that may be present in the blood by restricting the movement of molecules across capillary endothelial cells. To enter or leave capillaries, molecules must cross the endothelial cells themselves. Gases and other hydrophobic molecules penetrate through these cells relatively easily because they are able to move across cell membranes by simple diffusion through the lipid bilayer. As a consequence, these molecules are able to move freely between blood and brain tissue. One example of such a hydrophobic substance is ethanol (grain alcohol), which depresses CNS function by several mechanisms.

In contrast, hydrophilic substances such as ions, carbohydrates, and amino acids cannot cross the plasma membranes of any cells by simple diffusion, but instead must rely on mediated transport to cross capillary walls in the CNS. Because mediated transport utilizes transport proteins, which are selective to certain molecules, the blood-brain barrier is selectively permeable, allowing only certain compounds to move across it. Compounds such as glucose can penetrate the blood-brain barrier readily by facilitated diffusion because transport proteins for glucose are present in cell membranes. Glucose is transported across the blood-brain barrier by GLUT-1 carriers, which are distinct from the insulin-dependent GLUT-4 carriers found in the membranes of most cells. Other substances transported across the blood-brain barrier include insulin, amino acids, choline (needed for acetylcholine synthesis), and aspirin. The finding that insulin is transported across the blood-brain barrier might seem contradictory at first because glucose transport into neurons occurs through insulin-independent mechanisms. However, receptors for insulin are located on certain neurons in the CNS, including neurons in the satiety center of the hypothalamus. In fact, insulin actions in the CNS regulate food intake, not glucose metabolism. Specifically, insulin is considered a satiety hormone in that it decreases hunger. Thus, when insulin levels are low, such as in diabetes, a person is more likely to overeat and become obese. Other hydrophilic substances, such as catecholamines, inorganic ions (including H^+), and certain drugs (including several antibiotics), cannot penetrate the blood-brain barrier to any great degree because transport proteins for these compounds do not exist in the endothelial cells of CNS capillaries.

Health Fact

As a result of the selective permeability of the blood-brain barrier, the composition of cerebrospinal fluid differs from the composition of plasma (see Table 1). Therefore, certain CNS diseases can be adequately diagnosed only by sampling cerebrospinal fluid, usually by a procedure called a *spinal tap*.

Gray Matter and White Matter

The central nervous system has an orderly arrangement of neurons such that cell bodies, dendrites, and axon terminals tend to be found in clusters that appear gray and axons appear in clusters that appear white; these regions are called **gray matter** and **white matter,** respectively. The gray matter constitutes approximately 40% of the CNS and is the site of synaptic communication and neural integration (**Figure 5a**). The white matter constitutes approximately

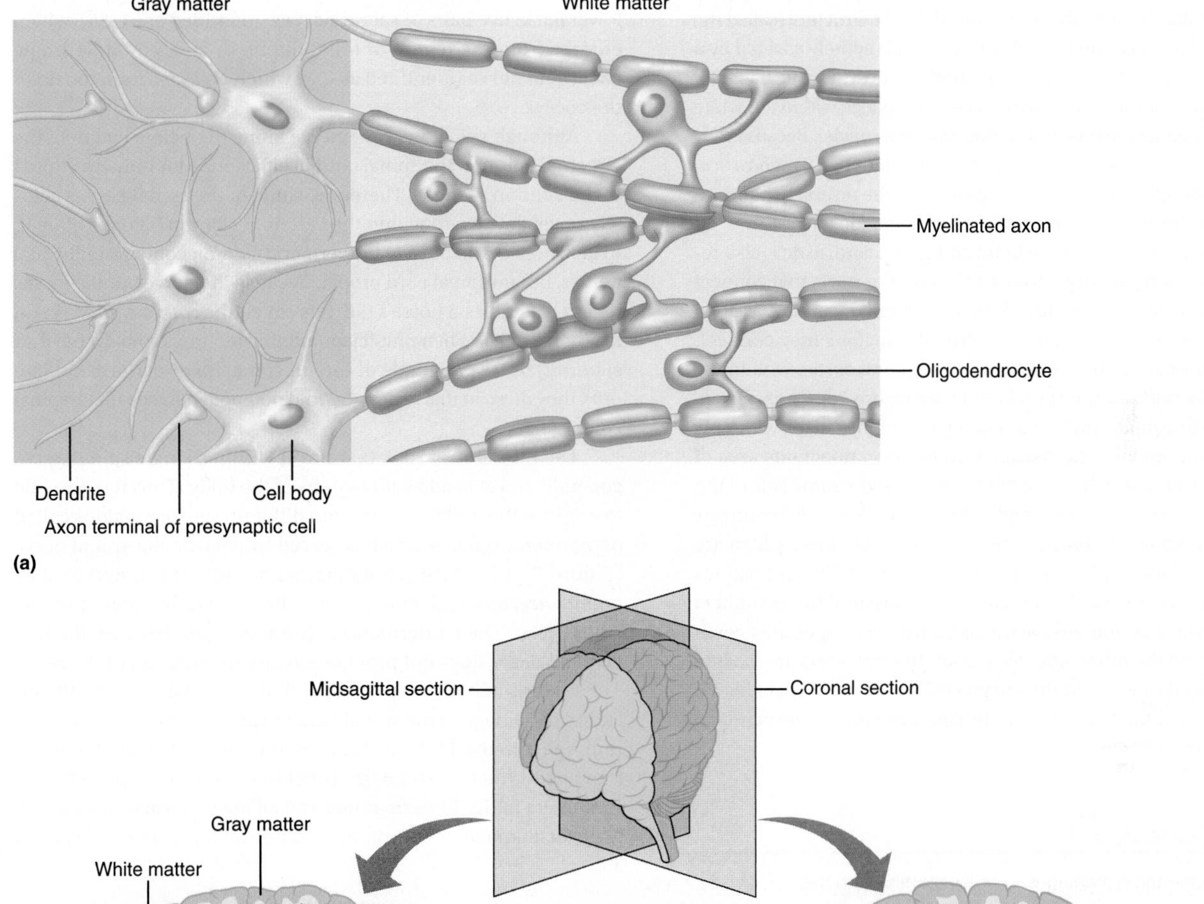

Figure 5 Makeup and arrangement of gray matter and white matter in the CNS. (a) Histology of gray matter and white matter. **(b, c)** Midsagittal and coronal sections of the brain, showing association fibers, commissural fibers, and projection fibers, tracts of white matter that connect different areas of the CNS.

Q On what part of the neuron are action potentials propagated?

60% of the CNS and appears white due to the presence of myelinated axons. (Myelin appears white because of its high lipid content.) Myelinated axons are specialized for the rapid transmission of information in the form of action potentials over relatively long distances. Glial cells are located throughout the CNS, in both gray matter and white matter.

When one looks at the external surface of the brain, only gray matter is visible, because the prominent globelike structure (called the *cerebrum*) that makes up the bulk of the brain is entirely covered by a thin layer of gray matter called the **cerebral cortex.** The white matter of the cerebrum is located beneath this layer. Embedded within this white matter are smaller areas of gray matter known as nuclei. Because they are found beneath the cortex, they are often referred to as *subcortical nuclei.* In the spinal cord, the arrangement is the reverse: The white matter is on the outside, whereas the gray matter is on the inside.

In the white matter of the brain and spinal cord, axons (also referred to as fibers) are organized into bundles or tracts that connect one region of gray matter with another. Different tracts are classified according to the locations of the particular regions they connect. **Projection fibers** or tracts connect the cerebral cortex with lower levels of the brain or the spinal cord (see Figure 5c). Thus, for example, *corticospinal tracts* connect regions of the cortex with gray matter in the spinal cord. **Association fibers** connect one area of cerebral cortex to another area of the cortex on the same side of the brain (see Figure 5c). For example, the *arcuate fasciculus* connects two regions known as *Broca's area* and *Wernicke's area,* which are important in language function and are located in different regions of the cerebral cortex on the same side. **Commissural fibers** connect cortical regions on one side of the brain with corresponding cortical regions on the other side. Most commissural fibers are located in a band of tissue called the **corpus callosum** (see Figure 5b, c), which connects the two halves of the cerebrum (called the **cerebral hemispheres**) together.

Quick Check 1

1. Name the three meninges in order, starting with the outermost.
2. What is cerebrospinal fluid? Where is it found? Where is it synthesized? Where is it absorbed?
3. What is the function of the blood-brain barrier? Which anatomical structure between endothelial cells limits the diffusion of water-soluble molecules?

2 The Spinal Cord

The **spinal cord** is a cylinder of nervous tissue that is continuous with the lower end of the brain and is surrounded by the vertebral column (**Figure 6**). The spinal cord is approximately 44 cm long in adults and ranges in diameter from 1 cm to 1.4 cm.

Spinal Nerves

Branching off the spinal cord at regular intervals are 31 pairs of **spinal nerves** (see Figure 6). Each pair of spinal nerves exits the vertebral column between two adjacent vertebrae. The spinal nerves and the region of spinal cord from which they originate are given designations according to where the nerves emerge from the vertebral column. There are eight pairs of **cervical nerves** (identified as C1–C8), which emerge from the vertebral column in the neck region; 12 pairs of **thoracic nerves** (identified as T1–T12), which emerge in the chest region; five pairs of **lumbar nerves** (identified as L1–L5), which emerge in the region of the lower back; five pairs of **sacral nerves** (identified as S1–S5), which emerge from the region of the tailbone or coccyx; and a single **coccygeal nerve** (identified as C_0), which emerges from the tip of the coccyx.

Although spinal nerves emerge along the whole length of the vertebral column, the spinal cord itself extends for only two-thirds of the column's length. Therefore, some of the spinal nerves actually travel downward within the vertebral column before exiting it. In fact, the bottom third of the vertebral column contains individual nerves, but no spinal cord proper. Because the nerve bundle in this region resembles a horse's tail, it is called the *cauda equina* (Latin for "horse tail"). When physicians administer drugs spinally (give an *epidural*) or take a sample of cerebrospinal fluid (perform a *spinal tap*), they do so in this region to avoid potential damage to the spinal cord itself.

The numerous axons that travel within a single spinal nerve generally travel to adjacent regions of the body. Thus it is possible to map out the body's surface into different sensory regions called *dermatomes*, each of which is served by a particular spinal nerve (**Figure 7**). (The face is not mapped because it is innervated by *cranial nerves*, which emerge from the brain rather than from the spinal cord. The C1 dermatome is not mapped because the first cervical nerve does not provide sensory innervation of the skin.) These kinds of body maps enable clinicians to determine the location of damage to the spinal cord or spinal nerves. For example, numbness in the T1 dermatome on the right side would indicate damage to the first thoracic spinal nerve on that side, whereas numbness in the T1 dermatome and all lower dermatomes would indicate damage to the spinal cord between the levels of C8 and T1.

Spinal Cord Gray and White Matter

The gray matter of the spinal cord is concentrated within a butterfly-shaped region in the interior of the cord, whereas the white matter is found in the surrounding outer region (**Figure 8**). The gray matter contains interneurons, cell bodies and dendrites of efferent neurons, and the axon terminals of afferent neurons. The efferent neurons travel in spinal nerves to effector organs; the afferent neurons travel in spinal nerves from sensory receptors in the body's periphery to the spinal cord.

The gray matter of the spinal cord is organized both anatomically and functionally into dorsal and ventral regions. The gray matter includes one *dorsal horn* and one *ventral horn* on each side (see Figure 8). The **dorsal horn** encompasses the dorsal (posterior) half of the gray matter on either side; the **ventral horn** encompasses the ventral (anterior) half. Afferent fibers originate in the periphery as sensory receptors and terminate in the dorsal horn, where they synapse on interneurons or efferent neurons. Note that the cell bodies of these afferent fibers are not located in the spinal

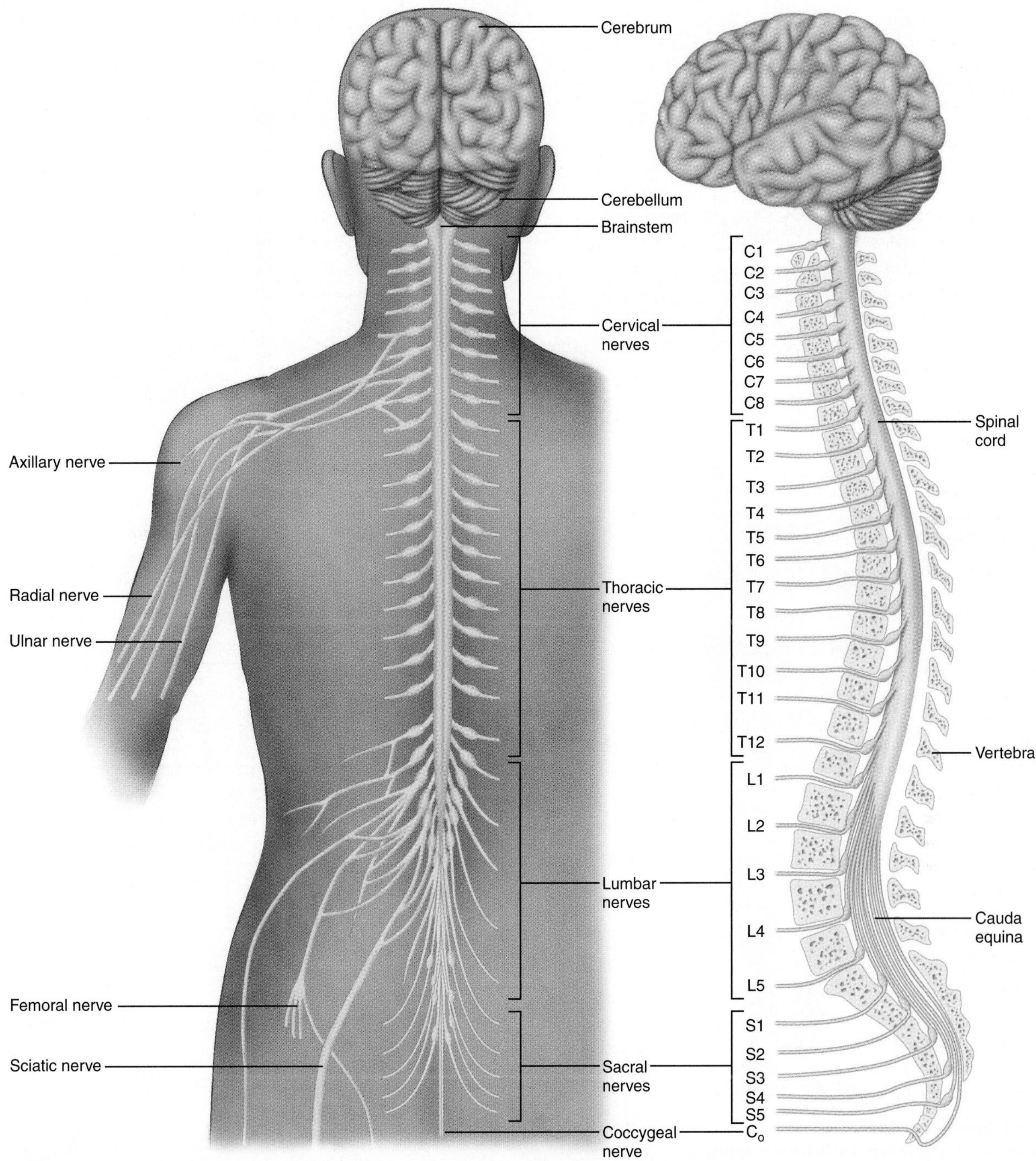

Figure 6 The spinal cord. (Left) Location of the spinal cord, as seen in a posterior view. (Right) Lateral view of the spinal cord, showing its position within the bony vertebral column. Also shown are the 31 pairs of spinal nerves, which leave the spinal cord between adjacent vertebrae.

cord itself; rather, they are located outside the spinal cord in clusters called **dorsal root ganglia.** (Recall that *ganglion* is a general term for any cluster of neuron cell bodies outside the CNS.) In contrast, the cell bodies of efferent neurons are located in the spinal cord. Efferent neurons originate in the ventral horn and travel to the periphery, where they form synapses with skeletal muscles.

In the thoracic and upper lumbar regions of the spinal cord, there is another area of gray matter called the *lateral horn,* or *intermediolateral cell column,* which is located between the dorsal and ventral horns on either side. The lateral horns are the origins of efferent neurons of the autonomic nervous system.

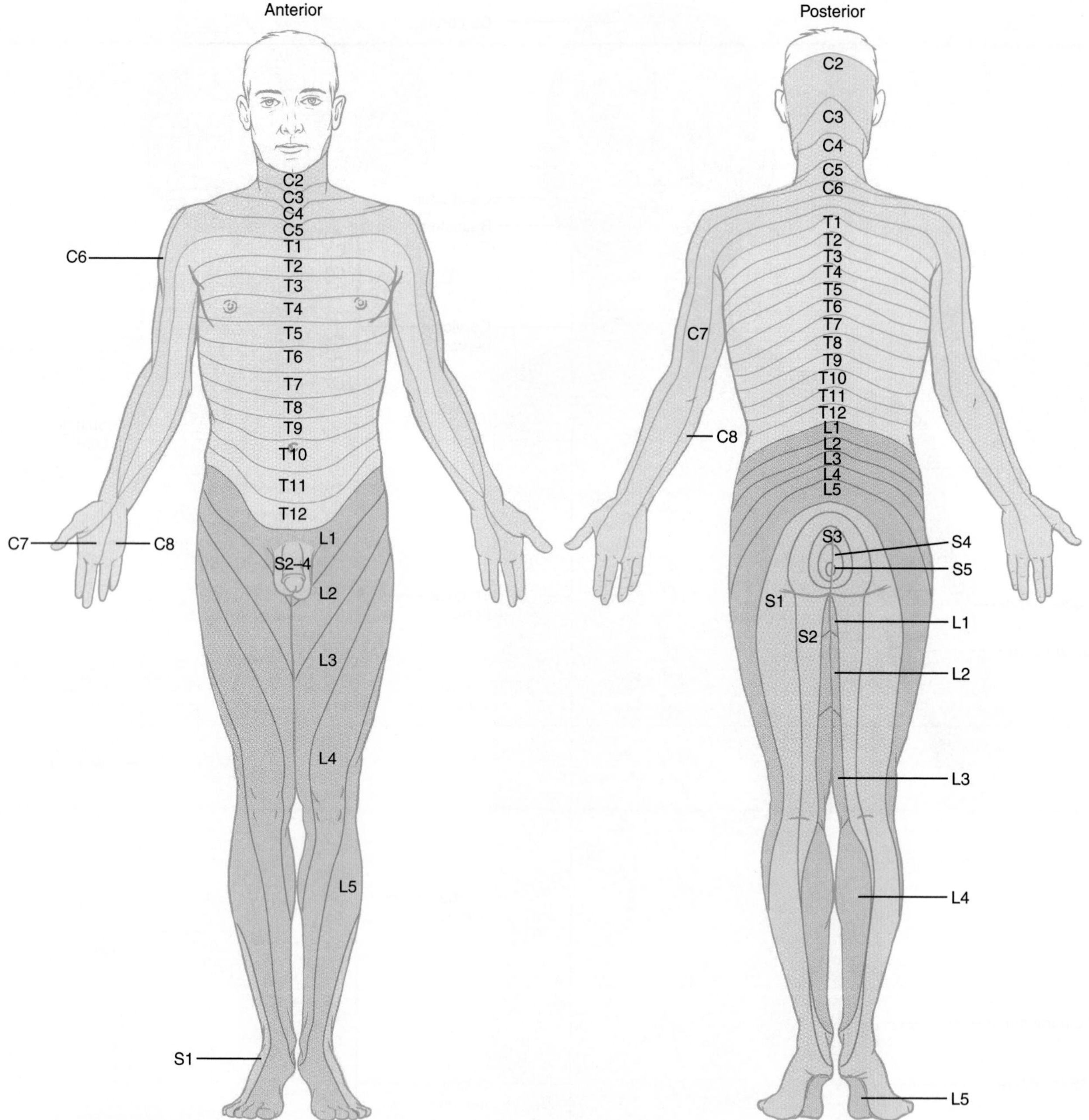

Figure 7 Dermatomes. Each dermatome is a sensory region on the surface of the body that is served by the spinal nerve indicated by the abbreviations.

Q Numbness in the left thumb indicates damage to which spinal nerve?

Apply Your Knowledge

Some people with peripheral neurological damage may lose sensory function from a specific spinal level, but not motor function. Where must the damage have occurred?

Afferent and efferent axons travel together in spinal nerves, but they separate into different bundles as they enter or leave the spinal cord. The bundles containing afferent axons are called **dorsal roots,** whereas the bundles containing efferent axons are called **ventral roots.** A short distance from the spinal cord, these dorsal and ventral roots come together to form the spinal nerves. All spinal nerves, therefore, are described as **mixed nerves** because they contain both afferent and efferent axons.

The white matter of the spinal cord consists of tracts that provide communication either between the different levels of the spinal cord or between the brain and various levels of the spinal

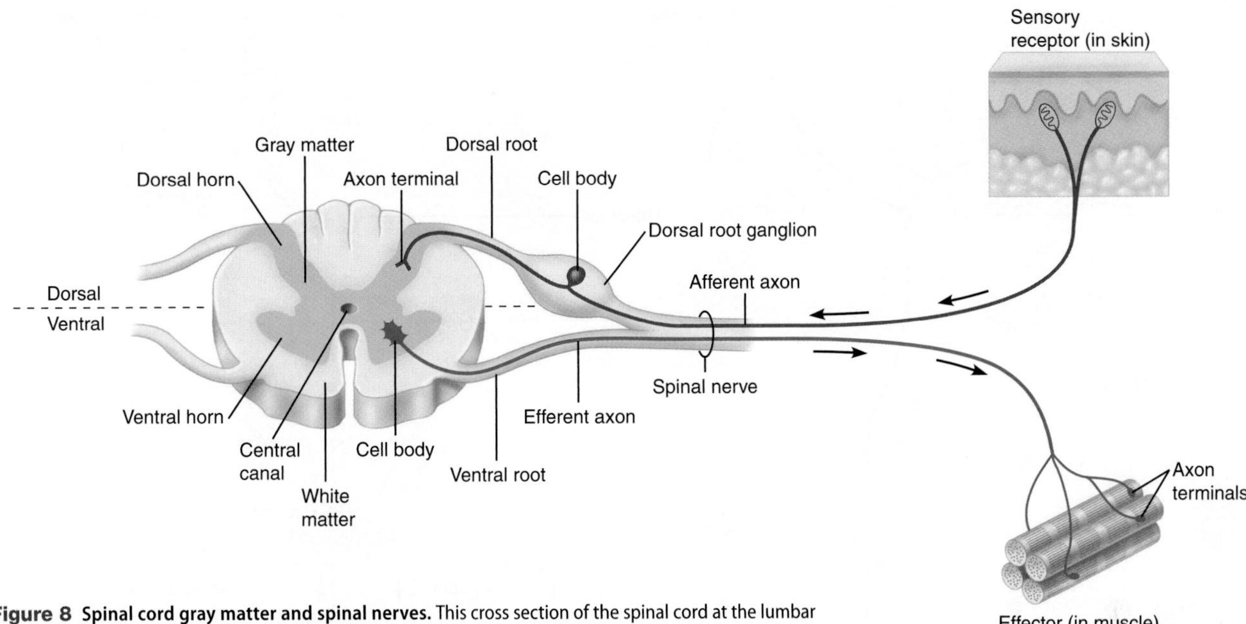

Figure 8 Spinal cord gray matter and spinal nerves. This cross section of the spinal cord at the lumbar level reveals the two functional halves of spinal cord grey matter: dorsal and ventral. Axons of afferent neurons enter the spinal cord through the dorsal roots and terminate in the dorsal horn; their cell bodies are located in dorsal root ganglia. Axons of different neurons originate in the ventral horn and exit through the ventral root. Because they contain axons of both afferent and efferent neurons, spinal nerves are considered mixed nerves.

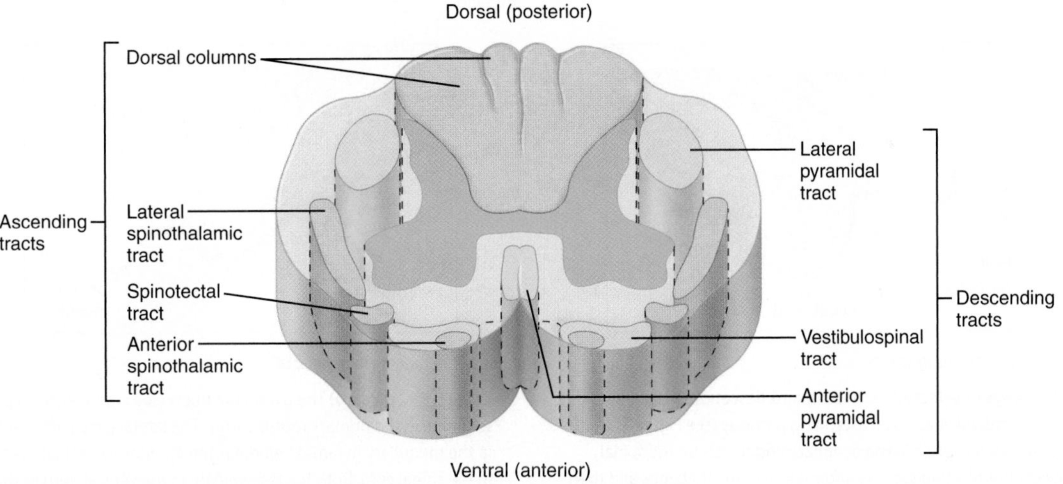

Figure 9 Cross section of white matter in the spinal cord. Spinal cord white matter consists of longitudinal tracts that run between the brain and the spinal cord, or between different spinal cord segments. Ascending tracts transmit information from spinal cord to brain, whereas descending tracts transmit information from brain to spinal cord. Only selected tracts are illustrated here.

cord (**Figure 9**). **Ascending tracts** transmit information from spinal cord to brain, whereas **descending tracts** transmit information from brain to spinal cord. Although the tracts in Figure 9 are labeled on only one side of the spinal cord for clarity, all tracts are *bilateral;* that is, any given tract is found on both sides of the spinal cord. For example, the *pyramidal tracts* are descending pathways that transmit motor commands to efferent neurons on both sides of the spinal cord. The *dorsal columns* are ascending tracts that transmit sensory information from the periphery to the brain.

The ascending and descending tracts effectively link the peripheral nerves to the brain. When afferent neurons are activated by a stimulus acting on a sensory receptor (for example, when a finger touches a thorn), action potentials travel along the nerve fiber from the sensory receptor to the axon terminal, usually in the dorsal horn of the spinal cord (**Figure 10a**). The axon terminal releases a neurotransmitter that transmits the signal to an interneuron (or, in rare cases, directly to an efferent neuron). Some of these interneurons form the ascending tracts, which

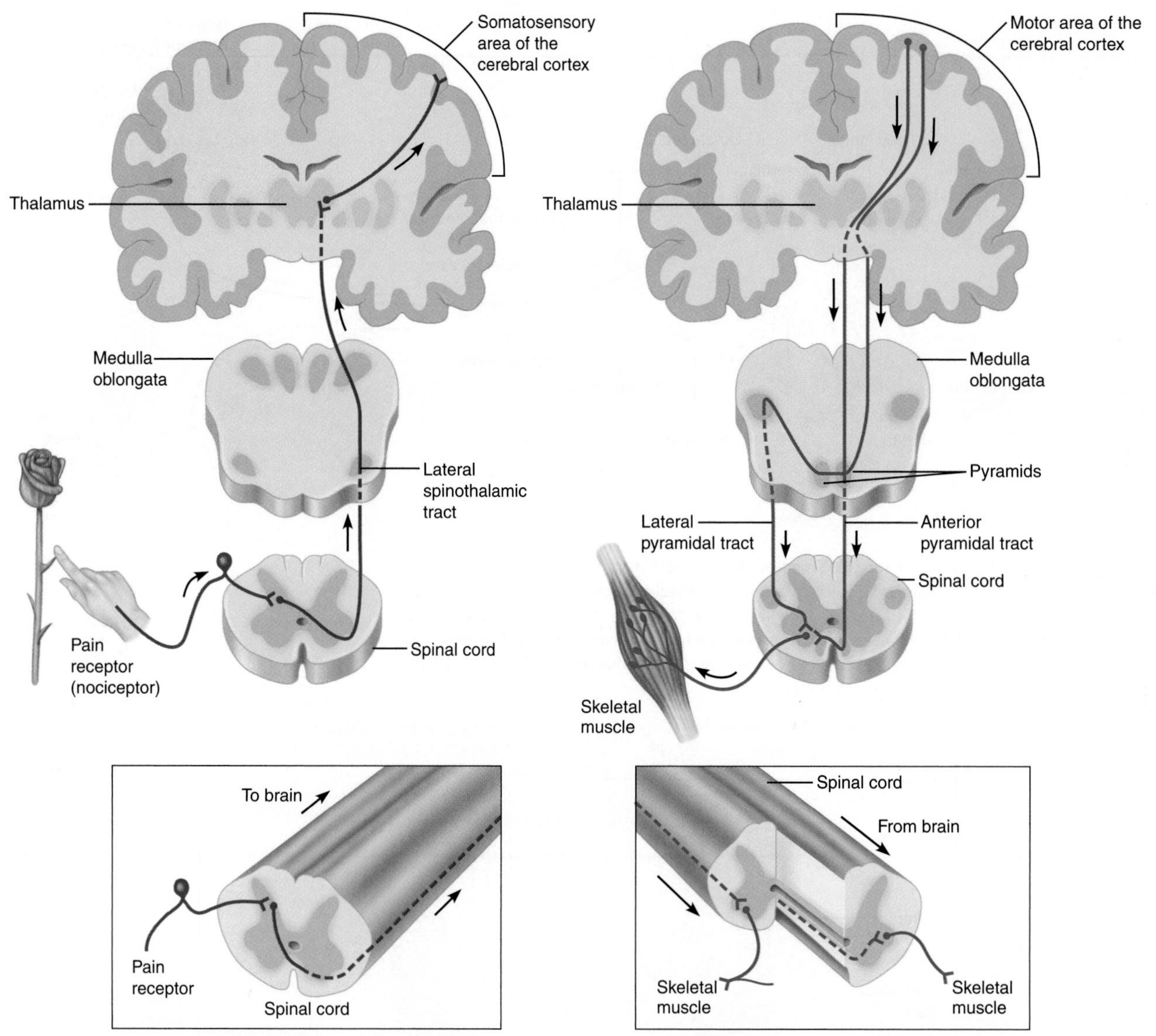

Figure 10 Pathways of selected ascending and descending tracts. **(a)** The lateral spinothalamic tract is an ascending pathway (see Figure 9) that originates with sensory receptors in the periphery and travels up the spinal cord, eventually communicating sensory information to the thalamus and then to the cerebral cortex. The spinothalamic tract crosses to the contralateral side in the spinal cord. **(b)** The pyramidal tracts (see Figure 9). Both pyramidal tracts originate in the primary motor cortex. The lateral pyramidal tract crosses over in the medullary pyramids, whereas the anterior pyramidal tract crosses over in the spinal cord. Both tracts terminate in the ventral horn of the spinal cord, where they communicate to motor neurons innervating skeletal muscle.

transmit the information to the brain so that perception of the stimulus can occur.

Signals from the brain travel along descending tracts to efferent neurons in the ventral horn (Figure 10b). For example, when you wish to wiggle your finger, the brain transmits commands through descending fibers to the efferent neurons that control the skeletal muscles that cause the finger to move. Controlling efferent neurons is not the only function that descending fibers perform, however. Some descending tracts modulate sensory information. For example, the body has pain-relieving *(analgesic)* systems that block the perception of pain during periods of stress. These systems work by blocking synaptic transmission between a pain-transmitting afferent neuron and interneurons in the spinal cord. By blocking the transmission of pain information to the brain, perception of a painful stimulus is prevented.

Notice that the ascending and descending tracts illustrated in Figure 10 generally cross over to the side opposite the side of origin. This is the norm, but some exceptions exist. When a pathway remains on the same side as its origin, it is called **ipsilateral.** When a pathway crosses to the side opposite its origin, it is called **contralateral.** Because most sensory and motor pathways cross to the opposite side within the CNS, sensory input to the right side of the body is transmitted to the left side of the brain for perception, and motor control of the right side of the body comes from the left side of the brain.

Quick Check 2

1. What is the functional difference between the dorsal and ventral horns of the spinal cord?
2. Why are spinal nerves called *mixed nerves*?
3. What is the functional difference between ascending and descending tracts in the spinal cord?

3 The Brain

The **brain** consists of three main parts: the forebrain, the cerebellum, and the brainstem (**Figure 11**). The **forebrain,** the largest and most superior part of the brain, is divided into left and right halves, or *hemispheres*, and consists of the cerebrum and diencephalon (Figure 11c). The **cerebrum** is a large, roughly C-shaped structure containing both gray and white matter. The gray matter areas

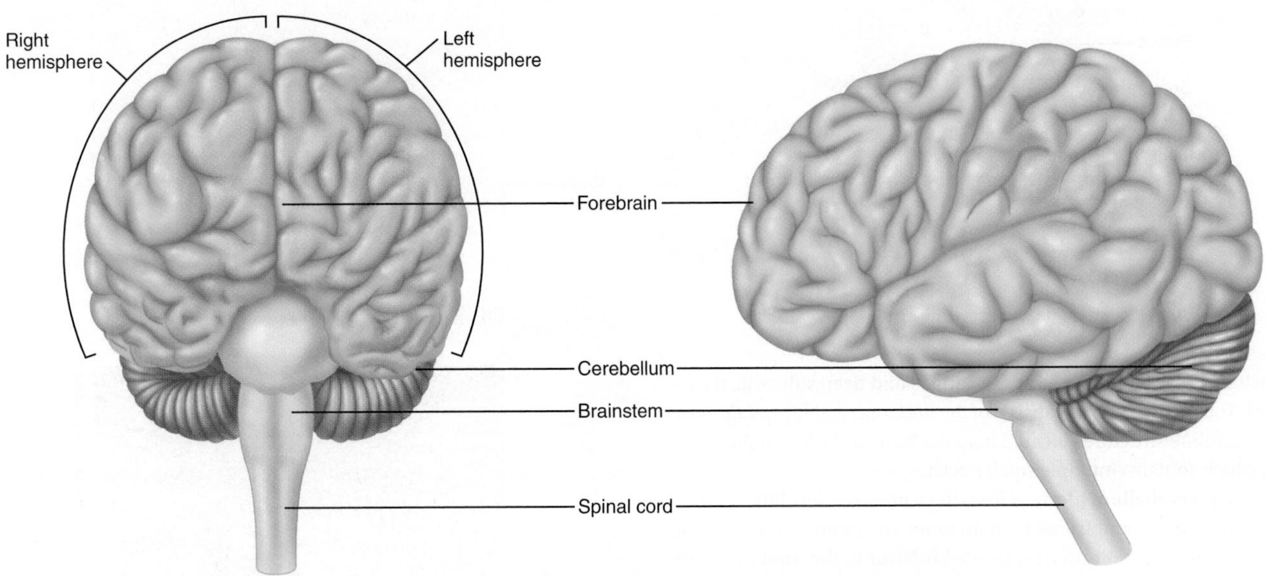

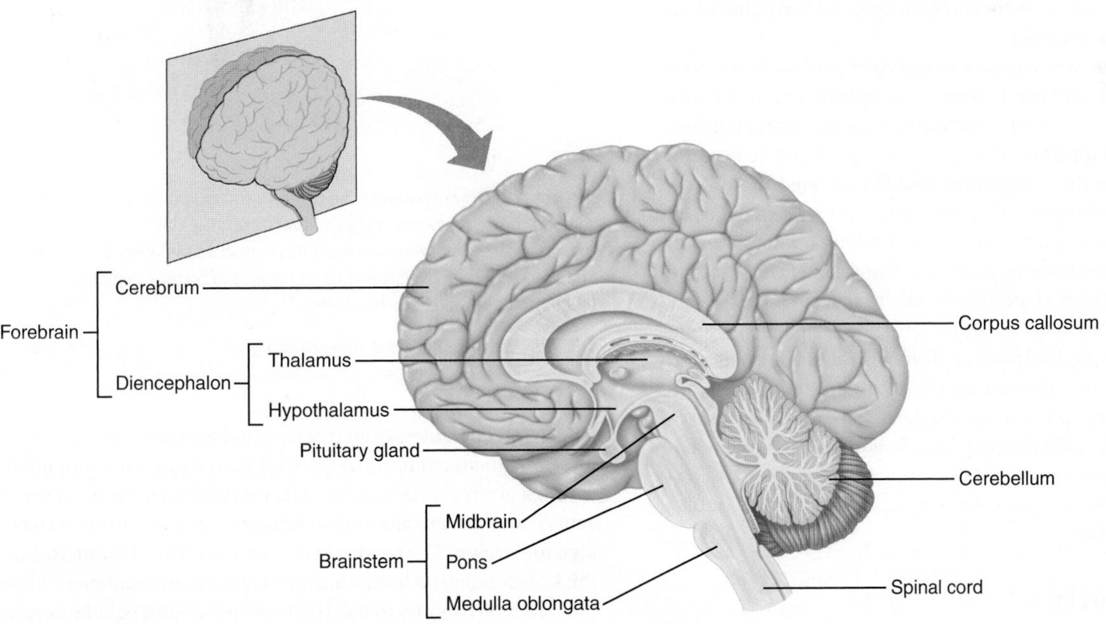

Figure 11 The brain. The brain is composed of three main parts: forebrain, cerebellum, and brainstem. **(a, b)** External views of the brain, showing its three main parts and their relation to the spinal cord. **(c)** A midsagittal section of the brain, showing the structures of the three main brain parts. Note the corpus callosum, the major fiber tract connecting the left and right cerebral hemispheres.

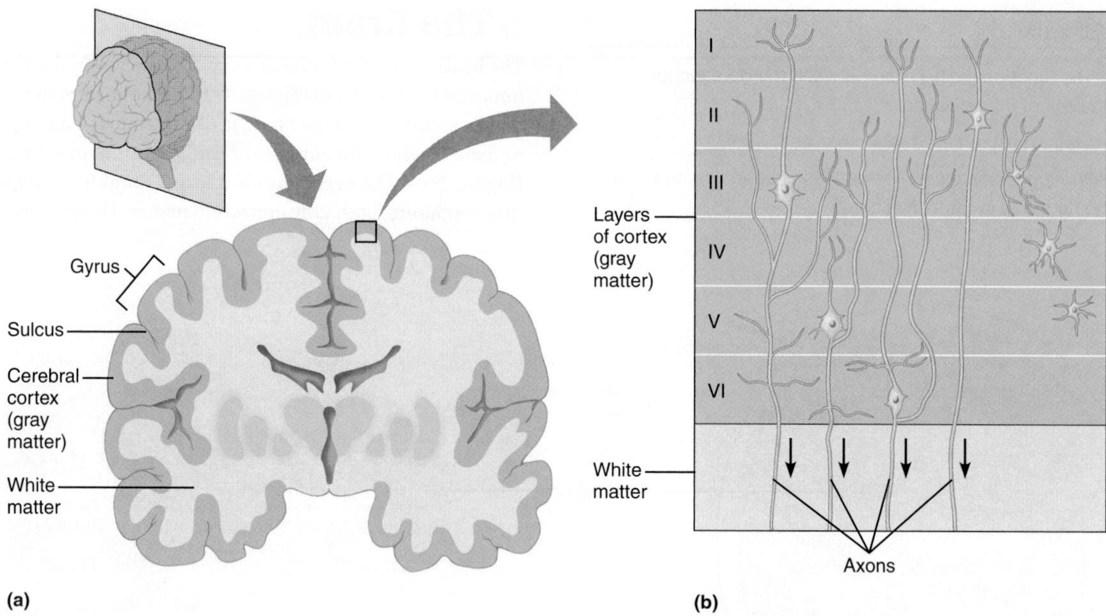

include the *cerebral cortex* at the surface and deep subcortical nuclei. The **diencephalon** consists of the *thalamus* and *hypothalamus*, two midline structures located near the base of the forebrain, each of which contains multiple small nuclei.

The **cerebellum** (derived from the Latin word for "little brain") is bilaterally symmetrical, with an outer cortex and inner nuclei, similar to the forebrain. It is located inferior to the forebrain and dorsal or posterior to the brainstem. The cerebellum functions in motor coordination and balance, providing feedback to motor systems to ensure smooth movements of the eyes and body (described in detail later in this chapter).

The **brainstem,** which is the caudalmost part of the brain, connects the forebrain and cerebellum to the spinal cord. It consists of three main regions: (1) the **midbrain,** the most rostral portion, which connects to the forebrain; (2) the **pons,** the middle portion, which connects to the cerebellum; and (3) the **medulla oblongata,** the most caudal portion, which connects to the spinal cord. The brainstem contains many nuclei that perform a wide variety of functions. Within the brainstem are processing centers for 10 of the 12 pairs of **cranial nerves,** peripheral nerves that emanate directly from the brain rather than the spinal cord. The 12 pairs of cranial nerves and their basic functions are listed in Table 2. Also located within the brainstem is the **reticular formation,** a diffuse network of nuclei that plays important roles in sleep-wake cycles, arousal of the cerebral cortex, and consciousness. In addition, the brainstem is important in the regulation of many involuntary functions controlled by the autonomic nervous system, such as cardiovascular function and digestion.

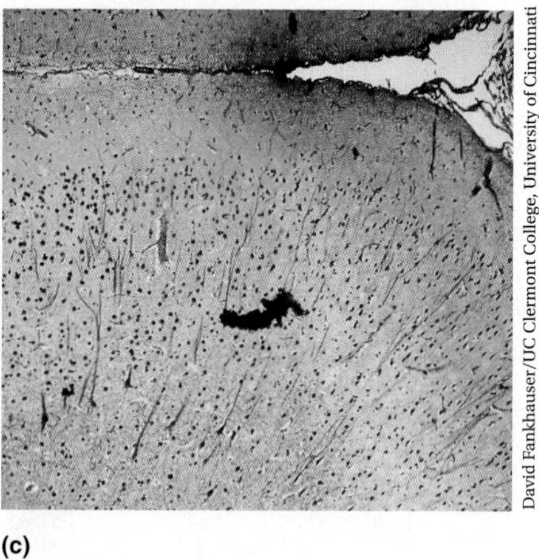

(c)

Figure 12 Organization of the cerebral cortex. (a) Convolutions of the cerebral cortex. **(b)** Layers of the cerebral cortex. The cell bodies are confined to a single layer, but axons and dendrites extend across layers. The actual arrangement of the layers depends on the area of cortex examined. **(c)** Photomicrograph of cortical layers.

Q What are the purposes of the sulci and gyri?

Cerebral Cortex

The cerebral cortex is the outermost portion of the cerebrum, comprising a thin, highly convoluted layer of gray matter (**Figure 12**). The convolutions, which consist of grooves called **sulci** (singular: *sulcus*) and ridges called **gyri** (singular: *gyrus*), allow for a greater volume of cerebral cortex to be accommodated within a given cranial volume. The cerebral cortex is the most advanced area of the brain and, in terms of evolution, the most recent area to develop. The cerebral cortex ranges from 1.5 mm to 4 mm thick, depending on location, and it contains an estimated 1 billion (10^9) neurons and 1 trillion (10^{12}) synapses. Although the cortex is thin, it is generally arranged in six functionally distinct layers (see Figure 12). The specific arrangement of layers varies, depending on the function of the particular cortical region.

TABLE 2 Cranial Nerves

Number	Name	Nerve class	Function
I	Olfactory	Sensory	Olfaction (smell)
II	Optic	Sensory	Vision
III	Oculomotor	Mixed	Eye movements; pupillary reflex; accommodation reflex; proprioception (position of muscles and joints)
IV	Trochlear	Mixed	Eye movements; proprioception
V	Trigeminal	Mixed	Motor control of chewing; somatic sensations of face, nose, mouth
VI	Abducens	Mixed	Eye movements
VII	Facial	Mixed	Motor control of facial muscles, salivary glands, tear glands; somatic sensations of face; taste
VIII	Vestibulocochlear	Sensory	Hearing; equilibrium
IX	Glossopharyngeal	Mixed	Motor control of swallowing and salivary glands; taste; visceral afferent from pharyngeal region; visceral afferent from baroreceptors
X	Vagus	Mixed	Motor and visceral afferent of thoracic and abdominal viscera; motor control of larynx and pharynx
XI	Accessory	Motor	Motor control of larynx and pharynx
XII	Hypoglossal	Mixed	Motor and somatic sensations of tongue

The cerebral cortex carries out the highest level of neural processing. It is in the cortex that we perceive our environment, formulate ideas, experience emotions, recall past events, and command our bodies to move. To carry out all of these complex functions, the cortex acts as an integrating center; that is, it receives many types of sensory input from different origins, consolidates this information, and uses it to formulate thoughts and actions.

Functional Organization of the Cerebral Cortex

Each of the cerebral hemispheres is divided into four regions known as *lobes* (**Figure 13**). The **frontal lobe** is the anterior part of the cerebrum. Immediately posterior to it is the **parietal lobe.** These two lobes are separated by the *central sulcus,* a prominent groove on each hemisphere of the brain. Located posterior and inferior to the parietal lobe is the **occipital lobe.** The **temporal lobe** is located inferior to the frontal and parietal lobes on the side of the cerebrum; it is separated from the frontal lobe by a deep groove called the *lateral sulcus* (or *Sylvian fissure*).

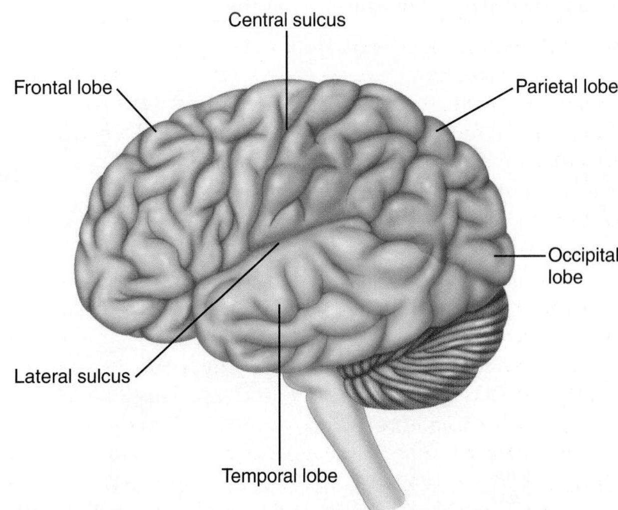

Figure 13 Lobes of the cerebrum. This lateral view of the left cerebrum shows its four distinct lobes: frontal, parietal, occipital, and temporal.

Within each lobe, the cerebral cortex can be subdivided into areas specialized for different functions (**Figure 14**). The occipital lobe is also called the *visual cortex* because the processing of visual information occurs there. Another example is the *auditory cortex,* an area located in the superior temporal lobe that functions in hearing. In the parietal lobe resides the **primary somatosensory cortex,** which is involved in the processing of *somatic* sensory information associated with both **somesthetic sensations** (surface sensations such as touch, itch, temperature, and pain) and **proprioception** (awareness of muscle tensions, and joint and limb positions). The frontal lobe contains the **primary motor cortex,** which initiates voluntary movement, and other areas involved in motor control. The frontal lobe also contains areas involved in language and planning, and it is important to the establishment of personality (see **Discovery: The Story of Phineas Gage**).

Several functional areas of the cerebral cortex are *topographically* organized, meaning that areas of the cerebral cortex can be mapped according to their function. The best examples of this topographical organization occur in the primary motor cortex in the frontal lobe, and in the primary somatosensory cortex in the parietal lobe. Maps of the *somatotopic organization* of these two cortical areas, in which body parts are mapped onto the cortical surface with which they correspond, are called the **motor homunculus** and **sensory homunculus** (plural: *homunculi; homunculus* = "little man"; **Figure 15**). In homunculi, each body part is shown next to the area of cerebral cortex devoted to it, and the relative size of the body part in the drawing represents the relative size of the cortical area devoted to that body part. Consider the fingers and thumb: Because they are capable of both delicate movements and fine tactile discrimination, the digits are greatly overrepresented in the cortex relative to their actual size. Thus homunculi show the extensive sizes of the area of the motor cortex containing neural circuits devoted to the movement of the digits (Figure 15a), and the area of the sensory cortex containing neural circuits that process sensory signals coming from the digits (Figure 15b). Other areas of cortex are also topographically organized. For example, one can construct a map of the visual field (what the eyes see) in the

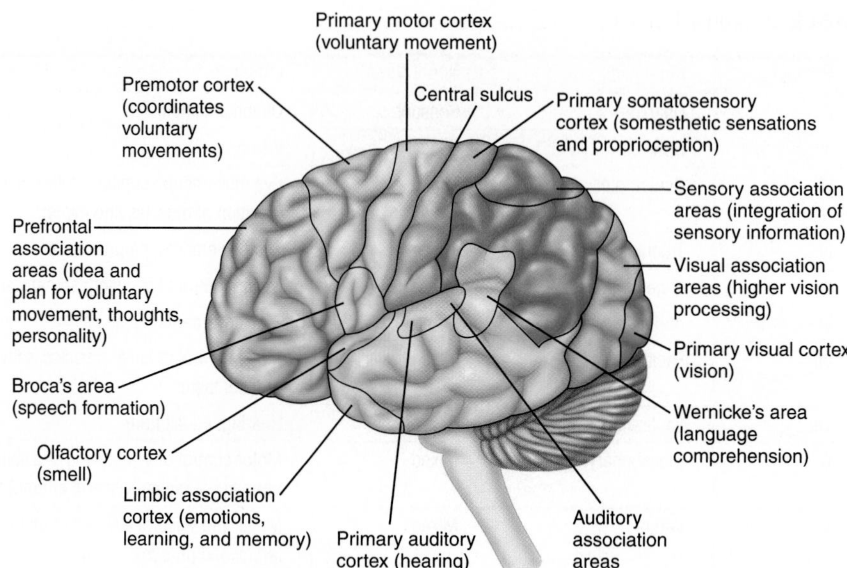

Figure 14 Functional areas of the cerebral cortex. Some selected areas of the cerebral cortex and the specific functions associated with them are illustrated.

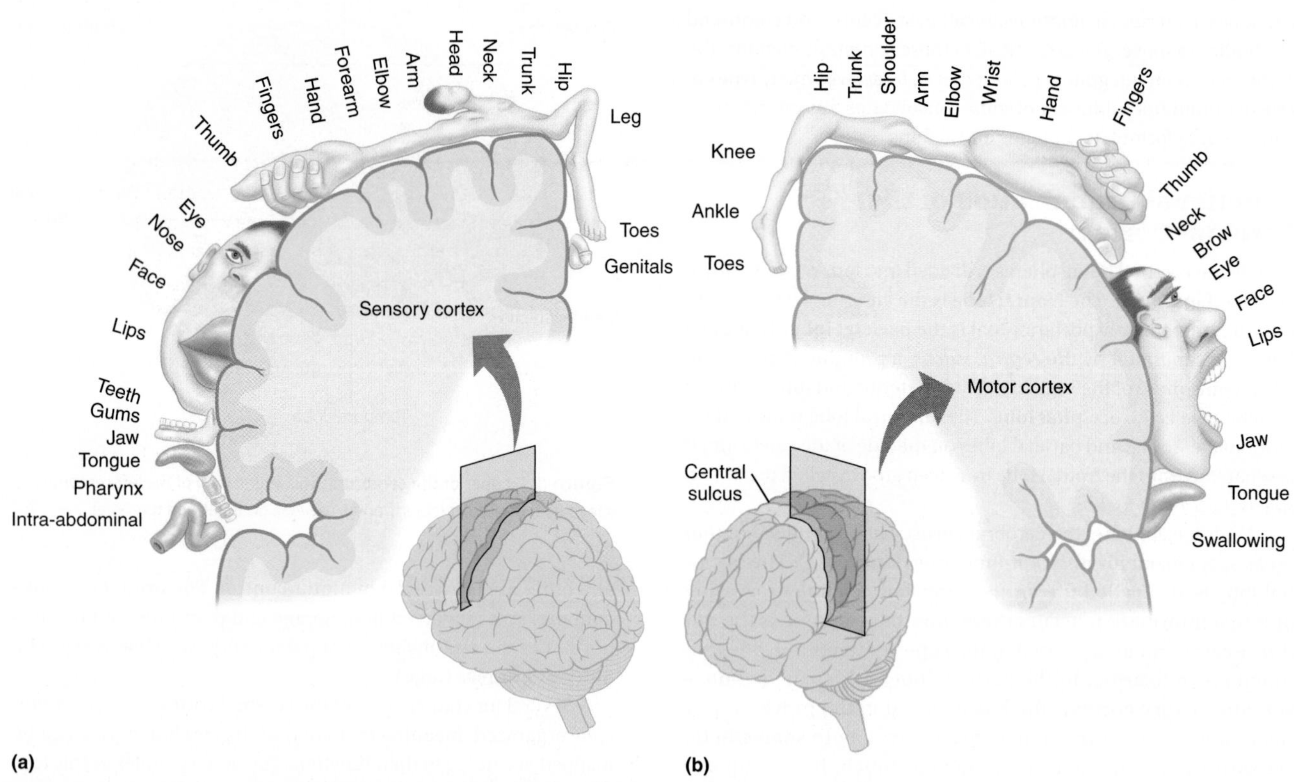

Figure 15 Motor and sensory homunculi. **(a)** Cross section of the primary somatosensory cortex, located just posterior to the central sulcus, and the corresponding somatotopic map of body parts.
(b) Cross section of the primary motor cortex, located just anterior to the central sulcus, and the corresponding somatotopic map of body parts.

primary visual cortex, or a frequency (tone or pitch of sound) map in the primary auditory cortex.

We have so far discussed the cortex as though each area handled one specific function. In reality, that is not precisely true. Certain areas of cortex, called **association areas,** are involved in more complex processing that requires integrating different types of information. For example, if you are awakened in the middle of the night by a loud bang outside your bedroom window, you must decide whether action is needed. You would try to determine what made the loud bang, whether it poses potential danger to you, and, if so, what you can do to avoid it. You might look outside and listen closely for any other unusual sounds. To make a decision, your brain must consolidate information from sensory systems (such as the eyes and ears) and from memory (what things make noises like that?). Much of this type of processing occurs in association areas.

Brain Lateralization

The division of the cerebrum into two hemispheres is indicated by functional differences in the two sections as well as by their anatomical separation. Certain brain functions are dominant on one side of the brain relative to the other, an organization called *brain lateralization*. Brain lateralization is not absolute; that is, not every person's brain is dominant for a particular function on the same side. For example, the left side of the brain controls movement of the right side of the body, whereas the right side of the brain controls movement of the left side of the body. In 90% of the population, the left brain is dominant for movement of the hand; thus 90% of the population is right-handed. As another example, approximately 95% of the population shows dominance for language on the left side of the brain.

Why brain lateralization exists is unknown, but some psychologists believe that it is an adaptive development to have certain survival functions be dominant on opposite sides of the brain, and that this adaptation contributes to survival of the species. In support of this theory, brain lateralization has been identified in many species of animals. Some of the differences in the left and right brains of humans include the following: The right brain is generally associated with movement of the left side of the body, sensory perception of stimuli to the left side of the body, spatial orientation, creativity, face recognition, music, dream imagery, philosophy, and intuition. The left brain is generally associated with movement of the right side of the body, sensory perception of stimuli to the right side of the body, logic and analytical processing, strong language capabilities, and math skills.

Indications of brain lateralization initially came from studies of stroke patients and patients who had the corpus callosum cut as a treatment for epilepsy so that their left and right hemispheres were literally separated (known as split-brain). In split-brain patients, the left and right brains function independently of each other. For

Functional Fact

example, if you show the numbers 1 and 2 in the left and right visual fields, respectively, to a split-brain patient and ask that person what number he or she sees, the individual would likely answer "two" in both instances, because the left brain is dominant for speech and perceives visual information from the right visual field.

Subcortical Nuclei

The subcortical nuclei are regions of gray matter located within the cerebrum. Among the more prominent subcortical nuclei are the basal nuclei (also called basal ganglia), which include the *caudate nucleus*, *globus pallidus*, and *putamen* (**Figure 16**). The **basal nuclei** are notable for their role in modifying movement and are described later in this chapter.

Diencephalon

The diencephalon lies inferior to the cerebrum and includes two midline structures: the thalamus and the hypothalamus (see Figure 16).

Thalamus

The **thalamus** is a cluster of nuclei that functions as a relay center. Except for stimuli related to the sense of smell, all sensory information follows a pathway that includes a direct relay through the thalamus to the cerebral cortex (see Figure 10a for two examples). Much sensory input is filtered by and refined in the thalamus before being transmitted to the cortex. In this manner, the thalamus seems to be important in directing attention, as when a mother is more attentive to her crying baby than to an airplane flying overhead even though the latter may be louder. The thalamus also relays information from the cerebellum and basal ganglia to the motor cortex to provide feedback in controlling movement (described later in this chapter).

Hypothalamus

The **hypothalamus,** which is located inferior to the thalamus, has many roles in regulating homeostasis. In particular, it serves as the major link between the two communication systems of the body, the endocrine and nervous systems. In response to neural and hormonal input, the hypothalamus releases tropic hormones that regulate the release of anterior pituitary hormones. In addition, the hypothalamus controls the release of hormones from the posterior pituitary, including antidiuretic hormone, which regulates plasma volume and osmolarity, and oxytocin, which regulates uterine contractions and milk ejection.

The hypothalamus also affects many behaviors. It contains satiety and hunger centers, which regulate eating behavior, and the thirst center, which regulates drinking behaviors. This portion of the diencephalon also contains the thermoregulatory center. In addition, because the hypothalamus is part of the limbic system (discussed shortly), it affects emotions and behaviors in response to emotions.

The hypothalamus exerts many of its responses through communication with the autonomic nervous system, the efferent branch of the peripheral nervous system that controls the internal organs. The hypothalamus has both direct and indirect inputs to the autonomic nervous system. For example, emotions can affect cardiovascular, respiratory, and digestive functions through hypothalamic input to autonomic control centers in the brainstem. The hypothalamus also coordinates the body's response to stress by triggering neural and hormonal changes that allow the body to adapt.

DISCOVERY

The Story of Phineas Gage

In 1848, while working as a railroad construction foreman, a 25-year-old man named Phineas Gage became the victim of a bizarre accident that yielded early insights into the brain's role in personality. As a result, his story will forever be a part of the history of science.

On the day of the accident, Gage was supervising workers who were laying a new section of track through the rocky hills of Vermont. To make way for the track, they were preparing to remove a large piece of rock by blasting. To do this, workers would bore a hole into the rock, fill it with gunpowder, cover the powder with a layer of sand to shield it from sparks, tamp down the powder and sand with a heavy metal rod (to make the explosion more powerful), and then ignite the powder with a fuse. Gage was in the middle of this procedure when he was distracted by a sound. After looking around to investigate, he returned to his task, unaware that he had forgotten to add the sand. When he dropped the tamping rod down the hole, it apparently created a spark that ignited the powder. The resulting explosion drove the rod upward into his face underneath the left cheekbone. The missile continued on its relentless course, tearing through bone, muscle, and brain tissue, until it shot from the top of his head with enough force to sail about 50 feet into the air. Gage flew backward, lay unconscious on the ground, and then began to convulse. Within minutes, however, he had regained consciousness and was able to move about (with assistance), despite the fact that a substantial mass of brain and skull tissue had been blown out of his head!

Amazed by what they were witnessing, Gage's coworkers transported him by carriage to a physician in a nearby town. For the next few days, Gage's life hung in the balance; the injury had caused the brain to swell, and infection had set in. Nevertheless, he passed through the crisis and within a few months was fully recovered and able to lead a nearly normal life.

Unfortunately, however, Gage did not emerge unchanged. Although the injury had no apparent effect on his ability to reason, speak, or move, it profoundly affected his personality. Prior to the accident he had been a calm, likable individual, but afterward he became quarrelsome and would fight with any man at the slightest provocation. His language became coarse and vulgar, and he could not restrain himself from using obscenities. Once a model citizen, even-tempered and sound in judgment, he was now impulsive, hot-headed, and irresponsible. In the words of his physician, he was "no longer Gage." Because he could not control his behavior, he lost his job and was forced to roam the countryside in search of employment. At one point, he took a job with the Barnum and Bailey Circus, displaying himself and his tamping rod as a sideshow attraction.

Phineas Gage died at age 37, 12 years after the accident that robbed him of his personality. His skull and the tamping rod that pierced it are currently kept at Harvard Medical School. By examining a sketch made by Gage's physician indicating the relative sizes of the tamping iron and Gage's skull, and a computer reconstruction of the probable path taken by the rod as it passed through his head, we can see that the accident likely obliterated a large portion of his frontal lobes, leaving the rest of the brain intact. Because the frontal lobes are important in goal-directed behavior, they exert a restraining influence on the more emotional, impulsive parts of the brain, such as the limbic system. After the accident, this restraining influence was no longer present, which allowed Gage's emotions to rule his behavior.

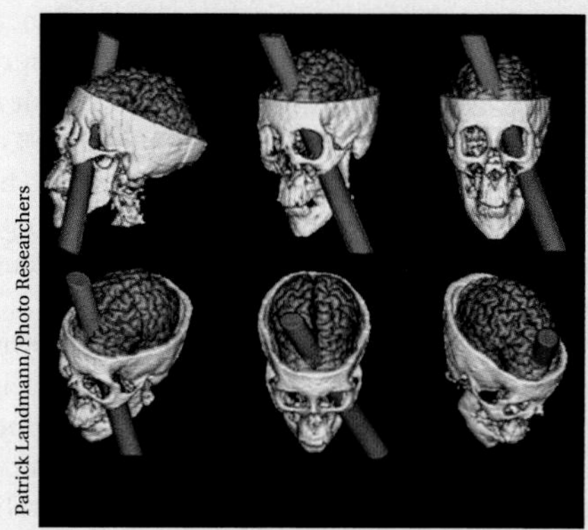

Various views of the approximate position of the tamping rod in Phineas Gage's head.

The suprachiasmatic nucleus of the hypothalamus generates and regulates the circadian rhythm, a series of endogenous fluctuations in body functions that occur on a 24-hour cycle. Extending from the diencephalon is an endocrine organ called the *pineal gland*; it secretes the hormone *melatonin*, which also plays a role in establishing circadian rhythms.

Limbic System

The **limbic system** is a diverse collection of brain areas that is found in all mammals and is associated with autonomic functions, motivation, memory, and emotions. It includes the *amygdala, hippocampus, fornix,* and *cingulate* and *parahippocampal gyri* of the cerebral cortex (**Figure 17**), as well as portions of the basal nuclei, prefrontal cortex, thalamus, and hypothalamus. The limbic system, as one of the more "primitive" areas of our brain, is involved in basic drives. For example, one of the oldest regions of the brain (in terms of evolution) is the amygdala, which is involved in aggression and fear. The hippocampus, a major component of the limbic system, is involved in learning and memory. The olfactory system provides important sensory input to the limbic system, especially in lower mammalian species.

Quick Check 3

1. Name the two structures that make up the diencephalon, and list their major functions.
2. Which three structures make up the brainstem?

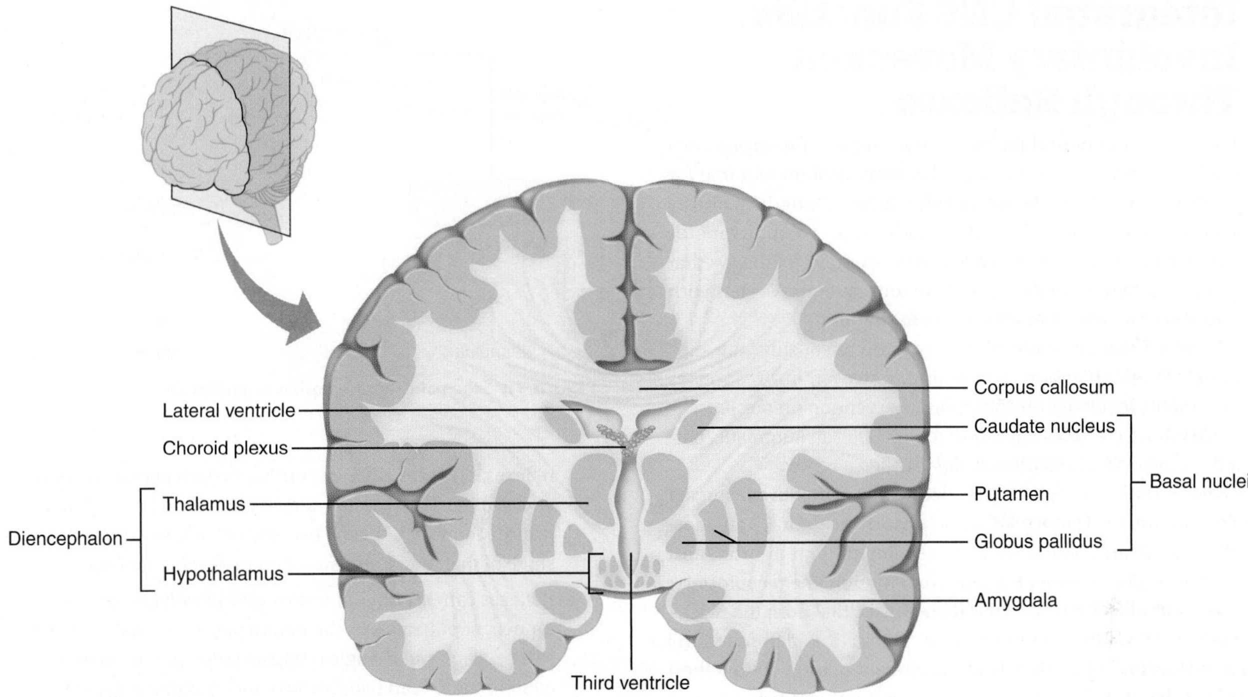

Figure 16 Subcortical gray matter. A coronal section of the cerebrum at the level indicated reveals gray matter areas: the basal nuclei (caudate nucleus, putamen, and globus pallidus), thalamus, hypothalamus, and the amygdala (part of the limbic system).

Q What is the function of the choroid plexus?

3. Name the four lobes of the cerebral cortex, and describe at least one function associated with each.
4. For each of the following functions, indicate whether the left or right cerebral hemisphere plays the predominant role: (a) creative activity, (b) logic, (c) motor control on the left side of the body, (d) sensory perception on the left side of the body, (e) language comprehension and expression.

Thus far we have focused on identifying specific areas of the CNS and describing their basic functions. Although it is true that different areas specialize in different functions, it is important to realize that most tasks carried out by the CNS require the coordinated action of many different areas working together. In addition, the CNS receives and transmits information through the peripheral nervous system in carrying out many tasks. For example, we can identify an object as an apple because we can see it, touch it, smell it, taste it, put those sensations together into a coherent "picture" of the object, and compare that picture with our concept of what an apple is, which derives from our remembered experiences with apples. Thus the seemingly simple task of identifying an apple involves the coordinated activity of several sensory areas, memory storage areas, and association areas. In the next section we examine several CNS functions that involve such coordinated actions, starting with *reflexes*. We then examine more complex tasks, including voluntary motor control, language, sleep and consciousness, emotions and motivation, and learning and memory.

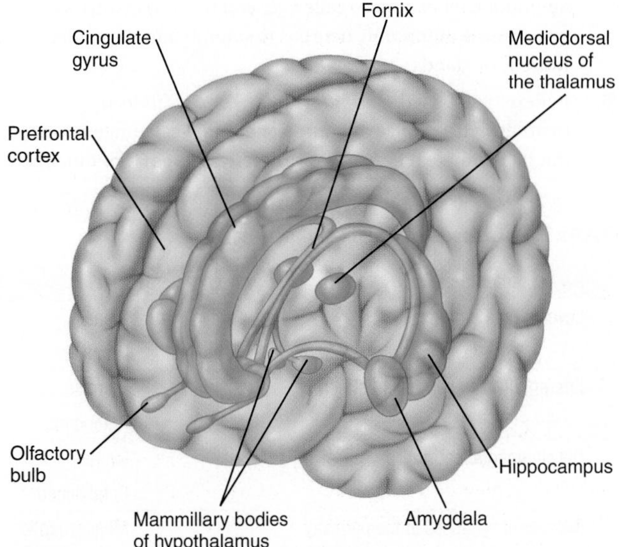

Figure 17 The limbic system.

4 Integrated CNS Function: Involuntary Movement Through Reflexes

We have seen that neural pathways are a series of neurons connected by synapses that form a line of communication required for a specific task. For example, we already learned about the pathway to perceive pain (Figure 10a) and the pathway for voluntary motor control (Figure 10b). We now move to more complex pathways that integrate information in the central nervous system. We start with the pathways for involuntary responses to stimuli.

When we react to some stimuli, we stop to consider the situation before taking action. Sometimes, however, our responses are automatic, involving no conscious intervention on our part, as when we jump in response to a loud sound. Such an automatic, patterned response to a sensory stimulus is called a **reflex.**

Neural pathways for reflexes, known as **reflex arcs,** consist of five components (**Figure 18**): ① a sensory receptor, ② an afferent neuron, ③ an integration center, ④ an efferent neuron, and ⑤ an effector organ. To set the reflex into motion, the receptor first detects a stimulus. Information is then transmitted from the receptor to the CNS via the afferent neuron. The CNS, which functions as the integrator, sends signals to the efferent neuron, which then transmits signals to the effector organ, stimulating it to produce a specific response. The integration can consist of a single neuron or a network of neurons.

Reflexes can be categorized into the following four groups, each of which contains two classes (Table 3):

1. Reflexes can be either *spinal* or *cranial*, based on the level of neural processing involved. In spinal reflexes, the highest level of integration occurs in the spinal cord; cranial reflexes require participation by the brain.
2. Reflexes can be either *somatic* or *autonomic,* depending on which efferent division controls the pathway. Somatic reflexes involve signals sent via somatic neurons to skeletal muscle; autonomic reflexes (also called *visceral reflexes*) involve signals sent via autonomic neurons to smooth muscle, cardiac muscle, or glands.
3. Reflexes can be either *innate* (inborn) or *conditioned* (learned). All people with normal neurological function have the same innate reflexes; we are born with them. Conditioned reflexes, however, differ according to each person's experiences in life. For example, after a person eats too much of a favorite food such that he becomes ill, future exposures to the smell of that food may cause him to develop nausea.
4. Reflexes can be either *monosynaptic* or *polysynaptic.* In monosynaptic reflexes, the neural pathway consists of only two neurons and a single synapse; polysynaptic reflexes contain more than two neurons and multiple synapses.

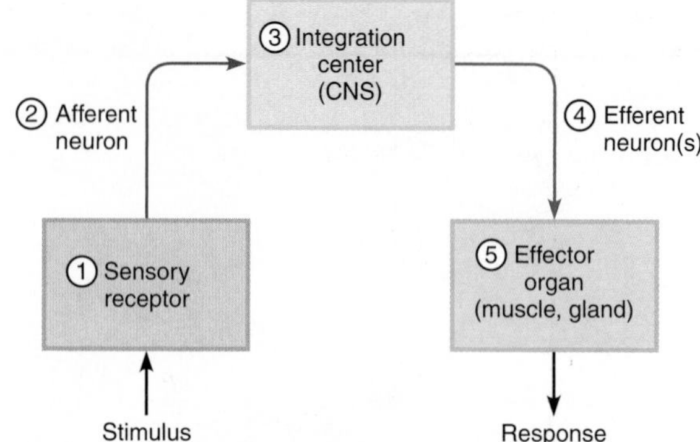

Figure 18 Schematic representation of a reflex arc.

Note that a given reflex is classified within all four groups. For example, consider what happens when the doctor taps your patellar tendon just below the kneecap. In response to the tap, the leg extends (straightens), kicking forward by the *knee-jerk reflex.* The knee-jerk reflex is a spinal reflex (the brain is not involved), a somatic reflex (because efferent signals are sent to skeletal muscles via somatic neurons), an innate reflex (we are born with it), and a monosynaptic reflex (the afferent neuron synapses directly with the efferent neuron).

Now consider what happens when the doctor shines a light into your left eye. Light induces the *pupillary light reflex,* during which the pupils of both eyes constrict even if the light is applied to only one eye. The pupillary light reflex is a cranial reflex (because it involves the brain), an autonomic reflex (because efferent signals are sent to the smooth muscle surrounding the pupil via autonomic neurons), an innate reflex, and a polysynaptic reflex (numerous synapses with interneurons occur between the afferent and efferent neurons).

TABLE 3 Classes of Reflexes

Basis of classification	Classes	Example
Level of neural processing	Spinal	Muscle spindle stretch reflex
	Cranial	Pupillary reflex
Efferent division controlling effector	Somatic	Muscle spindle stretch reflex
	Autonomic	Baroreceptor reflex to control blood pressure
Developmental pattern	Innate	Muscle spindle stretch reflex
	Conditioned	Salivation reflex of Pavlov's dogs
Number of synapses in the pathway	Monosynaptic	Muscle spindle stretch reflex
	Polysynaptic	All other reflexes

The knee-jerk and pupillary light reflexes are innate, as they need not be learned; we all have them as long as our nervous system is normal. In addition, however, we all have reflexes that are unique to us—reflexes we learned based on our life experiences. The classic example of a learned reflex is the *salivation reflex,* first described by Ivan Pavlov, a Russian physiologist who won the Nobel Prize in 1904 for his study on digestion in dogs. In his classic demonstration of the salivation reflex, over a period of time Pavlov rang a bell just before he fed the dogs. The dogs learned to associate the sound of the bell with the arrival of food, so that eventually they began to salivate upon hearing the bell, even if no food was forthcoming. Thus salivation in response to the sound of a bell was a conditioned (learned) reflex in these animals.

Stretch Reflex

The simplest example of a reflex is the **muscle spindle stretch reflex,** of which the knee-jerk reflex is an example. **Figure 19** depicts this reflex in detail. The stretch reflex is the only known monosynaptic reflex in the human body. The receptor in this reflex is a *muscle spindle,* a specialized structure found in skeletal muscles that detects lengthening of the muscle. In the knee-jerk reflex, tapping the patellar tendon below the kneecap stretches the quadriceps muscle in the upper thigh. This stretch excites muscle spindles in that muscle, thereby triggering action potentials that travel in afferent neurons to the spinal cord (integration center). In the spinal cord, the afferent neurons make direct excitatory synaptic connections (synapse 1 in Figure 19) with efferent neurons that innervate the quadriceps muscle, the same muscle that contains the muscle spindles, stimulating the quadriceps to contract and the leg to "kick" forward, or extend.

For extension of the leg to work effectively, the muscles causing extension (quadriceps) should not be opposed by the actions of other muscles that cause the leg to bend or flex (hamstrings). During the stretch reflex, contraction of the hamstrings could slow down or even prevent extension of the leg. This potential problem is avoided because the stretch reflex, once set into motion, triggers not only contraction of the quadriceps but also *inhibition* of the hamstrings. Afferent neurons from muscle spindles have collaterals that synapse with inhibitory interneurons innervating the motor neurons going to the hamstrings (synapse 2 in Figure 19). Simultaneous excitation of the quadriceps and inhibition of the hamstrings causes leg extension during the knee-jerk reflex.

Neural processing does not end here. Collaterals from muscle spindle afferent neurons also ascend to the brain, forming synapses with various interneurons. The brain uses information from muscle spindles (and other receptors from the muscles, joints, and skin) to monitor the contractile states of the various muscles and the positions of the limbs they control. Because the brain controls skeletal muscle contractions, this information provides feedback that enables the brain to adjust its commands to the muscles, a function that is necessary for the execution of smooth and accurate movements.

Apply Your Knowledge

Suppose a doctor taps the patellar tendon and the knee-jerk reflex does not occur. Which type of neurological damage can cause this reflex to disappear?

Withdrawal and Crossed-Extensor Reflexes

When a portion of the body is subjected to a painful stimulus, it withdraws from the stimulus automatically via a response called the **withdrawal reflex,** as occurs when a person steps on a tack (**Figure 20**). Stepping on a tack is perceived as painful because it activates special sensory receptors called *nociceptors,* which respond to intense stimuli that are damaging (or potentially damaging) to tissue. Afferent neurons from nociceptors transmit this information to the spinal cord, where they have excitatory synapses on interneurons. The interneurons then excite efferent neurons (synapse 1 in Figure 20) that innervate the skeletal muscles that cause withdrawal of the limb or other body part. In the example in Figure 20, the hamstring muscles contract to withdraw the leg.

For the withdrawal reflex to work effectively, the muscles that cause withdrawal should be excited, whereas the muscles that oppose withdrawal (or cause extension) should be inhibited. This is indeed the case because branches of the afferents activate inhibitory interneurons to the efferents innervating the quadriceps (synapse 2 in Figure 20). Therefore, the withdrawal reflex simultaneously triggers

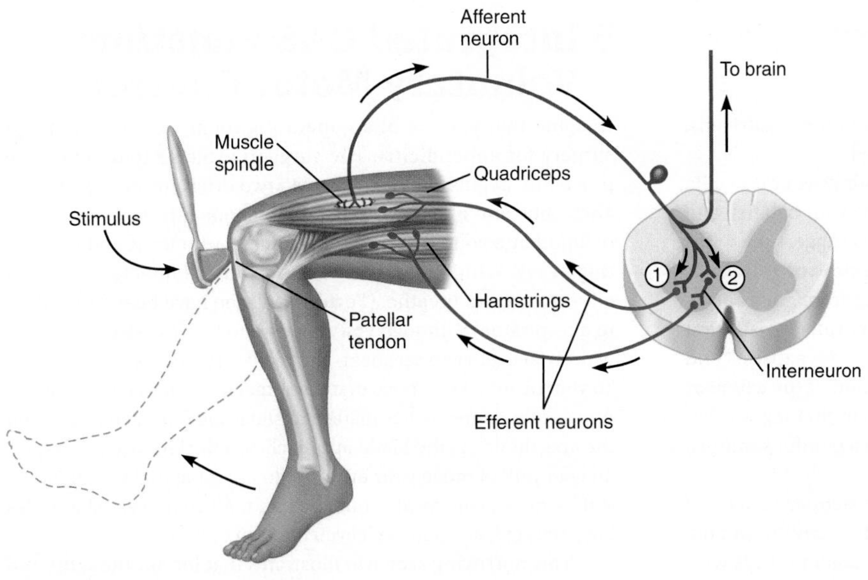

Figure 19 **The muscle spindle stretch reflex.**

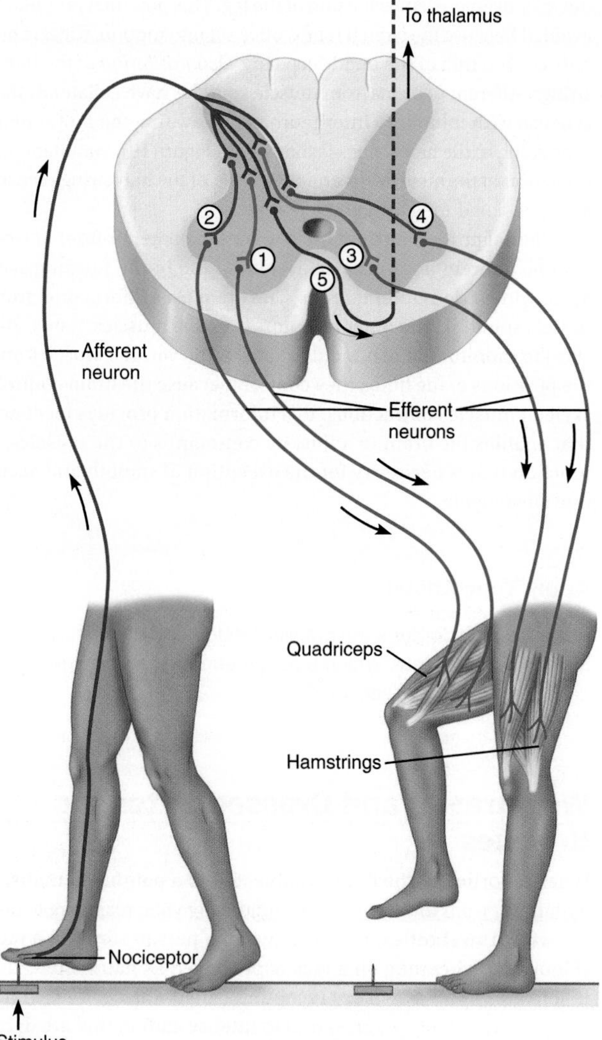

Figure 20 Withdrawal and crossed-extensor reflexes.

contraction of the hamstrings and relaxation of the quadriceps, allowing flexion of the leg to proceed unimpeded.

But another problem remains: If you withdraw your leg while it is supporting your weight, you may well lose your balance and fall over. Fortunately, when a painful stimulus triggers the withdrawal reflex, another reflex—the **crossed-extensor reflex**—is initiated simultaneously. The afferent neurons from nociceptors have branches that send signals (via interneurons) to efferent neurons controlling muscles on the opposite leg (synapses 3 and 4 in Figure 20). These signals trigger contraction of the extensor muscles and relaxation of the flexor muscles in that leg, so that when the first leg is withdrawn in response to a painful stimulus, the other leg is extended to support the body.

Would the person be aware that he or she stepped on a tack? Of course, because the same afferent neurons that activate both the withdrawal and crossed-extensor reflexes also communicate with ascending pathways that send information to the brain that a painful stimulus has been applied to the body (synapse 5 in Figure 20). Will the withdrawal reflex cause you to withdraw from every painful stimulus you encounter in life? Probably not, because sometimes reflexes are detrimental. Suppose you were to pick up a cup of tea served in your mother's favorite china and found the cup to be painfully hot. Knowing how much your mother likes her china, you might attempt to override the reflex and hold on to the cup, even if you experience pain by doing so. In this process, the brain activates inhibitory interneurons in the spinal cord that suppress activity in the neurons that control the withdrawal reflex. Remember, postsynaptic cells (in this case, the efferent neurons that control the muscles) sum information that comes from different synaptic inputs. If the inhibitory influence that descends from the brain is greater than the excitatory influence coming from the nociceptor afferents, then the withdrawal reflex will not occur.

Pupillary Light Reflex

The stretch reflex and withdrawal reflex are examples of somatic spinal reflexes. The pupillary light reflex, in contrast, is an example of an autonomic cranial reflex in which the stimulus is light entering one eye and activating photoreceptors. The photoreceptors activate afferent neurons that transmit signals to areas in the midbrain of the brainstem that function as the integration center. These midbrain areas then activate, through a polysynaptic pathway, autonomic efferents that innervate the smooth muscle surrounding the pupils of both eyes. The response is pupillary constriction (reduction in diameter of the pupils) in both eyes.

Quick Check 4

1. Is salivation in response to a bell an example of an innate reflex or a conditioned reflex? Why? What about salivation in response to the smell of food?
2. What are the five components of a reflex arc?

5 Integrated CNS Function: Voluntary Motor Control

Imagine that you are in an operating room, about to undergo surgery for appendicitis. The surgeon explains that before the procedure begins, you will be given two drugs intravenously—an anesthetic and a muscle relaxant. Someone injects a syringe full of liquid into your arm, and soon you feel your arms and legs getting heavy. Within minutes, you find yourself unable to move at all—not even to breathe. (Fortunately, you have been hooked up to a respirator.) Although you expected to become sleepy, you are wide awake and can see, hear, and feel everything that is going on. To your dismay, you notice that the surgeon has picked up a scalpel. Apparently, someone has made a mistake and forgotten to give you the anesthetic. As the blade approaches your skin, you try desperately to yell, to move your arms, to do anything that would let the staff know you are awake, but you can't. All your skeletal muscles are paralyzed, and you can't even blink an eye!

This horrifying scenario illustrates that for our thoughts and intentions to become actions, the nervous system must be able to communicate with muscles and trigger muscle contraction. The purpose of this section is to describe what happens when we control our muscles voluntarily—in other words, what happens when our intentions are translated into actions.

Neural Components for Smooth Voluntary Movements

The successful completion of a voluntary motor task involves four components: (1) developing the idea to move, (2) putting together a program of motor commands to carry out the movement, (3) executing the movement by activating the correct muscles at the correct time, and (4) constant feedback to ensure that the movement is carried out smoothly and successfully. Each of the steps is described next and illustrated in **Figure 21**.

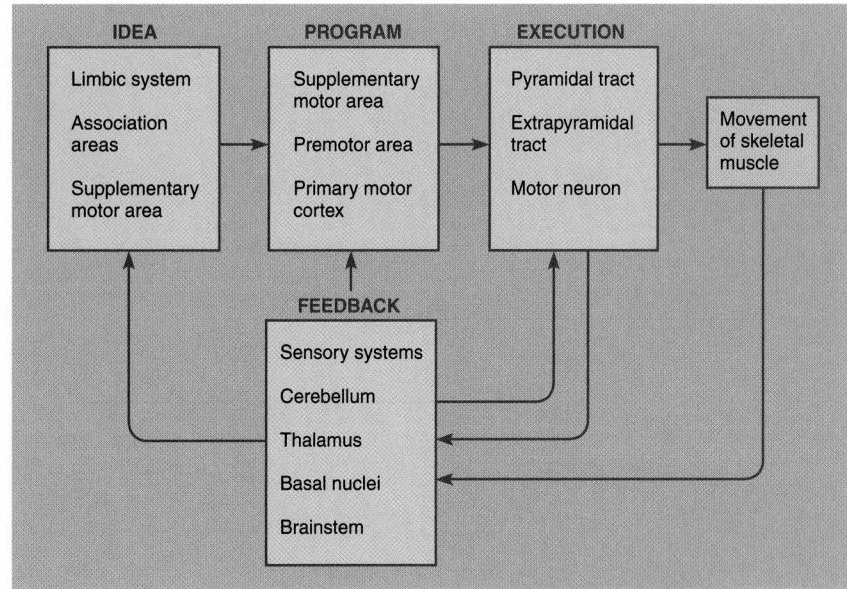

Figure 21 Steps in voluntary movement. Voluntary movement requires coordinated activity of several neural structures to ensure smooth skeletal muscle movement. The process begins with the idea to move.

The idea or intention to move must first be formulated. The formulation of intentions, and their translation into plans for movement, is exerted by several areas of the brain including the prefrontal cortex, association areas of the cerebral cortex, basal nuclei, and the limbic system. The idea is generally goal oriented, and is based on sensory input, memories, emotions, or other motivating factors.

A program of motor commands must be developed to ultimately direct the appropriate muscles to contract at the right time to produce the intended motion. Such development involves many areas, including the primary motor cortex, the prefrontal association area, and the premotor cortex. In addition to stimulating the correct muscles to contract at the appropriate time, it is just as important to inhibit contraction of muscles to inhibit motion.

The program of motor commands must be executed by sending commands to the muscles, telling them to contract. The factor that ultimately determines whether a certain muscle cell will contract is whether the efferent neuron innervating that muscle cell is active. Thus the execution of any motor program requires activation of the appropriate efferent neurons. The efferent neurons that control skeletal muscles originate in the ventral horn of the spinal cord and are called **lower motor neurons,** or simply **motor neurons.** (For those muscles that are controlled by cranial nerves, the motor neurons originate in certain cranial nerve nuclei of the brainstem.)

Input to motor neurons originates from many different levels in the nervous system. In the previous section, we saw that motor neurons can be excited or inhibited by input from afferent neurons, as is the case for spinal reflexes. We also saw that activity in motor neurons can be influenced by input descending from the brain, as when a person overrides a withdrawal reflex through conscious effort. In this section we are interested primarily in descending input from the brain to motor neurons. Several descending pathways are important in voluntary motor control that provide input to motor neurons; they can be divided into two categories: *lateral pathways* and *ventromedial pathways*.

As the motor task is being carried out, continual feedback to the CNS enables it to ascertain whether the task is being accomplished according to the intention and the program. Such feedback facilitates the CNS in making any needed adjustments in the program, and possibly even changing the original intention, so as to ensure smooth and reliable completion of the task. The cerebellum, thalamus, basal nuclei, certain brainstem nuclei, and various cortical association areas provide this feedback and assist with development of motor programming.

Lateral Pathways Control Voluntary Movement

The two types of tracts that descend from the brain to the spinal cord along lateral pathways and that control skeletal muscle activity are called *pyramidal tracts* and *rubrospinal tracts*. These pathways control movement of the distal extremities (portions of limbs located farthest from the trunk).

The **pyramidal tracts** (**Figure 22a**) are direct pathways from the primary motor cortex to the spinal cord. The axons of neurons in these tracts terminate in the ventral horn of the spinal cord and are called **upper motor neurons;** some synapse directly on motor neurons, whereas others synapse on interneurons. Most axons in the pyramidal tracts cross over to the opposite side of the CNS in an area of the medulla called the *medullary pyramids.* In addition, the axons have collaterals that travel to other motor areas, including brainstem nuclei. The pyramidal tracts are believed to primarily control the fine discrete movements of the distal extremities, especially the forearms, hands, and fingers. All descending motor pathways other than the pyramidal tracts are part of the extrapyramidal tracts.

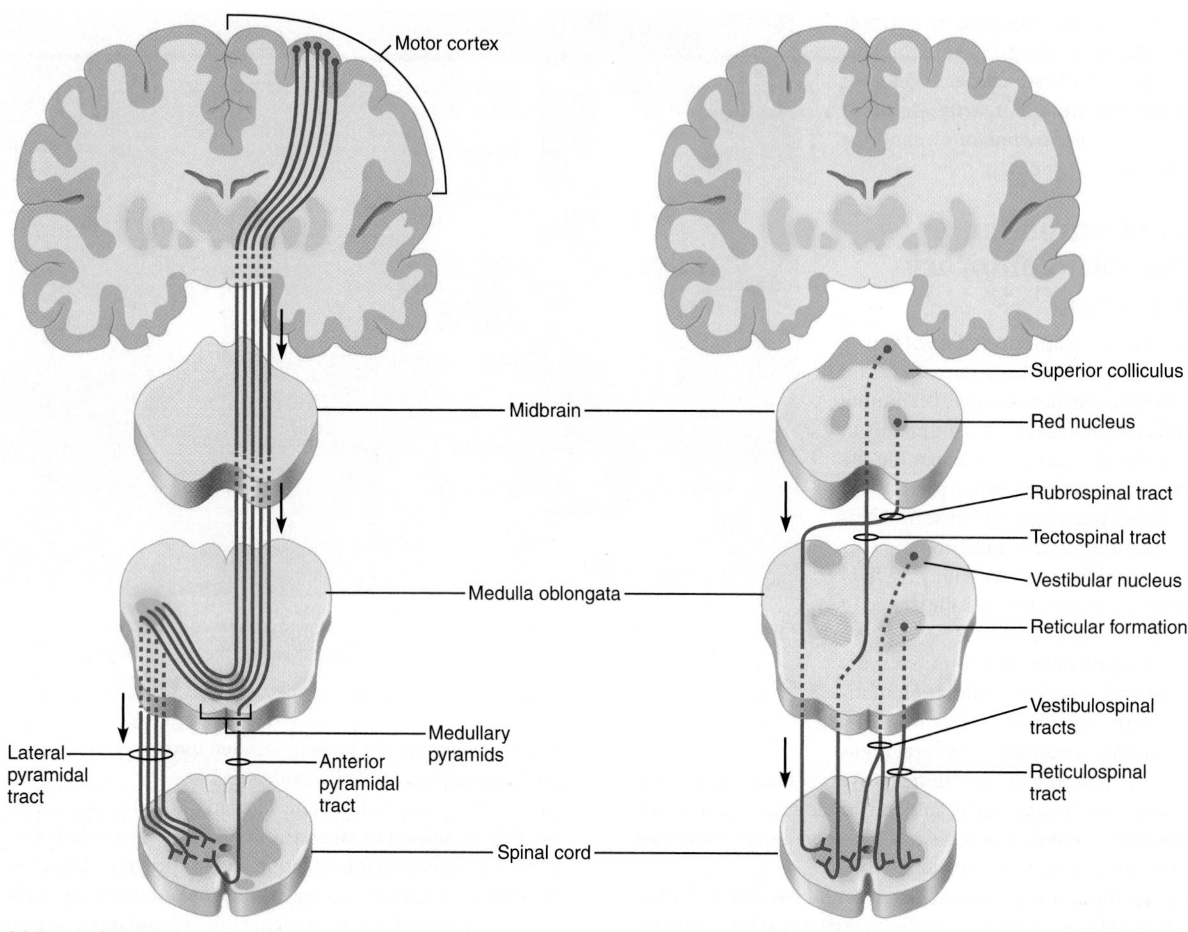

Figure 22 Pyramidal and extrapyramidal tracts. (a) The lateral pyramidal tract is primarily a crossed pathway, whereas the anterior pyramidal tract is primarily uncrossed until it reaches the spinal cord. **(b)** The extrapyramidal tracts include all motor tracts except the pyramidal tracts; some cross, some do not, and some are bilateral.

Q Where do the pyramidal tracts originate?

The **rubrospinal tracts** originate in the *red nuclei* of the midbrain. Axons of the rubrospinal tracts decussate (cross to the opposite side) and then join the axons of the pyramidal tract. Communication to the red nuclei is primarily from the primary motor cortex, making the rubrospinal tracts almost an extension of the pyramidal tracts, but with much less significance in motor control compared to the pyramidal tracts in humans.

Ventromedial Pathways Control Voluntary and Involuntary Movements

The ventromedial pathways include three descending tracts that originate in the brainstem (Figure 22b). Unlike the pyramidal tracts, the ventromedial pathways are indirect connections between the brain and spinal cord; that is, neurons of the ventromedial pathways do not form synapses on motor neurons. The primary influence of the ventromedial pathways is on the muscles of the trunk, neck, and proximal portions of the limbs (portions of the limbs closest to the trunk). Each of the three pathways is briefly described next.

The **vestibulospinal tracts** originate in the *vestibular nuclei* of the medulla, which receives information from the inner ear regarding movement of the head. The vestibular spinal tracts control the head, neck, and lumbar muscles so as to maintain posture and balance.

The **tectospinal tracts** originate in the *superior colliculi* of the midbrain, which receives input from the eyes and, to a lesser extent, from the skin and ears. The tectospinal tracts control head and eye movements that correctly position the eyes to look at a desired object and to follow that object if it is moving.

The **reticulospinal tracts** originate in the reticular formation of either the pons or the medulla. These pathways are important for balance, with the pontine reticulospinal tract stimulating extensors of the legs to maintain an upright posture against the force of gravity, and the medullary reticulospinal tract inhibiting those same extensors.

To carry out coordinated activity, the lateral and ventromedial pathways must work together. Generally speaking, the ventromedial pathways are involved in controlling large groups of muscles that contract together to maintain posture and balance, whereas the pyramidal tracts are more involved in the control of smaller groups of muscles that can contract independently. For example, when a seamstress threads a needle, the pyramidal system controls the fine movements in her arms, hands, and fingers, while the ventromedial pathways control the gross movement of her trunk, neck, and legs to maintain appropriate posture.

The Control of Posture by the Brainstem

The brainstem contains a number of nuclei that are part of the ventromedial pathways that project to the spinal cord. Even though the separate roles of these nuclei are not fully understood, it appears that they are generally involved in the involuntary control of posture. In addition, these nuclei cannot initiate voluntary movements.

Our muscles constantly adjust to changes in posture, whether we are moving or standing still. Even the simple act of reaching for a doorknob requires alteration of muscular tension in the legs and trunk because we change the position of our body's center of gravity when we extend an arm. To exert postural control, the brainstem uses information from sensory receptors in the skin, proprioceptors in muscles and joints, and receptors in the vestibular system. These inputs inform the brainstem about the body's position and the forces acting upon it. The brainstem also receives sensory information from the eyes and ears, so that it can make postural adjustments in response to visual or auditory stimuli. A good example is your reaction when you hear an unexpected sound. When you hear the snap of a twig in the forest, you may automatically stop walking and turn your head toward the source of the sound. In addition, this example illustrates how feedback to motor control systems can alter not only the programming of movement but also the intention or plan: The original plan may have been to take a long walk through the woods; depending on what snapped the twig, the new task may be to sprint out of the woods.

The Role of the Cerebellum in Motor Coordination

Before launching a spacecraft, mission controllers formulate a trajectory and then program the ship's computer with information regarding the planned route. After the launch, however, the ship never follows the plan exactly. Instead, an onboard guidance system continually monitors the ship's position and compares it with the planned route, so that it can make course corrections when necessary.

In many respects, the cerebellum acts much as a "guidance system" for movement of the human body. After the cortex formulates plans of action and commands the muscles to execute them, the cerebellum compares the actual movements as they are occurring with the plans, making corrections in the force and direction of movement whenever it detects a discrepancy. In addition, the cerebellum appears to play a role in maintaining muscle tone (a constant low level of tension that is present in muscles at rest). The cerebellum may also store memories of motor activities, so that previously executed movements can be refined and executed more smoothly when they are attempted again.

To do its job, the cerebellum must both receive information from the cortex regarding planned movements and be continually

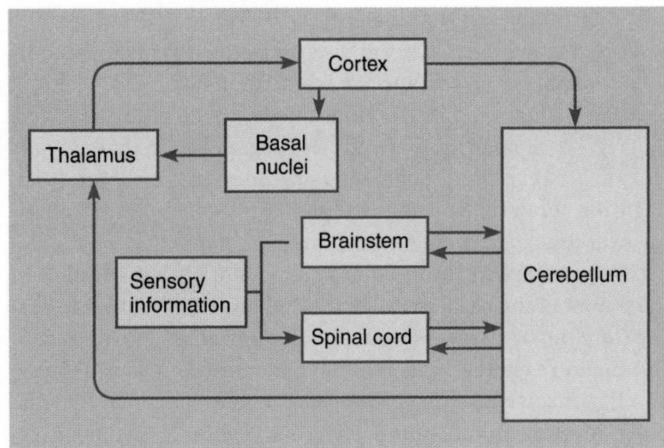

Figure 23 **Major pathways for information flow to and from the cerebellum.**

updated regarding how motions are actually being carried out. It receives input from sensorimotor areas of the cortex (areas involved in sensory and/or motor processing), the basal nuclei, the brainstem, and the spinal cord, as well as sensory information originating from all areas of the body (**Figure 23**). It then sends signals back to the cortex (via a relay in the thalamus), so that the cortex can adjust its output to complete the planned movement successfully.

A person with damage to the cerebellum retains the ability to carry out voluntary movement, but the resulting motions are clumsy and misdirected, and often made with inappropriate force. Consider the numerous intricate finger, wrist, and arm movements required to pick up a pencil—a task you can do without even thinking about it. What would ordinarily be a smooth sequence of movements, one leading seamlessly into the next, may be executed as a series of independent steps, each requiring conscious control, in an individual with cerebellar injury. A characteristic sign of cerebellar damage is an *intention tremor,* a sometimes-violent shaking motion that occurs only when a person makes a voluntary movement. When a person with cerebellar damage reaches for a light switch, for example, his or her hand does not head straight toward the switch, but instead begins to make erratic motions that become more pronounced as the hand nears the intended destination.

The Basal Nuclei in Motor Control

The basal nuclei are thought to have functions similar to the cerebellum, in that they provide feedback to the cortex for the development of motor strategies and smoothing out of movements. Some evidence also suggests that the basal nuclei are necessary for automatic performance of learned repetitive motions. The basal nuclei receive input from the cortex and send output back to the cortex via a relay in the thalamus. One of the functions of this "loop" is to assist the cortex in the selection and initiation of purposeful movement while inhibiting unwanted movements.

By studying two well-known diseases that affect motor control, *Huntington's chorea* and *Parkinson's disease,* we have learned that the effects of the basal nuclei on motor activity are generally inhibitory. Huntington's chorea is a progressive genetic disease characterized in its early stages by the loss of motor coordination and the appearance of exaggerated involuntary jerking or twitching motions involving the limbs and face. Loss of motor control is progressive, so that patients eventually become bedridden. At the same time, cognitive function (the ability to think) also declines, progressing over time to a complete deficit in mental function, or *dementia*. The damage in Huntington's chorea entails a disruption in the output of the basal nuclei to the thalamus. Normally, the output from the basal nuclei would inhibit unwanted movement; in Huntington's chorea, certain motor circuits become overactive.

Health Fact

Parkinson's disease is a progressive neurodegenerative disorder affecting motor control. Signs of this disease include involuntary tremors at rest; a stooped, shuffling gait; movements that are generally slow and stiff; and difficulty in initiating movement, which often requires careful planning and intense effort. Some disruption of cognitive function may occur as well. The dysfunction in Parkinson's disease involves a loss of input to the basal ganglia from dopaminergic neurons in an area of the midbrain called the *substantia nigra*.

Quick Check 5

1. What are the functional differences between the lateral and the ventromedial pathways for motor control?
2. Briefly describe the role of each of the following structures in motor control: the cerebellum, the basal nuclei, the brainstem nuclei, and the thalamus.

6 Integrated CNS Function: Language

Compared to the brains of any other species, the human brain excels in its ability to manipulate symbols that it uses to encode thoughts and ideas. Among these symbols are letters, words, and sounds, which we use to formulate *language*—the means by which we communicate our thoughts and ideas to others. Given the importance of language to humans, it should come as no surprise that our brains contain several areas specialized for this ability. In fact, two areas of the association cortex, which is located in the left hemisphere in 90%–95% of the population, are completely devoted to language: **Wernicke's area,** located in the posterior and superior portion of the temporal lobe and the inferior parietal lobe, and **Broca's area,** located in the frontal lobe (see Figure 14). Wernicke's area is involved in language comprehension—that is, the ability to understand language. Although close to the auditory cortex, Wernicke's area functions not in hearing language, but rather in comprehending language in many forms, such as sound, written words, or even hand signals such as those used in sign language. Broca's area, in contrast, is involved in language expression—that is, our ability to speak or write words. Although close to the primary motor cortex region that controls the mouth, it functions in forming oral or written words with appropriate syntax for comprehension by others. Several other areas of the brain also contribute to language.

Damage to either Wernicke's or Broca's areas produces *aphasia,* or language dysfunction, but the characteristics of the aphasia differ depending on which area is damaged. Damage to Wernicke's area produces *Wernicke's aphasia,* in which a person has difficulty understanding language, regardless of whether the words are spoken or written. Although a person suffering from Wernicke's aphasia can speak words, the words make no sense. Damage to Broca's area, by contrast, produces *Broca's aphasia.* In this case, affected individuals comprehend language and know what they want to say but cannot make the correct sounds or write the correct words. Thus, if you were to ask someone with Broca's aphasia, "How are you doing?" he or she might try to answer, but the response would come up slurred, stuttered, or grammatically incorrect. In contrast, a person with Wernicke's aphasia might respond with a fluent, but meaningless phrase such as "Sun school you see."

As stated earlier, Broca's and Wernicke's areas process sign language as well as spoken language. Surprisingly, the right auditory cortex, which cannot perform its normal function in deaf people, takes part in the processing of sign language. Damage to Broca's area in deaf people causes a deficit in the ability to form signs, even though the person can move his or her hands normally for other functions. Likewise, damage to Wernicke's area causes difficulty in interpreting signs, even though the person can form signs correctly with the hands. When put together, however, these signs would make little sense.

Some people who have learned two or more languages are bilingual or multilingual, respectively. Learning a new language is easier when one is younger because the brain is still developing and, therefore, is more flexible and adaptive. For example, the Japanese language does not include any sounds similar to the English "r." When an adult Japanese person tries to learn English, making the "r" sound is difficult because he never learned that particular motor skill; the muscle development and innervation may not be there to make the "r" sound.

When a person can speak more than one language, both languages are controlled by the same language centers, and there appears to be much overlap in neural processing. If a bilingual person has a stroke in the language centers, it may affect one or both languages. The most recently learned language will generally suffer the greatest deficit. However, if both languages were learned about the same time, then the two languages tend to be equally affected, and any recovery occurs simultaneously.

7 Integrated CNS Function: Sleep

We all know what sleep is, and we spend approximately one-third of our lives doing it. However, *why* we sleep is not clearly understood, and although the topic has been extensively studied, little is known about *how* we sleep. In this section we examine what is known about this familiar, yet mysterious phenomenon.

Functions of Sleep

Sleep is an active process in that it requires energy; it is also a necessary process that all mammals require. Sleep is defined as a cyclically occurring state of decreased motor activity and perception.

Note that activity is decreased compared to the awake state, but not altogether absent. Occasional bursts of motor activity occur, for example, when a sleeping person shifts position in bed. Perception also occurs because a person can be awakened by sensory stimuli such as sound or touch.

Researchers have proposed many theories to explain the biological usefulness of sleep. Some believe that its function is mainly restorative—that it provides a resting period that allows the body and the brain to fully recover from their daily activity. Others suggest that sleep allows us to conserve energy. Still others think that sleep may have evolved because dreaming is useful by providing, for example, an opportunity for mentally practicing and refining adaptive behaviors (escaping from predators, for instance) without actually having to perform them, which could place the body in danger. Some evidence suggests that sleep and dreaming may facilitate the storage of long-term memories. Another possibility—that sleep may be necessary for maintaining adequate immune system function—is supported by the observation that animals deprived of sleep for long periods often get infections.

Much of what we know about sleep and its effects on brain function comes from *electroencephalography,* a technique that uses electrodes affixed to the scalp to record electrical activity in the brain. Recordings of such activity, commonly known as *brain waves,* are called **electroencephalograms (EEGs).** Using this method, researchers have identified two different kinds of sleep: **slow-wave sleep (SWS),** which is characterized by multiple stages of low-frequency waves in the EEG, and **REM sleep,** which is characterized by high-frequency waves in the EEG and periodic episodes of rapid eye movement (REM).

In SWS, which is also called non-REM sleep (or NREM sleep), muscle tone is present but diminished compared to the awake state. Spinal reflexes are still present and occasional bursts of involuntary activity occur every 10–20 minutes.

Somnambulism (sleepwalking) is more likely to occur during the deeper stages of SWS than during lighter stages of SWS sleep and not at all during REM sleep. Sleepwalking occurs in approximately 40% of children but is rare in adults. Talking in your sleep (somniloquy) also occurs during SWS. Brain activity decreases in most regions during SWS, but activity in the parasympathetic nervous system and certain brainstem structures increases. During SWS, a person may have recurring thoughts and may dream, but dreaming is sparse and thoughts and dreams tend to be more logical and less emotional compared to those occurring in REM sleep. Snoring may occur during SWS.

In REM sleep, postural muscles lose their tone and become paralyzed. However, the muscles that control the face, eyeballs, and distal limbs twitch frequently. The jaw muscles relax, frequently causing the mouth to open. Snoring (if it occurs during SWS) stops. Compared to SWS, an overall increase in brain activity occurs, except in certain areas of the brainstem, where activity decreases. Breathing is generally faster and more variable. Activity in the sympathetic nervous system is increased, causing increases in blood pressure and heart rate. Loss of body temperature control causes body temperature to fluctuate more with environmental temperature. Dreams and thoughts in REM sleep are more elaborate and intense than in SWS and are more bizarre and narrative than in SWS. Nightmares may also occur during REM sleep. Children are more likely than adults to have nightmares, although some adults suffer from frequent nightmares (see **Clinical Connections: Post-Traumatic Stress Disorder**). Table 4 summarizes differences between SWS and REM sleep.

In comparison to sleep, anesthesia and coma are states of depressed activity in the central nervous system. Whereas a person can be awakened from the sleeping state, a person cannot be awakened from anesthesia or coma.

Sleep-Wake Cycles

The human body alternates between periods of wakefulness and periods of sleep, normally completing one complete sleep-wake cycle in a 24-hour period. Other body functions, such as body temperature, also vary with the sleep-wake cycle (**Figure 24**). Although we typically describe the average daily sleep time as 8 hours for adults, the length of the stages in this cycle actually varies with age. Infants tend to spend at least 17 hours per day sleeping, whereas teenagers spend approximately 9 hours sleeping. The natural biological rhythm for teenagers is also offset compared to adults in that teenagers are more alert later in the day and evening and less alert early in the morning. Adults spend less time sleeping as they age, with some adults needing only about 5 hours of sleep per day.

What causes our body to shift between these different levels of consciousness? Although the answer to this question is not completely known, certain areas of the brain appear to be involved in the regulation of sleeping and wakefulness. One of these areas is the reticular formation of the brainstem, a diffuse network of nuclei that form the *ascending reticular activating system* (*ARAS;* **Figure 25**). This area is critical in maintaining alert

TABLE 4 Characteristics of Slow-Wave Sleep and REM Sleep

	Slow-wave sleep	REM sleep
EEG	Slow; medium to high amplitude	Fast; low amplitude
Movement	Moderate muscle tone; some movement, as in shifting position	Little tone; no movement (paralysis) of postural muscles, twitches of distal muscles
Heart and respiration rates	Decreased relative to rest while awake	Increased relative to rest while awake
Dreams	Rare, logical, not detailed	Common, vivid, illogical
Rapid eye movements	Rare	Frequent
CNS site of induction	Forebrain	Pons

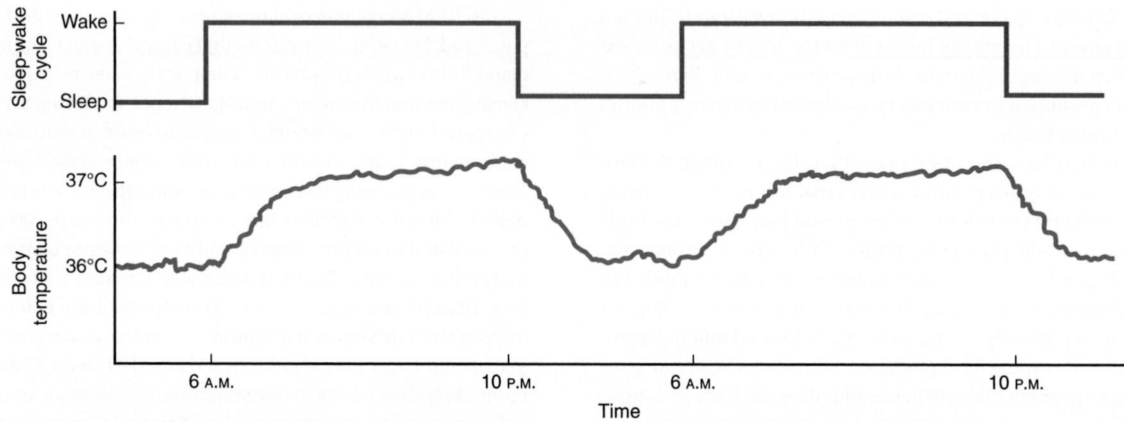

Figure 24 **Body temperature fluctuations during the sleep-wake cycle.**

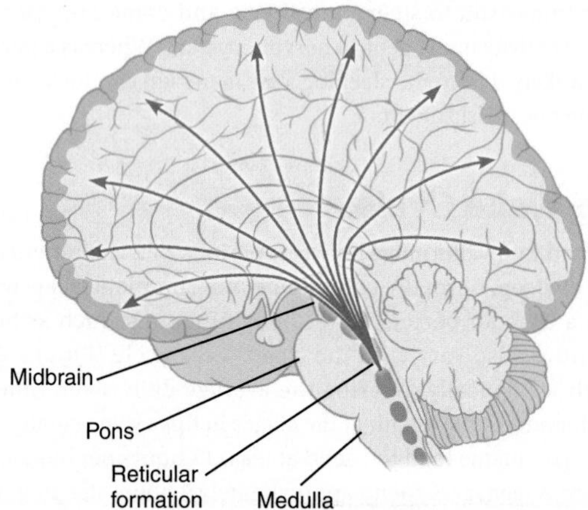

Figure 25 **Ascending reticular activating system.** The arrows indicate the spread of excitation that "arouses" the cortex.

wakefulness. Input from the ARAS projects to the cortex through relays in the thalamus, hypothalamus, and forebrain and "awakens" the cortex, making it more receptive to incoming signals from other pathways. Neurotransmitters associated with the ARAS, and therefore the awake state, include acetylcholine, norepinephrine, and dopamine. The role of acetylcholine in the alert awake state is mimicked by nicotine, a drug that binds nicotinic cholinergic receptors; likewise, the role of the catecholamines is mimicked by drugs such as amphetamines and cocaine. In another area involved in awaking the brain, the hypothalamus, neurotransmitters include histamine and orexin. The role of histamine in maintaining the awake state is evident when an individual takes an antihistamine, which makes the person drowsy. Orexin is a recently discovered peptide whose awakening actions are being explored by pharmaceutical companies in the development of drugs to keep people awake.

Other areas of the brain are involved in inducing sleep. SWS is induced by the forebrain. Recent research suggests that adenosine is a critical neurotransmitter for the induction of SWS. One of the current theories on the stimulatory effects of caffeine suggests that caffeine blocks adenosine receptors, thereby inhibiting sleep. REM sleep, in contrast, is induced by activity originating in the pons and is associated with release of acetylcholine. The precise mechanisms of sleep induction or maintenance of the awake state are not yet fully understood.

Electrical Activity During Wakefulness and Sleep

When you are awake and alert, an EEG shows a pattern of high-frequency, low-amplitude oscillations known as *beta waves,* which reflect electrical signals being generated by large numbers of neurons at different times (**Figure 26a**). When you are awake but resting, the EEG shifts to a new pattern of lower-frequency, higher-amplitude waves known as *alpha waves* (Figure 26b). Compared to beta waves, alpha waves reflect a greater degree of synchronization of electrical activity among neurons; that is, electrical signals are being generated in large groups of neurons at more or less the same time. This change in EEG pattern might be compared to changes in crowd noise in a stadium. When large numbers of people are talking at different times (desynchronized activity, similar to beta waves), the overall sound is a nearly constant din, and it is difficult to make out distinct patterns or words. When the crowd begins to chant in unison (synchronized activity, similar to alpha waves), the sound occurs in a series of discernible waves that can be recognized as words.

SWS occurs in four successive stages that are distinguishable by changes in both the EEG and the arousal threshold (the strength of a stimulus required to awaken a sleeper). Stage 1 SWS has the lowest arousal threshold, whereas stage 4 has the highest. In other words, sleep is "lightest" in stage 1 and "deepest" in stage 4. Stage 1 sleep is characteristic of the "drowsy" period between being asleep and being awake. Stage 2 sleep is light sleep, stage 3 is moderate sleep, and stage 4 is deep sleep. Stages 1–4 show successive increases in the degree of synchronization of the EEG, with stage 4 waves being the highest in amplitude and lowest in frequency, indicating a large degree of synchronization (see Figure 26c–e). The K-complexes seen in Figure 26c represent sleep spindles, which are periodic bursts of synchronized activity occurring in the cortex and thalamus. Sleep spindles commonly occur during sleep stages 2 and 3.

During REM sleep, the EEG shows a completely different pattern (Figure 26f). Instead of slow, high-amplitude waves, it contains

CLINICAL CONNECTIONS

Post-Traumatic Stress Disorder

When a person is subjected to a stressful circumstance, the hypothalamic-pituitary-adrenal axis becomes activated, causing cortisol release and activating the autonomic nervous system. Stress, which is defined as a strain or pressure on the body, comes in many forms, including exercise, changes in the environment, illness, injury, or fear. Cortisol helps the body adjust to stress by mobilizing energy stores, even at the expense of breaking down cell proteins to amino acids. The autonomic nervous system helps with the immediate adaptation to stress by increasing blood flow to the muscles and heart and by mobilizing energy stores.

Although our bodies are well suited for dealing with the daily stresses of life, chronic or traumatic stress can have long-lasting, sometimes devastating effects on the central nervous system. Post-traumatic stress disorder (PTSD) results when the body is traumatized by a horrific experience such as serving in war, being in a natural disaster, witnessing a crime, or being a victim of sexual assault. Approximately 30% of Vietnam War veterans and 8% of Gulf War veterans have been diagnosed with PTSD. Following the September 11, 2001, terrorist attacks on the World Trade Center and Pentagon, diagnosis of PTSD increased by 55% across the United States, and approximately 100% in New York City. Overall, more than 5 million Americans suffer from PTSD.

PTSD was not recognized as a brain disorder until 1980. Unlike the symptomatic diagnosis that is applied to most disorders, diagnosis of PTSD depends on the cause (exposure to a traumatic event) as well as the symptoms. Physiological effects include increased responsiveness of the sympathetic nervous system, anxiety, and sleep disorders (including recurring nightmares). Psychological effects include numbing of emotions, re-experiencing the traumatic event, and avoidance (phobias) of reminders of the traumatic event. PTSD is associated with decreased levels of serotonin in the brain, so it is treated with serotonin selective reuptake inhibitors (SSRIs), such as paroxetine (Paxil) and sertraline (Zoloft).

Black Star/Newscom.

The stress experienced by soldiers can lead to post-traumatic stress disorder.

Critical Thinking Questions

1. Describe how cortisol and the autonomic nervous system help our bodies deal with stress.
2. Describe both the physiological and the psychological effects of PTSD.
3. What do you suppose are the limitations of current treatments for PTSD? Can you anticipate new directions for future PTSD prevention or treatment?

fast, low-amplitude waves resembling those of the awake, alert state. Although the arousal threshold is higher during REM sleep than at any other time, a person is actually more likely to awaken spontaneously during REM sleep than at any other time. Because of this surprising fact, REM sleep is often referred to as *paradoxical* sleep.

During a typical night, a person progresses through the various stages of sleep in a fairly orderly, predictable way (**Figure 27**). When you first fall asleep, you go into stage 1 SWS. From there, you proceed through stages 2–4 in order. About an hour after sleep onset, you go in reverse order through these stages and then enter into your first period of REM sleep. At approximately 90-minute intervals, this pattern repeats a few times, except that you tend to spend less time in the deeper stages of SWS and more time in REM sleep as sleeping continues.

Notice in Figure 27 that REM sleep periods last longer and occur more closely together as sleep continues. In keeping with this pattern, when spontaneous awakenings occur, they occur much more frequently in the last few hours of sleep than in the beginning. As a rule, we remember only those dreams that we experience immediately before waking up. Because this is more likely to happen during REM sleep, we are most likely to recall REM-type dreams.

Quick Check 6

1. What is electroencephalography, and how is it useful in the study of sleep?
2. Which types of EEG changes are observed as a person progresses from stage 1 to stage 4 of SWS?
3. What is the ascending reticular activating system, and how is it thought to function in regulation of the sleep-wake cycle?

8 Integrated CNS Function: Emotions and Motivation

You know how you feel at this moment. You may be angry because your roommate is making too much noise while you are trying to study, or you may be excited about all the new physiology you are learning. Emotions are highly subjective; they differ among people

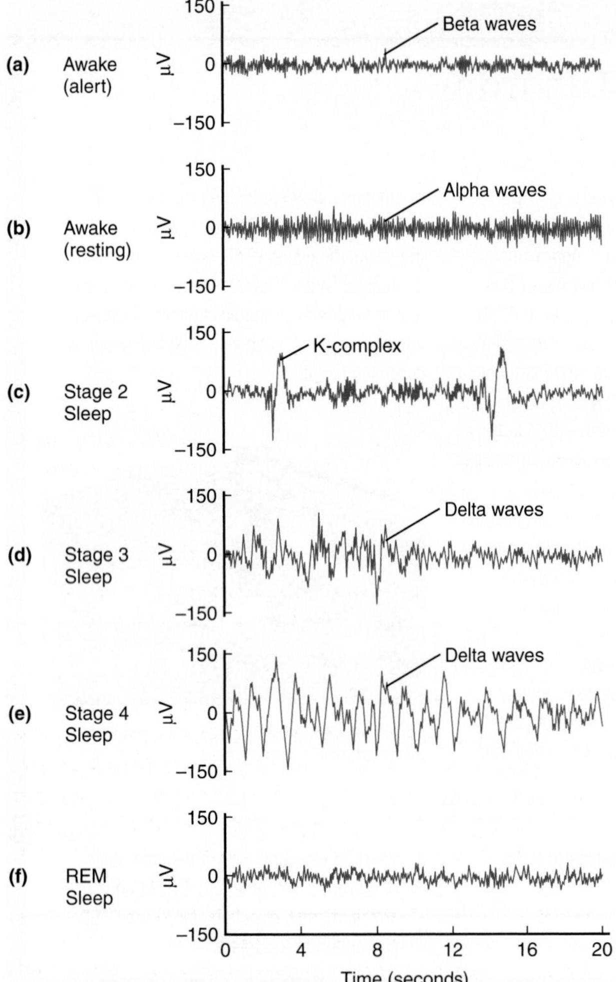

Figure 26 EEG recordings during wakefulness and sleep. These graphs are 20-second EEG samples recorded from a 21-year-old female subject. Amplitudes of EEG recordings are given in microvolts (μV), and frequencies are given in hertz (Hz). **(a)** Waking EEG is characterized by high-frequency, low-amplitude waves. **(b)** During relaxed waking, alpha waves of 8–12 Hz are present in the EEG. **(c)** The EEG during stage 2 sleep is characterized by high-amplitude K-complexes (seen here at 3 and 14 seconds) and sleep spindles (12- to 15-Hz waves seen here at 6, 10, and 12 seconds). **(d, e)** High-amplitude delta (0.3–3 Hz) waves are present in stage 3 sleep and occupy the entire 20-second period in the sample of stage 4 sleep. **(f)** EEG amplitude decreases and frequency increases in REM sleep. (Data provided courtesy of I. G. Campbell, University of California, Davis.)

Q During which phase of sleep is electrical activity of the brain most synchronized?

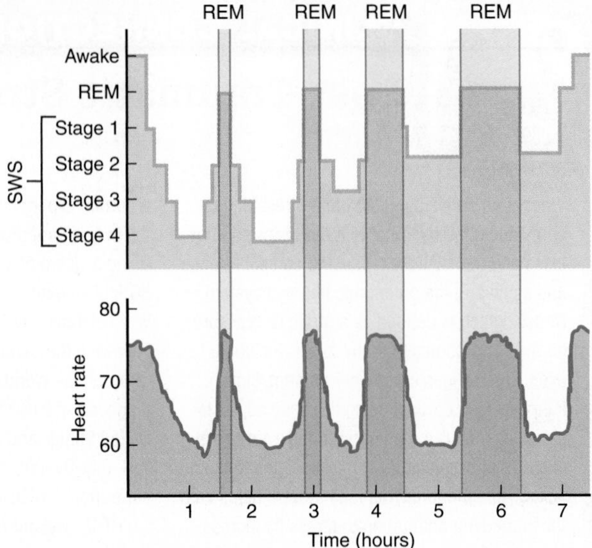

Figure 27 Stages of sleep. During 8 hours of sleep, a person moves among the different sleep stages. Initially, the person progresses from stage 1 SWS to stage 4 SWS, then returns to REM sleep. As the period of sleep continues, REM sleep becomes longer and more frequent. The lower panel shows the changes in heart rate during sleep. Notice that as a person goes into REM sleep, heart rate increases to near the levels in the awake state.

confronted with identical circumstances, and they also vary within an individual confronted with the identical set of events if they happen under different circumstances. For example, how you feel when your roommate gets an A on an exam depends on your own circumstances. If you are also doing well in school, you will probably feel happy; if you just did poorly on an exam, you might feel jealous. Because emotions are so varied, they are difficult to define and understand.

Emotions involve many diverse areas of the brain, including the cerebral cortex, limbic system, and hypothalamus (**Figure 28**). They are generally triggered by sensory input or memories. The "emotional response" encompasses a variety of autonomic (changes in heart rate, digestion, and so on), motor, and hormonal changes. To date, many theories have been proposed to explain the brain's role in emotions, but little strong evidence supports the primacy of any single theory. It does appear, however, that different neural pathways come into play for different emotions. For example, the amygdala is critical in feeling fear and anxiety; by contrast, the hypothalamus is associated with anger and aggression, although other structures (including the amygdala and midbrain) are involved in these emotions as well. The left and right hemispheres also seem to have different roles in emotions: The left hemisphere is more closely associated with positive emotions, whereas the right hemisphere is more closely associated with negative emotions.

Closely associated with emotions is *motivation,* the impulse that drives our actions. Motivations can be purely physiological; that is, they can arise simply because some need of the body is not being satisfied. For example, we eat because we feel hunger, which usually arises because the body is lacking certain nutrients. Motivations can also be driven by emotions acting in the absence of any obvious physiological drives. For example, some people eat ice cream when they feel sad, even when they are not hungry, because eating the ice cream brings a feeling of pleasure that may override the sadness.

Pleasure is a highly motivating emotion. The brain contains "pleasure centers" that can be acted on by many different stimuli. For example, the euphoria that people perceive when they consume alcohol, opioids, amphetamines, or nicotine may be due to the actions of these substances on pleasure centers. All of these substances are thought to produce euphoria by activating

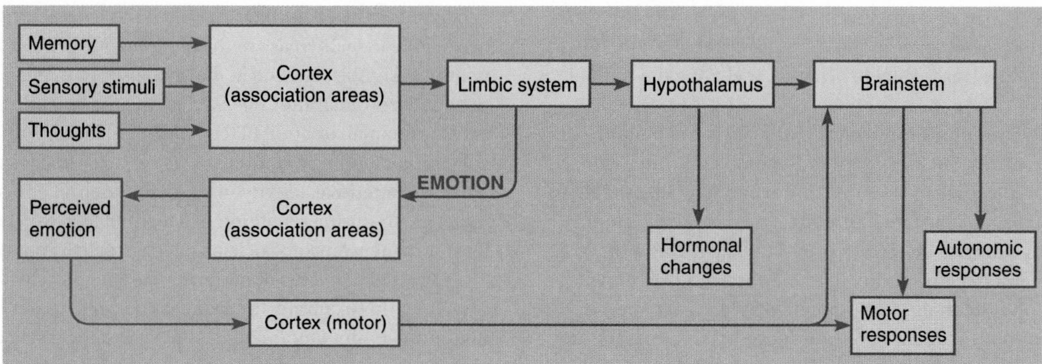

Figure 28 **The CNS structures involved in emotions.** Various structures of the brain are involved in producing emotions, both the "feeling" and the responses associated with them. Cortical association areas integrate thoughts, memory, and sensory information and communicate to the limbic system. The limbic system "creates" the emotion, but we are not aware of the emotion until it is transmitted back to the cortex for perception. Meanwhile, the limbic system also communicates the emotion to the hypothalamus, which is responsible for bodily responses coupled with the emotion, including hormonal changes (for example, adrenaline release), motor responses (for example, frowning), and autonomic responses (for example, changes in heart rate).

dopaminergic systems of the basal ganglia, which then project to the limbic system. The purpose of these systems is believed to be providing "rewards" for healthy behaviors such as seeking out food and water to satisfy body needs. The motivation that is elicited by these substances is so strong that animals, including people, can develop addictions to them. Sometimes these addictions override other motivations, even those closely connected to the body's needs. Addicted animals, for example, have been known to die of starvation because they habitually ignored food in favor of other pleasurable stimuli.

9 Integrated CNS Function: Learning and Memory

Whereas **learning** is the acquisition of new information or skills, **memory** is the retention of information, skills, or thoughts. Learning is an important part of life; we do it many times every day. In fact, every new experience is an opportunity for learning. Nevertheless, only a portion of what we experience is actually put into memory so that it can be recalled at some later time. If we were able to store and retrieve everything we ever experienced, then life would be much easier—we would never need to study for exams or look at a road map a second time! Learning and memory are closely linked, as are the brain areas involved in these processes.

Learning

Although learning has a simple definition, psychologists have long recognized several forms of learning that can be categorized into two major types: associative learning and nonassociative learning.

Associative learning is the type of learning that requires making connections between two or more stimuli. The salivation reflex in Pavlov's dogs is an example of associative learning, in that the dogs learned to "associate" the ringing of the bell with the arrival of food. If you were to try to teach your younger sister not to touch something that is hot, for example, she might not listen when you tell her it will hurt; however, when she touches a hot stove, the pain she feels will teach her not to do it again. In this case, your sister may learn two associative lessons: Do not touch something that is hot because it will hurt, and pay heed when your older sibling warns you about something. Using mnemonics in class to memorize a group of words is another example of associative learning.

Nonassociative learning occurs in response to repetition of a single stimulus (or similar stimuli) and includes the processes of habituation and sensitization. **Habituation** is a decrease in response to a repeated stimulus. For example, if construction workers were to start hammering outside your room while you are studying, you would first be distracted and likely unable to study. However, as the noise persists, you would likely find that you are no longer as aware of it. Studying would become possible, even though circumstances might not be ideal. In this case your brain has become *habituated* to a repetitive stimulus. In contrast, **sensitization** is an increase in response to repeated stimulus. Suppose, for example, that construction workers are hammering on your roof, directly over where you are sitting. All of a sudden you hear a creaking sound. You might jump out of concern that a piece of the roof is going to collapse. The hammering above has sensitized you to other sounds that could be indicative of danger.

Whether we become sensitized or habituated to repetitive stimuli depends on the circumstances. If we perceive a stimulus as important (such as a signal of impending danger), we tend to become sensitized to it; if we deem the stimulus to be unimportant, we tend to become habituated to it.

Memory

We saw how a younger sister learned to associate hot things with pain, but now we want her to remember that lesson so that she does not repeatedly burn herself in the future. To do so, she must put the information she learned into memory. Just as there are two types of learning, so there are two types of memory: procedural memory and declarative memory.

Procedural memory (or *implicit memory*) is the memory of learned motor skills and behaviors. Think back to when you first tried to play a musical instrument or a sport. At first, learning was difficult; each movement had to be thought through. Of course, now you are more proficient, because you have learned and memorized each skill. The type of memory involved when someone says, "It's like riding a bike—you never forget," is procedural memory. This type of memory involves several brain areas, including the cerebellum, basal nuclei, and pons.

Declarative memory (or *explicit memory*) is the memory of learned experiences, such as facts and events and other things that can be stated verbally ("declared"). You can state that the first three letters of the alphabet are A, B, and C (a declarative memory), but you cannot state the memories you use while riding a bicycle (procedural memory). The ability to remember and explain what generates action potentials is an example of declarative memory. Remembering your birth date and your parents' anniversary are other examples. Declarative memory is more closely associated with the everyday use of the word "memory." This type of memory involves the hippocampus.

Memory occurs at two levels, *short-term memory* and *long-term memory*. According to the currently accepted model of how information is processed and stored into memory, incoming information first enters the CNS and is stored as **short-term memory** (also called working memory), which is temporary storage lasting only a few seconds or minutes. Limited space is available for short-term memory, and information placed in short-term memory will be lost if it is not *consolidated* into **long-term memory,** which can last years or a lifetime. The mechanisms of consolidation are not well understood, but repetition appears to help (your little sister may not believe that something hot will always hurt her until she touches hot objects a few times), as does association. We tend to remember things that we can associate with items, people, or events that are important to us. Memorizing a friend's phone number is much easier than memorizing a sequence of seven random numbers. Likewise, most brides and grooms remember their wedding day, even though it occurred only once.

The processing of memories is complex and involves several, if not all, areas of the brain. The frontal lobe is critical for short-term memory, and the temporal lobe, including the hippocampus, is necessary for long-term memory. However, virtually the entire central nervous system can change with experiences, and these changes play a role in learning and memory, as described next.

Plasticity in the Nervous System

Learning and memory are able to occur because the nervous system is endowed with **plasticity,** the limited ability to alter its anatomy and function in response to changes in its activity patterns. This plasticity derives from both the fact that the function of existing synapses can be altered for long periods of time (hours to days), and the fact that new synaptic connections can develop. Recent studies show that our brains produce new neurons in areas involved in memory.

One example of plasticity in the nervous system is *long-term potentiation (LTP),* which in mammals was first discovered in the hippocampus and occurs at preexisting synapses. In LTP, repetitive stimulation of a particular synapse eventually leads to an increase in the *strength* of that synaptic connection; that is, repetition increases the likelihood that synaptic input will be able to trigger an action potential in the postsynaptic cell (assuming the synapse is excitatory). Such increases in synaptic strength occur because of an increase in the size of the excitatory postsynaptic potentials (EPSPs) that are generated in the postsynaptic cell. An increase in synaptic strength can be due to an increase in the postsynaptic cell's sensitivity to the neurotransmitter released at that synapse, an increase in the quantity of neurotransmitter released by the presynaptic cell with each action potential, or both. LTP is thought to be important in the consolidation of long-term memory because it provides a mechanism whereby repetitive activity in particular neural pathways, such as might occur during repetition of a learned fact, can leave a more or less permanent "record" of itself once the activity has ceased.

One mechanism of long-term potentiation is illustrated in **Figure 29**. In this example, the presynaptic cell releases the excitatory neurotransmitter glutamate. The postsynaptic cell has two types of glutamate receptors: AMPA (α-amino-3-hydroxy-5-methylisoxazoleproprionic acid) receptors and NMDA (*N*-methyl-D-aspartate) receptors. NMDA and AMPA are glutamate agonists that bind to glutamate receptors. When glutamate binds to NMDA receptors, calcium channels open; when this molecule binds to AMPA receptors, sodium channels open. When the presynaptic cell releases glutamate, it binds to both receptor types.

The concentration of glutamate in the synaptic cleft depends on the frequency of action potentials in the presynaptic cell. When low to moderate levels of glutamate are present in the synaptic cleft (Figure 29a), some AMPA receptor–dependent sodium channels open, and sodium diffuses into the cell, causing a depolarization. Some NMDA receptor–dependent calcium channels also open, but calcium cannot move through these channels because they are blocked by magnesium ions. The end result is an EPSP that may or may not reach threshold for an action potential.

When the action potential frequency increases in the presynaptic cell, large amounts of glutamate are released (Figure 29b), which produces a greater depolarization of the postsynaptic cell. The strong depolarization exerts a repulsive force on the magnesium ions that forces them out of the NMDA receptor channels. With the magnesium now gone and the channels open, calcium is able to flow into the postsynaptic cell, raising the cytosolic calcium concentration and triggering activation of other second messengers that activate protein kinases. These protein kinases phosphorylate several proteins that potentiate communication across the synapse by acting in two ways: AMPA receptor channels become more sensitive to glutamate, and the postsynaptic cell produces a paracrine that causes the presynaptic cell to release more glutamate. The net result is larger and more prolonged EPSPs, which increase the likelihood that action potentials will be generated in response to signals being transmitted at that synapse.

While brain plasticity is greater at young ages, it continues throughout life. Thus you can teach an old dog a new trick. Recent studies suggest that exercise may enhance brain function, as described in **Understanding Exercise: Can Exercise Affect the Brain?**

Quick Check 7

1. Which emotion leads to addiction? Which neurotransmitter is critical in the development of drug addiction?
2. When a dog is trained to sit by rewarding it with food each time the dog obeys a command, is associative learning or nonassociative learning involved?

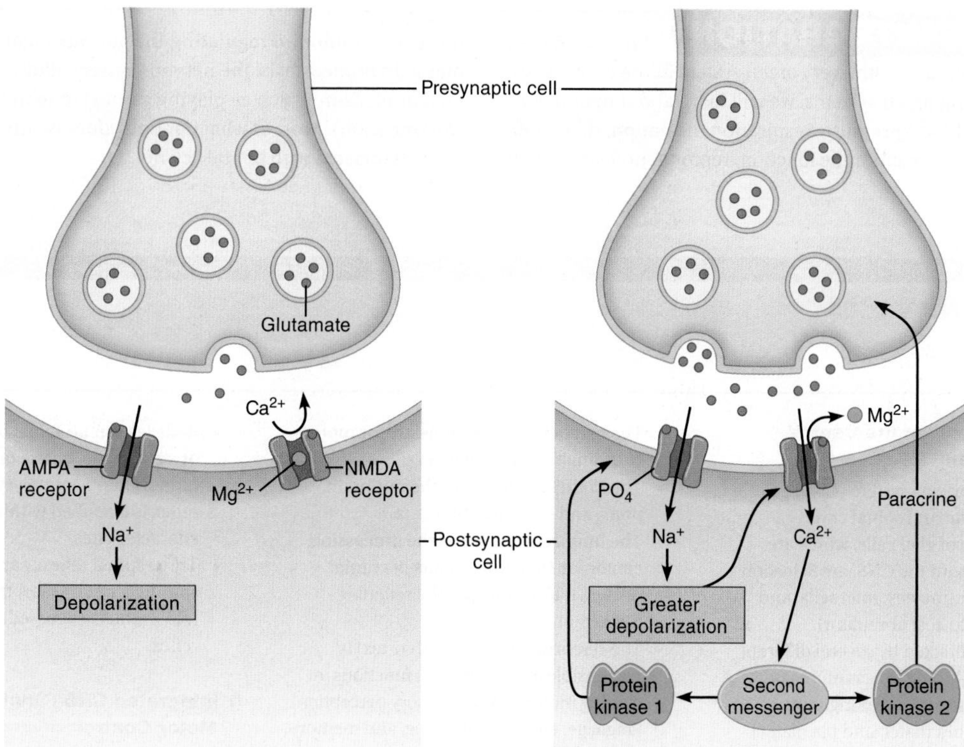

Figure 29 One mechanism for long-term potentiation. See the text for a detailed description.

Q What triggers exocytosis of glutamate?

Can Exercise Affect the Brain?

Performing physical activities such as exercise clearly involves the brain. You cannot run, throw, catch, or lift without sending neural signals to your skeletal muscles to cause a contraction that occurs at precisely the right time and at the appropriate strength. Exercise also benefits the body in many ways. Aerobic exercise (low- to moderate-intensity exercise that depends on adequate oxygen supply to the muscles) improves whole-body metabolism, body weight, and cardiovascular function, to name just a few examples. But what about the brain? Can exercise benefit the brain?

Several recent studies suggest that the answer is yes. For instance, aerobic exercise in elderly persons can improve cognitive function and may even slow the development of dementia in patients with Alzheimer's disease. Exercise also appears to alleviate mood disorders, such as depression, in both elderly and adult populations.

How does exercise do this? Several theories are being developed, but researchers have not fully elucidated the mechanism involved. One theory suggests that brain function is enhanced simply because exercise increases cardiovascular fitness. More efficient blood flow to the brain may enhance brain activity and decrease brain deterioration as one ages. Studies have shown that changes in brain function with exercise occur over the same time as improvement in cardiovascular function.

Other studies suggest that exercise enhances the metabolic activity of the brain, thereby improving brain function. In fact, the increase in whole-body metabolism from regular exercise is associated with changes in brain activity. Specifically, activation of a hypothalamic enzyme, AMP activated protein kinase (AMPK), enhances whole-body metabolism. Other types of metabolic changes occur in the brain as well. For example, exercise induces metabolic changes in the hippocampus, an area critical for memory. Although studies of these potential mechanisms of action are in their early stages, they show promise in explicating the brain-exercise link. Some researchers, for example, have shown that several proteins are expressed at higher levels in the hippocampus of exercising mice or rats compared to nonexercising animals. Given that some of these proteins are involved in synaptic communication, they may play a critical role in enhancing memory.

Although definitive conclusions about the neural benefits of exercise are a few years away, it's exciting to consider that exercise may help you maintain not only a healthy body, but also a healthy mind!

SYSTEMS INTEGRATION The central nervous system interacts with every organ system in the body. As we study the remaining organ systems, we will learn about neural control of heart rate, blood pressure, respiration, urination, digestion, absorption, metabolism, immune function, reproduction, and much more. In addition to regulating the various organ systems so as to maintain homeostasis, the nervous system allows us to elicit behaviors for pleasure (such as playing games) or to meet needs (such as working a job). Most of what our body does is either directly or indirectly associated with neural activity.

CHAPTER REVIEW

SUMMARY

1 General Anatomy of the Central Nervous System

- The central nervous system (CNS) consists of the brain and spinal cord.
- The four types of glial cells, which are found throughout the CNS, are Schwann cells, oligodendrocytes, microglia, and astrocytes (the most abundant).
- The CNS is protected by several different structures, including the cranium and vertebral column, the meninges (dura mater, arachnoid mater, and pia mater), and the blood-brain barrier.
- The cerebrospinal fluid also protects the brain by cushioning it from impacts.
- CNS tissue is organized into gray and white matter.
- Gray matter consists primarily of cell bodies, dendrites, and axon terminals; white matter consists mostly of myelinated axons.

iP Nervous I and II, Anatomy Review

2 The Spinal Cord

- Thirty-one pairs of spinal nerves come off the spinal cord, one pair at each vertebral level.
- All spinal nerves are mixed nerves, in that they all contain both afferent and efferent axons.
- The white matter of the spinal cord contains ascending and descending tracts that carry information to and from the brain, respectively.
- The gray matter is divided into ventral and dorsal horns; cell bodies and dendrites of efferents are located in the ventral horn, axon terminals of afferents are located in the dorsal horn, and small interneurons are scattered throughout.

3 The Brain

- The brain has three major parts: the forebrain, the cerebellum, and the brainstem.
- The forebrain includes the cerebrum (cerebral cortex and subcortical nuclei) and diencephalon (thalamus and hypothalamus).
- The cerebellum is responsible for motor coordination and balance.
- The brainstem includes the midbrain, pons, and medulla oblongata.
- The brainstem serves as the processing center for 10 of the 12 pairs of cranial nerves; it also contains the reticular formation.
- The cerebrum (the cerebral cortex) is responsible for higher brain functions including motor control, sensory perception, language, emotions, learning, and memory.
- The cerebral cortex is divided into four lobes: the frontal lobe, the parietal lobe, the occipital lobe, and the temporal lobe.
- The subcortical nuclei—regions of gray matter in the cerebrum—include the basal nuclei and nuclei of the limbic system.
- The basal nuclei are important in motor control, whereas the limbic system is involved with emotions, learning, and memory.
- The diencephalon, located beneath the cerebrum at the core of the forebrain, includes the thalamus, which serves as a relay station for information traveling to the cortex, and the hypothalamus, which regulates a number of body functions to maintain homeostasis.

4 Integrated CNS Function: Involuntary Movement Through Reflexes

- The simplest actions of the nervous system are reflexes, or automatic responses to a particular stimulus.
- Reflexes can be mediated spinally or cranially, may be either somatic or autonomic, and may be either conditioned or innate.
- Reflexes, or reflex arcs, are made up of a sensory receptor, an afferent neuron, an integration center, an efferent neuron, and an effector organ.
- Reflexes can be spinal or cranial, somatic or autonomic, innate or conditioned, and monosynaptic or polysynaptic; a given reflex is classified within all four groups of characteristics.
- Three spinal reflexes are the muscle spindle stretch reflex, the withdrawal reflex, and the crossed-extensor reflex.

5 Integrated CNS Function: Voluntary Motor Control

- Voluntary control of muscles involves many regions of the CNS.
- Several areas of the brain are involved in formulating plans for movement and setting up a command, including the prefrontal cortex, association areas of the cerebral cortex, basal nuclei, and the limbic system.
- The pyramidal and rubrospinal tracts transmit information about the plan to the motor neurons in the spinal cord, which are the neurons that stimulate the skeletal muscle to contract. The cerebellum and basal nuclei help make the movement smooth.
- The ventromedial pathways are more important for supportive movements such as maintaining posture than for carrying out precise voluntary movement. Involuntary contractions of skeletal muscles also occur.
- The ventromedial pathways include the vestibulospinal tracts, the tectospinal tracts, and the reticulospinal tracts.

6 Integrated CNS Function: Language

- The areas of the brain that play important roles related to language include Wernicke's area in the temporal lobe and Broca's area in the frontal lobe.
- An abnormality in language skills is called aphasia.

7 Integrated CNS Function: Sleep

- The purpose and mechanisms of sleep are not well understood.
- The stages of sleep can be monitored using an EEG.
- Slow-wave sleep is characterized by low-frequency, high-amplitude EEG waves, whereas REM sleep is characterized by high-frequency, low-amplitude EEG waves.
- The body moves through various stages throughout periods of sleep.

8 Integrated CNS Function: Emotions and Motivation

- Emotions are generated in the limbic system based on sensory input.
- The limbic system then transmits information to the cortex, where the emotion is actually perceived.
- Emotion-related responses also include changes in the autonomic nervous system.
- Motivation directs actions and is closely associated with pleasure.

9 Integrated CNS Function: Learning and Memory

- Learning is the acquisition of new skills or information, whereas memory is the retention of knowledge.
- Learning and memory require plasticity.
- One type of plasticity is long-term potentiation, in which communication across a synapse is prolonged.

EXERCISES

Multiple-Choice Questions

1. Touch perception requires that sensory neurons from the skin transmit information to
   a) The basal nuclei.
   b) The somatosensory cortex.
   c) The hypothalamus.
   d) The limbic system.
   e) The hippocampus.
2. Which of the meninges is closest to the neural tissue?
   a) Dura mater
   b) Arachnoid mater
   c) Pia mater
3. At rest, the brain accounts for how much of oxygen consumption within the body?
   a) 2%
   b) 5%
   c) 20%
   d) 40%
   e) 50%
4. Most of the fibers connecting the left and right cerebral hemispheres are located in the
   a) Hippocampus.
   b) Corpus callosum.
   c) Corticospinal tract.
   d) Central canal.
   e) Cerebral ventricles.
5. All spinal nerves contain
   a) Afferent fibers.
   b) Efferent fibers.
   c) Both afferent and efferent fibers.
   d) Neither afferent nor efferent fibers.
6. The origin of neurons of the corticospinal tract is the
   a) Ventral horn of the spinal cord.
   b) Somatosensory cortex.
   c) Primary motor cortex.
   d) Cerebellum.
   e) Thalamus.
7. Which of the following is *not* a component of the limbic system?
   a) Hippocampus
   b) Cerebral cortex
   c) Amygdala
   d) Midbrain
   e) Hypothalamus
8. The circadian rhythm is established by which brain area?
   a) Suprachiasmatic nucleus
   b) Amygdala
   c) Thalamus
   d) Occipital lobe of the cerebral cortex
   e) Pons
9. Parkinson's disease is caused by loss of dopaminergic innervation of the
   a) Cerebellum.
   b) Basal nuclei.
   c) Brainstem.
   d) Primary motor cortex.
   e) Thalamus.
10. Which of the following reflexes is monosynaptic?
    a) Withdrawal reflex
    b) Crossed-extensor reflex
    c) Muscle spindle stretch reflex
    d) Both a and c
    e) All of the above
11. The area of the brain important for the maintenance of posture is the
    a) Primary motor cortex.
    b) Basal nuclei.
    c) Brainstem.
    d) Cerebellum.
    e) Limbic system.
12. Long-term potentiation is
    a) A prolonged increase in synaptic activity at a given synapse.
    b) The formation of new synapses.
    c) Memory from more than 10 years ago.
    d) Activated through the ascending reticular activating system.
    e) All of the above.

Objective Questions

13. Where is cerebrospinal fluid synthesized?
14. Which type of junction between the endothelial cells of brain capillaries produces the blood-brain barrier?
15. Myelinated axons are found in (gray/white) matter.
16. Somatic efferents originate in the (dorsal/ventral) horn of the spinal cord.
17. The major function of the (cerebrum/cerebellum) is to coordinate body movements.
18. Which three structures make up the brainstem?
19. Which side of the brain is generally associated with creativity?
20. What are the two main structures of the diencephalon?
21. What is the major sensory relay nucleus to the cortex?
22. The area of the brain most closely associated with fear is the ________.
23. The ________ system is associated with emotions, learning, and memory.
24. The ability to recall information when taking physiology exams is an example of ________ memory.

25. The afferents that activate reflex motor actions are different from those that activate ascending tracts to communicate a sensory input to the brain. (true/false)

Essay Questions

26. Describe the different structures that protect central nervous system tissues.
27. Describe the energy supply to the brain.
28. Identify the four lobes of the cerebral cortex, and describe at least one function associated with each.
29. Describe the different components of the nervous system involved in accomplishing a voluntary motor task. Identify the structure(s) associated with each component.
30. Describe the different stages of sleep. During which sleep stage is a person most likely to roll over?
31. Describe the roles of NMDA and AMPA receptors in long-term potentiation.

Critical Thinking

32. Parkinson's disease is caused by a loss of dopaminergic function in the basal nuclei. To treat patients with Parkinson's disease, doctors try to elevate the dopamine levels in the brain. Dopamine, however, cannot cross the blood-brain barrier. Can you devise methods to elevate brain dopamine levels?
33. Suppose a person suffers a stroke that affects Wernicke's area. Describe what it would be like to carry on a conversation with this person.
34. Reflexes can be overridden by voluntary controls. Explain how this can occur, using your knowledge on synaptic transmission and neural integration.
35. Paralysis refers to the inability to voluntarily control movement and can include a loss of excitatory input to motor neurons (flaccid paralysis) or a loss of inhibitory input to motor neurons (spastic paralysis). In addition, some forms of paralysis are coupled with the absence of reflexes. Compare the types of paralysis that would be observed when the following areas of the nervous system are damaged: (1) primary motor cortex, (2) basal nuclei, and (3) motor neuron.

Photo credits for icons in feature boxes: Focus on Diabetes: Tanyae/Dreamstime; Clinical Connections: Iofoto/Dreamstime; Understanding Exercise: Pixattitude/Dreamstime; Discovery: Yuri Arcurs/Cutcaster.

Glossary

arachnoid mater (ah-rak-noyd) one of the three meninges, located between the dura mater and the pia mater

ascending tracts neural pathways that transmit information from the spinal cord or brainstem toward the brain

association areas areas of the cerebral cortex involved in complex processing that requires the integration of different types of information

association fibers neural pathways that connect one area of the cerebral cortex to another on the same side of the brain

associative learning learning by making connections between two or more stimuli

basal nuclei a particular group of nuclei located deep in the cerebrum that are important in regulating voluntary movements

blood-brain barrier physical barrier that exists between the blood and the interstitial fluid in the central nervous system (CNS), formed by tight junctions between endothelial cells of the cerebral capillaries

brain portion of the central nervous system that resides in the cranium

brainstem bottom-most part of the brain that connects the forebrain and cerebellum to the spinal cord; consists of three main regions: midbrain, pons, and medulla oblongata

Broca's area area of the association cortex devoted to language expression; located in the frontal lobe

central canal long, thin cylindrical canal that runs the length of the spinal cord and is continuous with the cerebral ventricles; contains cerebrospinal fluid

cerebellum (ser-ah-bel-um) a bilaterally symmetrical brain structure, with an outer cortex and inner nuclei; located below the forebrain and posterior to the brainstem

cerebral cortex thin layer of gray matter that covers the cerebrum

cerebral hemispheres the two halves of the cerebrum

cerebrospinal fluid (CSF) clear, watery fluid that surrounds and protects the central nervous system and is similar in composition to plasma

cerebrum (seh-ree-brum) largest structure of the brain, which contains both gray and white matter; gray matter areas include the cerebral cortex at the surface and deep subcortical nuclei

cervical nerves spinal nerves originating from the cervical spinal cord

choroid plexus (kor-oid) vascularized tissue lining the cerebral ventricles; synthesizes cerebrospinal fluid

coccygeal nerve spinal nerve originating from the coccyx

commissural fibers fibers that connect one region of the cerebral cortex to the corresponding region on the other side of the brain

contralateral referring to ascending and descending pathways that are on the side opposite their origin

corpus callosum (kor-pus kal-loh-sum) the primary band of nervous tissue that connects the two cerebral hemispheres

cranial nerves 12 pairs of peripheral nerves that emanate directly from the brain

cranium skull

crossed-extensor reflex reflex extension of a limb when a noxious stimulus is applied to the limb on the other side of the body

declarative memory memory of learned experiences, including facts and events

descending tracts pathways that transmit information from the brain or brainstem toward the spinal cord

diencephalon (dy-en-sef-ah-lon) the lowest portion of the forebrain; consists of the thalamus and hypothalamus

dorsal horn posterior half of the gray matter on either side of the spinal cord

dorsal root where a spinal nerve bifurcates before joining the spinal cord, the dorsalmost of the two resulting branches; contains afferent nerve fibers

dorsal root ganglia clusters of cell bodies of afferent neurons located outside the spinal cord

dura mater the outermost of the three meninges, closest to the bone

ependymal cells (ep-en-dee-mal) glial cells that line the cerebral ventricles of the brain and the central canal of the spinal cord

forebrain largest and uppermost part of the brain, divided into left and right halves, or hemispheres; consists of the cerebrum and diencephalon

frontal lobe one of the four lobes of the cerebrum, located in the anterior portion of the cerebrum; important in voluntary motor control, behavior, and personality traits

gray matter areas of the central nervous system (CNS) consisting primarily of cell bodies, dendrites, and axon terminals of neurons; where synaptic transmission and neural integration occur

gyri (jy-ri) ridges in the highly convoluted gray matter of the cerebral cortex; singular, *gyrus*

habituation decrease in response to a repeated stimulus; a form of learning

hypothalamus a region at the base of the brain that regulates autonomic functions and secretes several hormones, most of which regulate secretory activity of the pituitary gland

ipsilateral (ip-sih-lat-er-al) referring to ascending and descending pathways that are on the same side as their origin

learning acquisition of new information or skills

limbic system diverse collection of closely associated cortical regions, subcortical nuclei, and tracts in the forebrain; function in learning and emotions

long-term memory memory that is stored and can be recalled for years

lower motor neurons neurons that innervate skeletal muscle; also called *motor neurons*

lumbar nerves spinal nerves originating from the lumbar spinal cord

medulla oblongata (meh-duhl-ah ob-long-got-ah) lowest portion of the brainstem; connects to the spinal cord

memory retention of information, skills, or thoughts

meninges (men-in-jees) three membranes that separate the soft tissue of the central nervous system (CNS) from the surrounding bone; dura mater, arachnoid mater, and pia mater

midbrain uppermost portion of the brainstem; connects to the forebrain

mixed nerve bundle of axons containing both afferent and efferent fiber types

motor homunculus map that indicates which areas of the primary motor cortex in the frontal lobe are devoted to particular regions of the body

motor neuron efferent neuron of the somatic nervous system that synapses on skeletal muscle cells; originates in the spinal cord ventral horn or in analogous structures in the brainstem

nonassociative learning learning in association with repetition of a single stimulus

occipital lobe (ok-sip-ih-tul) one of the four lobes of the cerebrum, located in the posterior and inferior portion of the cerebrum; important in processing visual information

parietal lobe one of the four lobes of the cerebrum, located immediately behind the frontal lobe of the cerebrum; important in processing somatic sensations and sensory integration

pia mater innermost of the three meninges, adjacent to the nervous tissue

plasticity ability to change

pons middle portion of the brainstem; connects to the cerebellum

primary motor cortex region of the frontal lobe where voluntary movement is initiated

primary somatosensory cortex region of the parietal lobe of the cerebral cortex specialized for the processing of somatic sensory information

procedural memory memory of learned motor skills and behaviors

projection fibers axons in the central nervous system (CNS) that connect the cerebral cortex with lower levels of the brain or the spinal cord

proprioception (pro-pree-oh-cep-shun) the perception of the positions of the limbs and the body

pyramidal tract a direct pathway from the primary motor cortex to the spinal cord; controls fine voluntary movements

reflex automatic patterned response to a stimulus

reflex arc pathway by which a stimulus reflexively induces a response

REM sleep sleep characterized by high-frequency waves in the EEG and periodic rapid movements of the eyes; REM = rapid eye movement

reticular formation a diffuse network of nuclei located in the brainstem that is important in sleep-wake cycles, arousal of the cerebral cortex, and consciousness

sacral nerves spinal nerves originating from the sacral spinal cord

sensitization increased responsiveness to repetitive stimuli; a form of learning

sensory homunculus map that indicates which areas of the primary somatosensory cortex are devoted to particular regions of the body

slow-wave sleep (SWS) sleep characterized by multiple stages of low-frequency waves in the electroencephalogram (EEG)

somesthetic sensations sensations that arise from receptors in the skin

spinal cord portion of the central nervous system that travels in the vertebral column

spinal nerves a total of 31 pairs of nerves originating in the spinal cord and traveling to the periphery

subarachnoid space the space between the pia mater and the arachnoid mater, which is filled with cerebrospinal fluid

sulci (sul-sigh) grooves in the highly convoluted gray matter of the cerebral cortex; singular, *sulcus*

temporal lobe one of the four lobes of the cerebrum located at the side and separated from the frontal lobe by a deep groove; important in processing auditory information and language

thoracic nerves spinal nerves originating from the thoracic spinal cord

upper motor neurons neurons of the pyramidal tract that originate in the primary motor cortex and terminate in the spinal cord ventral horn

ventral horn anterior half of the gray matter on either side of the spinal cord

ventral root at the point at which a spinal nerve bifurcates before joining the spinal cord, the ventralmost of the two resulting branches; contains efferent nerve fibers

ventricles in the heart, the two lower chambers, which pump blood into the arteries; in the brain, chambers that contain cerebrospinal fluid

vertebral column bony structure that surrounds and protects the spinal cord

Wernicke's area area of the association cortex devoted to language comprehension; located in the posterior and superior portion of the temporal lobe and the inferior parietal lobe

white matter areas of the central nervous system (CNS) consisting primarily of myelinated axons; specialized for the rapid transmission of information over relatively long distances in the form of action potentials

withdrawal reflex reflex removal of a limb from a noxious stimulus

Answers

Figure Question Answers

Fig 5	Axons
Fig 7	C6 on the left side
Fig 12	To increase the volume of cerebral cortex
Fig 16	To synthesize CSF
Fig 22	In the primary motor cortex
Fig 26	Stage 4
Fig 29	Calcium entering the axon terminal after an action potential has opened voltage-gated calcium channels

Apply Your Knowledge Answers

Hydrocephalus is caused by an increase in cerebrospinal fluid, which can occur when too much of this fluid is produced by the choroid plexus or too little drainage of it occurs.

To block sensory function but not motor function at a specific spinal level, damage to the peripheral nervous system must be in the dorsal root or dorsal root ganglion—the only sites where motor and sensory axons travel in distinct paths.

If the knee-jerk reflex is absent, neurological damage can exist anywhere along the reflex pathway—for example, as receptor damage in the tissue, the afferent neuron, the spinal cord, or the efferent neuron. Further tests are needed to determine where the neurological damage occurred. Damage to the receptor would most likely be congenital and exist at birth. Damage to the afferent neuron would be accompanied by a loss of proprioception. Damage to the motor neuron would also cause lack of voluntary movement of the same muscle.

Answers to End-of-Chapter Multiple Choice and Objective Questions

Multiple Choice Questions

1. b 2. c 3. c 4. b 5. c 6. c 7. d 8. a 9. b 10. c 11. c 12. a

Objective Questions

13. choroid plexus
14. tight junction
15. white
16. ventral
17. cerebellum
18. midbrain, pons, medulla
19. right
20. thalamus, hypothalamus
21. thalamus
22. amygdala
23. limbic
24. declarative
25. false

The Nervous System: Autonomic and Motor Systems

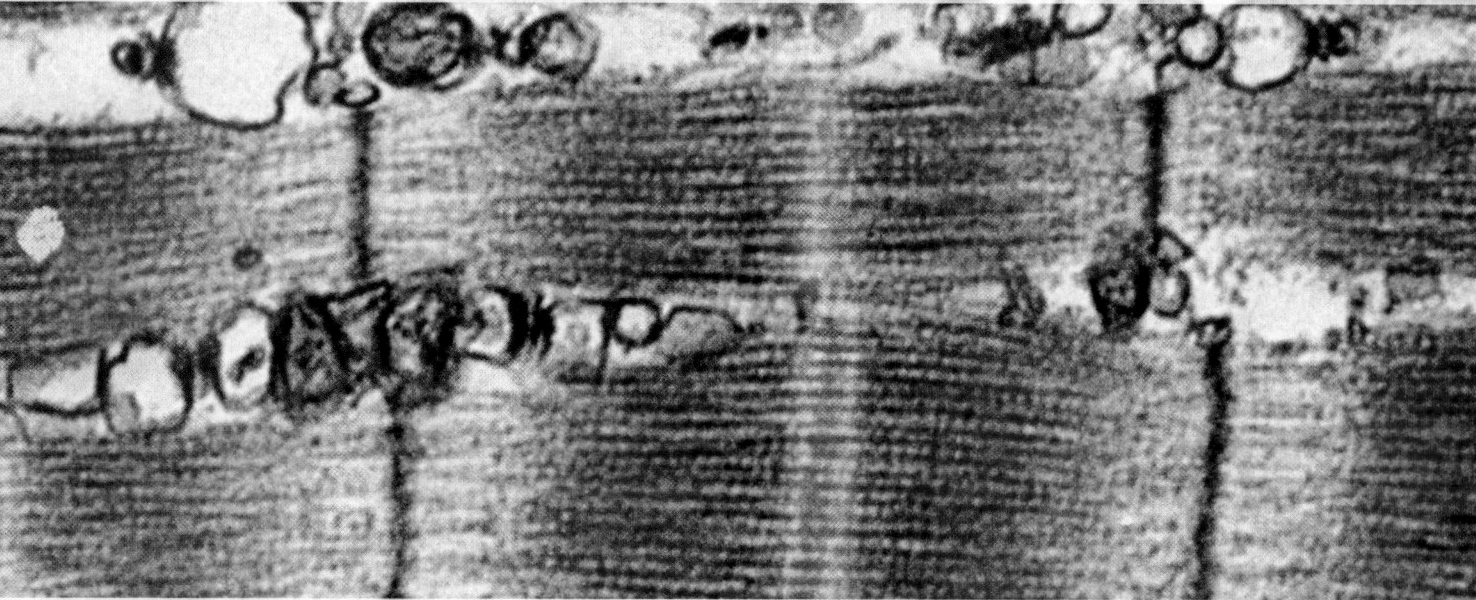

Biology Pics/Photo Researchers, Inc.

Colored transmission electron micrograph (TEM) of a longitudinal section through striated skeletal muscle.

Imagine it's a Friday evening and you are at the mall. You have just finished dinner and are at your favorite bookstore, browsing through the bestsellers. Your heart is beating slowly while your gastrointestinal organs are actively digesting and absorbing nutrients. All of a sudden, the fire alarm goes off and you smell smoke. Your heart starts pounding. Blood flow is diverted from your gastrointestinal organs to your skeletal muscles, and you flee the store with the other customers and employees. Shortly after you've reached the mall parking lot and safety, however, your body starts to return to its resting state.

How does your body shift so rapidly from a relaxed, resting state to one prepared for emergency action? Your central nervous system controls your muscles and other organs by way of signals sent through the efferent branch of the peripheral nervous system. When your brain detects impending danger, it can quickly communicate to the sympathetic branch of the autonomic nervous system that it should initiate the fight-or-flight response. In just a matter of seconds, your body adjusts to the crisis.

CHAPTER OUTLINE

LEARNING OUTCOMES *After studying this chapter you should be able to:*

- Describe the anatomy of the somatic nervous system and the two branches of the autonomic nervous system.
- Describe the chemical messengers and receptor types associated with the peripheral nervous system.
- Explain the basic function of the two branches of the autonomic nervous system and the concept of dual innervation.
- Describe how the central nervous system regulates or controls the autonomic branch of the peripheral nervous system.

Before You Begin

Make sure you have mastered the following topics:

- G proteins
- cAMP second messenger system
- Phosphatidylinositol second messenger system
- Adrenal medulla
- Acetylcholine
- Catecholamines
- Cholinergic receptors
- Adrenergic receptors
- Synapses

Many of the actions of the central nervous system are carried out by commands sent to muscles and glands by the efferent branch of the nervous system, which is the topic of this chapter. The efferent branch of the nervous system communicates with effector organs in much the same way that neurons communicate with other neurons at synapses. Thus knowledge of ion channels, electrochemical forces for ion movements, signal transduction mechanisms, graded potentials, and action potentials is critical to understanding the function of the efferent nervous system. In this chapter we will learn about the neurotransmitters of the efferent nervous system and their mechanisms of action.

The efferent nervous system consists of two branches: the autonomic and somatic nervous systems. We begin our study with the autonomic nervous system.

1 The Autonomic Nervous System

The autonomic nervous system innervates most effector organs and tissues in the body, including cardiac muscle, the smooth muscle found in blood vessels and various visceral organs (for example, the stomach and the respiratory airways), glands (for example, sweat glands, salivary glands, and some endocrine glands), and adipose tissue. It is called "autonomic" because its functions occur at a subconscious level. For example, a person does not consciously decide to increase his or her heart rate; instead, heart rate increases through the operation of subconscious neural mechanisms when the need arises, such as during exercise. For this reason, the autonomic nervous system is sometimes referred to as the *involuntary nervous system.*

Dual Innervation in the Autonomic Nervous System

Innervation of organs by the two branches of the autonomic nervous system is depicted in **Figure 1**; note that *both* branches of the autonomic nervous system innervate most organs, an arrangement called *dual innervation.* Even so, the functions of the two branches are generally opposite in nature, so their effects are also diametrically opposed to each other. This poses no conflict, however, because the two autonomic divisions are typically most active under different conditions. For instance, the parasympathetic nervous system is most active during resting conditions, when it both stimulates the digestive organs (enhancing the digestion and absorption of nutrients) and inhibits the cardiovascular system (decreasing heart rate); this state can be referred to as "rest and digest." In contrast, the sympathetic nervous system is most active during periods of excitation or physical activity, when it coordinates a group of physiological changes known as the **fight-or-flight response** that prepares the body to cope with threatening situations. During the fight-or-flight response, the rate and force of the heart's contractions increase, blood flow shifts from the gastrointestinal organs to skeletal and cardiac muscles, and energy stores are mobilized. These and other changes—all mediated by the sympathetic nervous system—prepare the body for intense physical exertion and otherwise adapt it for a possible response to a threatening situation.

The primary function of the autonomic nervous system is to regulate the function of effector organs so as to maintain homeostasis. At rest, both the sympathetic and parasympathetic branches are active, but the parasympathetic nervous system dominates. When the body becomes excited or stressed, however, the pattern shifts to feature less parasympathetic activity and more sympathetic activity. The existence of both systems working at cross purposes provides a push-pull mechanism that permits greater control over the effector organs. For example, both sympathetic and parasympathetic neurons control heart rate; to bring about an increase in heart rate, sympathetic activity increases (that is, the frequency of action potentials in sympathetic neurons increases) while parasympathetic activity decreases (that is, the frequency of action potentials in parasympathetic neurons decreases).

Anatomy of the Autonomic Nervous System

The autonomic nervous system consists of efferent pathways containing two types of neurons arranged in series that provide for communication between the CNS and the effector organ

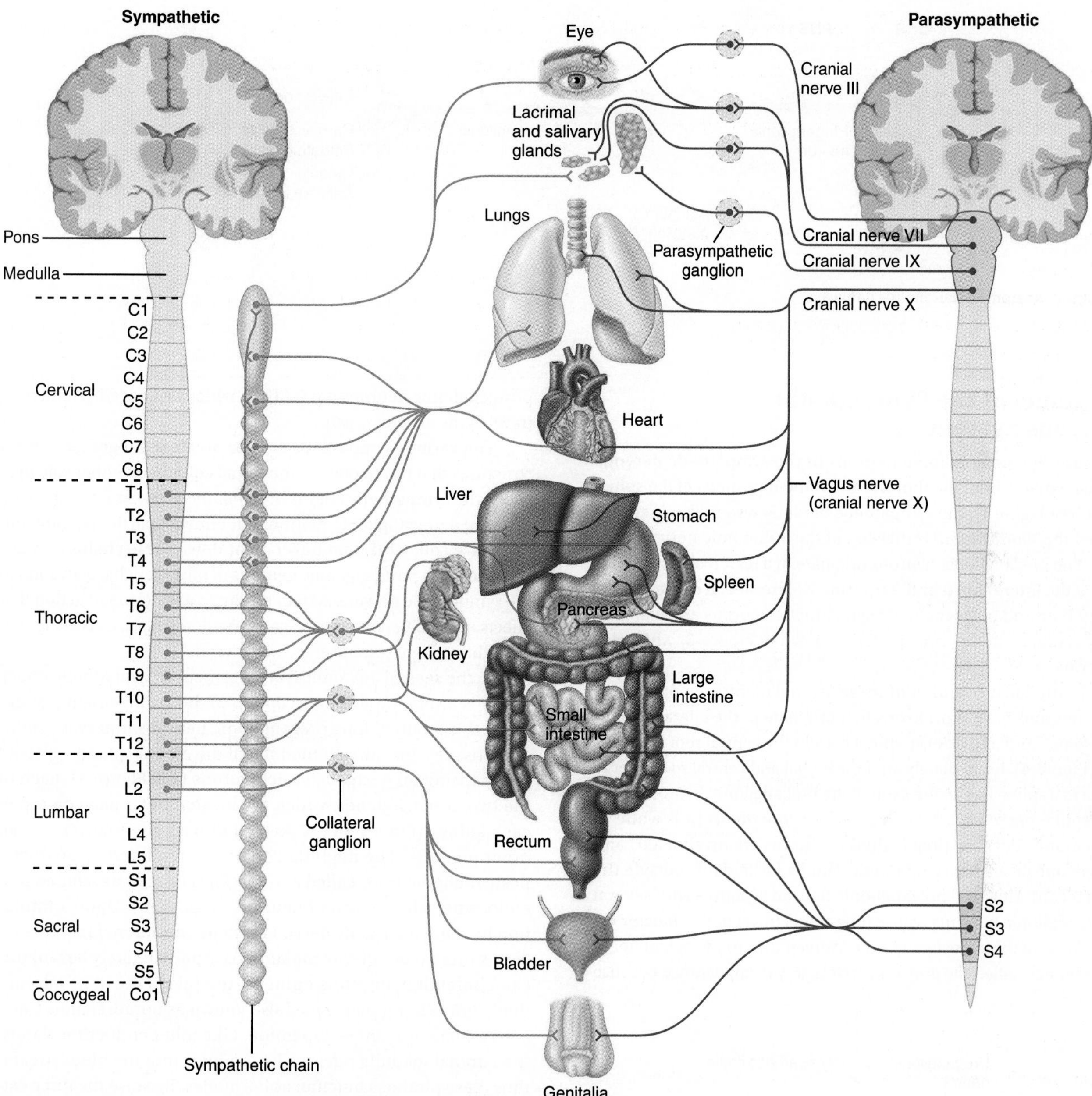

Figure 1 Dual innervation in the autonomic nervous system. In this detailed schematic diagram of the autonomic nervous system, parasympathetic pathways are shown in purple, and sympathetic pathways in green. Both branches of the autonomic nervous system generally innervate the same organs. (Only one side of the body is shown for each system; thus there are actually two sympathetic chains, two of each cranial nerve, and so on.)

(**Figure 2**). These neurons communicate with one another through synapses located in peripheral structures called **autonomic ganglia.** The neurons that travel from the CNS to the ganglia are called **preganglionic neurons;** the neurons that travel from the ganglia to the effector organs are called **postganglionic neurons.** Within each ganglion, therefore, are the axon terminals of preganglionic neurons and the cell bodies and dendrites of postganglionic neurons. A single preganglionic neuron generally synapses with several postganglionic neurons. In addition, other neurons located entirely within each ganglion, called *intrinsic neurons,* modulate the flow of information to the target organs. Although the two branches of the autonomic nervous system are similar in this regard, they show characteristic anatomical differences that are discussed in the following two sections.

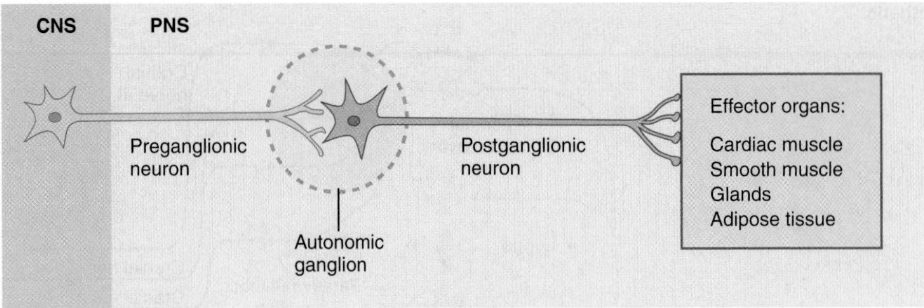

Figure 2 Anatomy of autonomic pathways.

Anatomy of the Sympathetic Nervous System

Because the preganglionic neurons in the sympathetic nervous system emerge from the thoracic and lumbar portions of the spinal cord (see Figure 1), the sympathetic nervous system is sometimes called the *thoracolumbar* division of the autonomic nervous system. The preganglionic neurons originate in a region of gray matter called the **lateral horn** or the **intermediolateral cell column.** Preganglionic and postganglionic sympathetic neurons are anatomically arranged in the three patterns shown in **Figure 3**, which are described next.

In the most common of these arrangements, the preganglionic neurons have short axons that originate in the lateral horn of the spinal cord and exit the spinal cord in the ventral root (Figure 3 and **Figure 4**). Immediately after the dorsal and ventral roots merge to form a spinal nerve, the axon of the preganglionic neuron leaves the spinal nerve via a branch called a *white ramus* (it is white in appearance because preganglionic neurons are myelinated) and enters one of several sympathetic ganglia located just outside the spinal cord. Here the preganglionic neuron synapses with several postganglionic neurons whose long axons travel to the effector organ. Most of these postganglionic axons return to the spinal nerve via a branch called the *gray ramus* (it is gray in appearance because postganglionic neurons are C fibers, which lack myelin), and then travel to the effector organ.

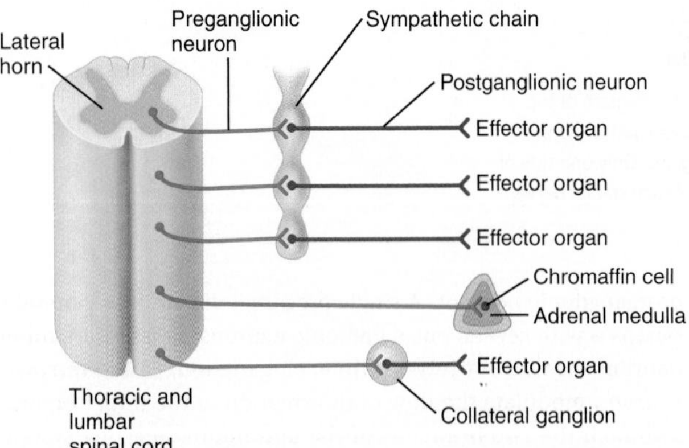

Figure 3 Anatomical pathways of preganglionic and postganglionic neurons in the sympathetic nervous system.

The various sympathetic ganglia are linked together to form structures that run parallel to the spinal column on either side of it, called the **sympathetic chains** or *sympathetic trunks* (see Figure 4). Because a preganglionic neuron that enters a particular ganglion may have collaterals that travel up or down these chains to make synapses with postganglionic neurons of other ganglia, activation of the sympathetic nervous system produces a widespread action that affects many different target organs simultaneously, as is evident in the fight-or-flight response.

In the second anatomical arrangement of sympathetic fibers, which is a very important exception to the arrangement just described, a group of long preganglionic neurons innervate endocrine tissue—the adrenal medulla of the adrenal gland—instead of synapsing on postganglionic neurons (see Figure 3). Each of the two adrenal glands, which are located in fat pads on top of each kidney (**Figure 5a**), is divided into an outer cortex and an inner medulla. The medulla consists of modified sympathetic postganglionic cells, called *chromaffin cells,* that developed into endocrine cells instead of neurons (Figure 5b). Upon stimulation by the sympathetic nervous system, the adrenal medulla releases three types of catecholamines. Approximately 80% of this catecholamine output is epinephrine (also known as adrenaline), 20% is norepinephrine (also known as noradrenaline), and a very small amount is dopamine. Like other endocrine glands, the adrenal medulla releases its products into the bloodstream; thus these products function as hormones. Because the three catecholamines travel throughout the body in the blood, this endocrine component of the sympathetic nervous system contributes to the widespread effects of sympathetic activation.

The third anatomical arrangement of sympathetic fibers includes preganglionic neurons that synapse with postganglionic neurons in structures called **collateral ganglia** (or prevertebral ganglia) situated somewhere between the CNS and the effector organ (see Figure 3). The collateral ganglia, which are found in pairs, include the celiac ganglia that innervate the stomach, liver, and spleen; the superior mesenteric ganglia that innervate the small intestine and upper part of the large intestine; and the inferior mesenteric ganglia that innervate most of the large intestine, the kidneys, urinary bladder, and reproductive organs.

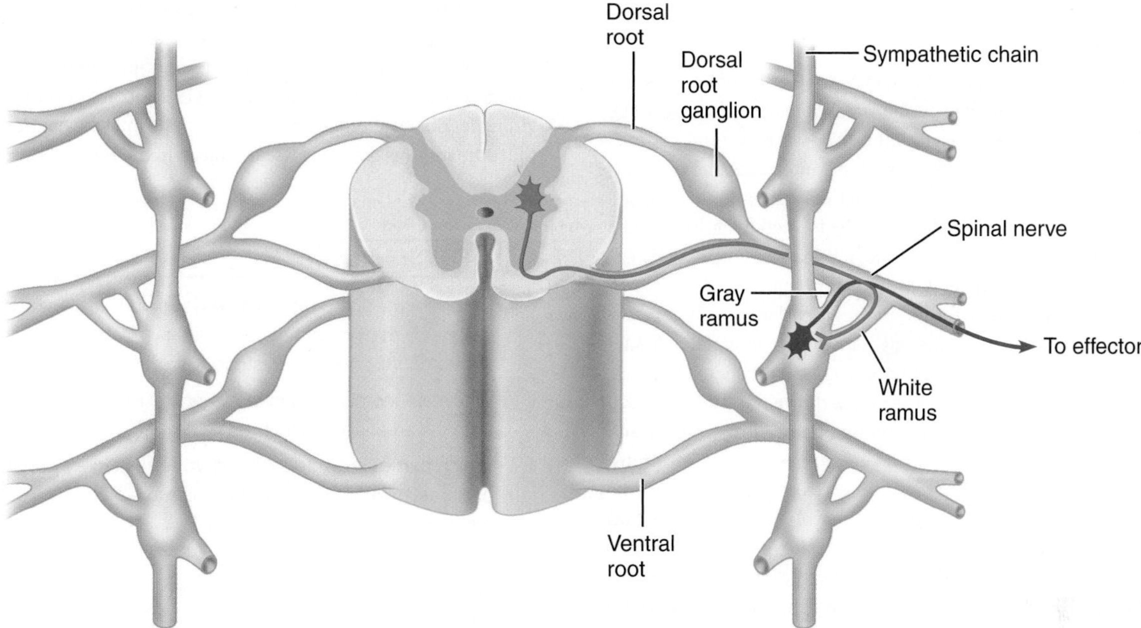

Figure 4 The most common pathway of sympathetic fibers.

Q In addition to sympathetic postganglionic axons, which types of axons travel in a spinal nerve?

The anatomical pathway for sympathetic neurons of the collateral ganglia includes a preganglionic neuron that exits the spinal cord in the ventral root and then enters the sympathetic chain via a white ramus, as previously described. The axon of the preganglionic neuron continues through this ganglion without forming a synapse, however; instead, it travels to a collateral ganglion via a sympathetic nerve. Within the collateral ganglion, the preganglionic neuron forms synapses with several postganglionic neurons that travel to target tissues. Because these ganglia are not interconnected, as are those of the sympathetic chain, they provide routes that enable the sympathetic nervous system to selectively target organs and thus exert more discrete effects.

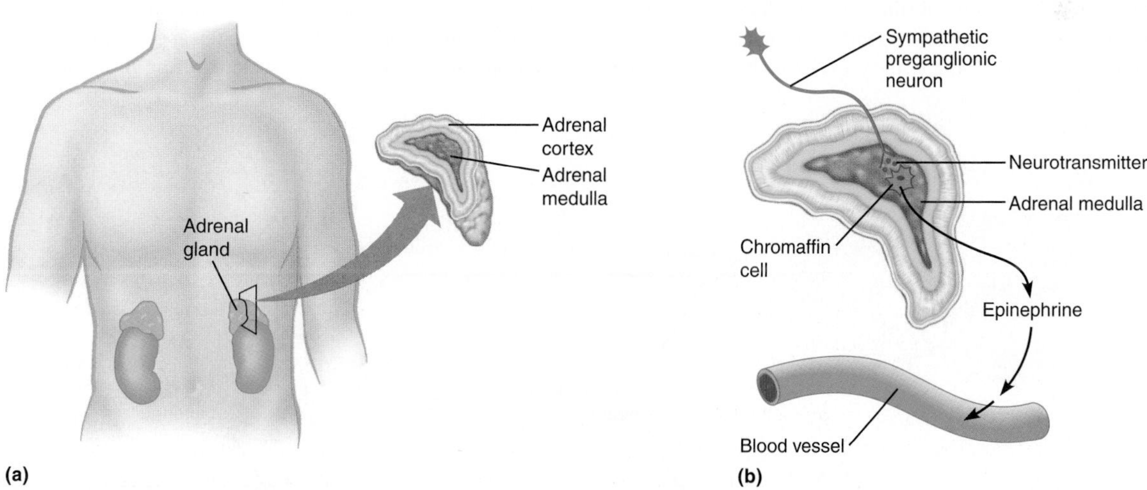

Figure 5 Sympathetic innervation of the adrenal gland. (a) The adrenal glands, located within fat pads on top of each kidney. **(b)** Sympathetic preganglionic neurons innervate chromaffin cells of the adrenal medulla, stimulating the release of epinephrine into the blood.

Q Which three hormones are secreted by chromaffin cells, and what are their proportions relative to the total amount secreted?

Anatomy of the Parasympathetic Nervous System

The preganglionic neurons of the parasympathetic nervous system originate in either the brainstem or the sacral spinal cord (see Figure 1), which explains why the parasympathetic nervous system is sometimes called the *craniosacral division* of the autonomic nervous system. Generally speaking, parasympathetic preganglionic neurons are relatively long and terminate in ganglia located near the effector organ (**Figure 6**); in the ganglia, they form synapses with short postganglionic neurons that travel to the effector organ.

In the cranial portion of the parasympathetic nervous system, axons of the preganglionic neurons originate from cranial nerve nuclei located in the brainstem, and then travel with axons in cranial nerves. One important cranial nerve is the **vagus nerve** (cranial nerve X), which originates in the medulla oblongata and innervates much of the viscera, including the lungs, heart, stomach, small intestines, and liver (see Figure 1). Other cranial nerves containing parasympathetic preganglionic axons are the *oculomotor nerve* (cranial nerve III), which innervates the smooth muscles of the eye that control pupil size and lens strength; the *facial nerve* (cranial nerve VII), which innervates the salivary glands; and the *glossopharyngeal nerve* (cranial nerve IX), which innervates the smooth muscle and glands of the throat and the viscera of the thorax and abdomen.

Unlike sympathetic preganglionic neurons, parasympathetic preganglionic neurons that originate in the spinal cord do not join with the spinal nerve. Instead, they join with other parasympathetic preganglionic neurons to form distinct pelvic nerves that innervate the colon, bladder, and reproductive organs (see Figures 1 and 6).

The Mixed Composition of Autonomic Nerves

Although it was long thought that nerves with fibers of the autonomic nervous system contained only efferent fibers, recent studies have shown that these nerves are generally mixed nerves. In fact, some of these nerves (for example, the vagus) actually contain more afferent fibers than efferent fibers. Only the efferent fibers in these nerves are part of the autonomic nervous system (which, by definition, is an *efferent* branch). The afferent fibers generally transmit signals from visceral receptors to the CNS and are important in providing feedback about the conditions of the internal environment so that adjustments can be made to maintain the body's homeostasis.

Quick Check 1

1. Compare the anatomical pathways for innervation of effector organs by the sympathetic and parasympathetic branches of the autonomic nervous system.
2. Which autonomic division regulates the release of epinephrine from the adrenal medulla?
3. Which part of the autonomic nervous system—sympathetic or parasympathetic—produces the more diffuse response? Why?

Autonomic Neurotransmitters and Receptors

The two primary neurotransmitters in the peripheral nervous system are acetylcholine and norepinephrine. Neurons that release the more common of the two, acetylcholine, are referred to as **cholinergic.** Acetylcholine is released by preganglionic neurons of both the sympathetic and parasympathetic branches of the autonomic nervous system, and by parasympathetic postganglionic neurons. The sympathetic preganglionic neurons that innervate the chromaffin cells of the adrenal medulla release acetylcholine, just like all other preganglionic neurons. In this case, however, acetylcholine acts on endocrine cells in the adrenal medulla to stimulate the release of epinephrine. Acetylcholine is also the sole neurotransmitter of the somatic branch

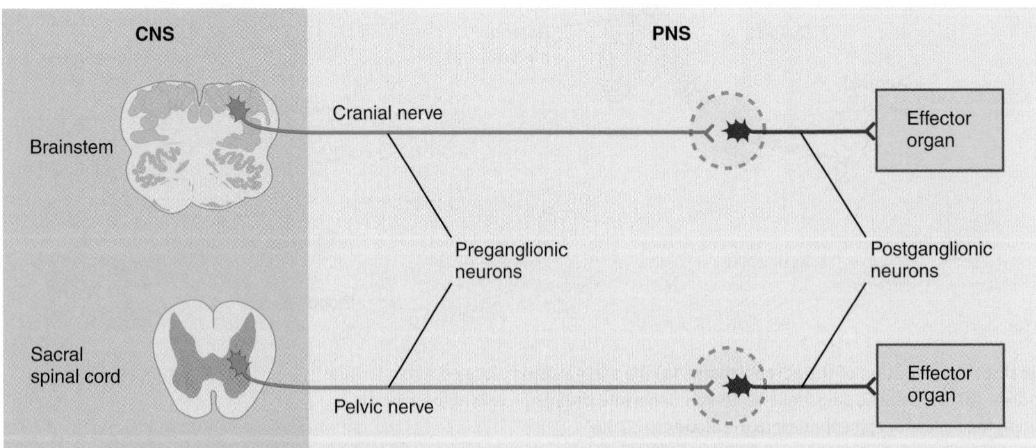

Figure 6 Parasympathetic nervous system pathways.

of the efferent nervous system (a topic that is discussed later in this chapter).

Norepinephrine is released by almost all sympathetic postganglionic neurons. Neurons that release norepinephrine are referred to as **adrenergic.** (Sympathetic postganglionic neurons that innervate the sweat glands release acetylcholine.)

The effects of a neurotransmitter depend on the type of postsynaptic receptors activated when the neurotransmitter binds to them. Acetylcholine and norepinephrine can each bind to different classes and subclasses of cholinergic and adrenergic receptors, respectively. We also consider receptors for the hormone epinephrine in this section because epinephrine is released from the adrenal medulla during activation of the sympathetic nervous system.

Types of Cholinergic Receptors

The two major classes of cholinergic receptors—**nicotinic receptors** and **muscarinic receptors**—are distinguished based on the findings from pharmacological studies using two acetylcholine agonists (chemicals that bind to a receptor and produce the biological effect): *nicotine* (found in tobacco) and *muscarine* (a toxin found in certain mushrooms). At least four subclasses of nicotinic receptors and five subclasses of muscarinic receptors exist. Nicotinic cholinergic receptors are located on the cell bodies and dendrites of sympathetic and parasympathetic postganglionic neurons (**Figure 7**), on chromaffin cells of the adrenal medulla, and on skeletal muscle cells. Muscarinic cholinergic receptors are found on effector organs of the parasympathetic nervous system, such as the heart, smooth muscles controlling the diameter of the pupil of the eye, and smooth muscles in the digestive tract (Figure 7).

The signal transduction mechanisms of cholinergic receptors are summarized here. All subclasses of nicotinic cholinergic receptors are associated with channels that allow both sodium and potassium ions to move through them. When acetylcholine binds to these receptors, the cation channels open, allowing sodium to diffuse into the cell and potassium to diffuse out of the cell. Because sodium is further from equilibrium, the flow of sodium into the postsynaptic cell exceeds the flow of potassium out of the cell, leading to the cell's depolarization. Thus nicotinic cholinergic receptors are associated with depolarization, or excitation, of the postsynaptic cell.

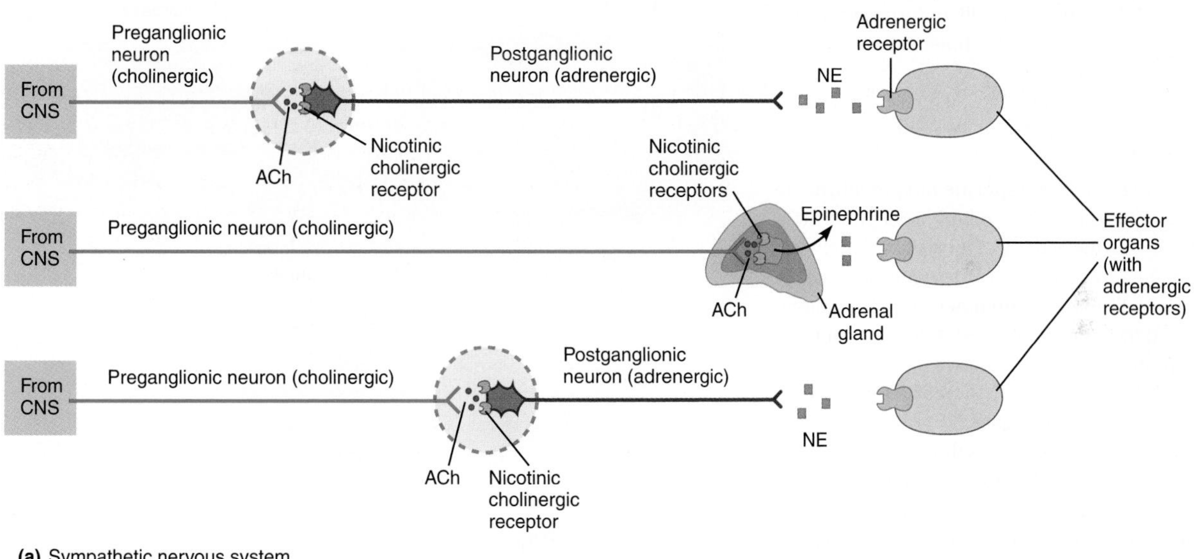

(a) Sympathetic nervous system

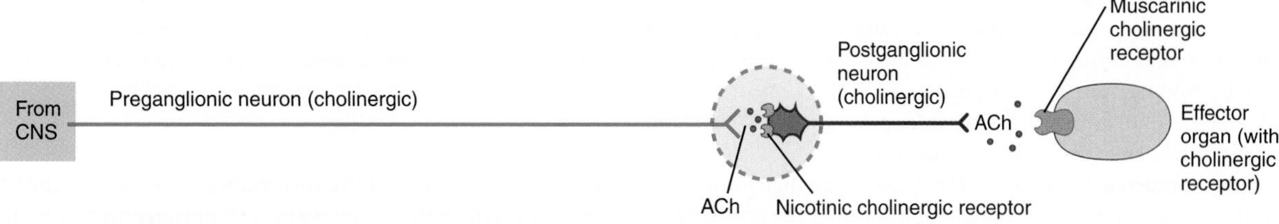

(b) Parasympathetic nervous system

Figure 7 Neurotransmitters and receptors in the autonomic nervous system. (a) Neurotransmitters and receptors for the three distinct anatomical pathways of the sympathetic nervous system. **(b)** Neurotransmitters and receptors in the parasympathetic pathway.

In contrast, all subclasses of muscarinic cholinergic receptors are coupled to G proteins and second messengers. The responses triggered by the binding of acetylcholine can be either excitatory or inhibitory, depending on the target cell in question and the nature of the signal transduction pathway.

Types of Adrenergic Receptors

Two major classes of adrenergic receptors are located in effector organs of the sympathetic nervous system: **alpha (α) receptors** and **beta (β) receptors.** Each of these is further divided into subclasses: α_1 and α_2, and β_1, β_2, and β_3.

Adrenergic receptors are coupled to G proteins that either activate or inhibit second messenger systems (**Figure 8**). Binding of norepinephrine or epinephrine to an α_1 receptor activates a G protein, which in turn activates the enzyme phospholipase C, which then catalyzes the conversion of phosphatidylinositol bisphosphate (PIP_2) to inositol triphosphate (IP_3) and diacylglycerol (DAG) (Figure 8a). IP_3 enters the cell and triggers the release of calcium from intracellular stores. Calcium then causes a response in the cell. DAG activates the enzyme protein kinase C, which catalyzes the phosphorylation of a protein, causing a response in the cell.

Binding of norepinephrine or epinephrine to an α_2 receptor activates an inhibitory G protein (G_i) that decreases the activity of the enzyme adenylate cyclase, thereby suppressing the synthesis of cAMP (Figure 8b). Binding of norepinephrine or epinephrine to a β receptor, by contrast, activates a stimulatory G protein (G_s) that increases the activity of the enzyme adenylate cyclase, thereby enhancing the synthesis of cAMP (Figure 8b).

Whereas acetylcholine is the only endogenous messenger that binds to cholinergic receptors, both norepinephrine and epinephrine bind to adrenergic receptors. The subclasses of adrenergic receptors have different affinities for these two messengers, however (**Table 1**). Notably, the α receptors have a greater affinity for norepinephrine than for epinephrine and are generally excitatory; that is, they stimulate muscle cells to contract or glands to secrete a product. By comparison, β_1 and β_3 receptors have approximately equal affinities for norepinephrine and epinephrine, and tend to be excitatory. β_2 receptors have a much greater affinity for epinephrine than for norepinephrine; as a consequence, norepinephrine does not normally function as a chemical messenger at these receptors. Activation of β_2 receptors generally produces an inhibitory response.

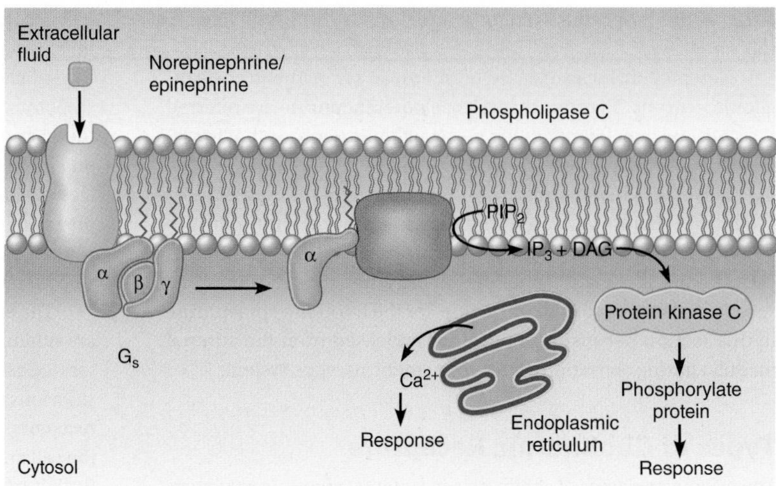

(a) Phosphatidylinositol bisphosphate (PIP_2) and α_1 receptors

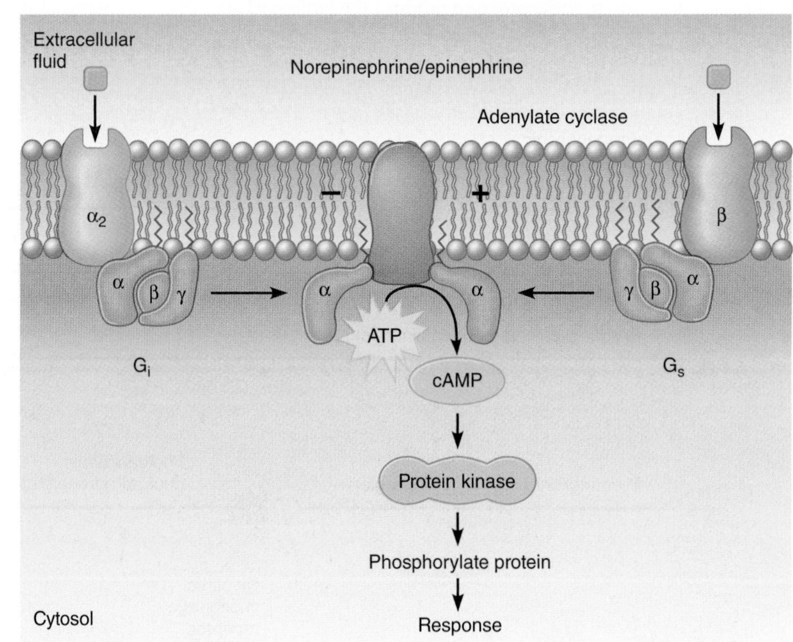

(b) cAMP and α_2 and β receptors

Figure 8 Signal transduction mechanism at effector organs of the sympathetic nervous system. (a) Binding of norepinephrine or epinephrine to an α_1 adrenergic receptor triggers activation of the phosphatidylinositol bisphosphate second messenger system. **(b)** Binding of norepinephrine or epinephrine to an α_2 adrenergic receptor triggers inhibition of the cAMP second messenger system, whereas binding to a β adrenergic receptor triggers excitation of the cAMP second messenger system.

Autonomic receptors are of great clinical significance. For example, people suffering from asthma or nasal congestion often take β_2 adrenergic agonists, such as epinephrine or ephedrine, to produce dilation of the respiratory airways; α_1 agonists can also be used to treat congestion. Hypertension can be treated

with drugs (such as propranolol) called beta-blockers, so named because they block the effects of norepinephrine or epinephrine at beta adrenergic receptors including those in the heart. Atropine, a muscarinic cholinergic antagonist, comes from the plant

TABLE 1 Adrenergic Receptors

Receptor type	Effector organ with receptor type	Relative affinities*	Signal transduction mechanism	Effect on effector organ†
α_1	Most vascular smooth muscle, pupils	NE > Epi	Activates IP_3	Excitatory
α_2	CNS, platelets, adrenergic nerve terminals (autoreceptors), some vascular smooth muscle, adipose tissue	NE > Epi	Inhibits cAMP	Excitatory
β_1	CNS, cardiac muscle, kidney	NE = Epi	Activates cAMP	Excitatory
β_2	Some blood vessels, respiratory tract, uterus	Epi >> NE	Activates cAMP	Inhibitory
β_3	Adipose tissue	NE = Epi	Activates cAMP	Excitatory

*NE = norepinephrine; Epi = epinephrine; > = greater than; >> = much greater than
†Effects are generalizations and not absolute.

Atropa belladonna; its name reflects the fact that extracts from the plant were once used to dilate the pupils of women's eyes for cosmetic purposes (*belladonna* means "beautiful woman"). Atropine is sometimes used to dilate the pupils of the eye before an eye exam and is used to treat intestinal spasms and nausea. To learn more about drugs that affect the autonomic nervous system, see Table 2.

Apply Your Knowledge

A person suffers from an overactive bladder. Based on information provided in Table 3, which type of drug (sympathetic or parasympathetic) could be given to treat this disorder? What might be some of the side effects of such a drug?

TABLE 2 Drugs Affecting the Autonomic Nervous System

Drug	Mechanism of action	Function
Pilocarpine	Muscarinic agonist	Treat dry mouth and glaucoma
Bethanechol	Muscarinic agonist	Treat non-obstructive urinary retention
Physostygmine	Inhibits acetylcholinesterase activity, can cross blood-brain barrier	Treat myasthenia gravis, Alzheimer's disease, and glaucoma
Neostygmine	Inhibits acetylcholinesterase activity; does not cross blood-brain barrier well	Treat myasthenia gravis
Atropine	Blocks muscarinic cholinergic receptors	Treat vagal-induced fainting, pupillary dilation for purposes of viewing retina, treat gastric ulcers
Scopolamine (usually applied as a patch)	Blocks muscarinic cholinergic receptors	Treat nausea and vomiting associated with motion sickness
Phenylephrine	α_1 agonist	Treat hypotension, local application to treat nasal congestion
Prazosin	α_1 antagonist	Treat hypertension and congestive heart failure
Clonidine	α_2 agonist	Treat cardiac arrest; terminate premature labor; through autoreceptors, decrease sympathetic outflow to treat hypertension
Yohimbine	α_2 antagonist	By blocking autoreceptors, increase sympathetic outflow to treat hypotension
Isoproterenol	β_1 and β_2 agonist	Bronchodilation for asthmatics
Propranolol	β_1 and β_2 antagonist	Treat ventricular arrhythmias and angina pectoris
Atenolol	β_1 antagonist that does not cross the blood-brain barrier	Treat hypertension
Timolol	β_1 antagonist given as eye drops	Treat glaucoma by reducing intraocular pressure
Ephedrine	Increases norepinephrine release and activity at target tissue	Decrease congestion
Phenylpropanolamine (Dexatrim)	Increases norepinephrine release	Promote weight loss

Autonomic Neuroeffector Junctions

The synapse between an efferent neuron and its effector organ is called a *neuroeffector junction.* Synapses between autonomic postganglionic neurons and their effector organs differ from ordinary neuron-to-neuron synapses in that the postganglionic neurons do not have discrete axon terminals. Instead, neurotransmitters are released from numerous swellings located at intervals along the axons of these neurons, called **varicosities** (**Figure 9a**). Within these varicosities, neurotransmitters are synthesized and then stored in vesicles (Figure 9b). The membrane of the axon contains the usual voltage-gated sodium and potassium channels that support propagation of action potentials. In addition, the membrane in the region of each varicosity contains voltage-gated calcium channels that open when an action potential reaches them.

The mechanism of neurotransmitter release from these varicosities is similar to the mechanism of transmitter release from an ordinary axon terminal (**Figure 10**). An action potential arriving at a varicosity opens voltage-gated calcium channels, which then allows calcium to enter the cytosol and stimulates the release of

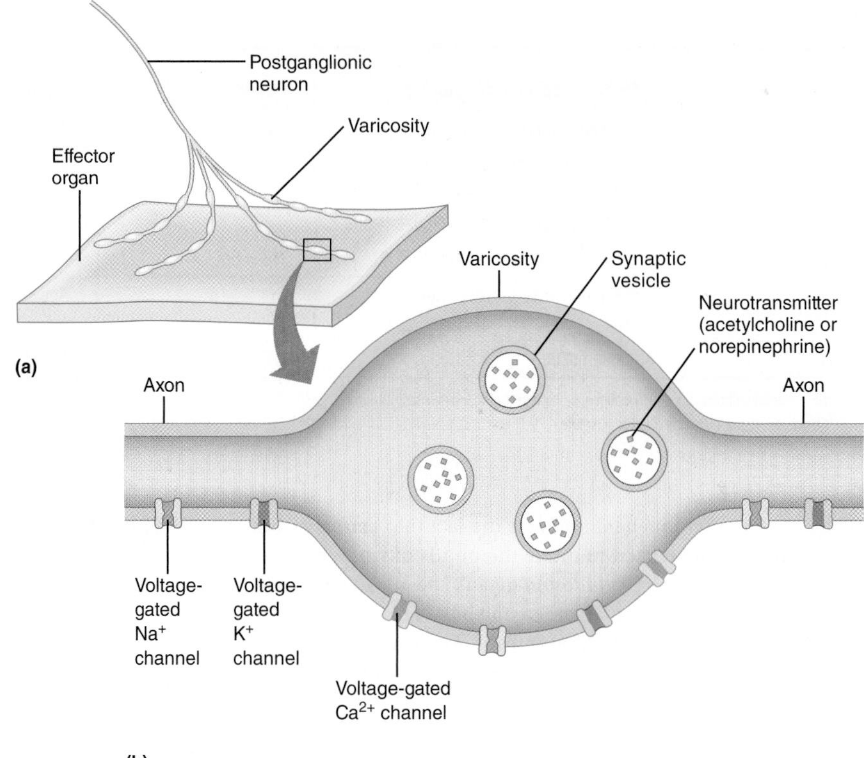

Figure 9 Neuroeffector junctions of the autonomic nervous system. (a) Near the effector organ, the axon of a postganglionic neuron splits into branches bearing swellings called varicosities. **(b)** An enlarged view of a varicosity.

Q Where, with respect to the varicosity depicted in Figure 9b, are action potentials propagated: at the left, at the right, or on both sides?

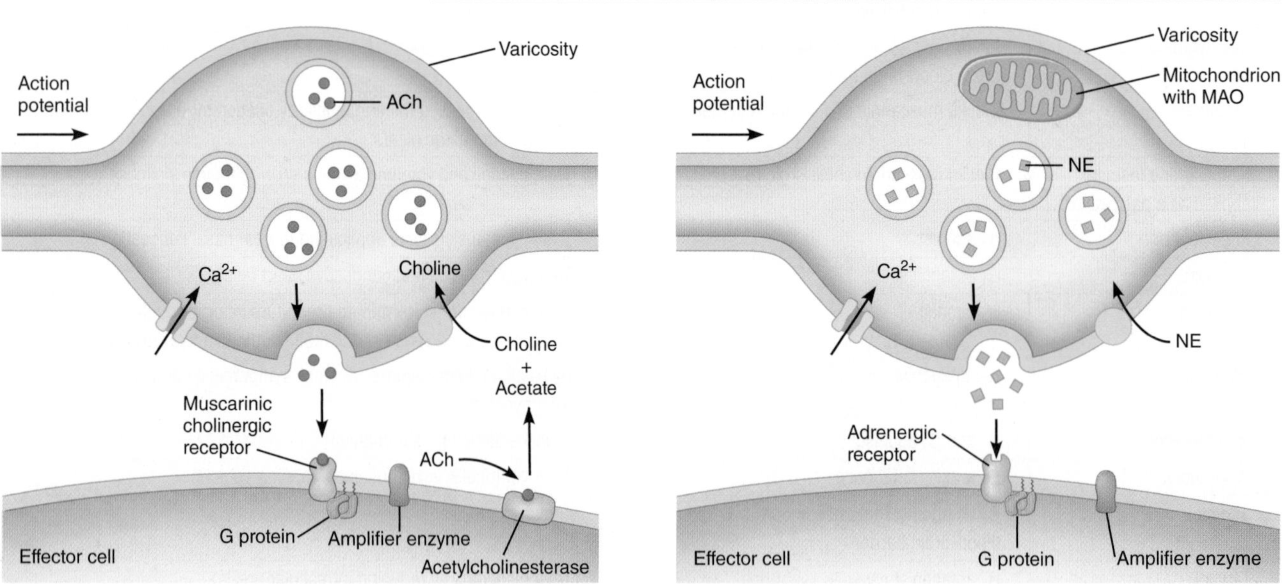

Figure 10 Neurotransmitter release from varicosities. (a) Release of acetylcholine from a parasympathetic varicosity. **(b)** Release of norepinephrine from a sympathetic varicosity.

neurotransmitter by exocytosis. However, unlike the case with ordinary synapses, no discrete anatomical arrangement exists between the varicosity and the effector organ. First, because a single postganglionic axon has several varicosities, an action potential propagated along the axon triggers the release of neurotransmitter from all of the varicosities. Second, because the distance between the varicosity and effector organ is greater than the width of a synaptic cleft, the neurotransmitter released from a varicosity diffuses over a greater area of the effector organ, binding to receptors on cells throughout the effector organ.

The effects of the neurotransmitter are terminated in the same way that they are in a typical neuron-to-neuron synapse: via either diffusion of the neurotransmitter away from the receptors, active reuptake of the neurotransmitter, or degradation of the neurotransmitter by enzymes. One such enzyme is *acetylcholinesterase,* which is located on the membrane of either the postganglionic neuron or the effector organ at cholinergic neuroeffector junctions. Following degradation of acetylcholine to acetate and choline by acetylcholinesterase, choline is actively transported back into the postganglionic varicosity and used to synthesize more acetylcholine. Monoamine oxidase, another degradative enzyme, is located at adrenergic neuroeffector junctions within mitochondria in postganglionic neurons, where it degrades catecholamines that have been actively transported back into the cells.

Regulation of Autonomic Function

To maintain homeostasis, the body must balance activity levels between the two branches of the autonomic nervous system. Under resting conditions, the body uses less energy than at other times; most of that energy is expended in the digestion and absorption of nutrients from recently ingested food. With low energy demands, the heart does not need to work as hard, because organs' demand for blood is low. Appropriately, the parasympathetic nervous system exerts dominant control over the body's organs under these conditions. By comparison, when the body is active, energy demands of the tissues increase. These demands must be met, even at the expense of suspending operation of the gastrointestinal organs. Under these conditions, parasympathetic activity decreases, and sympathetic activity increases. The result is increased activity of the heart, which helps increase blood flow to the skeletal muscles. But how does the brain regulate the balance between parasympathetic and sympathetic activity to meet the body's changing demands?

Most changes in the activity of the autonomic nervous system are accomplished through the operation of **visceral reflexes**—automatic changes in the functions of organs that occur in response to changing conditions within the body. Consider, for example, the autonomic response that operates when a person stands up rapidly (**Figure 11**). The drop in blood pressure that occurs because blood pools in the lower limbs due to the force of gravity is detected by receptors in some of the major arteries (aorta and carotid arteries); this information is then relayed to cardiovascular regulatory areas in the medulla oblongata through an afferent pathway. In response, these areas adjust their output to the sympathetic and parasympathetic nervous systems. Sympathetic output to the heart and blood vessels is

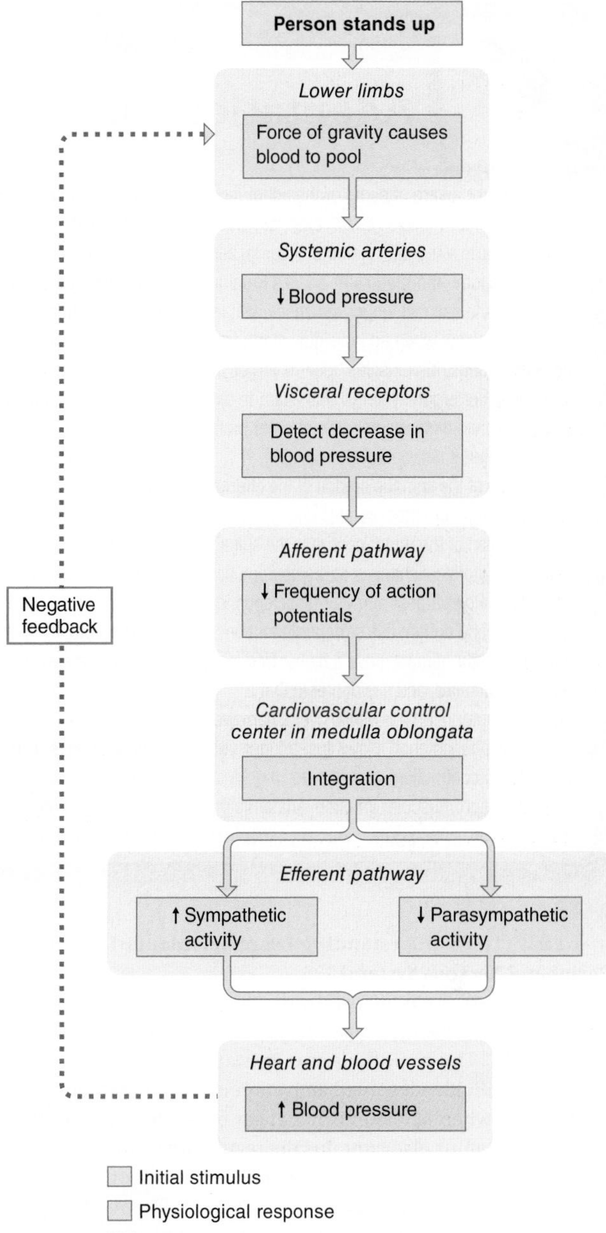

Figure 11 **Autonomic reflex response that controls blood pressure when a person stands up.**

enhanced, which brings blood pressure back to normal. At the same time, parasympathetic output to the heart decreases, which allows the blood pressure to increase unopposed. (Recall that parasympathetic activity tends to slow the heart—an action that tends to lower blood pressure.) Blood pressure typically returns to normal in just a few seconds, so the person is unaware of any changes that might have occurred. This reflex arc acts to prevent reduced blood flow to the brain and loss of consciousness when a person arises from lying down. The autonomic nervous system also adapts to long-term variations in activity, as for example

UNDERSTANDING EXERCISE

Adaptations of the Peripheral Nervous System

Most of us are aware of some of the adaptations that occur to our muscles as we exercise. When a person starts a regular routine of jogging, her muscles become more efficient and, as training continues, she can run longer distances with less fatigue. Likewise, when an individual starts training with weights, his muscles become larger and stronger. Other organ systems also adapt to exercise. Here we describe how the efferent branch of the peripheral nervous system adapts.

In terms of the somatic branch of the efferent nervous system, the most notable changes occur with activation of the motor neuron in the spinal cord and anatomical changes at the neuromuscular junction. Recruitment of motor units tends to be stronger and tighter: It is stronger in that once activated, motor neurons have a higher frequency of action potentials, and it is tighter in that a particular motor neuron is active over a narrower range. This phenomenon allows for stronger, yet more steady contractions of a muscle.

At the neuromuscular junction, anatomical changes occur with regular exercise, although the precise functional significance of these modifications remains unclear. Motor neurons, for example, develop a more extensive axonal branching at the neuromuscular junction. Thus, when an action potential arrives at the more numerous terminal boutons, more synaptic vesicles release their acetylcholine into the synaptic cleft, enhancing synaptic communication and possibly causing a stronger muscle contraction. In addition, regular exercise causes an increase in the concentration of acetylcholinesterase at the neuromuscular junction. Thus, while the muscle contracts more strongly due to enhanced acetylcholine release, it also relaxes more quickly due to a faster degradation of acetylcholine.

The autonomic nervous system also adapts to endurance exercise. For example, while a normal resting heart rate is in the range of 70–80 beats per minute, it is not unusual for an athlete to have a resting heart rate of 50 beats per minute. This is partially due to changes in the parasympathetic and sympathetic neural activity related to the heart. Parasympathetic neural activity decreases heart rate, whereas sympathetic neural activity increases heart rate. Under resting conditions, parasympathetic activity dominates and heart rate is relatively low. During exercise, sympathetic activity increases and parasympathetic activity decreases, causing an increase in heart rate. When a person engages in regular exercise, the resting heart rate is lowered due to neural adaptation: Parasympathetic activity is more active and sympathetic activity is less active during rest in an athlete compared to a non-exercising individual. Part of this adaptation reflects the greater efficiency of the heart and other organs in response to exercise. The decreased resting levels of sympathetic neural activity may have other benefits for the body as well, as the sympathetic nervous system is activated during the fight-or-flight response, which may cause additional wear on the body. For example, a high heart rate causes a higher blood pressure, which puts additional stress on the walls of the blood vessels.

in exercise (see **Understanding Exercise: Adaptations of the Peripheral Nervous System**).

The major areas of the brain that regulate autonomic function include the hypothalamus, pons, and medulla oblongata (**Figure 12**). The hypothalamus initiates the fight-or-flight response to elicit widespread activation of the sympathetic nervous system when a person is in danger or is otherwise excited. The hypothalamus also contains the regulatory centers for body temperature, food intake, and water balance—all of which are regulated in some way by autonomic neurons. The medulla oblongata and pons contain cardiovascular and respiratory regulatory centers that control the heart, blood vessels, and smooth muscle in the respiratory airways and regulate the automatic breathing patterns that do not require conscious thought. These areas of the brain receive input from other brain regions, including the hypothalamus, cerebral cortex, and limbic system. They also receive afferent information needed for reflex control of visceral function, such as that just described concerning blood pressure. Other autonomic reflexes involving the brainstem include the pupillary light reflex, the accommodation reflex, the vomiting reflex, and the swallowing reflex. Some visceral reflexes are also coordinated by neural circuits in the spinal cord and, therefore, are considered spinal reflexes. Examples include the reflexes that control urination, defecation, erection , and ejaculation.

One area of the brain that influences activity in autonomic control centers is the limbic system, which is involved in the development of emotions. Emotions have a strong effect on activity of the autonomic nervous system and, therefore, influence the functions of effector organs controlled

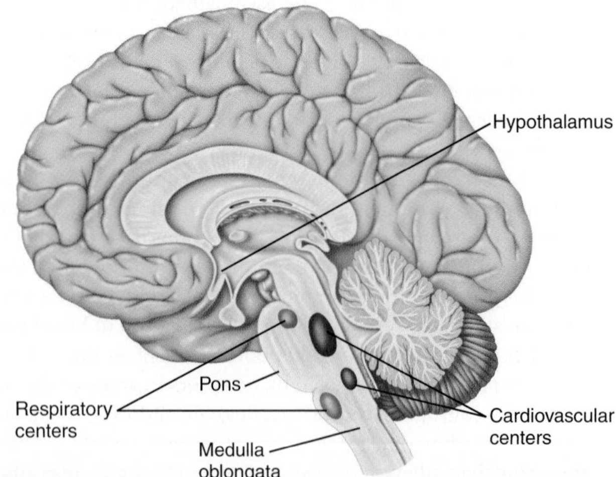

Figure 12 Areas of the brain that regulate autonomic function. This midsagittal section of the brain reveals the pons and medulla oblongata of the brainstem and the hypothalamus, all of which regulate autonomic function.

by it. We are all aware of some of the more common autonomic responses to emotions, including a racing heartbeat, an upset stomach ("butterflies"), blushing, fainting, and sweating.

Apply Your Knowledge

For each of the following emotional conditions—racing heartbeat, upset stomach, blushing, fainting, and sweating—identify whether it's the parasympathetic or the sympathetic branch of the nervous system that dominates the autonomic response. Which type of medication could be used to alleviate these emotional responses if they occur in excess?

The autonomic nervous system is also responsible for many of the symptoms associated with motion sickness, which is generally caused by a mismatch of sensory inputs, including those from the vestibular apparatus (which detects motion of the body), the visual system, and proprioceptors throughout the body. Common symptoms of this imbalance include nausea and sweating. The neural mechanisms mediating motion sickness are not fully understood, but communication between the vestibular system and autonomic control centers in the brainstem appears to be involved, with both branches of the autonomic nervous system becoming excited during this condition. A common treatment for motion sickness is the drug scopolamine, a muscarinic cholinergic antagonist that can be administered as a tablet or via a patch.

Table 3 summarizes the effects of the two branches of the autonomic nervous system. In the next section, we discuss the somatic nervous system.

Quick Check 2

1. What are the two primary neurotransmitters of the efferent branch of the peripheral nervous system, and which neuron(s) release them?
2. What is a neuroeffector junction? In which respects are autonomic neuroeffector junctions different from ordinary neuron-to-neuron synapses? In which respects are they the same?
3. What are visceral reflexes, and how are they involved in the control of autonomic functions?
4. Which areas of the CNS exert primary control over the autonomic nervous system?

2 The Somatic Nervous System

Unlike the autonomic nervous system, which controls the functions of many different types of effector organs, the somatic nervous system controls only one type of effector organ—skeletal muscle. Most skeletal muscles are connected to bones and, therefore, function in the support and movement of the body. In addition, the somatic nervous system has only a single type of efferent neuron: motor neurons, which innervate skeletal muscle. Most skeletal muscle is under voluntary control; that is, one can consciously decide to contract a muscle. For this reason, the somatic nervous system is also referred to as the *voluntary nervous system.*

Anatomy of the Somatic Nervous System

In the somatic nervous system, a single motor neuron travels from the central nervous system to innervate skeletal muscle cells (Figure 13); recall that two neurons are present in the autonomic nervous system pathway to an effector organ. Motor neurons originate in the ventral horn of the spinal cord (or the analogous brainstem nuclei) and receive input from multiple sources, including afferents (for spinal reflexes), the brainstem extrapyramidal tracts, and the cerebral cortex pyramidal tract. A single motor neuron innervates many muscle cells (called **muscle fibers**), but each muscle fiber is innervated by only one motor neuron. A motor neuron plus all the muscle fibers it innervates constitutes a **motor unit** (Figure 14). When a motor neuron is activated, it stimulates all the muscle fibers in its unit to contract.

The Neuromuscular Junction 

Each branch of a motor neuron synapses with a skeletal muscle fiber at a single highly specialized central region of the fiber called the **neuromuscular junction** (Figure 15a). The axon terminals of the motor neuron, called **terminal boutons,** store and release acetylcholine, the only neurotransmitter in the somatic nervous system. Opposite these terminal boutons is a specialized region of the muscle fiber's plasma membrane called the **motor end plate,** which has invaginations containing large numbers of nicotinic cholinergic receptors. Acetylcholinesterase, which is found between the invaginations of the motor end plate, terminates the excitatory signal, which allows the muscle fiber to relax.

The signal transmission mechanism at the neuromuscular junction is similar to that at excitatory neuron-to-neuron synapses (Figure 15b). When a motor neuron is activated by converging synaptic input, action potentials are propagated to the terminal boutons at the neuromuscular junctions of all muscle fibers in the motor unit. The resulting depolarization causes voltage-gated calcium channels in the boutons to open, thereby allowing calcium

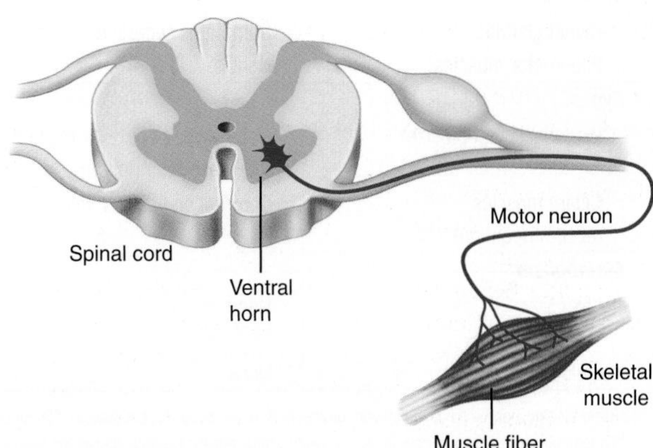

Figure 13 Anatomy of somatic nervous system pathway.

TABLE 3 Effects of Innervation by the Autonomic Nervous System

	Parasympathetic Nervous System*	Sympathetic Nervous System	
Organ system	**Effect**	**Effect**	**Adrenergic receptor class**
Heart			
SA node	Decreases heart rate	Increases heart rate	β_1
AV node	Decreases conduction velocity	Increases conduction velocity	β_1
Force of contraction	Decreases (small effect)	Increases	β_1
Blood vessels			
Arterioles to most of body	None	Vasoconstriction	α_1
Arterioles to skeletal muscle	None	Vasoconstriction	α_1
		Vasodilation (epinephrine)	β_2
Arterioles to brain	None	None	
Veins	None	Vasoconstriction	α_1
		Vasodilation (epinephrine)	β_2
Lungs			
Bronchial muscle	Contraction	Relaxation	β_2
Bronchial glands	Stimulates secretion	Inhibits secretion	α
Digestive tract			
Motility	Increased	Decreased	α_1, α_2, β_2
Secretions	Stimulated	Inhibited	α_2
Sphincters	Relaxation	Contraction	α_1
Pancreas			
Exocrine glands	Stimulates secretion	Inhibits secretion	α
Endocrine glands	Stimulates secretion	Inhibits secretion	α_2
Salivary glands	Stimulates watery secretion	Stimulates mucus secretion	α_1
Kidneys			
Renin release	None	Stimulated	β_1
Urinary bladder			
Bladder wall	Contraction	Relaxation (small effect)	β_2
Sphincter	Relaxation	Contraction	α_1
Male reproductive tract			
Blood vessels (erection)	Vasodilation	None	
Vas deferens and seminal vesicles (ejaculation)	None	Ejaculation	α_1
Female reproductive tract			
Uterus, nonpregnant	Unknown	Relaxation	β_2
Uterus, pregnant	Unknown	Contraction	α_1
Skin			
Sweat glands	Stimulates secretion	Stimulates secretion	α_1, muscarinic†
Piloerector muscles	None	Contraction (hairs stand up)	α_1
Eye			
Iris muscles (pupil size)	Contraction of circular muscle (pupillary constriction)	Contraction of radial muscle (pupillary dilation)	α_1
Ciliary muscles (accommodation)	Contraction for near vision	Relaxation for far vision (small effect)	β_2
Metabolism			
Liver	None	Stimulates glycogenolysis and gluconeogenesis	α_1, β_2
Adipose tissue	None	Stimulates lipolysis	β_3

*Receptor types for the parasympathetic nervous system are not given, because *all* effector organs have muscarinic cholinergic receptors.

†Sympathetic postganglionic neurons to the sweat glands release acetylcholine as the neurotransmitter.

to enter the cytosol and trigger the release of acetylcholine by exocytosis. Acetylcholine, in turn, diffuses across the synaptic cleft and binds to nicotinic cholinergic receptors at the motor end plate, causing cation channels to open. This allows sodium to flow into the muscle fiber and produces a depolarization called an **end-plate potential (EPP)** that is similar in many ways to the excitatory postsynaptic potentials (EPSPs) generated in neurons. The one major difference is that end-plate potentials are normally of sufficient magnitude to depolarize the muscle fiber to threshold, which generates an action potential in the muscle cell plasma membrane and triggers contraction of the muscle fiber. Whereas hormones regulate skeletal muscle *metabolism*, only innervation by a motor neuron controls skeletal muscle *contraction.* Therefore, all neural communication to skeletal muscle is excitatory; that is, it stimulates the muscle to contract. For skeletal muscle to relax, the neural stimulation must decrease.

Clearly, normal signal transmission at the neuromuscular junction is necessary for normal control of skeletal muscle contraction. If transmission at the neuromuscular junction is altered, such as occurs in the disease *myasthenia gravis,* then normal function of skeletal muscle is lost (see **Clinical Connections: Myasthenia Gravis**).

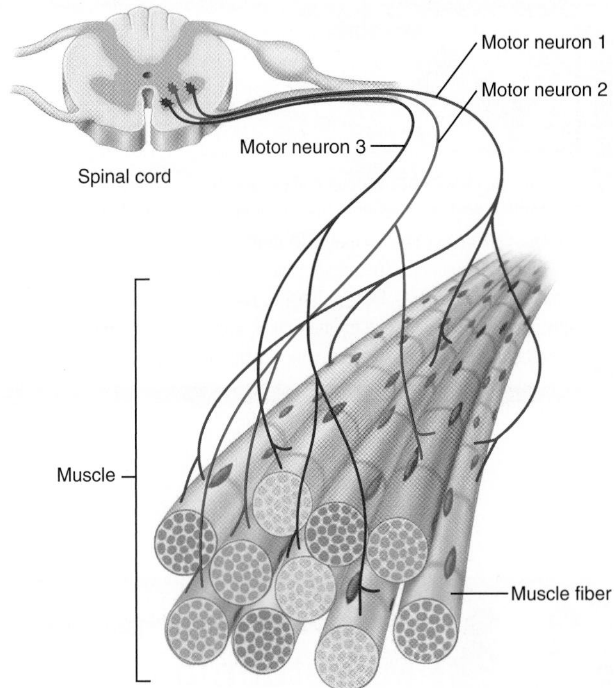

Figure 14 **Motor units.**

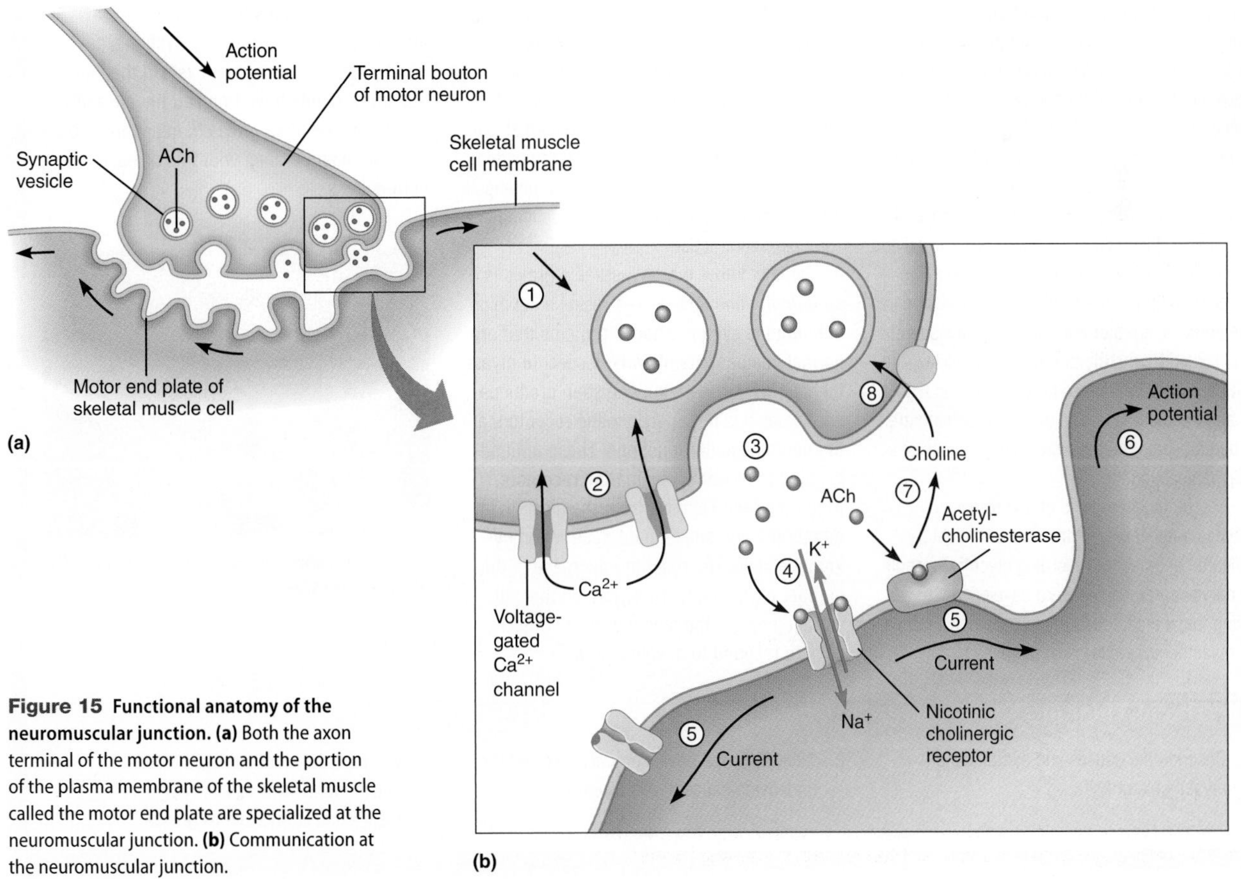

Figure 15 **Functional anatomy of the neuromuscular junction. (a)** Both the axon terminal of the motor neuron and the portion of the plasma membrane of the skeletal muscle called the motor end plate are specialized at the neuromuscular junction. **(b)** Communication at the neuromuscular junction.

FOCUS ON DIABETES

Diabetic neuropathy has widespread effects on the nervous system. Given that the autonomic nervous system controls so many different organ systems, autonomic neuropathy has a wide range of serious effects. First, autonomic neuropathy makes it more difficult for individuals with diabetes to control their disease, which requires controlling blood glucose levels. Many of the symptoms of hypoglycemia, such as sweating, shakiness, and palpitations, require autonomic activity to become manifest and, therefore, may go unnoticed if neuropathy is present. Damage to autonomic nerves to the heart also decreases the ability of the heart to adjust to changes in the demand for increased or decreased blood flow. Autonomic neuropathy affecting nerves to the gastrointestinal tract can cause nausea, vomiting, or constipation. This type of neuropathy can affect the kidneys and reproductive organs as well, causing symptoms such as urinary incontinence and erectile dysfunction.

CLINICAL CONNECTIONS

Myasthenia Gravis

Myasthenia gravis is a disease affecting signal transmission at neuromuscular junctions. Its victims are mostly women, and it usually strikes between the ages of 20 and 50 years. The defining characteristic of this disease is fatigue of unusually rapid onset and severity following the use of certain muscle groups. Because the muscles most frequently affected are those of the head, difficulties in speaking *(dysarthria)* and in swallowing *(dysphagia)* are common symptoms; drooping of the eyelids *(ptosis)*, shown in the photo, is another common sign. Myasthenia gravis also frequently targets limb muscles, with resultant weakness in the arms and legs. In some people, muscles used in breathing are affected, sometimes necessitating the use of a mechanical ventilator.

The underlying cause of this disease remained a mystery until the mid-20th century, when techniques for recording electrical signals in nerves and muscles were developed. Studies using these techniques revealed that symptoms of myasthenia gravis arose because motor neurons failed to excite muscle cells to contract. Even though motor neurons were capable of transmitting action potentials that were followed by end-plate potentials (EPPs) in muscle cells, these EPPs were often smaller than normal, particularly if neurons were repetitively stimulated. In such a case, muscle cell membrane potentials would often fail to reach threshold, resulting in a "dropped" contraction.

We now know that myasthenia gravis is an *autoimmune* disease—a disease in which the immune system attacks proteins that are normal components of body tissues. In myasthenia gravis, the immune system produces antibodies against acetylcholine receptors at the neuromuscular junctions. These antibodies bind to nicotinic cholinergic receptors, triggering their removal from the plasma membrane and subsequent destruction by immune cells. The resulting decrease in the number of functional receptors on the cell surface impairs the ability of these muscle cells to respond to acetylcholine.

Current therapies for myasthenia gravis attempt to blunt the immune system's action by reducing antibody levels in the blood. Such treatments have reduced the mortality rate for myasthenia gravis to near zero and enabled most persons with this disease to live normal lives.

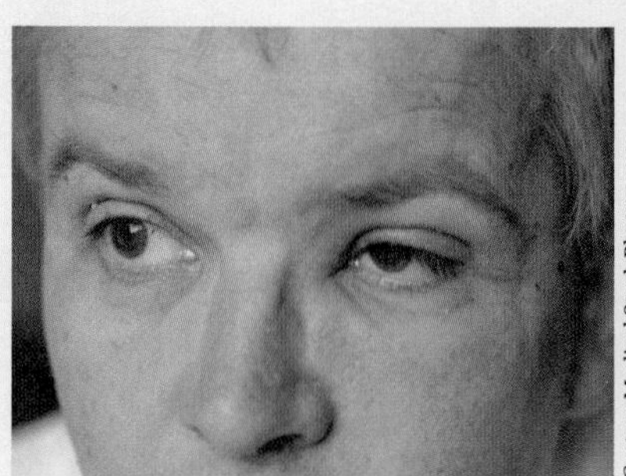

Custom Medical Stock Photo

This patient shows a drooping eyelid characteristic of myasthenia gravis.

Critical Thinking Questions

1. Describe the causes and symptoms of myasthenia gravis.
2. What technical advancements revealed the underlying cause of this disease?
3. What are the characteristics of an autoimmune disease, and how is it manifested in myasthenia gravis?

DISCOVERY

Curare

Curare is an extract of a plant *(Chondrodendron tomentosum)* found in South America. Indigenous people of that region crush and cook the roots and stems of the plant to produce a poison for the tips of their arrows and darts. (The word *curare* comes from the native word for "poison.") Traditionally, curare was used in hunting game. When an animal was struck by a curare-laced arrow or dart, it would become paralyzed and eventually die from respiratory failure.

The effective component of curare is a compound called *tubocurarine,* which blocks communication at the neuromuscular junction. Tubocurarine binds to nicotinic cholinergic receptors, thereby preventing acetylcholine from binding. When this occurs, skeletal muscles are unable to contract even when action potentials are transmitted in the motor neurons that innervate them.

In the late 19th and early 20th centuries, curare was studied for its possible pharmaceutical benefits. It was first used as a skeletal muscle relaxant to supplement general anesthesia in the 1930s. Today, curare has numerous clinical uses, including dilating hollow organs such as the rectum and relaxing the throat, which enables easier examination of these structures. It is also used for relief of *spastic paralysis,* a type of paralysis that can result from excessive skeletal muscle activity.

Bruce Coleman/Photoshot

A curare plant.

The neuromuscular junction is also targeted by many of the toxins present in animal venoms. The venom of the black widow spider, for example, contains the toxin *latroxin,* which by stimulating the release of acetylcholine at the neuromuscular junction induces muscle spasms and rigidity. Because the respiratory muscles (which are skeletal muscles) are affected, this venom can cause respiratory failure and death by inducing spastic contractions of these muscles. The venom of the rattlesnake, by contrast, contains the toxin *crotoxin,* which has the opposite effect of latroxin. Crotoxin inhibits the release of acetylcholine, which induces flaccid paralysis of skeletal muscles. Another toxin with paralyzing effects on skeletal muscle is *curare,* a poison used by native South Americans on the tips of the darts used in blow guns (see **Discovery: Curare**).

Functional Fact

Quick Check 3

1. Describe two differences between the somatic nervous system and the two branches of the autonomic nervous system with regard to the anatomical arrangement of their efferent neurons. What are the efferent neurons of the somatic nervous system called?
2. Which neurotransmitter is released by motor neurons at the neuromuscular junction? To which type of receptor does this neurotransmitter bind?
3. How many motor neurons innervate a single skeletal muscle fiber?

TABLE 4 Properties of the Autonomic and Somatic Nervous Systems

Property	Autonomic: parasympathetic	Autonomic: sympathetic	Somatic
Origin	Brainstem or lateral horns of the sacral spinal cord	Lateral horns of the thoracic and lumbar spinal cord	Ventral horns of the spinal cord
Neurons in pathway	Two (preganglionic and postganglionic)	Two (preganglionic and postganglionic)	One (motor neuron)
Effector organs	Cardiac muscle, smooth muscle, glands	Cardiac muscle, smooth muscle, glands, adipose tissue	Skeletal muscle
Neurotransmitters at neuroeffector junction	Acetylcholine	Norepinephrine	Acetylcholine
Receptor type at effector organ	Muscarinic cholinergic	Adrenergic (all classes)	Nicotinic cholinergic
Effects on effector organ	Either excitation or inhibition	Either excitation or inhibition	Excitation
Control	Primarily involuntary	Primarily involuntary	Primarily voluntary

SYSTEMS INTEGRATION As shown in Table 2, the peripheral nervous system provides communication to all the organs of the body, thereby coordinating actions to maintain homeostasis. For example, the autonomic nervous system innervates the cardiovascular system and stimulates adjustments when needed, such as decreasing sympathetic activity and increasing parasympathetic activity when one is resting to slow down heart rate and energy mobilization while increasing digestive processes. The opposite is triggered during times of arousal, when energy is mobilized, the heart beats faster and stronger, the pupils dilate, the liver produces glucose and releases it into the bloodstream, skeletal and cardiac muscles receive more blood, etc. so that the body can cope with the increased needs. There are many more examples of the coordinated actions of the autonomic nervous system that are beyond the scope of this chapter.

CHAPTER REVIEW

SUMMARY

1 The Autonomic Nervous System

- The efferent nervous system has two main branches: the autonomic nervous system and the somatic nervous system.
- Table 4 compares the properties of the two branches of the autonomic nervous system with those of the somatic nervous system.
- The autonomic nervous system includes the parasympathetic and sympathetic nervous systems, which innervate cardiac muscle, smooth muscle, glands, and adipose tissue.
- Effector organs have dual innervation; they are innervated by both the parasympathetic and sympathetic divisions.
- The parasympathetic nervous system is most active during rest, whereas the sympathetic nervous system is most active during periods of activity or excitation and is responsible for the fight-or-flight response.
- Pathways in the autonomic nervous system consist of two types of neurons that provide for communication between the CNS and the effector organ: preganglionic neurons and postganglionic neurons, which innervate the effector organ.
- The sympathetic nervous system also has an endocrine component, in that one set of preganglionic neurons innervates the adrenal medulla, stimulating the release of the hormone epinephrine.
- All preganglionic neurons contain the neurotransmitter acetylcholine.
- The parasympathetic postganglionic neurons also contain the neurotransmitter acetylcholine, but most sympathetic postganglionic neurons contain the neurotransmitter norepinephrine.
- The receptors for acetylcholine on postganglionic neurons are nicotinic cholinergic receptors, whereas the receptors for acetylcholine on effector organs in the parasympathetic nervous system are muscarinic cholinergic receptors.
- The receptors for norepinephrine and epinephrine on the effector organs in the sympathetic nervous system are adrenergic receptors.
- The synapse between an efferent neuron and its effector organ is called a neuro-effector junction.
- At the neuroeffector junctions between autonomic postganglionic neurons and their effector organs, neurotransmitter is released diffusely from varicosities and then binds to receptors on the effector organ.
- The mechanism of release of neurotransmitter is similar to that for neuron-to-neuron synapses.
- The autonomic nervous system is under involuntary control.
- Areas of the brain that influence autonomic activity include the brainstem, hypothalamus, and limbic system.

MAP iP Nervous II, Synaptic Transmission
iP Nervous II, Ion Channels
iP Nervous I, The Membrane Potential

2 The Somatic Nervous System

- The somatic division of the efferent nervous system consists of pathways composed of motor neurons that originate in the ventral horn of the spinal cord and innervate skeletal muscle cells.
- A motor unit is composed of a single motor neuron and the muscle cells it innervates.
- The synapse between a motor neuron and a skeletal muscle fiber is called a neuromuscular junction.
- The motor neuron contains the neurotransmitter acetylcholine.
- The receptors in skeletal muscle are nicotinic cholinergic.
- Binding of acetylcholine to nicotinic cholinergic receptors at the motor end plate produces an end-plate potential, which ultimately causes the skeletal muscle fiber to contract.

MAP iP Nervous II, Synaptic Transmission
MAP iP Muscular, The Neuromuscular Junction
iP Contraction of Motor Units

Check your understanding at MasteringA&P and apply your new knowledge to real-life case studies to ensure that you've mastered the content!

EXERCISES

Multiple-Choice Questions

1. Effector organs of the autonomic nervous system include all of the following *except*
   a) Heart muscle.
   b) Smooth muscle in the pupils of the eye.
   c) Respiratory muscles.
   d) Sweat glands.
   e) Salivary glands.
2. According to the concept of dual innervation by the autonomic nervous system, if sympathetic activity inhibits pancreatic secretions, then the parasympathetic nervous system should
   a) Inhibit pancreatic secretions as well.
   b) Stimulate pancreatic secretions.
   c) Have no effect on pancreatic secretions.
3. The adrenal medulla
   a) Contains sympathetic postganglionic neurons.
   b) Is part of the brainstem.
   c) Releases epinephrine into the blood.
   d) Is part of the parasympathetic nervous system.
   e) Is controlled by the somatic nervous system.
4. Which of the following cranial nerves does *not* contain parasympathetic preganglionic neurons?
   a) Oculomotor (cranial nerve III)
   b) Facial (cranial nerve VII)
   c) Glossopharyngeal (cranial nerve IX)
   d) Vagus (cranial nerve X)
   e) Hypoglossal (cranial nerve XII)
5. Which of the following receptor types does *not* activate G proteins?
   a) Nicotinic cholinergic
   b) Muscarinic cholinergic
   c) α_1 adrenergic
   d) β_1 adrenergic
   e) β_3 adrenergic
6. The origin of spinal preganglionic neurons is the
   a) Ventral horn of the spinal cord.
   b) Dorsal horn of the spinal cord.
   c) Lateral horn of the spinal cord.
7. The origin of motor neurons is the
   a) Ventral horn of the spinal cord.
   b) Dorsal horn of the spinal cord.
   c) Lateral horn of the spinal cord.
8. Which of the following second messengers stimulates the release of calcium from intracellular stores?
   a) cAMP
   b) Inositol triphosphate
   c) Diacylglycerol
9. Which of the following is the location of the cardiovascular regulatory centers?
   a) Thalamus
   b) Hypothalamus
   c) Limbic system
   d) Pons
   e) Medulla oblongata
10. How many motor neurons innervate a single skeletal muscle cell?
    a) Zero
    b) One
    c) Several
    d) Hundreds
    e) Millions
11. The motor end plate is
    a) The specialized synaptic terminal of the motor neuron.
    b) The specialized synaptic terminal of autonomic postganglionic neurons.
    c) The specialized region of skeletal muscle innervated by a motor neuron.
    d) The specialized region of an effector organ innervated by an autonomic postganglionic neuron.
12. Neurotransmitter is released from which portion of a postganglionic neuron?
    a) Terminal bouton
    b) Axon terminal
    c) Varicosity
    d) Cell body
    e) Dendrites

Objective Questions

13. Which branch of the autonomic nervous system has longer preganglionic neurons?
14. Which branch of the autonomic nervous system is most active when the body is at rest?
15. The communication between preganglionic neurons and postganglionic neurons in the autonomic nervous system is one-to-one. (true/false)
16. Which part of the adrenal gland secretes epinephrine?
17. Name the four cranial nerves that contain parasympathetic preganglionic neurons.
18. Autonomic nerves contain only efferent neurons. (true/false)
19. Which neurons in the peripheral nervous system are cholinergic?
20. Which neurons in the peripheral nervous system are adrenergic?
21. Which enzyme catalyzes the formation of diacylglycerol and inositol triphosphate?
22. A decrease in cAMP is associated with which class of adrenergic receptor?
23. Activation of α adrenergic receptors usually produces (excitation/inhibition).
24. β_2 adrenergic receptors have a greater affinity for (epinephrine/norepinephrine).
25. The motor end plate has (nicotinic/muscarinic) cholinergic receptors.
26. Skeletal muscle can be excited to contract only; that is, it cannot be inhibited to relax. (true/false)
27. The enyzme that degrades acetylcholine in the synaptic cleft is called ________.

Essay Questions

28. Describe the different anatomical arrangements found in the two branches of the autonomic nervous system and in the somatic nervous system.
29. Explain the concept of dual innervation.
30. Explain why sympathetic activation produces a more diffuse effect compared to activation of the parasympathetic nervous system.
31. Compare the signal transduction mechanisms for the different types of adrenergic receptors.
32. Which areas of the brain regulate autonomic function?

Critical Thinking

33. If a patient has such a slow heart rate that not enough blood circulates to the brain, which type of receptor agonist could be administered to increase heart rate? Which type of receptor antagonist could be administered? (See Table 3.)
34. Of the sympathetic preganglionic neurons and postganglionic neurons, one has myelinated axons whereas the other is unmyelinated. Based on the anatomy of the sympathetic nervous system in Figure 4, which neuron is myelinated and which is unmyelinated?
35. Explain how binding of norepinephrine to α_2 or β_1 receptors can produce an excitatory effect on the effector organ even though α_2 receptors inhibit production of cAMP whereas β_1 receptors stimulate production of cAMP.
36. Monoamine oxidase inhibitors are sometimes used to treat depression. What are some of the side effects you might expect with this drug class?

Photo credits for icons in feature boxes: Focus on Diabetes: Tanyae/Dreamstime; Clinical Connections: lofoto/Dreamstime; Understanding Exercise: Pixattitude/Dreamstime; Discovery: Yuri Arcurs/Cutcaster.

Glossary

alpha receptors class of adrenergic receptors

adrenergic pertaining to epinephrine (adrenaline) or norepinephrine (noradrenaline)

autonomic ganglia clusters of synapses between preganglionic and postganglionic neurons of the autonomic nervous system; singular, *autonomic ganglion*

beta receptors class of adrenergic receptors

cholinergic (koh-lin-er-jik) pertaining to acetylcholine

collateral ganglion sympathetic ganglion independent of the sympathetic chain

end-plate potential (EPP) a depolarization of the motor end plate of a skeletal muscle fiber caused by acetylcholine binding to nicotinic cholinergic receptors

fight-or-flight response the group of physiological changes coordinated by the sympathetic nervous system that prepares the body to cope with threatening situations

lateral horn a region of the gray matter of the spinal cord where certain autonomic preganglionic neurons originate; synonymous with *intermediolateral cell column*

motor end plate the specialized region of a skeletal muscle fiber's plasma membrane, located at the neuromuscular junction

motor unit a motor neuron and all the muscle fibers it innervates

muscle fibers muscle cells

neuromuscular junction the synapse between a motor neuron and a skeletal muscle cell

postganglionic neurons neurons of the autonomic nervous system that travel from autonomic ganglia to the effector organs

preganglionic neurons neurons of the autonomic nervous system that travel from the central nervous system to autonomic ganglia, where they communicate with postganglionic neurons

sympathetic chains structures parallel to the spinal column on either side in which the sympathetic ganglia are linked together in rows; also called *sympathetic trunks*

terminal bouton the axon terminal of a motor neuron, which stores and releases acetylcholine

vagus nerve (vay-gus) major parasympathetic nerve that originates in the medulla oblongata and innervates much of the viscera; cranial nerve X (CN X)

varicosities axon swellings of autonomic postganglionic neurons; store and release neurotransmitter

visceral reflexes automatic changes in the functions of organs that occur in response to changing conditions inside the body

Answers

Figure Question Answers

Fig 4	Axons of afferent neurons and motor neurons
Fig 5	Epinephrine (80%), norepinephrine (20%), dopamine (less than 1%)
Fig 9	On both sides

Apply Your Knowledge Answers

The parasympathetic nervous system stimulates bladder contraction, so a parasympathetic antagonist would decrease bladder contraction. Specifically, a muscarinic cholinergic antagonist could be used. However, such a drug would affect all effector organs of the parasympathetic nervous system, including the heart (heart rate would increase) and the gastrointestinal system (motility and secretion would decrease). The sympathetic nervous system inhibits bladder contraction at α_2 receptors, but this is a small effect. The sympathetic nervous system also stimulates contraction of the urethral sphincter, thereby inhibiting urination, through actions at α_1 receptors. Although an α_1 agonist could inhibit urination, it would not prevent the bladder contractions; thus the urge to urinate would still be present.

A racing heartbeat, upset stomach, blushing, and sweating can all occur with an excess in sympathetic activity and, therefore, can be blocked by a sympathetic antagonist. The type of antagonist would vary based on the receptor type activated. As indicated in Table 2, an α_1 antagonist could slow down heart rate, an α_1 antagonist could decrease blushing, and an α_1 antagonist could decrease sweating. Fainting is usually a result of increased parasympathetic activity, although an excess of sympathetic activity can also cause a person to faint. Thus a muscarinic cholinergic antagonist could be used for someone who tends to faint.

Answers to End–of-Chapter Multiple Choice and Objective Questions

Multiple Choice Questions

1. c 2. b 3. c 4. b 5. a 6. c 7. a 8. b
9. e 10. b 11. c 12. c

Objective Questions

13. parasympathetic
14. parasympathetic
15. false
16. medulla
17. oculomotor, glossopharyngeal, vagus, hypoglossal
18. false
19. parasympathetic preganglionic and postganglionic neurons, sympathetic preganglionic neurons, and motor neuron
20. sympathetic postganglionic neurons
21. phospholipase C
22. α_2 adrenergic
23. excitation
24. epinephrine
25. nicotinic
26. true
27. acetylcholinesterase

Answers to End-of-Chapter Multiple Choice and Objective Questions

Multiple Choice Questions

The Nervous System: Sensory Systems

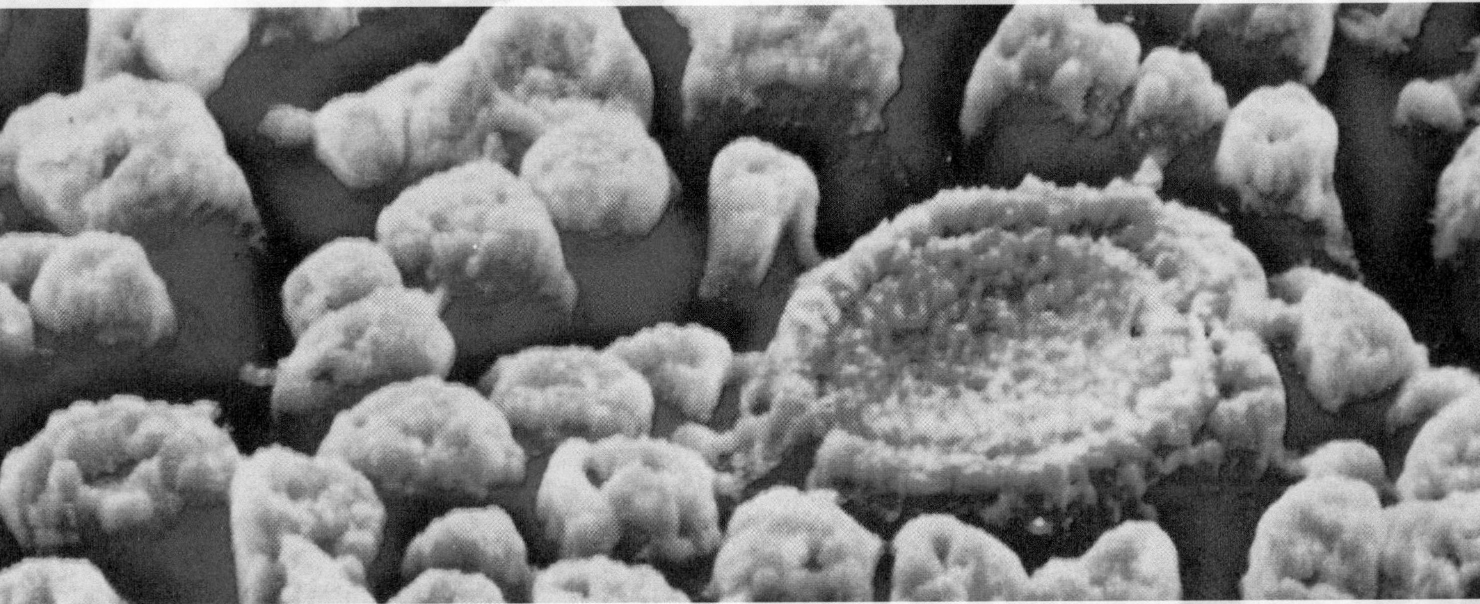

Science Source/Photo Researchers, Inc.

Colorized scanning electron micrograph (SEM) of taste buds.

Sit back for a moment and reflect on the world around you. Are you sitting or lying down? If you are indoors, what color are the walls? Are you warm or cold? Are the people near you noisy? Maybe someone has music on so loud that you can feel the vibrations. Your perception of the world around you is achieved through the functioning of your sensory systems. Even so, you cannot perceive everything in your environment. For example, if you've ever had an X-ray image taken, you could neither see nor hear the rays as they passed through your body; likewise, when you put some food in the microwave, you cannot see the waves that heat your food.

Why can we perceive some things in the world and not others? Humans have special sensory receptors that can detect specific forms of energy in the environment, such as light waves, sound waves, or vibrations, but we do not have receptors for X rays or microwaves. As a consequence, we cannot perceive these energy forms, even when they are present in our immediate environment. Furthermore, our perceptions of the stimuli that we *can* perceive are not absolute. When you jump into a lake, for example, you immediately perceive the cold water, but over time your body adapts to the chill: Within minutes the water does not feel as cold—even though the temperature of the water has not changed.

CHAPTER OUTLINE

LEARNING OUTCOMES *After studying this chapter, you should be able to:*

- Describe the function of sensory receptors, and explain how they perform that function.
- Describe the sensory transduction mechanism for each of the special senses.
- Describe the neural pathways of the different sensory systems, from sensory receptor to cerebral cortex.
- Describe how light is focused on the retina.
- Describe how the iris regulates the amount of light that enters the eye.
- Explain the differences between rods and cones in the eye.
- Describe the two sensory systems of the ear.
- Describe the two chemical senses.

Before You Begin

Make sure you have mastered the following topics:

- Ion channels
- Signal transduction mechanisms
- Cerebral cortex
- Spinal cord dorsal horn
- Ascending tracts

The peripheral nervous system consists of an afferent branch and an efferent branch, with the afferent branch sending information about the internal and external environments to the central nervous system (CNS) and the efferent branch sending information from the central nervous system to effector organs. Some of the structures of the central nervous system are involved in our perception of the world around us. This chapter focuses on the sensory systems involved in **perception**—the conscious interpretation of the world based on the sensory systems, memory, and other neural processes.

As we study the sensory systems, we will learn how the various signal transduction mechanisms detect and convey information about stimuli. These signal transduction mechanisms often depend on a receptor potential, a type of graded potential. Also, as with any branch of the nervous system, the importance of ion channels, electrochemical forces, and electrical signaling will be apparent.

In this chapter, we first describe general properties of sensory systems. Next, we discuss those senses that allow us to perceive touch and other stimuli associated with the skin surface, and that also allow us to perceive such things as the positions of our limbs and bodies. Finally, we discuss the special senses.

1 General Principles of Sensory Physiology

The afferent division of the peripheral nervous system transmits information from the periphery to the central nervous system. The information is detected by *sensory receptors* that respond to specific types of stimuli. Whereas some of these receptors detect stimuli in the external environment, others, called **visceral receptors,** detect stimuli that arise within the body. Visceral receptors transmit information to the CNS by a class of afferent neurons known as **visceral afferents.**

Examples of the many types of visceral receptors include *chemoreceptors* in major blood vessels that monitor O_2, CO_2, and H^+ levels in the blood; *baroreceptors* in certain blood vessels that monitor blood pressure; and *mechanoreceptors* in the gastrointestinal tract that monitor the degree of stretch or distension. Although the brain receives information from these receptors and utilizes this information for regulatory purposes, we are not consciously aware of these stimuli. For example, we cannot sense what the pH of our own blood is unless we take a blood sample and analyze it. Nevertheless, pH receptors in certain arteries and in the brainstem communicate to the brain whether the pH is normal, and this information in turn enables the brain to make regulatory decisions.

This chapter covers the sensory systems that detect information about the external environment. This information is transmitted to the cerebral cortex, reaches consciousness, and is perceived, making us aware of the world around us. As **Figure 1** shows, however, our perception of the world is not always fully accurate.

In Figure 1a, you perceive something that is not present—a square. Part of your perception is based on previous experiences. In this example, your brain perceives what it expects to see; that is, it fills in for perceived gaps in information. In Figure 1b, you perceive curved lines when the lines are actually straight, which shows that our perceptions may deceive us. In Figure 1c, some people perceive the image to be a rabbit, others a duck, others both, and still others neither. Individuals may perceive things differently, based in part on their individual past experiences.

The sensory systems that enable us to perceive the external environment include the somatosensory system and the special sensory systems. The **somatosensory system** is necessary for perception of sensations associated with receptors in the skin (**somesthetic sensations**) and for perception of the position of the limbs and the body (**proprioception**). Proprioception depends on specific proprioceptors in muscles and joints and on more generalized receptors in the skin. The **special senses** are necessary for senses of vision, hearing, balance and equilibrium, taste, and smell.

Receptor Physiology

Sensory receptors are specialized structures that detect a specific form of energy in the external environment. The energy form of a stimulus is called its **modality.** Examples of modalities include light waves, sound waves, pressure, temperature, and chemicals. The **law of specific nerve energies** states that a given sensory receptor is specific for a particular modality. Thus special cells in the eye called *photoreceptors* detect light waves, but not sound waves.

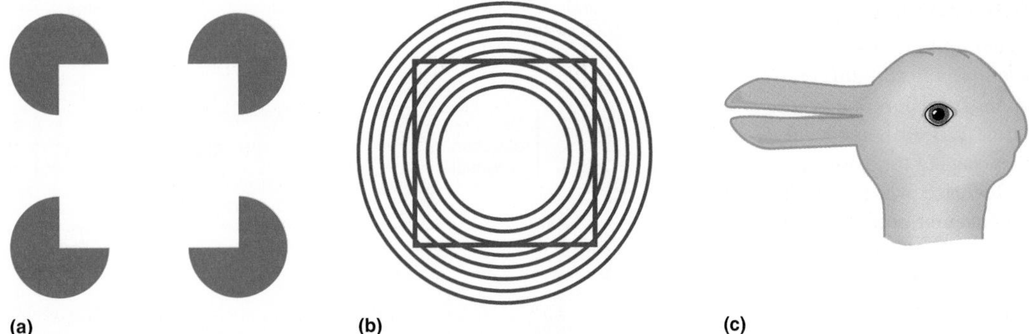

Figure 1 The fallibility of perception. These optical illusions demonstrate that our perception is not always accurate. **(a)** Where parts of the four green disks are missing, the brain perceives a square. **(b)** The effect of the concentric circles is to make straight lines appear curved. **(c)** This shape can resemble a duck or a rabbit—or both or neither.

The modality to which a receptor responds best is called the **adequate stimulus.** Table 1 lists the characteristics of several classes of sensory receptors.

Modalities other than the adequate stimulus may activate receptors, but only at relatively high energy levels. Moreover, a sufficiently strong "inadequate" stimulus will induce the perception of the adequate stimulus, as when a blow to the eye (pressure) activates photoreceptors, causing the person to perceive light.

Sensory Transduction

The function of sensory receptors is **transduction**—that is, the conversion of one form of energy into another. In **sensory transduction,** receptors convert the energy of a sensory stimulus into changes in membrane potential called **receptor potentials** or **generator potentials.** Receptor potentials resemble postsynaptic potentials in that they are graded potentials caused by the opening or closing of ion channels. The stronger the stimulus, the greater the change in membrane potential that occurs. However, unlike postsynaptic potentials, which are triggered by the binding of neurotransmitter to receptors, receptor potentials are triggered by sensory stimuli.

Sensory receptors exist in two basic forms, illustrated in Figure 2. In some cases, a sensory receptor is a specialized structure at the peripheral end of an afferent neuron (Figure 2a). When such a receptor is depolarized to threshold, an action potential is generated in the afferent neuron and propagated to the CNS, thereby transmitting information about the stimulus. In other cases, the sensory receptor is a separate cell that communicates through a chemical synapse with an associated afferent neuron (Figure 2b). When the sensory receptor is a separate cell, changes in the receptor cell's membrane potential cause the release of a chemical messenger or transmitter. (In some cases, a stimulus triggers a change in membrane potential that *inhibits* the release of a transmitter.) The greater the excitatory stimulus, the greater the amount of transmitter released. This transmitter then binds to receptors on the afferent neuron and causes changes in the membrane potential of that cell. In most instances, if the afferent neuron is depolarized to threshold, then an action potential is generated and transmitted by that cell to the CNS (photoreceptors are an exception). The mechanism of transduction varies for different types of receptors and is covered along with the individual sensory systems later in this chapter.

TABLE 1 Characteristics of Sensory Receptors

Receptor class	Sensation/visceral information	Modality
Photoreceptors	Vision	Photons of light
Chemoreceptors	Taste	Chemicals dissolved in saliva
	Smell	Chemicals dissolved in mucus
	Pain	Chemicals in extracellular fluid
Thermoreceptors		
Warm receptors	Warmth	Increase in temperatures between 30°C and 43°C
Cold receptors	Cold	Decrease in temperatures between 35°C and 20°C
Mechanoreceptors		
Pacinian corpuscle	Vibration	Pressure
Hair cells	Sound	Sound waves
	Balance and equilibrium	Acceleration

Receptor Adaptation

Some receptors continue to respond to a stimulus for as long as the stimulus is applied. However, most receptors *adapt* to the stimulus—that is, their response declines with the passage of time. **Receptor adaptation** is a decrease over time in the magnitude of the receptor potential in the presence of a constant stimulus. **Slowly adapting** or **tonic receptors** show little adaptation and, therefore, can function in signaling the intensity of a prolonged stimulus (**Figure 3a**). Examples of slowly adapting receptors include muscle stretch receptors, which detect muscle length, and *Merkel's disks,* which detect pressure applied to the skin (touch). **Rapidly adapting** or **phasic receptors** adapt quickly, and thus function best in detecting changes in stimulus intensity (Figure 3b). Rapidly adapting receptors respond at the onset of a stimulus, then adapt. Some rapidly adapting receptors also show a second, smaller response upon the termination of a stimulus, called the "off response." Examples of rapidly adapting receptors include *olfactory receptors,* which detect odors, and *Pacinian corpuscles,* which detect vibration in the skin.

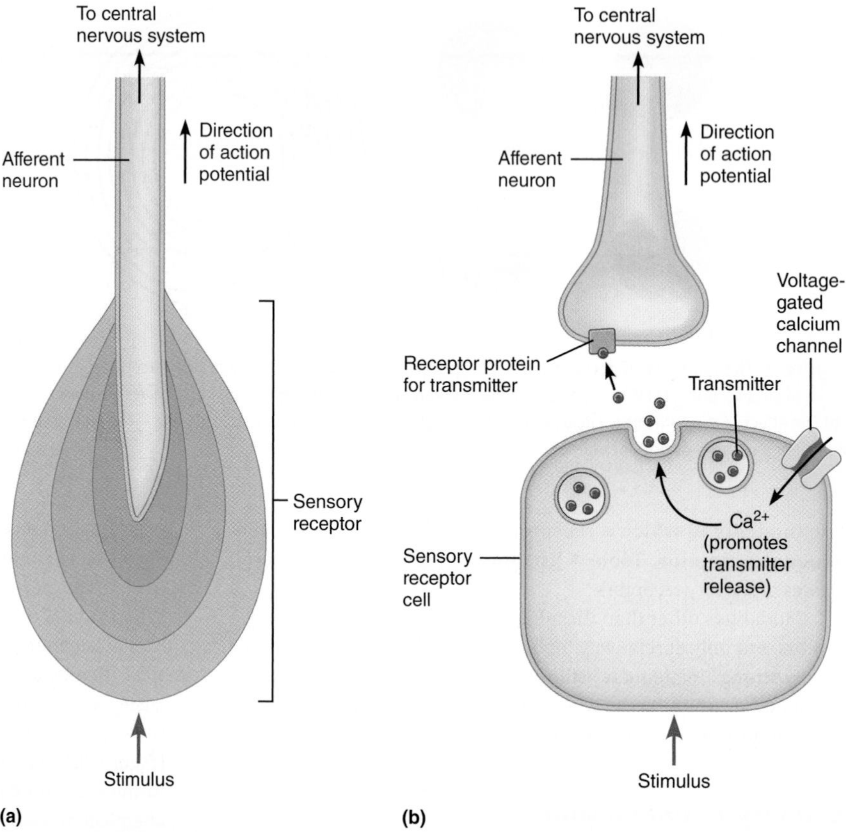

Figure 2 Structure and function of sensory receptors. (a) A sensory receptor that is a specialized ending of an afferent neuron. The stimulus acts on the sensory receptor by opening or closing ion channels, thereby producing a receptor potential. **(b)** A sensory receptor that is a separate cell from the afferent neuron. The stimulus changes the membrane potential of the receptor cell, which opens (or closes) a calcium channel, and cytosolic calcium concentration then increases (or decreases). Changes in calcium concentration trigger (or inhibit) the release of a chemical transmitter by exocytosis. The transmitter communicates to the afferent neuron by binding to receptors on the afferent ending and inducing a graded potential.

Sensory Pathways

Here we focus on the sensory pathways that relay sensory information to the appropriate area of the cortex.

The specific neural pathways that transmit information pertaining to a particular modality are referred to as **labeled lines,** and each sensory modality follows its own labeled line. Activation of a specific pathway causes perception of the associated modality, regardless of which stimulus actually activated the pathway. For example, pressure on the eye may trigger light flashes because the pressure generates signals in the pathway for perception of light. The same would be true if the neurons that transmit light information were activated by a blow to the back of the head where the visual cortex is located. The pathways for different modalities terminate in different sensory areas of the cerebral cortex (**Figure 4**). Although these pathways are distinct, we can make some generalizations concerning most pathways, starting with a sensory unit.

A **sensory unit** comprises a single afferent neuron and all the receptors associated with it (**Figure 5**). All of the

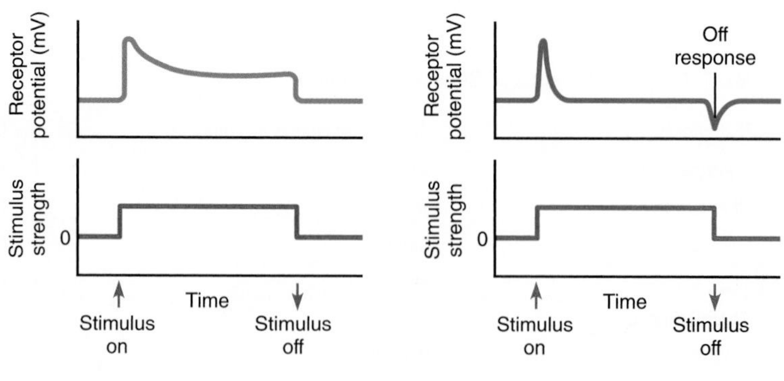

Figure 3 Responses of slowly adapting receptors and rapidly adapting receptors. (a) Slowly adapting receptors respond with a change in receptor potential that persists for the duration of the stimulus. **(b)** Rapidly adapting receptors respond with a change in receptor potential at the onset of a stimulus, but then adapt. The "off response" is a second, smaller response that occurs upon termination of a stimulus.

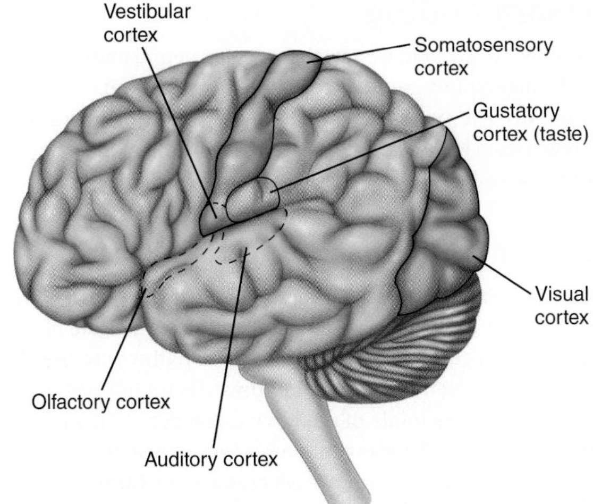

Figure 4 Sensory areas of the cerebral cortex. Note that the vestibular cortex actually lies on the underside of the somatosensory cortex in the parietal lobe, the auditory cortex lies on the internal side of the temporal lobe, and the olfactory cortex is on the inferior surface of the temporal lobe.

Q Name the lobe that corresponds to each of the following: visual cortex, auditory cortex, and somatosensory cortex.

receptors associated with a given afferent neuron are of the same type. Activation of any of the associated receptors may trigger action potentials in the afferent neuron. The area over which an adequate stimulus can produce a response (which can be either excitatory or inhibitory) in the afferent neuron is called the **receptive field** of that neuron; it corresponds to the region containing receptors for that afferent neuron. In the skin, for example, a single afferent neuron possessing Pacinian corpuscles branches in the periphery, with each branch terminating with a Pacinian corpuscle. Vibration on any area of the skin where these Pacinian corpuscles are located can depolarize the afferent neuron and, if strong enough, can generate action potentials. If vibration applied to any given area of the skin causes a response in a neuron, that area is included within that neuron's receptive field. Note, however, that receptive fields in the visual system are associated with areas in the visual field (the area of the external environment from which light rays can enter the eyes and be seen); afferent neurons in the visual pathway respond to light emanating from specific areas of the visual field. The other special senses do not have receptive fields.

Figure 6 illustrates a generalized pathway for the transmission of sensory information. Although not all sensory pathways follow this exact model, the model illustrates key similarities in sensory systems. The afferent neuron that transmits information from the periphery to the CNS is called the **first-order neuron.** A single first-order neuron may diverge within the CNS and communicate with several interneurons. In addition, interneurons may receive converging input from several first-order neurons. Some of these interneurons transmit the information to the thalamus, the

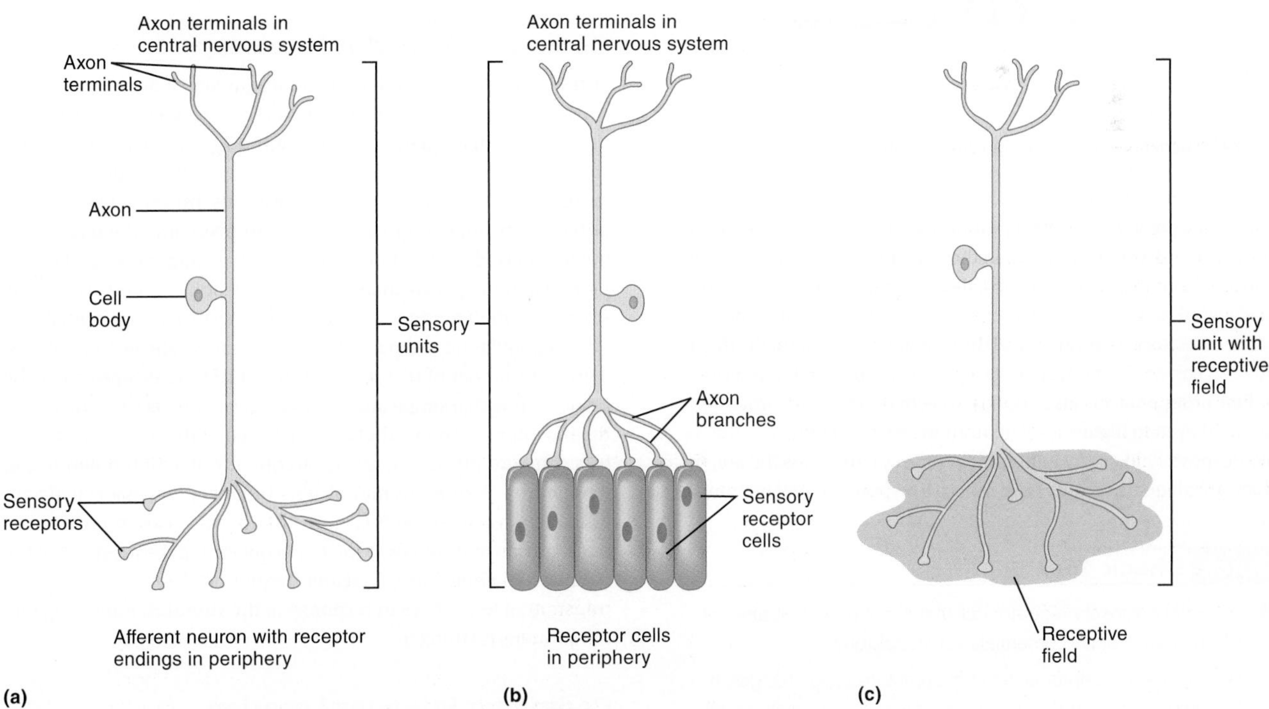

Figure 5 Sensory units and receptive fields. (a) A sensory unit in which the receptors are specialized endings of the afferent neuron. **(b)** A sensory unit in which the receptors are separate cells, each of which communicates to a single afferent neuron. **(c)** The receptive field for the sensory unit depicted in part (a).

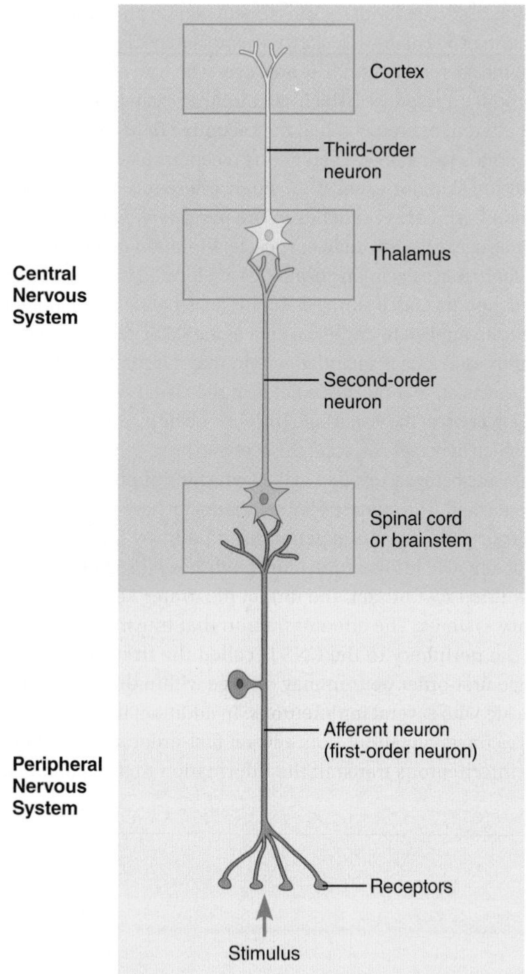

Figure 6 Generalized pathway for sensory systems.

major relay nucleus for sensory input; such interneurons are examples of **second-order neurons.** In the thalamus, these second-order neurons form synapses with **third-order neurons** that transmit information to the cerebral cortex, where sensory perception occurs. Different sensory pathways travel through different areas of the thalamus and cortex. The concept of receptive fields previously described for first-order neurons also applies to second- and third-order neurons, and even to higher-level neurons in sensory pathways—that is, the receptive field of any neuron in a sensory pathway is the area in which an adequate stimulus can induce a response in that neuron.

Quick Check 1

1. How can a sensory receptor cell that is not part of an afferent neuron cause action potentials in that neuron?
2. Which type of receptor is better at communicating changes in the intensity of a sensory input—rapidly adapting receptors or slowly adapting receptors?
3. Define *sensory unit.* How many different types of receptors are associated with a single sensory unit?

Sensory Coding

In the previous section, we saw that specific neural pathways transmit information from sensory receptors to the central nervous system. But when a stimulus acts on a sensory receptor, how does the nervous system identify the type, strength, and location of that stimulus? These tasks are made possible by *sensory coding.*

Coding for Stimulus Type

Stimulus type is coded by the receptor and pathway activated when the stimulus is applied. For example, light waves activate photoreceptors, which communicate via a specific pathway to the visual cortex. In fact, even if the visual pathway is inappropriately activated, as in the example of the blow to the eye, light will still be perceived because the visual pathway was activated. To see what happens when sensory pathways get crossed, see **Clinical Connections: Synesthesia**.

Our perceptions of stimuli are not necessarily based on a single sensory pathway; often the brain must integrate information from different sensory systems. For example, we perceive when our skin is wet, even though we do not have wetness receptors. This perception is possible because thermoreceptors and touch receptors, when activated appropriately, transmit a combination of signals interpreted by the brain as wetness. To demonstrate that there are no wetness receptors, do the following: Put a thin latex glove on your hand and then hold it under running water. You will be convinced that the glove is leaking water, even when it is not! If wet receptors existed, they would not have been activated by the inadequate stimulus in this example.

Coding for Stimulus Intensity

Stimulus intensity is coded by the frequency of action potentials *(frequency coding)* and the number of receptors activated *(population coding).* In frequency coding (**Figure 7**), a stronger stimulus results in a larger receptor potential. As long as the graded potential (in this case, the receptor potential) exceeds threshold for an action potential, stronger depolarizations can overcome the relative refractory period of an action potential and generate a second action potential more quickly than can weaker depolarizations. Therefore, stronger stimuli produce a higher frequency of action potentials.

In population coding, a stronger stimulus activates, or recruits, a greater number of receptors (**Figure 8**). These receptors may be associated with a single afferent neuron (Figure 8a), in which case the receptor potentials that are generated at the individual receptors sum and produce a greater frequency of action potentials in that neuron. A stimulus may also recruit receptors associated with different afferent neurons (Figure 8b), in which case more afferent neurons transmit signals to the CNS concerning the presence of the stimulus. In either case, a greater frequency of action potentials is transmitted to the CNS in response to the stimulus, indicating that the stimulus is stronger.

Coding for Stimulus Location

The coding of locations of tactile, proprioceptive, and visual stimuli is based on receptive fields. Recall that when an adequate stimulus

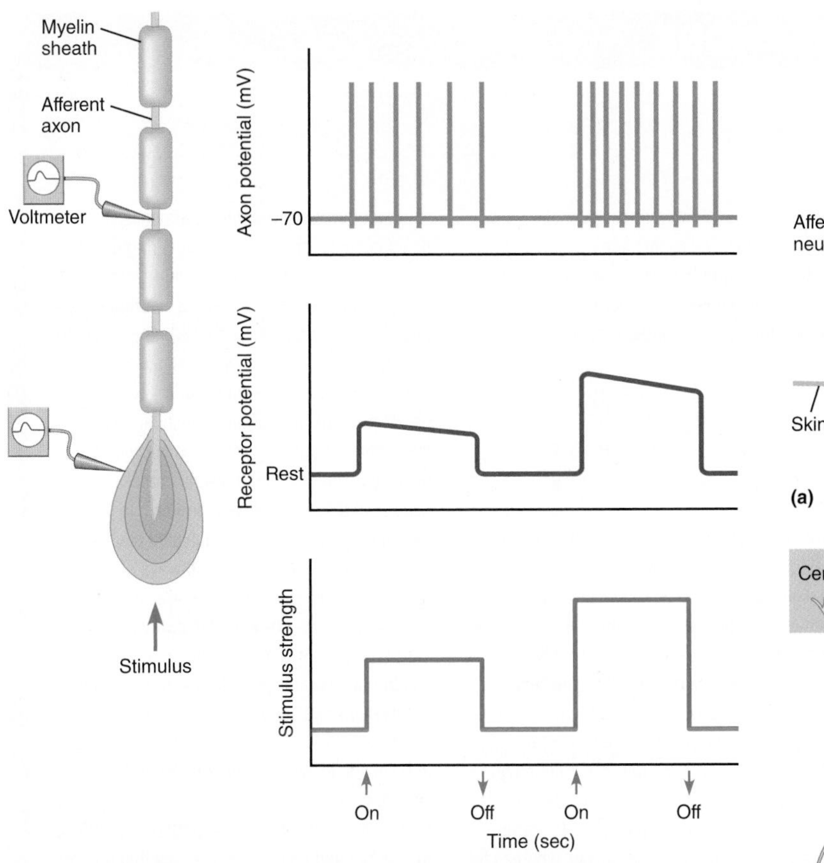

Figure 7 Coding of stimulus intensity. Changes in membrane potential along the axon (top) and at the receptor (middle) are plotted for two different stimulus intensities (bottom). Detection of a stronger stimulus by the receptor leads to more frequent action potentials along the axon.

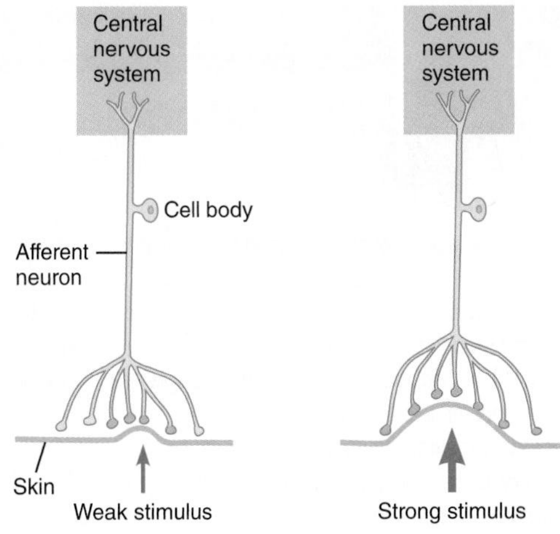

(a) Single sensory unit stimulated

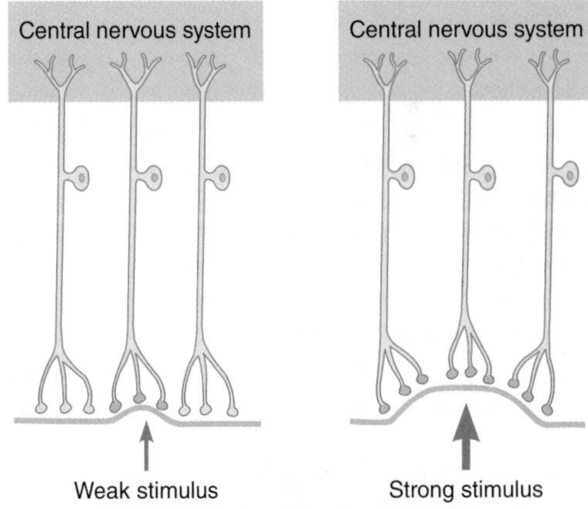

(b) Multiple sensory units stimulated

Figure 8 Coding of stimulus intensity by recruitment (population coding). (a) Receptor recruitment within a single sensory unit. **(b)** Receptor recruitment of additional sensory units.

is applied within a particular receptive field, it activates receptors associated with a specific afferent neuron. This concept is best illustrated with sensory receptors in the skin.

The precision with which the location of a stimulus is perceived is called **acuity.** In sensations associated with the skin, acuity depends on the size and number of receptive fields, the amount of overlap between the receptive fields, and a phenomenon called *lateral inhibition* (discussed shortly). If a specific afferent neuron is activated, then the stimulus must be located within that neuron's receptive field. However, the sizes of receptive fields vary considerably over the body. Localization of a stimulus is better in areas served by neurons with smaller receptive fields. Furthermore, information from one afferent neuron alone does not permit precise localization of a stimulus because the stimulus could be anywhere within its receptive field. Localization is improved by the overlapping of the receptive fields of different afferent neurons (**Figure 9**). Such overlapping improves localization via two phenomena: (1) the fact that any stimulus that occurs within the region of overlap between the receptive fields of two afferent neurons activates both neurons, and (2) lateral inhibition.

In **lateral inhibition,** a stimulus that strongly excites receptors in a given location inhibits activity in the afferent

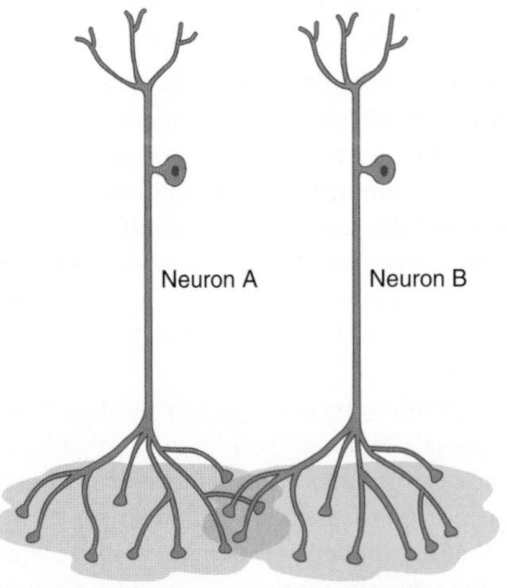

Figure 9 Overlapping receptive fields of two afferent neurons.

CLINICAL CONNECTIONS

Synesthesia

If you have ever noticed how food tastes blander when you have a cold, then you have experienced firsthand how the sense of smell is linked to the sense of taste. Our sensory systems do not exist in isolation, but rather interact with one another. In some individuals, however, the sensory systems seem to do more than merely interact; they actually seem to intertwine. Individuals with a condition called synesthesia (which literally means "joined perception") may "hear" colors, "see" sounds, or "taste" shapes. Synesthetes with "colored-hearing" may see green when they hear the word "dog," and see red when they hear the word "cat." They may see letters in the alphabet as always being a certain color—such as L appearing blue and N appearing orange—regardless of which color most other people perceive the letters to be. They may even taste steak when they see a triangle!

The following figure illustrates how a synesthete's perceptions might differ from typical perceptions. The top row illustrates the numbers 1–9 in black. A synesthete, however, may see each number as a different color, as shown in the middle row. The synesthete's association of a particular number with a particular color would remain consistent, regardless of the order in which the numbers appear, as shown in the bottom row.

1	2	3	4	5	6	7	8	9
1	2	3	4	5	6	7	8	9
3	5	1	6	9	7	2	4	8

Interestingly, synesthesia seems to be an involuntary automatic phenomenon that usually works in only one direction. For example, when a synesthete sees yellow when looking at the number 5, that person will always see the number 5 as yellow, but he will not necessarily see the number 5 when exposed to the color yellow. Synesthesia also seems to run in families, and synesthetes within a family tend to have the same form of synesthesia. Outside families, the synesthesia is highly variable—for some synesthetes the number 5 is yellow, whereas for others the number 7 is yellow.

Most forms of synesthesia are considered developmental anomalies in the wiring of the central nervous system. Magnetic resonance imaging (MRI) studies of colored-hearing synesthesia, for example, show that when synesthetes hear particular words, their visual association areas become active—an event that does not occur in normal individuals. This phenomenon suggests that auditory pathways in synesthetes developed connections to visual processing centers in these individuals' brains. It is thought that normal individuals also develop such sensory pathways but that the pathways undergo apoptosis (programmed cell death). In people with synesthesia, however, the pathways persist. That synesthesia is more common in young children than adults provides evidence that these "extra pathways" degenerate during development. Synesthesia can also be induced in adults in response to brain injury or epilepsy. Because synesthesia is rare and occurs in many variants, studies on its developmental patterns are difficult to conduct and tend to yield results that are far from definitive.

Critical Thinking Questions

1. Describe several specific examples of synesthesia.
2. What are the connections between the manifestation of synesthesia and the CNS?
3. Describe the differences in the auditory pathways of people with and without synesthesia. What do the differences suggest about brain function?

pathways of other nearby receptors. In the simplified pathways depicted in **Figure 10**, each afferent neuron synapses with only one second-order neuron. Note that the afferent neurons have collaterals that communicate with inhibitory interneurons in the CNS. In this case, the collaterals of afferent neuron Y_1 activate inhibitory interneurons that decrease the communication between afferent neurons with neighboring receptive fields (neurons X_1 and Z_1) and their second-order neurons (X_2 and Z_2, respectively).

Lateral inhibition increases acuity because it increases the contrast of signals in the nervous system—that is, it allows the transmission of strong signals in some neurons while suppressing the transmission of weaker signals arising in nearby neurons. When a stimulus is applied in the center of the receptive field for neuron Y_1, for example, that neuron is more strongly activated than are neurons X_1 or Z_1, as evidenced by a higher frequency of action potentials (see the lower graph in Figure 10). Because neuron Y_1 is the most highly activated, its activation of the inhibitory interneurons to neurons X_1 and Z_1 exceeds the inhibitory effect of the other interneurons, producing a greater inhibition of neurons X_2 and Z_2. Decreased activity occurs in all the second-order neurons, but the activities of X_2 and Z_2 are decreased more than that of neuron Y_2 (see the upper graph in Figure 10). The result is a greater difference in the frequency of action potentials among second-order neurons than in afferent neurons. More importantly, the frequency of action potentials in Y_2 is *much greater* than the frequencies in X_2 and Z_2. The resulting enhanced contrast between more important and less important neural signals allows for better localization of the stimulus, thereby increasing tactile acuity.

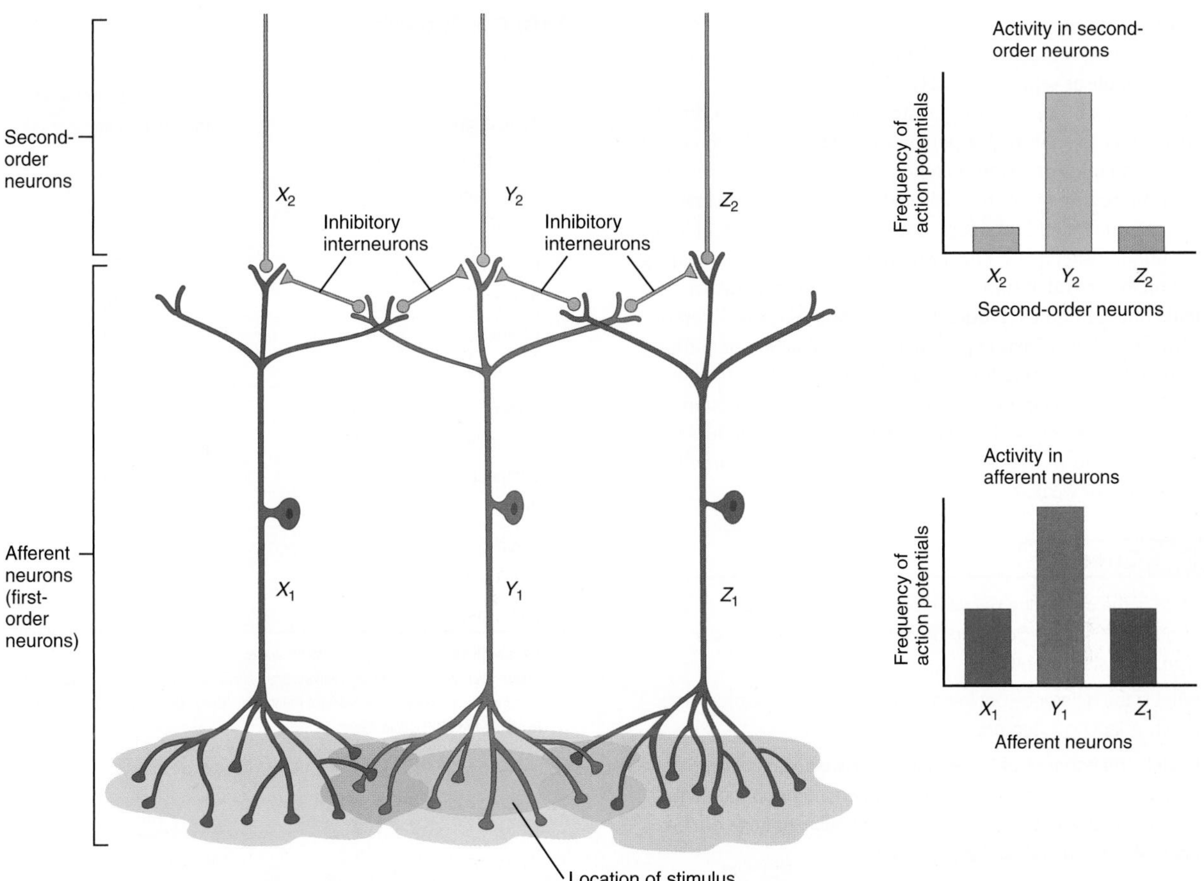

Figure 10 Lateral inhibition. Lateral inhibition enhances contrast between sites of strong and weak stimulation. In this example, the stimulus is applied in the center of the receptive field for afferent neuron Y_1, and in the periphery of the receptive fields for afferent neurons X_1 and Z_1. Lateral inhibition occurs when collaterals of one afferent neuron (here, Y_1) activate interneurons that inhibit communication between neighboring afferent neurons and their second-order neurons. The graphs show the frequencies of action potentials in the afferent and second-order neurons.

One measure of tactile acuity is **two-point discrimination,** the ability of a person to perceive two fine points pressed against the skin as two distinct points. The minimum distance that must exist between two points for them to be perceived as separate is termed the *two-point discrimination threshold;* points that are closer together than this distance are perceived as a single point. Two-point discrimination occurs only if the two points are applied to the receptive fields of two different afferent neurons (**Figure 11**). The smaller the receptive fields, the greater the ability to discriminate two points, and the greater the tactile acuity.

Tactile acuity varies over different regions of the body. The skin in some areas of the body is innervated by afferent neurons with few branches and, therefore, small receptive fields, whereas other areas are innervated by afferent neurons with extensive branching and large receptive fields. In addition, the smaller receptive fields overlap to a greater extent, which in turn contributes to greater tactile acuity. Table 2 lists two-point discrimination thresholds in several selected areas of the body. In the lips, the most sensitive areas, points as close together as 1 mm can be distinguished as two distinct points. The fingertips are also quite sensitive. Areas on the

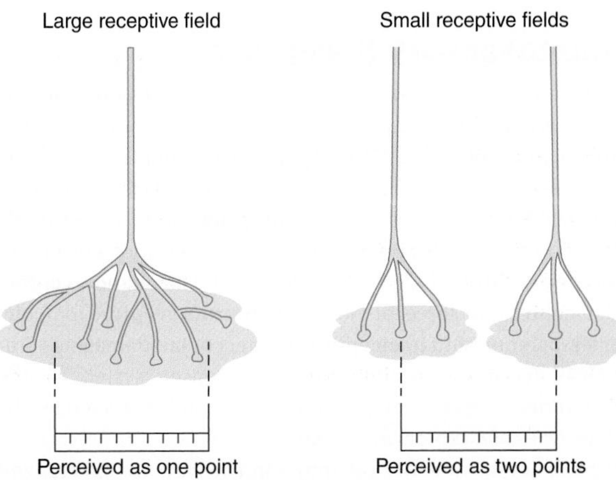

Figure 11 Two-point discrimination. The ability to discriminate between two separate points depends on the activation of separate receptive fields. The smaller the receptive field, the greater the two-point discrimination ability, and the greater the tactile acuity.

back, thigh, and upper arm, by contrast, are not very sensitive. In fact, two points as far apart as 40–50 mm (almost 2 inches!) may be indistinguishable as separate points in these areas.

Localization in some sensory systems is unrelated to coding by receptive fields. Even though a person can determine the direction from which an odor or a sound comes, neither olfactory nor auditory afferent neurons have receptive fields. Instead, these systems code for the quality and intensity of smell, and the pitch and loudness of sound, respectively. Localization in olfaction and in hearing is based on the arrival of stimuli at the two nostrils or at the two ears at slightly different times (**Figure 12**). The brain uses the difference in the time of arrival of action potentials at the olfactory or auditory cortex to determine where the stimulus originated. Localization in the auditory system is more acute than in the olfactory system, but both systems often rely on behavioral actions as well, such as your turning your head to determine whether the sound or smell becomes stronger or weaker.

Quick Check 2

1. What is the difference between population coding and frequency coding? Do these processes code a stimulus's type, intensity, or both?
2. What is the difference between rapidly adapting receptors and slowly adapting receptors?
3. Explain the concept of two-point discrimination. Is two-point discrimination better on the palm of the hand or on the thigh? Why? Where is the ability to locate a stimulus better—on the palm of the hand or on the thigh? Why?

TABLE 2 Two-Point Discrimination Thresholds for Selected Areas of the Body

Body region	Two-point discrimination threshold (mm)*
Lips (greatest acuity)	1
Index finger	2
Thumb	3
Palm of hand	10
Big toe	10
Forehead	18
Sole of foot	22
Breast	31
Abdomen	36
Shoulder	38
Back	42
Thigh	46
Upper arm	47
Calf (least acuity)	48

*Smaller distances indicate greater tactile acuity.
Source: Weinstein and Kenshalo, editors, *The Skin Senses*, copyright 1968. (Table: Two-Point Discrimination Thresholds for Selected Areas of the Body.) Courtesy of Charles C. Thomas Publisher, Ltd., Springfield, Illinois.

2 The Somatosensory System

The somatosensory system is involved with body sensations such as pressure, temperature, pain, and body position (*soma* = "body"). We first consider the types of receptors in this system.

Somatosensory Receptors

The somatosensory system responds to a variety of stimuli arising in many areas of the body, so it utilizes many receptor types. For example, proprioception of the body's position requires receptors in the muscles, tendons, ligaments, and joints, as well as receptors in the skin. Such receptors are called **proprioceptors,** one example of which is the muscle spindle. Somesthetic sensations of stimuli associated with the surface of the body require **mechanoreceptors** to detect pressure, force, or vibration; **thermoreceptors** to detect skin temperature; and **nociceptors** to detect tissue-damaging stimuli. Of all the sensory systems, the somatosensory system has the widest variety of receptor types; **Table 3** lists and characterizes the various types of receptors in the skin.

Most somatosensory receptors in the skin are specialized structures at nerve endings, which are easily identified under the light microscope. However, a few somatosensory receptor types lack identifiable specialized structures and, therefore, are called **free nerve endings.** The different structures are designed to respond to particular modalities of stimuli impinging on the skin.

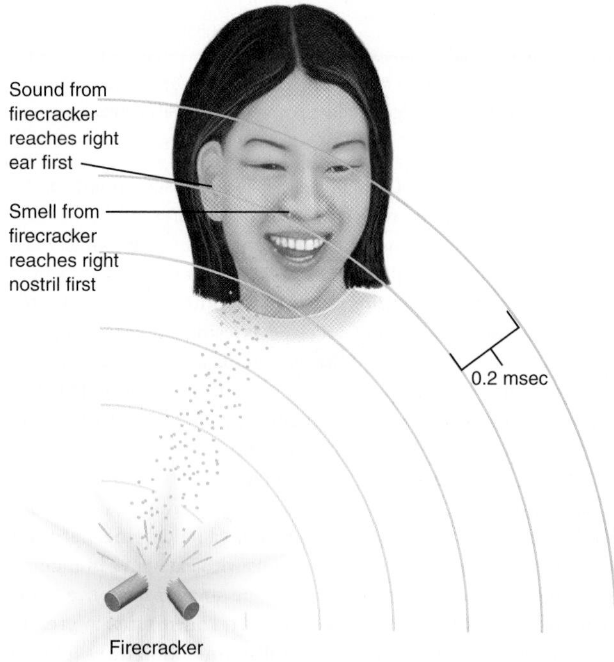

Figure 12 Localization in hearing and olfaction. Localization of a sound or an odor depends on the difference between the time the stimulus reaches the left and right ears or nostrils, respectively.

TABLE 3 Sensory Receptors in the Skin

Receptor class	Type	Associated afferent type	Location	Receptive field size	Adaptation	Modality
Mechanoreceptors	Free nerve ending	A-delta, C	Superficial, all skin	Small	Slow	Light touch
	Merkel's disk	A-beta	Superficial, all skin	Small	Slow	Pressure
	Pacinian corpuscle	A-beta	Deep, all skin	Large	Rapid	Vibration near 300 Hz
	Meissner's corpuscle	A-beta	Superficial, glabrous skin	Small	Rapid	Vibration near 50 Hz
	Hair follicle receptor	A-beta	Superficial, hairy skin	Small	Rapid	Bending of hair
	Ruffini's ending	A-beta	Deep, hairy skin	Large	Slow	Pressure
Thermoreceptors	Warm receptors (free nerve endings)	C	Superficial, all skin	Small	Rapid	Increase in skin temperature
	Cold receptors	A-delta	Superficial, all skin	Small	Rapid	Decrease in skin temperature
Nociceptors	Mechanical (free nerve endings)	A-delta	Superficial, all skin	Large	Slow	Intense mechanical stimulus
	Thermal (free nerve endings)	A-delta	Superficial, all skin	Small	Rapid	Intense hot or cold stimulus
	Polymodal (free nerve endings)	C	Superficial, all skin	Large	Slow	Intense mechanical or thermal stimulus; specific chemicals

Mechanoreceptors in the Skin

The various types of mechanoreceptors in the skin are depicted in **Figure 13**. Some receptors are found in superficial layers of the skin close to the *epidermis,* or outer layer of the skin. Superficial receptors include *Merkel's disks* and *Meissner's corpuscles.* Other receptors are located deeper, in the *dermis* or inner layer of the skin, including *hair follicle receptors, Pacinian corpuscles,* and *Ruffini's endings.* Meissner's corpuscles are found only in *glabrous* skin (hairless skin), whereas hair follicle receptors are found only in hairy skin. Note that slowly adapting mechanoreceptors respond to pressure (a sustained stimulus), whereas rapidly adapting receptors respond best to vibration (a constantly changing stimulus). Note as well that the sizes of the receptive fields for the various mechanoreceptors vary greatly, and that the smaller receptive fields provide better tactile acuity.

Thermoreceptors in the Skin

Thermoreceptors respond to the temperature of the receptor endings themselves and the surrounding tissue, rather than the temperature of the surrounding air. There are two types of thermoreceptors: warm receptors and cold receptors. **Warm receptors** respond to temperatures between 30°C and 45°C; the frequency of action potentials increases as skin temperature increases to 45°C, but then decreases rapidly (see **Figure 14a**). **Cold receptors** respond to temperatures between 35°C and 20°C;

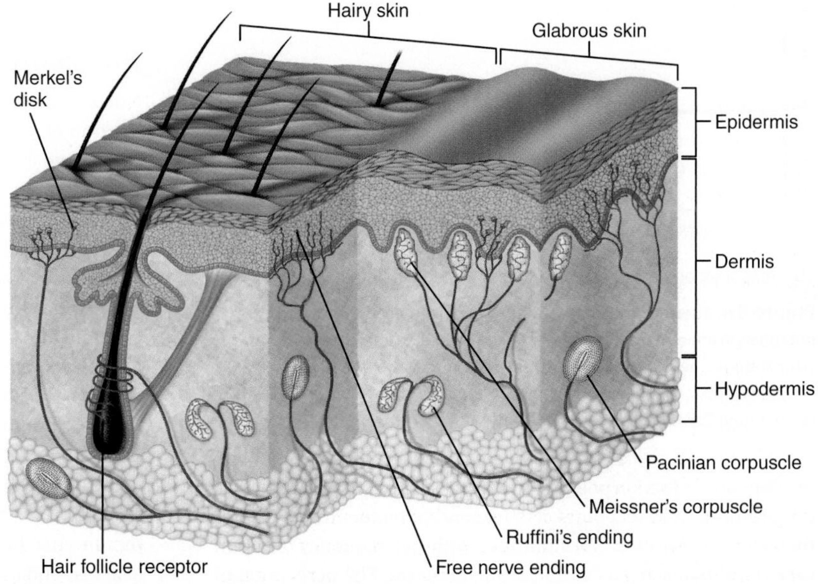

Figure 13 Sensory receptors in the skin.

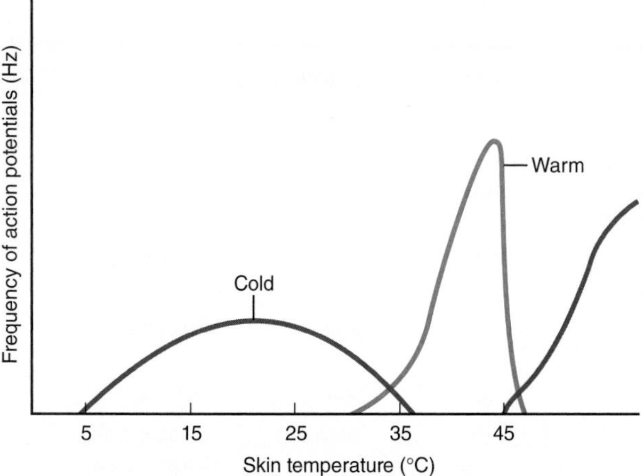

(a) Response of thermoreceptors to static temperature

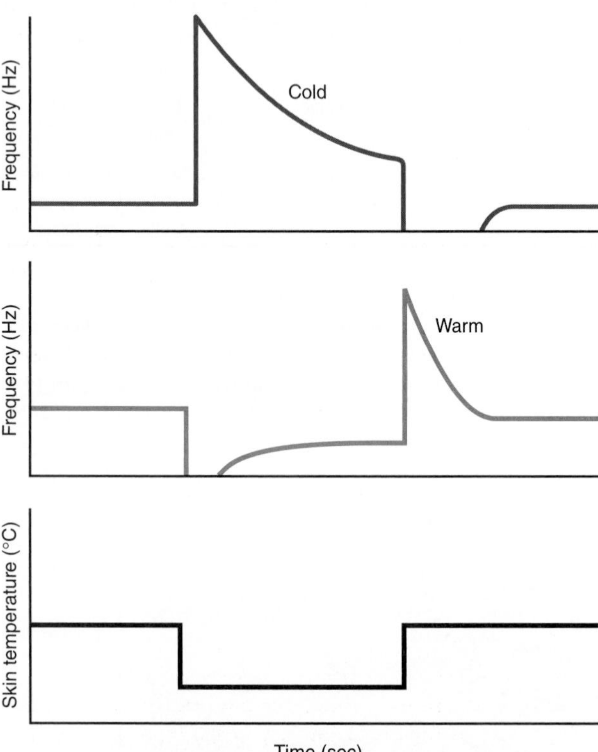

(b) Response of thermoreceptors to changes in temperature

Figure 14 Thermoreceptor responses. (a) The frequency of action potentials in afferents associated with warm or cold receptors when temperature is held at a set level. **(b)** The response of afferents of cold receptors (upper panel) and warm receptors (middle panel) to a decrease in temperature (lower panel).

the frequency of action potentials peaks at a skin temperature of 25°C (Figure 14a). Cold receptors also respond to temperatures greater than 45°C—a painfully *hot* stimulus—with the frequency of action potentials increasing as temperature increases. The perception of cold at these hot temperatures is referred to as *paradoxical cold.*

Thermoreceptors are free nerve endings that contain temperature-sensitive ion channels called transient receptor potential (TRP) channels. Some of these channels open or close solely in response to thermal stimuli, whereas others also respond to chemical stimuli and are involved in thermal transduction as well as transduction of painful stimuli, as described in the next section on nociceptors (Table 4). TRP channels are also involved in the senses of taste, olfaction, and vision , and visceral responses to various chemical stimuli that will not be discussed in this text. Seven subfamilies of TRP receptors have been identified, with receptors in the TRPV subfamily (TRPV1-4) being activated by heat, and receptors in both the TRPM subfamily (TRPM8) and the TRPA subfamily (TRPA1) being activated by cold. TRPV1 and TRPV2 respond to temperatures greater than 42°C and are found in nociceptors. TRPV3 and TRPV4 respond to temperatures in the range of 27°C to 42°C and are found in warm receptors. TRPV3 also responds to camphor, a natural chemical used topically that induces a warming sensation. The TRPM8 receptor responds to temperatures less than 25°C, and to menthol and eucalyptus oil, two compounds that induce a cooling sensation. TRPA1 receptors respond to temperatures less than 17°C and to mustard oil, garlic, and cinnamon.

Both warm and cold receptors are rapidly adapting, and they respond best when the temperature changes, as shown in Figure 14b. Notice that both types of thermoreceptors have tonic activity during the background (resting) temperature. When temperature decreases, the frequency of action potentials in axons associated with cold receptors increases sharply and then gradually declines (as the receptors adapt to the temperature change). Axons associated with warm receptors show the opposite response: The frequency of action potentials decreases sharply when the temperature drops and then increases rapidly when temperature returns to the resting state. In fact, it is the relative changes in responses of warm and cold receptors that the brain uses to interpret the temperature of the environment.

Nociceptors in the Skin

Nociceptors are the sensory receptors responsible for the transduction of *noxious stimuli* that we perceive as pain; that is, these free nerve endings respond to tissue-damaging (or potentially damaging) stimuli. Three types of nociceptors exist: **mechanical nociceptors,** which respond to intense mechanical stimuli, such as stubbing your toe; **thermal nociceptors,** which respond to intense heat (greater than 44°C), such as touching a hot stove; and **polymodal nociceptors,** which respond to a variety of stimuli, including intense mechanical stimuli, intense heat, intense cold, and chemicals released from damaged tissue. Chemicals that are released from damaged tissue and are capable of activating polymodal nociceptors include *histamine, bradykinin,* and *prostaglandins.*

Earlier we saw that certain TRP channels detect warm and cold stimuli, and now we learn that two of the TRP channels detect intense heat: TRPV1 and TRPV2 (Table 4). TRPV1 channels also respond to acid and to capsaicin, the chemical that makes chili peppers hot.

TABLE 4 Transient Receptor Potential Ion Channels in Thermoreceptors and Nociceptors

Channel	Receptor type	Temperature sensitivity	Chemical sensitivity
TRPV1	Nociceptors	≥42°C	Capsaicin, H^+ camphor
TRPV2	Nociceptors	≥52°C	
TRPV3	Warm receptors	33°C–38°C	Camphor
TRPV4	Warm receptors	27°C–42°C	
TRPM8	Cold receptors*	≤25°C	Menthol, eucalyptus oil
TRPA1	Cold receptors*	≤17°C	Cinnamon, garlic, mustard oil

*The presence of these channels in cold receptors has not been shown, but their response pattern is indicative of cold reception.

The Somatosensory Cortex

The perception of somatic sensations from all parts of the body begins in the primary somatosensory cortex (although studies suggest that some "crude" perception may occur in the thalamus). The somatosensory cortex is topographically oriented; that is, sensory information arising in neighboring areas of the body generally projects to neighboring areas of the cortex. Recall as well that the size of the cortex area devoted to somatic sensations from a specific area of the body is not proportional to the size of the body region, but rather to the sensitivity of the body region. The lips and fingertips, for example, are very sensitive areas because they have small two-point discrimination thresholds (see Table 2), and these body regions also have large areas of the primary somatosensory cortex devoted to them.

Recall also that the cerebral cortex has columnar organization. In the primary somatosensory cortex, vertical columns are organized according to sensory modality. For example, in the area of the primary somatosensory cortex for the thumb, one column is associated with pressure to the thumb, another column with vibration to the thumb, another column with cold, and so on. The next section describes the pathways along which information travels from the receptors to the primary somatosensory cortex.

Somatosensory Pathways

Two main pathways transmit information from peripheral somatosensory receptors to the central nervous system: the *dorsal column–medial lemniscal pathway* and the *spinothalamic tract.* These pathways transmit different types of sensory information to the thalamus, and then to the primary somatosensory cortex. In both cases, the pathways enter the spinal cord on one side and cross to the other side before reaching the thalamus. Therefore, somatosensory information from the right side of the body is perceived in the left somatosensory cortex, and vice versa.

The Dorsal Column–Medial Lemniscal Pathway

The **dorsal column–medial lemniscal pathway** transmits information from mechanoreceptors and proprioceptors to the thalamus; it crosses to the other side of the CNS in the medulla oblongata (**Figure 15a**). In this pathway, first-order neurons originate in the periphery and enter the dorsal horn of the spinal cord. Collaterals from the main axon may terminate in the spinal cord, communicating with interneurons such as those involved in spinal reflexes. However, the main branch of the axon ascends from the spinal cord to the ipsilateral (same side as the stimulus) brainstem in the *dorsal columns,* which are tracts of white matter that run dorsal and medial to the dorsal horn. The first-order neurons terminate in the medullary *dorsal column nuclei* (the medial gracile nucleus that receives input from the lower extremities and the lateral cuneate nucleus that receives information from the upper extremities), where they form synapses with second-order neurons. The second-order neurons then cross over to the contralateral side of the medulla in a tract called the *medial lemniscus,* and finally ascend to the thalamus. In the thalamus, the second-order neurons synapse with third-order neurons, which transmit information from the thalamus to the somatosensory cortex.

Apply Your Knowledge

A person is in a car accident and damages the dorsal column pathway on the left side of the thoracic spinal cord. Describe the changes this person would experience in sensory perception, paying particular attention to the modality of deficit(s), the side of deficit (left or right), and the level of deficit (arms and/or legs).

The Spinothalamic Tract

The **spinothalamic tract** transmits information from thermoreceptors and nociceptors to the thalamus; it crosses to the other side of the CNS within the spinal cord before it reaches the brain (Figure 15b). In this pathway, first-order neurons originate in the periphery at either thermoreceptors or nociceptors and enter the dorsal horn of the spinal cord. Here, first-order neurons may ascend or descend for a short distance (a few spinal segments) along *Lissauer's tract,* but eventually they form synapses with second-order neurons in the dorsal horn. The second-order neurons cross over to the contralateral side of the spinal cord, ascend in the anterolateral quadrant of the spinal cord through the brainstem, and then terminate in the thalamus. In the thalamus, second-order neurons form synapses with third-order neurons that ascend to the somatosensory cortex.

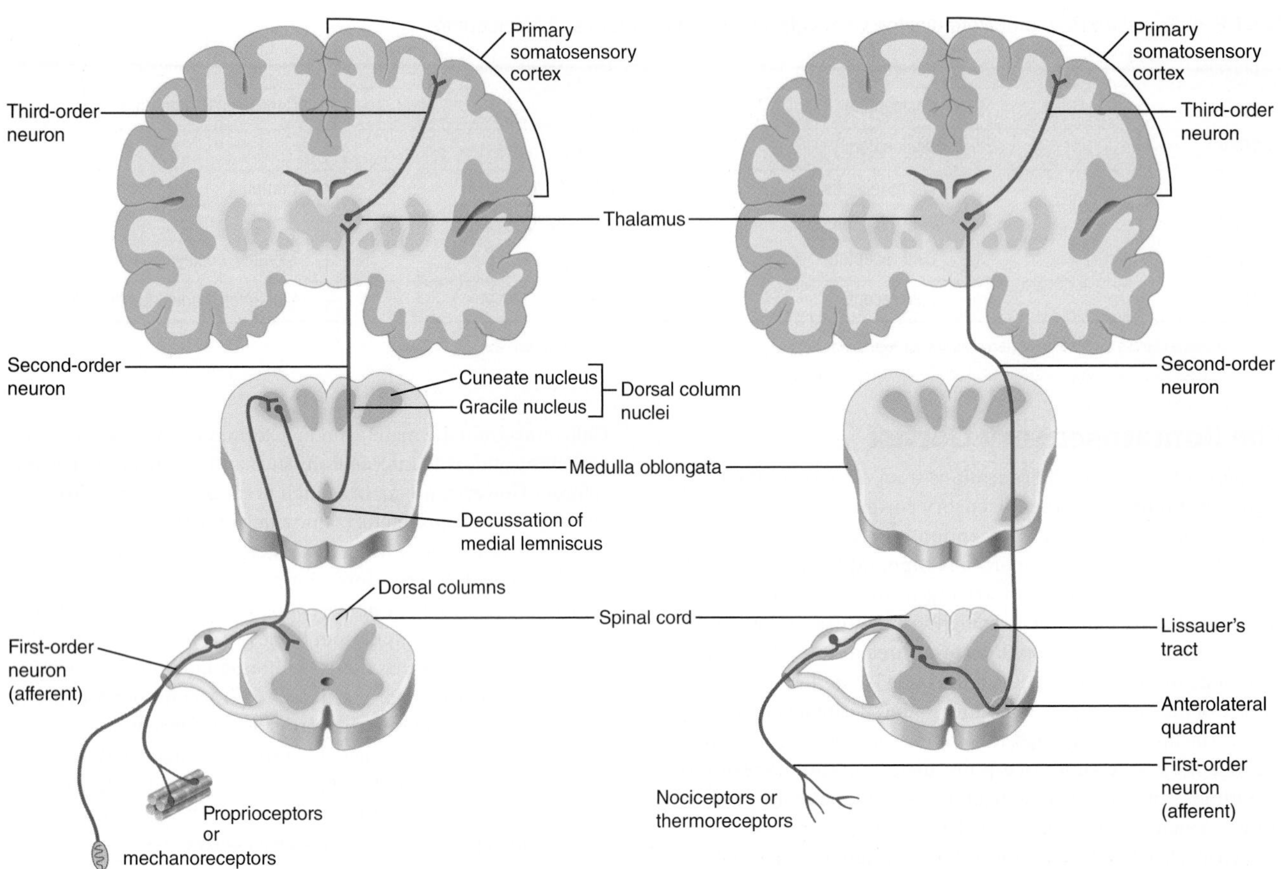

Figure 15 The two somatosensory pathways. (a) The dorsal column–medial lemniscal pathway transmits information from mechanoreceptors and proprioceptors to the CNS. **(b)** The spinothalamic tract transmits information from thermoreceptors and nociceptors to the CNS.

Apply Your Knowledge

A person is in a car accident and damages the spinothalamic tract on the left side of the thoracic spinal cord. Describe the changes this person would experience in sensory perception, paying particular attention to the modality of deficit(s), the side of deficit (left or right), and the level of deficit (arms and/or legs).

Pain Perception

Pain—one of the somesthetic sensations—is important because it teaches us to avoid subsequent encounters with potentially damaging stimuli. Pain is also important clinically, because it indicates that tissue damage may have occurred. Thus the mechanisms of pain perception warrant special consideration.

The Pain Response

The activation of nociceptors leads not only to the perception of pain but also to a variety of other body responses, including one or more of the following: (1) autonomic responses, such as increases in blood pressure and heart rate, increases in blood epinephrine levels, increases in blood glucose, dilation of the pupils of the eye, or sweating; (2) emotional responses, such as fear or anxiety; and (3) a reflexive withdrawal from the stimulus. The level of perceived pain varies considerably among individuals based on their past experiences and the circumstances under which the stimulus is applied. Thus a toothache that is barely noticeable during a busy day, for example, may become excruciating when the affected individual tries to fall asleep at night. To better understand how pain perception can vary, let's take a closer look at the mechanisms of pain perception.

Each of two types of pain—fast pain and slow pain—is perceived differently and is transmitted by a different class of afferent neurons. **Fast pain** is perceived as a sharp pricking sensation that can be easily localized; it is transmitted by *Aδ* (A-delta) *fibers,* thin, lightly myelinated axons with a conduction velocity of approximately 12–30 m/sec. **Slow pain,** in contrast, is perceived as a poorly localized, dull aching sensation; it is transmitted by *C fibers,* thin, unmyelinated axons with a conduction velocity of approximately 0.2–1.3 m/sec. To understand the differences between the two types of pain, think about the last time you stubbed your toe: An initial sharp pain (fast pain) was followed by a prolonged aching pain (slow pain).

The primary afferents, whether Aδ or C fibers, form synapses with second-order neurons in the dorsal horn of the spinal cord. Communication between these first- and second-order neurons involves different neurotransmitters, including two peptide transmitters, *substance P* and calcitonin gene-related peptide (CGRP) and glutamate. The role of substance P and glutamate in pain transmission is fairly well understood and will be described in more detail; CGRP is believed to function in a similar manner.

Substance P and glutamate are is released from primary afferent neurons and binds to receptors on second-order neurons. The second-order neurons ascend to the thalamus via the spinothalamic tract, the pathway involved in the perception and discrimination of pain. Nociceptive afferents also activate different ascending pathways that are necessary for interpreting the affective components of pain. The affective pathways ascend to the reticular formation of the brainstem, the hypothalamus, and the limbic system.

Visceral Pain

Pain is not limited to the body surface. For example, most people have suffered pain in their muscles after overexercising or while passing a kidney stone. The viscera are subject to tissue damage, and nociceptors in the organs detect this damage. But what does a person perceive when the appendix becomes inflamed or when suffering a heart attack? Generally, activation of nociceptors in the viscera produces a type of pain called **referred pain** (because it has been "referred" to the body surface). For example, a person having a heart attack generally complains about pain in the left chest, upper arm, and shoulder—not in the heart itself. Referred pain occurs because the second-order neurons that receive input from visceral afferents also receive input from somatic afferents (**Figure 16a**). According to one theory, the brain interprets information based on past experiences. Throughout a person's life, these second-order neurons are activated primarily by the somatic afferents, and thus the brain has learned that signals from these neurons are somatic in origin. When a person has a heart attack, then, the brain interprets these signals as a somatic disturbance because that is what such signals have meant in the past. Physicians use maps of surface locations of referred pain (Figure 16b) to ascertain which internal organ(s) may be causing a patient's pain.

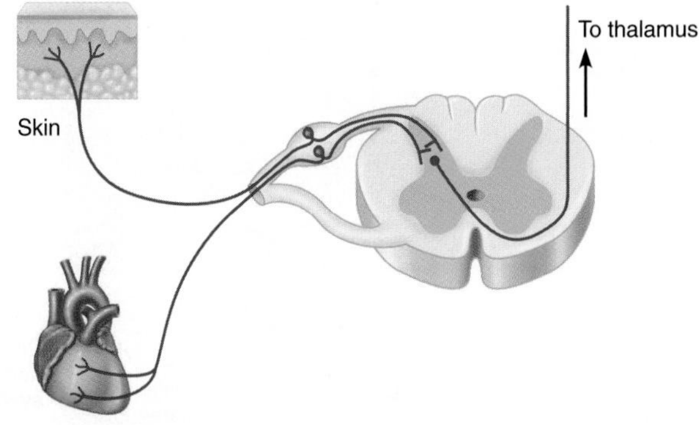

(a) Mechanism of referred pain

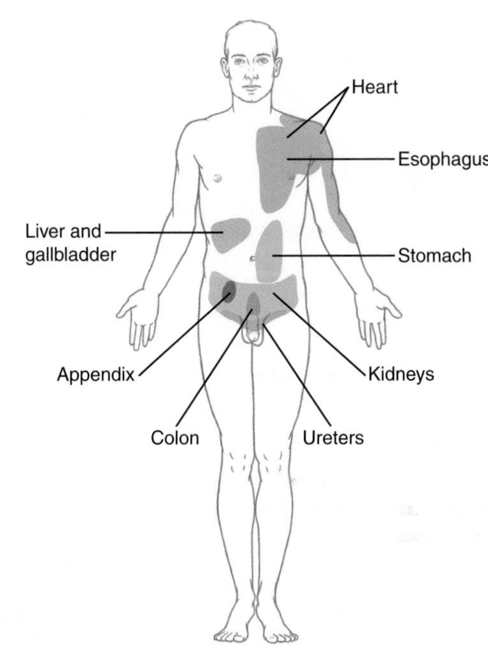

(b) Clinical map of referred pain

Figure 16 Mechanisms and sites of referred pain. (a) Referred pain occurs when visceral and somesthetic afferents converge on the same second-order neurons in the spinal cord. **(b)** Pain in specific visceral organs is generally referred to the areas of the body surface indicated in this map.

Modulation of Pain Signals

Signals about sensory information can be modulated as they are transmitted along sensory pathways; that is, facilitation or attenuation of signals can result in changes in the final perception of that information. Sensory signals can be modulated wherever there is a synapse in the pathway. The mechanisms involved in the modulation of pain have profound clinical significance.

The **gate-control theory** states that somatic signals of nonpainful sources can inhibit signals of pain at the spinal level (**Figure 17**). Among the various interneurons within the spinal cord are ones that inhibit the second-order neurons that transmit pain information. When these interneurons are active, the transmission of pain signals is suppressed, and the perception of pain is lessened. When information about a painful stimulus is being transmitted to the spinal cord by C fibers (Figure 17a), the collaterals of these C fibers *inhibit* the activity of the inhibitory interneuron, which allows transmission to proceed to the second-order neuron. Conversely, the same inhibitory interneuron is *stimulated* by collaterals from large-diameter myelinated afferents (Aβ fibers) associated with mechanical stimuli such as touch, pressure, and vibration (Figure 17b). If a nonpainful mechanical stimulus is applied simultaneously with a painful stimulus, the collaterals from the Aβ fibers stimulate the inhibitory interneuron, thereby decreasing the transmission of pain signals.

The gate-control theory describes why rubbing a painful area relieves the pain. It is also the basis for using *transcutaneous electrical nerve stimulation (TENS)* to treat pain. In TENS, a small current

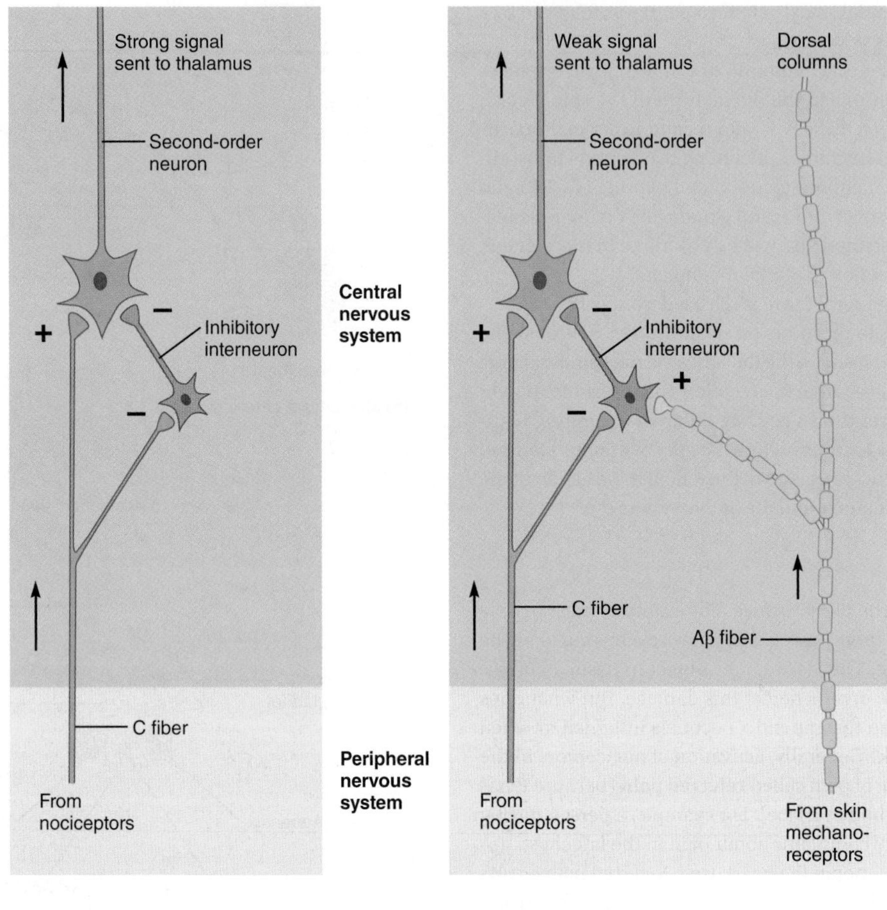

Figure 17 **Gate-control theory of pain. (a)** In unmodulated pain transmission, collaterals of the nociceptor afferents (C fibers) inhibit inhibitory interneurons, allowing transmission of pain signals to second-order neurons in the dorsal horn of the spinal cord and then to the thalamus. **(b)** In the modulation of pain transmission, collaterals of large-diameter afferents (Aβ fibers) branching from touch and pressure receptors excite the inhibitory interneuron, thereby decreasing the transmission of pain signals.

applied through the skin overlying a nerve activates large-diameter afferents, which relieves the pain. However, while nonpainful stimuli inhibiting pain can be of some therapeutic benefit, the brain also has the ability to block pain, or produce **analgesia,** through descending pathways that are part of the pain-blocking **endogenous analgesia systems.** Tapping into this system can produce remarkable clinical benefits.

Pain can be debilitating, and sometimes perceiving pain has distinct disadvantages. For example, seriously wounded soldiers are less likely to survive if they are immobilized by pain while their lives are at stake. Such wounded soldiers can often function without perceiving pain because the endogenous analgesia systems become active and block the pain, thereby allowing the body to cope with a more pressing need—survival.

Many brain areas are involved in the endogenous analgesia systems. One of the best-defined pathways is illustrated in **Figure 18**. Stressful situations can activate an area in the midbrain called the *periaqueductal gray matter.* This area communicates to areas in the medulla called the *nucleus raphe magnus* and the *lateral reticular formation.* Neurons from these areas descend to the spinal cord, where they block the communication between nociceptive afferent neurons and second-order neurons, as described next.

Recall that substance P is a neurotransmitter released from nociceptive afferents that communicate with second-order neurons. Inhibitory interneurons in the spinal cord form synapses on the cell body and dendrites of the second-order neurons as well as on the axon terminal of the nociceptive afferent neuron. These inhibitory interneurons release the endogenous opiate neurotransmitter *enkephalin,* which binds to opioid receptors on the second-order neuron and induces inhibitory postsynaptic potentials. Enkephalin also binds to opioid receptors on the axon terminal of the nociceptive afferent neuron, which inhibits the release of substance P, causing presynaptic inhibition. Both of these actions suppress signal transmission from the afferent neuron to the second-order neuron, thereby decreasing the transmission of pain signals to the brain. These inhibitory interneurons are activated by

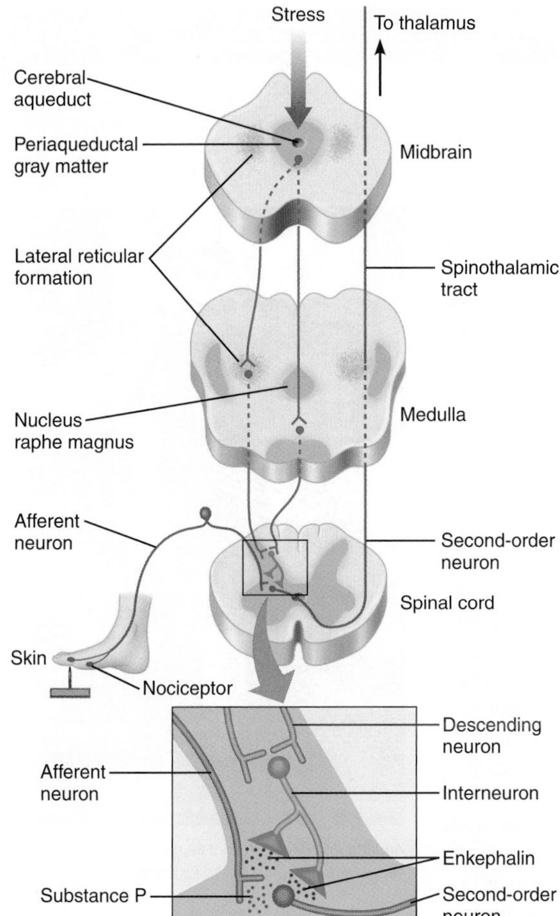

Figure 18 Endogenous analgesia systems. When a painful stimulus activates nociceptors, information is transmitted to the CNS via the spinothalamic tract. In the presence of stress, endogenous analgesia systems can block pain transmission at the level of the synapse between the nociceptive afferent neuron and the second-order neuron, as follows: The periaqueductal gray matter in the midbrain communicates to the lateral reticular formation and to the nucleus raphe magnus of the medulla. These regions have neurons that descend to the dorsal horn of the spinal cord and activate inhibitory interneurons that release the neurotransmitter enkephalin, which then blocks communication between the nociceptive afferent and the second-order neuron via two mechanisms: presynaptic inhibition of substance P release from the nociceptive afferent, and production of IPSPs on the second-order neuron.

descending neurons of the nucleus raphe magnus and lateral reticular formation.

More recently, descending modulation has been demonstrated to include pathways that facilitate pain. One example is the *hyperalgesia* that results when a person is ill. Several of the cytokines that induce symptoms of illness (for example, fever and fatigue) also act on areas of the brain to enhance the perception of pain.

Sometimes pain-transmitting or pain-modulating systems malfunction, resulting in chronic pain. One example is *phantom limb pain,* the perception of pain in an amputated limb (see **Clinical Connections: Phantom Limb Pain**).

Quick Check 3

1. Name the two types of thermoreceptors and describe the types of stimuli that excite them.
2. Name the three types of nociceptors and describe the types of stimuli that activate them.
3. Distinguish between fast pain and slow pain. Which class of afferent neurons is responsible for fast pain, and which is responsible for slow pain?
4. What kind of pain is experienced in the left shoulder by a person having a heart attack?

3 Vision

Much of what we learn about the world we learn through seeing. In this section we examine the visual system, which endows us with this important capability. We begin with the anatomy of the eye, the sensory organ of the visual system.

Anatomy of the Eye

The important structures of the eye are shown in **Figure 19**. The eye can be divided into three concentric layers. The outermost layer consists of the sclera and cornea. The **sclera,** a tough connective tissue, makes up the "white" of the eye. In the front of the eye, the sclera gives way to the **cornea,** a transparent structure that allows light to enter the eye.

The middle layer of the eye consists of the choroid, ciliary body, and iris. The **choroid** is located beneath the sclera and contains blood vessels that nourish the inner layer of the eye, which includes the photoreceptors. The **ciliary body** contains the **ciliary muscles,** which are attached to the lens by strands of connective tissue called **zonular fibers.** The **lens** focuses the light on the retina, the back of the eye where visual information is transduced. Contraction and relaxation of the ciliary muscles change the shape of the lens to focus light waves. The **iris** consists of two layers of pigmented smooth muscle located in front of the lens. The pigmentation of the iris determines eye color. The **pupil** is a hole in the center of the iris that allows light to enter the posterior part of the eye; it is not a structure. The iris regulates the diameter of the pupil, thereby regulating the amount of light that reaches the back of the eye.

The innermost layer of the eye is the **retina,** which consists of neural tissue and contains the **photoreceptors,** cells that detect the light waves. Photoreceptors are of two types, *rods* and *cones,* which detect dim light and bright light, respectively. Thus the retina functions in *phototransduction,* the conversion of light energy to electrical energy. Located just outside the retina and attached to the choroid is the **retinal pigmented epithelium.** This structure contains a high concentration of the black pigment melanin, which absorbs the light that strikes the back of the eye, preventing it from reflecting back across the retina.

Two areas of the retina are worth noting. One is the **fovea,** the central region on the retina, where light from the center of the visual field strikes. (The transparent lens focuses light waves on the retina, much as a camera lens focuses light waves on film.)

CLINICAL CONNECTIONS

Phantom Limb Pain 

When a person loses an arm or a leg, the damage can be psychologically devastating. To make matters worse, these patients often suffer chronic pain, seemingly coming from the limb that is no longer present!

An amputee often perceives other sensations coming from the lost limb (called *phantom sensations*), but the sensation of pain is the most debilitating. More than 50% of amputees perceive some sort of *phantom limb pain.* In some patients, this pain is relatively minor and subsides over time. In other patients, it is excruciating and persists almost constantly for the rest of their lives.

How can a person feel something in a missing body part? This question has baffled doctors for centuries. Early theories suggested that nerve endings in the stump can be activated, thereby producing the sensation of pain. However, the pain induced by stimuli to the stump produces *stump pain,* which is better characterized and more easily treated than true phantom limb pain. More recent theories suggest that changes occur within the central nervous system such that second- or third-order pain-transmitting neurons that no longer receive input from the missing limb become more sensitive to other types of synaptic input. Because neurons receive converging input from a variety of sources, other synapses become more effective in their ability to excite the second- or third-order neurons. Thus neurons that in the past transmitted information about painful stimuli in the limb now become excited by other synaptic input unrelated to the limb. The perception is still pain in the limb, however, because the pain pathway normally from the limb is activated.

Treatment of phantom limb pain has proved difficult, with only 25% of patients receiving long-term relief. Traditional chronic pain treatments, such as anti-inflammatory agents and opioids, have not been very successful in reducing phantom limb pain. Other treatments used include surgical manipulations, acupuncture, and transcutaneous electrical nerve stimulation. One of the more recent and fascinating techniques to treat phantom pain involves the use of mirrors. The patient uses a mirror to view his present limb but thinks that it is the missing limb. The patient can then exert some control over the mirrored missing limb, which helps relieve some pain.

An amputee may perceive sensations that seem to originate from the lost limb.

Critical Thinking Questions

1. Describe the similarities and differences between stump pain and phantom limb pain.
2. What are the recent theories explaining phantom limb pain?
3. Why has the treatment of this condition proved so difficult? What may hold promise for future treatment?

FOCUS ON DIABETES

The importance of the somesthetic sensations, and especially pain, to homeostasis is obvious when one looks at the effects of diabetes on foot health. Each year in the United States, approximately 80,000 to 90,000 individuals with diabetes undergo limb amputations resulting from the disease. The high blood glucose associated with uncontrolled diabetes damages the peripheral nerves (peripheral neuropathy), especially those supplying the feet. The precise mechanism by which this damage occurs is unknown, but the consequences are severe. Neuropathy can lead to a loss of sensation in the foot. Thus a person with diabetes may be unaware of sores on the feet—even a minor sore like a small blister. Being unaware is a hazard because the feet may not be cared for properly and the sores may become infected. A severe infection could lead to tissue death (gangrene), which might eventually progress to the point where amputation is necessary to prevent further spread. To avoid the risk of amputation, diabetics need to monitor and maintain their blood glucose levels and examine their feet frequently for visual signs of damage, as somatic signs could be absent.

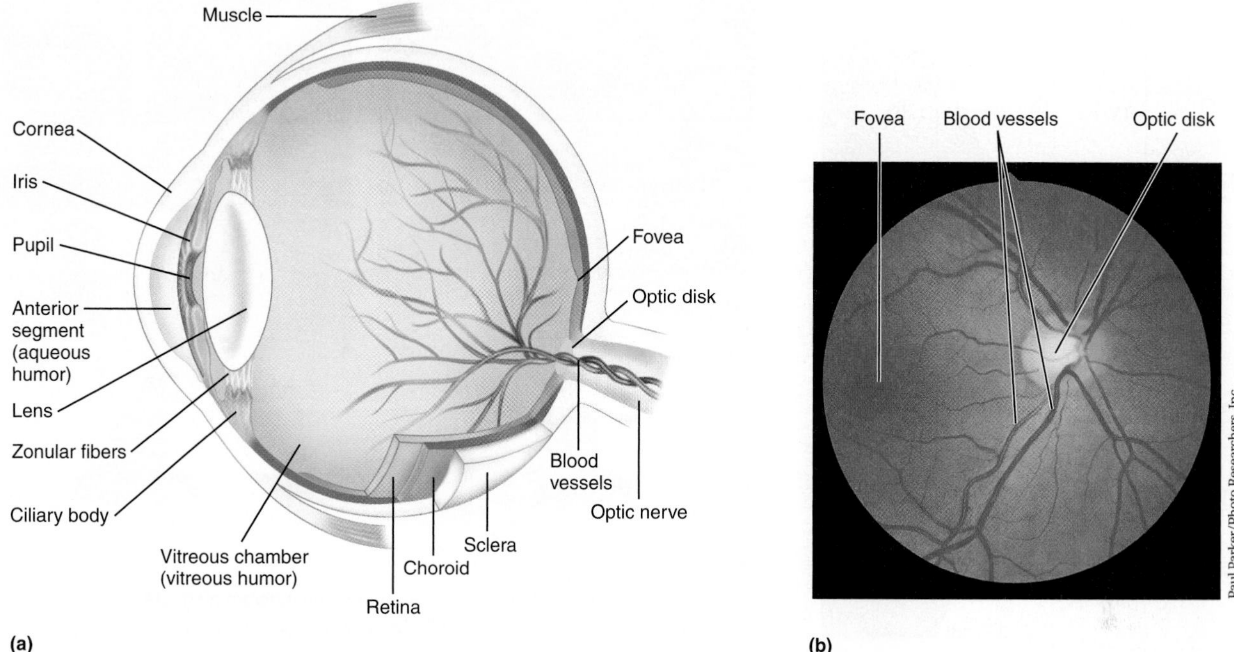

Figure 19 Anatomy of the eye. (a) The major structures of the eye as viewed in a horizontal section. **(b)** The retina as viewed through an ophthalmoscope.

For reasons described later, the fovea is the area of the retina with the greatest visual acuity. The other area of note is the **optic disk,** the portion of the retina where the optic nerve and blood vessels supplying the eye pass through the retina. Because there are no photoreceptors in the optic disk, this area is a **blind spot,** a region where light striking the retina cannot be transduced into neural impulses and, therefore, cannot be perceived (see **Figure 20** for a demonstration).

The lens and ciliary body separate the eye into two fluid-filled chambers. In front of these structures is the **anterior segment,** which is divided into an *anterior chamber* between the cornea and iris and a *posterior chamber* between the iris and lens. The anterior segment contains a clear, watery fluid called **aqueous humor,** which supplies nutrients to the cornea and lens. The cornea and lens are transparent, so light can pass through them easily. If these structures relied on blood to supply their nutrients, the presence of blood vessels would obstruct the light. Behind the lens and ciliary body is the **posterior segment** or **vitreous chamber** containing a firmer, jellylike material called **vitreous humor,** which maintains the spherical structure of the eye.

Figure 20 The blind spot. To demonstrate your blind spot, cover your right eye and look at the – sign with your left eye. Move the book close to your left eye while continuing to look at the – sign. When the book is a few inches from your face, the + sign should disappear because the light waves reflecting from it strike the optic disk.

The Nature and Behavior of Light Waves

Light is a form of energy—specifically, *electromagnetic waves.* Along with other forms of electromagnetic energy, including radio waves, television waves, X rays, and gamma rays, light is a part of the electromagnetic spectrum (**Figure 21**). Visible light includes those electromagnetic waves having wavelengths between about 350 nm and 750 nm; different colors correspond to different wavelengths within this range.

Because light is a wave, it exhibits the usual properties of waves, including reflection and refraction. **Reflection** is a phenomenon in which light waves strike and bounce off a surface. Reflection is important in vision because much of the light we perceive has reflected off the objects we are observing. (We also see emitted light, such as the light coming directly to the eye from the sun or a lightbulb.) Light that is absorbed by objects is not perceived; thus we perceive an object as green because it reflects to the eyes light of the wavelength corresponding to green (approximately 530 nm) while absorbing all other wavelengths. We see black when all the light waves are absorbed by the object we are looking at, and we see white when all the light waves are reflected from the object we are looking at to the eye.

Refraction refers to the bending of light waves as they pass through transparent materials of different densities. Refraction occurs because the speed of light is different in varying media: The greater the change in the speed, the greater the refraction. This property is important in vision because in its path from the visual field to the photoreceptors in the retina, light must pass through several different materials, including air, the cornea, the lens, and

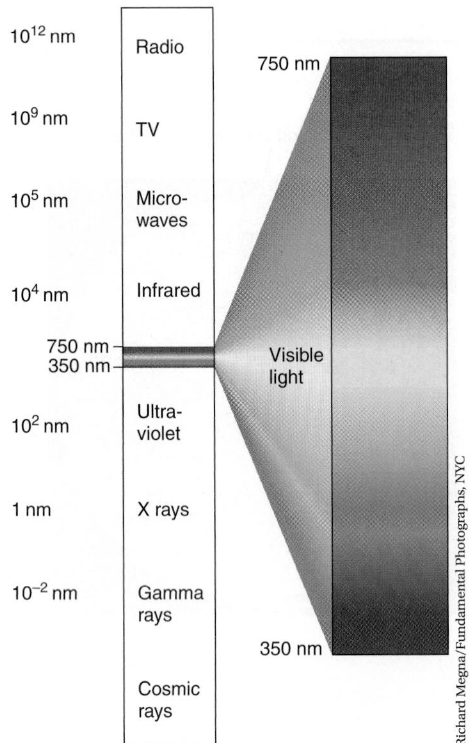

Figure 21 The electromagnetic spectrum. The numbers indicate wavelength in nanometers (1 nm = 1 × 10^{-9} meter). The band of visible light is highlighted.

(a) (b)

Figure 22 Refraction of light waves passing through different media. (a) Straw at an angle to the air-water interface. **(b)** Straw perpendicular to the air-water interface.

the vitreous and aqueous humors. Because refraction is important in the focusing of light waves on the retina, we consider it here in greater detail.

Try this experiment at home: Place a straw in a clear glass half full of water and let it rest at an angle. When so oriented, the straw looks broken at the air-water interface (**Figure 22a**). Now hold the straw straight up so that it is perpendicular to the air-water interface. The straw now appears to be in one piece again (Figure 22b). Light waves are refracted as they travel from one medium to another if they strike the second medium at an angle other than perpendicular. How much the light waves refract depends on the differences in the densities of the two media and the angle at which the light strikes them. Try holding the straw at different angles or putting the straw in different types of fluid.

Figure 23 illustrates the refraction of parallel light waves as they pass through the *concave* and *convex* surfaces often used in lenses, such as those in eyeglasses or telescopes. With either type of surface, light waves striking the surface perpendicular to it pass straight through. However, as the light strikes the surfaces at other angles, the concave lens causes the once-parallel light waves to diverge (move farther apart), whereas the convex lens causes the light waves to converge at a single point called the *focal point.* The distance from the long axis of the convex lens to the focal point is called the *focal length.*

Both the cornea and the lens have convex surfaces that function to converge the light waves entering the eye onto the retina, a process that is necessary if visual images are to be in focus. **Figure 24** shows how light waves from a viewed object are projected onto the retina. For us to see items in focus, light from a given point in the visual field must converge to a single point on the retina. Although the cornea has more than three times the refractive power of the lens, the refractive power of the cornea is constant, whereas that of lens can be varied as needed to focus light on the retina. For example, to view very near objects, the lens becomes rounder, increasing its refractive power to focus the image on the retina. The ability of the lens to increase its refractive power for viewing near objects is a process called **accommodation,** which we explore next.

Accommodation

If an object is to be seen clearly, the light reflected from any given point on the object must converge at a single point on the retina. When viewing something far away, the light waves enter the eye almost parallel to each other (**Figure 25a**), so little refractive power is needed to focus the light on the retina. In contrast, light waves from close-up objects are diverging as they enter the eye (Figure 25b), so the greater refractive power of a rounder lens is necessary to overcome this divergence and focus the light on the retina.

The shape of the lens is controlled by the circularly arranged ciliary muscle through the tension it applies to the zonular fibers, which attach the ciliary muscle to the lens. When muscle contracts, it gets shorter. Thus, when circular muscle contracts, the diameter of the circle becomes smaller. For viewing distant objects, the ciliary muscle relaxes, which increases the diameter of

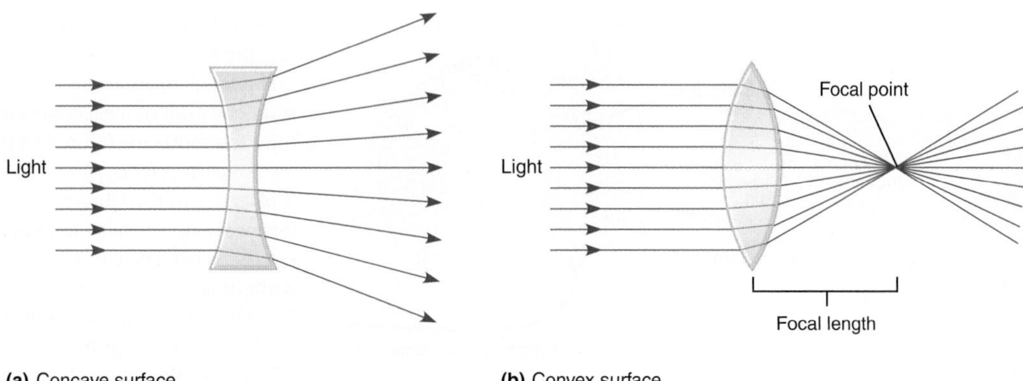

Figure 23 Refraction of light waves passing through curved surfaces. (a) Concave surfaces cause divergence of light waves. **(b)** Convex surfaces cause convergence of light waves to a focal point. The distance from the long axis of a convex lens to the focal point is called the focal length.

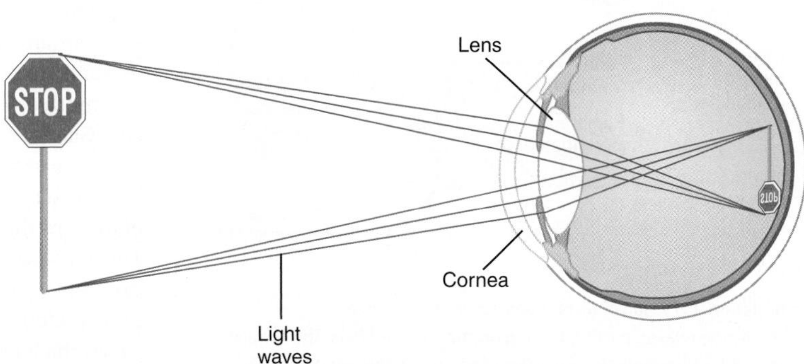

Figure 24 Refraction of light waves in the eye. A given point in the visual field comes to focus on a single point in the retina. Refraction of light waves as they pass through the convex cornea and lens of the eye causes the image to be inverted and reversed on the retina.

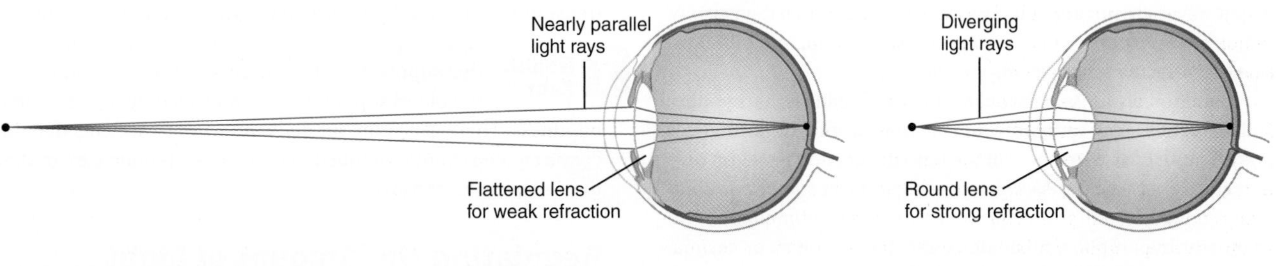

Figure 25 Focusing light from distant and near sources. (a) Light waves reflected from a distant object approach the lens parallel to one another. A relatively flat (weak) lens is sufficient to converge the light waves on the retina. **(b)** Light waves reflected from a near object diverge as they approach the lens. A rounder (strong) lens is needed to converge the light waves on the retina.

the circle and tightens the zonular fibers, pulling the lens into a flattened shape (**Figure 26a**). To achieve accommodation for viewing close-up objects (Figure 26b), the ciliary muscle contracts, reducing the diameter of the circle and lessening the tension on the zonular fibers. Because of its inherent elasticity, the lens becomes rounder when the tension on the zonular fibers is reduced. Accommodation is under control of the parasympathetic nervous system, which triggers contraction of the ciliary muscle for near vision. In the absence of parasympathetic activity, the ciliary muscle relaxes.

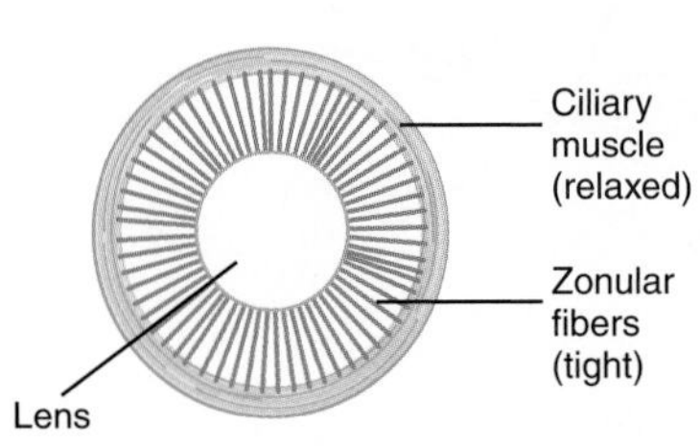

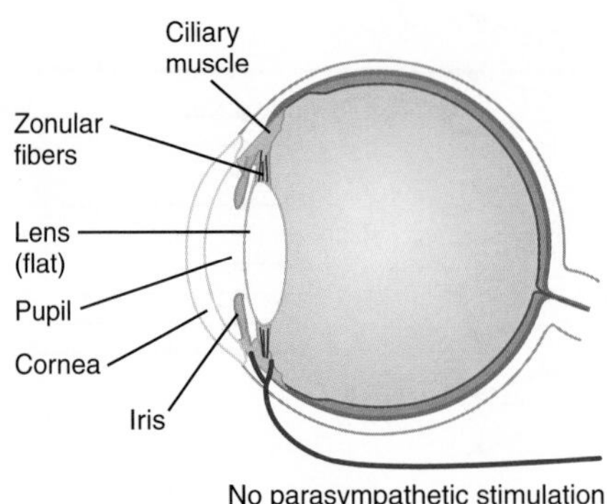

(a) Far vision of distant objects

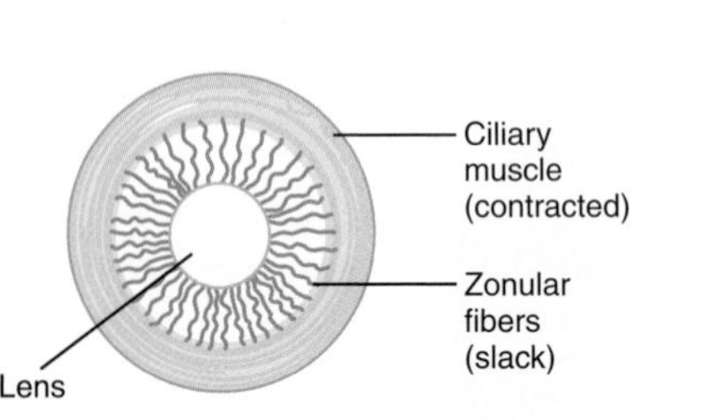

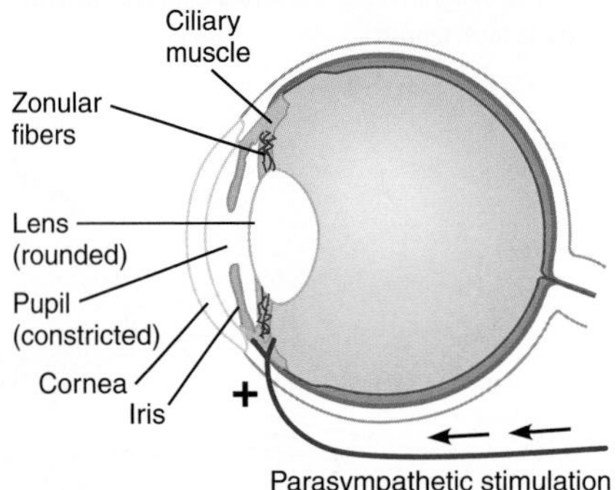

(b) Accommodation for near vision

Figure 26 Mechanism of accommodation. (a) Vision of distant objects. In the absence of parasympathetic stimulation, the ciliary muscle relaxes, putting tension on the zonular fibers. The zonular fibers pull on the lens, flattening the lens. **(b)** Accommodation for near vision. Under parasympathetic stimulation, the ciliary muscle contracts, reducing the tension on the zonular fibers and enabling the elastic lens to become rounder.

Clinical Defects in Vision

If light waves are not adequately focused on the retina, then vision is blurred. There are many causes of blurred vision, and different types of corrective lenses to improve vision.

Common visual defects include near-sightedness, or **myopia,** and far-sightedness, or **hyperopia.** In **emmetropia,** or normal vision (**Figure 27a**), a person can see both distant and close-up objects clearly because the eye can focus light from far sources without accommodation, and from near sources with accommodation. In myopia or hyperopia, a mismatch exists between lens or cornea strength and eyeball length.

A person with myopia (Figure 27b) can see near objects clearly, but not distant objects, because the lens or cornea is too strong for the length of the eyeball and bends light rays too much. In this situation, the eye can focus on close-up objects without accommodation, but light from distant objects is focused in front of the retina, resulting in a blurred image. To correct for myopia, a concave lens is placed in front of the eye. The lens causes light waves to diverge before reaching the eye. Under these conditions, the eye must accommodate to view close-up objects, and distant objects will be in focus without accommodation.

In hyperopia (Figure 27c), the lens or cornea is too weak for the length of the eyeball. Therefore, the eye can focus on distant objects only with accommodation, which means that the lens cannot increase accommodation enough to adjust for near vision. Light from close-up objects comes to focus behind the retina, resulting in a blurred image. To correct for hyperopia, a convex lens is placed in front of the eye. The lens causes light waves to converge before reaching the eye. Now the eye can see distant objects without accommodation, which gives the lens enough leeway to enable it to accommodate for near objects.

Many other clinical defects affect the ability to focus light on the retina. In **astigmatism,** irregularities on the surface of the cornea or lens cause erratic bending of light waves. Glasses or contact lenses can correct for astigmatism. **Presbyopia** is a hardening of the lens that occurs with aging; as the lens hardens, its loss of elasticity decreases its ability to become spherical, making accommodation for near vision more difficult. Glasses can correct for presbyopia to a certain extent, but the condition cannot be reversed. A **cataract** is an opacification of the lens that decreases its transparency. The opacification is caused by clumping of proteins in the eye as a person ages, creating a cloudy area in the lens. In severe cases, the lens can be removed and an artificial lens put in its place. In **glaucoma,** an increase in the volume of aqueous humor raises pressure in the anterior cavity of the eyeball, which can distort the shape of the cornea and shift the position of the lens. Shifting of the lens can transmit the increased pressure to the vitreous chamber, where it can compress optic nerve axons or the blood vessels that supply the retina. Excessive pressure can reduce the blood supply significantly, leading to permanent blindness. There is no cure for glaucoma, but treatments that decrease aqueous humor production or increase its drainage can slow the disease's progression.

Regulating the Amount of Light Entering the Eye

The eyes regulate the amount of light that enters them by varying the size of the pupils. In bright light, the pupils are small, or *constricted,* so that the photoreceptors do not become "bleached out" by too much light. Under bright light conditions, the pupils can constrict to approximately 2 mm in diameter. In dim light, by contrast, the pupils are large or *dilated* to allow more light in, which enhances the ability to see. Under conditions of dim light, the pupils can dilate to approximately 8 mm in diameter.

The size of the pupil is controlled by the iris. Recall that the iris consists of two layers of smooth muscle around the pupil. These two layers of smooth muscle are an inner **circular muscle** layer, also

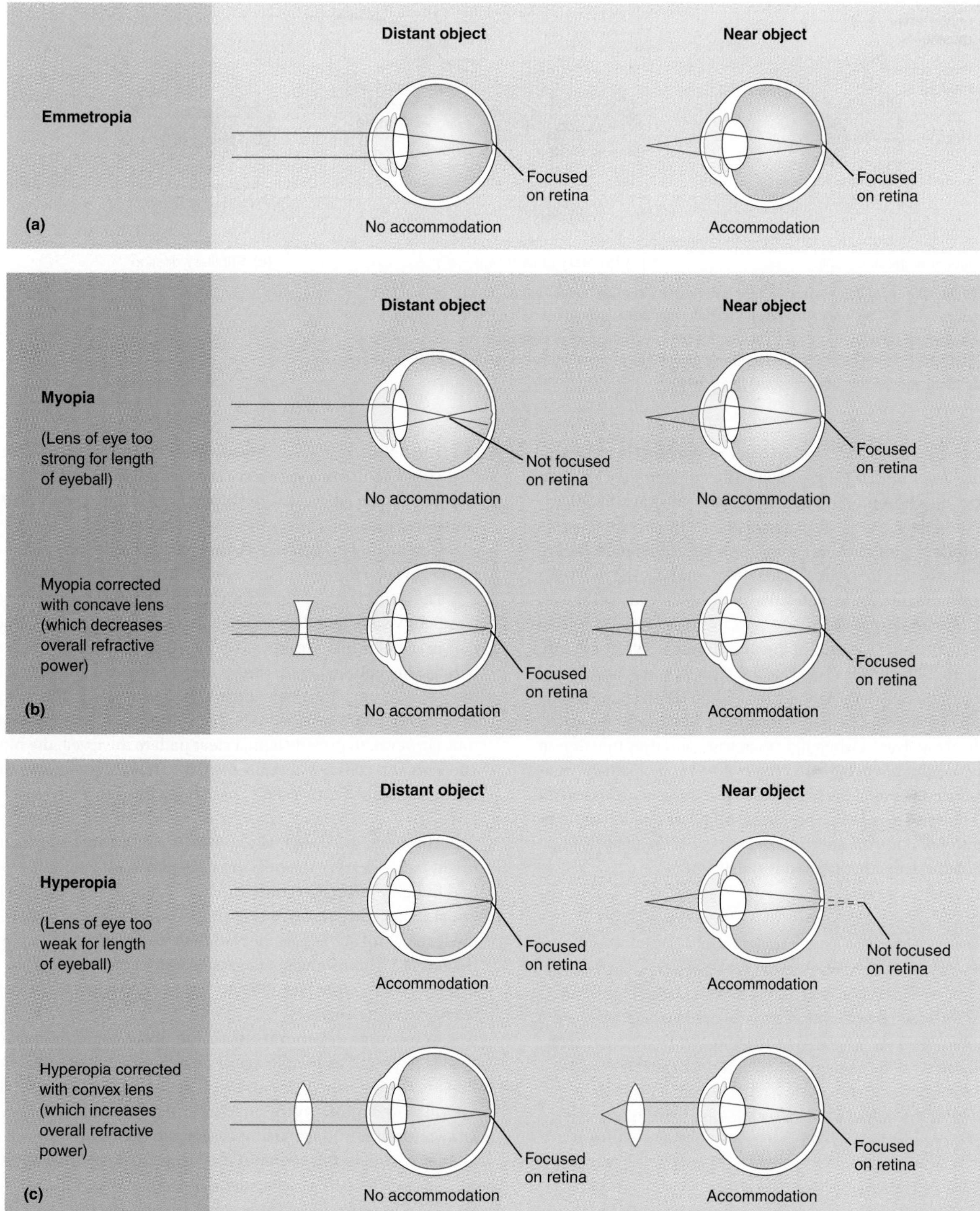

Figure 27 Normal, near-sighted, and far-sighted vision. **(a)** In emmetropia, or normal vision, distant objects are focused on the retina without accommodation, and near objects are focused with accommodation. **(b)** In myopia, or near-sightedness, the lens (or cornea) is too strong for the length of the eyeball. Near objects are focused without accommodation, and distant objects come into focus in front of the retina even without accommodation. Myopia can be corrected by using a concave lens to produce divergence of light waves before they enter the eye. **(c)** In hyperopia, or far-sightedness, the lens (or cornea) is too weak for the length of the eyeball. Distant objects are focused with accommodation, and near objects come into focus behind the retina, even with accommodation. Hyperopia can be corrected by using a convex lens to produce convergence of light waves that supplements the convergence produced in the eye.

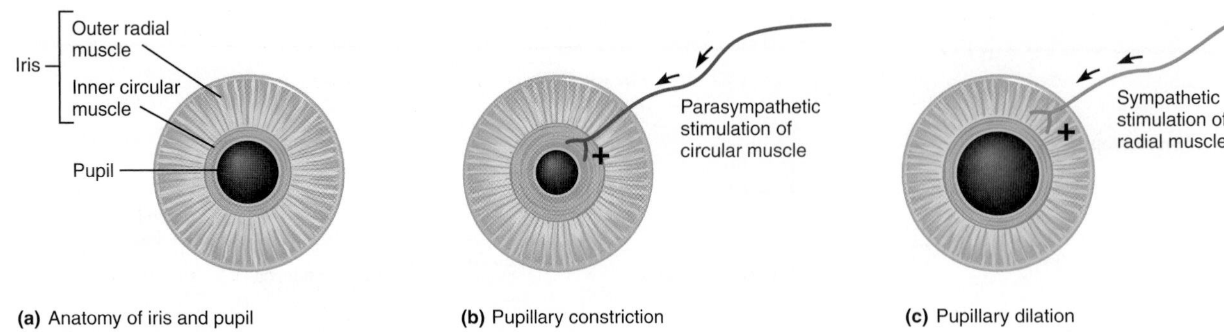

Figure 28 Regulation of the amount of light entering the eye. (a) The iris, which consists of two layers of smooth muscle—an inner circular layer and an outer radial layer—controls pupil size. The size of the pupil determines the amount of light that enters the eye. **(b)** Pupillary constriction, which is caused by parasympathetic stimulation of the circular muscle layer of the iris. **(c)** Pupillary dilation, which is caused by sympathetic stimulation of the radial muscle layer of the iris.

called the *constrictor muscle,* and an outer **radial muscle** layer, also called the *dilator muscle* (**Figure 28a**). The circular muscles form concentric rings around the pupil; when they contract, the diameter of the pupil decreases. Thus contraction of the circular muscles causes *pupillary constriction* (Figure 28b). The radial muscles are arranged like spokes in a wheel; when they contract, the diameter of the pupil increases. Thus contraction of the radial muscles causes *pupillary dilation* (Figure 28c).

The iris is under control of the autonomic nervous system. Parasympathetic neurons innervate the circular muscles; activity in these neurons causes the circular muscles to contract, producing pupillary constriction. The parasympathetic neurons to the circular muscle are activated when light increases, and thus function in regulating the amount of light that enters the eye. Sympathetic neurons innervate the radial muscles; activity in these neurons causes the radial muscles to contract, producing pupillary dilation. Sympathetic control of pupil diameter occurs as part of the fight-or-flight response of the sympathetic nervous system.

Apply Your Knowledge

"Red eye" occurs in pictures taken in dim light with a flash. In this situation, the eye is dilated prior to the flash, and the flash is so quick that the eye cannot compensate by constricting. When a lot of light enters the eye, some of it is reflected off the choroid and out of the eye (and to the camera). To protect against red eye, some cameras are equipped with a flash that emits a short burst of light prior to the actual flash for the picture. Explain how this pattern can prevent the red eye phenomenon in photographs.

The Retina

The retina, which is composed of neural tissue, is the location of photoreceptors—that is, the rods and cones. **Rods** provide the ability to see in black and white during relatively low light conditions, such as the light provided by the moon at night. **Cones** provide us with color vision, but they are active only in relatively bright light, such as the sunlight during the day.

The retina consists of three distinct layers (**Figure 29**): (1) an inner layer containing neurons called **ganglion cells,** (2) a middle layer containing neurons called **bipolar cells,** and (3) an outer layer containing rods and cones. Also present in the middle layer are *amacrine cells* and *horizontal cells,* neurons that modulate communication between the cells in the retina.

The retina is designed to provide the clearest image of the center of our visual field. Recall that light reflected from the center of the visual field strikes the fovea of the retina. As shown in Figure 29, because the photoreceptors are in the outer layer of the retina, light must pass through the inner and middle layers before striking them. In addition, blood vessels are in the light's path to the photoreceptors. However, to provide light a clear path to the fovea, the bipolar and ganglion cells are laterally displaced, creating a depression in the center of the retina called the **macula lutea** that surrounds the fovea.

The cones are designed to provide high-acuity color vision in bright light, whereas the rods are designed to provide some vision in dim light. The fovea contains cones only, contributing to high visual acuity. The ratio of rods to cones increases with distance from the fovea, until at the periphery of the retina only rods are present (**Figure 30**). This arrangement explains why we see objects better in dim light if we do not look directly at them, and why we see them in black and white only.

In **macular degeneration,** vision in the center of the visual field is impaired as people age. Two forms of macular degeneration are distinguished: dry and wet. In dry macular degeneration, the photoreceptors in the macula lutea degenerate for an unknown reason. Initially vision is blurred, but as the disease progresses, vision in the center of the visual field can be completely lost. In wet macular degeneration, new blood vessels develop (a process called *angiogenesis*) within and behind the retina. These new blood vessels are fragile and tend to burst. A person with macular degeneration progressively loses the high-acuity central vision but maintains peripheral vision.

One cause of macular degeneration is diabetic retinopathy. Diabetic retinopathy can be caused by development of new fragile blood vessels or by damage to existing blood vessels, which causes

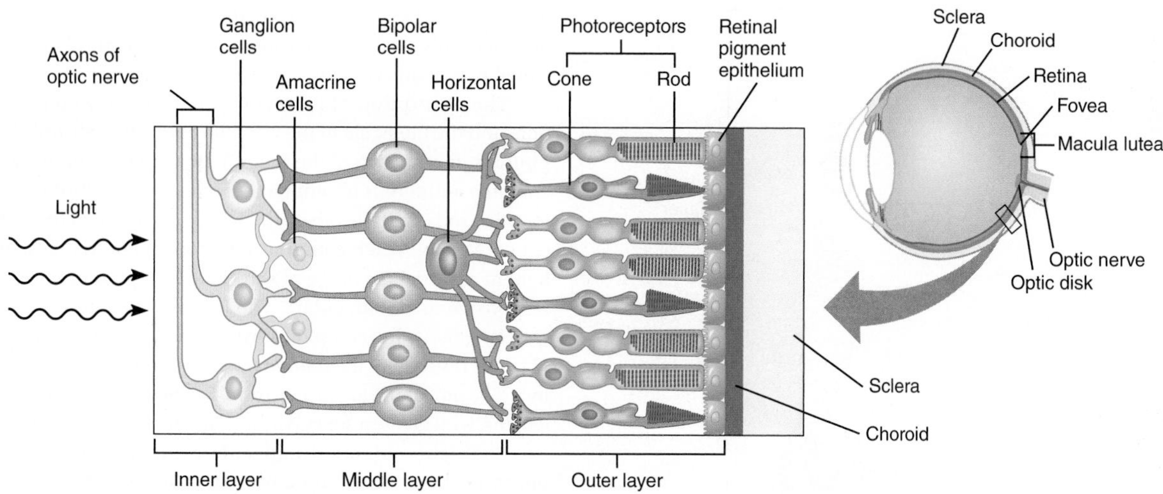

Figure 29 Anatomy of the retina. Located on the inner surface of the eye, the retina consists of three layers of neural tissue composed of the various types of cells depicted. Note that light must pass through the inner and middle layers of the retina before striking the photoreceptors in the outer layer. Deep within the retina is the retinal pigment epithelium, which absorbs light.

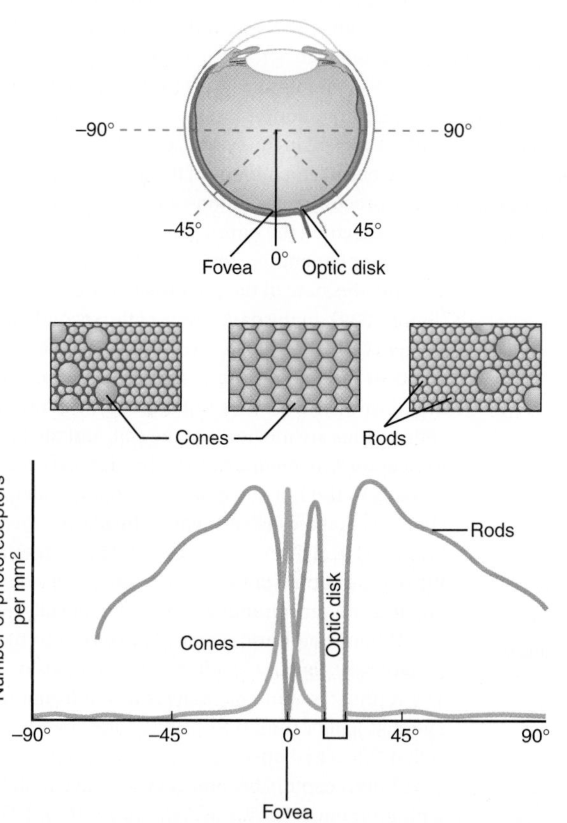

Figure 30 Distribution of rods and cones in the retina. The abundance of cones is greatest at the fovea but declines rapidly with distance from it. Rods are absent from the fovea but very abundant near it; they slowly decrease in abundance with distance from the fovea.

them to leak. In fact, diabetes is often first diagnosed by optometrists who notice excessive blood vessels in the retina.

Another disease of the retina is *retinitis pigmentosa*, an inherited disease characterized by progressive degeneration of photoreceptors in the peripheral retina. Loss of peripheral photoreceptors causes *tunnel vision*, where central vision through the macula is sustained but peripheral vision (and thus night vision) is lost.

Quick Check 4

1. What effect does contraction of the ciliary muscles have on the lens? Does this support vision of near objects or of distant objects?
2. What effects do the two branches of the autonomic nervous system have on the diameter of the pupil?
3. What are the locations of the vitreous humor and the aqueous humor? What is the function of each?
4. Which types of cells are found in each of the three layers of the retina?

Phototransduction

Phototransduction—the conversion of light energy into electrical signals—is carried out by the rods and cones. The basic morphology of the two types of photoreceptors is the same; each consists of two major portions referred to as *outer* and *inner segments* (**Figure 31**). The outer segment contains invaginations with membranous disks whose molecules absorb light waves, giving the photoreceptors the ability to respond to light. The inner segment contains the cell nucleus and various organelles and ends

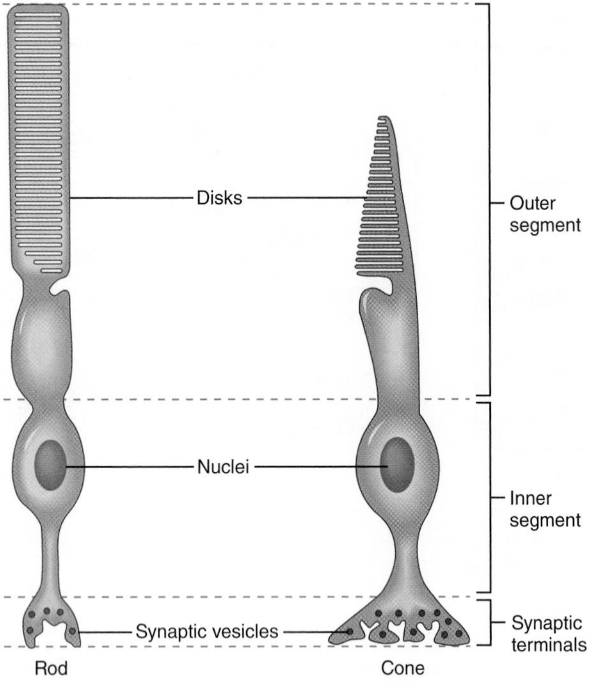

Figure 31 Morphology of the photoreceptors. Rods and cones have the same basic structural components: The outer segment consists of disks that contain the photopigment; the inner segment contains the nucleus and most of the organelles. The synaptic terminal contains the synaptic vesicles, which store a chemical transmitter used for communication.

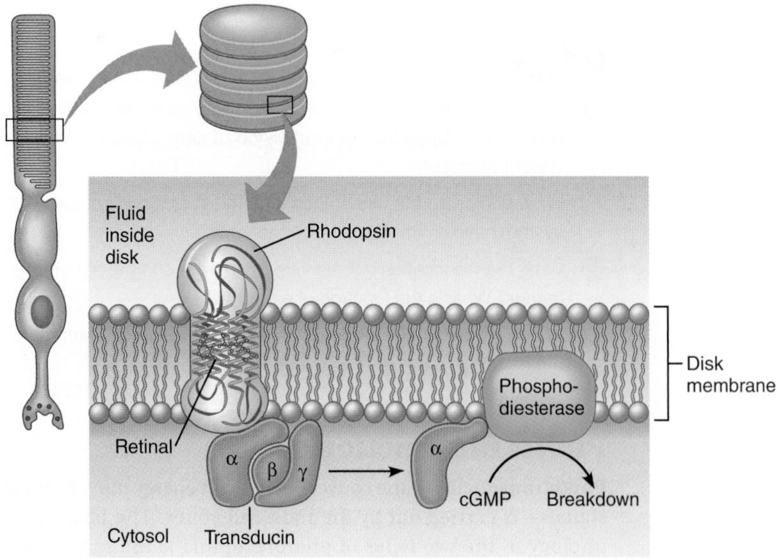

at the receptor's synaptic terminal, which is analogous to an ordinary neuron's axon terminal and is where a chemical messenger is stored in synaptic vesicles.

The absorption of light is the first step in phototransduction, and the molecule in the photoreceptors that absorbs light is a **photopigment.** Each of the four different types of photoreceptors contains a different photopigment. One type of photopigment is found in rods, and the other three types are found in three types of cones, each of which contains a photopigment that best absorbs light of a particular range of wavelengths and, therefore, is most responsive to certain colors. The L cones respond to the *longest* wavelengths, with greatest sensitivity to 560 nm, or the red spectrum. M cones respond to the *middle* wavelengths, with greatest sensitivity to 530 nm, or the green spectrum. S cones respond to *short* wavelengths, with greatest sensitivity to 420 nm, or the blue spectrum.

Each photopigment molecule contains a light-absorbing portion called *retinal* and a protein called an *opsin.* The retinal portion is the same in all photopigments, but the kind of opsin present determines which light wavelengths are absorbed by a given photopigment by altering the electromagnetic energies to which the retinal is sensitive. The four photopigments include **rhodopsin** in the rods and L opsin, M opsin, and S opsin in the long-, middle-, and short-wavelength cones, respectively.

The components of photoreceptors involved in phototransduction are shown in **Figure 32** using rods as an example. The photopigment of rods, rhodopsin, is located in the membrane of the disks. Also within the disk membrane is a G protein called **transducin** and the enzyme **phosphodiesterase,** which catalyzes the degradation of cGMP (if present in the cytosol, cGMP opens sodium channels located in the plasma membrane of the photoreceptor).

Let's examine the process of phototransduction as it occurs in rods. (Similar processes occur in the three types of cones, but different photopigments are involved.) We begin by considering the state of the photoreceptor in the dark (**Figure 33a**). In the dark, levels of the second messenger cGMP are high inside the outer segment ①, so cGMP opens sodium channels in the plasma membrane of the outer segment ②. Therefore, sodium ions are moving into the cell, and the photoreceptor is *depolarized* ③. This depolarization spreads to the inner segment and opens calcium channels that are also present in the plasma membrane ④. Calcium enters the cell ⑤, triggering the release of transmitter by exocytosis *in the dark.* The transmitter communicates to bipolar cells ⑥.

When the photoreceptor is exposed to light (Figure 33b), light is absorbed by the rhodopsin ①. The retinal component changes its conformation and dissociates from the opsin ②, leaving what is called "bleached opsin." (When opsin is bleached, the photoreceptors become less sensitive to light, a phenomenon known as *light adaptation.*) The bleached opsin activates transducin ③, which in turn activates the enzyme phosphodiesterase ④, which then catalyzes the breakdown of cGMP.

Figure 32 Components of rods. The photopigment, rhodopsin, is located in the membrane of the stacked disks located in the rod's outer segment. Rhodopsin is coupled to a G protein called transducin, which activates the enzyme phosphodiesterase, which in turn catalyzes the breakdown of cGMP.

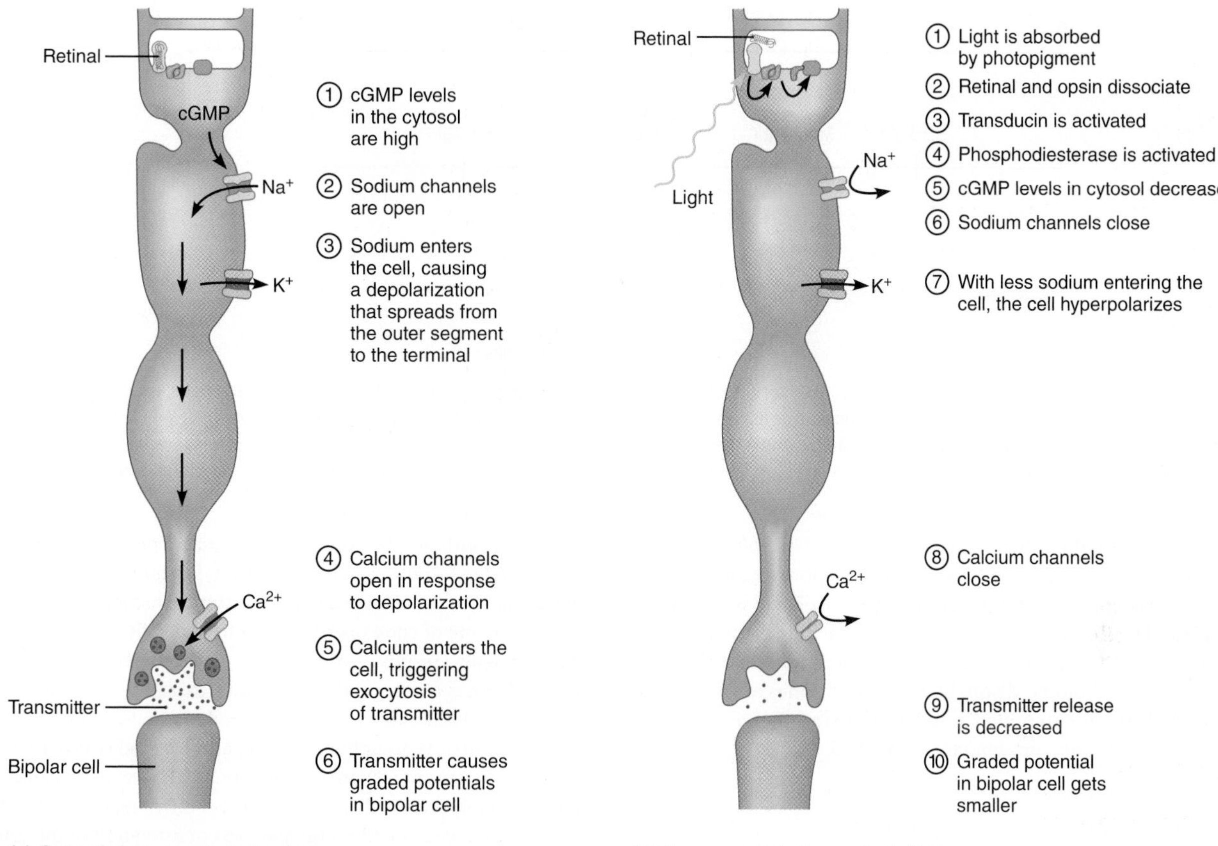

Figure 33 Phototransduction of light. (a) In the dark, photoreceptors release their chemical transmitter. **(b)** When light is present, this chemical transmitter is absorbed by the photopigment, initiating a sequence of events that decreases release of the transmitter.

With cGMP levels in the outer segment decreased ⑤, sodium channels close ⑥. Potassium leaking out of the cell causes a hyperpolarization that is no longer opposed by sodium movement into the cell ⑦. This hyperpolarization causes closing of calcium channels on the inner segment ⑧. With less calcium entering the cell, release of transmitter decreases ⑨. Thus, in the light, less transmitter is released from the photoreceptor terminal. Information about the presence of light is relayed, therefore, by a decrease in signaling to the next cells in the visual pathway, the bipolar cells ⑩.

Rods versus Cones

Rods and cones vary not only in the wavelength of light to which they respond (**Figure 34**), but also in the level or intensity of light that activates them. Rods are very sensitive to dim light (they can respond to a single photon of light) and are responsible for *scotopic vision*, or monochromatic (black-and-white) vision. In bright light, however, rods become saturated (completely bleached); that is, they become as hyperpolarized as possible and, therefore, cannot code for any additional brightness. Thus rods do not provide visual input in daylight. Cones, by comparison, require higher-intensity light to become activated. Because the three types of cones respond to

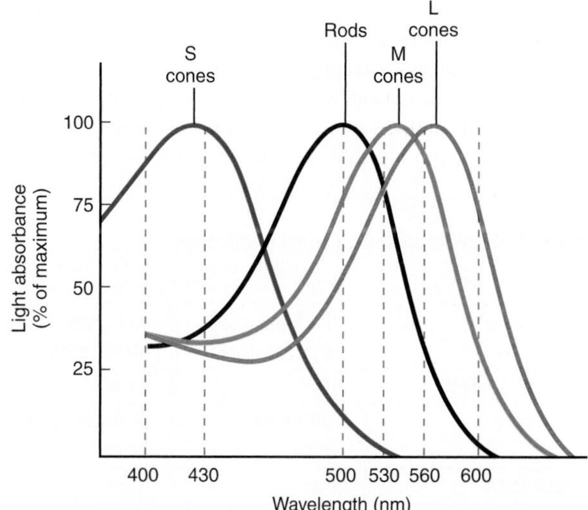

Figure 34 Absorbance spectra for the different photoreceptors. Rods can absorb light over the widest range of wavelengths. The absorbance spectra of the three types of cones overlap.

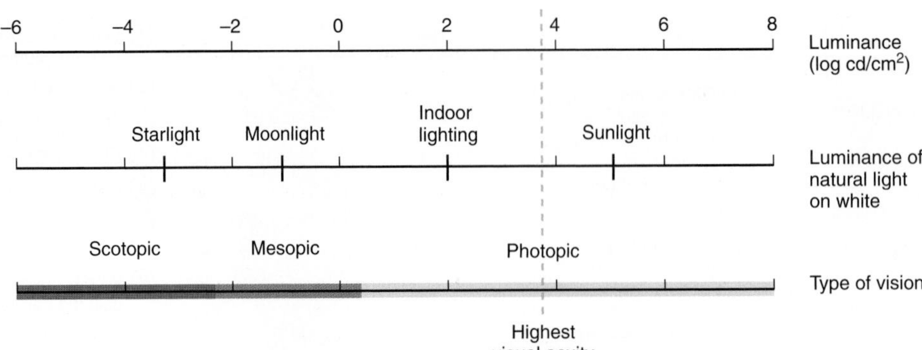

Figure 35 Rod and cone sensitivity. The luminance levels of natural light and normal indoor lighting and the types of vision under those luminance conditions are shown. Scotopic vision occurs under starlight and activates rods only, providing low-acuity, black-and-white images. Mesopic vision occurs under moonlight and activates rods and cones. Acuity and color vision are both limited, however. Photopic vision occurs at light levels at which rods become saturated and only cones are active, such as with indoor lighting and sunlight. Photopic vision provides the highest acuity and color vision.

different wavelengths of light, however, cones provide *photopic vision*, or color vision. **Figure 35** shows the response of rods and codes to varying levels of luminance (a measure of light emission). At low levels of luminance, only rods are active, so there is only scotopic vision. At high levels of luminance, only cones are active, so there is only photopic vision. Both rods and cones are active at intermediate levels of luminance and give rise together to *mesopic vision*. Rods and cones are compared in Table 5.

Color Vision

Color vision occurs only in bright light because the activation of cones requires a high intensity of light. The process of phototransduction in cones, however, is similar to that in rods and will not be described here.

The ability to perceive colors is based on the presence of three types of cones that respond best to light of different wavelengths. That more than one type of cone is necessary to distinguish colors is illustrated in **Figure 36**, which shows the absorbance spectra for two theoretical photoreceptors, P_1 and P_2. Consider two colors with wavelengths of 450 and 500 nm. Photoreceptor P_1 could not by itself distinguish these colors, because the absorbance of P_1 at 450 nm and 500 nm is equal. In contrast, with both P_1 and P_2 present, absorbance by both photoreceptors dictates that the wavelength had to be 500 nm and not 450 nm. But how do three types of cones allow us to perceive the full spectrum of colors?

Each type of cone responds best to a specific wavelength, but each cone also responds over a range of wavelengths, albeit to different degrees (Figure 34). Specifically, S or blue cones are most sensitive to light at a wavelength near 420 nm, M or green cones are most sensitive to light at 530 nm, and L or red cones are most sensitive to light at 560 nm (see Figure 34). However, because the absorbance spectra of the three types of cones overlap, many colors can be perceived based on the patterns of activation of the different cones. Whereas a given wavelength of light might elicit a response in more than one type of cone, the different cones generally respond to varying degrees. Thus our brains discern different colors by *comparing* the responses of the different cone types to each wavelength.

However, the relative responses of cones alone do not fully explain color perception. For example, why is there no such color

TABLE 5 Characteristics of Rods and Cones

	Rods	Cones
Types of vision	Black and white; night (dim light)	Color; day (bright light)
Sensitivity to light	High	Low
Abundance	100 million per retina	3 million per retina
Visual acuity	Low	High
Site of greatest concentration	Periphery of retina	Fovea
Degree of convergence with bipolar cells	High	Low

Figure 36 Theoretical state if only two types of cones existed. Notice that a cone with properties of P_1 could not distinguish between colors with wavelengths of 450 nm and 500 nm, and a cone with properties of P_2 could not distinguish between colors with wavelengths of 500 nm and 550 nm. With both cones present, however, P_1 and P_2 would both respond to the 500-nm color and the brain would be able to distinguish the color by integrating information from both cones, rather than depending on just one cone.

as reddish green or yellowish blue? Properties like these can be explained by the *opponent-process theory,* which states that red/green, blue/yellow, and black/white are opponent colors such that stimulation of one color in each pair inhibits the other. Therefore, we cannot see reddish green because the presence of green inhibits the perception of red. The opponent-process theory pertains to the level of ganglion cells, where some ganglion cells are excited by red in their visual fields and inhibited by green in the same regions.

The opponent-process theory also explains the concept of afterimages. To observe afterimages, perform the following: Stare at **Figure 37** for approximately 30 seconds, and then stare at a blank sheet of white paper for approximately 30 seconds. What do you see?

When you looked at the white paper, you should have seen the opponent colors because of adaptation to the original colors. Recall that most of the light we perceive is light waves reflected off objects. We see white when the full spectrum of light waves is reflected to the eye and all three cones are activated. To understand your perception, let's concentrate on what happened in the area of the visual field where you were looking at green. While you were looking at the picture, the green cones were activated, or bleached. When vision was shifted to the white paper, all wavelengths of light were reflected to the eye. However, in the visual field that originally detected green, the green cones were bleached and thus did not respond as strongly to the green wavelengths that were present in the white light. With less inhibition from green, red signals were transmitted to the CNS more strongly, and red was observed.

Light Input to Circadian Rhythms

Many systems of the body vary based on circadian rhythms. However, without any visual input, circadian rhythms would run longer than 24 hours. The photoreceptors that link the circadian rhythm to light-dark cycles are neither rods nor cones. These photoreceptors use a photopigment called *melanopsin.*

Figure 37 **Opponent-color adaptation.** Follow the instructions in the text to observe opponent colors.

Bleaching of Photoreceptors in Light

With small changes in light intensity, the eyes adapt simply through pupillary dilation and constriction, as described previously. But with more drastic changes in light intensity, dilation and constriction are not enough to manage the situation. Recall the last time you went to a matinee in a movie theater, where it was very dark inside. When you walked outside into the bright sunlight, your eyes were overwhelmed by the intensity of the light. Similarly, when a person moves from the sunny outdoors into a dark room, the ability to see is greatly impeded. In both cases, however, the eyes adapt in a matter of minutes to enable clear vision. What allows our eyes to adjust to these varying intensities of light?

When exposed to bright light, the rods become "bleached"; that is, most of the rhodopsin has absorbed light, and the opsin is in its active form. As a result, no more light can be absorbed until the rhodopsin has been returned to its original or "unbleached" state. (This is why rods function only in dim light.) Under these conditions, which correspond to when you first enter a dark room, rods are much less sensitive to light. Unbleaching of rods occurs in dim light, when the opsin returns to its inactive state. This reassociation of retinal and opsin takes several minutes, then retinal becomes sensitive to light again. By contrast, unbleached rods are extremely sensitive to light. Therefore, when you have been in the dark for a period of time and then emerge into the daylight, the bright light overwhelms the rods until they become bleached.

Neural Processing in the Retina

Communication between the different cells in the retina is quite complex and will be simplified here by describing one of the better-understood pathways. In this pathway, photoreceptors communicate to bipolar cells and horizontal cells by releasing the transmitter glutamate. Horizontal cells communicate to bipolar cells distinct from those receiving input from the same photoreceptors. Bipolar cells, in contrast, communicate to ganglion cells and amacrine cells. Amacrine cells communicate to ganglion cells.

Some degree of convergence generally exists between photoreceptors and bipolar cells; that is, more than one photoreceptor communicates to a single bipolar cell. Convergence also occurs between bipolar and ganglion cells. However, the extent of this convergence is considerably greater with rods than with cones. Thus, in the fovea and macula, where cones predominate, little convergence occurs; only a few photoreceptors converge onto one bipolar cell. In fact, in the fovea one cone communicates with one bipolar cell, which communicates with one ganglion cell, giving the highest degree of visual acuity. In contrast, in the periphery of the retina, where there are only rods, thousands of rods converge on one bipolar cell. Recall that in the somatosensory system, less convergence results in greater tactile acuity and two-point discrimination. Similarly, less convergence in the visual system provides greater visual acuity, because two separate light sources can be discriminated as distinct sources only if they trigger responses in separate cells along the visual pathway. Greater convergence, by comparison, provides greater sensitivity to light, because of spatial summation of inputs from several photoreceptors onto one bipolar cell.

To produce images, the visual cortex must process the information coming out of the retina as action potentials in ganglion cell axons. To understand the outgoing signals requires an understanding of the receptive fields of bipolar and ganglion cells, with the receptive field being the area of the visual field that, when stimulated with light, changes the cell's membrane potential. Bipolar cells and ganglion cells have complex receptive fields in that the responses of the cells varies based on whether the light stimulus is in the *center* of the receptive field or in the periphery or *surround* of the receptive field (**Figure 38**).

Bipolar Cell Receptive Fields

In the retina glutamate can produce both excitatory (depolarization) and inhibitory (hyperpolarization) effects on bipolar cells. Glutamate depolarizes bipolar cells when it binds to ionotropic receptors that open cation channels, thereby allowing sodium to enter the cell. Glutamate hyperpolarizes bipolar cells when it binds to metabotropic receptors that require G-protein activity.

Direct input from photoreceptor to bipolar cell is responsible for the response in the *center receptive field*. Recall that the release of glutamate from photoreceptors is high in the dark, but decreases in the light. Thus bipolar cells that are excited by glutamate are inhibited by light in its receptive field. Such bipolar cells are called OFF bipolar cells because they turn "off" to light. In contrast, bipolar cells that are inhibited by glutamate are excited by light in its receptive field. Such bipolar cells are called ON bipolar cells because they turn "on" to light.

OFF bipolar cells are excited by light in the *surround receptive field* due to complex interactions between photoreceptors, bipolar cells, and horizontal cells. Likewise, ON bipolar cells are inhibited by light in the surround receptive fields. Thus a more correct terminology for the two types of bipolar cells is ON-center, OFF-surround and OFF-center, ON-surround.

Ganglion Cell Receptive Fields

Bipolar cells are capable of transmitting graded potentials, but not action potentials. Nevertheless, when depolarized, they release transmitter to communicate with ganglion and horizontal cells. The receptive field properties of bipolar cells are continuous to ganglion cells; thus ganglion cells can be described as ON-center, OFF-surround, and OFF-center, ON-surround cells. The role of amacrine cells in lateral modulation of input is beyond the scope of this text.

The ganglion cells are the first neurons in the visual pathway that are capable of transmitting action potentials. The axons of these cells make up the **optic nerve** and serve as the output neurons of the visual pathway.

Figure 39 shows one simplified retinal pathway whereby ON-center, OFF-surround ganglion cells receive visual input. In this example, one cone is shown with its receptive field in the center and two cones are shown with receptive fields in the surround. The central cone releases an inhibitory neurotransmitter that communicates with a bipolar cell. Because the cone is inhibited by light, inhibition of an inhibitory signal (disinhibition) results in excitation of the bipolar cell. The bipolar cell communicates to the ganglion

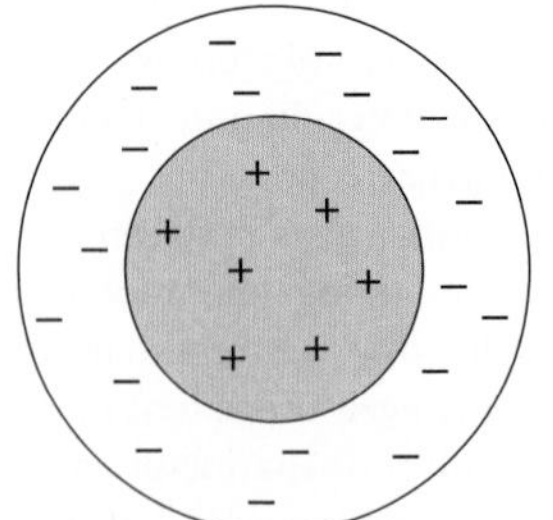

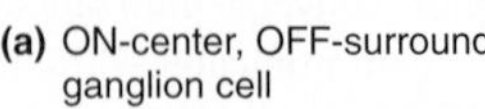

(a) ON-center, OFF-surround ganglion cell

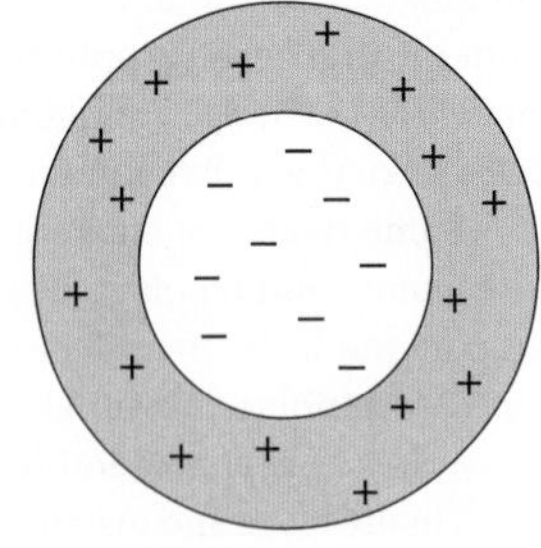

(b) OFF-center, ON-surround ganglion cell

Figure 38 Receptive fields of ganglion cells. Ganglion cells respond with both excitation (+) and inhibition (–) to light applied to the visual field. **(a)** ON-center, OFF-surround ganglion cells respond with excitation to light in the center of their receptive field and with inhibition to light in the surrounding area of their receptive field. **(b)** OFF-center, ON-surround ganglion cells respond with inhibition to light in the center of their receptive field, and with excitation to light in the surrounding area of their receptive field.

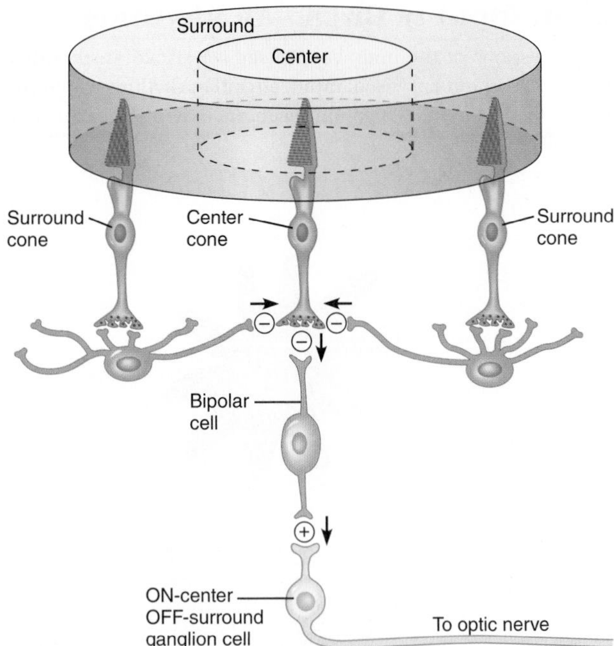

Figure 39 Neural circuitry resulting in an ON-center, OFF-surround ganglion cell. A single ganglion cell has different responses to light in the center of its visual field and light in the surround (outer edges) of its visual field. In this example, light in the center of the visual field excites the ganglion cell and light in the surround inhibits the ganglion cell. Horizontal cells provide lateral inhibition from cones receiving light in the surround to the synapse between cones receiving light in the center and a bipolar cell.

cell with an excitatory neurotransmitter. Thus the bipolar cell and ganglion cell are excited (or turned "on") by light impinging on the receptive field of this cone (the center).

The cones with receptive fields in the surround do not communicate directly with the bipolar cells. Instead, these cones communicate with horizontal cells that inhibit transmission between the center cone and the bipolar cell. The communication between the surround cones and the horizontal cells is excitatory; thus, when light strikes the surround and these cones are inhibited, the horizontal cell is also inhibited. When active, the horizontal cell inhibits the inhibition (disinhibits or excites) between the center cone and bipolar cell, thereby exciting the bipolar cell. However, when the horizontal cell is inhibited, the bipolar cell is not excited; likewise, the ganglion cell is not excited and it turns "off."

Receptive fields of ganglion cells become even more complex when color is considered. For example, some ganglion cells are excited by red in the center of the visual field and inhibited by red in the surround. These same ganglion cells are inhibited by green in the center of the visual field and excited by green in the surround. Green and red are *opponent* colors; that is, they cannot coexist. The presence of green inhibits red, and the presence of red inhibits green. Blue and yellow are also opponent colors, as are black and white. The opponent-process theory also explains why color blindness comes in pairs, such as red-green color blindness (see **Clinical Connections: Color Blindness**).

Neural Pathways for Vision

As previously described, the ganglion cells are the output neurons from the retina because they generate action potentials that are transmitted to the CNS. The axons of ganglion cells form the optic nerve (cranial nerve II). The two optic nerves exit each eye at the optic disk and combine at the base of the brain just in front of the brainstem to form the **optic chiasm.** In the optic chiasm, half the axons from each eye cross over to the other side of the brain (**Figure 40**). In Figure 40, note that input from the left visual field strikes the nasal retina (side closest to the nose) of the left eye and the temporal retina (side closest to the side of the head) of the right eye. Likewise, input from the right visual field strikes the nasal retina of the right eye and the temporal retina of the left eye. Therefore, both eyes receive information from both visual fields.

In the optic chiasm, axons originating from nasal ganglion cells cross to the opposite side, whereas axons originating from temporal ganglion cells stay on the side of origin. The result is that after the optic chiasm, all input from the right visual field travels in axons in the left side of the brain, and all input from the left visual field travels in axons in the right side of the brain. Although the axons are still those of ganglion cells, after the optic chiasm the axons travel in what is called the **optic tract.** The ganglion cells terminate in a nucleus in the thalamus called the **lateral geniculate body,** where they form synapses with neurons that ascend to the primary visual cortex in the occipital lobe. Pathways from the lateral geniculate body to the visual cortex on either side are called the **optic radiations.**

Apply Your Knowledge

Describe the visual defect that would occur following damage to each of the following on the right side: optic nerve, optic tract, and primary visual cortex.

In the somatosensory cortex, the topographic organization of the cortex is such that adjacent areas of the body are usually represented on adjacent areas of cortex. The visual cortex also has

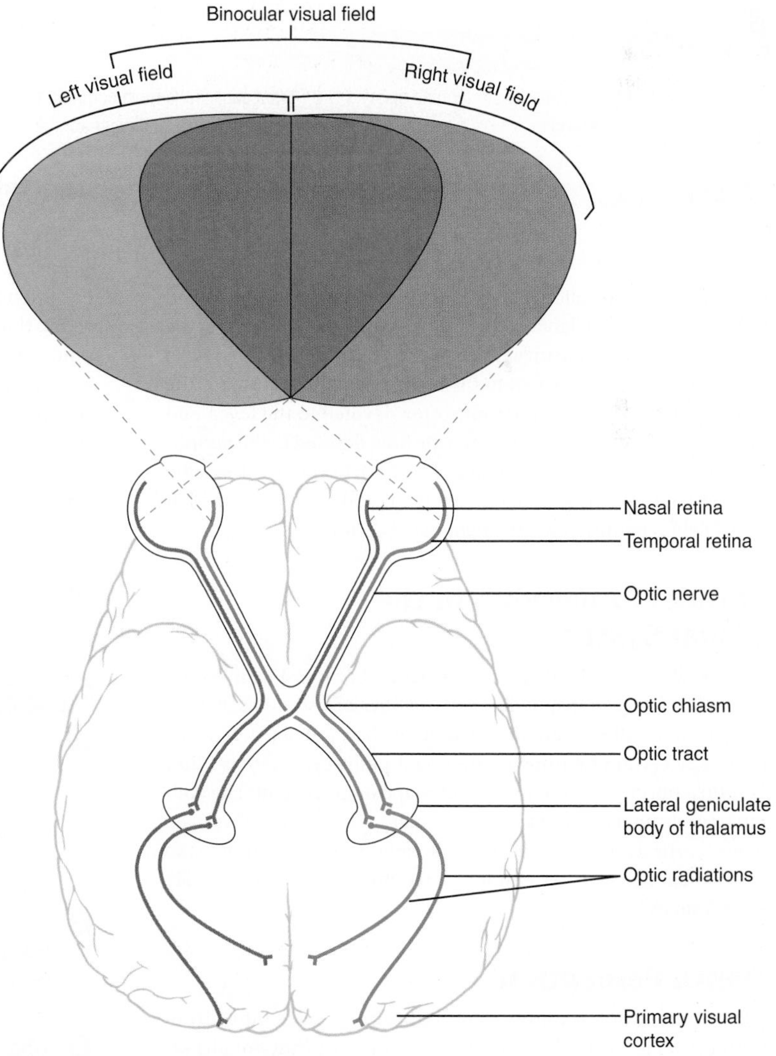

Figure 40 Neural pathways for vision.

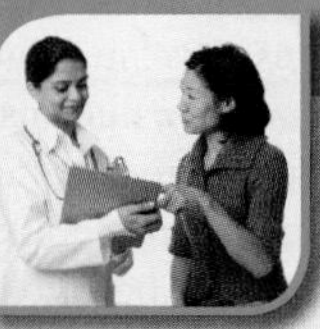

CLINICAL CONNECTIONS

Color Blindness

Our ability to perceive colors depends on the presence of three types of cones, each of which responds more preferentially to a particular wavelength of light. Lack of one of the types of cones impairs a person's ability to perceive different colors, which is the basis for color blindness.

The most common type of color blindness is red-green color blindness, in which red and green colors cannot be distinguished from each other. This type of color blindness is generally caused by a genetic defect in the photopigments of red or green cones. The genes that code for the red and green photopigments are recessive and are located on the X chromosome; because males have only one X chromosome, they are more likely to inherit this recessive trait. Blue color blindness also exists, but it is rarer and is not linked to the X chromosome.

A common test for color blindness employs what are known as *Ishihara charts.* In these charts, numbers are hidden within a pattern of colored dots. An example is shown in the figure. A person with normal vision can identify the number, but a person who is color blind is not able to see the number.

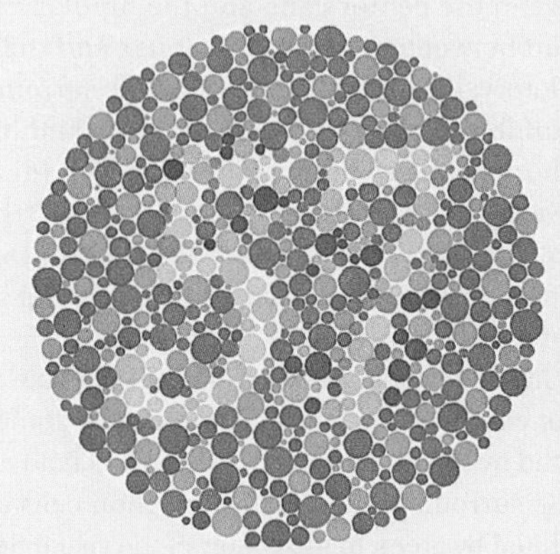

Critical Thinking Questions

1. Describe the elements that determine our ability to perceive color.
2. What causes the most common type of color blindness, and what are the genetic underpinnings for it?
3. Describe the nature of the test used to determine color blindness.

topographic organization, but in this case the *visual field* is mapped onto the cortex. Recall that in the somatosensory cortex, areas of the skin that are more sensitive (such as the lips and fingertips) have more area of cortex devoted to them. The same holds true for the visual system: The area of visual cortex devoted to the fovea and macula is considerably greater than the area devoted to the peripheral visual field. Due to the crossing of axons in the optic chiasm, the right visual field is mapped onto the left visual cortex, and the left visual field is mapped onto the right visual cortex.

Parallel Processing in the Visual System

The visual system clearly possesses parallel processing, in which parallel pathways transmit different qualities of a stimulus. For example, information about the color of an observed object is transmitted by certain neurons in the visual pathway, whereas other neurons transmit information about shape or movement. The coding stays distinct all the way to the primary visual cortex. However, higher cortical areas integrate the different qualities of a stimulus so that we can perceive, for example, a red fire truck moving quickly away from us.

Depth Perception

Depth perception requires that the brain receives input from both eyes. Figure 40 shows that images in most (but not all) areas of the left and right visual fields are detected by both eyes. For the portion of the visual fields that is detected by both eyes, called the *binocular visual field,* we are capable of depth perception. Depth perception depends on the fact that the left and right eyes, with their different positions in the head, see images from slightly different angles. The cortex uses these differences to construct a three-dimensional image of the world (that is, one that includes depth) rather than a flat, two-dimensional image. To test your depth perception, try the following: Hold two pencils at arm's length, one in each hand. With one eye open, try to touch the point of one pencil to the point of the other. Repeat with both eyes open.

Quick Check 5

1. Put the following components of the visual pathway in order such that they correctly reflect the path of transmission of visual information: optic tract, ganglion cell, photoreceptor, optic radiation, optic chiasm, bipolar cell, optic nerve, lateral geniculate nucleus, and visual cortex.
2. When are cGMP levels greatest in photoreceptors—when they are exposed to light, or when they are exposed to dark? What effect does cGMP have on sodium channels in the outer segment of the photoreceptors?
3. What are the two components of photopigment molecules? Which of these components absorbs light?

4 The Ear and Hearing

We now turn our attention to the sensory systems of the ear: first to the *auditory system,* which is responsible for hearing, and then to the *vestibular system,* which is responsible for balance or equilibrium. Although the stimuli involved vary, both of these sensory systems rely on hair cells to detect the movement of fluid within the cavities of the ear, which causes a receptor potential and release of a transmitter that communicates to an afferent neuron. We begin our examination of the auditory system by considering the anatomy of the ear.

Anatomy of the Ear

The ear can be divided into three parts: the *external ear,* the *middle ear,* and the *inner ear* (**Figure 41**). The external and middle ears are air-filled cavities, whereas the inner ear is filled with fluid.

The external ear includes the *pinna* and the **external auditory meatus,** or ear canal. The primary function of the external ear is to gather sound waves and conduct them to the **tympanic membrane** (or eardrum), which separates the external and middle ears.

The function of the middle ear is amplification of sound waves in preparation for the transmission of those waves from air to a fluid environment. Within the middle ear are three **ossicles,** or small bones, called the *malleus, incus,* and *stapes.* The three ossicles extend from the tympanic membrane to a thin membrane called the **oval window,** which is one connection between the middle and inner ears. The **round window** also connects the middle and inner ears.

The **eustachian tube,** which connects the middle ear with the *pharynx,* or throat, helps maintain normal pressure in the middle ear. Pressure changes, which may occur in the middle ear while flying or scuba diving or during ear infections, can be painful and can cause rupture of the tympanic membrane if they become large enough. Even mild pressure changes can change the ability to hear, because the pressure changes dampen sound vibrations. Opening the eustachian tube allows the pressure in the middle ear to equilibrate with the pressure in the pharynx, which alleviates any pressure difference across the eardrum. Swallowing or yawning can facilitate the opening of the eustachian tube.

The inner ear contains structures associated with both hearing and equilibrium. The **cochlea** is a spiral-shaped structure that contains the receptor cells for hearing. (The structures of the *vestibular apparatus,* including the *semicircular canals,* are discussed in the section on equilibrium.) The nerve that contains the afferents for both hearing and equilibrium, cranial nerve VIII, is the **vestibulocochlear nerve.**

The Nature of Sound Waves

Sound waves are mechanical waves caused by air molecules put into motion. **Figure 42a** shows the production of sound waves by a vibrating tuning fork. When the tines of a tuning fork vibrate, they generate waves in the air, similar to those produced when you touch your hand to the surface of a still pond. The waves consist of areas where the air molecules are closer together or *compressed,* and areas where the air molecules are farther apart or *rarefied.*

Figure 42b illustrates the properties of sound waves: loudness and pitch. The loudness (amplitude) of a sound is proportional to the difference in the densities of air molecules between the areas of compression and the areas of rarefaction: The greater the difference in densities, the louder the sound. The amplitude of a sound is most conveniently expressed in logarithmic units called *decibels (dB)* (see **Toolbox: Decibels**).

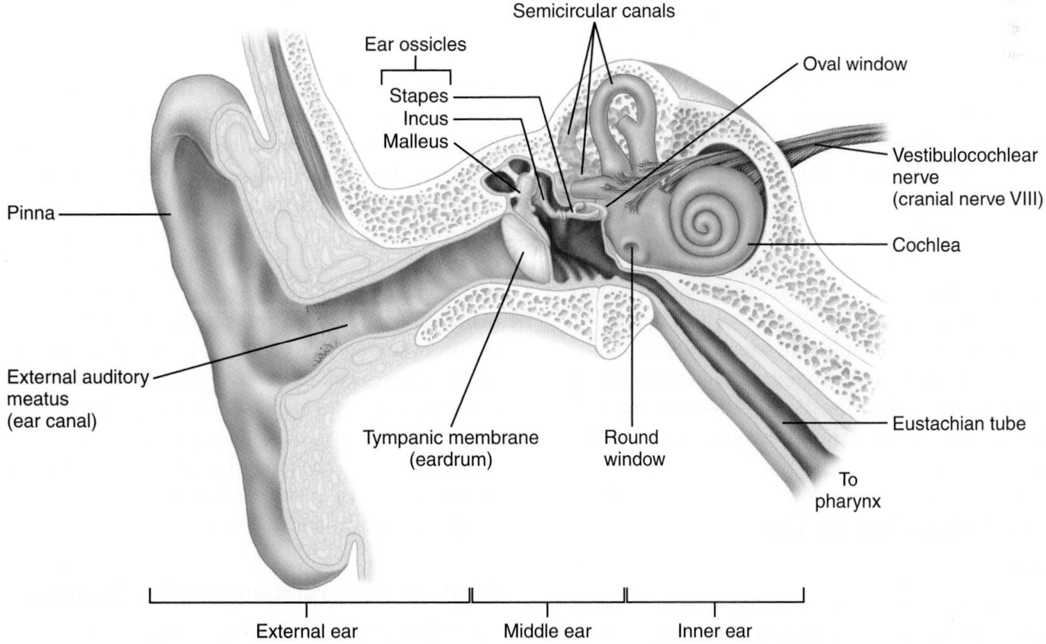

Figure 41 Anatomy of the ear. The major structures of the external ear, middle ear, and inner ear are labeled.

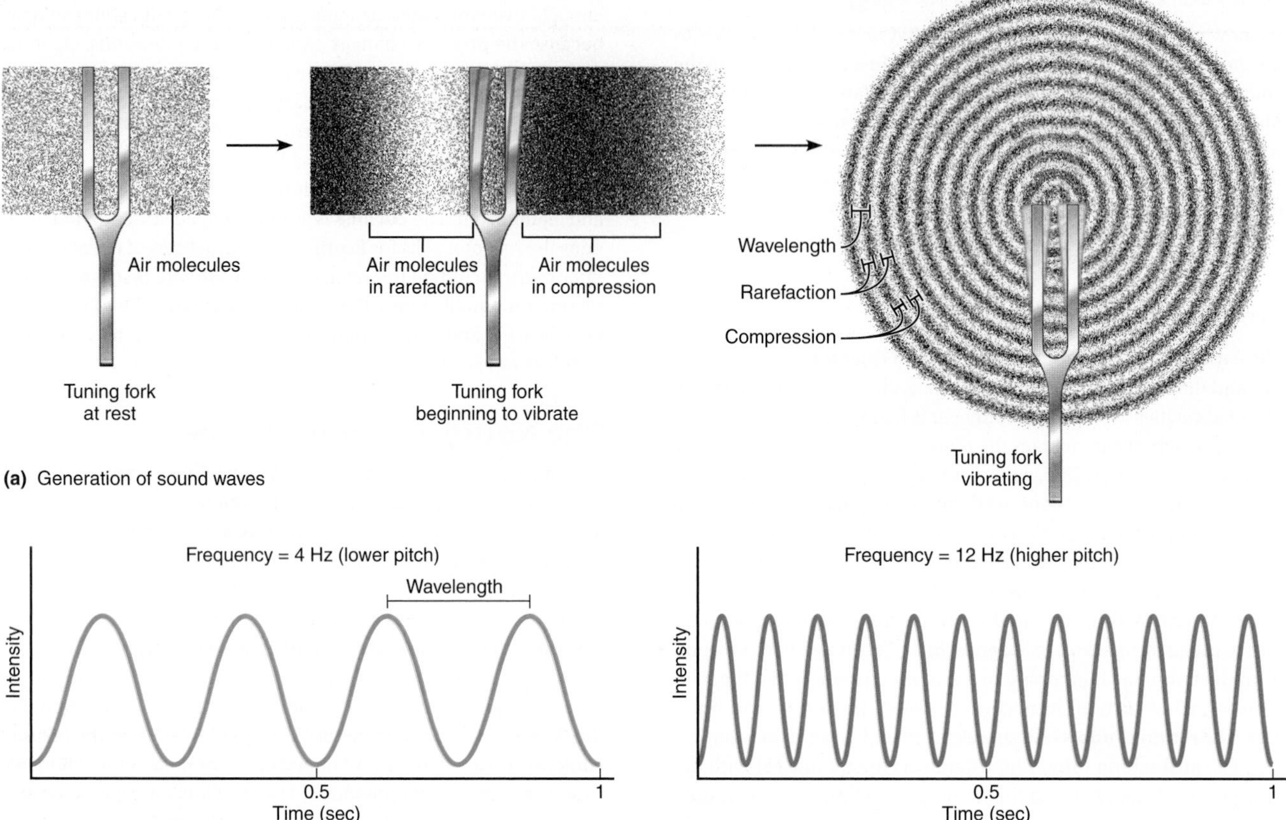

Figure 42 The nature of sound waves. (a) Vibrations produce sound waves, which consist of areas of high-density air molecules (compression) separated by areas of low-density air molecules (rarefaction). **(b)** Sound waves have an intensity (loudness) and a frequency (pitch).

The pitch of a sound is determined by the frequency of the sound waves. Low-frequency sound waves produce low-pitched sounds, such as those produced by a tuba. High-frequency sound waves produce high-pitched sounds, such as the squealing of car brakes. The frequency of sound waves is measured as the number of waves per second, or *hertz (Hz)*. The average person can hear sound waves with frequencies ranging from 20 to 20,000 Hz; the greatest auditory sensitivity occurs in the range between 1000 Hz and 4000 Hz.

The coding of amplitude and pitch of sound must be maintained both while sound waves move from the air-filled middle ear to the fluid-filled inner ear, and during the transduction process in the inner ear. The anatomy of the ear is exquisitely designed to sustain this coding.

Sound Amplification in the Middle Ear

When sound waves enter the ear, they strike the tympanic membrane, causing it to oscillate (vibrate) back and forth. These oscillations occur at the same frequency as those of the sound waves, and with an amplitude proportional to the amplitude of the sound waves. Because the malleus—the first of the small ossicles—is connected to the tympanic membrane (**Figure 43**), oscillations of the tympanic membrane cause the malleus to oscillate at the same frequency and with an amplitude reflecting that of the tympanic membrane's vibrations. The three ossicles are arranged in such a manner that they function as a series of levers; movement of the malleus causes a greater movement of the incus, which in turn causes an even greater movement of the stapes. The net effect is amplification of the motion produced by the initial sound waves. The stapes overlies the oval window to the fluid-filled cochlea, such that oscillations of the stapes generate waves in the fluid of the cochlea. Amplification of sound waves enhances the transmission of air-based waves to fluid-based waves. Amplification also occurs as sound waves are transmitted from the large-diameter tympanic membrane to the considerably smaller oval window, because a given force acting on a smaller surface produces a greater pressure.

Signal Transduction for Sound

The cochlea is the organ in which sound transduction occurs. Next we explore the anatomy of the cochlea and consider how it relates to the mechanism of sound transduction.

TOOLBOX

Decibels

The loudness (amplitude) of sound is based on the difference in density of air molecules in the areas of rarefaction and compression. Because the ear functions over a wide range of amplitudes, loudness is generally expressed in logarithmic units known as decibels (dB).

Measurements in decibels always compare the ratio of two intensities according to the following equation:

$$\text{intensity (dB)} = 20 \log \frac{\text{sound amplitude}}{\text{reference amplitude}}$$

In hearing, the reference amplitude is the threshold for normal hearing (that is, the minimum amplitude that can be perceived by humans), and is given a value of 1. Thus, with the denominator set at 1, the equation becomes

$$\text{intensity (dB)} = 20 \log (\text{sound amplitude})$$

At the threshold amplitude, the number of decibels is 0 [$20 \log(1) = 0$]. A sound that is 10 times the threshold for hearing is 20 dB [$20 \log(10) = 20$], whereas a sound that is 100,000 times the threshold is 100 dB [$20 \log(100{,}000) = 100$].

The intensities of various familiar sounds are listed in the table. Sounds approaching 100 dB create the potential for hearing loss; sounds at 130–140 dB approach the pain threshold.

Intensities of Some Familiar Sounds

Sound	Intensity (decibels)
Ticking watch	20
Elevator music	40
Conversational speech	50–60
Alarm clock	80
Live rock band	100
Jackhammer	110
Propeller airplane	120
Jet airplane	130

Functional Anatomy of the Cochlea

Understanding the mechanisms of sound transduction requires in-depth knowledge of the anatomy of the cochlea (**Figure 44**). From the outside (Figure 44a, b), the cochlea looks like a spiral seashell; the point at the end of the spiral is called the **helicotrema.** To depict the inside of the cochlea, the spiral in Figure 44d is partially uncoiled. A cross section of the cochlea is shown in parts (d) and (e) of Figure 44. Inside the cochlea are two membranes that separate it into three fluid-filled compartments: The **vestibular membrane** and the **basilar membrane** separate the cochlea into the **scala vestibuli,** the **scala tympani,** and the **scala media.** The vestibular and basilar membranes join at the helicotrema; thus an opening exists between the scala vestibuli and scala tympani at the helicotrema. The fluid in the scala vestibuli and scala tympani is called **perilymph;** it differs from the fluid in the scala media, which is called **endolymph.** Perilymph is similar in composition to cerebrospinal fluid, but endolymph is closer in composition to intracellular fluid, with a high concentration of potassium ions and a low concentration of sodium ions. In addition, there is an electrical potential in the endolymph relative to the perilymph called the *endocochlear potential,* which is approximately +80 mV.

The cochlea is a closed, fluid-filled structure separated from the middle ear at the oval and round windows. Because fluid is incompressible, to generate waves in the perilymph requires movement within the system without changing volume. When vibrations occur at the oval window, waves are generated in the perilymph of the scala vestibuli and travel through the scala tympani, where they cause motion of the round window. Thus the movement of the oval window, which would tend to change the volume of the perilymph, is countered by

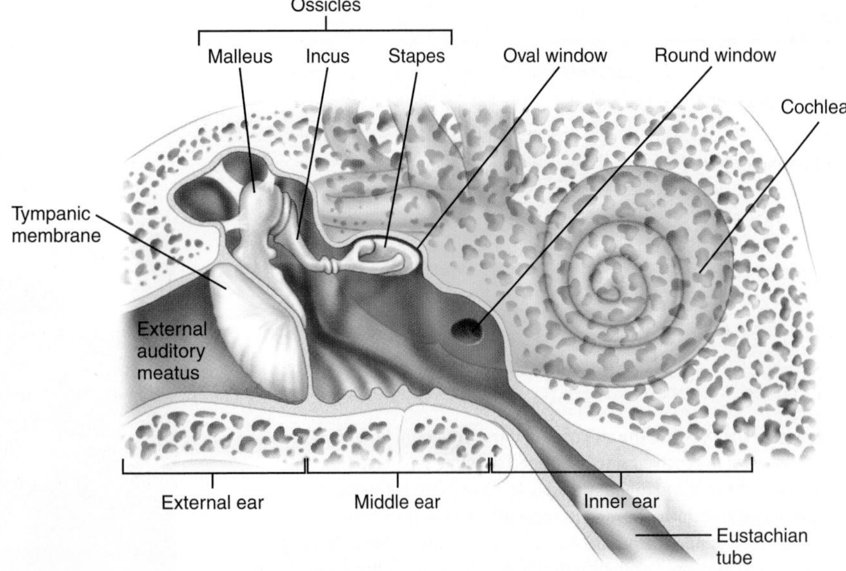

Figure 43 Structures that transmit sound waves in the middle ear. Sound waves hit the tympanic membrane, initiating vibrations in the ossicles. The stapes pushes the oval window, and vibrations are passed on to fluid in the cochlea.

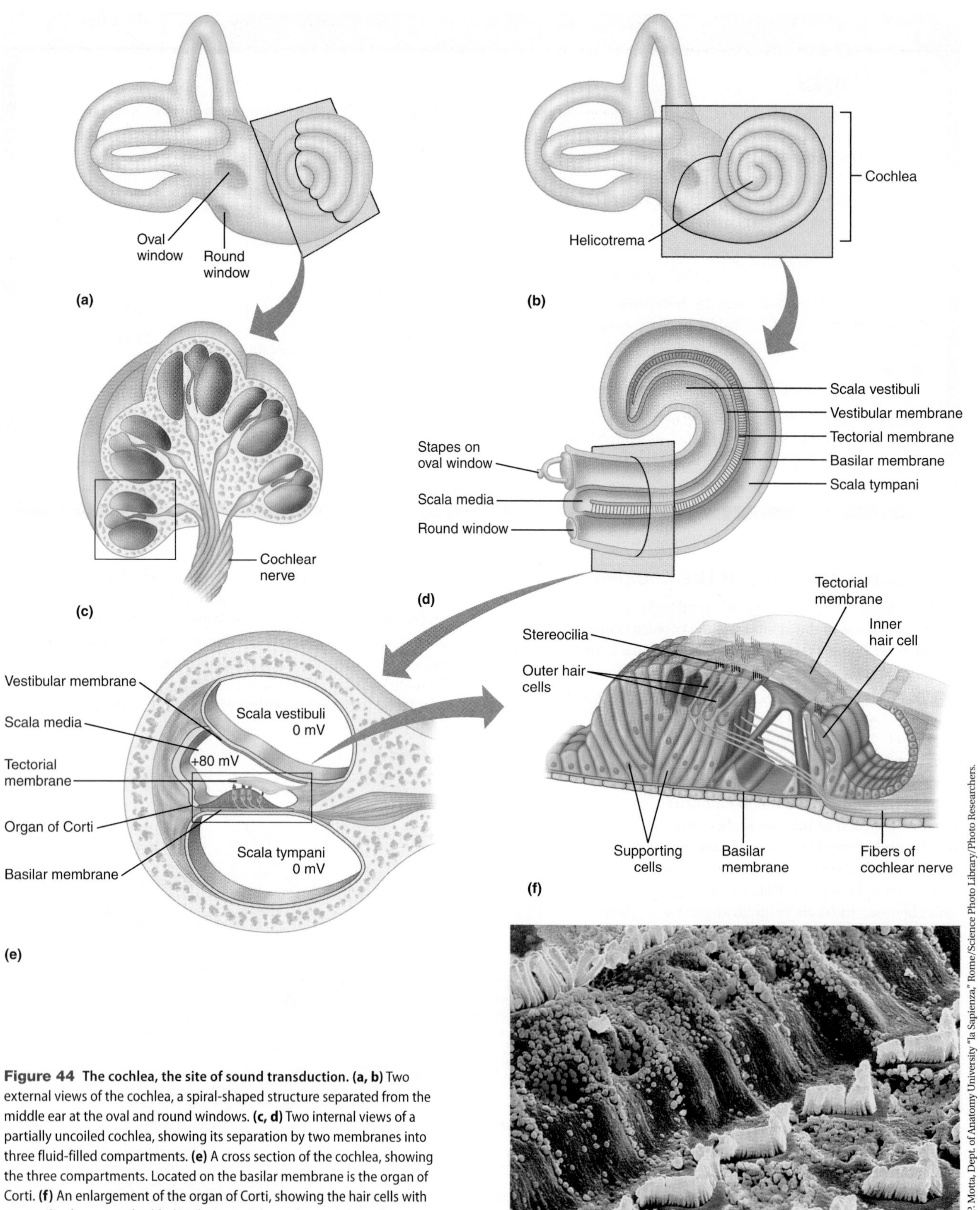

P. Motta, Dept. of Anatomy University "la Sapienza," Rome/Science Photo Library/Photo Researchers.

Figure 44 The cochlea, the site of sound transduction. (a, b) Two external views of the cochlea, a spiral-shaped structure separated from the middle ear at the oval and round windows. **(c, d)** Two internal views of a partially uncoiled cochlea, showing its separation by two membranes into three fluid-filled compartments. **(e)** A cross section of the cochlea, showing the three compartments. Located on the basilar membrane is the organ of Corti. **(f)** An enlargement of the organ of Corti, showing the hair cells with stereocilia that are embedded in the tectorial membrane. **(g)** An electron micrograph of hair cells in the cochlea.

movement of the round window, which dissipates the wave energy. Consider the following analogy: If you fill a syringe with water and cap off the opening, then you cannot move the plunger because the water can neither expand nor compress. However, if you connect two water-filled syringes with a tube, you can push one plunger down because the resulting force pushes the other plunger out. This arrangement allows you to move the plunger (the stapes) back and forth despite the fact that the water remains at a constant volume.

Functional Anatomy of the Organ of Corti

The **organ of Corti,** the sensory organ for sound, is located on top of the basilar membrane (Figure 44f). The organ of Corti contains **hair cells,** supporting cells, and an overlying membrane called the **tectorial membrane.** Hair cells were named for their hairlike projections called **stereocilia,** the tips of which are embedded in the tectorial membrane. Three rows of outer hair cells and one row of inner hair cells are present in the organ of Corti. Sound waves cause the bending of the stereocilia of the inner hair cells, which is the beginning of sensory transduction. The outer hair cells modulate the sensitivity of the inner hair cells and, therefore, receive primarily efferent innervation. Only the activity of the inner hair cells will be described here in detail.

The environment within hair cells and that surrounding them differ from the environments of most other cells of the body. Figure 44e shows the +80 mV potential found in the endolymph of the scala media in reference to the perilymph in the scala tympani and scala vestibuli. The hair cells are at a membrane potential of approximately –70 mV. Recall that the endolymph is also rich in potassium ions. The stereocilia of hair cells are surrounded by endolymph, whereas the remainder of the hair cell is surrounded by perilymph. Within the stereocilia are mechanically gated potassium channels. The presence of the endocochlear potential and the high concentration of potassium produce an electrochemical force that promotes the movement of potassium into the stereocilia. The role of these electrochemical forces for potassium movement in signal transduction will be described next.

Sound Transduction by Inner Hair Cells

Figure 45a shows the conduction of sound waves into the perilymph of the internal ear. The energy from waves in the perilymph causes the vestibular and basilar membranes to move relative to

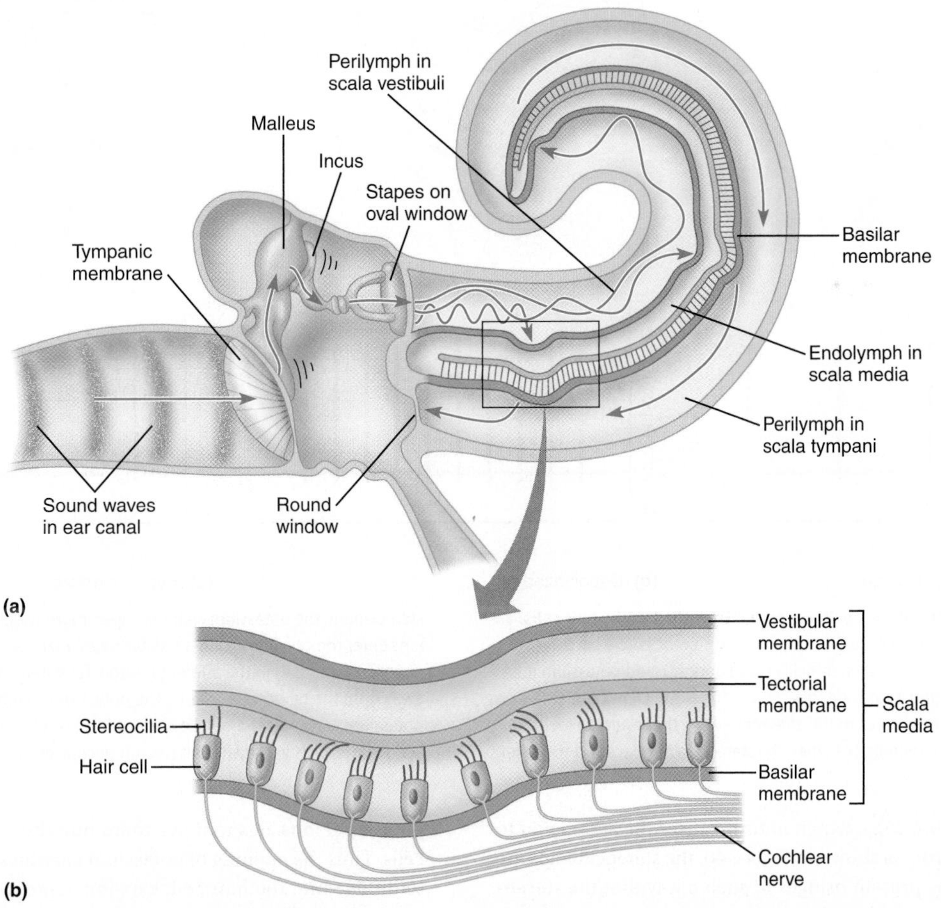

Figure 45 Conduction of sound waves in the ear. (a) Sound waves that enter the ear through the pinna and external auditory meatus strike the tympanic membrane, causing it to vibrate. The ossicles vibrate in response to the tympanic membrane and transmit the vibrations to the oval window. The vibrating oval window creates waves in the fluid (perilymph) of the cochlea. **(b)** Waves in the perilymph cause deflection of the membranes in the cochlea. When the membranes oscillate, the stereocilia of the hair cells bend, causing the opening or the closing of potassium channels.

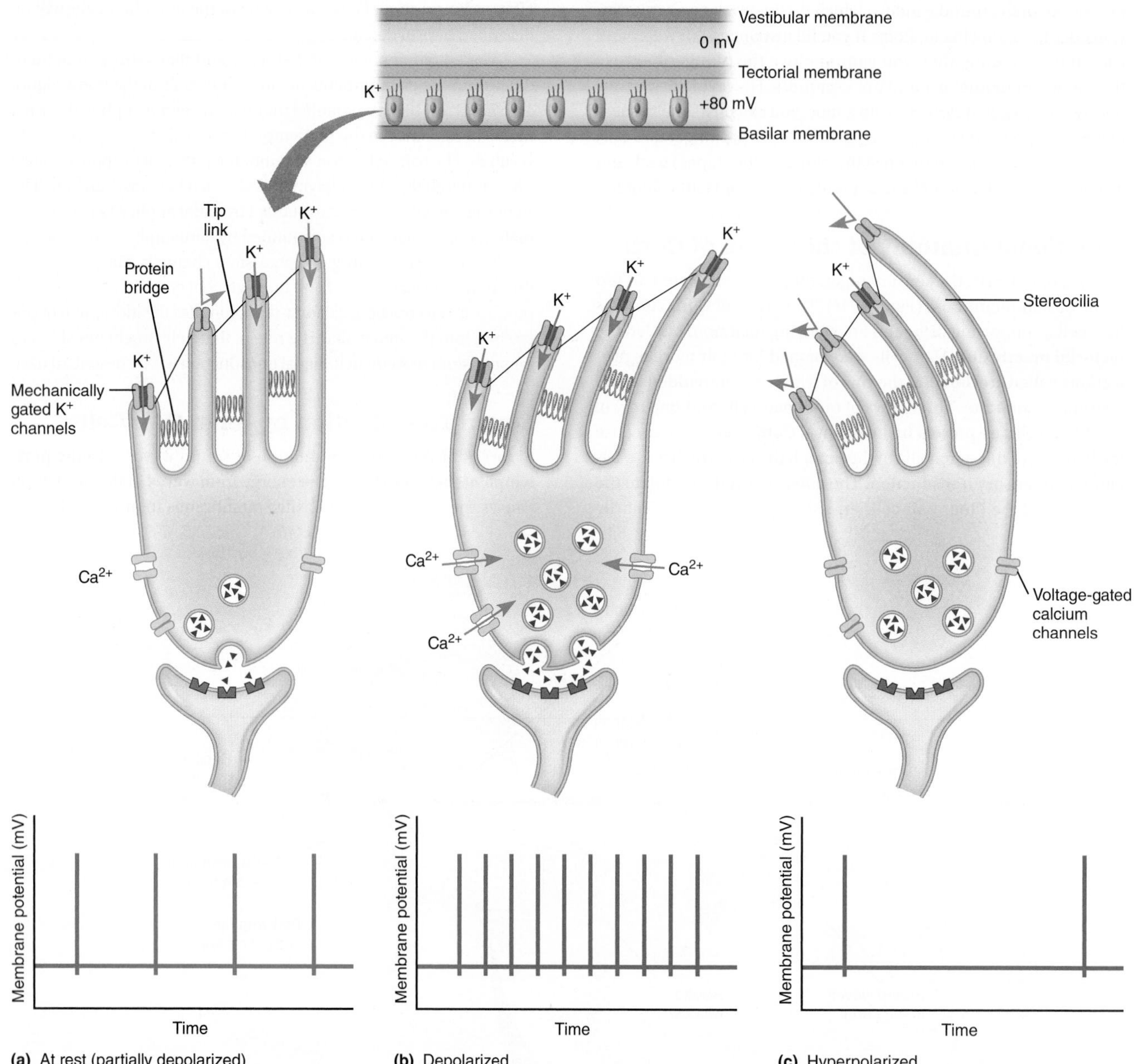

Figure 46 The roles of stereocilia in sound transduction by hair cells. **(a)** The mechanically gated potassium channels in stereocilia are partially opened when the cell is at rest (stereocilia stand upright), and potassium ions enter the cell, producing a small depolarization that is sufficient to release transmitter that communicates to the afferent neuron; the result is a low frequency of action potentials. **(b)** When the stereocilia bend toward the taller stereocilium, the potassium channels open more widely, and more potassium ions enter the cell, producing a greater depolarization and a higher frequency of action potentials in the afferent neuron. **(c)** When the stereocilia bend away from the taller stereocilium, the potassium channels close, and little potassium can enter the cell. Less transmitter is released, and the frequency of action potentials in the afferent neuron decreases.

one another (Figure 45b), which in turn causes the stereocilia to bend back and forth. As shown in **Figure 46**, the stereocilia tips are linked together by protein bridges in such a way that the stereocilia move together. Depending on the direction in which the stereocilia bend, potassium channels in the hair cells either open or close. Because of the cochlear potential and high concentration of potassium in the endolymph, the electrochemical force on potassium ions acts to move them into the stereocilia of the hair cells. Thus the opening of potassium channels prompts potassium to diffuse into the hair cell, causing depolarization; by contrast, closing of the potassium channels prevents potassium from diffusing into the hair cell, causing hyperpolarization.

Figure 46 shows how the bending of stereocilia causes potassium channels to open or close. Note that the stereocilia are linked together by elastic protein filaments and are of different sizes, such that one side of the hair cell has the tallest stereocilia. When the stereocilia extend straight up (no sound waves present), some tension on the elastic filaments holds the potassium channels in a partially opened state, allowing some potassium to diffuse into the hair cell and partially depolarize it (Figure 46a). This partial depolarization of the hair cell at rest causes the opening of some calcium channels in the hair cell, so calcium enters the cell and causes the release of transmitter by exocytosis. The transmitter communicates to an afferent neuron that is part of the cochlear nerve, resulting in a low frequency of action potentials (see Figure 45b).

When the stereocilia are bent in the direction of the tallest stereocilium (Figure 46b), then more tension is put on the elastic filaments and the potassium channels are opened to a greater extent. More potassium diffuses into the hair cell, producing a more depolarized state. More calcium channels open, resulting in the release of more transmitter, thereby increasing the frequency of action potentials in the afferent neuron.

When the stereocilia are bent away from the tallest stereocilium, the elastic filaments go slack and the potassium channels close (Figure 46c). With the potassium channels closed, potassium cannot diffuse into the cell, and the cell becomes hyperpolarized compared to the resting state. Under these conditions, calcium channels close and less transmitter is released, thereby decreasing the frequency of action potentials in the afferent neuron.

Coding of Sound Intensity and Pitch in the Cochlea

Given that receptor potentials are produced by bending of stereocilia, how are the intensity and pitch of sound coded? Louder sounds cause the stereocilia to bend farther in either direction, causing larger changes in the number of open potassium channels. This effect results in larger receptor potentials and larger variations in transmitter release.

The coding of sound frequency is based on the location of hair cells on the basilar membrane. Sound waves of different frequencies cause deflection of the basilar membrane at different regions because the structure of the basilar membrane varies over its length: The basilar membrane is stiff and narrow near the oval and round windows, but wide and flexible near the helicotrema (**Figure 47**). Therefore, high-frequency sound waves associated with high pitch cause the greatest deflection of the basilar membrane in the region closer to the oval and round windows, which activates hair cells located in this region. Low-frequency sound waves, by contrast, cause the greatest deflection of the basilar membrane in the region closer to the helicotrema, which activates hair cells located in this region of the basilar membrane. The frequency of sound, therefore, is coded by which hair cells are activated most strongly.

Neural Pathways for Sound

The transmitter released from hair cells binds to receptors on afferent neurons of the *cochlear nerve,* which is part of cranial nerve VIII. The hair cell transmitter depolarizes the afferent neuron:

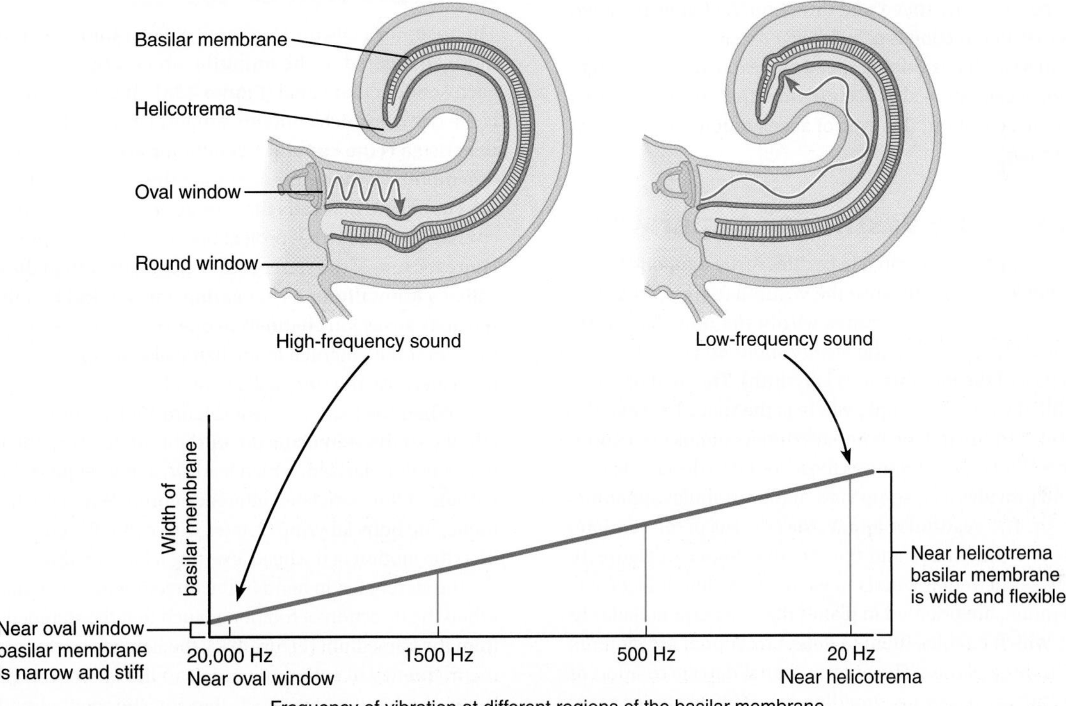

Figure 47 Coding for the frequency of sound. The frequency of sound is coded based on the location along the basilar membrane of activated hair cells. Variations in width and flexibility of the basilar membrane along its length dictate that high-frequency sounds cause deflections near the oval window, whereas low-frequency sounds cause deflections near the helicotrema.

The greater the degree of depolarization, the greater frequency of action potentials in the afferent neuron, which therefore codes for the intensity of the sound. The afferent neurons terminate in **cochlear nuclei** located in the medulla oblongata. Neurons of the cochlear nuclei travel both ipsilateral and contralateral to other brainstem nuclei that function as relays to the **medial geniculate nucleus** of the thalamus. Neurons from the medial geniculate travel to the **auditory cortex** in the temporal lobe. The brain compares the timing of the ipsilateral and contralateral signals to determine the direction of sound. Like other sensory systems, the auditory cortex has topographical organization. Its organization is *tonotopic;* that is, it maps the frequency of sound.

Quick Check 6

1. Explain how the pitch of a sound is coded. How is the loudness of sound coded?
2. When stereocilia of hair cells are bent in a certain direction, which ion moves into the hair cell to depolarize it?

5 The Ear and Equilibrium

The previous section discussed the anatomy of the ear as it relates to transduction of sound waves. This section presents ear anatomy as it pertains to the ability to detect acceleration of the body and position of the head—information required if balance and equilibrium are to be maintained. Acceleration is a change in an object's velocity, which depends on its speed and direction. A change in speed without a change in direction is called *linear acceleration;* it occurs, for example, in a car that is gaining speed while moving in a straight line. However, acceleration also occurs when the car turns, even if its speed remains constant; this type of acceleration is called *rotational acceleration.*

Anatomy of the Vestibular Apparatus

The vestibular apparatus is located in cavities of the temporal bones called the *bony labyrinth.* Because the vestibular apparatus consists of membrane-bound structures within the bony labyrinth, these structures are also called the *membranous labyrinth* (the cochlea is also part of the membranous labyrinth). The membranous labyrinth is filled with endolymph, whereas the space between the membranous labyrinth and the bony labyrinth contains perilymph; note that these fluids are the same as those found in the cochlea.

Figure 48 provides a close-up view of the vestibular apparatus of the inner ear. The **vestibular apparatus** consists of the **semicircular canals** and the **utricle** and the **saccule.** Notice in Figure 48 that the three semicircular canals in each ear, which detect rotational acceleration, are oriented in planes that are perpendicular to one another, which enables them to detect rotational movements of the head in three planes. The anterior canal detects rotation of the head up and down, as when nodding "yes." The posterior canal detects rotation of the head up and down to the side, as in moving the ear toward the shoulder. The lateral canal detects rotation of the head from side to side, as when shaking the head to indicate "no." The utricle and saccule together detect linear acceleration; the utricle detects acceleration forward and backward, whereas the saccule detects acceleration up or down.

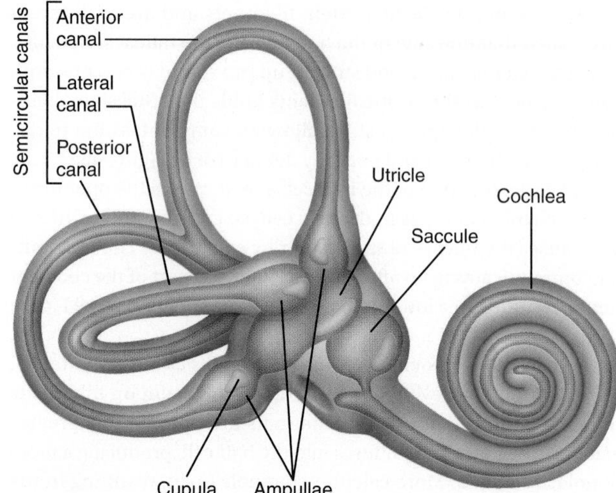

Figure 48 Anatomy of the vestibular apparatus in the inner ear. The vestibular apparatus includes the three semicircular canals, the utricle, and the saccule. The hair cells for detecting rotational acceleration are located in the ampullae of the semicircular canals.

The Semicircular Canals and the Transduction of Rotation

The receptor cells for rotational acceleration are another type of hair cell located in the **ampulla,** an enlarged area at the base of each semicircular canal (Figure 49a). At the base of each ampulla is the *cristae,* which contains support cells and hair cells. Overlying the cristae is the **cupula,** a gelatinous area separated from the endolymph by a membrane. Among the support cells at the base of the cupula are hair cells that are similar to those in the cochlea, including stereocilia that project upward into the cupula (Figure 49b). However, one of the stereocilia is much larger than the others and is called a **kinocilium.** As in hearing, mechanical bending of the stereocilia causes ion channels to open or close, resulting in a change in membrane potential in the hair cells. Next we examine how rotation can cause the stereocilia to bend.

When the head is at rest (Figure 49c), no force is acting on the cupula, so the stereocilia are upright. In this case the hair cells are partially depolarized, which leads to a low frequency of action potentials in the associated afferent neuron. When the head begins to rotate, the bony labyrinth rotates with it, but the endolymph lags behind the motion of the head, exerting a force on the cupula and causing the stereocilia to bend in the direction opposite that of rotation. When the direction of rotation is such that the stereocilia bend away from the kinocilium (Figure 49d), the hair cell is hyperpolarized, and the frequency of action potentials in the afferent neuron declines. When the head rotates such that the stereocilia bend toward the kinocilium (Figure 49e), the hair cells are depolarized, and the frequency of action potentials in the afferent neuron increases.

Note that when the head continues to rotate at a constant speed (see Figure 49c), the movement of fluid eventually catches

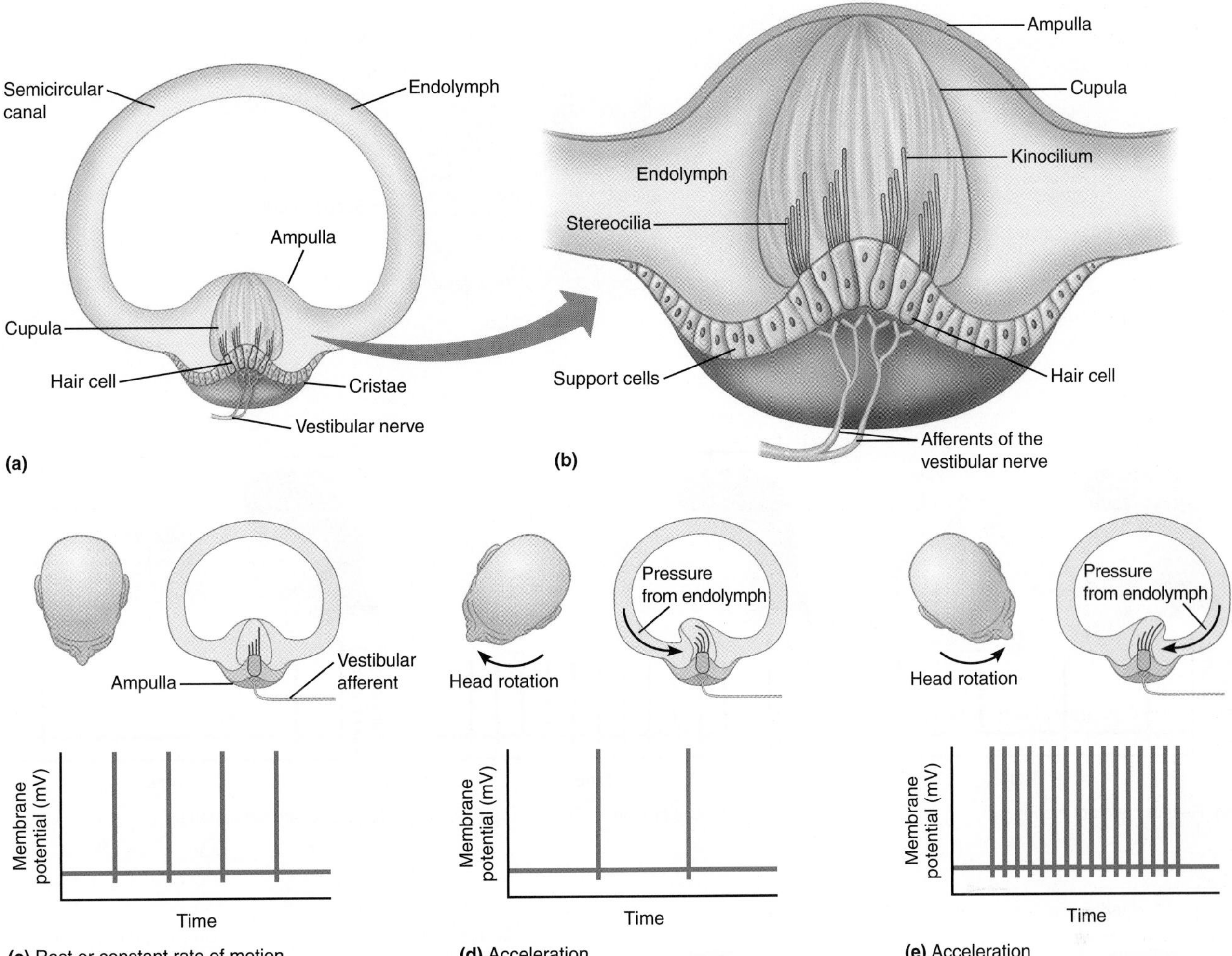

Figure 49 Functional anatomy of the semicircular canals.
(a) Within the ampulla at the base of each semicircular canal is the cupula, a gelatinous area separated from the endolymph by a membrane. **(b)** Hair cells—the receptor cells for acceleration—have stereocilia that extend into the gelatinous cupula; the longest stereocilium is the kinocilium. Hair cells communicate to afferent neurons. **(c)** When the head is still or moving at a constant rate of motion, no net force is acting on the cupula, and the stereocilia are upright. The hair cells are partially depolarized, releasing some chemical transmitter and causing a low frequency of action potentials in the afferent neuron. **(d)** During acceleration of the head, inertia of the fluid causes it to lag behind the motion of the head. As it does so, it pushes against the cupula in the direction opposite that of rotation, causing the stereocilia to bend. When the stereocilia bend away from the kinocilium, the hair cell is hyperpolarized, and the frequency of action potentials in the afferent neuron decreases. **(e)** During acceleration of the head in the opposite direction, the stereocilia bend toward the kinocilium, the hair cells are depolarized, and the frequency of action potentials in the afferent neuron increases.

Q Stereocilia are composed of which type of protein filament?

up to the movement of the bony labyrinth. In that case, no force is acting on the cupula, so the stereocilia are no longer bent, and the frequency of action potentials returns to that occurring when the head is at rest. Thus the semicircular canals detect only changes in the rate of rotation, not constant rotation.

What happened when as a child you spun around to make yourself dizzy and then suddenly stopped spinning? When the constant rotation suddenly stops, the head and bony labyrinth stop as well, but the endolymph keeps moving for a time. Thus, for a while, rotation is detected even though you are standing still, but it feels like it is occurring in the direction opposite that of the original rotation.

The Utricle and Saccule and the Transduction of Linear Acceleration

The utricle and saccule function in a manner similar to the semicircular canals in that they contain hair cells with stereocilia that bend, but their anatomy is somewhat different. The utricle and saccule, which are bulges located between the semicircular canal and cochlea of the inner ear (see Figure 48), have hair cells with stereocilia extending up into a gelatinous material (**Figure 50**). Within the upper edges of the gelatinous material are **otoliths,** small calcium carbonate crystals that add mass to the gelatinous material.

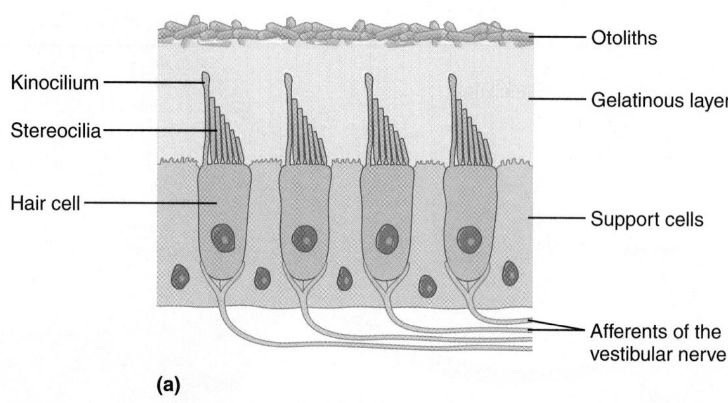

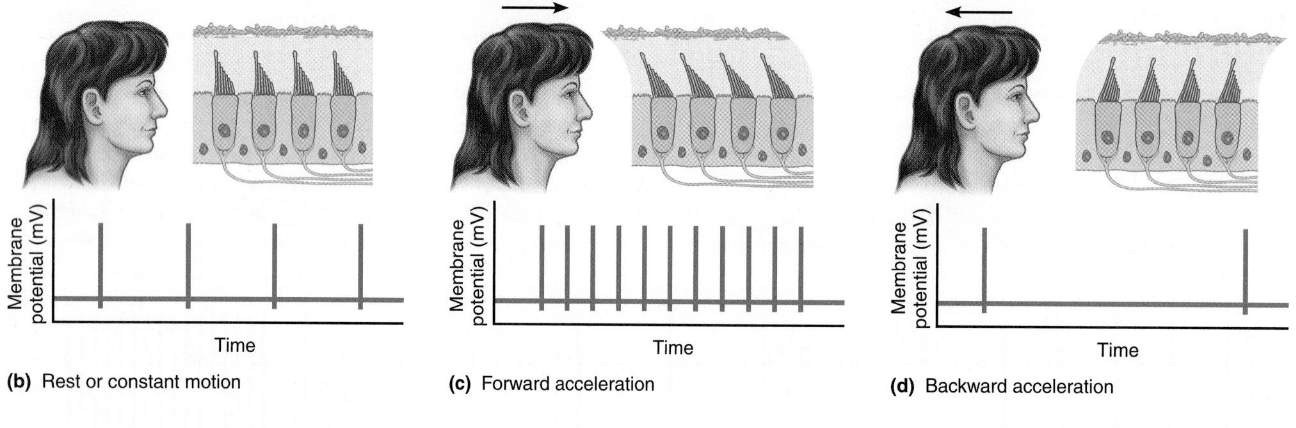

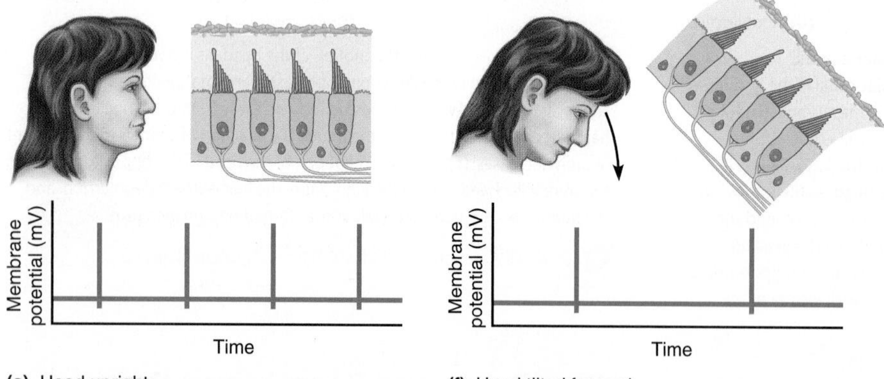

Figure 50 Functional anatomy of the utricle and saccule. (a) The receptor cells for linear acceleration in both the utricle and the saccule are hair cells with stereocilia extending into a gelatinous layer containing otolith crystals. Parts (b) through (f) depict events associated with the utricle only. **(b)** At rest or in constant motion, the stereocilia in the utricle stand erect, and the hair cell is partially depolarized. The afferent neuron has a low frequency of action potentials. **(c)** During forward acceleration, the otoliths cause a drag on the gelatinous mass of the utricle, causing the stereocilia to bend. In this example, the stereocilia are bending toward the kinocilium. The hair cell is depolarized, and the frequency of action potentials in the afferent neuron is increased. **(d)** During backward acceleration, the otoliths in the utricle cause a drag on the gelatinous mass in the opposite direction as in part (c). The stereocilia bend away from the kinocilium, causing a hyperpolarization of the hair cell and a decrease in the frequency of action potentials in the afferent neuron. **(e, f)** Effect of gravity on the utricle. In the example shown, when the head is tilted forward, gravity causes the otoliths to fall, pulling on the gelatinous mass. The stereocilia bend, causing a hyperpolarization and a decrease in the frequency of action potentials in the afferent neuron.

The hair cells of the utricle are oriented in horizontal rows in the head, with the stereocilia extending up vertically; the hair cells of the saccule are oriented in vertical rows in the head, such that the stereocilia are oriented horizontally. Because of these orientations, the utricle detects forward and backward linear acceleration, whereas the saccule detects up and down linear acceleration.

Figure 50b–d illustrates the bending of the stereocilia within the utricle during forward and backward linear acceleration. Although all kinocilia are shown oriented in the same direction (posterior) in this example, some kinocilia are located anteriorly and others posteriorly in the utricle. Hair cells with anterior kinocilia will behave in a manner opposite to that described here. When a person is at rest or moving forward or backward at a constant speed (Figure 50b), the hair cells are vertical, and the resulting depolarization results in moderately frequent action potentials in the afferent neuron. When the person begins to walk forward (Figure 50c), the otoliths cause the gelatinous mass to drag behind, bending the stereocilia toward the kinocilium and causing greater depolarization and more frequent action potentials. Backward linear acceleration (Figure 50d) causes bending of stereocilia away from the kinocilium, producing hyperpolarization and less frequent action potentials. When linear motion continues at a constant rate (see Figure 50b), the gelatinous material catches up with the movement of the body, and the stereocilia return to the at-rest position.

The utricle also plays a role in detecting the position of the head relative to gravity, a force that can cause acceleration. When the head is upright (Figure 50e), the stereocilia in the utricle are vertical. When the head is tilted forward (Figure 50f), however, gravity pulls the weighted otoliths downward, bending the stereocilia downward and producing hyperpolarization.

The saccule functions in the same way as the utricle, but because of its orientation when the head is held upright it detects up and down linear acceleration, as occurs when a person rides in an elevator.

Apply Your Knowledge

Ménière's disease is an inner ear disorder associated with changes in the composition of endolymph. Its exact cause is not certain, but one theory suggests that a rupture in the membranous labyrinth causes mixing of endolymph and perilymph, which diminishes the electrochemical gradient for potassium and other ions. Explain symptoms that might be associated with Ménière's disease.

Neural Pathways for Equilibrium

At rest, hair cells of the vestibular apparatus are partly depolarized and release some transmitter that communicates to afferents in the vestibular nerve. Therefore, at rest, a low frequency of action potentials is traveling along the vestibular afferents. When a hair cell becomes depolarized because of bending of the stereocilia toward the kinocilium, more transmitter is released, and the frequency of action potentials in the afferent neuron increases. Similarly, when a hair cell becomes hyperpolarized because of bending of the stereocilia away from the kinocilium, less transmitter is released, and the frequency of action potentials in the afferent neuron decreases.

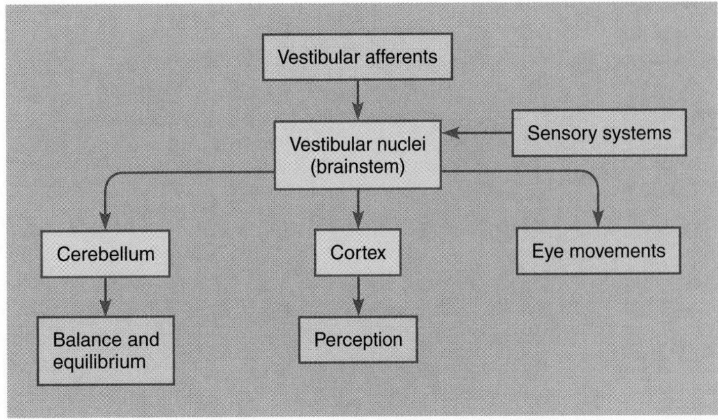

Figure 51 Neural pathways associated with the vestibular apparatus.

The vestibular afferents enter the brainstem as part of the vestibular nerve (**Figure 51**). Most terminate in the *vestibular nuclei*, although some travel directly to the cerebellum to provide immediate feedback for equilibrium and balance, which is important in motor coordination. The vestibular nuclei have projections to different areas of the cortex that allow perception of acceleration, but these pathways are not well understood. The vestibular nuclei are important in feedback control of movement for balance and in controlling eye movements. To carry out these functions, they receive input from various sensory systems, including those involved in vision, somesthetic sensations, and proprioception. This sensory information is used to provide output to motor neurons involved in balance and to match eye movements with body movements. If the eyes and head do not move together, vision will be blurred and motion sickness can occur.

Quick Check 7

1. Of the organs of the vestibular apparatus, which detect linear acceleration?
2. Why do the semicircular canals detect rotation when it starts or stops, but not when it is constant?

6 Taste

Although many of us think of chocolate as nectar of the gods, on another level chocolate is simply a bunch of chemicals mixed in just the right way to stimulate chemoreceptors in the mouth in a way that is pleasurable to most of us. This section describes how we detect the flavor of food.

Anatomy of Taste Buds

We can taste food because chemoreceptors in the mouth respond to certain chemicals in food. The chemoreceptors for taste are located in structures called **taste buds,** each of which contains

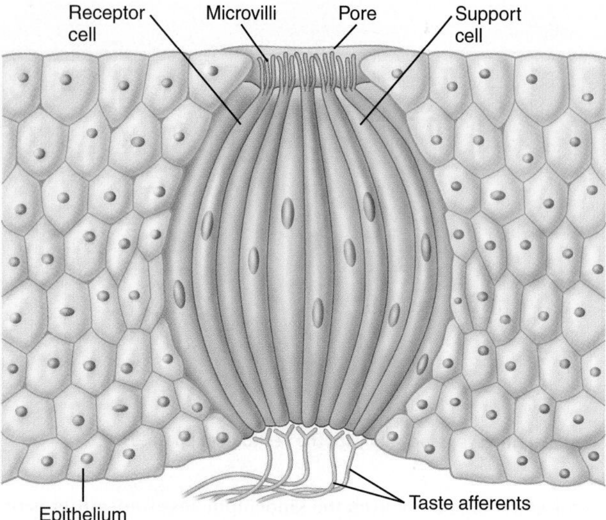

Figure 52 Taste receptor cells in a taste bud. Taste receptors contain microvilli that extend into a pore at the top of the taste bud, where they are exposed to food particles dissolved in saliva.

50–150 receptor cells and numerous support cells (**Figure 52**). At the top of each bud is a pore that allows receptor cells to be exposed to saliva and dissolved food molecules. Each person has more than 10,000 taste buds, located primarily on the tongue and the roof of the mouth, but also located in the pharynx.

Taste receptor cells are modified epithelial cells with microvilli that extend into the pore of a taste bud. Located on the plasma membrane of the microvilli are taste receptors that bind selectively to the different chemicals, called **tastants,** that give food its flavor. For a chemical to bind to the taste receptor, the chemical must dissolve in saliva, which is why a person does not taste food as well when the mouth is dry. The interaction of the tastant with the taste receptor changes the membrane potential of the taste receptor cell.

Signal Transduction in Taste

There are four primary tastes—sour, salty, sweet, and bitter—each requiring its own transduction mechanism. For all taste transductions, the tastant interacts with proteins on the receptor cells to activate a signal transduction mechanism that increases intracellular calcium levels either by depolarizing the cell to open voltage-gated calcium channels or by activating a second messenger system. Calcium triggers the release of a transmitter that communicates with taste afferents. Although not all transmitters released from taste receptor cells have been identified, ATP and serotonin are both known to be taste transmitters.

Salty tastes are caused by the presence of sodium ions in the food and, in turn, these ions' ability to cross the plasma membrane through *amiloride-sensitive sodium channels* (other cations such as lithium can also cross these channels and induce the salty flavor). When sodium levels outside the cell are increased by the consumption of salty foods, the inward electrochemical driving force on sodium ions increases, causing an increased flow of sodium into the cell, which depolarizes it. Voltage-gated calcium channels then open, and calcium triggers the release of a transmitter.

Sour tastes are caused by the presence of acid in the food. Sourness tends to be an unpleasant or aversive stimulus that may protect the body against the consumption of excessive acid. Whether taste receptor cells respond to hydrogen ions dissociated from acidic compounds or organic acids is uncertain; both responses may actually occur. Hydrogen ions are capable of crossing the membrane of taste receptor cells through the same amiloride-sensitive sodium channels responsible for salty tastes. Because hydrogen ions carry a positive charge, this movement would depolarize the receptor cell. However, this phenomenon cannot explain our ability to distinguish salty from sour. In addition, hydrogen ions block certain potassium channels in taste receptor cells, thereby decreasing potassium leakage out of the cell and in turn causing a depolarization that opens calcium channels.

Sweet tastes are caused by the presence of organic molecules such as sucrose. Sucrose and other sweet chemicals (such as fructose, maltose, aspartame, and sucralose [Splenda]) function as ligands that bind to G protein–coupled receptors; however, each binds to a different receptor. Sucrose binds to a G protein that activates the enzyme adenylate cyclase, which in turn catalyzes synthesis of cAMP. cAMP activates a protein kinase that catalyzes phosphorylation of potassium channels, thereby closing the channels. When potassium channels close and potassium leakage out of the cell decreases, the taste receptor cell becomes depolarized, causing calcium channels to open. In contrast, synthetic sweeteners activate a G protein that activates phospholipase C, which in turn catalyzes synthesis of inositol triphosphate. Inositol triphosphate releases calcium from intracellular stores, and the increased cytosolic calcium triggers transmitter release.

Bitter tastes are associated with a wide variety of nitrogen-containing compounds, including some that are toxic; that animals tend to avoid eating things that taste extremely bitter is, therefore, a protective mechanism. There are several different types of bitter substances and more than 30 different types of bitter taste receptors that work via different mechanisms. Some bitter foods contain molecules, such as quinine, that can block potassium channels, thereby decreasing potassium diffusion out of the cell, depolarizing the cell, and opening voltage-gated calcium channels. Calcium then enters the cell and triggers release of a transmitter. Yet other bitter tastes are caused by tastants binding to receptors coupled to G proteins, some of which activate phosphodiesterase, others adenylate cyclase, and yet others phospholipase C. Thus there are several mechanisms by which bitter-tasting substances increase calcium levels in the cytosol.

Researchers have identified another tastant that differs from the primary tastes in that it produces no taste sensation of its own, but instead enhances other tastes: *umami,* which is Japanese for "delicious." Chemicals associated with umami seem to be flavor enhancers. Umami is associated with amino acids—most commonly, glutamate; thus monosodium glutamate (MSG) is often added to food to enhance its flavor. MSG, and other ligands that induce umami, bind to receptors coupled to G proteins. Like bitter tastants, these ligands act through a wide array of responses.

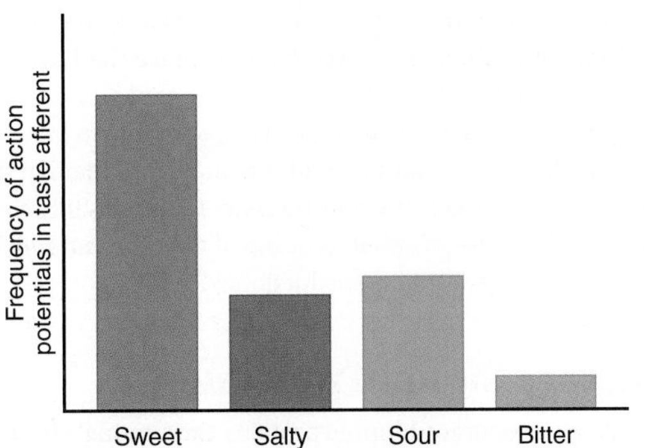

Figure 53 Neural coding of taste. Responses of a single taste afferent to the four primary tastes. This particular taste afferent responds best to sweet tastes.

Approximately 90% of taste receptors respond to more than one primary taste, and most respond to all four. Thus coding of qualities of taste is complex. Furthermore, several receptor cells communicate with a single afferent neuron, complicating the signals transmitted by taste afferents. For example, **Figure 53** shows a plot of the responses of an afferent neuron that responds most strongly to sweet molecules. Therefore, whereas most sensory systems code the quality of the stimulus based on receptor specificity and a labeled-line pathway, taste does not have that specificity. The perception of taste is complex and seems to depend on the pattern of activity in multiple afferent neurons, and it is further complicated by the fact that it depends on the sense of smell.

Neural Pathway for Taste

Taste receptor cells communicate to afferent neurons that travel in three cranial nerves: VII, IX, and X. These taste afferents terminate in the *gustatory nucleus* of the medulla, where they form synapses with second-order neurons. The second-order neurons travel to the ipsilateral thalamus, where they form synapses with third-order neurons that terminate in the *gustatory cortex,* which is near the region of the somatosensory cortex that corresponds to the mouth.

7 Olfaction

The sensation of *olfaction,* or smell, depends on many of the same mechanisms as the sensation of taste. In olfaction, however, specific chemical substances called **odorants** must dissolve in *mucus* if they are to bind to specific chemoreceptors and be smelled.

Anatomy of the Olfactory System

At the roof of the nasal cavity is the organ for smell, the **olfactory epithelium** (**Figure 54**), which has a surface area of approximately 1 cm^2 on each side of the nose. Two cell layers form the epithelium: the olfactory mucosa and the lamina propria. The olfactory mucosa contains several cell types, including the cell body of **olfactory receptor cells** (neurons), **basal cells** that are precursors for the development of new receptors cells, and **sustentacular cells** that maintain the extracellular environment surrounding the receptor cells. Olfactory receptor cells are the only neurons in the body currently known to be replaced regularly. The lamina propria

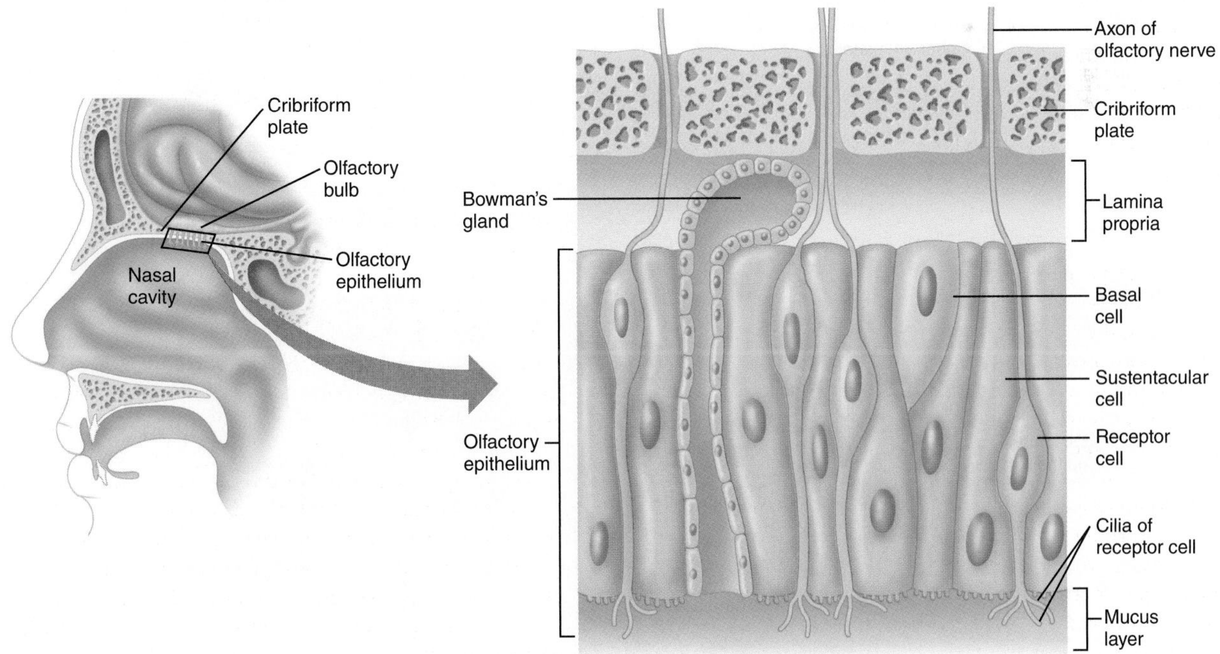

Figure 54 The olfactory epithelium. The olfactory epithelium, located in the nasal cavity, contains three types of cells: receptor cells, supporting cells, and basal cells. Cilia of the receptor cells extend into a layer of mucus in the nasal cavity, whereas axons of the receptor cells ascend to the olfactory bulb of the brain through holes in the cribriform plate of the skull.

contains Bowman's glands, which produce the mucus found in the nasal cavity, and olfactory neuron axons.

The olfactory receptor cells are bipolar neurons with cilia on the dendritic end that project down into the mucus lining the nasal cavity. The cilia, in turn, contain receptors that bind with specific odorant molecules. These cilia are immotile and increase the functional surface area of the olfactory epithelium about ten times on each side of the nose. The axons of olfactory receptor cells enter the CNS through holes (called foramina) in the *cribriform plate* of the skull and terminate in the olfactory bulb.

To perceive odors, odorants must traverse the mucus layer and bind to olfactory receptors. Hydrophilic molecules diffuse through the mucus, but hydrophobic molecules must bind to **olfactory binding proteins** to be transported through the mucus to the cilia with receptors. Odorant molecules can be detected at concentrations as low as 1 part per 10 million (1 odorant molecule per 10 million molecules). However, a higher concentration of odorants reaching the olfactory epithelium produces a stronger sense of smell. Behaviors such as sniffing both increase the flow rate and create turbulent flow that enhances movement of odorants into the mucus, thereby increasing the delivery of odorants to the olfactory epithelium. Some odorant molecules actually originate in the oral cavity and pass through the nasopharynx to the nasal cavity, contributing to the sense of smell. Binding of the odorant molecule to the receptor initiates signal transduction.

Olfactory Signal Transduction

Although many odorant-binding proteins exist, a single basic transduction process seems to apply universally in olfactory receptor cells. Binding of an odorant molecule to a membrane receptor activates a G protein called G_{olf}, which in turn activates the enzyme adenylate cyclase, which then catalyzes the formation of cAMP. Although cAMP typically functions as a second messenger by

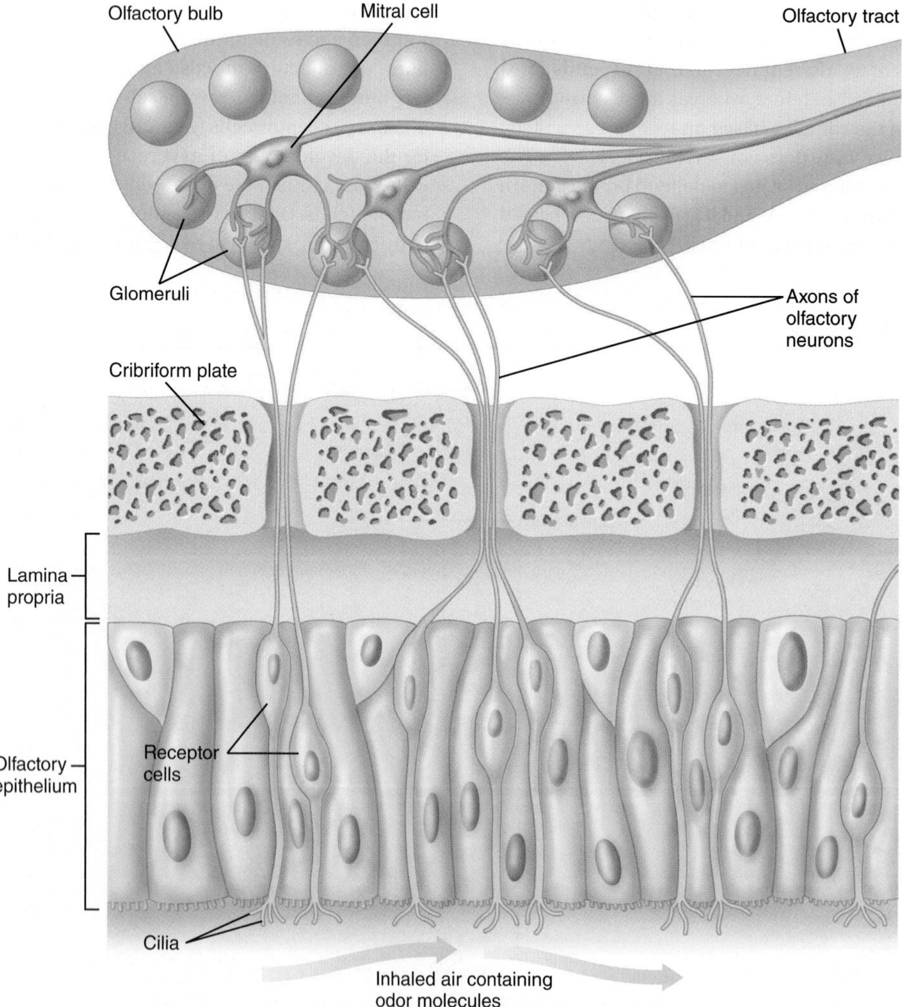

Figure 55 The initial portion of the neural pathway for olfaction. Communication between olfactory afferents and second-order neurons, called mitral cells, occurs in glomeruli of the olfactory bulb.

activating a protein kinase, in olfactory receptor cells cAMP instead binds to cation channels, opening them and allowing both sodium and calcium ions to enter the cell. The primary effect of sodium and calcium entry is depolarization. However, the entry of calcium into the cell also causes chloride channels to open, allowing chloride to move out of the cell and increasing depolarization of the receptor cell. If the depolarization is great enough, action potentials are generated on the axon of the receptor cell.

Although smell is one of the most sensitive sensations, it is also a rapidly adapting sensation. Current theories suggest that the same calcium used to depolarize the cell to generate action potentials also inhibits cAMP-gated cation channels.

Olfaction is a very acute sensation, such that most people have the ability to distinguish between millions of smells. Although the precise mechanism by which the olfactory system codes for different smells is unknown, at least 1000 genes have been identified that code for olfactory receptors; each olfactory receptor cell is believed to express only one of these genes. Thus one receptor cell responds to only one type of odorant. Given the anatomical differences in the neurons transmitting different qualities of olfaction, this separation can be maintained through the neural pathway for olfaction, described next.

Neural Pathway for Olfaction

Unlike other special sense receptors, olfactory receptors are specialized endings of afferent neurons, not separate cells. The axons of the afferent neurons make up the **olfactory nerve** (cranial nerve I). If a receptor cell is depolarized to threshold, action potentials are produced and transmitted to the axon terminal in the brain. **Figure 55** illustrates a portion of the neural pathway for olfaction. The axons terminate in an area of the brain called the olfactory bulb, where they communicate with second-order neurons called *mitral cells.* Numerous olfactory neurons communicate with a single mitral cell, and these synapses occur in clusters called *glomeruli.* There are approximately 8000 glomeruli in each olfactory bulb, and a single glomerulus receives all the olfactory input from a specific odorant. The smell ultimately perceived depends on the pattern of activation of olfactory receptor neurons.

The axons of the mitral cells form the *olfactory tract,* which is the neural pathway to an area of the brain called the *olfactory tubercle.* The neural pathways from the olfactory tubercle terminate in two areas of cerebral cortex: the olfactory cortex for the perception and discrimination of smells, and the limbic system for triggering olfactory-driven behaviors, such as sexual behavior. The pathway to the olfactory cortex starts with a synapse in the olfactory tubercle with another neuron. Most of these neurons travel directly to the olfactory cortex. Olfaction is the one sensory system whose pathway to the cortex does *not* include a synapse in the thalamus. However, recent studies have identified odor-responsive neurons in the thalamus that might function as a relay to the olfactory cortex. Other second-order neurons do not synapse in the olfactory tubercle but instead travel directly to the limbic system.

Quick Check 8

1. Name the four primary tastes.
2. What is the difference between a taste bud and a taste receptor?
3. Name the three types of cells found in the olfactory epithelium, and describe their functions.
4. Give three similarities between the systems responsible for taste and smell.

SYSTEMS INTEGRATION The body cannot function without constant feedback. That feedback comes in the form of information about the external environment transmitted by the sensory systems described in this chapter and information about the internal environment transmitted by visceral afferent systems that function in a manner similar to the sensory systems. For example, body temperature cannot be regulated without thermoreceptors to detect that body temperature is not normal. A person is less likely to protect himself or herself from tissue-damaging stimuli without the perception of pain. In addition, our ability to find food and water depends on our senses of vision and olfaction. Many more examples exist of how we use our sensory systems on a daily basis to keep our bodies healthy.

CHAPTER REVIEW

SUMMARY

1 General Principles of Sensory Physiology

- Our ability to perceive the world around us depends on the presence of sensory receptors and specific neural pathways to communicate information to the cerebral cortex.
- Stimulus type is coded by the receptor and the pathway activated.
- Stimulus intensity is coded by frequency coding and population coding.
- The ability to locate a stimulus depends on the size of receptive fields, the degree of overlap of receptive fields, and lateral inhibition.

2 The Somatosensory System

- The somatosensory system enables perception of stimuli associated with the body surface (somesthetic sensations) or the body position (proprioception).
- Some receptors in the somatosensory system are specialized nerve endings, whereas others are free nerve endings.
- Information about touch, pressure, vibration, and proprioception is transmitted to the thalamus through the dorsal column–medial lemniscal pathway.

- Information about pain and temperature is transmitted to the thalamus through the spinothalamic tract.
- Information from the thalamus is transmitted to the primary somatosensory cortex.

3 Vision

- The eyes function in transduction of light energy.
- To focus light on the retina, the eye can contract and relax the ciliary muscles to change lens shape, thereby altering its refractive power.
- Several clinical defects affect the ability of the eyes to focus light, including myopia, hyperopia, astigmatism, presbyopia, cataracts, and glaucoma.
- The pupils of the eye are capable of constricting or dilating to regulate the amount of light that enters the eye.
- Phototransduction occurs in the retina, which consists of neural tissue forming three layers: photoreceptors, bipolar cells, and ganglion cells.
- The photoreceptors—rods and cones—contain photopigments that absorb light.
- In the dark, photoreceptors are depolarized and release a chemical transmitter.
- In the light, the photopigments absorb the light and dissociate, causing a chain of reactions leading to hyperpolarization of the photoreceptor and decreased release of the transmitter.
- Photoreceptors communicate to bipolar cells and horizontal cells by releasing glutamate. Bipolar cells then communicate to ganglion cells and amacrine cells. If the ganglion cell is depolarized to threshold, then an action potential results.
- The axons of ganglion cells make up the optic nerve.
- Bipolar cells and ganglion cells have complex receptive fields; the responses of the cells vary based on whether the light stimulus is in the center of the receptive field or in the surround (outer edges) of the receptive field.
- Information is transmitted from the optic nerve to the optic chiasm, where half of the axons from each eye cross to the opposite side of the CNS, with all input from the right visual field ending up on the left side, and all information from the left visual field being sent to the right side.
- The axons of ganglion cells after the optic chiasm make up the optic tract.
- The optic tract terminates in the lateral geniculate body of the thalamus, where the ganglion cell axons communicate with neurons that transmit information to the visual cortex.

4 The Ear and Hearing

- The ear, which is made up of the external ear, middle ear, and inner ear, contains the receptor cells for two sensory systems, hearing and equilibrium.
- For hearing, sound waves must enter the external ear, be amplified in the middle ear, and then be transduced into neural impulses in the cochlea of the inner ear.
- Sound transduction occurs in hair cells of the organ of Corti, located in the cochlea.
- When sound waves reach the fluid-filled cochlea, they set up waves that cause movement of the basilar membrane. This movement causes the stereocilia to bend, which in turn causes the opening or closing of potassium channels and changes the electrical properties of the hair cells.
- When a hair cell is depolarized, a chemical transmitter is released and communicates with the afferent neurons in the cochlear nerve.
- Sound information is transmitted to the medial geniculate body of the thalamus and then to the auditory cortex.
- The auditory cortex contains a tonotopic map.
- The loudness of a sound is coded by the degree of bending of the stereocilia, whereas the pitch of sound is coded for by the location of hair cells on the basilar membrane.

5 The Ear and Equilibrium

- The vestibular apparatus of the inner ear includes the semicircular canals (for detecting rotation) and the utricle and saccule (for detecting linear acceleration).
- The vestibular apparatus contains hair cells with stereocilia that bend with acceleration of the head. In the semicircular canals, the hair cells are located in the ampulla.
- Hair cells in the utricle and saccule have stereocilia that extend into a gelatinous mass containing otoliths.
- The bending of the stereocilia opens or closes ion channels, thereby influencing the release of a chemical transmitter that communicates with afferent neurons in the vestibular nerve.
- Vestibular information is transmitted to the vestibular nuclei of the brainstem, which communicate to the thalamus and then the cortex for perception of equilibrium.
- Vestibular nuclei also communicate to the cerebellum for maintaining balance, and to brainstem nuclei that regulate eye movement.

6 Taste

- Both chemical senses—taste (gustation) and smell (olfaction)—depend on the binding of specific chemicals in food or the air to chemoreceptors on receptor cells.
- Taste receptor cells are located within taste buds.
- Molecules that bind to taste receptors are called tastants.
- Each of the four primary tastes utilizes a different transduction mechanism.
- A single taste receptor cell responds to all four primary tastes, but most strongly to only one such taste.
- Taste-related information is first transmitted via cranial nerves to the gustatory nucleus of the medulla, then relayed through the thalamus to the gustatory cortex.

7 Olfaction

- Olfactory receptors, located in the olfactory epithelium of the nasal cavity, respond to odorants dissolved in the mucus found there.
- Basal cells in the olfactory epithelium are precursors for receptor cells.
- An odorant must bind to olfactory-binding proteins in the mucus so that it can bind to olfactory receptors.
- Upon binding to receptors, odorants trigger the production of cAMP in the cytosol of the receptor cell, which through a series of steps depolarizes the cell.
- If depolarized to threshold, an action potential is transmitted along the axon of the receptor cell.
- Axons of the receptor cells, which together constitute the olfactory nerve, form synapses with mitral cells in glomeruli of the olfactory bulb.
- Mitral cells transmit information along two pathways, one terminating in the olfactory cortex and the other in the limbic system.

Check your understanding at MasteringA&P and apply your new knowledge to real-life case studies to ensure that you've mastered the content!

EXERCISES

Multiple-Choice Questions

1. The strength of a stimulus is coded by
   a) The size of the receptor potential.
   b) The size of the action potentials.
   c) The frequency of action potentials.
   d) Both a and c.
   e) All of the above.

2. The mechanism by which a receptor converts a stimulus into an electrical signal is called
   a) Conduction.
   b) Convection.
   c) Transduction.
   d) Modulation.
   e) Propagation.

3. In lateral inhibition,
   a) The nervous system produces contrast to emphasize more-important information over less-important information.
   b) Afferent neurons with neighboring receptive fields inhibit each other's communication to second-order neurons.
   c) The ability to locate the site of a stimulus is enhanced.
   d) Both a and c.
   e) All of the above.

4. Which of the following observations best illustrates the concept of the labeled line?
   a) When a boxer gets punched in the eye, he perceives light.
   b) Rotation of the head stimulates certain receptors in the vestibular system but not those in the visual system.
   c) Information from different photoreceptors converges on a single ganglion cell that projects to the lateral geniculate nucleus.
   d) Hair cells in the cochlea are stimulated by sound vibrations over a wide range of frequencies.

5. Which of the following best illustrates the concept of an adequate stimulus?
   a) When a boxer gets punched in the eye, he perceives light.
   b) Rotation of the head stimulates certain receptors in the vestibular system but not those in the visual system.
   c) Information from different photoreceptors converges on a single ganglion cell that projects to the lateral geniculate nucleus.
   d) Hair cells in the cochlea are stimulated by sound vibrations over a wide range of frequencies.

6. Rubbing a sore area can decrease the sensation of pain by
   a) Activating the endogenous analgesia systems.
   b) Referring the pain to another area of the body.
   c) Activating larger-diameter afferents, which activate an inhibitory interneuron, which inhibits the second-order neurons for pain.
   d) Decreasing the number of action potentials in nociceptor afferents.
   e) Presynaptic inhibition of substance P release.

7. In the dorsal column–medial lemniscal pathway,
   a) Proprioception information is transmitted to the brain.
   b) The first-order neuron communicates to the second-order neuron in the dorsal horn of the spinal cord.
   c) The pathway crosses to the contralateral side in the spinal cord.
   d) Both a and c.
   e) All of the above.

8. Which of the following is the correct name of the pathway from the retina to the optic chiasm?
   a) Optic tract
   b) Optic radiations
   c) Optic nerve
   d) Optic disk

9. Which of the following is the correct name of the pathway from the lateral geniculate nucleus of the thalamus to the visual cortex?
   a) Optic tract
   b) Optic radiations
   c) Optic nerve
   d) Optic chiasm
   e) Optic disk

10. Where would you expect to find the ascending tracts for somatosensory information?
    a) In the white matter of the spinal cord
    b) In a spinal nerve
    c) In the gray matter of the spinal cord
    d) None of the above

11. The ability to perceive different frequencies in sound vibrations is based on the fact that
    a) The stereocilia of any given hair cell respond to only one frequency.
    b) Different areas of the basilar membrane resonate at different frequencies, such that sound of a particular frequency causes only a certain region of the membrane to vibrate.
    c) The frequency of action potentials in the cochlear nerve varies in proportion to the frequency of a sound stimulus.

12. The stereocilia for hearing are exposed to
    a) Endolymph in the scala vestibuli.
    b) Perilymph in the scala vestibuli.
    c) Endolymph in the scala media.
    d) Perilymph in the scala media.
    e) Endolymph in the scala tympani.

13. The parasympathetic nervous system causes
    a) Contraction of the radial muscle of the iris.
    b) Contraction of the ciliary muscle.
    c) Pupillary dilation.
    d) Both a and c.
    e) All of the above.

Objective Questions

14. The two types of thermoreceptors are _______ and _______.
15. Receptors are most sensitive to energy from the _______ stimulus.
16. A phasic receptor adapts (quickly/ slowly) to a constant stimulus.
17. The three types of nociceptors are _______, _______, and _______.
18. Information about touch detected on the left side of the body is transmitted to the brain in the dorsal columns on the _______ side of the spinal cord.
19. When a photopigment absorbs light, cGMP levels (increase/decrease).
20. The first neurons that support production of action potentials in the visual pathway are (photoreceptors/bipolar cells/ganglion cells).
21. The pitch of sound vibration reflects its (amplitude/frequency).
22. A hair cell in the cochlea can be excited by sounds of different frequencies. (true/false)
23. The process by which the lens becomes stronger for close-up vision is called _______.
24. Rods and cones differ with regard to the type of (retinal/opsin) they contain.

25. A single ganglion cell will either be excited or inhibited by light applied to its visual field. (true/false)
26. The visual cortex on the left side of the brain receives information from the right eye only. (true/false)
27. Odorant molecules must be dissolved in mucus if they are to bind to olfactory receptors. (true/false)
28. A given taste receptor cell responds to only one of the four primary tastes. (true/false)

Essay Questions

29. Compare the response of rapidly and slowly adapting touch receptors when you place your hand on a vibrating speaker.
30. Explain the concepts of topographic organization of the cerebral cortex. Compare the topographic organization of the somatosensory cortex, the visual cortex, and the auditory cortex.
31. Explain how it is possible for one person's perception to differ from another person's perception.
32. Diagram the general sensory pathway for transmitting information from a receptor to the cortex. Try fitting each sensory system into the general pathway. Which sensory systems do not exactly fit this scheme?
33. Describe the sequence of events that occurs when photoreceptors are exposed to light.
34. Make a list of similarities between the olfactory (smell) and gustatory (taste) systems.
35. Make a list of the different types of sensory receptors in your body that are being activated at this moment.

Critical Thinking

36. A person goes to the hospital complaining of pain in the lower right abdomen. What might be the ailment? Which neural pathway is responsible for the sensation of pain?
37. Explain why you are normally unaware of your blind spot.
38. Stare at a bright light for 10 seconds and then stare at a white sheet of paper. What do you observe? Explain.
39. If a person suffers damage to the left side of the thoracic spinal cord (damaging all the white matter on that side), which somatic sensations would be lost and which limbs would be affected, and on which side of the body?
40. Which sensory inputs are generally affected by a severe cold? Explain.

Photo credits for icons in feature boxes: Focus on Diabetes: Tanyae/Dreamstime; Clinical Connections: Iofoto/Dreamstime; Understanding Exercise: Pixattitude/Dreamstime; Discovery: Yuri Arcurs/Cutcaster.

Glossary

accommodation reflex adjustments of the eye to view near objects; includes rounding of the lens, constriction of the pupil, and posterior movement of the lens

adequate stimulus the energy form or stimulus type to which a sensory receptor responds best

acuity in sensory systems, a measure of the precision of perception

ampulla enlargement at the base of each semicircular canal; contains the hair cells for detecting angular acceleration

analgesia absence of pain perception

anterior segment chamber of the eye in front of the lens and ciliary muscle

auditory cortex portion of the temporal lobe of the brain that processes auditory information

basal cells in the olfactory system, precursor cells for development of new olfactory receptor cells

basilar membrane membrane in the cochlea of the inner ear that separates the scala tympani from the scala media

blind spot anatomically, the optic disk of the retina where there are no photoreceptors; physiologically, the visual field where light strikes the optic disk and, therefore, cannot be detected

choroid middle layer of the posterior two-thirds of the eye; contains pigment that absorbs light

ciliary muscles smooth muscles in the eye attached to the lens by zonular fibers; regulate the curvature of the lens for focusing light

circular muscle inner layer of smooth muscle of the iris; also called *constrictor muscle*

cochlea a spiral-shaped structure in the inner ear that contains the receptor cells for hearing

cold receptors thermoreceptors that increase in responsiveness as the temperature decreases

cornea transparent structure at the front of the eye that allows light waves to enter

cones photoreceptors that enable visibility during relatively bright light and are responsible for color vision

cupula gelatinous area within the ampulla of the inner ear; contains hair cells that detect angular acceleration

dorsal column–medial lemniscal pathway an ascending tract that transmits sensory information from touch, pressure, vibration, and proprioceptors to the thalamus

endogenous analgesia systems neural pathways involved in blocking the transmission of pain signals to the brain, thereby producing analgesia

endolymph fluid found in the scala media of the cochlea in the inner ear

eustachian tube (you-stay-shun) a canal that connects the middle ear with the pharynx and allows equilibration of air pressure across the eardrum

emmetropia normal visual acuity

external auditory meatus ear canal

fast pain a sharp, pricking sensation that can be easily localized and is produced by activation of nociceptors; transmitted by AΔ fibers

free nerve endings somatosensory receptors in the skin that lack identifiable specialized sensory structures; includes some mechanoreceptors, thermoreceptors, and nociceptors

fovea (foh-vee-ah) central point on the retina of the eye, in which visual acuity is greatest

generator potential a change in membrane potential; also known as *receptor potential*

hair cells cells with stereocilia; receptor cells for hearing and equilibrium

hyperopia (hi-per-oh-pee-ah) common visual defect of the eye causing farsightedness

iris pigmented smooth muscle in the eye that sits in front of the lens and regulates the diameter of the pupil to control the amount of light entering the eye

kinocilium large stereocilia projecting from the receptor cells for equilibrium; the direction of bending in response to acceleration of the body determines the direction of receptor potentials in receptor cells

lateral geniculate body nucleus of the thalamus that transmits auditory information to the auditory cortex

lateral inhibition process during which a stimulus that strongly excites receptors in a certain location inhibits activity in the afferent pathways of other receptors located nearby

law of specific nerve energies states that a given sensory receptor is specific for a particular energy form or stimulus type

lens clear object in the eye that focuses light on the retina

macula lutea a depression in the center of the retina that contains the fovea; contains a high density of cones

mechanoreceptors sensory receptors that detect physical forces such as pressure or vibration

modality in the sensory system, the energy form of a stimulus

myopia (my-oh-pee-ah) common visual defect of the eye causing nearsightedness

nociceptors (noh-sih-sep-tors) sensory receptors on the surface of the body that detect tissue-damaging stimuli; include mechanical nociceptors that respond to intense mechanical stimuli, thermal nociceptors that respond to intense heat, and polymodal nociceptors

optic chiasm (ki-azm) portion of the visual neural pathway where some axons from both the right and left optic nerves cross to the opposite side of the brain

optic disk point on the retina where the optic nerve and blood vessels that supply the eye join the retina

optic nerve the nerve that transmits visual information from the eye to the optic chiasm; contains axons of retinal ganglion cells

optic tract tract containing axons of retinal ganglion cells that transmit visual information from the optic chiasm to the lateral geniculate nucleus of the thalamus

optic radiations pathways from the lateral geniculate body to the visual cortex on either side

ossicles three bones in the middle ear that transmit sound vibrations from the eardrum to the cochlea; include the malleus, incus, and stapes

oval window membrane between the middle and inner ear; transmits sound waves from the middle to inner ear

organ of Corti the sensory organ for sound; located on top of the basilar membrane in the cochlea of the inner ear

otoliths small calcium carbonate crystals within the gelatinous material found in the utricles and saccules

odorants chemical substances that must bind to specific chemoreceptors in order to be smelled

olfactory epithelium the organ for smell within the nasal cavity

olfactory receptor cells neurons that respond to olfactants

olfactory binding protein protein in the nasal mucus that binds olfactants and transports them to olfactory receptors

olfactory nerve the nerve that contains axons of olfactory receptor cells; cranial nerve I

perception conscious awareness

proprioception (pro-pree-oh-cep-shun) the perception of the positions of the limbs and the body

phasic receptors receptors that decrease in responsiveness to a continuing stimulus; also called *rapidly adapting receptors*

pupil hole through which light can enter the eye

perilymph fluid found in the scala vestibuli and scala tympani of the cochlea in the inner ear

photopigment the molecule in the photoreceptors that absorbs light, the first step of phototransduction

radial muscle outer layer of smooth muscle in the iris; also called *dilator muscle*

rapidly adapting receptors receptors that decrease in responsiveness to a continuing stimulus; also called *phasic receptors*

receptor potential graded potential caused by the opening or closing of ion channels on sensory receptors, and triggered by sensory stimuli

receptive field the area over which an adequate stimulus can produce a response, either excitatory or inhibitory, in an afferent neuron or higher-order neurons

refraction bending of light waves as they travel through media of different densities at an angle other than perpendicular

round window membrane between the middle and inner ear; dissipates sound waves in the inner ear

rods photoreceptors that enable visibility during relatively low light and are responsible for black-and-white vision

referred pain the perception of a painful stimulus as originating at a site on the body distinct from the location of the stimulus

somatosensory system (suh-mat-uh-sen-suh-ree) branch of the nervous system associated with perception of somatic sensations; associated with receptors in the skin and proprioception

somesthetic sensations sensations that arise from receptors in the skin

special senses senses of vision, olfaction, taste, hearing, and equilibrium

sensory receptors specialized neuronal structures that detect a specific form of energy in either the internal or external environment

slowly adapting receptors receptors that maintain responsiveness to a continuing stimulus; also called *tonic receptors*

sensory unit a single afferent neuron and all sensory receptors associated with it

spinothalamic tract somatosensory pathway that transmits information from thermoreceptors and nociceptors to the thalamus

slow pain a poorly localized, dull, aching sensation produced by activation of nociceptors; transmitted by C fibers

sclera (sklee-rah) tough connective tissue that makes up the white of the eye

scala vestibuli fluid-filled duct in the cochlea; also called *vestibular duct*

scala tympani fluid-filled duct in the cochlea; also called *tympanic duct*

scala media fluid-filled duct in the cochlea; also called *cochlear duct*

stereocilia (ster-ee-oh-sil-ee-ah) hairlike projections on the upper surfaces of hair cells in the inner ear that move in response to sound vibrations or acceleration of the head

semicircular canals structures in the inner ear that contain the receptor cells for rotational acceleration

saccule (sak-yool) structure of the inner ear; detects up and down linear acceleration

transduction conversion of the energy form of a stimulus into an electrical signal in the form of changes in membrane potential

tonic receptors receptors that maintain responsiveness to a continuing stimulus; also called *slowly adapting receptors*

thermoreceptors sensory receptors that detect temperature; include peripheral thermoreceptors that detect skin temperature and central thermoreceptors that detect core body temperature

tympanic membrane eardrum

taste buds structures on the tongue, roof of the mouth, and pharynx that contain taste receptor cells

tastants chemical substances that give foods their flavors

tectorial membrane (tek-tor-ee-al) membrane in the organ of Corti in which the tips of stereocilia are embedded

utricle structure of the inner ear; detects forward and backward linear acceleration

vestibulocochlear nerve nerve that contains the afferents for hearing and equilibrium; cranial nerve VIII

vestibular membrane membrane in the cochlea of the inner ear that separates the scala vestibuli from the scala media

vestibular apparatus structures of the inner ear that contain the receptor cells for equilibrium, including the semicircular canals, utricle, and saccule

vitreous chamber cavity of the eye posterior to the lens and ciliary muscle

vitreous humor jelly-like material found in the vitreous chamber of the eye; maintains the spherical structure of the eye

warm receptors thermoreceptors that increase responsiveness with an increase in temperature

zonular fibers strands of connective tissue in the eye that connect the ciliary muscles to the lens and are involved in adjusting the shape of the lens to focus light

Answers

Figure Question Answers

Fig 4 Visual cortex: occipital lobe; auditory cortex: temporal lobe; somatosensory cortex: parietal lobe

Fig 49 Microtubules

Apply Your Knowledge Answers

Damage of the dorsal columns affects the somatic sensations of touch, pressure, and proprioception. Because this pathway is ipsilateral until it reaches the brainstem, damage to the left spinal cord causes a lack of sensation on the left side of the body. Because input from the arms is above the level of damage (cervical spinal cord versus thoracic damage), only the legs are affected. Thus damage to the left thoracic dorsal columns of the spinal cord causes a lack of touch, pressure, and proprioception from the left legs.

Damage to the spinothalamic tract affects the somatic sensations of temperature and pain. Because this pathway crosses to the contralateral spinal cord with a couple of segments at the level of input to the spinal cord, damage to the left spinal cord causes lack of sensation on the right side of the body. Because input from the arms is above the level of damage (cervical spinal cord versus thoracic damage), only the legs are affected. Thus damage to the left thoracic spinothalamic tract of the spinal cord causes a lack of temperature and pain sensation from the right legs.

When a flash picture is taken with a camera that corrects for "red eye," the initial flash of light causes pupillary constriction. When the next flash occurs in sync with taking the picture, the now-constricted pupil allows less light to enter the eye and be reflected back to the camera, thereby minimizing the "red eye" phenomenon.

Damage to the right optic nerve would cause loss of vision from the right eye. Because the left eye perceives part of the right visual field, only the right temporal visual field would be lost. However, with no bilateral input going to the brain, depth perception would be decreased. Damage to the right optic tract would cause loss of input from the left nasal retina and right temporal retina, thereby resulting in loss of the left visual field. Damage to the right primary visual cortex would also result in loss of perception of the left visual field.

Because the membranous labyrinth is a functional part of both the cochlea and the vestibular apparatus, Meniere's disease can result in problems with hearing and equilibrium, often involving an excess of neural activity. Symptoms include tinnitus (ringing in the ears) and dizziness.

Answers to End-of-Chapter Multiple Choice and Objective Questions

Multiple Choice Questions

1. d 2. c 3. e 4. a 5. b 6. c 7. a 8. c 9. b 10. a 11. b 12. c 13. b

Objective Questions

14. warm, cold
15. adequate
16. quickly
17. mechanical, thermal, polymodal
18. left
19. decrease
20. ganglion cells
21. frequency
22. false
23. accommodation
24. opsin
25. false
26. true
27. true
28. false

Muscle Physiology

Muscle Physiology

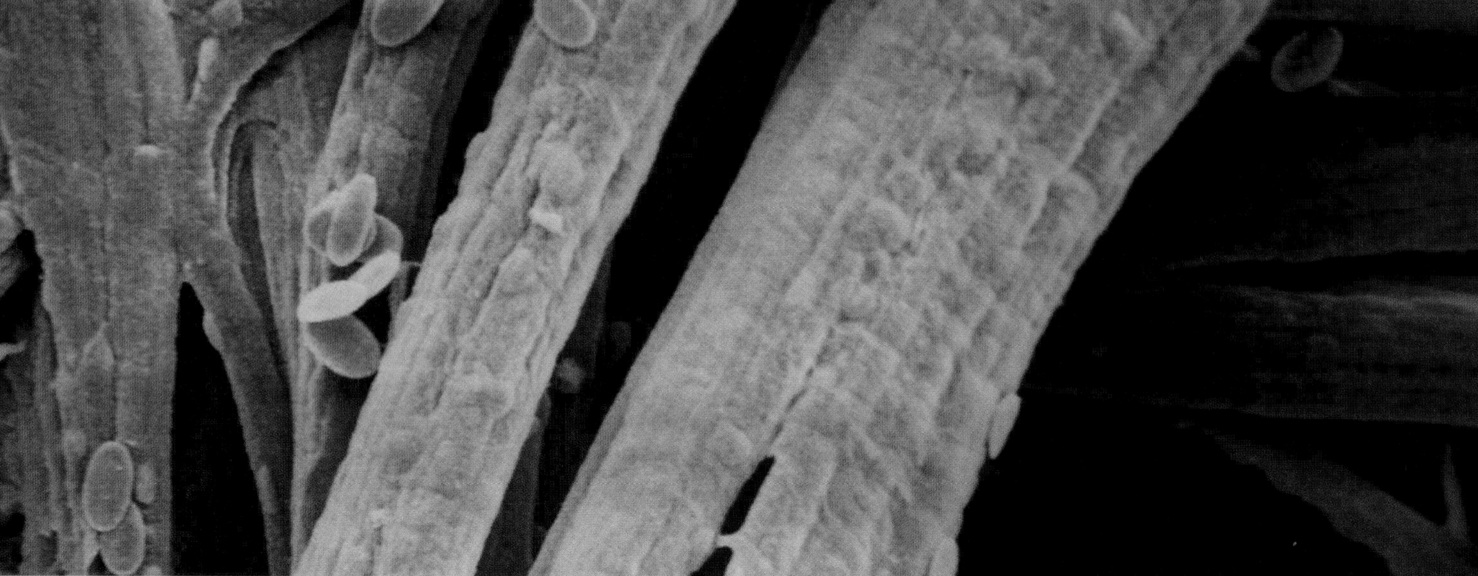

Asa Thoresen/Photo Researchers, Inc.

Scanning electron micrograph (SEM) of cardiac muscle cells.

A man named Luigi Galvani taught anatomy at the University of Bologna in Italy and served as professor of obstetrics in the Institute of Arts and Sciences from 1765 to 1797. In his teachings, he used frog dissections. In the process of preparing a dead frog, he would hang it on a copper hook and then suspend the frog on an iron wire. He did this during a thunderstorm and saw the freshly dead frog contract its muscles. This suggested to Galvani that frogs, and other animals, had their own form of electricity and that this "animal electricity" caused muscle contraction.

In further experiments, Galvani's assistant discovered that nerve tissue activated muscle tissue. While the assistant was dissecting a frog, he used a scalpel that was sitting near a machine that generated electricity. When he touched the scalpel to the sciatic nerve, the muscle contracted. Galvani concluded that electrostatic charge on the scalpel excited the nerve, and that nerves provide the electricity that causes muscle to contract. Although this is not precisely true, as you will learn in this chapter, Galvani's work led to the field of *electrophysiology,* the study of electrical activity in animals (or plants).

CHAPTER OUTLINE

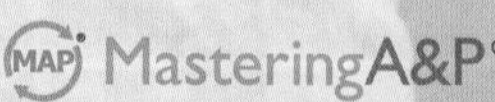

LEARNING OUTCOMES *After studying this chapter you should be able to:*

- Name the major structural features of a skeletal muscle cell, and briefly describe the relationship between the structure and the function of each structural feature.
- Describe the sequence of events that occurs in the crossbridge cycle, and relate this sequence to the sliding-filament model of muscle contraction.
- Identify the various factors that affect the force of muscle contraction.
- Name the three types of skeletal muscle fibers, and describe the major differences among them.
- Describe the major characteristics of smooth and cardiac muscle, and compare these muscle types to skeletal muscle.

Before You Begin

Make sure you have mastered the following topics:

- Neuromuscular junction
- Substrate-level and oxidative phosphorylation
- Lactic acid production
- Calmodulin

The effector organ of the somatic nervous system is skeletal muscle, whereas smooth and cardiac muscle are effector organs of the autonomic nervous system. In this chapter, you will learn about these muscle types and how they, like neurons, are excitable tissues that generate action potentials. However, action potentials in muscle tissue have a different effect from those in neurons: They induce contractions that generate a force.

The contractile machinery of muscle consists of proteins, and these proteins must interact with both ions and other proteins. Muscle is high-energy-demanding tissue, and the concepts of cell metabolism apply to muscle cells, with certain muscle cells depending more on oxidative phosphorylation for energy and others depending more on anaerobic glycolysis. Muscle must also respond quickly to commands to contract and, therefore, is under control of the nervous system.

To perform their jobs properly, the muscles that move the body must be able to respond faithfully and quickly to commands from the nervous system. Indeed, a typical skeletal muscle cell can activate its contractile machinery within milliseconds of receiving a neural signal, and it can turn off the machinery nearly as quickly. It is this quickness of response that enables us to perform complicated motions.

We begin our study of muscle physiology by examining muscle anatomy, moving from the gross anatomical level to the molecular level, with special emphasis on the structures that generate and regulate contractile force. Next, we investigate contractile and regulatory mechanisms to see how they work. Although we concentrate on *skeletal muscle,* most of the basic principles apply to each of the three muscle types found in the body. At the end of the chapter, we consider the special properties of the other two muscle types—*smooth muscle* and *cardiac muscle.*

1 Skeletal Muscle Structure

With few exceptions, **skeletal muscles,** such as the *biceps* of the arm, are connected to at least two bones. The exceptions to this rule include certain skeletal muscles that are connected to the skin (as is the case with some facial muscles), to cartilage (for example, muscles of the larynx), or to other muscles (the external anal sphincter, for instance). Muscles are connected to bones by **tendons,** cords of elastic connective tissue that transmit force from the muscle to the bone.

Structure at the Cellular Level

The part of the muscle that generates force is called the *body;* it constitutes the "meaty" part of the muscle (**Figure 1**). The connective tissue of the tendons is continuous with the connective tissue that surrounds the body of the muscle, called the *epimysium.* More connective tissue, called *perimysium,* extends into the body, dividing it into numerous bundles (called **fascicles**) of individual muscle cells. Each fascicle contains hundreds to thousands of muscle cells, which are called **muscle fibers** because of their elongated shape. Each muscle fiber runs the full length of the muscle (often positioned at a diagonal) and is encased in a thin sheath of connective tissue, called *endomysium.* Unlike most cells, which have a single nucleus, muscle fibers have many because each muscle fiber is formed during embryonic life from the fusion of several cells. These nuclei lie immediately below the muscle fiber's plasma membrane, which is called the **sarcolemma.**

A muscle fiber's semifluid cytoplasm, called sarcoplasm, is packed with mitochondria and hundreds of banded, rodlike elements called **myofibrils,** which contain the fiber's contractile machinery (**Figure 2**). Each myofibril comprises a bundle of overlapping thick and thin filaments made of the proteins *myosin* and *actin,* respectively. A saclike membranous network called the **sarcoplasmic reticulum** surrounds each of the myofibrils and is closely associated with other structures called **transverse tubules (T tubules),** which are continuous with the sarcolemma and penetrate into the cell's interior. Near the T tubule, the sarcoplasmic reticulum has enlargements called **lateral sacs** or **terminal cisternae,** which store calcium. Each T tubule is associated with two lateral sacs, forming a **triad.** The sarcoplasmic reticulum and T tubules play important roles in the activation of muscle contractions: They help transmit signals from the sarcolemma to the myofibrils, enabling a muscle cell to respond to neural input. The function of the sarcoplasmic reticulum is to store calcium ions (Ca^{2+}). As we will see, these calcium ions are released in response to electrical signals

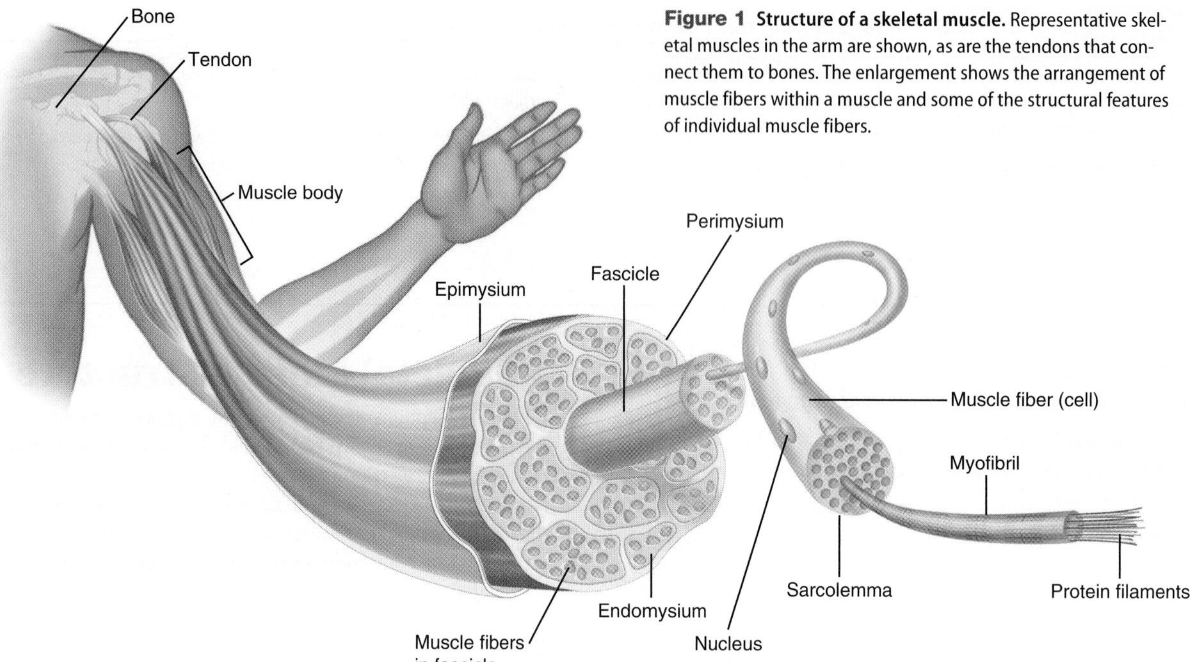

Figure 1 Structure of a skeletal muscle. Representative skeletal muscles in the arm are shown, as are the tendons that connect them to bones. The enlargement shows the arrangement of muscle fibers within a muscle and some of the structural features of individual muscle fibers.

that travel from the sarcolemma to the T tubules, and they serve as chemical messengers that carry these signals to the myofibrils to initiate contraction.

Structure at the Molecular Level

When viewed under a microscope, skeletal muscle cells have a striped appearance; for this reason, such muscle (and also cardiac muscle) is referred to as *striated muscle.* A close-up view shows that these striations are created by the orderly arrangement of protein fibers in the myofibrils called **thick filaments** and **thin filaments,** which run parallel to the muscle cell's long axis. The thick and thin filaments exist in a 2:1 ratio. Myofibrils are composed of a fundamental unit called a **sarcomere** that repeats over and over (**Figure 3**). Each sarcomere is bordered on either end by *Z lines,* which run perpendicular to the long axis and anchor the thin filaments at one end. The thick filaments in a sarcomere are connected by *M lines,* which also run perpendicular to the long axis.

Before the structure of the sarcomere was elucidated and its protein filaments were identified, early investigators identified certain regions of the muscle fiber based on the banding pattern as seen under a light microscope (Figure 3). The dark striation observed under the microscope is due to the presence of thick filaments and is called the **A band.** In the center of the A band is a region that is lighter than the sides because only thick filaments are present; that is, there are no thin filaments overlapping the thick filaments here. This region is called the **H zone.** The light striation comprises areas where there are thin filaments with no overlap with thick filaments; this region is called the **I band.** The Z line that connects thin filaments is located in the center of the I band.

The thin and thick filaments of the sarcomere are made up of two proteins called **actin** and **myosin,** respectively, which are referred to as *contractile proteins* because they constitute the machinery that generates contractile force. The thick and thin filaments are made up of structures arranged in an orderly, repeating fashion.

The basic components of each thin filament are actin monomers called G actin ("G" because they are globular proteins), each of which has a myosin-binding site. As we will see, the ability of actin and myosin to bind together under certain conditions is critical to a muscle's ability to generate force. G actins are linked together end to end, like pearls in a necklace, to form strands called F actin ("F" because they are fibrous proteins). Two F actins are arranged in a double helix to form the actin strands found in thin filaments (**Figure 4a**).

Also present in thin filaments are two *regulatory proteins* that enable muscle fibers to start or stop contracting: tropomyosin and troponin (Figure 4b). **Tropomyosin** is a long fibrous molecule that extends over numerous actin monomers in such a way that it blocks the myosin-binding sites in muscles at rest. **Troponin** is a complex of three proteins—one that attaches to the actin strand, another that binds to tropomyosin, and a third containing a site to which calcium ions can bind reversibly. As we will see, the binding of calcium to this site triggers muscle contraction by causing troponin to move tropomyosin aside, thereby exposing the myosin-binding sites on the actin molecules.

Each thick filament is made of hundreds of myosin molecules, each of which looks a bit like two golf clubs wrapped around each other (**Figure 5a**). Each myosin molecule is a dimer consisting of two intertwined subunits, each having a long tail and a fat, protruding head. These heads are called **crossbridges** because under certain conditions (discussed shortly) they bridge the gap between the thick and thin filaments. Within a thick filament, the myosin molecules bind to each other at their tail ends so that their heads

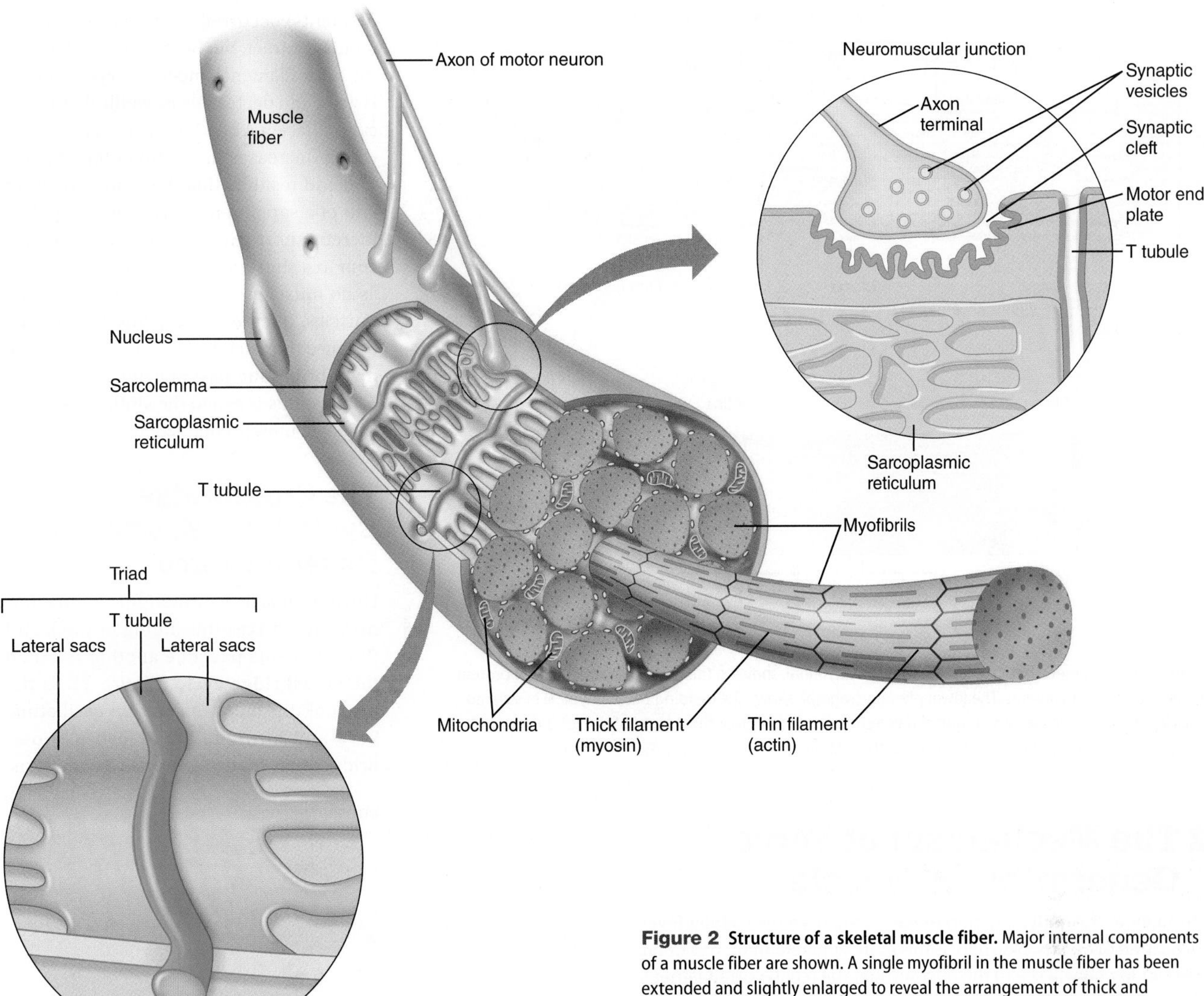

Figure 2 Structure of a skeletal muscle fiber. Major internal components of a muscle fiber are shown. A single myofibril in the muscle fiber has been extended and slightly enlarged to reveal the arrangement of thick and thin filaments within it. The enlarged views show magnified images of a neuromuscular junction and a triad.

extend in opposite directions away from the center (Figure 5b). The tails of adjacent myosin molecules are also arranged in a staggered fashion so that their heads protrude from the thick filament in an orderly helical pattern (Figure 5c). Because the middle of the thick filament is devoid of crossbridges, this region is appropriately called the *bare zone*.

The head is the "business end" of the myosin molecule—it is the part that actively generates a muscle's mechanical force. Each head possesses two sites that are critical to its force-generating ability: an *actin-binding site*, which is capable of binding to the actin monomers in the thin filaments, and an *ATPase site*, which has enzymatic activity and hydrolyzes ATP (see Figure 5a).

Like thin filaments, thick filaments have additional proteins associated with them—most notably *titin*, an extraordinarily elastic protein that can be stretched to more than three times its unstressed length (Figure 5d). Strands of titin extend along each thick filament from the M line to each Z line, anchoring the thick filaments in their proper positions relative to the thin filaments. When an external stretching force is applied to a muscle, titin strands (see Figure 5d) elongate as the sarcomeres lengthen, and these strands begin to exert an opposing force, just as a spring resists stretching. When the external force is removed, this opposing force pulls the Z lines and thick filaments closer together and causes the sarcomeres to shorten, allowing the titin strands to spring back to their original length. As this occurs, individual muscle fibers shorten, as does the whole muscle.

Quick Check 1

1. Name and describe the locations and general functions of the two contractile proteins and the two regulatory proteins present in sarcomeres.
2. Which two functions are performed by the heads of myosin molecules?

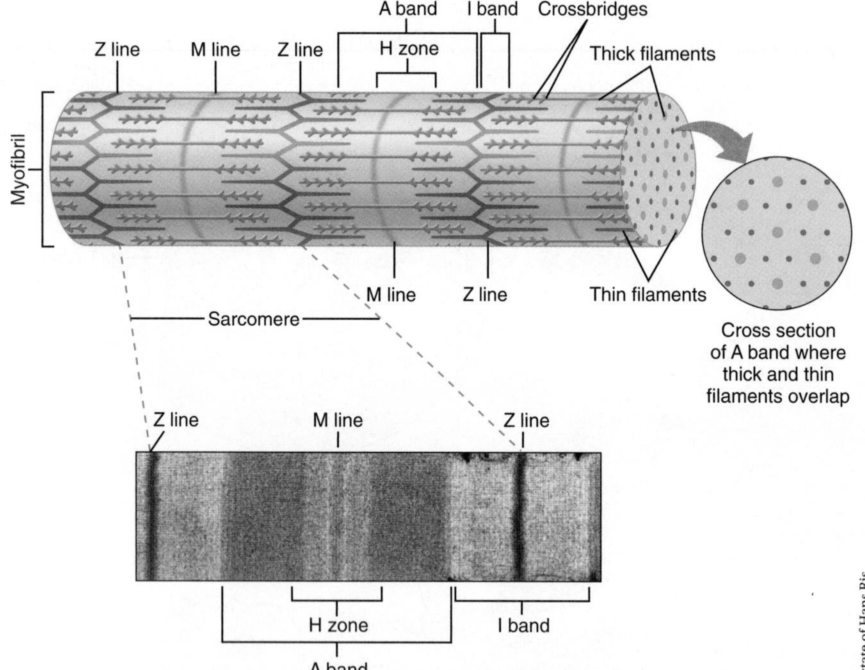

Estate of Hans Ris

Figure 3 Sarcomere structure. A drawing of a myofibril, showing the regular arrangement of protein filaments within sarcomeres. The lower photomicrograph shows the banding pattern typical of striated muscle; the photomicrograph at right shows a cross section through the A band of a myofibril, in which the three-dimensional arrangement of thick and thin filaments can be seen clearly.

2 The Mechanism of Force Generation in Muscle

Throughout the body, function follows form, from the cellular level to the level of organs. Nowhere is this more apparent than in skeletal muscle. As you will see in the next section, the regular arrangement of thick and thin filaments is the key to muscle contraction.

The Sliding-Filament Model

When physiologists first discovered the presence of actin and myosin in myofibrils, they thought that muscle contractions were caused by shortening of the proteins themselves. As advances in microscopy occurred, researchers discovered that during muscle-cell contraction the A band does not change in length, but the I bands and the H zone shorten (**Figure 6**). Given that the A band spans the length of the thick filaments in a sarcomere, this means the thick filaments do not change length when the muscle cell contracts. Researchers realized that the shortening of the I bands (which contain only thin filaments) occurred not because thin filaments contract, but because they slide past the thick filaments, moving deeper into the H zone and decreasing its width. As this occurs, adjacent A bands move closer together, which decreases the width of the I bands. The end result is that the Z lines at either end of a sarcomere move closer together, thereby shortening the sarcomere. As sarcomeres shorten, myofibrils also shorten, as do muscle fibers and ultimately whole muscles. In other words, muscles contract because the thick and thin filaments of the myofibrils slide past one another. Appropriately, this is called the **sliding-filament model** of muscle contraction.

The Crossbridge Cycle: How Muscles Generate Force

MAP

During muscle contraction, the mechanism that drives the sliding of thick and thin filaments past one another is called the **crossbridge cycle** (**Figure 7**). At the heart of this mechanism is an oscillating, back-and-forth motion of myosin crossbridges that is powered by ATP hydrolysis.

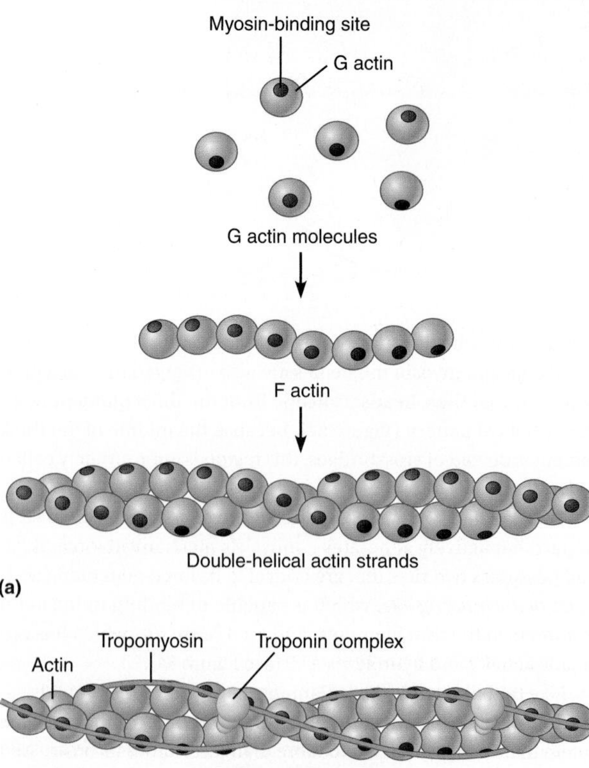

Figure 4 Structure of a thin filament. (a) The backbone of a thin filament consists of two strands of polymerized actin molecules wound together to form a double helix. Myosin-binding sites on individual actin molecules (G actin) are represented by dark dots. **(b)** A portion of a thin filament showing troponin and tropomyosin in their normal resting positions on the actin strands. Notice that actin's myosin-binding sites are covered by tropomyosin when a muscle cell is at rest.

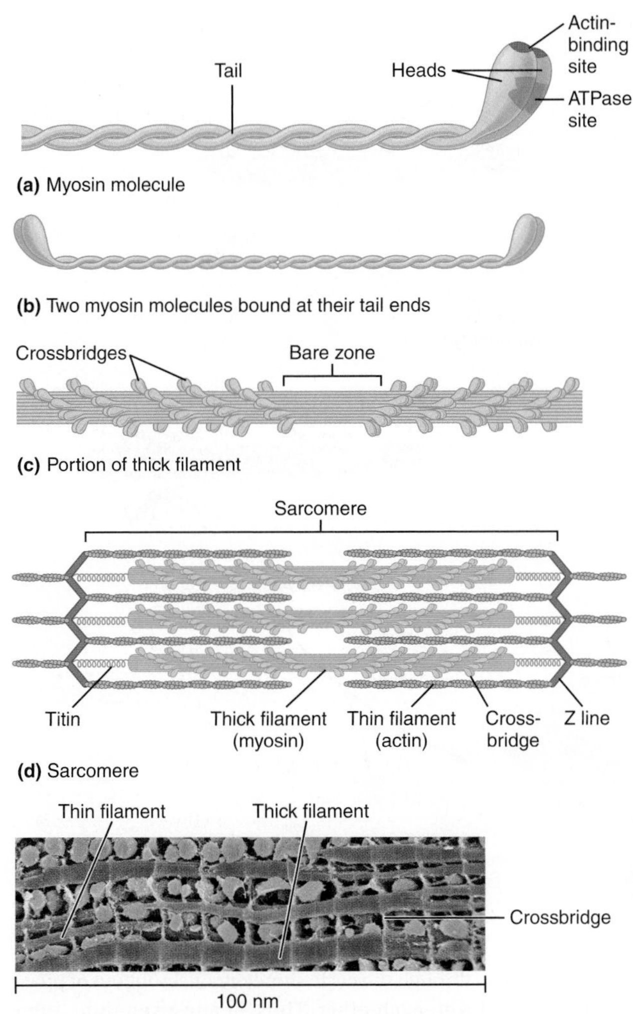

Figure 5 Structure of a thick filament. **(a)** A myosin molecule. Note the actin-binding and ATPase sites in the head region. **(b)** Two myosin molecules joined tail to tail. **(c)** A portion of a thick filament showing myosin heads (crossbridges) protruding at either end but not in the middle region (the bare zone). **(d)** A detailed view of a sarcomere showing the relative positions of thick and thin filaments and the protein titin, which anchors the thick filaments in place. **(e)** An electron micrograph of a sarcomere showing thick and thin filaments and crossbridges.

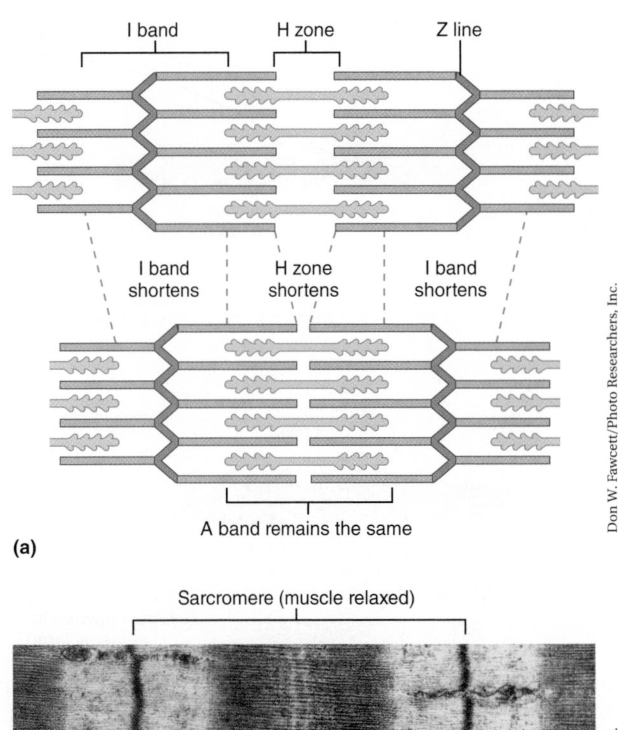

Figure 6 How changes in striation pattern are explained by the sliding-filament model of muscle contraction. **(a)** A schematic drawing and **(b)** photomicrographs showing the relative positions of the thick and thin filaments in sarcomeres in relaxed muscle (top) and contracted muscle (bottom).

Coupled with this activity is cyclic binding and unbinding of crossbridges to the thin filaments, which occurs in such a way that the motion of the crossbridges pulls the thin filaments toward the center of the sarcomere.

The back-and-forth movement of crossbridges is due to changes in the conformation of the myosin molecules. These conformational changes not only cause the heads to change position, but also alter both their ability to bind to actin monomers in the thin filaments and the *energy content* of the myosin molecules. One conformation of myosin is referred to as the *high-energy form*, which is indicated in step 5 of Figure 7. Myosin heads go into this conformation after they hydrolyze ATP. It is called the *high-energy form* because the myosin molecule stores energy that is released in the hydrolytic splitting of ATP. Myosin heads go into the other conformation, called the *low-energy form*, after the stored energy is released to drive the movement of the thin filaments (step 3).

Each crossbridge cycle involves the following five steps (see Figure 7):

① *Binding of myosin to actin.* We start with myosin in its energized form; that is, ADP and Pi (inorganic phosphate) are bound to the ATPase site of the myosin head. In this state, myosin has a high affinity for actin, and the myosin head binds to an actin monomer in the adjacent thin filament. This step can occur only in the presence of calcium, for reasons described below.

② *Power stroke.* The binding of myosin to actin triggers the release of the Pi from the ATPase site. During this process, the myosin head pivots toward the middle of the sarcomere, pulling the thin filament along with it.

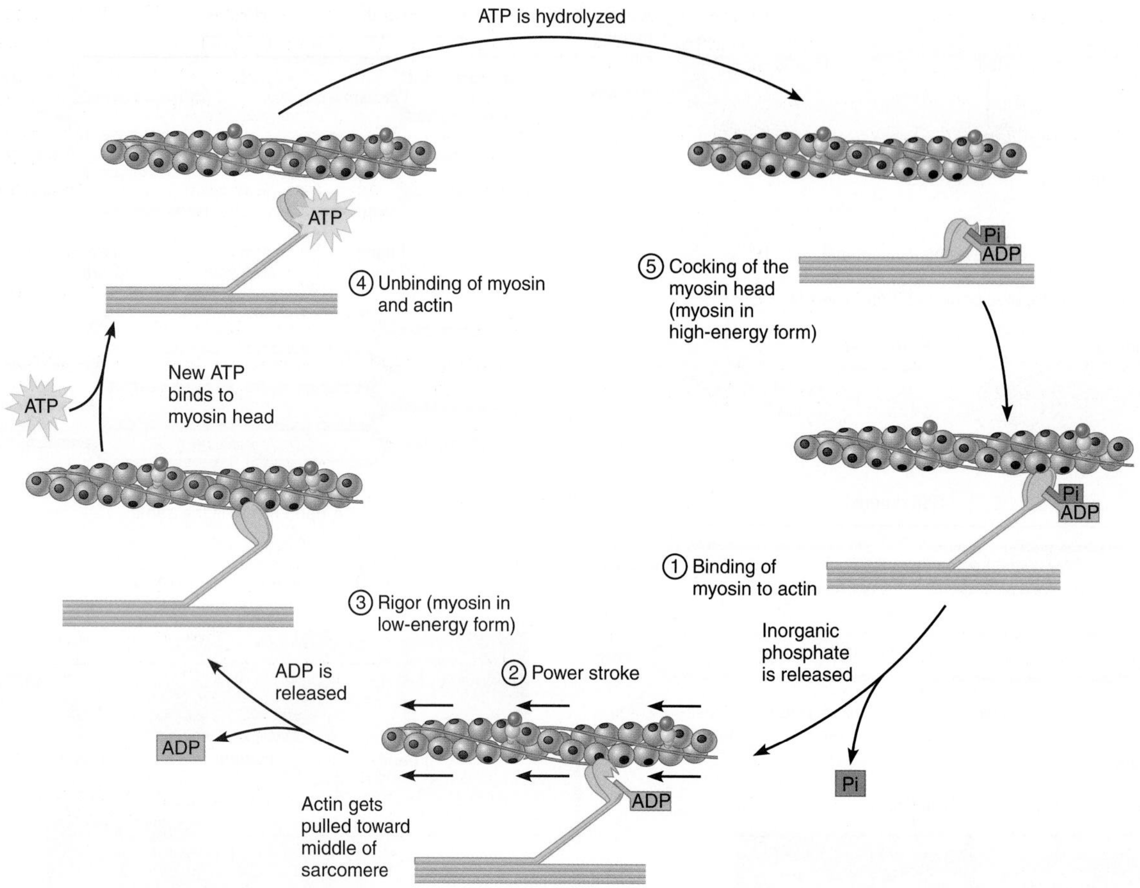

Figure 7 The crossbridge cycle.

③ *Rigor.* As the power stroke ends, ADP is released from the myosin head and the myosin molecule goes into its low-energy state. In this state, myosin and actin are tightly bound together, a condition called *rigor.* (*Rigor mortis*—the stiffening of the body that occurs after death—occurs because the crossbridge cycle gets stuck at this step due to (1) an excess of calcium when cell membranes are damaged and (2) a lack of ATP due to the termination of energy production. Rigor mortis continues until enzymes leaked by disintegrating cellular components begin to break down the myofibrils.)

④ *Unbinding of myosin and actin.* A new ATP enters the ATPase site on the myosin head, triggering a conformational change in the head, which decreases the affinity of myosin for actin, so the myosin detaches from the actin.

⑤ *Cocking of the myosin head.* Soon after it binds to myosin's ATPase site, ATP is split by hydrolysis into ADP and Pi, which releases energy. Some of the energy is captured by the myosin molecule as it goes into its high-energy conformation. Although ATP has been hydrolyzed at this point, the end-products of the reaction (ADP and Pi) remain bound to the ATPase site. If calcium is present, the cycle will continue by revisiting step 1.

Although a given crossbridge generates force only part of the time while it is active (during the power stroke), a muscle cell generates force continually during a contraction, because many crossbridges go through the cycle simultaneously but out of phase ("out of step") with each other. Thus, at any given time, some crossbridges are starting the cycle, others are finishing it, and still others are at various stages in between. If the cycles were in sync and all the crossbridges detached from actin at the same time, then the thin filaments would passively slide back to their original positions, making the crossbridge cycle inefficient. During actual contraction, there are always several crossbridges attached to actin at any given time—an arrangement that prevents this effect from occurring.

Because the crossbridges at opposite ends of a thick filament are oriented in opposite directions from each other (see Figures 3 and 5), the power strokes of crossbridges at opposite ends of the thick filament move in opposite directions, pulling the thin filaments on either side of the A band in toward the center and causing the sarcomere to shorten. When the crossbridge cycle stops and the contraction ends, the thin filaments passively slide back to their original position. During a contraction, each myosin head completes only about five crossbridge cycles per second, but because each thick filament has several hundred heads, thousands of power strokes can occur each second. For this reason, sarcomeres—and entire muscle fibers—can shorten very rapidly, in many cases taking less than a tenth of a second to contract fully. A skeletal muscle fiber is capable of shortening to approximately 60% of its resting length.

The crossbridge cycle could continue indefinitely, so long as there is ample ATP. To prevent this never-ending cycle from happening, the regulatory proteins troponin and tropomyosin interact with calcium, controlling the availability of myosin-binding sites on actin and thereby regulating the crossbridge cycle, as described next.

Excitation-Contraction Coupling: How Muscle Contractions Are Turned On and Off

Central nervous system ultimately controls skeletal muscle contractions, with *motor neurons* delivering to the muscles commands telling them when and when not to contract. Moreover, input from motor neurons always has an excitatory effect on muscle cells and serves to trigger contraction of those cells. Like neurons, muscle cells are *excitable*, meaning that they are capable of generating action potentials if their plasma membranes are depolarized to a sufficient degree. When a muscle cell receives input from a motor neuron, the cell depolarizes and fires an action potential that then stimulates contraction. The sequence of events that links the action potential to the contraction is referred to as **excitation-contraction coupling** (Figure 8).

The Role of the Neuromuscular Junction in Excitation-Contraction Coupling

The connection between a motor neuron and a muscle cell, referred to specifically as a *neuromuscular junction,* is fundamentally

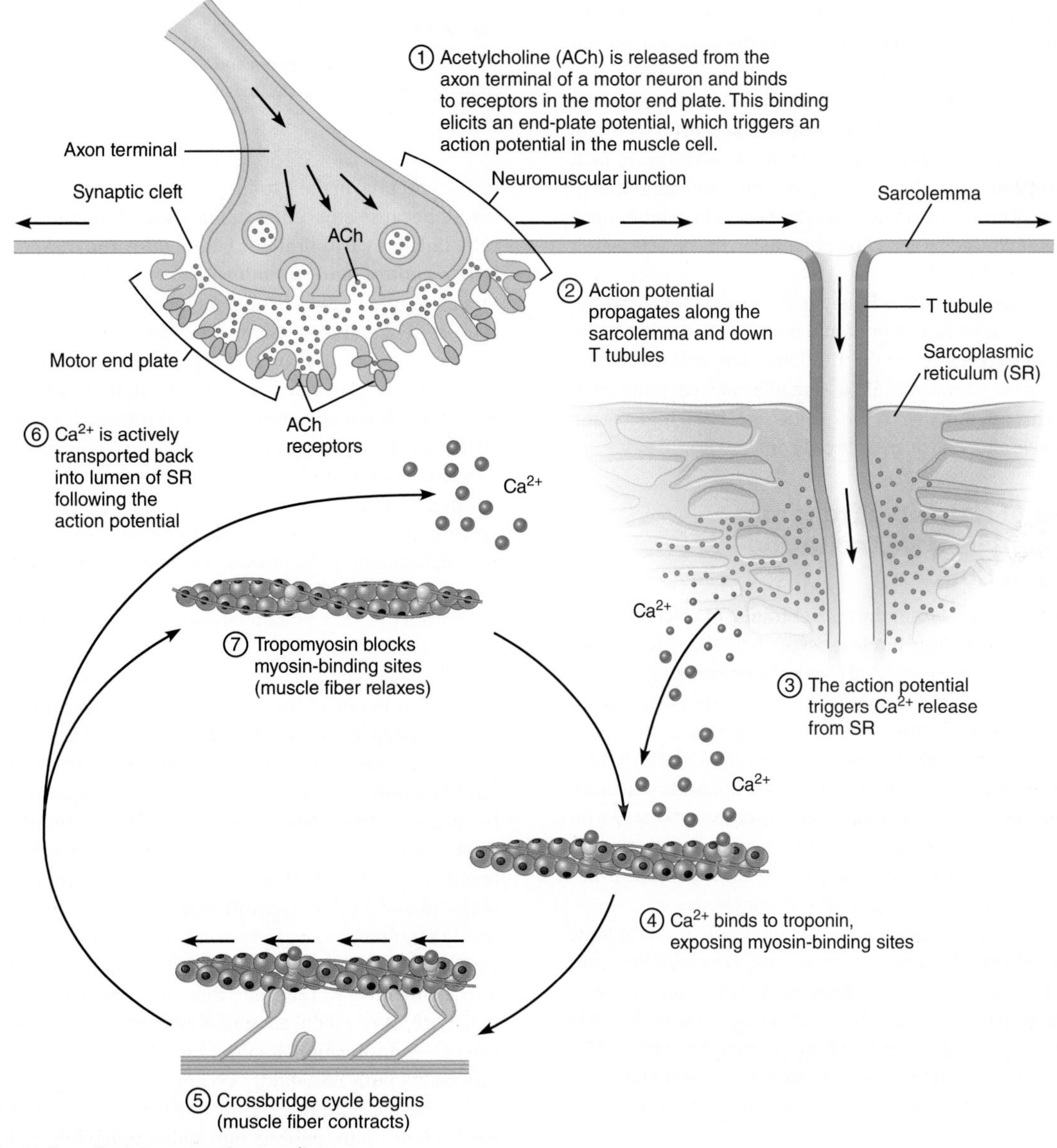

Figure 8 Events in excitation-contraction coupling.

Q Which enzyme breaks down acetylcholine after it is released?

similar to an "ordinary" synapse between two neurons in the nervous system. The motor neuron (the presynaptic cell) transmits an action potential and secretes the neurotransmitter *acetylcholine* upon its arrival at the axon terminal (step 1, Figure 8). After release, acetylcholine diffuses to the plasma membrane of the muscle cell (the postsynaptic cell), where it binds to specific receptors, triggering a change in ion permeability that results in a depolarization.

Despite its similarity to an ordinary synapse, a neuromuscular junction has many characteristics that make it special. Although a motor neuron typically branches and innervates more than one muscle cell, each muscle fiber receives input from only one motor neuron. At the neuromuscular junction, the motor neuron's terminal boutons fan out over a wide area of the sarcolemma. Opposite these boutons is a specialized region of the sarcolemma called the *motor end plate,* which is highly folded and contains a high density of acetylcholine receptors (see Figure 8). An action potential in the motor neuron triggers the release of acetylcholine from each of its many terminal boutons, which causes many acetylcholine receptors to become activated. As a consequence, the resulting depolarization (called an **end-plate potential**) is much larger than an ordinary postsynaptic potential—so large, in fact, that it is always above threshold and triggers an action potential in the muscle cell. Thus, if an action potential occurs in a motor neuron, it is always followed by an action potential in the muscle cells it innervates (step 2, Figure 8).

Once an action potential is initiated in a muscle cell, it propagates through the entire sarcolemma and down the T tubules. As the action potential travels through the T tubules, it triggers the release of calcium from the lateral sacs of the sarcoplasmic reticulum (step 3, Figure 8). This calcium then serves as the signal that initiates the crossbridge cycle and, hence, muscle cell contraction.

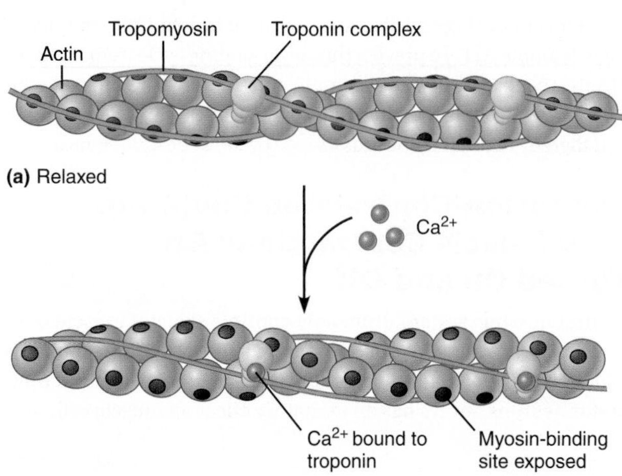

Figure 9 Actions of troponin and tropomyosin in excitation-contraction coupling. (a) Relaxed muscle. **(b)** Role of calcium in exposing the myosin-binding sites on actin.

The Roles of Calcium, Troponin, and Tropomyosin in Excitation-Contraction Coupling

When a muscle cell is relaxed, the concentration of calcium in the cytosol is very low, and little binding occurs between calcium and troponin. Troponin is in its normal (resting) conformation and, as a consequence, tropomyosin is positioned on the thin filaments in such a way that it blocks actin's myosin-binding sites, so the crossbridge cycle cannot occur (**Figure 9a**). The cytosolic calcium level is normally low because the membrane of the sarcoplasmic reticulum (SR) is equipped with calcium pumps that actively transport calcium ions from the cytosol into the SR. Due to the action of these pumps, the SR is able to accumulate calcium against a concentration gradient and, therefore, can function as a calcium storehouse.

In addition to calcium pumps, the membrane of the SR contains voltage-gated calcium channels that are normally closed, which prevents calcium inside the SR from leaking out. When an action potential travels through the T tubules, however, it causes these channels to open briefly, allowing calcium to flow out into the cytosol. The end result is an increase in the cytosolic calcium concentration.

Although the calcium channels that allow Ca^{2+} out of the SR are voltage-gated, an unusual kind of voltage-gating is present because the electrical signal that triggers gating occurs in the membrane of the T tubule—not the membrane of the SR itself. An action potential in the T tubule can trigger the release of Ca^{2+} from the SR because adjacent T tubules and SR membranes are physically linked by two proteins called *dihydropyridine receptors,* or DHP receptors, and *ryanodine receptors* (**Figure 10**). DHP receptors are found in the T tubule membrane and function as voltage sensors. Ryanodine receptors are found in the sarcoplasmic reticulum membrane and connected to the DHP receptors. Ryanodine receptors are also calcium channels that are gated by the DHP receptors. When an action potential travels down the T tubules, DHP receptors undergo a conformational change that transmits a signal to the ryanodine receptors, causing the calcium channels to open. Calcium then moves out of the sarcoplasmic reticulum and into the cytosol. Some calcium binds to specific sites on other SR calcium channels and causes them to open. In this manner, the initial release of calcium triggers the release of even more calcium from the SR, a phenomenon called *calcium-induced calcium release.*

As the cytosolic calcium concentration rises, some of the calcium binds to one of the three proteins making up each troponin complex (step 4 in Figure 8, and Figure 9b). This complex then undergoes a conformational change that causes tropomyosin to shift out of its normal resting position, thereby exposing the myosin-binding sites on the actin monomers. With the myosin heads of the thick filament now able to bind to actin, the crossbridge cycle can begin (step 5, Figure 8), and the sarcomere contracts. The design of the SR and T tubule network surrounding the myofibrils permits nearly simultaneous delivery of calcium to all sarcomeres of a muscle fiber, so the sarcomeres contract in unison, as does the entire muscle fiber. We will see later that the strength of contraction ultimately depends on the amount of calcium in the cytosol, with more calcium exposing more binding sites on actin for myosin, which in turn allows more crossbridge cycling to occur.

A muscle cell stops contracting when it no longer receives input from its motor neuron, and action potentials no longer occur in the sarcolemma. When an action potential triggers the release of calcium from the SR, this release does not continue indefinitely.

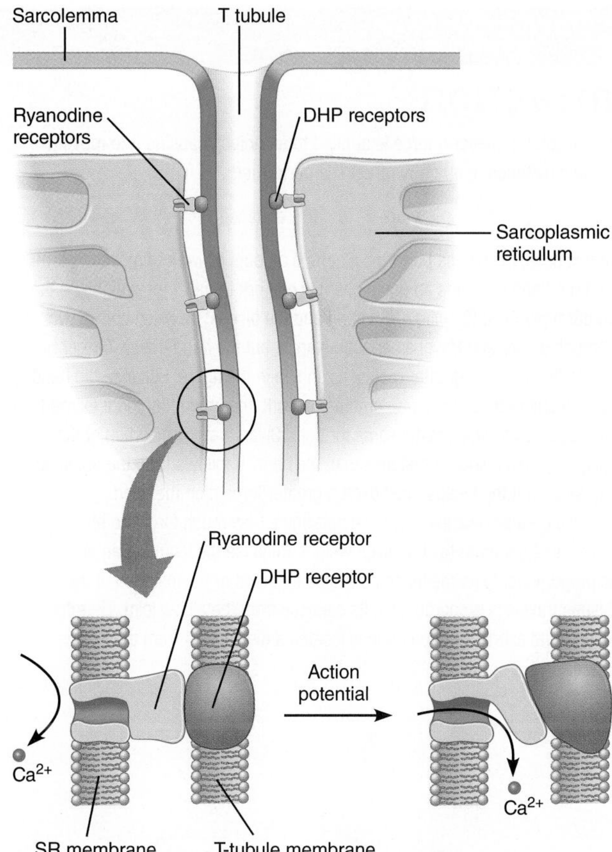

Figure 10 **Gating of sarcoplasmic reticulum calcium channels.**

The voltage-gated ryanodine receptors close when the membrane potential returns to normal. In addition, the calcium-induced calcium release is subjected to a sort of negative feedback: As the cytosolic calcium concentration increases, calcium ions begin to bind to certain sites on the SR calcium channels, causing them to close. (These sites are distinct from those that trigger channel opening and have a lower affinity for calcium, so they do not come into play until the cytosolic calcium level has become sufficiently high.) The closure of these channels turns off the release of calcium and enables the active transport of calcium back into the SR (an ongoing process) to clear calcium from the cytosol (step 6 in Figure 8), which causes the calcium concentration to fall. Because the binding of calcium to troponin is reversible, this concentration change causes calcium to dissociate from troponin, which allows both troponin and tropomyosin to revert to their original positions (step 7 in Figure 8). Subsequently, the number of exposed sites on the actin filament decreases, leading to a decline in the number of active crossbridges. Eventually, as calcium concentration returns to normal, the muscle contraction ends.

Apply Your Knowledge

Muscle cramps often result from spastic contraction of skeletal muscle. Based on what you know about muscle contraction, name some potential causes of muscle cramps.

Quick Check 2

1. Which of the following shortens during a muscle contraction: thick filaments, thin filaments, A bands, I bands, H zones, sarcomeres? (Choose all that apply.)
2. When ATP is hydrolyzed, the conformation of myosin crossbridges changes. Do these structures evolve into the high-energy form or the low-energy form?
3. What triggers the power stroke of the crossbridge cycle?
4. What is the function of the sarcoplasmic reticulum in skeletal muscle?
5. How does an increase or a decrease in the cytosolic calcium level cause the crossbridge cycle to turn on or off?

3 The Mechanics of Skeletal Muscle Contraction

When you use a given set of muscles in different activities, the contractions of those muscles vary in force and duration. As you lift a chair, for example, your arm muscles must exert more force than when you lift a paper clip. When you throw a baseball, muscle contractions are brief and explosive, but when you carry a suitcase, the contractions are smooth and sustained. In view of this great variability, you may be surprised to learn that when a muscle cell contracts in response to a single action potential, the result is always the same: Within a fraction of a second, the force rises to a maximum and then falls to zero nearly as quickly. This event, called a *twitch,* is like an action potential in that it is a reproducible, all-or-nothing event; if the cell is stimulated again, it gives the identical response. Surprising as it may seem, all skeletal muscle contractions—whether strong, weak, short, or long—are built on this simple muscle twitch, which we discuss next.

The Twitch

The axons of motor neurons typically branch before reaching their target cells, so that one neuron innervates several muscle fibers. Thus an action potential in a motor neuron triggers contraction of *all* the muscle cells that are connected to that neuron, and it is not possible to stimulate one cell without stimulating the others. A motor neuron and all the muscle fibers it innervates are collectively referred to as a **motor unit.** A **twitch** is the mechanical response of an individual muscle cell, a motor unit, or a whole muscle to a single action potential.

Phases of the Twitch

With the possible exception of an eye blink, a single isolated twitch exists only under artificial conditions produced in the laboratory. To observe twitches, a researcher can surgically remove a muscle from an animal and anchor it in an apparatus that permits contractile force to be measured and displayed on an oscilloscope, strip chart recorder, or other recording device. The muscle is then stimulated electrically, which causes individual muscle cells to have action potentials at the same time and contract together.

TOOLBOX

Physics of Skeletal Muscle Contraction

Bones function as levers. When you use your muscles to lift a weight or move a part of your body, the force your muscles must generate is actually much greater than you might think. For instance, to hold a 15-kg weight (about 33 pounds) in the palm of your hand, the biceps in your arm generates an impressive 105 kg (about 232 pounds) of force! The reason relates to the fact that the insertion of the biceps is located close to the elbow joint (approximately 5 cm away), whereas the hand is located farther from the joint (approximately 35 cm away); see figure **(a)**. To lift the weight, the biceps pulls upward on the bones of the lower arm, causing the hand to exert an upward force against the weight. (When the weight is being held steady or is being lifted at a constant speed, the magnitude of this force equals the downward force of gravity acting on the weight.) Because the insertion of the biceps is close to the elbow joint, which acts as a pivot, the muscle must exert a force significantly greater than the force exerted by the hand; see figure **(b)**.

To see why the biceps must exert more force, it is instructive to consider what happens when the weight is being lifted, as shown in figure **(c)**. Whenever a force is applied to an object, causing it to move a certain distance, a quantity of *work* is performed:

$$\text{work} = \text{force} \times \text{distance}$$

In this case, the biceps performs a certain amount of work to move the arm, and the hand performs an equal quantity of work to move the weight. As the forearm pivots at the elbow, the hand and the biceps' insertion both move through an arc and travel a certain distance, but the hand travels farther because its arc has a greater radius, as shown in the figure. Because the hand and biceps perform the same amount of work, the product of force exerted and distance traveled is the same for both. Given that the insertion of the biceps travels a shorter distance than the hand, it follows from the previous equation that the biceps must exert a greater force than the hand.

Using these principles, we can determine how much force the biceps must generate to lift a given weight. Because the length of an arc is proportional to its radius, the distance the hand or the insertion of the biceps travels is proportional to its distance from the elbow joint. Therefore, if the hand exerts a force F_1 and is located a distance R_1 from the elbow, it

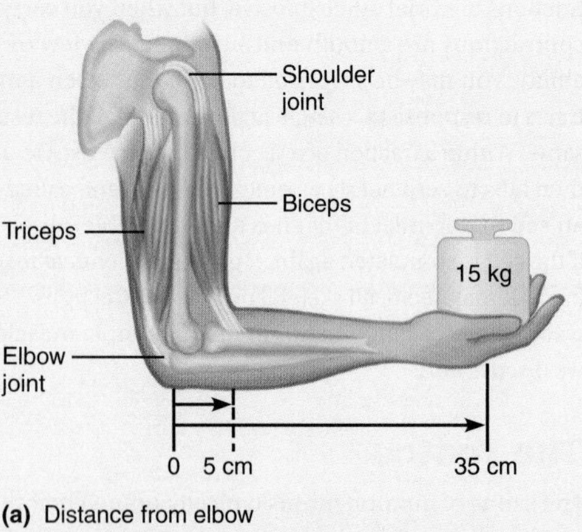

(a) Distance from elbow

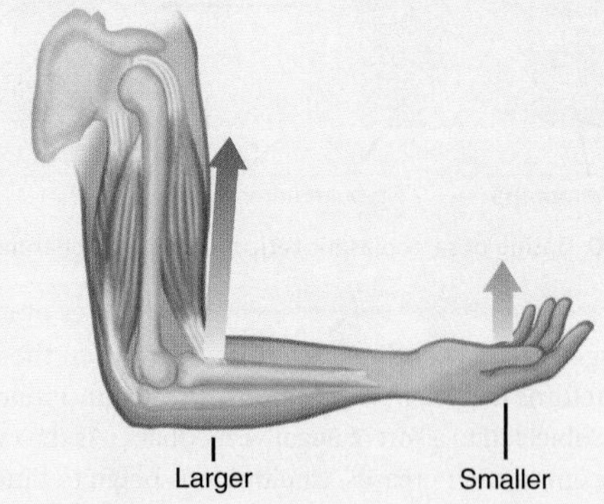

(b) Force exerted

When each end of a muscle is anchored in place, a single brief stimulus causes the contractile force to first rise, then fall. On the force tracing, a latent period is observed followed by a contraction phase and a relaxation phase (**Figure 11**). The **latent period** is the delay of a few milliseconds that occurs between the action potential in a muscle cell and the start of contraction, when the cell first begins to generate force. This time lag exists because the events of excitation-contraction coupling must occur before crossbridge cycling—and hence force generation—can begin. The *contraction phase,* which can range from 10 milliseconds to more than 100 milliseconds for different muscles, starts at the end of the latent period and ends when muscular tension peaks. (*Tension* is synonymous with *force* and is commonly expressed in units of mass, such as grams; muscle tension of 1 gram is equivalent to the force

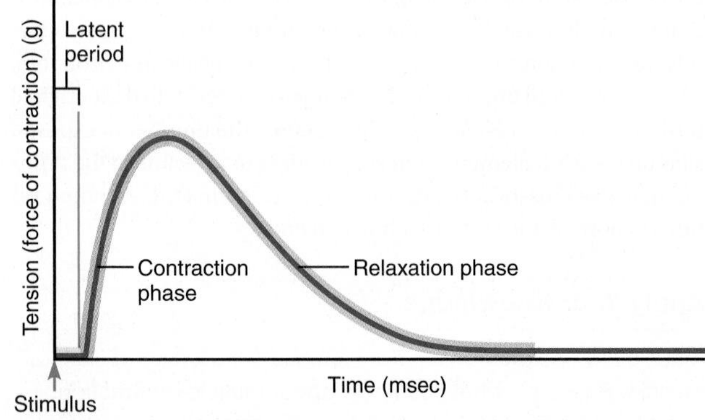

Figure 11 Phases of a twitch.

performs a quantity of work proportional to the product $F_1 \times R_1$; see figure **(d)**. Likewise, if the biceps exerts a force F_2 and its insertion is located a distance R_2 from the elbow, the work it performs is proportional to the product $F_2 \times R_2$. Because the hand and biceps perform equal amounts of work, the force of the hand and biceps are related by the following expression:

$$F_1 \times R_1 = F_2 \times R_2$$

or

$$F_2 = F_1 \times (R_1 / R_2)$$

Using the values given previously, we find that when the hand is lifting a 15-kg weight, the biceps exerts the following force:

$$F_2 = 15 \text{ kg} \times (35 \text{ cm} / 5 \text{ cm}) = 105 \text{ kg}$$

This is also true when the hand is holding the weight steady.

In the preceding example, the bones of the forearm act as a lever that pivots at the elbow. The biceps pulls upward on this lever, while the weight in the hand pushes down on it. To hold the weight steady (or lift it at constant speed), the biceps must exert a force greater than the downward force of the weight because its *lever arm*—the distance between the lever's pivot point and the point at which the force is applied—is shorter than that of the weight. As a reflection of this relationship, the shorter lever arm is said to put the biceps at a *mechanical disadvantage.* Consequences of the lever arm effect may be familiar to you: If you have ever had to change an automobile tire, for instance, you know that it is easier to loosen the lug nuts with a longer wrench than with a shorter wrench. The shorter wrench puts you at a mechanical disadvantage, such that you have to exert more force to loosen the nuts.

Even though the lever arm effect puts muscles at a mechanical disadvantage with respect to the force they must generate, this situation is actually advantageous in another way: As a muscle contracts, it shortens with a certain velocity, but the limb (or other body part) to which it is attached moves significantly faster. If, for example, the biceps shortens at a rate of 2 cm/sec, the hand moves at a rate seven times as great—14 cm/sec. Thus the lever arm effect makes it possible for the biceps and other muscles to move body parts more quickly than the muscles themselves can shorten. This behavior is important because it not only makes activities such as running and throwing possible but also enables us to move out of danger quickly in times of emergency.

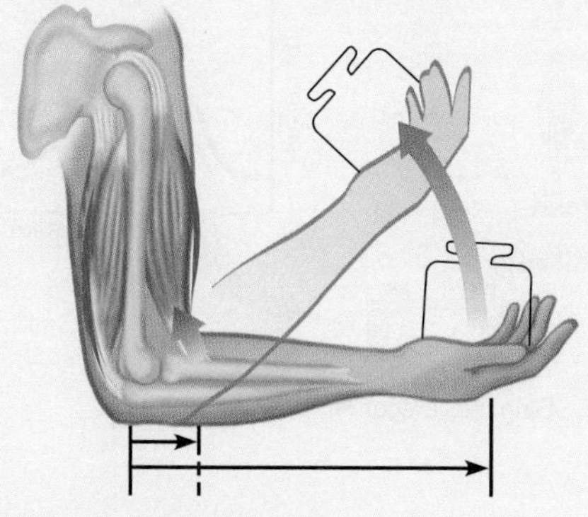

(c) Radius of arc (distance from elbow)

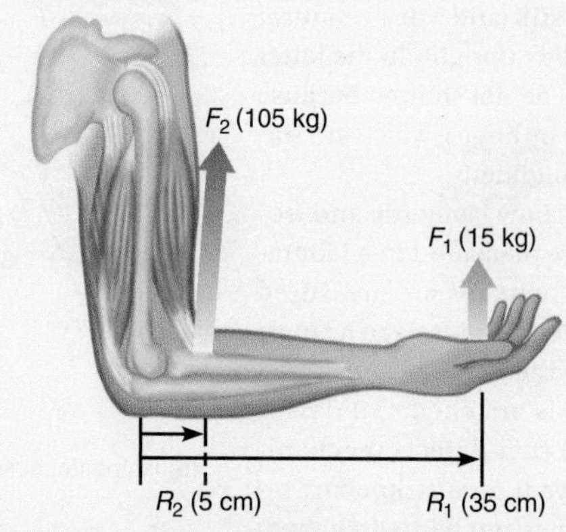

(d) Radius of arc and forces exerted

generated by a 1-gram weight hanging on a string.) During this time, cytosolic calcium levels are increasing as release of these ions exceeds their reuptake. The *relaxation phase,* which is typically the longest of the three phases, is the time between peak tension and the end of the contraction, when tension returns to zero. During this time, cytosolic calcium levels are decreasing (reuptake of these ions exceeds their release) and the number of active crossbridges is decreasing.

As mentioned, a characteristic feature of a muscle twitch is its reproducibility; that is, repetitive stimulation of a muscle produces several twitches in a row, each having the same magnitude and shape. (Later we will see that this is not true when the frequency of stimulation is high enough that the twitches follow one another closely.) The reason for this reproducibility relates to the all-or-nothing character of the muscle cell's action potentials: Assuming that over time no changes occur in the cell's properties (such as occurs with fatigue, discussed later), an action potential always triggers the same degree of calcium release from the SR, which causes the same rise in cytosolic calcium levels, exposes the same number of binding sites for the myosin crossbridges, and produces the same force.

Although twitches are reproducible for any given muscle or muscle fiber, twitches vary considerably from one muscle or muscle cell to another. One reason for inconsistency is that certain muscle fibers are inherently stronger than others—that is, they are capable of generating more force. Specifically, larger-diameter fibers can exert more force than smaller-diameter fibers (discussed later in this chapter). Another reason is that muscle fibers differ in the speed with which they can achieve peak force; differences in these so-called *fast-twitch fibers* and *slow-twitch fibers* are also discussed later in the chapter.

Isometric and Isotonic Twitches

There are two variations on the twitch, known as *isometric* and *isotonic* twitches, that differ not in how the basic mechanism of force generation works but rather in whether the muscle is allowed to shorten as it contracts. During an **isotonic contraction,** the muscle generates a constant (*iso*) tension (*tonic*) just greater than any forces opposing it (called *loads;* an example is the weight of a dumbbell), and the muscle length changes. Two types of isotonic contractions occur: concentric and eccentric. A muscle contraction involving a shortening of the muscle is called a **concentric contraction.** For example, when you lift an object with a curling motion at the elbows, your biceps are performing a concentric contraction. In contrast, during an **eccentric contraction,** the muscle length increases, such as when your biceps contract as you lower the object. During an **isometric contraction,** a muscle creates tension but maintains the same (iso) length (metric) because the load is greater than the force generated by the muscle. This occurs, for example, when you try to lift an object that is too heavy for you to move, or when you stand still and your postural muscles hold your body upright. In the latter instance, the muscles do not shorten because they are held in place by bones, which are stationary under those conditions.

Figure 12 shows how isometric and isotonic contractions are measured in a laboratory. In both cases, twitches are measured as the response of a whole muscle to a single stimulus; one end of the muscle is anchored while the other end is attached to a device (a force transducer) that detects mechanical force and displays it on a strip chart recorder. To measure an **isometric twitch,** both ends of the muscle are rendered immobile so that when the muscle is electrically stimulated, it develops tension but cannot shorten (Figure 12a). The force tracing shows that tension rises to a peak and then declines to the resting level. To measure an **isotonic twitch,** the lower end of the muscle is not anchored in place but is instead attached to a movable load, such that the muscle is able to shorten when it contracts (Figure 12b). Note that the curve for the isotonic twitch shows a distinct plateau, indicating that the force remains constant for a period of time. It is during this plateau phase that the muscle shortens and the load moves. Before the plateau phase, the force increases, but the muscle does not shorten because it is not yet generating enough force to lift the load. Only when the force becomes equal to the load does the muscle begin to shorten. This force remains constant so long as the load is moving, but eventually the muscle starts to relax and the load starts to fall, corresponding to an end in the plateau phase and a decline in the force.

Unlike an isometric twitch, an isotonic twitch is not an all-or-nothing event—its size and shape depend on the size of the load that is placed on the muscle. When the load is increased, for example, greater tension is needed to overcome it, and for this reason the force tracings show plateaus at higher tension levels (**Figure 13**). At the same time, the time lag (latent period) between the stimulus and the beginning of muscle shortening (the start of the plateau) also increases, because it takes the muscle longer to develop the force required to move the load. When the load exceeds the amount of force the muscle can generate, the muscle cannot move it and, therefore, contracts isometrically. Under these conditions, the force tracing has the rounded peak characteristic of an isometric twitch (Figure 13, far right).

When a muscle contracts isometrically, its sarcomeres shorten even though the whole muscle does not. This is possible because the sarcomeres (collectively referred to as the muscle's **contractile component [CC]**) do not extend the entire length of each muscle

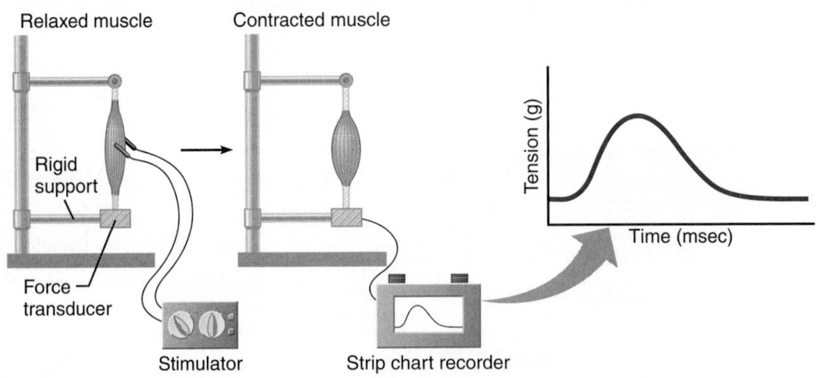

(a) Isometric muscle contraction

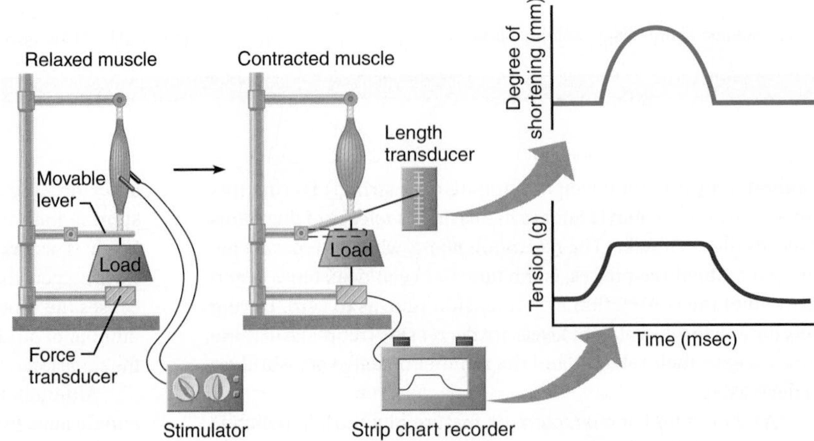

(b) Isotonic muscle contraction

Figure 12 Isometric and isotonic muscle contractions. (a) An isometric contraction, in which a muscle is mounted with both ends firmly anchored and thus cannot shorten on contraction. **(b)** An isotonic contraction, in which contraction of the muscle lifts a load.

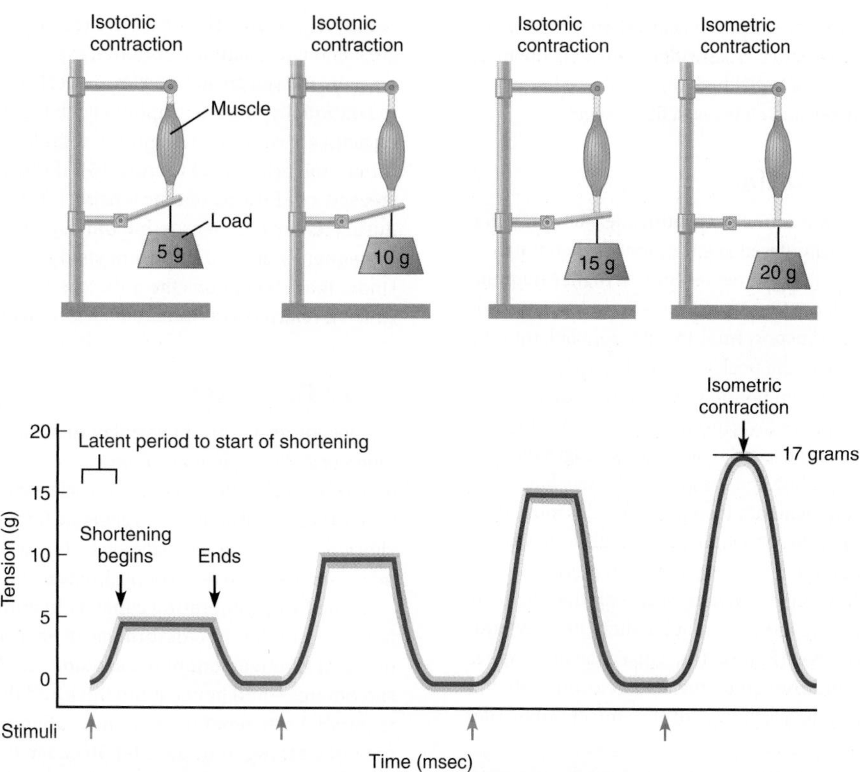

Figure 13 Effect of load on peak tension in an isotonic twitch. The responses to stimulation of a muscle subjected to four different loads are plotted. The muscle contracts isotonically when the load is 5, 10, or 15 grams, and in each case muscle tension plateaus just above the load. When the load is 20 grams, however, the muscle is unable to shorten because the load exceeds the maximum tension it can generate in a single twitch (17 grams); thus the muscle contracts isometrically. Note that as load increases, the duration of muscle shortening decreases, and the latent period before the onset of muscle shortening increases.

Q If a 16-gram load were placed on the muscle, would the contraction be isotonic or isometric?

fiber and, therefore, do not transmit force directly to the ends of the cells. Instead, the force is transmitted through certain cellular components that connect the myofibrils to the ends of the cells and then through connective tissue that anchors the ends of the cells in place and extends through the tendons. Those parts of a muscle (or muscle cell) that do not actively generate force but only serve to passively transmit force to the ends of the muscle are collectively referred to as the muscle's **series elastic component (SEC).** When a muscle contracts isometrically, the CC shortens and stretches the SEC, causing it to pull on the ends of the muscle. In so doing, the SEC lengthens as the CC shortens, giving an overall length change of zero.

Note that although muscles do contract isometrically in the body, they rarely (if ever) contract in a strictly isotonic fashion—that is, with a constant force as they shorten. When you walk, run, lift objects, or otherwise move such that contracting muscles shorten, your nervous system continually adjusts its input to the muscles to ensure that the muscular force is appropriate for the intended activity. This behavior is quite different from that seen in the laboratory situation depicted in Figures 12 and 13, in which a single invariant stimulus is delivered to a muscle. Furthermore, when a muscle contracts in the body, the load placed on the muscle is rarely constant (as it is in Figures 12 and 13), even when a constant weight is being lifted. This is the case because the positions of the joints change as you move, altering the loads placed on muscles. You can easily demonstrate this by attempting to hold a 10-pound weight steady in your hand under two conditions: with your elbow close to your body and bent at a 90-degree angle, and with your arm extended straight out in front of you. Without a doubt, you will find it more difficult to hold the weight steady in the second instance because the extended position of the arm puts greatly increased stress (that is, *load*) on certain muscles of the arm and shoulder. (To learn more about the physics of muscle activity, see **Toolbox: Physics of Skeletal Muscle Contraction**).

Factors Affecting the Force Generated by Individual Muscle Fibers

The force generated in a muscle depends on two factors: (1) the force generated in individual muscle fibers and (2) the number of muscle fibers contracting. The force generated in individual muscle fibers depends on the number of myosin-binding sites on actin that are exposed and, therefore, on the number of active crossbridges (those that can bind to actin). More active crossbridges generate more force, just as the force exerted on the rope in tug-of-war increases when more participants are added. In this section we examine factors affecting the number of active crossbridges and thus the force generated by the contraction of individual muscle fibers: frequency of stimulation, fiber diameter, and changes in fiber length.

Frequency of Stimulation

Isometric muscle twitches are actually reproducible, all-or-nothing events only if a muscle is stimulated at a frequency low enough to ensure that twitches are well separated in time. At higher frequencies, the rate of calcium release from the SR into the cytosol exceeds the rate of calcium active transport from the cytosol back into the SR, resulting in an increase in the peak tension. The more calcium in the cytosol, the more calcium bound to troponin, which moves the tropomyosin to expose more myosin-binding sites on actin. The more myosin-binding sites exposed, the more crossbridges that can participate in crossbridge cycling and the greater the tension development. As the stimulation frequency increases, muscles move from twitch contractions to *treppe,* to *summation,* and finally to *tetanus.*

Treppe occurs at a frequency of muscle stimulation where independent twitches follow one another closely such that the peak tension rises in a stepwise fashion with each twitch, until eventually it reaches a constant level (**Figure 14**). The cause of treppe is unknown, but it may result from an increase in cytosolic calcium between twitches or from a "warming" of the muscle fiber that occurs with work.

Summation and tetanus occur at greater frequencies of stimulation as twitches overlap in time. Compared to an action potential, which takes at most a few milliseconds to complete, a muscle twitch is fairly slow, taking anywhere from tens to hundreds of milliseconds to complete (**Figure 15**). For this reason, a muscle fiber can receive several action potentials in the time it takes to complete one twitch. When a muscle is stimulated repetitively such that additional action potentials arrive before twitches can be completed, the twitches become superimposed on one another, yielding a force greater than that of a single twitch; this process is called **summation** (**Figure 16**). Summation happens whenever twitches occur at a frequency such that calcium cannot be removed from the cytosol as rapidly as it is released from the SR. Calcium removal is necessary for relaxation; thus the muscle fiber cannot relax completely between twitches.

At higher frequencies of stimulation, summation reaches a peak called **tetanus.** (The term *tetanus* also refers to a disease in which toxins produced in a bacterial infection cause motor neurons to stimulate muscle contraction inappropriately; see **Clinical Connections: Tetanus**). In *unfused* (or incomplete) *tetanus,* the force demonstrates small oscillations with brief periods of relaxation between peaks (Figure 16). The peaks are reached when calcium levels are great enough to saturate troponin, exposing all myosin-binding sites on actin. At even greater frequencies, calcium levels are great enough to continually saturate troponin such that all myosin-binding sites on actin are continually exposed, resulting in a plateau called *fused* (or complete) *tetanus* (Figure 16). If the stimulus intensity is increased still further, tetanic tension (that is, the tension generated during tetanus) increases, but only up to a point; further increases in frequency beyond this point yield no further increases in force. Under these conditions the muscle is generating all the force it can, which is referred to as *maximum tetanic tension.*

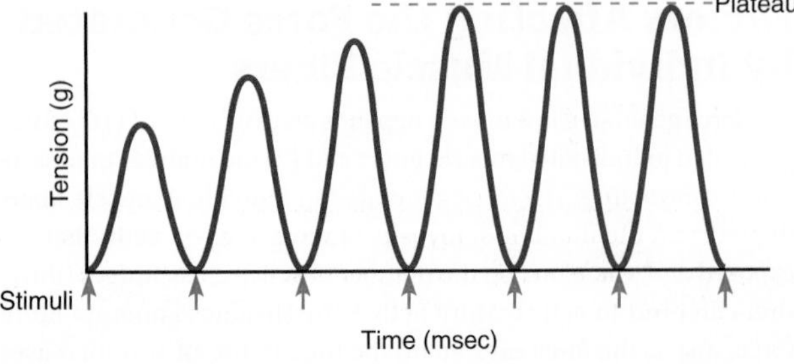

Figure 14 Treppe.

Fiber Diameter

We accept as a fact of everyday life that some muscles have an inherent ability to generate more force than others. Why else would we equate a weight lifter's bulging muscles with superior strength? The inherent ability of a muscle to generate force is referred to as the muscle's *force-generating capacity,* which is usually assessed by measuring maximum tetanic tension or peak tension in an isometric twitch.

The force-generating capacity of a muscle fiber depends on both the number of crossbridges in each sarcomere and the geometrical arrangement of the sarcomeres. A muscle that has more sarcomeres—and hence more thick and thin filaments—arranged in parallel can generate more force than a muscle with fewer sarcomeres arranged in parallel. Because the number of thick and thin filaments per unit of cross-sectional area does not vary significantly from one muscle to another, a fiber's *diameter* is the crucial variable that determines its force-generating capacity. The greater the fiber diameter, the greater its cross-sectional area, and the more force it can generate. This is why a weight lifter's bulging muscles are stronger than the average person's slimmer muscles.

Changes in Fiber Length

The length of a muscle fiber at the onset of contraction also affects the amount of tension that can be developed. However, it is the length of the sarcomeres that is important, not the number of sarcomeres in series. For each muscle fiber, there is an optimum length at which it can generate the most force due to maximum participation of myosin crossbridges in force generation. When a fiber

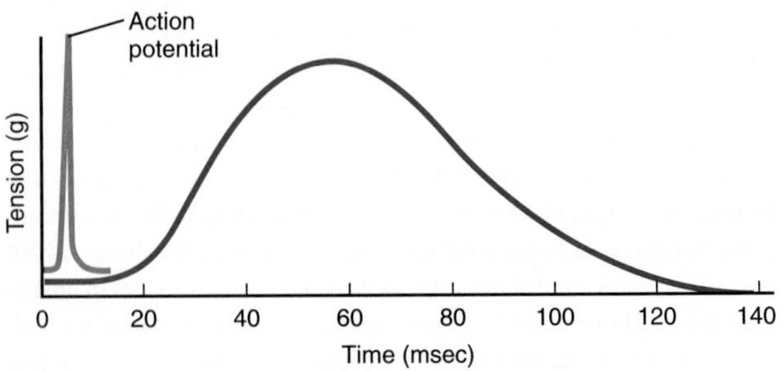

Figure 15 **Duration of an isometric twitch relative to an action potential.**

Q Which term is used to describe the series of events occurring in the time interval between an action potential and the rise in muscle tension?

CLINICAL CONNECTIONS

Tetanus

The disease known as tetanus results from infection with *Clostridium tetani*, an anaerobic bacterium whose spores are found in soil and animal feces. It most commonly arises from contamination of deep puncture wounds or burns, but can also occur after relatively minor wounds. Deep wounds are most typically involved because the anaerobic bacteria grow well in the oxygen-poor conditions found in such wounds. Symptoms of tetanus appear after an incubation period of between 2 days and 50 days, although a span of 5–10 days is most common.

Symptoms of the disease result from the action of tetanus neurotoxin secreted by the bacteria at the wound site. This toxin exerts its effects after reaching the CNS, where it binds to synaptic terminals and blocks the transmission of signals that normally inhibit the activity of motor neurons. As a consequence, motor neurons become hyperexcitable, leading to inappropriate stimulation of skeletal muscles. Depending on the extent of the toxin's spread, effects may be localized to muscles in the vicinity of the wound or may involve muscles all over the body. Local effects include soreness and increased muscle tone in the affected area, whereas systemic effects include spastic movements and intermittent convulsions.

The most frequent early systemic symptom of tetanus is stiffness of the jaw. Commonly, this stiffness then spreads to other areas including the neck, arms, or legs and may be accompanied by sore throat, headache, fever, restlessness, irritability, or difficulty in swallowing. As the disease progresses, muscle spasms may cause a patient to experience difficulty opening the jaw (trismus), which accounts for the disease's common name—lockjaw. Other facial muscles may become affected, causing the patient to take on a bizarre facial expression characterized by a fixed smile and raised eyebrows *(risus sardonicus)*. Death, if it comes, normally results from asphyxiation due to spasm of laryngeal or thoracic muscles. However, largely because of mass immunization programs, tetanus now afflicts fewer than 100 people in the United States each year.

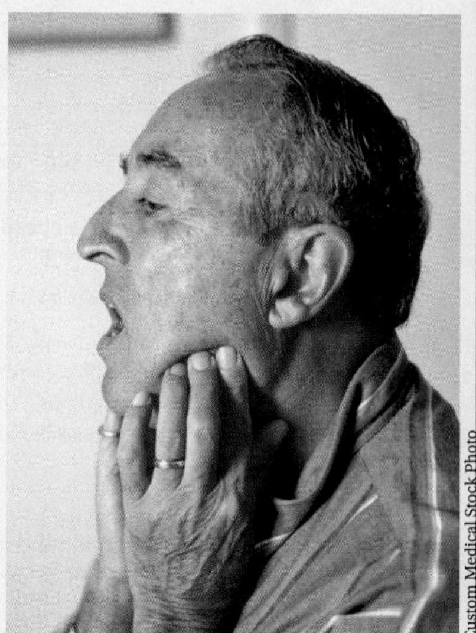

Stiffness of the jaw is an early systemic symptom of tetanus.

Critical Thinking Questions

1. Describe the causes and symptoms of tetanus. Why are deep wounds more typically involved in this disease?
2. What is the result of the tetanus toxin binding to synaptic terminals and blocking the transmission of inhibitory signals?
3. What are the early and later symptoms of this disease? What is the main cause of death from tetanus?

either shortens beyond or is stretched beyond this optimum range, its force-generating capacity decreases, because such changes in length alter the length of individual sarcomeres and reduce their ability to generate force.

Figure 17 shows how the force of contraction, measured as percentage of maximum tetanic tension or peak tension in an isometric twitch, varies with a muscle's length. The graph is an example of a *length-tension curve,* and its shape is a consequence of the sliding-filament model and the nature of the crossbridge cycle. We have seen that the force of a muscle contraction reflects the number of active crossbridges—the greater this number, the greater the force—and that crossbridge activity requires that myosin

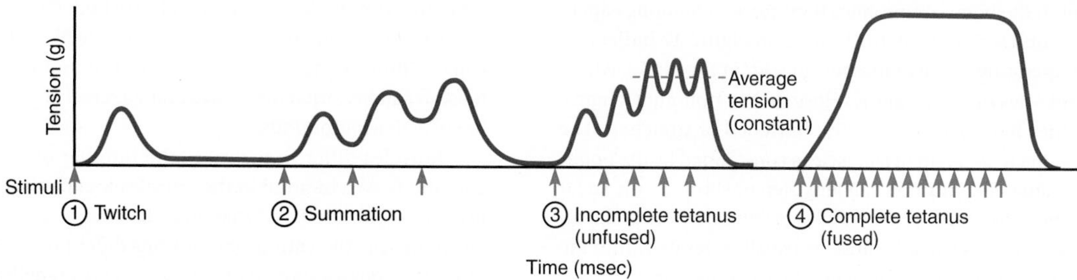

Figure 16 **Effects of high stimulus frequency: summation and tetanus.**

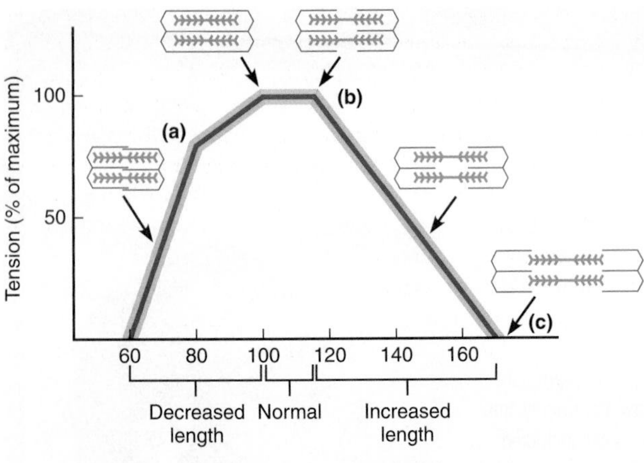

Figure 17 A length-tension curve. The curve plots the tension developed by a muscle as a function of the muscle's resting length. The center bracket indicates the normal range of muscle lengths *in situ*. Drawings of sarcomeres indicate changes in the relative positions of thick and thin filaments as muscle length changes.

crossbridges be able to bind to actin. If a muscle fiber is stretched beyond the optimum length, the tension generated by the fiber decreases linearly as its length increases (segment b–c in Figure 17). The reason is that the degree of overlap between thick and thin filaments decreases as sarcomeres lengthen, and crossbridges that are not overlapped by thin filaments cannot bind to actin and, therefore, cannot contribute to force generation.

If the muscle is significantly shorter than its optimum length, tension declines in two stages as the length decreases. At first tension decreases relatively slowly; once the length decreases beyond a certain point (to the left of point (a) in Figure 17), however, tension decreases more rapidly. As sarcomeres shorten beyond their optimum length, the thin filaments at opposite ends of the sarcomere begin to overlap each other, which interferes with their movement. Then, as the sarcomeres shorten beyond point (a) in Figure 17, the Z lines eventually come into contact with thick filaments, so most of the force generated by crossbridges is exerted on the sarcomere itself instead of being transmitted to the ends of the muscle fiber.

The constant force-generating capacity of a muscle fiber at the peak of the length-tension curve in Figure 17 occurs for two reasons: All crossbridges are overlapped by thin filaments and are capable of generating force, and sarcomeres are not short enough to allow thin filaments to come into contact, so interference between them does not occur.

The fact that muscles shorten when they contract does not mean that muscles in the body compromise their force-generating capacity during a contraction. Note the bracket in Figure 17 indicating the normal operating range of muscles *in situ* (in the body), which shows that muscles *in situ* are always close to their optimum lengths, even when they have been maximally shortened or stretched. The reason is that the movement of a muscle is constrained by the bones to which it is attached; muscles can lengthen or shorten only so far because the possible range of joint angles is limited by the architecture of the skeleton. As a result, muscles usually operate within the range of lengths in which they can generate maximum force.

Apply Your Knowledge

Athletes use the optimum muscle length to generate maximum force at the onset of activity. Can you think of any athletic events where an athlete positions himself or herself for a quick start?

Based on what we have seen concerning summation of contractions, we know that the action potential frequency in a muscle's motor units has a direct bearing on the force the muscle generates. The most force a single muscle fiber can develop by contracting isometrically is the maximum tetanic tension, which for most muscle fibers is only about five times greater than the peak tension in a single twitch. Given that muscular tension can vary over several orders of magnitude (for example, the force required to hold a paper clip compared to that required to hold a chair), it is clear that variation in action potential frequency can account for only a small fraction of the range of forces a muscle can generate. We see next that the force generated also depends on the number of muscle fibers contracting.

Regulation of the Force Generated by Whole Muscles

When a muscle contracts, only rarely do *all* of its fibers actively generate force. Although some motor units are active, the fibers in other motor units simply "go along for the ride," passively shortening in response to forces generated by actively contracting fibers. When larger forces are needed, the nervous system can activate some of these "extra" fibers, thereby increasing the total number of active fibers. Indeed, the nervous system exerts most of its control over muscular force by varying the number of active motor units; variation in the frequency of stimulation of individual fibers plays a secondary role. An increase in the number of active motor units is called **recruitment.**

Recruitment

We have seen that within a muscle, fibers belonging to a given motor unit are intermixed with fibers from other motor units. But not all motor units are created equal: They often differ in size, with some having relatively more fibers and others having relatively few. **Figure 18a** shows two motor units (X and Y) in a muscle, the fibers of which have identical force-generating properties and have been stimulated to give maximum tetanic tension. Motor unit X contains five fibers, whereas motor unit Y contains seven fibers. Thus, when motor unit X is stimulated to contract, it generates five times the force of a single fiber because its fibers are working together in parallel; likewise, stimulation of motor unit Y results in a force seven times that of a single fiber. Stimulation of both motor units produces a force 12 times greater than that of a single fiber (Figure 18b). Because a muscle may contain hundreds of motor units, muscular tension can be varied over a wide range merely by varying the number of active motor units.

Muscles differ in regard to the number of motor units they contain, from a handful in the muscles that control movements of the eyes to hundreds in larger muscles such as the biceps. Within a given muscle, the various motor units differ both in the number of fibers they possess and in the diameter and strength of those fibers.

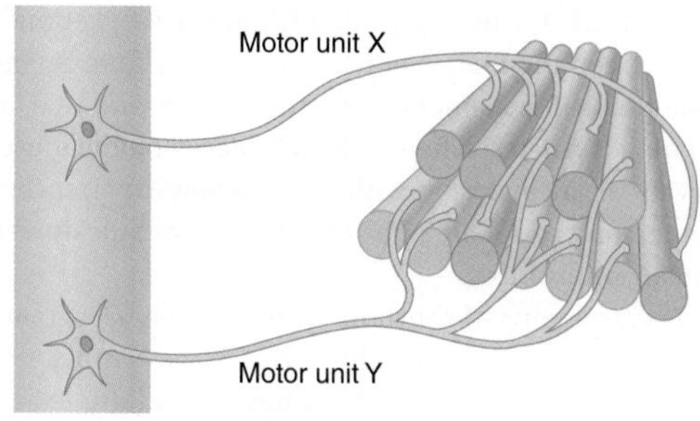

(a)

Tension (g)
12 fibers
7 fibers
5 fibers
Stimuli
X
Y
X + Y

(b)

Figure 18 Increases in force generation with recruitment of motor units. (a) Motor units X and Y, which possess five fibers and seven fibers, respectively. **(b)** Tension developed by motor unit X, by motor unit Y, and by motor units X and Y together.

The fibers within any given motor unit tend to fall within a narrow range of sizes, with some consisting mostly of small fibers and others consisting mostly of large fibers. Furthermore, motor units that have larger fibers also tend to have *more* fibers.

The Size Principle

When a muscle is called on to generate small forces, generally only the smaller motor units come into play; when larger forces are needed, larger motor units are recruited. This correspondence between the size of motor units and the order of recruitment is known as the **size principle.** In addition, when contractions are sustained over a long time, motor units are activated asynchronously—as one becomes active, another ceases its activity. In this manner, the total force of the muscle is maintained at a constant level without overworking any of the individual motor units.

The fact that motor units differ in size has practical implications for precise control of muscular force. As a general rule, fine control is easier when muscular forces are small, because only the smaller motor units are recruited. Smaller variations in force are possible under these conditions because the recruitment of additional motor units causes only small increases in the total number of active fibers. In contrast, when large forces are involved, only larger increments in force are possible because larger motor units are recruited.

The basis for the size principle is not only that motor units vary in size, but also that the *motor neurons* that control them vary in size. Larger motor units are controlled by motor neurons with larger-than-average cell bodies and axon diameters, whereas smaller motor units are controlled by neurons with smaller-than-average cell bodies and axon diameters. This distribution of neuron sizes has important consequences for muscular control. For complicated reasons, larger cells are more difficult to depolarize to threshold; more excitatory synaptic input (through spatial and temporal summation) is required to induce a larger neuron to fire. Thus, when gradually increasing synaptic input is delivered to a set of motor neurons, the small neurons will fire first and the large ones last.

This idea is illustrated schematically in **Figure 19**, in which small, medium, and large motor units (X, Y, and Z, respectively) are controlled by excitatory input from a neuron (W) located in

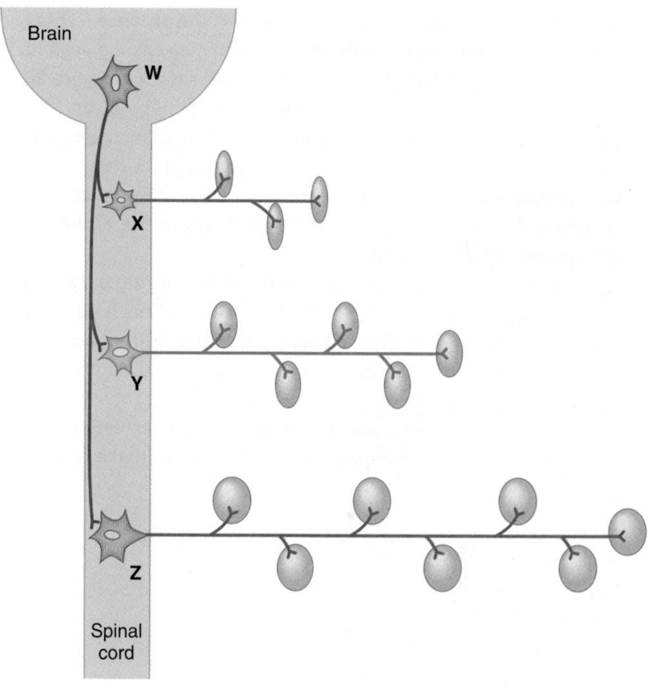

(a)

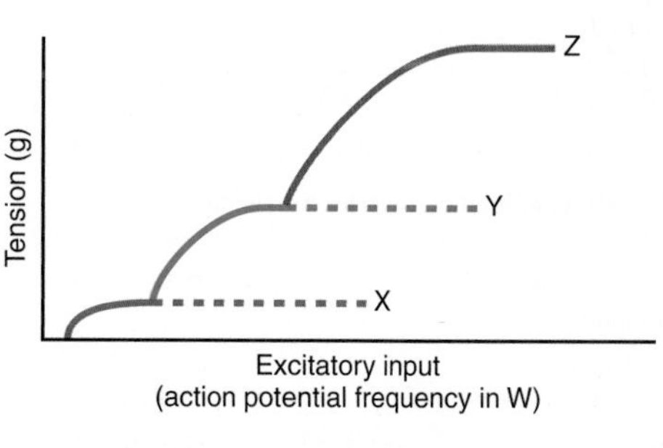

(b)

Figure 19 The size principle. (a) The anatomical relationship of three motor units (X, Y, and Z) of increasing size to an excitatory neuron W within the CNS. **(b)** As the frequency of action potentials in neuron W increases, the order of motor units activated proceeds from smallest (X) to largest (Z).

the CNS. Note that as the action potential frequency in neuron W increases, the motor units become active in order of increasing size. Note as well that as each motor unit is recruited, the force increases in stepwise increments that reflect both the increased number of fibers in the larger units and the larger size of those fibers.

Now we see that the force generated by whole muscles is affected by a combination of the factors acting on individual fibers and the number of fibers that are active. But more than just the force of muscle contraction is important in movement: The speed with which muscles contract is important as well.

Velocity of Shortening

To determine a muscle's velocity of shortening, the muscle is stimulated to contract isotonically; while it does so, the distance it shortens is plotted over time, usually while different loads are placed on the muscle. The results of these measurements, shown in **Figure 20**, reveal three effects: (1) The latent period of shortening (the time between the stimulus and the beginning of shortening) increases with increasing load (see Figure 13); (2) the duration of shortening (the time during which the muscle is shorter than its resting length) decreases with increasing load; and (3) the velocity of shortening decreases with increasing load.

The velocity of shortening is defined as the *rate of change* of the distance shortened, which is the slope of each curve in Figure 20. (Because the slope changes continually throughout the period of shortening, it is customary to use the initial slopes of the curves as the measures of shortening velocity.) When velocity of shortening is plotted as a function of the load, the result is a load-velocity curve like that in **Figure 21**. Note that as the load increases, the velocity of shortening gradually decreases, eventually reaching zero when the load is equal to or greater than the maximum tension that can be generated by the muscle, and that velocity of shortening is greatest when no load is placed on the muscle. These observations are in accord with our everyday experience: We know that a person cannot lift a box full of books as quickly as he or she can lift an empty box.

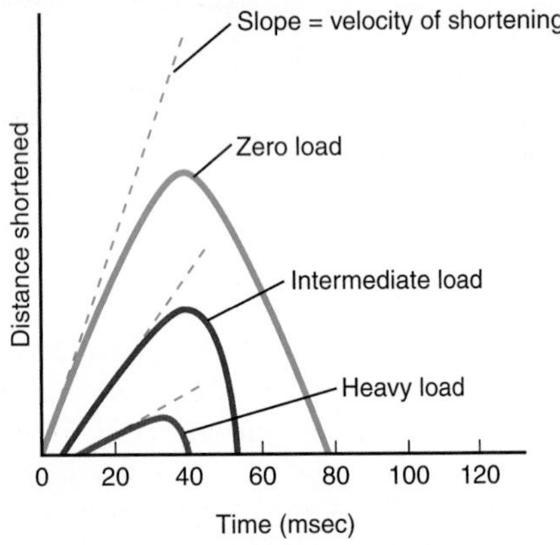

Figure 20 The effect of load on muscle shortening.

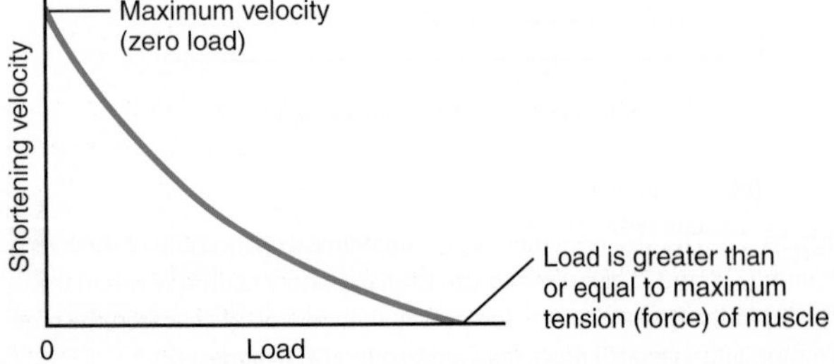

Figure 21 A load-velocity curve.

Measurements of shortening velocity tell us a great deal, including revealing that muscles differ in ways other than force-generating capacity. Researchers have found, for example, that certain types of muscle fibers can shorten faster than others. The bases for such differences are discussed in the next section.

Quick Check 3

1. What is a twitch? How does an isometric twitch differ from an isotonic twitch?
2. What is the size principle?
3. What is the meaning of the term *velocity of shortening*?

4 Skeletal Muscle Metabolism

Our muscles are capable of producing a variety of work, from the movement of our eyes to the intense work of moving furniture around. Our muscle cells require ATP to generate work, and the mechanism for producing ATP will vary based on the type of work needed. Furthermore, we have different types of muscle cells that specialize in metabolic machinery, velocity of shortening, amount of force generated, and degree of resistance to fatigue that enable us to accomplish these various tasks. We now look at muscle cell metabolism and muscle fiber type.

Muscle Cell Metabolism: How Muscle Cells Generate ATP

Muscles are able to begin contracting within a fraction of a second after receiving neural input and are capable of sustaining contractile activity for a time dependent on the intensity of activity. For example, think of a competitive sprinter, whose muscles begin to work near maximum capacity just after the starting gun goes off and continue to do so to the end of the race. For muscles to perform in this manner, muscle cells must have a readily available supply of ATP, even when the demand for ATP increases suddenly and rapidly. Energy sources for contracting muscle are shown in **Figure 22** and described next.

The Role of the Creatine/Creatine Phosphate System

When a muscle fiber is resting, its demand for ATP is small. When it is signaled to contract, however, the demand for ATP soars. At rest, a muscle cell contains a small store of ATP, but it cannot rely on this ATP for long once it begins contracting. To keep from depleting its ATP supply, it must gear up ATP production to keep pace with the increased rate of utilization.

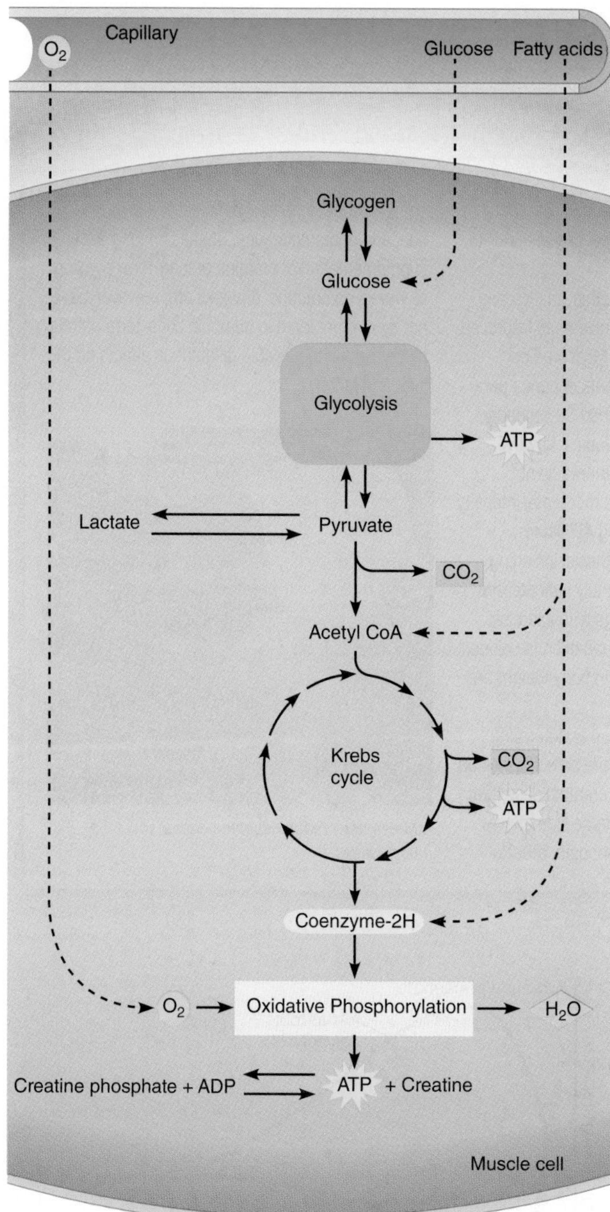

Figure 22 **Metabolic pathways in skeletal muscle.**

The ATP that powers muscle contraction is produced in muscle cells, as in other cells, by substrate-level phosphorylation and oxidative phosphorylation. When a cell's rate of ATP utilization increases, the concentration of ATP inside the cell falls and the concentration of ADP rises. These and other changes then stimulate the enzymes that control ATP-producing reactions, such that ATP is generated at a higher rate. Although this process occurs once a muscle cell begins to contract, the reactions need a few seconds to come up to speed. To ensure a steady supply of ATP in the meantime, muscles rely on an immediately available store of high-energy phosphate that is present in the form of **creatine phosphate,** a compound that donates its phosphate to ADP (which is always present) to form ATP.

The reaction of creatine phosphate with ADP is catalyzed by the enzyme *creatine kinase* and is reversible:

$$\text{Creatine phosphate} + \text{ADP} \xrightleftharpoons{\text{creatine kinase}} \text{creatine} + \text{ATP}$$

Thus this reaction generates ATP and creatine as end-products when it proceeds to the right, and it generates ADP and creatine phosphate when it proceeds to the left. When a muscle cell is at rest, the reaction is at equilibrium, and for every molecule of creatine phosphate formed, another is broken down to creatine. When muscle activity begins, however, ATP levels fall and ADP levels rise, which drives the reaction to the right by mass action. As a result, some ADP is converted to ATP that can be used in the crossbridge cycle, but creatine phosphate is also consumed. A resting cell contains enough creatine phosphate to supply four to five times the quantity of preformed ATP.

Although this reaction can produce ATP for only a short while, that is sufficient to give ATP-producing metabolic reactions time to gear up. (To learn about whether creatine supplements can improve muscle performance, see **Discovery: Creatine Supplements.**) When the muscle cell relaxes, the supply of creatine phosphate is replenished because the reduced demand for ATP causes the ATP concentration to rise and the ADP concentration to fall, which pushes the reaction to the left such that creatine is converted back to creatine phosphate. Thus creatine phosphate remains "on tap" for the next burst of contractile activity.

How Muscle Cell Metabolism Changes with Exercise Intensity

Exercise can occur at a variety of intensities from a walk (light), to a jog (moderate), to a sprint (high). All forms of exercise require a continued supply of ATP, but where that ATP comes from varies based on intensity of exercise and the blood flow's ability to supply enough oxygen to the muscle cells for oxidative phosphorylation. If oxygen levels are low, the muscle cells must rely on anaerobic metabolism to provide ATP.

When a muscle is exercised continuously at a light to moderate intensity (such that the oxygen supply is sufficient to keep up with demand), most of the needed ATP is supplied by oxidative phosphorylation (**Figure 23a**). Initially, muscle cells rely on their own glycogen stores to supply glucose as a substrate to generate ATP. As exercise continues, the muscles shift to begin utilizing glucose and fatty acids delivered to them by the bloodstream. For blood glucose to be used in this way, muscle cells must transiently increase the expression of insulin-independent glucose transporters, called GLUT-4, in their sarcolemmas. After approximately 30 minutes, glucose utilization decreases, and fatty acids become the dominant energy source.

In heavy exercise, oxidative phosphorylation becomes less important as a source of ATP, and substrate-level phosphorylation (particularly anaerobic glycolysis) becomes more important (Figure 23b). Even though glycolysis is capable of producing ATP all by itself, one result is the production of lactic acid, which accumulates in muscle tissue and can spill over into the bloodstream. (Recall that in glycolysis, glucose is converted to pyruvate, which normally undergoes further oxidation in the Krebs cycle and oxidative

DISCOVERY

Creatine Supplements

Use of creatine supplements by Olympic and professional athletes soared in the 1990s, and higher use of these compounds by recreational athletes soon followed. Athletes believed that the supplements made them bigger, stronger, and faster, with minimal side effects. Whether scientific research supports such beliefs remains debatable.

The body usually synthesizes all the creatine it needs on its own. The liver, pancreas, and kidneys synthesize creatine, which is then transported in the bloodstream to muscle cells (95% of all creatine in the body is found in skeletal muscle cells). The synthesis of creatine is subject to feedback inhibition when creatine levels are elevated. Thus consuming additional creatine decreases the endogenous synthesis of creatine, negating some of the effects of creatine supplements. Most studies indicate that creatine levels in muscle cells do, indeed, increase (by approximately 15–20%) with supplements, but the increase reaches a maximum within a couple of days. Studies also indicate that the degree of increase varies considerably between individuals, possibly because of differences in the test subjects' muscle profiles prior to creatine supplementation.

The evidence regarding whether an increase in the body's creatine levels translates to improved muscle activity is inconclusive. Most studies suggest that creatine supplements enhance performance in certain brief (less than 30 seconds), high-intensity activities. Most studies, however, also agree that creatine supplements do not improve performance in low- to moderate-intensity activities. In addition to buffering ATP levels, creatine stimulates protein synthesis, including actin and myosin. Studies generally indicate that combining creatine supplements with exercise increases muscle mass more than exercise alone, although most initial increases in body weight are due to water retention.

More research is also needed to determine whether creatine supplements are safe. Studies on short-term supplementation of creatine (less than 8 weeks) indicate no major adverse side effects. Minor side effects include weight gain, muscle cramps, gastrointestinal difficulties, dehydration, and heat intolerance. Studies on long-term supplementation of creatine or long-term "cycling" (8 weeks on creatine, 8 weeks off), however, have not been completed in humans. Long-term animal studies indicate possible deleterious effects on the liver and kidney.

Istvan Csak/Shutterstock

Athletes use creatine supplements to increase muscle mass.

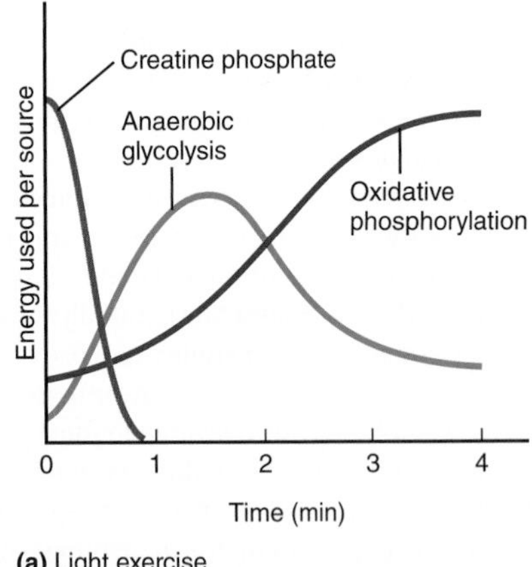

(a) Light exercise

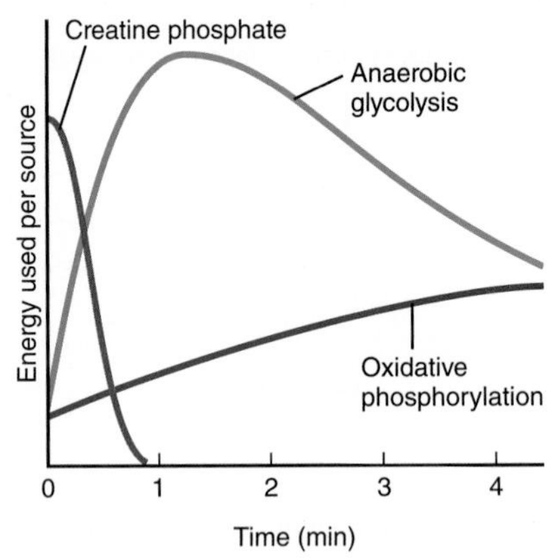

(b) Intense exercise

Figure 23 Comparison of energy sources during (a) light to moderate exercise and (b) intense exercise. During intense exercise, fatigue sets in within a couple of minutes.

phosphorylation; when pyruvate is produced at a rate exceeding that at which it is oxidized, it builds up and is converted to lactic acid.) This buildup of lactic acid is believed to be the reason for the burning sensation a person feels in his or her muscles after strenuous exercise.

So far, we have seen how muscle cells respond to neural signals, how they use ATP to power contractions, and how they ensure an adequate supply of ATP. But what makes one muscle stronger than another? How is it possible that we can use a given group of muscles, such as those of the arm and hand, to lift a chair at one moment and a paper clip immediately thereafter? We address these and related issues in the following section, which focuses on the *mechanics* of skeletal muscle contraction—the various factors that influence contractile speed and force.

Types of Skeletal Muscle Fibers

Although all skeletal muscle fibers are fundamentally alike with respect to the mechanisms of excitation-contraction coupling and force generation, they exhibit significant differences in terms of how quickly they can contract and how they produce most of their ATP. In this section we examine these differences and their effects on muscle performance in everyday life.

Differences in Speed of Contraction: Fast-Twitch Fibers and Slow-Twitch Fibers

When different muscles are stimulated to contract isometrically, some take longer to reach peak tension than others (**Figure 24**). The reason is these muscles contain different populations of fibers. Some muscles (such as the soleus muscle of the leg) contain mostly *slow-twitch fibers,* which contract relatively slowly. In other muscles (such as the extraocular muscles, which control eye movements), the predominant fibers are *fast-twitch fibers,* which contract relatively quickly. In still other muscles (such as the gastrocnemius of the leg), the proportion of slow-twitch and fast-twitch fibers is intermediate. Differences between the two fiber types are seen not only in isometric contractions but also in isotonic contractions. Fast-twitch fibers attain peak isometric tension sooner than slow-twitch fibers; moreover, they have higher maximum shortening velocities when they contract isotonically compared to slow-twitch fibers of similar length.

The difference between fast-twitch and slow-twitch fibers reflects the type of myosin present in their thick filaments. The velocity of muscle contraction depends on the rate of crossbridge cycling, which is limited by the rate-limiting step—ATP hydrolysis. So-called *fast myosin* has the inherent ability to hydrolyze ATP at a faster rate than *slow myosin,* and this ATPase rate has been found to correlate strongly with a fiber's speed of contraction. The higher ATPase rate of fast myosin implies that this form of myosin can complete more crossbridge cycles per second, which means that sarcomeres shorten faster, all else being equal.

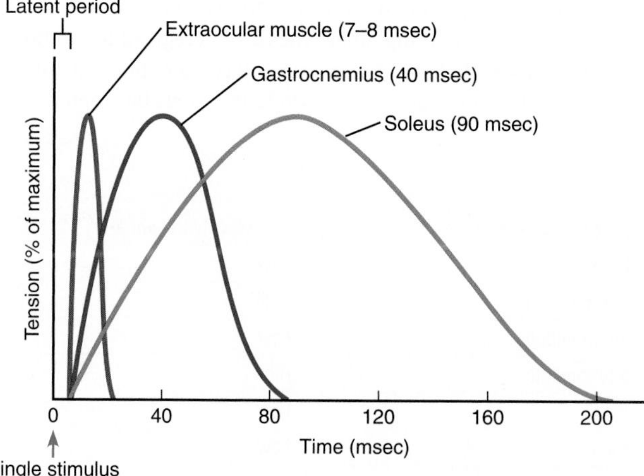

Figure 24 Differences in speed of contraction for three selected muscles. The curves represent isometric twitches; times in parentheses are those required for each muscle to develop maximum tension following a single stimulus (arrow). To facilitate comparison, the curves have been adjusted to make their peaks the same height, even though these muscles develop different maximum tensions.

Differences in the Primary Mode of ATP Production: Glycolytic Fibers and Oxidative Fibers

Even though all muscle fibers have the ability to produce ATP by both oxidative phosphorylation and substrate-level phosphorylation, they differ in their capacities for doing so and are grouped into two general categories on this basis. *Glycolytic fibers* have high cytosolic concentrations of glycolytic enzymes, so they can generate ATP rapidly via glycolysis (substrate-level phosphorylation); these fibers have a relatively low capacity for generating ATP via oxidative phosphorylation because they contain relatively few mitochondria, where oxidative phosphorylation occurs. By contrast, *oxidative fibers* are rich in mitochondria and have a high capacity for producing ATP via oxidative phosphorylation. However, these fibers contain relatively low concentrations of glycolytic enzymes and, therefore, have a low glycolytic capacity. Both fiber types are found in all muscles of the body, but their proportions vary among muscles.

The distinction between glycolytic and oxidative fibers goes beyond differences in glycolytic enzyme content or numbers of mitochondria. Oxidative fibers are generally of smaller diameter and well supplied with capillaries, whereas glycolytic fibers are of larger diameter and are surrounded by fewer capillaries (**Figure 25**). This arrangement makes sense given that oxidative fibers have a higher

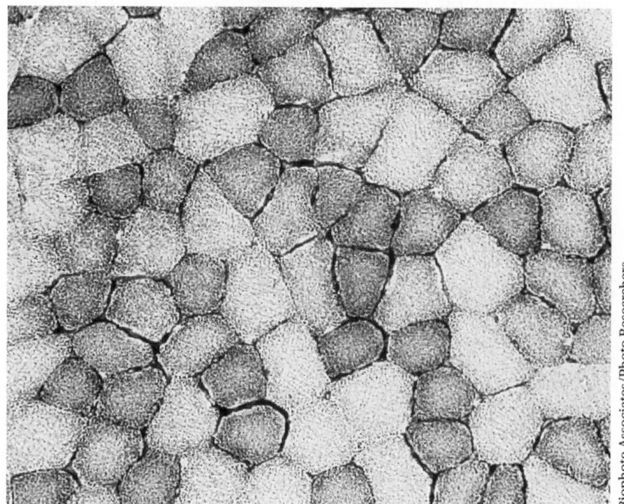

Figure 25 Oxidative and glycolytic muscle fibers. A micrograph of a cross section through a skeletal muscle showing numerous muscle fibers. The muscle has been prepared with a stain that causes mitochondria to appear purple and capillaries to appear black. Smaller fibers stain more deeply, indicating a higher density of mitochondria, which is typical of oxidative fibers. Note that capillary density is highest around these fibers.

capacity for utilizing oxygen and rely more heavily on rapid oxygen delivery for proper function. A rich capillary supply ensures rapid oxygen delivery to the interstitial fluid surrounding the fibers, and the fibers' small diameter minimizes the distance over which oxygen must diffuse to reach the mitochondria.

Another difference between the fiber types is that oxidative fibers contain an oxygen-binding protein known as **myoglobin,** whereas glycolytic fibers lack it. Myoglobin, like hemoglobin (the oxygen-carrying protein in red blood cells), is a reddish molecule that binds oxygen reversibly. It functions as an oxygen buffer by binding and thereby storing oxygen inside the cell; this oxygen can then be released whenever the oxygen concentration inside the cell declines (as can happen when a muscle contracts strongly and compresses nearby blood vessels, thereby interrupting the blood supply). Because this oxygen store is limited, however, it can supply adequate amounts of oxygen for only a short time before it must be replenished, which occurs when blood flow is restored and the oxygen concentration rises. Because myoglobin imparts a reddish brown color to oxidative fibers, these fibers are often referred to as *red muscle.* In contrast, glycolytic fibers, which lack myoglobin and hence this reddish color, are referred to as *white muscle.* Familiar examples of red and white muscle are the "dark" and "white" meat found in chicken.

Glycolytic fibers produce ATP less efficiently than oxidative fibers because fewer ATP molecules are synthesized per unit of fuel consumed. Nevertheless, glycolytic fibers are better able to produce ATP when oxygen availability is low because glycolysis does not require oxygen. When glycolytic fibers are active and producing ATP at a high rate, they must rely on anaerobic glycolysis for energy because these cells have a low oxidative capacity in addition to their high glycolytic capacity. One consequence of these differing capacities is that pyruvate, the end-product of glycolysis, is generated faster than it can be consumed, so it accumulates in these cells. As it accumulates, the pyruvate is converted to lactic acid, which has been implicated as a cause of muscle fatigue (discussed shortly). For this reason, glycolytic fibers fatigue more rapidly than oxidative fibers. In contrast, oxidative fibers produce little lactic acid so long as they are supplied with adequate amounts of oxygen, and as a consequence they are more resistant to fatigue. Lactic acid does not normally accumulate in these cells because their high oxidative capacity enables them to convert pyruvate to acetyl CoA as fast as it is produced.

Slow Oxidative, Fast Glycolytic, and Fast Oxidative Fibers

We have seen that skeletal muscle fibers can be classified as fast-twitch fibers or slow-twitch fibers on the basis of their contractile speeds, and as glycolytic fibers or oxidative fibers on the basis of their metabolic capacities. Not surprisingly, various combinations of contractile speed and oxidative or glycolytic capacity are possible. Indeed, three major classes of skeletal muscle fibers have been identified: slow oxidative fibers, fast glycolytic fibers, and the relatively rare fast oxidative fibers. Muscles generally contain all three fiber types, albeit in different proportions.

As their name implies, **slow oxidative fibers** contain slow myosin and have a high oxidative capacity, producing most of their ATP by oxidative phosphorylation. **Fast glycolytic fibers** contain fast myosin and have a high glycolytic capacity, producing most of their ATP through glycolysis. **Fast oxidative fibers** have a high oxidative capacity and contain fast myosin. (Actually, the myosin ATPase activity in these fibers is intermediate between the slowest and fastest myosin.)

Although there is no direct connection between a muscle fiber's type and its ability to generate force, an indirect connection exists in that the three fiber types also differ in diameter. Slow oxidative fibers are the smallest in diameter and are capable of generating only small forces. Fast glycolytic fibers have the largest diameter and generate the highest forces, whereas fast oxidative fibers are intermediate in terms of diameter and force-generating capacity. Properties of the three types of fibers are summarized in Table 1.

Response of the Three Fiber Types to Exercise Even though the different types of fibers are intermixed in skeletal muscles, all three types are not typically utilized every time a muscle contracts. For one thing, fiber types are segregated into different motor units, such that a given motor unit contains fibers of only one type. Furthermore, because a correlation exists between the size

TABLE 1 Properties of Skeletal Muscle Fiber Types

	Slow oxidative (red)	Fast oxidative (red)	Fast glycolytic (white)
Oxidative capacity	High	High	Low
Glycolytic capacity	Low	Intermediate	High
Speed of contraction	Slow	Intermediate	Fast
Myosin ATPase activity	Low	Intermediate	High
Mitochondrial density	High	High	Low
Capillary density	High	High	Low
Myoglobin content	High	High	Low
Resistance to fatigue	High	Intermediate	Low
Fiber diameter	Small	Intermediate	Large
Motor unit size	Small	Intermediate	Large
Force-generating capacity	Low	Intermediate	High

of a motor unit and the type of fiber it contains, the different fiber types are recruited in a specific order (in accordance with the size principle) as muscle tension increases.

Slow oxidative fibers are recruited first because they are located in the smaller motor units. Fast oxidative fibers are recruited next because they are found in intermediate motor units. Fast glycolytic fibers are the last to be recruited because they are found in the larger motor units. The fast glycolytic fibers are not usually recruited unless a muscle is generating a large amount of force—for example, during high-intensity exercises such as weight lifting or sprinting.

Resistance to Fatigue It is not uncommon for competitive cyclists to ride all day, covering considerably more than 100 miles in the process. By comparison, even Olympic-class weight lifters cannot lift at or near their maximum capacity for more than a few seconds. This difference in duration of exercise reflects the fact that muscles differ in their ability to resist **fatigue,** a decline in a muscle's ability to maintain a constant force of contraction in the face of long-term, repetitive stimulation. Although fatigue eventually sets in after any kind of muscular activity, it generally occurs more quickly when a muscle is stimulated at higher frequencies and when larger forces are generated (**Figure 26**).

Although the precise causes of muscle fatigue are not fully understood, different types of exercise are known to induce fatigue for different reasons. In high-intensity exercise, glycolytic muscle fibers are recruited and, as previously mentioned, have a tendency to generate lactic acid because of their low oxidative capacity. As a consequence, a rapid buildup of lactic acid in muscles occurs in high-intensity exercise. When contractions are strong and sustained, an additional factor can come into play: Strong contractions can compress vessels that supply blood to the muscles, which interrupts or reduces the muscles' blood supply. The resulting decrease in oxygen delivery causes muscle cells to produce larger amounts of lactic acid, which by lowering intracellular pH can alter enzyme activities and interfere in a variety of metabolic processes. In low-intensity exercise, by contrast, lactic acid accumulation is not generally a problem because few glycolytic fibers are recruited; the active fibers are mostly oxidative fibers, which have little tendency to produce lactic acid. The cause of fatigue in low-intensity exercise, which takes a longer time to develop, is thought to be linked to the depletion of energy reserves—glycogen, in particular. Full recovery from this type of fatigue requires approximately 24 hours, whereas recovery from the more rapid-onset type of fatigue requires only minutes or a few hours.

Very-high-intensity exercise can induce *neuromuscular fatigue,* which arises when motor units are stimulated to contract at high frequencies, such as occurs when large forces are generated. When cells are stimulated to contract strongly over long periods, repeated firing of motor neurons can deplete synaptic terminals of acetylcholine, which ultimately causes failure of neuromuscular transmission.

In addition to these mechanisms of fatigue, which affect either a muscle cell's ability to generate force or the nervous system's ability to trigger muscle contractions, fatigue has a psychological component that resists any purely physiological explanation. It is widely accepted among athletes that performance is influenced by mental state as well as by physical condition. Often, the athletes who run the fastest or the longest are the ones with the strongest "will to win." Competitors with less desire are more likely to succumb to discouragement when fatigue sets in and muscles begin to ache; as a consequence, they may put forth less effort.

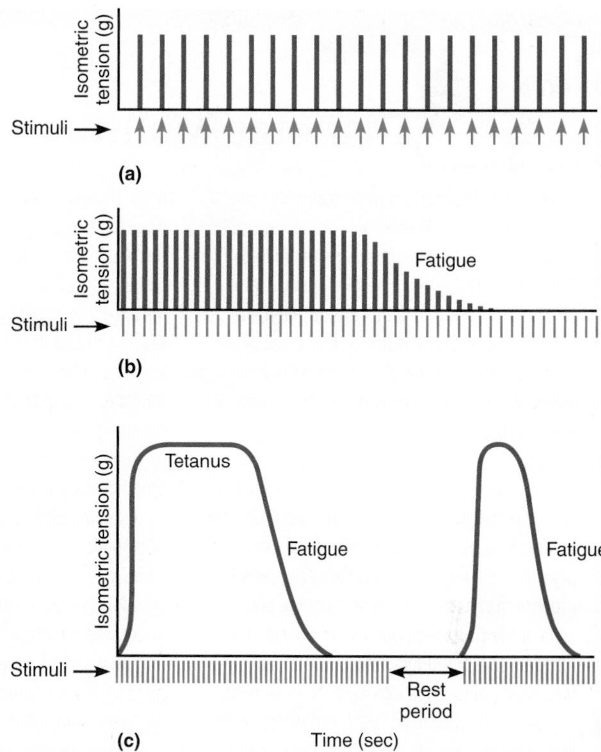

Figure 26 Muscle fatigue. In these records of tension developed by a muscle in response to repetitive stimuli (arrows), the time scale is compressed, so individual twitches appear as spikes. **(a)** Stimulation at a relatively low frequency produces little, if any, fatigue. Note that peak tension is constant over time. **(b)** Stimulation at a higher frequency produces fatigue, apparent in the decline in peak tension near the end of the record. **(c)** A muscle stimulated at a frequency high enough to yield maximum tetanic tension fatigues even more rapidly. After a recovery period, tension returns to previous levels upon resumption of stimulation, but fatigue sets in more rapidly than before.

Long-Term Responses of Muscles to Exercise Athletes go into training not simply to hone the skills needed in their particular sport, but also to change the body's physical condition so that it is fitter to perform those skills. Such changes involve, among other things, changes in muscles' cellular architecture that result from regular exercise over time. As a result of these changes, the muscles' capacity to generate force and resist fatigue is altered.

As every athlete knows, different types of exercise affect muscles in different ways—which explains why it is important to choose an exercise regimen that can achieve the desired results. For example, an aspiring marathon runner, who requires great endurance for her sport, should train using long-duration, low-intensity exercises ("aerobic" exercise) such as jogging. In contrast, an aspiring boxer, who desires bigger and stronger muscles, would do well to include short-duration, high-intensity exercise in his regimen.

FOCUS ON DIABETES

Diabetes Mellitus

Type 2 diabetes mellitus makes a person prone to many serious health issues, but following the recommended lifestyle regimen can decrease these risks. The most important lifestyle factors for controlling diabetes are nutrition and exercise. Appropriate nutrition and exercise help regulate body weight and blood glucose levels. We will describe some of the benefits of exercise here.

Exercise directly affects blood glucose levels, but the effect varies based on whether the exercise is light to moderate (aerobic) or intense (anaerobic). Normally, an increase in blood glucose (which typically occurs after a meal) triggers the release of insulin from the beta cells of the pancreas. The insulin travels in the blood to cells throughout the body, enhancing the uptake of glucose into most of these cells. The liver and muscle cells are stimulated to take in excess glucose, as these cells can convert glucose to glycogen for storage; in the case of the liver, glucose can be converted to fat for storage as well. Thus, by stimulating cellular uptake of glucose, insulin induces a decrease in blood glucose. When a person exercises, the demand of body cells for glucose, especially muscle cells, increases. At moderate levels of exercise, muscles take up glucose at a faster rate, thereby decreasing blood glucose levels. The uptake of glucose occurs through increased expression of insulin-independent glucose transporters called GLUT-4 transporters. In contrast, at intense exercise levels, the body often responds with the "stress response," a series of hormonal and neurological responses that promote mobilization of energy stores. Thus intense exercise increases blood glucose levels.

Exercise indirectly improves the health of diabetic individuals by reducing body fat. Obesity is one of the predisposing factors for type 2 diabetes mellitus, so reducing body fat decreases the chances of developing diabetes and decreases the severity of diabetes once it has developed. Aerobic exercise increases triglyceride catabolism, thereby promoting weight loss. Exercise also builds muscle, and muscle tissue burns more calories than fat under resting conditions. In this way, exercise increases the basal metabolic rate (the rate of whole-body metabolism when a person is awake, is lying down, and has fasted for 12 hours), further promoting weight loss.

Aerobic exercise can also produce a healthier heart and vasculature. Cardiac muscle, like skeletal muscle, can be conditioned based on use, leading the heart to become more efficient. Exercise also lowers the level of "bad" cholesterol (low-density lipoprotein [LDL]), which promotes heart and vasculature disease, and increases the level of "good" cholesterol (high-density lipoprotein [HDL]), which protects against heart and vasculature disease. Thus exercise can counter some of the deleterious effects of diabetes on the cardiovascular system. An especially important consideration is exercise's ability to improve blood flow to the nervous system, which may protect against the development of various forms of diabetic neuropathy.

Which type of exercise should a person with diabetes perform? Most doctors recommend slowly building up to moderate-intensity exercise, such as brisk walking, jogging, bicycling, swimming, playing basketball, or other activities that you can sustain for 20–30 minutes with your heart rate elevated by approximately 30–60%. It is better to exercise for short durations on several days of the week compared to engaging in long-duration exercise on only one day per week.

Aerobic exercise increases the oxidative capacity of muscle fibers, thereby increasing their resistance to fatigue. As a result of such training, some fast glycolytic fibers are effectively converted to fast oxidative fibers. (However, because exercise does not alter the type of myosin present in muscle fibers, slow-twitch fibers remain slow and fast-twitch fibers remain fast.) Other adaptations include increases in the size and number of mitochondria within the fibers, and an increase in the number of capillaries surrounding the fibers. In addition, the average diameter of the fibers decreases, which facilitates the movement of oxygen into the cells but also decreases the cells' force-generating capacity. Aerobic exercise also helps improve health in certain disease states such as diabetes (see **Focus on Diabetes: Diabetes Mellitus**).

In contrast, high-intensity exercise decreases the oxidative capacity of muscle fibers and increases their glycolytic capacity, thereby converting a portion of the fast oxidative fibers into fast glycolytic fibers. Changes include decreases in the size and number of mitochondria, increases in the concentration of glycolytic enzymes, and increases in average fiber diameter. However, the declining oxidative capacity of these fibers reduces their resistance to fatigue. As fibers grow, new myofibrils are synthesized, which enables the fibers to generate more force; reflecting the increase in average fiber diameter, entire muscles also become bulkier and more massive. Note that muscle growth is not due to the addition of new fibers, because muscle fibers are *postmitotic*—that is, they cannot divide to form new cells. Even though new fibers can be generated from immature precursors called *satellite cells,* this process normally happens only when fibers die and must be replaced. If muscle is not used, atrophy occurs. Atrophy also occurs in some disease states (see **Clinical Connections: Muscular Dystrophy**).

Quick Check 4

1. How does creatine phosphate help to maintain an adequate supply of ATP at the start of a muscle contraction?
2. What is the major difference between slow-twitch fibers and fast-twitch fibers?
3. Which three types of fibers are found in skeletal muscles?
4. How does aerobic exercise affect resistance to fatigue in muscles? How is resistance to fatigue affected by high-intensity exercise? Why?

CLINICAL CONNECTIONS

Muscular Dystrophy

Every Labor Day weekend from 1966 to 2010, comedian/actor Jerry Lewis hosted the Muscular Dystrophy Telethon to raise money for the Muscular Dystrophy Association. Muscular dystrophy (MD) is a group of more than 30 inherited diseases that cause progressive weakening of the muscles. The various forms are classified based on mode of inheritance, age of onset, distribution pattern of muscle weakness, and rate of progression. Most forms attack only skeletal muscle, but others affect smooth and/or cardiac muscle as well. Currently, there is no cure, and certain forms progress to an early death.

The most common form of MD in children is Duchenne MD, which primarily affects boys. Duchenne MD rapidly progresses from its onset, which tends to occur at 2–6 years of age; most patients depend on a respirator for breathing by age 20 and die in their early 20s. Becker MD is a milder form, whose onset often occurs in the late teens to early 20s. Both Duchenne and Becker MD are caused by a genetic mutation of the X-linked gene for dystrophin, a protein closely associated with the sarcolemma and cytoskeleton of cardiac and skeletal muscle cells. Dystrophin may function in maintaining the integrity of muscle tissue during contraction.

The mutation causing Becker MD is a deletion. Some nucleotides of the gene have been omitted, but the reading frame is intact, resulting in a modified dystrophin molecule. Some dystrophin activity is still maintained, so Becker MD is relatively mild. The mutation causing Duchenne MD is also a deletion, but in this instance the reading frame is altered, affecting a larger area of the code for amino acids. With little to no dystrophin activity, Duchenne MD progresses rapidly.

Early signs of Duchenne and Becker MD include frequent falling, difficulty standing, and weakness in the upper arm and leg muscles and pelvic muscles. Muscle weakness progresses with age and leads to an inability to walk, respiratory distress, and cardiac complications. Treatment includes physical therapy to delay the progression of *contracture* (shrinkage) of muscle tissue and, at later stages, respiratory therapy. Medications prescribed for these diseases include corticosteroids, whose anti-inflammatory actions may decrease progression of MD.

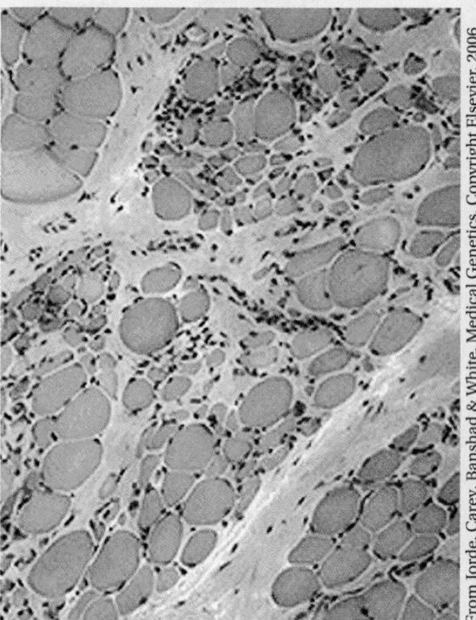

From Jorde, Carey, Banshad & White, Medical Genetics, Copyright Elsevier, 2006

Muscle biopsy from a patient with muscular dystrophy.

Critical Thinking Questions

1. What are the criteria used to classify the more than 30 forms of muscular dystrophy?
2. Describe the causes of Duchenne and Becker MD. How do these diseases affect the functionality of the dystophin gene?
3. Describe the symptoms of Duchenne and Becker MD. What treatments are currently available for these forms of MD?

5 Control of Skeletal Muscle Activity

Now we look at how selective activation of muscles allows for smooth, coordinated movement.

Muscle Activity Across Joints

Almost all skeletal muscles are connected to at least two bones by tendons. When a muscle contracts, one of these bones typically moves while the other remains relatively stationary. A muscle's point of attachment to the stationary bone is called the *origin*, whereas its point of attachment to the movable bone is the *insertion* (**Figure 27**). Depending on a muscle's function and location, its tendons may be short or long. The biceps muscle, for instance, is connected to the bones of the arm by short tendons, but the gastrocnemius muscle of the calf is connected to the heel bone by a long tendon (the *Achilles tendon*), an arrangement that frees the ankle of both bulk and weight. Similarly, long tendons permit the muscles that control the fingers to reside in the forearm, which is useful because a muscle-bound hand would be detrimental to finger dexterity.

A muscle can actively exert force only by contracting, which in the case of skeletal muscle means *pulling*, not pushing, on a bone. Nevertheless, it is clear that you can use muscular force to move the elbow joint, for example, in opposite directions: You can *flex* the forearm (that is, reduce the angle of the elbow joint) by contracting the biceps muscle, and you can *extend* the forearm (increase the angle of the elbow joint) by contracting the triceps muscle, which is

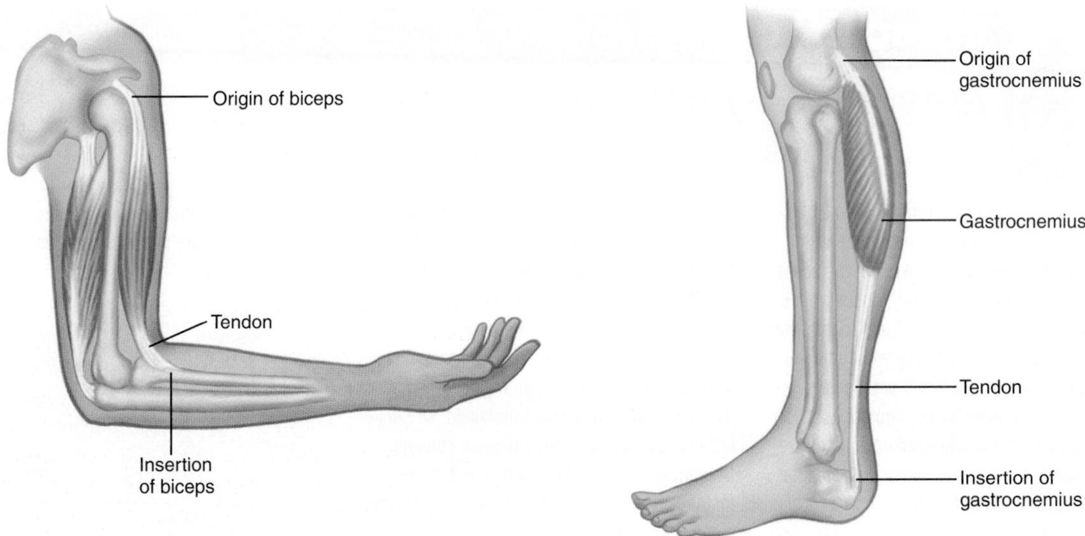

Figure 27 Attachment of muscle to bone.

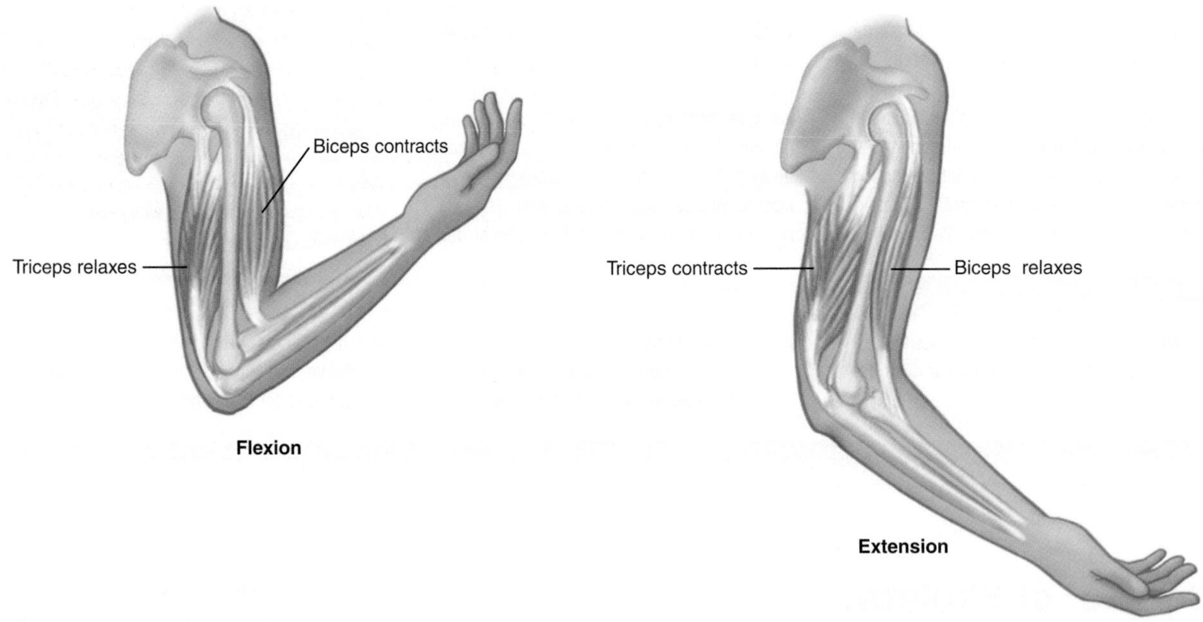

Figure 28 Antagonistic muscles at the elbow.

located opposite the biceps in the arm (**Figure 28**). The biceps and triceps are examples of muscles that are *antagonistic*—each exerts force in a direction that opposes the action of the other. (Most skeletal muscles across joints are arranged in antagonistic groups.) To flex the forearm, the biceps generates force actively while the triceps relaxes and stretches passively—that is, in response to forces exerted on it by the action of the biceps. To extend the forearm, the triceps contracts actively, and the biceps stretches passively. Thus most movements across joints involve simultaneous activation of motor neurons to one group of muscle fibers and inhibition of motor neurons to the antagonistic fibers. At times, however, simultaneous contraction of antagonistic muscle groups is useful. For example, when you "brace yourself" to receive a package whose weight you are unsure of, the biceps and triceps are stimulated to contract simultaneously to stiffen the elbow joint. Of course, how much tension is actually needed to support the package will depend on its weight, which you will not know until the package is in your hands. The next section discusses feedback for motor control.

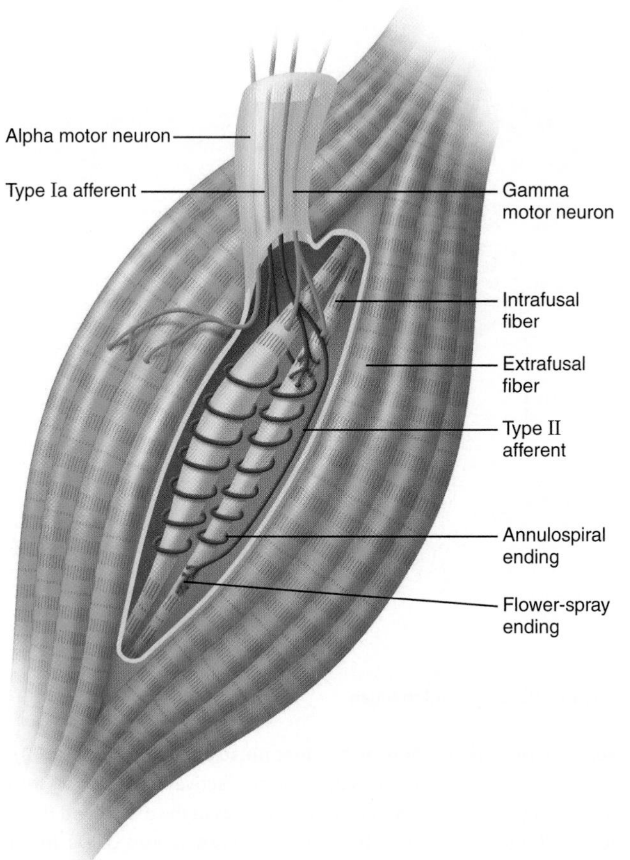

Figure 29 Muscle spindle.

Muscle Receptors for Coordinated Activity

Within skeletal muscle are sensory receptors and associated afferents that detect movement of the muscle and communicate this information to the central nervous system. There are two main receptor types: **muscle spindles,** which detect muscle length, and **Golgi tendon organs,** which detect muscle tension.

Muscle Spindles

Muscle spindles consist of 2–12 modified muscle fibers, called *intrafusal fibers,* enclosed in connective tissue (**Figure 29**). Intrafusal fibers run parallel with the muscle fibers for contraction, called *extrafusal fibers.* A muscle spindle has a central "bag" region that detects changes in muscle length. Two types of sensory endings are located around the central region: (1) an *annulospiral ending* that is wrapped around the central bag area and is connected to type Ia afferent fibers, and (2) flower-spray endings located to the side of the central bag of certain muscle spindles and connected to type II afferent fibers. On either side of the bag are striated contractile components that receive innervation from *gamma motor neurons* (the motor neurons that innervate extrafusal fibers are *alpha motor neurons*). As described next, contraction of the intrafusal fibers adjusts the sensitivity of the receptor.

When a muscle is stretched, the intrafusal fibers are stretched. This activates the receptors, causing the frequency of action potentials to increase in proportion to the degree of stretch (**Figure 30**). Likewise, a decrease in the length of the muscle, such as occurs during muscle contraction, decreases the frequency of action potentials in the afferent neurons. In this case, contraction of muscle fibers causes the muscle spindle to go slack, which would make

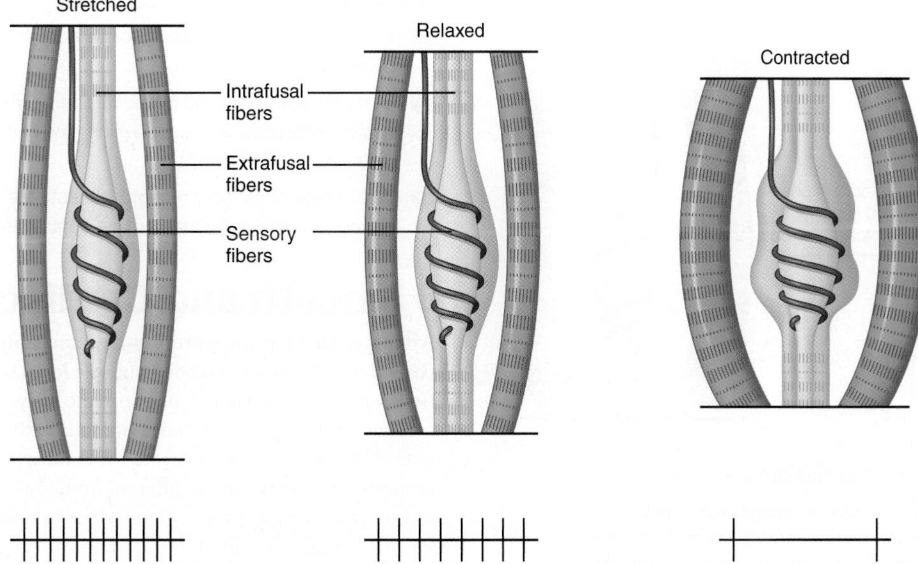

Figure 30 Response of muscle spindle afferents to changes in muscle length.

them inefficient detectors of any further changes in muscle length (**Figure 31a**). To maintain their effectiveness, gamma motor neurons stimulate contraction of intrafusal fibers.

During voluntary muscle contraction, alpha and gamma motor neurons are coactivated by upper motor neurons of the pyramidal tract to cause contraction of extrafusal and intrafusal muscle fibers *almost* simultaneously. Because axons of alpha motor neurons are larger in diameter than gamma motor neurons, action potentials will be conducted to the extrafusal fibers more quickly than to the intrafusal fibers. The extrafusal fibers will contract, creating slack on the muscle spindle, which in turn responds with a decrease in the frequency of action potentials in the afferent neuron. Within milliseconds, however, the intrafusal fibers contract, removing the slack in the muscle spindle. In this manner, the muscle spindle is reset to detect any further changes in muscle length (Figure 31b).

Golgi Tendon Organs

Golgi tendon organs (GTOs) are capsules of connective tissue intertwined with collagen fibers of tendons. They are located in series with muscle fibers and detect muscle tension (**Figure 32**). Muscle contraction stretches the tendons, activating the GTOs and increasing the frequency of action potentials in the associated Ib afferent fibers. Passive stretch of tendons also activates GTOs. Input from GTOs to the central nervous system provides reflex inhibition of the motor neuron controlling the muscle containing the GTOs. This *tendon reflex* protects muscle from excessive activity.

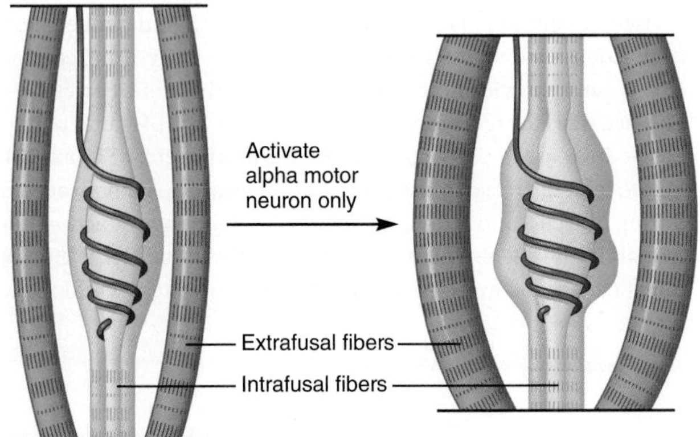

(a) No gamma activity

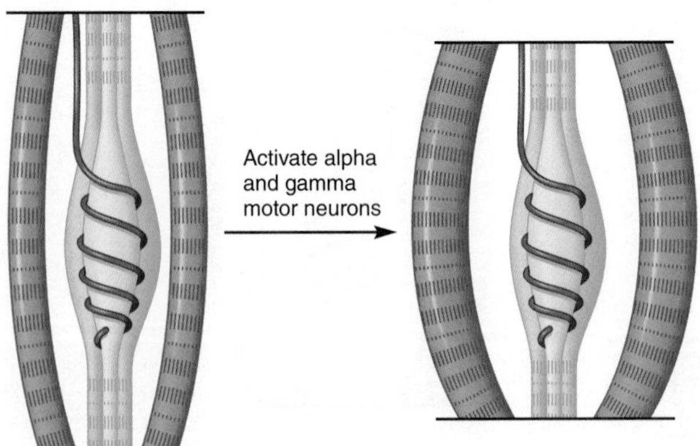

(b) Coactivation of alpha and gamma motor neurons

Figure 31 Coactivation of alpha and gamma motor neurons.
(a) Theoretical response of muscle spindle to activation of alpha motor neurons to the extrafusal fibers without gamma motor neuron activation. **(b)** How coactivation of alpha and gamma motor neurons resets the muscle spindle to detect further changes in muscle length.

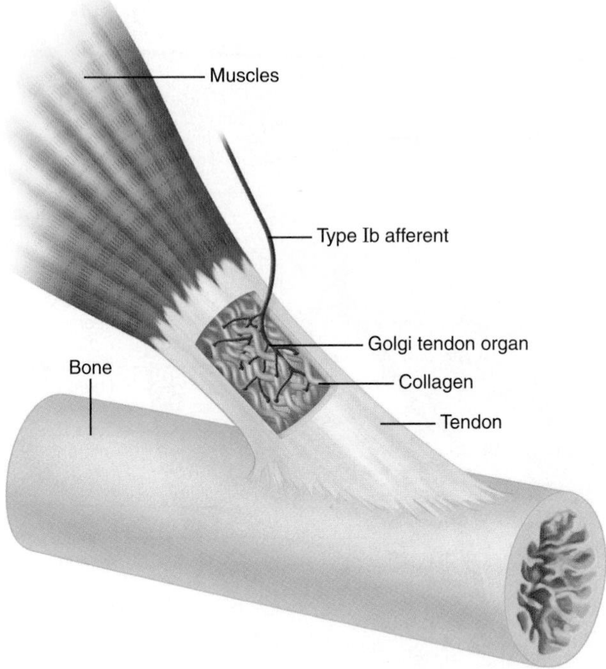

Figure 32 Golgi tendon organ.

Quick Check 5

1. Which type of sensory receptor detects the length of skeletal muscle?
2. Compare extrafusal and intrafusal fibers. Which type of motor neuron innervates each, and which type of motor neuron has the faster conduction velocity?
3. Which type of sensory receptor detects the tension generated during skeletal muscle contraction?

6 Smooth and Cardiac Muscle

When people hear the word *muscle,* most automatically think of *skeletal muscle,* which has been our sole focus so far in this chapter. However, two other types of muscle in the body—*smooth muscle* and *cardiac muscle*—perform their functions without attracting much notice. We now shift our focus to these muscles and to the special properties that make them different from skeletal muscle. For easy comparison, the three types of muscle are shown in **Figure 33**. The specific features of smooth muscle and cardiac muscle that are apparent in this diagram will be discussed in the next two sections.

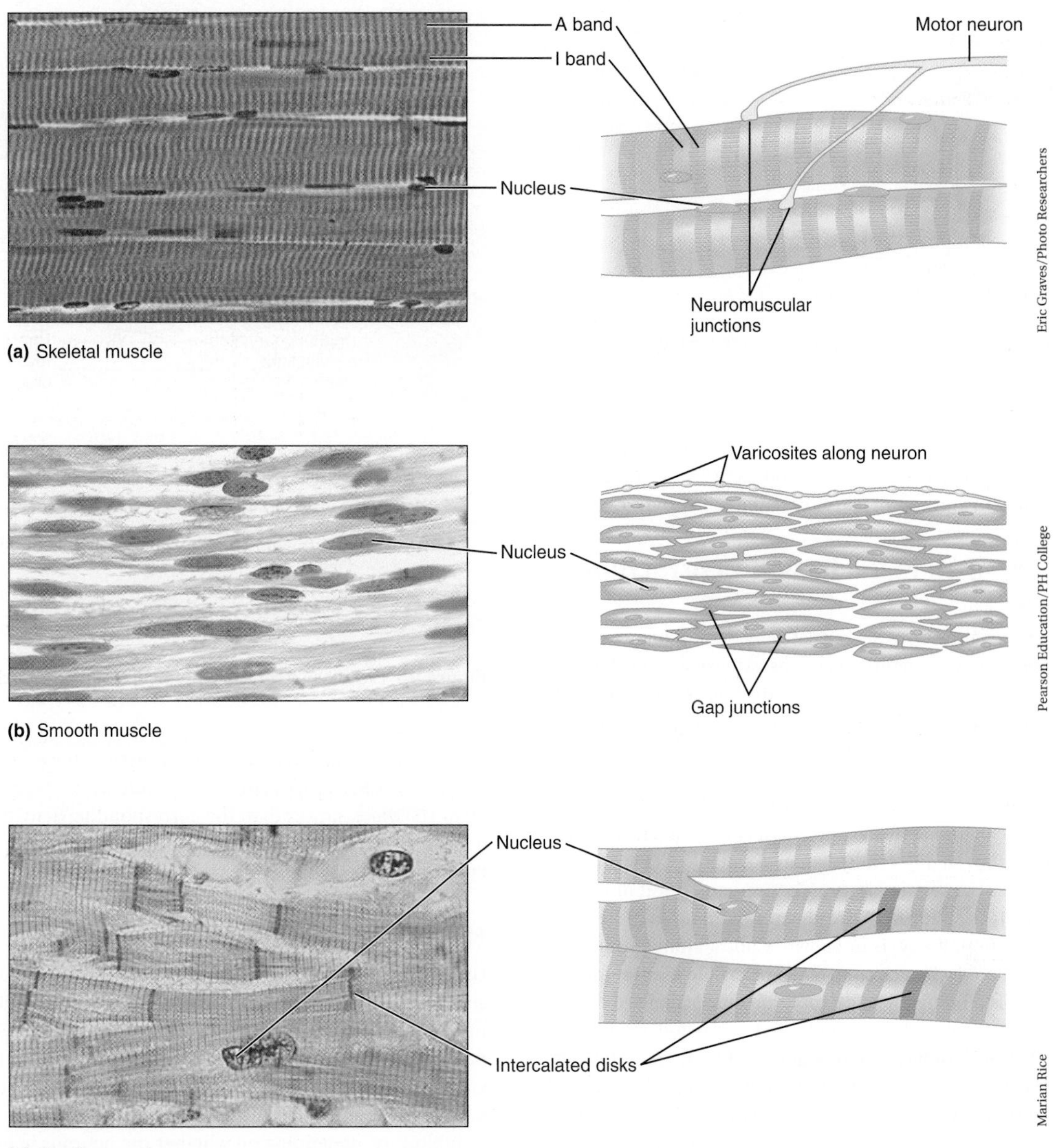

Figure 33 Photomicrographs (left) and schematic diagrams (right) of the three types of muscle in the body.

Q The neurons that control skeletal muscles and those that control smooth muscle belong to different branches of the efferent nervous system. What are these branches called?

Smooth Muscle

Smooth muscle—which gets its name from the fact that it lacks the striations characteristic of skeletal and cardiac muscle and, therefore, appears uniformly bright under the light microscope (Figure 33)—is the type of muscle found in internal organs, blood vessels, and other structures that are not under voluntary control. Functions performed by these muscles are many and varied and depend on the organ in which they are located. In the gastrointestinal tract, for example, smooth muscle contractions mix the ingested food with digestive secretions and propel it from one location to another. In blood vessels, smooth muscle regulates blood flow to organs and tissues by causing the vessels to constrict or dilate.

Like skeletal muscle, smooth muscle has thick and thin filaments and generates force through the crossbridge cycle. However, the filaments are not arranged in sarcomeres, which accounts for the lack of striations. Even though thick and thin filaments are arranged in parallel with each other, as in skeletal muscle, they tend to run obliquely in various directions, which means that contraction

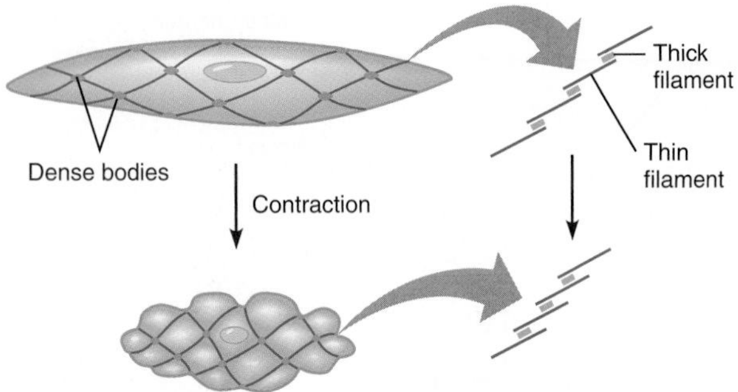

Figure 34 **Oblique arrangement of thick filaments and thin filaments in a smooth muscle cell.**

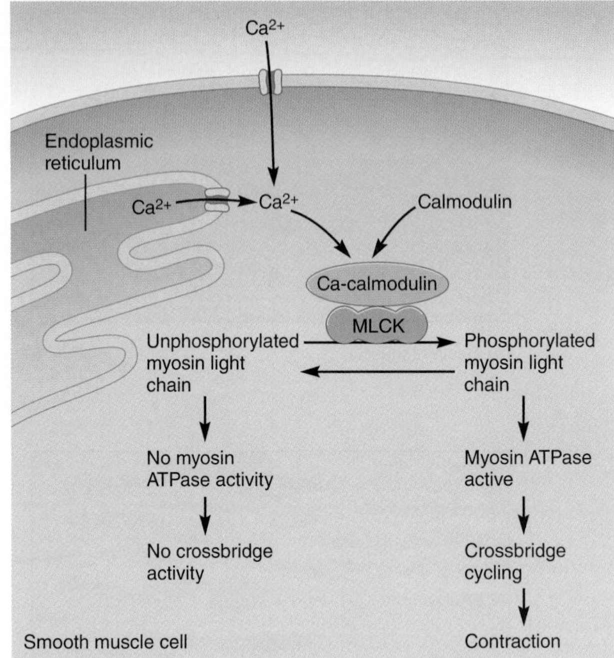

Figure 35 **Excitation-contraction coupling in smooth muscle.**

occurs along several axes (**Figure 34**). *Dense bodies*—points of attachment between these filaments and connective tissue inside the cells—serve to transmit contractile force to the cell's exterior.

The Mechanism of Excitation-Contraction Coupling

Smooth muscle contractions are regulated by intracellular calcium, but the sarcoplasmic reticulum is not as extensive as in skeletal muscle. Moreover, much of the calcium that triggers contraction comes from outside the cells, because when the cell is depolarized, voltage-gated calcium channels in the plasma membrane open and allow calcium to flow in. (Depolarization also triggers the release of calcium from the sarcoplasmic reticulum.)

Cytosolic calcium activates the crossbridge cycle in smooth muscle, but it does so differently than in skeletal muscle. Smooth muscle contains the regulatory protein troponin, but it does not contain tropomyosin. Furthermore, the role of troponin in smooth muscle contraction, if any, is unknown. Contractions in smooth muscle are triggered when calcium binds reversibly to **calmodulin** (**Figure 35**). This binding triggers a conformational change that enables the calcium-calmodulin complex to bind to and activate an enzyme called **myosin light chain kinase (MLCK).** The activated MLCK then catalyzes the phosphorylation of myosin crossbridges, activating the crossbridges and thereby initiating crossbridge activity. The resulting crossbridge cycling proceeds essentially as shown in Figure 7. Note that in smooth muscle the calcium signal that triggers crossbridge activity targets *myosin* filaments, whereas in skeletal muscle it targets *actin* filaments because that is where troponin and tropomyosin are located.

Termination of the crossbridge cycle in smooth muscle requires more than just removal of calcium from the cytosol, because the phosphate groups attached to myosin are covalently bound and, therefore, do not dissociate readily. For this reason, termination of the crossbridge cycle in smooth muscle requires the action of phosphatase enzymes, which, by removing phosphate groups, inactivate myosin. Because these phosphatases (which are continually active) compete with MLCK, activation of myosin occurs only when enough calcium is present to activate MLCK to a degree sufficient to overcome the phosphatase action.

Because the mechanism of excitation-contraction coupling is entirely different from that in skeletal muscle, it takes a lot longer both to initiate and to terminate smooth muscle contractions. This is no real handicap because smooth muscle is not exactly "built for speed": Myosin ATPase activity in smooth muscle is anywhere from 10 to 100 times slower than the corresponding activity in skeletal muscle, which means that smooth muscle contraction is an inherently slow process.

Neural Regulation of Contraction

Unlike skeletal muscle, which is regulated by motor neurons, smooth muscle is regulated by *autonomic neurons.* A given smooth muscle cell may be regulated by sympathetic neurons, parasympathetic neurons, or frequently both. Whereas motor neurons always exert excitatory effects on skeletal muscle cells, the effect of autonomic input to smooth muscle cells may be either excitatory or inhibitory, depending on whether the neurons in question are sympathetic or parasympathetic; these two types of neurons almost always affect a given smooth muscle cell in opposite ways. Furthermore, whether a muscle cell contracts or relaxes depends on where it is located. For instance, during urination, the parasympathetic nervous system causes contraction of the bladder wall smooth muscle and relaxation of urethral smooth muscle. In contrast, the sympathetic nervous system opposes urination by causing relaxation of the bladder wall and contraction of the urethra. The varying responses of smooth muscle to the same branch of the autonomic nervous system reflect differences in the type of neurotransmitter receptors found on the muscle cells rather than differences in the type of neurotransmitter released, because sympathetic or parasympathetic postganglionic neurons release the same neurotransmitters (norepinephrine and acetylcholine, respectively) at virtually all of their target tissues.

Another difference between skeletal and smooth muscle concerns the specificity of neural connections. In skeletal muscle, neural input is delivered to each cell individually because motor neurons are connected to specific cells via neuromuscular junctions. In contrast, smooth muscle cells do not receive direct synaptic connections to specific cells. Instead, neurotransmitter is released from varicosities (swellings) located at intervals along the axon and diffuses over a relatively long distance to large groups of cells (see Figure 33b). Consequently, neighboring muscle cells tend to contract or relax together. This synchronized activity is also promoted by the presence in most smooth muscle tissue of *gap junctions* that allow ions (and other small molecules) to move from one cell to another, so an electrical signal initiated in one cell spreads to neighboring cells.

Skeletal and smooth muscles also differ in regard to the electrical signal generated in response to neural input. Whereas skeletal muscle cells always respond to neural input with an action potential that triggers a reproducible twitch, smooth muscle cells may demonstrate slow but twitchlike contractions in response to action potentials (although this is not necessarily so). The membrane potential of most smooth muscle cells varies in a graded fashion in response to neural input, becoming depolarized if the input is excitatory and hyperpolarized if it is inhibitory, which causes the contractile force to increase or decrease in a graded fashion. Thus contractions in smooth muscle need not be elicited by action potentials. In fact, some smooth muscle cells do not have action potentials at all. When action potentials do occur, they are not usually followed by real twitches (that is, individual contractions, one in response to each action potential), but rather by greater tension.

Some types of smooth muscle cells are able to actively exert tension even in the absence of external stimulation, because resting calcium levels are high enough to maintain a constant low level of crossbridge activity. This resting tension is referred to as *tone.* (Skeletal muscles also exhibit some degree of basal tone, but in their case it is due to a constant low level of neural stimulation.) Many smooth muscle cells have the ability to contract in response to hormones and other chemical agents, independent of neural input. Furthermore, some smooth muscle cells exert active tension in response to mechanical stretch.

Multi-unit and Single-unit Smooth Muscle Smooth muscle tissue varies in both the degree to which muscle cells are connected by gap junctions and the pattern of innervation. In some places, most smooth muscle cells are not connected by gap junctions but instead are largely separate and richly supplied with neurons; smooth muscle of this type is referred to as **multi-unit smooth muscle** (Figure 36a). Multi-unit smooth muscle occurs in the large respiratory airways and large arteries, where the number of active smooth muscle cells may be large or small depending on the circumstances. In other locations, smooth muscle cells are extensively linked by gap junctions, such that electrical signals originating in a few cells are transmitted to the rest of the cells; this kind of smooth muscle is innervated by relatively few neurons and is referred to as **single-unit smooth muscle** (Figure 36b). Examples of organs containing single-unit smooth muscle are the gastrointestinal tract and the uterus, in which large groups of cells contract synchronously.

Spontaneous Changes in Membrane Potential In some cases, smooth muscle cells exhibit spontaneous depolarizations that occur on a regular basis and that may or may not be accompanied by action potentials. Two types of spontaneous depolarizations occur in certain smooth muscle cells: pacemaker potentials and slow-wave potentials (Figure 37). **Pacemaker potentials** are slow depolarizations caused by one or more of the following

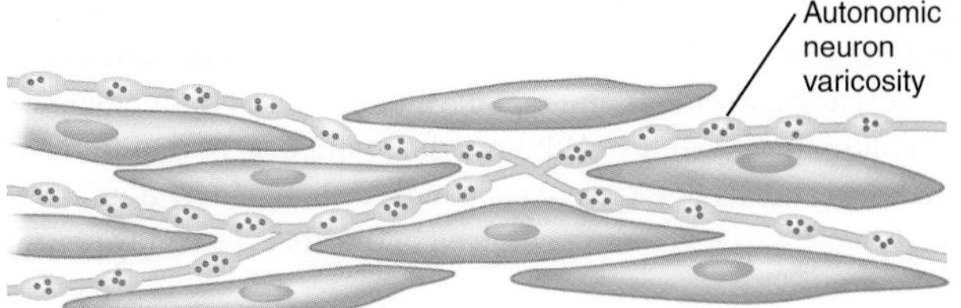

(a) Multi-unit smooth muscle

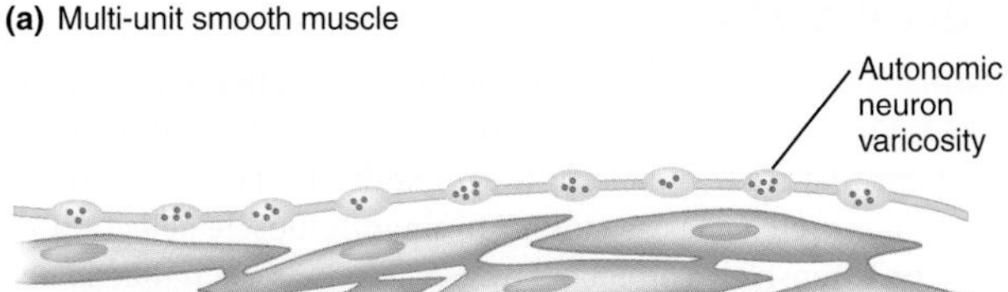

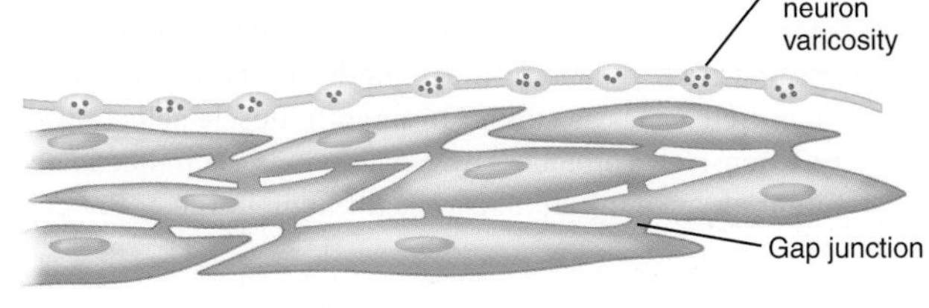

(b) Single-unit smooth muscle

Figure 36 **Multi-unit and single-unit smooth muscle.**

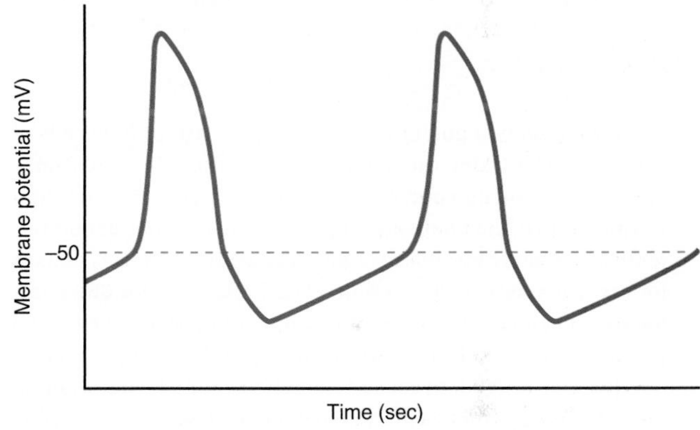

(a) Pacemaker potential

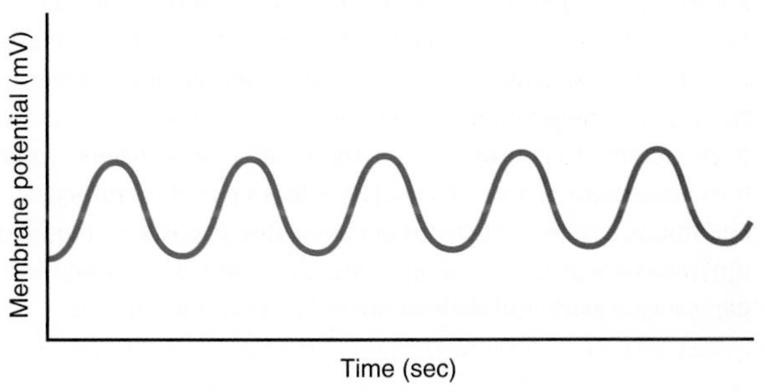

(b) Slow-wave potential

Figure 37 **Spontaneous depolarizations. (a)** Pacemaker potentials are slow depolarizations to threshold, resulting in an action potential. **(b)** Slow-wave potentials are cyclical fluctuations in membrane potential.

permeability changes: increases in sodium, increases in calcium, or decreases in potassium permeability. **Slow-wave potentials** are cyclical depolarizations and repolarizations caused by fluctuations in sodium permeability. Although the frequency or amplitude of electrical signals in pacemakers can be influenced by neural activity, pacemaker signals occur even in the absence of any neural influence.

Cardiac Muscle

Cardiac muscle is similar to skeletal muscle in that it is striated (see Figure 33c), has the same sarcomere structure, and has contractions that are regulated by the troponin-tropomyosin system. Cardiac muscle cells are similar to smooth muscle cells in that they are extensively connected by gap junctions, such that an action potential, once initiated, travels throughout the entire cell network.

Apply Your Knowledge

Cardiac muscle is most similar to slow oxidative skeletal muscle. Which type of exercise benefits cardiac function the most? Will the heart increase much in size as a result of such exercise?

Cardiac action potentials are broad and last for hundreds of milliseconds, making them quite different from the 1- to 2-millisecond, spiky action potentials that occur in skeletal muscle and most neurons. The relatively long duration of cardiac action potentials is significant: Because they last nearly as long as it takes for cardiac muscle cells to contract and relax (**Figure 38**), summation of cardiac muscle contractions cannot occur, even when the action potential frequency is high and the heart is beating rapidly. This is good, because summation would be detrimental to the heart's pumping action in that the heart would not be able to relax completely and fill with blood between contractions.

Certain heart muscle cells that are concentrated in two regions known as the *sinoatrial* and *atrioventricular nodes* exhibit pacemaker activity. The heartbeat is triggered by action potentials originating in pacemaker cells and does not depend on neural input. Because the signals that trigger the heartbeat originate within the heart muscle itself, the heart's contractile activity is said to be *myogenic,* whereas the contractile activity of skeletal muscle is said to be *neurogenic.* The autonomic nervous system regulates the heart muscle by modulating the frequency and force of heart muscle contractions. Table 2 compares the properties of skeletal, smooth, and cardiac muscle.

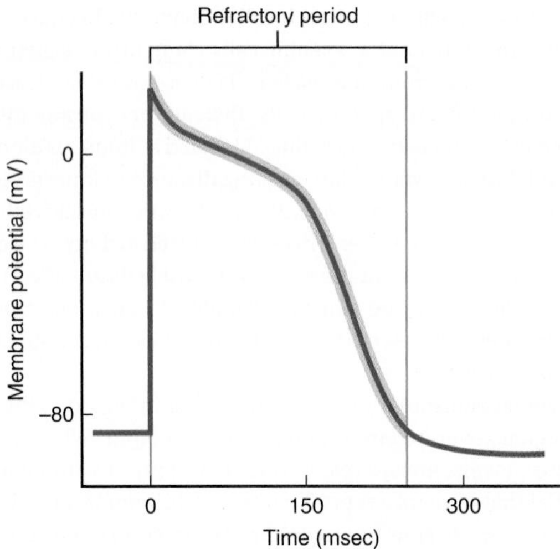

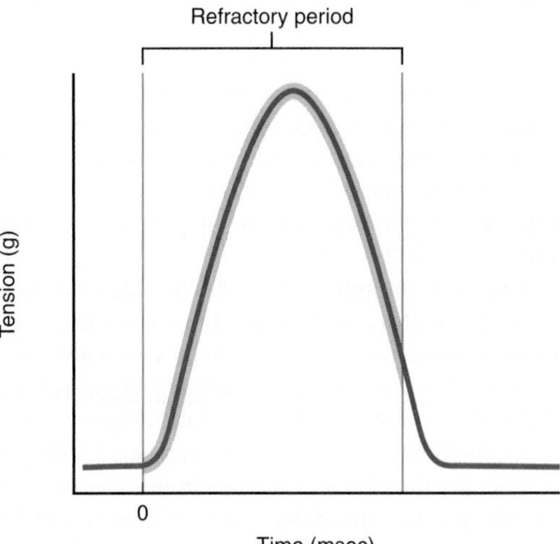

Figure 38 Relative durations of contractions and action potentials in cardiac muscle. The common time scale for the two graphs allows comparison of an action potential (above) with the tension developed in response to it (below) in a cardiac muscle cell.

Quick Check 6

1. How does smooth muscle differ from skeletal muscle with respect to the arrangement of thick and thin filaments? With respect to the mechanism of excitation-contraction coupling?
2. What is the difference between multi-unit smooth muscle and single-unit smooth muscle?
3. What does a pacemaker cell do, and in which types of muscle are these cells found?
4. What prevents summation of cardiac muscle contractions, and what is the significance of this phenomenon?

TABLE 2 Comparison of Skeletal, Smooth, and Cardiac Muscle

Property	Skeletal	Smooth (single-unit)	Smooth (multi-unit)	Cardiac
Striations (sarcomeres)	Yes	No	No	Yes
Actin and myosin	Yes	Yes	Yes	Yes
Level of control	Voluntary	Involuntary	Involuntary	Involuntary
Neural input	Somatic	Autonomic	Autonomic	Autonomic
Neuroeffector junction	Neuromuscular junction—specific	Varicosities—diffuse	Varicosities—diffuse	Varicosities—diffuse
Hormonal control	None	Several, depending on location	Several, depending on location	Epinephrine
Source of calcium	SR	SR and ECF	SR and ECF	SR and ECF
Regulatory protein that binds calcium	Troponin	Calmodulin	Calmodulin	Troponin
Gap junctions	No	Yes	No (or few)	Yes
Pacemaker activity	No	Yes	No	Yes
Myosin ATPase activity	Fastest	Slowest	Slowest	Intermediate
Recruitment	Yes	No	Yes	No

SYSTEMS INTEGRATION The activity of muscle has many diverse effects in the various body systems. Even skeletal muscle has an important role in homeostasis. Skeletal muscle, which includes the respiratory muscles, is responsible for shivering during thermoregulation, thereby directly contributing to maintaining homeostasis. Skeletal muscle also allows for behavior responses of the body to maintain homeostasis, such as putting on warm clothes when the weather is cold, or obtaining and consuming food in response to hunger. In addition, this type of muscle forms *sphincters* in the gastrointestinal tract and urinary system to regulate the flow of waste products out of the body.

Of course, cardiac muscle plays a key role in homeostasis, by pumping blood to the various organs. Smooth muscle has many different roles in homeostasis, as it is found in many different organs serving several functions. For example, smooth muscle in the vasculature regulates blood flow to organs and blood pressure. Smooth muscle in the airways regulates air flow into and out of the lungs. Smooth muscle in the wall of the gastrointestinal tract mixes food with digestive enzymes and fluids for adequate digestion and absorption, and then moves the food through the gastrointestinal tract from mouth to anus.

CHAPTER REVIEW

SUMMARY

1 Skeletal Muscle Structure

- Most skeletal muscles contain numerous elongated cells (muscle fibers) that generate contractile force using energy from ATP hydrolysis.
- Within muscle fibers are myofibrils that contain the contractile machinery.
- The sarcoplasmic reticulum surrounds the myofibrils, stores calcium ions, and is closely associated with transverse (T) tubules, which penetrate into the cell interior from the sarcolemma.
- Skeletal and cardiac muscle are striated, reflecting the orderly arrangement of thick and thin filaments in the myofibrils, which are made up of fundamental force-generating units (sarcomeres) that are joined in an end-to-end fashion.
- Thick and thin filaments contain the contractile proteins myosin and actin, respectively.
- The heads (crossbridges) of myosin molecules are responsible for generating the motion that drives contraction and possess two important sites: an actin-binding site and an ATPase site.
- Two regulatory proteins (troponin and tropomyosin) present on the thin filaments serve to initiate and terminate contractions.

MAP iP Muscular, Anatomy Review: Skeletal Muscle Tissue

MAP iP Muscular, Sliding-Filament Theory

2 The Mechanism of Force Generation in Muscle

- When a muscle contracts, thick and thin filaments slide past one another.
- This sliding is driven by the crossbridge cycle, in which the motion of crossbridges is coupled to their cyclic binding and unbinding to actin molecules in adjacent thin filaments.
- In skeletal muscle, each fiber receives input from one motor neuron, which branches and innervates more than one fiber.
- An action potential in a motor neuron triggers the release of acetylcholine, which binds to receptors in the muscle

fiber's motor end plate; the result is an electrical signal (end-plate potential) that triggers an action potential in the sarcolemma.
- The end-plate potential is followed by propagation of the action potential through the T tubules, release of Ca^{2+} from the sarcoplasmic reticulum, binding of Ca^{2+} to troponin, movement of tropomyosin away from actin's myosin-binding sites, and initiation of the crossbridge cycle.

3 The Mechanics of Skeletal Muscle Contraction

- A motor neuron plus the muscle fibers it innervates constitutes a motor unit.
- When a motor neuron fires an action potential, all fibers in the motor unit contract together.
- The mechanical response of a motor unit to a single action potential is a twitch, which is reproducible in size.
- Twitches can be isometric, in which case the muscle generates force but does not shorten, or isotonic, in which case the muscle shortens.
- Two types of isotonic contractions are possible: concentric and eccentric.
- The force generated by an entire muscle is determined by both the force generated by individual fibers (which depends on the frequency of stimulation, fiber diameter, and changes in fiber length) and the number of fibers that are contracting.
- Stimulation at high frequencies causes summation of twitches, such that the force eventually reaches a plateau (tetanus).
- The central nervous system regulates muscular force by varying both the action potential frequency in motor neurons and the number of active motor units (recruitment). As muscular force increases, motor units are recruited in order of increasing size, a phenomenon referred to as the size principle.

iP Muscular, Sliding-Filament Theory

iP Muscular, Contraction of Whole Muscle

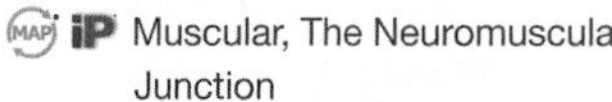

4 Skeletal Muscle Metabolism

- Muscles rely on creatine phosphate as their initial source of ATP.
- Skeletal muscles contain different types of fibers in various proportions.
- Fast-twitch fibers and slow-twitch fibers differ in their speed of contraction, which is related to the type of myosin they contain.
- Glycolytic fibers synthesize most of their ATP via glycolysis and generate lactic acid, which makes them fatigue rapidly.
- Oxidative fibers synthesize ATP mostly via oxidative phosphorylation and are more resistant to fatigue.
- The three major classes of skeletal muscle fibers are slow oxidative fibers, fast glycolytic fibers, and the somewhat rare fast oxidative fibers.

MAP iP Muscular, Muscle Metabolism

5 Control of Skeletal Muscle Activity

- Skeletal muscles are usually arranged in antagonistic pairs across joints.
- Movement of a joint generally involves contraction of one muscle group and relaxation of the antagonist.
- Control of muscle activity depends on feedback from muscle spindles and Golgi tendon organs to the central nervous system.
- Muscle spindles detect muscle length, and Golgi tendon organs detect tension.

6 Smooth and Cardiac Muscle

- Smooth muscle is found in internal organs and other structures that are not under voluntary control and is regulated by autonomic neurons.
- Contractions are triggered by the binding of Ca^{2+} to calmodulin, which activates myosin kinase, resulting in the phosphorylation of myosin crossbridges.
- Smooth muscle is regulated by autonomic neurons—sympathetic neurons, parasympathetic neurons, or both; the effect may be excitatory or inhibitory.
- Smooth muscle cells exhibit two types of spontaneous action potentials: pacemaker potentials and slow-wave potentials.
- In cardiac muscle, contractions are triggered by action potentials initiated in pacemaker cells.
- Action potentials travel from cell to cell through gap junctions, so the entire network of cells contracts as a unit.

iP Muscular, Anatomy Review: Skeletal Muscle Tissue

iP Cardiovascular, Anatomy Review: The Heart

Check your understanding at MasteringA&P and apply your new knowledge to real-life case studies to ensure that you've mastered the content!

EXERCISES

Multiple-Choice Questions

1. When a muscle cell is relaxed and intracellular ATP levels are normal, a crossbridge will remain in which of the following states?
   a) Bound to actin and in the low-energy form
   b) Bound to actin and in the high-energy form
   c) In the high-energy form, with ADP and P_i bound to it
   d) In the high-energy form, with ATP bound to it
   e) In the low-energy form with nothing bound to it
2. During a muscle contraction, which of the following does *not* change length?
   a) The distance between Z lines
   b) The width of I bands
   c) The width of A bands
   d) None of the above
3. Which of the following would tend to *reduce* the concentration of lactic acid that accumulates in a muscle cell as a result of contractile activity?
   a) Increasing the concentration of glycolytic enzymes
   b) Decreasing the oxygen supply to the cell
   c) Increasing the diameter of the cell
   d) Increasing the number of mitochondria in the cell
   e) All of the above

4. Which of the following statements is a valid generalization regarding the properties of smooth muscle?
   a) Neurotransmitters can either excite or inhibit smooth muscle contraction, but any given neurotransmitter is always excitatory or inhibitory, regardless of where the muscle is located.
   b) A given smooth muscle cell can respond to more than one type of neurotransmitter.
   c) Smooth muscle cells are generally unresponsive to neurotransmitters of all types.
   d) Smooth muscle cells can respond to neural input from the somatic or autonomic nervous systems.
   e) None of the above.

5. Which of the following is *not* a determinant of whole muscle tension?
   a) The number of muscle fibers contracting
   b) The tension produced by each contracting fiber
   c) The proportion of each motor unit that is contracting at any given time
   d) The extent of fatigue
   e) The frequency of action potentials in the motor neurons

6. In an isotonic contraction,
   a) Muscle length shortens.
   b) Muscle tension exceeds the force of the load.
   c) The load is moved.
   d) Both a and c.
   e) All of the above.

7. Which of the following is true for the excitation-contraction coupling of *all* muscle types (skeletal, cardiac, and smooth)?
   a) An action potential causes calcium levels in the cytosol to increase.
   b) Calcium binds to troponin.
   c) Thick and thin filaments slide past each other.
   d) Both a and c.
   e) All of the above.

8. During contraction of a skeletal muscle fiber,
   a) The thick filaments contract.
   b) The thin filaments contract.
   c) The A band becomes shorter.
   d) The I band becomes shorter.
   e) All of the above.

9. Which of the following statements concerning the characteristics of different types of muscle fibers is *false*?
   a) The higher the myosin ATPase activity, the faster the speed of contraction.
   b) Muscles that have high glycolytic capacity and large glycogen stores are more resistant to fatigue.
   c) Oxidative types of muscle fibers contain myoglobin.
   d) Oxidative fibers have a richer blood supply.
   e) Larger-diameter fibers can produce greater tension.

10. Which of the following muscle types contain gap junctions?
   a) Skeletal muscle
   b) Smooth muscle
   c) Cardiac muscle
   d) Both a and b
   e) Both b and c

Objective Questions

11. In skeletal muscle, when calcium is released from the sarcoplasmic reticulum, it binds to (troponin/tropomyosin) to initiate the crossbridge cycle.

12. When a muscle fiber contracts, the I bands shorten. (true/false)

13. Glycolytic fibers generate more force than oxidative fibers because they are larger in diameter. (true/false)

14. (Glycolytic/Oxidative) fibers contain high concentrations of the oxygen-binding protein myoglobin.

15. The plasma membrane of a muscle cell is also known as the ________.

16. During muscle contraction, ATP hydrolysis is catalyzed by (myosin head groups/actin monomers).

17. During an (isometric/isotonic) muscle contraction, a muscle develops contractile force but does not change in length.

18. The velocity of contraction of a muscle fiber is directly related to its (diameter/myosin ATPase activity).

19. A reduction in the number of active crossbridges is responsible for a decrease in force-generating capacity of a muscle fiber that is significantly (longer/shorter) than its optimum length.

20. (Oxidative/Glycolytic) muscle fibers are more resistant to fatigue.

21. According to the size principle, the force-generating capacity of a muscle fiber increases in direct proportion to its length. (true/false)

Essay Questions

22. Compare and contrast mechanisms of excitation-contraction coupling in striated muscle and smooth muscle. Be sure to include a description of the mechanisms responsible for termination of muscle contractions.

23. Compare and contrast skeletal muscle and smooth muscle with respect to regulation of contractile activity by the nervous system.

24. Describe the relationship between contractile force and action potential frequency in skeletal muscle, and explain how the situation differs in cardiac muscle.

25. Explain the size principle, including how it relates to the amount of tension developed by a skeletal muscle.

26. Discuss how oxidative muscle fibers differ from glycolytic fibers, and explain how each of these factors relates to the ability of a fiber to resist fatigue.

27. Describe the role of creatine kinase in muscle cell metabolism.

Critical Thinking

28. A physiologist measures peak twitch tension for a series of muscle contractions performed with different loads. If peak tension is plotted as a function of load, which of the following graphs would result?

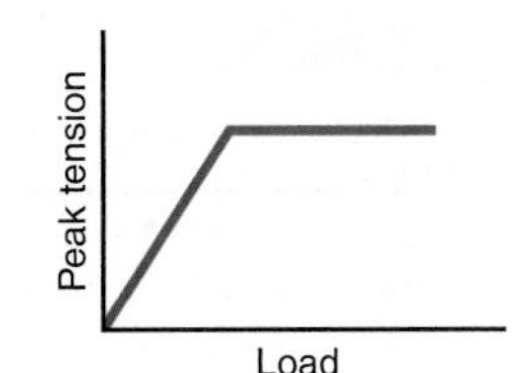

(a)

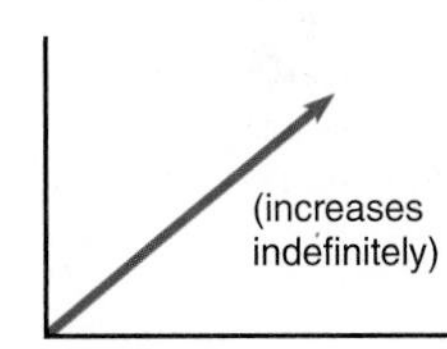

(b)

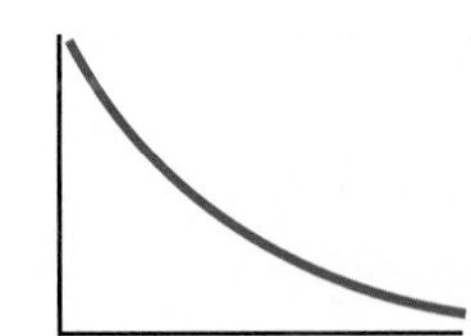

(c)

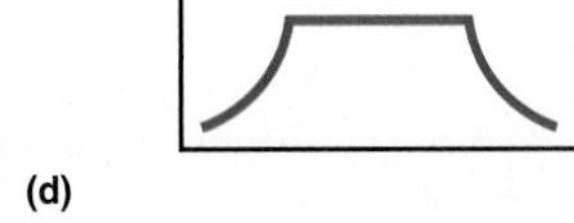

(d)

29. Following the contraction phase of a muscle twitch, cytoplasmic calcium concentration declines, and calcium ions dissociate from troponin. When the calcium concentration returns to its resting level, ATP hydrolysis ceases because
    a) ATP concentration goes to zero.
    b) Low calcium levels cause the energy of ATP hydrolysis to become positive, favoring ATP synthesis.
    c) ATP binding sites on myosin molecules lose their ability to catalyze ATP hydrolysis.
    d) The number of available (unoccupied) myosin ATPase sites declines to virtually zero.

30. Two muscle fibers (X and Y) generate the same isometric peak twitch tension, but fiber X contains myosin that hydrolyzes ATP at a high rate, whereas fiber Y has myosin that hydrolyzes ATP more slowly. Properties of these fibers are consistent with which of the following sets of curves?

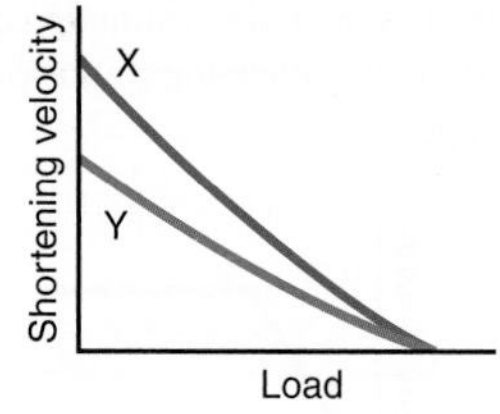

(a)

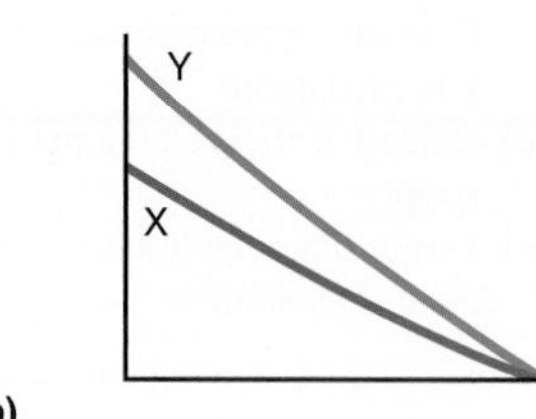

(b)

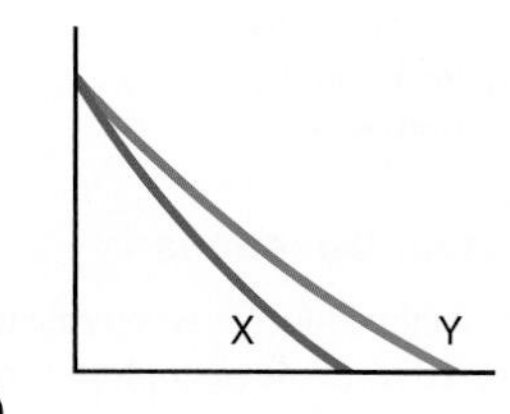

(c)

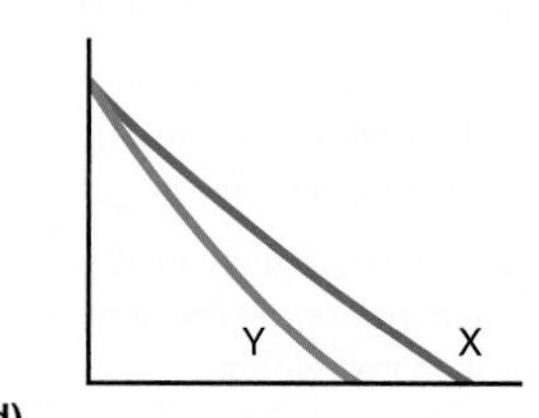

(d)

31. Suppose you have a beaker containing a suspension of purified myosin molecules and ATP in a saline solution. Based on your understanding of events occurring during the crossbridge cycle,
    a) ATP hydrolysis should occur.
    b) ATP hydrolysis will occur if actin is added to the solution.
    c) ATP hydrolysis will occur if calcium is added to the solution.
    d) ATP hydrolysis will occur if calcium and troponin are added to the solution.

32. During aerobic activities, glucose and fatty acids are delivered to skeletal muscle cells by the bloodstream. Based on what you already know about cell metabolism, explain the sources of glucose and fatty acids for contracting muscle and describe the metabolic pathways by which they provide energy.

33. Glycolysis under anaerobic conditions provides approximately 5% of the energy that could be produced under aerobic conditions. Thus one might assume that anaerobic exercises (such as weight training) would be a better way to lose weight than aerobic exercise. Explain why this is not true.

Photo credits for icons in feature boxes: Focus on Diabetes: Tanyae/Dreamstime; Clinical Connections: Iofoto/Dreamstime; Understanding Exercise: Pixattitude/Dreamstime; Discovery: Yuri Arcurs/Cutcaster.

Glossary

actin the most common microfilament; found in thin filaments in muscle fibers; provides structural support for microvilli

calmodulin (kal-mod-you-lin) cytosolic calcium-binding protein; modulates the activity of intracellular proteins

contractile component thick and thin filaments of muscle; compared to elastic components

creatine phosphate a compound in muscle cells that can donate a high-energy phosphate to adenosine diphosphate (ADP) to form adenosine triphosphate (ATP)

crossbridges protrusions on both ends of the thick filament that bind to actin and are responsible for generating the motion that causes muscle contraction

crossbridge cycle the mechanism that drives muscle contraction

end-plate potential (EPP) a depolarization of the motor end plate of a skeletal muscle fiber caused by acetylcholine binding to nicotinic cholinergic receptors

excitation-contraction coupling in a muscle cell, the sequence of events that links the action potential to the contraction

fast glycolytic fibers type of skeletal muscle with high myosin ATPase activity and high glycolytic capacity

fatigue a decline in a muscle's ability to maintain a constant force of contraction in the face of long-term, repetitive stimulation

Golgi tendon organ sensory receptor located near the junction between muscle and tendon that detects changes in muscle tension

isometric twitch (eye-soh-met-rick) a twitch during which a muscle generates force but does not shorten

isotonic twitch (eye-soh-tah-nik) a twitch during which a muscle shortens and lifts a constant load

latent period the time between a stimulus and the beginning of a response; in muscle physiology, the lag of a few milliseconds that occurs between the action potential in a muscle and when the muscle first begins to generate force

motor unit a motor neuron and all the muscle fibers it innervates

muscle fibers muscle cells

muscle spindle sensory receptor within skeletal muscle that detects changes in the length of the muscle

multi-unit smooth muscle smooth muscle lacking gap junctions; thus each fiber acts as a unit

myosin (my-oh-sin) the contractile protein found in thick filaments in striated muscle

myofibrils banded, rodlike elements that contain a muscle fiber's contractile machinery

myoglobin an oxygen-binding protein found in certain muscle cells

pacemaker potentials spontaneous depolarization in the resting membrane potential

recruitment an increase in the number of active neurons

skeletal muscle the type of muscle that is generally connected to bone

sarcomeres (sar-kuh-meers) the fundamental repeating units that make up myofibrils

sarcolemma (sar-co-lem-ah) a muscle fiber's plasma membrane

sarcoplasmic reticulum (SR) modified smooth endoplasmic reticulum in muscle cells that surrounds the myofibrils and stores calcium

series elastic component connective tissue that is in series with the sarcomeres of skeletal muscle

sliding-filament mechanism process of muscle contraction whereby thick and thin filaments slide past each other

size principle the correspondence between the size of motor units and the order of recruitment

slow oxidative fibers muscle cells with slow ATPase activity and high oxidative capacity

summation in neurophysiology, the adding together of graded potentials that occurs within a neuron; in muscle physiology, the adding together of twitches that occurs when a muscle is stimulated at high frequency

smooth muscle the type of muscle found in internal organs, blood vessels, and other structures that are not under voluntary control; lacks striations

single-unit smooth muscle smooth muscle with gap junctions linking the cells together so they function as a unit

tendons cords of elastic connective tissue that transmit force from skeletal muscles to bones

thick filaments filaments composed of myosin that form part of the contractile machinery of a muscle cell

thin filaments filaments composed of actin that form part of the contractile machinery of a muscle cell; also contain troponin and tropomyosin in striated muscle cells

transverse tubules (T tubules) structures that transmit action potentials from the sarcolemma into the cell's interior, triggering the release of calcium from the sarcoplasmic reticulum

tropomyosin (troh-poh-my-oh-sin) one of the two regulatory proteins in striated muscle; a long fibrous molecule that acts to block myosin-binding sites on thin filaments when a muscle is not contracting

troponin (troh-poh-nin) one of the two regulatory proteins in striated muscle; binds calcium reversibly and is responsible for starting the crossbridge cycle by moving tropomyosin out of its blocking position

twitch the mechanical response of a muscle to a single action potential

treppe increase in muscle tension with repeated twitch contractions

tetanus (tet-ah-nus) in a muscle being stimulated at high frequency, the plateau phase of the contraction, during which the tension is relatively constant

Answers

Figure Question Answers

Fig 8	Acetylcholinesterase
Fig 13	Isotonic
Fig 15	Excitation-contraction coupling
Fig 33	The somatic and autonomic nervous systems, respectively

Apply Your Knowledge Answers

Muscle cramps are usually caused by hyperexcitability of the motor neurons innervating the muscle. Nerve tissue is hyperexcitable when electrolytes are out of balance (such as occurs during dehydration) and during high levels of activity (for example, intense exercise). Spinal reflexes can also increase contraction of skeletal muscle during exercise when the muscle fatigues.

Sprinters use the optimal length concept when they position themselves in the runner's block. Baseball players use the optimal length in batting when they extend their arms prior to swinging. Many other examples exist.

Answers to End-of-Chapter Multiple Choice and Objective Questions

Multiple Choice Questions

1. c 2. c 3. d 4. b 5. c 6. e 7. d 8. d 9. b 10. e

Objective Questions

11. troponin
12. true
13. true
14. Oxidative
15. sarcolemma
16. myosin head groups
17. isometric
18. myosin ATPase activity
19. longer
20. Oxidative
21. false

The Cardiovascular System: Cardiac Function

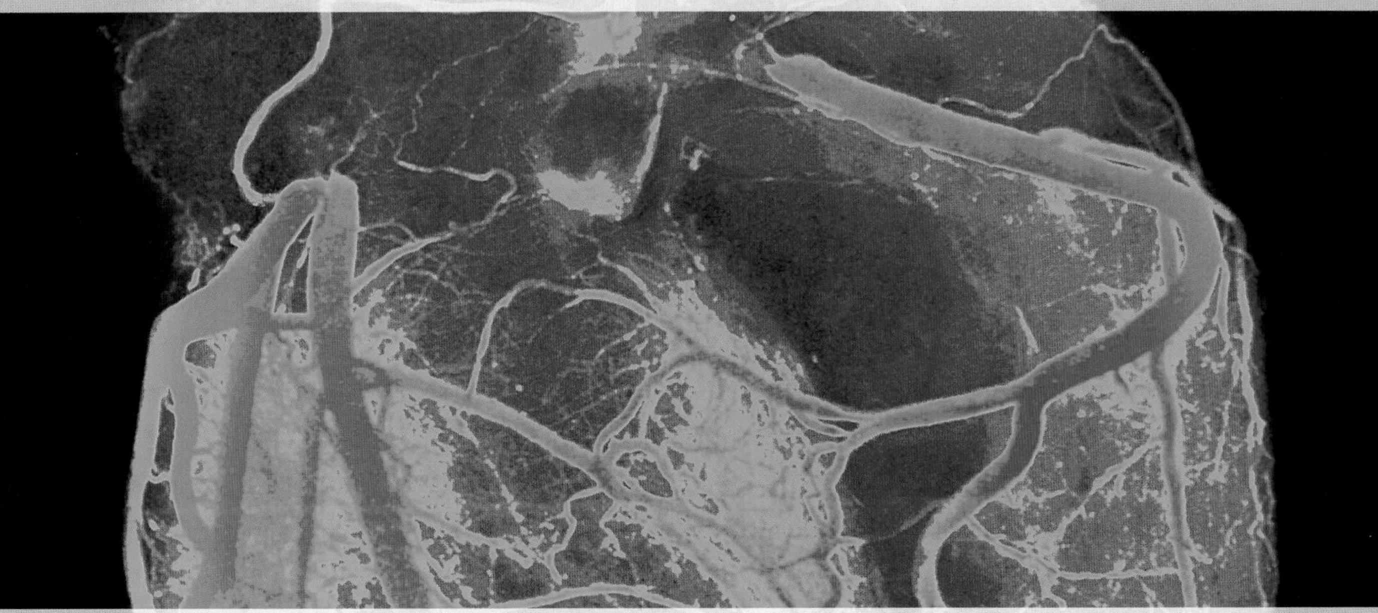

SPL/Photo Researchers, Inc.

Colorized microphotograph of a coronary angiogram.

The photo on this page shows a colored coronary angiogram of a healthy heart. The angiogram measures the blood flow through the coronary arteries, which supply the heart with blood. Because the heart beats approximately 70 times per minute when a person is at rest (and even faster when a person is active), adequate blood flow is critical in supplying the heart with sufficient amounts of the oxygen and nutrients it needs.

The ECG is a recording of the heart's electrical activity, which is responsible for the heartbeat. You have likely seen an ECG in the movies or on television hospital shows, with the ECG trace fluctuating violently when a patient is in trouble, or "flat lining" when the patient's heart stops beating. In this chapter, you will learn what the waves in the trace represent and how the electrical activity of the heart drives the rhythmic contraction that efficiently propels blood throughout the body. You will understand why the circulation of blood is so critical and why any interruption of that blood flow is potentially life-threatening. If blood flow were to stop suddenly, you would experience no more than a few seconds of consciousness before blacking out—and then you would have only a few minutes to live. Clearly, the heart must perform its functions continuously and nearly flawlessly for every minute of every day that you live.

This chapter focuses on the inner workings of the heart—a wondrous muscle that beats approximately 3 billion times in an average lifetime.

CHAPTER OUTLINE

MasteringA&P®

Go to MasteringA&P for helpful A&P Flix 3-D animations, chapter quizzes, pre-tests, Interactive Physiology tutorials, and more!

LEARNING OUTCOMES *After studying this chapter, you should be able to:*

- Identify the major components of the cardiovascular system, and briefly describe their functions.
- Identify the major structures of the heart, and describe the path of blood flow through the heart and vasculature.
- Explain the following events in the cardiac cycle: changes in ventricular, aortic, and atrial pressure; changes in ventricular volume; and heart sounds.
- Trace the path of action potentials through the conduction system of the heart, and relate the heart's electrical activity to its pumping action.
- Describe how the phases of the electrocardiogram relate to the events of the cardiac cycle.
- Explain the difference between extrinsic control and intrinsic control, and identify how these terms relate to the regulation of the heart.
- Explain how each of the following variables affects cardiac output: sympathetic and parasympathetic nervous activity, circulating epinephrine, afterload, preload, end-diastolic volume, ventricular contractility, and filling time.

Before You Begin

Make sure you have mastered the following topics:

- Membrane potentials
- Mechanism of muscle contraction
- Gap junctions
- Cardiac muscle
- Excitation-contraction coupling
- Autonomic nervous system

The internal environment in which cells survive must be maintained through integrated actions of the organ systems so as to maintain homeostasis. For the organ systems to do their jobs, they require energy in the form of ATP. The cardiovascular system is critical to the maintenance of homeostasis in two ways: It provides cells with the oxygen and nutrients necessary to generate ATP, and it delivers carbon dioxide and waste products to organ systems that eliminate them from the body. These nutrients, wastes, and gases must be exchanged between the cells and the external environment through a series of exchange mechanisms. The process of delivery and removal of molecules to meet the body's demands is ensured by the regulation of the cardiovascular system; this regulating function is performed by paracrines, the nervous system, and the endocrine system.

In this chapter, we begin with a description of the function of the heart, an organ consisting primarily of muscle, whose function is to pump blood (1) to the organs so as to provide nutrients and to eliminate wastes and (2) to the lungs so as to exchange gases. The heart muscle is under the control of the autonomic nervous system, and we will explore the specifics of how the parasympathetic and sympathetic branches of the autonomic nervous system control heart rate and the force with which the heart contracts.

1 An Overview of the Cardiovascular System

The ability of cells to exchange materials with their immediate environment is an absolute requirement of life. When exchanging materials with interstitial fluid, each cell relies on diffusion to bring needed materials, such as oxygen and nutrients, to it and to carry unwanted materials, such as carbon dioxide and other wastes, away from it. Because these materials ultimately come from or go into the *external* environment, which is at a considerable distance from most body cells, diffusion alone cannot provide all cells with what they need as quickly as they need it; that is, as a mechanism of transport, it is too slow. Ensuring that a much more efficient transport mechanism is available is the purpose of the cardiovascular system.

The **cardiovascular system** consists of three components: (1) the **heart**—a muscular pump that drives the flow of blood through blood vessels; (2) **blood vessels**—conduits through which the blood flows; and (3) **blood**—a fluid that circulates around the body, carrying materials to and from the cells.

At first glance, the function of the cardiovascular system seems simple: The heart pumps blood through blood vessels to various organs, and the blood carries oxygen and nutrients to tissues and removes carbon dioxide and other wastes. In reality, however, the heart is more than a pump: It performs sensory and endocrine functions that help regulate cardiovascular variables such as blood volume and pressure. The blood vessels are not just conduits for blood, but are also important sensory and effector organs that regulate blood pressure and the distribution of blood to various parts of the body. The blood not only carries nutrients and wastes, but also transports hormones from one part of the body to another, thereby serving as a communications link acting in conjunction with the nervous system. As we will see, it is not possible to fully understand how one part of the system works—the heart, for instance—without understanding how the rest of the system works. Furthermore, regulation of the cardiovascular system involves interactions with several other organ systems, including the nervous system, the endocrine system, and the kidneys.

Before we examine the workings of the heart—the central focus of this chapter—we introduce the major components of the cardiovascular system and their functions.

The Heart

The heart is a muscular organ whose function is to generate the force that propels blood through the blood vessels. The heart contains four chambers: two upper chambers, called **atria** (singular: *atrium*), which receive the blood that comes back to the heart from the vasculature, and two lower chambers, called **ventricles,** which receive blood from the atria and generate the force that pushes the blood away from the heart and through the blood vessels (**Figure 1**).

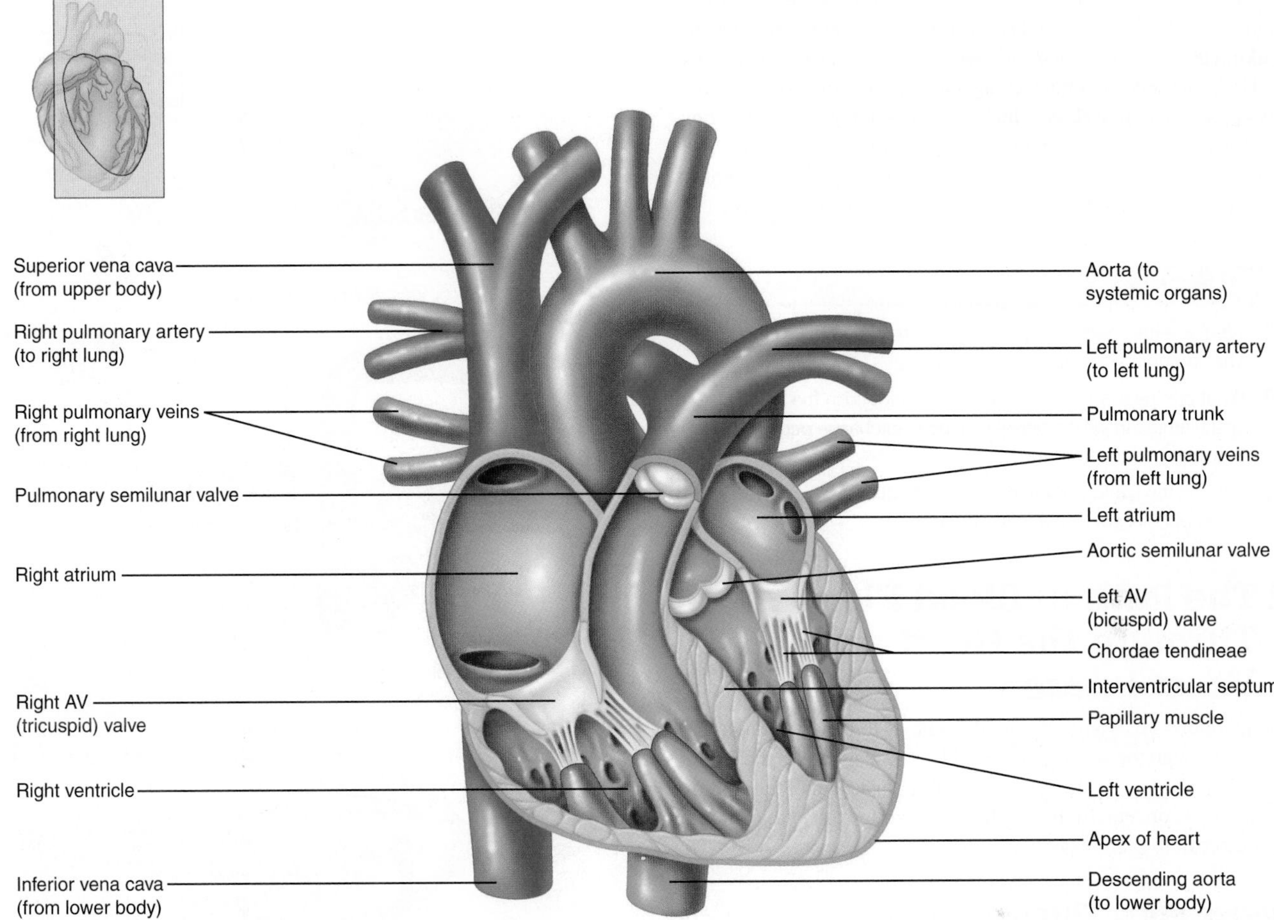

Figure 1 **A cutaway view of the heart showing the atria, ventricles, atrioventricular valves, and connections to major blood vessels.**

The heart can be functionally separated into left and right halves: The atrium and ventricle on the left side of the heart constitute the *left heart,* whereas the atrium and ventricle on the right side constitute the *right heart.* The atria and ventricles on either side of the heart are separated by a wall called the *septum* that prevents blood in the left heart from mixing with blood in the right heart. The portion separating the left and right atria is referred to as the *interatrial septum;* the portion separating the left and right ventricles is the *interventricular septum.* Just as the heart has left and right sides, so it also has a "top" and a "bottom." The wider upper pole (end) of the heart is known as the *base;* the narrower lower pole is the *apex.* The two sides of the heart work efficiently as two pumps to supply all organs with the nutrients and oxygen they need.

Blood Vessels

When blood moves through the body, it travels in a circular pattern through a system of blood vessels that carry it from the heart to the various organs and then back to the heart again. This system of blood vessels is called the **vasculature.** As blood flows away from the heart, the blood vessels branch repeatedly, becoming more numerous and smaller in diameter, just as the limbs of a tree become smaller and more numerous as you move from the trunk to the outer branches. The smallest blood vessels are the **capillaries,** which serve as the site of exchange between blood and interstitial fluid. After passing through the capillaries, blood flows back to the heart, where the vessels conveying it converge and become less numerous and larger in diameter.

When blood leaves the heart, it is transported to the body's organs and tissues in relatively large vessels called **arteries,** which branch repeatedly within the organs and tissues. The smallest arteries branch into still smaller vessels called **arterioles,** which carry blood to the capillaries. From the capillaries, blood moves to larger vessels called **venules,** which lead to still larger vessels called **veins,** which in turn carry blood back to the heart. Because blood moves from the heart into the vasculature and then from the vasculature back to the heart, the cardiovascular system is a *closed* system.

Blood

Although blood is a fluid, nearly half its volume is composed of cells. The most numerous cells are **erythrocytes,** also known as *red blood cells.* These cells contain *hemoglobin,* a protein that carries oxygen. The presence of hemoglobin in these cells gives them their characteristic red color. The remainder of the cells are **leukocytes,** or *white blood cells,* which come in a variety of types and help the body defend itself against invading microorganisms. Also present are **platelets,** which are not actually cells but rather cell fragments that play an important role in blood clotting. The liquid portion of the blood, called **plasma,** is made up of water containing dissolved proteins, electrolytes, and other solutes.

Quick Check 1

1. What are the "receiving chambers" of the heart called? The "pumping chambers"?
2. What are the five types of blood vessels found in the vasculature? In which vessel type does exchange occur between blood and tissue?
3. What is the liquid portion of the blood called? Which two types of cells are found in the blood?

2 The Path of Blood Flow Through the Heart and Vasculature

Blood follows an essentially circular path as it travels through the body. Although the sheer number of blood vessels makes the structure of the cardiovascular system complex, the layout of the system is simple in concept. In this section, we examine the path of blood flow through the cardiovascular system.

Series Flow Through the Cardiovascular System

The general pattern of blood flow through the cardiovascular system is shown in **Figure 2**. In the diagram you can see that the circulatory system consists of two divisions: the **pulmonary circuit,** which comprises all blood vessels within the lungs as well as those connecting the lungs with the heart, and the **systemic circuit,** which encompasses the rest of the blood vessels in the body. Note that these two divisions are supplied with blood by the different sides of the heart. The *right* heart supplies blood to the pulmonary circuit, whereas the *left* heart supplies blood to the systemic circuit. Notice that blood on one side of the heart never mixes with blood on the other side.

The pulmonary and systemic circuits both possess dense networks of capillaries called *capillary beds,* where exchange of nutrients and gases (oxygen and carbon dioxide) takes place. In pulmonary capillaries, oxygen (O_2) moves into the blood from air in the lungs while carbon dioxide (CO_2) leaves the blood. When it leaves pulmonary capillaries, the blood is relatively rich in oxygen and is called *oxygenated* blood. Capillary beds of the systemic circuit are located in all organs and tissues except the lungs. In these organs and tissues, cells consume oxygen and generate carbon dioxide; thus, as blood travels through systemic capillaries, oxygen leaves the blood and carbon dioxide enters it. Blood leaving the

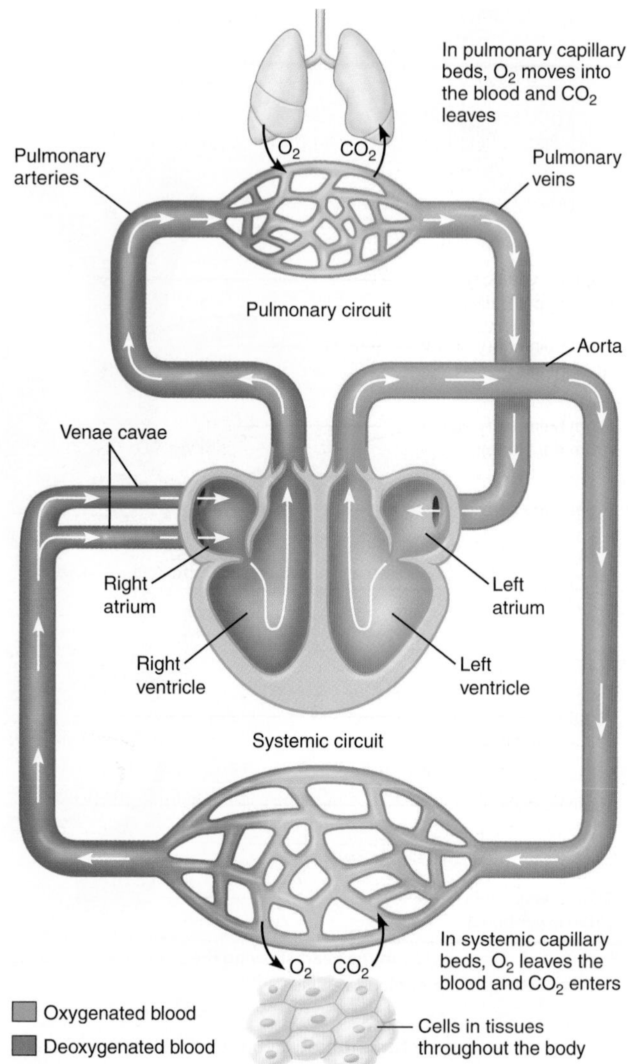

Figure 2 The path of blood flow through the cardiovascular system. The pulmonary and systemic circuits and major blood vessels connecting with the heart are shown. Arrows indicate the direction of blood flow.

systemic capillaries is called *deoxygenated* blood because it is relatively oxygen poor.

When oxygenated blood becomes deoxygenated, or vice versa, its color actually changes. Oxygenated blood is bright red, whereas deoxygenated blood is a darker red. Despite the fact that both forms of blood are red, the terms *red blood* and *blue blood* are frequently used to denote oxygenated blood and deoxygenated blood, respectively. Deoxygenated blood is called *blue* because it imparts a bluish color to veins visible beneath the skin. Oxygenated blood and deoxygenated blood are indicated by red and blue colors in Figure 2.

As blood flows within the cardiovascular system, it travels through the pulmonary and systemic circuits in an alternating fashion, returning to the heart each time (see Figure 2). Let us follow the path of blood flow step by step, starting in the left ventricle:

1. The left ventricle pumps oxygenated blood through the aortic valve into the **aorta,** a major artery whose branches carry blood to capillary beds of all organs and tissues in the systemic circuit.
2. Blood becomes deoxygenated in systemic capillaries and then travels back to the heart in the **venae cavae** (singular: *vena cava*), two large veins that carry blood into the right atrium. The *superior* vena cava carries blood from parts of the body above the diaphragm, whereas the *inferior* vena cava carries blood from parts below the diaphragm.
3. From the right atrium, blood passes through the tricuspid valve into the right ventricle.
4. The right ventricle pumps blood through the pulmonary semilunar valve into the pulmonary trunk, which almost immediately branches into the **pulmonary arteries,** which carry deoxygenated blood to the lungs. Note that the pulmonary arteries are the only arteries in the body carrying deoxygenated blood.
5. Blood becomes oxygenated in the lungs and then travels to the left atrium in the **pulmonary veins.** These are the only veins in the body carrying oxygenated blood.
6. From the left atrium, blood passes through the bicuspid valve into the left ventricle, which is where it started. The whole cycle then repeats.

The preceding path describes how blood flows through the pulmonary and systemic circuits *in series* with each other; that is, for the cardiovascular system as a whole, blood must pass through the two circuits in sequence before it can return to the starting point. However, it is important to realize that blood flows through both circuits simultaneously; that is, the right heart is pumping blood to the lungs at the same time that the left heart is pumping blood to the systemic organs. If we look at blood flow within either the systemic or pulmonary circuit, we see a different pattern—*parallel flow.*

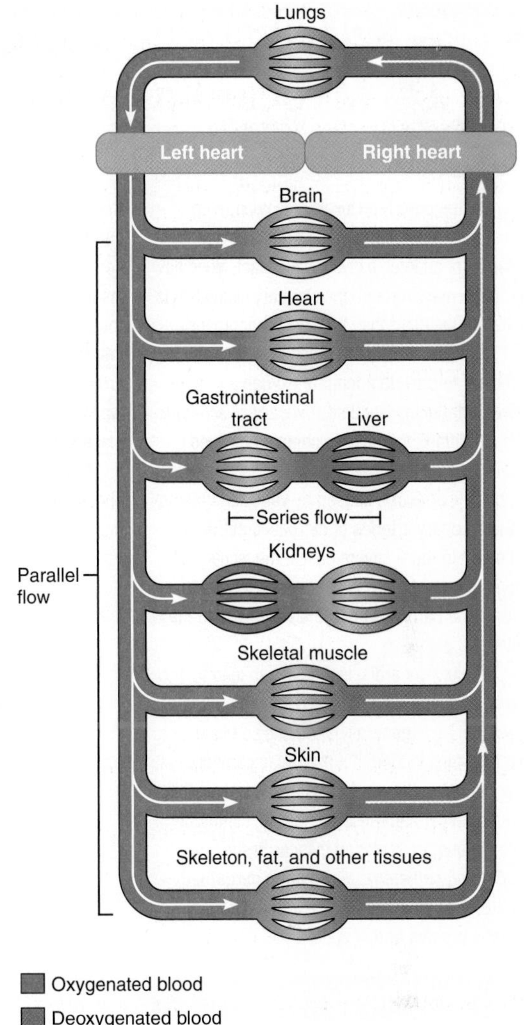

Figure 3 Blood flow patterns in the cardiovascular system. Blood flow through organs in the systemic circuit is called *parallel flow.*

Parallel Flow Within the Systemic or Pulmonary Circuit

Figure 3 shows why blood flow in the systemic circuit is called *parallel flow* (and why the systemic organs are said to be *in parallel* with one another). In the systemic circuit, blood does not flow from one organ directly to the next. Instead, blood travels through the aorta and into the arteries that branch off it to reach only one organ, before then flowing through veins that converge to either the superior or inferior vena cava. Moreover, the pattern of blood flow *within* organs, including the lungs in the pulmonary circuit, is also parallel because arteries branch to arterioles, which branch to capillaries, and so on.

As shown in Figure 3, the *heart* works in parallel with the other organs in the systemic circuit. Even though the heart pumps a large volume of blood, the blood within the heart's chambers does

FOCUS ON DIABETES

People with diabetes are more prone to heart attacks. Unfortunately, individuals with diabetes often do not feel the pain associated with a heart attack because of neuropathy of the nerves that transmit signals from the heart to the central nervous system. As a consequence, diabetics may suffer from *silent myocardial ischemia* and have little warning that a heart attack is occurring.

CLINICAL CONNECTIONS

Myocardial Ischemia

The heart requires high levels of oxygen, even at rest, because cardiac muscle has a limited capacity for anaerobic glycolysis. In fact, blood flow per tissue mass through the coronary circulation is higher for the heart than for most other organs.

A decrease in cardiac blood flow to levels insufficient to provide adequate oxygen and to remove metabolites is called *myocardial ischemia.* This condition can result in chest pain called *angina pectoris.* In some cases, the ischemia is temporary, or *acute,* caused by vascular spasms of the coronary arteries or increased activity of the heart. In more severe cases, the ischemia is prolonged, or *chronic,* often caused by *atherosclerosis,* a narrowing of the arteries due to the buildup of plaques.

Chronic myocardial ischemia can lead to *myocardial infarction (MI),* also known as *heart attack.* An MI is generally caused by a blood clot that becomes lodged in a narrowed coronary artery (see the figure). An MI causes irreversible damage to cardiac muscle—that is, contractile cells die and are replaced by scar tissue.

Cardiac catheterization is often performed to determine whether blockage of a coronary artery is present and, if so, to try to remove the obstacle using *balloon angioplasty.* During cardiac catheterization, a dye is injected into the coronary arteries, and its flow through the arteries is observed using X-ray imaging. To remove the blockage, a small deflated balloon is inserted into the artery; it is then inflated to expand the artery and push the plaque back against the walls of the artery. If necessary, a *stent* can be placed in the artery to hold it open.

A person at risk of having an MI is often put on a low-cholesterol, low-fat diet (to reduce the risk of atherosclerosis) and takes an aspirin each day (to decrease the likelihood of forming a blood clot). People who have already suffered an MI may be given a supply of nitroglycerin (a vasodilator) that they can take in the event of another episode. Other drugs used to help prevent a second MI include digitalis (which decreases heart rate so that the heart requires less energy), calcium channel blockers (which decrease the likelihood of coronary vascular spasms), and beta-blockers (which decrease heart rate and cardiac contractility).

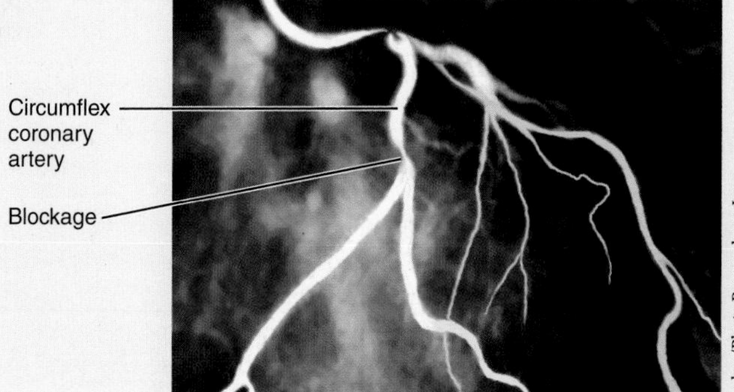

Angiogram of a heart showing blockage of the coronary artery.

Critical Thinking Questions

1. What are the differences between the causes of acute and chronic myocardial ischemia?
2. Explain the technique used to determine if a patient is at risk for a myocardial infarction.
3. Define the following terms: acute myocardial ischemia; chronic myocardial ischemia; angina pectoris; and myocardial infarction. In what key ways do they differ?

not supply the heart muscle with significant quantities of oxygen or nutrients. Instead, the heart muscle obtains most of its nourishment from blood via the **coronary arteries,** which branch off the aorta near its base and run through the heart muscle (**Figure 4**). A decrease in blood flow through the coronary arteries can lead to a heart attack (see **Clinical Connections: Myocardial Ischemia**).

The parallel arrangement of organs in the systemic circuit confers two distinct advantages. First, because each organ is fed by a separate artery, each receives fully oxygenated blood—that is, blood that has not been depleted of oxygen as a result of having already flowed through another organ. Second, because blood reaches the organs via parallel paths, blood flow to the organs can be independently regulated. Thus blood flow can be adjusted to match the constantly changing metabolic needs of organs.

Although parallel flow is the norm for the systemic circuit, in some exceptions to the rule, blood flows in series between two capillary beds (Figure 3). We know one of the exceptions, with the path of blood flow between the hypothalamus and the anterior pituitary. Capillary beds in the hypothalamus are in series with those in the anterior pituitary, connected together by *portal* veins. A *portal circulation* is one in which blood flows from one capillary bed to another before returning to the heart. Another portal circulation exists between the intestines and liver such that the liver processes absorb nutrients before they enter the general circulation. The kidneys have a special circulatory system with two capillaries in series, connected not by a portal vein but by an arteriole.

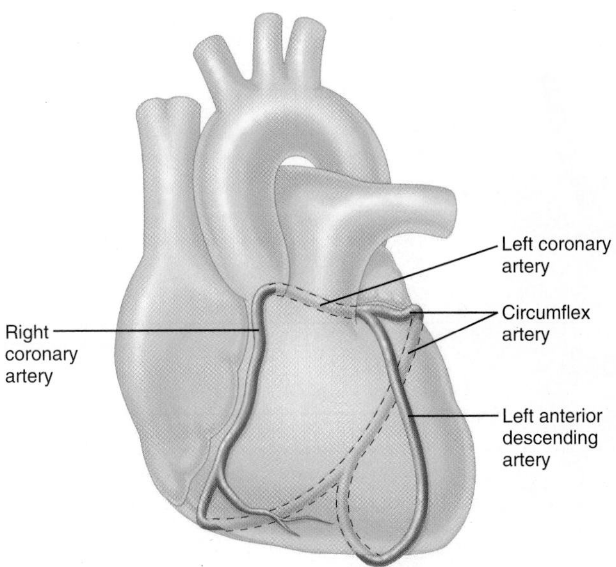

Figure 4 Coronary arteries.

Quick Check 2

1. Which organs receive blood flowing through the pulmonary circuit? Through the systemic circuit?
2. Compare the location and function of parallel and series blood flow.
3. Which blood vessels transport oxygenated blood? Deoxygenated blood?

3 Anatomy of the Heart

The heart is located centrally in the *thoracic cavity* (chest cavity) just above the *diaphragm*, a muscular partition that separates the thoracic cavity from the *abdominal cavity* (**Figure 5**). The heart is about the size of a fist, weighing approximately 300–350 grams in males and 250–300 grams in females. It is surrounded by a membranous sac called the *pericardium*, which contains pericardial fluid that lubricates the heart. Inflammation of the pericardium, a condition known as *pericarditis*, causes pain due to friction as the heart beats.

The heart generates the force that propels blood flow through the vasculature. To enable the heart to carry out this function, its walls consist primarily of cardiac muscle. To ensure that blood flows in the correct direction through the heart, four valves prevent backflow. In the next section, we look at the anatomy of the heart wall and the valves within the heart.

Myocardium and the Heart Wall

The heart wall consists of three layers: an outer layer of connective tissue called the **epicardium,** a middle layer of cardiac muscle called the **myocardium,** and an inner layer of epithelial cells called the **endothelium.** (The endothelial layer extends throughout the entire cardiovascular system.)

The heart's pumping action is conferred by the rhythmic contraction and relaxation of the myocardium. When muscle in the wall of an atrium or ventricle contracts, the wall moves inward and squeezes the blood in the chamber. This squeezing increases the pressure within the chamber and forces the blood out. When the muscle relaxes, the chamber expands and fills with blood.

Note in Figure 1 that ventricular muscle is substantially thicker than atrial muscle. This difference reflects the fact that the ventricles pump blood over relatively long distances through the vasculature (and not just into the next chamber, as the atria do), so they must work harder to pump a given volume of blood. Note as well that the ventricular muscle is much thicker on the left side than it is on the right side (**Figure 6**). The thicker muscle enables the left ventricle to develop greater pressure than the right ventricle, which is necessary because the left ventricle pumps blood to all the organs in the body except the lungs, whereas the right ventricle pumps blood only to the lungs.

Valves and Unidirectional Blood Flow

What we call the *heartbeat* is actually a wave of contraction that sweeps through heart muscle fibers (cells) in an orderly, coordinated fashion. Although the entire heart muscle functions as a unit, atrial muscle (the *atrial myocardium*) and ventricular muscle (the *ventricular myocardium*) are anchored to and separated by a layer of fibrous connective tissue called the *fibrous skeleton* of the heart. The fibrous skeleton also forms rings that anchor the *heart valves* in place.

The series of events that occur in a heartbeat is called the *cardiac cycle*, in which the atria contract first, followed by the ventricles. This cycling causes pressures in the chambers to fluctuate and, more importantly, changes the direction of pressure gradients for blood to flow. It is critical that blood flows through the heart in one direction. Thus, when the pressure gradients favor movement in the opposite direction, blood flow must be prohibited.

The heart has four valves that keep blood flowing in the proper direction within the heart itself and between the heart and the arteries connected directly to it (the *aorta* and the *pulmonary trunk*) (see Figure 1). The atrium and ventricle on each side are separated by **atrioventricular valves (AV valves),** which permit blood to flow from the atrium to the ventricle but not in the opposite direction. AV valves open or close passively in response to cyclic changes in pressure that occur with every heartbeat (**Figure 7**). When atrial pressure is higher than ventricular pressure, the valves open; when ventricular pressure becomes higher than atrial pressure, the valves close. The AV valve on the left consists of two flaps or *cusps* of connective tissue and, therefore, is called the **bicuspid valve;** it is also known as the **mitral valve,** because it looks like a bishop's hat (called a mitre). The right AV valve has three cusps and is called the **tricuspid valve.**

When a ventricle contracts, the increased ventricular pressure exerts an upward force against the AV valve. Because of this force, there is a potential danger that one or more valve cusps could be pushed into the atria, a condition called *prolapse.* If this were to happen, the edges of the cusps would no longer meet properly when the valve closes, and the valve would not be able to seal completely. Prolapse of the AV valves

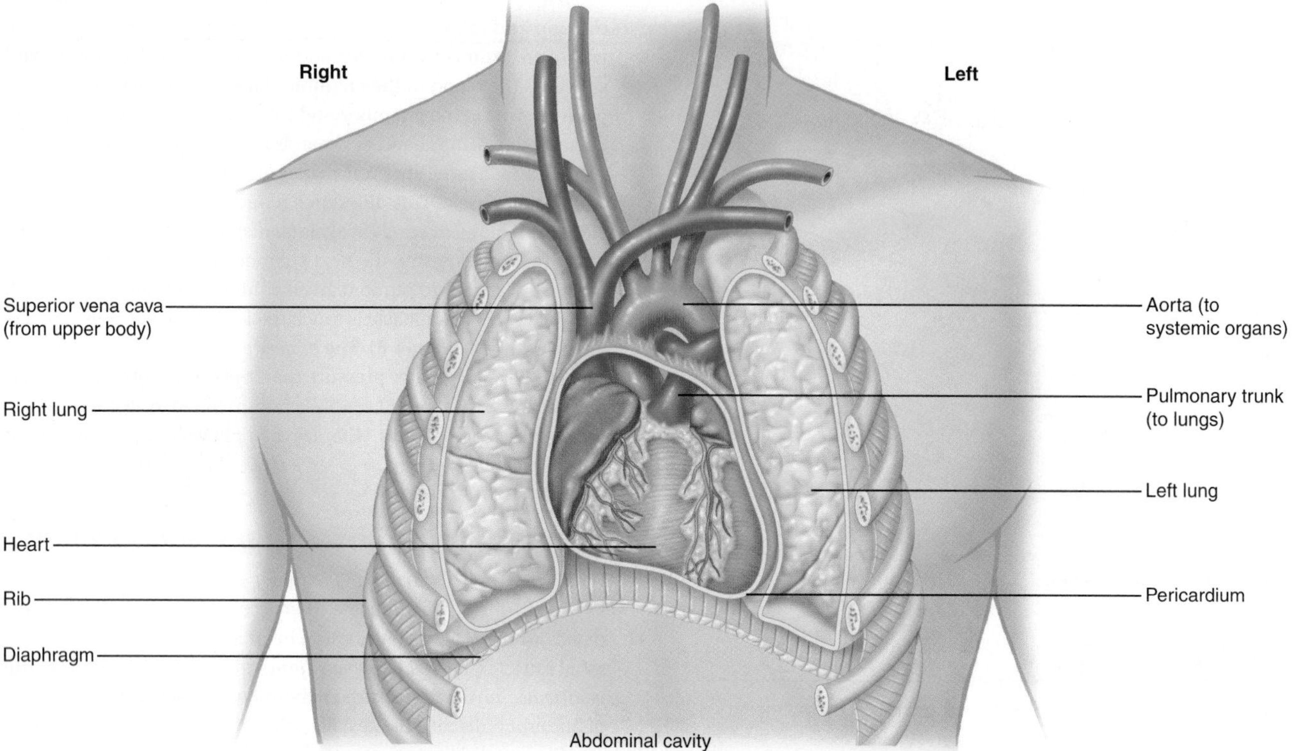

Figure 5 **Location of the heart in the thoracic cavity.**

is normally prevented because the valve cusps are held in place by strands of connective tissue (known as the **chordae tendineae**) that extend from the edges of the cusps to **papillary muscles,** which protrude from the ventricular wall. During ventricular contraction, the papillary muscles also contract, which exerts tension on the chordae tendineae. The chordae tendineae pull downward on the valve cusps, thereby enabling the AV valves to seal properly while resisting the upward force of ventricular pressure.

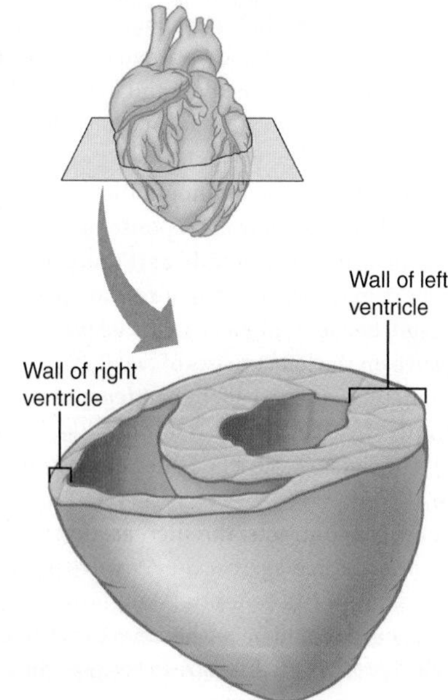

Figure 6 **Right and left ventricle muscle thickness.**

In addition to the AV valves, other valves, called **semilunar valves,** are found between the ventricles and arteries. The **aortic valve** is located between the left ventricle and the aorta, and the **pulmonary valve** is located between the right ventricle and the pulmonary trunk. The function of these valves is similar to that of the AV valves—to permit blood to flow forward while preventing it from flowing backward (**Figure 8**). The aortic and pulmonary valves open when ventricular pressure is greater than arterial pressure (when the ventricles contract). This mechanism allows blood to leave the ventricles and enter the arteries. When the ventricles relax and ventricular pressure becomes lower than arterial pressure, the valves close, thereby preventing blood from flowing back into the ventricles from the arteries.

Quick Check 3

1. What are the three layers of the heart wall? Which layer extends throughout the entire cardiovascular system, lining the lumen?
2. Which chamber of the heart has the thickest walls? Explain why.
3. Where are the atrioventricular valves located? What is their function? Where are the semilunar valves located? What is their function?
4. What is mitral valve prolapse?

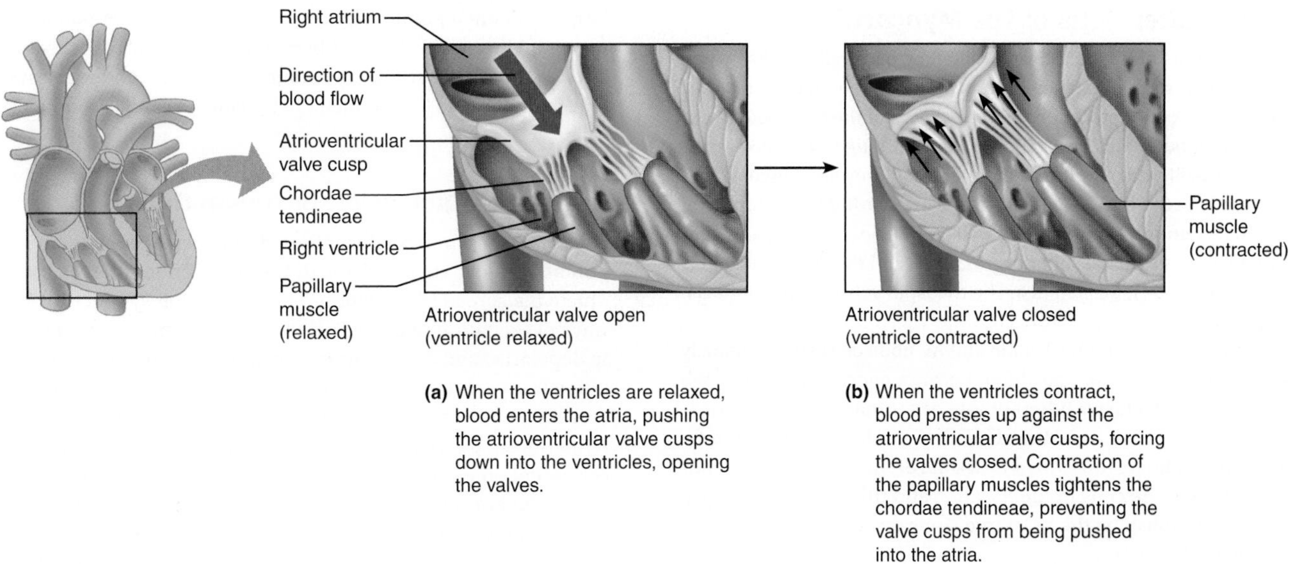

Figure 7 Action of the atrioventricular valves.

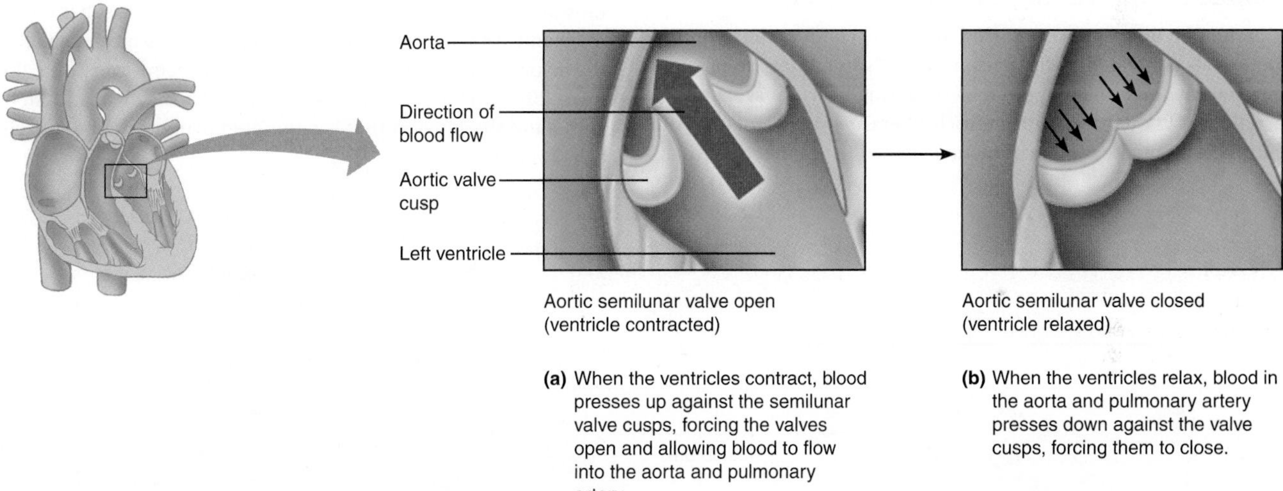

Figure 8 Action of the semilunar valves.

4 Electrical Activity of the Heart

For the heart to adequately pump blood through the circulatory system, the cardiac muscle must contract in a highly synchronized manner—first the contraction of both atria occurs, followed by the contraction of both ventricles. Cardiac contractions are coordinated by an elaborate conduction system that determines the sequence of excitation of cardiac muscle cells.

The Conduction System of the Heart

The cardiac muscle, unlike skeletal muscle, does not require commands from the central nervous system to contract. Instead, cardiac muscle contractions are triggered by signals originating from within the muscle itself. For this reason, the contractile activity of cardiac muscle is said to be *myogenic.* (The contractile activity of skeletal muscle is said to be *neurogenic* because the initial signal for contraction originates in neurons.) The ability of the heart to generate signals that trigger its contractions on a periodic basis—that is, to generate its own rhythm—is called **autorhythmicity.** The heart's autorhythmicity is due to the action of a small percentage of muscle cells, called *autorhythmic cells,* that generate little or no contractile force but are critical to the heart's pumping action because they coordinate and provide rhythm to the heartbeat. Two types of autorhythmic cells exist: (1) **pacemaker cells,** which initiate action potentials and establish the heart rhythm, and (2) **conduction fibers,** which transmit action potentials through the heart in a highly coordinated manner. Together, these cells make up the **conduction system** of the heart. The cells that generate the contractile force are called *contractile cells.*

Pacemaker Cells of the Myocardium

Contractions of the heart are initiated by pacemaker cells, which spontaneously generate action potentials. As their name suggests, pacemaker cells determine the rate or *pace* of the heartbeat by firing action potentials on a regular basis. Although pacemaker cells are located in nearly all parts of the heart, they are concentrated primarily in two specific regions of the myocardium: the **sinoatrial node (SA node),** located in the wall of the upper right atrium near where it joins with the superior vena cava, and the **atrioventricular node (AV node),** located near the tricuspid valve in the interatrial septum.

The rates at which SA node and AV node cells spontaneously generate action potentials differ. Because pacemaker cells in the SA node have a faster inherent rate of spontaneous depolarization, and because the SA node and AV node are connected by conduction fibers (described next), the SA node drives the depolarization of the cells in the AV node and throughout the heart, thereby establishing the heart rate. Thus the SA node is the pacemaker of the heart.

Conduction Fibers of the Myocardium

Conduction fibers are specialized to quickly conduct the action potentials generated by the pacemaker cells from place to place through the myocardium, thereby triggering heart muscle contractions. Although all cardiac muscle fibers are capable of transmitting action potentials, conduction fibers differ in that they are larger in diameter and, therefore, can conduct action potentials more rapidly than "ordinary" fibers. Action potentials can travel as fast as 4 meters per second in certain parts of the conduction system, as opposed to 0.3–0.5 meter per second in cardiac contractile cells.

Spread of Excitation Between Cells

Once action potential is initiated in pacemaker cells, action potentials move rapidly through the conduction fibers to coordinate the spread of excitation. The conduction system causes a wave of excitation to move first through the atria, causing them to depolarize and then contract as a unit. Next, the wave of excitation moves through the ventricles, causing them to depolarize and then contract as a unit. Rapid transmission of action potentials from pacemakers to conduction fibers to the contractile cells is possible because all cardiac muscle cells are connected to their neighbors by gap junctions, which permit electrical current to pass in the form of ions from one cell to another. In the heart, gap junctions are concentrated in structures called **intercalated disks,** which form the junctions between adjacent muscle fibers (**Figure 9**). Intercalated disks also contain large numbers of desmosomes, forming a physical bond between the disks that resists mechanical stress. This property is important because it enables the myocardium to resist stretching, which occurs every time the heart fills with blood, and because it enables the myocardium to withstand the tension that is generated every time the muscle cells contract.

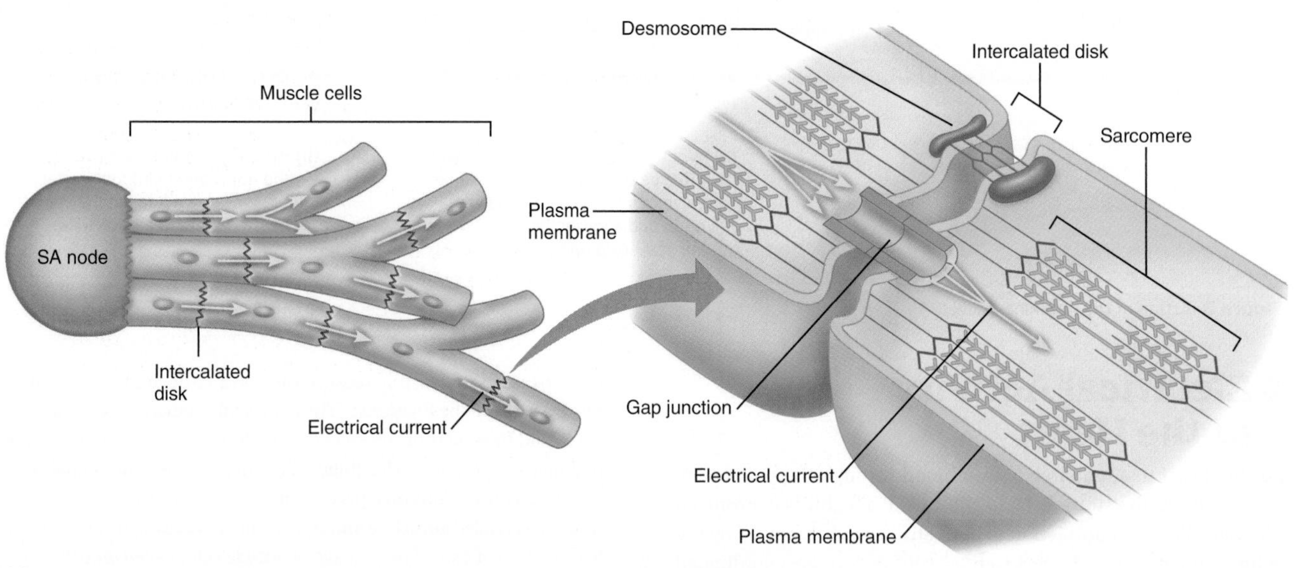

Figure 9 Electrical connections between cardiac muscle cells. (a) An action potential generated spontaneously in cells of the SA node spreads to adjacent muscle cells by means of electrical current passing through gap junctions in intercalated disks. **(b)** A schematic view of the junction between two adjacent muscle cells showing a gap junction and a desmosome.

Q Besides cardiac muscle, which other type of muscle possesses gap junctions?

Initiation and Conduction of an Impulse During a Heartbeat

The sequence of electrical events that normally triggers the heartbeat occurs as follows (**Figure 10**):

① An action potential is initiated in the SA node. From the SA node, impulses travel to the AV node by way of *internodal pathways*—systems of conduction fibers that run through the walls of the atria. As these signals move through the internodal pathways, they also spread through the bulk of the atrial muscle by way of *interatrial pathways.*

② The impulse is conducted to cells of the AV node, which transmit action potentials less rapidly than other cells of the conduction system. As a result, the impulse is momentarily delayed by about 0.1 second (called the **AV nodal delay**) before moving onward.

③ From the AV node, the impulse travels through the atrioventricular bundle, also known as the **bundle of His** (pronounced "hiss"), a compact bundle of muscle fibers located in the interventricular septum. The AV node and bundle of His are the only electrical connection between the atria and the ventricles, which are otherwise separated by the fibrous skeleton.

④ The signal travels only a short distance through the atrioventricular bundle before it splits into left and right **bundle branches,** which conduct impulses to the left and right ventricles, respectively.

⑤ From the bundle branches, impulses travel through an extensive network of branches referred to as **Purkinje fibers,** which spread through the ventricular myocardium from the apex upward toward the valves. From these fibers, impulses travel through the rest of the myocardial cells.

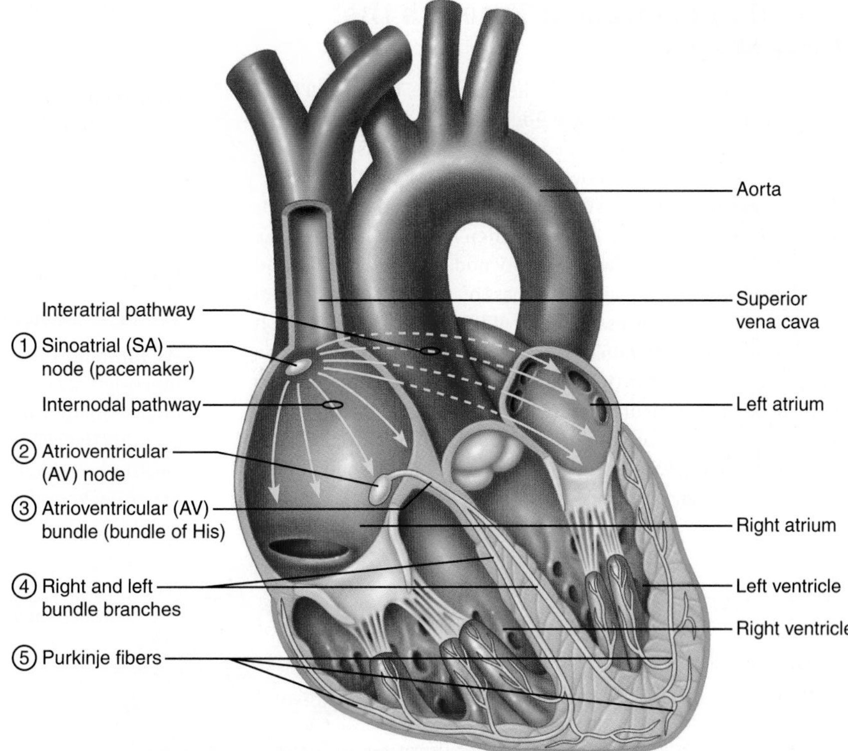

Figure 10 The conduction system of the heart.

Control of the Heartbeat by Pacemakers

Although the SA node and the AV node are both capable of generating spontaneous action potentials, the heartbeat is normally triggered by impulses originating from the SA node. The AV node rarely initiates contractions for two reasons. First, action potentials originating in the SA node travel through the AV node on their way to the ventricles. When this happens, cells in the AV node go into a refractory period, during which they cannot generate their own action potentials. Second, the SA node has a higher frequency of action potentials than the AV node—about 70 impulses per minute for the SA node, as opposed to 50 impulses per minute for the AV node. Thus the AV node rarely has a chance to fire an action potential because the SA node always "beats it to the punch."

However, if the SA node fails to fire an action potential or if it slows down dramatically, the AV node *will* initiate action potentials, which travel through the conducting system and trigger ventricular contraction in the normal manner. The AV node can also take over control of the heartbeat if conduction between the nodes becomes blocked or slows down for some reason. In these circumstances, the AV node functions as an emergency backup system that keeps the ventricles beating. If for some reason the AV node is unable to drive ventricular contraction, the heart has yet another backup system: Certain cells in the Purkinje fibers (sometimes referred to as *idioventricular pacemakers*) can take over. However, the firing frequency of these cells is only 30–40 impulses per minute.

Apply Your Knowledge

Sometimes a contraction is initiated outside the normal conduction pathway at a site called an *ectopic focus*. If this site is located in the atrium, it can cause a premature atrial contraction (PAC); current will then spread through gap junctions, followed by a ventricular contraction. If the ectopic focus is located in the ventricle, it can cause a premature ventricular contraction (PVC), with no atrial contraction being involved. This extra beat, called an *extrasystole*, is followed by a skipped beat, leaving a pause between ventricular contractions. Why would the heart skip its next regular beat following an extrasystole?

Spread of Excitation Through the Heart Muscle

As impulses propagate through the heart muscle, they travel in an orderly pattern as a kind of wave front—a "wave of excitation." As this wave of excitation spreads, contraction of the muscle follows. The pattern of excitation is shown in **Figure 11**.

The wave of excitation starts at the SA node and then spreads outward through the atria. The wave then "funnels" through the atrioventricular bundle by way of the AV node, which acts as a kind of bottleneck due to the relative slowness of impulse conduction in this region. This delay is essential for efficient cardiac function: It allows the wave of excitation to spread completely through the atria before it reaches the ventricles, ensuring that atrial contraction is complete before ventricular contraction starts. Given that the function of atrial contraction is to drive blood into the ventricles, ventricular contraction would work against the pumping action of the atria if no such delay occurred.

Once impulses reach the bundle branches and the Purkinje fibers, they are carried relatively quickly to the lower portion of the ventricles. From there, the wave of excitation fans out through the entire ventricular muscle. Thus ventricular contraction begins at the apex and spreads upward. This pattern makes sense when you consider that blood exits the ventricles from the top (see Figure 1). In this sense, ventricular contraction is reminiscent of how one should squeeze a tube of toothpaste—from the bottom up.

The Ionic Basis of Electrical Activity in the Heart

We now know that the heartbeat is triggered by action potentials that originate in pacemaker cells and propagate through the heart muscle in an orderly, predictable fashion. In this subsection, we examine the cellular mechanisms responsible for generating these electrical signals, beginning with events occurring in the membrane of pacemaker cells.

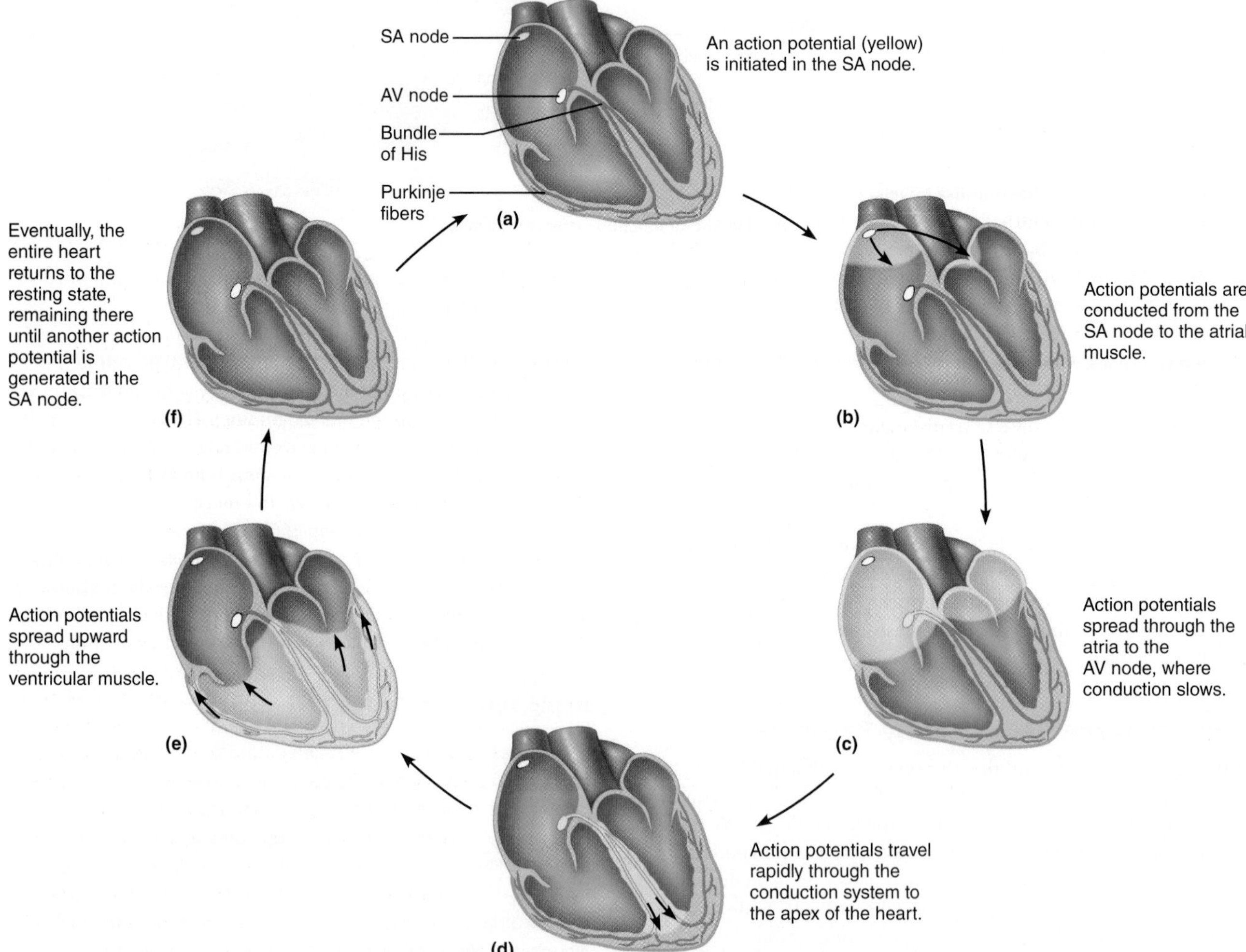

Figure 11 The spread of action potentials through the heart. The sequence of electrical excitation during a single heartbeat, starting with **(a)** depolarization of the SA node and ending with **(f)** the return of the heart to the resting state.

Electrical Activity in Pacemaker Cells

A cardiac contractile cell fires an action potential only when it is depolarized to threshold by a stimulus. Normally this stimulus is a circulating electrical current that originates in neighboring cells that are firing action potentials. We have seen that this current enters the cell through gap junctions that connect it with its neighbors. After entering, the current exits the cell by passing through the plasma membrane, and in doing so it triggers depolarization.

Recall that pacemaker cells are different because they can fire action potentials in the absence of any external stimulus, and they do so in a regular, periodic fashion. A pacemaker cell is able to fire action potentials spontaneously because it does not have a steady resting potential. After an action potential, a pacemaker cell immediately begins to depolarize slowly and continues to do so until its membrane potential reaches threshold, which triggers another action potential (**Figure 12**). Following this event, the membrane potential returns to about -60 to -70 mV and then begins another round of slow depolarization until another action potential is triggered. The slow depolarizations or "ramps" that lead up to each action potential are referred to as **pacemaker potentials.**

In pacemaker cells and other cardiac muscle cells, electrical signals are caused by changes in plasma membrane ion permeability brought about by the opening and closing of specific types of ion channels, just as in any other type of cell. To understand how these permeability changes affect the membrane potential, recall the following rule: *As a membrane's permeability to a particular ion increases relative to that of other ions, the membrane potential moves toward the equilibrium potential of that ion.* In cardiac muscle cells, the most important permeability changes involve sodium, potassium, and calcium ions (Na^+, K^+, and Ca^{2+}, respectively). Ion concentrations in cardiac muscle cells are similar to those in other cells—the intracellular fluid is rich in potassium but poor in sodium and calcium compared to extracellular fluid. Thus the equilibrium potential of potassium is negative, whereas the equilibrium potentials of sodium and calcium are both positive. (Approximate values are $E_K = -94$ mV, $E_{Na} = +60$ mV, and $E_{Ca} = +130$ mV.) Therefore, increased sodium or calcium permeability (P_{Na} or P_{Ca}) tends to make the membrane potential become more positive, whereas increased potassium permeability (P_K) tends to make it become more negative.

In pacemaker cells, electrical signals are triggered by changes in P_K, P_{Na}, and P_{Ca}, as shown in Figure 12. The slow depolarization that occurs in the early stages of the pacemaker potential is due to closing of potassium channels and opening of so-called *funny channels.* Potassium channels open during repolarization of the action potential, and then close when the membrane returns to its polarized state. Funny channels, so named because investigators noticed that they had some unusual characteristics, open after the cell repolarizes and allow sodium and potassium ions to cross the plasma membrane. The closing of potassium channels and the opening of funny channels during the early stages of the pacemaker potential have a net effect of decreasing potassium movement out of the cell and increasing sodium movement into the cell, causing the initial depolarization.

The funny channels are open for only a brief time, closing when the membrane potential approaches −55 mV, approximately 5 mV short of the threshold to generate an action potential. However, the initial depolarization triggers the opening of voltage-gated calcium channels called *T-type channels* (*T* stands for "transient"). This event raises P_{Ca}, which depolarizes the cell even further. Although the T-type channels stay open for only a short time before becoming inactivated, the resulting depolarization to threshold triggers the opening of a second population of voltage-gated calcium channels called *L-type channels* (*L* stands for "long-lasting"), which stay open longer and become inactivated only slowly. The result is a large increase in P_{Ca}, which produces the rapid depolarization characteristic of the upswing of the action potential. (As these calcium channels open, they also allow some sodium to flow into the cell, which increases P_{Na} and adds to the depolarizing effect.) This depolarization triggers the opening of potassium channels and, consequently, a rise in P_K that occurs shortly after the increase in P_{Ca}; the result is to pull the membrane potential back down. The subsequent fall in potential removes the stimulus for calcium channel opening, allowing these channels to begin closing. This event

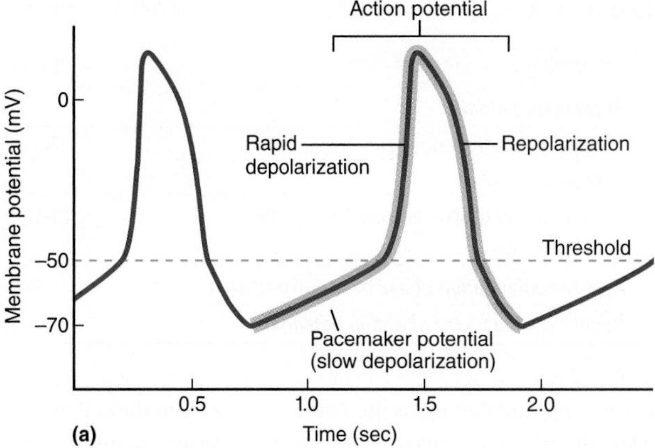

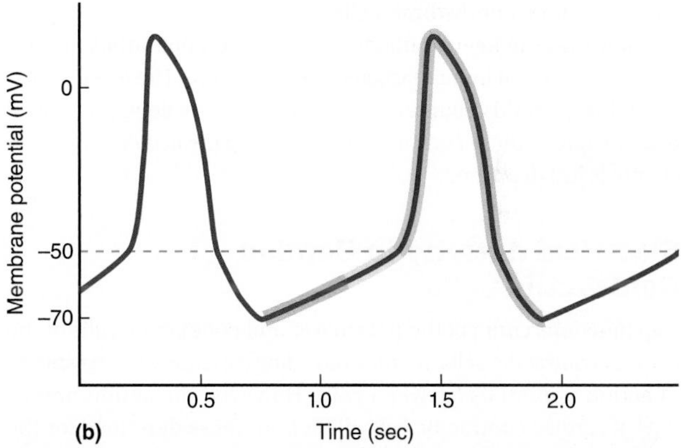

Figure 12 Electrical activity in a pacemaker cell. (a) A recording of the membrane potential showing action potentials and pacemaker potentials. **(b)** Changes in membrane permeability responsible for potential changes in pacemaker cells. During the initial spontaneous depolarization (orange), P_K is decreasing and P_{Na} is increasing. During the latter spontaneous depolarization (yellow), P_{Ca} is increasing and P_{Na} is decreasing. During the rapid depolarization (green), P_{Ca} is increasing more. During the repolarization phase (pink), P_{Ca} is decreasing and P_K is increasing.

TABLE 1 Ionic Bases of the Autorhythmic Cell Action Potential

Autorhythmic cell potential change	Ion channel gating	Ion movement
Pacemaker potential		
Initial period of spontaneous depolarization to subthreshold	Funny channels open	Sodium moves in, potassium moves out
Latter period of spontaneous depolarization to threshold	T-type calcium channels open	Calcium moves in
Rapid depolarization phase of action potential	L-type calcium channels open	Calcium moves in
Repolarization phase of action potential	Potassium channels open	Potassium moves out

reduces P_{Ca} and decreases the flow of calcium into the cell, which works along with the increase in P_K to repolarize the membrane and terminate the action potential. Table 1 summarizes the function of ion channels in autorhythmic cells.

Because the heart initiates its own action potentials, it does not require neural input to trigger its contractions. However, as discussed later in this chapter, autonomic neurons do exert control over the rate of these contractions by altering the activity of the ion channels just described.

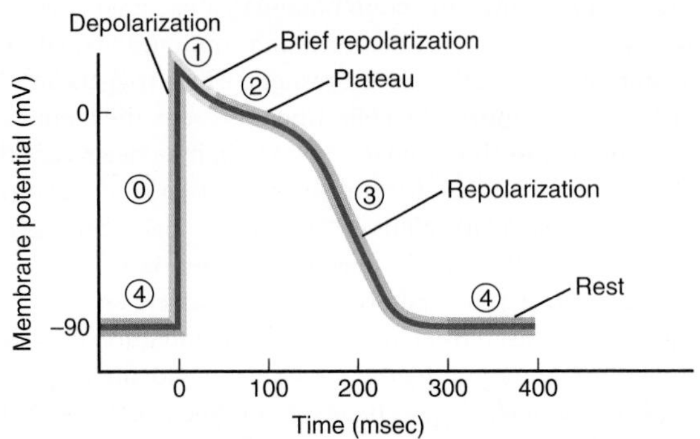

Figure 13 The cardiac action potential. An action potential recorded from a ventricular muscle cell. P_{Na} increases during phase 0 (green) and decreases during phase 1 (yellow). P_{Ca} increases and P_K decreases during phase 2 (orange), and then P_{Ca} decreases and P_K increases during phase 3 (purple). During phase 4 (blue), all ion channels are in their resting state (P_K high, P_{Ca} and P_{Na} low).

Electrical Activity in Cardiac Contractile Cells

Gap junctions connect the pacemaker and conducting cells of the heart to contractile cells, thereby providing a pathway for the spread of action potentials between cells. However, the action potentials of cardiac contractile cells differ from those described for the pacemaker cells. In addition, action potentials in cardiac contractile cells from different regions of the heart vary in regard to shape (time course and level of depolarization) and speed of propagation. These differences occur because contractile cells vary in the type and number of ion channels they possess, as well as in their physical dimensions. Despite these differences, two important events characterize most cardiac action potentials: (1) During a typical cardiac action potential, P_K *decreases* as a result of the action of a certain type of voltage-gated potassium channel that *closes* in response to depolarization. (Recall that in pacemaker cells and most other excitable tissues, P_K *increases* during an action potential because these tissues contain voltage-gated potassium channels that open in response to depolarization.) (2) During a cardiac action potential, depolarization causes the opening of voltage-gated *calcium* channels, which not only affects the membrane potential, but also is instrumental in triggering muscle cell contractions. These two events prolong the cardiac contractile cell action potential as described next for the ventricular contractile cell.

The majority of ventricular muscle cells, which make up the bulk of the myocardium, are unlike pacemaker cells in that they have stable resting potentials. They also have longer-lasting action potentials with a distinctive shape that can be divided into five phases (designated 0–4), as shown in **Figure 13** Permeability changes occurring during each of these phases are described next:

⓪ *Phase 0.* Phase 0 of the cardiac action potential is similar to the upswing of a neuronal action potential and is caused by similar events: Depolarization of the membrane triggers the opening of voltage-gated sodium channels, which raises P_{Na} and increases the flow of sodium ions into the cell. Consequently, the membrane potential becomes more positive, which triggers the opening of still more sodium channels, additional increases in P_{Na}, more depolarization, and so on. The result is a rapid rise in membrane potential that peaks between +30 and +40 mV.

① *Phase 1.* The sodium channels that were opened in phase 0 start to become inactivated, which reduces P_{Na}. This decreases the flow of sodium into the cell and causes the membrane potential to fall toward more negative values because of continued movement of potassium ions out of the cell. The membrane potential drops only a small amount, however, because the depolarization of the membrane that began in phase 0 has set into motion two additional events that are occurring at this time: (1) the closing of voltage-gated potassium channels (known as *inward rectifier channels*), which reduces P_K and decreases the flow of potassium out of the cell; and (2) the opening of L-type calcium channels, which raises P_{Ca} and increases the flow of calcium into the cell. Both of these changes act to depolarize the membrane, thereby counteracting the effect of sodium channel inactivation.

② *Phase 2.* During phase 2, which is also referred to as the *plateau phase,* most of the potassium channels that were closed in phase 1 stay closed, so that P_K remains lower than its resting value. At the same time, most of the calcium channels that opened in phase 1 remain open, and P_{Ca} remains elevated. The lowered P_K and elevated P_{Ca} both act to keep the membrane in its depolarized state.

③ *Phase 3.* During this phase, P_K increases, partly because of the action of a second population of potassium channels similar to those in neurons (called *delayed rectifier channels*), which open in response to depolarization. These channels begin to open during phases 1 and 2 but do not exert a significant influence on the membrane potential until phase 3 because they open slowly. As P_K rises, the flow of potassium out of the cell increases, which pulls the membrane potential down toward more negative values. Furthermore, this fall in potential removes the stimulus that kept the inward rectifier channels closed in phase 2; as a consequence, these channels begin to open, which raises P_K even further. The fall in potential also removes the stimulus that kept the calcium channels open during phase 2 and allows them to begin closing, which lowers P_{Ca} and reduces the flow of calcium into the cell. This phenomenon works hand in hand with the increase in P_K to repolarize the membrane, thereby terminating the action potential.

④ *Phase 4.* During this phase, which corresponds to the *resting potential,* P_K, P_{Na}, and P_{Ca} are at their resting values. Because P_K is much greater than P_{Na} or P_{Ca} under these conditions, the membrane potential is approximately −90 mV, which is close to the equilibrium potential of potassium.

Table 2 summarizes the role of ion channels in the contractile cell action potential.

Apply Your Knowledge

Verapamil is an L-type calcium channel blocker used to treat arrhythmias and hypertension (high blood pressure). How does verapamil affect pacemaker potentials and contractile cell action potentials?

Excitation-Contraction Coupling in Cardiac Contractile Cells

The mechanism by which a cardiac action potential stimulates contraction is similar to that for skeletal muscle in many respects, but also has similarities to smooth muscle (**Figure 14**). The stimulus that triggers an action potential in the cardiac muscle cell is current coming through gap junctions ①. An action potential spreads through the plasma membrane and down T tubules ②, causing voltage-sensitive calcium channels on the sarcoplasmic reticulum to open and release calcium into the cytosol ③. (Dihydropyridine [DHP] receptors of the T tubules are coupled to ryanodine receptors in the sarcoplasmic reticulum.) The action potential also triggers opening of voltage-gated calcium channels on the plasma membrane, allowing calcium to enter the cell. (This increased permeability to calcium is also responsible for the plateau phase of the action potential.) The calcium that enters during the plateau phase acts on the voltage-sensitive channels that trigger calcium release from the sarcoplasmic reticulum and stimulates them to stay open longer ④. As a result, the sarcoplasmic reticulum releases more calcium with each action potential. This phenomenon is known as *calcium-induced calcium release.*

Cytosolic calcium triggers contraction of cardiac muscle in the same way it triggers contraction of skeletal muscle: Calcium binds to troponin, shifting tropomyosin off of the myosin-binding sites on actin ⑤, and crossbridge cycling occurs ⑥. Most of the calcium that binds to troponin (95%) comes from the sarcoplasmic reticulum, with only 5% coming from the extracellular fluid. Relaxation of cardiac muscle requires removal of calcium from the cytosol ⑦, which occurs by three mechanisms: (1) As in skeletal muscle, a Ca^{2+}-ATPase unit located in the membrane of the sarcoplasmic reticulum actively transports calcium from the cytosol into the lumen of the sarcoplasmic reticulum. (2) Cardiac muscle also has a Ca^{2+}-ATPase unit located in the plasma membrane that actively transports calcium from the cytosol into the interstitial fluid. (3) Cardiac muscle also has a Na^+-Ca^{2+} exchanger in the plasma membrane that actively transports calcium out of the cell by countertransport with sodium. Without calcium bound to troponin, tropomyosin shifts back over the myosin-binding sites on actin and the muscle fiber relaxes ⑧.

Recording the Electrical Activity of the Heart with an Electrocardiogram

The **electrocardiogram** (**ECG** or EKG; the *K* is for the German form of the word, *elektrokardiograph*) is a noninvasive means of monitoring the electrical activity of the heart. To understand the interpretation of ECGs, it is important to remember that the recorded electrical events cause the contraction of cardiac muscle. Physicians use ECG recordings to determine whether problems exist in the electrical activity of the heart.

Functional Fact

TABLE 2 Ionic Bases of the Contractile Cell Action Potential

Phase of contractile cell action potential	Ion channel gating	Ion movement
0: Rapid depolarization	Sodium channels open	Sodium moves in
1: Small repolarization	Sodium channels become inactivated	Sodium movement in decreases
2: Plateau	Potassium inward rectifier channels close Calcium L-type channels open	Potassium movement out decreases Calcium moves in
3: Repolarization	Potassium delayed rectifier channels open Calcium L-type channels close	Potassium moves out Calcium movement in decreases
4: Resting potential	Potassium channels (both types) open Sodium and calcium channels remain closed	Potassium moves out Little sodium or calcium moves in

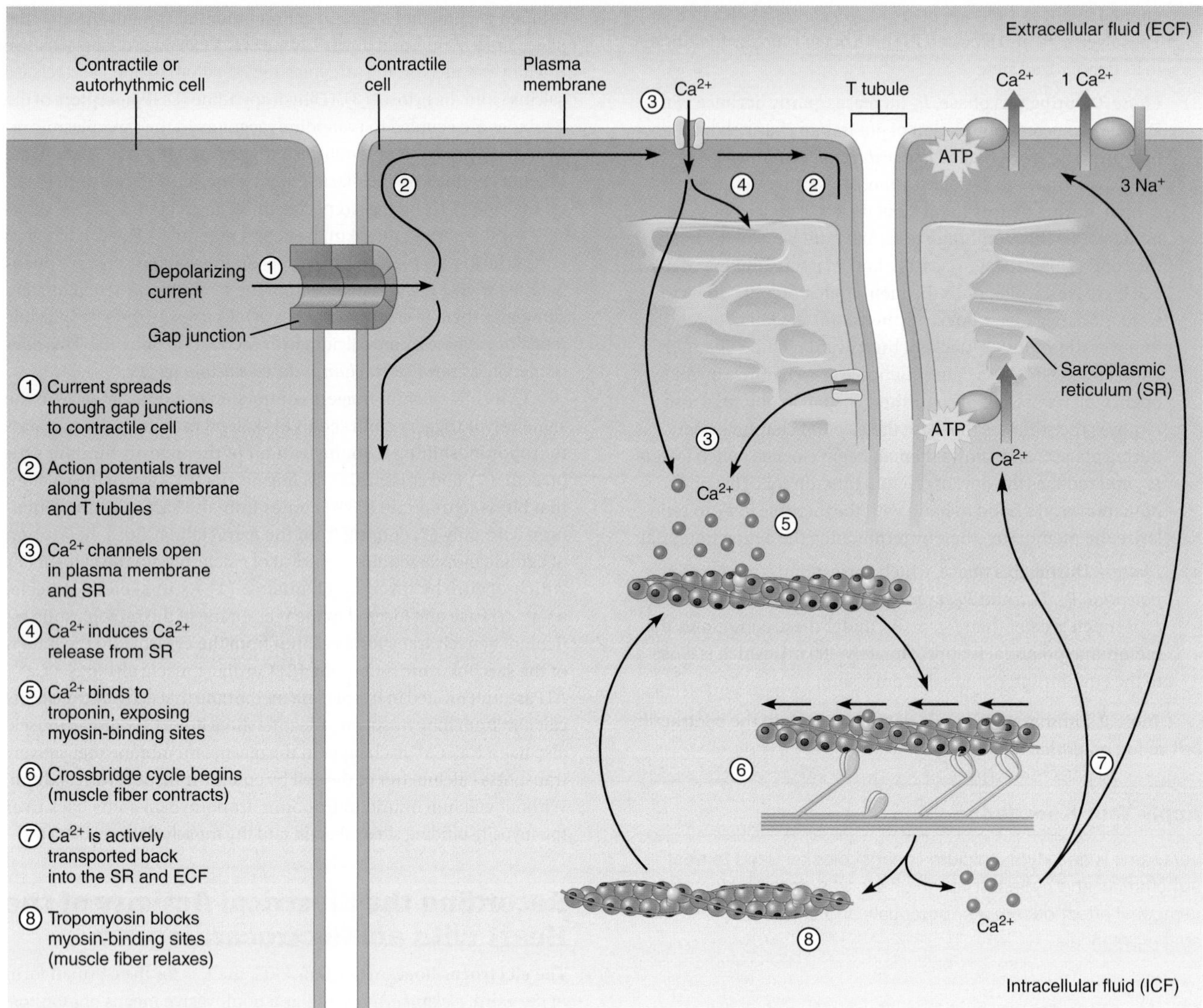

Figure 14 **Excitation-contraction coupling in cardiac muscle.**

Q In cardiac muscle, which of the transport mechanisms for calcium is passive? Which is primary active transport? Which is secondary active transport?

The ECG is a record of the overall spread of electrical current through the heart as a function of time during the cardiac cycle. The ECG is usually recorded by means of electrodes placed on the skin. The concept of recording ECGs is similar to that for recording EEGs; namely, electrical activity generated in nervous or muscle tissue spreads through the body because body fluids function as conductors. The more synchronized the activity, the larger the amplitude of signals that are recorded at a distance from the source. Because the electrical activity of the heart is highly synchronized, relatively large-amplitude electrical potentials that correspond to distinct electrical phases of the heart can be detected at the surface of the skin.

A Dutch physiologist, Willem Einthoven, developed the technique of ECG recordings. The procedure for standard ECG recording is based on an imaginary equilateral triangle surrounding the heart. The triangle is expanded until its corners fall on the right arm, left arm, and left leg—a pattern known as Einthoven's triangle (**Figure 15**). Electrodes placed on the skin at the corners of the triangle are connected in pairs to a voltage-measuring device such as an oscilloscope or chart recorder. Certain pairs of electrodes are referred to as *leads* and are designated by Roman numerals. One electrode in each lead is designated as the positive electrode, the other as the negative.

Each specific lead detects the difference in the surface electrical potential between the positive and negative electrodes. Lead I detects the potential at the left arm minus that at the right arm; lead II detects the potential at the left leg minus that at the right arm; lead III detects the potential at the left leg minus the left arm. The direction of the recorded waveforms (up or down) depends on

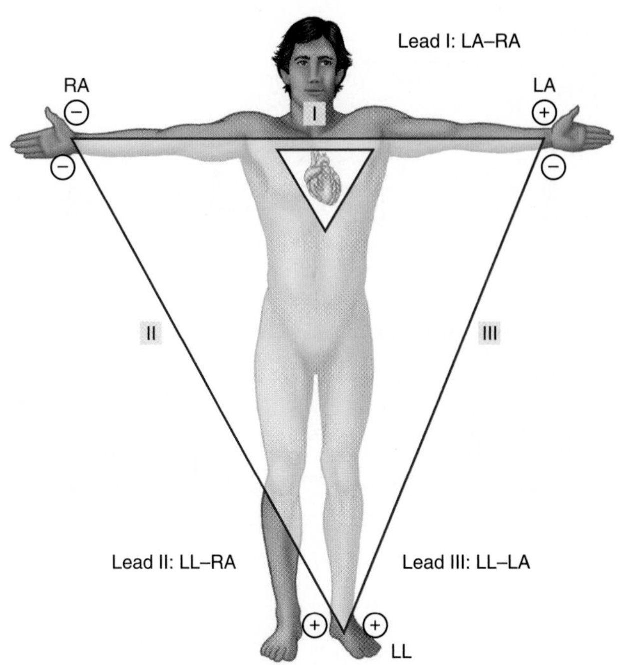

Figure 15 **Einthoven's triangle.**

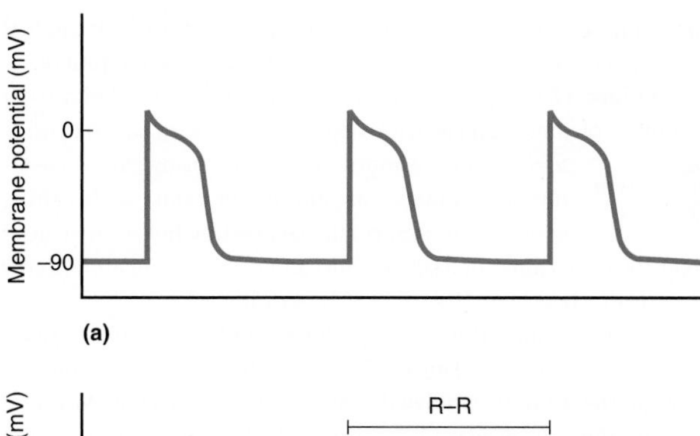

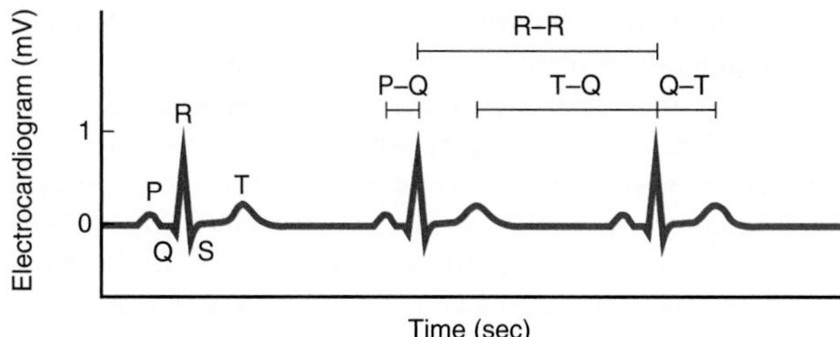

Component	Amplitude (mV)	Duration (sec)
P wave	0.2	0.10
QRS complex	1.0	0.08–0.12
T wave	0.2–0.3	0.16–0.27
P–Q interval	N/A	0.12–0.21
Q–T interval	N/A	0.30–0.43
T–Q segment	N/A	0.55–0.70
R–R interval	N/A	0.85–1.00

(b)

Figure 16 **Electrical activity of the heart. (a)** Recording of the membrane potential in a ventricle contractile cell. **(b)** Recording of a lead II ECG. The table gives normal values for ECG waves, intervals, and segments.

whether the difference in potential between the two electrodes is positive (which gives an upward deflection) or negative (which gives a downward deflection). For example, a depolarizing wave traveling toward the positive electrode causes an upward deflection, whereas a depolarizing wave traveling toward the negative electrode causes a downward deflection.

Clinical ECGs use chest electrodes as well as the limb electrodes, providing 12 different leads. Each lead gives a different electrical picture of the heart. Thus, even though each lead will show the same basic waveforms (P, QRS, and T waves; described next), the shapes of the waves differ.

The waveforms recorded with a standard lead II ECG and the action potential of a ventricular contractile cell are shown in **Figure 16**. ECGs are recorded on chart paper at a rate of 25 mm/sec with an amplitude of 1 mV/cm. The ECG normally shows three characteristic waveforms: (1) The **P wave** is an upward deflection that is due to atrial depolarization. (2) The **QRS complex** is a series of sharp upward and downward deflections due to ventricular depolarization; it is correlated with phase 0 of the ventricular contractile cell action potential. (3) The **T wave** is an upward deflection caused by ventricular repolarization; it is correlated with phase 3 of the ventricular contractile cell action potential. Atrial repolarization is generally not detected in an ECG recording because it occurs at the same time as the QRS complex. Between the waves, a normal ECG trace consists of a horizontal line, called the *isoelectric line,* indicating that no changes in electrical activity are occurring.

Although the phases of the ECG are associated with action potentials traveling through the heart muscle, the ECG is *not* simply a recording of an action potential. During the heartbeat, cells fire action potentials at different times, and the ECG reflects *patterns* of action potential firing in the entire *population* of cells that make up the heart muscle.

In addition to waves, certain intervals and segments can provide important information about the function of the heart. The *P-Q* or *P-R interval* occurs between the onset of the P wave and the onset of the QRS complex and is an estimate of the time of conduction through the AV node. The *Q-T interval* is the time from the onset of the QRS complex to the end of the T wave and is an estimate of the time the ventricles are contracting, called **ventricular systole.** The *T-Q segment* is the time from the end of the T wave to the beginning of the QRS complex and is an estimate of the time the ventricles are relaxing, called **ventricular diastole.** The *R-R interval* is the time between the peaks of two successive QRS complexes; it represents the time between heartbeats. Heart rate can be determined by dividing 60 seconds by the R-R interval. If the R-R interval is 1 second, for example, then the heart rate is 60 beats per minute. If the R-R interval is 0.5 second, then the heart rate is 120 beats per minute.

Figure 17 compares normal ECGs to those recorded during examples of abnormal electrical activity of the heart, called *cardiac arrhythmias*. Abnormal SA nodal firing can cause sinus arrhythmias, including either a sinus tachycardia, which is an abnormally fast resting heart rate (more than 100 beats/min), or a sinus bradycardia, which is an abnormally slow resting heart rate (less than 50 beats/min).

Altered conduction through the AV node can cause various degrees of heart block: During first-degree heart block, conduction through the AV node is slowed, causing a longer delay in AV nodal conduction (an increased P-Q interval). During second-degree heart block, conduction through the AV node does not always occur. If conduction does not occur, the ventricles do not depolarize (resulting in the absence of a QRS complex and T wave) and, therefore, do not contract. As a consequence, the 1:1 relationship between atrial contractions and ventricular contractions is lost. During third-degree heart block, conduction through the AV node does not occur at all, causing complete dissociation of atrial and ventricular contractions. The atria contract at the rate of SA nodal discharge, but the ventricles contract at the rate of bundle of His fiber discharge, which occurs only 30–40 times per minute. This slow rate of ventricular contraction is often insufficient to supply the body with the oxygen and nutrients required, which is why third-degree block can be deadly.

Sometimes the heart is depolarized by an electrical stimulus arising outside the normal conduction pathway. Because cardiac muscle cells are connected by gap junctions, an abnormal depolarization will spread throughout the heart, causing an extra contraction called an *extrasystole*. If the depolarization occurs in an atrium, then a premature atrial contraction (PAC) occurs, and conduction follows through the AV node, causing contraction of the ventricle. If the depolarization occurs in a ventricle, then a premature ventricular contraction (PVC) occurs, and no atrial contraction precedes it. PACs and PVCs are generally of little clinical significance unless they occur at high frequencies.

More serious arrhythmias are fibrillations, which occur when the heart muscle no longer demonstrates synchronized depolarization. In atrial fibrillation, atrial muscle fibers depolarize independently, so atrial contraction is inefficient in pumping blood to the ventricle. Atrial fibrillation results in weakness and light-headedness due to decreased blood flow, but as long as the ventricles still contract at a sufficient rate, atrial fibrillation is generally not deadly because contraction of the atria contributes little to ventricular filling (most ventricular filling is passive). During atrial fibrillation, however, blood churns in the atria, enhancing the likelihood that a blood clot will form; such a clot can move through the circulation and lodge in blood vessels, clogging them. Thus atrial fibrillation can cause pulmonary embolism, stroke, or heart attack, depending on where the clot lodges.

Ventricular fibrillation, by contrast, can cause death within minutes. When ventricular muscle cells depolarize independently, the ventricles can no longer efficiently pump the blood out to the tissues, including the brain. Clinicians must quickly *defibrillate* the ventricular muscle to keep the person alive. Defibrillation is often accomplished by passing a large current through the chest wall to

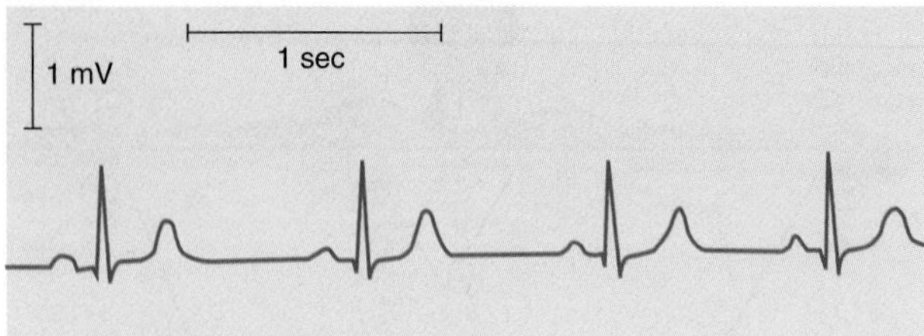

Sinus tachycardia (with inverted T wave)

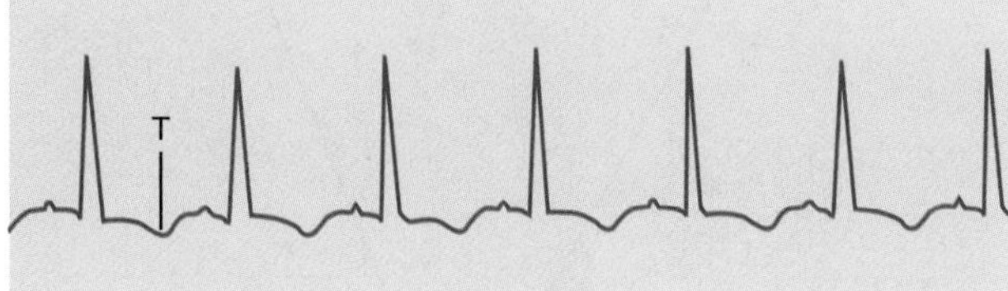

Sinus bradycardia

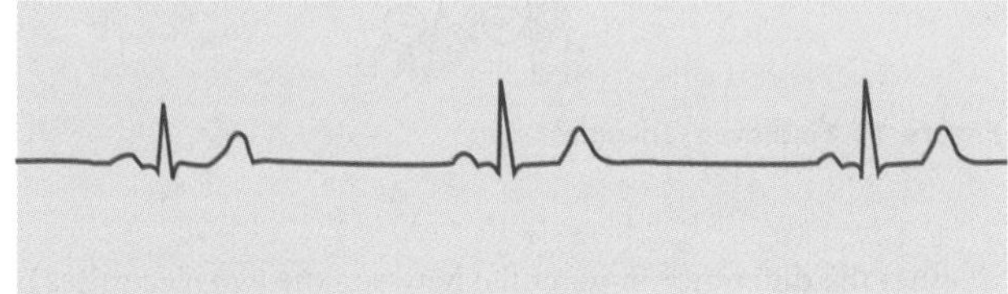

Heart block, third degree

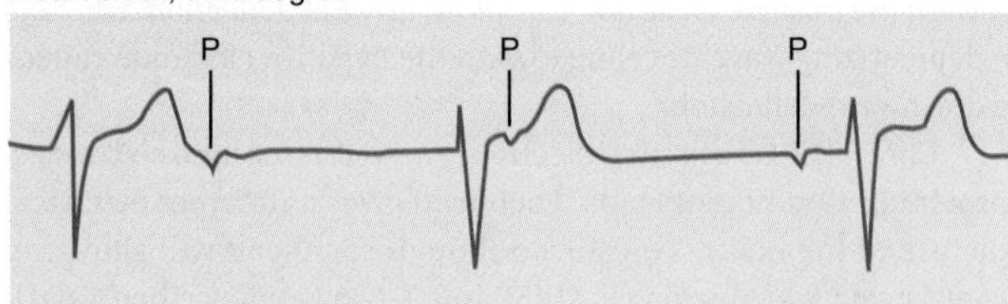

Premature atrial contraction (PAC)

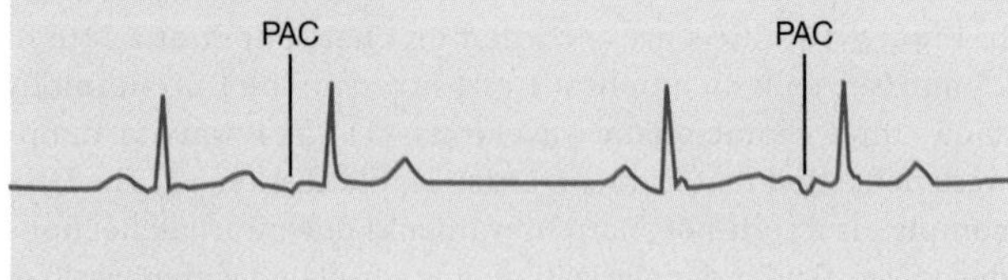

Ventricular fibrillation

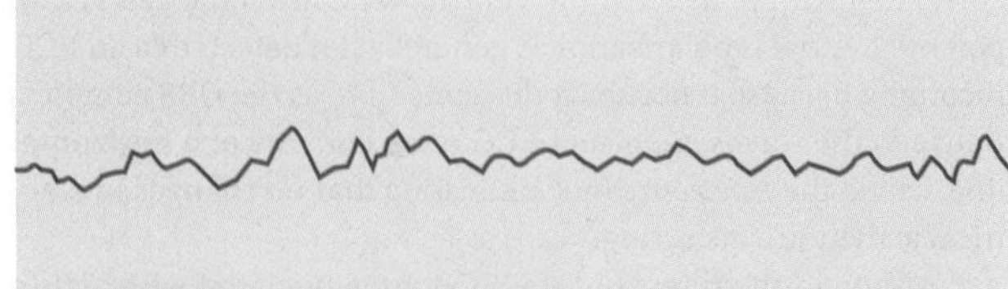

Figure 17 Lead II ECGs showing various arrhythmias. 25 mm on the horizontal axis = 1 sec; 1 cm on the vertical axis = 1 mV. Some data courtesy of C. V. Massey, University of South Alabama.

the heart such that the externally applied current depolarizes all the muscle cells at the same time, thereby returning synchronous electrical activity to the heart.

Quick Check 4

1. Under normal conditions, which controls the heartbeat—the SA node or the AV node? Explain why.
2. Arrange the order of the following terms so that they describe the normal path of electrical impulses through the heart: *Purkinje fibers, atrioventricular bundle, AV node, bundle branches, SA node, ventricular muscle, atrial muscle.*
3. The entry of calcium into a ventricular muscle cell helps to maintain depolarization of the membrane during the plateau phase of the action potential. Which other important function does this calcium perform?
4. Match the terms *P wave, QRS complex,* and *T wave* with the following events: ventricular depolarization, ventricular repolarization, atrial depolarization.

5 The Cardiac Cycle

The **cardiac cycle** includes all the events associated with the flow of blood through the heart during a single complete heartbeat. The discussion here concentrates on the following aspects of the cardiac cycle: (1) the various phases in the pumping action of the heart; (2) periods of valve opening and closure; (3) changes in atrial, ventricular, and aortic pressure, which reflect contraction and relaxation of the heart muscle; (4) changes in ventricular volume, which reflect the amount of blood entering and leaving the ventricle during each heartbeat; and (5) the two major heart sounds.

The relationships among the various aspects of the cardiac cycle are depicted in the *Wigger diagram* in **Figure 18**. (The pressure graphs pertain to the left heart only; the graphs for pressures in the right heart are similar, except that the peak pressures are lower.)

Phases of the Cardiac Cycle

Because the cardiac cycle involves the events of one heartbeat, a complete cycle involves both ventricular contraction and ventricular relaxation. As a result, the cycle can be divided into two major stages: **systole,** the period of ventricular contraction, and **diastole,** the period of ventricular relaxation. (Even though the atria also undergo periods of contraction and relaxation—termed *atrial systole* and *atrial diastole,* respectively—we use the terms *systole* and *diastole* to refer to ventricular events.)

We begin our examination of the cardiac cycle in the middle of diastole, a time at which the atria and ventricles are completely relaxed:

1. *Ventricular filling.* During mid-to-late diastole (phase 1 in Figure 18), blood returning to the heart via the systemic and pulmonary veins enters the relaxed atria and passes through the AV valves and into the ventricles under its own pressure. The return of blood from the veins to the heart, which is called **venous return,** occurs because the pressure in the veins is greater than that in the atria. During this time, the pulmonary and aortic (semilunar) valves are closed because ventricular pressure is lower than that in the aorta and pulmonary arteries.

   Late in diastole (at the end of phase 1), the atria contract, driving more blood into the ventricles. Shortly thereafter, the atria relax and systole begins. This entire phase of blood entering the ventricle is called **ventricular filling.**
2. *Isovolumetric contraction.* At the beginning of systole (phase 2), the ventricles contract, which raises the pressure within them. When ventricular pressure exceeds atrial pressure (which occurs very early in systole), the AV valves close; the semilunar valves remain closed because ventricular pressure is not yet high enough to force them open. At this point, no blood flows into or out of the ventricles because all the valves are closed. Thus, even though the ventricles are contracting, the volume of blood within them remains constant, so phase 2 is termed **isovolumetric contraction.** Phase 2 ends when the ventricular pressure is great enough to force open the semilunar valves so that blood can leave the ventricles.
3. *Ventricular ejection.* In the remainder of systole (phase 3), blood is ejected into the aorta and pulmonary arteries through the open semilunar valves, and ventricular volume falls. During the exit of blood from the ventricles, referred to as **ventricular ejection,** ventricular pressure rises to a peak and then begins to decline. When it falls below aortic pressure, the semilunar valves close, ending ejection (and systole) and marking the beginning of diastole.
4. *Isovolumetric relaxation.* At the onset of early diastole (phase 4), the ventricular myocardium is relaxing. Some blood is present in the ventricles, and it remains under pressure because it takes time for the tension in the ventricular muscle to wane. Ventricular pressure is simultaneously too low to keep the semilunar valves open and too high to allow the AV valves to open. Because all valves are closed and the volume of blood remains constant within the relaxing ventricles, phase 4 is referred to as **isovolumetric relaxation.**

   Once ventricular pressure decreases to less than atrial pressure, thereby permitting the AV valves to open again, blood enters the ventricles from the atria. This marks the beginning of phase 1, and the pump cycle begins once again.

The durations of systole and diastole are not equal. For a heart beating at the normal resting rate of 72 beats per minute (one beat every 0.8 second), most of the cardiac cycle (approximately 65%, or 0.5 second) is spent in diastole; systole lasts only about 0.3 second. This longer diastole gives the heart adequate time to fill with blood, which is essential for efficient pumping; it also gives the heart muscle more time to relax, which helps prevent fatigue.

Now that this overview of the cardiac cycle is complete, we can examine pressure and volume changes during this cycle.

Atrial and Ventricular Pressure

By convention, cardiovascular pressures (the pressure of blood in the chambers of the heart or in the vasculature) are given in millimeters of mercury (mm Hg). Atmospheric pressure is also measured in millimeters of mercury, with normal atmospheric pressure at sea

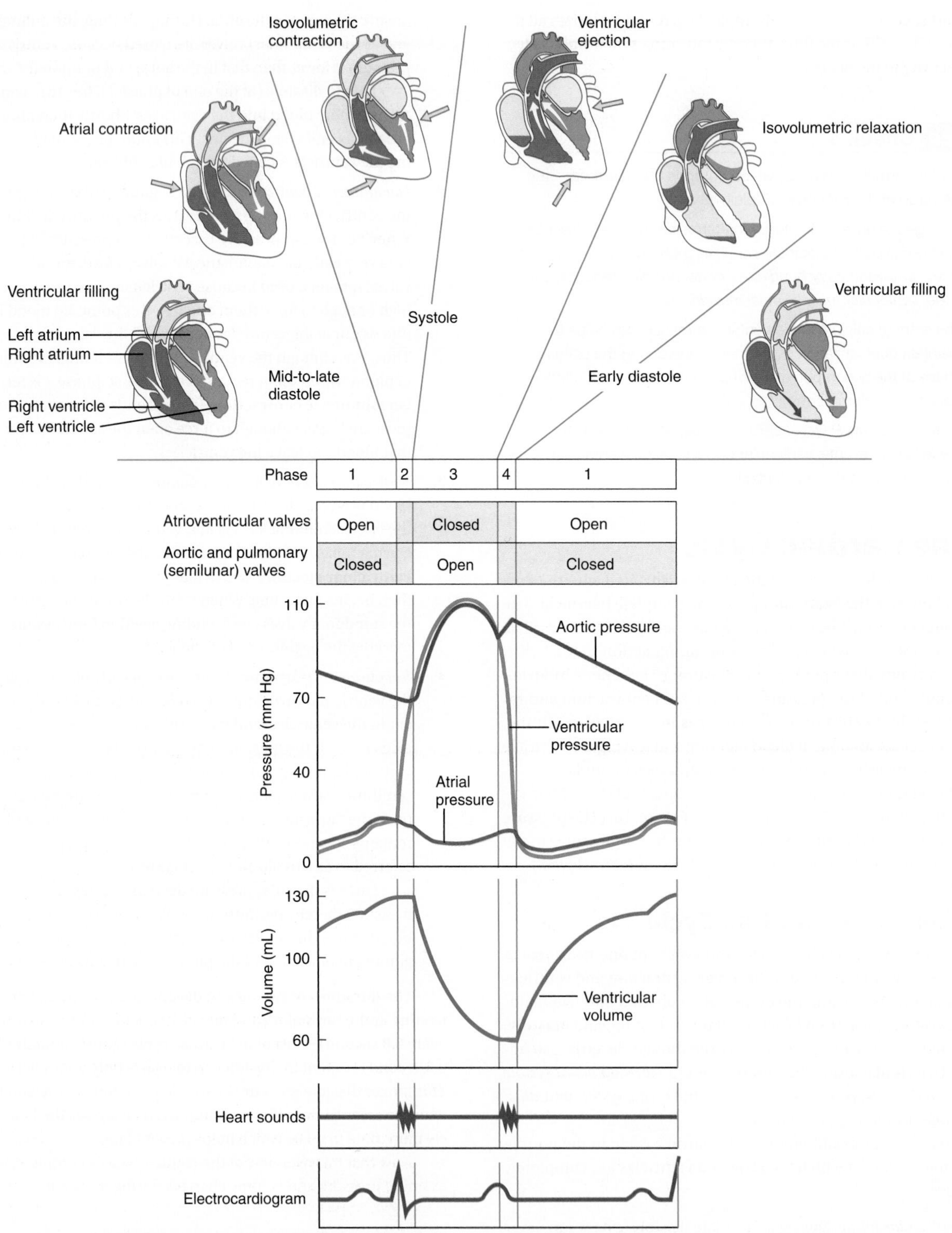

Figure 18 Cardiac cycle. All values are given in reference to the left heart. Diastole corresponds to phases 4 and 1, whereas systole corresponds to phases 2 and 3. Note the correlations between changes in pressure gradients across valves and the open/closed state of the valve and the changes in ventricular volume. Heart sounds are correlated with the closing of the valves. ECG waves are correlated with the mechanical events of the heart: The P wave precedes atrial contraction (evident as an increase in atrial pressure), while the QRS complex precedes ventricular contraction.

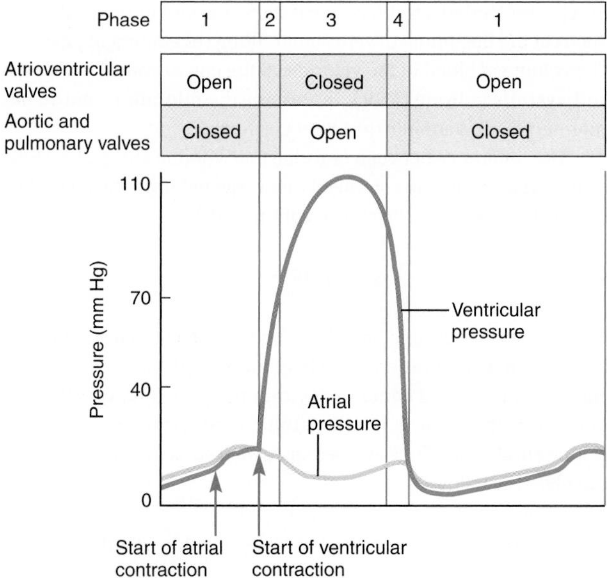

Figure 19 Ventricular pressure during the cardiac cycle. The AV valves close when ventricular pressure exceeds atrial pressure, which occurs immediately after the ventricles start to contract.

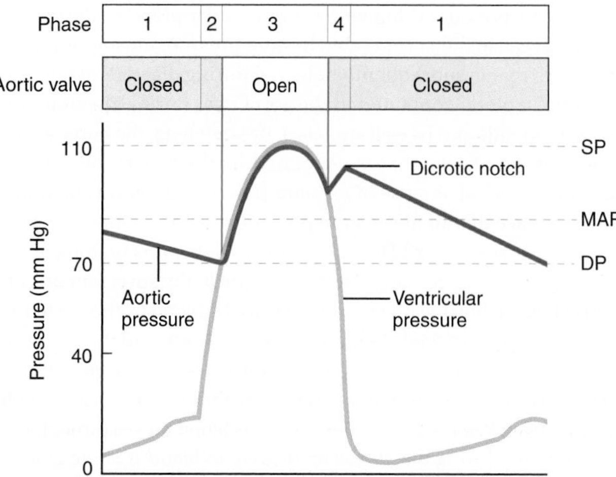

Figure 20 Aortic pressure during the cardiac cycle. The aortic valve opens when ventricular pressure exceeds aortic pressure, and then closes when aortic pressure exceeds ventricular pressure. The maximum pressure in the aorta during the cardiac cycle is the systolic pressure (SP), whereas the minimum pressure during the cardiac cycle is the diastolic pressure (DP). The average pressure during the cardiac cycle is the mean arterial pressure (MAP).

Q Why is the mean arterial pressure closer to diastolic pressure than to systolic pressure?

level being 760 mm Hg. In cardiovascular physiology, all pressures are given relative to atmospheric pressure, which is taken to be *zero.* Thus, when physiologists say that blood is at a pressure of 100 mm Hg, they mean that it is 100 mm Hg *above* atmospheric pressure.

Figure 19 shows the changes in ventricular pressure that occur during the cardiac cycle. In mid-diastole (phase 1), ventricular pressure stays very low until the end of the phase, when an abrupt but small rise occurs. This rise is due to atrial contraction, which adds a small volume of blood to the ventricle; the increase in volume causes ventricular pressure to rise. Shortly thereafter, a much larger increase in pressure that corresponds to ventricular systole occurs. In early ventricular diastole, the pressure falls to near zero, reflecting relaxation of the myocardium. Through the remainder of diastole, ventricular pressure slowly creeps upward as the ventricle passively fills with blood returning from the pulmonary circulation.

The increase in atrial pressure that occurs in late diastole indicates the beginning of atrial contraction (see Figure 18). This rise in pressure is small and short-lived, however, and is followed by a series of similar small increases at various times throughout the cardiac cycle. Because these changes in pressure are of little significance to overall cardiac function, we will not discuss them further.

Aortic Pressure

During diastole, no blood enters the aorta because the aortic valve is closed. However, blood is able to leave the aorta by flowing through blood vessels downstream in the systemic circuit. This continual exit of blood causes the volume of blood in the aorta to decrease during diastole, which leads to a slow decline in aortic pressure (**Figure 20**). (Imagine the pressure falling slowly in a leaky tire as air escapes.) At the end of this decline, aortic pressure reaches a minimum called the **diastolic pressure (DP).** As the next systolic period begins (phase 2), the aortic pressure continues to fall briefly because the aortic valve opens only when the ventricular pressure becomes high enough to force it open.

When the aortic valve opens and the ejection phase begins (phase 3), the aortic pressure rises quickly. This increased pressure reflects an increase in the volume of blood contained in the aorta, which occurs because blood is flowing into the aorta faster than it can flow out. (Imagine the pressure rising as you pump air into a leaky tire, even though air continues to escape from it.) Within a short time, however, the flow of blood from the heart begins to slow down. As a reflection of this condition, the aortic pressure does not continue to rise but instead reaches a maximum, called the **systolic pressure (SP),** and then starts to fall. At the end of systole the aortic valve closes, terminating the flow of blood into the aorta from the heart (beginning of phase 4). This event is marked by a "wiggle" in the aortic pressure curve called the *dicrotic notch.*

The average aortic pressure occurring during the cardiac cycle is known as the **mean arterial pressure (MAP).** This pressure is an extremely important variable in cardiovascular physiology because-MAP represents the driving force that pushes blood through the systemic circuit.

In Figure 20, it is apparent that the overall variation in aortic pressure that occurs during the cardiac cycle is much less than the variation in left ventricular pressure. During the ejection phase (phase 3), aortic and ventricular pressures are virtually identical because the aorta and ventricle are in communication through the aortic valve. Throughout the remainder of the cycle, however, the difference between aortic and ventricular pressure is substantial—ventricular pressure drops abruptly while aortic pressure remains elevated and decreases only slowly. This difference in pressure reflects the fact that the aorta and ventricle are no longer in communication because the aortic valve is closed.

Aortic pressure is higher than ventricular pressure during diastole because the aorta is able to store pressure during systole—pressure that is subsequently released during diastole. When the volume of blood contained in the aorta rises during ejection, the vessel expands and its wall stretches. By stretching, the aorta stores some of the energy that is generated by the heart; this energy becomes evident as a rise in *pressure.* (Because the aorta and other arteries have this ability to store pressure, they are said to function as *pressure reservoirs.*) During diastole, the heart is no longer actively generating pressure, but the pressure in the aorta remains elevated due to the inward force exerted on the blood by the stretched elastic tissue in its wall. As blood leaves the aorta and the pressure falls, the wall recoils. In doing so, the aorta releases the energy that it stored during systole. This energy drives the flow of blood through downstream vessels during diastole. Thus blood moves through the vasculature during diastole *even though no blood is being ejected from the heart at this time.* As a consequence, blood flows through the vasculature in a more or less continual fashion throughout the cardiac cycle, despite the fact that it exits the heart in spurts.

Ventricular Volume

Figure 21 shows a curve that traces changes in *ventricular volume*—in this case the volume of blood in the left ventricle—during the cardiac cycle. During mid-to-late diastole (phase 1), the volume of blood contained in the ventricle rises during ventricular filling. Volume increases quickly at first and then rises less rapidly as time goes on. Toward the end of diastole, a small but abrupt rise in volume reflects blood being pumped into the ventricle as a result of atrial contraction. The volume of blood in the ventricle at the end of diastole, referred to as the **end-diastolic volume (EDV),** represents the maximum ventricular volume attained during the cardiac cycle, which is reached just before the beginning of ejection. This volume does not change during the period of isovolumetric contraction (phase 2) but begins to diminish as soon as the aortic valve opens at the beginning of ejection (beginning of phase 3). It continues to decrease until the aortic valve closes at the end of the ejection period, at which time it stays constant until the AV valve opens at the beginning of ventricular filling (beginning of phase 1). The volume of blood in the ventricle at the end of systole, called the **end-systolic volume (ESV),** represents the minimum ventricular volume, which is attained just after ejection.

The difference between end-diastolic volume and end-systolic volume represents the volume of blood ejected from the heart during one beat, which is the **stroke volume (SV):**

$$SV = EDV - ESV$$

The average resting end-diastolic volume is 130 mL. Notice in Figure 21 that when the ventricle contracts (phase 3), it does not eject all of the blood contained within it; approximately 60 mL of blood remains in the ventricle at the end of systole under normal resting conditions. Thus an average resting stroke volume could be calculated as follows:

$$SV = EDV - ESV = 130 \text{ mL} - 60 \text{ mL} = 70 \text{ mL}$$

End-systolic volume is determined by a number of factors, including the force of ventricular contraction, which can be altered by autonomic neurons and hormones. An increase in the force of contraction, such as would occur during a fight-or-flight response, raises the stroke volume, so that a greater fraction of the end-diastolic volume is ejected. Conversely, a decrease in the force of contraction has the opposite effect. The fraction of end-diastolic volume ejected during a heartbeat is known as the **ejection fraction (EF):**

$$EF = SV/EDV$$

Using the normal values given, we obtain an ejection fraction of approximately 54% at rest:

$$EF = 70 \text{ mL}/130 \text{ mL} = 0.54$$

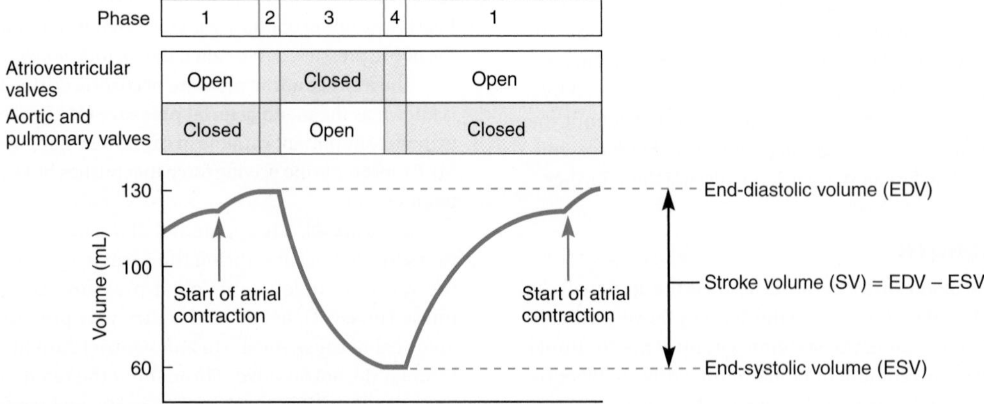

Figure 21 Changes in ventricular volume during the cardiac cycle.

Q This figure shows stroke volume for the left ventricle. Will stroke volume for the right ventricle be *higher*, *lower*, or the *same*?

Apply Your Knowledge

Given end-diastolic volume = 150 mL and end-systolic volume = 50 mL, calculate the stroke volume and the ejection fraction.

Pressure-Volume Curve

Now that we have an understanding of the pressure and volume changes in the heart, we can take a close look at the relationship between pressure and volume changes in the left ventricle. **Figure 22** shows a plot of left ventricular pressure versus left ventricular volume. The plot is a function of the cardiac cycle and, therefore, forms a loop in which each leg represents a particular phase of the cardiac cycle.

Phase 1 represents ventricular filling, when the mitral valve is open and the aortic valve is already closed. Ventricular volume increases from 60 to 130 mL, while pressure increases only slightly. Phase 2 represents isovolumetric contraction, with the volume equal to end-diastolic volume. At the beginning of phase 2, pressure in the ventricle goes up, exceeding that in the atria and causing closure of the mitral valve. Ventricular pressure continues to rise rapidly until it exceeds aortic pressure, at which point the aortic valve opens and phase 3, or ventricular ejection, starts. During ejection, the volume of blood in the ventricle decreases from 130 mL to 60 mL until the onset of diastole, which causes closure of the aortic valve as ventricular pressure falls below the corresponding pressure in the aorta. Phase 4 represents the isovolumetric phase of ventricular relaxation, with the volume equal to end-systolic volume. During

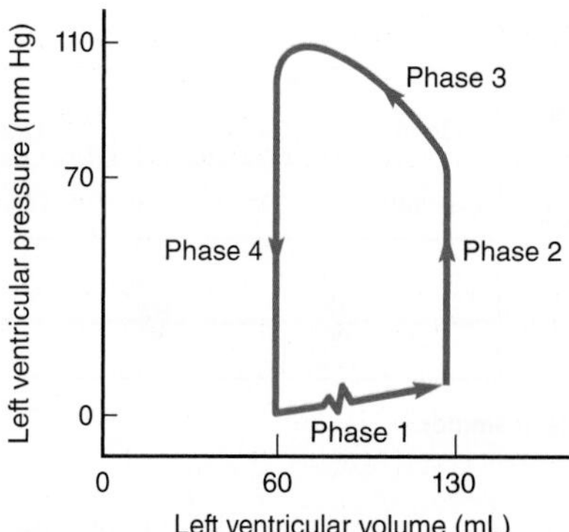

Figure 22 **Left ventricular pressure-volume curve.** Phases 1–4 are identical to those in Figure 18.

phase 4, pressure decreases rapidly until it drops below the atrial pressure and the mitral valve opens. At this point, the cycle begins anew at phase 1.

The pressure-volume curve provides information about the functioning heart. For example, the slope of the ventricular filling phase (phase 1) is a measure of the **compliance** of the left ventricle, or the ease with which the left ventricle can expand (increase in volume) with a given change in pressure: The greater the slope, the greater the compliance (see **Toolbox: Compliance**). The end-systolic volume is a measure of ventricular contractility, with a shift of the curve to the left indicative of greater contractility; that is, higher contractility will cause a greater ejection of

TOOLBOX

Compliance

A blood vessel (indeed, any hollow structure) tends to expand as the pressure inside it rises, and to contract as this pressure falls. Strictly speaking, it is not just the pressure *inside* the vessel that determines whether a vessel expands or contracts, but rather the *difference* between the pressure inside and the pressure outside. This difference is called the *distending pressure* (or *transmural pressure*). When the pressure inside a vessel is greater than the pressure outside it, the distending pressure is positive, and a net outward force acts on the wall of the vessel, tending to make it expand. As the distending pressure increases, the volume of the vessel increases. In contrast, if the pressure inside a vessel is less than the pressure outside it, the distending pressure is negative, and a net inward force acts on the wall and tends to compress the vessel.

The *compliance* of a vessel is strictly defined as the change in volume per unit change in distending pressure. Mathematically, this is expressed as follows:

$$\text{Compliance} = \Delta V / \Delta\,(P_{inside} - P_{outside})$$

where ΔV is the change in volume, and P_{inside} and $P_{outside}$ are the pressures inside and outside the vessel, respectively. The term within the parentheses is the distending pressure.

In simple terms, compliance is the ease with which a hollow vessel expands. A balloon with high compliance expands easily, whereas a balloon with low compliance requires greater effort to inflate it. That effort is used to create a pressure difference between the inside and outside of the balloon wall. The following diagram, which is a pressure-volume curve, graphically depicts the concept of compliance:

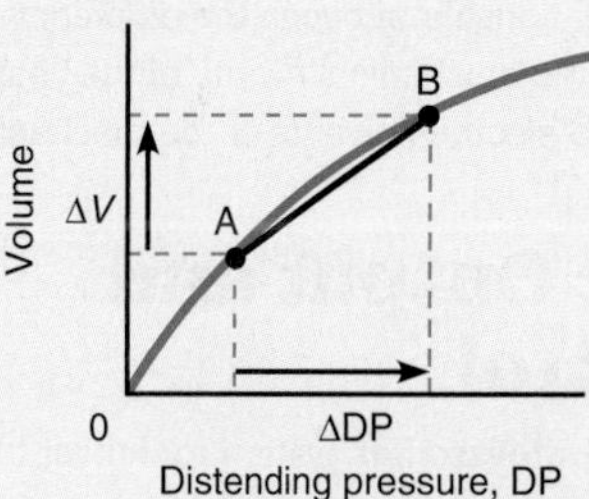

An increase in the distending pressure in a vessel, as when one goes from point A to point B in the diagram, produces a certain increase in the vessel's volume. The increase in volume is given by the vertical distance between the two points. The compliance is the change in vertical distance (volume) divided by the change in horizontal distance (distending pressure). In other words, the compliance of the vessel over the range between point A and B is the *slope* of the line joining the two points.

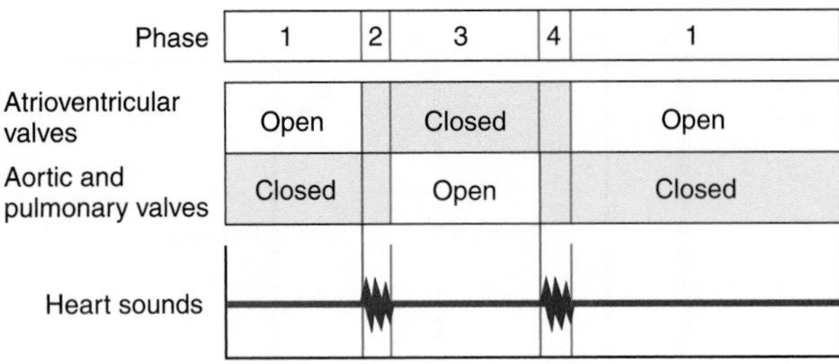

Figure 23 **Heart sounds.**

blood (higher stroke volume), leaving less blood in the heart at the end of systole. In contrast, a shift to the right could be indicative of heart failure.

Heart Sounds

The sounds of the beating heart that can be heard through a stethoscope are called *heart sounds.* In most people, each heartbeat comprises two distinct sounds: a soft, low-pitched "lubb," designated the "first sound," and a louder, sharper, higher-pitched "dupp," designated the "second sound." Thus the sounds of the beating heart are often described as lubb-DUPP, lubb-DUPP, lubb-DUPP, . . .

Comparing the timing of the heart sounds to the events in the cardiac cycle reveals that the heart sounds occur at the beginning of systole (phase 2), when the AV valves close, and at the beginning of diastole (phase 4), when the semilunar valves close (**Figure 23**). But contrary to what seems obvious, heart sounds are not caused by the valve cusps slapping together as they snap shut. Instead, the sounds are created by the turbulent rushing of blood through the valves as they are narrowing and about to close.

Quick Check 5

1. Which event is associated with the first heart sound? With the second heart sound?
2. Describe the relationship between the different waveforms of an ECG and the mechanical events of the heart. Which comes first—the electrical events or the mechanical events?

6 Cardiac Output and Its Control

The ability of the cardiovascular system to deliver blood to organs ultimately depends on the rate at which the heart's ventricles are able to pump blood. When a person is at rest, the left and right ventricles each pump a little more than 5 liters through the vasculature every minute. Because the total volume of blood in the body is around 5 liters, it takes only a minute for a ventricle to pump the equivalent of the entire blood volume. This translates into more than 2.6 million liters per year—and remember that this volume pertains to only *one* ventricle! The rate at which a ventricle pumps blood is called the **cardiac output (CO),** and it is usually expressed in liters per minute.

With each heartbeat, the left and right ventricles contract together. Thus the number of contractions per minute (called the *heart rate, HR*) is the same for both ventricles. The cardiac output is determined by the heart rate and the volume of blood that is pumped from each ventricle with every beat, which we know as the *stroke volume:*

$$\text{Cardiac output} = \text{heart rate} \times \text{stroke volume}$$

$$\text{CO} \quad = \quad \text{HR} \quad \times \quad \text{SV}$$

For an adult at rest, the normal resting heart rate is approximately 72 beats/min, and the stroke volume is 70 mL (0.07 liter); thus a normal resting cardiac output is

$$\text{CO} = 72 \text{ beats min} \times 0.07 \text{ L/beat} = 5.0 \text{ L/min}$$

Apply Your Knowledge

Given the following information, calculate the cardiac output: R-R interval = 0.6 second, end-diastolic volume = 160 mL, and end-systolic volume = 30 mL.

The cardiac output of the left ventricle equals the rate of blood flow through the systemic circuit; the cardiac output of the right ventricle equals the rate of blood flow through the pulmonary circuit. Over the long run, the left and right sides of the heart must have the same cardiac output, or else blood volume would shift from the pulmonary circuit to the systemic circuit, or vice versa. Because the heart rate *and* the cardiac output are the same for the right and left sides of the heart, both ventricles must also have the same average stroke volume.

Because we know that cardiac output is determined entirely by two variables, heart rate and stroke volume, our goal in this section is to examine the various factors that influence these two variables, so that we can develop an understanding of how these factors interact to determine cardiac output.

Although contractions of the heart are not *triggered* by input from the central nervous system, the nervous system does regulate various aspects of cardiac function, including the *rate* and *force* of heart muscle contraction. For this reason, neural input to the heart exerts a significant influence on cardiac output. In addition, cardiac output is influenced by hormones that circulate in the bloodstream. Regulation of the heart (or any other organ or tissue) by neural input, circulating hormones, or any other factor originating from outside the organ is referred to as **extrinsic control.** When the function of an organ or tissue is regulated by factors originating from within the organ or tissue itself, the function is said to be under **intrinsic control** (also known as *autoregulation*). Like most organs, the heart is regulated by extrinsic *and* intrinsic control, and we will see examples of both mechanisms in the following sections.

Autonomic Input to the Heart

Neural control of the heart is carried out by the autonomic nervous system. Fibers of the autonomic nervous system project to nearly every region of the heart, including the conduction system and the

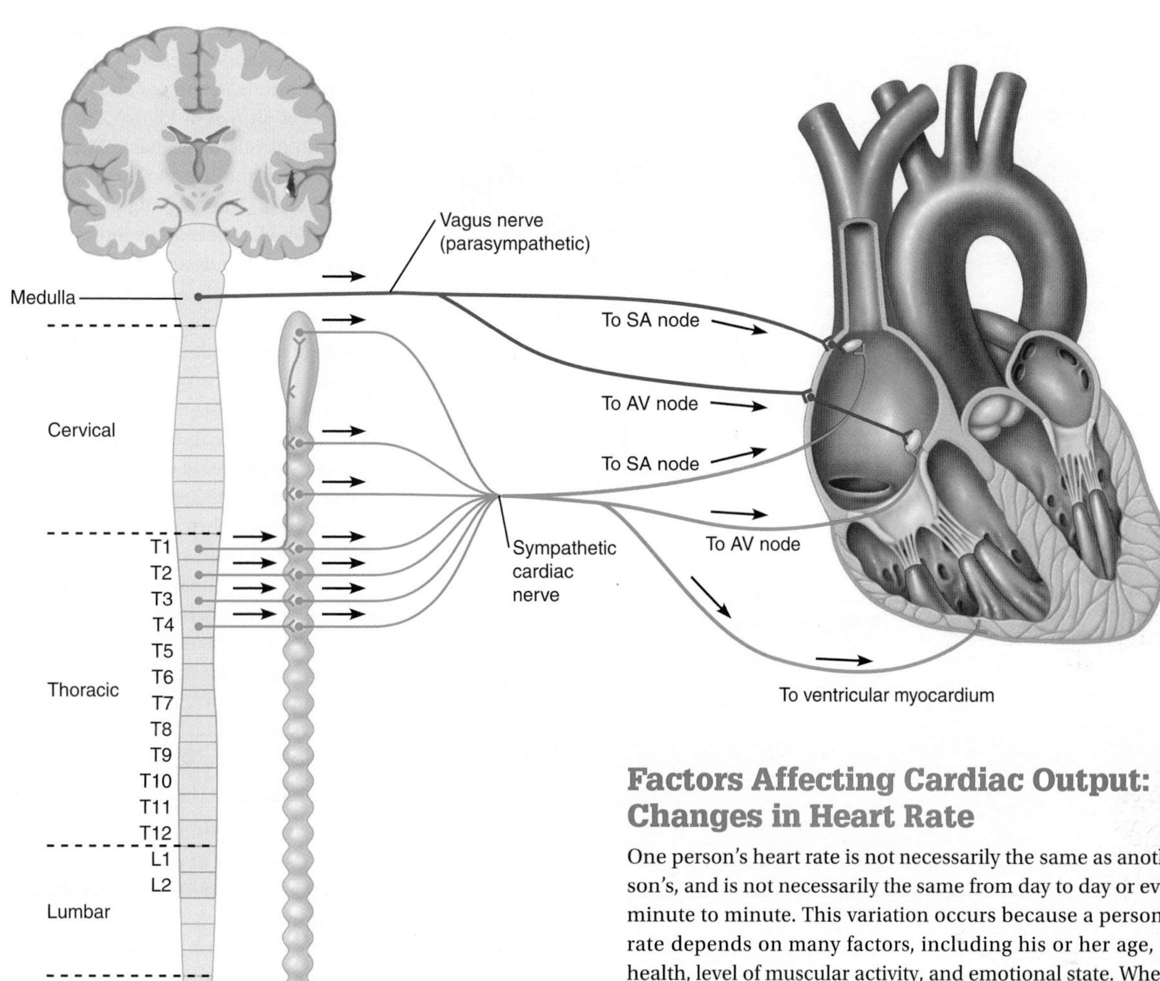

Figure 24 Major autonomic inputs to the heart.

Q Which neurotransmitter is secreted from axon terminals in the sympathetic chain?

bulk of the myocardium, and regulate both heart rate and stroke volume. These fibers include some belonging to the sympathetic nervous system and others belonging to the parasympathetic nervous system. At the various sites of action in the heart, sympathetic and parasympathetic neurons exert opposite effects (as is generally true throughout the body). However, the distribution of parasympathetic fibers is relatively sparse in the ventricles. As a consequence, the ventricular myocardium is regulated primarily by the sympathetic nervous system. Major autonomic inputs to the heart are shown in **Figure 24**.

Factors Affecting Cardiac Output: Changes in Heart Rate

One person's heart rate is not necessarily the same as another person's, and is not necessarily the same from day to day or even from minute to minute. This variation occurs because a person's heart rate depends on many factors, including his or her age, general health, level of muscular activity, and emotional state. When a person exercises or when he or she is excited, anxious, or frightened, the heart rate can increase from the normal resting value of around 70 beats per minute to more than 100 beats per minute, sometimes going as high as 180 beats per minute. In well-trained athletes, the resting heart rate is lower than average, typically around 50 beats per minute. In children, the resting heart rate is higher than that in adults. In this section, we focus on the various factors involved in the minute-by-minute regulation of heart rate, which is entirely under extrinsic control.

Neural Control of Heart Rate

Pacemaker cells of the SA node receive direct input from the autonomic nervous system. This relationship is important because input from these neurons alters the frequency of action potentials generated by these cells, which is normally the sole determinant of heart rate.

Increased activity in sympathetic neurons to the SA node increases the frequency of action potentials in the pacemaker cells by the mechanisms shown in **Figure 25a**. Sympathetic neurons release norepinephrine, which binds to β_1 adrenergic receptors on the SA nodal cells and activates the cAMP second messenger system. In turn, cAMP augments the opening of funny channels and T-type calcium channels. The net result is an increase in the slope

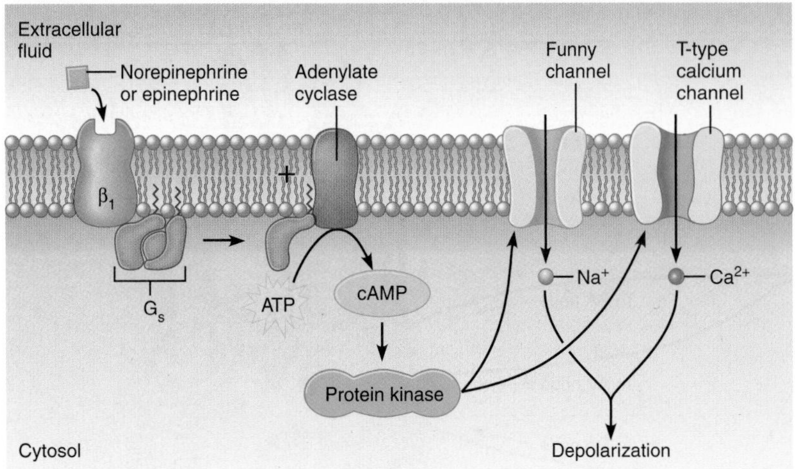

(a) Sympathetic

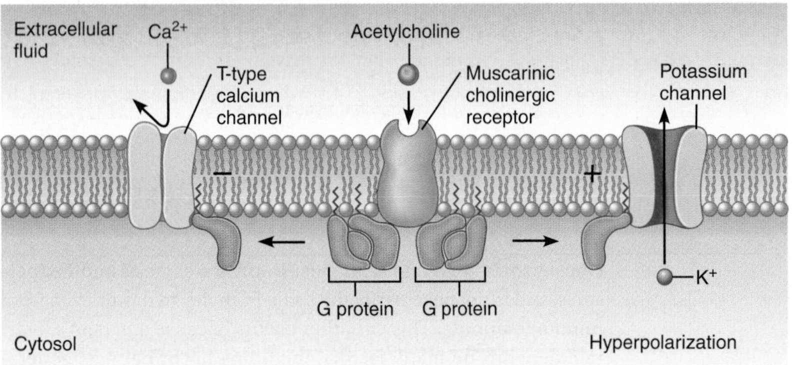

(b) Parasympathetic

Figure 25 Effects of autonomic nervous system on SA nodal cells. (a) The sympathetic nervous system enhances the rate of spontaneous depolarization through a cAMP-dependent mechanism that enhances movement of sodium and calcium into the pacemaker cell. **(b)** The parasympathetic nervous system decreases the rate of spontaneous depolarization through muscarinic receptors by decreasing the movement of calcium and sodium (not shown) into the cell and increasing the movement of potassium out of the cell.

of the spontaneous depolarization and a decrease in the level of repolarization such that threshold for an action potential is reached more quickly (**Figure 26**). The frequency of action potentials is thereby increased, causing an increase in heart rate, which tends to increase cardiac output.

Sympathetic neurons also project to the AV node and other parts of the conduction system, where they influence the speed with which action potentials are conducted. As sympathetic activity increases, action potentials move faster, which decreases the delay of impulse conduction between the atria and ventricles and shortens the time it takes for action potentials to travel through the ventricles. As a result, ventricular contraction starts sooner after atrial contraction and proceeds more quickly, which decreases the duration of systole. A decrease in systole is critical when heart rate increases, because the ventricles can become filled with blood only during diastole. Thus, even though the duration of diastole decreases more so than systole, there is ample time for the ventricles to fill.

Increased activity in parasympathetic neurons to the SA node decreases the frequency of action potentials in the pacemaker cells by the mechanisms shown in Figure 25b. Parasympathetic neurons release acetylcholine, which binds to muscarinic cholinergic receptors on the SA nodal cells; this binding augments the opening of potassium channels and suppresses the opening of funny channels and T-type calcium channels. The net result is a decrease in the slope of the spontaneous depolarization and a hyperpolarization of the membrane potential such that the threshold for an action potential is reached more slowly (Figure 26). The frequency of action potentials is thereby decreased, lowering the heart rate, which tends to decrease cardiac output.

Parasympathetic neurons also influence impulse conduction through the AV node and the rest of the conduction system. As activity in these neurons increases, the speed of impulse conduction decreases, which increases the delay of conduction between the atria and the ventricles as well as lengthens the time required for impulses to travel through the ventricles. As a result, the duration of systole increases.

Hormonal Control of Heart Rate

Although the function of the heart can be affected by a number of hormones, one hormone in particular—*epinephrine*—is significant in the minute-to-minute regulation of cardiac function. The effects of epinephrine, which is secreted by the adrenal medulla in response to increased sympathetic activity, are similar to those exerted by sympathetic neural activity: Epinephrine increases action potential frequency at the SA node and, therefore, increases heart rate. In addition, epinephrine increases the velocity of action potential conduction through cardiac muscle fibers.

Because increased sympathetic nervous activity is usually coupled with enhanced epinephrine secretion, the hormone's actions generally reinforce the effects of sympathetic neural input.

Other hormones that directly affect cardiac function include the *thyroid hormones,* which are secreted by the thyroid gland, and *insulin* and *glucagon,* which are secreted by the pancreas. These hormones primarily increase the force of myocardial contraction, but glucagon also promotes increased heart rate. The importance of these hormones, if any, in the short-term regulation of cardiac function is unclear.

Integration of Heart Rate Control

We have seen that the heart rate is determined entirely by the frequency of action potential firing by the SA node, which is in turn regulated primarily by three factors: (1) activity in *sympathetic neurons* projecting to the SA node, which tends to raise the heart rate; (2) activity in *parasympathetic neurons* projecting to the SA node,

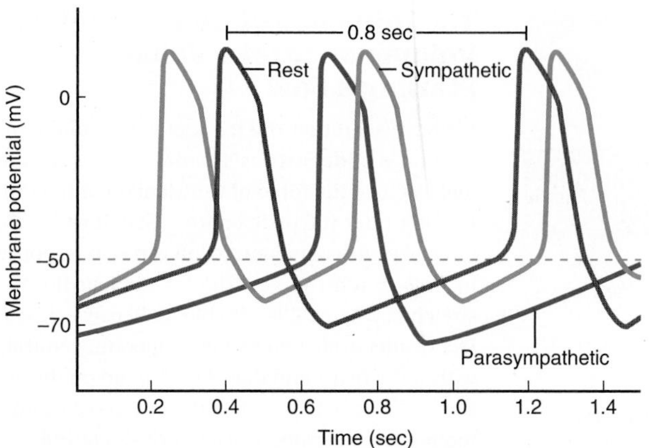

Figure 26 Effects of autonomic nervous system on SA nodal cell potentials. Activation of the sympathetic nervous system increases the rate of spontaneous depolarization, whereas activation of the parasympathetic nervous system decreases the rate of spontaneous depolarization.

which tends to lower the heart rate; and (3) levels of circulating *epinephrine,* which acts to raise the heart rate.

Both divisions of the autonomic nervous system are active at all times. As a result, the heart simultaneously receives signals from the sympathetic and parasympathetic nervous systems that act in opposite directions. Whether heart rate goes up or down depends on the relative rates of activity in the two branches. Activity varies in a "push-pull" manner, in which increases in sympathetic activity are usually accompanied by decreases in parasympathetic activity, and vice versa. Because the two divisions exert opposing effects, this push-pull arrangement ensures that changes in sympathetic and parasympathetic activity reinforce one another. For example, increases in sympathetic activity *and* decreases in parasympathetic activity both act to increase heart rate.

Laboratory experiments have shown that the SA node fires action potentials at a "natural" frequency of about 100 action potentials per minute in the absence of neural or hormonal influences. The fact that the heart rate in a person at rest is significantly lower (approximately 70 beats per minute) indicates that under normal resting conditions, the influence of parasympathetic input to the SA node predominates over the combined influence of sympathetic neurons and epinephrine, giving a net *suppressive* effect on the firing frequency. Any increase in firing frequency is usually triggered by an increase in sympathetic input (and a rise in epinephrine levels) in conjunction with a decrease in parasympathetic input. Any decrease in firing frequency is usually brought about by neural and hormonal changes in the opposite direction. As heart rate increases or decreases, cardiac output tends to increase or decrease, respectively.

Quick Check 6

1. Write the equation showing the relationship between heart rate, stroke volume, and cardiac output.
2. Which parts of the heart receive input from autonomic neurons? What effect does autonomic input have on each of these parts? (Include both sympathetic and parasympathetic influences.)
3. Which hormone reinforces the effects of the sympathetic nervous system on the heart?

Factors Affecting Cardiac Output: Changes in Stroke Volume

The second important determinant of cardiac output is stroke volume. Like heart rate, stroke volume can vary from moment to moment and depends on several factors. In the following sections we examine the primary factors that affect stroke volume: (1) *ventricular contractility,* a measure of the ventricles' capacity for generating force; (2) end-diastolic volume; and (3) *afterload,* the pressure that the ventricles have to work against as they pump blood out of the heart. As we will see, these variables are themselves influenced by a number of other factors, including neural input to the heart, hormones, and various physical variables affecting blood flow into or out of the heart.

The Influence of Ventricular Contractility on Stroke Volume

The first two factors just mentioned—ventricular contractility and end-diastolic volume—both affect stroke volume because they affect the force of ventricular contraction. What, then, is the difference between *contractility* and *contractile force*? The answer is that a change in *ventricular contractility* means a change in the force of ventricular contraction at any given end-diastolic volume. (The term **contractility** can be used to describe any muscle's capacity for generating force.) Generally speaking, any factor that causes the ventricles to contract with more force will tend to make stroke volume larger, which will in turn increase cardiac output. This is true regardless of whether the change in contractile force is due to a change in contractility or a change in end-diastolic volume.

Sympathetic Nervous Control of Ventricular Contractility Ventricular contractility, like heart rate, is regulated by the autonomic nervous system, but autonomic control of stroke volume is exerted almost entirely by the sympathetic nervous system. There is little or no parasympathetic influence on ventricular contractility because of the sparse distribution of parasympathetic fibers in the ventricular myocardium.

As mentioned previously, sympathetic neurons project not only to the heart's conduction system but also to the muscle cells that make up the bulk of the myocardium. Some of these neurons project to the atria and influence the force of atrial contraction. Increased sympathetic activity causes the atria to contract with more force, which raises atrial pressure and increases the volume of blood the atria pump into the ventricles. More importantly, sympathetic neurons project to the ventricular myocardium (see Figure 24), where they exert a direct influence on myocardial contractility. As sympathetic activity increases, ventricular contractility increases, which tends to raise cardiac output.

Sympathetic neurons exert their influence over ventricular contractility in the following manner: Action potentials in these neurons trigger the release of norepinephrine, which binds to β_1 adrenergic receptors on the contractile cells. Binding of neurotransmitter to these receptors activates the cAMP second messenger system. In turn, cAMP activates protein kinases that have four distinct effects in these cells that increase contractility (**Figure 27**): (1) They augment the open state of calcium channels in the plasma membrane, thereby increasing the flow of

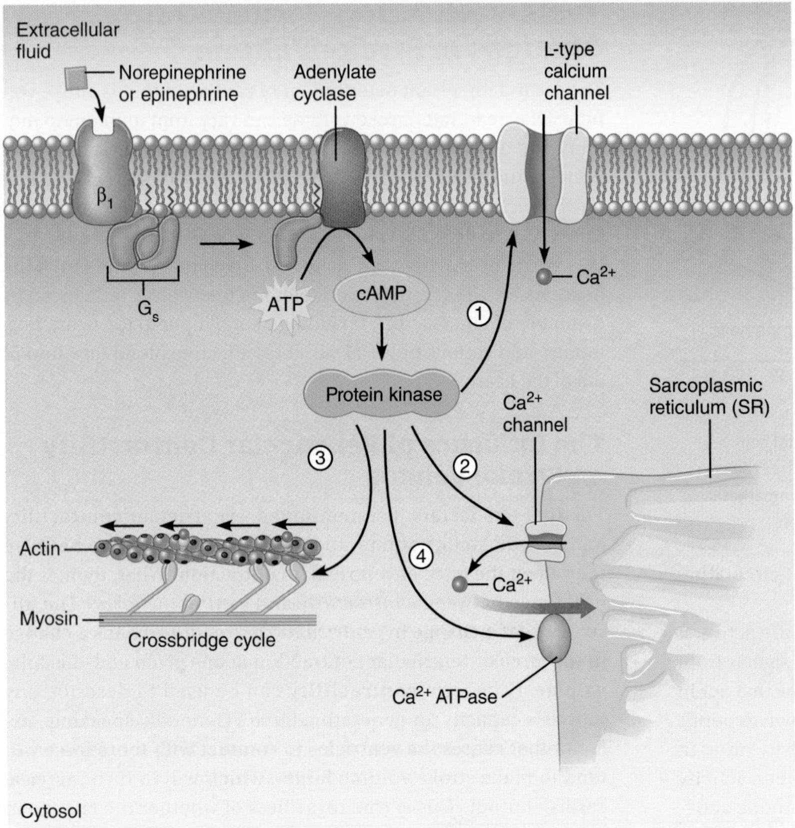

Figure 27 Effects of sympathetic activity on ventricular contractility. The sympathetic nervous system increases the strength and rate of contraction of ventricular contractile cells by (1) increasing calcium movement from the extracellular fluid to the cytosol, (2) increasing calcium movement from the sarcoplasmic reticulum to the cytosol, (3) increasing the rate of myosin ATPase activity, and (4) increasing the rate of calcium reuptake into the sarcoplasmic reticulum.

Q What is the rate-limiting step for crossbridge cycling?

calcium into the cell during an action potential; (2) they enhance the release of calcium from the sarcoplasmic reticulum; (3) they increase the rate of the myosin ATPase, thereby increasing the speed of crossbridge cycling; and (4) they enhance the rate of Ca^{2+}-ATPase activity on the sarcoplasmic reticulum, which increases calcium reuptake, thereby increasing the rate of relaxation of the contractile cell. The net result is that under sympathetic influence, contractile cells contract more rapidly and more strongly (**Figure 28**). To learn more about the activities of the sympathetic nervous system on cardiac function, see **Understanding Exercise: Sympathetic Activity**.

Hormonal Control of Ventricular Contractility Ventricular contractility is increased by a number of hormones, including insulin, glucagon, and thyroid hormones. Most importantly, it is regulated by epinephrine. Like norepinephrine, epinephrine binds to β receptors on cardiac muscle cells and increases intracellular cAMP levels. As a consequence, epinephrine increases myocardial contractility, thereby promoting increases in stroke volume and cardiac output.

The Influence of End-Diastolic Volume on Stroke Volume: Starling's Law

We have seen that the force of ventricular contraction is under extrinsic control. As this section makes clear, the force of ventricular contraction is also under intrinsic control. Specifically, the force of ventricular contraction varies in response to how much the ventricular myocardium is stretched when it fills with blood. Because this effect results from a mechanism operating entirely within the heart and does not depend on the actions of extrinsic factors such as nerves or hormones, it is an example of an intrinsic control.

Intrinsic control of cardiac function is exemplified by **Starling's law of the heart,** which can be stated as follows: *When the rate at which blood flows into the heart from the veins (that is, venous return) changes, the stretch on the ventricular myocardium changes, causing the ventricle to contract with greater or lesser force so that the stroke volume (output) matches the venous return (input).* The outcome of Starling's law is the following: if an increase in end-diastolic volume occurs, the force of ventricular contraction rises, producing an increase in stroke volume and cardiac output. Conversely, if the end-diastolic volume decreases, the force of ventricular contraction declines, producing a decrease in stroke volume and cardiac output.

The physiological basis for Starling's law is related to the fact that increases in end-diastolic volume cause muscle fibers in the ventricular myocardium to lengthen. Such stretching of the muscle fibers causes an increase in the force of contraction by two mechanisms. First, unlike skeletal muscle, cardiac muscle has an optimal length that is much greater than its resting length and that is never reached in a healthy heart. Therefore, increasing the length of the muscle by increasing the end-diastolic volume stretches the muscle fibers

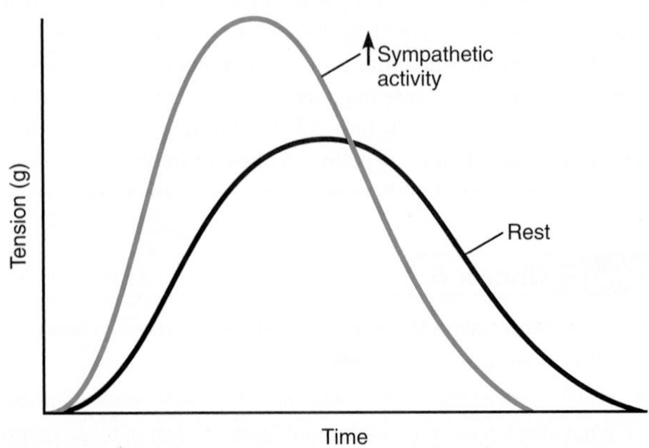

Figure 28 Changes in ventricular contractility induced by sympathetic activity.

UNDERSTANDING EXERCISE

Sympathetic Activity

During exercise, sympathetic nerve activity to the heart increases. The result is an increase in both heart rate and stroke volume. These major changes are coupled to some less well-known changes that are just as important to keep the heart functioning efficiently. As heart rate increases, the time span of the cardiac cycle shortens. Recall that the cardiac cycle includes systole and diastole. Although systole is essential to pump the blood out of the heart, diastole is essential to allow time for blood to return to the heart—with no blood in the heart, it would not matter how hard the muscle contracted. Thus the body is left with a dilemma.

Ideally, sympathetic activity (and a decrease in parasympathetic activity) would increase cardiac output by causing the heart to beat faster, yet still maintain as much filling time (diastole) as possible. The autonomic nervous system does so through two mechanisms: (1) Sympathetic activity increases not only heart rate but also the conduction time through the heart, so that less of a delay occurs between atrial and ventricular contraction, and (2) sympathetic contractility causes the ventricular muscle to contract not only more strongly but also faster—that is, both contraction time and relaxation time are decreased so that the time of systole declines.

closer to their optimal length for contraction, so that they contract with greater force. Second, stretching of the muscle fibers induces an increase in the affinity of troponin for calcium. As a consequence, binding between troponin and calcium is increased, which increases the number of crossbridges that are activated with each contraction.

The Starling Curve Figure 29 shows a graph referred to as a *Starling curve* or *cardiac function curve*. The Starling curve is basically a length-tension curve such as the one shown in Figure 17 for skeletal muscle. The end-diastolic volume is a measure of the length of the muscle fibers: As end-diastolic volume increases, the muscle fibers are stretched to a longer length. The stroke volume is a measure of cardiac muscle tension: As the force of contraction increases, stroke volume increases. The main difference between the length-tension curve for cardiac muscle and the corresponding curve for skeletal muscle is that cardiac muscle is always operating at lengths less than the optimum (thus the continuing positive slope to the curve).

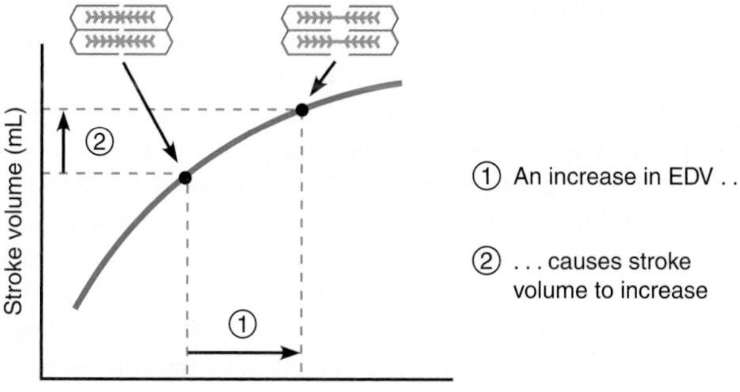

Figure 29 A Starling curve showing how stroke volume changes in response to changes in end-diastolic volume. The Starling curve shows the effect of muscle length (end-diastolic volume) on the tension produced (stroke volume).

When interpreting this graph and others like it, note that stroke volume depends on other factors besides end-diastolic volume. Earlier in the chapter we saw that stroke volume is influenced by sympathetic input to the ventricular myocardium. The curve in Figure 29 is drawn assuming that the degree of sympathetic input, along with all other factors that might affect stroke volume, is held constant. Further, note that such a curve pertains only to what happens in normal, healthy hearts. Normally, an increase in venous return causes an increase in end-diastolic volume, which triggers an increase in stroke volume according to Starling's law. In a person whose heart is chronically expanded, however, prolonged stretching of the heart muscle can weaken the connective tissue, which causes EDV to gradually increase. Eventually, the volume becomes so large that the slope of the cardiac function curve actually becomes *negative*—that is, stroke volume *decreases* as end-diastolic volume becomes larger. Under these conditions, the heart is capable of generating only weak contractions and is unable to adjust its volume to normal because it cannot expel the excess blood that it has accumulated.

Because changes in either sympathetic activity or end-diastolic volume affect the force of ventricular contraction, albeit by different mechanisms, it is possible to alter stroke volume either by changing sympathetic activity without changing end-diastolic volume or by changing end-diastolic volume without changing sympathetic activity. Thus cardiac function is described not by a single Starling curve but instead by a *family* of curves, each of which pertains to a given level of sympathetic activity (**Figure 30**). An increase in sympathetic activity shifts the Starling curve upward. As a result, stroke volume *at any given end-diastolic volume* increases, reflecting the fact that ventricular contractility has increased. A decrease in contractility means that the Starling curve shifts downward, so that stroke volume at any given end-diastolic volume decreases. In actuality, end-diastolic volume and contractility can change simultaneously.

Significance of Starling's Law On the face of it, Starling's law may seem a mere curiosity of no special significance. After all, if more blood flows into the heart, one would expect that more blood

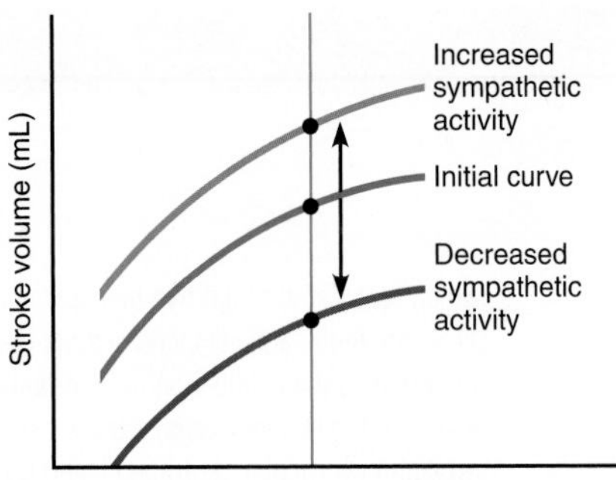

Figure 30 **A family of Starling curves, which shows the influence of sympathetic input on ventricular contractility.**

Q Does a shift from one Starling curve to another reflect intrinsic or extrinsic control of cardiac function?

should come out. But consider what would happen if venous return were suddenly to increase but stroke volume did not. (For simplicity, assume that heart rate is constant.) If cardiac output and venous return were equal to begin with, an increase in venous return with no change in stroke volume would cause the volume of blood remaining in the heart to increase because the amount flowing in would be greater than the amount flowing out. With each beat, the heart would accumulate more blood, which would eventually cause it to expand far beyond its normal size. From this viewpoint, Starling's law takes on a new significance: By adjusting stroke volume so that cardiac output matches venous return, the heart regulates its *size.*

Certain pathological conditions can lead to chronic enlargement of the heart, which can be detrimental to cardiac function for the following reasons: As the wall of a ventricle stretches, the muscle must develop more tension and, therefore, work harder just to maintain the same pressure on the blood contained within it (see **Toolbox: Laplace's Law**). If the ventricles get too big, they become unable to generate enough pressure to maintain adequate cardiac output; such an inability is called *heart failure.*

Factors Affecting End-Diastolic Volume According to Starling's law, the force of ventricular contraction (and hence stroke volume) rises or falls as the end-diastolic volume rises or falls, respectively. In turn, end-diastolic volume is determined by a number of factors, each of which indirectly influences the stroke volume through its effect on end-diastolic volume. Here we examine the various factors that influence end-diastolic volume.

End-diastolic volume is primarily determined by *end-diastolic pressure,* sometimes referred to as **preload.** Ventricular end-diastolic pressure is called preload because it places tension (or *load*) on the myocardium *before* it begins to contract. When a ventricle fills with blood during diastole, the process is similar to what happens when you blow up a balloon with air: As pressure inside rises, the balloon expands. Therefore, the final volume of a given balloon is determined by the final pressure of the air inside it. Likewise, the end-diastolic volume of a ventricle is determined by the pressure of the blood inside it at the end of diastole. As preload (end-diastolic pressure) increases, end-diastolic volume increases, and stroke volume increases according to Starling's law.

Preload is determined by a number of factors, including (1) filling time, which depends on heart rate, and (2) atrial pressure, which is determined by venous return and the force of atrial contraction. As heart rate decreases, filling time increases because diastole increases in duration. At a heart rate of 60 beats/min, diastole is approximately 0.6 second long; when the heart rate increases to 180 beats/min, the length of diastole decreases to a little more than 0.1 second. Because more time is allowed for the entry of blood into the ventricles when the heart rate is lower, a decrease in heart rate (increase in filling time) tends to increase both preload and end-diastolic volume. Preload and end-diastolic volume also tend to increase when atrial pressure rises, because atrial pressure is virtually identical to ventricular pressure when the AV valves are open (that is, during diastole). Atrial pressure, in turn, rises in response to increases in either venous return or the force of atrial contraction.

The most important factor influencing venous return is **central venous pressure,** the pressure of blood contained in the large veins that lead into the heart. As we will see, central venous pressure is affected by many variables, including changes in blood volume, muscular activity, and even posture (as when a person stands up or lies down). As central venous pressure rises, venous return increases because the increased pressure forces more blood to flow into the atria. This effect raises atrial pressure, which leads to an increase in preload. Consequently, end-diastolic volume increases, which produces an increase in stroke volume as described by Starling's law.

The Influence of Afterload on Stroke Volume

The previously mentioned factors influence stroke volume by altering the force of ventricular contraction. However, stroke volume depends not only on how much force the ventricular muscle develops, but also on how large a force it has to work against. (Consider a person attempting to push a wagon up a slope: The speed of the wagon depends not only on how much force the person exerts, but also on how much the wagon weighs.) When the heart ejects blood, the ventricular muscle works against arterial pressure in the same way that your muscles of respiration work against the pressure inside a balloon you are blowing up. For this reason, increases in arterial pressure tend to cause stroke volume to decrease. Because arterial pressure places a load on the myocardium *after* contraction starts, it is called **afterload.** For the left ventricle, afterload is determined by the pressure in the aorta *during the ejection period.* Generally speaking, afterload increases as mean arterial pressure rises.

TOOLBOX

Laplace's Law

The detrimental effect of increased size on the heart's pumping action has its basis in fundamental physical principles: In a vessel containing a gas or a liquid under constant pressure, such as a balloon or blood vessel, the outward force exerted by the pressure stretches the wall of the vessel, creating tension. This tension exerts an inward force that balances the outward force of the pressure, so that the system comes to equilibrium and the vessel neither expands nor contracts. For a spherical container, wall tension is proportional to the interior pressure (*P*) and to the radius of the container (*R*):

$$\text{Tension} = \frac{PR}{2}$$

This relationship is known as Laplace's law.

A consequence of Laplace's law is that if two vessels of different size contain a gas or a liquid at the same pressure, wall tension is greater in the larger vessel. Thus, if two different soap bubbles, one large and one small, contain air at the same pressure, the larger one is subject to more tension because its diameter is larger. This phenomenon explains why a larger bubble is more apt to break than a smaller one, all else being equal. Because the heart is not spherical, the relationship among the wall tension, pressure, and size is more complex for this organ than for a sphere, but the same basic idea applies: If a ventricle contains a large volume of blood, the muscle of the wall has to exert greater tension to generate a given pressure.

In a cylindrical vessel, wall tension is also proportional to the internal pressure and the radius:

$$\text{Tension} = PR$$

Thus a spherical vessel will have half the wall tension as a cylindrical vessel of the same radius.

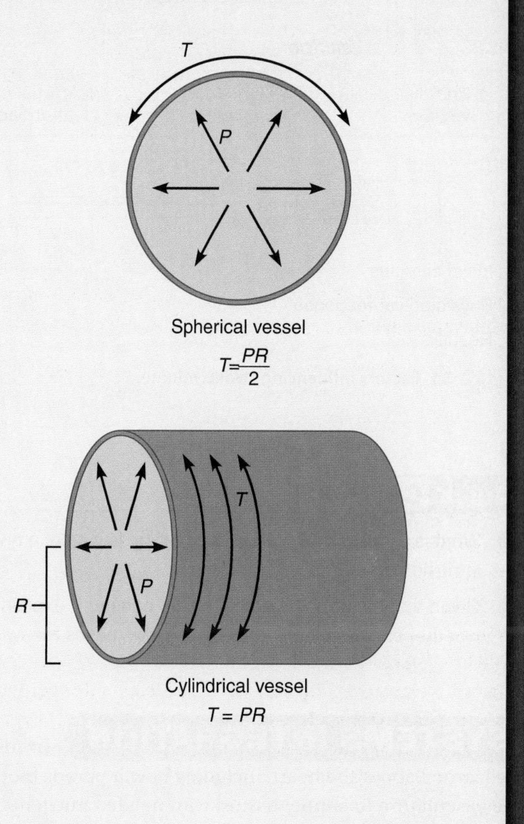

Summary of Factors Affecting Stroke Volume

We have seen that stroke volume is influenced by three major factors: (1) ventricular contractility, which is regulated by sympathetic nervous activity and circulating epinephrine; (2) end-diastolic volume, which is influenced primarily by end-diastolic pressure (preload); and (3) afterload, which depends on arterial pressure. Note that a change in stroke volume (and hence in cardiac output) can result from a change in just one variable (for example, an increase in sympathetic activity) or a simultaneous change in several variables (an increase in sympathetic activity coupled with an increase in end-diastolic volume, for instance). Because these variables interact and are themselves affected by several other factors, it may not always be possible to determine how the stroke volume behaves in every conceivable situation. Nevertheless, you should understand how each variable *tends* to affect the stroke volume. These factors are summarized in **Figure 31**.

Integration of Factors Affecting Cardiac Output

To this point, we have focused on factors that affect heart rate and stroke volume individually. In reality, both heart rate and stroke volume can change simultaneously, and it is the change in the *combination* of these two variables that determines whether cardiac output increases, decreases, or stays the same. In terms of extrinsic controls, the autonomic nervous system affects both heart rate and stroke volume (**Figure 32**). An increase in sympathetic activity, for instance, is usually coupled with a decrease in parasympathetic activity, both of which lead to an increase in heart rate; the rise in sympathetic activity also causes the stroke volume to increase. The net result is an increase in cardiac output. In contrast, a decrease in sympathetic activity and an increase in parasympathetic activity combine to produce decreases in heart rate, stroke volume, and cardiac output. In terms of intrinsic controls, stroke volume is affected by end-diastolic volume.

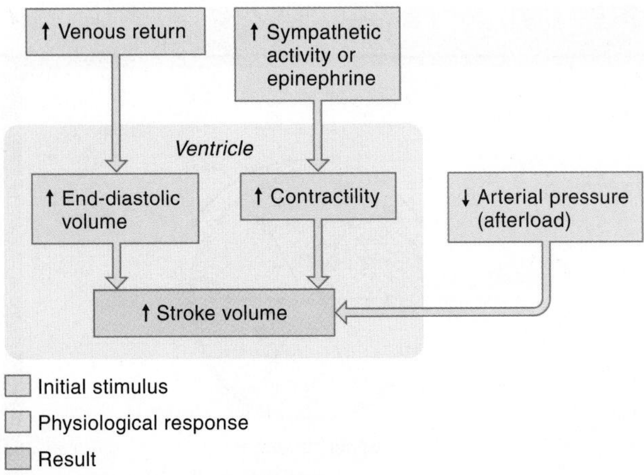

Figure 31 **Factors influencing stroke volume.**

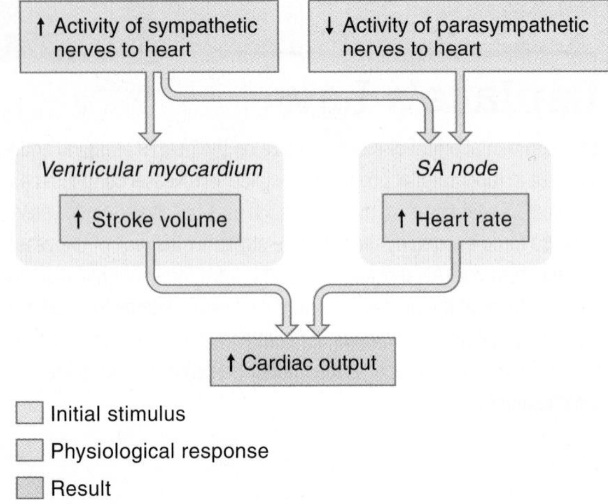

Figure 32 **Autonomic regulation of cardiac output.**

Quick Check 7

1. What is Starling's law of the heart? Why is it an example of intrinsic control?
2. Given that all other variables remain constant, explain whether stroke volume increases or decreases following an *increase* in each of the following variables: sympathetic activity, end-diastolic volume, afterload, preload, filling time.
3. What is preload? What is afterload? How does each affect stroke volume?

SYSTEMS INTEGRATION In this chapter, we learned about the heart, including how it pumps blood through the vasculature to supply tissues with needed nutrients and to remove waste products. The heart cannot carry out this function alone: It depends on the vasculature to provide the pathways for transport and the blood to do the transporting. In addition, the cardiovascular system depends on the respiratory system to provide oxygen to the blood and eliminate carbon dioxide from it; the gastrointestinal system, to provide nutrients to the blood; and the liver and kidneys, to remove waste products from the blood.

CHAPTER REVIEW

SUMMARY

1 An Overview of the Cardiovascular System

- The cardiovascular system includes the heart, blood vessels, and blood.
- The heart is a muscular organ possessing four chambers: the left and right atria, which receive blood as it returns to the heart from the vasculature, and the left and right ventricles, which pump blood away from the heart through the vasculature.
- The blood vessels (arteries, arterioles, capillaries, venules, and veins) function as conduits for blood flow.
- Blood consists of a liquid (plasma) in which the other components (erythrocytes, leukocytes, and platelets) are suspended.
- Blood acts as a medium that carries oxygen and nutrients to the body's cells while it carries away carbon dioxide and other waste products.

iP Cardiovascular, Anatomy Review: Blood Vessels

iP Cardiovascular, Anatomy Review: The Heart

MAP iP Cardiovascular, Cardiac Action Potential

MAP iP Cardiovascular, Cardiac Cycle

2 The Path of Blood Flow Through the Heart and Vasculature

- The vasculature is divided into a pulmonary circuit, which supplies blood to the lungs, and a systemic circuit, which supplies blood to all other organs and tissues of the body.
- In the pulmonary circuit, blood becomes oxygenated and gives up carbon dioxide; in the systemic circuit, it becomes deoxygenated and picks up carbon dioxide.

- Blood is ejected from the right ventricle through the pulmonary semilunar valve into the pulmonary trunk, which divides into left and right pulmonary arteries, carrying deoxygenated blood to the lungs.
- The pulmonary veins carry blood away from the lungs and deliver it to the left atrium, where the blood then moves into the left ventricle.
- The left ventricle pumps blood into the aorta, which delivers it to the systemic organs and tissues.
- Blood returns to the heart by way of the venae cavae, which carry it to the right atrium, where the blood then enters the right ventricle.
- In contrast to the series flow of blood through the right and left sides of the heart, blood flow through the systemic circuit takes the form of parallel flow, with different arteries supplying fresh blood to different organs.
- The branching of blood vessels ensures that each capillary bed receives fresh blood.

iP Cardiovascular, Anatomy Review: The Heart

MAP iP Pathway of Blood Through the Heart

3 Anatomy of the Heart

- The heart is located in the thoracic cavity and is surrounded by a pericardial sac.
- The heart wall is made up of the epicardium, the myocardium, and the endothelium; most of the heart consists of the myocardium.
- Valves in the heart ensure unidirectional flow of blood. Atrioventricular valves allow blood to flow from atrium to ventricle, whereas semilunar valves allow blood to flow from ventricle to artery (left ventricle to aorta and right ventricle to pulmonary trunk).

4 Electrical Activity of the Heart

- The heart muscle fibers that make up the heart's conduction system are specialized to initiate action potentials and conduct them rapidly through the myocardium.
- Contractions of the heart are triggered on a regular basis by action potentials initiated by pacemaker cells concentrated in certain regions of the myocardium.
- Normally the heartbeat is driven by pacemakers in the sinoatrial (SA) node, located in the upper right atrium.
- Following each action potential, pacemaker cells exhibit slow, spontaneous depolarizations (pacemaker potentials) that eventually depolarize the membrane to threshold and trigger the next action potential.
- In most cardiac contractile cells, action potentials are characterized by a broad plateau phase that largely results from an increase in the cell membrane's calcium permeability; the flow of calcium into the cells is important in triggering heart muscle contractions.
- The heart's electrical activity can be recorded using electrodes placed on the skin surface, yielding an electrocardiogram (ECG), which consists of three phases: a P wave, corresponding to atrial depolarization; a QRS complex, corresponding to ventricular depolarization; and a T wave, corresponding to ventricular repolarization.

iP Cardiovascular, Cardiac Action Potential

MAP iP Action Potentials in Autorhythmic Cells

MAP iP Intrinsic Conduction System of the Heart

MAP iP Generation of the Action Potential

MAP iP Propagation and Velocity of the Action Potential

5 The Cardiac Cycle

- The cardiac cycle is divided into two distinct periods: diastole (ventricular relaxation), during which ventricular filling occurs, and systole (ventricular contraction), during which the exit of blood from the ventricles (ejection) occurs.
- Aortic pressure varies throughout the cardiac cycle; it rises to a maximum (systolic pressure, SP) during systole and falls to a minimum (diastolic pressure, DP) during diastole.
- The average pressure throughout the cycle, which represents the driving force for blood flow through the systemic circuit, is the mean arterial pressure (MAP).
- Ventricular volume falls to a minimum at the end of systole (end-systolic volume, ESV) and rises to a maximum at the end of diastole (end-diastolic volume, EDV).
- The difference between these volumes is the stroke volume (SV), the volume pumped by each ventricle in a single heartbeat.
- The pressure-volume curve provides information about how well the heart is functioning.

MAP iP Cardiovascular, Cardiac Cycle

iP Cardiovascular, Factors That Affect Blood Pressure

MAP iP Cardiovascular, Cardiac Output

MAP iP Cardiovascular, Intrinsic Conduction System

6 Cardiac Output and Its Control

- The volume of blood pumped by each ventricle per minute is the cardiac output (CO), which depends on the heart rate (HR) and stroke volume: $CO = HR \times SV$.
- The heart is regulated by sympathetic and parasympathetic neurons and hormones (extrinsic control) as well as by factors operating entirely within the heart (intrinsic control).
- Heart rate, which is determined by the firing frequency of the SA node, is entirely under extrinsic control.
- Stroke volume is under extrinsic and intrinsic control and is affected by three major factors: ventricular contractility, end-diastolic volume, and afterload.
- The influence of end-diastolic volume on stroke volume is the basis of Starling's law of the heart, an example of intrinsic control of cardiac function.
- End-diastolic volume is primarily determined by end-diastolic pressure (preload).

iP Cardiovascular, Cardiac Output

iP Cardiovascular, Blood Pressure Regulation

MAP iP Regulation of Cardiac Output

MAP MasteringA&P®

Check your understanding at MasteringA&P and apply your new knowledge to real-life case studies to ensure that you've mastered the content!

EXERCISES

Multiple-Choice Questions

1. Minimum aortic pressure during the cardiac cycle is attained
   a) Immediately after closure of the aortic semilunar valve.
   b) Immediately before opening of the aortic semilunar valve.
   c) Immediately before opening of the atrioventricular valves.
   d) In mid-diastole.
   e) At the end of systole.
2. The first heart sound occurs when the atrioventricular valves close; thus it marks
   a) The end of the ejection period.
   b) The beginning of the ejection period.
   c) The beginning of systole.
   d) The beginning of isovolumetric contraction.
   e) Both c and d are true.
3. If you know end-diastolic volume, the only other thing you need to know to determine stroke volume is
   a) Afterload.
   b) Ventricular contractility.
   c) End-systolic volume.
   d) Heart rate.
   e) Cardiac output.
4. As a result of Starling's law, stroke volume should increase following an increase in
   a) Mean arterial pressure.
   b) Heart rate.
   c) Sympathetic activity.
   d) Afterload.
   e) Preload.
5. Sympathetic and parasympathetic input to the SA node influences
   a) Ventricular filling time.
   b) Ventricular contractility.
   c) Afterload.
   d) Atrial contractility.
   e) All of the above.
6. Which of the following contains deoxygenated blood?
   a) The right ventricle
   b) The left ventricle
   c) Pulmonary veins
   d) The aorta
   e) Both a and c are true
7. Which of the following is *not* normally apparent in the ECG?
   a) Atrial depolarization
   b) Atrial repolarization
   c) Ventricular depolarization
   d) Ventricular repolarization
   e) None of the above
8. The second heart sound occurs when the semilunar valves close; thus it marks
   a) The end of the ejection period.
   b) The beginning of the ejection period.
   c) The beginning of systole.
   d) The beginning of isovolumetric contraction.
   e) Both c and d are true.
9. The QRS complex of the ECG is due to
   a) Atrial depolarization.
   b) Atrial repolarization.
   c) Ventricular depolarization.
   d) Ventricular repolarization.
   e) Opening of the AV valves.
10. As a wave of action potentials travels from the atria to the ventricles, it is momentarily delayed by about 0.1 second as a result of slow conduction through
    a) The SA node.
    b) The AV node.
    c) The atrioventricular bundle.
    d) The left and right bundle branches.
    e) Purkinje fibers.
11. Which of the following is most likely to cause a *decrease* in the stroke volume of the left ventricle?
    a) An increase in mean arterial pressure
    b) An increase in end-diastolic pressure
    c) An increase in end-diastolic volume
    d) An increase in the activity of sympathetic nerves to the heart
    e) An increase in central venous pressure
12. Left ventricular pressure and aortic pressure are virtually identical during
    a) Isovolumetric contraction.
    b) Isovolumetric relaxation.
    c) Diastole.
    d) Systole.
    e) The ejection period.

Objective Questions

13. Heart rate is normally determined by the action potential frequency in the (SA/AV) node.
14. According to Starling's law, stroke volume should increase if end-diastolic volume (increases/decreases).
15. Heart rate is determined entirely by the inherent action potential frequency in cells of the SA node, with no external influences. (true/false)
16. Blood flow through the systemic circuit is driven by contractions of the (right/left) ventricle.
17. The valve located at the junction between the left ventricle and the aorta is an example of a(n) (atrioventricular/semilunar) valve.
18. (Isovolumetric contraction/Ejection) occurs immediately after diastole.
19. The maximum aortic pressure during the cardiac cycle is called (diastolic/systolic) pressure.
20. Under normal conditions, pressures in the left and right ventricles are equal during systole. (true/false)
21. Stroke volume and ________ completely determine cardiac output.
22. If end-diastolic volume does not change but end-systolic volume decreases, stroke volume (increases/decreases).
23. If end-diastolic volume does not change but end-systolic volume decreases, ejection fraction (increases/decreases).
24. If sympathetic and parasympathetic inputs are constant and end-diastolic volume increases, contractility of the ventricular myocardium increases. (true/false)
25. The period of relaxation of the heart muscle is known as ________.
26. The (P/T) wave of the ECG corresponds to ventricular repolarization.
27. Action potentials generated by pacemaker cells are called pacemaker potentials. (true/false)

Essay Questions

28. Discuss autonomic regulation of cardiac function. Include in your discussion a description of the effects of autonomic activity on the rate and force of ventricular contraction. Feel free to use Starling curves to clarify your discussion.
29. Describe the process of action potential propagation through the heart. Include a description of the role of pacemaker cells and gap junctions in cardiac electrical activity.
30. Discuss the interplay of the various influences on stroke volume. Be sure to include a discussion of Starling's law and the influence of autonomic neurons. Use graphs or charts in your explanation if you feel that it is appropriate.
31. Clarify the distinction between ventricular *contractility* and *force of contraction.* Use graphs or charts in your explanation if applicable.

Critical Thinking

32. Jane trained for a marathon for 6 months. Her resting heart rate is 50 beats/min and her blood pressure is 105/75 mm Hg. Jane's roommate, Sue, watches a lot of TV and has a resting heart rate of 85 beats/min and blood pressure of 135/85 mm Hg. Both women have the same resting cardiac output of 5.0 liters/min. Determine the resting stroke volume and mean arterial pressure for each woman. Based on your findings, whose heart is beating more efficiently?
33. Cardiac arrhythmias are often treated with verapamil, an L-type calcium channel blocker. Explain how such calcium blockers would affect the heart. Include effects on conduction and contraction. Which types of arrhythmias would these drugs be used to treat?
34. Damage to a valve can often be detected by auscultation with a stethoscope because valve damage causes murmurs. Mitral valve prolapse results from a defective valve and allows blood to flow backward. Where is the mitral valve? The murmur caused by mitral valve prolapse would be associated with which heart sound? Describe the changes in cardiac function that can result in severe cases of mitral valve prolapse.
35. In the ECG trace below, label all of the P, QRS, and T waves. Which abnormality is readily apparent? Based on your knowledge of the conduction pathway of the heart, determine where the abnormality is in the pathway.

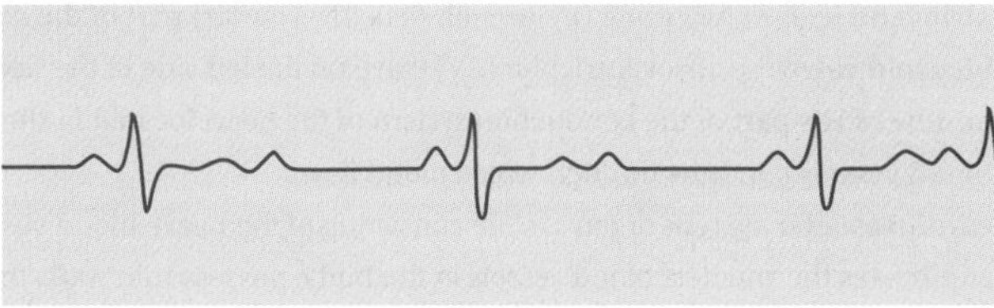

Photo credits for icons in feature boxes: Focus on Diabetes: Tanyae/Dreamstime; Clinical Connections: Iofoto/Dreamstime; Understanding Exercise: Pixattitude/Dreamstime; Discovery: Yuri Arcurs/Cutcaster.

Glossary

afterload the pressure that the ventricles have to work against as they pump blood

atria (ay-tree-ah) the heart's two upper chambers, which receive blood carried to the heart in veins; singular, *atrium*

arteries large vessels that carry blood away from the heart

arterioles small blood vessels that carry blood to the capillaries; walls contain smooth muscle that contracts and relaxes to regulate blood flow

aorta a major artery whose branches carry blood to all organs and tissues in the systemic circuit

atrioventricular (AV) valve (ay-tree-oh-ven-trik-you-lar) one of the valves that separate the atrium and the ventricle on either side of the heart

autorhythmicity the ability of the heart to generate signals that trigger its contractions on a periodic basis; the heart's ability to generate its own rhythm

atrioventricular (AV) node (ay-tree-oh-ven-trik-you-lar) part of the conduction system of the heart located near the tricuspid valve

bicuspid valve the atrioventricular (AV) valve on the left side of the heart, which possesses two cusps; also called the *mitral valve*

bundle of His part of the conduction system of the heart located in the interventricular septum

blood vessels conduits through which blood flows

cardiovascular system organ system consisting of the heart, blood vessels, and blood

capillaries the smallest blood vessels in the body; possess thin walls that permit material exchange between blood and tissues

conduction fibers specialized muscle cells that rapidly conduct action potentials through the heart

conduction system a set of specialized heart muscle cells that initiate and conduct action potentials

cardiac cycle a series of mechanical and electrical events occurring within the heart during a single beat

compliance a measure of the ability of blood vessels or other hollow structures to stretch as the pressure inside them rises

cardiac output (CO) the volume of blood ejected from each ventricle per minute

contractility capacity of a muscle to generate force

central venous pressure (CVP) the pressure in the large veins in the thoracic cavity that lead to the heart

diastole (dy-ass-toh-lee) the period of ventricular relaxation during a cardiac cycle

diastolic pressure (DP) the minimum aortic pressure attained during the cardiac cycle; occurs during diastole

erythrocyte (eh-rith-roh-site) red blood cell; functions in transport of oxygen and carbon dioxide

epicardium outer membrane surrounding the heart

endothelium (en-doh-thee-lee-um) a layer of epithelial cells that line the interior surface of the walls of all blood vessels and the heart

electrocardiogram (ECG) a recording of the heart's electrical activity obtained through electrodes placed on the body's surface

end-diastolic volume (EDV) the volume of blood contained within each ventricle at the end of diastole

end-systolic volume (ESV) the volume of blood in each ventricle at the end of systole

ejection fraction (EF) the fraction of end-diastolic volume ejected during a heartbeat

extrinsic control regulation of an organ or tissue by neural input, circulating hormones, or any other factor originating from outside the organ

isovolumetric contraction contraction of the ventricles with all heart valves closed, such that the volume of blood contained within the ventricles is constant; occurs early in systole

isovolumetric relaxation relaxation of the ventricles with all heart valves closed, such that the volume of blood contained within the ventricles is constant; occurs early in diastole

intrinsic control regulation of an organ or tissue by factors originating from within the organ or tissue itself; also known as *autoregulation* or *local regulation*

intercalated disk specialized junction between cardiac muscle cells that contains both desmosomes and gap junctions

leukocytes (loo-koh-sites) white blood cells; function in immune response

myocardium the entire cardiac muscle mass

mean arterial pressure (MAP) the average aortic pressure occurring during the cardiac cycle

mitral valve atrioventricular valve on the left side of the heart

platelets cell fragments that play an important role in blood clotting

plasma liquid in the blood made up of water and dissolved solutes, including proteins; represents approximately 20% of the total volume of extracellular fluid

pulmonary circuit the portion of the vasculature that encompasses all the blood vessels within the lungs and those connecting the lungs with the heart

pulmonary arteries arteries that carry blood to the lungs from the heart

pulmonary veins veins that carry blood from the lungs to the heart

pacemaker cells cells that are capable of generating pacemaker potentials; in cardiac or smooth muscle, they are responsible for triggering contractions

Purkinje fibers (purr-kin-gee) an extensive network of conducting fibers that spread through the ventricular myocardium

pacemaker potentials spontaneous depolarization in the resting membrane potential

P wave in an electrocardiogram (ECG), the waveform corresponding to atrial depolarization

preload ventricular end-diastolic pressure

QRS complex in an electrocardiogram (ECG), the waveform representing depolarization of the ventricles

systemic circuit the portion of the vasculature that encompasses all of the body's blood vessels, except those belonging to the pulmonary circuit

semilunar valves valves located between the ventricles and arteries on either side of the heart, which prevent blood from flowing back into the ventricles when the ventricles are relaxed

sinoatrial (SA) node (sy-noh-ay-tree-al) a region in the wall of the upper right atrium where pacemaker cells are concentrated; normally determines the heart rate

systole (sis-toh-lee) the period of ventricular contraction during a cardiac cycle

systolic pressure (SP) the maximum aortic pressure attained during the cardiac cycle; occurs during systole

stroke volume (SV) the volume of blood ejected from each ventricle during a single heartbeat

Starling's law of the heart states that when there is a change in the rate at which blood flows into the heart from the veins, the heart automatically adjusts its output to match the inflow

T wave in an electrocardiogram (ECG), the waveform representing ventricular depolarization

tricuspid valve the atrioventricular (AV) valve on the right side of the heart, which has three cusps

ventricles in the heart, the two lower chambers, which pump blood into the arteries; in the brain, chambers that contain cerebrospinal fluid

vasculature all the blood vessels in the body

venules blood vessels that carry blood from capillaries to veins

veins large blood vessels that carry blood toward the heart

venous return blood flow into the heart

ventricular ejection the exit of blood from the ventricles

Answers

Figure Question Answers

Fig 9 Smooth muscle

Fig 14 Passive = diffusion through calcium channels; primary active = Ca^{2+}-ATPase; secondary active = Na^{+}-Ca^{2+} exchanger

Fig 20 Because diastole lasts longer than systole

Fig 21 The same

Fig 24 Acetylcholine

Fig 27 Hydrolysis of ATP by the myosin ATPase

Fig 30 Extrinsic control

Apply Your Knowledge Answers

A premature ventricular contraction results from a depolarization of the ventricle musculature. This depolarization sends the ventricle into a refractory period, such that when the next wave of depolarization comes from the normal conduction pathway, the voltage-gated sodium channels are inactivated and the ventricle cannot be excited to contract. Thus the next beat is skipped.

By blocking L-type calcium channels in pacemaker cells, verapamil decreases the rate of the spontaneous depolarization, thereby decreasing the heart rate, which in turn decreases blood pressure. By blocking L-type calcium channels in cardiac contractile cells, verapamil decreases the flow of calcium into the cell during the plateau phase. Because this calcium contributes to excitation-contraction coupling, verapamil decreases the force of contraction, which decreases the stroke volume and, therefore, decreases blood pressure.

Stroke volume = end-diastolic volume − end-systolic volume = 150 mL − 50 mL = 100 mL
Ejection fraction = stroke volume/end-diastolic volume = 100 mL/150 mL = 0.67

Cardiac output = heart rate × stroke volume. From the data given, the heart rate can be calculated from the R-R interval:

$$\text{Heart rate} = \frac{60\ \text{sec/min}}{\text{R-R interval}} = \frac{60\ \text{sec/min}}{0.6\ \text{sec/beat}}$$

$$= 100\ \text{beats/min}$$

Stroke volume can be calculated from the end-diastolic and end-systolic volumes:
Stroke volume = end-diastolic volume − end-systolic volume = 160 mL − 30 mL = 130 mL/beat

Multiply heart rate and stroke volume to get cardiac output:

Cardiac output = heart rate × stroke volume = 100 beats/min × 130 mL/beat = 13,000 mL/min

Answers to End-of-Chapter Multiple Choice and Objective Questions

Multiple Choice Questions

1. b 2. e 3. c 4. e 5. a 6. a 7. b 8. a 9. c 10. b 11. a 12. e

Objective Questions

13. SA
14. increases
15. false
16. left
17. semilunar
18. Isovolumetric contraction
19. systolic
20. false
21. heart rate
22. increases
23. increases
24. false
25. diastole
26. T
27. false

The Cardiovascular System: Blood Vessels, Blood Flow, and Blood Pressure

The Cardiovascular System: Blood Vessels, Blood Flow, and Blood Pressure

SPL/Photo Researchers, Inc.

Colorized scanning electron micrograph of an arteriole.

Have you ever stood up too fast and felt dizzy or light-headed? Most people have experienced this sensation, called *orthostatic hypotension,* at one time or another. When a person stands up, the force of gravity pulls blood toward the legs and away from the upper parts of the body. This decreases the amount of blood returning to the heart, which decreases stroke volume and, therefore, cardiac output. With less blood flowing into the vasculature, blood pressure decreases, called *hypotension*. Because pressure is the driving force for blood flow, the amount of blood reaching the brain may be decreased, causing the symptoms of dizziness.

As you will learn in this chapter, arterial blood pressure and blood flow to the organs are regulated so as to deliver adequate supplies of oxygen and nutrients to organs and tissues under most circumstances. Thus, although you may sometimes feel dizzy when you stand up, usually the cardiovascular system rapidly adapts and maintains adequate blood flow to the brain so that your body can change position without detriment.

MasteringA&P®

Go to MasteringA&P for helpful A&P Flix 3-D animations, chapter quizzes, pre-tests, Interactive Physiology tutorials, and more!

CHAPTER OUTLINE

1. **Physical Laws Governing Blood Flow and Blood Pressure**
2. **Overview of the Vasculature**
3. **Arteries**
4. **Arterioles**
5. **Capillaries and Venules**
6. **Veins**
7. **The Lymphatic System**
8. **Mean Arterial Pressure and Its Regulation**
9. **Other Cardiovascular Regulatory Processes**

LEARNING OUTCOMES ***After studying this chapter, you should be able to:***

- Describe the physics of blood flow through blood vessels. Explain the concepts of pressure gradients and resistance.
- Describe the anatomy of the vasculature, and explain the basic functional properties of the different types of blood vessels.
- Explain the role of arterioles in varying resistance. Describe how intrinsic control of vascular resistance regulates blood flow to organs. Explain the role of extrinsic control of the arteriole radius in determining the mean arterial pressure.
- Explain how material is exchanged between blood and interstitium. Describe the forces that cause bulk flow of fluid across capillary walls.
- Explain how mean arterial pressure influences blood flow to individual organs and to the entire systemic circuit, and identify the factors that determine the mean arterial pressure.
- Describe what the arterial baroreceptor reflex is, and explain how it regulates the mean arterial pressure.
- Describe how changes in arterial carbon dioxide levels, body heat, and exercise affect cardiovascular function and mean arterial pressure.

Before You Begin

Make sure you have mastered the following concepts:

- Diffusion across membranes
- Osmosis
- Autonomic nervous system
- Smooth muscle
- Factors affecting cardiac output

Cardiac output —the rate at which the heart pumps blood into the vasculature—is regulated by controlling heart rate and stroke volume. The autonomic nervous system regulates heart rate through neural input to the sinoatrial (SA) node and stroke volume through neural input to the myocardium. Intrinsic regulation of stroke volume occurs as well. We now turn to the vasculature. Blood flow to the organs and blood pressure are regulated by both intrinsic and extrinsic controls. Smooth muscle can be regulated by intrinsic factors, the autonomic nervous system, and certain hormones. In this chapter, we see how the smooth muscle of arterioles is regulated by intrinsic controls to regulate blood flow to the organs and by extrinsic controls to regulate blood pressure.

We start with a discussion of the physical laws that govern the flow of blood through the vasculature. We then discuss the various types of blood vessels that make up the vasculature and their roles in cardiovascular function. In the last part of this chapter, we describe blood pressure, including how it is regulated by autonomic control of the vasculature and the heart.

1 Physical Laws Governing Blood Flow and Blood Pressure

The vasculature is much like an elaborate system of pipes that runs through the body, so the fundamental physical laws that describe the flow of any liquid through a system of pipes also pertain to blood flow in the cardiovascular system. The rule that is pertinent to our discussion here states that the flow rate of a liquid (the volume flowing per unit of time) through a pipe is directly proportional to the difference between the pressures at the two ends of the pipe (the *pressure gradient*) and inversely proportional to the resistance of the pipe:

$$\text{Flow} = \text{pressure gradient/resistance} = \Delta P/R$$

The quantity ΔP, the size of the pressure gradient, represents the driving force that *pushes* the flow of liquid through a pipe, with the fluid moving from high to low pressure; the quantity R, the resistance, is a measure of the various factors that *hinder* the flow of liquid through a pipe.

This rule is so crucial to our understanding of blood flow that it is the starting point for all our discussions pertaining to flow, pressure, and resistance in the cardiovascular system. It is so universally applicable that it applies to liquids flowing in a single pipe or blood vessel, or in a system of pipes or blood vessels, no matter how complicated. It even pertains to the flow of air into and out of the lungs.

In this section we examine the general principles that govern how pressure gradients and resistance affect blood flow in individual vessels and networks of vessels. First we consider pressure gradients.

Pressure Gradients in the Cardiovascular System

When you inflate a balloon, it expands because the pressure the air exerts on the inside of the balloon is greater than the pressure the air exerts on the outside. If you remove your fingers from the nozzle of the balloon, air rushes out for the same reason—air pressure is greater inside the balloon than outside it. Whenever there is a difference in pressure between two locations, the pressure gradient drives the flow from a region of higher pressure to one of lower pressure, or *down the pressure gradient.*

Air flowing out of a balloon, like the flow of blood through the cardiovascular system, is an example of *bulk flow*. Regardless of whether the flowing medium is a gas or a liquid, the driving force for bulk flow is always a pressure gradient, and the direction of flow is always down the gradient from a region of greater pressure to a region of lower pressure. This rule applies to blood flow and to all other examples of bulk flow that occur within the body, such as the flow of air into and out of the lungs.

The Role of Pressure Gradients in Driving Blood Flow

The primary function of the heart is to generate the pressure that drives the flow of blood through the vasculature. Strictly speaking, however, it is not absolute pressure that drives blood flow, but rather a pressure gradient. By pumping blood into the arteries, the heart increases the mean arterial pressure, which creates a difference in pressure between the arteries and veins that drives the flow of blood.

Figure 1 shows a useful model for explaining the relationship between pressure and flow in blood vessels. In the diagram, a tube or "blood vessel" connects two large reservoirs containing liquid. The vertical distance from the vessel to the surface of the liquid, the so-called *hydrostatic column*, determines the pressure at either end of the vessel. (The pressure also depends on the *density* of the liquid—its mass per unit volume—which we assume is constant.) A higher hydrostatic column corresponds to a greater pressure. (You can feel the effect of a hydrostatic column by diving underwater: The deeper you go, the more pressure the water exerts on your body.)

When the liquid level is different on the two sides (Figure 1a), a pressure gradient exists, and the liquid flows at a rate of 20 mL/min through the tube from the high-pressure side (100 mm Hg) to the low-pressure side (60 mm Hg), or down the pressure gradient ($\Delta P = 40$ mm Hg). As a result, the level on one side drops while that on the other side rises. Eventually, the levels become equal (Figure 1b), and flow stops because a pressure gradient no longer exists.

Figure 1c illustrates that the rate of flow through the tube depends only on the *difference* between the pressures at either end, not the absolute pressure. When the liquid level is raised on both sides, the pressure on both sides increases. However, if the difference between the levels remains constant (here, ΔP still equals 40 mm Hg), the flow does not change (flow still equals 20 mL/min).

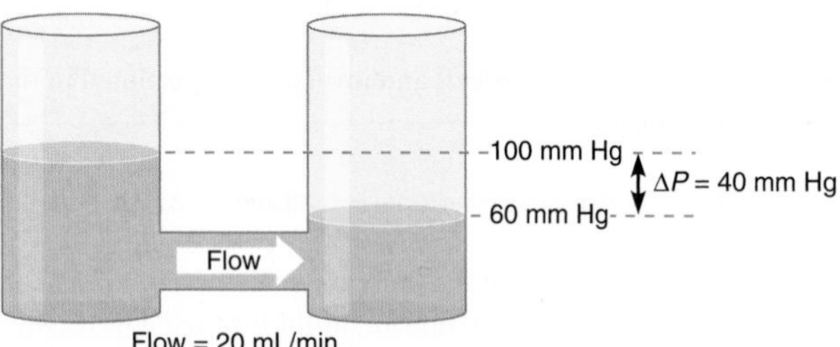

(a)

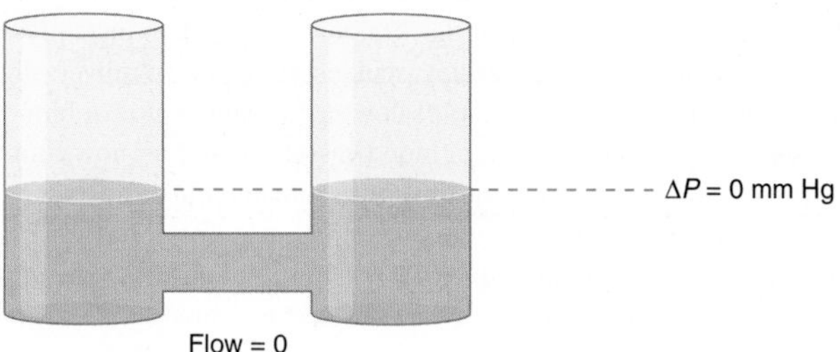

(b)

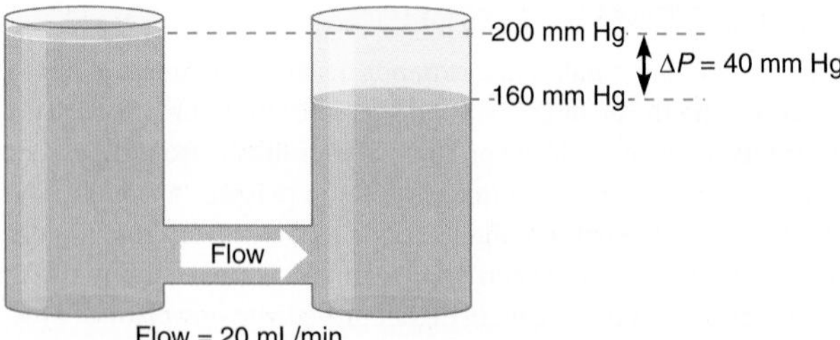

(c)

Figure 1 A model that relates blood flow to the pressure gradient. A single blood vessel is represented by a tube connecting two reservoirs, in which the depth of liquid determines the pressure. **(a)** The difference in the pressures in the two reservoirs produces a pressure gradient (ΔP) of 40 mm Hg, creating a flow of 20 mL/min, as indicated by the arrow. **(b)** When the levels of the liquid are the same in both reservoirs, the pressure gradient is zero, and hence flow is zero. **(c)** When the levels in both reservoirs are raised such that ΔP remains at 40 mm Hg, flow remains at 20 mL/min, indicating that the pressure gradient, not absolute pressure, determines flow.

Pressure Gradients Across the Systemic and Pulmonary Circuits

In the systemic circuit, *mean arterial pressure* (*MAP,* the average pressure in the aorta throughout the cardiac cycle) is approximately 85 mm Hg. At the other end of the circuit, in the large veins in the thoracic cavity that lead to the right atrium, the pressure—known as the **central venous pressure (CVP)**—is approximately 2-8 mm Hg, and the pressure in the vena cava just outside the right atrium is approximately 0 mm Hg. The difference between the MAP and the CVP is the pressure gradient that drives blood flow through the systemic circuit. Because CVP is so small, we simplify our description of flow through the systemic circuit by ignoring it (this omission decreases accuracy in determining the pressure gradient, but highly simplifies our descriptions of pressure gradients). Thus *the pressure gradient* (ΔP) *driving blood flow through the systemic circuit is equated to the mean arterial pressure* (**Figure 2**).

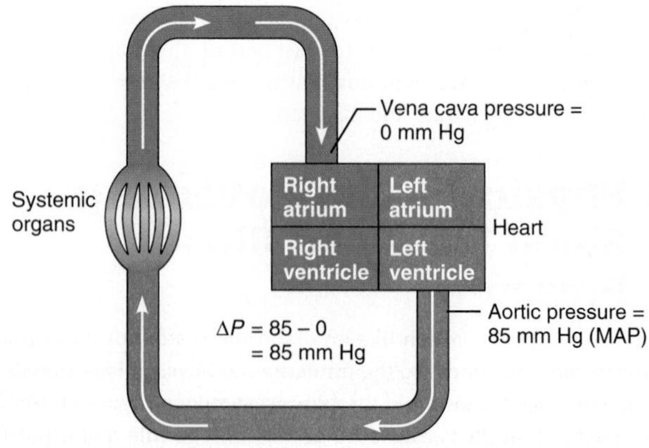

Figure 2 A pressure gradient is the driving force for blood flow. Aortic pressure averages about 85 mm Hg, whereas the pressure in the vena cava is close to zero where it joins the heart. This creates a pressure gradient of 85 mm Hg, which represents the overall driving force that pushes the flow of blood through the systemic circuit.

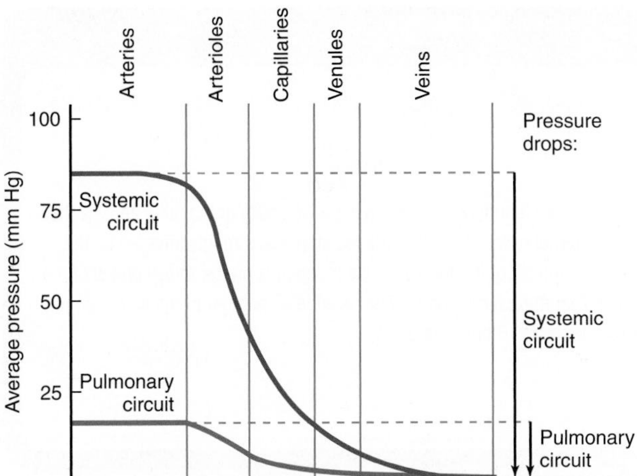

Figure 3 **Pressures and pressure drops in the pulmonary and systemic circuits.**

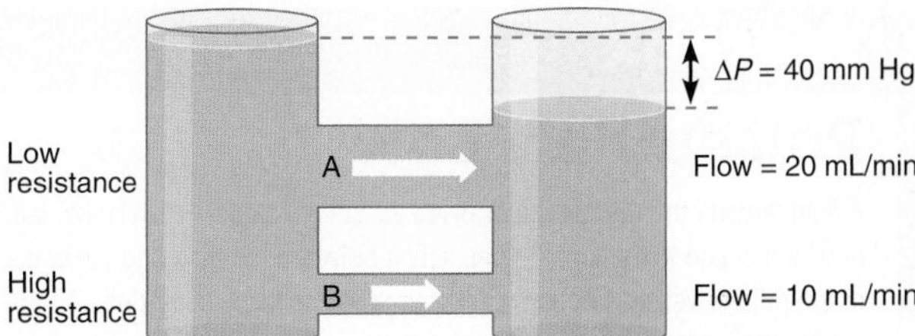

Figure 4 **The effect of resistance on flow.** In this model, two blood vessels are depicted as two tubes (A and B) connecting the reservoirs. Even though the pressure gradient is the same (40 mm Hg) for both vessels, blood flows through vessel B at a lower rate than through vessel A. Because it has a smaller diameter, vessel B has a higher resistance than vessel A.

Q If the pressure difference decreases to 20 mm Hg, what will be the flow rate in tube A? In tube B?

Blood flow through the pulmonary circuit is also driven by a pressure gradient—the difference between the pressure in the pulmonary arteries and the pressure in the pulmonary veins. However, this pressure gradient is smaller than the one that drives flow through the systemic circuit because pulmonary arterial pressure is lower than aortic pressure (**Figure 3**). During the cardiac cycle, pulmonary arterial pressure averages approximately 15 mm Hg, as opposed to approximately 85 mm Hg in the aorta. Pulmonary venous pressure, like CVP, is close to zero. Both the systemic and pulmonary circuits must circulate the same volume of blood. Based on our flow rule, if the pressure gradient to drive blood flow through the pulmonary circuit is relatively low, then the resistance must also be low, as described next.

Resistance in the Cardiovascular System

The blood flow through the pulmonary circuit is identical to that through the systemic circuit (approximately 5 liters/min at rest). But if the pressure gradient in the pulmonary circuit is lower than that in the systemic circuit, how is it possible to have the same blood flow in the two circuits? The answer can be deduced from the flow rule, $F = \Delta P/R$: The pulmonary circuit offers less *resistance* (because of its physical characteristics), so a smaller pressure gradient can achieve the same flow.

Here we examine the factors that determine the resistance of individual blood vessels and of networks of vessels such as the systemic and pulmonary circuits.

Resistance of Individual Blood Vessels

The resistance of any tube (including a blood vessel) is a measure of the degree to which the tube hinders or resists the flow of liquid through it. From the flow rule, it is apparent that for a given pressure gradient, a vessel with higher resistance yields a lower flow (**Figure 4**). Put another way, for a given pressure gradient, blood flow is greater when resistance is lower because it is easier for blood to flow.

Apply Your Knowledge

There are many similarities to resistance to blood flow described here. Compare Ohm's law to the flow rule, being sure to describe similarities as well as differences.

If you have ever drunk liquid through a drinking straw, you have experienced the effects of resistance. You likely noticed that it is easier to drink through a wide straw than through a narrow one, and that it is easier to drink through a short straw than through a long one. You also know that it is more difficult to drink a milkshake through a straw than it is to drink a soda. Resistance, then, depends both on the physical dimensions of the tube and on the properties of the fluid flowing through it—that is, on the tube's *radius* and *length*, and on the fluid's *viscosity* ("thickness" or "syrupiness"), respectively. The effects of these factors on blood flow are described next and mathematically in the **Toolbox: Poiseuille's Law:**

1. *Vessel radius.* Changes in resistance to blood flow in the cardiovascular system almost always result from changes in the radii of blood vessels: As radius decreases, resistance increases. A decrease in blood vessel radius is called **vasoconstriction;** an increase in vessel radius is called **vasodilation.**
2. *Vessel length.* Even though longer vessels have greater resistance than shorter ones (all else being equal), changes in vascular resistance are rarely attributable to changes in vessel length; vessels do not change length except as a person grows.
3. *Blood viscosity.* Vascular resistance increases as viscosity increases, but blood viscosity does not change appreciably under normal conditions. The major determinant of blood viscosity is the concentration of cells and proteins in the blood; as either concentration increases, blood viscosity increases. Neither the concentration of proteins nor the concentration of blood cells changes over the short term, but they may change during adaptations. For example, the concentration of red blood cells increases when a person is exposed to high altitude (because the air contains less oxygen).

TOOLBOX

Poiseuille's Law

A fluid flowing through a tube or blood vessel encounters resistance, some of which is due to frictional forces acting between the fluid and the walls of the tube or vessel. Friction within the fluid itself also contributes to the resistance, which explains why some fluids, such as molasses, flow more slowly than others, such as water. The speed at which a fluid moves varies from one location to another within the moving liquid. As a consequence, layers of fluid moving at different speeds rub against each other, which creates friction and dissipates energy. For a fluid moving smoothly through a cylindrical tube, the resistance *(R)* is given by the following equation, which is called *Poiseuille's law:*

$$R = \frac{8L\eta}{\pi r^4}$$

where L is the length of the tube, η is the viscosity of the fluid, and r is the tube's internal radius. Note that the resistance is strongly affected by the internal diameter of the tube because it depends on the *fourth power* of the radius. Therefore, if one tube is half as wide as another of the same length, its resistance is 16 times as great!

Resistance of Blood Vessel Networks: Total Peripheral Resistance

Although so far we have limited our consideration to the resistance of individual blood vessels, a *network* of blood vessels (such as the systemic or pulmonary circuits, the vasculature within an organ, or even a single capillary bed) also has a resistance. For blood vessel networks, the rules governing flow, pressure, and resistance are fundamentally the same as for individual vessels: For any network, the total flow increases in proportion to the pressure gradient along the network and decreases as the resistance of the network increases.

As we might expect, the resistance of a vascular network depends on the resistances of all the individual blood vessels it contains. Thus any factor that causes the resistance of individual vessels in a network to increase or decrease also tends to cause a corresponding increase or decrease in the resistance of the network as a whole. From this relationship, it follows that *vasoconstriction anywhere within a network of blood vessels tends to increase the resistance of the network, whereas vasodilation anywhere within a network tends to decrease the resistance of the network.* In the systemic circuit, the combined resistances of all the blood vessels within the circuit are known as **total peripheral resistance (TPR).**

Relating Pressure Gradients and Resistance in the Systemic Circulation

We can express the relationship among pressure, resistance, and flow in the systemic circuit by making some substitutions in the flow rule, $F = \Delta P/R$. First, because all the blood that flows from the heart goes through the systemic circuit, the flow is equal to the volume of blood flowing through the circuit each minute, or cardiac output (CO). We have already seen that this flow of blood is driven by the pressure gradient represented by the difference between MAP and CVP, and that this pressure gradient is virtually identical to MAP. In addition, we know that the resistance in the systemic circuit is TPR. Accordingly, substituting these variables into the flow rule yields the following equation:

$$\text{CO} = \text{MAP}/\text{TPR}$$

Now that we understand the physics governing blood flow, we describe the different components of the vasculature and their role in blood flow and exchange of material between the blood and interstitium.

Quick Check 1

1. What is a pressure gradient? Where are pressure gradients present in the vasculature?
2. What are some factors that affect resistance in the vasculature? How does resistance affect blood flow?

2 Overview of the Vasculature

Blood vessels are classified according to whether they carry blood away from or to the heart, and according to size. Arteries and smaller arterioles carry blood from the heart and to capillaries, which are drained by venules and then larger veins, which return the blood to the heart (**Figure 5**). The arterioles, capillaries, and venules can be seen only with the aid of a microscope and, therefore, are called the **microcirculation.** All blood vessels possess a hollow interior called the *lumen,* through which blood flows; the lumen of all blood vessels is lined by a layer of epithelium called the endothelium. Surrounding the lumen is a wall that varies in thickness and composition from one vessel type to another.

The smallest of all blood vessels, capillaries, consist of a layer of endothelial cells and a basement membrane; the walls of all other blood vessels contain various amounts of *smooth muscle* and fibrous and/or elastic connective tissue (**Figure 6**). Within the fibrous connective tissue are extracellular fibers made of a protein called *collagen,* which lends tensile strength to vessel walls, enabling them to withstand the pressure of blood within them without rupturing. Elastic connective tissue contains fibers of a highly stretchable extracellular protein called *elastin,* which enables blood vessels to expand or contract as the pressure of blood within them changes.

We next examine the structure and function of the various types of blood vessels, beginning with arteries.

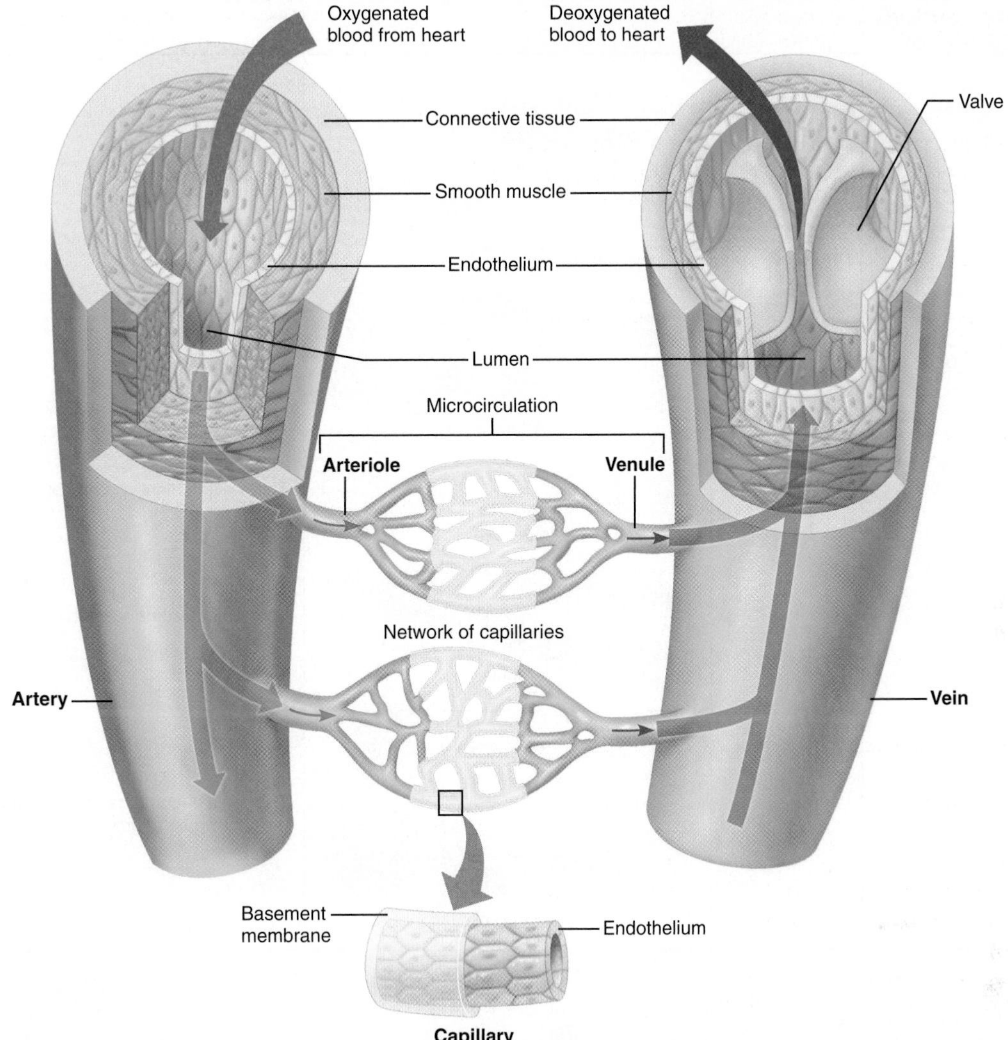

Figure 5 The relationships of blood vessels according to size and the direction of blood flow in the systemic circuit. The arrow at the upper left represents oxygenated blood (red) arriving from the heart, whereas the arrow at the upper right represents deoxygenated blood (blue) returning to the heart. Note the differences in lumen diameter and wall thickness between arteries and veins, and the presence of one-way valves in the vein.

3 Arteries

Arteries conduct blood away from the heart and toward the body's tissues. The largest artery, the aorta, has an internal diameter of about 12.5 mm and a wall that is 2 mm thick. The smaller arteries that branch off the aorta have internal diameters ranging from 2 mm to 6 mm and a wall thickness of about 1 mm, and these branch into yet smaller-diameter arteries. The larger arteries provide little resistance to blood flow, so they serve as a rapid conduit through which blood can travel. The walls of large arteries contain large amounts of elastic and fibrous tissue, enabling arteries to withstand the relatively high blood pressures present in these vessels compared to any other vessels in the vasculature.

As the arteries branch into smaller arteries, the amount of elastic tissue in the walls decreases while the amount of smooth muscle increases. Arteries less than 0.1 mm (100 microns) in diameter lose most of their elastic properties and are sometimes called *muscular arteries*. The smooth muscle enables regulation of the radius of small arteries in a manner similar to that for arterioles (described shortly).

Arteries: A Pressure Reservoir

The thickness of arterial walls, coupled with the relative abundance of elastic tissue, gives arteries both stiffness and the ability to expand and contract as the blood pressure rises and falls with each

Average internal diameter (mm)	Average wall thickness (mm)		Special features
4.0	1.0	Artery	Muscular, highly elastic
0.03	0.006	Arteriole	Muscular, well innervated
0.008	0.0005	Capillary	Thin-walled, highly permeable
0.02	0.001	Venule	Thin-walled, some smooth muscle
5.0	0.5	Vein	Thin-walled (compared to arteries), fairly muscular, highly distensible

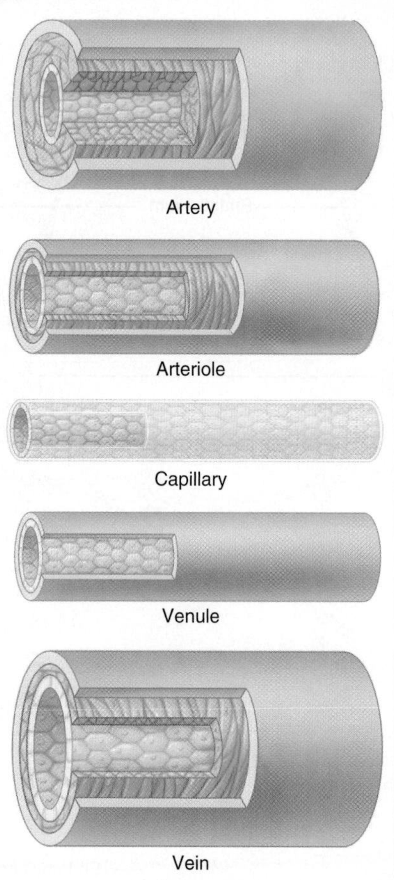

= Endothelium
= Smooth muscle
= Connective tissue
Wall thickness
Internal diameter

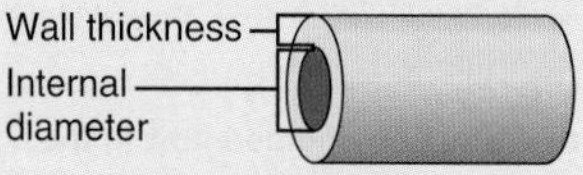

Figure 6 **Structural characteristics of the five blood vessel types.**

heartbeat. This combination of stiffness and flexibility enables arteries to perform one of their major functions—acting as *pressure reservoirs* (storage sites for pressure) to ensure a continual, smooth flow of blood through the vasculature even when the heart is not pumping blood (diastole).

Figure 7 illustrates the concept of a pressure reservoir. As the arterial walls are expanding due to increased volume during systole, the elastin fibers act much like a spring being stretched. This elastic force is stored such that during diastole, when no more blood is entering the arteries, the walls passively recoil inward, propelling blood forward. Therefore, blood moves through the vasculature continuously, propelled during systole by the ejection of blood from the heart and during diastole by the elastic recoil of the arterial walls. Although blood flow is continuous, palpitation of arteries, such as the radial artery in the wrist, reveals a pulse. The pulse is caused by a pressure wave that travels along the arteries in response to blood being pushed into the arteries during systole, causing the arterial walls to expand.

To serve as a pressure reservoir, arteries must have low compliance; it is a measure of the relationship between pressure and volume changes. In vessels with low compliance, such as arteries, a small increase in blood volume causes a large increase in blood pressure (or a large increase in pressure causes only a small degree of expansion of the blood vessel walls). Therefore, when the heart ejects blood into the arteries during systole and causes these vessels to expand, the resulting rise in pressure is greater than it would be if arteries' compliances were higher. The low compliance of arteries is a function of the elasticity of the vessel walls.

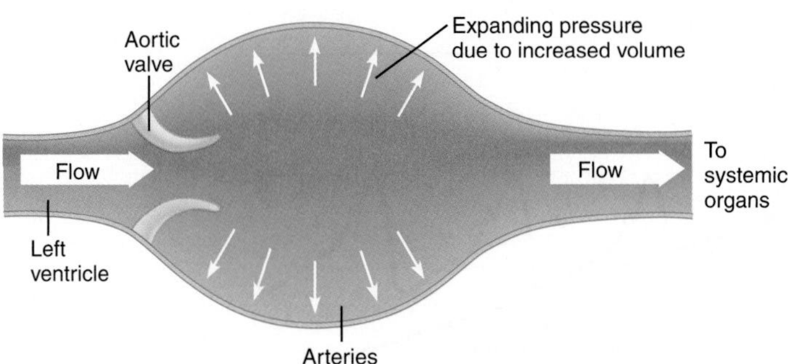

(a) Systole

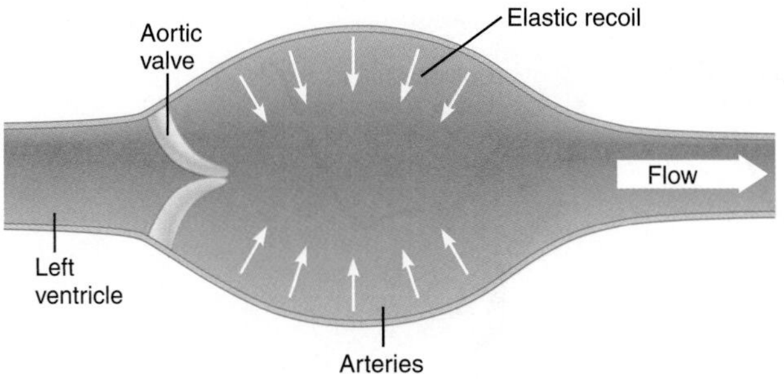

(b) Diastole

Figure 7 The role of arterioles as a pressure reservoir. Flow arrows indicate the movement of blood into and out of the arteries. Pressure arrows indicate the expansion and recoil of the arterial walls. **(a)** During systole, the pressure on the blood in the left ventricle is greater than that in the arteries, causing blood to flow through the aortic valve into the arteries. Blood entering the arteries causes their volume to increase, stretching the walls. **(b)** During diastole, the arterial walls recoil inward, pushing blood through the vasculature. Blood cannot flow back into the ventricle because the aortic valve is closed.

Arterial Blood Pressure

As blood is ejected from the ventricle into the aorta, pressure in the aorta increases to almost that of the ventricle. Pressure in the aorta does not stay elevated, however, because during diastole, blood quits flowing into the aorta yet continues to flow out, which causes a slow decline in arterial blood pressure to a minimum just prior to the next systole. The pressure in the aorta is called the **arterial blood pressure.** Ventricular pressure decreases to 0 mm Hg, whereas arterial pressure remains elevated. The elevated pressure in the arteries during diastole is due to the elastic recoil property described previously. As the arterial walls recoil inward, they exert a force on the blood, increasing the pressure. Because arterial blood pressure varies with the cardiac cycle, the maximum pressure that occurs during systole is called the systolic pressure, and the minimum pressure that occurs during diastole is called the diastolic pressure. The average arterial pressure during the cardiac cycle is the MAP.

Measuring Arterial Blood Pressure

When you go to a physician to have your blood pressure taken, the result is actually an *estimate* of the arterial pressure. Because there is no convenient way to measure aortic pressure directly, pressure is usually measured in the *brachial artery,* which runs through the upper arm. Pressure measured in this manner is close to aortic pressure because the brachial artery is not far from the heart and is also at about the same height as the aorta. (Blood pressure tends to be lower in upper regions of the body and higher in lower regions due to the force of gravity acting on blood.)

When a technician takes your blood pressure, he or she uses a device called a *sphygmomanometer,* which consists of an inflatable cuff and a pressure-measuring device that displays the air pressure inside the cuff; the technician also uses a stethoscope placed over the brachial artery to listen for sounds produced by turbulent blood flow (**Figure 8**).

To measure blood pressure, the technician places the cuff around the upper arm and inflates it to increase the cuff pressure. This pressure is transmitted through the tissue of the arm to the brachial artery, which runs to the lower arm. The technician increases cuff pressure until it is greater than systolic arterial pressure, which causes the artery to collapse, stopping blood flow. With no blood flowing through the artery, no sounds can be heard through the stethoscope.

Next, the technician opens a valve to slowly let air out of the cuff, allowing cuff pressure to fall. When cuff pressure drops to where it is just slightly below systolic arterial pressure, the artery opens briefly with each heartbeat when the pressure inside the artery is higher than that outside it, which forces the vessel open. Blood then flows through the artery in a turbulent fashion because it is forced through a narrow opening. This turbulence creates audible vibrations, called *Korotkoff sounds,* which can be heard through the stethoscope. When the Korotkoff sounds first appear, the technician notes the cuff pressure and records it as systolic arterial pressure.

The technician continues to let air out of the cuff while listening for sounds as blood moves through the brachial artery. Eventually, cuff pressure falls just below diastolic arterial pressure, from which point the artery stays open throughout the entire cardiac cycle because pressure inside the artery is always higher than that outside it. Under these conditions, blood flow returns to its laminar pattern and the Korotkoff sounds disappear. The technician notes the cuff pressure where these sounds first disappear and records it as the diastolic arterial pressure.

The blood pressure is recorded as systolic pressure (SP) over diastolic pressure (DP), SP/DP. Average normal values for a healthy individual are 110/70 mm Hg. From the blood pressure measurement, pulse pressure and MAP can be determined.

Pulse pressure (PP) is the difference between systolic pressure and diastolic pressure:

$$PP = SP - DP$$

Using average numbers for a healthy adult, the pulse pressure is

$$PP = 110 \text{ mm Hg} - 70 \text{ mm Hg} = 40 \text{ mm Hg}$$

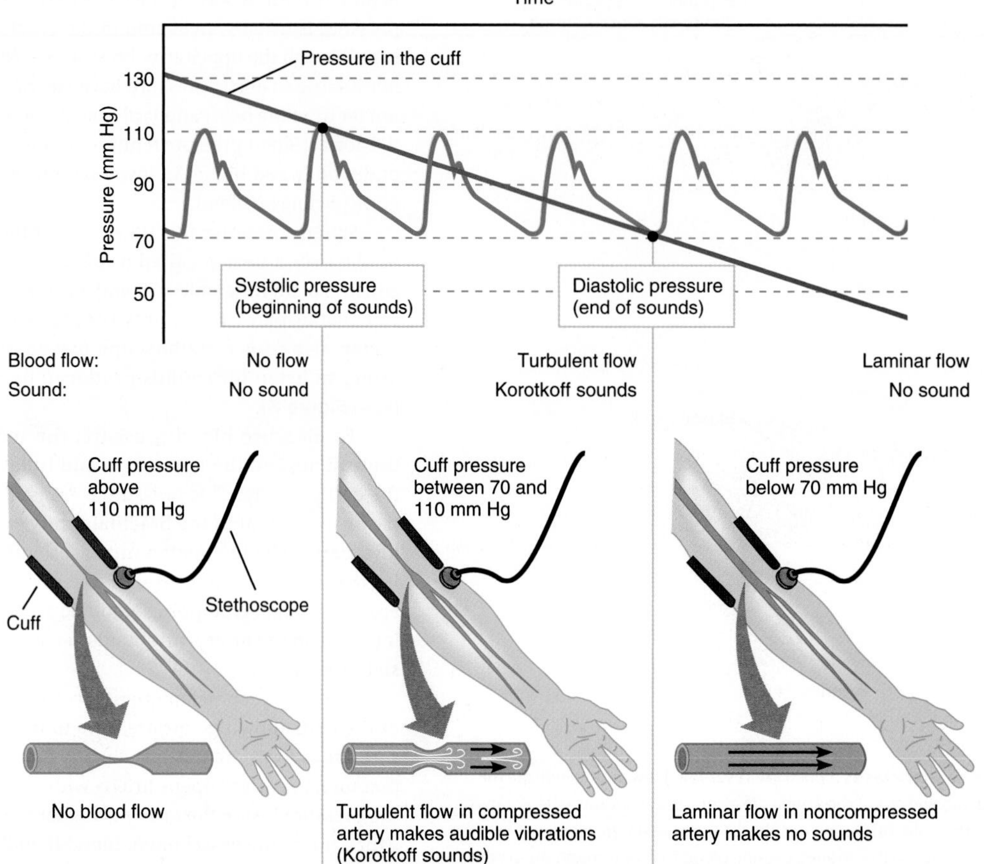

Figure 8 The events involved in blood pressure measurement. The straight line represents cuff pressure, and the wavy line represents arterial pressure.

In older people, a pulse pressure that is abnormally high may indicate *hardening of the arteries*, a condition in which the arteries become thickened and more rigid, which decreases their ability to stretch.

Health Fact

MAP, the average pressure occurring in the arteries during one cardiac cycle, is estimated by the following expression:

$$\text{MAP} = \frac{\text{SP} + (2 \times \text{DP})}{3}$$

For a normal healthy adult, this becomes

$$\text{MAP} = \frac{110 + (2 \times 70)}{3} = 83.3 \text{ mm Hg}$$

Notice that the MAP is *not* obtained simply by taking the average of the systolic and diastolic pressures, which is 90 mm Hg. Instead, MAP is a weighted mean, in which diastolic pressure is given twice the weight of systolic pressure. The reason has to do with the changes in aortic pressure as seen in the shape of the pressure wave: During a single cardiac cycle, aortic pressure is near its maximum for a relatively short period and is closer to the minimum about twice as long. Any change away from this pattern affects the MAP—even if systolic and diastolic pressure do not change. For this reason, the equation for MAP should be regarded as only an estimate.

Quick Check 2

1. Name the three types of tissue commonly found in blood vessel walls. Compare the wall thicknesses of the different blood vessels.
2. Define *compliance*. How does the compliance of arteries contribute to their function as a pressure reservoir?

4 Arterioles

The smallest arteries branch into even smaller arterioles, which lead either into a capillary bed or into metarterioles, which in turn lead into capillary beds. Thus arterioles serve as the passageway for blood to enter the capillaries, where exchange between blood and tissue takes place. The walls of arterioles contain little elastic material but have an abundance of circular smooth muscle that forms rings around the arterioles. Because this smooth muscle can contract or relax, thereby changing the diameter of arterioles, arterioles are best known as the site where resistance to blood flow can be regulated.

Arterioles and Resistance to Blood Flow

Blood flow depends on the pressure gradient and the resistance to blood flow ($F = \Delta P/R$). The arterioles are the blood vessels that provide the greatest resistance to blood flow; in fact, more than 60% of TPR is attributable to arterioles. (Although individual capillaries have a smaller radius than arterioles, the body contains so many capillaries that their total cross-sectional area is greater than that of arterioles; collectively, then, the capillaries have less resistance.) This system is illustrated in **Figure 9**, which shows that as blood flows from arteries to veins, pressure decreases gradually. A difference in pressure across any portion of the vasculature is called the *pressure drop* across that part. The largest pressure drop occurs along the arterioles: In the systemic circuit, blood enters the arterioles at an average pressure of 75–80 mm Hg and leaves them at a pressure of 35–40 mm Hg. The reason for this large pressure drop is related to the high resistance of arterioles. In addition, the pressure becomes less pulsatile as it moves through the vasculature. In contrast, the pressure drops in the larger vessels (the arteries and veins) are quite small, such that the pressure in these vessels is nearly uniform.

Not only do arterioles provide the greatest resistance to blood flow, but their resistance can also be regulated. In fact, the major function of arterioles is to act as points of control for regulating resistance to blood flow, which serves two functions: (1) controlling blood flow to individual capillary beds and (2) regulating MAP.

Resistance is regulated by the contraction and relaxation of the circular smooth muscle. Arteriolar smooth muscle is single-unit smooth muscle; thus it is partially contracted in the absence of any external factors, a state called **arteriolar tone** . External factors can either increase or decrease the contractile state. When arteriolar smooth muscle contraction increases, the radius of arterioles decreases (vasoconstriction) and resistance increases. When the arteriolar smooth muscle relaxes, the radius of arterioles increases (vasodilation) and resistance decreases (**Figure 10**).

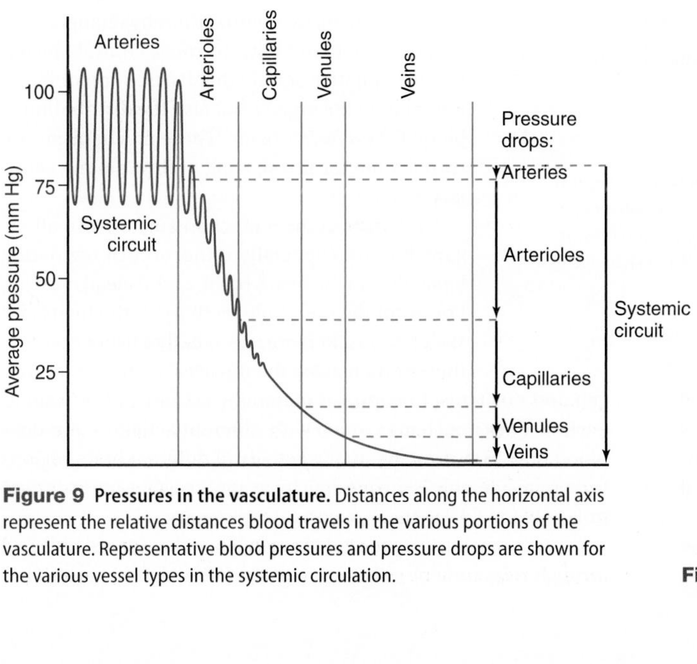

Figure 9 Pressures in the vasculature. Distances along the horizontal axis represent the relative distances blood travels in the various portions of the vasculature. Representative blood pressures and pressure drops are shown for the various vessel types in the systemic circulation.

Both intrinsic and extrinsic control mechanisms alter the contractile state of arteriolar smooth muscle, but for different reasons. Intrinsic controls include a variety of local metabolites that regulate blood flow to match the metabolic needs of the cells in the region. Extrinsic controls include both the autonomic nervous system and hormones; these controls regulate MAP.

Intrinsic Control of Blood Flow Distribution to Organs

Blood does not flow equally to all organs but instead gets distributed among the organs based on need. The distribution of blood flow is analogous to the problem of regulating water usage in a neighborhood: Whereas it is the job of the utility company to provide adequate water pressure for all houses in the system, the regulation of the flow of water to any individual house can be accomplished only by the people who live there, because they are the only ones who know of their moment-to-moment need for water. Similarly, extrinsic control mechanisms provide adequate arterial pressure for all

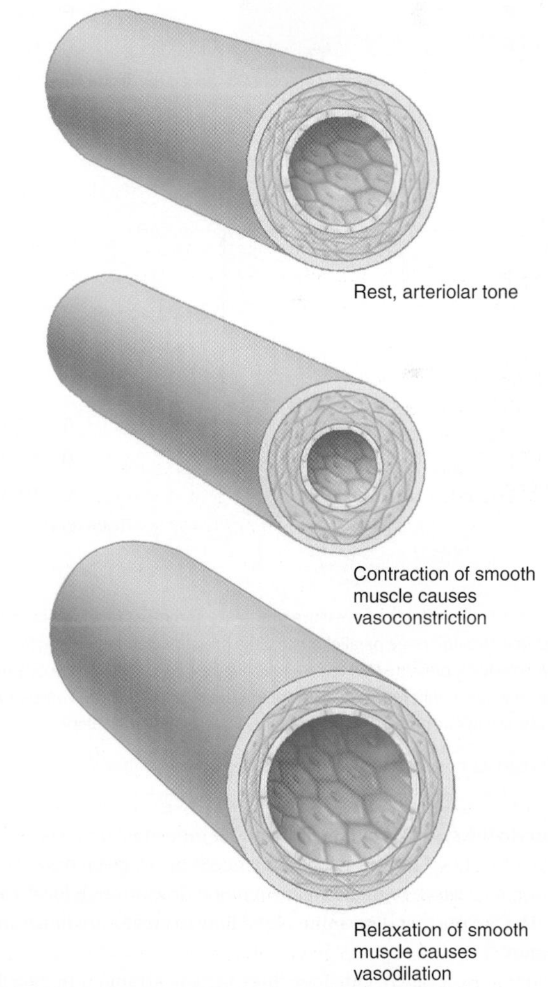

Figure 10 **Changes in the radius of arterioles.**

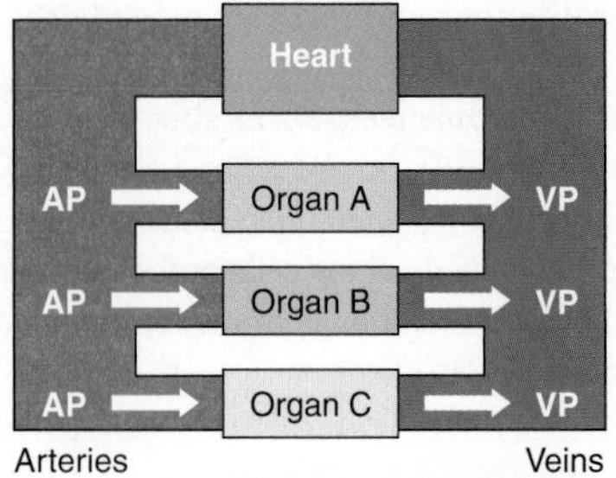

(a)

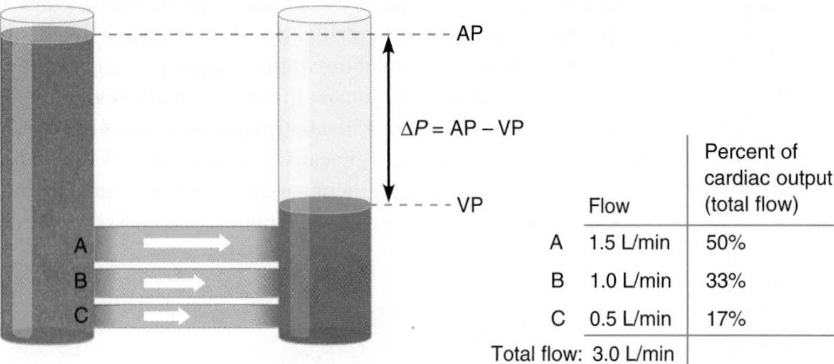

(b)

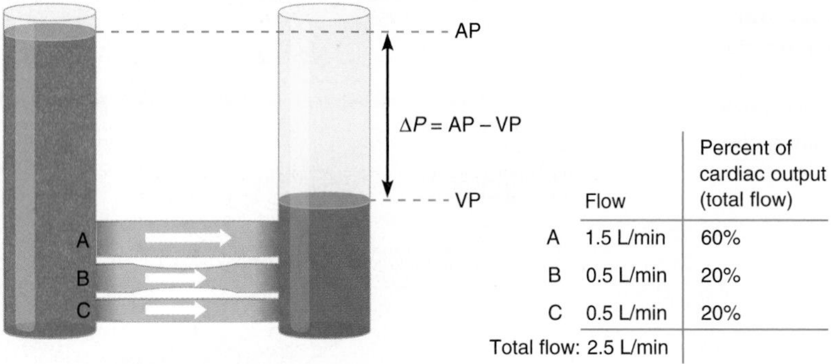

(c)

Figure 11 The effects of pressure gradients and resistance on blood flow to organs. (a) Blood flow through three parallel organs. The pressure gradient driving flow is the difference between arterial pressure (AP) and venous pressure (VP). **(b)** Differences among blood flows to organs are due to differences in resistance in the organs. **(c)** Given constant AP and VP, an increase in resistance in one organ (organ B) reduces flow to that organ alone.

Q In part (c), how do the resistances of tubes B and C compare?

organs in the systemic circuit, but only an individual organ or tissue can really "know" how much blood it needs at any given time. Thus the organ or tissue regulates its own blood flow through local control. The factors that determine blood flow to organs are illustrated in **Figure 11**.

In Figure 11a, we consider three organs arranged in parallel such that blood flow for each organ is driven by the same pressure gradient: the difference between arterial pressure and venous pressure. Suppose the flows through organs A, B, and C are 1.5 liters/min, 1.0 liter/min, and 0.5 liter/min, respectively (Figure 11b). Given that all these flows are driven by the same pressure gradient, the differences in flow must result from differences in resistance, according to the following rule:

$$\text{Organ blood flow} = \Delta P/\text{organ resistance}$$

On the basis of this relationship, the flows depicted in Figure 11b indicate that resistance is lowest in organ A, intermediate in organ B, and highest in organ C.

The effect of organ resistance on the *distribution* of blood flow is illustrated by comparing parts (b) and (c) of Figure 11. In both situations, we assume that the arterial pressure and venous pressure are the same across both organs, such that the pressure gradient is constant. The resistances, and thus the blood flows, of organs A and C are unchanged. The resistance of organ B, however, is higher in Figure 11c, which results in a decrease in blood flow compared to that in Figure 11b. Note that because the flow to organ B decreases while the flows to organs A and C remain unchanged, total flow declines, and organ B's *share* of CO decreases while the *shares* of organs A and C increase.

These findings have an important implication: Changes in the distribution of blood flow to organs—that is, changes in the percentage of CO supplied to each organ—are due primarily to changes in the vascular resistance of individual organs.

The vascular resistance of an organ (or tissue) is altered by the contraction or relaxation of smooth muscle in arterioles (and small arteries). Arterioles can vasoconstrict or vasodilate in response to intrinsic factors, thereby changing the resistance to blood flow. Intrinsic control mechanisms regulate not only the distribution of blood flow among the organs, but also the distribution of blood flow *within* organs. These mechanisms are responsible for the sharing of flow among capillary beds.

Intrinsic control mechanisms exist in all organs but are especially important in regulating blood flow to the heart, brain, and skeletal muscles. For example, metabolic activity in the heart and skeletal muscle increases greatly during exercise, thereby increasing the muscles' demands for oxygen and nutrients. In contrast, metabolic activity in the brain as a whole does not change much with different activities, nor does blood flow. However, metabolic activity of different brain regions varies considerably from one activity to another; thus intrinsic controls of blood flow to those regions are necessary.

Given that intrinsic control of organ blood flow is accomplished through relaxation or contraction of smooth muscle in arterioles, which control the flow through individual capillary beds within an organ or tissue, how do organs and tissues sense whether their blood flow is adequate? The answer is through *vascular smooth muscle* itself. In the following sections, we examine the responses of vascular smooth muscle in response to four factors: changes in metabolic activity, changes in blood flow, stretch of arteriolar smooth muscle, and local chemical messengers.

Regulation in Response to Changes in Metabolic Activity: Active Hyperemia

Vascular smooth muscle cells in arterioles are sensitive to conditions in extracellular fluid and respond to changes in the concentrations of a wide variety of chemical substances, including oxygen, carbon dioxide, potassium ions, and hydrogen ions, among others. These changes in concentration occur as a result of metabolic activity. Arteriolar smooth muscle either contracts or relaxes depending on whether concentrations of particular substances rise or fall. The general rule of thumb is this: *Changes associated with increased metabolic activity generally cause vasodilation, whereas changes associated with decreased metabolic activity induce vasoconstriction.* (An important exception to this rule occurs in the pulmonary vasculature.)

We can demonstrate how this rule works by considering the oxygen and carbon dioxide concentrations in tissue as an example (**Figure 12**). When blood flow is sufficient to satisfy the tissue's metabolic needs, the concentrations of oxygen and carbon dioxide in the tissue are in a steady state: The rate at which oxygen enters the tissue from the blood equals the rate at which it is consumed by the cells, and the rate at which carbon dioxide enters the blood equals the rate at which it is produced by the cells. Now suppose that the metabolic rate increases, such that the rates of oxygen consumption and carbon dioxide production rise, causing a decrease in tissue oxygen (called *hypoxia*) and an increase in tissue carbon dioxide. Initially, blood flow

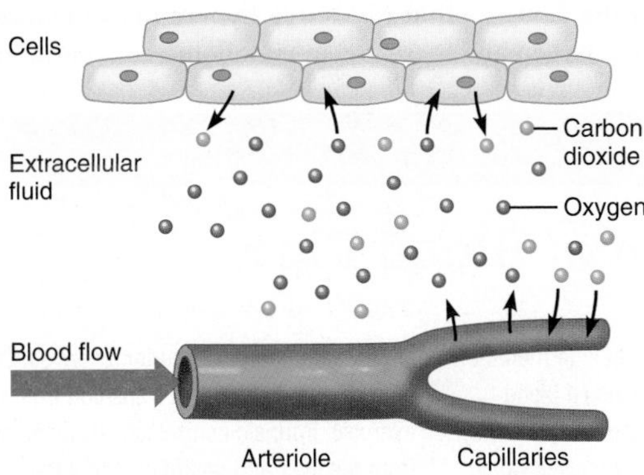

(a) Under normal steady-state conditions, oxygen (purple dots) is delivered to tissues by the blood as fast as it is consumed by cells, and carbon dioxide (green dots) is removed from tissues by the blood as fast as it is produced by cells.

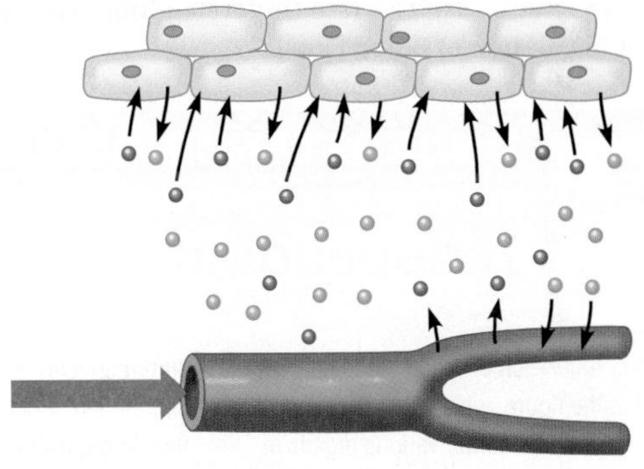

(b) An increase in the metabolic rate causes oxygen to be consumed faster than it is delivered and carbon dioxide to be produced faster than it is removed. The oxygen concentration in extracellular fluid decreases, while carbon dioxide concentration increases.

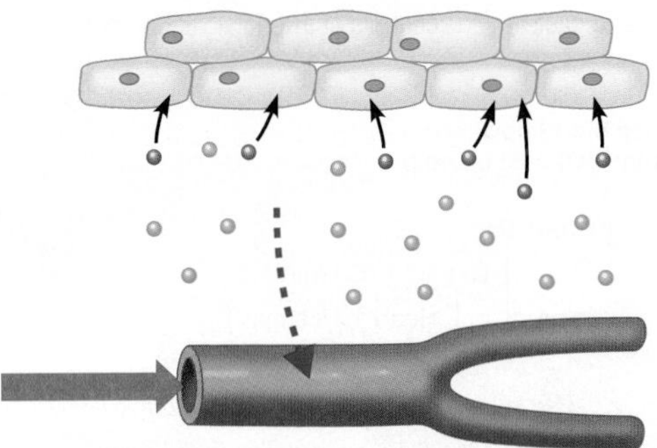

(c) The decreased oxygen concentration and increased carbon dioxide concentration act on arteriolar smooth muscle to promote vasodilation.

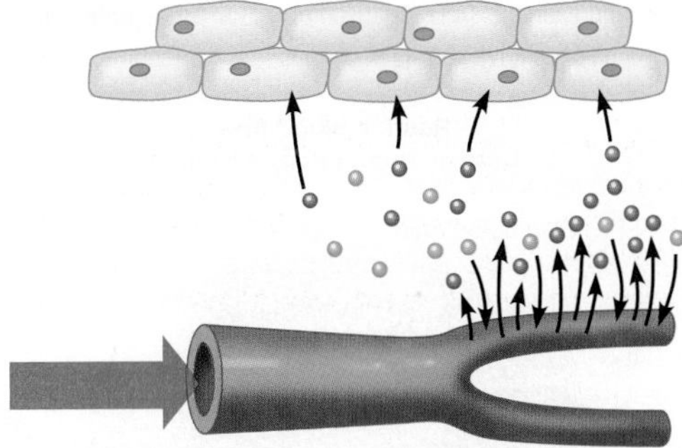

(d) Vasodilation promotes increased blood flow, which increases oxygen delivery to cells and carbon dioxide removal from cells.

Figure 12 Active hyperemia, a response to an increase in metabolic activity. (a) The delivery and consumption of oxygen and production and removal of carbon dioxide during the steady state. **(b–d)** Changes in local conditions in response to increased metabolic rate.

is insufficient to keep up with metabolic demand, a condition called *ischemia*. The decrease in oxygen and the increase in carbon dioxide both act on arteriolar smooth muscle, causing it to relax. When the muscle relaxes, arterioles dilate, vascular resistance in the tissue drops, and blood flow in that region increases. This increase in blood flow following an increase in metabolic activity is termed **active hyperemia** (*hyperemia* is a general term for a higher-than-normal rate of blood flow). As a result of this increase in blood flow, the oxygen delivery to the tissue and carbon dioxide removal from the tissue increase, and eventually a new steady state is achieved. Note that the changes in blood flow are a direct effect of the reduced oxygen and elevated carbon dioxide on the arterioles themselves and that no nerves or hormones are involved. Thus active hyperemia is an example of *intrinsic* control.

The same mechanism decreases blood flow if the local metabolic rate falls. As metabolic activity drops, oxygen levels tend to increase and carbon dioxide levels tend to decrease, which causes vasoconstriction of the arterioles, thereby increasing resistance to blood flow. A common example of active hyperemia is often observed during exercise (see **Understanding Exercise: Independent Regulation of Blood Flow**).

Regulation in Response to Changes in Blood Flow: Reactive Hyperemia

In the previous examples, tissue oxygen and carbon dioxide concentrations changed in response to changes in metabolic activity; tissue oxygen and metabolite concentrations can also change as a result of changes in *blood flow*. For example, if blood flow is blocked or reduced below adequate levels for any reason (such as occlusion of a blood vessel), the oxygen concentration falls and the carbon dioxide level rises because rates of oxygen consumption and carbon dioxide production exceed rates of delivery and removal, respectively. Both of these changes induce vasodilation and a reduction in vascular resistance, which tend to increase blood flow. Once the blockage is removed, the rate of flow will be higher than normal and will remain elevated until excess metabolites are removed and tissue oxygen concentration is restored to normal. Such an increase in blood flow in response to a previous reduction in blood flow is called **reactive hyperemia.**

Figure 13 compares reactive hyperemia to active hyperemia. Note that the basic mechanism underlying reactive hyperemia is the same as that for active hyperemia—decreases in tissue oxygen and increases in metabolite concentrations induce

UNDERSTANDING EXERCISE

Independent Regulation of Blood Flow

Independent regulation of organ blood flow is illustrated in the figure, which shows cardiac output and blood flow to the various organs at rest and during heavy exercise. At rest, cardiac output is 5 liters per minute; during exercise, it rises dramatically to 25 liters per minute, a fivefold increase.

We can see that blood flow to organs is independently regulated by comparing the proportions of total blood flow each organ receives at rest and during exercise. (A comparison of these proportions or relative shares of cardiac output enables us to identify what is termed the distribution of blood flow to organs.) If blood flow to organs were not regulated independently, the proportion of CO each organ receives would remain constant, and blood flow to every organ would rise as CO rises during exercise.

Whereas under resting conditions skeletal muscle receives 20–25% of the cardiac output, during exercise skeletal muscle can receive as much as 80–85% of the cardiac output. At the same time, blood flow to the liver and gastrointestinal tract declines from 20–25% of CO at rest to 3–5% of CO during exercise. Clearly, when the body makes a transition from rest to exercise, intrinsic controls divert blood away from the liver and gastrointestinal tract (where metabolic demand is low) and toward the muscles, which have high metabolic demands during exercise. The increased blood supply to muscles provides the oxygen and nutrients needed to generate contractile force. Blood flow is distributed according to need due to the local controls of arteriolar radius.

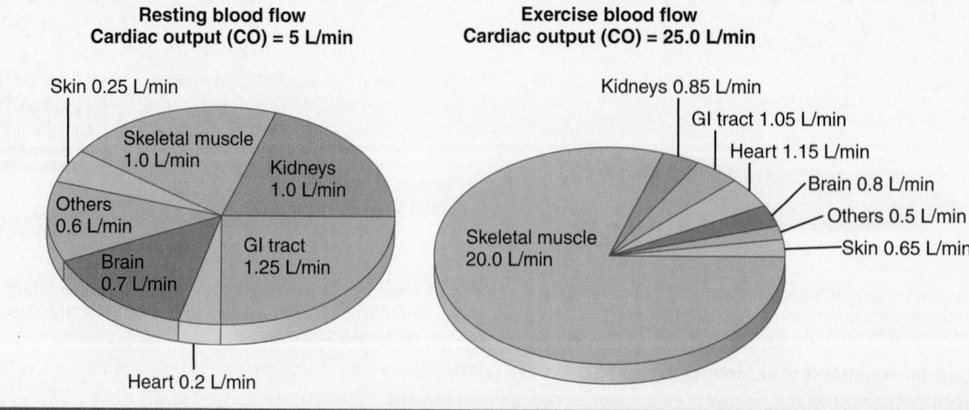

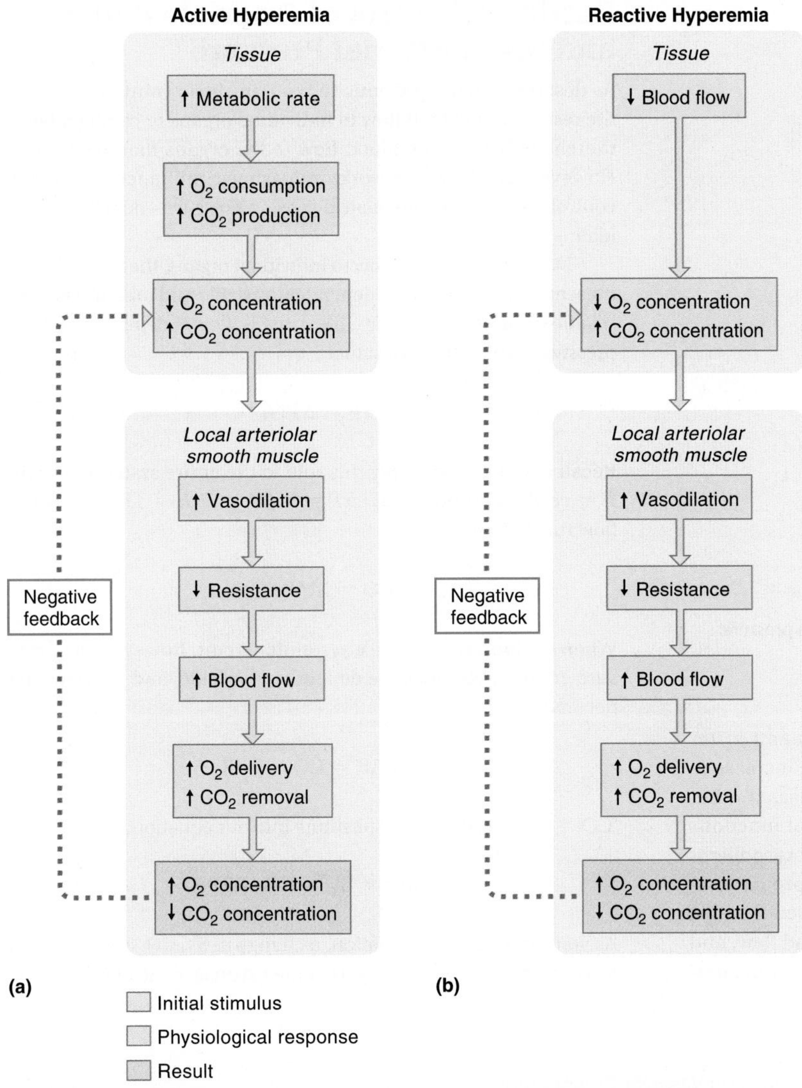

Figure 13 Comparison of active and reactive hyperemia. (a) Events in active hyperemia in response to increased utilization of oxygen and production of carbon dioxide. **(b)** Events in reactive hyperemia in response to a decrease in blood flow.

vasodilation and an increase in blood flow. The only difference is the *cause* of the changes in concentration. The sequence of events occurring in reactive hyperemia also works in reverse: If blood flow exceeds the level required for metabolic needs, intrinsic control mechanisms will induce vasoconstriction and a reduction in blood flow.

Regulation in Response to Stretch of Arteriolar Smooth Muscle: Myogenic Response

Certain tissues contain arteriolar smooth muscle called *stretch-sensitive fibers* because they are responsive to stretch, which occurs when the pressure of blood within the arterioles increases. When these fibers are stretched, they respond by contracting. A change in vascular resistance that occurs in response to stretch of blood vessels, and that does not require the action of sympathetic nerves, bloodborne hormones, or other chemical agents, is described as a **myogenic response.**

The pressure gradient that drives blood flow through a given organ or tissue is called the *perfusion pressure.* (In the case of a systemic organ, the perfusion pressure is virtually equivalent to MAP.) If perfusion pressure increases in an organ or tissue, blood flow increases and arteriolar pressure rises, which stretches the arteriolar walls because this condition increases the distending pressure across them. In arterioles containing stretch-sensitive smooth muscle, the muscle fibers then contract, which increases the arteriole's resistance and decreases the flow of blood through it (**Figure 14**). A decrease in perfusion pressure (which causes blood flow to decrease) brings about the opposite response—vasodilation and an increase in blood flow.

Myogenic control of vascular resistance in response to changes in perfusion pressure is similar to the other intrinsic control mechanisms we have discussed in that it involves a negative feedback loop. Unlike in either active or reactive hyperemia, however, the variable that remains constant in myogenic control is blood flow. Local regulation that tends to keep blood flow constant is called *flow autoregulation.*

It can be difficult to ascertain whether flow autoregulation is due to a myogenic response or to changes in metabolite concentrations because the two processes are interconnected and tend to go in the same direction. For example, an increase in perfusion pressure causes a rise in tissue oxygen levels and a fall in metabolite concentrations (assuming no change in metabolic activity) because the rise in pressure causes blood flow to increase. By themselves, these changes in concentration cause vasoconstriction, which reduces blood flow. The concentration changes also work synergistically: The increase in perfusion pressure increases stretch of arteriolar walls, which in turn induces vasoconstriction and a reduction in blood flow.

Regulation by Locally Secreted Chemical Messengers

Contractile activity of vascular smooth muscle is also affected by a variety of chemical substances, most of which are secreted by blood vessel endothelial cells or by cells in surrounding tissues (**Table 1**). One such substance is *nitric oxide,* which is released on a continual basis by endothelial cells in arterioles and acts on smooth muscle to promote vasodilation. Substances

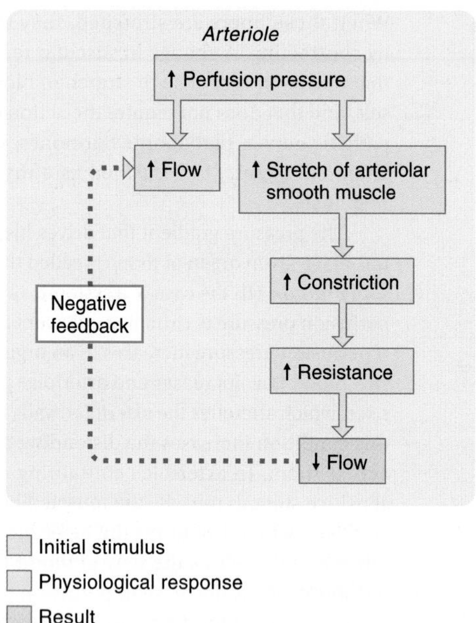

Figure 14 The myogenic response to changes in perfusion pressure.

produced by inflamed tissues, such as *bradykinin* and *histamine*, stimulate nitric oxide synthesis. The resulting increase in blood flow accounts for the redness of inflamed areas. Another potent vasodilator is *prostacyclin*, an eicosanoid that functions in preventing blood clots . *Adenosine* is an important vasodilator in the *coronary arteries*. Among the substances that promote vasoconstriction is *endothelin-1*, which is produced by endothelial cells. These intrinsic mechanisms regulate local blood flow, and extrinsic mechanisms regulate MAP, maintaining a pressure gradient that drives blood flow.

Extrinsic Control of Arteriole Radius and Mean Arterial Pressure

As described in the previous section, intrinsic controls can alter the resistance to blood flow to individual organs or capillary beds, thereby redistributing blood flow to the organs that need them. However, redistribution works only when simultaneous extrinsic controls maintain an adequate perfusion pressure—namely, by regulating MAP.

In describing blood flow to individual organs, the flow rule was presented with flow as the dependent variable and pressure and resistance as independent variables; that is, flow is dependent on the pressure gradient and resistance:

$$F = \Delta P / R$$

Recall that when we apply this rule to the entire systemic circuit, $F =$ cardiac output (CO), $\Delta P =$ MAP, and $R =$ TPR. Thus the flow rule becomes

$$\text{CO} = \text{MAP}/\text{TPR}$$

When we look at the entire systemic circuit, however, the pressure gradient becomes the dependent variable, and the flow rule becomes

$$\text{MAP} = \text{CO} \times \text{TPR}$$

$\text{CO} = \text{SV} \times \text{HR}$, we can substitute into our equation:

$$\text{MAP} = \text{SV} \times \text{HR} \times \text{TPR}$$

As you can see in this equation, a change in SV, CO, or TPR affects MAP. In this section, we describe the extrinsic control of arteriole

TABLE 1 Local Vasoactive Substances and Their Actions on Vascular Smooth Muscle

Substance	Source	Effect on vascular smooth muscle
Oxygen	Delivered to tissues by blood; consumed in aerobic metabolism	Vasoconstriction
Carbon dioxide	Generated in aerobic metabolism	Vasodilation
Potassium ions	Released from cells (particularly in muscle) as a result of repeated depolarization occurring during activity	Vasodilation (vasoconstriction at high concentrations)
Acids (hydrogen ions)	Generated during anaerobic metabolism (lactic acid) and by reaction of carbon dioxide with water (carbonic acid)	Vasodilation
Adenosine	Released by cells in certain tissues in response to hypoxia	Vasodilation
Nitric oxide	Released by endothelial cells on a continuous basis and in response to various chemical signals	Vasodilation
Bradykinin	Generated from a precursor protein (kininogen) by action of an enzyme (kallikrein) secreted by cells in certain tissues in response to various chemical signals	Vasodilation
Endothelin-1	Released by endothelial cells in response to various chemical signals and mechanical stimuli	Vasoconstriction
Prostacyclin	Released by endothelial cells in response to various chemical signals and mechanical stimuli	Vasodilation

radius as a factor that changes MAP. A later section in this chapter will more thoroughly cover the regulation of MAP.

Sympathetic Control of Arteriolar Radius

The sympathetic nervous system innervates the smooth muscle of most arterioles. During periods of increased sympathetic nerve activity, norepinephrine binds to α adrenergic receptors on arteriolar smooth muscle and activates the phosphatidylinositol bisphosphate second messenger system. The end result is vasoconstriction, which increases TPR and, in turn, MAP.

In addition to having α receptors, certain arteriolar smooth muscle (particularly that found within skeletal and cardiac muscle) has β_2 adrenergic receptors. Norepinephrine released from sympathetic postganglionic fibers does not normally bind to these receptors. However, epinephrine secreted from the adrenal medulla in response to sympathetic activity can bind to both α and β_2 receptors. Binding to α receptors, whether by norepinephrine or epinephrine, causes vasoconstriction, as described previously. However, binding of epinephrine to β_2 receptors activates the cAMP second messenger system, resulting in vasodilation and a decrease in resistance to blood flow.

Because epinephrine binds to both α and β_2 receptors, its effects on the vasculature are not readily apparent. When epinephrine is present at lower concentrations, it binds primarily to β_2 receptors and promotes vasodilation, because it has greater affinity for β_2 receptors than for α receptors. At higher concentrations, epinephrine binds to both types of receptors, and for this reason the hormone's effect on vascular resistance depends on which receptor type predominates. Because only α receptors are present on most arteriole smooth muscle, high concentrations of epinephrine usually promote vasoconstriction (the same effect as occurs with sympathetic neural input). However, because β_2 receptors predominate in the arteriole smooth muscle in cardiac and skeletal muscle, vasodilation occurs in these tissues in response to epinephrine. At the same time that this vasodilation promotes increased blood flow to these tissues, vasoconstriction decreases blood flow elsewhere.

The significance of this increased blood flow is clear given that large amounts of epinephrine are released during the fight-or-flight response, which prepares the body for vigorous physical exercise. In such exercise, the workload of the heart and skeletal muscles increases, and greater blood flow is required to meet the increased metabolic demand. The vasoconstriction in other vascular beds is also quite significant. In addition to decreasing blood flow to nonessential tissues, a generalized vasoconstriction promotes an increase in TPR, which tends to increase MAP, the driving force for blood flow.

The parasympathetic nervous system does not innervate arteriolar smooth muscle except that of the external genitalia, where it causes vasodilation.

Hormonal Control of Arteriolar Resistance

In addition to epinephrine, two other hormones regulate arteriolar resistance. These hormones cause vasoconstriction and increase MAP.

Vasopressin (ADH) Vasopressin is a hormone secreted by the posterior pituitary gland. Because it acts on the kidneys to limit urine output, vasopressin is also known as *antidiuretic hormone*. (Increased urine flow is called *diuresis*.) Along with this effect, and more to the point here, this hormone promotes vasoconstriction in most tissues—hence the name *vasopressin*. (Effects that tend to raise blood pressure are referred to as *pressor* effects.)

Angiotensin II Angiotensin II is a protein derived from a precursor called *angiotensinogen*, which is always present in the plasma. The generation of angiotensin II from angiotensinogen is a two-step process: Angiotensinogen is first converted to *angiotensin I* by *renin*, an enzyme secreted by the kidneys. Angiotensin I is then converted to angiotensin II by *angiotensin converting enzyme*, which is present on the inner surface of certain blood vessels. One of the many effects of angiotensin II is to promote vasoconstriction, thereby increasing TPR and MAP.

Table 2 summarizes extrinsic factors that regulate arteriole radius.

Quick Check 3

1. Define *vasodilation* and *vasoconstriction*. How do they affect resistance to blood flow?
2. Explain the different functions of intrinsic and extrinsic controls of arteriole radius. Name some intrinsic and extrinsic factors that affect radius.

5 Capillaries and Venules

Capillaries are the primary site where exchange of nutrients and waste products occurs between blood and tissue. In this section, we describe the functional anatomy of capillaries, the way in which blood flow into capillary beds is under local controls, and mechanisms of exchange between blood and tissue.

TABLE 2 Extrinsic Controls of Arteriole Radius

Extrinsic factor	Change in radius	Effect on MAP
Sympathetic nerves	Vasoconstriction	Increase
Epinephrine	Depends on receptor type α Adrenergic: vasoconstriction β_2 Adrenergic: vasodilation	Increase (dominant effect is at α receptors)
Vasopressin	Vasoconstriction	Increase
Angiotensin II	Vasoconstriction	Increase

Capillary Anatomy

Capillaries are the smallest blood vessels, measuring only 1 mm long and 5–10 μm in diameter, about the width of a red blood cell. Capillary walls are thin (0.5 μm), consisting of a single layer of endothelial cells surrounded by a basement membrane. The small diameter and thin wall provide a small diffusion distance between blood and the surrounding interstitial fluid.

Although capillaries are small, their extensive branching results in 10–40 billion capillaries in the human body, providing a surface area for exchange between blood and tissue of approximately 600 square meters. In fact, almost all cells of the body are within 1 mm of a capillary. In addition, capillaries do not exist in isolation, but rather occur in networks called *capillary beds*.

Because of their extensive branching, the total cross-sectional area of capillaries is considerably greater than that of the other blood vessels. Thus, as blood enters capillary beds, the velocity of blood flow decreases (**Figure 15**). This phenomenon is analogous to a river (small cross-sectional area) emptying into a lake (large cross-sectional area). The river current is fast, but as the water reaches the open lake, velocity of flow slows down as the water moves over a greater area. Velocity of blood flow through the capillaries is about 0.1 mm/sec, allowing approximately 1 second for exchange between blood and tissue to take place—ample time for simple diffusion to occur.

The size and surface area of capillaries are suitable for their primary function: facilitating exchange of substances between blood and tissue. Enhancing this exchange is the leakiness of the capillary wall, which varies in different tissues. For example, that tight junctions between endothelial cells of capillary walls within the central nervous system produce a functional barrier between blood and cerebrospinal fluid, the blood-brain barrier. Most capillaries, however, are considerably "leaky" and are classified as continuous, fenestrated, or discontinuous based on their degree of "leakiness."

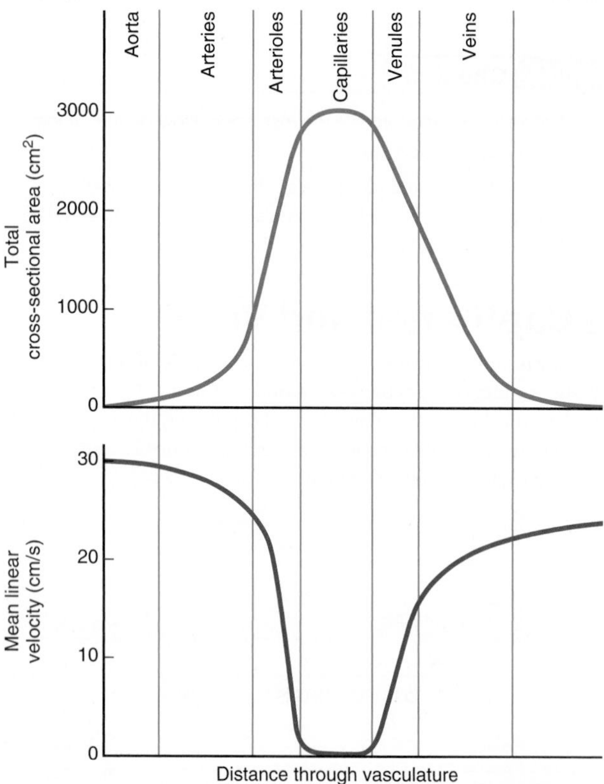

Figure 15 Total cross-sectional area and velocity of blood flow through the vasculature.

Continuous Capillaries

In *continuous capillaries* (**Figure 16a**), which are the most common type, the endothelial cells are joined together with tight junctions such that the spaces between them, called *intercellular clefts*, are relatively narrow; this arrangement limits the passage of some molecules. Continuous capillaries are highly permeable to water-soluble substances having small molecular sizes such that they can pass through the aqueous environment in the intercellular cleft, and to lipid-soluble substances that easily permeate the membranes of the endothelial cells (such as fatty acids and steroid hormones). In contrast, larger water-soluble molecules such as ions and amino acids encounter more difficulty when they attempt to pass through the narrow cleft. The permeability of continuous capillaries to proteins and other macromolecules is very low because these substances can neither readily cross membranes of endothelial cells nor easily penetrate the clefts between cells.

Fenestrated Capillaries

In *fenestrated capillaries* (Figure 16b), the endothelial cells possess relatively large pores *(fenestrations)* approximately 60–80 nm in diameter that allow for the rapid diffusion of small water-soluble substances. The size of fenestrations varies in different tissues, with some fenestrations large enough for proteins. Fenestrated capillaries are found primarily in organs whose functions depend on the rapid movement of materials across capillary walls, including the kidneys, intestines, and endocrine glands.

Discontinuous Capillaries and Sinusoids

Discontinuous capillaries serve as a transition from fenestrated capillaries to sinusoids. These capillaries are found in the spleen, liver, and bone marrow—areas where proteins and cells must cross the endothelium. The discontinuous capillaries form sinusoids in these organs. Sinusoids are basically large blood-filled spaces that function in the exchange of substances between blood and tissue. They are lined by highly fenestrated endothelium, but no basement membrane. In the liver, the presence of sinusoids allows newly synthesized proteins such as albumin or clotting factors to enter the plasma. In bone marrow and spleen, sinusoids enable blood cells to move into the spleen, where they are filtered, or to exit the bone marrow after the cells have been synthesized.

Local Control of Blood Flow Through Capillary Beds

The simplest way to regulate the exchange of material across capillary walls is to regulate the amount of blood flowing through a particular capillary bed. This is done through local control of smooth muscle

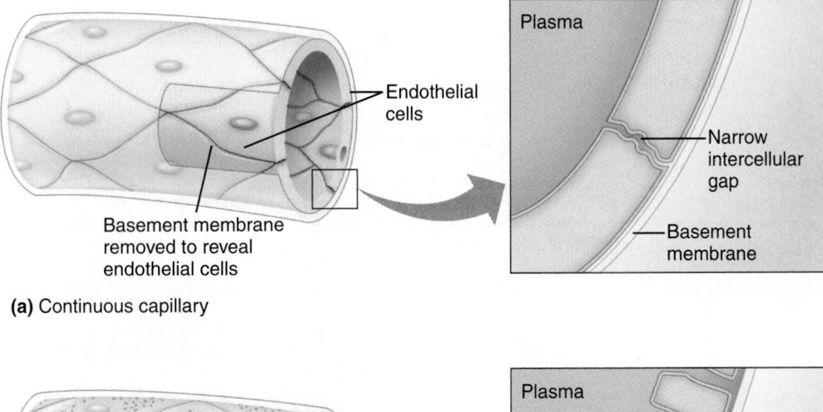

Figure 16 Two types of capillaries. (a) A continuous capillary, featuring narrow, water-filled gaps between endothelial cells. **(b)** A fenestrated capillary, which possesses pores (fenestrations) that penetrate through endothelial cells, in addition to intercellular gaps between endothelial cells.

located in various areas of the microcirculation (**Figure 17**). We already learned about the regulation of blood flow through arterioles. Next, we describe metarterioles and precapillary sphincters that regulate blood flow through certain capillaries.

Most tissues contain metarterioles, which are structurally intermediate between arterioles and capillaries; instead of the continuous layer of smooth muscle that surrounds arterioles, metarterioles possess isolated rings of smooth muscle that act as gatekeepers at strategic points. Unlike arterioles, which direct blood into the interbranching vessels in a capillary bed, metarterioles serve as bypass channels or *shunts* by directly connecting arterioles to venules. The presence of these shunts allows blood to continue flowing from arterioles to venules, bypassing capillaries.

Like the smooth muscle of arterioles, the smooth muscle of metarterioles can contract or relax to increase or decrease, respectively, resistance to blood flow. Because metarterioles serve as a bypass mechanism, when their resistance to blood flow is high, blood flow through the capillary beds increases, whereas when their resistance is low, blood flow through the capillary beds decreases. The smooth muscle of metarterioles is under local control by metabolites, similar to that of the arterioles. Whether metarterioles receive sympathetic innervation, as arterioles do, is still unclear.

Blood flow through capillaries is also regulated by smooth muscle that surrounds capillaries on the arteriole end, called **precapillary sphincters** (see Figure 17). Contraction of precapillary sphincters constricts the capillaries, increasing their resistance to blood flow. Precapillary sphincters are only affected by local controls—that is, metabolites produced based on the metabolic activity of the tissue around the sphincters. An increase in metabolites, such as carbon dioxide, causes relaxation of the sphincters and increases blood flow through the capillaries, whereas a decrease in carbon dioxide causes contraction of the sphincters and decreases blood flow. Because the precapillary sphincters contract and relax in response to levels of local metabolites, which are constantly changing, blood flow through capillaries is intermittent. This contraction/relaxation of precapillary sphincters is called *vasomotion*. Blood that enters the capillaries can participate in exchange with interstitial fluid through the mechanisms described next.

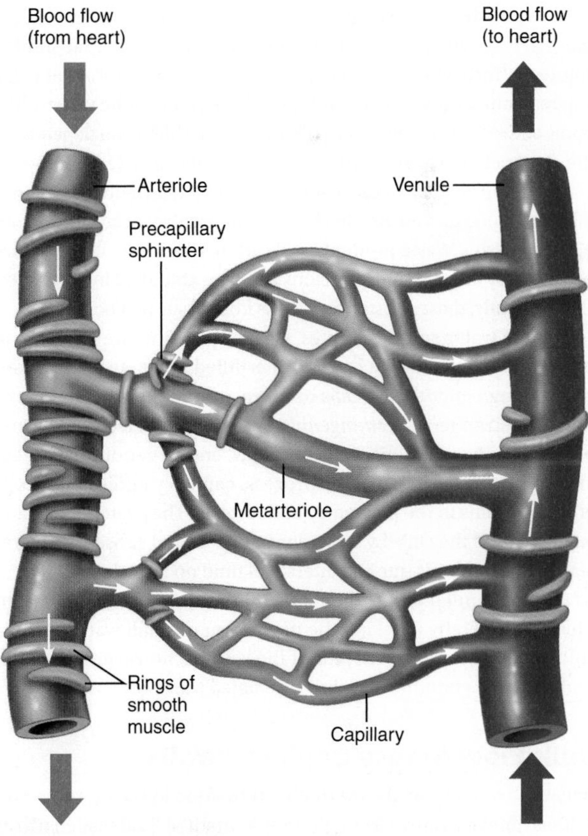

Figure 17 Microcirculation.

Movement of Material Across Capillary Walls

Movement of material across capillary walls serves two purposes: (1) exchange of material between blood and cells and (2) normal distribution of the extracellular fluid.

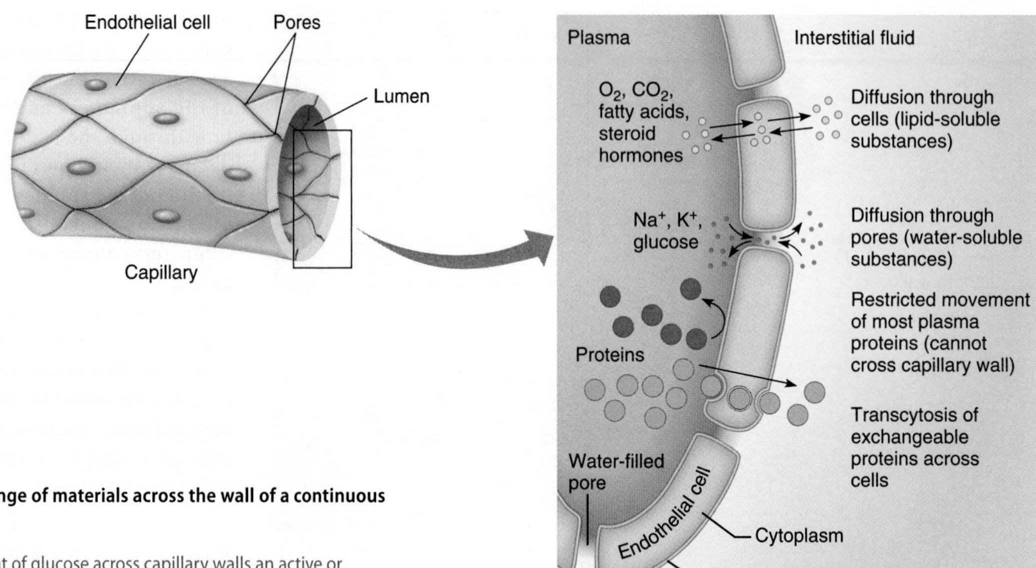

Figure 18 **Exchange of materials across the wall of a continuous capillary.**

Q Is the movement of glucose across capillary walls an active or passive process?

Exchange Across Capillary Walls

Materials are exchanged across capillary walls through different mechanisms, depending on the molecular size and lipid solubility of the substance and on the location of the capillary. Mechanisms of exchange include simple diffusion, transcytosis, and mediated transport (**Figure 18**). Most small solutes, whether lipid- or water-soluble, move across capillary walls by simple diffusion. Small lipid-soluble solutes diffuse across the plasma membranes of endothelial cells, whereas small water-soluble solutes diffuse through the water-filled pores between endothelial cells. The direction of diffusion depends on the electrochemical gradient for a specific substance (except the exchangeable proteins). Because nutrients and oxygen tend to be plentiful in the blood and are needed by cells, these substances diffuse from blood to tissue. Waste products and carbon dioxide, in contrast, are produced by the cells and, therefore, are more abundant in the tissues. Consequently, these substances diffuse from tissue to blood.

Large water-soluble solutes, such as proteins, are too large to pass through any but the largest water-filled pores, so they cannot pass through endothelial cells or move around them. Certain proteins (referred to as *exchangeable proteins*) are selectively transported across endothelial cells by a slow, energy-requiring process known as *transcytosis*. In this process, capillary endothelial cells engulf proteins in the plasma by endocytosis. The proteins are then ferried across the cells by vesicular transport and subsequently released by exocytosis into the interstitial fluid on the other side.

Brain capillary endothelial cells are linked together by tight junctions, thereby blocking the movement of small water-soluble solutes between cells. Certain of the solutes, however, are transported across endothelial cells by mediated transport.

Bulk Flow Across Capillary Walls

Because capillary walls are freely permeable to water and small solutes, fluid can move from blood to interstitial fluid (called **filtration**) or from interstitial fluid to blood (called **absorption**) based on pressure gradients. The purpose of this *bulk flow* is not to provide nutrients and remove waste products, but rather to maintain balance between the extracellular compartments—that is, between interstitial fluid and plasma. A shift in fluid from plasma to interstitial fluid causes a swelling of tissues, called **edema**.

The forces that drive the movement of fluid into and out of capillaries are called **Starling forces**. They include the following: (1) *capillary hydrostatic pressure* (P_{CAP}) due to the hydrostatic pressure of fluid inside the capillary, (2) *interstitial fluid hydrostatic pressure* (P_{IF}) due to the hydrostatic pressure of fluid outside the capillary, (3) capillary osmotic pressure (π_{CAP}) due to the presence of nonpermeating solutes inside the capillary, and (4) interstitial fluid osmotic pressure (π_{IF}) due to the presence of nonpermeating solutes outside the capillary. (Note that the Starling forces for movement across capillary walls are *not* related to the Starling law of the heart, which describes intrinsic regulation of stroke volume.) The four Starling forces are illustrated in **Figure 19a** and described next.

Hydrostatic Pressures Whenever a hydrostatic pressure exists across a semipermeable barrier such as a capillary wall, the pressure tends to move water across the wall. The capillary hydrostatic pressure favors filtration and is equal to the blood pressure in the capillaries, which varies along the length of the capillary, because the blood pressure decreases as blood flows from the arteriole end ($P_{CAP} = 38$ mm Hg) of the capillary to the venule end ($P_{CAP} = 16$ mm Hg) (Figure 18). The interstitial fluid hydrostatic pressure favors absorption and arises because fluid outside the capillary walls drives the fluid into the capillary. This pressure does not vary along the length of capillaries and is usually low ($P_{IF} = 1$ mm Hg).

Osmotic Pressures Recall that when a semipermeable barrier separates two solutions with different concentrations of nonpermeating solute, water tends to flow from the side where the solute

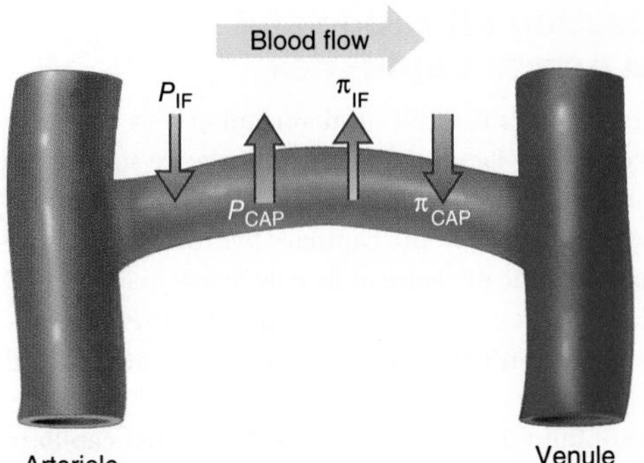

(a)

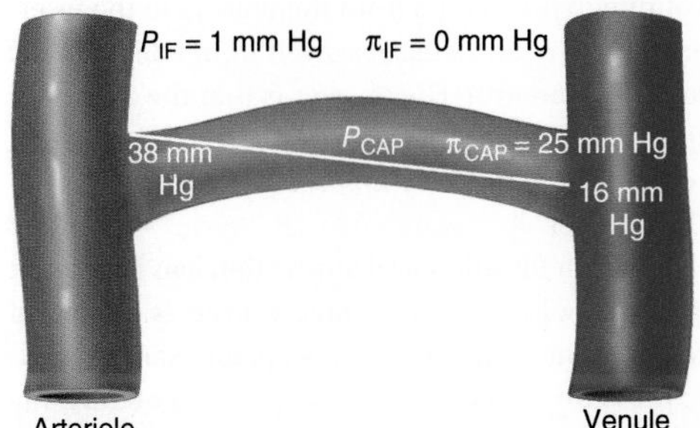

Arteriole end

Filtration pressure:	Absorption pressure:
P_{CAP} = 38 mm Hg	π_{CAP} = 25 mm Hg
π_{IF} = 0 mm Hg	P_{IF} = 1 mm Hg
38 mm Hg	26 mm Hg

NFP = Filtration pressure − Absorption pressure

= 38 mm Hg − 26 mm Hg = 12 mm Hg

Venule end

Filtration pressure:	Absorption pressure:
P_{CAP} = 16 mm Hg	π_{CAP} = 25 mm Hg
π_{IF} = 0 mm Hg	P_{IF} = 1 mm Hg
16 mm Hg	26 mm Hg

NFP = Filtration pressure − Absorption pressure

= 16 mm Hg − 26 mm Hg = −10 mm Hg

(b)

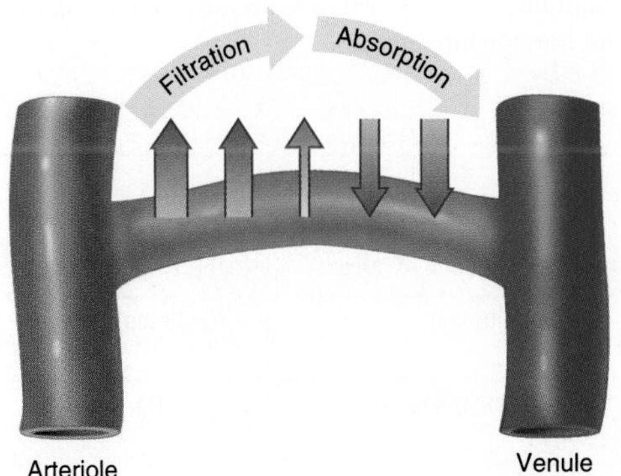

(c)

Figure 19 Starling forces across capillary walls. A schematic representation of an arteriole leading into a capillary bed (represented by a single tube) that drains into a venule is shown. **(a)** Of the four Starling forces, capillary hydrostatic pressure (P_{CAP}) and interstitial fluid osmotic pressure (π_{IF}) favor filtration, whereas interstitial fluid hydrostatic pressure (P_{IF}) and capillary osmotic pressure (π_{CAP}) favor absorption. **(b)** Average values for the Starling forces. Note that capillary hydrostatic pressure decreases as blood flows from the arteriole end of the capillary (P_{CAP} = 38 mm Hg) to the venule end (P_{CAP} = 38 mm Hg.) The diagonal line represents the decrease in P_{CAP}. **(c)** Net filtration pressure across the capillary wall is indicated by arrows. Note that filtration occurs near the arteriole end, whereas absorption occurs near the venule end.

Q Would an increase in the plasma protein concentration tend to favor increased filtration or absorption?

concentration is lower to the side where the solute concentration is higher. The osmotic pressure of a solution is a measure of its solute concentration, and it increases as the solute concentration increases. Put another way, the presence of a nonpermeating solute exerts an osmotic pressure that tends to draw water to the side where it is present in greater concentration. Only nonpermeating solutes exert an osmotic pressure because permeating solutes, such as glucose, are generally at equal concentrations across the capillary wall. The primary nonpermeating solutes in plasma are proteins, which are called *colloids*. Osmotic pressure that is exerted by proteins is referred to as **colloid osmotic pressure.**

The capillary osmotic pressure is the osmotic force exerted by proteins in the plasma. Plasma proteins draw water into the capillary, so they exert a force for absorption. Under normal conditions, the concentration of proteins in the plasma is 6–8 grams per 100 mL, which exerts an osmotic force of 25 mm Hg (Figure 19b). Although the capillary osmotic pressure is affected by any movement of water into or out of capillaries that might occur as blood flows through them, for the sake of simplicity we will ignore this factor and assume that the osmotic pressure of plasma does not change as blood flows from one end of the capillary to the other.

The interstitial fluid osmotic pressure is the osmotic force exerted by proteins in the interstitial fluid, which tends to draw fluid out of the capillary and into the interstitial fluid. Because relatively few proteins are present in the interstitial fluid, the osmotic pressure of interstitial fluid is approximately 0 mm Hg.

Net Filtration Pressure The direction of fluid flow across the wall of a capillary is determined by the **net filtration pressure** at any given location—that is, the difference in the filtration pressures and the absorption pressures. Of the four Starling forces, capillary hydrostatic pressure and interstitial fluid osmotic pressure favor filtration, whereas interstitial fluid hydrostatic pressure and capillary osmotic pressure favor absorption. The net filtration pressure (NFP) can be defined as follows:

$$\begin{aligned} \text{NFP} &= \text{filtration pressure} - \text{absorption pressure} \\ &= (P_{CAP} + \pi_{IF}) - (\pi_{CAP} + P_{IF}) \end{aligned}$$

When the sign of the NFP is positive, filtration occurs; when it is negative, absorption occurs.

Recall that while three of the Starling forces are relatively constant at any given location, the capillary hydrostatic pressure varies along the length of the capillary. Thus the NFP also varies along the length of the capillary, being higher at the arteriole end and lower at the venule end. Figure 19b shows the four Starling forces and the NFP at two points along the capillary: the arteriole end and the venule end. In this example, the capillary hydrostatic pressure is 38 mm Hg at the arteriole end and 16 mm Hg at the venule end. Interstitial fluid hydrostatic pressure is 1 mm Hg, capillary osmotic pressure is 25 mm Hg, and interstitial fluid osmotic pressure is 0 mm Hg. Based on these data, we can determine the NFP at the arteriole and venule ends of the capillary:

$$\text{Arteriole end: NFP} = (P_{CAP} + \pi_{IF}) - (\pi_{CAP} + P_{IF})$$
$$= (38 + 0) - (25 + 1)$$
$$= 12 \text{ mm Hg}$$

$$\text{Venule end: NFP} = (P_{CAP} + \pi_{IF}) - (\pi_{CAP} + P_{IF})$$
$$= (16 + 0) - (25 + 1)$$
$$= -10 \text{ mm Hg}$$

At the arteriole end, the NFP is 12 mm Hg, which favors filtration; at the venule end, the NFP is −10 mm Hg, which favors absorption. Thus, as blood enters a capillary bed, fluid moves out, but near the end of the capillary bed, *most* of the fluid returns to the blood (Figure 19c). Note that not all the fluid that leaves the blood by bulk flow is returned before the blood leaves the capillary bed.

Table 3 summarizes the forces for fluid movement across capillary walls.

Apply Your Knowledge

Jack has waited tables for several hours. The pressure in the veins of Jack's feet has increased to 26 mm Hg, and the average capillary hydrostatic pressure has increased to 32 mm Hg. Given an average capillary hydrostatic pressure of 32 mm Hg, interstitial fluid hydrostatic pressure of 1 mm Hg, capillary osmotic pressure of 25 mm Hg, and interstitial fluid osmotic pressure of 1 mm Hg, what is the net filtration pressure across the capillary walls of Jack's feet? What happens as a consequence of this pressure?

Factors Affecting Filtration and Absorption Across Capillaries

The rate at which fluid is filtered or absorbed across capillary walls is influenced by any factor that alters the relative sizes of the Starling forces and, hence, the NFP. Increased filtration, for example, is favored by an increase in the capillary hydrostatic pressure or interstitial fluid osmotic pressure or by a decrease in interstitial fluid hydrostatic pressure or capillary osmotic pressure. Changes in the opposite direction favor an increase in the rate of fluid absorption.

The volume of fluid that moves by bulk flow across capillary walls under normal conditions is considerably higher than total blood volume. Each day, approximately 20 liters of fluid is filtered and 17 liters is absorbed, giving a net volume of 3 liters per day that is filtered, which is roughly equal to an individual's entire plasma volume. If this volume of fluid shifts from the plasma to the interstitial space every day, why don't tissues swell from edema? Why doesn't blood volume decrease? The answer is that the 3 liters or so of filtered fluid is picked up from the interstitium and returned to the cardiovascular system by the *lymphatic system*, which is described later in this chapter.

The balance between filtration and absorption can be altered as a result of certain pathological conditions, or even as a result of everyday occurrences. When a person stands up, for example, capillary hydrostatic pressure increases in the lower parts of the body because the column of blood increases the hydrostatic pressure in the lower arterioles and veins. Any increase in pressure—whether it be at the arteriolar end of a capillary or at the venular end—tends to raise capillary blood pressure, which increases capillary hydrostatic pressure and, in turn, increases the rate of filtration.

An increase in capillary filtration, with accompanying tissue swelling, also occurs in response to certain types of injuries. When the skin is cut or abraded, for instance, the affected part becomes swollen within a few minutes. In part, this effect results from damage to capillaries, which allows protein-rich fluid to leak out. Consequently, the concentration of proteins in the interstitial fluid rises, which raises its osmotic pressure. In addition, certain cells in the injured area release a chemical called *histamine,* which increases the permeability of capillary walls to proteins, thereby increasing the leakage of proteins into the interstitium.

TABLE 3 Forces Affecting the Movement of Fluid Across Capillary Walls

Force	Definition	Direction of force	Approximate value
Capillary hydrostatic pressure, P_{CAP}	Hydrostatic pressure exerted by the presence of fluid inside the capillary	Filtration	16–38 mm Hg
Interstitial fluid hydrostatic pressure, P_{IF}	Hydrostatic pressure exerted by the presence of fluid outside the capillary	Absorption	1 mm Hg
Capillary osmotic pressure, π_{CAP}	Osmotic force due to presence of proteins in plasma	Absorption	25 mm Hg
Interstitial fluid osmotic pressure, π_{IF}	Osmotic force due to presence of proteins in interstitial fluid	Filtration	0 mm Hg
Net filtration pressure, NFP	Difference between forces for filtration and absorption	If positive: filtration; if negative: absorption	2 mm Hg

CLINICAL CONNECTIONS

Heart Failure

The term *heart failure* refers to any change in the heart that reduces its ability to maintain an adequate cardiac output. Sometimes heart failure is caused by a heart attack, or *myocardial infarction,* in which the death of muscle cells leads to weakening of the heart. Abnormalities in electrical conduction can also damage the heart.

Once the heart muscle weakens, it generates less contractile force, which reduces stroke volume and cardiac output. If this happens in the left heart, mean arterial pressure falls, and venous pressure rises. To understand why heart failure increases venous pressure, we must consider how venous pressure is affected by the pumping action of the heart. Imagine that the heart has stopped pumping blood. Only a small pressure (a few mm Hg) would exist in the cardiovascular system due to blood filling up blood vessels and stretching their walls. However, the pressure would be the same in the arteries, veins, and everywhere else, so no blood flow would occur. Now imagine that the heart were to suddenly resume pumping. In doing so, it would push blood into the arteries, which would effectively transfer a certain fraction of the total blood volume from the veins to the arteries. This transfer of blood would raise mean arterial pressure and lower venous pressure, as you can see in diagram (a), where the resumption of pumping occurs at the arrow. Arterial pressure rises more than venous pressure falls because the arteries have a lower compliance than the veins.

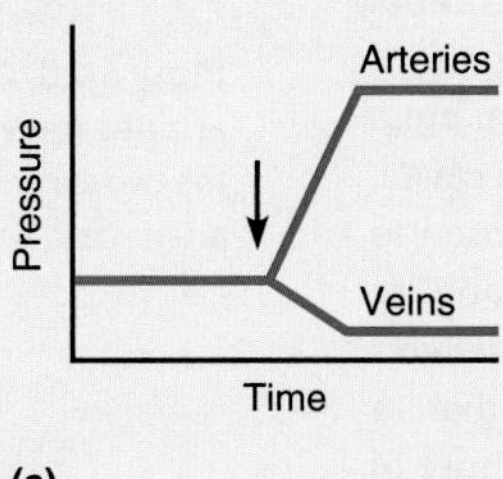

(a)

In heart failure, the heart's ability to pump blood is reduced, which causes a portion of the total blood volume to shift back into the veins from the arteries. This effect both reduces arterial pressure and raises venous pressure, as occurs at the arrow in diagram (b).

In heart failure, pressure can rise in either the systemic veins or the pulmonary veins, depending on which side of the heart is affected. Failure of the left ventricle *(left-sided failure)* causes pressure to rise in the pulmonary veins because they carry blood to the left side of the heart. In contrast, failure of the right ventricle *(right-sided failure)* causes pressure to rise in the vena cava and other systemic veins.

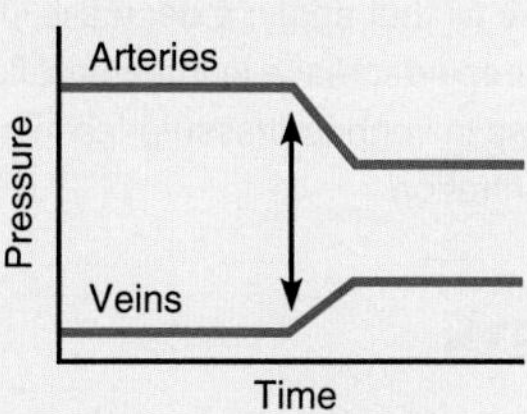

(b)

Edema—a common sign of heart failure—is a consequence of increased venous pressure. As venous pressure rises, the increase in pressure is transmitted to vessels upstream, including capillaries. When the left side of the heart fails, edema occurs in the lungs (called *pulmonary edema*) because the pressure increases in pulmonary capillaries. When the right side of the heart fails, it tends to cause edema in systemic tissues because pressure rises in systemic capillaries. When edema occurs, the condition is referred to as *congestive heart failure.*

Critical Thinking Questions

1. What is heart failure, and what causes it?
2. Explain the relationship between heart failure and venous pressure.
3. How does the side of the heart (right or left) affect differences in types of heart failure?

Other conditions that promote increased capillary filtration and edema include liver, kidney, and heart disease. Because most plasma proteins are manufactured by the liver, damage to the liver can result in a decrease in plasma protein concentration, which lowers plasma osmotic pressure. Certain forms of damage to the kidneys interfere with their ability to eliminate excess water (and solutes) in the urine, which results in the accumulation of excess fluid in the body. As a consequence, blood volume increases and blood pressure rises throughout the cardiovascular system. This increase in pressure raises capillary hydrostatic pressure, which increases filtration. Damage to the kidneys can also cause them to eliminate significant quantities of plasma proteins in the urine. (Normally, only negligible quantities of protein are voided in the urine.) This loss of protein triggers increased capillary filtration because it reduces plasma osmotic pressure. Damage to the heart can result in pulmonary edema, the accumulation of fluid in the lungs (see **Clinical Connections: Heart Failure**).

Venules

Capillaries come together to form vessels called *venules.* Venules are slightly smaller than arterioles, averaging about 20 μm in diameter, but their walls are much thinner than those of arterioles and contain little or no smooth muscle (see Figure 6). In fact, the smaller of the venules resemble capillaries more than arterioles in that they consist of a single layer of endothelium that is porous, allowing for exchange of materials between blood and interstitium. Thus exchange between blood and interstitium occurs in both capillaries and the small venules.

Quick Check 4

1. How do fenestrated capillaries differ from continuous capillaries?
2. Of the following changes, which would tend to cause an increase in the rate at which fluid is filtered from capillaries? Choose all that apply: a decrease in plasma osmotic pressure; a decrease in interstitial fluid osmotic pressure; an increase in venous pressure; an increase in plasma protein concentration.

6 Veins

Venules come together to form larger veins. On average, veins have slightly larger diameters than arteries, but have walls about half as thick; a typical vein has an internal diameter of 5 mm but a wall thickness of only 0.5 mm. The largest veins, the venae cavae, are even larger in diameter than the aorta (30 mm, as opposed to 12.5 mm) but have a wall thickness of only 1.5 mm compared to 2 mm for the aorta. The relative thinness of the walls of veins reflects the fact that blood pressure in the veins is significantly lower than in arteries. Although thinner, the walls of veins are similar to those of arteries in that they contain smooth muscle and elastic and fibrous connective tissue (see Figure 6).

Unlike any other blood vessels in the body, veins are equipped with one-way valves that permit blood to flow toward the heart but prevent it from flowing back toward organs and tissues (see Figure 5). These valves are present in veins located outside the thoracic cavity *(peripheral veins)* but are absent from veins located within the thoracic cavity *(central veins)* and in the central nervous system.

Veins: A Volume Reservoir

Unlike arteries, which function as pressure reservoirs, veins function as *volume reservoirs*, a property that is related to vessel compliance. Because veins are thin walled and easily stretched, they have high compliance—a relatively small increase in the pressure within veins causes a relatively large degree of expansion (increase in volume). Put another way, veins can accommodate a large increase in blood volume with little change in blood pressure, which makes them good at storing volume. Thus veins can hold a larger volume of blood than arteries can at a given pressure (**Figure 20**). In fact, the veins in the human body contain a substantially greater volume of blood than do the arteries (**Figure 21**), even though the pressure within veins is much lower than that within arteries. Under resting conditions, approximately 60% of the total blood volume is located within the systemic veins. This blood can be shifted to the arterial side of the circulation under more active conditions; in essence, it serves as a readily available blood reserve or volume reservoir. Blood is shifted by increasing venous return, which is described next.

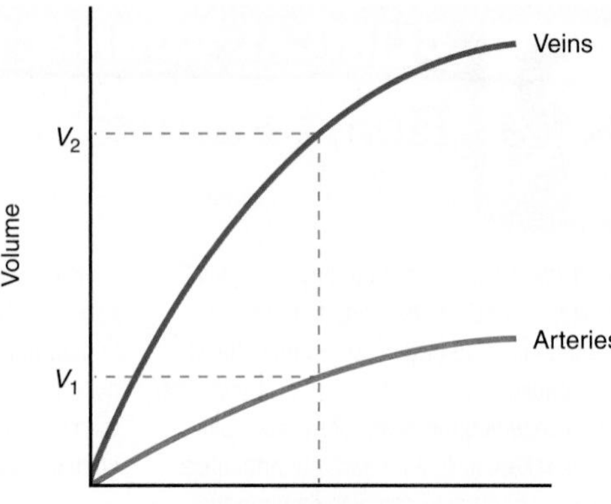

Figure 20 Curves showing how the volume of blood contained in arteries and veins varies with the pressure inside them. A comparison of the two curves shows that at a given pressure, veins hold more blood (V_2) than arteries do (V_1).

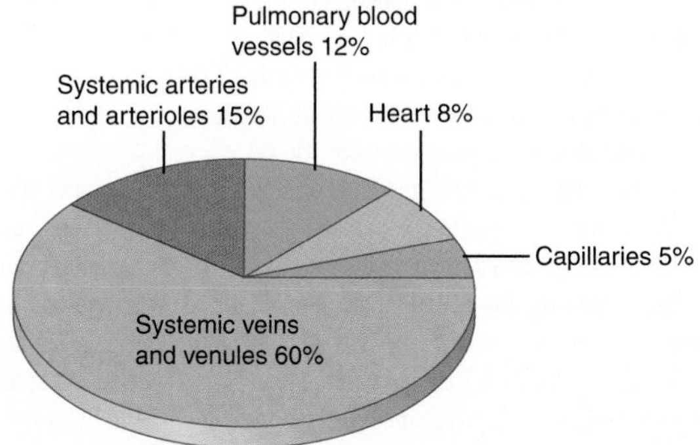

Figure 21 Distribution of blood volume in the various portions of the cardiovascular system. Percentages indicate proportions of total blood volume.

Factors That Influence Venous Pressure and Venous Return

Venous pressure has an important, though indirect, influence on MAP, so it affects the flow of blood to all systemic organs. The driving force for venous return is the pressure gradient between the peripheral veins and the right atrium—a gradient of approximately 15 mm Hg. Increases in venous pressure enhance venous return, which in turn increases end-diastolic volume and thus stroke volume, CO, and ultimately MAP.

In this section we examine four factors that affect venous pressure: the skeletal muscle pump, the respiratory pump, blood volume, and venomotor tone.

The Skeletal Muscle Pump

We have seen that peripheral veins contain one-way valves that allow blood to flow forward toward the heart but prevent it from flowing backward. When skeletal muscles contract, they press against veins traveling between them, which raises the venous pressure of blood. This increased pressure forces the more distal valves to close, preventing blood from flowing backward, and forces the more proximal valves to open, allowing blood to flow toward the heart (**Figure 22a**). When the muscles relax and the pressure drops, the reduced pressure allows the distal valves to open, so that blood can flow forward into the previously compressed vein; it also causes the proximal valve to close, thereby preventing blood from flowing away from the heart (Figure 22b). By alternately contracting and

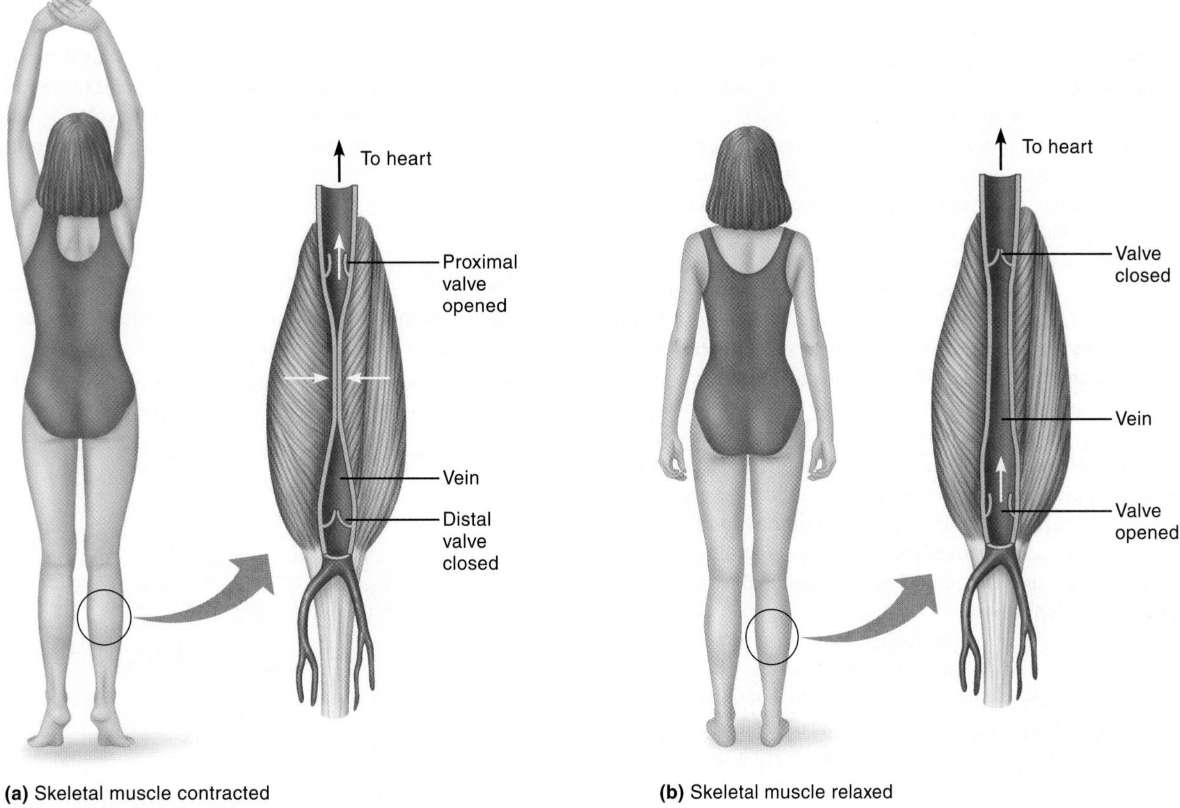

Figure 22 The skeletal muscle pump. (a) When a muscle contracts, it presses against veins, driving blood toward the heart (left). **(b)** When the muscle relaxes, backward flow is prevented by closure of one-way valves in the veins.

relaxing, muscles act as "pumps" that help drive blood toward the heart. For this reason, any exercise that involves rhythmic muscle contractions, such as walking or running, promotes an increase in venous return, increased stroke volume, and increased CO.

The Respiratory Pump

Just as exercise helps to move blood back to the heart through the action of the skeletal muscle pump, so the vigorous respiratory movements that accompany exercise help to move blood back to the heart. We call the effect of respiratory movements on venous return the *respiratory pump*.

The respiratory pump works in the following manner: When you inhale, your diaphragm pulls downward and your rib cage expands, which lowers pressure in the thoracic cavity and raises pressure in the abdominal cavity. This action creates a pressure gradient that promotes the movement of blood from abdominal veins to the central veins located in the thoracic cavity, thereby increasing blood flow toward the heart. When you exhale, thoracic pressure rises and abdominal pressure falls. This creates a pressure gradient that would tend to favor the backward movement of blood from the central veins to the abdominal veins, but such backward flow is prevented by the closure of valves in the abdominal veins. Instead, the rise in thoracic pressure drives the forward movement of blood from the central veins to the heart, thereby promoting increased venous return and CO.

Blood Volume

The body's total blood volume has an important influence on MAP through its effect on venous pressure. The relationship between blood volume and venous pressure is a simple one: An increase in blood volume produces an increase in venous pressure, and a decrease in blood volume produces a decrease in venous pressure. If blood volume falls as a result of bleeding or dehydration or for any other reason, venous pressure falls, as do venous return and end-diastolic volume. The resulting decrease in CO causes a corresponding decrease in MAP. Conversely, a rise in blood volume has the opposite effect, tending to increase MAP.

Because arterial pressure is strongly affected by blood volume, control of blood volume is an important part of blood pressure regulation. A decrease in blood volume activates reflex mechanisms that act to reduce the kidneys' output of water in the urine, which helps the body to conserve water, thereby maintaining blood volume and venous pressure. At the same time, activation of thirst centers in the brain induces a person to drink fluids. Absorption of these fluids by the gastrointestinal tract serves to restore lost blood volume, thereby maintaining adequate venous pressure. These reflex mechanisms that regulate blood volume generally involve the endocrine system and act relatively slowly; hence they regulate blood pressure over the long term (hours or days).

Certain forms of **hypertension** (elevated MAP), for example, are due to failure of the kidneys to excrete adequate amounts of salt and water, which results in the retention of excess fluid in the body. This excess fluid causes blood volume to increase, which in turn raises the MAP.

Health Fact

Under certain circumstances, the high compliance of veins actually works to the detriment of the heart's pumping action by leading to **venous pooling**—accumulation of blood in veins. When a person stands up, for instance, the force of gravity increases the pressure on the blood in the lower veins of the body, causing those veins to expand and enabling the volume of blood within them to increase. This pooling of venous blood is detrimental to the heart's pumping action because it reduces venous return; instead of returning to the heart, much of the blood entering the veins remains there. Thus venous pooling reduces arterial pressure in the same manner as does a reduction in blood volume.

A drop in MAP upon standing (referred to as *orthostatic hypotension*) may cause a person to feel dizzy because of a decrease in blood flow to the brain, but reflex mechanisms quickly compensate for this effect under normal conditions. The presence of aggravating factors, such as dehydration or a failing heart, may cause a person to faint upon standing, however. In such a case, fainting is actually advantageous because once a person has fallen over, blood that had pooled in the veins in the legs moves toward the central veins, just as water flows out of a glass when it is tipped over. This process increases venous return and thus increases CO and MAP, which helps restore blood flow to the brain.

Venomotor Tone

The smooth muscle in the walls of veins contracts or relaxes in response to input from the sympathetic nervous system and certain chemical agents. In terms of neural control, venous smooth muscle contains α adrenergic receptors and activity of the sympathetic nervous system triggers increased contractile activity, with a resulting rise in tension referred to as **venomotor tone**.

An increase in venomotor tone has two effects: (1) constriction of veins raises the venous pressure, which forces blood toward the heart, briefly increasing stroke volume, and (2) increased wall tension reduces venous compliance, which raises venous pressure and produces a sustained increase in stroke volume. Therefore, an increase in venomotor tone promotes a rise in CO and MAP. Changes in venomotor tone are an important component of the reflexes that regulate arterial pressure. When arterial pressure falls as a result of blood loss, for example, activity in sympathetic neurons increases, which acts to raise arterial pressure.

Figure 23 summarizes the factors that influence venous pressure. This diagram illustrates that increases in muscle pump activity, respiratory pump activity, blood volume, and venomotor tone all act to raise venous pressure and, therefore, tend to raise MAP.

Quick Check 5

1. Does an increase in venous pressure tend to increase or decrease cardiac output? Explain.
2. Compare the compliance of arteries and veins, and explain the functional significance of their difference.

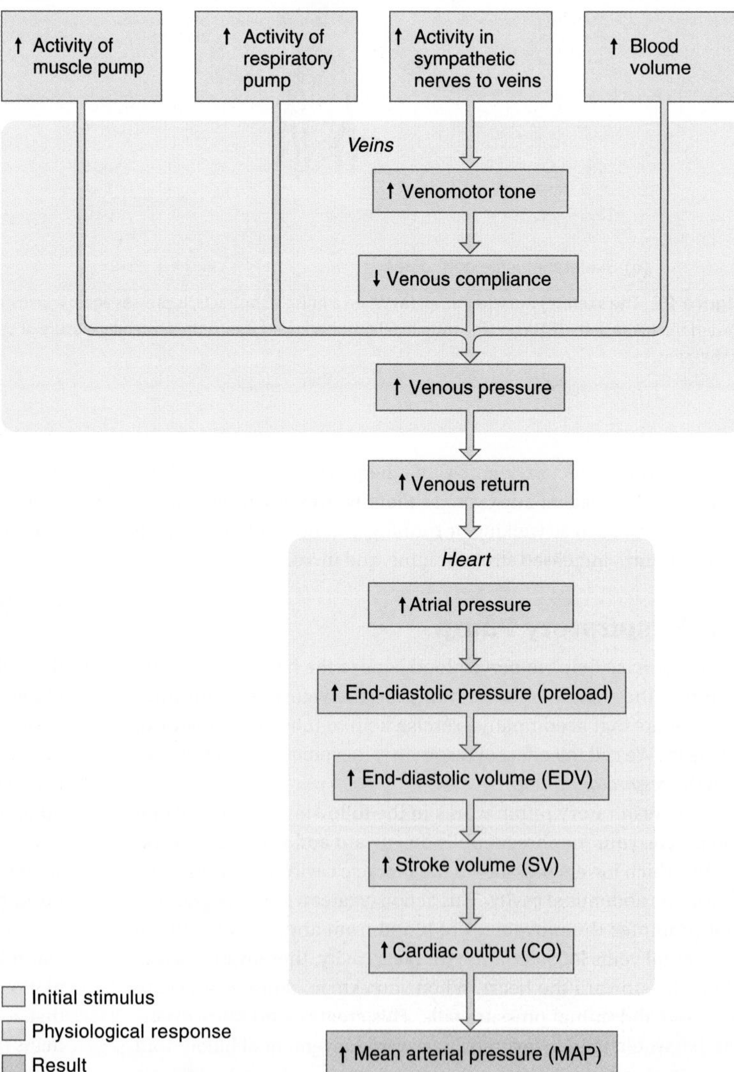

Figure 23 Factors affecting venous pressure and, therefore, mean arterial pressure.

7 The Lymphatic System

Even though approximately 3 liters of fluid leaks out of the capillaries each day, the tissues do not normally swell because this fluid enters the **lymphatic system,** a network of vessels (often referred to as *ducts*) that courses throughout the body (**Figure 24**). Once the fluid gets into the lymphatic system, it is carried through the ducts and eventually returned to the cardiovascular system. The lymphatic system is sort of a silent partner to the cardiovascular system: Although fluid continually moves through it, this fluid (called **lymph**) often goes unnoticed because it is clear, with only a slight yellow tinge to it. If you have ever scraped your knee slightly and noticed clear fluid oozing out of it, this is mostly lymph.

Fluid enters the lymphatic system by way of small, blind-ended ducts called **lymphatic capillaries,** the walls of which have large pores that allow water, small solutes, and even proteins and larger particles to pass through. As a consequence, any fluid that routinely leaks out of "ordinary" blood capillaries can easily move from the interstitium into the lymphatic system. From the lymphatic capillaries, fluid moves through a series of successively larger ducts called **lymphatic veins** until it eventually reaches one of the two ducts that drain into the bloodstream near the jugular veins, the *right lymphatic duct* and the *thoracic duct.* In this manner, filtered fluid is returned to the cardiovascular system. The lymphatic capillaries and veins have valves that allow fluid to enter the capillaries and the lymph to move through the veins in only one direction, toward the right lymphatic and thoracic ducts. The flow of lymph through the lymphatic veins to the ducts is similar to the flow of blood through "ordinary" veins, as previously described.

At certain points in the lymphatic system, lymph passes through structures known as **lymph nodes.** In lymph nodes, any particles that may be present in the lymph, including bacteria or other foreign matter, are filtered out and removed by phagocytic cells called **macrophages.** Lymphocytes and other cells of the immune system also congregate in the lymph nodes, making them important sites for the cellular interactions that are an integral part of the immune response.

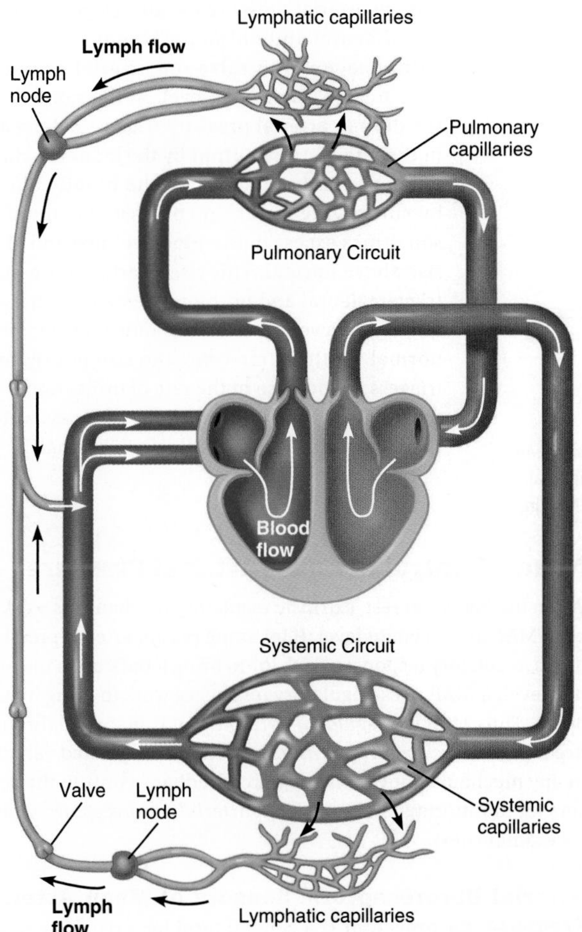

Figure 24 The lymphatic system.

8 Mean Arterial Pressure and Its Regulation

We have seen that two factors influence blood flow to an organ: MAP and the organ's resistance. When we consider that MAP influences blood flow to all organs in the systemic circuit, it becomes clear why matching blood flow to each organ's needs requires that the cardiovascular system maintain adequate MAP: Any decline in MAP tends to compromise blood flow to all the systemic organs. We now further explore the determinants of MAP and its regulation.

Determinants of Mean Arterial Pressure: Heart Rate, Stroke Volume, and Total Peripheral Resistance

One variation of the flow rule is $MAP = CO \times TPR$. CO is determined by heart rate (HR) and stroke volume (SV). Substituting these terms into the expression, we see that

$$MAP = HR \times SV \times TPR$$

Thus MAP is completely determined by three factors: (1) heart rate, (2) stroke volume, and (3) TPR.

The previous expression indicates that MAP should rise following an increase in heart rate or stroke volume (which tend to increase CO) or TPR. We can get an intuitive feeling for why increases in CO or TPR should cause MAP to rise by looking at **Figure 25**, which shows blood flow (wide arrows) into and out of the aorta in different circumstances. TPR is shown schematically as a constriction at the distal end of the aorta, where the blood flows out.

When MAP is steady (Figure 25a), blood flows into the aorta at the same rate as it flows out, such that the volume of blood contained within it does not change. Because the flow into the aorta consists of CO, the flow out of the aorta under these conditions equals CO. (The flows shown in Figure 25 represent *average* blood flows over several cardiac cycles. Within a single cardiac cycle, the flows into and out of the aorta differ from one another as aortic pressure cycles up and down.)

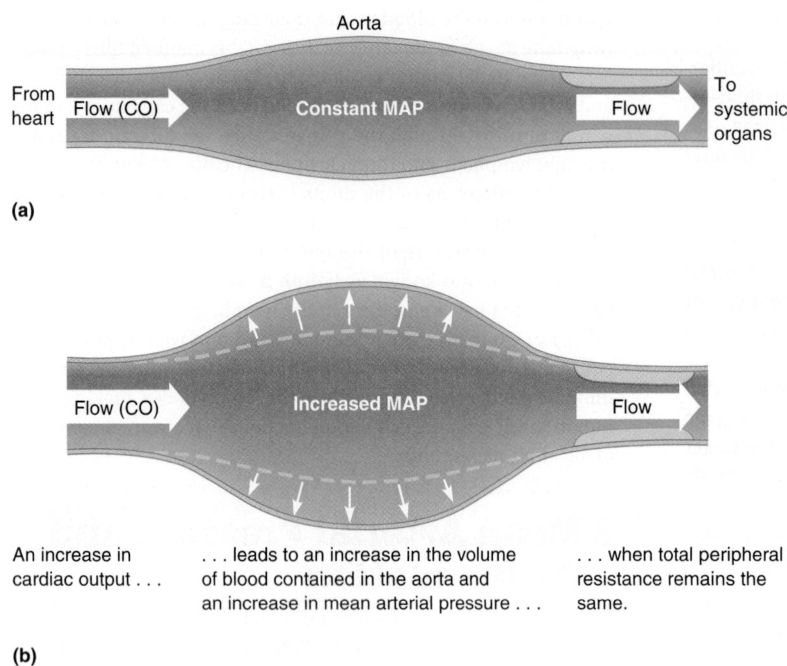

Figure 25 **How increases in cardiac output and total peripheral resistance increase mean arterial pressure. (a)** Constant MAP under conditions of constant cardiac output (CO) and total peripheral resistance (TPR). **(b)** Increased MAP in response to increased CO. **(c)** Increased MAP in response to increased TPR.

Figure 25b shows how MAP would change given a sudden increase in CO (due to an increase in heart rate and/or stroke volume), with TPR remaining constant. Under these conditions, blood flows into the aorta faster than it flows out, so that blood volume within the aorta increases and the vessel expands. This expansion stretches the wall of the aorta, causing it to exert a large inward force on the blood, so that the pressure of the blood rises. Thus an increase in heart rate or stroke volume causes MAP to rise. (If CO remains elevated, the rising aortic pressure causes the flow of blood out of the aorta to increase. Eventually, MAP reaches a steady level, with rates of blood flow into and out of the aorta being equal.)

Figure 25c shows how MAP would change when a sudden increase in TPR occurs while CO remains constant. The increase in TPR reduces the flow of blood out of the aorta, so that blood flows in faster than it flows out, and MAP rises.

We now look at how the body regulates MAP by regulating heart rate, stroke volume, and CO.

Regulation of Mean Arterial Pressure

Control of MAP is accomplished by extrinsic regulatory mechanisms—that is, mechanisms involving the control of organs and tissues by the nervous and endocrine systems. In this section we concentrate on understanding the *short-term* regulation of arterial pressure, which takes place over a time span of seconds to minutes. *Long-term* regulation of arterial pressure, involves control of the blood volume by the kidneys and takes place over periods ranging from minutes to days.

To see how short-term and long-term regulation work together, consider this example: When a person loses blood, the resulting decrease in blood volume causes a fall in venous pressure, which causes a decrease in venous return, end-diastolic volume, and CO, and ultimately a decrease in MAP. As we will see, this drop in arterial pressure usually triggers a number of neural and hormonal responses that act within seconds to raise the arterial pressure back toward its normal level. At the same time, the drop in arterial pressure triggers a decrease in the rate of urine output by the kidneys, which helps the body to maintain the blood volume by conserving water. In contrast, when a person drinks excess fluids, blood volume and MAP rise above normal. This rise in arterial pressure triggers neural and hormonal responses that act quickly to lower the pressure, returning it to near normal. At the same time, the rise in pressure triggers an increase in the rate of urine output by the kidneys, which rids the body of excess water, thereby returning blood volume to normal.

Neural Control of Mean Arterial Pressure

When the body is at rest, extrinsic regulatory mechanisms work to keep MAP at a constant level. If for some reason arterial pressure drops, regulatory responses "kick in" to bring it back up to the normal level; if MAP rises, regulatory responses work to bring it back down. Thus MAP is a *regulated variable* that is managed through *negative feedback control. Sensors* monitor the regulated variable in any mechanism involving negative feedback control; the sensors for monitoring MAP are called *arterial baroreceptors*, which we examine next.

Arterial Baroreceptors: Sensors of Mean Arterial Pressure **Baroreceptor** is a general term for a type of sensory receptor neuron in blood vessels and the heart that responds to changes in pressure within the cardiovascular system. (Recall that a *baro*meter is an instrument that monitors atmospheric pressure.) Arterial baroreceptors are found in two locations: the *aortic arch*, the curved portion of the aorta located close to where it emerges from the heart, and the *carotid sinuses* of the *carotid arteries*, which are located in the neck (**Figure 26**). Thus arterial baroreceptors are also called *sinoaortic baroreceptors*. These baroreceptors are strategically placed because pressure in the aorta affects blood flow to every organ in the systemic circuit, and because pressure in the carotid arteries affects blood flow to the brain, which is exceedingly sensitive to any reduction in its blood supply.

Arterial baroreceptors respond specifically to the stretching that occurs during pressure changes in arteries. The sensory endings of arterial baroreceptors are embedded within the walls of the aorta and carotid sinuses. When arterial pressure (or, more precisely, *distending pressure*) rises, the arteries expand, stretching the walls of the arteries and the sensory endings of the baroreceptors within them, and inducing depolarization. Depolarization triggers action potentials, which are then conducted to the central nervous system by the baroreceptors' axons. Increased pressure induces greater stretch of the sensory endings, which produces greater depolarization and an increase in action potential frequency (**Figure 27**).

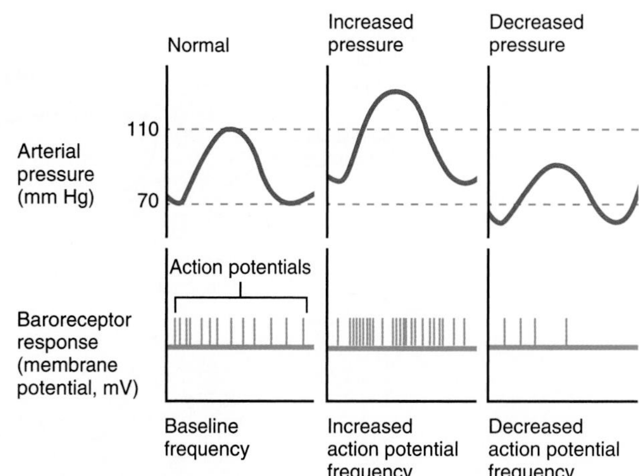

Figure 27 Response of arterial baroreceptors to changes in arterial pressure. The top trace in each panel shows an arterial pressure wave; vertical lines in the lower traces represent action potentials recorded from a carotid sinus baroreceptor.

Q Here, a change in the intensity of a stimulus (arterial pressure) triggers a change in the frequency of action potentials in afferent neurons. What is the general term for this phenomenon?

Arterial baroreceptors are important in the regulation of MAP because they relay pressure information to the central nervous system, which exerts control over cardiovascular function via autonomic neurons projecting to the heart and blood vessels. Baroreceptor input to the nervous system also triggers changes in the secretion of several hormones that target the heart and blood vessels.

Cardiovascular Control Center of the Medulla Oblongata Neural control of MAP is orchestrated primarily by the medulla oblongata. It possesses a diverse set of neural networks encompassing several nuclei, collectively called the *cardiovascular control center*, that regulate different aspects of cardiovascular function (**Figure 28**). The cardiovascular control center is able to assess various indicators of cardiovascular performance (such as arterial pressure) and decide whether this performance is sufficient to meet the body's current needs. If it is not, the cardiovascular control center instructs the cardiovascular system to make appropriate adjustments by sending output to effectors via autonomic nerves.

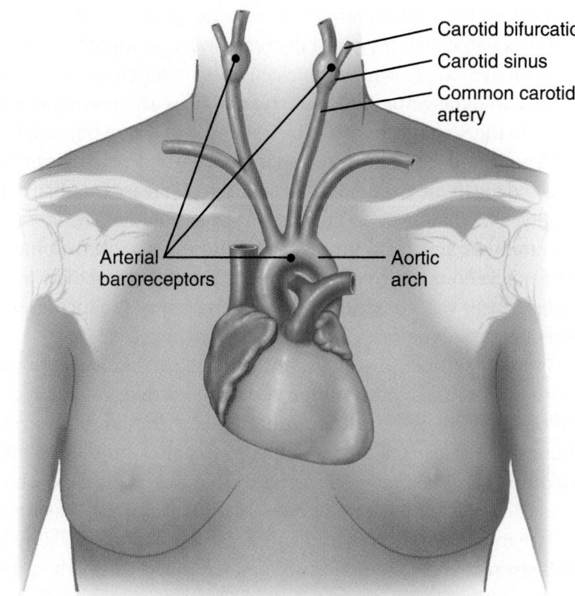

Figure 26 Arterial baroreceptors.

Information from a variety of sensory receptors projects to the cardiovascular control center. Foremost among these are the arterial baroreceptors, which inform the center about current pressures in the aortic arch and carotid sinus. Other receptors include *low-pressure baroreceptors* in the right atrium and large systemic veins (also called *volume receptors*), which monitor venous pressure, and *chemoreceptors* in the brain and carotid arteries, which monitor concentrations of oxygen, carbon dioxide, and hydrogen ions in arterial blood. (The functions of these receptors are discussed shortly.) Proprioceptors in skeletal muscle and joints, which sense body movement and position, and other receptors of various types in internal organs throughout the body also provide input; these receptors are important in the cardiovascular response to exercise.

The cardiovascular control center receives input not only from sensory receptors but also from higher brain areas, including the hypothalamus and cerebral cortex. The hypothalamus is important in orchestrating the *stress response*, a coordinated series of events that help the body cope with stimuli disrupting homeostasis; these reactions include the fight-or-flight response. The hypothalamus also regulates the resistance of blood vessels in the skin in response to changes in body temperature. While these changes in resistance control blood flow through the skin to regulate the rate of heat loss

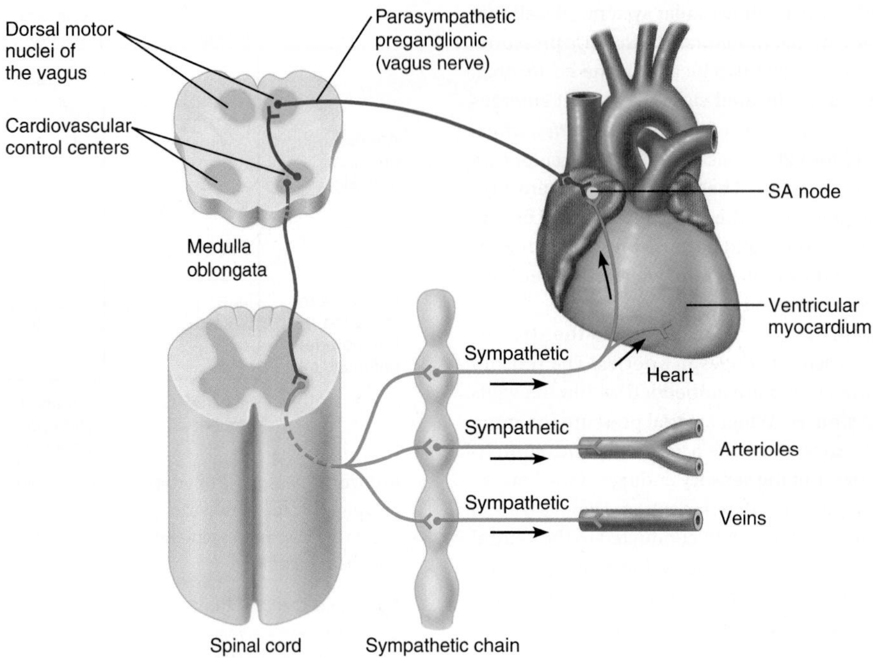

Figure 28 Major neural pathways in the control of cardiovascular function.

Q Does the sympathetic chain belong to the central nervous system or to the peripheral nervous system?

from the body, they also alter MAP. The precise nature of cortical influences on cardiovascular function is not fully understood, but the effects are thought to be wide ranging. The cortex is involved in the cardiovascular changes that occur in response to pain and emotional states (such as the rise in blood pressure that frequently accompanies anxiety) and in exercise. The cortex also exerts continual control over cardiovascular function by modulating the cardiovascular control center's responses to its sensory inputs.

Autonomic Inputs to Cardiovascular Effectors The cardiovascular control center integrates the information received from the various sources described earlier and determines which, if any, adjustments in the cardiovascular system are needed. To make these adjustments, the center communicates with the autonomic nervous system, thereby affecting the levels of activity in sympathetic and parasympathetic nerves to the heart and blood vessels.

Major autonomic innervations of the cardiovascular system include the following (Figure 28): (1) sympathetic and parasympathetic nerves to the sinoatrial node, which control heart rate; (2) sympathetic nerves to the ventricular myocardium, which control ventricular contractility; (3) sympathetic nerves to arterioles and other resistance vessels, which control vascular resistance, and (4) sympathetic nerves to veins, which control venomotor tone. Clearly, the primary neural influence on cardiovascular function is *sympathetic;* parasympathetic input comes into play only at the SA node.

The Baroreceptor Reflex

Most of us have probably felt the effects of baroreceptors in action. We have already seen that blood pressure can fall when a person stands, causing a sensation of dizziness. Within seconds, however, the dizziness goes away because the pressure has risen to nearly normal. This restoration of balance occurs because the fall in arterial pressure is detected by arterial baroreceptors, which trigger an increase in sympathetic activity and a decrease in parasympathetic activity, which brings about increases in heart rate, myocardial contractility, and vascular resistance. This sequence of events is called the **baroreceptor reflex**.

Figure 29 illustrates the baroreceptor reflex in response to a decrease in MAP. The drop in pressure is detected by the arterial baroreceptors and this information is transmitted to the cardiovascular control center, which triggers a decrease in parasympathetic activity and an increase in sympathetic activity to compensate for it. The decrease in parasympathetic activity and increase in sympathetic activity to the SA node increase the frequency of action potentials in the SA node and cause heart rate to increase. An increase in sympathetic activity to the ventricular myocardium increases cardiac contractility, which induces an increase in stroke volume. The increase in sympathetic activity to the veins also increases stroke volume by increasing venous return. An increase in sympathetic activity to arterioles causes vasoconstriction, which increases TPR. Increases in heart rate, stroke volume, and TPR all tend to increase MAP.

Figure 30 illustrates in graphical form the baroreceptor response to a drop in blood volume due to hemorrhage. When blood volume falls, the end-diastolic volume decreases, and the stroke volume and CO also fall. At the same time that CO drops, MAP falls. This drop in arterial pressure triggers the baroreceptor reflex, which drives the pressure back up to near its initial value by causing an increase in heart rate, TPR, and stroke volume. Notice in Figure 30 that heart rate and TPR increase to greater than resting values, but

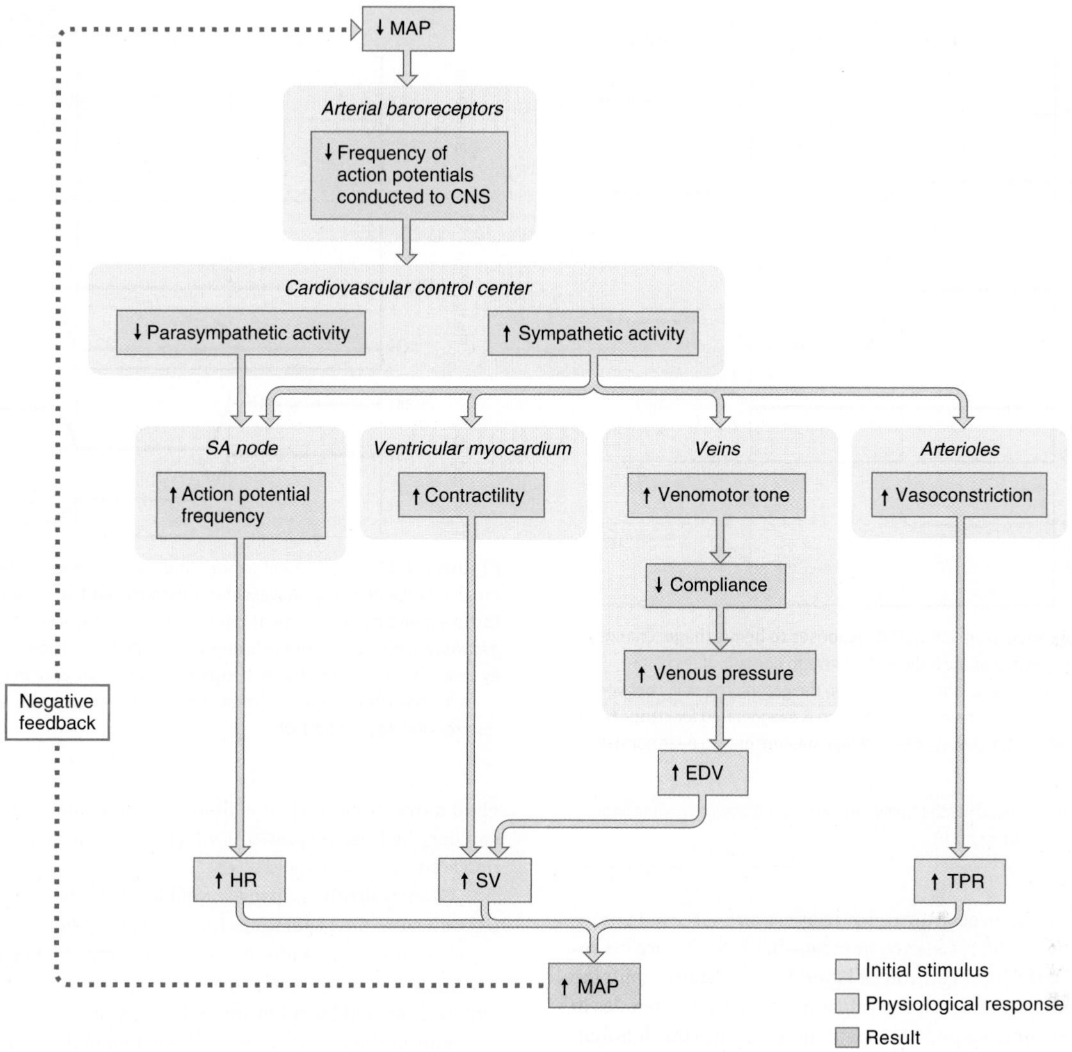

↓ MAP

Arterial baroreceptors

↓ Frequency of action potentials conducted to CNS

Cardiovascular control center

↓ Parasympathetic activity

↑ Sympathetic activity

SA node

↑ Action potential frequency

Ventricular myocardium

↑ Contractility

Veins

↑ Venomotor tone

↓ Compliance

↑ Venous pressure

Arterioles

↑ Vasoconstriction

↑ EDV

Negative feedback

↑ HR

↑ SV

↑ TPR

↑ MAP

Initial stimulus

Physiological response

Result

Figure 29 The events in the baroreceptor reflex in response to a drop in mean arterial pressure.

the stroke volume, which was decreased in the first place, returns to less than normal. The pressure does not return all the way to its original level because if it did, baroreceptors would relay signals to the cardiovascular control center, telling it that the pressure was normal; in that case the compensatory response to blood loss would shut down, and the pressure would drop again. Thus a small error signal is necessary for keeping the response activated and compensating for the blood loss.

Because a drop in arterial pressure triggers a subsequent compensatory rise, the baroreceptor reflex acts by negative feedback to keep pressure constant. The same principle holds when arterial pressure rises. In this case, the change in pressure triggers responses in the opposite direction, producing a decline in pressure.

The action of the baroreceptor reflex raises an interesting question: If baroreceptor reflexes work to keep MAP constant at normal levels, why do some people suffer hypertension? The reason is that hypertension is a chronic condition that develops slowly over long periods of time. The accompanying gradual rise in arterial pressure causes baroreceptors to lose their sensitivity such that they become reset at a new, higher pressure, which effectively becomes "normal."

Under these conditions, the baroreceptors continue to regulate arterial pressure, but they maintain it at a level higher than it would normally be. Once this reset has occurred, baroreceptors cannot correct the problem (see **Clinical Connections: Hypertension**).

When everything is considered, the *immediate* danger posed by low arterial pressure *(hypotension)* is far greater than that posed by hypertension, because a low MAP acts to reduce blood flow to all systemic organs, which can compromise their function and even permanently damage them. This is not to say that hypertension is not dangerous, but hypertension usually takes years to kill, whereas hypotension can kill in minutes. Thus the most important function of the baroreceptor reflex is to counteract potentially dangerous reductions in organ blood flow. However, we are now faced with a puzzle: If a fall in MAP triggers an increase in TPR, which tends to *reduce* organ blood flow, then how is blood flow maintained?

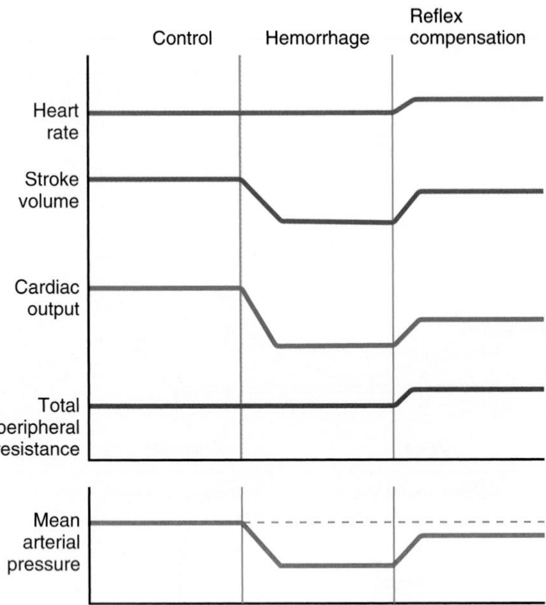

Figure 30 Baroreceptor-mediated responses to hemorrhage. Changes in cardiovascular variables are indicated relative to normal values before hemorrhage occurs (the "control" condition in the graph). The delay between hemorrhage and onset of reflex compensation is exaggerated for clarity. Note that reflex compensation raises mean arterial pressure only to near-normal levels.

Q In this example, which of the following decreases following hemorrhage: preload, afterload, or both?

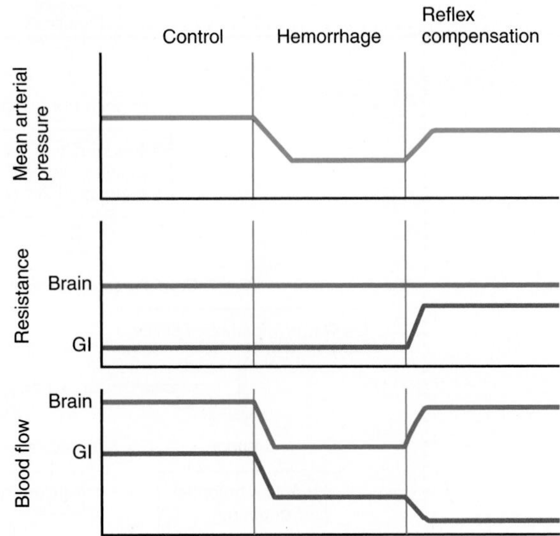

Figure 31 Effects of a hemorrhage-induced baroreceptor reflex on mean arterial pressure, on vascular resistance, and on blood flow in the brain and gastrointestinal tract. Increased vascular resistance in the gastrointestinal tract results in lower-than-normal blood flow to these organs. By contrast, virtually no change occurs in vascular resistance in the brain; as a result, blood flow to this vital organ returns to near normal as the reflex restores MAP to near normal.

When the baroreceptor reflex triggers an increase in sympathetic activity, the resistance in most—but not *all*—organs increases. First, the local controls of blood flow generally dominate extrinsic controls. For example, sympathetic activity to arterioles in active tissue cannot cause vasoconstriction, because the dominating effects of the local metabolites promote vasodilation. Furthermore, resistance in the brain and heart are affected very little by sympathetic influences, so they are not altered significantly by the baroreceptor reflex. When a baroreceptor reflex triggers a compensatory rise in MAP, it acts to maintain blood flow to the heart and brain—but it does so *at the expense of other organs*, whose blood flow actually decreases as a result of vasoconstriction. The end result is that blood flow shifts *away from* certain organs (such as the skin and gastrointestinal system) and *toward* other organs (such as the brain and heart) that are more vital in the short-term survival of the body (**Figure 31**).

Note that when a drop in arterial pressure is due to a loss of blood volume, the baroreceptor reflex is merely a "quick fix." As long as the drop in volume is not so great that the body cannot compensate for it, the baroreceptor reflex will keep arterial pressure high enough to ensure survival until the lost volume can be replaced by drinking fluids (or by blood transfusion, if necessary). Long-term regulation of arterial pressure is accomplished by controlling blood volume, which depends on adjusting the balance between fluid intake and excretion. (Temporary compensation for a decrease in blood volume can also occur through shifts in extracellular fluid: Fluid moves from interstitial fluid to plasma due to the decrease in capillary hydrostatic pressure, which results in a decrease in filtration across capillary walls.)

When individuals lose a lot of blood, the baroreceptor "quick fix" can sometimes get them into a "real fix" because the baroreceptor reflex reduces blood flow to most organs. Eventually these organs will require restoration of normal flow; otherwise, they will be damaged by either the reduced availability of oxygen and nutrients or the accumulation of toxic metabolic by-products. Because local controls also influence arteriolar smooth muscle, when tissues are deprived of an adequate blood supply for a prolonged period, local chemical changes cause vascular smooth muscle to relax. Unless lost blood volume is restored quickly (within one or two hours in the case of severe hemorrhage), these local influences will override the influence of sympathetic vasoconstrictor nerves on vascular smooth muscle, and vascular resistance will begin to decline. As a result, arterial pressure will fall, despite the actions of compensatory mechanisms to raise it. In this scenario, blood flows to the heart and brain decrease. This condition, known as *circulatory shock*, may become irreversible, in which case it progresses inexorably to death.

Hormonal Control of Mean Arterial Pressure

Arterial baroreceptors exert control over cardiovascular function not only via the baroreceptor reflex but also by regulating the secretion of the hormones epinephrine, vasopressin, and angiotensin II, which work hand in hand with the autonomic nervous system to regulate MAP.

CLINICAL CONNECTIONS

Hypertension

Hypertension (high blood pressure) is persistently elevated resting blood pressure greater than 120 mm Hg (systolic) over greater than 80 mm Hg (diastolic). This condition affects at least 60 million people in the United States (1 in 6 adults). Two main forms of hypertension are distinguished: *primary* and *secondary hypertension.*

Primary hypertension (or *essential hypertension*) accounts for 90–95% of all hypertension cases. The precise cause of primary hypertension in an individual cannot be established, but the condition is known to be associated with certain risk factors, including obesity, high cholesterol levels, smoking, and genetic disposition.

In secondary hypertension, the elevated blood pressure occurs secondary to another disease. Secondary hypertension includes *renal hypertension*, associated with kidney disease, and *endocrine hypertension*, associated with inappropriate secretion of a hormone.

In renal hypertension, the primary cause is a disorder of kidney function. Kidney disease may cause failure to excrete normal amounts of salt and water, leading to fluid retention and expansion of blood volume. Sometimes the problem is occlusion of blood flow to a kidney or within a kidney, which can trigger inappropriately high rates of renin release. Abnormally high levels of renin result in overproduction of angiotension II in the plasma, which stimulates vasoconstriction and thereby increases peripheral resistance. Angiotensin II also indirectly stimulates the kidneys to retain salt and water.

In endocrine hypertension, a hormone is inappropriately secreted by an endocrine gland. For example, a tumor of the adrenal medulla (called a *pheochromocytoma*) can oversecrete epinephrine, which stimulates increased cardiac output and total peripheral resistance, both of which elevate blood pressure.

Hypertension is closely associated with another disease, *atherosclerosis,* or hardening of the arteries. In atherosclerosis, a fatty plaque builds up in the walls of arteries, decreasing the elasticity of arterial walls and narrowing the lumen, as shown in the photograph. The decreased lumen increases the resistance of the blood vessel, thereby contributing to hypertension. However, hypertension also damages the walls of the arteries, predisposing them to atherosclerosis. Thus a vicious cycle ensues, in which hypertension and atherosclerosis each facilitate the development of the other.

In addition to producing atherosclerosis, hypertension has several adverse effects on the cardiovascular system. Because elevated arterial blood pressure increases the workload on the heart, it can increase the likelihood of a *myocardial infarction*, or heart attack. It can also lead to heart failure, because it increases afterload and can chronically elevate end-diastolic volume. Hypertension-induced damage to blood vessels may contribute to kidney failure or loss of vision as well.

Treatments for hypertension include diuretics, which promote increased excretion of salt and water by the kidneys, and specific antihypertensive drugs, such as beta-blockers and calcium channel blockers. *Beta-blockers* reduce cardiac output by interfering with the ability of epinephrine and norepinephrine to bind to beta receptors, thereby reducing the stimulatory influence of these agents on the heart. *Calcium channel blockers* reduce the flow of calcium into vascular smooth muscle cells across the plasma membrane, which reduces vasomotor tone and lowers peripheral resistance. Other drugs, called *angiotensin converting enzyme (ACE) inhibitors*, reduce plasma levels of angiotensin II by blocking the enzyme that catalyzes its formation from angiotensin I.

BSIP/Photo Researchers.

Plaque in the wall of an artery.

Critical Thinking Questions

1. What are the differences between primary and secondary hypertension, in terms of their causes and effects?
2. What is the relationship between the development of hypertension and atherosclerosis?
3. Describe the treatments for hypertension, as well as their relative benefits and limitations.

Recall that epinephrine is released in response to sympathetic nerve activity to the adrenal medulla. Low arterial pressure is a stimulus for epinephrine secretion, although evidence suggests that secretion is enhanced only when the drop in pressure is relatively severe. Epinephrine affects both CO and TPR. The effects of epinephrine on cardiac function mirror the actions of sympathetic input: At the SA node, epinephrine increases the action potential frequency of pacemaker cells, which increases heart rate. In the myocardium, epinephrine increases cardiac contractility, which increases stroke volume. In both cases, these effects are brought about as a result of epinephrine binding to receptors in cardiac tissue. The same receptors bind norepinephrine, which explains the similarity between epinephrine's actions and those of the sympathetic nervous system. At the vasculature, epinephrine has mixed effects: It produces vasoconstriction in most vascular beds, but it can cause vasodilation in skeletal and cardiac muscle. Under most circumstances, epinephrine causes an increase in TPR and increases blood pressure. Thus epinephrine tends to increase MAP by increasing heart rate, stroke volume, and TPR.

Vasopressin and angiotensin II both cause vasoconstriction, thereby increasing TPR and MAP. Vasopressin secretion is regulated by a variety of factors, including the level of activity in arterial baroreceptors. When arterial pressure falls, vasopressin release is enhanced, which promotes an increase in MAP. Vasopressin also reduces urine output by the kidneys to maintain plasma volume. Recall that angiotensin II is produced in response to renin secretion from the kidneys. When arterial pressure falls, the release of renin is stimulated both directly by reduced arterial pressure and indirectly through sympathetic activity to the kidneys. As a result of the increase in renin secretion, the plasma concentration of angiotensin I rises, an effect that is followed by an increase in the concentration of angiotensin II. Angiotensin II increases MAP in a number of different ways, including promoting vasoconstriction, reducing urine output by the kidneys, and stimulating thirst.

Extrinsic factors involved in the control of MAP are summarized in Table 4.

Control of Blood Pressure by Low-Pressure Baroreceptors (Volume Receptors)

In addition to the arterial baroreceptors, which monitor systemic arterial pressure, other baroreceptors monitor pressures elsewhere in the cardiovascular system, specifically on the low-pressure side of the circulation. Particularly important are baroreceptors in the walls of large systemic veins and in the walls of the right atrium. These receptors function in the same way as arterial baroreceptors in that they have receptor endings that respond to stretch; because of their locations, however, they monitor pressure on the *venous* side of the systemic circulation and, therefore, act directly to detect changes in *blood volume*. In recognition of the fact that the venous side of the circulation acts as a volume reservoir, low-pressure baroreceptors are frequently referred to as *volume receptors*. Additional baroreceptors located in the pulmonary vasculature act indirectly to monitor systemic venous pressure.

Because arterial pressure is influenced by venous pressure, the functions of low-pressure and arterial baroreceptors are intertwined; levels of activity in these receptors frequently vary in the same direction, because when venous pressure increases or decreases, arterial pressure also tends to increase or decrease, respectively. In addition, low-pressure baroreceptors exert many of the same actions as arterial baroreceptors. When venous pressure falls, for example, low-pressure baroreceptors trigger an increase in both sympathetic nerve activity and vasopressin secretion.

Quick Check 6

1. What is a baroreceptor? Where are arterial baroreceptors located? Where are low-pressure baroreceptors located? What is the baroreceptor reflex?
2. Indicate whether each of the following autonomic nervous activities increases or decreases when arterial pressure falls: sympathetic nervous activity, parasympathetic nervous activity, heart rate, myocardial contractility, vascular resistance (in most tissues), venomotor tone.
3. Name two organs whose vascular resistance is generally unaffected by baroreceptor reflexes. Why is this beneficial?

9 Other Cardiovascular Regulatory Processes

In addition to baroreceptor control of mean arterial blood pressure, other processes control cardiovascular function. Some of these, such as the thermoregulatory responses, alter the pattern of blood flow so that the body can adapt to a particular set of circumstances.

TABLE 4 Factors Involved in Extrinsic Control of Mean Arterial Pressure

Target organ or tissue	Neural or hormonal factor	Factor's effect on target	Influence on mean arterial pressure
Heart			
Sinoatrial node	Sympathetic nerves	↑HR	↑MAP
	Parasympathetic nerves	↓HR	↓MAP
	Epinephrine	↑HR	↑MAP
Ventricular myocardium	Sympathetic nerves	↑Contractility (↑SV)	↑MAP
	Epinephrine	↑Contractility (↑SV)	↑MAP
Arteriolar smooth muscle (most tissues)	Sympathetic nerves	Vasoconstriction (↑TPR)	↑MAP
	Epinephrine	Vasoconstriction or vasodilation, depending on concentration and location	Variable
	Vasopressin	Vasoconstriction (↑TPR)	↑MAP
	Angiotensin II	Vasoconstriction (↑TPR)	↑MAP
Venous smooth muscle	Sympathetic nerves	↑Venomotor tone	↑MAP
	Epinephrine	↑Venomotor tone	↑MAP

Other processes, such as respiratory sinus arrhythymia, have no known functional significance. What follows is a sampling of these other regulatory processes.

Respiratory Sinus Arrhythmia

Respiratory sinus arrhythmia is a rhythmic variation in heart rate in which inspiration is accompanied by increases in sympathetic activity and heart rate, whereas expiration is accompanied by increases in parasympathetic activity and a decrease in heart rate. Although not completely understood, the changes in sympathetic and parasympathetic activity influence the sinoatrial node. This type of arrhythmia is seen in only some people and appears to be more pronounced in children.

Respiratory sinus arrhythmia has at least one known practical consequence: Competitive archers and target shooters use it to aid their accuracy. If a person takes a deep breath and then breathes out while aiming, respiratory sinus arrhythmia slows the heart rate, enabling the shooter to fire between heartbeats and thus avoid even a slight movement of the body that could throw aim off.

Chemoreceptor Reflexes

Control of respiration is governed by *chemoreceptors* located in the carotid sinuses and the brain. These chemoreceptors are neurons specialized to monitor concentrations of carbon dioxide (and oxygen when it falls to extremely low levels) in the blood. When carbon dioxide rises, chemoreceptors stimulate breathing, so that more carbon dioxide is eliminated. These receptors also influence cardiovascular function, but the directions and sizes of these responses are difficult to predict, because some cardiovascular effects are primary (that is, they are due to stimulation of chemoreceptors themselves), whereas others are secondary (that is, they are due to changes in respiration triggered by stimulation of chemoreceptors).

One of the primary cardiovascular effects of chemoreceptor stimulation when arterial carbon dioxide levels rise is to trigger a decrease in heart rate and an increase in peripheral resistance. These responses seem counterproductive, because when the concentration of carbon dioxide in the blood rises, blood flow to tissues should be *increased* to maintain the same rate of carbon dioxide removal. However, when viewed from another perspective, these responses are entirely appropriate. Because of the decrease in heart rate, oxygen consumption by the heart muscle itself is diminished, which tends to conserve oxygen. Although the fall in heart rate also tends to reduce MAP, the increase in peripheral resistance offsets this effect. As a result, blood flow to the brain is maintained at normal or near-normal levels.

The chemoreceptor-mediated increase in peripheral resistance may also be adaptive in another way: When arterial carbon dioxide levels rise, the concentration of carbon dioxide rises in tissues all over the body. Given that vascular smooth muscle is sensitive to local carbon dioxide concentrations, such a rise in arterial carbon dioxide should trigger vasodilation in many tissues, leading to a potentially dangerous drop in peripheral resistance and MAP. By stimulating an increase in peripheral resistance, chemoreceptor reflexes work to protect against this possibility.

Thermoregulatory Responses

The ability to control heat loss through the skin is an essential component of the body's ability to regulate its own temperature. The body's response to changes in temperature is mediated by a *thermoregulatory center* in the hypothalamus that receives input from *thermoreceptors*, warm- or cold-sensitive neurons found at various locations throughout the body. The thermoregulatory center receives from these thermoreceptors information concerning skin and core temperatures.

Under normal conditions, a significant level of activity occurs in the sympathetic nerves that project to blood vessels in the skin. An increase in body temperature decreases the level of sympathetic activity in nerves supplying the skin, which induces relaxation of vascular smooth muscle. As a result, blood vessels dilate, the skin's vascular resistance decreases, and blood flow to skin increases. In addition, under conditions of heat stress, sweat glands produce bradykinin, whose vasodilatory effects tend to decrease vascular resistance in the skin. As a result of the increase in blood flow, the rate of heat loss through the skin increases, which helps to counteract the rise in body temperature. A decrease in body temperature has the opposite effect: Increased activity in the skin's sympathetic nerves stimulates contraction of smooth muscle in the blood vessels, which increases the resistance of blood vessels in the skin and decreases the blood flow through the skin. As a consequence, blood is diverted away from the skin and toward the deeper structures of the body, so that less heat is lost through the skin.

Because the influence of body temperature on skin blood vessels takes precedence over other reflex controls in most circumstances, changes in skin resistance can sometimes be dangerously maladaptive. For instance, when a person overexerts himself or herself on a very hot day, blood volume decreases due to excessive sweating, thereby decreasing blood pressure. The baroreceptor reflex triggers constriction of blood vessels throughout the body, including those in the skin. If that person then becomes overheated, the thermoregulatory center takes control, inducing dilation of skin blood vessels to increase heat loss. The resulting decrease in vascular resistance in skin vessels tends to cause a decrease in TPR. As a consequence, arterial pressure may fall, counteracting the baroreceptor reflex, which should be acting to raise TPR so as to maintain arterial pressure.

Cardio vascular function fluctuates considerably based on the demands placed on specific tissues. Nowhere is this more apparent than during exercise (see **Understanding Exercise: Cardiovascular Responses to Light Exercise**).

Quick Check 7

1. What is respiratory sinus arrhythmia?
2. What is a chemoreceptor? Which cardiovascular changes are triggered when chemoreceptors detect a drop in arterial oxygen levels?
3. When body temperature rises, does vascular resistance in the skin increase or decrease? How does this change help the body regulate its temperature?

UNDERSTANDING EXERCISE

Cardiovascular Responses to Light Exercise

Light exercise is accompanied by profound alterations in cardiovascular function. The graph shows that jogging increases cardiac output from the resting value of 5 liters per minute to more than 11 liters per minute; in highly trained athletes, cardiac output can reach 35 liters per minute. Heart rate also rises from the resting average of 72 beats per minute to about 135 beats per minute. Exercise causes dramatic changes in blood flow as well: Flow to skeletal muscles (and to cardiac muscle and skin as well) increases due to vasodilation in the vessels providing blood to these regions, while flow to the liver and the gastrointestinal tract decreases due to vasoconstriction in the vessels to these organs. Clearly, these changes are beneficial in several ways: (1) Delivery of oxygen and nutrients to cardiac muscle and active skeletal muscles is increased, which is appropriate given the increase in metabolic activity in these organs; (2) oxygen and nutrients are conserved by curtailing delivery to tissues for which the need for nutrients is not as acute; and (3) increased blood flow to the skin aids the body in getting rid of excess heat generated during exercise.

In exercise, the drop in vascular resistance in skin and muscle is not quite compensated for by the increase in the resistance of other organs, so total peripheral resistance drops. Blood pressure tends to rise slightly, however, due to the increase in cardiac output. Note in the figure that stroke volume increases markedly despite the fact that only a modest increase in end-diastolic volume occurs (in light exercise). This indicates that the increase in stroke volume is due not to the Starling effect but rather to an increase in ventricular contractility resulting from an increase in sympathetic nervous activity and increases in the levels of circulating epinephrine.

What triggers these responses to exercise? In the central nervous system, cortical and limbic regions of the brain exert a direct influence on the output of sympathetic and parasympathetic neurons, resulting in increased sympathetic activity and decreased parasympathetic activity to the heart that account for the increase in heart rate and ventricular contractility. In addition, sympathetic activity in the digestive system and other organs *increases* (resulting in vasoconstriction), while sympathetic activity to the skin *decreases* (which promotes vasodilation). This change in sympathetic input to the skin is a thermoregulatory response triggered by the rise in body temperature that accompanies exercise. Vasodilation in the skin is also promoted by bradykinin released from sweat glands, whose activity increases in exercise. The changes in sympathetic activity are facilitated by the cardiovascular control center, which receives input from various types of receptors in muscles, including chemoreceptors responsive to local chemical factors and mechanoreceptors and proprioceptors sensitive to muscular physical activity. The increase in blood flow to skeletal and cardiac muscles is largely a result of the action of local chemical factors on vascular smooth muscle causing vasodilation.

Other mechanisms also promote increased venous return in exercise, thereby facilitating the rise in cardiac output. One such mechanism is the skeletal muscle pump; another is the respiratory pump, whose activity increases in exercise because breathing becomes deeper and faster. A third mechanism is an increase in venomotor tone, which results from an increase in sympathetic activity to the veins.

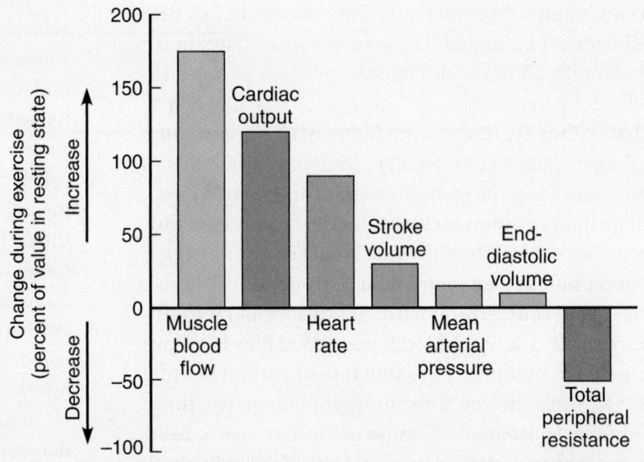

SYSTEMS INTEGRATION

In this chapter, we learned about the blood vessels, including how the body regulates blood flow to the various organs. Clearly, all organs are dependent on blood flow, but they do not always need the same amount of blood at all times. Reflecting this fact, blood flow is regulated through vasoconstriction or vasodilation of arterioles so as to supply blood to the tissue that needs it most at any given time. Not all arterioles are alike, however. Blood vessels to the brain lack receptors for extrinsic controls—as a consequence, the brain is never compromised in the body's attempts to maintain MAP. Blood vessels to the skin have a special innervation whereby sympathetic activity can cause vasoconstriction through unknown mechanisms. The purpose of this vasoconstriction is to conserve heat when body temperature decreases. Blood flow to the genitalia is under both sympathetic and parasympathetic control, with parasympathetic activity causing the vasodilation that is necessary to produce and maintain an erection.

In parallel flow of blood, each organ receives oxygenated blood. However, some exceptions to this rule exist. We have already encountered one such exception—the portal vein that connects capillary beds in the hypothalamus to capillary beds in the anterior pituitary. This arrangement allows for a large concentration of hypothalamic tropic hormones to affect the anterior pituitary. Another portal vein links capillary beds of the gastrointestinal tract to capillary beds of the liver. The absorbed molecules transported to the liver are processed before they enter the general circulation. Yet another exception is the connection of two capillary beds in the kidneys not by a portal vein, but rather by a second arteriole. Thus blood flows from (afferent) arterioles to (glomerular) capillaries to (efferent) arterioles to (peritubular) capillaries. This arrangement allows for finer control over the filtering of blood by the kidneys as well as a pathway for exchange between blood and kidneys after filtration has taken place.

CHAPTER REVIEW

SUMMARY

1 Physical Laws Governing Blood Flow and Blood Pressure

- The flow of blood through any vessel or network of vessels depends on the pressure gradient (ΔP) and the resistance (R) of the vessel or network: flow $= \Delta P/R$.
- The overall pressure gradient driving flow through the systemic circuit is the difference between MAP and CVP. Because CVP is virtually identical to MAP, the ΔP driving blood flow for the systemic circuit is equal to MAP.
- The main influence on vascular resistance is vessel radius, although vessel length and blood viscosity also play small parts in determining resistance.
- The combined resistances of all blood vessels in the systemic circuit is the total peripheral resistance (TPR).
- The flow rule for the systemic circuit can be written as CO $=$ MAP/TPR.

iP Cardiovascular, Factors That Affect Blood Pressure

iP Measuring Blood Pressure

2 Overview of the Vasculature

- All blood vessels possess a lumen that is lined by a layer of endothelial cells.
- The walls of all blood vessels, except for the capillaries, contain varying amounts of smooth muscle and connective tissue.
- The fibrous connective tissue within the blood vessels contains collagen; the elastic connective tissue contains elastin.

iP Cardiovascular, Anatomy Review: Blood Vessel Structure and Function

3 Arteries

- Arteries, which conduct blood away from the heart and toward the body's tissues, have thick walls that enable them to withstand the high pressure of blood within them, and have relatively low compliance.
- Arterial walls are elastic, a property that enables them to expand during systole and then recoil inward during diastole.
- Because of this elastic recoil, arteries function as a pressure reservoir that maintains blood flow throughout the cardiac cycle.
- Arterial blood pressure varies with the cardiac cycle. Systolic pressure is the maximum pressure that occurs during systole; diastolic pressure is the minimum pressure that occurs during diastole.
- MAP is the average arterial pressure during the cardiac cycle.

4 Arterioles

- Arterioles are the passageway for blood to enter the capillaries; they contain relatively large amounts of smooth muscle, which enables them to expand or contract, thereby regulating blood flow through capillary beds.
- Arterioles are important in regulating the distribution of CO to the organs and in controlling MAP.
- Regulation of the distribution of blood flow among the various systemic organs is achieved through intrinsic control of organ vascular resistance.
- Blood flow through any given systemic organ is determined by MAP and that organ's vascular resistance: Organ blood flow $=$ MAP/organ resistance.
- The resistance of an organ or tissue can change in response to variations in the metabolic activity of that organ or tissue because arteriolar smooth muscle is sensitive to local concentrations of chemicals produced or consumed in metabolism, including oxygen and carbon dioxide.
- Chemical changes associated with increased metabolic activity lead to vasodilation, decreased resistance, and increased blood flow (active hyperemia).
- The resistance of an organ can also change in response to local variations in blood flow.
- If blood flow becomes insufficient to meet metabolic demands (ischemia), local mechanisms induce vasodilation, with a resultant increase in blood flow (reactive hyperemia).
- In those tissues in which vascular smooth muscle is responsive to stretch, an increase in perfusion pressure causes arterioles to stretch, which stimulates vasoconstriction and a reduction in blood flow; this response is called a myogenic response.
- Extrinsic controls of arteriole radius (and therefore total peripheral resistance) regulate MAP: MAP $=$ CO $\times$ TPR.
- Extrinsic factors include the autonomic nervous system and hormones (epinephrine, vasopressin, and angiotensin II).

5 Capillaries and Venules

- Capillaries have the thinnest walls of all the blood vessels and are highly permeable to water and small solutes.
- The primary function of the capillaries is to permit exchange of materials between the blood and the tissues; they are classified as continuous, fenestrated, or discontinuous, depending on their degree of "leakiness."
- The movement of fluid across capillary walls is driven by the net filtration pressure, which depends on the Starling forces (capillary hydrostatic pressures, interstitial fluid hydrostatic pressures, capillary osmotic pressure, interstitial fluid osmotic pressure).
- Most of the fluid that is filtered from capillaries is returned to the cardiovascular system by absorption.
- Excess filtrate is returned to the cardiovascular system by the lymphatic system.
- Venules are thin-walled vessels that also participate in material exchange.

MAP iP Capillary Pressures and Capillary Exchange

6 Veins

- Veins are large thin-walled vessels.
- Most veins possess valves that permit blood to flow toward the heart but not away from it.
- Veins have a high compliance and function as volume reservoirs.
- Venous pressure influences arterial pressure because it affects venous return, end-diastolic volume, stroke volume, and CO.
- As venous pressure rises or falls, CO and MAP also tend to rise or fall, respectively.

- Factors affecting venous pressure include activity of the skeletal muscle pump, activity of the respiratory pump, blood volume, and venomotor tone (which is regulated by sympathetic input to the veins).

7 The Lymphatic System

- The lymphatic system is an open system of vessels starting with capillaries in the periphery.
- The capillaries pick up excess filtrate to form lymph.
- The lymph travels through a system of lymphatic veins to the thoracic duct, which returns the lymph to the blood in the right atrium.

MAP iP Interactive Physiology: Lymphatic Organs

8 Mean Arterial Pressure and Its Regulation

- To supply the organs and tissues with adequate blood flow, the driving force for flow (MAP) must be maintained.
- MAP is controlled by both short-term and long-term extrinsic regulatory mechanisms.
- Short-term regulation is achieved through neural and hormonal control of cardiovascular function; long-term regulation is achieved through control of blood volume, which involves the kidney.
- Short-term regulation includes the baroreceptor reflex, a negative feedback system whereby baroreceptors detect changes in MAP and relay this information to the cardiovascular control center, which then acts on the autonomic nervous system to exert appropriate control over cardiovascular function.
- Autonomic control of MAP is accomplished through (1) sympathetic and parasympathetic input to the SA node, which controls heart rate; (2) sympathetic input to the myocardium, which controls ventricular contractility and stroke volume; and (3) sympathetic input to arteriolar smooth muscle in most tissues, which regulates total peripheral resistance.

iP Cardiovascular, Factors That Affect Blood Pressure

iP Cardiovascular, Blood Pressure Regulation

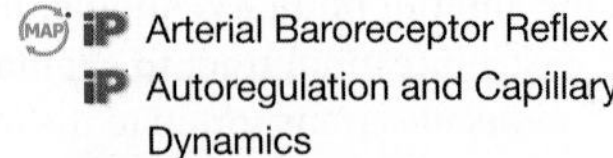

MAP iP Arterial Baroreceptor Reflex

iP Autoregulation and Capillary Dynamics

9 Other Cardiovascular Regulatory Processes

- The cardiovascular system is subject to a variety of regulatory processes that function in specific situations.
- Cardiovascular function is influenced by activity in arterial chemoreceptors, which monitor concentrations of oxygen and carbon dioxide in arterial blood.
- Regulation of blood flow to the skin, which is controlled by sympathetic nerves to skin blood vessels, is important in body temperature regulation.
- Cardiovascular responses to exercise are largely achieved through changes in the activity of autonomic nerves to the heart and blood vessels—changes that are orchestrated by cortical and limbic brain regions.
- Blood flow to the heart and skeletal muscle is also regulated by local factors operating within these tissues.

Check your understanding at MasteringA&P and apply your new knowledge to real-life case studies to ensure that you've mastered the content!

EXERCISES

Multiple-Choice Questions

1. Total peripheral resistance is
   a) The combined resistance of all organs in the body.
   b) The resistance of capillaries located in distal body parts.
   c) The combined resistance of all organs in the systemic circuit.
   d) The combined resistance of all the blood vessels within an organ or tissue.
   e) The resistance to blood flow through the heart.

2. Venous pressure increases
   a) When blood volume decreases.
   b) As a result of venous pooling.
   c) As a result of an increase in venomotor tone.
   d) When a person stands up.
   e) All of the above.

3. If total peripheral resistance is constant, an increase in MAP could be the result of
   a) Vasoconstriction in the systemic circuit.
   b) An increase in sympathetic input to the ventricular myocardium.
   c) A decrease in blood volume.
   d) Both a and b.
   e) None of the above.

4. Which of the following tends to promote edema in systemic tissues?
   a) A decrease in the concentration of plasma proteins
   b) An increase in pressure in the vena cava
   c) An increase in arterial pressure
   d) Leakage of proteins from capillaries into the interstitial fluid
   e) All of the above

5. Which of the following tends to cause a decrease in ventricular end-diastolic volume?
   a) An increase in venous pressure
   b) An increase in skeletal muscle pump activity
   c) A decrease in filling time
   d) An increase in blood volume
   e) An increase in venomotor tone

6. Knowing stroke volume, heart rate, and mean arterial pressure provides sufficient information to determine
   a) Total peripheral resistance.
   b) Cardiac output.
   c) Combined blood flows to all systemic organs.
   d) Both a and b.
   e) All of the above.

7. Which of the following blood vessels possess valves that prevent blood from flowing backward?
   a) Arteries
   b) Arterioles
   c) Capillaries
   d) Venules
   e) Veins

8. Exchange of oxygen and carbon dioxide between blood and respiring tissues occurs across the walls of
   a) Arteries.
   b) Arterioles.
   c) Capillaries.
   d) Veins.
   e) All of the above.

9. Where is the greatest proportion of total blood volume at rest?
   a) Heart
   b) Arteries
   c) Arterioles
   d) Capillaries
   e) Veins

10. Which of the following tends to cause a decrease in mean arterial pressure?
   a) A drop in total peripheral resistance
   b) An increase in stroke volume of the left ventricle

c) An increase in heart rate
d) An increase in venous return
e) An increase in sympathetic activity

11. Lymphatic capillaries differ from blood capillaries in that
   a) Lymphatic capillaries have a lower permeability to water.
   b) Lymphatic capillaries have a lower permeability to small solutes.
   c) Lymphatic capillaries are blind ended.
   d) Lymphatic capillaries are not connected to any other vessels.
   e) All of the above

12. Which of the following tends to cause increased venous pooling?
   a) A decrease in venomotor tone
   b) A decrease in the osmotic pressure of plasma proteins
   c) A decrease in mean arterial pressure
   d) Exercise
   e) Dehydration

13. Moment-to-moment changes in total peripheral resistance are normally due to changes in
   a) The lengths of blood vessels in the systemic circuit.
   b) The radius of certain blood vessels in the systemic circuit.
   c) The viscosity of blood.
   d) The mean arterial pressure.
   e) The cardiac output.

14. Which of the following does *not* tend to increase as a result of the action of the skeletal muscle pump?
   a) Venous pooling
   b) Venous return
   c) End-diastolic volume
   d) Stroke volume
   e) Cardiac output

15. If arterial pressure is elevated, baroreceptor signals trigger which of the following responses?
   a) A rise in vasopressin secretion
   b) A fall in plasma angiotensin II levels
   c) Increased activity in sympathetic vasoconstrictor nerves
   d) An increase in epinephrine secretion
   e) Increased total peripheral resistance

16. The baroreceptor reflex
   a) Is an example of intrinsic control of vascular resistance.
   b) Serves to maintain blood flows to all organs at nearly constant levels.
   c) Serves to maintain mean arterial pressure at a nearly constant level.
   d) Triggers a rise in arterial pressure following a sudden increase in cardiac output.
   e) All of the above

17. During exercise, carbon dioxide produced by muscle cells causes vasodilation in skeletal muscle. This is an example of
   a) Active hyperemia.
   b) Reactive hyperemia.
   c) Flow autoregulation.
   d) Extrinsic control of vascular resistance.
   e) Starling forces.

Objective Questions

18. An increase in total peripheral resistance tends to lower mean arterial pressure. (true/false)
19. Total peripheral resistance (increases/decreases) when general vasodilation of smaller arteries and arterioles occurs in systemic tissues.
20. Changes in blood flow distribution among organs are brought about by changes in mean arterial pressure. (true/false)
21. Veins have a (larger/smaller) compliance than arteries.
22. Blood flow to an organ is determined by both its vascular resistance and the ________.
23. (An increase/A decrease) in the radius of a blood vessel raises its resistance.
24. Over the course of a day, capillary filtration and absorption occur at the same rate, so that the net volume of fluid filtered from capillaries is zero. (true/false)
25. An increase in the hydrostatic pressure of blood in capillaries tends to increase the rate at which fluid is filtered across capillary walls. (true/false)
26. A drop in arterial blood pressure triggers (an increase/a decrease) in sympathetic nervous activity.
27. When venous baroreceptors detect a fall in pressure, arterial baroreceptors tend to detect a change in the opposite direction. (true/false)
28. An increase in blood flow in response to a rise in metabolic activity is referred to as active hyperemia. (true/false)

Essay Questions

29. Using your understanding of principles relating blood flow, pressure, and resistance, explain why mean arterial pressure depends only on heart rate, stroke volume, and total peripheral resistance.
30. Discuss the various factors that affect venous return to the heart.
31. Explain how blood flow to an organ can change even if mean arterial pressure does not.
32. Explain the concept of flow autoregulation, including how it differs from other forms of intrinsic control of vascular resistance. Explain how both the myogenic response and metabolic chemical factors can produce the phenomenon of flow autoregulation. Use the example of a compensatory decrease in flow following an increase in perfusion pressure.
33. Explain how vascular resistance is influenced by the actions of α (alpha) and β (beta) adrenergic receptors. On the basis of differences between these receptors, explain how it is possible for sympathetic nerve activity and circulating epinephrine to affect vascular resistance in opposite ways.
34. Describe the changes in cardiovascular function that occur in response to exercise. Explain how these changes help to adapt the body to the particular stresses that occur in exercise.

Critical Thinking

35. After working in a field for a whole day without drinking fluids, a man feels lightheaded when he stands up. Using your understanding of the factors influencing mean arterial pressure, explain why.
36. When the left side of the heart begins to fail, blood pressure tends to rise in the pulmonary veins and vessels upstream in the pulmonary circuit. Using your understanding of the factors affecting capillary filtration, explain why edema of the lungs is common in left heart failure.
37. Hypertension is dangerous because elevated blood pressure stresses the heart and blood vessels beyond their structural integrity. What are some potential drug types to treat hypertension?
38. Heat or cold is often applied to an injury. Explain the theory behind applying heat or cold and identify when each should be applied to an injury.

Photo credits for icons in feature boxes: Focus on Diabetes: Tanyae/Dreamstime; Clinical Connections: Iofoto/Dreamstime; Understanding Exercise: Pixattitude/Dreamstime; Discovery: Yuri Arcurs/Cutcaster.

Glossary

absorption movement of a substance from the external environment to the internal environment by transport across an epithelium; in the cardiovascular system, the bulk flow of fluid from the interstitial space into a capillary

active hyperemia (hy-per-ee-me-ah) the increase in blood flow occurring in response to an increase in metabolic activity

arterial blood pressure pressure exerted by the force of blood acting on the walls of arteries

arteriolar tone partial vasoconstriction of arterioles due to the inherent contractile activity of single-unit smooth muscle in arteriole walls

baroreceptor reflex negative feedback loop for regulating blood pressure

central venous pressure (CVP) the pressure in the large veins in the thoracic cavity that lead to the heart

edema swelling of tissue caused by bulk flow of fluid from the capillaries into the interstitial fluid

filtration the movement of fluid across capillary walls from plasma to the interstitium

hypertension higher than normal arterial blood pressure

lymph fluid that flows through the lymphatic system

lymph nodes peripheral lymphoid tissue

lymphatic capillaries open-ended vessels of the lymphatic system that absorb excess filtrate from the cardiovascular system

lymphatic system a network of vessels, or ducts, that courses throughout the body and contains lymphatic fluid

lymphatic veins ducts through which lymph flows toward the heart

macrophages phagocytic cells that filter and remove particles in the lymph and body tissues, including bacteria or other foreign matter

microcirculation vasculature that can be seen only with a microscope; includes arterioles, metarterioles, capillaries, and venules

myogenic response the inherent ability of smooth muscle to contract when stretched

net filtration pressure the sum of Starling forces that determines the direction of fluid movement across capillary walls

precapillary sphincter smooth muscle surrounding capillaries; regulates blood flow through capillary beds

reactive hyperemia a local increase in blood flow that occurs following termination of an occlusion of blood flow to the same area

Starling forces hydrostatic and osmotic pressures that exist across capillary walls and determine the direction of fluid movement

total peripheral resistance (TPR) in the systemic circuit, the combined resistance of all the blood vessels

vasoconstriction decrease in the radius of a blood vessel

vasodilation increase in the radius of a blood vessel

venomotor tone the degree of tension exerted by smooth muscle in the walls of veins; venous compliance decreases as venomotor tone increases

Answers

Figure Question Answers

Fig 4	10 mL/min; 5 mL/min
Fig 11	They are the same
Fig 18	Passive
Fig 19	Absorption
Fig 27	Frequency coding
Fig 28	The peripheral nervous system
Fig 30	Both

Apply Your Knowledge Answers

Ohm's law states that $I = E/R$, where I = current (charge flow), E = potential difference, and R = resistance. The flow rule states that $F = P/R$, where F = flow, P = pressure gradient, and R = resistance. For both current and fluid, flow is greater when there is a higher gradient (charge or pressure) and lower when there is a greater resistance. The types of gradient and the resistance vary, but the principles are the same.

The net filtration pressure = (32 mm Hg + 1 mm Hg) − (25 mm Hg + 1 mm Hg) = 7 mm Hg. Because of the increased filtration pressure (compared to normal), more fluid is filtered out of the blood into the interstitial fluid, causing edema in Jack's feet.

Answers to End-of-Chapter Multiple Choice and Objective Questions

Multiple Choice Questions

1. c 2. c 3. b 4. e 5. c 6. e 7. e 8. c 9. e 10. a 11. c 12. a 13. b 14. a 15. b 16. c 17. a

Objective Questions

18. false
19. decreases
20. false
21. larger
22. mean arterial pressure
23. decrease
24. false
25. true
26. increase
27. false
28. true

The Cardiovascular System: Blood

The Cardiovascular System: Blood

Science Source/Photo Researchers, Inc.

Colorized scanning electron micrograph of blood cells.

Anyone who has ever misjudged a school hallway corner and turned too soon has wound up with a bruise on the outer arm or hip. A bruise, or *contusion,* is caused by damage to blood vessels, which allows blood to escape into the tissues. The more blood that leaks into the tissues, the bigger the bruise; the closer the blood vessel damage is to the skin surface, the more colorful the bruise. The color of a bruise is due to the presence of *hemoglobin,* a protein in red blood cells that gives blood its red color. The metabolism of hemoglobin into bile pigments is responsible for the changing colors of a bruise over time.

Within the first couple of days, the site of injury appears red, with possible swelling and tenderness. As the bruise forms, the area takes on a reddish purple or blue color. As healing occurs, the color of the bruise changes to a greenish color and lastly to a yellowish color before the area returns to its normal color. These changes in color correspond to the color of the breakdown products of hemoglobin. Hemoglobin is broken down to form *biliverdin,* which is a green molecule. Biliverdin is then converted to *bilirubin,* which is a yellowish brown molecule. The injured area returns to its normal color as the biliverdin and bilirubin are absorbed into the bloodstream.

In this chapter, you will learn about the various components of blood, including hemoglobin. You will also learn about the formation of blood clots, which act to minimize the loss of blood when blood vessel damage occurs.

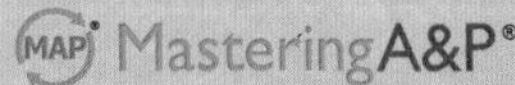

Go to MasteringA&P for helpful A&P Flix 3-D animations, chapter quizzes, pre-tests, Interactive Physiology tutorials, and more!

CHAPTER OUTLINE

LEARNING OUTCOMES ***After studying this chapter, you should be able to:***

- Identify the major components of blood, and describe their functions.
- Describe the life cycle of red blood cells.
- Identify the different classes of leukocytes, and explain their roles in the body's defense against pathogens.
- Describe the platelets and the mechanism of clot formation.

Before You Begin

Make sure you have mastered the following topics:

- Body fluids
- Phagocytosis
- Paracrines

We know the mechanisms by which the heart pumps blood to the tissues at a sufficient rate—the cardiac output—to provide a fresh supply of blood to the cells. We also know how this flow of blood could be regulated to distribute more blood to those tissues that need it by varying the resistance of blood vessels, and how the driving force for blood flow (the mean arterial pressure) is regulated. In this chapter, we examine the blood itself to see why it is so critical that our cells receive sufficient blood flow.

Now, we'll cover the different components of blood and describe which roles they play in the overall function of blood. We begin by discussing the liquid portion of blood, called **plasma,** and its role in the transport of proteins, hormones, electrolytes, organic nutrients, and waste products. Next, the cellular components of blood are described, including **erythrocytes** (red blood cells), which transport oxygen and carbon dioxide; **leukocytes** (white blood cells), which defend the body against pathogens; and **platelets** (cell fragments), which are critical in the formation of blood clots to prevent the loss of blood.

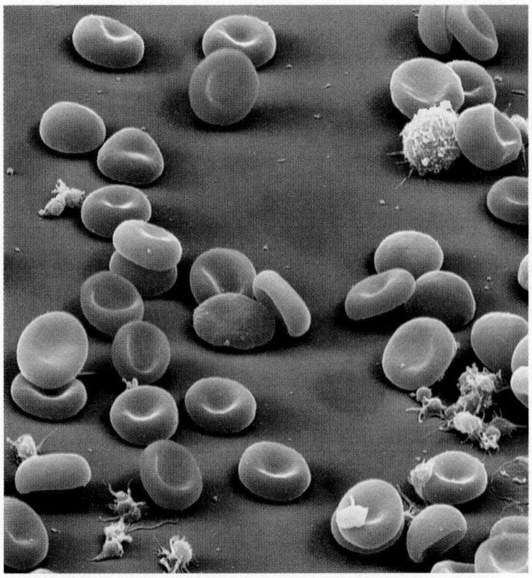

Oliver Meckes/Eye of Science/Photo Researchers

Figure 1 Cellular components of the blood. (Electron micrograph magnified approximately 3000 ×)

1 Overview of the Composition of Blood: The Hematocrit

The total volume of blood in a normal healthy adult is approximately 5.5 liters. It consists mostly of plasma (approximately 3 liters) and erythrocytes (approximately 2.5 liters), but also includes leukocytes and platelets (**Figure 1**). The fractional contribution of erythrocytes to the blood is called the **hematocrit** (abbreviated as *hct*), which is determined by centrifuging a sample of blood in a tube and is usually presented as a percentage (**Figure 2**). When blood is centrifuged, the elements of blood are separated based on density. Because erythrocytes are denser than other elements of the blood, they are pulled to the bottom of the tube. Plasma, the least dense component, remains at the top. Between these two layers rests a thin layer of leukocytes and platelets (accounting for less than 1% of the total blood volume) called the *buffy coat.*

To determine the hematocrit, the heights of the erythrocyte column and the whole blood column are measured, and then the

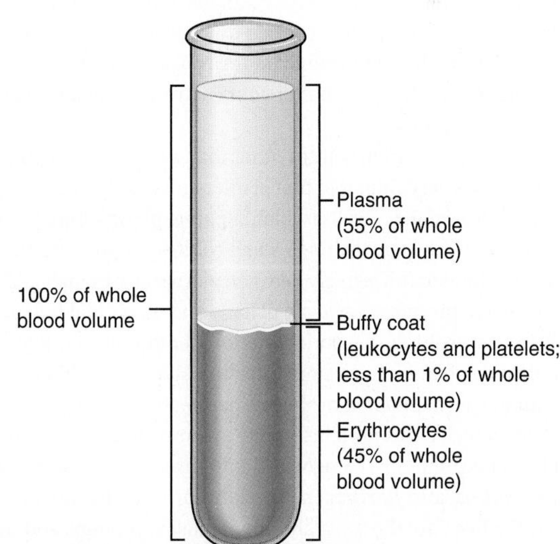

Figure 2 Determination of hematocrit. When a blood sample is centrifuged, erythrocytes are pulled to the bottom of the tube because they are denser than other blood elements.

hematocrit is calculated by determining the percentage of whole blood that consists of erythrocytes, as follows:

$$\text{Hematocrit} = \frac{\text{height of erythrocyte column}}{\text{height of whole blood column}} \times 100$$

The hematocrit is a useful clinical measure because it indicates whether a person has a normal complement of erythrocytes. For men, the normal range of the hematocrit is 42–52%, meaning that erythrocytes represent 42–52% of blood volume. For women, the normal range of the hematocrit is 37–47%. A low hematocrit indicates a lower than normal concentration of erythrocytes in the blood, whereas a high hematocrit indicates a higher than normal concentration of erythrocytes in the blood, called *polycythemia*. Polycythemia is a normal adaptive response in low-oxygen environments, such as occurs at high altitudes. For example, men living at an altitude of approximately 15,000 feet in the Andes often have hematocrits as high as 60. Further, the hematocrit increases within two weeks for people who travel to the Andes.

Apply Your Knowledge

What effect does polycythemia have on resistance to blood flow and thus on mean arterial blood pressure?

Now that we know the major components in blood, we will describe each component in detail. We start with the plasma, and then discuss erythrocytes, leukocytes, and lastly, platelets.

2 Plasma

Plasma is an aqueous solution in which a great variety of solutes are dissolved. These solutes include proteins, small nutrients (such as glucose, lipids, and amino acids), metabolic waste products (such as urea and lactic acid), gases (oxygen, carbon dioxide, nitrogen, and others), and electrolytes (such as sodium, potassium, and chloride). Although the proteins are the most abundant solutes in the plasma by weight, the smaller solutes are generally present in higher concentrations.

With respect to small solutes other than proteins, the composition of plasma is very similar to that of interstitial fluid. This similarity occurs because capillary walls (which separate plasma from interstitial fluid) are highly permeable to small solutes—a property that allows these solutes to move freely between plasma and interstitial fluid. With respect to proteins, however, plasma and interstitial fluid differ greatly in composition; the concentration of proteins in the plasma is significantly greater than that in interstitial fluid. This difference in concentration is maintained by the low permeability of capillary walls to proteins, which limits proteins' ability to move out of the plasma.

Plasma proteins are categorized into three main groups: *albumins, globulins,* and *fibrinogen.* The albumins, which are synthesized by the liver, are the most abundant plasma proteins and make a large contribution to the osmotic pressure of plasma, which affects the movement of fluid across capillaries. The globulins encompass a wide variety of proteins that transport lipids, steroid hormones, and other substances in the blood; play a critical role in the blood's ability to form clots; and are important in defending the body against foreign substances. Fibrinogen is synthesized by the liver and is a key substance in the formation of blood clots. **Serum** is plasma from which fibrinogen and other clotting proteins have been removed.

Major components of the plasma are listed in Table 1.

TABLE 1 Components of Plasma

Component	Description and importance
Water	Accounts for 90% of plasma volume; provides a dissolving and suspending medium for solutes and formed elements
Solutes	
Proteins	Accounts for 8% of plasma (by weight); most are synthesized by liver
Albumin	60% of plasma proteins; largely responsible for plasma osmotic pressure
Globulins	36% of plasma proteins; include clotting proteins, antibodies secreted by certain leukocytes during the immune response, and proteins that bind to lipids, fat-soluble hormones, and metal ions to transport these substances in the blood
Fibrinogen	Important in the formation of blood clots
Others	Enzymes, hormones, and antibacterial proteins
Nitrogenous waste products	By-products of metabolism, such as urea, uric acid, and creatinine
Organic nutrients	Materials absorbed from the intestines and used by cells throughout the body; include glucose and other simple sugars, amino acids, fatty acids, glycerol, triglycerides, cholesterol, and vitamins
Electrolytes	
Cations	Sodium, potassium, calcium, magnesium (important in neuromuscular signaling), and trace metals (important in normal enzyme activity)
Anions	Chloride (important in neuromuscular signaling), bicarbonate, and phosphate (important in maintenance of normal plasma pH)
Respiratory gases	Oxygen and carbon dioxide; most oxygen and some carbon dioxide is bound to hemoglobin in erythrocytes; a significant fraction of carbon dioxide is found in the plasma in the form of bicarbonate

3 Erythrocytes

Erythrocytes (red blood cells) are the most abundant cells in the blood, numbering about 5 million per cubic millimeter (mm^3) of blood. Among cells of the body, erythrocytes are unique in that they lack nuclei, mitochondria, and other organelles, such as ribosomes, that are necessary for manufacturing proteins. Erythrocytes are shaped like disks and are approximately 7.5 μm in diameter and a little more than 2 μm thick (see Figure 1). They are often described as *biconcave disks* because they are indented on both sides.

The shape of erythrocytes is due to the presence of a cytosolic protein called *spectrin*. Spectrin is a fibrous protein that forms a network linked to the plasma membrane. This *spectrin net* is flexible, giving erythrocytes the ability to bend and flex as necessary to move through capillaries that are sometimes smaller in diameter than the erythrocytes. In addition to its flexibility, the biconcave shape gives erythrocytes a large surface area, which makes them suitable for exchange. This ability is critical because, as we see next, the erythrocytes transport oxygen and carbon dioxide in blood for exchange with body cells and lung tissue.

Oxygen and Carbon Dioxide Transport

The major function of erythrocytes is to transport oxygen and carbon dioxide in the blood. Such transport is essential for the delivery of oxygen from the lungs to respiring cells, and for the delivery of carbon dioxide from respiring cells to the lungs, where it is eliminated from the body. Erythrocytes have a high capacity for carrying these gases because their cytoplasm contains two proteins: hemoglobin and carbonic anhydrase. **Hemoglobin** binds, and thus transports, oxygen and carbon dioxide, whereas **carbonic anhydrase** is essential for the transport of carbon dioxide only. In this chapter, we give a general description of these proteins and their importance to erythrocytes.

Hemoglobin's Reversible Binding of Oxygen and Carbon Dioxide

Hemoglobin is the most abundant protein in erythrocytes, with more than 250 million hemoglobin molecules available per erythrocyte. Hemoglobin is composed of four polypeptide chains of two types (two alpha chains and two beta chains), each of which has an iron-containing ring structure known as a *heme group* (**Figure 3**). The iron in hemoglobin is present in the ferrous form (Fe^{2+}), which imparts a red color to the erythrocytes and, hence, to blood. This iron is the site to which a molecule of oxygen binds. Because four heme groups are present in each hemoglobin molecule, each hemoglobin can bind four oxygens. Carbon dioxide binds reversibly to amino acids within the polypeptide chains; however, hemoglobin transports considerably less carbon dioxide than oxygen. Athletes often attempt to increase their hemoglobin content, as described in **Understanding Exercise: Effects of High Altitude.**

Carbonic Anhydrase and the Carbon Dioxide–Bicarbonate Reaction

Although less abundant, carbonic anhydrase is another erythrocyte protein that is critical in the transport of gases. Carbonic anhydrase is an enzyme that catalyzes the reversible conversion of carbon dioxide and water to carbonic acid:

$$CO_2 + H_2O \xrightleftharpoons{\text{carbonic anhydrase}} H_2CO_3$$

The pathway continues with the reversible dissociation of carbonic acid to yield a hydrogen ion and a bicarbonate ion:

$$H_2CO_3 \rightleftharpoons H^+ + HCO_3^-$$

The entire pathway can be written as follows:

$$CO_2 + H_2O \xrightleftharpoons{\text{carbonic anhydrase}} H_2CO_3 \rightleftharpoons H^+ + HCO_3^-$$

Thus carbon dioxide can be converted to free hydrogen ions, which affects the pH of blood. This reaction is critical to the transport of carbon dioxide in blood and the key role carbon dioxide has in maintaining the acid-base balance of the blood.

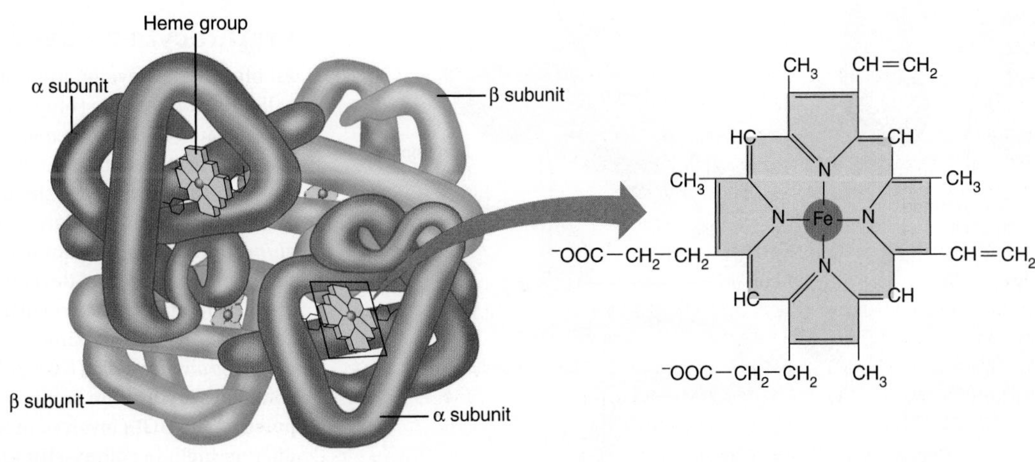

Figure 3 Hemoglobin. (a) The hemoglobin molecule consists of four polypeptide subunits, designated as either α or β, each covalently bound to a heme group containing iron. **(b)** Chemical structure of a heme group.

UNDERSTANDING EXERCISE

Effects of High Altitude

Because oxygen demands increase during exercise, some athletes try to increase their hemoglobin content, in order to increase the oxygen-carrying capacity of their blood. To do so, many athletes train at high altitudes, where the quantity of oxygen in the air is less than at sea level. When oxygen levels in the blood are low, more erythrocytes are produced; the larger numbers of erythrocytes, in turn, transport more oxygen to the tissues that need them. This is a valuable adaptation to high altitudes. When, after training, the athlete returns to sea level and everyday activities, his or her oxygen-carrying capacity is greater than needed. However, the increased number of erythrocytes may help with endurance during competition.

Some athletes try to increase their oxygen-carrying capacity through blood doping. In this practice, they increase their number of erythrocytes either by directly injecting blood cells into their system or by injecting *erythropoietin*, the chemical that stimulates erythrocyte synthesis. Blood doping is illegal in most professional sports.

There is also a downside to increasing the concentration of erythrocytes—namely, an increase in resistance to blood flow. A hematocrit greater than 60 indicates that impedance to blood flow is more detrimental than the increase in oxygen carrying capacity is beneficial. The increased hematocrit can also increase the friction between blood and the walls of the blood vessels, weakening the walls and making them more susceptible to atherosclerosis, which can lead to a number of health problems.

Life Cycle of Erythrocytes

Once erythrocytes are released into the bloodstream, they remain there for only about 120 days. They have neither a nucleus (no DNA) nor any organelles, so they cannot undergo cell division. Therefore, new erythrocytes are produced on a regular basis, at a rate of approximately 2–3 million per second or 200 billion per day! The bone marrow has the enormous task of producing these erythrocytes by a process called **erythropoiesis**, while the spleen removes the old erythrocytes from the blood. **Figure 4** shows the organs involved in the synthesis and breakdown of erythrocytes, described next.

Figure 4 Life cycle of erythrocytes. The production and breakdown of erythrocytes involves several organs. EPO = erythropoietin, T-iron = transferrin, and F-iron = ferritin.

Erythrocyte Production

All blood cells develop from precursor cells called *hematopoietic* (blood-forming) *stem cells,* located in the bone marrow (**Figure 5**). Whereas erythrocytes and most leukocytes come to full maturity in the bone marrow, T lymphocytes must migrate to the *thymus gland* (located in the thoracic cavity above the heart) before they develop to maturity. The development of a particular type of blood cell depends on cytokines called *hematopoietic growth factors (HGFs)*. The HGF that stimulates erythrocyte production is **erythropoietin.** The HGFs involved in leukocyte production include **colony-stimulating factors** and **interleukins.**

Erythropoietin is released from certain cells in the kidney in response to low oxygen levels in blood. Erythropoietin travels in the

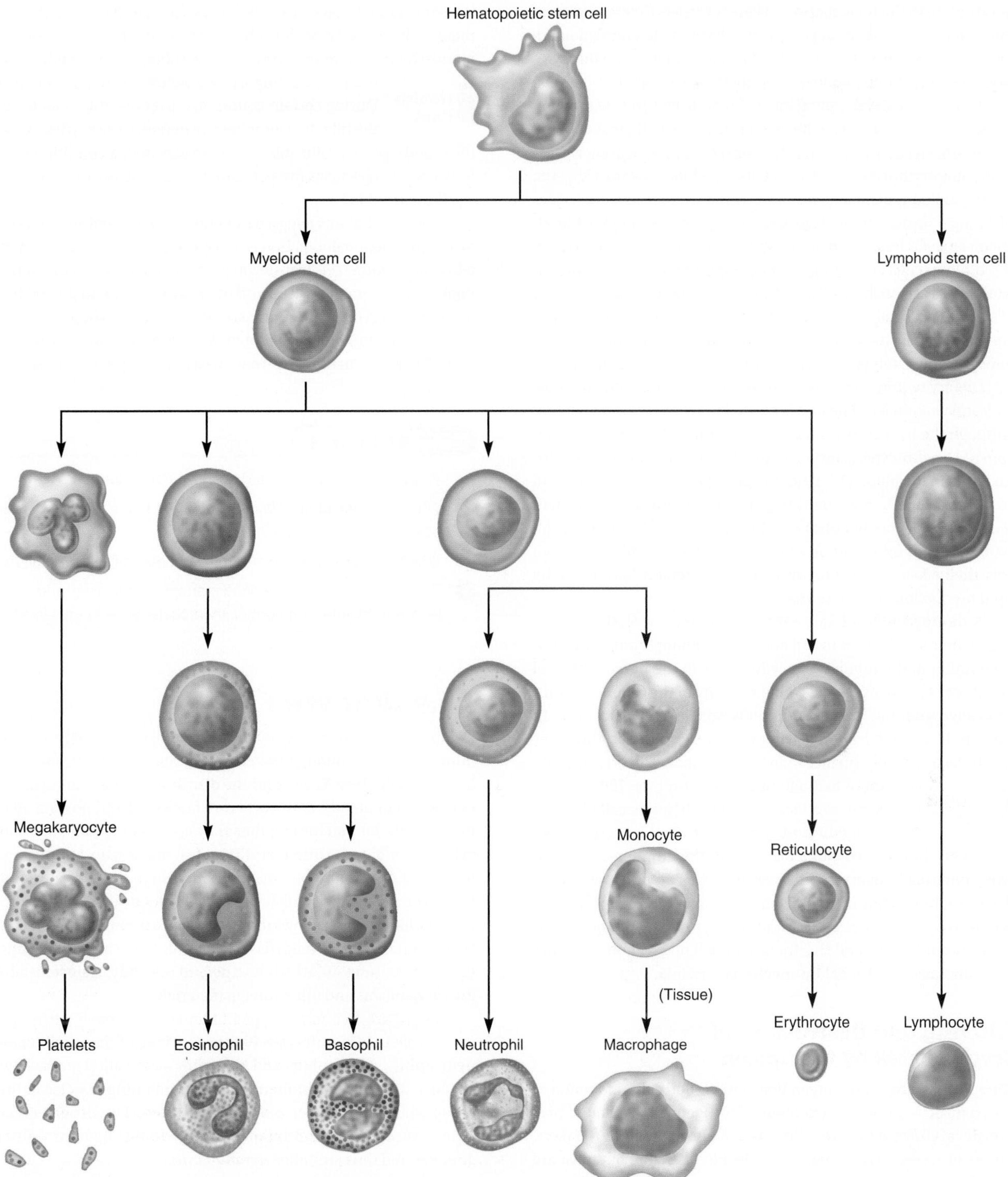

Figure 5 Production of blood cells and platelets. All blood cells are derived from hematopoietic stem cells. During the first level of differentiation, two cell lines are generated: myeloid stem cells, which are precursors for most blood cells, and lymphoid cells, which are precursors to lymphocytes.

bloodstream to the bone marrow, where it triggers differentiation of pluripotent cells to erythrocytes. Several levels of differentiation are needed to produce erythrocytes. During differentiation, erythrocytes produce hemoglobin and lose their nuclei and organelles. The last cell stage prior to development into the mature erythrocyte is the *reticulocyte*, a red blood cell with some ribosomes still present in the cytoplasm, giving the cell a weblike, or *reticular*, appearance. Normally, only erythrocytes are released into the bloodstream. However, under conditions of rapid erythrocyte synthesis, such as would occur following a severe hemorrhage, some reticulocytes enter the bloodstream as well. Under normal conditions, approximately 0.5% to 2.5% of red blood cells circulating in the peripheral blood are reticulocytes. In disease, such as sickle cell anemia, the retic count is typically 10% to 20%. If a manual retic count is performed, the blood smear is stained with new methylene blue to visualize the ribosomes. Some automated hematology analyzers can also perform retic counts.

Dietary requirements for erythrocyte production are iron, folic acid, and vitamin B_{12}. The iron is needed to synthesize hemoglobin. Although the liver stores some iron and some of the iron released from old erythrocytes is recycled, maintaining adequate hemoglobin levels in the blood (13–18 grams per 100 mL in men, and 12–16 grams per 100 mL in women) requires iron in the diet. Folic acid and vitamin B_{12} are necessary for the synthesis of DNA. Although deficiencies of folic acid or vitamin B_{12} affect cell division of all cells, their effects on erythropoiesis are most notable because of the rapid production of erythrocytes.

A shortage of the dietary elements for erythropoiesis can produce either a reduction in the amount of hemoglobin per cell or a reduction in the number of erythrocytes in the blood, either of which reduces the oxygen carrying capacity of the blood, a condition called **anemia.** If this condition is caused by a lack of iron in the diet, it is called *iron-deficiency anemia*, and the erythrocytes are characteristically smaller than normal due to a decrease in the amount of hemoglobin per erythrocyte. If the condition is due to a lack of vitamin B_{12}, it is called *pernicious anemia* and is characterized by larger than normal erythrocyte size but a decrease in the number of erythrocytes, and usually an overall increase in the hematocrit. Anemia can also result from hemorrhage (bleeding) or an abnormally high rate of erythrocyte destruction (hemolysis). Symptoms of anemia include chronic fatigue and shortness of breath on exertion. For more about anemia, see **Clinical Connections: Anemia**.

Filtering and Destruction of Erythrocytes by the Spleen

The spleen is a lymphoid organ that stores blood cells and removes old erythrocytes from the circulation (see Figure 4). As stated previously, erythrocytes have a life span of approximately 120 days. Some old erythrocytes hemolyze in the bloodstream, but most are engulfed by macrophages in the spleen and, to a lesser extent, in the liver. When macrophages destroy erythrocytes, the hemoglobin is also catabolized. After iron is removed, the resulting heme is converted to **bilirubin,** a yellow compound. Bilirubin is then released into the bloodstream, where it gives plasma its yellowish tinge. The bilirubin travels to the liver, where it is catabolized further. Most products of bilirubin catabolism are secreted in the bile to the small intestine and ultimately excreted in the feces, imparting a brownish tinge to the feces. Other bilirubin metabolites are released into the bloodstream, where they travel to the kidneys and are eliminated in the urine, imparting a yellowish color to the urine. During certain conditions (liver diseases, blockage of the bile duct, or excessive hemolysis of erythrocytes in the blood), plasma bilirubin levels are elevated, a condition called *jaundice,* which causes the skin and the whites of the eyes to take on a yellowish color.

The iron that was released by hemoglobin catabolism is recycled to form new hemoglobin. Iron is transported in blood bound to a protein called *transferrin.* Transferrin picks up iron from the gastrointestinal tract or from the spleen and transports the iron to the red bone marrow for erythrocyte synthesis, or to the liver where some iron can be stored bound to the protein *ferritin.* Some iron is also stored bound to ferritin in the spleen and in cells lining the small intestine.

Quick Check 1

1. What is the hematocrit, and how is it determined?
2. What is hemoglobin? Why is it important to the function of red blood cells?
3. Which reaction does the enzyme carbonic anhydrase catalyze?
4. Describe the roles of the bone marrow, spleen, liver, and kidney in maintaining normal erythrocyte levels in the blood.

4 Leukocytes

Compared to erythrocytes, leukocytes (white blood cells) are far less numerous in the blood, numbering only about 4000–10,000 per cubic millimeter (see **Table 2** for the distribution of erythrocytes and leukocytes in blood). Leukocytes are nucleated and possess all the normal cellular machinery; thus they are the only fully functional cells in the blood. Unlike erythrocytes, leukocytes are normally found not only in the bloodstream but also in other tissues of the body. The presence of leukocytes outside blood vessels results from their mobility, which allows them to squeeze through pores in capillaries and migrate through tissues. This ability to migrate is important to their function, which is to defend the body against invading microorganisms and other foreign materials.

There are five major types of leukocytes, each with a particular role in immunity (see **Figure 6**). Three of these cell types—neutrophils, eosinophils, and basophils—are called *granulocytes* because they have prominent protein-containing vesicles in their cytoplasm known as *cytoplasmic granules.* The other two cell types—monocytes and lymphocytes—do not have prominent granules and thus are called *agranulocytes.*

When your doctor orders a blood test of your leukocytes, two types of tests can be run: total white blood cells (usually included with a total of all cellular components of blood) and a differential white blood cell count. Both tests require a small sample of blood.

For the total white blood cell count, a hematology analyzer is used to determine the number of white blood cells in a specific volume of blood. (A hematology analyzer is a piece of automated

CLINICAL CONNECTIONS

Anemia 

Anemia—defined as a decrease in the oxygen carrying capacity of blood—is generally associated with a low hematocrit, which can result from either a decrease in the number of erythrocytes or a decrease in the size of erythrocytes (called *microcytosis*). A person can also develop anemia when the hematocrit is normal, or even elevated, but each red cell contains less than the normal concentration of hemoglobin. Because most of the oxygen transported in blood is bound to hemoglobin in erythrocytes, a decrease in erythrocyte abundance or size is associated with low levels of hemoglobin.

Six general categories of anemia are distinguished:

1. *Nutritional anemias* are caused by a dietary deficiency, most commonly an iron deficiency. Because iron is a component of hemoglobin, an iron deficiency decreases hemoglobin synthesis. The number of erythrocytes is normal, but less hemoglobin is present in each erythrocyte. With less hemoglobin, the erythrocytes are smaller, and the oxygen carrying capacity of the blood is diminished.

   Another essential nutrient for normal oxygen transport is folic acid, which is required for the synthesis of thymine, one of the bases in DNA. A deficiency of folic acid affects all cells of the body that undergo rapid cell division, because DNA replication is required for cell division. Given that each second the body produces 2–3 million erythrocytes by the rapid cell division of certain precursor cells, insufficient folic acid in the diet results in fewer erythrocytes, each of which is larger than normal (called *macrocytosis*). Large erythrocytes, however, tend to be more fragile and have a shorter life span.
2. *Pernicious anemia* is caused by a deficiency of intrinsic factor, which is required for the absorption of vitamin B_{12} in the intestinal tract. Because vitamin B_{12}, like folic acid, is required for the synthesis of thymine, this anemia has the same characteristics as anemia caused by a folic acid deficiency.
3. *Aplastic anemia* is caused by a defect in the bone marrow, the primary site of erythrocyte and leukocyte production; the result is a deficiency of both classes of blood cells. The availability of fewer erythrocytes causes anemia, and the presence of fewer leukocytes impairs the body's defense against pathogens such as bacteria and viruses.
4. *Renal anemia* is associated with decreased production of the hormone erythropoietin due to a pathological state in the kidneys. Because erythropoietin stimulates the synthesis of erythrocytes in the bone marrow, a decrease in erythropoietin production results in fewer erythrocytes in the blood.
5. *Hemorrhagic anemia* is caused by the rapid loss of blood. Within a few days following a severe hemorrhage, fluid intake and the shifting of body fluids between compartments increase the blood volume so that it returns to normal. Erythrocyte production takes a few weeks to rebound, however. Thus, even as blood volume returns to normal, the hematocrit remains low until a sufficient number of new erythrocytes can be produced.
6. *Hemolytic anemia* is caused by the rupture, or *hemolysis*, of an excessive number of erythrocytes. An increase in hemolysis can be caused by infections such as malaria, or by defects in the erythrocytes, such as occurs in *sickle cell anemia*. Sickle cell anemia is a hereditary disease caused by a defect in the gene that codes for hemoglobin. A single base in the gene coding for the beta chain is wrong, causing it to code for the wrong amino acid. This single amino acid substitution causes the hemoglobin molecules in an erythrocyte to polymerize (link to one another) when oxygen levels are low, making the erythrocyte sickle-shaped and fragile (see the photo).

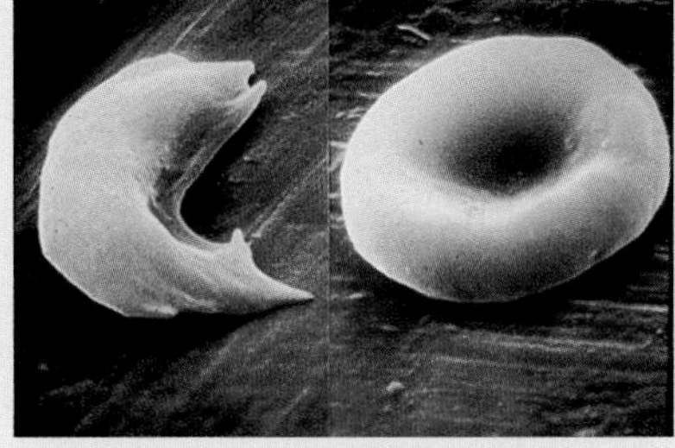

Oliver Meckes & Nicole Ottawa/Photo Researchers

A sickled (left) and normal (right) red blood cell.

Critical Thinking Questions

1. A patient has a normal hematocrit, but is anemic. Explain how the contrast can exist.
2. What is the nature of the genetic defect that causes hemolytic anemia?
3. Describe the role of folic acid in the development of nutritional and other types of anemias.

equipment that measures complete blood counts (CBCs) and differentials). Elevated leukocytes can be indicative of infection, inflammation, anemia, leukemia, bone marrow tumors, and stress. Decreased leukocytes can be caused by radiation therapy, or diseases of the liver, spleen, or bone marrow.

For the differential white blood cell count, a drop of blood is placed on a microscope slide and *smeared* across the slide into a thin layer. The blood smear is then stained (either manually or with an automated stainer) with Wright's stain, which reveals the distinctive nuclear shape and cytoplasmic color of the different leukocytes for easy identification. In the manual method, the laboratory scientist then counts 100 cells in a systematic manner, identifying each one. The number of each type of cell counted represents the percentage of leukocytes composed of that type of cell. Differential white blood cell counts can be used to identify a number of

TABLE 2 Cellular Composition of Blood

Component	Amount per microliter (mm^3)	Diameter (μm)	Anatomical features	Primary function
Erythrocytes	5,000,000	7–8	No nucleus; no organelles; biconcave disk	Transport O_2 and CO_2
Leukocytes	4000–10,000			Defend body against pathogens
Neutrophils	3000–7000	10–14	Multilobed nucleus; red- and blue-staining granules	Phagocytosis of foreign material
Eosinophils	100–400	10–14	Bilobed nucleus; red-staining granules	Kill parasites
Basophils	20–50	10–12	Multilobed nucleus; blue-staining granules	Secrete chemical mediators in inflammation and allergic reactions
Monocytes	100–700	14–24	Large kidney-shaped nucleus; no granules	Phagocytosis; mature into macrophages in tissue
Lymphocytes	1500–3000	5–17	Large round nucleus; little cytoplasm; no granules	B cells—secrete antibodies T cells—secrete cytokines that support immune response of other cells; secrete factors that kill infected or tumor cells
Platelets	100,000–500,000	2–4	Cytoplasmic fragments; granules	Hemostasis

infections and diseases based on changes in the normal distribution of leukocytes, described next. (Note that, in many larger laboratories, automated hematology analyzers calculate the differential cell counts, with programmed algorithms checking the accuracy of results. Abnormal results would get flagged as such, and a laboratory technologist would then perform a manual differential count.)

Neutrophils

Neutrophils constitute 50–80% of all leukocytes and are capable of one of the most important defense activities in the body: phagocytosis. As a phagocyte ("eating cell"), a neutrophil engulfs and digests microorganisms, abnormal cells, and foreign particles present in blood and tissues. Only a few other cell types (eosinophils, monocytes, and macrophages) are capable of phagocytosis. Newly produced neutrophils circulate in the blood for 7–10 hours and then migrate into the tissues, where they live for only a few days.

Elevated neutrophils in a differential white blood cell count, a condition called *neutrophilia,* generally indicate a bacterial or viral infection. Infectious agents cause mobilization of neutrophils from the bone marrow; neutrophils then travel to the infected tissue to phagocytose bacteria, affected cells, or debris. Early in an infection, some immature neutrophils, called *Band cells,* enter the blood. Neutrophil levels also increase during certain fungal infections, during inflammation, and following surgery or other trauma to the body.

Eosinophils

Approximately 1–4% of all leukocytes are **eosinophils.** Like neutrophils, eosinophils are phagocytic, but their main contribution in defense is in attacking parasitic invaders that are too large to be engulfed by neutrophils. Eosinophils mount an attack against these parasites by attaching to their bodies and discharging toxic molecules from their cytoplasmic granules. Unfortunately, eosinophil defense—and the overall immune response to such parasites—is weak. Eosinophil responses can sometimes be harmful because the toxic molecules they release can also damage normal tissues and may trigger allergic reactions.

An elevated eosinophil level usually indicates a parasitic infection or allergic reaction, but can also be caused by a number of diseases. Asthma, for example, is associated with long-term inflammation of the airways; eosinophils promote inflammation and their levels are elevated in asthmatics. In addition, levels of eosinophils can be elevated with autoimmune diseases and eczema.

Basophils

Basophils are nonphagocytic cells that are thought to defend against larger parasites; they may operate much like eosinophils by releasing toxic molecules that damage invaders. However, basophils also release histamine, heparin, and other chemicals that contribute significantly to allergic reactions, such as hay fever. Although basophils constitute less than 1% of all leukocytes, they have a significant effect in people with allergies. During an acute allergic reaction, basophil levels in the blood decrease as these cells move to the site of the reaction.

Monocytes

Monocytes, which account for 2–8% of the leukocytes, are important in phagocytic defense. Blood levels of monocytes increase during viral and parasitic infections, as well as with some forms of cancer. Monocyte levels decrease with HIV infection, rheumatoid arthritis, and exposure to steroids.

Health
Fact

New monocytes circulate in the blood for only a few hours before they migrate into tissues, where they become five to ten times larger and develop into very active phagocytic cells known as

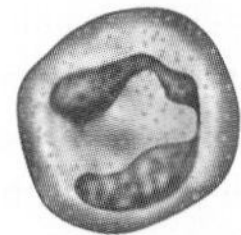
Neutrophil

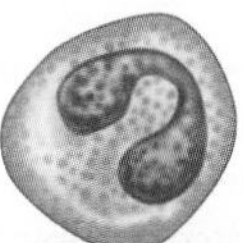
Eosinophil

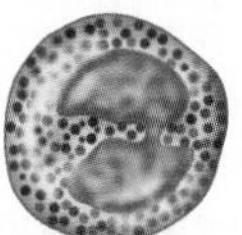
Basophil

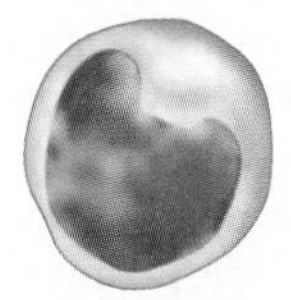
Monocyte

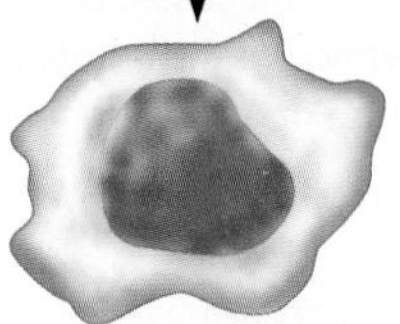
Macrophage

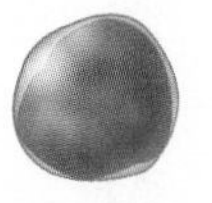
Lymphocyte

Figure 6 Leukocytes and their roles in immunity. See text for details.

macrophages ("big eaters"). Some macrophages, called "wandering macrophages," migrate throughout body tissues, whereas others, known as "fixed macrophages," remain at particular sites. Macrophages are especially abundant in connective tissue, in the wall of the gastrointestinal tract, in the alveoli of the lung, and in the walls of certain blood vessels in the liver (where they are known as Kupffer cells) and in the spleen. In the latter two sites, macrophages phagocytose abnormal, dead, and dying erythrocytes. Macrophages are also prevalent in regions of the spleen and lymph nodes where they are most likely to contact any infectious agents circulating in the blood and lymphatic fluid.

Macrophages can differentiate further into **dendritic cells**. Very few dendritic cells are found in the blood, and those that are tend to be immature. Dendritic cells are typically found in the tissue that separates the internal and external environments—that is, in epithelial tissue. Once dendritic cells respond to an infectious agent, however, they migrate to lymphoid tissue, where they present components of the infectious agent (antigens) to lymphocytes. Thus, although some dendritic cells derive from phagocytic macrophages, their function in the immune response is *antigen presentation,* rather than phagocytosis.

Lymphocytes

Lymphocytes constitute 20–40% of all leukocytes in the blood, and approximately 99% of all cells found in interstitial fluid. Lymphocytes are of three major types: *B lymphocytes (B cells), T lymphocytes (T cells),* and *null cells,* so called because they lack the cell membrane components that are characteristic of B cells and T cells. Most null cells are large, granular lymphocytes known as *natural killer (NK) cells.* Some dendritic cells derive from lymphocytes.

The various lymphocytes have complex functions in the immune system, which are presented briefly here. B lymphocytes differentiate into plasma cells, which secrete antibodies that target the immune response against a specific antigen. T lymphocytes come in a variety of forms that promote the immune response, including cytotoxic T cells that destroy abnormal cells, helper T cells that secrete chemicals promoting actions of other leukocytes, and suppressor T cells that decrease the immune response. Increased blood levels of lymphocytes tend to be associated with viral infections rather than with bacterial and parasitic infections. Lymphocyte levels also increase in Graves' disease. Decreased levels of lymphocytes occur with immunodeficiency diseases, lupus, and increased levels of steroids.

Quick Check 2

1. List the five major types of leukocytes, and describe one function of each type.
2. Which types of cells are capable of phagocytosis?
3. Name the different types of lymphocytes.

5 Platelets and Hemostasis

Platelets are colorless cell fragments that arise when portions of large bone-marrow cells called *megakaryocytes* break off. They are smaller than erythrocytes and contain mitochondria, smooth endoplasmic reticulum, and cytoplasmic granules, but no nucleus. Platelets can number from 100,000 to 500,000 per cubic millimeter and are important in triggering the sequence of events that leads to the formation of blood clots.

Blood vessels get damaged frequently, leading to internal and/or external bleeding. Although bleeding is usually minor, a lack of **hemostasis**—mechanisms to stop the bleeding—would mean even a superficial cut could cause a person to bleed to death. The process of hemostasis occurs in three steps: vascular spasm, formation of a platelet plug, and formation of a blood clot, or **thrombus**. Each of these steps is described next.

Apply Your Knowledge

An *embolus* is a blood clot that has broken loose and travels in the circulation. If an embolus lodges in an artery and obstructs blood flow, it forms an *embolism*. Explain some of the dangers of embolisms.

Vascular Spasm

When a blood vessel is damaged, intrinsic mechanisms trigger a constriction called a *vascular spasm,* which increases resistance to blood flow. Damage also tends to activate the sympathetic nervous system, which causes further vasoconstriction. With less blood flowing to the area of damage, blood loss is minimized. Decreasing blood loss alone is not sufficient, however; blood loss must be stopped altogether.

Platelet Plug

Platelets, also called *thrombocytes,* possess granules containing a variety of substances that can be secreted into the plasma, including ADP, serotonin, epinephrine, and a variety of chemicals that participate in the formation of a blood clot. Platelets also are "sticky" under certain circumstances, allowing them to adhere to surfaces, especially those of damaged blood vessels.

Both the formation of platelet plugs and the subsequent blood clot require the presence of platelets and a fairly large set of specific plasma proteins. In platelet plug formation, the key protein is

von Willebrand factor (vWf), which is secreted by megakaryocytes, platelets, and endothelial cells lining blood vessels. Although vWf is present in the plasma at all times, it accumulates at the site of vessel damage.

The first step in platelet plug formation is *platelet adhesion,* which occurs when blood vessel damage exposes the tissue underlying the endothelium (the layer of endothelial cells that line a vessel), called *subendothelial tissue.* When blood contacts subendothelial tissue, vWf binds to collagen fibers in the subendothelium, triggering the binding of platelets to vWf, which anchors platelets in place. Contact with vWf also activates platelets, changing their metabolism and surface properties (making them "sticky") and stimulating the secretion of certain products.

Two of the secretory products of activated platelets—serotonin and epinephrine—cause vasoconstriction, which increases resistance to blood flow and minimizes blood loss. A third secretory product of activated platelets, ADP, causes *platelet aggregation,* the second step in platelet plug formation. ADP stimulates morphological changes in the platelets that cause them to adhere to one another such that they form a mass, or aggregate. Aggregated platelets secrete more ADP, which stimulates further platelet aggregation, thereby providing a positive feedback loop that increases the rate of platelet plug formation (**Figure 7a**). ADP also stimulates the production of **thromboxane A_2 (TXA_2),** which further supports platelet aggregation.

TXA_2 is formed from a phospholipid, **arachidonic acid,** located in the membrane of platelets. TXA_2 has many roles in hemostasis, including stimulation of platelet aggregation, stimulation of ADP secretion (providing more positive feedback for platelet plug formation), and vasoconstriction (reducing blood flow to the area).

Platelet plug formation is limited to the area of blood vessel damage to prevent unnecessary blood clots that, by occluding blood flow, might deprive tissue of its needed nutrients and allow waste products to accumulate. Platelet plugs do not form on normal endothelium because healthy endothelial cells continuously release **prostacyclin** (also called prostaglandin I_2) and nitric oxide, both of which inhibit platelet aggregation (Figure 7b). Whereas aggregated platelets convert arachidonic acid to TXA_2 to facilitate platelet plug formation, healthy endothelial cells convert arachidonic acid to prostacyclin to inhibit platelet plug formation.

Platelets contain high concentrations of the contractile proteins actin and myosin. As platelets aggregate and form a plug, they contract to increase the "tightness" of the plug. Nevertheless, even a tight platelet plug cannot stop blood loss. The next step of hemostasis, forming the blood clot, is necessary.

Formation of a Blood Clot

Integral to the formation of a blood clot is a plasma protein called **fibrin,** which is necessary for the blood to coagulate or to be converted into a gel that traps erythrocytes and plugs the damage to the blood vessel and thus prevents blood loss. For this reason, blood clots are also called *fibrin clots* (**Figure 8**). Fibrin clot formation occurs secondary to platelet plug formation because it requires that a number of steps occur on phospholipids located on the surface of activated platelets, and because secretory products of aggregated platelets are necessary for clot formation.

The formation of a fibrin clot requires a sequence of reactions called the **coagulation cascade.** During this cascade, plasma proteins called **coagulation factors,** which are always present in the plasma in their inactive form, undergo a series of proteolytic activations resulting from hydrolysis of certain peptide bonds. Most coagulation factors are designated by Roman numerals, but the number of the factor provides no insight into its function or its location in

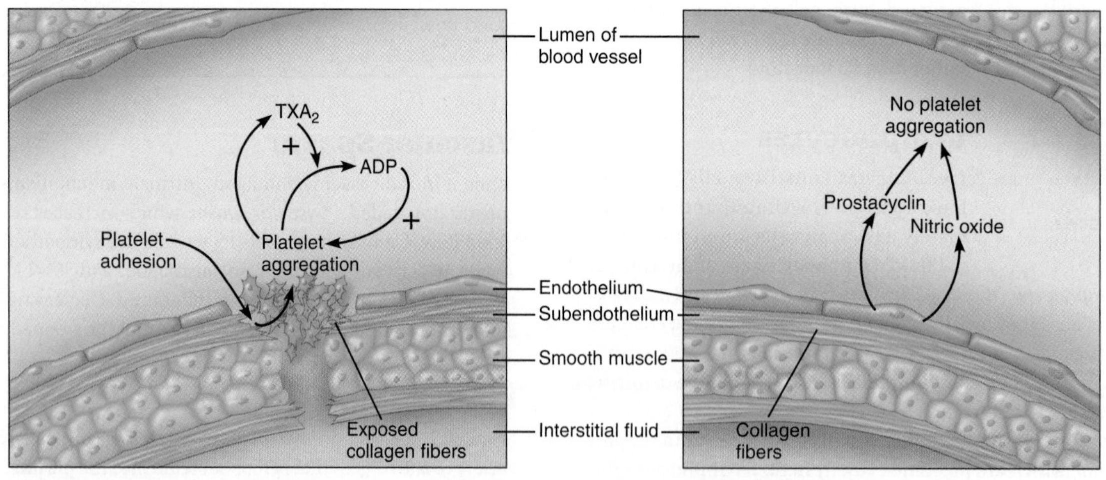

Figure 7 Formation of a platelet plug. (a) Platelets adhere to collagen fibers at the site of vessel damage. Adhered platelets secrete ADP to stimulate platelet aggregation at the site of adhesion, causing more platelets to secrete ADP. Aggregated platelets also release thromboxane A_2 (TXA_2) into the lumen of the blood vessel, which stimulates further platelet aggregation. **(b)** Healthy endothelial cells secrete nitric oxide (NO) and release prostacyclin, both of which inhibit platelet aggregation.

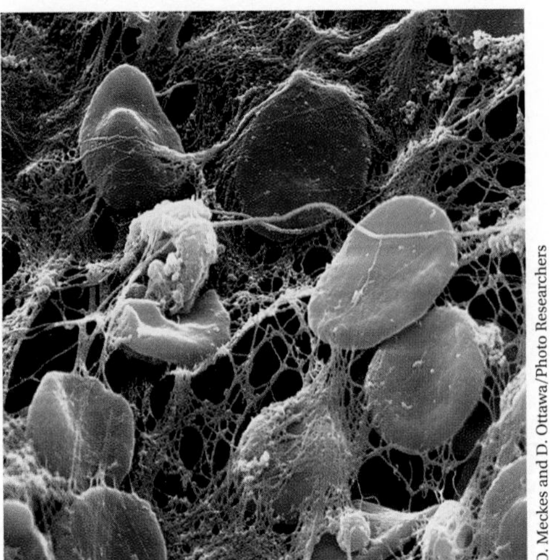

Figure 8 A fibrin clot.

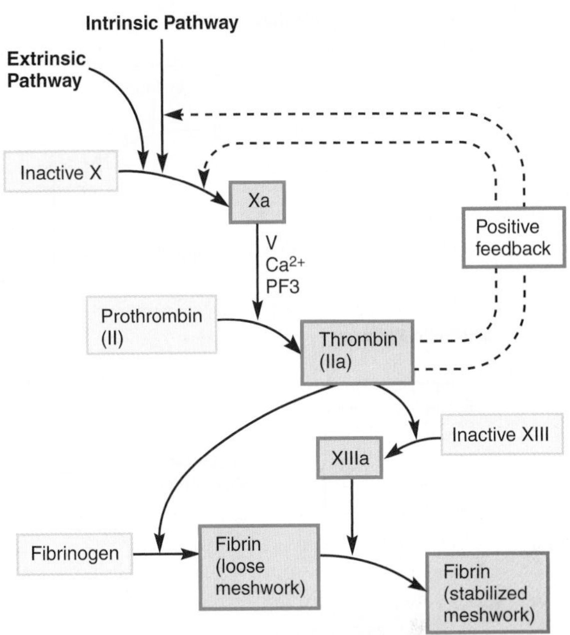

Figure 9 Role of thrombin in forming fibrin clot. Thrombin is activated by the action of factor Xa on prothrombin and provides positive feedback for its own synthesis. Once active, thrombin converts fibrinogen to fibrin and activates factor XIII.

the coagulation cascade; rather, these coagulation factors were numbered based on the order of their discovery. A lowercase "a" following the Roman numeral indicates activation of the factor. Most of the activated coagulation factors function as proteolytic enzymes for the next step of the cascade, although some serve as cofactors.

The ultimate step of clot formation is conversion of a filamentous plasma protein, *fibrinogen*, into its active form, fibrin. Conversion of fibrinogen to fibrin is a proteolytic reaction catalyzed by **thrombin,** the active form of another coagulation factor, prothrombin. Once formed, fibrin molecules adhere to one another, forming a loose meshwork of strands. This meshwork is then stabilized by formation of covalent linkages between strands, a reaction catalyzed by another coagulation factor, *factor XIII* (also called fibrin-stabilizing factor). Like fibrin and thrombin, factor XIII is present in the plasma in an inactive form and must be activated to XIIIa before participating in clot formation.

Central to formation of the stable fibrin meshwork is activation of thrombin (**Figure 9**). Factor Xa forms a *prothrombin activating complex* with factor V, Ca^{2+}, and PF3 (platelet factor 3—a phospholipid located on the surface of activated platelets). This complex converts prothrombin (factor II) to thrombin. Thrombin has several roles in clot formation, including converting fibrinogen to fibrin, activating factor XIII, and providing positive feedback for its own activation. Although not shown in Figure 9, thrombin contributes to additional platelet aggregation and stimulates platelets to secrete several products.

Two pathways lead to the activation of thrombin: an *intrinsic pathway* involving coagulation factors and other necessary chemicals that are already present in the plasma, and an *extrinsic pathway* involving some coagulation factors present in damaged tissue adjacent to the site of vessel damage. Both clotting pathways are generally activated simultaneously because rarely is vessel damage not accompanied by damage to other tissues. The stimuli that initiate the pathways, however, differ.

The intrinsic pathway starts when circulating factor XII (also called Hageman factor) is activated by contact with substances in the subendothelium, including collagen and phospholipids (**Figure 10**). Activation of factor XII triggers a cascade of reactions that leads to activation of factor X, to activation of thrombin, and ultimately to fibrin clot formation. The extrinsic pathway starts when tissue damage allows **tissue factor** (factor III) to contact the plasma and react with inactive factor VII to form a complex, which activates factor VII. The complex of factor VIIa and tissue factor then activates factor X, leading to the activation of thrombin. Even though the intrinsic and extrinsic pathways start from separate places, they eventually merge at activation of factor X to form a common pathway.

Only some of the several other chemicals in blood that participate in clot formation are shown in Figure 10. At several steps, Ca^{2+} (factor IV) and platelet factor 3 (PF3) are needed.

Factors Limiting Clot Formation

Like platelet aggregation, fibrin clot formation is limited to the immediate vicinity of the damaged area because certain proteins found in plasma and on the surface of endothelial cells act as **anticoagulants,** chemicals that inhibit blood clotting or coagulation (see **Discovery: Leeches and Bloodletting**,). During the initial phase of clotting, **tissue factor pathway inhibitor** is secreted by healthy endothelial cells and inhibits the extrinsic pathway. Additionally, healthy endothelial cells secrete a molecule called *thrombomodulin*, which binds to thrombin, forming a complex. When it is a part of the thrombomodulin-thrombin complex, thrombin cannot convert fibrinogen to fibrin; instead, it activates **protein C,** another plasma protein that continuously circulates in the plasma. Protein C is an anticoagulant that inhibits both the intrinsic and extrinsic pathways. Thus thrombin not only promotes clotting at injury sites, but also indirectly inhibits clotting in healthy tissue.

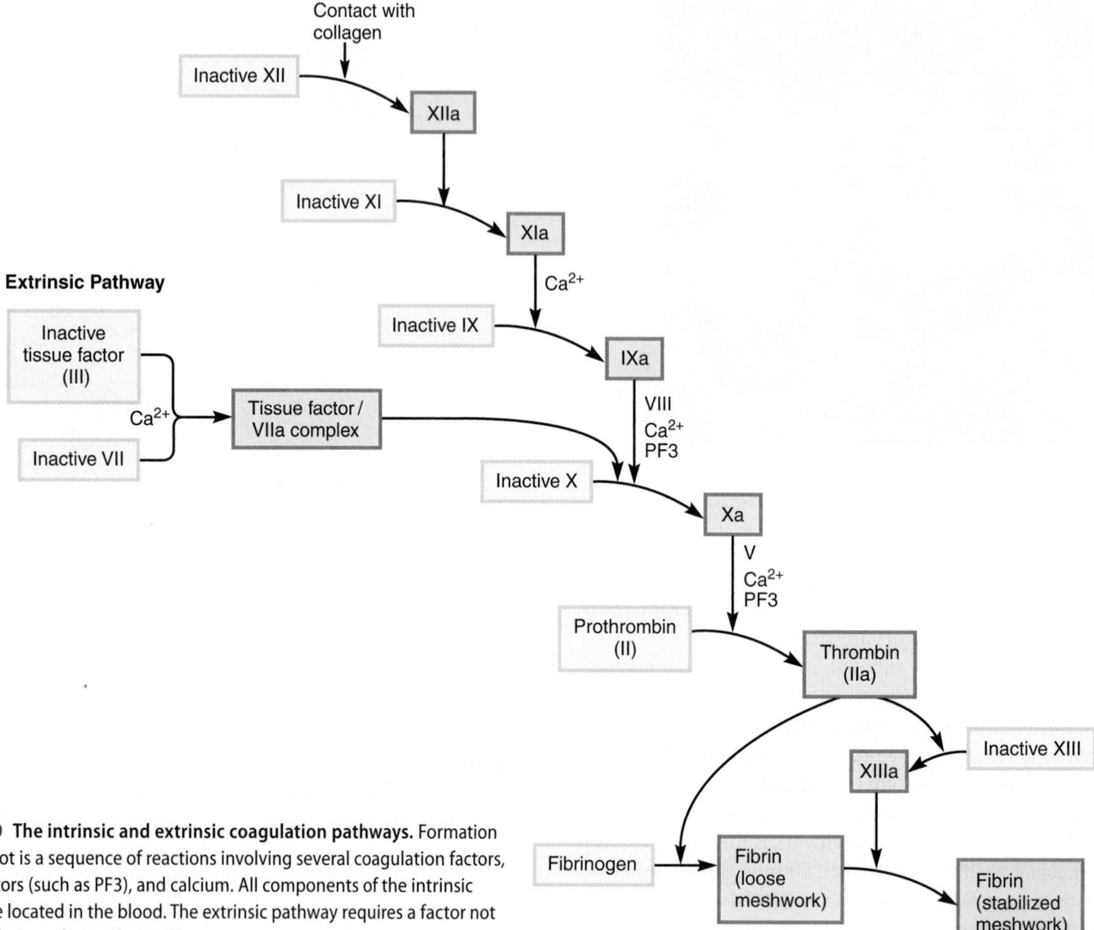

Figure 10 The intrinsic and extrinsic coagulation pathways. Formation of a fibrin clot is a sequence of reactions involving several coagulation factors, platelet factors (such as PF3), and calcium. All components of the intrinsic pathway are located in the blood. The extrinsic pathway requires a factor not in the blood: tissue factor (factor III).

Once formed, fibrin clots are eventually dissolved by **plasmin,** a protein derived from the plasma protein *plasminogen.* Plasmin dissolves clots by enzymatically breaking down fibrin. Plasminogen is converted to plasmin by *plasminogen activators* secreted by a variety of cell types. One example of a plasminogen activator is *tissue plasminogen activator (TPA),* which is secreted by endothelial cells during clot formation. Fibrin activates TPA, which subsequently converts plasminogen to plasmin.

The Role of Coagulation Factors in Clot Formation Disorders

Most coagulation factors are synthesized in the liver, which releases them into the plasma in their inactive form. *Serum* is plasma from which these coagulation factors have been removed.

Several coagulation factors are essential for the formation of a fibrin clot. The lack of any essential factor impairs clot formation and, therefore, results in excessive bleeding.

Hemophilia, a genetic disorder caused by a deficiency of the gene for a specific coagulation factor, is most commonly due to a deficiency of factor VIII in the blood. Another genetic bleeding disorder, von Willebrand's disease, is characterized by reduced levels of vWf, which interferes with platelet plug formation. However, because vWf also serves as a plasma carrier for factor VIII, the absence of vWf causes factor VIII to be less stable and to break down more quickly, leading to lower levels of factor VIII in the bloodstream. Excessive bleeding can also occur if vitamin K is lacking in the diet, because the liver requires vitamin K for the synthesis of many of the proteins necessary in blood clotting.

Aspirin as an Anticoagulant

Aspirin, one of the most commonly used analgesics, has received a great deal of attention lately for another of its actions: It prevents blood clotting. Aspirin is one of many anticlotting drugs given to people who are susceptible to stroke or coronary artery disease. Formation of blood clots in cerebral arteries (in the brain) or coronary arteries (in the heart) can have serious deleterious effects, including death.

When given at low dosages, aspirin acts as an anticoagulant by inhibiting the formation of TXA_2, thereby decreasing platelet aggregation and platelet plug formation. At high dosages, in contrast, aspirin decreases formation of prostacyclin, which actually increases the likelihood of clot

DISCOVERY

Leeches and Bloodletting

Leeches are invertebrates known for their ability to suck blood. From ancient times to the early 20th century, leeches (or lances) were used in *bloodletting,* a procedure involving the removal of blood from a patient in an attempt to treat disease. In 1094, some barbers adopted the practice of bloodletting and became known as "barber surgeons." Barber surgeons were trained not only to cut hair but also to perform surgery, extract teeth, and let blood. During the bloodletting procedure, the patient would hold onto a pole to make the veins more prominent. After the procedure, the bandages would be wrapped around the pole and placed outside the store for advertisement. This image of a red-striped pole is now familiarly known as a "barber's pole."

In the 1800s, bloodletting was used for surgery before the advent of anesthesia. If enough blood was removed, the patient would faint because of a low mean arterial pressure. In the 1900s, the practice of bloodletting came to be considered primitive and virtually came to a stop.

Today, however, the ancient practice of using leeches is being revived. Leeches are currently being used to increase blood flow to specific regions of the body in treatments of ischemia, pain, and inflammation. The medical value of a leech lies in its saliva, which contains an anesthetic, a vasodilator, and an anticoagulant. The anesthetic prevents the patient from feeling the presence of the leech, allowing the leech to stay attached longer. The vasodilator increases blood flow to the leech attachment site, while the anticoagulant prevents the formation of a blood clot.

Science Photo Library/Photo Researchers

Medicinal leeches drawing blood and injecting vasodilators and anticoagulants.

formation. Clinical studies have shown that low doses of aspirin both reduce the incidence of subsequent heart attacks and decrease the severity of damage when given after a heart attack.

Quick Check 3

1. What are platelets? What are some of their secretory products? Explain the role of platelets in forming a blood clot.
2. Name two chemicals that prevent formation of a platelet plug in healthy blood vessels.
3. Describe the role of thrombin in hemostasis.
4. Define *whole blood, plasma,* and *serum*.

6 Diabetes and Cardiovascular Disease

The cardiovascular system includes many intricate processes that must be performed with near-perfection for us to relax, exercise, be outside on cold or hot days, and just change from a sitting to a standing position. In such a complex and important system, things can—and do—go wrong. More than 80 million Americans suffer from some type of cardiovascular disease, which includes coronary artery disease, hypertension, stroke, and heart failure. Hypertension is by far the most prevalent of these conditions (see **Focus on Diabetes: Diabetes and Cardiovascular Disease**).

FOCUS ON DIABETES

Diabetes and Cardiovascular Disease

Cardiovascular diseases are twice as prevalent in patients with diabetes as in people who do not have diabetes; they also have a tendency to strike those individuals with diabetes at a younger age. Diabetes often increases the blood cholesterol level, predisposing a person to atherosclerosis, a condition in which cholesterol, fat, and other materials build up and form plaque in the artery walls. This plaque can eventually clog or block the arteries. Atherosclerosis is also known as *hardening of the arteries*.

Atherosclerosis can, in turn, lead to heart attack, stroke, and many other symptoms, depending on which blood vessels are affected. For example, a lack of blood flow can lead to severe foot disease, to the point that amputation may be required. A decrease in blood flow to the kidneys can lead to *end-stage renal disease*, a condition in which the patient requires regular dialysis to sustain life. Diabetes can also weaken blood vessels, making them fragile. This is the case in diabetic retinopathy, the leading cause of adult-onset blindness.

SYSTEMS INTEGRATION

The cardiovascular system is obviously essential for all other organ systems of the body to function properly, as it provides the nutrients and oxygen needed by those organs, while removing the waste products and carbon dioxide generated by them. The cardiovascular system is also an integral companion to the endocrine system: Movement of blood transports hormones to their target cells. In addition to fulfilling its circulatory role, the heart is an endocrine gland, secreting the hormone atrial natriuretic peptide. Likewise, the blood vessels contain angiotensin converting enzyme (ACE), an enzyme required for the activation of the hormone angiotensin II. The cardiovascular system is also part of the immune system; its leukocytes are delivered to the tissues to fight infections and remove debris.

Conversely, the cardiovascular system depends on other organs for its function. The lungs deliver oxygen to and remove carbon dioxide from the blood traveling through them. The gastrointestinal tract provides nutrients to the blood for delivery to the tissues. The urinary system filters the plasma, removing unnecessary chemicals from it. The spleen filters the blood and removes old red blood cells, which are replaced by new blood cells produced by the bone marrow. The two communication systems—the nervous and endocrine systems—regulate cardiovascular function to maintain mean arterial pressure, the driving force for blood flowing to the tissues.

CHAPTER REVIEW

SUMMARY

1 Overview of the Composition of Blood: The Hematocrit

- The hematocrit is determined by centrifuging blood to separate it into plasma, erythrocytes, leukocytes, and platelets.
- The hematocrit is the percentage of blood volume (approximately 42–45%) composed of erythrocytes.
- Plasma makes up approximately 55% of the total blood volume.
- Leukocytes and platelets together constitute less than 1% of the total blood volume.

2 Plasma

- Plasma is the liquid component of blood, and consists of water and dissolved solutes such as proteins and electrolytes.
- Plasma proteins include albumin, globulins, and fibrinogen.
- Plasma electrolytes include sodium, calcium, and chloride.
- Potassium and bicarbonate are also present in plasma, albeit to a lesser extent.
- Plasma is important in the transport of hormones, nutrients, waste products, clotting proteins, and antibacterial proteins.

3 Erythrocytes

- Erythrocytes (red blood cells) are small, disk-shaped cells that contain neither a nucleus nor any organelles.
- Erythrocytes transport oxygen and carbon dioxide between the lungs and tissues.
- Erythrocytes contain hemoglobin, a protein that binds and thus transports both oxygen and carbon dioxide in blood, and carbonic anhydrase, an enzyme that converts carbon dioxide to bicarbonate and hydrogen ions.
- Erythrocytes are produced in the bone marrow in response to erythropoietin secreted from the kidneys; this production requires iron, a necessary element of hemoglobin.
- Erythrocytes survive in the blood approximately 120 days and are then removed and broken down by the spleen.
- As the hemoglobin is catabolized, most of the iron is recycled for synthesis of new hemoglobin.
- Anemia, a decrease in the oxygen carrying capacity of blood, can be caused by a decrease in the amount of hemoglobin per erythrocyte or a decrease in the number of circulating erythrocytes.

4 Leukocytes

- Leukocytes (white blood cells) are larger than erythrocytes but much sparser in blood.
- Leukocytes function in the immune system, protecting the body against foreign matter.
- The five types of leukocytes are neutrophils, eosinophils, basophils, monocytes, and lymphocytes.
- Neutrophils, eosinophils and basophils are called granulocytes; monocytes and lymphocytes are called agranulocytes.
- There are two common types of blood tests that can be run to quantify leukocytes in the blood: a total white blood cell count (counts the total number of leukocytes in the blood) and a differential white blood cell count (counts the percentage of each type of leukocyte in the blood).
- Although leukocytes travel in blood, they migrate to the tissues where they carry out their defense functions; monocytes develop into macrophages in the tissues.
- Neutrophils, monocytes, and macrophages (and, to a lesser extent, eosinophils) are phagocytes—cells that engulf foreign matter or other debris to remove it from the body.
- Macrophages can differentiate into dendritic cells.

5 Platelets and Hemostasis

- Mechanisms of hemostasis include vascular spasm (blood vessel constriction), platelet plug formation, and blood clot formation.
- In platelet plug formation, platelets aggregate at the site of damage, forming a physical barrier to blood leakage. They then become activated, setting the stage for clot formation.
- In clot formation, fibrinogen (a soluble plasma protein), transformed by proteolytic cleavage into fibrin, forms a fibrous mesh around the platelet plug. This transformation follows the coagulation cascade.
- The coagulation cascade can be initiated by an intrinsic pathway, which involves only components present in plasma, or an extrinsic pathway, which involves factors present in tissues outside blood vessels.
- The blood clot is prevented from spreading beyond the site of damage by substances secreted by undamaged tissues and by other mechanisms.

6 Diabetes and Cardiovascular Disease

- Cardiovascular disease—including coronary artery disease, hypertension, stroke, and heart failure—affects approximately 80 million Americans, with hypertension by far the most prevalent among these conditions.
- Cardiovascular diseases are twice as prevalent in patients with diabetes as in people who do not have diabetes, and they tend to strike individuals with diabetes at a younger age.
- Diabetes increases the blood cholesterol level, predisposing a person to atherosclerosis (also known as *hardening of the arteries*). Atherosclerosis, in turn, can lead to heart attack, stroke, and other serious health symptoms.
- The decreased blood flow commonly associated with diabetes can lead to severe foot disease, sometimes requiring amputation, or to *end-stage renal disease,* a condition requiring regular dialysis treatments to sustain life. Weakened blood vessels, another common blood flow problem, can lead to diabetic retinopathy, the leading cause of adult-onset blindness.

MasteringA&P®

Check your understanding at MasteringA&P and apply your new knowledge to real-life case studies to ensure that you've mastered the content!

EXERCISES

Multiple-Choice Questions

1. Which of the following is *not* found in plasma?
   a) Albumin
   b) Sodium
   c) Glucose
   d) Hemoglobin
   e) Potassium
2. Which component of blood makes up most of the blood volume?
   a) Plasma
   b) Erythrocytes
   c) Leukocytes
   d) Platelets
3. Which organ synthesizes new erythrocytes?
   a) Liver
   b) Spleen
   c) Kidney
   d) Bone Marrow
   e) Heart
4. Which of the following classes of leukocytes function(s) in phagocytosis?
   a) Neutrophils only
   b) Basophils only
   c) Lymphocytes only
   d) Neutrophils and basophils
   e) Basophils and lymphocytes
5. Which class of leukocytes develops into macrophages in tissue?
   a) Neutrophils
   b) Eosinophils
   c) Basophils
   d) Monocytes
   e) Lymphocytes
6. Which class of leukocytes has granules that stain preferentially with a red dye?
   a) Neutrophils
   b) Eosinophils
   c) Basophils
   d) Monocytes
   e) Lymphocytes
7. Which cytokine stimulates erythrocyte production?
   a) ADH
   b) Erythropoietin
   c) Von Willebrand factor
   d) Epinephrine
   e) Renin
8. Which class of blood cell is the most abundant?
   a) Erythrocytes
   b) Neutrophils
   c) Basophils
   d) Monocytes
   e) Lymphocytes
9. Contact of blood with collagen triggers
   a) Platelet aggregation.
   b) Activation of the intrinsic clotting cascade.
   c) Activation of the extrinsic clotting cascade.
   d) Both a and b.
   e) All of the above.
10. What is the function of plasmin?
    a) Stimulate vasoconstriction of damaged blood vessels
    b) Inhibit platelet aggregation in healthy blood vessels
    c) Dissolve blood clots
    d) Activate thrombin
    e) Activate the extrinsic clotting pathway

Objective Questions

11. Plasma with clotting factors removed is called _______.
12. Iron can be stored in the liver bound to _______.
13. During the clotting cascade, _______ converts fibrinogen to fibrin.
14. (B lymphocytes/T lymphocytes) develop into antibody-secreting cells.
15. Catabolism of hemoglobin by the spleen produces (bilirubin/bile).
16. _______ is an enzyme that converts carbon dioxide and water into carbonic acid.
17. At low doses, aspirin inhibits production of _______, thereby decreasing platelet aggregation.

Essay Questions

18. Describe the components of erythrocytes, and explain how they are necessary for erythrocytes to carry out their function.
19. Describe the life cycle of erythrocytes. Be sure to include a description of the roles of the various organs needed for erythrocyte synthesis and breakdown.
20. Describe the body's mechanisms for preventing blood loss following damage to a blood vessel.

Critical Thinking

21. The percentage of blood that consists of erythrocytes affects blood viscosity. Explain how an increase in the hematocrit at higher altitudes can affect blood flow and blood pressure.
22. Aspirin is often prescribed at low doses to people at risk of suffering a heart attack or stroke. Explain the rationale for this prescription.

Photo credits for icons in feature boxes: Focus on Diabetes: Tanyae/Dreamstime; Clinical Connections: Iofoto/Dreamstime; Understanding Exercise: Pixattitude/Dreamstime; Discovery: Yuri Arcurs/Cutcaster.

Glossary

anticoagulant a substance that prevents the clotting of blood

arachidonic acid phospholipid found in the plasma membranes of cells and platelets; it is the precursor molecule for the synthesis of eicosanoids

basophils (bay-so-fils) leukocytes that defend against larger parasites; release histamine, heparin, and other chemicals

bilirubin product of hemoglobin catabolism

carbonic anhydrase enzyme that catalyzes the reversible reaction converting carbon dioxide and water to carbonic acid

coagulation factors plasma proteins that take part in the coagulation cascade

colony-stimulating factors substances that stimulate the production of a specific type of blood cell

eosinophils (ee-oh-sin-oh-fils) leukocytes that attack parasitic invaders too large to be engulfed

erythrocyte (eh-rith-roh-site) red blood cell; functions in transport of oxygen and carbon dioxide

erythropoiesis synthesis of new erythrocytes

erythropoietin hormone released from the kidney in response to low oxygen; stimulates the synthesis of erythrocytes

fibrin the last active clotting factor in the coagulation cascade, which forms polymers that make up the actual clot

hematocrit (heh-mat-oh-krit) the fraction of the blood volume that is occupied by red blood cells

hemoglobin (hee-moh-gloh-bin) a protein in red blood cells that carries oxygen and carbon dioxide

hemophilia genetic disorder affecting clotting factors and associated with excessive bleeding

hemostasis process of stopping bleeding; includes vascular spasms, platelet plug formation, and blood clot formation

interleukins cytokines that communicate messages between leukocytes

leukocytes (loo-koh-sites) white blood cells; function in immune response

monocytes leukocytes capable of phagocytosis; circulate in the blood for only a few hours, and then migrate into tissues where they enlarge and develop into phagocytic cells called macrophages

neutrophil (noo-troh-fil) most abundant leukocyte in the blood; capable of phagocytosis

plasma liquid in the blood made up of water and dissolved solutes, including proteins; represents approximately 20% of the total volume of extracellular fluid

plasmin plasma protein that dissolves a blood clot

platelets cell fragments that play an important role in blood clotting

prostacyclin eicosanoid released from healthy endothelial cells that inhibits platelet aggregation

protein C plasma protein that inhibits blood coagulation

serum plasma with coagulation factors removed

thrombin one of the active clotting factors in the coagulation cascade; reacts with fibrinogen to convert it to fibrin and activates factor XIII

thromboxane A_2(TXA_2) eicosanoid released from aggregated platelets that promotes hemostasis

thrombus blood clot

tissue factor a protein found in subendothelial tissues that initiates the extrinsic clotting pathway when exposed to blood

von Willebrand factor (vWF) protein that anchors platelets to collagen fibers during the platelet adhesion phase of hemostasis

Answers

Apply Your Knowledge Answers

Polycythemia increases resistance to blood flow, thereby increasing mean arterial pressure.

Embolisms are particularly dangerous if they become lodged in the blood vessels of the brain, heart, or lungs, where they may potentially cause stroke, heart attack, or pulmonary embolism, respectively. Each of these outcomes can be deadly.

Answers to End-of-Chapter Multiple Choice and Objective Questions

Multiple Choice Questions

1. d 2. a 3. d 4. d 5. d 6. b 7. b 8. b 9. d 10. b

Objective Questions

11. serum
12. ferritin
13. thrombin
14. B lymphocytes
15. bilirubin
16. Carbonic anhydrase
17. thromboxane A_2

The Respiratory System: Pulmonary Ventilation

The Respiratory System: Pulmonary Ventilation

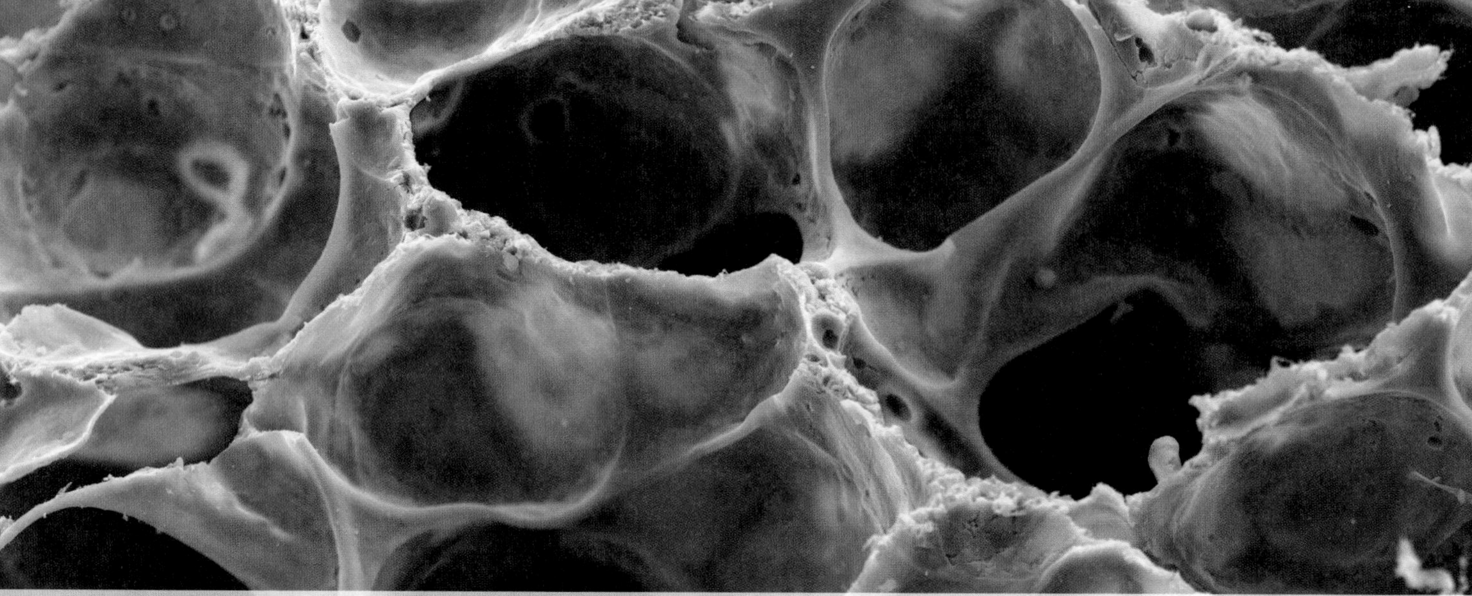

Scanning electron micrograph of (lung) alveoli.

There are many folk cures for *singultus*, including holding one's breath, rebreathing into a paper bag, drinking water while standing on one's head, or startling a person suffering from singultus. That's right: Singultus is the medical term for hiccups.

Hiccups are a spastic contraction of the *diaphragm,* the muscle of inspiration. During a hiccup, air rapidly enters a person's lungs and then rapidly stops as the *glottis* (a flap that closes the airways during swallowing to prevent food from entering the lungs) closes, producing the characteristic sound "hiccup." Most of us consider hiccups a nuisance. But many people suffer from intractable hiccups, which are bouts of hiccups that last more than 48 hours and could be indicative of an underlying disease. In pregnant women, hiccups may occur more frequently due to the pressure the fetus puts on the diaphragm.

Some of the folk cures are based on physiology of the respiratory system. For example, holding one's breath or rebreathing into a bag increases the amount of carbon dioxide in the blood, which may inhibit the neural pathway for hiccups (which is similar to the pathways that control breathing). Startling a person alters the individual's breathing pattern, which may also affect the neural pathway for hiccups. As for standing on one's head and drinking water, that feat could make a person choke, which would definitely alter respiration—and, possibly, hiccups.

CHAPTER OUTLINE

LEARNING OUTCOMES *After studying this chapter, you should be able to:*

- Compare internal respiration to external respiration, and describe the processes occurring in each.
- Describe the major structures of the respiratory system, and list the functions of each.
- Describe the anatomy of the respiratory membrane, and explain how its structure facilitates the exchange of gases between blood and air.
- Describe the anatomy of alveoli. Explain the roles of type I cells, type II cells, and alveolar macrophages in respiratory function.
- Explain the function of pulmonary surfactant.
- Describe the mechanics of breathing, and name the muscles of respiration. List the different pulmonary pressures, and explain their roles in ventilation.
- Describe the roles of lung compliance and airway resistance in ventilation.
- List the different lung volumes and capacities. Explain the clinical applications of lung volumes, forced vital capacity, and forced expiratory volume.

Before You Begin

Make sure you have mastered the following topics:

- Diffusion
- Pulmonary circulation
- Capillary anatomy
- Flow, pressure gradients, and resistance
- Compliance

We know that our cells produce most of the adenosine triphosphate (ATP) they need through oxidative phosphorylation, a process that requires oxygen. During energy metabolism, carbon dioxide is also produced. We are also aware of how blood flow to the tissues provides our cells with the nutrients they need, including oxygen, and removes waste products, including carbon dioxide. We now examine the respiratory system, which functions in delivering oxygen to blood and eliminating carbon dioxide from blood.

We all know that the depth (and rate) of breathing varies based on level of activity. When you exercise, for example, you breathe more deeply than while resting, and when something startles you, you might hold your breath for a moment before breathing a deep "sigh of relief." In either case, the deeper breath brings more air into the lungs, where oxygen in inhaled air moves into the blood, and where carbon dioxide moves out of the blood and leaves the body in exhaled air. In this chapter we explore *pulmonary ventilation*—the movement of air into and out of the lungs.

After a brief overview of respiratory function, we examine the functional anatomy of the respiratory system. We then explore how the muscles of respiration produce several different pulmonary pressures, and discover the roles of those pressures in pulmonary ventilation. Next we examine the factors that affect the rates of air flow during pulmonary ventilation. Finally, we conclude by considering the bases for and clinical significances of various lung volumes and capacities.

1 Overview of Respiratory Function

The respiratory system is so named because its function is **respiration,** the process of gas exchange. This exchange of gases occurs at two levels, termed *internal respiration* and *external respiration* (**Figure 1**). Whereas **internal respiration** (or cellular respiration) refers to the use of oxygen within mitochondria to generate ATP by oxidative phosphorylation, and the production of carbon dioxide as a waste product, **external respiration** refers to the exchange of oxygen and carbon dioxide between the atmosphere and body tissues, which involves both the respiratory and circulatory systems.

External respiration encompasses four processes:

① **Pulmonary ventilation,** the movement of air into the lungs (inspiration) and out of the lungs (expiration) by bulk flow.

② Exchange of oxygen and carbon dioxide between lung air spaces and blood by diffusion.

③ Transportation of oxygen and carbon dioxide between the lungs and body tissues by the blood.

④ Exchange of oxygen and carbon dioxide between the blood and tissues by diffusion.

This chapter focuses on the first process, pulmonary ventilation.

In addition to its main function—respiration—the respiratory system performs several other functions, including (1) contributing to the regulation of acid-base balance in the blood, (2) enabling vocalization, (3) participating in defense against pathogens and foreign particles in the airways, (4) providing a route for water and heat losses (via the expiration of air that was moistened and warmed during inspiration), (5) enhancing venous return (through the respiratory pump), and (6) activating certain plasma proteins as they pass through the pulmonary circulation.

2 Anatomy of the Respiratory System

The major organs of the respiratory system are the *lungs,* which are located in the thoracic cavity. Each lung is divided into lobes; the right lung has three lobes and the left lung has two lobes. Air gets into and out of the lungs by way of the *upper airways* and a network of tubes forming a system of passageways called the *respiratory tract* (**Figure 2**). The following sections describe how these various components of the respiratory system are specialized for certain functions.

Upper Airways

The term **upper airways** refers to air passages in the head and neck. Air enters the *nasal cavity* and/or the *oral cavity,* both of which lead to the **pharynx,** a muscular tube that serves as a common

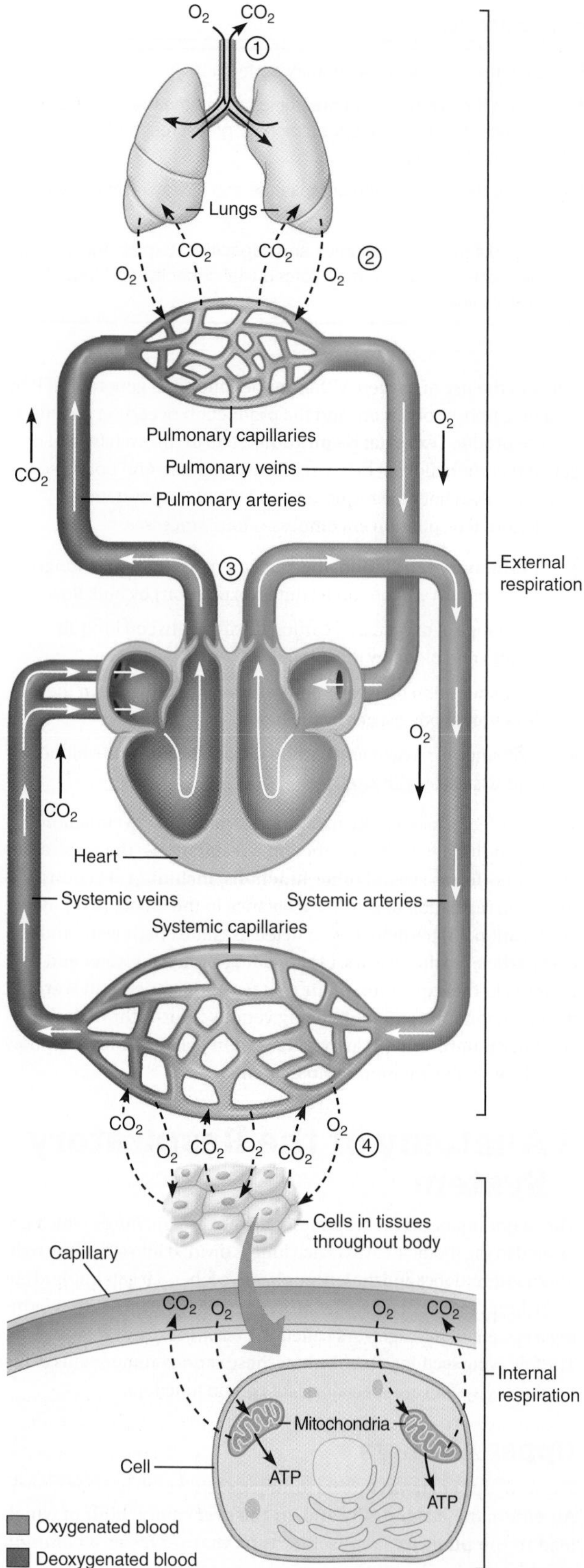

Figure 1 Relationship between external respiration and internal respiration. In external respiration, ① air moves between the atmosphere and the lungs, ② oxygen and carbon dioxide are exchanged between lung tissue and the blood, ③ oxygen and carbon dioxide are transported in the blood, and ④ oxygen and carbon dioxide are exchanged between systemic tissues and the blood. Internal respiration is the use of oxygen and production of carbon dioxide by cells, primarily within the mitochondria.

Q Which metabolic pathway in the mitochondria utilizes oxygen?

passageway for both air and food. After the pharynx, the passageways for food and air diverge: Food enters the esophagus, a muscular tube leading to the stomach, whereas air enters the first structure in the respiratory tract, the larynx. Occlusion of the upper airways can occur acutely when a person chokes on food or chronically during diseases such as sleep apnea (see **Clinical Connections: Sleep Apnea**).

The Respiratory Tract

The **respiratory tract** includes all air passageways leading from the pharynx to the lungs (including those present within the lungs themselves). The respiratory tract can be functionally divided into two components: a conducting zone and a respiratory zone (Figure 3). The **conducting zone,** the upper part of the respiratory tract, functions in conducting air from the larynx to the lungs. The **respiratory zone,** the lowermost part of the respiratory tract, contains the sites of gas exchange within the lungs. The primary anatomical difference between the conducting and respiratory zones that determines whether gas exchange occurs is in the *thickness* of the walls surrounding the air spaces; only air spaces with sufficiently thin walls can participate in gas exchange.

The Conducting Zone

The conducting zone starts with the **larynx,** a tube held open by *cartilage* (a dense connective tissue) in its walls. To keep food from entering the respiratory tract, the opening to the larynx, called the **glottis,** is covered by a flap of tissue called the **epiglottis,** which during swallowing is forced down over the glottis and prevents food or water from entering the larynx. The larynx houses the vocal cords (or vocal folds), which generate sounds by vibrating when air passes over them.

After the larynx, the next component of the respiratory tract is the **trachea,** a tube approximately 2.5 cm in diameter and 10 cm long that runs parallel with and anterior to the esophagus. Unlike the esophagus, which is collapsed except during swallowing, the trachea stays open because the front and sides of its wall contain 15–20 C-shaped bands of cartilage that provide structural rigidity. Without this rigidity, the trachea would collapse during inspiration as pressure drops. The absence of cartilage on the posterior wall of the trachea allows the esophagus behind it to expand with the passage of food.

After it enters the thoracic cavity, the trachea divides into left and right **bronchi** (singular: *bronchus*) that conduct air to each lung. Like the trachea, the bronchi contain cartilage; however, the cartilage forms rings around the entire circumference of the bronchus. Within each lung, the bronchi divide into smaller tubes called **secondary bronchi;** three secondary bronchi conduct air to the lobes of the right lung, and two secondary bronchi conduct air to

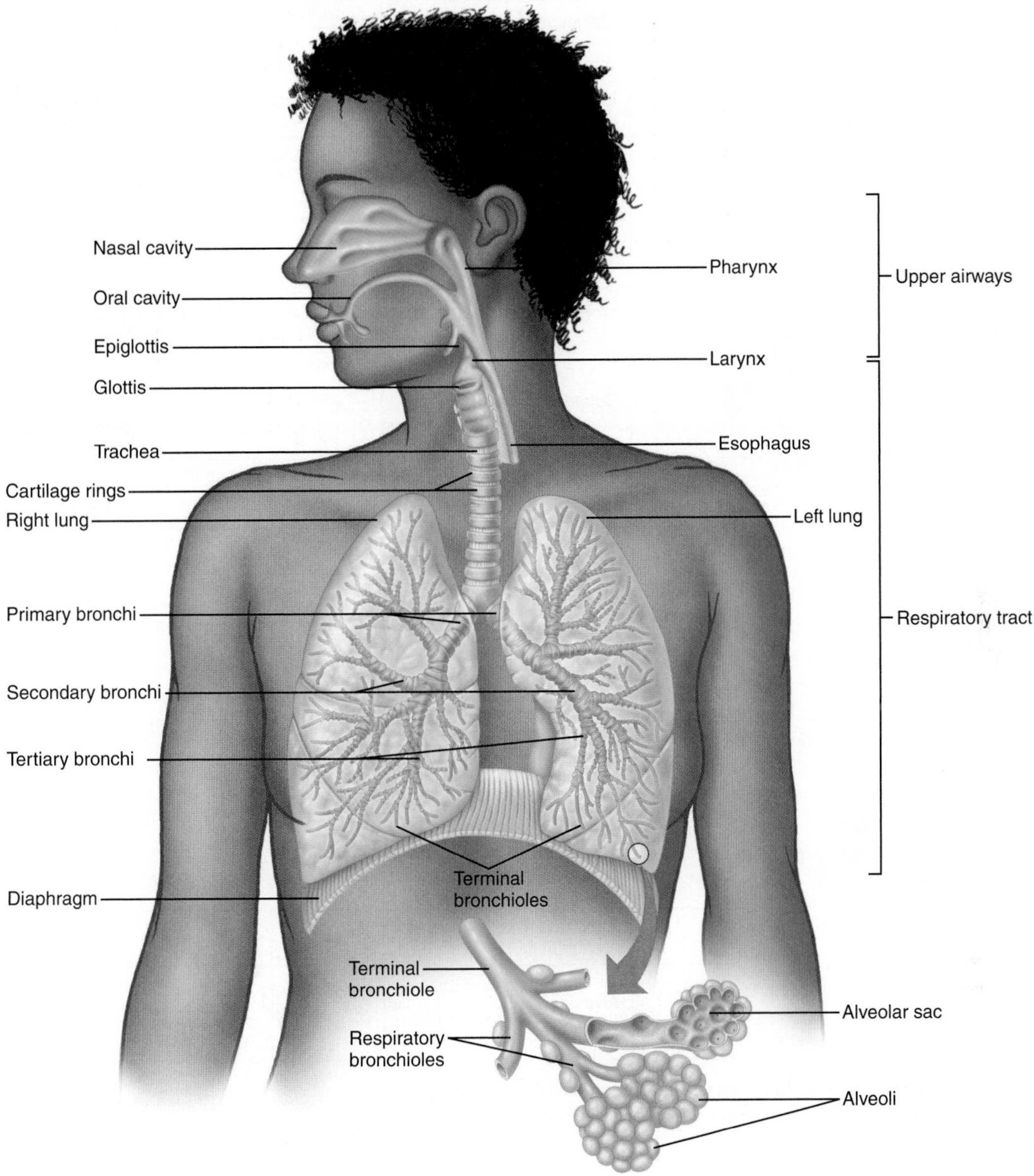

Figure 2 **Anatomy of the upper airways and the respiratory tract.**

the lobes of the left lung. The cartilage in secondary bronchi is less abundant than that in the primary bronchi and occurs as plates.

Each secondary bronchus divides into smaller *tertiary bronchi,* which in turn divide into successively smaller bronchi such that approximately 20–23 orders of branching occur. The extensive branching ultimately results in approximately 8 million tubules, the smallest less than 0.5 mm in diameter.

Once the tubules become less than 1 mm in diameter, they are called **bronchioles** ("little bronchi"). Unlike the larger bronchi, bronchioles have no cartilage and are capable of collapsing. To help prevent such collapse, the walls of bronchioles contain elastic fibers. The bronchioles divide further into **terminal bronchioles,** the final and smallest component of the conducting zone.

The primary function of the conducting zone is to provide a passageway through which air can enter and exit the respiratory zone, where gas exchange occurs. The conducting zone holds approximately 150 mL of air and is considered "dead space," because the air does not participate in gas exchange with blood in this region. The functional significance of the dead space is described later. As air travels through the conducting zone, its temperature is adjusted to body temperature and humidified to keep the respiratory tract moist.

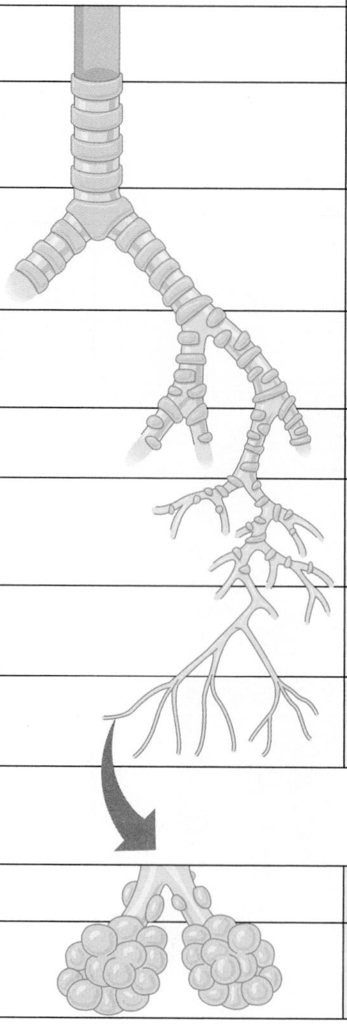

Conducting zone

Structure	Inner diameter (mm)	Cilia	Goblet cells	Cartilage	Smooth muscle
Larynx	35–45	+++	+++	+++	0
Trachea	20–25	+++	+++	+++ (C-shaped)	+
Primary bronchi	12–16	+++	++	+++ (rings)	++
Secondary bronchi	10–12	+++	++	+++ (plates)	++
Tertiary bronchi	8–10	+++	++	++ (plates)	++
Smaller bronchi	1–8	+++	+	+ (plates)	++
Bronchioles	0.5–1	++	+	0	+++
Terminal bronchioles	<0.5	++	0	0	+++

Respiratory zone

Structure	Inner diameter (mm)	Cilia	Goblet cells	Cartilage	Smooth muscle
Respiratory bronchioles	<0.5	+	0	0	+
Alveolar sacs	0.3	0	0	0	0

Figure 3 Anatomical features of the conducting and respiratory zones of the respiratory tract. 0 indicates not present, + indicates sparse, ++ indicates present, and +++ indicates abundant.

The conducting zone is lined by an epithelium that changes in composition as the tubules become smaller in diameter. The epithelium lining the larynx and trachea (and to a lesser extent, the bronchi) contains numerous goblet cells; also abundant in the epithelium throughout the conducting zone are ciliated cells (**Figure 4**). **Goblet cells** secrete a viscous fluid called *mucus,* which coats the airways and traps foreign particles in inhaled air; the **cilia** (hairlike projections) of the ciliated cells beat in a whiplike fashion to propel the mucus containing the trapped particles up toward the glottis and then into the pharynx, where the mucus is then swallowed. This process, called the *mucus escalator,* prevents mucus from accumulating in the airways and clears trapped foreign matter from this area. Accumulation of mucus in the airways increases the likelihood of infections such as bronchitis and pneumonia because it promotes retention and growth of bacteria. Because cilia are easily paralyzed by tobacco smoke, smoking disables the mucus escalator, so that mucus and trapped debris accumulate in the airways and can be cleared only by coughing; this is one reason for the familiar "smoker's cough." At levels below the bronchioles, phagocytic cells called macrophages engulf foreign matter in the interstitial space and on the surface of the epithelium.

Apply Your Knowledge

Mucus production in the conducting zone is essential to trap foreign particles and keep them from entering the lungs. Mucus production increases, however, when a person has a cold or other type of respiratory tract infection. What if any benefit does increased mucus production provide during such infections? We all know it is more difficult to breathe when we have a cold. Can you explain why?

CLINICAL CONNECTIONS

Sleep Apnea

Apnea is the cessation of breathing, and *sleep apnea* is a periodic cessation of breathing that occurs while sleeping. During sleep apnea, breathing may stop for just a few seconds or for several minutes, and these cessations may occur as often as 100 times per hour. A person with sleep apnea typically shifts from deep sleep to light sleep during episodes. Although the person often is unaware that these events are happening, several symptoms occur during sleep, including snoring, snorting or gasping sounds, and frequent awakenings. When awake, the sufferer often feels fatigued, sleepy, and irritable; suffers from morning headaches and dry mouth; and often has difficulty concentrating.

Two types of sleep apnea are distinguished: obstructive and central. Obstructive sleep apnea, the most common form, is often caused by excessive relaxation of muscles in the upper airways, causing them to collapse. Other possible causes include obstruction of the nasal passages or blockage of the air passages by the tongue, tonsils, and/or adenoids. Overweight people are more susceptible to obstructive sleep apnea, and this condition runs in families. Men are more likely to have sleep apnea than women. Central sleep apnea, a much rarer condition, is caused by neurological problems in the respiratory centers of the brain. Because the person does not attempt to breathe, snoring is absent in this type of sleep apnea.

Diagnosis of sleep apnea is difficult because the person is often unaware of the condition and there are no routine tests that can be performed, such as blood tests. Instead, diagnosis often requires polysomnography: the recording of electroencephalograms (EEGs), muscle tension, eye movements, respiration, blood oxygen levels, and sounds during sleep.

Treatments for sleep apnea range from behavioral modifications to surgical interventions for more serious cases. Behavioral modifications include sleeping on the side instead of the back, losing weight, and avoiding drugs with sleep-enhancing properties (such as alcohol and depressants). For moderate cases, patients may be put on continuous positive airway pressure machines (CPAPs), consisting of a mask that is placed over the nose and a pump that supplies air at a pressure that maintains open airways throughout the respiratory cycle. Surgical intervention involves physically removing the obstruction, which could include tonsillectomy, *uvulopalatoplasty* (removal of the uvula and part of the soft palate), and, in severe cases, *tracheostomy* (insertion of a tube into the trachea to bypass the upper airways).

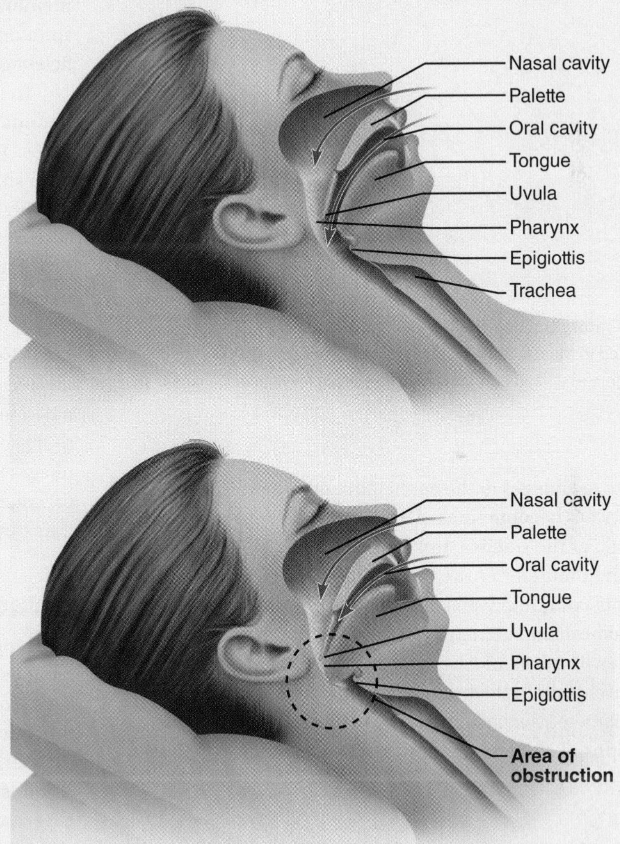

During snoring (top), airflow is blocked only partially; in sleep apnea (bottom), airflow is blocked completely.

Critical Thinking Questions

1. Describe the typical symptoms of sleep apnea, including symptoms experienced while awake.
2. What are the primary causes of the two types of sleep apnea?
3. Describe the behavioral modifications and surgical interventions for treating sleep apnea, and what you think their relative advantages and disadvantages might be.

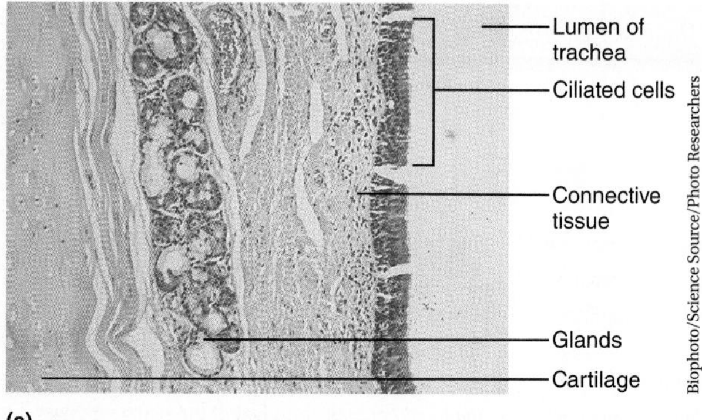

(a)

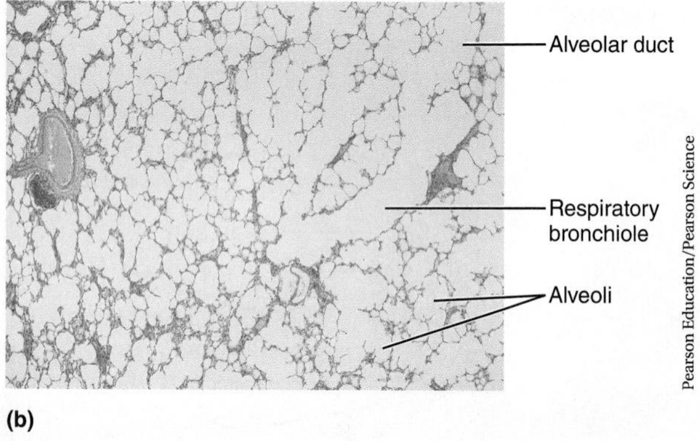

(b)

Figure 4 Respiratory tract epithelia. (a) A photomicrograph of tracheal epithelium, located in the conducting zone. **(b)** A photomicrograph of respiratory bronchioles and alveoli, located in the respiratory zone.

In addition to changes in the epithelium, other tissue changes occur as the airways become smaller. As we know, cartilage is plentiful in the walls of the trachea and bronchi, but it becomes less abundant as the diameter of the bronchi decreases, until in the bronchioles it is completely absent. Smooth muscle is sparse in the trachea and bronchi but increases in abundance as the airways become smaller. The lack of cartilage and the presence of circular smooth muscle within the bronchioles enable these airways to change their diameter; such changes alter the resistance to air flow, just as the actions of circular smooth muscle in arterioles enables them to alter resistance to blood flow.

The Respiratory Zone

The rate at which substances diffuse across a membrane increases as membrane surface area increases and as membrane thickness decreases. The arrangement of structures in the respiratory zone—the site of gas exchange—maximizes surface area and minimizes thickness, thereby facilitating the diffusion of oxygen and carbon dioxide between air and blood.

Past the site where the terminal bronchioles of the conducting zone branch, the respiratory zone begins (**Figure 5**). The first respiratory zone structures, **respiratory bronchioles,** terminate in *alveolar ducts,* which lead to **alveoli,** the primary structures where gas exchange occurs (Figure 5a). Most alveoli are organized into groups called **alveolar sacs,** which resemble clusters of grapes; some alveoli open off of respiratory bronchioles.

Adjacent alveoli are not completely independent structures. Because they are connected by *alveolar pores* (Figure 5b), air flows between alveoli, allowing for equilibration of pressure within the lungs.

Alveolar structure facilitates diffusion of gases between blood and air. The wall of an alveolus consists primarily of a single layer of epithelial cells called **type I alveolar cells** overlying a basement membrane (Figure 5c). We know that a capillary wall consists of a single layer of endothelial cells and an underlying basement membrane. In many places in the lungs, the alveolar epithelial cells and the endothelial cells of the nearby capillaries are so close together that their basement membranes are fused (Figure 5d). Together, the capillary and alveolar walls form a barrier, called the **respiratory membrane,** which separates air from blood. The thinness of the respiratory membrane—only about 0.2 μm thick—is essential for efficient gas exchange.

In addition to the thinness of the respiratory membrane, the abundance of alveoli and capillaries facilitates the diffusion of gases. The approximately 300 million alveoli in the lungs have a total surface area of approximately 100 square meters, which is roughly the size of a tennis court. The capillary supply to the alveoli is so rich that many physiologists think of pulmonary vasculature not in terms of arterioles and capillaries, but rather as a "sheet of blood" surrounding the alveoli. **Figure 6** shows a cast of the blood supply to the lungs, and of the alveoli.

Also located in the alveoli are *type II alveolar cells* and **alveolar macrophages,** which engulf foreign particles and pathogens inhaled into the lungs. These macrophages are cells that are free to roam around in the alveoli by amoeboid movements. Dead macrophages are moved from the alveoli into the conducting zone, where the mucus escalator carries the macrophages to the pharynx so that they can be swallowed with mucus.

Structures of the Thoracic Cavity

The lungs are located within the thoracic cavity. The **chest wall** is composed of structures that protect the lungs (**Figure 7**): the rib cage (consisting of 12 pairs of ribs), the sternum (breastbone), the thoracic vertebrae, and associated muscles and connective tissue (primarily hyaline cartilage). Muscles of the chest wall, which are responsible for breathing, include the **internal intercostals** and **external intercostals,** located between the ribs, and the dome-shaped **diaphragm,** which seals off the lower end of the chest wall and separates the thoracic and abdominal cavities. Muscles and connective tissue in the neck close off the chest wall at its upper end. The chest wall forms an airtight, continuous barrier around the lungs.

The interior surface of the chest wall and the exterior surface of the lungs are lined by a membrane called the **pleura,** which is composed of a layer of epithelial cells and connective tissue; each lung is surrounded by a separate **pleural sac** (see Figure 7). The side of the pleural sac attached to the lung tissue is called the *visceral pleura;* the side attached to the chest wall is called the *parietal pleura.* Between the two pleurae is a very thin compartment

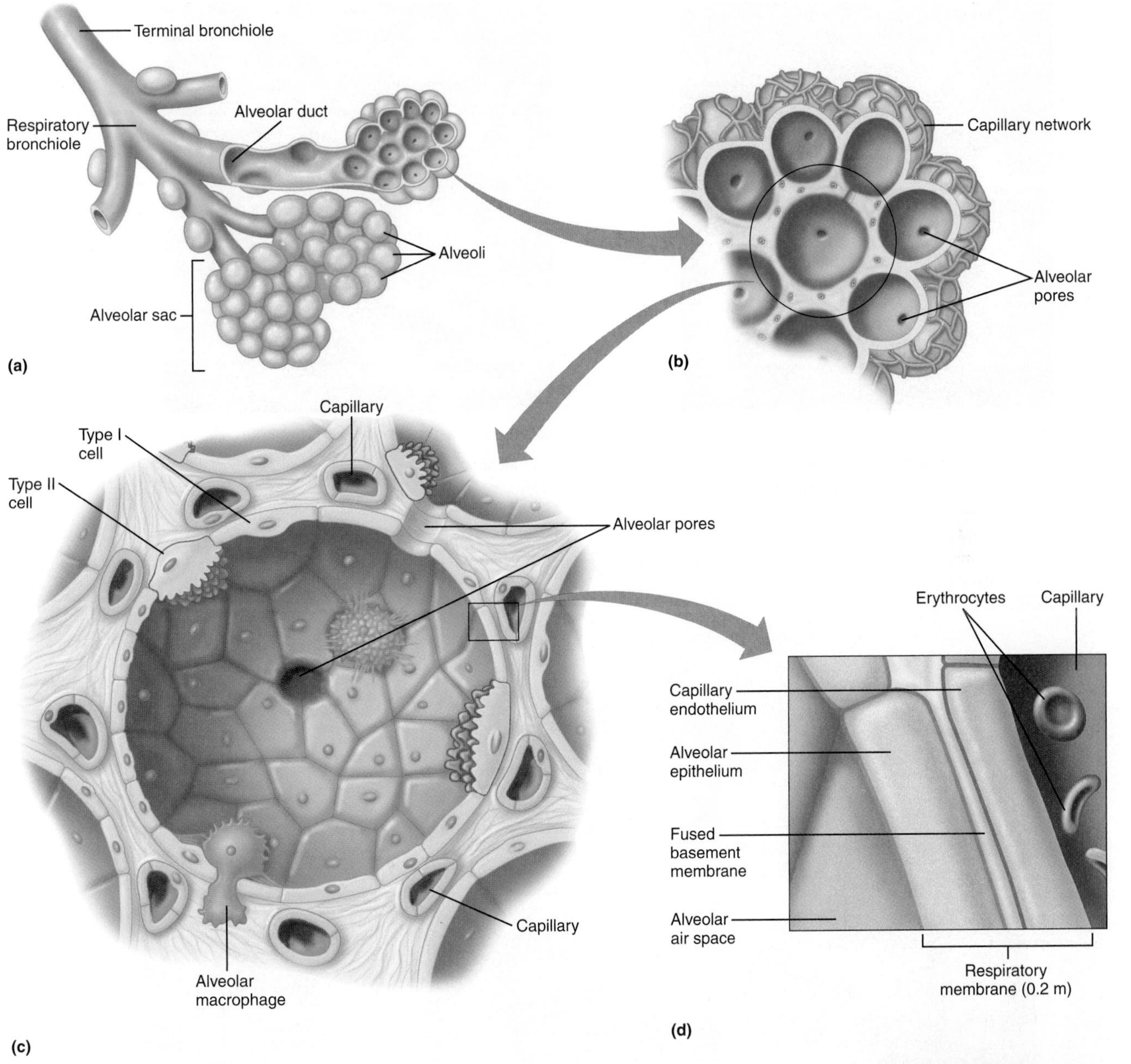

Terminal bronchiole
Respiratory bronchiole
Alveolar duct
Alveoli
Alveolar sac
(a)

Capillary network
Alveolar pores
(b)

Capillary
Type I cell
Type II cell
Alveolar pores
Capillary
Alveolar macrophage
(c)

Erythrocytes
Capillary
Capillary endothelium
Alveolar epithelium
Fused basement membrane
Alveolar air space
Respiratory membrane (0.2 m)
(d)

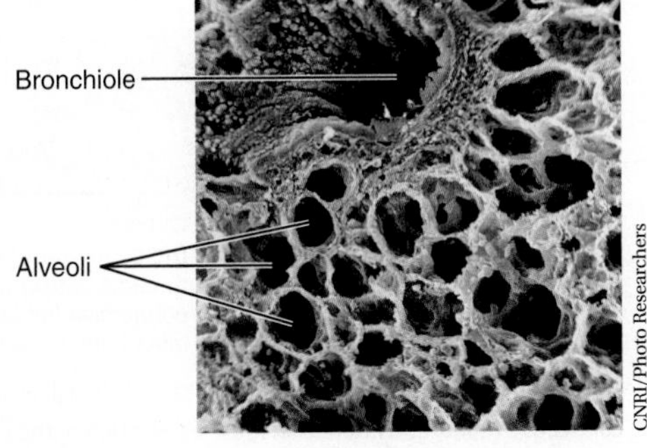

(e)

Figure 5 Anatomy of the respiratory zone. (a) Structures in the respiratory zone, which begins where terminal bronchioles branch into respiratory bronchioles. Alveoli are shown both in clusters called alveolar sacs at the ends of alveolar ducts, and associated with alveolar ducts and respiratory bronchioles. **(b)** The dense capillary network surrounding alveoli. **(c)** Wall of an alveolus in cross section. The alveolar wall contains type I cells, which make up the structure of the wall, and type II cells, which secrete surfactant. Also found in alveoli are macrophages. **(d)** Enlargement of the respiratory membrane showing the close association between alveolar and capillary walls. **(e)** Scanning electron micrograph of human lung showing a bronchiole and alveoli.

Q Macrophages are derived from which type of white blood cell?

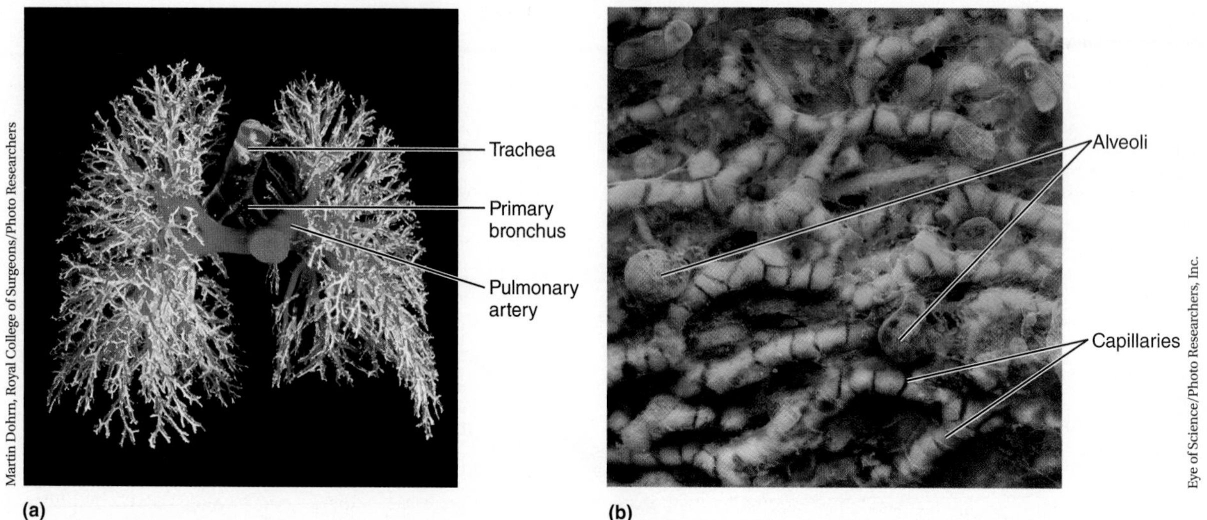

Figure 6 Blood supply to the lungs. (a) Resin cast of pulmonary arteries and bronchi. Blood vessels are shown in red, airways in white. **(b)** Scanning electron micrograph of capillaries around alveoli.

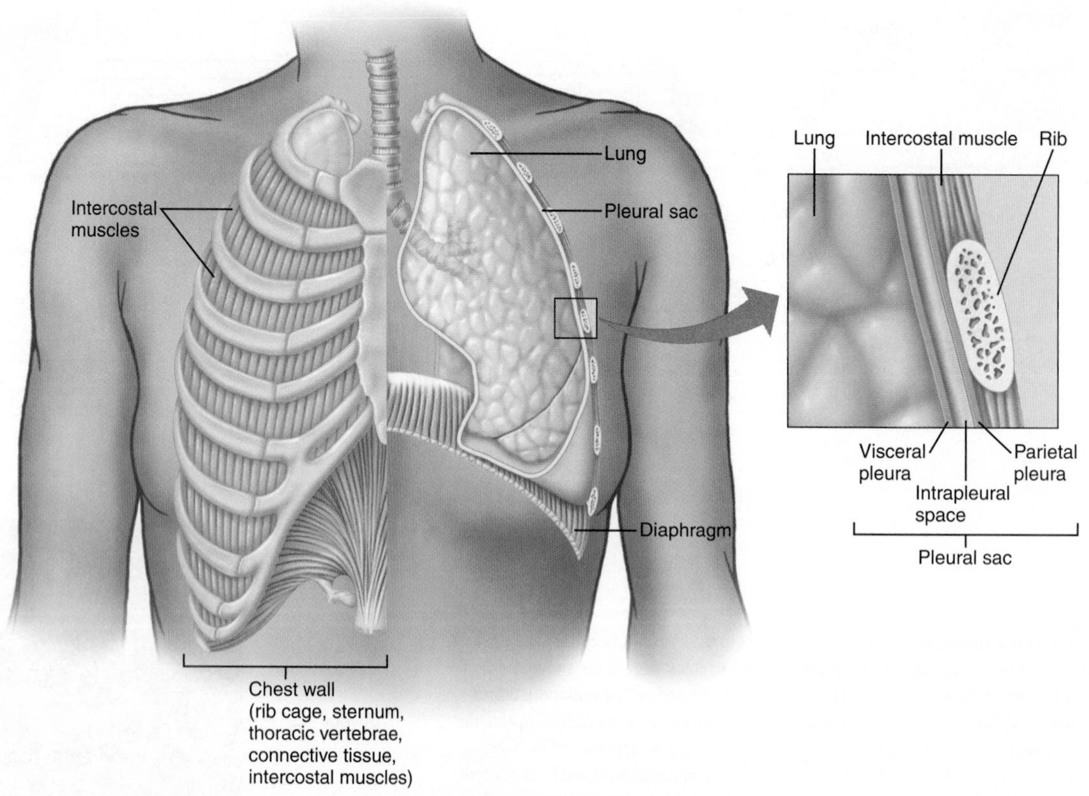

Figure 7 Chest wall and pleural sac. The chest wall includes the ribs, sternum, thoracic vertebrae, connective tissue, and intercostal muscles. The side of the pleural sac attached to the lung is called the visceral pleura; the side of the sac attached to the chest wall is called the parietal pleura. The fluid-filled intrapleural space is much thinner than shown here, with a total volume of approximately 15 mL.

called the **intrapleural space,** which is filled with a small volume (approximately 15 mL) of *intrapleural fluid.*

Quick Check 1

1. What are the anatomical and functional differences between the conducting zone and the respiratory zone of the respiratory tract?
2. Which structures make up the respiratory membrane?
3. Name the structures of the chest wall.

3 Forces for Pulmonary Ventilation

Air flow into or out of the lungs (*breathing* or *ventilation*) resembles blood flow through the vasculature in that both are examples of bulk flow driven by pressure gradients. Ventilation occurs because of the presence of pressure gradients between the alveoli and the outside air (atmosphere). Air moves down such a pressure gradient, from an area of high pressure to one of low pressure. Inspiration occurs when the pressure in the alveoli is less than the pressure in the atmosphere, creating a pressure gradient for air to move into the alveoli; expiration occurs when the pressure in the alveoli exceeds the pressure in the atmosphere, creating a pressure gradient for air to leave the alveoli. First we discuss the pressure gradients; then we discuss how these pressure gradients are created.

Pulmonary Pressures

Four primary pressures are associated with ventilation: atmospheric pressure, intra-alveolar pressure, intrapleural pressure, and transpulmonary pressure. **Figure 8a** illustrates these pressures when a lung is at rest (between breaths, at the end of a quiet expiration). The volume of air in the lungs between breaths is called the *functional residual capacity (FRC).* When the lungs are at the FRC, all forces acting across the lung and chest wall are balanced and the system is in a steady state.

Atmospheric pressure (P_{atm}) is the pressure of the outside air. At sea level, atmospheric pressure is normally 760 mm Hg, although it varies slightly with the weather. At altitudes higher than sea level, atmospheric pressure decreases. For simplicity, we consider only sea level and assume that atmospheric pressure is constant. All other lung pressures are expressed relative to atmospheric pressure.

Intra-alveolar pressure (P_{alv}) is the pressure of air within the alveoli. At rest (FRC), it is equal to atmospheric pressure, and thus is 0 mm Hg. However, intra-alveolar pressure varies during the phases of ventilation. In fact, the difference between intra-alveolar pressure and atmospheric pressure is the pressure gradient that drives ventilation. When atmospheric pressure exceeds intra-alveolar pressure (when intra-alveolar pressure is negative), inspiration occurs;

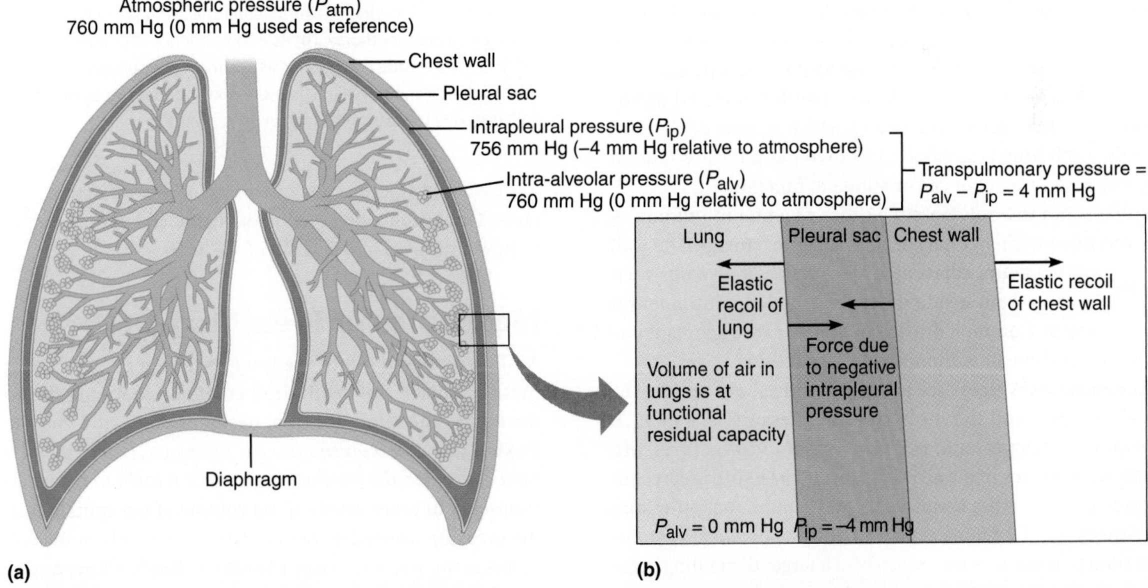

Figure 8 Pleural pressures at rest. (a) Pulmonary pressures for a lung at rest. Intra-alveolar pressure is the pressure within the alveoli; intrapleural pressure is the pressure in the pleural sac. Transpulmonary pressure is the difference between intra-alveolar pressure and intrapleural pressure. All pressures are given as absolute pressures and as pressures relative to atmospheric pressure. **(b)** Pressures and elastic forces when the lungs are at the functional residual capacity (between breaths). At the functional residual capacity, intra-alveolar pressure (P_{alv}) = atmospheric pressure = 0 mm Hg. The lung is distended, and an elastic recoil force tends to collapse it inward; the chest wall is compressed, and an elastic recoil force tends to expand it outward. The net force that these two opposing forces exert on the two sides of the pleural sac creates a negative intrapleural pressure (P_{ip}). The entire system is stable in that the elastic recoil forces of the lungs and chest wall are in balance, and no net change in size of the lung or chest wall occurs.

when intra-alveolar pressure exceeds atmospheric pressure (when intra-alveolar pressure is positive), expiration occurs.

Intrapleural pressure (P_{ip}) is the pressure inside the pleural space. Recall that the intrapleural space contains intrapleural fluid, not air. At rest, intrapleural pressure is −4 mm Hg. Although intrapleural pressure varies with the phase of ventilation, it is always negative during normal breathing (and always less than intra-alveolar pressure) because opposing forces exerted by the chest wall and the lungs tend to pull the parietal pleura and visceral pleura apart.

The lungs and the chest wall are both elastic, meaning that if they are stretched or compressed out of their natural positions, they tend to recoil, or spring back into those natural positions. At rest, the chest wall is compressed and tends to recoil outward (as would a compressed spring), whereas the lungs are stretched and tend to recoil inward, like an inflated balloon (Figure 8b). These forces tend to move the chest wall and lungs apart, but the two structures do not separate because the surface tension of the intrapleural fluid keeps the parietal pleura and the visceral pleura from pulling apart. (This phenomenon is similar to what happens when two wet microscope slides are stacked: The surface tension of the water holds the two slides tightly together, and it takes a lot of force to pull them apart.) Therefore, the chest wall pulls outward on the intrapleural space while the lungs pull inward (which tends to separate the visceral and parietal pleurae), but this in turn creates a negative intrapleural pressure that opposes the separation and thus opposes the outward and inward recoil forces of the chest wall and the lungs, respectively.

To maintain the negative intrapleural pressure, the pleural sac must be airtight. If the seal of the pleural sac is broken, as by a piercing wound to the chest, then the negative intrapleural pressure is lost as it equilibrates with atmospheric pressure. Without the negative intrapleural pressure, the lungs recoil and collapse while the chest wall recoils and expands (**Figure 9**). This results in a condition called **pneumothorax** (air in the intrapleural space), which is quite dangerous. Fortunately, each lung is isolated in its own pleural sac so that if one lung collapses, the other can continue to function.

Trauma is not the only possible cause of a pneumothorax. A *spontaneous pneumothorax* may occur if disease damages the wall of the pleura adjacent to a bronchus or alveolus such that air from inside the lungs enters the intrapleural space. Common diseases that may cause spontaneous pneumothorax include pneumonia and emphysema.

Transpulmonary pressure is the difference between the intrapleural pressure and the intra-alveolar pressure ($P_{alv} - P_{ip}$). The transpulmonary pressure is a *transmural pressure* (pressure difference on two sides of a wall). Transmural pressures represent the distending force across a wall, so that transpulmonary pressure indicates the distending force across the lungs. Specifically, an increase in transpulmonary pressure creates a larger distending pressure across the lungs, which is accompanied by the expansion of the lungs (alveoli).

When the lungs are at rest (between breaths), all breathing muscles are relaxed, and no air moves into or out of the lungs because there is no pressure gradient to drive air movement. To move air into or out of the lungs requires muscular force to create a pressure gradient. The generation of that muscular force is discussed in the following section on the mechanics of breathing—that is, how air moves into and out of the lungs.

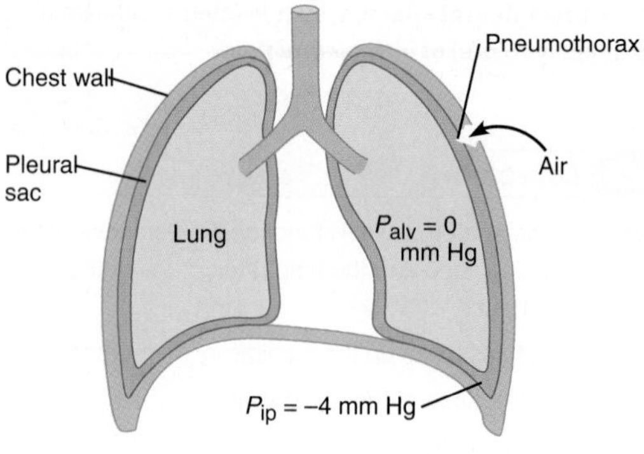

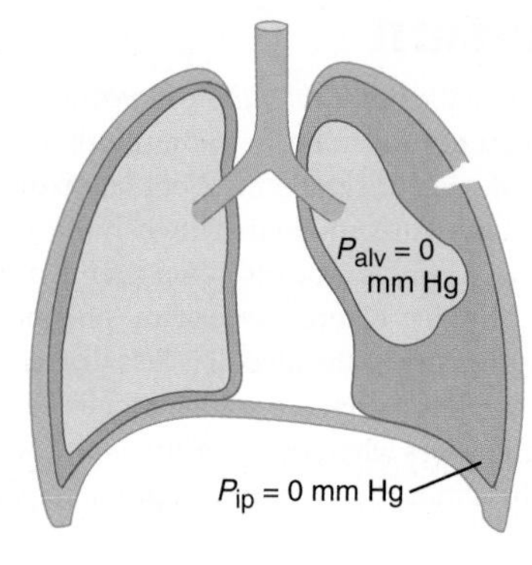

Figure 9 Pneumothorax. (a) Normal intrapleural pressure at rest is −4 mm Hg. Air entering the intrapleural space through a hole in the chest wall creates pneumothorax. **(b)** As a consequence of pneumothorax, intrapleural pressure equilibrates with atmospheric pressure. Without the negative force of intrapleural pressure drawing the lung outward, the lung collapses due to elastic recoil forces.

Mechanics of Breathing

Air flow into and out of the lungs is driven by pressure gradients that the muscles of respiration create by changing the volume of the lungs. The relationship between pressure and volume follows **Boyle's law:** *For a given quantity of any gas (such as air) in an airtight container, the pressure is inversely related to the volume of the container.* In other words, if the volume of the container increases, the pressure exerted by the gas falls; conversely, if the volume decreases, the pressure rises (**Toolbox: Boyle's Law and the Ideal Gas Law**). Although the lungs are not an airtight container and, therefore, Boyle's law is not strictly followed, pressure changes in the lungs occur in response to volume changes.

As we know that blood flow is determined by a pressure gradient and a resistance, as expressed by the following rule: Flow equals the pressure gradient divided by the resistance. Air flow into

TOOLBOX

Boyle's Law and the Ideal Gas Law

A fundamental relationship governing the behavior of gases is the *ideal gas law,* which is expressed as follows:

$$PV = nRT$$

where P is the gas pressure (in units of atmospheres; 1 atm = 760 mm Hg), V is the volume in which the gas is contained (liters), n is the quantity of gas (moles), R is the universal gas constant (0.083 liter-atm/mole-K), and T is the absolute temperature (kelvins, K). This equation can be rearranged such that pressure is given in terms of the other variables, as follows:

$$P = nRT/V$$

Given a fixed quantity of gas in a container at a given temperature (n and T are both constant), the pressure is inversely proportional to the volume of the container: $P \propto 1/V$. This is Boyle's law, illustrated in the figure. If instead the volume of the container is fixed but the quantity of gas is allowed to vary, the quantity RT/V becomes constant. Under these conditions, the pressure is proportional to the number of molecules in the container: $P \propto n$.

When we consider why gases exert pressure, we can see why the pressure varies with the number of gas molecules in a container. Gas molecules are normally in a state of constant thermal motion. Gas molecules inside a container continually collide with the wall and rebound off it. Every collision exerts a certain amount of force on the wall. The *pressure* is the total force exerted by all such collisions divided by the surface area of the wall (force per unit area). If the number of molecules is held constant but the volume of the container decreases, the density of molecules increases, so any given area of the wall experiences more collisions per second, and the pressure is higher. If more molecules are added to a container with a fixed volume, the same thing happens: Molecules are more densely spaced, which increases the number of collisions per unit area of wall, which raises the pressure.

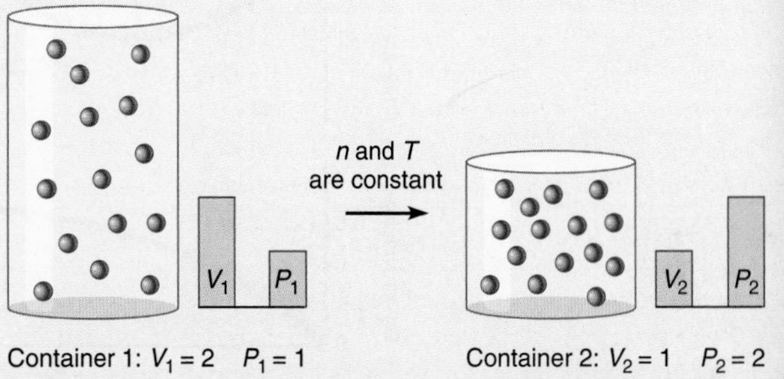

and out of the lungs also occurs by bulk flow, with the rate of flow determined by a pressure gradient ($P_{atm} - P_{alv}$) and resistance as follows:

$$\text{Flow} = \frac{P_{atm} - P_{alv}}{R}$$

Thus the driving force for air movement is the difference between atmospheric and intra-alveolar pressure. Because atmospheric pressure is constant, changes in alveolar pressure determine the direction of air movement.

Apply Your Knowledge

The flow rule can be applied to any section of the respiratory tract. Use this information to explain why the Heimlich maneuver (a sharp compression of the abdominal cavity just below the diaphragm) can dislodge food or other objects lodged in the trachea of a choking person.

Determinants of Intra-alveolar Pressure

Intra-alveolar pressure is determined by two factors: the quantity (moles) of air molecules in the alveoli, and the volume of the alveoli themselves. At rest, the alveoli contain a volume of air at atmospheric pressure that is equal to the functional residual capacity. At the beginning of inspiration, the lungs expand as a result of contraction of the inspiratory muscles and this expansion increases the volume of the alveoli, thereby lowering intra-alveolar pressure according to Boyle's law. The reduction in intra-alveolar pressure creates a pressure gradient that draws air into the lungs. During expiration, the reverse occurs: The lungs recoil, decreasing the volume of the alveoli and raising the intra-alveolar pressure. This effect creates a pressure gradient that drives air out of the lungs.

The pressure changes that occur during inspiration and expiration are shown in **Figure 10**. Note that during inspiration, intra-alveolar pressure falls at first but eventually rises to zero, because intra-alveolar pressure depends on both the quantity of alveolar air (that is, the number of air molecules) and its volume. Initially, the lungs expand and alveolar pressure decreases. However, as air flows into the alveoli during inspiration, the number of air molecules in the alveoli increases, so alveolar pressure increases (becomes less negative). Air stops flowing in when alveolar pressure rises to zero (which is atmospheric pressure); that is, air stops flowing in when there is no longer a pressure gradient to drive the flow. The same line of reasoning also explains why alveolar pressure rises and then falls during expiration. As the lung volume decreases, alveolar pressure increases, causing air to flow out. However, as air flows out, the quantity of air molecules in the alveoli decreases, which lowers the pressure toward zero. Air flow stops when alveolar and atmospheric pressures become equal.

The changes in the volume of the alveoli are produced by changes in the volume of the thoracic cavity, which involve the respiratory muscles (**Figure 11a**). The diaphragm and the external intercostal muscles are the primary inspiratory muscles, whereas

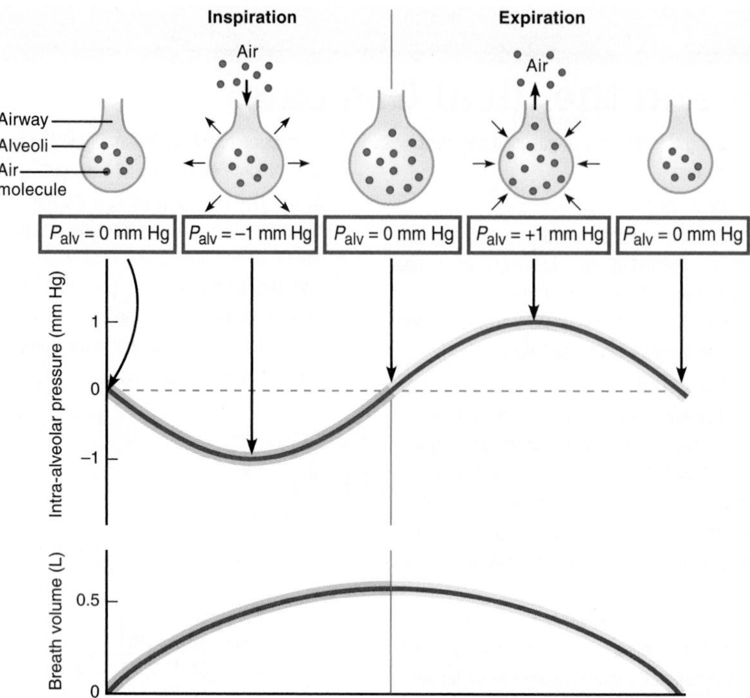

Figure 10 Changes in alveolar pressure and breath volume during inspiration and expiration. Before inspiration, intra-alveolar pressure is 0 mm Hg. During inspiration, expansion of the lungs causes intra-alveolar pressure to decrease. Air flow increases the quantity of gas in the lungs, which increases intra-alveolar pressure. At the end of inspiration, intra-alveolar pressure is equal to atmospheric pressure. During expiration, the lungs collapse inward, causing intra-alveolar pressure to increase. Air flows out of the lungs down a pressure gradient. At the end of expiration, intra-alveolar pressure is equal to atmospheric pressure, and air flow is zero.

Q If a person inhaled more than 0.5 liter of air in the same length of time as shown here, during inspiration would intra-alveolar pressure increase (become less negative or positive) or decrease (become more negative)?

the internal intercostals and abdominal muscles are the primary expiratory muscles, although expiration is often a passive process not requiring any muscle contraction.

Inspiration

The process of inspiration depicted in Figure 11b is initiated by neural stimulation of the inspiratory muscles (**Figure 12**). These skeletal muscles are stimulated to contract by the release of acetylcholine at the neuromuscular junction. Contraction of the diaphragm causes it to flatten and move downward; meanwhile, contraction of the obliquely oriented external intercostals causes the ribs to pivot upward and outward, expanding the chest wall. These combined actions increase the volume of the thoracic cavity. Other muscles of the neck (scalenes and sternocleidomastoids) and chest region (pectoralis minor) play subsidiary roles in inspiration, especially during forceful inspiration.

As the chest wall expands, it pulls outward on the pleura, causing the intrapleural pressure to decrease. This decrease in intrapleural pressure leads to an increase in the transpulmonary pressure, or the difference between the intrapleural pressure and the intra-alveolar pressure ($P_{alv} - P_{ip}$; **Figure 13**). An increase in transpulmonary pressure due to a decrease in P_{ip} creates a larger distending pressure across the lungs, so the lungs (alveoli) expand with the chest wall. When the lungs expand, pressure in the alveoli decreases to less than atmospheric pressure, so air flows into the alveoli by bulk flow; it continues to flow in until the pressure in the alveoli increases to atmospheric pressure. Stronger contractions of the inspiratory muscles produce a greater expansion of the thoracic cavity, making intrapleural pressure even more negative and creating a greater transpulmonary pressure. The result is greater lung expansion and a deeper inspiration—that is, a larger volume of air moves into the lungs.

Expiration

During quiet breathing, expiration is a *passive* process, in the sense that it does not require muscle contraction. At the end of an inspiration, the chest wall and lungs are expanded by muscle contraction. By simply relaxing these muscles, which occurs when motor neurons to the inspiratory muscles stop firing, the elastic chest wall and lungs recoil to their resting positions. As the chest wall and lungs recoil, the volume of the lungs decreases, causing alveolar pressure to increase to a value greater than atmospheric pressure. Air flows out (expiration occurs) due to the pressure gradient until the volume in the lungs equals the FRC.

A more forceful expiration can be produced by contraction of the expiratory muscles in a process called *active expiration*. Contraction of the expiratory muscles produces a greater and more rapid decrease in the volume of the thoracic cavity, which increases intra-alveolar pressure and causes a greater pressure gradient for air flow out of the alveoli.

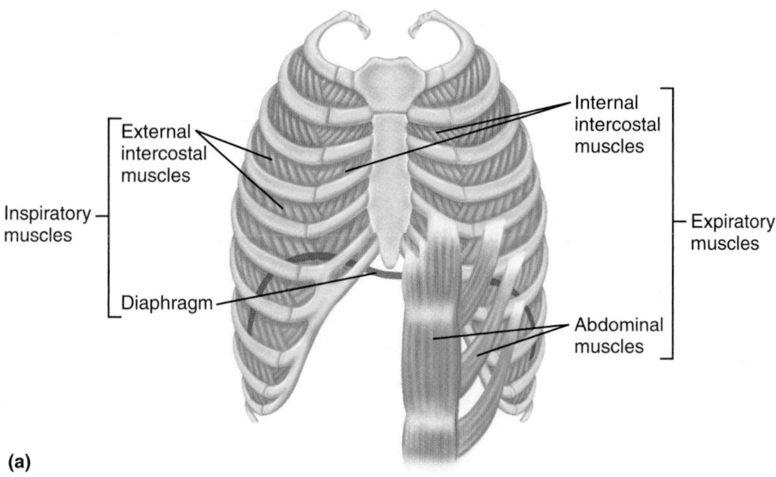

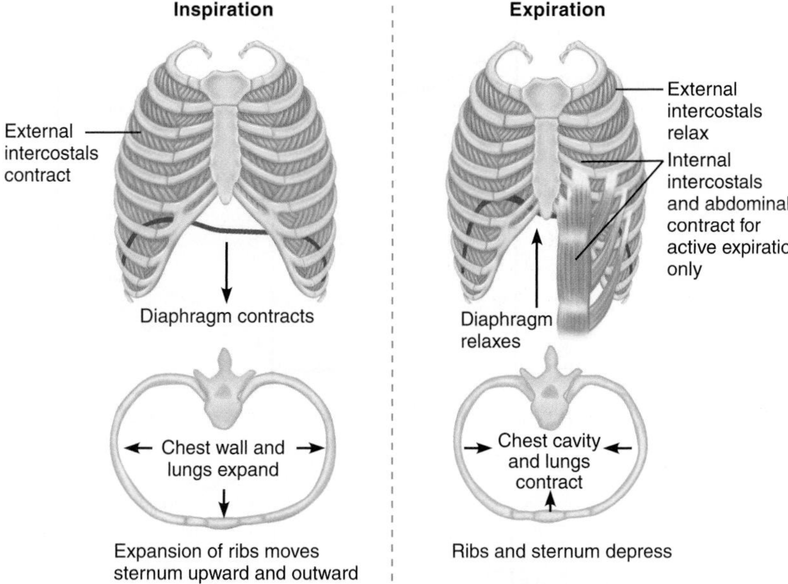

Figure 11 Respiratory muscles. (a) Locations of inspiratory and expiratory muscles. Notice the opposite orientation of the external and internal intercostal muscles. **(b)** Actions of respiratory muscles. When inspiratory muscles contract (at left), the chest wall expands, causing the lungs to expand. Quiet expiration (at right) occurs passively by relaxation of the muscles of inspiration, which allows the lungs and chest wall to recoil to their original positions. Active expiration requires contraction of the muscles of expiration, while the muscles of inspiration relax.

Quick Check 2

1. During inspiration, which is greater: intra-alveolar pressure or atmospheric pressure?
2. Name the muscles of inspiration. Which muscles are involved in active expiration?
3. At the end of an inspiration, in which direction do the lungs naturally recoil? The chest wall?

4 Factors Affecting Pulmonary Ventilation

The rate at which air flows into or out of the lungs is determined by two factors: the pressure gradient between the atmosphere and alveoli, and the airway resistance. In the previous section, we saw how pressure gradients produced by changes in the volume of the lungs drive inspiration and expiration. In this section, we consider the various factors that affect development of those pressure gradients and factors that affect airway resistance.

Lung Compliance

We know that arteries are elastic and recoil inward during diastole to keep blood flowing, and that the veins have high compliance, allowing them to hold large volumes of blood under low pressure. We have already seen that lungs are elastic and recoil after being stretched. Lungs also have relatively high compliance. Lung compliance is defined as the change in lung volume (ΔV) that results from a given change in transpulmonary pressure $\Delta(P_{alv} - P_{ip})$ (see **Toolbox: Compliance**):

$$\text{Lung compliance} = \frac{\Delta V}{\Delta(P_{alv} - P_{ip})}$$

A large lung compliance is advantageous, because a smaller change in transpulmonary pressure is needed to bring in a given volume of air and, therefore, less work or muscle contraction is required. Lung compliance depends on the elasticity of the lungs and on the surface tension of the fluid lining the alveoli. The lungs are elastic because of the presence of elastic fibers in the connective tissue. Forces exerted by these elastic fibers generally oppose lung expansion: As the lungs stretch, the fibers tend to recoil.

The surface tension of a liquid is a measure of the work required to increase the surface area of the liquid by a certain amount. A high surface tension is created by strong adhesive forces between the molecules of the liquid. The greater the surface tension, the more work needed to spread out the liquid. The surface tension of the lungs is caused by the air-liquid interface formed by the thin layer of fluid lining the internal surface of the alveoli (see **Toolbox: Pulmonary Surfactant and Laplace's Law**). As lung tissue expands, so does the fluid layer in the alveoli. Therefore, as the lungs expand, work is required not only to stretch the elastic tissue but also to increase the surface area of the fluid layer. Consequently, the surface tension of fluid acts to *decrease* lung compliance.

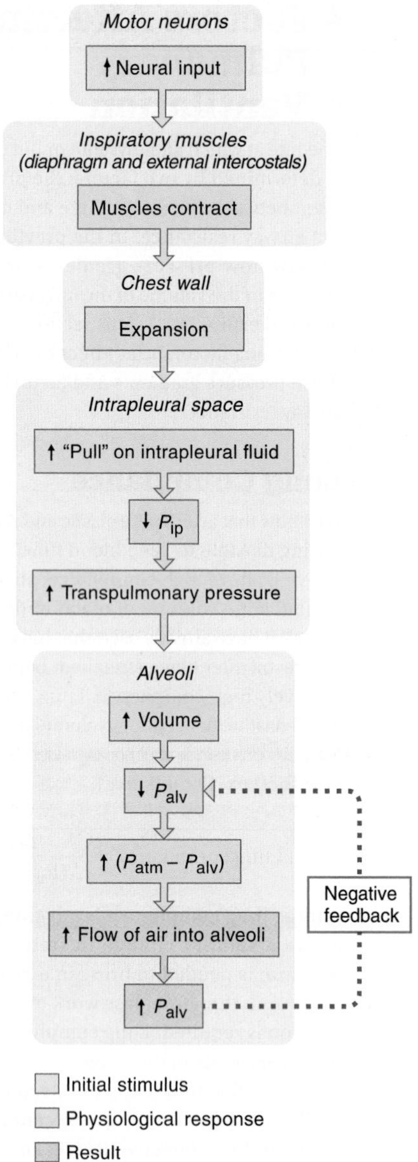

Figure 12 **Events in the process of inspiration.**

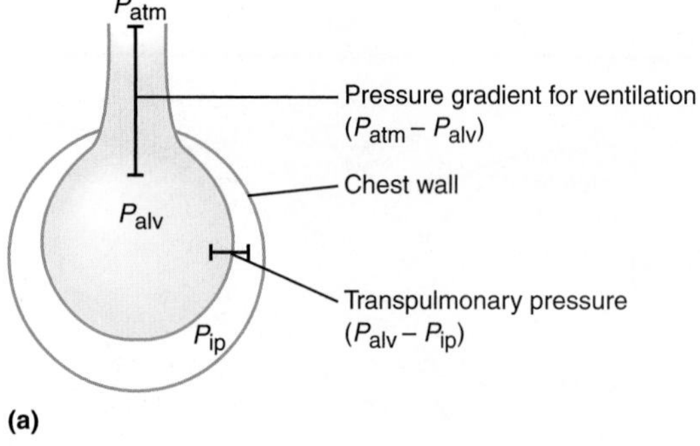

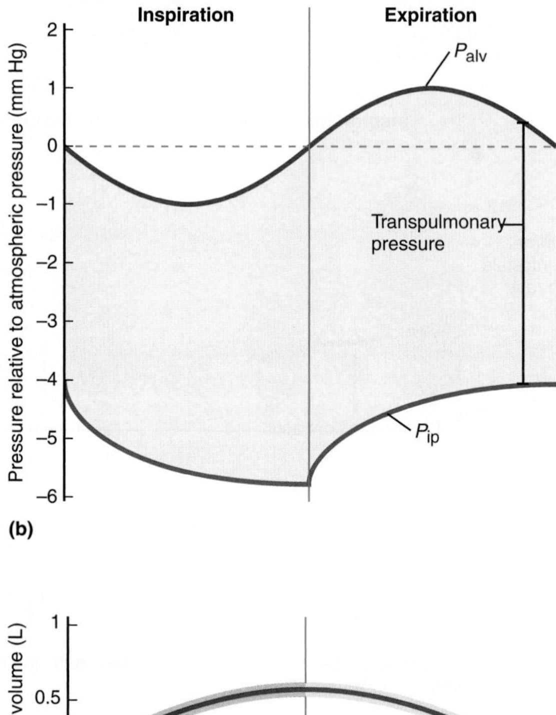

Figure 13 **Volume and pressure changes during inspiration and expiration. (a)** The difference between atmospheric and intra-alveolar pressures (the pressure gradient for ventilation) provides the force for moving air into or out of the lungs, and the transpulmonary pressure provides the force for the expansion of the lungs. **(b)** Changes in the intra-alveolar and intrapleural pressures that occur during breathing are such that the transpulmonary pressure increases during inspiration and during the beginning of expiration, and then decreases as expiration continues. **(c)** Changes in breath volume indicate that P_{ip} follows Boyle's law (intrapleural space is a closed system), but P_{alv} does not (alveoli are an open system due to movement of molecules in and out).

The presence of a detergent-like substance called **pulmonary surfactant** decreases the surface tension in alveoli. Pulmonary surfactant is secreted by **type II alveolar cells** located in the walls of alveoli. The surface tension of the fluid lining the alveoli is reduced (but not eliminated) by the action of pulmonary surfactant because surfactant interferes with the hydrogen bonding between water molecules. In this way, surfactant increases lung compliance and decreases the work of breathing.

Compliance is decreased if lung tissue becomes thickened, such as occurs with the formation of scar tissue in tuberculosis, or if surfactant production is decreased, such as occurs in *infant respiratory distress syndrome* (in which premature babies do not produce enough surfactant). When lung compliance decreases, the respiratory muscles must do more work to expand the lungs to a given volume.

TOOLBOX

Pulmonary Surfactant and Laplace's Law

The thin film of water that lines the alveoli increases the muscular work required to inflate the lungs because the water's surface tension decreases the lungs' compliance, making them harder to stretch. This surface tension creates another problem as well: Hydrogen bonding between adjacent water molecules tends to pull them into a round droplet, which in turn tends to pull the walls of an alveolus inward and make it collapse. What, then, prevents all of the alveoli from collapsing into one large "air bubble"?

To understand why the alveoli do not collapse, consider what happens in a soap bubble: The wall of a soap bubble contains water, which exerts surface tension and thus tends to pull the wall inward. As the bubble shrinks, however, it raises the pressure of the air contained within it, creating a distending pressure that opposes the tendency of the bubble to collapse. The bubble remains at a stable volume, neither collapsing nor expanding, as long as the distending pressure is just large enough to balance the inwardly directed forces created by the surface tension. When the lungs are not expanding or contracting, the volume of an alveolus remains stable for the same reason—the pressure of the air inside it balances the inward forces that would otherwise cause it to collapse.

According to *Laplace's law,* the air pressure (P) necessary to prevent the collapse of an alveolus (which is assumed to be spherical) is directly proportional to the surface tension (T) and inversely proportional to the alveolar radius (r):

$$P = \frac{2T}{r}$$

Here, P is the air pressure gradient across the alveolar wall, given as inside relative to the pressure outside, with outside pressure at zero. Thus, if two alveoli—one larger and one smaller—are subject to the same surface tension, the smaller one will require a greater pressure inside to keep from collapsing.

At the end of an inspiration or expiration, air pressure is the same inside all of the alveoli. (If air is not flowing, the pressure of the air must be the same everywhere.) The alveoli are not all the same size, however. This situation poses an interesting problem that is illustrated in the figure, which shows two adjacent alveoli of unequal sizes.

Let us assume for the moment that no surfactant is present. If the air pressure is just high enough to prevent the large alveolus from collapsing, then according to Laplace's law, it *cannot* be large enough to prevent the smaller alveolus from collapsing. Thus, if the two alveoli were at the same pressure initially, the smaller alveolus should collapse, which will raise the air pressure inside it (P_1), making it higher than the pressure in the larger alveolus (P_2). Air should then flow down the pressure gradient (from P_1 to P_2) and, therefore, from the smaller to the larger alveolus, as shown in the left part of the diagram.

In real lungs, however, surfactant is present and is actually more highly concentrated in the smaller alveoli. As a consequence, the surface tension in a small alveolus is lower than that in a large alveolus, which reduces the amount of pressure that must exist inside the small alveolus to prevent it from collapsing. For this reason, small and large alveoli can both have stable volumes at the same pressure, as shown in the right part of the diagram.

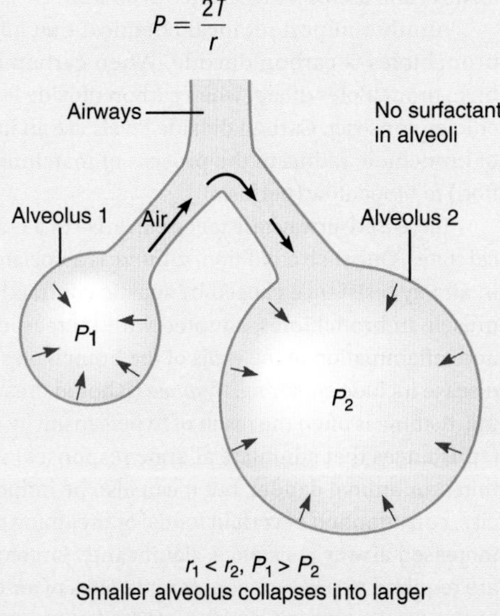

$r_1 < r_2$, $P_1 > P_2$
Smaller alveolus collapses into larger

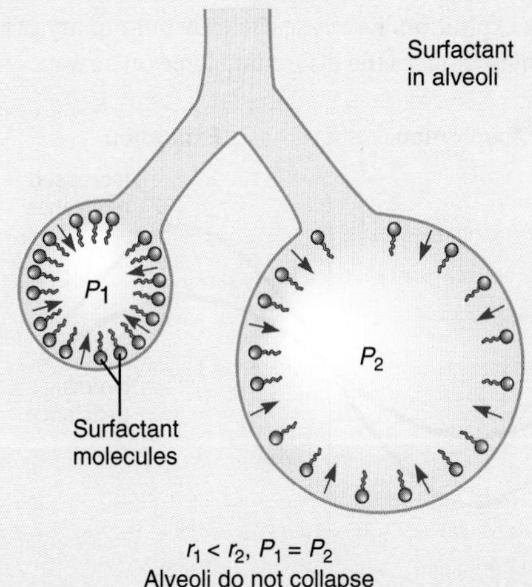

$r_1 < r_2$, $P_1 = P_2$
Alveoli do not collapse

Airway Resistance

In the context of this discussion, the term *airway resistance* refers to the resistance of the entire system of airways in the respiratory tract; it is analogous to *total peripheral resistance* in cardiovascular physiology. Airway resistance is determined primarily by the resistances of individual airways and is affected most strongly by changes in airway *radius:* As radius decreases, airway resistance increases.

In healthy lungs, resistance to air flow into and out of the lungs is low, because the radii of the tubes in the conducting zone are relatively large. Although the tubes decrease in radius as air moves down the conducting zone toward the alveoli, the total cross-sectional area of the smaller tubules increases due to the extensive branching. Therefore, overall resistance in the conducting zone is low. (This situation is similar to the low combined resistance of capillaries in the circulatory system.) This low resistance means that alveolar pressure need not differ much from atmospheric pressure to achieve normal rates of air flow under normal conditions. During quiet breathing *(eupnea),* the difference between alveolar and atmospheric pressure is generally less than 2 mm Hg. The effect of airway resistance on breathing is illustrated in **Figure 14**, which shows intra-alveolar pressure under normal conditions and when airway resistance is increased. (For purposes of illustration, we assume that the rates of air flow are the same in both conditions.) Note that when resistance increases, a larger pressure gradient is required to produce a given rate of air flow.

The resistance to air flow is affected by a number of factors, including passive forces exerted on the airways, contractile activity of smooth muscle in the bronchioles, and secretion of mucus into the airways. These passive forces are responsible for the changes in airway resistance that occur in a single breath, whereas changes in smooth muscle contractile activity are responsible for long-term variations in airway resistance.

The passive forces include changes in transpulmonary pressure that occur during inspiration and expiration, as well as *tractive forces* exerted on the airways by the pulling action of tissues surrounding them. Both of these forces decrease resistance during inspiration and increase resistance during expiration. During inspiration, the transpulmonary pressure increases because intrapleural pressure decreases more than intra-alveolar pressure decreases (see Figure 13). The increase in transpulmonary pressure that occurs during inspiration pulls outward on the airways, causing them to become distended (similar to what happens with the alveoli, only to a lesser extent), which decreases resistance as inspiration continues. During expiration, however, the transpulmonary pressure decreases, which reduces the distending force on airways and

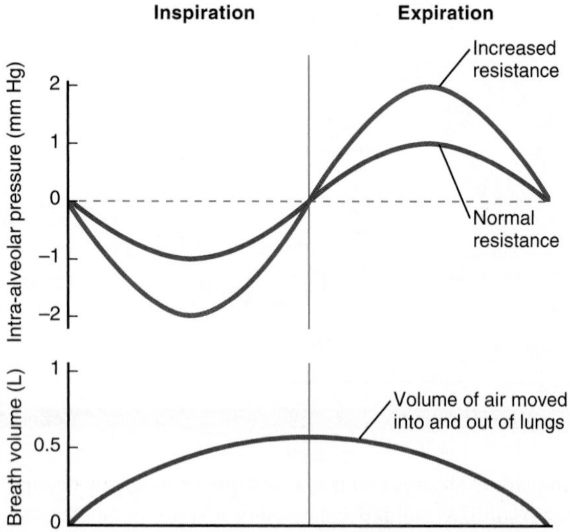

Figure 14 Effects of increasing airway resistance on the pressure changes required to move a fixed volume of air. When airway resistance is increased above normal, greater intra-alveolar pressure is required to move a given volume of air (in this case, 0.5 mL) into and out of the lungs over a given time period.

allows them to recoil inward, thereby increasing the resistance. As the surrounding tissue moves away from the airways during inspiration, tractive forces exerted by the tissues increase, which pulls the airways open and reduces resistance. During expiration, removal of these tractive forces reduces airway diameter and increases resistance.

Resistance to air flow in the bronchioles can also change as a result of contraction or relaxation of smooth muscle in the walls of the bronchioles. When this smooth muscle contracts, it decreases the radii of the bronchioles (called **bronchoconstriction**), which increases resistance. Conversely, relaxation of the smooth muscle causes the bronchiole radius to increase, a phenomenon known as **bronchodilation.** The contraction and relaxation of bronchiolar smooth muscle is subject to both extrinsic control (neural and hormonal signals) and intrinsic control (local chemical mediators).

The smooth muscle of the bronchioles is influenced by the autonomic nervous system: Sympathetic stimulation causes relaxation of the smooth muscles and increases the radius of bronchioles (called *bronchodilation*), whereas parasympathetic stimulation causes contraction of the smooth muscle and bronchoconstriction. The release of epinephrine from the adrenal medulla during sympathetic stimulation also causes bronchodilation.

Histamine, a chemical that is released locally during allergic reactions , causes contraction of the smooth muscle, resulting in bronchoconstriction. This effect only partially accounts for the difficulty in breathing during an allergic reaction, however, because histamine also stimulates mucus secretion, which builds up in the airways and increases resistance to air flow.

Another important local chemical that affects the radius of bronchioles is carbon dioxide. When carbon dioxide levels are high, bronchioles dilate; when carbon dioxide levels are low, bronchioles constrict. Carbon dioxide levels are an important regulator of bronchiole radius in the process of matching air flow (ventilation) to blood flow (perfusion).

Increased airway resistance can arise in a number of pathological states. One such condition, *asthma,* is associated with an increase in airway resistance caused by spastic contractions of the smooth muscle in bronchioles, coupled with increased mucus secretion and inflammation of the walls of the bronchioles. Symptoms of this disease include coughing, *dyspnea* (labored breathing), and wheezing. Asthma is often the result of hypersensitivity to certain *allergens* (substances that stimulate allergic responses) such as fungi, dust mites, or animal dander, but it can also be induced by stress, exercise, consumption of certain foods, or breathing cold air. Because of increased airway resistance, significantly larger pressure gradients are required to produce comparable rates of air flow, which greatly increases the work of breathing. The treatment for asthma varies for individuals, but can include *bronchodilators,* which induce relaxation of airway smooth muscle, and *corticosteroids,* which reduce inflammation. Other pathological states that increase airway resistance include *chronic obstructive pulmonary disease (COPD).* Whereas asthma involves acute (temporary) increases in airway resistance, COPD is associated with chronic (long-lasting) increases in airway resistance. (See **Clinical Connections: Chronic Obstructive Pulmonary Disease.**) Additionally, in some individuals, airway resistance can increase during sleep, resulting in apnea.

CLINICAL CONNECTIONS

Chronic Obstructive Pulmonary Disease 

The term *obstructive pulmonary disease* is an umbrella term for several disorders, all characterized by increased airway resistance. Three major disorders falling under this classification are emphysema, chronic bronchitis, and asthma. Whereas asthma is considered an acute condition, the other two disorders are chronic in nature. Together, asthma and chronic obstructive pulmonary disease (COPD) currently affect approximately 20 million people in the United States.

One disease that falls within the realm of COPD is *emphysema*, a permanent enlargement of air spaces in the respiratory zone accompanied by destruction of airway walls. The primary cause of emphysema is cigarette smoke. According to a widely accepted theory, the tissue destruction that occurs with emphysema is a result of the action of *proteases,* enzymes secreted by macrophages and other white blood cells during chronic inflammation. These proteases destroy tissue by breaking down proteins, including those in elastic connective tissue fibers. As a result, the lungs lose elasticity, and compliance increases. Loss of these fibers also diminishes tractive forces on airways, which reduces their diameter and makes them more likely to collapse, especially during expiration, when airway distending pressures are lower. The increase in compliance both increases functional residual capacity (that is, it reduces the volume that could be exhaled) and reduces the effort required to inhale. However, the loss of airway traction makes exhaling more difficult. As a consequence, people afflicted with emphysema tend to breathe at higher lung volumes (that is, the residual volume increases), resulting in decreased vital capacity. Destruction of tissue also reduces the combined surface area of alveolar walls, diminishing the capacity for gas exchange.

Chronic bronchitis, another disorder within the COPD umbrella, is inflammation of the airways lasting for at least three months a year for at least two consecutive years. It is characterized by inflammation and thickening of the airway lining, which reduces airway diameter and can lead to destruction of the normal tissue followed by *fibrosis* (thickening and scarring by the formation of connective tissue). Along with these symptoms comes an abnormally high rate of mucus secretion, which exacerbates the problem of airway obstruction and predisposes the lungs to infection.

Treatments for COPD include *bronchodilators* (such as β_2 adrenergic receptor agonists) and *anti-inflammatory drugs* (such as corticosteroids). Administration of oxygen is also helpful in combating hypoxemia, when it occurs.

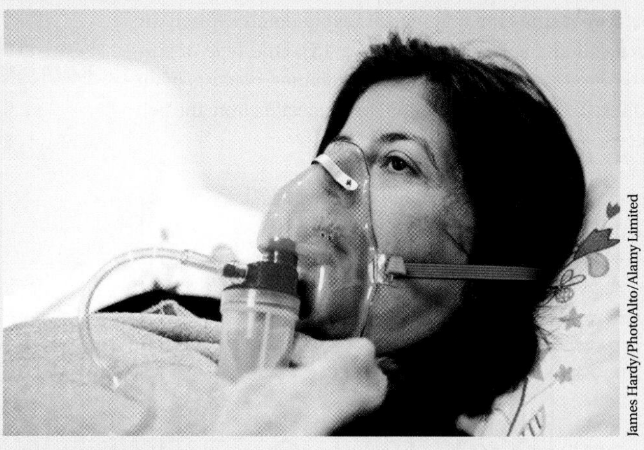
Oxygen administration.

James Hardy/PhotoAlto/Alamy Limited

Critical Thinking Questions

1. Describe the nature of the three COPD-related disorders described here. In what ways do they differ?
2. Explain the processes that lead to loss of elastic fiber in the lungs during emphysema.
3. What are proteases, and how do they affect respiratory tissue destruction?

Quick Check 3

1. If scarring of the lung resulted from some disease process, what would happen to lung compliance?
2. If surfactant secretion decreased, what would happen to lung compliance?
3. Will contraction of bronchiole smooth muscle cause the resistance to air flow to *increase* or *decrease*?

5 Clinical Significance of Respiratory Volumes and Air Flows

Certain pathological conditions affect the volume of air contained in the lungs or the rate of air flow into or out of the lungs. Clinicians measure lung volumes, calculate lung capacities (which are the sums of two or more measured lung volumes), and measure air flow rates

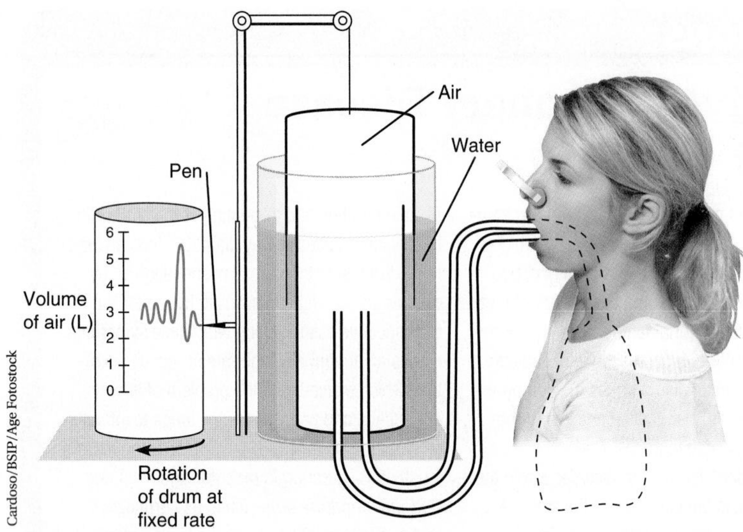

Cardoso/BSIP/Age Fotostock

Figure 15 Spirometry.

so as to gain information concerning pulmonary function. All of these values can be measured using a technique called *spirometry.*

Spirometry measures the volumes of inspired and expired air using a device called a **spirometer** (**Figure 15**). One type of spirometer, used primarily for measuring lung volumes, consists of an inverted, air-filled bell in a tub of water, plus a connection from the bell to an outlet via a hose that goes to the patient. The patient breathes air in and out of the bell. When the patient inhales, the volume of air in the bell decreases, and the bell descends deeper into the water. When the patient exhales, the volume of air in the bell increases, and the bell rises in the water. The bell is connected via a pulley system to a pen that moves up and down when the bell moves up and down. The pen is positioned to write on paper attached to a drum that rotates at a set speed. The movement of the pen up and down leaves marks on the paper that are calibrated to the volume of air that moved into or out of the lungs. The resulting trace is called a *spirogram* (**Figure 16**).

Today spirometers have become computerized, although the bell-and-pulley models are still sometimes used in teaching laboratories. In computerized systems, a person simply exhales as directed into a tube connected to a transducer that converts the volume of air into an electrical signal proportional to the volume. The computer translates the electrical signals and provides a readout of the various lung volumes. From the lung volume measurements, lung capacities can be calculated.

Although the rate of exhalation can be measured on the bell-and-pulley model spirometers, computer systems are used clinically. The maximum rate a person can exhale is critical in diagnosing certain pulmonary diseases.

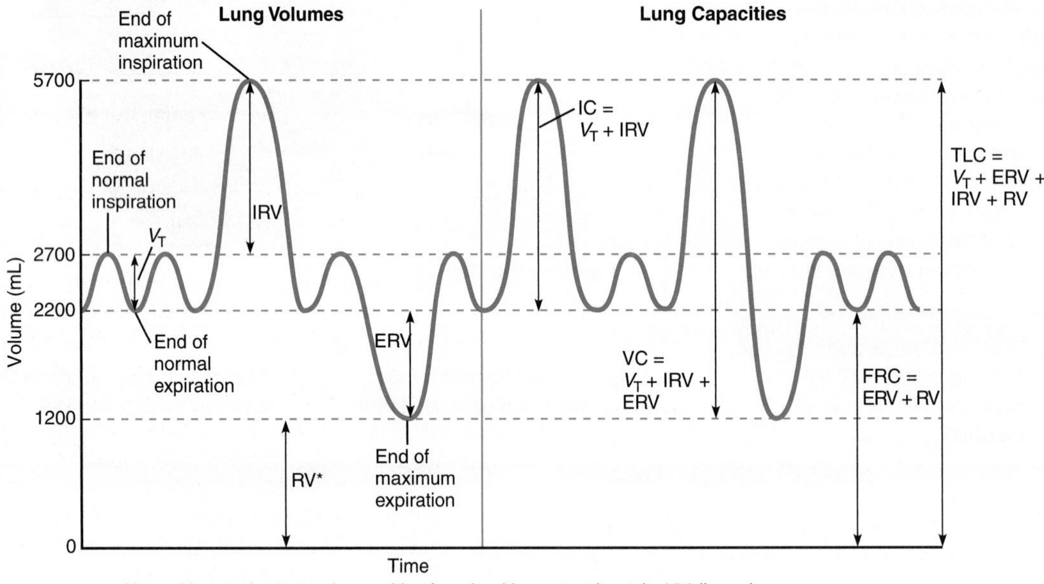

Normal lung volumes and capacities for a healthy, approximately 154-lb. male

Lung Volumes

V_T = Tidal volume = 500 mL
IRV = Inspiratory reserve volume = 3000 mL
ERV = Expiratory reserve volume = 1000 mL
RV = Residual volume* = 1200 mL

*Cannot be measured by spirometry

Lung Capacities

IC = Inspiratory capacity = V_T + IRV = 3500 mL
VC = Vital capacity = V_T + IRV + ERV = 4500 mL
FRC = Functional residual capacity = ERV + RV = 2200 mL
TLC = Total lung capacity = V_T + ERV + IRV + RV = 5700 mL

Figure 16 **Lung volumes and capacities measured using spirometry.** The curves shown were produced by spirometry (see Figure 15) and represent average values for an approximately 154-lb. male.

Lung Volumes and Capacities

Using spirometry, clinicians can measure three of the four nonoverlapping **lung volumes** that together make up the *total lung capacity.* The data in Figure 16 were obtained from a healthy 70-kg male sitting at rest. (Typically, the values would be smaller for a female and differ based on posture.) The volume of air that moves into and out of the lungs during a single, unforced breath is called the **tidal volume (V_T);** the average resting tidal volume is 500 mL. The maximum volume of air that can be inspired from the end of a normal inspiration is called the **inspiratory reserve volume (IRV)** and averages about 3000 mL. The maximum volume of air that can be expired from the end of a normal expiration is called the **expiratory reserve volume (ERV)** and averages about 1000 mL. Even after a maximum expiration, however, some air inevitably remains in the lungs and airways, because the negative intrapleural pressure prevents complete collapse of the lungs. The volume of air remaining in the lungs after a maximum expiration is called the **residual volume (RV);** it averages about 1200 mL.

The residual volume cannot be measured by spirometry, because spirometry measures only air that actually moves into or out of the lungs. One method of calculating residual volume is the *helium dilution method,* in which the subject breathes a gas mixture containing a known concentration of helium, which is not exchanged with the blood to any significant degree. During inspiration, helium in the inspired air mixes with all the air that is present in the lungs, which dilutes the helium. When the subject exhales, the concentration of helium in the expired air is measured and compared with the original concentration. With some simple calculations, the residual volume can be determined.

Lung capacities are sums of two or more of the lung volumes. The **inspiratory capacity (IC)** is the maximum volume of air that can be inspired at the end of a resting expiration; the sum of tidal volume and inspiratory reserve volume ($IC = V_T + IRV$) averages approximately 3500 mL. The **vital capacity (VC)** is the maximum volume of air that can be expired following a maximum inspiration; it is the sum of the tidal volume, inspiratory reserve volume, and expiratory reserve volume ($VC = V_T + IRV + ERV$) and averages approximately 4500 mL. The functional residual capacity (FRC) is the volume of air remaining in the lungs at the end of a resting tidal expiration; equal to the expiratory reserve volume and the residual volume ($FRC = ERV + RV$), it is approximately 2200 mL. The functional residual capacity is the volume of air in the lungs when the lungs are between breaths and the respiratory muscles are relaxed. Under these conditions, the elastic recoil of the lungs is balanced by the elastic recoil of the chest wall. The **total lung capacity (TLC)** is the volume of air in the lungs at the end of a maximum inspiration; it is the sum of the tidal volume, inspiratory reserve volume, expiratory reserve volume, and residual volume ($TLC = V_T + IRV + ERV + RV$) and averages approximately 5700 mL.

Pulmonary Function Tests

Spirometry is a simple method to test for respiratory disorders. A simple measure of lung volumes and calculations of lung capacities can help distinguish between *obstructive pulmonary diseases,* which involve increases in airway resistance, and *restrictive pulmonary disorders,* which involve decreases in lung compliance. In obstructive disorders, for example, the residual volume often increases because an increase in resistance makes it more difficult not only to inspire, but also to expire. The lungs become overinflated in this situation, and the functional residual capacity and total lung capacity are often increased. In contrast, restrictive disorders often involve structural damage to the lungs, pleura, or chest wall that decreases the total lung capacity and vital capacity.

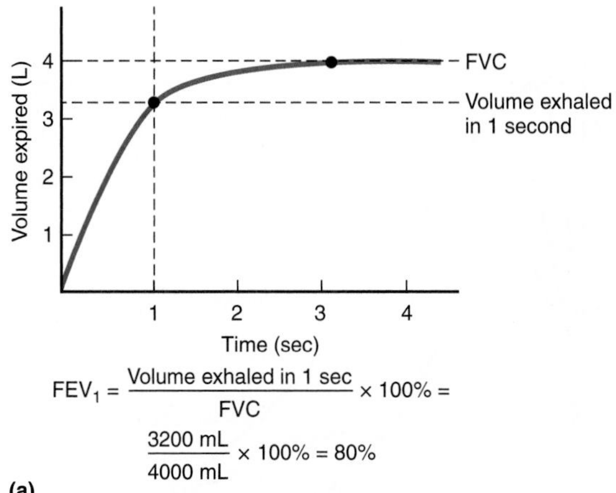

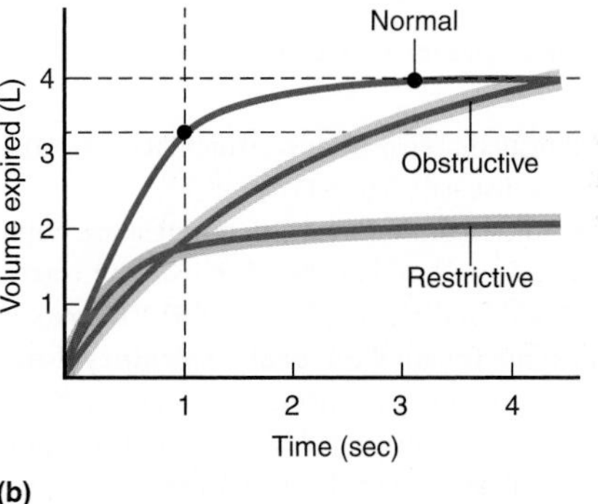

Figure 17 Pulmonary function test. Graphs illustrate how the Forced Expiratory Volume at 1 second (FEV_1) is measured **(a)**, and how it can be used to determine obstructive and restrictive pulmonary diseases **(b)**.

Even more useful than lung volume measurements in diagnosing pulmonary diseases are pulmonary function tests performed with computerized spirometry. Pulmonary function tests measure vital capacity and rates of inspiration and expiration. Specifically, measurements of **forced vital capacity (FVC)** and forced **expiratory volume (FEV)** provide information about lung capacity and compliance.

To test the forced vital capacity, the patient takes a maximum inspiration and then forcefully exhales as much and as rapidly as possible (**Figure 17**). The forced expiratory volume is a measure of the percentage of the FVC that can be exhaled within a certain length of time, most commonly 1 second (FEV_1). A normal FEV_1 is 80%, meaning that a person should be able to exhale 80% of the forced vital capacity within 1 second. If a person has a forced vital capacity of 4000 mL, for example, then following a maximum inspiration that person should be able to exhale 3200 mL in 1 second (Figure 17a). A low FVC is indicative of restrictive pulmonary disease, whereas a

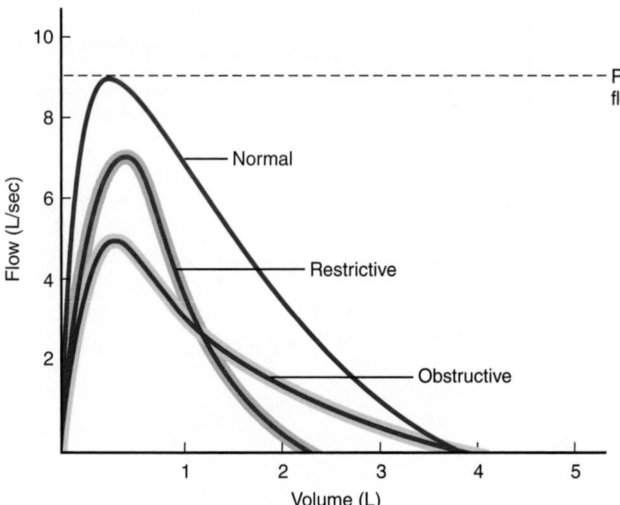

Figure 18 Flow-volume curve for expiration. Data are extrapolated from Figure 17.

low FEV_1 is indicative of increased resistance, which is characteristic of obstructive pulmonary disease (Figure 17b).

Clinically, data from a pulmonary function test are depicted in the form of a flow-volume curve, with volume on the *x*-axis and flow as liters per second on the *y*-axis (**Figure 18**). The maximum rate a person can exhale is called the **peak expiratory flow rate (PEFR),** which varies based on age and gender. For young adults, a normal PEFR for men is approximately 9 liters/second; for women, it is approximately 7 liters/second. Both obstructive and restrictive pulmonary diseases cause a decrease in the PEFR, but restrictive diseases also lead to a decrease in overall volume exhaled (FVC).

People with diabetes tend to have reduced pulmonary function with a decreased FVC. The reduced pulmonary function often manifests prior to other signs of diabetes, but is of unknown cause. Correlations with elevated blood glucose levels exist, but a cause-effect relationship has not been established.

Apply Your Knowledge

Based on the following information for a 70-kg male patient, determine whether he suffers from an obstructive or restrictive pulmonary disorder: vital capacity = 4800 mL, forced vital capacity = 5000 mL, and volume of air exhaled at 1 second of the forced vital capacity measurement = 3500 mL. Explain the basis of your diagnosis.

Alveolar Ventilation

Minute ventilation (V_E) is the total amount of air that flows into or out of the respiratory system in a minute. Minute ventilation can be calculated as the tidal volume times the number of breaths per minute, or **respiration rate (RR).** At rest, the average respiration rate is 12 breaths per minute. Therefore, the average resting minute ventilation is

$$\begin{aligned} V_E &= V_T \times \text{RR} \\ &= (500\ \text{mL/breath}) \times (12\ \text{breaths/min}) \\ &= 6000\ \text{mL/min} \end{aligned}$$

More important than the minute ventilation is the amount of fresh air that reaches the alveoli. Only a portion of the air breathed in actually participates in gas exchange, because a significant fraction of the air simply fills up the volume of airways in the conducting zone. For example, of a normal breath of 500 mL of air, only about 350 mL actually reaches the alveoli; the remaining 150 mL (30% of the tidal volume of 500 mL in this example) fills the trachea, the bronchi, and the bronchioles. The combined volume of these nonexchanging airways is referred to as the **anatomical dead space.** The effects of the anatomical dead space on the amount of "fresh air" reaching the lungs are illustrated in **Figure 19.** At the end of an expiration (Figure 19a), the conducting zone is full of "old" air—that is, air containing less oxygen and more carbon dioxide than atmospheric air because it once occupied the alveoli where oxygen and carbon dioxide were exchanged with blood. On the next inspiration (Figure 19b), that 150 mL of old air moves into the alveoli along with 350 mL of fresh air (assuming a tidal volume of 500 mL). Therefore, from a functional view, only 350 mL of fresh air enters the alveoli with each breath.

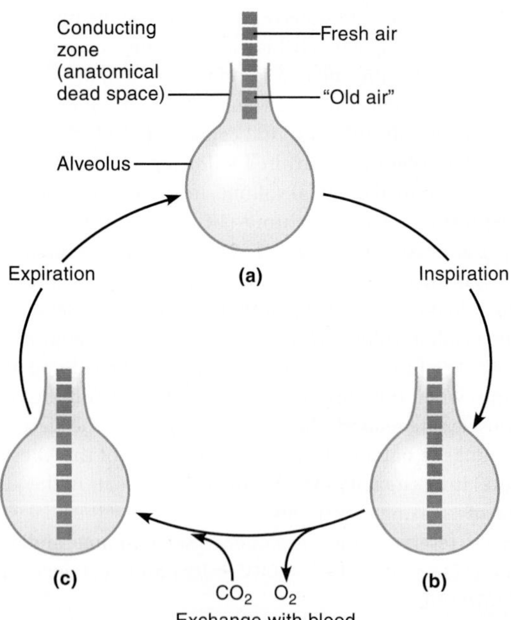

Figure 19 The effects of anatomical dead space on alveolar ventilation. **(a)** At the end of expiration, all of the air in the conducting and respiratory zones is stale air. **(b)** During inspiration, stale air from the conducting zone (anatomical dead space) enters the respiratory zone first, followed by atmospheric air. **(c)** During expiration, atmospheric air in the conducting zone is expired first, followed by stale air.

UNDERSTANDING EXERCISE

Effects of Exercise on Ventilation

During exercise, the body's demand for oxygen increases and more carbon dioxide must be eliminated from the body. To meet those needs, the respiratory system must increase minute alveolar ventilation. Ventilation increases during exercise as a consequence of both increased respiration rate and increased tidal volume. At low to moderate exercise intensities, the dominant adjustment is increased tidal volume. This outcome seems logical, because increasing tidal volume is more efficient than increasing respiration rate (the dead space volume is subtracted out of every breath). As exercise intensity increases further, tidal volume stabilizes at approximately 65% of vital capacity; any additional increases in ventilation are then attributable to increased frequency of respirations.

During long-term exercise regimens (training), many organs such as the heart and skeletal muscle hypertrophy and become more efficient. This is not true of the lungs. However, with training respiration can become a more efficient process, requiring less energy expenditure for the respiratory muscles.

Alveolar ventilation (V_A or minute alveolar ventilation) is a measure of the volume of fresh air reaching the alveoli each minute. It is similar to minute ventilation, except that the tidal volume has been corrected for the dead space volume (DSV):

$$\begin{aligned} V_A &= (V_T \times \text{RR}) - (\text{DSV} \times \text{RR}) \\ &= (500\ \text{mL/breath} \times 12\ \text{breaths/min}) \\ &\quad - (150\ \text{mL/breath} \times 12\ \text{breaths/min}) \\ &= 4200\ \text{mL/min} \end{aligned}$$

During times when oxygen demand of the tissues is increased, such as during exercise, alveolar ventilation must increase to meet those demands, either by increasing tidal volume, by increasing respiration rate, or both (see **Understanding Exercise: Effects of Exercise on Ventilation**). As illustrated in Table 1, it is more efficient to increase alveolar ventilation by increasing tidal volume than by increasing respiration rate. The reasoning is as follows: When the respiration rate is increased, the dead space volume is effectively subtracted out of each additional breath. In contrast, when tidal volume is increased, the total increase in volume in excess of the dead space volume adds to the fresh air reaching the alveoli.

Quick Check 4

1. Define the following terms: *tidal volume, minute ventilation, intra-alveolar pressure, anatomical dead space, alveolar ventilation.*
2. Following a maximum inspiration, a person expires maximally. Which lung capacity corresponds to the volume of air that has been expired? What is the volume of air left in the lungs called?
3. Given a tidal volume of 450 mL, a dead space volume of 150 mL, and a respiration rate of 15 breaths per minute, what is the minute ventilation? What is the minute alveolar ventilation?

TABLE 1 Effects of Respiration Rate and Tidal Volume on Minute Alveolar Ventilation

Tidal volume (mL)	Respiration rate (breaths/min)	Minute ventilation (mL/min)	Minute alveolar ventilation (mL/min)*
300	20	6000	3000
500	12	6000	4200
1000	6	6000	5100
500	24	12,000	8400
1000	12	12,000	10,200

*Assumes dead space volume = 150 mL.

SYSTEMS INTEGRATION

In this chapter, we learned how contraction of skeletal muscle creates pressure gradients that move air into and out of the lungs. These skeletal muscles are under the control of the somatic branch of the somatic nervous system. Inhaled air contains much-needed oxygen, which diffuses into pulmonary capillaries. At the same time, carbon dioxide in the pulmonary capillaries diffuses into the alveoli to be exhaled. Thus the lungs play a critical role in supplying oxygen and eliminating carbon dioxide—two functions that are essential for every organ system.

CHAPTER REVIEW

SUMMARY

1 Overview of Respiratory Function

- Respiration is the process of gas exchange; it includes both internal and external respiration.
- The four processes of external respiration are (1) pulmonary ventilation, (2) exchange of oxygen and carbon dioxide between lung air spaces and blood, (3) transport of oxygen and carbon dioxide in blood, and (4) exchange of oxygen and carbon dioxide between the blood and systemic tissues.
- Functions of the respiratory system include supply of oxygen to the tissues and elimination of carbon dioxide, management of the acid-base balance of the blood, vocalization, and protection against pathogens and irritants in the air.

2 Anatomy of the Respiratory System

- The lungs are the major organs of the respiratory system; the right lung is divided into three lobes and the left lung is divided into two lobes.
- The upper airways include the nasal cavity, oral cavity, and pharynx.
- After the pharynx, a common passageway for air and food, the passageways for food and air diverge.
- The respiratory tract is the pathway for air. It can be functionally divided into two components: a conducting zone and a respiratory zone.
- The conducting zone—larynx, trachea, bronchi, and bronchioles—functions in conducting air from the larynx to the lungs.
- The conducting zone is lined by epithelium that contains goblet cells and ciliated cells.
- The respiratory zone—respiratory bronchioles, alveolar ducts, alveoli, and alveolar sacs—is the site of gas exchange within the lungs; the alveoli are the primary sites of exchange.
- The wall of an alveolus contains type I cells and type II cells.
- The type I cells and endothelial cells of capillaries form the respiratory membrane, across which gas exchange occurs. The type II cells secrete pulmonary surfactant.
- Alveolar macrophages are also located in the alveoli.
- The chest wall consists of the rib cage, the sternum, the thoracic vertebrae, and associated muscles and connective tissue.
- Muscles of the chest wall include the internal and external intercostals and the diaphragm.
- The pleurae are membranes that line the chest wall and lung, forming a pleural sac around each lung.
- The space between the two membranes, called the intrapleural space, is filled with a thin layer of intrapleural fluid.

iP Respiratory, Anatomy Review: Respiratory Structures

MAP iP Respiratory, Control of Respiration

3 Forces for Pulmonary Ventilation

- The four primary pressures associated with ventilation are atmospheric pressure, intra-alveolar pressure, intrapleural pressure, and transpulmonary pressure.
- Atmospheric pressure is the pressure of the air outside the body.
- Intra-alveolar pressure is the pressure of air within the alveoli.
- Intrapleural pressure is the pressure of the intrapleural fluid. It is always negative during normal breathing, and always less than intra-alveolar pressure.
- Because the lungs and the chest wall are elastic, between breaths the chest wall tends to recoil outward, and the lungs to recoil inward. These forces tend to separate the chest wall from the lungs, thereby creating a negative intrapleural pressure.
- Transpulmonary pressure is the difference between the intrapleural pressure and the intra-alveolar pressure.
- Inspiration and expiration are driven by differences in atmospheric and intra-alveolar pressures.
- These pressure gradients are created when the volume of the lungs is changed. Inspiration is caused by contraction of the diaphragm and the external intercostal muscles; when these muscles contract, the volume of the thoracic cavity increases.
- As the thoracic cavity expands, the intrapleural pressure decreases, creating a force that expands the lungs as the chest wall expands.
- When intra-alveolar pressure decreases below atmospheric pressure, inspiration occurs.
- During quiet breathing, expiration occurs when the chest wall and lungs return passively to their original positions.
- Active expiration involves contraction of the internal intercostals and abdominal muscles.

MAP iP Respiratory, Pulmonary Ventilation

4 Factors Affecting Pulmonary Ventilation

- The rate of air flow into or out of the lungs is determined by the magnitude of the pressure gradient driving the flow and airway resistance.
- Lungs have a high compliance; they are easily stretched to increase lung volume for inspiration.
- Airway resistance depends primarily on the radius of the tubules of the respiratory tract.
- Airway resistance generally is low but can be affected by breathing mechanics, the autonomic nervous system, chemical factors, and pathological states.

MAP iP Respiratory, Pulmonary Ventilation

5 Clinical Significance of Respiratory Volumes and Air Flows

- Spirometry measures lung volumes and capacities.
- Lung volumes include tidal volume, inspiratory reserve volume, expiratory reserve volume, and residual volume.
- The various lung capacities include inspiratory capacity, vital capacity, functional residual capacity, and total lung capacity.
- Other lung measurements take into account the rate of air flow.
- The forced vital capacity is the amount of air a person can expire following a maximum inspiration, expiring as forcefully and rapidly as possible.

- The forced expiratory volume is a measure of the percentage of the forced vital capacity that can be exhaled within a certain time frame.
- Minute ventilation is the total amount of air that flows into or out of the respiratory system in a minute.
- Minute alveolar ventilation is a measure of the volume of fresh air reaching the alveoli each minute, calculated as minute ventilation corrected for dead space volume.
- To increase minute alveolar ventilation, it is more efficient to increase tidal volume than to increase respiration rate.
- The maximum rate a person can exhale is called the peak expiratory flow rate (PEFR); PEFR varies by age and gender.

Check your understanding at MasteringA&P and apply your new knowledge to real-life case studies to ensure that you've mastered the content!

EXERCISES

Multiple-Choice Questions

1. Which of the following is a component of internal respiration?
   a) Ventilation
   b) Transport of oxygen in the blood
   c) Diffusion of carbon dioxide from tissues to blood
   d) Diffusion of oxygen from blood to tissues
   e) Oxidative phosphorylation

2. Which of the following is *not* a function of the conducting zone of the respiratory system?
   a) Humidifying the air
   b) Adjusting the air to body temperature
   c) Exchanging gases between the respiratory system and the blood
   d) Secreting mucus
   e) Protecting the lungs from inhaled particles

3. The smallest airways in the conducting zone are
   a) Terminal bronchioles.
   b) Respiratory bronchioles.
   c) Alveolar ducts.
   d) Alveolar sacs.
   e) Bronchi.

4. Surfactant is secreted by
   a) Goblet cells.
   b) Alveolar macrophages.
   c) Type I cells.
   d) Type II cells.
   e) Ciliated cells.

5. The product of tidal volume and breathing frequency gives
   a) Respiration rate.
   b) Total lung capacity.
   c) Alveolar ventilation.
   d) Minute ventilation.
   e) Dead space volume.

6. When all muscles of respiration are relaxed and alveolar pressure is zero, lung volume is equal to
   a) Residual volume.
   b) Vital capacity.
   c) Functional residual capacity.
   d) Tidal volume.
   e) Total lung capacity.

7. Which of the following statements describes the lungs at the functional residual capacity?
   a) Atmospheric, intra-alveolar, and intrapleural pressures are all equal.
   b) The lungs tend to collapse due to their elastic properties.
   c) The chest wall tends to collapse due to its elastic properties.
   d) Both a and c are true.
   e) All of the above are true.

8. Which of the following factors decreases airway resistance?
   a) Activation of the parasympathetic nervous system
   b) Epinephrine
   c) Histamine

9. Pulmonary surfactant
   a) Prevents collapse of alveoli.
   b) Prevents small alveoli from joining with larger alveoli.
   c) Increases lung compliance.
   d) Both a and c.
   e) All of the above.

10. Which of the following muscles contract(s) during quiet expiration?
    a) Diaphragm
    b) Internal intercostals
    c) External intercostals
    d) None of the above
    e) All of the above

Objective Questions

11. Contraction of the diaphragm increases the rate of air flow during forced expiration. (true/false)
12. During inspiration, transpulmonary pressure (increases/decreases).
13. During inspiration, intrapleural pressure becomes (more/less) negative.
14. If airway resistance increases, a (higher/lower) transpulmonary pressure is required to produce a given rate of air flow during expiration.
15. Pulmonary surfactant (increases/decreases) the surface tension of water.
16. Pulmonary surfactant (increases/decreases) lung compliance.
17. Dead space volume is the volume of air in the (conducting zone/respiratory zone).
18. (Obstructive/Restrictive) lung diseases are characterized by increased airway resistance.
19. _______ cells secrete mucus.
20. The (internal/external) intercostals are muscles of inspiration.

Essay Questions

21. Describe the various ways in which the structure of the respiratory system is adapted to facilitate gas exchange.
22. Describe the balance between distending pressures and recoil forces acting on the lungs and chest wall at FRC. Include an explanation of why intrapleural pressure is subatmospheric.
23. Explain the difference between minute ventilation and minute alveolar ventilation. Include a discussion of the effect of tidal volume on alveolar ventilation.

24. Describe the role of surfactant in decreasing the surface tension in alveoli. What are some of the consequences of decreasing surface tension?
25. Describe the mechanics of ventilation. Which muscles contract for inspiration? Which muscles contract for expiration? How does the elasticity of the lungs and chest wall affect ventilation?

Critical Thinking

26. Which of the curves in the following graph illustrates a condition of greater compliance, A or B?

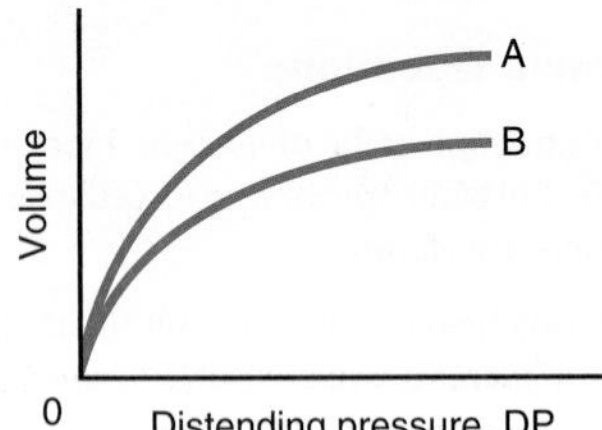

27. Determine your resting respiration rate by counting how many times you breathe in one minute. Assume a tidal volume of 500 mL and estimate your dead space volume as 1 mL per pound of body weight (that is, if you weighed 150 pounds, your dead space volume would be 150 mL). Calculate your minute ventilation and minute alveolar ventilation.

    During exercise, oxygen demand increases. Your respiratory system compensates by increasing minute alveolar ventilation. Calculate your minute alveolar ventilation assuming that your respiration rate increases 50% from what it was at rest. Then calculate your minute alveolar ventilation with your resting respiration rate but with an increase in tidal volume of 50% (from 500 mL to 750 mL). Does an increase in respiration rate or tidal volume more efficiently increase minute alveolar ventilation? Explain.
28. According to Boyle's law, pressure is inversely proportional to volume. In Figure 13, breath volume is shown to increase until the end of inspiration. Which pressure—intra-alveolar or intrapleural—follows Boyle's law? Explain.
29. Black widow spiders release a venom called latrotoxin, which enhances acetylcholine release. The venom is deadly because it interferes with breathing. Explain how this process works.

Photo credits for icons in feature boxes: Focus on Diabetes: Tanyae/Dreamstime; Clinical Connections: Iofoto/Dreamstime; Understanding Exercise: Pixattitude/Dreamstime; Discovery: Yuri Arcurs/Cutcaster.

Glossary

respiration the process of gas exchange within the body; includes internal respiration and external respiration

internal respiration cellular respiration that occurs in the mitochondria

external respiration the exchange of oxygen and carbon dioxide between the atmosphere and the tissues of the body; involves both the respiratory and cardiovascular systems

pulmonary ventilation the movement of air into and out of the lungs by bulk flow

upper airways air passages in the head and neck; include the nasal cavity, oral cavity, and pharynx

pharynx (fair-inks) a passageway leading from the mouth to the esophagus or larynx that serves as a common passageway for food and air

respiratory tract air passages leading from pharynx to lungs

conducting zone the upper part of the respiratory tract; conducts air from the larynx to the lungs; no gas exchange occurs

respiratory zone the lower part of the respiratory tract; the site of gas exchange within the lungs

larynx (lar-inks) the initial passageway of the respiratory tract, which contains the vocal cords

glottis opening to the larynx

epiglottis a flap of tissue over the glottis that prevents food or water from entering the larynx when swallowing

trachea (tray-key-ah) a cartilaginous tube of the respiratory tract, located between the larynx and the bronchi

bronchi branched tubes of the respiratory tract, located between the trachea and the bronchioles of the lungs; singular, *bronchus*

secondary bronchi branches off the bronchi leading to the lungs

bronchioles small tubules leading from the bronchi to the alveoli; less than 1 mm in diameter

terminal bronchioles bronchioles that lead directly to the airways of the respiratory zone of the respiratory tract; the last component of the conducting zone

goblet cells epithelial cells in the respiratory tract and gastrointestinal (GI) tract that secrete mucus

cilia hairlike processes found on certain epithelial cells in the respiratory tract and oviduct

respiratory bronchioles small tubules of the respiratory tract located between terminal bronchioles and alveolar ducts

alveoli (al-vee-oh-lie) terminal sacs of the respiratory tract, where most gas exchange occurs; usually grouped in clusters; singular, *alveolus*

alveolar sacs clusters of alveoli at the end of an alveolar duct

type I alveolar cells epithelial cells lining alveoli

respiratory membrane the structure across which gas exchange occurs in the lungs; a barrier between blood and air consisting of capillary endothelial cells and their basement membranes and alveolar epithelial cells and their basement membranes

chest wall structures that protect the lungs and form an airtight compartment around them; includes the rib cage, sternum, thoracic vertebrae, muscles, and connective tissue

internal intercostals muscles of expiration that are located between the ribs

external intercostals inspiratory muscles of the chest wall

diaphragm the primary inspiratory muscle for respiration; the muscular partition that separates the abdominal and thoracic cavities

pleura (plur-ah) the membrane that lines the chest wall and lung, forming a pleural sac around each lung

pleural sac membrane surrounding each lung

intrapleural space a fluid-filled compartment located between the lungs and chest wall; bounded by the visceral and parietal pleura

atmospheric pressure (P_{atm}) pressure of the outside air; at sea level, 760 mm Hg or 1 atmosphere

intra-alveolar pressure (P_{alv}) the pressure exerted by the air within the alveoli

intrapleural pressure (P_{ip}) the pressure of the fluid inside the pleural space

Answers

Figure Question Answers

Fig1	The electron transport chain
Fig 5	Monocytes
Fig 10	Decrease

Apply Your Knowledge Answers

The increase in mucus production during a cold helps prevent harmful substances from reaching the respiratory zone. However, it also increases resistance for air flow, making it more difficult to breathe.

Compression of the abdominal cavity during the Heimlich maneuver increases the pressure on the airways deep in the respiratory tract, thereby producing a pressure gradient for movement away from the lungs and toward the mouth. If enough pressure is produced, the object will be dislodged and ejected from the respiratory tract.

At 4800 mL, vital capacity is slightly elevated compared to normal. The forced vital capacity at 1 second is (3500 mL/5000 mL) $\times$ 100% $=$ 70%. This is considerably less than the normal 80%. The low forced vital capacity at 1 second is indicative of an obstructive pulmonary disease.

Answers to End-of-Chapter Multiple Choice and Objective Questions

Multiple Choice Questions

1. e 2. c 3. a 4. d 5. d 6. c 7. b 8. b 9. e 10. d

Objective Questions

11. false
12. increases
13. more
14. higher
15. decreases
16. increases
17. conducting zone
18. Obstructive
19. Goblet
20. external

The Respiratory System: Gas Exchange and Regulation of Breathing

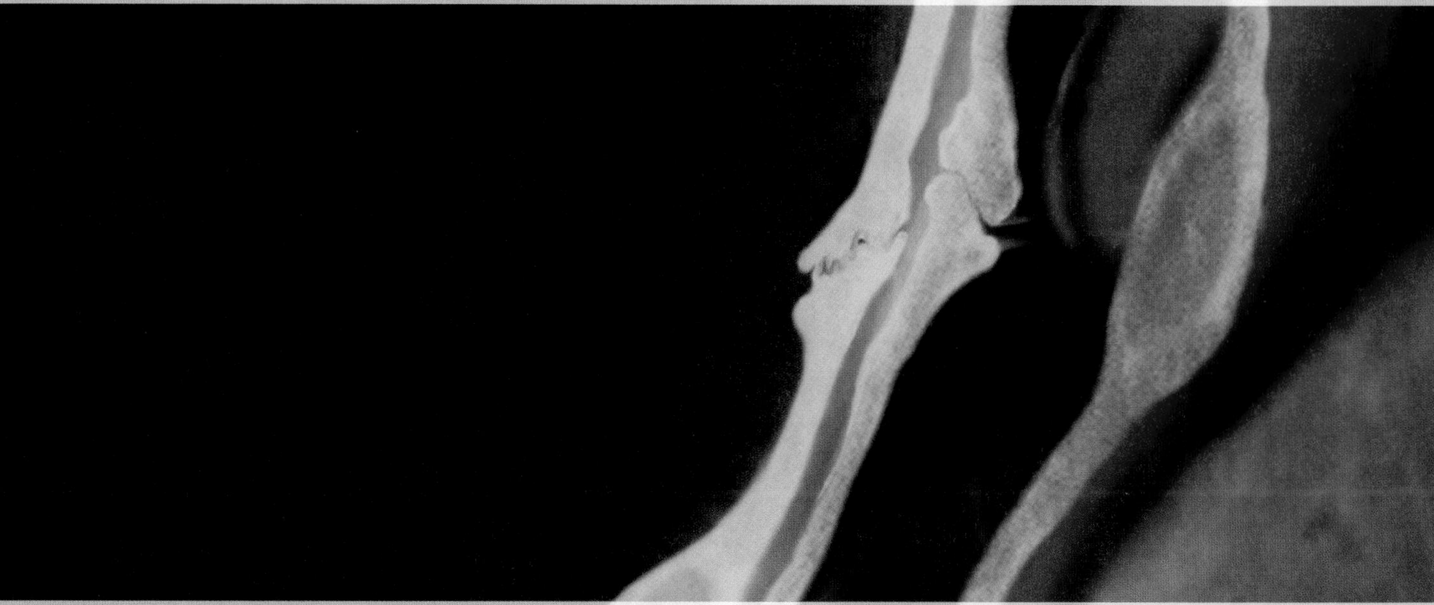

Anatomical Travelogue/Photo Researchers, Inc.

Drawing based on a scanning electron micrograph (SEM) of a respiratory membrane.

We've all held our breath at one time or another. Most people can hold their breath for a minute or less following a maximum inhalation under normal conditions. By comparison, following training, some individuals can hold their breath for more than six minutes!

Before the advent of scuba diving, pearls and sponges were harvested by "free-divers," people who would go underwater without the aid of an extra air supply. To help themselves stay underwater to harvest the oysters or sponges, the divers often decreased the amount of carbon dioxide in their blood by hyperventilating, breathing in excess of the body's needs at the time. In Japan, the amas (female pearl divers) still practice the art of free-diving. They can dive more than 100 feet and stay at the bottom harvesting oysters for up to two minutes before resurfacing. Today free-diving (or breath-hold diving) is a competitive sport in many areas of the world.

In this chapter, you will learn about the control of breathing and the effect of carbon dioxide on breathing. You will also learn that hyperventilating to hold your breath while you are underwater is a very dangerous practice, and one not to be attempted without extensive training. Many people drown because of the dizziness and confusion that can result from hyperventilation.

CHAPTER OUTLINE

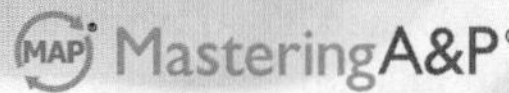

Go to MasteringA&P for helpful A&P Flix 3-D animations, chapter quizzes, pre-tests, interactive physiology tutorials, and more!

LEARNING OUTCOMES *After studying this chapter, you should be able to:*

- Describe the circulatory pathway for oxygenated and deoxygenated blood. Describe the exchange of gas in the lungs and in systemic tissues.
- List the normal partial pressures of oxygen and carbon dioxide in arterial and mixed venous blood, and explain how they contribute to the exchange of gases.
- Describe the mechanisms of transport of oxygen and carbon dioxide in the blood, including the role of hemoglobin in the transport of both gases.
- Explain the relationship between the P_{CO_2} of blood and the pH of blood. Describe the actions of carbonic anhydrase in erythrocytes as blood passes through the systemic and pulmonary circulations.
- Describe the neural mechanisms that establish the respiratory rhythm. Distinguish between the respiratory centers that establish the rhythm and those that regulate the rhythm.
- Explain the role of peripheral and central chemoreceptors in the control of ventilation.
- Describe how changes in P_{O_2} and P_{CO_2} in lung tissues can alter ventilation. Explain the ventilation-perfusion ratio.
- Explain how the respiratory system regulates acid-base balance of the blood by varying the rate of carbon dioxide expiration.

Before You Begin

Make sure you have mastered the following topics:

- Diffusion
- Law of mass action
- Affinity
- Structures of the brainstem
- Control of skeletal muscle contraction
- Pathway of blood flow
- Components of blood
- Hemoglobin

We know how air moves in and out of the lungs. But that is just the first step in respiration. Once in the lungs, the right gas—oxygen—has to diffuse into the bloodstream, and carbon dioxide must diffuse from the bloodstream into the lungs. The diffusion of oxygen and carbon dioxide follows Fick's law of diffusion.

In this chapter, we examine how gases in the blood can travel bound to the protein hemoglobin, and how they can be converted into another molecule by an enzyme. Both of these processes follow the law of mass action. The concentrations of oxygen and carbon dioxide in blood are tightly regulated by the autonomic nervous system and somatic nervous system. In this chapter, we learn how oxygen and carbon dioxide are exchanged efficiently between the alveoli and the blood, and how the blood transports oxygen and carbon dioxide to and from the tissues, respectively. We also learn how ventilation is regulated to maintain an adequate oxygen supply to the body's tissues, even when demands change.

1 Overview of the Pulmonary Circulation

The body may hunger for food and thirst for water, but it is absolutely voracious in its need for oxygen. The cells of the body consume approximately 250 mL of oxygen and produce approximately 200 mL of carbon dioxide each minute when at rest. To maintain homeostasis, the body must obtain the oxygen from the atmosphere and expel the carbon dioxide into the atmosphere.

To provide the cells with the oxygen they need, an average adult male inhales 6000 mL of air per minute, 4200 mL of which reaches the alveoli. Approximately 882 mL of this air (21%) is oxygen. Approximately 250 mL of this oxygen diffuses from the alveoli to the blood for subsequent consumption by the cells of the body; the remaining 632 mL is exhaled. With a consumption of 250 mL of oxygen per minute, the body consumes 360,000 mL of oxygen per day—nearly 100 gallons.

Under most conditions, concentrations of oxygen and carbon dioxide in systemic arterial blood are maintained at relatively constant levels for two reasons: (1) oxygen moves from alveolar air into the blood at the same rate it is consumed by the tissues; and (2) carbon dioxide moves into alveolar air from the blood at the same rate it is produced in the tissues (**Figure 1**). The ratio of the amount of carbon dioxide produced by the body to the amount of oxygen consumed is called the **respiratory quotient.** On average, cells consume 250 mL of oxygen per minute while producing 200 mL of carbon dioxide under resting conditions, so the average respiratory quotient at rest is 0.8 ($200/250 = 0.8$). The respiratory quotient varies with diet, because diet affects metabolism. For example, consuming a diet containing a large amount of lipids results in a relatively low respiratory quotient (near 0.7) because oxygen is consumed independent of carbon dioxide production during lipid catabolism. The respiratory quotient also varies with exercise and during conditions of hypoventilation and hyperventilation.

Figure 1 shows the movements of oxygen and carbon dioxide into and out of the lungs and tissues under resting conditions. Oxygen enters the alveoli and carbon dioxide leaves the alveoli by bulk flow of air during ventilation. Deoxygenated blood returns via systemic veins to the right atrium of the heart. The blood then enters the right ventricle, which pumps blood through pulmonary arteries to the lungs. Each minute the deoxygenated blood in the pulmonary arteries transports approximately 750 mL of oxygen and 2700 mL of carbon dioxide to the lungs. In the pulmonary capillaries, oxygen diffuses from the alveoli to the blood at a rate of 250 mL/min, and carbon dioxide diffuses from the blood to the alveoli at a rate of 200 mL/min. Therefore, the oxygenated blood in the pulmonary veins transports approximately 1000 mL of oxygen per minute (750 mL/min already in the deoxygenated blood plus 250 mL/min added from the alveoli) and 2500 mL of carbon dioxide per minute (2700 mL/min in the deoxygenated blood minus 200 mL/min lost to

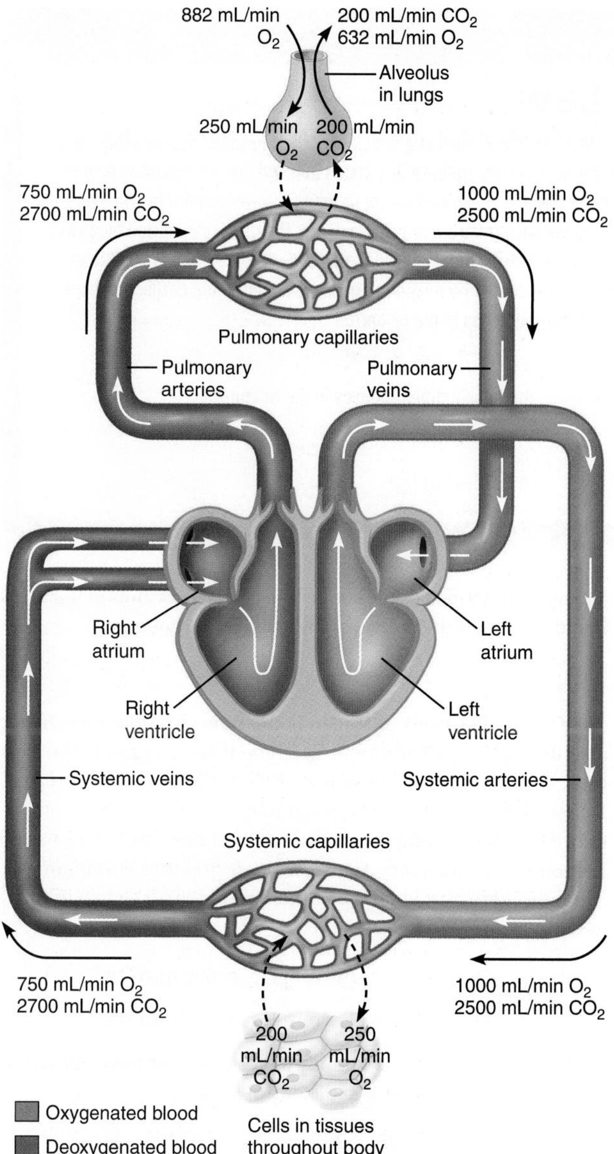

Figure 1 Movements of oxygen and carbon dioxide in pulmonary and systemic tissues during rest.

Q Under resting conditions, how much blood is pumped to the lungs each minute?

the alveoli). The oxygenated blood leaves the lungs and returns to the left atrium via pulmonary veins. Blood then moves into the left ventricle, which pumps the blood through the systemic arteries to the cells of the body. In the systemic capillaries, 250 mL of oxygen diffuses from the blood to the cells each minute, and 200 mL of carbon dioxide diffuses from the cells to the blood each minute. The now deoxygenated blood returns to the right atrium and the cycle begins again.

Oxygen and carbon dioxide move between alveolar air and blood by diffusion down concentration gradients. Oxygen is at a higher concentration in the alveoli and diffuses into the blood, whereas carbon dioxide is at a higher concentration in the blood

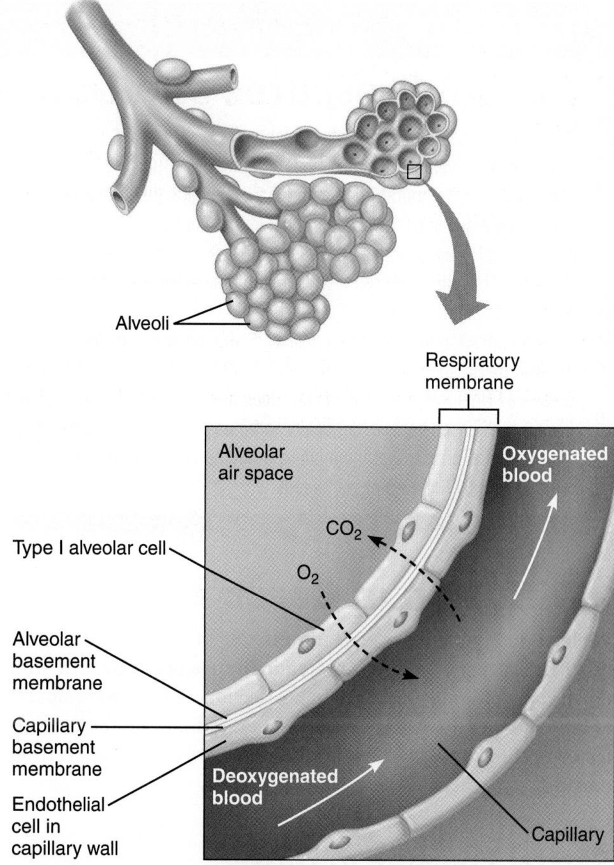

Figure 2 The respiratory membrane.

and diffuses into the alveoli. We know that oxygen and carbon dioxide are able to cross cell membranes by simple diffusion; we also know that the rate of transport in simple diffusion is proportional to the magnitude of the concentration gradient, the surface area of the membrane through which it is moving, and the permeability of the membrane to that particular substance. **Figure 2** shows the structure of the respiratory membrane, which is composed of three layers: type I epithelial cells in the alveolar wall and endothelial cells in the capillary wall "sandwiched" around their fused basement membranes. The respiratory membrane provides a large surface area of very thin membrane, favoring fast rates of diffusion for oxygen and carbon dioxide between alveolar air and blood.

Quick Check 1

1. Under normal conditions, if respiring tissues consume oxygen at a rate of 250 mL/min, at what rate does oxygen diffuse from the alveoli to blood?
2. If the cells of the body consume 300 mL of oxygen per minute and produce 270 mL of carbon dioxide per minute, then what is the respiratory quotient?
3. Which blood vessel contains blood with the greater concentration of oxygen—a pulmonary artery or a pulmonary vein?

TOOLBOX

Partial Pressures and Dalton's Law

According to the ideal gas law, for any pure gas, $PV = nRT$, where P = pressure, V = volume, n = number of moles, R = universal gas constant, and T = temperature. If V and T are constant, then the pressure exerted by a gas is directly proportional to the number of moles of the gas. If 1 mole of helium produces a pressure of 100 mm Hg, for example, then 2 moles of helium at the same volume and temperature would exert a pressure of 200 mm Hg.

What about a mixture of two gases? If we start with 1 mole of helium at 100 mm Hg and then add 1 mole of oxygen, what will happen to the pressure? Note that the ideal gas law does not depend on the kind of gas involved; all gases behave identically so far as pressure is concerned. Therefore, the pressure of the mixture would be 200 mm Hg, just as if we had added another mole of helium.

In 1801, the English chemist John Dalton formulated a law regarding mixtures of gases. Dalton's law states that *the pressure exerted by a mixture of gases is equal to the sum of the pressures exerted by the individual gases occupying the same volume alone.* The pressure exerted by an individual gas in a mixture is called the *partial pressure* of that gas. Therefore, the total pressure of a mixture of gases is a sum of the partial pressures exerted by each gas in the mixture:

$$P_{total} = P_1 + P_2 + P_3 + \cdots + P_n$$

where n is the total number of gases in the mixture.

2 Diffusion of Gases 

We have seen that the overall movements of oxygen and carbon dioxide between lungs and systemic tissues occur by diffusion down concentration gradients. In this section we look more closely at how these gradients are established and find that they are affected by the *partial pressures* and *solubilities* of oxygen and carbon dioxide.

Partial Pressure of Gases

We know that the pressure of a gas depends on its temperature and the number of gas molecules contained in a given volume. A gas (air, for instance) is often a mixture of more than one type of molecule. The total pressure of such a gas is the sum of the pressures of the individual gases that make up the mixture:

$$P_{total} = P_1 + P_2 + P_3 + \cdots + P_n$$

where n is the number of gases. In any gas mixture, the **partial pressure** of an individual gas is the proportion of the pressure of the entire gas that is due to the presence of the individual gas (see **Toolbox: Partial Pressures and Dalton's Law**). For example, if helium and nitrogen are mixed together in equal proportions and the pressure of the mixture is 500 mm Hg, then half the pressure (250 mm Hg) is exerted by helium and half by nitrogen. In this example, 500 mm Hg is the *total* pressure of the mixture, and 250 mm Hg is the *partial pressure* of helium or nitrogen.

The partial pressure of a gas mixture is determined by two factors: (1) the *fractional concentration* of that gas, which is the quantity of that gas (moles) relative to the total quantity of gas in the mixture, and (2) the total pressure exerted by the gas mixture. To find the partial pressure of a gas, we multiply these two factors together.

Air is composed almost entirely of two gases: nitrogen and oxygen. (Other gases such as carbon dioxide, argon, neon, helium, and methane are found in only minute amounts in air.) The amount of water vapor in air depends on the humidity. The total pressure of air can be described as the sum of the partial pressures of the gases found in air, plus the pressure of water vapor, as follows:

$$P_{air} = P_{N_2} + P_{O_2} + P_{H_2O}$$

On a molar basis, nitrogen is the most abundant gas in air, accounting for 79% of the molecules in air. Oxygen is the next most abundant gas, representing 21% of the molecules present. Depending on the humidity, water may become a critical component of air, decreasing the contribution of the other two gases to the total pressure of air. At sea level, air pressure is 760 mm Hg. Assuming zero humidity, the partial pressures of the two primary gases in air are

$$P_{N_2} = 0.79 \times 760 \text{ mm Hg} = 600 \text{ mm Hg}$$

$$P_{O_2} = 0.21 \times 760 \text{ mm Hg} = 160 \text{ mm Hg}$$

Carbon dioxide accounts for only 0.03% of the air molecules; thus the partial pressure of carbon dioxide in air is

$$P_{CO_2} = 0.0003 \times 760 \text{ mm Hg} = 0.23 \text{ mm Hg}$$

Recall that as air moves through the conducting zone, it is humidified to saturation (100% humidity). At 100% humidity and 37°C, the partial pressure of water is 47 mm Hg. With water contributing significantly to the pressure of air that enters the lungs, the partial pressures of the other gases in the air decrease to the following levels: $P_{N_2} = 563$ mm Hg, $P_{O_2} = 150$ mm Hg, and $P_{CO_2} = 0.21$ mm Hg.

Solubility of Gases in Liquids

Gas molecules can either exist in gaseous form or be dissolved in liquids. The ability of a gas to dissolve in water has physiological significance because oxygen and carbon dioxide are exchanged between air in the alveoli and blood (which is primarily water) and are then transported by the blood. Even when gas molecules are dissolved in a liquid, they have a certain partial pressure. When a gas mixture and a liquid are in contact with each other, gas

TOOLBOX

Henry's Law and Solubility of Gases

Gases can dissolve in liquids. The solubility of a gas in a liquid depends on temperature, the partial pressure of the gas over the liquid, the chemical properties of the gas, and the chemical properties of the liquid.

If a gas exists over a liquid, then the gas will dissolve in the liquid until equilibrium is reached. At equilibrium, the number of gas molecules dissolving in the liquid is equal to the number of gas molecules leaving the liquid. However, the concentrations of gas (moles/L) in the air and in the liquid are not equal at equilibrium. The concentration of gas in the liquid depends on the specific solubility for that gas in that particular liquid: The more soluble, the greater the concentration of gas in the liquid.

In 1803, William Henry developed a law to describe the solubility of gases in liquids. Basically, Henry's law states that when the temperature is constant, the concentration of a gas in a liquid is proportional to its partial pressure. In mathematical terms, Henry's law is

$$c = kP$$

where c is the molar concentration of the dissolved gas (moles/liter), P is the partial pressure of the gas, and k is the Henry's law constant at a specific temperature (determined experimentally; constants can be found in tables). The relationship between concentration and pressure can be written as

$$k = c/P$$

Because k is a constant, if the pressure changes, the relationship between the concentration of dissolved gas at the initial pressure (P_1) and at the subsequent pressure (P_2) can be described by

$$c_1/P_1 = c_2/P_2$$

Therefore, if pressure is doubled, then the concentration of gas dissolved in the liquid also doubles.

In the human body, gases such as nitrogen, oxygen, and carbon dioxide dissolve in plasma (and other body fluids) according to Henry's law. The partial pressure of the dissolved gas equals the partial pressure of the gas surrounding the liquid. At any given partial pressure, the concentration of carbon dioxide dissolved in plasma is greater than the concentration of oxygen in plasma, because carbon dioxide is more soluble. For either gas, as the partial pressure increases, the concentration also increases. Because the concentration of a gas in plasma is directly related to the partial pressure of the gas, movement of a gas by diffusion occurs from areas of high partial pressure (or high concentration) to areas of low partial pressure (low concentration).

---

molecules dissolve in the liquid until the system reaches equilibrium, at which point gas molecules dissolve in the liquid at the same rate that gas molecules move from the liquid to the gaseous state. At equilibrium, the dissolved gas molecules and those in the gaseous phase are said to be at the same partial pressure, and the concentration of gas molecules in the liquid is proportional to the partial pressure of the gas.

At a given partial pressure, the relative concentrations of the various dissolved gas molecules differ from one gas to another because some gases are more soluble in a given liquid than other gases; that is, some gases mix with the liquid more easily. For example, carbon dioxide is nearly 20 times more soluble in blood than is oxygen. Therefore, the concentration of dissolved gas molecules depends not only on the partial pressure but also on the solubility of the gas in that particular liquid. The relationship among the concentration, partial pressure, and solubility of a gas is described by Henry's law, mathematically expressed as

$$c = kP$$

where c is the molar concentration of the dissolved gas (moles of gas per liter of liquid), P is the partial pressure of the gas in atmospheres (1 atmosphere = 760 mm Hg), and k is the Henry's law constant, which varies based on the gas and the temperature (see **Toolbox: Henry's Law and Solubility of Gases**).

Let's look at a common example of Henry's law. Think about the last time that you opened a bottle of a carbonated beverage. The beverage was bottled under high pressure to force more gas, primarily carbon dioxide, to dissolve in the liquid. When the bottle was uncapped, the gas pressure in the bottle decreased as it equilibrated with atmospheric pressure; the pressure of carbon dioxide in the bottle decreased to approximately 0.23 mm Hg. When the partial pressure of carbon dioxide in the gas in the bottle decreased, carbon dioxide was no longer at equilibrium in the gas and in the liquid, so gas bubbles emerged from the liquid as they came out of solution, which decreased the partial pressure of carbon dioxide in the liquid until a new equilibrium was established. As a result of this process, the beverage went "flat." A similar phenomenon can occur within body fluids, as described in **Clinical Connections: The Bends**.

Figure 3a illustrates the solubility of oxygen in water. The system starts off by exposing water to air that has a partial pressure of 100 mm Hg of oxygen. Initially, no oxygen is dissolved in the water. Over time, however, oxygen dissolves in the water, and the concentration of oxygen in the water increases until equilibrium is reached—that is, until the number of oxygen molecules dissolving in water per unit time equals the number of oxygen molecules leaving the water. When a gas is in equilibrium with a liquid, the partial pressure of the gas in the liquid equals the partial pressure of the gas in the air; at equilibrium, the partial pressure of oxygen in the water is 100 mm Hg. (This analysis assumes that the volume of air is considerably greater than the volume of water, such that the amount of oxygen that dissolves in the water is small enough to not change the partial pressure of oxygen in the air.)

That the partial pressures of oxygen in air and water are equal at equilibrium does *not* mean that the *concentrations* of oxygen in the air and in the water are equal. In fact, the concentrations are far from equal because oxygen has low solubility in water. The concentration of oxygen in the air at 100 mm Hg and 37°C (as determined

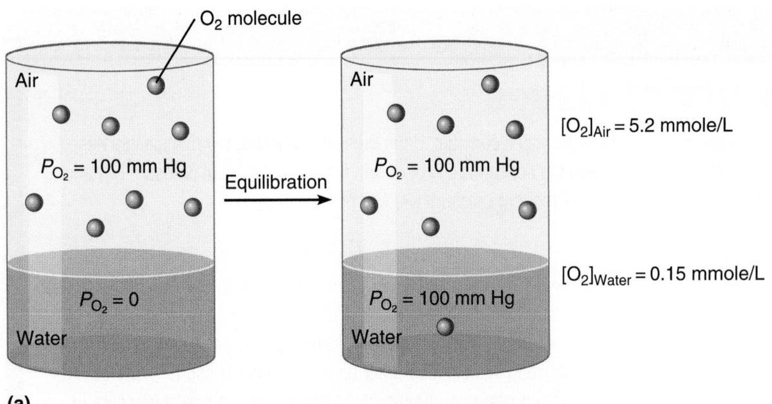

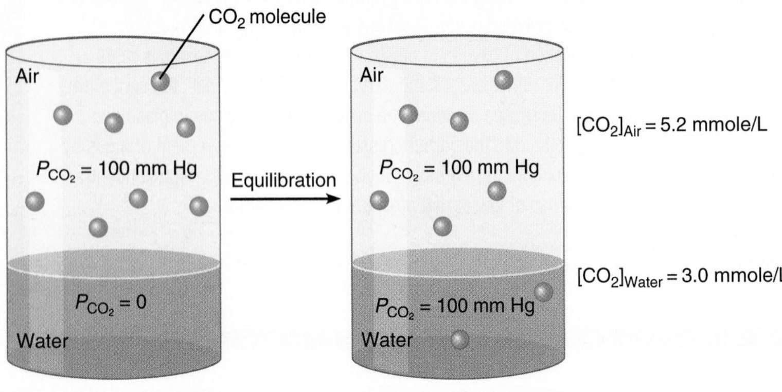

Figure 3 Solubilities of oxygen and carbon dioxide in water. The beakers at left depict the initial conditions in which water has just been exposed to air that has a partial pressure of gas of 100 mm Hg; the beakers at right show the conditions after equilibration, when the gas has dissolved in the water until its partial pressures in both media are equal. **(a)** After equilibration, the concentration of oxygen in water is much less than its concentration in air, indicating that the solubility of oxygen in water is low. **(b)** After equilibration, the concentration of carbon dioxide in water is greater than that of oxygen in water at the same partial pressure, indicating that carbon dioxide is more soluble in water than is oxygen.

by Boyle's law) is 5.2 mmole/liter, whereas the concentration of oxygen in the water at 100 mm Hg and 37°C (as determined by Henry's law) is 0.15 mmole/liter.

We can perform the same analysis with carbon dioxide, again starting with a partial pressure of 100 mm Hg (Figure 3b). When at equilibrium with water, the concentration of carbon dioxide in air is 5.2 mmole/liter, the same as that of oxygen at 100 mm Hg. However, the concentration of carbon dioxide in water is 3.0 mmole/liter—much greater than the concentration of oxygen in water at the same partial pressure. Thus we can see that carbon dioxide is much more soluble in water than is oxygen.

Apply Your Knowledge

At sea level, atmospheric pressure is 760 mm Hg. However, at the top of Mount Everest, atmospheric pressure is only 250 mm Hg. Given that the air in both places is 21% oxygen, what is the partial pressure of oxygen at the top of Mount Everest?

3 Exchange of Oxygen and Carbon Dioxide

In mixtures of gases, a particular gas will diffuse down its *partial pressure gradient.* Oxygen and carbon dioxide are exchanged between the alveoli and the blood, and between the blood and systemic tissues, by the same mechanism: Each gas diffuses down its own partial pressure gradient. This section describes the partial pressures that exist in the alveoli and blood, and the resulting exchange of gases that occurs.

Gas Exchange in the Lungs

Representative partial pressures of oxygen and carbon dioxide in alveolar air, blood, and respiring tissues are tabulated in **Table 1** and shown in **Figure 4**. Recall that in atmospheric air, the P_{O_2} is 160 mm Hg, and the P_{CO_2} is 0.23 mm Hg. Notice in Figure 4, however, that the P_{O_2} in the alveoli is only 100 mm Hg, whereas the P_{CO_2} is 40 mm Hg. Alveolar gas pressures differ from atmospheric pressures for three reasons: (1) Exchange of gases occurs continually between alveolar air and capillary blood; (2) upon inspiration, fresh atmospheric air mixes with air rich in carbon dioxide and relatively poor in oxygen in the dead space of the conducting zone; and (3) air in the alveoli is saturated with water vapor.

The blood entering the pulmonary capillaries is deoxygenated blood, with a P_{O_2} of 40 mm Hg and a P_{CO_2} of 46 mm Hg. As this blood passes by the alveoli, oxygen and carbon dioxide diffuse down their partial pressure gradients: Oxygen diffuses from alveoli to blood (P_{O_2} alveolar = 100 mm Hg; P_{O_2} blood = 40 mm Hg) while carbon dioxide diffuses from blood to alveoli (P_{CO_2} blood = 46 mm Hg; P_{CO_2} alveolar = 40 mm Hg). Eventually, diffusion reaches an equilibrium between alveolar air and capillary blood, such that the blood leaving the pulmonary capillaries and entering the pulmonary veins has a P_{O_2} of 100 mm Hg and a P_{CO_2} of 40 mm Hg (the same partial pressures as found in the alveoli).

Diffusion is a rapid process (taking approximately 0.25 second for the blood to equilibrate with alveolar air) that is nearly complete by the time blood has traveled only one-third the length of a capillary (**Figure 5**). The rapidity of gas exchange provides a margin of safety in that gases can still equilibrate between capillary blood and alveolar air even if blood is flowing at a rate as much as three times faster than the resting rate, as can occur during exercise. The diffusion rate is rapid because of the thinness of the respiratory membrane.

Whenever the respiratory membrane is effectively thickened, gas exchange is hampered. During pulmonary edema, for example, excess fluid builds up in the interstitial spaces of the lungs and

in the alveoli, effectively increasing the distance between alveoli and capillaries and, therefore, increasing the thickness of the diffusion

CLINICAL CONNECTIONS

The Bends

In 1943, Jacques Cousteau designed the first self-contained underwater breathing apparatus (SCUBA), called the Aqua-Lung, for use in studies of marine biology and oceanography. Since then, scuba diving has taken off as a recreational sport, with more than a million Americans registered as certified divers. Most recreational scuba divers breathe normal air from tanks that allow them to stay submerged for 30–60 minutes. When diving, scuba divers monitor the depth they are at because pressure increases with depth; for every 33 feet of depth, pressure increases by 760 mm Hg, or 1 atmosphere.

As a diver goes deeper, the increased pressure causes an increase in the concentration of gases, especially nitrogen, dissolved in the blood (according to Henry's law, $c = kP$, so as P increases, c increases). If a diver comes to the surface too quickly, nitrogen gas bubbles form in the blood as the pressure decreases (just as the carbon dioxide comes out of a carbonated liquid when the pressure is decreased by opening the bottle), causing *decompression sickness*. The nitrogen gas bubbles commonly lodge in joints and in the nervous system, causing problems as minor as discomfort upon the bending of joints (thus the name "the bends") or as severe as paralysis or even death. Gas bubbles can also clog blood vessels, producing *air embolisms*. By rising to the surface slowly (approximately 1 foot per second) and allowing the nitrogen gas to come out of solution slowly, divers can protect themselves from the bends.

A diver suffering from the bends is immediately put into a hyperbaric chamber for recompression. Recompression causes the nitrogen bubbles to go back into solution. Then the pressure is slowly decreased to atmospheric pressure, allowing the excess nitrogen to slowly come out of solution and be eliminated from the body during normal breathing.

Zoonar GmbH/Alamy

Critical Thinking Questions

1. What causes the increased gas concentration in the body as a person scuba dives?
2. Explain the sequence of events leading to decompression sickness.
3. Describe the symptoms of and treatment for decompression sickness.

barrier between air and blood. This thicker barrier decreases the rate of diffusion of gases, which reduces the exchange of gases between alveolar air and capillary blood, resulting in systemic arterial blood with a lower P_{O_2} and a higher P_{CO_2} than normal. The body will try to compensate for the decreased P_{O_2} and increased P_{CO_2} in arterial blood by increasing ventilation. (See **Clinical Connections: Pulmonary Edema**, for more details.)

TABLE 1 Typical Partial Pressures of Oxygen and Carbon Dioxide in Atmospheric Air and at Various Sites in the Body

	Oxygen	Carbon dioxide
Atmospheric air	160 mm Hg	0.3 mm Hg
Alveolar air	100 mm Hg	40 mm Hg
Pulmonary veins	100 mm Hg	40 mm Hg
Systemic arteries	100 mm Hg	40 mm Hg
Cells	≤40 mm Hg	≥46 mm Hg
Systemic veins	40 mm Hg	46 mm Hg
Pulmonary arteries	40 mm Hg	46 mm Hg

Gas Exchange in Respiring Tissue

Blood oxygenated in the pulmonary capillaries returns to the left atrium via the pulmonary veins. The blood then flows into the left ventricle, which pumps it out to the systemic capillaries, where gases are exchanged between tissue cells and blood. Blood entering the systemic capillaries has a P_{O_2} of 100 mm Hg and a P_{CO_2} of 40 mm Hg (see Figure 4). In the interstitial fluid surrounding the capillaries, the P_{O_2} is lower because respiring cells consume oxygen, and the P_{CO_2} is higher because the cells produce carbon dioxide. When oxygen and carbon dioxide diffuse down their pressure gradients, oxygen moves from blood to the tissue, and carbon dioxide moves from the tissue to blood.

The actual partial pressures of oxygen and carbon dioxide in the venous blood leaving the systemic capillaries depend on the metabolic activity of the tissue and the blood flow to the tissue. During intense exercise, for example, the metabolic activity of skeletal muscle is high relative to its blood flow, and thus the P_{O_2} in the interstitial fluid around the muscle tissue will be lower and the P_{CO_2} will be higher than they are in less active tissue. With larger pressure gradients between the capillary blood and the interstitial fluid, gas exchange occurs more rapidly, and more gas molecules are exchanged. More oxygen diffuses from the blood to the interstitial

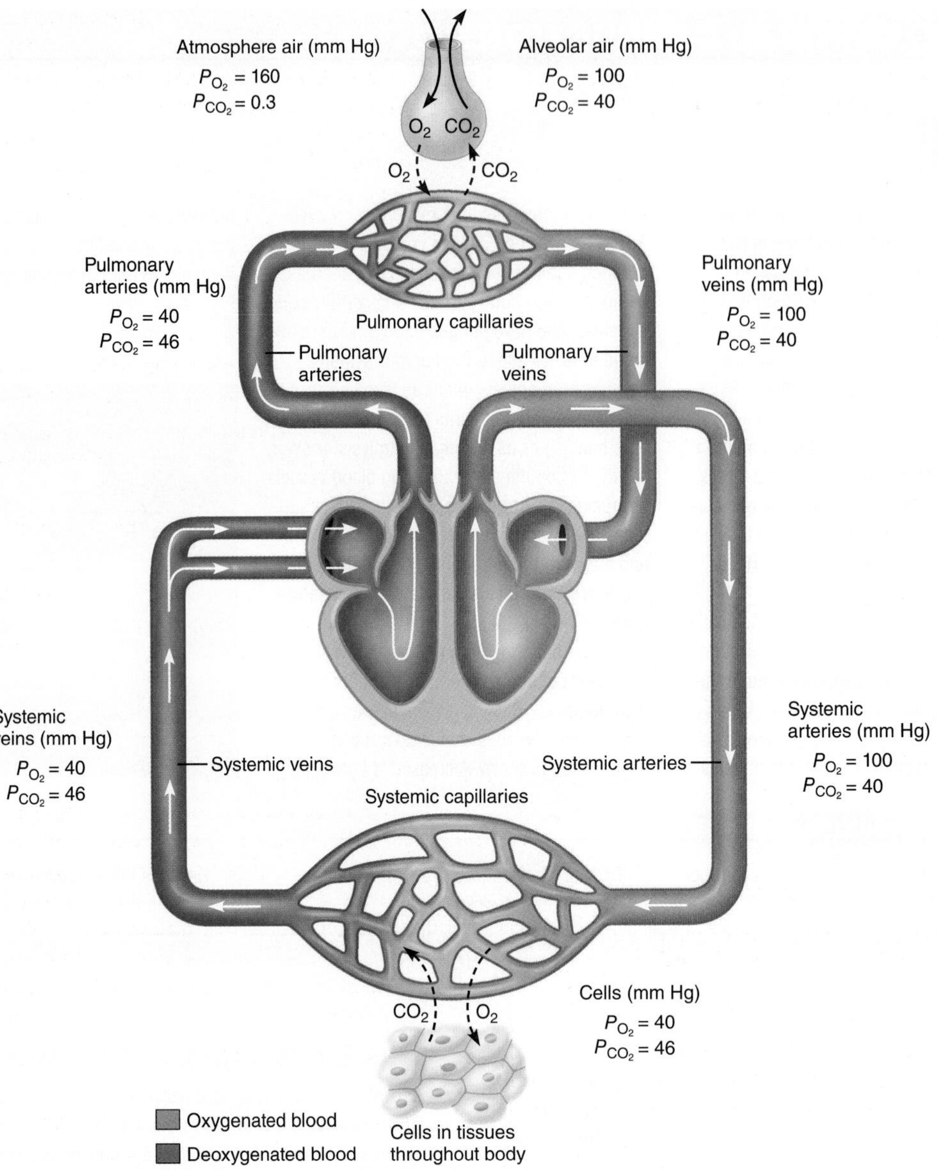

Figure 4 Partial pressures of oxygen and carbon dioxide in atmospheric air, in alveolar air, and at various sites in the body.

Q At which body sites does the partial pressure of carbon dioxide exceed the partial pressure of oxygen?

fluid, and more carbon dioxide diffuses from the interstitial fluid to the blood. Therefore, venous blood coming from capillaries that supply active tissue will have a lower P_{O_2} and a higher P_{CO_2} than venous blood coming from capillaries supplying less active tissue.

All venous blood returns to the right atrium and *mixes* together before being pumped by the right ventricle into the pulmonary artery. Therefore, the blood in the pulmonary artery is called **mixed venous blood.** Typically, the P_{O_2} of mixed venous blood is 40 mm Hg for a person at rest, and the P_{CO_2} is 46 mm Hg (see Figure 4). During strenuous exercise, however, the value for P_{O_2} decreases, and that for P_{CO_2} increases.

Determinants of Alveolar P_{O_2} and P_{CO_2}

Because alveolar P_{O_2} and P_{CO_2} determine systemic arterial P_{O_2} and P_{CO_2} it is important to understand the factors that affect alveolar partial pressures. Alveolar P_{O_2} and P_{CO_2} are normally determined by only three factors: (1) the P_{O_2} and P_{CO_2} of inspired air, (2) the minute alveolar ventilation (the volume of fresh air reaching the alveoli each minute), and (3) the rates at which respiring tissues consume oxygen and produce carbon dioxide.

Under most conditions, partial pressures in inspired air do not change appreciably and can be assumed to be constant (exceptions include changes in altitude and the use of artificial breathing

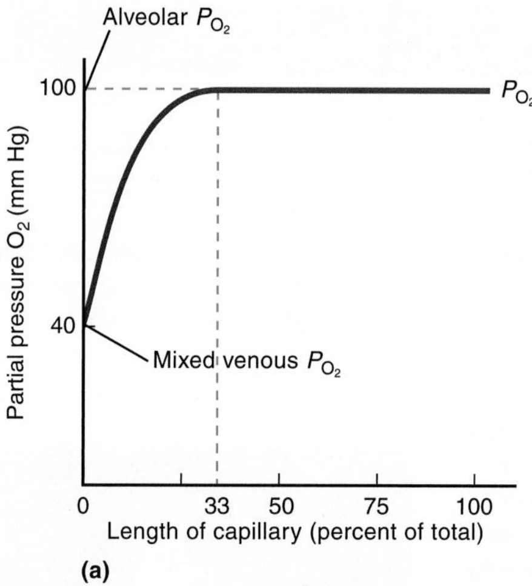

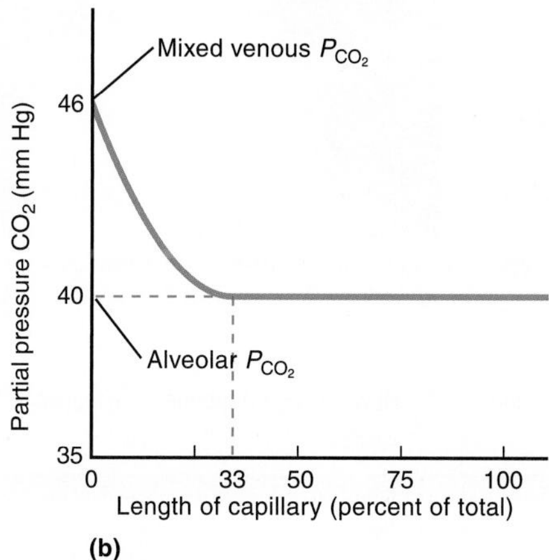

Figure 5 Gas exchange as a function of pulmonary capillary length. In the rapid exchange of gases between alveolar air and pulmonary capillary blood, equilibration of the partial pressures of **(a)** oxygen and **(b)** carbon dioxide occurs within the first 33% of a capillary's length. "Mixed venous blood" refers to the blood in the pulmonary artery that contains blood returned to the heart from the systemic veins.

mixtures such as 95% O_2 with 5% CO_2, which are administered for therapeutic purposes). Therefore, alveolar partial pressures are normally determined by alveolar ventilation and the rates of oxygen consumption and carbon dioxide production. In particular, the crucial factor is the alveolar ventilation *relative to* the rate of oxygen consumption or carbon dioxide production.

Under normal circumstances, alveolar ventilation matches the demands of the tissues; that is, if oxygen consumption and carbon dioxide production increase, then alveolar ventilation will increase as well. This appropriate increase in alveolar ventilation is called **hyperpnea,** and the mechanisms that trigger it are described later in this chapter. In hyperpnea, arterial P_{O_2} and P_{CO_2} do not change because alveolar ventilation increases "in step with" the increased rates of oxygen consumption and carbon dioxide production. In **hypoventilation,** alveolar ventilation is insufficient to meet the demands of the tissues; in this case, arterial P_{CO_2} rises above the normal value of 40 mm Hg, and arterial P_{O_2} decreases below the normal value of 100 mm Hg. In **hyperventilation,** alveolar ventilation exceeds the demands of the tissues; arterial P_{CO_2} decreases to less than 40 mm Hg, and P_{O_2} increases to greater than 100 mm Hg. Both hypoventilation and hyperventilation can alter normal functioning of the body, including the nervous system. (A review of terms used in respiratory physiology is provided in Table 2.)

Health Fact

Quick Check 2

1. List the normal resting values for P_{O_2} and P_{CO_2} (in mm Hg) for alveolar air, arterial blood, and mixed venous blood.
2. When oxygen diffuses from alveolar air to pulmonary capillary blood, what causes it to move? Why does carbon dioxide leave the blood and enter the alveoli?
3. When alveolar ventilation increases, what happens to alveolar P_{O_2} and P_{CO_2}?
4. What happens to alveolar P_{CO_2} when a person hyperventilates? What happens to alveolar P_{CO_2} when a person hypoventilates?

4 Transport of Gases in the Blood 

The previous section explained how gas is exchanged between air and blood, and between tissue and blood. This section describes how the gas is transported within the blood. Earlier, we saw that gases are not very soluble in blood. For example, when arterial P_{O_2} is at the normal value of 100 mm Hg, blood contains only about 3 mL of dissolved oxygen per liter of blood. To deliver dissolved oxygen to tissues at the normal resting rate of approximately 250 mL/min, cardiac output would have to be

TABLE 2 Some Terms Used in Respiratory Physiology

Term	Definition
Hyperpnea	An increase in ventilation to meet an increase in the metabolic demands of the body
Dyspnea	Labored or difficult breathing
Apnea	Temporary cessation of breathing
Tachypnea	Rapid, shallow breathing
Hyperventilation	A condition in which ventilation exceeds the metabolic demands of the body
Hypoventilation	A condition in which ventilation is insufficient to meet the metabolic demands of the body
Hypoxia	A deficiency of oxygen in the tissues
Hypoxemia	A deficiency of oxygen in the blood
Hypercapnia	An excess of carbon dioxide in the blood
Hypocapnia	A deficiency of carbon dioxide in the blood

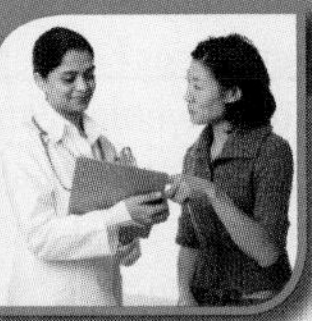

CLINICAL CONNECTIONS

Pulmonary Edema 

Pulmonary edema—the accumulation of excess fluid in the lungs—is a fairly common, yet dangerous, disorder. This condition is marked by signs of respiratory distress (including *tachypnea*, or rapid, shallow breathing) and by low oxygen levels in the tissues *(hypoxia)*, which may be evident as a bluish coloration *(cyanosis)* of the skin and mucous membranes.

Pulmonary edema occurs in two stages: *interstitial edema,* in which excess fluid accumulates in the interstitial spaces in lung tissue, and *alveolar edema,* in which fluid accumulates in the alveoli. In severe cases, fluid may even move into the airways, in which case the affected person may cough up a frothy foam. Pulmonary edema interferes with breathing in two ways: (1) by increasing the distance over which gases must diffuse to move between alveolar air and capillary blood, which impedes gas exchange, and (2) by interfering with the action of pulmonary surfactant, which causes a decrease in lung compliance and thus an increase in the work of breathing.

Pulmonary edema is similar to systemic edema. The most common cause of pulmonary edema is increased hydrostatic pressure in the pulmonary capillaries. Left heart failure can cause a backup of blood in the pulmonary circulation, which increases the pressure in the pulmonary veins. As pressure builds up in the pulmonary capillaries, fluid is pushed out of the capillaries and into the interstitial space of the lungs.

Treatment of pulmonary edema is critical because it is a life-threatening situation. The symptoms of pulmonary edema are treated by administering oxygen and diuretics (medications that increase fluid output by the kidneys). However, the cause of the pulmonary edema must be determined and treated appropriately once the symptoms are stabilized.

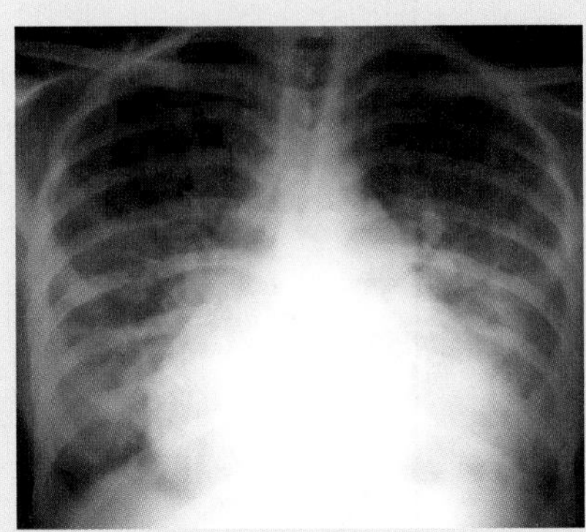

SPL/Photo Researchers

X-ray image of a person with pulmonary edema.

Critical Thinking Questions

1. Identify and describe the two stages of pulmonary edema.
2. Explain the cause of pulmonary edema, and how it interferes with breathing.
3. How are the symptoms of pulmonary edema treated?

approximately 83 liters/min if the total oxygen concentration in the blood were this low. Because the normal resting cardiac output is approximately 5 liters/min, oxygen must be transported in blood by a more efficient means than merely being dissolved in plasma. The same is true for carbon dioxide. First we examine how blood transports oxygen, and then how it moves carbon dioxide throughout the body.

Oxygen Transport in the Blood 

Transport of oxygen in the blood has a special need: The transport mechanism must be readily reversible such that oxygen enters the blood in the lungs and leaves the blood in other tissues of the body. Hemoglobin has a unique structure that allows oxygen to do just that.

Oxygen Transport by Hemoglobin

Every liter of arterial blood contains about 200 mL of oxygen. Approximately 3 mL of this oxygen (1.5%) is dissolved in the plasma or in the cytosol of erythrocytes, and only this dissolved oxygen contributes to the P_{O_2} in blood. The remaining 197 mL of oxygen (98.5%) is transported bound to hemoglobin. Although the bound oxygen does not contribute to the P_{O_2}, it is in equilibrium with the dissolved oxygen, and thus the amount of oxygen bound to hemoglobin is a function of the P_{O_2}.

Hemoglobin consists of four subunits, each of which contains a *globin* (globular polypeptide chain) and a *heme group* that contains iron. Each heme group is capable of binding one oxygen molecule, so each hemoglobin molecule can carry a total of four oxygen molecules. The complex of hemoglobin and bound oxygen is called *oxyhemoglobin;* a hemoglobin molecule without any oxygen is called *deoxyhemoglobin.*

In the lungs, as oxygen molecules move from alveolar air to capillary blood, they bind to hemoglobin (**Figure 6a**); when the blood reaches respiring tissues, oxygen molecules dissociate from the hemoglobin and diffuse to the cells (Figure 6b). For hemoglobin to function in oxygen transport, it is critical that it binds the oxygen *reversibly*—that is, tightly enough so that it can pick up large quantities of oxygen in the lungs, but not so tightly that it cannot release the oxygen into the respiring tissues later on.

The binding or release of oxygen depends on the P_{O_2} in the fluid surrounding hemoglobin. High P_{O_2} facilitates the binding of

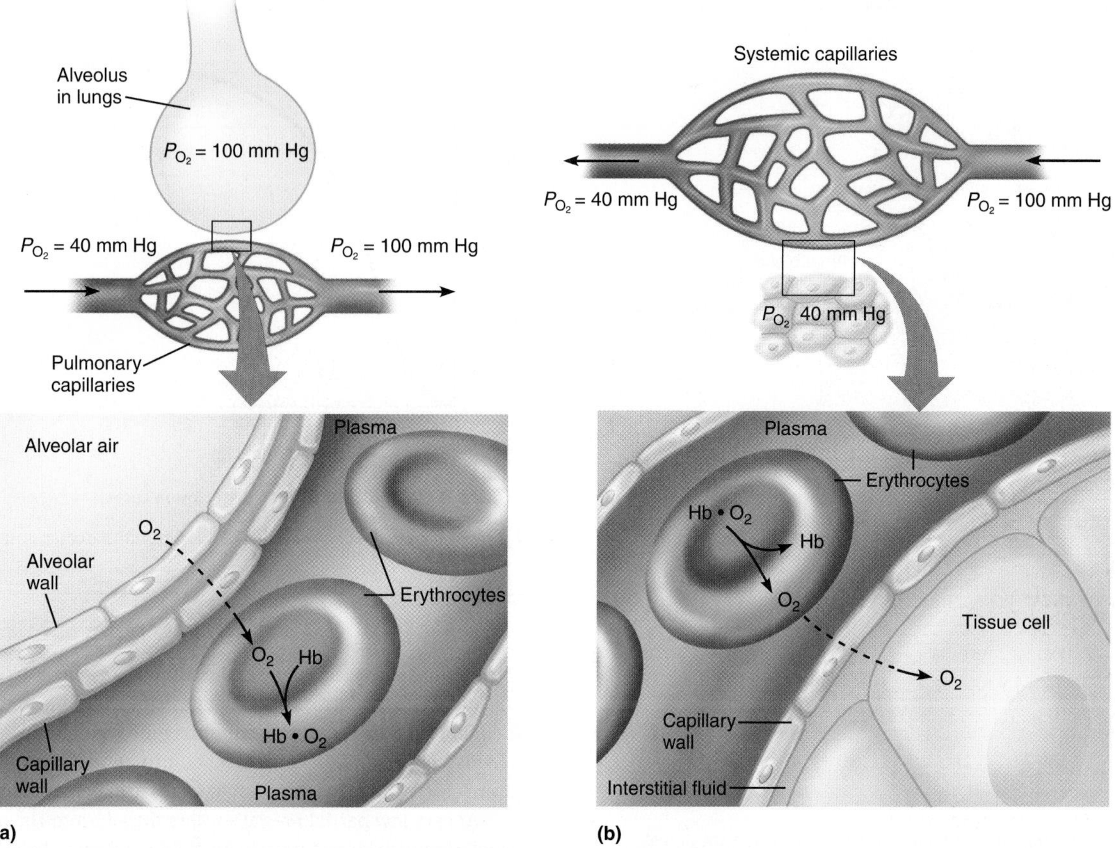

Figure 6 Transport of oxygen by hemoglobin. (a) Formation of oxyhemoglobin. Once oxygen diffuses from alveolar air to blood in pulmonary capillaries, it diffuses into erythrocytes and binds to hemoglobin for transport in the blood. **(b)** Dissociation of oxygen from hemoglobin. In systemic capillaries, hemoglobin in erythrocytes releases oxygen, which then diffuses from the blood into tissue cells.

Q What is the oxygen that moves into respiring tissue used for?

oxygen with hemoglobin, whereas low P_{O_2} facilitates release of oxygen from hemoglobin. The reaction of oxygen with hemoglobin can be written as

$$Hb + O_2 \rightleftharpoons Hb \cdot O_2$$

where Hb is deoxyhemoglobin, O_2 is the dissolved oxygen in blood, and Hb · O_2 is oxyhemoglobin. The law of mass action states that an increase in the concentration of the reactants drives the reaction to the right, resulting in the generation of more product. Therefore, as oxygen levels in the pulmonary capillaries increase, more oxyhemoglobin is formed. Conversely, as oxygen levels in the systemic capillaries decrease, the reaction is driven to the left to release oxygen from the hemoglobin.

Note, however, that because hemoglobin can bind up to four oxygen molecules, the previous equation could be written as follows:

$$Hb \overset{\overset{O_2}{\downarrow}}{\rightleftharpoons} Hb \cdot O_2 \overset{\overset{O_2}{\downarrow}}{\rightleftharpoons} Hb \cdot (O_2)_2 \overset{\overset{O_2}{\downarrow}}{\rightleftharpoons} Hb \cdot (O_2)_3 \overset{\overset{O_2}{\downarrow}}{\rightleftharpoons} Hb \cdot (O_2)_4$$

The law of mass action is still in effect in that the more oxygen available, the more oxyhemoglobin is formed. When all oxygen-binding sites on a hemoglobin molecule are occupied, the hemoglobin molecule is said to be 100% saturated.

When hemoglobin is 100% saturated, 1 gram (g) of hemoglobin carries 1.34 mL of oxygen. The normal concentration of hemoglobin in the blood is 12–17 g/dL, or an average of 150 g/liter. Therefore, the oxygen-carrying capacity of the hemoglobin in blood is about 200 mL of oxygen per liter of blood (1.34 mL/gm × 150 g/litter). At the normal arterial P_{O_2} of 100 mm Hg, hemoglobin is at approximately 98% of its oxygen-carrying capacity (is 98% saturated) (**Figure 7a**). When cardiac output is 5 liters per minute, the blood supplies almost 1000 mL of oxygen to respiring tissues each minute (5 liters of blood per minute times 200 mL of O_2 in a liter of blood equals 1000 mL of O_2 per minute). Because respiring tissues need only about 250 mL of O_2 per minute, just 25% of the oxygen diffuses into respiring cells; thus 75% of the binding sites on hemoglobin are still occupied when blood leaves the tissue at a P_{O_2} of 40 mm Hg (Figure 7b). Therefore, in mixed venous blood under resting conditions, hemoglobin is still 75% saturated.

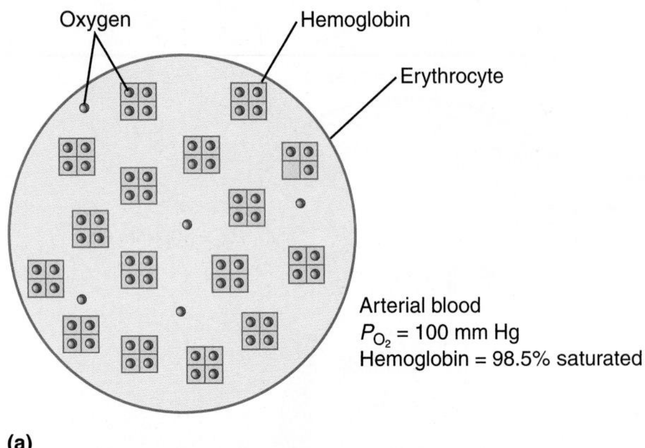

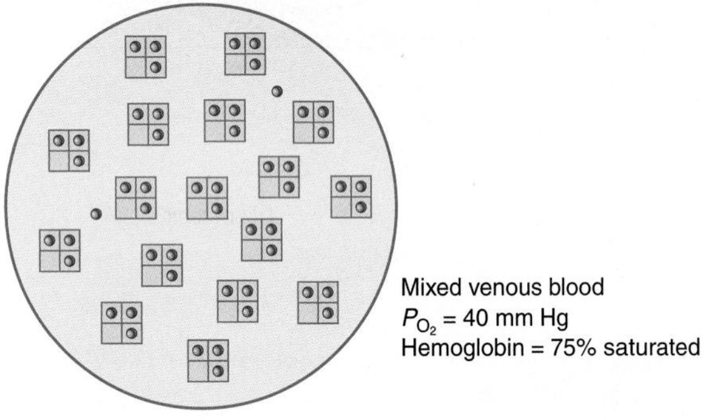

Figure 7 Saturation of hemoglobin. (a) In arterial blood, the partial pressure of oxygen is 100 mm Hg, and hemoglobin is 98.5% saturated; that is, almost all binding sites are occupied by oxygen. Very little oxygen is dissolved in the cytosol of the erythrocyte. **(b)** In mixed venous blood, the partial pressure of oxygen is 40 mm Hg, and hemoglobin is 75% saturated; that is, three of every four binding sites are occupied by oxygen.

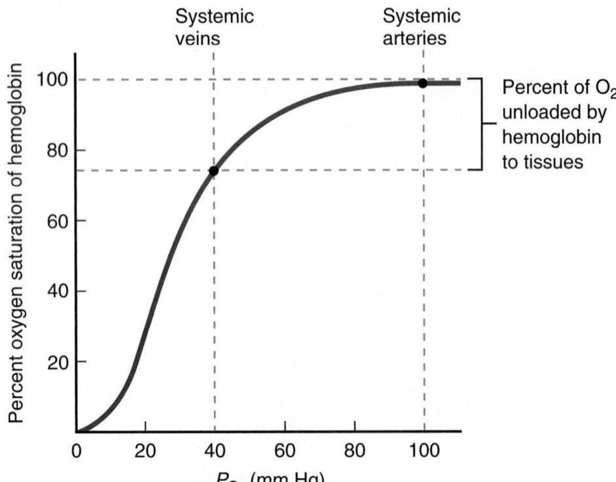

Figure 8 Hemoglobin-oxygen dissociation curve.

Q According to the graph, at what partial pressure of oxygen is hemoglobin 50% saturated?

Anemia is a decrease in the oxygen-carrying capacity of blood. It has many possible causes, including a deficiency or defect in hemoglobin (see Clinical Connections: Anemia). With less functioning hemoglobin in the blood, the oxygen-carrying capacity is decreased, and tissues may not be supplied with the oxygen they need, even when the P_{O_2} of blood is normal. As a consequence, people suffering from anemia tire more easily.

Health Fact

The Hemoglobin-Oxygen Dissociation Curve

The relationship between P_{O_2} and hemoglobin saturation just described can be summarized in the hemoglobin-oxygen dissociation curve, a plot of the percent saturation of hemoglobin as a function of the P_{O_2} (**Figure 8**). Although the percent saturation of hemoglobin increases as the P_{O_2} increases, the curve illustrating binding of oxygen to hemoglobin is not linear, but instead is S-shaped (sigmoidal), because the ability of hemoglobin to bind oxygen depends on how many oxygen molecules are already bound. Specifically, the binding of one oxygen molecule to hemoglobin increases the affinity of the hemoglobin molecule for oxygen and, therefore, increases the likelihood that another oxygen will bind with hemoglobin. The binding of oxygen to one of the subunits of a hemoglobin molecule induces a conformational change in the molecule that increases the affinity of the remaining subunits for oxygen (a process called *positive cooperativity*). Once this conformational change occurs, a given increase in P_{O_2} yields a larger increase in percent saturation.

At very low partial pressures (less than 15 mm Hg, a level not usually found in blood), most hemoglobin molecules have no oxygen bound to them. Under these conditions, the affinity of hemoglobin for oxygen is relatively low, and a given increase in P_{O_2} yields a small increase in percent saturation. As the P_{O_2} increases, more hemoglobin molecules have acquired at least one oxygen molecule, causing an increase in the affinity of hemoglobin for other oxygen molecules. This relationship is seen in the steep part of the hemoglobin-oxygen dissociation curve at P_{O_2} values between approximately 15 mm Hg and 60 mm Hg. At P_{O_2} values higher than approximately 60 mm Hg, the slope of the curve decreases because fewer binding sites are available as saturation increases. Above a P_{O_2} of approximately 80 mm Hg, the slope of the curve becomes nearly flat.

We can relate the hemoglobin-oxygen dissociation curve to events in the lungs and respiring tissues. Figure 8 indicates the P_{O_2} values and percent saturation of hemoglobin in the systemic arteries and veins under resting conditions. The P_{O_2} in the systemic arteries is approximately 100 mm Hg, and at this P_{O_2} hemoglobin is approximately 98% saturated. (To reach 100% saturation of hemoglobin would require a P_{O_2} of approximately 250 mm Hg.) In systemic veins, the P_{O_2} is approximately 40 mm Hg, and hemoglobin is approximately 75% saturated. Therefore, at rest, respiring tissues take up only 25% of the oxygen transported in blood, leaving a large reserve of oxygen available should the demands of respiring tissues increase.

Other Factors Affecting the Affinity of Hemoglobin for Oxygen

At least four other factors (temperature, pH, P_{O_2}, and 2,3-BPG) affect the affinity of hemoglobin for oxygen. Changes in the affinity

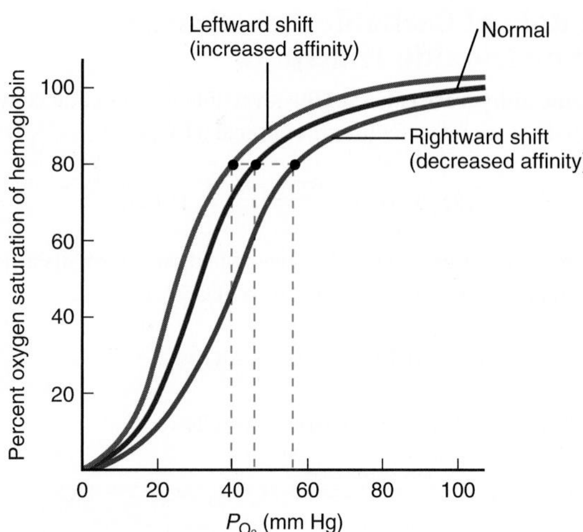

Figure 9 Effects of changes in the affinity of hemoglobin for oxygen on the hemoglobin-oxygen dissociation curve.

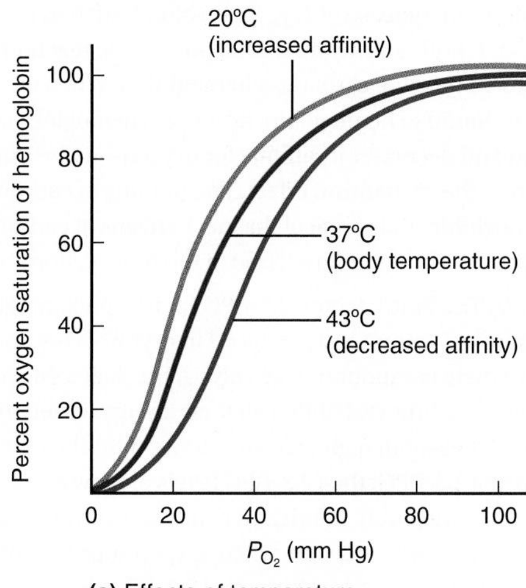

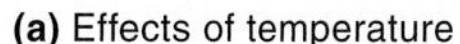

(a) Effects of temperature

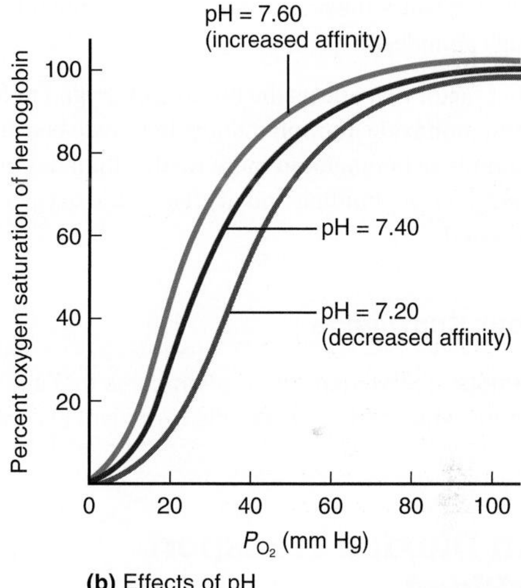

(b) Effects of pH

Figure 10 Effects of temperature and pH on the hemoglobin-oxygen dissociation curve. (a) Increases or decreases in temperature from normal body temperature (37°C) cause decreases or increases, respectively, in the affinity of hemoglobin for oxygen. **(b)** Increases or decreases in pH from normal arterial pH (7.4) cause increases or decreases, respectively, in the affinity of hemoglobin for oxygen.

of hemoglobin for oxygen are reflected in rightward or leftward shifts in the hemoglobin-oxygen dissociation curve (**Figure 9**). Decreases in the affinity cause the curve to shift rightward, indicating that a higher P_{O_2} is required to achieve any given level of saturation; a rightward shift also indicates that oxygen is *unloaded* more easily from hemoglobin, making it more available to the tissue. Increases in affinity cause leftward shifts, indicating that a lower P_{O_2} is required to achieve any given level of saturation; a leftward shift also indicates that oxygen is *loaded* more easily onto hemoglobin. Under normal conditions, for example, a P_{O_2} of approximately 45 mm Hg produces 80% saturation of hemoglobin. With a rightward shift, a P_{O_2} of greater than 45 mm Hg is required to produce the same 80% saturation; with a leftward shift, a P_{O_2} of less than 45 mm Hg produces 80% saturation.

Now we consider the factors that affect the affinity of hemoglobin for oxygen. The first three factors—temperature, pH, and P_{CO_2}—all work to promote oxygen unloading from hemoglobin in respiring tissues and oxygen loading onto hemoglobin in the lungs.

1. *Temperature.* Temperature affects the affinity of hemoglobin for oxygen by altering the structure of the hemoglobin molecule. This factor is nonspecific in that temperature affects the tertiary structure of all proteins. However, this structural change has important functional consequences. As metabolism of tissue increases, temperature increases, thereby decreasing the affinity of hemoglobin for oxygen (**Figure 10a**). As a result, more oxygen is unloaded in tissue that is highly active. Similarly, the decrease in temperature of blood as it travels to the lungs increases hemoglobin's affinity for oxygen, thereby promoting oxygen loading.
2. *pH.* The effect of pH on the hemoglobin-oxygen dissociation curve is known as the **Bohr effect.** It is based on the fact that when oxygen binds to hemoglobin, certain amino acids in the protein release hydrogen ions:

$$Hb + O_2 \rightleftharpoons Hb \cdot O_2 + H^+$$

By the law of mass action, an increase in hydrogen ion concentration (a decrease in pH) pushes the reaction to the left, causing some oxygen to dissociate from hemoglobin, even when P_{O_2} is kept constant. The Bohr effect is important because when hydrogen ions bind to hemoglobin, they decrease the affinity of hemoglobin for oxygen (Figure 10b), and oxygen is unloaded. The hydrogen ion concentration tends to increase in active tissue, which facilitates oxygen unloading.

3. P_{CO_2}. The P_{CO_2} affects the affinity of hemoglobin for oxygen because carbon dioxide reacts reversibly with certain amino groups in hemoglobin to form **carbaminohemoglobin** ($HbCO_2$):

$$Hb + CO_2 \rightleftharpoons HbCO_2$$

Therefore, an increase of P_{CO_2} in the blood, such as occurs when metabolic activity increases, pushes the reaction to the right, forming more carbaminohemoglobin. When carbon dioxide is bound to hemoglobin, it changes hemoglobin's conformation and decreases its affinity for oxygen—a phenomenon known as the **carbamino effect.** The binding of carbon dioxide to hemoglobin is also one of the mechanisms of carbon dioxide transport in the blood (discussed in the next section).

4. *2,3-BPG.* The fourth factor, 2,3-BPG (2,3-bisphosphoglycerate), is a chemical compound produced in erythrocytes from an intermediate compound in glycolysis, the anaerobic pathway by which erythrocytes obtain all their energy. When oxyhemoglobin is present in high concentrations, it inhibits the enzyme that forms 2,3-BPG; thus 2,3-BPG levels are low and have little effect on hemoglobin affinity. In contrast, if oxyhemoglobin levels are low, which occurs when oxygen supply is limited, then 2,3-BPG synthesis occurs, and 2,3-BPG decreases the affinity of hemoglobin for oxygen. This effect enhances the unloading of oxygen in the respiring tissues that need it. Conditions that enhance the production of 2,3-BPG include anemia and high altitudes.

Another factor that affects the binding of oxygen to hemoglobin is carbon monoxide. Carbon monoxide is toxic because, when present, it binds to hemoglobin more readily than oxygen, which prevents oxygen from binding and decreases the oxygen-carrying capacity of blood.

Apply Your Knowledge

Draw a hemoglobin-oxygen dissociation curve that indicates what happens when carbon monoxide is present in blood.

Carbon Dioxide Transport in the Blood 

Of the total amount of carbon dioxide in the blood, 5–6% is dissolved, 5–8% is bound to hemoglobin as carbaminohemoglobin (as previously described), and 86–90% is dissolved in the blood as bicarbonate ions (HCO_3^-) (**Table 3**).

Bicarbonate is formed from carbon dioxide within erythrocytes in systemic capillaries. Erythrocytes are able to form bicarbonate because they contain the enzyme carbonic anhydrase.

The Role of Carbonic Anhydrase in Carbon Dioxide Transport

Carbonic anhydrase catalyzes the reversible reaction that converts carbon dioxide and water to carbonic acid (H_2CO_3):

$$CO_2 + H_2O \xrightleftharpoons{\text{carbonic anhydrase}} H_2CO_3$$

The pathway continues with the reversible dissociation of carbonic acid, which yields a hydrogen ion and a bicarbonate ion:

$$H_2CO_3 \rightleftharpoons H^+ + HCO_3^-$$

The complete pathway can be written as follows:

$$CO_2 + H_2O \xrightleftharpoons{\text{carbonic anhydrase}} H_2CO_3 \rightleftharpoons H^+ + HCO_3^-$$

As shown in this final equation, carbon dioxide can be converted into H^+ and HCO_3^-. This reaction follows the law of mass action, meaning that an increase in carbon dioxide concentration (or an increase in P_{CO_2}) drives the reaction to the right, producing more hydrogen and bicarbonate ions; conversely, a decrease in carbon dioxide concentration (or a decrease in P_{CO_2}) drives the reaction to the left, producing carbon dioxide from hydrogen and bicarbonate ions. In addition to being important in the transport and exchange of carbon dioxide (as described in the next section), this reaction is important in acid-base balance, because an increase in P_{CO_2} causes an increase in the acidity of the blood (the concentration of hydrogen ions increases), and a decrease in P_{CO_2} causes a decrease in the acidity of the blood (the concentration of hydrogen ions decreases). Thus, in biological systems, CO_2 is considered an acid. The role of carbon dioxide in acid-base balance is described later in this chapter.

Carbon Dioxide Exchange and Transport in Systemic Capillaries and Veins

Respiring cells produce carbon dioxide at a rate of approximately 200 mL/min at rest, and this carbon dioxide must be removed by the circulatory and respiratory systems. The carbon dioxide produced in respiring cells diffuses, based on its partial pressure gradient, first into interstitial fluid and then into the plasma (**Figure 11a**). Once CO_2 is dissolved in the plasma, P_{CO_2} increases, creating a pressure gradient between plasma and erythrocytes such that the pressure in the plasma is greater. Thus carbon dioxide

TABLE 3 Carbon Dioxide Transport in Blood

	Systemic Arterial Blood		Systemic Venous Blood	
Form	CO_2 volume (mL/liter of blood)	% of total CO_2 in blood	CO_2 volume (mL/liter of blood)	% of total CO_2 in blood
Dissolved in blood	27	5.5	31	5.8
Dissolved as bicarbonate	439	89.6	470	87.0
Bound to hemoglobin	24	4.9	39	7.2
Total	490	100.0	540	100.0

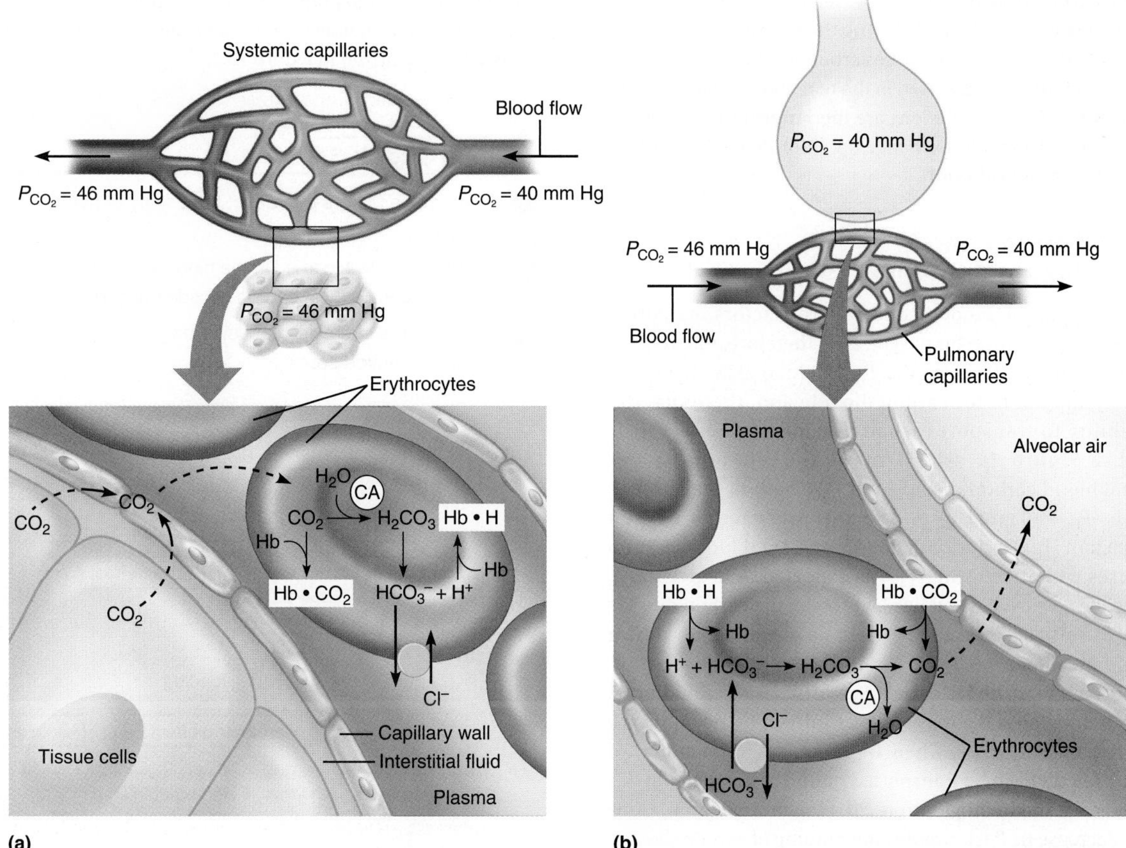

Figure 11 Carbon dioxide exchange and transport in blood. **(a)** Carbon dioxide produced in the tissues diffuses into the blood and then into an erythrocyte. The increased P_{CO_2} in the erythrocyte causes most carbon dioxide molecules to be converted to bicarbonate, some to bind to hemoglobin, and some to remain dissolved in the blood. The bicarbonate is transported out of the erythrocyte into the plasma. As bicarbonate is produced, the hydrogen ions produced are buffered by binding to hemoglobin. (CA = carbonic anhydrase.) **(b)** In the lungs, carbon dioxide diffuses from blood to alveolar air, decreasing the P_{CO_2} in blood. As P_{CO_2} in the erythrocytes decreases, bicarbonate enters the erythrocytes, and hydrogen ions are released from hemoglobin. The bicarbonate and hydrogen ions are then converted to carbon dioxide, which diffuses into the alveoli.

diffuses from plasma into the erythrocytes. Although some carbon dioxide remains dissolved in the blood and some binds to hemoglobin, most of the carbon dioxide is converted to bicarbonate and hydrogen ions by actions of carbonic anhydrase in the erythrocytes. This chemical reaction removes dissolved carbon dioxide from the blood and decreases the P_{CO_2}. The lower P_{CO_2} creates a greater gradient for diffusion of carbon dioxide from tissues to blood.

If bicarbonate and hydrogen ions were allowed to accumulate in erythrocytes, the reaction of carbon dioxide with water would slow down and eventually come to equilibrium. Under these conditions, carbon dioxide would no longer be converted to bicarbonate and hydrogen ions; instead, it would simply accumulate in the cells in its dissolved form, which could cause P_{CO_2} inside the cells to rise. As a result, the ability of the blood to transport carbon dioxide as bicarbonate would be hindered. To prevent this outcome from occurring, bicarbonate and hydrogen ions must be removed from the erythrocytes. As bicarbonate ion levels in the erythrocyte increase, bicarbonate ions are transported into the plasma and exchanged for chloride ions via a transport protein in the erythrocyte plasma membrane. The coupled movement of chloride ions into the erythrocyte and bicarbonate into the plasma is called the **chloride shift.** Meanwhile, many of the hydrogen ions produced by the reaction are buffered by binding to hemoglobin. Because of these chemical reactions and ion exchanges, most of the carbon dioxide in blood is transported as bicarbonate ions dissolved in the plasma (although some dissolved bicarbonate ions remain in the erythrocytes).

Carbon Dioxide Exchange and Transport in Pulmonary Capillaries and Veins

In the lungs, carbon dioxide diffuses down its pressure gradient from the blood in pulmonary capillaries to alveoli from where it will be exhaled, thereby decreasing blood carbon dioxide levels (Figure 11b). The loss of carbon dioxide causes bicarbonate and hydrogen ions in erythrocytes to combine and form carbonic acid, which is converted by carbonic anhydrase to carbon dioxide and water.

Most of the carbon dioxide created by this reaction diffuses into the alveoli to be exhaled, further driving the reaction in the direction that favors the formation of more carbon dioxide. As bicarbonate ions in the erythrocyte are used in the reaction, the levels inside the cell decrease. Bicarbonate ions are then transported from the plasma into erythrocytes in exchange for chloride ions by what is called the **reverse chloride shift.**

Effect of Oxygen on Carbon Dioxide Transport

Just as the P_{CO_2} of the blood is one of many factors that affect the affinity of hemoglobin for oxygen and, therefore, the oxygen-carrying capacity of blood, so the P_{O_2} of blood also affects the affinity of hemoglobin for carbon dioxide and, therefore, the blood's ability to transport carbon dioxide. This effect is illustrated in **Figure 12**, which shows the relationship between total blood carbon dioxide and P_{CO_2} both in the absence of oxygen (P_{O_2} = 0 mm Hg) and in the presence of oxygen (P_{O_2} = 100 mm Hg) Under both conditions, total blood carbon dioxide rises with increasing P_{CO_2}, as expected. At a given P_{CO_2} (shown by the vertical dashed line), however, the carbon dioxide content of the blood falls as the P_{O_2} rises—a phenomenon known as the **Haldane effect.**

One reason why the Haldane effect arises is that binding of oxygen to hemoglobin decreases the affinity of hemoglobin for carbon dioxide. Therefore, when P_{O_2} increases, the amount of carbon dioxide that can be transported bound to hemoglobin decreases because the hemoglobin is carrying more oxygen. Conversely, a decrease in P_{O_2} promotes the binding of carbon dioxide to hemoglobin.

In respiring tissues, where P_{O_2} is low and P_{CO_2} is high, the Haldane effect promotes the loading of carbon dioxide onto hemoglobin, while both the Bohr effect (effect of pH on hemoglobin's affinity for oxygen) and the carbamino effect (effect of carbon dioxide levels on hemoglobin's affinity for oxygen) work to promote oxygen unloading. In the lungs, where P_{O_2} is high and P_{CO_2} is low, the Haldane effect promotes unloading of carbon dioxide, while the Bohr effect and the carbamino effect promote oxygen loading. These interactions are summarized in **Figure 13.**

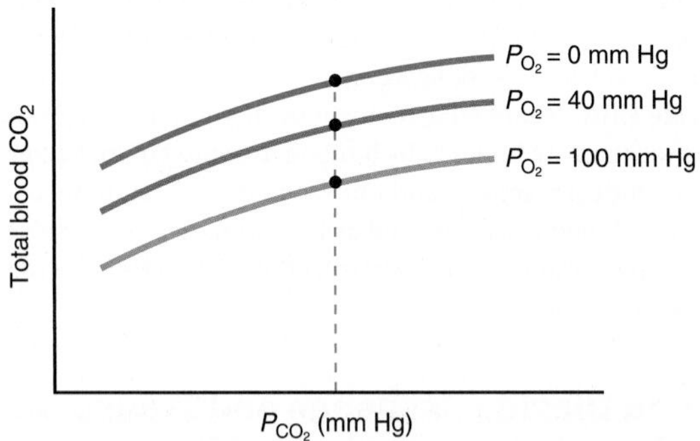

Figure 12 Effect of P_{O_2} on carbon dioxide transport. Increased P_{CO_2} in blood decreases the affinity of hemoglobin for carbon dioxide, which decreases the ability of the blood to transport carbon dioxide. Thus, at a given P_{CO_2} (vertical dashed line), more carbon dioxide is transported at a P_{O_2} = 0 mm Hg than at the venous P_{O_2} = 40 mm Hg, which is greater than the arterial P_{O_2} = 100 mm Hg.

Quick Check 3

1. Name three factors that affect the affinity of hemoglobin for oxygen.
2. Carbon dioxide is carried in the blood in which three forms? How is the majority of carbon dioxide transported?
3. Write out the chemical reaction catalyzed by the enzyme carbonic anhydrase.

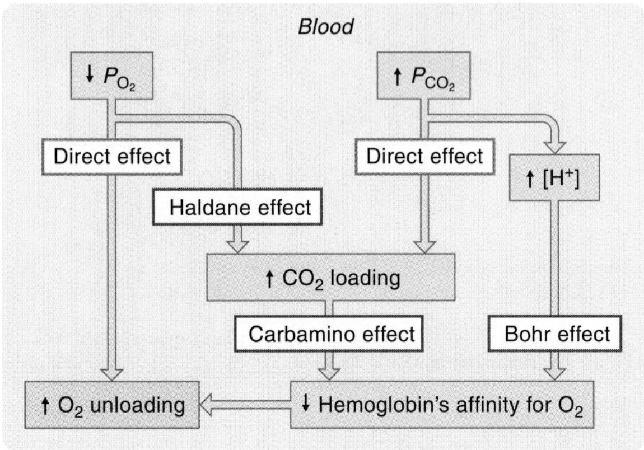

(a) CO_2 loading and O_2 unloading of hemoglobin in respiring tissue

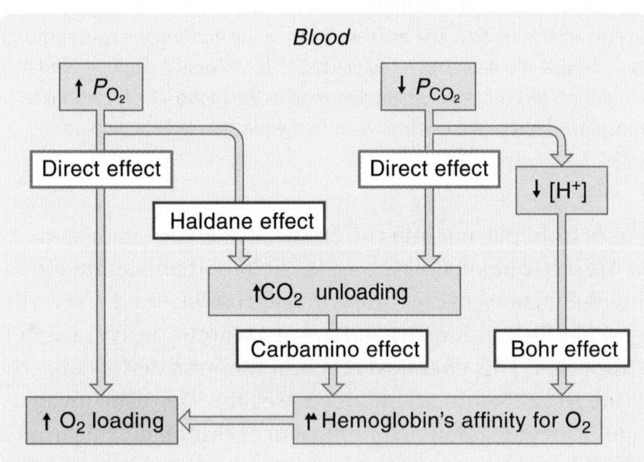

(b) CO_2 unloading and O_2 loading of hemoglobin in lungs

Initial stimulus
Physiological response
Result

Figure 13 Effects of P_{O_2} and P_{CO_2} on carbon dioxide and oxygen loading and unloading. (a) In systemic tissues, as P_{O_2} decreases and P_{CO_2} increases, carbon dioxide loading and oxygen unloading occur. **(b)** In the lungs, as P_{O_2} increases and P_{CO_2} decreases, carbon dioxide unloading and oxygen loading occur.

5 Central Regulation of Ventilation

In the simplest terms, the job of the respiratory system is to deliver oxygen to and remove carbon dioxide from cells at a rate sufficient to keep up with metabolic demands. The signals that indicate whether the respiratory system is doing this job adequately are the partial pressures of oxygen and carbon dioxide in systemic arterial blood. If the respiratory system can maintain P_{O_2} and P_{CO_2} at or near the levels that are normal for systemic arterial blood (100 mm Hg and 40 mm Hg, respectively), then it is delivering oxygen and removing carbon dioxide at adequate rates.

To maintain normal arterial partial pressures, the body must regulate minute alveolar ventilation (the volume of air reaching the alveoli each minute) such that it is neither too high nor too low. Recall that alveolar ventilation depends on the frequency and volume of breaths. In this section we will see how the breathing rhythm is generated, and how it is regulated to serve the metabolic demands of the body.

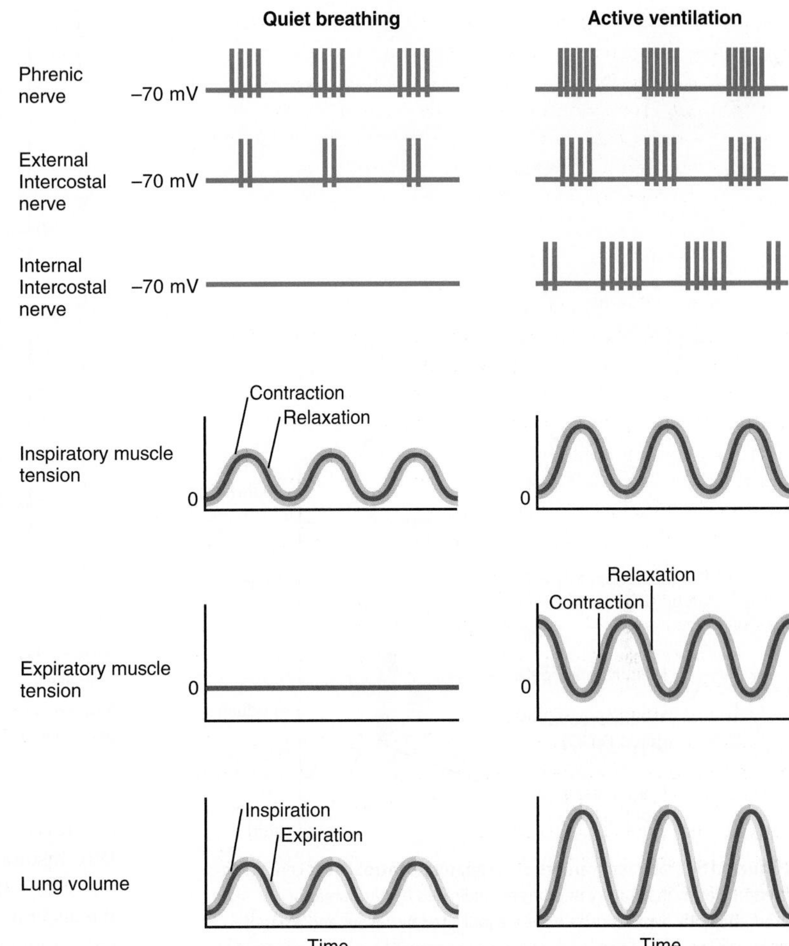

Figure 14 A comparison of quiet breathing and active ventilation. Ventilation involves cyclical changes in neural stimulation of respiratory muscles, which cause cyclical changes in lung volume. The nerve traces indicate the occurrences of action potentials over time.

Neural Control of Breathing by Motor Neurons

Breathing is a cycle of inspiration followed by expiration. Inspiration is an active process that requires contraction of the inspiratory muscles, including the diaphragm and external intercostals. During quiet breathing, in contrast, expiration is a passive process in which no muscle contraction is required. Thus, during quiet breathing, the breathing cycle consists of contraction of the inspiratory muscles during inspiration, followed by relaxation of the same muscles during expiration. During more active breathing, however, expiration becomes active and requires contraction of the expiratory muscles. Thus, during active breathing, the respiratory cycle consists of contraction of the inspiratory muscles and relaxation of the expiratory muscles during inspiration, followed by relaxation of the inspiratory muscles and contraction of the expiratory muscles during expiration.

Because the muscles of respiration are skeletal muscles, they are stimulated to contract by neural input from somatic motor neurons. The **phrenic nerve** innervates the diaphragm, whereas the internal and external **intercostal nerves** innervate the intercostal muscles. Figure 14 compares the peripheral events that occur during quiet breathing to those that occur during active breathing. During quiet breathing, action potential bursts occur in inspiratory motor neurons, triggering muscle contraction. During active breathing, action potential bursts in inspiratory motor neurons occur asynchronously with bursts in expiratory motor neurons. The neural signals that control the cyclical contractions of the respiratory muscles are generated in respiratory control regions located in the brainstem.

Generation of the Breathing Rhythm in the Brainstem

Breathing is under both voluntary and involuntary control. We alter our breathing rhythm voluntarily when we speak, sing, hold our breath, or sigh in dismay or exasperation. Most of the time, however, our breathing is automatic and requires no conscious effort.

Central control of respiration is not fully understood, but research indicates that respiratory control regions are present in the medulla and pons of the brainstem (Figure 15). Two general classes of neurons found in these regions, **inspiratory neurons** and **expiratory neurons,** generate action potentials during inspiration and expiration, respectively. Within these two general classes are subclasses of neurons that differ slightly in their patterns of activity and functions.

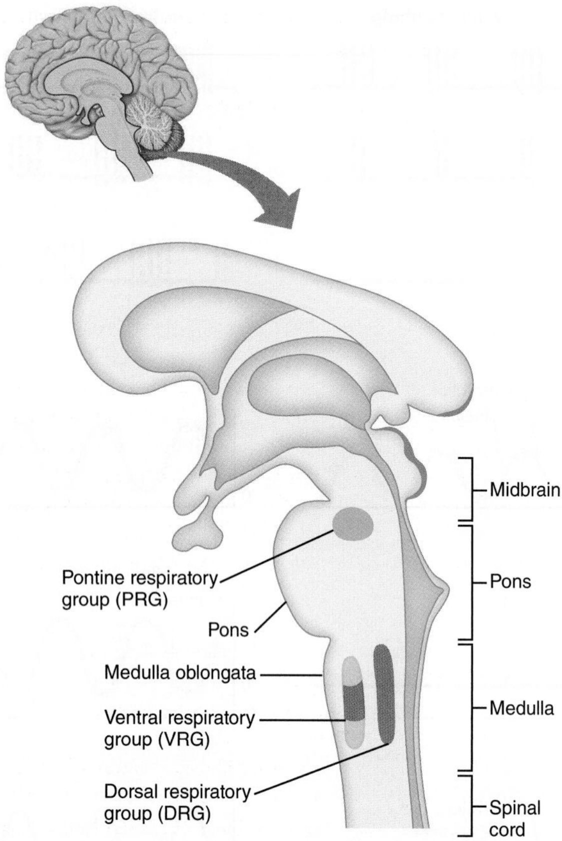

Figure 15 Brainstem centers of respiratory control. Areas containing predominantly inspiratory neurons are indicated by blue, areas of predominantly expiratory neurons are indicated by yellow, and areas of scattered inspiratory and expiratory neurons as well as mixed neurons are indicated by green.

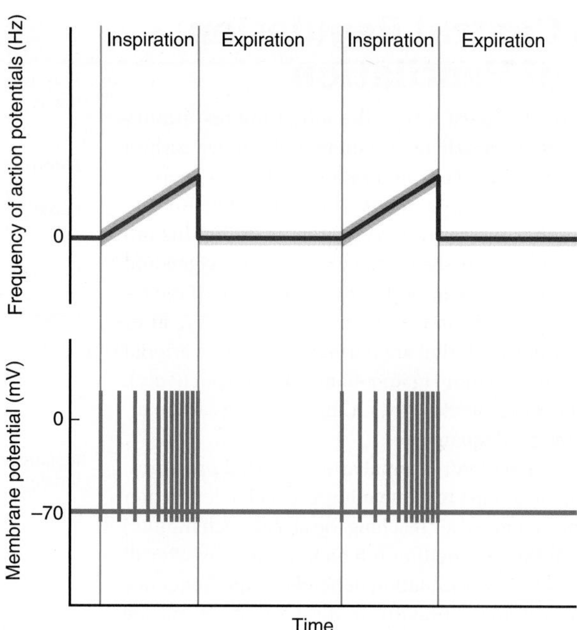

Figure 16 Activity of inspiratory neurons. Inspiratory neurons show a ramplike increase in the frequency of action potentials during inspiration, followed by a sudden termination of all activity at the end of inspiration and beginning of expiration.

Respiratory Control Centers of the Medulla

Two respiratory control centers are located on each side of the medulla: (1) an area encompassing the *nucleus ambiguus* and nearby structures in the ventral medulla called the **ventral respiratory group (VRG)** and (2) an area within the *nucleus tractus solitarius* in the dorsal medulla called the **dorsal respiratory group (DRG).** The VRG contains two regions of primarily expiratory neurons and one region of primarily inspiratory neurons (see Figure 15). The inspiratory neurons show a ramplike increase in activity during inspiration: The action potential frequency is low at the onset of inspiration but gradually increases until it reaches a maximum at the peak of inspiration, when it abruptly terminates and expiration begins (**Figure 16**). The DRG contains primarily inspiratory neurons, although some expiratory neurons are present as well (see Figure 15). The DRG inspiratory neurons have more complex patterns of activity than the VRG inspiratory neurons in that their firing pattern depends on the degree of stretch of the lungs.

Current theories suggest that inspiratory neurons in the VRG and DRG control the motor neurons in the cervical spinal cord that, in turn, control the inspiratory muscles. Put another way, VRG and DRG inspiratory neurons stimulate motor neurons of the phrenic and external intercostal nerves, which then cause contraction of the inspiratory muscles. Some inspiratory neurons of the VRG provide input to accessory respiratory muscles that do not participate in inspiration or expiration directly but instead provide supportive movements such as expansion of the opening to the larynx during inspiration. Some expiratory neurons of the VRG stimulate the motor neurons to expiratory muscles during active expiration. The role of other expiratory neurons is not as well understood, but one of their functions may be to suppress the activity of inspiratory neurons during expiration.

Respiratory Control Centers of the Pons

The respiratory center of the pons, called the **pontine respiratory group (PRG;** formerly called the *pneumotaxic center*), contains both inspiratory and expiratory neurons as well as *mixed neurons,* which have activity associated with both inspiration and expiration (see Figure 15). The PRG may facilitate the transition between inspiration and expiration.

Central Pattern Generator

We have seen that inspiratory neurons of the medulla control the motor neurons to the inspiratory muscles and that these neurons generate action potentials during inspiration but not during expiration. The source of this cycle of activity is called the central pattern generator.

The **central pattern generator (CPG)** is a network of neurons that generates a regular, repeating pattern of neural activity called the respiratory rhythm. The CPG is thought to be located in the pre-Botzinger complex in the ventrolateral medulla, although its mechanism of action is unknown. Two primary hypotheses have been proposed to explain how the CPG functions. The first hypothesis suggests that certain neurons in the CPG have pacemaker activity; that is, they spontaneously depolarize and generate action potentials in a cyclical manner similar to pacemaker cells of the heart. As yet, however, no studies have identified any cells with pacemaker activity in the respiratory centers of the brainstem. The second hypothesis suggests that complex interactions between networks of neurons are responsible for generating the breathing rhythm.

Model of Respiratory Control During Quiet Breathing

Figure 17 shows a simplified model for quiet breathing in which the breathing rhythm is produced by the CPG in the medulla. This rhythm is communicated to inspiratory neurons of the DRG and VRG, causing a ramplike increase in the frequency of action potentials in these cells, which communicate with motor neurons of the phrenic and external intercostal nerves. Early in inspiration—when the frequency of action potentials in the inspiratory neurons is relatively low—only a few motor neurons are recruited, and contraction of inspiratory muscles is weak. As inspiration continues, the greater frequency of action potentials in the inspiratory neurons causes recruitment of more motor neurons, leading to the recruitment of additional inspiratory muscle fibers and a stronger overall contraction, which in turn causes lung volume to increase. After a few seconds, activity in the inspiratory neurons terminates abruptly, causing a similar abrupt termination in motor neuron activity. As a result, the inspiratory muscles stop contracting, inspiration stops, and expiration begins. Following a brief period of quiescence, another burst of action potentials begins, marking the beginning of the next inspiration.

In this model, respiratory control areas of the medulla are primarily responsible for controlling breathing, but breathing is also affected by activity in other brain regions, including the pons,

cerebral cortex, cerebellum, limbic system, hypothalamus, and medullary cardiovascular regulatory areas. This web of relationships explains, for example, why breathing patterns change when a person experiences feelings of rage or fear.

Peripheral Input to Respiratory Centers

Several types of sensory input can alter respiration, presumably through indirect communication with the central pattern generator (see Figure 17). Particularly important in this regard are signals from central and peripheral **chemoreceptors** (chemically sensitive receptor cells) located in the brain and in systemic arteries. These chemoreceptors monitor chemical conditions in cerebrospinal fluid and arterial blood and are primarily responsible for regulating ventilation under resting conditions (as discussed in the next section).

Additional sensory inputs that affect breathing come from a variety of receptors located in the respiratory system and elsewhere. These sources of information include pulmonary stretch receptors in the smooth muscle of pulmonary airways, irritant receptors in the lining of the respiratory tract, proprioceptors in muscles and joints (which detect movement of the body), arterial baroreceptors (which detect changes in blood pressure), and nociceptors and thermoreceptors located throughout the body.

Pulmonary stretch receptors are excited by inflation of the lungs and do not appear to play a significant role in regulating breathing in humans. In other animal species, however, activation of these receptors inhibits inspiration and may act to protect the lungs against overinflation. *Irritant receptors* are stimulated by inhaled particulates, such as smoke or dust, and by some chemicals, such as sulfur dioxide. Stimulation of these receptors triggers bronchoconstriction, hyperpnea, and coughing and sneezing, which under certain circumstances help to clear these substances from the lungs. Coughing is initiated by irritant receptors in the trachea, whereas sneezing is triggered by irritant receptors in the nose and pharynx. Input to respiratory centers from muscle and joint proprioceptors plays a role in stimulating increased ventilation during exercise (see **Understanding Exercise: Role of Sensory Receptors**).

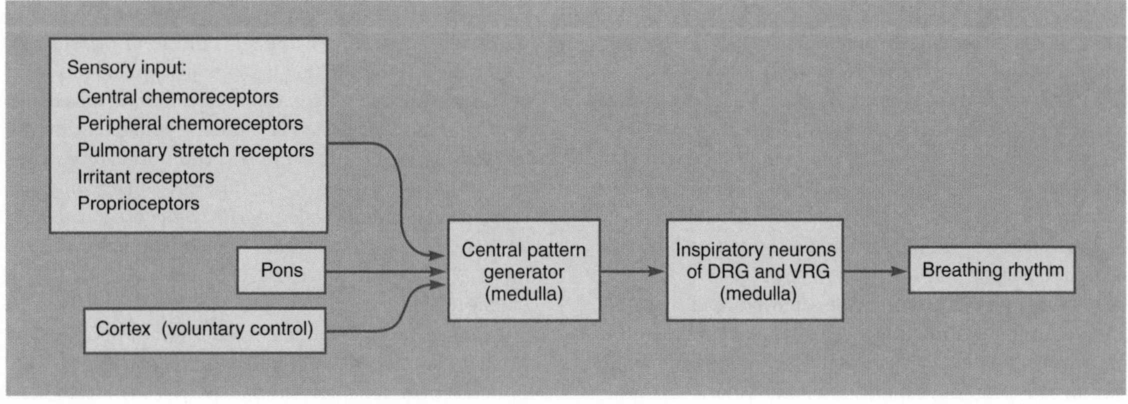

Figure 17 **Model of respiratory control during quiet breathing.**

UNDERSTANDING EXERCISE

Role of Sensory Receptors

As exercise begins, active skeletal muscles consume larger amounts of oxygen and produce larger amounts of carbon dioxide, but arterial concentrations of these gases change very little (at least during mild to moderate exercise). The reason is that feedforward control circuits increase ventilation before the gas concentrations in arterial blood can change. One of these feedforward circuits is the central command associated with the intention to perform exercise; another feedforward signal originates in mechanoreceptors and chemoreceptors in exercising skeletal muscles. Both of these feedforward signals stimulate ventilation.

At rest, changes in arterial oxygen and carbon dioxide concentrations regulate ventilation through negative feedback control. During exercise, however, these feedback mechanisms function primarily to prevent hyperventilation that might occur in response to the feedforward signals that *increased* ventilation during exercise.

The increase in ventilation during exercise, which is called *exercise hyperpnea,* is illustrated in the graph. At the onset of exercise, minute ventilation increases rapidly due to neural signals from the central command and due to feedback from mechanoreceptors and chemoreceptors in the muscle—a phenomenon referred to as the *neurogenic response.* As exercise progresses, minute ventilation tends to continue increasing, albeit at a much slower rate. Blood-borne signals trigger this more gradual increase, referred to as the *humoral response.* When exercise terminates, the reverse occurs: Ventilation initially decreases rapidly due to a decrease in neural signals from the central command and feedback from mechanoreceptors and chemoreceptors, but then continues to decrease more slowly until it returns to resting levels.

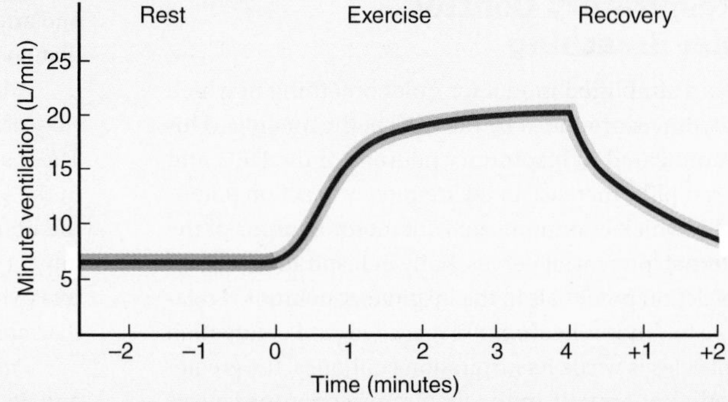

Quick Check 4

1. Where are inspiratory neurons located? Where are expiratory neurons located?
2. What is the function of the central pattern generator for breathing?
3. What is the function of irritant receptors?

6 Control of Ventilation by Chemoreceptors

Anyone who has ever experienced pneumonia, an asthma attack, or similar breathing difficulty can testify that few sensations are so unsettling as the feeling of lacking air, known as *air hunger.* You can demonstrate this effect to yourself by attempting to voluntarily slow down your own breathing while trying to keep the size of each breath the same. If you try this, you will probably begin to feel discomfort within the first seconds. Soon you will be completely unable to keep a steady breathing rhythm, and your breathing will become faster and deeper.

Those who attempt this demonstration inevitably fail, because altering the breathing pattern alters the chemical composition of systemic arterial blood. The reduction in breathing frequency, coupled with a constant tidal volume, causes alveolar ventilation to fall. This decrease in alveolar ventilation causes changes in the partial pressures of gases in alveolar air, which induce corresponding changes in arterial blood. Changes in chemical concentrations in the blood are detected by chemoreceptors located in major arteries and in the brain, which relay signals to the respiratory control center via afferent neurons. As a result of this neural input, the respiratory control center triggers an increase in the rate and depth of breathing to restore partial pressures in arterial blood to their normal values. The chemoreceptors also relay information to the cerebral cortex, giving rise to the conscious sensation of air hunger.

Chemoreceptors

Chemoreceptors monitor partial pressures of oxygen and carbon dioxide in arterial blood and relay this information to the respiratory control center, so that it can adjust ventilation in response to changes in these variables. Chemoreceptors involved in the control of breathing are classified as either peripheral or central, depending on their location. **Peripheral chemoreceptors** are located in the *carotid bodies* near the carotid sinus (**Figure 18**). (Other peripheral chemoreceptors called *aortic bodies* are located in the aortic arch and regulate respiration in many animal species, but not in humans.) The **central chemoreceptors** are located in the medulla oblongata.

Peripheral and central chemoreceptors differ not only in their location, but also in their structures and chemical sensitivities. Peripheral chemoreceptors are specialized chemically sensitive cells that are in direct contact with arterial blood and communicate

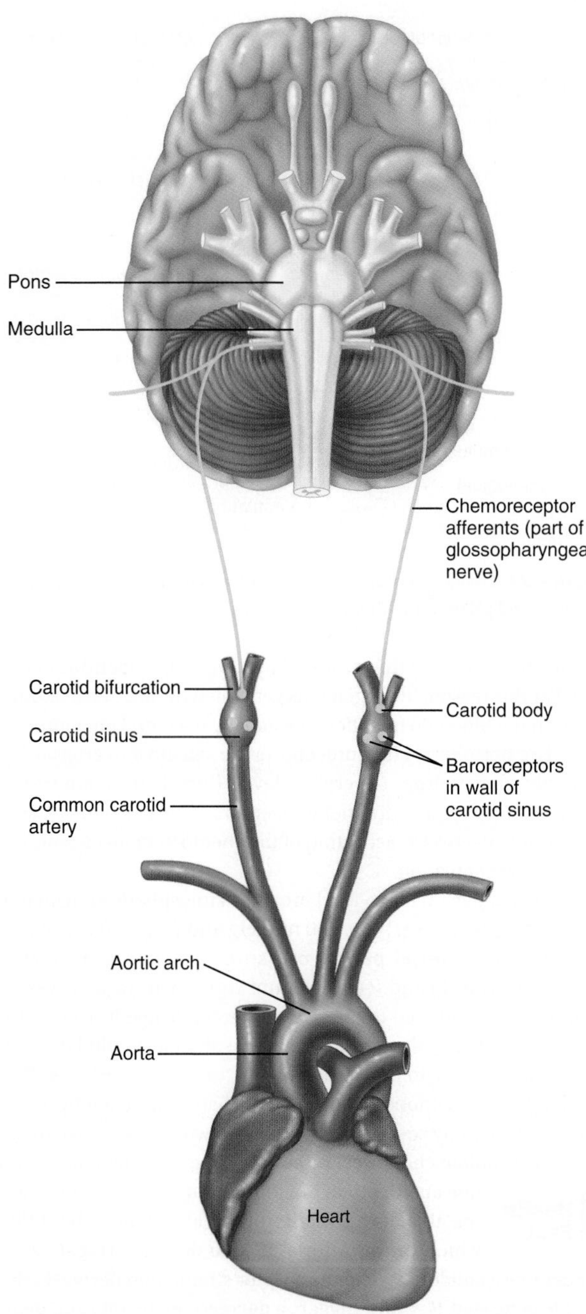

Figure 18 **Location of peripheral chemoreceptors in the carotid bodies.**

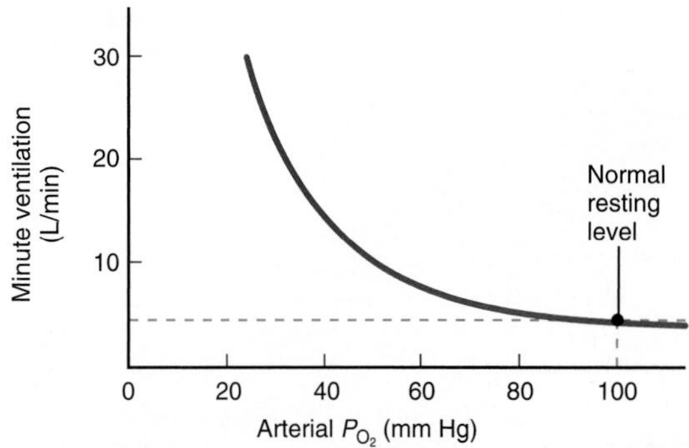

(a) Effects of arterial P_{O_2} on ventilation

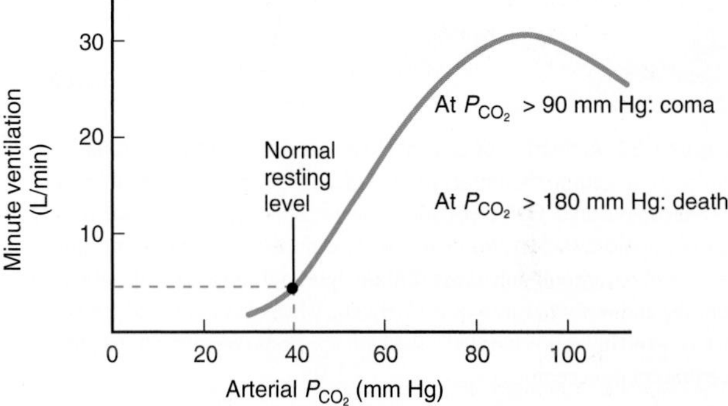

(b) Effects of arterial P_{CO_2} on ventilation

Figure 19 **Respiratory control by chemoreceptors. (a)** Declining arterial P_{O_2} has little effect on minute ventilation until the P_{O_2} drops to less than 60 mm Hg. **(b)** Increasing arterial P_{CO_2} has large effects on minute ventilation as P_{CO_2} increases above or decreases below normal. At P_{CO_2} greater than 90 mm Hg, coma and then death can occur.

Q Which class of chemoreceptors is responsible for the increased ventilation that occurs when P_{O_2} decreases to less than 60 mm Hg?

(through secretion of a chemical messenger) with afferent neurons projecting to medullary respiratory control regions. Peripheral chemoreceptors respond to changes in arterial P_{O_2}, P_{CO_2}, or pH (which changes when P_{CO_2} changes). Decreases in P_{O_2} can directly activate peripheral chemoreceptors, but only when the arterial P_{O_2} drops to below 60 mm Hg (**Figure 19a**). Because 60 mm Hg is an extreme drop in arterial P_{O_2}, oxygen is usually not a major factor in peripheral chemoreceptor activation. However, low P_{O_2} also increases the sensitivity of the peripheral chemoreceptors to carbon dioxide. In addition, peripheral chemoreceptors can directly respond to changes in blood pH (hydrogen ion concentration). In fact, changes in hydrogen ion concentration are the primary stimulus for the peripheral chemoreceptors. The hydrogen ions can come from many sources, but the main source is the reaction of carbon dioxide with water. Increases in arterial P_{CO_2} can also activate the peripheral chemoreceptors (Figure 19b), though mostly indirectly; first, the carbon dioxide must be converted to hydrogen ions and bicarbonate.

Central chemoreceptors are neurons in the medulla that respond directly to changes in hydrogen ion concentration in the cerebrospinal fluid surrounding this area. Because these chemoreceptors are located in the brain, they are protected by the blood-brain barrier, which separates the cerebrospinal fluid from the blood. Hydrogen ions cannot cross this barrier, but carbon dioxide can. The carbon dioxide that enters the brain does not affect the central chemoreceptors directly, but instead is converted to hydrogen ions and bicarbonate by carbonic anhydrase in the cerebrospinal fluid; the chemoreceptors are activated by the hydrogen ions that

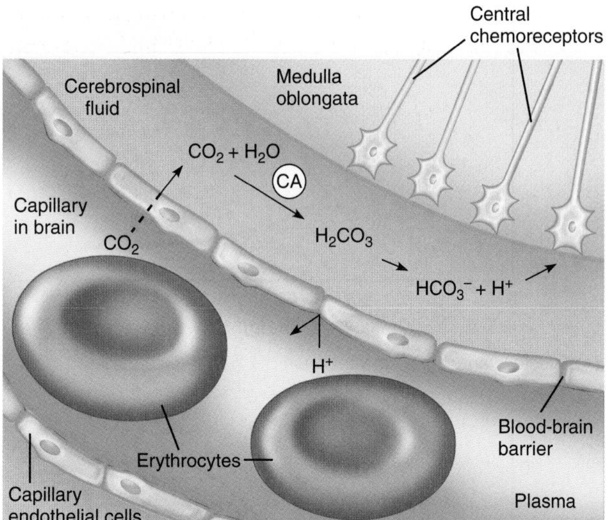

Figure 20 Activation of central chemoreceptors in the medulla oblongata. Central chemoreceptors respond best to changes in pH in the cerebrospinal fluid. However, hydrogen ions cannot cross the blood-brain barrier. Instead, carbon dioxide in the blood diffuses into the cerebrospinal fluid, where carbonic anhydrase (CA) catalyzes the conversion of carbon dioxide and water to carbonic acid (H_2CO_3), which dissociates to bicarbonate (HCO_3^-) and hydrogen ions (H^+). The hydrogen ions can then activate the central chemoreceptors.

Q Which type of junction links the endothelial cells of the capillary wall together to prevent hydrogen ions from passing between plasma and cerebrospinal fluid?

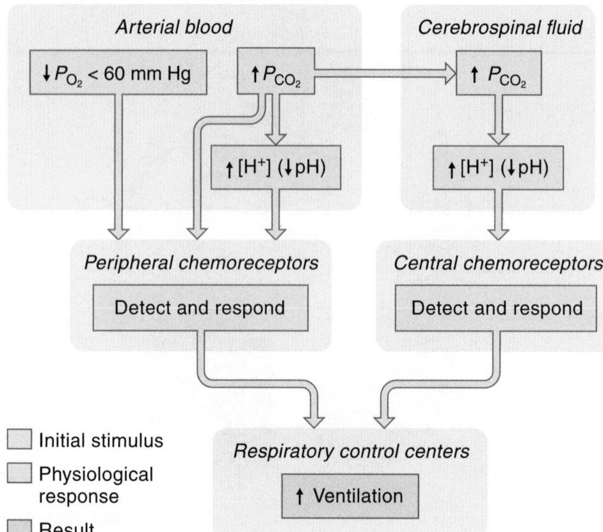

Figure 21 Chemoreceptor reflexes: The effects of changes in arterial P_{O_2}, P_{CO_2}, and pH on ventilation.

are generated as a result (**Figure 20**). Central chemoreceptors can also be activated by hydrogen ions generated by increased brain activity or decreased blood flow to the brain. Unlike the arterial chemoreceptors, central chemoreceptors are not sensitive to changes in P_{O_2}.

Although both peripheral and central chemoreceptors respond primarily to increases in arterial carbon dioxide levels, they are not a redundant system. Peripheral chemoreceptors respond more rapidly and are believed to be important for the moment-to-moment changes in ventilation to maintain normal arterial CO_2. The central chemoreceptors appear to be more important for maintaining the steady-state levels of CO_2.

Chemoreceptor Reflexes

Figure 21 summarizes the role of chemoreceptor reflex activity in maintaining normal arterial P_{O_2} and P_{CO_2}. Peripheral chemoreceptors respond directly and indirectly to changes in arterial P_{CO_2}, whereas central chemoreceptors respond indirectly to changes in arterial P_{CO_2}. In fact, changes in P_{CO_2} are the primary stimuli for changes in ventilation under normal conditions. The peripheral chemoreceptors respond to changes in arterial blood pH, which may come from carbon dioxide or other sources (such as lactic acid produced by cellular metabolism). The central chemoreceptors respond to changes in the concentration of hydrogen ions in the CSF but do not respond to changes in the hydrogen ion concentration of the blood because these ions cannot cross the blood-brain barrier. For this reason, hydrogen ions generated by lactic acid or other acids in the blood do not affect the activity of central chemoreceptors. The peripheral chemoreceptors are sensitive to arterial P_{O_2}, but only when it drops to very low levels (less than 60 mm Hg). In all cases, activation of the chemoreceptors results in an increase in ventilation; decreased activation of the chemoreceptors results in a decrease in ventilation.

When a person at sea level breathes atmospheric air with normal partial pressures (P_{O_2} = 160 mm Hg and P_{CO_2} = 0.2 mm Hg), alveolar and arterial partial pressures are also near normal (P_{O_2} = 100 mm Hg and P_{CO_2} = 40 mm Hg), so long as alveolar ventilation is matched to the body's metabolic needs. During hypoventilation, alveolar ventilation is less than it should be, which causes arterial P_{O_2} to decrease and P_{CO_2} to increase (**Figure 22a**). Under these conditions, chemoreceptors are stimulated by the increased P_{CO_2} (only severe hypoventilation will decrease P_{O_2} to levels that stimulate chemoreceptors) and trigger an increase in the rate and depth of breathing. During hyperventilation, the alveolar ventilation is greater than it should be, which causes arterial P_{CO_2} to decrease (Figure 22b). Under these conditions, chemoreceptor stimulation decreases due to the decreased P_{CO_2} and triggers a decrease in the rate and depth of breathing. In both conditions, the changes in ventilation should return the arterial P_{O_2} and P_{CO_2} toward normal values.

Quick Check 5

1. Describe the locations of the peripheral chemoreceptors and the central chemoreceptors.
2. Which stimuli activate the peripheral chemoreceptors? The central chemoreceptors?
3. How does carbon dioxide activate the peripheral and central chemoreceptors?

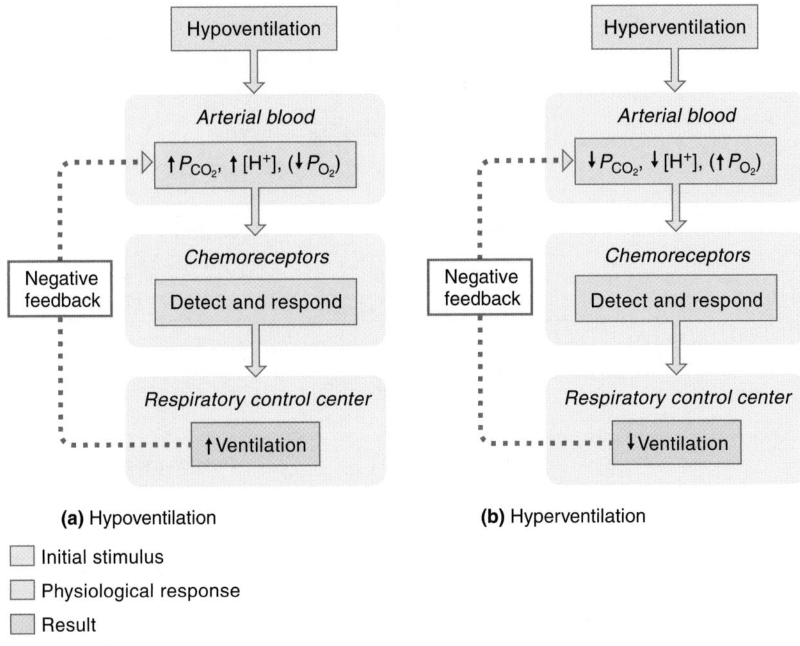

Figure 22 **The effects of hypoventilation and hyperventilation on minute ventilation.**

7 Local Regulation of Ventilation and Perfusion

In cases of obstructive lung diseases such as emphysema or bronchitis, it is not uncommon for airways in some regions of the lungs to become completely obstructed by mucus or other material that prevents air from flowing into alveoli. When such a blockage occurs, blood flowing to these alveoli is "wasted" in the sense that it cannot participate in gas exchange. In other types of lung diseases, the pulmonary capillaries may be damaged, leading to the opposite problem: Certain alveoli may receive air but not blood. In this case, air flowing to these alveoli is "wasted" in the sense that it does not participate in gas exchange.

Ventilation-Perfusion Ratios

In the normal lung, the rate of air flow to the alveoli (ventilation, $\dot{V}_A$) is matched to the rate of blood flow (perfusion, $\dot{Q}$); the dots over the abbreviations indicate that these are rates. The relationship of ventilation to perfusion is called the **ventilation-perfusion ratio** and is abbreviated $\dot{V}_A/\dot{Q}$. In the normal lung, $\dot{V}_A/\dot{Q}$ is approximately 1 (**Figure 23a**). Under these circumstances, the P_{O_2} and P_{CO_2} of the alveoli are at the normal values of 100 and 40 mm Hg, respectively. The P_{O_2} and P_{CO_2} in the blood emerging from the alveoli are also 100 and 40 mm Hg, respectively, because partial pressures in the capillary blood equilibrate with those in alveolar air.

The conditions mentioned in the introductory paragraph for this section illustrate extreme forms of ventilation-perfusion inequality. When airways are obstructed, for example, $\dot{V}_A$ in certain alveoli decreases, and the blood traveling in the capillaries to those alveoli does not undergo adequate gas exchange (Figure 23b). As a result, blood and air in these alveoli will have a lower P_{O_2} and a higher P_{CO_2} than normal. If ventilation of the alveoli is completely prevented, then the blood that emerges from the capillaries will have the same partial pressures of oxygen and carbon dioxide as the blood entering the capillaries. When pulmonary capillaries are damaged, perfusion is obstructed, causing a decrease in $\dot{Q}$ (Figure 23c). As a consequence, air in these alveoli will have a higher P_{O_2} and a lower P_{CO_2} than normal; however, little blood flows through these areas.

Ventilation-perfusion inequalities occur for other reasons. Under normal conditions, gravity affects the ventilation-perfusion ratio. Capillaries in the lungs can collapse if the pressure in them is not greater than the pressure of air in the alveoli. Gravity causes a greater capillary hydrostatic pressure, and thus greater blood flow, in the lower regions of the lungs. For example, in the uppermost regions of the lungs of a standing or sitting person, capillary pressure is greater than alveolar pressure at the arterial end of the capillary; in contrast, on the venous end, alveolar pressure is greater than capillary pressure. Blood flow is thus intermittent as the capillary alternately opens and closes. In the lower regions of the lungs, capillary pressure is always greater than alveolar pressure, so the capillary is always open and blood flow is continuous. Matching of ventilation to perfusion is especially important during exercise, when increased gas exchange between alveoli and capillaries is essential (see **Understanding Exercise: Recruiting Respiratory Reserve Capacities**).

Local Control of Ventilation and Perfusion

Given that differences in the ventilation or perfusion of certain alveoli can occur, whether because of diseases or gravity, local control mechanisms operate to match ventilation and perfusion to ensure efficient gas exchange. To obtain a ventilation-perfusion ratio of 1, areas with low ventilation require low perfusion, and areas with high ventilation require high perfusion. Changes in P_{O_2} and P_{CO_2} in the alveoli and the interstitial fluid of the lungs affect the contractile activity of smooth muscle in both the bronchioles, causing changes in bronchiolar diameter and thus in air flow, and the arterioles, causing changes in arteriole diameter and thus in perfusion of individual alveoli (**Figure 24**). The effects of changes in the partial pressures of oxygen and carbon dioxide on the radii of bronchioles and pulmonary arterioles are summarized in Table 4.

Oxygen acts primarily on the pulmonary arterioles; a low P_{O_2} causes a vasoconstriction (decrease in $\dot{Q}$). Carbon dioxide acts primarily on the bronchioles; a high P_{CO_2} causes bronchodilation (increase in $\dot{V}_A$). Therefore, in regions of the lungs with a high $\dot{V}_A/\dot{Q}$, the resulting increase in P_{O_2} and decrease in P_{CO_2} cause bronchoconstriction and vasodilation, thereby decreasing ventilation while increasing perfusion. In regions of the lungs with a low $\dot{V}_A/\dot{Q}$, the increase in P_{CO_2} causes bronchodilation and the decrease in P_{O_2} causes vasoconstriction. The net effect is an increase in ventilation and a decrease in perfusion. Note that the effects of oxygen and carbon dioxide on pulmonary arterioles are the opposite of the effects of these gases on systemic

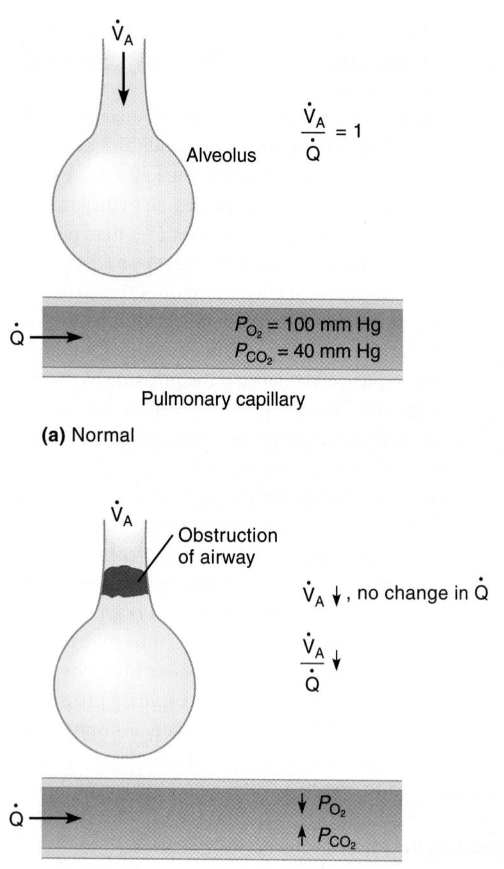

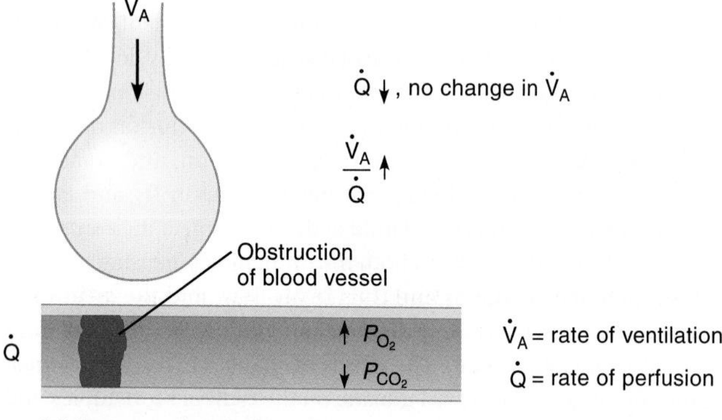

Figure 23 Ventilation-perfusion ratios. The dots over $\dot{V}_A$ and $\dot{Q}$ indicate that these quantities are rates. **(a)** Under normal conditions, ventilation and perfusion are matched ($\dot{V}_A/\dot{Q} = 1$). **(b)** If ventilation decreases and perfusion is normal, then $\dot{V}_A/\dot{Q} < 1$; arterial P_{O_2} decreases and P_{CO_2} increases. **(c)** If perfusion decreases and ventilation is normal, then $\dot{V}_A/\dot{Q} > 1$; arterial P_{O_2} increases and P_{CO_2} decreases.

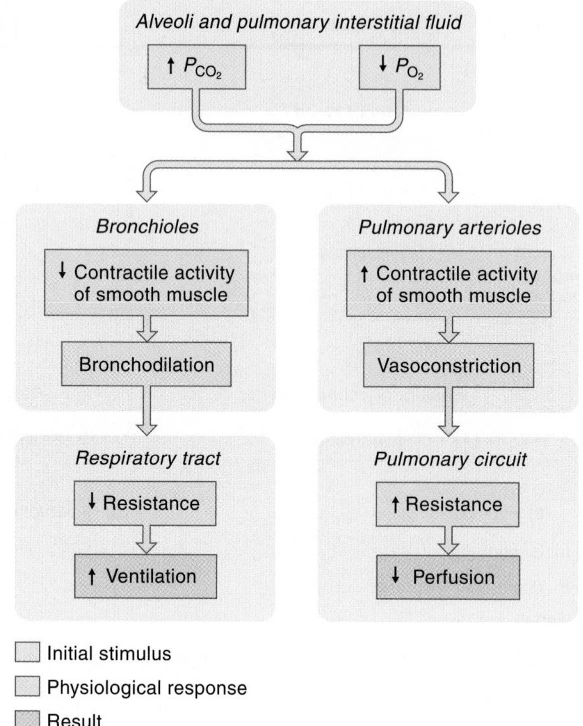

Figure 24 Local controls of ventilation and perfusion.

arterioles, where oxygen causes vasoconstriction and carbon dioxide causes vasodilation. The radii of arterioles and bronchioles can also be affected by other factors, such as epinephrine and the autonomic nervous system.

Quick Check 6

1. How is the P_{O_2} of alveolar air affected when the ventilation-perfusion ratio increases? How is P_{CO_2} affected?
2. What effect does a decrease in P_{CO_2} in the interstitial fluid around an alveolus have on the radii of arterioles and bronchioles in the region?
3. If blood flow to an alveolus is decreased, what should happen to the radius of the bronchiole leading to the alveolus if a normal ventilation-perfusion ratio is to be maintained?

8 The Respiratory System in Acid-Base Homeostasis

Although the primary function of the respiratory system is to control the oxygen and carbon dioxide content of arterial blood, it also plays an important role in regulating the blood's hydrogen ion concentration, or pH. We saw in previous sections that hydrogen ions can be buffered by hemoglobin, and that carbon dioxide is a source of hydrogen ions. In this section we focus on the role of hemoglobin and, more importantly, carbon dioxide in acid-base balance—that is, in maintaining normal blood pH.

UNDERSTANDING EXERCISE

Recruiting Respiratory Reserve Capacities

When the body is resting, the heart pumps 5 liters of blood per minute to both the systemic and pulmonary circuits, called the cardiac output. This 5-liter amount fills only about one-third of the pulmonary capillary capacity. Thus only one of every three capillaries is conducting blood at any given time, and more capillaries in the base of the lungs are carrying blood than in the apex. Alveoli near capillaries that are not conducting blood are collectively referred to as the *physiological dead space* because, like the air found in the anatomical dead space, air in these alveoli does not participate in gas exchange. As cardiac output increases during exercise, more of the capillaries begin carrying blood continuously, and perfusion becomes more uniform throughout the lungs. Thus the physiological dead space constitutes one type of pulmonary reserve capacity that can be accessed during exercise.

Once cardiac output exceeds three times the resting value, all of the reserve capillaries have been recruited; pulmonary pressure then rises modestly, and blood flows faster within individual capillaries. As a consequence, the amount of time each red blood cell spends exchanging gas with an alveolus is reduced. Transit time along the length of an alveolar capillary at rest is about 1 second, but a red blood cell can fully exchange oxygen and carbon dioxide in one-third of that time. Thus two-thirds of the time for red blood cells to travel through a pulmonary capillary constitutes a second type of pulmonary reserve capacity.

In combination, these two reserve mechanisms ensure that the pulmonary circuit can accommodate up to a sixfold increase in cardiac output during exercise, with no reduction in red blood cell oxygenation or carbon dioxide release.

Acid-Base Disturbances in Blood

The normal pH of arterial blood is 7.4, which is slightly more basic than pure water (pH = 7.0). Blood pH is tightly regulated by both the respiratory system and the kidneys, such that the pH rarely varies by more than a few hundredths of a unit in either direction; the normal pH range is 7.38–7.42. In fact, changes in blood pH of just a few tenths of a unit can have serious, even life-threatening consequences. If blood pH increases to 8.0 or decreases to 6.8 for even a few seconds, the result is death!

A change in arterial blood pH can have profound effects on body function because it alters the pH of fluids throughout the body. A change in pH alters the distribution of electrical charges in protein molecules, thereby altering their shape and interfering with their normal functions. Given that proteins have such a wide variety of functions in the body (as enzymes and receptors for messenger molecules, to name but two roles), changes in pH have widespread effects on body functions.

Arterial blood is considered to be excessively acidic if the pH is 7.35 or lower, a condition known as **acidosis.** (Note that acidosis occurs even when blood pH is, in fact, alkaline—that is, greater than 7.0—because acidosis refers to a pH that is *more acidic* than the normal blood pH of 7.4.) A pH of 7.45 or greater indicates that the blood is excessively alkaline or basic, a condition known as **alkalosis.** The danger of severe acidosis is that it depresses central nervous system activity. In severe cases, the depression of CNS activity can progress to coma and, ultimately, fatal respiratory failure. Severe alkalosis, in contrast, increases nervous system excitability, which can lead to uncontrollable muscle seizures and convulsions. Frequently, death occurs as a result of spasmodic contraction of respiratory muscles, which makes normal breathing movements impossible.

The Role of the Respiratory System in Acid-Base Balance

The respiratory system and kidneys work together to regulate blood pH;. In this section we concentrate on aspects of the respiratory system involved in acid-base balance.

Hemoglobin as a Buffer

Hemoglobin is considered to be a buffer because it can bind or release hydrogen ions. Deoxyhemoglobin's greater affinity for hydrogen ions compared to oxyhemoglobin's affinity is a component of

TABLE 4 Local Controls of the Radius of Bronchioles and Pulmonary Arterioles

Change in gas composition in lungs	Response of bronchioles	Response of pulmonary arterioles
Increased P_{CO_2}	Dilation (increased $\dot{V}_A$*)	Weak constriction (decreased $\dot{Q}$†)
Decreased P_{CO_2}	Constriction (decreased $\dot{V}_A$)	Weak dilation (increased $\dot{Q}$)
Increased P_{O_2}	Weak constriction (decreased $\dot{V}_A$)	Dilation (increased $\dot{Q}$)
Decreased P_{O_2}	Weak dilation (increased $\dot{V}_A$)	Constriction (decreased $\dot{Q}$)

*$\dot{V}_A$ = ventilation
†$\dot{Q}$ = perfusion

TOOLBOX

The Henderson-Hasselbalch Equation

The relationship between carbon dioxide and acidity can be described using the Henderson-Hasselbalch equation, which is based on the equilibrium for the dissociation of an acid (HA) into a free hydrogen ion (H^+) and a base (A^-):

$$HA \rightleftharpoons H^+ + A^-$$

The equilibrium constant (K) for this reaction is given by the following expression:

$$K = \frac{[H^+][A^-]}{[HA]}$$

which can be rearranged to give the hydrogen ion concentration under equilibrium conditions:

$$[H^+] = \frac{K[HA]}{[A^-]}$$

Because the hydrogen ion concentration is usually reported as pH, we can reformulate this equation by taking the negative log of both sides, which yields the Henderson-Hasselbalch equation:

$$pH = -\log K + \log\frac{[A^-]}{[HA]} = pK + \log\frac{[A^-]}{[HA]}$$

Note here that the negative log of the equilibrium constant is designated as p*K*, which is analogous to pH, the negative log of the hydrogen ion concentration.

We can apply the previous equation to the dissociation of carbonic acid (H_2CO_3) into bicarbonate (HCO_3^-) and hydrogen ions:

$$H_2CO_3 \rightleftharpoons HCO_3^- + H^+$$

by substituting $[HCO_3^-]$ for $[A^-]$ and $[H_2CO_3]$ for $[HA]$:

$$pH = pK + \log\frac{[HCO_3^-]}{[H_2CO_3]}$$

However, we can also express the equation in terms of the carbon dioxide concentration because carbonic acid is generated in the reaction of CO_2 with water:

$$CO_2 + H_2O \rightleftharpoons H_2CO_3$$

which is described by the following equilibrium constant:

$$K = \frac{[H_2CO_3]}{[CO_2][H_2O]}$$

Because $[H_2O]$ is virtually constant, this equation tells us that the ratio $[H_2CO_3]/[CO_2]$ is also constant. In other words, $[H_2CO_3]$ will always be proportional to $[CO_2]$. As a consequence, we can substitute $[CO_2]$ for $[H_2CO_3]$ to yield the Henderson-Hasselbalch equation in the following form (in which the appropriate p*K* value has been inserted):

$$pH = 6.1 + \log\frac{[HCO_3^-]}{[CO_2]}$$

This equation tells us that maintenance of a normal arterial pH of 7.4 requires that the ratio $[HCO_3^-]/[CO_2]$ remain constant at 20:1. If this ratio changes, blood pH must also change. The lungs and kidneys both play a role in helping to keep the ratio constant: The lungs regulate $[CO_2]$, while the kidneys regulate $[HCO_3^-]$.

Because blood carbon dioxide levels are commonly expressed in terms of the partial pressures, the previous equation can also be written in the following form:

$$pH = 6.1 + \log\frac{[HCO_3^-]}{(0.03)P_{CO_2}}$$

where 0.03 is a proportionality constant that converts the partial pressure of carbon dioxide (mm Hg) to a concentration (millimoles/liter).

the Bohr effect. In the tissues, hemoglobin unloads oxygen and binds hydrogen ions:

$$Hb \cdot O_2 \rightarrow O_2 + Hb \qquad Hb + H^+ \rightarrow HbH$$

This reaction is important because the tissues are producing carbon dioxide, which is quickly converted to bicarbonate and hydrogen ions. Some of these hydrogen ions can be buffered by hemoglobin, which helps to prevent the pH from becoming too acidic:

$$CO_2 + H_2O \rightarrow H_2CO_3 \rightarrow HCO_3^- + H^+$$

$$H^+ + Hb \rightarrow HbH$$

In the lungs, the reaction goes in reverse: Hemoglobin releases hydrogen ions and loads oxygen. The clearance of carbon dioxide tends to reduce the hydrogen ion concentration, an effect countered somewhat by the release of hydrogen ions from hemoglobin as it binds oxygen:

$$HbH \rightarrow H^+ + Hb \qquad Hb + O_2 \rightarrow Hb \cdot O_2$$

Bicarbonate Ions as a Buffer

Bicarbonate also serves as a major buffering system in blood. If the hydrogen ion concentration increases, hydrogen ions bind to bicarbonate to form carbon dioxide. Likewise, if carbon dioxide in the blood increases, it can be converted to bicarbonate and hydrogen ions. The production of hydrogen ions from carbon dioxide is important in acid-base balance: If carbon dioxide levels in the blood are allowed to increase, acidosis will result.

The relationship between carbon dioxide and acidity can be described using the Henderson-Hasselbalch equation, which can be applied to the equilibrium for the series of reactions that ultimately convert carbon dioxide and water to bicarbonate and hydrogen ions. The Henderson-Hasselbalch equation for this reaction is

$$pH = 6.1 + \log\frac{[HCO_3^-]}{[CO_2]}$$

(see **Toolbox: The Henderson-Hasselbalch Equation**).

This equation tells us that maintaining a normal blood pH of 7.4 requires that the ratio of bicarbonate to carbon dioxide remain

DISCOVERY

The Effects of High Altitude

Atmospheric pressure decreases at high altitudes, which means that the partial pressures of the gases in air at such locations decrease even though gas composition of the air does not change. For example, in Denver, Colorado (the "Mile High City"), the atmospheric pressure of air is 630 mm Hg, and at the top of Mount Everest, the atmospheric pressure of air is 250 mm Hg. The partial pressure of oxygen at these two locations would be as follows:

Denver: $P_{O_2} = (0.21) \times (630 \text{ mm Hg}) = 132.3 \text{ mm Hg}$

Mount Everest: $P_{O_2} = (0.21) \times (250 \text{ mm Hg}) = 52.5 \text{ mm Hg}$

Thus, as altitude increases, the air that reaches the alveoli contains less oxygen.

Because the alveolar P_{O_2} is decreased at high altitudes, the arterial P_{O_2} also decreases, a condition called *hypoxemia.* When oxygen levels in blood are low, less oxygen is available to the body's tissues, resulting in *hypoxia* (decreased oxygen in tissue). Initially the hypoxemia triggers compensatory responses in an attempt to raise arterial P_{O_2}. If the P_{O_2} drops below 60 mm Hg, activated peripheral chemoreceptors cause the respiratory center to increase ventilation. If ventilation increases relative to metabolic demands, however, both arterial P_{CO_2} and the hydrogen ion concentration in the blood will decrease. This suppresses the activation of both peripheral and central chemoreceptors, countering the stimulatory effect of low oxygen levels. In addition, a *respiratory alkalosis* occurs. With a decrease in the blood's acidity comes a shift in the hemoglobin-oxygen dissociation curve to the left as the affinity of hemoglobin for oxygen increases. An increase in affinity means that less oxygen is unloaded in the tissues, but it also means that more oxygen is loaded in the lungs.

Upon exposure to low oxygen levels, oxyhemoglobin levels decrease, causing an increase in erythrocytes' production of 2,3-BPG. The greater release of 2,3-BPG decreases the affinity of hemoglobin for oxygen, which increases oxygen unloading in the tissues, countering the effects of the alkalosis.

If exposed to high altitudes for a few days, the body begins to acclimate. The kidneys help maintain acid-base balance by excreting bicarbonate to match the loss of hydrogen ions that accompanies the reduction in arterial P_{CO_2}. If the body is exposed for longer periods, other acclimatizations occur. In response to the hypoxia, the kidneys secrete the hormone erythropoietin, which stimulates erythrocyte synthesis, causing a rise in the hematocrit up to 60%, a condition called *polycythemia.* With an increase in erythrocytes comes an increase in the concentration of hemoglobin in blood and, therefore, an increase in the oxygen-carrying capacity of the blood. However, polycythemia also increases blood viscosity, which increases the resistance to blood flow.

Karl Kummels/SuperStock

Mount Everest.

constant at 20:1. If this ratio changes, blood pH will also change. The lungs regulate the concentration of carbon dioxide, whereas the kidneys regulate the concentration of bicarbonate ions.

Respiratory disturbances in acid-base balance are due to changes in the concentration of carbon dioxide. **Respiratory acidosis** is an increase in the acidity of the blood due to increased carbon dioxide, which occurs, for example, during hypoventilation. **Respiratory alkalosis** is a decrease in the acidity of the blood due to decreased carbon dioxide, which occurs, for example, during hyperventilation or at high altitudes (see **Discovery: The Effects of High Altitude**). Other disturbances in acid-base balance are called metabolic acidosis and metabolic alkalosis.

Health Fact

Quick Check 7

1. Write out the chemical equations responsible for converting carbon dioxide to acid. Show all intermediate steps, indicate directions of reactions, and indicate the role of any enzymes.
2. What effect does hypoventilation have on P_{CO_2}, P_{O_2}, and blood pH?

SYSTEMS INTEGRATION In this chapter, we learn how the respiratory system provides oxygen to the blood and removes carbon dioxide (and hydrogen ions) from the blood. All systems of the body depend on oxygen for at least some of their energy, and carbon dioxide is a waste product that must be removed. The respiratory system works very closely with the cardiovascular system to carry out its function. In fact, the ventilation-perfusion ratio is regulated to ensure the most efficient gas exchange between blood and tissue. The respiratory system also helps defend against infectious agents through the use of the mucus escalator and alveolar macrophages. A function of the respiratory system that is often overlooked is its role in acid-base balance. The respiratory and renal systems work together to maintain normal blood pH.

CHAPTER REVIEW

SUMMARY

1 Overview of the Pulmonary Circulation

- To maintain relatively constant levels of oxygen in the systemic arterial blood, inspired oxygen moves from alveolar air into the blood at the same rate it is consumed by the tissues, just as carbon dioxide moves into alveolar air from the blood at the same rate it is produced by the tissues.
- The respiratory quotient is the ratio of the amount of carbon dioxide produced by the body to the amount of oxygen consumed.
- The right heart pumps deoxygenated blood to the pulmonary capillaries, where oxygen diffuses from the alveoli to blood, and carbon dioxide diffuses from blood to the alveoli.
- The respiratory membrane provides a large surface area and short distance for diffusion, so the diffusion rates are rapid.
- Once oxygenated in the lungs, blood returns to the left side of the heart, where it is pumped to the systemic capillaries in the tissues of the body.
- Oxygen diffuses from the blood to tissue, and carbon dioxide diffuses from tissue to the blood.
- After becoming deoxygenated, blood returns to the right side of the heart.

iP Respiratory, Anatomy Review: Respiratory Structures

iP Cardiovascular, Anatomy Review: The Heart

2 Diffusion of Gases

- The pressures of individual gases in a mixture are called partial pressures; they are equal to the fractional concentration of the gas multiplied by the total pressure.
- Gases can dissolve in liquids to varying degrees based on their solubility and partial pressure.
- The greater its solubility and its partial pressure, the larger the amount of a gas that dissolves in a liquid.
- Neither oxygen nor carbon dioxide is very soluble in water, but carbon dioxide is approximately 20 times more soluble than oxygen.

MAP iP Respiratory, Gas Exchange

3 Exchange of Oxygen and Carbon Dioxide

- Gas exchange occurs by diffusion down partial pressure gradients.
- In the lungs, oxygen diffuses from alveoli to blood, and carbon dioxide diffuses from blood to alveoli.
- In respiring tissues, oxygen diffuses from blood to tissue, and carbon dioxide diffuses from tissue to blood.
- The amount of oxygen and carbon dioxide diffusing across a particular systemic capillary depends on the activity of the tissue; more active tissues have greater partial pressure gradients, leading to greater diffusion rates.
- Alveolar P_{O_2} and P_{CO_2} are determined by the P_{O_2} and P_{CO_2} of inspired air, the alveolar ventilation, and the rates of oxygen consumption and carbon dioxide production in respiring tissue.
- Alveolar P_{O_2} and P_{CO_2} determine arterial P_{O_2} and P_{CO_2}.
- Normally, alveolar ventilation is matched to oxygen consumption and carbon dioxide production.
- Hyperpnea occurs when metabolic activity increases, causing ventilation to increase in an effort to match the oxygen demands of the tissue.

MAP iP Respiratory, Gas Exchange

4 Transport of Gases in the Blood

- Oxygen is transported in blood by being dissolved in blood (1.5%) or bound to hemoglobin (98.5%).
- The relationship between P_{O_2} and the amount of oxygen bound to hemoglobin is shown in the hemoglobin-oxygen dissociation curve.
- Factors in blood that influence the binding of oxygen to hemoglobin include temperature, pH, P_{CO_2}, 2,3-BPG, carbon monoxide, and P_{O_2}.
- The Bohr effect is the decrease in the affinity of hemoglobin for oxygen that occurs when hydrogen ions bind to hemoglobin.
- The carbamino effect is the decrease in the affinity of hemoglobin for oxygen that occurs when carbon dioxide binds to hemoglobin.
- The Haldane effect is the decrease in the affinity of hemoglobin for hydrogen ions and carbon dioxide that occurs when oxygen binds to hemoglobin.
- Carbon dioxide is transported in blood in three ways: dissolved in blood (5–6%), bound to hemoglobin (5–8%), or dissolved in blood as bicarbonate ions (86–90%).
- The conversion of carbon dioxide to bicarbonate plays a significant role in maintaining acid-base balance in the blood; bicarbonate is the main form in which carbon dioxide is transported between tissues and lungs.
- Carbonic anhydrase catalyzes the reversible reaction that converts carbon dioxide and water to carbonic acid, which then dissociates to hydrogen ions and bicarbonate.
- The chloride shift is the movement of chloride ions into erythrocytes and bicarbonate into plasma; the reverse chloride shift occurs when bicarbonate ions are transported from plasma into erythrocytes in exchange for chloride ions.

MAP iP Respiratory, Gas Transport

5 Central Regulation of Ventilation

- Breathing is a rhythmic process that is caused by cyclical neural excitation of respiratory muscles.
- The generation of the breathing rhythm requires the action of respiratory centers in the brainstem.
- The medullary respiratory control center includes the dorsal respiratory group and the ventral respiratory group.
- Inspiratory neurons in these regions activate the motor neurons that innervate the inspiratory muscles, causing inspiration.
- The pontine respiratory group may be involved in the transition between inspiration and expiration. Higher brain areas can influence respiration.
- Various stimuli affect ventilation, including changes in arterial P_{O_2} and P_{CO_2}, stretch of the lungs, irritants in the airways, proprioceptors, arterial baroreceptors, nociceptors, thermoreceptors, emotions, and voluntary controls.

MAP iP Respiratory, Control of Respiration

6 Control of Ventilation by Chemoreceptors

- Peripheral and central chemoreceptors detect changes in P_{O_2}, P_{CO_2}, and the pH of arterial blood.
- P_{CO_2} is the primary stimulus to the central chemoreceptors, but its effects are always indirect: CO_2 must first be converted to hydrogen ions (and bicarbonate).
- The peripheral chemoreceptors located in the carotid bodies respond directly to changes in pH and P_{CO_2} and to decreases in P_{O_2} to less than 60 mm Hg.
- Central chemoreceptors are located in the medulla oblongata and respond to changes in the pH of cerebrospinal fluid.

MAP iP Respiratory, Control of Respiration

7 Local Regulation of Ventilation and Perfusion

- The ventilation-perfusion ratio is the ratio between air flow to the alveoli and blood flow to the capillaries supplying those alveoli.
- In normal lungs, the ventilation-perfusion ratio is 1.
- If ventilation to a particular alveolus is decreased, perfusion will be decreased by vasoconstriction to sustain the normal ventilation-perfusion ratio.
- If perfusion to a particular alveolus is decreased, then air flow will be decreased by bronchoconstriction.

8 The Respiratory System in Acid-Base Homeostasis

- The pH of blood is tightly regulated between 7.38 and 7.42. The respiratory and urinary systems work together to maintain normal blood pH.
- Acidosis is a decrease in blood pH to 7.35 or less; alkalosis is an increase in blood pH to 7.45 or greater.
- The primary contribution of the respiratory system to acid-base balance is the regulation of arterial P_{CO_2}.
- Because carbon dioxide can be converted to carbonic acid, a change in P_{CO_2} can cause either respiratory acidosis or respiratory alkalosis.
- The respiratory system works in concert with the kidneys to maintain a ratio of bicarbonate to carbon dioxide of 20:1.

MAP iP Respiratory, Control of Respiration

MAP MasteringA&P®

Check your understanding at MasteringA&P and apply your new knowledge to real-life case studies to ensure that you've mastered the content!

EXERCISES

Multiple-Choice Questions

1. Under steady-state conditions, the rate at which oxygen enters pulmonary capillaries from alveolar air is equal to
   a) The rate at which oxygen is delivered to alveoli in inspired air.
   b) The rate at which oxygen is carried out of the alveoli in expired air.
   c) The rate at which oxygen is consumed in respiring tissues.
   d) The rate at which carbon dioxide is produced in respiring tissues.
   e) The rate at which carbon dioxide leaves the pulmonary capillaries and enters alveolar air.
2. At the normal resting P_{O_2} of mixed venous blood, hemoglobin is
   a) Nearly 100% saturated.
   b) Nearly 97% saturated.
   c) Nearly 75% saturated.
   d) Nearly 50% saturated.
   e) Nearly 25% saturated.
3. In respiring tissues, an increase in blood P_{CO_2} causes all of the following *except*
   a) An increase in the hydrogen ion concentration.
   b) An increase in the bicarbonate concentration.
   c) An increase in the carbaminohemoglobin concentration.
   d) An increase in the affinity of hemoglobin for oxygen.
   e) An increase in hemoglobin unloading of oxygen.
4. Which of the following does *not* affect alveolar P_{O_2}?
   a) The rate of oxygen consumption by respiring tissues
   b) Alveolar ventilation
   c) The P_{O_2} of inspired air
   d) The volume of air contained in the alveoli
   e) The humidification of air as it moves through the conducting zone
5. During hyperventilation, which of the following would be expected to happen?
   a) An increase in the P_{O_2} of arterial blood
   b) An increase in the P_{CO_2} of arterial blood
   c) An increase in the acidity of arterial blood
   d) An increase in the bicarbonate concentration of arterial blood
   e) A decrease in the pH of arterial blood
6. Which of the following would be expected to cause a decrease in the percent saturation of hemoglobin?
   a) An increase in P_{O_2}
   b) A decrease in blood pH
   c) A decrease in P_{CO_2}
   d) A decrease in temperature
   e) All of the above
7. Suppose that alveolar $P_{O_2} = 100$ mm Hg and $P_{CO_2} = 60$ mm Hg. Which of the following is true?
   a) pH will be less than normal.
   b) Percent saturation of hemoglobin by oxygen will be less than normal.
   c) Bicarbonate concentration will be greater than normal.
   d) Both a and c are true.
   e) All of the above are true.
8. Suppose a person's arterial P_{O_2} and P_{CO_2} are normal ($P_{O_2} = 100$ mm Hg; $P_{CO_2} = 40$ mm Hg). Which of the following would most likely stimulate an increase in ventilation?
   a) A decrease in P_{O_2} to 90 mm Hg
   b) A decrease in P_{CO_2} to 35 mm Hg
   c) An increase in P_{O_2} to 110 mm Hg
   d) An increase in P_{CO_2} to 45 mm Hg

9. A rise in arterial P_{CO_2} triggers an increase in ventilation by stimulating both central and peripheral chemoreceptors. The response of central chemoreceptors is due to
   a) Diffusion of carbon dioxide into brain extracellular fluid, which stimulates chemoreceptors directly.
   b) Diffusion of hydrogen ions into brain extracellular fluid, which stimulates chemoreceptors directly.
   c) Diffusion of carbon dioxide into brain extracellular fluid, which reacts with water to form hydrogen ions, which stimulate chemoreceptors directly.
   d) Diffusion of carbon dioxide into brain extracellular fluid, which reacts with water to form bicarbonate ions, which stimulate chemoreceptors directly.
   e) Direct stimulation by hydrogen ions in arterial blood.
10. When a person exercises, ventilation increases to meet the demands of more active tissues. This is an example of
    a) Hyperventilation.
    b) Hypoventilation.
    c) Hypoxia.
    d) Apnea.
    e) Hyperpnea.
11. The normal ratio of bicarbonate concentration to carbon dioxide concentration in arterial blood is
    a) 1:5.
    b) 5:1.
    c) 10:1.
    d) 20:1.
    e) 1:20.
12. Which of the following can hemoglobin bind and transport in blood?
    a) Oxygen
    b) Carbon dioxide
    c) Hydrogen ions
    d) Both a and c
    e) All of the above
13. Which of the following areas of the brain contain inspiratory neurons?
    a) The dorsal respiratory group only
    b) The ventral respiratory group only
    c) Both the dorsal and ventral respiratory groups
    d) Neither the dorsal nor ventral respiratory group

Objective Questions

14. Under normal conditions, the rate at which oxygen is brought into the alveoli in inspired air is (the same as/greater than/less than) the rate at which it is consumed in respiring tissues.
15. Under resting conditions, tissues normally extract (exactly half/more than half/less than half) of the oxygen that is delivered to them in arterial blood.
16. The amount of carbon dioxide in systemic arterial blood is less than 50% of that in mixed venous blood. (true/false)
17. When the P_{CO_2} of the blood increases, the concentration of bicarbonate (increases/decreases), and the concentration of hydrogen ions (increases/decreases).
18. The enzyme that catalyzes the conversion of carbon dioxide to carbonic acid is ________.
19. As the pH of the blood increases, the affinity of hemoglobin for oxygen (increases/decreases).
20. When a person hypoventilates, the P_{CO_2} of arterial blood (increases/decreases).
21. A decrease in alveolar ventilation would be expected to cause a(n) (increase/decrease) in arterial P_{O_2} and a(n) (increase/decrease) in arterial P_{CO_2}.
22. Hemoglobin with carbon dioxide bound to it is called ________.
23. In gas exchange in both the lungs and respiring tissues, oxygen and carbon dioxide always move down their partial pressure gradients. (true/false)
24. (Central/Peripheral) chemoreceptors respond directly to hydrogen ions produced during metabolism.
25. Coughing is triggered by stimulation of pulmonary ________ receptors.
26. In respiratory acidosis, arterial P_{CO_2} is (higher/lower) than normal.
27. An increase in the P_{CO_2} of alveolar air would be expected to trigger local (bronchoconstriction/bronchodilation) in airways.
28. An increase in the P_{O_2} of alveolar air would be expected to trigger local (vasoconstriction/vasodilation).

Essay Questions

29. Explain how changes in blood P_{CO_2} affect loading and unloading of oxygen in the lungs and in respiring tissues.
30. Sketch a hemoglobin-oxygen dissociation curve, and explain how it is affected by pH and P_{CO_2}. Include in your explanation changes in affinity and shifts in the hemoglobin-oxygen dissociation curve.
31. Describe what happens to oxygen and carbon dioxide as blood travels through the circulatory system. Start in the left ventricle and finish in the left atrium.
32. Suppose that alveolar P_{O_2} and P_{CO_2} are normal. If a sudden increase occurs in tissue metabolic activity and CO_2 production, but no change in minute alveolar ventilation occurs, then what would you expect to happen to arterial P_{CO_2}, mixed venous P_{CO_2}, and alveolar P_{CO_2}?
33. Describe how chemoreceptors work to keep arterial P_{CO_2} constant. Include an explanation of how arterial P_{CO_2} affects both central and peripheral chemoreceptors.

Critical Thinking

34. Carbon monoxide is a poison because it binds to hemoglobin at the heme group, preventing oxygen from binding. Draw a hemoglobin-oxygen dissociation curve for the normal condition and then with carbon monoxide present.
35. Chronic exposure to high altitude results in several adaptations to enhance oxygen delivery to tissues, one of which is an increase in the hematocrit or *polycythemia*.
36. Barbiturate overdoses cause respiratory depression. Describe what happens to the partial pressures of oxygen and carbon dioxide in arterial blood following an overdose. What happens to the pH? Will respiratory acidosis or alkalosis result? What would you do to treat someone following a barbiturate overdose?
37. Premature babies often do not secrete surfactant at adequate levels. Describe the consequences of low surfactant and appropriate treatments.

Photo credits for icons in feature boxes: Focus on Diabetes: Tanyae/Dreamstime; Clinical Connections: lofoto/Dreamstime; Understanding Exercise: Pixattitude/Dreamstime; Discovery: Yuri Arcurs/Cutcaster.

Glossary

acidosis condition in which arterial blood pH is 7.35 or lower

alkalosis condition in which arterial blood pH is 7.45 or greater

Bohr effect effect of hydrogen ions on the ability of hemoglobin to bind oxygen

carbamino effect (kar-bah-meen-oh) decrease in the affinity of hemoglobin for oxygen when carbon dioxide binds to hemoglobin

carbaminohemoglobin hemoglobin with carbon dioxide bound to it

central chemoreceptors chemoreceptors located in the medulla oblongata that respond directly to changes in hydrogen ion concentration in the cerebrospinal fluid and indirectly to arterial P_{CO_2}; function in regulating ventilation

chemoreceptors receptors that monitor the concentrations of certain chemicals in various locations in the body

chloride shift the movement of chloride ions into erythrocytes in exchange for the movement of bicarbonate into plasma

dorsal respiratory group (DRG) respiratory control center located on the dorsal side of the medulla; contains primarily inspiratory neurons

expiratory neurons neurons in the central nervous system that increase their firing rate during expiration

Haldane effect effect of oxygen on the binding of carbon dioxide to hemoglobin

hyperpnea (hy-perp-nee-ah) an increase in alveolar ventilation to match increased metabolic demands

hyperventilation an increase in alveolar ventilation such that metabolic demands of the tissue are exceeded

hypoventilation a decrease in alveolar ventilation such that metabolic demands of the tissue are not met

inspiratory neurons central nervous system (CNS) neurons that have an increased firing rate during inspiration

intercostal nerves spinal nerves that innervate the intercostal muscles of respiration

mixed venous blood blood in the pulmonary artery that includes a mixture of blood returning from all the systemic veins to the right atrium

partial pressure the proportion of the pressure of an entire gas mixture that is due to the presence of an individual gas

peripheral chemoreceptors chemoreceptors located in the carotid arteries that respond to changes in arterial P_{O_2}, P_{CO_2}, and pH, and are involved in regulating ventilation

phrenic nerve (fren-ik) nerve that innervates the diaphragm

respiratory acidosis decrease in blood pH caused by increases in plasma carbon dioxide levels

respiratory alkalosis increase in blood pH caused by decreases in plasma carbon dioxide levels

respiratory quotient the ratio of carbon dioxide produced by the body to the amount of oxygen consumed

ventilation-perfusion ratio relationship of ventilation to perfusion in alveoli; $\dot{V}_A/\dot{Q}$

ventral respiratory group (VRG) respiratory control center in the medulla; contains inspiratory and expiratory neurons

Answers

Figure Question Answers

Fig 1 5 liters
Fig 4 Cells, systemic veins, the right atrium and ventricle, and pulmonary arteries
Fig 6 To generate ATP through the electron transport chain and oxidative phosphorylation
Fig 8 26 mm Hg
Fig 19 Peripheral chemoreceptors
Fig 20 Tight junctions

Apply Your Knowledge Answers

Oxygen accounts for 21% of the air that is in the atmosphere. Because the total atmospheric pressure at the top of Mount Everest is only 250 mm Hg, the partial pressure of oxygen at this site can be determined as follows: 250 mm Hg $\times$ 0.21 = 52.5 mm Hg.

Drawing exercise. Students should refer to Figure 8 and the preceding section header The Hemoglobin-Oxygen Dissociation Curve for guidance.

Answers to End-of-Chapter Multiple Choice and Objective Questions

Multiple Choice Questions

1. c 2. c 3. d 4. d 5. a 6. b 7. e 8. d 9. c 10. e 11. d 12. e 13. c

Objective Questions

14. the same as
15. less than half
16. false
17. increases, increases
18. carbonic anhydrase
19. increases
20. increases
21. decrease, increase
22. carbaminohemoglobin
23. true
24. Peripheral
25. irritant
26. higher
27. bronchoconstriction
28. vasodilation

The Urinary System: Renal Function

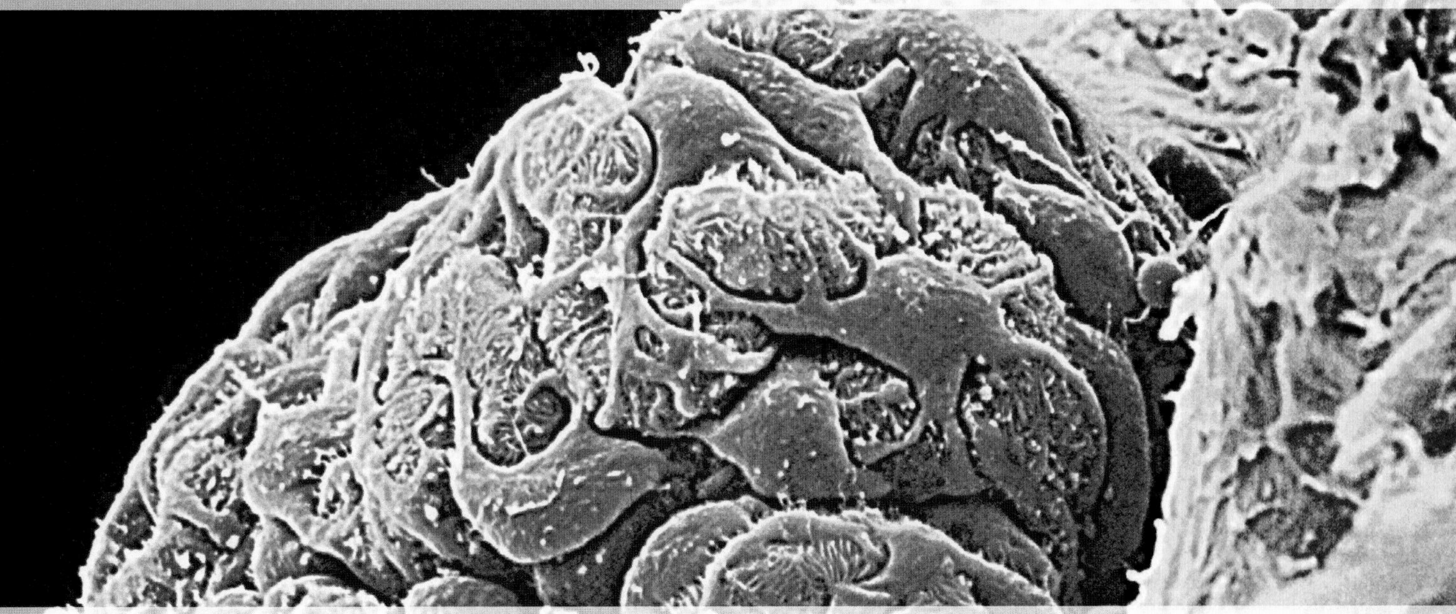

Susumu Nishinaga/Photo Researchers, Inc.

Colored scanning electron micrograph (SEM) of a glomerulus.

Marietta was born with a congenital kidney defect that left her with only one working kidney. A month before her 17th birthday, her remaining kidney starts to fail, and her doctors warn the family that complete renal failure is imminent. Marietta's mother knows that the chances of her daughter receiving a donor kidney in time are not favorable and that Marietta will soon be in a life-threatening situation. She volunteers to donate one of her kidneys to Marietta.

Unlike with most organ transplants, in which donor organs are removed from a deceased person, kidneys can be obtained from living donors. Because the kidneys normally function far below their maximum capacity, the loss of one kidney has little, if any, long-term effect on the body's ability to maintain homeostasis. Although this might seem to suggest that the kidneys do not work very hard, the reverse is true: The kidneys filter the entire plasma volume in the body (approximately 3 liters) every 22 minutes, and they are essential to maintaining the normal extracellular fluid environment that bathes the cells of the body. Without medical intervention, the loss of both kidneys would result in death within a couple of weeks.

CHAPTER OUTLINE

MasteringA&P®

Go to MasteringA&P for helpful A&P Flix 3-D animations, chapter quizzes, pre-tests, interactive physiology tutorials, and more!

LEARNING OUTCOMES ***After studying this chapter, you should be able to:***

- Identify and describe the functions of the following structures in the urinary system: nephron, glomerulus, renal tubule, collecting duct, ureter, bladder, and urethra.
- Describe how the urinary excretion of solutes and water influences the volume and composition of plasma, and identify other processes that affect plasma volume and composition.
- Explain how the basic renal exchange processes of filtration, secretion, and reabsorption affect the rate at which materials are excreted in the urine.
- Define the following terms and describe how they relate to renal function: *filtered load, glomerular filtration rate, clearance, transport maximum*, and *renal threshold*.
- Describe the events that occur during micturition.

Before You Begin

Make sure you have mastered the following topics:

- Movement of molecules across epithelia
- Movement of molecules across capillary walls
- Bulk flow across capillary walls due to Starling forces
- Components of blood
- Components of plasma
- Respiratory disturbances and compensations in acid-base balance

We know how the cardiovascular system circulates blood, providing cells with nutrients and removing waste products from the cells. We also know how the respiratory system eliminates one waste product, carbon dioxide. In this chapter, we examine how the kidneys regulate the composition of plasma by eliminating excess water and solutes (including certain waste products) and retaining needed water and solutes. By regulating the water volume within plasma, the kidneys also regulate mean arterial pressure.

In the process of regulating the solute content of plasma (and thereby the solute content of the extracellular fluid), the kidneys play a critical role in the maintenance of normal excitability of nerve and muscle tissues. Recall that the resting membrane potential depends on the electrochemical gradients for sodium and potassium ions across the cell membranes. The kidneys are necessary to maintain plasma levels of these ions. If these levels change, muscle and nerve tissue may become either hyperexcitable—causing muscle spasms or cramps, tingling sensations, or even cardiac arrhythmias—or hypoexcitable—causing numbness and weakness.

The kidneys also function in acid-base balance by regulating the rate of hydrogen ion excretion and the plasma concentration of bicarbonate. We know that the respiratory system regulates acid-base balance by regulating plasma levels of carbon dioxide. The kidneys and respiratory system work together to maintain a normal plasma pH of 7.4 by managing the ratio of bicarbonate to carbon dioxide to keep it at a ratio of 20:1.

For the kidneys to carry out all these functions (and more), they require a great deal of ATP to carry out active transport. The kidneys are under hormonal control and neural control by the autonomic nervous system.

We now turn to the kidneys and their role in filtering blood to eliminate wastes and any excess plasma water or solutes. In this chapter, we describe basic renal processes and the formation of urine. Homeostasis is maintained by the regulation of these processes.

1 Functions of the Urinary System

Of all the organs of the body, the kidneys are perhaps the most misunderstood and underappreciated. Because the kidneys filter the blood and produce **urine,** a fluid that is eliminated from the body, many people think that the sole function of the kidneys is to clear the blood of waste products. Such a view, however, overlooks the many other vital functions performed by these amazingly versatile organs. Although urine does contain metabolic by-products and other substances properly described as "wastes," it also contains water and solutes (such as sodium and potassium) that must be maintained at certain levels in the plasma and other body fluids. The significance of this fact is that the rate at which these materials are eliminated from the body, or *excreted,* by the kidneys has a significant impact on the volume and composition of the body fluids and, therefore, is highly regulated according to the body's needs.

The kidneys perform the following primary functions:

1. *Regulation of plasma ionic composition.* By increasing or decreasing the excretion of specific ions in the urine, the kidneys regulate the concentration of the following ions in the plasma: sodium (Na^+), potassium (K^+), calcium (Ca^{2+}), magnesium (Mg^{2+}), chloride (Cl^-), bicarbonate (HCO_3^-), hydrogen (H^+), and phosphates (HPO_4^{2-} and $H_2PO_4^-$).
2. *Regulation of plasma volume and blood pressure.* By controlling the rate at which water is excreted in the urine, the kidneys regulate plasma volume, which has a direct effect on total blood volume and, therefore, on blood pressure.
3. *Regulation of plasma osmolarity.* Because the kidneys vary the rate at which they excrete water relative to solutes, they have the ability to regulate the osmolarity (solute concentration) of the plasma.
4. *Regulation of plasma hydrogen ion concentration (pH).* By regulating the concentration of bicarbonate and hydrogen ions in the plasma, the kidneys partner with the lungs to regulate blood pH.

5. *Removal of metabolic waste products and foreign substances from the plasma.* Because the kidneys excrete wastes and other undesirable substances in the urine, they clear the plasma of waste products and eliminate them from the body. These materials include metabolic by-products such as urea and uric acid that are generated during protein and nucleic acid catabolism, respectively, as well as foreign substances such as food additives, drugs, or pesticides that enter the body from the external environment.

Because water and small solutes are freely exchanged between the plasma and interstitial fluid throughout most of the body, as the kidneys regulate the volume and composition of the plasma they likewise regulate the volume and composition of interstitial fluid. In addition, changes in the interstitial fluid affect the intracellular fluid. Thus the kidneys ultimately control the volume and composition of all the body's fluids. As we will see, the ability of the kidneys to form urine, and thus to perform their primary functions, hinges on their ability to filter and process large quantities of solutes and water.

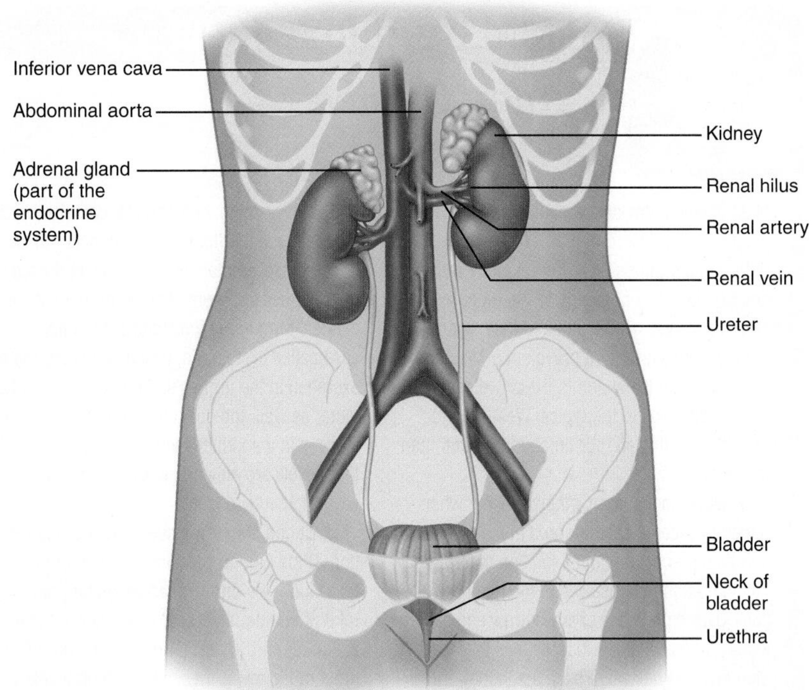

Figure 1 Structures of the urinary system.

The kidneys perform several secondary functions as well. They are considered endocrine organs because they secrete the hormone erythropoietin (which stimulates erythrocyte production by the bone marrow) and the enzyme renin (which is necessary for the production of angiotensin II, a hormone important in regulating salt and water balance for long-term control of blood pressure). The kidneys are also necessary for the activation of vitamin D_3 (ultimately to 1,25-$(OH)_2$ vitamin D_3), an important factor in regulating blood calcium and phosphate levels. Furthermore, the kidneys can function during periods of fasting to maintain a steady supply of plasma glucose by carrying out gluconeogenesis, the process by which molecules such as glycerol and certain amino acids are used to synthesize glucose. Our goal in this chapter is to gain an understanding of the basic process of urine formation.

2 Anatomy of the Urinary System

Although the kidneys produce urine, other organs are necessary for its excretion. We first take a brief look at the urinary system, including all the structures involved in urine formation and excretion. Then we describe the macroscopic and microscopic features of the kidneys that enable them to perform their crucial functions.

Structures of the Urinary System

The **urinary system** consists of two kidneys, two ureters, the urinary bladder, and the urethra (**Figure 1**). Once formed by the kidneys, the urine is conducted to the bladder by the ureters. The bladder stores the urine until it is time to excrete it; at this time the urine moves through the urethra and out of the body. Sometimes, crystals in the kidneys form into stones, which also may pass down the ureter to the urethra (see **Clinical Connections: Kidney Stones**).

The kidneys are paired organs lying at the rear wall of the abdominal cavity just above the waistline, at about the level of the 12th rib. Each kidney is roughly bean shaped and is about the size of a fist. Although most abdominal organs are enclosed within the *peritoneum,* a clear membrane that lines the abdominal cavity, the kidneys are located between the peritoneum and the wall of the abdominal cavity. Thus they are described as *retroperitoneal.*

The kidneys receive their blood supply from the **renal arteries,** which branch off the aorta and enter each kidney at a region called the *renal hilus.* Each kidney weighs only 115–170 grams (less than half a pound); their combined weight is less than 1% of the body weight of an average adult. Despite their small fraction of body weight, the kidneys receive about 20% of the cardiac output under normal resting conditions. This rich blood supply is crucial to the kidneys' function not only because it provides them with oxygen and nutrients (kidneys account for 16% of total body ATP usage), but also because it enables the kidneys to remove (or *clear*) unneeded solutes and water from the blood at a rapid rate and eliminate them as urine. Meanwhile, the blood (minus these cleared materials) returns to the general circulation via the **renal veins,** which run parallel to the renal arteries and drain into the inferior vena cava.

Macroscopic Anatomy of the Kidney

A cross section of a kidney reveals that it contains two major regions: a reddish-brown outer layer called the **cortex,** and an inner region

CLINICAL CONNECTIONS

Kidney Stones

Most people know of someone who has passed a kidney stone. More than 5% of the population of the United States will form a kidney stone, and 500,000 Americans go to the emergency room each year because of a kidney stone. Kidney stones are more prevalent in Western cultures (most likely due to the prevalence of processed foods in the typical Western diet), with males afflicted four times more often than females.

Kidney stones *(renal lithiasis)* occur when minerals become concentrated in the kidneys and then precipitate out of solution to form crystals that attach themselves to the renal epithelium. Over time, these crystals grow and become kidney stones. A stone (usually less than 1 cm in diameter) may dislodge from the kidney and enter a ureter, where it can cause debilitating pain in the side, back, and groin area. Hospitalization is often necessary for pain management.

Several factors increase a person's likelihood of developing kidney stones. The most common kidney stones (80%) consist of calcium oxalate crystals, but several other chemicals, including uric acid, cysteine, and struvite, can also be involved (see the figure). If the concentration of any of these substances increases in the kidneys, stones are more likely to form. Cysteine crystals form in persons with a hereditary disease called cysteinuria, which increases cysteine excretion in the urine. Fluid intake can be a key factor as well; the concentration of these substances in the kidney increases not only when their levels are elevated in the blood, but also when fluid intake is low.

Diet can also influence kidney stone formation. Diets high in proteins and low in fiber favor kidney stone formation, as can high intake of vitamin C or D. Some other substances, including citrate and magnesium, can help prevent kidney stones, as can high levels of fluid intake.

Once produced, kidney stones often are passed without medical intervention. Pain may or may not be present, but when it is, it can be excruciating, demonstrating that the intensity of pain does not always correlate with the level of tissue damage. To enhance natural passing of the stones, the person is advised to consume large volumes of water. If the stones are too large to be passed, they can be broken down by *extracorporeal shock wave lithotripsy* (ESWL), after which they are passed naturally. Very large stones may need to be surgically removed by either *nephrolithotomy* (removal of stones from the kidneys) or *uteroscopy* (removal of stones from the ureters).

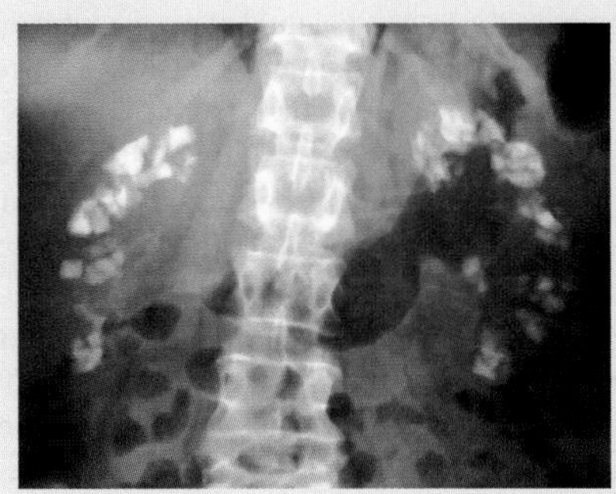

X-ray showing kidney stones.

Critical Thinking Questions

1. Describe the symptoms that indicate the presence of a kidney stone.
2. What treatments are available for removing or decreasing the size of kidney stones?
3. Describe the factors that increase a person's likelihood for developing kidney stones, including cultural ones.

called the **medulla,** which is darker and has a striped appearance (**Figure 2a**). The medulla is subdivided into a number of conical sections called *renal pyramids* (Figure 2b). At the tips of the renal pyramids, in the areas known as *papillae* (singular: *papilla*), tubules called *collecting ducts* drain into common passageways called *minor calyces* (singular: *calyx;* see Figure 2a). The minor calyces converge to form two or three larger passageways called *major calyces,* which drain into a single funnel-shaped passage called the **renal pelvis,** the initial portion of the ureter.

Within a kidney's many renal pyramids are more than a million microscopic subunits called **nephrons** (Figure 2c), which are the functional units of the kidneys; they do the work of filtering the blood and forming the urine. The most obvious feature of the nephron is a long, coiled tube (called a **renal tubule**) that forms a hairpin loop about midway along its length. During the process of urine formation, fluid flows through the renal tubules, during which time the fluid's composition is modified by exchange with the interstitial fluid. Fluid from individual tubules eventually drains into a set of common passageways called **collecting ducts**, where the composition of the fluid is further modified. The fluid that exits the collecting ducts and flows through the renal pelvis to the ureters is called *urine.*

Microscopic Anatomy of the Kidney

Each nephron is, in essence, a complete, self-contained "mini-kidney" that filters blood and forms urine. An individual nephron is composed of two parts: a *renal corpuscle* that filters the blood, and a renal tubule through which the filtrate travels and becomes modified to form urine (**Figure 3**).

Renal Corpuscle

A **renal corpuscle** consists of two parts: a spherical structure at the inflow end of the renal tubules called **Bowman's capsule,** and a tuft of capillaries called the **glomerulus.** The renal corpuscle is the site

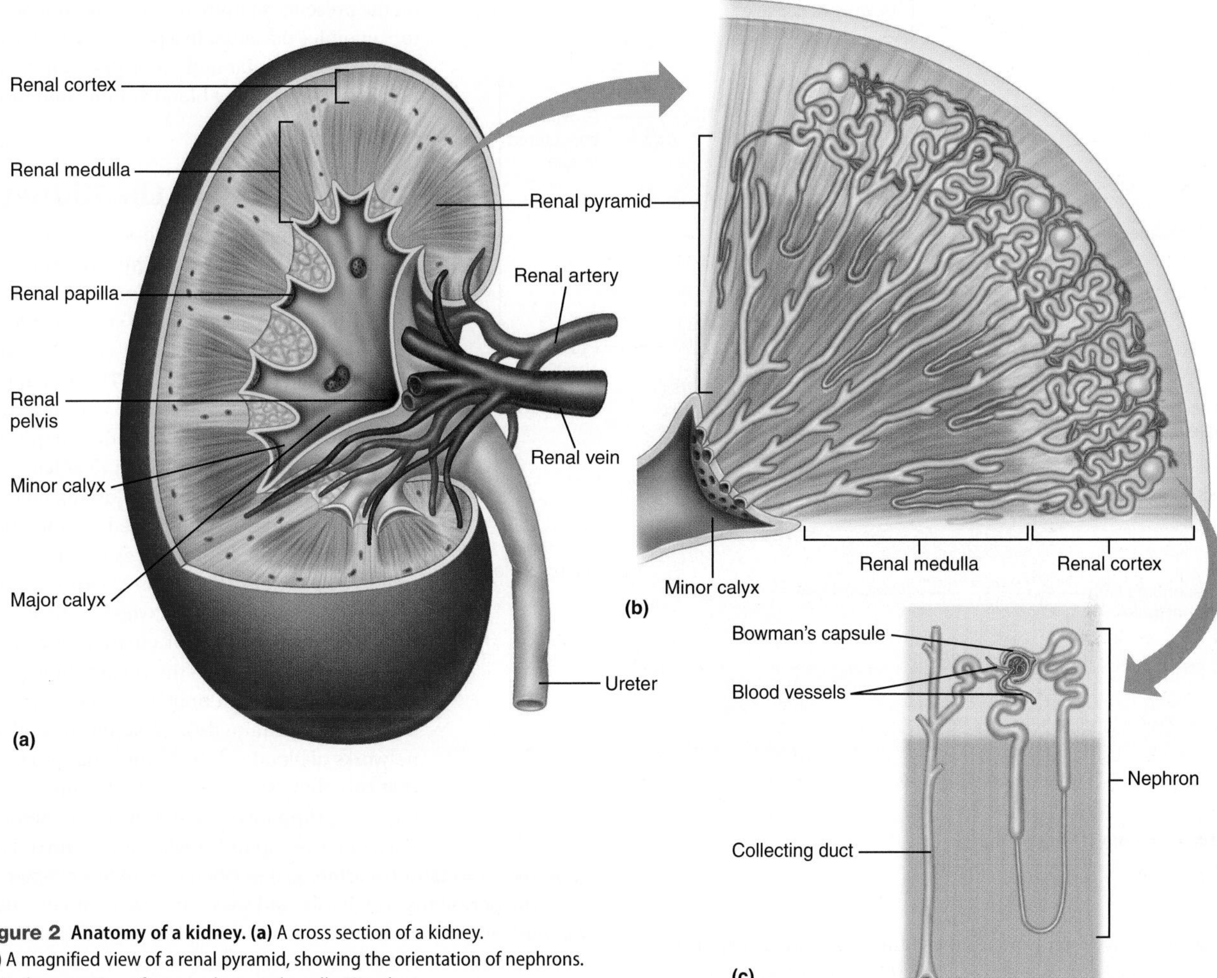

Figure 2 Anatomy of a kidney. (a) A cross section of a kidney. **(b)** A magnified view of a renal pyramid, showing the orientation of nephrons. **(c)** A close-up view of one nephron and a collecting duct.

where blood is filtered and where tubular fluid, or *filtrate,* has its origin. Before the blood is filtered, it enters the *glomerular capillaries* via an **afferent arteriole.** As the blood flows through the glomerular capillaries, protein-free plasma passes through the walls of the capillaries into Bowman's capsule by a process called **glomerular filtration.** The remaining blood leaves the glomerulus via an **efferent arteriole.** This arrangement of two arterioles in series with a capillary bed between them is unique to the renal corpuscle and allows greater regulation of glomerular filtration. The walls of the afferent and efferent arterioles contain smooth muscle that can contract or relax in response to input from paracrines and the sympathetic nervous system, thereby regulating their diameter and thus glomerular filtration (described in greater detail later in this chapter).

Renal Tubule

As the glomerular filtrate is formed, it flows from Bowman's capsule to the initial portion of the renal tubule, called the *proximal convoluted tubule* because of both its proximity to the capsule and its highly folded or *convoluted* structure; the fluid then moves to the *proximal straight tubule.* The two tubules together are called the **proximal tubule.**

The proximal tubule empties into the **loop of Henle,** the portion of the tubule that makes up the hairpin loop of the nephron. The loop of Henle is divided into three sections: (1) the descending limb, (2) the thin ascending limb, and (3) the thick ascending limb. The *descending limb* is a thin tubule leading from the proximal tubule and extending into the renal medulla. At the tip of the loop, the tubule reverses direction, becoming the *thin ascending limb,* which extends toward the cortex. As the tubule approaches the cortex, it widens into the *thick ascending limb.*

From the ascending limb of the loop of Henle, the fluid flows into the **distal convoluted tubule,** which resembles the proximal tubule in appearance but is considerably shorter. It next enters a short, straight terminal portion of the nephron, called the *connecting tubule,* which joins the nephron with the collecting duct. Several tubules empty their fluid into a single collecting duct. The collecting ducts then empty into the minor calyces, as previously described.

Cortical and Juxtamedullary Nephrons

Two classes of nephrons are distinguished based on their location: **cortical nephrons** and **juxtamedullary nephrons** (Figure 4).

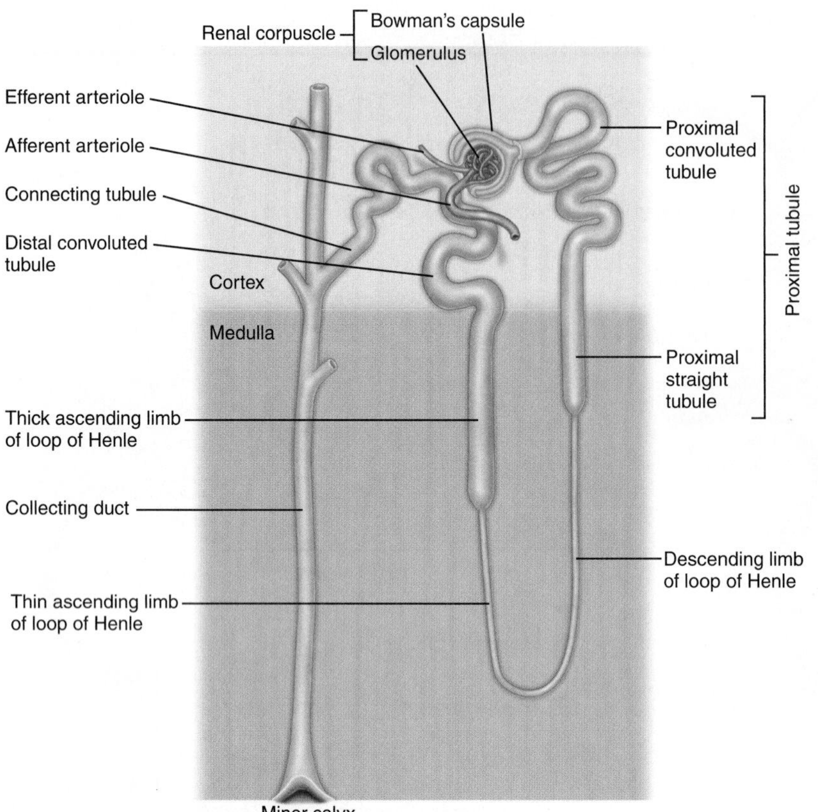

Figure 3 **Anatomy of a nephron.**

The vast majority of nephrons in the kidneys are cortical nephrons, which are located almost entirely within the renal cortex; only the tip of the loop of Henle dips into the renal medulla. In juxtamedullary nephrons, which constitute approximately 15–20% of all nephrons, the renal corpuscle is located near *(juxta)* the border between the cortex and medulla. The glomerulus, proximal convoluted tubule, and distal convoluted tubule are located in the cortex, whereas the loop of Henle dips deep into the renal medulla. Although the two types of nephrons are fundamentally similar, an important functional difference characterizes them: Whereas both cortical and juxtamedullary nephrons function directly in the processes involved in urine formation, juxtamedullary nephrons also function in maintaining an osmotic gradient in the renal medulla that is crucial to the kidneys' ability to produce highly concentrated urine and thus conserve water under certain conditions.

The Juxtaglomerular Apparatus

At a site where the initial portion of the distal tubule comes into contact with a nephron's afferent and efferent arterioles is a structure called the **juxtaglomerular apparatus** (Figure 5). The juxtaglomerular apparatus has two components: (1) a specialized cluster of the tubule's epithelial cells, called the **macula densa,** and (2) **granular cells** (or **juxtaglomerular cells**) in the wall of the afferent (and to a lesser extent the efferent) arterioles. Granular cells are named based on the presence of numerous cytoplasmic secretory granules that contain a product called *renin.* The juxtaglomerular apparatus plays an important role in regulating blood volume and blood pressure.

Blood Supply to the Kidney

The blood supply to the kidney is illustrated in **Figure 6a**. Within the kidney, the renal artery branches into *segmental arteries,* which in turn branch into a number of smaller *interlobar arteries,* which then feed into another set of arteries called *arcuate arteries.* The arcuate arteries branch into *interlobular arteries,* from which blood is carried to individual nephrons by the afferent arterioles, which next lead into the glomerular capillary beds. Coming off each of the glomerular capillary beds is the efferent arteriole, which then gives rise to one of two types of capillary beds (Figure 6b): (1) **peritubular capillaries,** which branch from the efferent arterioles of cortical nephrons and are located close to the renal tubules, and (2) **vasa recta,** which branch from the efferent arterioles of juxtamedullary nephrons and are networks of blood vessels forming hairpin loops that run along the loops of Henle and collecting ducts, dipping deep into the renal medulla. Each of these capillary beds has a distinct function in the formation of urine, as described later in this chapter.

The peritubular capillaries and vasa recta drain into the *interlobular veins.* From here, blood is carried away from nephrons by the *arcuate veins,* and then by the *interlobar veins,* which run parallel to their respective arterial counterparts, eventually draining into the renal vein.

Quick Check 1

1. Which structures make up the urinary system? What is the function of each?
2. What is a nephron? What are the two types of nephrons? How do they differ in location?
3. Which structures make up the juxtaglomerular apparatus? Where are they located?
4. Describe the path of blood flow into and out of the kidneys. In your description, be sure to identify where the blood is filtered.

3 Basic Renal Exchange Processes

In the kidneys, water and solutes are exchanged between plasma and fluid in the renal tubules to regulate the composition of plasma. Substances ultimately removed from the plasma are excreted in

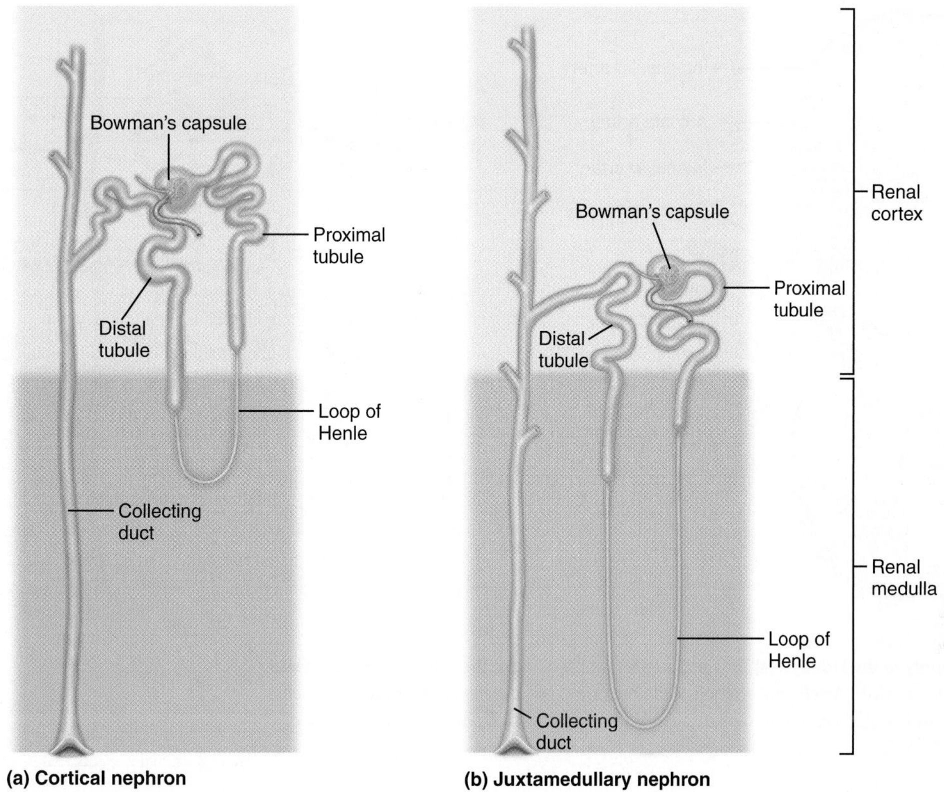

Figure 4 Locations of cortical and juxtamedullary nephrons. **(a)** A cortical nephron. **(b)** A juxtamedullary nephron.

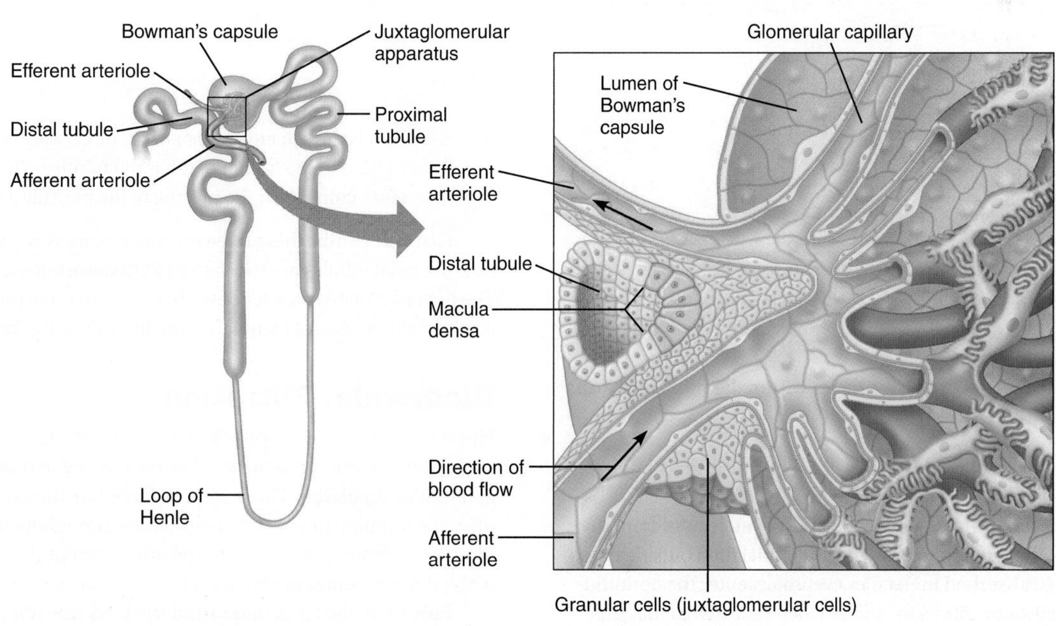

Figure 5 The juxtaglomerular apparatus.

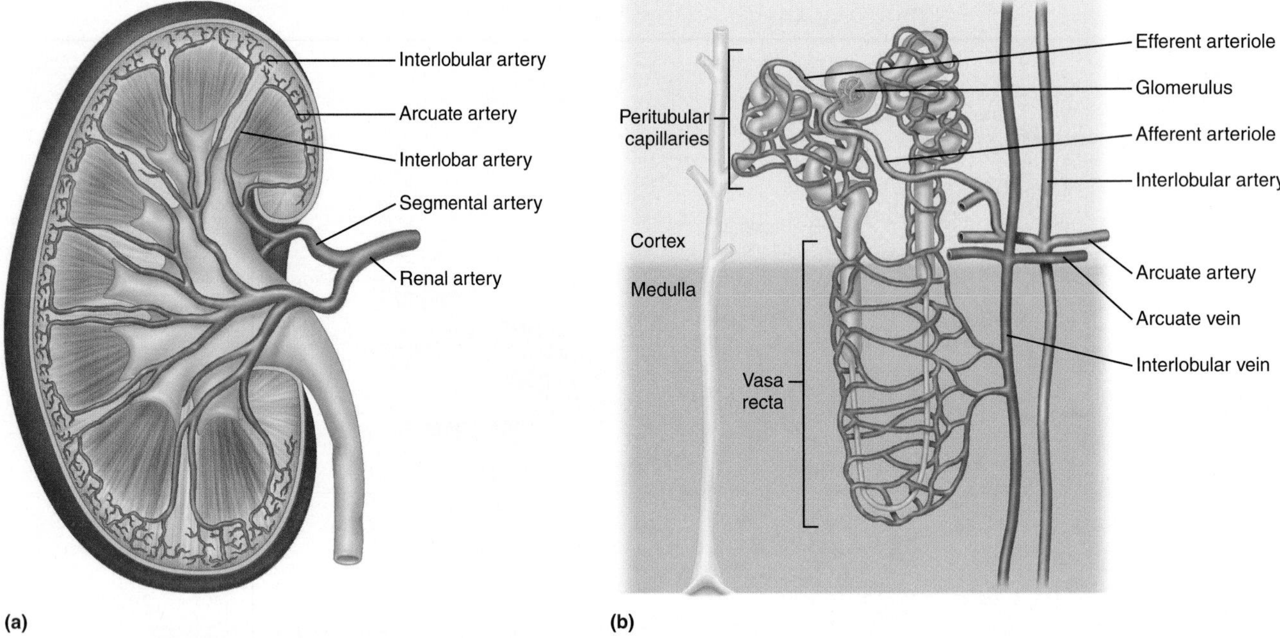

Figure 6 Blood supply to the kidneys. (a) The renal arteries supply blood to the kidneys and branch into the smaller arteries indicated. **(b)** An efferent arteriole leads into one of two different types of capillary beds: the peritubular capillaries located around the renal tubules, and the vasa recta located around the loops of Henle.

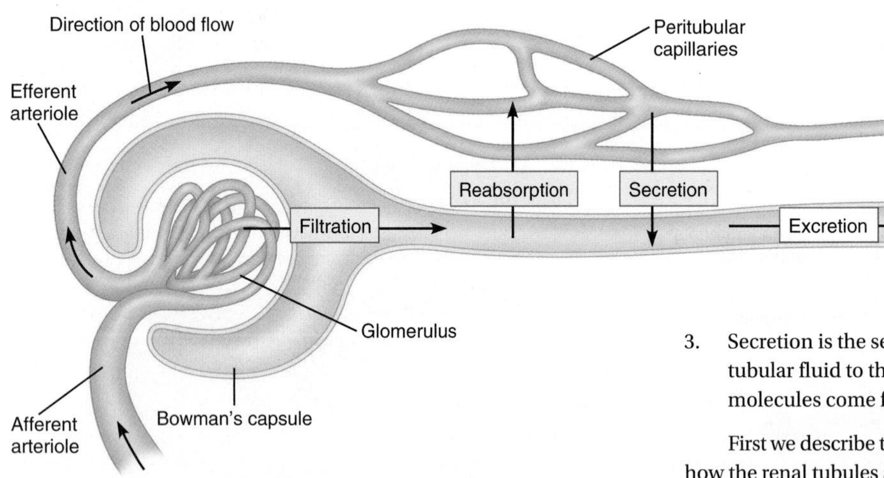

Figure 7 The three exchange processes in the renal tubules.

urine. Three exchange processes occur within the renal nephrons (**Figure 7**):

1. Glomerular filtration is the bulk flow of protein-free plasma from the glomerular capillaries into Bowman's capsule.
2. **Reabsorption** is the selective transport of molecules from the lumen of the renal tubules to the interstitial fluid outside the tubules. Reabsorbed molecules eventually enter the peritubular capillaries by diffusion, and are then returned to the general circulation.
3. Secretion is the selective transport of molecules from the peritubular fluid to the lumen of the renal tubules. These secreted molecules come from the plasma of the peritubular capillaries.

First we describe these three exchange processes; then we look at how the renal tubules are specialized for reabsorption and secretion. The final part of this chapter describes a fourth renal process: **excretion,** the elimination of materials from the body in the form of urine.

Glomerular Filtration

Filtration at the renal corpuscle is driven by Starling forces (hydrostatic and osmotic pressure gradients) existing across the walls of glomerular capillaries. These Starling forces are the same forces that drive the filtration of fluid from capillaries throughout the body. The filtrate resembles plasma in composition, except that it lacks most of the proteins found in plasma.

Figure 8 shows a magnified view of the renal corpuscle. The wall of Bowman's capsule and the wall of the renal tubule are

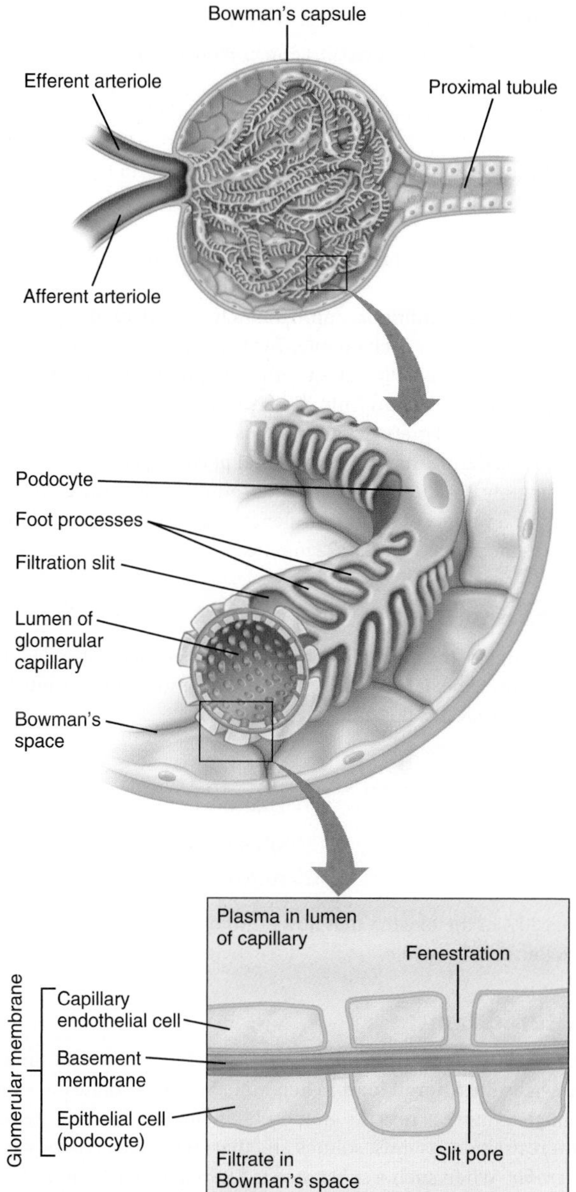

Figure 8 **Anatomy of the renal corpuscle.**

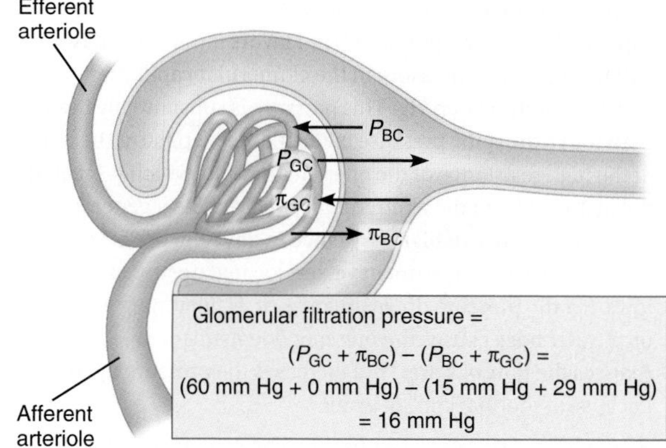

(a) Glomerular filtration pressure

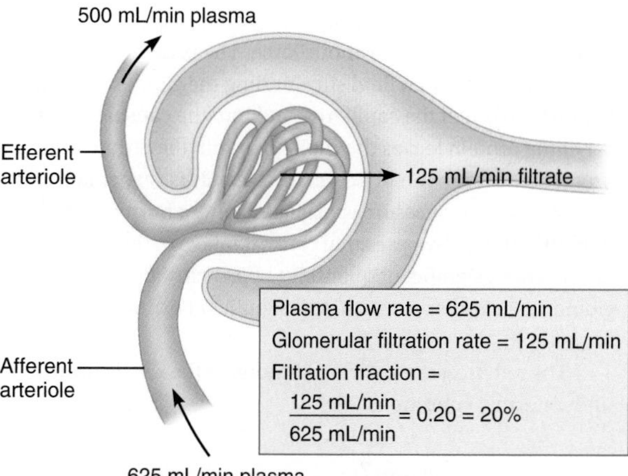

(b) Glomerular filtration rate and filtration fraction

Figure 9 **Glomerular filtration. (a)** Glomerular filtration pressure is the result of four Starling forces: (1) hydrostatic pressure in the glomerular capillaries (P_{GC}), (2) hydrostatic pressure in Bowman's capsule (P_{BC}), (3) osmotic pressure in the glomerular capillaries (π_{GC}), and (4) osmotic pressure in Bowman's capsule (π_{BC}). The net filtration pressure is 16 mm Hg. **(b)** Glomerular filtration rate and filtration fraction.

Q If proteins leaked out of the glomerular capillaries (which would decrease π_{GC} and increase π_{BC}), what would happen to the glomerular filtration pressure and to the glomerular filtration rate?

composed of a continuous layer of epithelial cells. In Bowman's capsule, this epithelium folds upon itself to envelop the glomerular capillaries. Below the epithelium is a basement membrane that acts as the primary filtration barrier for proteins. The glomerular filtrate must cross three barriers to enter Bowman's capsule: (1) the capillary endothelial cell layer, (2) the surrounding epithelial cell layer, and (3) the basement membrane that is sandwiched between them. The combination of these three layers makes up what is called the *glomerular membrane* or *filtration barrier.* The epithelial cells that cover the glomerular capillaries have special extensions or *foot processes,* giving them their name **podocytes.** As fluid moves out of the glomerular capillaries, it passes through gaps between the podocytes, called **slit pores.** The sizes of slit pores are regulated by *slit diaphragms.*

The presence of fenestrations (pores) in the capillary endothelium, the large number of slit pores in the surrounding capsule epithelium, and the large surface area of the filtration barrier combine to make the renal corpuscle favorable for the bulk flow of protein-free fluid between blood and the lumen of Bowman's capsule (called *Bowman's space*).

The sum of the Starling forces in the renal corpuscle is called the **glomerular filtration pressure,** which is analogous to the *net filtration pressure*. Four Starling forces play key roles in glomerular filtration (**Figure 9**):

1. *Glomerular capillary hydrostatic pressure.* The glomerular capillary hydrostatic pressure (P_{GC}) favors filtration and is equal to the blood pressure in the glomerular capillaries—approximately 60 mm Hg. This pressure is substantially higher than the hydrostatic pressure in most other capillaries because of the high resistance of the efferent arteriole, which is located downstream from the glomerular capillaries. As a general rule, the presence of high resistance in any network of vessels tends to raise the pressure in vessels located upstream while lowering the pressure downstream, just as tightening a clamp on a water hose causes the pressure downstream to fall (and reduces the flow of water) but increases the pressure upstream because of the "backup" of water.
2. *Bowman's capsule osmotic pressure.* The osmotic pressure in Bowman's capsule (π_{BC}) favors filtration. Recall that osmotic pressure of a fluid is created by the presence of nonpermeant solutes. Because proteins are generally the only solute that cannot move between plasma and Bowman's capsule, these molecules generate the osmotic force. The presence of proteins in the interstitial fluid surrounding the glomerulus tends to pull fluid out of the capillaries and into the capsule. Because very little protein leaves the capillaries with the filtrate, the protein concentration in Bowman's capsule is very small, and thus the osmotic pressure is negligible under normal conditions. (In certain diseases that cause damage to the glomerulus, however, significant quantities of protein can leak out of glomerular capillaries, creating a significant osmotic force for filtration.)

   The net pressure favoring filtration at the renal corpuscle under normal conditions is

$$P_{GC} + \pi_{BC} = 60 \text{ mm Hg} + 0 \text{ mm Hg} = 60 \text{ mm Hg}$$

3. *Bowman's capsule hydrostatic pressure.* Bowman's capsule hydrostatic pressure (P_{BC}) opposes filtration and is typically about 15 mm Hg. This pressure is considerably higher than the hydrostatic pressure in the interstitial fluid surrounding most capillary beds because the relatively large volume of fluid that filters out of the glomerular capillaries is "funneled" into the restricted space of Bowman's capsule.
4. *Glomerular osmotic pressure.* The glomerular osmotic pressure (π_{GC}) opposes filtration. The glomerular osmotic pressure opposes filtration because the presence of proteins in the plasma tends to draw filtrate back into the glomerulus. The osmotic pressure in the glomerulus is approximately 29 mm Hg, which is higher than the typical 25 mm Hg osmotic pressure found in most systemic capillaries, because the blood that flows through these capillaries loses a substantial fraction of its water as a result of glomerular filtration, and this loss of water causes the concentration of plasma proteins to increase.

   The net pressure opposing filtration at the renal corpuscle under normal conditions is

$$P_{BC} + \pi_{GC} = 15 \text{ mm Hg} + 29 \text{ mm Hg} = 44 \text{ mm Hg}$$

Glomerular Filtration Rate

The average net glomerular filtration pressure (GFP) is

$$\text{GFP} = (P_{GC} + \pi_{BC}) - (P_{BC} + \pi_{GC})$$

Substituting the values for the Starling forces into this equation gives us the glomerular filtration pressure:

$$(60 \text{ mm Hg} + 0 \text{ mm Hg}) - (15 \text{ mm Hg} + 29 \text{ mm Hg}) = 16 \text{ mm Hg}$$

Under normal conditions, approximately 625 mL of plasma flows through the kidneys each minute. The volume of the plasma filtered per unit time is called the **glomerular filtration rate (GFR),** and it is approximately 125 mL/min. Thus, over the course of a day, the kidneys filter 180 liters of plasma! This is an enormous volume of fluid, considering that the body's total plasma volume is only approximately 2.75 liters for an average adult. In fact, the GFR is so high that a volume of fluid equivalent to total plasma volume filters through the glomeruli every 22 minutes.

Filtration Fraction

The fraction of the renal plasma volume that is filtered is called the *filtration fraction.* It is equal to the glomerular filtration rate divided by the renal plasma flow rate (Figure 9b):

$$\text{Filtration fraction} = \frac{\text{GFR}}{\text{renal plasma flow}}$$

$$= \frac{125 \text{ mL/min}}{625\text{mL/min}} = 0.20 = 20\%$$

Thus 20% of the plasma that flows through the kidneys is filtered into Bowman's capsule.

Filtered Load

The quantity of a particular solute that is filtered per unit time is known as the **filtered load.** When molecules of a solute are small enough to move across the glomerular membrane without significant restriction, as most solutes are, that solute is said to be *freely filterable.* When such a substance is filtered, its concentration in the glomerular filtrate is virtually identical to its concentration in the plasma. Consequently, the filtered load, usually expressed in mole/min or the equivalent, equals the product of the GFR and the solute's plasma concentration (which is designated by the symbol *P*):

$$\text{Filtered load} = \text{GFR} \times \text{plasma concentration of X}$$

$$= \text{GFR} \times P_X$$

As an example, let's determine the filtered load of glucose. We already know that the normal GFR is 125 mL/min. Given that the normal plasma glucose concentration is 100 mg/dL = 1 mg/mL, then the filtered load of glucose under normal conditions is

$$\text{Filtered load} = 125 \text{ mL/min} \times 1 \text{ mg/mL} = 125 \text{ mg/min}$$

Note that the filtered load increases if either the plasma concentration of the solute or the GFR increases.

Apply Your Knowledge

Determine the filtration fraction and the filtered load for sodium given the following information: renal blood flow = 1.25 liter/min; hematocrit = 40; plasma sodium concentration = 140 mM; molecular weight of sodium = 23; and glomerular filtration rate = 125 mL/min.

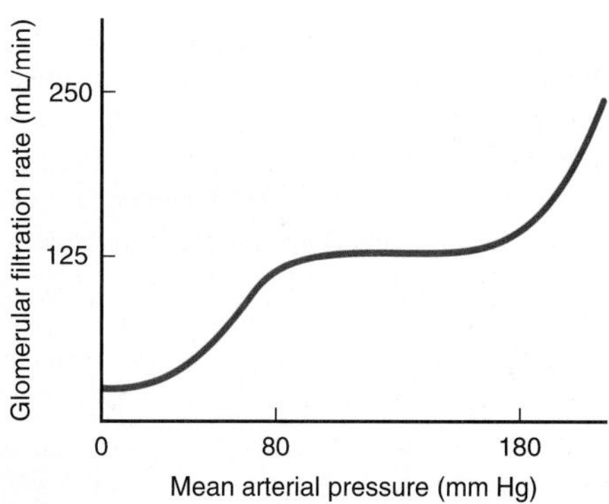

Figure 10 **Effect of mean arterial pressure on glomerular filtration rate.**

Q If the mean arterial pressure increased to 200 mm Hg, what would happen to the filtered load of any given solute?

Regulation of Glomerular Filtration Rate

Although 180 liters of fluid filters into the renal tubules every day, only about 1.5 liters of urine is normally excreted during the same period of time. The reason for this low excretion rate is that more than 99% of the fluid filtered out of the plasma is normally reabsorbed. Because the GFR is so large, however, even a small percentage change in its value will have an enormous effect on the volume of fluid filtered and, therefore, on the quantity of material that must be reabsorbed to maintain the same urinary output. A 10% increase in GFR, for example, would translate into an extra 18 liters of fluid entering the kidney tubules per day. Under most circumstances, however, GFR remains relatively constant because of the intrinsic and extrinsic mechanisms we discuss next.

Intrinsic Control of Glomerular Filtration Changes in mean arterial pressure (MAP) can potentially alter the GFR because the arterial pressure affects the glomerular capillary pressure, which in turn influences the glomerular filtration pressure. Although mean arterial pressure is regulated to stay constant by baroreceptor reflexes and other mechanisms, it does change in certain situations, such as during exercise. When arterial pressure increases, glomerular capillary pressure also tends to rise, which increases the glomerular filtration pressure and hence GFR. Conversely, when mean arterial pressure falls, the glomerular capillary pressure also tends to fall, which decreases the glomerular filtration pressure and GFR. Such changes in GFR are undesirable because they tend to make urine flow increase or decrease, respectively, which interferes with the kidneys' ability to regulate the volume and composition of the plasma.

Although variations in mean arterial pressure pose a potential problem, the kidneys can tolerate a change in mean arterial pressure over a fairly wide range (approximately 80–180 mm Hg) with very little change in GFR (**Figure 10**), because three intrinsic mechanisms regulate the GFR in the face of changes in arterial pressure. Two of these intrinsic mechanisms—*myogenic regulation* of afferent arteriolar smooth muscle and *tubuloglomerular feedback*—operate by changing the resistance of the afferent arteriole; the third factor, *mesangial cell contraction,* acts by changing the permeability of the filtration barrier.

Myogenic regulation of the GFR (**Figure 11a**) is similar to the myogenic regulation of blood flow that occurs in other parts of the body. The smooth muscle of the afferent arteriole is sensitive to stretch and responds to this condition by contracting. When mean arterial pressure rises, pressure in the afferent arteriole also rises, which causes its wall to stretch. Pressure also rises in the glomerular capillaries, which raises the glomerular filtration pressure and the GFR. However, in response to stretching, the afferent arteriole constricts, which increases its resistance to blood flow. As a result, the pressure in blood vessels downstream, including the glomerular capillaries, decreases. This decrease in pressure counteracts (but does not change) the initial rise in pressure that triggered constriction of the afferent arteriole. Thus, through this negative feedback mechanism, the glomerular capillary pressure, and hence the GFR, tends to stay nearly constant. As expected, a fall in mean arterial pressure has the opposite effects, triggering relaxation of the afferent arteriole and a subsequent rise in glomerular capillary pressure that counteracts the initial fall.

The smooth muscle of the afferent arteriole is sensitive not only to stretch but also to chemical agents that are secreted by cells of the macula densa, which are located in the nearby distal tubule. In **tubuloglomerular feedback,** a change in GFR causes a change in the flow of tubular fluid past the macula densa, which alters the secretion of certain (as yet unidentified) paracrines from the macula densa. These paracrines then trigger contraction or relaxation of the afferent arteriole, which causes a change in glomerular capillary pressure and GFR in the direction opposite to that of the original change. As a consequence of this negative feedback control of GFR, the flow of fluid past the macula densa changes such that it opposes the change in flow that triggered the response initially. For example, if an increase in GFR causes the flow of tubular fluid to increase, then the afferent arteriole constricts and the GFR decreases, thereby decreasing the flow. Thus the resistance of the afferent arteriole varies in such a way that the flow of fluid past the macula densa tends to be held constant. Figure 11b shows what happens when an increase in blood pressure causes an increase in GFR.

The third mechanism of GFR autoregulation is similar to myogenic regulation, except that it targets mesangial cells instead of blood vessels. Mesangial cells are modified smooth muscle cells located around glomerular capillaries that function much like

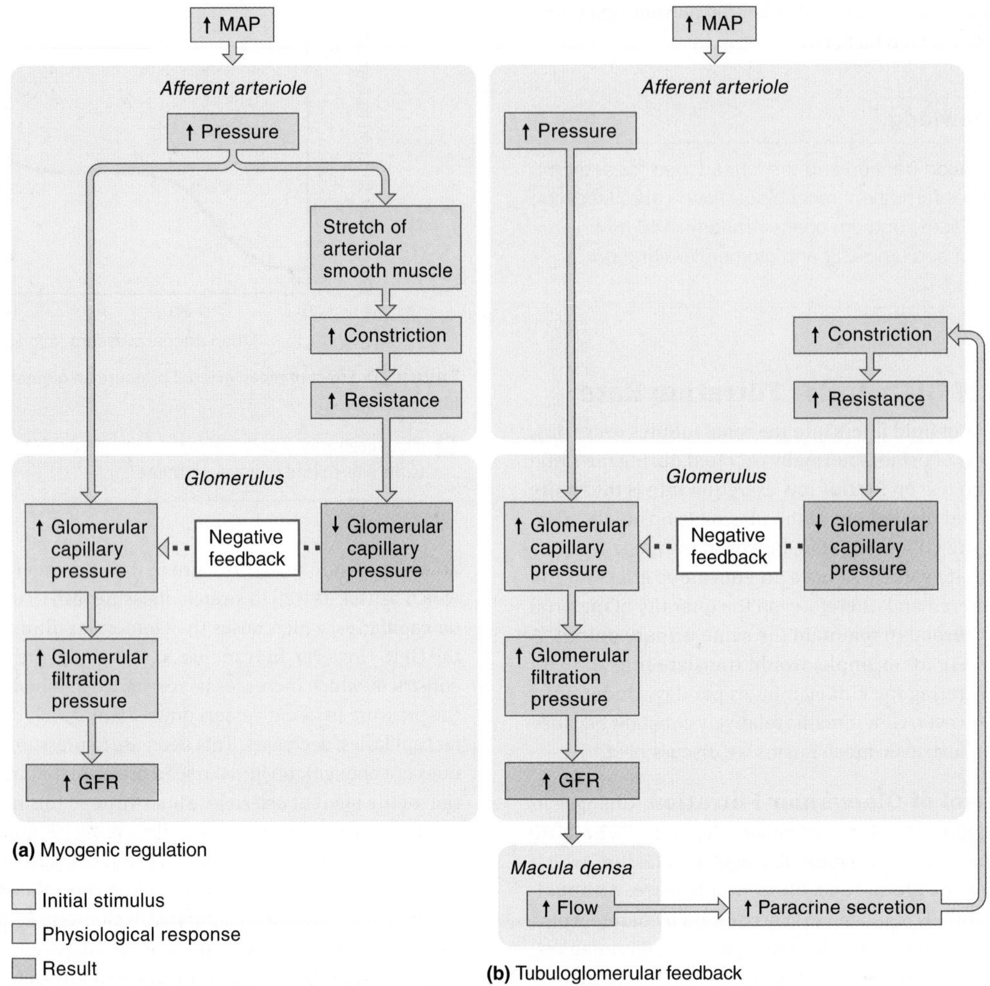

Figure 11 Intrinsic controls of glomerular filtration rate. (a) In response to a change in the mean arterial pressure, myogenic regulation of afferent arteriolar resistance prevents significant changes in GFR by affecting glomerular capillary pressure. **(b)** Tubuloglomerular feedback, triggered by chemical signals released from cells in the macula densa in response to increased GFR, prevents significant changes to GFR when mean arterial pressure changes by affecting glomerular capillary pressure.

precapillary sphincters. A contraction of the mesangial cells will decrease blood flow through some glomerular capillaries, thereby decreasing the surface area over which filtration can occur. The contractile state of mesangial cells depends on the pressure within the capillaries. An increase in blood pressure, which increases GFR, stretches mesangial cells. In response to the stretch, the mesangial cells contract, decreasing the surface area of capillaries available for filtration, which in turn decreases the GFR until it reaches the normal rate.

Extrinsic Control of Glomerular Filtration and Renal Blood Flow We just learned that glomerular filtration is regulated by intrinsic control mechanisms when mean arterial pressure varies between 80 and 180 mm Hg. However, when mean arterial pressure goes above or below this range, the GFR rises or falls, respectively, because intrinsic mechanisms are no longer able to prevent the glomerular capillary pressure from changing. **Figure 12** shows what happens when the mean arterial pressure falls due to hemorrhage or excessive sweating. When MAP falls to less than 80 mm Hg, GFR decreases directly due to the lowered filtration pressure. In addition, the fall in MAP triggers an increase in sympathetic nervous activity via baroreceptor reflexes. In response to increased input from sympathetic nerves, smooth muscle in the afferent and efferent arterioles contracts, and both afferent and efferent arterioles constrict, which increases the overall resistance of the renal vasculature and decreases the GFR. The increase in renal vascular resistance acts to decrease renal blood flow and also raises the total peripheral resistance, which in turn increases MAP. The decrease in GFR also decreases urine output, which helps the body conserve water. This process minimizes reductions in blood volume, which in turn counteracts further decreases in arterial pressure.

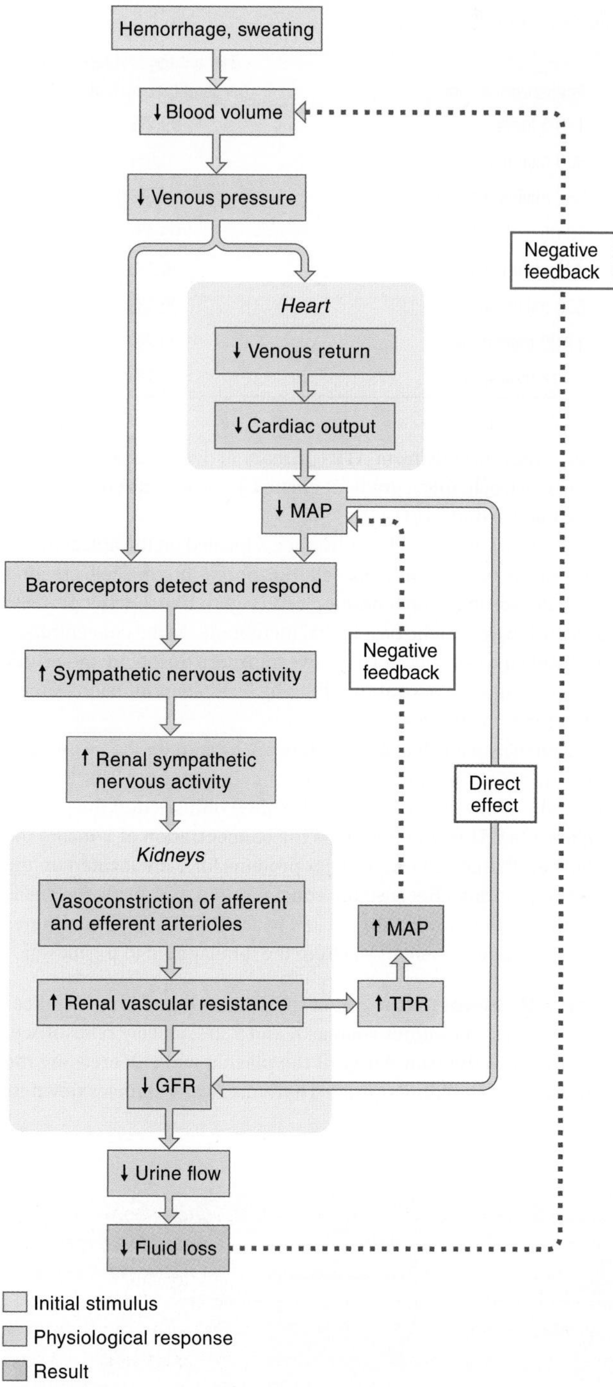

Figure 12 Extrinsic control of GFR and renal vascular resistance during fluid loss due to hemorrhage or sweating.

Q Which type of baroreceptors detects the decrease in venous pressure? Where are these baroreceptors located?

Quick Check 2

1. Indicate whether the GFR will tend to rise or fall in response to an increase in each of the following: the glomerular filtration pressure, the glomerular capillary hydrostatic pressure, the glomerular capillary osmotic pressure, the hydrostatic pressure in Bowman's capsule, the osmotic pressure in Bowman's capsule, the concentration of proteins in the plasma.
2. Name three intrinsic control mechanisms that regulate the GFR, and briefly explain how these mechanisms work to keep the GFR constant in the face of changes in the mean arterial pressure.
3. Is the control of afferent and efferent arteriolar resistance by renal sympathetic neurons an example of intrinsic control or extrinsic control? Why? How does an increase in sympathetic activity tend to affect the GFR?

Reabsorption

Reabsorption refers to movement of filtered solutes and water from the lumen of the tubules back into the plasma. If solutes and water were not reabsorbed, all filtered materials would be excreted, and it would take just eight minutes for a person to lose a liter of fluid in the urine (assuming that the GFR stays constant at its normal rate). Given that 1 liter of fluid is equivalent to 20% of total blood volume, the ability of the kidneys to reabsorb filtered solutes and water is an absolute necessity of life. In fact, 100% of many substances filtered at the glomerulus are reabsorbed by the renal tubules. The reabsorption of other substances is regulated to vary their excretion rate, which in turn regulates the concentration of these substances in the plasma. In this section, we first look at general properties of solute reabsorption; we then describe how water reabsorption across the tubular epithelium is coupled to solute reabsorption.

Solute and Water Reabsorption

Table 1 lists the quantities of several freely filtered substances that are filtered and reabsorbed in a single day under normal circumstances. The sheer mass of material that must be transported from tubular fluid back into the plasma by the renal tubule epithelium every day reveals how extensive the reabsorption process is. Many solutes are reabsorbed actively; that is, they are transported against their electrochemical gradients as they move from the tubular lumen to the plasma.

Reabsorption of most solutes occurs in the proximal and distal convoluted tubules. Recall that the efferent arteriole branches into peritubular capillary beds adjacent to the renal tubules in the renal cortex (see Figure 6b). The interstitial fluid between the epithelium of the renal tubules and the peritubular capillaries *(peritubular fluid)* fills the *peritubular space.*

When a substance is reabsorbed, it must move across two barriers: the tubule epithelium and the capillary endothelium (**Figure 13**). Capillaries are a barrier only to movement of macromolecules, such as proteins, and cells. Thus the epithelial cells lining the renal tubules form the primary barrier to reabsorption of the various solutes. Because tight junctions connect the epithelial

TABLE 1 Normal Rates of Filtration and Reabsorption for Water and Selected Solutes

Substance	Filtration rate	Reabsorption rate	Percentage of filtered load reabsorbed
Water	180 liters/day	178.5 liters/day	99.2%
Glucose	800 millimoles/day	800 millimoles/day	100%
Urea	933 millimoles/day	467 millimoles/day	50%
Na^+	25.20 moles/day	25.05 moles/day	99.4%
K^+	720 millimoles/day	620 millimoles/day	86.1%
Ca^{2+}	540 millimoles/day	530 millimoles/day	98.1%
Cl^-	18.00 moles/day	17.85 moles/day	99.2%
HCO_3^-	4.320 moles/day	4.318 moles/day	$>$99.2%

cells lining the renal tubules, movement of molecules between cells is restricted. The plasma membrane of the epithelial cells facing the tubule lumen is called the apical membrane and has microvilli; the plasma membrane facing the interstitial fluid is called the basolateral membrane. Microvilli on the apical membrane are abundant in the more proximal portions of the tubule but are sparse in the distal portions. Beneath the basolateral membrane is a basement membrane that does not contribute significantly as a barrier and thus will not be discussed further.

Some molecules are passively reabsorbed, whereas others require energy. **Figure 14** depicts the reabsorption of three uncharged solutes (X, Y, and Z) and water. (Note that the movement of charged substances is affected by electrical forces in addition to the forces described next.) Although each substance moves in the same direction, the actual mechanism of reabsorption differs.

Active Reabsorption of Solute In the examples shown in Figure 14a, substances X and Y are both actively transported, but through different mechanisms. Both substances are transported by mechanisms involving active transport across one membrane of the tubule in conjunction with passive movement across the other membrane. Recall that active transport requires energy to move a molecule against its electrochemical gradient. This energy can come either directly from ATP (primary active transport) or from an ion electrochemical gradient created at the expense of ATP (secondary active transport).

Active transporters for substance X located on the apical membrane of the tubule epithelial cell transport X into the cell, creating a high intracellular concentration of X (Figure 14a). Carrier proteins for X are located on the basolateral membrane. As the concentration of X inside the cell rises due to active transport, this solute moves out of the cell into the peritubular fluid by facilitated diffusion; X then diffuses into the plasma.

Active transporters for substance Y located on the basolateral membrane of the tubule epithelial cell transport Y out of the cell and into the peritubular fluid; Y then diffuses into the plasma (Figure 14a). This process keeps the concentration of Y inside the tubule epithelial cell low. Carrier proteins for Y are located on the apical membrane. Because the concentration of Y inside the cell is low, this solute moves into the cell by facilitated diffusion. Therefore, the net movement of Y is from the tubular fluid to the plasma.

Water Reabsorption Water diffusion is based on differences in osmolarity. As solutes such as X and Y are actively reabsorbed, they increase the osmolarity of the plasma while decreasing the osmolarity of the tubular fluid. Therefore, water diffuses down its

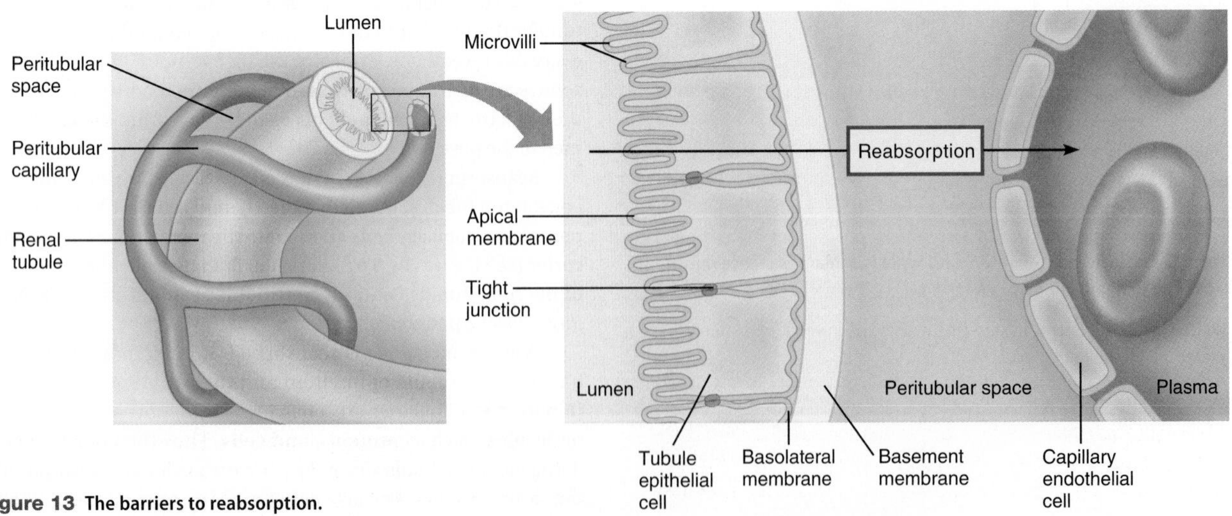

Figure 13 The barriers to reabsorption.

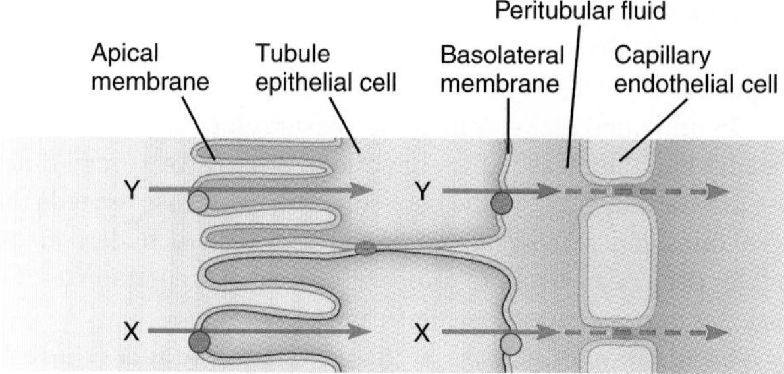

(a) Active solute reabsorption

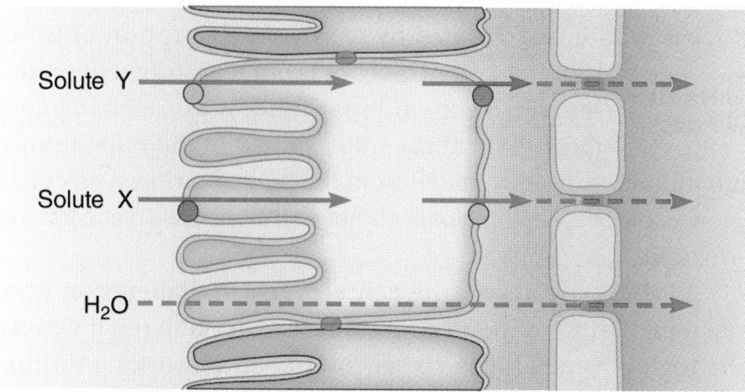

(b) Water reabsorption (passive)

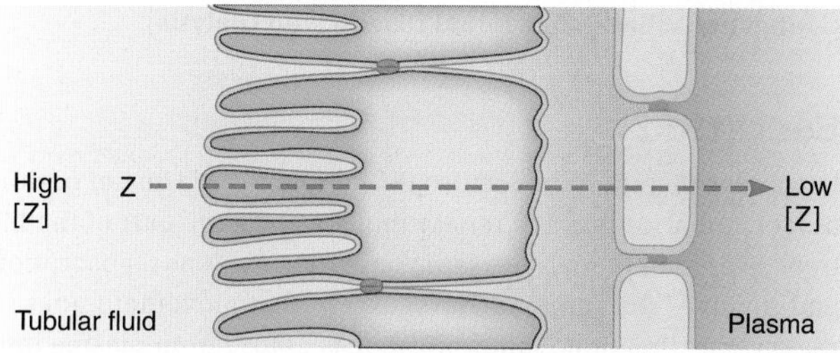

(c) Passive solute reabsorption via diffusion

Figure 14 Mechanisms of solute and water reabsorption. **(a)** Active reabsorption of solutes. **(b)** Passive reabsorption of water. **(c)** Passive reabsorption of a solute.

concentration gradient to a region of greater osmolarity (Figure 14b). Stated another way, water reabsorption follows the active reabsorption of solute. For water to follow solute, water must be able to permeate the tubule epithelium. Although water can permeate most plasma membranes in the body, some plasma membranes in the renal tubules are impermeable to water.

Passive Reabsorption of Solute Substance Z is passively reabsorbed via diffusion (Figure 14c). For this process to occur, two conditions must be satisfied: The concentration of Z must be greater in the tubular fluid than in the plasma, and Z must be able to permeate the plasma membranes of the tubule epithelium and the capillary endothelium. The tubular fluid, which came from plasma, has a higher concentration of Z than fluid in the capillary lumen, because most water that is filtered is reabsorbed as described earlier. As water leaves the lumen of the tubules and enters the plasma, the tubular concentration of Z increases, whereas the plasma concentration of this solute decreases. Therefore, Z diffuses down its concentration gradient from the tubular fluid to the plasma. Stated another way, reabsorption of molecules that can permeate membranes follows water reabsorption.

Transport Maximum

When solutes are transported from filtrate to plasma across the tubular epithelium by carrier proteins or pumps, those modes of transport can become saturated; that is, when solute concentration is high enough, all carrier proteins and pumps are occupied, and the system is operating at **transport maximum (Tm).**

The transport maximum is best understood for substances that are normally 100% reabsorbed in the renal tubules, such that none of the substance is excreted in the urine. When the plasma concentration of the solute increases, however, more solute filtered is at the glomerulus, causing an increase in the concentration of solute in the filtrate. At a certain plasma concentration of the solute, the corresponding filtrate concentration saturates the carrier proteins in the tubules, and some of the substance appears in the urine. The plasma concentration of solute at which "spillover" into the urine occurs is called the **renal threshold.** As the plasma concentration rises above the renal threshold, the rate at which the solute is excreted in the urine becomes progressively higher. A well-known example of transport maximum occurs in the active reabsorption of glucose.

Glucose, which is freely filtered from the glomerulus, is normally completely reabsorbed by active transport in the proximal tubule, so no glucose is excreted in the urine. **Figure 15** shows the mechanism of glucose reabsorption, which is similar to that for solute X in Figure 14a. Glucose is actively transported across the apical membrane by a sodium-linked active transport; that is, glucose is cotransported with sodium ions from the tubular fluid into the epithelial cell. Such cotransport concentrates glucose inside the epithelial cell. A carrier protein for glucose is located on the basolateral membrane. Because glucose is found in high concentration inside the epithelial cell, this solute is transported by this carrier down its concentration gradient into the peritubular fluid, where it can diffuse into the plasma.

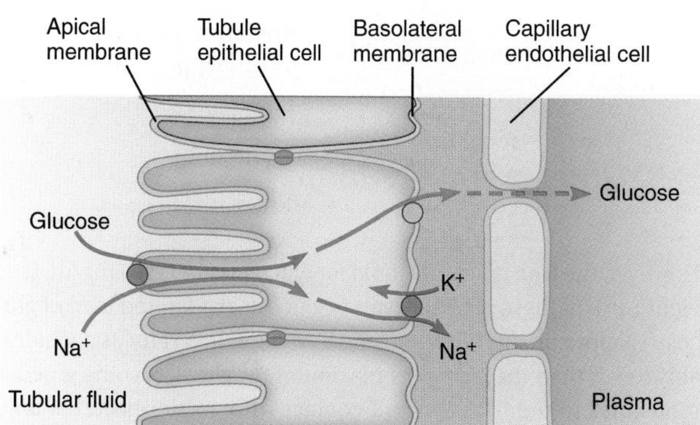

Figure 15 Mechanism of glucose reabsorption.

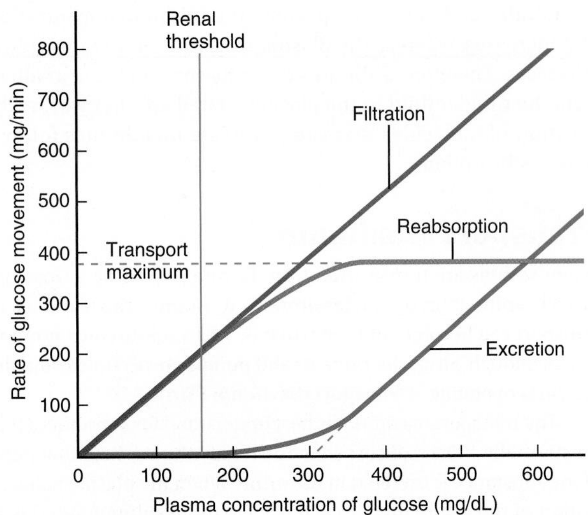

Figure 16 **Glucose filtration, reabsorption, and excretion as a function of plasma glucose concentration.**

Figure 16 shows the relationship between the plasma concentration of glucose and the handling of glucose by the kidneys. The transport maximum for glucose reabsorption is 375 mg/min. The normal plasma glucose level is 80–100 mg/dL. Given that GFR is 125 mL/min (=1.25 dL/min), the filtered load of glucose when its plasma concentration is 100 mg/dL is

$$\begin{aligned} \text{GFR} \times P_{\text{glucose}} &= 1.25\ \text{dL/min} \times 100\ \text{mg/dL} \\ &= 125\ \text{mg/min} \end{aligned}$$

which is well below the transport maximum of 375 mg/min. Therefore, all the glucose is reabsorbed, and none is excreted in the urine. However, if glucose levels increase, eventually the amount of glucose in the filtrate will exceed the capacity for reabsorption, and some glucose will be excreted in the urine.

The "theoretical" renal threshold for glucose—the plasma concentration at which the amount of glucose in the filtrate exceeds the transport maximum, and at which glucose appears in the urine—can be calculated as follows:

$$\begin{aligned} \text{GFR} \times \text{renal threshold} &= \text{transport maximum} \\ \text{renal threshold} &= \frac{\text{transport maximum}}{\text{GFR}} \\ \text{renal threshold} &= \frac{375\ \text{mg/min}}{1.25\ \text{dL/min}} \\ &= 300\ \text{mg/dL} \end{aligned}$$

However, the true renal threshold for glucose is 160–180 mg/dL (see Figure 16). At these plasma levels of glucose, the filtered load of glucose is approximately 225 mg/min. Although this value is considerably lower than the transport maximum for glucose, some glucose molecules in the filtrate avoid contact with carrier proteins and, therefore, are excreted in the urine even though the carrier proteins are not 100% saturated. (Notice the nonlinear relationship between glucose reabsorption and excretion rates and plasma concentration of glucose as the rates approach the maximum in Figure 16.)

In untreated diabetes mellitus, plasma glucose levels are elevated, a condition called hyperglycemia, and may be several times the normal value. When the concentration of glucose exceeds the renal threshold, glucose appears in the urine. Before modern methods for detecting glucose in urine were invented, a common test for diabetes mellitus was to taste the urine for sweetness.

Elimination of glucose in the urine also produces diuresis (increased water loss or urine volume). Recall that water moves across epithelium by osmosis, and generally water movement follows active transport of solute. Normally the actively reabsorbed glucose would be followed by passive reabsorption of water.

However, when glucose is not reabsorbed but rather remains in the tubular fluid, it exerts an osmotic force that causes water to stay in the renal tubules and ultimately be eliminated from the body in urine. As a consequence, early symptoms of diabetes include excessive thirst and urination.

As well as suffering the acute effects of diabetes on urine function, 20–30% of people with diabetes develop renal disease. The high plasma glucose levels of diabetics, working through mechanisms as yet unknown, damage the nephrons, inducing *diabetic nephropathy*. Over time, the disease may eventually progress to chronic renal failure and end-stage renal disease (see **Clinical Connections: End-Stage Renal Disease and Dialysis**).

Secretion

In tubular secretion, molecules move from the plasma of peritubular capillaries into the renal tubules to become part of the filtrate. Secretion follows the same basic processes as reabsorption and involves the same barriers, except that movement goes in the reverse direction. Some substances diffuse from plasma into the filtrate, whereas others are actively transported. Secretion by active transport requires either that proteins in the basolateral membrane actively transport the solute from interstitial fluid to inside the epithelial cell, or that proteins in the apical membrane actively transport the solute from inside the epithelial cell into the filtrate. Among the substances actively secreted by the renal tubules are ions, such as potassium ions and hydrogen ions; waste products, such as choline and creatinine; and foreign substances, such as the antibiotic penicillin. The end result of secretion is an increase in the quantity of solute excreted in the urine, which decreases the solute's plasma concentration.

Quick Check 3

1. Describe the barriers for reabsorption and secretion. Be sure to include all plasma membranes that must be crossed by the substance being reabsorbed or secreted.
2. Describe the general mechanism of water reabsorption.
3. Define the terms *transport maximum* and *renal threshold*.

CLINICAL CONNECTIONS

End-Stage Renal Disease and Dialysis 

Chronic renal failure is the progressive and irreversible loss of kidney structures, a condition that can either result from a variety of kidney diseases or occur secondary to another disease, such as diabetes or hypertension.

When the GFR decreases to less than 5% of normal, the result is *end-stage renal disease (ESRD).* In the United States, more than half a million people are treated for ESRD every year, 40% of whom developed the disease due to complications of diabetes.

ESRD is a fatal disease unless treated by kidney transplant or dialysis. Kidney transplant is the optimal treatment for a patient, but the number of kidneys available for transplant falls far short of the demand, with fewer than 20,000 kidney transplants begin performed each year. Thus patients must turn to dialysis, which provides for temporary relief of kidney failure. In this procedure, a semipermeable membrane is used that allows small solutes to exchange freely between two liquids. Although dialysis is not a cure for ESRD, it can prolong the life of the patient.

The most common form of dialysis is *hemodialysis,* in which a person's blood is pumped through a hemodialysis system, also called an artificial kidney (see the figure). The dialysis machine includes a membrane that separates the machine into two compartments, one containing dialysis fluid and the other containing blood. A catheter from an artery takes blood from the patient to the dialysis machine. As the blood moves through the machine, blood and dialysis fluid are exchanged across the membrane.

The membrane is semipermeable, allowing all molecules except blood cells and proteins to diffuse down their electrochemical gradients. The composition of the dialysate can be varied to favor movement of molecules in a particular direction. For example, if the patient is suffering from edema, then the solute concentration of the dialysate is made high to favor movement of water from blood to dialysate. After exchange, the blood leaves the machine and returns to the patient via a catheter connected to a vein.

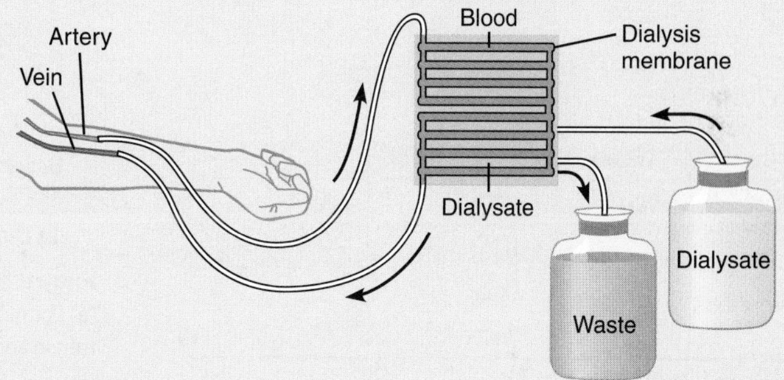

Critical Thinking Questions

1. Describe and distinguish between chronic renal failure and end-stage renal disease.
2. What are the current available treatments for renal failure?
3. Describe the process of hemodialysis.

4 Regional Specialization of the Renal Tubules

Because the properties of the tubule epithelium vary from region to region along the length of the tubule, both the substances transported and the mechanisms of transport differ in different regions of the tubule (Table 2). In the following sections, we look at transport in different regions of the renal tubules.

Nonregulated Reabsorption in the Proximal Tubule

In the proximal tubule, mechanisms of reabsorption are so efficient that 70% of the sodium and water that are filtered is reabsorbed by the time tubular fluid reaches the beginning of the loop of Henle. Solute and water reabsorption are coupled such that the process is iso-osmotic; that is, no change in solute concentration occurs in either the plasma or the filtrate. Some solutes (including glucose) are virtually 100% reabsorbed by this time. For this reason, the proximal tubule is said to function as a *mass absorber* whose function is to reabsorb the bulk of filtered solutes and water, thereby preventing their loss from the body. This reabsorption of solutes in the proximal tubule is an ongoing process that generally is *not* regulated.

The epithelium of the proximal tubule has a number of features that facilitate mass absorption (**Figure 17a**). First, the apical membrane is highly folded into many microvilli, a conformation known as a *brush border;* such folding increases the total surface area of the apical membrane, which facilitates transport. Second, transporting cells possess large numbers of mitochondria, which supply the large quantity of ATP necessary to drive active transport. Finally, the tight junctions between epithelial cells have a relatively high permeability to small solutes and water, making this a "leaky" epithelium. This high permeability at the tight junctions facilitates the diffusion of solutes and water between cells (called *paracellular transport*), further enabling the epithelium to transport large quantities of materials.

TABLE 2 Sites at Which Substances Are Reabsorbed and Secreted Across Renal Tubules

Tubule segment	Substances reabsorbed		Substances secreted
Proximal tubule	Na^+ Cl^- K^+ Ca^{2+} HCO_3^- Water	Glucose Amino acids Vitamins Urea Choline	H^+
Loop of Henle (descending limb)	Water		
Loop of Henle (ascending limb)	Na^+ Cl^- K^+	Mg^{2+} Ca^{2+}	
Distal tubule	Na^+ Ca^{2+} Cl^- Water		K^+ H^+
Collecting duct	Na^+ K^+ Cl^- Ca^{2+}	HCO_3^- H^+ Urea Water	K^+ H^+

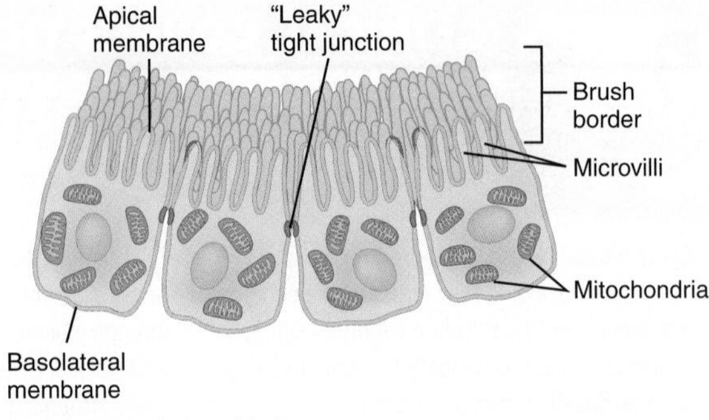

(a) Proximal tubule epithelium

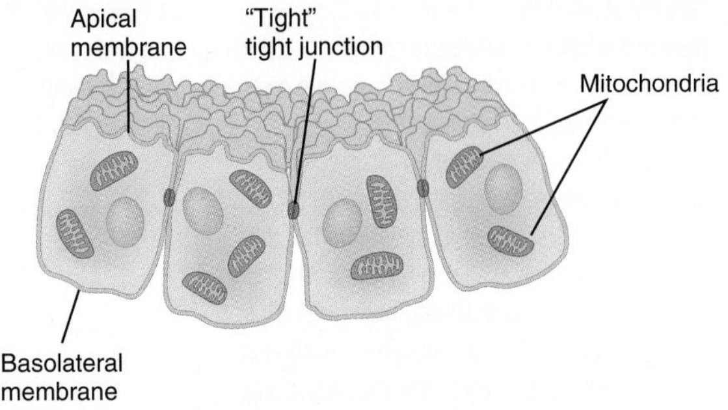

(b) Distal tubule and collecting duct epithelium

Figure 17 Epithelial cells in selected portions of a renal tubule. **(a)** Epithelial cells in the proximal tubule. **(b)** Epithelial cells in the distal tubule and collecting duct.

Q Is paracellular transport (between epithelial cells) more likely in the proximal tubule or in the distal tubule? Why?

Regulated Reabsorption and Secretion in the Distal Tubule and Collecting Duct

In contrast to the proximal tubule, the distal tubule and the collecting duct are specialized to allow *regulation* of reabsorption and secretion. In the epithelium of these tubules, the brush border is much less prominent than in the proximal tubule, or is even lacking altogether (Figure 17b). In addition, epithelial cells have fewer mitochondria, and tight junctions are far less permeable, making this an example of a "tight" epithelium. Furthermore, tubule epithelial cells have receptors for hormones that regulate the transport of water and several solutes, and water reabsorption does not always follow solute reabsorption. In the distal tubule and collecting duct, for example, sodium reabsorption is stimulated by aldosterone, a steroid hormone secreted by the adrenal cortex, and inhibited by atrial natriuretic peptide, a peptide hormone secreted by the atria of the heart. In these tubular segments, water reabsorption is stimulated by antidiuretic hormone (ADH), a peptide hormone secreted by the posterior pituitary.

Water Conservation in the Loop of Henle

The loop of Henle within juxtamedullary nephrons is specialized to create an osmotic gradient in the renal medulla, such that the fluid in the outer portion of the medulla (near the cortex) is at a lower osmolarity than the fluid in the inner portion of the medulla (near the renal pelvis). This gradient is critical to the kidneys' ability to conserve water. Both the loops of Henle within juxtamedullary nephrons and the collecting ducts are important in water reabsorption.

5 Excretion

Excretion by the kidneys entails the elimination of solute and water from the body in the form of urine. The rate at which a substance is excreted in the urine is an important quantity because it has a direct bearing on the volume and composition of the plasma. For any substance, the quantity that is excreted over a period of time is determined by a simple rule: *Material that enters the lumen of the renal tubules is excreted unless it is reabsorbed.* Because a substance can enter renal tubules either by filtration or by secretion, the quantity of a substance that is excreted in the urine over any given time period is given by the following expression:

$$\text{Amount excreted} = \text{amount filtered} + \text{amount secreted} - \text{amount reabsorbed}$$

or

$$E = F + S - R$$

Excretion Rate

The equation for calculating the amount of a substance excreted ($E = F + S - R$) indicates that the rate at which a solute is excreted in the urine (expressed in moles/min or the equivalent) depends entirely on three factors: (1) the filtered load, (2) the rate at which the solute is secreted, and (3) the rate at which solute is reabsorbed. **Figure 18** schematically depicts the renal processing of a hypothetical solute and water. In this figure, a single tubule represents all renal tubules combined, the GFR is 150 mL/min, the plasma concentration of the solute is 0.08 mmole/mL, and the solute is freely filterable. Therefore, the filtered load is calculated as follows:

$$\begin{aligned}\text{Filtered load} &= \text{GFR} \times \text{plasma concentration} \\ &= 150\ \text{mL/min} \times 0.08\ \text{mmole/mL} \\ &= 12\ \text{mmole/min}\end{aligned}$$

Figure 18 also shows that in one minute 3 mmole of solute is secreted and 6 mmole of solute is reabsorbed. Therefore, the excretion rate is calculated as

$$\begin{aligned}\text{Excretion rate} &= \text{filtered load} + \text{secretion rate} \\ &\quad - \text{reabsorption rate} \\ &= 12\ \text{mmole/min} + 3\ \text{mmole/min} \\ &\quad - 6\ \text{mmole/min} \\ &= 9\ \text{mmole/min}\end{aligned}$$

By calculating the filtered load of a specific solute and comparing this value to the amount of the solute excreted per minute, the net effect of renal processing (reabsorption or secretion) of the solute can be determined by two simple rules:

1. If the amount of solute excreted per minute is less than the filtered load, then the solute was reabsorbed in the renal tubules.
2. If the amount of solute excreted per minute is greater than the filtered load, then the solute was secreted in the renal tubules.

Note that only the *net* effect can be determined, because some substances are both reabsorbed and secreted. The amount excreted relative to the filtered load depends on which is greater, reabsorption or secretion. In the example depicted in Figure 18, both secretion and reabsorption of the solute occurred. Because the filtered load of the solute (12 mmole/min) exceeded the excretion rate (9 mmole/min), net reabsorption occurred.

Clearance

The rate at which a solute is excreted can be described in terms of **clearance,** a virtual measure of the volume of plasma from which a solute is completely removed or "cleared" by the kidneys per unit time (usually expressed in liters/hour). Given that we can measure how much of a solute is excreted per unit of time, the clearance depends on the volume of plasma that contained that amount of solute. The equation to calculate clearance is

$$\text{Clearance} = \frac{\text{excretion rate}}{\text{plasma concentration}}$$

To understand the meaning of clearance, let us return to the example in Figure 18. The excretion rate for this solute was

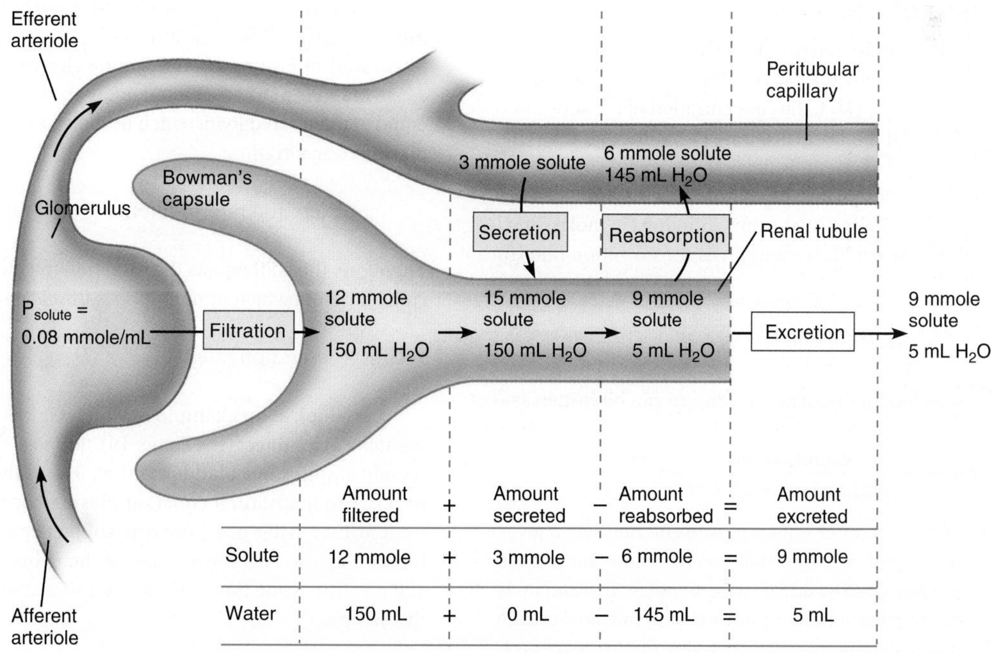

Figure 18 Schematic representation of the four basic renal processes.

9 mmole/min, or 540 mmole/hour. The plasma concentration of the solute was 0.08 mmole/mL, or 80 mmole/liter. Therefore, the clearance of this solute was

$$\text{Clearance} = \frac{540 \text{ mmole/hour}}{80 \text{ mmole/liter}} = 6.75 \text{ liters/hour}$$

This amount of 6.75 liters of plasma is called a "virtual" volume because the kidneys do not actually remove all the solute from 6.75 liters of plasma. (In fact, the body does not even contain this much plasma.) Instead, the kidneys excrete an amount of solute that equals the amount that would be contained in 6.75 liters of plasma. Note also that this amount, 6.75 liters of plasma per hour, is less than the amount of fluid filtered, the GFR (150 mL/min in the example, which equals 9 liters/hour). The clearance is a result of the combined actions of filtration, reabsorption, and secretion.

Clearance is an important concept because it tells us something about how the kidneys are handling one substance compared to another. If the kidneys simply excreted urine that had the same composition as the glomerular filtrate, for example, the clearance would be the same for all solutes despite the fact that the *excretion rate* (moles/min) would be different for each substance. If instead the kidneys excreted urine containing equal sodium and potassium concentrations, the excretion rate would be the same for the two solutes, but the potassium clearance would be greater than the sodium clearance because the plasma potassium concentration is lower than the plasma sodium concentration. The relative clearances of solutes tell us how urinary excretion affects the plasma concentration of one solute *relative to another.* Clearance is also used clinically to estimate GFR and renal blood flow rate, as described next.

Clinical Uses of Clearance

In renal physiology, the clearance of a substance is most often expressed in terms of three *measurable* variables: the concentration of the substance in the urine (U_X), the concentration of the substance in the plasma (P_X), and the urine flow rate (V). Although the symbol V is frequently used in scientific fields to designate volume, the use of V in this context designates the volume of urine produced *per unit time.* The product of the urinary concentration (in moles/liter) and the urine flow rate (in liters/min) gives the excretion rate (in moles/min):

$$\text{Excretion rate} = U_X \times V$$

Therefore, the equation for calculating clearance can be written as

$$\text{Clearance} = \frac{\text{excretion rate}}{\text{plasma concentration}} = \frac{U_X \times V}{P_X}$$

This form of the clearance equation is useful because it gives the clearance in terms of variables that can be readily measured. U_X and V can be determined by taking urine samples. To estimate V, it is necessary only to measure the volume of urine collected during a given period of time. For many substances, the urinary concentration U_X can be determined by standard laboratory tests. Similarly, the plasma concentration P_X can be determined by drawing blood samples and running standard tests.

To see how to calculate the clearance, suppose that a person produces 450 mL of urine in one hour. The urine flow rate is thus

$$\text{Flow rate} = 450 \text{ mL}/60 \text{ min} = 7.5 \text{ mL/min}$$

If the sodium concentration in the urine is determined to be 15 mM, and the plasma concentration is found to be the normal value of 145 mM, then the sodium clearance (C_{Na}) is calculated as follows:

$$C_{Na} = \frac{(15 \text{ mmole/liter})\,(7.5 \text{ mL/min})}{(145 \text{ mmole/liter})} = 0.78 \text{ mL/min}$$

Apply Your Knowledge

The plasma concentration of solute S is 1 mg/liter. A urine sample reveals that 0.5 liter of urine was excreted in 1 hour and had a concentration of S of 0.2 mg/L. What is the clearance of solute S?

Estimates of Glomerular Filtration Rate Clearance measurements also provide a relatively easy way of determining the GFR. If a substance is freely filtered and is neither reabsorbed nor secreted, then the amount of the substance in the urine must be equal to the amount filtered, or the filtered load. Under these conditions, the substance is entirely cleared from the volume of plasma that was filtered; in other words, the clearance of the substance is equal to the GFR. One substance that meets these requirements is *inulin,* a polysaccharide that is not produced in the body but can be injected into the bloodstream for the purpose of measuring the GFR. That inulin clearance is equal to the GFR is illustrated in **Figure 19** and can be shown mathematically. Because inulin is neither secreted nor reabsorbed, its excretion rate is equal to the filtered load, which by definition equals GFR times the plasma concentration:

$$\text{Excretion rate} = \text{filtered load} = \text{GFR} \times P_{\text{inulin}}$$

Therefore, the GFR equals the excretion rate divided by the plasma concentration, which by definition is the clearance:

$$\text{GFR} = \text{excretion rate}/P_{\text{inulin}} = (U_{\text{inulin}})(V)/P_{\text{inulin}} = C_{\text{inulin}}$$

Let's look at an example in which GFR is determined experimentally in a subject (see Figure 19). Suppose that the subject completely empties her bladder at time 0, and then inulin is infused into her to maintain a constant plasma concentration level of 4 mmole/liter. After one hour, the subject completely empties her bladder again, and the volume of the urine is determined to be 120 mL. The urine flow rate can be calculated from the volume of this sample:

$$V = 120 \text{ mL/hour} = 2.0 \text{ mL/min}$$

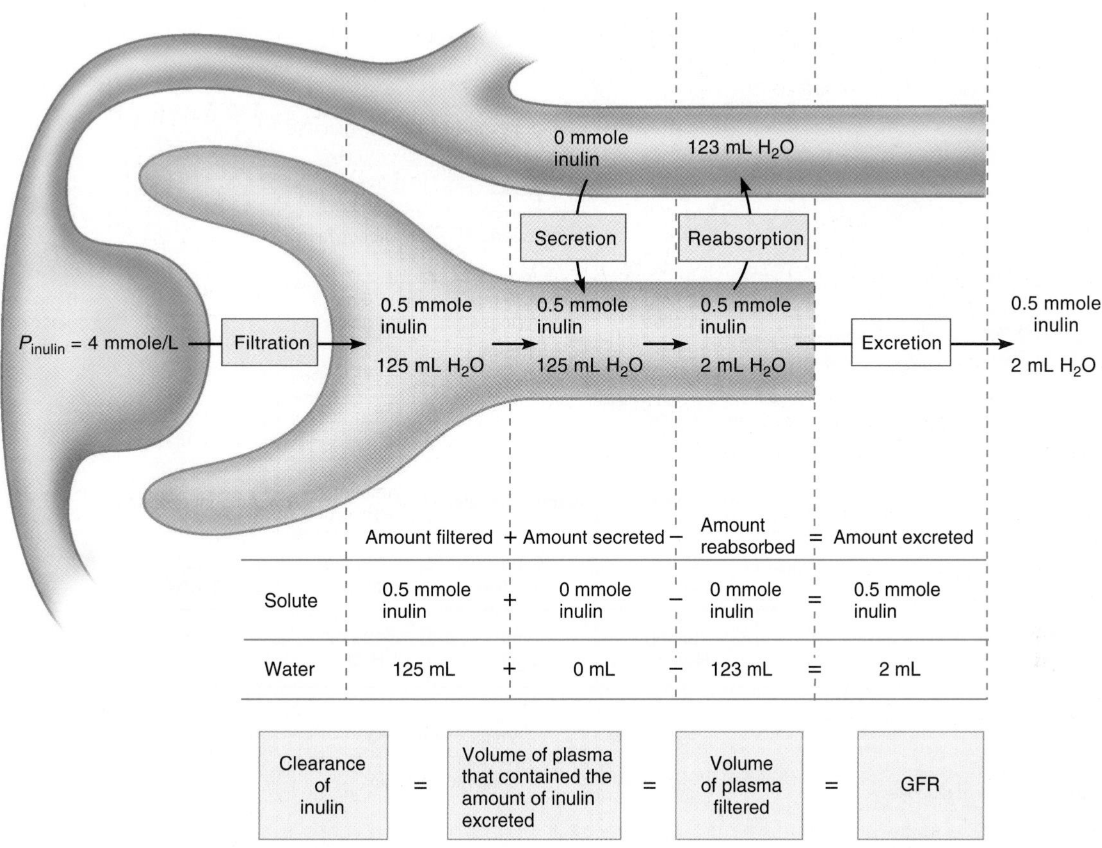

Figure 19 Clearance of inulin.

The concentration of inulin in the urine sample is measured to be 250 mmole/liter. Therefore, the clearance of inulin is

$$C_{\text{inulin}} = \frac{(U_{\text{inulin}})(V)}{(P_{\text{inulin}})} = \frac{(250\ \text{mmole/liter})(2.0\ \text{mL/min})}{(4\ \text{mmole/liter})}$$
$$= 125\ \text{mL/min}$$

The clearance of any substance that is freely filtered but neither reabsorbed nor secreted is equal to the GFR. Because the intravenous infusion of inulin is an invasive procedure, it is desirable to determine GFR using molecules that are normally present in the blood, even though none fits the requirements exactly. The chemical in the body that "best" meets the requirements to estimate GFR is creatinine, a waste product of muscle metabolism. Creatinine is freely filtered at the glomerulus and is not reabsorbed, but a small amount of it is secreted. Because the amount of creatinine excreted slightly exceeds the amount of creatinine filtered, the clearance of creatinine is slightly greater than the GFR. Even so, creatinine clearance provides a suitable clinical *estimate* of GFR.

Determining the Fates of Solutes in Renal Tubules

Clearance can be used to determine whether there has been a *net* reabsorption or *net* secretion of a certain substance in the renal tubules, but some substances are both reabsorbed and secreted. In such cases, the clearance depends on which was greater—reabsorption or secretion. The use of clearance in determining whether reabsorption or secretion of a substance occurred can be summarized by two rules:

1. If the clearance of a substance is greater than the GFR, then that substance was secreted in the renal tubules.
2. If the clearance of a substance is less than the GFR, then that substance was reabsorbed in the renal tubules.

We saw earlier that glucose is completely reabsorbed at normal plasma levels. Therefore, we should predict that the clearance of glucose would be zero: Because no glucose is excreted, no plasma has been cleared of glucose. This relationship is shown in **Figure 20a** and can be expressed mathematically as follows:

$$C_{\text{glucose}} = \frac{(0\ \text{mmole/liter})(V)}{(P_{\text{glucose}})} = 0\ \text{mL/min}$$

Another example involves para-aminohippuric acid (PAH), a foreign substance used clinically to measure renal blood flow. PAH is freely filtered at the glomerulus and is not reabsorbed. In addition, any PAH remaining in the plasma of renal capillaries following filtration is secreted into the renal tubules; that is, the plasma that enters the kidneys is completely cleared of PAH. Therefore, the clearance of PAH is equal to the flow of plasma that entered the kidney, or the *renal plasma flow* (Figure 20b).

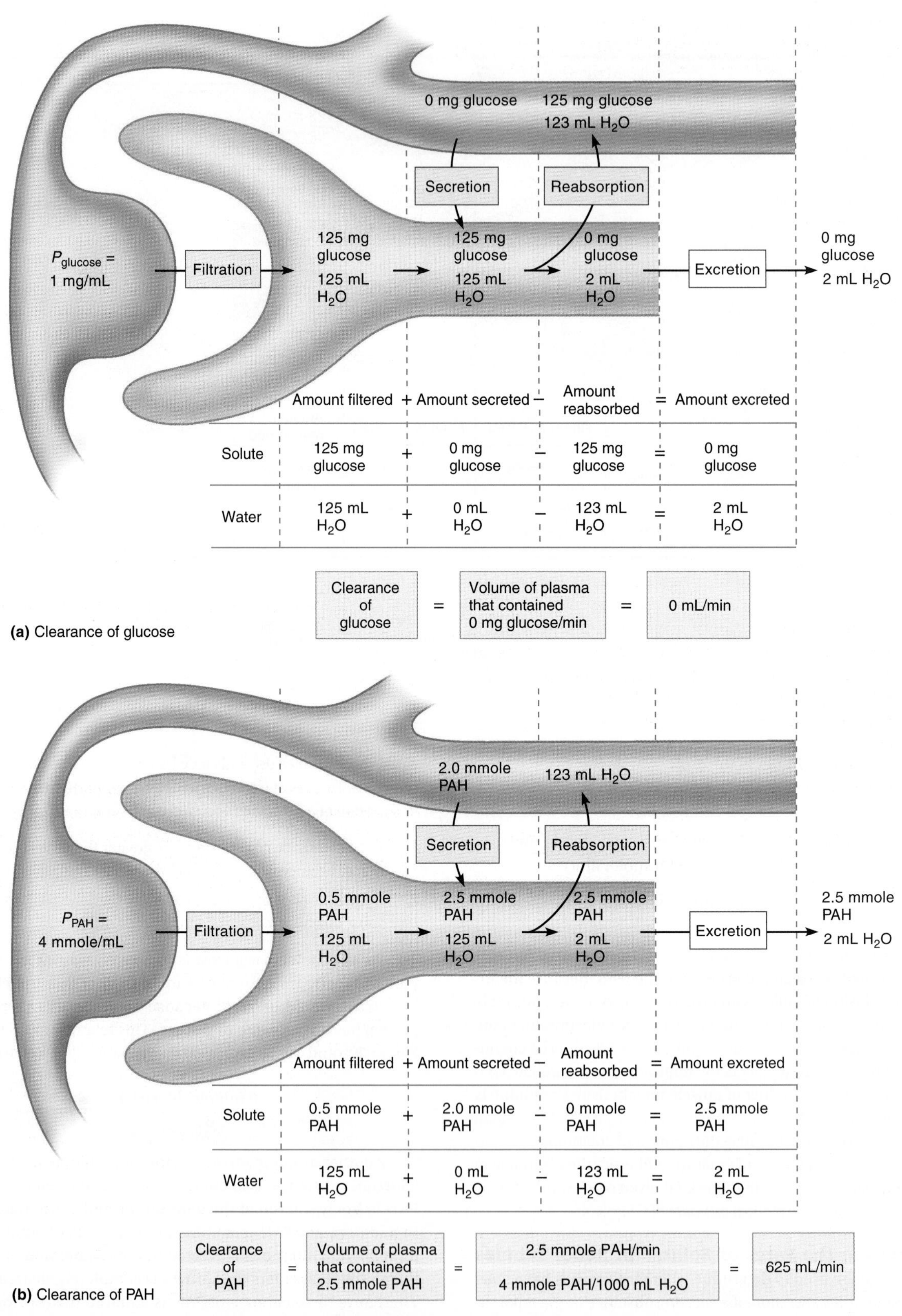

Figure 20 Examples of clearance. (a) Glucose clearance. **(b)** PAH clearance.

This can easily be converted to *renal blood flow* by measuring the hematocrit. For example, if the clearance of PAH is 625 mL/min, then renal plasma flow is 625 mL/min. With a normal hematocrit of 45, then

$$\text{Renal blood flow} = \frac{625 \text{ mL/min}}{1 - 0.45} = 1136 \text{ mL/min}$$

Table 3 lists the clearance rates for some of the common substances processed by the kidneys.

Apply Your Knowledge

Based on the data in Table 3, reabsorption of which solute is greater on a percentage basis, sodium or potassium?

Micturition

The fluid that remains in the renal tubules after filtration, reabsorption, and secretion is excreted as urine. This fluid drains from the collecting ducts into the renal pelvis and then into the ureter (see Figure 1). Wavelike contractions of the smooth muscle in the walls of the ureter propel the urine toward the bladder. The bladder stores the urine until it is eliminated in the process called **micturition,** or urination.

The anatomy of the bladder is shown in **Figure 21**. The wall of the bladder contains smooth muscle fibers that are connected by gap junctions and thus function as a unit. The smooth muscle fibers of the bladder are known collectively as the **detrusor muscle.** In the neck of the bladder, muscle fibers converge and overlap, forming a thickening of the wall referred to as the **internal urethral sphincter** that acts as a valve to regulate the flow of urine from the bladder. (A *sphincter* is any muscular structure that forms a ring around a hollow organ or tube and regulates the flow of material by contracting or relaxing.) The flow of urine from the bladder is also controlled by the contraction and relaxation of skeletal muscle that surrounds the urethra as it passes through the pelvic floor; this muscle is referred to as the **external urethral sphincter.** Sometimes when these sphincters do not operate properly, a person can excrete urine at inappropriate times due to urinary incontinence (see **Clinical Connections: Urinary Incontinence**).

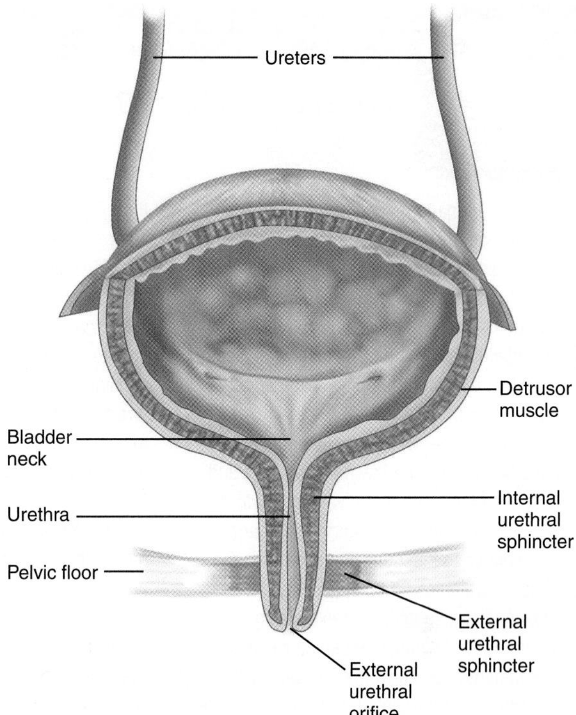

Figure 21 Anatomy of the urinary bladder and urethra. Structures are shown as they occur in females; in males, the urethra extends into the penis.

Q Which urethral sphincter is under voluntary control?

TABLE 3 Clearance of Some Common Substances Processed by the Kidneys

Substance	Clearance rate (mL/min)	Net renal processing (reabsorption or secretion)*
PAH	650	Secretion
Creatinine	140	Secretion
Inulin	125	None
Potassium	12.0	Reabsorption
Chloride	1.3	Reabsorption
Sodium	0.9	Reabsorption
Glucose	0	Reabsorption

*GFR = 125 mL/min. If clearance is greater than GFR, net secretion has occurred; if clearance is less than GFR, net reabsorption has occurred.

Micturition is under both voluntary and involuntary control. It occurs when the detrusor muscles contract, the smooth muscle of the internal urethral sphincter relaxes so that this sphincter opens, and the skeletal muscle of the external urethral sphincter relaxes so that this sphincter opens. The smooth muscle of the detrusor muscle and the internal urethral sphincter are under autonomic control, whereas the skeletal muscle is under voluntary control. Let's look at the activity of these muscles while the bladder fills with urine, and then during micturition.

Normally, the urine formed by the kidneys is delivered to the bladder on a continual basis. The internal and external sphincter muscles are contracted, so the sphincters remain closed; thus urine does not leave the bladder, and the volume of urine in the bladder increases. The detrusor muscles are relaxed and the bladder expands to accommodate the increasing volume. However, as expansion continues, stretch receptors in the bladder wall become activated. Eventually, activation of these receptors is sufficient to induce what is known as the *micturition reflex.*

Micturition is regulated by a spinal reflex that can be overridden by voluntary control in a trained individual (**Figure 22**).

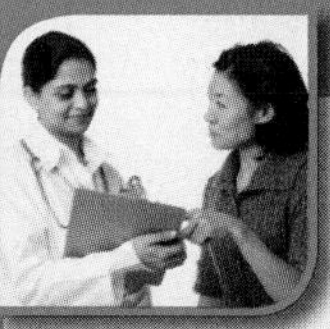

CLINICAL CONNECTIONS

Urinary Incontinence 

In infants, urination is an involuntary reflex in response to stretch of the bladder wall. Most children, by the age of 2 to 3 years old, learn to control this reflex until they reach a restroom. However, more than 25 million adults lose bladder control in this situation, resulting in unintentional loss of urine, called *urinary incontinence*.

Several types of urinary incontinence have been identified that vary in cause and severity. Most common in women is *stress incontinence,* the involuntary loss of urine in response to mechanical pressure on the bladder, which could be caused by sneezing, coughing, laughing, or exercising. Weak urethral sphincters can cause stress incontinence. Some women develop stress incontinence after giving birth to a child due to weakening pelvic muscles or shifts in the bladder position. Weakened pelvic muscles allow the neck of the bladder and the urethra to drop. Other women may develop stress incontinence after menopause, as the decrease in estrogen that occurs during this life stage is thought to weaken pelvic muscles. Men with enlarged prostate glands may develop stress incontinence due to pressure on the bladder.

Urge incontinence, also called *spastic bladder* or *overactive bladder,* is the frequent sudden urge to urinate. More common in the elderly population, urge incontinence is caused by factors such as urinary tract infections, damage to nerves innervating the bladder, diseases of the prostate in males, and hysterectomies in females.

Overflow incontinence occurs when a person has difficulty completely emptying the bladder, causing the bladder to always be partially full. Thus it takes less time to fill the bladder, causing frequent urges to urinate. Causes of overflow incontinence include weakened detrusor muscles, diabetic neuropathy, or blockage of the urethra.

Functional incontinence occurs secondary to other diseases that affect mobility or cognition. These diseases do not alter bladder function, but rather increase the awareness of the need to urinate and/or increase the time it takes to reach a restroom. For example, a decrease in mobility due to arthritis can cause a person to lose some urine prior to reaching the restroom.

Treatments of urinary incontinence vary from changes in diet or behavior, to medications, to surgery. Behavioral changes include decreasing fluid intake before bedtime and doing exercises that strengthen pelvic muscles. Medications prescribed for this condition include muscarinic cholinergic antagonists, which decrease detrusor muscle activity, and alpha adrenergic agonists, which stimulate contraction of the internal urethral sphincter. Surgery can resolve some forms of incontinence, but the type of surgery varies with the cause of incontinence. For example, blockage of the urethra may require surgical removal of the block (whether an internal or external structure is putting pressure on the urethra). An enlarged prostate gland can be surgically removed to decrease pressure on the bladder. In stress incontinence caused by weakened pelvic muscles, a sling can be used to raise the bladder.

Critical Thinking Questions

1. Describe the different types of urinary incontinence.
2. What treatments are available for urinary incontinence, and how do they differ?
3. What would you say are the relative advantages and disadvantages of the current available treatments?

In infants, the pathway is purely reflexive. As the bladder fills, expansion of the wall is detected by stretch receptors and this information is transmitted to the spinal cord. In the spinal cord, signals are relayed via interneurons to three sets of neurons that project back to the bladder and associated structures: (1) parasympathetic neurons, which innervate the detrusor muscle; (2) sympathetic neurons, which innervate the internal urethral sphincter; and (3) somatic motor neurons, which innervate muscles of the external urethral sphincter. Stimulation of stretch receptors excites parasympathetic neurons that travel in the pelvic nerve and stimulate the detrusor muscle to contract. Contraction of the detrusor muscle increases the pressure on the contents of the bladder, and when the contractions of the detrusor muscle become strong enough they cause the internal urethral sphincter to open. At the same time, stretch receptor activity inhibits sympathetic neurons projecting to the internal urethral sphincter and also inhibits somatic motor neurons projecting to the external urethral sphincter, allowing the sphincters to relax and open. The opening of both sphincters, coupled with the contraction of the detrusor muscle, allows the bladder to empty.

In older children and adults, the micturition reflex can be overridden by voluntary control. Signals from the stretch receptors that detect filling of the bladder are transmitted in ascending pathways to the cerebral cortex, giving rise to the conscious sensation of fullness of the bladder and serving as a signal to activate descending pathways that *inhibit* the parasympathetic neurons controlling the detrusor muscle, and to activate descending pathways that *excite* the motor neurons controlling the external urethral sphincter. This process allows postponement of the micturition reflex if needed. Such postponement cannot be continued indefinitely, however, because continued filling of the bladder leads to greater excitation of stretch receptors. Eventually the level of activity in these neurons becomes high enough to trigger the micturition reflex despite conscious efforts to the contrary, triggering uncontrollable urination.

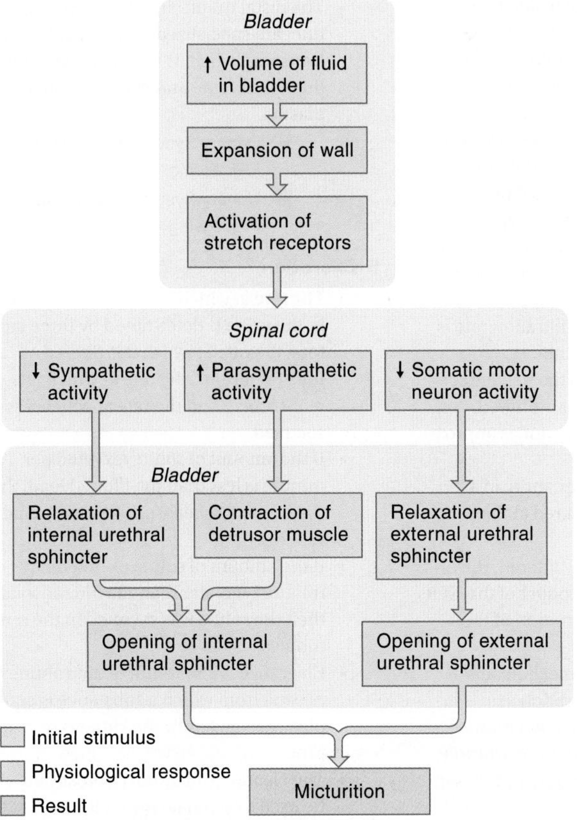

Figure 22 Elements of the micturition reflex.

In addition to exerting voluntary control to prevent micturition, a person can initiate micturition through voluntary relaxation of the external urethral sphincter and the lowering of the pelvic floor. This lowering both causes the bladder to drop, which pulls open the internal urethral sphincter, and stretches the bladder wall, which activates the stretch receptors and induces the micturition reflex. Contraction of the diaphragm and abdominal muscles can increase the volume of urine voided by increasing the pressure in the abdominal cavity, which increases the pressure on the bladder for micturition.

Quick Check 4

1. Name three structural differences between the epithelial cells of the proximal tubule and the distal tubule. The structure of which tubule is more favorable for exchange?
2. The clearance of which two substances can be used to estimate the GFR? The clearance of which substance can be used to estimate renal plasma flow?
3. If the clearance of molecule X is greater than the GFR, was X reabsorbed or secreted in the renal tubules?
4. Describe the innervation of the bladder, internal urethral sphincter, and external urethral sphincter. Voluntary inhibition of micturition is mediated through innervation to which of these structures?

SYSTEMS INTEGRATION

In this chapter, we learned how the kidneys form urine through the processes of filtration, reabsorption, and secretion. These processes are regulated to maintain normal fluid and electrolyte balance.

The importance of the kidneys in homeostasis becomes more apparent when we study the gastrointestinal system, which provides the major input of water and solutes to the body. The absorption of water and solutes across the gastrointestinal system is not regulated based on the body's needs, but rather is regulated to maximize absorption. This task puts a daunting demand on the kidneys, because at least the water and electrolytes that are consumed in excess must be eliminated from the body; most of that elimination occurs through the kidneys (some is lost through the skin and some by secretion in sweat). An accessory organ of the gastrointestinal system is the liver; the liver, like the kidney, eliminates some waste products. The liver and kidney also contribute to the maintenance of normal blood glucose levels through gluconeogenesis.

The kidneys play a role in acid-base balance. The kidneys cannot maintain normal plasma pH by themselves; instead, they rely on the respiratory system for assistance with this task. Thus the kidneys are dynamic organs that carry out a variety of functions, almost all of which require considerable interactions with the other organ systems.

CHAPTER REVIEW

SUMMARY

1 Functions of the Urinary System

- The primary function of the kidneys is to filter the blood to regulate the ionic composition, osmolarity, volume, and pH of plasma.
- The kidneys also remove metabolic waste products and foreign substances from the plasma and form urine.
  iP Urinary System, Anatomy Review

2 Anatomy of the Urinary System

- The urinary system consists of the kidneys (two), ureters, (two), urinary bladder, and urethra.

- The functional units of the kidneys are the nephrons, each of which consists of a Bowman's capsule, proximal tubule, descending loop of Henle, ascending loop of Henle, and distal tubule.
- The distal tubule drains into a collecting duct.
- Filtration occurs at the renal corpuscle, which includes Bowman's capsule and the glomerulus.
- The glomerular filtrate resembles plasma in composition except that it lacks proteins.
- As the filtrate moves through the nephron, its volume and composition change as a result of the reabsorption and secretion of water and solutes.
- Reabsorbed materials move from the tubular fluid in the lumen of the tubule to the peritubular fluid that surrounds the tubule, and then into the plasma of peritubular capillaries that surround the tubule.
- Secretion moves material from the plasma into the filtrate.
- The kidneys receive a large proportion of the cardiac output via the renal artery.
- An afferent arteriole leads into each glomerulus, and an efferent arteriole leaves the glomerulus.
- The efferent arteriole branches into either peritubular capillaries or vasa recta, which drain into veins that eventually lead to the renal vein.
- The juxtaglomerular apparatus consists of the macula densa cells in the distal tubule and granular cells in the walls of the afferent and efferent arterioles.
- The juxtaglomerular apparatus is important in the regulation of glomerular filtration and in salt and water reabsorption.

  MAP iP Urinary System, Glomerular Filtration

  iP Urinary System, Anatomy Review

3 Basic Renal Exchange Processes

- Glomerular filtration is driven by the four Starling forces that contribute to the glomerular filtration pressure: the glomerular capillary hydrostatic pressure, the hydrostatic pressure inside Bowman's capsule, the osmotic pressure of plasma in glomerular capillaries, and the osmotic pressure of fluid in Bowman's capsule.
- The glomerular filtration pressure and the presence of fenestrations in the glomerular capillaries and slit pores in the epithelium of Bowman's capsule favor the bulk flow of protein-free fluid between blood and the lumen of Bowman's capsule.
- The normal glomerular filtration rate is approximately 125 mL/min.
- The filtration fraction is the percentage of renal plasma flow that is filtered.
- The average filtration fraction is approximately 20%.
- The filtered load is the quantity of a certain solute that is filtered at the glomerulus.
- For a solute that is freely filtered, the filtered load equals the product of the GFR and the plasma concentration of the solute.
- Three intrinsic control mechanisms keep the glomerular filtration at a constant rate: myogenic regulation of smooth muscle in the afferent arteriole, tubuloglomerular feedback, and mesangial cell contraction.
- Extrinsic control of GFR includes sympathetic nervous control of smooth muscle in the afferent and efferent arterioles.
- The transport maximum occurs when the concentration of solute is great enough to saturate the carrier proteins during solute transportation.
- The plasma concentration at which the solute appears in the urine is called the renal threshold.

  MAP iP Urinary System, Glomerular Filtration

4 Regional Specialization of the Renal Tubules

- The proximal tubule, which is specialized to reabsorb large quantities of solutes and water, returns most filtered material to the bloodstream.
- The distal tubule and the collecting duct are specialized for the regulation of transport, which is important in controlling the volume and composition of the plasma.

  iP Urinary System, Early Filtrate Processing

  iP Urinary System, Late Filtrate Processing

5 Excretion

- The rate at which a substance is excreted in the urine is determined by three factors: the rate at which it is filtered at the glomerulus, the rate at which it is reabsorbed, and the rate at which it is secreted.
- If the amount of solute excreted per minute is less than the filtered load, then the solute was reabsorbed in the renal tubules.
- If the amount of solute excreted per minute is greater than the filtered load, then the solute was secreted in the renal tubules.
- Clearance is a measure of the volume of plasma from which a substance is completely removed by the kidneys per unit time.
- The clearance of inulin or creatinine can be used to estimate the GFR.
- The clearance of PAH can be used to estimate renal plasma flow and, therefore, renal blood flow.
- The fluid that remains in the renal tubules after filtration, reabsorption, and secretion is excreted as urine.
- Urine drains from the collecting ducts into the renal pelvis and then into the ureter.
- Wavelike contractions of smooth muscle in the wall of the ureter propel the urine toward the bladder.
- The bladder stores the urine until it is excreted during micturition.
- Micturition is under both reflex and voluntary control; the micturition reflex is triggered by stretching of the bladder wall.

Check your understanding at MasteringA&P and apply your new knowledge to real-life case studies to ensure that you've mastered the content!

EXERCISES

Multiple-Choice Questions

1. Which structure of the urinary system stores urine until it is excreted?
   a) Kidneys
   b) Bladder
   c) Ureter
   d) Urethra
   e) Gallbladder
2. Which of the following is *not* one of the mechanisms by which a solute can be exchanged between the plasma and the renal tubules?
   a) Glomerular filtration
   b) Secretion
   c) Excretion
   d) Reabsorption
3. Which type of specialized junction connects epithelial cells lining the renal tubules?
   a) Gap junctions
   b) Tight junctions
   c) Desmosomes
   d) Intercalated disks
   e) Slit pores
4. Which of the following does *not* favor a large glomerular filtration rate?
   a) Slit pores
   b) Fenestrations
   c) High glomerular hydrostatic pressure
   d) High resistance in the afferent arteriole
   e) High resistance in the efferent arteriole
5. Most reabsorption of water and solutes occurs in the
   a) Proximal tubule.
   b) Descending limb of the loop of Henle.
   c) Ascending limb of the loop of Henle.
   d) Distal tubule.
   e) Collecting duct.
6. In which of the following are microvilli most abundant?
   a) Bowman's capsule
   b) Glomerular capillaries
   c) Distal tubule
   d) Proximal tubule
   e) Collecting duct
7. Which of the following would occur if mean arterial pressure increased from 95 mm Hg to 125 mm Hg?
   a) Glomerular filtration rate would increase because of the increased glomerular capillary hydrostatic pressure.
   b) Glomerular filtration rate would decrease because of the increased Bowman's capsule hydrostatic pressure.
   c) Glomerular filtration rate would not change because of autoregulation.
   d) Glomerular filtration rate would not change because of activation of the sympathetic nervous system.
8. The normal fasting plasma glucose concentration is 100 mg/dL, and the renal threshold is 300 mg/dL. If the plasma concentration doubles to 200 mg/dL, then
   a) The rate at which glucose is reabsorbed will double.
   b) The capacity of the renal tubule for transporting glucose will be exceeded.
   c) Urinary water excretion will increase.
   d) Glucose clearance will increase.
   e) The filtered load of glucose is halved.
9. A substance S is freely filterable and is excreted at a rate (in moles/min) that is lower than the filtered load. On the basis of this information alone, which of the following is the *most precise* conclusion that can justifiably be drawn regarding the kidneys' processing of S?
   a) S is neither reabsorbed nor secreted.
   b) S is definitely reabsorbed and may be secreted.
   c) S is definitely secreted and may be reabsorbed.
   d) S is definitely both reabsorbed and secreted.
10. Which of the following observations would enable you to definitely conclude that a substance X is being secreted?
    a) The clearance of X is greater than the GFR.
    b) The concentration of X in the urine is greater than its concentration in the plasma.
    c) The concentration of X in the plasma is decreasing over time.
    d) Either a or c
    e) Any of the above
11. Micturition occurs in response to
    a) Relaxation of the detrusor muscle.
    b) Contraction of the internal and external urethral sphincters.
    c) Activation of parasympathetic neurons to the bladder.
    d) Activation of somatic motor neurons to the bladder.

Objective Questions

12. The (ureter/urethra) carries urine from the bladder to the outside of the body.
13. Urinary excretion is the elimination of urine from the bladder. (true/false)
14. The (afferent/efferent) arteriole carries blood toward the glomerulus.
15. The combination of a glomerulus and the surrounding Bowman's capsule is called a(n) ________ ________.
16. The hydrostatic pressure in glomerular capillaries is (higher/lower) than the corresponding pressure in most capillaries of the body.
17. The glomerular filtration rate tends to (increase/decrease) as the concentration of proteins in the plasma increases.
18. Autonomic neurons regulate contraction of the (internal/external) urethral sphincter.
19. The glomerular filtration pressure is synonymous with the hydrostatic pressure inside glomerular capillaries. (true/false)
20. The filtered load of a solute is determined by its plasma concentration and the (glomerular filtration rate/urine flow rate).
21. If the clearance of a substance is greater than the glomerular filtration rate, then that substance must have undergone (reabsorption/secretion) in the renal tubules.
22. The clearance of (PAH/creatinine) is approximately equal to the renal plasma flow rate.
23. Substances that are reabsorbed move into the (peritubular capillaries/tubule lumen).
24. An increase in the flow rate through the macula densa causes a(n) (increase/decrease) in the glomerular filtration rate.
25. Glucose reabsorption occurs primarily in the (proximal tubule/distal tubule).

Essay Questions

26. Describe the pathway of filtrate flow from the renal corpuscle to elimination from the body.
27. Explain why the glomerular filtration rate does not change with a moderate decrease in mean arterial pressure.
28. Compare the mechanisms of active and passive reabsorption of solute.
29. Explain the concept of transport maximum. Why does glucose appear in the urine of patients?
30. Explain why the proximal tubules are considered mass absorbers.
31. In your own words, describe what the term *clearance* means.

Critical Thinking

32. Suppose that sodium is being excreted in the urine at a rate of 0.5 mmole/min, and that sodium is present in the plasma at a concentration of 150 mM. If the GFR is 125 mL/min, what is the filtered load of sodium? What is the sodium clearance? Is sodium being reabsorbed or secreted by the renal tubules?

33. In many types of kidney disease, proteins can leak across the glomerular wall, increasing the colloid osmotic pressure of Bowman's capsule and decreasing the colloid osmotic pressure of the plasma. Calculate the glomerular filtration pressure given the following information from a patient with acute renal failure: glomerular hydrostatic pressure = 60 mm Hg, Bowman's capsule hydrostatic pressure = 20 mm Hg, glomerular osmotic pressure = 23 mm Hg, and Bowman's capsule osmotic pressure = 3 mm Hg. Is filtration across the kidneys greater or less than normal? Does urine flow rate increase or decrease?

34. Compare the responses of the smooth muscle in the afferent arterioles and the detrusor muscles to stretch.

Photo credits for icons in feature boxes: Focus on Diabetes: Tanyae/Dreamstime; Clinical Connections: Iofoto/Dreamstime; Understanding Exercise: Pixattitude/Dreamstime; Discovery: Yuri Arcurs/Cutcaster.

Glossary

Bowman's capsule in each nephron, a cup-shaped structure that surrounds the glomerulus and conducts filtrate into the renal tubule at the inflow end of the renal tubules; site where filtrate enters the renal tubules

clearance virtual measure of the volume of plasma from which a substance is completely removed or "cleared" by the kidneys per unit time

collecting ducts ducts that collect fluid from several different renal tubules and carry it to the renal pelvis for eventual elimination

cortex outer portion of an organ such as the cerebral cortex, adrenal cortex, and renal cortex

detrusor muscle smooth muscle fibers of the bladder

distal convoluted tubule the portion of a nephron farthest from the renal corpuscle

excretion elimination from the body through the kidneys (urine) or gastrointestinal (GI) tract (feces)

external urethral sphincter ring of skeletal muscle that regulates the excretion of urine

filtered load the quantity of a certain solute that is filtered at the glomerulus per unit time; equals the product of the glomerular filtration rate (GFR) and the plasma concentration of the solute

glomerular filtration the bulk flow of protein-free plasma from the glomerular capillaries into Bowman's capsule

glomerular filtration pressure sum of the Starling forces acting to move fluid across the capillary walls of the glomerulus

glomerular filtration rate (GFR) the volume of the plasma filtered per unit time from all renal glomeruli combined

glomerulus (gloh-mer-you-lus) in each nephron, a ball-like cluster of capillaries in the renal corpuscle; site of filtration; plural, *glomeruli*

granular cells specialized cells in the wall of the afferent and efferent arterioles that secrete renin; also called *juxtaglomerular cells*

internal urethral sphincter ring of smooth muscle that regulates the excretion of urine

juxtaglomerular apparatus (jux-tah-gloh-mer-you-lar) a collection of specialized cells in the distal tubules and the afferent and efferent arterioles near where the three structures come together in the kidney; regulates glomerular filtration and renin secretion

juxtaglomerular cells specialized cells in the wall of the afferent and efferent arterioles that secrete renin; also called *granular cells*

loop of Henle (hen-lee) the portion of the renal tubule that forms a hairpin loop that dips into the medulla, including the descending limb, the thin ascending limb, and the thick ascending limb

macula densa specialized cluster of epithelial cells found in the distal convoluted tubule in the region adjacent to the afferent and efferent arterioles

medulla an internal compartment of certain organs such as the adrenal medulla and renal medulla

micturition (mik-chur-rish-un) the elimination of urine from the body, or urination

myogenic regulation means of autoregulating the glomerular filtration rate that relies on the inherent ability of smooth muscle of the afferent arterioles to contract when stretched

nephrons functional units of the kidneys that filter the blood and form urine; each consists of a renal corpuscle (a glomerulus and Bowman's capsule) and a renal tubule

peritubular capillaries capillary bed that branches off the efferent arterioles of cortical nephrons and is located close to the renal tubules; functions in exchange with renal tubules during reabsorption and secretion

proximal tubule portion of the renal tubule nearest the renal corpuscle; includes the proximal convoluted tubule and the proximal straight tubule

reabsorption transport of a substance into blood

renal arteries arteries that branch off the aorta and provide the kidneys with their blood supply

renal corpuscle site of glomerular filtration; consists of glomerulus and Bowman's capsule

renal pelvis funnel-shaped passage forming the initial portion of the ureter

renal threshold the plasma concentration of solute at which the transport maximum is exceeded and excess solute appears in the urine

renal tubule a portion of a nephron, consisting of a long, coiled tube

renal veins blood vessels that transport blood from the kidneys back into the general circulation

transport maximum (Tm) rate of transport by carrier proteins when carriers are 100% saturated

tubuloglomerular feedback autoregulatory mechanism in which a change in the glomerular filtration rate is regulated by paracrines secreted from the macula densa, located downstream from the glomerulus

urinary system organ system that consists of two kidneys, two ureters, the urinary bladder, and the urethra

urine a fluid produced by the kidneys and eliminated from the body

vasa recta capillary bed that branches off the efferent arteriole of a juxtamedullary nephron and surrounds the loop of Henle; functions in maintaining medullary osmotic gradient

Answers

Figure Question Answers

Fig 9 Both would increase
Fig 10 The filtered load would increase because GFR increased
Fig 12 Low pressure or volume receptors located in the large systemic veins and right atrium
Fig 17 In the proximal tubule, because it has leakier tight junctions
Fig 21 The external urethral sphincter

Apply Your Knowledge Answers

The filtration fraction equals the glomerular filtration rate (GFR) divided by the renal plasma flow rate. Renal plasma flow rate is not given, but it can be calculated from the renal blood flow and hematocrit as follows:

Plasma flow = blood flow × (1 − hematocrit) = 1.25 liters/min × (1 − 0.4) = 1.25 liters/min × 0.6 = 0.75 liters/min = 750 mL/min

Filtration fraction = GFR/renal plasma flow = 125 mL/min/750 mL/min = 0.17 = 17%

The filtered load for sodium = GFR × plasma concentration of sodium:

Filtered load = 125 mL/min × 140 mmoles/1000 mL = 17.5 mmoles/min

To convert from millimoles to milligrams, 17.5 mmoles sodium × 23 mg/mmoles = 402.5 mg. Thus the filtered load of sodium is 402.5 mg/min.

The clearance of a substance can be calculated using the following equation:

Clearance = excretion rate/plasma concentration

Excretion rate of S = the urine concentration of the S times the urine flow rate = 0.2 mg/liter × 0.5 liter/hour = 0.1 mg/hour

Clearance of S = 0.1 mg/hour / 1 mg/liter = 0.1 liter/hour = 100 mL/hour

Because less blood is cleared of sodium than is cleared of potassium, more sodium is reabsorbed.

Answers to End-of-Chapter Multiple Choice and Objective Questions

Multiple Choice Questions

1. b 2. c 3. b 4. d 5. a 6. d 7. c 8. a 9. b 10. a 11. c

Objective Questions

12. urethra
13. true
14. afferent
15. renal corpuscle
16. higher
17. decrease
18. internal
19. false
20. glomerular filtration rate
21. secretion
22. PAH
23. peritubular capillaries
24. decrease
25. proximal tubule

The Urinary System: Fluid and Electrolyte Balance

Thomas Deerinck, NCMIR/SPL/Photo Researchers, Inc.

Colored scanning electron micrograph (SEM) of a renal tubule.

CHAPTER OUTLINE

We all have different eating and drinking habits. For example, some people need a large glass of water in the morning to quench their thirst; others need a few cups of coffee to get them going. The immediate outcome is the same—excretion of large volumes of water in the urine—but the consequences differ.

When you drink a large volume of water, it isn't long before your body eliminates the excess water consumed to maintain normal plasma volume and osmolarity. Likewise, when you consume salty food, such as potato chips, your body retains fluid and you feel thirsty. These are part of negative feedback systems to maintain normal fluid and electrolyte composition of the plasma. By comparison, the caffeine in coffee is a drug that disrupts your body's homeostasis by causing your body to eliminate water inappropriately, causing dehydration. Ultimately, drinking caffeine makes you thirsty to replace the water lost shortly after consumption.

In this chapter, you will learn about the mechanisms of water and solute movement into and out of the kidney tubules and the hormonal regulation of these movements. You will also learn why caffeine causes an increase in water excretion, even if plasma volume and osmolarity are normal.

MasteringA&P®

Go to MasteringA&P for helpful A&P Flix 3-D animations, chapter quizzes, pre-tests, interactive physiology tutorials, and more!

LEARNING OUTCOMES *After studying this chapter, you should be able to:*

- Explain the concept of balance.
- Describe the different sources of body water input and output.
- Explain the control of water balance and osmolarity by antidiuretic hormone.
- Describe how aldosterone and atrial natriuretic peptide regulate plasma sodium levels. Include an explanation of the stimuli that release these hormones.
- Describe the major mechanisms whereby water and sodium balance influence mean arterial pressure.
- Explain the role of aldosterone in potassium balance.
- Describe the major hormone systems that regulate calcium balance.
- Describe various factors that influence acid-base balance.
- Explain how buffers in the blood, actions of the respiratory system, and the kidneys compensate for acid-base disturbances.

Before You Begin

Make sure you have mastered the following topics:

- Active and passive transport mechanisms
- Transport across epithelia
- Osmolarity and osmosis
- Movement of ions across cell membranes
- The posterior pituitary and ADH
- The adrenal cortex and aldosterone
- Angiotensin II and blood pressure

We know that blood pressure is regulated through neural control of cardiac output and total peripheral resistance and through hormonal control of blood volume. Maintaining normal plasma composition and volume is a major facet of homeostasis. We know that kidneys filter blood and modify urine composition through reabsorption and secretion. In this chapter, we learn how the processes of reabsorption and secretion across renal tubules are regulated by the endocrine system to maintain normal plasma composition and volume. We then relate this control to long-term regulation of blood pressure.

1 The Concept of Balance

To maintain homeostasis, the human body must be kept in balance. To be "in balance" in this context means that what comes into the body and what is produced by the body must equal the sum of what is used by the body and what is eliminated from the body, as shown in the following equation:

$$\text{Input} + \text{production} = \text{utilization} + \text{output}$$

The kidneys play a key role in regulating fluid and electrolyte balance and acid-base balance.

Factors Affecting the Plasma Composition

The kidneys exert control over the volume and composition of plasma by regulating its solute and water content. The volume and composition depend on each other and must be maintained within narrow limits. The volume of plasma is determined almost entirely by its water content; solutes make only a negligible direct contribution to the volume. Nevertheless, the amount of solute in plasma indirectly affects plasma volume because changes in plasma osmolarity (solute concentration) can cause water to shift between the plasma and other body fluid compartments. Plasma volume has an important influence on the body's homeostasis because it affects mean arterial pressure. As we will see later in this chapter, mean arterial pressure is regulated over the long run by changes in plasma volume.

The solute and water content of the plasma is affected both by the movement of materials into and out of the body and by the movement of materials between different compartments within the body. **Figure 1** shows pathways for the movement of water and those solutes that are small enough to permeate capillary walls and cell membranes and, therefore, to move freely into and out of the plasma. The plasma and interstitial fluid compartments are combined in this diagram to emphasize their similarity in composition, which is due to the free exchange of water and small solutes between them.

As Figure 1 shows, the plasma can gain or lose materials by exchange with cells or with extracellular connective tissue, such as the bone matrix. When the bone is resorbed, for example, calcium and phosphates are released into the plasma. Their release raises the concentrations of these substances; conversely, the deposition of calcium and phosphates into bone lowers their plasma concentrations. Figure 1 shows three other routes by which the plasma and the external environment can exchange materials: The plasma can (1) either gain or lose materials by exchange with the lumen of the gastrointestinal tract, (2) either gain or lose materials by exchange with the lumens of the renal tubules, and (3) lose materials through sweating, hemorrhage, or respiration. (The plasma normally gains oxygen through respiration, however.)

Solutes and water are absorbed into the plasma from the gastrointestinal tract and also move from the plasma to the lumen of the tract in the form of saliva, bile, pancreatic juice, and other gastrointestinal secretions. Under normal conditions, the absorptive capacity of the gastrointestinal tract is sufficient to recover virtually 100% of all secreted materials as well as those materials that enter this tract via ingestion (eating or drinking). The rate at which solutes and water are lost by elimination from the gastrointestinal tract is very small compared to the rate at which these substances are lost through the excretion of urine by the kidneys. Consequently, *the transport of materials across the wall of the gastrointestinal tract normally amounts to a net gain of solutes and water by the body.*

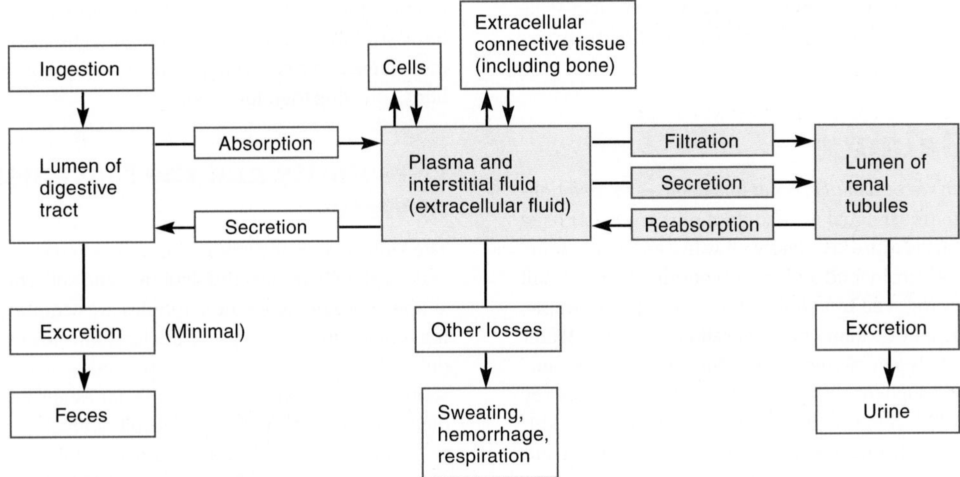

Figure 1 Material exchanges affecting plasma content. Materials (water and solutes) enter the body through ingestion into the digestive tract. Once in the body, materials are exchanged among the digestive tract, the plasma (and the interstitial fluid), cells of the body, connective tissue, and the renal tubules. The highlighted areas represent exchanges between plasma and the renal tubules.

Small solutes and water move from the plasma to the lumen of the renal tubules by glomerular filtration and secretion. Solutes and water are returned to the plasma from the renal tubules by reabsorption. Because not all this material is reabsorbed, *the transport of materials across the walls of the renal tubules amounts to a net loss of water and solutes by the body.* These lost materials are excreted in the urine.

Apply Your Knowledge

There are certain molecules that are not lost through renal function. Can you name two of these molecules?

Solute and Water Balance

When solutes and water enter and exit the plasma at the same rate, the plasma's volume and composition do not change, and the body is said to be in balance. Changes in volume and/or composition occur when materials enter the plasma faster than they exit, or vice versa (**Figure 2**). When a substance enters the body faster than it exits, the substance is said to be in a state of *positive balance;* under these conditions, the quantity of that substance in the plasma increases, unless the substance enters cells or is metabolized within the body. If a substance leaves the body faster than it enters, the substance is in a state of *negative balance;* under these conditions, the quantity of that substance in the plasma decreases.

For certain substances whose plasma concentrations are controlled by specific regulatory mechanisms, a state of positive balance or negative balance can exist with little or no change in the plasma concentration. When you eat a meal containing glucose, for example, your body goes into a state of positive glucose balance. The absorption of glucose from the lumen of the gastrointestinal tract increases the plasma glucose level, but this rise is transient because it triggers insulin secretion and other hormonal changes. Insulin causes an increase in cellular glucose uptake that, combined with other hormonally triggered responses, quickly lowers the blood glucose concentration, eventually restoring it to normal. Once the glucose enters cells, it is catabolized for energy or converted to glycogen or fats for storage.

We know that the kidneys filter 180 liters of plasma per day. Approximately 70% of the filtered water and sodium is reabsorbed in the proximal tubule in the absence of any regulation. However, the body can regulate the remaining filtered water and sodium to vary the amount excreted or retained based on the body's requirements for maintaining balance. The kidneys also function in the regulation of potassium and calcium levels in the blood and of acid-base balance. The regulation of renal excretion occurs primarily in the late distal tubules and collecting ducts. Two types of epithelial cells line these tubules: principal cells and intercalated cells. Water and electrolyte balance is regulated

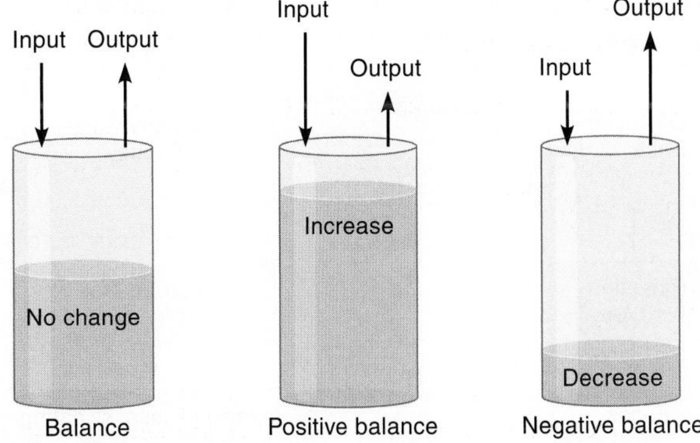

Figure 2 The concept of balance. A substance is in balance in the body when input equals output. A substance is in positive balance when input exceeds output, causing a net increase in the amount of that substance. A substance is in negative balance when output exceeds input, causing a net decrease in the amount of that substance.

through hormonal actions on **principal cells,** whereas acid-base balance is regulated through processes occurring within the **intercalated cells.**

2 Water Balance

For the body to be in water balance, water intake plus any metabolically produced water must equal water output plus any water used in chemical reactions (**Figure 3**). Nearly all water enters the body through the digestive tract in food and drinks consumed, but a small amount of water is produced by cellular metabolism (for example, during oxidative phosphorylation or condensation reactions). Water leaves the body through several routes, including *insensible loss* during respiration and through the skin, sweating, elimination of feces by the digestive tract, and excretion of urine by the kidneys. Only the kidneys, however, regulate the rate of water loss for the purpose of maintaining water balance. The importance of water balance is obvious when we consider what happens during exercise (see **Understanding Exercise: Sweating, Rehydration, and Water Balance**).

When over a given period a person drinks a quantity of water that exceeds the quantity of water lost, his or her body is said to be in a state of positive *fluid balance.* Plasma volume will rise above normal, a condition called *hypervolemia.* If instead the quantity of water lost is greater than the quantity gained in a given period, the body is in a state of negative fluid balance. Plasma volume will go below normal, a condition called *hypovolemia.* A state of normal blood volume is referred to as *normovolemia.*

The control of urinary water excretion is instrumental in regulating plasma volume and osmolarity. Plasma volume is directly related to blood pressure: Increases in plasma volume increase mean arterial pressure, whereas decreases in plasma volume decrease mean arterial pressure. Plasma volume can also affect osmolarity: An increase in plasma water with no increase in solute decreases osmolarity; a decrease in plasma water with no decrease in solute increases osmolarity. Changes in plasma osmolarity affect the movement of water between fluid compartments—in particular, between intracellular fluid and extracellular fluid. In turn, water movement can cause changes in the volume of cells throughout the body, affecting their function.

Osmolarity and the Movement of Water

The kidneys vary the volume of water excreted by regulating water reabsorption in the late distal tubules and collecting ducts. The forces acting on water across these tubules are osmotic gradients between the tubular fluid and the peritubular fluid. However, water can move only if the tubule membranes are permeable to water. This section describes the mechanisms responsible for establishing the osmotic gradients and regulating water permeability, and thus water reabsorption.

Under normal conditions, the total solute concentration of the plasma is approximately 300 mOsm (300 milliosmoles of solute per liter of plasma); this is also the normal osmolarity of interstitial fluid and intracellular fluid. Because the osmolarity is the same in both extracellular fluid (interstitial fluid and plasma) and intracellular fluid, the osmotic pressure is also the same, and there is no net force for water to either leave cells or enter them. Under these conditions, cell volumes do not change, and the intracellular fluid and extracellular fluid are said to be in *osmotic equilibrium.*

If a person whose plasma volume and osmolarity are normal drinks a significant quantity of pure water, the plasma volume will increase, and the concentration of solutes in the plasma will decrease. The absorption of water into the plasma will cause plasma osmolarity to fall below the normal level. Unless it is corrected, this decrease in osmolarity will cause cells to swell, because water will flow spontaneously from extracellular fluid, where the osmolarity is lower (and the concentration of water is higher), to intracellular fluid, where the osmolarity is higher (and water concentration is lower). Such swelling can have deleterious effects on the functioning of cells all over the body, but cells in the brain are particularly

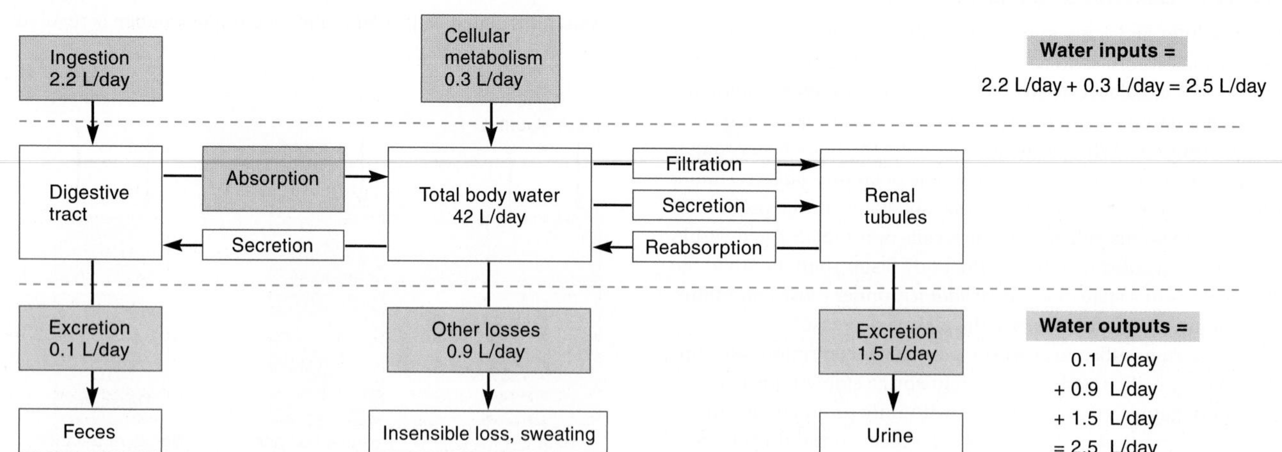

Figure 3 Factors affecting water balance. Body water input includes absorption across the digestive tract and production by cellular metabolism totaling approximately 2.5 L/day. Body water output includes excretion in the urine and feces and loss through sweat and respiration totaling approximately 2.5 L/day. Under these circumstances the body is in water balance.

UNDERSTANDING EXERCISE

Sweating, Rehydration, and Water Balance

Exercise requires catabolism to generate the energy needed for muscle contraction. Recall that energy catabolism generates heat. When the body is in the resting state, much of the heat generated by catabolism is used to maintain body temperature. During exercise, however, the excess heat generated must be dissipated. To eliminate excess heat, blood flow to the skin increases; heat can then be transferred to the environment by radiation, conduction, convection, and evaporation.

Radiation is the transfer of heat in the form of electromagnetic waves from a hot object to a cold object, such as the transfer of heat to your body when you stand near a fire or a hot stove. When your body is warmer than the environment, heat radiates from the body. Radiation is the primary mechanism for heat loss under resting conditions.

Conduction is the transfer of heat through direct contact between two objects. For example, when you touch cold metal, heat leaves your hand and moves to the metal; thus you feel cold. In the body, conduction occurs between adjacent tissues. For example, heat generated in exercising muscles can be transmitted to other tissues and ultimately to the skin, where it can then be transferred to the environment by radiation.

Convection is the movement of heat in currents, either in water (blood) or in air. With convection of heat in blood, blood vessels in the skin vasodilate, thereby increasing blood flow to the skin, where more heat can be lost by radiation. (When the temperature is cold, blood flow to the skin is decreased.) Convection also occurs in response to air flow. Your body generates a "blanket" of warm air around it. When you use a fan, you blow away this warm air, cooling your body. In contrast, on a cold day, the blanket of warm air helps maintain your body temperature. If the wind blows the warm air away, you feel colder than the actual temperature—a phenomenon known as the wind chill factor.

The last mechanism for dissipating heat is heat exchange through evaporation. When you sweat, heat is not lost in the sweating process but rather through the evaporation of the sweat from your body. Thus dripping sweat is ineffective in cooling the body. On humid days, evaporation of sweat is decreased because the water content of the air is high and, therefore, the ability to dissipate heat through sweat and evaporation is diminished. Sweat is the primary mechanism for dissipating heat during exercise.

Sweat is produced by sweat glands, which are located all over the skin. These sweat glands produce a primary secretion that contains a lot of water, sodium, and chloride. (Some sweat glands, called apocrine glands, also produce a sweat that contains fatty acids and proteins.) When activity of the body is low, most of the electrolytes secreted by the sweat glands are reabsorbed, such that few electrolytes actually leave the body in sweat. When the body is highly active, however, electrolyte loss is considerable, so the sodium and chloride need to be replaced.

Through sweating, exercise results in the loss of water and electrolytes, taking the body out of balance. To compensate for this effect, thirst stimulates the person to intake fluid, and hormonal control of the kidneys decreases water and electrolyte excretion to return the body to balance. Experts recommend that athletes hydrate *prior to* starting intense exercise and rehydrate frequently *during* exercise to prevent dehydration. Sport drinks have become the preferred drinks for hydration in athletes: They replenish lost electrolytes as well as lost fluids.

When a person does not rehydrate sufficiently during exercise, the individual becomes dehydrated. Although salt is lost in the sweat along with water, sweat is hypo-osmotic to body fluids; as a consequence, more water is lost than salt. As the body becomes dehydrated, the osmolarity of the body fluids increases, stimulating the release of antidiuretic hormone (ADH), which in turn increases water reabsorption in the kidneys, minimizing fluid loss.

Commonly a person overhydrates *following* exercise, which causes a decrease in the osmolarity of body fluids, inhibiting the release of ADH and thereby increasing water elimination in the urine. The hypo-osmolarity of the body fluids can cause neurons and glial cells in the brain to swell, putting pressure on the brain tissue in the fixed space of the cranium. Thus a person who overhydrates often suffers from headaches due to the swelling of brain cells.

sensitive to it. The swelling of brain cells can alter neurological function, producing headache, nausea, confusion, seizures, or coma. We normally do not worry about the possible adverse effects of drinking water because when excess water is ingested, the kidneys increase the rate at which they excrete water, which quickly restores the volume and osmolarity of the plasma to normal (however, see **Clinical Connections: Water Intoxication**). Ideally, the kidneys would excrete a volume of pure water equal to the volume of water that was drunk, but the kidneys cannot excrete pure water, so instead they excrete a large volume of urine with a low osmolarity.

Now suppose that a person whose plasma volume and osmolarity are normal eats a large quantity of salty pretzels without drinking any water. In this case, plasma volume will not change, but salt will be absorbed into the plasma, raising its osmolarity. If not corrected, the increased plasma osmolarity will cause cells to shrink because the osmolarity of the extracellular fluid exceeds that of intracellular fluid, resulting in water movement out of the cells. Such cell shrinkage induces many of the same neurological problems that cell swelling does. To avert this possibility, the kidneys excrete the excess solutes that were ingested while minimizing water loss by excreting a small volume of water. Ideally, the kidneys would excrete the excess solutes only, but this is not possible. Thus, when little or no water is ingested, the kidneys excrete a small volume of highly concentrated urine.

In these examples, the kidneys compensate for changes in plasma volume and osmolarity by adjusting the rate of water excretion. Because water is reabsorbed but not secreted, these adjustments are achieved through changes in the rate of water reabsorption.

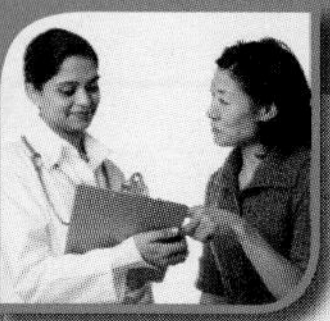

CLINICAL CONNECTIONS

Water Intoxication 

On January 12, 2007, 28-year-old Jennifer Strange died of water intoxication after participating in a radio contest, called "Hold Your Wee for a Wii." Contestants were instructed to drink a bottle of water every 15 minutes, with the volume of water in the bottles increasing over time. Whoever lasted the longest without urinating would win the Wii game console. Jennifer won second place, tickets to a Justin Timberlake concert, by drinking nearly two gallons of water in three hours. She complained of a headache after being interviewed, but then went home—where she subsequently passed out and died of water intoxication. The winner of the contest, Lucy Davidson, collapsed 15 minutes after leaving the station, and although feeling ill afterward, recovered fully.

It's difficult for many people to believe that water can be toxic. In fact, consumption of extreme volumes of water alters the osmolarity of the plasma, causing a shift of water from plasma to interstitial fluid and then into cells. Thus body cells swell, including brain cells that are held in a fixed space by the cranium. A minor amount of swelling of the brain cells can cause headaches and nausea, but severe swelling can interfere with brain function, resulting in death.

In October 2009, a jury found the radio station negligent in the death of Jennifer Strange and awarded her husband, Billy Strange, $16.5 million in damages.

Critical Thinking Questions

1. What effect does extreme water intake have on blood osmolarity?
2. What is the biggest health danger posed from the swelling of body cells?
3. What is the cause of death due to extreme water intake?

In the renal tubules, water reabsorption is passive and is coupled to the active reabsorption of solutes. Solute transport creates an osmotic gradient across the tubule epithelium. The precise mechanism responsible for creating the osmotic gradient varies in the different segments of the renal tubules. In the next sections we see that in the proximal and distal tubules, reabsorption of solute increases the osmolarity of the peritubular fluid (which drives reabsorption of water by osmosis), whereas in the collecting ducts, the *medullary osmotic gradient* drives reabsorption of water.

Water Reabsorption in the Proximal Tubule

Because the primary solute in extracellular fluid is sodium and most of the filtered sodium is reabsorbed in the proximal tubule, sodium is the primary solute responsible for producing the osmotic gradient that drives water reabsorption. Although the exact mechanism varies in different segments of the renal tubules, sodium reabsorption always involves the active transport of sodium across the basolateral membrane from the epithelial cell of the tubule into the peritubular fluid, where it can diffuse into the plasma of peritubular capillaries (**Figure 4**). Sodium crosses the apical membrane by a variety of mechanisms, including secondary active transport during which the movement of sodium is coupled to the movement of another molecule (such as glucose).

Water is reabsorbed by osmosis, as illustrated in **Figure 5**. Active reabsorption of sodium and other solutes in the proximal tubule creates an osmotic gradient, and therefore water follows the solutes. Thus fluid in the proximal tubule is iso-osmotic with the interstitial fluid of the renal cortex, with both fluids being at 300 mOsm. Because reabsorption of water creates a concentration gradient for permeating solutes to move from tubular fluid to the plasma in peritubular capillaries, the movement of permeating solutes follows water reabsorption.

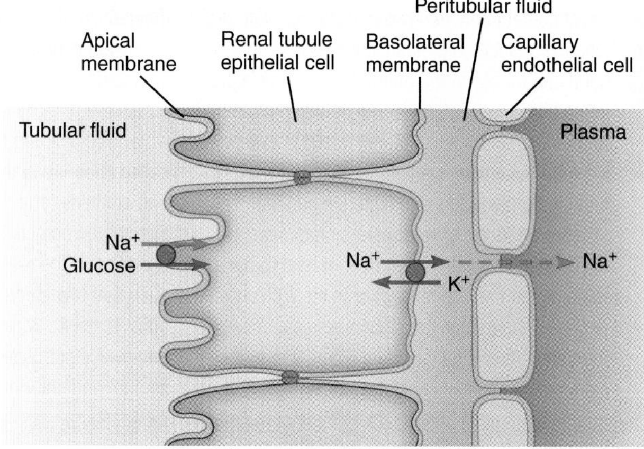

Figure 4 Mechanism of sodium reabsorption in the proximal tubule. Sodium is actively transported across the basolateral membrane into the peritubular fluid by the Na^+/K^+ pump. Sodium moves from the tubular fluid into the epithelial cell through sodium channels or via cotransport with other molecules, such as glucose.

Establishment of the Medullary Osmotic Gradient

Within the interstitial fluid of the renal medulla is the **medullary osmotic gradient;** the outer regions of the medulla have a lower osmolarity than the inner regions (**Figure 6**). The osmolarity varies

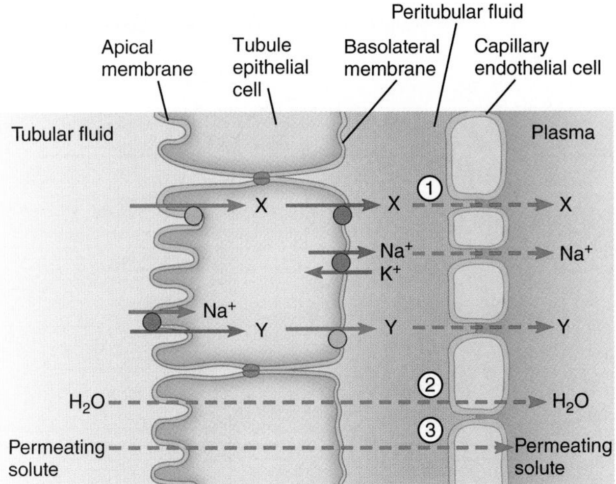

Steps for water and urea reabsorption:

① Solutes (Na^+, X, Y) are actively reabsorbed, increasing the osmolarity of peritubular fluid and plasma.

② Water is reabsorbed by osmosis.

③ Urea (permeating solute) is reabsorbed passively.

Figure 5 Mechanism of water reabsorption. When solutes, such as sodium ions or molecule X, are actively reabsorbed, the osmolarity of the peritubular fluid and plasma increases, such that water moves from the lumen of the renal tubules first into the peritubular fluid and then into plasma by osmosis. Permeating solutes follow water reabsorption.

from 300 mOsm at the cortical edge of the medulla to approximately 1200–1400 mOsm at the innermost portion of the medulla near the renal pelvis. This gradient, which is necessary for water reabsorption from the collecting duct, exists because of a mechanism known as the countercurrent multiplier and because of the facilitated diffusion of urea from the lumen of the collecting duct into the medullary interstitial fluid.

Countercurrent Multiplier

The properties of different portions of the loops of Henle of juxtamedullary nephrons are critical to the countercurrent multiplier and establishment of the medullary osmotic gradient. The descending limb is permeable to water, so water diffuses when an osmotic gradient exists. The thick ascending limb, by contrast, is impermeable to water; thus water does not diffuse even if an osmotic gradient exists. In addition, the thick ascending limb has sodium/potassium/chloride active transporters that the descending limb lacks. These transporters pump all three ions into the interstitial fluid, thereby increasing its osmolarity. The term *countercurrent* refers to the fact that fluid flowing through the descending and ascending limbs, which parallel each other, moves in opposite directions (see Figure 6).

Figure 7 illustrates how the countercurrent multiplier creates the medullary osmotic gradient. ① The figure starts with no osmotic gradient along the tubules or within the medullary interstitial fluid. The fluid that enters the descending limb from the proximal tubule is iso-osmotic with the interstitial fluid, at 300 mOsm. The fluid in the proximal tubule is iso-osmotic because water freely crosses the wall of the tubule and, therefore, is reabsorbed along with solutes. As fluid moves down the descending limb, there is no net movement of water across the tubule wall because there is no osmotic gradient. As the fluid begins to travel up the ascending limb of the loop of Henle, sodium, chloride, and potassium are actively transported from the tubule into the medullary interstitial fluid, ② increasing the osmolarity of the interstitial fluid from 300 mOsm to 400 mOsm and lowering the osmolarity of fluid in the ascending limb to 200 mOsm. When the osmolarity of the peritubular fluid increases, water moves out of the descending limb and into the peritubular fluid until ③ the two are iso-osmotic again at 400 mOsm. This creates a difference in osmolarity between the fluid in the descending limb and the fluid in the ascending limb, with the latter at a lower osmolarity (200 mOsm versus 400 mOsm). ④ As more fluid at 300 mOsm enters the loop of Henle from the proximal tubule, this fluid pushes the fluid ahead of it through the tubule, thereby pushing the higher-osmolarity fluid deeper into the medulla. ⑤ Active transport of sodium, chloride, and potassium in the ascending limb raises the osmolarity of deeper medullary interstitial fluid from 400 mOsm to 500 mOsm, which causes water movement into the medullary interstitial fluid from the descending limb. ⑥ The descending limb is now iso-osmotic with the medullary interstitial fluid. More fluid at 300 mOsm enters the loop of Henle from the proximal tubule, pushing the higher-osmolarity fluid toward the tip of the loop of Henle. ⑦ The process continues until the medullary osmotic gradient is created and the system is in a steady state.

At steady state, the fluid entering the loop of Henle from the proximal tubule is iso-osmotic to extracellular fluid, at 300 mOsm; however, the osmolarity of the tubular fluid within both limbs of the loop of Henle is greater in the deeper portions of the renal medulla. At the tip of the loop of Henle, the osmolarity of the tubular fluid is approximately 1200–1400 mOsm.

Note that at any given level in the medulla, the osmolarity of the fluid in the ascending limb is always lower than the osmolarity of fluid in the descending limb, because the ascending limb actively transports solutes out of the tubular fluid but prevents water from following them. As the tubular fluid leaves the loop of Henle and enters the distal tubule, it is hypo-osmotic to extracellular fluid at approximately 100–200 mOsm.

Role of Urea in the Medullary Osmotic Gradient

The countercurrent multiplier establishes the osmotic gradient, but additional solute is needed to maintain the gradient. That additional solute is **urea,** a waste product generated by the liver during catabolism of proteins, and the primary form by which nitrogen is eliminated from the body. In an average Western diet, urea makes up 40% of the solutes in the glomerular filtrate; consequently, it has profound effects on water movement across the renal tubules. As we will see, only approximately 40% of the filtered urea stays in the tubules, which is critical in preventing massive water loss through the osmotic effect of urea on water. Urea also decreases water loss in other ways, including strengthening the medullary osmotic gradient.

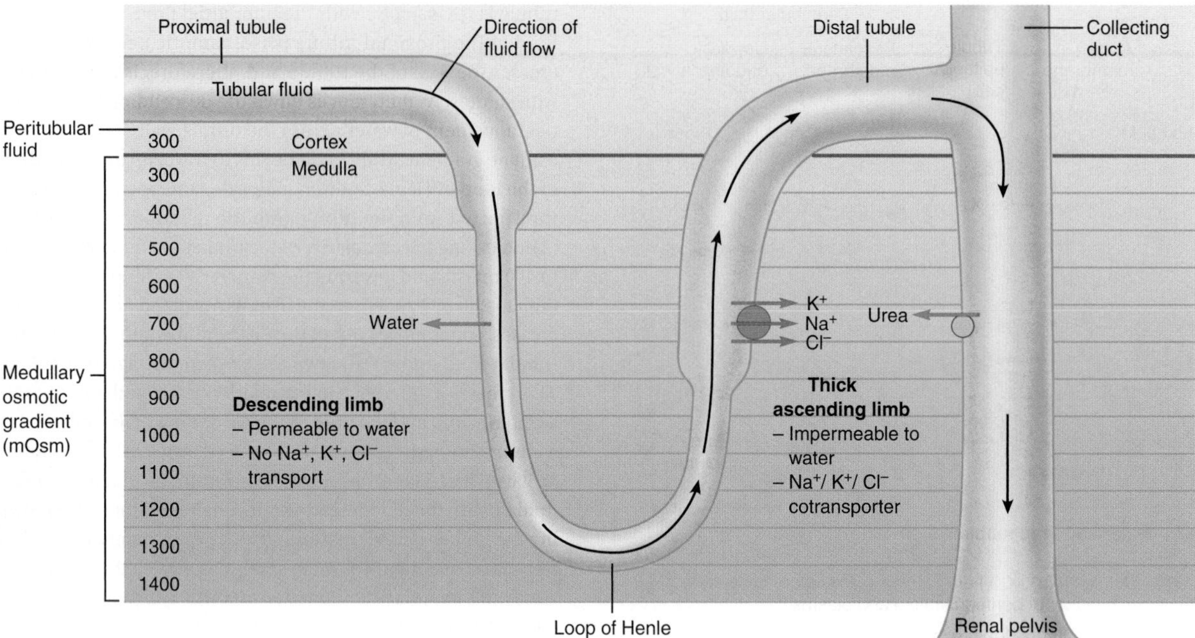

Figure 6 The medullary osmotic gradient. The descending limb of the loop of Henle is permeable to water, whereas the thick ascending limb is impermeable to water and contains transporters for Na^+, Cl^-, and K^+. Thus, differences in the transport of materials establish an osmotic gradient in the medullary interstitial fluid. The osmolarity of medullary interstitial fluid is 300 mOsm near the cortex, and increases continuously to a maximum of approximately 1400 mOsm near the renal pelvis.

Q Which of the two types of nephrons in the body is depicted in this figure?

Urea is highly water soluble and unable to permeate cell membranes in the absence of urea transporters. Such transporters are located in a number of cell types, including UT-B transporters in the endothelial cells of the descending limb of the vasa recta (discussed in the next subsection), and UT-A transporters of various types found in epithelial cells lining the renal tubules (**Figure 8**). In the portion of the collecting duct deep in the renal medulla, the epithelial cells contain UT-A_1 and UT-A_3 transporters that allow facilitated diffusion of urea from filtrate to the peritubular fluid, thereby contributing approximately 40% of the osmolarity of the medullary osmotic gradient (see Figure 6). Due to water reabsorption in the proximal tubule, some interstitial urea travels through UT-C transporters into the proximal tubule and can be used again to contribute to the medullary osmotic gradient when it reaches the collecting duct, thereby recycling the urea. (We will see later that active transporters for urea in the collecting duct are also critical in regulating water reabsorption.)

Role of the Vasa Recta in Preventing Dissipation of the Medullary Osmotic Gradient

As blood flows into the renal medulla to supply it with nutrients and oxygen, water tends to diffuse out of capillaries, and solutes tend to diffuse into them. However, the anatomical arrangement of the vasa recta capillaries, which accompany the loops of Henle as they dip into the medulla and return to the cortex (see Figure 1b), prevents the diffusion of water and solutes from dissipating the medullary osmotic gradient (**Figure 9**). Like the loop of Henle, the vasa recta can be described as having descending and ascending limbs based on the direction of fluid flow. As the descending limb of the vasa recta enters the renal medulla, the plasma has an osmolarity of 300 mOsm. As the descending limbs reach deeper into regions of the medulla with higher osmolarity, water leaves the capillaries by osmosis, and solutes enter the plasma by diffusion. This process, which would tend to reduce the osmolarity of the interstitial fluid if left unchecked, continues to the tip of the vasa recta because of the increasing osmolarity of the medullary interstitial fluid. However, as the blood flows back toward the cortex in the ascending limb, the direction of the osmotic gradient across the capillary walls reverses, so that water moves into the plasma and solutes move into the interstitial fluid. This movement tends to raise the osmolarity of the interstitial fluid. As a result, the osmolarity of the interstitial fluid stays relatively constant, and the osmolarity of plasma leaving the renal medulla in the vasa recta capillaries is just slightly hyperosmotic (325 mOsm) to that of the plasma entering the renal medulla.

Recall that the descending limb of the vasa recta has transporters for urea and that urea is concentrated deep in the medulla. Thus urea enters the descending limb increasing the osmolarity of the plasma, which would draw water into the descending limb, except that the medullary gradient is much stronger. More significant is

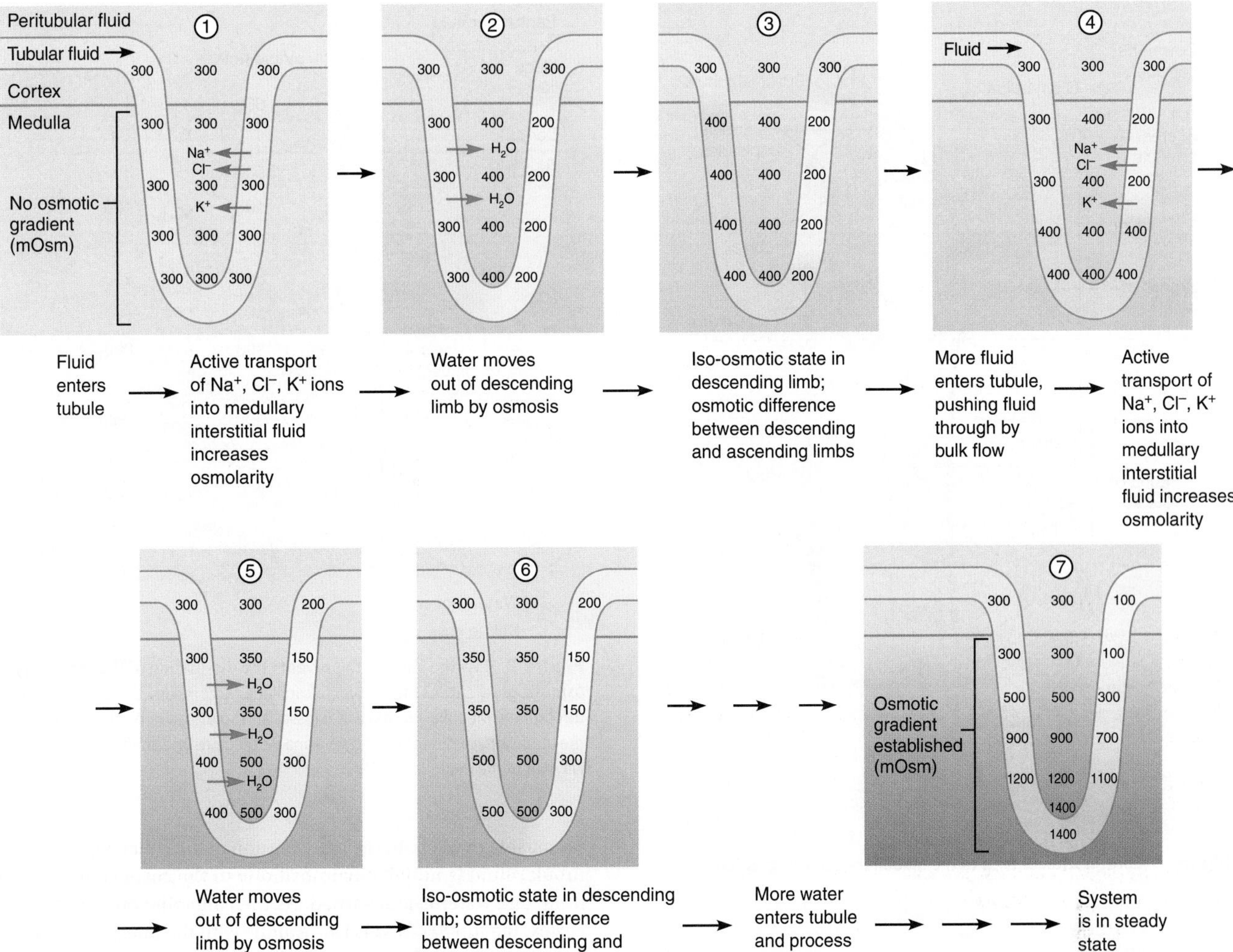

Figure 7 How the countercurrent multiplier establishes the medullary osmotic gradient. ① Initially all fluids are iso-osmotic at 300 mOsm. Active transport of solutes (Na^+, Cl^-, K^+) from the ascending limb of the loop of Henle into the medullary interstitial fluid increases the osmolarity of the interstitial fluid and decreases the osmolarity of the tubular fluid in the ascending limb. ② The increased osmolarity of the medullary interstitial fluid draws water from the lumen of the descending limb of the loop of Henle into the interstitial fluid, ③ increasing the osmolarity of the tubular fluid in the descending limb. ④ More tubular fluid then enters the loop of Henle, pushing the fluid farther into the renal tubules. The process of solute transport from the ascending limb ⑤ followed by water movement from the descending limb, ⑥ increasing the osmolarity of the tubular fluid in the descending limb, followed by more tubular fluid entering the loop of Henle, repeats until ⑦ the medullary osmotic gradient is established.

that the ascending limb of the vasa recta does not have urea transporters. Thus urea stays in the plasma (has been reabsorbed), contributing to the high osmolarity of plasma leaving the renal medulla.

Role of the Medullary Osmotic Gradient in Water Reabsorption in the Distal Tubule and Collecting Duct

Recall that 70% of the water filtered from plasma at the renal corpuscle is reabsorbed in the proximal tubule. Approximately 20% of the filtered water is reabsorbed in the distal tubule, and most of the remaining 10% is reabsorbed in the collecting ducts. In the initial portion of the distal tubule, the lumenal fluid (100–200 mOsm) is hypo-osmotic to the peritubular fluid (300 mOsm). As fluid moves down the collecting duct, the osmolarity of the lumenal fluid is always less than the increasing osmolarity of the medullary interstitial fluid, thereby increasing the osmotic force for water to move from the renal tubule into the interstitial fluid as tubular fluid moves down the collecting duct toward the renal pelvis. Thus, when water can permeate the wall of the collecting duct, it is reabsorbed.

Aquaporins and Water Permeability

The epithelial cells lining the late distal tubule and collecting duct are connected by tight junctions such that water cannot pass between cells from peritubular fluid to tubular fluid, or vice versa.

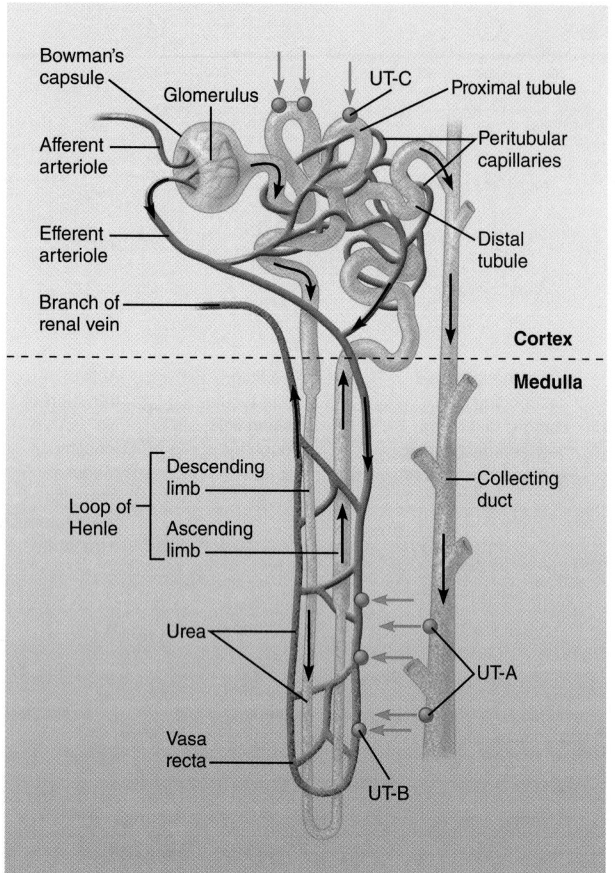

Figure 8 **Contribution of urea to the medullary osmotic gradient.**

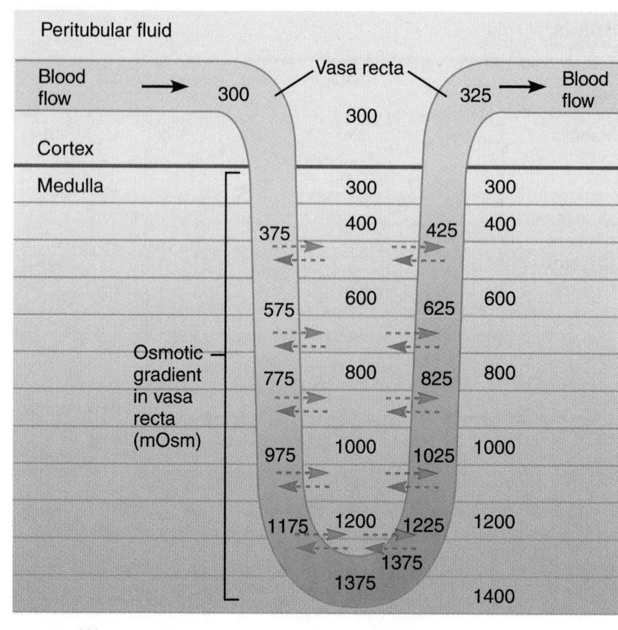

Figure 9 **How the vasa recta prevents the dissipation of the medullary osmotic gradient.** As the vasa recta capillaries accompany the loops of Henle through the medulla, plasma water losses and solute gains on the way into the medulla are counteracted by plasma water gains and solute losses on the way out of the medulla.

In addition, the lipid bilayers of these cells' plasma membranes are not permeable to water. The ability of water to cross the plasma membrane (and therefore the epithelial layer) depends on the presence of water channels or pores, called **aquaporins,** in the plasma membrane of principal cells. Aquaporin-3 is present in the basolateral membrane of principal cells at all times, whereas aquaporin-2 is present in the apical membrane only in the presence of antidiuretic hormone (ADH; discussed in the next section).

The effects of the medullary osmotic gradient and water permeability of the late distal tubule and collecting duct on water reabsorption are shown in **Figure 10**. Basically, the more permeable these tubules are, the greater the water reabsorption.

In Figure 10a, the walls of the late distal tubule and collecting duct are impermeable to water. As tubular fluid of 100 mOsm enters the late distal tubule, an osmotic gradient for water movement exists, but because the membrane is impermeable, water cannot move. The osmotic gradient gets larger as the tubular fluid at 100 mOsm travels down the collecting duct toward the renal pelvis, but water still cannot cross the impermeable membrane. The final result is excretion of a large volume of urine with low osmolarity.

Figure 10b shows how the kidneys can conserve water when the late distal tubule and collecting duct are made highly permeable to water. In the early portion of the collecting duct, the tubular fluid is initially hypo-osmotic to the cortical interstitial fluid, and water is reabsorbed. As the collecting duct leaves the cortex, the tubular fluid is iso-osmotic with the interstitial fluid at 300 mOsm. As the fluid moves down the collecting duct, water continues to be reabsorbed from the collecting duct into the medullary interstitial space, such that the fluid in the collecting duct always remains very nearly iso-osmotic with the medullary interstitial fluid. Eventually, the fluid reaches an osmolarity of 1400 mOsm at the end of the collecting duct. Tubular fluid osmolarity can never exceed the medullary interstitial fluid osmolarity because water will stop moving across the wall once the osmolarity inside the tubule equalizes with that outside the tubule. Therefore, the maximum osmolarity of urine is 1400 mOsm. Because those solutes that are not 100% reabsorbed must be excreted in the urine and because there is an upper limit on the osmolarity of urine, a minimum volume of water must be excreted to eliminate the solutes. This volume is the **obligatory water loss,** which is approximately 440 mL of water per day under normal conditions [see **Discovery: Don't Drink the (Sea)Water**].

The length of the loop of Henle determines the maximum concentration of urine. Longer loops of Henle can form a larger medullary osmotic gradient by the countercurrent multiplier, thereby facilitating greater water reabsorption. Camels, for example, have longer loops of Henle than humans and can generate urine with a concentration of 2800 mOsm. Australian hopping mice, which have the longest loops

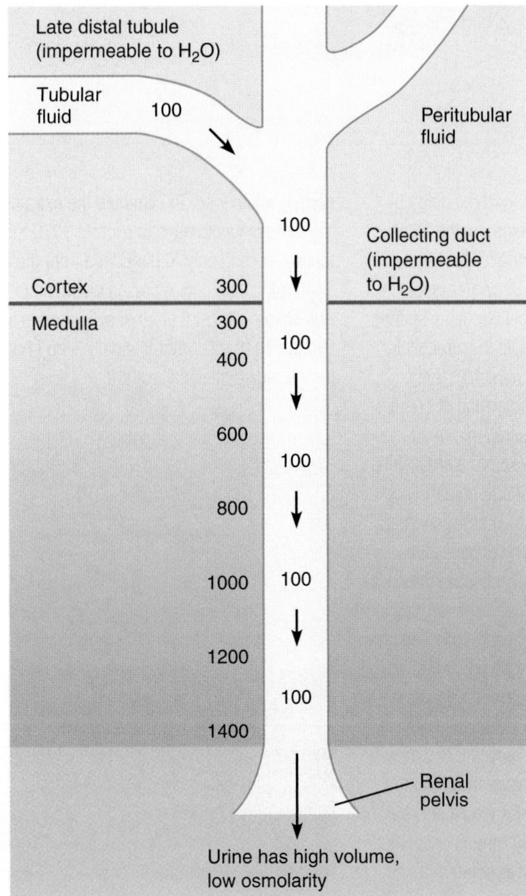

(a) Late distal tubule and collecting duct impermeable to water

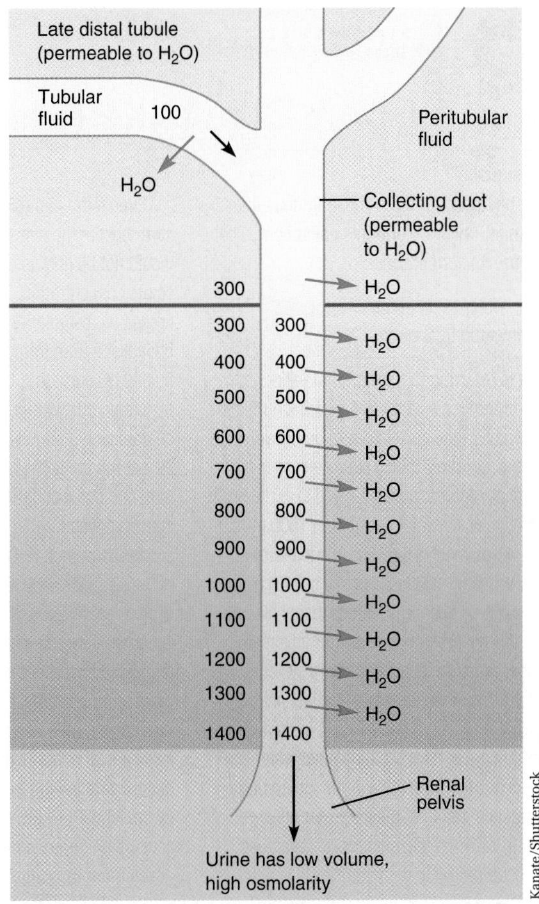

(b) Late distal tubule and collecting duct permeable to water

Figure 10 Water reabsorption across the late distal tubule and collecting duct. The fluid in the distal tubule and collecting duct is hypo-osmotic to the medullary interstitial fluid, creating an osmotic force for water to leave the tubules. **(a)** When the membrane of the late distal tubule and collecting duct is impermeable to water, water cannot leave the tubule and is excreted in the urine, producing a large volume of urine with low osmolarity. **(b)** When the membrane is permeable to water, water can leave the tubule. If the membranes are highly permeable, the final urine will be iso-osmotic with the deepest layers of the renal medulla, producing a small volume of urine with high osmolarity.

of Henle of any known species, can concentrate urine to 9800 mOsm. Because of their strong ability to conserve water, Australian hopping mice experience very small obligatory water loss and can survive on the water they consume in their food and produce metabolically.

Quick Check 1

1. Where does most water reabsorption occur in the renal tubules? Where does the regulation of water reabsorption in the renal tubules occur?
2. Compare the movements of solute and water across the epithelial cells lining the descending limb of the loop of Henle and those lining the ascending limb of the loop of Henle. Explain how this differential movement allows establishment of an osmotic gradient in the renal medulla.
3. Can pure water be excreted as urine? Explain.

Effects of ADH on Water Reabsorption

By varying the degree of permeability to water in the late distal tubules and collecting ducts, the kidneys can vary the concentration of the urine between 100 and 1400 mOsm. **Antidiuretic hormone (ADH;** also known as vasopressin) regulates the permeability of the late distal tubules and collecting ducts to water.

In the absence of ADH, the apical membrane of principal cells is impermeable to water, so water reabsorption cannot occur. ADH stimulates synthesis of aquaporin-2, along with its insertion into the apical membrane of principal cells in the late distal tubules and collecting ducts. In this way, water reabsorption and urine volume are regulated by variations in the plasma levels of ADH, which determines the number of aquaporin-2 pores, thereby establishing the permeability of principal cell membranes to water: The greater the permeability to water, the greater the water reabsorption.

Aquaporin-2 is stored in the membrane of cytoplasmic vesicles of principal cells, awaiting insertion into the apical membrane in

DISCOVERY

Don't Drink the (Sea)Water

You may have encountered these famous lines from Samuel Taylor Coleridge's epic poem, *The Rime of the Ancient Mariner:*

> Water, water everywhere
> Nor any drop to drink

The poet is referring to the fact that sailors cannot rely on the sea as a source of drinking water: If they do, they face certain death by dehydration.

The danger from drinking seawater lies in the concentration of the solutes dissolved in it. The solute concentration of seawater varies from 1000 mOsm to 1200 mOsm, most of which is due to dissolved sodium chloride. At 1200 mOsm, the solute concentration of seawater is four times that of plasma. Drinking this hyperosmotic seawater adversely affects the gastrointestinal and urinary systems.

Water movement in the gastrointestinal tract follows solute movement, as it does in the kidney tubules. When a person consumes seawater, the high concentration of solutes in the lumen of the gastrointestinal tract decreases water absorption, resulting in diarrhea and significant water loss, thereby contributing to dehydration.

Even if the gastrointestinal system could absorb most of the seawater consumed, the kidneys would not be able to compensate for the high solute concentration. Drinking seawater would increase the osmolarity of extracellular fluid, shifting body fluids from intracellular fluid to extracellular fluid. Cells, including the osmoreceptors in the hypothalamus, would shrink. Activation of osmoreceptors would stimulate the secretion of ADH, and as a result the kidneys would excrete a small volume of highly concentrated urine to help the body conserve water while ridding itself of the excess solutes. However, the maximum concentration of urine is 1200–1400 mOsm. The kidneys must excrete a minimum of 440 mL of water per day to accommodate the normal daily excretion of solutes (the obligatory water loss). In addition, the body loses approximately 1 liter of water daily through insensible losses and sweating. Thus the body loses a total of approximately 1440 mL of water daily. A person who tried to replace these fluids by drinking 1440 mL of seawater would also be consuming approximately 1500 mmole of solutes, which must then be eliminated in addition to the normal solute load. To eliminate the extra solutes, the kidneys would have to excrete 1070 mL urine, thereby netting only 370 mL fluid from the original 1440 mL consumed. A person would have to drink *and absorb* almost 6 liters of seawater daily to retain the 1440 mL water lost daily—an impossibility for the gastrointestinal system.

response to ADH. ADH acts on renal tubule cells by binding to receptors on the basolateral membrane (**Figure 11**). These receptors are coupled to a G protein that activates the enzyme adenylate cyclase, which then catalyzes the production of the second messenger cAMP. cAMP activates protein kinase A, which stimulates the insertion of aquaporin-2 pores into the apical membrane by exocytosis. As mentioned earlier, ADH also stimulates the synthesis of new aquaporin-2 molecules. At high concentrations of ADH, water reabsorption is high and urine output is low. (The word *antidiuretic* refers to something that counters the effects of *diuresis,* or increased urine flow; a diuretic is a drug that promotes urine flow.) At low concentrations of ADH, water reabsorption is low and urine output is high.

ADH also promotes water reabsorption indirectly. This hormone increases the concentration of active transporters that move urea from the interstitium into the collecting duct, thereby decreasing the medullary osmotic gradient and increasing the gradient for water reabsorption into the collecting duct.

Regulation of ADH Secretion

Recall that ADH is secreted from the posterior pituitary gland by neurosecretory cells that originate in the hypothalamus. Changes in the osmolarity of extracellular fluid are the strongest stimuli for ADH release. **Osmoreceptors**—cells located in the *organum vasculosum of the lamina terminalis* (OVLT) of the hypothalamus—monitor the osmolarity of extracellular fluid. Recall that the brain is protected from water-soluble substances in the extracellular fluid, which ensures that it is not exposed to normal extracellular fluid. However, the OVLT is one of a few regions of the brain not surrounded by the blood-brain barrier and, therefore, is well suited for its role in detecting osmolarity. An increase in osmolarity stimulates ADH secretion, which in turn increases water reabsorption (**Figure 12**). Conversely, a decrease in the osmolarity of extracellular fluid inhibits ADH secretion, which decreases water reabsorption and increases water excretion. Inadequate secretion of ADH leads to a form of diabetes called *diabetes insipidus* (see **Focus on Diabetes: Diabetes Insipidus**).

Plasma levels of ADH also depend on signals arising in baroreceptors that detect blood volume and blood pressure (**Figure 13**). Baroreceptors in the atria respond to changes in blood volume, whereas those in the aortic arch and carotid sinus (sinoaortic baroreceptors) respond to changes in blood pressure. As blood volume or blood pressure decreases, the frequency of action potentials in baroreceptor afferents also decreases. Decreased baroreceptor activity stimulates increased secretion of

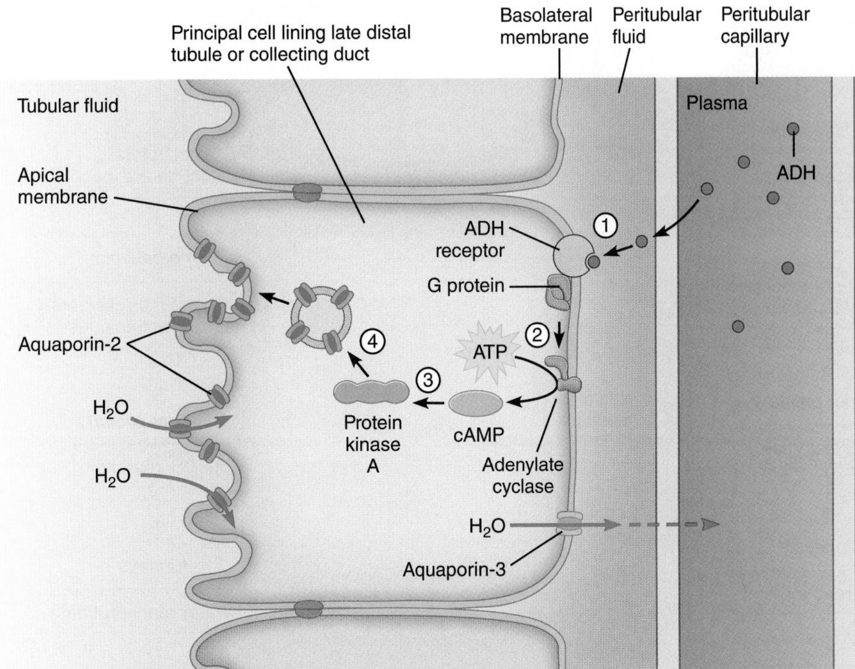

Figure 11 Effects of ADH on principal cells lining the late distal tubules and collecting ducts. (1) ADH binds to receptors on the membrane of principal cells. (2) Activation of a G protein activates the enzyme adenylate cyclase, which catalyzes the formation of cAMP, which then (3) activates protein kinase A. (4) The protein kinase A stimulates insertion of new water channels (aquaporin-2) into the apical membrane, which increases its water permeability.

Q Which type of hormone is ADH, and which endocrine gland secretes it?

ADH, which in turn increases water reabsorption and minimizes the stimuli for the release of ADH (see Figure 13). Note that if the plasma volume is smaller than normal, the kidneys cannot raise it back to the normal level; they can only minimize further fluid loss, which then minimizes any additional reductions in plasma volume. Achieving an increase in body water content requires behavioral changes—specifically, an input of water through drinking. If blood volume or blood pressure increases, then ADH secretion decreases, resulting in increased water excretion, which decreases blood volume and blood pressure.

Regulating Water Excretion by Changing GFR

We know that normally the glomerular filtration rate (GFR) is autoregulated to minimize increases in water and electrolyte loss when blood pressure is high. However, if blood pressure drops below normal levels (mean arterial pressure less than 80 mm Hg), then decreases in GFR play a significant role in water retention. When less water is filtered, less water will be excreted. These changes in GFR work hand in hand with other mechanisms to limit water loss when blood pressure is low. Likewise, if blood pressure increases to sufficiently high levels (mean arterial pressure greater than 180 mm Hg), then the GFR increases, and water loss increases. However, because blood pressures greater than 180 mm Hg occur only under pathological conditions, changes in GFR do not normally compensate for elevations in blood pressure.

Quick Check 2

1. Name two stimuli for ADH release from the posterior pituitary.
2. Describe the effects of ADH on principal cells.

3 Sodium Balance

Regulation of plasma sodium levels is critical because sodium is the primary solute in extracellular fluid, is critical to the function of excitable cells, and is used to transport other solutes. An increase in plasma sodium levels to greater than normal, called *hypernatremia,* is often accompanied by water retention and an increase in blood pressure. A decrease in plasma sodium levels to less than normal, called *hyponatremia,* is associated with low plasma volume and hypotension.

Sodium is freely filtered at the glomerulus and undergoes tubular reabsorption, but it is not secreted. Regulation of plasma sodium occurs at the level of reabsorption. In this section, we first look at renal handling of sodium. Next, we discuss the two hormones that function in the regulation of sodium reabsorption: aldosterone and atrial natriuretic peptide.

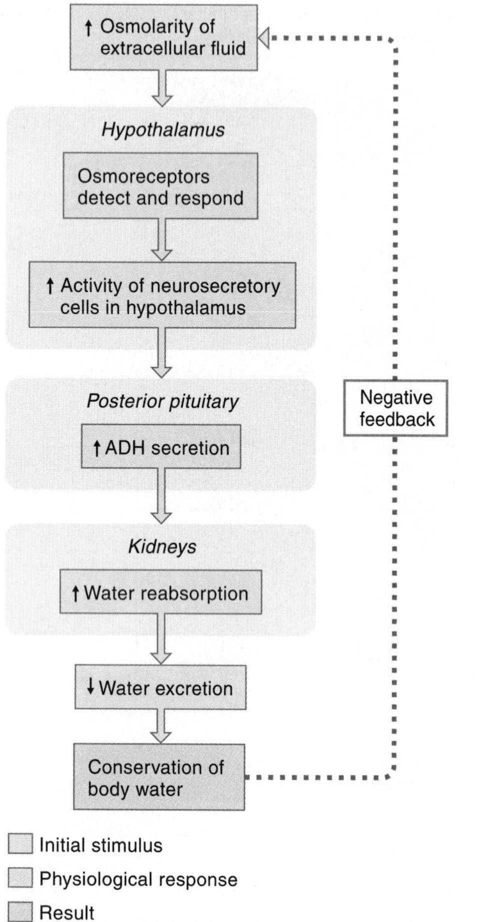

Figure 12 Pathway for extracellular fluid osmolarity and ADH secretion to interact. Increased osmolarity of extracellular fluid stimulates secretion of ADH, which increases water conservation by the kidneys and decreases extracellular fluid osmolarity.

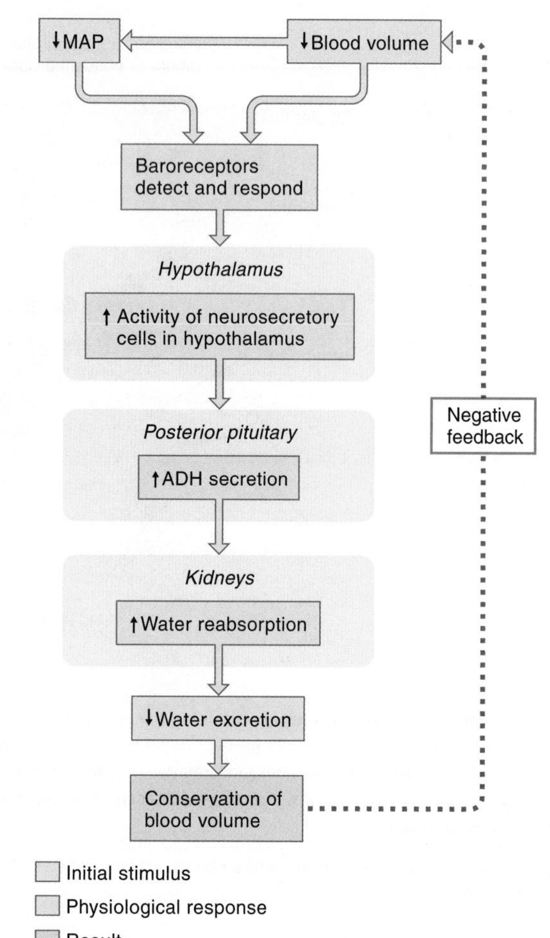

Figure 13 Effects of arterial and cardiac baroreceptors on ADH release. A decrease in mean arterial pressure and/or blood volume stimulates ADH release.

FOCUS ON DIABETES

Diabetes Insipidus

We most often think of diabetes as a disease associated with insulin and glucose. In reality, *diabetes* refers to any metabolic disorder causing an increase in thirst and the production of a large volume of urine (the word *diabetes* is derived from the Greek word for "siphon"). **Diabetes insipidus** occurs when ADH levels are low (central diabetes insipidus) or when the kidneys do not respond appropriately to ADH (nephrogenic diabetes insipidus). Because ADH acts on renal tubules to increase water reabsorption, a deficiency in ADH causes copious amounts of fluid loss in the urine.

The decrease in ADH characteristic of central diabetes insipidus can be caused by a head injury, inflammation of the hypothalamus, or tumors in the hypothalamus or posterior pituitary. Lack of response to ADH, which is characteristic of nephrogenic diabetes insipidus, can occur in renal diseases or can be hereditary. The most common hereditary form of diabetes insipidus is a mutation in the gene that codes for the ADH receptors found in the renal tubules. This gene is located on the X chromosome; thus hereditary diabetes insipidus occurs primarily in males. The symptoms of either form of diabetes insipidus include excessive urination *(polyuria)* and excessive fluid intake *(polydipsia)* as a compensatory response. People suffering from diabetes insipidus may lose as much as 20 liters of water per day! Excessive fluid loss results in high plasma sodium levels and increased plasma osmolarity. Treatment includes repairing the cause (for example, surgical removal of tumors) and fluid replacement.

Mechanisms of Sodium Reabsorption in the Renal Tubule

In all tubular segments where sodium is reabsorbed, sodium ions are actively transported. This active reabsorption is driven by Na^+/K^+ pumps located in the basolateral membrane of renal tubule epithelial cells. Because the active transport of sodium out of the epithelial cell keeps this ion's concentration low in the intracellular fluid, sodium passively enters the cell from the tubular lumen across the apical membrane. Even though this latter step is passive, the overall movement of sodium ions across the cell (that is, from tubular fluid to peritubular fluid) is active because it depends on the active transport of sodium across the basolateral membrane.

In the proximal tubule, the entry of sodium into the tubule epithelial cells is carried out by transport proteins in the apical membrane that couple sodium movement to the flow of other solutes. **Figure 14a** shows two such pathways for sodium entry into cells: (1) cotransport with solutes such as glucose and amino acids (designated by X in the figure), and (2) countertransport with hydrogen ions. In the first of these processes, energy released by the passive entry of sodium is harnessed to drive the flow of glucose or amino acids against their electrochemical gradients as they enter the epithelial cell. These solutes then exit the cell passively across the basolateral membrane. Given that the transport of glucose and amino acids from the tubular fluid to peritubular fluid requires an active step, the reabsorption of these solutes requires energy. The ultimate source of this energy is ATP hydrolysis, because the reabsorption of these solutes is coupled to the flow of sodium, which ultimately depends on the ATP-driven Na^+/K^+ pumps. Other transport proteins couple the passive entry of sodium to the active secretion of hydrogen ions into the tubule lumen, a process that is important in acid-base regulation. In addition, the active transport of sodium affects the movement of water across the renal tubules by osmosis.

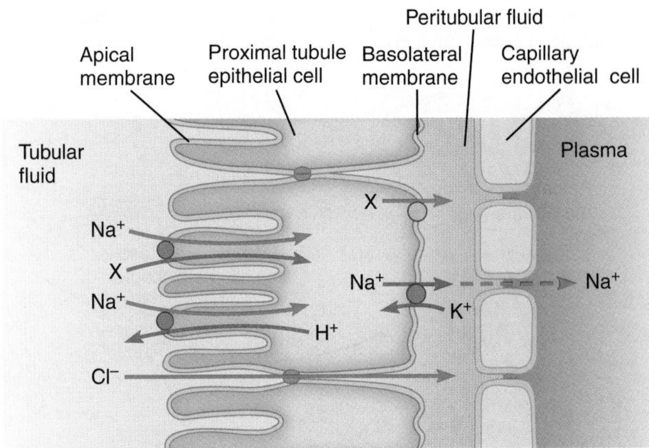

(a) Sodium reabsorption in the proximal tubule

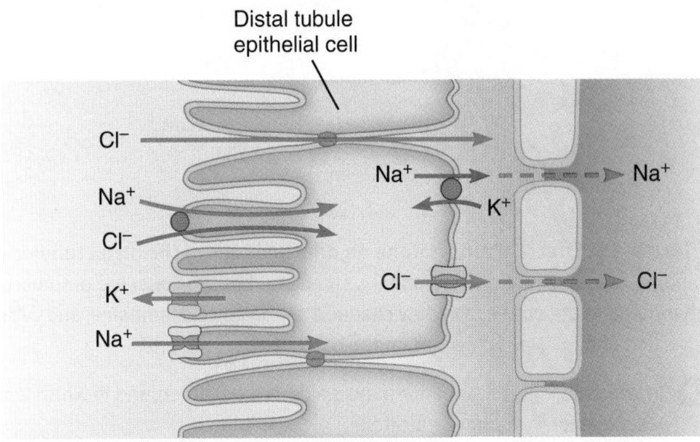

(b) Sodium reabsorption in the distal tubule

Figure 14 Mechanisms of sodium reabsorption in the proximal and distal tubules. (a) Sodium reabsorption in the proximal tubule. Sodium is actively transported across the basolateral membrane by the Na^+/K^+ pump. Sodium moves across the apical membrane either by cotransport with an organic molecule (X) such as glucose or an amino acid, or by countertransport with another ion such as hydrogen. Note that chloride follows sodium reabsorption. **(b)** Sodium reabsorption in the distal tubule. Again, sodium is actively transported across the basolateral membrane by the Na^+/K^+ pump. Sodium moves across the apical membrane either by cotransport with chloride ions or through sodium channels. Potassium secretion from peritubular fluid to the tubule lumen sometimes accompanies sodium reabsorption.

Apply Your Knowledge

Oubain blocks the Na^+/K^+ pump. Describe and explain the effects oubain would have on the renal transport of sodium and water.

Figure 14b shows the mechanism of active sodium reabsorption in the distal tubule. As in the proximal tubule, this process involves passive movement of sodium across the apical membrane into the tubule epithelial cell and active transport of sodium across the basolateral membrane out of the epithelial cell and into the peritubular fluid. In the distal tubule, however, sodium crosses the apical membrane by different mechanisms of passive transport. In the distal tubule, this ion enters the epithelial cell by two means: (1) by cotransport with other solutes such as chloride ions and (2) by diffusion through sodium channels.

Sodium reabsorption in the distal tubule is often coupled to potassium secretion, which, along with the cotransport of sodium with chloride, minimizes changes in the electrical potential that exists across the walls of the tubules. If the electrical potential is to be maintained, the reabsorption of a cation such as sodium must be balanced by reabsorption of anions (and to a lesser extent by secretion of other cations). Because chloride and bicarbonate are the most abundant anions in the tubular fluid, they account for the bulk of the anions that are reabsorbed with sodium. Potassium and hydrogen ions are the most common cations secreted as sodium ions are reabsorbed.

The Effects of Aldosterone

Aldosterone is a steroid hormone released from the adrenal cortex that regulates both the reabsorption of sodium and the secretion of potassium. Here we focus on the role of aldosterone in sodium reabsorption.

Aldosterone binds to cytosolic receptors in principal cells of the late distal tubules and collecting ducts, where it has several effects

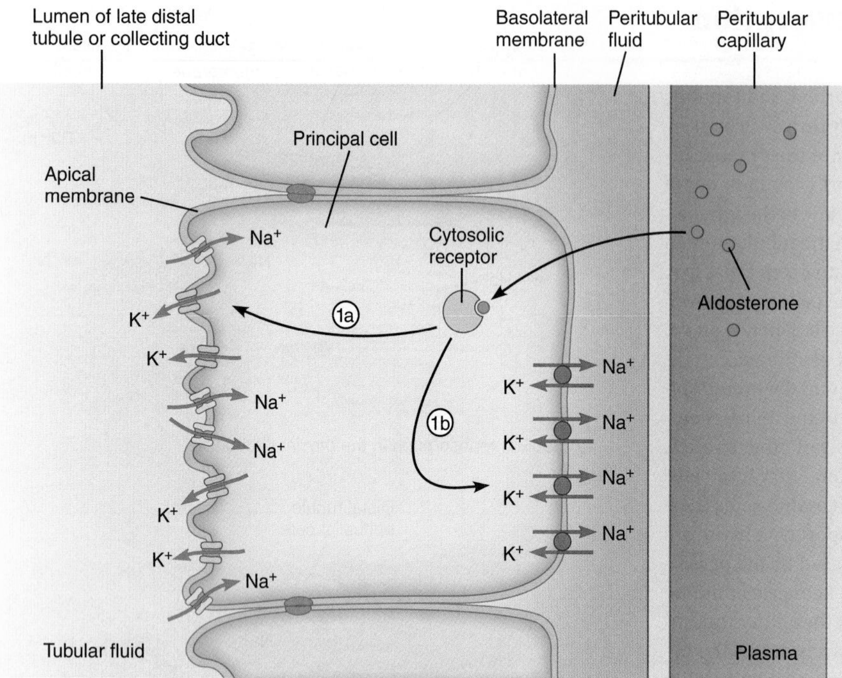

Figure 15 **Effects of aldosterone on principal cells of the distal tubules and collecting ducts.** After binding to its receptor, aldosterone (1a) stimulates both the opening of sodium channels and potassium channels and the synthesis of new channels on the apical membrane, and (1b) stimulates the synthesis and insertion of more Na^+/K^+ pumps on the basolateral membrane.

Q In which chemical class of hormone is aldosterone placed, and in which location in principal cells are receptors for the hormone located?

(**Figure 15**). Aldosterone increases the number of open sodium channels and potassium channels in the apical membrane, both by causing existing channels to open and by stimulating the synthesis of new channels. It also stimulates the synthesis of Na^+/K^+ pumps, which increases the concentration of Na^+/K^+ pumps in the basolateral membrane. Through these actions, aldosterone increases sodium reabsorption and potassium secretion simultaneously; it cannot affect one without affecting the other.

Of the various factors that control aldosterone release, the one that is most important in the control of sodium reabsorption is the renin-angiotensin-aldosterone system (RAAS).

The Renin-Angiotensin-Aldosterone System

Recall that where the distal tubule travels close to the afferent and efferent arterioles, these structures form the juxtaglomerular apparatus. Within the walls of the afferent arteriole are granular cells that secrete **renin.** Although often called a hormone, renin is actually a proteolytic enzyme. Within the walls of the distal tubule are the macula densa cells, which can detect changes in the sodium and chloride concentrations of the fluid and the flow rate of the tubular fluid. When sodium concentration in the tubular fluid decreases, renin secretion increases.

Once renin is released from the granular cells into the bloodstream, it initiates a series of reactions that lead to the release of aldosterone (**Figure 16**). Renin acts on another protein that is always present in the plasma, *angiotensinogen,* which (like most plasma proteins) is secreted by the liver. Renin cleaves off some amino acids from angiotensinogen, converting it to angiotensin I. As angiotensin I molecules circulate in the bloodstream, they encounter another proteolytic enzyme called **angiotensin converting enzyme (ACE),** which is bound to the inner surfaces of capillaries throughout the body and is particularly abundant in the capillaries of the lungs. ACE cleaves off some amino acids from angiotensin I, converting it to angiotensin II. In addition to acting as a vasoconstrictor that plays an important role in the regulation of mean arterial pressure , angiotensin II has another key function: It stimulates aldosterone release from the adrenal cortex. Angiotensin II also acts in the hypothalamus, where it stimulates ADH release and thirst.

Figure 17 summarizes the four mechanisms whereby angiotensin II increases mean arterial pressure:

1. Angiotensin II stimulates vasoconstriction of systemic arterioles, which by increasing the total peripheral resistance increases mean arterial pressure.
2. Angiotensin II stimulates the adrenal cortex to secrete aldosterone, which by increasing sodium reabsorption increases water reabsorption.
3. Angiotensin II stimulates the posterior pituitary to secrete ADH, which by increasing water reabsorption minimizes fluid loss and maintains plasma volume, thereby maintaining mean arterial pressure.
4. Angiotensin II activates hypothalamic neurons to stimulate thirst and fluid intake, which by increasing plasma volume increases mean arterial blood pressure.

Because the RAAS tends to increase blood pressure, a decrease in blood pressure serves as a primary stimulus for renin release through both intrinsic and extrinsic mechanisms (**Figure 18**). Two intrinsic controls of renin secretion exist. First, a decrease in afferent arteriolar pressure triggers renin release, because granular cells are directly sensitive to the degree of stretch of the afferent arteriole. Second, a large decrease in mean arterial pressure decreases the glomerular filtration rate, which, when coupled with the continual reabsorption of sodium and chloride in the proximal tubule and ascending loop of Henle, decreases sodium and chloride levels in the distal tubule. A decrease in sodium and chloride concentrations in the distal tubules is detected by macula densa cells of the tubules, which secrete a chemical signal that stimulates renin release from

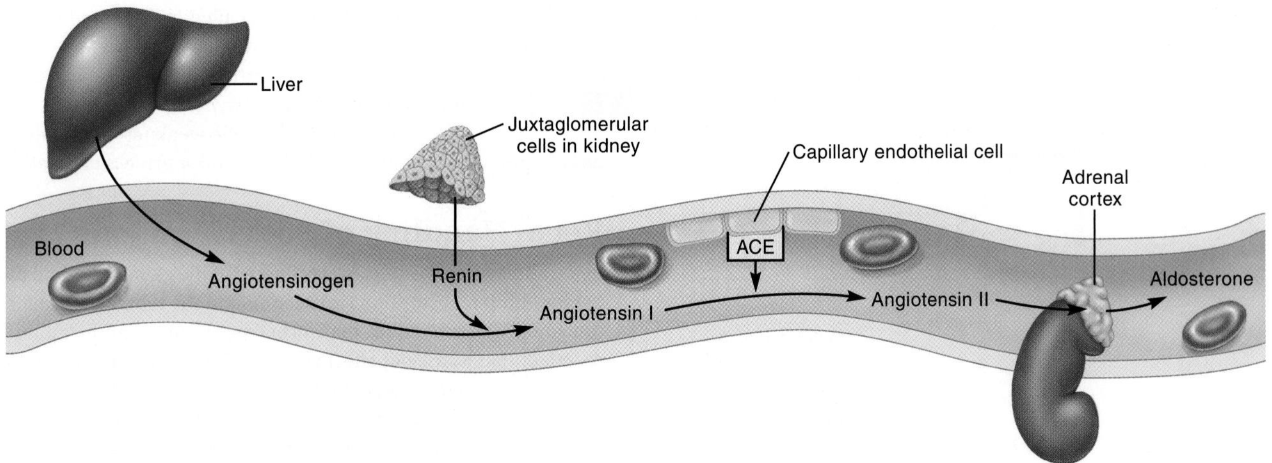

Figure 16 The renin-angiotensin-aldosterone system. The liver and the juxtaglomerular cells secrete and release angiotensinogen and renin, respectively, into the blood, where renin cleaves amino acids from angiotensinogen to form angiotensin I. Angiotensin converting enzyme (ACE), located on certain endothelial cells in capillaries, cleaves amino acids from angiotensin I to form angiotensin II, which travels in the bloodstream to the adrenal cortex, where it stimulates the release of aldosterone into the blood.

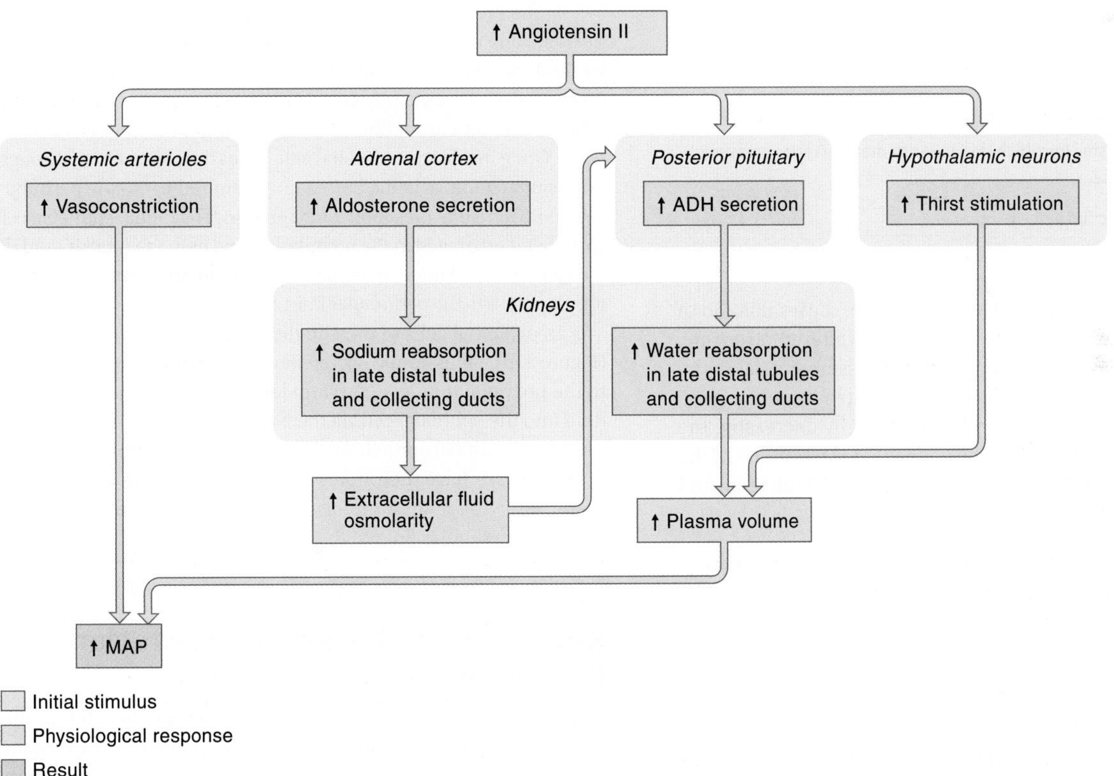

Figure 17 Mechanisms by which angiotensin II increases mean arterial pressure.

the juxtaglomerular cells. In terms of extrinsic control, an increase in renal sympathetic nerve activity stimulates renin release by direct input to granular cells. The sympathetic nervous system is activated during the baroreceptor reflex response to a decrease in blood pressure.

Atrial Natriuretic Peptide

Atrial natriuretic peptide (ANP) is secreted by cells in the atria of the heart in response to distension of the atrial wall, which occurs when plasma volume increases. ANP increases sodium excretion by increasing the glomerular filtration rate and by decreasing

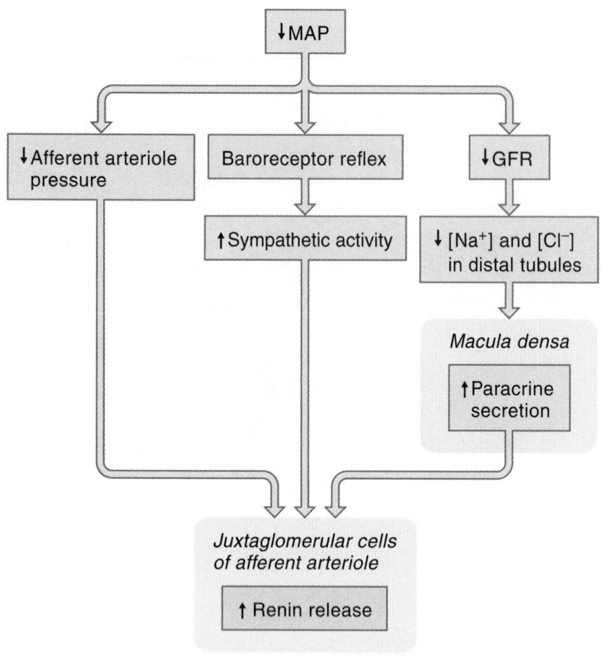

Figure 18 Mechanisms by which decreases in mean arterial pressure stimulate renin release.

Q What is renin's function in the plasma?

sodium reabsorption (**Figure 19**). This peptide causes dilation of the afferent arteriole and constriction of the efferent arteriole, which by increasing glomerular capillary pressure increases the glomerular filtration rate and increases the filtered sodium load. ANP decreases sodium reabsorption directly by decreasing the number of open sodium channels in the apical membrane of the principal cells. In addition, it decreases secretion of both renin and aldosterone.

Quick Check 3

1. Briefly explain how an increase in the secretion of renin stimulates the secretion of aldosterone. How does an increase in aldosterone secretion affect sodium reabsorption?
2. Where are the cells that secrete renin located? Name three stimuli for renin secretion.
3. How does atrial natriuretic peptide affect sodium reabsorption?

4 Potassium Balance

The gradient that results from high potassium concentrations in the intracellular fluid and low potassium concentrations in the extracellular fluid is critical to the function of excitable cells. An increase in plasma potassium levels is called *hyperkalemia;* it can cause cardiac arrhythmias, muscle weakness and cramps, dizziness, nausea, and diarrhea. A decrease in plasma potassium levels is called *hypokalemia;* it can cause cardiac arrhythmias, muscle weakness and tenderness, hypotension, confusion, alkalosis, and shortness of breath.

Renal Handling of Potassium Ions

In the kidneys, potassium is freely filtered at the glomerulus and undergoes both reabsorption and secretion in the tubules. Normally, the amount of potassium reabsorbed is greater than the amount secreted; that is, the net effect is reabsorption. In fact, most of the potassium filtered is reabsorbed.

Unlike with water and sodium, whose plasma levels are regulated by varying the amounts that are reabsorbed from the renal tubules, the plasma concentration of potassium is regulated by varying the amounts that are secreted into the renal tubules. As with sodium, renal handling of potassium ions varies within the renal tubules (**Figure 20**). Potassium ions are reabsorbed in the proximal tubule and secreted in the late distal tubule and collecting duct; the secretion is regulated.

In the proximal tubule (Figure 20a), potassium is reabsorbed by the following mechanisms: Potassium ions move from the peritubular fluid into the tubule epithelial cell via the Na^+/K^+ pump located on the basolateral membrane; potassium ions also move from the tubular fluid into the epithelial cell by some as yet unknown mechanism. Once inside the epithelial cell, potassium ions move through potassium channels in the basolateral membrane into the peritubular fluid. Therefore, most potassium entering the tubule epithelial cell (whether it originated in the peritubular fluid or in the tubular fluid) is reabsorbed. In addition, potassium ions can move between cells from tubule lumen to the peritubular fluid and then into the plasma.

In principal cells of the late distal tubule and collecting duct (Figure 20b), potassium is secreted by the following mechanism: As in the proximal tubule, potassium ions move from the peritubular fluid into the epithelial cell via the Na^+/K^+ pump in the basolateral membrane. Unlike epithelial cells in the proximal tubule, however, principal cells have their potassium channels in the apical membrane, allowing potassium ions to move out of the epithelial cell and into the tubular fluid of the distal tubule and collecting duct.

Regulation of Potassium Secretion by Aldosterone

Potassium secretion is regulated by aldosterone. Recall that this hormone increases the number of Na^+/K^+ pumps on the basolateral membrane in principal cells lining the late distal tubules and collecting ducts as well as the number of potassium channels in the apical membrane. The increase in Na^+/K^+ pumps causes greater potassium movement into the epithelial cells, which is followed by greater movement of potassium ions through apical potassium channels and into the lumen of the tubules, resulting in greater excretion of potassium in the urine.

As discussed previously, aldosterone secretion is regulated by the RAAS, whereby angiotensin II stimulates aldosterone release

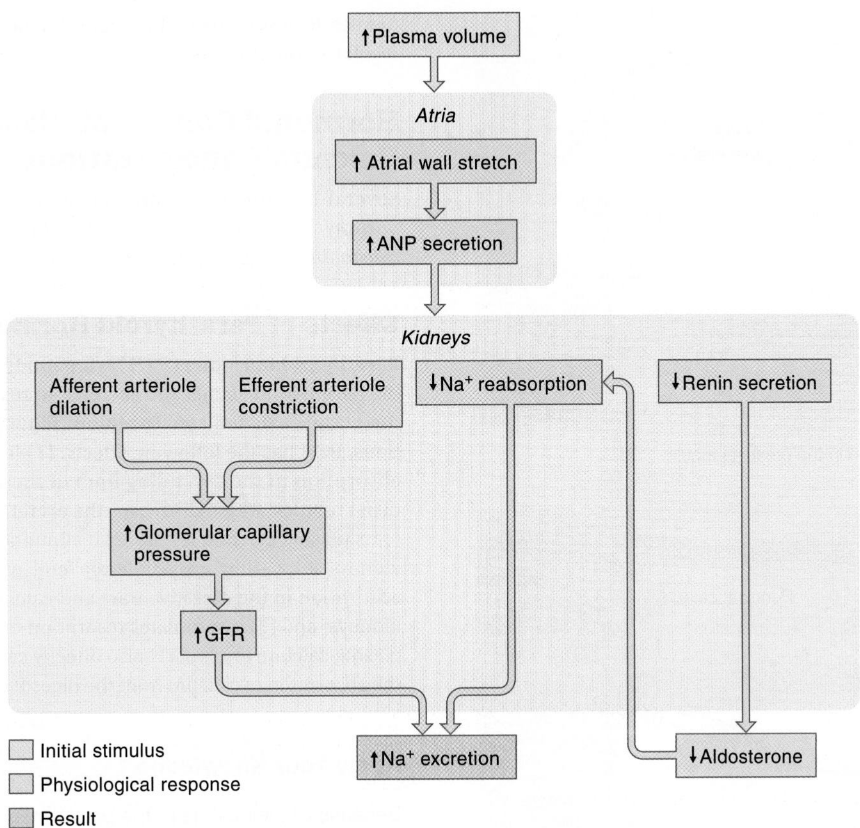

Figure 19 **Mechanisms by which secretion of atrial natriuretic peptide increases sodium excretion in response to increased plasma volume.**

from the adrenal cortex. However, high plasma potassium levels also stimulate aldosterone secretion directly, by acting on secretory cells in the adrenal cortex. The aldosterone released then increases potassium secretion, which brings plasma potassium levels toward normal.

5 Calcium Balance

Calcium is critical to the function of most cells: It triggers exocytosis of chemical messengers, stimulates secretion of various substances, stimulates muscle contraction, and increases the contractility of the heart and blood vessels. Calcium is also an important component of the bone and teeth. An increase

in plasma calcium, called *hypercalcemia,* has widespread effects on the body, including muscle weakness and atrophy, lethargy, behavioral changes, hypertension, constipation, and nausea. A decrease in plasma calcium, called *hypocalcemia,* causes numbness and tingling sensations, muscle cramps and spasms, exaggerated reflexes, and hypotension.

Plasma calcium concentration is regulated through the interaction of a number of organs, including the kidneys, digestive tract, bone, and skin (**Figure 21**). Calcium can be added to the plasma from bone and absorbed via the gastrointestinal tract, and it can be removed from the plasma by bone and the kidneys. Even though most of the calcium in the body (99%) is located in the bones, this calcium is not permanently fixed in the bone. The bone actually provides a reservoir of calcium such that when plasma calcium levels are low, the plasma can obtain calcium via a process called **resorption,** during which bone is broken down to liberate calcium ions. Conversely, when plasma calcium levels are high, calcium can be deposited into bone (see **Clinical Connections: Osteoporosis**). The body also obtains calcium from ingested food. Although the absorption of most substances by the gastrointestinal tract is not regulated, calcium absorption is regulated according to the needs of the body.

Renal Handling of Calcium Ions

Calcium is transported in blood both bound to carrier proteins and free in the plasma. Calcium that is free in the plasma is freely filtered at the glomerulus. Normally, 99% of the filtered calcium is reabsorbed as the tubular fluid moves through the renal tubules. Approximately 70% of the filtered calcium is reabsorbed in the proximal tubules, 20% is reabsorbed in the thick ascending limbs of the loops of Henle, and the remaining 10% is reabsorbed in the distal

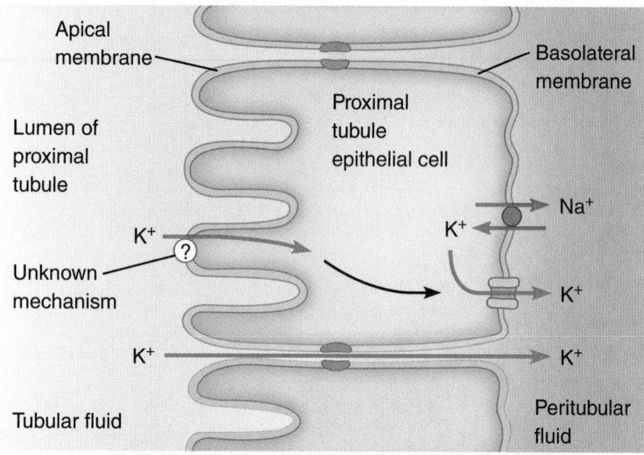

(a) Potassium reabsorption in the proximal tubule

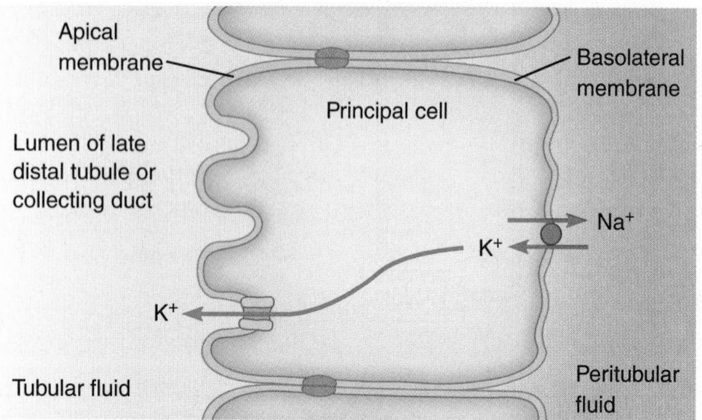

(b) Potassium secretion in the principal cells of the late distal tubule and collecting duct

Figure 20 Potassium transport in renal tubules. (a) In the proximal tubule, potassium is reabsorbed because of the presence of potassium channels in the basolateral membrane. **(b)** In principal cells of the distal tubule and collecting duct, potassium is secreted because of the presence of potassium channels in the apical membrane.

tubules. Reabsorption in the loops of Henle and the distal tubules is regulated by hormones.

Hormonal Control of Plasma Calcium Concentrations

Several hormones regulate plasma calcium levels, including *parathyroid hormone (PTH), 1,25-dihydroxycholecalciferol,* and *calcitonin.* PTH is the primary regulator of plasma calcium levels.

Effects of Parathyroid Hormone

Parathyroid hormone (PTH) is a peptide hormone produced in the parathyroid glands and secreted in response to a decrease in the plasma calcium concentration (**Figure 22**). Among its functions, PTH has the following effects: (1) It stimulates calcium reabsorption in the ascending limb of the loop of Henle and the distal tubules, which decreases the excretion of calcium and sustains plasma calcium levels; (2) it stimulates the activation in the kidneys of 1,25-dihydroxycholecalciferol, which stimulates calcium absorption in the digestive tract and calcium reabsorption in the kidneys; and (3) it stimulates resorption of bone, which increases plasma calcium levels. PTH also directly causes a small increase in the absorption of calcium from the digestive tract.

Apply Your Knowledge

Because of their location, the parathyroid glands were removed during thyroidectomy in the past. Which clinical symptoms would have indicated to early doctors that more than the thyroid gland was being removed during these procedures?

Effects of 1,25-Dihydroxycholecalciferol

1,25-Dihydroxycholecalciferol, or **1,25-(OH)$_2$D$_3$,** is a steroid hormone that acts to increase plasma calcium levels by stimulating calcium absorption from the digestive tract and calcium reabsorption in the distal tubules of the kidneys. 1,25-(OH$_2$)D$_3$ is synthesized

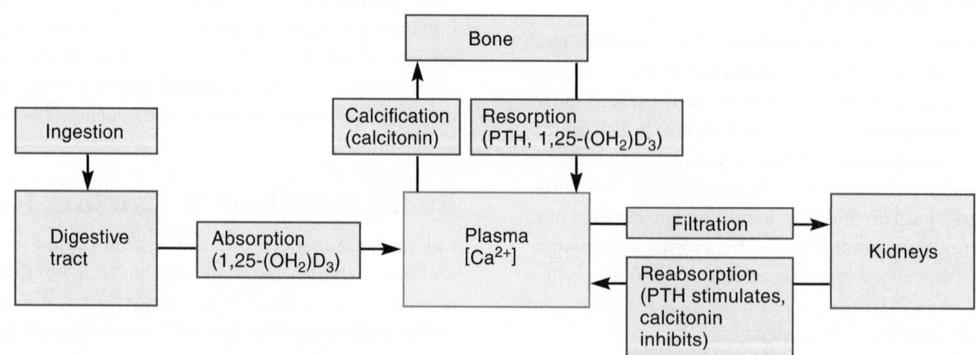

Figure 21 Routes of calcium exchange. Calcium can enter the plasma by absorption from the digestive tract or resorption of bone. Calcium leaves the plasma by calcification of bone or excretion in the urine. The amount of calcium excreted in the urine is regulated by varying the rate of calcium reabsorption.

CLINICAL CONNECTIONS

Osteoporosis 

Osteoporosis includes a group of disorders that result in a decrease in bone mass, which leads to bone fragility. Some forms of osteoporosis are localized to a specific bone. For example, immobilization of an extremity for a long period results in *disuse osteoporosis.* Most forms of osteoporosis, however, are more widespread and affect bones throughout the body. They include *primary osteoporosis,* in which the osteoporosis is the disorder, and *secondary osteoporosis,* in which the osteoporosis occurs secondarily to some other disorder, such as hyperthyroidism. Factors that increase the likelihood of developing primary osteoporosis include a genetic predisposition, obesity, and decreased physical activity.

Primary osteoporosis develops with age as bone undergoes changes in the rates of bone formation and resorption. Bone formation is carried out by cells called *osteoblasts*, whereas bone resorption is carried out by cells called *osteoclasts.* Bone formation exceeds bone resorption until a person reaches his or her early 30s, when bone reaches its maximum density. After the 30s, bone density decreases approximately 0.7% per year.

Certain hormones may be important in the development of osteoporosis. Estrogen has received considerable attention in this regard, because postmenopausal women are more subject to developing osteoporosis. A decrease in estrogen causes osteoblasts to increase production of a paracrine called *interleukin-6.* Interleukin-6 stimulates osteoclast activity, causing bone resorption. Estrogen may also stimulate osteoblast activity, such that a decrease in estrogen would result in a decrease in osteoblast activity and, in turn, less bone formation.

Diagnosis of osteoporosis is difficult during the early stages because bone degeneration must reach a level observable on X-ray examination before the disease's presence can be confirmed. Often, diagnosis is not made until after a bone fracture has resulted from the weakened state.

Treatment of osteoporosis often includes hormonal or dietary supplements. Estrogen supplements are commonly given to postmenopausal women to decrease the rate of bone resorption. Unfortunately, estrogen treatment cannot reverse any damage that has already occurred.

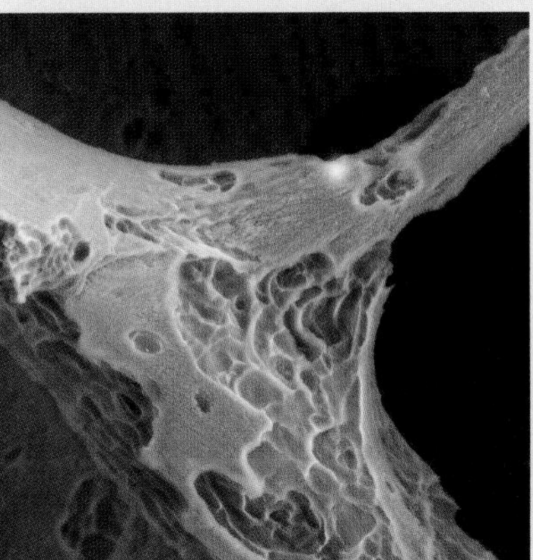

Dr. Tony Brain/Photo Researchers, Inc.

A scanning electron micrograph of osteoporosis in an 89-year-old female.

Critical Thinking Questions

1. What are the different types of osteoporosis, and what factors increase the likelihood of developing the condition?
2. How does estrogen influence the activity of osteoblast and osteoclast cells, and what is the role of osteoblasts and osteoclasts in osteoporosis development?
3. Describe the diagnosis process and the treatment options for osteoporosis.

in several steps from vitamin D_3, which can be either synthesized from 7-dehydrocholesterol in the skin upon exposure to sunlight or absorbed from the diet (**Figure 23**). Once in the plasma, vitamin D_3 travels to the liver, where it is converted to 25-hydroxyvitamin D_3, or 25-OH D_3. From the liver, 25-OH D_3 travels in the bloodstream to the kidneys, where, in response to low plasma calcium levels, PTH stimulates the conversion of 25-OH D_3 to 1,25-(OH_2) D_3.

Effects of Calcitonin

Unlike PTH and 1,25-$(OH_2)D_3$, calcitonin decreases plasma calcium levels. **Calcitonin** is a peptide hormone secreted from C cells of the thyroid gland, which are distinct from the cells that secrete thyroid hormone. Calcitonin secretion is triggered by increases in plasma calcium levels. Although the primary action of calcitonin is to increase bone formation by depositing calcium, it also decreases the reabsorption of calcium by the kidneys, which leads to an increase in urinary excretion of calcium; both of these actions decrease plasma calcium levels.

Quick Check 4

1. Which hormone regulates potassium secretion? What are some stimuli for the release of this hormone?
2. Name the three hormones that affect plasma calcium levels. Describe the effects of each.

6 Interactions Between Fluid and Electrolyte Regulation

The previous sections described fluid and electrolyte balance as though water and ions were independently regulated. In reality, that is not the case, because reabsorption of any ion increases the osmotic gradient for water reabsorption. In addition, hormones often affect multiple systems. Consider, for example, ADH: In addition to increasing water reabsorption, it stimulates production of sodium channels in the principal cells, thereby enhancing the movement of sodium ions from the tubule lumen into these cells.

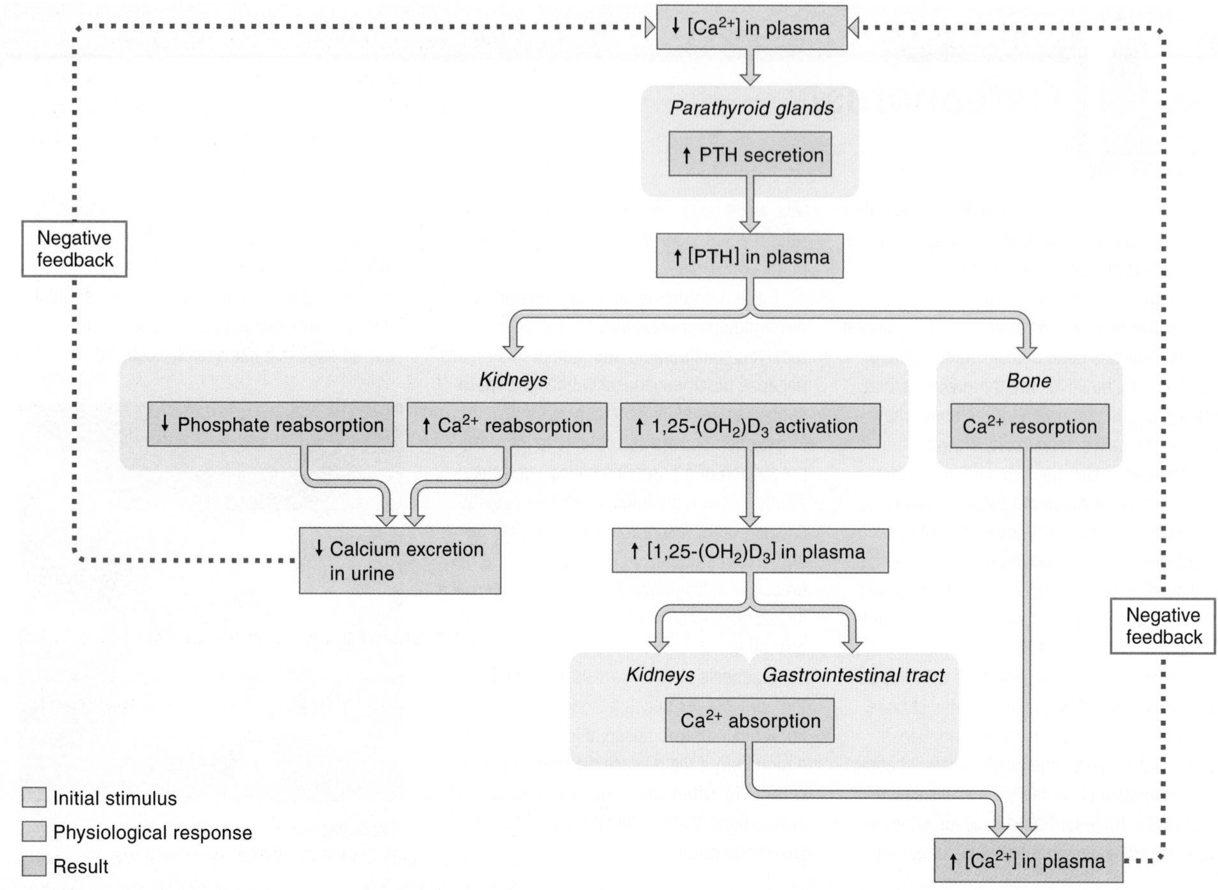

Figure 22 Role of parathyroid hormone in calcium balance. PTH release is stimulated by a decrease in plasma calcium levels. PTH then acts on the kidneys to increase calcium reabsorption and activate 1,25-(OH_2) D_3, and on the bone to increase calcium resorption, all of which leads to increases in plasma calcium levels.

As more sodium enters the cell, the Na^+/K^+ pump works faster to rid the cell of sodium, increasing sodium reabsorption and potassium secretion. ADH also decreases water flow to the late distal tubules and collecting ducts because less water is left following its reabsorption. The decrease in water increases solute concentration in the tubules, which affects the reabsorption or secretion of any solutes present in these tubules. In other examples, angiotensin II and ANP affect water reabsorption in addition to having their primary effects on sodium reabsorption. Whereas angiotensin II increases ADH secretion, which stimulates water reabsorption, ANP inhibits ADH secretion, which decreases water reabsorption.

We can illustrate the interactions among the various systems regulating fluid and electrolyte balance by considering the events following hemorrhage. We first used this example to demonstrate short-term neural control of mean arterial pressure. Here we focus on the restoration of blood volume following hemorrhage.

Hemorrhage results in a decrease in blood volume, which decreases venous pressure and, therefore, decreases venous return. With less blood returning to the heart, stroke volume is decreased, causing a decrease in mean arterial pressure. This decrease in mean arterial pressure is detected by arterial baroreceptors, which (through the baroreceptor reflex) activate the sympathetic nervous system and inhibit the parasympathetic nervous system. As a result, heart rate and stroke volume increase, and total peripheral resistance increases, all of which produce an increase in mean arterial pressure. This neural control of blood pressure occurs very rapidly, but it does not correct for the initial cause of the problem—the loss of blood volume.

Through its effects on blood volume, hormonal control of the kidneys is instrumental in long-term regulation of blood pressure. Several factors induced by hemorrhage contribute to renal control of water loss:

1. The decrease in mean arterial pressure stimulates renin secretion both directly and indirectly as follows: Pressure in the afferent arteriole is decreased, which stimulates renin secretion. The baroreceptor reflex increases sympathetic input to juxtaglomerular cells, which also stimulates renin secretion. Increased renin secretion from the kidneys causes an increase in plasma angiotensin II levels. Angiotensin II causes vasoconstriction, which increases the total peripheral resistance and, therefore, mean arterial pressure. Angiotensin II also stimulates the adrenal cortex to release aldosterone, which increases

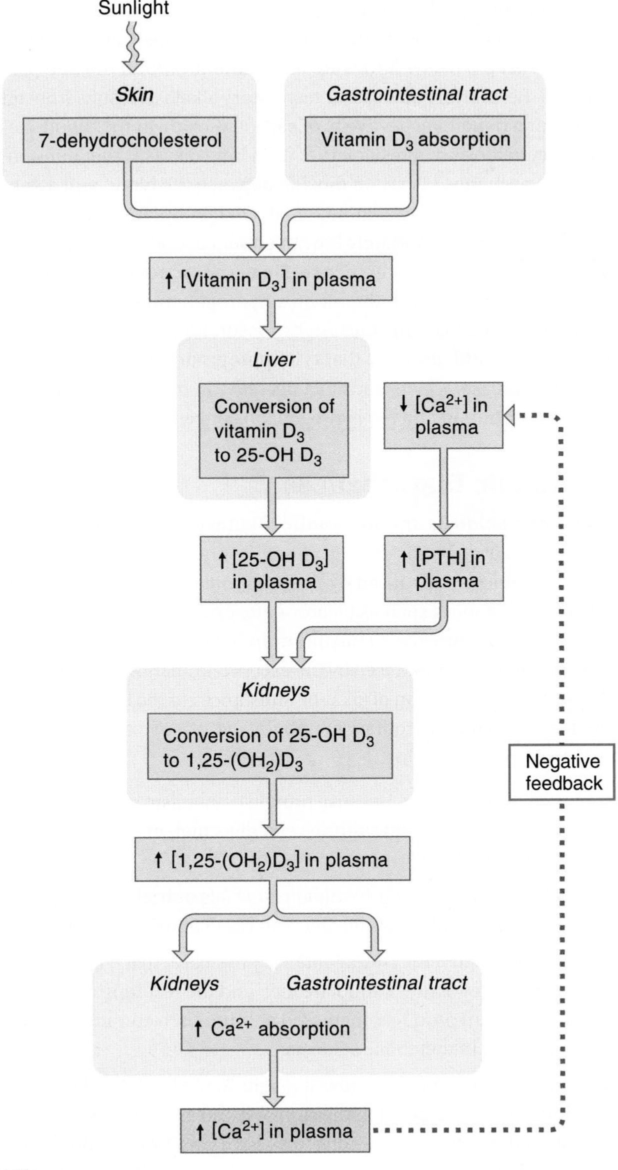

Figure 23 Activation of 1,25-$(OH_2)D_3$. Several steps are required to convert vitamin D_3 to 25-OH D_3 in response to low plasma calcium levels. PTH regulates the conversion of 25-OH D_3 to 1,25-(OH_2) D_3.

sodium reabsorption in the kidneys, which in turn increases water reabsorption. Moreover, angiotensin II acts in the hypothalamus to stimulate thirst, which increases fluid intake and, subsequently, plasma volume.

2. Decreased activity in venous and cardiac baroreceptors stimulates ADH release from the posterior pituitary gland. ADH increases water reabsorption in the kidneys, conserving plasma volume.
3. A decrease in mean arterial pressure decreases the glomerular filtration rate, which conserves water and sodium.
4. Hemorrhage is characterized by the loss of both plasma and blood cells. The blood cells are replaced through the synthesis of new cells, which takes days to be completed. The decrease in mean arterial pressure results in a decrease in blood flow to the kidneys, so the oxygen supply to the kidneys is decreased, and erythropoietin secretion is stimulated. Recall that following its release by the kidneys, erythropoietin travels in the blood to the bone marrow, where it stimulates the production of new erythrocytes.

Hemorrhage provides an example of how organ systems must work together to maintain homeostasis. In the next section we see how acid-base balance of the plasma is maintained by interactions between two organ systems: the lungs and the kidneys.

Quick Check 5

1. Describe the relationship between blood volume and blood pressure.
2. Compare the short-term and long-term regulation of blood pressure.
3. Name the hormones affecting blood volume, and describe their major actions.

7 Acid-Base Balance

The hydrogen ion concentration or pH of arterial blood is regulated by the combined actions of the lungs and the kidneys. It is essential that the arterial pH be tightly controlled within its normal range of 7.38 to 7.42. Note that this is a slightly alkaline pH. A decrease in pH (increase in hydrogen ion concentration) to less than 7.35 is called acidosis; an increase in pH (decrease in hydrogen ion concentration) to greater than 7.45 is called alkalosis; a pH range of 7.42 to 7.45 is considered borderline alkalosis. Changes in pH of as little as a few tenths of a unit in either direction can have profound effects on the body:

1. Interactions between hydrogen ions and certain amino acids result in conformational changes in proteins, and thus in changes in the proteins' functions. For example, enzyme activity may increase or decrease with pH changes.
2. Activity of the nervous system changes, although the precise mechanisms are not fully understood. Acidosis causes a decrease in the excitability of neurons, especially in the central nervous system (CNS); severe acidosis can lead to confusion, coma, and even death. Alkalosis, such as that produced by hyperventilation, causes an increase in the excitability of neurons.

3. Because of complex interactions in the movement of potassium and hydrogen ions across renal tubules (such as electrochemical interactions and competition for carrier proteins), acid-base disturbances are often linked to potassium imbalances. Acidosis results in potassium retention (hyperkalemia), whereas alkalosis results in potassium depletion (hypokalemia).

4. Acidosis causes cardiac arrhythmias and vasodilation of blood vessels to the skin due to impaired activity of catecholamines.

Apply Your Knowledge

A person hyperventilating often feels a tingling sensation in the hands and feet and may have muscle spasms and twitches. Explain the cause of these symptoms.

Sources of Acid-Base Disturbances

Figure 24 shows the various inputs and outputs of acid (free hydrogen ions) to the blood. Inputs include dietary and metabolic sources. Protein and fats in the diet provide hydrogen ions in the forms of amino acids and fatty acids. Cellular metabolism produces several acids, including carbon dioxide, lactic acid, and keto acids. Hydrogen ions can be removed from the blood by the kidneys or lungs. Whereas the lungs remove acid in the form of carbon dioxide during ventilation, the kidneys excrete hydrogen ions in the urine. As with many other systems in the body, these inputs and outputs must be balanced if normal blood pH is to be maintained. Because acid production occurs regularly in the body during metabolism, potential increases in blood acidity must be prevented.

Respiratory Disturbances

One source of acid is carbon dioxide. Recall that the enzyme carbonic anhydrase catalyzes the reaction converting carbon dioxide and water to carbonic acid, which dissociates to bicarbonate and hydrogen ions:

$$CO_2 + H_2O \xrightleftharpoons{\text{carbonic anhydrase}} H_2CO_3 \rightleftharpoons HCO_3^- + H^+$$

Carbon dioxide is produced during cellular metabolism—in particular, in the Krebs cycle. As the amount of carbon dioxide in the blood increases, it is converted to bicarbonate and hydrogen ions. When blood rich in carbon dioxide reaches the lungs, the reaction reverses, and carbon dioxide is exhaled. The respiratory system contributes to acid-base balance by regulating carbon dioxide levels in the blood.

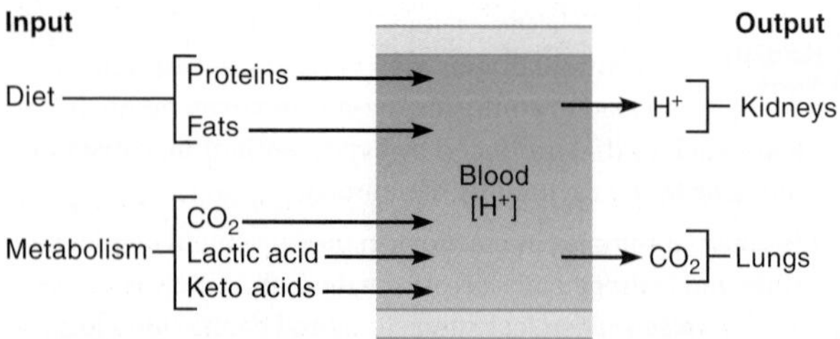

Figure 24 Inputs and outputs of acid to the blood. Diet is the primary contributor of acids to the blood, and cellular metabolism generates additional acids. To compensate for acid inputs, the kidneys excrete hydrogen ions and the respiratory system clears the blood of carbon dioxide.

Normally, the P_{CO_2} of arterial blood is maintained at 40 mm Hg by respiratory chemoreceptor reflexes, which adjust alveolar ventilation to keep pace with the rate of metabolic carbon dioxide production. Respiratory acidosis or respiratory alkalosis results from an excess or a deficit, respectively, of carbon dioxide in the blood. Respiratory acidosis can result either from lung diseases that interfere with the exchange of carbon dioxide between the blood and alveolar air, or from hypoventilation, an alveolar ventilation rate that is inappropriately low. In either case, arterial P_{CO_2} rises, leading to an increase in carbonic acid levels and an increase in the blood's hydrogen ion concentration. Respiratory alkalosis, in contrast, can result from hyperventilation, an alveolar ventilation rate that is inappropriately high; this causes P_{CO_2} to decrease, which leads to a decrease in the blood's carbonic acid concentration and the hydrogen ion concentration.

Metabolic Disturbances

Metabolic acidosis and **metabolic alkalosis** are disturbances in blood pH caused by something other than an abnormal P_{CO_2}. Metabolic acidosis can be caused by excess elimination from the body of alkaline substances such as bicarbonate, excess production of acid in metabolism, or excess consumption of acids in the diet. Metabolic alkalosis can be caused by the excess elimination of acids from the body or the addition of alkaline substances to the blood. Among the factors that can produce metabolic disturbances in acid-base balance are the following:

1. *A high-protein diet.* Because protein catabolism produces phosphoric acid and sulfuric acid, diets high in protein can result in metabolic acidosis.
2. *A high-fat diet.* Because catabolism of fats or triglycerides produces fatty acids, high-fat diets can result in metabolic acidosis.
3. *Strenuous exercise.* During intense physical activity, the oxygen demands of tissues cannot be met, and the resulting anaerobic metabolism produces increased amounts of lactic acid, which can result in metabolic acidosis.
4. *Excessive vomiting.* Because it results in the loss of hydrogen ions that are secreted in the stomach and are normally reabsorbed in the small intestine, excessive vomiting can produce metabolic alkalosis.
5. *Severe diarrhea.* Because it results in the loss of bicarbonate, which is produced in the upper small intestine and is normally reabsorbed in the lower small intestine, severe diarrhea can produce metabolic acidosis.
6. *Alterations in renal function.* Because the kidneys secrete hydrogen ions and reabsorb bicarbonate (as described shortly), changes in renal function can produce metabolic acidosis or metabolic alkalosis.

Defense Mechanisms Against Acid-Base Disturbances

Any alteration in the rate of acid or base production by the body changes the pH of arterial blood by upsetting the balance between the rates at which hydrogen ions are added to and removed from the blood. Fortunately, three "lines of defense" protect against such changes in pH: (1) *buffering of hydrogen ions,* which occurs through the binding or release of hydrogen ions by substances that are always present in the blood and other body fluid compartments; (2) *respiratory compensation,* which adjusts the rate at which carbon dioxide is cleared from the blood via elimination through the lungs; and (3) *renal compensation,* which adjusts the rate at which hydrogen ions are secreted and bicarbonate ions are reabsorbed by renal tubules. The term *compensation* is used because the respiratory and renal systems do not generally correct the cause of an acid-base disturbance; they simply compensate for it by activating mechanisms that work to restore normal arterial pH.

Buffering of Hydrogen Ions

Buffers—the first line of defense against changes in pH—act immediately to compensate for disturbances in pH. A chemical buffer is a compound that minimizes changes in pH when either an acid or a base is added to or removed from a solution (**Chemistry Review: Buffers**). The most important buffer in the extracellular fluid (which includes plasma) is bicarbonate. The buffering ability of bicarbonate is shown in the following equation:

$$HCO_3^- + H^+ \rightleftharpoons H_2CO_3$$

Important buffers located primarily in intracellular fluid include proteins and phosphates, whose buffering abilities are shown in the following equations:

$$\text{protein}^- + H^+ \rightleftharpoons H \cdot \text{protein}$$
$$HPO_4^{2-} + H^+ \rightleftharpoons H_2PO_4^-$$

The law of mass action determines whether a buffer binds or releases hydrogen ions. Consider the following example, which shows hydrogen ions (H^+) binding reversibly to a buffer (A^-):

$$H^+ + A^- \rightleftharpoons HA$$

When acid is added to a solution, the resulting increase in the hydrogen ion concentration pushes the reaction to the right, so that the ionized form of the buffer (A^-) combines with hydrogen ions to produce the protonated (or acid) form of the buffer (HA):

$$\overset{\text{add}}{\Downarrow}$$
$$H^+ + A^- \rightleftarrows HA$$

When acid is removed from a solution, the resulting decrease in hydrogen ion concentration pulls the equilibrium to the left, so that the protonated form of the buffer releases hydrogen ions:

$$\overset{\text{remove}}{\Uparrow}$$
$$H^+ + A^- \leftrightarrows HA$$

Of the three lines of defense against alterations in blood pH, buffering works the most rapidly; its response time is limited only by the time required for buffers to bind or release hydrogen ions. However, buffering can only *limit* changes in pH by binding or releasing hydrogen ions; it cannot *reverse* changes in pH unless more buffer molecules are added to or removed from the blood. Consider, for example, a situation in which pH has decreased. Once a buffer molecule has bound a hydrogen ion to remove it from solution and thus minimize the reduction in pH, that molecule cannot bind another hydrogen ion. The excess hydrogen ions bound to buffers must eventually be eliminated from the body, or the buffering capacity of blood will be exceeded. Therefore, buffering alone cannot defend against pH changes. Once arterial pH has deviated from its normal value, it can be returned to its norm only by respiratory or renal compensation.

Respiratory Compensation

The respiratory system—the second line of defense against changes in blood pH—usually acts within minutes. The respiratory system regulates pH by increasing or decreasing alveolar ventilation, which tends to raise or lower pH, respectively. An increase in alveolar ventilation lowers arterial P_{CO_2}. Given that carbon dioxide is in equilibrium with hydrogen ions and bicarbonate, the law of mass action dictates that hydrogen ions and bicarbonate ions decrease, as follows:

$$\overset{\text{increased alveolar ventilation removes } CO_2}{\Uparrow}$$
$$CO_2 + H_2O \leftrightarrows H_2CO_3 \leftrightarrows H^+ + HCO_3^-$$

Conversely, a decrease in alveolar ventilation raises the arterial P_{CO_2}. By the law of mass action, hydrogen ions and bicarbonate ions increase, as follows:

$$\overset{\text{decreased alveolar ventilation adds } CO_2}{\Downarrow}$$
$$CO_2 + H_2O \rightleftarrows H_2CO_3 \rightleftarrows H^+ + HCO_3^-$$

Unlike simple hydrogen ion buffering, which minimizes but cannot reverse changes in pH, respiratory compensation is a true homeostatic regulatory mechanism that can reverse pH changes. **Figure 25** shows what happens to ventilation when plasma pH decreases. The increase in hydrogen ions in the plasma activates the peripheral chemoreceptors, which reflexively increase ventilation. An increase in ventilation causes a decrease in arterial P_{CO_2}, which by the law of mass action induces the conversion of bicarbonate and hydrogen ions to carbon dioxide, thereby removing free hydrogen ions from solution and increasing the pH.

The respiratory system alone generally cannot completely restore the pH to its normal level. Usually, the last line of defense, renal compensation, must be called into action as well.

Renal Compensation

The third line of defense against changes in blood pH is the renal system, which takes hours or even days to compensate for changes in pH. The kidneys regulate the pH of arterial blood by regulating the renal excretion of hydrogen ions and bicarbonate, and by producing new bicarbonate, according to the following rules: *If the hydrogen ion concentration in the blood increases, the kidneys increase hydrogen ion secretion and bicarbonate reabsorption and synthesize new bicarbonate; if hydrogen ion concentration in the blood*

CHEMISTRY REVIEW

Buffers

A hydrogen ion *buffer* is an acid-base pair that has the ability to minimize changes in the proton concentration or pH by taking up hydrogen ions (H^+) from a solution when acid is added and by releasing hydrogen ions into the solution when acid is removed. The pH of a solution declines as the proton concentration increases and is defined as the logarithm of the inverse of the proton concentration: $pH = \log(1/[H^+])$. Because a buffer must be able to release and bind hydrogen ions, it functions as both an acid and a base. Substances that can release and take up hydrogen ions are defined as *weak acids* because they have a tendency to give up some, but not all, of their bound hydrogen ions when dissolved in water. In other words, they *partly dissociate* in solution. In contrast, a *strong acid,* such as hydrochloric acid (HCl), has little tendency to bind hydrogen ions and *completely dissociates* in solution.

To see how a buffer works, consider what happens when a weak acid (HA) is dissolved in water. The acid releases some of its hydrogen ions into solution, such that some molecules go into the base form (A^-). Over a given time period, some of these bases take up other hydrogen ions from solution. In so doing, they revert to the acid form, even as an equal number of acids release their hydrogen ions, thereby going to the base form. As a consequence, the concentration of hydrogen ions in solution does not change, and the concentrations of the acid and base forms of the buffer remain constant. Under these conditions, the system is in equilibrium, which is represented as follows:

$$HA \rightleftharpoons H^+ + A^-$$

When acid (hydrogen ions) is added to water, the pH decreases because the concentration of hydrogen ions increases. If a buffer is present, this increase in concentration pushes the equilibrium to the left by the law of mass action:

$$\overset{\text{add } H^+}{\Downarrow}$$
$$HA \;\overset{\longleftarrow}{\rightharpoondown}\; H^+ + A^-$$

The combination of hydrogen ions with the base form of the buffer (A^-) removes some of the added hydrogen ions from solution. Consequently, the pH still falls, but not as much as it would if the buffer were not present. When acid is removed from a solution, the pH rises because the hydrogen ion concentration falls. If a buffer is present, the decrease in hydrogen ion concentration pulls the equilibrium to the right:

$$\text{remove } H^+$$
$$\Uparrow$$
$$HA \;\overset{\longrightarrow}{\leftharpoondown}\; H^+ + A^-$$

The release of hydrogen ions from the acid form of the buffer molecule (HA) adds hydrogen ions to the solution, which replaces some of the hydrogen ions removed initially. Thus the pH rises, but not as much as it would have in the absence of the buffer.

From this discussion it is apparent that buffers limit increases or decreases in the pH of a solution when acid is added or removed. When acid is added to a given volume of pure water, the size of the resulting decrease in pH depends on the quantity of acid involved. When the same quantity of acid is added in the presence of a buffer, the resulting decrease in pH is smaller, as seen in the following figure:

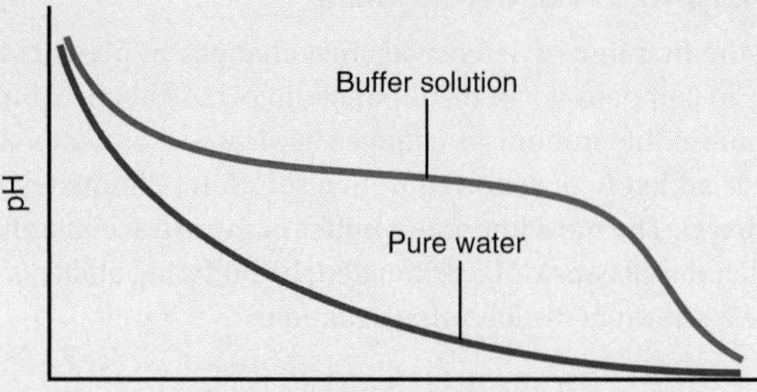

Notice that when the buffer is present, the pH is fairly constant over a certain range, but it changes more quickly as the pH becomes either very high or very low. As a result, a buffer's ability to protect against changes in pH is restricted to a certain range of pH values (the flat portion of the curve). Outside this range, a buffer is relatively ineffective at limiting changes in pH. Different buffers are most effective over different ranges of pH, depending on how strongly they tend to dissociate.

To be useful in protecting the blood against changes in pH, a buffer's effective range must be within the physiological pH range. The buffer should also be present at sufficient concentration to enable it to consume or release significant quantities of hydrogen ions. Important physiological buffers include bicarbonate ions (HCO_3^-) and monoprotonated phosphates (HPO_4^{2-}). These buffers take up and release hydrogen ions as follows:

$$H^+ + HCO_3^- \rightleftharpoons H_2CO_3$$

$$H^+ + HPO_4^{2-} \rightleftharpoons H_2PO_4^-$$

Proteins are also important physiological buffers because they contain amino acids that possess carboxyl groups ($—COOH$) and amino groups ($—NH_2$), which can accept or give up hydrogen ions as follows:

$$H^+ + —COO^- \rightleftharpoons —COOH$$
$$H^+ + —NH_2 \rightleftharpoons —NH_3^+$$

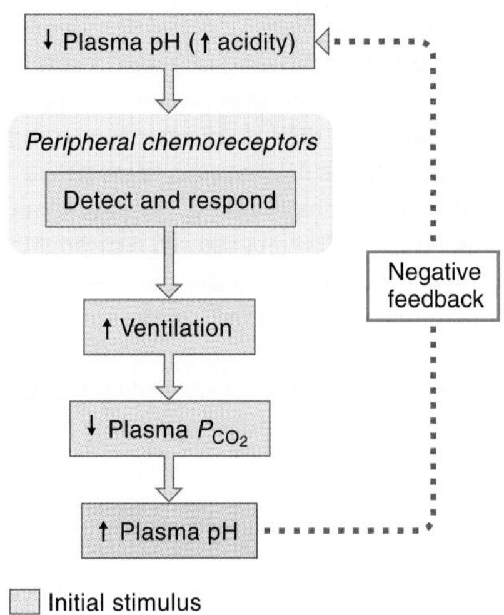

Figure 25 The mechanism by which decreases in plasma pH increase ventilation.

Q Where are the peripheral chemoreceptors located?

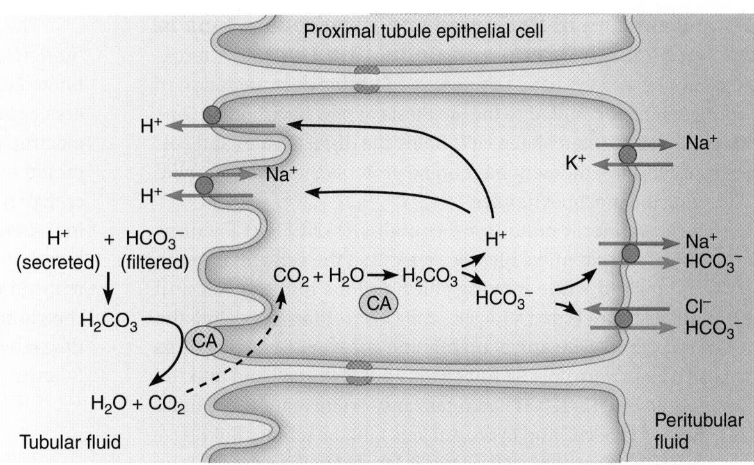

Figure 26 Bicarbonate reabsorption and hydrogen ion secretion in the proximal tubule. Filtered bicarbonate ions combine with secreted hydrogen ions to form carbonic acid, which is converted to water and carbon dioxide by carbonic anhydrase on the apical membrane. The carbon dioxide diffuses into the epithelial cell, where intracellular carbonic anhydrase catalyzes the conversion of carbon dioxide and water to carbonic acid; the carbonic acid then dissociates into bicarbonate and hydrogen ions. The hydrogen ions are secreted by countertransport with sodium ions, whereas the bicarbonate ions are reabsorbed by cotransport with sodium ions and by countertransport with chloride ions.

decreases, the kidneys decrease hydrogen ion secretion and bicarbonate reabsorption. The secretion of hydrogen ions is coupled to the reabsorption or synthesis of bicarbonate ions, as described shortly.

Several substances critical to renal compensation for acid-base disturbances are filtered at the glomerulus, including CO_2, H^+, HCO_3^-, and HPO_4^{2-}. The fates of these substances vary in the different segments of the renal tubules.

Renal Handling of Hydrogen and Bicarbonate Ions in the Proximal Tubule In the proximal tubule, bicarbonate reabsorption is coupled to hydrogen ion secretion (**Figure 26**). In epithelial cells of the proximal tubule, several carrier proteins required for movement of hydrogen or bicarbonate ions are located on either the basolateral or apical membrane. Which transporters are active depends on the pH of the extracellular fluid.

The basolateral membrane contains three transporters critical to acid-base balance: (1) Na^+/K^+ pumps that transport sodium ions out of the cell and into the peritubular fluid while transporting potassium ions into the cell, (2) Na^+/HCO_3^- cotransporters that transport both sodium and bicarbonate ions out of the cell and into the peritubular fluid, and (3) HCO_3^-/Cl^- countertransporters that transport chloride ions into the cell and bicarbonate ions into the peritubular fluid.

The apical membrane contains two transporters critical to acid-base balance: (1) Na^+/H^+ countertransporters that transport sodium ions into the cell and hydrogen ions out of the cell and into the tubular fluid, and (2) H^+ pumps that use ATP to transport hydrogen ions into the tubular fluid.

The enzyme carbonic anhydrase (CA), which is located in the cytosol and on microvilli of the apical membrane of the epithelial cell, catalyzes the following reversible reaction:

$$CO_2 + H_2O \rightleftharpoons H_2CO_3$$

The membrane-bound carbonic anhydrase converts carbonic acid (which comes from filtered bicarbonate ions) to carbon dioxide in the lumen of the proximal tubule. The carbon dioxide then diffuses into the epithelial cell, where it is converted back to carbonic acid by an intracellular carbonic anhydrase. The carbonic acid then dissociates by the following reversible reaction:

$$H_2CO_3 \rightleftharpoons H^+ + HCO_3^-$$

The hydrogen ion formed inside the epithelial cell through this reaction is secreted into the lumen of the tubules by either countertransport with sodium ions or active transport by the H^+ pumps. The intracellular concentration of sodium is kept low by the Na^+/K^+ pumps on the basolateral membrane. In the lumen of the tubule, hydrogen ions combine with filtered bicarbonate to form carbonic acid. The carbonic anhydrase located on microvilli catalyzes the conversion of carbonic acid to carbon dioxide and water. The carbon dioxide can then diffuse into the epithelial cell, as previously described.

The bicarbonate ion formed inside the epithelial cell by the carbonic anhydrase–catalyzed reaction moves from the epithelial cell into the peritubular fluid by either cotransport with sodium or countertransport with chloride. The Na^+/HCO_3^- cotransporter functions in reabsorption of both sodium and bicarbonate. The net effect for the bicarbonate ion is reabsorption, because bicarbonate is moved from the lumen of the tubules into the peritubular fluid, as follows: A bicarbonate ion in the lumen is converted to a carbon dioxide molecule, which then moves from the lumen into the epithelial cell, where it is converted back into a bicarbonate ion that moves into the peritubular fluid, where it can diffuse into the blood.

Overall, these actions in the proximal tubule produce three primary effects: (1) Under normal conditions, approximately 80–90% of the filtered bicarbonate is reabsorbed, (2) hydrogen ions are secreted, and (3) sodium is reabsorbed.

Renal Handling of Hydrogen and Bicarbonate Ions in the Late Distal Tubule and Collecting Duct In intercalated cells of the late distal tubule and collecting duct, secretion of hydrogen ions is coupled to the synthesis of new bicarbonate ions (**Figure 27**). The intercalated cells lining the distal tubules and collecting ducts have different membrane proteins than the epithelial cells lining the proximal tubules.

The basolateral membrane contains (1) HCO_3^-/Cl^- countertransporters that move bicarbonate out of the cell and into the peritubular fluid while moving chloride ions into the cell and (2) chloride channels that allow the chloride to diffuse back into the peritubular fluid. The apical membrane contains (1) H^+ pumps that use ATP to transport hydrogen ions out of the cell and into the tubular fluid and (2) K^+/H^+ countertransporters that move potassium ions into the cell and hydrogen ions into the tubular fluid.

The enzyme carbonic anhydrase is located in the cytosol of the intercalated cell. Carbon dioxide levels in the epithelial cell increase from either cellular metabolism or diffusion from the peritubular fluid into the cell. Inside the cell, carbonic anhydrase converts carbon dioxide and water to carbonic acid, which then dissociates into H^+ and HCO_3^-, as follows:

$$CO_2 + H_2O \rightleftharpoons H_2CO_3 \rightleftharpoons H^+ + HCO_3^-$$

This reaction is driven to the right by the removal of hydrogen ions and bicarbonate ions from the epithelial cell. Hydrogen ions are removed through transport across the apical membrane by the H^+ pump or in exchange for K^+ by the countertransporter, resulting in secretion of hydrogen ions. Bicarbonate ions are removed by countertransport with chloride ions across the basolateral membrane first into the peritubular fluid, and then into the plasma. This bicarbonate is not being reabsorbed, however, because it was never in the lumen of the renal tubules. Instead, it was produced by the epithelial cell and is considered *new* bicarbonate.

The secreted hydrogen ions decrease the pH of the tubular fluid. However, the pH of the tubular fluid, and therefore urine, is limited to a minimum of 4.5, at which point hydrogen ion transporters cease to move hydrogen ions into the lumen against a strong electrochemical gradient. To minimize decreases in urine pH, secreted hydrogen ions are buffered. Recall that hydrogen ions secreted in the proximal tubule were buffered by filtered bicarbonate ions. In the distal tubule and collecting ducts, in contrast, very little bicarbonate remains in the lumen of the tubules, as most of it was reabsorbed. Recall as well that phosphate ions are freely filtered by the glomerulus. In the lumen of the distal tubules and collecting ducts, hydrogen ions are buffered by phosphates according to the following equation:

$$HPO_4^{2-} + H^+ \rightleftharpoons H_2PO_4^-$$

Overall, these actions in the distal tubules and collecting ducts produce two primary effects: Newly formed bicarbonate ions are added to the plasma, and hydrogen ions are secreted into the tubular fluid.

Role of Glutamine in Renal Compensation During Severe Acidosis The mechanisms just described generally compensate for increases in hydrogen ion concentration produced by normal daily activities. They are not sufficient to compensate for large increases in plasma hydrogen ion concentration, however. Thus, under conditions of severe acidosis, a third renal mechanism contributes to compensation (**Figure 28**).

In the proximal convoluted tubule, glutamine is transported from both the tubular fluid and the peritubular fluid into the epithelial cells. Catabolism of glutamine in the epithelial cells yields bicarbonate ions and ammonia (NH_3):

$$\text{Glutamine} \rightarrow HCO_3^- + NH_3$$

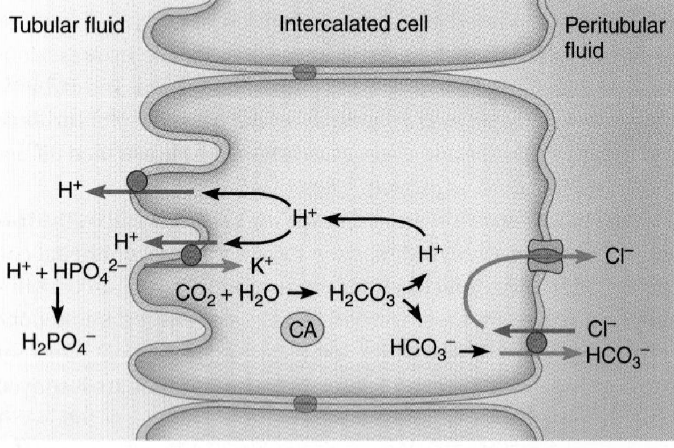

Figure 27 Bicarbonate synthesis and hydrogen ion secretion by intercalated cells of the distal tubule and collecting duct. Carbonic anhydrase in the cytosol converts carbon dioxide (which either is metabolically produced or diffuses into the cell from the peritubular fluid) into carbonic acid, which dissociates into hydrogen ions and bicarbonate ions. The hydrogen ions are secreted by an H^+ pump or a K^+/H^+ countertransporter; the bicarbonate ions are transported into the peritubular fluid by a HCO_3^-/Cl^- countertransporter.

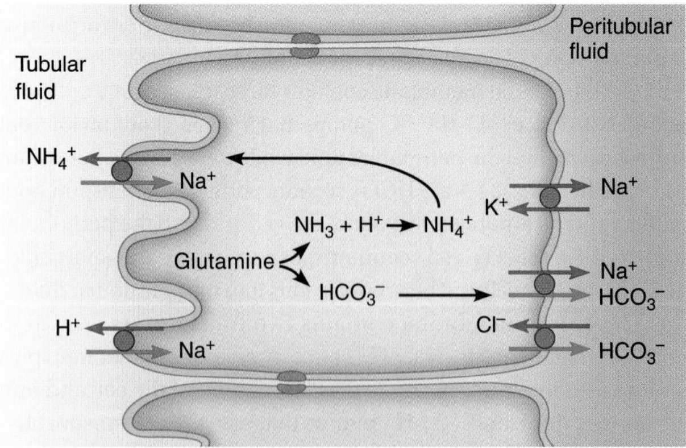

Figure 28 Bicarbonate production and hydrogen secretion by glutamine metabolism in the proximal tubule. Glutamine is catabolized to bicarbonate ions and ammonia. The bicarbonate ions are transported into the peritubular fluid by a Na^+/HCO_3^- cotransporter or a HCO_3^-/Cl^- countertransporter. The ammonia binds a hydrogen ion to form ammonium, which is secreted by countertransport with sodium ions.

The resulting bicarbonate moves into the peritubular fluid by either cotransport with sodium or countertransport with a chloride ion. This bicarbonate is not being reabsorbed, however; because it was never in the tubular fluid, *new* bicarbonate is being added to the blood. The ammonia is converted to ammonium (NH_4^+) by the following reaction:

$$NH_3 + H^+ \rightarrow NH_4^+$$

This ammonium is transported into the tubular fluid by countertransport with sodium ions and is eventually excreted.

The overall effect of these actions is that a new bicarbonate ion is added to the blood, and a hydrogen ion is secreted in the form of ammonium.

Compensation for Acid-Base Disturbances

Recall that the Henderson-Hasselbalch equation describes the relationship of plasma pH to the ratio of bicarbonate and carbon dioxide levels in blood:

$$pH = 6.1 + \log [HCO_3^-]/[CO_2]$$

Given that blood pH must be maintained at 7.4, this equation can be solved for the ratio of bicarbonate to carbon dioxide, as follows:

$$7.4 = 6.1 + \log [HCO_3^-]/[CO_2]$$
$$1.3 = \log [HCO_3^-]/[CO_2]$$
$$[HCO_3^-]/[CO_2] = 20$$

Therefore, for plasma pH to be normal, the ratio of bicarbonate to carbon dioxide must be 20:1. The respiratory and renal systems work together to control this ratio: The respiratory system controls carbon dioxide levels, and the kidneys regulate bicarbonate levels.

In acidosis, the ratio of bicarbonate to carbon dioxide decreases to less than 20:1, because of either a decrease in bicarbonate or an increase in carbon dioxide. In alkalosis, the ratio of bicarbonate to carbon dioxide is greater than 20:1, because of either an increase in bicarbonate or a decrease in carbon dioxide.

We turn now to a description of the four types of acid-base disturbances, and the ways in which the body compensates for them.

Respiratory Acidosis

During respiratory acidosis, carbon dioxide increases in the plasma, decreasing the pH:

$$\overset{\text{add}}{\Downarrow}$$
$$CO_2 + H_2O \rightleftharpoons \uparrow H_2CO_3 \rightleftharpoons \uparrow H^+ + \uparrow HCO_3^-$$

Hypoventilation can be caused by lung diseases, depression of the respiratory center in the brainstem, or diseases that affect respiratory muscles. As a result of hypoventilation, arterial P_{CO_2} increases, decreasing the ratio of bicarbonate to carbon dioxide. To bring the ratio (and therefore pH) back to normal, the kidneys increase the secretion of hydrogen ions and the reabsorption of bicarbonate ions. The lungs cannot compensate for the imbalance because that is where the problem developed initially (unless the hypoventilation was voluntary and not pathological, in which case the lungs can compensate by returning ventilation to normal).

Respiratory Alkalosis

Hyperventilation can be caused by fever or anxiety. Such hyperventilation decreases arterial P_{CO_2}, thereby increasing the ratio of bicarbonate to carbon dioxide:

$$\overset{\text{remove}}{\Uparrow}$$
$$CO_2 + H_2O \rightleftharpoons \downarrow H_2CO_3 \rightleftharpoons \downarrow H^+ + \downarrow HCO_3^-$$

To bring the ratio back to normal, the kidneys decrease the reabsorption of bicarbonate ions and secrete fewer hydrogen ions. The lungs cannot compensate for the imbalance because that is where the problem developed initially (unless the hyperventilation was voluntary).

Metabolic Acidosis

Metabolic acidosis is caused by an increase in acids in the plasma from sources other than carbon dioxide:

$$\overset{\text{add}}{\Downarrow}$$
$$CO_2 + H_2O \rightleftharpoons H_2CO_3 \rightleftharpoons \uparrow H^+ + \downarrow HCO_3^-$$

Causes of metabolic acidosis include diarrhea, which results in loss of bicarbonate through elimination of the intestinal contents; diabetes mellitus, which by increasing fat metabolism causes a buildup of keto acids; strenuous exercise, which increases lactic acid production; and renal failure. Compensation for metabolic acidosis includes an increase in ventilation and, if the kidneys are not part of the initial problem, an increase in production of new bicarbonate by the kidneys as well as excretion of H^+.

In respiratory compensation, an increase in the hydrogen ion level in the plasma activates peripheral chemoreceptors, which reflexively increase ventilation. The increase in ventilation decreases arterial P_{CO_2}, which drives the conversion of bicarbonate and hydrogen ions toward production of carbon dioxide, thereby decreasing the free hydrogen ion levels in blood. However, the lungs cannot completely compensate for the imbalance because the increased ventilation decreases arterial P_{CO_2}, which decreases the stimulatory influence of carbon dioxide on central chemoreceptors.

In renal compensation, the production of new bicarbonate is essential to replace bicarbonate that was lost via two processes: (1) use of bicarbonate to buffer excess acid and (2) decreases in plasma bicarbonate levels during respiratory compensation as described previously.

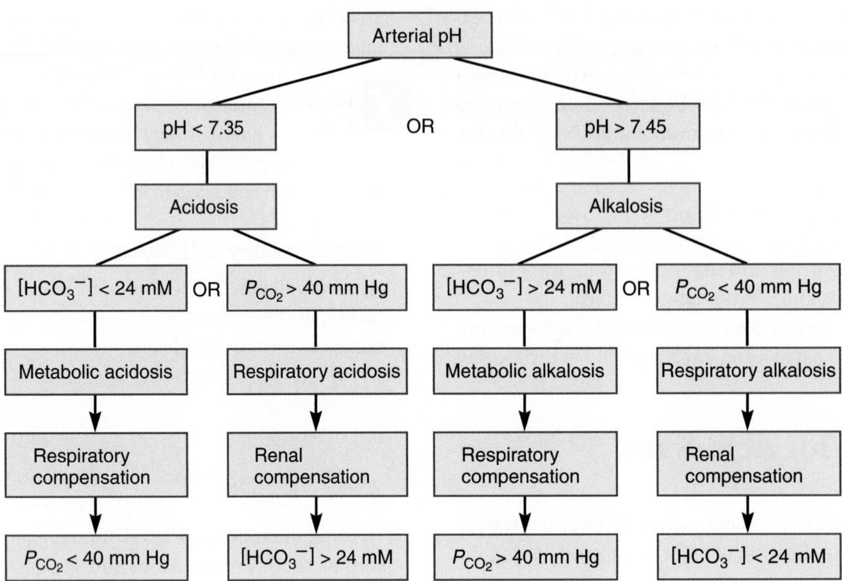

Figure 29 Summary of acid-base disturbances and compensation.

Metabolic Alkalosis

Metabolic alkalosis is caused by a decrease in acids in the plasma from sources other than carbon dioxide:

$$CO_2 + H_2O \rightleftharpoons H_2CO_3 \rightleftharpoons \downarrow H^+ + \uparrow HCO_3^- \quad (\text{remove } \Uparrow \text{ from } H^+)$$

Causes of metabolic alkalosis include vomiting (which results in a loss of acidic gastric contents) and ingestion of alkaline drugs such as sodium bicarbonate (baking soda) or antacids. Compensation for metabolic alkalosis involves both the lungs and the kidneys.

Health Fact

In respiratory compensation, decreases in hydrogen ion levels in the plasma remove a stimulatory effect on peripheral chemoreceptors, which reflexively decreases ventilation. The decrease in ventilation increases arterial P_{CO_2}, which combines with water to produce bicarbonate and hydrogen ions, increasing free hydrogen ion levels in blood. As in metabolic acidosis, the lungs cannot completely compensate for the imbalance because the decreased ventilation increases arterial P_{CO_2}, which activates the central chemoreceptors and increases ventilation.

In renal compensation, the kidneys excrete more bicarbonate ions and fewer hydrogen ions. Ridding the blood of bicarbonate shifts the equilibrium to the right, causing more carbon dioxide to react with water to form carbonic acid, which then dissociates and increases the plasma concentration of hydrogen ions.

Evaluation of Acid-Base Disturbances

Diagnosis of acid-base disturbances involves measuring plasma levels of hydrogen ion concentration (pH), carbon dioxide (P_{CO_2}), and bicarbonate. The different acid-base disturbances can be diagnosed as follows (**Figure 29**):

- A decrease in pH coupled with a decrease in plasma bicarbonate levels indicates that a metabolic acidosis is occurring. Because the lungs compensate for a metabolic acidosis by increasing ventilation, P_{CO_2} levels decrease.
- A decrease in pH coupled with an increase in P_{CO_2} indicates that a respiratory acidosis is occurring. Because the kidneys compensate for a respiratory acidosis by increasing bicarbonate reabsorption, plasma bicarbonate levels increase.
- An increase in pH coupled with an increase in plasma bicarbonate levels indicates that a metabolic alkalosis is occurring. P_{CO_2} levels increase as part of respiratory compensation.
- An increase in pH coupled with a decrease in P_{CO_2} indicates that a respiratory alkalosis is occurring. Plasma bicarbonate levels decrease as part of renal compensation.

Quick Check 6

1. Describe the three lines of defense against changes in acid-base balance. Which most rapidly corrects changes in blood pH? Which takes the longest to compensate for imbalances?
2. Explain how hydrogen ion secretion is coupled to bicarbonate reabsorption in the proximal tubule, and how it is coupled to new bicarbonate synthesis in the distal tubule and collecting duct.
3. What is the normal plasma ratio of bicarbonate ions to carbon dioxide? Why?

SYSTEMS INTEGRATION Just as the cardiovascular system is necessary to deliver blood to the organs, so the urinary system is necessary to make sure that the blood has the appropriate volume and content. However, the kidneys do much more than filter blood. Collectively, the kidneys and respiratory system regulate acid-base balance, which is necessary for normal protein function. Given that proteins carry out many functions of the cells and blood transports nutrients to all the organs, damage to or dysfunction of the urinary system affects every system of the body.

The kidneys have several interactions with the endocrine system. These organs secrete erythropoietin, which regulates red blood cell synthesis to ensure an adequate supply of oxygen is delivered to the cells. The kidneys are also necessary for activation of 1,25-dihyroxycholecalciferol, which regulates blood calcium levels to maintain normal muscle and nerve excitability and for normal bone growth. The kidneys also secrete renin, a proteolytic enzyme that is required for the activation of angiotensin II. Through activation of angiotensin II and regulation of water and electrolyte balance, the kidneys function in long-term regulation of blood pressure.

CHAPTER REVIEW

CHAPTER SUMMARY

1 The Concept of Balance

- To be in balance, the sum of the input and production of a substance must equal the sum of the output and utilization of that substance.
- The plasma can gain or lose materials by exchange with cells or extracellular connective tissue and/or as a result of exchange between it and the external environment.
- The plasma is in balance when solutes and water enter and exit the plasma at the same rate.
- A state of positive balance is said to exist when a substance enters the body faster than it exits.
- A state of negative balance exists when a substance leaves the body faster than it enters.

iP Fluid, Electrolyte, and Acid-Base Balance: Introduction to Body Fluids

2 Water Balance

- For water to be in balance, the input from consuming food and fluids and through cellular metabolism must equal the output in urine, feces, and insensible losses.
- The control of water excretion by the kidneys regulates plasma volume and osmolarity.
- Water reabsorption occurs in the renal tubules via osmosis that is coupled to the active reabsorption of solutes.
- The medullary osmotic gradient creates a force for water reabsorption via osmosis as tubular fluid moves through the distal tubule and collecting duct.
- The osmolarity in this gradient varies from 300 mOsm at the edge of the medulla to approximately 1200–1400 mOsm deep within the medulla; this gradient exists because of the countercurrent multiplier.
- The presence of urea—a waste product generated by the liver that is the primary form by which nitrogen is eliminated from the body—assists in maintaining the medullary osmotic gradient.
- Because the ascending limb of the vasa recta does not have urea transporters, the urea is reabsorbed in the plasma, contributing to the high osmolarity of plasma leaving the renal medulla.
- Most filtered water is reabsorbed in the proximal tubule.
- How much of the remaining 30% of filtered water can be reabsorbed in the late distal tubule and collecting ducts depends on the plasma levels of ADH.
- The tubular fluid in the late distal tubule and collecting ducts is hypo-osmotic to the interstitial fluid, creating an osmotic force for water reabsorption.
- ADH increases the permeability of these tubules to water, thereby allowing water reabsorption to occur.
- ADH is released from the posterior pituitary in response to either increases in the osmolarity of the extracellular fluid or decreases in blood pressure and blood volume.

MAP iP Reabsorption and Secretion in the Proximal Tubule

MAP iP Processing of Salt and Water in the Nephron

iP Fluid, Electrolyte, and Acid-Base Balance: Electrolyte and Water Homeostasis

MAP iP ADH and the Renin-Angiotensin-Aldosterone System

3 Sodium Balance

- Regulation of sodium reabsorption is critical to maintaining normal plasma sodium composition, which affects extracellular fluid osmolarity and tissue excitability.
- Sodium reabsorption also affects the reabsorption of other solutes and water and the secretion of some solutes.
- Sodium is actively reabsorbed across the renal tubules, and this reabsorption is driven by the Na^+/K^+ pumps located in the basolateral membrane of tubule epithelial cells.
- The hormones aldosterone and atrial natriuretic peptide regulate sodium reabsorption.
- Aldosterone release is controlled by plasma potassium levels and the renin-angiotensin-aldosterone system.
- Renin release is stimulated by an increase in sympathetic nerve activity, a decrease in afferent arteriole pressure, or a decrease in sodium and chloride levels in the distal tubules.
- Renin converts angiotensinogen to angiotensin I, which is converted by angiotensin converting enzyme to angiotensin II, which stimulates aldosterone release from the adrenal cortex.
- Aldosterone increases sodium reabsorption and potassium secretion.
- Atrial natriuretic peptide is secreted by cells in the atria in response to distension of the atrial wall caused by an increase in plasma volume.
- ANP decreases the glomerular filtration rate and sodium reabsorption, which increases sodium excretion.

iP Fluid, Electrolyte, and Acid-Base Balance: Electrolyte Homeostasis

4 Potassium Balance

- Potassium balance is critical to the normal function of excitable cells.
- Potassium undergoes both reabsorption and secretion in the renal tubules.
- Although the net effect of potassium movement across renal tubules is reabsorption, it is the secretion of potassium that is regulated.
- Potassium secretion is increased by aldosterone.
- High plasma potassium levels stimulate aldosterone release.

iP Fluid, Electrolyte, and Acid-Base Balance: Electrolyte Homeostasis

5 Calcium Balance

- Calcium is critical to the function of most cells in the body.
- Calcium can be added to the plasma from the bone and digestive tract, and removed from the plasma by the bone and kidneys.
- Parathyroid hormone (PTH) stimulates resorption of bone, calcium absorption in the digestive tract, calcium reabsorption in the kidneys, and activation of 1,25-$(OH_2)D_3$ in the kidneys.
- 1,25-$(OH_2)D_3$ stimulates calcium absorption in the digestive tract and calcium reabsorption in the kidneys.
- Calcitonin decreases plasma calcium levels by increasing bone formation and decreasing the reabsorption of calcium by the kidneys.

iP Fluid, Electrolyte, and Acid-Base Balance: Electrolyte Homeostasis

6 Interactions Between Fluid and Electrolyte Regulation

- Considerable overlap exists in the regulation of fluid and electrolytes, in that a single hormone often affects both water and electrolyte excretion by the kidneys.
- Movement of solute affects the forces acting on water molecules; movement of water affects the forces acting on solute molecules.
- Hemorrhage is an example of how systems interact to maintain homeostasis; interactions between the renal and cardiovascular systems regulate blood pressure in this situation in an attempt to return it to normal levels.

7 Acid-Base Balance

- Arterial pH is highly regulated to maintain the normal range of 7.38 to 7.42.
- A decrease in pH to less than 7.35 is called acidosis, whereas an increase in pH to greater than 7.45 is called alkalosis.
- The respiratory system contributes to acid-base balance by regulating carbon dioxide levels in the blood.
- Carbon dioxide can be converted to carbonic acid by the enzyme carbonic anhydrase.
- Respiratory acidosis is caused by an increase in P_{CO_2}, whereas respiratory alkalosis is caused by a decrease in P_{CO_2}.
- Metabolic acidosis and metabolic alkalosis are disturbances in blood pH caused by something other than an abnormally high or low P_{CO_2}, respectively.
- The three "lines of defense" that protect against changes in blood pH are buffering of hydrogen ions, respiratory compensation, and renal compensation.
- Buffering occurs immediately upon blood pH changes because chemical buffers are always present in the blood.
- The blood has a limited buffering capacity, so when excess hydrogen ions are added to the plasma, buffered hydrogen ions must eventually be eliminated by the body.
- The respiratory system acts within minutes to eliminate hydrogen ions in the form of carbon dioxide.
- The renal system takes hours to days to synthesize new bicarbonate and to eliminate hydrogen ions.

MAP iP Mechanisms to Control Acid/Base Homeostasis

MAP iP Acid/Base Problems

MAP MasteringA&P®

Check your understanding at MasteringA&P and apply your new knowledge to real-life case studies to ensure that you've mastered the content!

EXERCISES

Multiple-Choice Questions

1. Which of the following would be expected to trigger a decrease in the secretion of renin?
   a) A decrease in the mean arterial pressure
   b) A decrease in the pressure inside the afferent arteriole
   c) A decrease in the activity of renal sympathetic nerves
   d) A decrease in the concentration of sodium chloride in tubular fluid
   e) A decrease in GFR

2. Which of the following would be expected to trigger a decrease in the secretion of ADH?
   a) Ingestion of a large quantity of pure water
   b) Sweating
   c) Hemorrhage
   d) A decrease in mean arterial pressure
   e) Both a and c

3. Assume that fluid enters the distal tubule with an osmolarity of 100 mOsm, and that the maximum osmolarity of medullary interstitial fluid is 1100 mOsm. As plasma ADH levels rise, what happens to the osmolarity of the urine?
   a) It approaches 100 mOsm as a lower limit.
   b) It approaches 1100 mOsm as an upper limit.
   c) It eventually exceeds 1100 mOsm.
   d) It approaches 300 mOsm, the normal osmolarity of plasma.

4. Which of the following tends to be accompanied by an increase in the rate at which bicarbonate is excreted in the urine?
   a) A decrease in hydrogen ion secretion
   b) A decrease in the plasma bicarbonate concentration
   c) An increase in the production of ammonia by renal tubule epithelial cells
   d) Both a and b
   e) All of the above

5. Assuming that arterial P_{CO_2} is normal, metabolic acidosis promotes which of the following?
   a) Increased hydrogen ion secretion by the renal tubule
   b) Decreased alveolar ventilation
   c) Increased bicarbonate reabsorption
   d) Both a and c
   e) All of the above

6. Which of the following tends to promote an increase in sodium excretion?
   a) An increase in the glomerular filtration rate
   b) An increase in plasma renin concentration
   c) An increase in the secretion of atrial natriuretic peptide
   d) Both a and c
   e) All of the above

7. The osmolarity of tubular fluid increases as it flows through the descending limb of the loop of Henle because
   a) Solutes are passively transported into the descending limb.
   b) Solutes are actively transported into the descending limb.
   c) Solutes are passively transported out of the ascending limb.
   d) Water moves passively into the descending limb.
   e) Water moves passively out of the descending limb.

8. Because sweat is essentially a salt solution with an osmolarity lower than that of plasma, severe sweating leads to a reduction in plasma volume and an increase in plasma osmolarity. How do these changes affect the secretion of ADH?
   a) The decrease in plasma volume inhibits ADH secretion, but the increase in osmolarity stimulates it.
   b) The decrease in plasma volume stimulates ADH secretion, but the increase in osmolarity inhibits it.
   c) Both the decrease in plasma volume and the increase in osmolarity stimulate ADH secretion.
   d) Both the decrease in plasma volume and the increase in osmolarity inhibit ADH secretion.

9. In the cytosol of intercalated cells, the carbon dioxide that is converted to carbonic acid can come from
   a) The lumen of the distal tubule and the collecting duct.
   b) Metabolism inside the intercalated cell.
   c) Catabolism of glutamine.
   d) Both a and b
   e) All of the above

10. In the lumen of the proximal tubule, secreted hydrogen ions are primarily buffered by
    a) Bicarbonate.
    b) Phosphates.
    c) Proteins.
    d) Sulfates.
    e) Carbon dioxide.

11. An increase in mean arterial pressure stimulates which of the following?
    a) ADH release
    b) Angiotensin II production
    c) Aldosterone release
    d) Renin release
    e) Increased water excretion in urine

12. Which of the following does *not* stimulate aldosterone release?
    a) Atrial natriuretic peptide
    b) An increase in plasma potassium
    c) An increase in renin secretion
    d) An increase in angiotensin II production

13. Where is the Na^+/K^+ pump located?
    a) On the basolateral membrane of proximal tubule cells, and on the apical membrane of principal cells
    b) On the apical membrane of proximal tubule cells, and on the basolateral membrane of principal cells
    c) On the basolateral membrane of both proximal tubule cells and principal cells
    d) On the apical membrane of both proximal tubule cells and principal cells

14. In epithelial cells lining the proximal tubules, carbonic anhydrase is located
    a) On the basolateral membrane only.
    b) On the apical membrane only.
    c) Inside the cell only.
    d) Inside the cell and on the apical membrane.
    e) Inside the cell and on the basolateral membrane.

15. A person has the following symptoms: arterial pH = 7.48, P_{CO_2} = 44 mm Hg, plasma bicarbonate concentration = 27 mM What is the diagnosis?
    a) Respiratory acidosis
    b) Respiratory alkalosis
    c) Metabolic acidosis
    d) Metabolic alkalosis

Objective Questions

16. An increase in the reabsorption of solutes (increases/decreases) water reabsorption.

17. Most solutes and water are reabsorbed in the (proximal tubule/distal tubule and collecting duct).

18. Epithelial cells of the descending limb of the loop of Henle actively transport solutes from tubular fluid to the surrounding peritubular space. (true/false)

19. Urine flow rate increases as the plasma ADH level (increases/decreases).

20. Stretching of the atria of the heart promotes the secretion of ________ ________ ________, a hormone that promotes sodium excretion.

21. If the plasma volume is below normal, an increase in water reabsorption will be sufficient to restore it to normal. (true/false)

22. ________ stimulates the insertion of Na^+/K^+ pumps into the plasma membrane of principal cells of the distal tubules and collecting ducts.

23. ________ stimulates the insertion of water pores into the plasma membrane of epithelial cells of the distal tubules and collecting ducts.

24. ADH increases water permeability of the loop of Henle. (true/false)

25. Potassium secretion is (stimulated/inhibited) by aldosterone.

26. Calcitonin (increases/decreases) plasma calcium levels.

27. Resorption of bone (increases/decreases) plasma calcium levels.

28. By adding new bicarbonate to the blood, the kidneys can bring about a compensatory (increase/decrease) in the plasma pH.

29. The kidneys can excrete urine that is pure water. (true/false)

30. There is no limit to the amount of hydrogen ions the kidneys can excrete. (true/false)

Essay Questions

31. Describe the cellular effects of ADH on principal cells in the distal tubule and collecting duct.

32. Describe how each of the following hormones affects water reabsorption, either directly or indirectly: ADH, angiotensin II, aldosterone, and ANP.

33. Describe the effects of aldosterone on renal handling of sodium and potassium.

34. What are the stimuli for renin release? Describe the pathway by which renin leads to aldosterone release.

35. Explain how the following hormones affect blood pressure: ADH, angiotensin II, aldosterone, and ANP.

36. Describe the changes in arterial P_{CO_2} and bicarbonate levels that occur during metabolic acidosis. Which change is compensatory in nature?

Critical Thinking

37. You know that potassium plays a role in establishing the resting membrane potential. Cardiac arrhythmias are irregularities in the electrical activity of the heart. On the basis of this information, explain how both hyperkalemia and hypokalemia can cause cardiac arrhythmias.
38. ACE inhibitors are commonly used to treat hypertension. Explain how they work.
39. Potassium imbalances are often linked to acid-base disturbances. Explain how regulation of potassium and regulation of hydrogen ions influence each other. Can you think of other instances where regulating the balance of one substance in blood may come at the expense of the balance of another substance?

Photo credits for icons in feature boxes: Focus on Diabetes: Tanyae/Dreamstime; Clinical Connections: Iofoto/Dreamstime; Understanding Exercise: Pixattitude/Dreamstime; Discovery: Yuri Arcurs/Cutcaster.

Glossary

aquaporins water channels located in the plasma membrane of most cells

obligatory water loss minimum volume of water that must be excreted in urine to eliminate solutes

antidiuretic hormone (ADH) hormone secreted by the posterior pituitary that regulates water reabsorption by the kidneys; also called *vasopressin*

osmoreceptors (oz-moh-ree-sep-tors) receptors that detect the osmolarity of various body fluids

aldosterone (al-dos-stir-own) a steroid hormone released from the adrenal cortex that regulates the reabsorption of sodium and secretion of potassium

renin enzyme released by the kidney; converts angiotensinogen to angiotensin I

angiotensin converting enzyme (ACE) enzyme that converts angiotensin I to angiotensin II

atrial natriuretic peptide (ANP) hormone secreted from the atrium that regulates plasma sodium levels

resorption the breakdown of bone tissue

parathyroid hormone (PTH) peptide hormone released from the parathyroid glands that regulates plasma calcium levels

1,25-dihydroxycholecalciferol or 1,25-$(OH_2)D_3$ steroid hormone derived from vitamin D that regulates plasma calcium levels; also called *calcitriol*

calcitonin peptide hormone released from C cells of the thyroid gland that regulates plasma calcium levels

metabolic acidosis decrease in blood pH in response to changes other than carbon dioxide levels

metabolic alkalosis increase in blood pH in response to changes other than carbon dioxide levels

Answers

Figure Question Answers

Fig 6 A juxtamedullary nephron
Fig 11 A peptide hormone; the posterior pituitary gland
Fig 15 A steroid hormone; in the cytosol
Fig 18 It cleaves amino acids off angiotensinogen to form angiotensin 1
Fig 25 Carotid bodies

Apply Your Knowledge Answers

Proteins are not filtered so they are not eliminated by the kidneys, and glucose is 100% reabsorbed, so it is not eliminated by the kidneys.

The Na^+/K^+ pump is located on the basolateral membrane of epithelial cells lining the proximal and distal tubules. It pumps sodium ions into the interstitial space to be picked up by blood, and it pumps potassium ions into the cell. The fate of the sodium is critical in this question. Because of the pump's action, sodium levels in the cell are low, creating an electrochemical gradient for sodium to move from the tubule lumen across the apical membrane into the epithelial cell. This movement is often coupled to the transport of other molecules, such as glucose and amino acids. Reabsorption of sodium, and the solutes transported with sodium, create an osmotic gradient for reabsorption of water from the proximal tubule.

Oubain blocks the Na^+/K^+ pump, which will dissipate the electrochemical gradient for sodium reabsorption. When there is no sodium reabsorption, glucose and certain amino acids will also not be reabsorbed. Thus the osmotic gradient for water reabsorption across the proximal tubule will be low and less water will be reabsorbed.

The parathyroid glands secrete parathyroid hormone, which increases blood calcium levels. Without parathyroid glands, calcium levels would be low, possibly causing arrhythmias, hypotension, osteoporosis, muscle weakness, and cramping.

Hyperventilation causes a decrease in blood carbon dioxide levels. Because carbon dioxide functions as an acid in biological systems (it can be converted to carbonic acid by the enzyme carbonic anhydrase), a decrease in carbon dioxide causes a decrease in hydrogen ions, or an alkalosis. Alkalosis increases the excitability of neurons, especially peripheral nerves. Afferent neurons become active and transmit action potentials to the central nervous system, giving the sensation of tingling.

Answers to End–of-Chapter Multiple Choice and Objective Questions

Multiple Choice Questions

1. c 2. a 3. b 4. d 5. d 6. d 7. e 8. c 9. b 10. a 11. e 12. a 13. c 14. d 15. d

Objective Questions

16. increases
17. proximal tubule
18. false
19. decreases
20. atrial natriuretic]peptide
21. false
22. Aldosterone
23. Antidiuretic hormone (ADH)
24. false
25. stimulated
26. decreases
27. increases
28. increase
29. false
30. false

The Gastrointestinal System

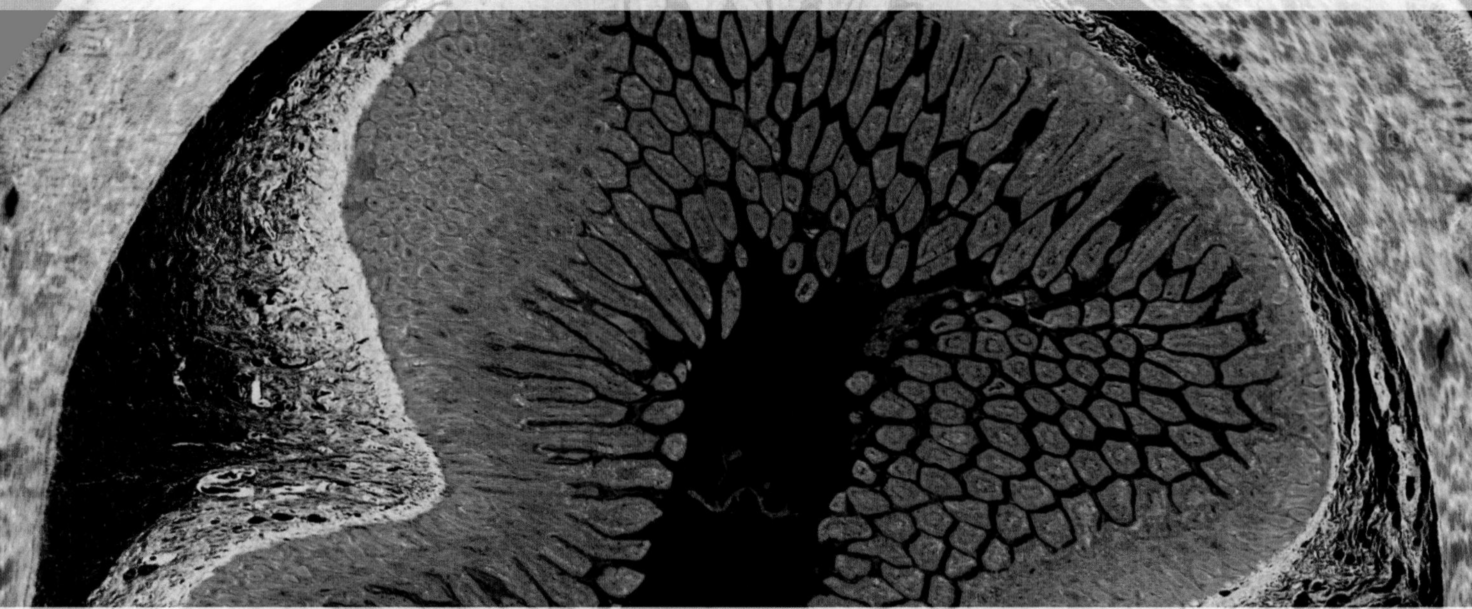

Steve Gschmeissner/Photo Researchers, Inc.

Colored light micrograph of the small intestine (cross-section).

Marjorie was attending a family reunion picnic. Upon swallowing a bite of barbeque ribs, she realized that the food was "stuck" in her throat. Marjorie knew this feeling—it was happening to her more and more often. No amount of swallowing water, bread, or anything else would make it budge. After what seemed like an hour, she could feel her throat relaxing and the lump of food slowly slide down the rest of the way.

The next day Marjorie went to see her doctor to find out why she had been having difficulty swallowing (called *dysphagia*). Her doctor ordered upper gastrointestinal tract radiography. The results showed that Marjorie had ulcers in her esophagus caused by *gastroesophageal reflux disease (GERD).* During and after a meal, the stomach lining secretes acid into the lumen to facilitate the digestion of proteins. A ring of smooth muscle called the *lower esophageal sphincter* normally closes off the opening to the stomach so that the contents cannot go backward into the esophagus, but this sphincter opens during swallowing. If the lower esophageal sphincter does not close completely, some of the acidic content of the stomach escapes into the esophagus (an event called *gastric reflux*), and the acid irritates the esophageal lining and activates nociceptors. The damage to the esophageal tissues caused by this irritation can lead to spasm, scarring, or narrowing of the esophagus—all of which may cause dysphagia. In this chapter, we examine the organs and functions of the gastrointestinal system.

CHAPTER OUTLINE

LEARNING OUTCOMES ***After studying this chapter, you should be able to:***

- Identify the major organs of the gastrointestinal system, and describe the functions of each.
- Identify the various tissue layers that make up the wall of the gastrointestinal tract.
- Describe the fundamental mechanisms involved in the absorption of carbohydrate, protein, and lipid digestion products. Explain how the mechanism of lipid absorption is related to the hydrophobic nature of fats.
- In general terms, describe the role of short reflex pathways, long reflex pathways, and gastrointestinal hormones in the control of digestive function.
- Describe the functions of saliva, stomach acid, pancreatic juice, and bile. Explain how the secretion of each of these substances is regulated.
- Define peristalsis, segmentation, migrating motility complex, haustration, mass movement, and basic electrical rhythm. Describe the role of each in digestion.

Before You Begin

Make sure you have mastered the following topics:

- Biomolecules
- Microvilli
- Epithelial transport
- Enzymes
- Hydrolysis
- Smooth muscle
- Autonomic nervous system

We know that the kidneys influence the composition of the extracellular fluid by regulating the output of water and electrolytes. In this chapter, you will learn how the *gastrointestinal system* is necessary for the input of water, electrolytes, and nutrients. The renal and gastrointestinal systems share many similarities; both involve massive transport across epithelium, for example. However, there is one major difference between the two systems: The renal system is regulated to maintain normal composition of the extracellular fluid, whereas the gastrointestinal system is regulated to absorb all of the nutrients we consume, whether we need them or not. As we learn about the gastrointestinal system, we will revisit several cell components and functions, including biomolecules, enzyme activity, and transport mechanisms. The smooth muscle in the wall of the gastrointestinal tract is excited and contracts by several mechanisms. We will apply what we learned about the nervous and endocrine systems to the enteric nervous system and the gastrointestinal hormones. We begin this chapter with an overview of the basic functions of the gastrointestinal system.

1 Overview of Gastrointestinal System Processes

Most of the nutrient molecules found in food are too large to be transported into the bloodstream intact. For this reason, these molecules must be chemically broken down to smaller molecules by enzymes in the lumen of the gastrointestinal tract, a process called **digestion.** This process is aided by the mechanical breakdown of food, which both enables it to move more easily through the tract and renders it more susceptible to the action of digestive enzymes.

Once the larger nutrient molecules have been reduced to smaller digestive end-products, these molecules are transported into the bloodstream via a process called **absorption.** To aid in digestion and absorption, fluids containing enzymes and other substances are transported into the lumen of the gastrointestinal tract via a process called **secretion.** As these processes are occurring, the contents of the lumen are mixed and slowly propelled from one end of the tract to the other by the contractile activity of muscle located in the wall of the tract itself, which gives digestive organs the ability to move, called **motility.** The four basic processes of digestion, absorption, secretion, and motility are summarized in **Figure 1**.

In the next section, we explore the details of the gastrointestinal system in terms of how its anatomy relates to its function. In subsequent sections, we focus on the four basic digestive processes, including how they are regulated.

2 Functional Anatomy of the Gastrointestinal System

The gastrointestinal system (digestive system) comprises two major divisions: (1) the **gastrointestinal tract** (also known as the **GI tract** or *digestive tract*), which consists of several organs joined in series to form a passageway through which food and digestion products are conducted, and (2) **accessory glands,** a number of glands located outside the GI tract that secrete various fluids and enzymes into the lumen of the tract to aid the digestive process.

The Gastrointestinal Tract

The GI tract *in vivo* is essentially a hollow tube approximately 4.5 meters (15 feet) long that runs through the body and opens to the outside at either end. (When removed from the body, the GI tract is twice as long, 30 feet, due to the relaxation of smooth muscle found in the wall of the tract.) The tract begins at the *mouth,* where food enters, and ends at the *anus,* where unabsorbed material exits. These and the other organs of the GI tract—the *pharynx, esophagus, stomach, small intestine, colon,* and *rectum*—are shown in **Figure 2** and discussed later in the chapter. Although each organ has its own unique structure and function, the wall of the GI tract maintains a relatively uniform structure throughout most of its length, which we examine next.

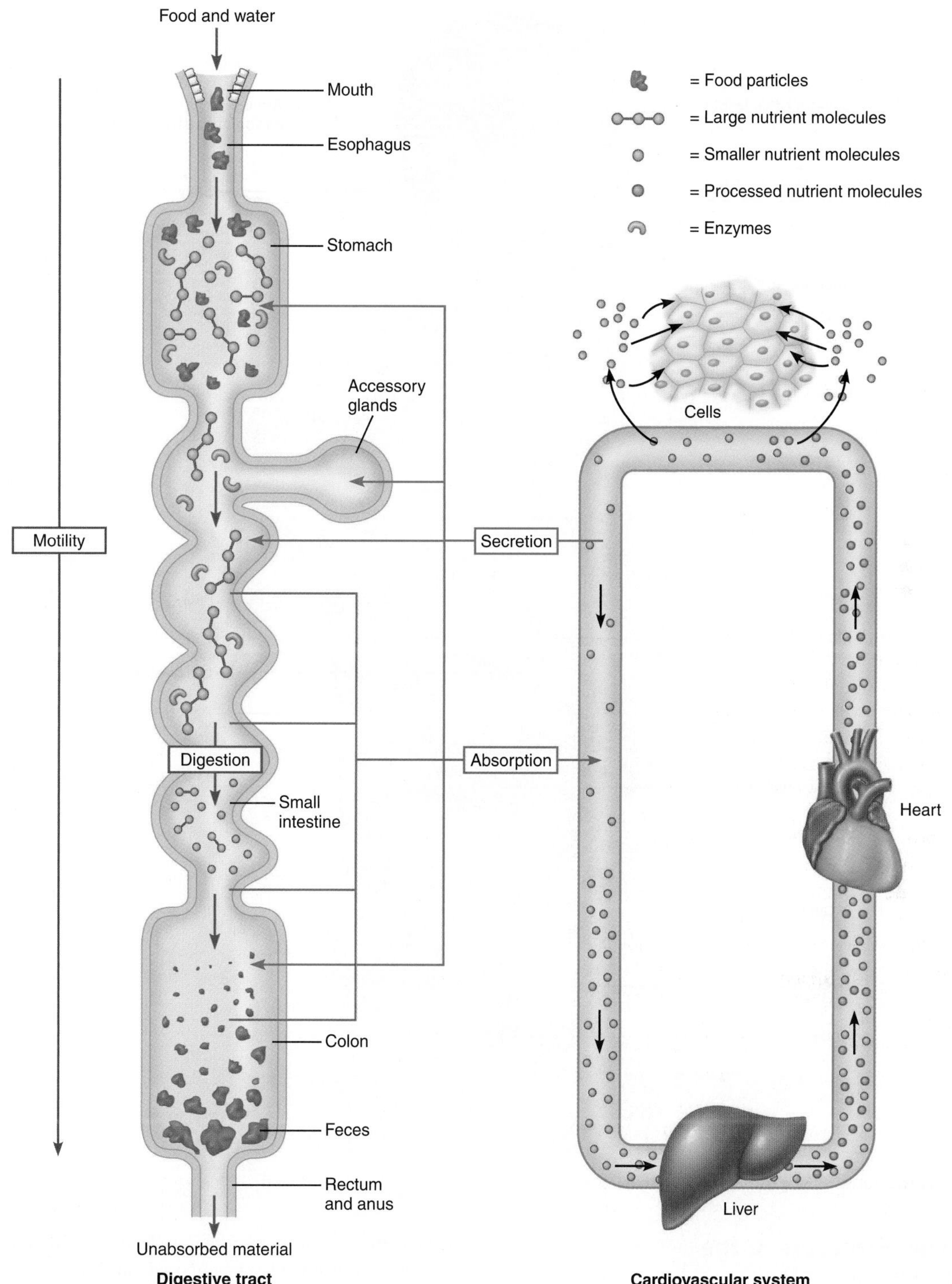

Figure 1 Overview of the four basic digestive processes: digestion, absorption, secretion, and motility.

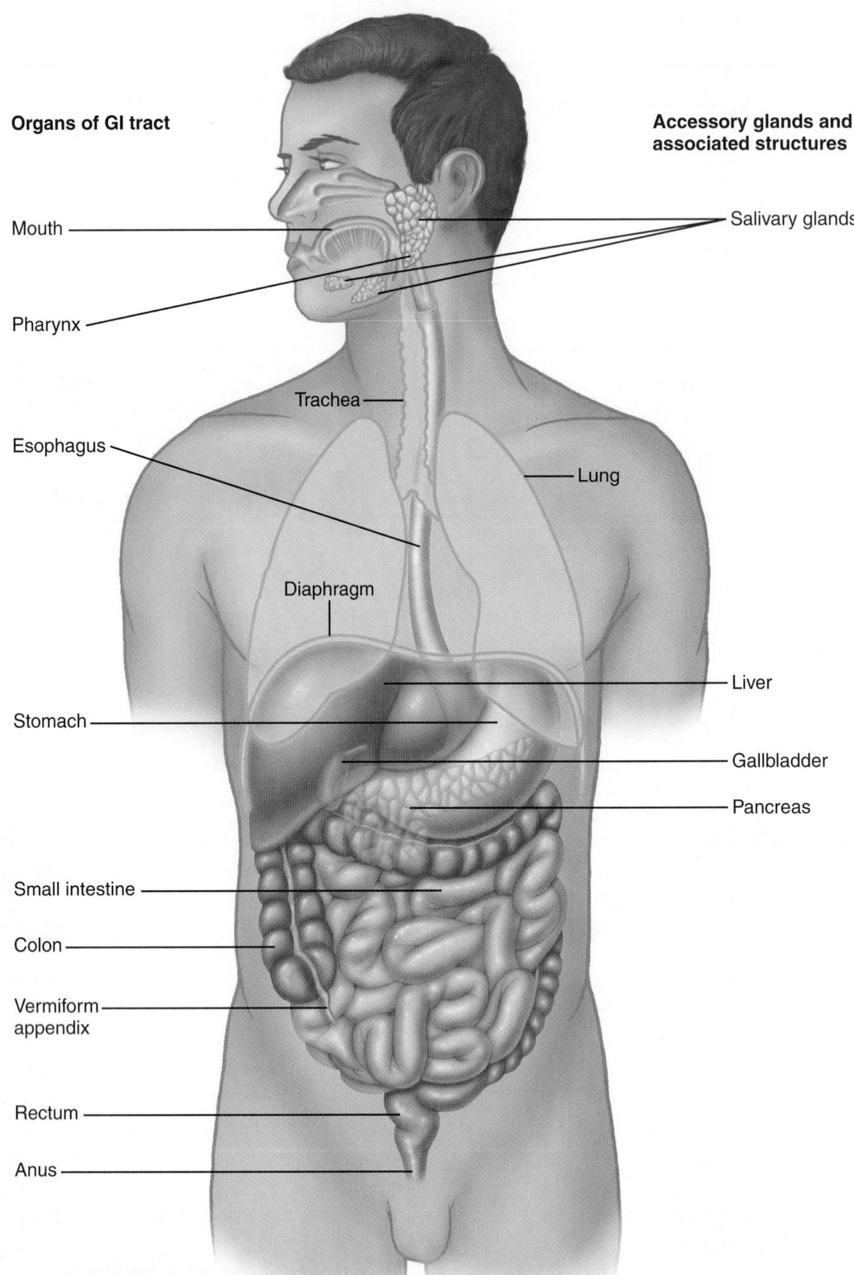

Figure 2 Major structures of the gastrointestinal system.

Generalized Structure of the Gastrointestinal Wall

The wall of most of the GI tract (exceptions being the mouth, pharynx, upper third of the esophagus, and external-most portion of the anus) has the same general features shown in **Figure 3**. Four distinct layers are found in the GI tract wall: (1) the **mucosa,** which lines the lumen of the GI tract; (2) the **submucosa,** an underlying layer of connective tissue; (3) the **muscularis externa,** a layer made up primarily of smooth muscle fibers; and (4) the **serosa** (or *adventitia*), an outer layer composed mostly of connective tissue.

The Mucosa The mucosa is composed of three layers: (1) an innermost layer of cells called the *mucous membrane,* (2) a middle layer called the *lamina propria,* and (3) an outer layer of smooth muscle called the *muscularis mucosae.*

The **mucous membrane** is a layer of epithelial cells of various types (collectively referred to as **enterocytes**) that lines the inside

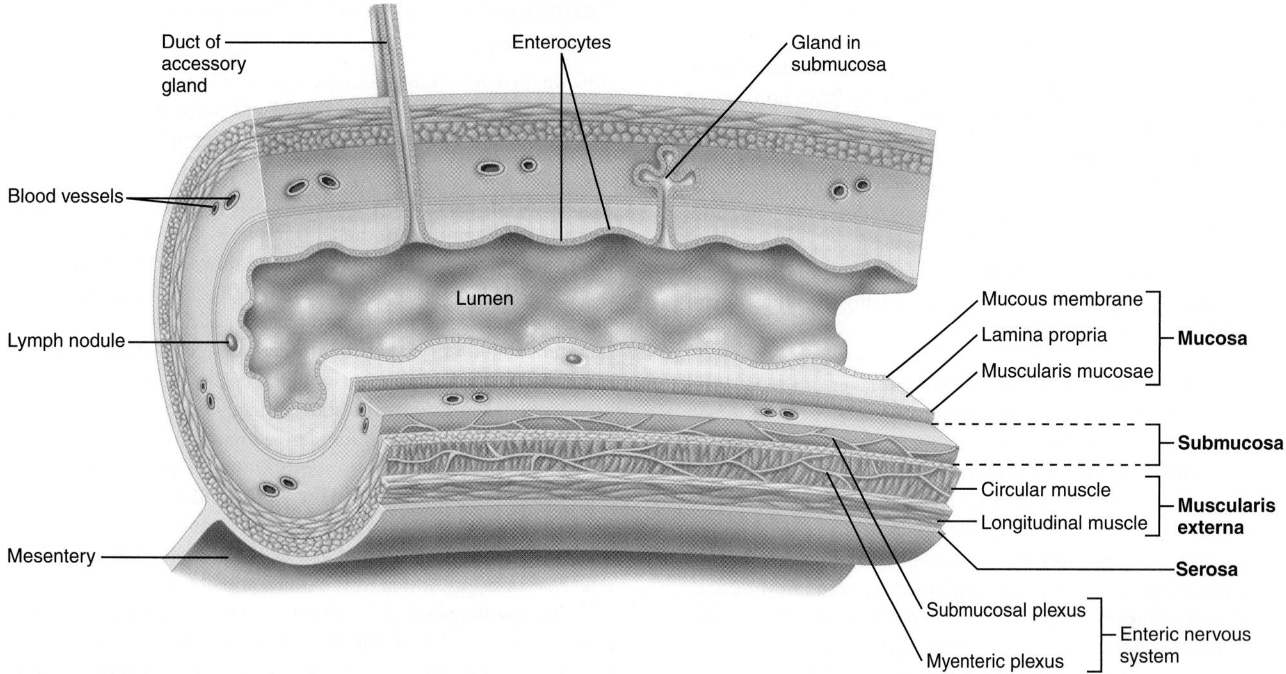

Figure 3 Generalized structure of the gastrointestinal wall, depicting the four major layers and the tissues within them.

of the GI tract, forming a continuous barrier separating the lumen from the body's internal environment. Some enterocytes are classified as *absorptive cells* because they are specialized for absorption, the transport of nutrients and other materials from the lumen to the bloodstream. Other enterocytes are classified as *exocrine cells* because they secrete materials such as fluids and enzymes into the lumen (which is outside the body). Exocrine cells include **goblet cells,** which secrete **mucus** (a sticky, viscous fluid containing glycoproteins called *mucins*) throughout the length of the GI tract; mucus forms a coating that protects the lining against abrasion and substances in the lumen that may attack tissue. Still other enterocytes are *endocrine cells,* which secrete hormones into the bloodstream. For example, *G cells* in the stomach mucous membrane secrete the hormone *gastrin* into the bloodstream. As we will see, gastrin and several other hormones play an important role in the regulation of digestive function.

The **lamina propria** is a layer of connective tissue underlying the mucous membrane. Contained within this layer are small blood vessels, nerves, and lymphatic vessels that communicate with larger nerves and vessels in still deeper tissue layers. The lamina propria also contains lymphoid tissue, including *lymph nodules* and *Peyer's patches,* that is important in defending the body against bacteria, which are plentiful in the lumen of the intestines.

The **muscularis mucosae** is a thin layer of smooth muscle that serves primarily to contract the mucosa into folds, which stirs the lumenal contents and promotes contact with the mucosal surface. Within this layer are *longitudinal* muscle fibers, which run parallel to the tract's long axis, and *circular* muscle fibers, which run around the tract's circumference.

The Submucosa The submucosa is a thick layer of connective tissue that provides the GI tract with much of its distensibility and elasticity, enabling it to tolerate a large degree of stretch without sustaining damage. This layer also contains many of the tract's larger blood and lymphatic vessels. At its outer border is a network of nerve cells known as the **submucosal plexus** *(Meissner's plexus),* which communicates with another nerve cell network in the muscularis externa called the **myenteric plexus** *(Auerbach's plexus)* (see Figure 3); together these nerve plexuses make up the **enteric nervous system** (or *intrinsic nervous system*).

The enteric nervous system comprises an elaborate network of sensory neurons, motor neurons, and interneurons located within the wall of the GI tract. It is capable of regulating many GI functions independent of external influences. Input to the enteric nervous system comes from receptors located in the GI tract and autonomic neurons. Output from the enteric nervous system goes to effector cells (smooth muscle, exocrine glands, and endocrine glands) located within the GI tract and plays an important role in the control of digestive function.

The Muscularis Externa The muscularis externa, which is largely responsible for the motility of the GI tract, contains two separate layers of smooth muscle: an inner layer of circular muscle and an outer layer of longitudinal muscle. The circular muscle layer consists of single-unit smooth muscle capable of generating spontaneous depolarizations, called *slow-wave potentials.* The circular muscle makes up the bulk of the muscularis externa. Contractions of the circular muscle layer decrease the diameter of the lumen of the GI tract. The longitudinal muscle layer consists of multi-unit smooth muscle that depends on neural input for contraction; contraction of the longitudinal muscle layer shortens the GI tract. Coordinated contractions of the circular and longitudinal muscles propel the lumenal contents through the GI tract and mix the contents with secretions that help digest food particles. Motility also promotes contact between the digestive end-products and the mucosal epithelium, thereby increasing the efficiency of absorption.

The Serosa The serosa—the outermost layer of the GI tract wall—consists of an inner layer of fibrous connective tissue, which provides structural support, and an outer layer of epithelial tissue called the **mesothelium,** which secretes a watery lubricating fluid that makes it easier for organs to slide past one another. The mesothelium (along with a layer of underlying connective tissue) is continuous with the **mesenteries,** a system of clear, thin membranes that interconnects most of the abdominal organs and houses nerves and blood vessels running to them (**Figure 4**). The mesenteries help to anchor the organs in place and are continuous with the **peritoneum,** a membrane lining the inside of the abdominal cavity.

Quick Check 1

1. Name the four major layers that make up the wall of the GI tract.
2. Name the three major cell types in the mucous membrane, and briefly describe the function of each.
3. Name the two major divisions of the enteric nervous system, and give their locations.

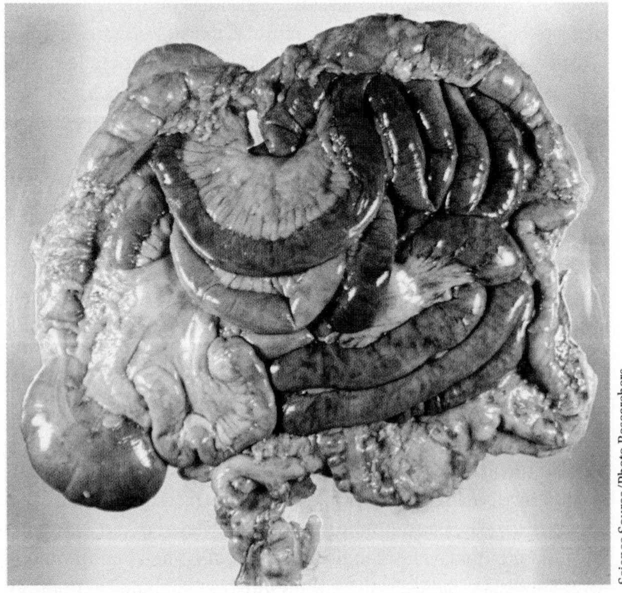

Figure 4 Mesenteries.

Functional Anatomy of Gastrointestinal Tract Organs

Now we turn our attention to the individual organs of the GI tract, progressing in order from the upper end of the tract to the lower end. We begin with the mouth, pharynx, and esophagus.

The Mouth, Pharynx, and Esophagus The *mouth* or *oral cavity* is the beginning of the GI tract; it is where food enters and where the processes of mechanical breakdown and digestion begin (**Figure 5**). In the mouth, food is chewed (a process called *mastication*) to decrease the size of food particles and to mix them with a secretion called *saliva.* Saliva lubricates the food and contains an enzyme called *salivary amylase,* which begins the digestion of carbohydrates by breaking down starch and glycogen.

From the mouth, the food-saliva mixture is propelled by the tongue into the pharynx (commonly known as the throat), a common passageway for food and air. At the end of the pharynx, the passageways for food and air diverge. Whereas air enters the larynx and trachea via the glottis and proceeds toward the lungs, food enters the esophagus, which runs parallel and dorsal to the trachea.

The **esophagus** is a muscular tube whose primary function is to conduct food from the pharynx to the stomach. Unlike the trachea, it is thin walled and pliant, so that it can easily stretch to accommodate food as it is swallowed; when food is not present, however, it is normally collapsed. The esophagus is unusual among the organs of the GI tract in that its wall contains both skeletal muscle (in the upper third of its length) and smooth muscle (in the bottom two thirds).

The movement of food from the pharynx to the esophagus is regulated by the **upper esophageal sphincter** (see Figure 5), a ring of skeletal muscle surrounding the esophagus at its upper end. (A **sphincter** is generally defined as a ring of muscle that surrounds an orifice and regulates the passage of material through it by altering its diameter.) At the esophagus's lower end is the **lower esophageal sphincter,** a ring of smooth muscle that regulates the flow of food from the esophagus to the stomach. Both of these sphincters are normally closed; they open only when food is being swallowed. The lower esophageal sphincter prevents the contents of the stomach, which are acidic, from entering the esophagus. However, backflow of stomach contents into the esophagus *(gastric reflux)* can occasionally occur and produce *heartburn,* a burning sensation in the chest caused by irritation of the esophageal lining.

The Stomach An important function of the **stomach,** a J-shaped sac located beneath the diaphragm, is to store food after it is swallowed and to release it into the small intestine. The lining of the stomach contains glands (called **gastric glands**) that secrete a watery fluid called **gastric juice** into the lumen. Contractile activity of smooth muscle in the stomach's wall pulverizes food into smaller particles and mixes it with gastric juice, forming a mixture called **chyme.**

The stomach is divided into three major anatomical regions (see Figure 5): a domed upper portion called the *fundus,* which extends above the lower esophageal sphincter; a middle region called the *body,* which accounts for the bulk of the stomach's volume; and a lower region called the *antrum,* which is narrower and smaller in volume. Contractions of the antrum propel chyme from

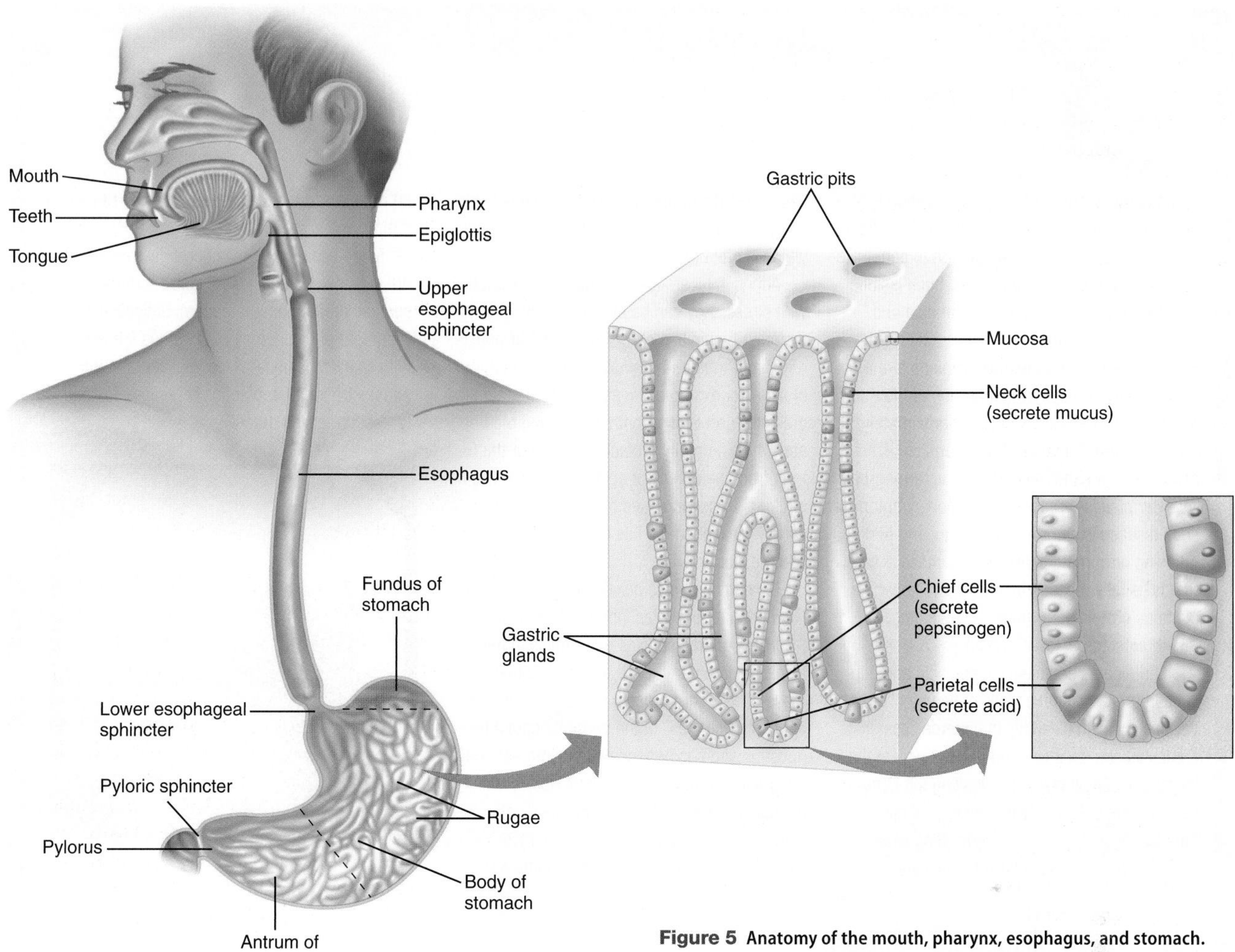

Figure 5 Anatomy of the mouth, pharynx, esophagus, and stomach. The enlarged views show gastric pits and gastric glands.

the stomach into the small intestine, a process called *gastric emptying*. As chyme exits the stomach, it passes through a narrow passage called the **pylorus** on its way to the small intestine. The flow of chyme through the pylorus is regulated by a surrounding ring of smooth muscle called the **pyloric sphincter,** which opens and closes with each cycle of stomach contraction, such that chyme exits the stomach in spurts.

The stomach wall differs anatomically and functionally in the three regions of this organ. In the fundus, the wall is thin and easily expands to accommodate increases in volume following a meal. In the body of the stomach, the gastric mucosa folds longitudinally into rugae, which flatten as the stomach expands. These two anatomical specializations allow the stomach to expand approximately 20 times from its empty volume (50 mL) to its full volume (1000 mL). The antrum has the thickest muscle layer and produces the strong contractions responsible for gastric mixing and emptying. Both the fundus and the body contain **gastric pits,** which are populated by cells that secrete the products found in gastric juice.

Gastric pits contain a variety of secretory cells, including both exocrine and endocrine cells. In the neck, or upper region, of the pits, **neck cells** secrete mucus. Deeper in the pits are gastric glands, which contain the following cells: (1) **chief cells,** which secrete **pepsinogen,** the precursor for a proteolytic enzyme called **pepsin;** (2) **parietal cells,** which secrete both hydrogen ions (as HCl) to acidify the stomach contents and **intrinsic factor,** which is necessary for the absorption of vitamin B_{12}; and (3) **G cells,** which secrete the hormone **gastrin.** Whereas parietal and chief cells secrete their products into the lumen of the stomach, G cells secrete gastrin into the bloodstream.

The lumen of the stomach is the only locale in the GI tract where the contents are acidic. In fact, the pH of stomach contents can be as low as 2, which is equivalent to a 10 mM solution of hydrochloric acid! This acidity is necessary for converting pepsinogen into its active form, pepsin; it is also useful because it helps denature proteins in the food and kills many foodborne bacteria, thereby protecting the body against certain infections.

The stomach's lining is protected against the potentially harmful effects of acid and pepsin by a surface covering of mucus and bicarbonate, called the **gastric mucosal barrier.** Neck cells in the gastric pits secrete mucus, whereas surface epithelial cells secrete both mucus and bicarbonate. If the gastric mucosal barrier is

CLINICAL CONNECTIONS

Ulcers 

Approximately 10% of the U.S. population is affected by *ulcers*, which are erosions of the lining of the GI tract that are deep enough to penetrate through the muscularis mucosae. Such lesions result from the combined action of acid and pepsin, which break down tissues. The most vulnerable areas are the lower regions of the stomach and the uppermost portion of the duodenum; duodenal ulcers occur approximately 10 times more frequently than stomach ulcers. Affliction with either type of ulcer is termed *peptic ulcer disease.* The most common symptom of such disease is a chronic, rhythmic, and periodic gnawing or burning pain in the stomach area that is usually relieved by drinking milk (which contains proteins that buffer acid), eating, or taking antacids.

If an ulcer is sufficiently deep, acid and pepsin can break down the walls of nearby blood vessels, causing bleeding. In extreme cases, the lesion can penetrate through the wall of the stomach or small intestine, allowing the lumenal contents to leak into the abdominal cavity. This condition, known as a *perforating ulcer,* is extremely serious because the leakage of bacteria-laden material almost invariably leads to inflammation of the peritoneum *(peritonitis),* which is often fatal.

Although ulcers are commonly associated with stress, they can be brought on by other risk factors. Chronic use of aspirin and other nonsteroidal anti-inflammatory drugs increases the risk of ulcer, for example, because these agents suppress the secretion of both mucus and bicarbonate, which normally protect the lining of the GI tract from the effects of acid and pepsin. The risk of ulcer is also increased by chronic alcohol use and by the leakage of bile from the duodenum into the stomach.

Researchers have discovered that many ulcers result from infection with *Helicobacter pylori*, a bacterial species that thrives in the acidic environment of the stomach. Although these bacteria are present in the stomachs of approximately 40% of the general population (most of whom experience no symptoms), they are present in more than 90% of people with duodenal ulcers, and in 70% of those patients with gastric ulcers. Infection with *H. pylori* bacteria triggers ulcers by precipitating an immune response that destroys tissue, rendering it more vulnerable to attack by acid and pepsin.

Most ulcers can be successfully treated by drugs that suppress gastric acid secretion, such as proton pump inhibitors. Successful treatment of ulcers in those persons infected with *H. pylori* is usually accomplished by the administration of antibiotics in conjunction with acid-suppressing drugs.

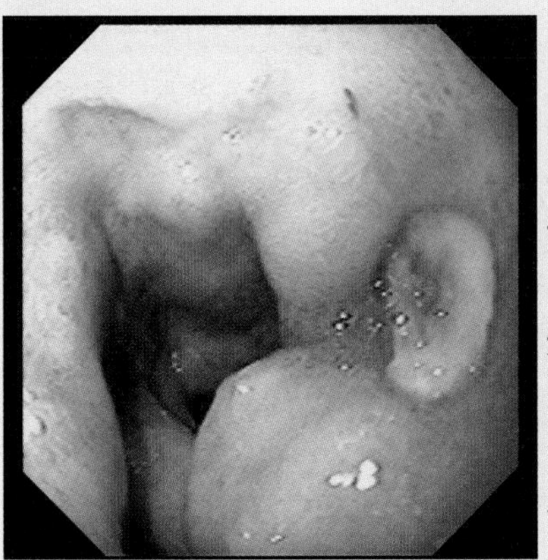

David M. Martin, M. D./Photo Researchers, Inc.

Duodenal ulcer viewed through an endoscope.

Critical Thinking Questions

1. What are ulcers, what causes them, and what are risk factors for developing them?
2. Describe how a bacterial infection can result in an ulcer.
3. What treatments are available for ulcers and how do they work?

penetrated by acids, such as salicylic acid (aspirin), an ulcer can result (**Clinical Connections: Ulcers**).

The Small Intestine From the stomach, chyme travels to the **small intestine,** a coiled tube about 2–3 cm in diameter and 2.5–3 meters long *in vivo* (on a postmortem basis, the small intestine is 6–8 meters in length due to loss of smooth muscle contraction) that is the primary site for the digestion of all nutrients in food. The small intestine is also where most of the ingested nutrients, water, vitamins, and minerals (inorganic ions such as sodium, potassium, and calcium) are absorbed.

On the basis of subtle anatomical distinctions, the small intestine is divided into three major regions (**Figure 6**): an initial relatively short portion called the **duodenum,** which begins at the pylorus and extends for approximately 30 cm; a middle portion called the **jejunum,** which extends for about another 1 meter; and a terminal portion called the **ileum,** which extends approximately 1.5 meters and joins the colon.

In the duodenum, chyme is mixed with a watery secretion from the pancreas called **pancreatic juice,** which contains a wide variety of digestive enzymes and is rich in bicarbonate. Bicarbonate neutralizes the acid in the chyme when it exits the stomach, which is necessary because the enzymes in pancreatic juice function at the normal pH of the small intestine (which is slightly basic) but not at an acidic pH. In addition to pancreatic juice, the duodenum receives **bile,** a fluid secreted by the liver that contains bicarbonate and *bile salts,* which aid in the digestion of fats, as described later.

As enzymes break down nutrients in the chyme, digestive end-products are released into solution and absorbed by cells in the

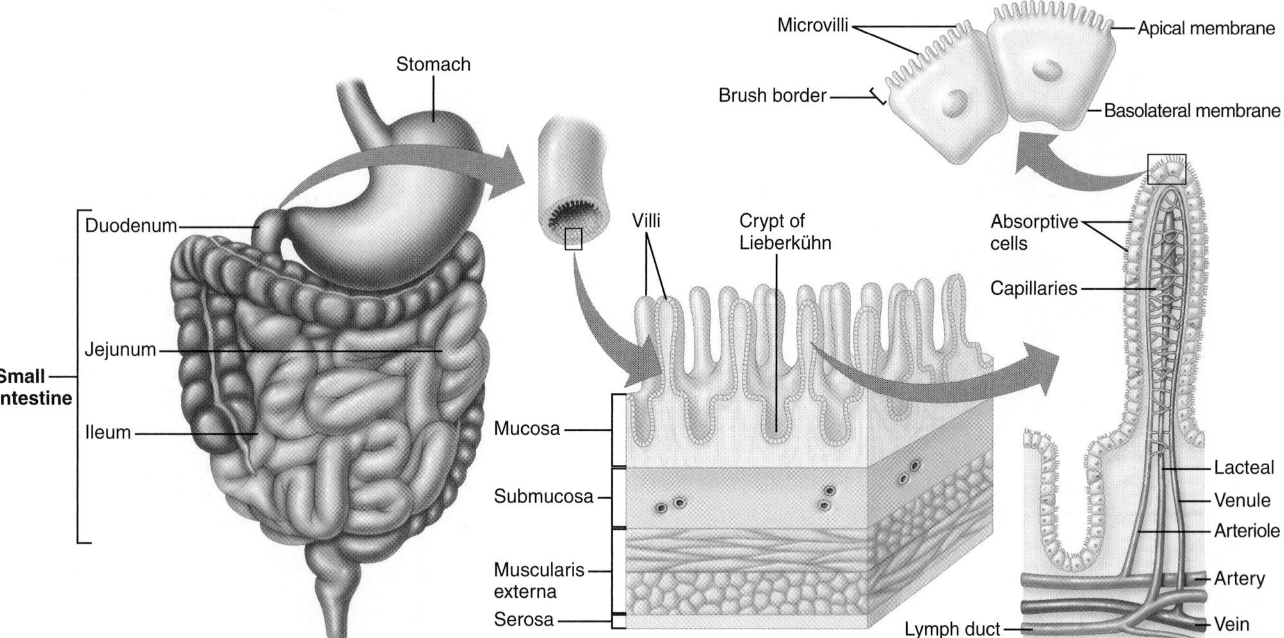

Figure 6 Anatomy of the small intestine. The duodenum, jejunum, and ileum are shown in relation to other GI tract organs; the enlarged views show the structure of villi and microvilli in the duodenum.

mucosal epithelium. These simultaneous processes of digestion and absorption begin in the duodenum and continue to completion in the remainder of the small intestine. Unless an unusually large quantity of food has been ingested, absorption is typically completed within the first 20% of the small intestine's length, or before the chyme has reached the ileum. Thus the small intestine has a large excess capacity for absorbing nutrients, indicating that its absorptive mechanisms are highly efficient.

The small intestine's absorptive efficiency is attributable in part to the fact that the mucosal surface consists of finger-like projections called **villi** (singular: *villus*) that increase the surface area by roughly a factor of 10, compared to what the surface area would be if the inside of the small intestine was simply a smooth cylinder (see Figure 6). Each villus houses structures crucial to the absorption of nutrients, including a network of *capillaries* and a blind-ended lymphatic vessel called a **lacteal.** After nutrients are absorbed from the lumen of the intestine into the cells lining it, they are transported across the mucosal epithelium into the interstitial fluid on the other side. From there, most absorbed nutrients diffuse into the capillaries and then are carried away from the intestine in the general circulation. (Absorbed *fats* are an exception to this rule, as they diffuse into lacteals, as described shortly.)

Another factor that increases the absorptive efficiency of the small intestine is the presence of a brush border on the mucosal surface, which increases the surface area by a factor of 20. This **brush border** is made up of microvilli, which are located on the apical surface of epithelial cells and which number from 3000 to 6000 per cell. The brush border and villi are most prominent in the proximal portions of the small intestine (the duodenum, for instance), but become less pronounced as one progresses toward more distal portions of the small intestine.

The epithelium of the brush border is designed for digestion and absorption. Several digestive enzymes and carrier proteins are located on the microvilli. Because of their location, these enzymes are called **brush border enzymes.** Other areas of the intestinal epithelium are designed for secretion.

Secretory cells of the small intestine are located in pits known as *crypts of Lieberkühn* (see Figure 6). These cells secrete copious amounts of bicarbonate-rich fluid, called *succus entericus,* into the lumen. This fluid is secreted mostly in the more proximal portions of the small intestine and is almost completely absorbed (along with fluid that is ingested) before the chyme reaches the colon.

Water-soluble materials that are absorbed from the intestines are carried by the bloodstream to the liver, which extracts certain nutrients for further processing. (Nutrients that remain in the blood are carried into the general circulation.) Blood from intestinal capillaries drains into the *mesenteric veins* and is carried to the liver by the **hepatic portal vein** (**Figure 7**). Blood delivered by the hepatic portal vein is deoxygenated, but a supply of oxygenated blood (which is necessary for the liver's proper function) is delivered to the liver by the *hepatic artery.* Blood is carried from the liver to the general circulation by the *hepatic vein,* which drains into the inferior vena cava.

The Colon The small intestine empties into the **colon,** which is larger in diameter (6 cm) than the small intestine, but considerably shorter (1.5 meters long). Like the small intestine, the colon is divided into regions on the basis of its anatomy (**Figure 8**). The four regions of the colon include the following

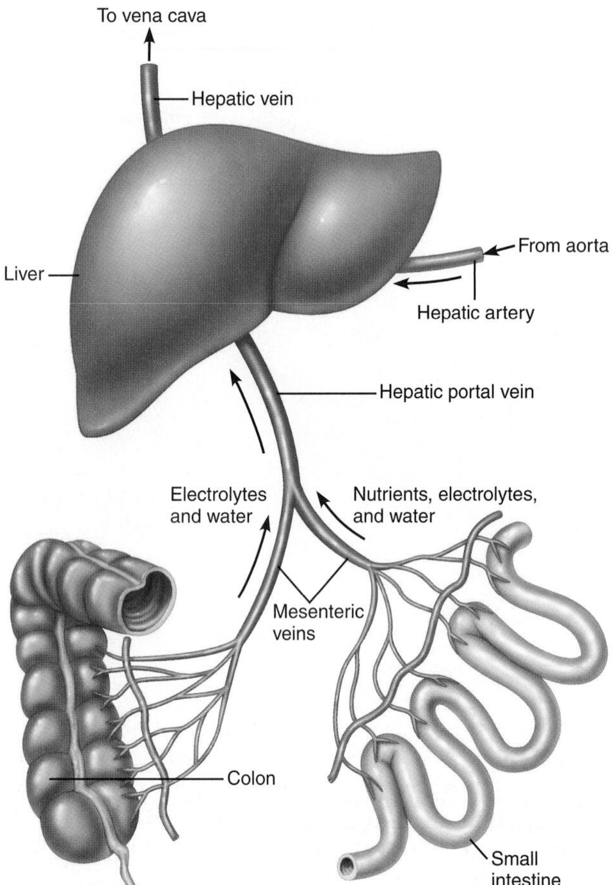

Figure 7 Circulatory route of materials absorbed in the small intestine and colon. Materials absorbed in the intestines travel to the liver via the mesenteric veins and the hepatic portal vein, and then into the general circulation via the hepatic vein.

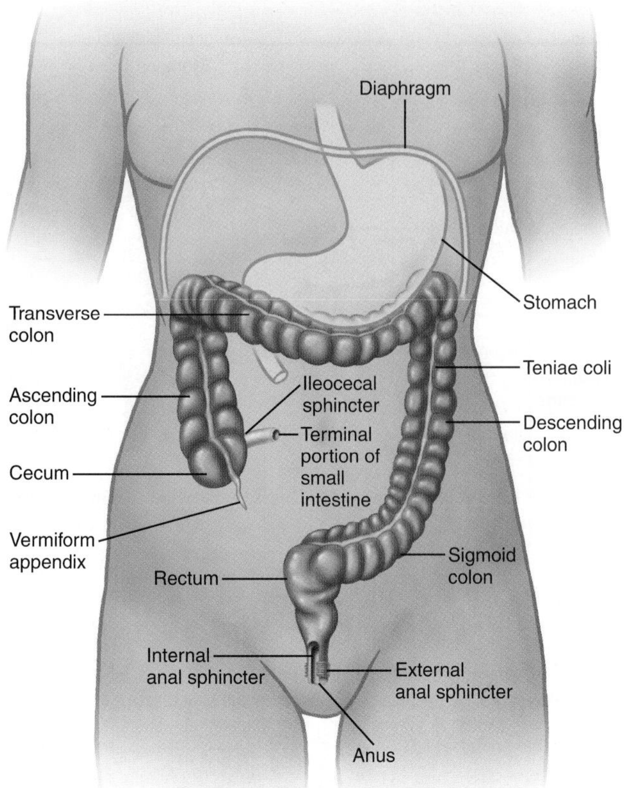

Figure 8 Major anatomical features of the colon.

segments: (1) the *ascending colon,* which runs upward on the right side of the body from the end of the small intestine toward the diaphragm; (2) the *transverse colon,* which runs across the abdominal cavity; (3) the *descending colon,* which runs downward on the left side; and (4) the *sigmoid colon,* an S-shaped segment leading to the rectum. The first three segments are specialized for absorbing water and ions from the chyme; the sigmoid colon serves primarily as a storage depot for whatever material remains in the lumen after absorption has occurred. Although the wall of the colon has the same fundamental structure as other parts of the GI tract, the longitudinal muscle layer of the muscularis externa is not continuous but is instead compressed into three relatively narrow bands called *teniae coli,* which run the colon's length.

At the junction between the ileum and the colon, the flow of material is regulated by a ring of smooth muscle called the **ileocecal sphincter** (see Figure 8). Below this junction is a blind-ended bulb called the *cecum,* to which is attached the *vermiform appendix,* a wormlike appendage having no known function.

Health Fact

On rare occasions, the opening of the appendix can become blocked and then inflamed, a condition known as *appendicitis.* If this condition persists, the appendix may rupture, spilling the lumenal contents into the abdominal cavity. Rupture is a dangerous situation because it almost always leads to *peritonitis* (inflammation of the peritoneum), which is fatal in most cases if left untreated.

By the time chyme reaches the colon, it contains very few digestible nutrients because most of the materials present in the chyme have already been absorbed. What remains consists mainly of water, inorganic ions, indigestible material from food, and bacteria. The colon's primary function is to reduce the volume of the chyme by absorbing most of the water remaining in it, thereby transforming it into a semisolid material called *feces.* The colon then stores the feces until it is ready to be eliminated from the body. Together, the cecum, colon, and rectum constitute what is often referred to as the *large intestine.*

More than 700 species of bacteria live in the lumen of the colon, forming what is called the *intestinal flora.* These bacteria have several functions, including the digestion of fiber and the production of vitamin K and vitamin B_7 (biotin). By-products of bacterial metabolism include gases *(flatus)* such as nitrogen and carbon dioxide, and smaller amounts of hydrogen, methane, and hydrogen sulfide.

Functional Fact

Antibiotics may attack the intestinal flora as well as the invading pathogen, thereby causing intestinal discomfort. Food products containing bacteria, such as yogurt, can be consumed to supplement the intestinal flora.

The Rectum and Anus Intermittently, the colon contracts strongly, pushing fecal material into the rectum. This material does not exit the body immediately because the movement of material through the anus is controlled by two sphincters: the **internal anal sphincter,** which is composed of smooth muscle, and the **external anal sphincter,** a ring of skeletal muscle that controls the opening to the outside. Relaxation of both sphincters, which are normally closed, allows fecal material to be eliminated from the body, a process called *defecation.*

Quick Check 2

1. Arrange the following terms in order such that they correctly describe the path of ingested material as it travels through the GI tract: *descending colon, esophagus, rectum, stomach, lower esophageal sphincter, pharynx, ascending colon, ileum, upper esophageal sphincter, transverse colon, duodenum, sigmoid colon, ileocecal sphincter, pyloric sphincter, jejunum.*
2. Describe the brush border. Where is it located, and what is its function?
3. Where does most digestion and absorption occur in the GI tract?

The Accessory Glands

The accessory glands of the digestive system include the **salivary glands,** which secrete *saliva;* the pancreas, which secretes *pancreatic juice;* and the **liver,** which secretes *bile.* Although these accessory glands look very different superficially and perform different functions, they share many structural and functional similarities. Secretions of these glands are carried to the GI tract via ducts lined by epithelial cells, and within the body of a gland the ducts branch extensively, terminating in an enclosed space that is completely surrounded by a layer of specialized secretory epithelial cells (**Figure 9**). In the salivary glands and pancreas, these epithelial cells are arranged in ball-like clusters called **acini** (singular: *acinus*).

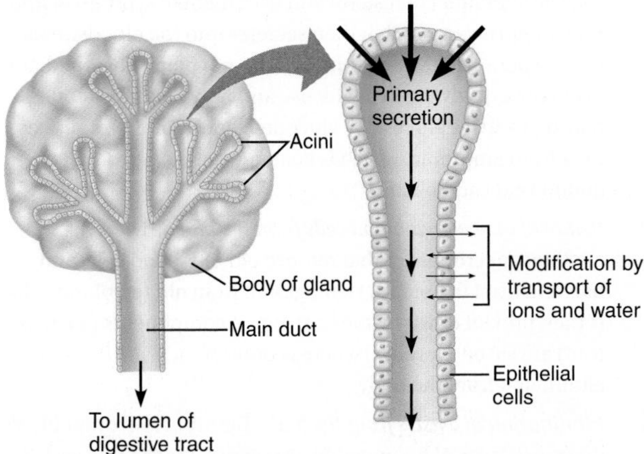

Figure 9 Generalized structure of an accessory gland. The gland depicted here is typical of the salivary glands or the pancreas. Heavy arrows at right indicate the formation of the primary secretion in an acinus; lighter arrows indicate modification of this secretion by the transport of ions and water within the ducts.

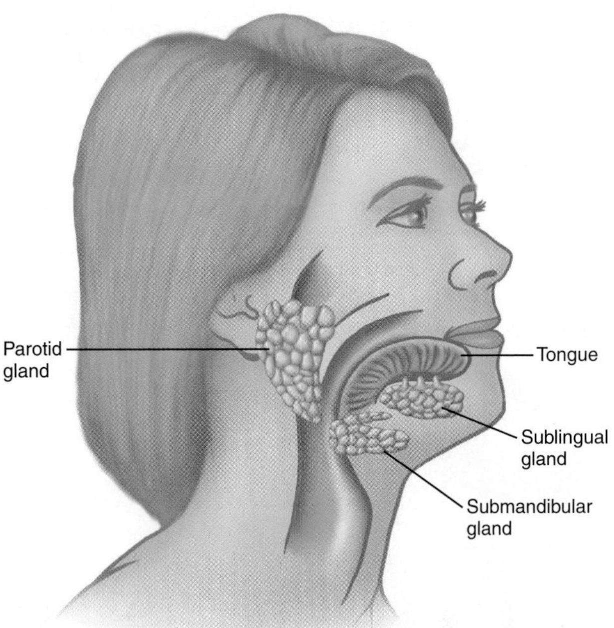

Figure 10 Location of the major salivary glands.

(The arrangement of liver secretory cells is described later in this section.) The acinar cells secrete a fluid (referred to as the *primary secretion*) containing water, inorganic ions, and other solutes whose nature depends on the gland in question. As this fluid flows through the ducts, the epithelial cells lining the ducts secrete or absorb ions and/or water, thereby modifying the fluid's composition.

The Salivary Glands

Saliva is produced by three pairs of major salivary glands (**Figure 10**): the *parotid glands,* located on both sides of the head at approximately ear level; the *sublingual glands,* located beneath the tongue on either side; and the *submandibular glands,* located on each side of the head beneath the lower jaw. Other smaller salivary glands are located in the wall of the mouth and pharynx.

Among the components of saliva are (1) *bicarbonate,* which makes the saliva alkaline and helps to neutralize acid; (2) *mucus,* which lubricates the food and protects the lining of the mouth from abrasion; (3) *salivary amylase,* a digestive enzyme that breaks down starch and glycogen; and (4) *lysozyme,* an enzyme that by destroying or *lysing* certain bacteria helps to prevent tooth decay.

The Pancreas

The pancreas, located behind and beneath the stomach (see Figure 2), is not only an exocrine organ of the digestive system but also an endocrine organ that secretes hormones important in the regulation of metabolism. The **exocrine pancreas** comprises the numerous acini and their associated ducts, whereas the **endocrine pancreas** consists of *pancreatic islets* scattered among the acini and ducts (**Figure 11**). Ducts from the acini converge to larger ducts, which eventually converge to the *pancreatic duct,* the main duct that carries pancreatic juice to the duodenum.

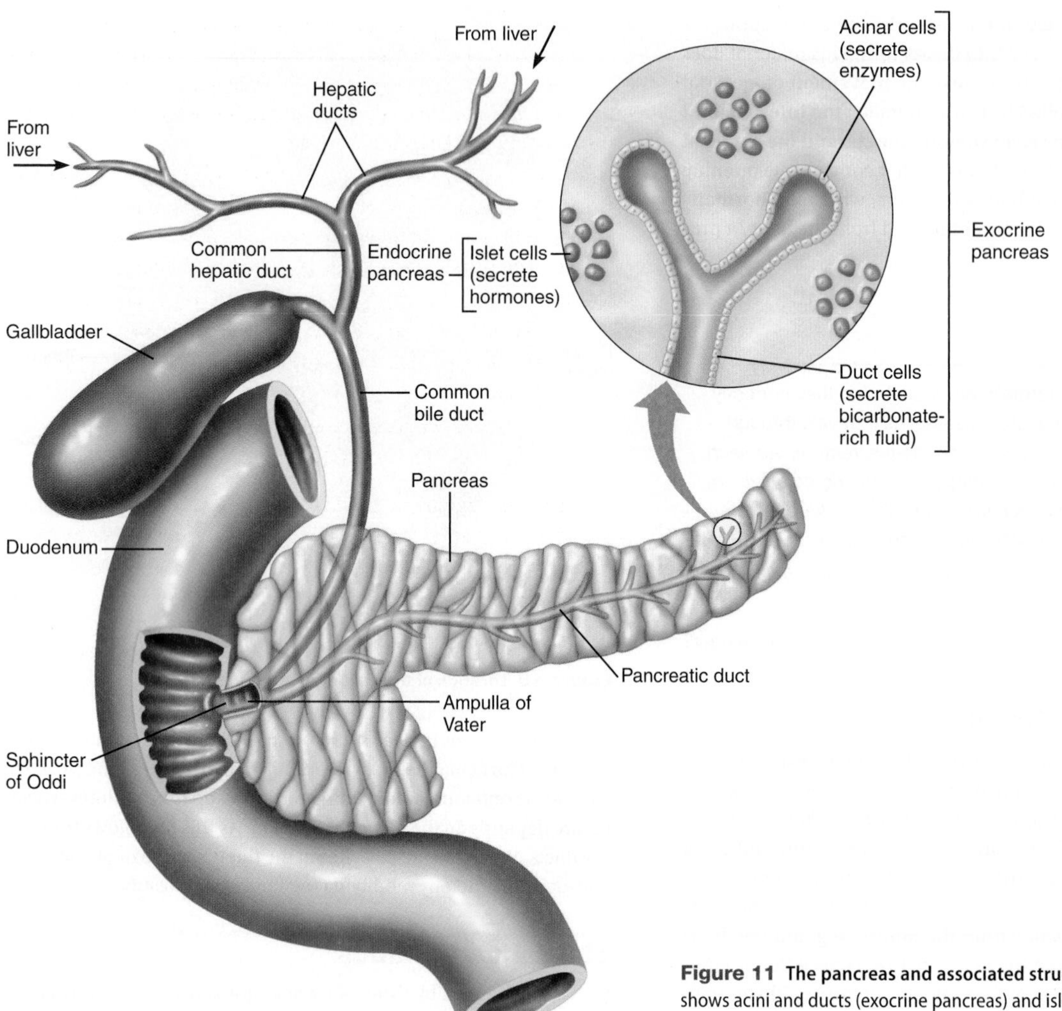

Figure 11 The pancreas and associated structures. The enlarged view shows acini and ducts (exocrine pancreas) and islets (endocrine pancreas).

Pancreatic juice is rich in bicarbonate and also contains several digestive enzymes, including *pancreatic amylase,* which is similar to salivary amylase and breaks down starch and glycogen, and *pancreatic lipases,* which break down fats. Also present are a number of *proteases,* which break down proteins, and *nucleases,* which break down nucleic acids. Relative to its weight, the pancreas secretes more protein than any other tissue in the body, most of it in the form of digestive enzymes.

The Liver

The liver—the largest organ in the abdominal cavity—is amazingly versatile. Among its more important functions are the following:

1. *Secretion of bile.* As mentioned previously, the liver secretes bile, which contains bicarbonate, phospholipids, inorganic ions, and bile salts.
2. *Metabolic processing of nutrients.* Following a meal, the liver converts some of the absorbed glucose to glycogen, and some absorbed amino acids to fatty acids; the liver also synthesizes triglycerides and cholesterol and uses them to synthesize *lipoprotein particles,* which it then secretes into the bloodstream. During periods in which nutrients are not being absorbed, the liver converts glycogen to glucose and fatty acids to ketones. It also produces glucose by gluconeogenesis and synthesizes urea from ammonia, which is generated as a by-product of amino acid catabolism.
3. *Removal of aged red blood cells from the blood.* The liver contains macrophages that remove old red blood cells and bacteria from the blood. Hemoglobin from old red blood cells is then broken down by the liver; some components (such as iron) are saved for reuse, whereas others (such as *bilirubin*) are eliminated from the body.
4. *Elimination of wastes from the body.* Bilirubin and other breakdown products of hemoglobin are secreted in the bile and eliminated from the body in the feces. (Bilirubin and similar breakdown products are referred to as *bile pigments* because in combination they impart a greenish color to the bile.) Other substances eliminated in the bile include excess cholesterol,

insulin, foreign compounds such as drugs or poisons, and trace metals. The liver also chemically transforms many hydrophobic compounds (including toxins) into more hydrophilic forms so that they can be more readily dissolved in the plasma and eliminated by the kidneys.

5. *Synthesis of plasma proteins.* The liver synthesizes most of the proteins that are present in the plasma, including albumin, steroid-binding and thyroid-hormone-binding proteins, clotting proteins, angiotensinogen, and thrombopoietin.
6. *Secretion and modification of hormones.* The liver participates with the kidney in the activation of vitamin D, and it secretes *insulin-like growth factors.* The liver also helps to clear many hormones from the body by metabolizing them.
7. *Storage of essential molecules.* The liver stores certain vitamins (A, D, and B_{12}) and metals (iron and copper).

Because it manufactures and secretes bile, the liver is considered to be part of the *biliary system,* which comprises all structures involved in synthesizing or storing bile and delivering it to the GI tract (**Figure 12**). Bile is manufactured and secreted by the liver continually, but is released into the GI tract only when food is

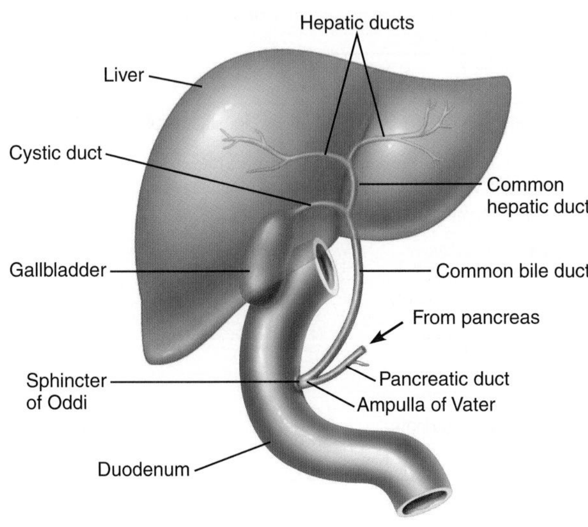

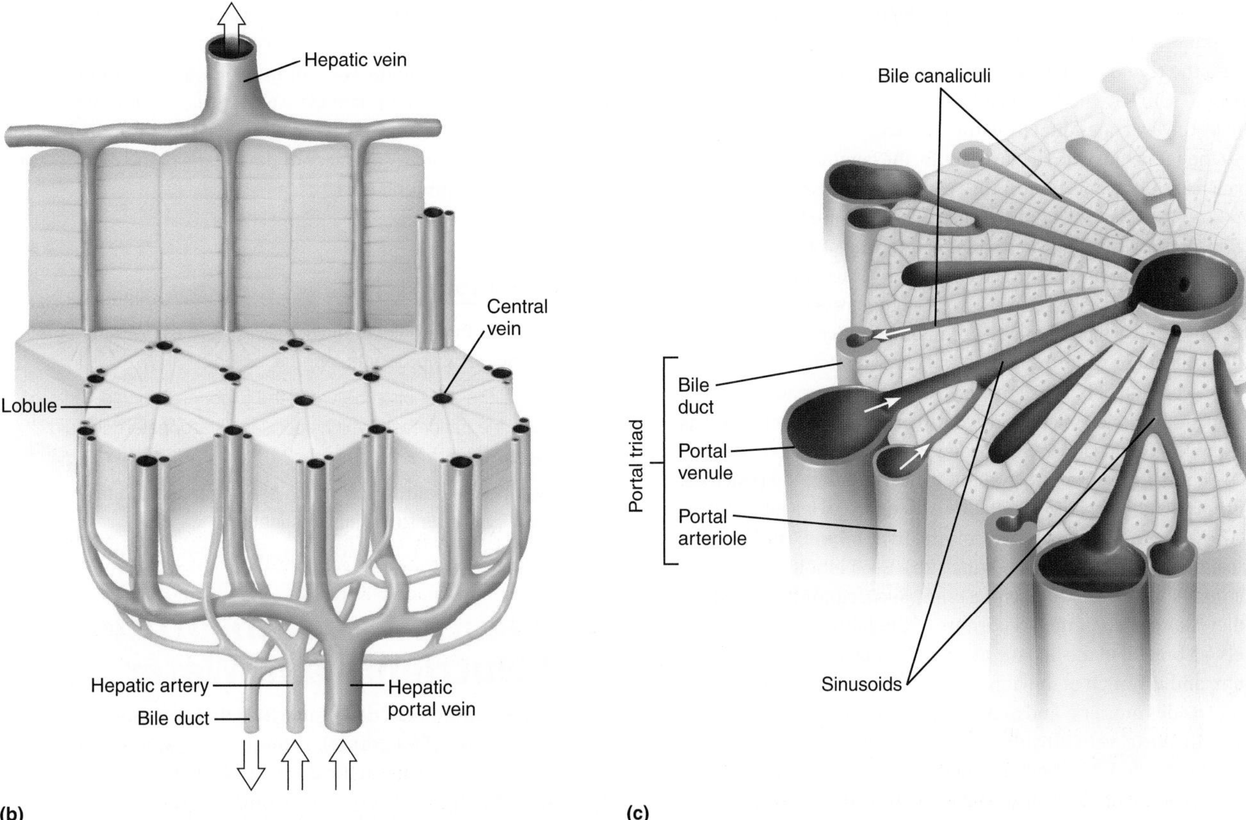

Figure 12 Structures of the biliary system. (a) Gross anatomy of the liver, gallbladder, and their associated ducts. **(b)** Microscopic anatomy of the liver, showing hepatocytes, sinusoids, and bile canaliculi. **(c)** Lobule.

Q Nutrients are carried from the small intestine to the sinusoids via branches of which blood vessel?

TABLE 1 Gastrointestinal System Organs and Their Functions

Organ	Functions
Mouth	Mechanical breakdown of food; mixing of food with saliva; initiation of chemical digestion of carbohydrates by salivary amylase
Pharynx	Conduction of food to esophagus
Esophagus	Conduction of food to stomach
Stomach	Mechanical breakdown of food; secretion of acid, pepsinogen, and intrinsic factor; initiation of chemical digestion of proteins by pepsin; secretion of gastrin into bloodstream; transformation of food into chyme
Small intestine	Chemical digestion of all nutrient classes by pancreatic enzymes and brush border enzymes; absorption of digestive end-products, water, ions, and vitamins; secretion of enterogastrones into bloodstream; secretion of bicarbonate-rich fluid
Colon	Absorption of ions and water; transformation of chyme into feces; storage of feces
Rectum	Storage of feces prior to elimination
Anus	Control of defecation
Salivary glands	Secretion of saliva (contains amylase, mucus, bicarbonate, and lysozyme)
Pancreas	Secretion of pancreatic juice (contains digestive enzymes and bicarbonate)
Liver	Secretion of bile (contains bile salts and bicarbonate); processing of absorbed nutrients
Gallbladder	Storage and concentration of bile

present. Between meals, the bile secreted by the liver is stored in a small muscular sac called the **gallbladder,** which is located immediately adjacent to and beneath the liver (Figure 12a). During meals, the gallbladder is stimulated to contract, which forces the stored bile into the *common bile duct,* which carries it to the duodenum. For people who have had their gallbladders removed, any secretion of bile by the liver goes directly to the GI tract.

At their juncture with the duodenum, the common bile duct and the pancreatic duct converge to form a common passageway (called the *ampulla of Vater*) for the flow of bile and pancreatic juice (see Figure 11). The flow of these two fluids is regulated by a ring of smooth muscle called the **sphincter of Oddi,** which is normally closed but opens when food is present. Between meals, when the sphincter is closed, the bile secreted by the liver backs up in the common bile duct and "spills over" into the gallbladder, where it is stored and concentrated.

The secretion of bile, like that of saliva and pancreatic juice, is a two-step process involving the formation of a primary secretion and its subsequent modification in ducts. However, the spongelike liver differs substantially in structure from the salivary glands and pancreas. The liver is a highly organized structure with right and left lobes.

The functional unit of the liver is the **liver lobule,** consisting of liver cells called **hepatocytes** (Figure 12b). Each lobule is a roughly hexagonal-shaped structure with a *central vein* running through its middle. Along each of the six corners of the lobule runs a triad of vessels—a branch of the hepatic artery, a branch of the hepatic portal vein, and a branch of the hepatic duct (Figure 12c). The hepatic artery provides nutrient- and oxygen-rich blood to the liver, whereas the hepatic portal vein provides recently absorbed nutrients to the liver for processing. The branches of the hepatic artery and hepatic portal vein drain into liver **sinusoids,** blood-filled cavities that function much like veins except that exchange occurs within them. The sinusoids drain into branches of the hepatic vein. A single layer of hepatocytes forms the walls of the sinusoids and can take up necessary substrates from the blood to produce bile. On the side of the hepatocytes opposite the sinusoid are **bile canaliculi.** Hepatocytes synthesize bile from substrates taken up from the sinusoids and then secrete the bile into the bile canaliculi, which drain into **bile ducts** that eventually converge to form the **common hepatic duct;** this duct carries bile away from the liver to the gallbladder or small intestine. Also found in the walls of sinusoids are **Kupffer cells,** hepatic macrophages that phagocytose bacteria, old red blood cells, and other debris that may be in the blood passing through the liver.

Table 1 provides a summary of the organs of the digestive system and their major functions.

Quick Check 3

1. Name the three pairs of major salivary glands. What are the main secretory products found in saliva?
2. What is the distinction between the exocrine pancreas and the endocrine pancreas? Which secretes bicarbonate-rich fluid? Digestive enzymes? Hormones?
3. Describe the pathway by which bile is transported from its site of synthesis to its final destination.

3 Digestion and Absorption of Nutrients and Water

A typical recommended daily American diet consists of a mixture of carbohydrates (500 grams), proteins (125 grams), and lipids (50 grams). (To learn about the diet of athletes, see **Understanding Exercise: The Role of Diet**) Most of these nutrients are consumed as macromolecules that must be enzymatically broken down into smaller molecules before they can be absorbed, a process called *chemical digestion* or simply *digestion.* Digestion generally involves the hydrolysis of macromolecules to smaller subunits. In the next three subsections, we see how molecules in each of these major

UNDERSTANDING EXERCISE

The Role of Diet

When a person exercises, he or she uses energy from previously consumed nutrients. The type of exercise determines the immediate source of energy used. Weight training or other intense exercises utilize primarily anaerobic glycolysis to generate the ATP needed for contracting muscles, whereas jogging or other moderate exercises utilize primarily oxidative phosphorylation to generate the needed ATP. The primary substrate for anaerobic glycolysis is glucose that was stored in the muscle cells as glycogen. The primary substrates for oxidative phosphorylation include blood glucose (provided by the liver) and fatty acids (provided by adipose tissue).

People who take exercise seriously also take their diets seriously. Low fat is a healthy choice for any person, exerciser or not. Weight trainers tend to favor a high-protein diet to provide the amino acids necessary for muscle growth. Joggers tend to favor a high-carbohydrate diet, especially the day before a long-distance run. But what do the experts recommend?

Athletes, like any other people, need a well-balanced diet that includes carbohydrates (up to 400 grams/day), proteins (up to 84 grams/day), and fats (approximately 25% of energy can come from fat—an essential nutrient—without any detriment to athletic performance). The overall caloric intake of athletes needs to be higher than that of sedentary persons to maintain energy balance while calories are being burned during exercise. Diets of athletes also need to contain vitamins and minerals, especially the B-complex vitamins and the antioxidant nutrients (vitamins A, E, and C, and beta-carotene and selenium). Female athletes, especially, must have adequate levels of calcium, iron, and zinc.

Fluids are critical before, during, and after exercise to maintain body water balance. Fluid balance also enhances athletic performance. Athletes often drink water *before* performance to prepare for the loss of fluid from sweating. They may then consume fluids *during* exercise to maintain fluid balance, as dehydration can interfere with exercise performance. Although sports drinks are popular with athletes, the hypertonicity of these beverages can decrease the rate of gastric emptying, thereby decreasing the rate of fluid absorption. The electrolytes in sports drinks are needed only with intense exercise or long-duration exercise, and the carbohydrates are generally not needed unless the exercise occurs long after the last meal or continues for several hours.

What should one eat, if anything, before exercise? Eating one hour before exercise tends to improve performance, but it is critical that the right nutrients are consumed and in a relatively small quantity (more of a snack than a meal). The pre-exercise meal should be high in fluids and carbohydrates, and low in fats and fibers. The low fiber is necessary to minimize gastric motility.

Today's world is full of supplements designed to improve human performance. The most popular choices among weight trainers are creatine and protein shakes. Creatine is believed to increase the creatine phosphate levels in muscle cells to generate more power at the onset of exercise. Consuming creatine on a regular basis can increase cellular content by 15% to 30%. Data suggest that creatine increases lean body mass, and most studies show an improvement in exercise capacity with the use of the supplement. Caffeine is another supplement commonly used by athletes to enhance performance. Studies suggest that it enhances long-duration exercise by increasing fat metabolism, thereby sparing glycogen for later use.

In summary, athletes require more fluids and carbohydrates in their diet compared to non-athletes. More protein is desirable for those trying to build muscle mass.

nutrient classes are digested and absorbed. Then we see how the digestive system handles vitamins, minerals, and water—components of food that are not digested but instead are simply absorbed in their original forms.

Carbohydrates

Most carbohydrates in the diet are in the form of polysaccharides, such as starch and cellulose from plants and plant products (pasta and breads, for example) and glycogen from animals and animal products. Although all polysaccharides are polymers of glucose, they differ in structure. Cellulose cannot be digested, because it is not a substrate of digestive enzymes. Furthermore, the intestinal flora break down some cellulose into absorbable units, although the amount transformed in this way is negligible. Therefore, most cellulose passes through the intestinal tract and is eliminated from the body in feces. Cellulose is a major component of what nutritionists call dietary fiber and is considered an important part of the diet because it aids in intestinal motility. Other consumed carbohydrates include disaccharides such as sucrose (table sugar) and lactose (milk sugar), and to a lesser extent, monosaccharides such as glucose and fructose. The absorbable forms of carbohydrates are monosaccharides. Thus most consumed carbohydrates must be digested before they can be absorbed.

Digestion of Carbohydrates to Monosaccharides

Polysaccharides are digested by **amylases** located in saliva and pancreatic juice; thus digestion begins in the mouth. Salivary amylase can digest polysaccharides for only a short time, however, before it is inactivated by the acidic environment of the stomach. Pancreatic amylase continues the digestion of polysaccharides in the small intestine.

Amylases are limited in their ability to digest polysaccharides because they cannot break bonds at branch points or between glucose monomers at the end of polysaccharide chains. Thus salivary and pancreatic amylases reduce starch or glycogen to either maltose (disaccharides consisting of two glucose monomers) or short, branched polysaccharides called *limit dextrins* (**Figure 13**).

Digestion of carbohydrates to monosaccharides is completed by a variety of brush border enzymes bound to the apical

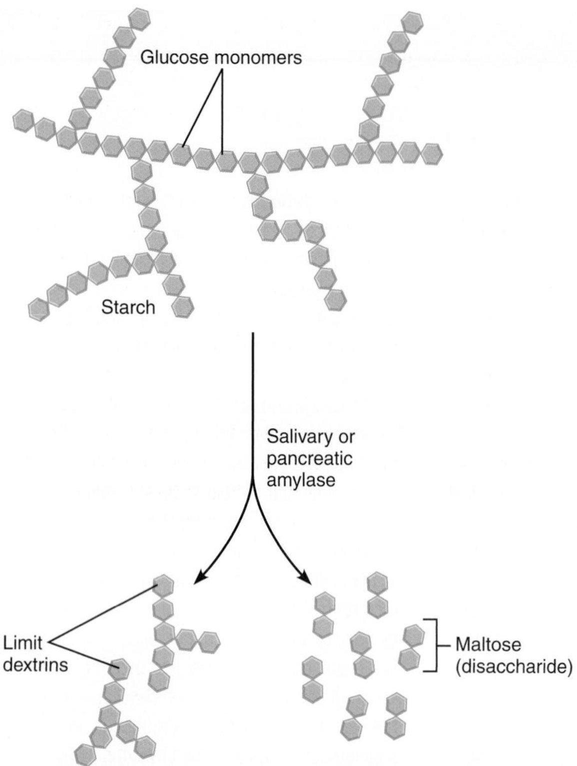

Figure 13 Digestion of starch by salivary or pancreatic amylase. Complete digestion of starch by amylases produces maltose and limit dextrins.

Q Is starch digestion an *anabolic* reaction or a *catabolic* reaction?

membranes of absorptive cells lining the small intestine. These enzymes include *dextrinase* and *glucoamylase,* which break down limit dextrins and straight-chain glucose polymers to glucose monomers; *sucrase,* which hydrolyzes sucrose to glucose and fructose; *lactase,* which hydrolyzes lactose to glucose and galactose; and *maltase,* which hydrolyzes maltose to two glucose molecules. (To see what happens when one of these enzymes is deficient, see **Clinical Connections: Lactose Intolerance**.) These enzymes are in close proximity to other membrane-bound proteins that transport the monosaccharides into the epithelial cells, the first step of absorption. To illustrate the action of brush border enzymes, the digestion and absorption of maltose is shown in **Figure 14**.

Absorption of Carbohydrates

Once carbohydrates have been digested to monosaccharides, they are absorbed by carrier-mediated transport across epithelial cells lining intestinal villi. Glucose and galactose enter the epithelial cells via cotransport with sodium across the apical membrane and then cross the basolateral membrane by facilitated diffusion (see Figure 14). The same cotransporter on the apical membrane is used for both monosaccharides. Fructose is absorbed by facilitated diffusion, so no energy is required. Following transport across the epithelium, these molecules then diffuse into capillaries and travel to the liver for processing.

Proteins

Proteins enter the GI tract through ingestion, secretion (mucins and enzymes, for example), and the turnover of enterocytes, which are sloughed into the lumen of the GI tract. Proteins must be digested to tripeptides, dipeptides, and amino acids prior to absorption. Two

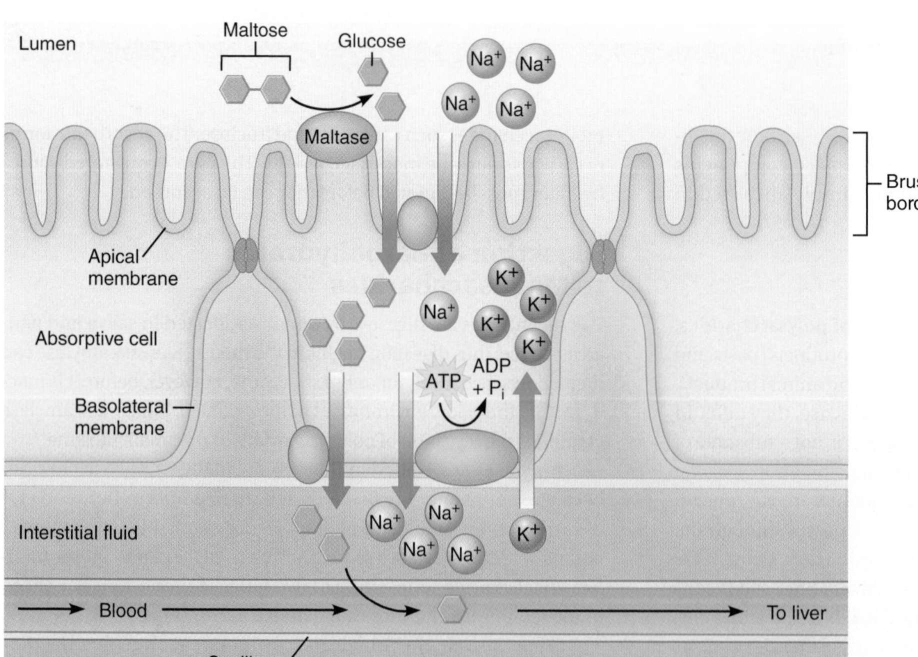

Figure 14 Digestion and absorption of maltose. Maltose, a disaccharide consisting of two glucose monomers (hexagons), and an end-product of starch digestion by amylase, is digested by maltase, a brush border enzyme. The end-products—glucose molecules—are transported across the apical membrane of mucosal epithelial cells by sodium-linked secondary active transport. Energy for this process is derived from the action of Na^+/K^+ pumps in the basolateral membrane, which create an inwardly-directed electrochemical gradient for sodium ions. Glucose molecules exit cells by facilitated diffusion across the basolateral membrane and then diffuse into capillaries, which carry them into hepatic circulation.

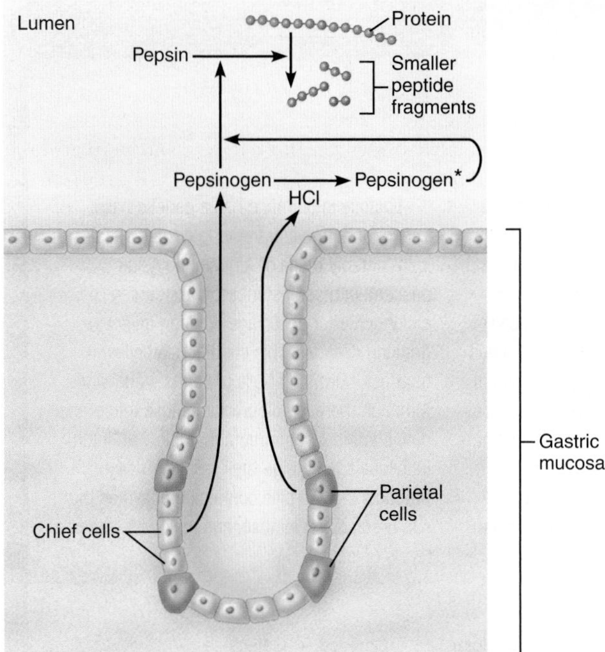

Figure 15 Activation and activity of pepsin in the stomach. Chief cells secrete pepsinogen, which is partially activated by hydrochloric and (HCl) secreted by parietal cells. Partially-activated pepsinogen molecules (pepsinogen) then fully activate other pepsinogen molecules by cleaving off amino acids to form pepsin, which digests proteins to smaller fragments.

types of enzymes digest proteins: (1) **endopeptidases,** which split polypeptides at interior peptide bonds, thereby producing smaller peptides, and (2) **exopeptidases,** which cleave off amino acids from one end of the polypeptide, thereby producing amino acids.

To protect the secretory cells from the proteolytic action of these enzymes, the enzymes are stored in these cells in an inactive form, called **zymogens.** The zymogens are sequestered in vesicles called *zymogen granules* and are secreted by exocytosis. Once they are secreted into the lumen of the GI tract, other enzymes convert zymogens into active enzymes by **proteolytic activation** (cleaving of amino acids off a precursor to form the active protein).

Digestion of Proteins to Small Peptides and Amino Acids

The digestion of proteins begins in the stomach with the action of *pepsin,* an endopeptidase. *Pepsinogen,* the precursor for pepsin, is secreted by chief cells of the stomach (**Figure 15**). Pepsinogen is partially activated by interaction with hydrogen ions secreted by the parietal cells of the stomach. The partially active pepsinogen can then fully activate other pepsinogens by proteolytic activation. At that point, pepsin can start digestion of proteins in the lumen of the stomach, but it cannot complete their digestion to individual amino acids because its actions are limited to splitting bonds between certain amino acids only. In addition, after gastric emptying, pepsin becomes inactive in the alkaline lumen of the small intestine.

The pancreas secretes several zymogens into the duodenum, including trypsinogen, chymotrypsinogen, and procarboxypeptidase. *Enterokinase,* a brush border enzyme, proteolytically activates trypsinogen by converting it to *trypsin* (**Figure 16**). Trypsin then activates other zymogens through proteolytic activation, including *chymotrypsin* and *carboxypeptidase.* Like pepsin, trypsin and chymotrypsin are endopeptidases that break peptide bonds between certain amino acids only, which limits their ability to fully digest proteins. Carboxypeptidase and a brush border enzyme, *aminopeptidase,* are exopeptidases that finish protein digestion by cleaving amino acids from the carboxyl and amino ends of a polypeptide chain, respectively. The final digestion products include amino acids, dipeptides, and tripeptides.

Absorption of Amino Acids and Small Peptides

Once proteins are digested, amino acids are actively transported into intestinal epithelial cells by cotransport with sodium across the apical membrane. There are at least four distinct amino acid carriers that can transport the 20 amino acids. Dipeptides and tripeptides are also actively transported across the apical membrane, but their transporters are distinct from those used for amino acids. Once inside epithelial cells, the dipeptides and tripeptides are broken down by intracellular proteases to amino acids. Amino acids are then transported across the basolateral membrane by facilitated diffusion and diffuse into the blood.

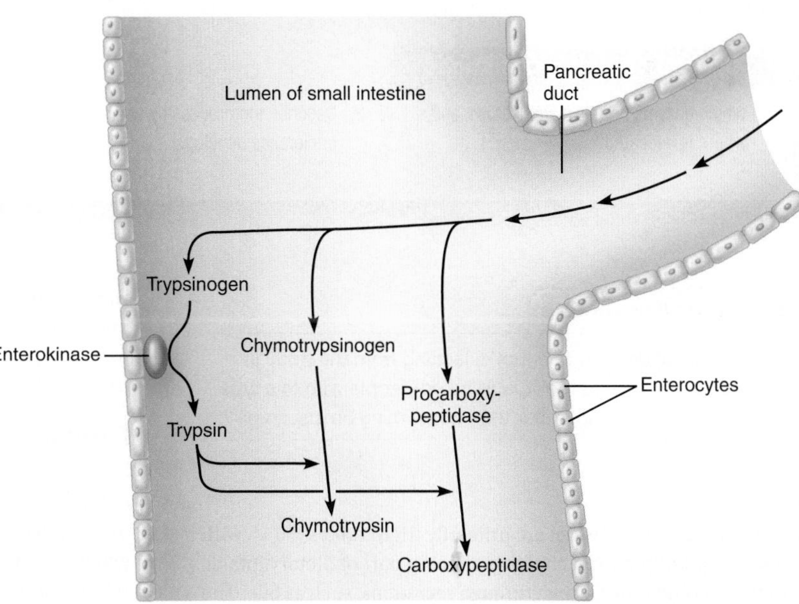

Figure 16 Activation of proteases in the small intestine. Trypsinogen, a zymogen secreted by the pancreas, is converted to trypsin by the brush border enzyme enterokinase. Trypsin then converts other zymogens to their active forms in the lumen of the small intestine and also converts other trypsinogen molecules to trypsin (not shown).

CLINICAL CONNECTIONS

Lactose Intolerance

Most nutrient molecules in the human diet must be broken down by enzymes in the digestive system before they can be absorbed. For instance, disaccharides such as sucrose (table sugar) and lactose (milk sugar) must be broken down to monosaccharides, which are then absorbed by the small intestine. The enzymes that perform this task are bound to the plasma membrane of intestinal epithelial cells and, therefore, are well positioned to come into contact with nutrients in the lumen of the intestine. After a disaccharide is reduced to monosaccharides, these monosaccharides are absorbed and transported into the bloodstream.

In the epithelium of the small intestine, the lactose present in milk is broken down by the membrane-bound enzyme *lactase* to glucose and galactose, which are then absorbed (see the figure). Normally, the processes of enzymatic breakdown and absorption are so efficient that virtually 100% of the lactose molecules are cleared from the lumen in this manner. In certain individuals with a condition known as *lactose intolerance,* however, this is not the case.

Lactose intolerance develops when cells stop producing lactase, such that the enzyme concentration falls. When this occurs, a person begins to have trouble digesting lactose and, therefore, cannot absorb it. The problem arises because lactose then remains in the lumen of the intestine, serving as a nutrient for the bacteria that normally live there. Lactose stimulates bacterial growth and the resultant production of gas and other waste products. These substances irritate the lining of the intestine and cause bloating, discomfort, and diarrhea. Lactose-intolerant individuals experience these symptoms after drinking milk or eating other dairy products and eventually learn to avoid them.

Lactose intolerance has a genetic basis and is more prevalent among certain peoples, most notably those of African and Asian descent. In most instances, it appears in childhood after about age 6, when milk drinking declines. This condition is believed to affect more than 50% of adults worldwide. Although there is no cure for lactose intolerance, new commercial products that allow afflicted individuals to indulge their taste for dairy products include pills containing enzymes that can break down lactose, and milk with reduced levels of lactose.

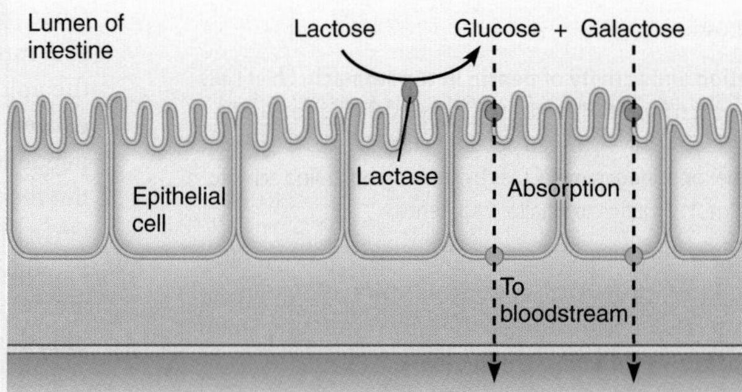

Critical Thinking Questions

1. Which enzyme breaks down lactose, and where is it located?
2. Describe the process by which lactose intolerance develops.
3. List the symptoms of lactose intolerance, and explain when lactose intolerance is most likely to appear.

Apply Your Knowledge

Insulin is a peptide hormone that is lacking from the body in some forms of diabetes mellitus. Why do people with this type of diabetes inject insulin rather than consuming an insulin pill?

Lipids

Lipids consumed in the diet are primarily triglycerides (90%), with some phospholipids and cholesterol being part of dietary intake. Other lipids enter the GI tract through secretions, such as bile. The digestion of lipids follows the same basic pattern as that for carbohydrates and proteins—large molecules are broken down by enzymes to smaller molecules before they are absorbed. However, carbohydrates and proteins are water soluble, whereas lipids are water insoluble. Therefore, the mechanism of lipid absorption is markedly different from what we have seen so far. In this section, we place particular emphasis on the digestion and absorption of triglycerides.

Digestion of Triglycerides

Lipids are digested by a class of enzymes called **lipases.** Digestion of lipids begins in the mouth with *lingual lipase,* an enzyme found in saliva. This process continues in the stomach with lingual lipase, which remains active in the acidic stomach environment, and *gastric lipase,* an enzyme secreted into the gastric lumen. Most digestion of lipids, however, does not take place until the lipids reach the small intestine, where the chyme is mixed with *pancreatic lipases.*

Because lipids are hydrophobic, ingested fats do not mix readily with the rest of the stomach contents; instead, they coalesce to form large droplets that float on top of the chyme. As these fats leave the stomach, they do so as large globules that are practically indigestible because lipases are water-soluble (like other digestive enzymes) and can act only on molecules near a globule's surface. Thus the vast majority of the lipid molecules in these globules cannot be reached by the enzymes.

The Action of Bile Salts Efficient digestion of lipids is made possible by the action of bile, which first comes into contact with fat globules in the duodenum. The bile does not actually digest lipids, because it contains no enzymes; it simply facilitates the action of lipases by breaking fat globules down into smaller droplets, a process called **emulsification.** By increasing the total surface area of the droplets, emulsification increases the amount of lipid that is exposed to water and, therefore, is susceptible to enzymatic digestion.

The emulsification of fats is due primarily to the action of **bile salts,** cholesterol derivatives synthesized by hepatocytes and secreted in the bile. Although cholesterol itself is very hydrophobic because it is composed almost entirely of nonpolar hydrocarbons, bile salts possess a number of oxygen-containing polar groups (such as hydroxyl groups [—OH] and carboxyl groups [—COOH]) that are hydrophilic (**Figure 17a**). Because all of these polar groups are located on one side of the molecule (the opposite side is nonpolar), bile salts are amphipathic.

When bile salts come into contact with a fat globule, their hydrophobic sides face inward (toward the hydrophobic droplet), and their hydrophilic sides face outward (toward the water), as shown in **Figure 18**. In so doing, bile salts endow the droplets with a polar "coating" that allows them to mix more readily with water and break down into smaller droplets.

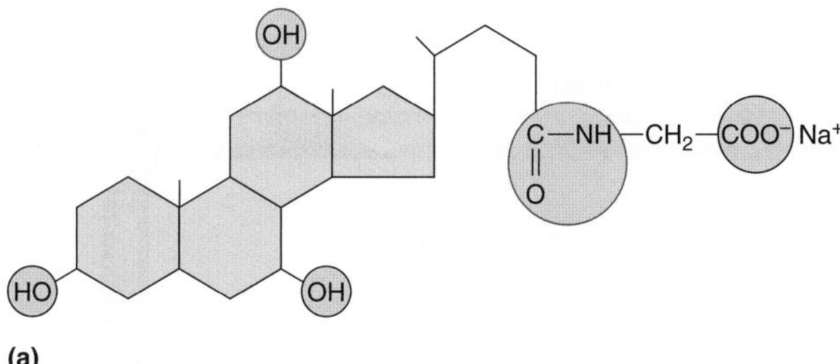

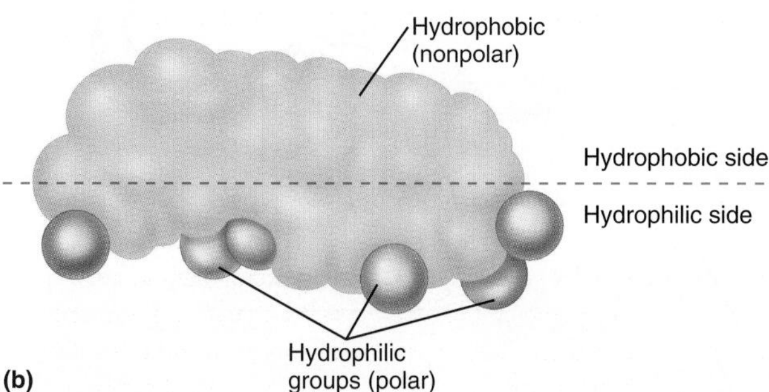

Figure 17 **A representative bile salt. (a)** Structural formula, in which hydrophilic groups are highlighted by shading. **(b)** A space-filling model illustrating the amphipathic nature of the molecule.

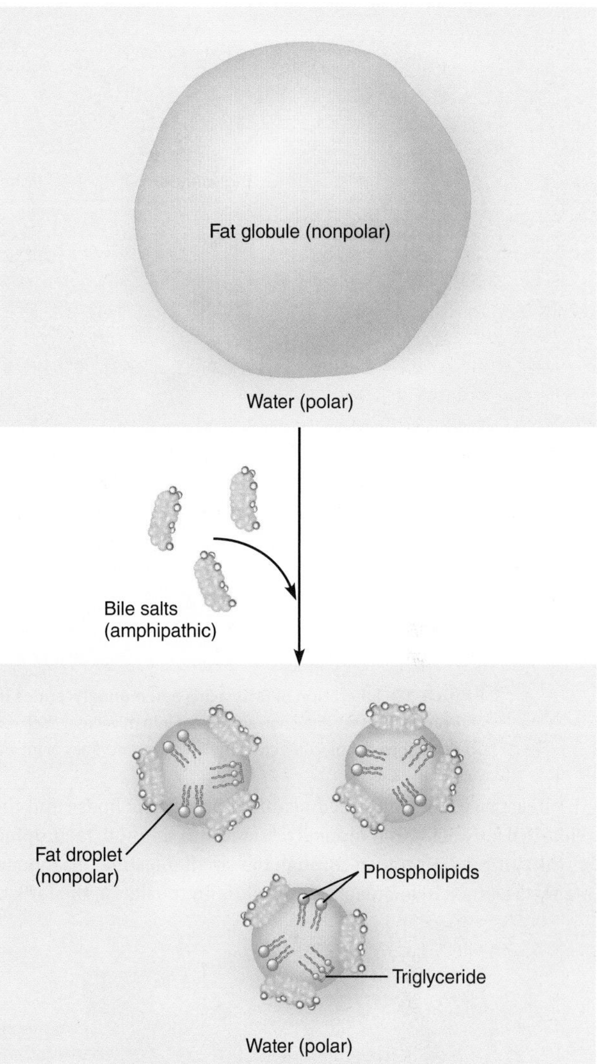

Figure 18 **Emulsification of a fat globule by bile salts.**

The Action of Pancreatic Lipase As fat is being emulsified in the duodenum, it also mixes with pancreatic lipase. For pancreatic lipase to interact with the hydrophobic fat globule, it must bind with *colipase,* a peptide secreted in pancreatic juice. A colipase molecule binds to a lipase molecule in a one-to-one fashion, allowing the lipase to interact with the fat globule and digest outer layers of lipids (**Figure 19**).

Lipases act on triglycerides to break the bonds linking fatty acids to the two carbons on either end of the glycerol backbone. As a result, the end-products of triglyceride digestion are two free fatty acids and a monoglyceride (a glycerol molecule to which a single fatty acid is attached). Some of these dissolved end-products are quickly absorbed into epithelial cells lining the small intestine; others remain in the chyme, aggregating with bile salts, cholesterol, and other lipid-soluble substances into small particles called *micelles,* which readily exchange lipids with the surrounding solution. As more dissolved lipids are absorbed, the micelles release more products to be absorbed.

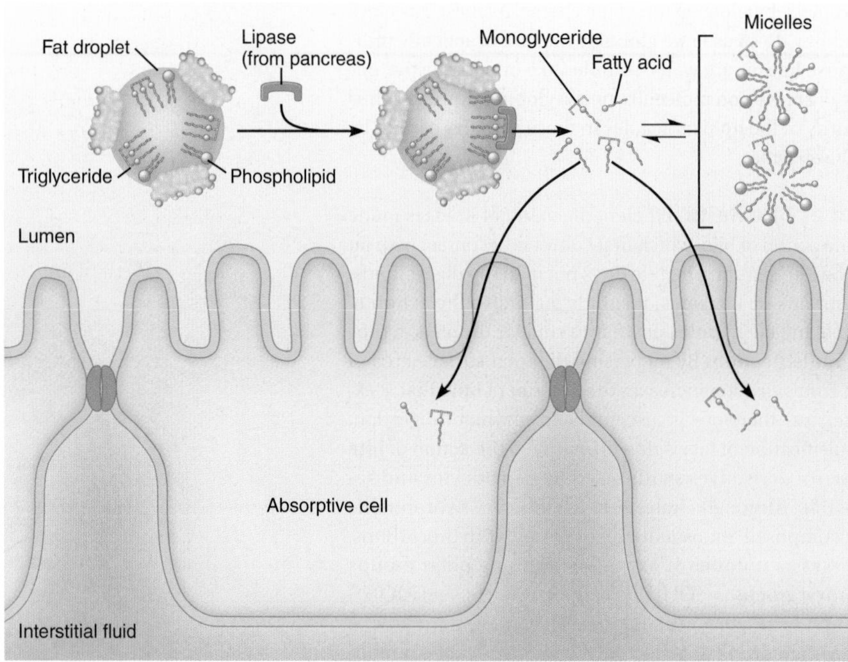

Figure 19 **Liberation of fatty acids and monoglycerides from fat droplets by lipases.** Note that these end-products of fat digestion aggregate into micelles, which exist in equilibrium with dissolved molecules. Some dissolved molecules cross the apical membranes of intestinal epithelial cells by simple diffusion.

Because of the continued action of pancreatic lipase and the release of fatty acids and monoglycerides into solution, fat droplets shrink during their transit through the small intestine, disappearing by the time chyme reaches the colon. In the ileum, the bulk of the bile salts that were secreted into the duodenum (approximately 95%) are absorbed into the circulation; these bile salts are eventually recycled by the liver and secreted again in the bile via a pathway referred to as the *enterohepatic circulation* (**Figure 20**).

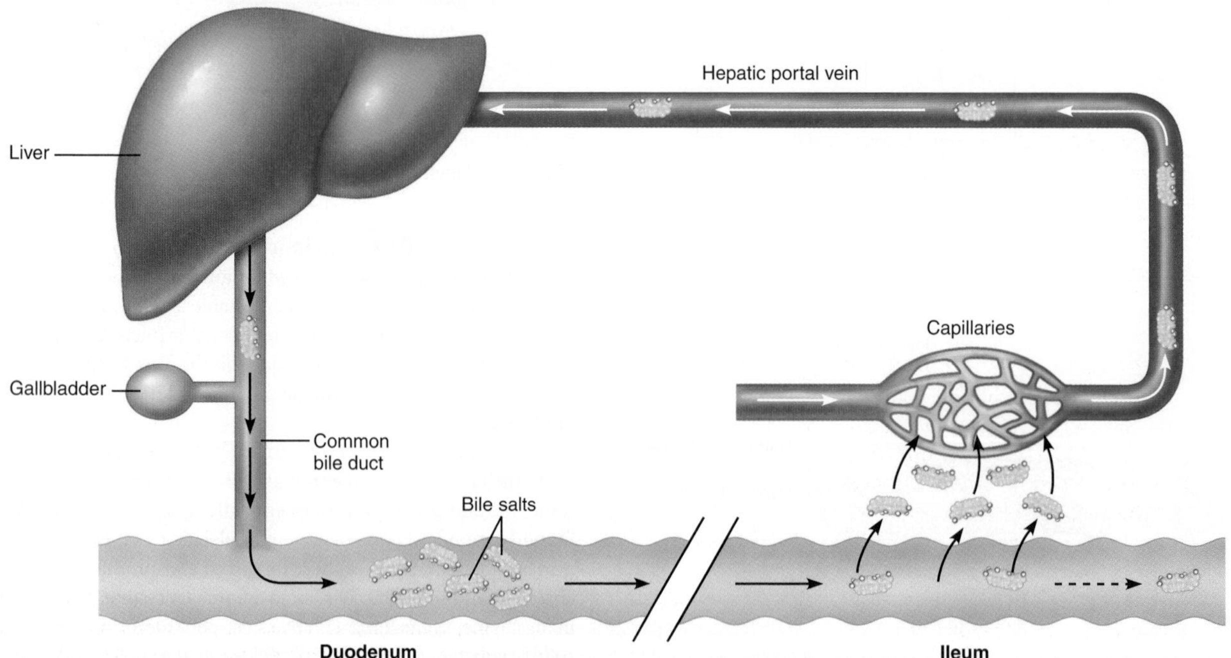

Figure 20 **The enterohepatic circulation.** Bile salts are absorbed in the ileum, return to the liver via the hepatic portal circulation, and are eventually resecreted in the bile.

Absorption of Lipids

The first step in the absorption of lipids is the entry of fatty acids and monoglycerides into absorptive intestinal epithelial cells (enterocytes), which occurs by simple diffusion. Inside the cells, these molecules enter the smooth endoplasmic reticulum, where they are acted upon by enzymes that *reassemble* them into triglycerides (**Figure 21**). (Other end-products of lipid digestion are also reassembled in a similar fashion.) These lipids are then packaged by the Golgi apparatus into large particles called **chylomicrons,** which belong to a general class of particles known as **lipoproteins** (see **Discovery: Lipoproteins and Plasma Cholesterol**).

Following their synthesis, chylomicrons are secreted by exocytosis across the basolateral membrane and into the interstitial fluid (see Figure 21). After reaching the interstitial fluid, they enter the lymphatic system via the lacteals, which, like other lymphatic capillaries, have openings in their walls that are large enough to allow such particles to pass through. (Chylomicrons cannot enter the bloodstream directly because they are too large to cross capillary walls.) The flow of lymphatic fluid eventually carries the chylomicrons to the bloodstream, where much of the lipid contained within them is released to cells for use, including cells in adipose tissue.

Quick Check 4

1. The digestion of carbohydrates is complete when they have been reduced to which form? When are proteins digested completely?
2. Name the substrate for each of the following enzymes: sucrase, lactase, dextrinase, trypsin, chymotrypsin, pepsin, amylase, carboxypeptidase, and enterokinase.
3. Are absorbed fats transported directly into the bloodstream? Why or why not? What are chylomicrons, and what is their role in the absorption of fats?
4. What role do bile salts play in the digestion of fats?

Absorption of Vitamins

Vitamins are organic molecules needed in small amounts for the normal physiological function of cells. They are consumed in their absorptive form, and do not undergo digestion. The mechanism of absorption of vitamins depends on whether the vitamin is hydrophobic or hydrophilic.

Hydrophobic or *fat-soluble vitamins* (A, D, E, and K) are absorbed by epithelial cells in parallel with lipids because they dissolve readily in lipid droplets, micelles, and chylomicrons.

Water-soluble vitamins are absorbed by special transport proteins, with some requiring active transport, and others facilitated diffusion. One water-soluble vitamin, vitamin B_{12}, cannot be absorbed by itself; it can be absorbed only when bound to *intrinsic factor,* which is secreted into the lumen of the stomach by parietal cells. Intrinsic factor binds to vitamin B_{12} to form a complex that is subsequently absorbed in the ileum. Because vitamin B_{12} is necessary for the synthesis of hemoglobin, its deficiency leads to a particular form of anemia called *pernicious anemia,* which can result from a lack of vitamin B_{12} in the diet or a lack of intrinsic factor (see Clinical Connections: Anemia).

Apply Your Knowledge

Olestra, a fat substitute, consists of a six-carbon backbone with six fatty acids attached to it. Olestra is neither digested nor absorbed. Explain why olestra would not be absorbed and what effect its consumption could have on the absorption of fat-soluble vitamins.

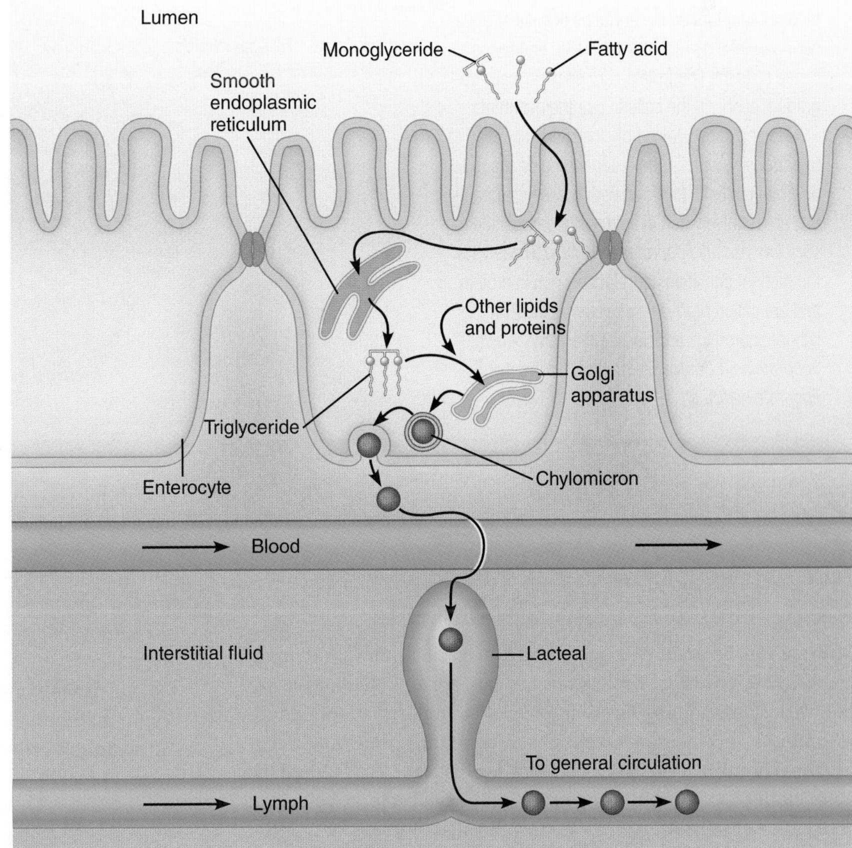

Figure 21 Events in the absorption of fats. Fatty acids and monoglycerides absorbed into enterocytes are synthesized into triglycerides, which are then packaged into chylomicrons. After exocytosis into interstitial fluid, chylomicrons enter the lymphatic system via lacteals and eventually reach the general circulation.

Q What prevents chylomicrons from entering the bloodstream directly via capillaries in intestinal villi?

DISCOVERY

Lipoproteins and Plasma Cholesterol

Chylomicrons belong to a general class of particles known as *lipoproteins,* which contain both lipids and proteins. Lipoproteins range from 75 to 1500 μm in diameter and have an inner core of hydrophobic lipids surrounded by a hydrophilic coat, as shown in the figure. Lipoproteins are useful for transporting lipids in blood because their hydrophilic surfaces allow them to mix with water, which lipids cannot do.

The inner core of a lipoprotein contains triglycerides, cholesterol, and other hydrophobic lipids and fat-soluble compounds. The outer coat contains phospholipids, other amphipathic lipids, and proteins called *apolipoproteins,* which both increase the particles' water solubility and serve as cofactors for various enzymes. Certain apoproteins bind to receptors on the surfaces of cells, triggering uptake by receptor-mediated endocytosis. Inside cells the lipoproteins are enzymatically degraded, enabling the cells to use their contents.

Plasma lipoproteins are classified according to their density, which differs depending on the ratio of protein to lipid. Particles with the lowest densities are relatively rich in lipid and are classified as *very low density lipoproteins (VLDLs).* Those with the highest densities are relatively rich in protein and are called *high-density lipoproteins (HDLs).* Other lipoproteins are classified as *low-density lipoproteins (LDLs)* or as *intermediate-density lipoproteins (IDLs).*

LDL particles have been implicated in the development of *atherosclerotic plaques,* cholesterol-rich deposits on the linings of blood vessels that can cause coronary artery disease. A person's risk of heart disease in such a case depends on the relative quantities of cholesterol carried by HDLs and LDLs. LDL cholesterol is often called "bad cholesterol" because higher concentrations of LDLs in the blood are associated with a higher risk for atherosclerosis. In contrast, HDL cholesterol is referred to as "good cholesterol" because higher HDL levels are associated with lower risks for the disease. As the ratio of HDL cholesterol to LDL cholesterol increases, the risk of atherosclerosis decreases.

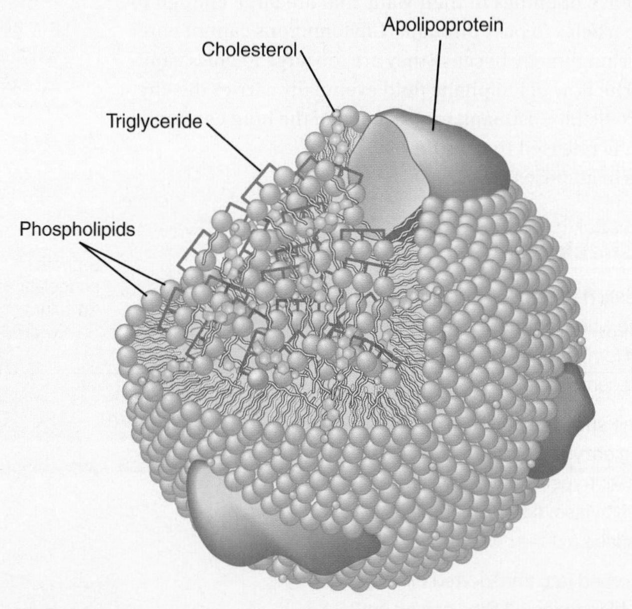

Absorption of Minerals

Absorption of Sodium and Chloride

Most sodium enters the GI tract through secretions (80%), although some is ingested daily. The precise mechanism for sodium absorption varies between regions of the intestine. In the duodenum and jejunum, the epithelium is "leaky," similar to the epithelium in the proximal tubules of the kidney. Thus, as some water is absorbed by paracellular transport, it brings with it dissolved solutes by a process called *solvent drag.* Active transport of sodium occurs in the jejunum, ileum, and colon, with the ileum and colon having "tight" epithelium. Active absorption depends on the action of a basolateral Na^+/K^+ pump, which keeps sodium levels inside enterocytes low. Sodium enters enterocytes across the apical membrane by a variety of mechanisms, including cotransport with other molecules and ions (cotransport is more important in the proximal intestine, where most solutes are absorbed), and countertransport with hydrogen ions.

Absorption of sodium is generally coupled to chloride absorption to maintain electroneutrality. In the leaky jejunum, chloride simply follows the positively charged sodium. In the ileum and colon, chloride is actively absorbed by countertransport with bicarbonate across the apical membrane and by facilitated diffusion across the basolateral membrane. The bicarbonate transported into the lumen buffers hydrogen ions secreted into the lumen and forms carbonic acid in the lumen. Carbonic anhydrase, located in the lumen of the intestinal tract, catalyzes the breakdown of carbonic acid to water and carbon dioxide.

Absorption of Potassium

Potassium is passively absorbed in the small intestine, but it may be absorbed or secreted in the colon, depending on electrochemical gradients. When the concentration in the lumen is approximately 25 mM or less, potassium is secreted. When the lumenal concentration is greater than 25 mM, potassium is absorbed. Because increased volume of water in the chyme (which occurs in diarrhea) results in a low concentration of potassium in the chyme, less potassium is absorbed. Severe diarrhea, therefore, can cause hypokalemia.

Absorption and Secretion of Bicarbonate

In the jejunum, bicarbonate ions are passively absorbed. As stated previously, hydrogen ions in the lumen interact with bicarbonate ions that are in the lumen (from the pancreatic juice) to form carbonic acid, which then dissociates to carbon dioxide and water. Some of the carbon dioxide diffuses into the enterocyte and reacts with water inside the cell to form carbonic acid. The intracellular carbonic acid dissociates to hydrogen ions and bicarbonate. The hydrogen ions are secreted in exchange for sodium, and the bicarbonate is transported across the basolateral membrane and then diffuses into the bloodstream. In the ileum and colon, bicarbonate is secreted in exchange for chloride ions. The bicarbonate buffers the contents of the intestine.

Regulated Absorption of Calcium

Unlike most substances consumed, the absorption of calcium is regulated based on the body's needs. Calcium is actively absorbed in the duodenum and jejunum via two steps: (1) Calcium binds to a brush border protein, called *calcium-binding protein,* and is then taken into the cell by an unknown mechanism; and (2) calcium is transported out of the cell by a Ca^{2+} pump on the basolateral membrane. Calcium absorption is enhanced by the hormone $1, 25\text{-}(OH)_2D_3$ which increases the amount of calcium-binding protein.

Absorption of Iron

The absorption of iron, like that of calcium and vitamin B_{12}, requires the aid of a protein. In this case, the protein is *transferrin,* which is secreted by enterocytes into the lumen of the small intestine. In the lumen, transferrin binds to iron ions to form a complex, which then binds to a receptor on the apical membrane. Both the receptor and the complex are taken into the cell by receptor-mediated endocytosis, and the iron is either stored in the enterocyte as *ferritin* or transported across the basolateral membrane to the blood, where it is carried by another form of transferrin. The absorption of iron from enterocytes is closely matched to the body's needs. As plasma levels of iron decrease, ferritin releases more iron, which then crosses the apical membrane and enters the blood.

Absorption of Water

Although we consume only some 2 liters of water per day, secretions from the stomach, intestine, and accessory glands result in another 7 liters of water entering the GI tract. Therefore, absorption of water is critical to maintain normal fluid balance. The absorption of water is passive and is driven by an osmotic gradient across the mucosal epithelium—a gradient created by the transport of solutes from the lumen to the interstitial fluid. One of these solutes, sodium, is especially important in the creation of this gradient: It is the most abundant of all solutes in the lumen and in the interstitial fluid and, therefore, makes the largest contribution to the osmotic pressure. Approximately 95% of the water that is initially present in the duodenum is typically absorbed by the time chyme reaches the colon.

4 General Principles of Gastrointestinal Regulation

The GI system is unlike the other organ systems we have studied in that, for the most part, it does not act to maintain constancy of conditions in the body's internal environment—at least not directly—because generally the amount of material absorbed by the GI tract is not determined by conditions inside the body. Most of the nutrients ingested are absorbed completely, so virtually nothing with nutritive value is eliminated in the feces.

Although GI functions are not regulated according to the conditions in the internal environment, they are influenced by an impressive array of neural and hormonal regulatory mechanisms, many of which involve negative feedback control. Many of these mechanisms work to control conditions in the lumen of the GI tract to maximize absorption. Now we take a brief look at general principles pertaining to the regulation of GI function.

Neural and Endocrine Pathways of Gastrointestinal Control

Regulation of gastrointestinal function involves the autonomic nervous system, enteric nervous system, and GI hormones secreted from the stomach and small intestine. The enteric nervous system consists of its own sensory neurons, motor neurons, and interneurons. Influences of the autonomic nervous system on GI functions are generally indirect, mediated via communication to the enteric nervous system.

The function of GI organs is influenced by stimuli arising from within the GI tract, such as the presence or absence of food or changes in the acidity of the lumenal contents. Conditions in the lumen of the tract are monitored by three types of receptor neurons located within the GI tract wall: mechanoreceptors, which detect the degree of distension of the wall; chemoreceptors, which monitor the concentrations in the lumen of specific substances such as hydrogen ions and fats; and osmoreceptors, which monitor the osmolarity of the lumenal contents. Afferents from these receptors project to both the enteric nervous system and the central nervous system (CNS).

The enteric nervous system and CNS exert their control over GI function via neurons projecting to various types of effector cells

TABLE 2 Gastrointestinal Hormones and Their Actions

Hormone	Site of secretion	Stimuli for secretion	Actions
Gastrin	Stomach	Proteins and protein digestion products in stomach; distension of stomach; parasympathetic input to stomach	Stimulates gastric secretion and motility; stimulates ileal motility and relaxes ileocecal sphincter; stimulates mass movement of colon
Cholecystokinin (CCK)	Duodenum and jejunum	Fat or protein digestion products in duodenum	Inhibits gastric secretion and motility; potentiates actions of secretin on pancreatic bicarbonate secretion; stimulates pancreatic enzyme secretion; stimulates bile secretion by liver; stimulates gallbladder contraction and relaxation of sphincter of Oddi
Secretin	Duodenum and jejunum	Acid in duodenum	Inhibits gastric secretion and motility; stimulates pancreatic bicarbonate secretion; potentiates actions of CCK on pancreatic enzyme secretion; stimulates bile secretion by liver
Glucose-dependent insulinotropic peptide (GIP)	Duodenum and jejunum	Glucose, fats, or acid in duodenum; distension of duodenum	Inhibits gastric secretion and motility; stimulates insulin secretion by pancreas

in GI organs, usually smooth muscle cells or secretory cells (which may be exocrine or endocrine). Neural input to smooth muscle cells controls GI motility, whereas input to secretory cells regulates release of chemicals. Whereas exocrine cells secrete their product into the lumen of the GI tract, endocrine cells secrete their product, hormones, into the bloodstream, where the hormones circulate and eventually return to the GI system to exert their effects. The actions of four GI hormones are well established—gastrin, secreted by the stomach, and **cholecystokinin (CCK), secretin,** and **glucose-dependent insulinotropic peptide (GIP),** secreted by cells in the duodenum and jejunum and referred to collectively as *enterogastrones.* Table 2 summarizes the sites of and stimuli for the release of these hormones, and their actions.

Short and Long Reflex Pathways

Neural and hormonal control of GI function is summarized in **Figure 22**, which shows that a stimulus in the lumen of the GI tract can trigger a response without any involvement of the CNS. In this case, signals may travel from receptors to the intrinsic nerve plexuses and then directly to the effectors, following what is called a *short reflex pathway;* alternatively, signals may follow a *long reflex pathway,* traveling from receptors to the CNS and then to the intrinsic nerve plexuses, which relay information to the effectors. (In some cases the CNS may relay signals directly to the effectors.)

As a general rule, long reflex pathways involve either the sympathetic nervous system or the parasympathetic nervous system, or both. Increased parasympathetic activity usually enhances GI

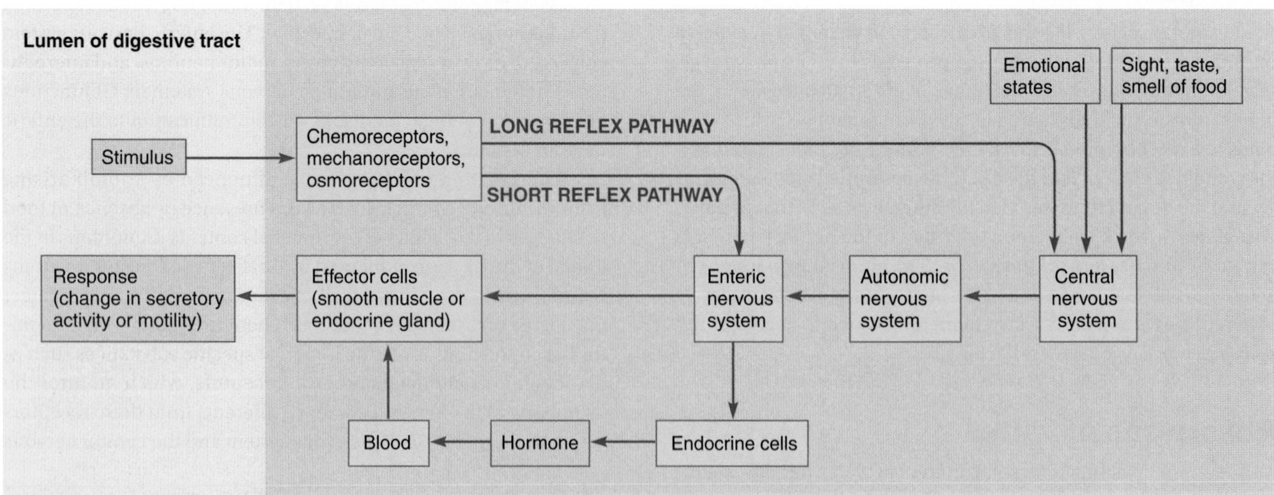

Figure 22 Summary of neural and endocrine control of gastrointestinal function.

activity, which is manifested as an increase in muscle activity or product secretion; conversely, sympathetic activity generally has the opposite effect, promoting a reduction in GI activity. (There are some exceptions to this rule, however.)

Phases of Gastrointestinal Control

A given region of the GI tract can respond to stimuli arising either within that region or in more remote regions. For example, the rate of stomach acid secretion is influenced both by the degree of acidity of the stomach contents (secretion decreases with increasing acidity) and by the level of acidity in the duodenum. The function of GI organs can also be affected by stimuli arising from outside the GI tract; for instance, stomach acid secretion is influenced by the smell of food.

Control of GI function by stimuli arising in the head (such as the smell, taste, or thought of food) is referred to as **cephalic-phase control,** which is always affected by input from the CNS. Control of GI function by stimuli arising in the stomach or the small intestine is referred to as **gastric-phase control** or **intestinal-phase control,** respectively. Gastric-phase and intestinal-phase stimuli exert their effects via long or short reflex pathways or by altering the secretion of GI hormones.

Regulation of Food Intake

Physiological regulation of food intake generally involves the hypothalamus and can occur over either the short term or the long term. Eating a meal because you are hungry and then stopping because your hunger has disappeared is an example of short-term regulation; increases in your daily food consumption over the weeks after you begin an exercise program is an example of long-term regulation. One factor that is important in long-term regulation is **leptin,** a hormone secreted by adipose tissue (**Figure 23**).

Adipocytes secrete leptin at levels proportional to the amount of adipose tissue. When the dietary intake of calories in nutrients exceeds the body's demands, fat is deposited in adipose tissue. This action promotes the release of leptin, which then acts on appetite-control centers in the arcuate nucleus of the hypothalamus to reduce the sensation of hunger. Physiological signals that have hunger-suppressing effects, such as leptin, are referred to as *satiety signals.* Leptin stimulates the release of other satiety factors from the arcuate nucleus, including *alpha melanocyte stimulating hormone* (αMSH) and *cocaine and amphetamine related transcript* (CART), which coordinate responses to reduce fat stores. In addition to inducing satiety to reduce food intake, αMSH and CART increase sympathetic activity and stimulate the release of the anterior pituitary tropic hormones known as thyroid stimulating hormone (TSH) and adrenocorticotropic hormone (ACTH), thereby promoting a general increase in the body's metabolic rate and reducing fat storage. Some people are born with congenital deficiencies in leptin. These individuals tend to eat excessively and nondiscriminatively—a condition called *hyperphagia*—which results in severe obesity. Administration of leptin over time can reverse their severe hunger, with the treated patients

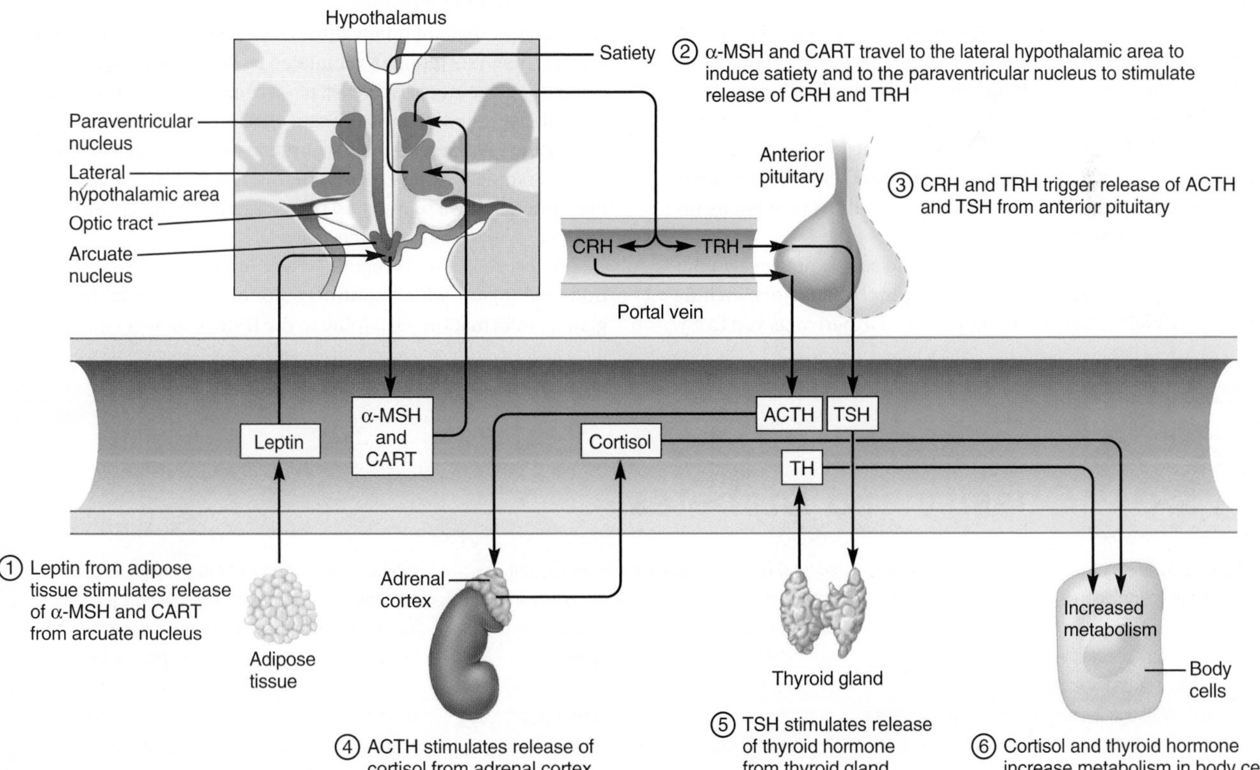

Figure 23 **Effects of satiety and orexigenic factors on food intake.**

then losing weight quickly. However, administration of leptin in obese individuals without a genetic defect in leptin has little effect on their obesity.

Opposing the actions of satiety factors are *orexigenic factors*, including neuropeptide Y (NPY) and agouti-related peptide (AgRP), both of which are released from the arcuate nucleus and act on the lateral hypothalamus to promote eating behavior and decrease metabolism. Orexigenic factors achieve this effect by increasing parasympathetic activity, decreasing sympathetic activity, and inhibiting the secretion of ACTH and TSH.

Food intake is also influenced by a number of other physiological variables that are important in short-term regulation. The absorption of nutrients from a typical meal, for example, stimulates pancreatic islet cells to release the hormone insulin, which acts on the hypothalamus to induce satiety. Another hunger-suppressing hormone, CCK, is released in response to the presence of food in the duodenum. Neural signals from certain mechanoreceptors and chemoreceptors in the wall of the GI tract signal the presence of food or digestion products in the lumen, and these signals also suppress hunger.

Just as there are short-term satiety factors, so there are also short-term orexigenic factors, including the gastric hormone *ghrelin*, which is released from the stomach when it is empty. Ghrelin promotes the release of NPY and AgRP, which then promote feeding behavior. The removal of ghrelin-secreting cells during gastric bypass surgeries may contribute significantly to the effectiveness of this procedure in producing weight loss.

Functional Fact

People become obese when they consume, on a regular basis, more calories than their bodies use. Most of the excess calories are stored as triglycerides in adipose tissue. Four major factors contribute to development of obesity: genetic factors, environmental factors, excessive food intake, and insufficient exercise. Regardless of its cause, obesity predisposes a person to many diseases, including atherosclerosis, heart disease, liver disease, and type 2 diabetes mellitus (see **Focus on Diabetes: Obesity and Diabetes**).

Health Fact

Even though food intake is necessary to supply nutrients to the body, food intake is strongly affected by unrelated factors, including psychosocial influences. From your own experience, you likely know that even when you are not particularly hungry, you can be persuaded to eat when you see an appealing dish, when you have a rare opportunity to eat a favorite food, or when you are with a group of friends who are eating.

Quick Check 5

1. Name three major types of receptors that monitor conditions in the lumen of the GI tract.
2. What is the distinction between a short reflex pathway and a long reflex pathway?
3. To what do the terms cephalic phase, gastric phase, and intestinal phase refer?
4. Name the various satiety and orexigenic factors, the sites that release them, and the stimuli for their release.

5 Gastrointestinal Secretion and Its Regulation

So far, we have learned that several substances are secreted into the lumen of the GI tract. In this section, we look at the mechanisms of exocrine secretion, including how these mechanisms are regulated.

Saliva Secretion

The secretion of saliva is controlled by neural input to the salivary glands via both branches of the autonomic nervous system. Unlike most tissues that are dually innervated by the autonomic nervous system, saliva secretion is stimulated by both parasympathetic and sympathetic neurons. However, parasympathetic activity stimulates copious secretion of a watery saliva, whereas sympathetic activity stimulates minute secretion of a protein-rich, viscous saliva. (A nervous person giving a presentation often suffers from "dry mouth" even though saliva production is increased with the increased activity of the sympathetic nervous system.)

Saliva secretion is regulated by the *salivary center* in the medulla oblongata, which controls autonomic output to the salivary glands. Information pertaining to the taste of food is relayed to the

FOCUS ON DIABETES

Obesity and Diabetes

Type 2 diabetes mellitus is associated with tissue resistance to insulin. Tissue resistance is a failure of target tissues to respond normally to insulin and can occur anywhere along the signaling pathway for insulin, starting with insulin's binding to its receptor. Although the way insulin resistance arises is not fully understood, obesity is its primary cause.

Obesity is defined as an amount of body fat 20% or greater than what is considered normal for a particular gender and height. Although adipose tissue is traditionally thought of as an energy storage depot, it secretes hormones that may promote some of the detrimental effects of obesity. We saw in the previous section that adipose tissue secretes leptin to decrease food intake; however, leptin may promote insulin resistance as well. In addition, adipose tissue secretes *resistin*, a cytokine also secreted by macrophages during inflammation; like leptin, resistin may promote insulin resistance. Better understanding of the adipose tissue chemical mediators that induce insulin resistance could lead to the development of pharmaceutical agents that would decrease the likelihood of an obese person developing type 2 diabetes mellitus.

salivary center by chemoreceptors in taste buds. The presence of food in the mouth stimulates these receptors, resulting in activation of the parasympathetic nervous system and increased saliva secretion. Activity of the salivary center is also affected by input from the cerebral cortex, which relays information pertaining to the sight and smell of food and other cephalic-phase stimuli.

Acid and Pepsinogen Secretion in the Stomach

The mechanism of acid secretion in the stomach is in some ways reminiscent of the mechanism of hydrogen ion secretion in the renal tubules. The acid secreted by the stomach is generated inside parietal cells by the *carbonic anhydrase (CA)*-catalyzed reaction, converting the reaction of carbon dioxide and water into hydrogen ions and bicarbonate (**Figure 24**). A proton pump that uses ATP transports hydrogen ions out of the parietal cell and into the lumen of the stomach in exchange for potassium ions, which move from the lumen into the cell. Bicarbonate ions exit the parietal cell across the basolateral membrane in exchange for chloride ions, which then move into the lumen through channels in the apical membrane. The net result is that hydrogen and chloride ions are transported into the lumen while bicarbonate is transported into the interstitial fluid.

The secretion of acid by parietal cells is stimulated by parasympathetic nervous activity, gastrin, and histamine, which is secreted by cells in the stomach lining and acts as a paracrine agent. Acid secretion is controlled by cephalic-phase, gastric-phase, and intestinal-phase stimuli and is generally stimulated when food is present in the stomach and suppressed when the food leaves. Because the stimuli that affect acid secretion also tend to affect the secretion of pepsinogen in the same manner, the secretion of pepsinogen generally rises and falls in parallel with changes in acid secretion.

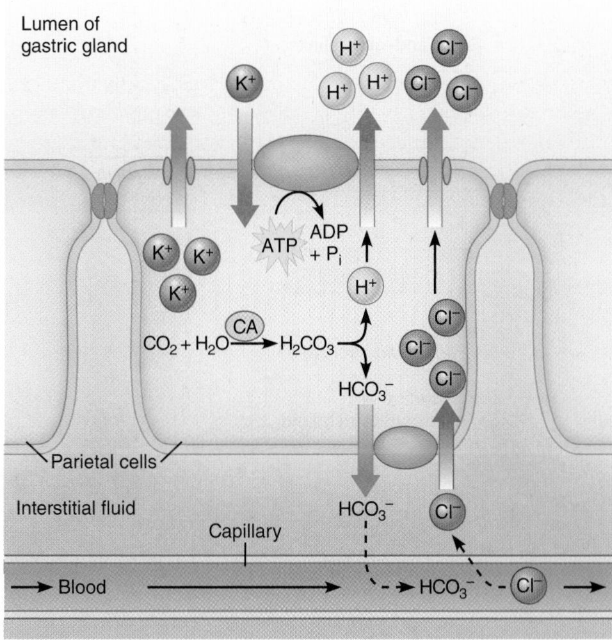

Figure 24 Mechanism of gastric acid secretion. Note that because potassium ions are actively transported from the lumen into parietal cells and then passively leak back into the lumen, these ions are simply recycled across the apical membrane.

Q Is the electrochemical driving force on hydrogen ions directed into or out of the cell?

As shown in **Figure 25a**, cephalic-phase stimuli arising either from the sight, taste, and smell of food or from the acts of chewing and swallowing trigger increased activity in parasympathetic nerves to the stomach, which in turn stimulates both parietal cells and chief cells to secrete acid and pepsinogen, respectively, and G cells to secrete gastrin. The gastrin enters the bloodstream and stimulates the parietal and chief cells, reinforcing the effects of parasympathetic input.

Once food reaches the stomach, gastric-phase stimuli come into play (Figure 25b). The presence of proteins and protein digestion products in the lumen stimulates chemoreceptors in the wall of the stomach, whereas the presence of food distends the stomach, which activates mechanoreceptors. As a result, signals are relayed to parietal cells and chief cells via short reflex and long reflex pathways, triggering the release of acid and pepsinogen; signals are also sent to G cells, triggering the release of gastrin. In addition, the proteins present in the lumen exert a direct effect on G cells that stimulates them to secrete gastrin.

As food leaves the stomach, both gastric-phase and intestinal-phase stimuli work to *reduce* acid and pepsinogen secretion. The exit of food from the stomach reduces the degree of distension and lowers the concentration of proteins and protein digestion products in the lumen, causing a *withdrawal* of the stimuli that previously stimulated gastric secretion. The exit of proteins also tends to increase gastric acidity (because proteins normally buffer some of the secreted hydrogen ions); the greater acidity acts directly on G cells to suppress gastrin secretion, thereby causing the withdrawal of another stimulus for acid secretion.

The reduction in gastric secretion that occurs during gastric emptying is also triggered by signals arising from the entry of food into the duodenum. As chyme leaves the stomach, the osmolarity of the duodenal content rises, the concentrations of fats and acid in the content rise, and the duodenum becomes more distended. These changes stimulate chemoreceptors, osmoreceptors, and mechanoreceptors, which relay signals via long reflex and short reflex pathways to parietal and chief cells in the stomach, inducing a decrease in acid and pepsinogen secretion. In addition, signals relayed to the endocrine cells in the small intestine increase the secretion of CCK, secretin, and GIP, all of which suppress the secretory activity of parietal and chief cells.

Secretion of Pancreatic Juice and Bile

The secretion of pancreatic juice begins in the pancreatic acini, where cells produce a relatively small volume of fluid containing water, electrolytes, and digestive enzymes. As this fluid flows through the ducts leading from the acini, duct cells secrete a larger volume of bicarbonate-rich fluid that is added to it. Even though both components of the pancreatic juice—enzyme-rich fluid and bicarbonate-rich fluid—are secreted together during a meal, the regulatory mechanisms that control their secretions are somewhat

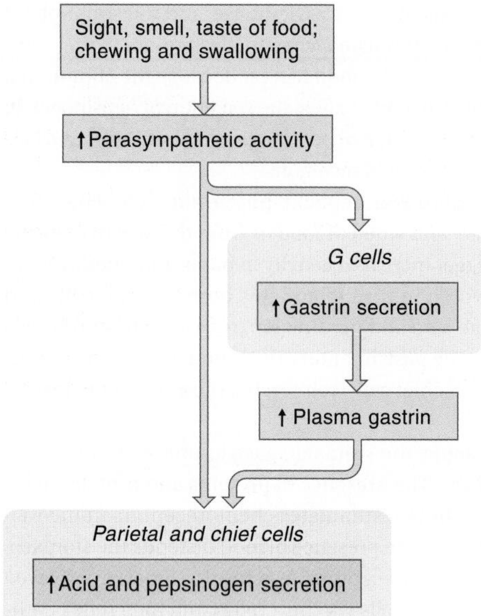

(a) Cephalic-phase control of gastric secretion

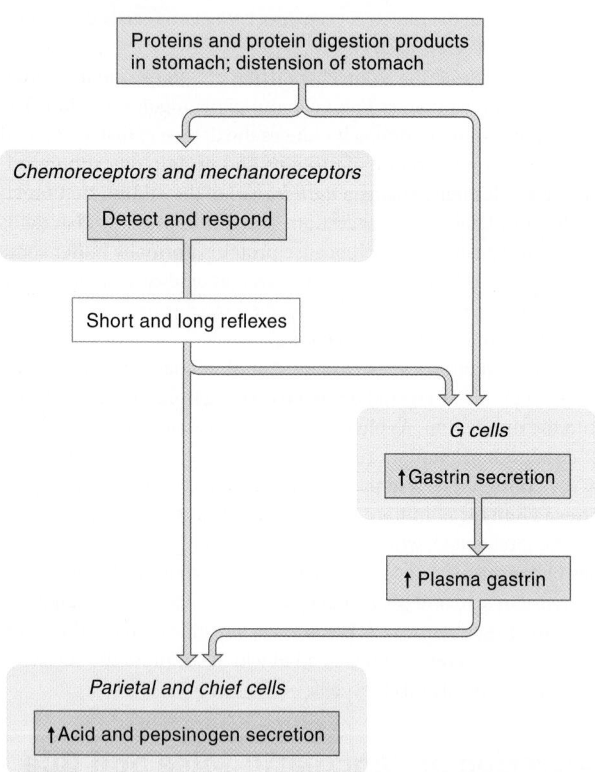

(b) Gastric-phase control of gastric secretion

Initial stimulus
Physiological response
Result

Figure 25 Regulation of gastric secretion. Shown are the pathways by which acid and pepsinogen are secreted in response to **(a)** cephalic-phase stimuli and **(b)** gastric-phase stimuli when food is being consumed and is present in the stomach.

separate, and for this reason the composition of the pancreatic juice can vary.

The secretion of pancreatic juice is influenced by cephalic-phase, gastric-phase, and intestinal-phase stimuli, although the last predominate. The strongest influences on pancreatic secretion are the hormones CCK and secretin, which are released in response to the presence of food in the duodenum. CCK acts primarily on acinar cells to stimulate enzyme secretion, whereas secretin acts primarily on duct cells to stimulate the secretion of bicarbonate-rich fluid. Although CCK by itself is only a weak stimulus for bicarbonate secretion, its effect becomes stronger when secretin is present. Likewise, secretin by itself exerts only a weak stimulatory effect on enzyme secretion, but this effect becomes much stronger when CCK is present. Put another way, when CCK and secretin are both present, these hormones amplify each other's effects, a phenomenon known as **potentiation.**

The release of secretin and the subsequent secretion of bicarbonate-rich fluid are strongly affected by the acidity of the duodenal contents (**Figure 26**). The flow of chyme from the stomach

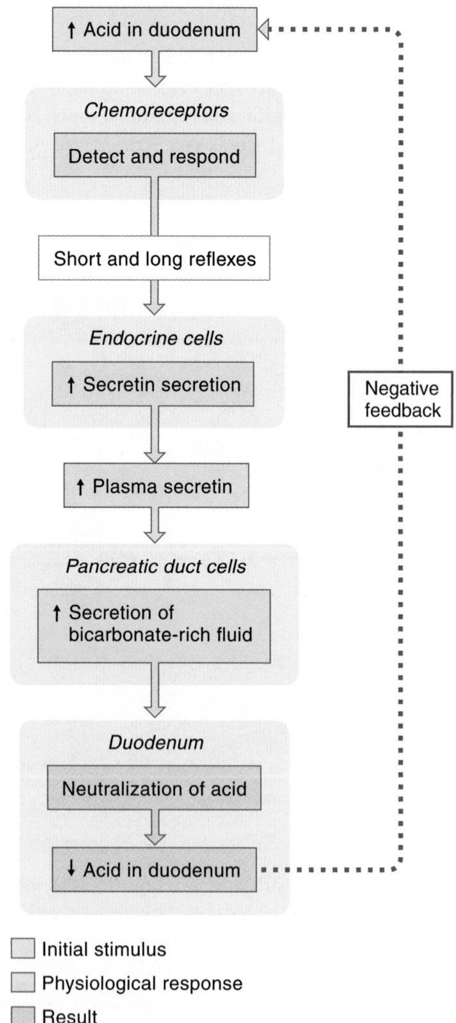

Initial stimulus
Physiological response
Result

Figure 26 Stimulation of the secretion of bicarbonate-rich fluid by acidity in the duodenum.

into the duodenum raises the acidity of the duodenal contents, which stimulates chemoreceptors in the wall. As a result, signals are relayed via short and long reflex pathways to endocrine cells in the duodenum and jejunum, which then release secretin in response. The resulting rise in plasma secretin levels acts on pancreatic duct cells to stimulate the secretion of bicarbonate-rich fluid, which combines with hydrogen ions in the duodenum to lower the acidity of the contents. (Note that these events constitute negative feedback control, because a rise in acidity triggers a series of events that culminate in a fall in acidity.) This reduction of acidity is necessary for the proper activity of pancreatic and other enzymes that work in the small intestine.

The secretion of CCK is regulated primarily by the concentrations of protein digestion products and fat in the duodenum, which increase as chyme leaves the stomach. The increased levels of these products activate chemoreceptors, which relay signals via short reflex and long reflex pathways to endocrine cells, triggering the release of CCK, which acts on acinar cells to stimulate enzyme secretion. Later, once digestion products have left the duodenum, these signals are turned off, so CCK secretion falls.

CCK and secretin are also responsible for regulating the entry of bile into the duodenum (**Figure 27**). In response to increased acidity in the duodenum, secretin acts on the liver to stimulate bile secretion; the presence of protein digestion products and fat in the duodenum stimulates the secretion of CCK, which by promoting gallbladder contraction and relaxation of the sphincter of Oddi, allows bile to flow into the duodenum.

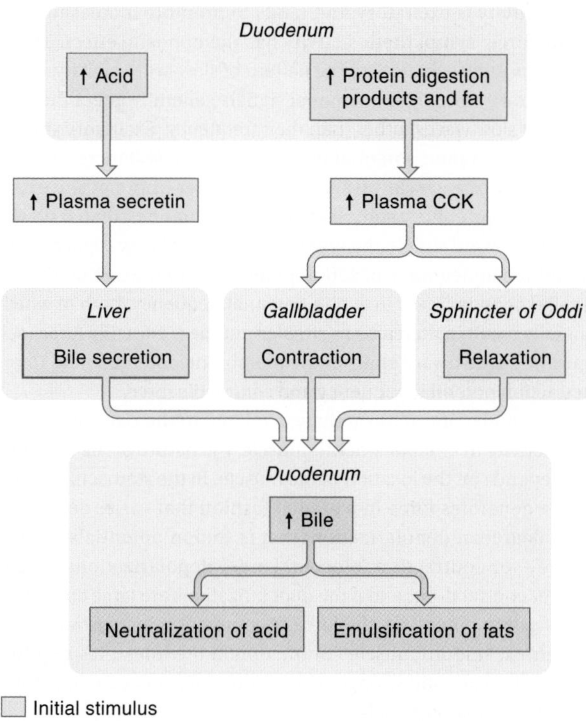

Figure 27 **The mechanisms by which secretin and CCK regulate the entry of bile into the duodenum.**

Rates of Fluid Movement in the Digestive System

The volume of fluid that moves into and out of the digestive system each day is impressive, amounting to several times the normal plasma volume (**Figure 28**). The liver and pancreas combined secrete into the GI tract an average of 2 liters of fluid per day, and the salivary glands secrete another 1.5 liters. An additional 3.5 liters is secreted by glands in the wall of the stomach and small intestine, and approximately 2 liters enters the GI tract via ingested water. The great majority of the secreted and ingested fluid—some 8.5 liters each day—is absorbed by the small intestine; the colon also absorbs some fluid. When the movements of fluid into and out of the GI tract

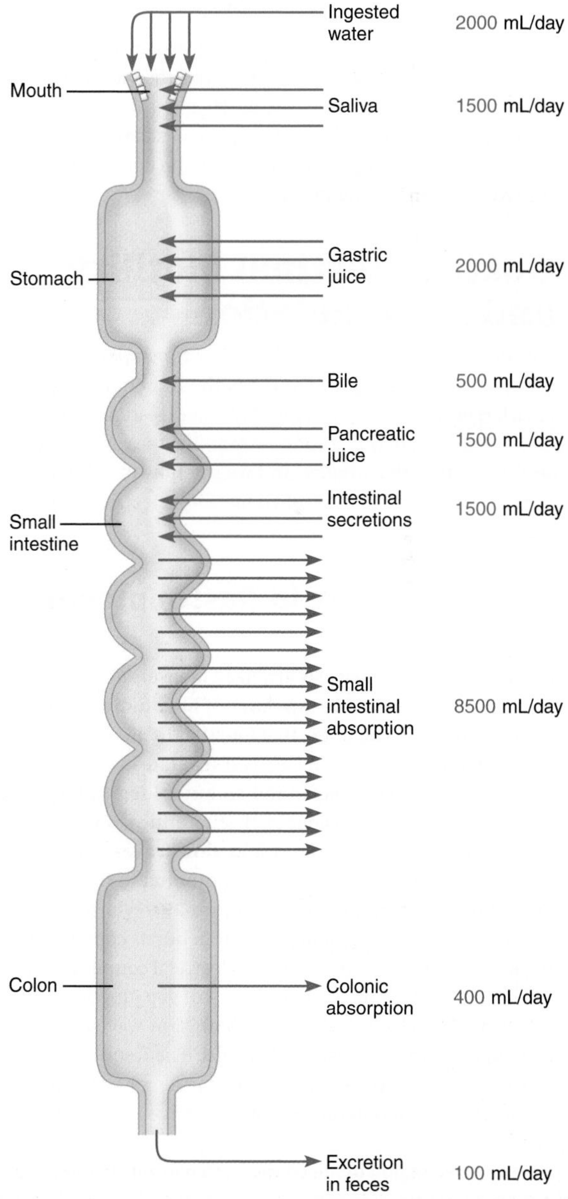

Figure 28 **Approximate daily fluid flows in the gastrointestinal system.**

have been accounted for, the difference between the amount that enters and the amount that is absorbed is quite small—only about 100 mL per day. This unabsorbed water is eliminated in the feces.

Quick Check 6

1. What is the role of carbonic anhydrase in stomach acid secretion? Which other ions are secreted along with hydrogen ions?
2. Of the following stimuli, which would be expected to stimulate gastric acid secretion: distension of the stomach, distension of the duodenum, increased protein digestion products in the stomach, increased acidity in the duodenum, increased osmolarity in the duodenum?
3. Of the following stimuli, which would be expected to stimulate the secretion of enzymes or bicarbonate-rich fluid by the pancreas: increased acidity in the duodenum, a rise in plasma CCK levels, a rise in plasma secretin levels, decreased protein digestion products in the duodenum, increased fat in the duodenum?

6 Gastrointestinal Motility and Its Regulation

The GI tract is capable of generating finely tuned patterns of motion that both propel forward and mix the lumenal contents. In fact, it is largely because of this motility that digestion and absorption are so efficient. In this section we examine the various patterns of motility in digestive organs, and we see how motility is regulated. We begin with the basis of GI motility: electrical activity in smooth muscle.

Electrical Activity in Gastrointestinal Smooth Muscle

Recall that GI smooth muscle is arranged in layers at two levels. In the muscularis mucosae is a thin layer of longitudinal muscle fibers and circular muscle fibers that function to mix lumenal contents. In the muscularis externa are two separate layers of smooth muscle: an inner circular layer and an outer longitudinal layer. The inner circular layer is capable of generating slow, spontaneous, graded depolarizations known as **slow waves,** which when large enough to bring the membrane potential to threshold are accompanied by action potentials (**Figure 29**). When action potentials are generated, they tend to occur in bursts coinciding with the peaks of the slow waves. Extensive electrical coupling between smooth muscle cells, due to the presence of gap junctions, allows propagation of this electrical activity (both slow waves and action potentials) to nearby nonpacemaker cells. The frequency of slow waves varies from region to region along the length of the tract because muscle cells in different regions are driven by different sets of pacemaker cells.

In any given region of the GI tract, circular muscle slow waves occur at regular intervals at a fairly constant frequency. This pattern, which is referred to as the **basic electrical rhythm (BER),** is affected by neural activity and hormones. In general, parasympathetic activity is excitatory and tends to promote increased contractile force; sympathetic activity has the opposite effect. These influences reflect changes in the pattern of slow-wave activity.

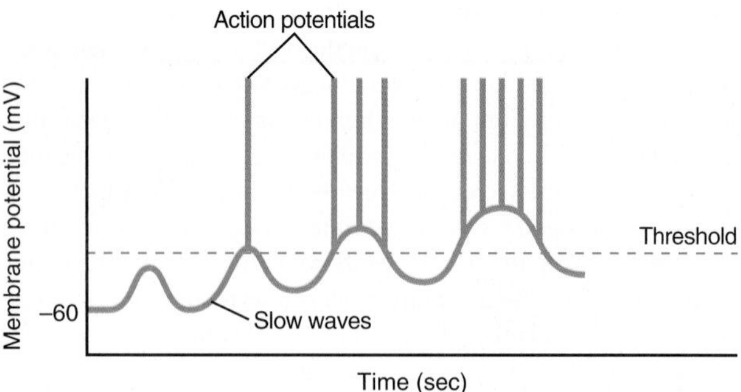

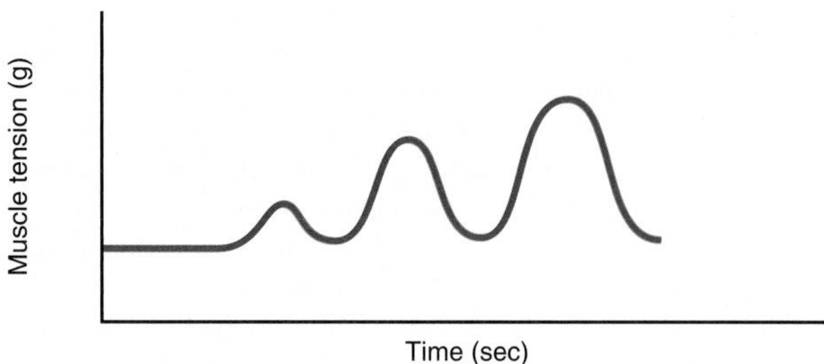

Figure 29 Slow-wave electrical activity in gastrointestinal smooth muscle. Depolarizations that exceed threshold result in action potentials (top), inducing gastrointestinal smooth muscle contractions whose strengths vary with the frequency of the action potentials (bottom).

Neural activity and hormonal signals generally affect the *amplitude* of slow waves rather than their frequency. Excitatory stimuli shift the membrane potential upward, so that it still increases and decreases with each slow wave, but at a higher (more positive) average value; inhibitory stimuli shift the membrane potential downward. An upward shift in the membrane potential may trigger either the sudden appearance of action potentials (if none are occurring initially) or an increase in action potential frequency, both of which generally result in increases in circular muscle contractile force (see Figure 29). Downward shifts in the membrane potential tend to reduce action potential frequency and contractile force.

Although contractile activity is related to the pattern of electrical activity in circular smooth muscle, the nature of the relationship depends on the location of the muscle. In the stomach, smooth muscle generates force in a graded fashion that varies according to the degree of depolarization; that is, action potentials are *not* required for contraction to occur. Larger depolarizations trigger stronger contractions, and if the depolarizations are large enough to trigger action potentials, the force of contraction increases further. In contrast, smooth muscle contraction in the intestines requires action potentials, and stronger contractions occur with a higher frequency of action potentials.

In the sections that follow, we examine motility in greater detail, beginning with some generalized motility patterns observed in the smooth muscle of the GI tract, followed by motility patterns in the different organs of the GI tract.

Peristalsis and Segmentation

The basic electrical rhythm sets up a wave of contraction in the muscularis externa that travels longitudinally down the intestinal tract. These waves, called **peristalsis,** propel the contents forward at a rate dependent on the basic electrical rhythm (**Figure 30**). Peristalsis is caused by the coordinated activity between the circular and longitudinal muscle layers in adjoining segments of the GI tract. In the proximal GI segment, the circular muscle contracts while the longitudinal muscle relaxes. This effect decreases the diameter of the tract. In the distal GI segment, the circular muscle relaxes while the longitudinal muscle contracts, causing an increase in the diameter of the tract that prepares it to receive the contents from the proximal segment. Activity in the distal segment actually precedes activity in the proximal segment as it prepares for the arrival of contents.

In the small intestine, the muscularis externa is also capable of **segmentation,** alternating contractions between intestinal segments that mix the chyme. In segmentation, while one segment of circular muscle in the intestine contracts, segments on either side relax, thereby allowing contents to move in both directions from the contraction (as opposed to the unilateral propulsion of contents during peristalsis). The result is that chyme is shuttled back and forth, mixing the contents and bringing digestive products into close contact with the mucosa.

Chewing and Swallowing

Chewing is like breathing, in that it is controlled both consciously and unconsciously; we can control the rate and force of chewing motions voluntarily, or we can chew without even thinking about it. Unconscious chewing is orchestrated by a *chewing reflex* that is activated by the presence of food in the mouth as follows: When food is not present in the mouth, the muscles that hold the lower jaw closed are tonically active and constantly exert force. However, the presence of food in the mouth stimulates pressure receptors and triggers inhibition of the jaw-closing muscles, allowing the jaw to drop in response to the pull of gravity. Opening of the jaw relieves the pressure of food against the receptors, which removes the inhibitory stimulus from the jaw-closing muscles, allowing them to close the jaw once again. When this happens, however, the pressure is restored, which triggers inhibition of the jaw-closing muscles and allows the jaw to open, and so on. Thus the operation of this reflex induces the alternating opening and closing motion of the jaw that constitutes chewing.

Chewing and the associated tongue movements both reduce the food mass to smaller particles and ensure that food is mixed thoroughly with saliva, transforming the food into a semisolid mass called a *bolus.* When the bolus becomes soft and moist enough to swallow, the tongue propels it to the back of the mouth and into the pharynx, where it stimulates mechanoreceptors that initiate the *swallowing reflex,* a series of muscle contractions coordinated by the *swallowing center* in the medulla. The swallowing reflex involves the following series of steps (**Figure 31**):

① As the bolus begins to descend from the pharynx, it presses downward on the epiglottis (a flap of tissue guarding the glottis), causing it to cover the glottis and preventing the bolus from entering the larynx and trachea. Closure of the glottis is aided by muscles of the neck, which raise the larynx. Reflex mechanisms also inhibit inspiratory muscles, which suppresses breathing motions.

② The upper esophageal sphincter relaxes, enabling the entry of the bolus into the esophagus. Once the bolus passes through, the sphincter closes again.

③ Entry of the bolus into the esophagus stimulates stretch receptors, triggering peristalsis. The peristaltic wave propels the bolus toward the stomach, which it reaches in about 9 seconds.

④ Upon arrival of the bolus at the lower end of the esophagus, the lower esophageal sphincter relaxes momentarily to allow the bolus to enter the stomach.

⑤ In the event that the initial or primary peristaltic wave is unsuccessful in delivering the bolus to the stomach, it is followed by additional secondary waves initiated by stimulation of mechanoreceptors in the esophagus.

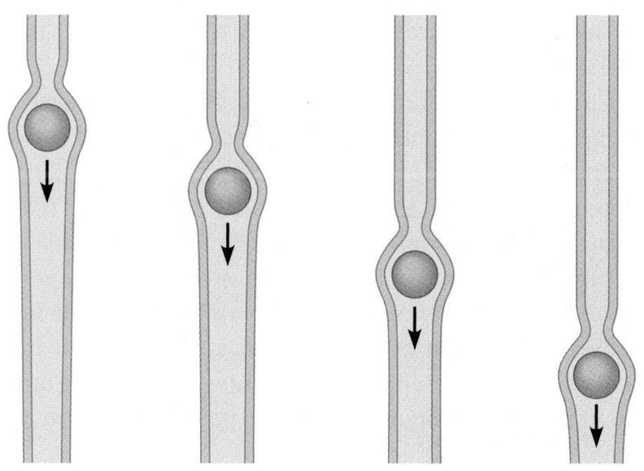

(a) Peristalsis

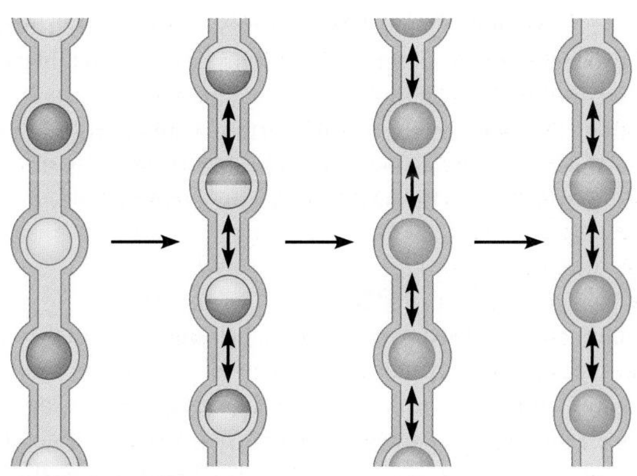

(b) Segmentation

Figure 30 Comparison of peristalsis and segmentation. (a) In peristalsis, waves of contraction travel a short distance and then die out, propelling the chyme along the GI tract's length. **(b)** In segmentation, waves of contraction cause neighboring segments of intestine to alternately contract and relax, mixing the chyme by shuttling it back and forth.

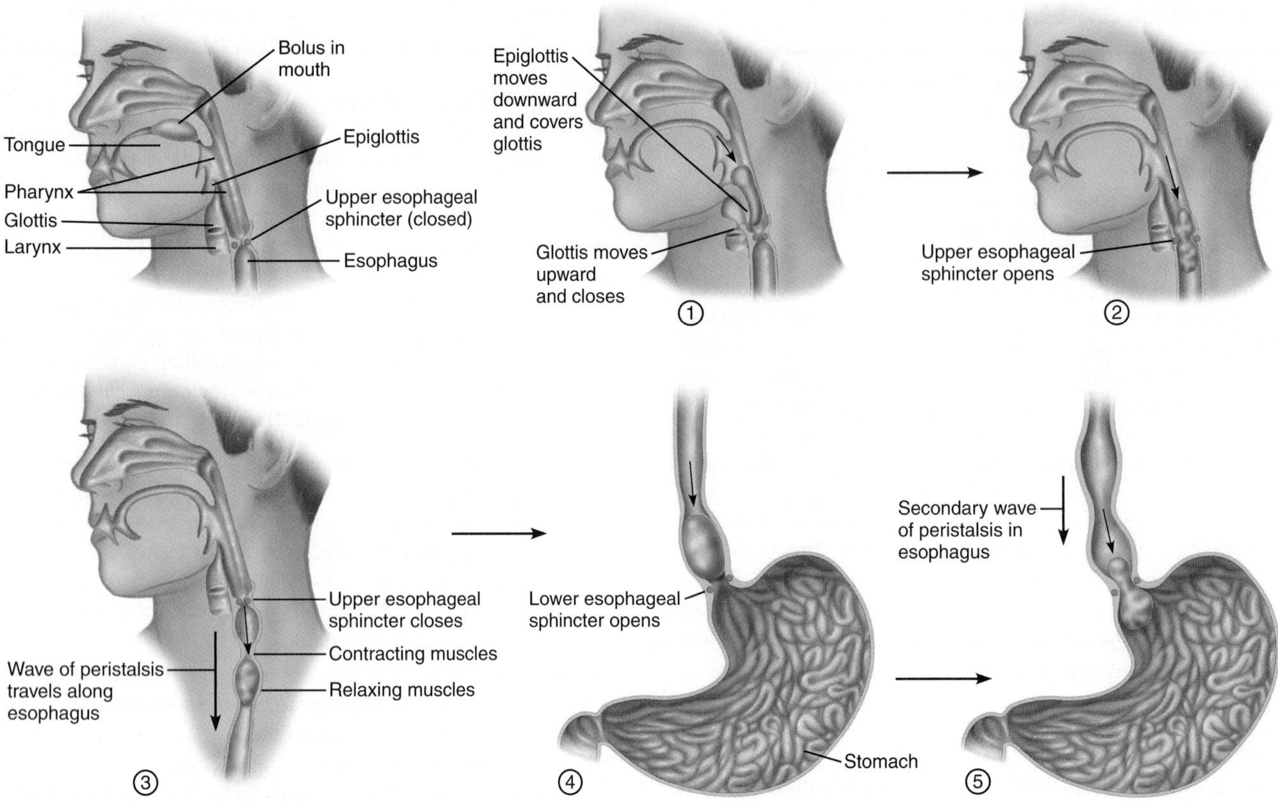

Figure 31 **Events of the swallowing reflex.**

Prior to the arrival of the bolus, the stomach makes preparations to accommodate it through a process called *receptive relaxation.* In this process, the swallowing center in the medulla triggers relaxation of smooth muscles in the upper portion of the stomach, which increases its volume.

Gastric Motility

Following ingestion of a meal, the muscles of the stomach accomplish two tasks: mixing of the chyme to ensure that food and gastric juice are thoroughly combined, and regulation of gastric emptying such that chyme enters the small intestine at an appropriate rate. Both of these tasks are achieved by peristaltic waves in the stomach, which are coordinated by the enteric nervous system.

Gastric Motility Patterns

Peristaltic waves travel downward from the upper body of the stomach to the pylorus, normally at a rate of about three per minute. Each wave begins as a weak contraction but progressively increases in force as it advances toward the pylorus. Because the pyloric sphincter is closed, the wave of contraction does not push the bulk of the chyme forward; instead, most of it is forced to flow backward, which mixes it.

However, as peristaltic waves increase in strength, they force the pyloric sphincter open as chyme is propelled forward, spurting into the duodenum. The rate of gastric emptying depends on several factors, including the composition of the chyme, the volume of chyme in the stomach, and the force of gastric contractions. In general, emptying is faster when the volume of chyme is larger and gastric contractions are stronger. However, only liquid and small particles in the chyme can move through the pyloric sphincter. Liquids enter the duodenum approximately 3 minutes after ingestion. Smaller particles (less than 1 mm in diameter) follow 20–30 minutes later. In contrast, large particles may stay in the stomach up to 9 hours before they are broken down to a size that can move through the pyloric sphincter.

During periods of fasting, peristaltic contractions eventually cease, and the stomach becomes quiescent for an hour or two, after which time another pattern of activity appears. First, the antrum begins to undergo a series of intense contractions (called a **migrating motility complex**) that are accompanied by relaxation of the pyloric sphincter. After a while, these contractions stop and the stomach goes into another period of quiescence, which is followed by another burst of contractions, and so on. This pattern of activity sweeps the stomach of its contents, including any particles that might have been too large to pass through the pyloric sphincter.

Regulation of Gastric Motility

Control of gastric motility is achieved primarily through changes in the force of smooth muscle contractions. (Changes in frequency do occur but are relatively minor.) The force of gastric contraction increases in response to gastrin and decreases in response to CCK, secretin, and GIP. (In fact, GIP was originally called *gastric inhibitory peptide* because of its inhibitory influence on gastric motility.)

The control of gastric motility is similar to the control of gastric secretion, in that motility is influenced by cephalic-phase, gastric-phase, and intestinal-phase stimuli. Effective cephalic-phase stimuli include pain, fear, and depression, which normally inhibit gastric motility, and anger and aggression, which stimulate it. Gastric-phase stimuli (including distension of the stomach) and intestinal-phase stimuli (including distension of the duodenum and changes in lumenal acidity, osmolarity, and fat concentration) regulate gastric emptying such that chyme exits the stomach at a rate that is well matched to the small intestine's ability to process it. Specifically, if gastric emptying exceeds the rate at which the small intestine can process contents, then distension of the duodenum occurs, and increases in lumenal acidity, osmolarity, and fat concentration all occur, thereby decreasing gastric motility and thus gastric emptying.

Vomiting

On occasion, certain conditions can cause the contents of the stomach and sometimes of lower portions of the GI tract to be forcefully expelled through the mouth, a phenomenon known as *vomiting*. Vomiting can be triggered by a variety of stimuli, including illness (such as influenza), strong emotional states, severe pain, severe distension of the stomach or small intestine, rotational motion of the head (as in motion sickness), or the ingestion of certain substances (such as copper sulfate). Substances that stimulate vomiting are known as *emetics*. Given that many emetics are poisonous, vomiting acts as a protective mechanism that removes these substances from the GI tract before substantial quantities can be absorbed into the blood.

Vomiting involves a complex sequence of events (called the *vomiting reflex*) that is coordinated by a region in the medulla called the *vomiting center*. Prior to the actual act of vomiting, a person usually experiences a sensation of nausea, the skin becomes pallid, and heart rate and sweating increase. Eventually, a series of deep inspirations are followed by closure of the glottis. Abdominal muscles begin to contract strongly, causing the abdominal wall to move inward as the inspiratory movements cause the diaphragm to move downward. This combination of motions raises the abdominal pressure substantially and effectively squeezes the stomach, raising the pressure inside. Finally, the lower esophageal sphincter relaxes, allowing stomach contents to enter the esophagus. If contractions are not strong enough to eject this material into and out of the mouth, it returns to the stomach; this can occur several times before abdominal contractions become strong enough to push the material through the upper esophageal sphincter and out through the mouth.

Motility of the Small Intestine

Like the stomach, the small intestine exhibits different patterns of motility depending on whether food is present or absent. Motility patterns also differ in the muscularis externa and muscularis mucosae.

Motility Patterns in the Small Intestine

When chyme is present in the small intestine, brief periods of peristalsis that propel the chyme forward are interspersed with relatively longer periods of segmentation that mix the chyme. During fasting, peristalsis and segmentation cease and are replaced by migrating motility complexes that periodically move through the intestine, sweeping it of its contents.

Segmentation, peristalsis, and migrating motility complexes result from the contractile activity of the external muscle layers, but the innermost layer (the thin muscularis mucosae) also undergoes regular contractions when food is present. These contractions mix chyme near the surface of the mucosa and may exert a "massaging" effect on the lacteals, which helps propel the lymphatic fluid forward.

Regulation of Motility of the Small Intestine

Contractions of the small intestine are influenced both by distension and by input from extrinsic nerves and hormones. During intestinal phase control, moderate distension of the intestine triggers an increase in contractile force, a response mediated by both short and long reflex pathways that tends to relieve the distension by propelling the intestinal contents onward. During gastric-phase control, gastrin both stimulates motility in the ileum and promotes relaxation of the ileocecal sphincter to increase the rate at which these contents move from the small to large intestine, thereby making room for more chyme to enter the duodenum.

The small intestine also exhibits a number of specialized reflexes that are orchestrated by the CNS and come into play only in certain situations. In the *intestino-intestinal reflex*, severe distension or injury to any portion of the small intestine inhibits contractile activity throughout the rest of the intestine, which helps protect the injured part from further stretching and additional injury. In the *ileogastric reflex*, distension of the ileum causes inhibition of gastric motility, which decreases the rate at which chyme enters the duodenum. In the *gastroileal reflex*, the presence of chyme in the stomach triggers increased motility in the ileum.

Quick Check 7

1. Describe the motility patterns responsible for gastric emptying, and explain how gastric emptying is regulated.
2. Describe the steps of the swallowing reflex.
3. Compare the mechanics and functions of peristalsis and segmentation in the small intestine.

Motility of the Colon

In the colon, contractile activity serves to mix the chyme, to expose it to the mucosal surface (which facilitates the absorption of minerals and water), and to propel the lumenal contents toward the rectum for storage and eventual elimination.

Motility Patterns in the Colon

The more proximal portions of the colon exhibit a pattern of motility called **haustration,** which is similar to segmentation in the small intestine, except that the segments *(haustra)* are delineated by permanent folds in the intestinal wall and are regular in appearance (**Figure 32**). Haustration is also significantly slower than segmentation; its contractions occur at a rate of about two per hour.

About three or four times a day, a different pattern of activity begins. This activity, called **mass movement,** is like a peristaltic wave, except that after a given portion of the intestine contracts, it remains contracted for a longer time before relaxing. These waves of contraction propel the lumenal contents forward rapidly and sweep the colon clean.

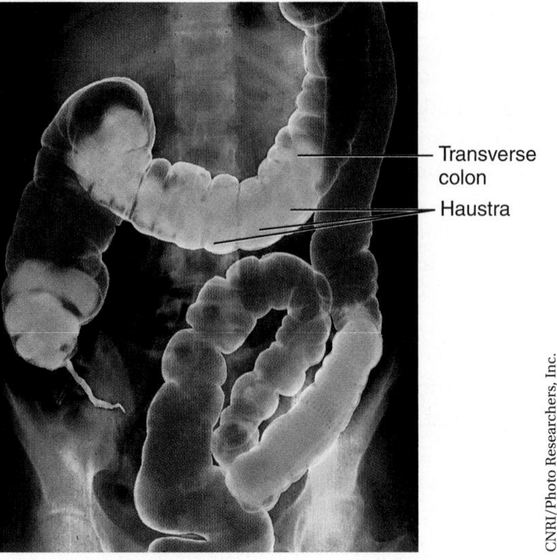

CNRI/Photo Researchers, Inc.

Figure 32 An X-ray image of a human colon showing haustra.

Regulation of Motility of the Colon

Like the small intestine, the colon exhibits specialized reflexes that act only in certain circumstances. In the *colonocolonic reflex,* distension of one part of the colon induces relaxation of the remaining parts. In the *gastrocolic reflex,* the presence of a meal in the stomach triggers an increase in colonic motility and an increase in the frequency of mass movements.

Defecation

The elimination of feces from the body is controlled both unconsciously by a *defecation reflex* and voluntarily. The defecation reflex is triggered by distension of the rectum, which occurs as fecal material begins to enter the rectum from the colon, usually during mass movements of the colon. This distension stimulates stretch receptors and initiates several events. First, smooth muscle in the wall of the rectum is stimulated to contract, which raises the pressure inside. Peristaltic contractions of the sigmoid

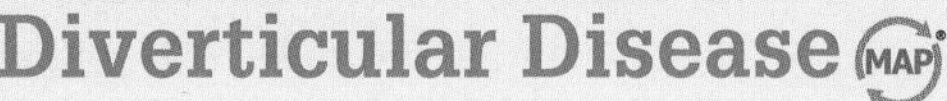

Diverticular Disease

Between 10% and 30% of U.S. adults older than age 40 suffer from diverticular disease; 40% of them don't know that they have this disorder. Diverticular disease occurs when the wall of the large intestine weakens over time, an effect likely exacerbated by the excess pressure of frequent constipation. *Diverticula* are outpockets, approximately 5–15 mm in diameter, that occur in the mucosa and submucosa. The exact cause of weakening of the colon wall is not known, but there is a genetic predisposition to this condition in many cases. Diet also plays a role; low-fiber diets and traditionally Western diets have been implicated in the formation of diverticula. Defects in GI motility or wall strength also contribute to the disease.

Diverticula occur mostly in the sigmoid colon, but they can occur in all areas of the large and small intestine. Although diverticula in themselves are usually asymptomatic, they can lead to *diverticulitis,* an inflammatory disease that causes severe pain. The diverticula allow chyme to become lodged in, and irritate the walls of, the colon. Symptoms of diverticulitis include pain, fever, and chills.

Treatments of mild diverticular disease include a high-fiber diet and medications to control bowel movements. Nuts are excluded from the diet because they have a tendency to lodge themselves in the diverticula. Treatments of diverticulitis include anti-inflammatory agents, antibiotics, and pain medication. Surgical resection of the colon can be done in severe cases.

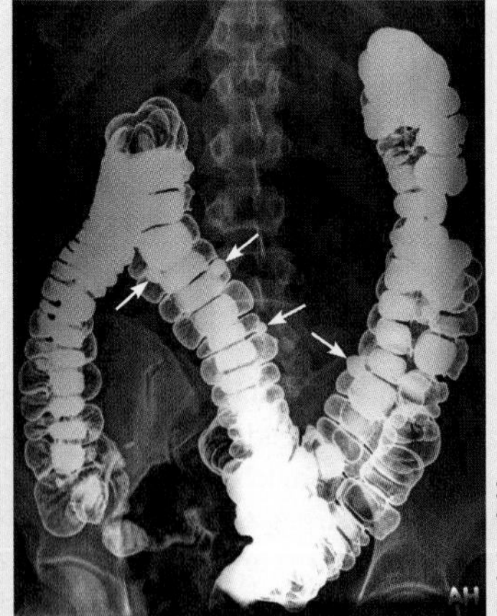

Cindy Stanfield

X-ray of colon showing several diverticula (arrows).

Critical Thinking Questions

1. What is diverticular disease and where in the body can diverticula occur?
2. Explain the causes and describe the symptoms of diverticulitis.
3. Describe the various treatment options available for diverticular disease.

colon are also stimulated, which propels more fecal material into the rectum, further raising the pressure. At the same time, the internal anal sphincter relaxes while the external anal sphincter contracts, preventing the material from exiting the body. If the colonic contractions are strong enough to raise the pressure in the rectum to a certain critical level, the external anal sphincter relaxes, allowing defecation to proceed. In adults and children who are toilet trained, defecation may be postponed by voluntary contraction of the external anal sphincter. Constipation, caused by insufficient defecation, can lead to increased pressure in the colon, which can cause diverticula (see **Clinical Connections: Diverticular Disease**).

SYSTEMS INTEGRATION As you learned in this chapter, the GI system is regulated to maximize absorption of nutrients, not to regulate the amount and types of nutrients absorbed so as to maintain homeostasis. Thus one could question the role of the GI system in homeostasis. However, if the GI system did not carry out its function efficiently, then none of the organ systems could survive.

Our cells need nutrient molecules every second of every day to survive. The GI tract provides those nutrient molecules. Of course, our bodies could not perform other necessary functions if they devoted each moment to bringing in new nutrients through the GI system. Thus the GI system evolved to efficiently absorb nutrients when nutrients were available.

In addition to providing nutrients, the GI system secretes hormones that regulate its own function as well as food intake. The GI tract can absorb and store iron to later be used in the synthesis of hemoglobin. The liver has several functions besides the processing of nutrients, including storing carbohydrates that can later be released into the bloodstream in the form of glucose to provide energy for the nervous system, synthesizing ketones for energy use by the nervous system, removing old erythrocytes and some waste products (such as ammonia, which the liver converts to urea) from the blood, and synthesizing plasma proteins such as clotting proteins needed for homeostasis. In all of these ways, the GI system interacts with the other organ systems to maintain homeostasis.

CHAPTER REVIEW

SUMMARY

1 Overview of Gastrointestinal System Processes

- The primary purpose of the gastrointestinal system is to extract needed materials from ingested food and deliver them to the bloodstream for distribution to cells throughout the body.
- Nutrient molecules are too large to be transported in the bloodstream; they must be broken down into smaller molecules by enzymes in the lumen of the GI tract (digestion).
- The resultant digestion products are absorbed into the bloodstream.
- To aid in these processes, fluids and enzymes are secreted into the lumen of the GI tract, and motility propels the lumenal contents from one digestive organ to the next.

  iP Digestive: Orientation

2 Functional Anatomy of the Gastrointestinal System

- The gastrointestinal system consists of the GI tract and accessory glands.
- Four layers make up the wall of the GI tract: the mucosa, the submucosa, the muscularis externa, and the serosa.
- The mucosa, which lines the lumen of the GI tract, contains the mucous membrane, an epithelial layer populated by secretory, absorptive, and endocrine cells.
- The outer border of the submucosa is made up of the enteric nervous system, which consists of the submucosal plexus and the myenteric plexus.
- The muscularis externa contains circular and longitudinal muscles that propel the lumenal contents through the GI tract.
- The serosa is composed of an inner layer of connective tissue and the mesothelium.
- Organs of the GI tract include the mouth, esophagus, stomach, small intestine, colon, rectum, and anus.
- The primary function of the esophagus is to conduct food from the mouth to the stomach.
- The flow of material between organs is regulated by sphincters.
- The stomach's important function is to store food after it is swallowed and release it into the small intestine.
- The small intestine is the primary site for the digestion and absorption of all nutrients.
- The colon's primary function is to transform the chyme (food mixed with gastric juice) into feces.
- Fecal matter is pushed from the colon into the rectum, where it is expelled from the body through the anus.
- Accessory glands include the salivary glands (which secrete saliva), the pancreas (which secretes pancreatic juice containing enzymes and bicarbonate), and the liver (which secretes bile and processes absorbed nutrients).

  iP Digestive: Anatomy Review

3 Digestion and Absorption of Nutrients and Water

- Digestion of starch and glycogen begins in the mouth with the action of salivary amylase and is continued in the small intestine by pancreatic amylase.
- Brush border enzymes in the small intestine reduce carbohydrates to monosaccharides, which are transported across the mucosal epithelium and diffuse into the bloodstream.
- Protein digestion begins with the action of pepsin in the stomach, and continues in the small intestine with the action of both pancreatic enzymes (including trypsin, chymotrypsin, and carboxypeptidase, secreted as inactive zymogens that are activated upon entry into the GI tract) and membrane-bound enzymes (including enterokinase and aminopeptidase).

- Most proteins are reduced completely to amino acids, which are transported into the bloodstream.
- Dietary fats (mostly triglycerides) are reduced to fatty acids and monoglycerides by pancreatic lipases and by bile salts, which emulsify fat droplets.
- Products of fat digestion enter epithelial cells by simple diffusion and then are reassembled into triglycerides, which are transported (along with other lipids) into the lymphatic system in the form of chylomicrons, a type of lipoprotein.
- Vitamins and minerals are absorbed chemically unaltered.
- Water absorption occurs secondary to solute absorption and is driven by an osmotic gradient.

iP Digestive: Digestion and Absorption

MAP iP Enzymatic Digestion and Absorption

4 General Principles of Gastrointestinal Regulation

- Gastrointestinal regulatory mechanisms maximize the efficiency of digestion and absorption but generally do not act to maintain homeostasis.
- Gastrointestinal function is regulated by short reflex and long reflex pathways involving the enteric nervous system, autonomic nervous system, and hormones (including gastrin, secretin, cholecystokinin, and glucose-dependent insulinotropic peptide).
- The enteric nervous system receives input both from the autonomic nervous system and from mechanoreceptors, chemoreceptors, and osmoreceptors that monitor conditions in the GI tract.

MAP iP Digestive: Control of the Digestive System

5 Gastrointestinal Secretion and Its Regulation

- Saliva secretion is controlled by autonomic input to the salivary glands and is coordinated by the medullary salivary center.
- Gastric secretion of acid and pepsinogen is influenced by cephalic-phase, gastric-phase, and intestinal-phase stimuli and controlled by neural and hormonal reflexes.
- Pancreatic secretion is also controlled by neural and hormonal signals (primarily secretin and cholecystokinin).
- Bile secretion by the liver is stimulated by secretin and by cholecystokinin, which also stimulates contraction of the gallbladder.

MAP iP Digestive: Secretion

6 Gastrointestinal Motility and Its Regulation

- Gastrointestinal smooth muscle contractions are triggered by slow waves generated by pacemaker cells.
- Nerves and hormones generally influence the strength of contractions, but not their frequency.
- The stomach and intestines exhibit motility patterns that change depending on conditions in the lumen.

iP Digestive: Motility

MAP MasteringA&P®

Check your understanding at MasteringA&P and apply your new knowledge to real-life case studies to ensure that you've mastered the content!

EXERCISES

Multiple-Choice Questions

1. What do sodium, fatty acids, and vitamin A have in common?
   a) They are not enzymatically modified prior to their absorption into the bloodstream.
   b) They cross the apical membranes of enterocytes by simple diffusion.
   c) They are transported into blood capillaries in the villi.
   d) They are all hydrophilic.
   e) All of the above
2. Blockage of the flow of bile into the duodenum interferes with the digestion of which of the following?
   a) Carbohydrates only
   b) Lipids only
   c) Proteins only
   d) Carbohydrates and lipids only
   e) Carbohydrates and proteins only
3. Which of the following is an accurate statement regarding the various phases of gastrointestinal control?
   a) The autonomic nervous system is involved in cephalic-phase regulation only.
   b) The autonomic nervous system is involved in gastric-phase regulation only.
   c) The autonomic nervous system is involved in intestinal-phase regulation only.
   d) The autonomic nervous system is involved in gastric-phase and intestinal-phase regulation only.
   e) The autonomic nervous system is involved in all three phases of regulation.
4. Failure of the salivary glands to secrete amylase would make it impossible to digest which of the following?
   a) Proteins
   b) Fats
   c) Disaccharides
   d) Starch
   e) None of the above
5. Which of the following tends to inhibit acid secretion by the stomach?
   a) An increase in the osmolarity of duodenal contents
   b) Entry of stomach acid into the duodenum
   c) The arrival of food in the stomach
   d) Both a and b
   e) All of the above
6. Which of the following best illustrates the phenomenon of potentiation?
   a) Bile secretion is normally stimulated by both secretin and CCK.
   b) Secretin and CCK act primarily on different parts of the hepatic secretory apparatus.
   c) Secretin and CCK both stimulate bile duct cells to secrete fluid.
   d) The secretion of fluid by bile duct cells is greater when both CCK and secretin are present, compared to when either hormone is present alone (at the same concentration).
7. The enzyme enterokinase is directly or indirectly responsible for the proper functioning of
   a) Bile salts.
   b) Lipases.
   c) Trypsin.
   d) Chymotrypsin.
   e) Both trypsin and chymotrypsin.

8. Which of the following is an example of a zymogen?
   a) Enterokinase
   b) Chymotrypsinogen
   c) Salivary amylase
   d) Cholecystokinin
   e) All of the above
9. Increases in gastric motility are generally accompanied by increases in ileal motility because of the
   a) Gastroileal reflex.
   b) Ileogastric reflex.
   c) Gastrocolic reflex.
   d) Colonocolonic reflex.
   e) Intestino-intestinal reflex.
10. Which of the following is a digestion product of amylase?
   a) Maltose
   b) Glucose
   c) Sucrose
   d) Lactose
   e) Fructose
11. Increases in which of the following can induce satiety?
   a) Alpha melanocyte stimulating hormone
   b) Thyroid stimulating hormone
   c) Leptin
   d) Both a and c
   e) All of the above

Objective Questions

12. Through the action of pancreatic amylase alone, starch can be broken down to a form that can be absorbed completely. (true/false)
13. The glands that secrete acid in the stomach are examples of accessory glands. (true/false)
14. Receptive relaxation is an example of (cephalic-phase/gastric-phase) control of stomach function.
15. The (submucosal/myenteric) nerve plexus is located within the muscularis externa.
16. The _______ is the outermost layer of the gastrointestinal wall, consisting of the mesothelium and an underlying layer of connective tissue.
17. When food is present in the small intestine, contraction of the gallbladder is stimulated by (secretin/cholecystokinin).
18. Bile salts are necessary for the proper functioning of pancreatic (proteases/lipases).
19. The lamina propria is located within the (mucosa/submucosa).
20. Gastrin (stimulates/inhibits) the secretion of acid by the stomach.
21. In the stomach, pepsinogen is secreted by (chief cells/parietal cells).
22. The small intestine is periodically swept of its contents by (segmentation/migrating motility complexes).
23. The term *enterohepatic circulation* refers to the conduction of blood from the intestine to the liver via the hepatic portal vein. (true/false)
24. Nerves and hormones exert their effects on gastrointestinal motility primarily by altering the (frequency/amplitude) of slow waves.
25. Chewing is under voluntary control but is also controlled by reflex neural pathways. (true/false)
26. Disaccharides are broken down to monosaccharides by (pancreatic enzymes/brush border enzymes).

Essay Questions

27. Describe the process by which ingested triglycerides are digested, absorbed, and transported into the bloodstream. Indicate which of the steps are adaptations to triglycerides' hydrophobic nature, and explain why. Compare these steps to the corresponding processes pertaining to the digestion and absorption of hydrophilic substances such as proteins or carbohydrates.
28. When chyme moves from the stomach to the small intestine, several stomach functions are altered. Identify these functions, describe the changes that occur, and explain the regulatory mechanisms that bring about these changes. Explain why these changes make sense in light of the gastrointestinal system's ultimate function.
29. Describe the structural adaptations of the small intestine that increase its capacity for absorbing nutrients.
30. Describe the processes of segmentation, peristalsis, haustration, and mass movement, and explain their roles in digestive function. What is the apparent function of the muscularis mucosae?
31. Describe the various mechanisms that regulate the secretion of acid and pepsinogen by the stomach. Compare these mechanisms to those that regulate gastric motility.

Critical Thinking

32. Pantoprazole (Protonix) is a fairly new drug that blocks the active transport of hydrogen ions in the stomach. Explain the consequences of taking this drug and which disorder(s) it might be useful to treat. What would be some of the drug's negative side effects?
33. The body composition (fat mass versus fat-free mass) of Americans includes a greater proportion of fat than ever before, and the trend is toward continued increases in fat. The medical community is concerned with this trend because obesity predisposes people to many diseases (hypertension and diabetes, for example). New diets are proposed on a regular basis. Describe the physiological basis behind some of these new diets.
34. Based on what you learned about long-term regulation of food intake, explain why recreational drugs such as cocaine can cause weight loss.

Photo credits for icons in feature boxes: Focus on Diabetes: Tanyae/Dreamstime; Clinical Connections: Iofoto/Dreamstime; Understanding Exercise: Pixattitude/Dreamstime; Discovery: Yuri Arcurs/Cutcaster.

Glossary

acini small sacs of secretory cells leading to a duct; singular, *acinus*

amylase enzyme that digests starch and glycogen; includes salivary amylase and pancreatic amylase

basic electrical rhythm (BER) a pattern of electrical activity in gastrointestinal smooth muscle, in which waves of depolarization occur at regular intervals

bile fluid secreted by the liver that contains bicarbonate and bile salts

bile salts cholesterol derivatives manufactured by the liver and found in bile; their function is to emulsify fats in the small intestine

brush border the collection of microvilli that are located on the apical membranes of epithelial cells lining the small intestine or renal tubule

cephalic-phase control in the gastrointestinal system, the regulation of gastrointestinal (GI) function by stimuli originating in the head

chief cells specialized cells in the gastric glands that secrete pepsinogen

cholecystokinin (CCK) a hormone secreted by the duodenum and jejunum that inhibits gastric secretion and motility, and stimulates pancreatic enzyme secretion, gallbladder contraction, and relaxation of the sphincter of Oddi

chylomicrons lipoprotein particles formed during the process of lipid absorption in the small intestine

chyme a mixture of food particles and gastrointestinal secretions; found in the stomach and intestines

colon an organ of the gastrointestinal (GI) tract that absorbs water and ions from the chyme, and stores feces; comprises the ascending colon, descending colon, transverse colon, and sigmoid colon

digestion the breakdown of nutrient molecules that are present in food to smaller molecules by enzymes in the lumen of the digestive tract

duodenum initial portion of the small intestine

emulsification the process by which the action of bile salts breaks fat globules down into smaller droplets

endopeptidase enzyme that digests proteins by breaking peptide bonds in the middle of the protein molecule

enteric nervous system a system of neural networks in the wall of the gastrointestinal (GI) tract that regulates many digestive functions; also called the *intrinsic nervous system*

enterocytes epithelial cells of various types found in the mucous membrane of the stomach or intestine

esophagus a muscular tube whose primary function is to conduct food from the pharynx to the stomach

exopeptidase enzyme that digests proteins by cleaving off amino acids from the end of the polypeptide

external anal sphincter ring of skeletal muscle that regulates the excretion of feces

G cells endocrine cells of the stomach that secrete gastrin

gallbladder small muscular sac located immediately adjacent to the liver; stores bile in between meals

gastric glands glands in the lining of the stomach that secrete gastric juice

gastric juice watery secretion of the stomach containing hydrochloric acid, pepsinogen, and mucus

gastric mucosal barrier mucus layer that protects the lining of the stomach from the effects of acid and pepsin

gastric-phase control regulation of gastrointestinal function by stimuli originating in the stomach

gastric pits openings in the lining of the stomach that lead to gastric glands

gastrin hormone secreted by the stomach that regulates many gastrointestinal functions, including gastric acid secretion

glucose-dependent insulinotropic peptide (GIP) hormone secreted by the duodenum and jejunum that stimulates insulin secretion by pancreas

haustration mixing movements of the colon

hepatocytes liver cells

ileocecal sphincter ring of smooth muscle that regulates movement of chyme from the ileum to the cecum

ileum terminal portion of the small intestine

internal anal sphincter ring of smooth muscle that regulates the excretion of feces

intestinal-phase control regulation of gastrointestinal (GI) function by stimuli originating in the small intestine

intrinsic factor a protein secreted by parietal cells of the stomach that binds to vitamin B_{12} and that is necessary for the absorption of this vitamin in the small intestine

jejunum middle portion of the small intestine

lacteal blind-ended lymphatic vessel found within each villus in the lining of the small intestine

lamina propria in the wall of the gastrointestinal (GI) tract, a layer of connective tissue within the mucosa

leptin hormone secreted by adipose cells that is important in physiological regulation of food intake

lipases enzymes present in pancreatic juice that digest lipid substrates

lipoproteins protein- and lipid-containing particles that possess a hydrophobic interior surrounded by a hydrophilic coat; the primary vehicle of lipid transport in the blood or lymphatic system

liver accessory gland in the gastrointestinal (GI) system that secretes bile; also processes certain absorbed nutrients and performs many other functions

lower esophageal sphincter a ring of smooth muscle at the lower end of the esophagus that regulates the flow of food from the esophagus to the stomach

mass movement peristaltic-like wave in the colon that propels its contents forward

mesenteries a system of clear thin membranes that interconnects most of the abdominal organs and houses nerves and blood vessels running to them; helps anchor organs in place

mesothelium a layer of epithelial tissue found on the outer surfaces of the organs of the gastrointestinal (GI) tract

migrating motility complex cyclical patterns of intense motility followed by relaxation in the stomach and small intestine that clears the lumen of its contents

motility in the gastrointestinal (GI) tract, patterns of smooth muscle contraction

mucosa the innermost of the four major layers that make up the wall of the gastrointestinal (GI) tract; composed of three layers including the mucous membrane, the lamina propria, and the muscularis mucosae

mucous membrane a layer of epithelial cells that lines the inside of the gastrointestinal (GI) tract and other hollow cavities in the body

mucus sticky, viscous fluid containing glycoproteins called mucins

muscularis externa one of the four major layers of the wall of the gastrointestinal (GI) tract, located between the submucosa and serosa; composed primarily of smooth muscle fibers

muscularis mucosae thin layer of smooth muscle located in the mucosa of the gastrointestinal (GI) tract wall

myenteric plexus one of the two neural networks that make up the enteric nervous system; located in the muscularis externa

neck cells cells in gastric glands that secrete mucus

pancreatic juice watery secretion from the pancreas that contains digestive enzymes and bicarbonate

parietal cells cells in the gastric glands that secrete hydrogen ions and intrinsic factor

pepsin the fully active form of pepsinogen; acts in the lumen of the stomach to begin the digestion of proteins

pepsinogen an inactive precursor of the enzyme pepsin, secreted by specialized chief cells

peristalsis a wave of momentary contraction that travels along the length of a hollow vessel or organ, such as the gastrointestinal (GI) tract

peritoneum a membrane lining the inside of the abdominal cavity

potentiation in endocrinology, the amplification of one hormone's effects by another hormone

proteolytic activation rendering an enzyme active by cleaving off certain amino acids

pyloric sphincter ring of smooth muscle that regulates gastric emptying

pylorus a narrow passage between the stomach and the duodenum

salivary glands glands that secrete saliva, located in the mouth and pharynx

secretin a hormone secreted by the duodenum and jejunum that inhibits gastric secretion and motility and stimulates pancreatic bicarbonate secretion

secretion movement of substance from the internal environment to the external environment by transport across an epithelium; movement of substance from inside a cell to outside the cell by movement across the plasma membrane

segmentation the mixing motility pattern of the small intestine

serosa the outermost of the four major layers of the wall of the digestive tract; composed mostly of connective tissue

slow waves cyclical fluctuations in the resting membrane potential

small intestine an organ of the gastrointestinal (GI) tract, consisting of a coiled tube 8–10 feet long; the primary site for the digestion and absorption of all the nutrients in food

sphincter a ring of muscle that surrounds an orifice and regulates the passage of material through it by altering its diameter

sphincter of Oddi a ring of smooth muscle that regulates the flow of bile and pancreatic juice into the duodenum

stomach an organ of the gastrointestinal (GI) tract that stores food and releases it into the small intestine; major anatomical regions are the fundus, body, and antrum

submucosa one of the four major layers of the wall of the gastrointestinal (GI) tract, located between the mucosa and the muscularis externa; composed mostly of connective tissue

submucosal plexus one of the two neural networks that make up the enteric nervous system; located in the submucosa

upper esophageal sphincter ring of skeletal muscle that controls the movement of food from the pharynx to the esophagus

villi folds in the mucosal surface of the small intestine that facilitate the transport of materials by increasing the surface area of the epithelium; singular, *villus*

zymogens inactive precursor forms of pancreatic digestive enzymes that are stored in secretory cells and released by exocytosis

Answers

Figure Question Answers

Fig 12 The hepatic portal vein
Fig 13 A catabolic reaction
Fig 21 Chylomicrons are too large
Fig 24 Into the cell

Apply Your Knowledge Answers

Because insulin is a peptide, if it is consumed in the diet, it will be digested by proteases in the stomach and small intestine. Thus it must be injected into the bloodstream directly.

Olestra is a synthetic molecule, and human gastrointestinal systems do not have the enzymes necessary to digest it to absorbable units. Olestra will, therefore, be eliminated in the feces. Because it is fat soluble, it can draw other fat-soluble substances with it, including the fat-soluble vitamins.

Answers to End-of-Chapter Multiple Choice and Objective Questions

Multiple Choice Questions

1. a 2. b 3. e 4. e 5. d 6. d 7. e 8. b 9. a 10. a 11. d

Objective Questions

12. false
13. false
14. cephalic-phase
15. myenteric
16. serosa
17. cholecystokinin
18. lipases
19. mucosa
20. stimulates
21. chief cells
22. migrating motility complexes
23. false
24. amplitude
25. true
26. brush border enzymes

The Endocrine System: Regulation of Energy Metabolism and Growth

The Endocrine System: Regulation of Energy Metabolism and Growth

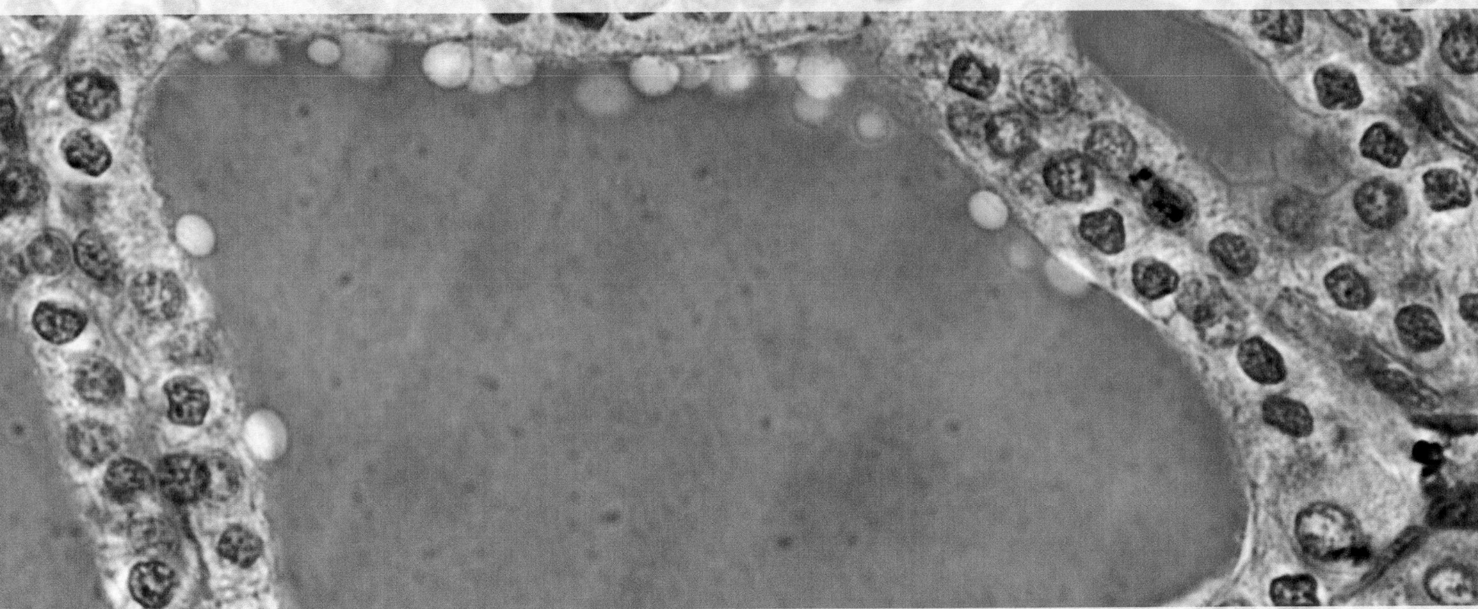

Michael Ross/Photo Researchers, Inc.

Colorized light micrograph of a follicle in the thyroid gland.

Have you ever been stuck in classes from 10:00 until 1:00? If so, you probably had difficulty concentrating part of the time because you started feeling hungry before that last class let out. At 1:30, however, you replenished your energy reserves by eating a late lunch. That allowed you to study until 6:00 before the next wave of hunger set in.

Energy balance differs from other body balances because energy input from meals is intermittent. This uneven input creates a challenge for our organ systems, but the challenge is met through the actions of primarily two hormones, insulin and glucagon. You will learn in this chapter how insulin and glucagon regulate energy metabolism to maintain adequate glucose for our neurons to function. You will also learn how other hormones affect energy metabolism as the body grows or copes with stress.

CHAPTER OUTLINE

Go to MasteringA&P for helpful A&P Flix 3-D animations, chapter quizzes, pre-tests, interactive physiology tutorials, and more!

LEARNING OUTCOMES *After studying this chapter, you should be able to:*

- Compare the metabolic pathways operating during energy mobilization to those operating during energy utilization.
- Explain the concepts of negative energy balance and positive energy balance.
- Describe the hormonal control of metabolism during absorptive and postabsorptive states.
- Describe how the body regulates body temperature.
- Explain how growth hormone regulates growth.
- Describe the synthesis and secretion of thyroid hormones. Distinguish between direct and permissive actions of thyroid hormones.
- Describe the effects of glucocorticoids on whole-body metabolism. Compare the physiological effects of glucocorticoids to their pharmacological effects.
- Describe the stress response.

Before You Begin

Make sure you have mastered the following topics:

- Enzymes
- Metabolic pathways
- Glycolysis
- Krebs cycle
- Electron transport chain
- Pancreas
- Thyroid gland
- Hypothalamus-anterior pituitary tropic hormones
- Adrenal glands
- Muscle metabolism

Whole-body metabolism is regulated primarily by hormones. In this chapter, we examine the coordinated regulation of metabolic pathways in different organs to maintain adequate energy supplies for all the body's cells, focusing on hormonal control of energy balance.

As we study these hormones and their actions, we will find that their effects are varied and often overlap. To help you understand these hormones, the first section of this chapter presents general principles relating to energy metabolism and energy balance. A comparison of energy metabolism during and between meals follows. The chapter then describes the hormones that regulate blood glucose level, especially insulin and glucagon. The next section covers thermoregulation, a process closely linked to energy metabolism. The hormones that regulate growth and the primary hormones that regulate whole-body metabolism—thyroid hormones—are then discussed. The final section of the chapter describes the glucocorticoids and their role in adapting to stress.

1 An Overview of Whole-Body Metabolism

The next time you find yourself sitting down to eat a meal, ask yourself this question: Why am I eating? If you do this on several occasions, you will find that many possible reasons for eating exist. Perhaps you smelled something cooking that stimulated your appetite, or you saw an advertisement for a food you like. Perhaps you decided to join some friends who were on their way to dinner, or maybe you just looked at your watch and decided it was mealtime. The number of possibilities is large. In fact, we have so many different motivations for eating that it is easy to forget that the ultimate reason—obtaining nutrition—is a biological necessity, because food is our sole source of energy and the raw materials from which our bodies are made.

Although biological necessity drives our need for food, our eating patterns vary and are influenced by other factors. (You have probably skipped meals to study for exams or have overeaten during the holidays.) In most cases, this inconsistency is no cause for concern because the body has ways of maintaining the steady supply of energy that cells need despite changes in the pattern of food intake. Between meals the body converts energy stores (including large carbohydrates, proteins, and lipids) into smaller molecules that cells can use for energy. When you eat, the body replenishes these stores by converting nutrients into energy storage molecules.

The way the body stores and utilizes energy—*energy metabolism*—is influenced not only by eating patterns but also by such factors as growth, stress, and metabolic rate. In all cases, whether the body stores or utilizes energy is controlled primarily by endocrine signals. Two critical concepts drive the control of energy metabolism:

1. Because food intake is intermittent, the body must store nutrients during periods of intake and then break down these stores during periods between meals.
2. Because the brain depends on glucose as its primary energy source, blood glucose levels must be maintained at all times, even between meals.

To fully appreciate the control of metabolism, we need to review some key concepts of cellular metabolism and relate them to the whole body.

Anabolism

An interesting aspect of metabolism is that the same small biomolecules that provide energy are also used to synthesize larger biomolecules. A good example is acetyl CoA, which can be catabolized in the Krebs cycle for energy and also serves as a substrate for triglyceride and cholesterol synthesis. Thus, because carbohydrates, lipids, and proteins can all be catabolized to acetyl CoA, they can all eventually be converted to lipids. Many other metabolic intermediates of glycolysis and the Krebs cycle can be used to synthesize larger biomolecules. For example, some metabolic intermediates can be converted to amino acids and used for protein synthesis, whereas other intermediates can be used in the synthesis of phospholipids.

Regulation of Metabolic Pathways

If anabolic and catabolic pathways have several of the same intermediates, what governs the direction of metabolism? The most important factors in determining which metabolic pathways are in operation are the number and activity of the enzymes involved in the pathways. The activity of enzymes can be regulated by changing their concentration through synthesis or degradation, or by changing the activity of individual enzyme molecules through allosteric or covalent regulation. Hormones that regulate metabolic pathways do so by regulating the activity of enzymes in one or more of these ways.

Metabolic pathways are also controlled by compartmentation. Whereas glycolysis occurs in the cytosol, the Krebs cycle occurs in the mitochondrial matrix. Compartmentation also occurs on the tissue level, because some enzymes are found in the cells of only certain tissues. In addition, hormones differentially affect tissues based on the types of receptors on the cells in that tissue. Tissues or organs that have special metabolic activities include the brain, skeletal muscle, the liver, and adipose tissue. We will explore how these tissues and organs affect whole-body metabolism shortly; first we consider how the body handles different classes of biomolecules, from absorption, to cellular uptake, to utilization by the cell.

2 Energy Intake, Utilization, and Storage

When we eat, digestion in the gastrointestinal (GI) tract breaks down the large molecules in food into smaller molecules, which are then absorbed into the bloodstream. Of our three main nutrient classes, carbohydrates are transported in the blood as glucose, proteins are transported as amino acids, and lipids are transported in lipoproteins. The blood flow distributes these nutrients to tissues throughout the body, where they are eventually taken up by cells. Inside cells, these molecules undergo one of three possible fates:

1. Biomolecules can be broken down into smaller molecules, in the process releasing energy that can be used for driving various cellular processes such as muscular contraction, transport, secretion, or anabolism.
2. Biomolecules can be used as substrates to synthesize other molecules needed by cells and tissues for normal function, growth, and repair.
3. Biomolecules in excess of those required for energy and synthesis of essential molecules are converted to energy storage molecules that provide energy during periods between meals. The two primary energy storage molecules are glycogen and triglyceride (fat).

The ultimate fate of consumed molecules depends on their chemical nature and the body's needs at the time of consumption, as described next.

Uptake, Utilization, and Storage of Energy in Carbohydrates

Although carbohydrates are consumed in a variety of forms, monosaccharides—especially glucose—are the forms found in the bloodstream. **Figure 1a** illustrates the fate of glucose in the blood. Molecules of glucose are transported into cells throughout the body by *glucose transporters* ①. Inside cells, glucose can be oxidized for energy ②, which generates carbon dioxide as a waste product; can provide substrates for other metabolic reactions ③; or can be converted to glycogen for storage ④. If glucose levels in the cell decrease, glycogen can be broken down to glucose by glycogenolysis ⑤.

Although this series of steps accurately describes what happens to glucose *in the body as a whole,* it may or may not describe what happens in individual cells. Most cells, for example, can oxidize glucose but have a limited ability to synthesize and store glycogen.

Uptake, Utilization, and Storage of Energy in Proteins

As depicted in Figure 1b, amino acids rather than whole proteins are transported in the bloodstream. Following uptake into cells ①, amino acids are used for the synthesis of proteins ②, or catabolized for energy by proteolysis ③. Because proteins consist of amino acids, protein catabolism produces amino acids ④, which can then be catabolized for energy or released into the bloodstream for use by other cells. Cells utilize protein catabolism for energy less so than carbohydrates and lipids, but when proteins are used, ammonia (NH_3) and carbon dioxide are produced. The highly toxic ammonia is converted by the liver to *urea,* which is eventually eliminated in the urine.

Uptake, Utilization, and Storage of Energy in Fats

When the body uses dietary carbohydrates and proteins, they are taken into cells in the form of smaller components (glucose or amino acids), which can either be catabolized for energy or assembled into larger molecules. The same process occurs for fats, although the process is a little more complicated.

Figure 1c illustrates the body's handling of triglycerides, the predominant form in which fats are present in the diet. Triglycerides are transported in the bloodstream in **lipoproteins,** small protein- and lipid-containing particles. An assortment of lipoproteins of varying densities transports lipids to various target cells throughout the body. Once they reach their target cells, lipids must leave the lipoprotein before they can enter the cells.

To facilitate entry into cells, triglycerides at the outer surface of lipoproteins are broken down by the enzyme **lipoprotein lipase** ①, which is located on the inside surface of capillaries throughout the body and is particularly dense in capillaries running through

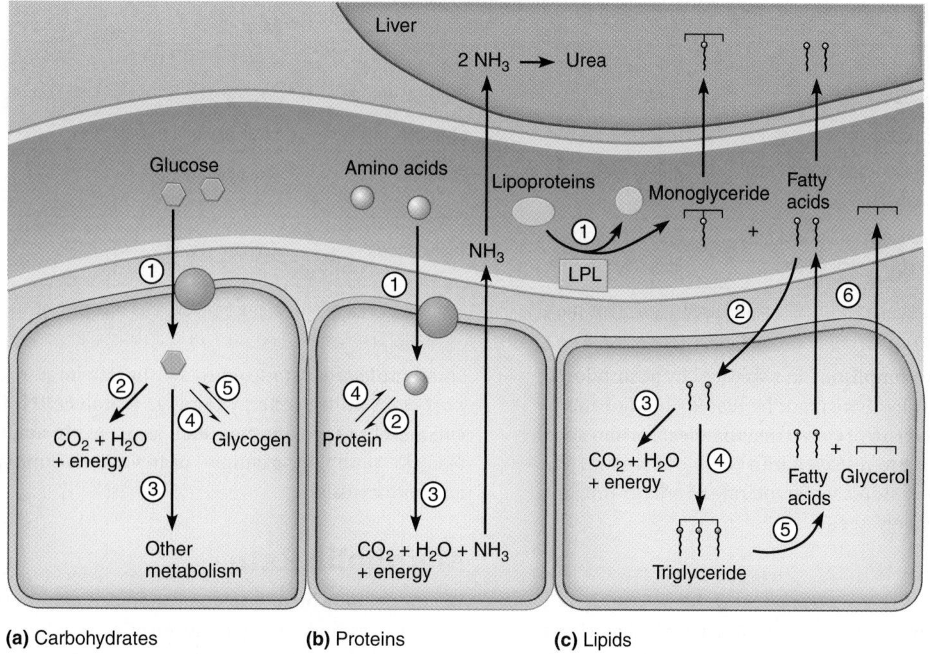

Figure 1 Transport, uptake, and cellular fates of biomolecules. Numbered steps are described in the text for the handling of **(a)** carbohydrates, **(b)** proteins, and **(c)** lipids. LPL is lipoprotein lipase.

Q Which three metabolic pathways are necessary for the handling of glucose shown in part (a), step 2?

adipose tissue (body fat). This enzyme breaks down triglycerides into fatty acids and monoglycerides; the fatty acids are then taken up by nearby cells ②, while the monoglycerides remain in the bloodstream and are eventually metabolized in the liver.

After entering cells, fatty acids may be oxidized for energy ③ or combined with glycerol to form new triglycerides ④, which are stored in fat droplets in the cytosol. This storage occurs mainly in **adipocytes,** adipose tissue cells that are specialized for fat storage. (The glycerol used in triglyceride synthesis is not derived from absorbed triglycerides, but instead is synthesized within adipocytes.) Stored triglycerides can subsequently be broken down into glycerol and fatty acids ⑤, which can be catabolized for energy or released into the bloodstream for use by other cells ⑥. The catabolism of glycerol and fatty acids produces carbon dioxide as a waste product. The breakdown of triglycerides to fatty acids and glycerol, such as occurs in step 1 or step 5, is called *lipolysis.*

Table 1 summarizes the body's processing of carbohydrates, proteins, and lipids. Fatty acids are included among the smaller nutrient molecules (even though they are not absorbed in this form) because they are the form in which fats are made available to most cells. Some small nutrients can be *interconverted;* for example, glucose can be synthesized from amino acids, and fatty acids can be synthesized from glucose or amino acids. These interconversions have a significant role in whole-body metabolism, as we will see shortly.

3 Energy Balance

To maintain homeostasis, the human body must be kept in balance. To be "in balance" in this context means that what comes into the body and what is produced by the body equal the sum of what is used by the body and what is eliminated by the body. This relationship is summarized in the following equation:

$$\text{Input} + \text{production} = \text{utilization} + \text{output}$$

In terms of energy balance, the body does not produce energy (an impossibility according to the laws of thermodynamics) and so this equation becomes

$$\text{Energy input} = \text{energy utilization} + \text{energy output}$$

Energy is used by the body to perform work, and energy output exists as heat released. Therefore, the balance equation becomes

$$\text{Energy input} = \text{work performed} + \text{heat released}$$

The endocrine system regulates the body's energy balance to ensure that a steady supply of small nutrients is always available to all cells to meet their energy demands. As cells expend energy, they draw on stores of nutrients both within cells and in the bloodstream to obtain more energy. This pool of nutrients must be continually replenished if uninterrupted energy expenditure is to occur. This

TABLE 1 Summary of Carbohydrates, Protein, and Lipid Processing

	Form absorbed across GI tract	Form circulating in blood	Form stored	Storage site	Percentage of total energy stored
Carbohydrates	Glucose	Glucose	Glycogen	Liver, skeletal muscle	1%
Proteins	Amino acids, some small peptides	Amino acids	Proteins	Skeletal muscle*	22%
Lipids	Monoglycerides and fatty acids (in chylomicrons)	Free fatty acids, lipoproteins	Triglycerides	Adipose tissue	77%

* Even though proteins are found in all cells of the body, most of the proteins that are mobilized for energy come from skeletal muscle cells.

replenishment can be accomplished in two ways: by absorption of more nutrients into the bloodstream or by *mobilization* of energy stores—that is, the catabolism of stored macromolecules into small nutrient molecules that are released into the bloodstream. The body mobilizes its energy stores when the rate of energy intake is insufficient to meet its energy needs.

Energy Input

Energy input into the body arrives in the form of absorbed nutrients. When a particular nutrient molecule (such as glucose) is oxidized in the body, a certain quantity of energy is liberated; this quantity represents the *energy content* of the molecule. A person's *energy intake* is the total energy content of all nutrients absorbed.

The energy content of a nutrient is generally determined by burning a known quantity of the substance in an instrument called a *calorimeter,* which measures the total amount of energy released in the form of heat as the substance burns. This amount is usually expressed in kilocalories per gram of the substance burned. Kilocalories can also be written as Calories, with a capital C (1 kilocalorie = 1 Calorie = 1000 calories). The energy content of nutrient molecules varies for the three nutrient classes, averaging 4 Calories per gram of carbohydrates, 4 Calories per gram of protein, and 9 Calories per gram of fat. Thus fat, which has a higher energy content than carbohydrates and proteins, is the most efficient form in which to store energy.

Energy Output

In a calorimeter, all the energy contained in a molecule is converted to heat during oxidation, so that the energy content of the molecule (input) is equal to the quantity of energy in the heat (output). Although the body also relies on the oxidation of molecules to liberate energy, the energy released during oxidation or other catabolic reactions always takes two forms: heat and work (**Figure 2**). Approximately 60% of the energy in consumed nutrients goes to heat production, which is necessary to maintain body temperature. Most of the remaining 40% of energy is used to synthesize ATP, which is used to perform cellular work (a process that releases still more heat).

The energy-requiring processes of cells are classified into three main categories: mechanical, chemical, and transport. *Mechanical work* uses intracellular protein filaments to generate movement, such as occurs in muscle contraction or the beating of cilia lining the respiratory tract. *Chemical work* is used to form bonds during chemical reactions, such as the bond formation that occurs when small molecules are used to synthesize large molecules. *Transport work* utilizes energy to move a molecule from one side of a cell membrane to another, such as occurs in active transport (the Na^+/K^+ pump, for example) or in vesicular transport (exocytosis and endocytosis).

Metabolic Rate

When the body breaks down nutrients, it either releases energy as heat or uses that energy to perform work. The amount of energy so expended per unit time is the body's **metabolic rate.** A person's metabolic rate is influenced by a number of factors, including muscular activity, age, gender, body surface area, and environmental temperature. When sitting still, for example, the rate of energy expenditure is approximately 100 kilocalories per hour; when riding a bicycle at a moderate rate, it tops 300 kilocalories per hour.

The **basal metabolic rate (BMR)** is the rate of energy expenditure of a person who has fasted for at least 12 hours and is awake, lying down, and physically and mentally relaxed; under these conditions, both metabolic rate and work performed are minimal. The BMR is usually estimated by measuring a person's rate of oxygen consumption, which correlates with the rate at which nutrients are oxidized in the body.

The BMR represents the energy requirement of performing such necessary tasks as pumping blood and transporting ions. This rate generally increases as body weight increases, because a larger mass of tissue requires greater energy expenditure for its upkeep; therefore, the BMR is usually expressed as the rate of energy expenditure per unit body weight. For adults, it averages 20–25 kilocalories per kilogram of body weight per day. Most of this expenditure is due to activity in the nervous system and skeletal muscles, which

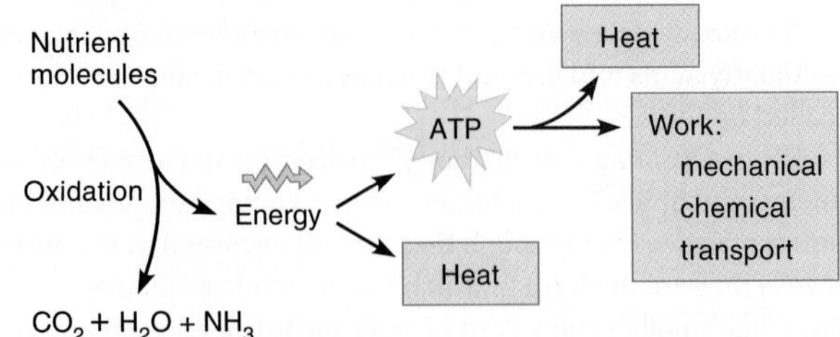

Figure 2 The forms of energy produced by the oxidation of nutrient molecules. Some of the energy liberated during oxidation is used to generate ATP, which can perform various types of work within cells; the rest of the energy takes the form of heat.

account for 40% and 20–30% of the BMR, respectively. The BMR varies from tissue to tissue; muscle tissue, for example, has a higher resting metabolic rate than adipose tissue. The BMR (per unit body weight) also varies with age: It is greater in growing children because of the energy expended in the synthesis of new tissue, and it is usually lower in the elderly than in young adults.

Apply Your Knowledge

Notice that the BMR is lowest when a person is awake, not asleep.

Negative and Positive Energy Balance

In today's society, virtually everyone is aware that body weight, diet, and exercise are interrelated. When people eat a lot of food but do not get much exercise, their body weight tends to increase; when they eat less and exercise more, their body weight decreases. These changes in body weight occur when energy input and output are not balanced.

If the body is not in energy balance—that is, if energy input and energy output are not equal—then the difference between energy input and output determines whether the amount of stored energy increases or decreases. From our balance equation, we can determine the amount of energy that is stored:

$$\begin{aligned}\text{Energy stored} &= \text{energy input} - \text{energy output}\\ &= \text{energy input} - (\text{work performed}\\ &\quad + \text{heat released})\end{aligned}$$

When a person takes in energy at a rate greater than he or she expends energy as heat and work, the quantity of stored energy increases. This condition, called *positive energy balance,* tends to be associated with increases in body weight; a net synthesis of macromolecules from absorbed nutrients occurs. (Later in this chapter we see that most excess nutrients are converted to lipids for storage.) When the rate of energy intake is less than the rate at which energy is expended as heat or work, the quantity of stored energy decreases. This condition, known as *negative energy balance,* tends to be associated with decreases in body weight. Under these conditions, a net breakdown of macromolecules (including lipid stores) provides energy for body functions.

When someone goes on a diet to lose weight, the idea is to decrease food intake and achieve negative energy balance. The same result can be attained through exercise, which increases work and heat production (energy output). To learn more about how energy is metabolized during exercise, see **Understanding Exercise: Energy Metabolism.**

Although the concept of energy balance is useful for explaining why diet and exercise affect body weight, note that a change in the body's energy content is not necessarily equivalent to a change in mass. If during a given time span 100 grams of glucose is absorbed and 100 grams of triglycerides is oxidized, then body weight does not change; however, the body's energy content decreases because 1 gram of glucose contains less energy than 1 gram of triglycerides.

Quick Check 1

1. What are the storage forms of carbohydrates and lipids?
2. Once energy is taken into the body, either it is stored or else it appears in which two other forms?

4 Energy Metabolism During the Absorptive and Postabsorptive States

We have seen that maintaining energy balance requires that energy input equal energy output. However, the body is generally not in energy balance at any given time, because the rate of energy input is determined by feeding, which is intermittent. For approximately 3–4 hours after a typical meal, nutrients are absorbed during the **absorptive state,** after which time absorption stops until the next meal. During this time, the rate of energy input generally exceeds energy output, putting the body in positive energy balance. Nutrients in the blood are plentiful. Glucose serves as the primary energy source for cells, while fats, amino acids, and excess glucose are taken up by liver, muscle, and fat cells and converted to energy storage molecules. The **postabsorptive state** corresponds to the time between meals, when nutrients are not being absorbed; during this time, the rate of energy expenditure is greater than the rate of energy intake. Energy stores are mobilized to provide the energy cells need. Whereas glucose serves as the energy source for cells in the central nervous system, other cells in the body utilize other energy sources (such as fatty acids), thereby *sparing* glucose for the central nervous system.

Energy metabolism during the absorptive and postabsorptive states can be summarized by the following rule: *During the absorptive state, energy is stored in macromolecules; during the postabsorptive state, these energy stores are mobilized.* This rule is important because it means that although we might eat different amounts of nutrients at different times, the body provides a constant supply of nutrients to cells—a supply that cannot be interrupted for even a minute because the body must expend energy continuously just to stay alive. In the following sections we see how the body performs this necessary function.

Metabolism During the Absorptive State

The absorptive state is primarily an anabolic state—that is, the majority of reactions involve synthesis of macromolecules. However, cellular metabolism differs among cell types. In this subsection we explore the typical absorptive state metabolic responses in body cells in general, in skeletal muscle cells, in liver cells, and in adipocytes (**Figure 3**).

Body Cells in General

The body's energy needs are supplied primarily by absorbed glucose, which is plentiful after a typical meal. Glucose is taken into the cells and catabolized as the body's primary fuel. Absorbed fatty acids and amino acids can also be catabolized for energy, particularly if the diet is rich in these nutrients but poor in carbohydrates. Fatty acids

UNDERSTANDING EXERCISE

Energy Metabolism

How carbohydrate, fat, or protein supplies ATP to fuel exercise depends on the intensity and duration of the exercise. We have seen that ATP is broken down to ADP during the crossbridge cycling that pulls thin filaments along thick filaments and causes muscle contraction. The instantaneous recharging of ADP back to ATP is accomplished by the donation of high-energy phosphate from creatine phosphate. By itself, however, this "energy reserve" can supply only a few seconds of energy.

Glucose (whether free in the blood or stored as glycogen in muscle and liver) is another energy source that can respond quickly because it can be broken down to produce ATP through anaerobic glycolysis. Recall that although anaerobic glycolysis liberates energy quickly, it is not a very efficient way of generating ATP. Exercising while this metabolic pathway is dominant (before oxygen delivery can be increased) means that lots of glucose will be broken down to produce rather limited amounts of usable energy. As a consequence, this metabolic pathway is appropriate for sprinting, but it cannot support long-duration exercise such as long-distance running.

The metabolic pathways that use oxygen (aerobic metabolism), such as fatty acid oxidation and the Krebs cycle, respond more slowly to the changes in metabolic rate during exercise as the activities of rate-limiting enzymes in these pathways change with the buildup or depletion of metabolic reaction products. Certain reaction products (including heat, H^+, and CO_2) cause vasodilation, which increases blood flow to exercising muscles. These products also decrease the affinity of hemoglobin for oxygen, resulting in greater oxygen unloading in the muscle tissue. Thus, during long-duration exercise, the body initially utilizes creatine phosphate for energy; then, as exercise continues, the body switches from an anaerobic/glycolysis-dominant metabolism to an aerobic/lipolysis-dominant metabolism.

If a person begins exercising relatively gradually (at a low intensity), a greater proportion of ATP is generated through aerobic lipolytic metabolism. This conserves both blood glucose—whose conservation is important because glucose is, with rare exceptions, the only fuel used by the brain—and muscle glycogen—whose conservation seems to be important in preventing fatigue during aerobic exercises. Thus, for a person who wants to burn fat during an exercise session, longer-duration, light-to-moderate exercise is better than short-term, more-intense exercise. However, building muscle mass through intense exercise helps burn fat as well, as more muscle mass equates to a higher BMR (recall that muscle tissue has a higher resting metabolism than other types of tissue).

undergo oxidation to provide acetyl CoA subunits for the Krebs cycle, and amino acids are converted to keto acids (organic acids with a carbonyl group, C=O), many of which serve as intermediates for the Krebs cycle. Amino acids can also be used to synthesize proteins.

Note, however, that proteins are not synthesized as "storage molecules." Instead, most body proteins have important structural and functional roles in cells and are continuously turned over; that is, old proteins are degraded and replaced with new ones. For this reason, the body's protein mass remains relatively stable and does not increase simply in response to the absorption of an excess of amino acids. Body cells catabolize proteins for energy only under extreme conditions, because doing so comes at the expense of losing functioning molecules.

Skeletal Muscle Cells

Like body cells in general, skeletal muscle cells take up glucose and amino acids from the blood for their own needs. However, unlike most body cells, skeletal muscle cells can convert glucose to

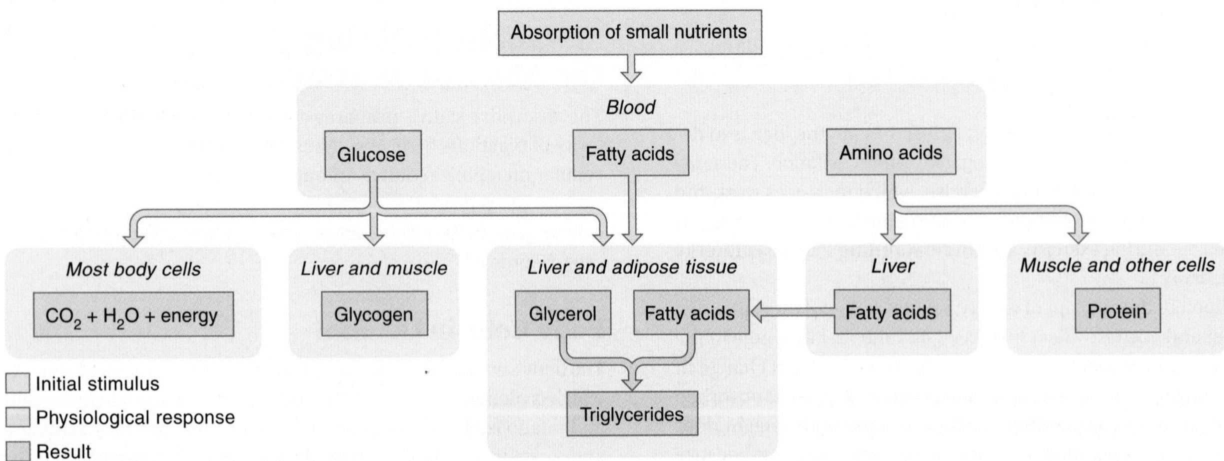

Figure 3 Major metabolic reactions of the absorptive state.

glycogen for storage. Within individual muscle cells, these glycogen stores are limited, but taken together they constitute the majority (approximately 70%) of the body's total stored glycogen.

Liver Cells

The liver converts nutrient molecules to energy stores that can subsequently be mobilized to supply energy to most cells in the body. The liver converts glucose to glycogen or fatty acids, and fatty acids to triglycerides. The glycogen is stored in the liver (which contains approximately 24% of the body's glycogen stores), whereas the triglycerides are transported to adipose tissue for storage. Between liver and skeletal muscle, the body can store a maximum of approximately 500 grams of glycogen, an amount that is sufficient to meet the body's energy demands for only a few hours. Any absorbed glucose that exceeds the quantity that is needed for energy or that can be stored as glycogen is converted first to fatty acids and then to triglycerides.

The liver also takes up amino acids. Although the liver uses some amino acids to synthesize proteins (including plasma proteins), most amino acids are converted to keto acids, many of which act as intermediates in glycolysis or the Krebs cycle and can be used for energy. Most of the keto acids, however, are used to synthesize fatty acids; thus they ultimately end up as triglycerides.

The triglycerides synthesized in the liver must be transported to adipose tissue, which is achieved by packaging the triglycerides into particles called **very low density lipoproteins (VLDLs).** Briefly, VLDLs transport triglycerides to the cells of the body. The plasma membranes of most cells contain the enzyme lipoprotein lipase, which catabolizes triglycerides at the outer surface of the VLDLs to fatty acids and monoglycerides. The fatty acids then diffuse into cells, where they can be used for energy (by most body cells) or converted back into triglycerides for storage (in adipocytes). Adipocytes have a high concentration of lipoprotein lipase on their plasma membranes and, therefore, take up most of the fatty acids transported in VLDLs.

Adipocytes

Adipocytes store energy in the form of triglycerides, or fat. Absorbed triglycerides are transported to adipocytes by chylomicrons, the smallest of the lipoproteins. Lipoprotein lipase catabolizes the triglycerides in chylomicrons in the same manner as just described for VLDLs. Excess absorbed glucose enters the adipocytes and is converted to triglycerides. In addition, triglycerides synthesized in the liver are transported to adipocytes by VLDLs for storage.

Energy Reserves

Whereas the body is limited in its ability to store energy in the form of glycogen or protein, it is practically unlimited in its ability to store energy as fats. As a consequence, triglyceride synthesis represents the final common pathway for all nutrients that are absorbed in excess of the body's needs. In most people, fat accounts for 20–30% of total body weight, but in very overweight individuals it can account for as much as 80%. To people who are weight conscious, the body's propensity for storing nutrients as fats is at best an annoyance. However, given that triglycerides contain more energy than carbohydrates or proteins, triglyceride synthesis is clearly the best way to store the most energy in the least weight.

Under normal circumstances, glycogen stores account for 1% or less of the body's total energy reserves and can supply a person's energy needs for only a few hours of quiet activity. Proteins account for 20–25% of the total energy reserves. Although large amounts of protein can be mobilized for energy without serious consequences, particularly from skeletal muscle, continual use of protein for energy is harmful and potentially fatal because it eventually compromises cellular function. Thus a significant portion of the energy contained in protein stores is, in reality, unavailable for use. Fat stores represent 75–80% of the total energy reserves and contain enough energy to sustain human life for approximately two months. For this reason, fats are absolutely essential to the body's ability to withstand prolonged periods of fasting. Table 2 tabulates the average energy stores of a healthy man weighing about 154 lbs.

Metabolism During the Postabsorptive State

Within a few hours after a typical meal, absorption of nutrients ceases. Without absorbed nutrients for energy, the body catabolizes glycogen, proteins, and fats to meet its energy needs (**Figure 4**). Thus the postabsorptive state is primarily a catabolic state. In addition, unlike most body cells, central nervous system cells rely on glucose as their sole energy source (cells of the central nervous system can also obtain energy from ketone bodies during extreme conditions). Therefore, a primary function of the postabsorptive state is to maintain plasma glucose levels. Too large a

reduction in plasma glucose can result in serious impairment of brain function, loss of consciousness, and even death.

Given the importance of maintaining a steady supply of glucose during the postabsorptive state, a question naturally arises: Because glucose is derived from the breakdown of glycogen, which is in relatively short supply (enough to last only a few hours), how is glucose made available for longer periods? The body synthesizes new glucose from amino acids, glycerol, and other breakdown products of catabolism, a process known as **gluconeogenesis.** In addition, most tissues turn almost exclusively to other energy sources, primarily fatty acids, thereby conserving glucose for use by the central nervous system; this process is called **glucose sparing.**

As in the absorptive state, cellular metabolism during the postabsorptive state differs among the types of cells.

TABLE 2 Energy Stores (as a Percentage of Total Energy Reserves) in a Healthy, Approximately 154-lb. Man

	Glycogen	Triglycerides	Proteins (mobilizable)
Skeletal muscle	71	<1	98
Liver	24	<1	2
Adipose tissue	5	99	<1
Brain	<1	0	0

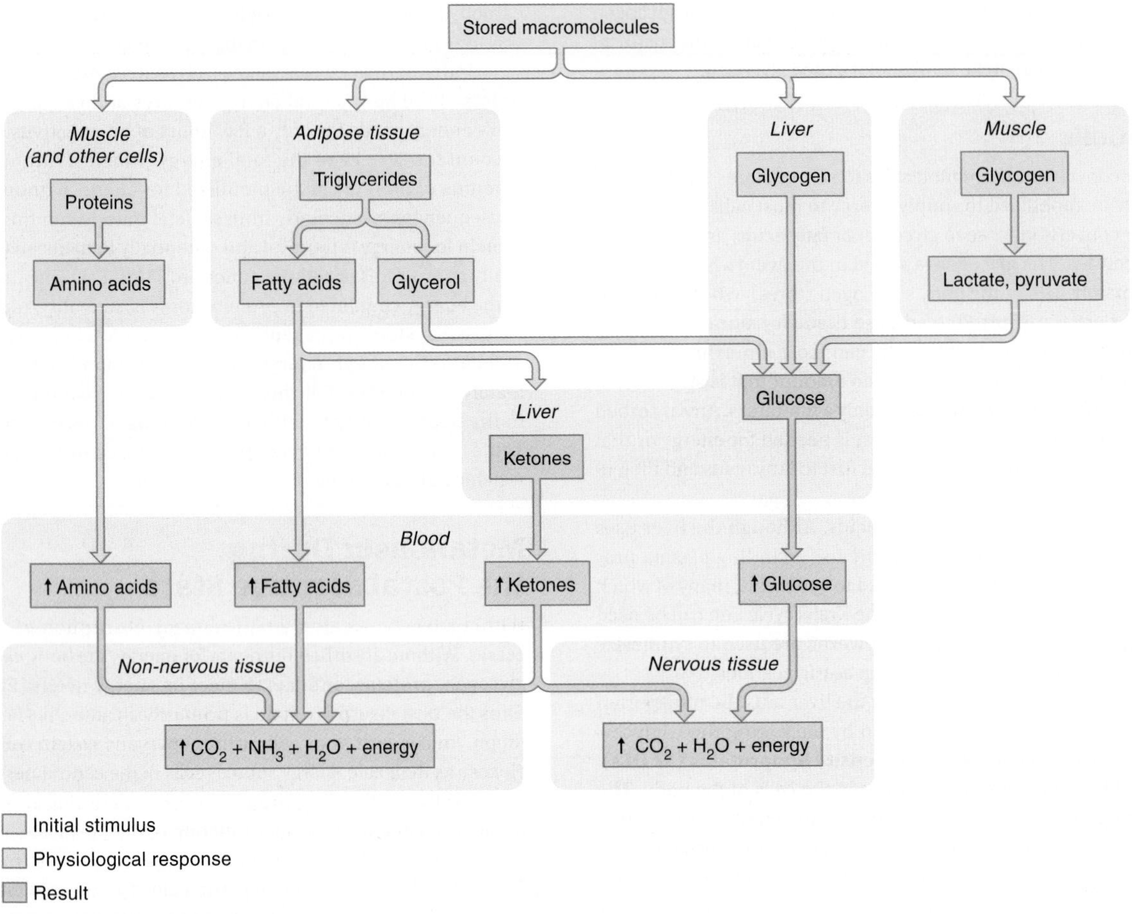

Figure 4 **Major metabolic reactions of the postabsorptive state.**

Q When the liver converts amino acids, lactate, or pyruvate to glucose, which type of process is occurring?

Body Cells in General

Most cells utilize fatty acids instead of glucose for energy, sparing the glucose for the central nervous system.

Skeletal Muscle Cells

In a skeletal muscle cell, any glucose formed from glycogen during glycogenolysis can be used for energy only within that muscle cell. Glycogen is catabolized to glucose-6-P (a glucose molecule with a phosphate group attached to the sixth carbon; it is an intermediate in glycolysis). The phosphate group cannot be removed from the glucose because skeletal muscle cells lack the enzyme (glucose-6-phosphatase) that catalyzes its removal. For glucose to be transported out of a cell, it must be in its unphosphorylated form. Thus the glucose formed by glycogenolysis in skeletal muscle cells remains in the cell and is catabolized by glycolysis to form pyruvate or lactate. Any lactate produced then travels to the liver for further processing, as described shortly.

Skeletal muscle cells can also catabolize proteins to amino acids, which are then transported into the bloodstream and carried to the liver for further processing.

Liver Cells

The liver is the primary source of plasma glucose during the postabsorptive state. Its glycogen stores can be broken down by glycogenolysis to glucose-6-P, and its supply of glucose-6-phosphatase then catalyzes the conversion of glucose-6-P to glucose. This glucose can, in turn, be transported out of the liver cells and into the bloodstream. Therefore, liver glycogen stores, unlike skeletal muscle glycogen stores, can be mobilized to provide glucose to the blood.

The liver is also the primary site of gluconeogenesis. (Some gluconeogenesis does occur in the kidneys, which also produce the enzyme glucose-6-phosphatase.) Like the glucose produced by glycogenolysis, the newly synthesized glucose is transported from the liver into the bloodstream for use by other cells in the body.

During the postabsorptive state, **ketogenesis** occurs in the liver; during this process, some of the fatty acids are converted to ketone bodies, which are then released into the bloodstream and eventually catabolized by most tissues. The production of ketones is important because during prolonged fasting, the central nervous system acquires the ability to use ketones for energy, thereby freeing it from some of its dependence on glucose.

Adipocytes

In the postabsorptive state, adipose tissue supplies fatty acids to the bloodstream as energy sources for body cells, thereby sparing glucose for the central nervous system. Adipose tissue achieves this feat by catabolizing stored triglycerides into glycerol and free fatty acids. The glycerol is also released into the bloodstream, where it travels to the liver and is catabolized by glycolysis.

Quick Check 2

1. When glucose or amino acids are absorbed in excess of the quantities oxidized or stored as glycogen or proteins, what happens to them?
2. Where is most of the body's glycogen stored? What is the storage site of most of the glycogen that can supply glucose for cells throughout the body?
3. During the postabsorptive state, most tissues use fatty acids instead of glucose as their primary energy source. Why is this preference important in whole-body metabolism?

5 Regulation of Absorptive and Postabsorptive Metabolism

The transitions between the absorptive and postabsorptive states are marked by profound alterations in the metabolic activity of tissues throughout the body. In this section we see how these metabolic changes are triggered primarily by endocrine signals involving the pancreatic hormones insulin and glucagon. In addition, epinephrine and sympathetic nerve activity play a role in this form of metabolism.

The Role of Insulin

The metabolic adjustments that occur as the body switches between the postabsorptive and absorptive states are largely triggered by changes in the plasma concentration of **insulin,** a peptide hormone secreted by beta cells located in the pancreatic islets of Langerhans. Even though insulin and the factors that influence its secretion exert numerous effects, they all share a common thread: *Insulin promotes the synthesis of energy storage molecules and other processes characteristic of the absorptive state* (**Figure 5**). In other words, insulin is an anabolic hormone. Accordingly, its secretion is stimulated by signals associated with feeding and the absorption of nutrients into the bloodstream.

Factors Affecting Insulin Secretion

During the absorptive period, insulin secretion by beta cells increases, causing plasma insulin levels to increase, and promoting many of the metabolic processes characteristic of the absorptive state. During the postabsorptive state, insulin secretion decreases, causing a decrease in the plasma concentration of insulin, which helps turn off the absorptive processes. This raises a key question: How do beta cells know when to increase or decrease insulin secretion?

Table 3 shows that several factors influence insulin secretion. Particularly important among these is plasma glucose concentration. During the absorptive period, plasma glucose levels increase as glucose is transported into the bloodstream from the GI tract. This increase stimulates insulin secretion through a direct effect of glucose on beta cells, which are sensitive to the concentration of glucose in the fluid surrounding them (**Figure 6**). Glucose enters beta cells by facilitated diffusion utilizing GLUT2 transporters. Glucose is catabolized in a process that generates ATP, which then closes potassium channels in the beta cell membrane. With less potassium moving out, the beta cell becomes

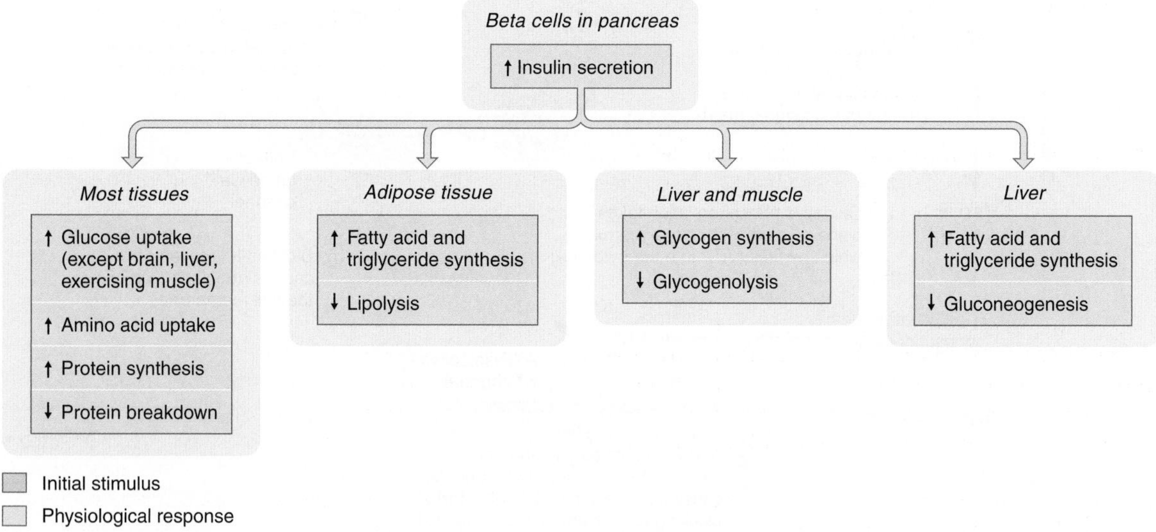

Figure 5 Actions of insulin on target tissues.

TABLE 3 Factors Affecting Insulin and Glucagon Release

Factor	Effect on insulin secretion	Effect on glucagon secretion
↑ Plasma [glucose]	Increase	Decrease
↑ Plasma [amino acids]	Increase	Increase
↑ Plasma [GIP]	Increase	Increase
↑ Parasympathetic activity	Increase	Increase
↑ Sympathetic activity	Decrease	Increase
↑ Plasma [epinephrine]	Decrease	Increase

depolarized, which opens calcium channels in the plasma membrane. Calcium then enters the cell and triggers exocytosis of insulin. During the postabsorptive period, plasma glucose levels decrease, which in turn decreases insulin secretion. Insulin secretion is influenced in a similar fashion by the plasma amino acid concentration: Increases in plasma amino acid levels cause an increase in uptake into beta cells. The amino acids generate ATP through the Krebs cycle and oxidative phosphorylation, and ATP closes potassium channels.

Hormones and input from the autonomic nervous system also influence insulin secretion. Secretion is stimulated by glucose-dependent insulinotropic peptide (GIP) and glucagon-like polypeptide 1 (GLP-1), both hormones secreted by cells in the wall of the small intestine. Parasympathetic nerve activity also increases insulin release. This factor is significant because parasympathetic activity, GIP production, and GLP-1 production all increase in response to the presence of food in the GI tract. Given that feeding occurs prior to the absorption of nutrients, these feedforward signals prepare the body for transitioning to the absorptive state by triggering insulin secretion in advance of rising plasma glucose levels. Insulin secretion is inhibited by sympathetic nervous activity and circulating epinephrine.

Actions of Insulin

Through its actions on a variety of target tissues, insulin influences almost every major aspect of energy metabolism (see Figure 5). It promotes energy storage by stimulating the synthesis of fatty acids and triglycerides in the liver and adipose tissue, glycogen in liver and skeletal muscle, and proteins in most tissues. At the same time, it opposes the catabolism of energy stores by inhibiting the breakdown of proteins, triglycerides, and glycogen, and by suppressing gluconeogenesis by the liver. In short, insulin promotes reactions associated with the absorptive state and suppresses reactions linked to the postabsorptive state.

Along with its metabolic actions, insulin affects the transport of nutrients across the membranes of all body cells except those in the liver and central nervous system. In most tissues, insulin stimulates the uptake of amino acids by cells, which facilitates the hormone's stimulatory effect on protein synthesis. There are at least four insulin-sensitive active transporters that move at least eight of the nonessential amino acids from plasma into cells, thereby decreasing amino acid levels in the plasma. Uptake of amino acids into cells varies based on cell type, with uptake into skeletal muscle cells occurring to the greatest extent.

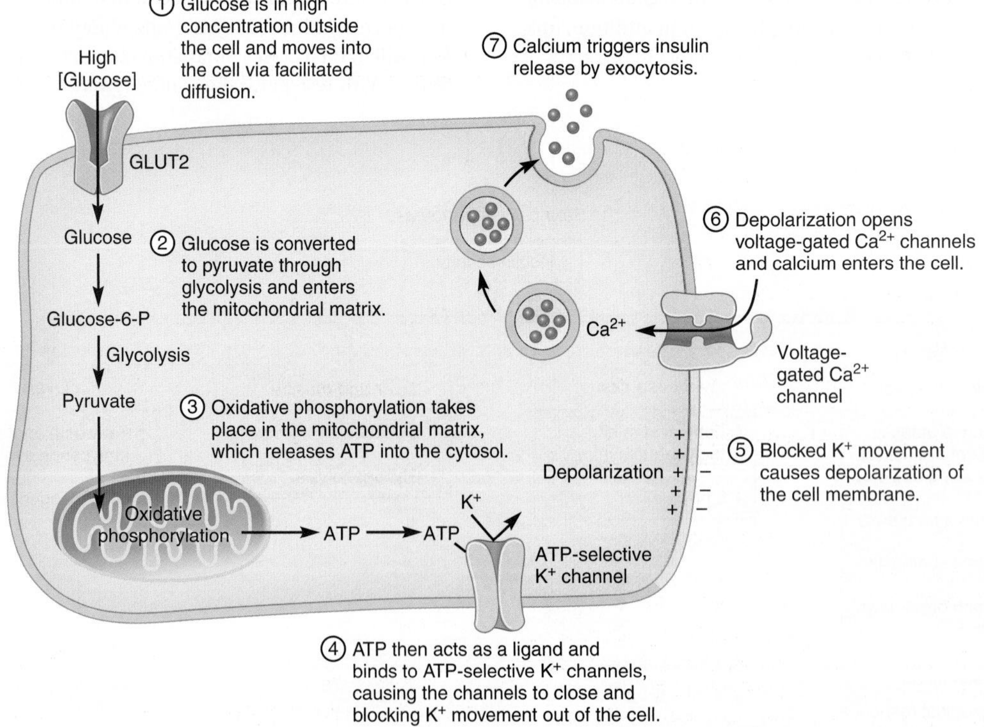

Figure 6 Actions of glucose on insulin secretion.

Insulin also stimulates the uptake of glucose in some tissues by increasing the number of glucose transport proteins in cell plasma membranes. Thirteen different glucose transporters are coded in the human genome, called GLUT1 through GLUT13. (We saw earlier that glucose entered pancreatic beta cells through GLUT2 transporters.) These transporters differ in their location, sensitivity to hormones, and specificity (some are more specific for transporting other monosaccharides, such as fructose, than for transporting glucose). Only one of these transporters, GLUT4, is sensitive to insulin. GLUT4 is a facilitated diffusion transporter for glucose and is the most important glucose transporter for whole-body metabolism. Skeletal muscle and adipose tissue cells have pools of GLUT4 transporters that are stored in vesicles in the cytoplasm. Insulin can either trigger insertion of these *stored* transporters into the plasma membrane by exocytosis or stimulate the synthesis of *new* transporters. Exercise also triggers expression of GLUT4 on the plasma membrane of skeletal muscle cells and adipocytes.

The transport of glucose in the central nervous system and the liver is not affected by insulin. The isolation of this process is critical, because when insulin levels are low during the postabsorptive state, glucose uptake by most cells is decreased, sparing the glucose for use by cells of the central nervous system, where glucose transport is not affected by insulin. In addition, during the postabsorptive state, the liver produces glucose that must be transported into the blood.

In addition to the actions shown in Figure 5, insulin has important growth-promoting effects. Although insulin by itself does not stimulate growth, it must be present in the blood for growth hormone to exert its normal effects, a form of permissiveness. This need results, at least in part, from insulin's role in promoting protein synthesis, DNA synthesis, and cell division, all of which are essential to tissue growth.

The Role of Glucagon

Insulin's actions to bring about the body's metabolic adaptations to the absorptive and postabsorptive states are reinforced by contrary changes in **glucagon,** a peptide hormone secreted by alpha cells of pancreatic islets of Langerhans. Put another way, insulin and glucagon are antagonists, hormones whose actions oppose each other: Insulin promotes processes of the absorptive state; glucagon promotes processes of the postabsorptive state.

Glucagon secretion decreases during the absorptive state and increases during the postabsorptive state. Because insulin levels are also changing with these states, the metabolic adjustments from one state to the other are orchestrated by contrary changes in plasma levels of insulin and glucagon.

Factors Affecting Glucagon Secretion

Most of the signals that stimulate the secretion of glucagon are the same signals that inhibit the secretion of insulin (see Table 3). For instance, decreases in blood glucose both stimulate glucagon secretion and inhibit insulin secretion. Likewise, glucagon secretion is stimulated by sympathetic nervous activity and epinephrine, which have a suppressive effect on insulin secretion. Some studies suggest that glucagon and insulin function as paracrines in the islets of Langerhans, with insulin inhibiting the secretion of glucagon from alpha cells, and glucagon inhibiting the secretion of insulin from beta cells. Because of the opposite controls of these hormones, plasma glucagon levels tend to rise as insulin levels fall, and vice versa.

Actions of Glucagon

Figure 7 shows that the actions of glucagon, though more limited than those of insulin, oppose insulin's actions. In the liver, glucagon promotes glycogenolysis and gluconeogenesis (which increase blood glucose levels), ketone synthesis, and breakdown of proteins, while inhibiting the opposing processes of glycogen and protein synthesis. In adipose tissue, glucagon stimulates lipolysis and suppresses triglyceride synthesis. These actions lead to glucagon's classification as a catabolic hormone. The overall effect of glucagon promotes mobilization of energy stores and synthesis of "new" energy sources (glucose and ketone bodies) that can be used by tissues; all of these actions are characteristic of the postabsorptive state.

Negative Feedback Control of Blood Glucose Levels by Insulin and Glucagon

Plasma glucose levels are normally tightly regulated by the antagonistic actions of insulin and glucagon to maintain stability. (Other hormones, described below, also regulate plasma glucose levels.) This stability is important because deviations too far from normal in either direction can have serious adverse effects on health. The

normal fasting level of glucose in the blood is 70–110 mg/dL (clinical measures usually assess blood—not plasma—glucose levels). Fasting blood glucose levels greater than 140 mg/dL constitute *hyperglycemia,* which is often indicative of **diabetes mellitus,** a serious and increasingly common

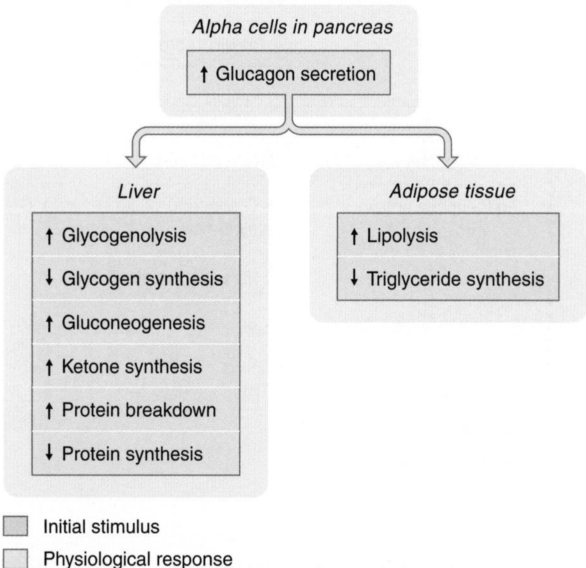

Figure 7 Actions of glucagon on target tissues.

disease involving defects in insulin production or signaling. Fasting blood glucose levels below 60 mg/dL constitute *hypoglycemia,* which has widespread deleterious effects on nervous system function because the nervous system uses glucose almost exclusively as its source of energy.

Insulin and glucagon together control plasma glucose concentration through negative feedback (**Figure 8**). An increase in plasma glucose concentration increases insulin secretion and decreases glucagon secretion from the pancreas, both of which cause a decrease in plasma glucose. Similarly, a decrease in plasma glucose concentration decreases insulin secretion and increases glucagon secretion, both of which cause an increase in plasma glucose.

Figure 8a diagrams insulin's negative feedback control of plasma glucose concentration. An increase in plasma glucose stimulates insulin secretion from pancreatic beta cells, and the actions of insulin decrease plasma glucose. Insulin decreases plasma glucose concentration in three ways: (1) by promoting the uptake of glucose into cells by increasing the expression of GLUT4 transporters in the plasma membrane; (2) by reducing the concentration of free glucose within cells by converting it to glycogen, which promotes glucose uptake by increasing the size of the glucose concentration gradient; and (3) by suppressing gluconeogenesis, thereby reducing the rate at which new glucose is released into the bloodstream. If plasma glucose concentration decreases, insulin secretion decreases, causing an increase in plasma glucose.

Figure 8b illustrates glucagon's negative feedback control of plasma glucose concentration. A decrease in plasma glucose stimulates glucagon secretion from alpha cells of the pancreas, and the actions of glucagon increase plasma glucose. Glucagon increases plasma glucose concentration by promoting gluconeogenesis and glycogenolysis in the liver, which directly increases plasma glucose concentration, and by stimulating lipolysis in adipose tissue, which provides fatty acids as an alternate energy source to glucose.

Stimulation of Insulin and Glucagon Secretion by Amino Acids

Although insulin secretion and glucagon secretion are usually affected in opposite ways by a given stimulus, some exceptions to this pattern have been noted (see Table 3). For example, an increase in plasma amino acids stimulates the release of both insulin and glucagon. Although this effect might seem counterproductive, it is not. Consider what happens following a person's consumption of a meal rich in proteins but low in carbohydrates: When the nutrients are absorbed, plasma amino acid levels rise significantly, but glucose levels either do not change or rise only slightly. The rise in amino acids stimulates insulin secretion, which promotes increases in amino acid uptake by cells. At the same time, the rise in insulin tends to promote a decrease in plasma glucose. Because the plasma glucose level was already near normal, this change is inappropriate and potentially dangerous. However, the rise in plasma amino acids also stimulates the secretion of glucagon, which tends to promote an *increase* in plasma glucose. (Note that glucagon does not affect amino acid uptake.) When amino acids are absorbed with significant amounts of glucose, as occurs in a typical diet, the effect of insulin prevails over that of glucagon because insulin secretion is stimulated by amino acids and glucose, whereas these two stimuli affect glucagon secretion in opposite ways.

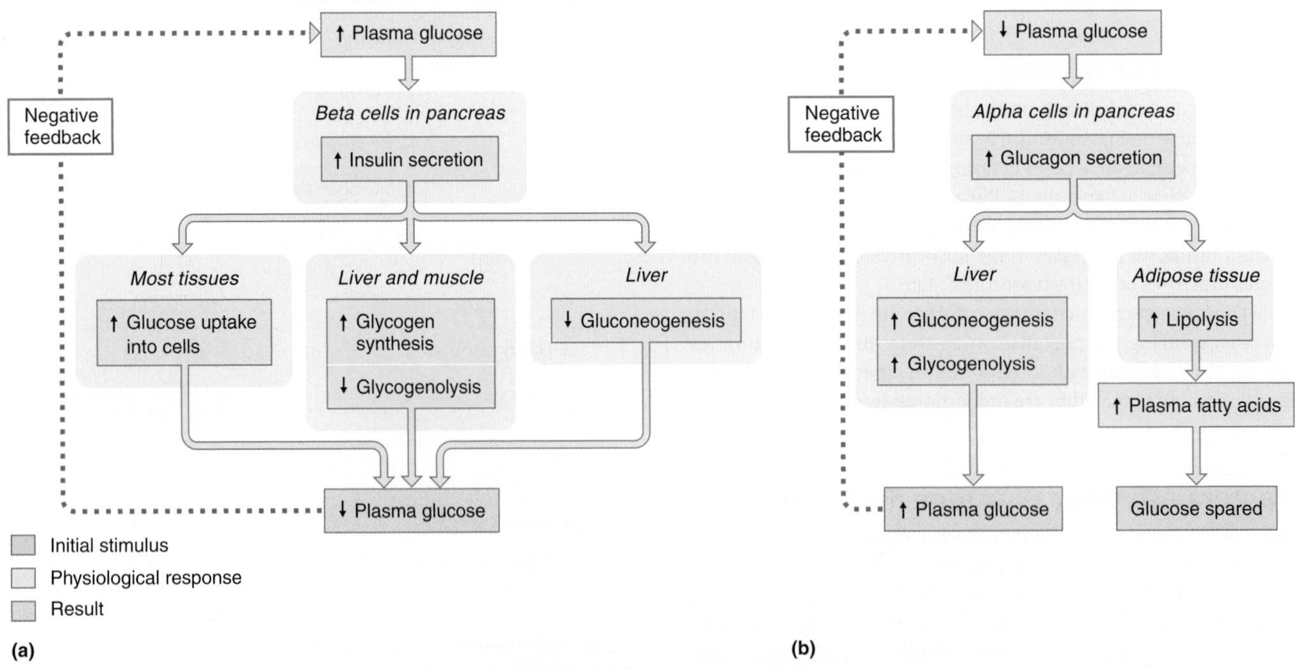

Figure 8 Regulation of plasma glucose concentration. (a) Negative feedback control of plasma glucose by insulin. **(b)** Negative feedback control of plasma glucose by glucagon.

Effects of Epinephrine and Sympathetic Nervous Activity on Metabolism

Table 3 indicates that the sympathetic nervous system and epinephrine suppress insulin secretion and stimulate glucagon secretion, thereby indirectly promoting metabolic adjustments to the postabsorptive state. Sympathetic neurons and epinephrine also directly affect metabolism of certain target tissues.

The postabsorptive period is characterized by decreased plasma glucose levels, which act directly on alpha and beta pancreatic cells to increase glucagon secretion and decrease insulin secretion, respectively. In similar fashion, a decrease in plasma glucose acts directly on *glucose receptors* in the central nervous system to raise the level of activity in sympathetic neurons, which triggers a rise in epinephrine secretion by the adrenal medulla; the resulting increase in plasma epinephrine acts on the liver to increase glycogenolysis and gluconeogenesis, on skeletal muscle to increase glycogenolysis, and on adipose tissue to increase lipolysis. Similar actions are promoted by sympathetic neural input to the liver and adipose tissue. (Skeletal muscle cells do not receive input from sympathetic neurons.)

Although sympathetic control of metabolism plays a role in adapting the body to the postabsorptive state, under normal circumstances its influence is relatively minor compared to that of insulin and glucagon. The major importance of the sympathetic influence on metabolism becomes apparent in the body's reaction to *stress*—a general term for any condition that actually or potentially poses a serious challenge to the body's ability to maintain homeostasis. (Stress includes both physical conditions, such as dehydration, hemorrhage, infection, exposure to temperature extremes, trauma, or severe exercise, and psychological states, such as pain, fear, or anxiety.) Activation of the sympathetic nervous system by *stressors* (things that cause stress) triggers the familiar constellation of fight-or-flight responses (accelerated heart rate, generalized vasoconstriction, dilation of respiratory airways, and so on); it also

FOCUS ON DIABETES

Diabetes Mellitus

In diabetes mellitus, insulin regulation of energy metabolism and blood glucose levels is either impaired or altogether absent. The disease has two basic forms: (1) *type 1 diabetes mellitus,* previously known as *insulin-dependent diabetes mellitus* (IDDM), or *juvenile-onset diabetes,* which usually appears before age 20 and accounts for 5–10% of all cases, and (2) *type 2 diabetes mellitus,* previously known as *non-insulin-dependent diabetes mellitus* (NIDDM), or *adult-onset diabetes,* which usually appears after age 40 and accounts for the vast majority of cases. In type 1 diabetes mellitus, insulin secretion is reduced or absent, usually because of a reduction in the number of active pancreatic beta cells; in type 2 diabetes mellitus, the primary defect is a reduction in target cell responsiveness to insulin.

The hallmark of both forms of diabetes mellitus is a persistent hyperglycemia, which is an expected consequence of reduced insulin activity. Hyperglycemia arises in part due to reduced glucose uptake and utilization by many tissues, and in part due to increased glucose output by the liver, which results from increased gluconeogenesis and glycogenolysis. Frequently these effects are exacerbated by abnormally high plasma glucagon levels. Although hyperglycemia normally has a *suppressive* effect on glucagon secretion, glucagon secretion is often *elevated* in diabetes because the glucose permeability of alpha cells in the pancreas (which secrete glucagon) is insulin dependent. A lack of insulin hampers the ability of glucose to enter these cells, which "tricks" them into behaving as if the glucose level is lower than it actually is. (Recall that glucagon secretion is stimulated when plasma glucose levels fall.)

Some other metabolic abnormalities are usually more pronounced in type 1 diabetes mellitus than in type 2 diabetes mellitus. For example, overstimulation of lipolysis and suppression of triglyceride synthesis (due to a lack of insulin or an excess of glucagon) can result in *hyperlipidemia,* an excess of fatty acids and other lipids in the blood. Excess utilization of fatty acids for energy can also lead to *ketosis,* elevated ketone levels in the blood. A lack of insulin also interferes with protein synthesis, resulting in excessive protein catabolism. This change hampers normal tissue repair and causes muscle weakness and retardation of growth in children.

Adverse consequences of diabetes are many and varied, and occur secondary to the hyperglycemia and metabolic disturbances associated with the disease. Elevation of blood glucose results in *glucosuria* (the presence of glucose in urine) and excessive urine output caused by osmotic forces exerted by glucose in kidney tubules. In fact, the very name of the disease refers to these symptoms; the Greek roots of *diabetes mellitus* mean "sweet siphon." The high rate of water loss via the urine predisposes individuals with diabetes to dehydration and loss of electrolytes (Na^+, K^+, and others) from the plasma. Unless this loss is compensated for by water intake (which also must be abnormally high) or other measures, dehydration can quickly lead to circulatory collapse and death. Electrolyte disturbances can result in neuromuscular problems. Excess ketone production, which is a greater problem in type 1 diabetes mellitus than in type 2 diabetes mellitus, is also potentially dangerous because some ketones are acids and can cause a drop in blood pH—a particular form of metabolic acidosis called *ketoacidosis.* This condition depresses central nervous system function and can ultimately lead to coma and death.

Because beta cells are unable to secrete insulin in type 1 diabetes mellitus, the only treatment available for this form of the disease is administration of insulin, which must be given by injection. In contrast, type 2 diabetes mellitus can usually be controlled or even reversed through weight reduction and exercise, which increases tissues' insulin responsiveness. Use of drugs that stimulate insulin secretion may also be beneficial.

elevates plasma glucose levels (due to increased gluconeogenesis and glycogenolysis in the liver) and plasma levels of fatty acids and glycerol (due to increased lipolysis in adipocytes). These actions make fuel more readily available to cells, in the process helping prepare the body for the strenuous physical activity inherent in the fight-or-flight response. The increased availability of fuel also prepares the body for other activities requiring energy, such as tissue repair or fighting infections. Other components of the body's response to stress are described later in this chapter.

Clearly, whole-body metabolism is highly regulated, for important reasons. Notably, diseases of body metabolism affect all organ systems, as is most apparent in *diabetes mellitus* (see **Focus on Diabetes: Diabetes Mellitus**).

Quick Check 3

1. The concentration of which hormone, insulin or glucagon, is increased during the absorptive period?
2. Sympathetic nervous activity and epinephrine promote metabolic reactions characteristic of which state, the absorptive state or the postabsorptive state?
3. For each of the following processes, indicate whether it is promoted by insulin or by glucagon: gluconeogenesis, glucose uptake by cells, glycogenolysis, glycogen synthesis, catabolism of energy stores, protein synthesis, a decrease in blood glucose levels, triglyceride synthesis, lipolysis.

6 Thermoregulation

When nighttime comes to the desert, the snakes, lizards, and insects that were active in the noonday sun begin to sink into a state of relative torpor; many are barely able to move at all. This change occurs because the falling temperature causes these animals' bodies to cool, which slows down biochemical reactions and other metabolic processes. By contrast, humans and other mammals are less affected by changes in the ambient temperature because they have the ability to maintain their body temperatures within a fairly narrow range through **thermoregulation.** Animals with this ability are said to be *homeothermic*, whereas those lacking this ability are called *poikilothermic*.

Temperature Balance

All living things produce heat as a by-product of metabolism, but humans (like all other homeothermic animals) are able to control body temperature by regulating the rates at which heat is produced and lost from their bodies. To maintain a normal body temperature, the rate of heat production must be balanced against the rate of heat loss. In negative heat balance, heat loss exceeds heat produced; as a result, body temperature falls below normal, a condition called **hypothermia.** People who are trapped in snowstorms or swept from a boat into icy waters, for example, are vulnerable to hypothermia. Such misfortunes can quickly lead to stupor, loss of consciousness, multiple-organ failure, and ultimately death. In positive heat balance, heat produced exceeds heat loss and body temperature increases to levels above normal; this condition, called **hyperthermia,** can lead to loss of consciousness, convulsions, respiratory failure, and death. Adverse effects begin to appear when body temperature approaches 41°C (106°F); a temperature of 43°C (109°F) or higher is usually fatal. The ability to maintain normal body temperature depends on the adequacy of heat transfer with the environment.

Mechanisms of Heat Transfer Between the Body and the External Environment

Under most conditions, the body loses heat to the environment because the surrounding temperature is often lower than body temperature. When the rate of heat loss equals the rate of heat generation, body temperature does not change. Generally speaking, heat is lost by four mechanisms: (1) radiation, (2) conduction, (3) evaporation, and (4) convection.

In **radiation,** thermal energy is transferred from the body to the environment in the form of electromagnetic waves. A general law of physics states that all objects emit and absorb these waves, albeit to varying degrees. When an object is warmer than its surroundings, it loses heat by emitting more energy than it absorbs. For example, when you are outside in the cool fall air, you emit radiant energy to the air molecules around you, making your body colder. By contrast, if an object is cooler than its surroundings, it gains heat by absorbing more energy than it emits. For example, if you are sitting by a campfire, your body absorbs radiant heat and becomes warmer.

Conduction is the transfer of thermal energy between objects that are in direct contact with each other. As in radiation, heat is always transferred from the warmer object to the cooler object. When you touch cold metal, for example, you feel colder because thermal energy is transferred directly from your skin to the metal.

In **evaporation,** heat is lost from an object through the evaporation of water from its surface. For water to evaporate from your body, the water must be converted from liquid form to gaseous form, a process that requires thermal energy obtained from the body. Water evaporates from the skin, the lining of the airways, and other moist surfaces such as the lining of the mouth. This kind of **insensible water loss** occurs continually without your being aware of it (hence the name). Your body also loses water through the evaporation of **sweat,** a salt-containing solution secreted by numerous small **sweat glands** in the skin. Unlike insensible water loss, which occurs continuously, sweat production is regulated according to the body's needs. When increased heat loss is desirable, sweat production increases. As a result, more water evaporates from the skin surface, carrying thermal energy away from the body.

When the environmental temperature is warmer than the body temperature, radiation and conduction transfer heat *into* the body. This transfer adds to the heat generated by the body itself, boosting the need for heat loss. During such circumstances, the body must rely on *evaporation* to carry heat away by increasing the production of sweat. Sweating cools the body under these conditions because water continues to evaporate even when it is cooler than its surroundings, assuming that the humidity of the surroundings is not too high. In a humid environment, sweating is not as efficient as in drier air because water cannot evaporate into already water-saturated air. To facilitate thermoregulation, sweat requires the process of evaporation; that is, dripping sweat does not cool the body. Thus humidity contributes to the *heat index*, when meteorologists report that the air feels hotter than it is.

Convection—the transfer of heat from one place to another by a moving gas or liquid—contributes to heat loss on a windy day. On a still day, the air that is closer to your skin warms up as it absorbs heat from your body's surface. This warmer air forms a kind of "blanket" around you that slows down the rate of heat loss by conduction. Because the air in this protective layer contains moisture that has evaporated from your skin, it tends to have a higher humidity than the surrounding air. The presence of this moisture near the skin reduces the rate of evaporative heat loss. In contrast, when the surrounding air is moving, as on a windy day, the thickness of the protective "blanket" of air is reduced, and conductive and evaporative heat loss both increase. Thus convection is responsible for the *wind chill factor* reported by meteorologists on cold, windy days.

Regulation of Body Temperature

The body's thermoregulatory efforts maintain the **core temperature,** the temperature within internal structures, including those of the central nervous system and abdominal and thoracic cavities (**Figure 9**). This temperature is normally regulated at approximately 37°C (98.6°F). Within the hypothalamus is the body's **thermoregulatory center,** which contains both heat-losing and heat-promoting centers. Input about the core temperature is transmitted to the thermoregulatory center from **central thermoreceptors** that include temperature-sensitive neurons within this region of the brain, in other areas of the central nervous system, and in other internal organs. Input of changes in core temperature is necessary to initiate the thermoregulatory responses that return the core temperature to normal. Other thermoreceptors, called **peripheral thermoreceptors,** are located in the skin; they detect the temperature of the skin, which is usually well below the core temperature and is more variable. Our bodies do not regulate skin temperature, but information about skin temperature enables us to compensate for changes in environmental temperatures by making behavioral adjustments, such as dressing appropriately or avoiding extreme temperatures altogether.

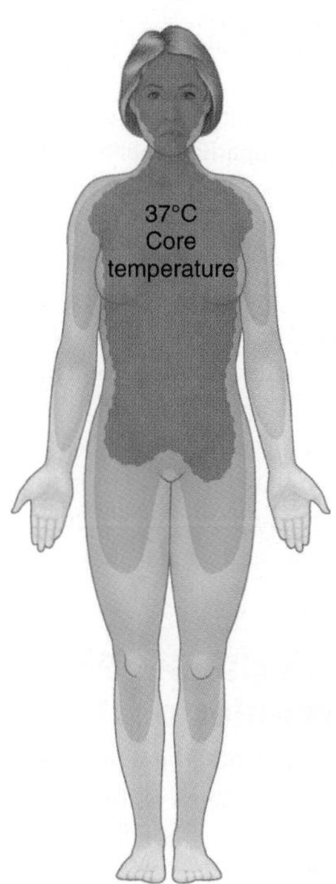

Figure 9 Core body temperature. Body temperature is maintained near 37°C in core regions of the body, including the thoracic and abdominal cavities, cranium, and most proximal areas of the limbs. Note that temperature decreases from the 37°C when moving from the core to the skin.

Thermoregulation in the Thermoneutral Zone

The primary mechanism for regulating body temperature is to vary the amount of blood flowing to the skin, where thermal energy in the blood can be exchanged with that in the environment (the heat actually moves from blood to cutaneous tissue and then out of the body). When body temperature decreases, blood flow to the skin decreases, so that the blood loses less of its heat to the environment. Likewise, when body temperature increases, blood flow to the skin increases, so that the blood can lose more of its heat to the environment. Alterations in blood flow to the skin are sufficient to maintain body temperature when the environmental temperature is maintained within a narrow range called the **thermoneutral zone** (25–30°C).

The sympathetic nervous system regulates cutaneous blood flow based on input from the thermoregulatory center. Arteriole radius is regulated by sympathetic activity, with increases in sympathetic activity causing vasoconstriction. The arterioles to skin follow this same rule. Thus, when body temperature decreases, sympathetic activity to the arterioles increases, causing vasoconstriction and thereby decreasing cutaneous blood flow to conserve body heat. In contrast, when body temperature increases, sympathetic activity decreases, causing vasodilation and an increase in blood flow to the skin, thereby transferring heat to the environment. When the environmental temperature is outside the thermoneutral zone, other thermoregulatory mechanisms must be activated to maintain body temperature.

Heat Generation in a Cold Environment

When the environmental temperature drops to less than 25°C, decrease in cutaneous blood flow alone cannot prevent body temperature from falling. In this circumstance, the heat-promoting center of the hypothalamus communicates to effector organs to stimulate shivering and decrease sweat production (**Figure 10**).

Shivering is the primary mechanism by which our bodies can increase heat production for the purpose of increasing body temperature, a process called **shivering thermogenesis.** Shivering involves the rapid rhythmical contraction of skeletal muscles and is controlled by a spinal reflex. As the muscles contract, they generate heat. Human infants and hibernating mammals are also capable of generating heat through **non-shivering thermogenesis.** Infants have a special form of adipose tissue, called *brown adipose tissue,* that generates heat through the uncoupling of the electron transport

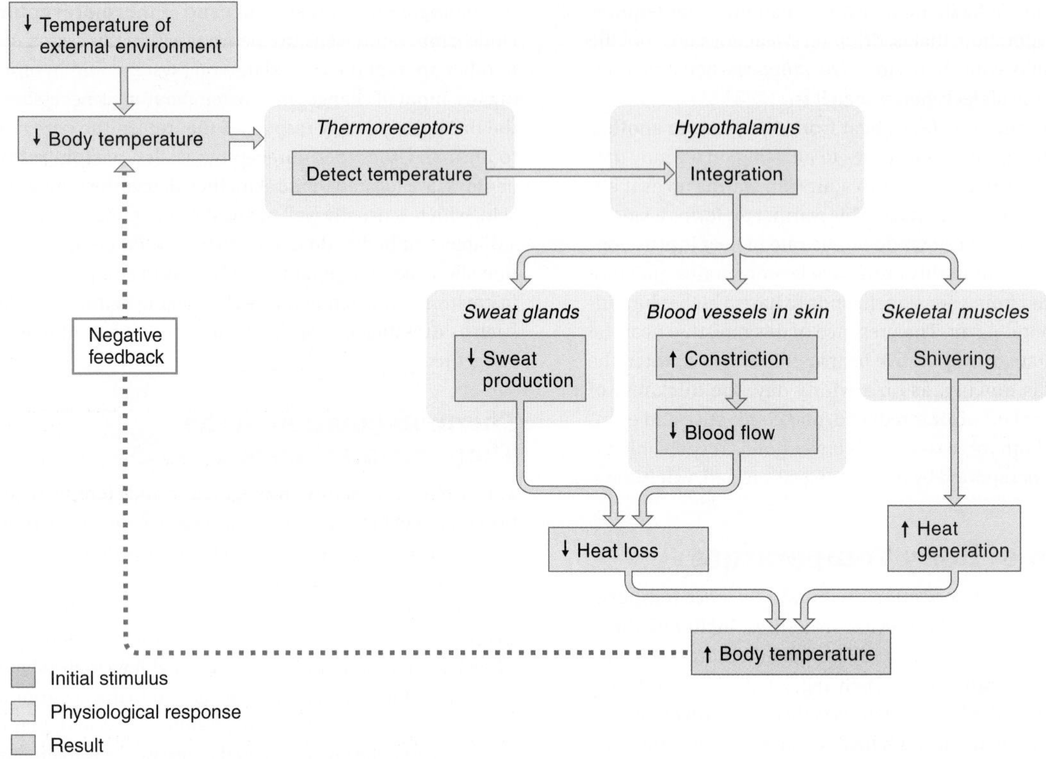

Figure 10 Events occurring in the body during thermoregulation.

chain from oxidative phosphorylation. Thus the energy released by electrons is "lost" as heat instead of being harnessed to synthesize ATP. By comparison, adult humans have little, if any, brown adipose tissue.

Heat Loss in a Warm Environment

When environmental temperature increases above the thermoneutral zone, the body must respond with more than cutaneous vasodilation if it is to remove enough heat to bring body temperature back toward normal. Specifically, the body produces sweat for evaporative heat loss.

The average person has approximately 2.5 million sweat glands located in skin throughout the body, except in the lips, nipples, and external genitalia. Two types of sweat glands are distinguished: (1) **eccrine glands** (the more common), which are located all over the body but especially in the forehead, palms of the hands, and soles of the feet, and (2) **apocrine glands,** which are located primarily in the axillary region (arm pits) and the anal-genital region. Eccrine glands empty into pores at the surface of the skin, whereas apocrine glands empty into hair follicles. Eccrine glands are active at birth, but apocrine glands do not become active until puberty. The amount of sweat produced by both types of glands depends on the body temperature and level of sympathetic activity.

Eccrine sweat glands produce a primary secretion of water, sodium, and chloride, plus a trace amount of potassium (**Figure 11**). As this fluid makes its way through the duct leading to the skin surface, the sodium and chloride are actively reabsorbed, and water follows the same path by osmosis. However, the quantity of water and electrolytes reabsorbed depends on the rate of sweat production; as the rate increases, less is reabsorbed and more is secreted outside the body. The sympathetic nervous system controls the rate of sweat production; as sympathetic activity increases (whether due to a warm environment or activation of the fight-or-flight response), sweat production increases. Sympathetic innervation of sweat glands is atypical, in that the sympathetic postganglionic neurons secrete acetylcholine instead of norepinephrine.

Apocrine sweat glands produce sweat in a similar manner, but proteins and fatty acids are also found in the sweat produced via this mechanism. Proteins and fatty acids secreted on the skin surface provide nutrition that allows bacteria to grow. As the bacteria grow, they generate waste products that create the odor associated with sweat in the axillary and genital areas.

Alterations in the Set Point for Thermoregulation: Fever

During an infection, certain white blood cells produce cytokines that function as **pyrogens,** chemicals that induce fever. A normal response of the immune system, this effect promotes several immune responses that fight the infection. Thus fever is actually considered beneficial because it enhances the body's ability to defend itself.

Pyrogens induce a fever through actions on the thermoregulatory center, adjusting the temperature to be maintained by thermoregulatory processes to a higher level. The body responds by

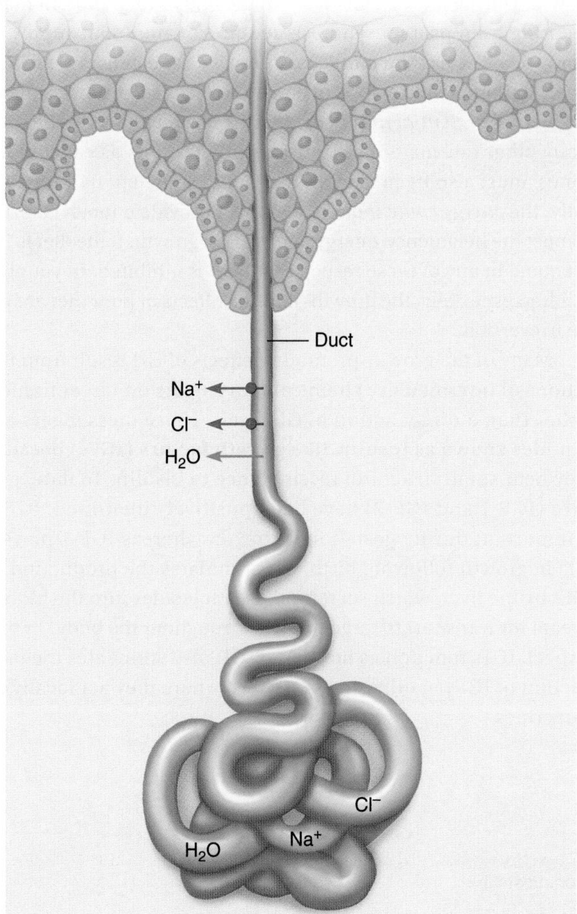

Figure 11 Eccrine sweat gland.

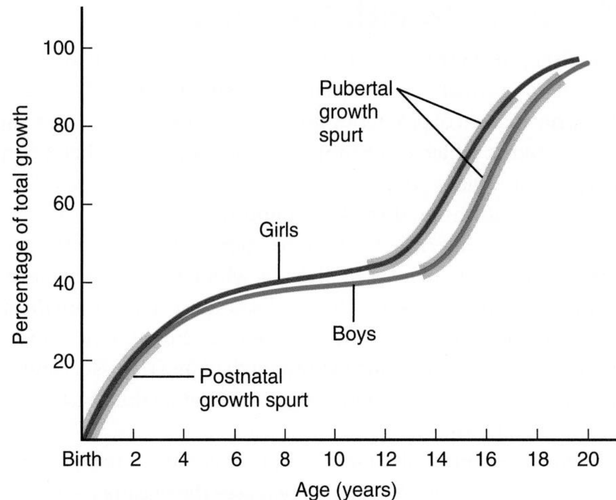

Figure 12 **A representative human growth curve.** Note the postnatal and pubertal growth spurts.

Q According to this graph, who enters puberty at an earlier age—girls or boys?

increasing heat production and minimizing heat loss. As a result, a person with an infection tends to appear white (due to decreased blood flow to the skin) and to shiver. The person may also feel cold even though he or she is warm because of the contradictory neural signals sent during this situation.

7 Hormonal Regulation of Growth

Although feeding (or the lack of it) is certainly a significant factor in the regulation of the body's overall metabolism, it is not the only one. In this section we explore the actions of hormones that play little (if any) role in everyday adjustments to feeding and fasting but nevertheless exert important influences on energy metabolism as they promote growth.

Body Growth

During their first two years of life, humans experience a dramatic increase in height and body weight, a phenomenon called the *postnatal growth spurt* (**Figure 12**). After age 2 years, growth continues at a slower rate until the beginning of adolescence (about age 11 years for girls and age 13 years for boys), at which time another period of rapid growth, known as the *pubertal growth spurt,* begins. At the end of adolescence, which occurs in the late teens, growth stops, and individuals attain their full adult stature. Thereafter, no further increase in height is possible, although obviously a person's weight may increase. Unless otherwise noted, in this text the term *growth* refers to the bodily changes that normally accompany increases in height only.

During periods of growth, both the size and number of cells in the body's soft tissues (nonbony tissues such as skin and muscle) increase, as does the length and thickness of bones. Observed increases in height are mostly attributable to increases in the length of bones in the legs and vertebral column. Lengthening of long bones in the limbs (the femur of the thigh or the humerus of the arm, for instance) is largely responsible for the changes in body proportion that accompany growth.

Body growth during childhood is regulated primarily by hormones, but it is also influenced by a person's genetic makeup, diet, and other factors such as disease or stress. Many of the bodily changes occurring during growth are attributable to the actions of **growth hormone (GH),** a peptide hormone secreted by the anterior pituitary. Other hormones that are essential for normal growth include insulin, the thyroid hormones, and the sex hormones (androgens and estrogens), which are especially important during the pubertal growth spurt. In addition, growth of various organs and tissues is influenced by numerous *growth factors* and *growth-inhibiting factors* that are usually specific to certain types of tissues and act locally as paracrine or autocrine agents. *Nerve growth factor,* for example, promotes elongation and proliferation of the axons and dendrites of neurons.

In the following subsection, we concentrate on the actions of growth hormone and the factors influencing its secretion; the influence of other hormones on growth is discussed shortly.

Effects of Growth Hormone

In children, GH exerts several effects on bones and soft tissues that result in body growth. In adults, it demonstrates many of the same effects, but instead of promoting growth it maintains bone mass and *lean body mass,* which is the proportion of body weight that is contributed by muscle (as opposed to fat).

Growth hormone directly promotes growth in two ways: It stimulates protein synthesis, thereby increasing cell size **(hypertrophy),** and it stimulates cell division, which results in increased cell number **(hyperplasia).** The results of these actions are *linear growth* (increased height) due to the elongation of bones, an increase in lean body mass due to the growth of muscle tissue, and an increase in the size of individual organs, including the heart, lungs, kidneys, and intestines.

Growth hormone also indirectly promotes a number of actions that affect growth (**Figure 13**). GH increases the plasma concentrations of glucose, fatty acids, and glycerol by inhibiting glucose uptake in adipose tissue and skeletal muscle, by stimulating lipolysis in adipose tissue, and by stimulating gluconeogenesis in the liver. Increasing plasma levels of these nutrients make energy more readily available to tissues, which must expend energy to grow. GH also promotes increased uptake of amino acids by cells in muscle and other tissues, which facilitates protein synthesis.

Growth hormone alone cannot ensure normal growth; an adequate diet is clearly essential in providing the raw materials for growth. During growth, increases in the body's total protein mass resulting from protein synthesis require an abundant supply of amino acids, which are most easily provided by eating protein-rich foods. Certain amino acids can be synthesized if they are lacking in the diet, but *essential amino acids* must be consumed in the diet. Many other raw materials for tissue growth, such as calcium for bones, must also be present in the diet in sufficient quantity. Finally, the *energy content* of the diet must provide enough calories to meet the heightened energy demands of growth. If the diet is inadequate in any of these respects, growth is inhibited. In younger children especially, the growth-stunting effects of poor diet are often irreversible.

Many of the growth-promoting effects of GH result from the actions of intermediary chemical messengers on target tissues, rather than a direct action of GH itself. These messengers are peptides known as **insulin-like growth factors (IGFs)** because they bear some structural resemblance to insulin. To date, two IGFs (IGF-1 and IGF-2) have been positively identified. IGF-2 is important during gestational growth, whereas IGF-1 plays a role in growth following birth. GH stimulates the production of IGFs by the liver, which secretes these molecules into the bloodstream for transport to target tissues throughout the body. In this respect, IGFs function as hormones. GH also stimulates the production of IGFs in other target tissues, where they act locally as paracrines.

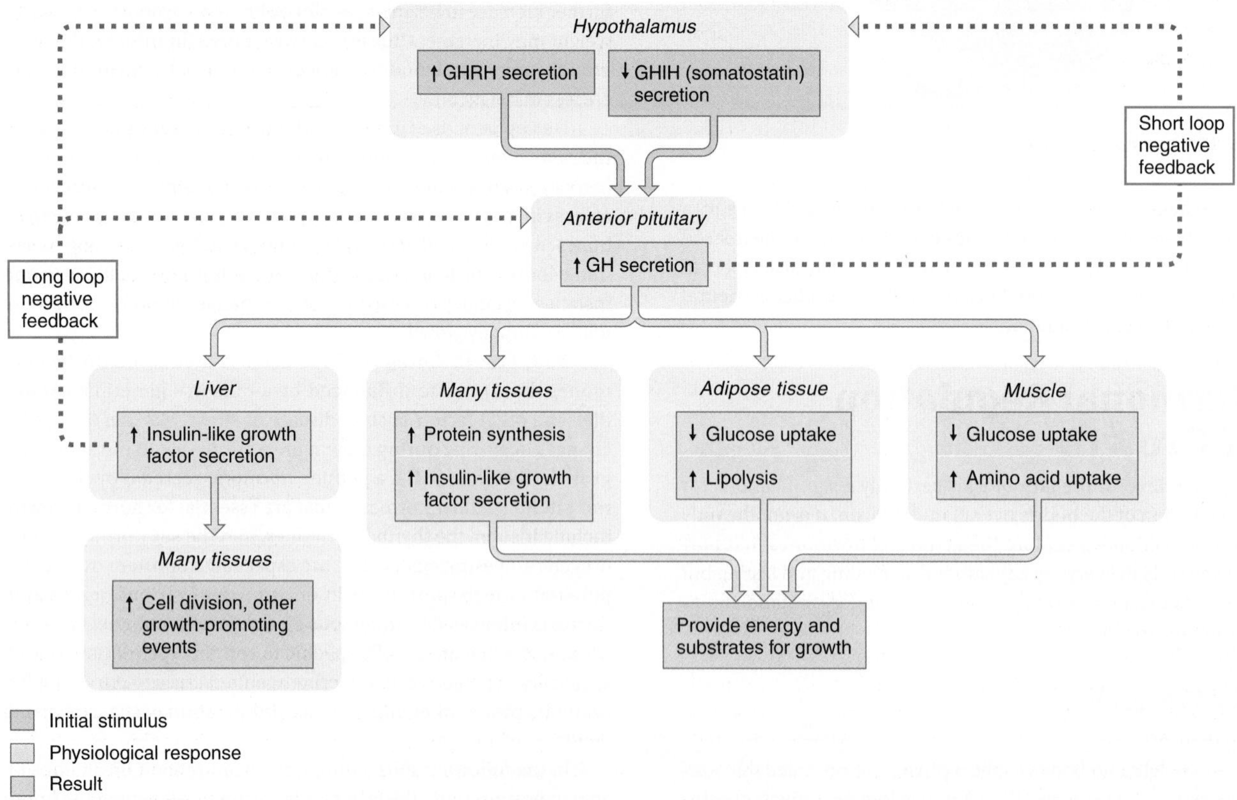

Figure 13 **Growth hormone secretion and actions on target tissues.**

Factors Affecting Growth Hormone Secretion

Secretion of GH by the anterior pituitary is regulated by two hypothalamic hormones: **growth hormone releasing hormone (GHRH),** which stimulates GH secretion, and **growth hormone inhibiting hormone (GHIH;** also known as **somatostatin),** which inhibits GH secretion (see Figure 13). Although the relative importance of these two hormones is unclear, variations in GH secretion are likely triggered primarily by GHRH, with GHIH playing a relatively minor role. As is true with other anterior pituitary hormones, GH secretion is regulated through negative feedback loops. GH limits its own secretion via short loop negative feedback on the hypothalamus. Plasma IGFs also exert long loop negative feedback controls on the hypothalamus and anterior pituitary to inhibit GHRH and GH secretion, respectively.

GHRH secretion is regulated by neural input of various types to the hypothalamus. Secretion is affected by plasma nutrient concentrations; specifically, decreases in plasma glucose or fatty acid levels, or increases in plasma amino acid levels, stimulate GHRH secretion. Because it promotes changes in the opposite direction (by reducing glucose uptake and increasing lipolysis and amino acid uptake), GH acts by negative feedback to limit variations in these nutrient concentrations in plasma. GH secretion is also stimulated in response to exercise, stress, or sleep. A boost in GH during exercise or stress is useful because it tends to counteract reduced plasma levels of glucose and fatty acids, thereby helping to maintain a steady supply of these much-needed energy sources. The significance of heightened GH secretion during sleep is not understood. GH secretion is also subject to a circadian rhythm that is mediated by neural input to the hypothalamus, with GH levels increasing at night and decreasing by day. (Secretion reaches its peak about 1–2 hours after the onset of sleep.)

Average daily plasma levels of GH also vary with a person's age. These levels generally reach a maximum during puberty, but then decline with age. Decreased GH levels are thought to be at least partially responsible for some signs of aging, such as decreased muscle mass and increased body fat.

Bone Growth

Because stimulation of bone growth is an important part of growth hormone's actions, it is appropriate to consider the nature of bone in some detail here. Bone is also an important reserve for calcium, which can be liberated and moved into the bloodstream when plasma calcium levels decrease.

To support the weight of the body and to withstand the forces placed on it by contracting muscles, bone must be strong but not brittle. Crystals of calcium phosphate in a form known as *hydroxyapatite* $[Ca_{10}(PO_4)_6(OH)_2]$ give bone a mineral component that helps it withstand compressive forces (that is, "squeezing" or "crushing" forces). **Osteoid,** an organic component that consists of collagen fibers embedded in a gel-like substance, gives bone its ability to withstand tensile or stretching forces, making it less prone to fracture.

Despite its nonliving organic and mineral components, bone is a dynamic living tissue that contains cells. The dynamic nature of bone is evident not only in its ability to grow during childhood but also in the fact that bone can heal following a fracture and adapt its structure in response to forces placed on it. In a person who engages regularly in heavy lifting, for instance, the weight-bearing bones gradually increase in thickness and strength. In a person who is sedentary or bedridden, bone mass diminishes over time. Such restructuring of bone is called *remodeling,* and it is critical to the body's ability to regulate plasma calcium levels.

Central to the remodeling of bone are mobile cells known as osteoblasts and osteoclasts, both of which are found on the outer surfaces and inner cavities of bone tissue (**Figure 14**). **Osteoblasts,** or "bone makers," are responsible for building up the mass of bone tissue, a process called **deposition. Osteoclasts,** or "bone breakers," are responsible for breaking down bone tissue, a process called **resorption.** When the activity of osteoblasts exceeds that of osteoclasts—that is, when deposition exceeds resorption—bone growth occurs.

Osteoblasts initiate bone deposition by laying down the osteoid, which is followed by deposition of calcium phosphate, a process called **calcification.** The mechanism of calcification, which takes several days, is poorly understood, but it seems that interstitial calcium is attracted to the osteoid and calcifies it. As an osteoblast works to build bone, it eventually becomes immobilized within the surrounding tissue and is transformed into an **osteocyte,** another type of cell that maintains the surrounding matrix but no longer actively lays down new bone tissue. An osteocyte is distinguishable from other bone cells in that it possesses long, filamentous processes that extend through channels in the bone tissue called *canaliculi.* These cell processes come into contact with processes belonging to other osteocytes or with nearby osteoblasts, such that the cells can communicate with one another through gap junctions. This communication is important because it allows cells in the interior of the bone tissue to exchange nutrients, wastes, and other materials with blood vessels, which run through bone cavities.

Osteoclasts affect resorption of bone tissue by secreting both acid that dissolves calcium phosphate crystals, and enzymes that break down osteoid. Resorption releases calcium and phosphate into the bloodstream. Recall that bone deposition and resorption are also important for calcium balance.

Apply Your Knowledge

Rickets is a disease, primarily affecting children, in which the bones are weak because of a deficiency in vitamin D, calcium, or phosphate. Explain how each of these deficiencies can result in weak bones.

Bone deposition is also necessary for increases in height, which occur through the growth of long bones. **Figure 15** shows the structure of a typical long bone, which consists of a long, nearly cylindrical *shaft* capped at either end by a knob called an *epiphysis.* In the bones of growing children, the epiphyses are separated from the shaft by a thin layer of tissue called the **epiphyseal plate,** which is composed of **cartilage,** a soft material similar to uncalcified osteoid. The epiphyseal plate plays a key role in the elongation of bones during growth. Inside most bones is a central cavity containing red

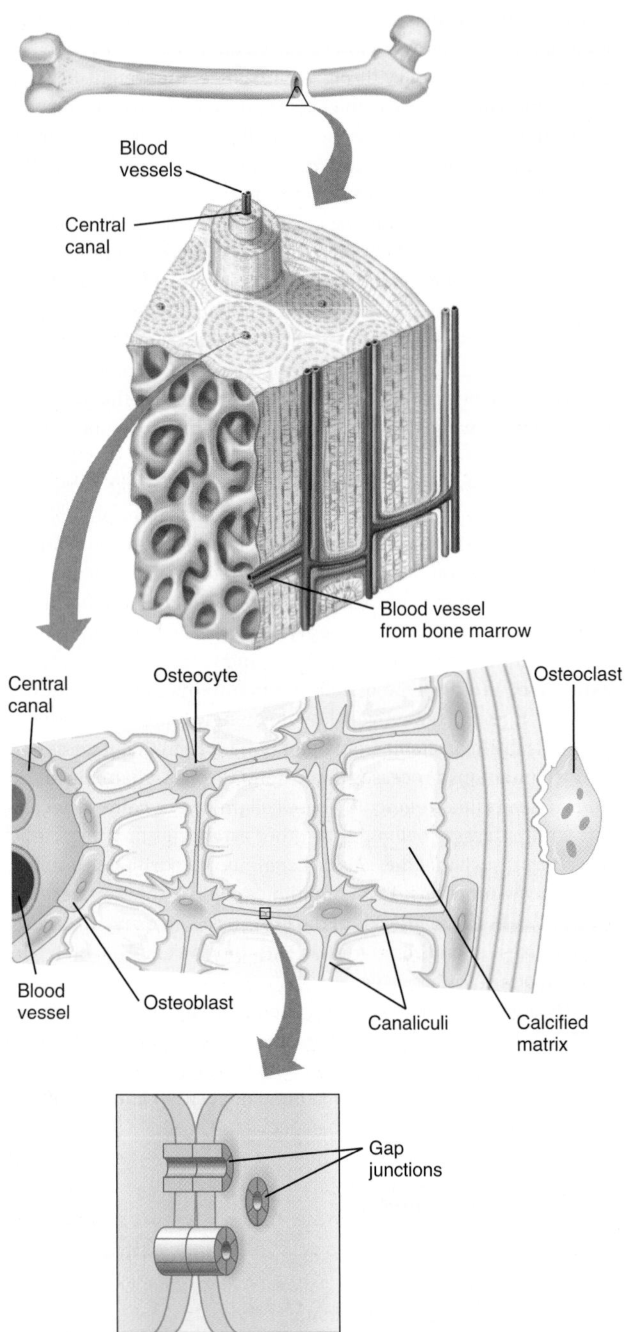

Figure 14 **The structure of bone.**

Q Which type of cell breaks down bone?

bone marrow and yellow bone marrow. *Red bone marrow* produces red and white blood cells; *yellow bone marrow* contains primarily adipocytes.

Under the influence of GH, bones increase in both circumference and length. The increase in circumference is brought about through the action of osteoblasts, which lay down new tissue on the outer bone surfaces. This process is accompanied by resorption of bone by osteoclasts at the inner surface of the marrow cavity. As a consequence, the diameter of the marrow cavity increases as the outer diameter of the bone increases, which minimizes weight gain while increasing strength.

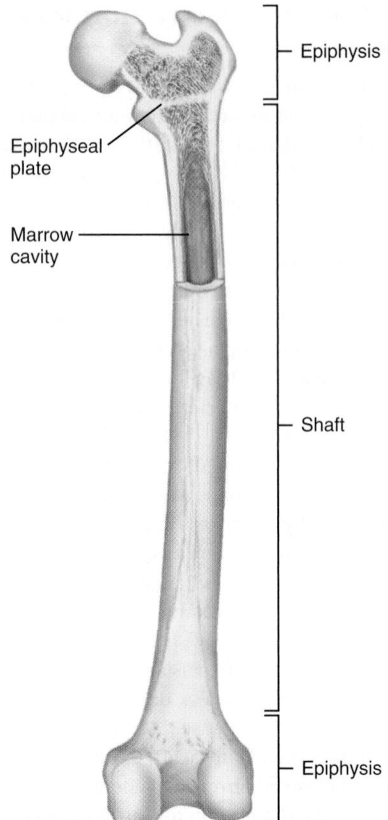

Figure 15 **Structure of a long bone (the femur).**

Increases in the length of a bone result from the addition of new bone tissue to either end of the bone shaft (**Figure 16**). This process begins with the activity of cells in the epiphyseal plate called **chondrocytes,** which are similar to osteoblasts except that they produce cartilage rather than bone. Under the influence of GH, chondrocytes increase in size and number, forming a layer of cartilage that causes the epiphyseal plate to become wider (elongating the bone). As new cartilage forms, chondrocytes located in the region adjacent to the shaft die, and osteoblasts from the nearby bone tissue replace them and begin converting the cartilage to bone. In this manner, new bone is added to the end of the bone shaft.

In late adolescence, the epiphyseal plates become completely filled in with bone tissue, a process called *epiphyseal plate closure.* When closure is complete, GH can no longer stimulate bone elongation, and bones stop lengthening. This is why further increases in height are generally not possible after adolescence (even though changes in bone circumference and remodeling of bone continue). Plate closure is influenced by sex hormones—that is, androgens and estrogens—during puberty.

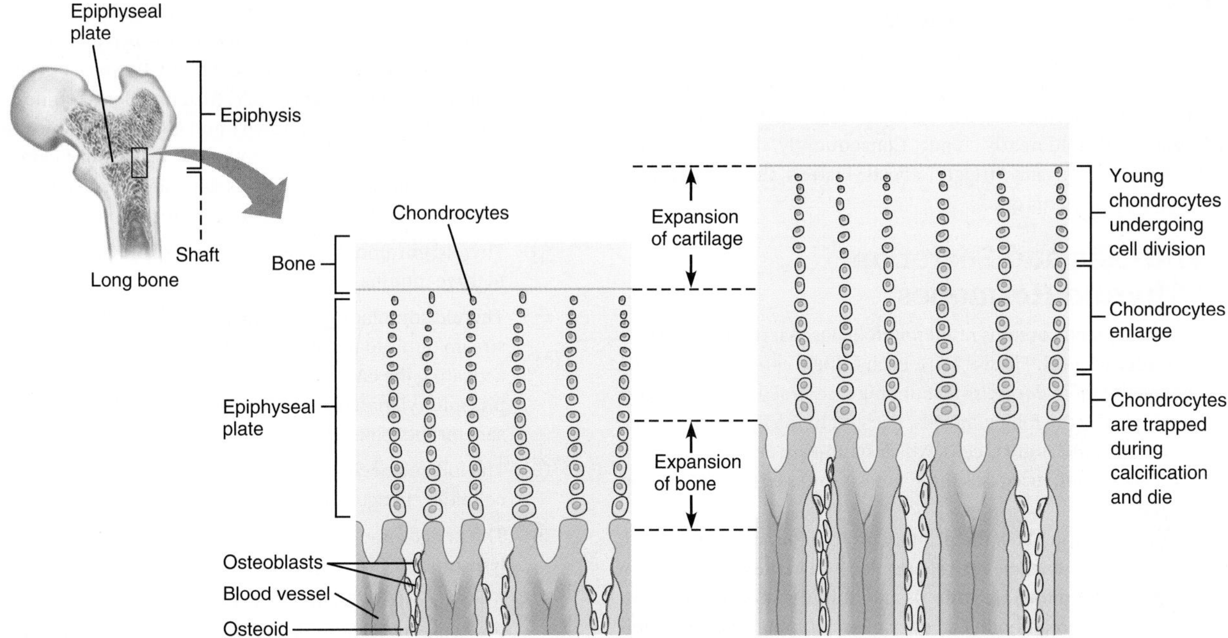

Figure 16 Long bone elongation. Growth occurs at the epiphyseal plate. Chondrocytes lay down cartilage, which is invaded by osteoblasts. The osteoblasts cause calcification or bone formation.

Effects of Abnormal Growth Hormone Secretion

Deficient GH secretion during childhood leads to a condition known as *dwarfism,* an irreversible stunting of growth, poor muscle development, and higher-than-normal amounts of body fat. Dwarfism is sometimes also caused by deficient tissue responsiveness to GH, which can result from defective GH receptors, insufficient production of IGFs, or failure of tissues to respond to IGFs. Abnormally low GH secretion in adults produces few noticeable signs other than decreased muscle or bone mass.

When excessive secretion of GH occurs before closure of the epiphyseal plates, the result can be *gigantism,* a condition in which stature is abnormally large but the body is normally proportioned. An excess of GH that occurs following plate closure results in *acromegaly*. In this syndrome, no change in height occurs, but the overgrowth of soft tissues and an increase in bone circumference produce a characteristic pattern of disfiguration in certain body parts. Individuals with untreated acromegaly typically have an overly wide, protruding jaw (sometimes referred to as a "lantern jaw") and overly long limbs.

Other Hormones That Affect Growth

Normal body growth requires the actions of other hormones besides GH, including thyroid hormones, sex hormones, and insulin. Thyroid hormones are required for the synthesis of GH and are generally permissive for its actions; that is, thyroid hormones are required for GH to exert its effects on target tissues. (Other actions of thyroid hormones are described later in this chapter.) Insulin is also permissive for growth because it is required for secretion of IGF-1 and for normal protein synthesis in general.

Sex hormone levels remain low until a few years before puberty and play little (if any) role in early childhood growth. In puberty, however, the dramatic increase in sex hormone secretion is essential for the growth spurt that normally occurs during this period. In contrast to insulin and the thyroid hormones, which are permissive for growth, the sex hormones actively promote growth by stimulating the secretion of GH and IGF-1. In addition, the sex hormones stop bone elongation by virtue of their role in promoting epiphyseal plate closure. Androgens (testosterone in men and adrenal androgens in women) exert an additional growth-promoting effect by directly stimulating protein synthesis in many tissues, including skeletal muscle. The marked rise in muscle mass that occurs in boys during puberty is largely due to the increased testosterone levels observed during this period.

In contrast to the growth-promoting hormones mentioned so far, high concentrations of the glucocorticoids (such as cortisol) secreted from the adrenal cortex inhibit growth, in part because they promote bone resorption and protein catabolism. It is also worth noting that glucocorticoid secretion is stimulated by stress, which is one possible explanation for the observation that illness and other forms of stress can have a growth-retarding effect. Glucocorticoids and their role in stress are described later in this chapter.

Quick Check 4

1. Describe the roles of osteoblasts, osteoclasts, and osteocytes in bone remodeling.
2. What are IGFs? What is their role with respect to body growth?
3. Why are further increases in body height not possible after adolescence?

8 Thyroid Hormones

Unlike the hormones we have studied so far, which show large changes in their rates of secretion throughout the course of a normal day, the thyroid hormones show little variation and their plasma levels hold nearly steady. Consequently, the thyroid hormones do not normally "trigger" effects; instead, they simply work to maintain the status quo.

Synthesis and Secretion of Thyroid Hormones

The thyroid gland contains numerous follicles that produce the thyroid hormones (TH; **Figure 17a**). Each follicle consists of a single outer layer of *follicular cells* surrounding a central protein-rich colloid secreted by the follicular cells. Located in the interstitial space between the follicles are C cells, which synthesize and secrete calcitonin. This section discusses the synthesis and secretion of thyroid hormones.

Thyroglobulin (TG), the primary protein found in the colloid, is the precursor molecule for TH. Also located in the colloid are the enzymes required for TH synthesis and iodide (ionized form of iodine, I^-). The thyroglobulin and enzymes are synthesized in the follicular cells and secreted into the colloid by exocytosis; the iodide is actively transported by follicular cells from the blood into the colloid and is a necessary component of TH. Thus all components of TH synthesis are located in the colloid.

The steps of TH synthesis and secretion proceed as follows (Figure 17b):

① Tyrosine residues of TG are iodinated. Addition of one iodide forms *monoiodotyrosine (MIT),* whereas addition of a second iodide to the same tyrosine residue forms *diiodotyrosine (DIT).*

② Two iodinated tyrosine residues (MIT or DIT) on the same TG molecule are coupled, meaning that the two tyrosine residues are linked together by a covalent bond. If two DIT groups combine, the final product is 3,5,3′, 5′-tetraiodothyronine or $\mathbf{T_4}$ (also called *thyroxine*); if a DIT and an MIT combine, the final product is 3,5,3′-triiodothyronine or $\mathbf{T_3}$. T_3 and T_4 are the thyroid hormones, although at this step they are still attached to TG. Note that two MIT groups cannot combine.

③ Thyroid hormones are stored in the colloid bound to TG for up to three months.

④ Thyroid stimulating hormone (TSH) arriving via the bloodstream binds to receptors on the membrane of follicular cells, activating the cAMP second messenger system. The result is phosphorylation of a variety of follicular cell proteins necessary for the release of thyroid hormones.

⑤ The follicular cells take in iodinated TG molecules from the colloid by phagocytosis.

⑥ The phagosome containing the iodinated TG fuses with a lysosome.

⑦ Exposure of the TG molecule to lysosomal enzymes that break down the thyroglobulin triggers the release of free T_3 and T_4 into the follicular cell. Because T_3 and T_4 are lipophilic, they can diffuse across the plasma membrane and into the bloodstream, where they are selectively bound by protein carriers that include *thyroxine-binding globulin* and *transthyretin,* and nonselectively bound by albumin.

T_4 is normally produced and secreted at a rate approximately 10 times greater than the rate for T_3. However, T_3 is approximately 4 times more potent at the target tissues. Most of the T_4 that is secreted into the plasma is eventually converted by the liver, kidneys,

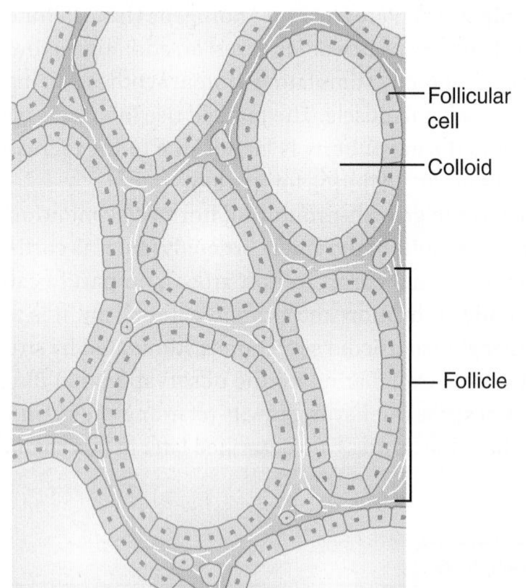

(a) Thyroid follicles

(b) Synthesis and secretion of thyroid hormones

Figure 17 Synthesis and secretion of thyroid hormones. (a) Thyroid follicles, the sites where thyroid hormones are synthesized. **(b)** Steps of thyroid hormone synthesis, which are described in the text in detail. Thyroid hormones can be stored in the colloid for months after their formation there, until binding of TSH stimulates the endocytosis into follicular cells of TG-thyroid hormone complexes, which are enzymatically degraded to release the thyroid hormones into the cells and ultimately into the bloodstream.

or target tissues to the more active form, T_3; in fact, the majority of T_3 in the plasma is synthesized from circulating T_4. Conversion of T_4 to T_3 is called *activation.*

TH levels remain virtually constant under normal conditions because the primary control of this hormone's secretion occurs via negative feedback (**Figure 18**). As we have seen, TH secretion is stimulated by TSH from the anterior pituitary. Secretion of TSH is, in turn, stimulated by **thyrotropin releasing hormone (TRH)** from the hypothalamus. Once released into the bloodstream, TH feeds back to the hypothalamus and the anterior pituitary to limit the secretion of TRH and TSH. Interestingly, T_4 provides stronger negative feedback than T_3.

The only known stimulus for TRH secretion, and thus for TH secretion, is exposure to cold temperatures. This action is more pronounced in infants than in older children and is virtually absent in adults. In infants, the cold-stimulated TRH secretion is thought to promote heat production as the infant adapts to a colder environment outside the mother's body. TRH secretion, and thus TH secretion, is inhibited by stress through neural input to the hypothalamus, although the significance of this inhibition is unknown.

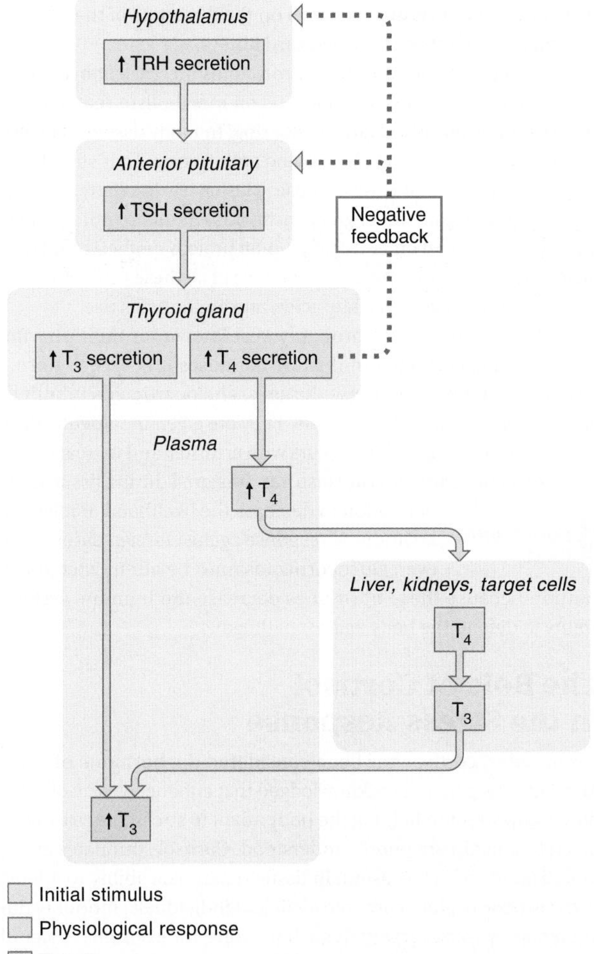

Figure 18 Factors affecting the secretion of thyroid hormones. T_3 also provides negative feedback (not shown), but to a lesser extent than T_4.

Apply Your Knowledge

A *goiter,* or enlargement of the thyroid gland, occurs for many reasons, one of which is a deficiency of iodine in the diet. Explain how iodine deficiency can cause the thyroid gland to enlarge.

Actions of Thyroid Hormones

Thyroid hormones are lipophilic and thus easily cross membranes; the receptors for these hormones are found in the nuclei of target cells. Binding of TH to receptors alters the rate of transcription of mRNA from DNA, thereby altering protein synthesis in the target cell. Such alterations take hours to days to exert an observable effect in the target cell; once induced, however, the effect generally lasts for days.

The primary action of the TH is to raise the body's BMR—that is, to increase the rate of oxygen consumption and energy expenditure at rest. As a result, heat generation also increases, a phenomenon termed a *calorigenic* effect. TH-stimulated increases in metabolism occur in most (but not all) tissues of the body; notable exceptions include the brain, spleen, and gonads. One way in which TH increases metabolism is by increasing the rate of Na^+/K^+ pump activity in cells. As ATP is hydrolyzed during activity of the Na^+/K^+ pump, heat is liberated. Meanwhile, ATP is used up, necessitating higher rates of fuel oxidation and ATP production, which generates even more heat. TH also promotes increased numbers of mitochondria in cells and increases in the concentrations of certain enzymes involved in oxidative phosphorylation.

In addition to stimulating energy *utilization,* TH promotes increased energy *mobilization* when present in *higher-than-normal* concentrations by enhancing glycogenolysis, conversion of muscle proteins to amino acids, and lipolysis. Elevated TH levels also promote gluconeogenesis and ketone synthesis. Conversely, at *lower-than-normal* concentrations, TH has the opposite effect, promoting glycogenesis and protein synthesis.

Many effects of TH are permissive. This hormone promotes the synthesis of beta adrenergic receptors, for example. Recall that adrenergic receptors bind epinephrine and norepinephrine, the chemical messengers of the sympathetic nervous system. Thus TH *permits* many tissues to respond to sympathetic neural input and to circulating epinephrine.

Thyroid hormones are necessary for normal growth and development of many other tissues, and for maintaining normal function after growth has ceased. Many of these effects are mediated through stimulation of GH release (in synergism with glucocorticoids, discussed shortly) and permissiveness to GH in target tissues. Developmental actions of TH are especially important with respect to the nervous system. In infants, TH deficiency can lead to a form of irreversible brain damage called *cretinism,* in which mental development is retarded and growth is stunted. In this condition, axons and dendrites of nerve cells are poorly developed, and myelination of axons is defective. Cretinism can be prevented by early diagnosis of hypothyroidism and initiation of T_3 replacement therapy. In the fully developed

nervous system, TH is essential for normal function. In adults, TH deficiency can lead to impairment of mental function, but defects are fully reversible upon restoration of normal TH levels.

9 Glucocorticoids

At normal plasma concentrations, the glucocorticoids, which are steroid hormones secreted by the adrenal cortex, maintain a wide variety of essential body functions. At higher concentrations they play a crucial role in the body's adaptation to stress.

Factors Affecting Secretion of Glucocorticoids

Secretion of glucocorticoids by the adrenal cortex is stimulated by **adrenocorticotropic hormone (ACTH)** from the anterior pituitary, which in turn is stimulated by **corticotropin releasing hormone (CRH)** from the hypothalamus (**Figure 19**). Because glucocorticoids are steroid hormones and, therefore, are lipophilic, glucocorticoids diffuse out of the adrenal cortex and into the bloodstream immediately after synthesis. Plasma glucocorticoid levels are normally regulated by negative feedback on the hypothalamus and anterior pituitary, which limits the secretion of CRH and ACTH, respectively.

Cortisol is the primary glucocorticoid released from the adrenal cortex. Like growth hormone, it is secreted in bursts and exhibits a circadian rhythm. Although the amount of hormone secreted per burst is virtually constant, the burst *frequency* varies with the time of day: It is higher in the morning and lower at night. This pattern is tied to the sleep-wake cycle and reverses in people who are awake at night and sleep during the day.

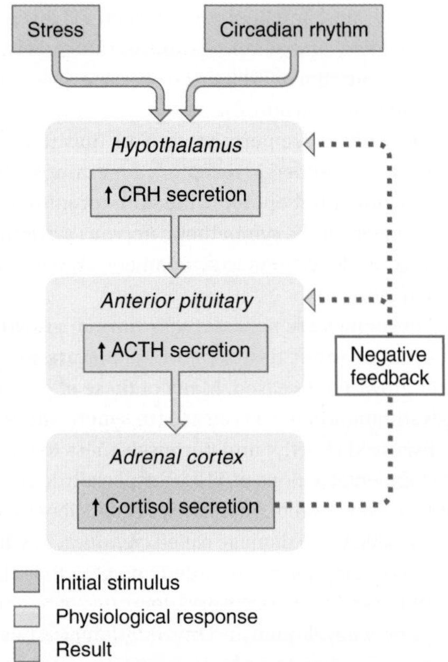

Figure 19 Factors affecting the secretion of cortisol.

Stress, whether physical or emotional, is an important stimulus for cortisol secretion. Those stressors that are most effective in stimulating cortisol secretion are usually noxious stimuli such as surgery, trauma, burns, infection, shock, and pain; other stressors include exposure to temperature extremes, strenuous exercise, and anxiety.

Actions of Glucocorticoids

Although the glucocorticoids do not *trigger* normal adjustments to the postabsorptive state, their presence is essential to the body's ability to mobilize fuels in response to signals from other hormones (for example, insulin and glucagon). The primary actions of glucocorticoids are to maintain normal concentrations of enzymes necessary both for the breakdown of proteins, fats, and glycogen, and for the conversion of amino acids to glucose in the liver. For this reason, the glucocorticoids are necessary for survival during prolonged fasting. In their absence, the resulting deficiencies in gluconeogenesis can lead to death by hypoglycemia once glycogen stores have been depleted.

Glucocorticoids are also required for GH secretion (in synergism with TH) and for maintaining the normal responsiveness of blood vessels to vasoconstrictive stimuli such as sympathetic nervous activity, epinephrine, and angiotensin II. In addition, glucocorticoids exert a variety of effects on the functions of the immune system, the nervous system, and the kidneys.

When plasma levels of glucocorticoids increase above resting levels, they exert a number of effects on metabolism that enhance energy mobilization and glucose sparing. In many tissues, they promote decreased uptake of glucose and amino acids. They stimulate lipolysis in adipose tissue, which raises plasma levels of fatty acids and glycerol. At the same time, glucocorticoids stimulate protein breakdown in muscle and other tissues, inhibit protein synthesis, and stimulate gluconeogenesis. As a consequence of all these actions, plasma concentrations of glucose, fatty acids, and amino acids rise.

Glucocorticoids are probably best known for their pharmacological effects when administered at doses that exceed normal physiological levels. At these dosages, glucocorticoids inhibit inflammation and allergic reactions. They are given therapeutically to treat inflammation, such as occurs with arthritis, and to treat certain allergies. Glucocorticoids are also administered during tissue transplantation to decrease the likelihood of rejection, an immune response against foreign tissue. However, glucocorticoids must be administered with caution because these hormones decrease the immune system's ability to defend the body against pathogens.

The Role of Cortisol in the Stress Response

For decades, cortisol has been considered the hormone of stress. Although it is generally acknowledged that enhanced cortisol secretion is important in helping the body adapt to stress, the reasons for this relationship are poorly understood. Cortisol stimulates energy mobilization, which is useful in tissue repair. The ability to tolerate stress is poor in glucocorticoid-deficient individuals; mortality during recovery from surgery in such persons, for example, is significantly higher than the corresponding rate in normal individuals.

In reality, cortisol is only one facet of the body's response to stress. In general, if a stimulus is effective in triggering increased cortisol secretion, it also triggers a consistent pattern of other neural and hormonal responses; for example, stress tends to promote increased activity of the sympathetic nervous system and secretion of epinephrine. These activities elicit the familiar fight-or-flight responses and stimulate gluconeogenesis, glycogenolysis, and lipolysis, which augment cortisol's energy-mobilizing action. Other changes generally associated with stress include increased secretion of antidiuretic hormone by the posterior pituitary, increased renin release by the kidneys, and elevated plasma levels of angiotensin II. These responses help maintain blood pressure, and thus adequate blood flow to the heart and brain. This generalized, stereotypical pattern of stress responses is referred to as *general adaptation syndrome.*

Effects of Abnormal Glucocorticoid Secretion

An excess or deficiency of glucocorticoid secretion can result from either a defect originating in the adrenal cortex (a primary disorder) or a defect in the secretion of the tropic hormones CRH or ACTH (a secondary disorder).

Hypersecretion of cortisol is associated with a characteristic pattern of signs known as *Cushing's syndrome.* These signs include hyperglycemia (due to stimulation of gluconeogenesis and inhibition of glucose uptake) and protein depletion, which results in muscle wasting, weakness, and fragility in many tissues due to the breakdown of connective tissue. A frequent consequence of Cushing's syndrome is a tendency to bruise easily, which indicates weakened blood vessels. Although cortisol generally has a stimulatory effect on lipolysis, it also stimulates fat synthesis and proliferation of adipocytes in certain regions of the body, which promotes an unusual pattern of body

fat distribution: Fat is deposited in the abdomen and above the shoulder blades, giving patients a protruding stomach and a hump-backed appearance; fat is also deposited in the face. Other regions, however, are not affected.

Hyposecretion of cortisol, known as *Addison's disease,* is characterized by hypoglycemia and poor tolerance of stress. In the primary form of this disease, which is usually a result of destruction of the adrenal cortex, there is often a defect in the secretion of aldosterone. Because aldosterone normally promotes sodium retention and potassium secretion by the kidneys, Addison's disease is marked by excess sodium excretion and potassium retention, which by altering plasma sodium and potassium levels results in cardiac arrhythmias and other neuromuscular signs.

Quick Check 5

1. What does it mean to say that thyroid hormones have a calorigenic effect?
2. Describe the roles of thyrotropin releasing hormone and thyroid stimulating hormone in the regulation of thyroid hormone secretion.
3. Indicate which of the following actions are promoted by glucocorticoids: glycogen synthesis, glycogenolysis, an increase in plasma glucose levels, gluconeogenesis, protein synthesis.
4. How does stress affect glucocorticoid secretion?

The metabolic effects of all the hormones discussed in this chapter are summarized in Table 4.

TABLE 4 Summary of the Metabolic Effects of the Hormones Discussed in This Chapter

	Site of secretion	Primary stimuli for secretion (indirect stimuli in parentheses)	Net effect on carbohydrate metabolism	Effect on plasma glucose	Net effect on lipid metabolism	Net effect on protein metabolism
Insulin	Beta cells of islets of Langerhans in pancreas	↑ Plasma glucose ↑ Plasma amino acids	↑ Glucose uptake into cells ↑ Glycogen stores	↓ Plasma glucose	↑ Triglyceride stores	↑ Amino acid uptake into cells ↑ Protein synthesis
Glucagon	Alpha cells of islets of Langerhans in pancreas	↓ Plasma glucose ↑ Plasma amino acids	↑ Glycogenolysis ↑ Gluconeogenesis	↑ Plasma glucose	↑ Lipolysis	↑ Proteolysis
Epinephrine	Adrenal medulla	Sympathetic nerve activity (stress, exercise)	↑ Glycogenolysis	↑ Plasma glucose	↑ Lipolysis	None
Growth hormone	Anterior pituitary	GHRH from hypothalamus (↓ plasma glucose, ↑ plasma amino acids, ↓ fatty acids, sleep, stress, exercise)	↓ Glucose uptake into cells	↑ Plasma glucose	↑ Lipolysis	↑ Amino acid uptake into cells ↑ Protein synthesis
Thyroid hormones (T_3 and T_4)	Thyroid gland	TSH from anterior pituitary (TRH from hypothalamus, cold temperatures in infants)	↑ Glycolysis	None	↑ Lipolysis	↑ Protein synthesis
Cortisol	Adrenal cortex	ACTH from anterior pituitary (CRH from hypothalamus, stress)	↓ Glucose uptake into cells ↑ Gluconeogenesis	↑ Plasma glucose	↑ Lipolysis	↓ Amino acid uptake into cells ↑ Proteolysis

SYSTEMS INTEGRATION Nowhere is systems integration more apparent than with the endocrine system, and especially the control of whole body metabolism. We learned that the nervous system depends on a constant supply of glucose, but too much glucose causes many complications as evident in diabetes. Glucose enters the blood through absorption from the gastrointestinal tract during the absorptive state and mostly from synthesis by the liver (gluconeogenesis and glycogenolysis) during the postabsorptive state. The pancreatic hormones insulin and glucagon regulate whole body metabolism, but so do cortisol, epinephrine, growth hormone, and insulin-like growth factor during periods of stress and growth. The sympathetic nervous system contributes to regulation of blood glucose during stress and the postabsorptive state and the parasympathetic nervous system promotes absorption of nutrients during the absorptive state.

For the endocrine system to regulate metabolism, growth, and coping with stress requires that the cardiovascular system does its job of delivering blood (containing these hormones) to the target tissues. The respiratory system functions in providing oxygen needed to generate most of the body's ATP and removing carbon dioxide produced during this process. Skeletal muscle's high metabolic rate helps maintain energy balance by utilizing nutrients, and smooth muscle functions indirectly through its role in other systems such as the cardiovascular, respiratory, and gastrointestinal systems.

The endocrine system affects every organ system, by altering its metabolism as described in this chapter or by altering function of the various organ systems as described throughout the book. For example, antidiuretic hormone regulates water reabsorption by the kidneys and total peripheral resistance, both of which are important in controlling blood pressure. The gastrointestinal hormone CCK regulates not only gastrointestinal function but also food intake. And as we learned in this chapter, glucocorticoids suppress the immune system and growth, both detrimental effects observed when a person is subjected to chronic stress.

CHAPTER REVIEW

SUMMARY

1 An Overview of Whole-Body Metabolism

- Whole-body metabolism requires the coordination of cellular metabolic activities.
- Cells use energy in the form of ATP, which they obtain from the oxidation of small nutrient molecules such as glucose, fatty acids, and amino acids.
- Cellular metabolism must be coordinated so that nutrients are provided to the appropriate cells when needed.

2 Energy Intake, Utilization, and Storage

- Energy is released in cells by the breakdown of nutrients into smaller molecules, as when glucose, amino acids, or fatty acids are oxidized to yield waste products.
- Energy mobilization comprises the breakdown of macromolecules into small nutrient molecules that are released into the bloodstream.
- Energy is stored by converting small nutrient molecules into macromolecules.
- Glucose is stored as glycogen in skeletal muscle and liver.
- Fatty acids and glycerol are stored as triglycerides in adipose tissue.
- Amino acids are stored as proteins in all cells, but especially in skeletal muscle cells.

3 Energy Balance

- Energy input must equal energy output to maintain energy balance.
- Energy input comes from ingested nutrients, whereas energy output is the energy expended as heat or work.
- Positive energy balance occurs when energy input exceeds energy output; negative energy balance occurs when energy output exceeds energy input.
- The body's metabolic rate is the total amount of energy released per unit time as a result of nutrient oxidation.
- The metabolic rate at rest is the basal metabolic rate (BMR).

4 Energy Metabolism During the Absorptive and Postabsorptive States

- When the body is in an absorptive state, glucose is used by most tissues as the primary fuel.
- Absorbed nutrients are also converted to glycogen, triglycerides, and proteins.
- Excess amino acids and glucose are mostly converted to fatty acids and stored as triglycerides.
- When the body is in a postabsorptive state, stored glycogen, triglycerides, and proteins are catabolized for energy.
- Fatty acids are used by most tissues as the primary fuel.
- The nervous system, unlike most other tissues, relies on a steady supply of glucose for its energy.
- Utilization of nonglucose fuels conserves glucose for use by the nervous system, a phenomenon called glucose sparing.
- The liver is able to produce more glucose by gluconeogenesis.

5 Regulation of Absorptive and Postabsorptive Metabolism

- Metabolic adjustments to the absorptive state are promoted by insulin; they include synthesis of energy stores and uptake of glucose and amino acids by cells in many tissues.
- Insulin also suppresses gluconeogenesis and regulates plasma glucose levels via negative feedback control.
- Metabolic adjustments to the postabsorptive state promoted by glucagon include glycogenolysis, protein breakdown by the liver, lipolysis, gluconeogenesis, and ketone synthesis.
- Glucagon also helps to regulate blood glucose levels.
- Other postabsorptive metabolic adjustments are promoted by increased epinephrine secretion and sympathetic nervous activity.

iP Endocrine: Actions of Hormones on Target Cells

6 Thermoregulation

- The thermoregulatory system acts to maintain a constant core temperature through thermoregulatory centers in the hypothalamus; central and peripheral thermoreceptors that detect core and skin temperature, respectively; and effector organs including cutaneous blood vessels, sweat glands, and skeletal muscle.
- Changes in cutaneous blood flow regulate body temperature when environmental temperature is within the thermoneutral zone.
- When the environmental temperature is lower than the thermoneutral zone, shivering is also needed to generate heat.
- When the environmental temperature is higher than the thermoneutral zone, sweating contributes to heat loss through evaporation.

7 Hormonal Regulation of Growth

- Body growth during childhood is promoted by the actions of growth hormone (GH), which is secreted by the anterior pituitary and acts to promote the growth of soft tissues and bones.
- In adulthood, GH acts to maintain bone mass and lean body mass.
- Actions promoted by GH include hypertrophy, hyperplasia, protein synthesis, lipolysis, gluconeogenesis, and amino acid uptake by cells.
- GH also inhibits glucose uptake by adipose tissue and muscle.
- Combined metabolic actions work to raise plasma levels of glucose, fatty acids and glycerol, thereby making energy more readily available to growing tissues.
- Many GH actions are mediated by IGFs synthesized by the liver and other tissues.

MAP iP Bone

8 Thyroid Hormones

- Thyroid hormone (TH) is normally secreted by the thyroid gland at near-constant rates and increases the metabolic rate in most tissues of the body.
- At high concentrations, TH mobilizes energy stores.
- TH is necessary for normal growth, development, and maintenance of normal function in many tissues, particularly the nervous system.
- TH is secreted in two forms: T_3, the more active form, and T_4, the more abundant form.

MAP iP Endocrine: The Hypothalamic-Pituitary Axis

9 Glucocorticoids

- Glucocorticoids are released by the adrenal cortex and are important in the body's response to stress.
- Glucocorticoids are also required so that the body can mobilize energy stores during postabsorptive periods.

MAP iP Endocrine: Response to Stress

Check your understanding at MasteringA&P and apply your new knowledge to real-life case studies to ensure that you've mastered the content!

EXERCISES

Multiple-Choice Questions

1. Which of the following is an example of a permissive effect of a hormone?
   a) The effect of thyroid hormones on growth
   b) The effect of insulin on glucose uptake by cells
   c) The effect of sex hormones on the secretion of growth hormone
   d) Both a and c
   e) All of the above
2. Which of the following is an example of a glucose-sparing effect of cortisol?
   a) Inhibition of ACTH release
   b) Stimulation of gluconeogenesis by the liver
   c) Stimulation of lipolysis
   d) Stimulation of glycogen breakdown
   e) Stimulation of protein synthesis
3. Which of the following cells of the pancreas secrete insulin?
   a) Alpha cells
   b) Beta cells
   c) Delta cells
   d) Exocrine cells
   e) Duct cells
4. Stress stimulates secretion of which of the following hormones?
   a) Growth hormone
   b) Epinephrine
   c) Thyroid hormones
   d) ACTH
   e) All of the above
5. Hypoglycemia inhibits secretion of which of the following?
   a) Growth hormone
   b) Insulin
   c) Epinephrine
   d) Glucagon
   e) All of the above
6. In the postabsorptive state, the central nervous system uses which of the following as its *primary* source of energy?
   a) Fatty acids
   b) Amino acids
   c) Glucose
   d) Glycerol
   e) Ketones
7. Which of the following cell types is directly responsible for building new bone material?
   a) Osteoblasts
   b) Osteoclasts
   c) Osteocytes
   d) Chondrocytes
8. Which of the following is true of adulthood?
   a) Growth hormone exerts no effects on body tissues.
   b) The secretion of growth hormone ceases altogether.
   c) Growth hormone cannot stimulate increases in the length of long bones.
   d) The structure of bone becomes permanently fixed.
   e) None of the above
9. Which form of thyroid hormone has greater activity at target cells?
   a) T_3
   b) T_4
   c) Neither; T_3 and T_4 have equal activity
10. Which of the following hormones is a steroid?
    a) Thyroid hormones
    b) Insulin
    c) Glucagon
    d) Growth hormone
    e) Cortisol

Objective Questions

11. Energy mobilization is promoted by (insulin/glucagon).
12. Secretion of (insulin/glucagon) is increased during the absorptive period.
13. Insulin and glucagon both help regulate the plasma glucose concentration. (true/false)
14. Breakdown of triglycerides yields fatty acids and ________, which can be used by cells for energy.
15. Conversion of amino acids to fatty acids is more likely to occur in the (absorptive/postabsorptive) state.
16. Conversion of amino acids to glucose is more likely to occur in the (absorptive/postabsorptive) state.
17. An increase in plasma thyroid hormone levels tends to make the body's energy balance more (positive/negative).
18. Energy that is taken into the body is either stored or appears as work or ________.
19. Stress tends to (stimulate/inhibit) GHRH secretion.
20. Many of growth hormone's effects on target tissues are due to it triggering release of other chemical messengers called ________.
21. Closure of the epiphyseal plates is promoted by (growth hormone/sex hormones).
22. Thyroid hormones promote increased responsiveness of target tissues to (sympathetic/parasympathetic) nerve activity.
23. Glucocorticoids promote (increased/decreased) plasma glucose levels.
24. Stimulation of gluconeogenesis by glucagon is an example of a glucose-sparing effect. (true/false)
25. Plasma glucocorticoids have a(n) (stimulatory/inhibitory) effect on the secretion of ACTH.

Essay Questions

26. Describe the regulation of plasma glucose by insulin and glucagon. Include a description of the role of negative feedback.
27. Describe how insulin, glucagon, and the sympathetic nervous system work together to maintain adequate plasma glucose levels during fasting. Why is this coordination important?
28. Describe the various factors that determine the body's energy balance. Be sure to describe what happens to energy that is liberated as a result of fuel oxidation.
29. Describe the similarities between the metabolic actions of thyroid hormones and glucocorticoids.
30. Describe the metabolic actions of growth hormone, and explain how these actions promote growth.

Critical Thinking

31. Some athletes have used anabolic steroids in hopes of enhancing their athletic performance. Which type of steroids might they use? Based on what you read in this chapter, what could be some of the deleterious effects of taking such steroids?
32. Many hormone deficiencies are genetic in origin. Explain how a steroid hormone deficiency can be caused by a mutant gene.
33. Physicians often recommend that people with hypoglycemia consume several small meals per day, as opposed to the typical three large meals per day. Explain how this dietary pattern can affect insulin and glucagon levels in the blood, and ultimately blood glucose levels.
34. An obese patient goes to his doctor complaining of lethargy. The doctor suspects that the patient has hypothyroidism, but does not know whether the disorder would be primary or secondary. Which tests should the doctor run to determine whether hypothyroidism exists, and if it does, whether it is primary or secondary? Explain the expected results if the patient has a tumor in the hypothalamus that is affecting the release of TRH.

Photo credits for icons in feature boxes: Focus on Diabetes: Tanyae/Dreamstime; Clinical Connections: Iofoto/Dreamstime; Understanding Exercise: Pixattitude/Dreamstime; Discovery: Yuri Arcurs/Cutcaster.

Glossary

absorptive state period following a meal during which nutrients are absorbed

adipocytes fat cells that store triglycerides

basal metabolic rate (BMR) the slowest metabolic rate of the body, indicative of the energy necessary to sustain vital functions

calcification formation of bone

cartilage a connective tissue secreted by chondrocytes that is similar to uncalcified bone

chondrocytes (kon-droh-sites) cells that produce cartilage

conduction the transfer of thermal energy from one object to another that occurs when the objects are in direct contact with each other; in the nervous system, the propagation of action potentials along an axon

core temperature temperature deep within the body

corticotropin releasing hormone (CRH) (kor-tih-koh-troh-pin) tropic hormone released by the hypothalamus that stimulates secretion of adrenocorticotropic hormone (ACTH) from the anterior pituitary

deposition formation of bone

diabetes mellitus (die-ah-bee-teez) a disease involving defects in insulin production or signaling

epiphyseal plate (ep-ih-fiz-ee-al) a thin layer of tissue that separates the epiphyses of a long bone from the shaft; plays a key role in the elongation of bones during growth

evaporation conversion of a liquid to a gas; one of the mechanisms of thermoregulation

glucagon a peptide hormone secreted by alpha cells of the pancreas; promotes metabolic processes of the postabsorptive state

gluconeogenesis (gloo-koh-nee-oh-jen-ih-sis) process during which new glucose molecules can be synthesized from noncarbohydrate precursors by the liver

glucose sparing process by which non-nervous tissues convert to using fatty acids for energy rather than glucose, which is spared for use by the central nervous system

growth hormone (GH) peptide hormone secreted by the anterior pituitary; essential for normal growth

growth hormone releasing hormone (GHRH) tropic hormone released by the hypothalamus that stimulates growth hormone secretion from the anterior pituitary

hyperplasia increase in cell number

hyperthermia a condition of higher than normal body temperature

hypertrophy increase in cell size

hypothermia a condition of lower than normal body temperature

insulin a peptide hormone secreted by beta cells of the pancreas; promotes metabolic processes of the absorptive state

lipoprotein lipase enzyme found on the inside surface of capillaries throughout the body that breaks down triglycerides

metabolic rate amount of energy expended by the body per unit time

osteoblasts mobile cells within bone that secrete the extracellular matrix during bone formation

osteoclasts mobile cells that are responsible for breaking down bone tissue

osteocyte type of bone cell that maintains the extracellular matrix

osteoid (os-tee-oyd) organic gel-like substance secreted by osteoblasts that forms the matrix of bone

postabsorptive state period between meals during which stored nutrients are mobilized

radiation transfer of thermal energy in the form of electromagnetic waves

resorption the breakdown of bone tissue

somatostatin tropic hormone released by the hypothalamus that inhibits growth hormone secretion from the anterior pituitary; also known as *growth hormone inhibiting hormone (GHIH)*

thermoregulation utilization of feedback mechanisms to maintain normal body temperature

thyrotropin releasing hormone (TRH) tropic hormone secreted from the hypothalamus that stimulates secretion of thyroid stimulating hormone (TSH) from the anterior pituitary

very low density lipoproteins (VLDLs) one of a group of lipoprotein particle types consisting of lipids and proteins in various ratios; VLDLs contain a high ratio of lipids to proteins, which results in their low density

Answers

Figure Question Answers

Fig 1	Glycolysis, the Krebs cycle, and the electron transport chain
Fig 4	Gluconeogenesis
Fig 12	Girls
Fig 14	Osteoclast

Apply Your Knowledge Answers

Certain areas of the brain show increased activity during sleep compared to rest. Thus more energy is needed during sleep.

Bone formation requires calcium and phosphate as major components. Vitamin D is the precursor for 1,25-dihydroxycholrcalciferol, a hormone necessary for bone deposition.

When iodine levels in the diet are low, thyroid hormone synthesis and secretion are decreased. A lack of thyroid hormone removes inhibitory input (loss of negative feedback) to the hypothalamus and anterior pituitary to secrete TRH and TSH, respectively. Tropic hormones stimulate growth of the glandular tissue as well as the secretion of the target hormones. In this case, TSH stimulates growth of the thyroid gland in an attempt to increase thyroid hormone synthesis.

Answers to End-of-Chapter Multiple Choice and Objective Questions

Multiple Choice Questions

1. d 2. c 3. b 4. e 5. b 6. c 7. a 8. c 9. a 10. e

Objective Questions

11. glucagon
12. insulin
13. true
14. glycerol
15. absorptive
16. postabsorptive
17. negative
18. heat
19. inhibit
20. (IGFs) insulin-like growth factors
21. sex hormones
22. sympathetic
23. increased
24. false
25. inhibitory

Diabetes Mellitus

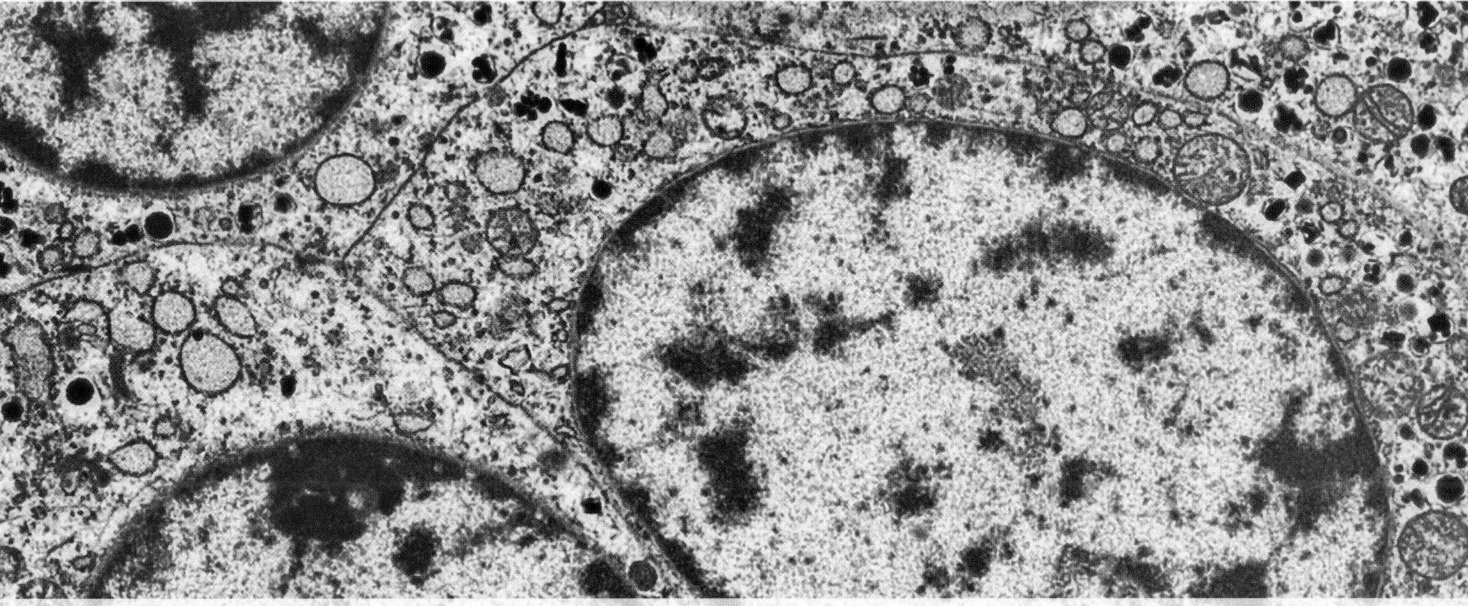

Steve Gschmeissner/Photo Researchers

Colorized transmission electron micrograph of the pancreas showing several cells in an islet of Langerhans.

Aiden received his first insulin shot when he was 10 years old. He was a thin boy, and his mother, Janice, noticed that Aiden was urinating frequently and drinking fluids excessively. His level of physical activity was also declining. Janice took Aiden to the doctor, who immediately noted the fruity odor of Aiden's breath—a sign of ketoacidosis. The doctor ordered urine and blood tests. Aiden's urine tested positive for glucose and ketones, and his blood test showed elevated levels of glucose and glycated hemoglobin. His blood also tested positive for human leukocyte antigen. Aiden was diagnosed with diabetes mellitus type 1.

Janice was terrified at the prospect of her son growing up with this lifelong disease. However, the doctor reassured her that with careful management of his condition, Aiden could live a happy, normal life. Now, at age 15, Aiden manages his diabetes with a "pod"—an insulin pump that monitors his blood glucose and automatically administers insulin for him. He plays on the high school baseball team, has good grades, and is hoping for a college scholarship.

CHAPTER OUTLINE

MasteringA&P®

Go to MasteringA&P for helpful A&P Flix 3-D animations, chapter quizzes, pre-tests, interactive physiology tutorials, and more!

LEARNING OUTCOMES *After studying this chapter, you should be able to:*

- Identify the different causes of type 1 and type 2 diabetes mellitus.
- Describe the initial manifestations of diabetes mellitus.
- Identify the causes and effects of ketoacidosis.
- Describe the different mechanisms through which diabetes may lead to a coma.
- Explain how hyperglycemia causes many of the adverse effects in diabetes.
- Explain how damage to the microvasculature leads to adverse effects in diabetes.
- Define macrovascular damage, and identify its serious consequences in diabetes.
- Describe the normal wound healing process and the way in which it is impaired in diabetes.
- Describe current treatments of diabetes.
- Describe some of the areas of new research on diabetes.

Before You Begin

Make sure you have mastered the following concepts:

- Hormone mechanism of actions
- Insulin release
- Insulin actions
- Renal handling of glucose
- Regulation of blood glucose
- Endocrine pancreas
- Osmolarity
- Autoimmune disease
- Microvasculature
- Peripheral nervous system

Today, 24 million people in the United States have diabetes, a 13.5% increase from the number who had this disease just a few years ago. Diabetes affects the various organ systems, as summarized in **Figure 1**. We now look more closely at the progression of this disease and its attack on the organ systems.

1 Classification of Diabetes Mellitus

Currently, two categories of diabetes mellitus are distinguished: type 1, previously called insulin-dependent or juvenile-onset diabetes mellitus; and type 2, previously called insulin-independent or adult-onset diabetes mellitus. As the previous nomenclatures suggest, these diseases vary in both their causes and their ages of onset.

Type 1 Diabetes Mellitus

Type 1 diabetes mellitus (type 1 DM) is divided into two subclasses, both of which involve destruction of pancreatic beta cells and, therefore, a decrease in insulin secretion. Type 1a DM is an autoimmune disease in which the immune system attacks the beta cells of the pancreas, causing a loss of insulin secretion from these cells. The antigens most likely to be attacked include proinsulin, glutamic acid decarboxylase, tyrosine phosphorylase-like protein, zinc transporter 8, and other proteins in the islets of Langerhans. Type 1b DM, also called idiopathic type 1 DM, is a rarer (5–10% of type 1 DM cases) form of diabetes that involves destruction of pancreatic beta cells. In this form, the mechanism of destruction is unknown, but most likely does not involve autoimmune dysfunction.

In type 1a DM, the cause of the erroneous immune response appears to have at least a partial genetic basis, as a family history of diabetes indicates a predisposition for developing the disease. Research made possible by completion of the Human Genome Project identified mutations at more than 20 different loci as causes for predisposition toward diabetes mellitus. One of those loci codes is for human leukocyte antigen proteins of the major histocompatibility complex, which is the basis of self versus non-self recognition. In other cases of type 1 DM, however, genetics alone is not the cause; environmental factors also play a role. For example, if

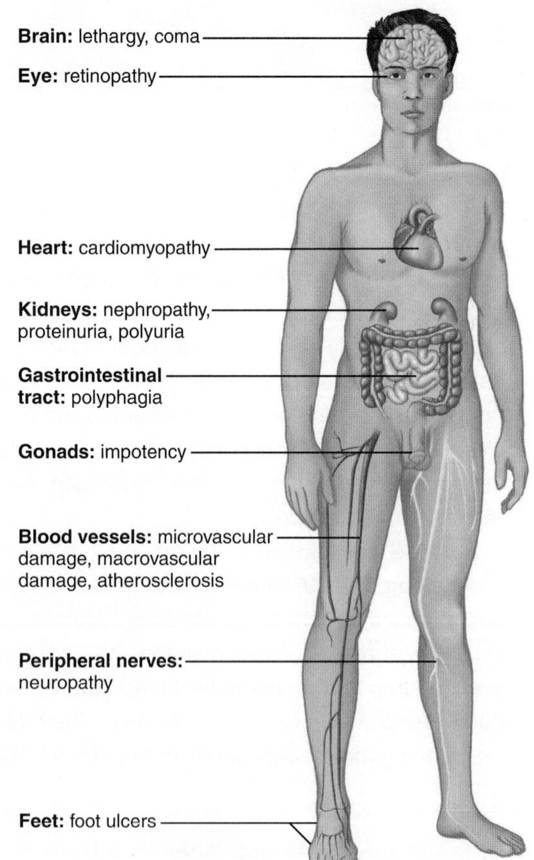

Figure 1 **Effects of diabetes on organs throughout the body.**

an identical twin has type 1 DM, there is only a 30–50% chance that his or her twin will also have the disease. Genetics also cannot account for the increasing rate of type 1 DM, such as the 50% increase in the number of cases in the United States over the last 20 years.

Viral infections may be another cause of type 1a DM; alternatively, viral infections may unmask a genetically predisposed form of type 1 DM. Viral infections of the gastrointestinal tract are particularly suspect, with coxsackie B the most likely virus to cause type 1 DM. These viruses are thought to infect beta cells, inducing an immune attack against these cells. Other viruses, such as those responsible for rubella, mumps, and Epstein-Barr viral infection, do not infect beta cells but may induce an immune response that attacks beta-cell antigens.

Whether the type 1 diabetes is genetically or virally induced, the immune system destroys beta cells such that insulin secretion is insufficient to regulate blood glucose. At approximately 7 years of age through early adulthood (when 80% or more of the beta cells are destroyed), symptoms begin to appear. These symptoms, which include hyperglycemia and lethargy, are mild at first, but progress for the remainder of the person's life. An early diagnosis will help control the disease, but because the majority of beta cells have been destroyed by the time symptoms appear, the disease is not reversible.

Type 2 Diabetes Mellitus

Type 2 diabetes mellitus (type 2 DM) is associated with insulin resistance; that is, target cells to insulin show a decreased response to insulin. Although the cause(s) of this insulin resistance is (are) unknown, several factors predisposing individuals to this condition have been identified. For example, genetics seems to play a bigger role in type 2 DM compared to type 1 DM. More than half of all children who develop type 2 DM have a family member who has type 2 DM. Lifestyle also plays a major role, with a sedentary lifestyle and obesity promoting development of the disease.

The increased incidence of type 2 DM is the primary cause of the diabetes epidemic. Although type 2 DM generally develops after age 45 (with the incidence increasing with age), the age at time of diagnosis is decreasing. In 1988, the average age of diagnosis was approximately 52 years, but by 2000 that age had decreased to 46 years. Also, the number of children with type 2 DM is increasing. Table 1 summarizes the main differences between type 1 and type 2 DM.

Quick Check 1

1. Which is more prevalent, type 1 or type 2 DM?
2. Which form of diabetes depends more on a familial history of the disease?

TABLE 1 Type 1 and Type 2 Diabetes Mellitus

	Type 1 diabetes mellitus	Type 2 diabetes mellitus
Prevalence as a percentage of all DM cases in United States	5%–10%	90%–95%
Previous nomenclature	Insulin-dependent DM (IDDM) Juvenile-onset DM	Insulin-independent DM (IIDM) Adult-onset DM Non-insulin-dependent DM (NIDDM)
Pathology	Autoimmune disease of pancreatic beta cells causing insufficient release of insulin into blood	Tissue resistance to insulin
Age of onset	Adolescence to young adulthood	Adulthood, but occurrence in children is increasing
Clinical diagnostics	Decreased blood insulin Elevated blood glucose Impaired glucose tolerance Elevated hemoglobin A_{1c} Presence of antibodies to beta cell components	Normal or elevated blood insulin (may decrease as the disease progresses) Elevated blood glucose Impaired glucose tolerance Elevated hemoglobin A_{1c} No antibodies to beta cells
Familial	30%–50% concordance in identical twins Family history in 10%	60%–80% concordance in identical twins Family history in more than 50%
Complications	Hyperglycemia, causing increased glycosylation of proteins Ketoacidosis is common	Hyperglycemia, causing increased glycosylation of proteins Ketoacidosis is rare
Management	Insulin is mandatory in all cases Diet Exercise	Insulin is used occasionally Diet Exercise

2 Acute Effects of Diabetes Mellitus

Diabetes mellitus is a long-term disease associated with many chronic ailments. However, acute changes in blood glucose cause short-term symptoms and alter whole-body metabolism.

Acute Hyperglycemia 

Diabetes mellitus predisposes a person to hyperglycemia, or elevated blood glucose levels. Several factors may increase blood glucose in diabetics, including forgetting to take blood-lowering medication, eating a high-carbohydrate meal, decreased physical activity or increased intense physical activity, stress, or illness. Hyperglycemia may cause relatively harmless symptoms such as dry mouth, thirst, and increased urination; it can also cause more severe symptoms such as hypotension, headaches, fatigue, blurred vision, difficulty thinking, and mood changes. Severe hyperglycemia can lead to **ketoacidosis,** a dangerous metabolic acidosis that arises when cells use fats for energy instead of glucose.

Diabetic Ketoacidosis

Ketoacidosis occurs primarily in type 1 DM (although it can occur in type 2 DM) during an episode of hyperglycemia triggered by low insulin levels (**Figure 2**). The ketoacidosis results from hormone-induced alterations in metabolism.

When hyperglycemia occurs in type 1 DM, for the reasons described previously, most cells of the body cannot use the plentiful blood glucose without insulin to induce glucose transporters into the cells' plasma membranes. Thus the cells must rely on another energy source. During hyperglycemia, glucagon, cortisol, and growth hormone levels all increase, triggering lipolysis, which in turn releases fatty acids into the blood. Fatty acids diffuse into cells, which then use them to produce ATP by beta oxidation and oxidative phosphorylation. The by-products of this beta oxidation process travel to the liver, where they are converted to ketones, causing ketoacidosis.

Hyperglycemia indirectly induces epinephrine release in response to low blood pressure. One effect of hyperglycemia is an increase in urine output due to the osmotic pull of excess filtered glucose on water. The resulting hypovolemia decreases blood pressure, and baroreceptors in the carotid sinus and aortic arch and volume receptors in the right atrium respond to and communicate these decreases to cardiovascular control centers in the medulla oblongata. To restore blood volume and pressure to normal values, the sympathetic nervous system is activated and epinephrine is released from the adrenal medulla.

Collectively, epinephrine, glucagon, cortisol, and growth hormone act on the liver to promote gluconeogenesis and glycogenolysis, which mobilizes glucose for use as an energy source, thereby adding to the hyperglycemia. As mentioned earlier, these hormones also promote lipolysis in adipose tissue, increasing free fatty acids at a rate up to ten times the normal levels in an effort to provide adequate energy to the cells. In the liver, incomplete oxidation of fatty acids yields acetyl CoA, which can be used for gluconeogenesis or **ketogenesis,** the production of ketones. This ketogenesis, in turn, increases blood ketone levels. Meanwhile, elevated fatty acids decrease ketone oxidation in peripheral tissues, further adding to the blood ketone levels. Ketones are acids—thus ketoacidosis ensues.

Symptoms of ketoacidosis vary, but involve several organ systems. High levels of ketones produce acetone breath (a fruity odor). Ketones are excreted in the urine of diabetics along with glucose, adding to the osmotic force to excrete more water, which pulls sodium with it by solvent drag. As a consequence, ketoacidosis can lead to dehydration, sodium imbalance, and hypotension. Corrections for acidosis occur through buffers in the blood and through respiratory and renal compensation. Thus ketoacidosis causes tachypnea (rapid breathing) as the respiratory system tries to compensate for the metabolic acidosis. Acidosis also causes movement of hydrogen ions into cells in exchange for potassium ions, producing electrolyte imbalances that affect excitable cells. Furthermore, ketones induce nausea and vomiting.

Ketoacidosis may be severe when diabetes is coupled with such precipitating conditions as infection, trauma, emotional stress, and alcohol consumption. Failure to treat ketoacidosis may lead to cardiovascular failure, coma, and even death. Treatment includes rehydration with fluids and electrolytes, insulin administration, and treatment of the cause. Bicarbonate can also be administered in severe cases. Because of the serious risks associated with ketoacidosis, a person with diabetes mellitus should control his or her blood glucose levels with appropriate medications, diet, and lifestyle to prevent the development of this imbalance, and should seek medical attention when early signs of acidosis appear.

Hyperosmolar Nonketotic Coma

Elderly patients with diabetes may suffer from hyperosmolar nonketotic coma when their glucose levels reach extremely high levels, in the range of 600 to 2400 mg/dL. This marked imbalance causes more glucose, and thus more water, to be lost in the urine, severely dehydrating the person. Severe dehydration leads to a decrease in blood volume, which in turn decreases blood flow to the tissues. If tissues do not get the blood and oxygen they need, the cells will turn to anaerobic glycolysis as a means of energy production. This alternative means of generating energy may produce a slight acidosis. If enough fluid is lost such that blood osmolarity increases to a level greater than 325 mOsm, the person is likely to go into a coma.

Hyperosmolar nonketotic coma progresses more slowly than ketoacidosis. Because the person is elderly, the sympathetic response to decreased blood volume and blood pressure is relatively weak; for this reason, epinephrine levels and thus fatty acids are not elevated as dramatically as they are in ketoacidosis. Instead, the primary danger in hyperosmolar nonketotic coma is blood clotting. Many elderly persons have atherosclerosis, which means their platelets tend to coagulate more easily, increasing the likelihood of thrombus formation.

Treatment of hyperosmolar nonketotic coma is similar to that of ketoacidosis: Fluids and electrolytes must be replaced and insulin injections are used to decrease blood glucose.

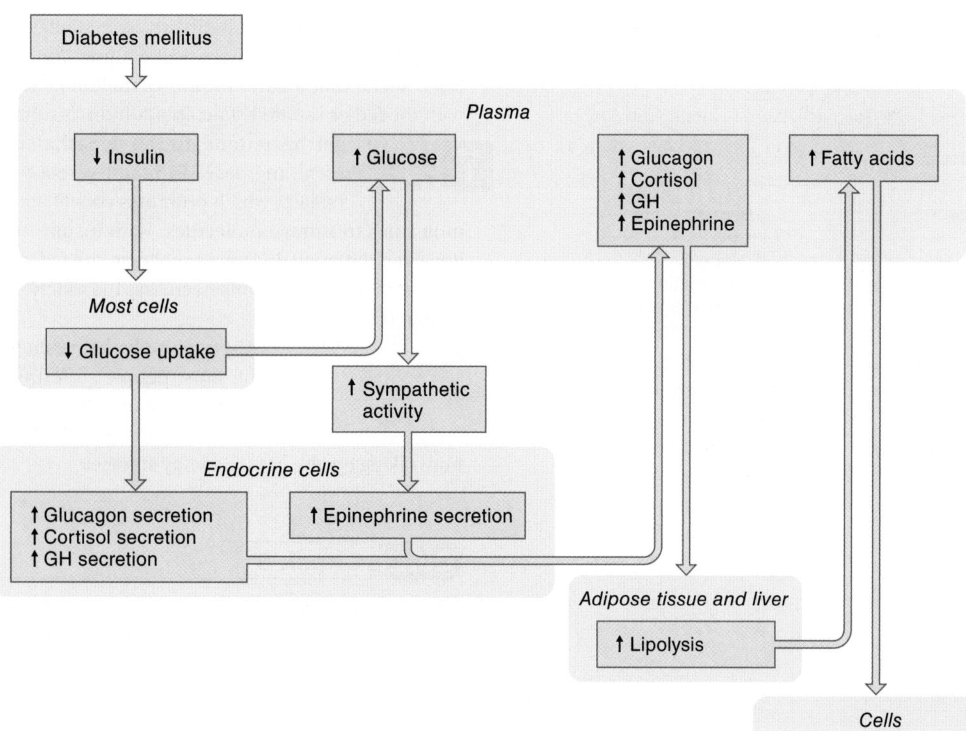

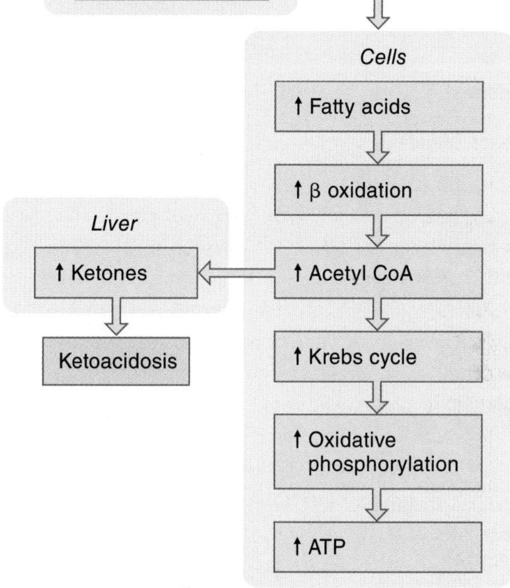

Figure 2 Ketoacidosis. The pathways by which diabetes can produce ketoacidosis.

Hypoglycemic Coma

As noted earlier, hyperglycemia can result in a coma, but hypoglycemic (low blood glucose) coma also can occur with mismanagement of diabetes. The hyperglycemic coma described previously was caused by too little insulin or a lack of insulin—that is, diabetes caused the coma. In contrast, *treatment* of diabetes—specifically, the taking of too much insulin—causes hypoglycemic coma.

Health Fact

Hypoglycemia, defined as a blood glucose level less than 50 mg/dL, can be induced by inappropriate use of medications for diabetes mellitus. Because the nervous system relies on glucose as its energy source, symptoms of this imbalance affect most body systems and vary with the degree of the hypoglycemia. These symptoms include nausea, hunger, cold or wet skin, tingling, trembling, and tachycardia during mild stages, and blurred vision, confusion, dizziness, anxiety, poor coordination, lethargy, and sleepiness in advanced stages. If allowed to progress, hypoglycemia can cause seizures and loss of consciousness, which may eventually lead to a coma. A person showing early signs of hypoglycemia should slowly consume simple carbohydrates in fruit juices, soda, hard candies, or other quick sources of sugar.

Functional Fact

Quick Check 2

1. How are ketones produced, and what are their adverse effects on the body?
2. What causes hypoglycemic coma?

3 Chronic Complications of Diabetes Mellitus: Early Stages

Early in the development of diabetes, symptoms are mild and the disease often goes undiagnosed. The early symptoms include the "three polys": **polyuria** (excessive urination), **polydipsia** (excessive thirst), and **polyphagia** (excessive eating) (**Figure 3**).

Recall that water moves passively in and out of the renal tubules based on osmotic gradients. As solute excreted in the urine increases, it tends to pull more water with it by osmosis. Normally, no glucose is

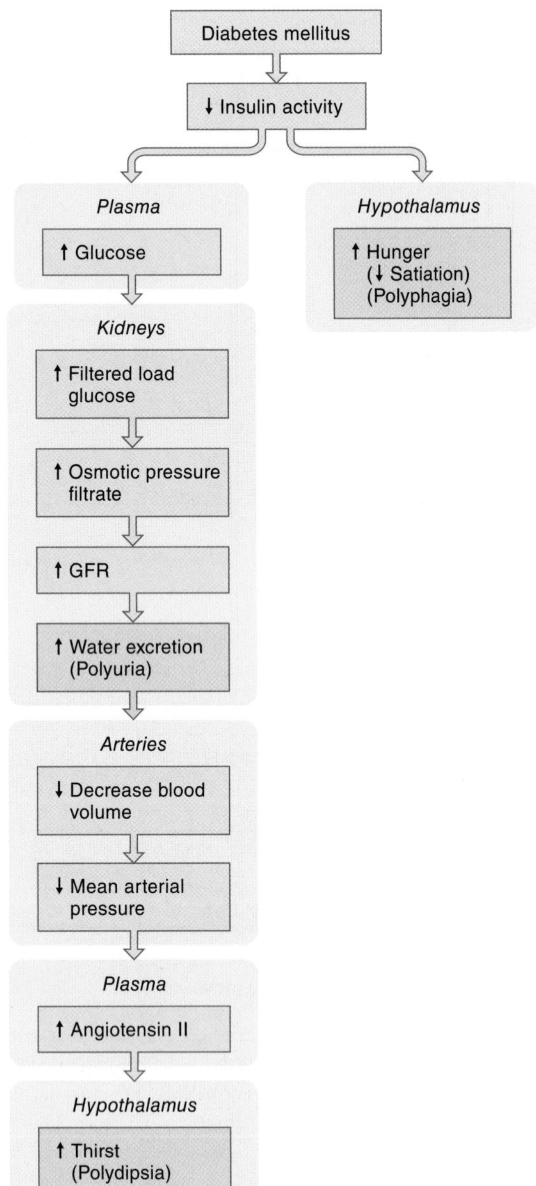

Figure 3 Effects of diabetes on eating, drinking, and urination. The flowchart indicates the pathways by which diabetes can produce the "three polys": polyuria, polydipsia, and polyphagia.

excreted in the urine; instead, all of the glucose is reabsorbed in the proximal tubules. In diabetes mellitus, however, the elevated blood glucose levels exceed the renal threshold for glucose reabsorption, causing some glucose to remain in the renal tubules and thereby increasing the osmotic force for water movement into the lumen. In this way, polyuria results from hyperglycemia.

Health Fact

Polyuria increases water loss, which tends to decrease both plasma volume and blood pressure. Recall that a decrease in blood pressure elevates the level of angiotensin II, the most potent vasoconstrictor in the blood. In addition to its vasoconstriction effects, angiotensin II acts on thirst centers in the anterior hypothalamus to trigger polydipsia.

Health Fact

Polyphagia can result directly from diabetes mellitus or from the hyperglycemia that occurs in conjunction with diabetes mellitus. Recall that insulin promotes metabolic changes associated with the fed or satiated state. In addition, insulin suppresses hunger by actions on the hypothalamus. Following a meal, increases in blood glucose stimulate insulin release, which promotes conversion of small nutrient molecules to storage molecules. With insulin activity decreased, however, the body does not recognize the fed state. Some studies suggest that hyperglycemia decreases the satiety effects of cholecystokinin (CCK) as well.

Functional Fact

Although polyphagia should increase body weight, most people suffering from type 1 DM tend to be malnourished and to have decreased body weight. Inadequate insulin prevents the liver and adipocytes from taking up glucose for energy storage.

Quick Check 3

1. Do levels of angiotensin II increase or decrease during hyperglycemia? What effect does this trend have on thirst?
2. Which two hormones induce satiety?

4 Chronic Complications of Diabetes Mellitus: Advanced Stages

Diabetes mellitus can be controlled to a large extent through diet and other lifestyle changes. Insulin and other medications are required to control type 1 DM; they may also be used to control advanced-stage type 2 DM. Maintaining normal blood glucose levels prevents or slows the progression of advanced symptoms. When reading about the progression of diabetes in this section, understand that many people with diabetes can avoid the more severe symptoms through proper blood glucose management.

Adverse Effects of Hyperglycemia

Hyperglycemia has many adverse effects on the body. First, it results from the inability of most cells to take up glucose and use it for energy. Without the use of glucose, cells are constrained in their energy resources, which in turn limits their activities. Although most cells can utilize fatty acids for energy production, to do so the fatty acids must be liberated from adipose tissue (and to a lesser extent from the liver) in lipoproteins that deliver lipids (including cholesterol) to the cells. This process takes time, and it also increases the cholesterol level in the blood, promoting atherosclerosis. Thus energy metabolism is slower and fatigue sets in more rapidly.

In contrast, most tissue damage occurring in diabetes results from elevated levels of glucose in certain cells due to the inability of these cells to decrease glucose uptake under conditions of hyperglycemia, such as retinal endothelial cells, mesangial cells in the renal corpuscle, neurons, and Schwann cells. High *intra*cellular glucose damages cells through the various mechanisms shown in **Figure 4**. Specifically, glucose can follow several metabolic pathways inside cells leading to different products.

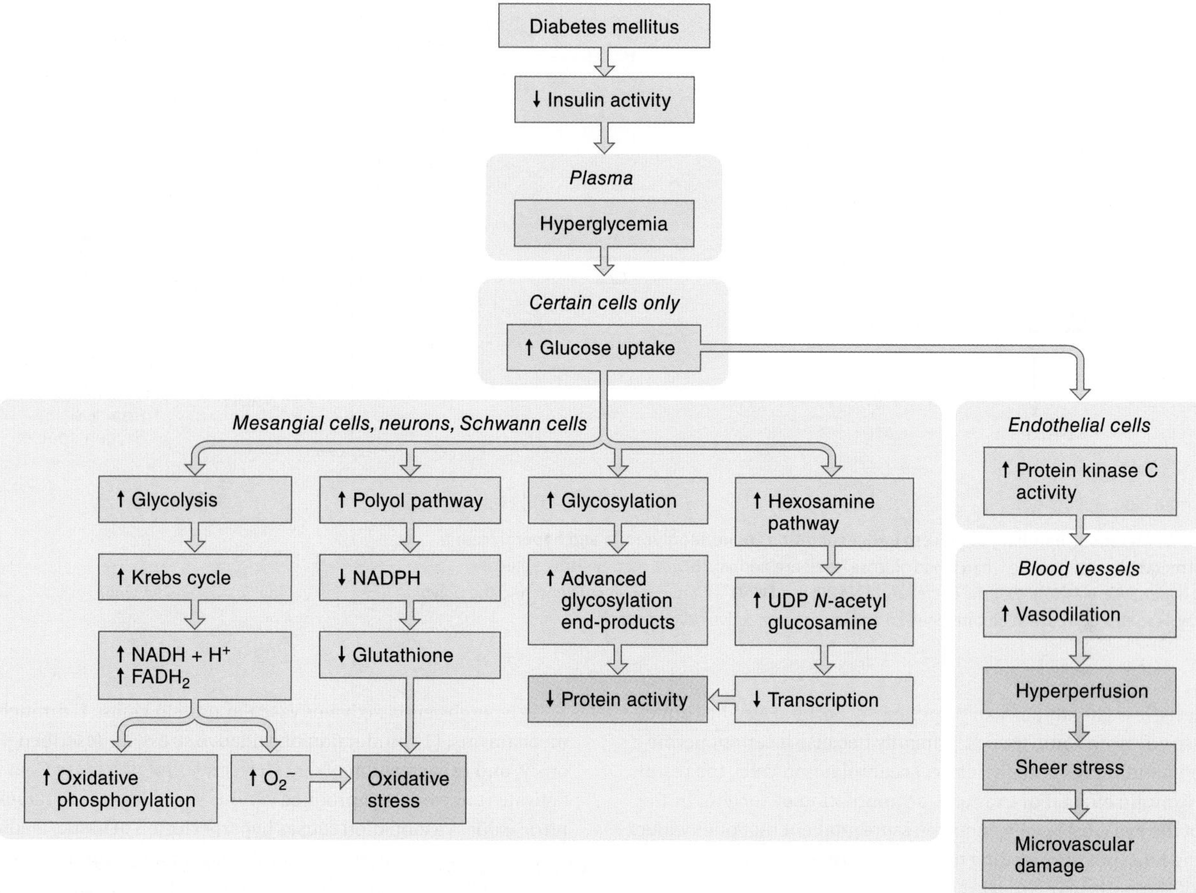

Figure 4 Pathways by which diabetes mellitus damages tissue.

Glucose reacts with several proteins in the body through **glycosylation,** the non-enzymatic addition of glucose to a protein, to form *advanced glycosylation end-products (AGEs)*. We know that glycosylation of hemoglobin to form hemoglobin A_{1c} (HbA_{1c}) provides a marker of diabetes. Almost any protein in the body can undergo glycosylation, which interferes with the protein's activity. For example, glycosylation of *fibronectin*, a protein that acts in wound healing, interferes with the wound-healing process. Other deleterious effects of AGEs in diabetes include increased capillary permeability, impaired vasodilatory responses mediated through nitric oxide, atherosclerosis, nephropathy, neuropathy, elevated levels of the low-density lipoproteins (LDLs) that transport cholesterol to body cells, enhanced oxidative stress, induction of cytokines that cause angiogenesis, and increased DNA mutation. Glycosylation is a reversible process, so it is subject to the law of mass action. With high glucose levels, the law of mass action pushes reactions toward glycosylation.

When normal levels of glucose are present in the blood, cells utilize glucose to generate ATP through glycolysis or convert it to glycogen for storage. In contrast, when present in high intracellular concentrations (hyperglycemic conditions), glucose can enter the polyol pathway, which converts glucose to sorbitol, which is then converted to fructose. That is, the polyol pathway converts monosaccharides such as glucose to their alcohol form, called a *polyol*—sorbitol is the glucose polyol (**Figure 5**). The rate-limiting enzyme in the polyol pathway is *aldose reductase*. Glucose normally does not enter the polyol pathway because hexokinase (the first enzyme of glycolysis) has a much greater affinity for glucose than does aldose reductase. High concentrations of glucose, however, can overcome the low affinity of aldose reductase. Once synthesized, sorbitol can be converted to fructose by the enzyme *sorbitol dehydrogenase*.

Much evidence supports the theory that glucose's entry into the polyol pathway contributes to diabetic symptoms. For example, aldose reductase and sorbitol dehydrogenase are found in certain tissues damaged by diabetes (including the retina, lens, cornea, kidney, Schwann cells, and endothelium); aldose reductase inhibitors relieve some complications of diabetes in humans (discussed later in this chapter); and sorbitol dehydrogenase inhibitors have been shown to reduce diabetes complications in animal studies. The polyol pathway contributes to adverse effects of diabetes by introducing oxidative stress. Both enzyme-catalyzed steps of the polyol pathway generate intermediates that increase production of reactive oxygen species. In the first step of the polyol pathway, NADPH is oxidized to NADP. Synthesis of glutathione, an antioxidant, depends on NADPH levels; thus, when NADPH is utilized in sorbitol

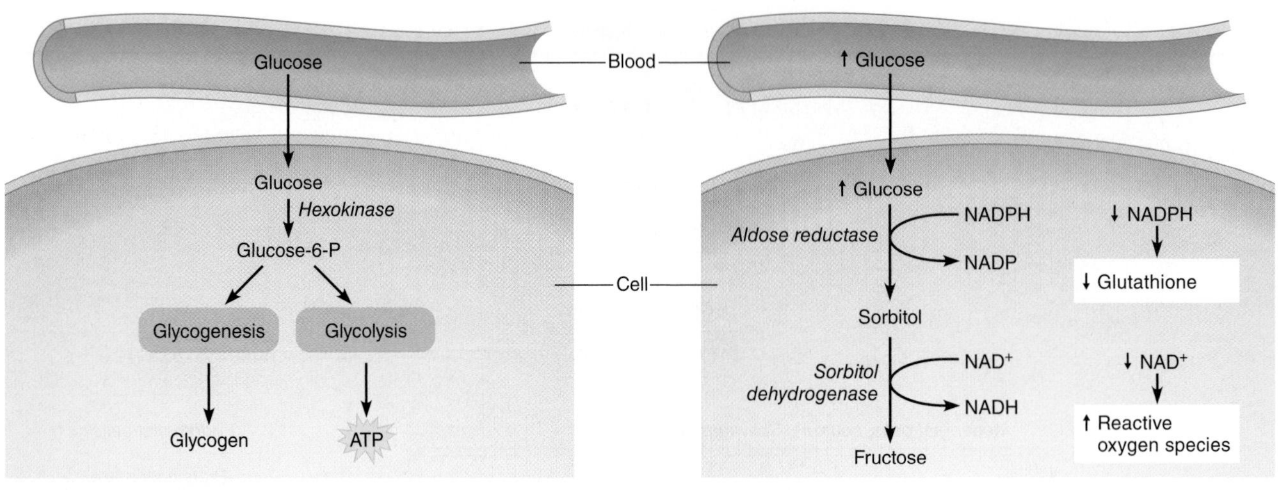

(a) Normal blood glucose

(b) Hyperglycemia

Figure 5 Metabolic handling of glucose under conditions of normoglycemia and hyperglycemia. **(a)** Cell metabolism of glucose when blood glucose levels are normal. **(b)** Cell metabolism of glucose during hyperglycemia. Note that the decreases in NADPH and NAD^+, caused by glucose entering the polyol pathway, lead to a decrease in glutathione and an increase in reactive oxygen species.

synthesis, less glutathione can be produced. Sorbitol also has direct adverse effects in some tissues, primarily because it cannot permeate cell membranes and, therefore, accumulates in cells; the result is an osmotic effect. For example, accumulation of sorbitol in the lens of the eye creates an osmotic pressure gradient that pulls water into the lens, possibly leading to cataract formation. In the second step of the pathway, NADH + H^+ is generated and can then be converted by *NADH oxidase* to reactive oxygen species.

Under hyperglycemic conditions, glucose may enter the *hexosamine pathway*, which ultimately interferes with gene expression. One of the intermediates of glycolysis, *fructose-6-phosphate,* can continue on through glycolysis or be converted to glucosamine-6-phosphate by the enzyme *glutamine-fructose-6-phosphate amidotransferase*, more simply known as GFAT. Glucosamine-6-phosphate is then converted to UDP (uridine diphosphate) *N*-acetyl glucosamine, which is inserted into threonine and serine residues of transcription factors, thereby interfering with gene expression.

Elevated intracellular glucose also increases production of free radicals, contributing to oxidative stress. High levels of glucose entering first glycolysis and then the citric acid cycle produce an excess of reduced coenzymes, NADH + H^+ and $FADH_2$. These coenzymes donate their electrons to the electron transport chain in the inner mitochondrial membrane. Recall that movement of these electrons down the chain is normally coupled to ATP synthesis by oxidative phosphorylation. The presence of too many donated electrons, however, causes the electron transport chain to back up. In turn, one of the components, coenzyme Q, is prompted to donate its electrons to an oxygen molecule, producing superoxides, O_2^- (**Figure 6**). The mitochondrial enzyme *superoxide dismutase* degrades the superoxide to hydrogen peroxide, H_2O_2, which can then be converted to water and oxygen. Oxidative stress has many adverse effects on the body, including activation of vascular protein kinase C.

Hyperglycemia activates vascular protein kinase C through two mechanisms: (1) production of oxidative stress, as described previously, and (2) activation of diacylclycerol. Protein kinase C, in turn, activates enzymes that produce various vasodilatory agents such as nitric oxide. Vasodilation causes hyperperfusion of tissue, leading to microvascular damage, and promotes development of atherosclerosis. In fact, all of the effects described previously in the discussion of hyperglycemia can induce microvasculature damage (**Figure 7**).

Effects of Diabetes Mellitus on the Microvasculature

Diabetes mellitus affects blood vessels throughout the body, thereby contributing to the widespread effects of this disease. The precise mechanism of vasculature pathology varies in type 1 DM versus type 2 DM, although hyperglycemia contributes to development of this complication in both.

In type 1 DM, hyperglycemia increases blood flow and pressure in the microcirculation, which causes shear stress to the blood vessels. Endothelial cells become damaged, scarring the microvasculature and thickening the basement membrane. This scarring, or sclerosis, limits the ability of arterioles to dilate and induces a rigidity that hinders local control of blood flow, including active and reactive hyperemia and autoregulation. As a consequence, tissues may be deprived of increased blood flow during times of need, leading to tissue damage.

Type 2 DM decreases the vasodilatory responses of the microvasculature, probably through decreased nitric oxide production. Recall that nitric oxide is a vasodilator released from endothelial cells in response to a variety of stimuli. Current theories propose that nitric oxide release is impaired in type 2 DM. A lack of nitric oxide release will leave arterioles constricted, thereby decreasing local blood flow to active tissue.

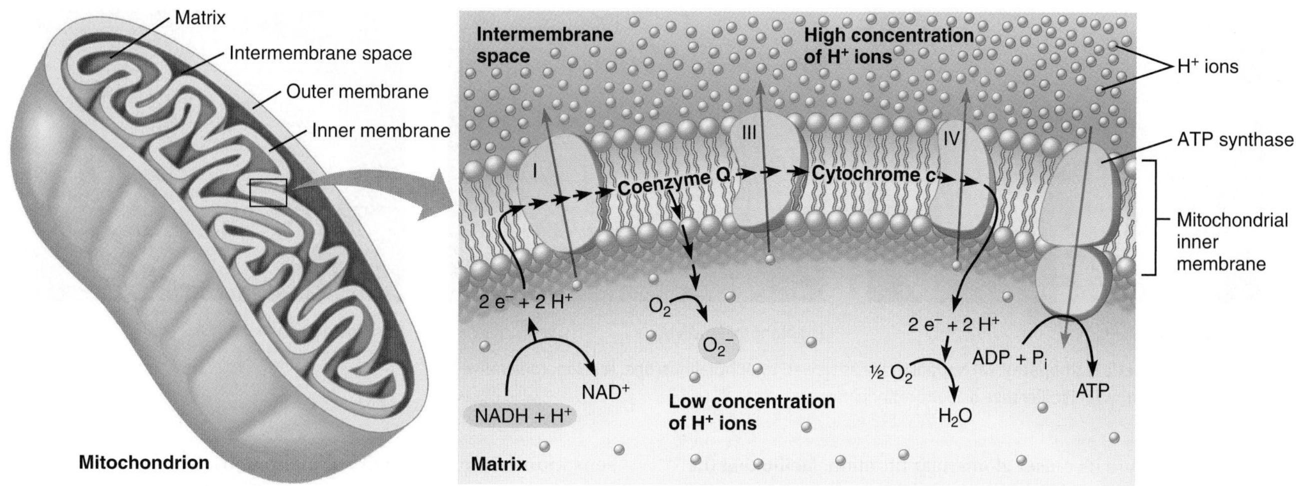

Figure 6 Generation of supraoxide through electron transport. Supraoxide is produced when electrons back up in the transport chain.

Damage to the microvasculature contributes to several diabetic conditions, including retinopathy, nephropathy, neuropathy, and delayed wound healing.

Retinopathy

Diabetic retinopathy—damage to the blood vessels of the eye—is the leading cause of adult blindness in the United States. There are two stages of diabetic retinopathy: nonproliferative and proliferative.

The disease progresses from the less severe nonproliferative stage, which starts with microaneurysms (weak bulges in the walls of blood vessels) in the retina and leads to blockage of blood vessels, to severe proliferative diabetic retinopathy, in which many new blood vessels form (angiogenesis) to compensate for the decrease in blood flow due to blockage of blood vessels (**Figure 8**). However, these new blood vessels are weak and leaky, causing blurred vision and possibly blindness.

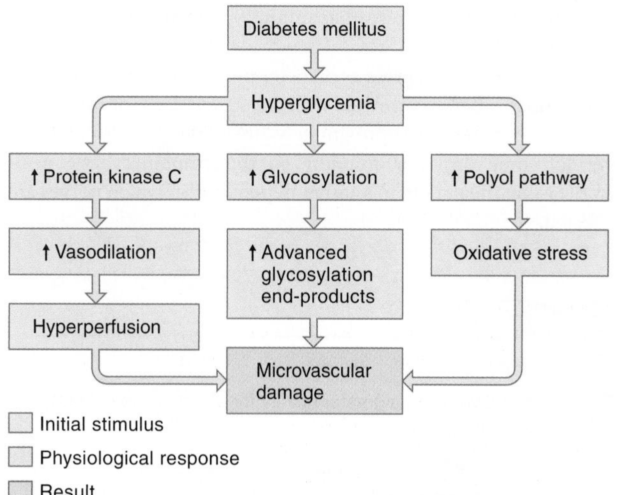

Figure 7 Diabetes, hyperglycemia, and microvascular disease.

Nephropathy

Approximately 30% of people with diabetes will develop some level of nephropathy, and diabetic nephropathy is the number one cause of end-stage renal disease in the United States. Diabetic nephropathy—damage to the kidneys—can be caused by the hyperglycemia, hypertension, and hyperlipidemia that occur with diabetes. Some individuals also appear to have a genetic predisposition to develop diabetic nephropathy; the rate of disease progression and severity appear to have genetic components as well.

Hyperglycemia causes kidney damage through glycosylation, activation of protein kinase C, release of several cytokines, and increased activity of the polyol pathway. Glycosylation of amino acids in blood and in renal tissue increases glomerular leakiness, production of free radicals, and damage to the microvasculature. Hyperglycemia and the polyol pathway cause oxidative stress, which in turn increases protein kinase C activity. Protein kinase C promotes nephropathy in two ways: (1) by triggering release of paracrines that cause vasodilation, which increases blood flow into the glomerulus, and (2) by increasing secretion of extracellular matrix by mesangial cells. Hyperglycemia triggers the release of several inflammatory cytokines, including vascular endothelial growth factor (VEGF), transforming growth factor beta (TGF-β), tumor necrosis factor (TNF), and several interleukins. These cytokines have a variety of negative effects in the kidney, including thickening the glomerular basement membrane and making it more leaky, promoting apoptosis of normal renal cells, and interfering with tight junctions between tubule cells.

Hypertension, which is one of the more serious complications of diabetes mellitus, increases the glomerular hydrostatic pressure and produces shear stress on the vasculature. The afferent arteriole, and to a lesser extent the efferent arteriole, dilates in response to paracrines released as a consequence of hypertension, thereby contributing to the elevated glomerular hydrostatic pressure as well. One such paracrine is nitric oxide. Nitric oxide levels tend to increase in the early stages of nephropathy, but decrease as the severity of the disease increases. Over time, the increased glomerular

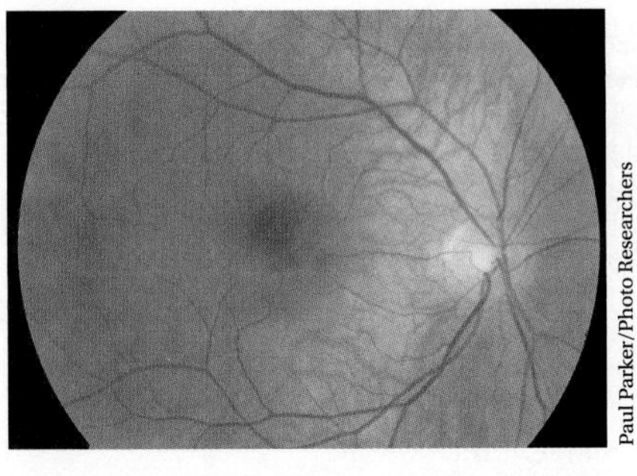
Paul Parker/Photo Researchers

(a) Normal retina

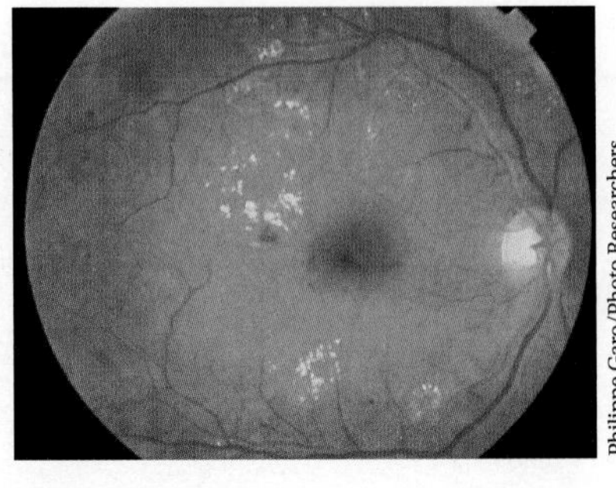
Philippe Garo/Photo Researchers

(b) Nonproliferative

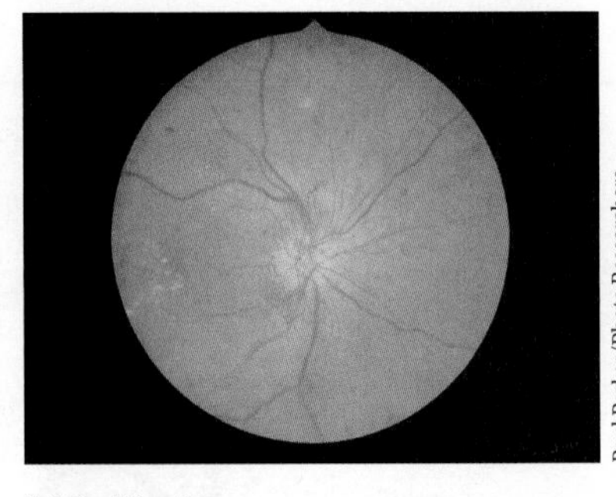
Paul Parker/Photo Researchers

(c) Proliferative

Figure 8 Diabetic retinopathy. (a) Normal retina as viewed by ophthalmoscope. **(b)** Nonproliferative diabetic retinopathy. **(c)** Proliferative diabetic retinopathy.

hydrostatic pressure increases glomerular filtration, facilitating the development of nephropathy. Increased filtration across a leakier capillary wall allows proteins to filter into the renal tubules and appear in the urine. In fact, one of the earliest signs of nephropathy is proteinuria, the presence of proteins (especially albumin) in urine. Treatment of hypertension, such as with ACE inhibitors, helps alleviate some of the problems associated with nephropathy and may even reverse some of the damage.

The symptoms of diabetic nephropathy vary depending on the severity and cause of this complication. Because the kidneys have a functional reserve (less than 20% of kidney capacity is needed for homeostasis), symptoms of diabetic nephropathy do not appear until late stages of the disease. The most common early symptom, which is used for early monitoring and diagnosis, is proteinuria. Recall that proteins cannot cross the basement membrane in the renal corpuscle; as a consequence, they normally escape glomerular filtration. When the renal corpuscle is damaged, however, filtered proteins create an osmotic force that pulls more water into Bowman's space, thereby increasing the glomerular filtration rate. Symptoms that develop later resemble those observed in other forms of kidney failure, such as fatigue, systemic edema, nausea, and headache. When the nephropathy progresses to end-stage renal disease, the glomerular filtration rate decreases as the kidneys fail to carry out their function. In this scenario, dialysis is required to sustain life.

Neuropathy

One of the most debilitating complications of diabetes mellitus is diabetic neuropathy. In severe cases, it can result in pain, autonomic dysfunction, and even amputation. Diabetic neuropathy occurs in the peripheral nervous system, affecting afferents and efferents of both the somatic and autonomic nervous systems. Glial cells are affected as well. Uncontrolled hyperglycemia leads to nerve and glial cell damage by many of the same mechanisms through which hyperglycemia causes kidney damage, including glycosylation, oxidative stress, and increased activity of the polyol pathway.

Health Fact

Because the peripheral nervous system has so many different functions, diabetic neuropathy presents with a wide range of symptoms. Neuropathies associated with sensory afferents can induce *paresthesia* (abnormal sensation of the skin such as tingling, pricking, or numbness), *dysesthesia* (unpleasant sensation produced by normally innocuous stimuli), or pain. The abnormal sensation is often perceived across a diffuse area of skin in the extremities, most often the legs and feet, although occasionally it remains limited to a single area. Altered sensation to the feet may contribute to diabetic foot ailments, as a person may be unaware of a sore on the foot. If the sore becomes infected, gangrene can set in, and amputation may ultimately be required. In fact, distal sensory diabetic neuropathy is the number one cause of nontraumatic lower extremity amputation in the United States. Neuropathy of somatic efferents (motor neurons to skeletal muscles) causes muscle weakness.

Neuropathies of the autonomic nervous system can affect almost any organ in the body. Table 2 lists some of the common symptoms of neuropathies affecting organ systems. Notably, reflex control of autonomic functions can be impaired, interfering with the body's ability to maintain homeostasis. For example, if nerves in the baroreceptor reflex (on either the afferent or efferent side) are damaged, then the body cannot respond appropriately to changes in blood pressure, such as the drop in blood pressure that occurs on standing. Such damage can cause frequent bouts of orthostatic hypotension and can lead to fainting. Cardiac autonomic neuropathy may lead to cardiac infarction, arrhythmias, and cardiac arrest.

Neuropathies progress as diabetes progresses, with the degree of neuropathy being correlated with the effectiveness of controlling blood glucose levels. Approximately 70% of people with diabetes develop some form of neuropathy, but the symptoms vary considerably from one person to another depending on which nerves are affected.

Quick Check 4

1. Define *glycosylation*.
2. Which enzyme has a greater affinity for glucose: hexokinase or aldose reductase?
3. During the early stages of nephropathy, is the glomerular filtration rate increased or decreased?
4. Which nerves may be affected by diabetic neuropathy?

TABLE 2 Diabetic Neuropathies of the Autonomic Nervous System

Organ system	Complications	Symptoms
Cardiovascular	Orthostatic hypotension Resting tachycardia	Dizziness and fainting on standing
Urinary	Loss of bladder sensation Incontinence	Overfilling of bladder and inability to completely empty, leading to urinary tract infections Uncontrolled urination
Gastrointestinal	Gastroparesis (delayed gastric emptying) Esophageal atony (decreased muscle tone) Gastric atony Colonic atony	Loss of appetite due to inappropriate feeling of fullness Gastric reflux and heartburn; difficulty swallowing Nausea; vomiting; anorexia Constipation
Skeletal muscle	Neuropathy of somatic efferents	Atrophy of muscles in hands (if cervical motor neurons are affected) or legs (if lumbar motor neurons are affected)
Reproduction	Men: erectile dysfunction, impotence Women: vaginal dryness, lack of sexual arousal	
Thermoregulation	Decreased sweating (except excessive sweating after a meal) Dampened control of cutaneous blood flow	Decreased heat tolerance Decreased ability to regulate body temperature in the thermoneutral zone as well as outside this zone
Vision	Sluggish pupil reflex, including a decreased resting pupil diameter in the dark and delayed pupillary constriction in the light	Poor vision in the dark
Cardiovascular during exercise	Decreased tolerance to exercise	Autonomic nervous system cannot adjust heart and blood vessels to meet the extra demands created by exercise
Visceral afferents	Decreased sensitivity to changes in the internal environment such as hypoglycemia	Body does not detect or adjust to hypoglycemic symptoms such as increased heart rate, palpitations, shakiness, and sweating

5 Progression of Diabetes Mellitus to Critical States

Once started, diabetes progresses for the rest of a person's life. Even when blood glucose levels are carefully controlled, some progression inevitably occurs. Many people with diabetes mellitus die because of complications of the disease. In this section we describe some of the more serious consequences of diabetes that are often encountered later in life.

Effects of Diabetes on the Macrovasculature

Macrovascular diseases affect the large blood vessels, including the coronary arteries, aorta, and large arteries in the brain and limbs.

Macrovascular disease in diabetes develops secondary to atherosclerosis, which is promoted by lipid transport, microvascular disease, and glycosylation.

Diabetes increases lipid transport to cells as well as the production of reactive oxygen species and free radicals that contribute to the development of atherosclerosis. Specifically, diabetes both increases the number of LDLs that transport lipids (including cholesterol) to the cells and decreases the number of high-density lipoproteins (HDLs) that transport lipids and cholesterol to the liver.

Macrovascular disease makes a person more prone to heart attacks, strokes, and peripheral artery disease. Such disease accounts for the deaths of 70% of people with diabetes, although generally these people have been diagnosed with diabetes for at least 15 years.

Glycosylation contributes to atherosclerosis in a number of ways—for example, through elevation of LDL levels, suppression of HDL function, and secretion of cytokines that promote cellular proliferation and plaque formation in the walls of the blood vessels. The danger of macrovasculature disease in diabetes is so severe that in cardiac risk assessments, diabetes mellitus is considered equivalent to a person having suffered a heart attack.

Diabetic Cardiomyopathy

Another serious consequence of diabetes is *cardiomyopathy*, a disease of cardiac muscle that weakens the heart. With this condition,

contractility of the cardiac muscle is reduced, which decreases stroke volume and, in turn, cardiac output. People with diabetes tend to have enlarged hearts to compensate for the weaker muscle, and they are more prone to myocardial infarction. Recent studies suggest that better control of blood glucose levels can decrease the incidence of cardiomyopathy in type 1 DM. Whether diabetic cardiomyopathy exists in the absence of macrovascular disease remains the subject of debate.

A phenomenon more commonly observed in people with diabetes than in the general population is asymptomatic ("silent") myocardial infarction. If diabetes has caused an autonomic neuropathy as well as macrovascular disease and cardiomyopathy, the lack of visceral afferent output from the heart during an ischemic episode may prevent signals from traveling to the central nervous system to communicate a painful stimulus. As a consequence, a heart attack is more likely to go unnoticed and untreated.

Quick Check 5

1. What is the relationship between atherosclerosis and macrovascular disease?
2. Macrovascular disease predisposes a person to which potential deadly complications?

6 Delayed Wound Healing

One of the more disturbing and surprising complications of diabetes is amputation. Many people with diabetes lose sensation in their feet and, therefore, do not notice wounds to them. Although normally wounds tend to heal on their own, wound healing is impaired in diabetes. To better understand abnormal wound healing, we first describe the normal wound-healing process for skin damage. (Organ damage is more complicated and will not be described here.)

Wound Healing

Wound healing is a synchronized procession of three phases following injury and blood clot formation: inflammation, proliferation, and remodeling (**Figure 9**). Although one phase leads into the next phase, some overlap occurs as healing takes place.

Phase 1: Inflammation

Within seconds following infliction of a wound, hemostasis begins as the body attempts to minimize blood loss. Blood vessels constrict, platelets aggregate and form a plug, and a fibrin clot forms. Platelets activated during hemostasis secrete cytokines and eicosanoids necessary for the progression of inflammation. The most important of these cytokines are transforming growth factor β (TGF-β) and platelet-derived growth factor (PDGF); these two growth factors stimulate chemotaxis of neutrophils, monocytes, and fibroblasts to the injured area.

Macrophages in the area of the wound immediately start the inflammatory process, which typically lasts from 24 to 72 hours. Macrophages phagocytose foreign matter, including bacteria that have infiltrated the wound. Within approximately 10 hours, neutrophils and monocytes migrate to the wound area. The monocytes mature into macrophages that, along with the neutrophils, continue the phagocytosis of debris. These activated neutrophils and macrophages also secrete cytokines, such as *basic fibroblast growth factor*, which attract fibroblasts and other cells to the wound area to initiate the next phase of wound healing.

Phase 2: Proliferation

During proliferation, the number of cells in the wound area increases and new extracellular matrix is laid down. The primary cells that lay down extracellular matrix are fibroblasts. The fibroblasts migrate to the center of the wound area, where they synthesize and secrete collagen around the fibrin meshwork to form a scaffold around which new tissue will grow. Fibroblasts also produce proteoglycans and glycosaminoglycans that form the ground substance around which granular tissue will form. Granular tissue develops deep in the wound as new capillaries grow up around the collagen scaffold. Fibroblasts also secrete cytokines, including PDGF, TGF-β, keratinocyte (epidermal cell) growth factor, and insulin-like growth factor.

While a collagen scaffold is forming in the interior of the wound, keratinocytes proliferate and migrate up to 3 cm across the opening of the wound to seal it from the exterior. Endothelial cells proliferate as part of angiogenesis, because an enhanced blood supply is needed to provide oxygen and nutrients supporting the growth of the new tissue. The edges of the wound may move toward each other (contracture), although the extent of contracture in humans is limited. The proliferative phase starts in the middle of the inflammation and can last for as long as three weeks.

Phase 3: Remodeling

The last phase of wound healing entails formation of new tissue, or remodeling. Fibroblasts continue to lay down collagen, but now the collagen becomes organized and forms cross-links so that the wound can withstand mechanical stress. The new *scar* tissue is weaker than the original tissue. Remodeling can take anywhere from three weeks to a year.

Altered Wound Healing in Diabetes

Diabetes slows down wound healing, and several factors contribute to this impairment. For example, both the inflammatory and proliferative phases of wound healing are prolonged in diabetes. Glycosylation decreases the ability of neutrophils and macrophages to phagocytose bacteria, slowing down the inflammatory phase. Diabetes causes erythrocytes to lose their flexibility, which decreases oxygen delivery to the wound area. Angiogenesis is also decreased in diabetes, further limiting blood flow to the wound. Because the remodeling phase of wound healing is an energy-intensive process, the lack of oxygen and nutrients slows the healing. Diabetes also hinders wound healing by decreasing secretion of collagen, cytokines, and growth factors, and by decreasing keratinocyte and fibroblast proliferation and migration.

Foot injuries in people with diabetes often result in *diabetic foot ulcers* (**Figure 10**). These ulcers look similar to a healing wound, but because neuropathy may suppress the sensation of pain, the person may be unaware that the wound is not healing properly. For 84% of people with diabetes who ultimately require leg amputation, the problem started with a diabetic foot ulcer.

Quick Check 6

1. What are the three phases of wound healing?
2. Name the various cell types involved in the wound-healing process, and identify the phases in which they play a role.
3. What do fibroblasts secrete?

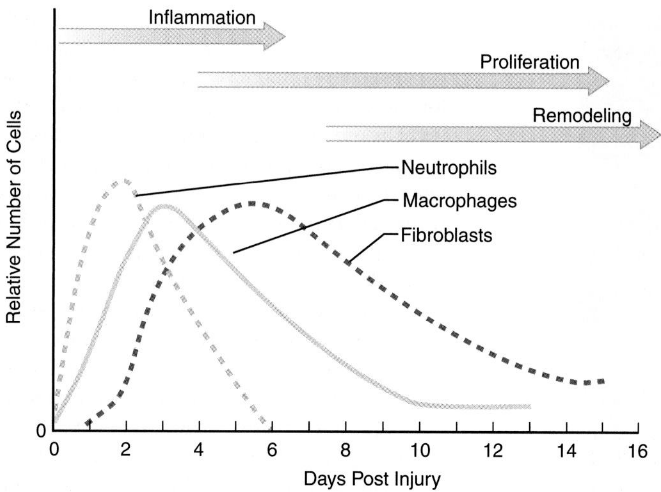

(a) Time frame of wound healing

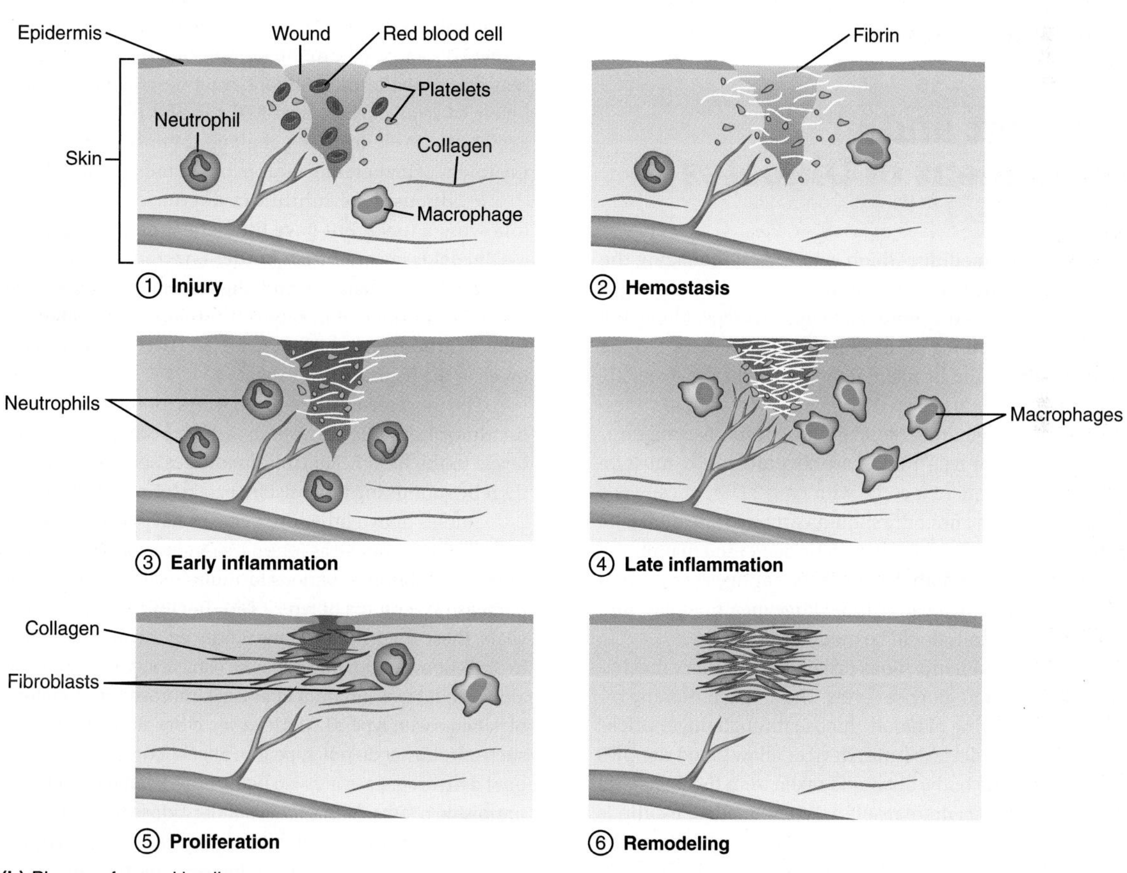

(b) Phases of wound healing

Figure 9 Normal wound healing. (a) Time frame of the different phases of wound healing. The times shown are approximate, as they vary with the severity and type of wound. **(b)** Phases of wound healing.

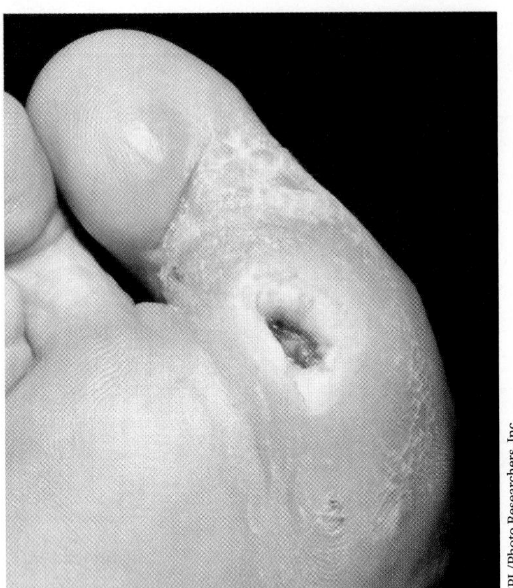

Figure 10 Diabetic foot ulcer.

7 Treatment and Management of Diabetes Mellitus

Treatment of diabetes mellitus, which is based on managing the disease (there is no cure), has been touched on briefly throughout this chapter, but deserves more thorough coverage. Managing diabetes requires maintaining normal blood glucose levels. Fasting blood glucose levels should be maintained at less than 100 mg/dL, and hemoglobin A_{1c} should be less than 7.

Fundamental to regulating blood glucose levels is consumption of a healthy diet. For type 1 DM, carbohydrate intake must be regulated. However, for many people with type 2 DM, obesity is a contributing factor. Thus, not only should carbohydrate intake be regulated, but overall intake also needs to be decreased to promote weight loss. In all cases of diabetes mellitus, eating several small meals each day is better than eating three large meals per day, and carbohydrate consumption should be moderate and slow.

For type 1 DM, insulin injections provide the primary mechanism for regulating blood glucose levels. Use of such therapy requires frequent monitoring of blood glucose through finger pricks and glucose monitoring devices (most devices allow blood samples from other regions of the body, but blood taken from the finger still tends to yield the most accurate results). When needed, insulin is injected subcutaneously, as proteins cannot be absorbed from the intestinal tract. Three types of insulin preparations—rapidly acting, intermediate acting, and prolonged acting—provide different therapeutic approaches to regulating blood glucose. In addition, a subcutaneous insulin pump can be used to administer insulin on a more regular basis. However, insulin injections themselves can be dangerous in that if a person takes too much, the individual will experience a rapid and large drop in blood glucose, which can lead to hypoglycemic shock or coma.

For type 2 DM, four classes of oral hypoglycemic agents are used to regulate blood glucose levels: sulfonylureas, biguanides, α-glucosidase inhibitors, and thiazolidinediones. The use of sulfonylureas to treat diabetes came about accidentally during studies of its use in treating typhoid fever during World War II. In 1942, Dr. Marcel Janbon and colleagues found that sulfonylureas stimulated insulin release from beta cells of the pancreas. More recently, investigators have shown that sulfonylureas bind to receptors on ATP-sensitive potassium channels in the membrane of beta cells. Sulfonylureas close these potassium channels, causing depolarization of the beta cells, which opens voltage-gated calcium channels. Calcium then enters the beta cells and triggers insulin release by exocytosis. More than 60 years after Janbon et al.'s discovery, sulfonylureas remain the drugs most commonly used to treat type 2 DM.

Biguanides act by reducing glucose production in the liver and increasing the sensitivity of receptors to insulin. Because these medications lower blood glucose without altering insulin secretion, they are prescribed to treat both type 1 DM and type 2 DM. Biguanides decrease gluconeogenesis in the liver and increase glucose uptake into cells by as yet unknown mechanisms. The most commonly used agent, metformin (Glucophage), appears to activate AMP-activated protein kinase, which regulates the switching between anabolic and catabolic reactions. Biguanides also decrease levels of triglycerides and LDLs in the blood while producing a slight increase in HDLs, making them suitable options for the treatment of diabetes mellitus in obese individuals.

α-Glucosidase inhibitors decrease blood glucose levels following a meal, but have little effect on fasting glucose levels. α-Glucosidases are brush border enzymes that hydrolyze disaccharides, trisaccharides, and oligosaccharides to glucose or other monosaccharides. Blockage of these enzymes decreases absorption of carbohydrates. Thus, to be effective, α-glucosidases must be taken at the beginning of a meal.

The most recently approved drugs for treatment of diabetes are the thiazolidinediones, which reduce both glucose and fatty acid levels in the blood. Thiazolidinediones sensitize tissue to insulin, thereby countering the resistance to insulin that causes type 2 DM.

Studies of two other drug classes have shown that these agents have limited promise as future treatments for type 2 DM. Aldose reductase inhibitors of various formulas have been tested as means to decrease symptoms of type 2 DM. In clinical trials over the last ten years, these medications have produced inconsistent improvements in diabetic symptoms. Six separate clinical studies have shown that the aldose reductase inhibitor epalrestat improves neuropathy symptoms of patients with type 2 DM. However, other aldose reductase inhibitors such as tolrestat do not appear to affect neuropathy. Inhibitors of sorbitol dehydrogenase are also being tested, but have yet to yield highly promising results. At this time, neither aldose reductase inhibitors nor sorbitol dehydrogenase inhibitors have been approved by the U.S. Food and Drug Administration (FDA) for use in the treatment of type 2 DM.

Quick Check 7

1. Which drug is essential for the treatment of type 1 DM?
2. Which four classes of drugs are used to treat type 2 DM?
3. Why are the sulfonylureas unsuitable for treatment of type 1 DM?

8 Current Research on Diabetes Mellitus

As with most diseases, research on diabetes mellitus tends to focus on either better treatment mechanisms or a more thorough understanding of the disease, which could lead to better treatment mechanisms. Thousands of published articles report on diabetes research each year, and PubMed (a database maintained by the U.S. National Library of Medicine and the National Institutes of Health) lists 26 journals with "diabetes" in their title. This list does not include magazines designed for the layperson. A thorough review of diabetes research is far beyond the scope of this text, but this section provides an overview of some of the ongoing diabetes research.

New Techniques for Insulin Administration

For virtually all cases of type 1 DM and for many cases of type 2 DM, regular subcutaneous injections of insulin are necessary to control the disease. Much research on diabetes is directed toward discovery of more convenient mechanisms of insulin administration. Insulin cannot be administered orally because it is broken down to amino acids in the gastrointestinal tract. Various alternative routes of delivery have been explored, including rectal, ocular, nasal, and oral administration in an encapsulated form that protects insulin from digestion. To date, these routes have had limited success.

Pharmaceutical companies continue exploring the pulmonary route, in recognition of the large surface area in the lungs available for absorption. Inhalers with various forms of insulin (powder, mist, and nebulizers) are currently being studied, and in 2006 the FDA approved the first inhalable form of insulin, Exubera. The company marketing Exubera inhalers, Pfizer, later withdrew the product due to poor sales. Some studies also suggest that inhaled insulin promotes tumor growth in the lungs of smokers and can lead to death.

Oral forms of insulin are the preferred choice, but research is far from finalizing a form that allows adequate absorption across the gastrointestinal tract. The rectal and ocular routes of administration are no more convenient than injections.

Development of New Medicines for the Treatment of Diabetes Mellitus

Although several medications exist to treat diabetes, no one medication has been able to cure diabetes or eradicate all the symptoms. Researchers continue to develop and test drugs, such as aldose reductase inhibitors and sorbitol dehydrogenase inhibitors, to better manage the disease (and possibly even cure it). In addition, both long-term and short-term glucagonlike peptide-1 (GLP-1) inhibitors are currently being tested as diabetes therapies, and some have been approved for clinical use. GLP-1 inhibitors block the actions of glucagon, the pancreatic hormone that promotes gluconeogenesis and glycogenolysis. In 2010, the FDA approved a short-term GLP-1 inhibitor, Victoza, which has to be injected once a day to regulate blood glucose. A long-term GLP-1 inhibitor, Bydureon, received FDA approval in 2012; one injection of this drug regulates blood glucose levels for a week. Both of these drugs are still being studied for potentially dangerous side effects.

Obtaining FDA approval to market a new drug takes several years and requires studies of the proposed therapy's efficacy and toxicity. Unfortunately, once approved, some drugs need to be withdrawn from the market when new problems are discovered. Thus getting new medicines to the patients who need them sometimes takes longer than desired. The U.S. Congress passed the Orphan Drug Act in 1983 to expedite research and development of drugs that treat a rare disease or condition. In May 2012, the FDA granted orphan status to DiaPep277, a drug that may prevent the autoimmune response against beta cells in type 1 DM. DiaPep277 is in late-stage clinical trials and could prevent the development of type 1 DM.

Nonpharmaceutical Therapies for the Treatment of Diabetes Mellitus

As described earlier, there is no simple way to manage diabetes and there is no cure for this disease. Nevertheless, scientists have not given up the search for a cure, which would eliminate the need for continual treatment. Type 1 DM is a more homogeneous disease than type 2 DM, making it a better candidate for such research. To cure type 1 DM, the body must be supplied with a mechanism for producing and releasing insulin when blood glucose levels are elevated, and this system must turn itself off when blood glucose levels return to normal. Such a cure would also benefit many people with type 2 DM, as beta cells tend to stop functioning in that disease as well.

Much research has focused on increasing the number of pancreatic beta cells in people with diabetes. Some researchers are trying to stimulate beta cells (as well as all islet of Langerhans cells) to replicate and repopulate the pancreas. Other researchers are studying the prospect of transplanting islet cells from a donor into a diabetic recipient. Most recent studies propose transplantation of stem cells into the pancreas, followed by stimulation to encourage these precursor cells to develop into beta cells. Gene therapy may also be used in the treatment of diabetes mellitus. For example, genes that produce insulin might be injected into other cells of the body, or the genes responsible for the autoimmune response that causes diabetes mellitus might be manipulated in some way. Most studies addressing these potential routes of therapy are still in early stages, with the treatments being tested in animal models. All of the techniques have met with some success in these animal studies, however.

Another strategy for treating type 1 DM is preventing the loss of beta cells. Beta cells undergo apoptosis in response to a genetic abnormality or viral infection. This programmed cell death is triggered by cytokines. If these cytokines can be identified, development of inhibitors to them might be able to stop the development of type 1 DM. Such treatment has some challenges, however. For example, it would require detection earlier than current diagnostic capabilities allow.

Because type 1 DM accounts for only 5–10% of all cases of diabetes, even if the previously described cures were developed and adopted worldwide, the diabetes epidemic would still exist. Curing type 2 DM will be much more complicated and will involve manipulating the biochemical pathways by which insulin acts. Gene therapy may prove beneficial if a common receptor or enzyme is found to be malfunctioning in a large percentage of patients with type 2 DM. Enzymes currently being studied for this possibility include

UNDERSTANDING EXERCISE

Glucose Transport in Exercising Muscle and Diabetes

The benefits of exercise in controlling diabetes have long been recognized, but the mechanisms underlying these effects remain poorly understood. Both aerobic and anaerobic exercise have been shown to be beneficial in persons with diabetes. Recent studies suggest that exercise improves glucose metabolism in diabetes both immediately and with a regular exercise program. Following a single bout of activity, exercise decreases blood glucose levels. With regular exercise, insulin sensitivity of tissues increases.

Skeletal muscle uses glucose as its primary energy source during exercise, turning to fatty acids and amino acids only when glucose availability is limited. Glucose availability depends not only on the plasma concentration of glucose, but also on the transport of glucose from blood into the muscle cells. Skeletal muscles have three types of glucose transporters: GLUT1, GLUT4, and GLUT5. GLUT4 is much more prevalent than the other forms. Although insulin regulates GLUT4 levels in muscle during rest, these levels increase during muscle contraction through an insulin-independent mechanism. GLUT4 transporters are located in cytosolic vesicle membranes and are inserted into the plasma membrane by exocytosis. Insulin triggers insertion of GLUT4 transporters into cell membranes by binding to insulin-specific receptors with tyrosine kinase activities. The tyrosine kinase catalyzes phosphorylation of another component of the insulin receptor, which then activates phosphatidylinositol 3-kinase. Phosphatidylinositol 3-kinase triggers exocytosis. Muscle contraction also triggers insertion of GLUT4 carriers by exocytosis, but the signaling mechanism, although unknown, differs from that for insulin. Several studies have suggested that separate pools of vesicles containing GLUT4 exist—one responsive to insulin, the other to contracting muscle. Studies also suggest increases in intracellular calcium levels during contraction serve as a signal for exocytosis, albeit not directly. Protein kinase C has also been implicated in contraction-induced increases in glucose transport.

In diabetes mellitus, although glucose transport into skeletal muscle at rest is impaired, glucose transport increases during exercise. Thus the hyperglycemia of diabetes mellitus can be immediately decreased by a single bout of exercise. For as yet unknown reasons, blood glucose levels remain decreased for several hours following exercise as well.

Regular exercise promotes an increase in sensitivity to insulin, thereby countering the insulin resistance observed in type 2 DM and creating more sensitive tissue to the decreased insulin in type 1 DM. Some of the beneficial changes with regular exercise may arise secondary to weight loss and increased muscle mass. In addition, weight-independent changes are thought to occur.

Further studies on the benefits of exercise in controlling diabetes may lead to new treatments for this chronic disease. For example, if the biochemical pathway for insulin-independent expression of GLUT4 transporters could be determined, then new drugs targeting this pathway might be used to increase GLUT4 expression. An understanding of the pathways by which regular exercise increases insulin sensitivity may also lead to the development of new drugs.

AMP-activated protein kinase. With a better understanding of type 2 DM, other proteins may emerge as targets of gene therapy in the future.

Because of the association of type 2 DM with obesity, one treatment that seems to be quite successful is surgery for weight loss (**Clinical Connections: Bariatric Surgery as a Treatment for Diabetes**). In one study, surgery performed on obese people with type 2 DM resolved the diabetes in 75% of the cases. Other mechanisms of treating obesity have also led to reductions in type 2 DM. Research continues on the associations between obesity, diabetes, exercise, and aging, and seeks to explain how exercise may be used to prevent obesity and diabetes even as a person ages. Learn more about the benefits of exercise in **Understanding Exercise: Glucose Transport in Exercising Muscle and Diabetes**.

Quick Check 8

1. What are some different mechanisms for insulin administration that are being studied?
2. What are some mechanisms for potentially increasing the number of beta cells in type 1 DM?

SYSTEMS INTEGRATION

As we've seen, diabetes mellitus affects every body system. It does so because our organ systems must work together to maintain homeostasis. Although we often learn about physiology one system at a time, systems do not work in isolation in real life; instead, they depend on one another to maintain homeostasis. A disturbance in one system often causes disturbances in other systems, and nowhere is this more apparent than in diabetes mellitus. This chapter has sought to demonstrate the integrated actions of the organ systems by seeing what happens when disease attacks one system—in this case, the endocrine system. To fully appreciate and understand physiology, you must look at the body as a whole, rather than as just a collection of organ systems.

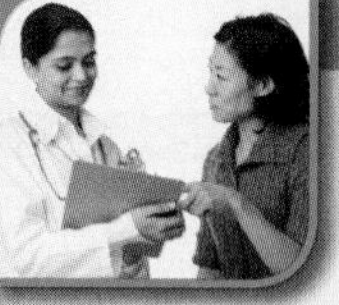

CLINICAL CONNECTIONS

Bariatric Surgery as a Treatment for Diabetes

Obesity predisposes a person to many diseases, including hypertension, heart disease, osteoarthritis, sleep apnea, gastroesophageal reflux, and type 2 DM. Some people who have difficulty losing weight through traditional methods often also have genetic predispositions toward obesity. However, they have an alternative therapy available to them: bariatric surgery.

Bariatric surgery promotes weight loss by decreasing the amount of food intake (restrictive) and the amount of nutrients absorbed (malabsorptive). Two major types of bariatric surgeries are performed today: (1) gastric banding or stapling and (2) gastric bypass. Both are effective in weight loss and control, although gastric bypass is considered the superior option.

Gastric banding (or stapling) is a purely restrictive procedure involving separation of the stomach into two parts, limiting the volume of food that can enter the stomach. An inflatable band is placed around the upper part of the stomach and inflated until only a small opening exists between the upper stomach and the rest of the stomach (Figure a). This band can be inflated or deflated, making this an adjustable and reversible procedure. In gastric stapling, surgical staples are used to divide the stomach into upper and lower partitions with a band placed between the two partitions to create a small pouch (Figure b). In both gastric banding and stapling, the small upper portion of the stomach fills more quickly than it can empty into the lower portion, generating the sensation of fullness with less food intake.

Gastric bypass, a more extensive surgical procedure, is both restrictive and malabsorptive. Several types of gastric bypass surgeries exist, the most common being roux-en-Y gastric bypass (RGB). In this procedure, stapling creates a small pouch about the size of a walnut in the upper stomach. A bypass is then created, whereby the pouch empties into the jejunum of the small intestine, skipping the lower stomach and duodenum (Figure c).

More extensive gastric bypass procedures include biliopancreatic diversion (BPD) alone and biliopancreatic diversion with duodenal switch (BPD-DS). In BPD, approximately 75% of the lower stomach is removed and a bypass is created from the remaining stomach to the ileum. In BPD-DS, 75% of the stomach is removed, leaving a vertical sleeve from the esophagus to the duodenum. A bypass is formed from the initial section of the duodenum to the ileum, eliminating most of the small intestine from the gastrointestinal tract. A portion of the duodenum is left intact because most iron and calcium are absorbed there. Because most nutrients are absorbed in the small intestine, even if a person eats the same quantity of food as prior to surgery, most of the calories avoid absorption and are eliminated in the feces, causing weight loss.

Bariatric surgery is not without risks, which is why it is used only in cases of chronic and severe obesity. In past years, the requirement for bariatric surgery was a body mass index (BMI) of 40 (approximately 80 pounds overweight for women and 100 pounds overweight for men). Today, as a result of the surgery's success rate, a BMI between 35 and 40—coupled with a secondary disease or predisposition to a secondary disease such as DM—qualifies people for surgery. The amount of weight loss varies considerably between individuals and sustained success depends to a great extent on lifestyle changes, including a healthy diet and regular exercise. With BPD-DS, patients lose on average 75–80% of their excess weight; with RGB, they lose on average 60–70% of their excess weight. The incidence of type 2 DM is reduced when these procedures are performed, and the progression of the disease can be reversed following surgery. Bariatric surgery also reduces blood pressure in people with hypertension and decreases the incidence and complications of sleep apnea, osteoarthritis, and gastroesophageal reflux.

The risks of bariatric surgery include infection, herniation, and malnutrition. Infection is a risk following any surgery, but bariatric surgeries have the added danger of gastric contents leaking out of the gastrointestinal tract, through incisions or staples, into the abdominal cavity. Infections caused by such leakage are treated with antibiotics and may require surgery. Hernias occur in approximately 3% of people who undergo laparoscopic bariatric surgery and in approximately 25% of individuals who require abdominal incisions for the surgery; such complications require surgical correction. Malnutrition—especially vitamin and iron deficiencies—occurs in approximately 40% of patients who undergo bariatric surgery, requiring the use of dietary supplements. A common side effect following surgery, caused by consumption of a high-carbohydrate meal, is *dumping syndrome,* which results in nausea, abdominal pain, sweating, and diarrhea. More severe but infrequent risks include blood clots in the legs and even death (the incidence of death is less than 1% in those who undergo RGB surgery and approximately 3% in those who receive BPD-DS surgery).

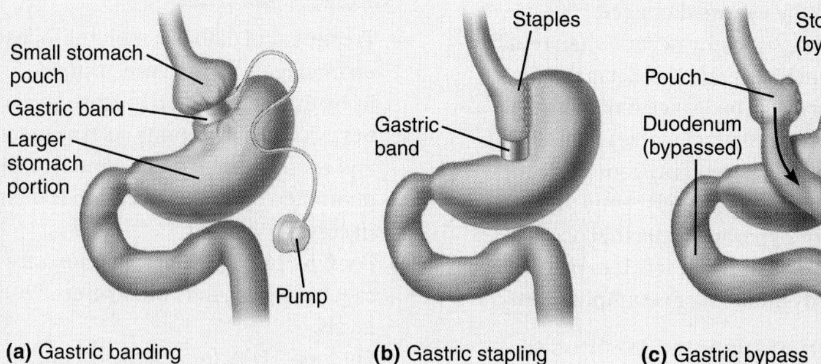

(a) Gastric banding **(b)** Gastric stapling **(c)** Gastric bypass

Surgical gastric banding options.

Critical Thinking Questions

1. Explain how bariatric surgery can cure some cases of type 2 DM.
2. Gastric banding is classified as a restrictive procedure, whereas gastric bypass is classified as both restrictive and malabsorptive. Explain the distinctions between these classifications.
3. Describe the risks and benefits of bariatric surgery procedures.

CHAPTER REVIEW

SUMMARY

1 Classification of Diabetes Mellitus

- The two categories of diabetes mellitus (DM) are type 1 and type 2.
- Type 1 DM is further subdivided into type 1a DM and type 1b DM.
- Type 1a DM is an autoimmune disease whereby the immune system destroys pancreatic beta cells causing the body to produce insufficient amounts of insulin.
- Type 1b DM (idiopathic type 1 DM), which is rarer than type 1a DM, involves the destruction of pancreatic cells, but the mechanism of this destruction is not known.
- In type 2 DM, target cells show reduced sensitivity to insulin.
- The number of people with both types of diabetes mellitus is increasing, although type 2 DM is by far the more prevalent, accounting for 90% of all diabetes mellitus cases.

2 Acute Effects of Diabetes Mellitus

- Acute or episodic events of diabetes mellitus vary from relatively mild to severe and life-threatening.
- Acute hyperglycemia, or extremely elevated blood glucose levels, can lead to ketoacidosis.
- Ketoacidosis, a more common complication in type 1 DM, results from increased lipolysis that leads to the production of ketones, a type of organic acid.
- Mild ketoacidosis causes dehydration and electrolyte imbalances; severe ketoacidosis can lead to coma and death.
- Elderly patients with diabetes may suffer from hyperosmolar nonketotic coma, which arises when loss of glucose in the urine creates a severe dehydration such that blood osmolarity increases to 325 mOsm or greater.
- Hypoglycemia symptoms vary from confusion and dizziness to seizures and loss of consciousness.
- Hypoglycemic coma can occur when a person with diabetes takes too much insulin, causing blood glucose to drop to dangerously low levels, resulting in coma and possibly death.

3 Chronic Complications of Diabetes Mellitus: Early Stages

- The "three polys"—polyuria, polydipsia, and polyphagia—are the earliest symptoms of diabetes mellitus.
- Polyuria, or excessive urination, is a direct result of the hyperglycemia.
- Polydypsia, or the sensation of thirst, arises from the water loss and eventual dehydration from polyuria.
- Polyphagia occurs when insulin does not signal satiety and the body does not shift to the fed (satiated) state.

4 Chronic Complications of Diabetes Mellitus: Advanced Stages

- The chronic complications from diabetes mellitus are mostly related to hyperglycemia or damage to the microvasculature.
- Even though glucose is plentiful in the blood when a person has diabetes, cells cannot use it for energy and must turn to fatty acids and amino acids instead, thereby inducing lethargy in the person with DM.
- Hyperglycemia leads to glycosylation of various proteins throughout the body, and promotes the conversion of glucose to sorbitol.
- Diabetes mellitus weakens blood vessels throughout the body, causing sclerosis of the microvasculature, which minimizes the ability of local factors to regulate blood flow based on demand.
- Damage to the microvasculature may contribute to retinopathy, nephropathy, neuropathy, and delayed wound healing.
- Diabetic retinopathy occurs when blood vessels in the eye are damaged.
- Diabetic nephropathy occurs when renal epithelium is damaged, and it is the leading cause of end stage renal disease.
- Diabetic neuropathy involves damage to the peripheral nervous system and can be caused by the hyperglycemia, hypertension, or hyperlipidemia that occurs with diabetes, and can result in pain, autonomic dysfunction, and amputation.

5 Progression of Diabetes Mellitus to Critical States

- Once started, diabetes progresses for the rest of a person's life; many people with DM die because of complications from the disease.
- One complication of DM is macrovascular disease, which affects the large blood vessels and makes a person more prone to heart attack, stroke, peripheral artery disease, and cardiomyopathy.
- Diabetes increases lipid transport to cells as well as the production of reactive oxygen species and free radicals that contribute to the development of atherosclerosis.

6 Delayed Wound Healing

- In diabetes, the process of wound healing is slowed considerably.
- When a person suffers a wound, macrophages are recruited to the area of damage and initiate the inflammatory process.
- Neutrophils and monocytes follow macrophages to the wound area and secrete chemicals to attract fibroblasts to the area.
- Fibrobasts start the proliferative phase by laying down new extracellular matrix; they also secrete chemicals that stimulate keratinocyte proliferation.
- During remodeling, the collagen laid down by fibroblasts becomes cross-linked to form scar tissue.
- In diabetes, the wound healing process is delayed during both the inflammatory and proliferative phases, and secretion of collagen and several chemical mediators of the wound-healing process is decreased.
- Hyperglycemia decreases the activity of neutrophils and macrophages.

7 Treatment and Management of Diabetes Mellitus

- Treatment of diabetes mellitus focuses on managing the disease, primarily by managing blood glucose levels; behavioral adjustments such as diet and exercise help in this endeavor, but pharmacological intervention is often necessary.
- For type 1 DM, insulin injections are required to regulate blood glucose levels.
- For type 2 DM, four classes of oral hypoglycemic agents are used to regulate blood glucose levels: sulfonylureas, biguanides, α-glucosidase inhibitors, and thiazolidinediones.

8 Current Research on Diabetes Mellitus

- Research on DM focuses on better treatment mechanisms, new medications for treatment, or a more thorough understanding of the disease.

- Some of the current research areas associated with better treatment include better routes for insulin administration, beta cell transplantation, and genetic modification of cells to produce insulin-secreting cells.
- Research on new medicines, including aldose reductase inhibitors and sorbitol dehydrogenase inhibitors, is ongoing.
- Much research on the mechanisms of disease progression in diabetes focuses on chemical mediators that cause the disease or cause symptoms of the disease.
- Once the mediators have been identified, drugs can be developed to target those chemicals.

Check your understanding at MasteringA&P and apply your new knowledge to real-life case studies to ensure that you've mastered the content!

EXERCISES

Multiple-Choice Questions

1. Which of the following best describes type 1 DM?
   a) Bacterial infection of the adrenal medulla
   b) Viral infection of the adrenal medulla
   c) Autoimmune disease of the endocrine pancreas
   d) Autoimmune disease of the exocrine pancreas
   e) Disease of chronic obesity
2. Which of the following is *not* a symptom of ketoacidosis?
   a) Acetone breath
   b) Increased ventilation
   c) Increased lipolysis
   d) Muscle tremors
   e) Increased sympathetic activity
3. What causes a diabetic coma?
   a) Severe hyperglycemia
   b) Too much insulin, which causes severe hypoglycemia
   c) Acidosis
   d) Increased sympathetic activity to blood vessels of the brain
   e) Vasoconstriction of blood vessels to the brain
4. What are the "three polys" of diabetes mellitus?
   a) Polydipsia, polyphagia, polyuria
   b) Polyglycemia, polylipemia, polyuria
   c) Polydipsia, polyglycemia, polycythemia
   d) Polycythemia, polyphagia, polyuria
   e) Polyphagia, polyinsulinemia, polydipsia
5. Glycosylation of which protein is used as a marker for diagnosing diabetes?
   a) Insulin
   b) Hexokinase
   c) Hemoglobin
   d) Aldose reductase
   e) Protein kinase C
6. Proliferative retinopathy is caused by which of the following?
   a) Microaneurysms causing blood vessels in the retina to leak
   b) Extensive angiogenesis in the retina resulting in weak, fragile blood vessels that leak
   c) Increased production of rods and cones such that they overlap and interfere with each other's function
   d) Extensive development of the choroid, putting pressure on the optic nerve
   e) Extensive development of the retinal pigmented epithelium, putting pressure on the optic nerve
7. What is an early symptom of nephropathy used for diagnostic purposes?
   a) Glucose in the urine
   b) Albumin in the urine
   c) Blood in the urine
   d) Excessive thirst
   e) Dehydration
8. Which of the following is *not* a phagocyte?
   a) Fibroblasts
   b) Neutrophils
   c) Macrophages
   d) Monocytes
9. Which of the following is *not* used to treat type 2 DM?
   a) Sulfonylureas
   b) Biguanides
   c) α-Glucosidase inhibitors
   d) Thiazolidinediones
   e) Phenylephrine
10. Which of the following is an enzyme currently being studied as a target for gene therapy of diabetes mellitus?
    a) Hexokinase
    b) Protein kinase C
    c) Aldose reductase
    d) AMP-activated protein kinase
    e) Tyrosine kinase

Objective Questions

11. Which is the more prevalent, type 1 or type 2 diabetes mellitus?
12. Which is an autoimmune disease, type 1 or type 2 diabetes mellitus?
13. The non-enzymatic addition of glucose to a protein is called ________.
14. Acidosis occurs in diabetes due to the production of ________.
15. Define polyuria, polydipsia, and polyphagia.
16. Amputation in diabetes occurs secondary to ________.
17. Name the three phases of wound healing. Which phases are delayed in diabetes?
18. How do sulfonylureas increase insulin secretion?
19. True or False: Weight-loss surgery can reverse some cases of type 2 DM.
20. What is a diabetic coma?

Essay Questions

21. Compare type 1 and type 2 DM with respect to their causes, effects, and treatment.
22. Describe the mechanisms by which researchers hope to increase the number of active beta cells in the pancreas in persons with diabetes.
23. Describe the direct adverse effects of hyperglycemia.

Critical Thinking

24. Using information from ***The Endocrine System: Regulation of Energy Metabolism and Growth,*** describe the different whole-body metabolic changes that occur in diabetes mellitus due to the decreased activity of insulin.
25. Describe some behavioral modifications that should improve the health of a person with diabetes mellitus.

Photo credits for icons in feature boxes: Focus on Diabetes: Tanyae/Dreamstime; Clinical Connections: Iofoto/Dreamstime; Understanding Exercise: Pixattitude/Dreamstime; Discovery: Yuri Arcurs/Cutcaster.

Glossary

ketoacidosis the buildup of ketones in blood due to breakdown of stored fats for energy

Answers

Answers to End-of-Chapter Multiple Choice and Objective Questions

Multiple Choice Questions

1. c 2. d 3. b 4. a 5. c 6. b 7. b 8. a 9. e 10. c

Objective Questions

11. Type 2 diabetes mellitus
12. Type 1 diabetes mellitus
13. glycosylation
14. ketones
15. Polyuria = excessive urination; polydipsia = excessive thirst; polyphagia = excess eating
16. Acceptable answers include delayed wound healing; diabetic neuropathy; gangrene; or diabetic foot ulcer.
17. Three phases: inflammation, proliferation, remodeling; all these phases are delayed in diabetes.
18. They cause closing of potassium channels on the beta cell membrane, which depolarizes the cell.
19. True
20. A coma resulting from taking an excess of insulin, thereby dropping blood glucose levels to dangerously low levels.

The Reproductive System

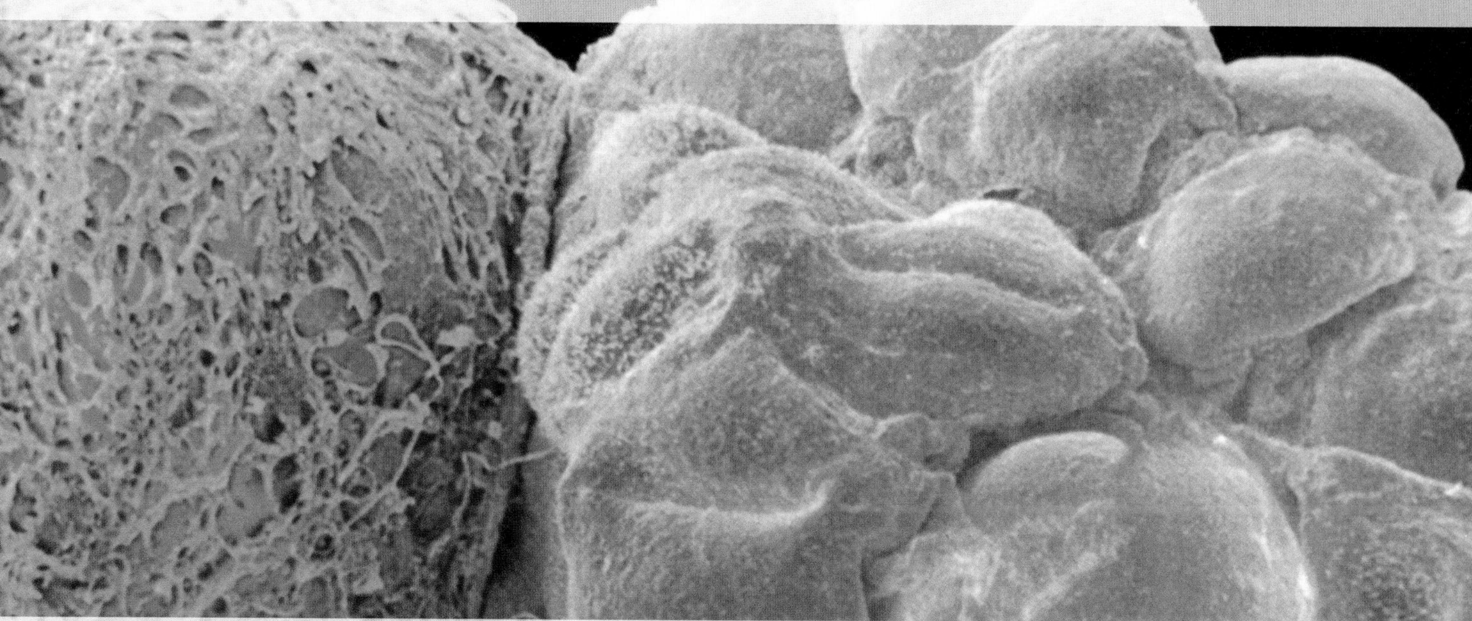

Dr. Yorgos Nikas/Photo Researchers, Inc.

Colorized scanning electron micrograph (SEM) of a human embryo at the blastocyst stage, five days after fertilization.

Jared and Jamal are fraternal twins. Jared is tall, with light brown hair and blue eyes, and wears contact lenses because he is near-sighted. Jamal is average height, has black hair and brown eyes, and enjoys perfect vision. Given that Jared and Jamal have the same biological parents, how can the two brothers exhibit such different characteristics? Answering this question requires an understanding of the reproductive process and the way in which genetic makeup is determined. In this chapter, we examine the processes of reproduction, from copulation and fertilization through fetal growth to parturition and lactation. We also explore what differentiates females from males, and discover how these differences contribute to the reproductive process.

CHAPTER OUTLINE

Go to MasteringA&P for helpful A&P Flix 3-D animations, chapter quizzes, pre-tests, interactive physiology tutorials, and more!

LEARNING OUTCOMES *After studying this chapter, you should be able to:*

- Describe the events in meiosis, and explain its significance in gametogenesis and in the promotion of genetic diversity.
- Describe the role of sex chromosomes and sex hormones in the development of sexual characteristics.
- Describe the process of spermatogenesis and its hormonal regulation.
- Explain the cyclic variations in plasma hormone levels that occur during the menstrual cycle. Describe how these hormones regulate the ovarian and uterine changes that occur during this cycle.
- Describe the events that occur during fertilization, implantation, and early embryonic development.
- Describe the regulation of estrogen and progesterone secretion during pregnancy, and explain how these hormones help maintain pregnancy and prepare the body for parturition.

Before You Begin

Make sure you have mastered the following topics:

- Hypothalamic tropic hormones
- Negative feedback
- Mitosis

Throughout this text, we have emphasized the concept of *homeostasis*: the maintenance of a constant internal environment by the interworking of the organ systems. In this chapter, we diverge from that concept. The reproductive system's primary function is not focused on survival of the individual, but rather survival of the species. We begin our study of the reproductive system with a general overview of reproductive physiology, emphasizing similarities between the male and female systems and the means by which they work together to create offspring. In subsequent sections we concentrate on sex-specific aspects of reproductive function.

1 An Overview of Reproductive Physiology

Reproduction in humans is *sexual,* meaning that offspring are produced as a result of the mating of parents of different sexes. Each offspring inherits a unique combination of genes from both parents and, therefore, develops its own unique set of characteristics. This ability to create new combinations of genes is a hallmark of sexual reproduction, and it is one reason for the great diversity of living things in nature.

The Role of Gametes in Sexual Reproduction

In sexual reproduction, each parent produces cells called **gametes,** each of which contains a copy of half of that parent's genetic material. Male gametes are known as **spermatozoa** (or more commonly, *sperm*); female gametes are called **ova** (singular: *ovum*) or *eggs.* Gametes have half the number of chromosomes that most cells of the body have.

Most cells of the body are described as **diploid** because they contain two sets of 23 chromosomes (designated $2n$). In contrast, gametes are described as **haploid** because they contain only a single set of 23 chromosomes (designated as n). As a result of mating, gametes from each parent may fuse together to produce a new cell, a phenomenon known as **fertilization** (**Figure 1**). (Fertilization in humans is frequently referred to as *conception.*) This new cell, known as a **zygote,** has the potential to develop into a completely new individual and is diploid, having received one half of its chromosomes from each parent.

Gametes are generated from a pool of specialized, relatively undifferentiated precursor cells known as *germ cells* in a process called **gametogenesis.** In this process, the diploid germ cells undergo a series of cell divisions that ultimately reduces the number of chromosomes in each cell by half (that is, from $2n$ to n).

A single sperm or ovum contains a total of 23 chromosomes: 22 autosomes plus one sex chromosome. Each chromosome is distinguishable from the others by its shape; it also possesses its own characteristic set of genes. **Sex chromosomes** are the chromosomes that determine an individual's sex. They are of two types—the Y or male chromosome and the X or female chromosome. **Autosomes** are the chromosomes other than the sex chromosomes.

During fertilization, the genetic materials from a sperm and an ovum combine to yield a total of 46 chromosomes (44 autosomes plus two sex chromosomes), the normal complement of chromosomes found in most cells of the body. Unlike the sex chromosomes, autosomes always come in matching pairs, the members of which are similar in size and shape and possess genes governing the

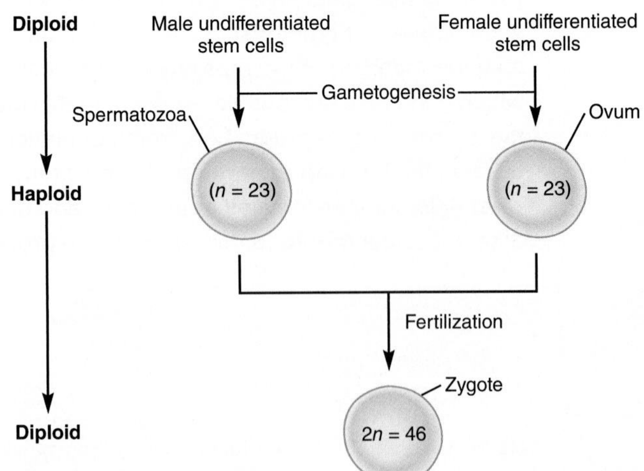

Figure 1 The process of fertilization from undifferentiated stem cells to the zygote. The process of going from diploid to haploid to diploid is shown on the left side with vertical arrows.

same characteristics. Thus the members of each pair of autosomes are said to be *homologous.* In the zygote and all other diploid cells that arise from it, one member of each pair of homologous chromosomes is inherited from the egg and, therefore, is of *maternal* origin (that is, comes from the mother), whereas the other member of the pair is inherited from the sperm and, therefore, is of *paternal* origin (that is, comes from the father).

The fact that autosomes come in pairs means that for every autosomal gene, there is a corresponding gene on the homologous chromosome that governs the same trait or characteristic. Thus autosomal *genes* occur in pairs, one each from the father and the mother. The genes in each pair, though similar in many important respects, are not necessarily identical because genes generally come in a variety of different versions, or **alleles.** The allele inherited from the mother may be the same as or different from the allele inherited from the father. A familiar example involves a gene that determines eye color. One version of the gene (one allele) codes for blue eyes, whereas another allele codes for brown eyes. Because each person has two genes for eye color, it is possible for a person to have two "blue" alleles, two "brown" alleles, or one "blue" allele and one "brown" allele.

The rule that chromosomes come in matching pairs does not apply to the sex chromosomes. The Y chromosome is significantly

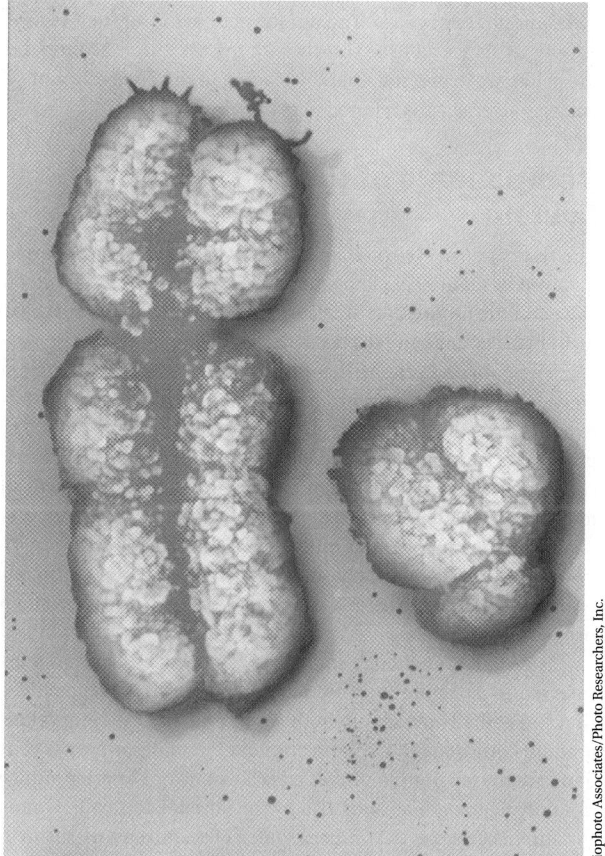

Figure 2 **Electron micrograph of X (left) and Y (right) chromosomes.**

smaller than the X chromosome and has a different shape (**Figure 2**). Furthermore, most of the genes present on the X chromosome are not matched by corresponding genes on the Y chromosome, and vice versa. Because many genes that are present on the X chromosome but absent from the Y chromosome are necessary for life, every individual must inherit at least one X chromosome (see **Clinical Connections: X-Linked Genes**). Females inherit two X chromosomes, whereas males inherit one Y and one X chromosome.

Gene Sorting and Packaging in Gametogenesis: Meiosis

In both males and females, all the gametes an individual will ever produce are ultimately derived from a relatively small set of diploid germ cells. At some point in life these cells undergo mitosis to produce a colony or *clone* of daughter cells, each possessing an exact copy of all 46 of the individual's chromosomes. Subsequently these cells undergo **meiosis,** a series of two cell divisions following only a single replication of DNA, thereby generating daughter cells with half the normal chromosome number. Eventually these cells become mature sperm or ova.

In meiosis, an individual's chromosomes are sorted and packaged in such a way that each of the final daughter cells receives one sex chromosome plus one chromosome from each of the 22 pairs of autosomes. The sorting and packaging that occurs in meiosis proceeds in the following series of steps (**Figure 3**):

① The process begins with a diploid germ cell containing 46 chromosomes ($2n$). In Figure 3, only 4 chromosomes are shown (2 homologous pairs). The color indicates whether the chromosome is of maternal or paternal origin and the length indicates homologous pairs (a pair will have the same length). (For simplicity, the germ cell's nuclear membrane is not shown.)

② In a process similar to what occurs prior to mitosis, the nuclear membrane breaks down, and the original DNA is replicated, yielding exact copies of all 46 chromosomes ($2n \times 2$). The two copies of each chromosome (called *sister chromatids*) remain joined together at a structure called the centromere.

③ Homologous chromosomes begin to group together in pairs such that genes on paternal chromosomes line up opposite the corresponding genes on maternal chromosomes.

④ Portions of homologous chromosomes begin to overlap and exchange segments with one another, a phenomenon known as *crossing over.*

⑤ Crossing over results in new chromosomes that contain mixtures of maternal and paternal genes. Because crossing over occurs randomly, the number of possible new chromosomes is extremely large.

⑥ Pairs of homologous chromosomes line up along a plane bisecting the cell in such a way that a random mixture of maternal and paternal chromosomes is present on either side of the cell. This random grouping of maternal and paternal chromosomes is known as *independent assortment.*

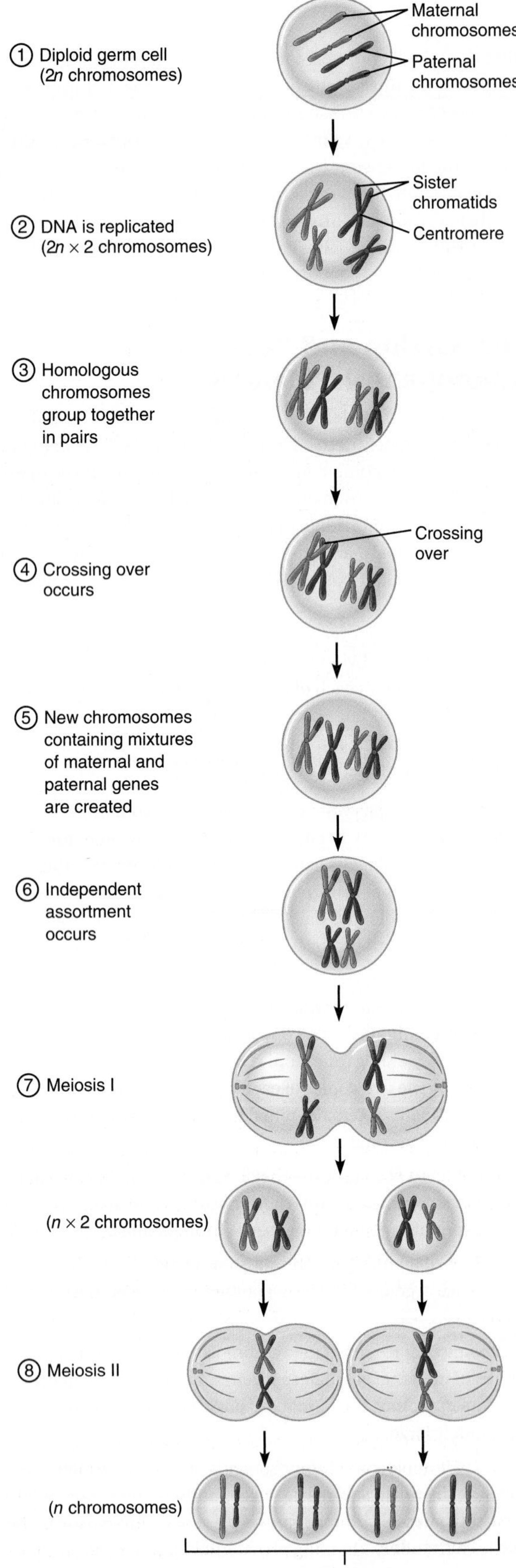

Figure 3 Meiosis. Details of the steps of meiosis are described in the text.

⑦ The first meiotic division *(meiosis I)* occurs. The cell divides in two, and each daughter cell receives one chromosome from each homologous pair. As a result of the independent assortment, each daughter cell receives a different combination of maternal and paternal chromosomes. Because the chromosomes are duplicates consisting of two sister chromatids, each cell receives half the normal chromosome number but two copies of every chromosome ($n \times 2$). Thus each cell receives two copies of each autosome (either the maternal or paternal version) and two copies of either an X or a Y chromosome.

⑧ The second meiotic division *(meiosis II)* occurs. Each of the cells generated in meiosis I divides in two, yielding a total of four cells. In this process, sister chromatids separate, and each of the new haploid daughter cells receives one sister chromatid from each chromosome. Consequently, each of the four cells receives single copies of 23 chromosomes ($n \times 1$), which are eventually enclosed within a new nuclear membrane (not shown).

Because of crossing over and independent assortment, a given parent can produce a large number of genetically different gametes, even though all are derived from the same gene pool. Accordingly, each child of a given set of parents normally inherits his or her own unique set of genes. The unique gene set is referred to as the **genotype.** As we will see shortly with the sex chromosomes, how the genes are expressed, thereby leading to the emergence of certain characteristics, is referred to as the **phenotype.**

Components of the Reproductive System

The *reproductive system* of an individual encompasses all organs involved in mating, gametogenesis, or other functions directly involved in the production of offspring. Reproductive system organs include the gonads and the accessory reproductive organs.

The **gonads** (*testes* in the male and *ovaries* in the female) are considered the primary reproductive organs because they perform two functions that ultimately govern all reproductive activity: They produce gametes, and they secrete sex hormones. Sex hormones include a variety of steroids that promote gametogenesis, growth and maintenance of reproductive organs, development of secondary sex characteristics, and various other effects throughout the body. The testes secrete a class of sex hormones known as **androgens** (notably *testosterone*); the ovaries secrete **estrogens** (such as *estradiol*) and **progesterone.**

Note that even though androgens and estrogens are commonly referred to as "male" and "female" sex hormones, respectively, they are actually present in both sexes, albeit in different relative amounts. Androgens are more abundant in males, whereas estrogens are more abundant in females. Sex hormones are produced not only by the gonads but also by the adrenal cortex. In women, the adrenal cortex is the primary source of androgens, which are responsible for promoting the sex drive in either sex.

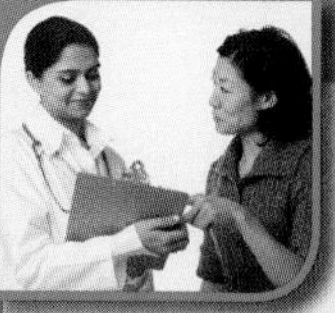

CLINICAL CONNECTIONS

X-Linked Genes

Gametogenesis results in germ cells with a haploid number of chromosomes, including one sex chromosome. In the ovum, the sex chromosome is always an X, but the spermatozoa can have an X or a Y. After fertilization, the resulting zygote, and all cells produced from it, will have either two Xs (female) or an X and a Y (male). The X chromosome is considerably larger than the Y chromosome, meaning that it contains more base pairs and, therefore, more genes. In fact, the X chromosome contains more than 1400 genes, whereas the Y chromosome contains approximately 200 genes.

Inheritance of specific traits is primarily due to the inherited genome. In the case of the inherited autosomes, each cell has two sets of homologous chromosomes with the same genes, but different alleles. For most of these chromosomes, both alleles are expressed, although one may be dominant relative to the other. Thus, if a mutated gene is passed on in one set of chromosomes (from the ovum, for example), a normal gene in the other set of chromosomes (from the sperm) can decrease the significance of the genetic abnormality. For example, sickle cell anemia is an inherited disease in which the gene for the beta globulin of hemoglobin is mutated (HbS). A person who inherits HbS from both parents will have sickle cell *disease*, whereas a person who inherits HbS from only one parent will have the less severe sickle cell *trait*. In this way, homologous chromosomes protect individuals, to some extent, against hereditary diseases.

Because the X chromosome contains essential genes, mutations can lead to deleterious effects. However, in the female the likelihood of expressing the mutated phenotype is lessened by the fact that she has two X chromosomes. In contrast, because the male inherits only one X chromosome, mutated genes are expressed in males, leading to an increased likelihood of X-linked diseases. For example, X-linked genes include the genes that code for the photopigments sensitive to green and red light wavelengths; these two genes are very similar and are located near each other on the X chromosome. Males are, therefore, more likely to be red-green colorblind when they have a mutation in these genes. More than 150 such X-linked diseases have been identified, including hemophilia, impaired mental development, and Duchenne muscular dystrophy.

Interestingly, the gene that codes for androgen receptors is located on the X chromosome. Thus the testosterone produced by males in response to having a Y chromosome, and therefore testes, can produce responses on target cells only because of the expression of the gene on the X chromosome. A mutation of this X-linked gene can lead to androgen insensitivity syndrome, in which case testes develop because of the presence of the srY gene (but they generally do not descend), and the reproductive tract and external genitalia develop into the female phenotype.

Critical Thinking Questions

1. Why are men more likely to demonstrate X-linked diseases than women?
2. Explain how homologous chromosomes can protect individuals against some hereditary diseases.
3. Explain why the gene that codes for androgen receptors, located on the X chromosome, is an interesting phenomenon.

The *accessory reproductive organs* include structures specialized to perform additional functions required for proper reproductive activity, such as transporting gametes from one place to another and providing nourishment for gametes once they are produced. Accessory organs include (1) the organs of the **reproductive tract,** a system of interconnecting passageways through which gametes are transported after leaving the gonads, and (2) various glands that secrete fluids into the reproductive tract.

Events Following Fertilization

Sperm are transferred from the male to the female during the act of mating, known as **copulation.** Ova, in contrast, remain in the female's body, where they are fertilized by the arriving sperm. After fertilization, the ovum undergoes many cell divisions, eventually giving rise to billions of cells. As the cells increase in number, they also differentiate and begin to organize into distinct tissues according to instructions encoded in their genes. Over a period of approximately nine months, these cells and tissues gradually develop into a functional human organism. During this time the developing human is carried within its mother's body (a condition referred to as **gestation** or *pregnancy*), an arrangement that provides it with nourishment and protection from the outside environment. In the first two months after conception, the developing human is called an **embryo;** from that point on, it is referred to as a **fetus.** As the fetus develops, it eventually acquires the capacity to live outside its mother's body and separates from it in a process known as **parturition** (birth). Next we examine the events in embryonic and fetal development that determine the offspring's gender.

Sex Determination

Because a female's germ cells possess two X chromosomes (and thus no Y chromosome), every egg she ever produces possesses an X chromosome. In contrast, a male produces some sperm with an X chromosome and other sperm with a Y chromosome. For this reason, the sex of the offspring is determined entirely by the genetic makeup of the sperm, not the egg: When the sperm bears an X chromosome, the fertilized egg inherits two Xs (one from the mother and one from the father) and develops into a female; when the sperm bears a Y chromosome, the egg inherits one Y and one X and develops into a male. Thus XX is the genotype for females and XY is the genotype for males. That males and females are born in approximately equal numbers is a consequence of the fact that, on average, approximately 50% of all sperm cells possess a Y chromosome. Although a person's gender is normally determined by the genotype inherited, the genes themselves do not confer the full complement of male or female traits; instead, they simply determine whether a fetus will develop ovaries or testes, a process referred to as **sex determination.**

In its early life, the embryo possesses primitive gonads that have the potential to become either ovaries or testes. Whether a fetus develops ovaries or testes is determined by the presence or absence in the embryo of a gene located on the Y chromosome—the **srY gene**—which codes for a protein called **testis-determining factor.** (srY stands for *s*ex-determining *r*egion of the *Y* chromosome.) When the embryo inherits a Y chromosome, the srY gene instructs the primitive gonads to become testes; when the embryo inherits two X chromosomes, the srY gene is absent, so the primitive gonads develop into ovaries. Thus the primitive gonads develop into ovaries "by default"—that is, unless they are instructed by the srY gene to do otherwise.

Sex Differentiation

In the first few weeks of development, the embryo is said to be *sexually indifferent* and possesses both (rudimentary) female and male reproductive systems, called **Wolffian ducts** *(mesonephric ducts)* and **Müllerian ducts** *(paramesonephric ducts),* respectively (**Figure 4**). These structures have the potential to give rise to all reproductive organs except the gonads, the *external genitalia* (sex organs that are externally visible), and, in females, the vagina (the initial portion of the reproductive tract, which leads to the outside of the body).

The development of gonads in the embryo (sex determination) sets the stage for the development of other sexual characteristics (**sex differentiation**). Whether these characteristics are male or female depends on the presence or absence of two hormones

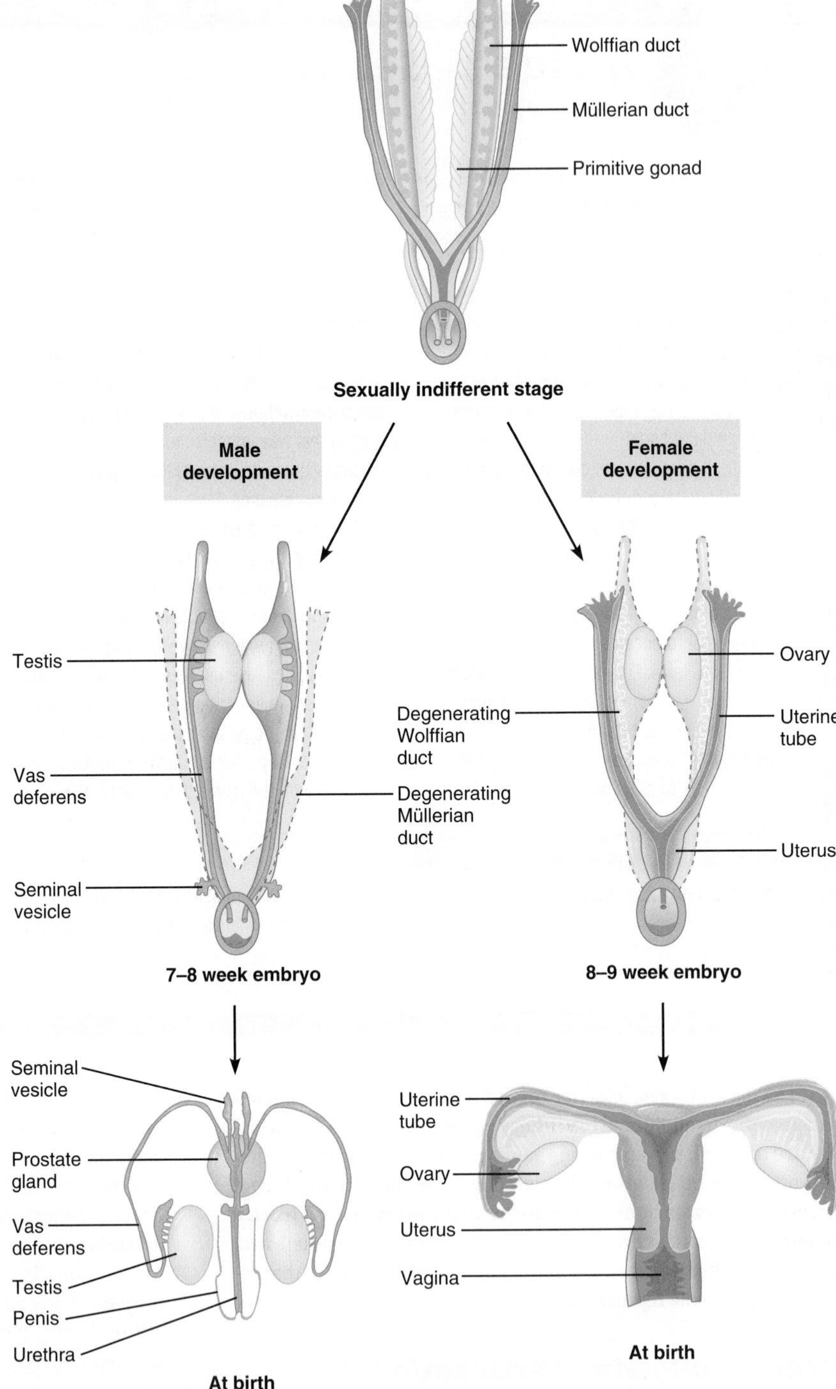

Figure 4 Development of male and female internal reproductive organs during embryonic life. Events shown include sex determination (development of primitive gonads into either testes or ovaries) and sex differentiation (development of either Wolffian ducts or Müllerian ducts into male or female reproductive structures, respectively). The penis and vagina do not develop from Wolffian or Müllerian ducts but are included for purposes of orientation.

normally secreted by the testes—**testosterone** and **Müllerian inhibiting substance (MIS).** If these hormones are present (and target tissues are able to respond to them normally), the fetus develops as a male; if they are absent, the fetus develops as a female. Thus this hormone-driven process of sex differentiation establishes the phenotypic sex.

When testis-determining factor is present, functioning testes develop by approximately 40 days post-fertilization. Testosterone and MIS secretion begin approximately 8 weeks post-fertilization, with testosterone secretion peaking at 20 weeks. Testosterone acts on the Wolffian ducts to promote the development of male reproductive organs, whereas MIS promotes the regression and eventual disappearance of the Müllerian ducts, thereby preventing the development of female organs. At the same time, testosterone acts on certain other embryonic tissues to promote the development of male external genitalia (such as the penis and scrotum, which are discussed shortly). In many target tissues, however, it is not testosterone that directly triggers these changes; instead, the enzyme 5α-reductase in these tissues converts testosterone to dihydrotestosterone, which then binds to receptors and triggers the target tissue response.

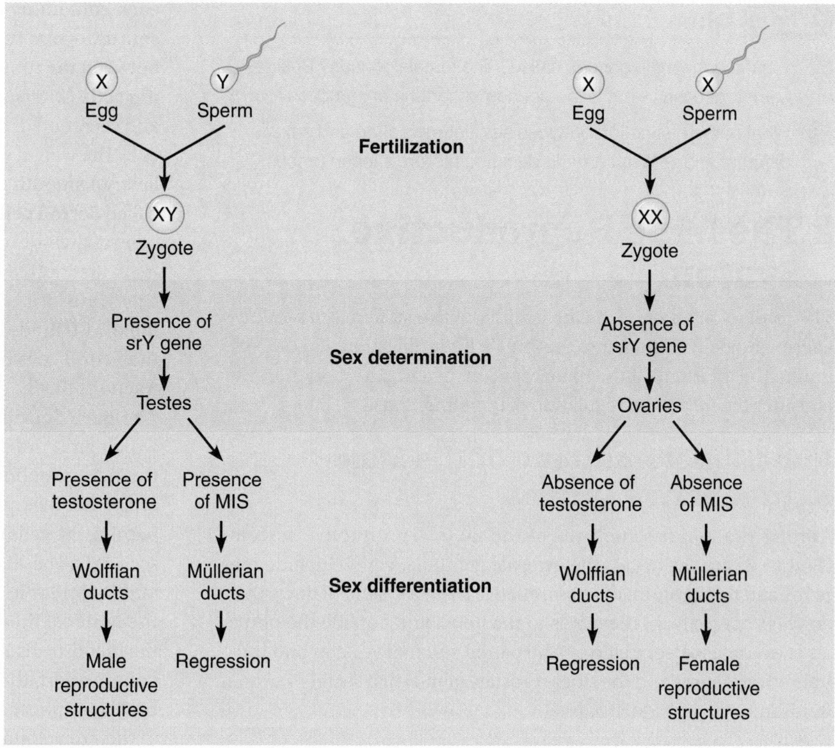

(a) Pivotal events in male development **(b)** Pivotal events in female development

Figure 5 The roles of sex determination and sex differentiation in the development of males and females.

When the embryo develops ovaries instead of testes, the absence of testosterone causes the Wolffian ducts to regress, thereby preventing the development of male organs. In addition, the absence of MIS allows the Müllerian ducts to develop into female organs. The lack of testosterone also eliminates the influences that would otherwise stimulate the development of male external genitalia and allows female external genitalia to develop instead. The roles of sex determination and sex differentiation in the development of males and females are depicted in **Figure 5**.

Apply Your Knowledge

Sometimes, more than one sex chromosome is passed on to the zygote by the sperm or the ova. Assuming normal autosomes, given the following sex chromosome genotypes, describe the resulting phenotype: XXY, XXXY, XYY, XO (O = absent chromosome), XY with mutation of the srY gene.

Patterns of Reproductive Activity over the Human Life Span

Humans are not born with the ability to reproduce but instead acquire this ability during **puberty,** a period of sexual maturation that typically begins sometime between the ages of 10 and 14 and extends to the mid- to late teens. In the years before puberty, the reproductive organs are immature and incapable of generating sperm or ova. In addition, a person's outward appearance is childlike, with little obvious distinction between the bodies of boys and girls. During puberty, the reproductive organs mature, gametogenesis begins, and other physiological changes occur throughout the body. In the process, males and females develop various *secondary sex characteristics*—the external features that distinguish the sexes from each other, such as the growth of facial hair in men and the widening of the hips relative to the shoulders in women.

Although both men and women acquire full reproductive function during puberty, the subsequent patterns of reproductive activity are very different for men and women. Activity of the female reproductive system exhibits cyclic variations known as the *menstrual cycle,* which averages about one month (28 days) in length, during which usually only a single ovum is produced. The male reproductive system, however, remains continuously active and normally produces millions of sperm every day. Furthermore, females produce only a limited number of ova in their lifetime and lose the ability to reproduce in middle age (around the age of 45 or 50). This loss of reproductive capacity marks the beginning of a period known as **menopause,** which is characterized by changes in hormonal secretory patterns and a host of other physiological changes. In contrast, men retain the ability to produce sperm—and thus the ability to reproduce—throughout their adult lives.

Quick Check 1

1. What are the male gonads called? The female gonads? Besides gametogenesis, what is the gonads' other primary function?
2. What is the distinction between sex determination and sex differentiation? What are the deciding factors in each process?

2 The Male Reproductive System

The primary function of the male reproductive system is to produce sperm and deliver them into the female for fertilization. First we examine the anatomy of the male reproductive organs, and then we consider the hormonal regulation of their functions.

Functional Anatomy of the Male Reproductive Organs

Among the major structures of the male reproductive system (**Figure 6**) are the testes; the external genitalia, which include the penis and the scrotum; the reproductive tract, a system of ducts that conveys sperm from the testes to the penis and outside the body; and several accessory glands (the paired seminal vesicles and bulbourethral glands and the single prostate gland) that secrete various fluids into the reproductive tract.

The Testes

The **testes,** the male gonads, are a pair of small egg-shaped organs that are encapsulated in a layer of fibrous connective tissue. Each testis (**Figure 7a**) is divided internally into 250–300 compartments, each containing a set of thin, highly coiled hollow tubes known as **seminiferous tubules,** where sperm are produced. In the spaces between the tubules (Figure 7b) are clusters of cells known as **Leydig cells** *(interstitial cells),* which are responsible for the synthesis and secretion of testosterone and other androgens.

The wall of a seminiferous tubule (Figure 7c) contains an outer layer of smooth muscle cells and an inner layer of epithelial cells called **Sertoli cells,** whose primary function is to nurture sperm and control their development. Sperm cells in various stages of development are located in the spaces between the Sertoli cells, and mature sperm are located in the lumen of the seminiferous tubules, which is filled with fluid. The smooth muscle layer, which is separated from the Sertoli cells by a basement membrane, exhibits peristaltic contractions that help propel the sperm and fluid through the tubule.

Each Sertoli cell is joined to its neighbors by tight junctions, which limit the diffusion of materials between cells. The cells and the tight junctions that connect them form a barrier (called the *blood-testis barrier*) that isolates the lumenal fluid from the fluid bathing the cells on the other side (referred to as the *basal compartment*). The basal compartment communicates freely with the blood and is similar in composition to normal interstitial fluid, whereas the lumenal fluid has a different composition. The significance of the blood-testis barrier is that by maintaining this difference in fluid composition, the barrier ensures that when sperm move from the basal compartment to the lumenal compartment at a certain stage of their development, the new fluid environment they encounter in the lumen fulfills their developmental requirements. In addition, the blood-testis barrier prevents the male's immune system from attacking the sperm cells.

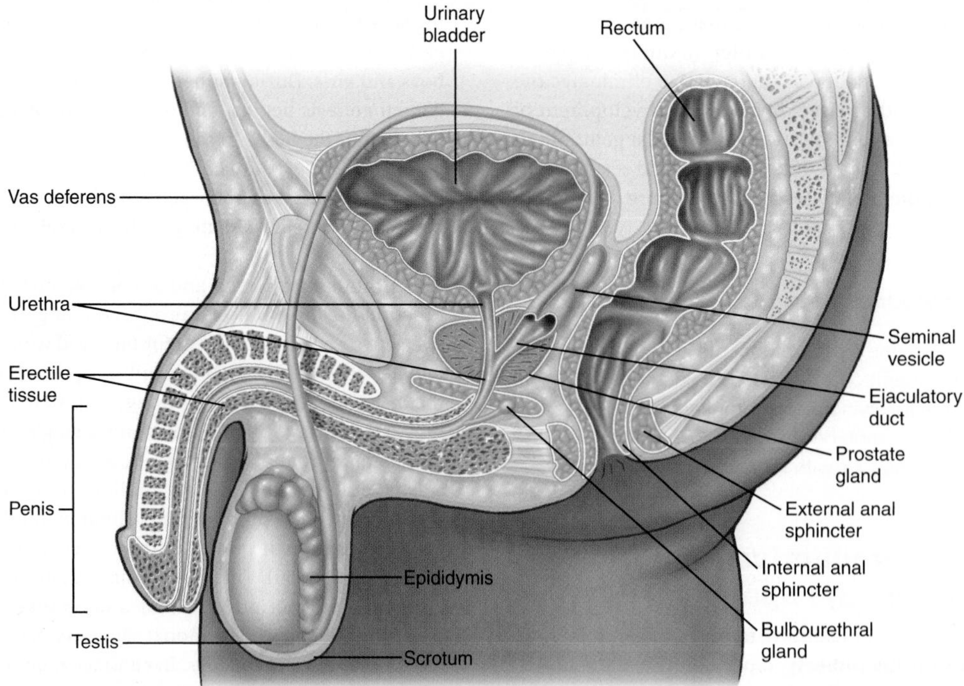

Figure 6 The male reproductive system, as seen in a sagittal section.

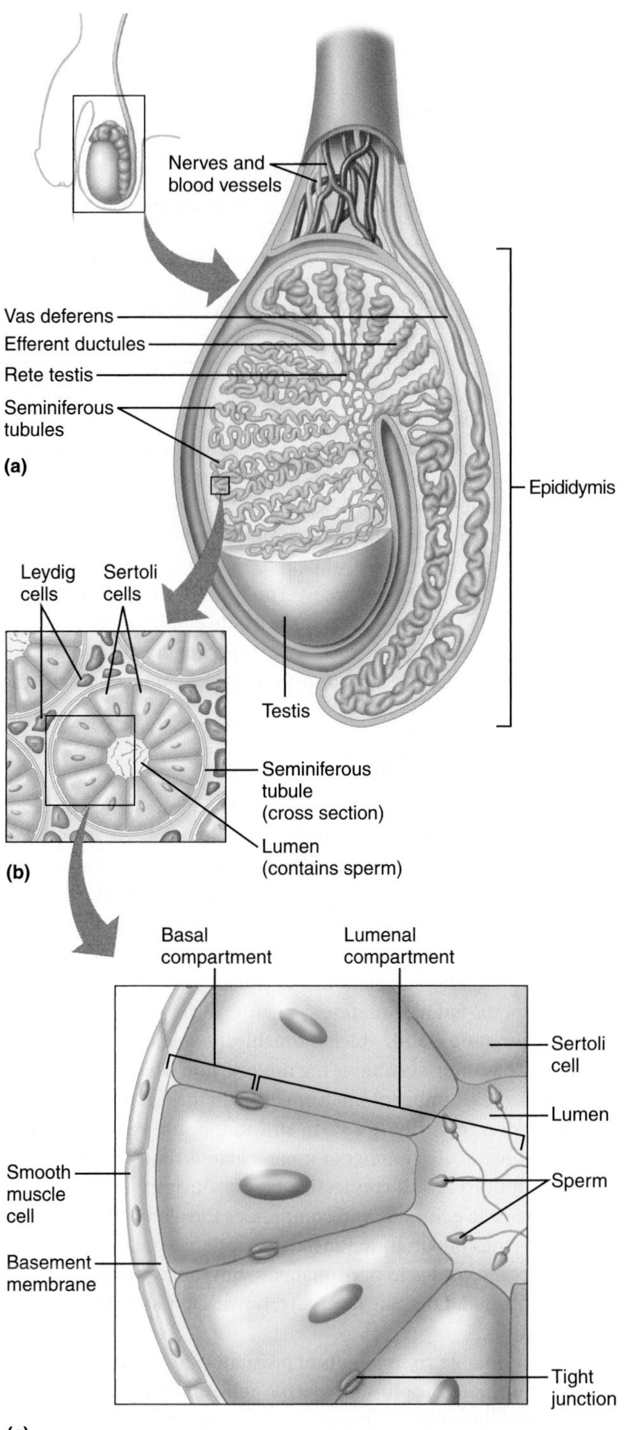

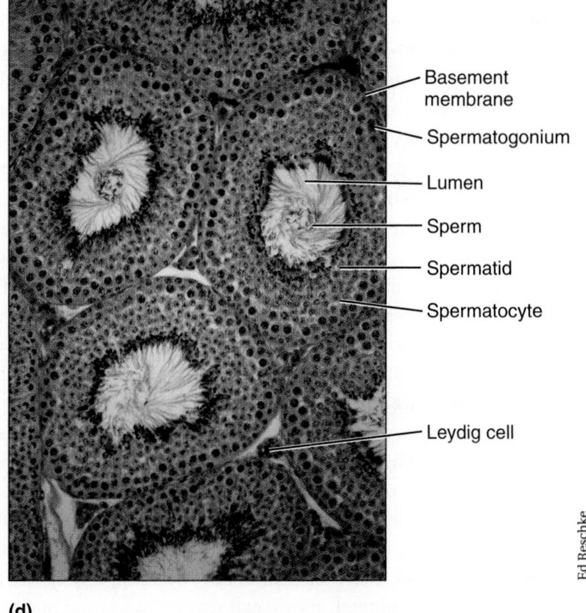

Figure 7 The testis and associated structures. (a) A sagittal section of the testis and epididymis showing the seminiferous tubules and other ducts that carry sperm away from the testis. **(b)** An enlargement showing seminiferous tubules in cross section and Leydig cells located in the spaces between the tubules. **(c)** A portion of a seminiferous tubule showing the structure of the wall. This diagram is highly schematized to emphasize the essential features of this structure. **(d)** Photomicrograph of a cross section through seminiferous tubules.

Sertoli cells perform a number of other functions crucial to sperm development and transport. They are responsible for secreting the lumenal fluid, which serves as a medium for sperm development and transport, and for transporting nutrients to the developing sperm cells. The Sertoli cells also manufacture and secrete **androgen-binding protein,** which by binding androgens reversibly acts as an "androgen buffer" that helps maintain in the lumenal fluid a steady concentration of androgens, which must be present for sperm development to take place. Additionally, Sertoli cells play a critical role in orchestrating the development of sperm cells, which is under the control of testosterone and *follicle stimulating hormone (FSH)*. Sertoli cells are the target cells of testosterone

TABLE 1 Functions of Sertoli Cells in Males and Granulosa Cells in Females

Sertoli cells	Granulosa cells
Secrete fluid into the lumen of seminiferous tubules to support sperm development and transport	Secrete antral fluid
Provide nutrients for the developing sperm	Provide nutrients for the developing oocyte
Secrete paracrines that stimulate spermatogenesis	Secrete paracrines that support follicle development
Secrete inhibin	Secrete inhibin
Secrete Müllerian inhibiting substance in fetal life	Secrete estrogens
Form blood-testis barrier	Secrete a substance that forms the zona pellucida
Secrete androgen-binding protein into the lumen of seminiferous tubules	

and FSH; they secrete chemical messengers that act directly on the immature sperm cells, stimulating growth and development. Moreover, Sertoli cells are responsible for the secretion of a hormone called *inhibin* (whose function is discussed later), and in embryonic life they secrete MIS. The functions of Sertoli cells are summarized in Table 1.

The External Genitalia

The reproductive function of the **penis,** the male copulatory organ, is to penetrate into the vagina during copulation and deposit sperm inside the female reproductive tract, thereby initiating the steps that lead to fertilization. The penis is ordinarily flaccid and hangs downward (see Figure 6), but during sexual arousal it swells, elongates, and becomes much firmer and straighter, a process known as **erection.** These changes enable the penis to penetrate into the vagina so that sperm can be deposited as close to the site of fertilization as possible. At the height of sexual arousal, sperm are forcibly ejected through the urethra (an event called **ejaculation**), which runs the length of the penis and opens to the outside at its tip. The urethra also serves as the passageway through which urine moves from the urinary bladder to the outside of the body.

Changes in the penis's shape during erection are possible because much of its volume is composed of **erectile tissue** (see Figure 6), spongy masses of connective tissue and smooth muscle containing numerous interconnected vascular spaces. Ordinarily only a small volume of blood is present within the erectile tissue, but during erection the volume and pressure of this blood increase, producing the swelling and elongation characteristic of erection.

The **scrotum,** a sac of skin and underlying connective tissue that is suspended beneath the penis, houses the two testes in separate right and left compartments divided by a septum. The scrotum provides an environment for the testes in which the temperature is slightly below core body temperature, which is important because even a small rise in testicular temperature can suppress sperm production. (Once produced, sperm can survive at normal body temperature, however.)

The Reproductive Tract

Within each testis, the seminiferous tubules converge to form a network of short interconnected tubules (called the *rete testis*) that lead to a set of small tubes *(efferent ductules)* that penetrate the fibrous outer covering of the testis (see Figure 3). Outside the testis, these tubes lead to the **epididymis,** which is loosely attached to the testis's outer surface and contains a single highly coiled duct. This duct, in turn, leads to a larger and more muscular thick-walled tube called the **vas deferens,** which enters the pelvic cavity.

In the pelvic cavity, the vas deferens connects with a duct from one of the seminal vesicles, at which point it becomes known as an *ejaculatory duct* (see Figure 6). The two ejaculatory ducts (one from each testis) penetrate into the prostate gland, where they join the urethra. The urethra then emerges from the prostate, joins with ducts leading from the bulbourethral glands, and traverses the length of the penis.

Accessory Glands

The **seminal vesicles** (see Figure 6) are two elongated glands approximately 5 cm long that produce a viscous yellowish fluid forming approximately 70% of what becomes **semen,** the fluid ejaculated from the penis. Seminal vesicle fluid is alkaline, which serves to neutralize the acidic environment of the vagina. It also contains fructose to provide energy for the sperm, proteins to facilitate semen's clotting after ejaculation, and prostaglandins to protect the sperm from the female immune system.

The **prostate gland,** which is roughly the shape of an onion and the size of a walnut, surrounds the urethra, into which it empties its secretions via a number of small openings in the urethral wall. Prostate secretions contribute 25% to 30% of the semen. The prostate actually consists of a number of smaller glands that are surrounded by smooth muscle and connective tissue. During ejaculation, the smooth muscle contracts and compresses the glands, forcing their secretions into the urethra. Prostate secretions contain high levels of citrate, zinc, and acid phosphatases, but their functions are not well established. Low zinc levels are, however, associated with low sperm counts.

Other notable components of prostate secretions include various enzymes, such as prostate specific antigen (PSA). PSA functions in breaking down clotted semen to free the sperm inside the female reproductive tract. The larger the prostate gland, the more PSA it produces. Initial screenings for prostate cancer include measuring blood levels of PSA—if PSA is elevated, then further tests are recommended. Unfortunately, the PSA test results in a large number of *false positives*—that is, only 30% of men with high PSA levels actually have prostate cancer. Given this fact, scientists continue to look for other

Functional Fact

possible markers of prostate cancer. Low levels of both citrate and zinc have been identified as having some correlation with prostate cancer, but this relationship has not been firmly established.

The **bulbourethral glands**—a pair of round, pea-sized glands located beneath the prostate—secrete a small volume of fluid containing a viscous, sticky mucus that is released prior to ejaculation. This fluid functions primarily as a lubricant during copulation.

In the following sections we take a closer look at how the male reproductive system performs its functions, and how these functions are regulated.

Hormonal Regulation of Reproductive Function in Males

Reproductive function in males is largely governed by the gonadotropins, androgens (including testosterone), and gonadotropin releasing hormone (GnRH) via the processes diagrammed in **Figure 8**.

The **gonadotropins** are two protein hormones that are secreted by the anterior pituitary and act on the gonads: **follicle stimulating hormone (FSH)** and **luteinizing hormone (LH).** (Later in the chapter we will see that these hormones are also involved in controlling reproductive function in females.) The primary function of FSH in males is to act on Sertoli cells to stimulate **spermatogenesis,** or sperm production. LH, by contrast, stimulates the secretion of androgens by Leydig cells. Because androgens are steroids, they are lipophilic and, therefore, readily diffuse from the Leydig cells to all testicular tissues, including the Sertoli cells, where androgens function as paracrines. The androgens also enter the bloodstream, which delivers them to tissues all over the body where they function as hormones.

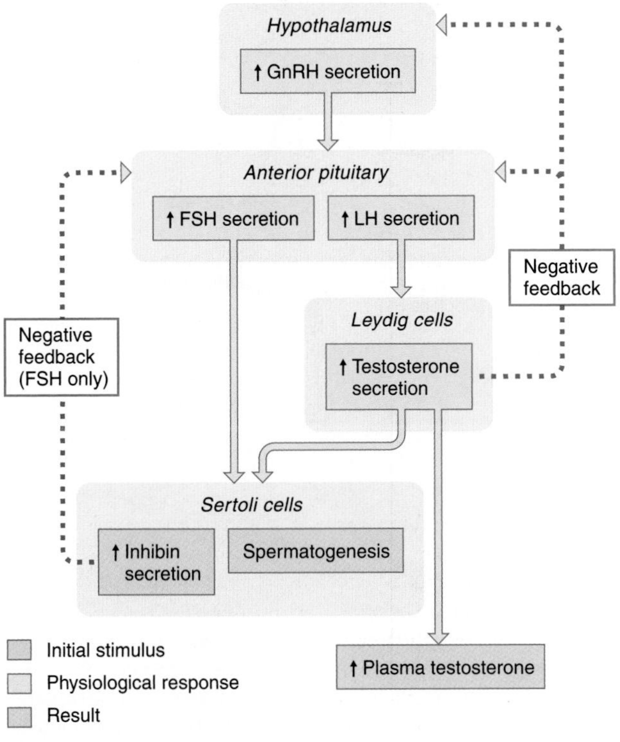

In the testes, testosterone works with FSH to promote spermatogenesis. Testosterone also promotes the development and growth of accessory reproductive organs during puberty and is necessary for the maintenance and continued function of these structures in adult life. For this reason, a lack of testosterone leads to a decrease in the size of accessory organs and a decrease in smooth muscle and glandular activity. In addition, testosterone is responsible for the development and maintenance of secondary sex characteristics and acts on the brain to promote the sex drive.

Gonadotropin releasing hormone (GnRH) is a hypothalamic tropic hormone that stimulates the secretion of gonadotropins and, therefore, tends to promote androgen secretion. (GnRH also stimulates gonadotropin secretion in females, as we will see.) Most hypothalamic tropic hormones stimulate the release of one particular hormone from the anterior pituitary, but GnRH is different because it stimulates the release of two hormones—FSH and LH—that are produced and secreted by anterior pituitary cells called *gonadotrophs*. Most gonadotrophs secrete either LH or FSH, allowing for the differential regulation of these hormones.

Testosterone and other androgens normally limit their own secretion through negative feedback. Specifically, they act both on the hypothalamus, to suppress the secretion of GnRH, and on the anterior pituitary, making it less responsive to GnRH. As a result, testosterone tends to suppress the release of FSH and LH, which in turn tends to suppress subsequent testosterone secretion. Sertoli cells also secrete a protein hormone called **inhibin,** which suppresses the release of FSH (but not the release of LH) by the anterior pituitary.

Rates of sex hormone secretion in males are fairly constant and are not subject to the cyclic variations that occur in females during the menstrual cycle. Throughout a male's reproductive life, the hypothalamus releases GnRH in bursts occurring at approximately two-hour intervals, which causes rates of FSH and LH secretion to rise during bursts and fall between bursts. Despite these fluctuations, *average* plasma levels of FSH and LH (and hence levels of testosterone) remain fairly constant from day to day. Significant variations over the long term do occur, however. Notably, testosterone levels rise dramatically during puberty, reach a peak sometime in the third decade of life, and subsequently decline slowly.

In prepubescent males, testosterone levels are relatively low, reflecting low rates of GnRH and gonadotropin secretion. At the beginning of puberty, however, GnRH secretion rises dramatically in response to a change in brain activity that alters neural input to the hypothalamus. (The precise nature of this change in brain activity is presently unknown.) The increase in GnRH triggers a similar rise in gonadotropin secretion, which stimulates growth and maturation of the testes as well as testosterone secretion.

The rise in plasma testosterone levels that occurs during puberty stimulates further maturation of the reproductive system and other widespread changes throughout the body. For example, testosterone stimulates the secretion of growth hormone by the anterior pituitary, which in turn stimulates the bone growth responsible for the dramatic increase in stature that occurs during this period.

Figure 8 **Hormonal regulation of reproductive function in males.**

Q How does GnRH travel from the hypothalamus to the anterior pituitary?

UNDERSTANDING EXERCISE

Gender Differences

Males and females differ not only in their primary reproductive functions, but also in their secondary sexual characteristics. The latter include both the obvious physical differences (facial hair in men, for example) and more subtle differences that can affect how the body responds to physical activity.

The male sex hormones—the androgens—promote muscle growth, so males tend to be stronger than females. Before puberty, little gender difference in muscle strength exists. In contrast, after the onset of puberty, boys rapidly surpass girls in muscle strength, and adult females tend to have less than 75% of the strength of adult males. Although training can increase muscle strength in both sexes, this increase is proportional to initial strength; thus even women who exercise regularly tend to have lower muscle strength than men, especially in the upper body. The strength-promoting effect of androgens has led to their use by both males and females who seek to gain an advantage in athletic competition.

In contrast to the muscle-promoting effects of androgens, the female sex hormones—the estrogens—tend to promote storage of triglycerides in adipose tissue. As a consequence, males have a higher lean mass (muscle-to-fat ratio) than females.

Men and women also appear to differ in terms of their energy metabolism during exercise. Studies suggest that during aerobic exercise, women tend to mobilize fat stores for energy sooner than males, thereby maintaining muscle glycogen stores for a longer period of time. This phenomenon may help women resist fatigue. In contrast, men tend to have higher levels of epinephrine, which enhances glycogenolysis in liver and muscle cells.

Although data suggest that males and females differ in strength and energy utilization, the overall benefits of exercise appear to be the same for both sexes. Exercise benefits the cardiovascular and respiratory systems of both sexes. A person who performs regular aerobic exercises has a lower resting heart rate, enhanced coronary circulation, increased blood volume, and increased metabolic rate. In addition, trained individuals have a higher capacity for increasing cardiac output and oxygen catabolism during exercise.

Testosterone's actions also trigger the development of secondary sex characteristics during puberty, including growth of facial hair; growth of coarse body hair in the pelvic region and armpits; increased protein synthesis in muscle that leads to increased muscle size and strength, particularly in the upper body; growth of the larynx, which causes the voice to deepen; and secretion of a thick oil by *sebaceous glands* (oil glands) in the skin. Table 2 summarizes the actions of androgens in males. Differences in androgen levels in males and females are responsible for many of the differences observed in physical capacities during exercise (see **Understanding Exercise: Gender Differences**).

Sperm and Their Development

Fully developed sperm cells possess three distinct segments (Figure 9): a bulblike *head*, a short cylindrical *midpiece*, and a long, slender *tail* (or *flagellum*). Within the head are the chromosomes and a large vesicle called an **acrosome,** which contains enzymes and other proteins that enable the sperm to fuse with the egg

TABLE 2 Selected Actions of Androgens in Males

Stimulate spermatogenesis
Promote development of secondary sex characteristics during puberty and maintenance of these characteristics in adult life
Increase sex drive
Promote protein synthesis in skeletal muscle
Stimulate growth hormone secretion, which permits bone growth during adolescence
Promote development of male reproductive structures during embryonic life

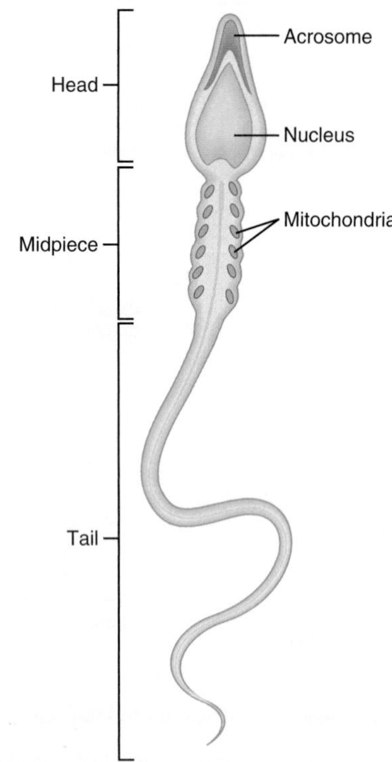

Figure 9 Anatomy of a sperm cell.

during fertilization. The tail contains complex machinery made up of proteins that hydrolyze ATP and convert the released chemical energy into motion. The resulting whiplike movements of the tail propel the sperm forward in a swimming motion. The midpiece anchors the tail to the head and contains mitochondria, which produce the ATP required for motility.

Spermatogenesis starts near the basement membrane in the seminiferous tubules with undifferentiated germ cells called **spermatogonia** (**Figure 10**). As spermatogenesis proceeds, germ cells gradually migrate toward the lumen of the tubule in the spaces between Sertoli cells. At a certain point the developing germ cells move from the basal compartment to the lumenal compartment by passing through the tight junctions, which open up temporarily to let the cells pass through. As a result of this migration, cells at different stages of development are characteristically found at certain locations in the tubule; the older, more mature cells are nearer the lumen, whereas the newer, less differentiated cells are nearer the basement membrane.

Each male is born with a finite number of spermatogonia, but these cells undergo mitosis repeatedly, giving males the capacity to produce sperm for an indefinite period following puberty. One of the daughter cells produced by each mitotic division differentiates further and eventually becomes a mature sperm; the other daughter cell does not differentiate and remains a spermatogonium. Therefore, the total number of spermatogonia does not change, ensuring that this pool of potential sperm is never depleted.

To become a mature sperm, a spermatogonium goes through the following stages of development (see Figure 10):

① The spermatogonium's chromosomes are replicated, and the cell differentiates to become a *primary spermatocyte,* which has 46 duplicated chromosomes possessing two sister chromatids apiece. (In Figure 10, the chromosome number is designated as $2n \times 2$ to indicate that the chromosomes are duplicated.)

② The primary spermatocyte goes through meiosis I to yield two *secondary spermatocytes,* which possess 23 duplicated chromosomes ($n \times 2$).

③ Secondary spermatocytes undergo meiosis II to become *spermatids,* which have 23 single chromosomes (n) but have not developed the mature sperm's characteristic features.

④ Spermatids differentiate to become spermatozoa (n), which possess the characteristic head, midpiece, and tail. After spermatozoa are formed, they are released into the lumen of the seminiferous tubule.

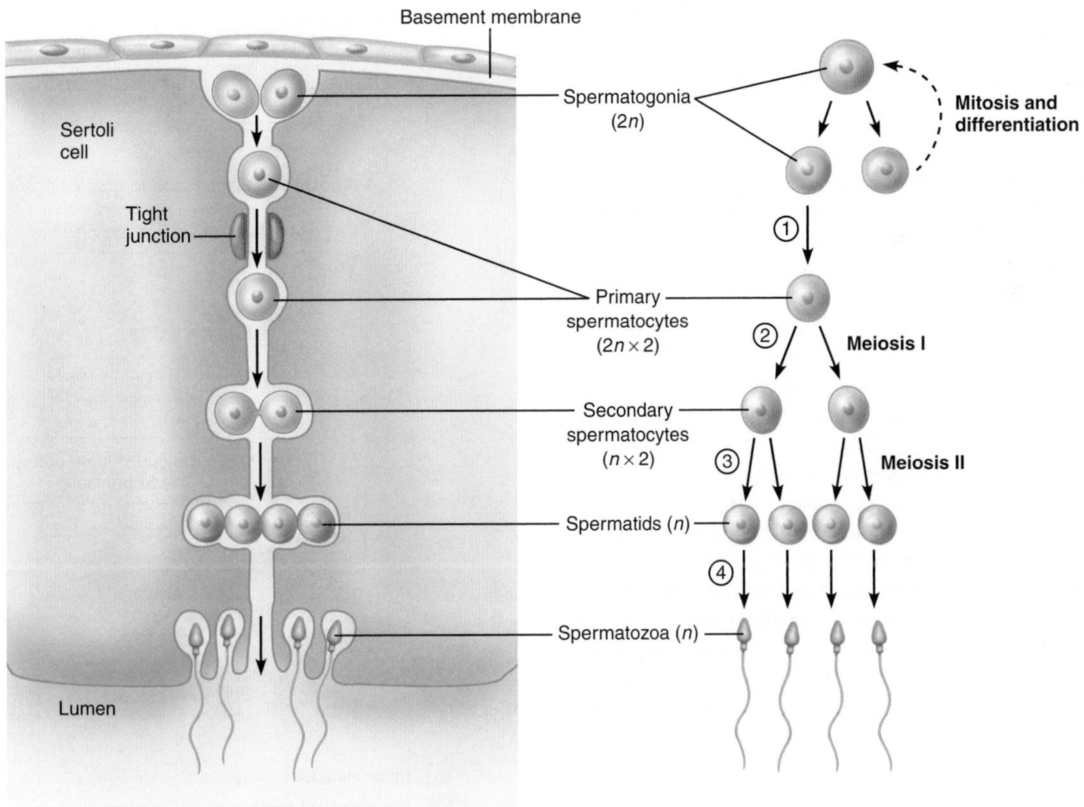

Figure 10 Spermatogenesis. As germ cells migrate from the basal compartment to the lumen of a seminiferous tubule (depicted at left), they undergo various stages of development (depicted at right). The dashed arrow indicates that some spermatogonia do not differentiate following mitosis and remain as spermatogonia.

Q Over a male's lifetime, does the number of spermatogonia normally increase, decrease, or stay the same?

At this stage, the sperm are immotile (incapable of self-propulsion), and they remain so for approximately 20 days. Sperm acquire motility only after they have moved from the seminiferous tubules to the epididymis to undergo further maturation. Transport of sperm to the epididymis is driven by peristaltic contractions in the seminiferous tubules and by the flow of lumenal fluid, a natural consequence of its continual secretion by Sertoli cells. Much of this fluid is absorbed in the epididymis, causing the sperm to become more concentrated. Peristalsis propels the sperm within the epididymis and to the vas deferens, where they are held until sexual arousal triggers ejaculation.

The Sexual Response in Males

The sexual response in males is controlled by the autonomic nervous system and consists of two reflex phases: erection and ejaculation. Activity in parasympathetic neurons stimulates erection, whereas activity in sympathetic neurons triggers ejaculation. Inputs to these autonomic neurons include afferent signals from mechanoreceptors in the penis and other sensitive parts of the body (called *erogenous zones*) and descending signals from higher brain centers. We now discuss these two reflexes and consider how the level of input to the spinal cord causes a shift in the output from parasympathetic control of the penis to sympathetic control of the reproductive tract and accessory organs.

Erection

The events leading to erection are depicted in **Figure 11**. When a male is not sexually aroused, very little blood is in the erectile tissue of the penis, and the penis is flaccid. However, mechanical stimulation of the penis or descending input from the brain acts on interneurons in the spinal cord that orchestrate changes in the activity of sympathetic and parasympathetic neurons that project to small arterioles supplying the erectile tissue with blood. Activity in sympathetic neurons to these arterioles decreases, whereas activity in parasympathetic neurons increases. These parasympathetic neurons are unusual not only because they control arteriolar resistance, but also because they release nitric oxide, which has potent vasodilatory effects. (Vascular resistance is controlled in most regions of the body by sympathetic—not parasympathetic—neurons.)

The combined result of the decrease in sympathetic activity and the increase in parasympathetic activity is vasodilation, which increases blood flow to the erectile tissue. As a result, the vascular spaces inside the tissue fill with blood and expand, which brings about erection (**Clinical Connections: Erectile Dysfunction**). The expansion of erectile tissue due to vasodilation also compresses veins in the penis, which by reducing the outflow of blood from the vascular spaces promotes the accumulation of blood in the erectile tissue. As more blood accumulates, it further compresses the veins, leading to the accumulation of still more blood in a positive feedback loop that promotes expansion of the erectile tissue.

Ejaculation

As the level of mechanical stimulation of the penis increases in conjunction with psychological stimuli, the output shifts from parasympathetic control of the penis to sympathetic control of the reproductive tract and accessory glands, which triggers ejaculation. During ejaculation, the male experiences intense psychological stimulation accompanied by increases in heart rate and blood pressure and a host of other physiological changes, a phenomenon called *orgasm.*

Ejaculation, like erection, is controlled by a spinal reflex, but in this case the reflex causes an increase in the activity of sympathetic neurons. This sympathetic activity triggers strong contractions in the epididymis, vas deferens, and ejaculatory ducts, and it also stimulates the secretion of fluids by the seminal vesicles and prostate gland. As a result, a mixture of sperm and fluid (called **semen**) moves into the urethra, an event called **emission.** Following emission, smooth muscle in the urethra and skeletal muscle at the base of the penis undergo a series of strong contractions that expel the semen through the urethral opening. The reflex also triggers closure of the urethral sphincter at the base of the bladder, which prevents urine from mixing with the semen during ejaculation.

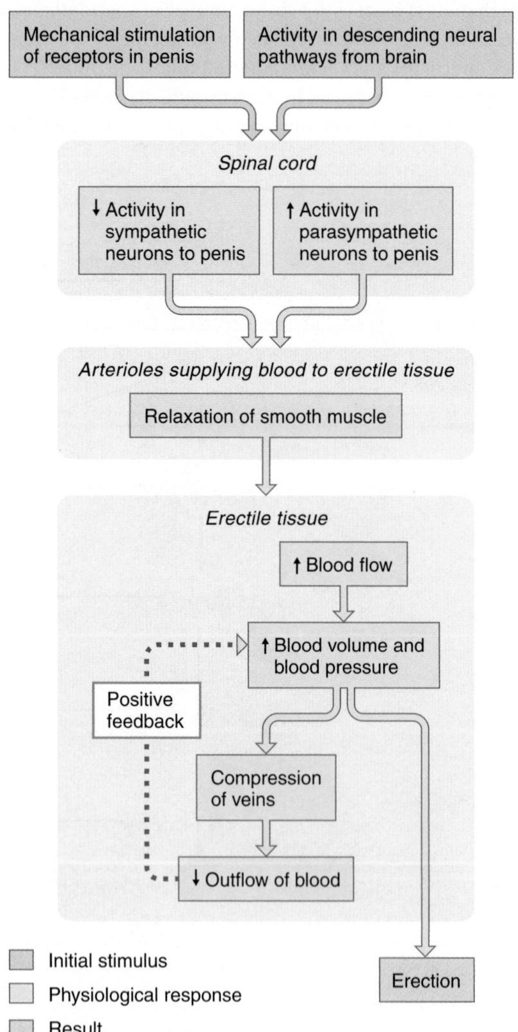

Figure 11 Events leading to erection.

Q When the activity of parasympathetic neurons increases, does the secretion of nitric oxide increase or decrease?

CLINICAL CONNECTIONS

Erectile Dysfunction

Erectile dysfunction (ED), commonly known as *impotence,* is the inability to attain or maintain an erection that is sufficiently rigid to permit sexual intercourse. Most often, ED results from damage to vascular or connective tissues in the penis, occurring as a normal consequence of aging, but it can also result from damage either to spinal tracts involved in the control of blood flow to erectile tissue or to afferent or efferent nerves to the penis. ED can be triggered by endocrine dysfunction (a lack of testosterone, for example) or other diseases, including kidney disorders, multiple sclerosis, atherosclerosis, or diabetes mellitus.

The risk of this condition increases with age and with use of a variety of substances, including cigarettes/cigars, alcohol, and various drugs such as antihypertensive agents, antidepressants, and tranquilizers. ED doesn't require a purely physical basis, however; indeed, in 10% to 20% of cases, the cause is thought to be purely psychological. Feelings of stress, anxiety, guilt, or depression are all known to increase the risk of ED.

Treatments for ED include oral testosterone replacement (if the primary problem is a lack of testosterone) and the injection into the penis of drugs that increase blood flow, a method that has the disadvantage of inducing erection almost immediately. Other treatments include surgery to rebuild damaged arteries or to insert into the penis prosthetic devices that allow simulation of naturally occurring erections.

In recent years the treatment of ED has been revolutionized by a drug called Viagra (generic name: seldenafil) that enables erection to occur if taken an hour or so before intercourse. This drug works by mimicking the effects of nitric oxide, which induces arterial dilation in the penis. Nitric oxide normally works by stimulating guanylate cyclase in smooth muscle cells, which converts GTP to cyclic GMP (cGMP). cGMP promotes relaxation of arterial smooth muscle and an increase in blood flow to erectile tissue. Viagra mimics the effect of nitric oxide by inhibiting the phosphodiesterase enzyme that breaks down cGMP, thereby promoting a rise in intracellular cGMP levels.

Since the discovery of Viagra, two other drugs have been approved by the FDA for the treatment of ED: tadalafil (brand name: Cialis) and vardenafil (brand names: Levitra and Staxyn). All three drugs work by increasing blood flow to the penis, but the time frame during which the increased blood flow occurs varies. Vardenafil and Viagra take about 30 minutes to have an effect, with the effects lasting approximately 4 hours with Viagra and 5 hours with vardenafil. Cialis is a faster-acting drug, with initial effects occurring 15 minutes after ingestion, and its effects last much longer (up to 36 hours in some cases).

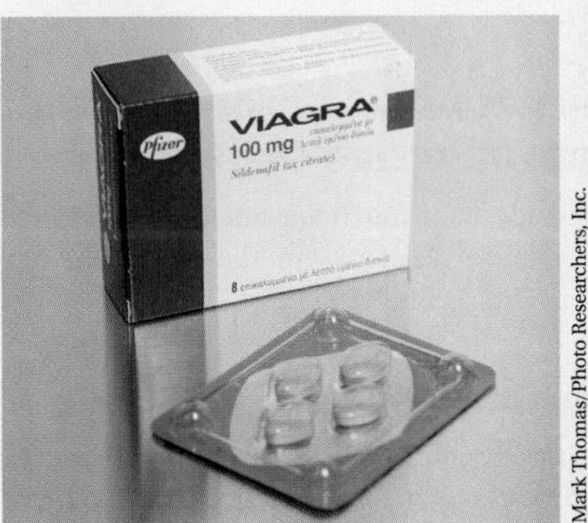

Mark Thomas/Photo Researchers, Inc.

Critical Thinking Questions

1. Describe the various factors that lead to erectile dysfunction.
2. What role does a man's age and mental state play in the development of ED?
3. What are the similarities and differences between the three medications commonly used to treat erectile dysfunction?

Quick Check 2

1. For each of the following descriptions, indicate whether it applies to Sertoli cells or to Leydig cells: are present in seminiferous tubules; are present outside seminiferous tubules; secrete testosterone; are target cells of FSH; are target cells of LH; secrete inhibin.
2. Arrange the following structures so that they correctly describe the path of sperm transport from the testes to the penis: ejaculatory duct, vas deferens, seminiferous tubule, urethra, epididymis, efferent ductule, rete testis.
3. Describe the role of the autonomic nervous system in the male sexual response.

3 The Female Reproductive System

Whereas the male reproductive system has the relatively uncomplicated job of producing sperm and delivering them to the female, the female reproductive system is responsible not only for the production and transport of ova, but also for everything else that must occur to enable a new human being to be brought into the world. The female reproductive system exhibits the following general characteristics:

1. *Cyclic changes in activity.* Each month the female reproductive system undergoes a series of dramatic structural and functional changes called the **menstrual cycle,** which is

accompanied by regular, periodic changes in the secretion of hypothalamic, anterior pituitary, and ovarian hormones. The beginning of each cycle, which lasts for about 28 days, is marked by **menstruation,** the shedding of blood and tissue from the surface of the uterine lining.

2. *Restricted periods of fertility.* Ova mature at different rates and are typically released singly from the ovaries at intervals of approximately 28 days—once per menstrual cycle. The release of the egg, known as **ovulation,** is a prerequisite for fertilization and usually occurs in the middle of the cycle. A female is fertile (capable of having an egg fertilized) for only the few days per cycle that roughly coincide with ovulation.
3. *Limited gamete production.* Ova develop from a pool of germ cells whose number is fixed at birth at approximately 2–4 million. The number of potential ova declines steadily throughout life because most germ cells degenerate (a phenomenon called *atresia*) at some point in their development. Only some 400 eggs are released by ovulation over a woman's lifetime.

Functional Anatomy of the Female Reproductive Organs

Our study of the functional anatomy of the female reproductive system (**Figure 12**) centers on the ovaries, the reproductive tract, and the external genitalia.

The Ovaries

The **ovaries,** the female gonads, are a pair of slightly flattened oval structures, each about the size of a walnut, located in the pelvic region on either side of the uterus (see Figure 12a). Each ovary is composed mostly of dense connective tissue that is well supplied with blood vessels and enclosed within a layer of fibrous connective tissue. Embedded within this connective tissue at the beginning of puberty are more than 300,000 spherical structures called **follicles,** each of which contains a single developing ovum. As the ovum develops, the rest of the follicle develops along with it, increasing in size and structural complexity. Among the follicles present in an ovary at any given time, most are in the earliest stages of development.

A follicle in the earliest stage of development (called a **primordial follicle**) is a simple structure consisting of a developing ovum or *oocyte* surrounded by a single layer of specialized epithelial cells called follicle cells (see Figure 12a). As the follicle develops, the epithelial cells increase to more than one layer and the cells are then called **granulosa cells.** In later stages of development, the granulosa cells proliferate, and the outermost layer is transformed into another cell type known as **theca cells.**

Granulosa cells perform a number of important functions that are analogous in many ways to those performed by Sertoli cells in males (see Table 1). Specifically, granulosa cells function as intermediaries between the oocyte and the hormones that control its development, in that both estrogen and FSH stimulate granulosa cells to secrete chemical messengers that target the oocyte. In addition, they secrete inhibin, which suppresses FSH secretion (as it does in males). Granulosa cells also transport nutrients to the oocyte's interior via cytoplasmic bridges that communicate with the oocyte's interior through gap junctions.

Granulosa cells are responsible for manufacturing and secreting the female sex hormones, estrogens and progesterone. Estrogen synthesis requires androgens that are first synthesized in theca cells. After synthesis, these androgens diffuse from the theca cells to the granulosa cells, where they are enzymatically modified to form estrogens, which are then secreted.

The Reproductive Tract

The female reproductive tract is a system of interconnected passages comprising the internal cavities of the uterus, uterine tubes, and vagina. The centerpiece is the **uterus,** a hollow, pear-shaped organ located in the center of the pelvic cavity that functions to house and nourish the developing fetus (see Figure 12a). During pregnancy the fetus is held in the upper portion of the uterus, called the *body.* In the lower, narrower portion of the uterus (called the **cervix**), the *cervical canal* leads directly to the vagina. During parturition, the infant passes through this canal and then through the vagina to exit the mother's body. For this reason, the cervical canal and vagina are collectively referred to as the *birth canal.*

To accommodate the growing fetus, the wall of the uterus must be strong and flexible. Beneath the outer layer of epithelial cells and connective tissue (called the *perimetrium*) is a layer of smooth muscle called the **myometrium,** which occupies most of the wall's thickness. Contractions of the myometrium help to expel the fetus from the uterus during birth. The innermost layer, called the **endometrium,** comprises a layer of epithelial cells and an underlying layer of thick connective tissue. The endometrium contains numerous glands that secrete fluid that bathes the uterine lining; it also provides nutritional support to the embryo following fertilization.

The **vagina,** a canal approximately 8 to 10 cm long leading from the cervix to the outside of the body, receives the penis during mating and, therefore, is considered the female organ of copulation (Figure 12). Its thin wall contains a middle layer of smooth muscle, and its inner surface is bathed by fluid that leaks from the uterus or is secreted by glands in the cervix. Because of the action of bacteria that normally live in the vagina, this fluid is acidic. The acidic environment, although potentially harmful to sperm, limits the growth of certain pathogenic microorganisms that could easily enter the vagina from the external environment.

Extending from the upper part of the uterus on either side are the **uterine tubes** (also called *fallopian tubes* or *oviducts*), which function in transporting the egg from the ovary to the uterus (see Figure 12a). The uterine tubes are also the site where fertilization normally takes place. Each tube, which is approximately 10 cm long, terminates in a funnel-like structure (the *infundibulum*) fringed by fingerlike projections *(fimbriae)* that partially envelop the ovary on that side. When an ovum is released from the ovary, it is swept into the uterine tube by the motion of the surrounding fluid, which is drawn into the tube by the beating of cilia lining the tube's inner surface. For the first few minutes after ovulation, peristaltic contractions of the tube propel the ovum toward the uterus; thereafter, the ovum is propelled solely by ciliary motion. The entire trip from the ovary to the uterus takes roughly four days.

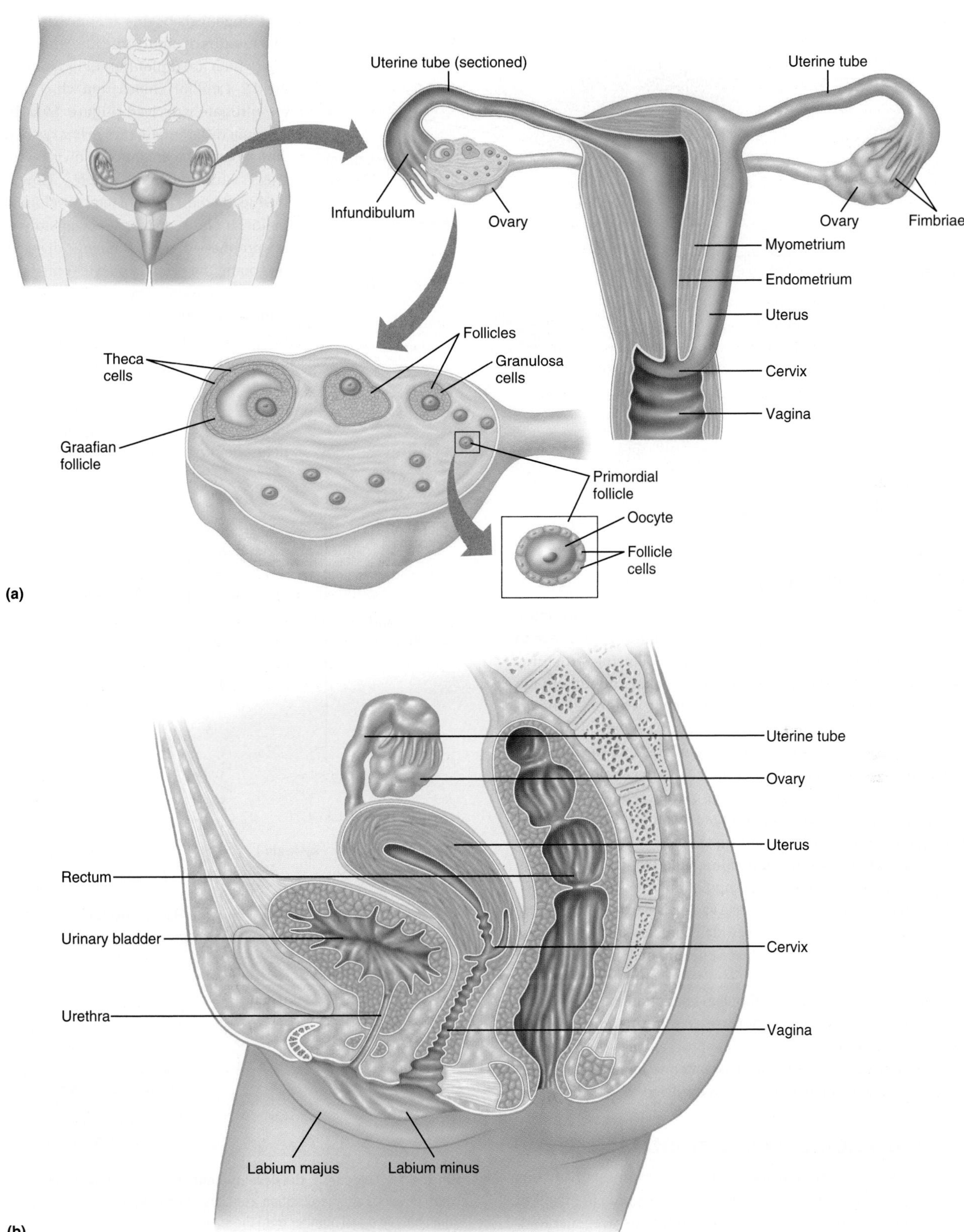

Figure 12 The female reproductive system. (a) A frontal section of the uterus and associated structures; the enlargement of an ovary shows ovarian follicles in different stages of development. **(b)** A sagittal section showing the major reproductive organs.

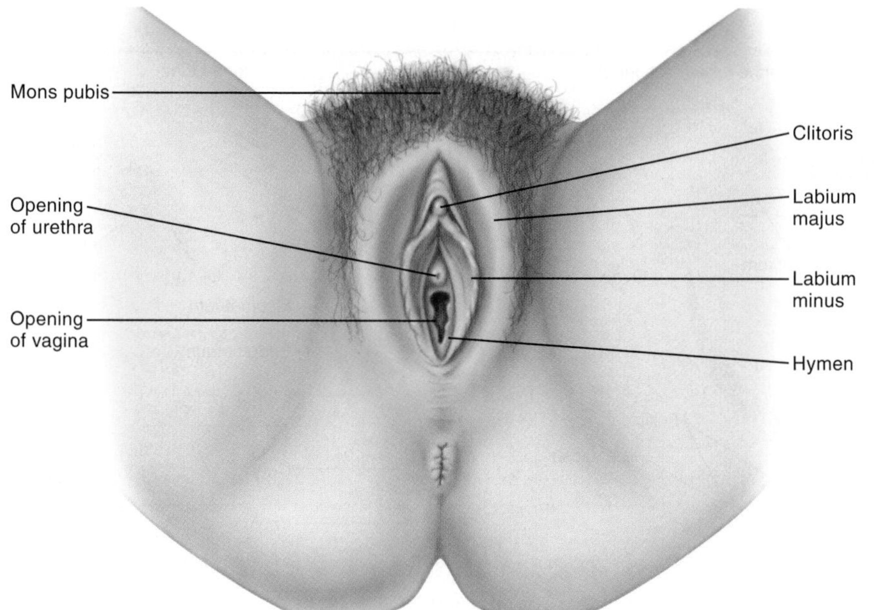

Figure 13 **Female external genitalia.**

Oogenesis begins at approximately 2 to 3 months of embryonic life, when oogonia (which are diploid) undergo mitosis to yield 2–4 million clones from which all ova are ultimately derived (**Figure 14**). These oogonia differentiate into *primary oocytes,* which begin the first meiotic division (meiosis I) with replication of DNA. However, during the remainder of fetal development, meiosis of most primary oocytes stops during prophase of the first meiotic division; at this point, the primary oocytes enter a state known as *meiotic arrest* and remain there until just before ovulation. Primary oocytes that do not enter meiotic arrest undergo apoptosis, leaving 1–2 million primary oocytes at birth in meiotic arrest with 46 chromosomes that have two sister chromatids apiece ($2n \times 2$). Meiosis stays in arrest until just before ovulation.

The External Genitalia

The external genitalia of females are located in the single structure called the **vulva** (or *pudendum*), which surrounds the entrance to the reproductive tract and encompasses the following structures (**Figure 13**): (1) the *mons pubis,* a mound of skin and underlying fatty tissue located centrally in the lower pelvic region; (2) the *labia majora* (singular: *labium majus*) and *labia minora* (singular: *labium minus*), a set of outer and inner skin folds that surround the entrance to the vagina; (3) the *vestibule,* an area enclosed by the labia minora that leads to the openings of the vagina and the urethra; (4) the *clitoris,* a small erectile organ located at the front of the vestibule; and (5) *vestibular glands* (not shown in Figure 13), a pair of small glands that secrete a lubricating fluid that facilitates entry of the penis into the vagina during copulation. Prior to the first episode of copulation, the opening to the vagina is partially covered by a thin membrane called the *hymen,* which contains small blood vessels. This membrane is often ruptured by entry of the penis into the vagina during the first episode of copulation, although it can also rupture from physical trauma without any penetration.

In the remainder of this section, we examine in more detail the day-to-day functioning of the female reproductive system that occurs in the absence of fertilization. Thereafter we focus on processes that occur only when an ovum becomes fertilized.

Ova and Their Development

In the process called **oogenesis,** ova develop from relatively undifferentiated germ cells called **oogonia.** The oogonia are fixed in number prior to birth, and they are not continuously regenerated, as are spermatogonia. Furthermore, the process of meiosis, which transforms oogonia into fully mature ova, begins in fetal life but is not completed until fertilization has occurred.

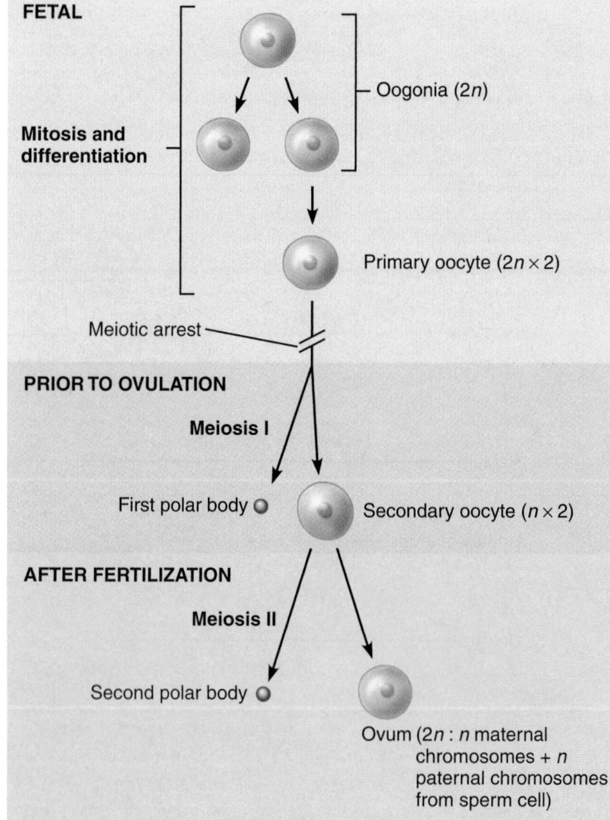

Figure 14 **Oogenesis, the development of ova.** Note that the ovum, which develops after fertilization, contains 23 chromosomes of maternal origin plus an equal number from the sperm.

Q How many sex chromosomes are present in an oogonium, and which chromosomes are they?

By the time a female reaches puberty, only some 400,000 primary oocytes still exist; the other primary oocytes have undergone degeneration *(atresia)*. Once puberty starts, one primary oocyte per month continues meiosis I, completing the process just prior to ovulation, producing two daughter cells possessing 23 replicated chromosomes each ($n \times 2$). One of the daughter cells, called a *secondary oocyte,* receives most of the cytoplasm during the cell division and continues to develop further; the other daughter cell, called the *first polar body,* degenerates and is lost.

Only in the event that the secondary oocyte is fertilized does meiosis II occur. This second meiotic division yields an ovum, which receives most of the cytoplasm, and a *second polar body,* which degenerates. The fertilized ovum contains 23 single chromosomes that it inherits from the secondary oocyte plus an equal number that it inherits from the sperm, giving a total of 46 chromosomes ($2n$).

The Sexual Response in Females

The female sexual response is similar to the male response in many respects. Mechanical stimulation of erogenous zones and psychological stimuli activate the same autonomic pathways in females as in males. In females, sexual excitement produces a characteristic set of changes, including vascular congestion (increased blood volume) in certain tissues and muscular contraction in many areas of the body. For example, the clitoris begins to swell and elongate due to engorgement of its erectile tissue with blood. The breasts also become engorged, the nipples become erect, and the vagina begins to secrete a viscous lubricating fluid. The clitoris, which is equipped with numerous sensory endings, is stimulated by friction during sex acts, and this induces pleasurable sensations and raises the level of sexual arousal. When arousal reaches a certain level, it can trigger orgasm, which is accompanied by rhythmic contractions of the vagina and uterus and many other responses, including intense pleasure, a rise in blood pressure and heart rate, and widespread skeletal muscle contractions.

Quick Check 3

1. How does oogenesis differ from spermatogenesis with regard to the number of gametes produced? The timing of gamete production?
2. When do meiosis I and meiosis II occur in oogenesis?

The Menstrual Cycle

Once a female reaches puberty and then for her entire reproductive life, widespread changes in body function occur during the menstrual cycle. These changes include the following: cyclic changes in ovarian structure and function (referred to as the **ovarian cycle**), cyclic changes in uterine structure and function (referred to as the **uterine cycle**), and cyclic changes in the secretion of ovarian, hypothalamic, and pituitary hormones. In the next two sections we examine the ovarian and uterine cycles to see how they set the stage for possible fertilization and pregnancy, which are necessary for fulfillment of the reproductive system's ultimate function, the creation of new life. We then link these events to the hormonal changes that cause them.

The Ovarian Cycle

The ovarian cycle is divided into two phases: the **follicular phase,** which is 14 days long on average, and the **luteal phase,** which is 14 days long (**Figure 15**). The follicular phase commences with the start of menstruation (day 1 of the menstrual cycle) and ends with ovulation; the luteal phase coincides with the remainder of the menstrual cycle.

The Follicular Phase Within an ovary at any phase of the ovarian cycle, follicles exist at different stages of development, as shown in Figure 15a and b. Most of the follicles are in the primordial phase and about 2 mm in diameter ①. A small fraction of follicles begin to develop, with each follicle's development proceeding independently of the others. As a follicle begins to develop, its cells proliferate into multiple layers around the oocyte and differentiate into granulosa cells ②. At this point, the primordial follicle becomes a primary follicle; this stage of development is called the *preantral stage* (for reasons described shortly). During the preantral stage, granulosa cells secrete noncellular material that forms a thick membrane between them and the oocyte called the **zona pellucida.** Because the zona pellucida separates the oocyte from other tissue, including blood, nourishment to an oocyte is provided by gap junctions located in cytoplasmic extensions between granulosa cells and the oocyte. Also at this stage, certain connective tissue cells differentiate to form the outer layer of theca cells. Some follicles do not develop beyond this stage and undergo atresia. In fact, at any stage of development, such a high percentage of follicles undergo atresia that 1000 or more follicles are lost every month.

Follicles that continue developing form a fluid-filled cavity called an **antrum,** which increases in size as development continues ③. At this point, the follicle is called a secondary follicle; this stage of development is called the *early antral stage.* Some follicles will stop developing at this point and undergo atresia.

At the beginning of the follicular phase, approximately 10 to 25 follicles from the pool of preantral or early antral follicles in each ovary are recruited to develop further. After about seven days, one of these follicles (referred to as the **dominant follicle**) is selected to develop to full maturity, setting the stage for ovulation. The ovary containing the dominant follicle is the *dominant ovary* for that cycle. Most studies indicate that the dominant ovary occurs randomly, although a few studies suggest an alternating pattern may occur. The remaining follicles undergo atresia and are lost.

Follicular growth and development are stimulated by both FSH and the estrogens secreted by the follicles themselves. During the follicular phase, plasma FSH levels gradually fall (for reasons explained later), which tends to cause rates of estrogen secretion to decrease. Those follicles that cannot sustain adequate rates of estrogen secretion undergo atresia. Selection of the dominant follicle depends on its ability to secrete adequate levels of estrogen in the face of falling FSH levels. The dominant follicle is more sensitive to FSH than are the other follicles, and it also shows some responsiveness to LH, which does not appear until a follicle has reached a certain stage in development. Thus the dominant follicle ensures its dominance by developing more quickly than the other follicles.

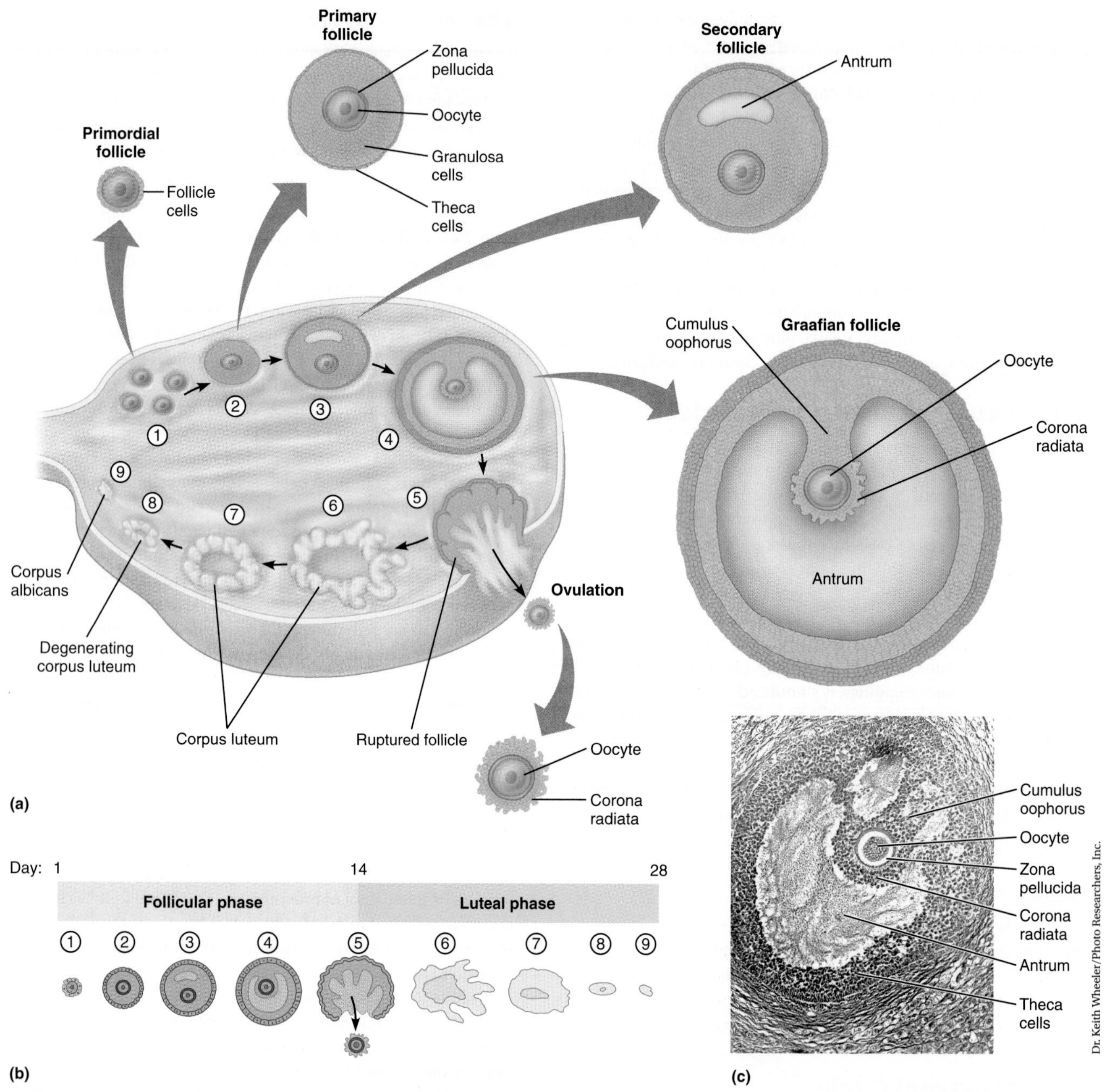

Figure 15 **The ovarian cycle. (a)** Cross section of an ovary depicting the various stages of the ovarian cycle; events occurring during these stages are described in detail in the text. **(b)** A diagram relating the stages depicted in (a) to the follicular and luteal phases of the ovarian cycle. Note that ovulation marks both the end of the follicular phase and the beginning of the luteal phase. **(c)** Photomicrograph of a Graafian follicle.

As the dominant follicle continues to develop into the *late antral stage,* the antrum grows and displaces some of the cellular tissue around the oocyte. Several layers of granulosa cells continue to surround the oocyte, forming the *corona radiata,* while other granulosa cells form the *cumulus oophorus,* a bridge of cells that attaches the oocyte and corona radiata to the wall of the follicle, now called a **Graafian follicle** ④. Eventually, meiosis I occurs, and the oocyte (now a *secondary oocyte*) detaches from the follicle wall and floats freely in the antral fluid, along with the surrounding corona radiata. The antrum continues to expand and the follicle eventually grows to 2.0 to 2.5 cm in diameter just before ovulation. Sometimes ovulation does not occur, however, and a cyst may form (see **Clinical Connections: Ovarian Cysts**).

CLINICAL CONNECTIONS

Ovarian Cysts 

Each month during a woman's reproductive years, a follicle develops in the ovary and releases an oocyte at ovulation. Following ovulation, the follicle develops into the corpus luteum, which ultimately degenerates into the corpus albicans if pregnancy does not occur. Sometimes during this process, a cyst, or fluid-filled sac, may form.

Ovarian cysts can be either functional or abnormal. Functional cysts include follicular cysts and luteal cysts, named for the phase of the ovarian cycle during which they develop. A *follicular cyst* occurs when the follicle does not release its oocyte and grows to larger than 2 cm. *Luteal cysts* develop after ovulation and occur when the corpus luteum does not degenerate. Sometimes during ovulation, a blood vessel ruptures and the luteal cyst fills with blood, forming a *hemorrhagic cyst.* Both follicular and luteal cysts are benign and usually disappear on their own within 90 days. Nevertheless, some functional cysts may continue to grow until they induce pain or burst, which induces intense pain.

Abnormal ovarian cysts occur in many forms, some more severe than others. *Endometriomas* occur as a result of endometriosis, a condition whereby endometrial tissue grows outside of the uterus—in this case, on the ovary. Endometriomas are also called *chocolate cysts* because they contain old blood with a dark brown appearance. If these cysts burst, the contents often cause secondary problems in the pelvic region that can affect fertility. *Dermoid cysts* are benign tumors that generally occur in young females and resemble tumors on the skin. These cysts tend to grow quite large, up to 6 inches, and may contain hair and bone tissue. In *polycystic ovary disease,* several cysts form as a result of gonadotropin imbalances. If symptomatic, benign ovarian cysts can be treated with progesterone, which suppresses ovulation, or can be surgically removed.

Malignant ovarian cysts are rare, but their diagnosis frequently occurs during late stages of their development because they remain relatively asymptomatic until excessive growth has occurred. Cysts that occur before adolescence or after menopause are more likely to be malignant. Malignant cysts require immediate treatment, which includes surgical removal followed by chemotherapy and/or radiation treatment.

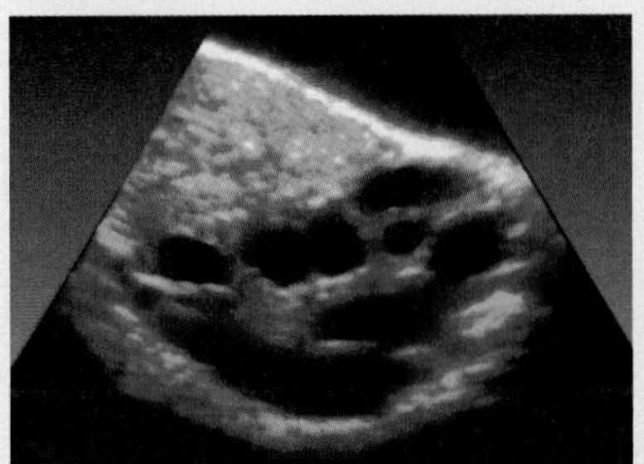

Ultrasound showing an ovarian cyst.

Critical Thinking Questions

1. Describe the different types of functional ovarian cysts and their main characteristics.
2. Why are endometriomas potentially severe? Describe the effects of dermoid cysts.
3. What are the treatment options for benign and malignant ovarian cysts?

The Luteal Phase In ovulation, which marks the beginning of the luteal phase, the wall of the Graafian follicle ruptures ⑤, causing a flow of antral fluid that carries the oocyte (with its surrounding cells) to the ovary's surface. The ruptured follicle is then transformed into a gland called the **corpus luteum** ⑥, which secretes estrogens and progesterone. Both ovulation and formation of the corpus luteum are triggered by the same event—an abrupt rise in plasma LH levels. In fact, its role in the formation of the corpus luteum is why this hormone is called luteinizing hormone.

After ovulation, the released oocyte enters the uterine tube, and its fate ultimately determines that of the corpus luteum. If the oocyte is not fertilized, the corpus luteum reaches its maximum activity within 10 days of its formation ⑦ and then begins to degenerate ⑧ to scar tissue called the *corpus albicans* ⑨. This degeneration causes a decline in plasma estrogen and progesterone levels that sets the stage for menstruation and the beginning of the next follicular phase. If the oocyte is fertilized and implants in the uterus (that is, if pregnancy occurs), then hCG is released first from the endometrium and then from the placenta. HCG sustains the corpus luteum such that it persists well into the gestation period.

On rare occasions (1% to 2% of all ovarian cycles), two or more follicles are selected to become dominant. When this happens, two (or more) oocytes are released at ovulation. In the rare event that both are fertilized, the result will be two offspring referred to as *fraternal twins,* who inherit different sets of genes and, therefore, are not identical. These offspring are also called *dizygotic twins* because they originate from different zygotes.

The Uterine Cycle

The uterine cycle, which occurs in concert with the ovarian cycle, is divided into three phases: the **menstrual phase,** which begins on day 1 and lasts for three to five days corresponding to the first few days of the ovarian follicular phase; the **proliferative phase,** which lasts for the remainder of the follicular phase (typically from the end of menstruation to day 14); and the **secretory phase,** which coincides with the ovarian luteal phase (**Figure 16**).

The Menstrual Phase The menstrual phase of the uterine cycle is so named because it corresponds to the period of menstruation—that is, the shedding of the uterine lining.

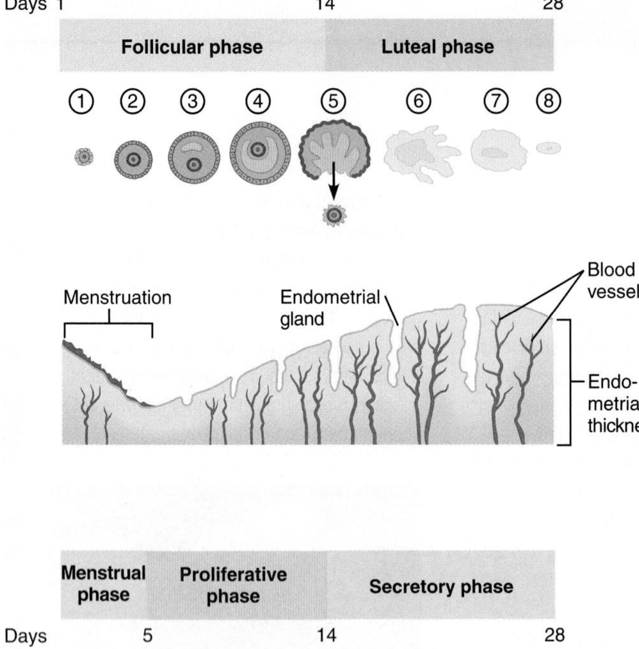

Figure 16 The uterine cycle. Changes in endometrial thickness that occur during the three phases of the uterine cycle are illustrated beneath the events of the ovarian cycle; the shared timeline enables a comparison of events in the two cycles. Note that the beginning of the uterine cycle is marked by shedding of endometrial tissue and blood, and that increased endometrial thickness is accompanied by growth of endometrial glands and proliferation of blood vessels.

Menstruation is triggered in response to the fall in plasma estrogen and progesterone that occurs as the corpus luteum degenerates. At first, blood vessels in the outermost layer of the endometrium begin to constrict, which reduces blood flow to the tissue. As a result, these tissues die and start to separate from the underlying endometrial tissues, which remain intact. The dead tissue is gradually shed from the endometrial surface, which causes rupture of blood vessels and bleeding. Over the next few days a mixture of blood and sloughed tissue seeps into the vagina from the uterus and exits the body, a phenomenon called *menstrual flow.*

The Proliferative Phase During the proliferative phase of the uterine cycle, which begins with the end of menstruation, the uterus renews itself in preparation for possible pregnancy, which might occur following the next ovulation. The endometrial tissues that were spared from destruction in the menstrual phase begin to grow anew, and the smooth muscle in the underlying myometrium thickens. The endometrial glands enlarge, and blood vessels increase in abundance. In the cervical canal, glands begin to secrete a thin mucus that bathes the inner surface; should sperm be deposited in the vagina, this mucus will facilitate their migration through the uterus, a necessary prelude to fertilization. Uterine changes in the proliferative phase are promoted by estrogens, whose plasma levels rise due to the growth and increasing secretory activity of the dominant follicle.

The Secretory Phase During the secretory phase of the uterine cycle, the endometrium (which has been rebuilt during the proliferative phase) is transformed in such a way as to make it a favorable environment for implantation and subsequent housing and nourishment of the developing embryo. The blood supply of the endometrium becomes enriched as arteries branch. The endometrial glands enlarge further and begin to secrete fluids rich in glycogen, which the embryo uses as an energy source in its early stages of growth. Furthermore, the secretions of the cervical glands become stickier and more viscous, eventually blocking the cervical canal by forming a "plug" that effectively isolates the uterus from microorganisms in the outside environment that could possibly harm a developing embryo. These uterine changes are promoted by progesterone, whose plasma levels rise during the secretory phase (as do estrogen levels) due to the action of the corpus luteum.

As the end of the secretory phase approaches, the corpus luteum degenerates, causing plasma estrogen and progesterone levels to fall. This decline causes a withdrawal of these hormones' growth-promoting influences on the endometrium, which triggers the previously described events of menstruation. With the onset of menstruation, the secretory phase ends and the next menstrual phase (and follicular phase of the ovarian cycle) begins. If fertilization and implantation have occurred, the corpus luteum does not degenerate, and estrogen and progesterone levels remain elevated. As a result, secretory-phase uterine conditions are maintained into pregnancy.

Hormonal Changes During the Menstrual Cycle

In this portion of our discussion, we examine the hormonal changes that occur in the menstrual cycle relative to the phases of the ovarian cycle. Not only do hormones affect the cycle, but the cycle also affects plasma levels of the hormones. We begin by examining events occurring in the early and middle portions of the follicular phase.

Hormonal Changes in the Early to Mid-follicular Phase The hormonal changes occurring during the menstrual cycle are plotted in the graphs in **Figure 17**. As is apparent in Figure 17a, the early follicular phase is marked by short-lived declines in plasma estrogen and progesterone levels, which are actually the final stages of declines that began in the previous luteal phase as a result of degeneration of the corpus luteum. Because estrogens and progesterone tend to suppress the secretion of gonadotropins by the anterior pituitary, plasma levels of LH and FSH show a slight increase at this time, reflecting a withdrawal of this suppressive action (see Figure 17b). The rising FSH level stimulates a number of follicles to develop and grow.

FSH binds to receptors on granulosa cells and promotes their growth and proliferation, which causes follicles to increase in size (**Figure 18**). Under the influence of FSH, the outer layer of granulosa cells differentiates into theca cells, which possess LH receptors. LH stimulates the theca cells to secrete androgens, which then travel to the granulosa cells and are converted to estrogens. Eventually, a dominant follicle emerges that secretes estrogens at a high rate, causing plasma levels of these hormones to rise rapidly. These estrogens feed back on the hypothalamus and anterior pituitary, suppressing LH and FSH secretion and halting the initial

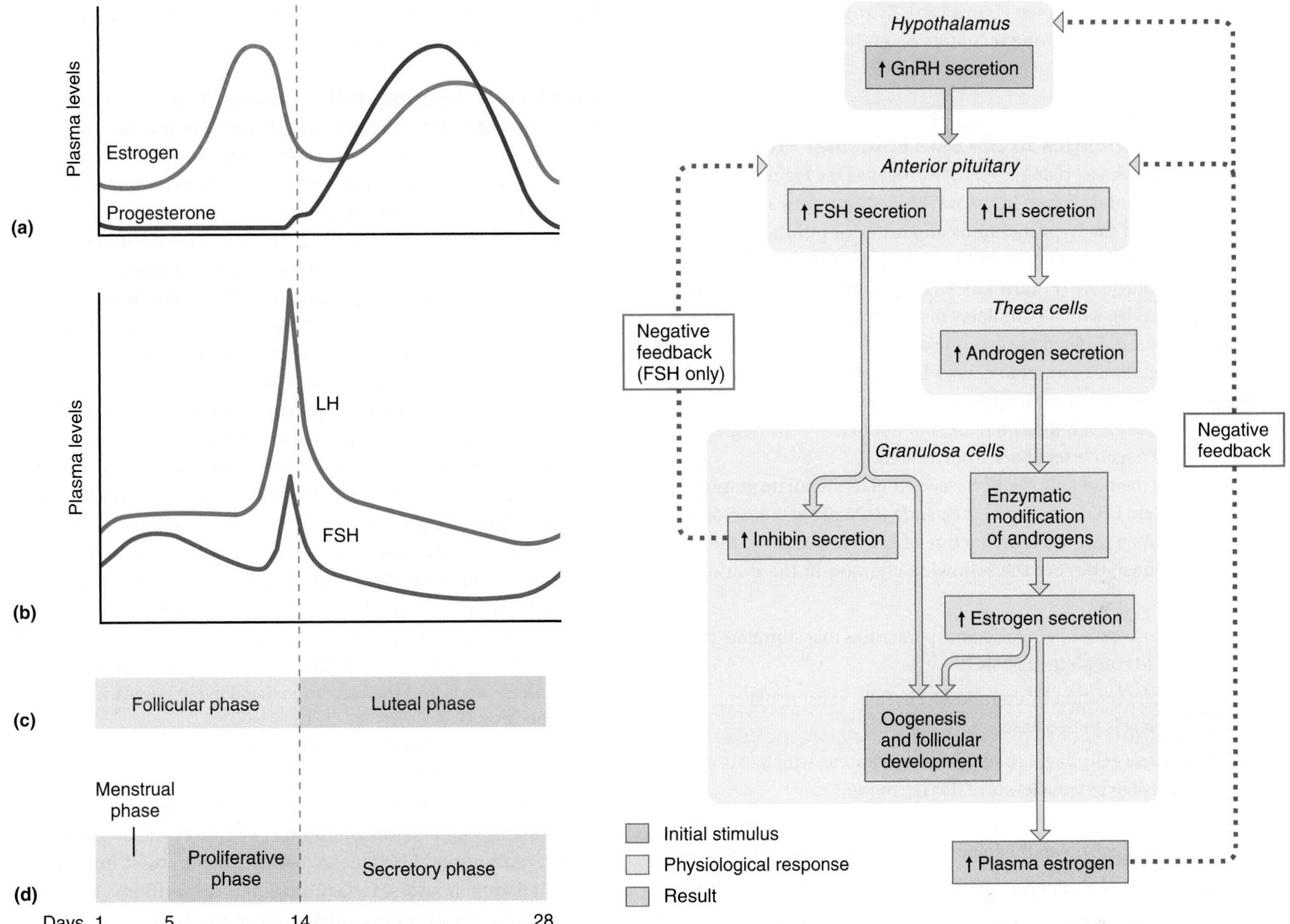

Figure 17 Summary of hormonal changes in the menstrual cycle. (a) Plasma estrogen and progesterone levels. **(b)** Plasma gonadotropin levels. **(c)** Phases of the ovarian cycle. **(d)** Phases of the uterine cycle. The vertical dashed line indicates the occurrence of ovulation.

Figure 18 Regulation of hormone secretion during the early to mid-follicular phase. Notice that at relatively low plasma levels, estrogen provides negative feedback for the release of GnRH, FSH, and LH.

rise in plasma levels of these hormones. The LH level remains fairly steady, but FSH level falls due to the influence of inhibin, which is secreted at ever-increasing rates by granulosa cells of the growing follicle. This drop in FSH is largely responsible for triggering atresia of nondominant follicles, as previously described.

During the early to mid-follicular phase, estrogens promote a variety of physiological changes throughout the body, including uterine changes characteristic of the proliferative phase. Estrogens also work in conjunction with FSH to promote oogenesis and follicular growth (Table 3). In addition, these hormones promote changes that prepare the body for subsequent events of the menstrual cycle. For instance, estrogens induce the expression of LH receptors on granulosa cells, which renders these cells responsive to LH in preparation for the rise in LH that occurs just before the end

TABLE 3 Actions of Estrogens

Follicular phase	Pregnancy (and luteal phase)
Promote endometrial growth	Promote growth of duct tissue and deposition of fat in the breasts
Promote oogenesis and follicle development	Suppress lactation
↑ LH receptors on granulosa cells	↑ Prolactin secretion
↑ Progesterone receptors in the uterus	↑ Growth and contractile activity of uterine smooth muscle
↑ LH secretion (late follicular phase only)	↑ Oxytocin receptors in the uterus
	Promote fluid retention

of the follicular phase (see Figure 17b). Estrogens also induce the expression of progesterone receptors on endometrial cells, which primes the endometrium to respond to progesterone during the luteal phase.

Hormonal Changes in the Late Follicular Phase In the late follicular phase, rising levels of estrogens (see Figure 17a) trigger a fundamental change in the way these hormones affect secretory activity of the hypothalamus and anterior pituitary: Instead of suppressing the secretion of LH, estrogens *stimulate* the secretion of this hormone (**Figure 19**). As a consequence, LH levels rise (see Figure 17b), which stimulates more estrogen secretion, which stimulates more LH secretion, and so on. These events constitute a positive feedback loop that causes LH secretion to rise abruptly, bringing about a dramatic upswing in plasma LH levels called the **LH surge.** (FSH levels also increase, but not to the same degree because of the negative feedback from inhibin.)

During the mid-follicular phase, estrogens stimulate granulosa cells to express LH receptors on their surfaces, making them responsive to LH. As a result, the rising tide of LH that occurs in the late follicular phase triggers the following changes in the dominant follicle:

1. Granulosa cells begin to secrete paracrines that stimulate the oocyte to complete meiosis I.
2. Estrogen secretion by granulosa cells falls, causing plasma estrogen levels to decrease.
3. Granulosa cells begin to secrete progesterone, which causes a small rise in plasma levels of this hormone.
4. Granulosa cells begin to secrete enzymes and paracrines that begin to break down the follicular wall. About 18 hours following the peak of the LH surge, the wall ruptures, triggering ovulation.
5. Granulosa cells and theca cells begin to differentiate into the cells forming the corpus luteum.

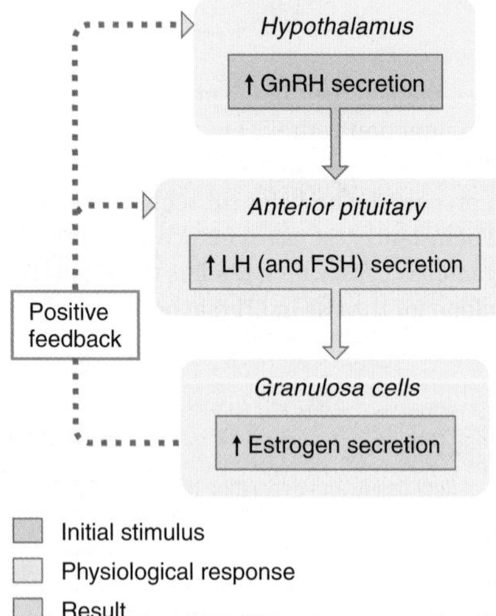

Figure 19 Regulation of hormone secretion during the late follicular phase. Notice that at relatively high plasma levels, estrogen provides positive feedback for the release of GnRH, FSH, and LH.

The appearance of the corpus luteum sets the stage for other events of the luteal phase, which are described next.

Hormonal Changes in the Luteal Phase Early in the luteal phase, the corpus luteum is growing but not yet fully functional as an endocrine organ. Plasma estrogen levels continue to fall (see Figure 17a) because the rate of estrogen secretion by the corpus luteum is not yet high enough to compensate for the loss of the estrogen-secreting dominant follicle. This decline in the estrogen levels removes the stimulus for LH secretion, thereby terminating the LH surge (see Figure 17b). Progesterone secretion by the corpus luteum causes plasma levels of this hormone to begin rising. As time passes, the corpus luteum grows and eventually attains full maturity. Estrogen and progesterone secretion continue to rise, causing a corresponding dramatic increase in plasma levels that peak at about the middle of the luteal phase. Estrogens work to maintain many of the same effects exerted during the follicular phase, whereas progesterone promotes secretory-phase uterine changes and other changes that generally adapt the body for possible pregnancy (see Table 3 and Table 4). After about the tenth day of the luteal phase (in the absence of an implanted embryo), the corpus luteum begins to degenerate and its secretory activity declines, causing a drop in estrogen and progesterone levels that soon triggers menstruation. With the beginning of menstruation, the luteal phase ends and the next menstrual cycle begins with a new follicular phase.

As estrogen levels rise during the luteal phase, possible stimulatory effects of these hormones on LH secretion are blocked by progesterone, which strongly suppresses secretory activity of the hypothalamus and anterior pituitary (**Figure 20**). Thus progesterone prevents further LH surges, which might otherwise be triggered by high estrogen levels. In fact, plasma LH (and FSH) levels decrease throughout the luteal phase due to this inhibitory influence of progesterone.

We have seen that the female sex hormones are responsible for regulating many of the events that occur during the menstrual cycle. In the next section we consider the long-term effects of these hormones—estrogens, in particular.

Long-Term Hormonal Regulation of Female Reproductive Function

Estrogen and progesterone levels are low in early childhood but increase dramatically in puberty. The rising estrogen levels promote the development of female secondary sex characteristics, including widening of the hips relative to the shoulders; deposition of fat in the breasts, hips, and buttocks; secretion of skin oil (which is thinner than that of males); and growth of coarse body hair in the

TABLE 4 Actions of Progesterone

Luteal phase and pregnancy
Promotes secretory-phase uterine conditions
Suppresses uterine contractile activity
Promotes growth of glandular tissue in breasts
Suppresses milk production

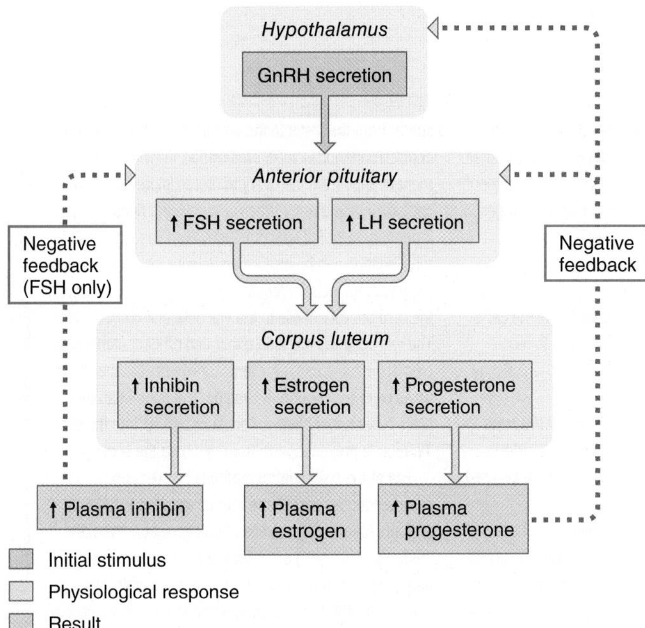

Figure 20 **Regulation of hormone secretion during the luteal phase.**

pattern characteristic of females. (Elongation of the hairs themselves is promoted by androgens, as it is in males.) Estrogens also promote the growth of bones during adolescence, which is similar to the action of testosterone in males.

During a woman's reproductive life, average estrogen levels remain high and help maintain the secondary sex characteristics; when she reaches menopause, however, these levels decline as menstrual cycling comes to a halt. Over time, this relative lack of estrogens causes a reversal of many of the changes that occurred during puberty and triggers many other changes throughout the body. For example, the breasts and reproductive organs, which grow during puberty, gradually shrink in size during menopause. Many women also experience so-called hot flashes—increases in body temperature and blood flow to the skin of the face and neck followed by profuse sweating—which appear suddenly for no apparent reason and last for a few seconds to minutes. The decrease in estrogen levels during and after menopause causes an increased risk for coronary heart disease and bone fractures; the latter effect results from osteoporosis, a decrease in the mineral content of bones. Some women take estrogen and progesterone supplements to minimize the health risks following menopause. However, recent studies suggest that these supplements may increase the likelihood of developing breast cancer or cardiovascular diseases.

Quick Check 4

1. Identify the two phases of the ovarian cycle, and briefly describe the events that occur in each.
2. Identify the three phases of the uterine cycle, and briefly describe the events that occur in each.

4 Fertilization, Implantation, and Pregnancy

For fertilization to occur, sperm must be introduced into the female reproductive tract from about five days before ovulation to one day after ovulation, because sperm can live in the tract for a maximum of five days and because the oocyte remains viable for only 12 to 24 hours after its release from the ovary.

When first deposited in the female reproductive tract, sperm are unable to fertilize an oocyte; they gain this capacity only after they have been in the tract for several hours, during which time they undergo a process called *capacitation.* One result of capacitation is a change in the pattern of tail movements, which permits sperm to move faster; another is alteration of the plasma membrane so that the sperm can fuse with the oocyte, which is necessary for fertilization to occur.

Sperm deposited in the vagina move by self-propulsion through the cervical canal into the uterus, where they move along the uterine wall and eventually enter the opening to the uterine tube. Sperm migration is also aided by vaginal and uterine contractions that occur during sexual arousal. Of the millions of sperm that are deposited, only a few hundred typically reach the uterine tube; the others die along the way. In the tube, sperm move toward the oocyte as it approaches them from the opposite direction. Fertilization, when it occurs, usually takes place in the uterine tubes. The fertilized egg undergoes several cell divisions and moves on to the upper portion of the uterus, where it adheres to and then embeds in the uterine wall (an event called **implantation**). If implantation is successful, further development of the fertilized egg continues. If fertilization does not occur, implantation does not occur, and the egg is eventually expelled from the uterus during menstruation. In addition, even if fertilization occurs, implantation might not occur, such that the fertilized oocyte leaves the body during menses. Preventing implantation is the basis of the so-called *morning-after* (see **Discovery: Birth Control Methods**).

Events of Fertilization

During sperm migration, several sperm may reach the oocyte and work together to break down the barriers to fertilization—namely, the corona radiata and zona pellucida (**Figure 21**). Once sperm penetrate the corona radiata, they bind to sperm-binding proteins in the zona pellucida. This binding then triggers an event called the **acrosome reaction,** which results in release of the contents of the acrosome to the outside. Among the acrosome's contents are enzymes capable of digesting the substance of the zona pellucida, thereby allowing the sperm access to the oocyte's surface.

The first sperm to reach the oocyte binds to a receptor on the plasma membrane, which precipitates transport of the head of the sperm into the egg's cytoplasm. This fusion also stimulates the oocyte to complete meiosis II and become an ovum. Once inside the ovum, the sperm's plasma membrane disintegrates and its chromosomes migrate toward the center of the cell, as do the chromosomes that were originally present in the oocyte. (At this stage the chromosomes are contained in their nuclear membranes, which later

DISCOVERY

Birth Control Methods

Many sexually active couples use various methods of birth control to postpone or prevent pregnancy. Some birth control methods are reversible, whereas others are permanent. Most birth control methods work by preventing fertilization (a process called *contraception*); others work by preventing implantation; and a few terminate pregnancy (called *abortion*.)

Reversible contraceptives can be categorized into behavioral, barrier, or chemical mechanisms. Behavioral mechanisms include *abstinence, withdrawal,* and the *rhythm method* (or natural family planning). Abstinence is the only fail-proof birth control mechanism. In withdrawal, the male withdraws the penis from the vagina prior to ejaculation; use of this method is effective only 75% to 80% of the time, because some leakage of semen occurs prior to ejaculation and withdrawal depends on accuracy in timing. The rhythm method requires that a couple predict when the female is fertile and refrain from sexual intercourse for a short time during each menstrual cycle—namely, for a few days around the time of ovulation. Successful use of this method depends on an accurate prediction of when ovulation occurs. Predictions are often based on the calendar (a woman's cycle pattern), measurements of body temperature (body temperature increases approximately 1°C following ovulation), and monitoring of changes in vaginal secretions. Because these predictions are prone to error, a woman using the rhythm method faces a 20% chance of becoming pregnant over a year's time.

Barrier methods include devices that block sperm from leaving the male and devices that block sperm from entering the cervical canal of the female. The male *condom*, a thin latex rubber sheath that is placed over the penis, decreases the amount of sperm that leave the male. There is also a female condom that is inserted into the vagina. The likelihood of getting pregnant with either type of condom is greater than 20%. Other female barrier devices include the *diaphragm, cervical cap,* and *sponge,* which are placed in the vagina prior to intercourse. Often, spermicidal creams or jellies, which inhibit sperm motility, are used in conjunction with diaphragms and caps; the sponge contains spermicide. These barrier methods have a relatively high failure rate due to faulty structure or inappropriate use; 15% to 20% of couples using these birth control methods will face an unexpected pregnancy over the course of a year.

Chemical methods of birth control include spermicides, hormones, and copper. Spermicides are best used in conjunction with other birth control methods; when used alone, they are associated with a 35% chance of pregnancy per year. Hormonal control comes in a variety of forms that vary in effectiveness from less than 1% to 10% chance of pregnancy in a year, with the effectiveness being highly dependent on correct utilization of the hormones. Hormonal control comes in two primary forms: (1) estrogen and a progesterone-like substance (progestin) or (2) progestin alone. When the combination of hormones is used, treatment generally occurs for three weeks to prevent ovulation followed by one week without hormones to allow menses to occur. In contrast, progestin alone can be used for longer periods of time to prevent ovulation. However, a woman's menses will be attenuated or absent during continued treatment with progestin. Other effects of progestin include thickening of the cervical mucus, which blocks the ability of sperm to reach the oocyte, and a thinning of the uterine wall, which may prevent implantation.

Hormonal controls include *oral contraceptives* (the "Pill"), *patches* worn on the skin, *intramuscular (IM) injections*, and *vaginal contraceptive rings* that are inserted for three weeks and then removed for one week.

Oral contraceptives contain a combination of estrogens and progestin, or just the latter hormone. These pills must be taken daily to be effective. The combination pill is used more commonly and varies in terms of the doses of the two hormones to match more natural changes in the body; it functions by preventing ovulation. The fourth-week pill contains no hormone, thereby allowing menses to occur. The progestin-only pill, also called the *minipill*, comes in two doses that vary in their method of action. Low-dose progestin pills thicken the cervical mucus to prevent the sperm from reaching the egg. Higher-dose progestin pills thicken the cervical mucus but also prevent ovulation. It is important to note that oral contraceptives, while relatively effective when used correctly, lose their effectiveness when taken in combination with antibiotics.

Intramuscular administration of high-dose progestin is used to produce long-term inhibition of ovulation, lasting up to three months. These shots tend to be more effective than the pill because of behavioral issues: A woman is more likely to miss one day taking a pill than having a shot once every three months.

Patches and vaginal rings function by releasing estrogens and progestin to prevent ovulation. Both birth control methods are applied by the woman using them. Each patch is effective for one week, and patches are used for three weeks and then removed for the fourth week, during which the woman has menses.

An *intrauterine device (IUD)* is a long-term but not permanent method of birth control. This small T- or Y-shaped plastic device is inserted by a physician into the uterus, where it releases either copper or progestin, with copper being toxic to sperm and progestin preventing implantation. The copper IUD can be effective for as long as 10 years, and the progestin IUD is effective for as long as 5 years. The failure rate of many IUDs is less than 1%.

A more reliable, and generally permanent, method of contraception is *surgical sterilization*, which has a failure rate near zero. In men, the most common method of sterilization is *vasectomy*, in which the vas deferens is cut and tied or cauterized to prevent the passage of sperm from the testis to the urethra. In women, the most common method is *tubal sterilization*, in which the movement of eggs from the ovary to uterus is prevented by cauterizing the uterine tubes or by placing rings or clips around them *(tubal ligation)*. Today both vasectomy and tubal ligation can be performed without incisions. Vasectomy without incision requires a small puncture wound in the scrotum for access to the vas deferens on both sides. The vas deferens are then either tied off or cauterized. In women, doctors use a *hysteroscope* to access the uterine tubes by entry through the vagina. The doctor can then place a permanent insert, Adiana or Essure, into the uterine tube to prevent sperm from reaching the oocytes.

All of the birth control methods just described are intended for use before intercourse, but RU 486 (mifepristone) or high doses of oral contraceptives (the "morning-after pill") are effective in preventing pregnancy when taken within one to five days after intercourse. RU 486 binds to progesterone receptors, thereby blocking progesterone's ability to promote secretory-phase uterine conditions and suppress uterine contractions. If fertilization has occurred, RU 486 can prevent implantation, or if implantation has occurred, can trigger detachment of the embryo from the endometrium and its subsequent expulsion from the uterus. High doses of oral contraceptives, by comparison, can inhibit ovulation for three to five days after unprotected sex, thereby preventing fertilization from ever occurring.

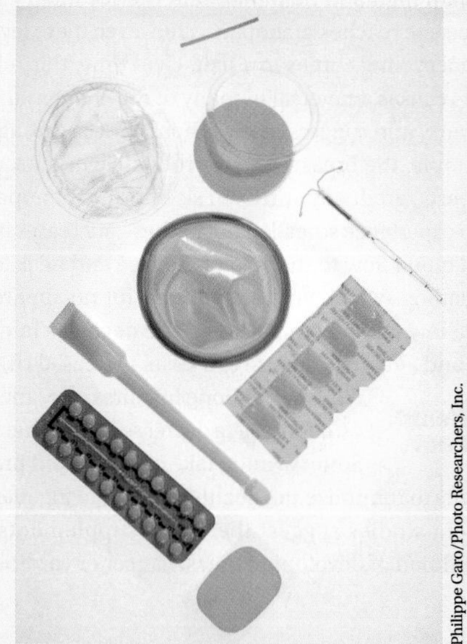

Philippe Garo/Photo Researchers, Inc.

Various birth control devices including a condom, a contraceptive implant, a contraceptive patch, contraceptive pills, a diaphragm, an intrauterine device (IUD), spermacide cream, contraceptive sponges, and a vaginal ring.

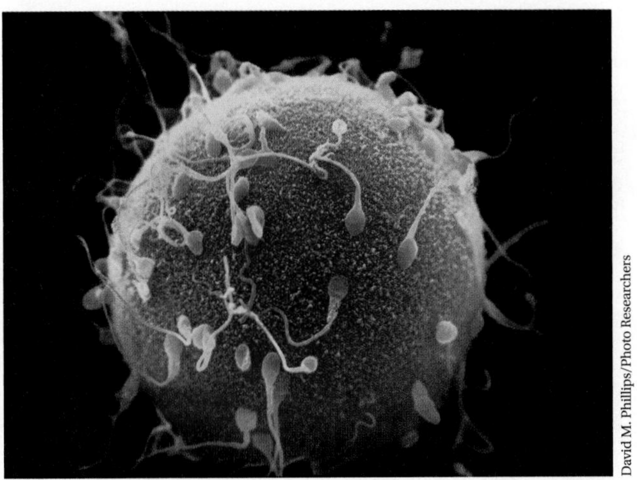

Figure 21 Scanning electron micrograph of sperm trying to fertilize an oocyte.

disintegrate.) The chromosomes of the sperm and ovum combine, and the DNA is then replicated in preparation for the first mitotic division of the cell (now a *zygote*).

The fact that hundreds of sperm typically reach the uterine tube raises the distinct possibility that more than one sperm could fertilize an oocyte, a phenomenon called *polyspermy.* If this were to happen, the fertilized ovum would contain more than the normal number of chromosomes, and it would most likely fail to develop. Fortunately, polyspermy is averted by a number of mechanisms set in motion when a sperm first fuses with the oocyte's plasma membrane. Such fusion triggers the exocytosis of vesicles near the plasma membrane, which releases enzymes into the space between it and the surrounding zona pellucida. As a result, sperm-binding proteins in the zona pellucida become inactivated, and the zona pellucida hardens and pulls away from the plasma membrane. Because these changes create a barrier around the oocyte that prevents its fertilization by other sperm, they are referred to as the *block to polyspermy.*

Early Embryonic Development and Implantation

After fertilization has occurred, the ovum undergoes several mitotic divisions (**Figure 22a**), transforming within the next few days into a round ball of cells called a **morula.** Cell divisions at this stage are unlike subsequent cell divisions in that the total volume of cytoplasm does not increase. For this reason the number of cells increases with each division, but the size of each cell decreases. Cell division of this type is often referred to as *cell cleavage.* Three to four days after fertilization, repeated cell cleavages yield a total of 16 to 32 cells, each of which is *totipotent*—that is, has the potential to develop into a complete human being. By this time the morula has reached the uterus; for the next three to four days it floats in the intrauterine fluid and undergoes more cell divisions.

On rare occasions, the morula splits into two (or occasionally more) parts. When this occurs, each part can develop into a complete individual. The resulting offspring are *identical twins,* who look alike and have identical genes. They are also called *monozygotic twins* because they originate from the same zygote.

Eventually the morula develops into a more complex, hollow structure called a **blastocyst,** which has shed the zona pellucida and possesses the following components (see Figure 22a): a spherical outer cell layer called the **trophoblast;** a cluster of cells on the inside called the **inner cell mass,** which is attached to the trophoblast on one side and eventually gives rise to the embryo; and a fluid-filled cavity, called the **blastocoele,** adjacent to the inner cell mass. At this stage the cells have differentiated from one another; that is, they have lost their totipotency and are committed to eventual development into different tissues.

About six or seven days after fertilization, the blastocyst attaches to the uterine wall (Figure 22b). For the next few weeks, cells of the trophoblast secrete enzymes that digest the adjacent endometrial cells, from which the embryo draws nourishment. The trophoblast also secretes paracrines that promote local changes in the endometrial tissue. Among these changes, collectively known as the *decidual response,* is an increase in the number of capillaries, which helps increase the delivery of oxygen and nutrients to the area. At the point of contact with the uterine wall, trophoblastic cells proliferate, differentiate, and begin to infiltrate into the endometrial tissue (Figure 22c). As described shortly, the trophoblast and underlying endometrium eventually develop into the **placenta,** a structure that allows efficient exchange of gases, nutrients, and wastes between the mother's circulatory system and the circulatory system of the developing embryo.

Apply Your Knowledge

An ectopic pregnancy occurs when the embryo implants somewhere other than the uterus (usually in the uterine tube). Why do ectopic pregnancies fail to go to full term, and why are they dangerous for the mother?

Later Embryonic and Fetal Development

After implantation, the embryo continues to grow in size and complexity. At a certain stage, it becomes a fetus, which depends on the placenta for nourishment. In describing embryonic and fetal development, time is usually identified relative to the first day of the woman's last menses (*gestational age*), but it can also be given relative to fertilization.

Development from Embryo to Fetus

In the region of contact between the embryo and the uterine wall at about seven weeks, the trophoblast thickens and develops into the **chorion,** a tissue that eventually grows into a tough envelope that encapsulates the embryo, isolating it from its surroundings (**Figure 23**). Inside the embryo, other changes are occurring: A fluid-filled cavity called the *amniotic cavity* begins to form and grow in size within the inner cell mass (see Figure 22c). As the amniotic cavity expands, the surrounding cells develop into an epithelial

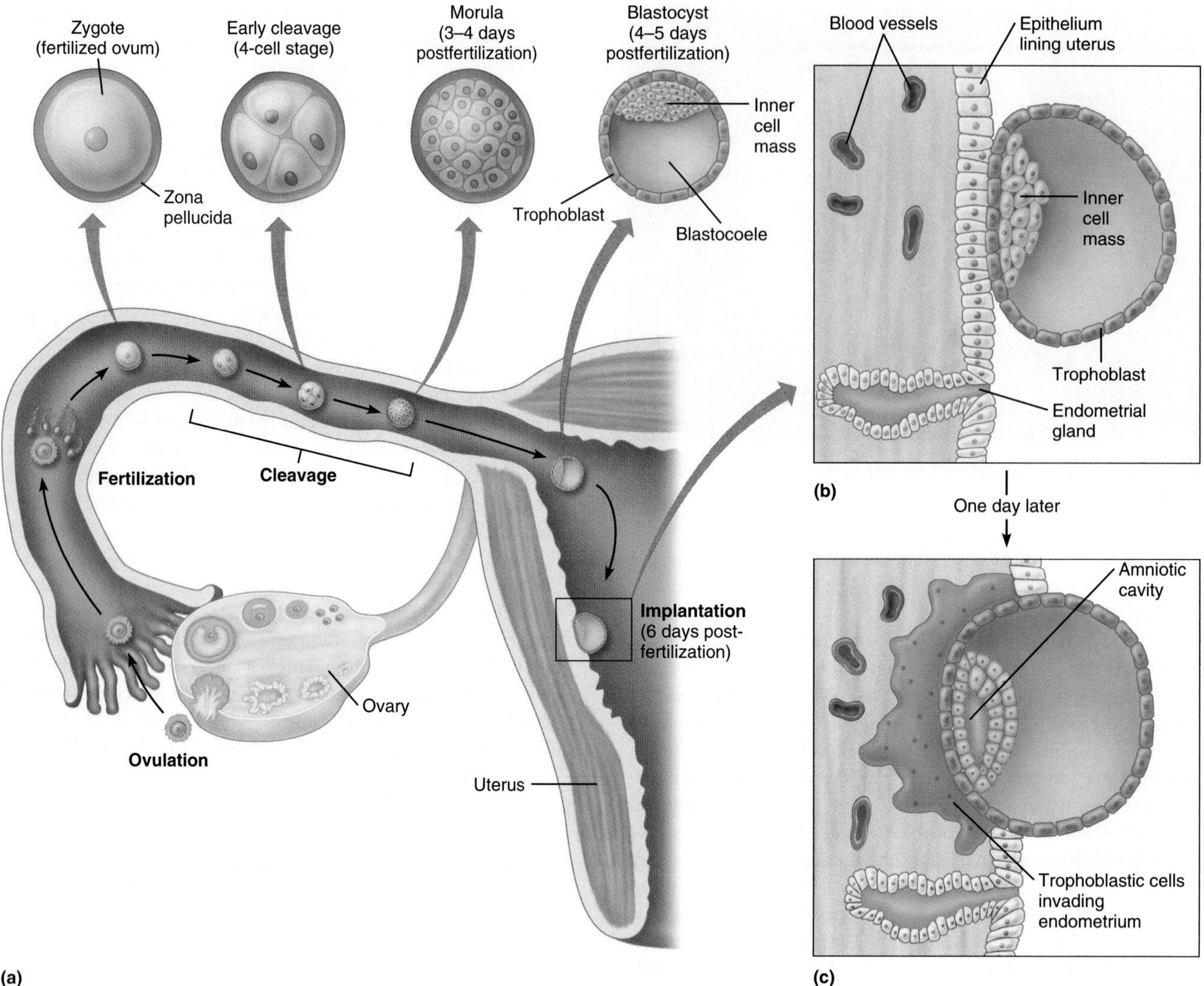

Figure 22 **Early embryonic development and implantation.** **(a)** Stages of early embryonic development. Events shown begin with ovulation and proceed through implantation. **(b)** Early implantation, showing the blastocyst adhering to the surface of the endometrium. **(c)** A later stage of implantation, showing embryonic cells penetrating into the endometrial tissue. Note the appearance of the fluid-filled amniotic cavity within the inner cell mass.

tissue called the **amnion** (or *amniotic sac;* see Figure 23), which eventually fuses with the chorion to form a single membrane that surrounds the developing embryo. The fluid contained within the amnion (called *amniotic fluid*) is similar in composition to normal extracellular fluid; it bathes the embryo and forms a physical barrier between it and the uterine wall that cushions the developing embryo against possible physical trauma and protects it against sudden changes in temperature.

As development proceeds, the embryo undergoes a series of dramatic transformations in structure and function. Within the inner cell mass, cells begin to differentiate into distinct tissues that will ultimately give rise to all the organ systems found in adults. By week 5, the heart, brain, spinal cord, and gastrointestinal tract begin to develop. At week 6, the heart is beating. By 8 weeks, the lungs begin to develop. By 9 weeks, all organ systems have begun to develop. Over time, these primitive organ systems gradually acquire more of their adult characteristics; different systems develop at different rates. As this development is occurring, the embryo's external appearance is also rapidly changing. By the end of week 10, although it is about the size of a kidney bean, the embryo has a recognizably human form, complete with clearly discernible limbs as well as a head and face. From this point onward, the developing human is described as a *fetus.*

Formation of the Placenta

In its earlier stages of development, the embryo is able to obtain nutrients through the breakdown of endometrial tissue and to obtain

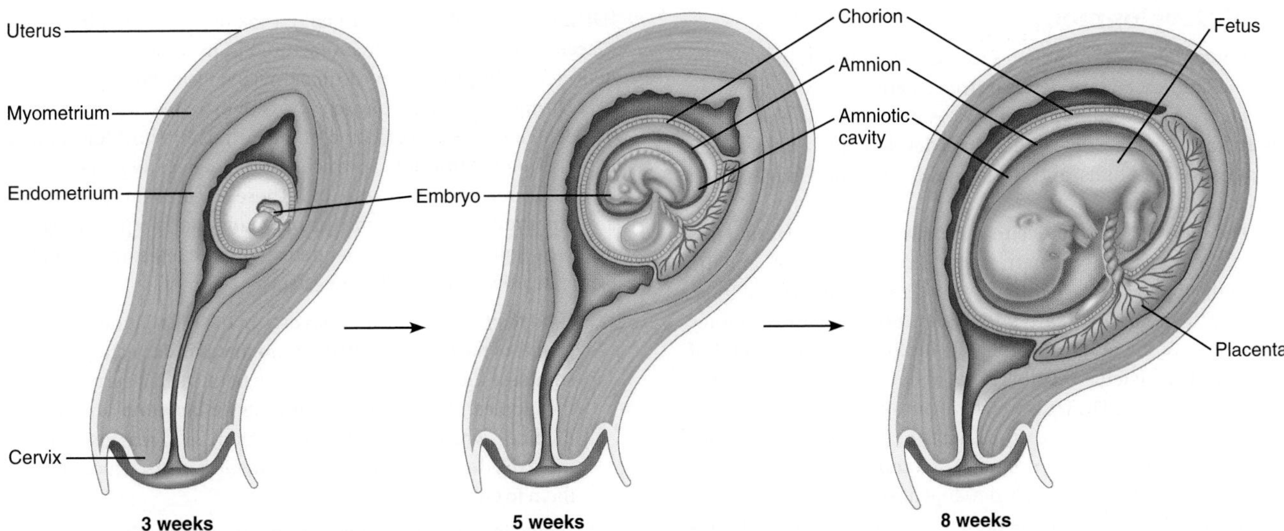

Figure 23 Growth and development of the embryo and the structures that support it. At approximately eight weeks after the first day of the woman's last menses, the developing human is termed a fetus.

sufficient quantities of oxygen by diffusion from nearby blood vessels. However, as the embryo grows, its need for oxygen increases such that it can no longer rely on diffusion from the mother's tissues as a mechanism of delivery; likewise, its demand for nutrients increases such that it can no longer obtain sufficient amounts from the surrounding tissue.

To obtain essential materials (and to rid itself of carbon dioxide and other wastes), the growing embryo or fetus relies on the placenta, a structure that permits the ready exchange of gases, nutrients, and other materials between the fetus's bloodstream and the maternal circulatory system. The placenta, which develops in the first few weeks after implantation in the region of contact between the embryo and the endometrium, is made of specialized, closely interwoven fetal and maternal tissues (**Figure 24**).

Placental development begins as the chorion sends out fingerlike projections called *chorionic villi.* The villi contain both capillaries of the fetal circulation and cells that secrete paracrines that alter the structure and function of the surrounding endometrial tissue, such that the villi become surrounded by sinuses containing maternal blood. As a result, maternal blood and fetal blood come into close proximity (which facilitates material exchange) without actually mixing. (Separation of maternal and fetal blood is important because it prevents the mother's immune system from attacking the fetal tissues.) Maternal blood is delivered to the placental sinuses by the *uterine artery* and is returned to the mother's general circulation by the *uterine vein.* Fetal blood is carried to the placenta by the paired *umbilical arteries* and back to the fetus's general circulation by the *umbilical vein,* both of which are housed in a ropelike structure called the **umbilical cord,** which extends between the placenta and the fetus. As we will see shortly, the placenta is not just an organ of exchange; it also secretes several hormones that are important in maintaining pregnancy and preparing the mother's body for the eventual birth of the child.

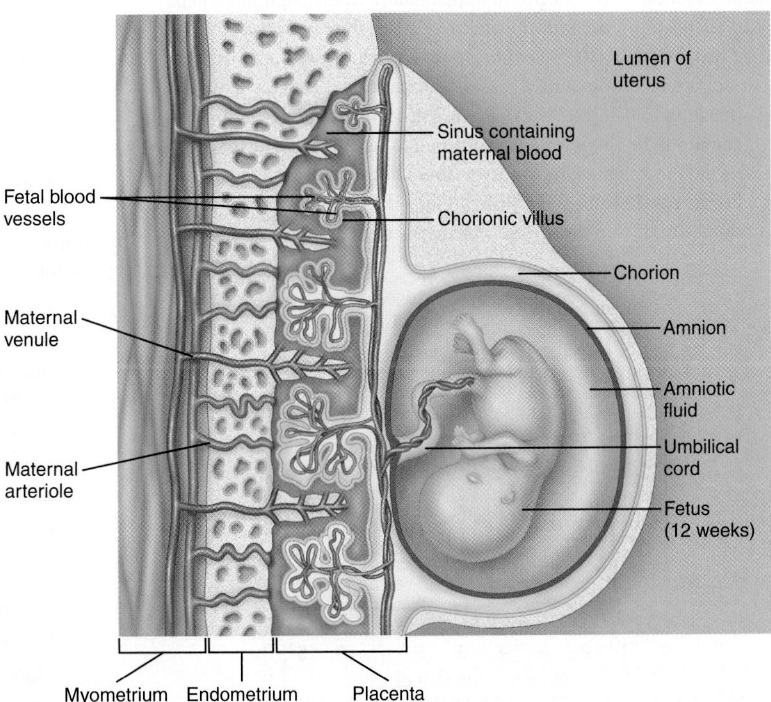

Figure 24 The placenta. The interweaving of fetal and maternal tissues enables the exchange of materials between the circulatory systems of mother and fetus.

Fetal Development

While a complete description of fetal development is beyond the scope of this text, an overview is provided here.

The first trimester of pregnancy is defined as weeks 1–12 of gestation. During this time, the embryo develops into a fetus, which has gonads, tooth buds, and closed eyelids. The total length of the fetus is 1 to 2 inches.

The second trimester occurs during weeks 13–28. During this period, the liver and bone marrow produce red blood cells and the liver and pancreas secrete products into the gastrointestinal tract. The heartbeat can be heard with a stethoscope, and voluntary movements of the fetus are felt by the mother. Toward the end of this trimester, the fetus's eyelids begin to open and close and some reflexes are present. The fetus weighs about 1 to 2 pounds.

At the beginning of the third trimester (weeks 29–40), the fetus's lungs function in gas exchange even though surfactant is not yet secreted. The bones are fully developed. The body starts to store ions and nutrients. Later in the third trimester, the fetus continues rapid development of all systems and gains considerable weight as it prepares for parturition. Surfactant production in the lungs begins at approximately week 36.

Hormonal Changes During Pregnancy

During the first three months of pregnancy, the corpus luteum is maintained by the actions of a hormone called **human chorionic gonadotropin (hCG),** which is secreted by the chorionic portion of the placenta and exerts many of the same effects as LH. Under the influence of hCG, the corpus luteum continues to secrete large amounts of estrogens and progesterone, which exert several actions described later. Secretion of hCG begins at implantation, when trophoblast cells start to infiltrate the endometrial tissue, but then increases for the next two months as the placenta grows (**Figure 25**). At the end of this period, the secretion of hCG falls off dramatically, and as a result the corpus luteum begins to degenerate. Estrogen and progesterone levels remain high, however, and continue to rise until parturition because the placenta begins to secrete these hormones just as secretion by the corpus luteum is declining. Placental estrogens are actually manufactured from androgens (primarily dehydroepiandrosterone [DHEA]) secreted by the adrenal cortex of the developing fetus. Placental progesterone is manufactured from cholesterol, which is delivered to the placenta by the maternal bloodstream.

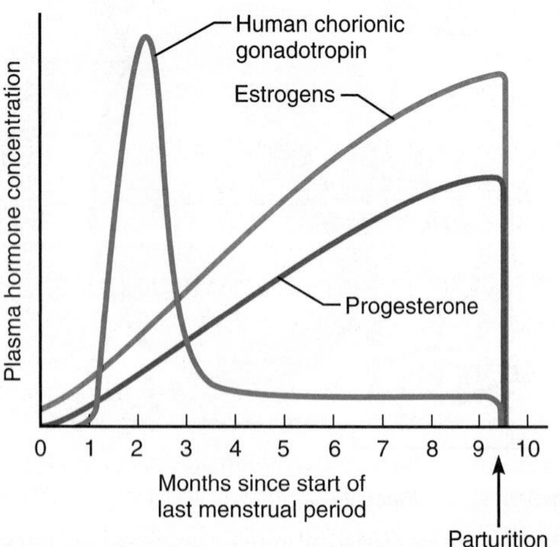

Figure 25 **Plasma concentrations of estrogen, progesterone, and human chorionic gonadotropin during pregnancy.**

During pregnancy, plasma levels of estrogen and progesterone, which rise after formation of the corpus luteum, continue to increase until parturition, at which time estrogen and progesterone levels fall precipitously (Figure 25). These elevated hormone levels are important for maintaining the pregnancy and for preparing the mother's body for the eventual delivery of the fetus. In particular, estrogen promotes the following effects:

1. *Growth of duct tissue in the breasts.* In the milk-producing glands of the breasts (called **mammary glands**), estrogens stimulate the development of tissue in ducts that carry milk to the nipples. This prepares the breasts for **lactation** (milk production), which begins after parturition. Estrogens also promote the deposition of fatty tissue in the breasts, which causes them to enlarge.
2. *Prolactin secretion.* Estrogen stimulates the anterior pituitary's secretion of prolactin, which promotes breast growth and prepares the mammary glands for lactation. Although prolactin and estrogens both work to build up the milk-producing apparatus during pregnancy, actual milk production does not occur yet because the high plasma levels of estrogens and progesterone during pregnancy suppress lactation.
3. *Growth and enhanced contractile responsiveness of uterine smooth muscle.* Estrogens stimulate the growth of uterine smooth muscle, thereby enhancing the ability of the uterus to contract and expel the fetus during parturition. Estrogens also work to promote increased contractile activity in uterine smooth muscle and to increase its responsiveness to *oxytocin,* which stimulates uterine contractions during birth. Despite the fact that estrogens enhance contractile activity by themselves, the uterine contractions that occur during pregnancy are infrequent and weak. The reason is that elevated levels of progesterone during this period block the contraction-promoting actions of estrogens and oxytocin.

While estrogens are working to promote these effects, progesterone is also working to maintain the pregnancy and prepare the mother's body for parturition. In particular, this hormone promotes the following effects:

1. *Growth of glandular tissue in the breasts.* In the breasts, progesterone stimulates the growth of glandular tissue (as opposed to ductal tissue, whose growth is stimulated by estrogens). This glandular tissue is responsible for secreting the milk that is produced after parturition.
2. *Suppression of contractile activity in uterine smooth muscle.* As previously mentioned, progesterone suppresses uterine contractions, thereby counteracting the effects of estrogens and oxytocin on uterine smooth muscle. Because contractions could prematurely expel an embryo or fetus from the uterus, this suppressive action is necessary for successful implantation and for the maintenance of pregnancy.
3. *Maintenance of secretory-phase uterine conditions.* Progesterone works during pregnancy to maintain the same secretory-phase uterine conditions that it promotes during the luteal phase of the ovarian cycle. This ensures hospitable conditions for embryonic and fetal growth.

In pregnancy, progesterone exerts negative feedback on the hypothalamus and anterior pituitary that serves to keep rates of LH and FSH secretion low. Consequently, no new dominant follicles appear and no LH surges occur, despite the fact that estrogen levels are elevated at this time. In addition, the placenta secretes a new hormone called **placental lactogen,** which has effects similar to some of the actions of growth hormone and prolactin. Placental lactogen works hand in hand with estrogens and prolactin to promote not only breast growth but also the mobilization within the mother of energy stores needed to meet the increased demands arising from the growth of breasts and fetal tissue.

While placental hormones induce some insulin resistance in the mother that slightly elevates blood glucose levels, in some pregnant women the level of insulin resistance continues to increase in response to the continued release of these hormones, thus elevating glucose levels even more (see **Focus on Diabetes: Gestational Diabetes**).

Quick Check 5

1. Define the following terms: *capacitation, acrosome reaction, polyspermy, morula, blastocyst, inner cell mass, trophoblast, implantation, placenta, chorion, chorionic villi, amnion, amniotic fluid, umbilical cord.*
2. Where does fertilization occur? Implantation? What is the primary function of the placenta?
3. Estrogens and progesterone are secreted by which structure during the first three months of pregnancy? During the final six months?
4. For each of the following hormones, name the site from which it is secreted, and give a brief description of its function: human chorionic gonadotropin, prolactin, oxytocin, placental lactogen.

5 Parturition and Lactation

Pregnancy normally lasts about nine months (40 weeks), during which time the fetus gains in size, weight, and maturity such that it acquires the ability to live outside its mother's body. Once fetal maturation is complete, parturition occurs.

Events of Parturition

In the last few weeks of pregnancy, weak and infrequent uterine contractions begin to occur. Over time, these contractions gradually increase in strength and frequency. Because estrogens are known to promote contractile activity in uterine smooth muscle, it is likely that these contractions are triggered, at least in part, by the increased levels of estrogens relative to progesterone that exist in late pregnancy. Moreover, uterine smooth muscle is capable of initiating its own contractions in response to stretch, and as the fetus grows and the uterus expands, the resulting stretch of the smooth muscle may initiate contractile activity.

To facilitate delivery, the cervix undergoes a process called *ripening* in the weeks before parturition. During this process, the cervix becomes softer and more flexible as a result of the enzymatic breakdown of collagen fibers in its connective tissue. These changes make it easier for the cervical canal to enlarge so that the fetus can pass through it.

Shortly before parturition, the fetus, which is normally positioned upside down in the mother's body, moves downward and comes into contact with the closed cervix (**Figure 26a**), so that the strong uterine contractions of parturition will push the fetus through the cervical canal headfirst. Moreover, the fetus's headfirst orientation enables the head, which is narrower than the rest of the body, to act as a wedge to force the cervical canal open. (Opening of the canal is called *cervical dilation.*) In a small fraction of cases, the fetus's legs or hips pass through the cervical canal first, a phenomenon called *breech birth;* because the hips are wider than the head, such a delivery is more difficult than normal and increases

FOCUS ON DIABETES

Gestational Diabetes

Approximately 4% of pregnant women in the United States develop *gestational diabetes*, a condition of elevated blood glucose that first appears during pregnancy. Traits that make a woman more likely to develop gestational diabetes include being overweight, older than 25 years of age, prediabetic, and of African American, Native American, Asian, or Hispanic descent.

Symptoms of gestational diabetes are few, but include increased thirst and urination. Diagnosis is often made during routine prenatal care.

The consequences of gestational diabetes can be severe for both mother and baby. The mother is likely to develop high blood pressure and urinary tract infections during untreated gestational diabetes. She is also more likely to develop type 2 diabetes mellitus. Because glucose crosses the placenta, the fetus's blood glucose levels will be elevated, which will increase insulin production by the fetal pancreas. Excess insulin promotes growth (insulin is permissive to growth hormone), so the fetus is likely to be larger than normal. Following birth, the baby may be hypoglycemic due to excess insulin and insufficient glucose; hypoglycemia can lead to seizures and other health problems for the infant. Babies born to a mother with gestational diabetes also have an increased likelihood of developing type 2 diabetes later in life.

Treatment of gestational diabetes centers on maintaining normal blood glucose levels. Thus the mother needs to eat a well-balanced diet and monitor her blood glucose levels; she is encouraged to exercise lightly. Insulin injections can be used if behavioral adjustments do not keep blood glucose levels low enough. Following delivery, both mother and child need to maintain a healthy lifestyle to minimize their chances of developing type 2 diabetes mellitus.

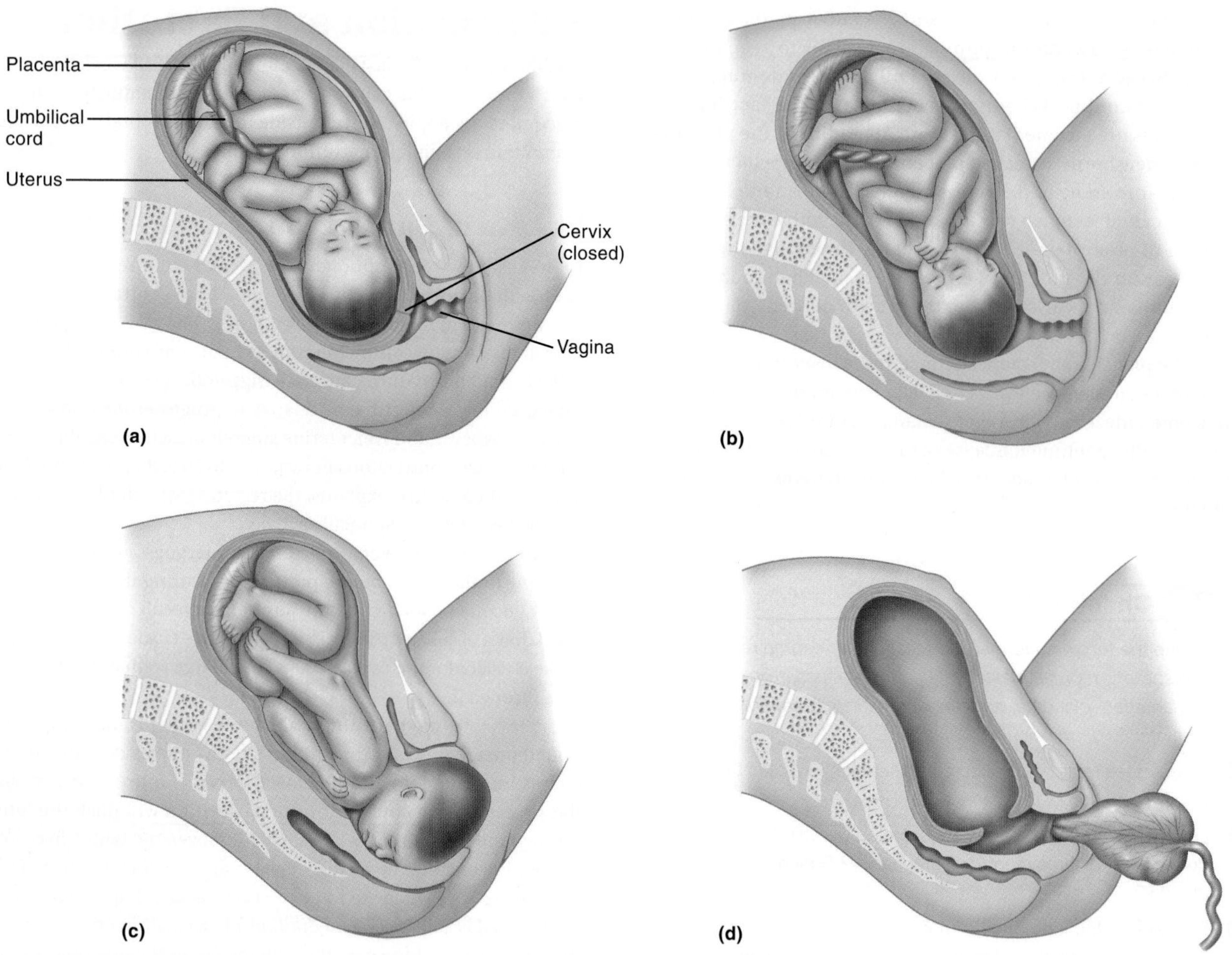

Figure 26 **Events of parturition. (a)** Position of the fetus in the uterus before parturition. **(b)** Dilation of the cervix. **(c)** Movement of the fetus through the birth canal (delivery). **(d)** Expulsion of the placenta following delivery.

the time required for parturition. Breech birth also exposes the fetus to greater danger because it increases the risk that the umbilical cord may become compressed between the fetus and the wall of the birth canal, which chokes off the fetus's blood supply. For these reasons, *cesarean section* (surgical delivery through the abdominal wall) is often recommended when breech birth is imminent.

In the hours just before parturition, the amniotic sac typically ruptures, causing the leakage of amniotic fluid to the outside of the body. This event is often the first reliable signal to the mother that parturition is imminent. Soon afterward, a series of strong uterine contractions (known as *labor*) begins. Each contraction starts at the top of the uterus and then travels downward through the smooth muscle as a wave. (Contractions are triggered by electrical signals that move from cell to cell through gap junctions.) The downward movement of these contractions pushes the contents of the uterus against the cervix, which eventually forces the cervical canal open. In labor, uterine contractions may be assisted by voluntary contractions of the mother's abdominal muscles, which increase pressure on the uterine contents.

Initially the contractions of labor are separated by intervals of 10 to 15 minutes, but contractions come closer together as labor progresses. The rising intrauterine pressure causes dilation of the cervical canal (Figure 26b), which eventually reaches a maximum diameter of approximately 10 centimeters. At that point the fetus moves into the canal and within a few minutes emerges from the mother's body (Figure 26c). Blood vessels in the umbilical cord begin to constrict, and the placenta then separates from the endometrium. (The umbilical cord is usually tied off by birth attendants and then cut on the maternal side of the knot.) Finally, a wave of strong uterine contractions expels the placenta (now referred to as the *afterbirth*) outside the mother's body (Figure 26d).

Once the contractions of labor begin, the stretch sensitivity of uterine smooth muscle tends to perpetuate them. As the contractions push the fetus downward, the lower portion of the uterus is stretched, which stimulates more contractile activity. Uterine contractions are also promoted by oxytocin, whose release is stimulated in labor. The pressure exerted by the fetus against the uterine wall excites stretch receptors located in the lower portion of the uterus, which triggers the release of oxytocin from the posterior pituitary. Oxytocin enters the bloodstream and then stimulates uterine contractions both by direct action on muscle cells and by stimulation of other cells in the myometrium to secrete prostaglandins, which act locally to promote contractile activity. The contraction-promoting effects of oxytocin are normally blocked during pregnancy by progesterone, which helps to keep uterine contractile activity low.

Although the precise nature of the signals that trigger parturition are not fully understood, evidence suggests that they may originate from the fetus itself. The fetal portion of the placenta is known to secrete corticotropin releasing hormone (CRH), the same hormone that is secreted by the hypothalamus in adults and stimulates the release of adrenocorticotropic hormone (ACTH) from the anterior pituitary. In the fetus, placental CRH stimulates the release of ACTH, which in turn stimulates the adrenal gland to secrete DHEA. This hormone is then converted by the placenta to estrogens, which enter the maternal bloodstream. The higher rates of placental CRH secretion are believed to be largely responsible for the rapidly rising estrogen levels that occur toward the end of pregnancy and help to initiate the events of labor. Because some placental CRH leaks into the maternal blood, it is also possible that rising CRH levels act as a direct trigger for parturition. Other possible triggers of parturition include increases in local oxytocin levels, increases in prostaglandins, and aging of the placenta.

Apply Your Knowledge

Today, almost 20% of births are induced. Prostaglandins are usually given to ripen the cervix, and oxytocin is often given to stimulate uterine contractions. When would inducing labor be beneficial, and what are some potential problems with this practice?

Lactation

The decrease in estrogens and progesterone that occurs in parturition is important not only because it helps to promote uterine contractions in labor, but also because it allows lactation to begin. In the first months after birth, the infant has no teeth and cannot eat solid foods. For this reason, the milk produced by the mother's breasts can serve as a valuable source of early nutrition. For the first one or two days after birth, this milk is little more than a watery fluid (called *colostrum*) containing a vast array of proteins but few other nutrients. Subsequently, however, the milk becomes enriched with a variety of additional constituents, including fat, lactose (the same sugar present in cow's milk), growth factors and hormones (which promote tissue development in the infant), and antibodies, which confer on the infant some degree of immunity against bacteria and other pathogens. (In early life, the infant is not able to make sufficient quantities of its own antibodies because its immune system is not yet fully developed.)

Milk is produced in the mammary glands by clusters of round, saclike glands called *alveoli* and is delivered by milk ducts to openings in the nipples (**Figure 27a**). To obtain this milk, the infant simply sucks on the nipples (an act known as *suckling*), which triggers the flow of milk through the ducts (sometimes referred to as "let-down"). The alveoli are surrounded by *myoepithelial cells* (Figure 27b), which are epithelial cells with the ability to contract. Suckling stimulates these cells to contract, which compresses the alveoli and forces the milk to flow through the ducts, a phenomenon called *milk ejection*.

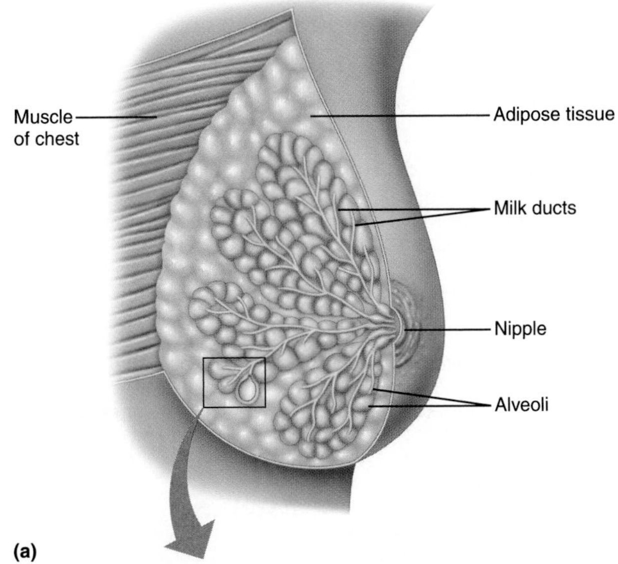

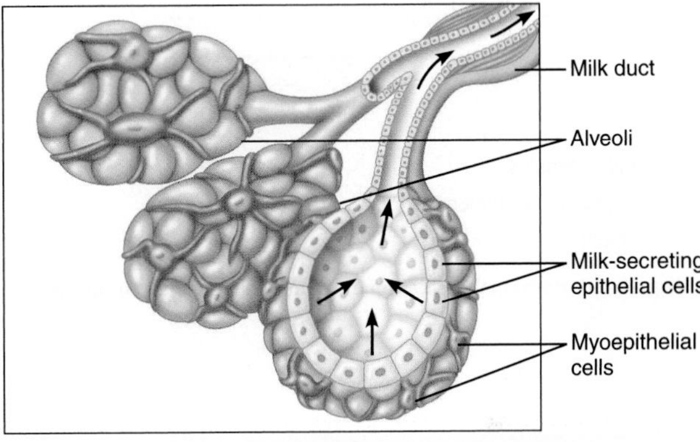

Figure 27 **Mammary glands. (a)** A cutaway view of a breast, showing the alveoli and milk ducts that constitute the mammary glands. **(b)** An enlargement showing milk-secreting epithelial cells and myoepithelial cells in alveoli.

The events through which suckling induces milk ejection are diagrammed in **Figure 28**. Suckling stimulates tactile receptors in the nipples, which project to the hypothalamus and excite neurosecretory cells that extend to the posterior pituitary and secrete oxytocin. Oxytocin travels in the bloodstream and stimulates myoepithelial cells to contract. Suckling also compels the actual production of milk by stimulating the release of prolactin releasing hormone (PRH) by the hypothalamus and inhibiting the release of prolactin inhibiting hormone (PIH). Both of these changes stimulate the anterior pituitary to secrete prolactin, which induces milk production by cells in the alveoli.

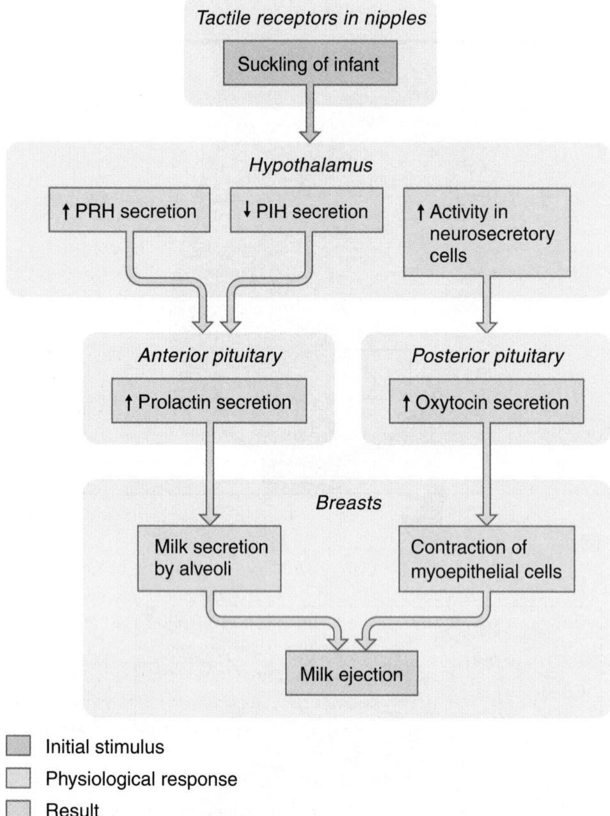

Figure 28 **Hormonal regulation of lactation in response to suckling.**

During puberty, rising estrogen and progesterone levels stimulate growth of the breasts and development of the milk-producing apparatus. Estrogens promote not only the development of tissue in the ducts but also the deposition of fatty tissue in the breasts, which results in visible breast enlargement; progesterone, by contrast, promotes development of the alveoli. In pregnancy, the breasts become enlarged and the milk-producing apparatus becomes fully developed as a result of the combined actions of increased plasma levels of estrogens, progesterone, prolactin, and placental lactogen. (Recall that prolactin release is stimulated by estrogen at this time.) No milk is actually produced, however, because the high estrogen and progesterone levels block the secretory activity of cells in the alveoli. After parturition, the decrease in estrogens and progesterone levels unblocks the stimulatory effect of prolactin on milk production and allows lactation to proceed.

Quick Check 6

1. For each of the following descriptive phrases, indicate whether it pertains to oxytocin, prolactin, estrogens, or progesterone: (a) inhibits uterine contractions during pregnancy, (b) secretion is stimulated by estrogens, (c) promotes development of milk ducts and deposition of fat in breasts during puberty, (d) promotes development of alveoli during puberty, (e) is secreted in response to excitation of uterine stretch receptors, (f) is secreted in response to tactile stimulation of nipples, (g) stimulates milk production in response to suckling, (h) stimulates milk ejection in response to suckling.

SYSTEMS INTEGRATION Although the reproduction system does not contribute directly to homeostasis, it can alter homeostasis and it depends on the other body systems to function appropriately, especially during pregnancy. The reproductive system, like all systems, depends upon a constant supply of energy and oxygen, delivered by the respiratory, gastrointestinal, and cardiovascular systems. During pregnancy and lactation, this demand goes up considerably. However, even in the male and the nonpregnant female, the reproductive system depends on all the systems, in particular the endocrine system that controls most aspects of reproduction. The cardiovascular system, for example, not only provides nutrients to the reproductive organs and the developing gametes but also provides the blood that engorges the penis or clitoris during sexual arousal. The nervous system provides short-term regulation of reproductive function, such as the autonomic nervous system's control of the sexual response. Smooth muscle is found within the reproductive organs, and skeletal muscle contracts during sexual arousal, especially during an orgasm. In males, the urinary and reproductive systems share a common pathway to the external environment, the urethra.

Sexual activity influences many organ systems through neural and endocrine pathways. For example, sympathetic activity increases during the sexual response causing a wide array of changes in the body including increased heart rate and blood glucose levels. Androgens promote skeletal muscle growth and both androgens and estrogens promote growth and development of secondary sexual traits. During pregnancy, more nutrients must be consumed and water retention occurs to support the developing fetus. Often blood pressure and blood glucose levels increase during pregnancy and gestational diabetes can develop.

CHAPTER REVIEW

SUMMARY

1 An Overview of Reproductive Physiology

- Human reproduction involves the fundamental processes of gametogenesis, fertilization, pregnancy, and parturition.
- Reproductive ability is acquired during puberty, a time during which the reproductive organs mature, gametogenesis begins, and secondary sex characteristics develop.
- Males are able to reproduce continuously through adulthood; in contrast, the reproductive capacity in females is cyclic and is lost at menopause.
- The male and female reproductive systems include the gonads (testes in males and ovaries in females) and the accessory reproductive organs.
- The gonads carry out gametogenesis and secrete sex hormones (androgens in males and estrogens and progesterone in women).
- The accessory reproductive organs include the organs of the reproductive tract and various glands that secrete fluids into the tract.

2 The Male Reproductive System

- The male reproductive system includes the testes, external genitalia (penis and scrotum), the reproductive tract (epididymis, vas deferens, ejaculatory duct, and urethra), and accessory glands (seminal vesicles, bulbourethral glands, and prostate gland).
- Sperm are formed in the testes in seminiferous tubules, which are lined by Sertoli cells.
- During sexual arousal, a mixture of sperm and fluids (semen) is forcibly ejected from the penis through the urethra, an event called ejaculation.
- Reproductive function in males is controlled by the gonadotropins—specifically, androgens (including testosterone) and GnRH.
- GnRH stimulates the secretion of gonadotropins and tends to promote androgen secretion.
- FSH and LH are gonadotropins—that is, protein hormones secreted by the anterior pituitary.
- Spermatogenesis and other Sertoli cell functions are stimulated by FSH.
- LH stimulates androgen secretion by Leydig cells.
- During males' reproductive life, androgen levels remain fairly steady because they limit their own secretion through negative feedback control of GnRH and gonadotropin secretion.

3 The Female Reproductive System

- The female reproductive system includes the ovaries, the reproductive tract (uterus, uterine tubes, and vagina), and the external genitalia (mons pubis, labia majora, labia minora, vestibule, clitoris, and vestibular glands).
- The female reproductive system exhibits the following three characteristics: cyclic changes in activity, restricted periods of fertility, and limited gamete production.
- The menstrual cycle begins with menstruation, the shedding of tissue and blood from the endometrium.
- The menstrual cycle lasts approximately 28 days and is marked by cyclic changes in pituitary and ovarian hormone secretion.
- Ova develop from a set of germ cells whose number is fixed at birth but that do not become fully mature until fertilization has occurred.
- Each ovary contains numerous follicles, each of which contains one oocyte.
- Follicles also contain granulosa cells, which nourish the oocyte, regulate its development, and secrete estrogens.
- In later stages of development, the granulosa cells proliferate and the outermost layer becomes transformed into theca cells.
- The ovarian cycle is divided into a follicular phase and a luteal phase.
- The follicular phase begins with the recruitment of preantral follicles for further development and ends with release of the oocyte (ovulation), which then enters the nearby uterine tube.
- The luteal phase begins with ovulation and ends with either fertilization of the oocyte and pregnancy or degeneration of the corpus luteum and menstruation.
- Coinciding with the ovarian cycle is a uterine cycle consisting of a menstrual phase, a proliferative phase, and a secretory phase.
- In the follicular phase, FSH targets granulosa cells to stimulate growth and estrogen secretion; LH stimulates theca cells to secrete androgens, which granulosa cells then convert to estrogens.
- The estrogens promote oogenesis (along with FSH) and proliferative-phase uterine changes.
- In the late follicular phase, rising estrogen levels trigger an LH surge, which is responsible for ovulation and development of the corpus luteum.
- In the luteal phase, the corpus luteum secretes estrogen and progesterone, which suppresses LH and FSH secretion and promotes secretory-phase uterine changes.
- In the absence of fertilization, the corpus luteum degenerates, causing a drop in estrogen and progesterone levels that then triggers menstruation.

4 Fertilization, Implantation, and Pregnancy

- After fertilization, which normally occurs in the uterine tube, the zygote is eventually transformed into a blastocyst, which implants in the endometrium.
- At the point of implantation in the uterus, embryonic and endometrial tissues develop into a placenta, which permits exchange of materials between the mother and the developing embryo.

- During pregnancy, estrogens and progesterone (secreted by the corpus luteum at first and then by the placenta) promote many effects, including growth and development of the mammary glands, secretion of prolactin by the anterior pituitary, and maintenance of secretory-phase uterine conditions.

5 Parturition and Lactation

- Parturition typically occurs 40 weeks after fertilization and is accompanied by a wave of strong uterine contractions (labor), dilation of the cervix, expulsion of the fetus from the uterus, and separation of the placenta from the uterine wall.
- After delivery, nourishment of the infant is by milk secreted by the mammary glands.
- Suckling of the infant triggers secretion of prolactin, which promotes milk production, and of oxytocin, which promotes milk ejection.

EXERCISES

Multiple-Choice Questions

1. In both males and females, gonadotropin secretion by the anterior pituitary is stimulated by which of the following?
   a) Inhibin
   b) Androgens
   c) GnRH
   d) FSH
   e) GHRH
2. In the embryo, which of the following effects is promoted by testosterone?
   a) Development of the primitive gonads into testes
   b) Regression of Müllerian ducts
   c) Development of Müllerian ducts into male reproductive organs
   d) Development of Wolffian ducts into male reproductive organs
   e) Expression of the srY gene
3. The testes are housed in a structure called the
   a) prostate gland.
   b) scrotum.
   c) penis.
   d) epididymis.
   e) vas deferens.
4. In the first step of spermatogenesis, spermatogonia differentiate into cells called
   a) spermatids.
   b) primary spermatocytes.
   c) secondary spermatocytes.
   d) spermatozoa.
   e) spermatophytes.
5. Cells in the ovaries secrete all of the following hormones *except*
   a) estrogens.
   b) progesterone.
   c) androgens.
   d) luteinizing hormone.
   e) inhibin.
6. In oogenesis, when does meiosis I occur?
   a) In early fetal life
   b) At birth
   c) Just before ovulation
   d) After ovulation but before fertilization
   e) After fertilization
7. During the early to mid-follicular phase of the ovarian cycle, which hormone(s) stimulate(s) granulosa cell functions?
   a) Progesterone
   b) FSH
   c) LH
   d) GnRH
   e) Estrogens
8. In the late luteal phase, why do estrogen and progesterone levels fall?
   a) Rupture of the dominant follicle
   b) Degeneration of the corpus luteum
   c) An inhibitory effect of LH on secretory activity of the corpus luteum
   d) The inhibitory effect of inhibin on the secretory activity of granulosa cells
   e) All of the above
9. In the uterine cycle, which phase immediately precedes the proliferative phase?
   a) Menstrual phase
   b) Secretory phase
   c) Luteal phase
   d) Follicular phase
10. The placenta not only serves as an organ of exchange but also secretes all of the following hormones *except*
    a) prolactin.
    b) chorionic gonadotropin.
    c) placental lactogen.
    d) progesterone.
    e) estrogens.

Objective Questions

11. In meiosis I, maternal and paternal chromosomes are segregated into separate daughter cells. (true/false)
12. The srY gene codes for (testosterone receptors/testis-determining factor), which determine(s) whether an embryo develops testes or ovaries.
13. In the absence of testosterone and MIS, the Müllerian ducts (persist/degenerate) in the embryo, and female structures eventually develop.
14. FSH and LH are classified as (sex hormones/gonadotropins).
15. GnRH, which is secreted by the hypothalamus, stimulates the secretion of both FSH and LH from the anterior pituitary. (true/false)
16. In the testes, androgens are secreted by (Sertoli cells/Leydig cells).
17. Spermatogenesis is stimulated by testosterone and (FSH/LH), which targets Sertoli cells.
18. The head of a sperm contains chromosomes and a(n) ________, a vesicle containing enzymes needed for fertilization.
19. Erection is accompanied by a(n) (increase/decrease) in the activity of sympathetic neurons projecting to blood vessels in the penis.
20. Once sperm are deposited in the female reproductive tract, they cannot fertilize the oocyte until they have undergone a process called ________.
21. Fertilization usually occurs in the (uterus/uterine tube).

22. The second half of the ovarian cycle is called the (luteal/follicular) phase.
23. In a follicle, the oocyte is surrounded by a layer of (granulosa/theca) cells that provide it with nourishment and regulate its development.
24. (FSH/LH) stimulates theca cells to secrete androgens, which are converted to estrogens by granulosa cells.
25. The inner layer of the uterine wall is called the (endometrium/myometrium).
26. During the (proliferative/secretory) phase of the uterine cycle, the lining of the uterus thickens under the influence of rising estrogen levels.
27. Ovulation is triggered by (FSH/LH).
28. The corpus luteum secretes estrogens and (LH/progesterone).
29. In the late follicular phase, LH secretion is stimulated by (estrogens/progesterone).
30. Degeneration of the corpus luteum causes hormonal changes that trigger (ovulation/menstruation).
31. Before implantation, the morula develops into a(n) ________, which consists of an inner cell mass contained within a hollow, fluid-filled outer cell layer.
32. Secretory-phase uterine conditions are promoted by (estrogen/progesterone), which inhibits gonadotropin secretion during the last half of the ovarian cycle.
33. During the last six months of pregnancy, estrogens and progesterone are secreted by the (ovaries/placenta).
34. During labor, strong uterine contractions are induced by (prolactin/oxytocin).
35. Suckling stimulates the release of ________, which promotes milk production by the breasts.

Essay Questions

36. Compare the hormonal regulation of reproductive function in males with that in females during the early to mid-follicular phase, drawing as many parallels as possible. Be sure to consider the actions of hormones as well as the mechanisms that regulate their secretion.
37. Compare the steps in spermatogenesis with those of oogenesis, drawing as many parallels as possible.
38. Describe all the events that must occur before a sperm deposited in the female reproductive tract can fertilize an oocyte released from an ovary.
39. Describe the events that occur during the LH surge, including the actions of LH on target cells. Be sure to include a description of the events that trigger the surge and prime the target tissue's responses to it.
40. Describe the processes that give rise to elevated rates of estrogen and progesterone secretion during pregnancy. Explain how these hormones promote the maintenance of pregnancy and prepare the body for the eventual birth and nourishment of the infant.

Critical Thinking

41. After fertilization, some zygotes may have the following three sex chromosomes: XXY. Will this zygote develop? If so, will it develop into a male or female? Explain.
42. Endometriosis is a disease in which some of the endometrial tissue of the uterus grows outside the uterus and attaches to organs in the abdominal cavity. What are two ways by which endometrial tissue can get into the abdominal cavity? The endometrial tissue located outside the uterus responds to hormones just like normal endometrial tissue does. Which changes occur to this abnormal tissue during a female cycle, and which symptoms develop from these changes?
43. Some males are sterile because their sperm cannot fertilize an oocyte. What are some possible defects in sperm that would prevent them from fertilizing an oocyte?
44. The genes for red photopigments and green photopigments are located on the X chromosome. Explain how this makes males more prone to red-green color blindness.

Photo credits for icons in feature boxes: Focus on Diabetes: Tanyae/Dreamstime; Clinical Connections: Iofoto/Dreamstime; Understanding Exercise: Pixattitude/Dreamstime; Discovery: Yuri Arcurs/Cutcaster.

Glossary

acrosome large vesicle within the head of the sperm; contains enzymes and other proteins that allow the sperm to fuse with the egg during fertilization

acrosome reaction process whereby acrosome releases its enzymes

amnion (am-nee-on) a membrane that forms a fluid-filled sac around the embryo; also called *amniotic sac*

androgens a class of sex hormones secreted by the testes

antrum fluid-filled cavity in the developing follicle

autosome any chromosome other than the sex chromosomes

blastocyst early stage of embryo development recognizable by the presence of an inner cell mass

bulbourethral glands (bul-bo-you-reeth-ral) accessory glands of the male reproductive system

cervix the lower, narrower portion of the uterus, containing a central canal that leads directly to the vagina

chorion (kor-ee-on) the outermost membrane that encapsulates the embryo, forming a tough envelope that isolates it from its surroundings

copulation the act of mating

corpus luteum (kor-pus loo-tee-um) a gland that develops from the ruptured follicle following ovulation and secretes progesterone and estrogen

diploid in reference to a cell's chromosome number, possessing the full complement of chromosomes

dominant follicle a follicle that has been selected to go to full maturation during the course of a menstrual cycle

ejaculation the discharge of semen from the penis

embryo the developing human within the first two months after conception

emission the movement of semen into the penis

endometrium (en-doh-mee-tree-um) the innermost layer of tissue in the uterus, made up of epithelial cells and an underlying layer of thick connective tissue

epididymis (ep-ih-did-eh-mus) duct loosely attached to the testes' outer surface; serves as a site for sperm maturation

erectile tissue spongy masses of connective tissue and smooth muscle containing numerous interconnected vascular spaces; found in the penis, clitoris, and certain other structures that expand when filled with blood

erection process occurring in male sexual arousal, during which the penis swells, elongates, and becomes firmer and straighter to facilitate its entry into the vagina during copulation; also occurs in the clitoris during female sexual arousal

estrogens a class of sex hormones secreted by the ovaries

fertilization the process by which two gametes (one from each parent) fuse together to produce a new cell; known as *conception* in humans

fetus the developing human after two months following conception

follicle spherical structure in the ovary containing a single developing ovum

follicle stimulating hormone (FSH) a gonadotropic hormone that stimulates gametogenesis and regulates other gonadal functions in either sex

follicular phase (fuh-lik-you-lar) the initial phase of the ovarian cycle, during which follicular maturation occurs; terminates with ovulation

gametes (gam-eets) cells of reproduction produced in meiosis that contain half of a person's genetic material; examples are sperm cells in males and ova in females

gametogenesis (gah-mee-toh-jen-ih-sis) the production of gametes from a pool of germ cells

gestation pregnancy

gonadotropin releasing hormone (GnRH) a hypothalamic tropic hormone that stimulates the secretion of gonadotropins by the anterior pituitary

gonadotropins two hormones—follicle stimulating hormone (FSH) and luteinizing hormone (LH)—that are secreted by the anterior pituitary and regulate gonadal function in either sex

gonads the primary reproductive organs; serve as sites of gametogenesis and sex hormone secretion

Graafian follicle a fully matured follicle, just prior to ovulation

granulosa cells specialized cells that surround a developing ovum and support its development

haploid in reference to a cell's chromosome number, possessing half the number of chromosomes found in most other cells of the body

human chorionic gonadotropin (hCG) a hormone secreted by the chorion that maintains the corpus luteum during the first three months of pregnancy

implantation process by which an embryo adheres to and becomes embedded in the wall of the uterus

inhibin a protein hormone secreted by the gonads that suppresses the release of follicle stimulating hormone (FSH) from the anterior pituitary

lactation milk production that takes place in the mammary glands

Leydig cells the cells in the testes that secrete androgens

LH surge a rapid increase in the secretion of luteinizing hormone that triggers ovulation

luteal phase the final phase of the ovarian cycle during which the corpus luteum is active

luteinizing hormone (LH) a gonadotropic hormone that stimulates sex hormone secretion and regulates other gonadal functions in either sex

mammary glands milk-producing glands in the female breasts

meiosis (my-oh-sis) a series of two cell divisions that generate daughter cells (gametes) with half the normal chromosome number; involved in gametogenesis

menopause cessation of reproductive capacity that occurs in females in middle age

menstrual cycle a sequence of events involving changes in reproductive function that occurs in females on a regular basis with a period of about one month

menstrual phase phase of the uterine cycle during which the uterine lining is shed

menstruation shedding of blood and tissue from the uterine lining

morula earliest stage of an embryo after cell division, consisting of a ball of identical cells

Müllerian ducts rudimentary precursor to the female reproductive system

Müllerian inhibiting substance (MIS) a hormone secreted by the testes that inhibits development of the Müllerian ducts

myometrium a layer of smooth muscle in the uterine wall

oogenesis (oh-uh-jen-ih-sis) the production of ova

oogonia undifferentiated germ cells that are precursors for ova

ovarian cycle cyclic changes in ovarian structure and function; consists of two phases, the follicular phase and the luteal phase

ovaries female gonads

ovulation release of an oocyte from an ovarian follicle

parturition the process of birth

penis the male copulatory organ

placenta a structure consisting of maternal and fetal tissues that allows exchange of gases, nutrients, and wastes between the mother's circulatory system and the circulatory system of the fetus

placental lactogen a hormone secreted by the placenta that promotes breast growth and the mobilization of energy stores in the mother

primordial follicle follicle at the earliest stage of development

progesterone a sex hormone secreted by the ovaries, primarily during the luteal phase of the menstrual cycle

proliferative phase phase of the uterine cycle during which the uterus renews itself after the menstrual phase

prostate gland an accessory gland of the male reproductive system that secretes fluid into the urethra

puberty period of sexual maturation

reproductive tract system of passageways through which gametes are transported

scrotum a sac that houses the testes

secretory phase phase of the uterine cycle during which the uterine endometrium prepares for the possible arrival and subsequent implantation of the fertilized egg

seminal vesicles accessory glands of the male reproductive system; the first to secrete fluid into the reproductive tract during sperm transport

seminiferous tubules (sem-ih-nif-er-ous) thin, highly coiled hollow tubes in the testes where sperm are produced

sense strand the strand of DNA that has the same code as the transcribed mRNA by virtue of its complementary nature to the template strand

Sertoli cells epithelial cells lining the wall of a seminiferous tubule whose primary function is to support sperm development

sex chromosomes X and Y chromosomes that determine a person's sex

sex determination the role of genes in deciding a person's sex

sex differentiation the role of testosterone and Müllerian inhibiting substance in determining the sexual characteristics of a person

spermatogenesis (sper-mah-toh-jen-ih-sis) sperm production

spermatogonia undifferentiated germ cells that are precursors for sperm

spermatozoa (spur-ma-toh-zoh-ah) newly formed sperm cells with a characteristic head, midpiece, and tail

srY gene gene located on the Y chromosome that codes for testis-determining factor

testes male gonads

theca cells (thee-ca) cells that surround the granulosa cells in a follicle

umbilical cord a rope-like structure that extends from the placenta to the fetus; contains vessels that provide blood to the fetus

uterine cycle cyclic changes in uterine structure and function that occur during the menstrual cycle

uterine tubes tubes protruding from the upper part of the uterus on either side that function in egg transport and fertilization; also called *fallopian tubes* or *oviducts*

uterus a hollow, pear-shaped organ located in the center of the pelvic cavity that functions to house and nourish the developing human

vagina a canal leading from the cervix to the outside of the body; receives the penis during intercourse; the female organ of copulation

vas deferens (vas def-er-enz) a duct leading from each testis to the ejaculatory duct; carries sperm and fluid

vulva external genitalia of the female; includes the mons pubis, labia majora, labia minora, the vestibule, and vestibular glands

Wolffian ducts rudimentary precursor to the male reproductive system

zona pellucida a thick layer of noncellular material that forms between an oocyte and the surrounding granulosa cells

zygote (zy-goat) fertilized ovum before cleavage begins

Answers

Figure Question Answers

Fig 8	Via hypothalamic-pituitary portal vessels
Fig 10	Stays the same
Fig 11	It increases
Fig 14	Two X chromosomes

Apply Your Knowledge Answers

Presence of a Y chromosome results in development of testes, and thus a male, regardless of how many X chromosomes are present, so long as there is at least one X. Thus, XY, XXY, and XXXY would all result in a male phenotype, although there may be some decreased fertility. XYY would also result in a male. XO would result in a female, where OY would not be viable. XY with a defect in the srY gene would be a female.

During an ectopic pregnancy, the embryo becomes implanted outside the uterus, often in the uterine tube. Only the wall of the uterus is designed to support a growing fetus; the wall of the remainder of the reproductive tract is designed to facilitate transport of sperm or ova and for parturition. The wall of the reproductive tract will not be able to provide the nutrition needed by the growing embryo, nor will there be sufficient space for the embryo to grow. This latter fact often causes pain to the pregnant woman, and the embryo may be spontaneously aborted. The abortion can cause excessive bleeding for the woman, which can be dangerous.

Inducing labor would naturally be beneficial for a woman who is going beyond term with a child. Inducing labor is also used with multiple births if the multiple babies are taking up too much room and interfering with one another's survival capacity. The induction in this case will result in premature babies, but each baby will have a better chance of survival. Inducing labor can cause premature births, which brings up a number of complications for the child, such as those associated with insufficient surfactant.

Answers to End–of-Chapter Multiple Choice and Objective Questions

Multiple Choice Questions

1. c 2. d 3. b 4. b 5. d 6. c 7. b 8. b 9. a 10. a

Objective Questions

11. false
12. testis-determining factor
13. persist
14. gonadotropins
15. true
16. Leydig cells
17. FSH
18. acrosome
19. decrease
20. capacitation
21. uterine tube
22. luteal phase
23. granulosa
24. LH
25. endometrium
26. proliferative
27. LH
28. progesterone
29. estrogens
30. menstruation
31. blastocyst
32. progesterone
33. placenta
34. oxytocin
35. prolactin

The Immune System

The Immune System

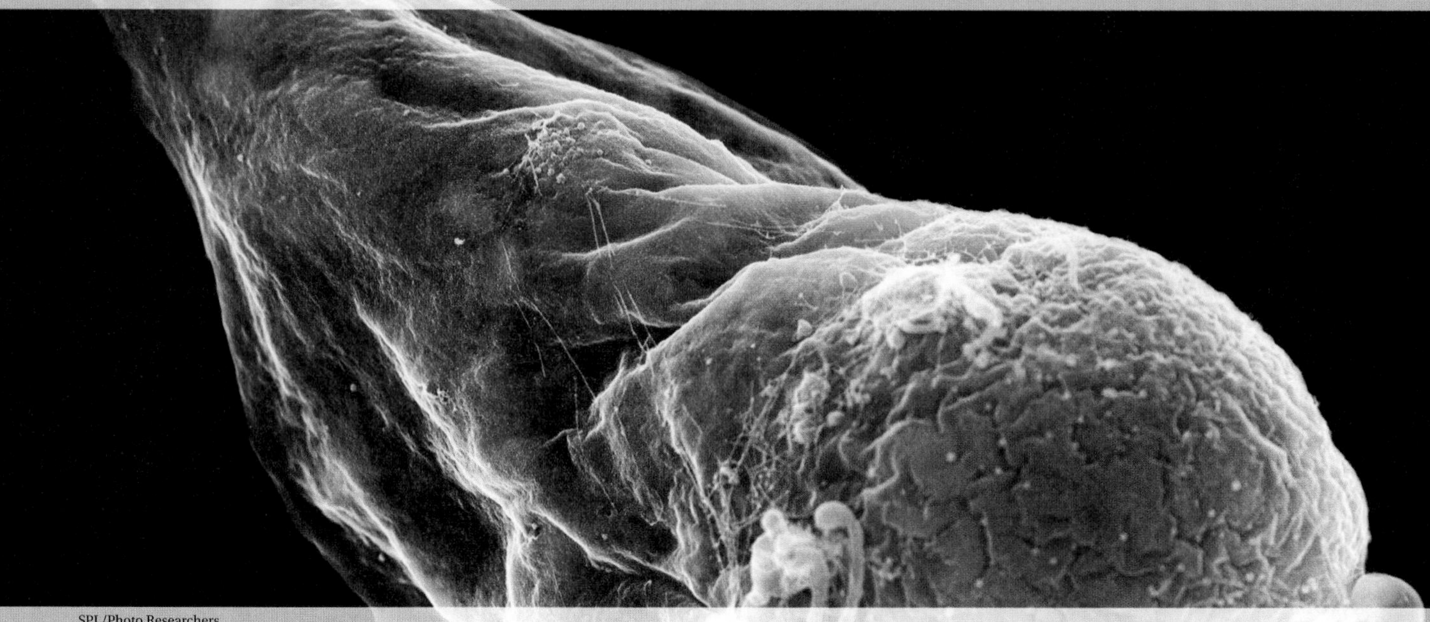

SPL/Photo Researchers

Colorized scanning electron micrograph (SEM) of a neutrophil.

Ming is a bright first-year law student who is a champion speed skater. Usually in excellent health, Ming is surprised and dismayed when she develops a sore throat and a fever and feels so tired that she can't gather enough energy to so much as look at a textbook—or even at her skates! Her physician at the campus health center tells her she has a viral disease called *mononucleosis,* also known as "mono." Ming learns that there is no specific treatment for mono, but that in time she will feel better thanks to her *immune system*—the organs, tissues, circulating cells, and secreted molecules that resist and defeat infections. Like all of us, Ming understands the power of this system: When she had an infected cut, it healed; when she was ill with the flu, she got better; and when she was immunized against the bacterial infection *diphtheria*, she became protected from that disease for life.

CHAPTER OUTLINE

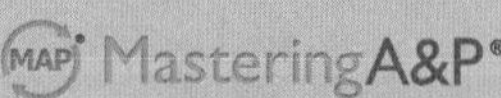

Go to MasteringA&P for helpful A&P Flix 3-D animations, chapter quizzes, pre-tests, interactive physiology tutorials, and more!

LEARNING OUTCOMES *After studying this chapter, you should be able to:*

- Identify the lymphoid organs, and briefly describe their functions.
- Identify the different pathogens that can invade the body causing disease.
- Explain events that occur during inflammation. Describe how the skin and mucous membranes, inflammation, interferons, natural killer cells, and the complement system contribute to the body's nonspecific defenses.
- Describe the complement system and its role in both innate and adapted immunity.
- Describe humoral immunity—that is, how B cells, through the production of antibodies, contribute to immune responses.
- Describe cell-mediated immunity—that is, how helper T cells and cytotoxic T cells contribute to immune responses.
- Explain how immunization can lead to protection from infectious disease.
- Discuss the major immunological issues regarding blood transfusion and organ transplantation.
- Explain how immune dysfunction can result in allergy, autoimmunity, or immunodeficiency.

Before You Begin

Make sure you have mastered the following topics:

- Phagocytosis
- Paracrines
- Histamine
- Composition of blood
- Vasodilation
- Hematopoiesis
- Leukocytes
- Lymphatic system

In this chapter we learn how the immune system protects the body against disease-causing organisms and nonliving substances that gain access to it via air, food, and water. **Immunity** refers to the immune system's capacity to protect individuals from disease by recognizing and eliminating potentially **pathogenic** (disease-generating) agents, including bacteria, bacterial toxins, viruses, parasites, and fungi. In addition to resisting foreign agents, the immune system disposes of unneeded components of the body, including aging cells and cellular debris present in diseased tissue; participates in wound healing; and sometimes recognizes and eliminates mutant cells that may develop into cancer. As it seeks out and recognizes materials that are foreign to the body, the immune system also rejects tissues and cells that are not identical to "self" (that is, to the cells and tissues of the individual in question); such rejection is the primary obstacle to organ transplantation.

The presence in the body of such foreign and abnormal substances induces the immune system to develop an **immune response,** a complex series of physiological events that culminates in the destruction and elimination of these substances. The immune response is divided into two basic aspects: **nonspecific** (or **innate**) **immunity,** which provides the first defenses against pathogens regardless of the type of pathogen, and **specific** (or **adaptive**) **immunity,** which represents a stronger defense against a specific pathogen(s) that takes longer to develop.

Sometimes, inappropriate immune responses occur and can lead to disease. People with allergies, for example, experience exaggerated immune responses to otherwise benign foreign materials such as dust, pollen, or peanuts. In addition, some people suffer from *autoimmune diseases* in which a person's immune system attacks his or her own tissues and cells. Rheumatoid arthritis, diabetes mellitus, and multiple sclerosis are just a few examples of the many diseases that can result from autoimmune responses.

1 Anatomy of the Immune System

The immune system consists of three components: (1) physical barriers between the external and internal environments, including epithelium (skin and the mucous membranes lining the respiratory, renal, and gastrointestinal systems) and chemicals such as the acidic environment of the stomach; (2) leukocytes (white blood cells), which are responsible for producing a wide range of immune responses; and (3) **lymphoid tissues,** such as the bone marrow, thymus, spleen, lymph nodes, and tonsils, in which leukocytes develop, reside, and come into contact with foreign materials. We discuss each of these components in detail in the following sections.

Physical Barriers

The skin and mucous membranes that separate the internal from the external environment prevent the entry of most microorganisms into the body, thereby providing nonspecific protection against foreign matter. The skin is made up of an outer *epidermis* and an inner *dermis* (**Figure 1**). The epidermis consists of tightly packed epithelial cells and lacks blood vessels. The outermost layer of the epidermis is composed of dead epithelial cells and keratin, a protein that keeps water from entering the body through the skin. This outer layer is shed continuously, requiring replacement of the epidermis every 15–30 days. Located within the dermis are sebaceous glands. These glands secrete an acidic oily substance, sebum, into the hair follicle where it lubricates the hair and inhibits bacterial growth. Some bacteria, however, thrive in sebum and are partly responsible for acne.

The mucous membranes form a continuous barrier with the skin. They also produce viscous mucus, which bathes the surfaces of exposed epithelia and can trap foreign matter and potential pathogens. Microorganisms that enter the upper respiratory tract, for example, are often caught in mucus and are then transported out of the

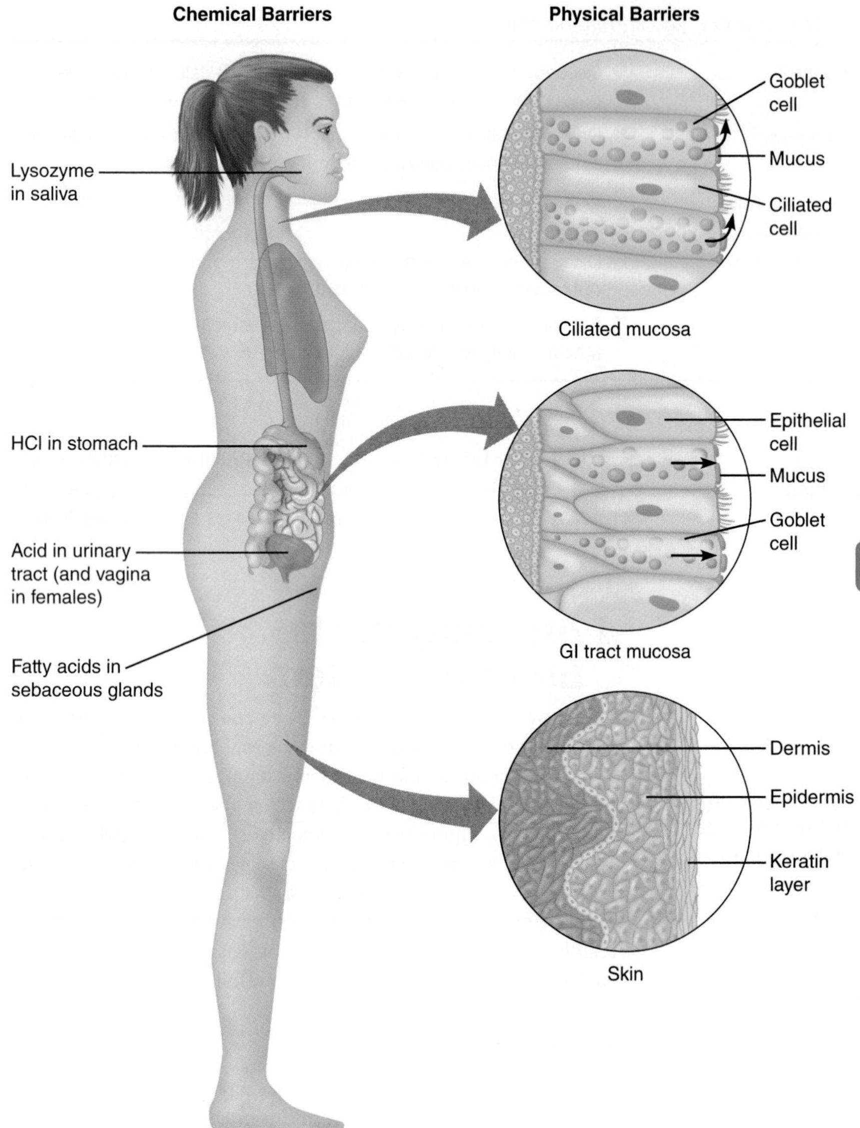

Figure 1 **Chemical and physical barriers to pathogens.**

lungs toward the oropharynx by the mucociliary (mucus) escalator, or coughed into the mouth and swallowed or spit out. In the stomach, mucous membranes produce acid that kills most of the microorganisms that enter the lumen of the gastrointestinal tract.

These physical barriers represent the beginning of the nonspecific defense against pathogens. We will learn shortly how the other two components of the immune system contribute to both nonspecific and specific defenses.

Leukocytes

The body manufactures five major types of leukocytes, all of which derive from the same hematopoietic stem cells in the bone marrow (**Figure 2**). Neutrophils, eosinophils, and basophils are granulocytes; they have cytoplasmic granules that contain secretory products—the release of these products is called *degranulation.* Monocytes and lymphocytes, the agranulocytes, lack granules. In addition to the leukocytes, three other types of tissue cells—*macrophages, mast cells,* and *dendritic cells*—are derived from the hematopoietic stem cells. Unlike the leukocytes, however, these cells mature in the tissues.

Phagocytes

Neutrophils, eosinophils, monocytes, macrophages, and dendritic cells are all phagocytes—they can engulf foreign particles and microorganisms, thereby removing them from blood and tissues. Neutrophils are the most abundant of the leukocytes and are particularly important in fighting bacterial infections. An elevation in the number of neutrophils in blood is used clinically as a determinant of infection. In addition to phagocytosis, neutrophils release several cytokines involved in inflammation. Although eosinophils are also phagocytes, their primary role in the immune system is to defend against parasites by releasing toxic substances. (Learn about the effects of exercise on the number of circulating leukocytes in **Understanding Exercise: Leukocyte Mobilization.**)

Functional Fact

Monocytes are phagocytes in the blood, but they differentiate into macrophages in the tissue to phagocytose foreign matter. Macrophages are five to ten times larger than monocytes and have greater activity (both phagocytic and secretory activity are enhanced). Macrophages are found in almost all tissues of the body, with some localized to certain tissue only *(fixed macrophages)* and others able to migrate to different tissues (*free* or *wandering macrophages*). Fixed macrophages include microglia, alveolar macrophages, *Kupffer cells* in the liver, and *histiocytes* in connective tissue.

Lymphocytes

Lymphocytes are unique in that they provide the immune system with diversity, specificity, memory, and the ability to distinguish between self and nonself. Lymphocytes are of three major types: **B lymphocytes (B cells), T lymphocytes (T cells),** and **null cells,** so called because they lack cell membrane components that are characteristic of B cells and T cells. Most null cells are large, granular lymphocytes known as **natural killer (NK) cells** (described shortly).

When B cells contact foreign or abnormal molecules known as **antigens,** they develop into **plasma cells,** which secrete antibodies. **Antibodies,** also known as **immunoglobulins,** are proteins present in the plasma and interstitial fluid that target specific antigens for

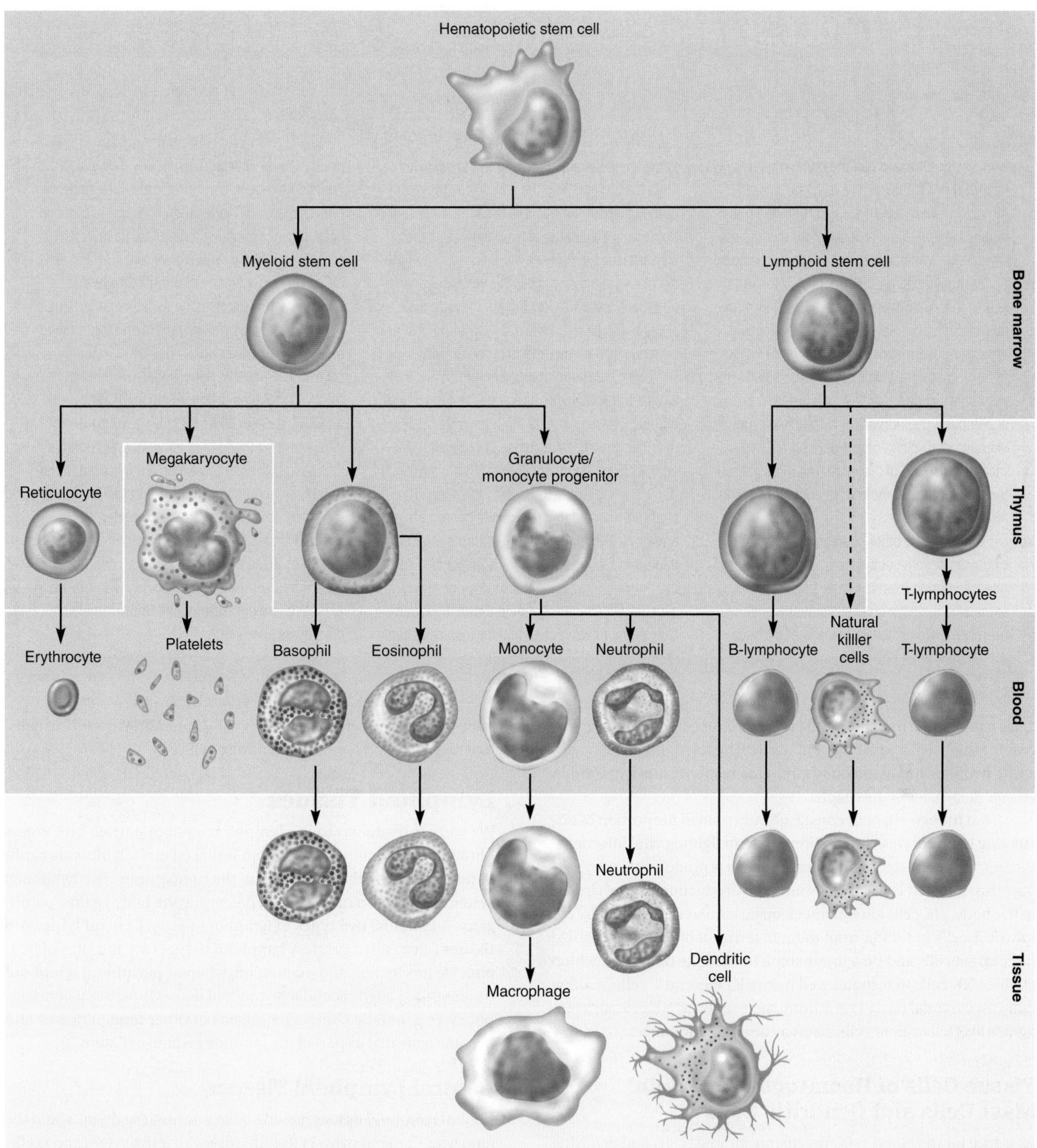

Figure 2 Immune cell production.

destruction. If, for example, a B cell contacts the bacterium *Staphylococcus aureus,* it develops into a plasma cell that secretes antibodies that bind only to *S. aureus*. These antibodies do not damage the bacteria by themselves but instead mark them for destruction by various mechanisms that are described later in the chapter.

T cells, in contrast, do direct damage to foreign or abnormal cells. T cells contact infected cells, mutant cells, and transplanted cells and then take several days to develop into active **cytotoxic T cells** that destroy the infected or abnormal cells. They kill the target cells by secreting molecules that form pores in the target cell's membrane; once

UNDERSTANDING EXERCISE

Leukocyte Mobilization

Exercise promotes a healthy body in many ways, but what does it do for the immune system? Almost any exercise (even just running up a flight of stairs) leads to a transient increase in circulating white blood cells, primarily because of the mobilization of neutrophils. Approximately half of the neutrophils in the vascular system do not circulate freely, but rather are loosely attached (electrostatically) to the inner walls of large-diameter veins, especially veins in the pulmonary circuit. The fluid shear stresses that accompany the increased cardiac output of exercise strip these cells free; epinephrine also triggers this kind of "release." As a result, neutrophil counts can more than double following moderate exercise.

Likewise, the other populations of leukocytes can become activated by stress hormones released during exercise or by fragments of extracellular matrix and cells damaged during exercise. Once activated, these cells produce adhesion molecules that bind to the vascular endothelium and help the leukocytes move out of the bloodstream and into the interstitial space. Some subpopulations of lymphocytes may exhibit transient decreases of 50% or more of the circulating number of cells via this mechanism; other leukocyte subpopulations may increase as a result of increased lymphatic flow to the thoracic duct due to skeletal muscle pump activity.

These transient changes in numbers of circulating leukocytes and anecdotal reports of chronic illnesses in elite athletes have been (mis)interpreted as decreases in overall readiness to resist infection. In reality, little scientific evidence exists to support this view. The changes in circulating leukocyte counts are small and brief. The main battlefield during infection is usually in the interstitial space, especially near the barriers to the external environment (e.g., skin, respiratory tract), not in the bloodstream. Only a tiny fraction of all leukocytes are actually present in the bloodstream at any moment; a twofold change in the number of circulating cells represents a shift in location of only 1% of these cells. Most controlled studies that have actually measured infection rates have failed to find any difference in incidence of infection between individuals who exercise and sedentary individuals.

The resting (basal) numbers and functioning of leukocytes generally do not differ in highly trained and untrained individuals. The spectrum of cytokines produced by the leukocytes of active people, however, may be somewhat different than those of inactive people. Given that some of these cytokines are involved in the pathological mechanisms leading to atherosclerosis and other inflammatory diseases, such small, training-induced changes in the concentration of these cytokines over long periods of time may contribute mechanistically to the well-established exercise-induced improvement of cardiovascular health.

it has a perforated membrane, the target cell succumbs to lysis, a process in which it fills with fluid and bursts. The roles of B lymphocytes and T lymphocytes in the body's immune response are discussed in greater detail later in the chapter.

Even though NK cells constitute only a small proportion of circulating lymphocytes, they are important in fighting viral infections. Viruses, unlike bacteria, must enter cells to reproduce; thus, by killing virus-infected cells NK cells limit the production of new viruses in the body. NK cells kill by a mechanism similar to that used by cytotoxic T cells but differ from them in terms of how they recognize their target cells and how they exhibit immediate readiness, which enables NK cells to respond well before B cells and T cells, making them an essential part of early immune responses. NK cells also recognize and kill mutant cells that may develop into cancer.

Tissue Cells of Hematopoietic Origin: Mast Cells and Dendritic Cells

As is true for the blood cells, precursors for mast cells and dendritic cells are formed in the bone marrow from hematopoietic stem cells (see Figure 2). These cells circulate for a brief time before entering the tissues where they mature. **Mast cells** are found in skin and mucous membranes, where they secrete histamine along with other substances. **Dendritic cells,** named for their long processes that look like dendrites in neurons, are similar to macrophages in their ability to phagocytose or endocytose pathogens and, as we will see later, to activate certain types of T cells. Four types of dendritic cells have been identified: Langerhans cells, interstitial dendritic cells, myeloid cells, and lymphoid dendritic cells.

Lymphoid Tissues

We were introduced to the lymphatic system earlier. The excess filtrate from the circulatory system is picked up by lymphatic capillaries and returned to the blood in the right atrium. The lymphatic system also plays a major role in defending the body against pathogens. It contains two types of lymphoid tissue: **Central lymphoid tissues** (also called primary lymphoid tissues) are the sites of lymphocyte production and maturation, whereas **peripheral lymphoid tissues** (also called secondary lymphoid tissues) are sites where lymphocytes generally contact pathogens or other foreign matter and become activated as part of the immune response (**Figure 3**).

Central Lymphoid Tissues

Central lymphoid tissues include bone marrow, the thymus, and the fetal liver. Bone marrow is the site where all leukocytes (and erythrocytes as well) develop from hematopoietic stem cells. Whereas B lymphocytes and most other leukocyte types come to full maturity in the bone marrow, T lymphocytes must migrate to the thymus gland (located in the thoracic cavity above the heart) before they develop to maturity. After B cells and T cells reach maturity, they migrate from central lymphoid tissues to the peripheral lymphoid tissues where they are most likely to contact foreign substances.

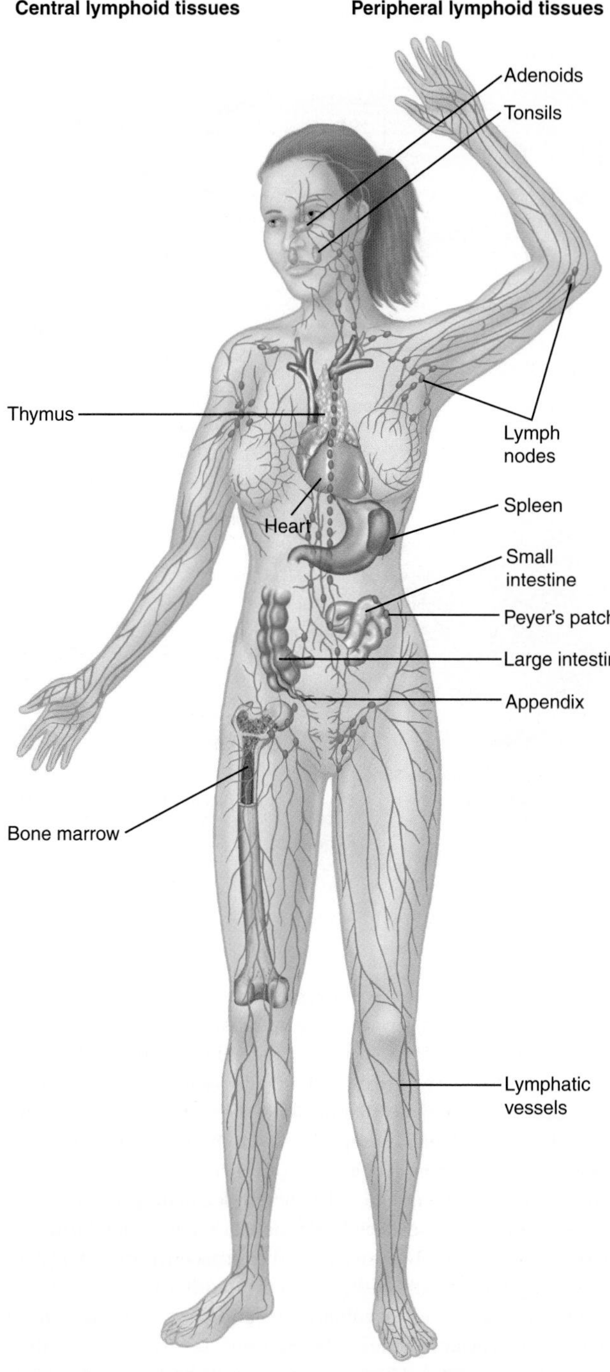

Figure 3 Lymphoid tissues. Lymphocytes reach maturity in the central lymphoid tissues and interact with foreign antigens in the peripheral lymphoid tissues. These lymphoid tissues are interconnected by blood vessels (not shown) and lymphatic vessels through which lymphocytes circulate.

Q A few days after infection with chicken pox, a person experiences viremia, a high level of viruses in the blood. In which lymphoid tissue are these viruses most likely to contact lymphocytes?

Peripheral Lymphoid Tissues

Peripheral lymphoid tissues are essentially collections of B cells, T cells, and macrophages; they include the spleen, lymph nodes, tonsils, adenoids, appendix, lymph nodules of the gastrointestinal tract, and regions in the lining of the gastrointestinal tract called *Peyer's patches.*

Each of the peripheral lymphoid tissues contains a dense network of cells that trap microorganisms and foreign particles, and each is located where it can ensnare invaders soon after they enter the body. While the spleen is collecting worn-out erythrocytes from the blood, it also filters out bloodborne microorganisms and foreign particles. Once trapped in the spleen, these particles are eventually cleared by the actions of macrophages and lymphocytes. Similarly, microorganisms and particles carried in lymph are trapped by lymph nodes, which are found throughout the body where lymphatic vessels converge. Whereas the macrophage and lymphocyte networks of the spleen and lymph nodes filter blood and lymph, respectively, those of the tonsils and adenoids trap inhaled particles and microorganisms, and those of the appendix, lymph nodules, and Peyer's patches trap substances that enter the body in ingested food or water.

Quick Check 1

1. Which types of leukocytes are phagocytes? Which types of tissue cells of hematopoietic origin are phagocytes?
2. Name the two central lymphoid tissues. Why are they known as *central* lymphoid tissues?
3. List six peripheral lymphoid tissues. What general role do these tissues play in immunity?

2 Pathogens That Activate the Immune Response

The field of immunology started in 1796 when Edward Jenner discovered that administration of *cowpox*, also called vaccinia, to humans could protect against smallpox infections. What caused the smallpox infections, however, was unknown; microorganisms were not discovered until 1882, when Robert Koch described his discovery of *Mycobacterium tuberculosis*, the bacterium that causes tuberculosis. Since then, four different classes of **pathogens** (microorganisms that cause specific diseases) have been identified: viruses, bacteria, fungi, and parasites (**Figure 4**). The ability of a specific pathogen to cause a disease is called **virulence.** In this section, the various pathogens and their mechanisms of infection are briefly described.

Viruses

Viruses are the smallest of pathogens, only 20–400 nm (nanometer, 10^{-9} m) in size, and consist of either single- or double-stranded nucleic acid (either RNA or DNA) surrounded by a protein coat. Viruses cannot survive without a *host*, which is another organism that the virus invades. Once a virus invades a cell of the host, it uses that host cell's metabolic machinery to reproduce itself, with

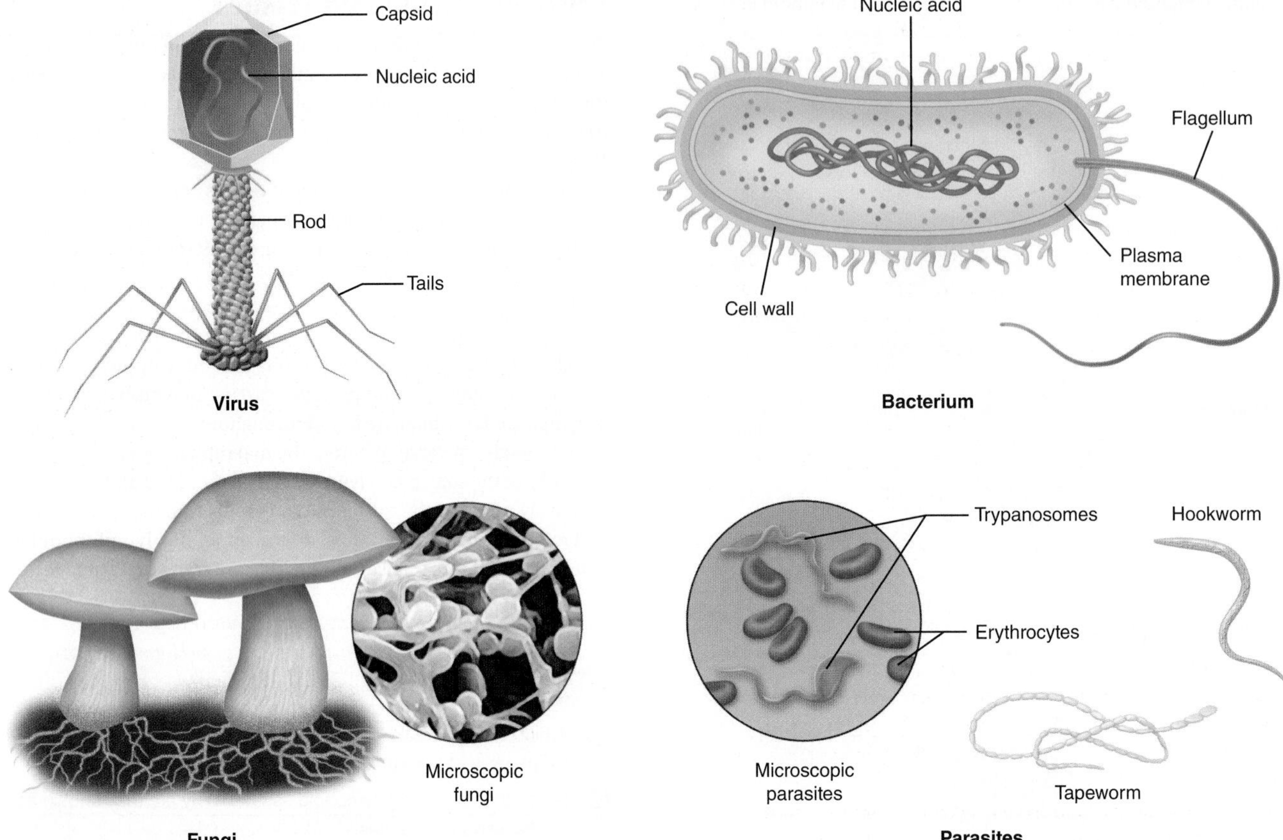

Figure 4 **Classes of pathogens:** viruses, bacteria, fungi, and parasites.

the ability to double the number of viruses every hour. Thus 1000 viruses will become 2000 in 1 hour, 4000 in 2 hours, and 8000 in 3 hours. Eventually, the cell lyses, releasing the virus to invade other cells.

Almost all viral infections result in disease. Viruses are specific in terms of which host cells they invade and which diseases they cause. These pathogens attack a particular cell type based on interactions between the virus and a receptor on the cell. For example, the acquired immunodeficiency syndrome (AIDS) virus binds to *CD4 receptors* on *helper T cells*, which are lymphocytes that contribute to the immune response (described in greater detail later in this chapter). As a consequence, people infected with the AIDs virus have weakened immune systems.

Health Fact

Bacteria

Bacteria are small prokaryotic cells, approximately 1–5 microns in size. Unlike eukaryotic cells, which are found in multicelled organisms, prokaryotic cells have neither a nucleus nor organelles, although they do have a cell wall. Most bacteria contain a single circular chromosome that contains the entire bacterial DNA. Bacteria are classified in two groups, based on whether they stain with cresyl violet. Gram-positive bacteria take up the cresyl violet, whereas gram-negative bacteria have enzymes that break up the cresyl violet, preventing it from accessing the interior of the cell.

Bacteria cause diseases by two different mechanisms: invasion of tissue and release of toxins. To invade tissue, these pathogens must produce chemicals that facilitate their entry into tissue and their ability to *colonize*, or reproduce. On their outer surface, bacteria expose a large molecule called **adhesin** that can bind to certain receptors on eukaryotic cells, thereby facilitating the bacteria's adhesion to tissues—especially epithelial tissues, which form barriers between the body's external and internal environments (skin, respiratory tract, gastrointestinal tract, and urinary system). Adhesion begins the process of colonization and prevents bacteria from being washed away by fluids. After adhesion, the bacteria must then penetrate the epithelial barrier protecting the body's internal environment, which requires a break in the epithelium.

Bacteria also release toxins—*endotoxins* or *exotoxins*—to cause disease. **Endotoxins** are located on the bacterial cell walls of all gram-negative bacteria, where they activate certain aspects of the immune system to induce fever and other symptoms of disease. **Exotoxins** are proteins secreted primarily by gram-positive bacteria. Three classes of exotoxins exist: *cytotoxins* that attack host cells, *neurotoxins* that interfere with neural signaling, and *enterotoxins* that affect the lining of the gastrointestinal tract.

Transmission of bacteria to another host is required for survival of the bacteria. Such transmission often occurs through body secretions and excretions. For example, sexually transmitted diseases, such as *Chlamydia trachomatis* infection, cause the urethra or cervix to produce a discharge that increases the likelihood of the bacteria being passed to another host during a sexual encounter. In contrast, bacteria that affect the gastrointestinal tract, such as *Salmonella enterica*, are excreted in the feces and enter the environment. Lack of proper hygiene may result in *S. enterica* being consumed by the next host. *S. enterica*, like many bacteria, has an incredible adaptation response that ensures the pathogen can survive under many different conditions, including anaerobic or acidic environments.

Fungi

Fungi (singular *fungus*) are neither plants nor animals; they are in their own kingdom consisting of single-celled and multicellular eukaryotic organisms such as yeast, molds, and mushrooms. Fungi do not consume nutrients, but instead absorb nutrients from the environment in which they live. Some fungi provide a service to the environment by absorbing debris from dead organic material, whereas others have a symbiotic relationship with other organisms, such as the fungi that inhabit the roots of vascular plants and provide them essential nutrients. Yet another fungus, *Penicillium fungi*, provides us with the first-discovered antibiotic, penicillin. Other fungi, however, can cause disease in humans.

Fungal infections in animals, or *mycoses* (singular: *mycosis*), occur frequently and include infections such as athlete's foot and ringworm. Many fungi live in the soil; when they reproduce, they produce spores that either stay in the original environment or enter the air. Thus people tend to get fungal infections on the skin (from the soil or air) or in the respiratory tract (from the air). Some fungi are normally found on the skin or other epithelium that separates the internal and external environments. Fungi grow slowly; those found on the epithelium are usually restrained from extensive growth by bacteria in the same areas. For example, a woman may contract a vaginal fungal infection ("yeast infection") as a response to taking an antibiotic for another bacterial infection. The antibiotic kills off the "good" bacteria that normally exist in the vagina, thereby allowing the fungi to grow unchecked.

Some fungi cause severe infections in humans, such as histoplasmosis, a respiratory disease caused by inhaling the fungus *Histoplasma capsulatum*, which often enters the soil in bird and bat droppings. This fungus is often found on the floor of caves, leading to its other names—spelunker's lung and caver's disease. The symptoms of histoplasmosis are similar to those associated with the flu, and include fever, chills, cough, chest pain, joint pain, and mouth sores. In addition, the disease may leave the infected person with scarred lung tissue. Histoplasmosis can also start in other areas of the body exposed to the environment, such as the eyes. Histoplasmosis of the eyes can produce permanent retinal damage and, therefore, blindness.

Parasites

Parasites are organisms that invade a host in their quest to obtain nutrients, in the process harming the host. Two main classes of parasites are distinguished: *protozoans* and *metazoans*. Protozoans are unicellular eukaryotic organisms that vary considerably in size from 10 μm to 20 cm, with most less than 50 μm. An example of a protozoan is *Trypanosoma brucei*, the pathogen that causes African sleeping sickness. Metazoans are multicellular animals that have different tissue types, and include humans. Metazoan parasites include worms (e.g., pinworms, tapeworms) and arthropods (e.g., ticks, lice).

Parasitic infections are often spread through a *vector*, an organism that transmits a parasite from one host to another, or via the *fecal-oral route*, in which eggs are excreted in the host's feces and enter the soil, where they are ultimately picked up and consumed either in water or food products. Parasitic infections are more pronounced in the tropics, where mosquitoes are a primary vector for diseases such as malaria and African sleeping sickness, and in poverty-stricken areas characterized by poor sanitation and hygiene.

Now that we have a basic understanding of the pathogens that can cause disease, we turn to how the body defends itself against such pathogens.

Quick Check 2

1. Define the following terms: pathogen, virulence, virus, bacteria, endotoxin, exotoxin, fungus, parasite, and vector.
2. Which pathogens are unicelluar organisms?
3. Which pathogens have no organelles?

3 Organization of the Body's Defenses

The body's defenses are categorized as either *nonspecific* or *specific*, depending on their selectivity for the invader and the point at which the response occurs. The mechanisms that constitute **nonspecific defenses** (or **innate defenses**) defend against potentially harmful substances without regard to their precise identity. These mechanisms operate even before foreign material enters the body, in the form of the skin and mucous membranes—the body's first line of defense against infection. If those barriers are broken, internal nonspecific defenses immediately begin to operate. Nonspecific mechanisms also clear wounds and damaged tissue of debris and contribute to healing. **Specific immune responses** (or **adaptive immunity**), in contrast, are highly selective (meaning they target specific substances) and come into play after nonspecific responses have already begun. These responses are specific because they are mediated by lymphocytes, which are uniquely designed to recognize particular substances and aid in their destruction. Unlike nonspecific responses, which always operate with about the same speed and effectiveness, specific responses become stronger with each exposure to a particular offending agent.

Nonspecific Defenses

Nonspecific defenses rapidly protect the body against infectious agents because, while they have to recognize a pathogen as such, they do not have to recognize the specific pathogen. Other than the physical barriers, the other nonspecific defenses described in this subsection distinguish self from pathogens through common chemical structures on the surface of many pathogens called **pathogen-associated molecular patterns** (PAMPs). Proteins associated with nonspecific defense mechanisms act as **pattern-recognition receptors** that recognize and bind the PAMPs, thereby initiating the defense process. These receptors include *macrophage mannose receptors, scavenger receptors, CD14, toll-like receptors, mannose-binding lectin,* and *complement receptors* (*CR3* and *CR4*), all of which are described here.

The body's nonspecific defenses include (1) physical barriers (described previously); (2) **inflammation,** a complex series of events causing accumulation of proteins, fluid, and phagocytic cells in an area of tissue that has been injured or invaded by microorganisms; (3) **interferons,** a family of related proteins that can induce virus resistance to other cells; and (4) the **complement system,** a group of plasma proteins that act to lyse foreign cells, especially bacteria. We consider each of these nonspecific defense mechanisms next.

Inflammation

Microbial invasion or damage to tissue triggers a complex series of events that rapidly lead to inflammation of the affected tissue, which in turn recruits cells and soluble factors to the site of infection. Five major events occur in inflammation, typically in the following order: (1) nearby macrophages engulf debris and foreign matter, (2) nearby capillaries dilate and become more permeable to proteins and fluid, (3) foreign matter is contained, (4) additional leukocytes migrate into the region, and (5) recruited leukocytes continue to help clear the infection, mainly by phagocytosis. Note that these events are nonspecific; in other words, inflammation proceeds in much the same way regardless of which bacteria, virus, or injury triggered it. The coordination of these events is not fully understood, but appears to involve the interaction of several chemical mediators coming from the invading pathogen, release by damaged cells, release by leukocytes, and movement from blood to the infected tissue. In the following subsections we examine the major events in inflammation by considering the body's response to an injury in which the skin is cut and abraded.

Phagocytosis of Pathogens and Debris by Nearby Macrophages Macrophages already present in affected tissues can quickly detect bacteria introduced into a cut by using the mannose, scavenger, CD14, and complement receptors found on the macrophages' surface. These receptor proteins cause attachment of the macrophage to the bacterium, thereby initiating phagocytosis (**Figure 5a**), in which the macrophages begin to engulf the bacteria. Attaching to bacteria in this way also stimulates the macrophages to secrete cytokines, small proteins secreted by cells in response to a stimulus (in this case, contact with bacteria) that act as autocrine or paracrine agents. The various kinds of cytokines exert a wide range of effects on the body. *Interleukins,* for example, earned their name because they act as chemical messengers that send signals between (*inter-*) leukocytes (*-leukins*). These cytokines act individually and collectively to induce a number of changes,

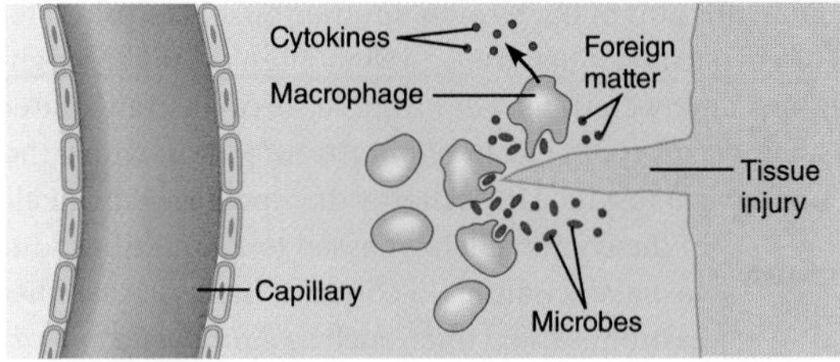

(a) Phagocytosis by nearby macrophages

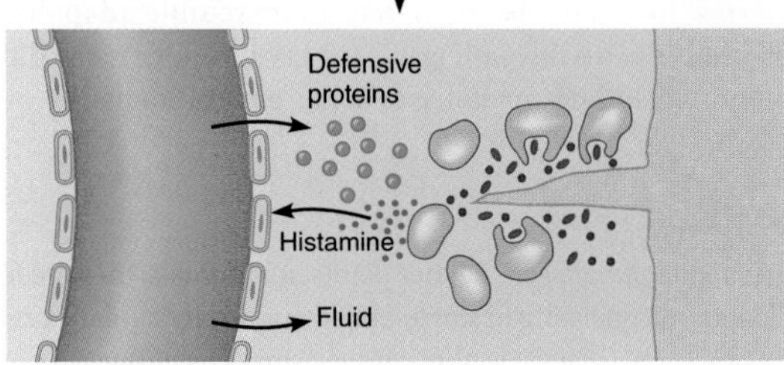

(b) Dilation and increased permeability of capillary

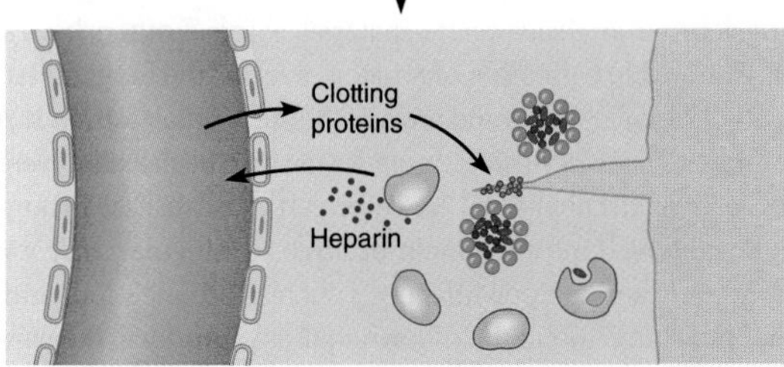

(c) Containment of bacteria and foreign matter

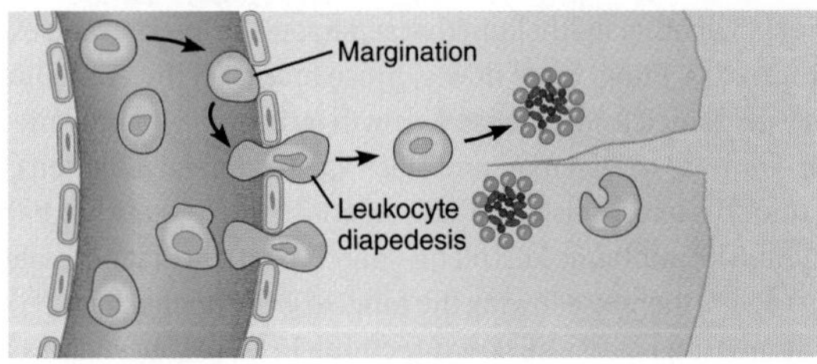

(d) Leukocyte proliferation and migration

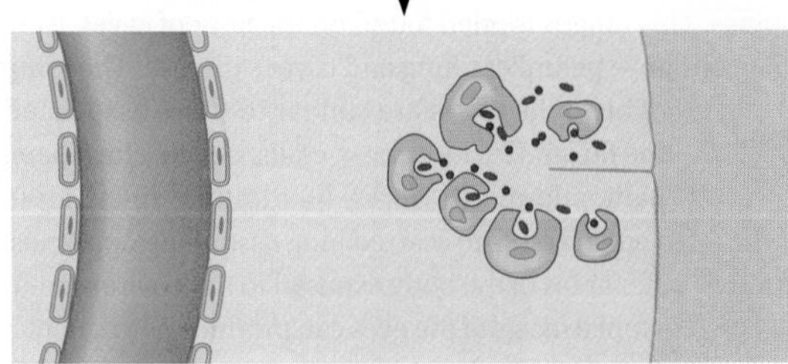

(e) Continued activity of recruited leukocytes

Figure 5 Major events in local inflammation. Events are described in detail in the text.

Q Drugs known as antihistamines block the action of histamine. Explain how antihistamines diminish the symptoms of inflammation.

including synthesis of adhesion molecules by blood vessel endothelial cells and release of more neutrophils from bone marrow, as we will see shortly. The phagocytic activity of these macrophages limits the spread of bacteria early in infection, but these cells are usually too few in number to eliminate all the foreign cells and debris present at an injury site.

Dilation and Increased Permeability of Capillaries Within minutes of bacterial invasion, nearby blood vessels dilate (Figure 5b), thereby increasing local blood flow. The capillary walls also become more permeable, allowing proteins and fluid normally contained in the plasma to move into tissue spaces. The increased blood flow brings additional leukocytes and defensive proteins into the local circulation, and the increased capillary permeability allows these proteins to move into the tissues where they are needed. In addition, the leukocytes that gather in these dilated vessels migrate from the blood into the tissue spaces by a process we examine shortly.

Both the vasodilation and the increased capillary permeability are induced by *histamine* released from a variety of cells in response to injury. The vascular changes mediated by histamine are ultimately responsible for the four characteristic symptoms of inflammation (from the Latin *inflammo,* "to set on fire"), which were described as early as the first century A.D.: redness *(rubor)*, swelling *(tumor)*, heat *(calor)*, and pain *(dolor)*. The increased blood flow both reddens the tissue (which is more or less apparent depending on skin tone) and makes it warmer. Both the histamine-induced increase in blood flow and capillary permeability contribute to the edema that follows. As the capillaries become engorged with blood, the resulting increase in hydrostatic pressure plus the increased interstitial osmotic pressure that accompanies the leakage of plasma proteins causes fluid to filter out of the capillaries and into the tissue spaces, resulting in edema. The edema exerts pressure against the surrounding tissue and skin, which contributes to pain, as does the production of pain-inducing chemicals such as *bradykinin* that stimulate nearby sensory neurons. Thus, even though these vascular changes bring some discomfort, they also help to gather nonspecific defenses to the site of injury.

Containment of Foreign Matter Early in the process, mast cells and basophils release the anticoagulant heparin, which temporarily suspends blood clotting and allows leukocytes unimpeded access to the area of tissue injury. In time, however, clotting factors that have leaked from the plasma into the tissue become active and form clots around clusters of bacteria, thereby inhibiting their spread within the body (Figure 5c). With the aid of other plasma proteins as well as proteins released from damaged tissues, the process of clot formation continues and effectively walls off the region of damage and infection. Eventually the portion of the clot that is at the skin's surface dries and hardens, forming a scab. Over time, the scab will be replaced with new cells as wound healing takes place.

Leukocyte Migration and Proliferation As previously mentioned, macrophages already present in the damaged tissue are the first to phagocytose bacteria and debris, but soon they are joined by other phagocytic cells. About an hour after the injury, neutrophils accumulate in great numbers within the affected tissue; about ten hours later, monocytes begin to move to the tissue, where they develop into large, active macrophages. To arrive at the infection site, leukocytes must leave the blood close to the site of injury and then move through the tissue to the injury site.

The signaling that tells leukocytes to move through blood vessel walls at the right location in the body is achieved by a process of regulated transit involving four events: margination, attachment, diapedesis, and chemotaxis (Figure 5d). *Margination* is the weak interaction of leukocytes with endothelial cells lining the blood vessel near the site of injury; it depends on the expression of adhesion molecules in both endothelial cells and leukocytes. Cytokines, including tumor necrosis factor alpha (TNF-α) and interleukin-1 (IL-1), released at the site of injury signal endothelial cells to express adhesion molecules called *selectins*, which loosely bind leukocytes in the blood. This interaction gives leukocytes the chance to receive activating signals from another cytokine, interleukin-8 (IL-8), which in turn signals the leukocytes to form other adhesion molecules called *integrins* that bind the cells tightly to the blood vessel wall. *Attachment* is soon followed by the cell's transit across the wall, known as *diapedesis* ("jump across"). In this step the leukocyte essentially "crawls" between endothelial cells of the blood vessel wall and through the basement membrane beneath. Once in the tissues, leukocytes move steadily and directly toward the point of injury by a process called *chemotaxis*, as the leukocytes are attracted by IL-8, which is at its highest concentration at the site of injury, as well as chemicals released from bacteria and injured tissues themselves. Through this process, phagocytic leukocytes are delivered to the place where they are most needed: the original site of injury and bacterial invasion. Once at the site, phagocytes contact and engulf additional bacteria and debris.

Apply Your Knowledge

One difficulty in vaccinations is the mutability of pathogens—bacteria and virus can develop resistant strains. This is especially true of the influenza (flu) virus. How are current flu vaccines prepared, and what are some of the challenges facing microbiologists today?

Note that the mass movement of neutrophils and monocytes from blood to tissues during inflammation does not diminish the number of circulating leukocytes. In fact, a common sign of bacterial infection is *leukocytosis,* a fourfold to fivefold increase in the number of circulating neutrophils. This increase occurs because the same cytokines that stimulated leukocyte migration signal the bone marrow to stimulate the proliferation and release of neutrophils and, later, monocytes into the circulation.

Continued Clearing of Infection by Recruited Leukocytes The inflammatory processes described so far have brought defensive proteins (such as clotting factors) and phagocytic leukocytes (neutrophils and macrophages) to the site of injury. We have seen that phagocytes perform two important functions: They engulf bacteria and debris, and they secrete cytokines that further regulate inflammation and other defensive mechanisms. Let's take a closer look at these two important tasks.

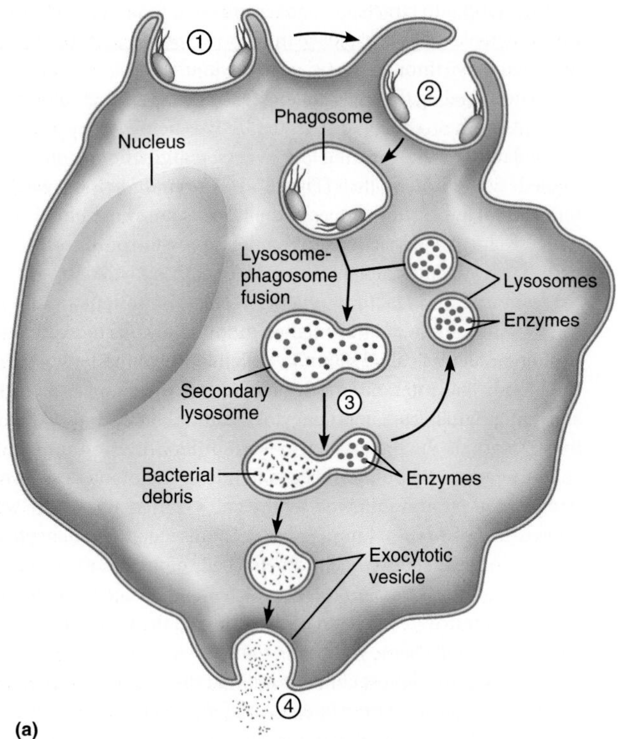

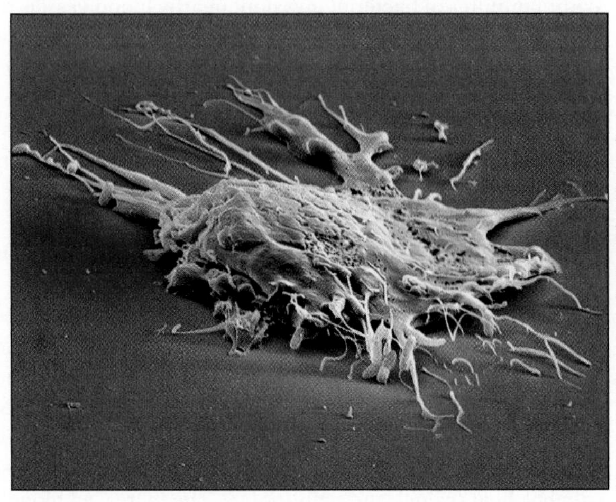

Figure 6 Phagocytosis. (a) The steps of phagocytosis: ① attachment, ② internalization, ③ degradation, and ④ exocytosis. **(b)** Electron photomicrograph of a macrophage phagocytosing bacteria.

Phagocytosis is achieved in four steps: attachment, internalization, degradation, and exocytosis (**Figure 6**). The first step, *attachment,* enables the phagocyte to distinguish between those substances that should be engulfed and those that should not. The selectivity of a phagocyte for its target material is mainly determined by how well the two adhere. Phagocytes tend to bind to damaged and dead cells with irregular, rough surfaces, but not to healthy cells. Phagocytes also attach to many types of bacteria through pattern-recognition receptors, including toll-like receptors (named for their similarity to toll receptors identified in *drosophila*). Ten toll-like receptors are found in humans, each of which binds to surface structures of bacteria. For example, toll-like receptor 1 (TLR-1) binds to lipopolysaccharides and lipopeptides on bacteria's surfaces. TLR-5, in contrast, binds to flagellin, the tail-like structure of some bacteria.

Sometimes, attachment is enhanced by **opsonins,** proteins (including antibodies) that bind tightly to the foreign material and make it easier for the phagocyte to engulf it. Phagocytosis is greatly enhanced by this binding of opsonins, a process known as *opsonization* ("to make tasty"); a macrophage can engulf and destroy a substance approximately 4000 times faster when it is coated with antibodies.

As soon as the phagocyte has attached to its prey, *internalization* occurs. In less than 0.01 second, the phagocyte's plasma membrane extends outward around the site of attachment and surrounds the material, enclosing it within a large intracellular vesicle called a phagosome. The phagosome moves toward the cell interior and fuses with a lysosome, which contains a variety of digestive enzymes. *Degradation* of the phagocytosed material then occurs within the enlarged, enzymatically active organelle, now called a secondary lysosome. Subsequently, the phagocyte uses some of the harmless bacterial debris (amino acids, for example) and eliminates others by exocytosis.

In addition to stimulating phagocytosis, attachment of a phagocyte to foreign cells and debris trigger the secretion of cytokines, including those mentioned previously. Once these cytokines enter the bloodstream, they have access to all body tissues, including the brain. IL-1 and TNF-α act on the hypothalamus to raise body temperature; that is, they function as *endogenous pyrogens*. They achieve this feat by stimulating the hypothalamus to release prostaglandins, which in turn adjust the body's "thermostat" setting, raising the temperature above normal. Although a high fever can be dangerous, a moderate one is generally thought to benefit the host's defenses; elevated temperatures are thought to decrease the rates of bacterial and viral replication, and to accelerate phagocytosis and other defensive reactions.

The cytokines IL-1, IL-6, and TNF-α also stimulate liver cells to produce *acute-phase proteins,* a group of proteins having a wide range of antibacterial and inflammatory effects and whose plasma concentrations increase soon after an infection begins. One of these is *C-reactive protein,* which binds to the surface of many types of bacteria. A C-reactive protein molecule acts as an opsonin, marking the bacteria for phagocytosis, and activates the complement system, described shortly.

Finally, IL-1 helps induce the proliferation and differentiation of B lymphocytes and T lymphocytes. B cell differentiation leads to the production of antibodies that mark selected foreign substances for destruction, whereas T cell differentiation leads to *cell-mediated immunity,* in which certain types of T cells kill specific abnormal or infected body cells.

Interferons

Another nonspecific defense mechanism prevents the spread of viruses within the body by interfering with virus replication. This defense is provided by a group of related proteins appropriately named interferons. The secretion of two interferons, called interferon-α and interferon-β, from virus-infected cells signals to the surrounding cells the presence of the virus and induces the cells' resistance.

The viral nucleic acid that accumulates in a virus-infected cell during viral replication stimulates the cell to synthesize and secrete interferons. Even though the infected cell will likely die, the interferons it secretes bind to nearby healthy cells, initiating in them a series of intracellular changes that cause the cells to become more resistant to the virus. This virus-resistant state is conferred by the presence of RNA-degrading enzymes and protein synthesis inhibitors in the cytoplasm. Because these enzymes and inhibitors are potentially dangerous to the cell itself, they come into play only after viral infection ensues and last only a short time. Interferon-induced cells are thus poised to block the production of new viruses, but they can do so for only a limited amount of time. Both the production of interferon and the virus-resistant state it induces are nonspecific; virtually any viral nucleic acid can induce interferon production, and interferon-induced resistance can defeat virtually any viral infection.

A third type of interferon, *interferon-γ*, is secreted not by virus-infected cells but rather by active T cells and NK cells. It has a broader range of effects. In addition to inhibiting viral replication, interferon-γ enhances phagocytosis in macrophages, boosts antibody production in B cells, and helps to activate NK cells and cytotoxic T cells, both of which kill virus-infected cells and cancer cells. Likewise, by inhibiting cell division, interferon-γ has a direct effect on cancer cells, suppressing the growth of tumors.

Natural Killer Cells

NK cells function in early nonspecific defense by recognizing the general features of infected or abnormal cells and releasing **perforins,** proteins that form pores in the infected or abnormal cell's membrane. The presence of such pores increases the permeability of the membrane to ions and water. As fluid moves into the cell, the cell swells and eventually lyses.

NK cells are activated by interferons and by cytokines released by activated macrophages. These cells defend the body against viral infection, tumors, bacteria, and protozoans. In the case of virus-infected cells, NK cells can act on them without detecting the virus itself, and they stand ready to act immediately. NK cells thus provide an essential, general defense in the early stages of both viral infection and tumor growth. These cells also come into play during specific immune responses because their activities can be enhanced by the cytokine *interleukin-2 (IL-2)* and by antibodies, which are produced by T cells and plasma cells, respectively, during specific responses.

The Complement System

The fourth and final nonspecific defense mechanism, the complement system, can act specifically or nonspecifically depending on whether antibodies are present (antibodies give specificity to the defense mechanism). The complement system is so named because it completes or fulfills ("complements") the actions of specific antibodies. The complement system consists of approximately 30 plasma proteins that act to destroy invading microorganisms, especially bacteria. The first in a series of complement protein reactions occurs in association with a bacterium and leads to a cascade of activation steps in which each component activates the next in the series, similar to what occurs in the coagulation cascade.

The complement cascade can be activated in three ways (**Figure 7**): (1) by binding to antibodies that are already attached to bacterial cells (known as the *classical pathway*); (2) by binding of a serum protein, mannose-binding lectin, to mannose-containing carbohydrates on bacteria or viruses (known as the *lectin pathway*); or (3) by binding directly to carbohydrates present on the surface of a broad range of bacterial cells (known as the *alternative pathway*). These pathways converge to generate the same set of complement proteins (**Figure 8**), which protects against infections by four different mechanisms: (1) covalently bonding to pathogens and functioning as opsonins, (2) recruiting more phagocytes to the site of infection, (3) triggering histamine secretion by mast cells, and (4) development of a membrane attack complex (MAC), a collection of pore-forming proteins that pierce the bacterial membrane (**Figure 9**). The insertion of MAC into the membrane causes the cell to lose its integrity, such that it fills with fluid, swells, and then bursts (Figure 7). Such complement-mediated lysis is the primary way by which antibody-coated bacteria are killed.

Classical Pathway The classical pathway is part of the body's specific defenses, as its activity depends on the presence of antibodies. The main proteins of the classical pathway are named C1 through C9, and most consist of subunits. Activation of the classical pathway requires C1 to bind to antibodies that are interacting with a pathogen; C1 then cleaves another complement protein, C2, starting the cascade. The completion of the cascade produces the MAC.

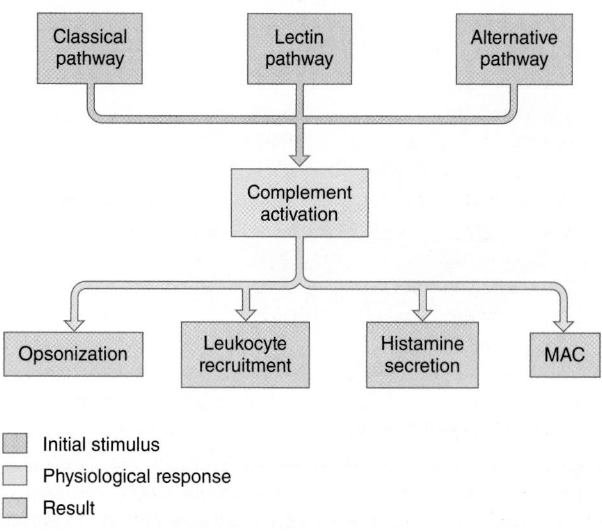

Figure 7 **Complement activation and actions.**

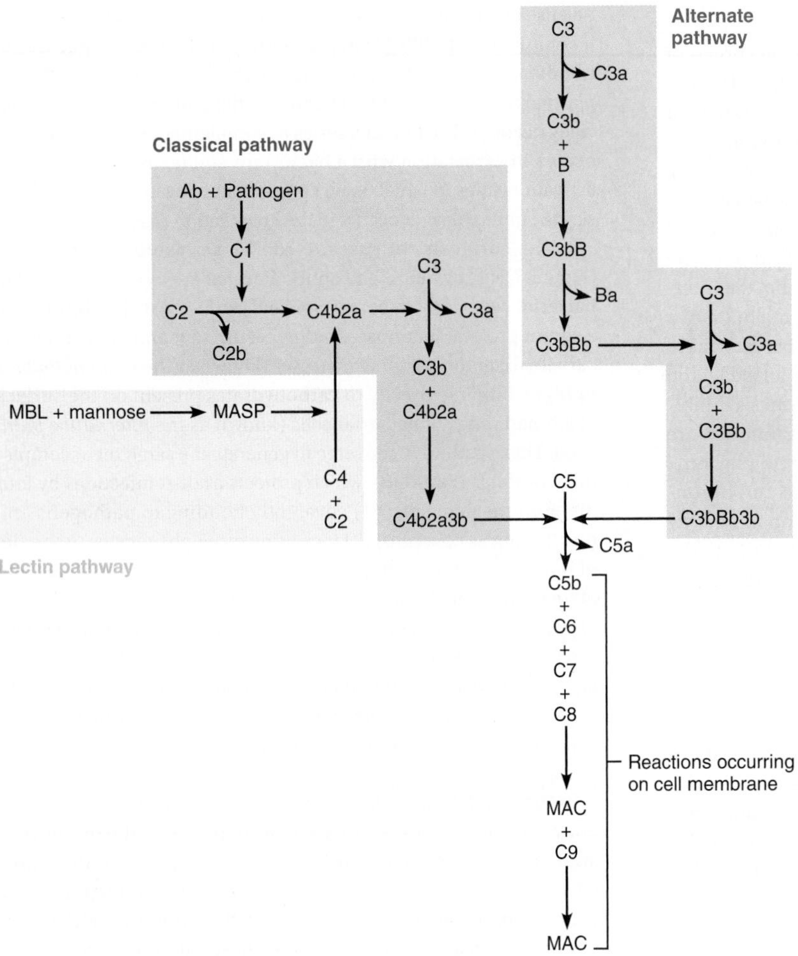

Figure 8 Complement pathways.

Lectin Pathway The lectin pathway differs from the classical pathway in its lack of specificity, speed of activation, and mechanism of activation. Nonspecific pathways tend to mount a faster response because they do not rely on the production of antibodies. Lectins are proteins that bind specific carbohydrates and are critical in cell recognition. One lectin in particular, *mannose binding lectin* (MBL, also known as mannose binding protein), functions in identifying pathogens and activating the complement system. In the lectin pathway, MBL binds to mannose residues in glycoproteins or polysaccharides located on the surface of bacteria. In the plasma are MBL-associated proteins (MASPs) that form complexes with MBL. The MBL-MASP complexes activate the complement pathway through proteolytic activation of C4 and C2. Once the complement cascade is activated, it proceeds in the same way as in the classical pathway.

Alternative Pathway The alternative pathway, like the lectin pathway, functions in the body's nonspecific defenses and, therefore, provides a faster response than the classical pathway. The alternative pathway activates the complement system at a point further along in the cascade of reactions than either the classical or lectin pathway—that is, it activates C3, which then activates C5. From this point forward, the three pathways are identical.

Actions of Complements Earlier, we learned about how activation of the complement

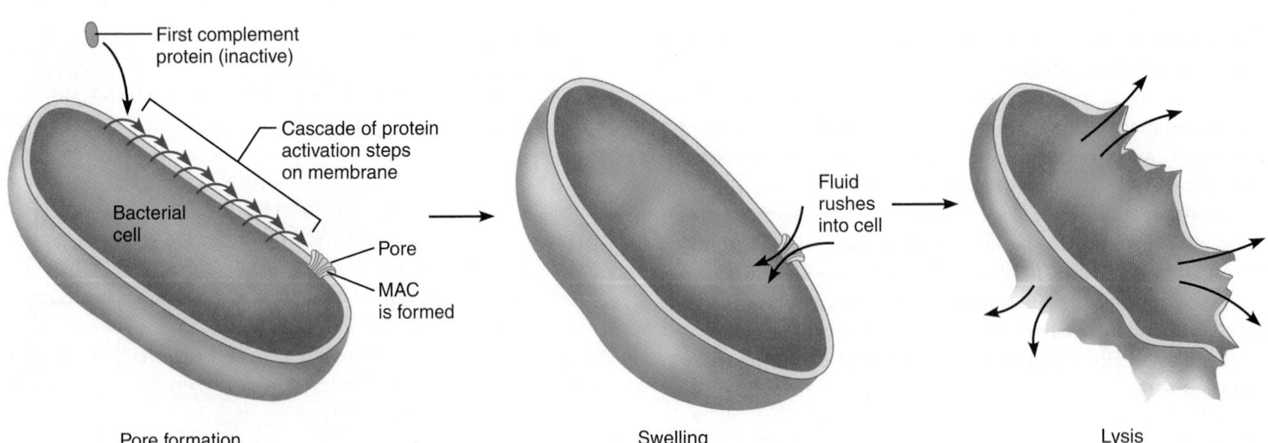

Figure 9 **Formation of the MAC complex and cell lysis.** After a series of activation steps, some of the complement proteins form a membrane attack complex (MAC), which inserts itself into the cell membrane as a protein-lined channel or pore. Fluid rushes into the cell, causing it to swell and burst (undergo lysis).

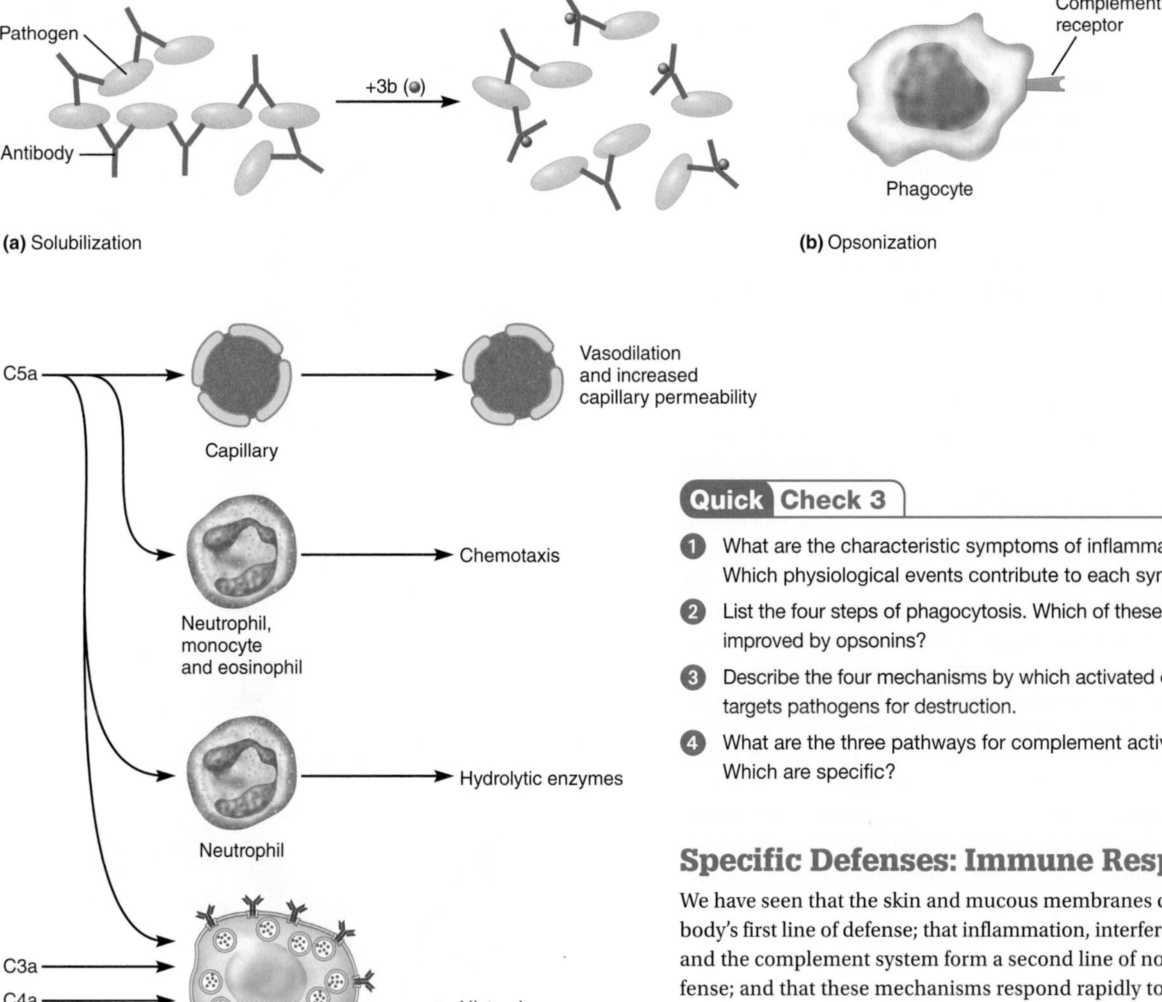

Figure 10 Actions of complement proteins independent of MAC. **(a)** Solubilization by C3b. **(b)** Opsonization by C3b. **(c)** Enhancing of inflammation by C3a, C4a, and C5a.

system can lead to formation of the MAC, which consists of C5–C9. However, other complement proteins also function directly in defending the body against pathogens (**Figure 10**). For example, C3b is produced in very large quantities and binds to antibody-laden cells, thereby preventing accumulation of too many antibodies on one pathogen; this serves to keep the pathogen-antibody complex soluble and prevent blockage of capillaries (Figure 10a). C3b also functions as an opsonin (Figure 10b). C3a, C4a, and C5a promote inflammation (Figure 10c). Specifically, these subunits all trigger release of vasoactive amines and activate mast cells. C5a also causes chemotaxis, vasodilation, increased vascular permeability, and activates neutrophils.

Quick Check 3

1. What are the characteristic symptoms of inflammation? Which physiological events contribute to each symptom?
2. List the four steps of phagocytosis. Which of these steps is improved by opsonins?
3. Describe the four mechanisms by which activated complement targets pathogens for destruction.
4. What are the three pathways for complement activation? Which are specific?

Specific Defenses: Immune Responses

We have seen that the skin and mucous membranes constitute the body's first line of defense; that inflammation, interferons, NK cells, and the complement system form a second line of nonspecific defense; and that these mechanisms respond rapidly to injury or exposure to foreign material or infective agents, even if such exposure is the initial one. Nevertheless, nonspecific defenses are not always completely successful in eliminating foreign materials. Fortunately, the body has an exquisite and powerful mechanism in reserve called the immune response.

Suppose that the body's nonspecific defenses were not completely effective in responding to the cut and abrasion we previously discussed. While these defenses continue to operate, bacteria, bacterial fragments, and other foreign molecules gain access to the bloodstream and become trapped in the netlike architecture of the spleen. Likewise, bacteria present in interstitial fluid are carried into lymphatic vessels and eventually into the netlike lymph nodes, which swell and become tender. In the spleen, lymph nodes, and other lymphoid tissues, the bacteria come into contact with B lymphocytes and T lymphocytes, thereby inducing these cells to generate efficient and selective immune responses that work throughout the body to eliminate the invaders.

B lymphocytes and T lymphocytes each generate a particular kind of immune response. B lymphocytes develop into plasma cells that secrete antibodies, the actions of which bring about **humoral immunity,** so called because antibodies circulate in the blood and lymph, body fluids long ago called "humors." In contrast, certain T lymphocytes develop into active cytotoxic T cells, which bind to and kill abnormal body cells. This type of immune response constitutes **cell-mediated immunity,** so called because cytotoxic T cells must come into direct contact with their targets to act on them.

The circulating antibodies of the humoral response defend mainly against bacteria, toxins, and viruses present in body fluids. In contrast, the T cells of the cell-mediated response are active against bacteria and viruses that are hidden within infected body cells. Moreover, the cell-mediated response operates in the body's reaction to transplanted tissue and cancer cells, both of which are perceived as foreign.

Features of Immune Responses

Approximately 2400 years ago, Thucydides of Athens described how people who were sick and dying of plague were attended by others who had recovered, "for no one was ever attacked a second time." We have all made similar observations: Those of us who experienced many childhood diseases do not worry about getting them again because exposure to each disease confers lifelong immunity to that disease. In this section we explore the nature of specific immune responses by addressing the following questions: Why does a measles infection generate a response to that disease but to no other disease (a phenomenon known as specificity)? How is it that the immune system is already poised to defend against a first exposure to chicken pox (which relates to a property known as diversity)? How do we acquire our lifelong immunity to chicken pox after the first exposure (which depends on the immune system's memory)? Finally, why does the immune system respond against foreign microbes but not against the body's own tissues (a phenomenon known as self-tolerance)? As we will see, these special features of the immune response—specificity, diversity, memory, and self-tolerance—derive from the nature of B lymphocytes and T lymphocytes.

Specificity B cells and T cells bind and respond to foreign or abnormal molecules known as *antigens.* **Antigens** ("*anti*body *gen*erators") are typically complex protein or polysaccharide components of viruses, bacteria, fungi, protozoa, parasitic worms, pollen, transplanted tissue, and tumor cells. Each antigen has a unique structure and contains different recognition sites called **epitopes** or *antigenic determinants.* These epitopes can be detected by specific lymphocytes, which then target the invaders bearing that epitope for destruction.

B cells and T cells are able to recognize specific antigens because they possess antigen-binding proteins called **antigen receptors** on their surfaces. All of the receptors on a particular B cell or T cell bind to a specific antigen. The antigen receptors on B cells and T cells are similar to receptors on antibodies.

A typical antibody is a Y-shaped molecule consisting of four protein chains: two identical *heavy chains* and two identical *light chains* that are joined by disulfide bridges (**Figure 11a**). Each antibody molecule possesses both *variable regions (V)* and *constant regions (C).* The constant regions consist of portions of the heavy and light chains that make up the tail of the Y. Five types of heavy chains exist, which serve as the basis of five classes of antibodies, each with its own mechanism of eliminating the antigen. The classes of antibodies and their mechanisms of action are described later in the text.

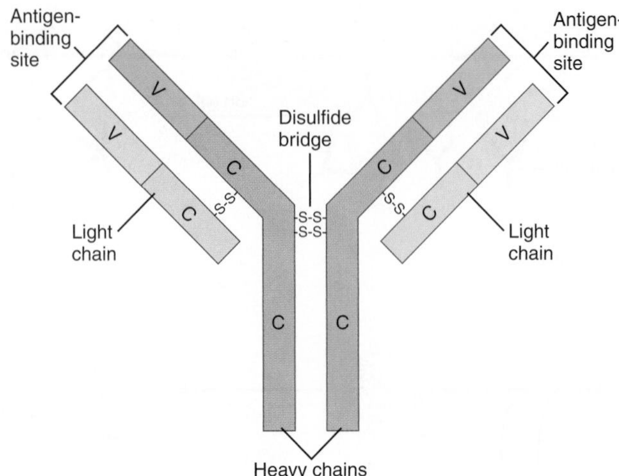

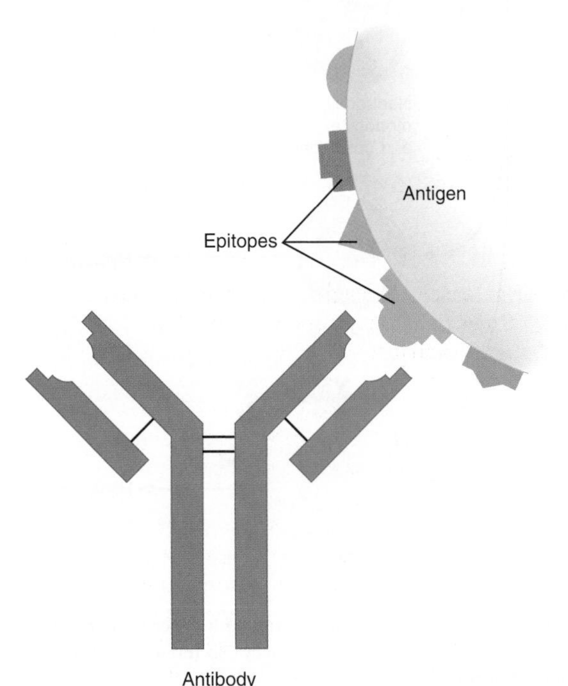

Figure 11 The basis of antigen-antibody specificity. (a) Each antibody molecule consists of two identical heavy chains and two identical light chains linked by disulfide bridges. The combined structures of the variable (V) regions form the antigen-binding sites of the molecule, which vary from antibody to antibody. **(b)** A schematic representation of how the complementary physical structures of an antigen's epitope and the antibody's antigen-binding site constitute antigen-antibody specificity.

The variable regions of an antibody, so called because the amino acid sequences in these regions vary extensively from antibody to antibody, consist of portions of heavy and light chains that make up the top of the Y. The two identical variable regions form

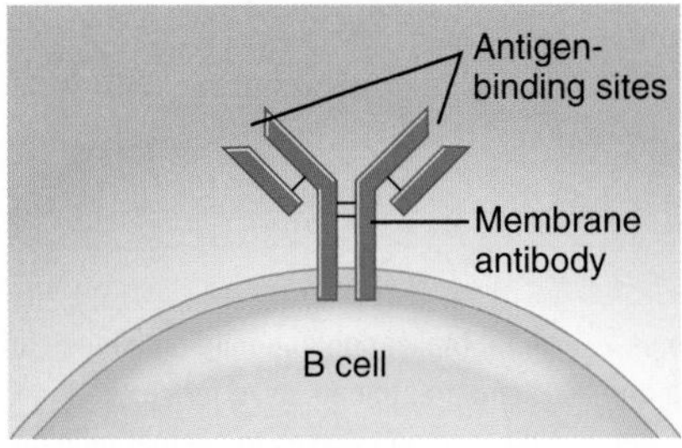

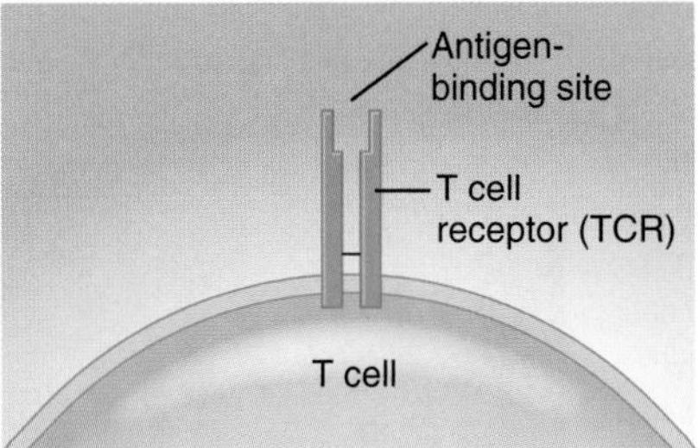

Figure 12 B cell and T cell antigen receptors. Antigen receptors on B cells are called membrane antibodies (or membrane immunoglobulins); antigen receptors on T cells are called T cell receptors (TCRs). Each lymphocyte has approximately 100,000 identical receptors specific for a particular antigen.

two identical antigen-binding sites. Thus each antibody molecule is capable of binding two epitopes of the same kind.

The interaction between the antigen-binding site and its antigen is similar to that between an enzyme and its substrate, which also shows specificity: The unique shape of the binding site allows for a close fit between the antibody and its antigen (Figure 11b), such that multiple noncovalent bonds can form between chemical groups on the respective molecules.

The antigen receptors on B cells are similar to antibody molecules except that the receptors are bound to the plasma membrane (**Figure 12**), whereas antibodies are secreted into the extracellular fluid. For this reason, B cell antigen receptors are often called *membrane antibodies* or *membrane immunoglobulins.* A B cell's antigen receptors have the same specificity as the antibodies it later secretes as a plasma cell.

The antigen receptors on T cells, called *T cell receptors (TCRs),* come from the same family of proteins that includes antibodies, but are different in structure from antibodies (Figure 12). Moreover, unlike antibodies, T cell receptors act only as cell surface receptors for antigen. They are never secreted.

Diversity A single T lymphocyte or B lymphocyte has approximately 100,000 antigen receptors, all with the same specificity. Therefore, each B cell and T cell can detect just a few of the millions of possible antigens that might gain entry to the body. The particular antigen receptor molecules produced by a given lymphocyte are determined by random genetic events that occur early in the development of the lymphocyte. These early events, which are unique to each lymphocyte and occur prior to any contact with foreign antigen, generate a phenomenal array of B lymphocytes and T lymphocytes in the body, each bearing antigen receptors of a particular specificity. With this diversity of lymphocytes, the immune system has the capacity to recognize and respond to millions of different antigens—even to antigens as yet unseen in the universe!

The specificity of B cell and T cell receptors explains the specificity of the B cell and T cell responses. When a particular microorganism invades the body, it interacts with and activates only those lymphocytes that have receptors specific for the antigens the microorganism possesses. When the virus that causes chicken pox invades the body, for example, only lymphocytes specific for chicken pox antigens are activated to proliferate (by successive cell divisions) and differentiate. In other words, the foreign antigen triggers an immune response against itself. This antigen-driven activation of lymphocytes is called **clonal selection,** and it is absolutely necessary for immune responses (**Figure 13**). Lymphocyte differentiation gives rise to two populations of identical cells (clones): **effector**

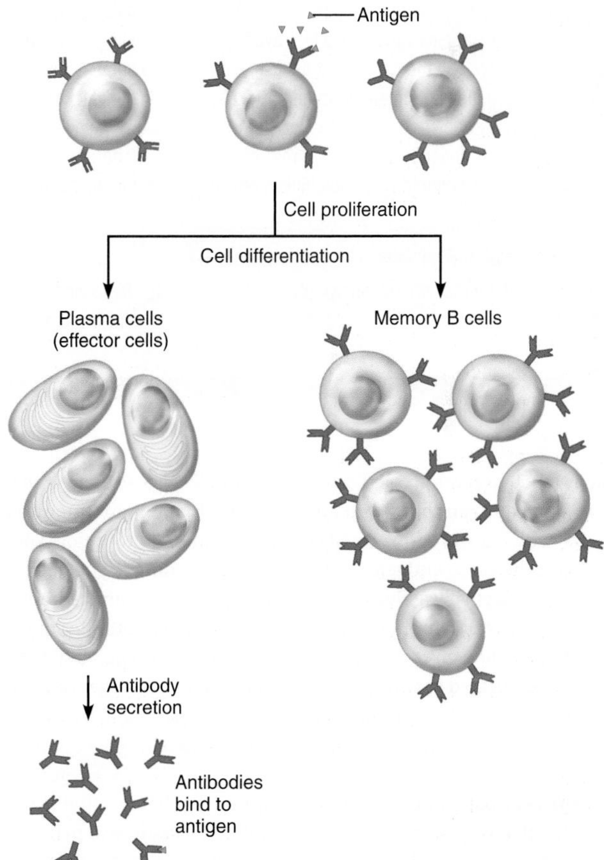

Figure 13 Clonal selection in B lymphocytes. Each lymphocyte produces a randomly generated set of identical antigen receptors. When a foreign antigen and a lymphocyte's receptors are sufficiently complementary in structure, binding induces the lymphocyte to proliferate and differentiate into a population (clone) of short-lived effector lymphocytes (in this case, plasma cells) and a clone of long-lived memory lymphocytes. Note that clonal selection also occurs in T lymphocytes.

CLINICAL CONNECTIONS

Shingles

Chicken pox was a common childhood disease until 1995, when a vaccine against the *varicella zoster* virus was approved in the United States. Chicken pox is highly contagious, and before the vaccine became available more than 10 million children developed this infection each year. That number is now down to only 2 million children annually. The primary symptom of chicken pox is an itchy red rash with spots, or *pox*, that can form blisters. Sore throat, fever, and loss of appetite are also commonly observed with this disease.

Although the immune system generally relieves the body of chicken pox within 10–14 days, some viruses can survive by traveling in sensory axons to the dorsal root ganglia, where the pathogens can remain dormant for years. As the body ages and the immune system weakens, or under conditions of stress or injury, one out of ten people who had chicken pox as children will develop shingles (also known as herpes zoster) as the virus reactivates in a dorsal root ganglion and travels back through the axon to the skin. Shingles begins with a tingling or painful sensation localized to one side of the body or face, basically limited to a single dermatome (area innervated by a nerve). In severe cases, such as in immune-deficient individuals, more than one dermatome can be affected. Next, painful blisters and a rash develop in the same area. Within the blisters of the rash is the virus, which can infect other people not previously exposed to chicken pox. Although shingles generally clears in 2–3 weeks, in some people it develops into *postherpetic neuralgia*, a debilitating, painful disease that lasts for years. Unlike chicken pox, shingles can recur in an individual who had it previously.

There is no cure for shingles. Treatment includes antiviral medication that decreases viral replication and mild pain medication, the same treatments used for chicken pox. If the disease progresses to postherpetic neuralgia, stronger pain medications and treatments are required. Because of the chicken pox vaccine, however, fewer adults will develop shingles, and this disease may eventually be eradicated.

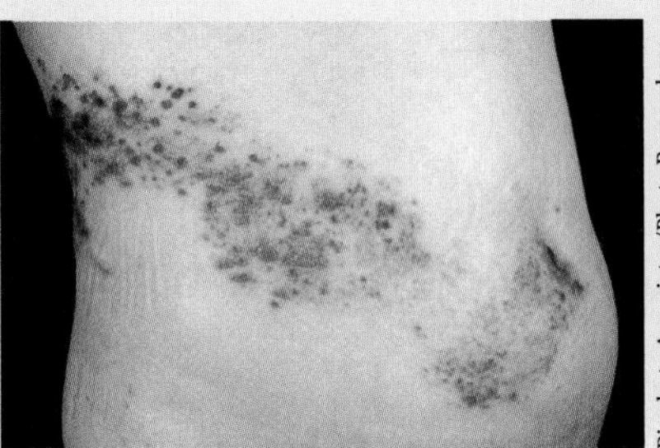

Biophoto Associates/Photo Researchers

Shingles (herpes zoster) rash on the abdomen.

Critical Thinking Questions

1. What are the causes and symptoms of chicken pox?
2. Describe the factors that increase the likelihood an individual will develop shingles.
3. Where do some viruses (such as herpes zoster) "hide" in the human body, and under what circumstances can they reactivate?

cells (such as plasma cells), which are short-lived cells that combat the same antigen that stimulated their production, and **memory cells,** which are long-lived cells bearing membrane receptors specific for the same antigen. Each antigen, by binding to specific receptors, selectively activates a tiny fraction of cells from the body's diverse array of lymphocytes. This relatively small number of selected cells then gives rise to clones of thousands of cells, all specific for and dedicated to eliminating that particular microorganism. (To learn about how the chicken pox virus avoids destruction by the immune system, see **Clinical Connections: Shingles**).

Memory The antigen-induced lymphocyte changes that occur when a person is first exposed to an antigen constitute a **primary immune response.** In the primary response, antigen-selected B cells and T cells proliferate and differentiate into effector cells (antibody-producing plasma cells and cytotoxic T cells, respectively) about 10–17 days after exposure to the antigen (**Figure 14**). Often, a person becomes ill during this time, because it takes only a few days for most viruses or bacteria to cause symptoms. Eventually, however, the symptoms of illness diminish and disappear as antibodies and cytotoxic T cells help clear the offending agent from the body.

Once someone has suffered through an infection, he or she is likely to be immune to further infection by that same microorganism. The basis of this so-called **acquired immunity** is that upon subsequent exposures to that same antigen, the response (called a **secondary immune response**) occurs much more rapidly (only 2–7 days), is greater in magnitude, and is more prolonged than a primary response. The secondary response results from the existence of immunological memory, which is due to the fact that each exposure to an antigen gives rise not only to effector cells but also to clones of long-lived memory T cells and memory B cells. Upon subsequent exposure to the same antigen, these memory cells are poised to quickly proliferate and differentiate into still more memory and effector cells.

Self-Tolerance Given that the diverse repertoire of lymphocyte specificities is randomly generated, how is it that an individual's B cells and T cells do not react to the body's proteins and polysaccharides? The answer is that as B cells and T cells mature in the bone marrow and thymus, their antigen receptors are, in effect, tested for their potential to recognize and react against self. In general, those lymphocytes whose receptors have the potential to react

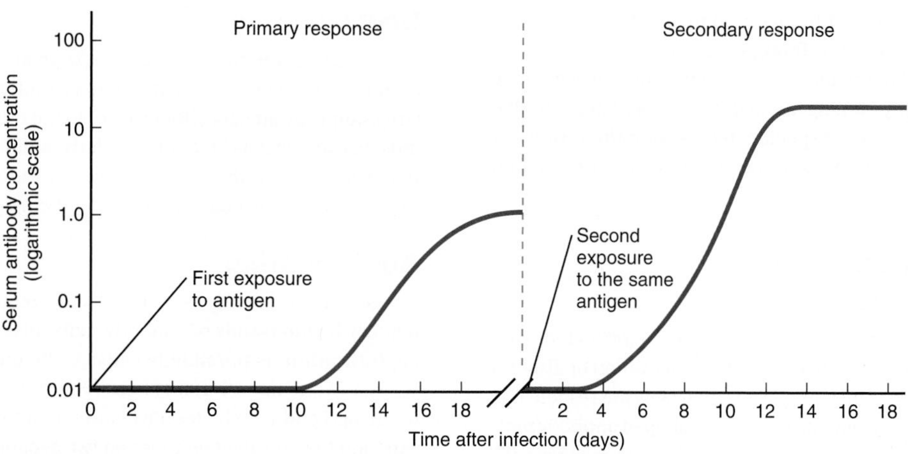

Figure 14 Primary and secondary responses to a given antigen. In secondary responses, serum antibody concentrations rise sooner, reach a higher level, and remain elevated longer than those stimulated by the initial exposure to the antigen.

Q If a person's initial exposure to antigen X coincided with a secondary exposure to antigen Y, what kind of response to antigen X would occur?

with self-molecules either are rendered nonfunctional or undergo *apoptosis,* the self-destructive events that accompany *programmed cell death.* As a result of these processes, only those lymphocytes that are reactive against foreign *(nonself)* molecules remain. This capacity to distinguish self from nonself, known as *self-tolerance,* continues to develop even as the lymphocytes migrate to lymphatic organs. Failure of self-tolerance can lead to various autoimmune diseases (discussed later).

Quick Check 4

1. What is humoral immunity? Which lymphocyte type is responsible for this kind of immunity?
2. What is cell-mediated immunity? Which lymphocyte type is responsible for this kind of immunity?
3. List the four key features of immune responses, and provide a brief explanation of each.
4. What is the basis for the stronger and more rapid immune response that occurs upon secondary exposure to an antigen?

4 Humoral Immunity

Now that we have discussed the characteristics of immune responses, we examine in greater detail the first type of immunity: humoral immunity. We begin by considering the role of B lymphocytes.

The Role of B Lymphocytes in Antibody Production

Millions of B lymphocytes with different specificities normally circulate in the blood and lymph or reside in peripheral lymphoid tissues. When one of the B cells binds with an antigen, the cell responds in the two ways depicted in Figure 13. First, the B cell is stimulated to proliferate, which increases the number of B cells with the same specificity; second, the cells differentiate, such that some become long-lived memory B cells and others become short-lived antibody-synthesizing effector cells called **plasma cells.** A plasma cell secretes approximately 2000 antibody molecules per second over its life span of 4–7 days. These antibodies then circulate throughout the blood and lymph for several weeks, binding to the same antigens that stimulated their production, thereby marking them for destruction by phagocytosis or complement-mediated lysis.

The antigens that evoke the production of both plasma cells and memory B cells are known as *T-dependent antigens* because they do so only with help from a special kind of T cell called a **helper T cell.** We examine the precise functions of helper T cells shortly; we mention them here because they influence B cell activation. When helper T cells respond to specific antigens, they secrete many cytokines, including interleukin-2 (IL-2). Together, IL-2 and T-dependent antigens (mainly proteins) induce the proliferation of B cells. By contrast, polysaccharide antigens, such as those found on many bacterial surfaces, can activate B cells without T cell help; these antigens are known as *T-independent antigens.* Because they contain long arrays of repeated subunits, these polysaccharides bind with many antigen receptors on the B cell surface, providing a strong enough stimulus to induce the B cell to proliferate even in the absence of IL-2. However, without IL-2, all of the proliferating cells differentiate into antibody-secreting plasma cells; none develop into memory B cells. Even repeated exposures to a particular T-independent antigen produce primary responses only, because the antigen alone never generates immunological memory. Thus B cell responses to T-independent antigens take 10–17 days to occur, and the quantity of antibodies produced is smaller than that produced in a secondary response. Specific events in B cell activation and the development of humoral immunity are discussed in the following sections.

Next we will see how antibodies bind to and mediate the disposal of microorganisms and foreign materials. Although antibodies are generated in the immune response, they also enhance nonspecific defenses by binding to—and thereby marking—specific microorganisms to focus nonspecific defenses on the invader at hand. A typical bacterium can be coated with as many as 4 million antibody molecules!

Antibody Function in Humoral Immunity

We have seen that an antibody has two functions: First it binds specifically to an antigen, and then it aids in the inactivation or disposal of that antigen. Whereas the antigen-binding sites are responsible for the recognition of antigen, the tail of the Y-shaped antibody molecule is responsible for the mechanisms by which it mediates antigen disposal. Five classes of antibodies are distinguished based on the structure of this tail: IgG, IgM, IgA, IgE, and IgD. (*Ig* stands for *immunoglobulin.*)

The way in which an antibody aids in antigen disposal depends on its class. The structures and functions of each of the five immunoglobulin classes are summarized in **Figure 15.** All classes of immunoglobulins can mediate the simplest forms of antigen attack—neutralization and agglutination—whereas particular classes specialize in opsonization of the antigen, activation of the complement system, and stimulation of NK cells. The functions of IgG are shown in **Figure 16.**

Neutralization

In **neutralization,** the antibody blocks an antigen's activity just by binding to it (Figure 16a). For example, antibodies can neutralize a virus simply by attaching themselves to the molecules that the virus must use to infect its host cell. Similarly, antibodies that coat bacterial toxins (such as the toxin produced by *Clostridium tetani,* which causes the disease tetanus) can render them inactive.

Agglutination

Antigens are frequently neutralized and clumped together simultaneously by thousands of antibody molecules. This process, called **agglutination,** is possible because each antibody molecule has at least two antigen-binding sites (Figure 16b). (Those belonging to the IgM and IgA classes have more than two.) IgG, for example, can bind to equivalent epitopes on two separate pathogens, linking them together. When numerous IgG molecules likewise bind to two microbes apiece, a large complex forms.

Opsonization

Once bound to antibodies, an antigen is effectively opsonized and, therefore, rendered more susceptible to phagocytosis (Figure 16c). As mentioned previously, antibodies act as opsonins because they can bind to both antigens and phagocytic cells. IgG antibodies are specialized for opsonization in that their tails bind to specific surface receptors on phagocytic cells. This binding triggers the phagocytes to engulf both the antibodies and their targeted prey.

Class of antibody IgM	Structure	Features	Roles in antigen disposal
IgM	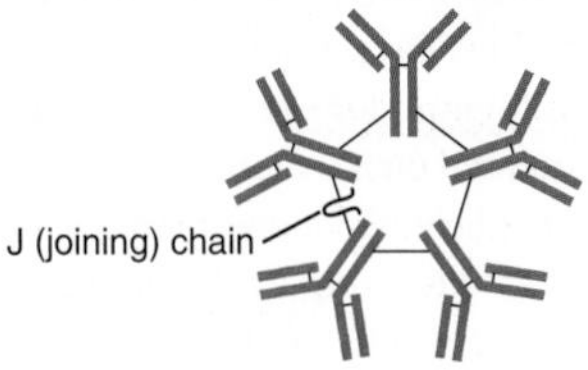 J (joining) chain	The most common class of antibody produced in the primary response to antigen	Neutralizes antigen Agglutinates antigen Activates complement
IgD		Important as an antigen receptor on B cells	Neutralizes antigen Agglutinates antigen
IgG		The most common class of antibody in the blood, and the major class of antibody produced in secondary responses. Crosses the placenta, so it is important in fetal and newborn immunity	Neutralizes antigen Agglutinates antigen Activates complement Opsonizes antigen Enhances NK cell activity
IgE		Involved in allergies such as hay fever	Neutralizes antigen Agglutinates antigen Binds to mast cells and basophils, causing them to release histamine
IgA	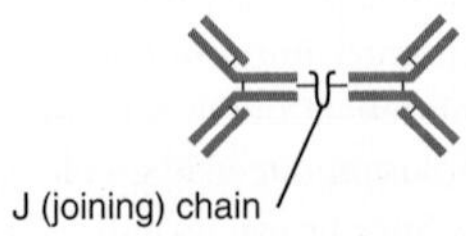 J (joining) chain	Crosses epithelial cells, so is present on mucosal surfaces and in breast milk; is important in immunity in newborns	Neutralizes antigen Agglutinates antigen

Figure 15 **The five major classes of antibodies.**

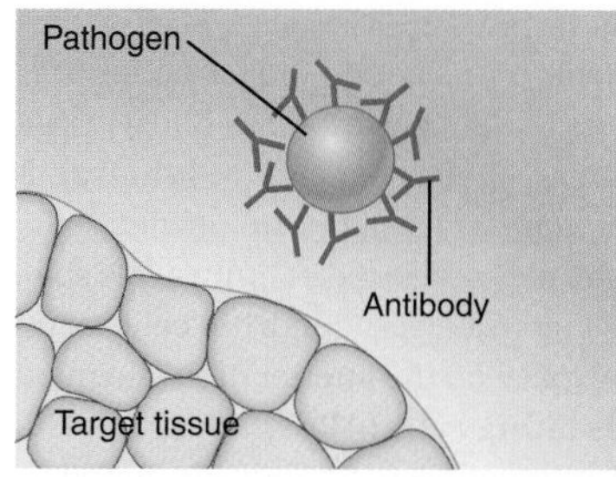

(a) Neutralization: Antibodies block the activity of a pathogen.

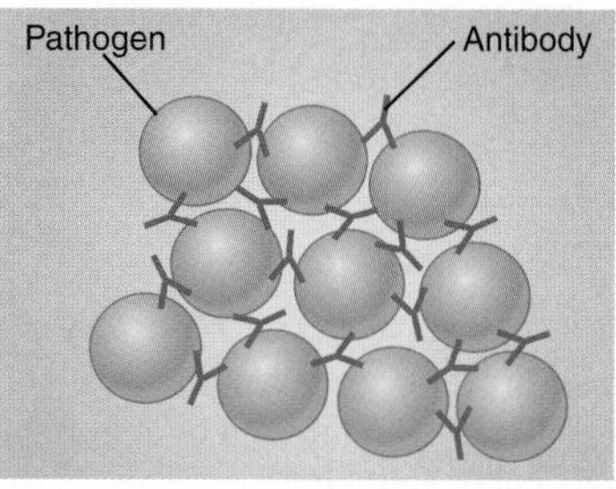

(b) Agglutination: Multiple pathogens are aggregated by antibody molecules.

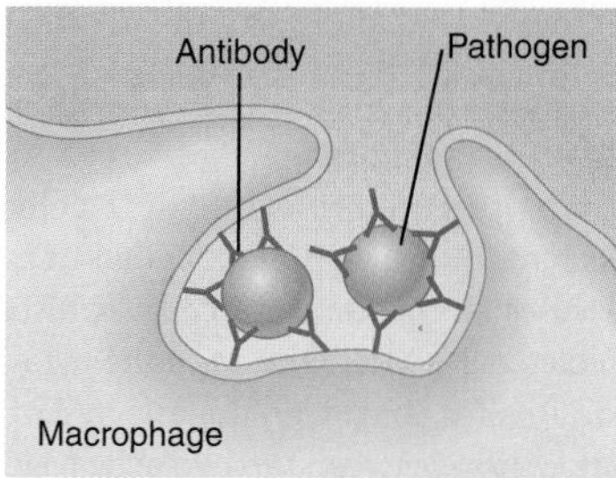

(c) Opsonization: Pathogens bound by antibodies are more efficiently engulfed by phagocytes.

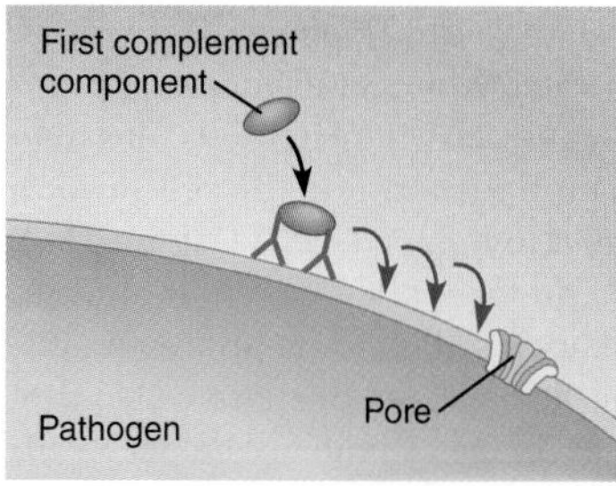

(d) Complement activation: Antibodies bound to pathogens activate the complement cascade, resulting in lysis of the cell.

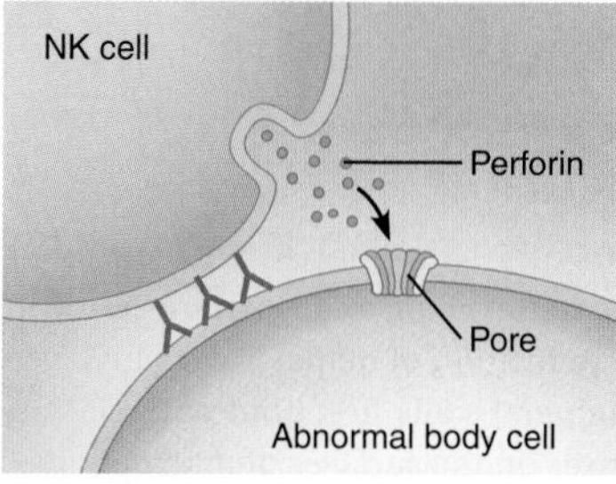

(e) Enhanced NK cell activity: Abnormal body cells that are bound by antibodies are recognized by NK cells and are subsequently lysed.

Figure 16 **Antibody-mediated mechanisms of antigen disposal.**

Complement Activation

Both IgM and IgG antibodies can activate the complement system, which brings about the lysis of bacteria to which the antibodies are bound (Figure 16d). Although the complement system can also be activated when certain complement proteins bind directly to some kinds of bacteria (the nonspecific, lectin, and alternative pathways), the specific, classical pathway is effective against almost any bacterial cell that has been marked with antibodies. Regardless of the initiating event, not only do activated complement proteins bring about the destruction of bacteria, but many of these proteins also help to advance and regulate inflammation.

Enhanced Activation of Natural Killer Cells

Finally, IgG antibodies can enhance the nonspecific killing action of NK cells (Figure 16e). Recall that NK cells broadly recognize abnormal features of tumor cells and virus-infected cells and then produce membrane-perforating molecules that lead to the lysis of these cells. Often these abnormal body cells also possess abnormal surface molecules that can stimulate the production of antibodies. The antibodies mark the cells for death—in this case, death by NK cells. Like phagocytes, NK cells have surface receptors that bind with the constant regions of IgG. In this way, antibodies provide a link between a specifically targeted, abnormal cell and the NK cell, serving once again to focus nonspecific responses upon a particular foreign substance.

Quick Check 5

1. Explain two differences between plasma cells and memory B cells.
2. What is the difference between T-dependent antigens and T-independent antigens? Why are no memory B cells generated in response to a T-independent antigen?
3. Draw an IgG molecule, and list the four other classes of antibodies.
4. State five ways in which antibodies help eliminate antigen.

5 Cell-Mediated Immunity

Antibodies defend against invaders and other antigens floating free in the blood and lymph, but they cannot access antigens located inside cells, such as pathogen-infected cells or tumor cells. T cells, in contrast, make contact with and respond to body cells that are infected or otherwise abnormal. Because T cells require direct contact between them and their targets, their responses are described as being *cell-mediated.* We begin our discussion of cell-mediated responses by examining the roles of T cells in this type of immunity.

Roles of T Lymphocytes in Cell-Mediated Immunity

Three major types of T lymphocytes have been identified: helper T cells, cytotoxic T cells, and suppressor T cells. **Helper T cells,** the primary regulators of immune responses, operate indirectly by secreting cytokines that enhance the activity of B cells, cytotoxic T cells, suppressor T cells, and helper T cells themselves. In addition, helper T cells produce cytokines that enhance the actions of macrophages and NK cells, which are essential to nonspecific defenses. **Cytotoxic T cells,** in contrast, are directly responsible for cell-mediated immunity in that they kill cells infected by viruses or intracellular bacteria and cells that are otherwise abnormal (such as cancer cells and transplanted cells). **Suppressor T cells** are not well understood, but they are thought to produce cytokines that suppress the activity of B cells, helper T cells, and cytotoxic T cells.

All three types of T cells have antigen receptors (T cell receptors [TCRs]) that detect foreign antigens on body cells, but they do so only when these antigens are associated with a special class of normal self-identifying proteins known as **major histocompatibility complex (MHC) molecules.** This type of recognition differs from the way in which B cells recognize antigens, in that B cells and the antibodies they secrete are able to bind to epitopes in their native forms (for instance, as they exist on the surface of a bacterium). For T cells to be activated, however, their antigen receptors must contact an MHC molecule on the surface of some other body cell that is bound to a small fragment of antigen. Thus MHC molecules must first bind to a foreign antigen that is present within a body cell; they must then transport the antigen to the cell surface. This process, which is called **antigen presentation,** is necessary for T cells to interact with the antigen. Before we examine T cell responses in greater detail, let's take a closer look at MHC molecules and their roles in T cell maturation, antigen presentation, and T cell activation.

MHC Molecules: Markers of Self

Major histocompatibility complexes consist of more than 100 genes (located on chromosome 6) that code for MHC molecules. Although MHC molecules have the same immune function in all people, each person's MHC molecules are unique to himself or herself; it is virtually impossible for the tissues of any two people (except those of identical twins) to have the same set of MHC molecules, also called *human leukocyte antigens* (HLA) or *tissue type.* That MHC molecules differ from person to person makes them markers of an individual, or *self.* While MHC molecules are critical in preventing a person's immune system from attacking his or her own cells, these molecules are also responsible for rejection of transplanted skin grafts and organs that are foreign to the recipient; specifically, the MHC molecule has antigens that are attacked by the recipient's immune system. In fact, the existence of MHC molecules was discovered during investigations of graft rejection. These proteins play a major part in determining whether transplanted tissue is accepted as self (histocompatible: *histo* = "tissue") or rejected as foreign. Blood compatibility is more likely to be observed than tissue compatibility, because erythrocytes have no MHC molecules.

Two classes of MHC molecules that differ in structure and the cells that express them allow the immune system to identify "self" from foreign matter, but by different mechanisms involving antigen presentation to the immune system. **Class I MHC** molecules are glycoproteins found on the surfaces of all nucleated cells—that is, on almost every cell of the body, although they are found in greater levels on the membranes of leukocytes. Class I MHC molecules present endogenous (intracellular) antigens to cytotoxic T cells; such antigens develop in tumor cells and virus-infected cells. In contrast, **class II MHC** molecules are found on only a few specialized cell types, including macrophages, dendritic cells, activated B cells, and the cells that make up the interior of the thymus, all of which are **antigen-presenting cells** (APCs). Class II MHC molecules present to helper T cells endogenous antigens (found in the extracellular fluid) that are taken into the cell by phagocytosis. We look next more closely at how the MHC molecules carry out their functions.

The Role of MHC Molecules in Antigen Presentation and T Cell Activation

Each MHC molecule possesses a binding site that can bind to a variety of foreign antigen fragments. Thus, as a newly synthesized MHC molecule makes its way to the surface of an infected or abnormal cell, it can capture an antigen fragment in its binding site and carry it out to the surface of the cell. How the MHC molecule captures and presents the antigen and which T cell class recognizes the antigen vary for the two classes of MHC molecules.

Class I MHC molecules capture antigens synthesized within infected cells or tumor cells and transport these antigens to the cell surface, where they are presented to cytotoxic T cells (**Figure 17a**). Cytotoxic T cells have a surface protein called CD8 that binds to class I MHC molecules; thus cytotoxic T cells are also called CD8 cells. With the infected cell's MHC bound to the cytotoxic T cell's CD8, the cytotoxic T cell is able to kill the infected or tumor cell. In addition, the interaction between the two cells triggers cytotoxic T cell proliferation (with the help of IL-2 secreted by activated macrophages).

Class II MHC molecules are found in phagocytes, which take foreign antigens into the cell through phagocytosis or receptor-mediated endocytosis. The class II MHC molecules bind to the internalized antigen and transport it to the cell surface, thereby presenting the antigen to helper T cells (Figure 17b). Helper T cells have a surface protein called CD4 that binds to class II MHC molecules, enhancing the interaction with antigens presented on APCs. Thus helper T cells are also called CD4 cells. Exposure to the presented antigen stimulates helper T cells to secrete cytokines that induce and regulate other immune responses. Importantly, helper T cells do not kill the APCs, which are immune cells.

Helper T Cell Activation

Like B lymphocytes, millions of T lymphocytes with different specificities normally circulate in the blood and lymph or reside in peripheral lymphoid tissues. Activation of helper T cells involves two simultaneous events: Helper T cells first bind with class II MHC–foreign antigen complexes on the surfaces of macrophages, dendritic cells, and B cells, and then helper T cells receive from

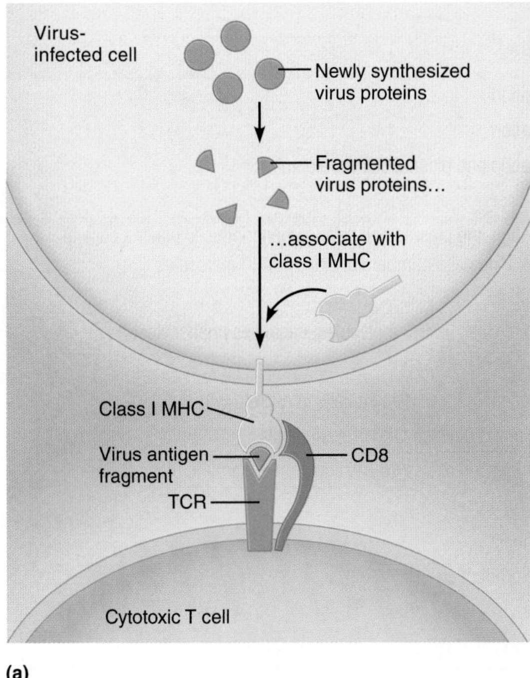

(a)

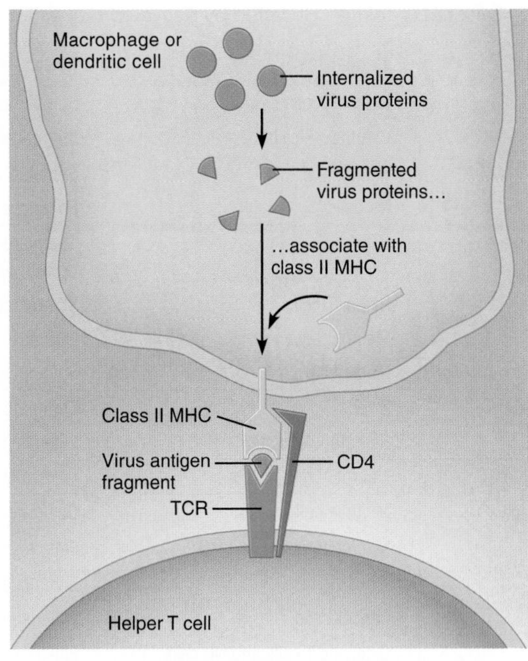

(b)

Figure 17 Presentation of antigens to T cells by major histocompatibility complex (MHC) molecules. (a) Class I MHC molecules, which are made by all nucleated cells of the body, capture fragments of viral (or bacterial) antigens synthesized within an infected cell and transport them to the cell surface. A cytotoxic T cell then binds to the infected cell through its T cell receptors (TCRs) and CD8 molecules. **(b)** Class II MHC molecules, which are made by macrophages, dendritic cells, and activated B cells, capture fragments of foreign antigens internalized by phagocytosis or receptor-mediated endocytosis, respectively, and transport them to the cell surface. A helper T cell then binds to the presenting cell through its TCR and CD4 molecules.

these cells an inducing signal in the form of IL-1. As a result, the helper T cells proliferate and differentiate. Some of the daughter cells begin to secrete cytokines, while a small proportion of them become long-lived memory T cells. Like humoral responses, cell-mediated responses can take up to 17 days to develop, especially upon first exposure to an antigen. Once memory T cells are present in the body, however, subsequent responses to that same antigen occur more quickly and are much more vigorous.

Activated helper T cells secrete several types of cytokines that help stimulate and regulate the activities of other helper T cells, B cells, cytotoxic T cells, macrophages, mast cells, NK cells, and hematopoietic stem cells (Table 1). Thus helper T cells appear to be the central coordinators of immune responses—which is why the depletion of helper T cells, a primary characteristic of *acquired immune deficiency syndrome (AIDS),* is so devastating to the immune system (see **Clinical Connections: AIDS**).

Cytotoxic T Cell Activation: The Destruction of Virus-Infected Cells and Tumor Cells

A cytotoxic T cell becomes activated to kill when two events occur simultaneously: The cytotoxic T cell binds with a class I MHC–foreign antigen complex on the surface of a virus-infected cell and it receives an inducing signal (in the form of IL-2) from a helper T cell. Once activated, the cytotoxic T cell releases perforins, which cause the target cell to lyse. The secretion of such lysis-inducing proteins is a common phenomenon in immune reactions; we previously encountered it in the complement cascade and in the way NK cells destroy their target cells. Cytotoxic T cells also release proteins called **fragmentins,** which first gain entry into target cells through the perforin-induced pores and then work their way inside the cells to bring about their death through apoptosis. Pathogens released from the dead cell are rapidly eliminated by nearby phagocytic cells or are targeted for destruction by antibodies.

In similar fashion, cytotoxic T cells also defend against tumor cells. These cells sometimes possess distinctive molecules, called *tumor antigens,* that are not present on normal cells. The class I MHC molecules that normally exist on tumor cells present fragments of these antigens to cytotoxic T cells, thereby initiating their killing response. Note that some cancers and viruses (Epstein-Barr virus, for example) actively inhibit the production of class I MHC molecules on affected cells, which enables such cancers and viruses to escape detection by cytotoxic T cells.

As we have seen, effective immune responses arise from multiple direct and indirect interactions among macrophages, helper

TABLE 1 Selected Cytokines Secreted by Activated Helper T Cells

Cytokine	Target cells	Effects on target cells
Interleukin-2	Helper T cells	Stimulates proliferation
	Cytotoxic T cells	Stimulates proliferation
	B cells	Stimulates proliferation and plasma cell development
	NK cells	Enhances activity
Interleukin-4	B cells	Stimulates proliferation and plasma cell development; induces plasma cells to secrete IgE and IgG; increases numbers of class II MHC molecules
	Helper T cells	Stimulates proliferation
	Macrophages	Increases numbers of class II MHC molecules; enhances phagocytosis
	Mast cells	Stimulates proliferation
Interleukin-5	B cells	Stimulates proliferation; induces plasma cells to secrete IgA
	Hematopoietic stem cells	Induces proliferation and development of eosinophils
Interleukin-10	Macrophages	Inhibits cytokine production (helps down-regulate the immune response)
Interferon-γ	Multiple cell types	Confers resistance to viruses
	Macrophages	Enhances phagocytosis
	B cells	Enhances antibody production
	Cancer cells	Inhibits proliferation
	Cytotoxic T cells and NK cells	Enhances killing capacity of cytotoxic T cells and NK cells

T cells, B cells, and cytotoxic T cells. These collective actions are depicted in **Figure 18**.

Quick Check 6

1. List the three major types of T cells, and explain their functions.
2. What are the two classes of MHC molecules? On which types of cells is each class found?
3. Which type of T cell has CD4 on its surface, and with which class of MHC molecules does this cell associate?
4. List four cytokines that are produced by helper T cells, and describe how each helps regulate immune responses.

6 Immune Responses in Health and Disease

We have seen that effective immune responses depend on the interplay of many kinds of cells and molecules. In this section, we explore how the immune system generates immunity and how it responds to transplantation and transfusion. We also see how disease can result from malfunctioning of the immune system.

Generating Immunity: Immunization

In 1798, a physician named Edward Jenner was investigating the spread of smallpox when he noticed that milkmaids in Gloucestershire, England, who had previously been ill with cowpox (a mild disease contracted from cows) usually escaped smallpox infection, even when this disfiguring, often fatal disease was running rampant in their community. Suspecting that exposure to cowpox conferred some sort of protection against smallpox, Jenner began to deliberately inoculate people with cowpox, boldly predicting that this action would bring an end to the terrible smallpox scourge. This treatment came to be known as *vaccination* after the Latin word *vacca,* meaning "cow." Indeed, thanks to intensive worldwide vaccination programs in the 20th century, smallpox has been eradicated.

In vaccination, also known as *immunization,* a safe form of a microorganism or a collection of its components that are not expected to cause disease is introduced into the body, where the inoculum both stimulates an immune response to that pathogen and—even more significantly—induces immunological memory (**Figure 19**). When a vaccinated person subsequently encounters the natural pathogen, his or her body mounts a strong immune response similar to a naturally occurring secondary immune response; a successfully vaccinated person is immune to that pathogen. Beginning in the 20th century, routine immunization of infants and children has proved extremely effective in preventing many infectious diseases (**Table 2**). The term *immunization* is also used to refer to the conferral of immunity that results from a natural infection. Both artificial immunization and natural immunization confer a type of protection referred to as **active immunity,** reflecting the fact that it depends on the ability of the immunized person's immune system to mount a response. Unfortunately, not all infectious agents (for example, the viruses that cause the common cold) are easily managed by vaccination.

Ready-made antibodies to a particular antigen can also be introduced into the body to provide another kind of protection called **passive immunity.** This type of immunization is termed "passive" because it does not require a response from the immunized person's

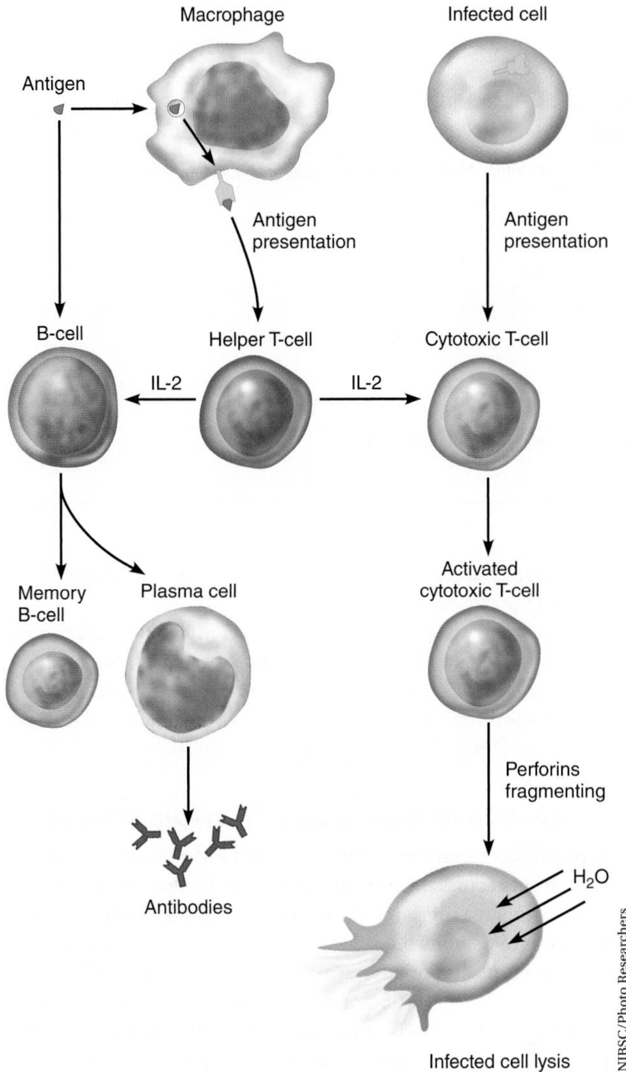

Figure 18 The role of helper T cells in activation of B cells and cytotoxic T cells. A helper T cell is activated by specific contact with a macrophage-presenting viral antigen. The activated helper T cell secretes IL-2, which activates B cells and cytotoxic T cells. B cells proliferate and differentiate into plasma cells that secrete antibodies. Cytotoxic T cells secrete perforins and fragmentins that cause lysis of an infected cell.

immune system. Because no foreign antigens are introduced into the body in this procedure, the person's B cells do not make antibodies, and no memory B cells are made. Although the introduced antibodies help fight an ongoing infection, passive immunization does not induce long-term immunity. The antibodies typically used for passive immunization are first isolated from people who are already immune to a particular disease; these antibodies are then injected into another person's body, conferring in the process a short-lived but immediate protection from that disease. A person who has been bitten by a rabid animal, for example, may be injected with antibodies collected from other people who have been vaccinated against rabies. This measure is important because rabies may progress more rapidly than the time it takes for a person to mount an active immune response. Most individuals exposed to rabies are given both passive

and active immunizations; the injected antibodies fight the virus for a few weeks, and then the person's own immune response, which is induced by both the immunization and the infection itself, takes over.

Passive immunity also occurs naturally when IgG antibodies in the blood of a pregnant woman cross the placenta and reach the fetus. In addition, IgA antibodies (see Figure 15) are passed from a mother to her nursing infant in breast milk, especially in the earliest form of milk, known as colostrum. Passive immunity persists only as long as the transferred antibodies do—a few weeks to a few months—but it provides the infant with at least some protection from infections until the baby's own immune system has had a chance to mature.

Apply Your Knowledge

Leukocyte adhesion deficiency (LAD) is a hereditary disease characterized by a lack of functional integrins in leukocytes. What effect does LAD have on inflammation, and which symptoms would you expect to see from a person with LAD?

Roles of the Immune System in Transfusion and Transplantation

The immune system's ability to distinguish self from nonself, although essential to a healthy immune system, limits our ability to transfuse blood or to transplant tissue between individuals. For this reason, material from the donor must be matched, as closely as possible, to the recipient so as to minimize immune reactions. In addition, transplant recipients are given medicines that suppress the ability of their immune system to react against the foreign tissue. Note that the body's hostile response to an incompatible transfusion or transplant is not a disorder of the immune system, but rather a normal response to foreign antigens.

Blood-Group Compatibility

Blood is classified into different types (designated by the letters A, B, AB, or O) according to the presence or absence of certain antigens on the surface of a person's red blood cells. An individual who has type A blood possesses red blood cells with surface antigens known as A antigens. These antigens are perceived as self by the individual's immune system; they are not antigenic to their "owner." Similarly, B antigens are found on type B red blood cells, and both A antigens and B antigens are found on type AB red blood cells. Type O red blood cells possess neither antigen.

An individual with type B blood does not produce antibodies to B antigen because that antigen is recognized as "self." This person will, however, have antibodies to A antigen (called anti-A antibodies), *even if the individual has never been exposed to type A blood.* The presence of circulating antibodies specific for A antigen seems to suggest that the individual's B lymphocytes have detected and responded against type A blood. In fact, what has occurred is that antibodies to the A antigen have been produced in response to bacteria that are normally present in the body and possess epitopes that are

TABLE 2 Recommended 2012 Vaccinations by the Centers for Disease Control and Prevention

Vaccine	Regimen
Hepatitis B (HepB)	Sequence of 3 shots at birth, 1–2 months, and 6–18 months
Rotavirus (RV)	Sequence of 3 shots at 2, 4, and 6 months
Diphtheria, tetanus, pertussis (DTaP)	Sequence of 6 shots at 2 months, 4 months, 6 months, 15–18 months, 4–6 years, and 11–12 years, then tetanus every 10 years
Haemophilus influenzae type b (Hib)	Sequence of 4 shots at 2, 4, 6, and 12–15 months
Pneumococcal	Sequence of 4 shots at 2, 4, 6, and 12–15 months
Inactivated polio virus	Sequence of 4 shots at 2 months, 4 months, 6–18 months, and 4–6 years
Influenza	Annually starting at 6 months
Measles, mumps, and rubella (MMR)	Two shots at 12–15 months and 4–6 years
Varicella (chicken pox)	Two shots at 12–15 months and 4–6 years
Hepatitis A (HepA)	Two shots at 12–24 months and then 6–18 months after the first short
Meningococcal (MCV)	One shot at 11–12 years and a booster at 16 years (can be used in at-risk infants)

very similar to the blood group antigens A and B. A person with type B blood does not make antibody to the B-like bacterial antigens—those are too much like self—but the individual does make antibodies to A-like bacterial antigens, which are perceived as foreign by his or her immune system. Thus, in an individual with type B blood, the anti-A antibodies that are constantly circulating in the blood induce an immediate and devastating *transfusion reaction* in the event that this individual receives a transfusion with type A or type AB blood, both of which contain the A antigen. For these reasons, a person with type O blood is considered to be a *universal donor* because any anti-A or anti-B antibodies that may be present in the recipient will find no target on the type O donor cells. In contrast, a person with type AB blood is considered to be a *universal recipient;* because such an individual lacks antibodies against A antigen or B antigen, transfusion of blood of any blood group into this individual will not induce a transfusion reaction.

One more issue must be considered with blood donations: When whole blood is transfused, it contains antibodies that could, in a mismatched transfusion, attack the *recipient's* red blood cells. This usually is not a problem because *packed* red blood cells, rather than whole blood, are typically used in transfusions.

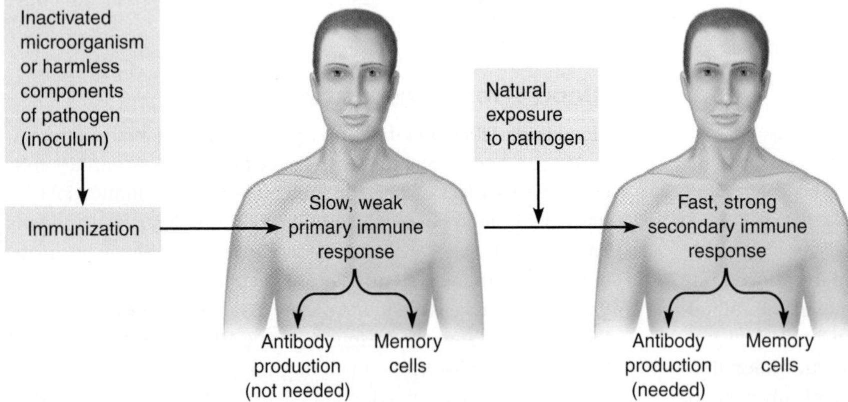

Figure 19 Acquisition of long-term immunity through vaccination (immunization). Introduction into the body of the inoculum stimulates immune responses that both target the introduced materials for destruction and generate memory cells. On subsequent natural exposure to the pathogen, memory cells mount a rapid secondary response that prevents or diminishes the symptoms of the disease.

Q Five-year-old Ivan receives an immunization against chicken pox and develops antibodies and memory cells specific for the virus. Which of these two components of his immune response is more important for Ivan, and why?

Tissue Grafts and Organ Transplantation

As previously described, MHC molecules are present in many different forms across the human population. In turn, two people are unlikely to have the same HLA profile. Thus HLA molecule mismatches are responsible for stimulating the rejection that occurs in tissue grafts and organ transplants. Note that MHC molecules do not play a role in transfusion reactions because red blood cells do not have MHC molecules.

To minimize the chance of rejection, physicians attempt to match the HLA molecules of the donor and the recipient as closely as possible, using a procedure called *tissue typing.* When a recipient has no identical twin (who would have exactly the same tissue type), siblings usually provide the closest HLA match.

In addition to testing and matching HLA molecules, physicians attempt to minimize the chance of rejection of a graft or transplant by prescribing drugs that suppress the recipient's immune responses. The complication with this strategy is that it renders the recipient more susceptible to infections and cancer during the course of treatment. The main effect of drugs such as cyclosporin A, FK506, and sirolimus, which have greatly improved the success of organ transplants, is to inhibit production of IL-2, which in turn inhibits

CLINICAL CONNECTIONS

Aids

In 1981, an outbreak of *Kaposi sarcoma* occurred among gay men in California and New York. That same year, a specific form of pneumonia, *pneumocystis,* broke out in New York City. In 1982, the Centers for Disease Control and Prevention (CDC) linked the outbreaks to blood and called the disease AIDS (acquired immune deficiency syndrome). The virus that caused AIDS, human immunodeficiency virus (HIV), was first discovered in France in 1984, and in 1985 it was confirmed by researcher Robert Gallo to be the cause of AIDS.

In 1982, 1600 people in the United States had AIDS and approximately 700 of them died. In 1995, AIDS was the number one cause of death in people ages 25–44. The World Health Organization (WHO) estimates that 25 million people worldwide have died of AIDS since 1981. In 2008, 33.4 million people in the world had AIDS or were infected with HIV, including 2.1 million children younger than the age of 15. Each year, approximately 2.7 million more people are infected with HIV, and 1.8 million people die of AIDS. Clearly, the AIDS epidemic threatens people worldwide.

HIV is a retrovirus composed of RNA, a viral core consisting of protein p24, and a fatty coat called a viral membrane. Inside the viral membrane is a protein matrix consisting of protein p17. Extending out from the viral membrane are spikes formed from proteins gp120 and gp41. The RNA contains nine genes that code for viral proteins, including those necessary for HIV to infect cells and to replicate.

HIV can replicate only when it is inside human cells. Current theories propose that this virus contacts and infects dendritic cells in mucous membranes. The dendritic cells then move to the lymphatic organs, where they infect other immune cells, such as helper T cells. HIV spikes bind to helper T cell CD4 receptors, allowing the virus to enter the cell, minus the viral membrane. Once inside the cell, an HIV enzyme, *reverse transcriptase,* synthesizes DNA from the viral RNA. This new DNA then enters the cell nucleus, where another HIV enzyme, *integrase,* splices the viral DNA into the human DNA.

A virus-infected cell may stay dormant for years before showing signs of disease. In the case of helper T cells, only when the cell is activated will it start transcribing the viral DNA into mRNA, and then translating the mRNA to produce more viral particles and HIV proteins. Most translated viral proteins are long, so another HIV enzyme, *protease,* breaks these proteins into smaller polypeptides that come together and form new viral particles in a process called *assembly.*

As long as the virus stays in one cell, it does not harm the body. Following assembly, however, the viral particles go through a process called *budding,* during which viral particles push their way through the helper T cell membrane, and in the process, take some membrane with it; this material then becomes the fatty coat for this virus. Within the fatty coat are the components necessary to bind to another CD4 receptor on another helper T cell. In this way, the virus spreads within the body.

Transmission of the virus to other individuals may occur through sexual contact, blood transfusions, and needle sharing. The virus can also be transmitted from mother to child through exchange across the pregnant woman's placenta to her fetus or through breast milk given to an infant. HIV infection is not spread by donating blood, casual contact, or mosquito bites.

Once a person is infected with HIV, he or she usually shows acute symptoms shortly after infection, but then may go into a latent period in which no symptoms are apparent for several years. Even though the person may be asymptomatic, he or she is still contagious, however. Once the disease starts to develop, HIV infection progresses from symptomatic HIV infection to advanced HIV infection or AIDS. Diagnosis of AIDS includes a helper T cell count of fewer than 200 cells/mm^3 (normal helper T cell levels are in the range of 500–1600 cells/mm^3).

AIDS compromises the immune system, making an individual more susceptible to infectious diseases. The precise symptoms experienced vary based on the stage of the disease. During acute HIV infection, a person may show flu-like symptoms, including fever, muscle aches, sore throat, headache, swollen lymph nodes, and rash. The acute phase is followed by an asymptomatic latent period. Symptoms then reemerge in the form of fever, fatigue, swollen lymph nodes, diarrhea, weight loss, and shortness of breath. At this point in the disease, a person is more susceptible to opportunistic infections including herpes simplex virus (cold sores on mouth or blisters in genital areas), herpes zoster (shingles), *Kaposi sarcoma* (herpes virus-induced cancer of the skin, lungs, or bowel), and tuberculosis. Once AIDS develops (which is almost certain to occur within about 10 years of initial infection if left untreated), symptoms include chills and fevers that last several weeks, night sweats, shortness of breath, chronic diarrhea, white lesions on the tongue, persistent fatigue, blurred vision, weight loss, and skin rashes. Opportunistic infections that often strike at this stage include pneumonia, meningitis, and encephalitis.

There is no cure for AIDS, but treatments have been developed that can slow its progression. Most of these measures focus on stopping viral replication in the body through antiretroviral therapy (ART) medicines. Usually a combination of ARTs are used, each working through a different mechanism. For example, ART drugs include non-nucleotide reverse transcriptase inhibitors, nucleotide reverse transcriptase inhibitors, protease inhibitors, integrase inhibitors, and entry or fusion inhibitors. Although these treatments can decrease the symptoms of HIV infection, a person still remains contagious. Studies are ongoing regarding whether or not ART can decrease the risk of transmission.

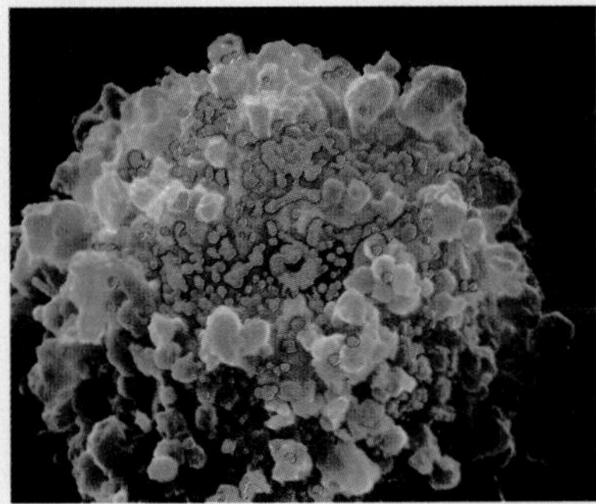

Colored scanning electron micrograph (SEM) of a T-lymphocyte blood cell (green) infected with the Human Immunodeficiency virus (HIV)(red), the causative agent of AIDs.

Critical Thinking Questions

1. Explain the processes by which a person becomes infected with HIV and the spread of the virus throughout the body.
2. List the stages of HIV infection and with their symptoms.
3. What are the treatments available for AIDS? How do they impair the spread of the virus?

B cell and T cell activation and the mounting of immune responses. Because of their immune-suppressing effects, antirejection drugs have been used to treat autoimmune diseases, albeit with only limited success. Unfortunately, one of the unsuccessful cases has been the treatment of diabetes mellitus. These antirejection drugs, when used in transplant patients, actually increase the likelihood that the transplant recipient will develop type 2 diabetes mellitus.

Transplantation of bone marrow, which is often used successfully to treat leukemia and other cancers, as well as various blood-cell diseases, has a different type of rejection risk. Before receiving the transplant, the recipient is typically treated with radiation to eliminate his or her own bone marrow cells. Such treatment both eliminates the abnormal cells and effectively inactivates that person's immune system, which minimizes the likelihood that the recipient will reject the graft. The great danger in this procedure is that the donated marrow, which contains lymphocytes, will mount immune responses against the recipient, an example of what is known as a *graft versus host reaction.* The intensity of such a reaction can be minimized if the HLA molecules of the donor and the recipient are well matched, which is why bone marrow donor programs continually seek volunteer donors the world over. Because of the diverse array of HLA molecules within the human population, a diverse pool of potential donors is needed.

Quick Check 7

1. Which cell surface molecules are tested in the process of tissue typing?
2. What is the name of the phenomenon in which a grafted tissue mounts an immune response against a recipient?

Immune Dysfunctions

The complex, highly regulated interplay of foreign substances with lymphocytes and other cells that make up the immune system provides us with extraordinary protection from infections. In addition, a growing body of evidence suggests that immune function is intimately associated with nervous system and endocrine system functions. When this delicately balanced network of interacting cells and molecules is disrupted, the effects on the individual can range from the minor inconvenience of some allergies to the serious and sometimes devastating consequences of autoimmune and immunodeficiency diseases. In the following sections we examine what can happen when immune function goes awry.

Allergy

Allergies (also known as hypersensitivity reactions) are exaggerated responses to certain environmental antigens known as allergens. The most common allergies involve antibodies of the IgE class (see Figure 15). Allergies occur in some people who are genetically predisposed to produce more than the usual amount of IgE when they are exposed to allergens. If an individual produces high levels of IgE in response to pollen, the result is an allergic reaction commonly known as hay fever. As shown in **Figure 20**, some of these IgE antibodies do not bind to the pollen but instead attach by their tails to mast cells. When the person is subsequently exposed to pollen, the pollen grains bind to the antigen-binding sites of these IgE antibodies, causing adjacent antibody molecules to become cross-linked and inducing the mast cell to degranulate—that is, to release histamine and other inflammatory agents into the surrounding fluid. Following its release, histamine causes dilation and increased permeability of small blood vessels in the immediate vicinity. These inflammatory events lead to the typical symptoms of hay fever: sneezing, runny nose, tearing eyes, and breathing difficulty, which can result from histamine-induced contraction of smooth muscle in respiratory airways. Drugs that act as antihistamines diminish allergy symptoms by blocking histamine receptors located on blood vessel endothelial cells and smooth muscle cells.

The most serious consequence of an acute allergic response is *anaphylactic shock,* a life-threatening reaction to injected or ingested allergens. Anaphylactic shock results from widespread degranulation of mast cells throughout the body, which triggers abrupt dilation of peripheral blood vessels. This dilation is not just confined to a restricted area, as in hay fever, but is widespread and, therefore, causes a precipitous drop in total peripheral resistance and mean arterial pressure. As a consequence of the drop in pressure, death may occur within a few minutes. Bee venom and penicillin are examples of allergens that can trigger anaphylactic shock in people who are extremely allergic to them. Some susceptible individuals carry syringes containing the hormone epinephrine as a prophylactic measure. When injected into the body, epinephrine counteracts the allergic response by stimulating increased cardiac output and constriction of blood vessels, both of which tend to raise blood pressure back toward normal levels.

Autoimmune Diseases

When the immune system loses tolerance to self and begins to react against normal molecules of the body, autoimmune disease can result. In a disease called *systemic lupus erythematosus* (often called lupus), for example, the immune system generates antibodies (known as auto-antibodies) against all sorts of self molecules, resulting in a widespread array of signs including skin rashes, fever, arthritis, and kidney dysfunction. *Rheumatoid arthritis*, another antibody-mediated autoimmune disease, causes painful inflammation of and eventual damage to the cartilage and bone of joints. In *insulin-dependent diabetes mellitus*, another autoimmune disease, the insulin-producing beta cells of the pancreas are targeted by cell-mediated immune responses. Another autoimmune disease is *multiple sclerosis* (MS), the most common chronic neurological disease in developed countries. In this disease, T cells that have a propensity to attack normal myelin are thought to infiltrate the central nervous system and cause demyelination of nerve fibers, thereby precipitating a number of serious neurological abnormalities (**Clinical Connections: Multiple Sclerosis**).

The causes of autoimmunity are varied and complex. Although much remains to be learned about these diseases, we know that people who inherit particular MHC molecules also are more likely to develop certain autoimmune diseases. For example, individuals who inherit certain class II MHC molecules are at higher risk of

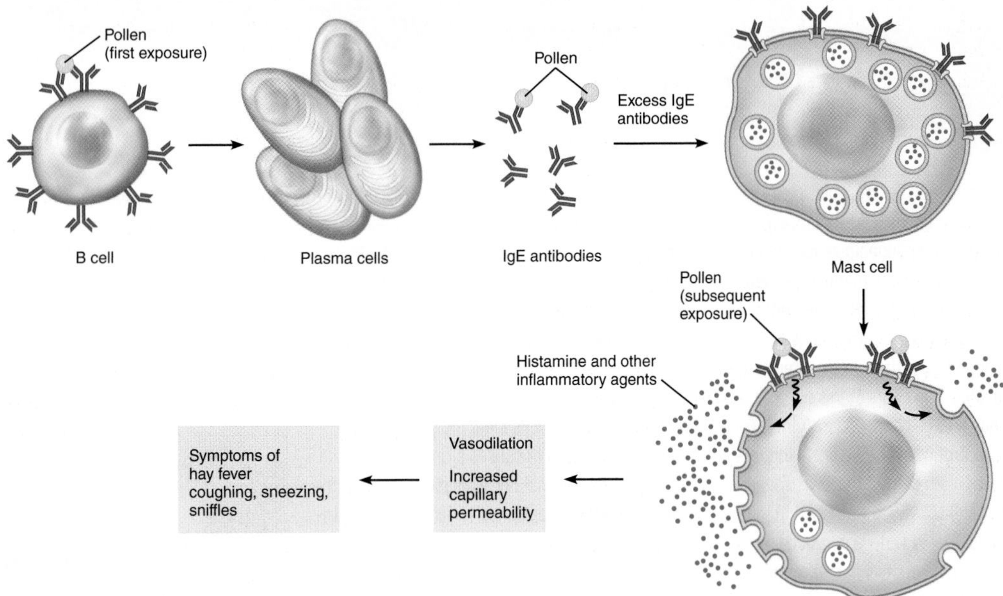

Figure 20 Events in hay fever, an allergic response. In response to exposure to allergens such as pollen, an allergic person produces abundant IgE molecules; excess IgE binds to the surface of mast cells. On subsequent exposure to the same kind of pollen, the allergens and bound IgE bind, cross-linking the IgE and inducing rapid degranulation (histamine release). Histamine induces increased vascular permeability and the resulting symptoms of hay fever.

Q Anaphylactic shock is characterized by a drop in blood pressure due, in part, to widespread histamine release. Which actions of histamine contribute to this decrease in blood pressure?

developing insulin-dependent diabetes mellitus than are members of the general population.

Immunodeficiency Diseases

In allergies or autoimmune diseases, problems arise due to overactivity in the immune system; conversely, immunodeficiency diseases arise as a result of underactivity in the immune system. There are almost as many immunodeficiency diseases as there are functions within the immune system. Many congenital immunodeficiencies affect the function of either humoral or cell-mediated immunity, but in *severe combined immunodeficiency disease* (SCID), both branches of the immune system fail to function. For people with this genetic disease, long-term survival may require a transplant of healthy bone marrow that will continue to supply functional B cells and T cells. One type of SCID is caused by deficiency of the enzyme adenosine deaminase (ADA), which plays a role in the breakdown of the DNA building blocks known as purines. ADA is particularly active in lymphocytes, and its deficiency results in the accumulation of lethal by-products. ADA deficiency has been treated with some success by gene therapy (**Clinical Connections: Gene Therapy for Severe Combined Immunodeficiency Disease**); in this procedure, an individual's own bone marrow cells are removed, provided with a functional ADA gene, and then returned to the body.

Immunodeficiency is not always a congenital condition; an individual may also develop immune dysfunction later in life. For example, certain cancers can cause immunodeficiency by suppressing the immune system. A prime example is *Hodgkin's disease,* which damages the lymphatic system. Another well-known and devastating acquired immunodeficiency is AIDS (see Clinical Connections: AIDS).

The Role of Stress in the Immune Response

Clinical and personal observations have long suggested a correlation between psychosocial factors and immune function. Indeed, several studies have suggested that a positive outlook is associated with improved health in both cancer and AIDS survivors; conversely, people hospitalized for depression have been found to exhibit diminished immunity. The mind-immunity relationship arises from multiple complex interactions that exist among the immune system, the nervous system, and the endocrine system.

Many of these interactions are well established. Some steroid hormones are known to suppress immune responses; corticosteroids, for example, reduce the number and activity of immune cells and are potent anti-inflammatory agents. Likewise, research on wild animals has demonstrated that chemicals such as PCB and DDT, which mimic naturally occurring endocrine hormones, reduce the effectiveness of immune responses and increase the

CLINICAL CONNECTIONS

Multiple Sclerosis 

Autoimmune diseases such as multiple sclerosis (MS) occur when the immune system wrongly identifies normal body components as foreign and mounts an attack against them. In multiple sclerosis, the tissue under attack is myelin, the material that wraps around and insulates the axons of neurons and is abundant in the white matter of the brain and spinal cord. The name of this disease refers to the many lesions or *scleroses* (from the Greek word for "scarring" or "hardening") that result from the destruction of myelin *(demyelination).*

The symptoms of MS range from mild to severe and may appear in various combinations, depending on where the demyelination occurs. Symptoms typically include blurred vision (or blindness in one eye), muscle weakness, and trouble maintaining balance while walking (ataxia). These symptoms may worsen, leading to a complete inability to walk or stand. Other symptoms include muscle spasticity; tremors; impairment of pain, temperature, and touch sensations; speech disturbances; vertigo; and fatigue. MS is usually diagnosed between the ages of 20 and 40 and occurs more often in women than in men. Approximately 250,000 to 350,000 people are currently living with this disease in the United States.

Because the cerebrospinal fluid of MS patients has been shown to contain activated cytotoxic T cells (which are absent from the CSF of healthy people), it is likely that these cells are primarily responsible for the demyelination that occurs in MS. Despite intensive research, it is still not clear *why* the body's cytotoxic T cells turn against its own myelin. One possibility is that an infection of some sort either induces inappropriate immune responses to myelin or causes a breakdown in the blood-brain barrier that normally prevents lymphocytes from leaving capillaries within the CNS. Lymphocytes have no opportunity to develop tolerance to self-antigens (including myelin proteins) that are located in and around the brain and spinal cord; when lymphocytes are subsequently exposed to these antigens, they perceive them as foreign and mount defensive responses, causing irreparable CNS damage. Scientists have studied a number of infectious agents (including measles, mumps, and rubella viruses) that may induce the autoimmune responses typical of MS but have been unable to implicate any particular causative agent.

Treatment and perhaps even prevention of this and other devastating autoimmune diseases require better understanding of the precise mechanisms by which B cells and T cells respond against foreign antigens while remaining tolerant of the body's own tissues.

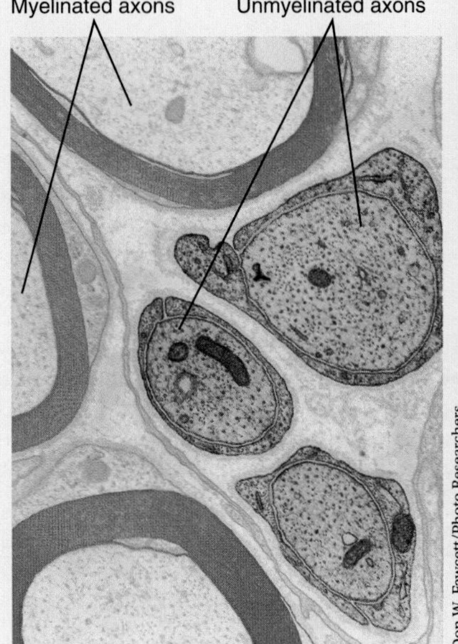

Micrograph of myelinated and unmyelinated axons.

Critical Thinking Questions

1. What is the cause of multiple sclerosis? What aspects make this an autoimmune disease?
2. Describe the symptoms and disease progression of multiple sclerosis.
3. Explain the current thinking on possible reasons why the immune system attacks the myelin sheaths of the nervous system.

incidence of infection. By contrast, both growth hormone and thyroid hormone play important roles in T cell development and function.

The nervous, endocrine, and immune systems interact in a variety of ways. For example, autonomic nerves innervate lymphoid organs such as the bone marrow and the spleen, often terminating at B cell and T cell clusters, and lymphocytes bear receptors for epinephrine and acetylcholine. Additionally, lymphocytes and macrophages secrete cytokines that affect both the endocrine and central nervous systems. For example, IL-1 secreted by macrophages induces fever, suppresses appetite, inhibits thyroid function, and stimulates the release of pituitary hormones.

The correlation between stress and immune function continues to be demonstrated in various ways. One study that tested college students for immune functions just after a vacation and again

 **Functional Fact** during final exams found that their immune systems were impaired in various ways (for example, plasma interferon levels were lower) during exam week.

Quick Check 8

1. Which class of antibody is induced in allergies such as hay fever? How do these antibodies trigger the symptoms of allergy?
2. Does rheumatoid arthritis result from a humoral autoimmune response or from a cell-mediated autoimmune response?
3. Name two lymphoid organs that are innervated by the autonomic nervous system.

CLINICAL CONNECTIONS

Gene Therapy for Severe Combined Immunodeficiency Disease

Severe combined immunodeficiency disease (SCID) is an inherited disorder caused by mutations in genes coding for proteins critical to both humoral and cell-mediated immunity. It affects both males and females, with a frequency of about three per million live births. People with this disorder exhibit increased susceptibility to all types of infections, requiring their isolation from all infectious agents.

The most common form of SCID is X-SCID, also known as *"bubble-boy" disease.* X-SCID is caused by mutation of a gene coding for the interleukin-2 receptor (IL2RG), which is located on immature blood cells (hematopoietic stem cells) and is necessary for development of B cells, T cells, and natural killer cells. IL2RG is located on the X chromosome. Because females receive two X chromosomes, both chromosomes would have to be mutated to cause SCID. Males, however, receive only one X chromosome, making them more susceptible to X-SCID.

The most commonly used treatment for SCID is bone marrow transplant. (Bone marrow contains hematopoietic stem cells.) Antibiotics can also be administered, and in some forms of SCID, the deficient protein can be injected on a regular basis. These treatments, however, have shown only limited success.

The advent of *gene replacement therapy,* or gene therapy, has offered new hope for those born with genetic diseases such as SCID. The concept is simple: If a gene is defective (that is, if it produces an aberrant product or inadequate levels of it), then introducing a functional copy of the gene—one that directs normal synthesis of the product—should cure the problem. For gene therapy to be successful in treating X-SCID, a normal IL2RG must be introduced into hematopoietic stem cells, and the gene must also be replicated normally when the cells divide. Ideally, cells carrying the new gene should persist in the individual, so that repeated treatments are not needed.

Gene therapy for X-SCID is accomplished in four steps: (1) Hematopoietic stem cells are collected from the blood, bone marrow, or umbilical cord of an infant or child with the disease; (2) the harvested cells are treated with a chemical that causes them to proliferate, thereby generating additional target cells for gene insertion; (3) a healthy gene for the IL-2 receptor is introduced into the cells, often using a nonpathogenic virus as a carrier; and (4) the cells, now producing normal IL-2 receptors, are infused back into the bloodstream of the affected child.

The use of gene therapy to treat SCID is still in the experimental phase and has encountered several setbacks. Early results from clinical trials treating X-SCID were promising, allowing some children to live outside of their protected "bubble" environments. However, scientists remain concerned about the viruses used to carry genes into human cells. In 2003, four of 11 children in a clinical trial in France developed T cell leukemia, and two died from that disease. As a result, the U.S. Food and Drug Administration put a temporary hold on clinical trials of gene therapy utilizing retroviral vectors (retroviruses—the type used in X-SCID gene therapy—produce double-stranded DNA from their single-stranded RNA). After the publication of the French studies, 10 children were treated for X-SCID in London, and all had some temporary relief, but one eventually died of T cell leukemia; as a consequence, these clinical trials were terminated in 2007. While gene therapy seems to hold promise in treating X-SCID, concerns remain about its potential to induce other illnesses, especially cancer, and the inserted genes seem to be short-lived. Scientists are currently working on other mechanisms of inserting genes into human cells.

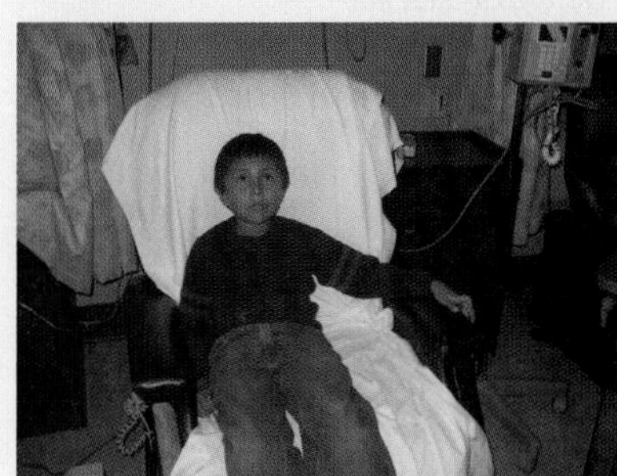

Felicia Fonseca/AP Images

A young boy in Tuba City, AZ, receiving a treatment infusion for SCID.

Critical Thinking Questions

1. What is the cause of X-SCID? Why are men more susceptible than women?
2. Describe the current available treatments for X-SCID.
3. Explain the four steps of gene therapy for X-SCID. How effective has gene therapy been for this disease?

SYSTEMS INTEGRATION

Every cell of every tissue of every organ can be injured or infected, and thus depends on the immune system for protection from disease; this is nowhere more apparent than in immune deficiency disorders such as AIDs. But the immune system also depends on other organ systems to function properly.

We already learned about the widespread locations of immune tissue and cells. From bone marrow, to blood, to lymphatic tissue, to intercellular spaces, immune cells are everywhere a pathogen is present. To get where they are needed requires the coordinated activities of the cardiovascular system, bone marrow, the spleen, lymph nodes, and lymph.

Numerous cytokines, including interleukins, provide local signals to coordinate activity near the site of injury or infection; but cytokines also travel long distances in the blood to areas such as the hypothalamus, signaling it to elevate body temperature or induce fever. The epithelium separating the internal from the external environment, forms both physical and chemical barriers to infectious agents. The liver produces proteins critical to various stages of the immune response and, along with the kidney, degrades toxins for elimination from the body.

In summary, while the immune system actively protects our body, the other organ systems enable it to do so.

CHAPTER REVIEW

SUMMARY

- The components of the immune system—physical barriers, lymphoid tissues, leukocytes and the molecules they produce—work together to generate and regulate immune responses.
- The function of the immune response is twofold: to clear foreign materials from the body and to bring about long-term immunity to infectious diseases.

1 Anatomy of the Immune System

- The physical barriers of the immune response are the skin and the mucous membranes.
- There are five major types of leukocytes: neutrophils, eosinophils, and basophils (the granulocytes), plus monocytes and lymphocytes (the agranulocytes).
- Neutrophils, monocytes, and macrophages (which arise from monocytes) are phagocytic; they engulf and destroy foreign matter and debris.
- Eosinophils and basophils defend against large parasites and are also involved in allergic reactions.
- The lymphocytes consist of B lymphocytes (B cells), and T lymphocytes (T cells), which exhibit specificity, and null cells, which are nonspecific.
- B cells differentiate into plasma cells that secrete antibodies, or immunoglobulins, when they come in contact with antigens.
- Some T cells develop into cytotoxic T cells that destroy infected or abnormal cells by destroying the infected cell's plasma membrane.
- Most null cells are large granular lymphocytes known as natural killer (NK) cells, which mount an important, early defense against viral infections.
- Leukocytes develop to maturity in the central lymphoid tissues (the bone marrow and, in the case of T cells, the thymus).
- The central lymphoid tissues include bone marrow, the thymus, and the fetal liver; once the B and T cells reach maturity, they migrate from the central lymphoid tissues to the peripheral lymphoid tissues.
- The peripheral lymphoid tissues exhibit a netlike architecture that traps foreign matter present in the blood (spleen), lymphatic fluid (lymph nodes), air (tonsils, adenoids), and food and water (appendix and Peyer's patches).

iP Immune: Immune System Overview

iP Immune: Anatomy Review

iP Immune: Common Characteristics of B and T Lymphocytes

2 Pathogens That Activate the Immune Response

- Four classes of pathogens exist: viruses, bacteria, fungi, and parasites.
- Viruses invade certain cells and use them to create more viruses.
- Bacteria cause infection by invasion of tissue and subsequent colonization, and by release of endotoxins or exotoxins.
- Fungal infections occur when the fungi release spores into the air that are either inhaled or contact the skin.
- Parasites invade a host to obtain nutrients, a process that harms the host; the parasites spread infection either through a vector or via the fecal-oral route.

3 Organization of the Body's Defenses

- When an infectious agent breaks through the body's first line of defense (the skin and mucous membranes, for example), it faces both nonspecific defenses and immune responses.
- Nonspecific (innate) defenses provide the body's most rapid defenses against infection or injury.
- The nonspecific defense mechanisms are physical barriers, inflammation, interferons, and the complement system.
- In inflammation, oxygen, nutrients, defensive molecules, and phagocytic cells are drawn to the affected region.
- Interferons, which are secreted from virus-infected cells, protect the surrounding healthy cells from infection, whereas NK cells can recognize and kill virus-infected cells.
- The complement system can be activated by three different pathways: the classical pathway, the lectin pathway, or the alternative pathway.
- These pathways merge to generate the same set of complement proteins.
- B cells and T cells provide for the features of immune responses: specificity, diversity, memory, and self-tolerance.
- There are two types of immune responses: the humoral response and the cell-mediated response.

iP Immune: Humoral Immunity

iP Immune: Cellular Immunity

iP Immune: Innate Host Defenses

MAP iP Overview of Innate and Adaptive Body Defenses

4 Humoral Immunity

- The humoral response is the result of B cell activation.
- After making contact with specific antigen, B cells proliferate and develop into long-lived memory B cells and short-lived plasma cells.
- Whereas memory cells provide for long-term immunity to the antigen, plasma cells secrete antibodies that bind to and target the antigen, and then recruit other defenses (such as phagocytic cells) to destroy it.

iP Immune: Humoral Immunity

5 Cell-Mediated Immunity

- The cell-mediated response occurs when cytotoxic T cells detect a specific antigen presented by a class I MHC molecule (on a virus- or bacteria-infected cell or a tumor cell) and develop into active killers.
- Cytotoxic T cells destroy their targets in two ways: by releasing perforins, which form pores in the stricken cell membrane, and by releasing fragmentins, which enter the cell and induce apoptosis.
- Both the humoral and cell-mediated responses are supported and regulated by cytokines secreted by activated helper T cells.
- Helper T cells are activated to proliferate and to secrete cytokines when they contact a specific antigen presented by a class II MHC molecule (on a macrophage, dendritic cell, or a B cell); at the same time, long-lived memory T cells are generated.
- The responses of helper T cells, cytotoxic T cells, and B cells coordinate the specific disposal of, and long-term immunity to, offending antigens.

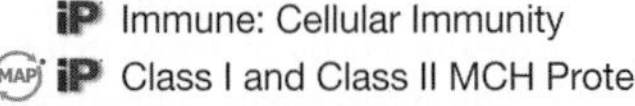

iP Immune: Cellular Immunity

iP Class I and Class II MCH Proteins

iP Cells and Proteins in the Innate Internal Defense System

6 Immune Responses in Health and Disease

- The aim of vaccination, or immunization, is to provide protection from infection.
- Both immunization and natural infection induce active immunity because they depend on the response of a person's own immune system and generate memory against the agent.
- Passive immunity is generated when antibodies are transferred from one person to another.
- These ready-made antibodies immediately target and mediate the disposal of antigens for which they are specific.
- Passive immunization is used when a dangerous bacterium or virus has entered the body of a person who is not already immune to it.
- The immune system's capacity to distinguish self from nonself limits our ability to share tissues through blood transfusion and transplantation.
- The survival of a graft requires making the best possible HLA match and using immunosuppressive drugs to diminish the recipient's immune responses.
- Bone marrow transplantation poses a particular problem when tissues are not well matched: The donated marrow, which contains lymphocytes, can mount immune responses against the recipient, resulting in a type of rejection known as a graft versus host reaction.
- Immune dysfunction can result in allergies, autoimmune diseases, or immunodeficiency diseases.
- An allergy results from an exaggerated response to environmental antigens (allergens).
- Autoimmune diseases occur when the immune system reacts against self, as in rheumatoid arthritis or multiple sclerosis.
- An immunodeficiency disease can result when a component of immunity is incapacitated by an inherited or acquired condition.
- Immunodeficiency diseases may affect humoral or cell-mediated immune function, or both (as in SCID).
- Evidence suggests that the immune system, the nervous system, and the endocrine system are physiologically linked.
- Neuroendocrine mechanisms have been shown to regulate immune responses, and immune responses in turn can induce changes in both endocrine and neural function.

iP Endocrine System: Response to Stress

MasteringA&P®

Check your understanding at MasteringA&P and apply your new knowledge to real-life case studies to ensure that you've mastered the content!

EXERCISES

Multiple-Choice Questions

1. Which of the following class of pathogens consists of only unicellular organisms?
   a) Virus
   b) Bacteria
   c) Fungi
   d) Parasites
   e) None of the above

2. Which of the following conditions would lead to the most serious immunodeficiency disease?
   a) Lack of IgG
   b) Lack of neutrophils
   c) Lack of B cells
   d) Lack of cytotoxic T cells
   e) Lack of helper T cells

3. Which of the following molecules can opsonize antigen?
   a) A T cell receptor
   b) Interferon
   c) An antibody
   d) A perforin
   e) Interleukin-2

4. Lymphocytes contact foreign antigen in all of the following tissues *except*
   a) bone marrow.
   b) the spleen.
   c) lymph nodes.
   d) the appendix.
   e) Peyer's patches.

5. Which of the following is *not* true about humoral immunity?
   a) It involves B cells.
   b) It involves antibodies.
   c) It involves cytotoxic T cells.
   d) It can provide passive immunity when transferred from one person to another.

6. Macrophages
   a) are derived from monocytes.
   b) have class II MHC molecules.
   c) are phagocytic.
   d) are indirectly involved in specific immunity.
   e) all of the above.

7. Activated cytotoxic T cells release pore-forming molecules called
   a) histamines.
   b) complement proteins.
   c) perforins.
   d) immunoglobulins.
   e) ready-porins.

8. An individual with type AB blood
   a) is considered a universal blood donor.
   b) is considered a universal blood recipient.
   c) produces antibodies to the B antigen.
   d) produces antibodies to the A antigen.
   e) is Rh-positive.

9. Which of the following events can result in lifelong immunity?
   a) Passage of maternal antibodies to a developing fetus
   b) An inflammatory response to a splinter
   c) Phagocytosis of bacteria by a neutrophil
   d) Administration of the polio vaccine
   e) Administration of antibodies against the rabies virus

10. Foreign antigens phagocytosed by macrophages are presented by
    a) class I MHC molecules to cytotoxic T cells.
    b) class II MHC molecules to helper T cells.
    c) class I MHC molecules to helper T cells.
    d) class II MHC molecules to CD8-bearing cells.
    e) class II MHC molecules to cytotoxic T cells.

11. Of the following events, which occurs earliest in the process of local inflammation?
    a) Increased capillary permeability
    b) Fever
    c) Attack by cytotoxic T cells
    d) Release of histamine
    e) Lysis of microbes mediated by antibodies and complement

12. Which of the following is *not* true about helper T cells?
    a) They function in both the cell-mediated and humoral immune responses.
    b) They secrete antibody.
    c) They bear surface CD4 molecules.
    d) They are subject to infection by HIV.
    e) When activated, they secrete IL-2 and other cytokines.
13. Which of the following is *not* an antigen-presenting cell?
    a) Macrophage
    b) Dendritic cell
    c) Mast cell
    d) Activated B cells
    e) Thymus cells
14. IL-2 is important for the activation of all of the following cell types *except*
    a) B cells.
    b) cytotoxic T cells.
    c) NK cells.
    d) T helper cells.
    e) macrophages.
15. Which of the following is an autoimmune disease in which myelinated neurons become the target of the immune response?
    a) Myasthenia gravis
    b) Multiple sclerosis
    c) Diabetes mellitus
    d) Rheumatoid arthritis

Objective Questions

16. Fill in the blank with the abbreviation for the cell type that mediates the stated function: helper T cell (T_H), cytotoxic T cell (T_C), B cell (B), or macrophage (M).
    a) ________ Phagocytosis
    b) ________ Secretion of cytokines such as IL-2
    c) ________ Killing of virus-infected cells
    d) ________ Specific binding to free virus
    e) ________ Differentiation into antibody-secreting plasma cells
17. Fill in the blank with the letter that applies to the stated situation: humoral immune response (H), cell-mediated response (CM), both (B), or neither (N).
    a) ________ Occurs in a viral infection
    b) ________ Involves the production of antibodies
    c) ________ Involves the phagocytic activity of neutrophils
    d) ________ Involves killing of virus-infected cells
    e) ________ Involves T cells bearing CD8
18. A person who experiences life-threatening allergic reactions to bee sting venom might be given an experimental drug designed to (block/enhance) the binding of IgE to mast cells.
19. When a macrophage is infected by a virus, viral antigen will be presented by (class I/class II MHC molecules) to a (helper T cell/cytotoxic T cell).
20. Evidence exists of interactions among the immune system, the nervous system, and the endocrine system. (true/false)
21. A young girl who has never been immunized against tetanus cuts her foot on a rusty nail. In the emergency room her wound is cleaned, and she is given an injection of tetanus antitoxin (antibody to tetanus toxin). This is considered (active/passive) immunization.
22. Macrophages internalize foreign antigens by (endocytosis/phagocytosis), whereas B cells do so by (endocytosis/phagocytosis).
23. A person who has had a thymectomy as a treatment for a thymic tumor will likely experience a diminished (T cell/B cell) count.

Essay Questions

24. You notice that the area surrounding a paper cut on your hand has become red, warm, and swollen. Briefly describe the processes that lead to these symptoms of inflammation.
25. Through which lymphoid organs will a cell destined to be an active, mature T cell travel during its lifetime? Briefly state the function of each organ you mention.
26. Immune responses exhibit four features, all of which are attributed to lymphocyte function. List these four features and discuss how they arise.
27. Draw a diagram showing the central role of helper T cells in both the humoral and cell-mediated immune responses.

Critical Thinking

28. Consider a simple artificial antigen containing two distinct epitopes. Describe and diagram the process of clonal selection that results as B cells specific for these epitopes are selected by the antigen and are stimulated to proliferate and to differentiate into memory B cells and plasma cells. Also diagram how secreted antibodies interact with this antigen.
29. An infant who has experienced multiple bacterial infections since birth has been diagnosed with a macrophage deficiency. Describe how this deficiency affects the baby's nonspecific defenses and immune responses. Do you think that a neutrophil deficiency would present with a more severe or a less severe state of immunodeficiency?
30. A child needs a kidney transplant. Both of his parents and his two older siblings are willing to donate the needed organ. Tissue typing reveals that his father has the HLA genotype "8, 9" (one HLA gene copy is "8," the other is "9"), and his mother has an HLA genotype "2, 6." Considering that a child inherits one copy of the HLA gene from each parent, which is the better potential donor—a parent or a sibling? Is it possible to have a 100% donor-recipient match of the HLA type?

Photo credits for icons in feature boxes: Focus on Diabetes: Tanyae/Dreamstime; Clinical Connections: Iofoto/Dreamstime; Understanding Exercise: Pixattitude/Dreamstime; Discovery: Yuri Arcurs/Cutcaster.

Glossary

active immunity a type of immune protection that depends on the ability of the immune system to mount a response

agglutination the linking together of many antigens through the specific binding of many antibodies

antibodies proteins present in the blood and interstitial fluids that target particular antigens for destruction; also known as *immunoglobulins*

antigen presentation the movement of an antigen from inside the infected cell to the cell surface by MHC molecules

antigens protein or polysaccharide components of viruses, bacteria, fungi, protozoa, parasitic worms, pollen, transplanted tissue, and tumor cells that can be identified by immune cells

cell-mediated immunity reaction of certain types of T cells in which they seek to kill abnormal or infected body cells

central lymphoid tissues tissues where leukocytes develop to maturity, including the bone marrow and thymus

class I MHC molecules found on the surfaces of all nucleated cells; identify cells as self

class II MHC molecules found on specialized cell types, including macrophages, activated B and T cells, and the cells of the interior of the thymus

clonal selection antigen-driven proliferation of lymphocytes necessary for specific immune responses

complement system 30 proteins that act to destroy invading microorganisms, especially bacteria

cytotoxic T cells (sy-toh-tox-ik) T cells that contact virus-infected cells, mutant cells, and transplanted cells, and gain the ability to destroy the infected or abnormal cells

effector cells in the immune system, short-lived cells that work to combat the antigen that triggered the immune response

epitopes (ep-ih-topes) sites on an antigen that are recognized by antibodies

fragmentin protein released from cytotoxic T cells that enters target cells and induces apoptosis

helper T cells T cells that secrete many different cytokines, thereby influencing the actions of other lymphocytes

humoral immunity a specific immune response generated by B lymphocytes and conferred by antibodies that circulate in the blood and lymph

immune response series of events that destroy and eliminate foreign and abnormal substances from the body

immunity capacity of the immune system to defend against disease by recognizing and eliminating disease-causing agents from the body

inflammation a complex series of events that culminate in the accumulation of proteins, fluid, and phagocytic cells in an area where tissue has been injured or invaded by microorganisms

interferons a group of related proteins that interfere with virus replication

lymphoid tissue sites where leukocytes are exposed to foreign substances in blood or lymph

mast cells cells dispersed throughout the body's connective tissues that help signal the early events of inflammation by releasing histamine

memory cells in the immune system, long-lived cells that bear membrane receptors specific for the antigen that triggered the immune response

natural killer (NK) cells large, granular lymphocytes; secrete chemicals that kill tumor cells and infected cells; also known as *null cells*

neutralization in the immune system, an antibody's blockage of an antigen's activity by binding to it

nonspecific defense mechanism that defends against potentially harmful substances without regard to their precise identity

null cells lymphocytes that lack certain membrane components, including antigen binding sites

opsonins (op-son-ins) proteins generated in immune responses that bind tightly to foreign materials and make it easier for a phagocyte to engulf them later

passive immunity immunity passed on to the host in the form of ready-made antibodies

perforin protein that forms pores in a cell's membrane, causing lysis

peripheral lymphoid tissues tissues that trap foreign matter present in the blood, including the spleen, lymph nodes, tonsils, adenoids, appendix, and Peyer's patches

plasma cells mature form of B cells that secrete antibodies

primary immune response antigen-induced lymphocyte response that occurs when a person is exposed to an antigen for the first time

secondary immune response antigen-induced lymphocyte response caused by subsequent exposure to an antigen

specific immune response mechanism that targets and eliminates specific substances

Answers

Figure Question Answers

Fig 3 The spleen

Fig 5 Inhibition of capillary dilation reduces redness and heat; inhibition of increased capillary permeability reduces edema

Fig 14 A primary response

Fig 19 Memory cells. Antibodies function in the short term; they will help clear the harmless vaccine from Ivan's body. Memory cells, in contrast, can survive a lifetime and are poised to combat the chicken pox virus whenever he is exposed to it.

Fig 20 Histamine induces increased capillary permeability and vasodilation, resulting in the loss of fluid from blood vessels and decreased peripheral resistance, respectively. Both of these events contribute to a drop in mean arterial pressure.

Apply Your Knowledge Answers

Because so many different strains of flu viruses are circulating in the world, scientists make an educated guess as to which strain is most likely to be active in a particular season. The vaccine is then made up of antigens associated with approximately 200 different flu strains to increase the likelihood of immunizing against the correct strain.

Integrins are important adhesion molecules that bind leukocytes tightly to the blood vessel wall so that the leukocytes can eventually move across the wall. Because LAD causes a lack of functional integrins, the inflammatory response would be weakened, as blood cells cannot leave the blood as easily to move to the site of damage/infection in the tissue. The healing process would proceed more slowly in a person suffering from LAD.

Answers to End-of-Chapter Multiple Choice and Objective Questions

Multiple Choice Questions

1. b 2. e 3. c 4. a 5. c 6. e 7. c 8. b 9. d 10. b 11. d 12. b 13. e 14. e 15. b

Objective Questions

16. (a) M (b) T_H (c) T_C (d) B (e) B
17. (a) B (b) H (c) N (d) CM (e) CM
18. block
19. class II MHC molecules, helper T cell
20. true
21. passive
22. phagocytosis, endocytosis
23. T cell

Index

Page references followed by "f" indicate illustrated figures or photographs; followed by "t" indicates a table.

C

D

E

G

H

J

K

L

O

P

Q

R

S

T

U

V

W

Y

Z